THIRD EDITION

Fundamentals of Anatomy and Physiology

Frederic H. Martini, Ph.D.

with

William C. Ober, M.D.
Art coordinator and illustrator

Claire W. Garrison, R.N.
Illustrator

Kathleen Welch, M.D.
Clinical consultant

Ralph T. Hutchings
Biomedical photographer

D1306933

Prentice Hall, Upper Saddle River, New Jersey 07458

Library of Congress Cataloging-in-Publication Data

MARTINI, FREDERIC
Fundamentals of anatomy and physiology/Frederic H. Martini; with William C. Ober, art coordinator and illustrator, Claire W. Garrison, illustrator, Kathleen Welch, clinical consultant, Ralph T. Hutchings, biomedical photographer—3rd ed.
p. cm.

Includes index.
ISBN 0-13-298952-2 (hard cover)
1. Human physiology. 2. Human anatomy. I. Title.
[DNLM: 1. Anatomy. 2. Physiology. QS 4 M3855f 1995]
QP34.5.M27 1995
612—dc20
DNLM/DLC 94-37668
for Library of Congress CIP

Dedication

To the students who suggested that I start this project, and to my wife, family, and friends whose support has enabled me to continue to improve it.

Executive Editor: *David Kendric Brake*
Development Editor: *Dan Schiller*
Production Editor: *Debra A. Wechsler*
Managing Editor: *Linda Schreiber*
Director, Production & Manufacturing: *David W. Riccardi*
Editorial Director: *Tim Bozik*
Editor in Chief, Development: *Ray Mullaney*
Managing Editor, Production: *Kathleen Schiaparelli*
Marketing Manager: *Gary June*
Copy Editor: *Margo Quinto*
Interior Designer: *Lee Goldstein, Lorraine Mullaney*
Cover Designer: *Tom Nery*
Creative Director: *Paula Maylahn*
Buyer: *Trudy Pisciotti*
Page Layout: *John Jordan, Richard Foster, Karen Noferi, Eric Hulsizer, Shari Toron, David Tay*
Photo Research: *Stuart Kenter*
Photo Editor: *Lorinda Morris-Nantz*
Illustrators: *William C. Ober, M.D.; Claire W. Garrison, R.N.; Ron Ervin; Tina Sanders; and Craig Luce*
Technical Art: *Karen Noferi; Vantage Art*
Production Assistance: *Lorraine Patsco*
Art Coordinators: *Liz Nemeth, Veronica Czinege-Schwartz*
Editorial Assistant: *Carrie Brandon*

Cover Photograph: *Lois Greenfield* ©

A Simon & Schuster Company
Upper Saddle River, New Jersey 07458

Printed in the United States of America
10 9 8 7 6 5 4

ISBN 0-13-298952-2

Prentice-Hall International (UK) Limited, *London*
Prentice-Hall of Australia Pty. Limited, *Sydney*
Prentice-Hall Canada Inc., *Toronto*
Prentice-Hall Hispanoamericana, S.A., *Mexico*
Prentice-Hall of India Private Limited, *New Delhi*
Prentice-Hall of Japan, Inc., *Tokyo*
Simon & Schuster Asia Pte. Ltd., *Singapore*
Editora Prentice-Hall do Brasil, Ltda., *Rio de Janeiro*

-ia, state or condition: insomnia
idi-, *idios,* one's own: idiopathic
ile-, *ileum:* ileocolic sphincter
ili-, ilio-, *ilium:* iliac
in-, in, within, or denoting negative effect: inactivate
infra-, *infra,* beneath: infraorbital
inter-, *inter,* between: interventricular
intra-, *intra,* within: intracapsular
ipsi-, *ipse,* itself: ipsilateral
iso-, *isos,* equal: isotonic
-itis, *-itis,* inflammation: dermatitis
karyo-, *karyon,* body: megakaryocyte
kerato-, *keros,* horn: keratin
kino-, -kinin, *kinein,* to move: bradykinin
lact-, lacto-, -lactin, *lac,* milk: prolactin
lapar-, *lapara,* flank or loins: laparoscopy
-lemma, *lemma,* husk: plasmalemma
leuko-, *leukos,* white: leukocyte
liga-, *ligare,* to bind together: ligase
lip-, lipo-, *lipos,* fat: lipoid
lith-, *lithos,* stone: cholelithiasis
lyso-, -lysis, -lyze, *lysis,* dissolution: hydrolysis
macr-, *makros,* large: macrophage
mal-, *mal,* abnormal: malabsorption
mamilla-, *mamilla,* little breast: mamillary
mast-, masto-, *mastos,* breast: mastoid
mega-, *megas,* big: megakaryocyte
melan-, *melas,* black: melanocyte
men-, *men,* month: menstrual
mero-, *meros,* part: merocrine
meso-, *mesos,* middle: mesoderm
meta-, *meta,* after, beyond: metaphase
micr-, *mikros,* small: microscope
mono-, *monos,* single: monocyte
morpho-, *morphe,* form: morphology
multi-, many: multicelllular
-mural, *murus,* wall: intramural
myelo-, *myelos,* marrow: myeloblast
myo-, *mys,* muscle: myofilament
narc-, *narkoun,* to numb or deaden: narcotics
nas-, *nasus,* nose: nasolacrimal duct
natri-, *natrium,* sodium: natriuretic
necr-, *nekros,* corpse: necrosis
nephr-, *nephros,* kidney: pronephros
neur-, neuro-, *neuron,* nerve: neuromuscular
oculo-, *oculus,* eye: oculomotor
odont-, *odontos,* tooth: odontoid process
-oid, resembling: odontoid process
oligo-, *oligos,* little, few: oligopeptide
-ology, *logos,* the study of: physiology
-oma, *-oma,* swelling: carcinoma
onco-, *onkos,* mass, tumor: oncology
oo-, *oon,* egg: oocyte
ophthalm-, *ophthalmos,* eye: ophthalmic nerve
-opia, *ops,* eye: optic
orb-, *orbita,* a circle: orbicularis oris
orchi-, *orchis,* testis: orchiectomy
orth-, *orthos,* correct, straight: orthopedist
-osis, *-osis,* state, condition: neurosis
osteon, osteo-, *os,* bone: osteocyte
oto-, *otikos,* ear: otoconia
para-, *para,* beyond: paraplegia
patho-, -path, -pathy, *pathos,* disease: pathology
pedia-, *paidos,* child: pediatrician
per-, *per,* through, throughout: percutaneous
peri-, *peri,* around: perineurium
phag-, *phagein,* to eat: phagocyte
-phasia, *phasis,* speech: aphasia
-phil, -philia, *philus,* love: hydrophilic
phleb-, *phleps,* a vein: phlebitis
-phobe, -phobia, *phobos,* fear: hydrophobic
phot-, *phos,* light: photoreceptor
-phylaxis, *phylax,* a guard: prophylaxis
physio-, *physis,* nature: physiology
-plasia, *plasis,* formation: dysplasia
platy-, *platys,* flat: platysma
-plegia, *plege,* a blow, paralysis: paraplegia
-plexy, *plessein,* to strike: apoplexy
pneum-, *pneuma,* air: pneumotaxic center
podo, *podon,* foot: podocyte
-poiesis, *poiesis,* making: hemopoiesis
poly-, *polys,* many: polysaccharide
post-, *post,* after: postanal
pre-, *prae,* before: precapillary sphincter
presby-, *presbys,* old: presbyopia
pro-, *pro,* before: prophase
proct-, *proktos,* anus: proctology
pterygo-, *pteryx,* wing: pterygoid
pulmo-, *pulmo,* lung: pulmonary
pulp-, *pulpa,* flesh: pulpitis
pyel-, *pyelos,* trough or pelvis: pyelitis
quadr-, *quadrans,* one quarter: quadriplegia
re-, *back,* again: reinfection
retro-, *retro,* backward: retroperitoneal
rhin-, *rhis,* nose: rhinitis
-rrhage, *rhegnymi,* to burst forth: hemorrhage
-rrhea, *rhein,* flow, discharge: amenorrhea
sarco-, *sarkos,* flesh: sarcomere
scler-, sclero-, *skleros,* hard: sclera
-scope, *skopeo,* to view: colonoscope
-sect, *sectio,* to cut: transect
semi-, *semis,* half: semitendinosus
-septic, *septikos,* putrid: antiseptic
-sis, state or condition: metastasis
som-, -some, *soma,* body: somatic
spino-, *spina,* spine, vertebral column: spinodeltoid
-stalsis, *staltikos,* contractile: peristalsis
sten-, *stenos,* a narrowing: stenosis
-stomy, *stoma,* mouth, opening: colostomy
stylo-, *stylus,* stake, pole: styloid
sub-, *sub,* below; subcutaneous
super-, *super,* above or beyond: superficial
supra-, *supra,* on the upper side: supraspinous fossa
syn-, *syn,* together: synthesis
tachy-, *tachys,* swift: tachycardia
telo-, *telos,* end: telophase
tetra-, four: tetralogy of Fallot,
therm-, thermo-, *therme,* heat: thermoregulation
thorac-, thorax, chest: thoracentesis
thromb-, *thrombos,* clot: thrombocyte
-tomy, *temnein,* to cut: appendectomy
tox-, *toxikon,* poison: toxemia
trans-, *trans,* through: transudate
tri-, three: trimester
-trophic, -trophin, -trophy, *trophikos,* nourishing: adrenocorticotrophic
tropho-, *trophe,* nutrition: trophoblast
tropo-, *tropikos,* turning: troponin
uni-, one: unicellular
uro-, -uria, *ouron,* urine: glycosuria
vas-, *vas,* vessel: vascular
zyg-, *zygotos,* yoked: zygote

Text and Illustration Team

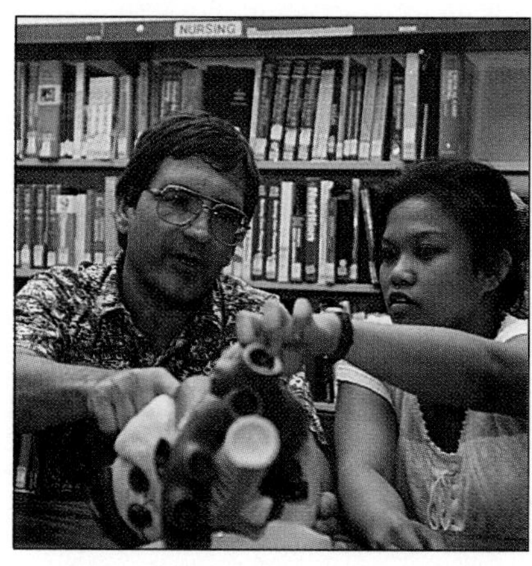
NURSING

Frederic H. Martini received his Ph.D. from Cornell University in Comparative and Functional Anatomy. He has broad interests in vertebrate biology, with special expertise in anatomy, physiology, histology, and embryology. Dr. Martini's publications include journal articles, technical reports, magazine articles, and a book for naturalists on the biology and geology of tropical islands. Working with Professor Michael J. Timmons, he has coauthored an undergraduate textbook on *Human Anatomy* (Prentice Hall, 1995).

Dr. Martini has been involved in teaching undergraduate courses in anatomy and physiology (comparative and/or human) since 1970. During the 1980s he spent his winters teaching courses, including human anatomy and physiology, at Maui Community College, and his summers teaching an upper-level field course in vertebrate biology and evolution for Cornell University, at the Shoals Marine Laboratory (SML). Dr. Martini now teaches part-time in the winters, and devotes most of his attention to developing new approaches to A&P education. His primary interest is in the use of appropriate technologies in creating an integrated learning system. Dr. Martini is a member of the Human Anatomy and Physiology Society, the National Association of Biology Teachers, the American Society of Zoologists, the Society for College Science Teachers, the Western Society of Naturalists, and the National Association of Underwater Instructors.

Dr. William C. Ober (art coordinator and illustrator) received his undergraduate degree from Washington and Lee University and his M.D. from the University of Virginia in Charlottesville. While in medical school he also studied in the Department of Art as Applied to Medicine at Johns Hopkins University. After graduation Dr. Ober completed a residency in family practice, and is currently on the faculty of the University of Virginia as a Clinical Assistant Professor in the Department of Family Medicine. He is also part of the Core Faculty at Shoals Marine Laboratory, where he teaches biological illustration in the summer program. Dr. Ober now devotes his full attention to medical and scientific illustration.

Claire W. Garrison, R.N., (illustrator) practiced pediatric and obstetric nursing for nearly 20 years before turning to medical illustration as a full-time career. Following a five-year apprenticeship, she has worked as Dr. Ober's associate since 1986. Ms. Garrison is also a Core Faculty member at Shoals. Texts illustrated by Dr. Ober and Ms. Garrison have received national recognition and awards from the Association of Medical Illustrators (Award of Excellence), American Institute of Graphics Arts (Certificate of Excellence), Chicago Book Clinic (Award for Art and Design), Printing Industries of America (Award of Excellence), and Bookbuilders West. They are also recipients of the Art Directors Award.

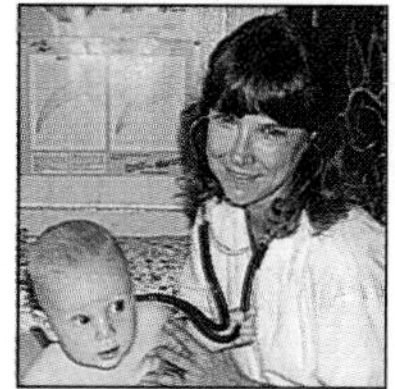

Dr. Kathleen Welch (clinical consultant) received her M.D. from the University of Washington in Seattle, and did her residency at the University of North Carolina in Chapel Hill. For two years she served as Director of Maternal and Child Health at the LBJ Tropical Medical Center in American Samoa, and subsequently was a member of the Department of Family Practice at the Kaiser Permanente Clinic in Lahaina, Hawaii. She has been in private practice since 1987. Dr. Welch is a Fellow of the American Academy of Family Practice. She is also a member of the Hawaii Medical Association and the Human Anatomy and Physiology Society.

Ralph T. Hutchings is a biomedical photographer who was associated with the Anatomy Department of the Royal College of Surgeons for 20 years. An engineer by training, Ralph has focused his attention for years now on photographing the structure of the human body. The result has been a series of color atlases, including the *Color Atlas of Human Anatomy*, *The Color Atlas of Surface Anatomy* and *The Human Skeleton* (all published by Mosby-Yearbook Publishing, St. Louis, Missouri, USA). Ralph makes his home in North London where he tries to balance the demands of his photographic assignments with his hobbies of early motor cars and airplanes.

Contents in Brief

Contents

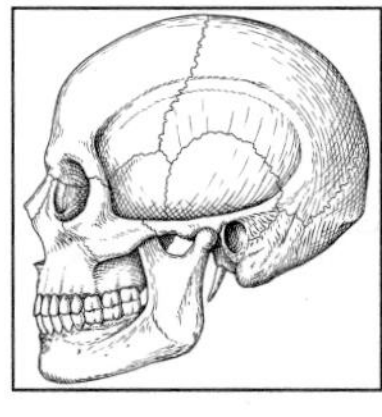

6 Osseous Tissue and Skeletal Structure *174*

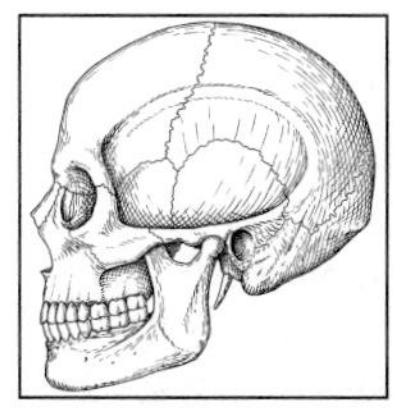

7 The Axial Skeleton *200*

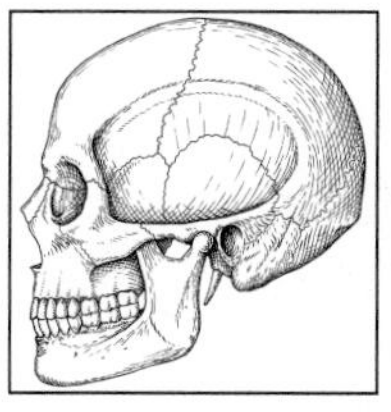

8 The Appendicular Skeleton *238*

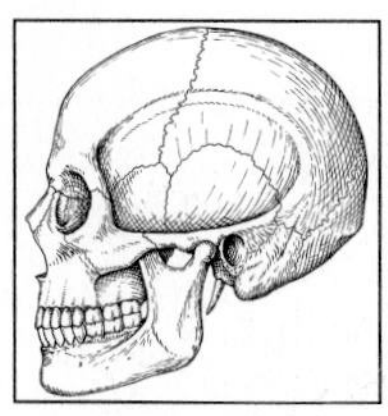

9 Articulations 260

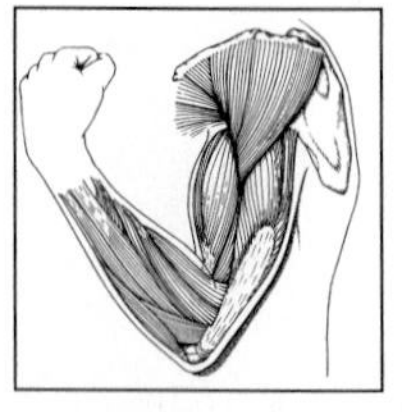

10 Muscle Tissue 284

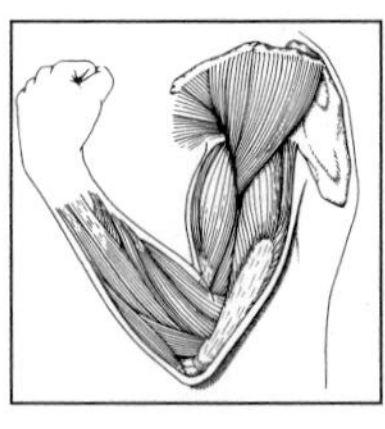

11 The Muscular System 324

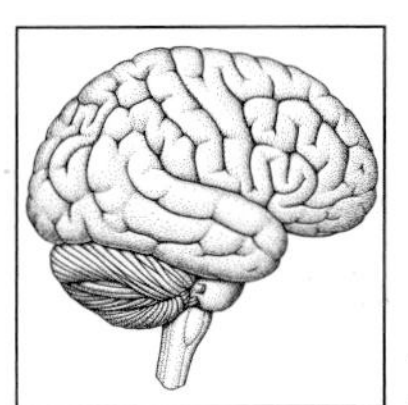

12 Neural Tissue *378*

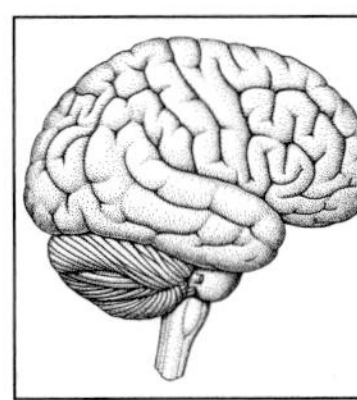

13 The Spinal Cord and Spinal Nerves *426*

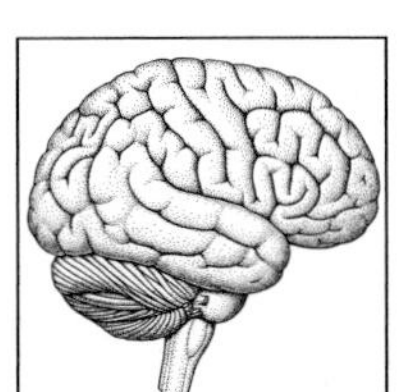

16 The Autonomic Nervous System *526*

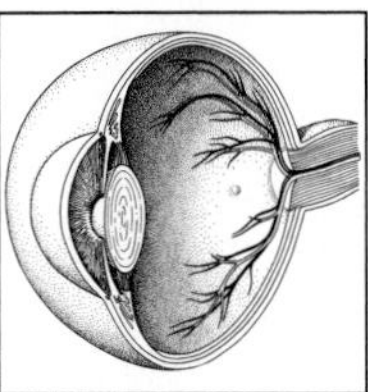

17 Sensory Function *550*

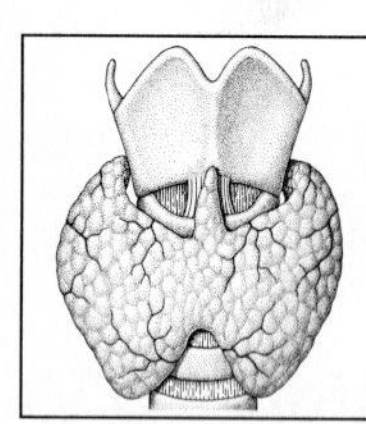

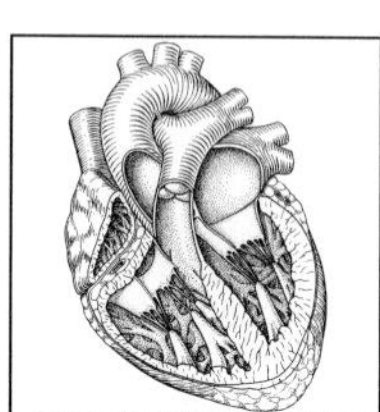

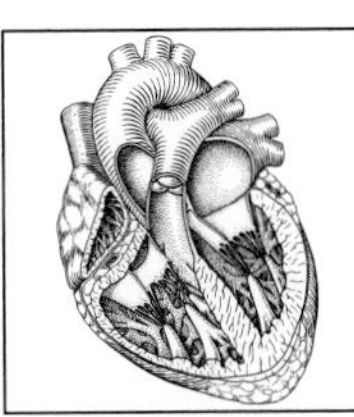

20 The Heart 680

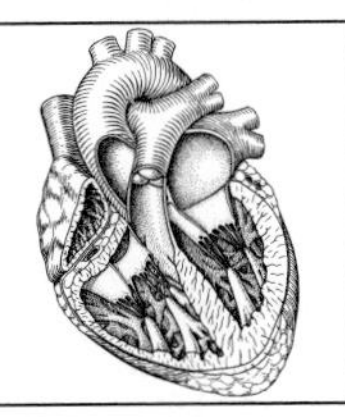

21 Blood Vessels and Circulation 716

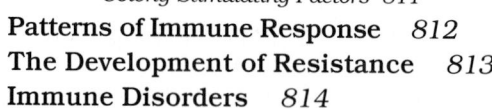

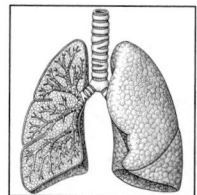

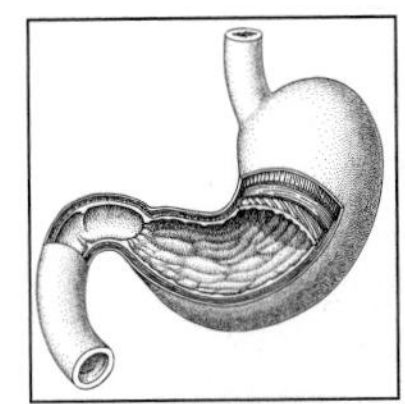

24 The Digestive System 876

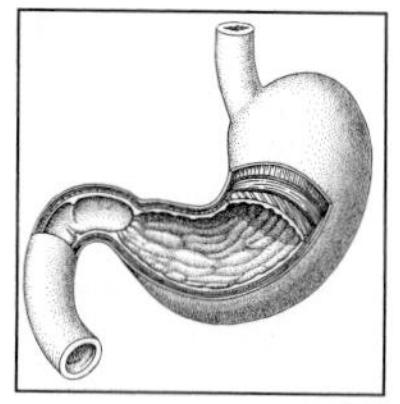

25 Metabolism and Energetics *934*

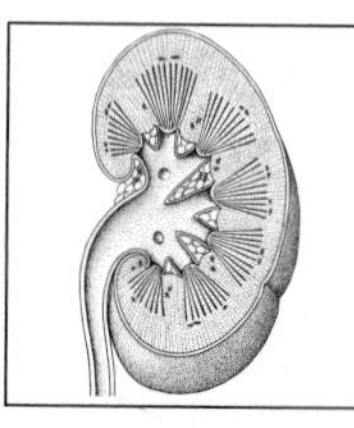

26 The Urinary System *978*

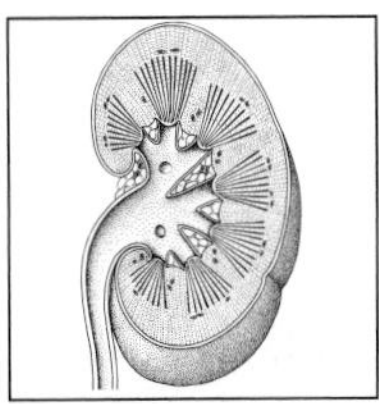

27 Fluid, Electrolyte, and Acid-Base Balance *1026*

28 The Reproductive System *1058*

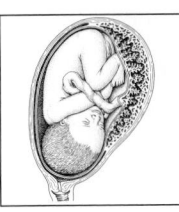

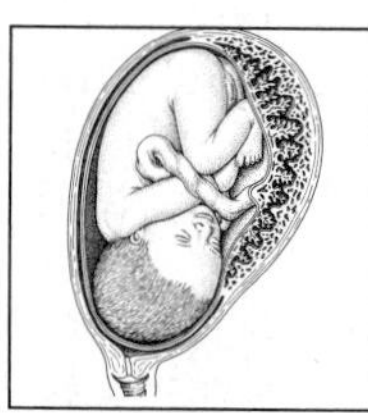

29 Development and Inheritance *1108*

Appendixes *A-1*

Glossary of Key Terms *G-1*

Index *I-0*

Preface

THE CHALLENGE

Nearly everyone today knows that medical knowledge is expanding at an unprecedented rate; that interest in health and fitness are at an all-time high; that the population as a whole is aging. Students of anatomy and physiology realize that never before have there been more opportunities for employment in health-related fields, from nursing to sports training, from occupational safety to dietetics They therefore typically begin this course with high hopes.

All too often, however, these hopes quickly turn to frustration. For many students this will be their first college-level science course, and they must not only learn a new vocabulary but learn new ways of studying and organizing information. Moreover, returning students must also balance the demands of their academic work with outside responsibilities, such as jobs and families. Not surprisingly, therefore, nearly all students suffer from a lack of time. Time to go to class, time to spend in lab, time to study at home. Time to memorize several thousand new technical terms that must be correlated with hundreds of anatomical illustrations; time to develop a thorough understanding of complex physiological processes. Above all, time to integrate what they learn in various chapters, so that they can see how the body functions as a whole rather than as a collection of separate organs or systems.

No textbook can give students more time. But a carefully crafted text can help students make better use of the time they do have. The key to producing such a book is understanding how students learn, and what impediments stand in the way of their learning.

THE THIRD EDITION

Instructors naturally share the frustrations of their students— multiplied by the number of students in their courses. This text has evolved from edition to edition in response to the concerns and suggestions from instructors as well as students. Many instructors have been happy to offer advice, ideas, and comments for the preparation of the third edition. Day-by-day notes of instructors using the second edition, on-campus meetings, focus groups, and extensive reviewing of several drafts all played important roles in the development of the book. The ultimate goal, however, has remained the same since the planning of the first edition nearly a decade ago: to give students every possible assistance in surmounting what may be the greatest challenge of their educational career.

The second edition of this text represented a major advance in the direction mapped out in the first edition. The goal was to create, not merely a superior textbook, but a complete *learning system* in which text, art, pedagogical apparatus, and supplements all worked together as an integrated whole. That second edition saw the introduction of many innovative pedagogical features: **concept check questions, concept links, system integrators, boxed embryology summaries,** a **three-level review system** for the end-of-chapter material, a companion ***Clinical Manual*** containing supplemental clinical and other material cross-referenced to the text. The reception of these features has been highly enthusiastic; they have been retained, and in some instances refined, in the third edition. A number of new features have been added:

- The Objectives that precede each chapter are now integrated into an outline of the chapter's major sections, complete with page numbers. This feature enables students to preview the contents of the chapter, provides focus for their study, and facilitates review.
- The Selected Clinical Terminology section at the end of each chapter has been expanded to include definitions of terms discussed in the *Applications Manual* as well as those that occur in the text.
- Red dots have been appended to text figure references so that students can more easily shift their attention from the text to the art and back again without losing their place in the narrative.
- Many figures in the text have been cross-referenced to X-ray or MRI images in the Scanning Atlas section of the *Applications Manual* or to color plates of dissection or surface anatomy photographs in the Cadaver Atlas section of the *Applications Manual*. Cross-references to related animations on the Laser Disk or Video Tutor have also been marked with icons.
- A new Focus format, designed for easier integration of text, art, and tabular material, has been introduced.

These features, along with others retained from the second edition, are not only explained but shown in the following Student Preface, which I urge you to consult. A few of the most important features of this revision, however, demand some additional explanation.

The Art Program

The art in the second edition was so well received and highly praised by users of the text that it may seem odd to find it listed under the heading of "new and improved." But art is crucially important to anatomy and physiology— there are few other courses in which so much depends on seeing. Both anatomical structures and physiological processes must be visualized if students are to attain true understanding. Thus there is always room for improvement in even the finest art program. The major changes in the art program for this edition fall into the following categories:

- New figures to clarify physiological processes, such as the creation of the resting potential and the generation and propagation of action potentials. (See for example Figures 12-9, 12-13, 12-15, 12-17, and 12-18.)
- New anatomical art and flow charts for the circulatory system. (See Chapter 21, pp. 750-769.)
- Clearer, more graphic diagrams showing pathways of homeostatic regulation. (See for example Figures 18-16, 21-17, 21-19, and 26-11.)
- More compound figures showing the relationship between macro and micro levels of structure. (See for example Figures 24-13, 24-16, and 24-23.)
- New, revised, and/or enlarged anatomical art for many organs and systems. (See for example Figures 12-2, 23-6, 23-9, 28-17, and 28-18.)
- The use of anatomical photographs by Ralph Hutchings, an internationally-renowned biomedical photographer, to supplement the anatomical paintings. (See for example Figure 24-12.)

Enhanced Coverage of Physiological Concepts

The discussions of muscle physiology, neurophysiology, sensory mechanisms, cardiovascular physiology, immunology, urinary function, and acid-base balance have been updated and expanded. Additional illustrations have been provided to support the enhanced coverage.

A New Approach to the Treatment of Clinical Topics

Instructors vary widely in the amount of clinical material they choose to cover in their courses. Clearly, it is not possible to add clinical material indiscriminately to the text without making the book both inordinately long and extremely cumbersome to use for those whose orientation is not clinical. One aim of this revision has been to provide even greater flexibility in the handling of clinical and applied topics. The text contains all pedagogically essential clinical material while making additional clinical topics accessible for instructors who require them and unobtrusive for those who do not.

- Clinical and applied subjects of general interest and pedagogical importance are treated in Clinical Discussions. These brief sections occur at relevant points within the context of the chapter, set off only by an icon, title, and vertical colored bar. The topics covered in this way are those that shed the greatest light on normal function; hence their integration into the flow of the narrative.
- Major diseases, syndromes, or other health-related topics that require more extended treatment, such as AIDS, coronary artery disease, and the abuse of steroids, are discussed in boxed essays set off from the main text. These essays can be read in context, studied separately, or bypassed entirely, as instructor preference or student interest dictate.
- Much additional clinical and applied material is contained in the *Applications Manual* that accompanies the text. Some of the discussions in the *Applications Manual* provide more detailed information about clinical conditions introduced in Clinical Discussions or Boxes in the text, such as AIDS. Others deal with topics that are not covered in the text.

All discussions in the *Applications Manual* are clearly referenced at the relevant point in the text by an icon title, in blue.

The Applications Manual

The *Applications Manual*, a greatly expanded and enhanced descendant of the *Clinical Manual* that accompanied the second edition of this text, contains a wealth of applied material that can broaden and deepen the student's experience of anatomy and physiology. The *Applications Manual* includes:

- An introduction to diagnostics, discussions of the history and language of anatomy and physiology, an account of the scientific method, and a selection of research topics that highlight the application of chemistry and cell biology to the understanding, diagnosis, or treatment of homeostatic disorders.
- A comprehensive compendium of clinical and diagnostic topics (described above), organized to parallel the text chapter by chapter and system by system. Related Critical Thinking questions following each system and Clinical Problems following groups of related systems extend the three-level learning system featured in the text, helping students to develop their powers of integration and analysis.
- *Case Studies* that provide further opportunities for problem-solving in realistic clinical situations. The studies presented, based on actual case histories, require the application of material from many different parts of the text. They are accompanied by questions that focus attention on the relevant facts, guiding the student through the diagnostic process toward the formation of plausible hypotheses. These questions can also be used as the basis for classroom discussions or collaborative learning exercises.
- A *Scanning Atlas* of photographs produced by various modern imaging techniques. The Scanning Atlas includes a number of unlabeled images that students can label to test their knowl-

edge of anatomical structure and develop their powers of visualization.

• A *Cadaver Atlas* of full-color anatomical photographs by Ralph Hutchings. These photographs supplement the illustrations in the text, to which they are fully cross-referenced. A selection of surface anatomy photographs of live models is included for comparison with the dissection views.

The text and *Applications Manual* are tied together by a comprehensive set of cross-references, allowing students to locate related material with ease.

THE SUPPLEMENTS PACKAGE

The ancillaries for this edition of *Fundamentals of Anatomy and Physiology* reflect the same concern for the needs of students and instructors that characterize the text.

For the Instructor

• **The Instructor's Resource Guide** (written by Lucia Tranel and Alice Mills) is a unique supplement designed to be useful to the new instructor as well as the one who has been teaching Anatomy and Physiology for many years. Not only have the authors created many lively and original analogies and teaching tips, but they have also collected suggestions from other instructors across the country. The result is a new kind of instructor's manual, one that provides a platform of exchange for those teaching this challenging course.
ISBN: 0-13-340894-9

• **The Test Item File** (written by Edwin Bartholomew and George Karleskint) features over 3000 questions and parallels the three-level learning system of the text. This test bank also includes 200 pieces of unlabeled text art that can be used to create labeling exercises for exams.
ISBN: 0-13-340902-3

• **The Prentice Hall Test Manager Plus** is an easily accessible software version of the above Test Item File and is available for IBM and Mac. This program will allow an instructor to add, delete, and edit questions to customize tests and quizzes.
ISBN: (IBM) 0-13-341025-0
(MAC) 0-13-340985-6

• **The Prentice Hall Transparency Set for Anatomy and Physiology** combines 450 color acetates from both the text and other sources and 200 masters. This collection includes 50 cadaver images.
ISBN: 0-13-341017-X

• **The Prentice Hall Laserdisc for Anatomy and Physiology** and its **Bar Code Manual** encourage instructors to bring a wealth of images into the classroom and lab. The disc features text art, quiz frames (text figures with labels removed), histology slides, 300 cadaver images, pictures of common lab models and equipment and a complete set of cat dissection photographs. The video portion of the disc has high-quality three-dimensional physiology animations and footage demonstrating lab experiments that are inaccessible to many faculty members (i.e. those utilizing computer analysis, animal specimens, and human body fluids). The disc can also be accessed using the **Prentice Hall Multimedia Presentation Manager for Anatomy and Physiology,** software that allows instructors to organize visual presentations and create custom overlays for images that accompany lectures. The Presentation Manger has its own **User's Guide.**
Laserdisc ISBN: 0-13-341033-1
Multimedia Presentation Manager ISBN:
(IBM) 0-13-340795-7 (MAC) 0-13-340977-5

For the Student

• Each new copy of the text comes packaged with an ***Applications Manual*** (written by Frederic Martini, Kathleen Welch, and Martha Newsome). This unique supplement provides students with access to interesting and relevant clinical and diagnostic information. Critical thinking questions, clinical problems, and case studies offer students the opportunity to think analytically. Historical background and information on recent research give them a frame of reference for their study of Anatomy and Physiology. Two atlases, one of color cadaver photographs and one of sectional scans, are included as additional image resources.
ISBN: 0-13-341009-9

• **The Study Guide** (written by Charles Seiger) is a thorough revision of the successful Study Guide that accompanied the second edition of the text. The three-level learning system has been retained, and the number of labeling exercises and Concept Maps has been significantly increased.
ISBN: 0-13-340936-8

• **The Anatomy and Physiology Video Tutor** gives students the opportunity to see the dynamics of the most difficult physiological processes, even without access to computers. This seventy-five minute video focuses on the concepts that instructors across the country have consistently identified as the most challenging. Physiological processes are demonstrated through the use of top-quality three-dimensional animations and video footage. On-camera narration and the accompanying frame-referenced study booklet promote repeated concept review.
ISBN: 0-13-340969-4

• **The Audio Study Tapes** (written by Dan Mark) teach students the pronunciation of difficult terms while reinforcing chapter objectives. Anatomy and Physiology challenges students with as many new vocabulary terms as a first course in a foreign language. These tapes offer students the same advantages that students of foreign languages have traditionally had available to them.
ISBN: 0-13-340951-1

• **The Interactive Computerized Study Guide** (written by Jean Helgeson) is available in both software and CD platforms. It offers over 1500 interactive exercises in seven different question types including labeling exercises and matrices linking structure to function. Immediate feedback accompanies all answers, and incorrect answers generate additional reading assignments from the text.
ISBN: (IBM) 0-13-340944-9
(MAC) 0-13-34888-9-6

• **The Laboratory Manual** to accompany *FUNDAMENTALS OF ANATOMY AND PHYSIOLOGY*, 3/E (written by Roberta Meehan) is page-referenced to the text and includes over seventy different exercises to provide maximum flexibility. Thoroughly revised and reorganized form the last edition, this lab manual includes coverage of cat and fetal pig dissections. It also features a unique organizational system that enables students to place the lab exercises in context. A companion **Instructor's Manual** is also available.
Laboratory Manual ISBN: 0-13-341910-4

• **The New York Times Contemporary View** is a program sponsored jointly by Prentice Hall and *The New York Times* and is designed to enhance student access to current, relevant information. Articles are selected by the text author and compiled into a free supplement that helps students make the connection between the classroom and the outside world.

ACKNOWLEDGMENTS
This textbook is not the product of any single individual; rather, it represents a group effort. Foremost on the list stand the faculty and reviewers whose advice, comments, and collective wisdom helped to shape the text into its final form. Their interest in the subject, concern for the accuracy and method of presentation, and experience with students of widely varying abilities and backgrounds made the review process an educational experience. To these individuals, who carefully recorded their comments, opinions, and sources, I would like to express my sincere thanks and best wishes. The following individuals devoted large amounts of time to reviewing drafts of the third edition:

Maxine A'Hearn, *Prince George's Community College*
CeCe Barto, *Tomball College*
Steven Bassett, *Southeast Community College*
Robert Bauman, Jr., *Amarillo College*
Latsy Best, *Palm Beach Community College-North*
Charles Biggers, *University of Memphis*
Mimi Bres, *Prince George's Community College*
Wayne Carley, *Lamar University*
Suzzette Chopin, Texas A&M *University at Corpus Christi*
Richard Coppings, *Chattanooga State College*
Gene Carella, *Niagara Community College*
Charles Daniels, *Kapiolani Community College*
Ellen Dupre, *Indian River Community College*
Lee Famiano, *Cuyahoga Community College*
Marion Fintel, *Jefferson State Community College*
Ruby Fogg, *New Hampshire Tech*
Sharon Fowler, *Dutchess Community College*
Mildred Fowler, *Tidewater Community College*
Mildred Galliher, *Cochise College*
Mac E. Hadley, *University of Arizona*
William Hairston, *H.A.A.C.*
Cecil Hampton, *Jefferson College*
Ernest Harber, *San Antonio College*
Ann Harmer, *Orange Coast Community College*
Linden Haynes, *Hinds Community College*
Mary Healey, *Springfield College*
Jean Helgeson, *Collin County Community College*
James Horwitz, *Palm Beach Community College*
Beth Howard, *Rutgers University*
Yvette Huet-Hudson, *University of North Carolina Charlotte*
Angie Huxley, *Pima Community College*
Aaron James, *Gateway Community College*
Desiree Jett, *Essex County College*
Drusilla Jolly, *Forsyth Technical Community College*
David Kalichstein, *Ocean County College*
George Karleskint, *St. Louis Community College*
William Kleinelp, *Middlesex County College*
Mary Lockwood, *University of New Hampshire*
Greg Maravellas, *Bristol Community College*
Dan Mark, *Penn Valley Community College*
Jane Marks, *Scottsdale Community College*
Mike McCusker, *Eastern College*
Roberta Meehan, *University of Norther Colorado*
Alice Mills, *Middle Tennesee State University*
Ron Mobley, *Wake Technical Community College*
Rose Morgan, *Minot State University*
Aubrey Morris, *Pensacola Junior College*
Elizabeth Naugle, *Lane Community College*
Martha Newsome, *Tomball College*
Bill Nicholson, *University of Arkansas*
Mary Theresa Ortiz, *Kinsborough Community College*
David Parker, *Northern Virginia Community College*
Lois Peck, *Philadelphia College of Pharmacy*
Philip Penner, *Borough of Manhattan Community College*
Clifford Pohl, *Duquesne University*
Anil Rao, *Metro State College of Denver*
Kevin Ryan, *Stark Technical College*
Charles Seiger, *Atlantic Community College*
Mark Shoop, *Macon College*
Jeffery Smith, *Delgado Community College*
Eric Sun, *Macon College*
Caryl Tickner, *Stark Technical College*
Lucia Tranel, *St. Louis Community College and St. Louis College*
Steve Trautwein, *Southeast Missouri State University*
Pat Turner, *Howard Community College*
Jane Wallace, *Chattanooga State Technical Community College*

Eva Weinreb, *Community College of Philadelphia*
Rosamund Wendt, *Community College of Philadelphia*
Stephen Williams, *Glendale Community College*
Eric Wise, *Univeristy of California at Santa Barbara*
Jamie Young, *Chattanooga State Technical Community College*

Participants in the focus groups included:

Maxine A'Hearn, *Prince George's Community College*
Charles Biggers, *University of Memphis*
Gene Carella, *Niagara County Community College*
Robert M. Carey, *University of Arizona*
Kim Cooper, *Arizona State University*
Grant Dahmer, *University of Arizona*
Lee Farello, *Niagara County Community College*
Delaine Gilcrease, *Mesa Community College*
Aaron James, *Gateway Community College*
C. Ward Kischer, *University of Arizona*
Jane Marks, *Scottsdale Community College*
Melvin Mills, *Scottsdale Community College*
Robert L. Moskowitz, *Community College of Philadelphia*
David Parker, *Northern Virginia Community College*
Gary Quick, *Paradise Valley Community College*
Jean Rigden, *Scottsdale Community College*
Charles Seiger, *Atlantic Community College*
Philip Sokolove, *University of Maryland-Baltimore County*
Eric Sun, *Macon College*
Kathy Taylor, *University of Arizona*
Pat Turner, *Howard Community College*
Eva Weinreb, *Community College of Philadelphia*
Rosamund Wendt, *Community College of Philadelphia*

My gratitude is also extended to the many faculty and students at campuses across the country (and outside the country) who have provided suggestions and comments while the third edition was under development. The third edition was built on the solid foundation established by reviewers of earlier editions. I would like to acknowledge their assistance as well. Reviewers, focus group participants, and others involved with the preparation of the second edition included:

Maxine A'Hearn, *Prince George's Community College*
Debra Joan Barnes, *Contra Costa College*
Edwin Bartholomew, *Maui Community College*
Steven Bassett, *Southeast Community College*
Charles Biggers, *University of Memphis*
Cynthia Bottrell, *Scott Community College*
C. David Bridges, *Purdue University*
William M. Chamberlain, *Indiana State University*
William D. Chapple, *University of Connecticut*
O. D. Cockrum, *Texas State Technical College*
Gerald R. Dotson, *Front Range Community College*
John Dziak, *Community College of Allegheny*
Ann Funkhouser, *University of the Pacific*
Jeff Gerst, *North Dakota State University*
Kathleen M. Gorczyca, *North Shore Community College*
Ernest Joe Harber, *San Antonio College*
Ann Harmer, *Orange Coast Community College*
Jean A. S. Helgeson, *Collin County Community College*
Vickie S. Hennessy, *Sinclair Community College*
Donna Hoel, *Stark Technical College*
Elvis J. Holt, *Purdue University*
George Karleskint, *St. Louis Community College*
Susan Lustick, *San Jacinto College North*
Bob McDonough, *DeKalb College*
Roberta Meehan, *University of Northern Colorado*
Ann Miller, *Middlesex Community College*
Richard Mostardi, *Akron University*
Elizabeth Naugle, *Mission College*
Mark Paulissen, *Slippery Rock University*
Robert L. Preston, *Illinois State University*
Carolyn C. Robertson, *Tarrant County Junior College*
Charles Seiger, *Atlantic Community College*
Robert A. Sinclair, *San Antonio College*
David S. Smith, *San Antonio College*
Thomas S. Spurgeon, *Colorado State University*
Jay Templin, *Widener University*
Michael Wood, *Del Mar College*

Reviewers of the first edition included:

Paul Anderson, *Massachusetts Bay Community College*
Dean Beckwith, *Illinois Central College*
Doris Benfer, *Montgomery Community College*
Spencer R. Bowers, *Oakton Community College*
Sandra Bruner, *Polk Community College*
Ray Bruns, *Mesa Community College*
Anthony Chee, *Houston Community College*
Charles Daniels, *Kapiolani Community College*
Louis Giacinti, *Milwaukee Area Technical College*
Bonnie Gordon, *Memphis State University*
John P. Harley, *Eastern Kentucky University*
Jerri K. Lindsey, *Tarrant County Junior College*
Thomas W. McCort, *Cuyahoga Community College*
Ruth McFarland, *Mt. Hood Community College*
Richard Northrup, *Delta College*
Betty Orr, *Sinclair Community College*
David L. Parker, *Northern Virginia Community College*
Robert Pollack, *Nassau Community College*
Kristi Sather-Smith, *Hinds Junior College*
David Smith, *San Antonio College*
Richard Symmons, *California State University at Hayward*
Dennis Taylor, *Hiram College*
Kent M. Van De Graaff, *Brigham Young University*

Dan Schiller, Senior Development Editor at Prentice Hall, also played a vital role in the shaping of the third edition. His keen eye for detail was a great help in keeping the text organization, general tone, and level of presentation consistent throughout. The accuracy and currency of the clinical material in this edition and in the *Applications Manual* in large part reflects the detailed clinical reviews performed by my wife, Kathleen Welch, M.D. Virtually without exception, reviewers stressed the importance of accurate, integrated, and visually attractive illustrations in aiding the students to understand essential material. The revision of the art program was primarily directed by Bill Ober, M.D. and Claire Garrison, R.N. In many respects Dr. Ober also served as a reviewer, for he took the time to discuss the manuscript goals before making decisions about the accompanying figures. His suggestions concerning topics of clinical importance, presentation sequence, and revisions to the proposed art were of incalculable value to me and to the project in general.

Many of these figures include color photographs or micrographs collected from a variety of sources. Much of the work in tracking down these materials was performed by Stuart Kenter, whose efforts are greatly appreciated. Many of the light micrographs prepared by the author used commercially available slides and photomicrographic equipment generously provided by the Shoals Marine Laboratory of Cornell University. Mr. Howard Edgerton, Manager, Histology Slide Department, Carolina Biological Supply, earned my heartfelt thanks by allowing me to explore his department's extensive slide collection in Burlington, NC. Thanks are also due to Wards Scientific, for allowing me to use selected slides from their reference collection. Dr. Eugene C. Wasson, III and the staff of Maui Radiology Consultants, Inc. provided valuable assistance in selecting and printing the CT and MRI scans found in the Applications Manual. A special word of gratitude to Tobi Zausner for her fine eye and pertinacious nature, without which we would be lacking many of our best chapter opening photographs.

The author also wishes to express his appreciation to the editors and support staff at Prentice Hall who made the entire project possible. Special thanks are owed to Tim Bozik, Editor in Chief for Science and Mathematics, for his support of the project; David K. Brake, Biology Editor, for being the driving force and project coordinator; Ray Mullaney, Editor in Chief of ESM Development, for his calm and patient supervision; Linda Schreiber, Managing Editor, for her outstanding job in initiating and overseeing a highly complex supplements program; David Riccardi, Production Director, and Kathleen Schiaparelli, Managing Editor for ESM Production, for providing the extraordinary resources (human as well as technological) required to produce a book of this size and complexity; Debra Wechsler, Senior Production Editor, for somehow managing to keep people, text, and art moving in the proper direction at appropriate times; Commander John Jordan and his intrepid crew—Karen Noferi, Richard Foster, Shari Toron, and Eric Hulsizer in particular —for their extraordinary dedication and craftsmanship in shaping these pages; Paula Maylahn, Creative Director, for her supervision of the design, the art program, and much else; Barbara Marttine and Adrian Clapp, for their help with the art; Lorraine Patsco, for important insights, moral support, and the loan of key personnel at crucial junctures; Naomi Sysak, for keeping the supplements moving smoothly through production; and John Nestor, Veronica Schwartz, and Liz Nemeth for their indefatigable work on the *Applications Manual* (not just another supplement but a book in itself).

No one person could expect to produce a flawless textbook of this scope and complexity. Any errors or oversights are strictly my own, rather than those of the reviewers, artists, or editors. To help improve future editions, readers with pertinent information, suggestions, or comments concerning the organization or content of this textbook should send their remarks to: David Brake, Executive Editor for Biology, Prentice Hall, Inc., 1208 East Broadway, Suite 200, Tempe AZ 85282. Any and all comments and suggestions will be deeply appreciated, and carefully considered in the preparation of the fourth edition.

Frederic Martini

To the Student

Dear Student,

This text was designed to help you master the terminology and basic concepts of human anatomy and physiology. It should also make it possible for you to begin applying what you learn to every-day problems and situations. These aims have helped to shape every aspect of the book.

I have no doubt that you will find the study of anatomy and physiology to be one of the most interesting, challenging, and satisfying of all your educational experiences. Because the subject is so broad, extra care has been taken to give you every possible assistance in organizing new information. Many learning aids are built into the format of the text to make your study of this material easier and more rewarding. This book is meant to be a "machine for learning," one that can help you to focus your efforts and get the most from the time and energy you invest. I encourage you to examine this overview carefully and to consult your instructor if you have further questions about how to use this textbook.

Best wishes!

Frederic H. Martini

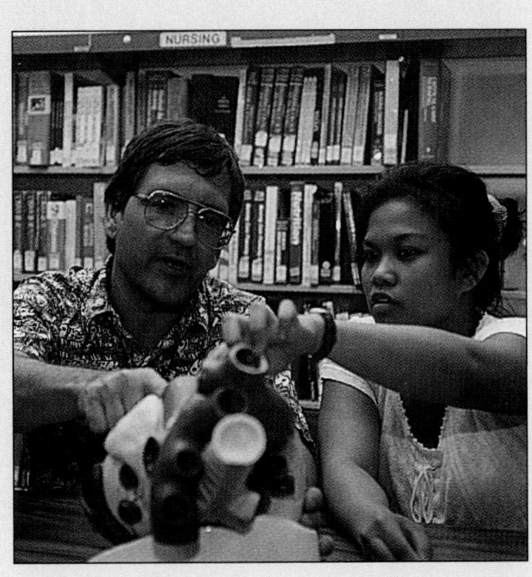
NURSING

Chapter Opening Photograph and Caption

Each chapter begins with an arresting photograph that dramatizes some aspect of the subject discussed in the chapter.

The accompanying narrative relates the photograph to themes that will be developed in the course of the chapter, providing you with a preview of the chapter's main subject matter.

CHAPTER

25

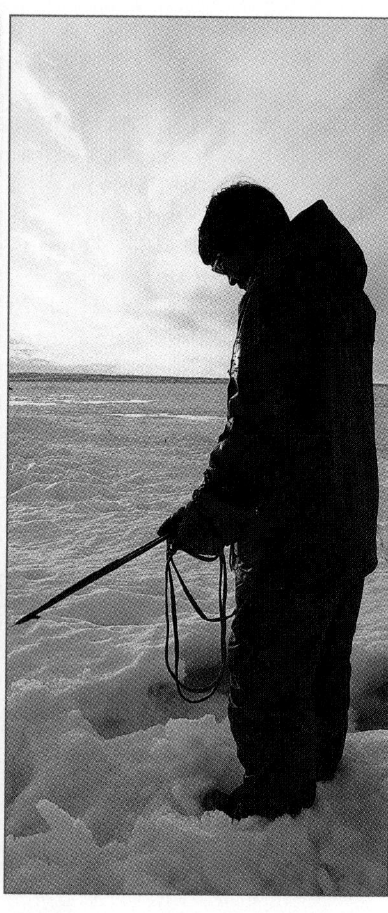

Human beings are remarkably adaptable animals— they can maintain homeostasis near the equator, where the year-round temperature averages 27°C, and near the poles, where the temperature remains below 0°C most of the time. The key to survival under such varied environmental conditions is the preservation of normal body temperature. In part this can be accomplished by shedding or adding layers of clothing, but hormonal and metabolic adjustments are also crucially important. Thus, the metabolic rate rises in cold climates, generating additional heat that warms the body. This chapter examines several aspects of metabolism, including nutrition, energy use, heat production, and the regulation of body temperature under normal and abnormal conditions.

Metabolism and Energetics

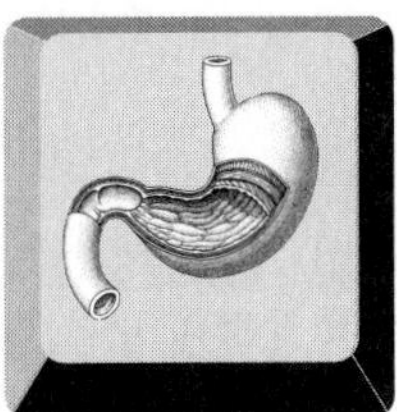

Chapter Outline and Objectives

Chapter Outline and Objectives

Before you begin using a chapter, it is important to know what it will cover and what you will be expected to learn from it. For this reason, every chapter opens with an **outline** that gives you an overview of its contents. Each entry in this outline corresponds to a major heading in the text, with a page number provided for easy reference.

The basic themes of the chapter are set forth in a series of **learning objectives.** These objectives will help you to structure your reading of the chapter and keep your attention focused on the key points. Each objective is placed under the heading where the relevant text material can be found, creating a "framework for learning" that can guide your study right from the start.

Clinical Discussions

Important **clinical topics** are presented in context, set off by an icon, title, and vertical color bar alongside the discussion. These topics have been selected not only for their medical importance but for their value in helping you to learn essential principles of anatomy and physiology. Many of them deal with diseases or therapies; they are designed to show how an understanding of abnormal conditions can shed light on normal function, and vice versa. (Discussions that focus on **personal health and fitness** rather than disease, diagnosis, or clinical practice are identified by a green jogger icon rather than a red caduceus.)

Key Terms

Throughout the text, the most important new terms are presented in **boldface** type. Whenever the **pronunciation** of a new term is not obvious, a pronunciation guide is presented in parentheses. In addition, the **derivations** of many terms are given, as an aid to memory and to help you recognize word roots that occur again and again in the language of anatomy and physiology.

Applications Manual References

Many additional topics of clinical significance are treated in the *Applications Manual* that accompanies each copy of this text. The *Applications Manual* also contains extensive and detailed coverage of clinical topics that are introduced in the text. Topics in the *Applications Manual* are identified in the text by the **AM icon,** followed by the title of the relevant discussion (in blue).

Concept Links

To understand anatomy and physiology you must remember many terms and be able to relate concepts introduced in different chapters. Concept links provide a quick visual signal that the new material being presented is related to or builds on earlier discussions. Each **"links" icon** is followed by a page reference to help you find the relevant material from an earlier section of the book.

The Epidermis 153

addition to forming the protective barrier in the stratum corneum, keratin is the basic structural component of hair and nails.

Stratum Corneum

As true keratin fibers are developing, the cells become thinner and flatter. The cell membranes thicken and become less permeable. The nuclei and organelles disintegrate, the cells die, and their subsequent dehydration creates a tightly interlocked layer of keratin sandwiched between phospholipid cell membranes. These layers of flattened and dead cells covering the epidermis form the **stratum corneum** (KOR-nē-um; *cornu*, horn). There are normally 15–30 layers of keratinized cells in the stratum corneum. *Keratinization* occurs everywhere on exposed skin surfaces except over the anterior surfaces of the eyes.

It takes 15–30 days for a cell to move from the stratum germinativum to the stratum corneum. The dead cells usually remain in the exposed stratum corneum layer for an additional two weeks before they are shed or washed away. This arrangement places the deeper portions of the epithelium and underlying tissues beneath a protective barrier composed of dead, durable, and expendable cells.

PERMEABILITY OF THE STRATUM CORNEUM Although the stratum corneum is water-resistant, it is not waterproof, and water from the interstitial fluids slowly penetrates the surface, to be evaporated into the surrounding air. Roughly 500 ml (about 1 pt) of water is lost in this way each day. This process is called **insensible perspiration** to distinguish it from the **sensible perspiration** produced by active sweat glands. Damage to the stratum corneum, as in a serious burn or *xerosis* (excessively dry skin) can drastically increase the rate of insensible perspiration and produce a potentially dangerous fluid loss. AM *Disorders of Keratin Production*

When the skin is immersed in water, osmotic forces may move water into or out of the epithelium. ∞ [p. 75] Sitting in a freshwater bath causes water to move into the epidermis because fresh water is hypotonic (has fewer dissolved materials) compared with body fluids. The epithelial cells may swell to four times their normal volumes, a phenomenon particularly noticeable in the thickly keratinized areas of the palms and soles. Swimming in the ocean reverses the direction of osmotic flow, because the ocean is a hypertonic solution. As a result, water leaves the body, crossing the epidermis from the underlying tissues. The process is slow, but long-term exposure to seawater endangers survivors of a shipwreck by accelerating dehydration.

DRUG ADMINISTRATION THROUGH THE SKIN Drugs in oils or other lipid-soluble carriers can penetrate the epidermis. The movement is slow, particularly through the layers of cell membranes in the stratum corneum, but once a drug reaches the underlying tissues it will be absorbed into the circulation. Placing a drug in a solvent that is lipid-soluble can assist its movement through the lipid barriers. Drugs can also be administered by packaging them in *liposomes*, artificially produced lipid droplets. Liposomes containing DNA fragments have been used experimentally to introduce normal genes into abnormal human cells. For example, genes carried by liposomes have been used to alter the cell membranes of skin cancer cells so that they will be attacked by the immune system. Another experimental procedure involves creating a transient change in skin permeability by administering a brief pulse of electricity. This electrical pulse temporarily changes the positions of the cells in the stratum corneum, creating channels that allow the drug to penetrate.

A useful technique for long-term drug administration involves placing a sticky patch containing a drug over an area of thin skin. To overcome the relatively slow rate of diffusion, the patch must contain an extremely high concentration of the drug. This procedure, called *transdermal administration*, has the advantage that a single patch may work for several days, making daily pills unnecessary. AM *Transdermal Medications*

- ✓ Excessive shedding of cells from the outer layer of skin in the scalp causes dandruff. What is the name of this layer of skin?
- ✓ As you pick up a piece of lumber, a splinter pierces the palm of your hand and lodges in the third layer of the epidermis. Identify this layer.
- ✓ Why does swimming in fresh water for an extended period cause epidermal swelling?
- ✓ Some criminals sand the tips of their fingers so as not to leave recognizable fingerprints. Would this practice permanently remove fingerprints? Why or why not?

❑ Skin Color

The color of the skin is due to an interaction between: (1) pigment composition and concentration and (2) the dermal blood supply.

Skin Pigmentation

Figure 5-4

The epidermis contains variable quantities of two pigments, *carotene* and *melanin*.

- **Carotene** (KAR-ō-tēn) is an orange-yellow pigment that normally accumulates inside epidermal cells; it is most apparent in cells of the stratum corneum of light-skinned individuals. This pigment is also distributed in fatty tissues

Concept Check Questions

These questions, located at the ends of major sections in each chapter, encourage you to stop and assess your understanding of the basic concepts presented in the preceding pages. Because concepts build upon one another, you need to be sure you have mastered important material before going on to the next section. These questions are designed to be answered quite quickly and easily if you have grasped the preceding discussion. If the questions seem difficult, you probably need to review that material before moving on. **Answers** to the Concept Check Questions are provided in Appendix I at the end of the text.

Figure Numbers Accompanying Major Text Headings

To enhance the coordination between text and art, each of the major headings includes a **list of figures discussed** in the text that follows. This feature will be especially important as you both preview and review the material in a chapter.

Boxes

Clinical or health-related topics of particular importance are presented in Boxes set off from the main text. These essays cover major diseases, such as breast cancer, and other subjects of special interest, such as the abuse of steroid drugs. Discussions that are primarily clinical in their focus are identified by a red **caduceus icon;** those dealing with personal health and fitness are marked by a green **jogger icon.**

Drugs and Synaptic Function

Many drugs interfere with key steps in the process of synaptic transmission. These drugs may (1) interfere with transmitter synthesis, (2) alter the rate of transmitter release, (3) prevent transmitter inactivation, or (4) prevent transmitter binding to receptors. The discussion that follows is limited to clinically important compounds that exert their effects at cholinergic synapses. Their sites of activity are indicated in Figure 12-22●.

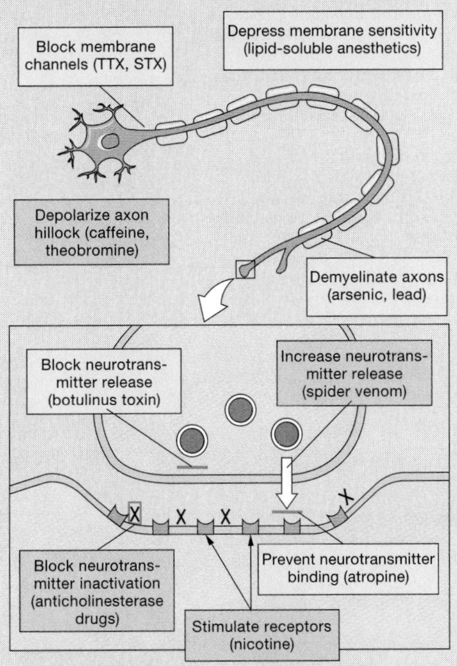

Block membrane channels (TTX, STX)
Depress membrane sensitivity (lipid-soluble anesthetics)
Depolarize axon hillock (caffeine, theobromine)
Demyelinate axons (arsenic, lead)
Block neurotransmitter release (botulinus toxin)
Increase neurotransmitter release (spider venom)
Block neurotransmitter inactivation (anticholinesterase drugs)
Prevent neurotransmitter binding (atropine)
Stimulate receptors (nicotine)

● **FIGURE 12-22**
Mechanism of Drug Action at a Cholinergic Synapse. Factors that facilitate neural function and make neurons more excitable are shown in violet. Factors that inhibit or depress neural function are shown in blue.

Botulinus toxin is responsible for the primary symptom of *botulism*, a widespread paralysis of skeletal muscles. [p. 295] Botulinus toxin blocks the release of ACh at the presynaptic membrane of cholinergic neurons. The venom of the black widow spider has the opposite effect. It causes a massive release of ACh that produces intense muscular cramps and spasms.

Anticholinesterase drugs, sometimes called *cholinesterase inhibitors*, block the breakdown of ACh by acetylcholinesterase. The result is an exaggerated and prolonged stimulation of the postsynaptic membrane. At the neuromuscular junctions, this abnormal stimulation produces an extended and extreme state of contraction. Military nerve gases block cholinesterase activity for weeks, although few persons exposed are likely to live long enough to regain normal synaptic function. Most animals utilize ACh as a neurotransmitter, and anticholinesterase drugs, such as *malathion*, are in widespread use in pest-control projects.

Drugs such as **atropine** or **d-tubocurarine** prevent ACh from binding to the postsynaptic receptors. The latter compound is a derivative of *curare*, a plant extract used by certain South American tribes to paralyze their prey. Curare and related compounds induce paralysis by preventing stimulation of the neuromuscular junction by ACh. Atropine can also be administered intentionally to counteract the effects of anticholinesterase poisoning. Other compounds, including **nicotine,** an active ingredient in cigarette smoke, bind to the receptor sites and stimulate the postsynaptic membrane. There are no enzymes to remove these compounds, and the effects are relatively prolonged. *Neuroactive Drugs*

serotonin, histamine, GABA, and any of the endorphins bind to membrane receptors, the receptors activate an intracellular enzyme, **adenylate cyclase,** also known as *adenylyl* (or *adenyl*) *cyclase*. This enzyme catalyzes the formation of **cyclic-AMP** from ATP at the inner surface of the cell membrane. Cyclic-AMP, or **cAMP,** is a **second messenger** that may open membrane channels and/or activate intracellular enzymes, depending on the nature of the postsynaptic cell. Other second messengers, including cGMP, calcium ions, and *inositol trisphosphate* will be encountered in later chapters.

3. Gases such as NO and CO are lipid-soluble so that they can diffuse across the cell membrane and bind to enzymes inside the cell. These enzymes then promote the appearance of second messengers that can have various effects on cellular activity.

It can be very difficult to distinguish between neurotransmitters and neuromodulators on either biochemical or functional grounds. In general, neuromodulators (1) have long-term effects that are relatively slow to appear, (2) trigger responses that involve second messengers, (3) may affect the presyn-

Focus Features

The focus format is designed to present text, art, and tabular material in an integrated manner. The aim is to help you assimilate complex information as efficiently as possible.

Figure Reference Locators

Studying a highly visual subject like anatomy and physiology involves a great deal of looking back and forth between the text and the numerous figures. **Red dots** next to figure references in the text serve as place markers, making it easy to return to your spot in the narrative after consulting a figure. The same red dots precede each figure caption, tying the art and text together.

488

The Facial Nerve (N VII)

Figure 14-22

Primary function: Mixed (sensory and motor) to face

Origin: Sensory from taste receptors on anterior two-thirds of tongue; motor from motor nuclei of pons

Passes through: Internal acoustic canal of temporal bone, along internal acoustic canal and facial canal to reach stylomastoid foramen [p. 210]

Destination: Sensory to sensory nuclei of pons. *Somatic motor:* muscles of facial expression. [p. 334] *Visceral motor:* lacrimal (tear) gland and nasal mucous glands via sphenopalatine ganglion; submandibular and sublingual salivary glands via submandibular ganglion

The **facial nerve** is a mixed nerve. The cell bodies of the sensory neurons are located in the **geniculate ganglion,** and the motor nuclei are in the pons. The sensory and motor roots emerge from the side of the pons and enter the internal acoustic canal of the temporal bone (Figure 14-22●). The nerve then passes through the facial canal to reach the face via the stylomastoid foramen. [p. 210] The sensory neurons monitor proprioceptors in the facial muscles, provide deep pressure sensations over the face, and receive taste information from receptors along the anterior two-thirds of the tongue. Somatic motor fibers control the superficial muscles of the scalp and face and deep muscles near the ear.

The facial nerve carries preganglionic autonomic fibers to the *sphenopalatine* and *submandibular ganglia.* Postganglionic fibers from the **sphenopalatine ganglion** innervate the lacrimal gland and small glands of the nasal cavity and pharynx. The **submandibular ganglion** innervates the *submandibular* and *sublingual* (*sub,* under + *lingua,* tongue) salivary glands.

Bell's palsy is a cranial nerve disorder that results from an inflammation of the facial nerve. The condition probably results from a viral infection. Symptoms include paralysis of facial muscles on the affected side and loss of taste sensations from the anterior two-thirds of the tongue. The condition is usually painless, and in most cases, Bell's palsy "cures itself" after a few weeks or months.

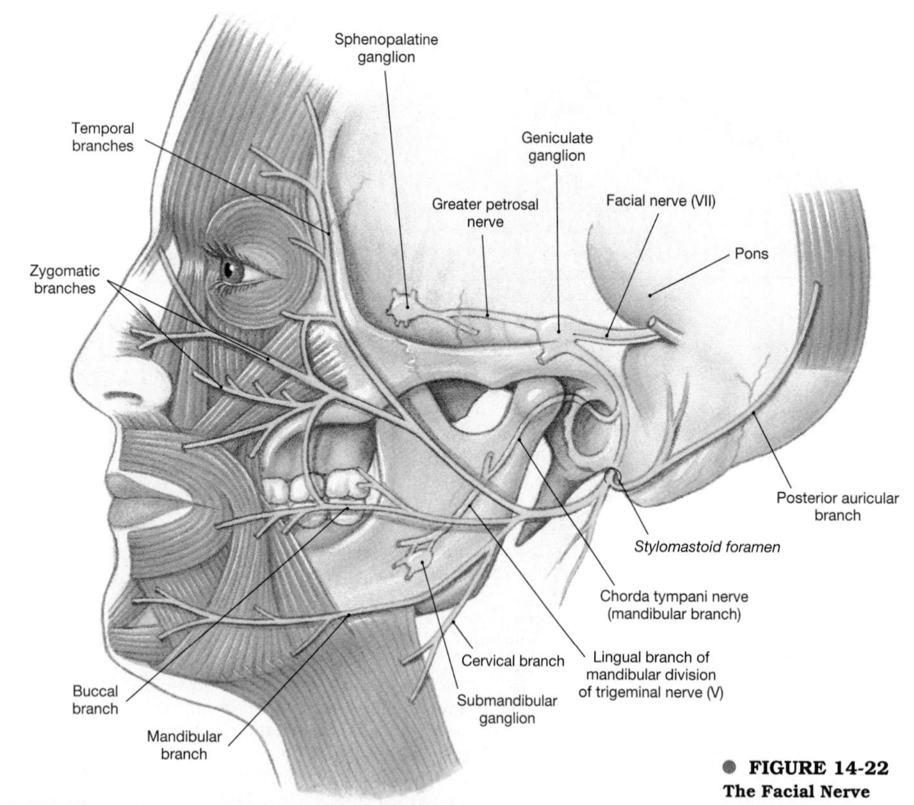
Sphenopalatine ganglion
Temporal branches
Geniculate ganglion
Greater petrosal nerve
Facial nerve (VII)
Pons
Zygomatic branches
Posterior auricular branch
Stylomastoid foramen
Chorda tympani nerve (mandibular branch)
Cervical branch
Lingual branch of mandibular division of trigeminal nerve (V)
Buccal branch
Submandibular ganglion
Mandibular branch

● **FIGURE 14-22**
The Facial Nerve

A **demonstration icon** is used to identify figures for which there is a related animation on the **Laser Disc** or **Video Tutor.**

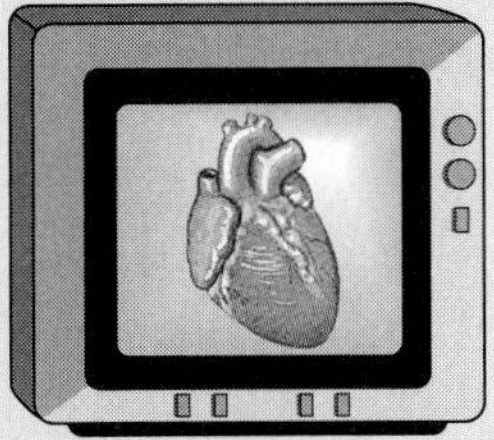

The **Video Tutor**, a 75-minute VCR tape containing numerous three-dimensional animations, allows you to visualize complex physiological processes as they unfold in time. Simultaneous narration and an accompanying frame-referenced booklet assist you in study and review. (Order ISBN 0-13-340969-4 or ask your instructor.)

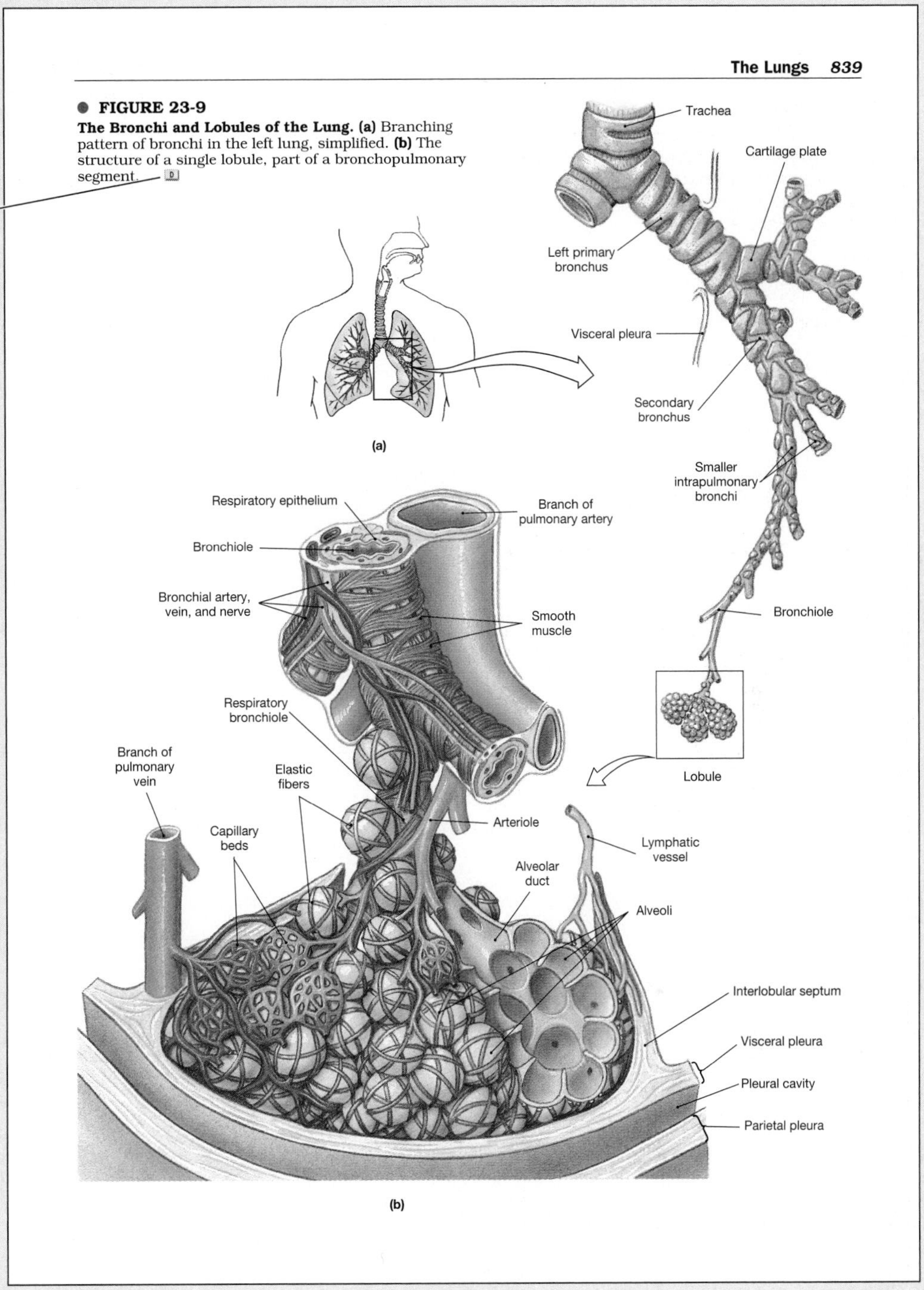

The Lungs 839

● **FIGURE 23-9**
The Bronchi and Lobules of the Lung. (a) Branching pattern of bronchi in the left lung, simplified. **(b)** The structure of a single lobule, part of a bronchopulmonary segment.

Trachea
Cartilage plate
Left primary bronchus
Visceral pleura
Secondary bronchus
Smaller intrapulmonary bronchi
Bronchiole
Lobule
(a)

Respiratory epithelium
Branch of pulmonary artery
Bronchiole
Bronchial artery, vein, and nerve
Smooth muscle
Respiratory bronchiole
Branch of pulmonary vein
Elastic fibers
Capillary beds
Arteriole
Lymphatic vessel
Alveolar duct
Alveoli
Interlobular septum
Visceral pleura
Pleural cavity
Parietal pleura
(b)

Art

To understand anatomy and physiology you must be able to visualize structures in the human body at several different scales. Many figures in the text are designed help you bridge the gap between the relatively familiar macroscopic world and the much less familiar microscopic world. Multiple **breakouts** show the relationship between different levels of structure. Cells, for example, are shown as parts of tissues, and tissues as parts of organs. **Orientation figures** are used extensively to show the location of important bones and organs in the body.

Embryology Summaries

Located throughout the book, these features use a combination of art and text to present the formation of significant organs, structures, and systems during **embryological and fetal development.** The embryology summaries are entirely self-contained so that they can be integrated with the chapter, studied separately, or omitted as you and your instructor see fit.

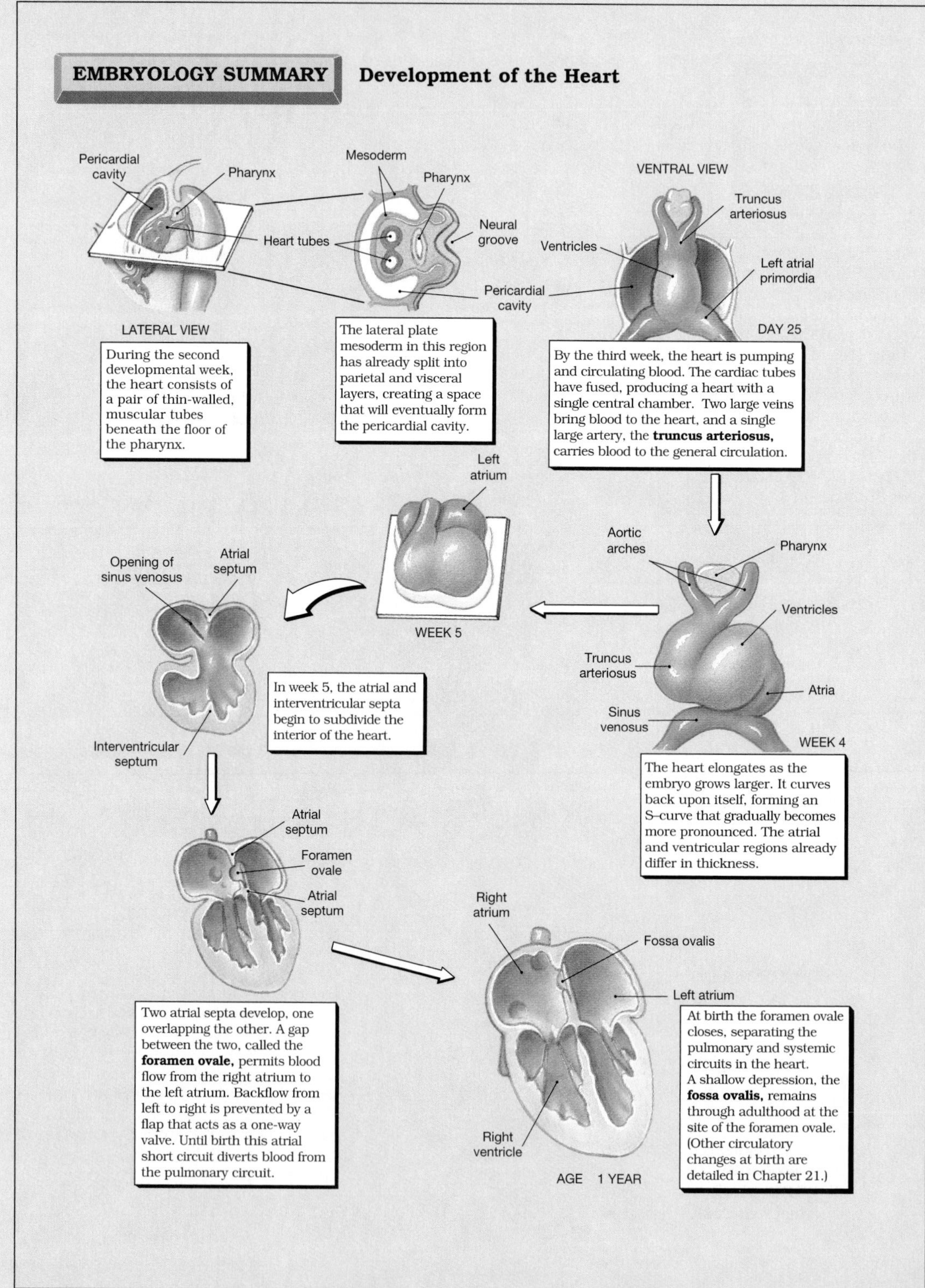

EMBRYOLOGY SUMMARY
Development of the Heart
Pericardial cavity
Pharynx
Mesoderm
Pharynx
Heart tubes
Neural groove
Pericardial cavity
VENTRAL VIEW
Truncus arteriosus
Ventricles
Left atrial primordia
DAY 25
LATERAL VIEW
During the second developmental week, the heart consists of a pair of thin-walled, muscular tubes beneath the floor of the pharynx.
The lateral plate mesoderm in this region has already split into parietal and visceral layers, creating a space that will eventually form the pericardial cavity.
By the third week, the heart is pumping and circulating blood. The cardiac tubes have fused, producing a heart with a single central chamber. Two large veins bring blood to the heart, and a single large artery, the **truncus arteriosus,** carries blood to the general circulation.
Aortic arches
Pharynx
Ventricles
Truncus arteriosus
Atria
Sinus venosus
WEEK 4
The heart elongates as the embryo grows larger. It curves back upon itself, forming an S-curve that gradually becomes more pronounced. The atrial and ventricular regions already differ in thickness.
Left atrium
WEEK 5
Opening of sinus venosus
Atrial septum
Interventricular septum
In week 5, the atrial and interventricular septa begin to subdivide the interior of the heart.
Atrial septum
Foramen ovale
Atrial septum
Two atrial septa develop, one overlapping the other. A gap between the two, called the **foramen ovale,** permits blood flow from the right atrium to the left atrium. Backflow from left to right is prevented by a flap that acts as a one-way valve. Until birth this atrial short circuit diverts blood from the pulmonary circuit.
Right atrium
Fossa ovalis
Left atrium
Right ventricle
AGE 1 YEAR
At birth the foramen ovale closes, separating the pulmonary and systemic circuits in the heart.
A shallow depression, the **fossa ovalis,** remains through adulthood at the site of the foramen ovale. (Other circulatory changes at birth are detailed in Chapter 21.)

System Integrators

These figures, found at the end of each systems chapter, show how the system under consideration interacts with each of the body's other systems. They thus reinforce the concept that the body functions as an integrated unit.

The interactions among systems are color-coded:

- tan is used to highlight the contributions of the system in question to each of the body's other systems
- blue is used for the contributions of other systems to the one being studied.

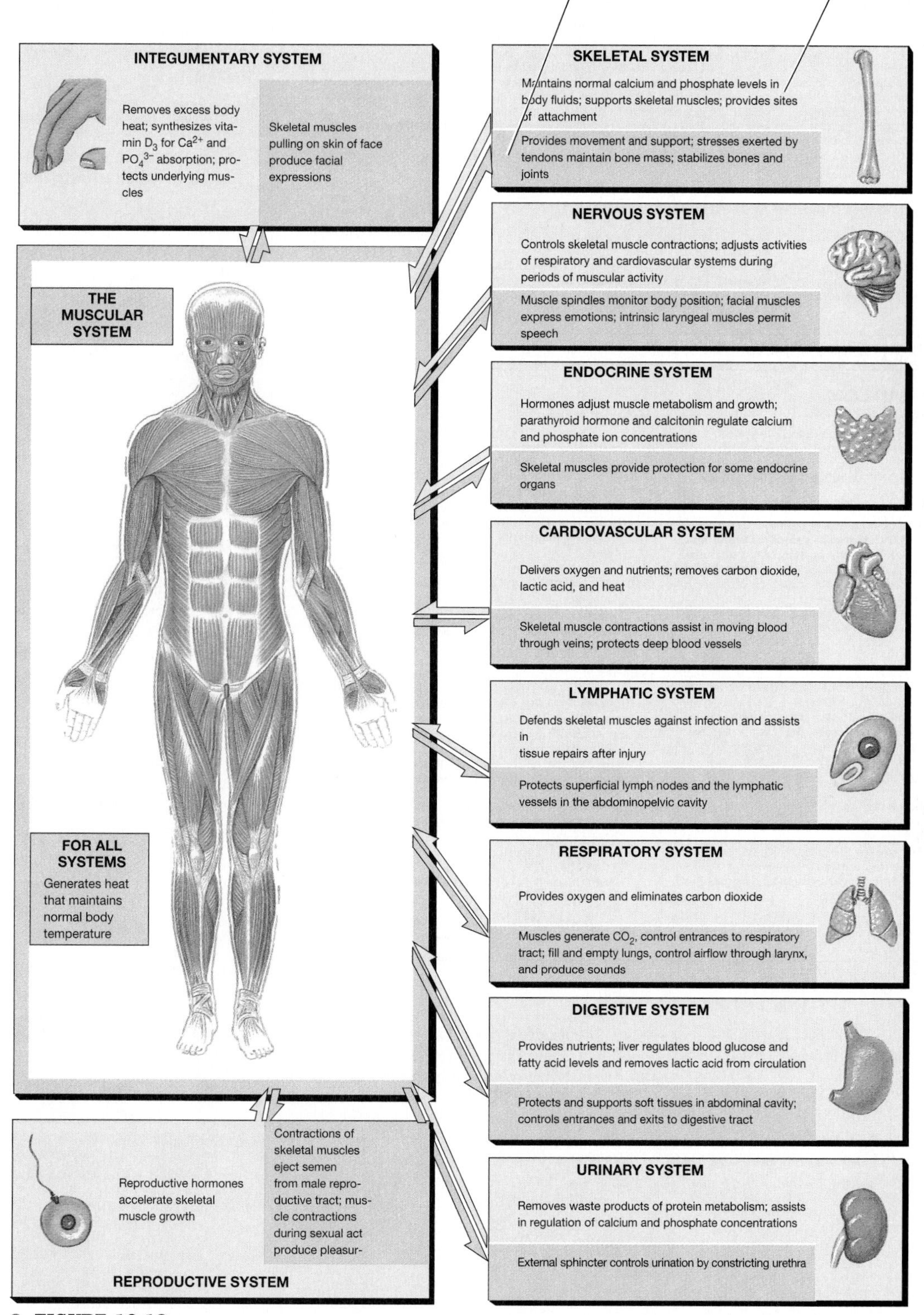

INTEGUMENTARY SYSTEM
Removes excess body heat; synthesizes vitamin D_3 for Ca^{2+} and PO_4^{3-} absorption; protects underlying muscles
Skeletal muscles pulling on skin of face produce facial expressions
THE MUSCULAR SYSTEM
FOR ALL SYSTEMS
Generates heat that maintains normal body temperature
SKELETAL SYSTEM
Maintains normal calcium and phosphate levels in body fluids; supports skeletal muscles; provides sites of attachment
Provides movement and support; stresses exerted by tendons maintain bone mass; stabilizes bones and joints
NERVOUS SYSTEM
Controls skeletal muscle contractions; adjusts activities of respiratory and cardiovascular systems during periods of muscular activity
Muscle spindles monitor body position; facial muscles express emotions; intrinsic laryngeal muscles permit speech
ENDOCRINE SYSTEM
Hormones adjust muscle metabolism and growth; parathyroid hormone and calcitonin regulate calcium and phosphate ion concentrations
Skeletal muscles provide protection for some endocrine organs
CARDIOVASCULAR SYSTEM
Delivers oxygen and nutrients; removes carbon dioxide, lactic acid, and heat
Skeletal muscle contractions assist in moving blood through veins; protects deep blood vessels
LYMPHATIC SYSTEM
Defends skeletal muscles against infection and assists in tissue repairs after injury
Protects superficial lymph nodes and the lymphatic vessels in the abdominopelvic cavity
RESPIRATORY SYSTEM
Provides oxygen and eliminates carbon dioxide
Muscles generate CO_2, control entrances to respiratory tract; fill and empty lungs, control airflow through larynx, and produce sounds
DIGESTIVE SYSTEM
Provides nutrients; liver regulates blood glucose and fatty acid levels and removes lactic acid from circulation
Protects and supports soft tissues in abdominal cavity; controls entrances and exits to digestive tract
URINARY SYSTEM
Removes waste products of protein metabolism; assists in regulation of calcium and phosphate concentrations
External sphincter controls urination by constricting urethra
Reproductive hormones accelerate skeletal muscle growth
Contractions of skeletal muscles eject semen from male reproductive tract; muscle contractions during sexual act produce pleasur-
REPRODUCTIVE SYSTEM

● **FIGURE 10-19**
Functional Relationships between the Muscular System and Other Systems

Selected Clinical Terminology

Terms Discussed in This Chapter

carpal tunnel syndrome: An inflammation of the sheath surrounding the flexor tendons of the palm. *(p. 360)*
compartment syndrome: Ischemia resulting from accumulated blood and fluid trapped within a musculoskeletal compartment. *(p. 374)*
diaphragmatic hernia (hiatal hernia): A hernia that occurs when abdominal organs slide into the thoracic cavity. *(p. 347)*
hernia: A condition involving an organ or body part that protrudes through an abnormal opening. *(p. 347)*
inguinal hernia: A condition in which the inguinal canal enlarges and abdominal contents are forced into it. *(p. 347)*
intramuscular (IM) injection: Administering a drug by injecting it into the mass of a large skeletal muscle. *(p. 369)*
ischemia (is-KĒ-mē-a): A condition of "blood starvation" resulting from compression of regional blood vessels. *(p. 374)*
rotator cuff: The muscles that surround the shoulder joint; a frequent site of sports injuries. *(p. 352)*

AM ***Additional Terms Discussed in The Application Manual***

bone bruise: Bleeding within the periosteum of a bone.
bursitis: Inflammation of the bursae around one or more joints.
muscle cramps: Prolonged, involuntary, painful muscular contractions.
sprains: Tears or breaks in ligaments or tendons.
strains: Tears or breaks in muscles.
stress fractures: Cracks or breaks in bones subjected to repeated stresses or trauma.
tendinitis: Inflammation of the connective tissue surrounding a tendon.

CHAPTER REVIEW

STUDY OUTLINE

INTRODUCTION, p. 326

1. The **muscular system** includes all the skeletal muscle tissue that can be controlled voluntarily.

BIOMECHANICS AND MUSCLE ANATOMY, p. 326

Organization of Skeletal Muscle Fibers, p. 326

1. A muscle can be classified according to the arrangement of fibers and fascicles as a **parallel muscle, convergent muscle, pennate muscle,** or **circular muscle (sphincter).** A pennate muscle may be **unipennate, bipennate,** or **multipennate.** *(Figure 11-1)*

Skeletal Muscle Length-Tension Relationships, p. 327

Levers, p. 328

2. A **lever** can change the direction, speed, or distance of muscle movements and modify the force applied to muscles.
3. Levers may be classified as **first-class, second-class,** or **third-class levers;** the last are the most common type of lever in the body. *(Figure 11-2)*

MUSCLE TERMINOLOGY, p. 329

Origins and Insertions, p. 329

Actions, p. 329

1. Each muscle may be identified by its **origin, insertion,** and **primary action.** A muscle may be classified as a **prime mover** or **agonist,** a **synergist,** an **antagonist,** or a **fixator.**

Names of Skeletal Muscles, p. 330

2. The names of muscles often provide clues to their location, orientation, or function. *(Table 11-1)*

Axial and Appendicular Muscles, p. 330

3. The **axial musculature** arises on the axial skeleton; it positions the head and spinal column and moves the rib cage. The **appendicular skeleton** stabilizes or moves components of the appendicular skeleton. *(Figure 11-3)*

THE AXIAL MUSCULATURE, p. 334

1. The axial muscles fall into logical groups based on location or function or both.
2. **Innervation** refers to the identity of the nerve that controls a muscle.

Muscles of the Head and Neck, p. 334

3. The muscles of facial expression are the **orbicularis oris, buccinator, epicranius (frontalis** and **occipitalis),** and **platysma.** *(Figure 11-4; Table 11-2)*
4. Six **extrinsic eye muscles (oculomotor muscles)** control external eye movements: the **inferior** and **superior rectus, lateral** and **medial rectus,** and **superior** and **inferior obliques.** *(Figure 11-5; Table 11-3)*
5. The muscles of mastication (chewing) are the **masseter, temporalis,** and **pterygoid.** *(Figure 11-6; Table 11-4)*
6. The muscles of the tongue are necessary for speech and swallowing and assist in mastication. They are the **genioglossus, hyoglossus, palatoglossus,** and **styloglossus.** *(Figure 11-7; Table 11-5)*

Selected Clinical Terminology

The first part of this section reviews the definitions and (where necessary) the pronunciations of the most important medical terms discussed **in the chapter.** Page references are provided for easy review.

The second part of this section gives definitions of medical terms that appear in related discussions **in the *Applications Manual.***

Study Outline

The Study Outline provides you with a concise but detailed **summary** of the chapter's major facts and concepts.

The short numbered statements are organized under the **main chapter headings,** which appear with their **page references.** Thus you can easily review the organization of the chapter as well as its contents.

Terms that are boldfaced in the text appear in **boldface** in the summary.

The **figures and tables** relevant to each section are also listed.

Audio Study Tapes help you learn the pronunciation of difficult terms while reinforcing chapter objectives. (Order ISBN 0-13-34095-1 or ask your instructor.)

Level 1 questions allow you to test your recall of the chapter's basic information and terminology.

Review Questions

Questions are divided into **three learning levels** to guide your progress toward increasing mastery of the material.

424 Chapter 12 Neural Tissue

REVIEW QUESTIONS

LEVEL 1 **Reviewing Facts and Terms**

1. Regulation by the nervous system provides:
   (a) relatively slow but long-lasting responses to stimuli
   (b) swift, long-lasting responses to stimuli
   (c) swift but brief responses to stimuli
   (d) relatively slow, short-lived responses to stimuli
2. The efferent division of the PNS:
   (a) brings sensory information to the CNS
   (b) carries motor commands to muscles and
   (c) processes and integrates sensory data
   (d) is the seat of higher functions in the b
3. The part of the nervous system that prov untary control over skeletal muscle contraction
   (a) somatic nervous system
   (b) autonomic nervous system
   (c) visceral motor system
   (d) sympathetic division of the ANS
4. Smooth muscle, cardiac muscle, and gla among the targets of the:
   (a) somatic nervous system
   (b) sensory neurons
   (c) afferent division of the PNS
   (d) autonomic nervous system
5. The cellular layer that lines the ventricle brain and the central canal of the spinal cord
   (a) microglia (b) ganglia
   (c) ependyma (d) oligodendrocyt
6. Glial cells responsible for maintaining th brain barrier are the:
   (a) microglia (b) ependymal cell
   (c) astrocytes (d) oligodendrocyt
7. Phagocytic cells found in neural tissue of are called:
   (a) astrocytes (b) ependymal cell
   (c) oligodendrocytes (d) microglia
8. Substances being transported from an axo nal to the soma of the same neuron are deliver
   (a) axoplasmic transport
   (b) synaptic conduction
   (c) retrograde flow
   (d) active transport
9. All the motor neurons that control skeletal mu
   (a) multipolar neurons
   (b) myelinated bipolar neurons
   (c) unipolar, unmyelinated sensory neuro
   (d) anaxonic neurons
10. The type of neural cells responsible for the of sensory inputs and coordination of motor out
   (a) neuroglia (b) interneurons
   (c) sensory neurons (d) motor neurons
11. Depolarization of a neuron cell membrane the membrane potential toward:
   (a) 0 mV (b) –70 mV
   (c) –90 mV (d) a, b, and c are
12. The primary determinant of the resting me potential is the:
   (a) membrane permeability to sodium
   (b) membrane permeability to potassium
   (c) intracellular negatively charged proteins
   (d) negatively charged chloride ions in the ECF
13. Receptors that bind acetylcholine at the motor end plate are:
   (a) chemically regulated channels

Level 2 questions help you to check your grasp of concepts and your ability to integrate ideas presented in different parts of the chapter.

Chapter Review 425

27. What two integrated steps are necessary for transfer of nerve impulses from neuron to neuron?
28. What is the *functional* difference between voltage-regulated, chemically regulated, and mechanically regulated channels?
29. What four basic characteristics are associated with graded potentials?
30. State the all-or-none principle regarding action potentials.
31. Describe the four steps involved in the generation of an action potential.
32. Describe the four steps that take place at a typical cholinergic synapse.
33. What environmental factors play a role in altering the function of neurons in the CNS and PNS?

LEVEL 2 **Reviewing Concepts**

34. If the resting membrane potential is –70 mV and the threshold is –55 mV, a membrane potential of –60 mV will:
   (a) produce an action potential
   (b) make it easier to produce an action potential
   (c) make it harder to produce an action potential
   (d) hyperpolarize the membrane
35. A graded potential:
   (a) decreases with distance from the point of stimulation
   (b) spreads passively because of local currents
   (c) may involve either depolarization or hyperpolarization
   (d) a, b, and c are correct
36. For an action potential to begin, an area of excitable membrane must:
   (a) have its voltage-regulated gates inactivated
   (b) be hyperpolarized
   (c) be depolarized to threshold level
   (d) not be in a relative refractory period
37. During an absolute refractory period, the membrane:
   (a) continues to hyperpolarize
   (b) cannot respond to further stimulation
   (c) can respond to a larger-than-normal depolarizing stimulus
   (d) will respond to summated stimulation
38. Action potentials are restricted to areas of excitable membranes that contain voltage-regulated channels.
   (a) true (b) false
39. The all-or-none principle states that:
   (a) the properties of an action potential are independent of the strength of the depolarizing stimulus
   (b) all stimuli will produce action potentials
   (c) all graded potentials will generate action potentials
   (d) any cell membrane can generate and propagate an action potential if stimulated to threshold
40. The loss of positive ions from the interior of a neuron produces:
   (a) depolarization (b) threshold
   (c) hyperpolarization (d) an action potential
41. Continuous propagation of an action potential cannot occur in:
   (a) myelinated axons (b) unmyelinated axons
   (c) Type A fibers (d) Type B fibers
42. Why can't neurons in the CNS be replaced when they are lost to injury or disease?
43. What purpose do collaterals serve in the nervous system?
44. What is the difference between axoplasmic transport and retrograde flow?
45. What is the difference between continuous propagation and saltatory propagation?
46. How does an action potential in a skeletal muscle fiber differ from one in a neuron?
47. How does the action of a neurotransmitter differ from that of a neuromodulator?
48. What is the difference between temporal summation and spatial summation?
49. How does the neuronal activity of divergence differ from that of convergence?
50. What functions of neurons necessitate the support of energy from ATP?
51. Why is an electrical synapse a more efficient carrier of nerve impulses from cell to cell than a chemical synapse?
52. When a runner experiences the "runner's high," why is the suppression of pain common?
53. Multiple sclerosis (MS) is a demyelination disorder. How does this condition produce muscular paralysis and sensory losses?

LEVEL 3 **Critical Thinking and Clinical Applications**

54. If neurons in the central nervous system lack centrioles and are unable to divide, how can a person develop brain cancer?
55. Harry suffers from a kidney condition that causes changes in his body's electrolyte levels (concentration of ions in the extracellular fluid). As a result of this problem, he is exhibiting tachycardia, an abnormally fast heart rate. What ion is involved, and how does a change in its concentration cause Harry's symptoms?
56. Twenty neurons synapse with a single receptor neuron. Fifteen of these neurons release neurotransmitters that produce EPSPs at the postsynaptic membrane, and the other five release neurotransmitters that produce IPSPs. Each time one of the neurons is stimulated, it releases enough neurotransmitter to produce a 2-mV change in potential at the postsynaptic membrane. If the threshold of the postsynaptic neuron is 10 mV, how many of the excitatory neurons must be stimulated to produce an action potential in the receptor neuron if all five inhibitory neurons are stimulated? (Assume that spatial summation occurs.)
57. Myelination of peripheral neurons occurs rapidly through the first year of life. How can this process explain the increased abilities of infants during their first year?
58. A drug that blocks ATPase is introduced into an experimental neuron preparation. The neuron is then repeatedly stimulated and recordings are made of the response. What effect would you expect to observe?

Level 3 questions let you develop your powers of reasoning and analysis by applying chapter material to plausible real-world and clinical situations.

A comprehensive **Study Guide** is available to help you master the material in each chapter. Utilizing the same three-level system employed in the text, the *Study Guide* includes questions in a variety of formats, including matching, multiple choice, completion, essay, labeling, and concept mapping. (Order ISBN 0-13-340936-8 or ask your instructor.)

A unique **Interactive Computerized Study Guide,** available as a CD-ROM, offers over 1500 **interactive exercises** in 7 different formats, including labeling exercises and structure/ function matching. All answers generate immediate feedback, including page references for reviewing relevant information in the text in response to wrong answers. (Order ISBN 0-13-340944-9 or ask your instructor.)

Applications Manual

The *Applications Manual* that accompanies each copy of this text provides a wealth of supplemental material to enrich your learning experience. Although all essential principles of anatomy and physiology are covered in the text, the *Applications Manual* allows you to explore many clinical and diagnostic topics in greater depth.

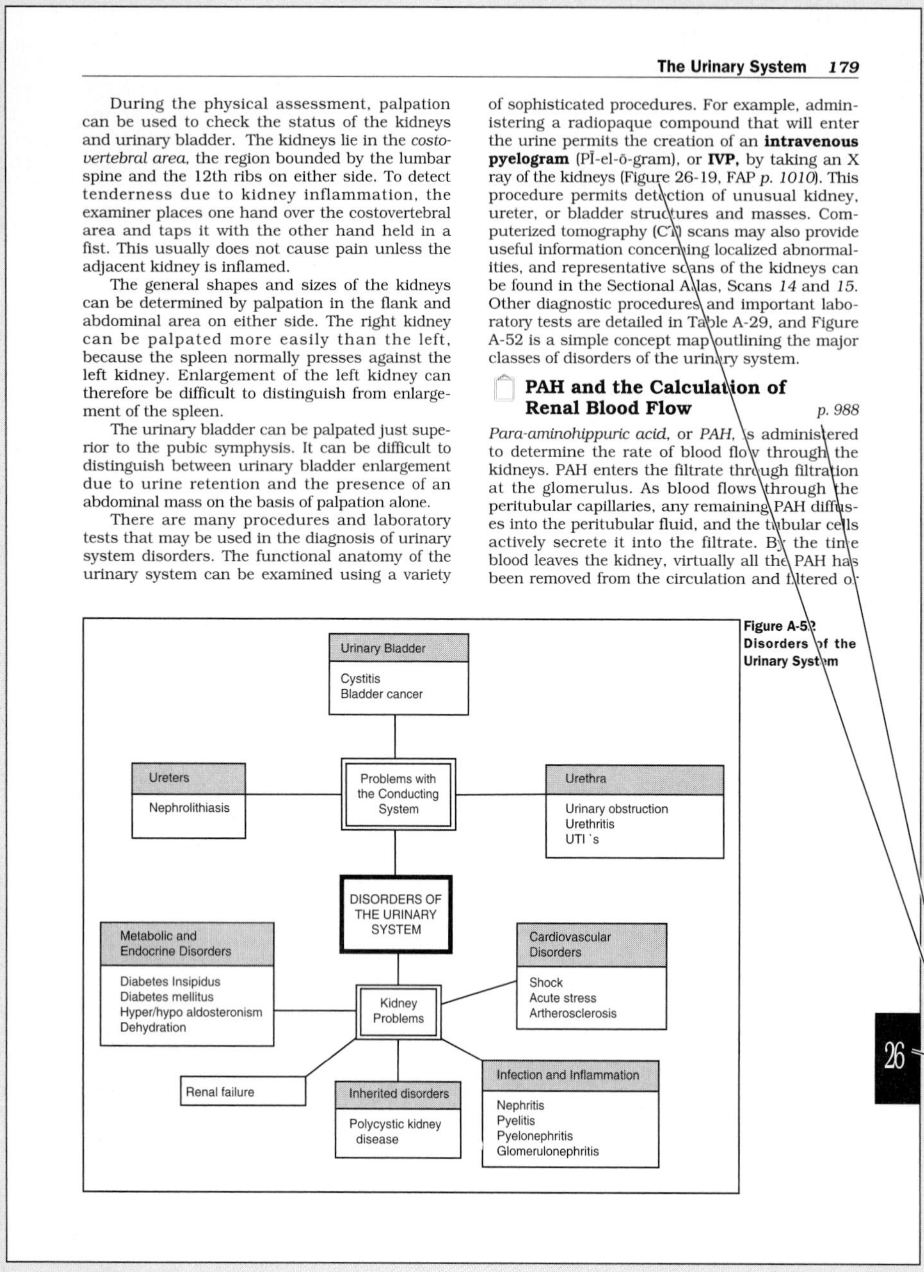

During the physical assessment, palpation can be used to check the status of the kidneys and urinary bladder. The kidneys lie in the *costovertebral area*, the region bounded by the lumbar spine and the 12th ribs on either side. To detect tenderness due to kidney inflammation, the examiner places one hand over the costovertebral area and taps it with the other hand held in a fist. This usually does not cause pain unless the adjacent kidney is inflamed.

The general shapes and sizes of the kidneys can be determined by palpation in the flank and abdominal area on either side. The right kidney can be palpated more easily than the left, because the spleen normally presses against the left kidney. Enlargement of the left kidney can therefore be difficult to distinguish from enlargement of the spleen.

The urinary bladder can be palpated just superior to the pubic symphysis. It can be difficult to distinguish between urinary bladder enlargement due to urine retention and the presence of an abdominal mass on the basis of palpation alone.

There are many procedures and laboratory tests that may be used in the diagnosis of urinary system disorders. The functional anatomy of the urinary system can be examined using a variety of sophisticated procedures. For example, administering a radiopaque compound that will enter the urine permits the creation of an **intravenous pyelogram** (PĪ-el-ō-gram), or **IVP**, by taking an X ray of the kidneys (Figure 26-19, FAP *p. 1010*). This procedure permits detection of unusual kidney, ureter, or bladder structures and masses. Computerized tomography (CT) scans may also provide useful information concerning localized abnormalities, and representative scans of the kidneys can be found in the Sectional Atlas, Scans *14* and *15*. Other diagnostic procedures and important laboratory tests are detailed in Table A-29, and Figure A-52 is a simple concept map outlining the major classes of disorders of the urinary system.

PAH and the Calculation of Renal Blood Flow *p. 988*

Para-aminohippuric acid, or *PAH*, is administered to determine the rate of blood flow through the kidneys. PAH enters the filtrate through filtration at the glomerulus. As blood flows through the peritubular capillaries, any remaining PAH diffuses into the peritubular fluid, and the tubular cells actively secrete it into the filtrate. By the time blood leaves the kidney, virtually all the PAH has been removed from the circulation and filtered or

Figure A-52 Disorders of the Urinary System

DISORDERS OF THE URINARY SYSTEM

Problems with the Conducting System

Urinary Bladder: Cystitis; Bladder cancer

Ureters: Nephrolithiasis

Urethra: Urinary obstruction; Urethritis; UTI 's

Kidney Problems

Metabolic and Endocrine Disorders: Diabetes Insipidus; Diabetes mellitus; Hyper/hypo aldosteronism; Dehydration

Cardiovascular Disorders: Shock; Acute stress; Artherosclerosis

Renal failure

Inherited disorders: Polycystic kidney disease

Infection and Inflammation: Nephritis; Pyelitis; Pyelonephritis; Glomerulonephritis

26

More than a collection of information on clinically-related topics, the *Applications Manual* is designed to supplement and extend the text's three-level learning system. It includes many features that can help you strengthen your grasp of the subject, improve your powers of analysis, and develop the ability to apply what you have learned.

- The **Critical Thinking questions** at the end of each system help you to sharpen your ability to think analytically. (These are comparable to the Level 3 questions in the text).

- The **Clinical Problems,** placed after groups of related systems, assist you in integrating information about several different body systems and enable you to practice making reasonable inferences in realistic situations.

- The **Case Studies** that follow the Body Systems section provide further opportunities for you to develop your powers of analysis, integration, and problem-solving. The studies presented, based on actual case histories, draw on material from the entire text (as they would in real life). The questions, keyed to crucial points in each presentation, help you to identify the relevant facts and form plausible hypotheses.

- The **introductory sections** of the *Applications Manual* include an overview of basic diagnostic principles and procedures, an explanation of the scientific method, and discussions of the application of chemistry and cell biology to the diagnosis and treatment of disease.

- The **Body Systems** section is organized to parallel the text chapter by chapter and system by system. Included are more detailed discussions of many clinical topics introduced in the text, as well as discussions of additional diseases, syndromes, and diagnostic techniques not covered in the text.

- **Critical Thinking questions, Clinical Problems,** and **Case Studies,** tied to the discussions, help you develop your powers of integration and analysis.

- A **Scanning Atlas** of photographs produced by various modern imaging techniques lets you view the interior of the human body section by section. These images will help you to develop a better understanding of three-dimensional relationships within the body.

- A full-color **Cadaver Atlas** of dissection photographs allows you to visualize the internal structure of all major body regions and organs. A selection of surface anatomy photographs of live models is included for comparison with the dissection views.

Extensive **cross-referencing** integrates the text and *Applications Manual,* making it easy to move back and forth between related material in the two books.

- The Scanning Atlas includes a number of **unlabeled images.** By labeling them yourself, you can test your knowledge of anatomical structure while developing your powers of visualization in three dimensions.

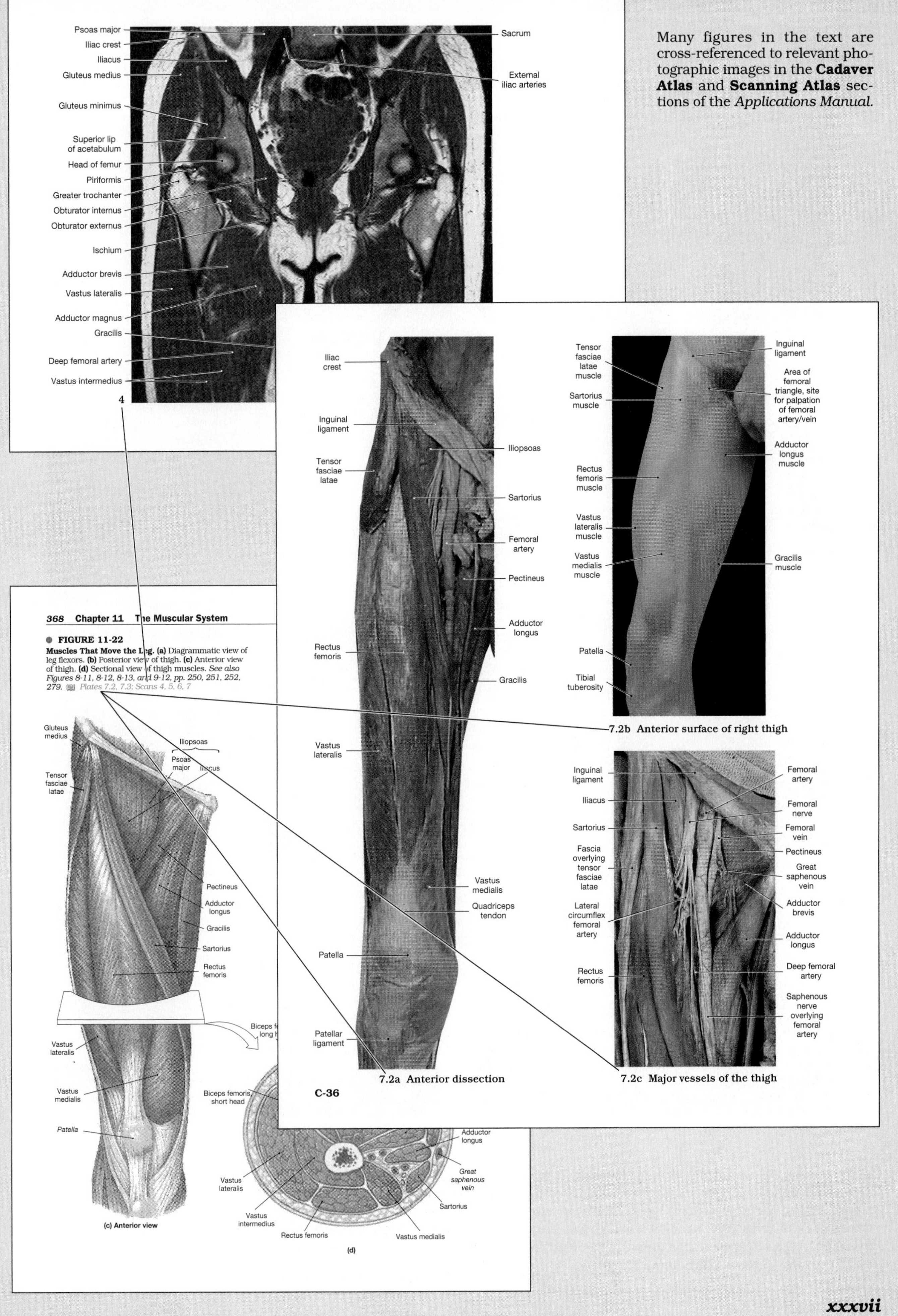
Psoas major
Iliac crest
Iliacus
Gluteus medius
Gluteus minimus
Superior lip of acetabulum
Head of femur
Piriformis
Greater trochanter
Obturator internus
Obturator externus
Ischium
Adductor brevis
Vastus lateralis
Adductor magnus
Gracilis
Deep femoral artery
Vastus intermedius
Sacrum
External iliac arteries
4
368 Chapter 11 The Muscular System
FIGURE 11-22
Muscles That Move the Leg. (a) Diagrammatic view of leg flexors. (b) Posterior view of thigh. (c) Anterior view of thigh. (d) Sectional view of thigh muscles. See also Figures 8-11, 8-12, 8-13, and 9-12, pp. 250, 251, 252, 279. Plates 7.2, 7.3; Scans 4, 5, 6, 7
Gluteus medius
Tensor fasciae latae
Iliopsoas
Psoas major
Iliacus
Pectineus
Adductor longus
Gracilis
Sartorius
Rectus femoris
Vastus lateralis
Vastus medialis
Patella
(c) Anterior view
Biceps femoris, short head
Adductor longus
Great saphenous vein
Sartorius
Vastus lateralis
Vastus intermedius
Rectus femoris
Vastus medialis
(d)
Iliac crest
Inguinal ligament
Tensor fasciae latae
Iliopsoas
Sartorius
Femoral artery
Pectineus
Adductor longus
Gracilis
Rectus femoris
Vastus lateralis
Vastus medialis
Quadriceps tendon
Patella
Patellar ligament
7.2a Anterior dissection
C-36
Tensor fasciae latae muscle
Sartorius muscle
Rectus femoris muscle
Vastus lateralis muscle
Vastus medialis muscle
Patella
Tibial tuberosity
Inguinal ligament
Area of femoral triangle, site for palpation of femoral artery/vein
Adductor longus muscle
Gracilis muscle
7.2b Anterior surface of right thigh
Inguinal ligament
Iliacus
Sartorius
Fascia overlying tensor fasciae latae
Lateral circumflex femoral artery
Rectus femoris
Femoral artery
Femoral nerve
Femoral vein
Pectineus
Great saphenous vein
Adductor brevis
Adductor longus
Deep femoral artery
Saphenous nerve overlying femoral artery
7.2c Major vessels of the thigh

Many figures in the text are cross-referenced to relevant photographic images in the **Cadaver Atlas** and **Scanning Atlas** sections of the *Applications Manual.*

CHAPTER 1

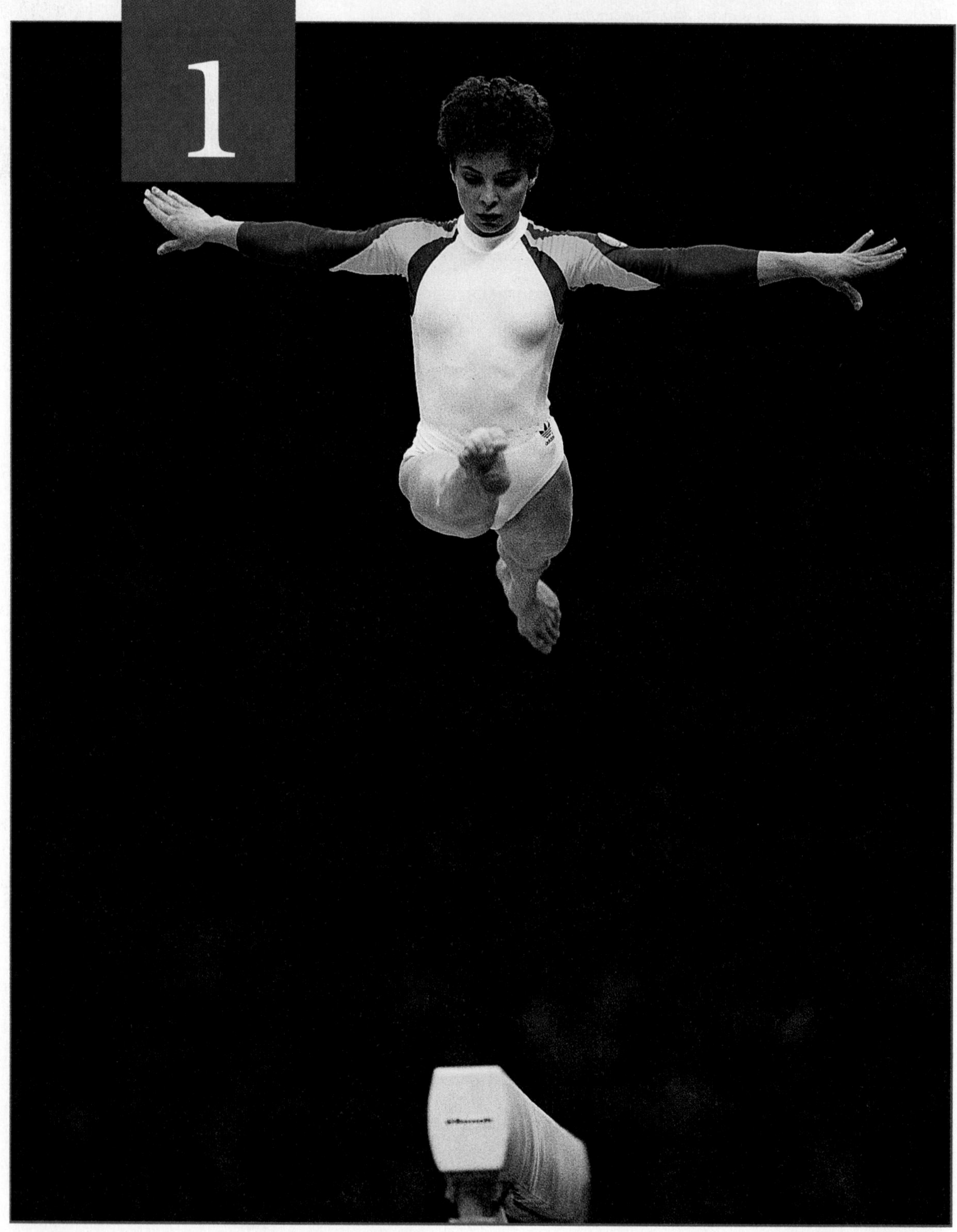

Every day we perform a balancing act. It is an amazingly complex performance, and it goes on continuously, without a break or intermission. Every part of our physical being is involved in this balance, from individual atoms to the largest organs and organ systems. Our bodies face constant challenges: injury, disease, and stresses both physical and mental. Every change, whether it occurs inside us or outside, threatens this delicate balance. Adjusting to these changes is crucial, for if the body loses its balance it may plunge out of control, into dysfunction, illness, and even death.

An Introduction to Anatomy and Physiology

Chapter Outline and Objectives

The world around us contains an enormous diversity of living organisms that vary widely in appearance and lifestyle. One aim of **biology**, the study of life, is to discover the unity and pattern that underlie this diversity. Biologists have found that all living things, despite their obvious differences, share certain basic characteristics:

- **Responsiveness:** Organisms respond to changes in their immediate environment; this property is also called irritability. You move your hand away from a hot stove, your dog barks at approaching strangers, fish are scared by loud noises, and tiny amoebas glide toward potential prey. Organisms also make longer-term changes as they adjust to their environments. For example, an animal may grow a heavier coat as winter approaches, or migrate to a warmer climate. The capacity to make such adjustments is termed *adaptability*.
- **Growth:** Over a lifetime, organisms grow larger, increasing in size through an increase in the size or number of their cells. In multicellular organisms, the individual cells become specialized to perform particular functions. This specialization is called *differentiation*. Growth and differentiation often produce changes in form and function. For example, the anatomical proportions and physiological capabilities of an adult human are quite different from those of an infant.
- **Reproduction:** Organisms reproduce, creating subsequent generations of similar organisms.
- **Movement:** Organisms are capable of producing movement, which may be internal (transporting food, blood, or other materials inside the body) or external (moving through the environment).
- **Metabolism:** Organisms rely on complex chemical reactions to provide the energy for responsiveness, growth, reproduction, and movement. They must also synthesize complex chemicals, such as proteins. The term *metabolism* refers to all the chemical operations under way in the body. Normal metabolic operations require the *absorption* of materials from the environment. To generate energy efficiently, most cells require various nutrients, as well as oxygen, an atmospheric gas. The term *respiration* refers to the absorption, transport, and use of oxygen by cells. Metabolic operations often generate unneeded or potentially harmful waste products that must be eliminated through the process of *excretion*.

These are the basic characteristics of living things, both plant and animal. Several additional functions can be distinguished when you consider animals as complex as fish, cats, or human beings. Whereas very small organisms can absorb materials across exposed surfaces, creatures larger than a few millimeters seldom absorb nutrients directly from their environment. For example, human beings cannot absorb steaks, apples, or ice cream without processing them first. That processing, called *digestion*, occurs in specialized areas where complex foods are broken down into simpler components that can be absorbed easily. Respiration and excretion are also more complicated for large organisms. Humans have specialized structures responsible for gas exchange (lungs) and waste elimination (kidneys). Finally, because absorption, respiration, and excretion are performed in different portions of the body, there must be an internal transportation system, or *circulation*.

Biology includes subspecialties, each with a slightly different perspective. This text considers two biological subjects, *anatomy* (a-NAT-o-mē) and *physiology* (fiz-ē-OL-ō-jē). In the course of the next 28 chapters you will become familiar with the basic anatomy and physiology of an unusually large and interesting type of animal, the human being.

The Sciences of Anatomy and Physiology

The word *anatomy* has its origins in the language of ancient Greece, as do many other anatomical terms and phrases. A literal translation would be "a cutting open." **Anatomy** is the study of internal and external structure and the physical relationships between body parts. For example, someone studying anatomy might examine how a particular muscle attaches to the skeleton. **Physiology**, another word adopted from the Greek, is the study of how living organisms perform the vital functions listed above. Someone studying physiology might consider how a muscle contracts or what forces a contracting muscle exerts on the skeleton.

Anatomy and physiology are closely integrated both theoretically and practically. Anatomical information provides clues about probable functions, and physiological mechanisms can be explained only in terms of the underlying anatomy. This observation leads to a very important concept: *All specific functions are performed by specific structures.*

Anatomists and physiologists approach the relationship between structure and function from different perspectives. To understand the difference, consider a simple nonbiological analogy. Suppose that an anatomist and a physiologist were asked to examine and report on a pickup truck. The anatomist might begin by measuring and photographing the various parts of the truck and, if possible, taking it apart and putting it back together. The anatomist would then be able to explain its key structural relationships—for example, how the movement of the pistons in the engine cylinders caused the drive shaft to rotate, and how this motion was then conveyed by the transmission to the wheels. The physiologist would of course note the relationships between the components, but the primary focus would be on functional characteristics, such as the ideal ratio of gasoline to air for most efficient combustion, the amount of power that the engine could generate, the amount of force transmitted to the wheels in different gears, and so on.

The link between structure and function is always present, but not always understood. For example, the superficial anatomy of the heart was clearly described in the fifteenth century, but almost 200 years passed before the pumping action of the heart was demonstrated.

This text will provide a familiarity with anatomical structures and an appreciation of the physiological processes that make human life possible. This information should enable you to understand many kinds of disease processes. It will also provide a perspective that should prove useful for making intelligent decisions about your personal health.

❑ Anatomy

Anatomy can be divided into *microscopic anatomy* and *macroscopic (gross) anatomy*, based on the degree of structural detail under consideration. Other anatomical specialties focus on specific processes or medical applications.

Microscopic Anatomy

Microscopic anatomy deals with structures that cannot be seen without magnification. The boundaries of microscopic anatomy are established by the limits of the equipment used. A light microscope can show you basic details about cell structure, whereas an electron microscope enables you to see individual molecules only a few nanometers across. As we proceed through the text, we will be considering details at all levels, from macroscopic to microscopic. (Readers unfamiliar with the terms used to describe measurements and weights over this size range should consult the reference tables in Appendix II.)

Microscopic anatomy can be subdivided into specialties that consider features within a characteristic range of sizes. **Cytology** (sī-TOL-o-jē) analyzes the internal structure of individual cells, the simplest units of life. Living cells are composed of chemical substances in various combinations, and our lives depend on the chemical processes occurring in the trillions of cells that form our body. For this reason we will consider basic chemistry (Chapter 2) before examining cell structure (Chapter 3).

Histology (his-TOL-o-jē) takes a broader perspective and examines **tissues,** groups of specialized cells and cell products that work together to perform a particular function. The trillions of cells in the human body can be assigned to four *tissue types* that are the focus of Chapter 4. Tissues in combination form **organs** such as the heart, kidney, liver, and brain. Many organs are easily examined without a microscope, and at the organ level we cross the boundary into gross anatomy.

Gross Anatomy

Gross anatomy, or **macroscopic anatomy,** considers features visible with the unaided eye. There are many ways to approach gross anatomy:

- **Surface anatomy** is the study of general form and superficial markings.
- **Regional anatomy** focuses on all the superficial and internal features in a specific area of the body, such as the head, neck, and trunk. Advanced courses in anatomy often stress a regional approach because it emphasizes the spatial relationships between structures already familiar to students.
- **Systemic anatomy** is concerned with the structure of *organ systems,* such as the skeletal and muscular systems. Organ systems are groups of organs that function together in a coordinated manner. For example, the heart, blood, and blood vessels form the *cardiovascular system,* which distributes oxygen and nutrients throughout the body. There are 11 organ systems in the human body, and they will be introduced later in the chapter.
- **Developmental anatomy** examines the changes in form that occur during the period between conception and physical maturity. Because it considers anatomical structures over such a broad range of sizes (from a single cell to an adult human), techniques used in developmental anatomy are similar to those used in both microscopic and gross anatomy. Developmental anatomy is important in medicine because many structural abnormalities can result from errors in development. The most extensive structural changes occur during the first two months of development. The study of these early developmental processes is called **embryology.**

There are a number of anatomical specialties that are most important in a clinical setting. Examples include *medical anatomy* (anatomical features that change during illness), *radiographic anatomy* (anatomical structures as seen using specialized imaging techniques, discussed later in this chapter), and *surgical anatomy* (anatomical landmarks important in surgery). Physicians normally use a combination of anatomical, physiological, and behavioral information when evaluating a patient. AM *Steps in the Diagnostic Process*

Physiology

Physiology considers the function of anatomical structures. A physiologist attempts to determine the physical and chemical processes responsible for vital functions. **Human physiology** is the study of the functions of the human body. These functions are complex and much more difficult to examine than most anatomical structures. As a result, there are even more specialties in physiology than in anatomy.

- **Cell physiology,** the study of the functions of living cells, is the cornerstone of human physiology. Cell physiology deals with events at the chemical and molecular levels: both chemical processes within cells and chemical interactions between cells. Chapters 2–4 focus on the chemical structure, internal organization, and control mechanisms of living cells and tissues.
- **Special physiology** is the study of the physiology of specific organs. For example, *cardiac physiology* is the study of heart function.
- **Systemic physiology** considers all aspects of the function of specific organ systems. *Cardiovascular physiology, respiratory physiology,* and *reproductive physiology* are examples of systemic physiology.
- **Pathological physiology** studies the effects of diseases on organ or system functions. (The Greek word *pathos* refers to disease.) Modern medicine depends on an understanding of both normal and pathological physiology; the practitioner must know not only what is wrong but how to correct it. You will find extensive information on clinically important topics in subsequent chapters and in the *Applications Manual.*

There are also special topics in physiology that address specific functions of the human body as a whole. These specialties focus on physiological interactions between multiple organ systems. For example, *exercise physiology* studies the physiological adjustments to exercise.

Levels of Organization

Figure 1-1

Our study of the human body will begin with an overview of microscopic anatomy and then proceed to the gross and microscopic anatomy of each organ system. When considering events from the microscopic to the macroscopic scale, we are examining several interdependent *levels of organization.* Figure 1-1 ● presents an example of the relationships among the various levels of organization.

- *Atoms,* the smallest stable units of matter, combine to form *molecules* with complex shapes. This is the *chemical,* or *molecular, level of organization.*
- Molecules interact to form *organelles,* such as the protein filaments found in cardiac muscle cells. Organelles are structural and functional components of *cells,* the smallest living units in the body. This is the *cellular level of organization.*
- Heart muscle cells are components of one form of *muscle tissue.* Muscle tissue, one of the four tissue types, is an example of the *tissue level of organization.*
- Layers of muscle tissue form the bulk of the wall of the heart, a hollow, three-dimensional organ. We are now at the *organ level of organization.*
- Each time it contracts, the heart pushes blood into a network of blood vessels. Together the heart, blood, and blood vessels form the *cardiovascular system,* an example of the *organ system level of organization.*
- All the organ systems of the body work together to maintain life and health. This brings us to the highest level of organization, that of the *organism,* in this case a human being.

The organization at each level determines the characteristics and functions of higher levels. For example, the arrangement of atoms and molecules at the molecular level creates the protein filaments that, at the cellular level, give cardiac muscle cells the ability to contract powerfully. Because at the tissue level these cells are linked together, forming cardiac muscle tissue, their contractions are coordinated, producing a heartbeat. When that beat occurs, the internal anatomy of the heart, an organ, enables it to function as a pump. The heart is filled with blood and connected to the blood vessels, and the pumping action circulates blood through the cardiovascular system. Through interactions with the respiratory, digestive, urinary, and other systems, the cardiovascular system performs a variety of functions essential to the survival of the individual.

It should also be noted that something that affects the *system* will ultimately affect all of its components. For example, the heart cannot pump blood effectively after a massive blood loss if there is not enough blood to fill the circulatory system. If the heart cannot pump and blood cannot flow, oxygen and nutrients cannot be dis-

● **FIGURE 1-1**
Levels of Organization.
Interacting atoms form molecules that combine in the protein fibers of heart muscle cells. These cells interlock, creating heart muscle tissue that constitutes most of the walls of a three-dimensional organ, the heart. The heart is one component of the cardiovascular system, which also includes the blood and blood vessels. All of the organ systems combine to create an organism, a living human being.

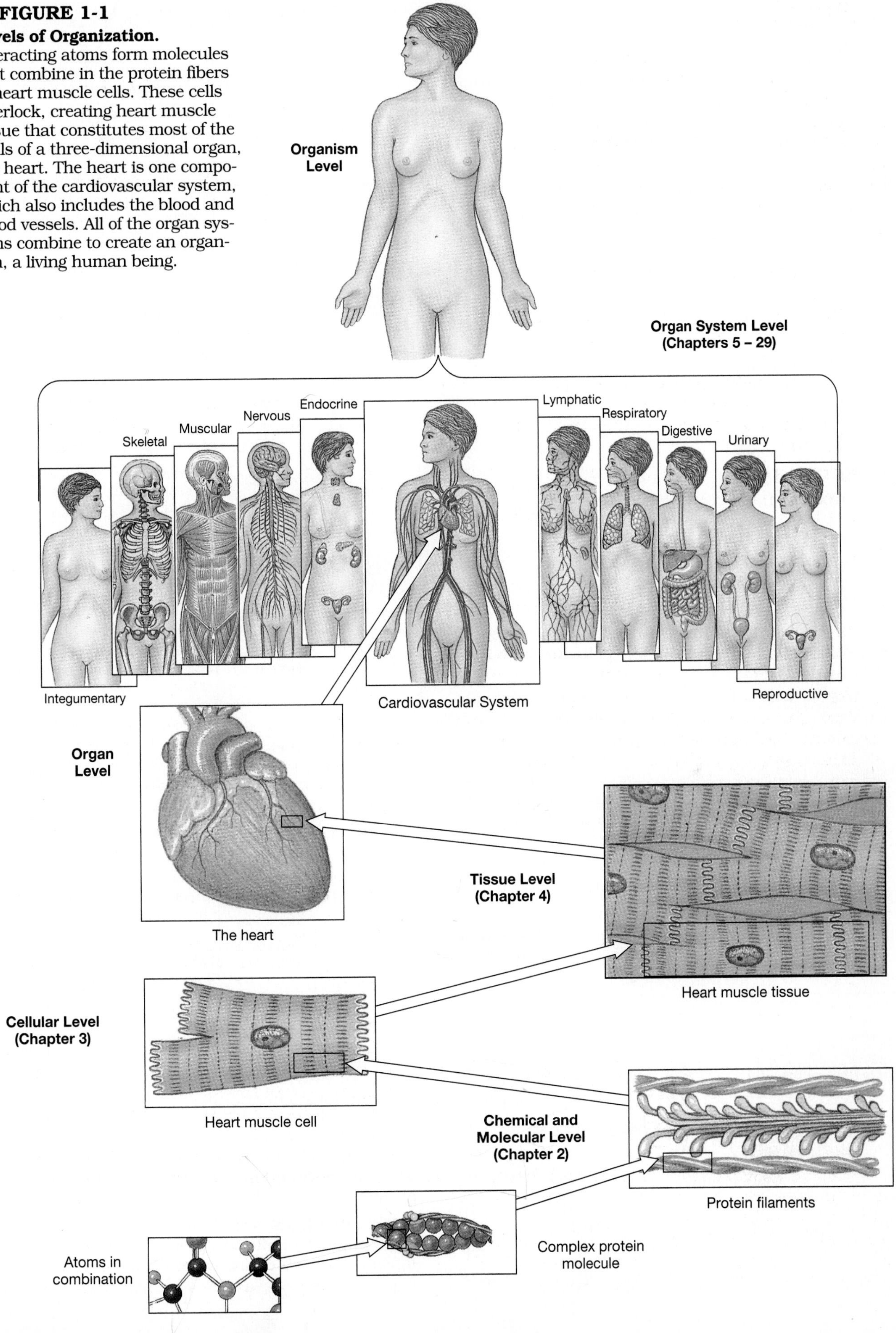

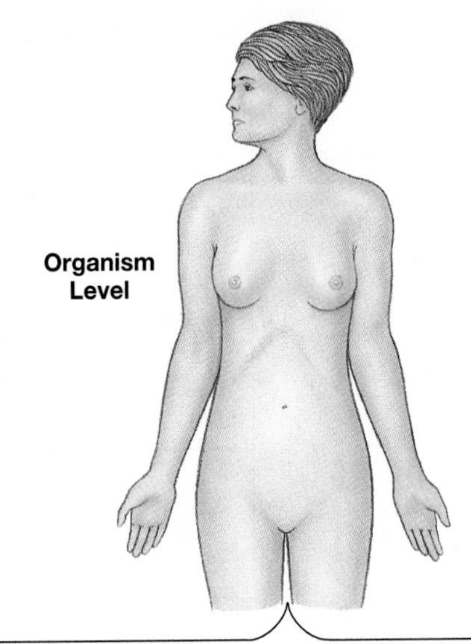
Organism Level
Organ System Level (Chapters 5 – 29)
Integumentary
Skeletal
Muscular
Nervous
Endocrine
Cardiovascular System
Lymphatic
Respiratory
Digestive
Urinary
Reproductive
Organ Level
The heart
Tissue Level (Chapter 4)
Heart muscle tissue
Cellular Level (Chapter 3)
Heart muscle cell
Chemical and Molecular Level (Chapter 2)
Protein filaments
Complex protein molecule
Atoms in combination

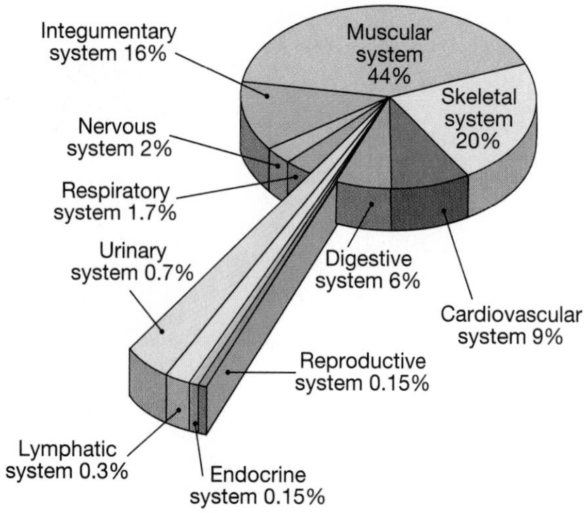

Integumentary system 16%
Muscular system 44%
Skeletal system 20%
Nervous system 2%
Respiratory system 1.7%
Urinary system 0.7%
Digestive system 6%
Cardiovascular system 9%
Reproductive system 0.15%
Lymphatic system 0.3%
Endocrine system 0.15%

(a) Organ system composition of the human body by weight

Organ System		Major Functions
	Integumentary system	Protection from environmental hazards, temperature control
	Skeletal system	Support, protection of soft tissues, mineral storage, blood formation
	Muscular system	Locomotion, support, heat production
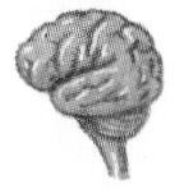	Nervous system	Directing immediate responses to stimuli, usually by coordinating the activities of other organ systems
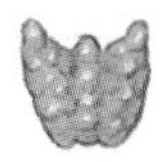	Endocrine system	Directing long-term changes in the activities of other organ systems
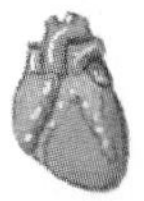	Cardiovascular system	Internal transport of cells and dissolved materials, including nutrients, wastes, and gases
	Lymphatic system	Defense against infection and disease
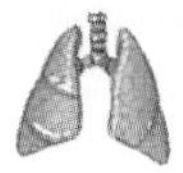	Respiratory system	Delivery of air to sites where gas exchange can occur between the air and circulating blood
	Digestive system	Processing of food and absorption of nutrients, minerals, vitamins, and water
	Urinary system	Elimination of excess water, salts, and waste products
	Reproductive system	Production of sex cells and hormones

(b)

● **FIGURE 1-2**
An Introduction to Organ Systems

tributed. In a very short time, the tissue begins to break down as heart muscle cells die from oxygen and nutrient starvation. Of course these changes will not be restricted to the cardiovascular system; all the cells, tissues, and organs in the body will be damaged.

✔ What characteristics of life are white blood cells exhibiting when they migrate to an injury site in response to chemicals released from damaged cells?

✔ A histologist investigates structures at what level of organization?

✔ A researcher studies the factors that cause heart failure. What names might be used for this specialty?

An Introduction to Organ Systems

Figures 1-2, 1-3

Figure 1-2● introduces the 11 organ systems in the human body. This figure indicates the major functions of each system and its relative contribution to the total body weight. Figure 1-3● provides an overview of the individual systems and their major components.

✔ The ability to move about and produce heat are major functions made possible by which body system?

✔ Individuals suffering from AIDS exhibit an impaired ability to defend themselves against infection. Which organ system is affected?

● FIGURE 1-3
The Organ Systems of the Human Body

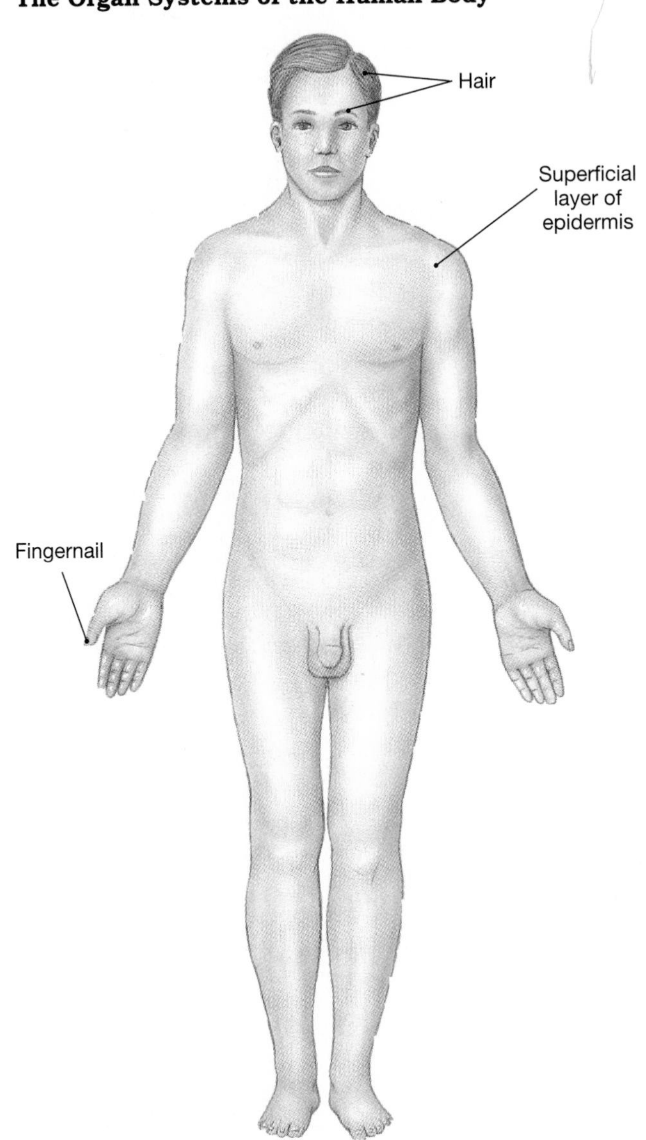
Hair
Superficial layer of epidermis
Fingernail

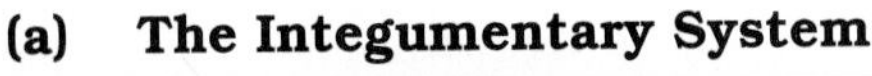
(a) The Integumentary System

Organ	Primary Functions
EPIDERMIS	Covers surface, protects underlying tissues
DERMIS	Nourishes epidermis, provides strength, contains glands
HAIR FOLLICLES	Produce hair
Hairs	Provide sensation, provide some protection for head
Sebaceous glands	Secrete lipid coating that lubricates hair shaft
SWEAT GLANDS	Produce perspiration for evaporative cooling
NAILS	Protect and stiffen distal tips of digits
SENSORY RECEPTORS	Provide sensations of touch, pressure, temperature, pain
SUBCUTANEOUS LAYER	Stores lipids, attaches skin to deeper structures

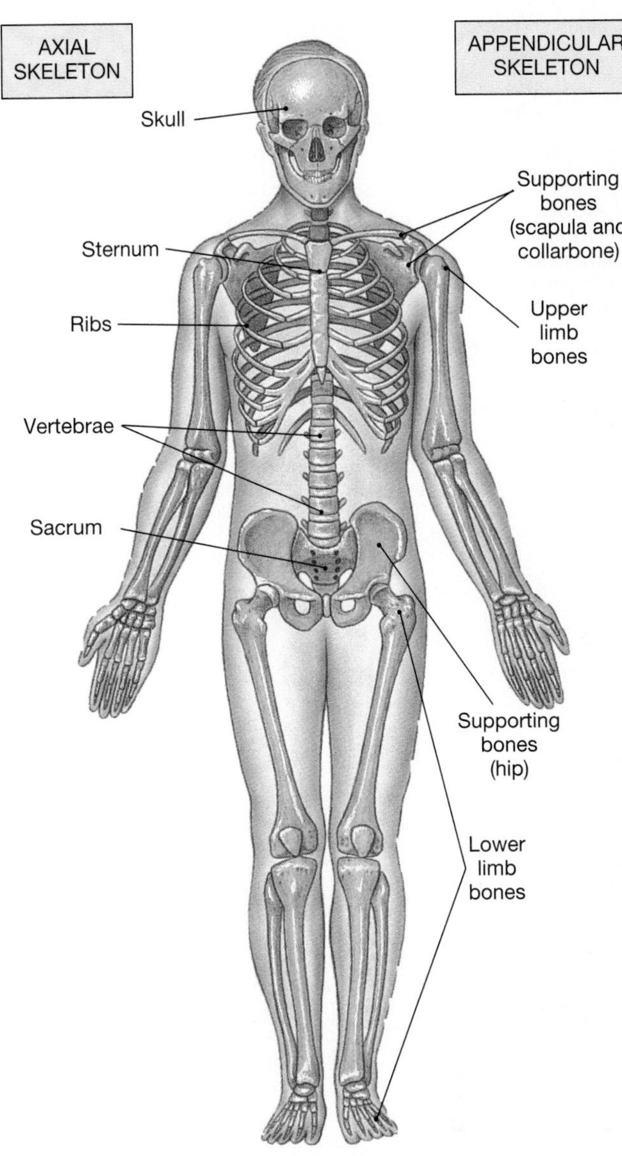
AXIAL SKELETON
APPENDICULAR SKELETON
Skull
Supporting bones (scapula and collarbone)
Sternum
Ribs
Upper limb bones
Vertebrae
Sacrum
Supporting bones (hip)
Lower limb bones

(b) The Skeletal System

Organ	Primary Functions
BONES (206), CARTILAGES, AND LIGAMENTS	Support, protect soft tissues; store minerals
Axial skeleton (Skull, vertebrae, sacrum, ribs, sternum)	Protects brain, spinal cord, sense organs, and soft tissues of chest cavity; supports the body weight over the lower limbs
Appendicular skeleton (Limbs and supporting bones)	Provides internal support and positioning of limbs; supports and moves axial skeleton
BONE MARROW	Primary site of blood cell production

● **FIGURE 1-3 continued**

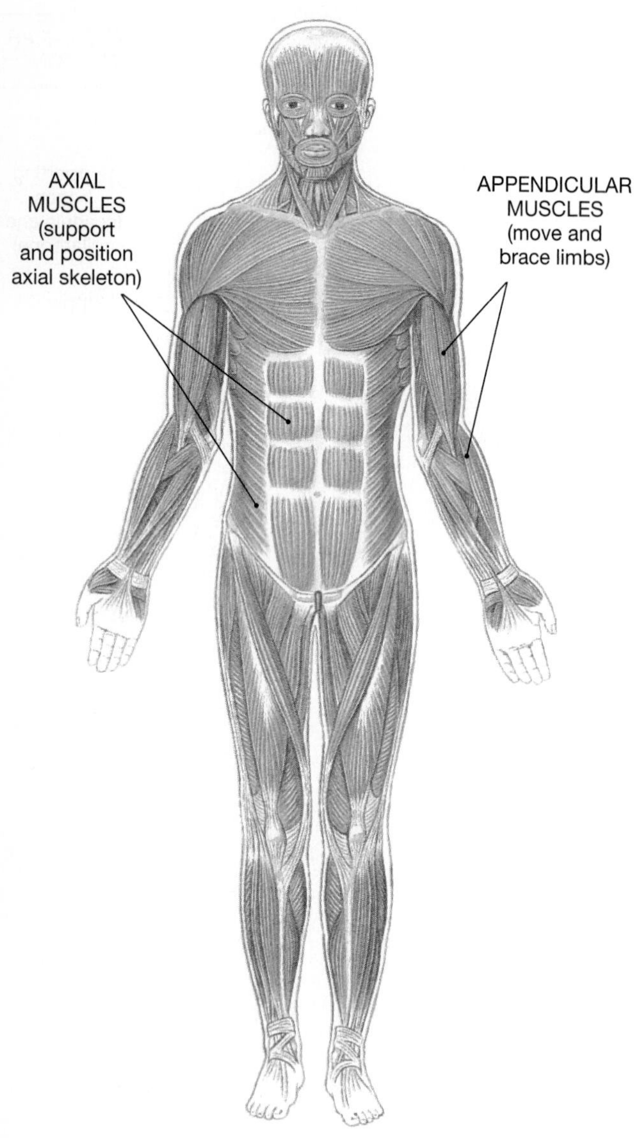
AXIAL
MUSCLES
(support
and position
axial skeleton)
APPENDICULAR
MUSCLES
(move and
brace limbs)

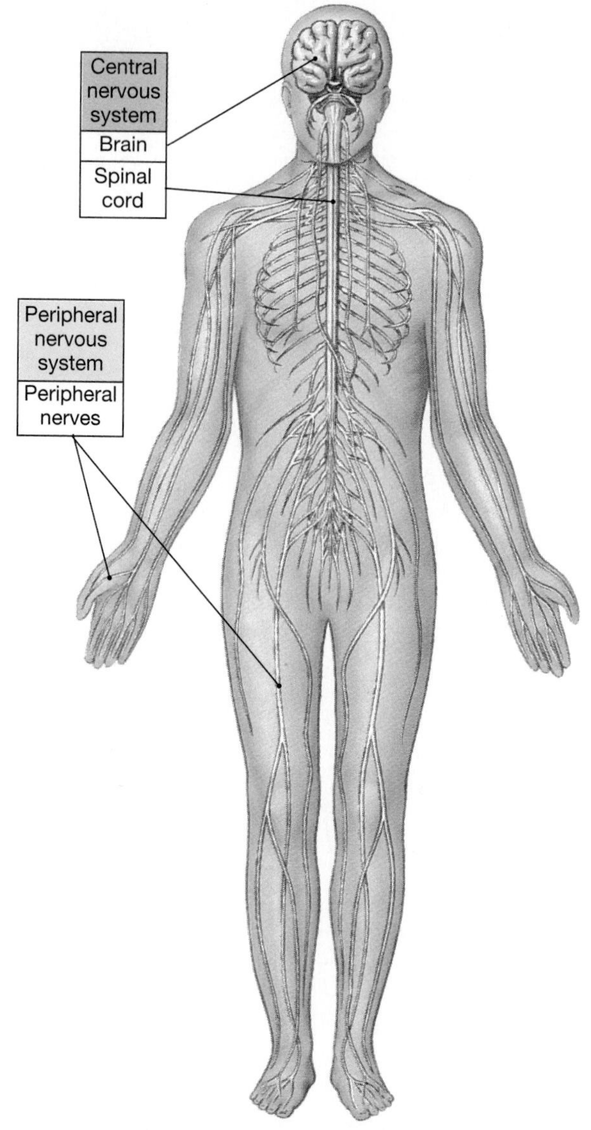
Central
nervous
system
Brain
Spinal
cord
Peripheral
nervous
system
Peripheral
nerves

(c) The Muscular System

Organ	*Primary Functions*
SKELETAL MUSCLES (700)	Provide skeletal movement, control entrances and exits of digestive tract, produce heat, support skeletal position, protect soft tissues
TENDONS, APONEUROSES	Harness forces of contraction to perform specific tasks

(d) The Nervous System

Organ	*Primary Functions*
CENTRAL NERVOUS SYSTEM (CNS)	Control center for nervous system: processes information, provides short-term control over activities of other systems
Brain	Performs complex integrative functions, controls voluntary activities
Spinal cord	Relays information to and from the brain, performs less complex integrative functions; directs many simple involuntary activities
PERIPHERAL NERVOUS SYSTEM (PNS)	Links CNS with other systems and with sense organs

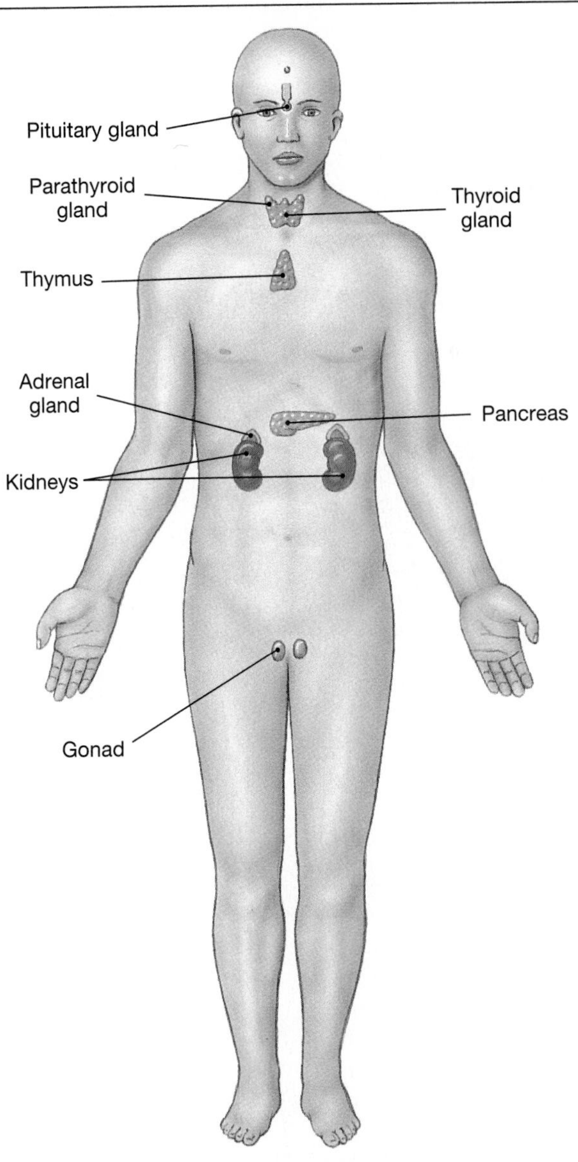

Pituitary gland
Parathyroid gland
Thyroid gland
Thymus
Adrenal gland
Pancreas
Kidneys
Gonad

(e) The Endocrine System

Organ	Primary Functions
PITUITARY GLAND	Controls other glands, regulates growth and fluid balance
THYROID GLAND	Controls tissue metabolic rate and regulates calcium levels
PARATHYROID GLAND	Regulates calcium levels (with thyroid)
THYMUS	Controls lymphocyte maturation
ADRENAL GLANDS	Adjust water balance, tissue metabolism, cardiovascular and respiratory activity
KIDNEYS	Control red blood cell production and elevate blood pressure
PANCREAS	Regulates blood glucose levels
GONADS	
Testes	Support male sexual characteristics and reproductive functions *(see Figure 1-3k)*
Ovaries	Support female sexual characteristics and reproductive functions *(see Figure 1-3k)*

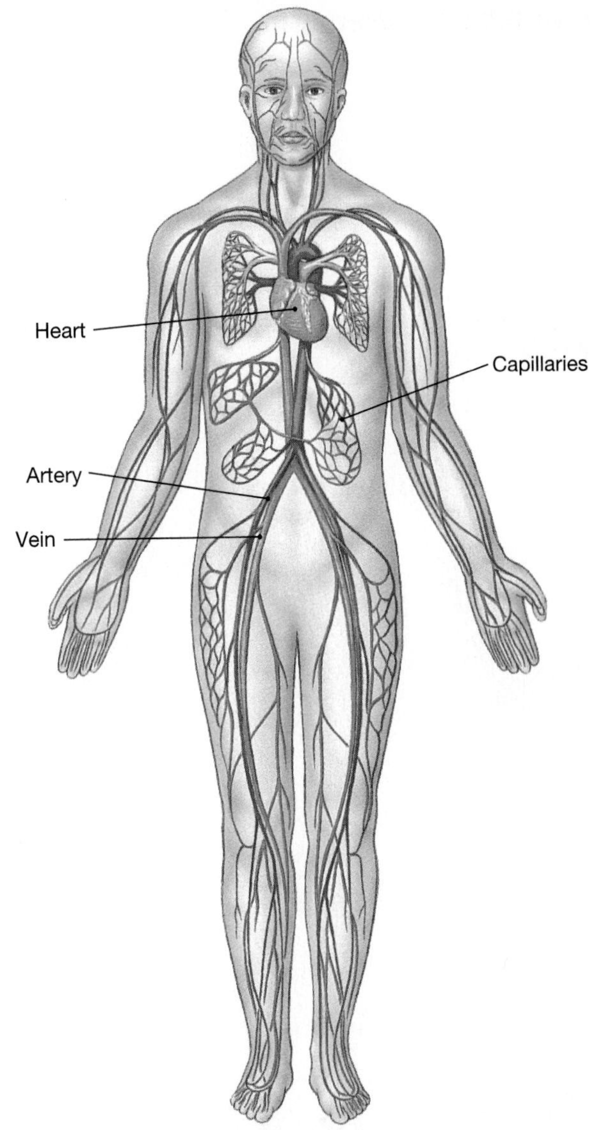

Heart
Capillaries
Artery
Vein

(f) The Cardiovascular System

Organ	Primary Functions
HEART	Propels blood, maintains blood pressure
BLOOD VESSELS	Distribute blood around the body
Arteries	Carry blood from heart to capillaries
Capillaries	Site of diffusion between blood and interstitial fluids
Veins	Return blood from capillaries to the heart
BLOOD	Transports oxygen and carbon dioxide, delivers nutrients, removes waste products, assists in defense against disease

FIGURE 1-3 continued

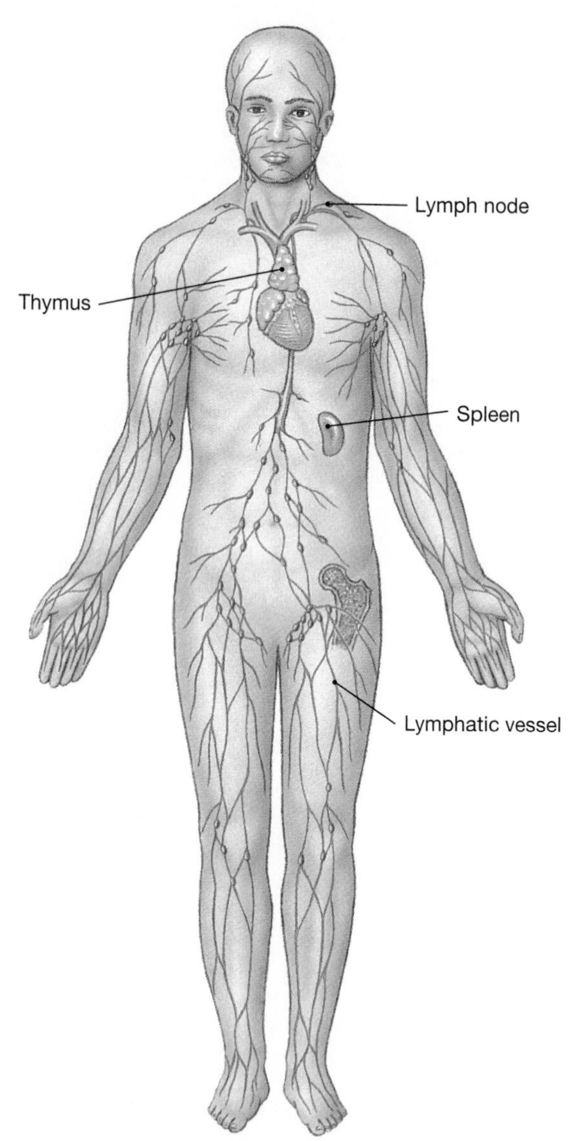
Lymph node
Thymus
Spleen
Lymphatic vessel

(g) The Lymphatic System

Organ	*Primary Functions*
LYMPHATIC VESSELS	Carry lymph (water and proteins) from peripheral tissues to the veins of the cardiovascular system
LYMPH NODES	Monitor the composition of lymph, engulf pathogens, stimulate immune response
SPLEEN	Monitors circulating blood, engulfs pathogens, stimulates immune response
THYMUS	Controls development and maintenance of one class of lymphocytes (T cells); immature lymphocytes and other blood cells are produced in the bone marrow

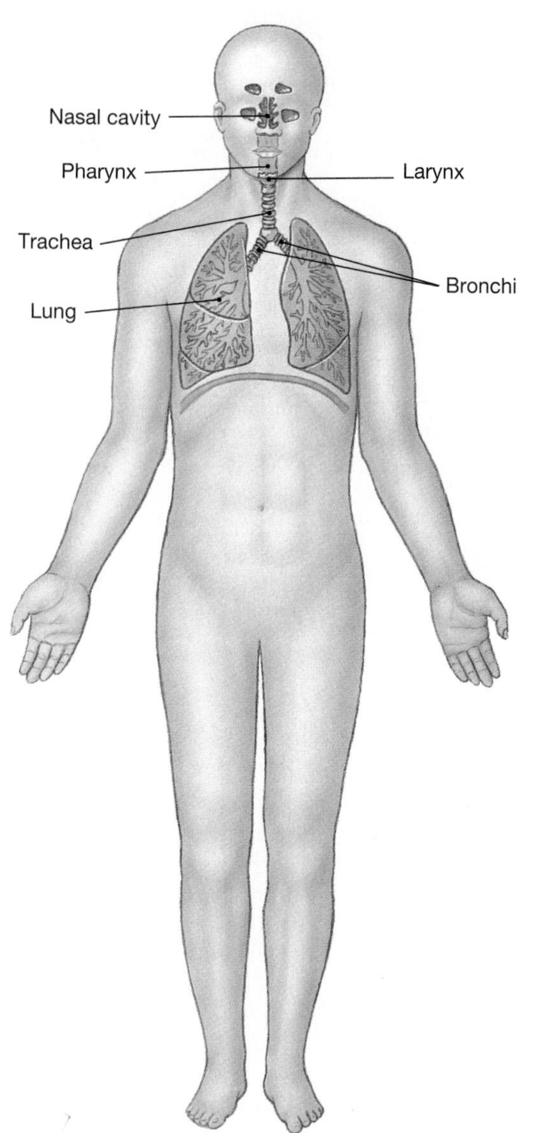
Nasal cavity
Pharynx
Larynx
Trachea
Bronchi
Lung

(h) The Respiratory System

Organ	*Primary Functions*
NASAL CAVITIES	Filter, warm, humidify air; detect smells
PHARYNX	Chamber shared with digestive tract; conducts air to larynx
LARYNX	Protects opening to trachea and contains vocal cords
TRACHEA	Filters air, traps particles in mucus; cartilages keep airway open
BRONCHI	Same as trachea
LUNGS	Include airways and alveoli; volume changes responsible for air movement
ALVEOLI	Sites of gas exchange between air and blood

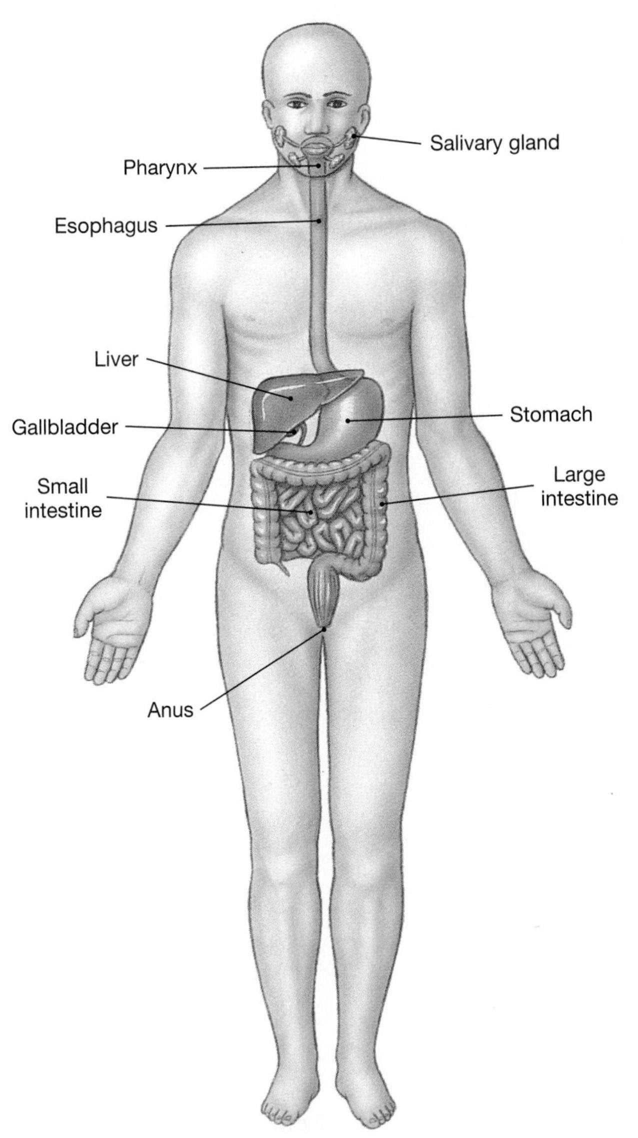

Salivary gland
Pharynx
Esophagus
Liver
Gallbladder
Stomach
Small intestine
Large intestine
Anus

(i) The Digestive System

Organ	Primary Functions
SALIVARY GLANDS	Provide lubrication, produce buffers and enzymes that begin digestion
PHARYNX	Passageway connected to esophagus
ESOPHAGUS	Delivers food to stomach
STOMACH	Secretes acids and enzymes
SMALL INTESTINE	Secretes digestive enzymes, absorbs nutrients
LIVER	Secretes bile, regulates blood composition of nutrients
GALLBLADDER	Stores bile for release into small intestine
PANCREAS	Secretes digestive enzymes and buffers; contains endocrine cells *(see Figure 1-3e)*
LARGE INTESTINE	Removes water from fecal material, stores wastes

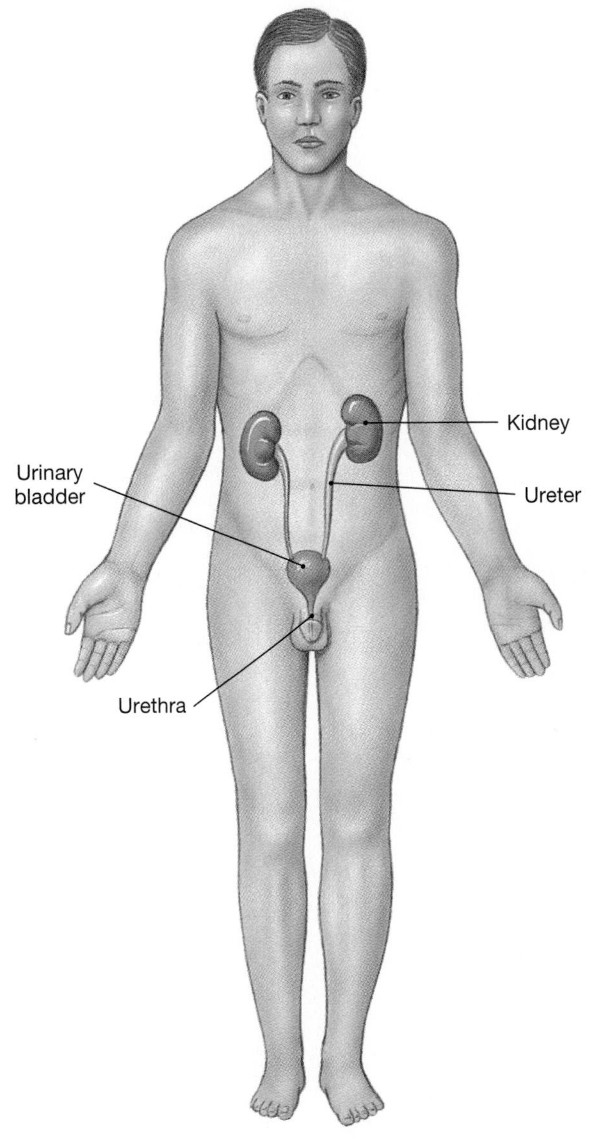

Kidney
Urinary bladder
Ureter
Urethra

(j) The Urinary System

Organ	Primary Functions
KIDNEYS	Form and concentrate urine, regulate blood pH and ion concentrations; endocrine functions noted in *Figure 1-3e*
URETERS	Conduct urine from kidneys to urinary bladder
URINARY BLADDER	Stores urine for eventual elimination
URETHRA	Conducts urine to exterior

● **FIGURE 1-3 continued**

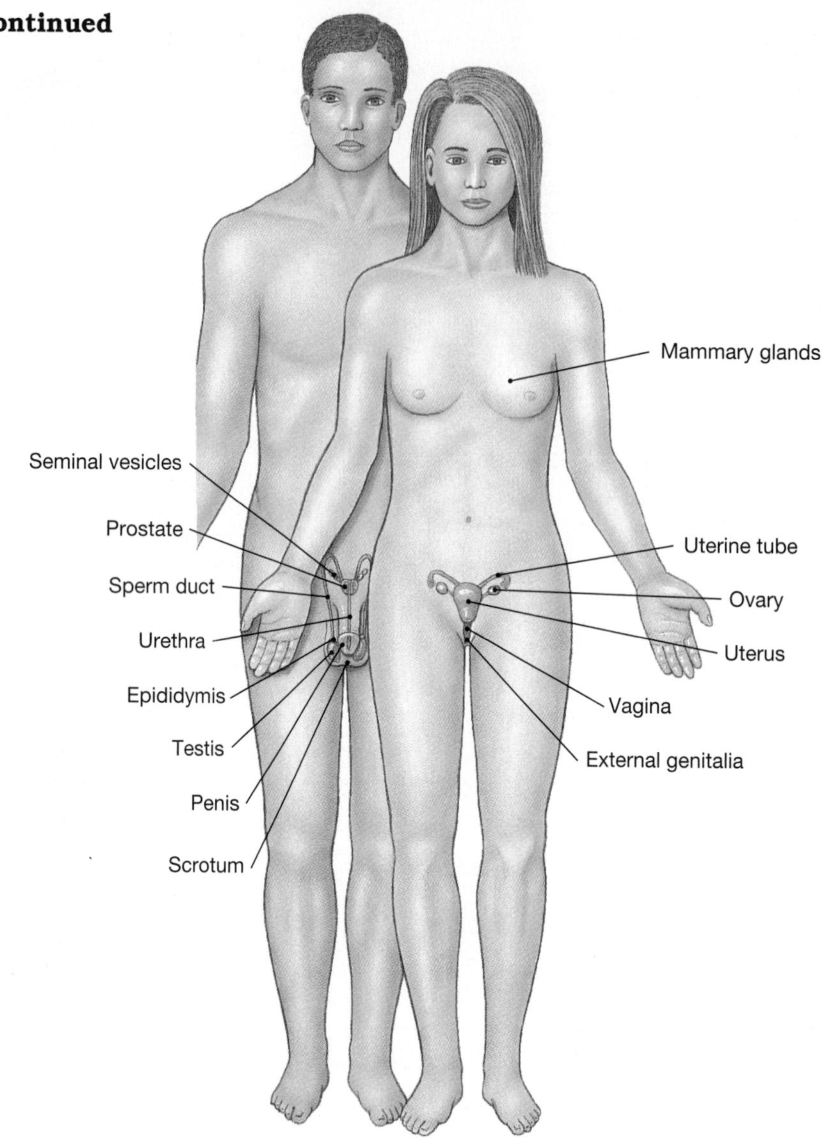
Mammary glands
Seminal vesicles
Prostate
Uterine tube
Sperm duct
Ovary
Urethra
Uterus
Epididymis
Vagina
Testis
External genitalia
Penis
Scrotum

(k) The Reproductive System of the Male and Female

Organ	*Primary Functions*
TESTES	Produce sperm and hormones *(see Figure 1-3e)*
ACCESSORY ORGANS	
Epididymis	Site of sperm maturation
Ductus deferens (sperm duct)	Conducts sperm between epididymis and prostate
Seminal vesicles	Secrete fluid that makes up much of the volume of semen
Prostate	Secretes buffers and fluid
Urethra	Conducts semen to exterior
EXTERNAL GENITALIA	
Penis	Erectile organ used to deposit sperm in the vagina of a female; produces pleasurable sensations during sexual act
Scrotum	Surrounds and controls temperature of the testes

Organ	*Primary Functions*
OVARIES	Produce ova (eggs) and hormones *(see Figure 1-3e)*
UTERINE TUBES	Deliver ova or embryo to uterus; normal site of fertilization
UTERUS	Site of embryonic development and diffusion between maternal and embryonic bloodstreams
VAGINA	Site of sperm deposition; birth canal at delivery; provides passage of fluids during menstruation
EXTERNAL GENITALIA	
Clitoris	Erectile organ, produces pleasurable sensations during sexual act
Labia	Contain glands that lubricate entrance to vagina
MAMMARY GLANDS	Produce milk that nourishes newborn infant

Homeostasis and System Integration

Organ systems are interdependent, interconnected, and packaged together in a relatively small space. The cells, tissues, organs, and systems of the body live together in a shared environment, like the inhabitants of a large city. City dwellers breathe the city air and drink the water provided by the local water company; cells in the human body absorb oxygen and nutrients from fluids that surround them. If a city is blanketed in smog, or the water is contaminated, the inhabitants will become ill. Similarly, if body fluid composition becomes abnormal, cells will be injured or destroyed. For example, suppose there are changes in the temperature or salt content of the blood. The effect on the heart could range from a minor adjustment (heart muscle tissue contracts more often, and the heart rate goes up) to a total disaster (the heart stops beating altogether, and the individual dies).

Homeostatic Regulation

Figure 1-4

A variety of physiological mechanisms act to prevent potentially disruptive changes in the environment inside the body. **Homeostasis** (*homeo*, unchanging + *stasis*, standing) refers to the existence of a stable internal environment. To survive, every living organism must maintain homeostasis. The term *homeostatic regulation* refers to the adjustments in physiological systems that are responsible for the preservation of homeostasis.

Homeostasis is absolutely vital—a failure of homeostatic regulation means illness and, in many cases, death within a relatively short period. The principle of homeostasis is the central theme of this text, and the foundation for all modern physiology. An understanding of homeostatic regulation will help you to make accurate predictions about the body's responses to normal and abnormal conditions.

Two general mechanisms are involved in homeostatic regulation: *autoregulation* and *extrinsic regulation.*

- **Autoregulation** occurs when the activities of a cell, tissue, organ, or system change automatically when faced with some environmental variation. For example, when the cells within a tissue need more oxygen, they release chemicals that dilate blood vessels in the area. This dilation accelerates the rate of blood flow and provides more oxygen to the region.
- **Extrinsic regulation** results from the activities of the nervous or endocrine system, organ systems that can control or adjust the activities of many different systems simultaneously. For example, when you are exercising, the nervous system issues commands that increase the heart rate so that blood will circulate faster. The nervous system also reduces blood flow to organs, such as the digestive tract, that are relatively inactive. The oxygen in the circulating blood is thus saved for the active muscles.

In general, the nervous system performs crisis management by directing rapid, short-term, and very specific responses. When you accidentally set your hand on a hot stove, the rising temperature produces a painful, localized disturbance of homeostasis. The nervous system responds by ordering the contraction of specific muscles that will pull the hand away from the stove. The effects last only as long as the neural activity continues, usually a matter of seconds.

By contrast, the endocrine system releases chemical messengers, called *hormones*, that affect tissues and organs throughout the body. The responses may not be immediately apparent, but when the effects appear they often persist for days or weeks. Examples of endocrine function include the long-term regulation of blood volume and composition and the adjustment of organ system function during starvation or stress.

Regardless of which system is involved, homeostatic regulation always focuses on *the stabilization of a particular aspect of the internal environment.* The regulatory mechanism is activated by a *stimulus* that appears or changes when conditions become abnormal. The mechanism itself consists of a **receptor** sensitive to that particular stimulus and an **effector** whose activity has an effect on the same stimulus. A **control center,** or *integration center,* is placed between the receptor and the effector. You are probably already familiar with several examples of homeostatic regulation, although not in those terms. As an example, consider the operation of the thermostat in a house or apartment (Figure 1-4●).

The thermostat is a control center that monitors room temperature. The dial on the thermostat establishes the **set point,** or optimal level for the controlled condition—in this case, the temperature you find most comfortable. In our example, the set point is at 22° C (about 72° F). The function of the thermostat is to keep room temperature within acceptable limits, usually within a degree or so of the set point. This thermostat receives information from a receptor (a thermometer exposed to the air in the room), and it controls an effector (an air conditioner).

When the temperature at the thermometer (a receptor) increases outside the normal range, the

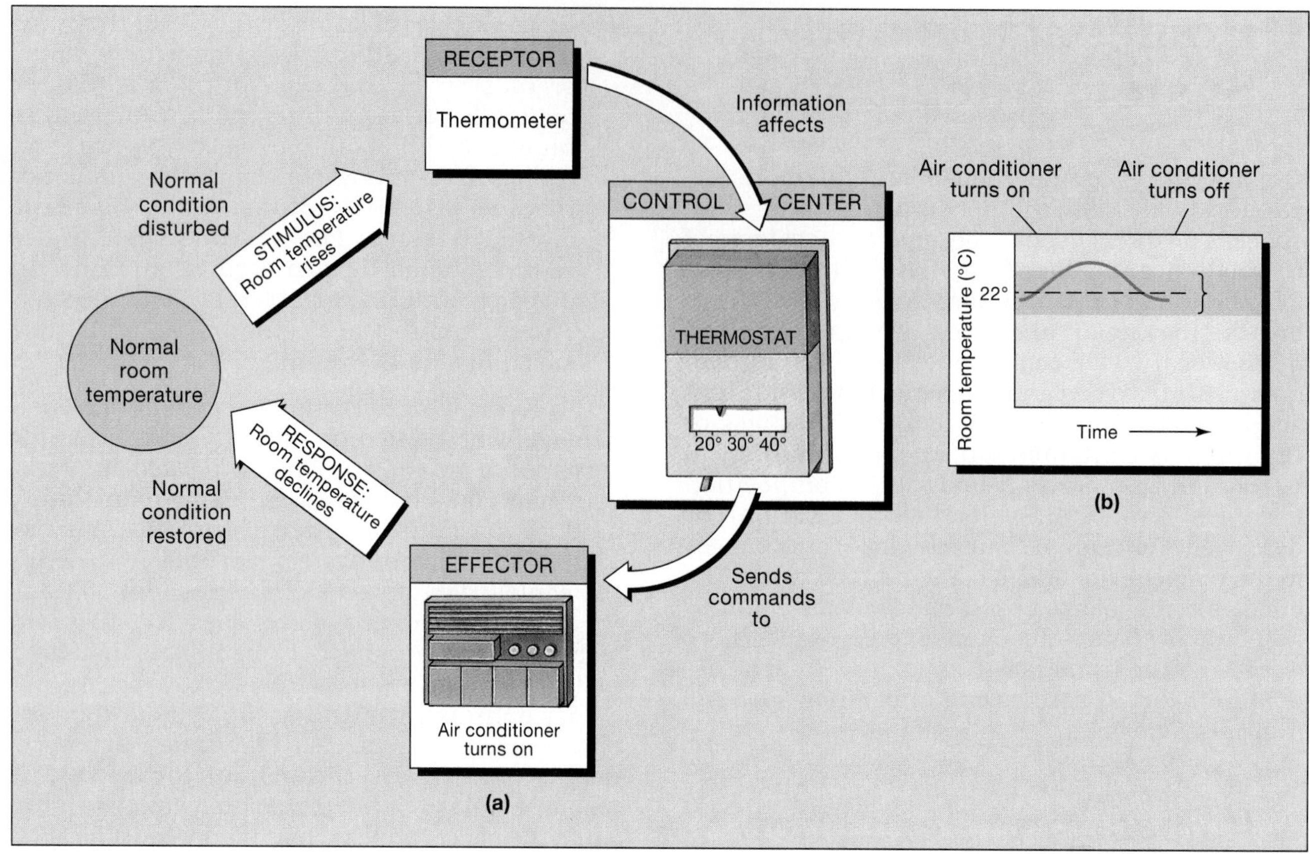
RECEPTOR
Thermometer
Information affects
CONTROL CENTER
THERMOSTAT
20° 30° 40°
Normal condition disturbed
STIMULUS: Room temperature rises
Normal room temperature
RESPONSE: Room temperature declines
Normal condition restored
EFFECTOR
Air conditioner turns on
Sends commands to
(a)
Air conditioner turns on
Air conditioner turns off
Room temperature (°C)
22°
Time
(b)

● **FIGURE 1-4**

Negative Feedback in Room Temperature Control. (a) A thermostat stabilizes room temperature by turning on an air conditioner or heater as needed to keep room temperature within acceptable limits. Whether the room temperature rises or falls, the thermostat (a control center) triggers an effector response that restores normal temperature. When room temperature rises, the thermostat turns on the air conditioner, and the room temperature returns to normal levels. **(b)** With this regulatory system, room temperature oscillates near the thermostatic set point.

thermostat (a control center) turns on the air conditioner (an effector). The air conditioner then cools the room (Figure 1-4a●). When the temperature at the thermometer approaches the set point, the thermostat turns off the air conditioner (Figure 1-4b●).

Negative Feedback

Figure 1-5

The essential feature of the example we have been discussing can be summarized very simply: *A variation outside of normal limits triggers an automatic response that restores homeostasis.* This method of homeostatic regulation is called **negative feedback** because an effector activated by the control center acts to eliminate or reduce the magnitude of the stimulus.

Most homeostatic mechanisms in the body involve negative feedback. For example, consider the control of body temperature, a process called *thermoregulation.* Thermoregulation involves altering the relationship between heat loss, which occurs primarily at the body surface, and heat production, which occurs in all active tissues.

The thermoregulatory control center is located in the brain. It receives information from two sets of temperature receptors, one in the skin and the other in the control center in the brain. At the set point, the general body temperature will be approximately 37° C (98.6° F) (Figure 1-5a●). If body temperature rises above 37.2° C, activity in the control center targets two different effectors: (1) smooth muscles in the walls of blood vessels supplying the skin and (2) sweat glands. The blood vessels dilate, increasing blood flow through vessels near the body surface, and the sweat glands accelerate their secretion. The skin then acts like a radiator, losing heat to the environment, and the evaporation of sweat speeds the process. When body temperature returns to normal, the control center becomes inactive, and superficial blood flow and sweat gland activity decrease to previous levels.

Negative feedback is the primary mechanism for homeostatic regulation and provides long-term control over internal conditions and systems. Homeostatic mechanisms using negative feedback normally ignore minor variations, and they maintain a normal *range* rather than a fixed value. In

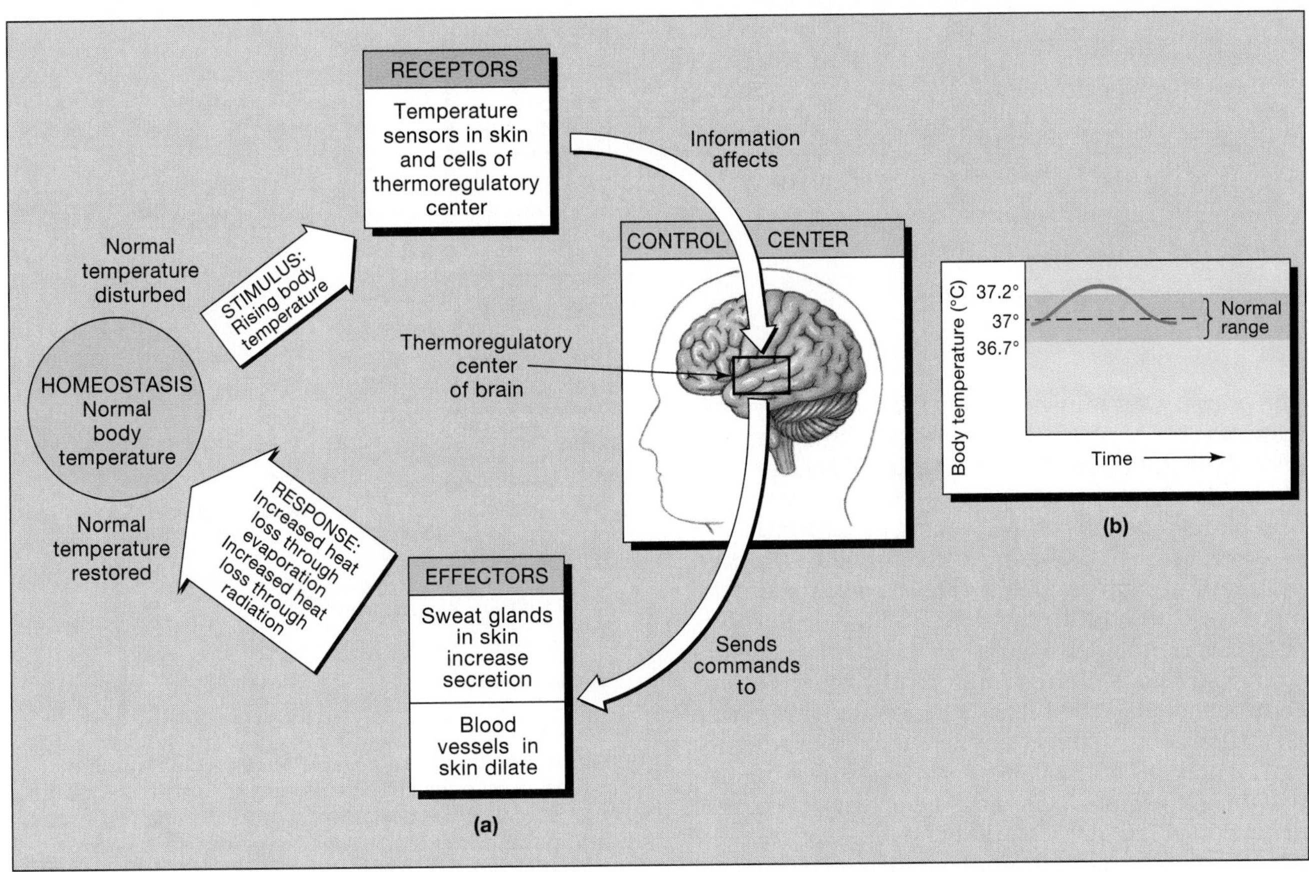
RECEPTORS
Temperature sensors in skin and cells of thermoregulatory center
Information affects
CONTROL CENTER
Normal temperature disturbed
STIMULUS: Rising body temperature
HOMEOSTASIS Normal body temperature
Thermoregulatory center of brain
Normal temperature restored
RESPONSE: Increased heat loss through evaporation Increased heat loss through radiation
EFFECTORS
Sweat glands in skin increase secretion
Blood vessels in skin dilate
Sends commands to
(a)
37.2°
37°
36.7°
Normal range
Body temperature (°C)
Time
(b)

● **FIGURE 1-5**
Negative Feedback in Body Temperature Control. (a) Comparable events occur in the regulation of body temperature. A control center in the brain functions as a thermostat with a set point of 37° C. If body temperature climbs above 37.2° C, heat loss is increased through enhanced blood flow to the skin and increased sweating. **(b)** The thermoregulatory center keeps body temperature oscillating within an acceptable range, usually between 36.7° and 37.2° C.

the example above, body temperature oscillates around the "ideal" set-point temperature (Figure 1-5b●). Thus any measured value, such as body temperature, can vary from moment to moment or day to day for any single individual.

The variability *between* individuals is even greater, for each person has homeostatic set points that may differ due to genetic factors, age, gender, or environmental conditions. It is therefore impractical to define "normal" homeostatic conditions very precisely. By convention, physiological values are reported either as averages, the average value obtained by sampling a large number of individuals, or as a range that includes 95 percent or more of the sample population. For instance, 5 percent of normal adults have a body temperature outside the normal range (below 36.7° C or above 37.2° C). But these temperatures are perfectly normal *for them,* and the variations have no clinical significance. This variability should be kept in mind when reviewing a lab report or clinical discussion, since an unusual value—even one outside the normal range—may represent individual variation rather than a homeostatic malfunction.

Positive Feedback

Figure 1-6

In **positive feedback,** the *initial stimulus produces a response that exaggerates the stimulus.* For example, suppose that the thermostat was wired so that when the temperature rose it would turn on a *heater* rather than the air conditioner. In that case the initial stimulus (rising room temperature) would cause a response (heater turns on) that would exaggerate the stimulus. The room temperature will continue to rise until some external factor switches off the thermostat, unplugs the heater, or intervenes in some other way.

Control mechanisms relying on positive feedback are not nearly as common as those involving negative feedback. However, *positive feedback is important in controlling physiological processes that, once initiated, must be completed quickly.* For example, consider the process of labor that results in the delivery of a newborn infant. Once labor begins, it is important that it be completed as soon as possible to avoid unduly stressing both mother and infant.

Homeostasis disturbed
STIMULUS: Extreme distortion of uterus
RECEPTORS
Stretch receptors of uterine wall
Information affects
CONTROL CENTER
Endocrine center in brain
Sends commands to
EFFECTORS
Muscles in uterine wall contract
RESPONSE: Contraction and further distortion of uterus
Positive feedback loop

● **FIGURE 1-6**
Positive Feedback. In positive feedback, a stimulus produces a response that reinforces the original stimulus. Positive feedback is important in accelerating processes that must proceed to completion rapidly. In this example, positive feedback enhances labor contractions that continue until delivery has been completed.

The primary stimulus initiating labor and delivery is distortion of the uterus by the growing fetus. The trigger for the initiation of labor contractions is a rise in the uterine level of *oxytocin* (ok-si-TO-sin), a hormone that stimulates the contractions of uterine muscles. As we shall see in Chapter 29, the oxytocin is derived from several sources. One source of oxytocin, and the regulatory mechanism that controls its secretion, has been diagrammed in Figure 1-6●. Stretch receptors in the uterine wall are monitored by an endocrine control center in the brain. When sufficient uterine distortion occurs, the control center releases oxytocin. As uterine contractions occur, the fetus is pushed toward the vagina. This movement causes extreme distortion of the lower portion of the uterus, triggering the release of additional oxytocin. The uterine contractions then become more forceful, leading to greater movement and distortion. Each time the control center responds, the action of the effectors causes an increase in receptor stimulation. This kind of cycle, a *positive feedback loop*, can be broken only by some external force or process—in this instance, the delivery of the newborn infant, which eliminates the uterine distortion. Labor and delivery will be examined more carefully in Chapter 29; blood clotting, another important example of positive feedback, will be considered in Chapter 19.

 Why is homeostatic regulation important to human beings?

 What happens to the body when homeostasis breaks down?

 Why is positive feedback helpful in blood clotting but unsuitable for regulation of body temperature?

❑ Homeostasis and Disease

Physiological mechanisms do a remarkably good job of maintaining a constant internal environment, regardless of our ongoing activities. But when homeostatic regulation fails, as a result of infection, injury, or genetic abnormality, organ systems begin to malfunction and the individual experiences the symptoms of illness, or **disease.** The chapters that follow devote considerable attention to the mechanisms responsible for a variety of human diseases. Specific conditions are presented throughout the text. An understanding of normal homeostatic mechanisms can usually enable you to draw conclusions about what might be responsible for observed symptoms. Major organizational and functional patterns relevant to clinical practice are discussed in the *Applications Manual.* AM Homeostatic Failure

A Frame of Reference for Anatomical Studies

Early anatomists faced serious problems in communication. Stating that a bump is "on the back" does not give very precise information about its location. So anatomists created maps of the human body. The landmarks are prominent anatomical structures, distances are measured in centimeters or inches, and specialized directional terms are used. In effect, anatomy uses a special language that must be learned almost at the start.

A familiarity with Latin roots and patterns makes anatomical terms more understandable. As new terms are introduced, notes on pronunciation and relevant word roots will be provided. Additional information on foreign word roots, prefixes, suffixes, and combining forms can be found on the front endpapers.

Latin and Greek terms are not the only foreign terms imported into the anatomical vocabulary over the centuries, and the vocabulary continues to expand. Many anatomical structures and clinical conditions were initially named after either the discoverer or, in the case of diseases, the most famous victim. Over the last 100 years most of these commemorative names, or *eponyms*, have been replaced by more precise terms. Those interested in historical details should refer to the relevant sections of the *Applications Manual.* AM *Historical Perspectives in Anatomy and Physiology; The Language of Anatomy and Physiology.*

Superficial Anatomy

A familiarity with anatomical landmarks and directional references will make subsequent chapters more understandable because none of the organ systems except the integument can be seen from the body surface. You must create your own mental maps and extract information from the anatomical illustrations that accompany this discussion.

Anatomical Landmarks

Figure 1-7

Important anatomical landmarks are presented in Figure 1-7●. The anatomical terms are given in boldface, the common names in plain type, and the anatomical adjectives in parentheses. You should become familiar with all three terms. For example, the term *brachium* refers to the arm, and later chapters will discuss the brachial artery, brachial nerve, and so forth. Understanding the terms and their origins will help you to remember the location of a particular structure, as well as its name.

Standard anatomical illustrations show the human form in the **anatomical position.** In this position the hands are at the sides with the palms facing forward. Figure 1-7 shows an individual in the anatomical position as seen from the front (Figure 1-7a●) and back (Figure 1-7b●). Unless otherwise noted, all the descriptions given in this text refer to the body in the anatomical position. A person lying down in the anatomical position is said to be **supine** (SŪ-pīne) when lying face up and **prone** when lying face down.

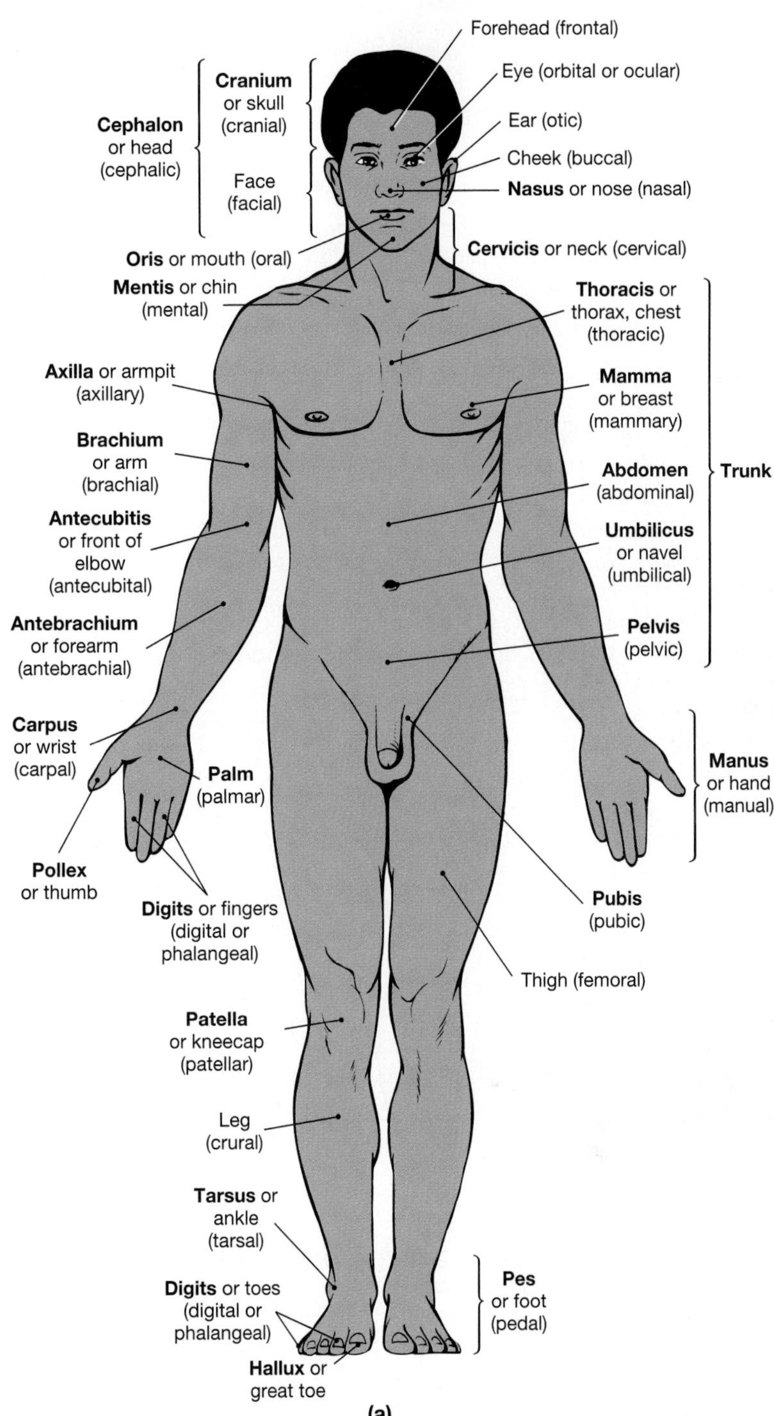

Forehead (frontal)
Cranium or skull (cranial)
Eye (orbital or ocular)
Cephalon or head (cephalic)
Ear (otic)
Face (facial)
Cheek (buccal)
Nasus or nose (nasal)
Oris or mouth (oral)
Cervicis or neck (cervical)
Mentis or chin (mental)
Thoracis or thorax, chest (thoracic)
Axilla or armpit (axillary)
Mamma or breast (mammary)
Brachium or arm (brachial)
Abdomen (abdominal)
Trunk
Antecubitis or front of elbow (antecubital)
Umbilicus or navel (umbilical)
Antebrachium or forearm (antebrachial)
Pelvis (pelvic)
Carpus or wrist (carpal)
Manus or hand (manual)
Palm (palmar)
Pollex or thumb
Pubis (pubic)
Digits or fingers (digital or phalangeal)
Thigh (femoral)
Patella or kneecap (patellar)
Leg (crural)
Tarsus or ankle (tarsal)
Pes or foot (pedal)
Digits or toes (digital or phalangeal)
Hallux or great toe
(a)

● **FIGURE 1-7**

Anatomical Landmarks. The anatomical terms are shown in boldface type, the common names are in plain type, and the anatomical adjectives are in parentheses.

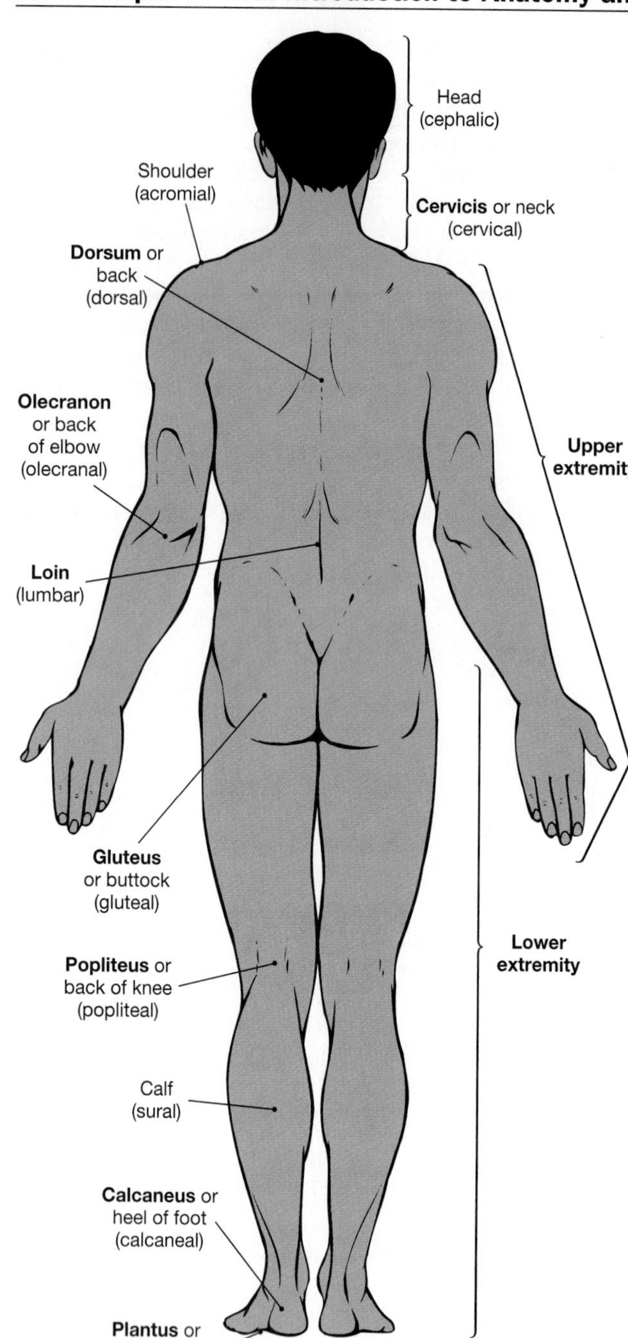

Head
(cephalic)
Shoulder
(acromial)
Cervicis or neck
(cervical)
Dorsum or
back
(dorsal)
Olecranon
or back
of elbow
(olecranal)
Upper
extremity
Loin
(lumbar)
Gluteus
or buttock
(gluteal)
Lower
extremity
Popliteus or
back of knee
(popliteal)
Calf
(sural)
Calcaneus or
heel of foot
(calcaneal)
Plantus or
sole of foot
(plantar)
(b)

● **FIGURE 1-7 continued**

Anatomical Regions

Figures 1-7, 1-8

Major regions of the body are listed in Table 1-1 and shown in Figure 1-7●. In addition to specific landmarks, anatomists and clinicians often need to use regional terms to describe a general area of interest or injury. Two approaches have developed, both concerned with mapping the surface of the abdominopelvic region. Clinicians refer to the **abdominopelvic quadrants.** The region is divided into four segments using a pair of imaginary lines that intersect at the *umbilicus* (navel). This simple method, shown in Figure 1-8a●, provides useful references for the description of aches, pains, and injuries. The location can assist the doctor in deciding the possible cause; for example, tenderness in the right lower quadrant (RLQ) is a symptom of appendicitis, whereas tenderness in the right upper quadrant (RUQ) may indicate gallbladder or liver problems.

TABLE 1-1 Regions of the Human Body (see Figure 1-7)

Structure	*Area*
Cephalon (head)	Cephalic region
Cervicis (neck)	Cervical region
Thoracis (chest)	Thoracic region
Abdomen	Abdominal region
Pelvis	Pelvic region
Loin (lower back)	Lumbar region
Buttock	Gluteal region
Pubis (anterior pelvis)	Pubic region
Brachium (arm)	Brachial region
Antebrachium (forearm)	Antebrachial region
Manus (hand)	Manual region
Thigh	Femoral region
Leg (anterior)	Crural region
Calf	Sural region
Pes (foot)	Pedal region

Anatomists like to use more precise regional distinctions to describe the location and orientation of internal organs. They recognize nine **abdominopelvic regions** (Figure 1-8b●). Figure 1-8c● shows the relationship among quadrants, regions, and internal organs.

Anatomical Directions

Figure 1-9

Table 1-2 and Figure 1-9● show the principal directional terms and examples of their use. There are many different terms, and some can be used interchangeably. For example, *anterior* refers to the front of the body, when viewed in the anatomical position; in human beings, this term is equivalent to *ventral*, which actually refers to the belly. Although your instructor may have additional recommendations, the terms that appear frequently in later chapters have been emphasized. Before continuing, take the time to review Table 1-2 in detail, and practice using these terms at every opportunity. If you are familiar with the basic vocabulary, you will find all the descriptions in future chapters easier to follow. When following subsequent anatomical descriptions, you will find it useful to remember that the terms *left* and *right* always refer to the left and right sides of the *subject*, not of the observer.

❑ Sectional Anatomy

A presentation in sectional view is sometimes the only way to illustrate the relationships among the parts of a three-dimensional object. An understanding of sectional views has become increasingly important since the development of electronic imaging techniques that enable us to see inside the living body without resorting to surgery.

Planes and Sections

Figures 1-10, 1-11

Any slice through a three-dimensional object can be described with reference to three **sectional planes,** indicated in Table 1-3 and Figure 1-10●. The **transverse plane** lies at right angles to the long axis of the body, dividing it into **superior** and **inferior** sections. A cut in this plane is called a **transverse section,** or *cross section.* The **frontal,** or **coronal, plane** and the **sagittal plane** parallel the long axis of the body. The frontal plane extends from side to side, dividing the body into **anterior** and **posterior** sections. The sagittal plane extends from front to back, dividing the body into *left* and *right* sections. A cut that passes along the midline and divides the body into left and right halves is a **midsagittal section;** a cut parallel to the midsagittal line is a **parasagittal section.**

Sometimes it is helpful to compare the information provided by sections made along different planes. You can experiment with this procedure by mentally sectioning this book, as in Figure 1-11a●. (Performing this experiment is not recommended, unless you are dropping the course.) Each sectional plane provides a different perspective on the structure of the book; when combined with observations of the external anatomy, they create a reasonably complete picture.

Obtaining a more accurate and detailed picture would entail choosing one sectional plane and making a series of sections at small intervals. This process, called **serial reconstruction,** permits the analysis of

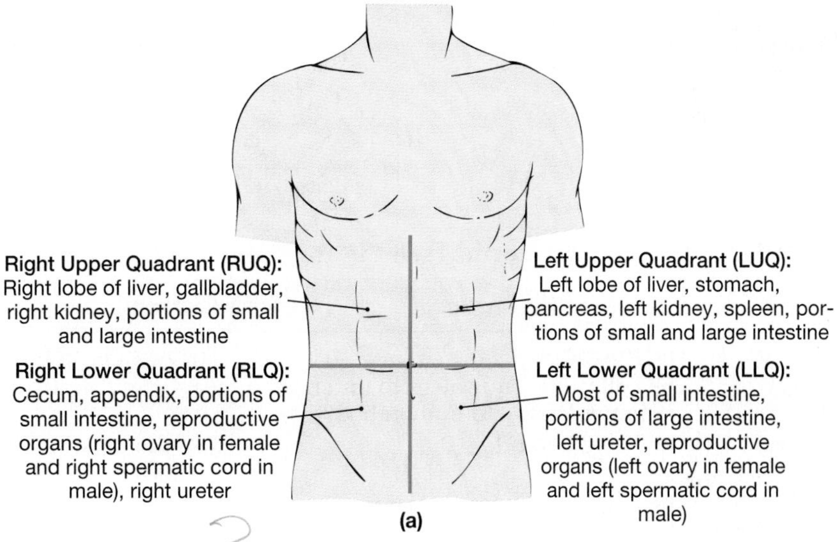

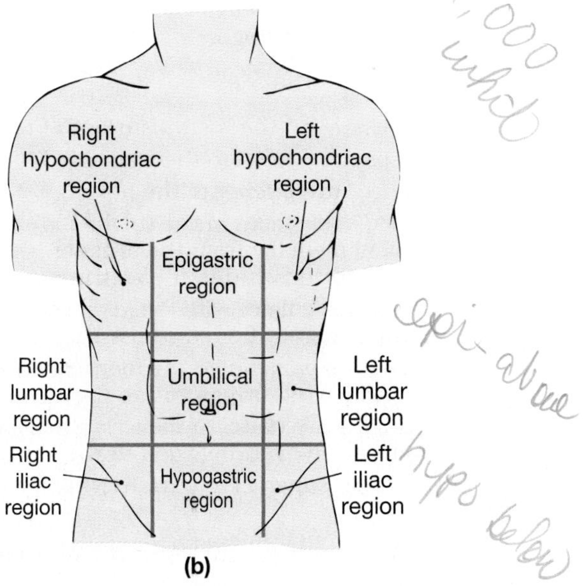

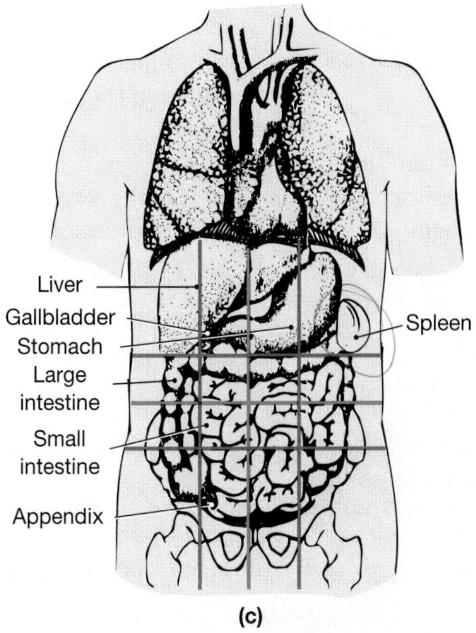

Right Upper Quadrant (RUQ):
Right lobe of liver, gallbladder, right kidney, portions of small and large intestine
Left Upper Quadrant (LUQ):
Left lobe of liver, stomach, pancreas, left kidney, spleen, portions of small and large intestine
Right Lower Quadrant (RLQ):
Cecum, appendix, portions of small intestine, reproductive organs (right ovary in female and right spermatic cord in male), right ureter
Left Lower Quadrant (LLQ):
Most of small intestine, portions of large intestine, left ureter, reproductive organs (left ovary in female and left spermatic cord in male)
(a)
Right hypochondriac region
Left hypochondriac region
Epigastric region
Right lumbar region
Umbilical region
Left lumbar region
Right iliac region
Hypogastric region
Left iliac region
(b)
Liver
Gallbladder
Stomach
Large intestine
Small intestine
Appendix
Spleen
(c)

● **FIGURE 1-8**
Abdominopelvic Quadrants and Regions. (a) Abdominopelvic quadrants divide the area into four sections. These terms, or their abbreviations, are most often used in clinical discussions. **(b)** More precise regional descriptions are provided by reference to the appropriate abdominopelvic region. **(c)** Quadrants or regions are useful because there is a known relationship between superficial anatomical landmarks and underlying organs.

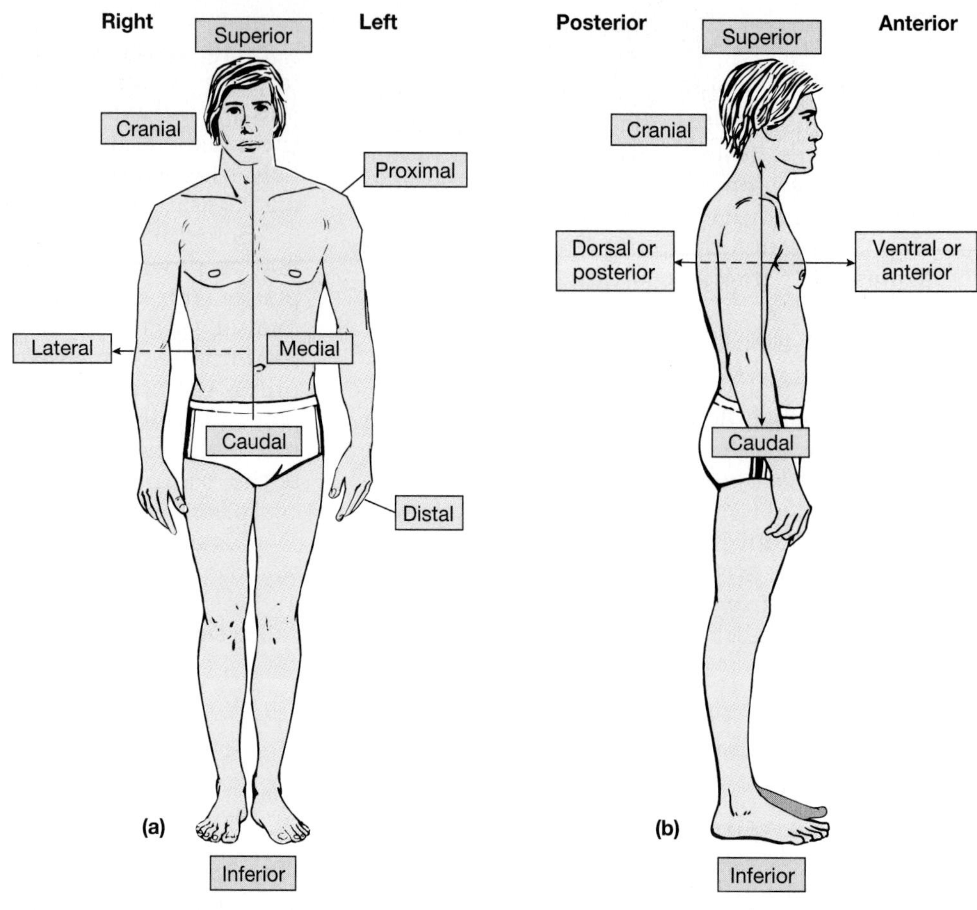
Right
Left
Superior
Cranial
Proximal
Lateral
Medial
Caudal
Distal
(a)
Inferior
Posterior
Anterior
Superior
Cranial
Dorsal or posterior
Ventral or anterior
Caudal
(b)
Inferior

FIGURE 1-9 Directional References. Important directional terms used in this text are indicated by arrows; definitions and descriptions are included in Table 1-2.

TABLE 1-2 Directional Terms (see Figure 1-9)

Term	*Region or Reference*	*Example*
Anterior	The front; before	The navel is on the *anterior (ventral)* surface of the trunk.
Ventral	The belly side (equivalent to anterior when referring to human body)	
Posterior	The back; behind	The shoulder blade is located posterior *(dorsal)* to the rib cage.
Dorsal	The back (equivalent to posterior when referring to human body)	The *dorsal* body cavity encloses the brain and spinal cord.
Cranial or cephalic	The head	The *cranial,* or *cephalic,* border of the pelvis is *superior* to the thigh.
Superior	Above; at a higher level (in human body, toward the head)	
Rostral	The nose	The eyes are on the *rostral* surface of the head.
Caudal	The tail (coccyx in humans)	The hips are *caudal* to the waist.
Inferior	Below; at a lower level	The knees are *inferior* to the hips.
Medial	Toward the body's longitudinal axis	The *medial* surfaces of the thighs may be in contact; moving medially from the arm across the chest surface brings you to the sternum.
Lateral	Away from the body's longitudinal axis	The thigh articulates with the *lateral* surface of the pelvis; moving laterally from the nose brings you to the eyes.
Proximal	Toward an attached base	The thigh is *proximal* to the foot; moving proximally from the wrist brings you to the elbow.
Distal	Away from an attached base	The fingers are *distal* to the wrist; moving distally from the elbow brings you to the wrist.
Superficial	At, near, or relatively close to the body surface	The skin is *superficial* to underlying structures.
Deep	Farther from the body surface	The bone of the thigh is *deep* to the surrounding skeletal muscles.

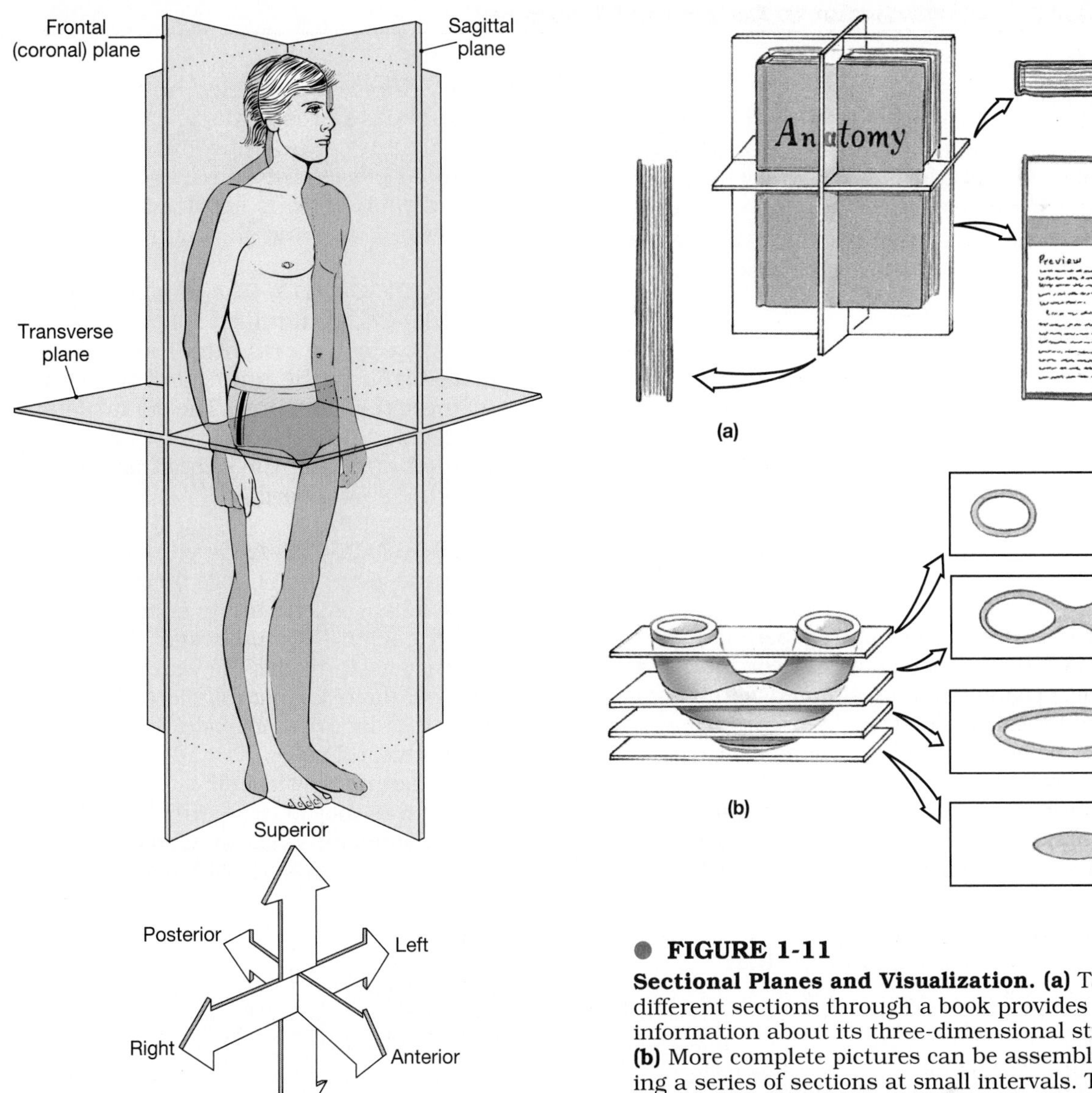

Frontal (coronal) plane
Sagittal plane
Transverse plane
Superior
Posterior
Left
Right
Anterior
Inferior
Anatomy
(a)
(b)

● **FIGURE 1-10**

Planes of Section. The three primary planes of section are indicated here. Table 1-3 defines and describes them.

● **FIGURE 1-11**

Sectional Planes and Visualization. (a) Taking three different sections through a book provides detailed information about its three-dimensional structure. **(b)** More complete pictures can be assembled by taking a series of sections at small intervals. This process is called serial reconstruction. Notice how the sectional views change; although it is a simple tube, a piece of elbow macaroni can look like a pair of tubes, a dumbbell, an oval, or a solid, depending on where the section was taken. The effects of sectional plane should be kept in mind when looking at slides under the microscope.

TABLE 1-3 Terms That Indicate Planes of Section (see Figure 1-10)

Orientation of Plane	*Adjective*	*Directional Reference*	*Description*
Parallel to long axis	Sagittal	Sagittally	A *sagittal section* separates right and left portions. You examine a sagittal section, but you section sagittally.
	Midsagittal		In a *midsagittal section* the plane passes through the midline, dividing the body in half and separating right and left sides.
	Parasagittal		A *parasagittal section* misses the midline, separating right and left portions of unequal size.
	Frontal or coronal	Frontally or coronally	A *frontal*, or *coronal, section* separates anterior and posterior portions of the body; coronal usually refers to sections passing through the skull.
Perpendicular to long axis	Transverse or horizontal	Transversely or horizontally	A *transverse*, or *horizontal, section* separates superior and inferior portions of the body.

relatively complex structures. Figure 1-11b● shows the serial reconstruction of a simple bent tube. The same procedure could be used to visualize the path of a small blood vessel or to follow a loop of the intestine. Serial reconstruction is an important method for studying histological structure and for analyzing the images produced by sophisticated clinical procedures.

Body Cavities

Figures 1-8, 1-12, 1-13

Viewed in sections, the human body is not a solid object, like a rock, in which all of the parts are fused together. Many vital organs are suspended in internal chambers called *body cavities.* These cavities have two essential functions: (1) They protect delicate organs, such as the brain and spinal cord, from accidental shocks and cushion them from the thumps and bumps that occur during walking, jumping, and running; and (2) they permit significant changes in the size and shape of visceral organs. For example, because they are situated within body cavities, the lungs, heart, stomach, intestines, urinary bladder, and many other organs can expand and contract without distorting surrounding tissues and disrupting the activities of nearby organs.

A **dorsal body cavity** surrounds the brain and spinal cord, and a much larger **ventral body cavity,** or **coelom** (SĒ-lom; *koila,* cavity), surrounds developing organs of the respiratory, cardiovascular, digestive, urinary, and reproductive systems. Relationships between the dorsal and ventral body cavities and their various subdivisions are indicated in Figure 1-12● and shown in Figure 1-13●.

DORSAL BODY CAVITY The dorsal body cavity (Figure 1-13a●) is a fluid-filled space whose limits are established by the **cranium,** the bones of the skull that surround the brain, and the *vertebral arches* of the spinal vertebrae. The dorsal body cavity is subdivided into the **cranial cavity,** which encloses the brain, and the **spinal cavity,** which surrounds the spinal cord.

VENTRAL BODY CAVITY As development proceeds, internal organs grow and change their relative positions. These changes lead to the subdivision of the ventral body cavity. The **diaphragm** (DĪ-a-fram), a flat muscular sheet, divides the ventral body cavity into a superior *thoracic cavity,* enclosed by the chest wall, and an inferior *abdominopelvic cavity,* enclosed by the abdomen and pelvis.

Many of the organs within these cavities change size and shape as they perform their functions. For example, the stomach swells at each meal, and the heart contracts and expands with each beat. These

● **FIGURE 1-12**
Relationships of the Various Body Cavities

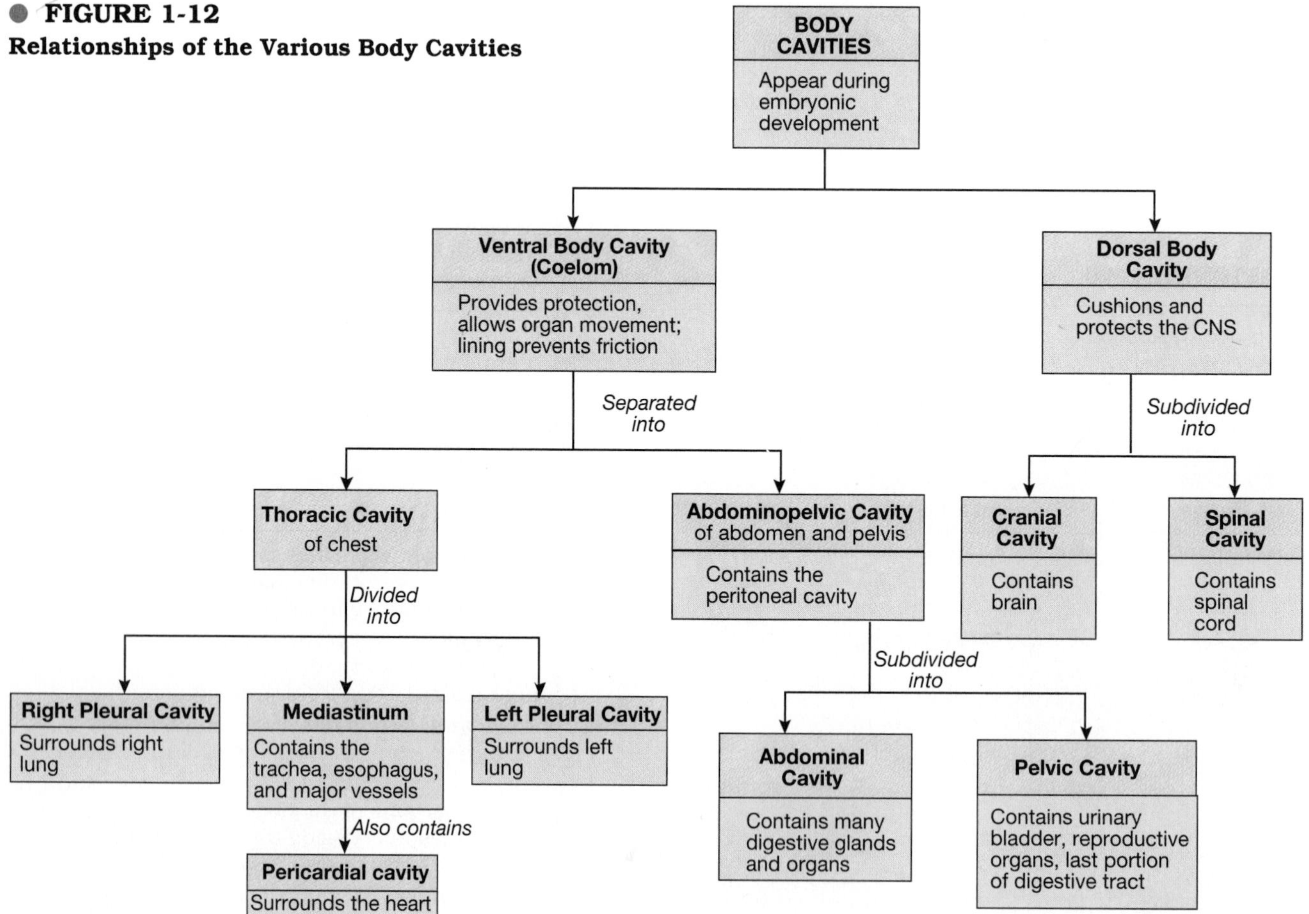
BODY CAVITIES
Appear during embryonic development
Ventral Body Cavity (Coelom)
Provides protection, allows organ movement; lining prevents friction
Dorsal Body Cavity
Cushions and protects the CNS
Separated into
Subdivided into
Thoracic Cavity of chest
Abdominopelvic Cavity of abdomen and pelvis
Contains the peritoneal cavity
Cranial Cavity
Contains brain
Spinal Cavity
Contains spinal cord
Divided into
Subdivided into
Right Pleural Cavity
Surrounds right lung
Mediastinum
Contains the trachea, esophagus, and major vessels
Left Pleural Cavity
Surrounds left lung
Also contains
Pericardial cavity
Surrounds the heart
Abdominal Cavity
Contains many digestive glands and organs
Pelvic Cavity
Contains urinary bladder, reproductive organs, last portion of digestive tract

● **FIGURE 1-13**

Body Cavities. (a) The dorsal body cavity is bounded by the bones of the skull and vertebral column. The muscular diaphragm divides the ventral body cavity into a superior thoracic (chest) cavity and an inferior abdominopelvic cavity. The pericardial cavity is located inside the chest cavity. **(b)** The heart is suspended within the pericardial cavity like a fist pushed into a balloon. The attachment site, corresponding to the wrist of the hand in the model, lies at the connection between the heart and major blood vessels. **(c)** An anterior view of the ventral body cavity, showing the central location of the pericardial cavity within the chest cavity. **(d)** The relationships are seen more clearly in this sectional plane, which shows how the mediastinum divides the thoracic cavity into two pleural cavities.

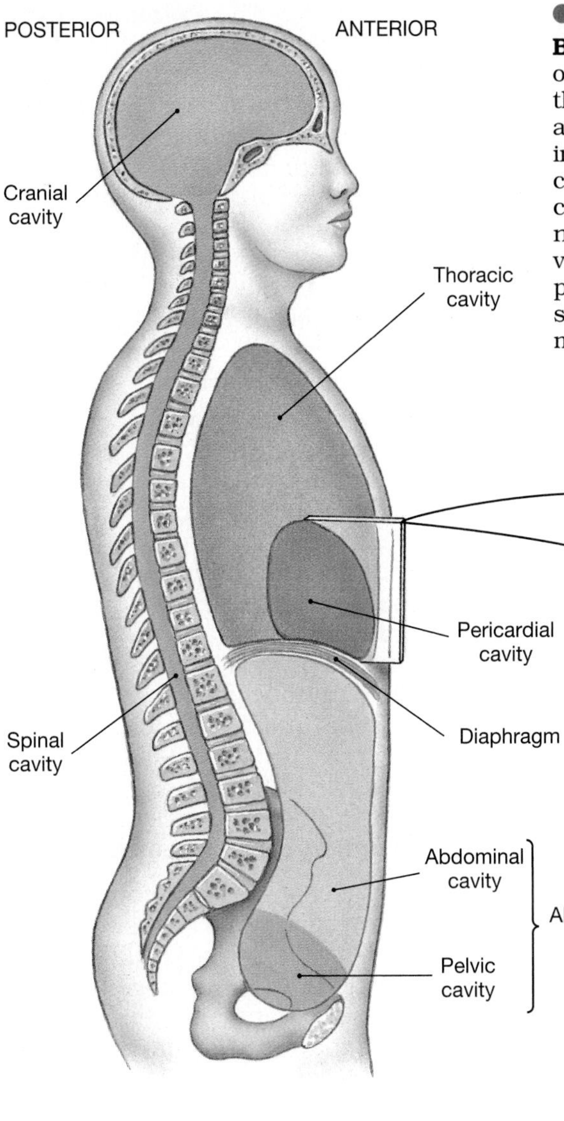

POSTERIOR
ANTERIOR
Cranial cavity
Thoracic cavity
Pericardial cavity
Diaphragm
Spinal cavity
Abdominal cavity
Pelvic cavity
Abdominopelvic cavity

(a)

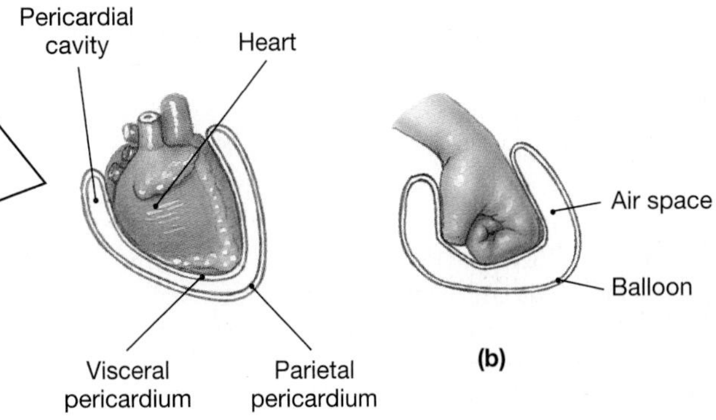

Pericardial cavity
Heart
Visceral pericardium
Parietal pericardium
Air space
Balloon

(b)

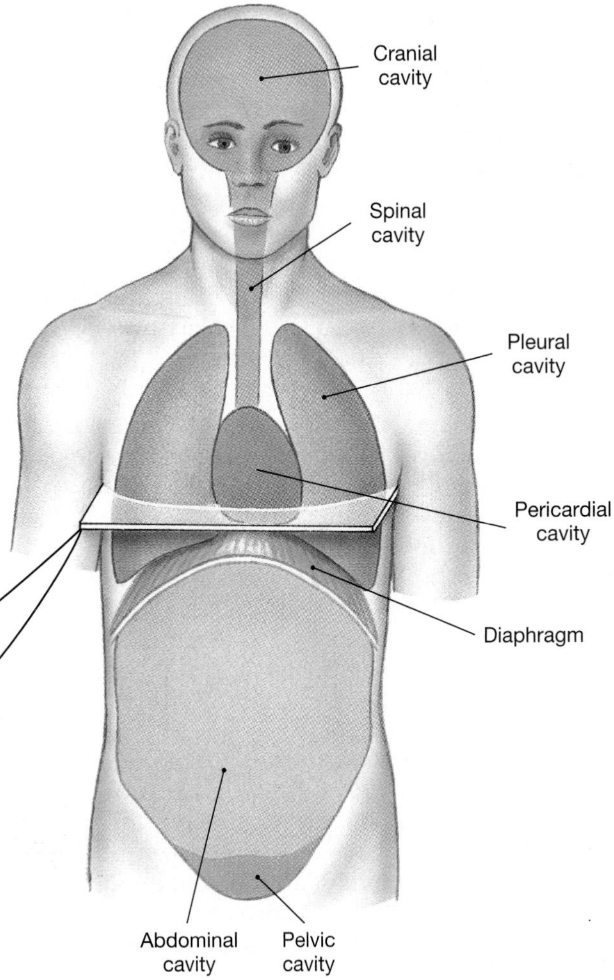

Cranial cavity
Spinal cavity
Pleural cavity
Pericardial cavity
Diaphragm
Abdominal cavity
Pelvic cavity

(c)

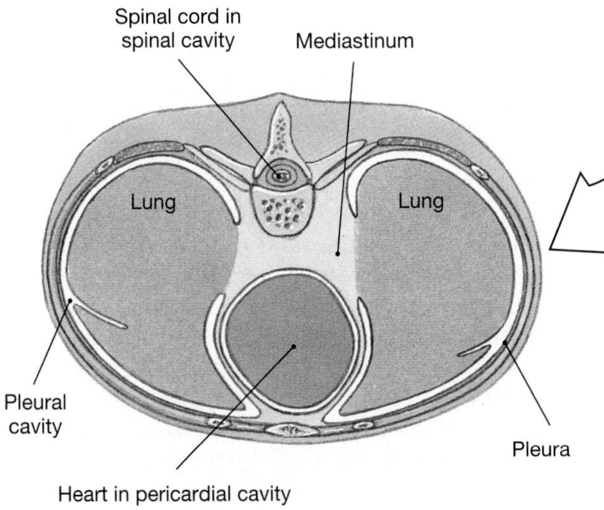

Spinal cord in spinal cavity
Mediastinum
Lung
Lung
Pleural cavity
Pleura
Heart in pericardial cavity

(d)

organs project into moist internal spaces that permit expansion and limited movement, but prevent friction. There are three such chambers in the thoracic cavity and one in the abdominopelvic cavity. The internal organs that project into these cavities are called **viscera** (VIS-e-ra).

The Thoracic Cavity. The walls of the **thoracic cavity** surround the lungs and heart, associated organs of the respiratory, cardiovascular, and lymphatic systems, the inferior portions of the esophagus, and the thymus gland. The thoracic cavity contains three internal chambers: a single *pericardial cavity* and a pair of *pleural cavities.* These cavities are lined by shiny, slippery, and delicate *serous membranes.*

- The heart projects into the **pericardial cavity.** The relationship between the heart and the cavity resembles that of a fist pushing into a balloon (Figure 1-13b•). The wrist corresponds to the base of the heart, and the balloon corresponds to the serous membrane lining the pericardial cavity. During each beat, the heart changes in size and shape. The pericardial cavity permits these changes, and the slippery lining of the cavity prevents friction between the heart and adjacent structures. The serous membrane is called the **pericardium** (*peri-*, around + *kardia,* heart). The layer covering the heart is the *visceral pericardium,* and the opposing surface is the *parietal pericardium.*

  The pericardium lies within the **mediastinum** (mē-dē-as-TĪ-num or mē-dē-AS-ti-num). The mediastinum is the portion of the thoracic cavity that lies between the left and right *pleural cavities.* The connective tissue of the mediastinum surrounds the pericardial cavity and heart, the large arteries and veins attached to the heart, and the thymus, trachea, and esophagus.
- There is one **pleural cavity** on each side of the mediastinum. Each pleural cavity encloses a lung, and the relationship between a lung and a pleural cavity is the same as that between the heart and the pericardial cavity. The serous membrane lining a pleural cavity is called a **pleura** (PLOO-ra). The outer surfaces of the lungs are covered by the *visceral pleura,* and the *parietal pleura* covers the opposing mediastinal surfaces and the inner body wall.

The Abdominopelvic Cavity. The abdominopelvic cavity can be divided into a superior *abdominal cavity* and an inferior *pelvic cavity.* The abdominopelvic cavity contains the **peritoneal** (per-i-tō-NĒ-al) **cavity,** an internal chamber lined by a serous membrane known as the **peritoneum** (per-i-tō-NĒ-um). The *parietal peritoneum* lines the body wall. A narrow, fluid-filled space separates the parietal peritoneum from the *visceral peritoneum* that covers the enclosed organs.

- ***The abdominal cavity.*** The **abdominal cavity** extends from the inferior surface of the diaphragm to an imaginary plane extending from the inferior surface of the lowest spinal vertebra to the anterior and superior margin of the pelvic girdle. The abdominal cavity contains the liver, stomach, spleen, kidneys, pancreas, small intestine, and most of the large intestine. (The positions of these organs can be seen in Figure 1-8c•, p. 19). Many of these organs project partially or completely into the peritoneal cavity, much as the heart or lungs project into the pericardial or pleural cavities.
- ***The pelvic cavity.*** The portion of the ventral body cavity inferior to the abdominal cavity is the **pelvic cavity.** The pelvic cavity, enclosed by the bones of the pelvis, contains the last segments of the large intestine, the urinary bladder, and various reproductive organs. For example, the pelvic cavity of a female contains the ovaries, uterine tubes, and uterus; in a male, it contains the prostate gland and seminal vesicles. The inferior portion of the peritoneal cavity extends into the pelvic cavity. The superior portion of the urinary bladder in both sexes, and the uterine tubes, the ovaries, and the superior portion of the uterus in females are covered by peritoneal membrane.

This chapter provided an overview of the locations and functions of the major components of each organ system. It also introduced the anatomical vocabulary needed to follow more detailed anatomical descriptions in later chapters. Modern methods of visualizing anatomical structures in living individuals are summarized in Figures 1-14• and 1-15•. Many of the figures in later chapters contain images produced by the procedures outlined in that section.

What type of section would separate the two eyes?

If a surgeon makes an incision just inferior to the diaphragm, what body cavity will be opened?

This concludes our preview of the major topics covered in this text and the underlying concepts that will guide our discussions. The next three chapters will take you on a tour of the principal levels of organization, from individual atoms to individual human beings. As we proceed through the text, we will emphasize major structural and functional patterns. To sharpen your analytical skills, we have included critical thinking questions at the ends of subsequent chapters. An appreciation for how science "works" may help you to solve these problems, and the *Applications Manual* includes an essay on that topic. AM *The Scientific Method*

FOCUS Sectional Anatomy and Clinical Technology

The term **radiological procedures** includes not only those scanning techniques that involve the use of radiation to create a photographic or computer-generated image of internal structures, but also methods that employ radiation sources outside the body. Physicians who specialize in the performance of these procedures and the analysis of the images they produce are called **radiologists.** Radiological procedures can provide detailed information about internal systems. Figures 1-14 and 1-15● compare the views provided by several different techniques used by radiologists and other clinicians. These figures include examples of *X-rays, CT scans, MRI scans,* and *ultrasound images.* Other examples of clinical technology will be found in later chapters and in the *Applications Manual.* As you encounter them, it will be helpful to remember that when anatomical diagrams or clinical procedures present cross-sectional views of the body, the sections are presented as though the observer were standing at the feet and looking toward the head of the subject.

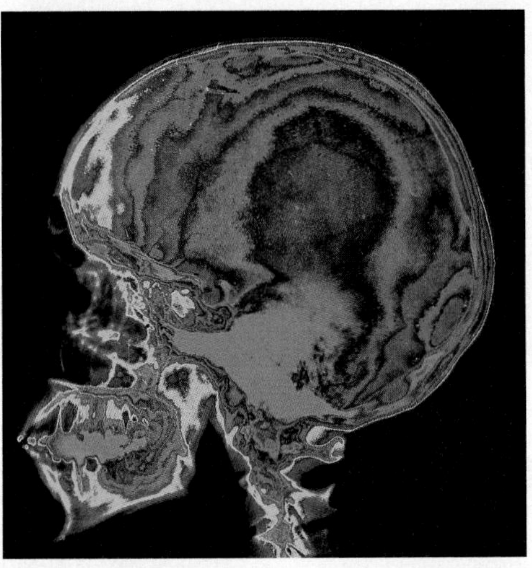

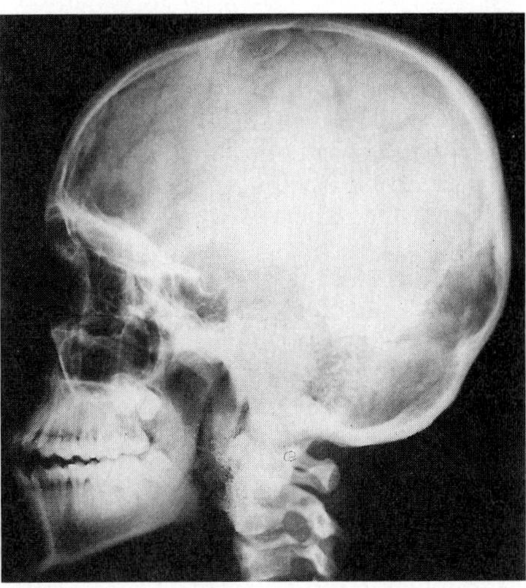

● **FIGURE 1-14**

X-rays. (a) An X-ray of the skull, taken from the left side. **X-rays** are a form of high-energy radiation that can penetrate living tissues. In the most familiar procedure, a beam of X-rays travels through the body and strikes a photographic plate. All of the projected X-rays do not arrive at the film; some are absorbed or deflected as they pass through the body. The resistance to X-ray penetration is called **radiodensity.** In the human body, the order of increasing radiodensity is as follows: air, fat, liver, blood, muscle, bone. The result is an image with radiodense tissues, such as bone, appearing in white, and less dense tissues in shades of gray to black. The picture is a two-dimensional image of a three-dimensional object; in this image it is difficult to decide whether a particular feature is on ⁺he left side (toward the viewer) or on the right side (away from the viewer).

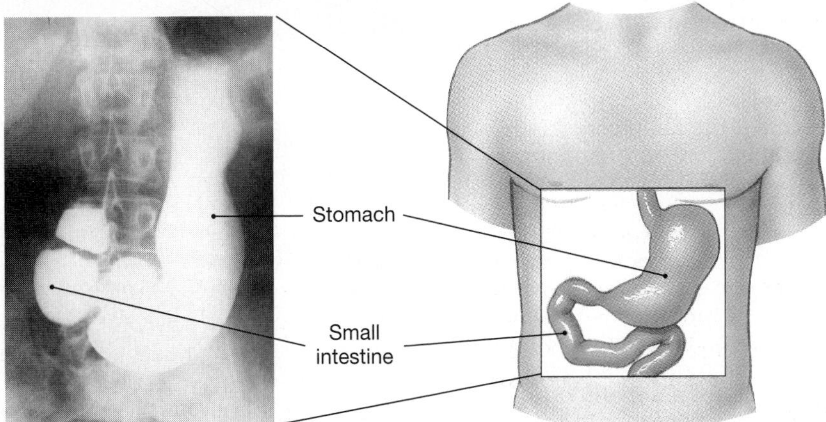
Stomach
Small intestine

(b) A barium-contrast X-ray of the upper digestive tract. Barium is very dense, and the contours of the gastric and intestinal lining can be seen outlined against the white of the barium solution.

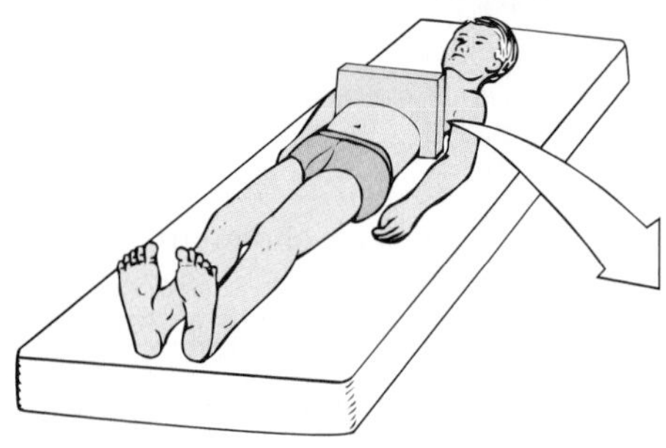

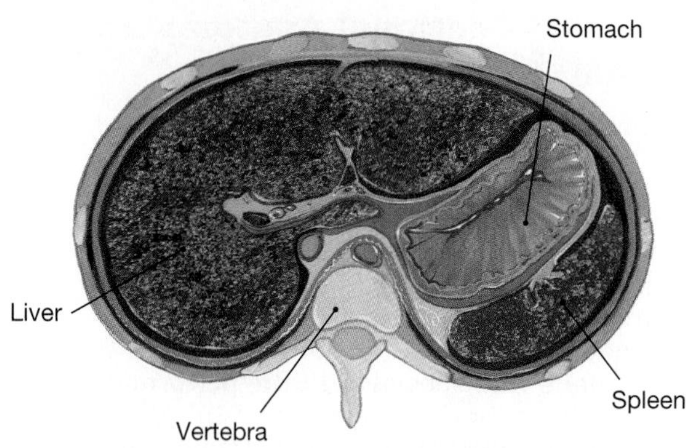
Stomach
Liver
Spleen
Vertebra

Diagrammatic view

● FIGURE 1-15

Scanning Techniques

(a) A color-enhanced CT scan of the abdomen. **CT** (**C**omputerized **T**omography), formerly called **CAT** (**C**omputerized **A**xial **T**omography), uses computers to reconstruct sectional views. A single X-ray source rotates around the body and the X-ray beam strikes a sensor monitored by the computer. The source completes one revolution around the body every few seconds; it then moves a short distance and repeats the process. The result is usually displayed as a sectional view in black and white, but it can be colorized for visual effect. CT scans show three-dimensional relationships and soft tissue structure more clearly than standard X-rays.

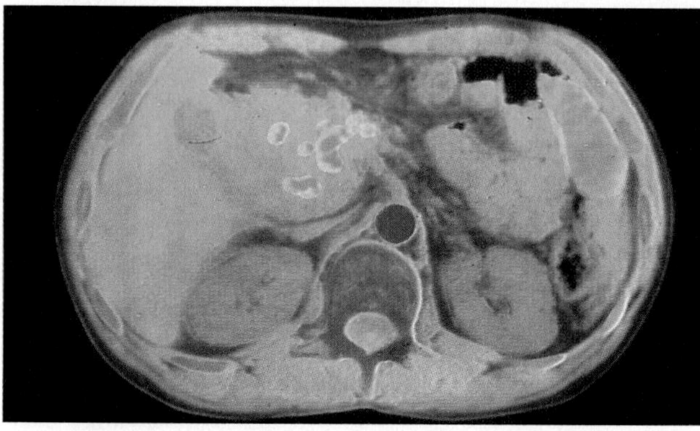

(a)

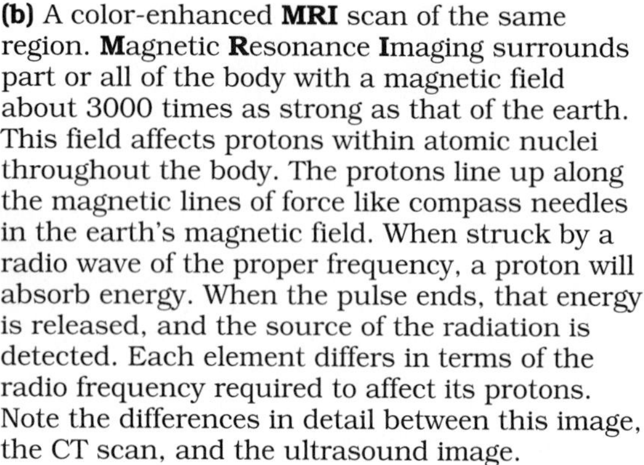

(b) A color-enhanced **MRI** scan of the same region. **M**agnetic **R**esonance **I**maging surrounds part or all of the body with a magnetic field about 3000 times as strong as that of the earth. This field affects protons within atomic nuclei throughout the body. The protons line up along the magnetic lines of force like compass needles in the earth's magnetic field. When struck by a radio wave of the proper frequency, a proton will absorb energy. When the pulse ends, that energy is released, and the source of the radiation is detected. Each element differs in terms of the radio frequency required to affect its protons. Note the differences in detail between this image, the CT scan, and the ultrasound image.

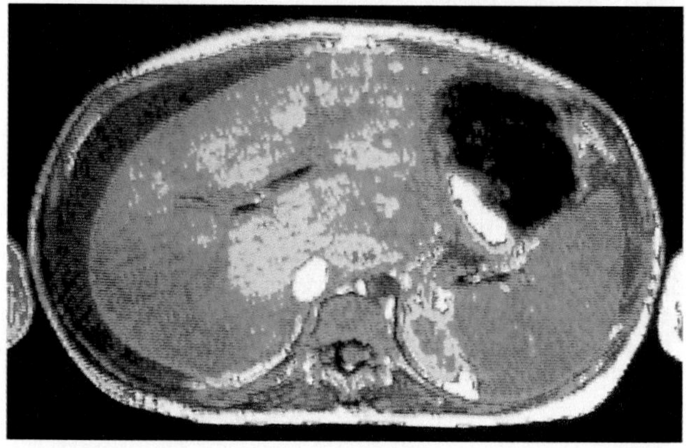

(b)

(c) An ultrasound scan of the abdomen. In **ultrasound** procedures, a small transmitter contacting the skin broadcasts a brief, narrow burst of high-frequency sound and then picks up the echoes. The sound waves are reflected by internal structures, and a picture, or **echogram,** can be assembled from the pattern of echoes. These images lack the clarity of other procedures, but no adverse effects have been reported, and fetal development can be monitored without a significant risk of birth defects. Special methods of transmission and processing permit analysis of the beating heart, without the complications that can accompany dye injections.

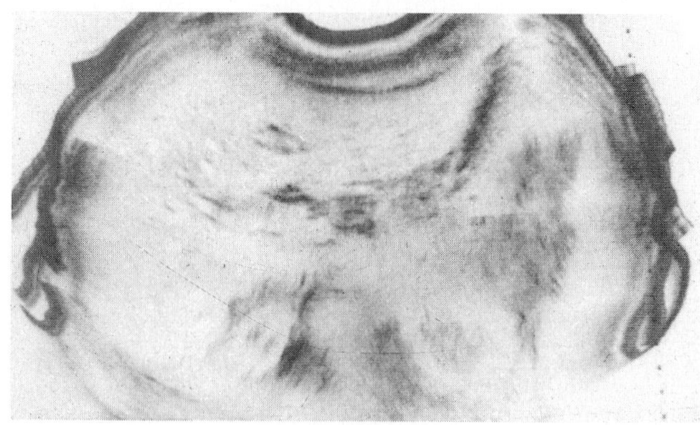

(c)

Selected Clinical Terminology

Terms Discussed in This Chapter

abdominopelvic quadrant: One of four regions of the anterior abdominal surface. *(p. 18)*
abdominopelvic region: One of nine regions of the anterior abdominal surface. *(p. 18)*
CT, CAT (computerized [axial] tomography): An imaging technique that reconstructs the three-dimensional structure of the body. *(p. 26)*
disease: A malfunction of organs or organ systems resulting from failure of homeostatic regulation. *(p. 16)*
echogram: An image created by ultrasound. *(p. 26)*
embryology: The study of structural changes during the first two months of development. *(p. 3)*
histology (his-TOL-o-jē): The study of tissues. *(p. 3)*
MRI (magnetic resonance imaging): An imaging technique that employs a magnetic field and radio waves to portray subtle structural differences. *(p. 26)*
radiologist: A physician who specializes in performing and analyzing radiological procedures. *(p. 25)*
ultrasound: An imaging technique that uses brief bursts of high-frequency sound reflected by internal structures. *(p. 26)*
X-rays: High-energy radiation that can penetrate living tissues. *(p. 25)*

AM Additional Terms Discussed in the Applications Manual

auscultation (aws-kul-TĀ-shun): Listening to a patient's body sounds using a stethoscope.
inspection: A careful observation of a patient's appearance and actions.
palpation: Using hands and fingers to feel the patient's body as part of a physical exam.
percussion: Tapping with the fingers or hand to obtain information about the densities of a patient's underlying tissues.

CHAPTER REVIEW

STUDY OUTLINE

Introduction, p. 2

1. **Biology** is the study of life; one of its goals is to discover the unity and patterns that underlie the diversity of living organisms.
2. All living things have certain common characteristics: They exhibit **responsiveness** to changes in their environment and show *adaptability;* they are capable of **growth,** and create future generations through **reproduction;** they can produce **movement;** they are supported by the chemical processes of **metabolism.**

THE SCIENCES OF ANATOMY AND PHYSIOLOGY, p. 2

1. **Anatomy** is the study of internal and external structure and the physical relationships between body parts. **Physiology** is the study of how living organisms perform vital functions. All specific functions are performed by specific structures.

Anatomy, p. 3

2. The boundaries of **microscopic anatomy** are established by the equipment used. **Cytology** analyzes the internal structure of individual cells. **Histology** examines **tissues**, groups of cells that have specific functional roles. Tissues combine to form **organs,** anatomical units with multiple functions.
3. **Gross (macroscopic) anatomy** considers features visible without a microscope. It includes **surface anatomy** (general form and superficial markings); **regional anatomy** (superficial and internal features in a specific area of the body); and **systemic anatomy** (structure of major organ systems).
4. **Developmental anatomy** examines the changes in form that occur between conception and physical maturity. **Embryology** studies processes that occur during the first two months of development.

Physiology, p. 4

5. **Human physiology** is the study of the functions of the human body. It is based on **cell physiology,** the study of the functions of living cells. **Special physiology** studies the physiology of specific organs. **Systemic physiology** considers all aspects of the function of specific organ systems. **Pathological physiology** studies the effects of diseases on organ or system functions.

LEVELS OF ORGANIZATION, p. 4

1. Anatomical structures and physiological mechanisms are arranged in a series of interacting levels of organization. *(Figure 1-1)*

AN INTRODUCTION TO ORGAN SYSTEMS, p. 6

1. The 11 organ systems of the body are the **integumentary, skeletal, muscular, nervous, endocrine, cardiovascular, lymphatic, respiratory, digestive, urinary,** and **reproductive systems.** *(Figures 1-2, 1-3)*

HOMEOSTASIS AND SYSTEM INTEGRATION, p. 13

Homeostatic Regulation, p. 13

1. **Homeostasis** is the presence of a stable environment within the body; through **homeostatic regulation,** physiological systems preserve homeostasis.
2. **Autoregulation** occurs when the activities of a cell, tissue, organ, or system change automatically in response to an environmental change. **Extrinsic regulation** results from the activities of the nervous or endocrine systems.
3. Homeostatic regulation usually involves a **receptor** sensitive to a particular stimulus and an **effector** whose activity affects the same stimulus.
4. **Negative feedback** is a corrective mechanism involving an action that directly opposes a variation from normal limits. *(Figures 1-4, 1-5)*
5. In **positive feedback** the initial stimulus produces a response that exaggerates the stimulus. *(Figure 1-6)*

Homeostasis and Disease, p. 16

6. Symptoms of **disease** appear when failure of homeostatic regulation causes organ systems to malfunction.

A FRAME OF REFERENCE FOR ANATOMICAL STUDIES, p. 16

Superficial Anatomy, p. 17

1. Standard anatomical illustrations show the body in the **anatomical position.** If the figure is shown lying down, it can be either **supine** (face up) or **prone** (face down). *(Figure 1-7; Table 1-1)*
2. **Abdominopelvic quadrants** and **abdominopelvic regions** represent two different approaches to describing anatomical regions of the body. *(Figure 1-8)*
3. The use of special directional terms provides clarity when describing anatomical structures. *(Figure 1-9; Table 1-2)*

Sectional Anatomy, p. 19

4. The three **sectional planes** (**frontal** or **coronal plane, sagittal plane,** and **transverse plane**) describe relationships between the parts of the three-dimensional human body. *(Figure 1-10; Table 1-3)*
5. **Serial reconstruction** is an important technique for studying histological structure and analyzing images produced by radiological procedures. *(Figure 1-11)*
6. **Body cavities** protect delicate organs and permit changes in the size and shape of visceral organs. The **dorsal body cavity** contains the **cranial cavity** (enclosing the brain) and **spinal cavity** (surrounding the spinal cord). The **ventral body cavity**, or **coelom,** surrounds developing respiratory, cardiovascular, digestive, urinary, and reproductive organs. *(Figure 1-12)*
7. During development the **diaphragm** divides the ventral body cavity into the superior **thoracic** and inferior **peritoneal cavities.** By birth the thoracic cavity contains two **pleural cavities** (each containing a lung) and a **pericardial cavity** (which surrounds the heart). The **abdominopelvic cavity** consists of the **abdominal cavity** and the **pelvic cavity.** It contains the **peritoneal cavity,** an internal chamber lined by a serous membrane, the **peritoneum.** *(Figure 1-13)*
8. Important **radiological procedures** (which can provide detailed information about internal systems) include **X-rays, CT, MRI,** and **ultrasound.** Each technique has its advantages and disadvantages. *(Figures 1-14, 1-15)*

REVIEW QUESTIONS

LEVEL 1 Reviewing Facts and Terms

Matching: Match each item in Column A with the most closely related item in Column B. Use letters for answers in the spaces provided.

	Column A	Column B
___	1. cytology	a. study of tissues
___	2. physiology	b. constant internal environment
___	3. histology	c. face up
___	4. metabolism	d. study of functions
___	5. homeostasis	e. positive feedback
___	6. muscle	f. system
___	7. heart	g. study of cells
___	8. endocrine	h. negative feedback
___	9. temperature regulation	i. brain and spinal cord
___	10. labor and delivery	j. all chemical activity in body
___	11. supine	k. thoracic and abdominopelvic
___	12. prone	l. tissue
___	13. ventral body cavity	m. serous membrane
___	14. dorsal body cavity	n. organ
___	15. pericardium	o. face down

16. The process by which an organism increases the size and/or number of cells is called:
(a) reproduction (b) adaptation
(c) growth (d) metabolism

17. The processes of absorption, respiration, and excretion are usually part of which basic life function?
(a) responsiveness (b) adaptability
(c) metabolism (d) movement

18. The study of internal and external structure and the physical relationships between body parts is called:
(a) histology (b) anatomy
(c) physiology (d) embryology

19. The specialist who attempts to determine the physical and chemical processes responsible for vital functions is a:
(a) physiologist (b) physician
(c) pathologist (d) cytologist

20. The relative stability of an organism's internal environment is called:
(a) homeopathy (b) uniformity
(c) equilibrium (d) homeostasis

21. The automatic change that occurs when the activities of a cell, tissue, organ, or system are faced with some environmental variation is called:
(a) autoregulation (b) extrinsic regulation
(c) deregulation (d) disease

22. Crisis management by directing rapid, short-term, specific responses is a function of the:
(a) endocrine system
(b) nervous system
(c) hormones
(d) autoregulatory system

23. Growth and sexual maturation are regulated primarily by the:
(a) nervous system (b) circulatory system
(c) endocrine system (d) digestive system

24. When a variation outside of normal limits triggers a response that restores the normal condition, the type of regulation involved is termed:
(a) negative feedback (b) positive feedback
(c) compensation (d) adaptation

25. In positive feedback the initial stimulus produces a response that:
(a) suppresses the stimulus
(b) has no effect on the stimulus
(c) interferes with the process being completed
(d) exaggerates the stimulus

26. Failure of homeostatic regulation in the body results in:
(a) autoregulation
(b) extrinsic regulation
(c) disease
(d) positive feedback

27. The terms that apply to the front of the body when in anatomical position are:
(a) posterior, dorsal (b) back, front
(c) medial, lateral (d) anterior, ventral

28. A cut through the body that passes perpendicular to the long axis of the body and divides the body into a superior and inferior section is known as a:
(a) sagittal section (b) transverse section
(c) coronal section (d) frontal section

29. The cranial and spinal cavity are found in the:
(a) ventral body cavity (b) thoracic cavity
(c) dorsal body cavity (d) abdominopelvic cavity

30. The diaphragm, a flat muscular sheet, divides the ventral body cavity into a superior __________ cavity and an inferior __________ cavity.
(a) pleural, pericardial
(b) abdominal, pelvic
(c) thoracic, abdominopelvic
(d) cranial, thoracic

31. The mediastinum is the region between the:
(a) lungs and heart
(b) two pleural cavities
(c) thorax and abdomen
(d) heart and pericardium

LEVEL 2 Reviewing Concepts

32. What basic functions are performed by all living things?

33. (a) Define the term anatomy.
(b) Define the term physiology.

34. How does the study of cytology differ from the study of histology?

35. Beginning with the molecular level, list in correct sequence the levels of organization from the simplest level to the most complex level.

36. What is homeostatic regulation and what is its physiological importance?

37. What distinguishes autoregulation from extrinsic regulation?

38. What necessary components are involved in homeostatic regulation? What is the function of each?

39. How does negative feedback differ from positive feedback?

40. Describe the position of the body when it is in the anatomical position.

41. What is the role of serous membranes in the body?

42. As a surgeon you perform an invasive procedure that necessitates cutting through the peritoneum. Are you more likely to be operating on the heart or on the stomach?

43. In which body cavity would each of the following organs or systems be found?
(a) brain and spinal cord
(b) cardiovascular, digestive, and urinary systems
(c) heart, lungs
(d) stomach, intestines

44. What are the names of the serous membranes that line each of the body cavities containing the following organs?
(a) heart (b) lungs
(c) intestines

LEVEL 3 Critical Thinking and Clinical Applications

45. A hormone called calcitonin from the thyroid gland is released in response to increased levels of calcium ions in the blood. If the control of this hormone is by negative feedback, what effect would it have on blood calcium levels?

46. During exercise, blood flow to skeletal muscles increases. The initial response that increases blood flow is automatic and independent of the nervous and endocrine systems. What type of homeostatic regulation is this? Why?

47. An anatomist wishes to make detailed comparisons of medial surfaces of the left and right sides of the brain. This work requires sections that will show the entire medial surface. What kind of sections should be ordered from the lab for this investigation?

CHAPTER 2

Water is so common, and seems so ordinary, that it is often taken for granted. Yet all living things, including human beings, are totally dependent on water. You can survive for weeks without food, but only for a few days without water. The basic reason may surprise you: in large part, you are water. Water makes up roughly 66 percent of your body weight.

Chemically, water is a simple substance: two atoms of hydrogen joined to one of oxygen. Yet when these three atoms are linked by chemical bonds they produce a substance with many unusual properties. In this chapter you will learn how atoms are bound together to form molecules, the building blocks of living cells. You will also meet the larger molecules that form the structural framework of the body's cells and tissues, and enable our cells to grow, divide, and perform the many other functions that make life possible.

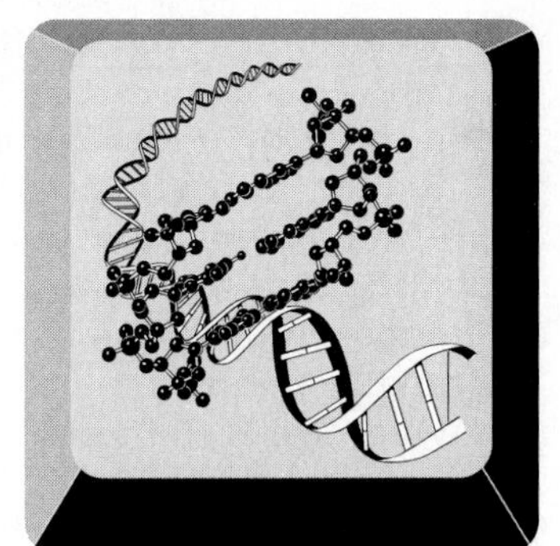

The Chemical Level of Organization

Chapter Outline and Objectives

1 Describe an atom and how atomic structure affects interactions between atoms.

2 Compare the different ways in which atoms combine to form molecules and compounds.

3 Use chemical notation to symbolize chemical reactions.

4 Distinguish among the major types of chemical reactions that are important for studying physiology.

5 Describe the crucial role of enzymes in metabolism.

6 Distinguish between organic and inorganic compounds.

7 Explain how the chemical properties of water make life possible.

8 Discuss the importance of pH and the role of buffers in body fluids.

9 Describe the physiological roles of inorganic compounds.

10 Discuss the structure and functions of carbohydrates, lipids, proteins, nucleic acids, and high-energy compounds.

Our study of the human body begins at the most fundamental level of organization, that of individual atoms and molecules. Air, elephants, oranges, oceans, rocks, and people are all composed of atoms in varying combinations. The unique characteristics of each object, living or nonliving, result from the types of atoms involved and the ways those atoms combine and interact. **Chemistry** is the branch of science that concerns itself with the structure of matter, including the interactions between atoms. A familiarity with chemical principles is essential to an understanding of the anatomy and physiology of the cells, tissues, organs, and organ systems of the human body.

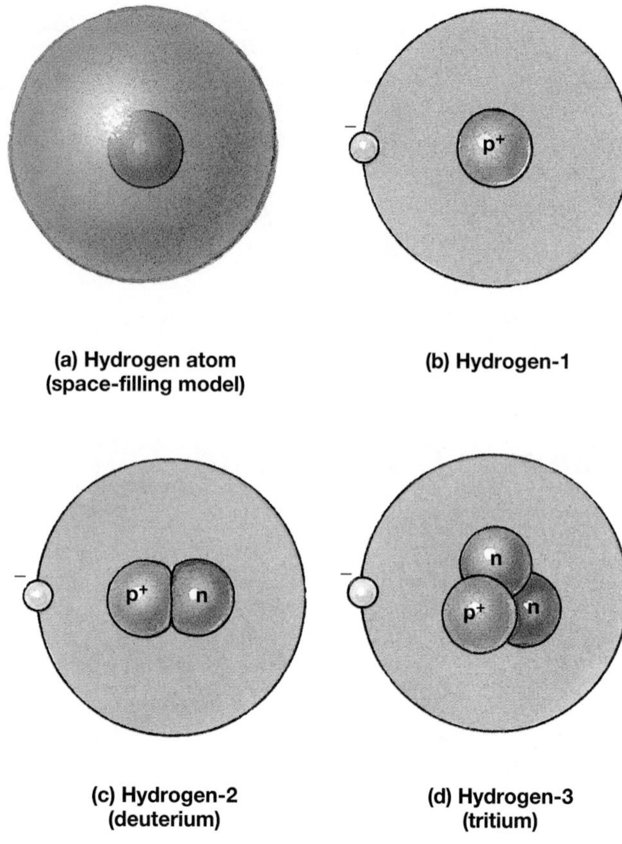
p+
(a) Hydrogen atom (space-filling model)
(b) Hydrogen-1
p+
n
n
p+
n
(c) Hydrogen-2 (deuterium)
(d) Hydrogen-3 (tritium)

● **FIGURE 2-1**

Hydrogen Atoms. (a) The electron cloud of a hydrogen atom is formed by the orbiting of an electron around the nucleus. **(b)** A two-dimensional model depicting the electron in an electron shell makes it easier to visualize the components of the atom. A typical hydrogen nucleus contains a single proton and no neutrons. **(c)** A deuterium (^2H) nucleus contains a proton and a neutron. **(d)** A tritium (^3H) nucleus contains a pair of neutrons in addition to the proton.

Atoms and Molecules

Atoms are the simplest chemical units of matter—no chemical change can alter their identities. Although there are dozens of different *subatomic particles*, only three are important for understanding chemical properties. These three *fundamental particles* are **protons, neutrons,** and **electrons.** Protons and neutrons are similar in size and mass, but *pro*tons bear a *positive* electrical charge whereas *neut*rons are *neutral*—that is, uncharged. Electrons are lighter than protons, only 1/1836th as massive, and they bear a negative electrical charge.

Structure of the Atom

Figure 2-1

All atoms contain protons and electrons, normally in equal numbers. The number of protons in an atom is known as its **atomic number. Hydrogen** is the simplest atom, with an atomic number of 1. An atom of hydrogen contains one proton and one electron. The proton is located in the center of the atom and forms the **nucleus.** The electron whirls around the nucleus at high speed, forming an *electron cloud* (Figure 2-1a●). It orbits the nucleus because of the electrical attraction between the proton (+) and the electron (–). This attraction is an example of an *electrical force.*

For convenience, atomic structure is often illustrated in the simplified form shown in Figure 2-1b●. In this representation the electrons are shown in a circular **electron shell.**

The dimensions of the electron cloud determine the overall size of the atom. To get an idea of the scale involved, consider that if the nucleus were the size of a tennis ball the electron cloud would have a radius of six *miles*. In reality, atoms are so small that atomic measurements are most conveniently reported in terms of *nanometers* (NA-nō-mē-terz) (nm) or *angstroms* (Å). A nanometer is 10^{-9} meters (0.000000001 m), and an angstrom is one-tenth of that size (10^{-10} m). The very largest atoms approach 0.5 nm in diameter (0.00000005 cm, or 0.00000002 in.).

When neutrons are present in an atom they are found in the nucleus together with the protons. Most hydrogen nuclei lack neutrons, but a small number of hydrogen nuclei and the nuclei of all other atoms contain neutrons. The number of neutrons present does not affect the properties of an atom, other than its mass.

Elements and Isotopes

Figures 2-1, 2-2

The basic components of matter are the chemical **elements,** substances that cannot be broken down into simpler substances by ordinary chemical processes. Only 92 different elements occur in nature (although 17 more have been made by

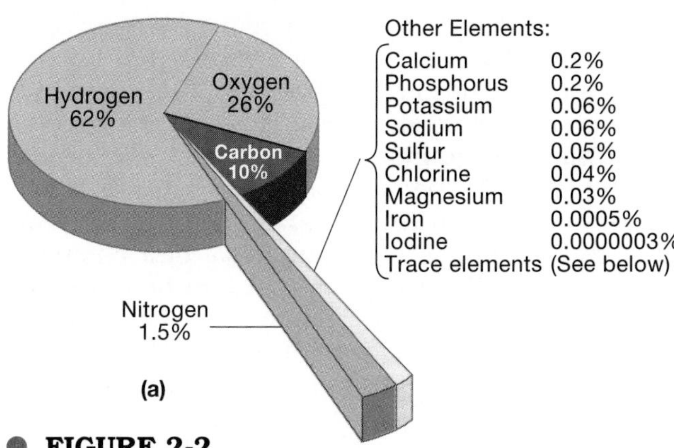
Other Elements:
Calcium 0.2%
Phosphorus 0.2%
Potassium 0.06%
Sodium 0.06%
Sulfur 0.05%
Chlorine 0.04%
Magnesium 0.03%
Iron 0.0005%
Iodine 0.0000003%
Trace elements (See below)
Hydrogen 62%
Oxygen 26%
Carbon 10%
Nitrogen 1.5%
(a)

Element (% of body weight)	Significance
Hydrogen (9.7)	A component of water and most other compounds in the body.
Oxygen (65)	A component of water and other compounds; oxygen gas essential for respiration.
Carbon (18.6)	Found in all organic molecules.
Nitrogen (3.2)	Found in proteins, nucleic acids, and other organic compounds.
Calcium (1.8)	Found in bones and teeth; important for membrane function, nerve impulses, muscle contraction, and blood clotting.
Phosphorus (1)	Found in bones and teeth, nucleic acids, and high-energy compounds.
Potassium (0.4)	Important for proper membrane function, nerve impulses, and muscle contraction.
Sodium (0.2)	Important for membrane function, nerve impulses, and muscle contraction.
Chlorine (0.2)	Important for membrane function and water absorption.
Magnesium (0.06)	A cofactor for several enzymes.
Sulfur (0.04)	Found in many proteins.
Iron (0.007)	Essential for oxygen transport and energy capture.
Iodine (0.0002)	A component of hormones of the thyroid gland.

(b)

● **FIGURE 2-2**

Principal Elements in the Human Body. (a) The percentages given are estimates of the contribution made by each element to the total number of atoms in the body. Note that just four elements (C, H, O, and N) contribute over 99 percent to the total. Among the so-called trace elements present in quantities too small to show are silicon (Si), fluorine (F), copper (Cu), manganese (Mn), zinc (Zn), selenium (Se), cobalt (Co), molybdenum (Mo), cadmium (Cd), chromium (Cr), tin (Sn), aluminum (Al), and boron (B). The functions of some of these in the body are still poorly understood. **(b)** Representative functions of each major element in the body; the percentages indicate the contribution of each element to total body weight.

means of nuclear reactions in the laboratory). The atoms of an element all have the same atomic number. Each element has a chemical symbol, an abbreviation recognized by scientists everywhere. Most of the symbols are easily connected with the English names of the elements, but a few, such as *Na* for sodium, are abbreviations of their Latin names. Figure 2-2 indicates the 13 most abundant elements in the human body and their relative contributions to the total number of atoms in the body (Figure 2-2a●) and to the total body weight (Figure 2-2b●). This list is incomplete, for the human body contains atoms of another 13 elements that are found in such small amounts that their percentage values are meaningless.

The number of neutrons in the nucleus can vary among the atoms of a single element. For example, although most hydrogen nuclei consist of a single proton, 0.015 percent also contain one neutron, and a very small percentage contain two. **Isotopes** of an element are atoms whose nuclei contain different numbers of neutrons. Because the presence or absence of neutrons has very little effect on the chemical properties of an atom, isotopes are usually indistinguishable except on the basis of weight. The **mass number**—the total number of protons and neutrons in the nucleus—is used to designate a particular isotope. Thus, the three isotopes of hydrogen are hydrogen-1, or ^{1}H; hydrogen-2, or ^{2}H, also known as *deuterium;* and hydrogen-3, or ^{3}H, also known as *tritium* (Figure 2-1c,d●).

Radioisotopes are isotopes with unstable nuclei that emit subatomic particles in measurable amounts. Weakly radioactive isotopes are sometimes used in diagnostic procedures that assess the structural and functional state of internal organs. AM *Radioisotopes and Clinical Testing; Radiopharmaceuticals*

Atomic Weights

A typical atom of oxygen, which has an atomic number of 8, contains eight neutrons. The mass number of this isotope is therefore 16. The mass numbers of other isotopes of oxygen will be different, depending on the number of neutrons present. Such mass numbers, however, simply tell us the number of subatomic particles in the nuclei of different atoms. They do not express the actual mass or weight of the atoms. (For example, they do not take into account the slight difference between the mass of a proton and that of a neutron.)

The unit used to measure the **atomic weight** of atoms is the **dalton** (also known as the *atomic mass unit* or *amu*). The dalton is very close to the weight of a single proton. Thus the atomic weight of the most common isotope of hydrogen is very close to 1, and that of the most common isotope of oxygen is very close to 16.

The atomic weight of an *element* is the *average* weight of an atom of that element, taking the presence of isotopes into account. For example, the atomic weight of hydrogen is more than 1 (1.0079) because some hydrogen atoms (0.015 percent) have a mass number of 2, and an even smaller percentage have a mass number of 3. In most instances, the atomic weight of an element is very close to the mass number of the most common isotope of that element. Atomic weights are included in Appendix III.

For every element, a quantity that has a mass in grams equal to its atomic weight will contain the same number of atoms. That quantity[1] has been given a special name: a **mole.** Expressing relationships in moles rather than grams makes it much easier to keep track of the relative numbers of atoms in chemical samples and processes. Most chemical analyses and laboratory reports record concentrations in terms of moles or fractions of moles. AM *Solutions and Concentrations*

Electrons and Energy Levels

Figure 2-3

Atoms are electrically neutral because every positively charged proton is balanced by a negatively charged electron. Note that each increase in the atomic number is accompanied by a comparable increase in the number of electrons orbiting the nucleus. These electrons occupy an orderly series of electron shells, or *energy levels.*

The number and arrangement of electrons in an atom's outer energy level determine the chemical properties of that element. Each energy level can accommodate a specific number of electrons. For example, the innermost shell can hold only two electrons. As indicated in Figure 2-3●, a hydrogen atom has one electron in this energy level, but a helium atom has two. Because it has a full outer energy level, a helium atom is very stable, and it will not ordinarily react with other atoms. Lithium has three electrons, so in a lithium atom the first level is filled and the third electron occupies a second energy level. The second level can hold up to eight electrons, and this level is filled in a neon atom with an atomic number of 10. Neon atoms, like helium atoms, are thus very stable. (To see how other elements are related to those discussed here, consult Appendix III, the periodic table.)

Chemical Bonds and Chemical Compounds

Helium, neon, and argon are called **inert gases** because their atoms, having full outer energy levels, neither react with one another nor combine with atoms of other elements. Atoms with unfilled outer energy levels are relatively unstable. Such atoms can achieve stability by gaining, losing, or sharing electrons. These processes create **chemical bonds** that hold the atoms together. When such chemical bonding occurs between atoms of different elements, the result is a chemical **compound.** A compound is a new chemical substance with properties that can be quite different from those of its component elements. For example, a mixture of hydrogen and oxygen is highly flammable, but combining hydrogen and oxygen atoms produces water, a compound used to put out fires.

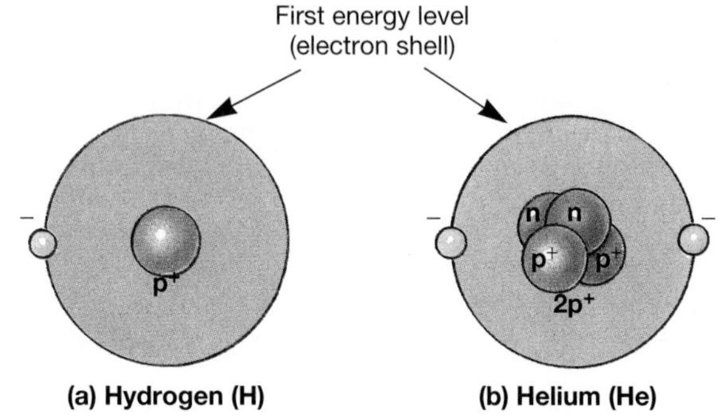
First energy level
(electron shell)
p+
n n
p+ p+
2p+
(a) Hydrogen (H)
(b) Helium (He)

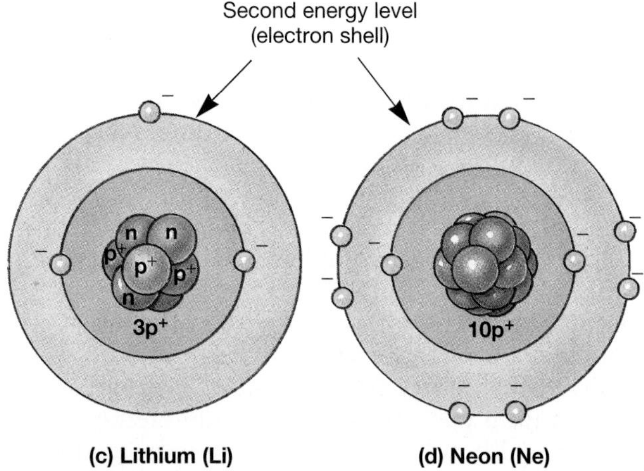
Second energy level
(electron shell)
n n
p+ p+ p+
n
3p+
10p+
(c) Lithium (Li)
(d) Neon (Ne)

● **FIGURE 2-3**

Atoms and Energy Levels. (a) A typical hydrogen atom has one proton and one electron. The electron orbiting the nucleus occupies the first energy level, diagrammed as an electron shell. **(b)** An atom of helium has a pair of protons and two electrons. The two electrons orbit in the same energy level. **(c)** The first energy level can hold only two electrons. In a lithium atom, with three protons and three electrons, the third electron occupies a second energy level. **(d)** The second level can hold up to eight electrons. A neon atom has 10 protons and 10 electrons; thus both the first and second energy levels are filled. Note that helium, lithium, and neon atoms contain neutrons as well as protons in their nuclei.

[1]The number of atoms, ions, or molecules in a mole—called *Avogadro's number*—is 6.023×10^{23}, or about 600 billion trillion.

Ionic Bonds

Figure 2-4

Ionic bonds are created by the electrical attraction between atoms that have gained or lost electrons, and thus carry an electrical charge. If we assign a value of +1 to the charge on a proton, the charge on an electron is −1. As long as the number of protons is equal to the number of electrons, an atom will be electrically neutral. If an atom loses an electron, it will exhibit a charge of +1 because there will be one proton without a corresponding electron. Losing a second electron would leave the atom with a charge of +2. Adding an extra electron to the atom will give it a charge of −1; adding a second electron would produce a charge of −2. Atoms or molecules that have + or − charges are called **ions.** Ions with a positive charge are **cations** (KAT-ī-onz); those with a negative charge are **anions** (AN-ī-onz).

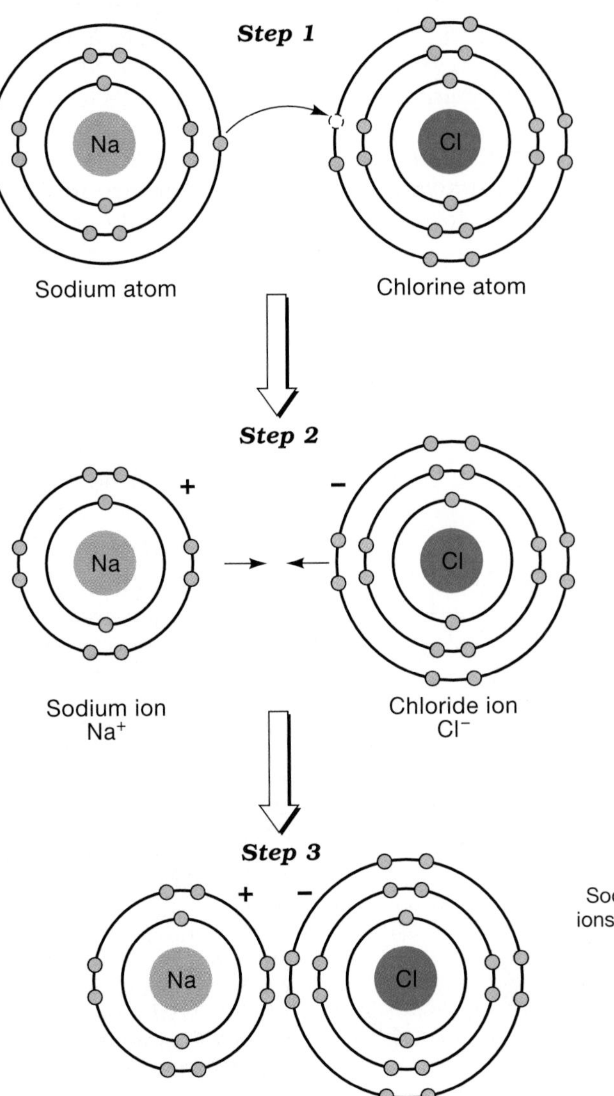
Step 1
Na
Cl
Sodium atom
Chlorine atom
Step 2
+
−
Na
Cl
Sodium ion
Na^+
Chloride ion
Cl^-
Step 3
+
−
Na
Cl
Sodium chloride
NaCl
(a)

In an ionic (ī-ON-ik) bond, anions and cations are held together by the attraction between positive and negative charges. In forming an ionic bond:

- One atom loses one or more electrons, becoming a cation with a + charge.
- Another atom gains those electrons, becoming an anion with a − charge.
- The opposite charges draw the two ions together.

The formation of a representative ionic bond is illustrated in Figure 2-4a●. The sodium atom diagrammed in Figure 2-4a (Step 1) has an atomic number of 11, so this atom normally contains 11 protons and 11 electrons. Electrons fill the first and second energy levels, and a single electron occupies level 3. Losing that "extra" electron would give the sodium atom a full outer energy level and produce a sodium ion with a +1 charge. But the electron cannot simply be thrown away; it must be donated to another atom. A chlorine atom has seven electrons in its outer energy level. An additional electron would fill this energy level, and a sodium atom can provide it. In the process (Step 2) the chlorine atom becomes a *chloride ion* with a −1 charge.

Both atoms have now become stable ions with filled outer energy levels. But the two ions do not move apart after the electron transfer, because the positively charged sodium ion is attracted to the negatively charged chloride ion (Step 3). The combination of oppositely charged ions forms the ionic compound *sodium chloride,* otherwise known as table salt. Large numbers of sodium and chloride ions interact to form highly structured crystals (Figure 2-4b●), held together only by the electrical attraction of oppositely charged ions.

Covalent Bonds

Figures 2-3a, 2-5, 2-6

Another way that atoms can complete their outer electron shells is by sharing electrons with other

Chloride ions (Cl^-)

Sodium ions (Na^+)

(b)

● **FIGURE 2-4**

Ionic Bonding. (a) Step 1: A sodium atom loses an electron, which is accepted by a chlorine atom. Step 2: Because the sodium (Na^+) and chloride (Cl^-) ions have opposite charges, they are attracted to one another. Step 3: The association of sodium and chloride ions forms the ionic compound sodium chloride. **(b)** Large numbers of sodium and chloride ions form a crystal of sodium chloride.

atoms. The result is a **molecule**—two or more atoms held together by **covalent** (KŌ-vā-lent) **bonds.**

Individual hydrogen atoms, as diagrammed in Figure 2-3a●, are not found in nature. Instead, we find hydrogen molecules (Figure 2-5a●). Molecular hydrogen is a gas present in the atmosphere in very small quantities. The two hydrogen atoms share their electrons, with each electron whirling around both nuclei. The sharing of one pair of electrons creates a **single covalent bond.**

Notice that three different methods can be used to show the structure of a hydrogen molecule: (1) the electron-shell model, which diagrams the positions of the electrons in concentric shells that represent the energy levels; (2) the space-filling model, which shows the molecule in three dimensions; and (3) the structural formula, which uses a solid line to indicate a shared electron pair.

Oxygen, with an atomic number of 8, has two electrons in its first energy level and six in the second. The oxygen atoms diagrammed in Figure 2-5b● attain a stable electron configuration by sharing two pairs of electrons, forming a **double covalent bond.** Molecular oxygen is an atmospheric gas that is very important to living organisms; our cells would die without a relatively constant supply of oxygen.

The chemical reactions in our bodies that consume oxygen also produce a waste product, **carbon dioxide.** The oxygen atoms in a carbon dioxide molecule form double covalent bonds with the carbon atom, as indicated in Figure 2-5c●.

Covalent bonds are very strong because the electrons hold the atoms together. In typical covalent bonds the atoms remain electrically neutral because each shared electron spends just as much time at home as away. (If two people were tossing a pair of baseballs back and forth as fast as they could, on the average, each person would have just one.) Covalent bonds, especially between carbon atoms, create the stable framework of the large molecules that make up most of the structural components of the human body.

Covalent bonds usually form molecules that complete the outer electron shells of the atoms involved. An atom or molecule that contains unpaired electrons in its outer shell is called a *free radical.* Free radicals are highly reactive and persist only for a very short time before additional reactions occur. Free radicals sometimes form as intermediaries in chemical reactions. Living cells usually have mechanisms to tie up or eliminate free radicals, to prevent the damage or destruction of vital compounds, such as proteins. However, **nitric oxide** (Figure 2-5d●) is a free radical that has important functions in the body. It is involved in chemical communication in the nervous system, the control of blood vessel diameter, blood clotting, and the defense against bacteria and other pathogens.

POLAR COVALENT BONDS Many covalent bonds involve relatively equal sharing of electrons. Some, however, do not, because elements differ in how strongly they hold shared electrons. An unequal sharing of electrons creates a **polar covalent bond.** For example, in a molecule of water (Figure 2-6●), an oxygen atom forms covalent bonds with two hydrogen atoms. The oxygen atom has a much stronger attraction for the shared electrons than the hydrogen atoms do, so the electrons spend most of their time orbiting around the oxygen

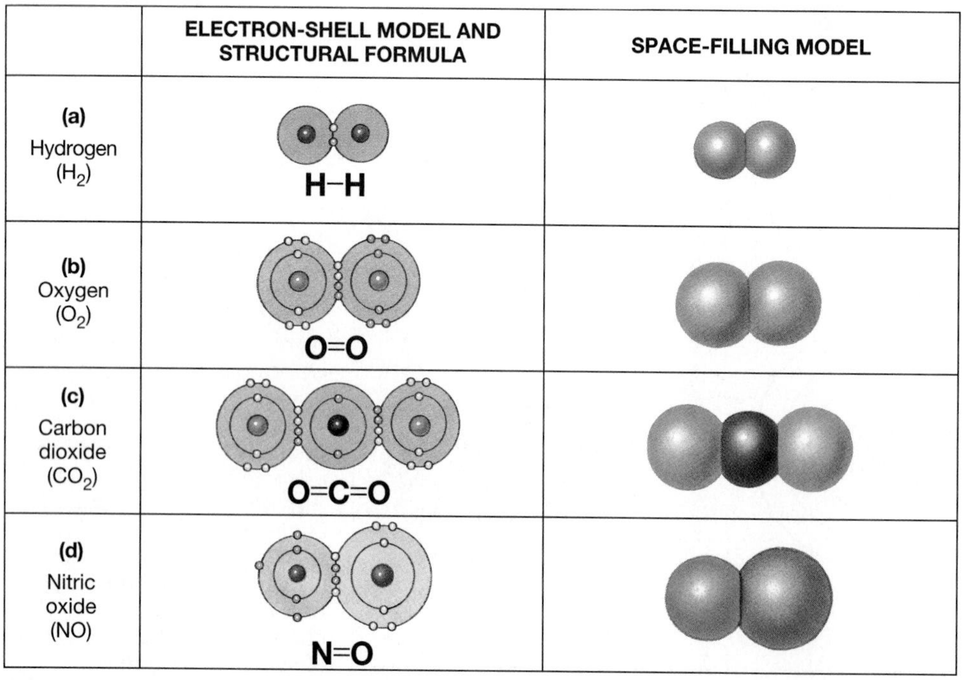

	ELECTRON-SHELL MODEL AND STRUCTURAL FORMULA	SPACE-FILLING MODEL
(a) Hydrogen (H_2)	H–H	
(b) Oxygen (O_2)	O=O	
(c) Carbon dioxide (CO_2)	O=C=O	
(d) Nitric oxide (NO)	N=O	

● **FIGURE 2-5**
Covalent Bonds. (a) In a molecule of hydrogen, two hydrogen atoms share their electrons so that each has a filled outer electron shell. This sharing creates a single covalent bond. **(b)** A molecule of oxygen consists of two oxygen atoms that share two pairs of electrons. The result is a double covalent bond. **(c)** In a molecule of carbon dioxide, a central carbon atom forms double covalent bonds with a pair of oxygen atoms. **(d)** A molecule of nitric oxide, NO; this molecule is held together by a double covalent bond, but the outer electron shell of the nitrogen atom requires an additional electron.

nucleus. Because it has two extra electrons part of the time, the oxygen atom develops a slight negative charge, indicated by δ^-. At the same time, the hydrogen atoms develop slight positive charges (δ^+), for their electrons are away part of the time. This unequal sharing makes polar covalent bonds somewhat weaker than other covalent bonds.

Hydrogen Bonds

Figure 2-7

Covalent and ionic bonds tie atoms together in a relatively stable framework. Comparatively weak attractive forces act between atoms in different parts of a large molecule as well as between adjacent molecules. *Hydrogen bonds* are the most important of these attractive forces. A **hydrogen bond** is the attraction between a δ^+ hydrogen atom involved in a polar covalent bond and an atom of either oxygen or nitrogen that is either part of another molecule or located at a distant site on the same molecule. The oxygen or nitrogen atom must also be involved in a polar covalent bond, giving it a δ^- charge. The hydrogen bond that is created is a very weak attractive force that reflects the attraction between opposite charges.

Hydrogen bonds do not create molecules, but they can change molecular shapes or pull molecules together. For example, hydrogen bonding occurs between water molecules (Figure 2-7a●). At the water surface the attraction between molecules slows the rate of evaporation and creates the phenomenon known as *surface tension* (Figure 2-7b●). Surface tension acts as a barrier that keeps small objects from entering the water; that is how small insects can "walk" across the surface of a pond or puddle. Surface tension in a layer of tears keeps small objects such as dust particles from touching the surface of the eye. At the cellular level, hydrogen bonds affect the shapes and properties of complex molecules, such as proteins; they may also determine the three-dimensional relationships between molecules.

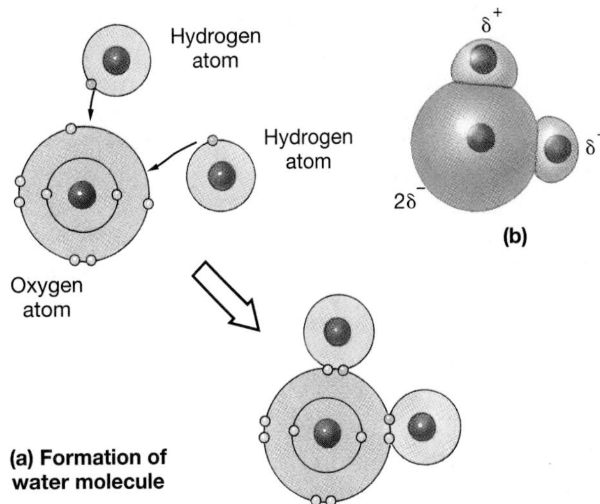
Hydrogen atom
Hydrogen atom
Oxygen atom
(a) Formation of water molecule
δ+
δ+
2δ−
(b)

● **FIGURE 2-6**

Polar Covalent Bonds and the Structure of Water. **(a)** In forming a water molecule, an oxygen atom completes its outer energy level by sharing electrons with a pair of hydrogen atoms. The sharing is unequal because the oxygen atom holds the electrons more tightly than do the hydrogen atoms. **(b)** Because the oxygen atom has two extra electrons much of the time, it develops a slight negative charge; the hydrogen atoms become weakly positive. The bonds in a water molecule are called polar covalent bonds.

✔ Oxygen and neon are both gases at room temperature. Oxygen combines readily with other elements but neon does not. Why?

✔ How is it possible for two samples of hydrogen to contain the same number of atoms but have different weights?

✔ What kind of bond holds atoms in a water molecule together? What attracts water molecules to one another?

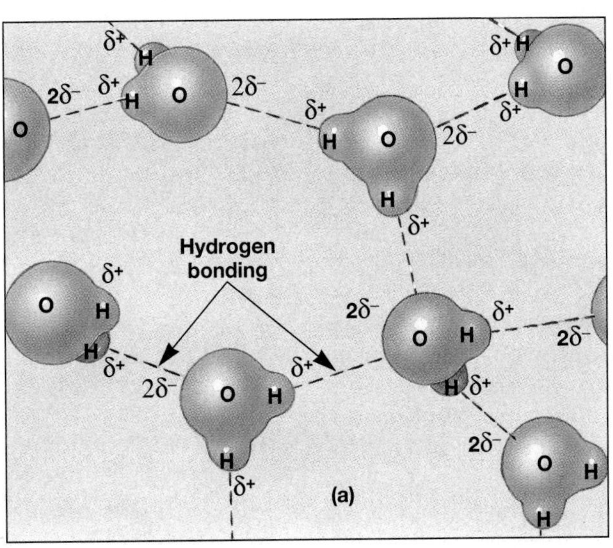
Hydrogen bonding
O
H
δ+
2δ−
(a)

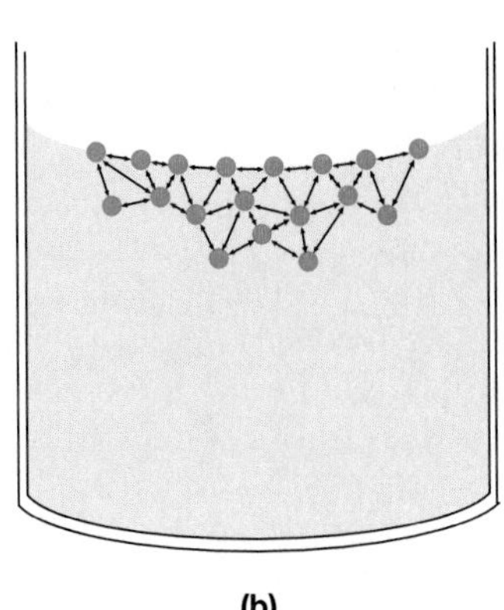
(b)

● **FIGURE 2-7**

Hydrogen Bonds. (a) The hydrogen atoms of a water molecule have a slight positive charge, and the oxygen atom has a slight negative charge (see Figure 2-6). Attraction between the hydrogen atom of one water molecule and the oxygen atom of another represents a hydrogen bond (indicated by dashed lines). **(b)** Hydrogen bonding between water molecules at a free surface restricts evaporation and creates surface tension.

TABLE 2-1 Rules of Chemical Notation

1. The abbreviation of an element indicates one atom of that element:

   H = an atom of hydrogen; O = an atom of oxygen

2. A number preceding the abbreviation of an element indicates more than one atom:

   2 H = two individual atoms of hydrogen

   2 O = two individual atoms of oxygen

3. A subscript following the abbreviation of an element indicates a molecule with that number of atoms:

   H_2 = a hydrogen molecule composed of two hydrogen atoms

   O_2 = an oxygen molecule composed of two oxygen atoms

4. In a description of a chemical reaction, the interacting participants are called *reactants*, and the reaction generates one or more *products*. An arrow indicates the direction of the reaction, from reactants (usually on the left) to products (usually on the right). In the reaction below two atoms of hydrogen combine with one atom of oxygen to produce a single molecule of water.

   $$2\,H + O \rightarrow H_2O$$

5. A superscript plus or minus sign following the abbreviation for an element indicates an ion. A single plus sign indicates an ion with a charge of +1 (loss of one electron). A single minus sign indicates an ion with a charge of –1 (gain of one electron). If more than one electron has been lost or gained, the charge on the ion is indicated by a number preceding the plus or minus.

   Na^+ = one sodium ion (has lost 1 electron)

   Cl^- = one chloride ion (has gained 1 electron)

   Ca^{2+} = one calcium ion (has lost 2 electrons)

6. Chemical reactions neither create nor destroy atoms—they merely rearrange them into new combinations. Therefore, the numbers of atoms of each element must always be the same on both sides of the equation. When this is the case, the equation is *balanced.*

   Unbalanced: $H_2 + O_2 \rightarrow H_2O$

   Balanced: $2\,H_2 + O_2 \rightarrow 2\,H_2O$

Chemical Notation

Before we can consider the specific compounds found in the human body, we must be able to describe chemical compounds and reactions effectively. Using sentences to describe chemical structures and events often leads to confusion, and a simple form of "chemical shorthand" makes communication much more efficient.

The chemical shorthand we will use is known as *chemical notation.* Chemical notation enables us to describe complex events in a brief and precise fashion. The rules of chemical notation are summarized in Table 2-1.

You cannot actually handle individual molecules, nor could you easily count the billions of molecules that take part in ordinary chemical processes in the lab or in the body. As an example, suppose you wanted to create water from hydrogen and oxygen, using the balanced equation

$$2\,H_2 + O_2 \rightarrow 2\,H_2O$$

The first step in performing such an experiment would be to calculate the **molecular weights** involved. The molecular weight is equal to the sum of the atomic weights of the components. The atomic weight of hydrogen is close to 1, so one hydrogen molecule (H_2) would have a *molecular* weight of 2. Oxygen has an atomic weight of around 16, so the molecular weight of an oxygen molecule (O_2) is roughly 32. In practical terms, if you wanted to perform the experiment you would take 4g of hydrogen, combine it with 32g of oxygen, and produce 36g of water. You could also work with ounces, pounds, or tons, as long as the proportions remained the same.

Chemical Reactions

Living cells remain alive and functional by controlling internal *chemical reactions*. In a **chemical reaction,** chemical bonds between atoms are broken as atoms in the reacting substances, or **reactants,** are rearranged in different combinations to form new substances, the **products.** In effect, each cell is a chemical factory. For example, growth, maintenance, and repair, secretion, and contraction all involve complex chemical reactions. The term **metabolism** (meh-TAB-o-lizm) refers to all the chemical reactions in the body. Living cells use chemical reactions to provide the energy needed to maintain homeostasis and perform essential functions.

❑ Basic Energy Concepts

Before we can consider various types of reactions, we must understand some basic principles concerning the relationships between matter and energy. **Work** is movement or a change in physical structure. **Energy** is the capacity to perform work; movement or physical change will not occur unless energy is provided. There are two major types of energy: *kinetic energy* and *potential energy.*

- **Kinetic energy** is the energy of motion. When a car hits a tree, it is kinetic energy that does the damage.
- **Potential energy** is stored energy. It may result from the position of an object (a book on a high shelf) or its physical or chemical structure (a stretched spring or a charged battery).

Kinetic energy must be used to lift a book, stretch a spring, or charge a battery. The potential energy is converted back into kinetic energy when the book falls, the spring recoils, or the battery discharges, and the kinetic energy can be used to perform work.

A conversion between potential energy and kinetic energy is never 100 percent efficient. Each time an energy exchange occurs, some of the energy is released as **heat.** Heat is an increase in random molecular motion, and the temperature of an object is directly related to its heat content.

Living cells perform work in many forms. For example, a resting skeletal muscle contains potential energy in the form of the positions of protein filaments and the covalent bonds between molecules inside the cell. When a muscle contracts, it performs work; potential energy is converted to kinetic energy, and heat is released. The amount of heat is proportional to the amount of work done. As a result, when we exercise, our body temperature rises.

❑ Types of Reactions

Three types of chemical reactions are important to the study of physiology: *decomposition reactions, synthesis reactions,* and *exchange reactions.*

Decomposition

A **decomposition** reaction breaks a molecule into smaller fragments. You could diagram a typical decomposition reaction as:

$$AB \rightarrow A + B$$

Decomposition reactions occur outside cells as well as inside them. For example, a typical meal contains molecules of fats, sugars, and proteins that are too large and too complex to be absorbed and used by our bodies. Decomposition reactions in the digestive tract break these down into smaller fragments before absorption begins.

Catabolism (kah-TAB-o-lizm; *katabole,* a throwing down) is the decomposition of molecules within cells. A covalent bond is a form of potential energy. When such a bond is broken, it releases kinetic energy that can perform work. By harnessing the energy released in this way, living cells perform vital functions such as growth, movement, and reproduction.

Synthesis

Synthesis (SIN-the-sis) is the opposite of decomposition. A synthesis reaction assembles larger molecules from smaller components. Relatively simple synthetic reactions could be diagrammed as:

$$A + B \rightarrow AB$$

Synthesis reactions may involve individual atoms or the combination of molecules to form even larger products. The formation of water from hydrogen and oxygen molecules is an example of synthesis. Synthesis always involves the formation of new chemical bonds, whether the reactants are atoms or molecules.

Anabolism (a-NAB-o-lizm; *anabole,* a throwing upward) is the synthesis of new compounds within the body. Because it takes energy to create a chemical bond, anabolism is usually an "uphill" process. Living cells must "balance their energy budgets," with catabolism providing the energy to support anabolism as well as other vital functions.

Exchange

In an **exchange reaction** parts of the reacting molecules are shuffled around, as in:

$$AB + CD \rightarrow AD + CB$$

You will notice that there are two reactants and two products. Although the reactants and products contain the same components (A, B, C, and D), those components are present in different combinations. In an exchange reaction, the reactant molecules AB and CD break apart (a decomposition) before they interact with one another to form AD and CB (a synthesis). If breaking the old bonds releases more energy than it takes to create the new ones, the exchange reaction will release energy, usually in the form of heat. Reactions that release energy are said to be **exergonic** (*exo-,* outside). If the energy required for synthesis exceeds the amount released by the associated decomposition reaction, additional energy must be provided. Reactions that absorb energy are termed **endergonic** (*endo-*, inside).

❑ Reversible Reactions

Chemical reactions are at least theoretically reversible, so that if A + B → AB, then AB → A + B. Many important biological reactions are freely reversible. Such reactions can be represented as an equation:

$$A + B \rightleftharpoons AB$$

This equation reminds you that there are really two reactions occurring simultaneously, one a synthesis (A + B → AB) and the other a decomposition (AB → A + B). At **equilibrium** (ē-kwi-LIB-rē-um) the two rates are in balance. As fast as a molecule of AB forms, another degrades into A + B.

It is possible to predict the result of a disturbance in the equilibrium condition. When the concentration of a reactant rises, for example, the rate of reaction will increase. In the example above, adding additional AB molecules will increase the rate of conversion to A and B. The concentrations of A and B then rise, and as these concentrations increase so does the rate of the reverse reaction—the formation of AB. Eventually an equilibrium is again established.

Not all chemical reactions are easily reversed. The requirements for the two reactions may differ, so that at any given time and place the reaction will proceed chiefly in one direction. For example, the synthesis reaction may occur when the A and B molecules are heated, and the decomposition reaction when AB molecules are placed in water. In this case the reaction would be written as:

$$A + B \underset{H_2O}{\overset{\text{heat}}{\rightleftharpoons}} AB$$

Decomposition reactions involving water are important in the breakdown of complex molecules in the body; several examples of this process, called **hydrolysis** (hī-DROL-i-sis; *hydro-*, water + *lysis*, dissolution), will be encountered later in the chapter. In hydrolysis, one of the bonds in a complex molecule is broken and the components of a water molecule (H and OH) are added to the resulting fragments:

$$A–B–C–D–E + H_2O \rightarrow A–B–C–H + HO–D–E$$

❑ Enzymes and Chemical Reactions

Most chemical reactions do not occur spontaneously, or they occur so slowly that they would be of little value to living cells. Before a reaction can proceed, enough energy must be provided to *activate* the reactants. **Activation energy** is the amount of energy required to start a reaction. Although many reactions can be activated by changes in temperature or acidity, such changes are deadly to cells. For example, to break down a complex sugar you must boil it in an acid solution.

Enzymes are special proteins that promote chemical processes by lowering the activation energy requirements. Enzymes belong to a class of substances called **catalysts** (KAT-uh-lists; *katalysis*, dissolution), compounds that accelerate chemical reactions without themselves being permanently changed. Enzymes are organic catalysts, and each enzyme is made by a living cell to promote a specific reaction. These reversible reactions can be written as:

$$A + B \overset{\text{enzyme \#1}}{\rightleftharpoons} AB$$

Although the presence of an appropriate enzyme can accelerate a reaction, an enzyme affects only the *rate* of a reaction, not the direction of the reaction or the products that will be formed. *An enzyme cannot bring about a reaction that would otherwise be impossible.*

The complex reactions that support life proceed in a series of interlocking steps, each step controlled by a different enzyme. Such a reaction sequence is called a **pathway.** A synthetic pathway could be diagrammed as:

$$A \underset{\text{Step 1}}{\overset{\text{enzyme \#1}}{\rightleftharpoons}} B \underset{\text{Step 2}}{\overset{\text{enzyme \#2}}{\rightleftharpoons}} C \underset{\text{Step 3}}{\overset{\text{enzyme \#3}}{\rightleftharpoons}} \text{etc.}$$

Sometimes the steps in the synthetic pathway differ from those of the decomposition pathway, and separate enzymes are involved.

✔ In living cells, glucose, a six-carbon molecule, is converted into two three-carbon molecules by a reaction that releases energy. How would you classify this reaction?

✔ If the product of a reversible reaction is continuously removed, what effect do you think this will have on the equilibrium?

■ Inorganic Compounds

Figure 2-2

Although the human body is very complex, it contains a relatively small number of elements, as indicated in Figure 2-2•, p. 33. But knowing the identity and quantity of each element will not help you to understand a human being any more than studying the alphabet will help you to understand this text. The rest of this chapter focuses attention on **nutrients** and **metabolites** (me-TAB-o-lits; *metabole*, change). Nutrients are the essential elements and molecules obtained from the diet. Metabolites include all the molecules synthesized or broken down by chemical reactions inside our bodies. Nutrients and metabolites can be broadly categorized as **inorganic** or **organic.** Inorganic compounds usually do not contain carbon and hydrogen atoms as the primary structural ingredients, whereas carbon and hydrogen always form the basis for organic compounds.

The most important inorganic substances in the body are (1) *carbon dioxide*, a byproduct of cell metabolism; (2) *oxygen*, an atmospheric gas required in important metabolic reactions; (3) *water*, which accounts for most of the body weight; and (4) *inorganic acids, bases*, and *salts*, compounds held together partially or completely by ionic bonds.

❑ Water and Its Properties

Water, H_2O, is the single most important constituent of the body, accounting for almost two-thirds of its total weight. A change in body water content can have fatal consequences because virtually all physiological systems will be affected.

Although familiar to everyone, water really has some very unusual properties. The properties of water are a direct result of the hydrogen bonding that occurs between adjacent water molecules.

1. ***Water is a medium for most chemical reactions and a participant in many reactions.*** Most chemical reactions in the body take place in solution. Water molecules participate in some of these reactions, especially catabolic reactions that break large molecules into smaller fragments. Specific examples will be encountered later in the chapter.
2. ***Water has a very high heat capacity,*** and its freezing and boiling points are far apart. Heat capacity is the ability to absorb and retain heat. Water has an unusually high heat capacity because water molecules in the liquid state are attracted to one another through hydrogen bonding. Important consequences of this attraction include:

   - The temperature must be quite high before individual molecules have enough energy to break free and become water vapor. Consequently water stays in the liquid state over a broad range of environmental temperatures.
   - Water carries a great deal of heat away with it when it finally does change from a liquid to a vapor. This feature accounts for the cooling effect of perspiration on the skin.
   - It takes an unusually large amount of heat energy to change the temperature of 1 g of water by 1°C. Thus, once a volume of water has reached a particular temperature, it will change temperature only slowly, a property called *thermal inertia.* Because water accounts for roughly 66 percent of the weight of the human body, thermal inertia helps stabilize body temperature. Similarly, water's high heat capacity allows the blood plasma to transport and redistribute large amounts of heat as it circulates within the body. For example, heat absorbed as the blood flows through active muscles will be released when the blood reaches vessels in the relatively cool body surface.
3. ***Water is an effective lubricant.*** There is little friction between water molecules. Thus, if two opposing surfaces are separated by even a thin layer of water, friction will be greatly reduced. (That is why driving on wet roads can be tricky; your tires may start sliding on a layer of water, rather than maintaining contact with the road.) Liquid within joints prevents friction between opposing bony surfaces, and a small amount of water in the ventral body cavities prevents friction between internal organs, such as the heart or lungs, and the body wall. ∞ [p. 24]
4. ***Water is an excellent solvent.*** Water will dissolve a remarkable number of inorganic and organic molecules, creating a *solution.* As they dissolve, the molecules break apart, releasing ions or molecules that become uniformly dispersed throughout the solution. The chemical reactions within living cells occur in solution, and the watery component, or *plasma,* of blood distributes dissolved nutrients and waste products throughout the body. The solvent properties of water are so important that we will consider them in detail in the next section.

Solutions

Figure 2-8

A remarkable number of inorganic and organic molecules will dissolve in water, creating a solution. Every **solution** consists of a medium, or **solvent,** in which atoms, ions, or molecules of another substance, or **solute,** are dispersed. In an **aqueous solution** water is the solvent. Water is particularly effective as a solvent because of its chemical structure. Because the hydrogen atoms are attached to the oxygen atom by polar covalent bonds, the hydrogen atoms have a slight positive charge and the oxygen atom has a slight negative charge (Figure 2-8a●). The bonds in a water molecule are oriented so as to place the hydrogen atoms relatively close together. As a result, the water molecule has positive and negative poles. A water molecule is therefore called a **polar molecule.**

Many inorganic compounds held together by ionic bonds will undergo **ionization** (ī-on-i-ZĀ-shun), or **dissociation** (di-sō-sē-Ā-shun), in water. In this process, shown in Figure 2-8b•, ionic bonds are broken as the individual ions interact with the positive or negative ends of polar water molecules. The result is a mixture of cations and anions surrounded by water molecules held by the attraction between opposite charges. The water molecules form a **hydration sphere** that completely surrounds each ion.

A solution containing anions and cations will conduct an electrical current; cations (+) move toward the negative terminal, and anions (–) move toward the positive terminal. Electrical forces across cell membranes affect the functioning of all cells, and small electrical currents carried by ions are essential to muscle contraction and nerve function. Chapters 10 and 12 will discuss these events in more detail.

Soluble inorganic molecules whose ions will conduct an electrical current in solution are called **electrolytes** (e-LEK-trō-līts). Sodium ions (Na^+), potassium ions (K^+), calcium ions (Ca^{2+}), and chloride ions (Cl^-) are released by the dissociation of electrolytes in blood and other body fluids. Table 2-2 contains a more complete listing of important electrolytes and the ions released by their dissociation. Alterations in the body fluid concentrations of these ions will disturb almost every vital function. For example, declining potassium levels will lead to a general muscular paralysis, and rising concentrations will cause weak and irregular heartbeats. The concentrations of ions in body fluids are carefully regulated, primarily by coordinating activities at the kidneys (excretion), the digestive tract (absorption), and the skeletal system (storage or release).

TABLE 2-2 Important Electrolytes That Dissociate in Body Fluids

Compound		Ion		Ion
NaCl (sodium chloride)	→	Na^+	+	Cl^-
KCl (potassium chloride)	→	K^+	+	Cl^-
$CaCl_2$ (calcium chloride)	→	Ca^{2+}	+	2 Cl^-
$NaHCO_3$ (sodium bicarbonate)	→	Na^+	+	HCO_3^-
$MgCl_2$ (magnesium chloride)	→	Mg^{2+}	+	2 Cl^-
Na_2HPO_4 (disodium phosphate)	→	2 Na^+	+	HPO_4^{2-}
Na_2SO_4 (sodium sulfate)	→	2 Na^+	+	SO_4^{2-}

Organic molecules often contain polar covalent bonds, which also attract water molecules. The hydration spheres that form then carry these molecules into solution, as shown in Figure 2-8c•. The molecule in this figure is an important soluble sugar, *glucose*. Molecules that readily associate with water molecules are called **hydrophilic** (hī-drō-FI-lik; *hydro-*, water + *philos*, loving).

Molecules that do not readily associate with water are called **hydrophobic** (hī-drō-FŌ-bik; *hydro-*, water + *phobos*, fear). Hydrophobic molecules have few if any polar covalent bonds. When placed in contact with water molecules, hydration spheres do not form and the molecules do not dissolve. Body fat deposits consist of large, insoluble droplets of hydrophobic molecules, trapped in the watery interior of cells. Gasoline, heating oil, and diesel fuel are examples of hydrophobic molecules not found in the body.

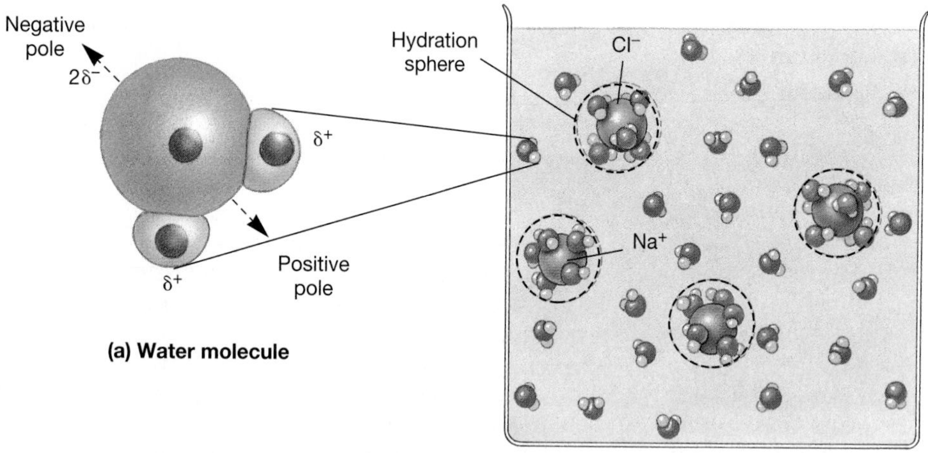

Negative pole
2δ⁻
δ⁺
δ⁺
Positive pole
(a) Water molecule
Hydration sphere
Cl⁻
Na⁺
(b) Sodium chloride in solution
Glucose molecule
(c) Glucose in solution

• **FIGURE 2-8**
Water Molecules and Solutions.
(a) In a water molecule, oxygen forms polar covalent bonds with two hydrogen atoms. Because the hydrogen atoms are positioned toward one end of the molecule, the molecule has an uneven distribution of charges. This creates positive and negative poles. **(b)** Ionic compounds dissociate in water as the polar water molecules disrupt the ionic bonds. The ions in solution are surrounded by water molecules, creating hydration spheres. **(c)** Hydration spheres will also form around an organic molecule containing polar covalent bonds. If the molecule is small it will be carried into solution, as shown here with glucose.

SOLUTE CONCENTRATIONS Physiologists and clinicians often monitor inorganic and organic solute concentrations in body fluids such as blood or urine. Data may be reported in several different ways, and two methods will be introduced here. The first method is to give the number of atoms, molecules, or ions in a specific volume of solution. Values are reported in terms of moles (mol, M) or millimoles (mmol, mM) per liter. Extending the definition given earlier in the chapter, a *mole* is the number of atoms or molecules in a sample that has a weight in grams equal to the atomic weight (for an element) or the molecular weight (for a molecule). Expressing relationships in moles rather than units of weight makes it much easier to keep track of the relative numbers of atoms or molecules involved. Physiological concentrations are often stated in millimoles per liter (mmol/l).

Concentrations can be reported in terms of millimoles only when dealing with a specific ion or molecule of known molecular weight. When you cannot be very specific about the chemical substance involved, a second method is used. This method reports the weight of material dissolved in a unit volume of solution. Values are most often expressed in terms of milligrams (mg) per deciliter (dl, or 100 ml). Appendix VII contains a detailed analysis of blood and other body fluids.

Colloids and Suspensions

Figure 2-8c

Body fluids may contain large and complex organic molecules, such as proteins and protein complexes, that are held in solution by their association with water molecules, as in Figure 2-8c●. A solution containing dispersed proteins or other large molecules is called a **colloid.** Liquid jello is a familiar example of a rather viscous (thick) colloid.

The particles or molecules in a colloid will remain in solution indefinitely. A **suspension** contains even larger particles that will, if undisturbed, settle out of solution because of the force of gravity. Stirring beach sand into a bucket of water creates a temporary suspension that ends as the sand settles to the bottom. Whole blood is an example of a suspension, because the cells circulating in the bloodstream are suspended in the plasma. If clotting is prevented, the cells in a blood sample will settle to the bottom of the container.

Hydrogen Ions in Body Fluids

Figure 2-9

The concentration of hydrogen ions in body fluids is precisely regulated. Hydrogen ions are extremely reactive in solution; in excessive numbers they can break chemical bonds, change the shapes of complex molecules, and disrupt cell and tissue functions.

A few hydrogen ions are normally present, even in a sample of pure water, because some of the water molecules dissociate, releasing cations and anions. The dissociation of water is a reversible reaction that can be represented as:

$$H_2O \rightleftharpoons H^+ + OH^-$$

The dissociation of a water molecule yields a hydrogen ion (H^+) and a **hydroxide** (hī-DROK-sīd) **ion,** OH^-. Very few water molecules ionize in pure water, and the number of hydrogen and hydroxide ions is very small. The quantities are usually reported in moles (p. 34), making it easy to keep track of the relative numbers of hydrogen and hydroxide ions. A liter of pure water contains around 0.0000001 moles of hydrogen ions and an equal number of hydroxide ions. In other words, the **concentration** of hydrogen ions in that solution is 0.0000001 moles per liter. This can be written:

$$[H^+] = 1 \times 10^{-7} \text{ mol/l}$$

The brackets around H^+ indicate "the concentration of," another example of chemical notation.

pH The hydrogen ion concentration is so important to physiological processes that a special shorthand is used to express it. The **pH** of a solution is defined as the negative logarithm of the hydrogen ion concentration in moles per liter. Thus, instead of using the above expression, we could state that the pH of pure water is –(–7), or 7. This saves space, but you must always remember that the pH number is an exponent. Thus a pH of 6, for example ($[H^+] = 1 \times 10^{-6}$, or 0.000001), means that the concentration of hydrogen ions is 10 times as great as it is at a pH of 7 ($[H^+] = 1 \times 10^{-7}$, or 0.0000001). For common liquids the pH scale, included in Figure 2-9●, ranges from 0 to 14.

Although pure water has a pH of 7, solutions display a wide range of pH values, depending on the nature of the solutes involved. A solution with a pH of 7 is **neutral** because it contains equal numbers of hydrogen and hydroxide ions. A solution with a pH below 7 is **acidic** (a-SI-dik), meaning that hydrogen ions predominate. A pH above 7 is **basic,** or **alkaline** (AL-kuh-lin), with hydroxide ions in the majority.

The pH of the blood normally ranges from 7.35 to 7.45. Abnormal fluctuations in pH can damage cells and tissues by breaking chemical bonds, changing the shapes of proteins, and altering cellular function. The human body generates significant quantities of acids that threaten homeostasis and must be neutralized. Under unusual circumstances, the loss of acids in body fluids may promote an equally disruptive increase in pH. The term **acidosis** refers to an abnormal physiological state caused by low blood pH (below 7.35); a pH below 7 can produce coma. **Alkalosis** results from an abnormally high pH (above 7.45); a blood pH above 7.8 usually causes uncontrollable, sustained skeletal muscle contractions.

Inorganic Acids and Bases

The body contains both inorganic and organic acids and bases. We will discuss acids and bases here because the inorganic acids and bases generated in the body are often responsible for producing acidosis or alkalosis.

An **acid** is any solute that dissociates to release hydrogen ions and shift the pH toward acidity. (Because a hydrogen atom that loses its electron consists solely of a naked proton, hydrogen ions are often referred to simply as protons, and acids as "proton donors.")

A **strong acid** ionizes completely, and the reaction is essentially one-way. Hydrochloric acid (HCl) is an excellent example, for in water it ionizes as:

$$HCl \rightarrow H^+ + Cl^-$$

The stomach produces this powerful acid to assist in the breakdown of food. Hardware stores sell HCl, as "muriatic acid," for cleaning sidewalks and swimming pools.

A **base** is a solute that removes hydrogen ions from a solution. Many common bases are compounds that dissociate in solution to liberate a hydroxide ion. Hydroxide ions have a strong affinity for hydrogen ions and quickly react with them, tying them up in water molecules. For example, sodium hydroxide, NaOH, is a **strong base,** because in solution it ionizes completely. The reaction that releases sodium and hydroxide ions can be written as:

$$NaOH \rightarrow Na^+ + OH^-$$

Strong bases have a variety of industrial and household uses; drain openers and lye are two familiar examples.

Weak acids and **weak bases** fail to ionize completely, and a significant number of molecules remain intact. For the same number of molecules in solution, weak acids and weak bases therefore have less of an impact on pH than strong acids and strong bases. *Carbonic acid* (H_2CO_3) is an example of a weak acid found in body fluids.

Salts

A **salt** is an ionic compound consisting of a cation that is not a hydrogen ion and an anion that is not a hydroxide ion. Because they are held together by ionic bonds, many salts ionize completely in water, releasing cations and anions. For example, sodium chloride, or table salt, immediately ionizes into Na^+ and Cl^- ions. Sodium and chloride are the most abundant ions in body fluids. However, many other ions are present in lesser amounts, released by the ionization of various other compounds. Ionic concentrations in the body are regulated by mechanisms described in Chapters 26 and 27.

The ionization of sodium chloride does not affect the local concentrations of hydrogen or hydroxide ions, so NaCl, like many salts, is a "neutral" solute. Other salts, however, indirectly affect the concentration of H^+ and OH^- ions through their interactions with water molecules. Thus their presence may make a solution slightly acidic or basic.

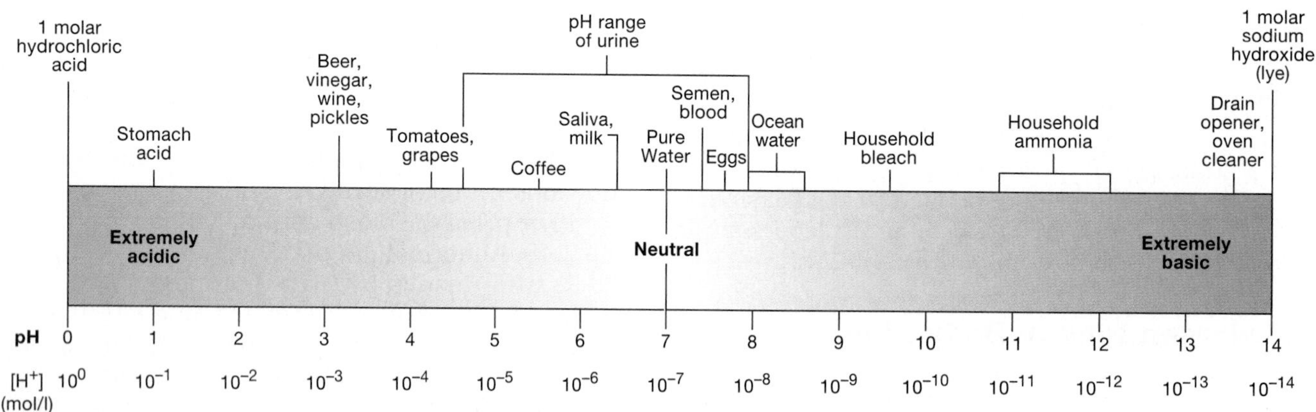
1 molar hydrochloric acid
Stomach acid
Beer, vinegar, wine, pickles
Tomatoes, grapes
pH range of urine
Coffee
Saliva, milk
Pure Water
Semen, blood
Eggs
Ocean water
Household bleach
Household ammonia
Drain opener, oven cleaner
1 molar sodium hydroxide (lye)
Extremely acidic
Neutral
Extremely basic
pH 0 1 2 3 4 5 6 7 8 9 10 11 12 13 14
[H+] (mol/l) 10^0 10^-1 10^-2 10^-3 10^-4 10^-5 10^-6 10^-7 10^-8 10^-9 10^-10 10^-11 10^-12 10^-13 10^-14

FIGURE 2-9

pH and Hydrogen Ion Concentration. Note that the scale is logarithmic; an increase or decrease of one unit corresponds to a tenfold change in H^+ concentration.

☐ Buffers and pH Control

Figure 2-9

Buffers are compounds that stabilize the pH of a solution by removing or replacing hydrogen ions. Buffers and **buffer systems** in body fluids maintain pH within normal limits; Figure 2-9● includes the pH of several body fluids. Buffer systems typically involve a weak acid and its related salt, which functions as a weak base. For example, the *carbonic acid–bicarbonate buffer system*, detailed in Chapter 27, consists of carbonic acid and *sodium bicarbonate*, $NaHCO_3$, otherwise known as baking soda. Antacids such as Alka-Seltzer® and Rolaids® use sodium bicarbonate to neutralize excess hydrochloric acid in the stomach. The effects of **neutralization** are most evident when you add a strong acid to a strong base. For example, adding hydrochloric acid to sodium hydroxide results in the neutralization of both the strong acid and the strong base:

$$HCl + NaOH \rightarrow H_2O + NaCl$$

This reaction produces water and a salt, in this case sodium chloride.

Why does a solution of table salt conduct electricity but a sugar solution does not?

How does an antacid help to decrease stomach discomfort?

Organic Compounds

Organic compounds contain the elements carbon and hydrogen, and usually oxygen as well. Organic molecules often contain long chains of carbon atoms linked by covalent bonds. These carbon atoms typically form additional covalent bonds with hydrogen or oxygen atoms, and less often, with nitrogen, phosphorus, sulfur, iron, or other elements.

Many organic molecules are soluble in water. Although the discussion above focused on inorganic acids and bases, there are also important organic acids and bases. For example, *lactic acid* is an organic acid, generated by active muscle tissues, that must be neutralized by the buffers in body fluids.

There are four major classes of organic compounds: **carbohydrates, lipids, proteins,** and **nucleic acids.** In addition, the human body contains small quantities of many other organic compounds whose structures and functions will be considered in later chapters.

☐ Carbohydrates

A **carbohydrate** (kar-bō-HĪ-drāt) molecule has carbon, hydrogen, and oxygen in a ratio near 1:2:1. Familiar carbohydrates include the sugars and starches that make up roughly half of the typical American diet. Our tissues can break down most carbohydrates, and, although they may have other functions, carbohydrates are most important as sources of energy. We will focus attention on *monosaccharides, disaccharides*, and *polysaccharides.*

Monosaccharides

Figure 2-10

A **simple sugar,** or **monosaccharide** (mon-ō-SAK-uh-rīd; *mono-*, single + *sakcharon*, sugar), is a carbohydrate containing from three to seven carbon atoms. A simple sugar may be called a triose (three-carbon), tetrose (four-carbon), pentose (five-carbon), hexose (six-carbon), or heptose (seven-carbon). The hexose **glucose** (GLOO-kōs), $C_6H_{12}O_6$, is the most important metabolic "fuel" in the body. Figure 2-10● diagrams the structure of a glucose molecule in several ways. The straight-chain form (Figure 2-10a●) shows the component atoms clearly, but in the body the ring form (Figure 2-10b●) is more common. The actual arrangement of atoms in three dimensions is shown most realistically in a space-filling model (Figure 2-10c●).

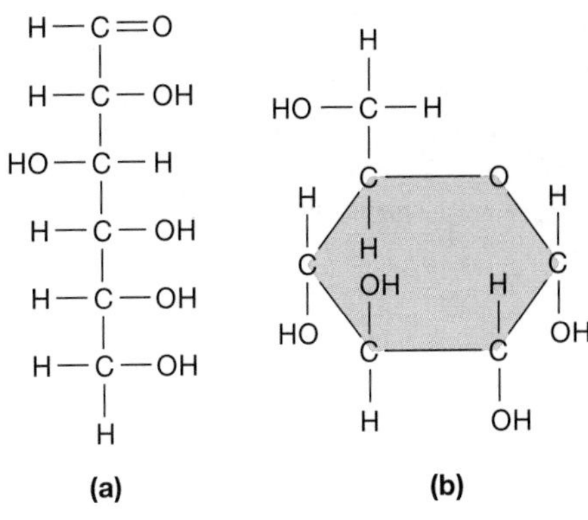
H—C=O
H—C—OH
HO—C—H
H—C—OH
H—C—OH
H—C—OH
H
(a)
H
HO—C—H
C
O
H
H
H
C
C
OH
H
HO
OH
C
C
H
OH
(b)

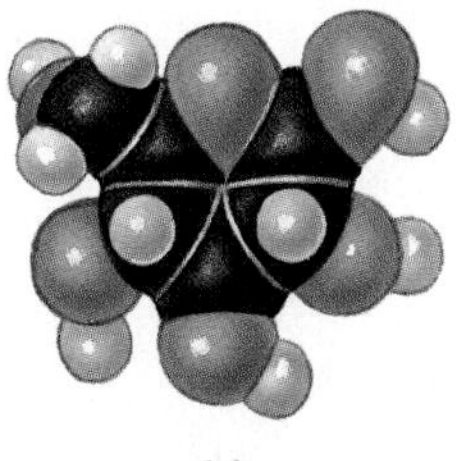
(c)

● **FIGURE 2-10**

Glucose. (a) The straight-chain structural formula. **(b)** The ring form that is most common in nature. **(c)** A space-filling model that shows the actual three-dimensional organization of the atoms in the ring.

It is important to note that each organic molecule has a specific three-dimensional shape, and that molecular shape is an important characteristic. **Isomers** are molecules that have the same molecular formula but different shapes. That is, they contain the same atoms, but those atoms are bonded together in different arrangements. The body generally treats different isomers as distinct molecules, and they have different properties and fates. For example, the simple sugars glucose and *fructose* are isomers. Fructose is found in many fruits and in secretions of the male reproductive tract. Although its chemical formula is the same as that of glucose, separate enzymes and reaction sequences control its breakdown and synthesis. Simple sugars such as glucose dissolve readily in water, and they are rapidly distributed throughout the body by the blood and other body fluids. AM *The Pharmaceutical Use of Isomers*

Disaccharides and Polysaccharides

Figures 2-11, 2-12

Carbohydrates other than simple sugars are complex molecules composed of monosaccharide building blocks. Two simple sugars joined together form a **disaccharide** (dī-SAK-uh-rīd; *di-*, two). Disaccharides such as **sucrose** (table sugar) have a sweet taste, and they are also quite water-soluble. Figure 2-11a● diagrams the reaction responsible for the formation of sucrose. This is an example of **dehydration synthesis,** the linking together of chemical units by the removal of water to create a more complex molecule. The reverse of this process, shown in Figure 2-11b●, is an example of *hydrolysis,* a process introduced on p. 40.

Many foods contain disaccharides, but they must be disassembled through hydrolysis before they can be broken down to provide useful energy. Most popular "junk foods," such as candies or sodas, abound in simple sugars (often fructose) and disaccharides such as sucrose. Some people cannot tolerate sugar for medical reasons; others avoid it in an effort to control their weight. (Excess sugars are converted to fat for long-term storage.) Many of these people are using *artificial sweeteners* in their foods and beverages. These compounds have a very sweet taste, but they either cannot be broken down in the body, or they are used in insignificant amounts. AM *Artificial Sweeteners*

Dehydration synthesis can continue adding monosaccharides or combining disaccharides to create very complex molecules. These large molecules are called **polysaccharides** (pol-ē-SAK-uh-rīdz; *poly-*, many). Polysaccharide chains may be straight or highly branched. **Starches** are glucose-based polysaccharides. Most starches are manufactured by plants. The digestive tract can break these molecules into simple sugars, and starches such as those found in potatoes and grains represent a major dietary energy source. *Cellulose,* a structural component of many plants, is a polysaccharide that our bodies cannot digest at all.

Glycogen (GLĪ-ko-jen), or *animal starch,* is a branched polysaccharide composed of interconnected glucose molecules (Figure 2-12●). Like most other large polysaccharides, glycogen will not dissolve in water or other body fluids. Liver and mus-

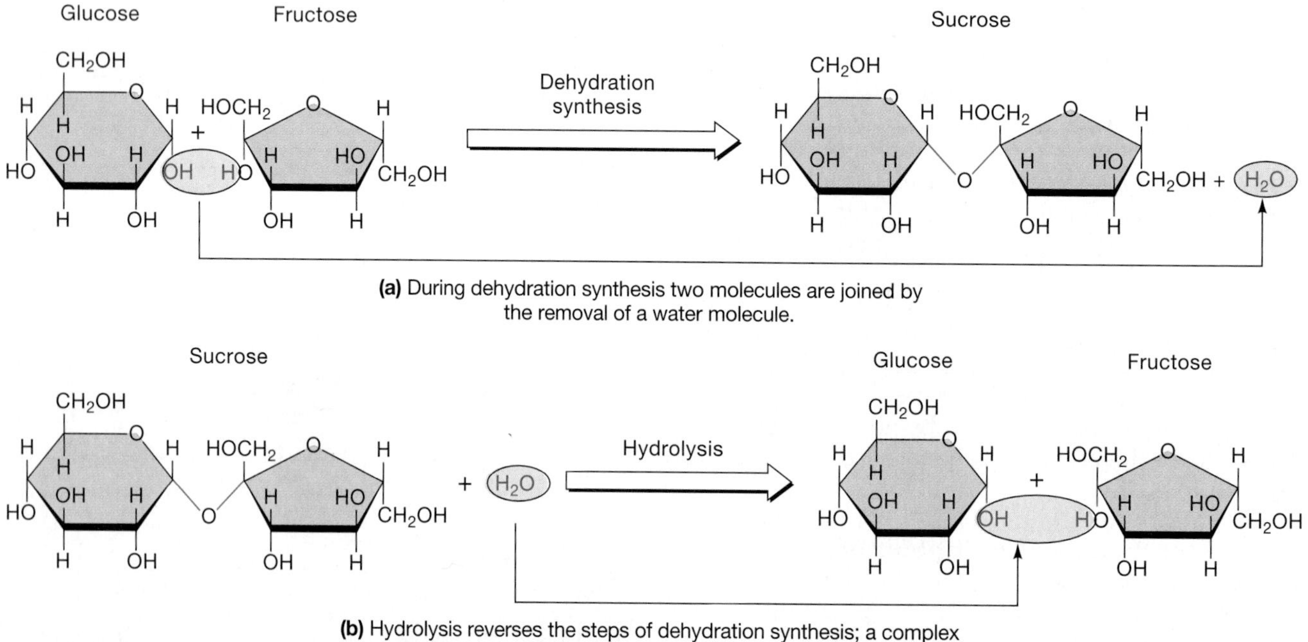

(a) During dehydration synthesis two molecules are joined by the removal of a water molecule.

(b) Hydrolysis reverses the steps of dehydration synthesis; a complex molecule is broken down by the addition of a water molecule.

● **FIGURE 2-11**
The Formation and Breakdown of Complex Sugars

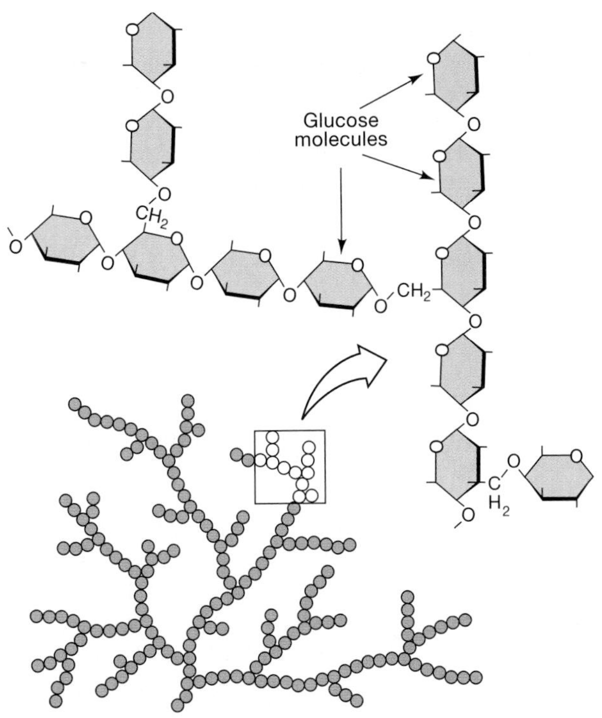
Glucose molecules
O
CH_2
CH_2
C
H_2

● **FIGURE 2-12**
Structure of a Polysaccharide. Liver and muscle cells store glucose as the polysaccharide glycogen, a long, branching chain of glucose molecules.

cle tissues manufacture and store large quantities of glycogen. When these tissues have a high demand for energy, glycogen molecules are broken down into glucose; when demands are low, the tissues absorb glucose and rebuild glycogen reserves.

Despite their metabolic importance, carbohydrates account for less than 3 percent of our total body weight. Table 2-3 summarizes information concerning the carbohydrates.

❑ Lipids

Lipids (*lipos*, fat) also contain carbon, hydrogen, and oxygen, but the ratios do not approximate 1:2:1. In general, a lipid molecule contains much less oxygen than a carbohydrate having the same number of carbon atoms. In addition to carbon, hydrogen, and oxygen, lipids may contain small quantities of other elements, such as phosphorus, nitrogen, or sulfur. Familiar lipids include fats, oils, and waxes. Most lipids are insoluble in water, but special transport mechanisms carry them in the circulating blood.

Lipids form essential structural components of all cells. In addition, lipid deposits are important as energy reserves, for on the average lipids provide roughly twice as much energy as carbohydrates, gram for gram, when broken down in the body. When the supply of lipids exceeds the demand for energy, the excess is stored in fat deposits. For this reason there has been great interest in developing *fat substitutes* that provide less energy but have the same taste and texture as lipids. AM *Fat Substitutes*

All together, lipids normally account for 10–12 percent of our total body weight. There are many different kinds of lipids in the body. Major lipid types are presented in Table 2-4 and diagrammed in Figures 2-13, 2-14, 2-15, and 2-16. We will consider six classes of lipid molecules: *fatty acids, glycerides, prostaglandins, steroids, phospholipids,* and *glycolipids.*

Fatty Acids

Figure 2-13

Fatty acids are long carbon chains with hydrogen atoms attached. One end of the carbon chain always bears a **carboxylic** (kar-bok-SIL-ic) **acid group,** a structure that can be represented as:

$$R\ldots -\overset{\displaystyle OH}{\overset{/}{C}}=O$$

The carbon chain is abbreviated by "R" and the carboxylic acid group by COOH. The name *carboxyl* group should help you remember that a *carbon* and a hydr*oxyl* (—OH) group are the important structural features. Figure 2-13a● shows a representative fatty acid, *lauric acid.*

In a fatty *acid* the polar covalent bond between the oxygen and hydrogen atoms of the

TABLE 2-3 Carbohydrates in the Body

Structure	*Examples*	*Primary Functions*	*Remarks*
Monosaccharides (simple sugars)	Glucose, fructose	Energy source	Manufactured in the body and obtained from food; found in body fluids.
Disaccharides	Sucrose, lactose, maltose	Energy source	Sucrose is table sugar, lactose is present in milk; all must be broken down to monosaccharides before absorption.
Polysaccharides	Glycogen	Storage of glucose molecules	Glycogen is found in animal cells, other starches and cellulose in plant cells.

TABLE 2-4 Representative Lipids and Their Functions in the Body

Lipid Type	*Examples*	*Primary Functions*	*Remarks*
Fatty acids	Lauric acid	Energy sources	Absorbed from food or synthesized in cells; transported in the blood for use in many tissues.
Glycerides	Monoglycerides, diglycerides, triglycerides	Energy source, energy storage, insulation, and physical protection	Stored in fat deposits; must be broken down to fatty acids and glycerol before they can be used as an energy source.
Prostaglandins		Chemical messengers coordinating local cellular activities	Produced in most body tissues.
Steroids	Cholesterol	Structural component of cell membranes, hormones, digestive secretions in bile	All have the same carbon-ring framework.
Phospholipids, glycolipids		Structural components of cell membranes	

carboxylic acid group breaks down in solution, releasing a hydrogen ion. This reaction can be summarized as:

$$\ldots - \overset{\text{OH}}{\overset{/}{\text{C}}}=\text{O} \longrightarrow \ldots - \overset{\text{O}^-}{\overset{/}{\text{C}}}=\text{O} + \text{H}^+$$

When placed in solution, only the carboxyl end of a fatty acid associates with water molecules, for this is the only hydrophilic portion of the molecule. The rest of the carbon chain is hydrophobic, so fatty acids have a very limited solubility in water.

In a **saturated** fatty acid each carbon atom in the hydrocarbon tail has four single covalent bonds. If some of the carbon-to-carbon bonds are double covalent bonds, the fatty acid is **unsatu-**

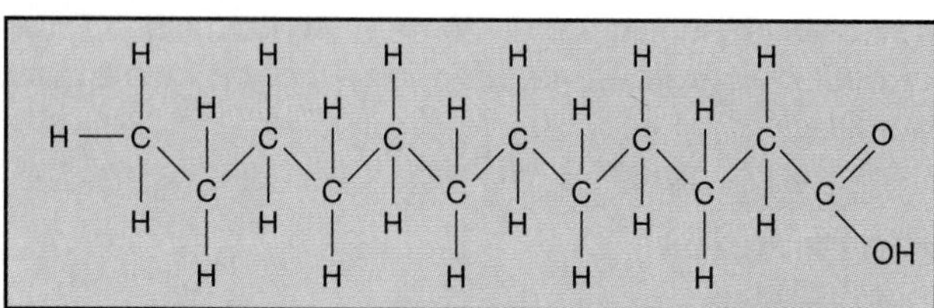

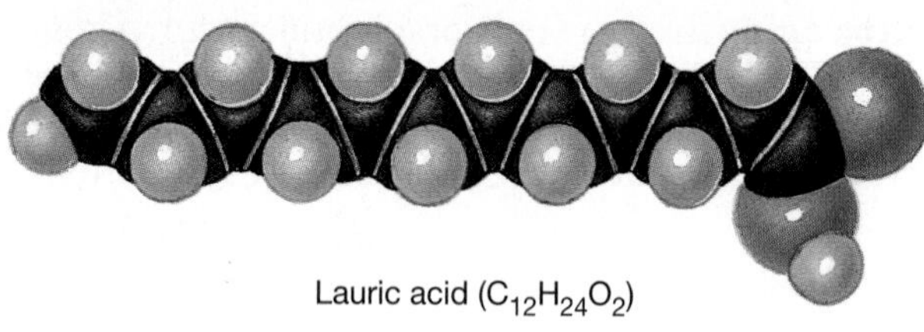

Lauric acid ($C_{12}H_{24}O_2$)

(a)

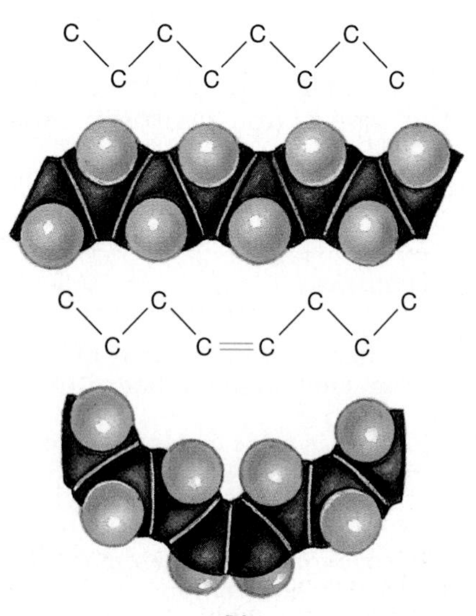

(b)

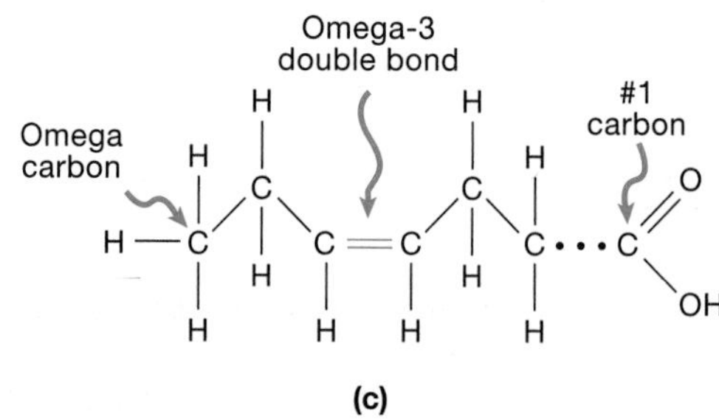
Omega-3
double bond
Omega
carbon
#1
carbon

(c)

● **FIGURE 2-13**
Fatty Acids. (a) Lauric acid demonstrates two structural characteristics common to all fatty acids: a long backbone of carbon atoms and a carboxylic acid group (— COOH) at one end. **(b)** A fatty acid may be saturated or unsaturated. Unsaturated fatty acids have double covalent bonds, and their presence causes a sharp bend in the molecule. **(c)** An omega-3 fatty acid has an unsaturated bond three carbons before the end of the carbon chain.

rated. These terms refer to the number of hydrogen atoms in the fatty acid. Replacing a double bond between carbon atoms with a single bond allows the molecule to accept two more hydrogens; hence a double-bonded chain is said to be unsaturated (see Figure 2-13b●). A **monounsaturated fatty acid** has a single unsaturated bond in the carbon tail. A **polyunsaturated** fatty acid contains multiple unsaturated bonds.

FATTY ACIDS AND HEALTH Both saturated and unsaturated fatty acids can be broken down for energy, but a diet containing large amounts of saturated fatty acids increases the risk of heart disease and other circulatory problems. Butter, fatty meat, and ice cream are popular dietary sources of saturated fatty acids. Vegetable oils such as olive oil or corn oil contain a mixture of monounsaturated and polyunsaturated fatty acids. Current research indicates that the monounsaturated fats may be more effective than polyunsaturated fats in lowering the risk of heart disease. In fact, the manufacture of margarine and vegetable shortening from polyunsaturated fats produces compounds called *trans* fatty acids that may actually *increase* the risk of heart disease. Increasing the proportion of *oleic acid*, an 18-carbon monounsaturated fatty acid, in cooking oils and other products could therefore yield health benefits.

The carbons in a fatty acid molecule are numbered beginning at the carboxylic acid end; the last carbon in the chain is called the *omega* carbon. Eskimos have lower rates of heart disease than other populations, despite the fact that the Eskimo diet contains high quantities of fats and cholesterol. The fatty acids in the Eskimo diet have an unsaturated bond three carbons before the omega carbon, a position known as "omega minus 3," or *omega-3* (Figure 2-13c●). *Omega-3 fatty acids* are found in fish flesh and fish oils. For unknown reasons the presence of omega-3 fatty acids in the diet reduces the risks of heart disease, rheumatoid arthritis, and other inflammatory diseases.

Glycerides

Figures 2-11, 2-14

Individual fatty acids cannot be strung together in a chain by dehydration, as simple sugars can. But they can be attached to another compound, **glycerol** (GLI-se-rol), through a similar reaction. As illustrated in Figure 2-14●, dehydration synthesis can produce a **monoglyceride** (mo-nō-GLI-se-rīd) consisting of glycerol plus one fatty acid. Subsequent reactions can yield a **diglyceride** (glycerol + two fatty acids), and then a **triglyceride** (glycerol + three fatty acids). Hydrolysis breaks the glycerides into fatty acids and glycerol. A comparison of Figure 2-14 with Figure 2-11● demonstrates the fact that dehydration synthesis and hydrolysis operate in the same way, whether the molecules involved are carbohydrates or lipids.

Triglycerides, otherwise known as **neutral fats,** have several important functions:

1. Fatty deposits in specialized sites represent a significant energy reserve. In times of need the triglycerides are disassembled to yield fatty acids that can be broken down to provide energy.
2. Fat deposits under the skin serve as insulation, preventing heat loss to the environment. Many of these deposits lie just under the skin, and heat loss across a layer of lipids is only about one-third of that through other tissues.
3. A fat deposit around a delicate organ such as a kidney provides a cushion that protects against shocks or blows.

Triglycerides are stored in the body as lipid droplets within cells. These lipids absorb and accumulate lipid-soluble vitamins, drugs, or toxins that appear in body fluids. This has both positive and negative effects. For example, the body's lipid reserves retain both valuable lipid-soluble vitamins (A, D, E, K) and potentially dangerous lipid-soluble pesticides, such as DDT. AM *Drugs and Lipids*

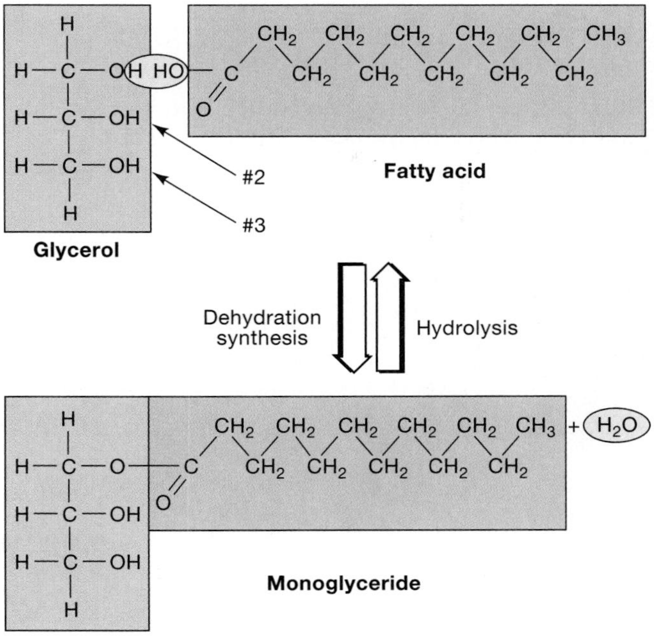

H—C—OH HO—C(=O)—CH_2—CH_2—CH_2—CH_2—CH_2—CH_2—CH_2—CH_2—CH_2—CH_2—CH_3
H—C—OH
H—C—OH
Glycerol
#2
#3
Fatty acid
Dehydration synthesis
Hydrolysis
H—C—O—C(=O)—CH_2—CH_2—CH_2—CH_2—CH_2—CH_2—CH_2—CH_2—CH_2—CH_2—CH_3 + H_2O
H—C—OH
H—C—OH
Monoglyceride

● **FIGURE 2-14**

Glyceride Formation. The formation of a glyceride involves the attachment of fatty acids to the carbons of a glycerol molecule. This example shows the formation of a monoglyceride (glycerol + one fatty acid). Attachment of a second fatty acid at site #2 would create a diglyceride, and addition of a third at site #3 would produce a triglyceride.

Prostaglandins

Figure 2-15a

Prostaglandins (pros-tuh-GLAN-dinz) are fatty acids that have five of their carbon atoms joined in a ring (Figure 2-15a●). Cells release prostaglandins to coordinate or direct local cellular activities. They are extremely powerful and effective in minute quantities. Almost every tissue in the body contains prostaglandins; the effects vary depending on the nature of the prostaglandin and the site of release. Consider three examples:

1. Prostaglandins released by damaged tissues stimulate nerve endings and produce the sensation of pain (Chapter 17).
2. Prostaglandins released in the uterus help trigger the start of labor contractions (Chapter 29).
3. Prostaglandins released in the stomach may help prevent stomach ulcers (Chapter 24).

Prostaglandins are often called "local hormones" because hormones are chemical messengers produced at one site to regulate activities under way in a *different* portion of the body. There are many other types of local hormones in the body that are lipid-based. Examples encountered in later chapters include the *thromboxanes* involved in blood clotting (Chapter 19) and the *leukotrienes* involved in inflammation (Chapter 22).

Steroids

Figure 2-15b,c

Steroids are large lipid molecules that share a distinctive carbon framework (Figure 2-15b●). They differ in the carbon chains attached to this basic structure. Figure 2-15c● shows the structural formula for the steroid **cholesterol** (ko-LES-ter-ol; *chole-*, bile + *stereos*, solid). Cholesterol and related steroids are important for the following reasons:

- ***All animal cell membranes contain cholesterol.*** Cells need cholesterol to maintain their cell membranes as well as for cell growth and division.
- ***Steroid hormones are involved in the regulation of sexual function.*** Examples include the sex hormones, such as *testosterone* and the *estrogens*.
- ***Steroid hormones are important in the regulation of tissue metabolism and mineral balance.*** Examples include the hormones of the adrenal cortex, called *corticosteroids*, and *calcitriol*, a hormone important in calcium ion regulation.
- ***Steroids called bile salts are required for the normal processing of dietary fats.*** Bile salts are produced in the liver and secreted in bile. These steroids interact with lipids in the intestinal tract and facilitate the digestion and absorption of lipids.

Cholesterol is obtained from two sources: (1) by absorption from animal products in the diet and (2) by synthesis within the body. Meat, cream, and egg yolks are especially rich dietary sources of cholesterol. A diet high in cholesterol can be harmful, because a strong link exists between high blood cholesterol levels and heart disease. Current nutritional advice suggests reducing cholesterol intake to under 300 mg per day; this represents a 40 percent reduction for the average American adult. Unfortunately, because the body can synthesize cholesterol as well, it is sometimes

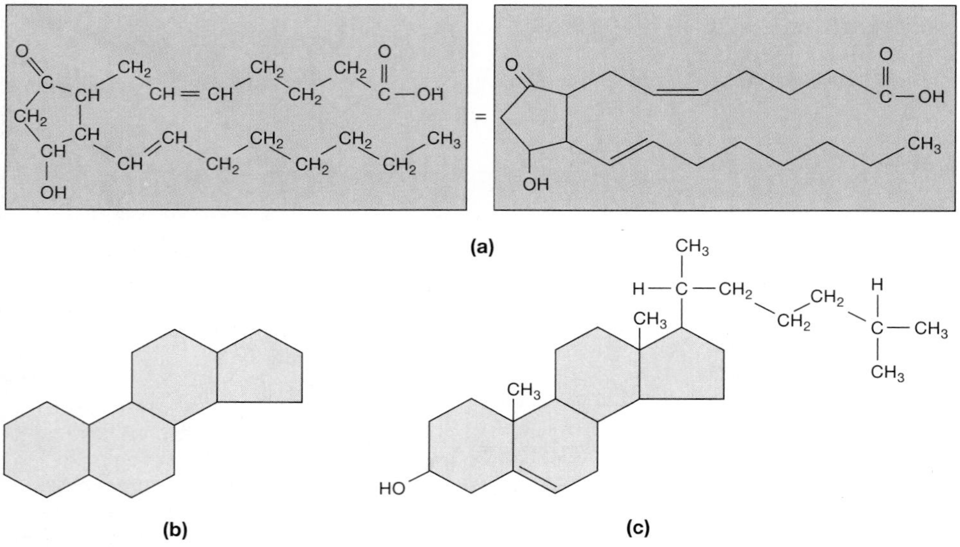

O
CH2
CH2
CH2
O
C
CH
CH = CH
CH2
C—OH
CH2
CH
CH
CH2
CH2
CH3
CH
CH
CH2
CH2
CH2
OH
=
O
O
C—OH
CH3
OH
(a)
(b)
CH3
H—C—CH2
CH2
H
CH3
CH2
C—CH3
CH3
CH3
HO
(c)

● **FIGURE 2-15**
Prostaglandins and Steroids. (a) Prostaglandins are unusual short-chain fatty acids. To appreciate the chain structure, compare the complete structural formula for a typical prostaglandin (left) with the abbreviated formula (right), where the molecular framework is shown as a solid line. Each change in the direction of the line indicates the presence of a carbon atom. This diagrammatic method simplifies the chemical formula and makes it easier to visualize the complete molecule. **(b)** All steroids share this complex four-ring structure. **(c)** Individual steroids differ in the side chains attached to the carbon rings. This is a molecule of cholesterol.

difficult to control blood cholesterol levels by dietary restriction alone. The connection between blood cholesterol levels and heart disease will be examined more closely in later chapters.

Phospholipids and Glycolipids

Figure 2-16

Phospholipids (FOS-fō-lip-idz) and **glycolipids** (GLĪ-cō-lip-idz) are structurally related, and both can be synthesized from fatty acids. In a *phospholipid* (Figure 2-16a●) a phosphate group $(PO_4)^{3-}$ serves as a link between a diglyceride and a nonlipid group. In a *glycolipid* (Figure 2-16b●) a carbohydrate is attached to a diglyceride. Note that placing *-lipid* last in these names indicates that the molecule consists primarily of lipid.

The long fatty acid "tails" of these molecules are hydrophobic, but the nonlipid "heads" are hydrophilic. In water, large numbers of these molecules tend to form droplets, or **micelles** (mī-SELLZ), with the hydrophilic portions on the outside (Figure 2-16c●). Most meals contain a mixture of lipids and other organic molecules, and micelles

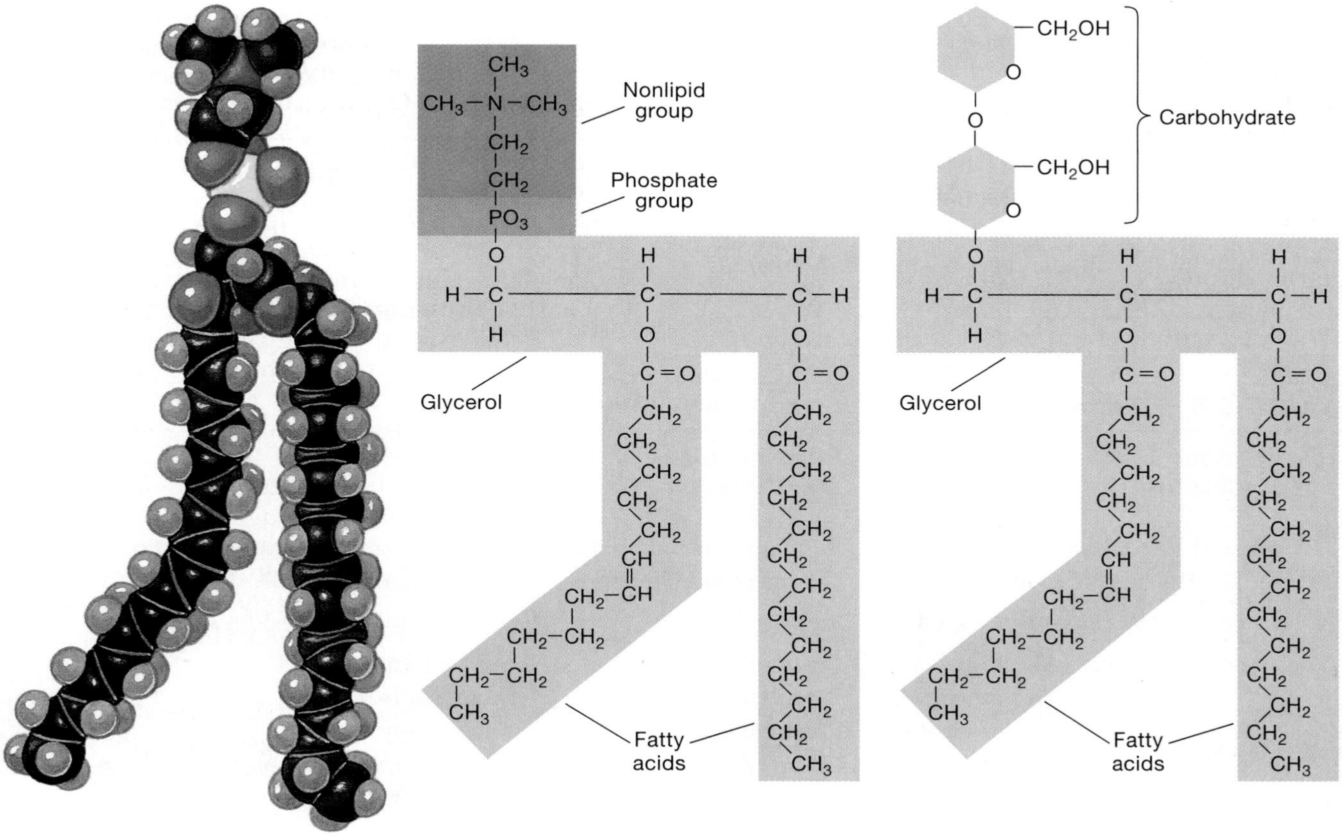

(a) Phospholipid

(b) Glycolipid

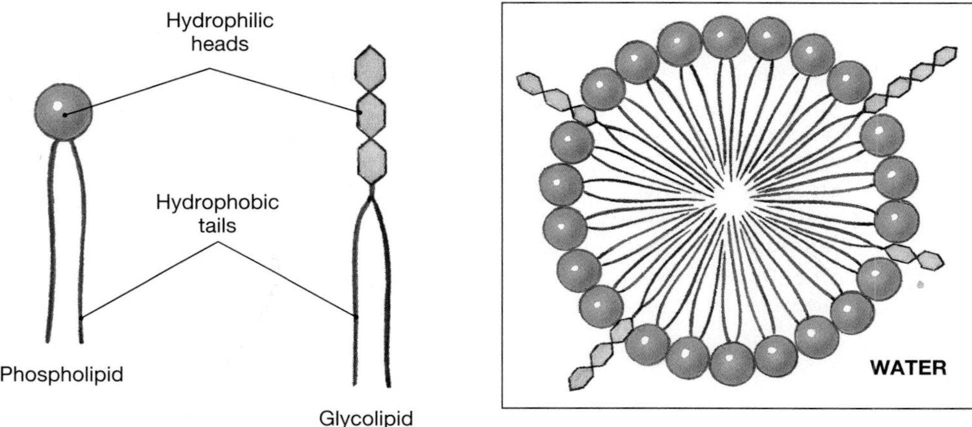
Hydrophilic heads
Hydrophobic tails
Phospholipid
Glycolipid
WATER

(c) Micelle structure

● **FIGURE 2-16 Phospholipids and Glycolipids. (a)** In a phospholipid a phosphate group links a nonlipid molecule to a diglyceride. This is a molecule of *lecithin*, a representative phospholipid. **(b)** In a glycolipid a carbohydrate is attached to a diglyceride. **(c)** When present in large numbers these molecules will form micelles, with the hydrophilic heads facing the water molecules and the hydrophobic tails on the inside of the droplet.

form as the food breaks down in the digestive tract. In addition to phospholipids and glycolipids, micelles may contain other insoluble lipids, such as steroids, glycerides, and long-chain fatty acids.

Cholesterol, phospholipids, and glycolipids are called **structural lipids** because they form and maintain cellular membranes. Membranes composed of hydrophobic lipids surround each cell, and internal membranes subdivide the interior of the cell. Thus the watery interior of the cell is relatively isolated from the watery extracellular environment, and the intracellular compartments, separated from each other by hydrophobic membranes, can maintain distinctive chemical compositions.

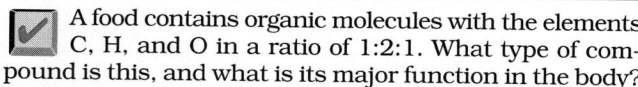

A food contains organic molecules with the elements C, H, and O in a ratio of 1:2:1. What type of compound is this, and what is its major function in the body?

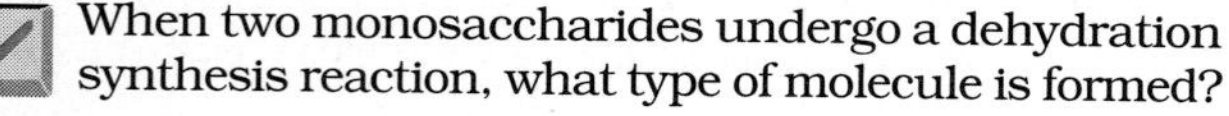

When two monosaccharides undergo a dehydration synthesis reaction, what type of molecule is formed?

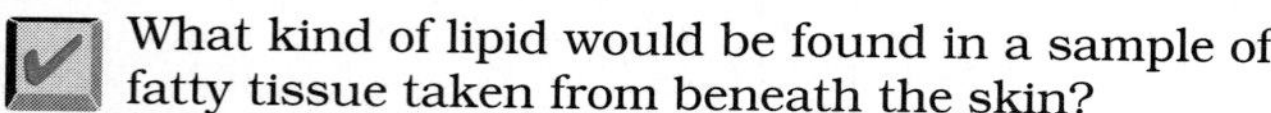

What kind of lipid would be found in a sample of fatty tissue taken from beneath the skin?

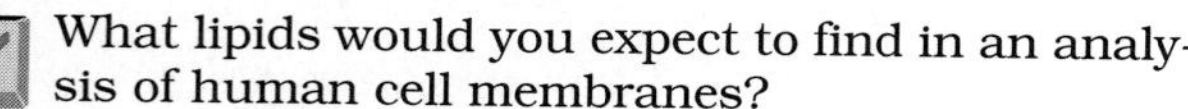

What lipids would you expect to find in an analysis of human cell membranes?

❑ Proteins

Proteins are the most abundant organic components of the human body, and in many ways the most important. There are roughly 100,000 different kinds of proteins, and they account for about 20 percent of the total body weight. All proteins contain carbon, hydrogen, oxygen, and nitrogen; smaller quantities of sulfur may also be present. Table 2-5 identifies the major types of proteins in the body.

Proteins perform a variety of essential functions. These include seven major functional categories.

1. *Support.* **Structural proteins** create a three-dimensional framework for the body, providing strength, organization, and support for cells, tissues, and organs.
2. *Movement.* **Contractile proteins** are responsible for muscular contraction; related proteins are responsible for the movement of individual cells.
3. *Transport.* Insoluble lipids, respiratory gases, special minerals, such as iron, and several hormones are carried in the blood attached to special **transport proteins.** Other specialized

TABLE 2-5 Representative Proteins and Their Functions

Functional Classification	*Examples*	*Representative Location*	*Function*
Structural proteins	Keratin	Skin surface, hair, nails	Provide strength and waterproofing
	Collagen	Dermis of skin, tendons	Provide tensile strength
Contractile proteins	Actin, myosin	Muscle cells	Perform contraction and movement
Transport proteins	Albumin	Circulating blood	Transports fatty acids
	Transferrin	Circulating blood	Transports iron
	Apolipoproteins	Circulating blood	Transport glycerides
	Hemoglobin	Circulating blood	Transports oxygen in blood (within red blood cells)
Buffers	Intracellular and extracellular proteins	Within cells and body fluids	Stabilize pH
Enzymes	Hydrolases	All cells	Catalyze hydrolysis of organic molecules
	Kinases	All cells	Attach phosphate groups to organic substrates
	Proteases	All cells; digestive secretions of stomach, pancreas	Break down proteins
	Carbohydrases	All cells; digestive secretions of salivary glands, pancreas	Break down carbohydrates
	Lipases	All cells; digestive secretions of pancreas	Break down lipids
Antibodies (Immunoglobulins)	Gamma-globulins	Circulating blood	Attack foreign proteins and pathogens
Hormones	Insulin, glucagon	Circulating blood	Coordinate and/or control metabolic activities

proteins transport materials from one part of a cell to another.

4. ***Buffering.*** Proteins provide a considerable buffering action, helping to restrict alterations in pH.
5. ***Metabolic regulation.*** Enzymes accelerate chemical reactions in living cells. The sensitivity of enzymes to environmental factors is extremely important in controlling the pace and direction of metabolic operations.
6. ***Coordination and control.*** Protein *hormones* can influence the metabolic activities of every cell in the body or affect the function of specific organs or organ systems.
7. ***Defense.*** The tough, waterproof proteins of the skin, hair, and nails protect the body from environmental hazards. The proteins of the **immune system,** known as **antibodies,** protect us from disease. Special clotting proteins restrict bleeding after an injury to the circulatory system.

Structure of Proteins

Figure 2-17

Proteins consist of chains of small organic molecules called **amino acids.** The human body contains significant quantities of 20 different amino acids. A typical protein contains 1000 amino acids, but the largest protein complexes may have a hundred thousand or more. Each amino acid consists of a central carbon atom to which four groups are attached: (1) a hydrogen atom; (2) an **amino group** (— NH_2); (3) a carboxylic acid group (— COOH); and (4) a variable group, known as an **R group,** or *side chain* (Figure 2-17a●). Different R groups distinguish one amino acid from another, giving each its individual chemical properties. The name *amino acid* refers to the presence of the *amino* group and the carboxylic *acid* group. These are hydrophilic groups, and amino acids are relatively small, soluble molecules. Figure 2-17b● shows two representative amino acids, **glycine** and **alanine.** (All 20 amino acids commonly found in proteins are diagrammed in Appendix IV.) In the normal pH range of body fluids, the carboxylic acid groups on many amino acids give up their hydrogens, going from — COOH to — COO^-, and these amino acids are negatively charged.

As indicated in Figure 2-17b●, dehydration synthesis can link two amino acids. This process creates a covalent bond between the carboxylic acid group of one amino acid and the amino group of another. Such a bond is known as a **peptide bond,** and the molecule created is a **dipeptide.**

The chain can be lengthened by the addition of more amino acids. Attaching a third produces a **tripeptide,** and there are tetrapeptides, pentapeptides, and so forth. Tripeptides and larger peptide chains are called **polypeptides.** Polypeptides containing more than 100 amino acids are usually called *proteins.* You are probably already familiar with the names of several important proteins, including *hemoglobin* in red blood cells and *keratin* in fingernails and hair. Because each protein contains amino acids that are negatively charged, the entire protein has a net negative charge. For this reason, proteins are often indicated by the abbreviation **Pr⁻**.

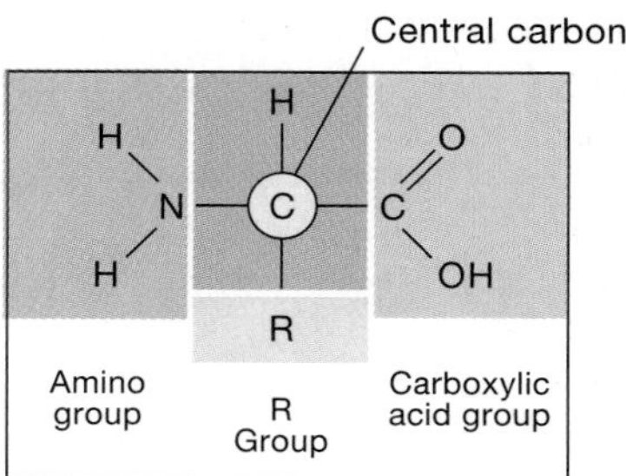

Central carbon
H
H
N
C
C
O
H
OH
R
Amino group
R Group
Carboxylic acid group

(a) Structure of an amino acid

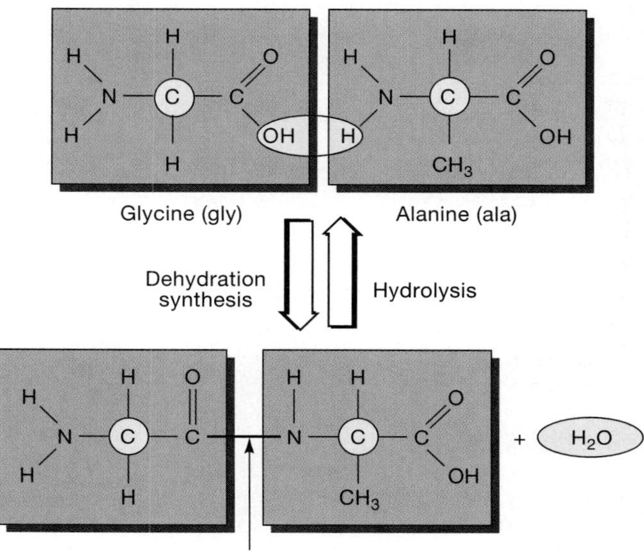

Glycine (gly)
Alanine (ala)
OH
H
CH3
Dehydration synthesis
Hydrolysis
Peptide bond
+ H2O

(b) Peptide bond formation

● **FIGURE 2-17**

Amino Acids and Peptide Bonds. (a) Each amino acid consists of a central carbon atom to which four different groups are attached: a hydrogen atom, an amino group (— NH_2), a carboxylic acid group (— COOH), and a variable group generally designated R. **(b)** Peptides form as dehydration synthesis creates a peptide bond between the carboxyl group of one amino acid and the amino group of another. In this example glycine and alanine are linked to form a dipeptide.

Protein Shape

Figure 2-18

The characteristics of a particular protein are determined in part by the R groups on its component amino acids. But the properties of a protein are more than

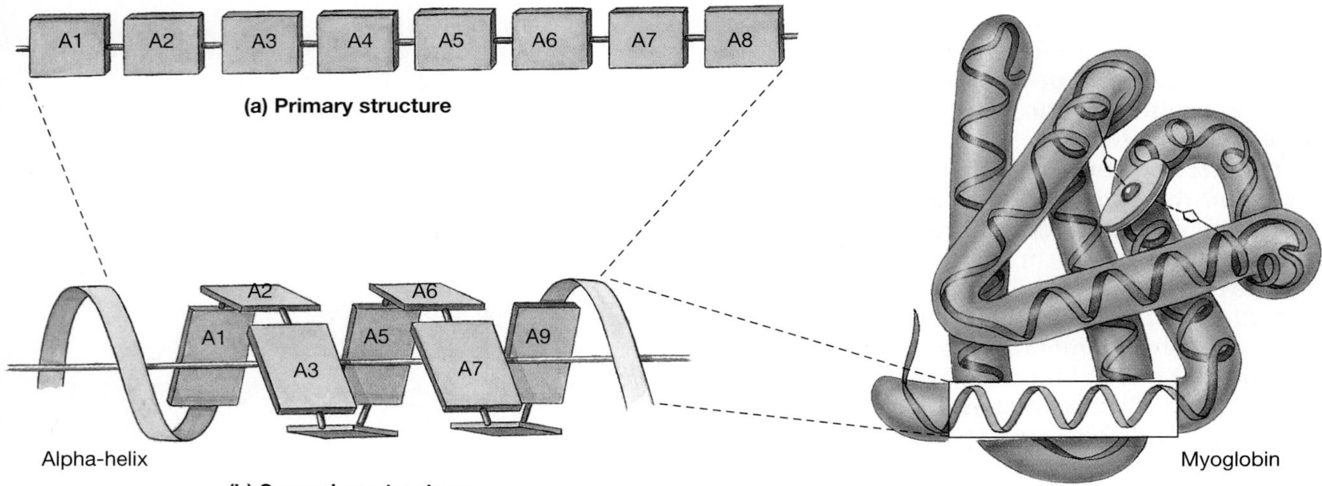
A1
A2
A3
A4
A5
A6
A7
A8
(a) Primary structure
A2
A6
A1
A3
A5
A7
A9
Alpha-helix
(b) Secondary structure
Myoglobin
(c) Tertiary structure

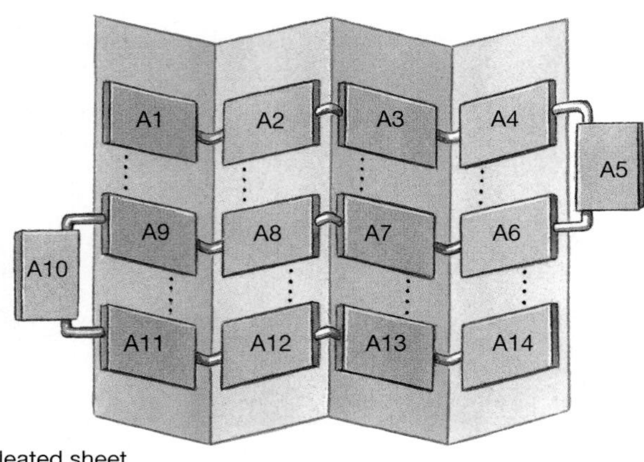
A1
A2
A3
A4
A5
A9
A8
A7
A6
A10
A11
A12
A13
A14
Pleated sheet

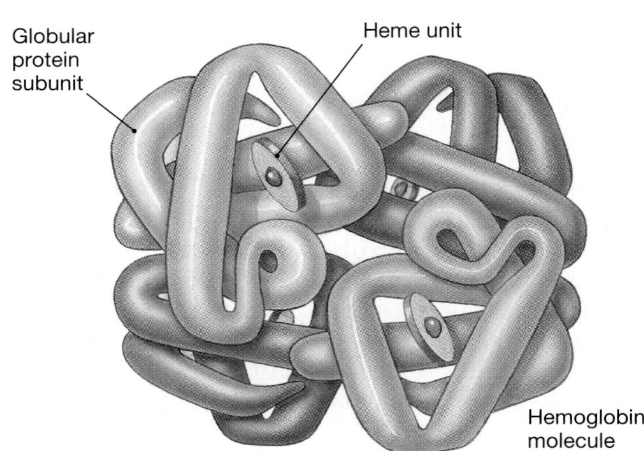
Globular protein subunit
Heme unit
Hemoglobin molecule
Keratin fiber
(d) Quaternary structure

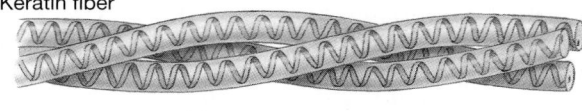

just the sum of the properties of its parts, for polypeptides and proteins can have very complex shapes. There are four levels of structural complexity:

1. The **primary structure** of a protein is the sequence of amino acids along its length. The primary structure of a short peptide chain is diagrammed in Figure 2-18a●.
2. The **secondary structure** appears as hydrogen bonding occurs between parts of the polypeptide chain. This bonding often creates a simple spiral, known as an **alpha-helix,** shown in Figure 2-18b●. Less often, hydrogen bonding produces a flat **pleated sheet.** A single polypeptide chain may have both helical and pleated sections.
3. The **tertiary structure** of a protein is the complex coiling and folding that gives the protein its unique shape. Tertiary structure results primarily from interactions between the polypeptide chain and the surrounding water molecules and partly from interactions between the R groups of amino acids in different parts of the molecule. Figure 2-18c● shows the tertiary structure of **myoglobin,** a protein found in muscle cells. Myoglobin is an example of a **globular protein.** Globular proteins are compact, usually rounded, and water-soluble.

● **FIGURE 2-18**
Protein Structure. (a) The primary structure of a polypeptide is the sequence of amino acids along its length. **(b)** The secondary structure is created by hydrogen bonding along the length of the polypeptide chain. Such bonding often produces a simple spiral (an alpha-helix) or a flattened arrangement known as a pleated sheet. **(c)** Tertiary structure. This is the structure of myoglobin, a globular protein involved in the storage of oxygen in muscle tissue. In the cylindrical segments the polypeptide chain is arranged in an alpha-helix. **(d)** The quaternary structure develops when separate polypeptide subunits interact to form a larger molecule. A single hemoglobin molecule contains four globular subunits, each structurally similar to myoglobin. Hemoglobin transports oxygen in the blood; the oxygen binds reversibly to the heme units. In keratin, three fibrous subunits intertwine like the strands of a rope. Keratin is a tough, water-resistant protein found in skin, hair, and nails.

4. Some proteins contain several polypeptide chains, each with its own secondary and tertiary structure. Interactions between these polypeptides determine the **quaternary structure** of the functional protein. The protein **hemoglobin** (Figure 2-18d●) contains four globular subunits, each structurally similar to myoglobin. Hemoglobin is found within red blood cells, where it binds and transports oxygen. In **keratin** (Figure 2-18d●) three alpha-helical polypeptides are wound together like the strands of a rope. Keratin is the tough, water-resistant protein found in skin, nails, and hair. Keratin is an example of a **fibrous protein.** Fibrous proteins are tough, durable, and generally insoluble. **Collagen** (COL-a-jen), another fibrous protein composed of helical polypeptide chains, forms a strong supporting framework in tissues and organs throughout the body.

SHAPE AND FUNCTION Proteins are very versatile, and they have a variety of different functions. The shape of a protein determines its functional properties, and the primary determinant of shape is the sequence of amino acids. The 20 major amino acids can be linked in an astonishing number of different combinations, creating proteins of enormously varied shape and function. Changing the identity of a single amino acid out of the 10,000 or more in a protein may significantly alter its functional properties. For example, several cancers and *sickle cell anemia,* a blood disorder, result from single changes in the amino acid sequences of complex proteins. AM *Metabolic Anomalies*

The tertiary and quaternary shapes of complex proteins depend not only on their amino acid sequence, but also on the local environmental characteristics. Small changes in the ionic composition, temperature, or pH of their surroundings can thus affect the function of proteins. Protein shape can also be affected by hydrogen bonding to other molecules in solution. The significance of these factors is most striking when one considers the function of enzymes, for these proteins are essential to the metabolic operations under way in every cell in the body.

Enzyme Function

Figure 2-19

The reactants in enzymatic reactions, called the **substrates,** interact to yield a specific product. Before the enzyme can function as a catalyst, the substrate must bind to a special region of the enzyme called the **active site.** The tertiary or quaternary structure of the enzyme molecule determines the shape of the active site, typically a groove or pocket into which the substrate nestles. This physical fit is reinforced by weak electrical attractive forces such as hydrogen bonding.

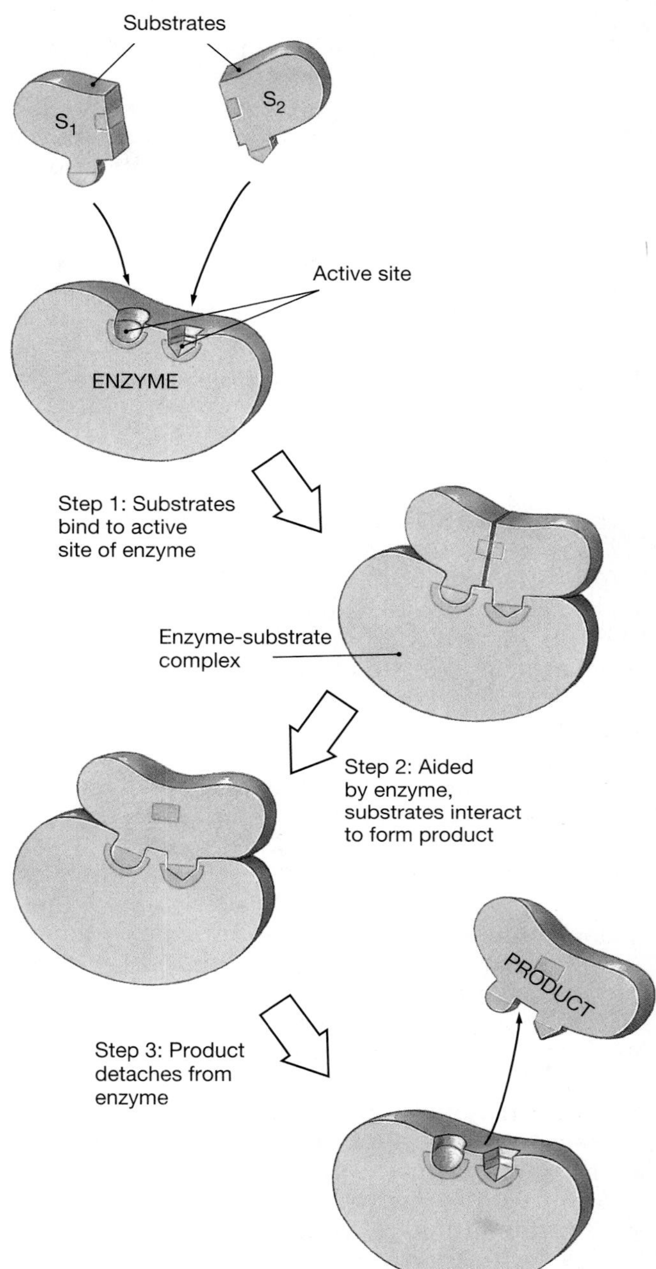
Substrates
S_1
S_2
Active site
ENZYME
Step 1: Substrates bind to active site of enzyme
Enzyme-substrate complex
Step 2: Aided by enzyme, substrates interact to form product
PRODUCT
Step 3: Product detaches from enzyme

● **FIGURE 2-19**
Enzyme Structure and Function. Each enzyme contains a specific active site somewhere on its exposed surface. In Step 1 a pair of substrate molecules (S_1 and S_2) bind to the active site. In Step 2 the substrates interact, forming a product that detaches from the active site (Step 3). Because the structure of the enzyme has not been affected, the entire cycle can be repeated.

Figure 2-19● diagrams enzyme function in a synthesis reaction. Substrate binding occurs at the active site (Step 1). The enzyme then promotes product formation (Step 2). It appears likely that substrate binding results in a change in the shape of the protein, and that this shape change promotes the reaction by placing physical stresses on the substrate molecules. The completed product then detaches from the active site, and the enzyme is ready to repeat the cycle.

Our example was an enzyme that catalyzed a synthesis reaction. Other enzymes may catalyze decomposition reactions or exchange reactions. The specificity of an enzyme is determined by the ability of a particular substrate to bind to the active site, a relatively small portion of the entire protein. Differences in the quaternary structure of the enzyme that do not affect the active site or change the response of the enzyme to substrate binding have no effect on enzyme function. **Isozymes** are enzymes that differ in structure but catalyze the same reaction. AM *Isozymes and Clinical Testing*

COFACTORS AND ENZYME FUNCTION **Cofactors** are ions or molecules that must bind to the enzyme before substrate binding can occur. Examples of such cofactors would include ions such as calcium (Ca^{2+}) or magnesium (Mg^{2+}), which bind at the enzyme's active site. Cofactors may also bind at other sites, as long as they produce the necessary change in the shape of the active site.

Coenzymes are nonprotein organic molecules that function as enzyme cofactors. The body converts many **vitamins** into essential coenzymes. Vitamins are structurally related to lipids or carbohydrates, but they have unique functional roles. The human body cannot synthesize most vitamins, so they must be obtained from the diet. (Chapter 25 will present more information concerning nutrition and vitamins.)

The relationship between enzymes and cofactors is one example of how enzyme activity can be controlled. Without a cofactor, the enzyme is intact but nonfunctional; with the cofactor, the enzyme can catalyze a specific reaction. In this case binding the cofactor makes the active site functional. A variety of other factors can activate or inactivate an enzyme. In fact, virtually any factor that changes the tertiary or quaternary shape of an enzyme may switch it "on" or "off." AM *The Control of Enzyme Activity*

TEMPERATURE AND pH Each enzyme works best at an optimal combination of temperature and pH. As temperatures rise, proteins change shape, and enzyme function deteriorates. Very high body temperatures (over 43° C, or 110° F) cause death because at these temperatures proteins undergo **denaturation,** an irreversible change in tertiary or quaternary shape. Denatured proteins are nonfunctional, and the loss of structural proteins and enzymes causes irreparable damage to organs and organ systems. You see denaturation in progress each time you fry an egg, for the clear egg white contains abundant dissolved proteins. As the temperature rises, the protein structure changes. Eventually the egg proteins become completely denatured, forming an insoluble white mass.

Enzymes are equally sensitive to pH changes. *Pepsin*, an enzyme that breaks down proteins in the stomach contents, works best at a pH of 2.0 (strongly acid). The small intestine contains *trypsin*, another enzyme that attacks proteins. Trypsin works only in an alkaline environment, with an optimum pH of 7.7 (somewhat basic).

ENZYMES AND HOMEOSTASIS Each cell contains an assortment of enzymes, and any particular enzyme may be active under one set of conditions and inactive under another. Because the change is immediate, enzyme activation or inactivation is an important method for short-term control over reaction rates and pathways. These control mechanisms will be considered in greater detail in later chapters.

Glycoproteins and Proteoglycans

Glycoproteins (GLĪ-kō-prō-tēnz) and **proteoglycans** (prō-tē-o-GLĪ-kanz) are combinations of protein and carbohydrate molecules. Glyco*protein* molecules are large proteins with small carbohydrate groups attached. Glycoproteins may function as enzymes, antibodies, hormones, or protein components of cell membranes. Glycoproteins in cell membranes play a major role in the identification of normal and abnormal cells, as well as in the initiation and coordination of the immune response. These mechanisms will be detailed in Chapter 22. Proteo*glycans* are large polysaccharide molecules connected to short polypeptide chains. Proteoglycan secretions coat the surfaces of the respiratory and digestive tracts, providing lubrication. Proteoglycans dissolved in tissue fluids give them a syrupy consistency.

Proteins are chains of what small organic molecules?

Which level of protein structure would be affected by an agent that breaks hydrogen bonds?

Why does boiling a protein affect its structural and functional properties?

How might a change in an enzyme's active site affect its function?

❑ Nucleic Acids

Nucleic (noo-KLĀ-ik) **acids** are large organic molecules composed of carbon, hydrogen, oxygen, nitrogen, and phosphorus. Nucleic acids store and process information at the molecular level, inside living cells. There are two classes of nucleic acid molecules, **deoxyribonucleic** (dē-ok-sē-rī-bō-noo-KLĀ-ik) **acid,** or **DNA,** and **ribonucleic** (rī-bō-noo-KLĀ-ik) **acid,** or **RNA.**

The DNA in our cells determines our inherited characteristics, such as eye color, hair color, blood type, and so on. It affects all aspects of body structure and function because DNA molecules encode the information needed to build proteins. By directing the synthesis of structural proteins, DNA controls the shape and physical characteristics of our bodies. By controlling the manufacture of enzymes, DNA regulates not only protein synthesis but all aspects of cellular metabolism, including the creation and destruction of lipids, carbohydrates, and other vital molecules.

Several different forms of RNA cooperate to manufacture specific proteins using the information provided by DNA. The functional relationships between DNA and RNA will be detailed in Chapter 3.

Structure of Nucleic Acids

Figure 2-20

A nucleic acid consists of a series of **nucleotides** linked by dehydration synthesis. A single nucleotide has three basic components: a sugar, a phosphate group, and a **nitrogenous base** (Figure 2-20●). The

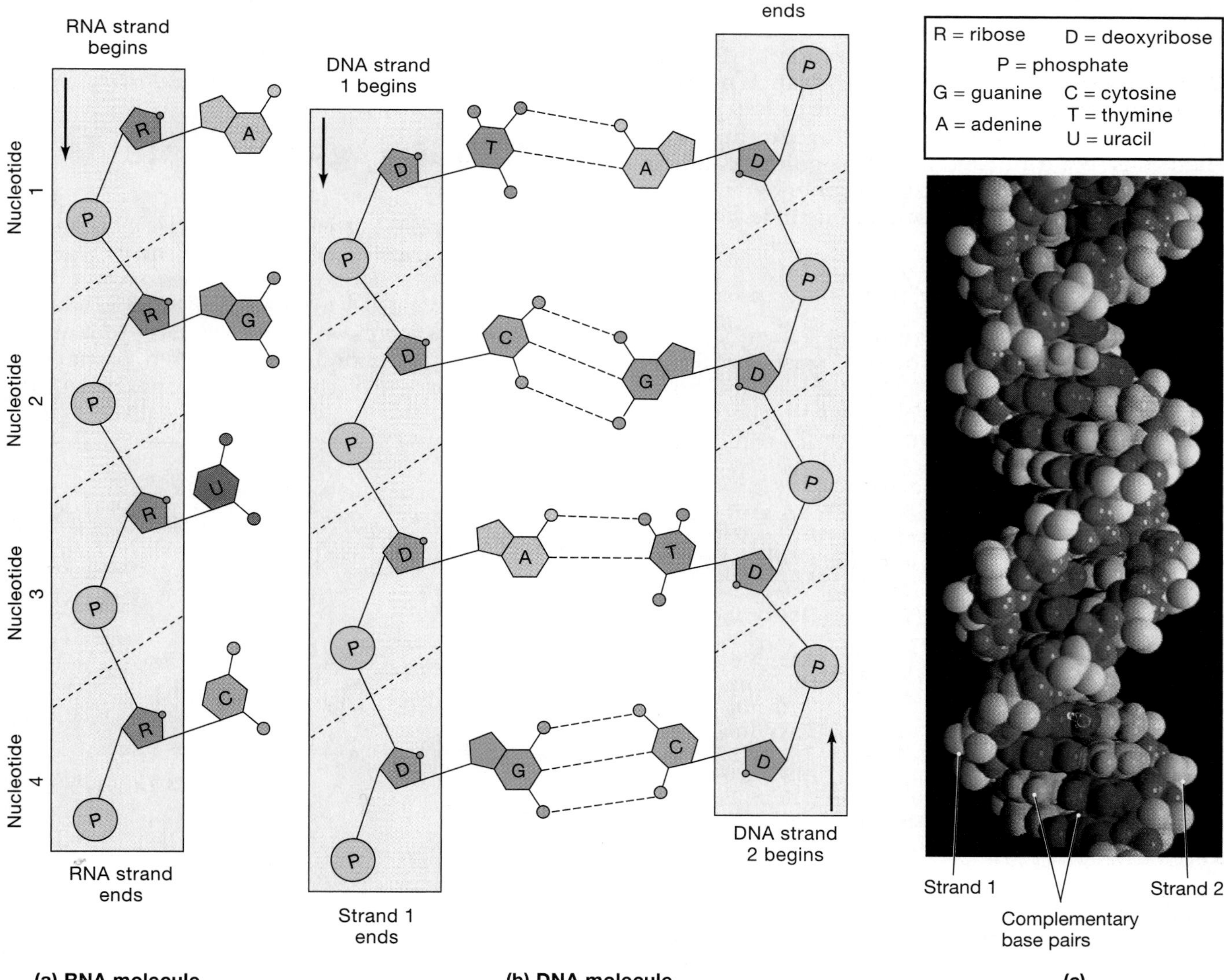
RNA strand begins
RNA strand ends
Nucleotide 1
Nucleotide 2
Nucleotide 3
Nucleotide 4
DNA strand 1 begins
Strand 1 ends
Strand 2 ends
DNA strand 2 begins
R = ribose
D = deoxyribose
P = phosphate
G = guanine
C = cytosine
A = adenine
T = thymine
U = uracil
Strand 1
Strand 2
Complementary base pairs
(a) RNA molecule
(b) DNA molecule
(c)

● **FIGURE 2-20**

Nucleic Acids: RNA and DNA. Nucleic acids are long chains of nucleotides. Each molecule starts at the sugar-nitrogenous base of the first nucleotide and ends at the phosphate group of the last member of the chain. An RNA molecule **(a)** consists of a single nucleotide chain. Its shape is determined by the sequence of nucleotides and the interactions between them. A DNA molecule **(b)** consists of a pair of nucleotide chains linked by hydrogen bonding between complementary base pairs. **(c)** A three-dimensional model of a DNA molecule shows the double helix formed by the two DNA strands.

sugar is always a **pentose** (five-carbon sugar), either **ribose** (in RNA) or **deoxyribose** (in DNA). Each pentose is attached to a phosphate group $(PO_4)^{3-}$ and a nitrogenous (nitrogen-containing) base. There are five different nitrogenous bases: **adenine** (A), **guanine** (G), **cytosine** (C), **thymine** (T), and **uracil** (U). Adenine and guanine are two-ringed structures called **purines**; the others are single-ringed **pyrimidines.** Both RNA and DNA contain adenine, guanine, and cytosine. Uracil is found only in RNA, and thymine only in DNA.

In the formation of a nucleic acid, attachment of a nitrogenous base to a carbohydrate creates a **nucleoside** that is named after the nitrogenous base. For example, attaching a guanine molecule produces a guanine nucleoside. A **nucleotide** forms when a phosphate group binds to the carbohydrate of the nucleoside. Dehydration synthesis then attaches the phosphate group of one nucleotide to the carbohydrate of another. The "backbone" of a nucleic acid molecule is a linear sugar-to-phosphate-to-sugar sequence with the nitrogenous bases projecting to one side.

RNA and DNA

Figure 2-20

There are important structural differences between RNA and DNA. A molecule of RNA consists of a single chain of nucleotides (Figure 2-20a●). Its shape depends on the order of the nucleotides and the interactions between them. There are three types of RNA in our cells: *messenger RNA (mRNA), transfer RNA (tRNA),* and *ribosomal RNA (rRNA).* These types have different shapes and functions, but all three are required for the synthesis of proteins, a process that will be detailed in Chapter 3. A DNA molecule consists of a pair of nucleotide chains (Figure 2-20b●). Hydrogen bonding between opposing nitrogenous bases holds the two strands together. Because of their shapes, adenine can bond only with thymine, and cytosine only with guanine. As a result, adenine-thymine and cytosine-guanine are known as **complementary base pairs.**

The two strands of DNA twist around one another in a **double helix** that resembles a spiral staircase. The stair steps correspond to the nitrogenous base pairs. Figure 2-20c● presents a three-dimensional view of a DNA molecule. Table 2-6 summarizes the differences between DNA and RNA.

TABLE 2-6 A Comparison of RNA and DNA

Characteristic	*RNA*	*DNA*
Sugar	Ribose	Deoxyribose
Nitrogenous bases	Adenine Guanine Cytosine Uracil	Adenine Guanine Cytosine Thymine
Number of nucleotides in typical molecule	Varies from under 100 nucleotides to around 50,000	Always over 45 million nucleotides
Shape of molecule	Varies, depending on hydrogen bonding along the length of the strand; 3 main types (mRNA, rRNA, tRNA)	Paired strands coiled in a double helix
Function	Performs protein synthesis as directed by DNA	Stores genetic information that controls protein synthesis by RNA

❑ High-Energy Compounds

Figure 2-21

Catabolism releases energy, and living cells can use that energy in constructive ways. The energy released by catabolic reactions is first harnessed to create high-energy bonds. A **high-energy bond** is a covalent bond that stores an unusually large amount of energy. When that bond is later broken, perhaps in a distant portion of the cell, the energy will be

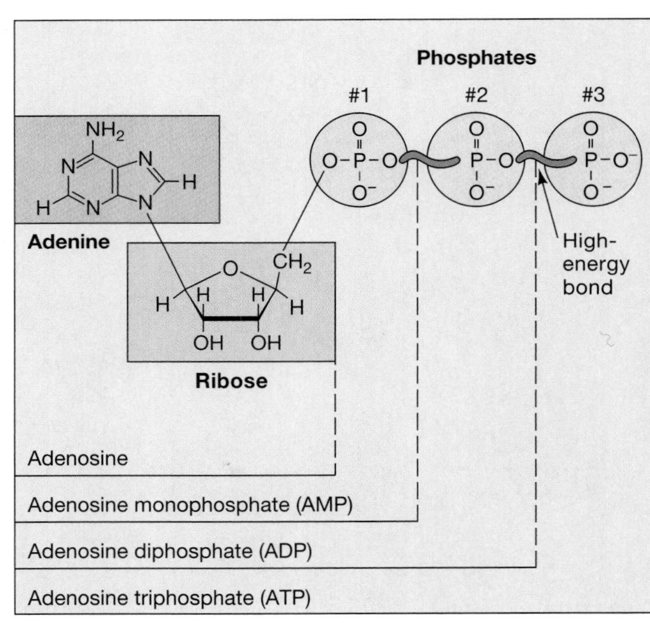
Phosphates
#1
#2
#3
NH_2
Adenine
CH_2
OH OH
Ribose
High-energy bond
Adenosine
Adenosine monophosphate (AMP)
Adenosine diphosphate (ADP)
Adenosine triphosphate (ATP)

● **FIGURE 2-21**
The Structure of ATP. A molecule of ATP consists of an adenine nucleoside to which three phosphate groups have been joined. Both the second and third phosphates are bound to the molecule by high-energy bonds. Cells most often store energy by attaching a third phosphate group to ADP. Removing the phosphate group releases the energy for cellular work, including the synthesis of other molecules.

released under controlled conditions. In our cells a high-energy bond usually connects a phosphate group (PO_4^{3-}) to an organic molecule. The resulting complex is called a **high-energy compound.**

Phosphorylation (fos-for-i-LĀ-shun) is the attachment of a phosphate group to another molecule, and this process does not necessarily produce high-energy bonds. The creation of a high-energy compound requires (1) a phosphate group, (2) appropriate enzymes, and (3) suitable organic substrates. The most important substrate is an adenine nucleoside, *adenosine,* with two phosphate groups attached. This combination is called **adenosine diphosphate,** or **ADP.** ADP is created by the phosphorylation of the nucleotide **adenosine monophosphate,** or **AMP,** one of the building blocks of nucleic acids (p. 56). It takes a significant energy input to convert AMP to ADP, and the second phosphate is attached by a high-energy bond. An even greater amount of energy is required to add a third phosphate and create the high-energy compound **adenosine triphosphate,** or **ATP.**

TABLE 2-7 Structure and Function of Biologically Important Compounds

Class	*Building Blocks*	*Sources*	*Functions*
INORGANIC			
Water	Hydrogen and oxygen atoms	Absorbed as liquid water or generated via metabolism	Solvent; transport medium for dissolved materials and heat; cooling through evaporation; medium for chemical reactions; reactant in hydrolysis
Acids, bases, salts	H^+, OH^-, various anions and cations	Obtained from the diet or generated via metabolism	Structural components; buffers; sources of ions
Dissolved gases	Oxygen, carbon, nitrogen, and other atoms	Atmosphere	O_2 required for normal cellular metabolism CO_2 generated by cells as a waste product NO chemical messenger involved in cardiovascular, nervous, and lymphatic systems
ORGANIC			
Carbohydrates	C, H, O, sometimes N; CHO in a 1:2:1 ratio	Obtained in diet or manufactured in the body	Energy source; some structural role when attached to lipids or proteins; energy storage
Lipids	C, H, O, sometimes N or P; CHO not in 1:2:1 ratio	Obtained in diet or manufactured in the body	Energy source; energy storage; insulation; structural components; chemical messengers; physical protection
Proteins	C, H, O, N, often S	20 common amino acids; roughly half can be manufactured in the body, others must be obtained in the diet	Catalysts for metabolic reactions; structural components; movement; transport; buffers; defense; control and coordination of activities
Nucleic acids	C, H, O, N, and P; nucleotides composed of phosphates, sugars, and nitrogenous bases	Obtained in diet or manufactured	Storage and processing of genetic information
High-energy compounds	Nucleotides joined to phosphates by high-energy bonds	Synthesized by all cells	Storage or transfer of energy

Figure 2-21● details the structure of ATP. Within our cells the conversion of ADP to ATP represents the primary method of energy storage, and the reverse reaction provides a mechanism for controlled energy release. The arrangement can be summarized as:

$$ATP + H_2O \rightleftharpoons ADP + \text{phosphate group} + \text{energy}$$

When energy sources are available, our cells make ATP from ADP; when energy is required, the reverse reaction occurs.

Although ATP is the most abundant high-energy compound, there are others. These compounds are typically other nucleotides that have undergone phosphorylation. For example, *guanosine triphosphate* (**GTP**) and *uridine triphosphate* (**UTP**) are nucleotide-based high-energy compounds that transfer energy in specific enzymatic reactions.

Table 2-7 summarizes information concerning the inorganic and organic compounds considered in this chapter.

TABLE 2-8 Turnover Rates

Cell Type	*Component*	*Half-Life (Days)**
Liver	Total protein	5–6
	Enzymes	1 hour to several days, depending on the enzyme
	Glycogen	1–2
	Cholesterol	5–7
Muscle cell	Total protein	30
	Glycogen	0.5–1
Neuron	Phospholipids	200
	Cholesterol	100+
Fat cell	Triglycerides	15–20

*Times are average life expectancies. Most values were obtained f m studies on mammals other than humans.

Chemicals and Living Cells

The human body is more than a collection of chemicals. Biochemical building blocks form functional units called **cells.** Each cell behaves like a miniature organism, responding to internal and external stimuli. A lipid membrane separates the cell from its environment, and internal membranes create compartments with specific functions. Proteins form an internal supporting framework and act as enzymes to accelerate and control the chemical reactions that maintain homeostasis. Nucleic acids direct the synthesis of all cellular proteins, including the enzymes that enable the cell to synthesize a wide variety of other substances. Carbohydrates provide energy for vital activities and form part of specialized compounds, such as proteoglycans or glycolipids.

Cells are dynamic structures that adapt to changes in their environment. That adaptation may involve changes in the chemical organization of the cell. Such changes are easily made because organic molecules other than DNA are temporary rather than permanent components of the cell. Their continual removal and replacement are part of the process of **metabolic turnover.**

Most of the organic molecules in the cell are replaced at intervals ranging from hours to months. The average time between synthesis and recycling is known as the *turnover rate.* Table 2-8 indicates the turnover rate for the organic components of representative cells.

✔ Analysis of a large organic molecule indicates that it is composed of the sugar ribose, nitrogenous bases, and phosphate groups. Which nucleic acid is this?

What molecule is produced by the phosphorylation of ADP?

Selected Clinical Terminology

Terms Discussed in This Chapter

cholesterol: A steroid important in the structure of cellular membranes, which in high concentrations increases the risks of heart disease. *(p. 50)*

mole (M, mol), **millimole** (mM, mmol): The number of atoms or molecules in a sample that has a weight in grams equal to the atomic weight (for an element) or the molecular weight (for a molecule). 1 mmol = 0.001 M. These units are often used to measure solute concentrations. *(p. 34)*

omega-3 fatty acids: Fatty acids, abundant in fish flesh and fish oils, that have a double bond three carbons away from the end of the hydrocarbon chain. Their presence in the diet has been linked to reduced risks of heart disease and other conditions. *(p. 49)*

radioisotopes: Isotopes with unstable nuclei that emit subatomic particles in measurable amounts. *(p. 33)*

AM Additional Terms Discussed in the Applications Manual

nuclear imaging: A procedure in which an image is created on a photographic plate or video screen by the radiation emitted by injected radioisotopes.

PET (positron emission tomography) **scan:** A nuclear imaging technique in which the emitted radiation is analyzed and the image is created by a computer.

tracer: A compound labeled with a radioisotope that can be tracked within the body by the radiation it releases.

CHAPTER REVIEW

STUDY OUTLINE

ATOMS AND MOLECULES, p. 32

1. **Atoms** are the smallest units of matter; they consist of **protons, neutrons,** and **electrons.** (*Figure 2-1*)

Structure of the Atom, p. 32

2. An **element** consists of atoms that have (usually) equal numbers of protons (**atomic number**) and electrons. Within an atom, an **electron cloud** surrounds the **nucleus.** (*Figures 2-1, 2-2, 2-3*)
3. The **mass number** of an atom indicates the total number of protons and neutrons in its nucleus. **Isotopes** are atoms of the same element whose nuclei contain different numbers of neutrons.
4. Electrons occupy a series of *energy levels* that are often illustrated as **electron shells.** The electrons in the outermost energy level determine an atom's chemical properties. (*Figure 2-3*)

Chemical Bonds and Chemical Compounds, p. 34

5. Atoms can combine to form a **molecule;** combinations of atoms of different elements form a **compound.**
6. An **ionic bond** results from the attraction between **ions,** atoms that have gained or lost electrons. **Cations** are positively charged, and **anions** are negatively charged. (*Figure 2-4*)
7. Some atoms share electrons to form a molecule held together by **covalent bonds.** Sharing one pair of electrons equally creates a **single covalent bond;** sharing two pairs equally forms a **double covalent bond.** An unequal sharing of electrons creates a **polar covalent bond.** (*Figures 2-5, 2-6*)
8. A **hydrogen bond** is a weak but important force that can affect the shapes and properties of molecules. (*Figure 2-7*)

CHEMICAL NOTATION, p. 38

1. *Chemical notation* allows us to describe reactions between **reactants** that generate one or more **products.** (*Table 2-1*)

CHEMICAL REACTIONS, p. 38

1. **Metabolism** refers to all the chemical reactions in the body. Through metabolism cells capture, store, and use energy to maintain homeostasis and to support essential functions.

Basic Energy Concepts, p. 39

2. **Work** involves movement of an object or a change in its physical structure, and **energy** is the capacity to perform work. There are two major types of energy: kinetic and potential.
3. **Kinetic energy** is the energy of motion. **Potential energy** is stored energy that results from the position or structure of an object. Conversions from potential to kinetic energy are not 100 percent efficient; every such energy exchange releases **heat.**

Types of Reactions, p. 39

4. Cells gain energy to power their functions by breaking down organic molecules, a process called **catabolism.** Much of this energy supports **anabolism,** the synthesis of new organic molecules.
5. A **chemical reaction** may be classified as a **decomposition, synthesis,** or **exchange reaction. Exergonic** reactions release energy; **endergonic** reactions absorb energy.

Reversible Reactions, p. 40

6. At **equilibrium** the rates of two opposing reactions are in balance.

Enzymes and Chemical Reactions, p. 40

7. **Activation energy** is the amount of energy required to start a reaction. **Enzymes** control many chemical reactions within our bodies. Enzymes are organic **catalysts**—substances that accelerate chemical reactions without themselves being permanently changed.

INORGANIC COMPOUNDS, p. 40

1. **Nutrients** and **metabolites** can be broadly classified as **organic** (carbon-based) or **inorganic.**

Water and Its Properties, p. 41

2. Water is the most important inorganic component of the body.
3. Many inorganic compounds, called **electrolytes,** will undergo **ionization** in water to form ions. (*Figure 2-8; Table 2-2*)
4. The **pH** of a solution indicates the concentration of hydrogen ions it contains. Solutions can be classified as **neutral, acidic,** or **basic (alkaline)** on the basis of pH. (*Figure 2-9*)

Inorganic Acids and Bases, p. 44

5. An **acid** releases hydrogen ions, and a **base** removes hydrogen ions from a solution.

Salts, p. 44

6. A **salt** is an electrolyte whose cation is not H^+ and whose anion is not OH^-.

Buffers and pH Control, p. 45

7. **Buffers** maintain pH within normal limits in body fluids.

ORGANIC COMPOUNDS, p. 45

1. **Organic compounds** contain carbon and hydrogen, and usually oxygen as well. They can be classified as carbohydrates, lipids, proteins, or nucleic acids.

Carbohydrates, p. 45

2. **Carbohydrates** are most important as an energy source for metabolic processes. The three major types are **monosaccharides (simple sugars), disaccharides,** and **polysaccharides.** (*Figures 2-10, 2-11, 2-12; Table 2-3*)

Lipids, p. 47

3. **Lipids** are water-insoluble molecules that include fats, oils, and waxes. There are six important classes of lipids: **fatty acids, glycerides, prostaglandins, steroids, phospholipids,** and **glycolipids.** (*Figures 2-13, 2-14, 2-15, 2-16; Table 2-4*)
4. **Triglycerides (neutral fats)** consist of three fatty acid molecules attached to a molecule of **glycerol.**
5. **Cholesterol** is a precursor of steroid hormones and is an important component of membranes.

Proteins, p. 52

6. **Proteins** perform a great variety of functions in the body. Important types of proteins include **structural proteins, contractile proteins, transport proteins, enzymes,** and **antibodies.** (*Table 2-5*)
7. Proteins are chains of **amino acids** linked by **peptide bonds.** (*Figures 2-17, 2-18*)
8. The reactants in an enzymatic reaction, called **substrates,** interact to yield a **product** by binding to the enzyme at the **active site.** (*Figure 2-19*)
9. The shape of a protein determines its functional characteristics. Each protein works best at an optimal combination of temperature and pH.

Nucleic Acids, p. 56

10. **Nucleic acids** store and process information at the molecular level. There are two kinds of nucleic acids: **deoxyribonucleic acid (DNA)** and **ribonucleic acid (RNA).** (*Figure 2-20; Table 2-6*)
11. Nucleic acids are chains of **nucleotides.** Each nucleotide contains a sugar, a phosphate group, and a **nitrogenous base.** The sugar is always **ribose** or **deoxyribose.** The nitrogenous bases found in DNA are **adenine, guanine, cytosine,** and **thymine.** In RNA, **uracil** replaces thymine.

High-Energy Compounds, p. 58

12. Cells store energy in **high-energy compounds** for later use. The most important high-energy compound is **ATP (adenosine triphosphate)**. When energy is available, cells make ATP by adding a phosphate group to ADP (**phosphorylation**). When energy is needed, ATP is broken down to ADP and phosphate. (*Figure 2-21; Table 2-7*)

CHEMICALS AND LIVING CELLS, p. 60

1. Biochemical building blocks form functional units called **cells.**
2. The continual removal and replacement of cellular organic molecules (other than DNA), a process called **metabolic turnover,** allows cells to change and adapt to their environments. (Table 2-8)

REVIEW QUESTIONS

LEVEL 1 Reviewing Facts and Terms

Match each item in Column A with the most closely related item in Column B. Use letters for answers in the spaces provided.

	Column A	Column B
___	1. atomic number	a. sum of atomic weights
___	2. mass number	b. water-insoluble
___	3. covalent bond	c. synthesis
___	4. ionic bond	d. catalyst
___	5. molecular weight	e. sharing of electrons
___	6. catabolism	f. $A + B \rightleftharpoons AB$
___	7. anabolism	g. stabilize pH
___	8. exchange reaction	h. number of protons
___	9. reversible reaction	i. decomposition
___	10. hydrophobic	j. carbohydrates, lipids, proteins
___	11. acid	k. loss or gain of electrons
___	12. enzyme	l. water, salts
___	13. buffer	m. H^+ donor
___	14. organic compounds	n. number of protons and neutrons
___	15. inorganic compounds	o. $AB + CD \rightarrow AD + CB$

16. In atoms, protons and neutrons are found:
    (a) only in the nucleus
    (b) outside the nucleus
    (c) inside and outside the nucleus
    (d) in the orbitals

17. Isotopes of an element differ from each other in the number of:
    (a) protons in the nucleus
    (b) neutrons in the nucleus
    (c) electrons in the outer shells
    (d) protons and electrons in the nucleus

18. The number and arrangement of electrons in an atom's outer energy level determines its:
    (a) atomic weight
    (b) atomic number
    (c) electrical properties
    (d) chemical properties

19. The bond between sodium and chlorine in the compound sodium chloride (NaCl) is:
    (a) an ionic bond
    (b) a single covalent bond
    (c) a nonpolar covalent bond
    (d) a double covalent bond

20. The oxygen atoms in a molecule of oxygen are held together by:
(a) a single covalent bond
(b) a double covalent bond
(c) a polar covalent bond
(d) an ionic bond

21. All of the chemical reactions that occur in the human body are collectively referred to as:
(a) anabolism (b) catabolism
(c) metabolism (d) homeostasis

22. Which of the following equations illustrates a typical decomposition reaction?
(a) $A + B \rightarrow AB$ (b) $AB + CD \rightarrow AD + CB$
(c) $2\ A_2 + B_2 \rightarrow 2\ A_2B$ (d) $AB \rightarrow A + B$

23. Of the following selections, the pH of the *least* acidic solution is:
(a) 6.0 (b) 4.5
(c) 2.3 (d) 1.0

24. A pH of 7.8 in the human body typifies a condition referred to as:
(a) acidosis (b) alkalosis
(c) metabolic acidosis (d) homeostasis

25. A(n) ____________ is a solute that dissociates to release hydrogen ions, and a(n) ____________ is a solute that removes hydrogen ions from solution.
(a) base, acid (b) salt, base
(c) acid, salt (d) acid, base

26. Carbohydrates, lipids, and proteins are formed from their basic building blocks by:
(a) removal of a water molecule between the building blocks
(b) adding a water molecule between the building blocks
(c) adding carbon to each molecule
(d) adding oxygen to each molecule

27. Complementary base pairing in DNA includes:
(a) adenine-uracil; cytosine-guanine
(b) adenine-thymine; cytosine-guanine
(c) adenine-guanine; cytosine-thymine
(d) guanine-uracil; cytosine-thymine

28. What are the three stable fundamental particles found in atoms?

29. What is the difference between potential and kinetic energy?

30. What is the role of enzymes in chemical reactions?

31. List six inorganic molecules found in abundance in the human body.

32. What four major classes of organic compounds are found in the body?

33. List three important functions of triglycerides (neutral fats) in the body.

34. List seven major functions performed by proteins.

35. (a) What three basic components make up a nucleotide of DNA? (b) What three basic components make up a nucleotide of RNA?

36. What three components are required to create the high-energy compound ATP?

LEVEL 2 Reviewing Concepts

37. If an isotope of oxygen has 8 protons, 10 neutrons, and 8 electrons, its mass number is:
(a) 26 (b) 16
(c) 18 (d) 8

38. *Hydrophilic* molecules will readily associate with:
(a) lipid molecules
(b) insoluble droplets of hydrophobic molecules
(c) water molecules
(d) a, b, and c are all correct

39. The speed or rate of a chemical reaction is influenced by:
(a) the presence of catalysts
(b) the temperature
(c) the concentration of the reactants
(d) a, b, and c are all influencing factors

40. Using the periodic table of the elements as a reference, determine the following information about *calcium*.
(a) number of protons
(b) number of electrons
(c) number of neutrons
(d) atomic number
(e) atomic weight
(f) number of electrons in each shell

41. Explain the differences between nonpolar covalent bonds, polar covalent bonds, and ionic bonds.

42. How does thermal inertia help to stabilize body temperature?

43. Why does pure water have a neutral pH?

44. What role do buffer systems play in the human body?

45. A biologist analyzes a sample that contains an organic molecule and finds the following constituents: carbon, hydrogen, oxygen, nitrogen, and phosphorus. On the basis of this information, is the molecule a carbohydrate, a lipid, a protein, or a nucleic acid?

LEVEL 3 Critical Thinking and Clinical Applications

46. The element sulfur has an atomic number of 16 and an atomic weight of 32. How many neutrons are in the nucleus of a sulfur atom? Assuming that sulfur forms covalent bonds with hydrogen, how many hydrogen atoms could bond to one sulfur atom?

47. An important buffer system in the human body involves carbon dioxide (CO_2) and bicarbonate ion (HCO_3^-) as shown below.

$$CO_2 + H_2O \rightleftharpoons H_2CO_3 \rightleftharpoons H^+ + HCO_3^-$$

If a person becomes excited and exhales large amounts of CO_2, how will this affect his body's pH?

48. A student needs to prepare 1 liter of a solution of NaCl that contains 1.2 moles of ions per liter of solution. How many moles of NaCl should the student use to prepare the solution?

CHAPTER 3

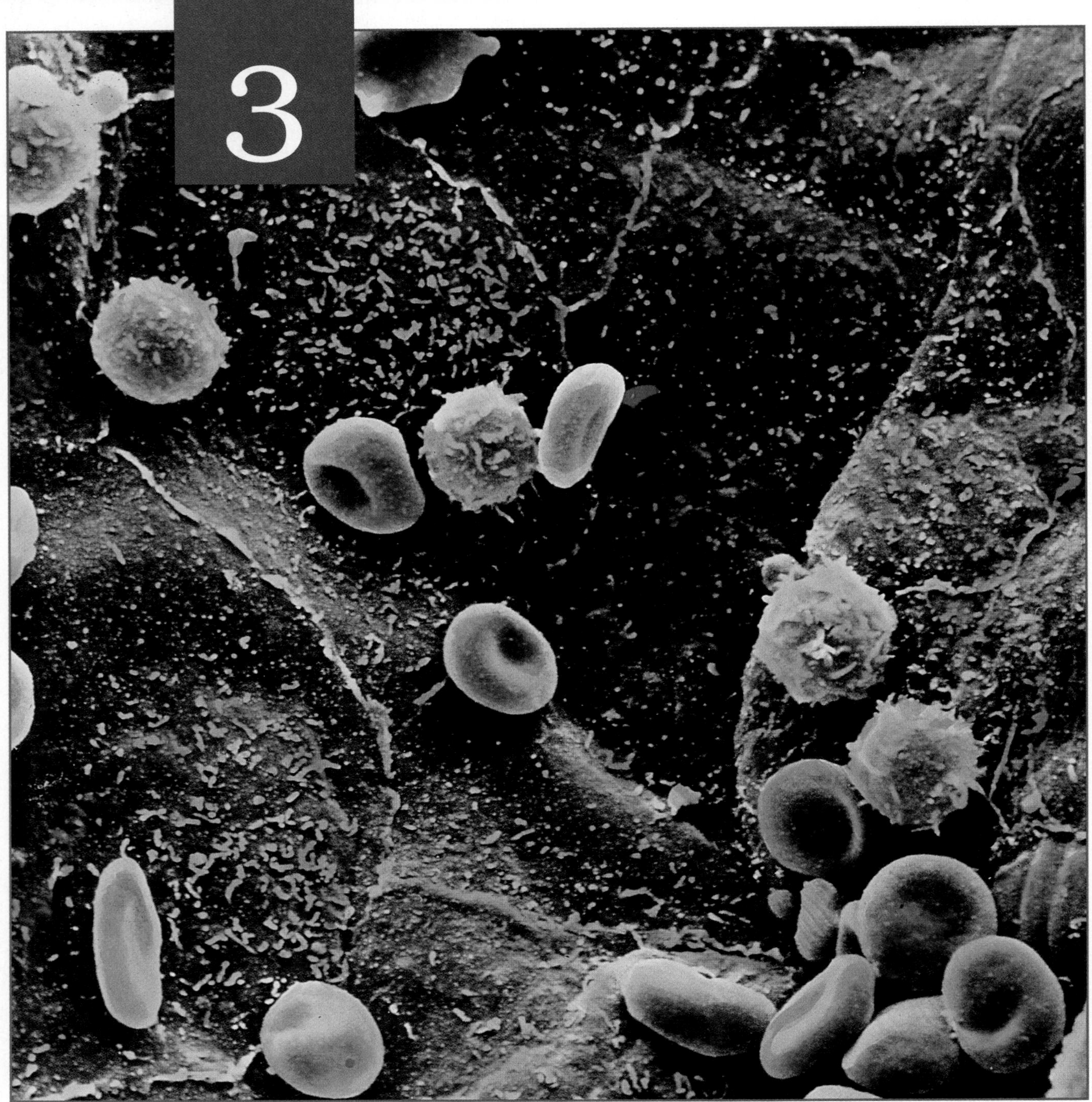

Many familiar structures, from pyramids to patchwork quilts, are made up of numerous small, similar components. The human body is built on the same principle, for it is made up of several trillion tiny units, called cells. Cells, however, differ from many other building blocks in a number of ways. For one thing, they vary widely in size and appearance. In this photo, for example, several relatively small blood cells lie on a layer of large, flattened cells that line a blood vessel.

An individual brick can perform few (if any) of the functions of a building. But a cell can perform many of the functions of the body of which it is a part. Indeed, cells are the smallest entities that can perform all the basic life functions discussed in Chapter 1. In addition, each cell in the body performs specialized functions and plays a role in the maintenance of homeostasis. As you can imagine, cells are complex structures, with many specialized components. In this chapter we will explore the functional organization of a representative cell.

The Cellular Level of Organization

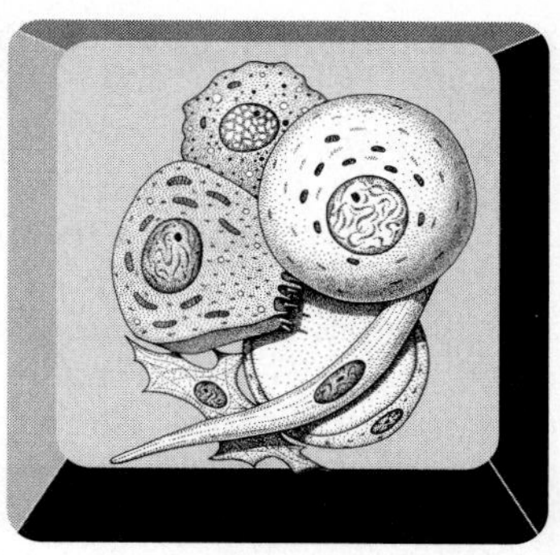

Chapter Outline and Objectives

1 List the functions of the cell membrane and the structural features that enable it to perform those functions.

2 Specify the routes by which different ions and molecules can enter or leave a cell and the factors that may restrict such movement.

3 Describe the various transport mechanisms that cells use to facilitate the absorption or removal of specific substances.

4 Explain the origin and significance of the transmembrane potential.

5 Describe the organelles of a typical cell and indicate the specific functions of each.

6 Explain the functions of the cell nucleus.

7 Discuss the nature and importance of the genetic code.

8 Summarize the process of protein synthesis.

9 Describe the stages of the cell life cycle.

10 Describe the process of mitosis and explain its significance.

11 Define differentiation and explain its importance.

Atoms are the building blocks of molecules; cells are the building blocks of the human body. Cells were first described by the English scientist Robert Hooke, around 1665. Hooke used an early light microscope to examine dried cork. He observed thousands of tiny empty chambers, which he named *cells*. Later that decade other scientists, observing the structure of living plants, realized that in life these spaces were filled with a gelatinous material. Research over the next 175 years led to the *cell theory*, the concept that cells are the fundamental units of all plant and animal tissues. Since that time the cell theory has been expanded to incorporate several basic concepts relevant to our discussion of the human body.

1. Cells are the building blocks of all plants and animals.
2. Cells are produced by the division of preexisting cells.
3. Cells are the smallest units that perform all vital physiological functions.
4. Each cell maintains homeostasis at the cellular level.
5. Homeostasis at the tissue, organ, system, and organism levels reflects the combined and coordinated actions of many cells.

Cells have a variety of forms and functions. Figure 3-1 ● gives examples of the range of cell sizes and shapes found in the human body. The relative proportions of the cells in this figure are correct, but all have been magnified roughly 500 times. Together, these and other types of cells create and maintain all anatomical structures and perform all vital physiological functions.

The human body contains trillions of cells, and all our activities, from running to thinking, result from the combined and coordinated responses of millions or even billions of cells. Yet each cell also functions as an individual entity, responding to a variety of environmental cues. As a result, anyone interested in understanding how the human body functions must first become familiar with basic concepts in cell biology. AM *The Nature of Pathogens*

Studying Cells

Cytology (sī-TOL-o-jē; *cyto-*, cell + *-ology*, the study of) is the study of the structure and function of cells. What we have learned since the 1950s has given us new insights into cellular physiology and the mechanisms of homeostatic control. Today, cytology is part of the broader discipline **cell biology,** which incorporates aspects of biology, chemistry, and physics. The two most common methods used to study cell and tissue structure are *light microscopy* and *electron microscopy*. (For information on the limitations of these techniques, see Appendix II.)

This chapter describes the structure of a typical animal cell, some of the ways in which cells interact with their environment, and how cells reproduce. Later chapters consider the coordination of cellular activities in physiological systems.

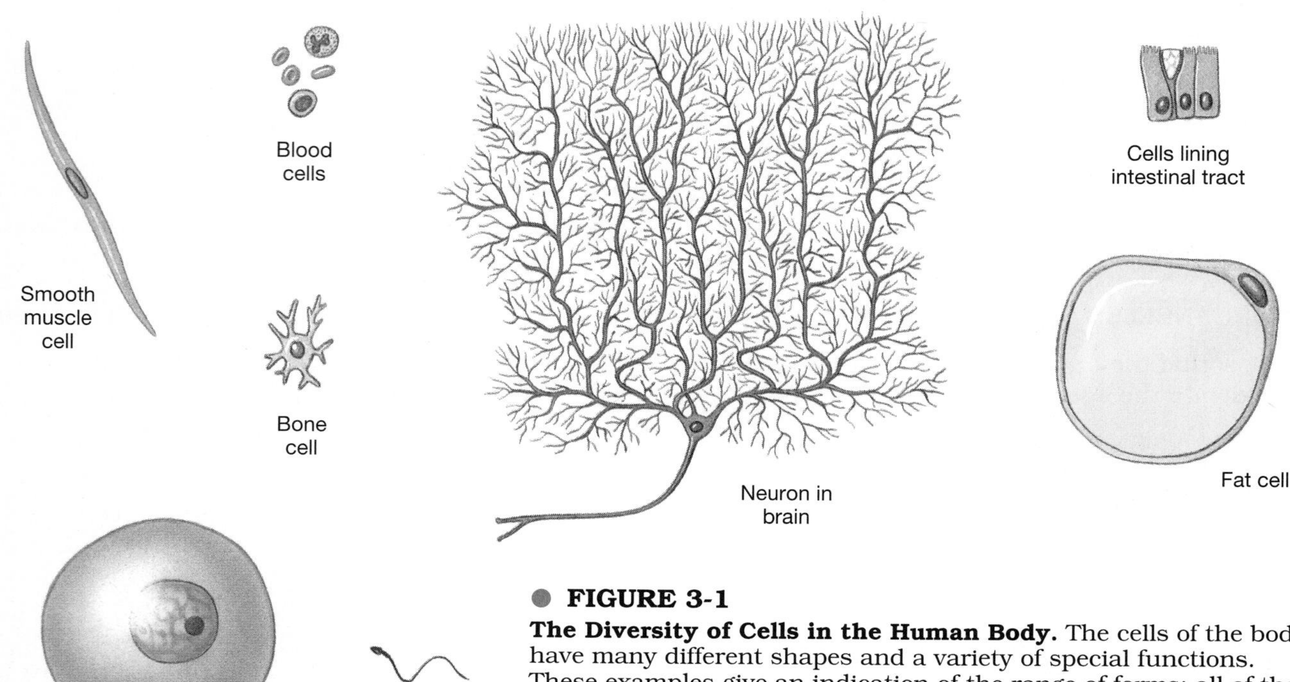
Smooth muscle cell
Blood cells
Bone cell
Neuron in brain
Cells lining intestinal tract
Fat cell
Ovum
Sperm

● **FIGURE 3-1**

The Diversity of Cells in the Human Body. The cells of the body have many different shapes and a variety of special functions. These examples give an indication of the range of forms; all of the cells are shown with the dimensions they would have if magnified approximately 500 times.

❑ An Overview of Cellular Anatomy

Figure 3-2

The human body contains two classes of cells: *somatic cells* and *sex cells* (also called *germ cells* or reproductive cells). There are only two types of **sex cells:** (1) the sperm of a male and (2) the ova of a female. **Somatic cells** (*soma*, body) include all of the other cells in the human body. This chapter will examine somatic cells; sex cells will be considered in the chapters dealing with the reproductive system.

The "typical" somatic cell is like the "average" person. Any description masks enormous individual variations. Our model cell will share features with most cells of the body without being identical to any. Figure 3-2● shows such a cell, and Table 3-1 summarizes the structures and functions of its parts.

Our cell is surrounded by a watery medium known as the **extracellular fluid.** The extracellular fluid found in most tissues is called **interstitial** (in-

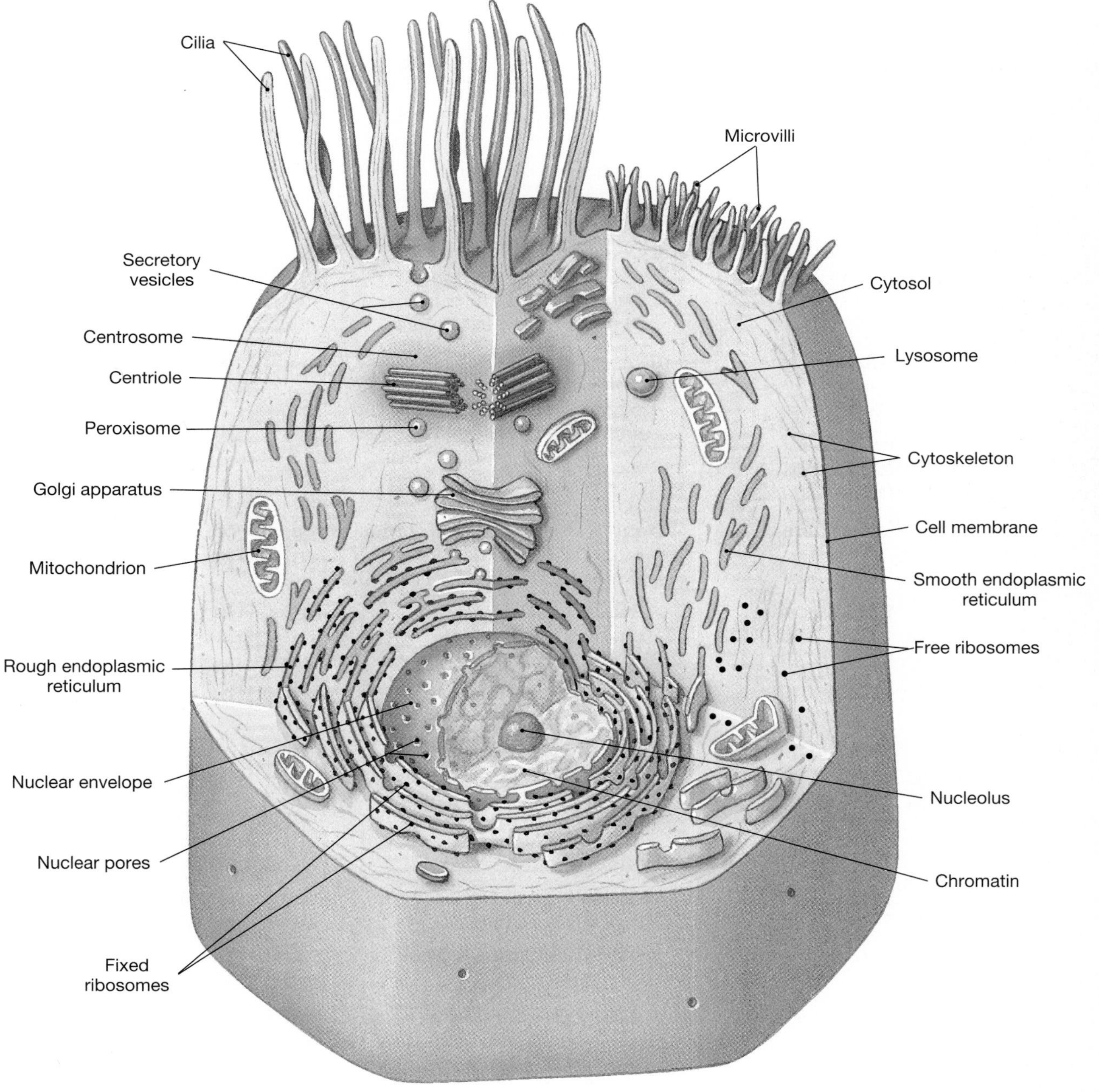
Cilia
Microvilli
Secretory vesicles
Centrosome
Centriole
Peroxisome
Golgi apparatus
Mitochondrion
Rough endoplasmic reticulum
Nuclear envelope
Nuclear pores
Fixed ribosomes
Cytosol
Lysosome
Cytoskeleton
Cell membrane
Smooth endoplasmic reticulum
Free ribosomes
Nucleolus
Chromatin

● **FIGURE 3-2**

The Anatomy of a Typical Cell. See Table 3-1 for a summary of the functions associated with the various cell structures.

TABLE 3-1 Organelles of a Typical Cell

Appearance	Structure	Composition	Function
	CELL MEMBRANE	Lipid bilayer, containing phospholipids, steroids, and proteins	Isolation, protection, sensitivity, support; controls entrance/exit of materials
	CYTOSOL	Fluid component of cytoplasm	Distributes materials by diffusion
	NONMEMBRANOUS ORGANELLES		
	Cytoskeleton: Microtubule Microfilament	Proteins organized in fine filaments or slender tubes	Strength, movement of cellular structures and materials
	Microvilli	Membrane extensions containing microfilaments	Increase surface area to facilitate absorption of extracellular materials
	Cilia	Membrane extensions containing 9 microtubule doublets + a central pair	Movement of materials over surface
Centrioles	**Centrosome**	Cytoplasm containing two centrioles, at right angles: Each centriole is composed of 9 microtubule triplets	Essential for movement of chromosomes during cell division; organization of microtubules in cytoskeleton
	Ribosomes	RNA + proteins; fixed ribosomes bound to endoplasmic reticulum, free ribosomes scattered in cytoplasm	Protein synthesis
	MEMBRANOUS ORGANELLES		
	Mitochondria	Double membrane, with inner folds (cristae) enclosing important metabolic enzymes	Produce 95% of the ATP required by the cell
	Endoplasmic reticulum (ER)	Network of membranous channels extending throughout the cytoplasm	Synthesis of secretory products; intracellular storage and transport
	Rough ER	Has ribosomes attached to membranes	Secretory protein synthesis
	Smooth ER	Lacks attached ribosomes	Lipid and carbohydrate synthesis
	Golgi apparatus	Stacks of flattened membranes (saccules) containing chambers (cisternae)	Storage, alteration, and packaging of secretory products and lysosomes
	Lysosomes	Vesicles containing powerful digestive enzymes	Intracellular removal of damaged organelles or of pathogens
	Peroxisomes	Vesicles containing degradative enzymes	Neutralization of toxic compounds
	NUCLEUS	Nucleoplasm containing nucleotides, enzymes, and nucleoproteins; surrounded by double membrane (nuclear envelope)	Control of metabolism; storage and processing of genetic information; control of protein synthesis
	Nucleolus	Dense region in nucleoplasm containing DNA and RNA	Site of rRNA synthesis and assembly of ribosomal subunits

ter-STISH-ul) **fluid** (*interstitium*, something standing between). A *cell membrane* separates the cell contents, or *cytoplasm*, from the interstitial fluid. The cytoplasm, which surrounds the membranous *nucleus*, can be further subdivided into a liquid, the *cytosol*, and intracellular structures collectively known as *organelles* (or-gan-ELZ; "little organs").

The Cell Membrane

The **cell membrane,** also called the **plasma membrane,** or *plasmalemma* (*lemma*, husk), forms the outer boundary of the cell. The general functions of the cell membrane include:

1. ***Physical isolation.*** The cell membrane is a physical barrier that separates the inside of the cell from the surrounding extracellular fluid. Conditions inside and outside the cell are very different, and those differences must be maintained to preserve homeostasis. For example, the cell membrane is a barrier that keeps enzymes and structural proteins inside the cell.
2. ***Regulation of exchange with the environment.*** The cell membrane controls the entry of ions and nutrients, the elimination of wastes, and the release of secretions.
3. ***Sensitivity.*** The cell membrane is the first part of the cell affected by changes in the composition, concentration, or pH of the extracellular fluid. It also contains a variety of receptors that allow the cell to recognize and respond to specific molecules in its environment. For example, the cell membrane may bind chemical signals from other cells. A single alteration in the cell membrane may trigger the activation or deactivation of enzymes that affect many cellular activities.
4. ***Structural support.*** Specialized connections between cell membranes or between membranes and extracellular materials give tissues a stable structure. For example, the cells at the surface of the skin are bound to one another, and those in the deepest layers are attached to extracellular protein fibers in underlying tissues.

Membrane Structure

The cell membrane is extremely thin and delicate, ranging from 6 to 10 nm in thickness. The cell membrane contains lipids, proteins, and carbohydrates. In terms of relative abundance, phospholipids are the largest contributors to membrane structure, followed by proteins, glycolipids, and cholesterol.

Membrane Lipids

Figure 3-3

Figure 3-3• presents a diagrammatic view of cell membrane structure. The cell membrane is called a **phospholipid bilayer** because the phospholipids form two distinct layers. In each layer the phospholipid molecules lie so that the hydrophilic heads are at the surface and the hydrophobic tails are on the inside, in association with cholesterol and small quantities of other lipids.

Note the similarities in lipid organization between a micelle, described in Chapter 2 (Figure 2-16c•, p. 51), and the cell membrane. Ions and water-soluble compounds cannot enter the interior of a micelle because the lipid tails of the phospholipid molecules are highly hydrophobic and will not associate with water molecules. For the same reason, these solutes cannot cross the lipid portion of a cell membrane. This feature makes the membrane very effective in isolating the cytoplasm from the surrounding fluid environment. Such isolation is important because the composition of the cytoplasm is very different from that of the extracellular fluid, and those differences must be maintained.

Membrane Proteins

Figures 3-3, 3-4

Several types of proteins are associated with the membrane. **Integral proteins** are part of the membrane structure. Most integral proteins span the entire width of the membrane one or more times and are therefore known as *transmembrane proteins.* **Peripheral proteins** are bound to the inner or outer surfaces of the membrane and are easily separated from it. Integral proteins greatly outnumber peripheral proteins. Membrane proteins (Figure 3-4•, p. 72) may function as *anchors, identifiers, enzymes, receptors, carriers,* or *channels.*

1. ***Anchors.*** Membrane proteins may attach the cell membrane to other structures and stabilize its position. Inside the cell, membrane proteins are bound to the *cytoskeleton,* a network of supporting filaments within the cytoplasm. Outside the cell, other membrane proteins may attach the cell to extracellular protein fibers or to another cell.
2. ***Identifiers.*** The cells of the immune system recognize other cells as normal or abnormal on the basis of the presence or absence of characteristic **recognition proteins.** Many important recognition proteins are glycoproteins; one group, the *major histocompatibility complex* (MHC), will be discussed in Chapter 22.
3. ***Enzymes.*** Enzymes in cell membranes may be integral or peripheral proteins. These enzymes catalyze reactions in the extracellu-

● FIGURE 3-3
The Cell Membrane 

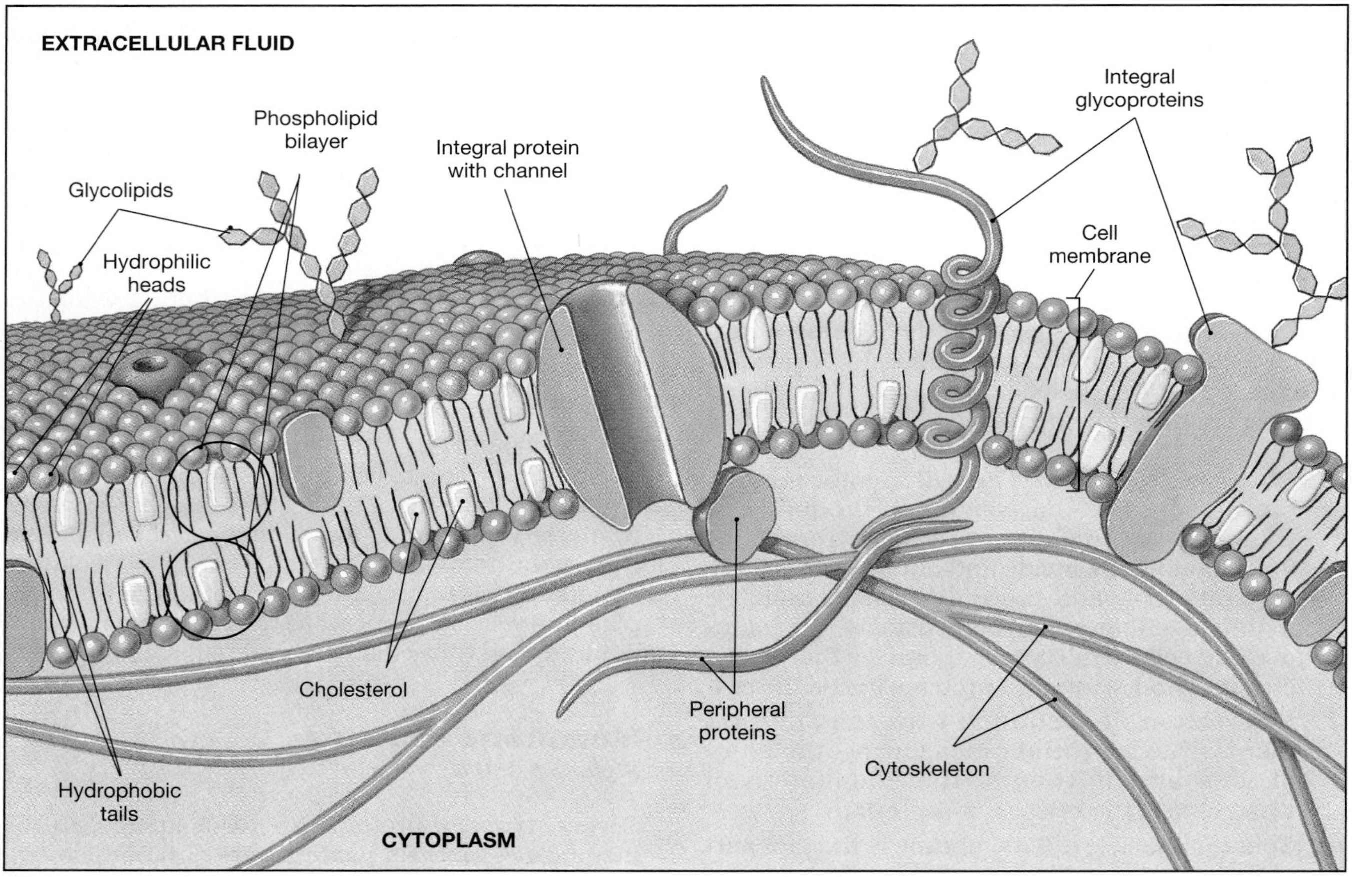

EXTRACELLULAR FLUID
Phospholipid bilayer
Integral protein with channel
Integral glycoproteins
Glycolipids
Hydrophilic heads
Cell membrane
Cholesterol
Peripheral proteins
Cytoskeleton
Hydrophobic tails
CYTOPLASM

lar fluid or within the cytosol, depending on the location of the protein and its active site. For example, dipeptides are broken down into amino acids by enzymes on the exposed membranes of cells lining the intestinal tract.

4. *Receptors.* **Receptor proteins** in the cell membrane are sensitive to the presence of specific extracellular materials, called *ligands.* A receptor protein exposed to an appropriate ligand will bind to it, and that binding may trigger changes in the activity of the cell. For example, binding of the hormone *insulin* to a specific membrane receptor is the key step that leads to an increase in the rate of glucose absorption by the cell. Cell membranes differ in terms of the type and number of receptor proteins they contain, and these differences account for their differing sensitivities to hormones and other solutes.
5. *Carriers.* **Carrier proteins** bind solutes and transport them across the cell membrane. The transport process involves a change in the shape of the carrier protein—the shape change occurs when solute binding occurs and reverses when the solute is released. Carrier proteins may or may not require ATP as an energy source. For example, virtually all cells have carrier proteins that can bring glucose into the cytoplasm without expending ATP, but these cells must expend ATP to transport ions such as sodium and calcium across the cell membrane and out of the cytoplasm.
6. *Channels.* Some integral proteins contain a central pore, or **channel,** that forms a passageway that permits the movement of water and small solutes across the cell membrane. Ions will not dissolve in lipids, and they cannot cross the phospholipid bilayer; ions and other small water-soluble materials can cross the membrane only by passing through a channel. Ion movements through channels are involved in a variety of physiological mechanisms. Physiologists speak of sodium channels, calcium channels, potassium channels, and so forth, to refer to channels with specific permeability properties. There are two major kinds of channels: (1) **leak channels,** which permit water and ion movement at all times (although the rate may vary from moment to moment); and (2) **gated channels,** which can open or close to regulate ion passage. Channels account for around 0.2 percent of the total membrane surface area.

Membrane structure is not rigid, and the embedded proteins drift from place to place across the sur-

Figure 3-3 (continued) Composition of the Cell Membrane

Compound	*Percentage of Weight*	*Function*	*Remarks*
Lipids	42	Form barrier to prevent free exchange of water-soluble materials between the intracellular and extracellular fluids	Double layer of phospholipids covers most of the cell surface; cholesterol is dissolved in the phospholipid layer
Phospholipids	25		
Cholesterol	13		
Other lipids	4		
Proteins	55	Form channels that regulate the passage of ions, perform active transport, facilitated transport, and act as receptors for specific extracellular materials	May be partially or totally embedded in the phospholipid layer
Carbohydrates (glycoproteins, proteoglycans)	3	Act as receptors that bind specific extracellular materials, provide lubrication and protection, identify the cell as "self" to cells responsible for immune defenses	Found as components of glycoproteins, glycolipids, and proteoglycans of the glycocalyx

face of the membrane like ice cubes in a punch bowl. In addition, the composition of the cell membrane can change over time, because components of the membrane surface are removed and replaced through processes described later in the chapter.

The inner and outer surfaces of the cell membrane differ in protein and lipid composition. For example, some cytoplasmic enzymes are found only on the inner surface of the membrane, and some receptors are found exclusively on its outer surface.

Membrane Carbohydrates

The carbohydrates in the cell membrane are found as components of complex molecules such as *proteoglycans, glycoproteins,* and *glycolipids,* introduced in Chapter 2. ∞ *[pp. 51,56]* The carbohydrate portions of these large molecules extend away from the outer surface of the cell membrane, forming a layer known as the **glycocalyx** (*calyx,* cup). The glycocalyx has a variety of important functions:

1. The glycoproteins and glycolipids form a viscous layer that lubricates and protects the cell membrane.
2. Because the components are sticky, the glycocalyx can help anchor the cell in place. It also participates in the locomotion of specialized cells.
3. Glycoproteins and glycolipids can function as receptors, binding specific extracellular compounds. Such binding can alter the properties of the cell surface and indirectly affect the behavior of the cell.
4. The characteristics of the glycocalyx are genetically determined. For example, your blood type (A, B, AB, or O) is determined by the presence or absence of membrane glycoproteins. Normal glycoproteins are recognized by the body's immune system as "self" rather than "foreign," and this recognition system keeps the immune system from attacking the tissues of the body.

What component of the cell membrane is primarily responsible for its ability to form a physical barrier between the cell's internal and external environments?

What kind of integral proteins allow water and small ions to pass through the cell membrane?

Membrane Permeability

Precisely which substances can enter or leave the cytoplasm is determined by the **permeability** of the cell membrane. If nothing could cross the cell membrane, it would be described as **impermeable.** If any substance at all could cross without difficulty, the membrane would be **freely permeable.** Cell membranes actually fall somewhere in between and are thus said to be **selectively permeable.** A selectively permeable membrane permits the free passage of some materials and restricts the passage of others. The distinction may be on the basis of size, electrical charge, molecular shape, lipid solubility, or some combination of factors.

The permeability of a cell membrane varies depending on the organization and identity of membrane lipids and proteins. Passage across the membrane may be passive or active. *Passive processes* move ions or molecules across the cell membrane without any energy expenditure by the cell. *Active processes* require that the cell expend energy, usually in the form of ATP.

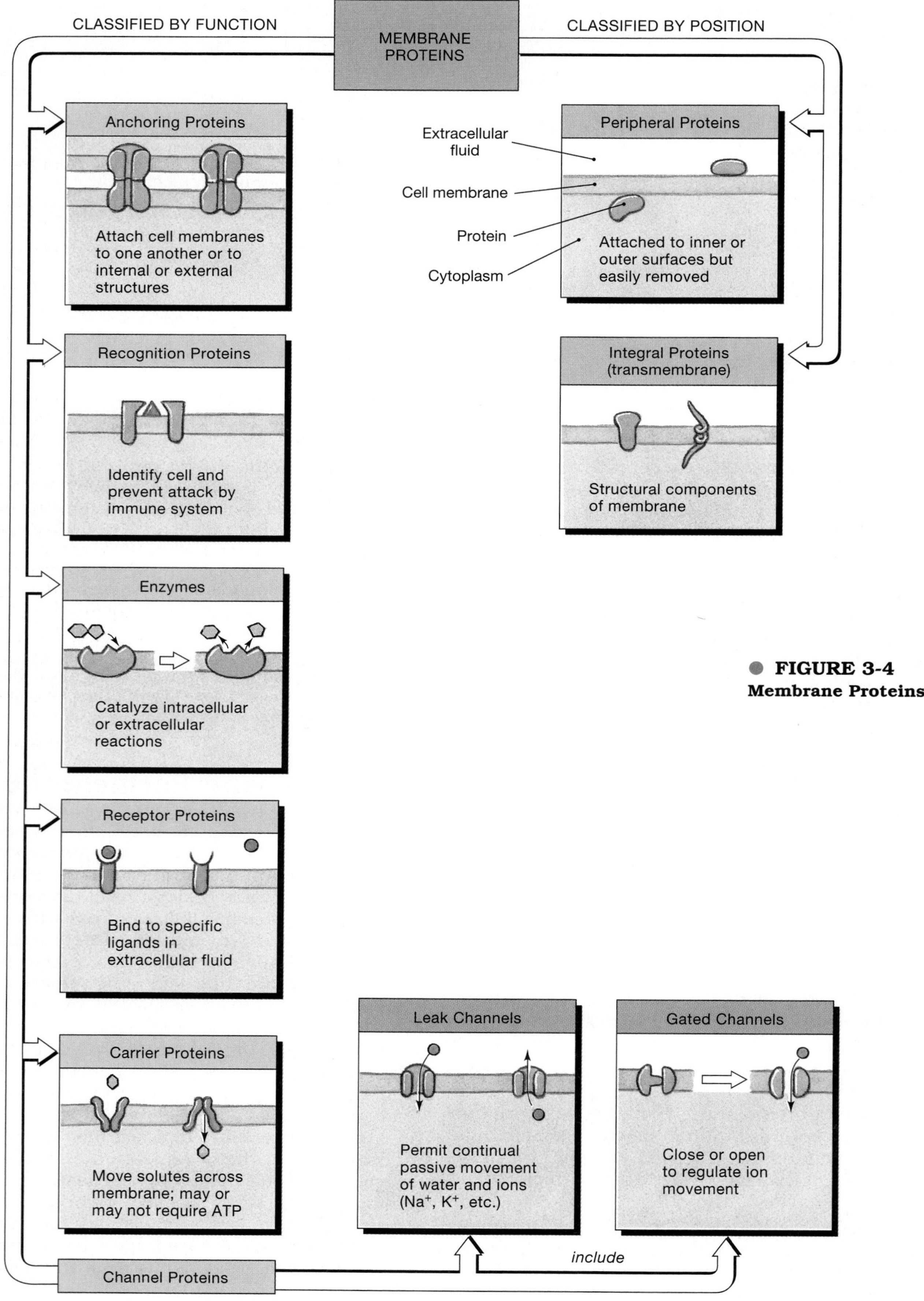
CLASSIFIED BY FUNCTION
MEMBRANE PROTEINS
CLASSIFIED BY POSITION
Anchoring Proteins
Attach cell membranes to one another or to internal or external structures
Peripheral Proteins
Extracellular fluid
Cell membrane
Protein
Cytoplasm
Attached to inner or outer surfaces but easily removed
Recognition Proteins
Identify cell and prevent attack by immune system
Integral Proteins (transmembrane)
Structural components of membrane
Enzymes
Catalyze intracellular or extracellular reactions
Receptor Proteins
Bind to specific ligands in extracellular fluid
Carrier Proteins
Move solutes across membrane; may or may not require ATP
Leak Channels
Permit continual passive movement of water and ions (Na^+, K^+, etc.)
Gated Channels
Close or open to regulate ion movement
include
Channel Proteins

● **FIGURE 3-4**
Membrane Proteins

Transport processes may also be categorized by the nature of the mechanism involved. The major categories are:

- *Diffusion*, which results from the random motion and collisions of ions and molecules. Diffusion is a passive process.
- *Filtration*, which occurs when hydrostatic pressure forces fluid and solutes across a membrane barrier. Filtration is also a passive process.
- *Carrier-mediated transport*, which requires the presence of specialized integral membrane proteins. Carrier-mediated transport may be passive or active, depending on the substance transported and the nature of the transport mechanism.
- *Vesicular transport*, which involves the movement of materials within small membranous sacs, or *vesicles*. Vesicular transport is an active process.

Diffusion

Figures 3-5, 3-6, 3-7, 3-8

Ions and molecules are in constant motion, colliding and bouncing off one another and off any obstacles in their paths. The movement is random: A molecule can bounce in any direction. One result of this continual random motion is that, over time, the molecules in any given space will tend to become evenly distributed. As this occurs, there will be a net movement of material from areas of relatively high concentration to areas of relatively low concentration. This distribution process is known as **diffusion.** The difference between the high and low concentrations represents a **concentration gradient,** and diffusion tends to eliminate that gradient. After that gradient has been eliminated, diffusion continues, but there is no longer net movement in any particular direction. For convenience, we restrict use of the term *diffusion* to the directional movement that eliminates concentration gradients; this process is sometimes called *net diffusion.* Because diffusion spreads materials from a region of high concentration to one of relatively lower concentration, it is often described as proceeding "down a concentration gradient" or "downhill."

We all have experienced the effects of diffusion, which occurs in air as well as water. The smell of fresh flowers in a vase can sweeten the air in a large room, just as a drop of ink spreads to color an entire glass of water. In each case you begin with an extremely high concentration of molecules in a very localized area. Consider ink dropped in a water glass (Step 1, Figure 3-5●). Placing that drop in a large volume of clear water establishes a steep concentration gradient for the dye: The dye concentration is high at the drop and negligible everywhere else. As diffusion proceeds, the dye molecules spread through the solution (Step 2) until they are distributed evenly (Step 3).

Diffusion is important in body fluids because it tends to eliminate local concentration gradients. For example, an active cell generates carbon dioxide, and the intracellular concentration is relatively high. Carbon dioxide concentrations are lower in the surrounding interstitial fluid, and lower still in the circulating blood. Because cell membranes are freely permeable to carbon dioxide, it can diffuse down its concentration gradient, traveling from the interior of the cell into the interstitial fluid, and from the interstitial fluid into the bloodstream, for eventual delivery to the lungs.

To be effective, diffusion of nutrients, waste products, and dissolved gases must be able to keep pace with the demands of active cells. Important factors that influence diffusion rates include:

Step 1

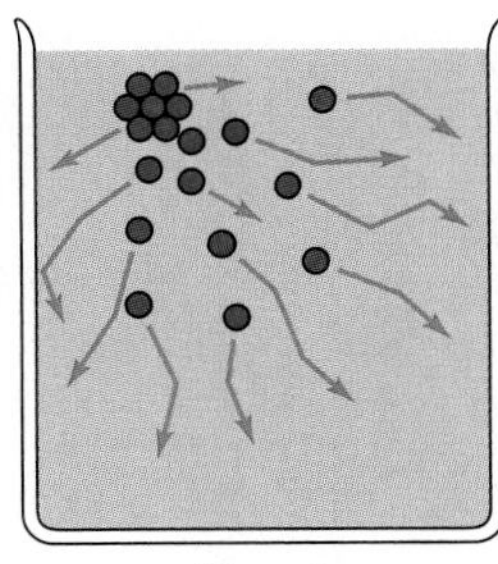
Step 2

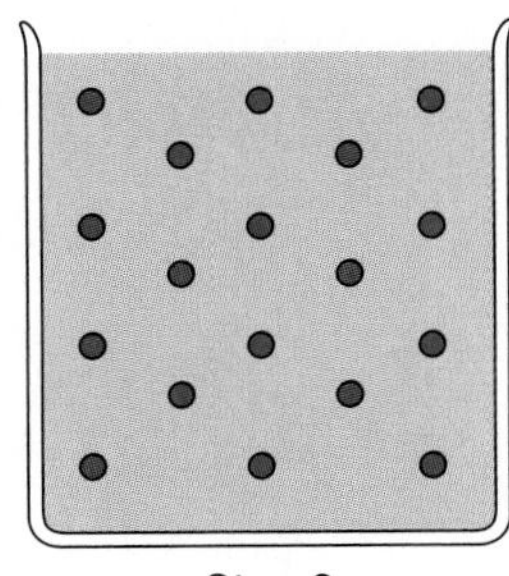
Step 3

● **FIGURE 3-5**

Diffusion. Step 1: Placing an ink drop in a glass of water establishes a strong concentration gradient because there are many ink molecules in one location and none elsewhere. Step 2: As diffusion occurs, the ink molecules spread through the solution. Step 3: Eventually diffusion eliminates the concentration gradient, and the ink molecules are distributed evenly. Molecular motion continues, but there is no directional movement.

1. ***Distance.*** Concentration gradients are eliminated quickly over short distances. The greater the distance, the longer the time required. In the human body, diffusion distances are usually small. For example, few living cells are farther than 125 μm from a blood vessel.
2. ***Size of the gradient.*** The larger the concentration gradient, the faster diffusion proceeds. When cells become more active, the intracellular concentration of oxygen declines. This change increases the concentration gradient for oxygen between the inside of the cell (relatively low) and the interstitial fluid outside (relatively high). The rate of oxygen diffusion into the cell then increases.
3. ***Molecule size.*** Ions and small organic molecules such as glucose diffuse more rapidly than large proteins.
4. ***Temperature.*** The higher the temperature, the faster the diffusion rate. The human body maintains a temperature of around 37° C, and diffusion proceeds more rapidly at this temperature than at normal environmental temperatures.
5. ***Electrical forces.*** For reasons that will be explored in a later chapter, the interior of the cell membrane has a net negative charge relative to the exterior surface. Opposite charges (+ and –) are attracted to one another, whereas similar charges (+ and + or – and –) repel one another. The negative charge on the inside of the cell membrane will tend to pull positive ions from the extracellular fluid into the cell and oppose the entry of negative ions. For example, compared with the cytosol, interstitial fluid contains relatively high concentrations of sodium ions (Na^+) and chloride ions (Cl^-). The concentration gradient, or *chemical gradient,* and the electrical gradient both favor the diffusion of the positively charged sodium ions into the cell. By contrast, diffusion of the negatively charged chloride ions is favored by the chemical gradient but *opposed* by the electrical gradient. For any given ion, the net result of the chemical and electrical forces acting on it is called the **electrochemical gradient.**

DIFFUSION ACROSS CELL MEMBRANES In the extracellular fluids of the body, water and dissolved solutes diffuse freely. A cell membrane, however, acts as a barrier that selectively restricts diffusion: Some substances can pass through easily, whereas others cannot penetrate the membrane at all. There are only two ways for an ion or molecule to diffuse across a cell membrane: (1) pass through one of the membrane channels or (2) move across the lipid portion of the membrane (Figure 3-6•). If there is an electrochemical gradient for a particular ion or molecule, whether that substance moves across the cell membrane will depend on two major factors: its lipid solubility and its size relative to the sizes of the membrane channels.

1. ***Lipid solubility.*** Alcohol, fatty acids, and steroids can enter cells easily because they can diffuse through the lipid portions of the membrane. Dissolved gases, such as oxygen and carbon dioxide, and lipid-soluble drugs also enter

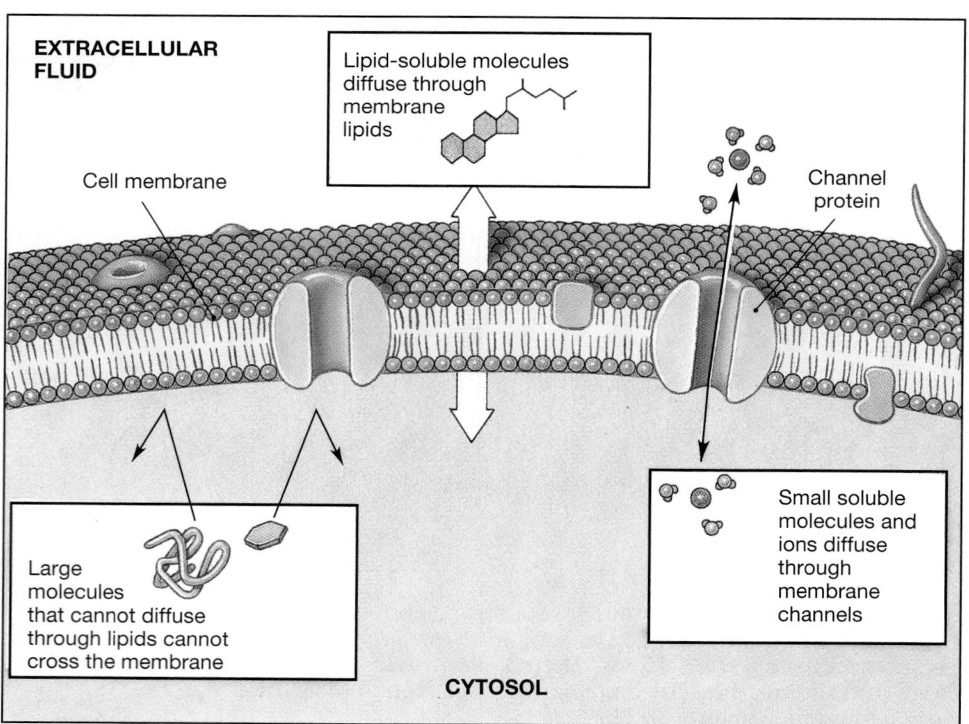
EXTRACELLULAR FLUID
Lipid-soluble molecules diffuse through membrane lipids
Cell membrane
Channel protein
Large molecules that cannot diffuse through lipids cannot cross the membrane
Small soluble molecules and ions diffuse through membrane channels
CYTOSOL

• **FIGURE 3-6**
Diffusion across Cell Membranes. Small ions and water-soluble molecules diffuse through membrane channels. Lipid-soluble molecules can cross the membrane by diffusing through the phospholipid bilayer. Large molecules that are not lipid-soluble, such as proteins and carbohydrates, cannot diffuse through the cell membrane.

and leave our cells by diffusion through the lipid bilayer. AM *Drugs and the Cell Membrane*

2. ***Size.*** To diffuse into the cytoplasm, a water-soluble compound must pass through a membrane channel. These channels are very small, averaging about 0.8 nm in diameter. Water molecules can enter or exit freely, as can ions such as sodium, potassium, calcium, hydrogen, and chloride, but even a small organic molecule, such as glucose, is too big to fit through the channels.

OSMOSIS: A SPECIAL CASE OF DIFFUSION The net diffusion of water across a membrane is so important that it is given a special name, **osmosis** (oz-MŌ-sis; *osmos,* thrust). For convenience, we will always use the term *osmosis* when considering water movement and restrict use of the term *diffusion* to the movement of solutes.

Intracellular and extracellular fluids are solutions that contain a variety of dissolved materials. Each solute tends to diffuse as if it were the only material in solution. For example, the diffusion of sodium ions occurs in response to the existence of a concentration gradient for sodium. Thus, a change in the concentration of other ions will have no effect on the rate or direction of sodium ion diffusion. Some solutes diffuse into the cytoplasm, others diffuse out, and a few, such as proteins, are unable to diffuse across the cell membrane at all. But if we ignore the individual identities, and simply count ions and molecules, we find that the *total* concentration of dissolved ions and molecules on either side of the cell membrane stays the same.

This state of equilibrium persists because the entire cell membrane is freely permeable to water. Whenever a solute concentration gradient exists, there is also a concentration gradient for water. Because dissolved solute molecules occupy space that would otherwise be taken up by water molecules, the higher the solute concentration, the lower the water concentration. As a result, *water molecules will tend to diffuse across a membrane toward the solution containing a higher solute concentration,* because this movement is down the concentration gradient for water.

Three characteristics of osmosis should be remembered:

1. Osmosis is the diffusion of water molecules across a membrane.
2. Osmosis occurs across a selectively permeable membrane that is freely permeable to water but not freely permeable to solutes.
3. In osmosis, water will flow across a membrane *toward the solution that has the higher concentration of solutes,* because that is where the concentration of water is lowest.

OSMOSIS AND OSMOTIC PRESSURE Figure 3-7● diagrams the process of osmosis. Step 1 shows two solutions (A and B) with differing solute concentrations separated by a selectively permeable mem-

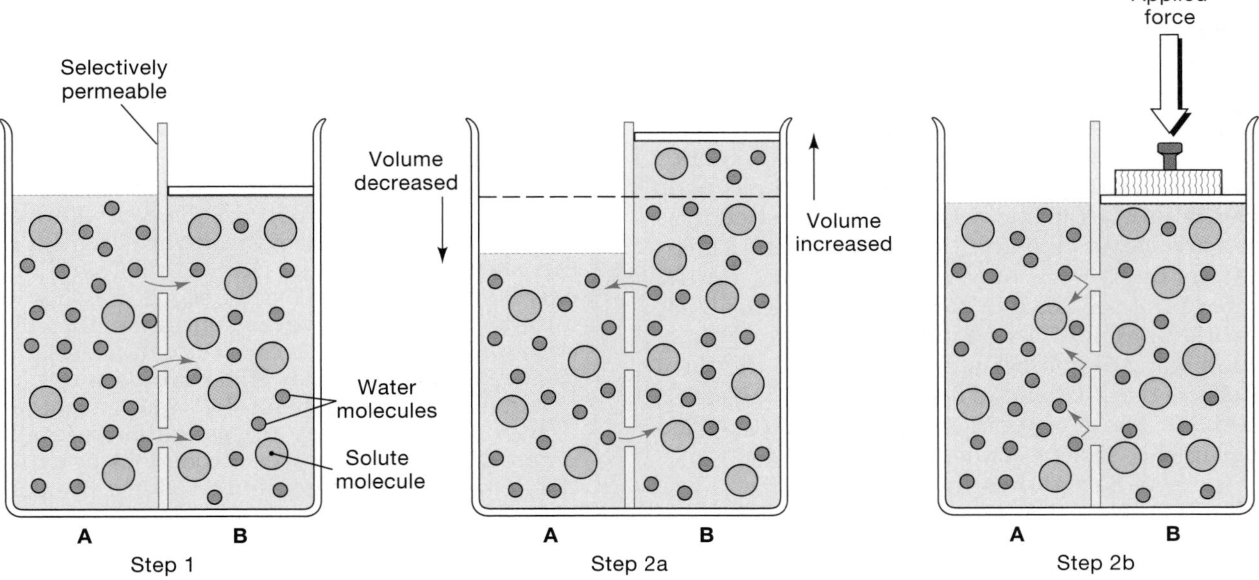

Selectively permeable
Volume decreased
Volume increased
Applied force
Water molecules
Solute molecule
A
B
Step 1
A
B
Step 2a
A
B
Step 2b

● **FIGURE 3-7**

Osmosis. Step 1: Two solutions containing different solute concentrations are separated by a selectively permeable membrane. Water molecules (small dots) begin to cross the membrane toward the solution with the higher concentration of solutes (larger circles) (solution B). Step 2a: At equilibrium the solute concentrations on the two sides of the membrane are equal. The volume of solution B has increased at the expense of solution A. Step 2b: Osmosis can be prevented by resisting the volume change. The osmotic pressure of solution B is equal to the amount of hydrostatic pressure required to stop the osmotic flow. D

brane. As osmosis occurs, water molecules cross the membrane until the solute concentrations in the two solutions are identical (Step 2a). Thus the volume of solution B increases at the expense of solution A. The greater the initial difference in solute concentrations, the stronger the osmotic flow. The **osmotic pressure** of a solution is an indication of the force of water movement *into that solution* as a result of its solute concentration. The osmotic pressure of a solution can be measured in several ways. For example, a strong enough opposing pressure can prevent the osmotic flow of water into the solution. Pushing against a fluid generates **hydrostatic pressure;** in Step 2b, hydrostatic pressure opposes the osmotic pressure of solution B, and no net osmotic flow occurs.

Osmosis eliminates solute concentration differences much more quickly than one would predict on the basis of diffusion rates for other molecules. When water molecules cross a membrane they move in groups held together by hydrogen bonding. So, whereas solute molecules usually diffuse through membrane channels one at a time, water molecules move together in large numbers. This phenomenon is called **bulk flow.**

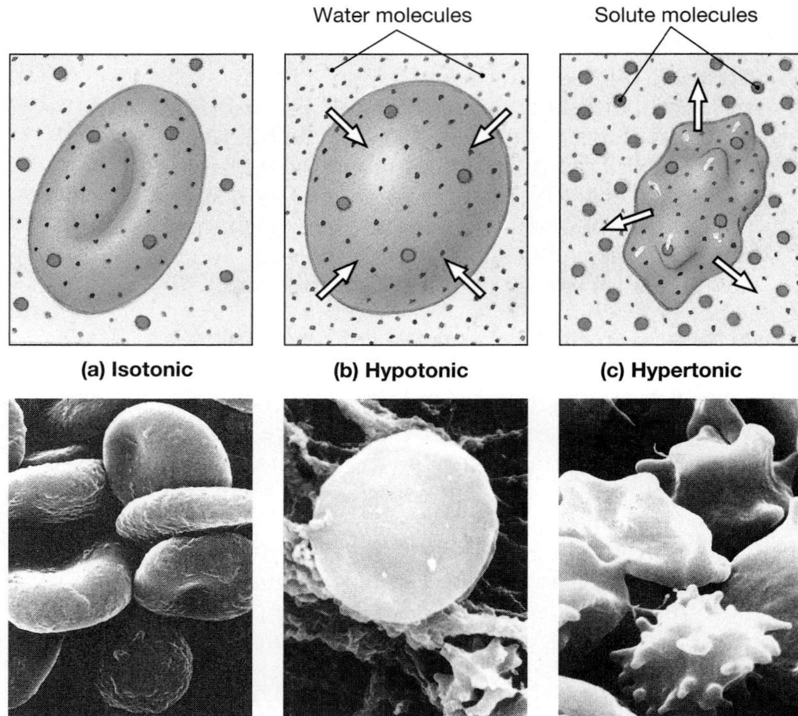
Water molecules
Solute molecules
(a) Isotonic
(b) Hypotonic
(c) Hypertonic

● **FIGURE 3-8**
Osmotic Flow across Cell Membranes. White arrows indicate the direction of osmotic water movement. **(a)** Because these red blood cells are immersed in an isotonic saline solution, no osmotic flow occurs and the cells have their normal appearance. **(b)** Immersion in a hypotonic saline solution results in the osmotic flow of water into the cells. The swelling may continue until the cell membrane ruptures. **(c)** Exposure to a hypertonic solution results in the movement of water out of the cells. The red blood cells shrivel and become crenated. (SEM × 833)

OSMOLARITY AND TONICITY The total solute concentration in a solution is its **osmotic concentration,** or **osmolarity.** If a given solution contains solutes in the same concentration as the cytosol, it is said to be *isosmotic.* If a solution has a higher solute concentration, it is termed *hyperosmotic* (*hyper,* above); if it has a lower solute concentration, it is *hyposmotic* (*hypo,* below). The effect of such solutions on the cell may vary, depending on the nature of the solutes (see the following clinical discussion). Therefore, when describing the effects of various osmotic solutions on living cells, the term **tonicity** is used instead. If a solution does not cause an osmotic flow of water in or out of a cell, it is called **isotonic** (*iso,* same + *tonos,* tension).

Figure 3-8a● shows the appearance of a red blood cell immersed in an isotonic solution. If a red blood cell is immersed in a **hypotonic** solution, water will flow into the cell, causing it to swell up like a balloon (Figure 3-8b●). Ultimately the cell may rupture, releasing the intracellular contents. This event is known as **hemolysis** (*hemo-,* blood + *lysis,* dissolution). A cell placed in a **hypertonic** solution will lose water through osmosis. As this occurs the cell shrivels and dehydrates. The shrinking of red blood cells, shown in Figure 3-8c●, is called **crenation.**

TONICITY AND NORMAL SALINE Although often used interchangeably, the terms *osmolarity* and *tonicity* do not always mean the same thing. For example, consider a solution that has the same osmolarity as the intracellular fluid but a higher concentration of one or more individual ions. If any of those ions can diffuse into the cell, the osmolarity of the intracellular fluid will increase and that of the extracellular solution will decrease. Osmosis will then occur, moving water into the cell. If the process continues, the cell will eventually burst. In this case the extracellular solution and the intracellular fluid were isosmotic (equal in osmolarity), but the solution was *not* isotonic.

It is often necessary to give patients large volumes of fluid after a severe blood loss or dehydration. One fluid often administered is a 0.9 percent (0.9 g/dl) solution of sodium chloride (NaCl). This solution, which approximates the normal osmotic concentration of the extracellular fluids, is called **normal saline.** It is used because sodium and chloride are the most abundant ions in the extracellular fluid. There is little net movement of either ion across cell membranes; thus normal saline is essentially isotonic with respect to body cells. An alternative treatment involves the use of an isotonic solution containing **dextran,** a carbohydrate that cannot cross cell membranes.

FLUID SHIFTS Minor fluctuations in intracellular and extracellular solute concentrations are eliminated in a matter of seconds by *fluid shifts,* the osmotic movement of water in or out of cells. It can take considerably longer for fluid shifts to compensate for systemwide changes in solute concentrations. For example, after you drink a large glass of pure water, it may take a half-hour for your intracellular and extracellular fluids to become isotonic again. Severe alterations in body water content, such as those occurring in dehydration, are extremely dangerous. Fluid shifts and water balance will be considered in later chapters dealing with metabolism and kidney function.

Filtration

Figure 3-9

In **filtration,** hydrostatic pressure forces water across a membrane, and solute molecules are selected on the basis of size. If the membrane pores are large enough, molecules of solute will be carried along with the water. We can see filtration in action in a coffee machine (Figure 3-9a●). Gravity forces hot water through the filter, and the water carries with it a variety of dissolved compounds. The large coffee grounds never reach the pot because they cannot fit through the fine pores in the filter.

In the body, the heart pushes blood through the circulatory system and generates hydrostatic pressure. Filtration occurs across the walls of small blood vessels, pushing water and dissolved solutes into the tissues of the body (Figure 3-9b●). Filtration across specialized blood vessels in the kidneys is an essential step in the production of urine. Because the filtration pores are very large, the clusters of water molecules carry dissolved ions with them, thus accelerating the filtration process (Figure 3-9c●). This process is called *bulk flow.*

✓ How would a decrease in the concentration of oxygen in the lungs affect the diffusion of oxygen into the blood?

✓ Some pediatricians recommend using a 10 percent salt solution to relieve congestion for infants with stuffy noses. What effect would such a solution have on the cells lining the nasal cavity, and why?

Carrier-Mediated Transport

Figures 3-10, 3-11

In **carrier-mediated transport,** integral proteins bind specific ions or organic substrates and facilitate their movement across the cell membrane. All forms of carrier-mediated transport share several characteristics in common with enzymes:

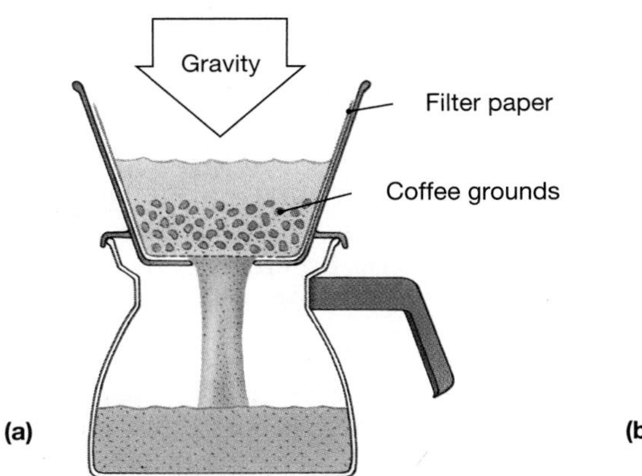
Gravity
Filter paper
Coffee grounds
(a)

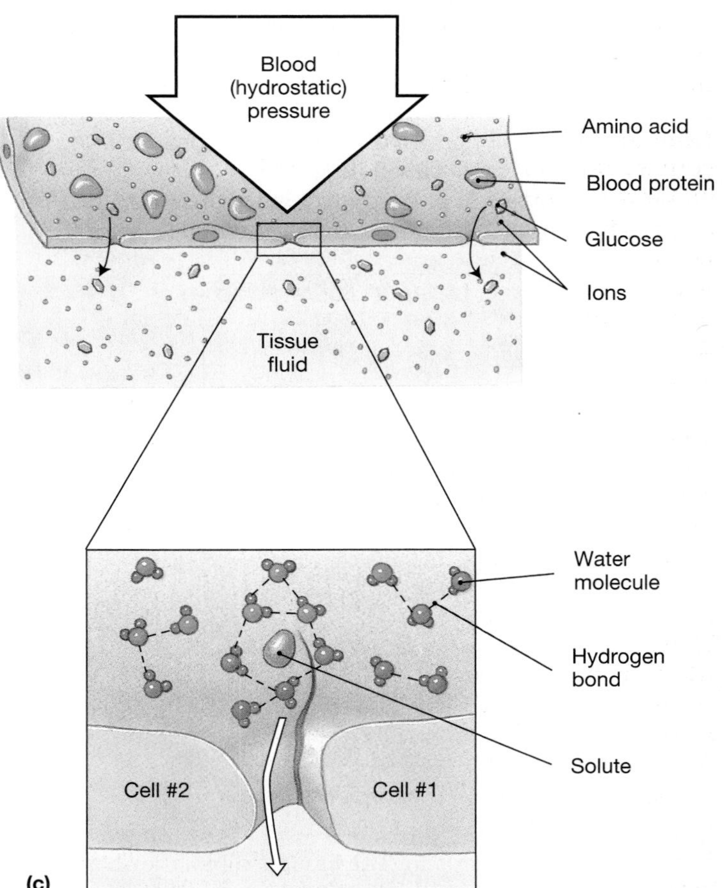
(b)
Blood (hydrostatic) pressure
Amino acid
Blood protein
Glucose
Ions
Tissue fluid
Water molecule
Hydrogen bond
Solute
Cell #2
Cell #1
(c)

● **FIGURE 3-9**

Filtration. (a) In filtration, materials are removed from a solution on the basis of size. In a coffee maker, the paper filter keeps the coffee grounds trapped, but lets smaller, dissolved molecules pass through. Gravity provides the hydrostatic pressure needed to force the coffee water through the filter. **(b)** In the most delicate blood vessels, blood pressure forces water and dissolved nutrients out of the bloodstream and into the interstitial fluid. Blood proteins are too large to pass through the openings between the cells that line the vessel. **(c)** Because adjacent water molecules are attracted to one another through the formation of hydrogen bonds, they cross a membrane in groups. Dissolved solutes small enough to fit through the pores will be carried along.

1. ***Specificity.*** Carrier proteins show *specificity.* That is, each carrier protein in the cell membrane is selective about what substances it will bind and transport. For example, the carrier protein that transports glucose will not transport other simple sugars.
2. ***Saturation.*** The rate of transport into or out of the cell is limited by the availability of carrier proteins, just as enzymatic reaction rates are limited by enzyme concentrations. When all the available carrier molecules are operating at maximum speed, the carrier system is said to be **saturated.** Because no more carrier proteins are available, and the existing ones cannot work any faster, the rate of transport cannot increase, regardless of the size of the concentration gradient.
3. ***Regulation.*** Carrier protein activity can be altered by the binding of other molecules, such as hormones. Hormones thus provide an important means of coordinating carrier protein activity throughout the body. The interplay between hormones and cell membranes will be examined further in chapters dealing with the endocrine system (Chapter 18) and metabolism (Chapter 25).

Two major examples of carrier-mediated transport are *facilitated diffusion* and *active transport.*

FACILITATED DIFFUSION Many essential nutrients, such as glucose and amino acids, are insoluble in lipids but too large to fit through membrane channels. These compounds can be passively transported across the membrane by carrier proteins in a process called **facilitated diffusion**. The molecule to be transported first binds to a **receptor site** on the protein. It is then moved to the inside of the cell membrane and released into the cytoplasm. The process is diagrammed in Figure 3-10●.

As in the case of simple diffusion, no ATP is expended in facilitated diffusion, and the molecules move from an area of higher concentration to one of lower concentration. However, facilitated diffusion differs from ordinary diffusion because once the carrier proteins are saturated, the rate of transport cannot increase, regardless of further changes in the concentration gradient.

ACTIVE TRANSPORT In **active transport** the high-energy bond in ATP (or another high-energy compound) provides the energy needed to move ions or molecules across the membrane. Although it has an energy cost, active transport offers one great advantage: It is not dependent on a concentration gradient. As a result, the cell can import or export specific materials *regardless of their intracellular or extracellular concentrations.*

All living cells contain carrier proteins called **ion pumps** that actively transport the cations sodium (Na^+), potassium (K^+), calcium (Ca^{2+}), and magnesium (Mg^{2+}) across their cell membranes. Specialized cells can transport additional ions such as iodide (I^-), chloride (Cl^-), and iron (Fe^{2+}). Many of these carrier proteins move a specific cation or anion in one direction only, either into or out of the cell. In a few instances, one carrier protein will move more than one kind of ion at the same time. If one kind of ion moves in one direction and the other moves in the opposite direction, the carrier protein is called an **exchange pump.**

THE SODIUM-POTASSIUM EXCHANGE PUMP Sodium and potassium ions are the principal cations in body fluids. Sodium ion concentrations are high in the extracellular fluids but relatively low in the

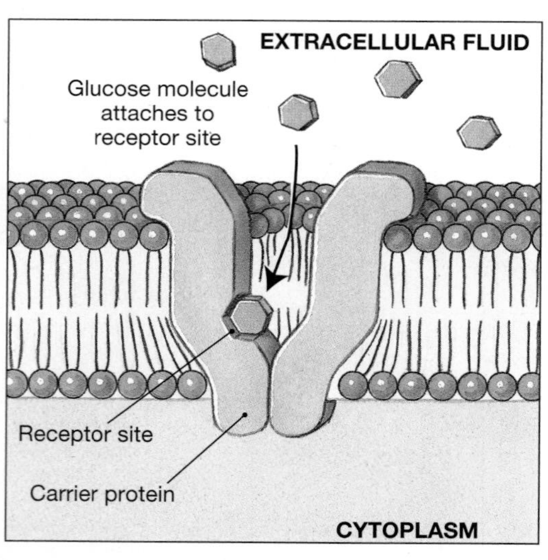
EXTRACELLULAR FLUID
Glucose molecule attaches to receptor site
Receptor site
Carrier protein
CYTOPLASM

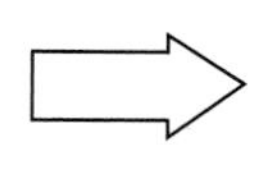
Change in shape of carrier protein

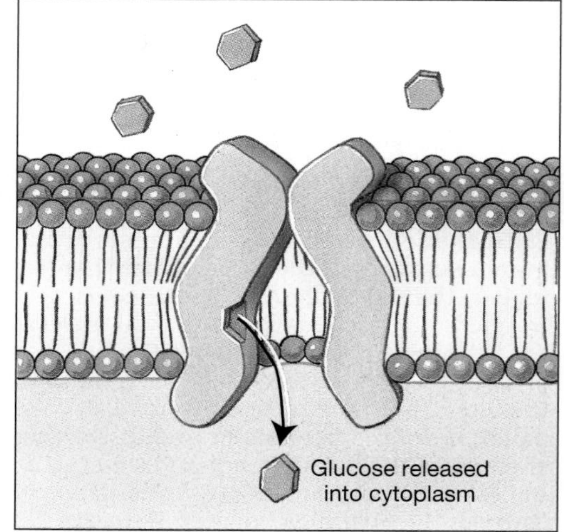
Glucose released into cytoplasm

● **FIGURE 3-10**

Facilitated Diffusion. In this process an extracellular molecule, such as glucose, binds to a receptor site on a carrier protein. The binding alters the shape of the protein, which then releases the molecule to diffuse into the cytoplasm.

cytoplasm. The distribution of potassium in the body is just the opposite—low in the extracellular fluids and high in the cytoplasm. As a result, sodium ions slowly diffuse into the cell, and potassium ions diffuse out through leak channels. Homeostasis within the cell depends on ejecting sodium ions and recapturing lost potassium ions. This exchange is accomplished through the activity of a **sodium-potassium exchange pump**. The protein involved in this process is called *sodium-potassium ATPase*.

As indicated in Figure 3-11●, the sodium-potassium exchange pump exchanges intracellular sodium for extracellular potassium. On the average, for each ATP molecule consumed, three sodium ions are ejected and two potassium ions are reclaimed by the cell. If ATP is readily available, the rate of transport depends on the concentration of sodium ions in the cytoplasm. When that concentration rises, the pump becomes more active. The energy demands are impressive; sodium-potassium ATPase may consume up to 40 percent of the ATP produced by a resting cell.

SECONDARY ACTIVE TRANSPORT In **secondary active transport,** the transport mechanism itself does not require energy, but the cell often needs to expend ATP at a later time to preserve homeostasis. Like facilitated transport, a secondary active transport mechanism moves a specific substance down a concentration gradient. Unlike the proteins in facilitated transport, however, these carrier proteins can also move another substance at the same time, without regard to its concentration gradient. In effect, the concentration gradient for one substance provides the driving force needed by the carrier protein, and the second substance gets a free ride. In **cotransport,** or *symport,* the carrier transports both substances in the same direction, either into or out of the cell. In **countertransport**, or *antiport,* one substance moves into the cell and the other moves out.

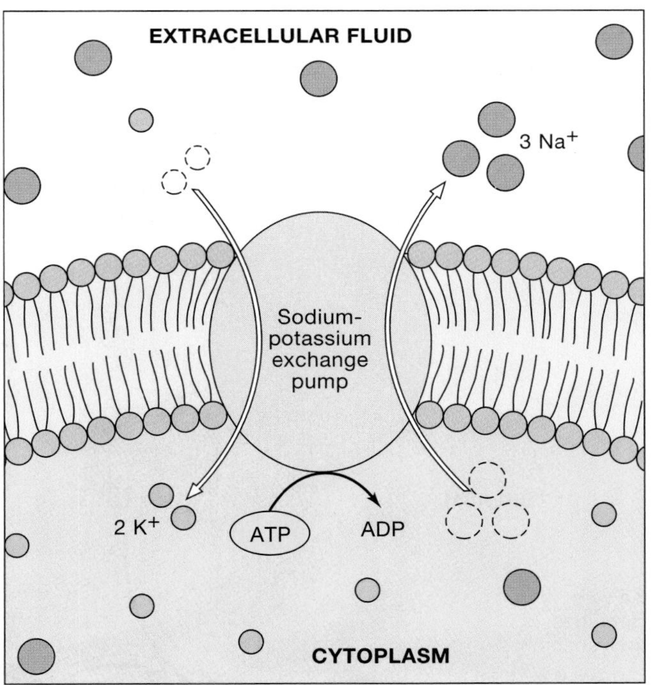
EXTRACELLULAR FLUID
3 Na+
Sodium-potassium exchange pump
2 K+
ATP
ADP
CYTOPLASM

● **FIGURE 3-11**

The Sodium-Potassium Exchange Pump. The operation of the sodium-potassium exchange pump is an example of active transport. For each ATP molecule converted to ADP, this protein, called *sodium-potassium ATPase,* carries three Na^+ ions out of the cell and two K^+ ions into the cell.

The concentration gradient for sodium ions most often provides the driving force for cotransport mechanisms that move materials into the cell. For example, sodium-linked cotransport is important in the absorption of glucose and amino acids along the intestinal tract. Sodium ions are also involved with many countertransport mechanisms. Sodium-calcium countertransport is responsible for keeping intracellular calcium ion concentrations very low.

Although the transport activity proceeds without a direct energy expense, the cell must expend ATP to pump the arriving sodium ions out of the cell, via the sodium-potassium exchange pump.

Vesicular Transport

Figures 3-12, 3-13, 3-19

In **vesicular transport,** materials move in or out of the cell through the formation or breakdown of **vesicles,** small membranous sacs. Because large volumes of fluid and solutes are transported in this way, this process is also known as *bulk transport.* There are two major categories of vesicular transport, *endocytosis* and *exocytosis.*

ENDOCYTOSIS Endocytosis is the packaging of extracellular materials in a vesicle at the cell surface for import into the cell. This process involves relatively large volumes of extracellular material. There are three major types of endocytosis: *receptor-mediated endocytosis, pinocytosis,* and *phagocytosis.* All three are active processes that require energy in the form of ATP.

All forms of endocytosis produce cytoplasmic vesicles whose contents remain isolated within the vesicle. Movement of materials into the surrounding cytoplasm may involve active transport, simple or facilitated diffusion, or the destruction of the vesicle membrane.

Receptor-Mediated Endocytosis. **Receptor-mediated endocytosis** involves the formation of small vesicles at the membrane surface. Receptor-mediated endocytosis is a selective process; it produces vesicles that contain a specific target molecule in high concentrations. The process begins when materials in the extracellular fluid bind to receptors on the membrane surface (Figure 3-12●). The receptor molecules are usually glycoproteins, and each binds to a specific target, or *ligand,* such as a transport protein or hormone.

Receptors bound to ligands cluster together in pits on the cell surface, creating areas of cell membrane coated with ligands. The coated pits then form grooves or pockets that pinch off to produce **coated vesicles.** Inside the cell the coated vesicles fuse with *lysosomes,* vesicles containing digestive enzymes. The fusion of one or more lysosomes with another vesicle creates a *secondary lysosome.* Lysosomal enzymes then free the ligands from their receptors, and the ligands enter the cytosol by diffusion or active transport. The vesicle membrane then pinches off from the secondary lysosome and returns to the cell surface, its receptors ready to bind more ligands.

Many important substances, including cholesterol and iron ions (Fe^{2+}), are distributed through the body attached to special transport proteins. The transport proteins are too large to pass through membrane pores, but they can enter the cell through receptor-mediated endocytosis.

Pinocytosis. **Pinocytosis** (pi-no-si-TŌ-sis), or "cell drinking," is the formation of small vesicles filled with extracellular fluid. This process is not as selective as receptor-mediated endocytosis because there are no receptors involved. The target appears to be the fluid contents in general, rather than specific bound ligands. In pinocytosis, a deep groove or pocket forms in the cell membrane and then pinches off. The fate of a pinocytotic vesicle is the same as that of a vesicle formed through receptor-mediated endocytosis:

1. The vesicle usually fuses with one or more lysosomes.
2. Lysosomal enzymes break down any organic molecules inside the vesicle.
3. Nutrients, such as lipids, sugars, or amino acids, enter the cytoplasm by diffusion or active transport.
4. The membrane of the pinocytotic vesicle then returns to the cell surface.

Phagocytosis. **Phagocytosis** (fa-go-si-TŌ-sis), or "cell eating," produces vesicles containing *solid objects* that may be as large as the cell itself. This process is diagrammed in Figure 3-13a●. Cytoplasmic extensions, called **pseudopodia** (sū-dō-PŌ-dē-a; *pseudo-*, false + *podon,* foot), surround the object, and their membranes fuse to form a vesicle. The vesicle will then fuse with many lysosomes, whereupon its contents are digested by lysosomal enzymes. Figure 3-13b● gives a three-dimensional view of phagocytosis in action.

Most cells display pinocytosis, but phagocytosis is performed only by specialized cells, such as the *macrophages* that protect tissues by engulfing bacteria, cell debris, and other abnormal materials.

EXOCYTOSIS **Exocytosis** (ek-sō-sī-TŌ-sis) is the functional reverse of endocytosis. In exocytosis, seen

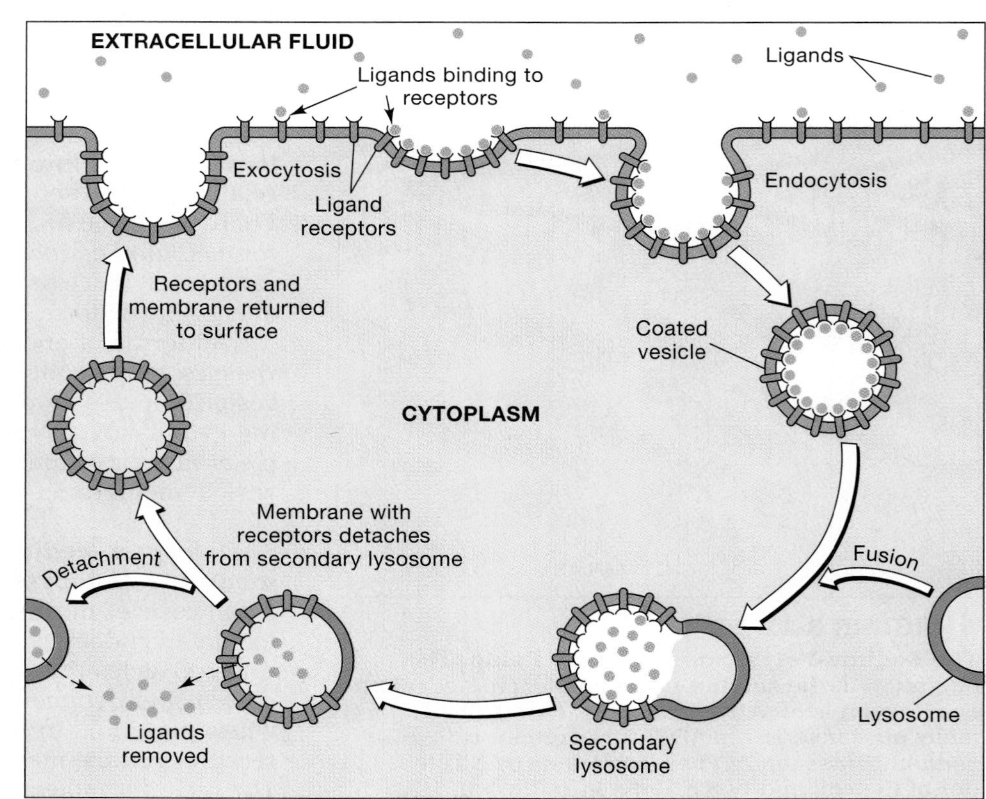
EXTRACELLULAR FLUID
Ligands binding to receptors
Ligands
Exocytosis
Ligand receptors
Endocytosis
Receptors and membrane returned to surface
Coated vesicle
CYTOPLASM
Membrane with receptors detaches from secondary lysosome
Detachment
Fusion
Ligands removed
Secondary lysosome
Lysosome

● **FIGURE 3-12**
Receptor-Mediated Endocytosis. In this process, specific target molecules called ligands bind to receptors, usually glycoproteins, in the membrane surface. Membrane areas coated with ligands pinch off to form vesicles that fuse with primary lysosomes. The ligands are freed from the receptors and, if necessary, broken down by enzymes before diffusing or being transported into the surrounding cytoplasm. The membrane containing the receptor molecules separates from the membrane of the lysosome and returns to the cell surface to bind additional ligands.

in Figures 3-12 and 3-19●, p. 90), a vesicle created inside the cell fuses with the cell membrane and discharges its contents into the extracellular environment. The ejected material may be a secretory product, such as mucus, or waste products accumulating within endocytic vesicles (Figure 3-12●) or secondary lysosomes (Figure 3-19●).

In a few specialized cells, endocytosis produces vesicles on one side of the cell that are discharged through exocytosis on the opposite side. This method of bulk transport is common in cells lining *capillaries*, the most delicate blood vessels. These cells use a combination of pinocytosis and exocytosis to transfer fluid and solutes from the bloodstream into the surrounding tissues.

A Review of Membrane Permeability

Many different mechanisms can be moving materials into and out of the cell at any given moment. Before proceeding further, review and compare the mechanisms summarized in Table 3-2.

The Transmembrane Potential

Figure 3-14

As noted above (p. 74), the inside of the cell membrane has a slight negative charge with respect to the outside. The cause is a slight excess of positively charged ions outside the cell membrane and a slight excess of negatively charged ions inside the cell membrane. This unequal charge distribution is created by differences in the permeability of the membrane to various ions, as well as by active transport mechanisms.

Although the positive and negative charges are attracted to one another, they are separated by the lipid membrane. When positive and negative charges are held apart, there is a **potential difference.** We will use the term **transmembrane potential** when referring to the potential difference across a cell membrane.

The **volt** (V) is the unit of measurement for potential difference; most cars have 12-volt batteries. The transmembrane potentials of living cells are much smaller, averaging about 0.07 V for a neuron. This value is usually reported as –70 mV, with mV indicating millivolts (thousandths of a volt) and the minus sign signifying that the inside of the cell membrane contains an excess of negative charges as compared to the outside.

SIGNIFICANCE OF THE TRANSMEMBRANE POTENTIAL If the lipid barrier were removed, the positive and negative charges would move together. The cell membrane thus acts like a dam across a stream. A dam resists the water pressure building up on the upstream side; a cell membrane resists electrochemical forces that would otherwise drive ions into or out of the cell. The water retained behind a dam and the ions held on either side of the cell membrane have *potential energy.* People have designed many ways to make use of the potential energy stored behind a dam—for example, turning a mill wheel or a turbine. Similarly, cells have ways of utilizing the potential energy stored in the transmembrane potential. For example, it is the transmembrane potential that makes possible the

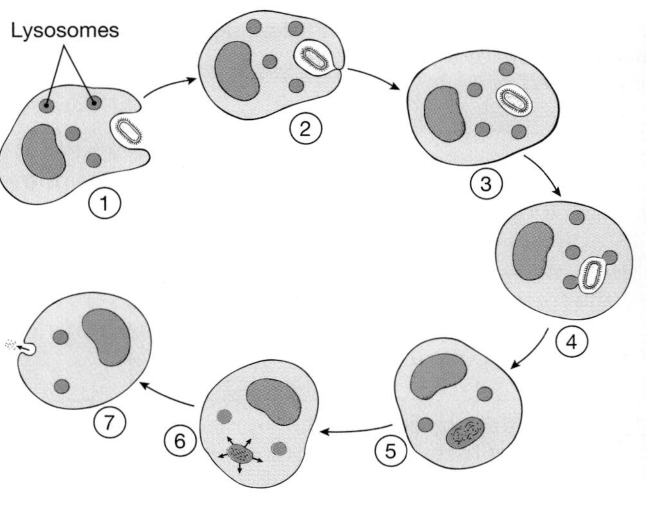
Lysosomes
1
2
3
4
5
6
7

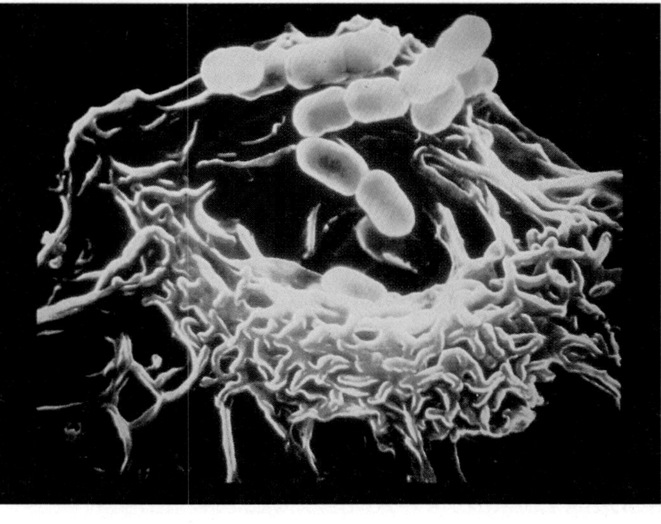

(a) (b)

● **FIGURE 3-13**

Phagocytosis. (a) A phagocytic cell first comes in contact with the foreign object and sends cytoplasmic extensions around it (1). The extensions approach one another (2) and then fuse to trap the material within an endocytic vesicle (3). Lysosomes fuse with this vesicle, activating digestive enzymes that gradually break down the structure of the phagocytized material (4–6). Undissolved residue is then ejected by the cell (7). **(b)** SEM showing a phagocyte engulfing several bacteria.

TABLE 3-2 Summary of Mechanisms Involved in Movement across Cell Membranes

Mechanism	*Process*	*Factors Affecting Rate*	*Substances Involved*
DIFFUSION	Molecular movement of solutes; direction determined by relative concentrations	Size of gradient, molecular size, charge, lipid solubility	Small inorganic ions, lipid-soluble materials (all cells)
Osmosis	Movement of water molecules toward solution containing relatively higher solute concentration; requires membrane	Concentration gradient, opposing osmotic or hydrostatic pressure	Water only (all cells)
FILTRATION	Movement of water, usually with solute, by hydrostatic pressure; requires filtration membrane	Amount of pressure, size of pores in filter	Water and small ions (blood vessels)
CARRIER-MEDIATED TRANSPORT			
Facilitated diffusion	Carrier molecules passively transport solutes down a concentration gradient; requires membrane	As above, plus availability of carrier protein	Glucose and amino acids (all cells)
Active transport	Carrier molecules actively transport solutes regardless of any concentration gradients	Availability of carrier, substrate, and ATP	Na^+, K^+, Ca^{2+}, Mg^{2+} (all cells); other solutes by specialized cells
VESICULAR TRANSPORT			
Endocytosis	Creation of vesicles containing fluid or solid material	Stimulus and mechanics incompletely understood; requires ATP	Fluids, nutrients (all cells); debris, pathogens (specialized cells)
Exocytosis	Fusion of vesicles containing fluids and/or solids with the cell membrane	Stimulus and mechanics incompletely understood; requires ATP	Fluids, debris (all cells)

transmission of information in the nervous system, and thus our perceptions and thoughts. As we shall see in later chapters, changes in the transmembrane potential also trigger the contraction of muscles and the secretion of glands.

THE RESTING POTENTIAL The **resting potential** is the transmembrane potential in an undisturbed cell. Each cell type has a characteristic resting potential between –10 mV and –100 mV. Examples include fat cells (–40 mV), thyroid cells (–50 mV), neurons (–70 mV), skeletal muscle cells (–85 mV), and cardiac muscle cells (–90 mV). The negativity of the normal resting potential primarily reflects the fact that the interior of the cell contains an abundance of negatively charged proteins, which cannot cross the cell membrane.

The cell must expend energy to maintain the resting potential because leak channels permit the entry of sodium ions and the departure of potassium ions. The sodium-potassium exchange pump stabilizes the resting potential by ejecting sodium ions from the cytosol and reclaiming potassium ions from the extracellular fluid. Figure 3-14● provides a diagrammatic view of a cell membrane at its resting potential.

✔ During digestion in the stomach, the concentration of hydrogen (H^+) ions rises to many times the concentration found in the cells of the stomach. What type of transport process could produce this result?

✔ If the cell membrane were freely permeable to sodium (Na^+), how would the transmembrane potential be affected?

✔ When certain types of white blood cells encounter bacteria, they are able to engulf them and bring them into the cell. What is this process called?

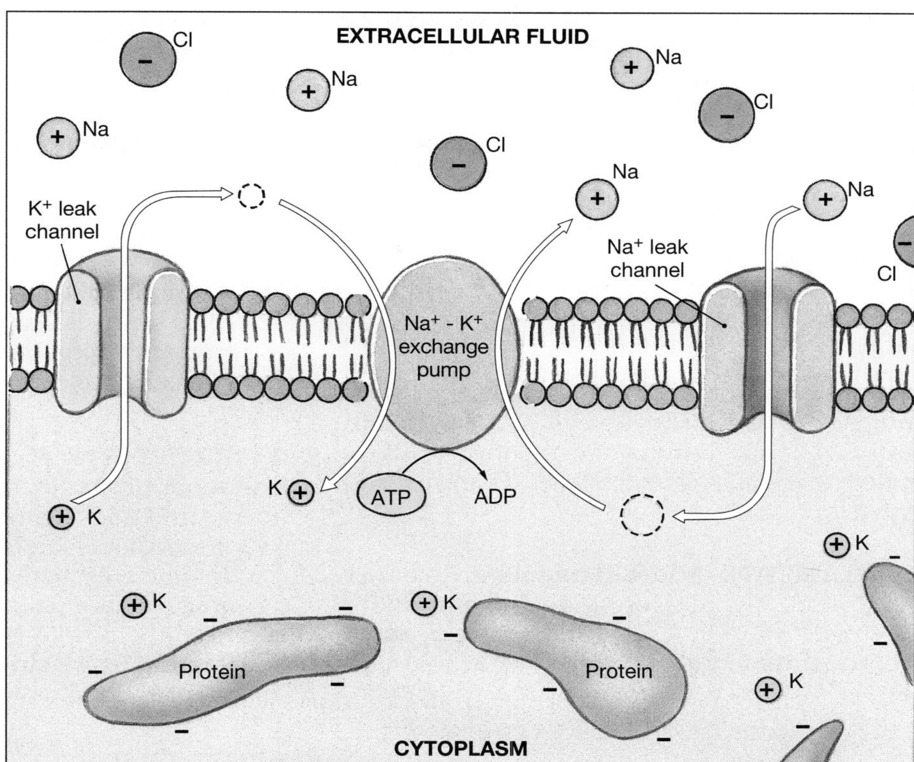
EXTRACELLULAR FLUID
Cl
Na
Na
Na
Cl
Cl
Na
Na
K+ leak channel
Na+ leak channel
Cl
Na+ - K+ exchange pump
K
ATP
ADP
K
K
K
K
K
Protein
Protein
CYTOPLASM

FIGURE 3-14 Ion Distribution across the Cell Membrane. Potassium ions are continually leaking out of the cell through potassium leak channels, and sodium ions are entering through sodium leak channels. At the resting potential, three sodium ions enter the cell for every two potassium ions that leave the cell. These movements are counteracted by the activities of the sodium-potassium exchange pump, which removes three sodium ions from the cytoplasm in exchange for two potassium ions from the extracellular fluid (Figure 3-11). The net result is that the exchange pump maintains a stable ion distribution across the cell membrane.

The Cytoplasm

Cytoplasm is a general term for the material located inside the cell membrane and outside the membrane surrounding the nucleus. Cytoplasm contains many more proteins than does the extracellular fluid, and proteins account for 15–30 percent of the weight of the cell. The cytoplasm includes two major subdivisions:

1. **Cytosol,** or **intracellular fluid.** The cytosol contains dissolved nutrients, ions, soluble and insoluble proteins, and waste products. The cell membrane separates the cytosol from the surrounding extracellular fluid.
2. **Organelles.** Organelles are structures that perform specific functions within the cell.

The Cytosol

The three most important differences between cytosol and extracellular fluid are:

1. The cytosol contains a relatively high concentration of potassium ions, whereas extracellular fluid contains a high concentration of sodium ions.
2. The cytosol contains a relatively high concentration of suspended proteins. Many of the proteins are enzymes that regulate metabolic operations, while others are associated with the various organelles. The cytosol is a *colloid* with a consistency that varies between that of thin maple syrup and almost-set gelatin. [p. 43]
3. The cytosol usually contains relatively small quantities of carbohydrates and large reserves of amino acids and lipids. Carbohydrates are broken down to provide energy, and the amino acids are used to manufacture proteins. Lipids are primarily used as an energy source when carbohydrates are unavailable.

The cytosol may also contain masses of insoluble materials known as **inclusions.** Among the most common inclusions are stored nutrients. For example, glycogen granules in liver or in skeletal muscle cells, and lipid droplets in fat cells.

Organelles

Each organelle performs specific functions that are essential to normal cell structure, maintenance, and metabolism. Cellular organelles can be divided into two broad categories. **Nonmembranous organelles** are always in direct contact with the cytosol. **Membranous organelles** are surrounded by lipid membranes that isolate them from the cytosol, just as the cell membrane isolates the cytosol from the extracellular fluid.

The cell's nonmembranous organelles include the *cytoskeleton, microvilli, centrioles, cilia, flagella,* and *ribosomes.* Membranous organelles include the *mitochondria, the endoplasmic reticulum, the Golgi apparatus, lysosomes,* and *peroxisomes.* (The *nucleus,* which is also surrounded by a membranous envelope and therefore sometimes included with the membranous organelles, will be the focus of a separate section.)

The Cytoskeleton

Figure 3-15

The **cytoskeleton** (Figure 3-15●) is an internal protein framework that gives the cytoplasm strength and flexibility. It has four major components: *microfilaments, intermediate filaments, thick filaments,* and *microtubules.*

MICROFILAMENTS Microfilaments are slender protein strands, usually less than 6 nm in diameter. The most abundant microfilaments are composed of the protein **actin.** In most cells, microfilaments are scattered throughout the cytoplasm, and they also form a dense layer just under the cell membrane.

Microfilaments have two major functions:

1. Microfilaments anchor the cytoskeleton to integral proteins of the cell membrane. They thus provide additional mechanical strength to the cell and attach the cell membrane to the enclosed cytoplasm.
2. Actin microfilaments can interact with intermediate or thick filaments composed of the protein **myosin** to produce active movement of a portion of a cell or to change the shape of the entire cell.

INTERMEDIATE FILAMENTS Intermediate filaments are intermediate in size between microfilaments and thick filaments. The protein composition of intermediate filaments, which range from 7 to 11 nm in diameter, varies from one cell type to another. Intermediate filaments (1) provide strength, (2) stabilize the positions of organelles, and (3) transport materials within the cytoplasm. For example, many cells contain intermediate filaments composed

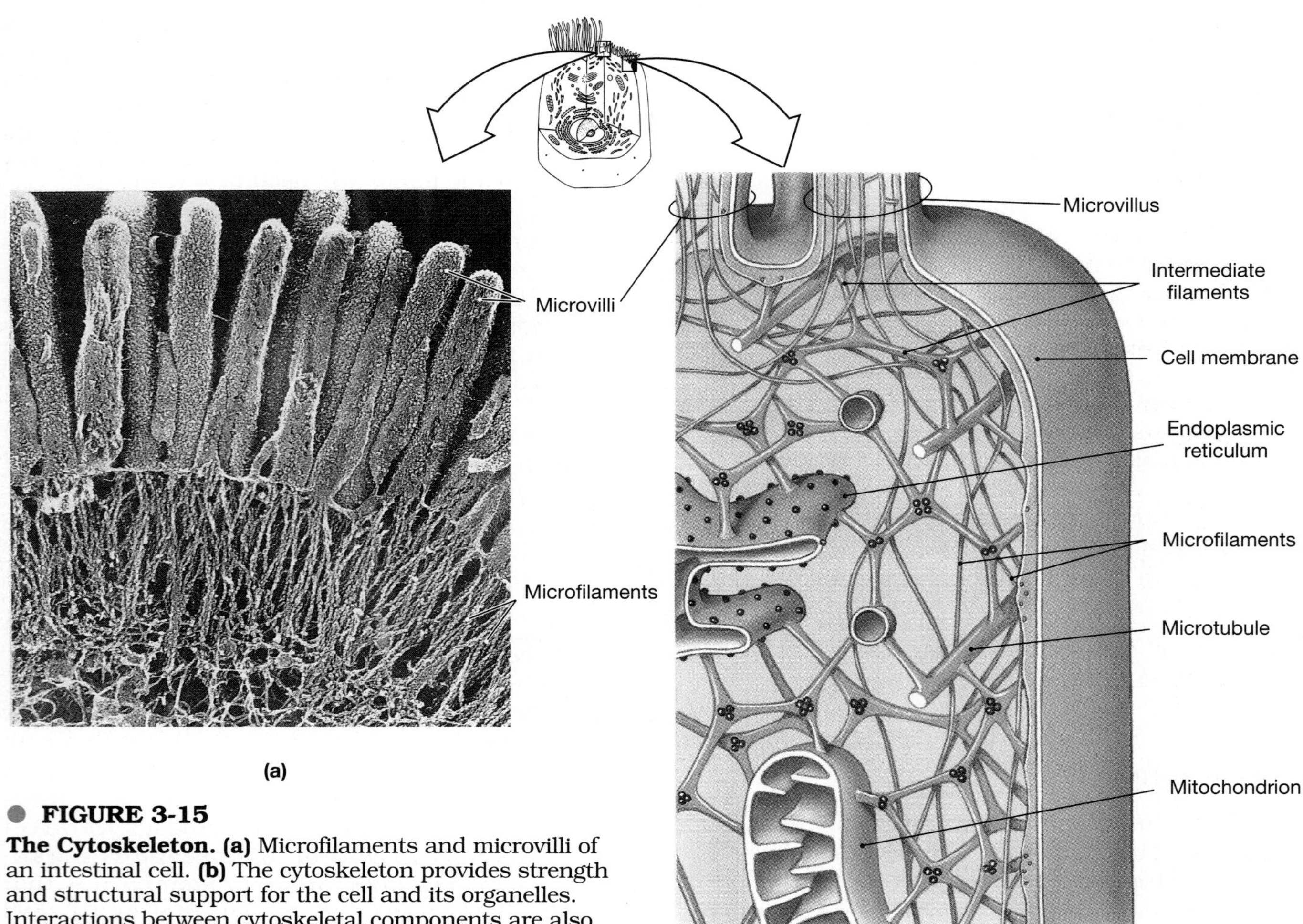
Microvilli
Microfilaments
(a)
Microvillus
Intermediate filaments
Cell membrane
Endoplasmic reticulum
Microfilaments
Microtubule
Mitochondrion
(b)

● **FIGURE 3-15**

The Cytoskeleton. (a) Microfilaments and microvilli of an intestinal cell. **(b)** The cytoskeleton provides strength and structural support for the cell and its organelles. Interactions between cytoskeletal components are also important in moving organelles and changing the shape of the cell.

of the protein **myosin.** Interactions between these filaments and actin filaments can change the shapes of these cells. Neurons contain *neurofilaments,* specialized intermediate filaments that provide structural support and assist in the movement of materials within the cytoplasm.

THICK FILAMENTS Thick filaments are relatively massive bundles of myosin protein subunits up to 15 nm in diameter. Thick filaments are abundant in skeletal and cardiac muscle cells, where they interact with actin filaments to produce powerful contractions.

MICROTUBULES Microtubules, found in all our cells, are hollow tubes built from the globular protein **tubulin.** Microtubules are the largest components of the cytoskeleton, with diameters of around 25 nm. The number and distribution of microtubules within a cell change over time. A microtubule forms through the aggregation of tubulin molecules. It persists for a time and then disassembles into individual tubulin molecules once again.

Microtubules have a variety of functions:

1. Microtubules form the primary components of the cytoskeleton, giving the cell strength and rigidity and anchoring the position of major organelles.
2. Disassembly of microtubules provides a mechanism for changing the shape of the cell, perhaps assisting in cell movement.
3. Microtubules can attach to organelles and other intracellular materials and move them around within the cell.
4. During cell division, microtubules form the *spindle apparatus* that distributes the duplicated chromosomes to opposite ends of the dividing cell. This process will be considered in more detail in a later section.
5. Microtubules form structural components of organelles such as *centrioles, cilia,* and *flagella.*

The cytoskeleton as a whole incorporates microfilaments, intermediate filaments, and microtubules into a network that extends throughout the cytoplasm. The organizational details are as yet poorly understood, because the network is extremely delicate and thus hard to study in an intact state. Figure 3-15● is based on our current knowledge of cytoskeletal structure.

Microvilli

Figure 3-15

Microvilli are small, finger-shaped projections of the cell membrane (Figure 3-15●). Microvilli greatly increase the surface area of the cell exposed to the extracellular environment. They cover the surfaces of cells that are actively absorbing materials from the extracellular fluid, such as the cells lining the digestive tract. A network of microfilaments stiffens each microvillus and anchors it to the underlying cytoskeleton. Interactions between these microfilaments and the cytoskeleton can produce a waving or bending action. Their movements help to circulate fluid around the microvilli, bringing dissolved nutrients into contact with receptors on the membrane surface.

Centrioles, Cilia, and Flagella

Figure 3-16

The cytoskeleton contains numerous microtubules that function individually. Microtubules can also interact to form more complex structures known as *centrioles, cilia,* and *flagella.*

CENTRIOLES A **centriole** (Figure 3-16a●) is a cylindrical structure composed of short microtubules. There are nine groups of microtubules, with three in each group. Each of these nine triplets is connected to its nearest neighbors on either side.

All animal cells that are capable of reproducing themselves contain a pair of centrioles, which direct the movement of DNA strands during cell division (discussed later in this chapter). Centrioles are not found in mature red blood cells, skeletal muscle or cardiac cells, or typical neurons.

The **centrosome,** the cytoplasm surrounding the centrioles, is the heart of the cytoskeletal system. Microtubules of the cytoskeleton usually begin at the centrosome and radiate through the cytoplasm.

CILIA Cilia (singular *cilium*) contain nine pairs of microtubules surrounding a central pair (Figure 3-16b●). Cilia are anchored to a compact **basal body** situated just beneath the cell surface. In section, the organization of microtubules in the basal body resembles that of a centriole. The exposed portion of the cilium is completely covered by the cell membrane. Cilia "beat" rhythmically, as depicted in Figure 3-16c●, and their combined efforts move fluids or secretions across the cell surface. For example, cilia lining the respiratory tract beat in a synchronized manner to move sticky mucus and trapped dust particles toward the throat and away from delicate respiratory surfaces. If the cilia are damaged or immobilized by heavy smoking or some metabolic problem, the cleansing action is lost, and the irritants will no longer be removed. As a result, chronic respiratory infections develop.

FLAGELLA Flagella (fla-JEL-uh; singular *flagellum,* whip) resemble cilia, but they are much longer. Flagella move a cell through the surrounding fluid, rather than moving the fluid past a stationary cell. The sperm cell is the only human cell that has a fla-

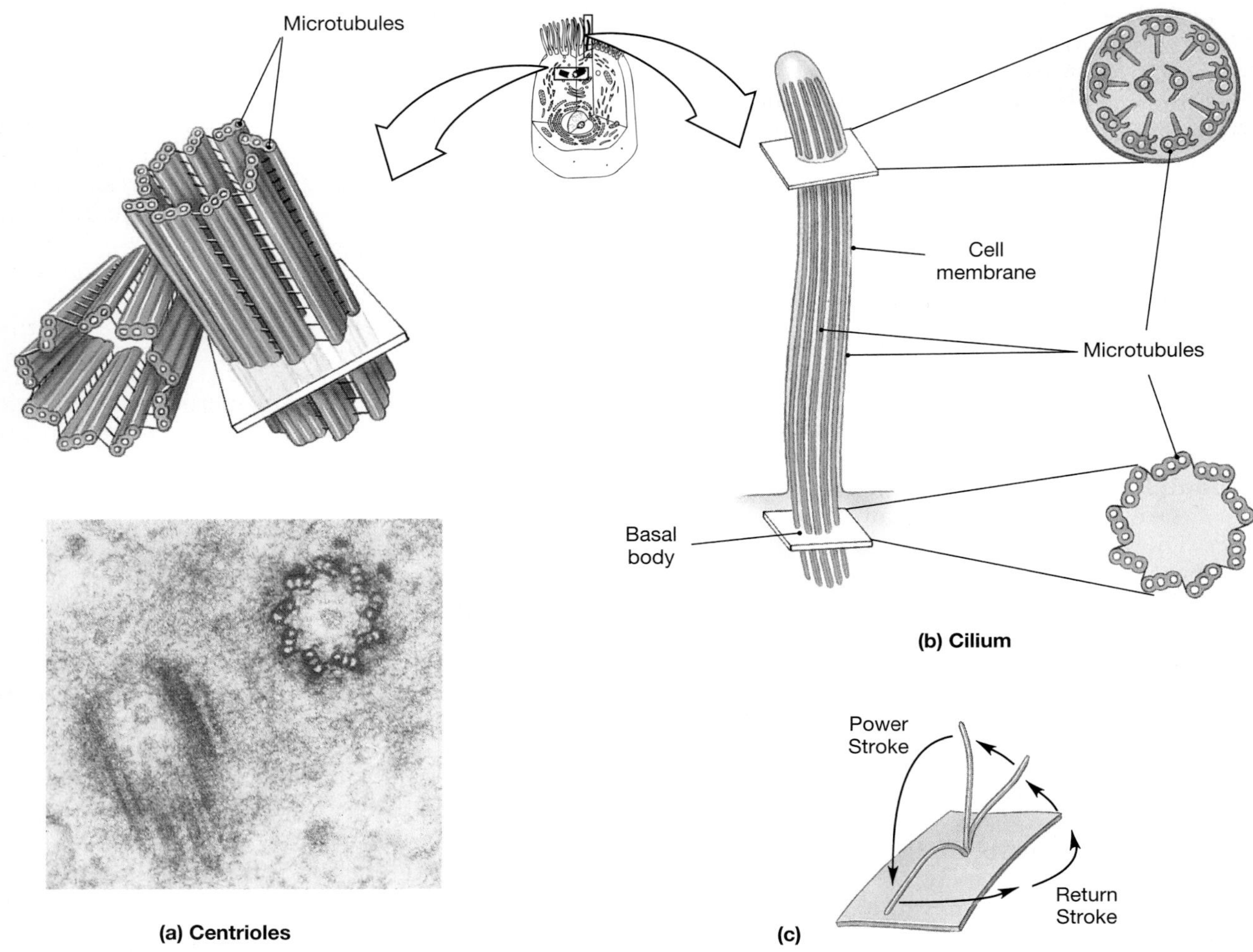
Microtubules
Cell
membrane
Microtubules
Basal
body
(b) Cilium
Power
Stroke
Return
Stroke
(a) Centrioles
(c)

● **FIGURE 3-16**

Centrioles and Cilia. (a) The centrosome contains a pair of centrioles oriented at right angles to one another. Note the correlation between the three-dimensional structure and the sectional view. A centriole consists of nine microtubule triplets (known as a 9 + 0 array). **(b)** A cilium contains nine pairs of microtubules surrounding a central pair (9 + 2 array). The basal body to which the cilium is anchored has a structure similar to that of a centriole. **(c)** A single cilium swings forward and then returns to its original position. During the power stroke the cilium is relatively stiff, but during the return stroke it bends and moves parallel to the cell surface.

gellum, used to move the cell along the female reproductive tract. If sperm flagella are paralyzed or otherwise abnormal, the individual will be sterile because immobile sperm cannot reach and fertilize an egg.

Ribosomes

Figures 3-2, 3-25

Ribosomes are organelles that manufacture proteins, using information provided by the DNA of the nucleus. The number of ribosomes within a particular cell varies depending on the type of cell. For example, liver cells, which manufacture blood proteins, contain far more ribosomes than do fat cells, which synthesize triglycerides.

Ribosomes cannot be seen clearly with the light microscope. In an electron micrograph, ribosomes are dense granules roughly 25 nm in diameter. Each ribosome consists of roughly 60 percent RNA and 40 percent protein.

A functional ribosome consists of two subunits that are normally separate and distinct. The subunits differ in size; one is a **light ribosomal subunit** and the other a **heavy ribosomal subunit** (Figure 3-25●, p. 97). These subunits contain special proteins and **ribosomal RNA,** one of the RNA types introduced in Chapter 2. ∞ [p. 58] Before protein synthesis can begin, a light and a heavy ribosomal subunit must join together with a strand of *messenger RNA* to create a functional ribosome.

There are two major types of functional ribosomes: *free ribosomes* and *fixed ribosomes* (Figure 3-2●, p. 67). **Free ribosomes** are scattered throughout the cytoplasm; the proteins they manufacture enter the cytosol. **Fixed ribosomes** are attached to the *endoplasmic reticulum* (*ER*), a membranous organelle. Proteins manufactured by fixed ribosomes enter the ER, where they are modified and packaged for secretion. Ribosomal structure and functions will be examined in later sections dealing with the endoplasmic reticulum and protein synthesis.

 Cells lining the small intestine have numerous fingerlike projections on their free surface. What are these structures, and what is their function?

How would the absence of a flagellum affect a sperm cell?

Mitochondria

Figure 3-17

Mitochondria (mī-tō-KON-drē-uh; singular *mitochondrion; mitos,* thread + *chondros,* cartilage) are small organelles that can have a variety of shapes, from long and slender to short and fat. The number of mitochondria in a particular cell varies depending on its energy demands. Red blood cells have none, but mitochondria may account for 20 percent of the volume of an active liver cell.

Mitochondria have an unusual double membrane (Figure 3-17a●). An outer membrane surrounds the entire organelle. The inner membrane contains numerous folds, called **cristae,** which increase the surface area exposed to the fluid con-

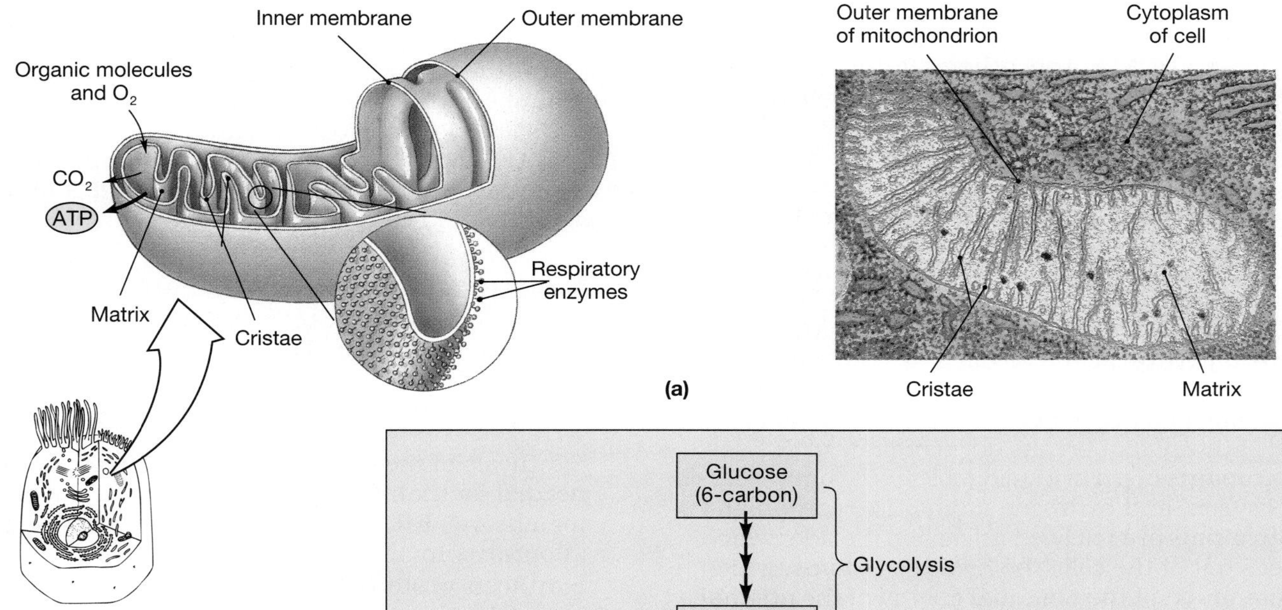

Inner membrane
Outer membrane
Organic molecules and O_2
CO_2
ATP
Matrix
Cristae
Respiratory enzymes
(a) 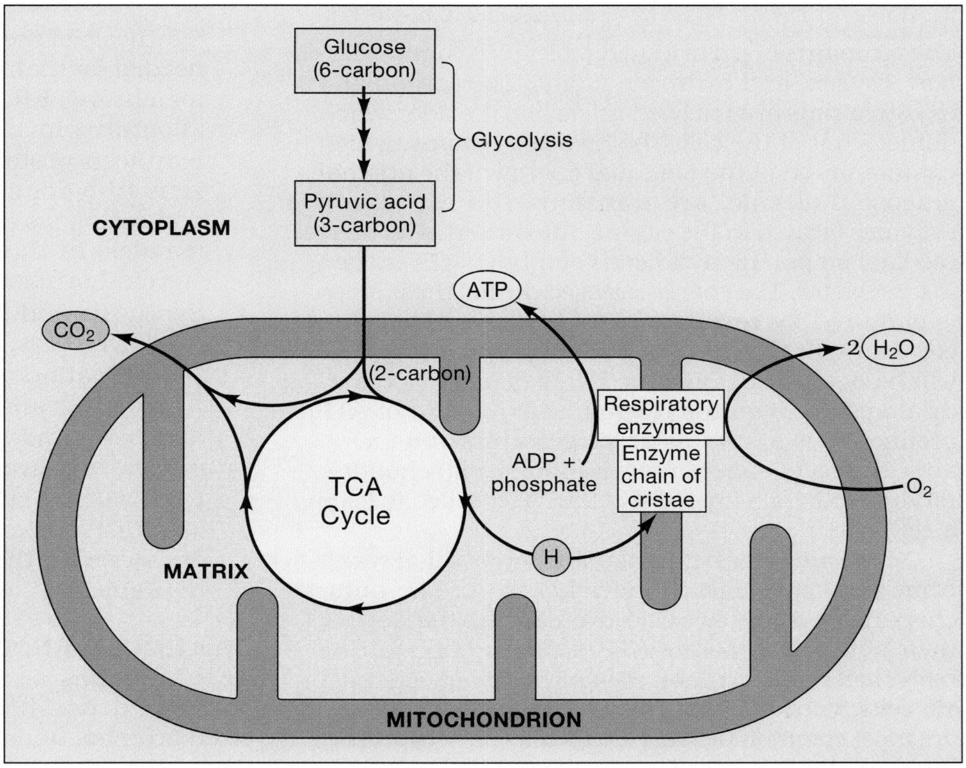
Outer membrane of mitochondrion
Cytoplasm of cell
Cristae
Matrix
Glucose (6-carbon)
Glycolysis
Pyruvic acid (3-carbon)
CYTOPLASM
ATP
CO_2
2 H_2O
(2-carbon)
Respiratory enzymes
ADP + phosphate
Enzyme chain of cristae
O_2
TCA Cycle
H
MATRIX
MITOCHONDRION
(b)

● **FIGURE 3-17**
Mitochondria. (a) This TEM shows a typical mitochondrion in section, and the sketch details its three-dimensional organization. (TEM × 46,332) **(b)** Overview of the role of mitochondria in energy production. Mitochondria absorb short carbon chains (such as pyruvic acid) and oxygen, and generate carbon dioxide and ATP.

tents, or **matrix,** of the mitochondrion. Metabolic enzymes in the matrix perform the reactions that provide energy for cellular functions.

Most of the chemical reactions that release energy occur in the mitochondria, but most of the cellular activities that require energy occur in the surrounding cytoplasm. Cells must therefore store energy in a form that can be moved from place to place. Energy is stored and transferred in the form of *high-energy bonds.* Such a bond usually attaches a phosphate group (PO_4^{3-}) to a suitable molecule, creating a *high-energy compound.* [p. 58] The most important high-energy compound is *adenosine triphosphate,* or *ATP.* Living cells break the high-energy phosphate bond under controlled conditions, reconverting ATP to ADP and phosphate and releasing energy for the cell's use.

MITOCHONDRIAL ENERGY PRODUCTION Most cells generate ATP and other high-energy compounds through the breakdown of carbohydrates, especially glucose. Although most of the ATP production occurs inside mitochondria, the first steps take place in the cytosol. In this reaction sequence, called **glycolysis,** 6-carbon glucose molecules are broken down into 3-carbon molecules of *pyruvic acid.* These molecules are then absorbed by the mitochondria.

Once inside the mitochondrial matrix, the absorbed pyruvic acid molecules are broken into 2-carbon fragments for entry into the **tricarboxylic acid cycle,** or **TCA cycle.** The TCA cycle is an enzymatic pathway that breaks down organic molecules. The remnants of pyruvic acid molecules contain carbon, oxygen, and hydrogen atoms. The carbon and oxygen atoms are released as carbon dioxide, which diffuses out of the cell. The hydrogen atoms, which contain much of the potential energy of the original glucose molecule, are transferred to a chain of enzymes located in the cristae. After a series of intermediate steps, the hydrogens combine with oxygen to form water. The energy released during these steps is captured by **respiratory enzymes** and used to convert ADP to ATP. The entire reaction sequence will be detailed in Chapter 25. Because mitochondrial activity requires oxygen, this method of ATP production is known as **aerobic respiration** (*aero-*, air + *bios,* life). Aerobic respiration in mitochondria produces about 95 percent of the ATP needed to keep a cell alive.

There are several inheritable disorders that result from abnormal mitochondrial activity. The mitochondria involved have defective enzymes that reduce their ability to generate ATP. Cells throughout the body may be affected, but symptoms involving muscle cells, neurons, and the receptor cells in the eye are most common because these cells have especially high energy demands. AM Mitochondrial DNA, Disease, and Evolution

The Endoplasmic Reticulum

Figures 3-18, 3-19

The **endoplasmic reticulum** (en-dō-PLAZ-mik re-TIK-ū-lum), or **ER,** is a network of intracellular membranes that is connected to the *nuclear envelope* surrounding the nucleus. The endoplasmic reticulum has four major functions:

1. ***Synthesis.*** The membrane of the ER synthesizes proteins, carbohydrates, and lipids.
2. ***Storage.*** The ER can hold synthesized molecules or materials absorbed from the cytosol without affecting other cellular operations.
3. ***Transport.*** Materials can travel from place to place in the ER.
4. ***Detoxification.*** Drugs or toxins can be absorbed by the ER and neutralized by enzymes within it.

The endoplasmic reticulum (Figure 3-18•) forms hollow tubes, flattened sheets, and round chambers. The chambers are called **cisternae** (sis-TUR-nē; singular *cisterna,* a reservoir for water). There are two distinct types of endoplasmic reticulum, **smooth endoplasmic reticulum (SER)** and **rough endoplasmic reticulum (RER).**

THE SMOOTH ENDOPLASMIC RETICULUM There are no ribosomes associated with the smooth endoplasmic reticulum. The SER has a variety of functions that center around the synthesis of lipids and carbohydrates. Those functions include:

1. Synthesis of the phospholipids and cholesterol needed for maintenance and growth of the cell membrane, ER, nuclear membrane, and Golgi apparatus in all cells.
2. Synthesis of steroid hormones, such as androgens (the dominant sex hormones in males) and estrogens (the dominant sex hormones in females) in the reproductive organs and the steroid hormones of the adrenal glands.
3. Synthesis and storage of glycerides, especially triglycerides, in liver and fat cells.
4. Detoxification or inactivation of drugs in the SER of liver and kidney cells.
5. Synthesis and storage of glycogen in skeletal muscle and liver cells.
6. Removal and storage of calcium ions (Ca^{2+}) or larger molecules from the cytosol. Calcium ions are stored in the SER of skeletal muscle cells, neurons, and many other cell types.

THE ROUGH ENDOPLASMIC RETICULUM The RER functions as a combination workshop and shipping depot. It is where many newly synthesized proteins undergo chemical modification and where they are packaged for export to their next destination, the *Golgi apparatus.*

The outer surface of the rough endoplasmic reticulum contains fixed ribosomes (Figure 3-18b●). Ribosomes synthesize proteins using instructions provided by a strand of mRNA; the mechanism will be detailed in a later section. ∞ [p. 58] As the polypeptide chains grow, they enter the cisternae of the endoplasmic reticulum. Inside the ER, each protein assumes its secondary and tertiary structure. Some of the proteins are enzymes that will function inside the ER. Other proteins are chemically modified within the ER by the attachment of carbohydrates, creating glycoproteins. Most of the proteins and glycoproteins produced by the RER are packaged into small membranous sacs that pinch off the tips of the cisternae. These **transport vesicles** deliver the products to the Golgi apparatus (Figure 3-19c●).

THE RER AND SER IN SPECIALIZED CELLS The amount of endoplasmic reticulum and the proportion of RER to SER vary depending on the type of cell and its ongoing activities. For example, pancreatic cells that manufacture digestive enzymes contain an extensive RER, but the SER is relatively small. The situation is just the reverse in the cells that synthesize steroid hormones in the reproductive system.

The Golgi Apparatus

Figure 3-19

The **Golgi** (GŌL-jē) **apparatus** (Figure 3-19a,b●) consists of flattened membrane discs, called *saccules.* A typical Golgi apparatus consists of five to six saccules; a single cell may contain several sets, each resembling a stack of dinner plates. Most often these stacks lie near the nucleus of the cell. The Golgi saccules communicate with the ER and with the cell surface through the formation, movement, and fusion of vesicles.

The major functions of the Golgi apparatus are:

1. Synthesis and packaging of secretions, such as mucus or enzymes, for release through exocytosis.
2. Renewal or modification of the cell membrane.
3. Packaging of special enzymes within vesicles for use in the cytosol.

SECRETORY VESICLES AND EXOCYTOSIS Figure 3-19c● diagrams the role of the Golgi apparatus in packaging secretions. Protein and glycoprotein synthesis occur in the RER, and transport vesicles then move these products to the Golgi apparatus. The vesicles usually arrive at a convex saccule known as the *forming face,* which is usually directed toward the nucleus. The transport vesicles then fuse with the Golgi membrane, emptying their contents into the cisterna. Inside the Golgi, enzymes modify the arriving proteins and glycoproteins. For example, the carbohydrate structure of a glycoprotein may be changed, or a phosphate group, sugar, or fatty acid may be attached to a protein.

Material moves from saccule to saccule by means of small **transfer vesicles.** Ultimately the product arrives at the *maturing face,* which is usually oriented toward the cell surface. At the maturing face, vesicles form that carry materials away from the Golgi apparatus. Three different types of vesicles are produced at the Golgi apparatus:

- *Secretory vesicles.* **Secretory vesicles** are vesicles containing secretions that will be discharged from the cell through exocytosis (Figure 3-19d●).

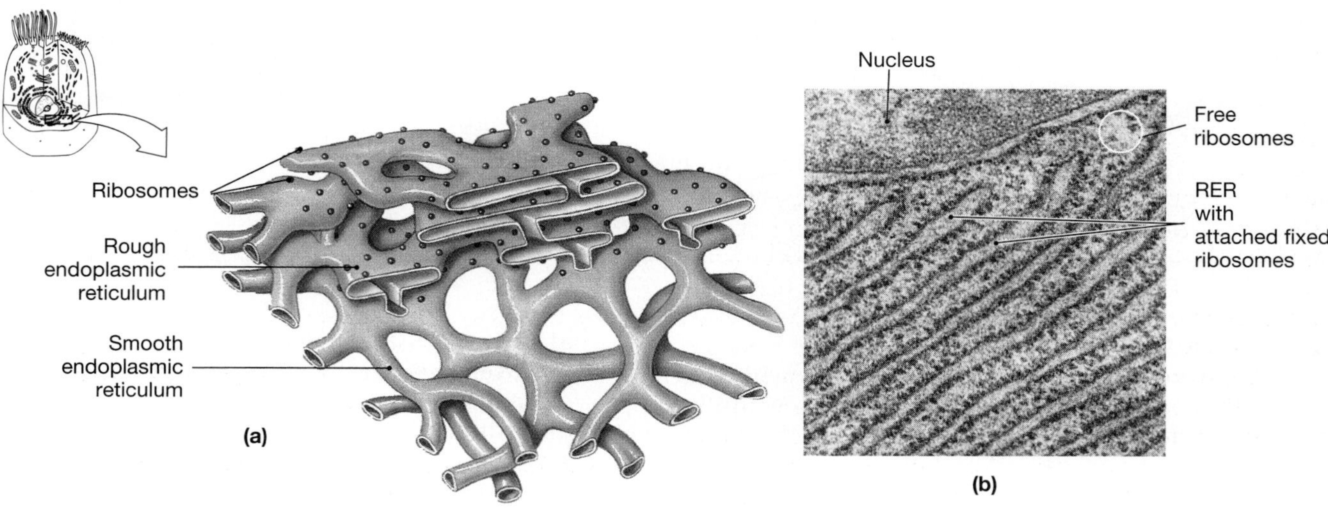
Ribosomes
Rough endoplasmic reticulum
Smooth endoplasmic reticulum
(a)
Nucleus
Free ribosomes
RER with attached fixed ribosomes
(b)

● **FIGURE 3-18**

The Endoplasmic Reticulum. (a) This diagrammatic sketch indicates the three-dimensional relationships between the rough and smooth endoplasmic reticula. **(b)** Rough endoplasmic reticulum and free ribosomes in the cytoplasm of a cell. (TEM × 73,600)

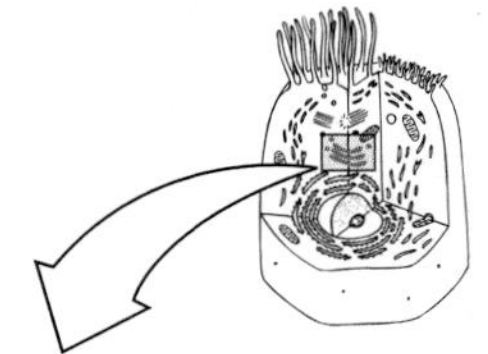

FIGURE 3-19

The Golgi Apparatus. (a) A sectional view of the Golgi apparatus of an active secretory cell. (TEM × 60,280) **(b)** A three-dimensional view of the Golgi apparatus with a cut edge corresponding to part (a). **(c)** The functional link between the ER and the Golgi apparatus. Golgi structure has been simplified to clarify the relationships between the membranes. Transport vesicles carry the secretory product from the endoplasmic reticulum to the Golgi apparatus, and transfer vesicles move membrane and materials between the Golgi saccules. At the maturing face, three functional categories of vesicles develop. Secretory vesicles carry the secretion from the Golgi to the cell surface, where exocytosis releases the contents into the extracellular fluid. Other vesicles add surface area and integral proteins to the cell membrane. Lysosomes and peroxisomes, which remain in the cytoplasm, are vesicles filled with enzymes. **(d)** Exocytosis at the surface of a cell. D

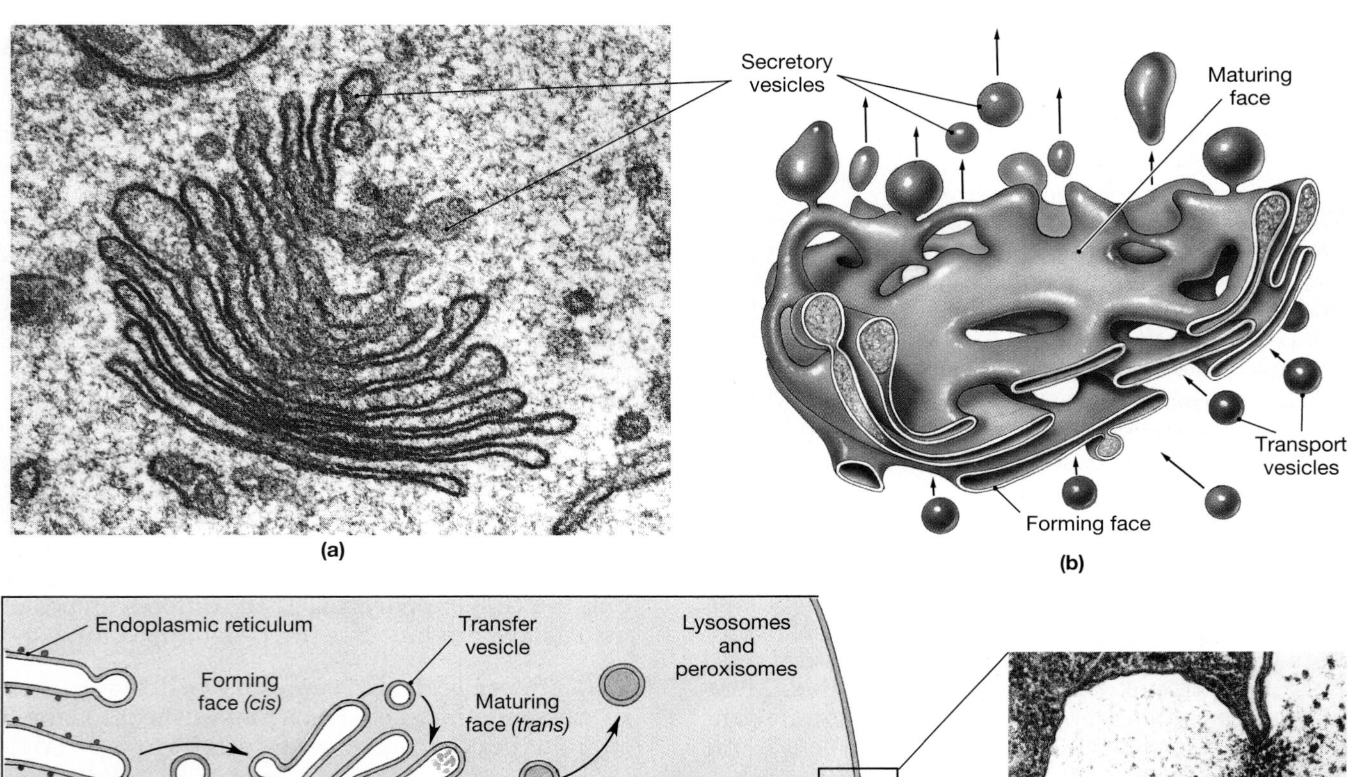
Secretory vesicles
Maturing face
Transport vesicles
Forming face
(a)
(b)

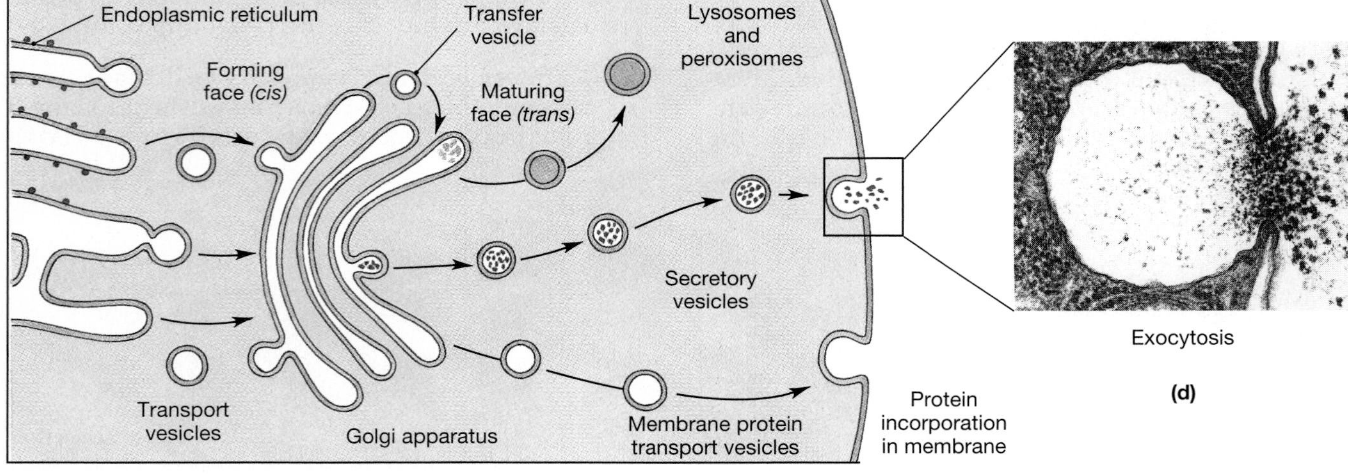
Endoplasmic reticulum
Forming face (cis)
Transfer vesicle
Maturing face (trans)
Lysosomes and peroxisomes
Secretory vesicles
Transport vesicles
Golgi apparatus
Membrane protein transport vesicles
Protein incorporation in membrane
Exocytosis
(c)
(d)

- ***New cell membrane components.*** As vesicles produced at the Golgi apparatus fuse with the surface of the cell, they are adding new lipids and proteins to the cell membrane. At the same time, other areas of the cell membrane are being removed as endocytosis occurs. The Golgi apparatus can thus change the properties of the cell membrane over time. For example, new glycoprotein receptors can be added, making the cell more sensitive to a particular stimulus; alternatively, receptors can be removed, making the cell less sensitive. Such changes can profoundly alter the sensitivity and functions of the cell.
- ***Packaging of intracellular enzymes.*** A third class of vesicles produced at the Golgi apparatus never leaves the cytoplasm. These vesicles, called *lysosomes*, contain digestive enzymes. Their varied functions will be detailed in the next section.

Lysosomes

Figures 3-12, 3-19c, 3-20

Lysosomes (LĪ-sō-sōmz; *lyso-*, dissolution + *soma*, body) are vesicles filled with digestive enzymes. They are produced at the Golgi apparatus (Figure 3-19c•). *Primary lysosomes* contain inactive enzymes. Activation occurs when the lysosome fuses with the membranes of damaged organelles, such as mitochondria or fragments of the endoplasmic reticulum. This fusion creates a *secondary lysosome*, which contains active enzymes. These enzymes then break down the lysosomal contents. Nutrients reenter the cytosol, and the remaining material is eliminated by exocytosis.

Lysosomes also function in the defense against disease. Cells may remove bacteria, as well as liquids and organic debris through endocytosis. (The role of lysosomes in receptor-mediated endocytosis was detailed in Figure 3-12•, p. 80.) Lysosomes fuse with the vesicles created through endocytosis, and the digestive enzymes then break down the contents and release usable substances, such as sugars or amino acids. In this way, the cell both protects itself against pathogenic organisms and obtains valuable nutrients.

Figure 3-20• summarizes lysosomal functions. Lysosomes perform essential cleanup and recycling functions inside the cell. For example, when muscle cells are inactive, lysosomes gradually break down their contractile proteins; if the cells become active once again, this destruction ceases. This regulatory mechanism fails in a damaged or dead cell. Lysosomes then disintegrate, releasing active enzymes into the cytosol. These enzymes rapidly destroy the proteins and organelles of the cell, a process called **autolysis** (aw-TAH-li-sis; *auto-*, self). We do not know how to control lysosomal activities, or why the enclosed enzymes do not digest the lysosomal walls unless the cell is damaged.

Problems with lysosomal enzyme production cause more than 30 serious diseases affecting children. In these conditions, called *lysosomal storage diseases*, the lack of a specific lysosomal enzyme results in the buildup of waste products and debris normally removed and recycled by lysosomes. Affected individuals may die when vital cells, such as those of the heart, can no longer continue to function. AM *Lysosomal Storage Diseases*

Peroxisomes

Peroxisomes are smaller than lysosomes and carry a different group of enzymes. In contrast to lysosomes, which are produced at the Golgi apparatus, peroxisomes probably originate at the RER. Peroxisomes absorb and neutralize toxins, such as alcohol or hydrogen peroxide (H_2O_2), that are absorbed from the interstitial fluid or generated by chemical reactions in the cytoplasm. Peroxisomes are most abundant in liver cells, which are responsible for removing and neutralizing toxins absorbed by the digestive tract. Peroxisomes protect all cells from the potentially damaging effects of *free radicals* produced during normal metabolic reactions. [p. 36] It has been suggested that the cumulative damage produced by free radicals inside and outside our cells is a major factor in the aging process.

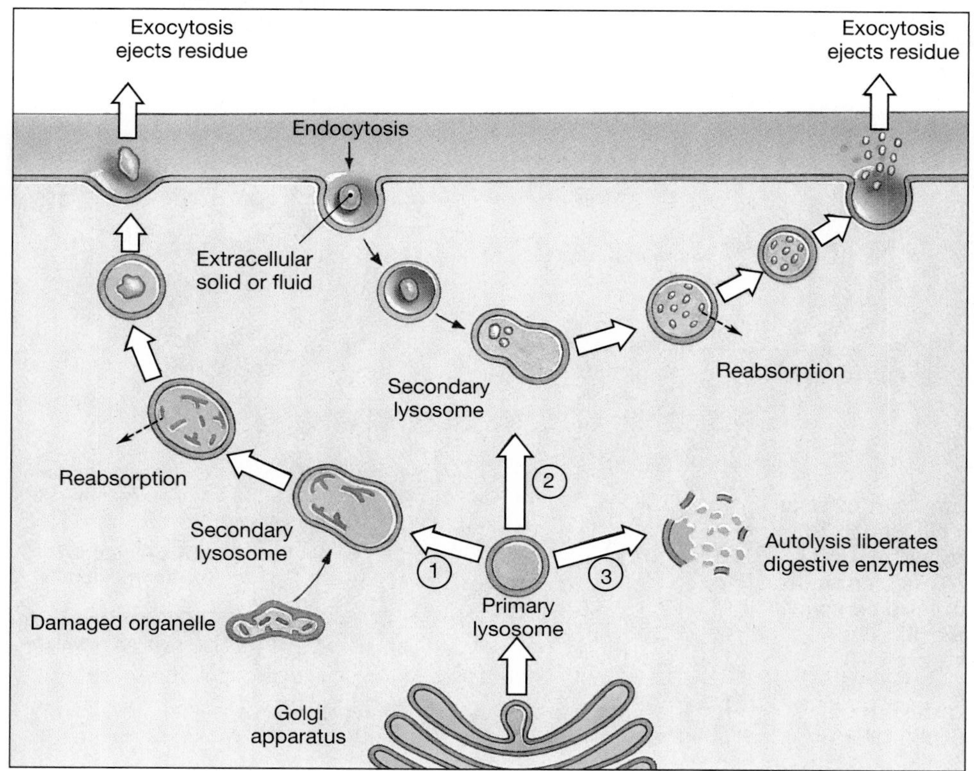

Exocytosis ejects residue
Endocytosis
Exocytosis ejects residue
Extracellular solid or fluid
Secondary lysosome
Reabsorption
Reabsorption
Secondary lysosome
2
1
3
Autolysis liberates digestive enzymes
Primary lysosome
Damaged organelle
Golgi apparatus

• **FIGURE 3-20**
Lysosomal Functions. Primary lysosomes, formed at the Golgi apparatus, contain inactive enzymes. Activation may occur under three basic conditions: (1) when the primary lysosome fuses with the membrane of another organelle, such as a mitochondrion; (2) when the primary lysosome fuses with an endocytic vesicle containing fluid or solid materials from outside the cell; or (3) in autolysis, when the lysosomal membrane breaks down following death or injury to the cell.

Membrane Flow

With the exception of the mitochondria, all the membranous organelles in the cell are either interconnected or in communication through the movement of vesicles. The RER and SER are continuous and connected to the nuclear envelope. Transport vesicles connect the ER with the Golgi apparatus, and secretory vesicles link the Golgi apparatus with the cell membrane. Finally, vesicles forming at the exposed surface of the cell remove and recycle segments of the cell membrane. This continual movement and exchange is called **membrane flow.** In an actively secreting cell, an area equal to the entire membrane surface may be replaced each hour.

Membrane flow is another example of the dynamic nature of cells. It provides a mechanism for cells to change the characteristics of their cell membranes—lipids, receptors, channels, anchors, and enzymes—as they grow, mature, or respond to a specific environmental stimulus.

✔ Microscopic examination of a cell reveals that it contains many mitochondria. What does this observation imply about the cell's energy requirements?

✔ Certain cells in the ovaries and testes contain large amounts of smooth endoplasmic reticulum (SER). Why?

The Nucleus

Figures 3-2, 3-21

The **nucleus** is the control center for cellular operations. A single nucleus stores all the information needed to control the synthesis of the approximately 100,000 different proteins in the human body. The nucleus determines the structural and functional characteristics of the cell by controlling what proteins are synthesized, and in what amounts. A cell without a nucleus could be compared to a car without a driver. However, a car can sit idle for years, but a cell without a nucleus will disintegrate within 3–4 months.

Most cells contain a single nucleus, but there are exceptions. For example, skeletal muscle cells have many nuclei, and mature red blood cells have none. Figure 3-21● details the structure of a typical nucleus. A **nuclear envelope** surrounds the nucleus and separates it from the cytosol. The nuclear envelope is a double membrane containing a narrow **perinuclear space** (*peri-*, around). At several locations, the nuclear envelope is connected to the rough endoplasmic reticulum, as shown in Figure 3-2●, p. 67.

The nucleus directs processes that take place in the cytosol and must in turn receive information about conditions and activities in the cytosol. Chemical communication between the nucleus and

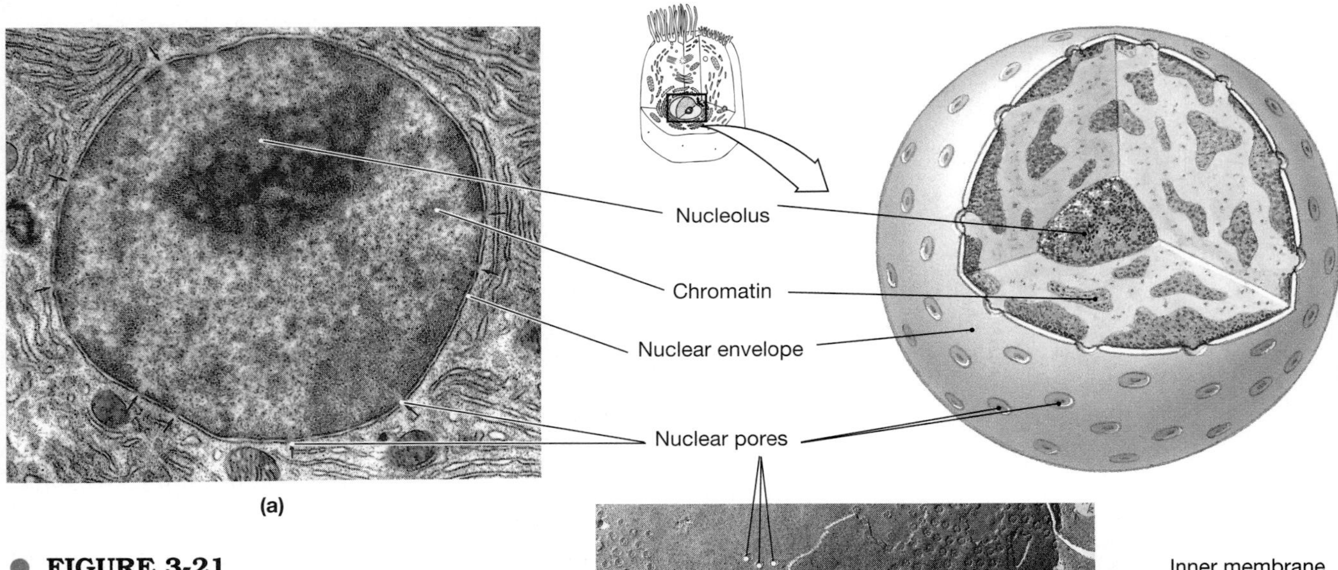
Nucleolus
Chromatin
Nuclear envelope
Nuclear pores
(a)
Inner membrane of nuclear envelope
Broken edge of outer membrane
Outer membrane of nuclear envelope
(b)

● **FIGURE 3-21**
The Nucleus. (a) TEM showing important nuclear structures. (TEM × 4828) **(b)** The cell seen in this SEM was frozen and then broken apart so that internal structures could be seen. This technique, called freeze-fracture, provides a unique perspective on the internal organization of cells. The nuclear envelope and nuclear pores are visible; the fracture broke away part of the outer membrane of the nuclear envelope, and the cut edge can be seen crossing the center of the nucleus. (SEM × 9240)

cytosol occurs through **nuclear pores.** These pores, which cover about 10 percent of the surface of the nucleus, are large enough to permit the movement of ions and small molecules, but too small for the free passage of proteins or DNA.

The term *nucleoplasm* refers to the fluid contents of the nucleus. The nucleoplasm contains ions, enzymes, RNA and DNA nucleotides, proteins, small amounts of RNA, and DNA. The DNA strands form complex structures known as *chromosomes* (*chroma*, color). The nucleoplasm also contains a network of fine filaments, the **nuclear matrix,** that provides structural support and may be involved in the regulation of genetic activity.

❑ Chromosome Structure

Figure 3-22

It is the DNA in the nucleus that stores the vital information, and this DNA is found in complex structures called **chromosomes.** Each chromosome contains DNA strands bound to special proteins called **histones.** The DNA strands coiling around the histones help package the DNA in a small space, and interactions between the DNA and the histones help determine what information is available to the cell at any given moment.

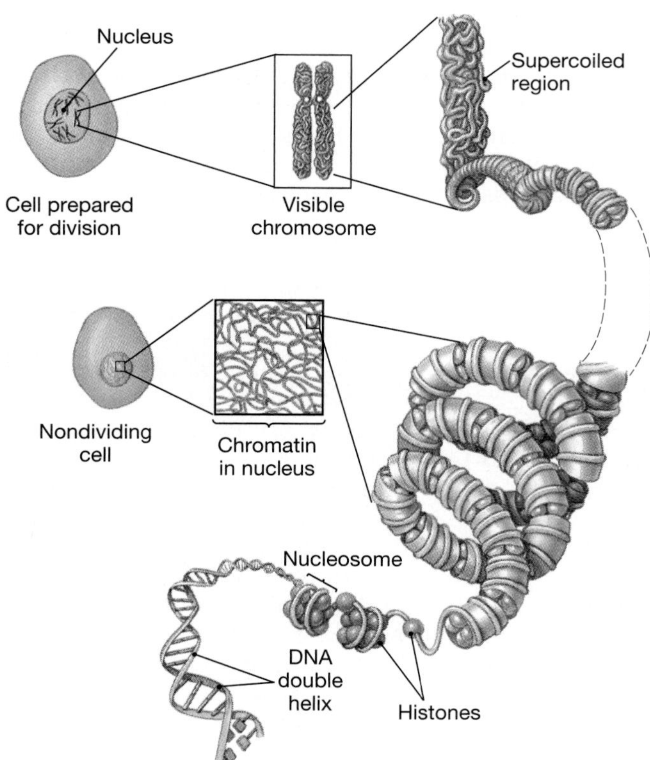
Nucleus
Supercoiled region
Cell prepared for division
Visible chromosome
Nondividing cell
Chromatin in nucleus
Nucleosome
DNA double helix
Histones

● **FIGURE 3-22**
Chromosome Structure. DNA strands are coiled around histones to form nucleosomes. Nucleosomes form coils that may be very tight or rather loose. In cells that are not dividing, the DNA is loosely coiled, forming a tangled network known as chromatin. When the coiling becomes tighter, as it does in preparation for cell division, the DNA becomes visible as distinct structures called chromosomes.

Somatic cell nuclei of humans contain 23 pairs of chromosomes; one member of each pair is derived from the mother and one from the father. The structure of a typical chromosome is diagrammed in Figure 3-22●.

At intervals the DNA strands wind around the histones, forming a complex known as a **nucleosome.** The entire chain of nucleosomes may coil around other proteins. The degree of coiling determines whether the chromosome is long and thin or short and fat. Chromosomes in a dividing cell are very tightly coiled, so they can be seen clearly as separate structures in light or electron micrographs. In cells that are not dividing, the chromosomal material is loosely coiled, forming a tangle of fine filaments known as **chromatin** that gives the nucleus a clumped, grainy appearance.

Most nuclei contain one to four dark-staining areas called **nucleoli** (noo-KLĒ-o-li); singular *nucleolus*). Nucleoli are nuclear organelles that synthesize rRNA and assemble the ribosomal subunits. [p. 58] A nucleolus contains histones and enzymes as well as RNA, and it forms around a chromosomal region that includes the DNA bearing instructions for producing ribosomal proteins and RNA. Nucleoli are most prominent in cells that manufacture large amounts of proteins, such as liver cells and muscle cells, because these cells need large numbers of ribosomes.

❑ The Genetic Code

The **genetic code** is the method of information storage within the DNA strands of the nucleus. An understanding of the genetic code has enabled us to determine how cells build proteins and how various structural and functional characteristics are inherited from generation to generation. We are now beginning to experiment with the manipulation of the genetic information in human cells, and the techniques may revolutionize the treatment of a variety of serious diseases.

The basic structure of nucleic acids was described in Chapter 2. [pp. 56–58] A single DNA molecule consists of a pair of strands held together by hydrogen bonding between complementary nitrogenous bases. *Information is stored in the sequence of nitrogenous bases along the length of the DNA strands.* Those nitrogenous bases are *adenine*, A; *thymine*, T; *cytosine*, C; and *guanine*, G.

The genetic code is called a *triplet code* because a sequence of three nitrogenous bases can specify the identity of a single amino acid. Each **gene** consists of all the triplets needed to produce a specific peptide chain. The number of triplets in a gene varies, depending on the size of the peptide represented. A relatively short peptide chain might require fewer than a hundred triplets, whereas the instructions for building a large protein might involve a thousand or more.

Each gene also contains segments responsible for regulating its activity. In effect these are triplets that say "do (or do not) read this message," "message starts here," and "message ends here." The "read me," "don't read me," and "start here" signals form a special control segment at the start of each gene.

DNA FINGERPRINTING Every nucleated somatic cell in the body carries an identical set of 46 chromosomes. Not all of the DNA of these chromosomes codes for proteins, however, and a significant percentage of DNA segments have no known function. Some of the "useless" segments contain the same nucleotide sequence repeated over and over. The number of segments and the number of repetitions vary from individual to individual. The chances of any two individuals, other than identical twins, having the same pattern of repeating segments is less than one in 9 billion. In other words, it is extremely unlikely that you will ever encounter someone else who has the same pattern of repeating nucleotide sequences present in your DNA.

Individual identification can therefore be made on the basis of a DNA pattern analysis, just as it can on the basis of a fingerprint. Skin scrapings, blood, semen, hair, or other tissues can be used as a sample source. Information from **DNA fingerprinting** has already been used to convict (and to acquit) persons accused of committing violent crimes, such as rape or murder.

❑ Gene Activation and Protein Synthesis

Each DNA strand contains thousands of individual genes. These genes are normally tightly coiled, and histones bound to the control segments prevent their activation. Before a specific gene is activated, enzymes must temporarily disrupt the weak bonds between the nitrogenous bases and detach the histone guarding the control segment. The regulation of this process is only partially understood. Another enzyme, **RNA polymerase,** then binds to the initial segment of the gene.

The events that follow can be divided into two stages:

1. ***Transcription.*** RNA polymerase uses the genetic information to assemble a strand of mRNA.
2. ***Translation.*** Ribosomes use the information carried by the mRNA strand to assemble functional proteins.

❑ Transcription

Figure 3-23

Activated genes do not leave the nucleus, nor do they lose their connections with other genes along the length of the DNA strand. Instead, a messenger carries the instructions from the nucleus to the cytoplasm. The carrier is a single strand of **messenger RNA (mRNA).** The process of mRNA formation is called **transcription,** because the mRNA is in a sense "taking notes" from the gene. Figure 3-23• details this process.

Step 1. Once the DNA strands have separated and the control segment has been exposed, transcription can begin. The key event is the attachment of RNA polymerase to the gene. RNA polymerase promotes hydrogen bonding between the nitrogenous bases of the gene and complementary nucleotides in the nucleoplasm. The nucleotides involved are those characteristic of RNA, not of DNA; RNA polymerase may attach adenine, guanine, cytosine, or uracil, but never thymine. Thus, wherever an A occurs in the DNA strand, the polymerase will attach a U rather than a T.

Step 2. The RNA polymerase interacts with only a small portion of the gene at any one time as it travels along the DNA strand. Moving from triplet to triplet, the enzyme collects additional nucleotides and attaches them to the growing chain. In this way it assembles a complete strand of mRNA.

Step 3. At the "stop here" command, the enzyme and the mRNA strand detach from the DNA strands, and transcription ends. The complementary DNA strands now reassociate.

Each gene includes a number of noncoding, or "nonsense," triplets that do not contain information needed to build a functional protein. As a result, the mRNA strand assembled during transcription, sometimes called immature mRNA or *pre-mRNA,* must be "edited" before it leaves the nucleus to direct protein synthesis. In this **RNA processing,** the nonsense regions, called *introns,* are snipped out, and the remaining segments, or *exons,* are spliced together. This process creates a much shorter, functional strand of mRNA that then enters the cytoplasm via one of the nuclear pores.

❑ Translation

Figures 3-24, 3-25

Translation is the construction of a functional polypeptide using the information provided by an mRNA strand. To understand this process you must follow the "message" from the gene to the mRNA strand to the finished peptide chain. The message is spelled out in triplets: A sequence of three DNA nucleotide bases stands for one amino acid in the

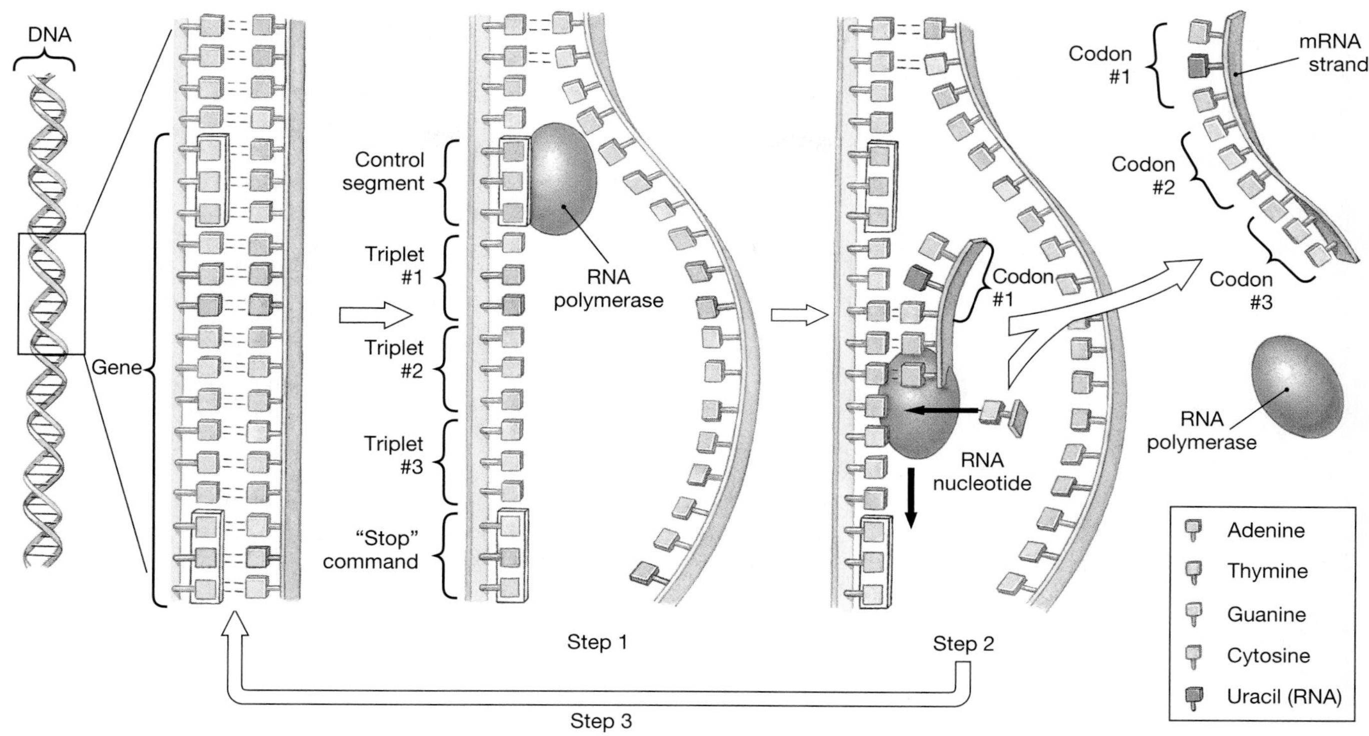
DNA
Gene
Control segment
Triplet #1
Triplet #2
Triplet #3
"Stop" command
RNA polymerase
Step 1
Codon #1
RNA nucleotide
Step 2
Codon #1
Codon #2
Codon #3
mRNA strand
RNA polymerase
Step 3
Adenine
Thymine
Guanine
Cytosine
Uracil (RNA)

● **FIGURE 3-23**

Transcription. This diagram shows a small portion of a single DNA molecule, containing a single gene available for transcription. Step 1: The two DNA strands separate, and RNA polymerase binds to the control segment of the gene. Step 2: The RNA polymerase moves from one nucleotide to another along the length of the gene. At each site, complementary RNA nucleotides form hydrogen bonds with the DNA nucleotides of the gene. The RNA polymerase then strings the arriving nucleotides together into a strand of mRNA. Step 3: Upon reaching the stop signal at the end of the gene, the RNA polymerase and the mRNA strand detach, and the two DNA strands reassociate.

finished polypeptide. In transcription, the nucleotide sequence of the gene determines the nucleotide sequence of the mRNA strand. Thus each DNA triplet will correspond to a sequence of three nucleotide bases in the mRNA strand. Such a three-base sequence is called a **codon** (KŌ-don). Codons contain nitrogenous bases that are complementary to those of the triplets in the gene. For example, if the DNA triplet is TCG, the corresponding codon on the mRNA strand will be AGC.

Each mRNA codon designates a particular amino acid to be incorporated into the polypeptide chain; Figure 3-24● includes several examples. *During translation the sequence of codons will determine the sequence of amino acids in the polypeptide.* The amino acids are provided by a relatively small and mobile type of RNA, called **transfer RNA (tRNA)** (Figure 3-24●). Each tRNA binds and delivers an amino acid of a specific type. There are more than 20 different kinds of transfer RNA, at least one for each of the amino acids used in protein synthesis.

At one place in its structure, each kind of tRNA molecule has a different trio of nitrogenous bases, known as an **anticodon.** The base sequence of the anticodon indicates the type of amino acid carried by the tRNA. For example, a tRNA with the anticodon GGG always carries the amino acid *proline,* whereas a tRNA with the anticodon CGG carries *alanine.* Figure 3-24● includes examples of other codons and anticodons that specify individual amino acids, and summarizes the relationships among triplets, codons, and anticodons.

The tRNA molecules thus provide the physical link between codons and amino acids. During translation, each codon along the mRNA strand will bind a complementary anticodon on a tRNA molecule. Thus, if the mRNA has the codons (CCC)-(GCC)-(UUA) it will bind to tRNAs with anticodons (GGG)-(CGG)-(AAU). *The amino acid sequence of the peptide chain created is determined by the sequence delivered by the tRNAs.* So in this case the amino acid sequence in the peptide would be proline-alanine-leucine.

Translation comprises three phases: *initiation,* which begins the process; *elongation,* which produces the peptide chain; and *termination,* which ends the process and releases the completed peptide. These phases are diagrammed in Figure 3-25●.

Initiation

Step 1. Initiation begins as the mRNA strand binds to a light ribosomal subunit. The first codon of

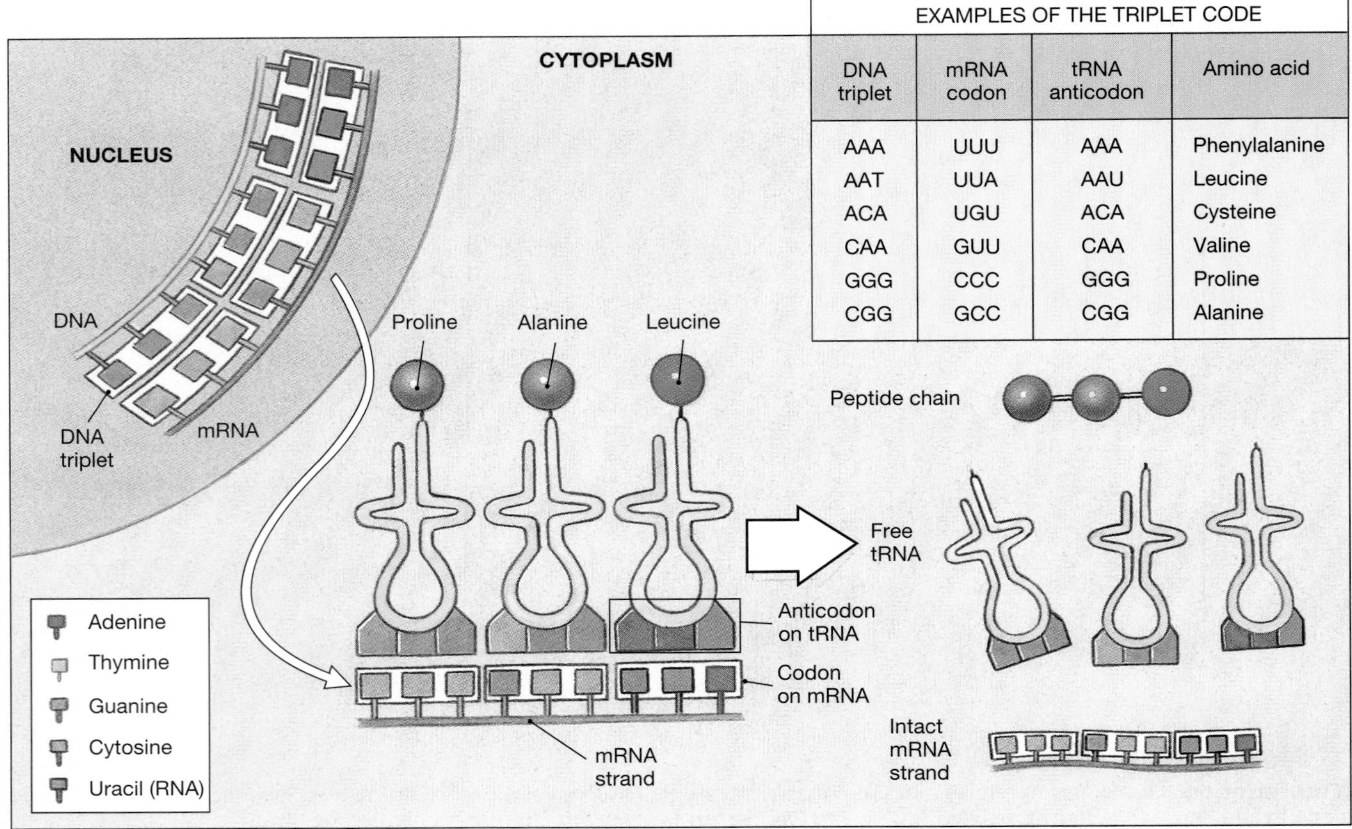
CYTOPLASM
NUCLEUS
DNA
DNA triplet
mRNA
Proline
Alanine
Leucine
Peptide chain
Free tRNA
Anticodon on tRNA
Codon on mRNA
mRNA strand
Intact mRNA strand
Adenine
Thymine
Guanine
Cytosine
Uracil (RNA)

EXAMPLES OF THE TRIPLET CODE

DNA triplet	mRNA codon	tRNA anticodon	Amino acid
AAA	UUU	AAA	Phenylalanine
AAT	UUA	AAU	Leucine
ACA	UGU	ACA	Cysteine
CAA	GUU	CAA	Valine
GGG	CCC	GGG	Proline
CGG	GCC	CGG	Alanine

● **FIGURE 3-24**
An Overview of Protein Synthesis. An mRNA strand contains codons that are complementary to triplets on the DNA strand. Molecules of tRNA contain anticodons that are complementary to the codons. Different amino acids are delivered by tRNAs with different anticodons, and examples are indicated. The sequence of amino acids in the completed peptide will reflect the sequence of tRNA arrival.

the mRNA strand, or *start codon*, always has the base sequence AUG. It therefore binds a tRNA with the complementary anticodon sequence UAC. This tRNA carries the amino acid *methionine*. (The methionine will ultimately be removed from the finished protein.)

Step 2. When this binding occurs, a heavy ribosomal subunit joins the complex to create a complete ribosome. The mRNA strand nestles in the gap between the light and heavy subunits.

Elongation

Step 3. A second tRNA now arrives at the adjacent site of the ribosome, and its anticodon binds to the next codon of the mRNA strand.

Step 4. Enzymes of the heavy ribosomal subunit then break the linkage between the tRNA molecule and its amino acid. At the same time, they attach the amino acid to its neighbor by means of a peptide bond. The ribosome then moves one codon down the mRNA strand.

Step 5. The cycle is then repeated with the arrival of another molecule of tRNA. The tRNA already stripped of its amino acid drifts away. It will soon bind to another amino acid and be ready to participate in protein synthesis at a later time.

Termination

Step 6. Elongation continues, adding amino acids to the growing polypeptide chain, until the ribosome reaches a "stop here" signal, or *stop codon*, at the end of the mRNA strand. The ribosomal subunits now detach, leaving an intact strand of mRNA and a completed polypeptide.

Translation proceeds swiftly, producing a typical protein in around 20 seconds. The mRNA strand remains intact, and it can interact with other ribosomes to create additional copies of the same peptide chain. The process does not continue indefinitely, however, because mRNA strands are broken down and the nucleotides recycled after a few minutes, or at most a few hours.

Polyribosomes

Figure 3-26

During translation, only two mRNA codons are "read" by a ribosome at any one time, although the entire

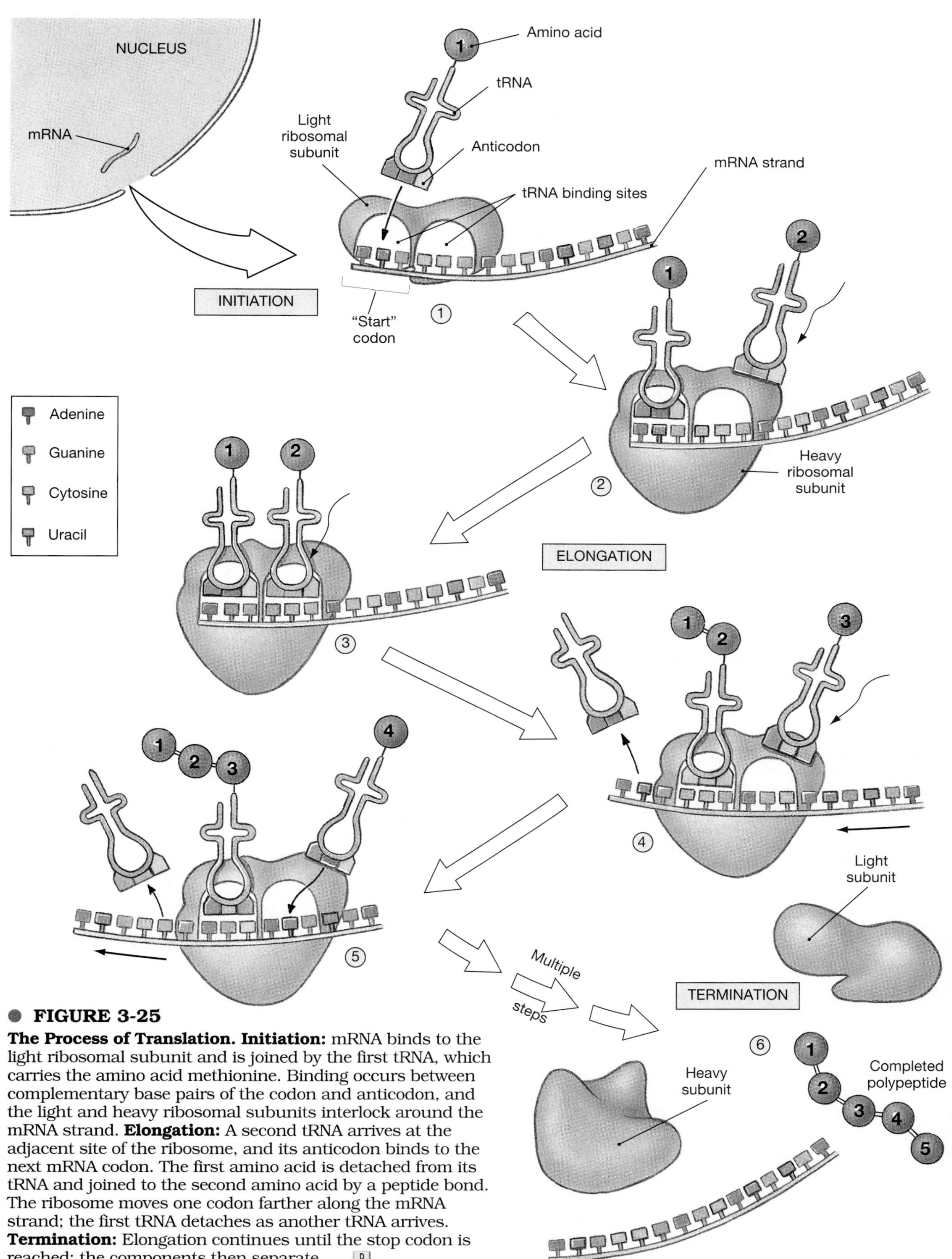

NUCLEUS
mRNA
Amino acid
tRNA
Light ribosomal subunit
Anticodon
mRNA strand
tRNA binding sites
INITIATION
"Start" codon
1
Adenine
Guanine
Cytosine
Uracil
2
Heavy ribosomal subunit
ELONGATION
3
4
5
Light subunit
Multiple steps
TERMINATION
6
Heavy subunit
Completed polypeptide

FIGURE 3-25

The Process of Translation. Initiation: mRNA binds to the light ribosomal subunit and is joined by the first tRNA, which carries the amino acid methionine. Binding occurs between complementary base pairs of the codon and anticodon, and the light and heavy ribosomal subunits interlock around the mRNA strand. **Elongation:** A second tRNA arrives at the adjacent site of the ribosome, and its anticodon binds to the next mRNA codon. The first amino acid is detached from its tRNA and joined to the second amino acid by a peptide bond. The ribosome moves one codon farther along the mRNA strand; the first tRNA detaches as another tRNA arrives. **Termination:** Elongation continues until the stop codon is reached; the components then separate.

strand may contain thousands of codons. As a result, there is ample room for many ribosomes to bind to a single mRNA strand. At any given moment, each will be reading a different part of the same message, but each will end up constructing a copy of the same protein. The arrangement is similar to a line of people at a buffet lunch—each person will assemble the same meal, but always a step behind the person ahead. A series of ribosomes attached to the same mRNA strand is called a **polyribosome,** or *polysome* (Figure 3-26•). Polyribosome formation greatly increases the rate of protein synthesis.

MUTATIONS **Mutations** are permanent alterations in a cell's DNA that affect the nucleotide sequence of one or more genes. The simplest is a **point mutation,** a change in a single nucleotide that affects one codon. The triplet code has some flexibility because several different codons can specify the same amino acid. But a point mutation that produces a codon that specifies a different amino acid will usually change the structure of the completed protein. A single change in the amino acid sequence of a structural protein or enzyme can prove fatal. Several cancers and two potentially lethal blood disorders, *thalassemia* and *sickle cell anemia,* result from variations in a single nucleotide. [p. 57]

More than 100 inherited disorders have been traced to abnormalities in enzyme or protein structure that reflect single alterations in nucleotide sequence. More elaborate mutations, such as additions or deletions of nucleotides, can affect multiple codons within a gene, several adjacent genes, or the structure of one or more chromosomes.

Because mutations most often occur during DNA replication, they are most likely to involve cells undergoing cell division. A single cell, a group of cells, or an entire individual may be affected. This last will be the case if the changes occur early in development. For example, a mutation affecting the DNA of an individual's gametes (sperm or eggs) will be inherited by that individual's children. Our understanding of genetic structure is opening the possibility for diagnosing and correcting some of these problems. For a discussion of the principles and technologies involved, see the *Applications Manual.* AM *Genetic Engineering and Gene Therapy*

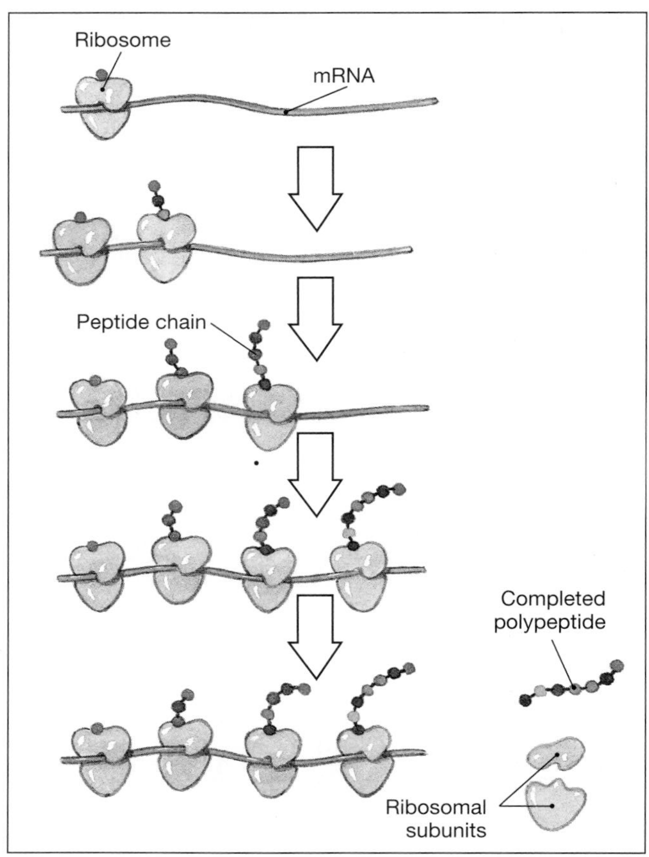
Ribosome
mRNA
Peptide chain
Completed polypeptide
Ribosomal subunits

● **FIGURE 3-26**
Polyribosomes. Each polyribosome consists of a strand of mRNA that is being read simultaneously by a number of ribosomes. In this way a single strand of mRNA can produce many polypeptide molecules in a short time.

❑ DNA and the Control of Cell Structure and Function

The DNA of the nucleus controls the cell by directing the synthesis of specific proteins. Through control of protein synthesis virtually every aspect of cell structure and function can be regulated. There are two levels of control involved:

1. The DNA of the nucleus has *direct* control over the synthesis of structural proteins, such as cytoskeletal components, membrane proteins (including receptors), and secretory products. By issuing appropriate orders, the nucleus can alter the internal structure of the cell, its sensitivity to substances in its environment, or its secretory functions to meet changing circumstances.
2. The DNA of the nucleus has *indirect* control over all other aspects of cellular metabolism because it regulates the synthesis of enzymes. By ordering the production of appropriate enzymes, the nucleus can regulate all the metabolic activities and functions of the cell. For example, the nucleus can accelerate the rate of glycolysis by increasing the number of needed enzymes in the cytoplasm.

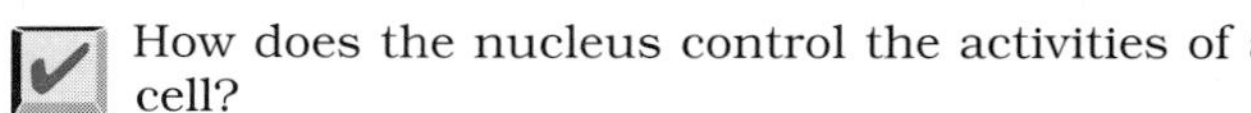
How does the nucleus control the activities of a cell?

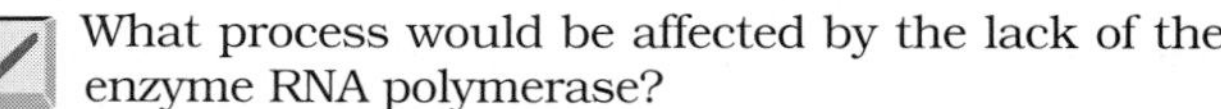
What process would be affected by the lack of the enzyme RNA polymerase?

During the process of DNA replication, a nucleotide was deleted from a sequence that coded for a polypeptide. What effect would this deletion have on the amino acid sequence of the polypeptide?

The Cell Life Cycle

Figure 3-27

Between fertilization and physical maturity a human being goes from a single cell to roughly 75 trillion cells. This amazing increase in size and complexity involves a form of cellular reproduction called **cell division.** The division of a single cell produces a pair of daughter cells, each half the size of the original. These cells will grow until each reaches the size of the original cell, when they will usually divide in turn.

Even when development has been completed, cell division continues to be essential to survival. Although cells are highly adaptable, they can be damaged by physical wear and tear, toxic chemicals, temperature changes, or other environmental stresses. In addition, cells, like individuals, are subject to aging. The life span of a cell varies from hours to decades, depending on the type of cell and the environmental stresses involved. Many cells appear to be "programmed" to self-destruct after a certain period of time. Their destruction results from the activation of specific "suicide genes" in the nucleus. The genetically controlled death of cells is called **apoptosis** (ap-op-TŌ-sis; *ptosis*, a falling away). A gene involved in the regulation of this process has been identified. This gene, called *bcl-2*, appears to prevent apoptosis and keep a cell alive and functional. If something interferes with the function of this gene, the cell self-destructs.

Because a typical cell does not live nearly as long as a typical person, cell populations must be maintained over time by cell division. Central to cell reproduction is the accurate duplication of the cell's genetic material and its distribution to the two new daughter cells formed by division. This process of nuclear division is called *mitosis* (mī-TŌ-sis). Mitosis occurs during the division of somatic cells. Production of sperm and ova involves a distinct process, *meiosis* (mī-Ō-sis), that will be described in Chapter 28.

Figure 3-27• presents the life cycle of a typical cell in greater detail. That life cycle includes a relatively brief period of mitosis preceded and followed by an *interphase* period of variable duration.

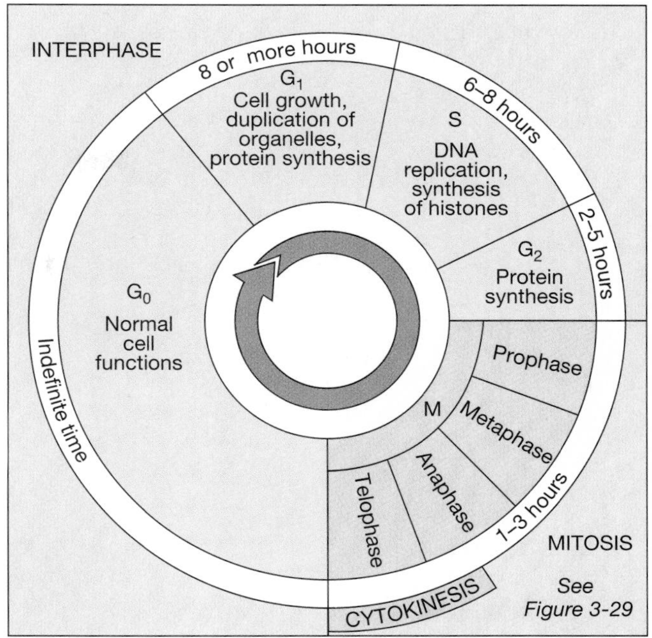
INTERPHASE
8 or more hours
G_1
Cell growth, duplication of organelles, protein synthesis
6–8 hours
S
DNA replication, synthesis of histones
2–5 hours
G_2
Protein synthesis
G_0
Normal cell functions
Indefinite time
Prophase
M
Metaphase
Anaphase
Telophase
1–3 hours
MITOSIS
See Figure 3-29
CYTOKINESIS

● **FIGURE 3-27**
The Cell Life Cycle

Interphase

Most cells spend only a small part of their time actively engaged in cell division. Somatic cells spend the majority of their functional lives in a state known as **interphase.** During interphase a cell performs all of its normal functions plus, if necessary, it makes preparations for cell division. Interphase can be divided into the G_0, G_1, S, and G_2 phases.

The G_0 Phase

An interphase cell in the **G_0 phase** is not preparing for mitosis, but is performing all other normal cell functions. Some mature cells, such as skeletal muscle cells and many neurons, remain in G_0 indefinitely and may never undergo mitosis. In contrast, *stem cells*, which divide repeatedly with very brief interphase periods, bypass the G_0 phase.and proceed directly to the G_1.

The G_1 Phase

A cell that is going to divide first enters the **G_1 phase.** In this phase the cell manufactures enough mitochondria, cytoskeletal elements, centrioles, endoplasmic reticulum, ribosomes, Golgi membranes, and cytosol to make two functional cells. In cells dividing at top speed, G_1 may last as little as 8–12 hours. Such cells pour all of their energy into mitosis, and all other activities cease. If G_1 lasts for days, weeks, or months, preparation for mitosis occurs as the cells perform their normal functions.

The S Phase and DNA Replication

Figure 3-28

When the activities of G_1 have been completed, the cell enters the **S phase.** Over the next 6–8 hours the cell duplicates its chromosomes. In this process, it copies its DNA and combines it with histones and other proteins in the nucleus.

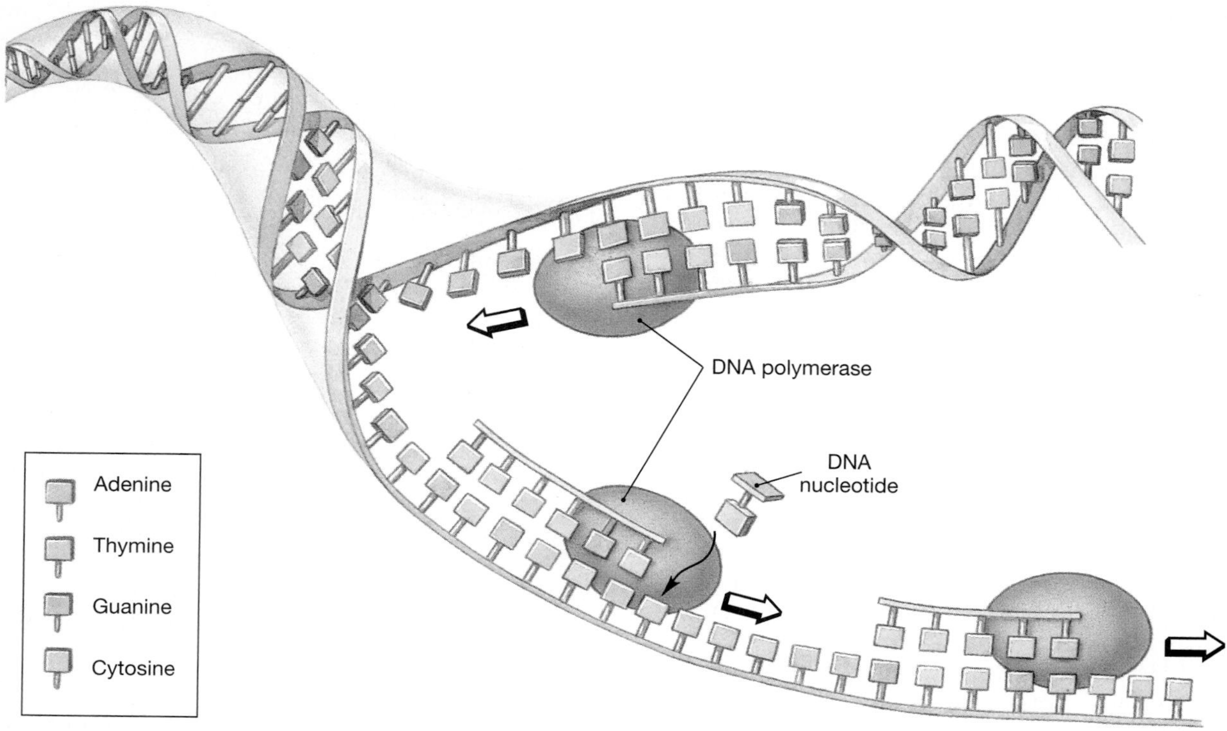
DNA polymerase
DNA nucleotide
Adenine
Thymine
Guanine
Cytosine

● **FIGURE 3-28**
DNA Replication. In replication, the DNA strands unwind, and DNA polymerase begins attaching complementary DNA nucleotides along each strand. This process produces two identical copies of the original DNA molecule.

Throughout the life of a cell, the DNA strands in the nucleus remain intact. DNA synthesis, or **DNA replication,** occurs during S phase in cells preparing to undergo mitosis or meiosis. The goal of this replication is to copy the genetic information in the nucleus so that one set of chromosomes can be given to each of the two cells produced.

DNA REPLICATION A DNA molecule consists of a pair of DNA strands held together by hydrogen bonding between complementary nitrogenous bases. Figure 3-28● diagrams the process of DNA replication. It begins when various enzymes unwind the strands and disrupt the weak bonds between the nitrogenous bases. As they do so, molecules of the enzyme **DNA polymerase** bind to the exposed nitrogenous bases. This enzyme promotes bonding between the nitrogenous bases of the DNA strand and complementary DNA nucleotides dissolved in the nucleoplasm.

Many molecules of DNA polymerase are working simultaneously along the DNA strands. The two DNA strands are oriented in opposite directions (Figure 2-20●, p.57), but DNA polymerase can work in only one direction. On one strand DNA polymerase moves forward as the original strands unwind, producing an intact complementary DNA strand (Figure 3-28●). On the other strand, DNA polymerase works *away* from the point of unwinding, creating short nucleotide chains that must be linked together by enzymes called **ligases** (LI-gās-ez; *liga*, to tie). The final result is a pair of identical DNA molecules.

The G_2 Phase

Once DNA replication has been completed, there is a brief (2–5 hours) **G_2 phase** devoted to last-minute protein synthesis. The cell then enters the **M phase,** and mitosis begins.

❑ Mitosis

Figure 3-29

Mitosis is a process that separates the duplicated chromosomes of the original cell into two identical nuclei. Mitosis specifically refers to the division and duplication of the *nucleus* of the cell. Division of the cytoplasm to form two distinct new cells involves a separate but related process known as **cytokinesis** (sī-tō-ki-NĒ-sis; *cyto-*, cell + *kinesis*, motion). Figure 3-29● summarizes the four stages of mitosis: *prophase, metaphase, anaphase,* and *telophase.*

Stage 1: Prophase

Prophase (PRŌ-fāz; *pro*, before) begins when the chromosomes coil so tightly that they become visible as individual structures. As a result of DNA replication during the S phase, there are now two copies of each chromosome. Each copy, called a **chromatid** (KRŌ-ma-tid), is connected to its duplicate at a single point, the **centromere** (SEN-trō-mēr). The region of the centromere is surrounded by a protein complex called a **kinetochore** (ki-NE-tō-kōr).

As the chromosomes appear, the two pairs of centrioles move toward opposite poles of the nucleus. An array of microtubules, called **spindle fibers,** extends between the centriole pairs. Smaller microtubules, called *astral rays,* radiate into the surrounding cytoplasm. Prophase ends with the disappearance of the nuclear envelope.

Stage 2: Metaphase

Metaphase (MET-a-fāz; *meta,* after) begins after the disintegration of the nuclear envelope. The spindle fibers now enter the nuclear region, and the chromatids become attached to spindle fibers called *chromosomal microtubules*. The attachment occurs at the kinetochore on opposite sides of the centromere linking each pair of chromatids. Once attachment has been completed, the chromatids move to a narrow central zone called the **metaphase plate.** Metaphase ends when all of the chromatids are aligned in the plane of the metaphase plate.

Stage 3: Anaphase

Anaphase (AN-a-fāz; *ana,* apart) begins when the kinetochore of each chromatid pair splits and the chromatids separate. The two **daughter chromosomes** are now pulled toward opposite ends of the cell along the chromosomal microtubules. This movement involves an interaction between the kinetochore and the microtubule. Anaphase ends when the daughter chromosomes arrive near the centrioles at opposite ends of the cell.

Stage 4: Telophase

During **telophase** (TĒL-ō-fāz; *telo,* end) the cell prepares to return to the interphase state. The nuclear membranes form, the nuclei enlarge, and the chromosomes gradually uncoil. Once the chromosomes have relaxed, and the fine filaments of chromatin become visible, nucleoli reappear and the nuclei resemble those of interphase cells. This stage marks the end of the process of mitosis.

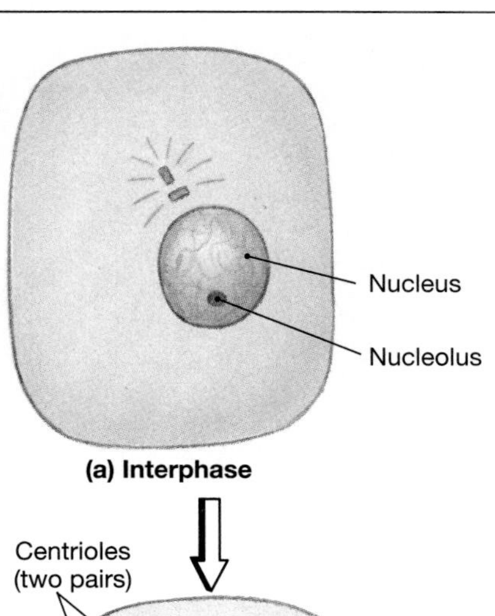
Nucleus
Nucleolus
(a) Interphase

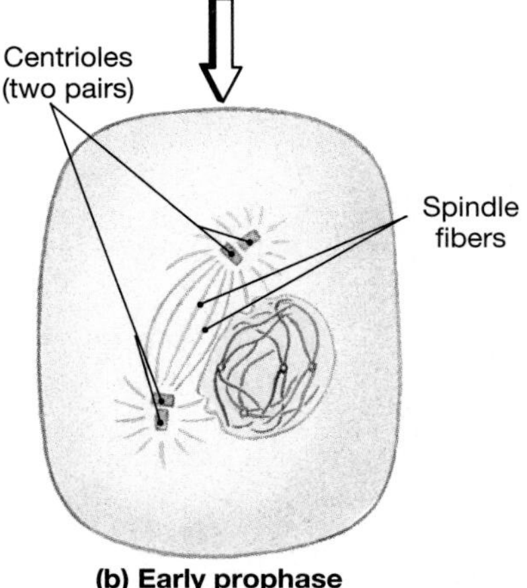
Centrioles (two pairs)
Spindle fibers
(b) Early prophase

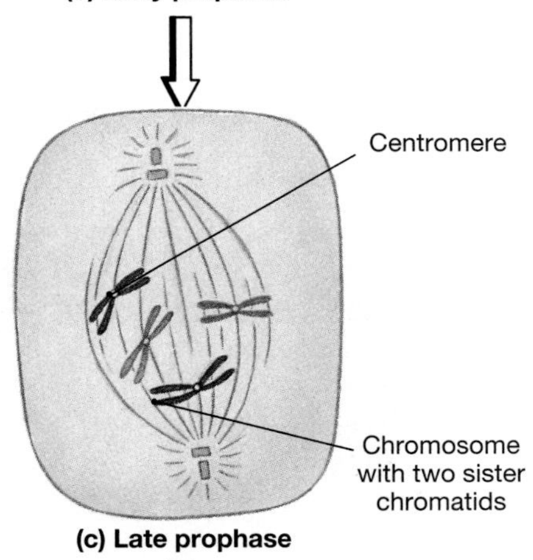
Centromere
Chromosome with two sister chromatids
(c) Late prophase

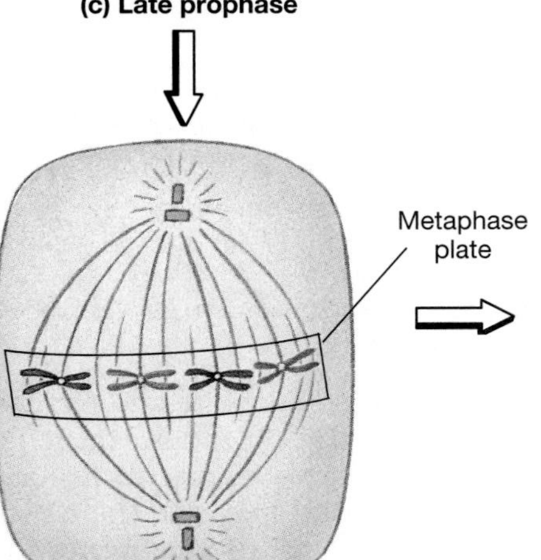
Metaphase plate
(d) Metaphase

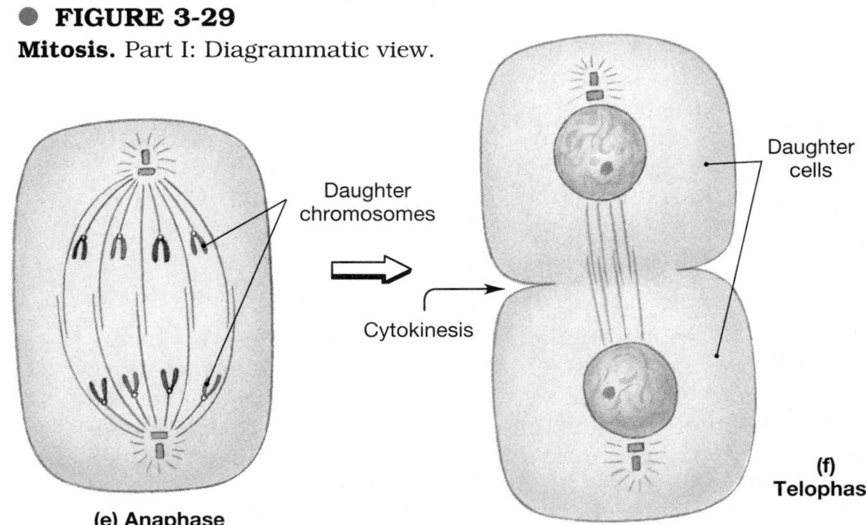
Daughter chromosomes
(e) Anaphase
Cytokinesis
Daughter cells
(f) Telophase

● **FIGURE 3-29**
Mitosis. Part I: Diagrammatic view.

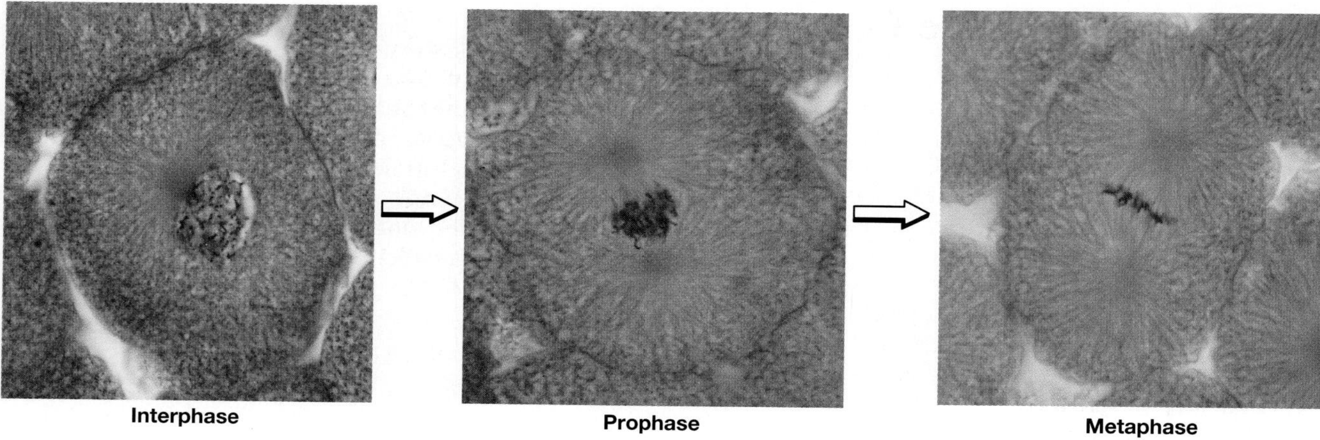
Interphase
Prophase
Metaphase

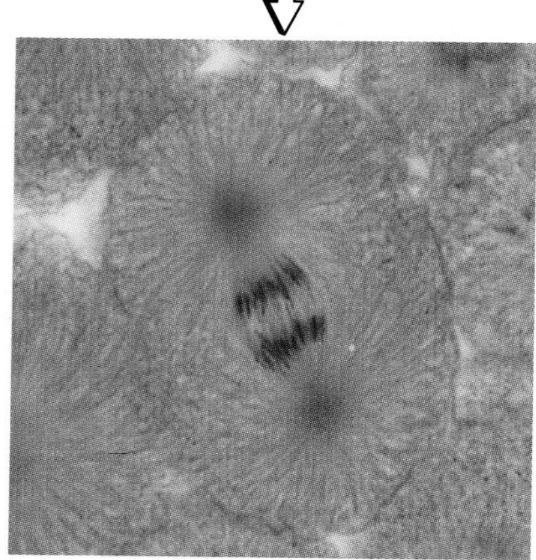
Anaphase

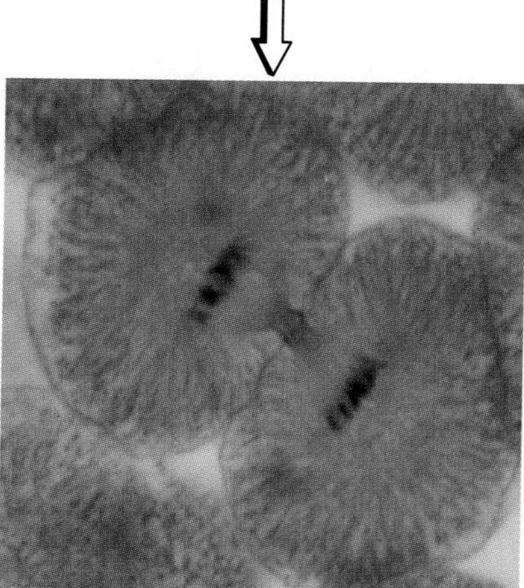
Telophase

● **FIGURE 3-29**
Mitosis. Part II: Light micrographs (× 581).

❑ Cytokinesis

Cytokinesis usually begins in late anaphase. As the daughter chromosomes approach the ends of the spindle apparatus, the cytoplasm constricts along the plane of the metaphase plate. This process continues throughout telophase and is usually completed sometime after a nuclear membrane has reformed around each daughter nucleus. The completion of cytokinesis marks the end of the process of cell division.

❑ The Mitotic Rate

The preparations for cell division that occur between G_1 and the end of the S phase are difficult to recognize in a light micrograph. However, the start of mitosis is easy to recognize, because the chromosomes become condensed and highly visible. The frequency of cell division can thus be estimated by the number of cells in mitosis at any given time. As a result, the term **mitotic rate** is often used when discussing rates of cell division. In general, the longer the life expectancy of a cell type, the slower the mitotic rate. Relatively long-lived cells, such as muscle cells and neurons, either never divide or do so only under special circumstances. Other cells, such as those covering the surface of the skin or the lining of the digestive tract, are subject to attack by chemicals, pathogens, and abrasion. They survive for only days or even hours. Special cells called **stem cells** maintain these cell populations through repeated cycles of cell division.

Stem cells are relatively unspecialized, and their only function is the production of daughter cells. Each time a stem cell divides, one of the daughter cells develops functional specializations while the other prepares for further divisions. The rate of stem cell division can vary, depending on the tissue and the demand for new cells. In heavily abraded skin, stem cells may divide more than once a day, but stem cells in adult connective tissues may remain inactive for years.

❑ Energetics and Cell Division

Dividing cells use an unusually large amount of energy. For example, they must synthesize new organic materials and move organelles and chromosomes within the cell. All these processes require ATP in substantial amounts. Cells that do not have adequate energy sources cannot divide, and in starvation normal cell growth and maintenance grind to a halt. For this reason, prolonged starvation stunts growth, slows wound healing, lowers resistance to disease, thins the skin, and changes the lining of the digestive tract. The same changes are seen in the late stages of many cancers, because the cancer cells are "stealing" the nutrients that would otherwise be used to support normal cell growth and maintenance.

❑ Regulation of the Cell Life Cycle

Mitotic rates are usually well controlled, and in normal tissues the rates of cell division balance cell loss or destruction. The rates are genetically controlled, and many different stimuli may be responsible for activating genes that promote cell division. The most important of these stimuli appear to be extracellular compounds, usually peptides, that stimulate the division of specific cell types. These compounds include several hormones and a variety of **growth factors.** Table 3-3 lists several of these stimulatory compounds and their target tissues; these hormones and factors will be discussed in later chapters. Each of these compounds appears to exert its effects by binding to receptors on the cell membrane. This binding initiates a series of biochemical events that ultimately trigger cell division. The effects of these stimulatory factors may be opposed by a poorly understood class of peptides called *chalones.*

Many of the peptide growth factors bind to membrane receptors at the cell surface. Binding activates enzymes inside the cell, starting a chain reaction leading to activation of genes that prepare the cell to undergo cell division. The presence or absence of the appropriate binding protein determines whether or not a particular cell will respond to a particular growth factor. The mechanism inside the cell appears to be very similar, however, regardless of the type of membrane receptor. Binding at the membrane surface triggers the activation and release of intermediaries known as *Ras proteins.* These proteins in turn activate intracellular enzymes and promote gene activation. Growth factors that do not work via Ras proteins may use other intermediaries or may enter the cell and exert their effects directly on the nucleus.

Genetic mechanisms for inhibiting cell division have recently been identified. The genes involved are known as *repressor genes.* One gene, called *p53,* controls a protein that resides in the nucleus and activates genes that direct the production of growth-inhibiting factors inside the cell. Roughly half of all cancers are associated with abnormal forms of the *p53* gene.

Cell Division and Cancer

When the balance between cell division and growth versus cell death breaks down, a tissue begins to enlarge. A **tumor,** or **neoplasm,** is a mass or swelling produced by the abnormal cell growth and division. In a **benign tumor** the cells usually remain within a connective tissue capsule. Such a tumor seldom threatens an individual's life. Surgery can usually remove the tumor if its size or position disturbs tissue function.

Cells in a **malignant tumor** are no longer responding to normal control mechanisms. These cells spread into the surrounding tissues, a process called **invasion**. The cancer cells may also spread to other tissues and organs. This dispersion is called **metastasis** (me-TAS-ta-sis; *meta,* after + *stasis,* standing still). Metastasis is dangerous and difficult to control. Once in new locations, the metastatic cells produce secondary tumors.

The term **cancer** refers to an illness characterized by malignant cells. Cancer cells gradually lose their resemblance to normal cells. They change size and shape, often becoming unusually large or abnormally small. Organ function begins to deteriorate as the number of cancer cells increases. The cancer cells may not perform their original func-

TABLE 3-3 Representative Chemical Factors Affecting Cell Division

Factor	*Source*	*Effect*	*Target*
Prolactin	Anterior pituitary gland	Stimulation of growth, cell division, development	Gland and duct cells of mammary glands
Nerve growth factor (NGF)	Salivary glands, other sources suspected	Stimulation of nerve cell repair and development	Neurons and glial cells
Epidermal growth factor (EGF)	Duodenal glands, other sources suspected	Stimulation of stem cell divisions and epithelial repairs	Epidermis of skin
Fibroblast growth factor (FGF)	Unknown	Division and differentiation of fibroblasts and related cells	Connective tissues
Erythropoietin	Kidneys (primary source)	Stimulation of stem cell divisions and maturation of red blood cells	Bone marrow
Thymosins and related compounds	Thymus	Stimulation of division and differentiation of lymphocytes (especially T cells)	Thymus and other lymphoid tissues and organs
Chalones	Many tissues	Inhibition of cell division	Cells in the immediate area

tions at all, or they may perform normal functions in an unusual way. For example, endocrine cancer cells may produce normal hormones, but in abnormally large amounts. Cancer cells do not use energy very efficiently, and they grow and multiply at the expense of healthy tissues, competing for space and nutrients with normal cells. This accounts for the starved appearance of many patients in the late stages of cancer. We will return to the subject of cancer in later chapters dealing with specific systems, and the *Applications Manual* contains detailed information on cancer formation, development, and treatment. AM *Cancer*

Cell Diversity and Differentiation

The liver cells, fat cells, and neurons of an individual contain the same chromosomes and genes, but in each case a different set of genes has been turned *off.* In other words, liver cells differ from other cells because they have different genes available for transcription.

When a gene is functionally eliminated, the cell loses the ability to create a particular protein, and thus to perform any functions involving that protein. Each time another gene switches off, the cell's functional abilities become more restricted. This specialization process is called **differentiation.**

Fertilization produces a single cell with all of its genetic potential intact. There follows a period of repeated cell divisions, and differentiation begins as the number of cells increases. Differentiation produces specialized cells with limited capabilities. These cells form organized collections known as *tissues,* each with discrete functional roles. The next chapter examines the structure and function of tissues and considers the role of tissue interactions in the maintenance of homeostasis.

✔ Analysis of a living cell indicates that it is actively manufacturing enough organelles to serve two functional cells. This cell is probably in what phase of its life cycle?

What would happen if spindle fibers failed to form in a cell during mitosis?

Selected Clinical Terminology

Terms Discussed in This Chapter

benign tumor: A mass or swelling in which the cells usually remain within a connective tissue capsule; rarely life-threatening. *(p. 103)*
cancer: An illness characterized by the metastasis of malignant tumor cells. *(p. 103)*
dextran: A carbohydrate that cannot cross cell membranes; often administered to patients following blood loss or dehydration. *(p. 76)*
DNA fingerprinting: Identifying an individual on the basis of repeating nucleotide sequences in his or her DNA. *(p. 94)*
invasion: The spread of cancer cells from a primary tumor into surrounding tissues *(p. 103)*
malignant tumor: A mass or swelling in which the cells no longer respond to normal control mechanisms, but divide rapidly. *(p. 103)*
metastasis (me-TAS-ta-sis): The spread of malignant cells into distant tissues and organs. *(p. 103)*
normal saline: A solution that approximates the normal osmotic concentration of extracellular fluids. *(p. 76)*
tumor (neoplasm): A mass or swelling produced by abnormal cell growth and division. *(p. 103)*

AM *Additional Terms Discussed in the Applications Manual*

carcinogen (kar-SIN-o-jen): An environmental factor that stimulates the conversion of a normal cell to a cancer cell.
genetic engineering: The popular term for research and experiments related to changing the genetic makeup of an organism.
karyotyping (KAR-ē-ō-tī-ping): The determination of an individual's chromosome complement.
mutagen (MŪ-ta-jen): A factor that can damage DNA strands and sometimes cause chromosomal breakage, stimulating the development of cancer cells.
oncogene (ON-kō-jēn): A cancer-causing gene created by a somatic mutation in a normal gene involved with growth, differentiation, or cell division.
recombinant DNA: DNA created by splicing a specific gene from one organism into the DNA strand of another organism.

CHAPTER REVIEW

STUDY OUTLINE

INTRODUCTION, p. 66

1. Contemporary cell theory incorporates several basic concepts: (1) Cells are the building blocks of all plants and animals; (2) cells are produced by the division of preexisting cells; (3) cells are the smallest units that perform all vital physiological functions; (4) each cell maintains homeostasis at the cellular level; and (5) homeostasis at the tissue, organ, system, and individual levels reflects the combined and coordinated actions of many cells. *(Figure 3-1)*

STUDYING CELLS, p. 66

An Overview of Cellular Anatomy, p. 67

1. A typical cell is surrounded by the **extracellular fluid,** specifically the **interstitial fluid** of a tissue. The cell's outer boundary is the **cell membrane,** or **plasma membrane,** which is a **phospholipid bilayer.** *(Figure 3-2, Table 3-1)*

THE CELL MEMBRANE, p. 69

Membrane Structure, p. 69

1. The cell membrane contains lipids, proteins, and carbohydrates. Phospholipids are the largest contributors to membrane structure. *(Figure 3-3)*
2. **Integral proteins** are part of the membrane itself; **peripheral proteins** are attached but can separate from it. **Channels** allow water and ions to move across the membrane; some channels are called **gated** because they can open or close. *(Figure 3-4)*

Membrane Permeability, p. 71

3. Cell membranes are **selectively permeable.**
4. **Diffusion** is the net movement of material from an area where its concentration is relatively high to an area where its concentration is lower. Diffusion occurs until the **concentration gradient** is eliminated. *(Figures 3-5, 3-6)*
5. Diffusion of water across a membrane in response to differences in solute concentration is **osmosis.** The force of movement is **osmotic pressure.** *(Figures 3-7, 3-8)*
6. In **filtration,** hydrostatic pressure forces water across a membrane; if membrane pores are large enough, molecules of solute will be carried along. *(Figure 3-9)*
7. **Facilitated diffusion** is a type of **carrier-mediated transport.** *(Figure 3-10)*
8. **Active transport** mechanisms consume ATP but are independent of concentration gradients. Some **ion pumps** are **exchange pumps.** *(Figure 3-11)*
9. In **vesicular transport,** material moves into or out of the cell in membranous sacs. Movement into the cell is accomplished through **endocytosis,** an active process that can take three forms: **pinocytosis, receptor-mediated endocytosis,** and **phagocytosis.** Movement out of the cell is accomplished through **exocytosis.** *(Figures 3-12, 3-13, 3-19; Table 3-2)*
10. The **potential difference** between the two sides of a cell membrane is a **transmembrane potential.** The transmembrane potential in an undisturbed cell is its **resting potential.** *(Figure 3-14)*

THE CYTOPLASM, p. 83

The Cytosol, p. 83

1. The **cytoplasm** contains a fluid **cytosol** and surrounds **organelles.**

Organelles, p. 83

2. **Membranous organelles** are surrounded by lipid membranes that isolate them from the cytosol. They include the mitochondria, endoplasmic reticulum, Golgi apparatus, lysosomes, and peroxisomes. *(Table 3-1)*
3. **Nonmembranous organelles** are always in contact with the cytosol. They include the cytoskeleton, microvilli, centrioles, cilia, flagella, and ribosomes. *(Table 3-1)*
4. The **cytoskeleton** gives the cytoplasm strength and flexibility. It has four components: **microfilaments, intermediate filaments, thick filaments,** and **microtubules.** *(Figure 3-15)*
5. **Microvilli** are small projections of the cell membrane that increase the surface area exposed to the extracellular environment. *(Figure 3-15)*
6. **Centrioles** direct the movement of chromosomes during cell division and organize the cytoskeleton. *(Figure 3-16)*
7. **Cilia** beat rhythmically to move fluids or secretions across the cell surface. *(Figure 3-16)*
8. **Flagella** move a cell through surrounding fluid, rather than moving fluid past a stationary cell.
9. **Ribosomes** are intracellular factories that manufacture proteins. There are **free ribosomes** in the cytoplasm and **fixed ribosomes** attached to the ER.
10. **Mitochondria** are responsible for 95 percent of the ATP production within a typical cell. *(Figure 3-17)*
11. The **endoplasmic reticulum (ER)** is a network of intracellular membranes. There are two types: rough and smooth. **Rough endoplasmic reticulum (RER)** contains ribosomes; **smooth endoplasmic reticulum (SER)** does not. *(Figure 3-18)*
12. The **Golgi apparatus** packages lysosomes and **secretory vesicles.** Secretions are discharged from the cell via *exocytosis. (Figure 3-19)*
13. **Lysosomes** are vesicles filled with digestive enzymes. The process of endocytosis is important in ridding the cell of bacteria and debris. *(Figure 3-20)*
14. **Peroxisomes** also carry enzymes; they absorb and neutralize toxins. *(Figure 3-19)*

THE NUCLEUS, p. 92

1. The **nucleus** is the control center for the cellular operations. It is surrounded by a **nuclear envelope,** through which it communicates with the cytosol by way of **nuclear pores.** *(Figure 3-21)*

Chromosome Structure, p. 93

2. The nucleus controls the cell by directing the synthesis of specific proteins using information

stored in the DNA of **chromosomes.** *(Figures 3-22, 3-23)*

The Genetic Code, p. 93

3. The cell's information storage system, the **genetic code,** is called a **triplet code** because a sequence of three nitrogenous bases identifies a single amino acid. Each **gene** consists of all the triplets needed to produce a specific polypeptide chain. *(Figure 3-24)*

Gene Activation and Protein Synthesis, p. 94

Transcription, p. 94

4. **Transcription** is the process of forming a strand of **messenger RNA (mRNA),** which carries instructions from the nucleus to the cytoplasm. *(Figures 3-23, 3-24)*

Translation, p. 94

5. During **translation** a functional polypeptide is constructed using the information from an mRNA strand. Each trio of nitrogenous bases along the mRNA strand is a **codon;** the sequence of codons determines the sequence of amino acids in the polypeptide. Translation proceeds in three steps: **initiation, elongation,** and **termination.** *(Figures 3-25, 3-26)*
6. **Ribosomal RNA (rRNA)** forms part of the structure of the ribosomes involved in translation. Molecules of **transfer RNA (tRNA)** bring amino acids to the active site of the ribosomal complex. *(Figure 3-25)*

DNA and the Control of Cell Structure and Function, p. 98

THE CELL LIFE CYCLE, p. 99

1. **Mitosis** refers to the nuclear division of somatic cells. Sex cells (sperm and ova) are produced by **meiosis.**

Interphase, p. 99

2. Most somatic cells spend most of their time in **interphase.** *(Figures 3-27, 3-28)*

Mitosis, p. 100

3. Mitosis proceeds in four stages: **prophase, metaphase, anaphase,** and **telophase.** *(Figure 3-29)*

Cytokinesis, p. 102

4. During the process of **cytokinesis,** the cytoplasm is divided, ususally producing two identical daughter cells.

The Mitotic Rate, p. 102

5. In general, the longer the life expectancy of a cell type, the slower the **mitotic rate. Stem cells** undergo frequent mitoses to replace other, more specialized cells.

Energetics and Cell Division, p. 102

6. Cell division is genetically controlled. A variety of **growth factors** can stimulate cell division and growth. *(Table 3-3)*

Regulation of the Cell Life Cycle, p. 103

CELL DIVERSITY AND DIFFERENTIATION, p. 104

1. **Differentiation** is the process of specialization that produces cells with limited capabilities. These specialized cells form organized collections called tissues, each of which has certain functional roles.

REVIEW QUESTIONS

LEVEL 1 Reviewing Facts and Terms

1. All of the following membrane transport mechanisms are passive processes *except:*
   (a) diffusion
   (b) facilitated diffusion
   (c) vesicular transport
   (d) filtration

2. The principal *cations* in body fluids are:
   (a) calcium and magnesium
   (b) chloride and bicarbonate
   (c) sodium and potassium
   (d) sodium and chloride

3. _______________ ion concentrations are high in the extracellular fluids, and _______________ ion concentrations are high in the cytoplasm.
   (a) Calcium, magnesium
   (b) Chloride, sodium
   (c) Potassium, sodium
   (d) Sodium, potassium

4. In a resting transmembrane potential, the inside of the cell is _______________ and the cell exterior is _______________.
   (a) slightly negative, slightly positive
   (b) slightly positive, slightly negative
   (c) slightly positive, neutral
   (d) slightly negative, neutral

5. The process that generates carbon dioxide while removing hydrogen atoms from organic molecules is called:
   (a) the tricarboxylic acid cycle
   (b) active transport
   (c) glycolysis
   (d) phosphorylation

6. The reaction sequence in which glucose is broken down into pyruvic acid is:
   (a) aerobic respiration
   (b) the TCA cycle
   (c) mitochondrial energy production
   (d) glycolysis

7. The construction of a functional polypeptide using the information provided by an mRNA strand is:
   (a) translation (b) transcription
   (c) replication (d) gene activation

8. Our somatic cell nuclei contain _______________ pairs of chromosomes.
   (a) 8 (b) 16
   (c) 23 (d) 46

9. Termination is the final stage in the production of an:
   (a) DNA molecule (b) mRNA molecule
   (c) tRNA molecule (d) protein

10. The term *differentiation* refers to:
   (a) the loss of genes from cells
   (b) the acquisition of new functional capabilities by cells
   (c) the production of functionally specialized cells
   (d) the division of genes among different types of cells

11. The interphase of the cell life cycle is divided into the following phases:
   (a) prophase, metaphase, anaphase, and telophase
   (b) G_0, G_1, S, and G_2
   (c) mitosis and cytokinesis
   (d) a, b, and c are correct

12. List the five basic concepts that make up the modern-day cell theory.

13. What are the four general functions of the cell membrane?

14. What are the primary functions of the membrane proteins?

15. By what four major transport mechanisms do things get into and out of cells?

16. What four important factors influence diffusion rates?

17. State three general characteristics of osmosis that define its role in the body.

18. List (a) the nonmembranous organelles and (b) the membranous organelles of a typical cell.

19. What are the three major functions of the endoplasmic reticulum?

20. List the four stages of mitosis in their correct sequence.

LEVEL 2 Reviewing Concepts

21. Diffusion is important in body fluids because it tends to:
   (a) increase local concentration gradients
   (b) eliminate local concentration gradients
   (c) move substances against concentration gradients
   (d) create concentration gradients

22. Osmotic pressure differs from hydrostatic pressure because the osmotic pressure of a solution is an indication of the force of water movement resulting from:
   (a) its solute concentration
   (b) the volume of water
   (c) the permeability of the membrane
   (d) a, b, and c are correct

23. When a cell is placed in a ______________ solution it will lose water through osmosis. The process results in the ____________ of red blood cells.
   (a) hypotonic, crenation
   (b) hypertonic, crenation
   (c) isotonic, hemolysis
   (d) hypotonic, hemolysis

24. Suppose that a DNA segment has the following nucleotide sequence: CTC ATA CGA TTC AAG TTA. Which of the following nucleotide sequences would be found in a complementary mRNA strand?
   (a) GAG UAU GAU AAC UUG AAU
   (b) GAG TAT GCT AAG TTC AAT
   (c) GAG UAU GCU AAG UUC AAU
   (d) GUG UAU GGA UUG AAC GGU

25. How many amino acids are coded in the DNA segment in the previous question?
   (a) 18 (b) 9
   (c) 6 (d) 3

26. How does hydrostatic pressure differ from osmotic pressure?

27. What general characteristics are important in carrier-mediated transport mechanisms?

28. What are the similarities between facilitated diffusion and active transport? What are the differences?

29. What role does the sodium-potassium pump play in stabilizing the resting membrane potential?

30. How does the cytosol differ in composition from the interstitial fluid?

31. Differentiate between transcription and translation.

32. List the phases of the interphase state in the cell life cycle in sequence, and briefly describe what happens in each.

33. List the stages of mitosis, and briefly describe the events that occur in each.

34. What is cytokinesis, and what role does it play in the cell cycle?

LEVEL 3 Critical Thinking and Clinical Applications

35. Experimental evidence shows that the transport of a certain molecule exhibits the following characteristics: (1) The molecule moves along its concentration gradient; (2) at concentrations above a given level there is no increase in the rate of transport; and (3) cellular energy is not required for transport to occur. What type of transport process is at work?

36. Two solutions, A and B, are separated by a semipermeable barrier. Over a period of time, the level of fluid on side A increases. Which solution initially had the higher concentration of solute?

37. In kidney dialysis, a person's blood is passed through a bath that contains a mixture of several ions and molecules. The blood is separated from the dialysis fluid by a membrane that allows water, small ions, and small molecules to pass, but does not allow large proteins or blood cells to pass. What should the composition of the dialysis fluid be for it to remove urea (a small molecule) without changing blood volume (removing water from the blood)?

CHAPTER 4

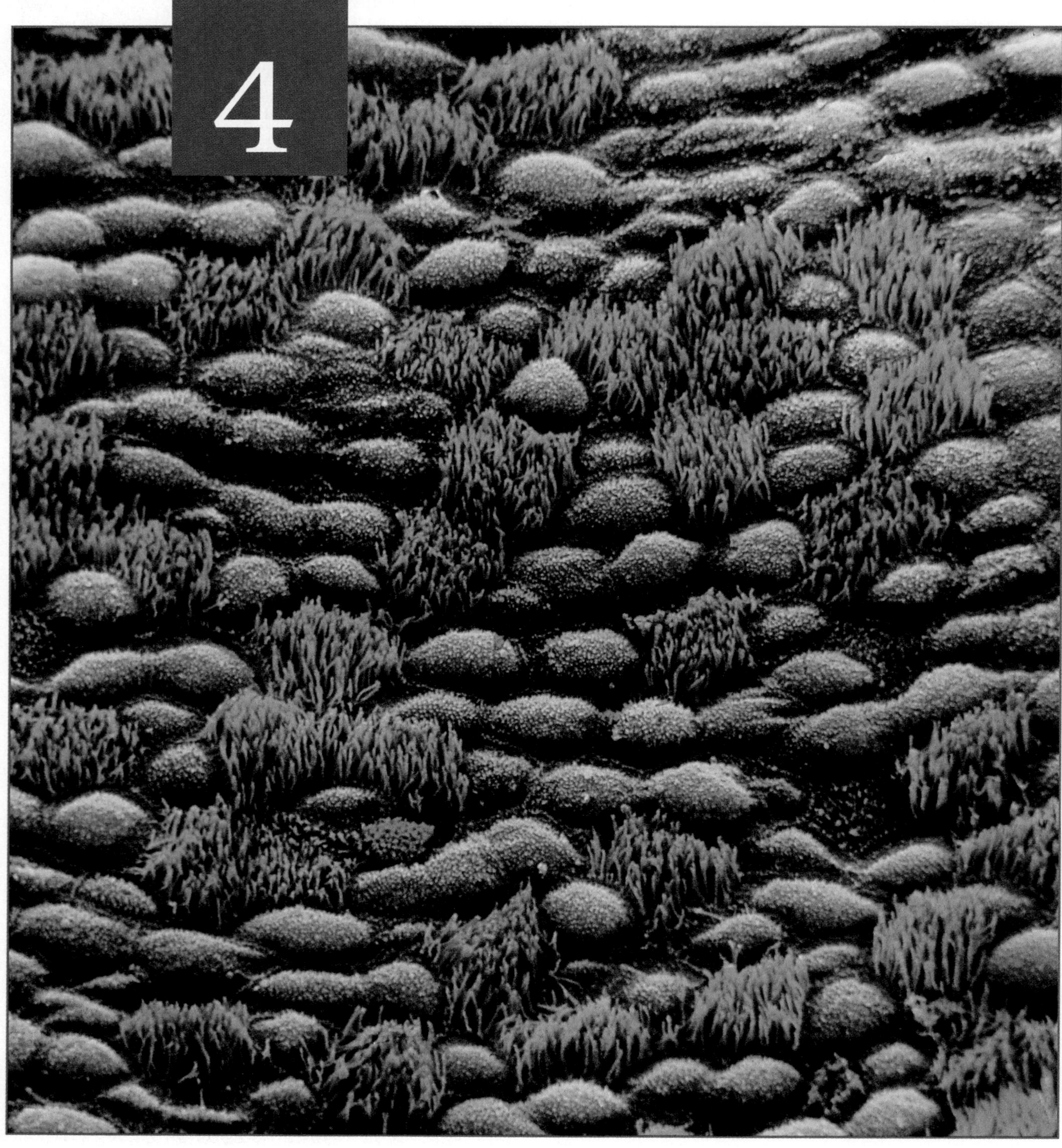

Exotic creatures on the deep sea floor?

Well, no. The "creatures" that you see here are much smaller than the strange life forms photographed in the ocean depths, and much closer to home. These are cells that line the trachea, the airway leading to your lungs. The "tentacles" are cilia that help to remove dirt and harmful microbes from inhaled air.

Notice that the surface is formed by more than one type of cell. Groups of cells specialized to perform a particular set of functions are found in tissues. Each of the body's several different tissues has its own distinctive structure and its own role to play in maintaining homeostasis. We will meet them all in this chapter.

The Tissue Level of Organization

Chapter Outline and Objectives

No single cell contains the metabolic machinery and organelles needed to perform all the many functions of the human body. Instead, through the process of differentiation, each cell develops a characteristic set of structural features and a limited number of functions. These structures and functions can be quite distinct from those of nearby cells. Nevertheless, cells in a given location all work together.

A detailed examination of the body reveals a number of patterns at the cellular level. Although there are trillions of cells in the human body, there are only about 200 types of cells. These cell types combine to form **tissues,** collections of specialized cells and cell products that perform a relatively limited number of functions. There are four basic **tissue types:** *epithelial tissue, connective tissue, muscle tissue,* and *neural tissue.*

- *Epithelial tissue* covers exposed surfaces, lines internal passageways and chambers, and forms glands.
- *Connective tissue* fills internal spaces, provides structural support for other tissues, and stores energy reserves.
- *Muscle tissue* contracts to perform specific movements and in the process generates heat that warms the body.
- *Neural tissue* carries information from one part of the body to another in the form of electrical impulses.

Histology, the study of tissues, provides beautiful examples of the interplay between form and function. This chapter will discuss the characteristics of each major tissue type, focusing on the relationship between cellular organization and tissue function. Later chapters will consider the patterns of tissue interaction in various organs and systems in greater detail.

Epithelial Tissue

Epithelial tissue includes *epithelia* and *glands,* secretory structures derived from epithelia. An **epithelium** (e-pi-THĒ-lē-um) is a layer of cells that forms a barrier with specific properties. Epithelia cover every exposed body surface. The surface of the skin is a good example, but epithelia also line the digestive, respiratory, reproductive, and urinary tracts—passageways that communicate with the outside world. Epithelia also line internal cavities and passageways, such as the chest cavity, fluid-filled chambers in the brain, eye, and inner ear, and the inner surfaces of blood vessels and the heart.

Important characteristics of epithelia include:

- Epithelia consist mainly of cells, rather than extracellular materials.
- An epithelium may consist of a single layer of cells (a *simple epithelium*) or multiple cell layers (a *stratified epithelium*).
- Although firmly attached to underlying connective tissues, an epithelium always has a free surface exposed to the external environment or to some internal chamber or passageway.
- There are no blood vessels in epithelia. Because of this **avascular** (ā-VAS-kū-lar; *a-*, without + *vas,* vessel) condition, epithelial cells must obtain nutrients by diffusion and/or absorption from deeper tissues or from their exposed surfaces.
- The mitotic rate in epithelia can be very high, especially in situations where the epithelial cells are exposed to harsh physical or chemical environments.

Functions of Epithelial Tissue

Epithelia perform essential functions that can be summarized as follows:

1. ***Provide physical protection.*** Epithelia protect exposed and internal surfaces from abrasion, dehydration, and destruction by chemical or biological agents.
2. ***Control permeability.*** Any substance that enters or leaves the body has to cross an epithelium. Some epithelia are relatively impermeable, whereas others are easily crossed by compounds as large as proteins. Many epithelia contain the molecular "machinery" needed for selective absorption or secretion. The epithelial barrier can be regulated and modified in response to various stimuli. For example, hormones can affect the transport of ions and nutrients through epithelial cells. Even physical stress can alter the structure and properties of epithelia—think of the calluses that form on your hands when you do rough work for a period of time.
3. ***Provide sensations.*** Most epithelia are extensively innervated by sensory nerves. Specialized epithelial cells can detect changes in the environment they confront and convey information about such changes to the nervous system. For example, touch receptors in the deepest epithelial layers of the skin respond to pressure by stimulating adjacent sensory nerves. A **neuroepithelium** is an epithelium containing sensory cells providing sensations of smell, taste, sight, equilibrium, and hearing.
4. ***Produce specialized secretions.*** Epithelial cells that produce secretions are called *gland cells.* Individual gland cells are often scattered among other cell types in an epithelium. In a **glandular epithelium** most or all of the epithelial cells produce secretions.

❑ Specializations of Epithelial Cells

Figure 4-1

Epithelial cells have several specializations that distinguish them from other body cells. Many epithelial cells are specialized for (1) the production of secretions, (2) the movement of fluids over the epithelial surface, or (3) the movement of fluids through the epithelium itself. These specialized epithelial cells usually show a definite **polarity** along the axis that extends from the basement membrane to the exposed surface of the epithelium. In other words, the organelles are distributed unevenly along this axis. The actual arrangement varies depending on the functional activities of the individual cells. The cells shown in Figure 4-1a● show a common type of polarity.

Many epithelial cells lining internal passageways have microvilli on their exposed surfaces; there may be just a few, or the entire surface may be carpeted by them. Microvilli are especially abundant on epithelial surfaces where absorption and secretion take place, such as along portions of the digestive and urinary tracts. The epithelial cells in these locations are transport specialists, and a cell with microvilli has at least 20 times the surface area of a cell without them. Microvilli are shown in Figure 4-1b●. **Stereocilia** are structurally similar to microvilli, but they are very long (up to 250 μm) and incapable of active movement. These structures are found in two places only: (1) along portions of the male reproductive tract, and (2) on receptor cells of the inner ear.

Figure 4-1b● shows the surface of a **ciliated epithelium.** A typical ciliated cell contains about 250 cilia that beat in a coordinated fashion. Substances are moved over the epithelial surface by the synchronized beating of cilia, like a continuously moving escalator. For example, the ciliated epithelium that lines the respiratory tract moves mucus up from the lungs and toward the throat. The mucus traps foreign particles such as dust, pollen, and pathogens, and carries them away from more delicate surfaces deeper in the lungs. Injury to the cilia or to the epithelial cells in general can stop ciliary movement and block the protective flow of mucus. This is one effect of smoking; other unpleasant effects will be considered in later sections and chapters.

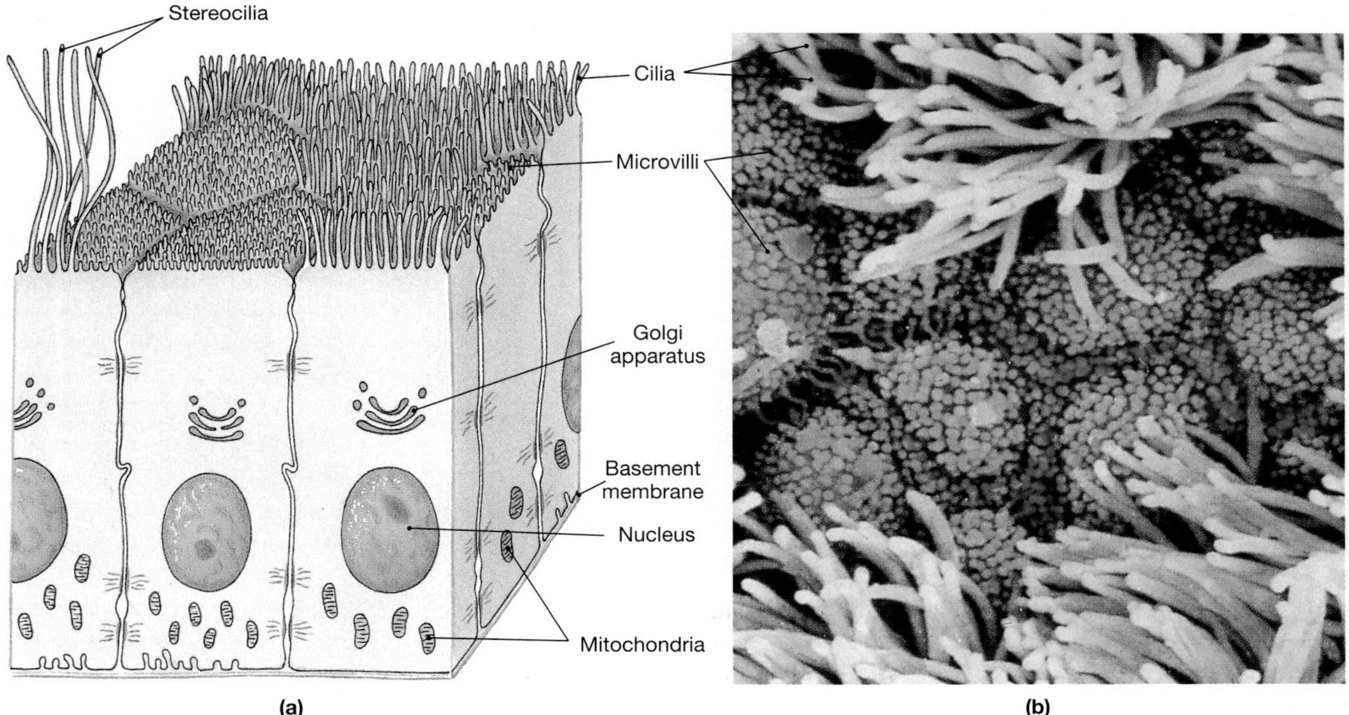
Stereocilia
Cilia
Microvilli
Golgi apparatus
Basement membrane
Nucleus
Mitochondria
(a)
(b)

● **FIGURE 4-1**

Polarity of Epithelial Cells. (a) Many epithelial cells differ in internal organization along an axis between the free surface and the basement membrane. The free surface frequently bears microvilli; less often, this surface may have cilia or (very rarely) stereocilia. (All three would not normally be found on the same group of cells but are depicted here for purposes of illustration.) In some epithelia, such as the lining of the kidney tubules, mitochondria are concentrated near the base of the cell, probably to provide energy for the cell's transport activities. **(b)** An SEM showing the surface of a ciliated epithelium that lines most of the respiratory tract. The small, bristly areas are microvilli found on the exposed surfaces of mucus-producing cells that are scattered among the ciliated epithelial cells. (SEM × 15,846)

Maintaining the Integrity of Epithelia

Three factors are involved in maintaining the physical integrity of an epithelium: (1) intercellular connections, (2) attachment to the basement membrane, and (3) replacement of damaged epithelial cells.

Intercellular Connections

Figures 4-2, 4-3

Cells in epithelia are firmly attached to one another, and the epithelium as a unit is attached to extracellular fibers of the underlying connective tissues. Because epithelial cells are specialists in intercellular connection, we will consider the topic at this time. However, many other cells in the body form permanent or temporary bonds to other cells or extracellular materials.

Intercellular connections may involve extensive areas of opposing cell membranes, or they may be concentrated at specialized attachment sites called *cell junctions.*

- Large areas of opposing cell membranes may be interconnected by binding between transmembrane proteins called **cell adhesion molecules,** or CAMs. For example, CAMs on the attached base of an epithelium lock them onto the underlying basement membrane.
- There are four types of specialized **cell junctions:** *gap junctions, tight junctions, intermediate junctions,* and *desmosomes.*

GAP JUNCTIONS At a **gap junction** (Figure 4-2a●) two cells are held together by an interlocking of membrane proteins. Because these are channel proteins, the result is a narrow passageway that lets small molecules and ions pass from cell to cell. Gap junctions are most common in cardiac muscle and smooth muscle tissue.

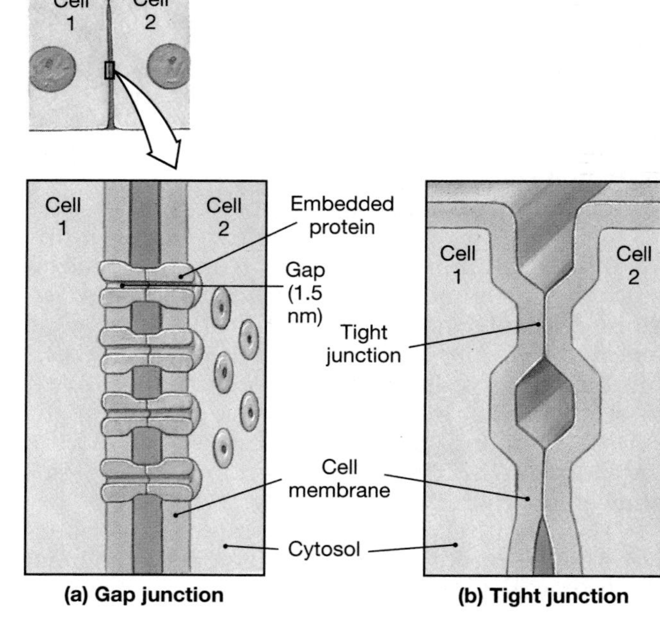
Cell 1
Cell 2
Cell 1
Cell 2
Embedded protein
Gap (1.5 nm)
Tight junction
Cell membrane
Cytosol
Cell 1
Cell 2
(a) Gap junction
(b) Tight junction

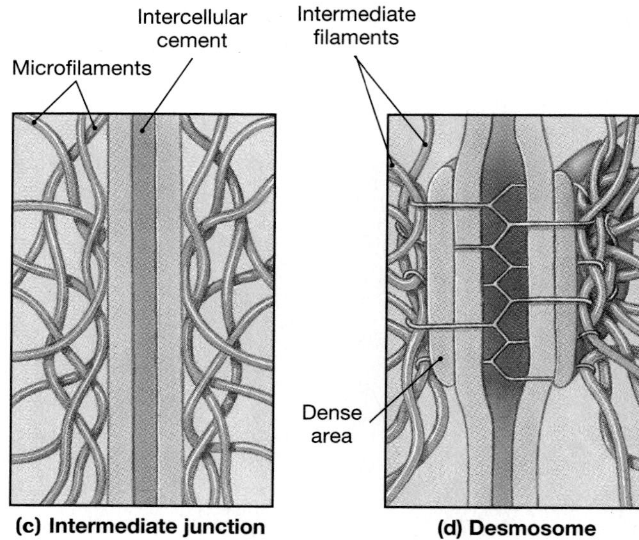
Intercellular cement
Intermediate filaments
Microfilaments
Dense area
(c) Intermediate junction
(d) Desmosome

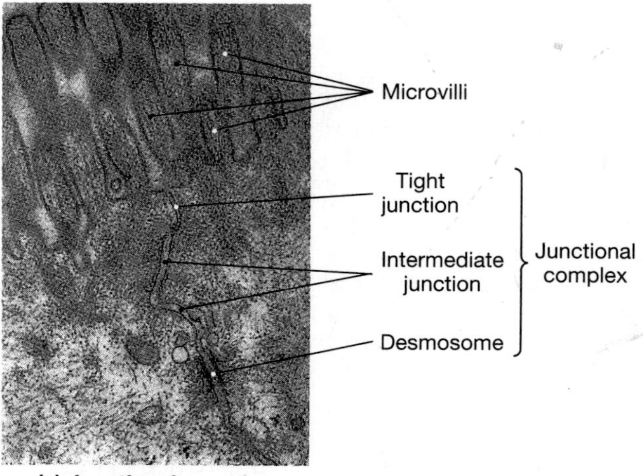
Microvilli
Tight junction
Intermediate junction
Desmosome
Junctional complex
(e) Junctional complex

● FIGURE 4-2
Cell Attachments. **(a)** At a gap junction, binding of membrane proteins creates a cytoplasmic connection between two cells. **(b)** A tight junction is formed by fusion of the outer layers of two cell membranes. **(c)** At an intermediate junction the membranes are held together by intercellular cement. A network of microfilaments strengthens the area of attachment. **(d)** A desmosome has a more organized network of intermediate filaments. Desmosomes attach one cell to another or attach a cell to extracellular structures, such as the protein fibers in connective tissues. **(e)** A junctional complex consists of a tight junction, an intermediate junction, and a desmosome. The tight junction is closest to the cell surface and forms a barrier to the diffusion of fluids and solutes between the cells.

TIGHT JUNCTIONS At a **tight junction,** shown in Figure 4-2b●, there is a partial fusion of the lipid portions of the two cell membranes. Because the membranes are fused together, tight junctions block the passage of water or solutes between the cells. Tight junctions are found near the exposed surfaces of cells lining the digestive tract, thereby keeping enzymes, acids, and wastes from damaging delicate underlying tissues.

INTERMEDIATE JUNCTIONS At an **intermediate junction** (Figure 4-2c●) the opposing cell membranes, while remaining distinct, are held together by a thick layer of proteoglycans. This proteoglycan layer is called **intercellular cement;** its primary component is the polysaccharide **hyaluronic acid.** The cytoplasm at an intermediate junction contains a dense network of microfilaments that anchor the junction to the cytoskeleton. This arrangement adds strength and helps to stabilize the shape of the cell.

DESMOSOMES At **desmosomes** (DEZ-mo-sōmz; *desmos*, ligament + *soma*, body) there is a very thin proteoglycan layer between the opposing cell membranes, reinforced by a network of intermediate filaments that lock the two cells together (Figure 4-2d●). Desmosomes are very strong, and they can resist stretching and twisting. These connections are most abundant between cells in the superficial layers of the skin. The desmosomes create links so strong that even dead skin cells are usually shed in thick sheets, rather than individually. Desmosomes are also found in cardiac muscle tissue, interconnecting the muscle cells.

JUNCTIONAL COMPLEXES Cells lining the digestive tract, respiratory tract, or other passageways are held together by **junctional complexes.** A single junctional complex consists of a tight junction, an intermediate junction, and a desmosome, with the tight junction closest to the cell surface. Figure 4-2e● details the structure of a typical junctional complex.

INTERLOCKING MEMBRANES In addition to junctional complexes, adjacent epithelial cells are often locked together by extensive folding of opposing cell membranes (Figure 4-3c●). The combination of junctional complexes, intercellular cement, and physical interlocking gives the epithelium strength and stability. The extensive connections between cells hold them together and may deny access to chemicals or pathogens that may cover their free surfaces. If the epithelium is damaged or the connections broken, infection can easily occur. For this reason infections are a serious risk after a severe burn or abrasion.

The tight interconnection of epithelial cells has one noteworthy disadvantage: Epithelia are avascular, and the interconnections that tie cells together also slow or stop diffusion between cells. However, in epithelia containing many layers of cells, tight junctions often create a network of narrow passageways that branch throughout the epithelium. These *interfacial canals* provide a route for the distribution of nutrients throughout the epithelium.

The Basement Membrane

Figure 4-3

Epithelial cells not only hold onto one another, they remain firmly connected to the rest of the body. The inner surface of each epithelium is attached to a special two-part **basement membrane** (Figure 4-3b●). The layer closer to the epithelium, called the **basal lamina** (LA-mi-nuh; *lamina*, thin layer), contains glycoproteins and a network of fine protein filaments. The basal lamina provides a barrier that restricts the movement of proteins and other large molecules from the underlying connective tissue into the epithelium. The deeper portion of the basement membrane, the **reticular lamina,** contains bundles of coarse protein fibers produced by connective tissue cells. The reticular lamina gives the basement membrane its strength. Attachments between the fibers of the basal lamina and those of the reticular lamina hold the two together.

Epithelial Maintenance and Repair

An epithelium must continually repair and renew itself. Epithelial cells lead hard lives, for they may be exposed to disruptive enzymes, toxic chemicals, pathogenic bacteria, or mechanical abrasion. Under severe conditions, such as those encountered inside the small intestine, an epithelial cell may survive for just a day or two before it is lost or destroyed. The only way the epithelium can maintain its structure over time is through the continual division of stem cells, introduced in Chapter 3. ∞ [p. 102] These stem cells, also known as **germinative cells,** are usually found in the deepest layers of the epithelium, close to the basement membrane.

 Microscopic examination of an epithelial surface reveals the presence of many microvilli. What is the probable function of this epithelium?

 In what kinds of tissue would you be most likely to find gap junctions?

 Why do you find the same epithelial organization in the pharynx, esophagus, anus, and vagina?

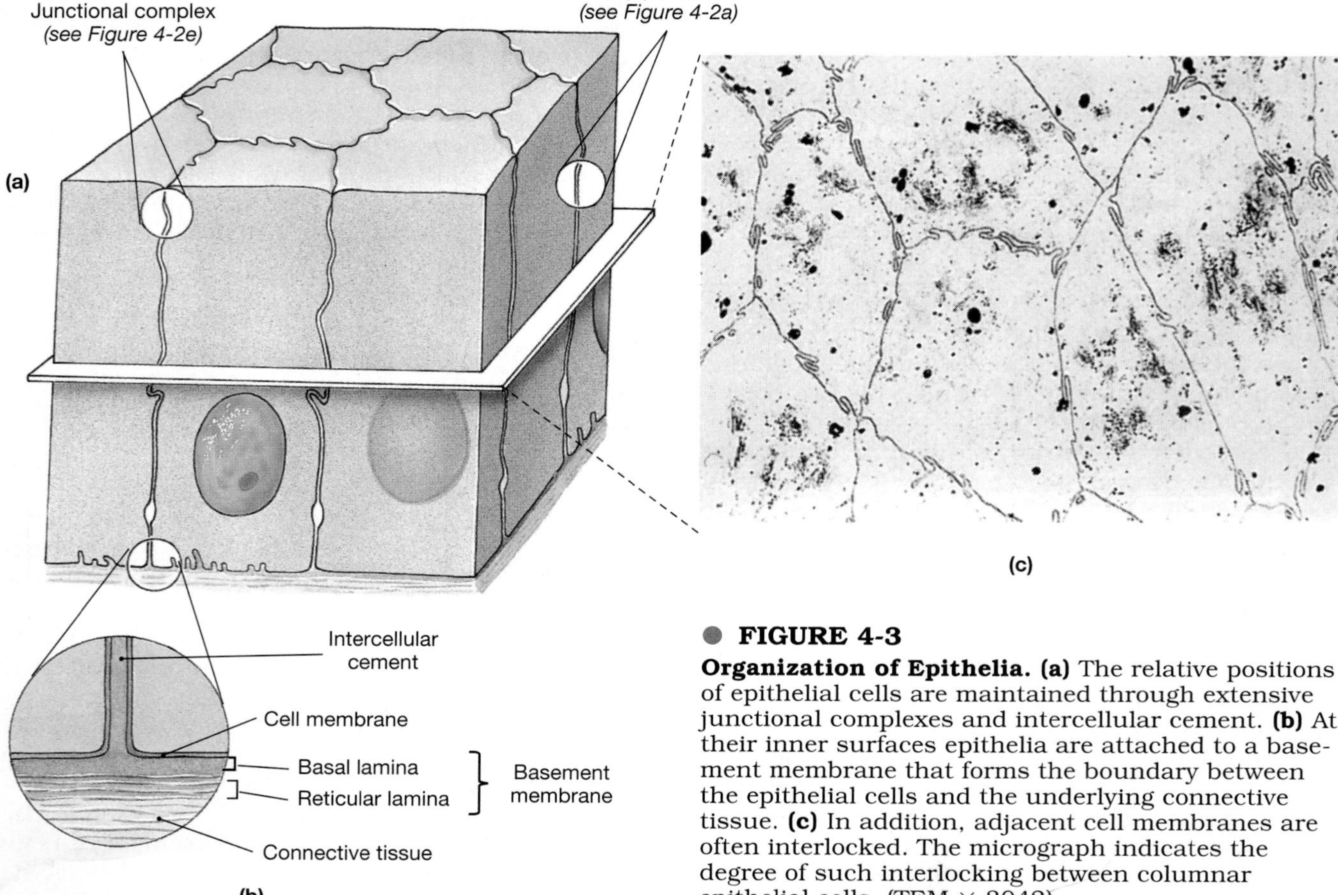
Junctional complex
(see Figure 4-2e)
Gap junction
(see Figure 4-2a)
(a)
(c)
Intercellular
cement
Cell membrane
Basal lamina
Reticular lamina
Basement
membrane
Connective tissue
(b)

● **FIGURE 4-3**
Organization of Epithelia. (a) The relative positions of epithelial cells are maintained through extensive junctional complexes and intercellular cement. **(b)** At their inner surfaces epithelia are attached to a basement membrane that forms the boundary between the epithelial cells and the underlying connective tissue. **(c)** In addition, adjacent cell membranes are often interlocked. The micrograph indicates the degree of such interlocking between columnar epithelial cells. (TEM × 3042)

❑ Classification of Epithelia

Epithelia are classified according to the number of cell layers and the shape of the exposed cells. The classification scheme recognizes two types of layering—*simple* and *stratified*—and three cell shapes—*squamous, cuboidal,* and *columnar.* The shape classification is based on the appearance of the cell when seen in a section that is perpendicular to the epithelial surface and the basement membrane.

If there is only a single layer of cells covering the basement membrane, the epithelium is termed **simple epithelium.** Simple epithelia are relatively thin, and because all the cells have the same polarity, the nuclei form a row above the basement membrane. Because they are so thin, simple epithelia are also relatively fragile. A single layer of cells cannot provide much mechanical protection, and simple epithelia are found only in protected areas inside the body. They line internal compartments and passageways, including the ventral body cavities, the chambers of the heart, and all blood vessels.

Simple epithelia are also characteristic of regions where secretion or absorption occurs, such as the lining of the intestines and the gas-exchange surfaces of the lungs. In these places the thinness of simple epithelia is an advantage, for it reduces the time required for materials to cross the epithelial barrier.

A **stratified epithelium** has several layers of cells above the basement membrane. Stratified epithelia are usually found in areas subject to mechanical or chemical stresses, such as the surface of the skin and the lining of the mouth. Combining the two basic epithelial layouts (simple and stratified) and the three possible cell shapes (squamous, cuboidal, and columnar) enables one to describe almost every epithelium in the body. In most stratified epithelia, the shapes of the cells change as they approach the exposed surface. As a result, the classification scheme for stratified epithelia is based only on the appearance of the cells in the superficial layer.

Squamous Epithelia

Figure 4-4

In a **squamous epithelium** (SKWĀ-mus; *squama,* plate or scale) the cells are thin, flat, and somewhat irregular in shape, like puzzle pieces (Figure 4-4a●). In a sectional view the nucleus occupies

SIMPLE SQUAMOUS EPITHELIUM

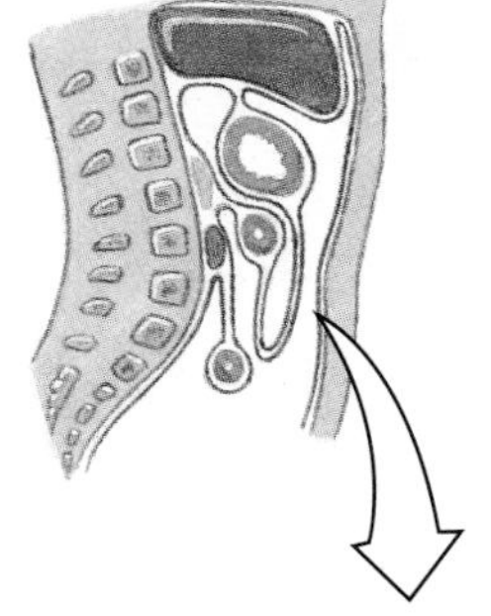

LOCATIONS: Mesothelia lining ventral body cavities; endothelia lining heart and blood vessels; portions of kidney tubules (thin sections of loop of Henle), inner lining of cornea, alveoli of lungs

FUNCTIONS: Reduces friction, controls vessel permeability, performs absorption and secretion

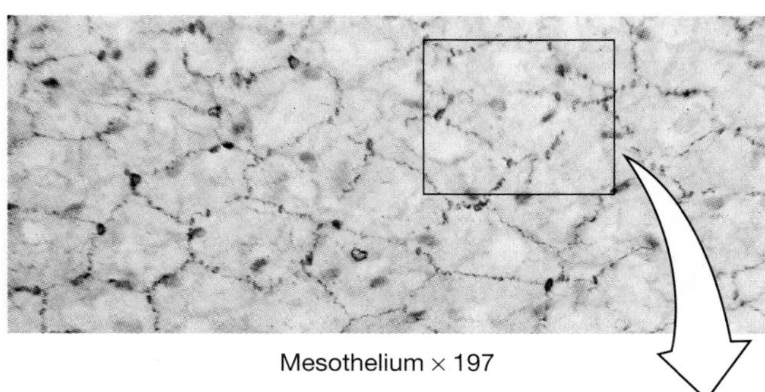

Mesothelium × 197

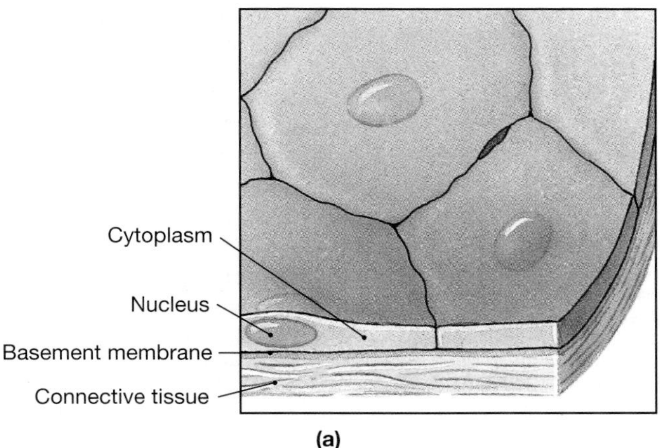
Cytoplasm
Nucleus
Basement membrane
Connective tissue

(a)

STRATIFIED SQUAMOUS EPITHELIUM

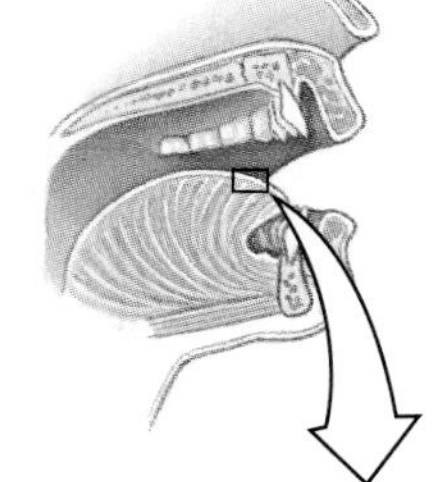

LOCATIONS: Surface of skin; lining of mouth, throat, esophagus, rectum, anus, and vagina

FUNCTIONS: Provides physical protection against abrasion, pathogens, and chemical attack

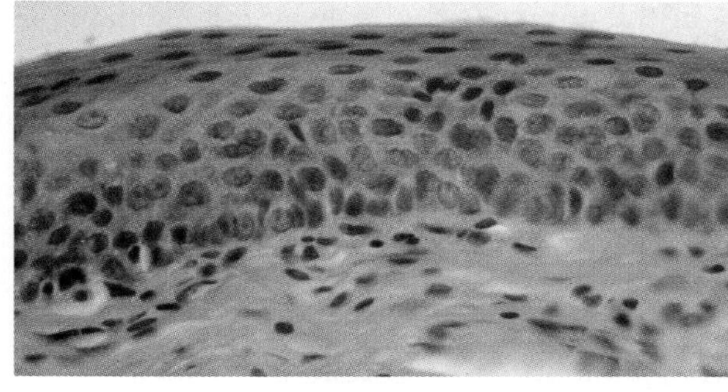

Stratified squamous epithelium × 310

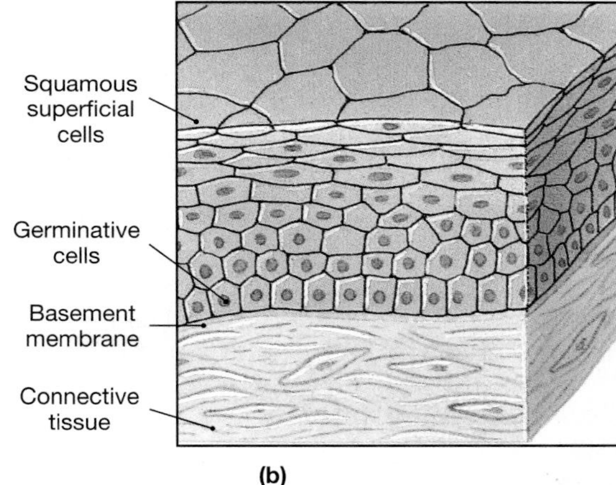
Squamous superficial cells
Germinative cells
Basement membrane
Connective tissue

(b)

● **FIGURE 4-4**

Squamous Epithelia. **(a)** A superficial view of the simple squamous epithelium (mesothelium) that lines the peritoneal cavity. The three-dimensional drawing shows the epithelium in superficial and sectional view. **(b)** A sectional view of the stratified squamous epithelium that covers the tongue.

the thickest portion of each cell; from the surface, the cells look like fried eggs laid side by side. A **simple squamous epithelium** is the most delicate type of epithelium in the body. This type of epithelium is found in protected regions where absorption takes place or where a slick, slippery surface reduces friction. Examples include the respiratory exchange surfaces *(alveoli)* of the lungs, the lining of the ventral body cavities, and the inner surfaces of the circulatory system.

Special names have been given to simple squamous epithelia that line chambers and passageways that do not communicate with the outside world. The simple squamous epithelium that lines the ventral body cavities is known as a **mesothelium** (mez-ō-THĒ-lē-um; *mesos*, middle). The pleura, peritoneum, and pericardium each contain a superficial layer of mesothelium. The simple squamous epithelium lining the heart and all blood vessels is called an **endothelium** (en-dō-THĒ-lē-um).

A **stratified squamous epithelium** (Figure 4-4b●) is usually found where mechanical stresses are severe. Note how the cells form a series of layers, like a stack of plywood sheets. The surface of the skin and the lining of the mouth, esophagus, and anus are areas where this epithelial type provides protection from physical and chemical attack.

Cuboidal Epithelia

Figure 4-5

The cells of a **cuboidal epithelium** resemble little hexagonal boxes; they appear square in typical sectional views. The nuclei are near the center of each cell, with the distance between adjacent nuclei roughly equal to the height of the epithelium. A **simple cuboidal epithelium** provides limited protection and occurs in regions where secretion or absorption takes place. Such an epithelium lines portions of the kidney tubules, as seen in Figure 4-5a•. In the pancreas and salivary glands, simple cuboidal epithelia secrete enzymes and buffers and line the ducts that discharge those secretions. The thyroid gland contains chambers, called *thyroid follicles,* that are lined by a cuboidal secretory epithelium. Thyroid hormones accumulate within the follicles before being released into the bloodstream.

Stratified cuboidal epithelia are relatively rare; they are found along the ducts of sweat glands (Figure 4-5b•) and in the larger ducts of the mammary glands. A **transitional epithelium,** shown in Figure 4-5c•, is unusual because, unlike most epithelia, it tolerates considerable stretching. It is called transitional because the appearance of the epithelium changes as the stretching occurs. A transitional epithelium is found in regions of the urinary system, such as the urinary bladder, where large changes in volume occur. In an empty urinary bladder the epithelium seems to have many layers, and the outermost cells are typically plump cuboidal cells. The layered appearance results from overcrowding; the actual structure of the epithelium can be seen in the full urinary bladder, when the pressure of the urine has stretched the lining to its natural thickness.

Columnar Epithelia

Figure 4-6

Columnar epithelial cells are also hexagonal in cross-section, but they are taller and more slender than cuboidal epithelial cells. The nuclei are crowded into a narrow band close to the basement membrane, and the height of the epithelium is several times the distance between two nuclei (Figure 4-6a•). A **simple columnar epithelium** is often encountered where absorption or secretion is under way, such as inside the small intestine. Inside the stomach and large intestine, the secretions of simple columnar epithelia provide protection from chemical stresses.

Portions of the respiratory tract contain a columnar epithelium that includes several different cell types with varying shapes and functions. Because the cell nuclei are situated at varying distances from the surface, the epithelium appears to be layered or stratified. But it is not truly stratified, because all of the epithelial cells contact the basement membrane. Because it looks stratified but isn't, it is known as a **pseudostratified columnar epithelium** (Figure 4-6b•). Pseudostratified

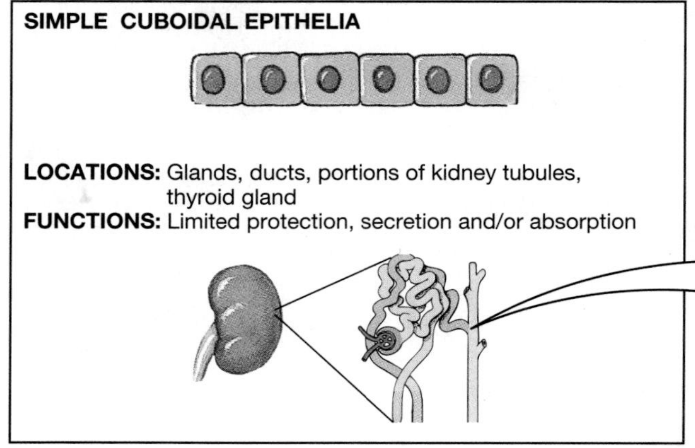
SIMPLE CUBOIDAL EPITHELIA
LOCATIONS: Glands, ducts, portions of kidney tubules, thyroid gland
FUNCTIONS: Limited protection, secretion and/or absorption

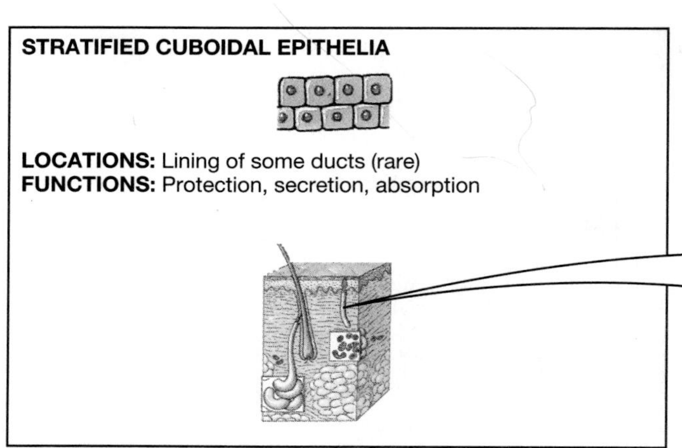
STRATIFIED CUBOIDAL EPITHELIA
LOCATIONS: Lining of some ducts (rare)
FUNCTIONS: Protection, secretion, absorption

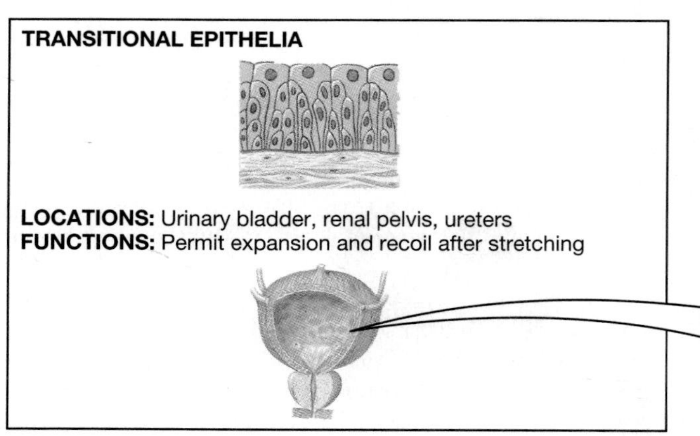
TRANSITIONAL EPITHELIA
LOCATIONS: Urinary bladder, renal pelvis, ureters
FUNCTIONS: Permit expansion and recoil after stretching

• **FIGURE 4-5**

Cuboidal Epithelia. (a) A section through the cuboidal epithelial cells of a kidney tubule. The diagrammatic view emphasizes structural details that permit the classification of an epithelium as cuboidal. **(b)** Sectional view of the stratified cuboidal epithelium lining a sweat gland duct in the skin. **(c)** At left, the lining of the empty urinary bladder, showing transitional epithelium in the relaxed state. At right, the lining of the full bladder, showing the effects of stretching on the arrangement of cells in the epithelium. (LM × 408)

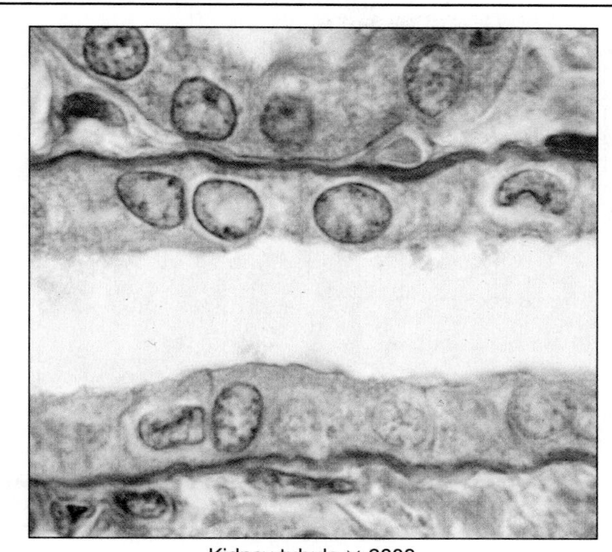
Kidney tubule × 2000

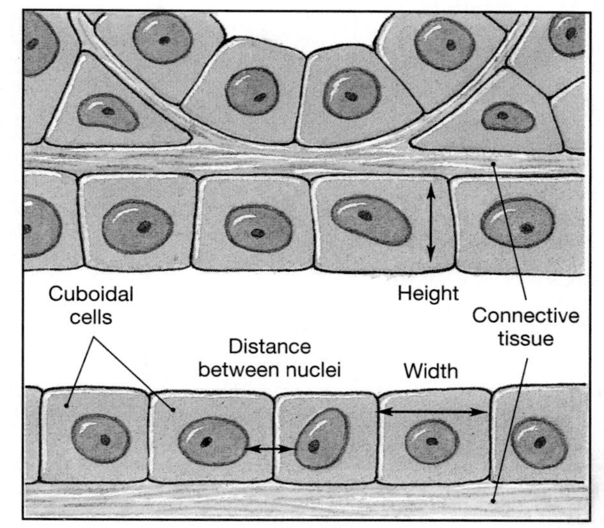
Cuboidal cells
Height
Connective tissue
Distance between nuclei
Width

(a)

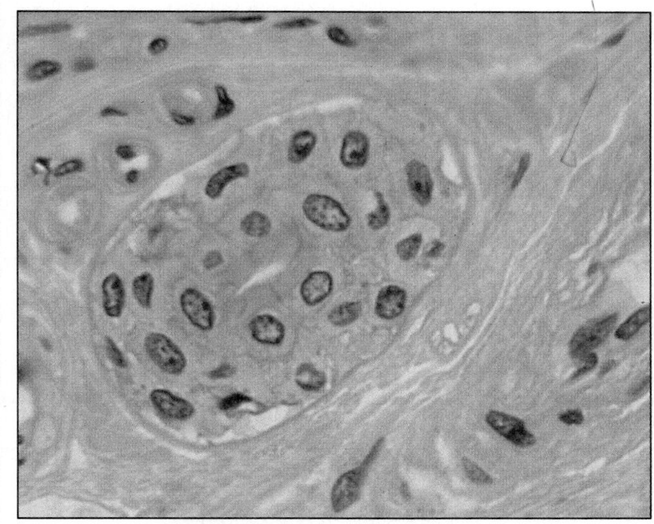
Sweat gland duct × 802

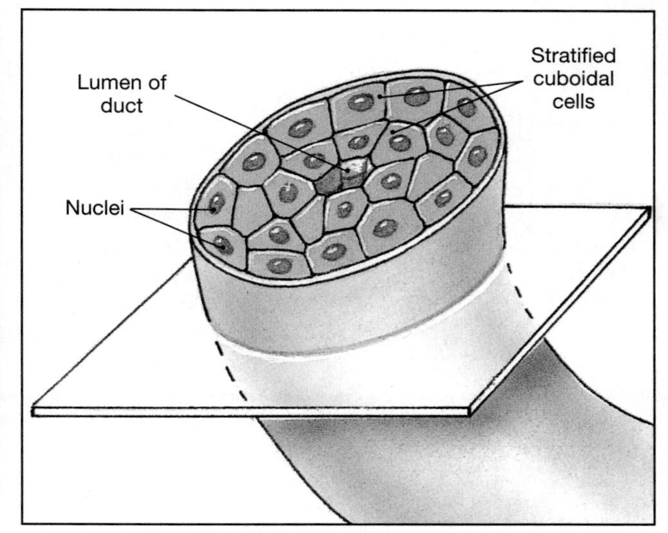
Lumen of duct
Stratified cuboidal cells
Nuclei

(b)

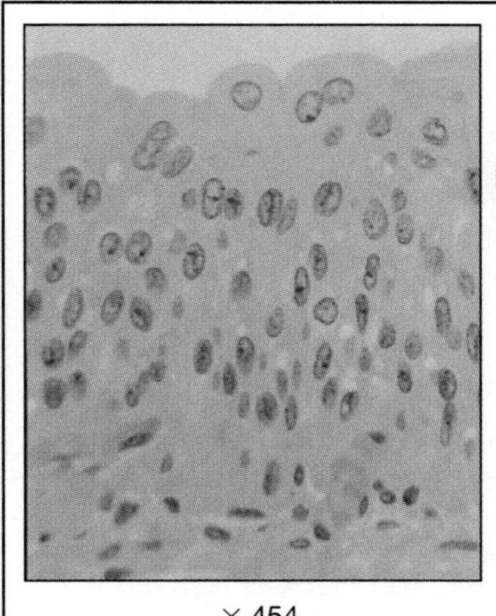
× 454

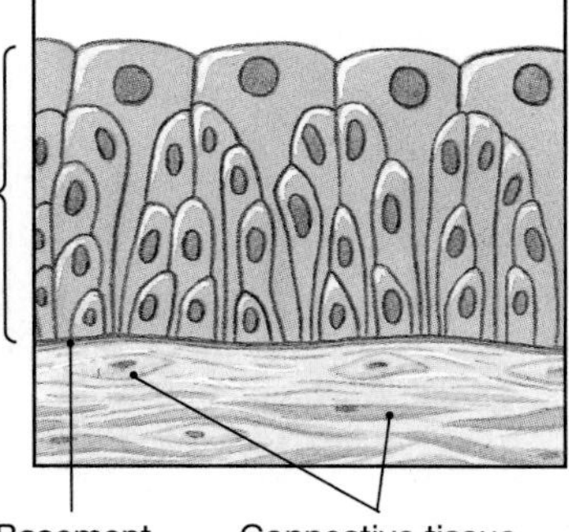
Epithelium (relaxed)
Basement membrane
Connective tissue and smooth muscle layers

Epithelium (stretched)
Basement membrane
Connective tissue and smooth muscle layers

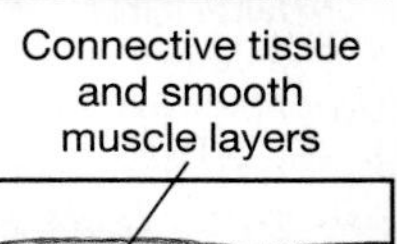

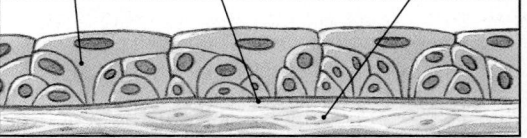

Empty bladder (c) Full bladder

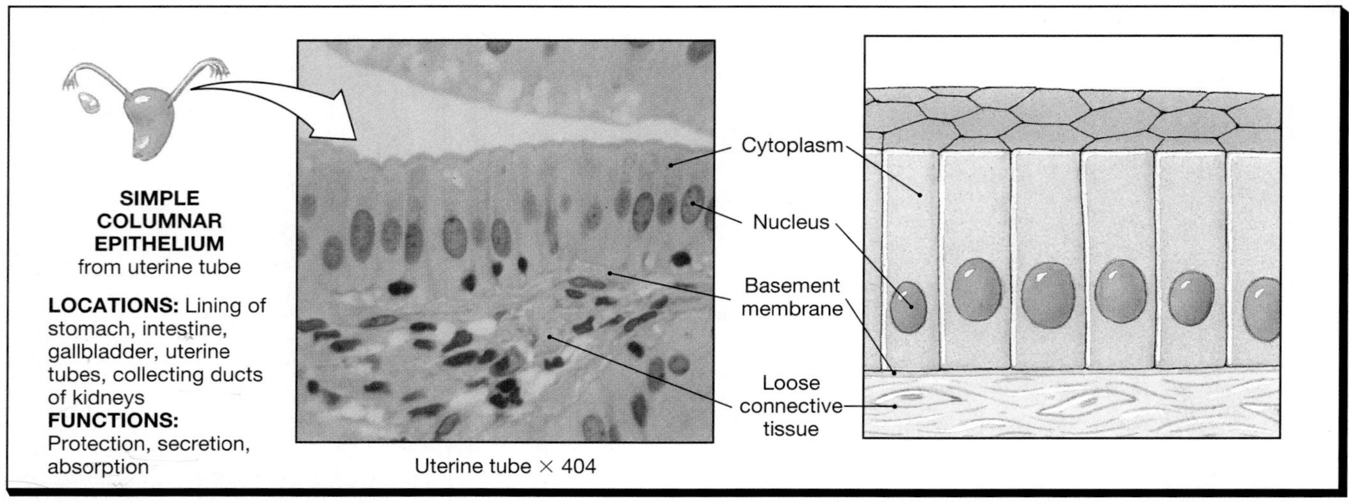
SIMPLE COLUMNAR EPITHELIUM
from uterine tube
LOCATIONS: Lining of stomach, intestine, gallbladder, uterine tubes, collecting ducts of kidneys
FUNCTIONS: Protection, secretion, absorption
Uterine tube × 404
Cytoplasm
Nucleus
Basement membrane
Loose connective tissue

(a)

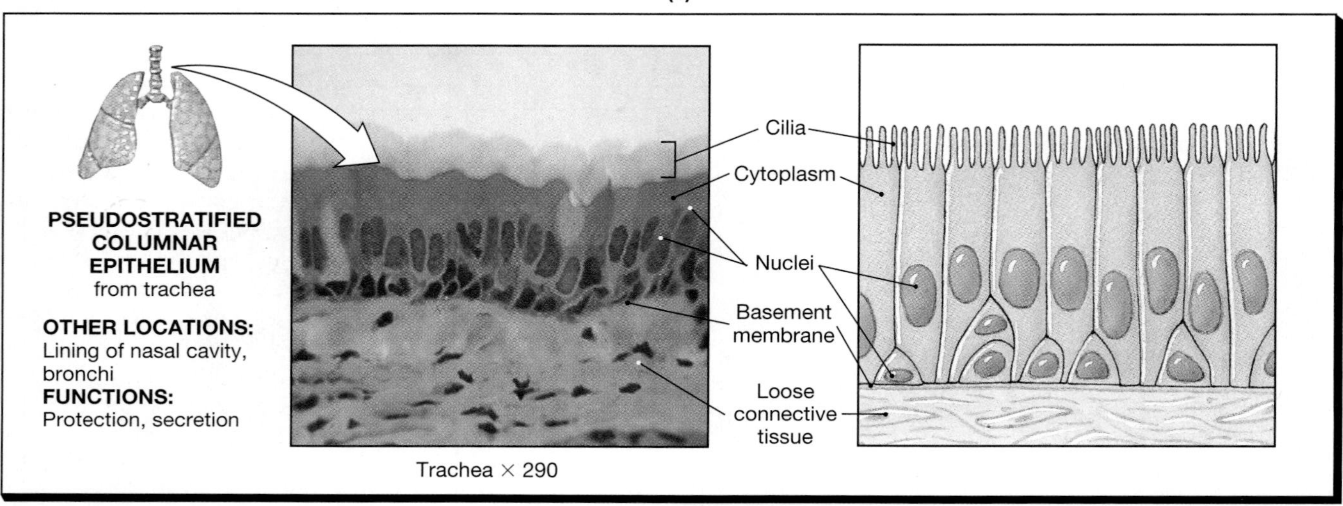
PSEUDOSTRATIFIED COLUMNAR EPITHELIUM
from trachea
OTHER LOCATIONS: Lining of nasal cavity, bronchi
FUNCTIONS: Protection, secretion
Trachea × 290
Cilia
Cytoplasm
Nuclei
Basement membrane
Loose connective tissue

(b)

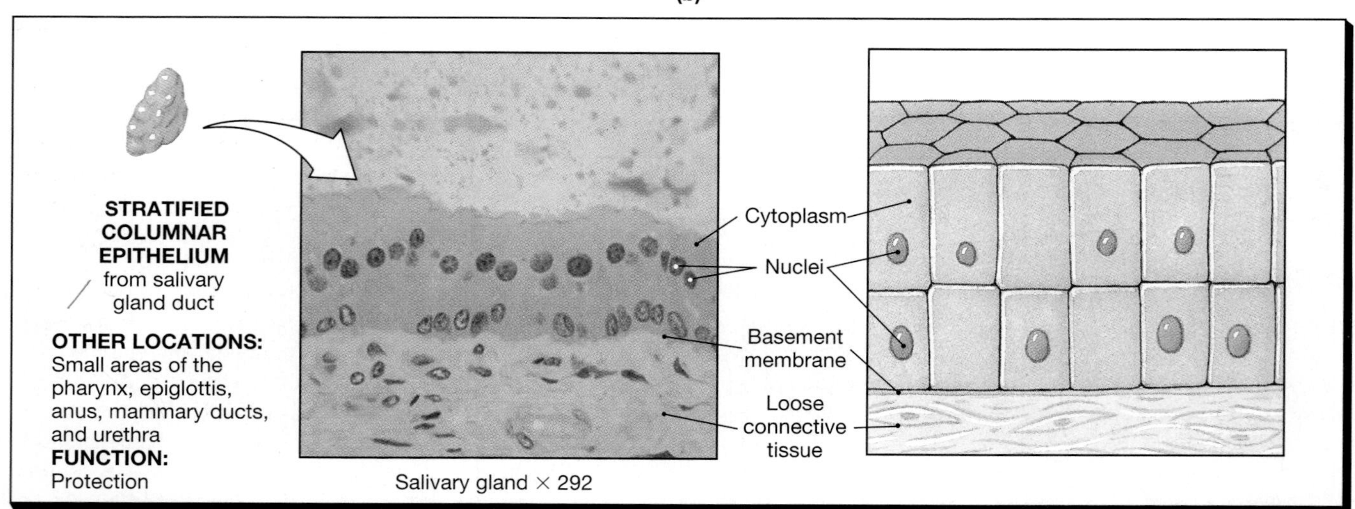
STRATIFIED COLUMNAR EPITHELIUM
from salivary gland duct
OTHER LOCATIONS: Small areas of the pharynx, epiglottis, anus, mammary ducts, and urethra
FUNCTION: Protection
Salivary gland × 292
Cytoplasm
Nuclei
Basement membrane
Loose connective tissue

(c)

FIGURE 4-6

Columnar Epithelia. (a) Micrograph showing the characteristics of simple columnar epithelium. In the diagrammatic sketch, note the relationships between the height and width of each cell; the relative size, shape, and location of nuclei; and the distance between nuclei. Contrast these observations with the corresponding characteristics of simple cuboidal epithelia (Figure 4-5a). **(b)** The pseudostratified, ciliated, columnar epithelium of the respiratory tract. Note the uneven layering of the nuclei. **(c)** A stratified columnar epithelium is sometimes found along large ducts, such as this salivary gland duct. Note the thickness of the epithelium and the location and orientation of the nuclei.

columnar epithelial cells always possess cilia. This type of epithelium lines most of the nasal cavity, the trachea (windpipe), bronchi, and portions of the male reproductive tract.

Stratified columnar epithelia are relatively rare, providing protection along portions of the pharynx, urethra, and anus, as well as along a few large excretory ducts. The epithelium may have only two layers (Figure 4-6c●) or multiple layers. When multiple layers exist, only the superficial cells are columnar.

EXFOLIATIVE CYTOLOGY **Exfoliative cytology** (eks-FŌ-lē-a-tiv; *ex,* from + *folium,* leaf) is the study of cells shed or collected from epithelial surfaces. The cells may be examined for a variety of reasons: for example, to check for cellular changes that indicate cancer formation or to identify the pathogens involved in an infection. The cells are collected either by sampling the fluids that cover the epithelia lining the respiratory, digestive, urinary, or reproductive tracts or by removing fluid from one of the ventral body cavities. The sampling procedure is often called a *Pap test,* named after Dr. George Papanicolaou. Probably the most familiar Pap test is the test for cervical cancer, which involves scraping a small number of cells from the tip of the **cervix,** a portion of the uterus that projects into the vagina.

Amniocentesis is another important test that relies on exfoliative cytology. In this procedure, exfoliated epithelial cells are collected from a sample of *amniotic fluid,* the fluid that surrounds and protects a developing fetus. Examination of these cells can determine if the fetus has a genetic abnormality, such as *Down syndrome,* that affects chromosomal number or structure.

❑ Glandular Epithelia

Many epithelia contain gland cells that produce secretions. In general, glands and gland cells are classified as *endocrine* or *exocrine* according to the final distribution of their secretions.

Endocrine Glands

Endocrine glands (*endo,* inside) release their secretions into the surrounding interstitial fluid. Because there are no endocrine ducts, endocrine glands are often called *ductless glands.* From the interstitial fluids, these secretions, called *hormones,* enter the circulation for distribution throughout the body. Hormones regulate or coordinate the activities of other cells that may reside in the same tissue or in other tissues and organs.

Endocrine cells may be part of an epithelial surface, such as the lining of the digestive tract, or they may be separate, as in the pancreas, thyroid gland, thymus, and pituitary gland. Endocrine cells, organs, and hormones are considered further in Chapter 18.

Only a few complex glands produce both exocrine and endocrine secretions. For example, the pancreas contains endocrine cells that secrete hormones as well as exocrine cells that produce digestive enzymes and buffers.

Exocrine Glands

Figures 4-7, 4-8

Exocrine glands (*exo-,* outside) produce secretions that are discharged onto epithelial surfaces. Most exocrine glands release their secretions into tubular passageways, called **ducts,** that empty onto the surface of the skin or onto an epithelial surface lining one of the internal passageways that communicates with the exterior. There are many kinds of exocrine secretions, all performing a variety of functions.

Exocrine glands may be classified by their *mode of secretion,* the *type of secretion,* and the *structure of the gland.*

MODES OF SECRETION A glandular epithelial cell may use one of three methods to release its secretions: *merocrine secretion, apocrine secretion,* or *holocrine secretion.* In **merocrine secretion** (MER-ō-krin; *meros,* part + *krinein,* to separate) the product is released through exocytosis (Figure 4-7a●). This is the most common mode of secretion. One merocrine secretion, called **mucus,** is an effective lubricant, a protective barrier, and a sticky trap for foreign particles and microorganisms. The mucous secretions of merocrine glands coat the passageways of the digestive and respiratory tracts. In the skin, merocrine sweat glands produce the watery perspiration that helps cool the body on a hot day.

Apocrine secretion (AP-ō-krin; *apo-,* off) involves the loss of cytoplasm as well as the secretory product (Figure 4-7b●). The outermost portion of the cytoplasm becomes packed with secretory vesicles before it is shed. The thick, sticky underarm perspiration targeted by the deodorant industry results from apocrine secretion. Milk production in the female breasts involves a combination of merocrine and apocrine secretion.

Merocrine and apocrine secretions leave a cell intact and able to continue secreting. **Holocrine secretion** (HOL-ō-krin; *holos,* entire) destroys the gland cell. During holocrine secretion, the entire cell becomes packed with secretory products and then bursts apart (Figure 4-7c●), releasing the secretion but killing the cell. Further secretion depends on the lost gland cells being replaced by the division of stem cells. Sebaceous glands, associated with hair follicles, produce an oily hair coating by means of holocrine secretion.

TYPES OF SECRETIONS Exocrine glands may be categorized according to the nature of the secretion produced:

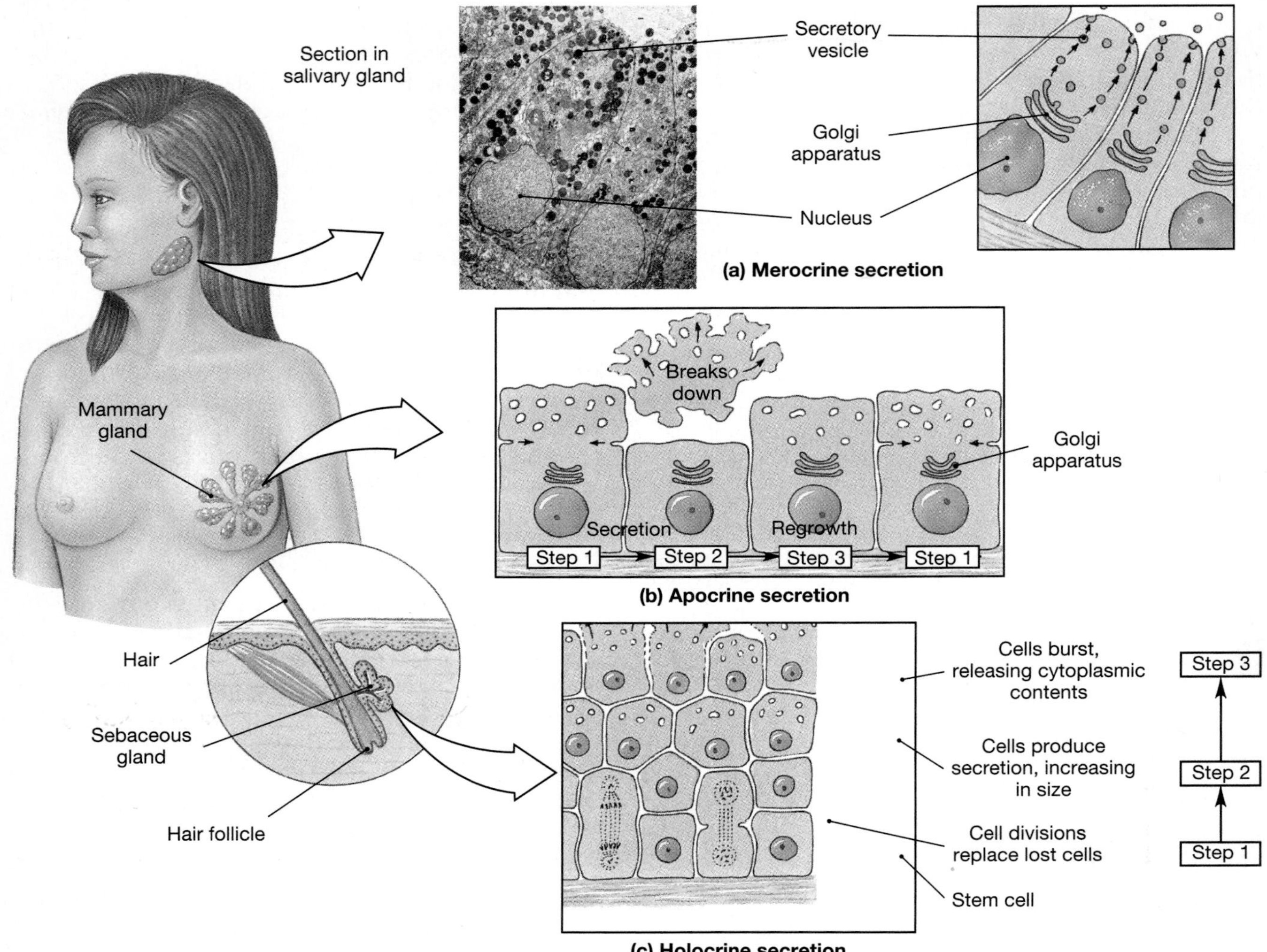

Section in salivary gland
Secretory vesicle
Golgi apparatus
Nucleus
(a) Merocrine secretion
Mammary gland
Breaks down
Golgi apparatus
Secretion
Regrowth
Step 1
Step 2
Step 3
Step 1
(b) Apocrine secretion
Hair
Sebaceous gland
Hair follicle
Cells burst, releasing cytoplasmic contents
Cells produce secretion, increasing in size
Cell divisions replace lost cells
Stem cell
Step 3
Step 2
Step 1
(c) Holocrine secretion

● **FIGURE 4-7**

Mechanisms of Glandular Secretion. (a) In merocrine secretion, secretory vesicles are discharged at the surface of the gland cell through exocytosis. **(b)** Apocrine secretion involves the loss of cytoplasm. Inclusions, secretory vesicles, and other cytoplasmic components are shed in the process. The gland cell then undergoes a period of growth and repair before releasing additional secretions. **(c)** Holocrine secretion occurs as superficial gland cells break apart. Continued secretion involves the replacement of these cells through the mitotic divisions of underlying stem cells.

- **Serous glands** secrete a watery solution that usually contains enzymes.
- **Mucous glands** secrete a viscous mucus.
- **Mixed exocrine glands** contain more than one type of gland cell and may produce two different exocrine secretions, one serous and the other mucous. The *submandibular gland,* one of the salivary glands, is an example of a mixed exocrine gland.

GLAND STRUCTURE In epithelia that contain scattered gland cells, the individual secretory cells are called **unicellular glands. Multicellular glands** include glandular epithelia and aggregations of gland cells that produce exocrine or endocrine secretions.

Unicellular Exocrine Glands. The only examples of **unicellular exocrine glands** in the body are **goblet cells.** Goblet cells, which secrete mucus, are scattered among other epithelial cells. For example, the pseudostratified columnar epithelium that lines the trachea and the columnar epithelium of the small and large intestines both contain an abundance of goblet cells.

Multicellular Exocrine Glands. The simplest **multicellular exocrine gland** is called a **secretory sheet.** In a secretory sheet, glandular cells dominate the epithelium and release their secretions into an inner compartment. The mucus-secreting cells that line the stomach are an example. Their continual secretion protects the stomach from the acids and enzymes it contains.

Most other multicellular glands are found in pockets set back from the epithelial surface; their secretory products travel through one or more

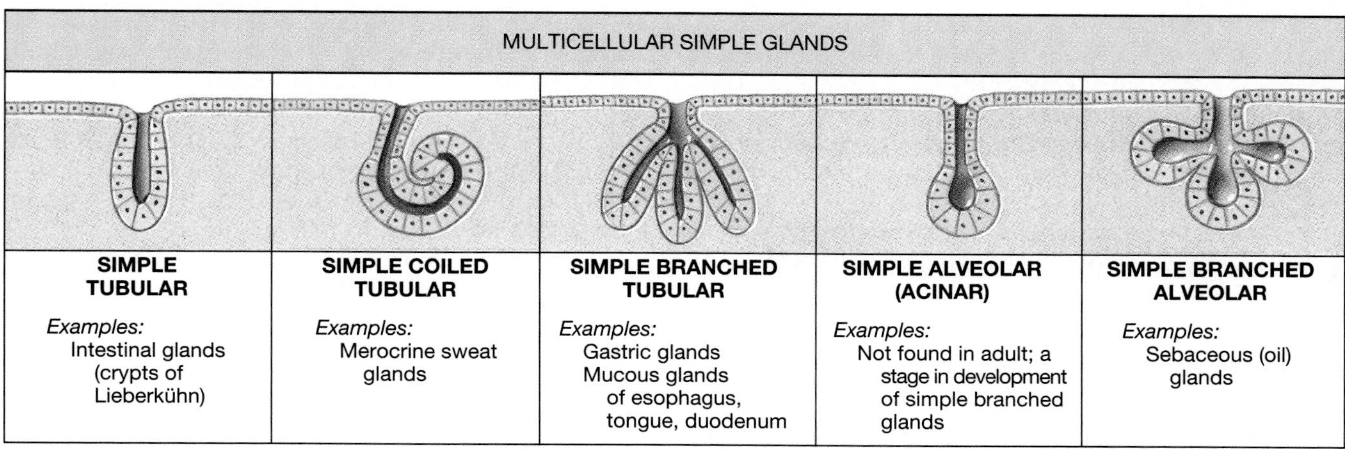
MULTICELLULAR SIMPLE GLANDS
SIMPLE TUBULAR
Examples: Intestinal glands (crypts of Lieberkühn)
SIMPLE COILED TUBULAR
Examples: Merocrine sweat glands
SIMPLE BRANCHED TUBULAR
Examples: Gastric glands Mucous glands of esophagus, tongue, duodenum
SIMPLE ALVEOLAR (ACINAR)
Examples: Not found in adult; a stage in development of simple branched glands
SIMPLE BRANCHED ALVEOLAR
Examples: Sebaceous (oil) glands

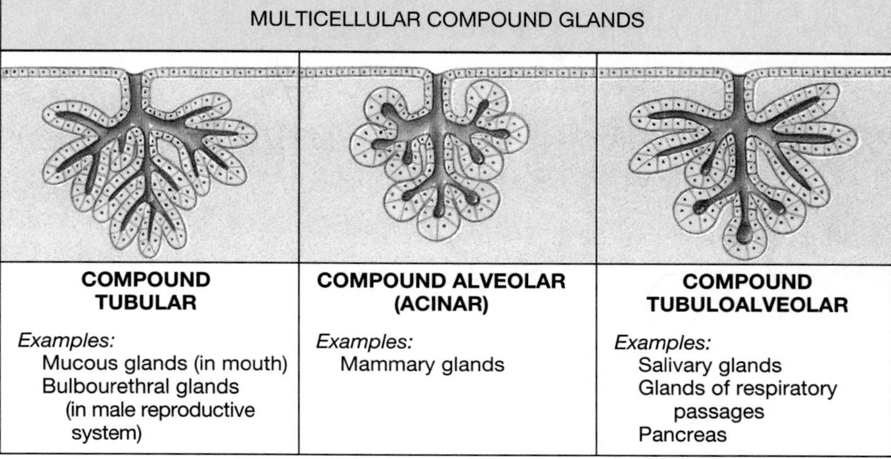
MULTICELLULAR COMPOUND GLANDS
COMPOUND TUBULAR
Examples: Mucous glands (in mouth) Bulbourethral glands (in male reproductive system)
COMPOUND ALVEOLAR (ACINAR)
Examples: Mammary glands
COMPOUND TUBULOALVEOLAR
Examples: Salivary glands Glands of respiratory passages Pancreas

● **FIGURE 4-8**
A Structural Classification of Exocrine Glands

ducts to reach the epithelial surface. Examples include the salivary glands that produce mucus and digestive enzymes.

Multicellular exocrine glands may be complex in structure. Three characteristics are used to describe the organization of a multicellular gland: (1) the shape of the secretory portion of the gland, (2) the branching pattern of the duct, and (3) the relationship between the ducts and the glandular areas.

1. Glands whose glandular cells form tubes are called **tubular.** Those that form blind pockets are called **alveolar** (al-VĒ-ō-lar; *alveolus,* sac), or **acinar** (A-si-nar; *acinus,* chamber). Glands whose gland cells form both tubes and pockets are called **tubuloalveolar** or tubuloacinar.
2. A gland is called **simple** if there is a single duct that does not divide on its way to the gland cells. The gland is called **compound** if the duct divides repeatedly.
3. A gland is called **branched** if several glandular areas (tubular or acinar) share a common duct. Note that the term *branched* always refers to the *glandular areas* not to the duct.

Figure 4-8● diagrams this method of classification based on gland structure. Specific examples of each gland type will be discussed in later chapters.

✔ You examine a tissue with a light microscope and see a simple squamous epithelium on the outer surface. Can this be a sample of the skin surface?

✔ The secretory cells of sebaceous glands fill with secretions and then rupture, releasing their contents. What type of secretion is this?

✔ Close examination of a gland indicates that there are no ducts to carry the glandular secretions. Chemical tests reveal that the gland's secretions are released directly into the extracellular fluid. What type of gland is this?

Connective Tissues

Connective tissues include tissues such as bone, fat, and blood, which are quite different in appearance and function. Nevertheless, all connective tissues have three basic components: (1) specialized cells, (2) extracellular protein fibers, and (3) a fluid known as the **ground substance.** The extracellular fibers and ground substance constitute the **matrix** that surrounds the cells. Whereas epithelial tissue consists almost entirely of cells, the extracellular matrix typically accounts for most of the volume of connective tissues.

Connective tissues are found throughout the body but are never exposed to the environment outside the body. Many connective tissues are highly vascular and contain sensory receptors that provide pain, pressure, temperature, and other sensations.

The functions of connective tissues include:

1. Establishing a structural framework for the body.
2. Transporting fluids and dissolved materials from one region of the body to another.
3. Providing protection for delicate organs.
4. Supporting, surrounding, and interconnecting other tissue types.
5. Storing energy reserves, especially in the form of lipids.
6. Defending the body from invading microorganisms.

❑ Classification of Connective Tissues

Connective tissue can be classified into three categories: (1) *connective tissue proper*, (2) *fluid connective tissues*, and (3) *supporting connective tissues.*

1. **Connective tissue proper** refers to connective tissues with many types of cells and extracellular fibers in a syrupy ground substance. These connective tissues differ in the number of cell types they contain and the relative properties and proportions of fibers and ground substance. For example, *adipose* (fat) *tissue* and *tendons* are both connective tissue proper, but they have very different structural and functional characteristics. Connective tissue proper is divided into *loose connective tissues* and *dense connective tissues* on the basis of the relative proportions of cells, fibers, and ground substance.
2. **Fluid connective tissues** have a distinctive population of cells suspended in a watery matrix that contains dissolved proteins. There are two fluid connective tissues, *blood* and *lymph.*
3. **Supporting connective tissues** are of two types, *cartilage* and *bone.* These tissues have a less diverse cell population than connective tissue proper and a matrix that contains closely packed fibers. The matrix of cartilage is a gel whose characteristics vary depending on the predominant fiber type. The matrix of bone is said to be **calcified** because it contains mineral deposits, primarily calcium salts. These minerals give the bone its rigidity.

❑ Connective Tissue Proper

Figure 4-9

Connective tissue proper contains extracellular fibers, a viscous ground substance, and a varied cell population. Some of these cells are involved with local maintenance, repair, and energy storage. These cells include *fibroblasts, fixed macrophages, adipocytes, mesenchymal cells,* and, in a few locations, *melanocytes.* Other cells are responsible for the defense and repair of damaged tissues. These cells include *free macrophages, mast cells, lymphocytes, plasma cells,* and *microphages.*

The number of cells and cell types within a tissue at any given moment varies depending on local conditions. Refer to Figure 4-9a● as we describe the cells and fibers of connective tissue proper.

The Cell Population

The major cell types of connective tissue proper include the following.

- **Fibroblasts** (FĪ-brō-blasts) are the most abundant fixed cells in connective tissue proper. These slender or *stellate* (star-shaped) cells are responsible for the production and maintenance of the connective tissue fibers. Each fibroblast manufactures and secretes protein subunits that interact to form large extracellular fibers. In addition, fibroblasts secrete hyaluronic acid, a proteoglycan that gives the ground substance its syrupy consistency.
- **Fixed macrophages** (MAC-rō-fā-jez; *phagein,* to eat) are large, amoeboid cells that are scattered among the fibers. These cells engulf damaged cells or pathogens that enter the tissue. Although they are not abundant, they play an important role in mobilizing the body's defenses. When stimulated they release chemicals that activate the immune system and attract large numbers of wandering cells involved in tissue defense.
- **Adipocytes** (AD-i-pō-sīts) are also known as fat cells, or *adipose cells.* A typical adipocyte contains a single, enormous lipid droplet. The nucleus and other organelles are squeezed to one side, making the cell in section resemble a class ring. The number of fat cells varies from one type of connective tissue to another, from one region of the body to another, and from individual to individual.
- **Mesenchymal cells** are stem cells that are present in many connective tissues. These cells respond to local injury or infection by dividing to produce daughter cells that differentiate into fibroblasts, macrophages, or other connective tissue cells.
- **Melanocytes** (me-LAN-ō-sīts) synthesize and store a brown pigment, **melanin** (ME-la-nin), that gives the tissue a dark color. Melanocytes are common in the epithelium of the skin, where they play a major role in determining skin color. However, melanocytes are also abundant in con-

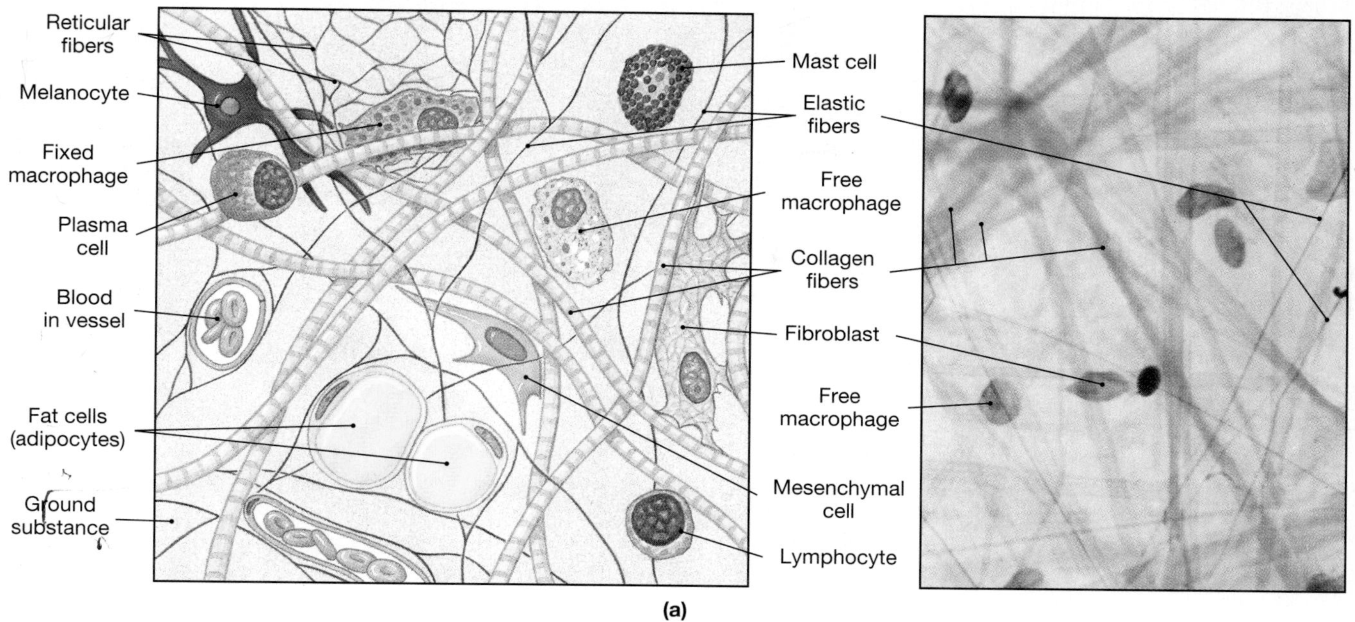
Reticular fibers
Melanocyte
Fixed macrophage
Plasma cell
Blood in vessel
Fat cells (adipocytes)
Ground substance
Mast cell
Elastic fibers
Free macrophage
Collagen fibers
Fibroblast
Free macrophage
Mesenchymal cell
Lymphocyte
(a)

● **FIGURE 4-9**

The Cells and Fibers of Connective Tissue Proper. (a) A summary of the cell types and fibers of connective tissue proper. (LM × 384) **(b)** Mesenchyme, the first connective tissue to appear in the embryo. (LM × 136) **(c)** Mucous connective tissue. This sample was taken from the umbilical cord of a fetus. Mucous connective tissue in this location is also known as *Wharton's jelly.* (LM × 136)

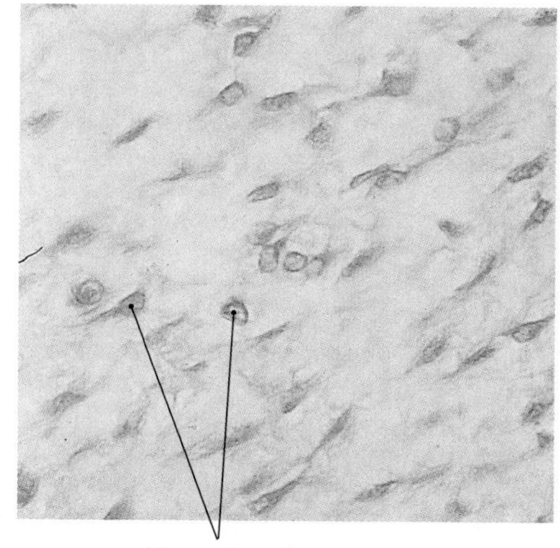
Mesenchymal cells
(b)

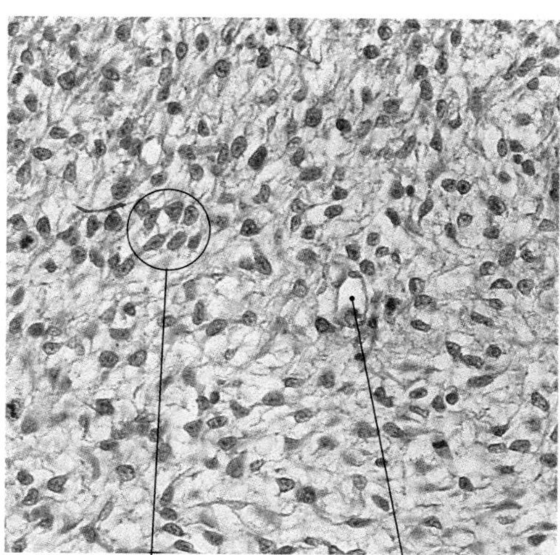
Mesenchymal cells
Blood vessel
(c)

nective tissues of the eye, and they are present in the dermis of the skin, although there are regional and individual differences in the number present.

- **Free macrophages** wander throughout the body. When an infection occurs, these phagocytic cells are drawn to the affected area. In effect, the few fixed macrophages in a tissue provide a "front-line" defense that will be reinforced by the arrival of free macrophages and other specialized cells in massive numbers.
- **Mast cells** are small, mobile connective tissue cells often found near blood vessels. The cytoplasm of a mast cell is filled with granules of **histamine** (HIS-tuh-mēn) and **heparin** (HEP-uh-rin). These chemicals, released following injury or infection, stimulate local inflammation. Histamine and heparin granules are also found in *basophils,* blood cells that enter damaged tissues and enhance the inflammation process.
- **Lymphocytes** (LIM-fō-sīts), like free macrophages, migrate throughout the body. Their numbers increase markedly wherever tissue damage occurs, and some of the lymphocytes may then develop into **plasma cells.** Plasma cells are responsible for the production of *antibodies,* proteins involved in defending the body against disease.
- **Microphages** are phagocytic blood cells that move through connective tissues in small numbers. When an infection or injury occurs, chemicals released by fixed macrophages and mast cells attract them in large numbers.

EMBRYOLOGY SUMMARY The Development of Tissues

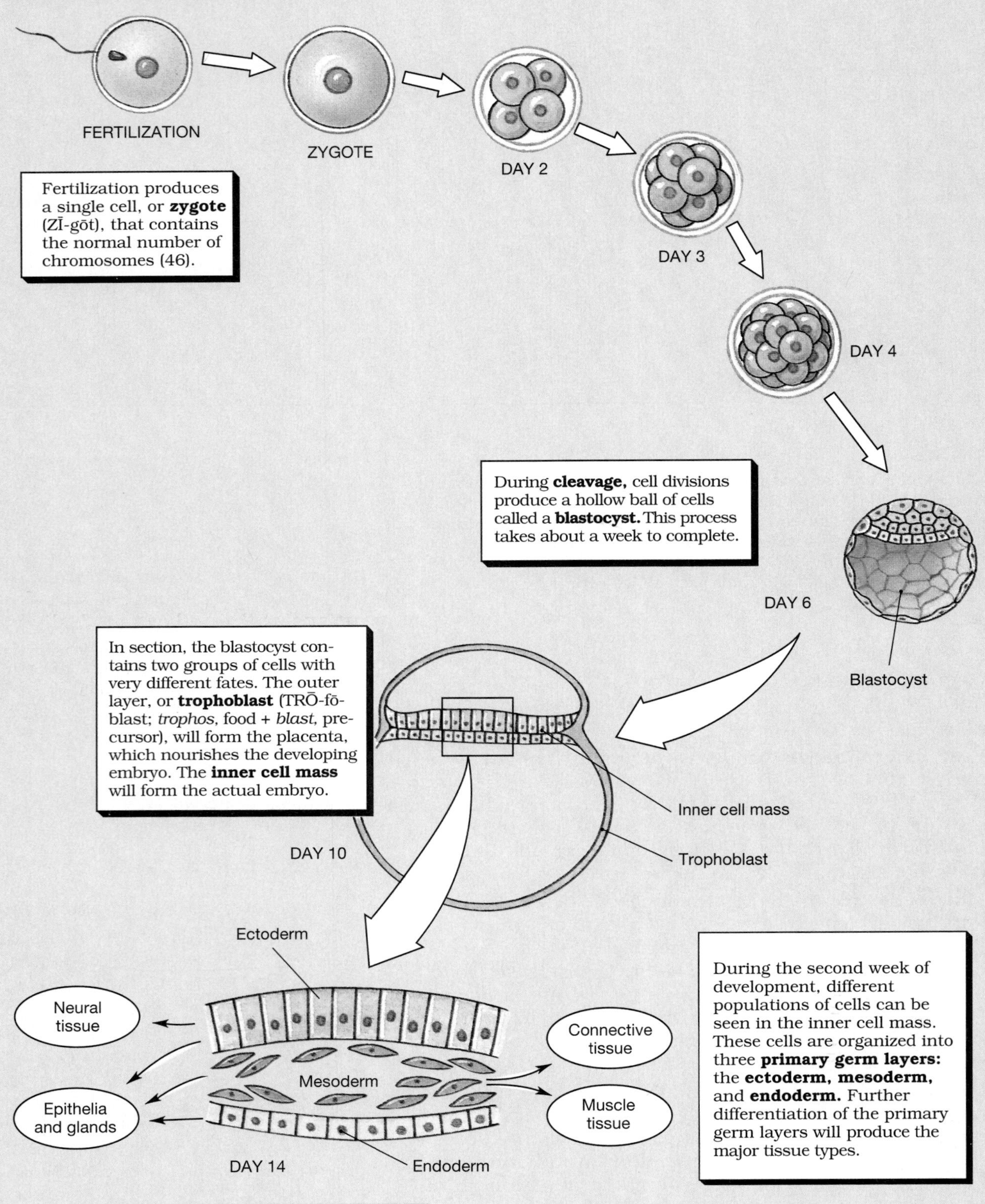
FERTILIZATION
ZYGOTE
DAY 2
DAY 3
DAY 4
Fertilization produces a single cell, or **zygote** (ZĪ-gōt), that contains the normal number of chromosomes (46).
During **cleavage,** cell divisions produce a hollow ball of cells called a **blastocyst.** This process takes about a week to complete.
DAY 6
Blastocyst
In section, the blastocyst contains two groups of cells with very different fates. The outer layer, or **trophoblast** (TRŌ-fō-blast; *trophos,* food + *blast,* precursor), will form the placenta, which nourishes the developing embryo. The **inner cell mass** will form the actual embryo.
Inner cell mass
DAY 10
Trophoblast
Ectoderm
Neural tissue
Mesoderm
Connective tissue
Epithelia and glands
Muscle tissue
DAY 14
Endoderm
During the second week of development, different populations of cells can be seen in the inner cell mass. These cells are organized into three **primary germ layers:** the **ectoderm, mesoderm,** and **endoderm.** Further differentiation of the primary germ layers will produce the major tissue types.

All three germ layers participate in the formation of functional organs and organ systems. Their interactions will be detailed in later Embryology Summaries dealing with specific systems.

Connective Tissue Fibers

Three basic types of fibers are found in connective tissue: *collagen fibers, reticular fibers,* and *elastic fibers.* Fibroblasts form all three types of fibers through the secretion of protein subunits that interact within the matrix.

1. **Collagen fibers** are long, straight, and unbranched. These are the most common fibers in connective tissue proper. Each collagen fiber consists of a bundle of fibrous protein subunits wound together like the strands of a rope. Like a rope, a collagen fiber is flexible, but it is stronger than steel when pulled from either end. *Tendons,* which connect skeletal muscles to bones, consist almost entirely of collagen fibers. Typical *ligaments* are similar to tendons, but they connect one bone to another. Tendons and ligaments can withstand tremendous forces; uncontrolled muscle contractions or skeletal movements are more likely to break a bone than snap a tendon or ligament.
2. **Reticular fibers** (*reticulum,* network) contain the same protein subunits as collagen fibers, but they are arranged in a different way. These fibers are thinner than collagen fibers, and they form a branching, interwoven framework that is tough but flexible. Because they form a network, rather than sharing a common alignment, reticular fibers can resist forces applied from many different directions. This interwoven network stabilizes the relative positions of the organ's cells, blood vessels, and other structures despite changing positions and the pull of gravity.
3. **Elastic fibers** contain the protein *elastin.* Elastic fibers are branched and wavy, and after stretching they will return to their original length. *Elastic ligaments* are dominated by elastic fibers. They are relatively rare, but have important functions, such as interconnecting the vertebrae.

Ground Substance

Figure 4-9a

Ground substance fills the spaces between cells and surrounds the connective tissue fibers (Figure 4-9a●). Ground substance in normal connective tissue proper is clear and colorless. It often has a consistency similar to that of maple syrup, caused by the presence of proteoglycans and glycoproteins. ∞ [p. 56] The ground substance is dense enough that bacteria have trouble moving through it (imagine swimming in molasses). This density slows the spread of pathogens through the tissue and makes them easier for phagocytes to catch.

MARFAN'S SYNDROME **Marfan's syndrome** is an inherited condition caused by the production of an abnormal form of *fibrillin,* a glycoprotein important to normal connective tissue strength and elasticity. Because connective tissues are found in most organs, the effects of this defect are widespread. The most visible sign of Marfan's syndrome involves the skeleton; individuals with this syndrome are usually tall, with abnormally long arms, legs, and fingers. But the most serious consequences involve the cardiovascular system. Roughly 90 percent of the people with Marfan's syndrome have abnormal cardiovascular systems. The most dangerous result is that the weakened connective tissues in the walls of major arteries, such as the aorta, may burst, causing a sudden, fatal loss of blood.

Embryonic Connective Tissues

Figure 4-9

Mesenchyme, or *embryonic connective tissue,* is the first connective tissue to appear in the developing embryo. Mesenchyme contains star-shaped cells that are separated by a matrix that contains very fine protein filaments. This connective tissue (Figure 4-9b●) gives rise to all other forms of connective tissue, including fluid connective tissues, cartilage, and bone. **Mucous connective tissue** (Figure 4-9c●) is a loose connective tissue found in many portions of the embryo, including the umbilical cord.

Neither of these embryonic forms of connective tissue is found in the adult. However, many adult connective tissues contain scattered mesenchymal cells that can assist in tissue repairs after injury.

Loose Connective Tissues

Figures 4-9, 4-10

Loose connective tissue, or **areolar tissue** (*areola,* little space), is the "packing material" of the body. It fills spaces between organs, provides cushioning, and supports epithelia. Loose connective tissue also surrounds and supports blood vessels and nerves, stores lipids, and provides a route for the diffusion of materials.

Typical loose connective tissue (Figure 4-9a●) is the least specialized connective tissue in the adult body. This tissue contains all of the cells and fibers found in any connective tissue proper. Loose connective tissue has an open framework, and ground substance accounts for most of its volume. This syrupy fluid cushions shocks and, because the fibers are loosely organized, loose connective tissue can distort without damage. The presence of elastic fibers makes it fairly resilient, so this tissue returns to its original shape after external pressure is relieved.

Loose connective tissue forms a layer that separates the skin from deeper structures. In addition to providing padding, the elastic properties of this layer allow a considerable amount of independent movement. Thus, pinching the skin of the arm does not affect the underlying muscle. Conversely, contractions of the underlying muscles do not pull against the skin—as the muscle bulges, the loose connective tissue stretches. Because this tissue has an extensive circulatory supply, drugs are typically injected into the loose connective tissue layer under the skin.

In addition to delivering oxygen and nutrients and removing carbon dioxide and waste products, the capillaries in loose connective tissue carry wandering cells to and from the tissue. Epithelia usually cover a layer of loose connective tissue, and fibroblasts are responsible for maintaining the reticular lamina of the basement membrane. The epithelial cells rely on diffusion across that membrane, and the capillaries in the underlying connective tissue provide the necessary oxygen and nutrients.

There are two specialized varieties of loose connective tissue that are relatively common: *adipose tissue* and *reticular tissue.*

ADIPOSE TISSUE The distinction between loose connective tissue and fat, or **adipose tissue**, is usually somewhat arbitrary. Adipocytes account for most of the volume of adipose tissue (Figure 4-10a●), but only a fraction of the volume of loose connective tissue. Adipose tissue provides padding, cushions shocks, acts as an insulator to slow heat loss through the skin, and serves as packing or filler around structures. Adipose tissue is common under the skin of the groin, sides, buttocks, and breasts. It fills the bony sockets behind the eyes, surrounds the kidneys, and dominates extensive areas of loose connective tissue in the pericardial and abdominal cavities.

Brown Fat. In infants and young children, the adipose tissue between the shoulder blades, around the neck, and possibly elsewhere in the upper body is different from most of the adipose tissue in the adult. It is highly vascularized, and the individual adipocytes contain numerous mitochondria. Together these characteristics give the tissue a deep, rich color responsible for the name **brown fat**. When these cells are stimulated by the nervous system, lipid breakdown accelerates. The cells do not capture the energy released, and it is absorbed by the surrounding tissues as heat. This heat warms the circulating blood, which distributes it throughout the body. In this way an infant can accelerate metabolic heat generation by 100 percent very quickly. There is little if any brown fat in the adult; with increased body size, skeletal muscle mass, and insulation, shivering is significantly more effective in elevating body temperature.

Adipose Tissue and Weight Loss. Adipocytes are metabolically active cells—their lipids are continually being broken down and replaced. When nutrients are scarce, adipocytes deflate like collapsing balloons. This is what occurs during a weight-loss program. Because the cells are not killed, merely reduced in size, the lost weight can easily be regained in the same areas of the body.

Adipocytes in the adult are incapable of dividing, and the number of fat cells in peripheral tissues is established in the first few weeks of a newborn's life, perhaps in response to the amount of fats in the diet. However, this is not the end of the story, because loose connective tissues also contain mesenchymal cells. If circulating lipid levels are chronically elevated, the mesenchymal cells will divide, giving rise to cells that differentiate into fat cells. As a result, areas of loose connective tissue can become adipose tissue in times of nutritional plenty, even in the adult. In the procedure known as *liposuction,* unwanted adipose tissue is surgically removed. Because adipose tissue can regenerate through differentiation of mesenchymal cells, liposuction provides only a temporary solution to the problem.

RETICULAR TISSUE **Reticular tissue** is found in organs such as the spleen and liver, where reticular fibers create a complex three-dimensional network, or *stroma,* that supports the functional cells of these organs. This fibrous framework is also found in the lymph nodes and bone marrow (Figure 4-10b●). Fixed macrophages and fibroblasts are associated with the reticular fibers, but these cells are seldom visible because the organs are dominated by specialized cells with other functions.

Dense Connective Tissues

Figure 4-11

Most of the volume of **dense connective tissues** is occupied by fibers. Dense connective tissues are often called **collagenous** (ko-LA-jin-us) **tissues** because collagen fibers are the dominant fiber type. Two types of dense connective tissues are found in the body: (1) *dense regular connective tissue* and (2) *dense irregular connective tissue.*

DENSE REGULAR CONNECTIVE TISSUE In **dense regular connective tissue** the collagen fibers are arranged parallel to each other, packed tightly, and aligned with the forces applied to the tissue. Four major examples of this tissue type are *tendons, aponeuroses, elastic tissue,* and *ligaments.*

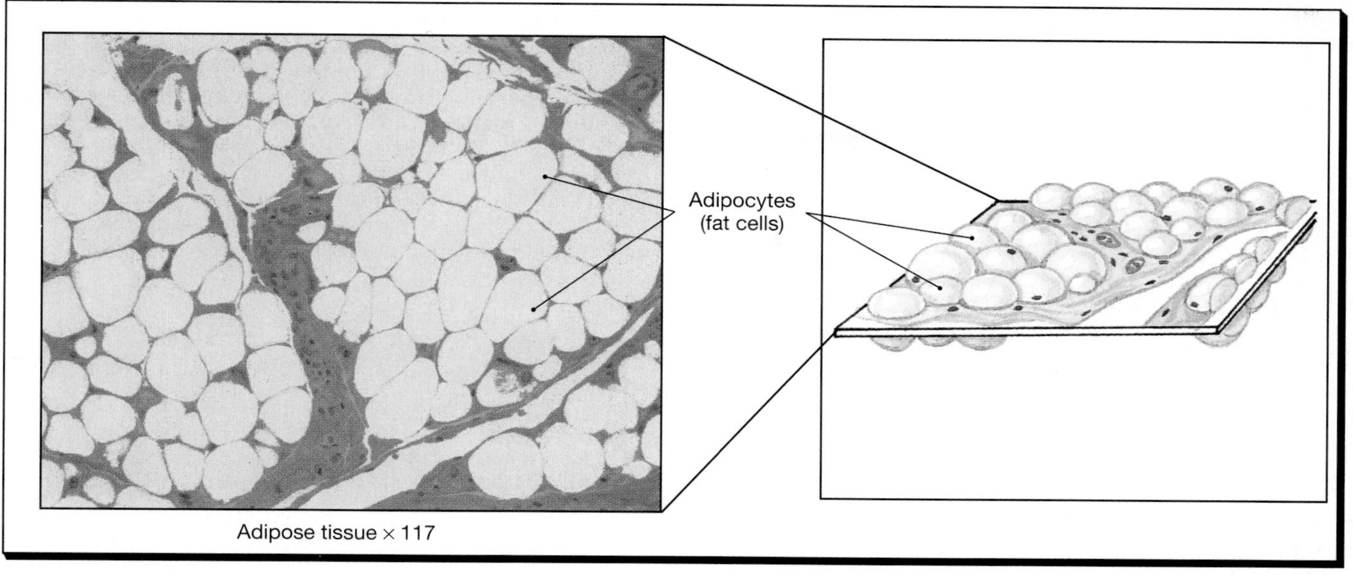
Adipocytes (fat cells)
Adipose tissue × 117

(a)

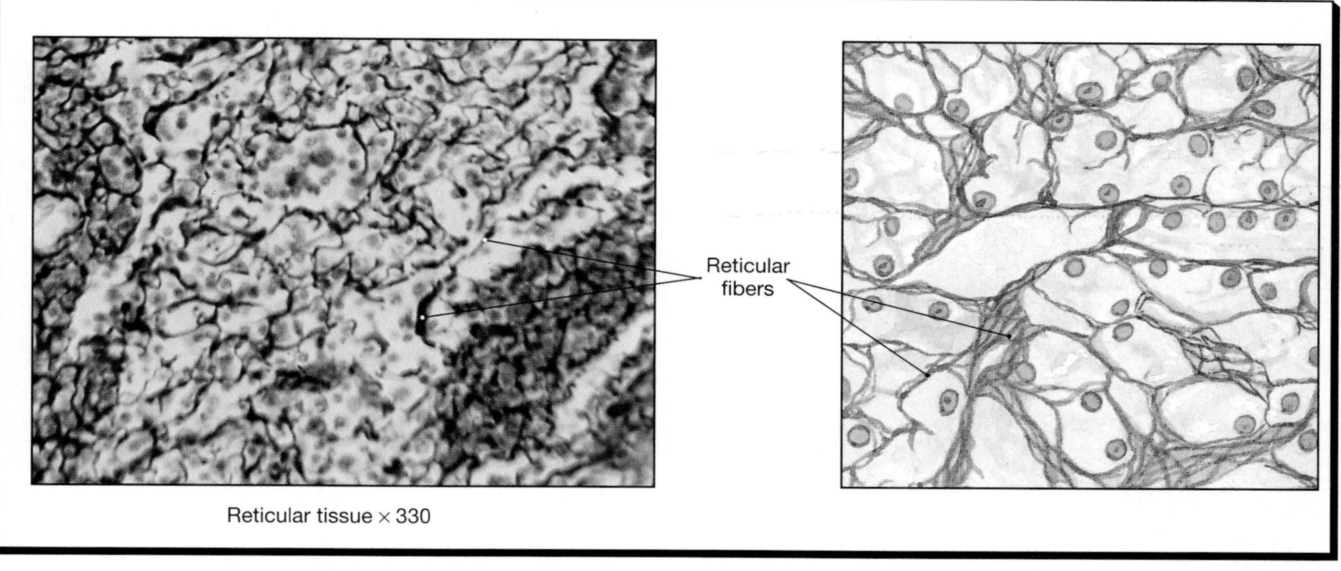
Reticular fibers
Reticular tissue × 330

(b)

● **FIGURE 4-10**

Adipose and Reticular Tissues. (a) Adipose tissue is a loose connective tissue that is dominated by adipocytes. In standard histological preparations, the tissue looks empty because the lipids in the fat cells dissolve in the alcohol used in tissue processing. **(b)** Reticular tissue has an open framework of reticular fibers. These fibers are usually very difficult to see because of the large numbers of cells around them.

1. **Tendons** (Figure 4-11a●) are cords of dense regular connective tissue that attach skeletal muscles to bones. The collagen fibers run along the longitudinal axis of the tendon and transfer the pull of the contracting muscle to the bone. Large numbers of fibroblasts are found between the collagen fibers.
2. **Aponeuroses** (ap-ō-noo-RŌ-sēz) are collagenous sheets or ribbons that resemble flat, broad tendons. Aponeuroses may cover the surface of a muscle and assist in attaching superficial muscles to another muscle or separate structure.
3. **Elastic tissue** contains large numbers of elastic fibers. Because elastic fibers outnumber collagen fibers, the tissue has a springy, resilient nature. This ability to stretch and rebound allows it to tolerate cycles of expansion and contraction. Elastic tissue often underlies transitional epithelia; it is also found in the walls of blood vessels, and it surrounds the respiratory passageways.
4. **Ligaments** (LIG-uh-ments) resemble tendons, but they extend from one bone to another. Ligaments often contain significant numbers of elas-

tic fibers as well as collagen fibers, and they can tolerate a modest amount of stretching. An even higher proportion of elastic fibers is found in **elastic ligaments,** which resemble tough rubber bands. Although uncommon elsewhere, elastic ligaments along the spinal column are very important in stabilizing the positions of the vertebrae (Figure 4-11b●).

DENSE IRREGULAR CONNECTIVE TISSUE The fibers in **dense irregular connective tissue** form an interwoven meshwork and do not show any consistent pattern (Figure 4-11c●). These tissues provide strength and support to areas subjected to stresses from many directions. A layer of dense irregular connective tissue gives skin its strength; a piece of cured leather (animal skin) provides an excellent illustration of the interwoven nature of this tissue. Dense irregular connective tissue also forms a thick fibrous layer, called a **capsule,** that surrounds internal organs, such as liver, kidneys, and spleen, and encloses the cavities of joints.

✔ Lack of vitamin C in the diet interferes with the ability of fibroblasts to produce collagen. What effect might this interference have on connective tissue?

✔ Many allergy sufferers take antihistamines to relieve their allergy symptoms. What type of cell produces the molecule that this medication blocks?

✔ Chemical analysis of a connective tissue reveals that the tissue contains primarily triglycerides. What type of connective tissue is this?

● **FIGURE 4-11**

Dense Connective Tissues. (a) The dense regular connective tissue in a tendon. Notice the densely packed, parallel bundles of collagen fibers. The fibroblast nuclei can be seen flattened between the bundles. **(b)** An elastic ligament from between the vertebrae of the spinal column. The bundles are fatter than those of a tendon or ligament composed of collagen. **(c)** The deep dermis of the skin contains a thick layer of dense irregular connective tissue.

DENSE REGULAR CONNECTIVE TISSUE

LOCATIONS: Between skeletal muscles and skeleton (tendons and aponeuroses); between bones (ligaments); covering skeletal muscles (deep fasciae)
FUNCTIONS: Provides firm attachment; conducts pull of muscles; reduces friction between muscles; stabilizes relative positions of bones

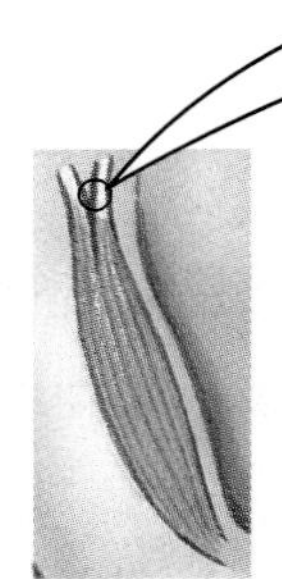

ELASTIC TISSUE

LOCATION: Between vertebrae of the spinal column (ligamentum flava and ligamentum nuchae)
FUNCTIONS: Stabilizes positions of vertebrae; cushions shocks

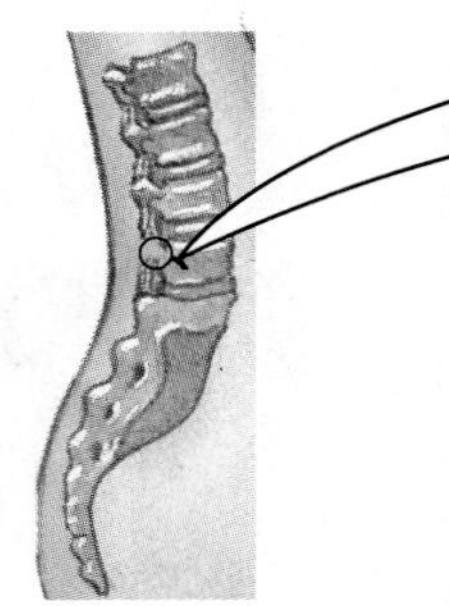

DENSE IRREGULAR CONNECTIVE TISSUE

LOCATIONS: Capsules of visceral organs; deep dermis of skin; periostea and perichondria; nerve and muscle sheaths
FUNCTIONS: Provides strength to resist forces applied from many directions; helps prevent overexpansion of organs such as the urinary bladder

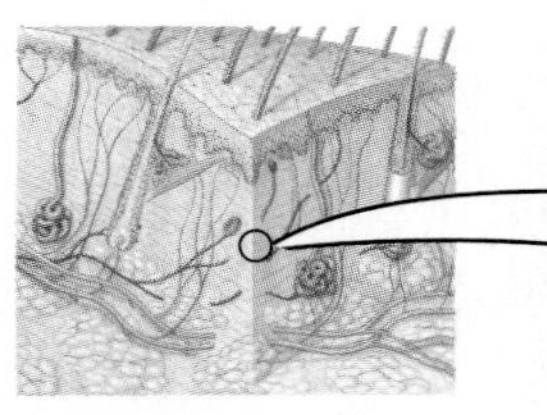

❑ Fluid Connective Tissues

Figure 4-12

Blood and **lymph** are connective tissues that contain distinctive collections of cells in a fluid matrix. The watery matrix of blood and lymph contains cells and many different types of suspended proteins that do not form insoluble fibers under normal conditions.

Blood contains blood cells and fragments of cells, collectively known as *formed elements* (Figure 4-12●). A single cell type, the **red blood cell,** or **erythrocyte** (e-RITH-rō-sīt; *erythros,* red), accounts for almost half of the volume of blood. Red blood cells are responsible for the transport of oxygen and, to a lesser degree, of carbon dioxide in the blood. The watery ground substance, called **plasma,** also contains small numbers of **white blood cells** or **leukocytes** (LOO-kō-sīts; *leuko,* white). White blood cells include the phagocytic microphages (*neutrophils* and *eosinophils*), basophils, *lymphocytes,* and macrophages called **monocytes.** The white blood cells are important components of the immune system, which protects the body from infection and disease. Tiny packets of cytoplasm, called **platelets,** contain enzymes and special proteins. Platelets function in the clotting response that seals breaks in the endothelial lining.

The extracellular fluid of the body includes three major subdivisions: *plasma, interstitial fluid,* and *lymph.* Plasma is normally confined

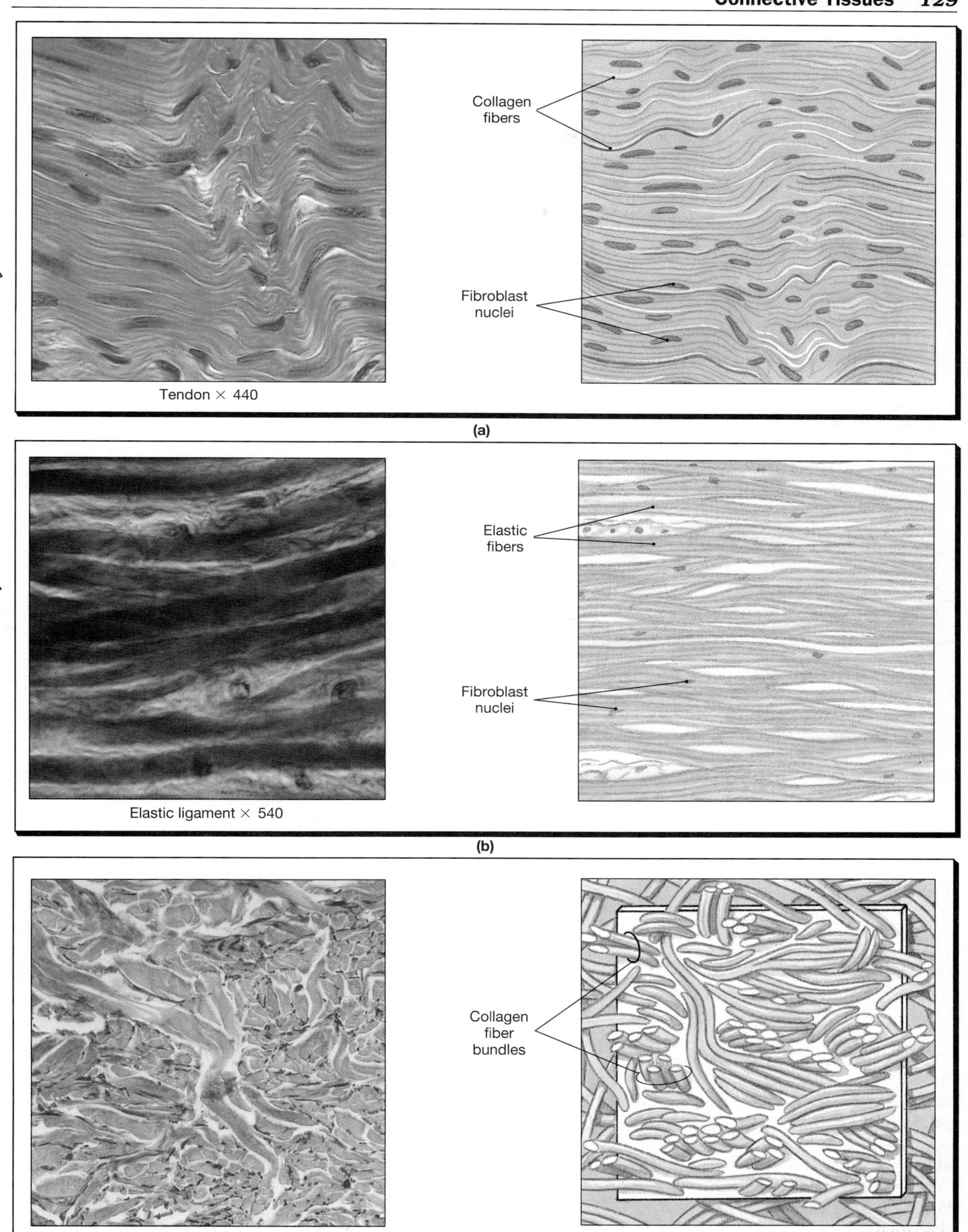
Collagen fibers
Fibroblast nuclei
Tendon × 440
(a)
Elastic fibers
Fibroblast nuclei
Elastic ligament × 540
(b)
Collagen fiber bundles
Deep dermis × 111
(c)

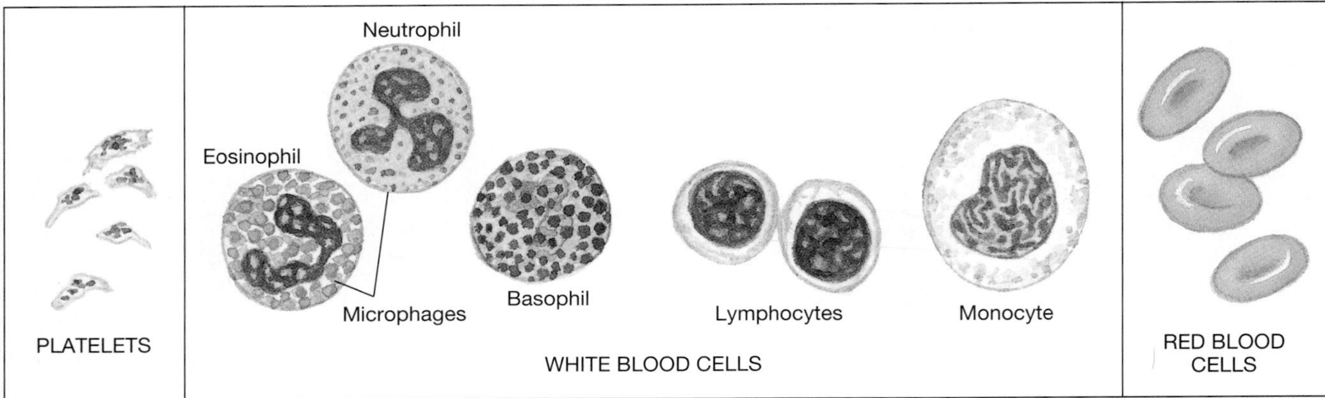
Neutrophil
Eosinophil
Microphages
Basophil
Lymphocytes
Monocyte
PLATELETS
WHITE BLOOD CELLS
RED BLOOD CELLS

FIGURE 4-12
Formed Elements of the Blood

to the vessels of the circulatory system, and the contractions of the heart keep it in constant motion. *Arteries* carry blood away from the heart toward fine, thin-walled vessels called **capillaries.** *Veins* drain the capillaries and return blood to the heart, completing the circuit. In the tissues, filtration moves water and small solutes out of the capillaries and into the interstitial fluid that bathes the cells of the body. The major difference between plasma and interstitial fluid is that plasma contains large numbers of suspended proteins.

Lymph forms as interstitial fluid enters small passageways, or **lymphatics,** that return it to the circulatory system. Along the way, cells of the immune system monitor the composition of the lymph and respond to signs of injury or infection. The number of cells in lymph may vary, but ordinarily 99 percent of them are lymphocytes, and the rest wandering macrophages or microphages.

Supporting Connective Tissues

Cartilage and bone are called supporting connective tissues because they provide a strong framework that supports the rest of the body. In these connective tissues the matrix contains numerous fibers and, in some cases, deposits of insoluble calcium salts.

Cartilage

Figure 4-13

The matrix of **cartilage** is a firm gel that contains proteoglycans called **chondroitin sulfates** (kon-DROY-tin; *chondros*, cartilage). Cartilage cells, or **chondrocytes** (KON-drō-sīts), are the only cells found within the matrix. These cells live in small pockets known as **lacunae** (la-KOO-nē; *lacus*, pool). The physical properties of cartilage depend on the type and abundance of extracellular fibers, as well as the proteoglycan content.

HYALINE CARTILAGE

LOCATIONS: Between tips of ribs and bones of sternum; covering bone surfaces at synovial joints; supporting larynx (voicebox), trachea, and bronchi; forming part of nasal septum
FUNCTIONS: Provides stiff but somewhat flexible support; reduces friction between bony surfaces

ELASTIC CARTILAGE

LOCATIONS: Pinna of external ear; tip of nose; epiglottis
FUNCTIONS: Provides support, but tolerates distortion without damage and returns to original shape

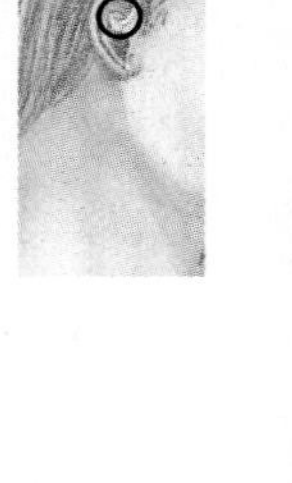

FIBROCARTILAGE

LOCATIONS: Intervertebral discs separating vertebrae along spinal column; pads within knee joint; between pubic bones of pelvis
FUNCTIONS: Resists compression; prevents bone-to-bone contact; limits relative movement

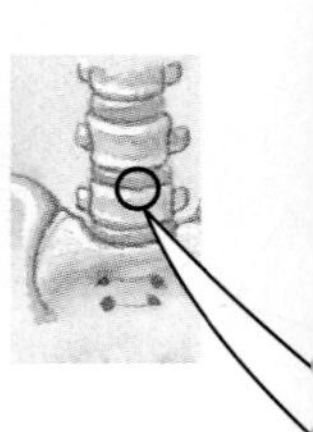

FIGURE 4-13
Types of Cartilage. (a) Hyaline cartilage. Note the translucent matrix and the absence of prominent fibers. **(b)** Elastic cartilage. The closely packed elastic fibers are visible between the chondrocytes. **(c)** Fibrocartilage. The collagen fibers are extremely dense, and the chondrocytes are relatively far apart.

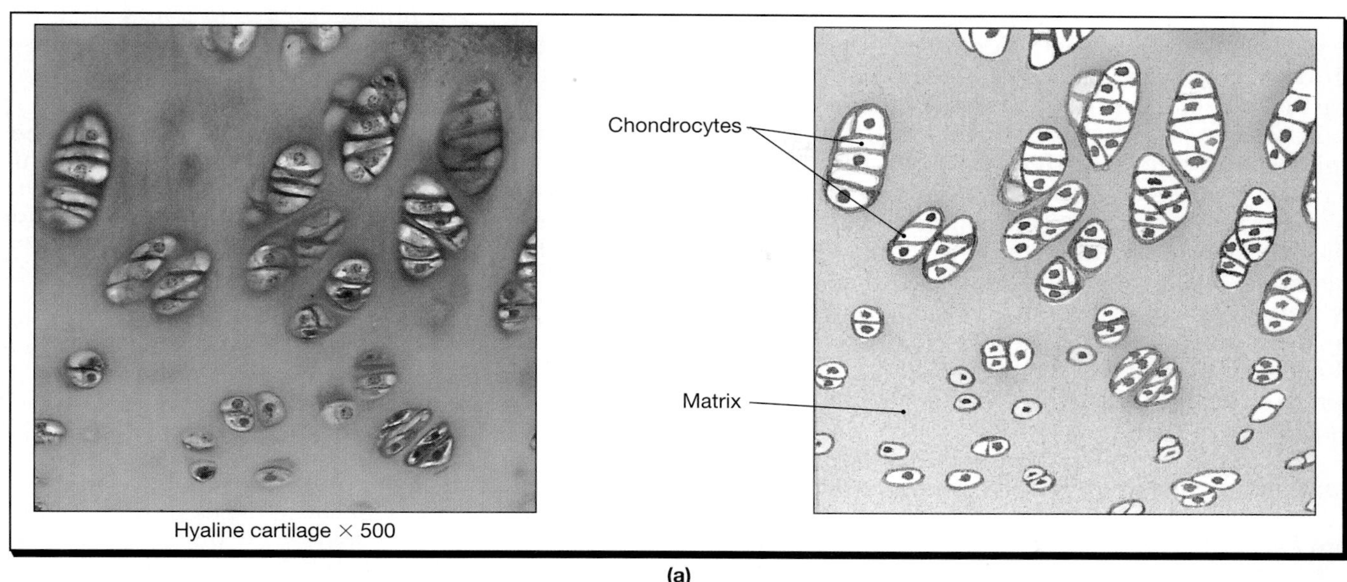
Chondrocytes
Matrix

Hyaline cartilage × 500

(a)

Chondrocyte

Elastic fibers in matrix

Elastic cartilage × 358

(b)

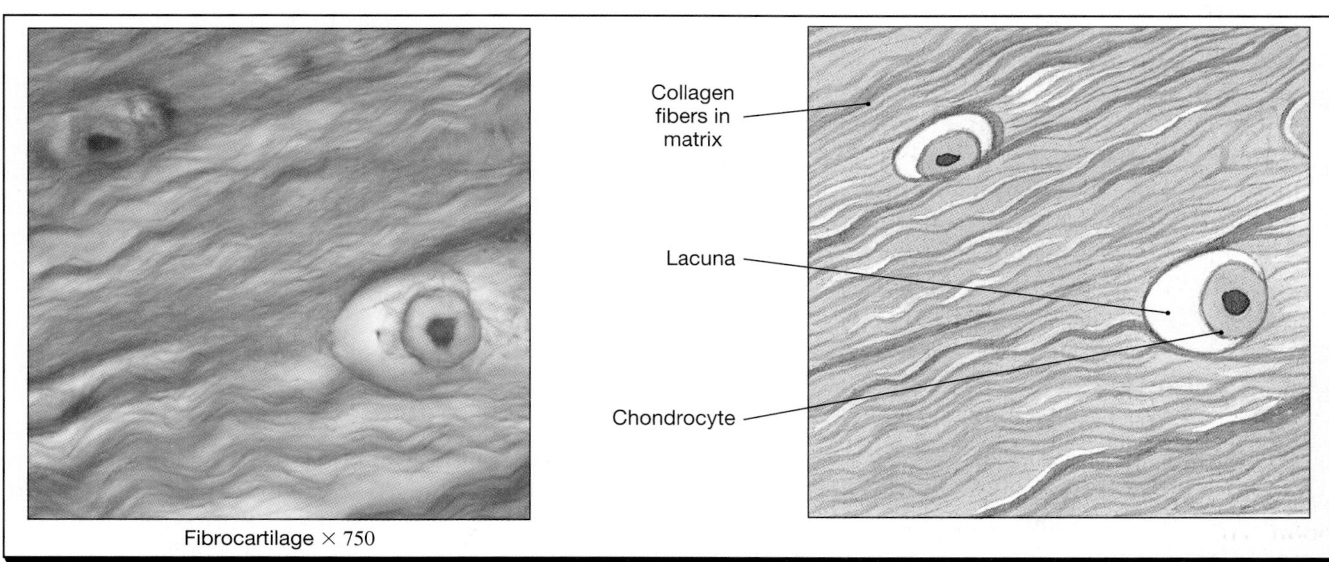
Collagen fibers in matrix
Lacuna
Chondrocyte

Fibrocartilage × 750

(c)

Cartilage is avascular, and all nutrient and waste product exchange must occur by diffusion through the matrix. Blood vessels do not grow into cartilage because chondrocytes produce a chemical that discourages their formation. This chemical has been named **antiangiogenesis factor** (*anti-*, against + *angeion*, vessel + *genno*, to produce). There is now interest in this compound as a potential anticancer agent. Cancers enlarge rapidly because blood vessels usually grow into areas where cells are crowded and active, improving nutrient and oxygen delivery. This growth could theoretically be prevented by antiangiogenesis factor, but the quantities produced in normal human cartilage are extremely small.

A cartilage is usually set apart from surrounding tissues by a fibrous **perichondrium** (pe-re-KON-drē-um; *peri-*, around). The perichondrium contains two distinct layers, an outer, fibrous region of dense irregular connective tissue and an inner, cellular layer. The fibrous layer provides mechanical support and protection and attaches the cartilage to other structures. The cellular layer is important to the growth and maintenance of the cartilage.

TYPES OF CARTILAGE Three major types of cartilage are found in the body: (1) *hyaline cartilage*, (2) *elastic cartilage*, and (3) *fibrocartilage*.

1. **Hyaline cartilage** (HĪ-uh-lin; *hyalos*, glass) is the most common type of cartilage. The matrix of hyaline cartilage contains closely packed collagen fibers, making it tough but somewhat flexible. Because the fibers do not stain well, they are not always apparent in light microscopy (Figure 4-13a•). Examples of this type of cartilage in the adult body include: (1) the connections between the ribs and the sternum, (2) the supporting cartilages along the conducting passageways of the respiratory tract, and (3) the cartilages covering opposing bone surfaces within many joints, such as the elbow or knee.
2. **Elastic cartilage** (Figure 4-13b•) contains numerous elastic fibers that make it extremely resilient and flexible. Elastic cartilage forms the external flap (*pinna*) of the outer ear, the epiglottis, and the tip of the nose.
3. **Fibrocartilage** has little ground substance, and the matrix is dominated by collagen fibers (Figure 4-13c•). The collagen fibers are densely interwoven, making this tissue extremely durable and tough. Fibrocartilaginous pads lie between the spinal vertebrae, between the pubic bones of the pelvis, and around or within a few joints and tendons. In these positions they resist compression, absorb shocks, and prevent damaging bone-to-bone contact. Cartilages in general heal poorly, and damaged fibrocartilages in joints such as the knee can interfere with normal movements.

CARTILAGES AND KNEE INJURIES The knee is an extremely complex joint that contains both hyaline and fibrocartilage. The hyaline cartilage covers bony surfaces, and fibrocartilage pads within the joint prevent bone contact when movements are under way. Many sports injuries result in tearing of the fibrocartilage pads. This tearing places more strain on the articular cartilages and leads to further joint damage. Because cartilages are avascular, they heal poorly, and joint cartilages heal even more slowly than other cartilages. Surgery usually produces only a temporary or incomplete repair.

Recent advances in tissue culture have enabled researchers to grow fibrocartilages in the laboratory. Chondrocytes removed from the knees of injured dogs are cultured in an artificial framework of collagen fibers. They eventually produce masses of fibrocartilage that can be inserted into the damaged joints. Over time the pads change shape and grow, restoring normal joint function. In the future this technique may be used to treat severe knee injuries in humans.

CARTILAGE GROWTH Cartilages grow by two mechanisms: *interstitial growth* and *appositional growth*. In **interstitial growth,** chondrocytes within the cartilage matrix undergo division and the daughter cells produce additional matrix. This process enlarges the cartilage, rather like inflating a balloon. In **appositional growth,** cells of the inner layer of the perichondrium undergo repeated cycles of division. The innermost cells begin producing cartilage matrix and differentiate into chondrocytes. This process gradually increases the size of the cartilage by adding to its surface. Interstitial growth is most important during embryonic development. Appositional growth begins early in development and continues through adolescence. Neither interstitial nor appositional growth occurs in the cartilages of normal adults. However, appositional growth may occur in unusual circumstances, such as after cartilage damage or under excessive stimulation with *growth hormone* from the pituitary gland. Minor damage to cartilage can be repaired by appositional growth at the damaged surface. After more severe damage, the injured portion of the cartilage will be replaced by a dense fibrous patch.

Bone

Figure 4-14

Because the detailed histology of **bone,** or **osseous** (OS-ē-us) **tissue** (*os*, bone), will be considered in Chapter 6, this discussion will focus on significant differences between cartilage and bone. Roughly one-third of the matrix of bone consists of collagen fibers. The balance is a mixture of calcium

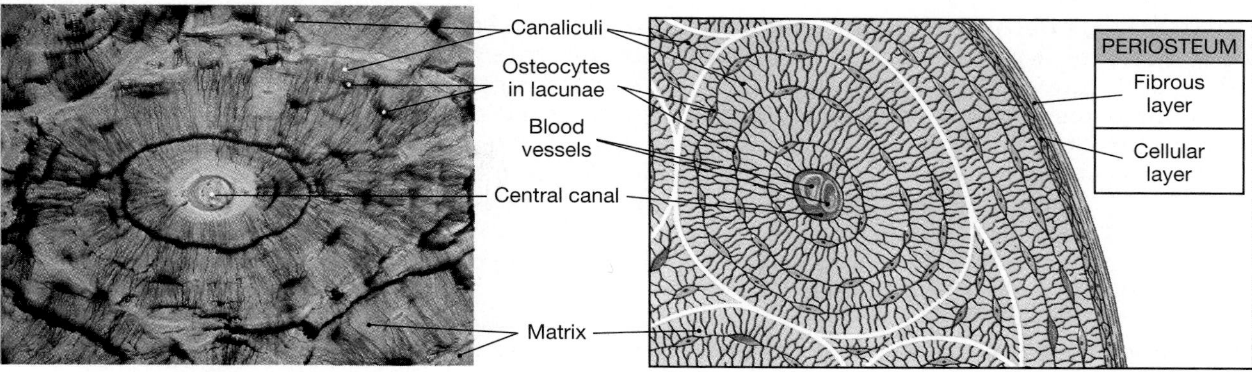
Canaliculi
Osteocytes in lacunae
Blood vessels
Central canal
Matrix
PERIOSTEUM
Fibrous layer
Cellular layer

● **FIGURE 4-14**

Bone. The osteocytes in bone are usually organized in groups around a central space that contains blood vessels. Bone dust produced during grinding fills the lacunae and the central canal, making them appear dark. (LM × 362)

salts, primarily calcium phosphate with lesser amounts of calcium carbonate. This combination gives bone truly remarkable properties. By themselves, calcium salts are hard but rather brittle. Collagen fibers are stronger, but relatively flexible. In bone, the minerals are organized around the collagen fibers. The result is a strong, somewhat flexible combination that is very resistant to shattering. In its overall properties, bone can compete with the best steel-reinforced concrete.

The general organization of osseous tissue can be seen in Figure 4-14●. Lacunae within the matrix contain bone cells, or **osteocytes** (OS-tē-ō-sīts). The lacunae are often organized around blood vessels that branch through the bony matrix. Although diffusion cannot occur through the calcium salts, osteocytes communicate with the blood vessels and with one another through slender cytoplasmic extensions. These extensions run through long, slender passages in the matrix. These passageways, called **canaliculi** (kan-a-LIK-ū-lē; little canals), form a branching network for the exchange of materials between the blood vessels and the osteocytes.

Except within joint cavities, where they are covered by a layer of hyaline cartilage, bone surfaces are sheathed by a **periosteum** (pe-rē-OS-tē-um) composed of fibrous (outer) and cellular (inner) layers. The periosteum assists in the attachment of a bone to surrounding tissues and to associated tendons and ligaments. The cellular layer functions in bone growth and participates in repairs after an injury. Unlike cartilage, bone undergoes extensive remodeling on a regular basis, and complete repairs can be made even after severe damage has

TABLE 4-1 A Comparison of Cartilage and Bone

Characteristic	*Cartilage*	*Bone*
STRUCTURAL FEATURES		
Cells	Chondrocytes in lacunae	Osteocytes in lacunae
Ground substance	Chondroitin sulfate (proteoglycan) in water	Insoluble crystals of calcium phosphate and calcium carbonate
Fibers	Collagen,elastic, and reticular fibers (proportions vary)	Collagen fibers predominate
Vascularity	None	Extensive
Covering	Perichondrium, two-part	Periosteum, two-part
Strength	Limited: bends easily but hard to break	Strong: resists distortion until breaking point is reached
METABOLIC FEATURES		
Oxygen demands	Relatively low	Relatively high
Nutrient delivery	By diffusion through matrix	By diffusion through cytoplasm and fluid in canaliculi
Growth	Interstitial and appositional	Appositional only
Repair capabilities	Limited ability	Extensive ability

occurred. Bones also respond to the stresses placed upon them, growing thicker and stronger with exercise and thin and brittle with inactivity. Table 4-1 summarizes the similarities and differences between cartilage and bone.

 Why does cartilage heal so slowly?

 If a person has a herniated intervertebral disc, what type of cartilage has been damaged?

 Which two connective tissues have a matrix that is fluid?

Membranes

Figure 4-15

Epithelia and connective tissues combine to form membranes that cover and protect other structures and tissues in the body. There are four such membranes: (1) *mucous membranes,* (2) *serous membranes,* (3) the *cutaneous membrane (skin),* and (4) *synovial membranes.*

Mucous Membranes

Figure 4-15a

Mucous membranes line cavities that communicate with the exterior, including the digestive, respiratory, reproductive, and urinary tracts (Figure 4-15a●). The epithelial surfaces are kept moist at all times; they may be lubricated, either by mucus produced by goblet cells or multicellular glands or by exposure to fluids such as urine or semen. The loose connective tissue component of a mucous membrane is called the **lamina propria** (PRŌ-prē-uh). We will consider the organization of specific mucous membranes in greater detail in later chapters.

Many mucous membranes are lined by simple epithelia that perform absorptive or secretory functions, such as the simple columnar epithelium of the digestive tract. Other types of epithelia may be involved, however. For example, a stratified squamous epithelium is part of the mucous membrane of the mouth, and the mucous membrane along most of the urinary tract has a transitional epithelium.

Serous Membranes

Figure 4-15b

Serous membranes line the sealed, internal cavities of the body. There are three serous membranes, each consisting of a *mesothelium* supported by loose connective tissue (Figure 4-15b●). These membranes were introduced in Chapter 1 ∞ *[p. 24]*: (1) The *pleura* lines the pleural cavities and covers the lungs; (2) the *peritoneum* lines the peritoneal cavity and covers the surfaces of the enclosed organs; and (3) the *pericardium* lines the pericardial cavity and covers the heart. Serous membranes are very thin, and they are firmly attached to the body wall and to the organs they cover. When you are looking at an organ, such as the heart or stomach, you are really seeing the tissues of the organ through a transparent serous membrane.

The opposing surfaces of a serous membrane are in close contact at all times. Minimizing friction between these opposing surfaces is the primary function of serous membranes. Because the mesothelia are very thin, serous membranes are relatively permeable, and tissue fluids diffuse onto the exposed surface, keeping it moist and slippery.

The fluid formed on the surfaces of a serous membrane is called a **transudate** (TRAN-sū-dāt; *trans,* across). Specific transudates are called **pleural fluid, peritoneal fluid,** or **pericardial fluid,** depending on their source. In normal healthy individuals the total volume of transudate is extremely small, just enough to prevent friction between the walls of the cavities and the surfaces of internal organs. But after an injury or in certain disease states, the volume of transudate may increase dramatically, complicating existing medical problems or producing new ones. Examples of conditions characterized by accelerated transudate production include *pleural effusions, pericardial effusions,* and *ascites.* AM *Problems with Serous Membranes*

The Cutaneous Membrane

Figure 4-15c

The **cutaneous membrane** of the skin covers the surface of the body. It consists of a stratified squamous epithelium and a layer of loose connective tissue (Figure 4-15c●) that is reinforced by underlying dense connective tissue. In contrast to serous or mucous membranes, the cutaneous membrane is thick, relatively waterproof, and usually dry.

Synovial Membranes

Figure 4-15d

Bones of the skeleton contact one another at joints, or *articulations.* Each joint is surrounded by a fibrous capsule, and the inner surfaces of the joint cavity are lined by a **synovial membrane** (sin-Ō-vē-ul). Such a membrane consists of extensive areas of loose connective tissue bounded by a superficial layer of squamous or cuboidal cells (Figure 4-15d●). Although usually called an epithelium, it differs from other epithelia in two respects: (1) There is no basement membrane, and (2) the cellular layer is incomplete, and gaps exist between adjacent cells. Some of the lining cells are

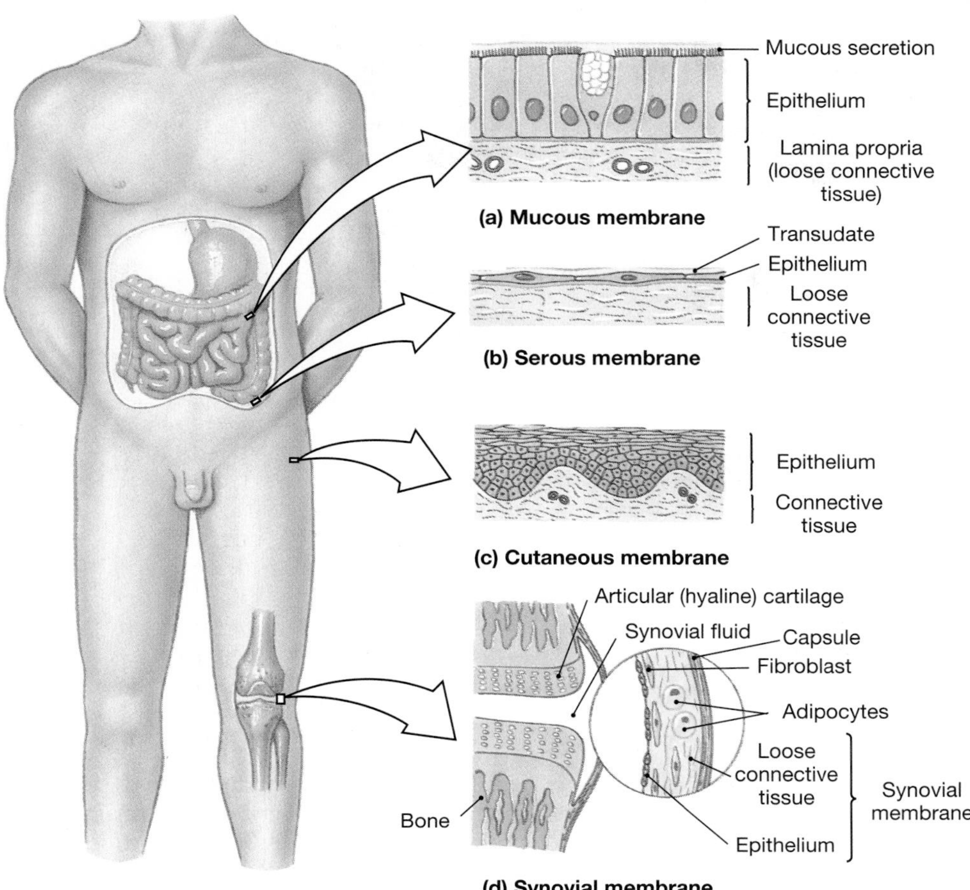
Mucous secretion
Epithelium
Lamina propria (loose connective tissue)
(a) Mucous membrane
Transudate
Epithelium
Loose connective tissue
(b) Serous membrane
Epithelium
Connective tissue
(c) Cutaneous membrane
Articular (hyaline) cartilage
Synovial fluid
Capsule
Fibroblast
Adipocytes
Loose connective tissue
Synovial membrane
Bone
Epithelium
(d) Synovial membrane

● **FIGURE 4-15**
Membranes. (a) Mucous membranes are coated with the secretions of mucous glands. Mucous membranes with a simple columnar epithelium line most of the digestive and respiratory tracts and portions of the urinary and reproductive tracts. **(b)** Serous membranes line the ventral body cavities (the peritoneal, pleural, and pericardial cavities). **(c)** The cutaneous membrane of the skin covers the outer surface of the body. **(d)** Synovial membranes line joint cavities and produce the fluid within the joint.

phagocytic and others are secretory. The secretory cells regulate the composition of the **synovial fluid** within the joint cavity.

The Connective Tissue Framework of the Body

Figure 4-16

Connective tissues create the internal framework of the body. Layers of connective tissue connect the organs within the dorsal and ventral body cavities with the rest of the body. These layers (1) provide strength and stability, (2) maintain the relative positions of internal organs, and (3) provide a route for the distribution of blood vessels, lymphatics, and nerves. The connective tissue layers and wrappings can be divided into three major components: the *superficial fascia*, the *deep fascia*, and the *subserous fascia*. The functional anatomy of these layers is illustrated in Figure 4-16●.

- The **superficial fascia,** or **subcutaneous layer** (*sub*, below + *cutis*, skin) is also termed the **hypodermis** (*hypo*, below + *dermis*, skin). This layer of loose connective tissue separates the skin from underlying tissues and organs. It provides insulation and padding and lets the skin or underlying structures move independently.
- The **deep fascia** consists of dense connective tissue. The organization of the fibers resembles that of plywood: All the fibers in an individual layer run in the same direction, but the orientation of the fibers changes from one layer to another. This arrangement helps the tissue to resist forces applied from many different directions. The tough *capsules* that surround most organs, including the kidneys and the organs in the thoracic and peritoneal cavities, are components of the deep fascia. The perichondrium around cartilages, the periosteum around bones and the ligaments that interconnect them, and the connective tissues of muscle, including tendons and aponeuroses, are all part of the deep fascia. The dense connective tissue components are interwoven; for example, the deep fascia around a muscle blends into the tendon, whose fibers intermingle with those of the periosteum. This arrangement creates a strong, fibrous network for the body and ties structural elements together.

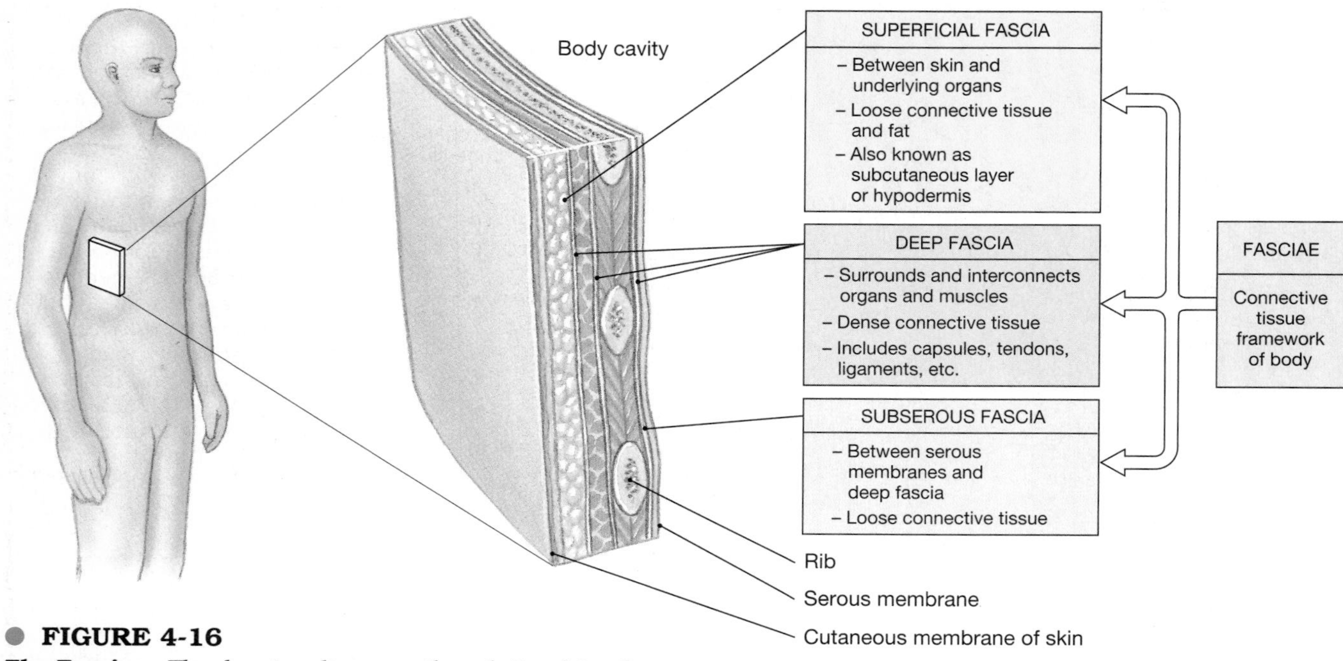
Body cavity
SUPERFICIAL FASCIA
– Between skin and underlying organs
– Loose connective tissue and fat
– Also known as subcutaneous layer or hypodermis
DEEP FASCIA
– Surrounds and interconnects organs and muscles
– Dense connective tissue
– Includes capsules, tendons, ligaments, etc.
SUBSEROUS FASCIA
– Between serous membranes and deep fascia
– Loose connective tissue
FASCIAE
Connective tissue framework of body
Rib
Serous membrane
Cutaneous membrane of skin

● **FIGURE 4-16**
The Fasciae. The drawing diagrams the relationship of connective tissue elements in the body.

- The **subserous fascia** is a layer of loose connective tissue that lies between the deep fascia and the serous membranes that line body cavities. Because this layer separates the serous membranes from the deep fascia, movements of muscles or muscular organs do not severely distort the delicate lining.

What cavities in the body are lined by serous membranes?

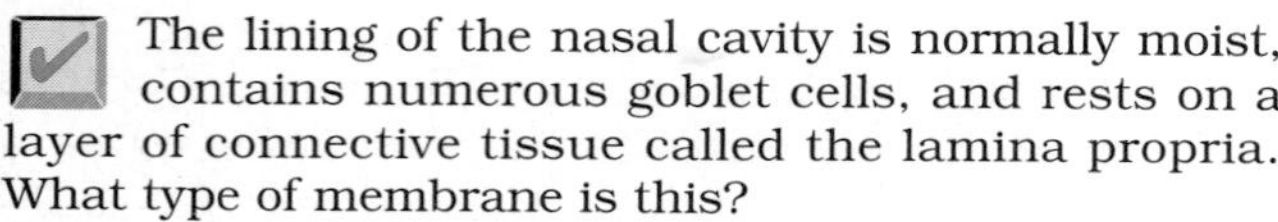
The lining of the nasal cavity is normally moist, contains numerous goblet cells, and rests on a layer of connective tissue called the lamina propria. What type of membrane is this?

A sheet of tissue has many layers of collagen fibers that run in different directions in successive layers. What type of tissue is this?

Muscle Tissue

Muscle tissue is specialized for contraction. Muscle cells possess organelles and properties distinct from those of other cells. They are capable of powerful contractions that shorten the fiber along its longitudinal axis. Because they are different from "typical" cells, the term **sarcoplasm** is used to refer to the cytoplasm of a muscle fiber, and one refers to the **sarcolemma** rather than the cell membrane.

Three types of muscle tissue are found in the body: (1) *skeletal muscle,* (2) *cardiac muscle,* and (3) *smooth muscle.* The contraction mechanism is similar in all three, but they differ in their internal organization. We will describe each muscle type in greater detail in Chapter 10. This discussion will focus on general characteristics rather than specific details.

❑ Skeletal Muscle Tissue

Figure 4-17a

Skeletal muscle tissue contains very large muscle fibers. Skeletal muscle fibers are very unusual because they may be 0.3 meters (1 ft) or more in length. Because the individual muscle cells are relatively long and slender, they are usually called **muscle fibers.** Each cell is **multinucleated,** containing hundreds of nuclei. Nuclei lie just under the surface of the sarcolemma (Figure 4-17a●). Skeletal muscle fibers are incapable of dividing, but new muscle fibers can be produced through the division of **satellite cells,** mesenchymal cells that persist in adult skeletal muscle tissue. As a result, skeletal muscle tissue can at least partially repair itself after an injury.

Because the actin and myosin filaments are arranged in organized groups, skeletal muscle fibers have a banded or *striated* appearance. Skeletal muscle fibers will not usually contract unless stimulated by nerves, and the nervous system provides voluntary control over their activities. Thus, skeletal muscle is called **striated voluntary muscle.**

Skeletal muscle tissue is tied together by loose connective tissue. The collagen and elastic fibers

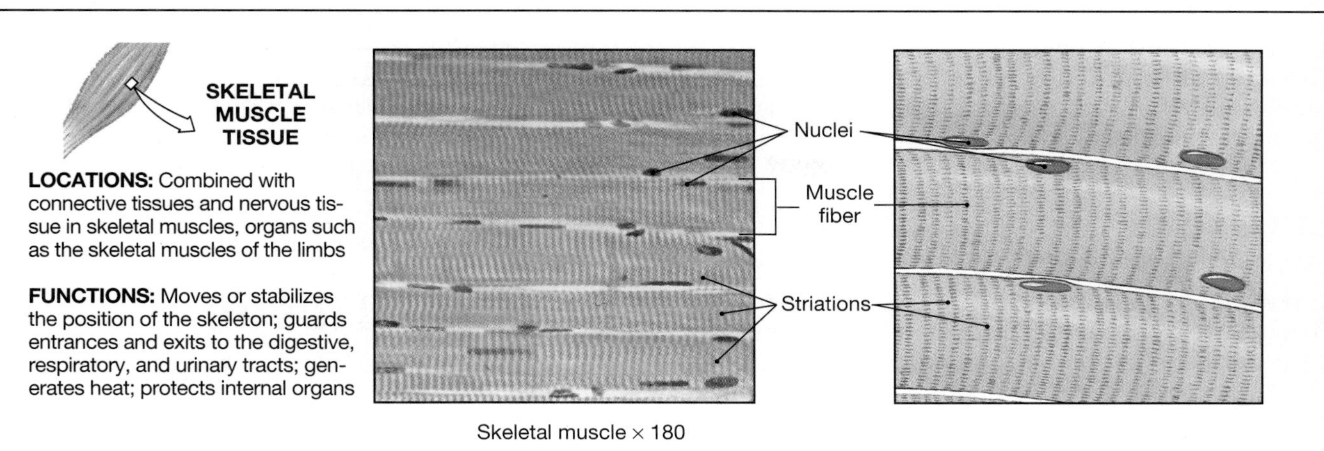
SKELETAL MUSCLE TISSUE
LOCATIONS: Combined with connective tissues and nervous tissue in skeletal muscles, organs such as the skeletal muscles of the limbs
FUNCTIONS: Moves or stabilizes the position of the skeleton; guards entrances and exits to the digestive, respiratory, and urinary tracts; generates heat; protects internal organs
Nuclei
Muscle fiber
Striations

Skeletal muscle × 180

(a)

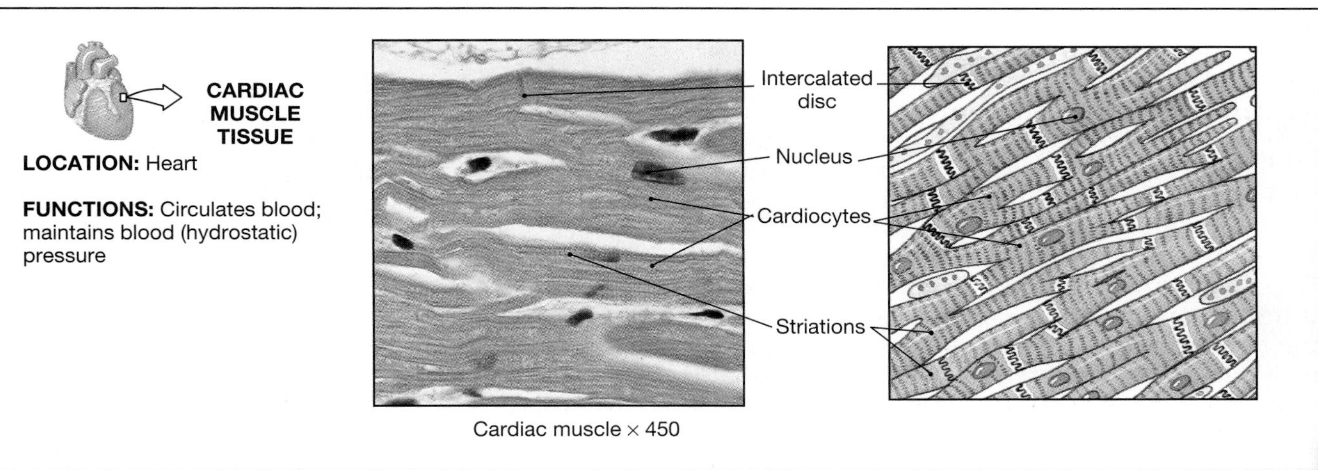
CARDIAC MUSCLE TISSUE
LOCATION: Heart
FUNCTIONS: Circulates blood; maintains blood (hydrostatic) pressure
Intercalated disc
Nucleus
Cardiocytes
Striations

Cardiac muscle × 450

(b)

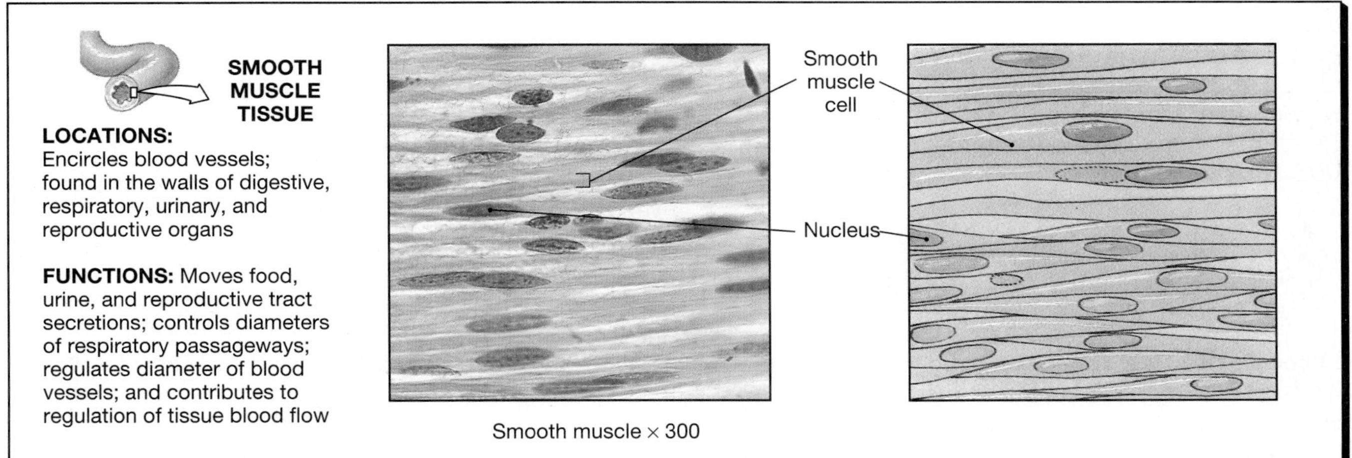
SMOOTH MUSCLE TISSUE
LOCATIONS:
Encircles blood vessels; found in the walls of digestive, respiratory, urinary, and reproductive organs
FUNCTIONS: Moves food, urine, and reproductive tract secretions; controls diameters of respiratory passageways; regulates diameter of blood vessels; and contributes to regulation of tissue blood flow
Smooth muscle cell
Nucleus

Smooth muscle × 300

(c)

FIGURE 4-17

Muscle Tissue. (a) Skeletal muscle fibers. Note the large fiber size, prominent banding pattern, multiple nuclei, and unbranched arrangement. **(b)** Cardiac muscle cells. Cardiac muscle cells differ from skeletal muscle fibers in three major ways: size (cardiac muscle cells are smaller), organization (cardiac muscle cells branch), and number and location of nuclei (a typical cardiac muscle cell has one centrally placed nucleus). Both contain actin and myosin filaments in an organized array that produces striations. **(c)** Smooth muscle cells. Smooth muscle cells are small and spindle-shaped, with a central nucleus. They do not branch, and there are no striations.

surrounding each cell and group of cells blend into those of a tendon or aponeurosis that conducts the force of contraction, usually to a bone of the skeleton. When the muscle tissue contracts, it pulls on the attached bone.

❑ Cardiac Muscle Tissue

Figure 4-17b

Cardiac muscle tissue is found only in the heart. A typical cardiac muscle cell, also known as a **cardiocyte**, or *cardiac myocyte*, is smaller than a skeletal muscle fiber. A typical cardiac muscle cell usually has one centrally placed nucleus, although cardiocytes may occasionally have as many as five. The prominent striations visible in Figure 4-17b● resemble those of skeletal muscle because the actin and myosin filaments are arranged the same way in both fiber types.

Cardiac muscle cells form extensive connections with one another. The connections occur at specialized regions known as **intercalated discs.** As a result, cardiac muscle tissue consists of a branching network of interconnected muscle fibers. The interconnections help channel the forces of contraction and coordinate the activities of individual cardiac muscle cells. Like skeletal muscle fibers, cardiac muscle cells are incapable of dividing, and because this tissue lacks satellite cells, cardiac muscle tissue damaged by injury or disease cannot regenerate.

Cardiac muscle cells do not rely on nerve activity to start a contraction. Instead, specialized cardiac muscle cells, called **pacemaker cells,** establish a regular rate of contraction. Although the nervous system can alter the rate of pacemaker activity, it does not provide voluntary control over individual cardiac muscle cells. Therefore, cardiac muscle is called **striated involuntary muscle.**

❑ Smooth Muscle Tissue

Figure 4-17c

Smooth muscle tissue can be found (1) in the walls of blood vessels, (2) around hollow organs such as the urinary bladder, and (3) in layers around the respiratory, circulatory, digestive, and reproductive tracts. A smooth muscle cell is a small, spindle-shaped cell with tapering ends, containing a single oval nucleus (Figure 4-17c●). Smooth muscle cells can divide, and smooth muscle tissue can regenerate after an injury. The actin and myosin filaments in smooth muscle cells are organized differently from those of skeletal and cardiac muscle, and there are no striations. Smooth muscle cells may contract on their own, or their contractions may be triggered by nerve cell activity. When areas of smooth muscle are innervated, the nerve activity is involuntarily controlled. Because the nervous system usually does not provide voluntary control over smooth muscle contractions, smooth muscle is known as **nonstriated involuntary muscle.**

Neural Tissue

Figure 4-18

Neural tissue, also known as *nervous tissue* or *nerve tissue,* is specialized for the conduction of electrical impulses from one region of the body to another. Most of the neural tissue in the body (98 percent) is concentrated in the brain and spinal cord, the control centers for the nervous system. Neural tissue contains two basic types of cells: **neurons** (NOO-ronz; *neuro,* nerve), and several different kinds of supporting cells, collectively called **neuroglia** (noo-RŌG-lē-uh; *glia,* glue). Our conscious and unconscious thought processes reflect the communication between neurons in the brain. Such communication involves the propagation of electrical impulses in the form of changes in the transmembrane potential. Information is conveyed by the frequency and pattern of impulse generation. Neuroglia have different functions, such as providing a supporting framework for neural tissue and helping to supply nutrients to neurons.

Neurons are the longest cells in the body, many reaching a meter (39 inches) in length. Most neurons are incapable of dividing under normal circumstances, and they have a very limited ability to repair themselves after injury. A typical neuron has a cell body, or **soma,** that contains a large prominent nucleus (Figure 4-18●). Extending from the soma are various branching processes termed **dendrites** (DEN-drīts; *dendron,* a tree) and a single **axon.** Dendrites receive information from other cells, and axons carry information to other cells. Because axons are often very long and slender, they are also called **nerve fibers.** Chapter 12 will discuss the properties of neural tissue in greater detail.

✔ What type of muscle tissue has small, tapering cells with single nuclei and no obvious striations?

✔ Microscopic examination of a tissue reveals irregularly shaped cells with many fibrous projections, some several centimeters long. These are probably what type of cell?

 If skeletal muscle fibers are incapable of dividing, how is new skeletal muscle formed?

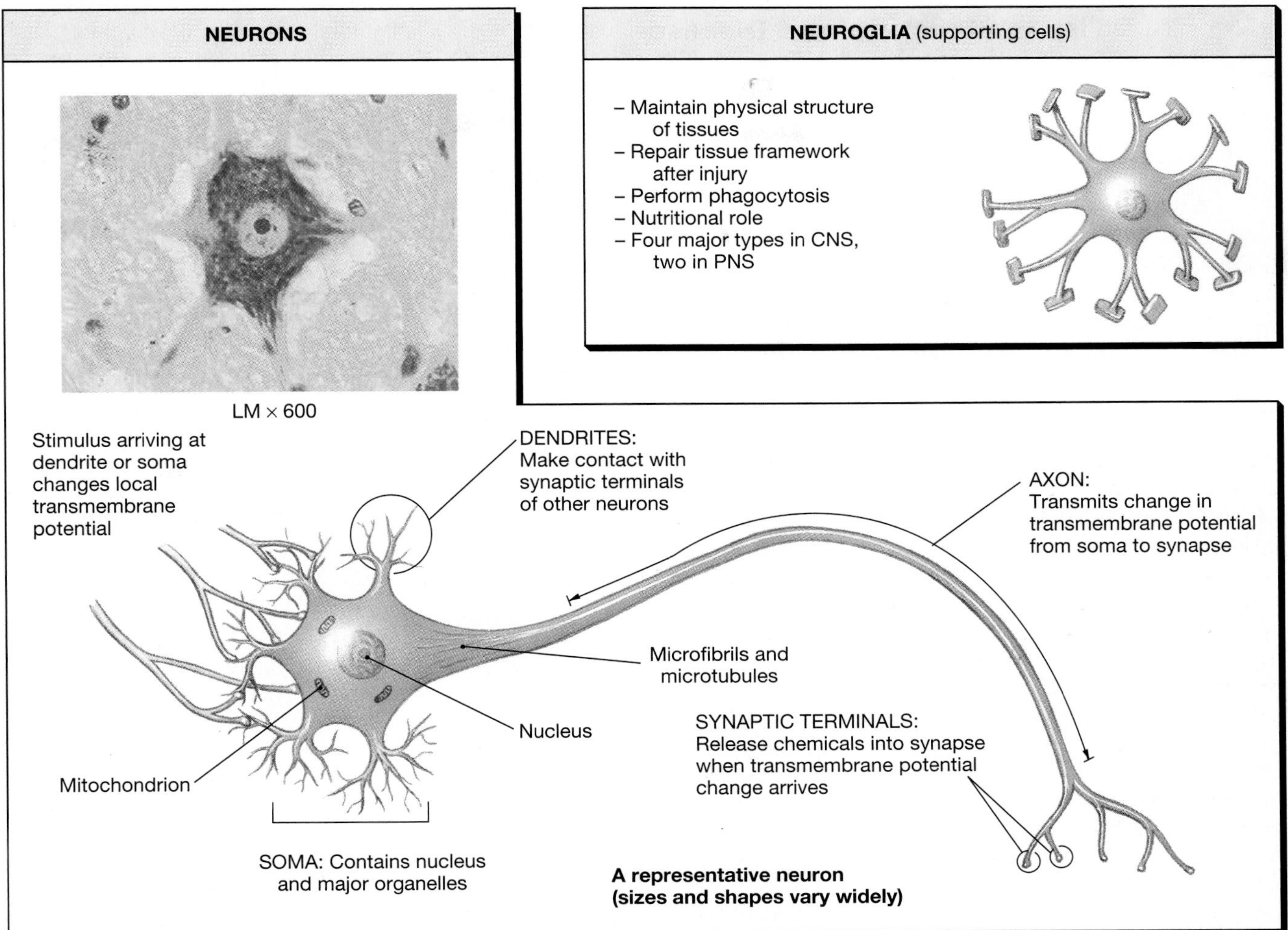
NEURONS
LM × 600
NEUROGLIA (supporting cells)
– Maintain physical structure of tissues
– Repair tissue framework after injury
– Perform phagocytosis
– Nutritional role
– Four major types in CNS, two in PNS
Stimulus arriving at dendrite or soma changes local transmembrane potential
DENDRITES: Make contact with synaptic terminals of other neurons
AXON: Transmits change in transmembrane potential from soma to synapse
Microfibrils and microtubules
Nucleus
SYNAPTIC TERMINALS: Release chemicals into synapse when transmembrane potential change arrives
Mitochondrion
SOMA: Contains nucleus and major organelles
A representative neuron (sizes and shapes vary widely)

FIGURE 4-18
Neural Tissue

Tissue Injuries and Aging

Tissues in the body are not isolated; they combine to form organs with diverse functions. Any injury to the body affects several different tissue types simultaneously, and these tissues must respond in a coordinated manner to preserve homeostasis.

Inflammation and Regeneration

The restoration of homeostasis after a tissue injury involves two related processes. First the area is isolated while damaged cells, tissue components, and any dangerous microorganisms are cleaned up. This phase, which coordinates the activities of several different tissues, is called **inflammation,** or the **inflammatory response.** Inflammation begins immediately after an injury and produces several familiar sensations, including swelling, redness, warmth, and pain. An **infection** is an inflammation resulting from the presence of pathogens, such as bacteria.

Second, the damaged tissues are replaced or repaired to restore normal function. This repair process is called **regeneration.** Inflammation and regeneration are controlled at the tissue level. The two phases overlap; isolation establishes a framework that guides the cells responsible for reconstruction, and repairs are under way well before cleanup operations have ended. Later chapters, especially Chapters 5 and 22, will examine inflammation and regeneration in more detail.

Aging and Tissue Repair

Tissues change with age, and there is a decrease in the speed and effectiveness of tissue repairs. In

Tissue Structure and Disease

Pathologists (pa-THOL-o-jists) are physicians who specialize in the study of disease processes. Diagnosis, rather than treatment, is usually the main focus of their activities. In their analyses, pathologists integrate anatomical and histological observations to determine the nature and severity of the disease. Because disease processes affect the histological organization of tissues and organs, tissue samples, or **biopsies,** often play a key role in their diagnoses.

Figure 4-19 diagrams the histological changes induced by one relatively common irritating stimulus, cigarette smoke. The first abnormality to be observed is **dysplasia** (dis-PLĀ-zē-uh), a change in the normal shape, size, and organization of tissue cells. It is usually a response to chronic irritation or inflammation, and the changes are reversible. The normal trachea (windpipe) is lined by a pseudo-stratified, ciliated, columnar epithelium. The cilia move a mucous layer that traps foreign particles and moistens incoming air. The drying and chemical effects of smoking first paralyze the cilia, halting the movement of mucus (Figure 4-19a●) As mucus builds up, the individual coughs to dislodge it (the well-known "smoker's cough").

Epithelia and connective tissues may undergo more radical changes in structure, caused by the division and differentiation of stem cells. **Metaplasia** (me-tuh-PLĀ-zē-uh) is a structural change that dramatically alters the character of the tissue. In our example, heavy smoking first paralyzes the cilia, and over time the epithelial cells lose their cilia altogether. As metaplasia occurs, the epithelial cells produced by stem cell divisions no longer differentiate into ciliated columnar cells. Instead, they form a stratified squamous epithelium that provides a greater resistance to drying and chemical irritation (Figure 4-19b●). This epithelium protects the underlying tissues more effectively, but it completely eliminates the moisturization and cleaning properties of the epithelium. The cigarette smoke will now have an even greater effect on more delicate portions of the respiratory tract. Fortunately, metaplasia is reversible, and the epithelium gradually returns to normal once the individual quits smoking.

During **anaplasia** (a-nuh-PLĀ-zē-uh) tissue organization breaks down. Tissue cells change size and shape, often becoming unusually large or abnormally small. In anaplasia (Figure 4-19c●), which occurs in smokers developing one form of lung cancer, the cells divide more frequently, but not all divisions proceed in the normal way, and many of the tumor cells have abnormal chromosomes. Unlike dysplasia and metaplasia, anaplasia is irreversible.

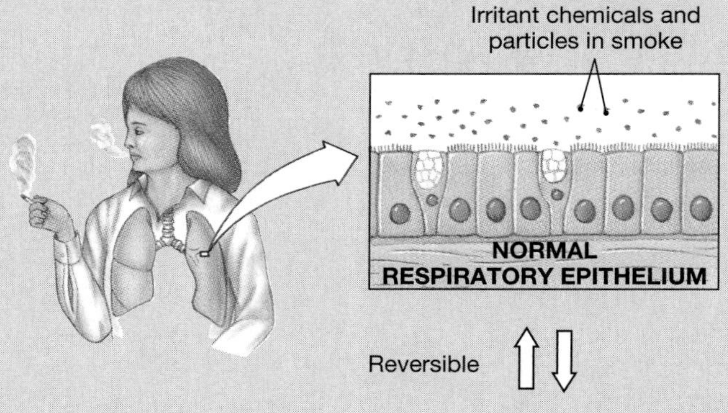
Irritant chemicals and particles in smoke
NORMAL RESPIRATORY EPITHELIUM

Reversible

(a) The cilia of respiratory epithelial cells are damaged and paralyzed by exposure to cigarette smoke. These changes cause the local buildup of mucus and reduce the effectiveness of the epithelium in protecting deeper, more delicate portions of the respiratory tract.

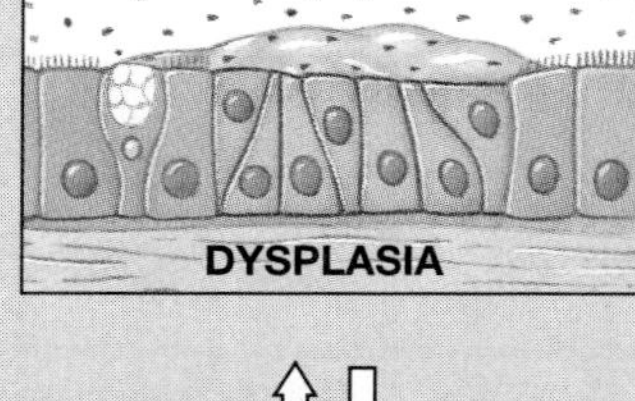
DYSPLASIA

Reversible

(b) In metaplasia, a tissue changes its structure. In this case the stressed respiratory surface converts to a stratified epithelium that protects underlying connective tissues but does nothing for other areas of the respiratory tract.

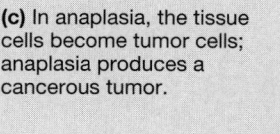

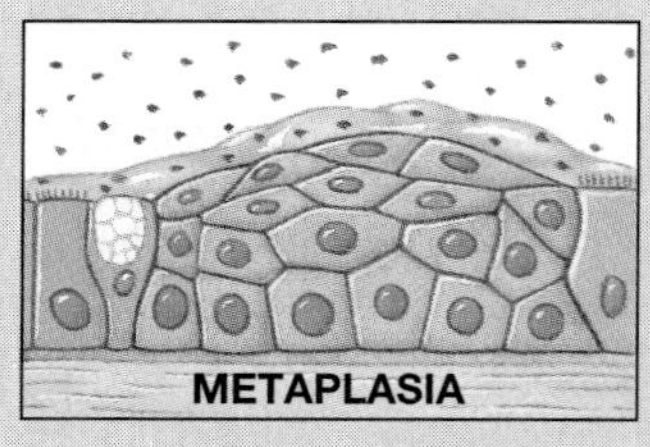
METAPLASIA

Irreversible

(c) In anaplasia, the tissue cells become tumor cells; anaplasia produces a cancerous tumor.

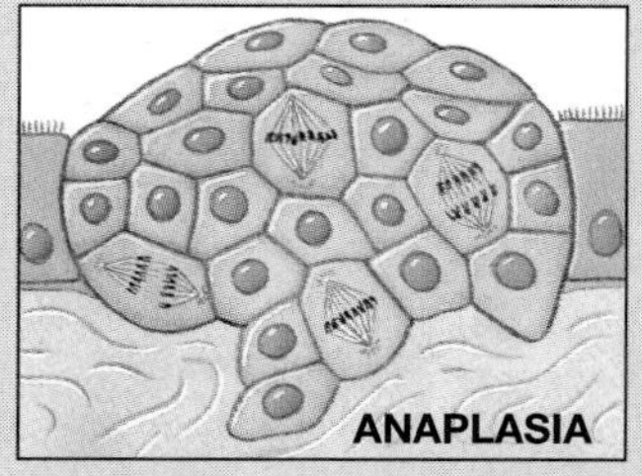
ANAPLASIA

● **FIGURE 4-19**
Changes in a Tissue under Stress

general, repair and maintenance activities throughout the body slow down, and a combination of hormonal changes and alterations in lifestyle affect the structure and chemical composition of many tissues. Epithelia get thinner and connective tissues more fragile. Individuals bruise easily and bones become brittle; joint pains and broken bones are common complaints. Because cardiac muscle cells and neurons cannot be replaced, cumulative damage can eventually cause major health problems such as cardiovascular disease or deterioration in mental function.

In future chapters we will consider the effects of aging on specific organs and systems. Some of these changes are genetically programmed. For example, the chondrocytes of older individuals produce a slightly different form of proteoglycan than those of younger people. The difference probably accounts for the thinner and less resilient cartilage of older people. In some cases the tissue degeneration can be temporarily slowed or even reversed. The age-related reduction in bone strength in women, a condition called *osteoporosis*, often results from a combination of inactivity, low dietary calcium levels, and a reduction in circulating *estrogens* (sex hormones). A program of exercise, calcium supplements, and hormonal replacement therapies can usually maintain normal bone structure for many years.

❑ Aging and Cancer Incidence

Cancer rates increase with age, and roughly 25 percent of all Americans develop cancer at some point in their lives. It has been estimated that 70–80 percent of these cases result from chemical exposure, environmental factors, or some combination of the two, and 40 percent of these cancers are caused by cigarette smoke. Each year over 500,000 Americans are killed by cancer, making this Public Health Enemy Number 2, second only to heart disease. The *Applications Manual* contains a detailed discussion of cancer development, growth, and treatment. AM Cancer

This chapter has introduced the four basic tissue types found in the human body. In terms of their contribution to total body weight, muscle tissue is most important (around 50 percent), followed by connective tissues (45 percent), epithelium and glands (3 percent), and neural tissue (2 percent). In combination these tissues form all of the organs and systems discussed in subsequent chapters. The next chapter considers the skin, or *integument*, an accessible structure that has structural characteristics intermediate between those of organs and organ systems.

Selected Clinical Terminology

Terms Discussed in This Chapter

anaplasia (a-nuh-PLĀ-zē-uh): An irreversible change in the size and shape of tissue cells. *(p. 140)*
antiangiogenesis factor: A secretion produced by chondrocytes that inhibits the growth of blood vessels. *(p. 132)*
dysplasia (dis-PLĀ-zē-uh): A change in the normal shape, size, and organization of tissue cells. *(p. 140)*
liposuction: A surgical procedure to remove unwanted adipose tissue by sucking it through a tube. *(p. 126)*
metaplasia (me-tuh-PLĀ-zē-uh): A structural change that alters the character of a tissue. *(p. 140)*
pathologists (pa-THOL-ō-jists): Physicians who specialize in the study of disease processes. *(p. 140)*
regeneration: The process of repairing injured tissues that follows inflammation. *(p. 139)*

AM *Additional Terms Discussed in the Applications Manual*

adhesions: Restrictive fibrous connections that can result from surgery, infection, or other injuries to serous membranes.
ascites (a-SĪ-tēz): Accumulation of fluid in the peritoneal cavity, usually caused by liver disease, kidney disease, or heart failure.
chemotherapy: Administration of drugs that either kill cancerous tissues or prevent mitotic divisions.
immunotherapy: Administration of drugs that help the immune system recognize and attack cancer cells.
oncologists (on-KOL-o-jists): Physicians who specialize in identifying and treating cancers.
pericarditis: An inflammation of the pericardial lining that may lead to the accumulation of pericardial fluid (a *pericardial effusion*).
peritonitis: Inflammation of the peritoneum following infection or injury.
pleural effusion: Accumulation of fluid within the pleural cavities as the result of chronic infection or inflammation of the pleura.
pleuritis (pleurisy): An inflammation of the pleural cavities. This condition may cause the production of a sound known as the *pleural rub*.
primary tumor (primary neoplasm): The site at which a cancer initially develops.
remission: A stage in which a tumor stops growing or grows smaller; the goal of cancer treatment.
secondary tumor: A colony of cancerous cells formed by metastasis, the spread of cells from a primary tumor.

EMBRYOLOGY SUMMARY Development of Organ Systems

Many different organ systems show similar patterns of organization. For example, the digestive, respiratory, urinary, and reproductive systems each include passageways lined by epithelia and surrounded by layers of smooth muscle. These patterns are the result of developmental processes under way in the first 2 months of embryonic life.

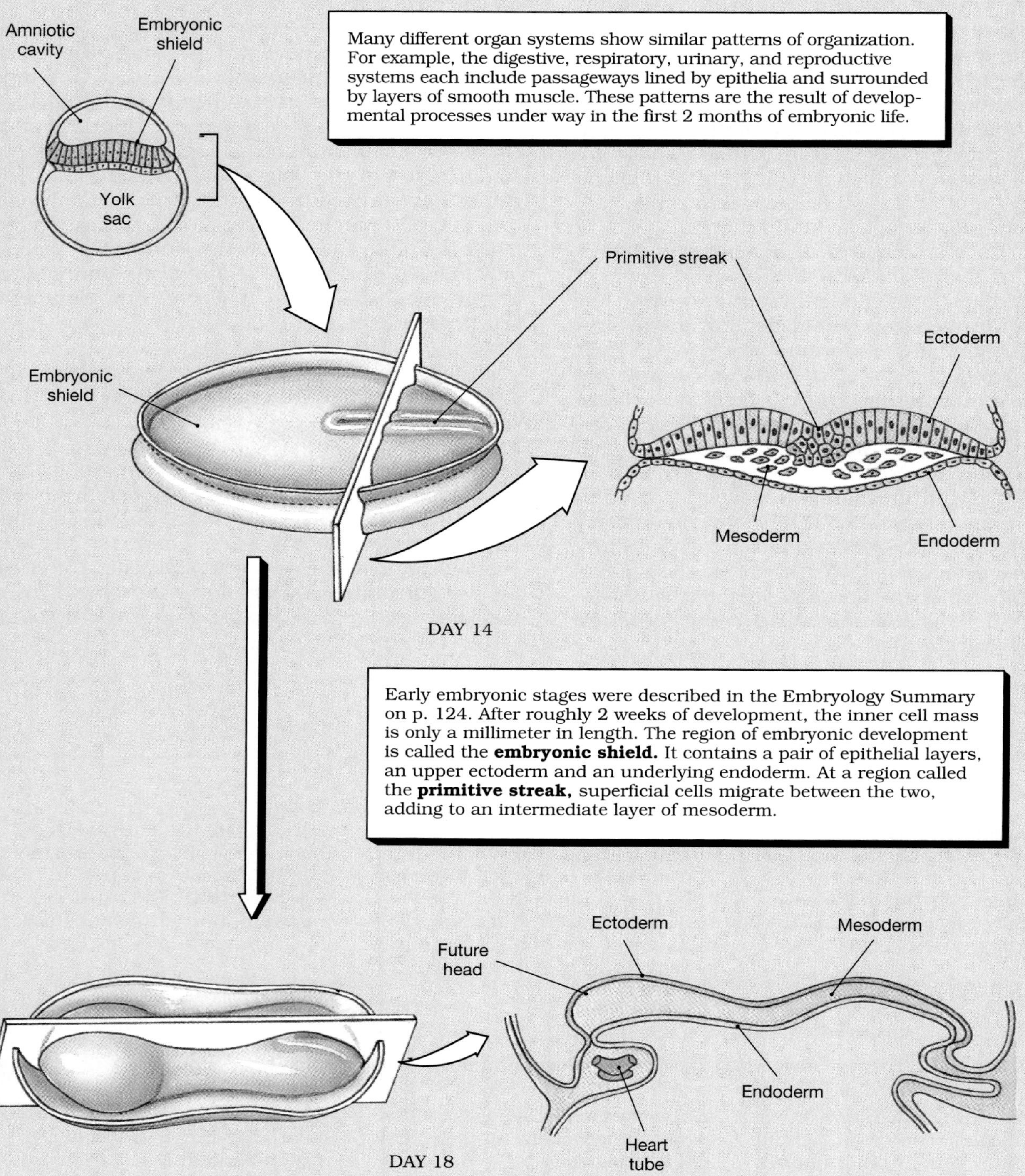
Amniotic cavity
Embryonic shield
Yolk sac
Embryonic shield
Primitive streak
Ectoderm
Mesoderm
Endoderm
DAY 14
Ectoderm
Mesoderm
Future head
Endoderm
Heart tube
DAY 18

Early embryonic stages were described in the Embryology Summary on p. 124. After roughly 2 weeks of development, the inner cell mass is only a millimeter in length. The region of embryonic development is called the **embryonic shield.** It contains a pair of epithelial layers, an upper ectoderm and an underlying endoderm. At a region called the **primitive streak,** superficial cells migrate between the two, adding to an intermediate layer of mesoderm.

By day 18 the embryo has begun to lift off the surface of the embryonic shield. The heart and blood vessels have already formed, well ahead of the other organ systems. Unless otherwise noted, discussions of organ system development in later chapters will begin at this stage.

DERIVATIVES OF PRIMARY GERM LAYERS	
Ectoderm forms:	Epidermis of integumentary system and the hair follicles, nails, and glands (sweat, milk, and sebum) communicating with the skin surface Lining of the mouth, salivary glands, nasal passageways, and anus Nervous system, including brain and spinal cord Portions of endocrine system (parts of pituitary and adrenal glands) Pharyngeal cartilages and related musculature
Mesoderm forms:	Lining of the body cavities (pleural, pericardial, peritoneal) Muscular, skeletal, cardiovascular, and lymphatic systems Kidneys and part of the urinary tract Gonads and most of the reproductive tract Connective tissues supporting all organ systems Portions of endocrine system (parts of adrenal and endocrine tissues of reproductive tract)
Endoderm forms:	Most of the digestive system: epithelium (except mouth and anus), exocrine glands (except salivary glands), the liver and pancreas Most of the respiratory system: epithelium (except nasal passageways) and mucous glands Portions of urinary and reproductive systems (ducts and sex cells) Portions of endocrine system (thymus, thyroid, parathyroids, pancreas, part of pituitary)

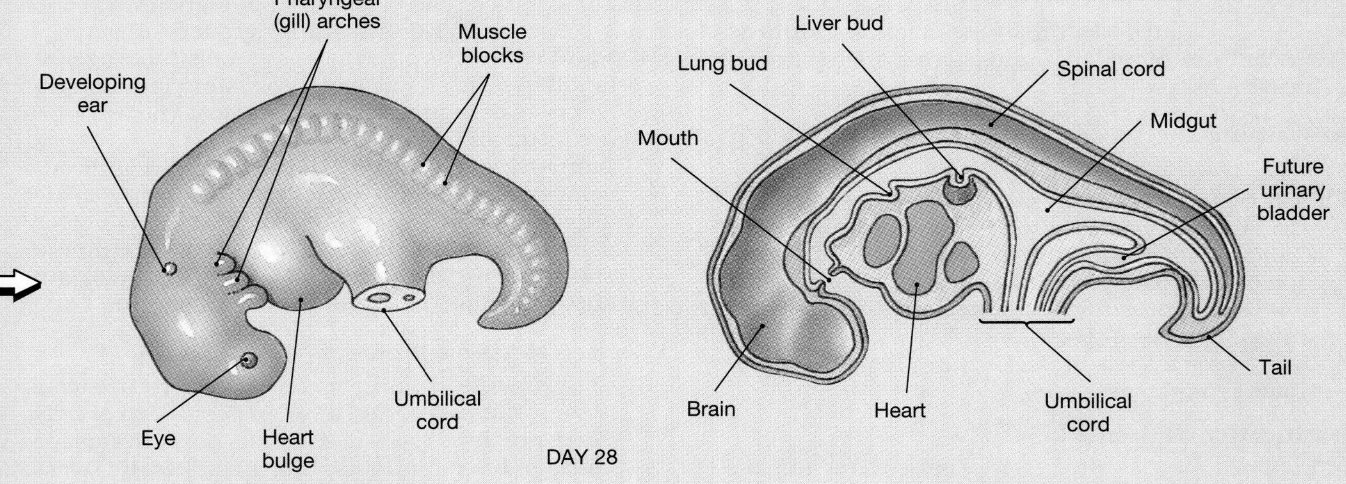
Pharyngeal (gill) arches
Muscle blocks
Developing ear
Eye
Heart bulge
Umbilical cord
DAY 28
Liver bud
Lung bud
Spinal cord
Midgut
Mouth
Future urinary bladder
Brain
Heart
Umbilical cord
Tail

After 1 month major organ systems have begun to form. The role of each of the primary germ layers in the formation of organs is summarized in the accompanying table; details will be found in later chapters.

CHAPTER REVIEW

STUDY OUTLINE

INTRODUCTION, p. 110

1. **Tissues** are collections of specialized cells and cell products that are organized to perform a relatively limited number of functions. There are four **tissue types:** epithelia, connective tissue, muscle tissue, and neural tissue.

EPITHELIAL TISSUE, p. 110

1. An **epithelium** is an **avascular** layer of cells that forms a barrier with certain properties.

Functions of Epithelial Tissue, p. 110

2. Epithelia provide physical protection, control permeability, provide sensation, and produce specialized secretions. Gland cells are epithelial cells that produce secretions.
3. **Exocrine** secretions are discharged onto the body surface or into passageways that communicate with the exterior. **Endocrine** secretions, known as *hormones*, are released by gland cells into the surrounding tissues.

Specializations of Epithelial Cells, p. 111

4. Epithelial cells are specialized to allow them to maintain the physical integrity of the epithelium and to perform secretory or transport functions. *(Figure 4-1)*
5. The inner surface of each epithelium is connected to a two-part **basement membrane** consisting of a **basal lamina** and a **reticular lamina.** Divisions by **germinative cells** continually replace the short-lived epithelial cells.
6. Epithelial cells may show **polarity,** and often have microvilli.
7. The coordinated beating of the cilia on a **ciliated epithelium** moves materials across the epithelial surface.

Maintaining the Integrity of Epithelia, p. 112

8. Cells can attach to other cells or to extracellular protein fibers by means of **cell adhesion molecules** (CAMs) or at specialized attachment sites. There are four types of **cell junctions: gap junctions, tight junctions, intermediate junctions,** and **desmosomes.** *(Figures 4-2, 4-3)*
9. Cells in some areas of the body are linked by **junctional complexes.**

Classification of Epithelia, p. 114

10. Epithelia are classified on the basis of the number of cell layers and the shape of the exposed cells.
11. A **simple epithelium** has a single layer of cells covering the basement membrane, while a **stratified epithelium** has several layers. In a **squamous epithelium** the cells are thin and flat. Cells in a **cuboidal epithelium** resemble little hexagonal boxes; those in a **columnar epithelium** are taller and more slender. *(Figures 4-4, 4-5, 4-6)*

Glandular Epithelia, p. 119

12. A glandular epithelial cell may release its secretions through merocrine, apocrine, or holocrine mechanisms. *(Figure 4-7)*
13. In **merocrine secretion,** the most common method of secretion, the product is released through exocytosis. **Apocrine secretion** involves the loss of both secretory product and cytoplasm. Unlike the first two methods, **holocrine secretion** destroys the cell; it becomes packed with secretions and finally bursts.
14. In epithelia that contain scattered gland cells, individual secretory cells are called **unicellular glands. Multicellular glands** are organs that contain glandular epithelia that produce exocrine or endocrine secretions.
15. Exocrine glands can be classified on the basis of structure. *(Figure 4-8)*

CONNECTIVE TISSUES, p. 121

1. Connective tissues are internal tissues with many important functions: establishing a structural framework; transporting fluids and dissolved materials; protecting delicate organs; supporting, surrounding, and interconnecting tissues; storing energy reserves; and defending the body from microorganisms.
2. All connective tissues have specialized cells, extracellular protein fibers, and a **ground substance.**

Classification of Connective Tissues, p. 122

3. **Connective tissue proper** refers to connective tissues that contain varied cell populations and fiber types surrounded by a syrupy ground substance.
4. **Fluid connective tissues** have a distinctive population of cells suspended in a watery ground substance containing dissolved proteins. The two types are blood and lymph.
5. **Supporting connective tissues** have a less diverse cell population than connective tissue proper and a dense ground substance that contains closely packed fibers; the combination of fibers and ground substance is called a **matrix.** The two types of supporting connective tissues are cartilage and bone.

Connective Tissue Proper, p. 122

6. Connective tissue proper contains fibers, a viscous ground substance, and a varied population of cells.
7. There are three types of fibers in connective tissue: **collagen fibers, reticular fibers,** and **elastic fibers.**
8. Connective tissue proper is classified as **loose** or **dense connective tissue.** There are two specialized types of loose connective tissues: **adipose tissue** and **reticular tissue.** Most of the volume in dense connective tissue consists of fibers. There are two types of dense connective tissue: **dense regular connective tissue** and **dense irregular connective tissue.** *(Figures 4-9, 4-10, 4-11)*

Fluid Connective Tissues, p. 128

9. **Blood** and **lymph** are connective tissues that contain distinctive collections of cells in a fluid matrix.
10. Blood contains **formed elements: red blood cells, white blood cells,** and **platelets.** Its watery ground substance is called **plasma.** *(Figure 4-12)*
11. **Arteries** carry blood from the heart and toward **capillaries,** where water and small solutes move into the **interstitial fluid** of surrounding tissues. **Veins** return blood to the heart.
12. Lymph forms as interstitial fluid enters the **lymphatics,** which return lymph to the circulatory system.

Supporting Connective Tissues, p. 130

13. Cartilage and bone are called supporting connective tissues because they support the rest of the body.
14. The matrix of **cartilage** is a firm gel that contains **chondroitin sulfates** (proteoglycans) and cells called **chondrocytes.** A fibrous **perichondrium** separates cartilage from surrounding tissues. There are three types of cartilage: **hyaline cartilage, elastic cartilage,** and **fibrocartilage.** *(Figure 4-13)*
15. Cartilage grows by two different mechanisms, **interstitial growth** and **appositional growth.**
16. Chondrocytes rely upon diffusion through the avascular matrix to obtain nutrients.
17. **Bone,** or **osseous tissue,** has a matrix consisting of collagen fibers and calcium salts, giving it unique properties. *(Figure 4-14; Table 4-1)*
18. **Osteocytes** depend on diffusion through **canaliculi** for nutrient intake.
19. Each bone is surrounded by a **periosteum** with fibrous and cellular layers.

MEMBRANES, p. 134

1. Membranes form a barrier or interface. Epithelia and connective tissues combine to form membranes that cover and protect other structures and tissues. There are four types of membranes: **mucous, serous, cutaneous,** and **synovial.** *(Figure 4-15)*

Mucous Membranes, p. 134

Serous Membranes, p. 134

The Cutaneous Membrane, p. 134

Synovial Membranes, p. 134

THE CONNECTIVE TISSUE FRAMEWORK OF THE BODY, p. 135

1. Internal organs and systems are tied together by a network of connective tissue proper that includes the **superficial fascia** (separating the skin from underlying tissues and organs), the **deep fascia** (dense connective tissue), and the **subserous fascia** (the layer between the deep fascia and the serous membranes that line body cavities). *(Figure 4-16)*

MUSCLE TISSUE, p. 136

1. **Muscle tissue,** consisting of **muscle fibers,** is specialized for contraction. There are three types of muscle tissue: **skeletal muscle, cardiac muscle,** and **smooth muscle.** *(Figure 4-17)*

Skeletal Muscle Tissue, p. 136

Cardiac Muscle Tissue, p. 138

Smooth Muscle Tissue, p. 138

NEURAL TISSUE, p. 138

1. **Neural tissue** is specialized to conduct electrical impulses that convey information from one area of the body to another.
2. Cells in neural tissue are either neurons or neuroglia. **Neurons,** whose axons are often called **nerve fibers,** transmit information as electrical impulses. There are several kinds of **neuroglia,** but their basic functions include: isolating and protecting the cell membranes of neurons; supporting neural tissue; regulating the composition of the interstitial fluid; and defending neural tissue from pathogens and repairing injuries. *(Figure 4-18)*
3. A typical neuron has a **soma, dendrites,** and an **axon.** The axon carries information in the form of changes in the transmembrane potential. At *synaptic terminals,* the neuron communicates with other cells.

TISSUE INJURIES AND AGING, p. 139

Inflammation and Regeneration, p. 139

1. Any injury affects several tissue types simultaneously, and they respond in a coordinated manner. Homeostasis is restored in two processes: inflammation and regeneration.
2. **Inflammation,** or the **inflammatory response,** isolates the injured area while damaged cells, tissue components, and any dangerous microorganisms are cleaned up. **Regeneration** is the repair process that restores normal function.

Aging and Tissue Repair, p. 139

3. Tissues change with age. Repair and maintenance grow less efficient, and the structure and chemical composition of many tissues are altered.

Aging and Cancer Incidence, p. 141

4. Cancer incidence increases with age, with roughly three-quarters of all cases caused by exposure to chemicals or environmental factors.

REVIEW QUESTIONS

LEVEL 1 Reviewing Facts and Terms

Match each item in Column A with the most closely related item in Column B. Use letters for answers in the spaces provided.

	Column A	Column B
___	1. histology	a. hyaluronic acid
___	2. microvilli	b. lines heart and all blood vessels
___	3. gap junction	c. repair and renewal
___	4. tight junction	d. ligament
___	5. ground substance	e. endocrine secretion
___	6. basal and reticular laminae	f. merocrine secretion
___	7. germinative cells	g. absorption and secretion
___	8. mesothelium	h. fat cells
___	9. endothelium	i. holocrine secretion
___	10. mucus	j. study of tissues
___	11. destroys gland cell	k. unicellular exocrine glands
___	12. hormones	l. tendon
___	13. goblet cells	m. intercellular connection
___	14. adipocytes	n. histamine and heparin
___	15. wandering cells	o. interlocking of membrane proteins
___	16. mast cells	p. lines ventral body cavities
___	17. bone-to-bone attachment	q. intercalated discs
___	18. muscle-to-bone attachment	r. striated, voluntary
___	19. skeletal muscle	s. defense and repair
___	20. cardiac muscle	t. basement membrane

21. The four basic tissue types found in the body are:
(a) epithelial, connective, muscle, neural
(b) simple, cuboidal, squamous, stratified
(c) fibroblasts, adipocytes, melanocytes, mesenchymal
(d) lymphocytes, macrophages, microphages, adipocytes

22. Long microvilli incapable of movement are called:
(a) cilia (b) flagella
(c) stereocilia (d) a, b, and c are correct

23. A type of junction common in cardiac and smooth muscle tissue is the:
(a) desmosome (b) intermediate junction
(c) tight junction (d) gap junction

24. The most abundant connections between cells in the superficial layers of the skin are:
(a) intermediate junctions
(b) gap junctions
(c) desmosomes
(d) tight junctions

25. The bulk movement of fluids across the epithelial lining of a capillary is called:
(a) active transport (b) epithelial transport
(c) facilitated diffusion (d) simple diffusion

26. Mucous secretions that coat the passageways of the digestive and respiratory tracts result from ________________ secretion.
(a) apocrine (b) merocrine
(c) holocrine (d) endocrine

27. The type of tissue that contains a fluid known as the ground substance is
(a) epithelial (b) neural
(c) muscle (d) connective

28. The three basic types of fibers found in connective tissue are:
(a) tendons, ligaments, aponeuroses
(b) loose, dense, irregular
(c) cartilage, bone, collagen
(d) collagen, reticular, elastic

29. The three types of loose connective tissue include:
(a) collagen, reticular, elastic
(b) areolar, adipose, reticular
(c) collagen, bone, cartilage
(d) fluid, supporting, connective tissue proper

30. Four major examples of dense regular connective tissue are:
(a) cartilage, bone, tendons, ligaments
(b) aponeuroses, elastic, cartilage, bone
(c) tendons, aponeuroses, elastic tissue, ligaments
(d) collagen, reticular, bone, tendons

31. The three major types of cartilage found in the body are:
(a) collagen, reticular, elastic
(b) areolar, adipose, reticular
(c) hyaline, elastic, fibrocartilage
(d) aponeuroses, reticular, elastic

32. The primary function of serous membranes in the body is:
(a) to minimize friction between opposing surfaces
(b) to line cavities that communicate with the exterior
(c) to perform absorptive and secretory functions
(d) to cover the surface of the body

33. The layer of loose connective tissue that separates the skin from underlying tissues and organs is the:
(a) superficial fascia (b) subcutaneous layer
(c) hypodermis (d) a, b, and c are correct

34. Large muscle fibers that are multinucleated, striated, and voluntary are found in:
(a) cardiac muscle tissue
(b) skeletal muscle tissue
(c) smooth muscle tissue
(d) a, b, and c are correct

35. Intercalated discs and pacemaker cells are characteristic of:
(a) smooth muscle tissue
(b) cardiac muscle tissue
(c) skeletal muscle tissue
(d) a, b, and c are correct

36. Axons, dendrites, and a soma are characteristics of cells found in:
(a) neural tissue (b) muscle tissue
(c) connective tissue (d) epithelial tissue

37. An inflammation resulting from the presence of pathogens, such as bacteria, is a(n):
(a) contusion (b) ascites
(c) angiogenesis (d) infection

38. A secretion produced by chondrocytes that inhibits the growth of blood vessels is:
  (a) chondroitin
  (b) proteoglycan
  (c) antiangiogenesis factor
  (d) chondromycin

39. The repair process necessary to restore normal function in damaged tissues is called:
  (a) isolation (b) regeneration
  (c) reconstruction (d) a, b, and c are correct

40. What four major characteristics are typical of epithelial tissue?

41. What are the four essential functions of epithelial tissue?

42. What four types of cell junctions serve as cellular attachments in the body?

43. What three types of layering make epithelial tissue recognizable?

44. What three cell shapes describe almost every epithelium in the body?

45. What three methods do various glandular epithelial cells use to release their secretions?

46. What three basic components are found in connective tissues?

47. What six basic functions characterize the role of connective tissue in the body?

48. What fluid connective tissues and supporting connective tissues are found in the human body?

49. What three types of fibers are found in connective tissues?

50. Give four major examples of dense regular connective tissue.

51. What three major types of cartilage are found in the body?

52. What four kinds of membranes composed of epithelial and connective tissue cover and protect other structures and tissues in the body?

53. What are the names of the transudates formed on the surfaces of the serous membranes of the (a) lungs; (b) heart; and (c) organs in the abdomen?

54. What three types of muscle tissue are found in the body?

55. What two cell populations make up neural tissue? What is the function of each?

LEVEL 2 Reviewing Concepts

56. In surfaces of the body where mechanical stresses are severe the dominant epithelium is:
  (a) stratified squamous epithelium
  (b) simple cuboidal epithelium
  (c) simple columnar epithelium
  (d) stratified cuboidal epithelium

57. Why is the presence of cilia on the respiratory surfaces important?

58. Why does holocrine secretion require continuous cell division?

59. What is the difference between an exocrine and an endocrine secretion?

60. A significant structural feature in the digestive system is the presence of tight junctions located near the exposed surfaces of cells lining the digestive tract. Why are these junctions so important?

61. Why are dead skin cells usually shed in thick sheets, rather than individually?

62. Why are infections always a serious threat after a severe burn or abrasion?

63. Why can't adults generate metabolic heat as readily as infants and young children?

64. After a weight-loss program, why is the lost weight often regained quickly in the same areas of the body?

65. Cartilage heals poorly and in many instances does not heal or recover after a severe injury. Why not?

66. What characteristics make the cutaneous membranes different from the serous and mucous membranes?

67. Why is cardiac muscle tissue that has been damaged by injury or disease incapable of regeneration?

68. Where are you most likely to find elastic ligaments in the body and why?

LEVEL 3 Critical Thinking and Clinical Applications

69. Assuming that you had the necessary materials to perform a chemical analysis of body secretions, how could you determine whether a secretion was merocrine or apocrine?

70. After many years of chronic smoking, Mr. Butts finds himself plagued by a hacking cough. Explain the causes of this cough.

71. A biology student accidentally loses the labels of two prepared slides she is studying. One is a slide of animal intestine and the other of animal esophagus. You volunteer to help her sort them out. How would you decide which slide is which?

72. You are asked to develop a scheme that can be used to identify the three different types of muscle tissue in two steps. What would the two steps be?

CHAPTER 5

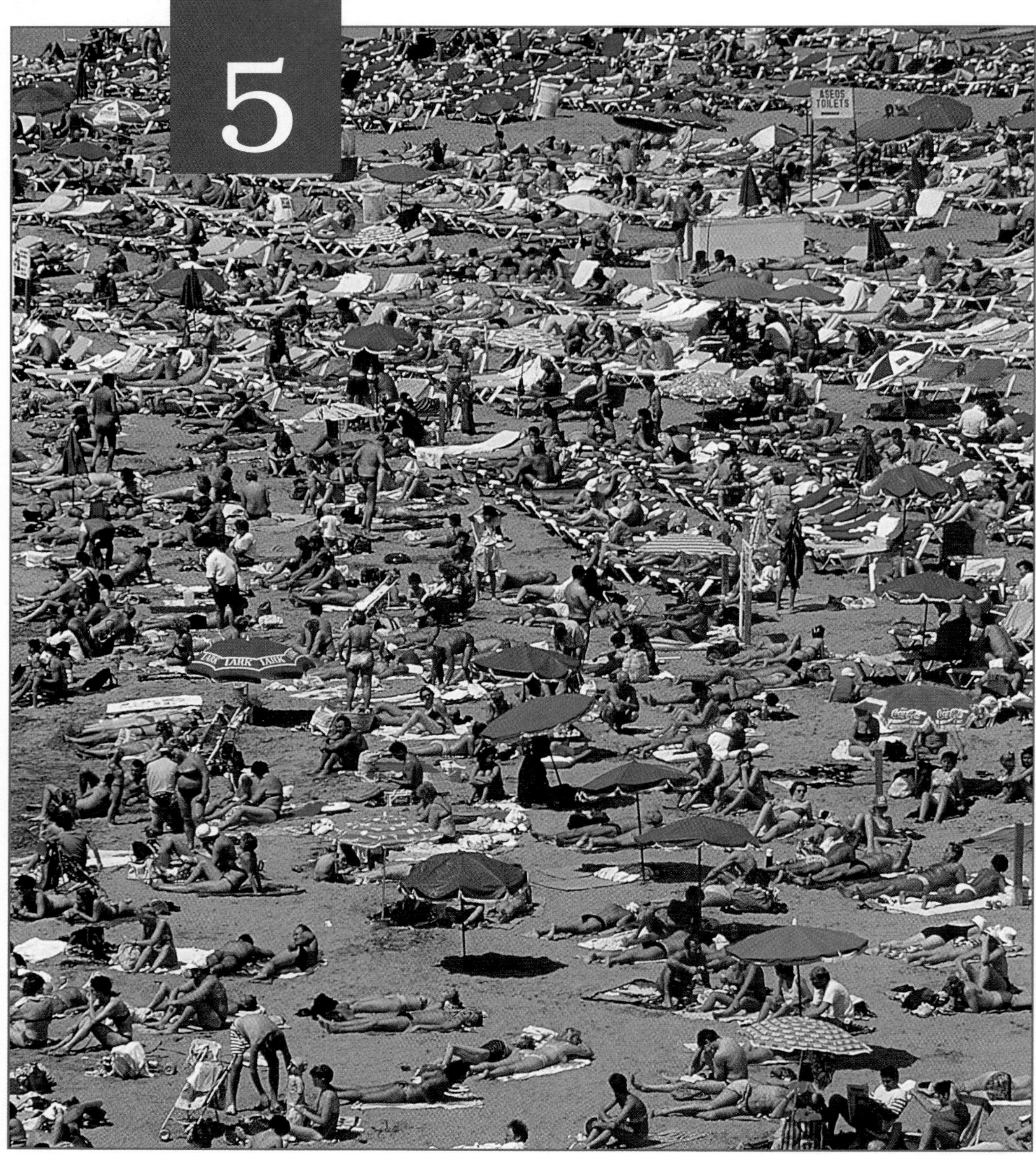
ASEOS
TOILETS
LARK

How many of the sun-lovers who crowd the beaches on a warm day ever stop to consider the contributions and sacrifices made by their skin? This remarkable structure absorbs ultraviolet radiation, prevents dehydration, preserves normal body temperature, and tolerates the chafing and abrasion of the sand. Although few people think of it in these terms, the skin is actually an organ—the largest organ of the body. In this chapter we will examine its varied functions.

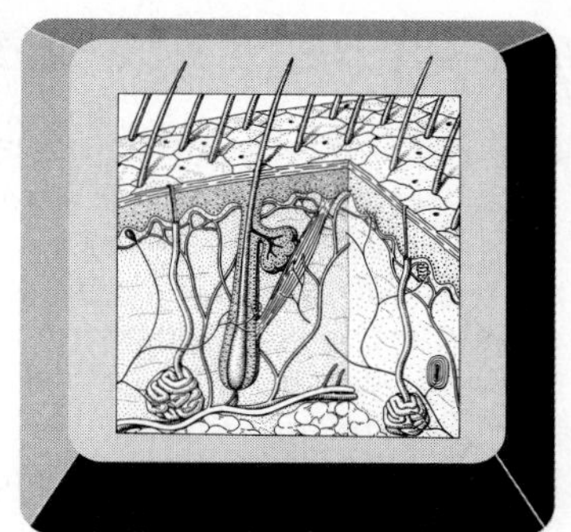

The Integumentary System

Chapter Outline and Objectives

1 Describe the main structural features of the epidermis and explain their functional significance.

2 Explain what accounts for individual and racial differences in skin, such as skin color.

3 Discuss the effects of ultraviolet radiation on the skin and the role played by melanocytes.

4 Describe the structure and functions of the dermis.

5 Explain the mechanisms that produce hair and determine hair texture and color.

6 Discuss the various kinds of glands found in the skin and their secretions.

7 Explain how the sweat glands of the integumentary system play a major role in regulating body temperature.

8 Describe the anatomical structure of nails and how they are formed.

9 Explain how the skin responds to injury and repairs itself.

10 Summarize the effects of the aging process on the skin.

The skin, or **integument** (in-TEG-ū-ment), can be considered as either a large, highly complex organ or a structurally integrated organ system. Most people recognize both possibilities and consider the terms *integument* and **integumentary system** as interchangeable. The components of the integumentary system include the epithelium and underlying connective tissue of the skin and the associated hairs, nails, and exocrine glands.

Few organs are as accessible, sizable, varied in function, and as underappreciated. Your integument accounts for around 16 percent of your total body weight, and its surface is continually abused, abraded, attacked by microorganisms, irradiated by sunlight, and exposed to environmental chemicals. Of all the body systems, this is the only one we see every day.

Because others see this system as well, we devote a lot of time to improving the appearance of the skin and associated structures. Washing the face, brushing and trimming hair, taking showers, and applying deodorant are activities that modify the appearance or properties of the skin. And when something goes wrong with the skin, the effects are immediately apparent. Even a relatively minor skin condition or blemish will be noticed at once, whereas more serious problems in other systems are often ignored. In addition to watching for abnormal *skin signs*, physicians note the general appearance of the skin because changes in skin color, flexibility, elasticity, or sensitivity may indicate a dysfunction in another system. AM *Examination of the Skin; Diagnosing Skin Conditions*

The integumentary system has two major components, the *cutaneous membrane* and the *accessory structures*.

1. The **cutaneous membrane** has two components, the superficial epithelium, or **epidermis** (*epi*, above), and the underlying connective tissues of the dermis.
2. The **accessory structures** include hair, nails, and multicellular exocrine glands. These structures are located in the dermis and protrude through the epidermis to the skin surface.

The integument does not function in isolation. An extensive network of blood vessels branches through the dermis, and sensory receptors that monitor touch, pressure, temperature, and pain provide valuable information to the central nervous system about the state of the body. The general structure of the integument is shown in Figure 5-1●. Beneath the dermis, the loose connective tissue of the **subcutaneous layer,** also known as the *superficial fascia,* or *hypodermis,* separates the integument from the deep fascia around other organs, such as muscles and bones. ∞ [p. 135] Although it is often not considered a part of the integument, we will consider the subcutaneous

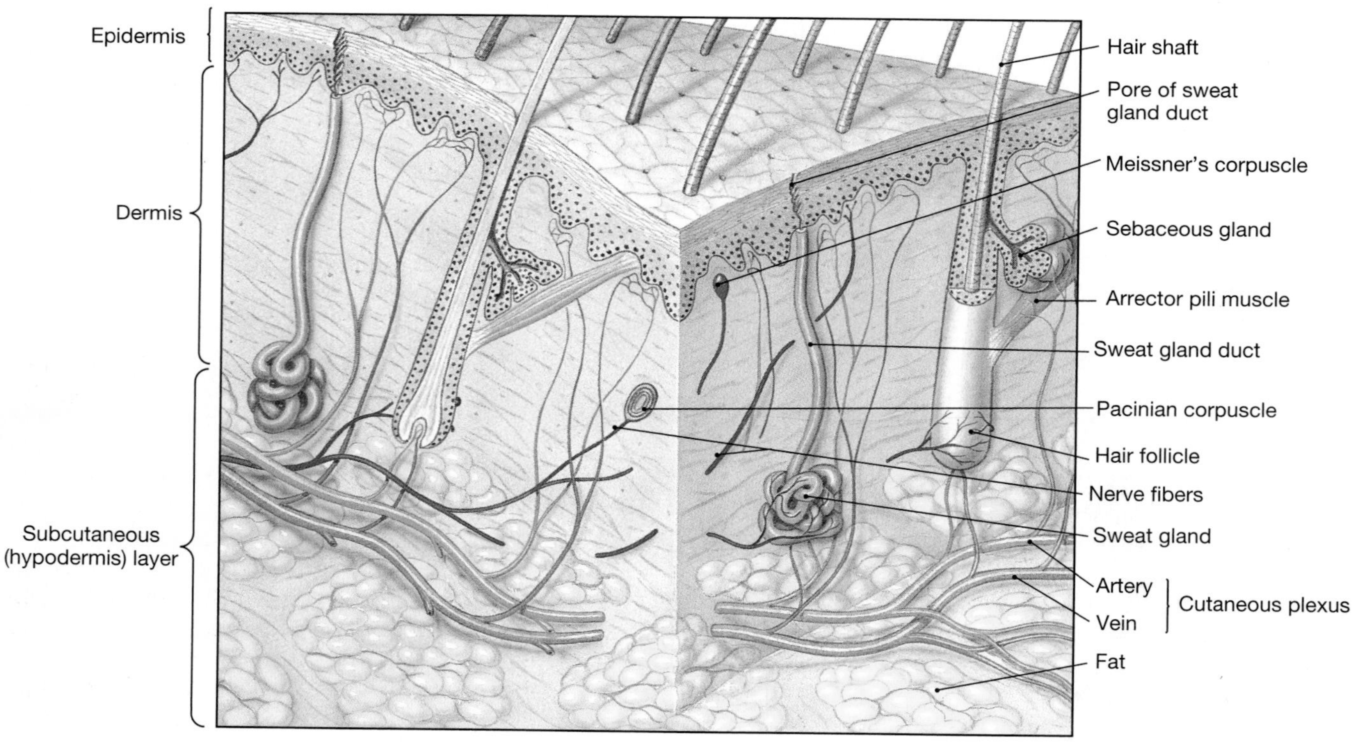
Epidermis
Dermis
Subcutaneous (hypodermis) layer
Hair shaft
Pore of sweat gland duct
Meissner's corpuscle
Sebaceous gland
Arrector pili muscle
Sweat gland duct
Pacinian corpuscle
Hair follicle
Nerve fibers
Sweat gland
Artery
Vein
Cutaneous plexus
Fat

● **FIGURE 5-1**
Components of the Integumentary System. Relationships among the major components of the integumentary system (with the exception of nails, shown in Figure 5-10).

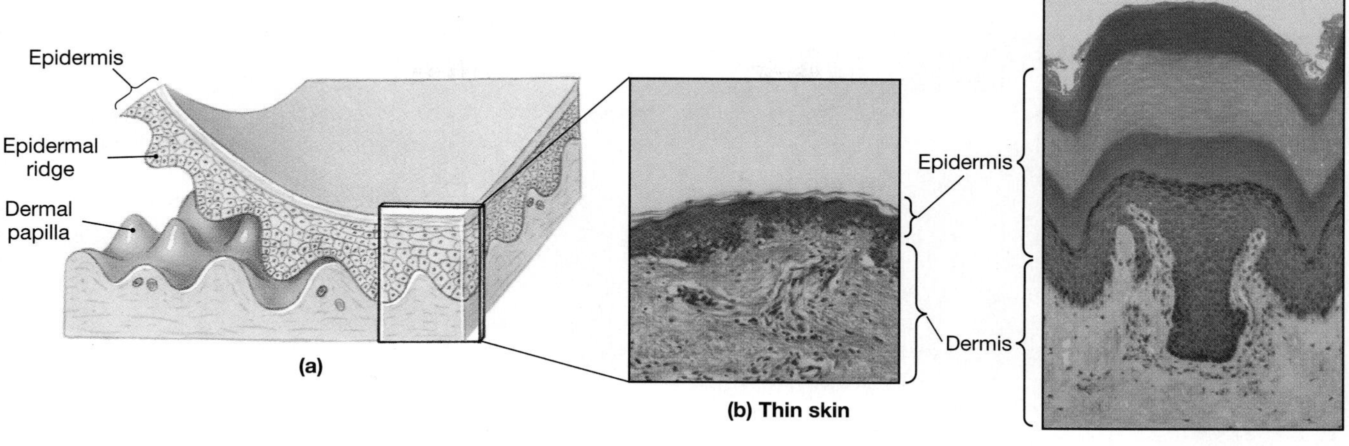
Epidermis
Epidermal ridge
Dermal papilla
(a)
Epidermis
Dermis
(b) Thin skin
(c) Thick skin

● **FIGURE 5-2**
Thin and Thick Skin. (a) The basic organization of the epidermis. The proportions of the various layers change depending on the location sampled. **(b)** Thin skin covers most of the exposed body surface. (LM × 122) **(c)** Thick skin covers the surfaces of the palms and soles. (LM × 165)

layer in this chapter because of its extensive interconnections with the dermis.

The general functions of the skin include:

1. ***Protection*** of underlying tissues and organs.
2. ***Excretion*** of salts, water, and organic wastes.
3. ***Maintenance*** of normal body temperature.
4. ***Synthesis*** of a steroid, *vitamin* D_3, that is subsequently converted to a hormone, *calcitriol*, important to normal calcium metabolism.
5. ***Storage*** of nutrients.
6. ***Detection*** of touch, pressure, pain, and temperature stimuli and the relay of that information to the nervous system.

Each of these functions will be explored more fully as we discuss the individual components of the integument.

The Epidermis

Figures 5-1, 5-2

The epidermis consists of a stratified squamous epithelium (Figures 5-1 and 5-2●).

The most abundant epithelial cells, called **keratinocytes,** form several different layers, but the precise boundaries between these layers are often difficult to see in a light micrograph. In **thick skin,** found on the palms of the hands and soles of the feet, five layers can be distinguished. Only four layers can be distinguished in the **thin skin** that covers the rest of the body. The terms *thick* and *thin* refer to the relative thickness of the epidermis, not to the integument as a whole.

The majority of the body is covered by thin skin. In a sample of thin skin (Figure 5-2a,b●), the epidermis is a mere 0.08 mm thick. The epidermis in a sample of thick skin (Figure 5-2c●) may be as much as six times as thick.

❑ Layers of the Epidermis

Figure 5-3

Figure 5-3● shows the layers in a section of thick skin. Beginning at the basement membrane and traveling toward the free surface, we find the *stratum germinativum,* the *stratum spinosum,* the *stratum granulosum,* the *stratum lucidum,* and the *stratum corneum.*

Stratum Germinativum

Figures 5-2, 5-3

The innermost epidermal layer is the **stratum germinativum** (STRA-tum jer-mi-na-TĒ-vum), or *stratum basale* (Figure 5-3●). This layer is 3–5 cells thick. The cells closest to the underlying loose connective tissue are firmly attached to the basement membrane of the epidermis. The stratum germinativum forms **epidermal ridges** that extend into the dermis, increasing the area of contact between the two regions. Dermal projections called **dermal papillae** (singular *papilla*, a nipple-shaped mound) extend between adjacent ridges (Figure 5-2a●).

The contours of the skin surface follow the ridge patterns, which vary from small conical pegs (in thin skin) to the complex whorls seen on the thick skin of the palms and soles. Ridges on the palms and soles increase the surface area of the skin and increase friction, ensuring a secure grip.

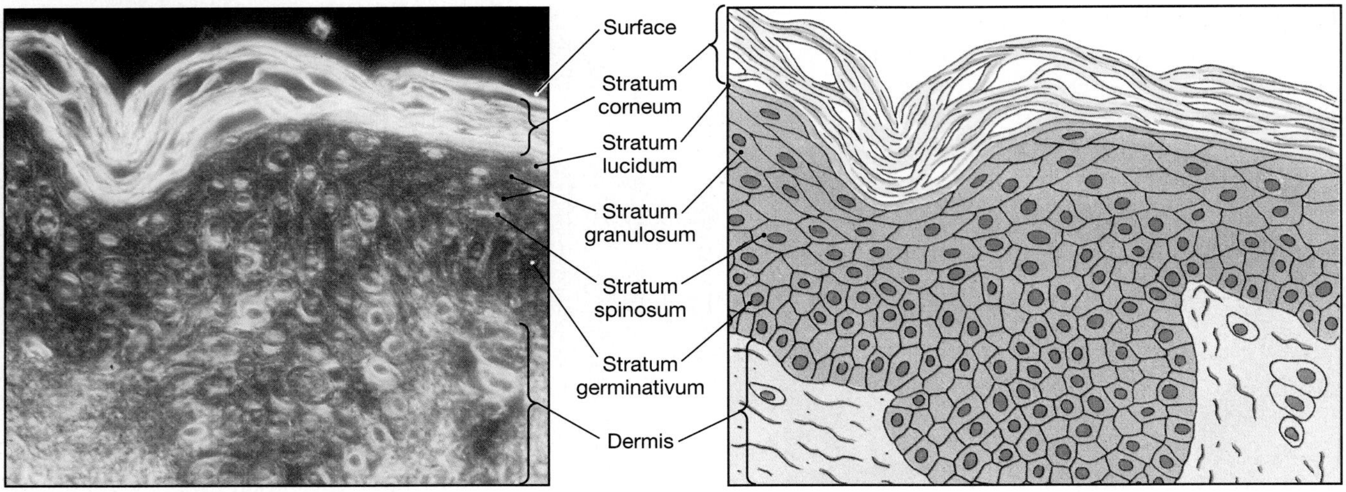

DARK-FIELD LIGHT MICROGRAPH OF EPIDERMIS

LAYER

Surface

Stratum corneum

Stratum lucidum

Stratum granulosum

Stratum spinosum

Stratum germinativum

Dermis

FIGURE 5-3
The Structure of the Epidermis. A light micrograph through a portion of the epidermis, showing the major stratified layers of epidermal cells. (LM × 128)

Ridge shapes are genetically determined: Those of each person are unique and do not change in the course of a lifetime. Fingerprints are ridge patterns on the tips of the fingers that can be used to identify individuals; they have been used in criminal investigations for over a century.

CELLS OF THE STRATUM GERMINATIVUM Large stem cells dominate the stratum germinativum. The divisions of these cells replace the more superficial keratinocytes that are lost or shed at the epithelial surface. Skin surfaces that lack hair contain specialized epithelial cells known as **Merkel cells.** These cells are found among the deepest cells of the stratum germinativum. They are sensitive to touch—when compressed or disturbed, Merkel cells release chemicals that stimulate sensory nerve endings. (There are many other kinds of sensory receptors in the dermis, and these will be described in later sections.)

The brown tones of the skin result from the synthetic activities of *melanocytes*, pigment cells introduced in Chapter 4. ∞ [p. 122] These cells are scattered throughout the stratum germinativum, with cell processes extending into more superficial layers. Skin color will be discussed in more detail in a later section.

Stratum Spinosum

Each time a stem cell divides, one of the daughter cells is pushed above the germinativum into the next layer, the **stratum spinosum** (spiny layer). The stratum spinosum is 8–10 cells thick, and the keratinocytes are bound together by desmosomes. Standard histological procedures shrink the cytoplasm, but the cytoskeletal elements and desmosomes remain intact, making the cells look like miniature pincushions. Some of the cells entering this layer from the stratum germinativum continue to divide, further increasing the thickness of the epithelium. The stratum spinosum also contains **Langerhans cells**, cells involved with the immune response. Langerhans cells are responsible for stimulating a defense against (1) microorganisms that manage to penetrate the superficial layers of the epidermis and (2) superficial skin cancers. Langerhans cells cannot be seen in standard histological preparations.

Stratum Granulosum

The layer of cells above the stratum spinosum is the **stratum granulosum** (grainy layer). The stratum granulosum consists of 3–5 layers of keratinocytes displaced from the stratum spinosum. By the time cells reach this layer, most have stopped dividing, and they begin manufacturing large quantities of the protein **keratohyalin** (ker-a-tō-HĪ-a-lin). Near the boundary with the stratum lucidum, the nuclei and other organelles start to degenerate.

Stratum Lucidum

In the thick skin of the palms and soles, a glassy **stratum lucidum** (clear layer) covers the stratum granulosum. The 3–5 layers of cells are flattened, densely packed, and filled with **eleidin** (e-LĒ-i-din), a protein derived from keratohyalin. Keratohyalin and eleidin represent preliminary steps in the production of the fibrous protein **keratin** (KER-a-tin; *keros*, horn), introduced in Chapter 2. ∞ [p. 55] Keratin is tough, durable, and water-resistant. In addition to forming the protective barrier in the stratum corneum, keratin is the basic structural component of hair and nails.

Stratum Corneum

As true keratin fibers are developing, the cells become thinner and flatter. The cell membranes thicken and become less permeable. The nuclei and organelles disintegrate, the cells die, and their subsequent dehydration creates a tightly interlocked layer of keratin sandwiched between phospholipid cell membranes. These layers of flattened and dead cells covering the epidermis form the **stratum corneum** (KOR-nē-um; *cornu*, horn). There are normally 15–30 layers of keratinized cells in the stratum corneum. *Keratinization* occurs everywhere on exposed skin surfaces except over the anterior surfaces of the eyes.

It takes 15–30 days for a cell to move from the stratum germinativum to the stratum corneum. The dead cells usually remain in the exposed stratum corneum layer for an additional two weeks before they are shed or washed away. This arrangement places the deeper portions of the epithelium and underlying tissues beneath a protective barrier composed of dead, durable, and expendable cells.

PERMEABILITY OF THE STRATUM CORNEUM Although the stratum corneum is water-resistant, it is not waterproof, and water from the interstitial fluids slowly penetrates the surface, to be evaporated into the surrounding air. Roughly 500 ml (about 1 pt) of water is lost in this way each day. This process is called **insensible perspiration** to distinguish it from the **sensible perspiration** produced by active sweat glands. Damage to the stratum corneum, as in a serious burn or *xerosis* (excessively dry skin) can drastically increase the rate of insensible perspiration and produce a potentially dangerous fluid loss. AM *Disorders of Keratin Production*

When the skin is immersed in water, osmotic forces may move water into or out of the epithelium. *[p. 75]* Sitting in a freshwater bath causes water to move into the epidermis because fresh water is hypotonic (has fewer dissolved materials) compared with body fluids. The epithelial cells may swell to four times their normal volumes, a phenomenon particularly noticeable in the thickly keratinized areas of the palms and soles. Swimming in the ocean reverses the direction of osmotic flow, because the ocean is a hypertonic solution. As a result, water leaves the body, crossing the epidermis from the underlying tissues. The process is slow, but long-term exposure to seawater endangers survivors of a shipwreck by accelerating dehydration.

DRUG ADMINISTRATION THROUGH THE SKIN Drugs in oils or other lipid-soluble carriers can penetrate the epidermis. The movement is slow, particularly through the layers of cell membranes in the stratum corneum, but once a drug reaches the underlying tissues it will be absorbed into the circulation. Placing a drug in a solvent that is lipid-soluble can assist its movement through the lipid barriers. Drugs can also be administered by packaging them in *liposomes*, artificially produced lipid droplets. Liposomes containing DNA fragments have been used experimentally to introduce normal genes into abnormal human cells. For example, genes carried by liposomes have been used to alter the cell membranes of skin cancer cells so that they will be attacked by the immune system. Another experimental procedure involves creating a transient change in skin permeability by administering a brief pulse of electricity. This electrical pulse temporarily changes the positions of the cells in the stratum corneum, creating channels that allow the drug to penetrate.

A useful technique for long-term drug administration involves placing a sticky patch containing a drug over an area of thin skin. To overcome the relatively slow rate of diffusion, the patch must contain an extremely high concentration of the drug. This procedure, called *transdermal administration*, has the advantage that a single patch may work for several days, making daily pills unnecessary. AM *Transdermal Medications*

Excessive shedding of cells from the outer layer of skin in the scalp causes dandruff. What is the name of this layer of skin?

As you pick up a piece of lumber, a splinter pierces the palm of your hand and lodges in the third layer of the epidermis. Identify this layer.

Why does swimming in fresh water for an extended period cause epidermal swelling?

Some criminals sand the tips of their fingers so as not to leave recognizable fingerprints. Would this practice permanently remove fingerprints? Why or why not?

Skin Color

The color of the skin is due to an interaction between: (1) pigment composition and concentration and (2) the dermal blood supply.

Skin Pigmentation

Figure 5-4

The epidermis contains variable quantities of two pigments, *carotene* and *melanin*.

- **Carotene** (KAR-ō-tēn) is an orange-yellow pigment that normally accumulates inside epidermal cells; it is most apparent in cells of the stratum corneum of light-skinned individuals. This pigment is also distributed in fatty tissues in the dermis. Carotene is found in a variety of orange-colored vegetables, such as carrots and squashes. Caucasian vegetarians with a special fondness for carrots can actually turn orange from an overabundance of carotenes in the skin. Carotene can be converted to vitamin A, which is required for (1) the normal maintenance of epithelia and (2) the synthesis of photoreceptor pigments in the eye.

- **Melanin** is a brown, yellow-brown, or black pigment produced by melanocytes.

MELANOCYTES Melanocytes (Figure 5-4●) are found in the stratum germinativum, squeezed between the epithelial cells. Melanocytes manufacture melanin pigments from molecules of the amino acid *tyrosine.* The melanin is stored in intracellular vesicles called *melanosomes.* These vesicles are "injected" into keratinocytes of the stratum germinativum and stratum spinosum. This transfer of pigmentation colors the entire epidermis. Within these epidermal cells, the melanin pigments become concentrated in the region around the nucleus. In this location the melanin pigments protect the DNA of the nucleus from the damaging effects of *ultraviolet radiation.*

Sunlight contains significant amounts of **ultraviolet** (UV) **radiation.** A small amount of UV radiation is beneficial, for it stimulates synthetic activity in the epidermis (a process discussed in a later section). However, too much ultraviolet radiation produces immediate effects of mild or even serious burns. Long-term damage can result from repeated exposure. For example, UV radiation damages fibroblasts, leading to impaired maintenance of the dermis. The resulting structural alterations lead to premature wrinkling. In addition, skin cancers can result from chromosomal damage in germinative cells or melanocytes. One of the major consequences of the global depletion of the ozone layer in the upper atmosphere will be a sharp increase in the rate of skin cancers, such as *malignant melanoma.* AM Skin Cancers

Melanocytes respond to UV exposure by increasing their activity. Unfortunately, the response is not rapid enough to prevent a sunburn the first day at the beach. Melanin synthesis accelerates slowly, peaking around ten days after the initial exposure.

The ratio between melanocytes and germinative cells ranges between 1:4 and 1:20, depending on the region of the body surveyed. For example, there are about 1000 melanocytes per square millimeter in the skin covering most areas of the body. Higher concentrations (around 2000/mm^2) are found in the cheeks and forehead, the nipples, and in the genital region (the *scrotum* of males and the *labia majora* of females). Observed differences in skin color between individuals and even races do not reflect different *numbers* of melanocytes, but merely different levels of synthetic activity. Even the melanocytes of *albino* individuals are distributed normally, although they are incapable of producing melanin.

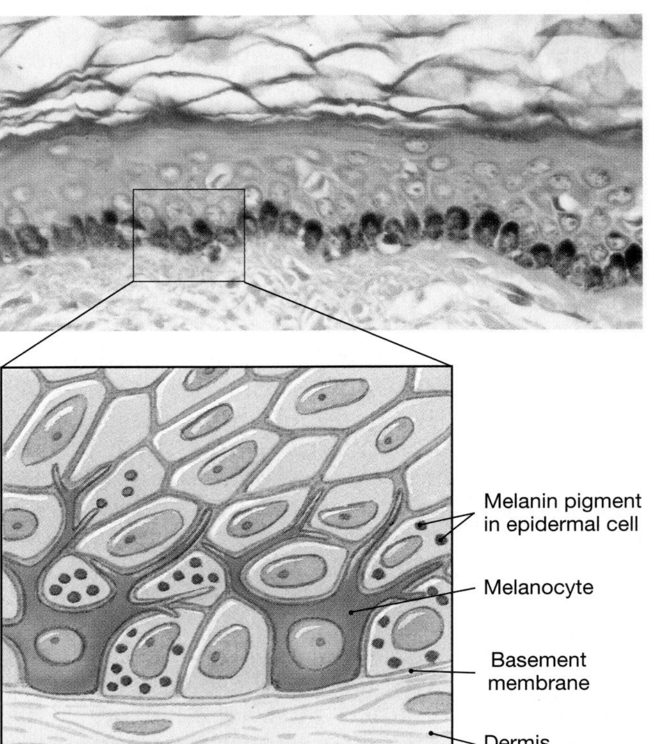
Melanin pigment in epidermal cell
Melanocyte
Basement membrane
Dermis

● **FIGURE 5-4**
Melanocytes. The micrograph and accompanying drawing indicate the location and orientation of melanocytes in the stratum germinativum of a black person.

ABNORMAL SKIN PIGMENTATION Several diseases that have primary impacts on other systems may have secondary effects on skin color and pigmentation. Because the skin is easily observed, these color changes can be useful in diagnosis. For example:

- In *jaundice* (JAWN-dis) the liver is unable to excrete bile, and a yellowish pigment accumulates in body fluids. In advanced stages, the skin and whites of the eyes turn yellow.
- Some tumors affecting the pituitary gland result in the secretion of large amounts of *melanocyte-stimulating hormone* (MSH). This hormone causes a darkening of the skin, as if the individual has an extremely deep bronze tan.
- In *Addison's disease* the pituitary gland secretes large quantities of *ACTH,* a hormone that is structurally similar to MSH. The result of ACTH on the skin coloration is also similar to that of MSH.
- In *vitiligo* (vi-ti-LĪ-gō) individuals lose their melanocytes. The condition develops in about 1 percent of the population, and the incidence increases among individuals with thyroid gland disorders, *Addison's disease,* and several other disorders. It is suspected that this disorder develops when the immune defenses malfunction, and antibodies attack normal melanocytes. The primary problem with vitiligo is cosmetic, especially for individuals with darkly pigmented skin. Michael Jackson is said to suffer from vitiligo. AM Changes in Skin Color

Dermal Circulation

Blood with abundant oxygen is bright red, and blood vessels in the dermis give the skin a reddish tint that is most apparent in light-pigmented individuals. When those vessels are dilated, as during inflammation, the red tones become much more pronounced.

When the circulatory supply is temporarily reduced, the skin becomes relatively pale; a frightened Caucasian may "turn white" because of a sudden drop in blood supply to the skin. During a sustained reduction in circulatory supply, the blood in the superficial vessels loses oxygen and changes color to a much darker red tone. Seen from the surface, the skin takes on a bluish coloration called **cyanosis** (sī-uh-NŌ-sis; *kyanos*, blue). In individuals of any skin color, cyanosis is most apparent in areas of thin skin, such as the lips or beneath the nails. It can be a response to extreme cold or a result of circulatory or respiratory disorders, such as heart failure or severe asthma.

An **ulcer** is a localized shedding of an epithelium. **Decubitis ulcers,** also known as bedsores, may affect bedridden or mobile patients with circulatory restrictions, especially when splints, casts, or bedding continually presses against superficial blood vessels. Such sores most often affect the skin near joints or projecting bones, where the dermal blood vessels are pressed against underlying structures. The chronic lack of circulation kills epidermal cells, removing a barrier to bacterial infection, and eventually the dermal tissues deteriorate as well. (A comparable degeneration, or *necrosis*, will occur in any tissues deprived of adequate circulation.) Bedsores can be prevented or treated by frequent changes in body position that vary the pressures applied to specific blood vessels.

❑ The Epidermis and Vitamin D_3

Although strong sunlight can damage epithelial cells and deeper tissues, limited exposure to sunlight is very beneficial. When exposed to ultraviolet radiation, epidermal cells in the stratum spinosum and stratum germinativum convert a steroid related to cholesterol into **vitamin D_3**, or **cholecalciferol** (kō-le-kal-SIF-e-rōl). The liver converts cholecalciferol to an intermediary product used by the kidneys to synthesize the hormone **calcitriol** (kal-si-TRĪ-ōl). Calcitriol is essential for normal calcium and phosphorus absorption by the small intestine, and an inadequate supply leads to impaired bone maintenance and growth.

Cholecalciferol can also be absorbed by the digestive tract if it is present in the diet. This fact accounts for the use of the term *vitamin*, even though the body can synthesize cholecalciferol. Children who live in areas with overcast skies, and whose diets lack cholecalciferol, can have abnormal bone development. This condition has largely been eliminated in the United States because dairy companies add cholecalciferol, usually identified on the label as "vitamin D," to the milk sold in grocery stores. Chapter 6 will consider the hormonal control of bone growth in greater detail.

❑ Epidermal Growth Factor

Epidermal growth factor, or **EGF,** is one of the peptide growth factors introduced in Chapter 3. ∞ [p. 103] Produced by glands of the digestive tract, EGF has widespread effects on epithelia, especially the epidermis. Its effects include:

- Promoting the divisions of germinative cells in the stratum germinativum and stratum spinosum.
- Accelerating the production of keratin in differentiating epidermal cells.
- Stimulating epidermal development and epidermal repair after injury.
- Stimulating synthetic activity and secretion by epithelial glands.

Epidermal growth factor has such a pronounced effect that it can be used in tissue culture to stimulate the growth and division of epidermal cells outside of the body. Sheets of epidermal cells can be produced in this fashion. This procedure has been used in the treatment of extensive burns; the burned areas can be covered by epidermal sheets "grown" from a small sample of intact skin from the burn victim. (This treatment will be considered in the discussion of burns on p. 170.)

Why does exposure to sunlight or sunlamps cause the skin to become darker?

Why does the skin appear red when it is warm?

In some cultures, women are required to be completely covered except for their eyes when they go outside. These women exhibit a high incidence of problems with their bones. Why?

The Dermis

Figure 5-1

The dermis lies beneath the epidermis (Figure 5-1●, p. 150). It has two major components, a superficial *papillary layer* and a deeper *reticular layer*.

❑ Dermal Organization

Figures 5-1, 5-2

The **papillary layer** consists of loose connective tissue. This region contains the capillaries and the sensory neurons supplying the surface of the skin. The papillary layer derives its name from the dermal papillae that project between the epidermal ridges, as shown in Figure 5-2●, p. 151.

The deeper **reticular layer** consists of an interwoven meshwork of dense irregular connective tissue. Bundles of collagen fibers leave the reticular layer to blend into those of the papillary layer above, so the boundary line between these layers is indistinct. Collagen fibers of the reticular layer also extend into the underlying subcutaneous layer.

In addition to extracellular protein fibers, the dermis contains all the cells of connective tissue proper. ∞ [p. 122] Accessory organs of epidermal origin, such as hair follicles and sweat glands, extend into the dermis. In addition, the reticular and papillary layers of the dermis contain networks of blood vessels, lymph vessels, and nerve fibers (Figure 5-1●, p. 150).

Wrinkles and Stretch Marks

The interwoven collagen fibers of the reticular layer provide considerable tensile strength, and the extensive array of elastic fibers enables the dermis to stretch and recoil repeatedly during normal movements. The water content of the skin also helps maintain its flexibility and resilience, properties known as *skin turgor*. As a result, dehydration causes a temporary decline in skin flexibility. Age, hormones, and the destructive effects of ultraviolet radiation permanently reduce the amount of elastin in the dermis, producing wrinkles and sagging skin. The extensive distortion of the dermis that occurs over the abdomen during pregnancy or after a substantial weight gain often exceeds the elastic capabilities of the skin. The resulting damage to the dermis prevents it from recoiling to its original size after delivery or a rigorous diet. The skin then wrinkles and creases, creating a network of **stretch marks.**

Tretinoin (*Retin-A*) is a derivative of vitamin A that can be applied to the skin as a cream or gel. This drug was originally developed to treat acne, but it also increases blood flow to the dermis and stimulates dermal repairs. As a result, the rate of wrinkle formation decreases, and existing wrinkles become smaller. The degree of improvement varies from individual to individual.

Lines of Cleavage

Figure 5-5

At any one location, the majority of the collagen and elastic fibers are arranged in parallel bundles. The orientation of these bundles depends on the stress placed on the skin during normal movement; the bundles are aligned to resist the applied forces. The resulting pattern of fiber bundles establishes the **lines of cleavage** of the skin. Lines of cleavage, shown in Figure 5-5●, are clinically significant because a cut parallel to a cleavage line will usually remain closed, whereas a cut at right angles to a cleavage line will be pulled open as cut elastic fibers recoil. Surgeons choose their incision patterns accordingly, for a neatly closed incision will heal faster and with less scarring.

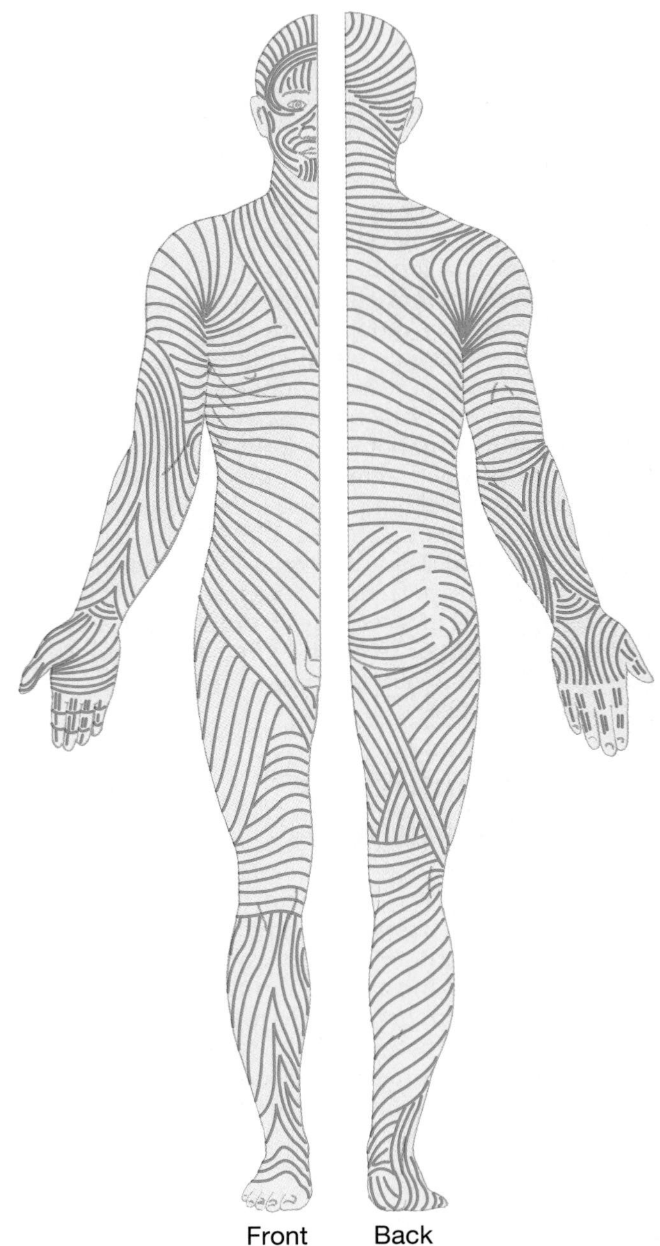
Front
Back

● **FIGURE 5-5**
Lines of Cleavage of the Skin. Lines of cleavage follow lines of tension in the skin. They reflect the orientation of collagen fiber bundles in the dermis.

❑ Dermal Circulation and Innervation

The Dermal Blood Supply

Arteries supplying the skin form a network in the subcutaneous layer along the border with the reticular layer. This network is called the *cutaneous plexus.* Tributaries of these arteries supply the adipose tissues of the subcutaneous layer and the tissues of the integument. As small arteries travel toward the epidermis, branches supply the hair follicles, sweat glands, and other structures in the dermis. On reaching the papillary layer these small arteries form another branching network, the *papillary plexus,* which provides arterial blood to capillary loops that follow the contours of the epidermal-dermal boundary. These capillaries empty into a network of venules that in turn form small veins that descend through the dermis to reach larger veins in the subcutaneous layer. Tumors that form in the dermal blood vessels during development create temporary or permanent *birthmarks.* Many of these birthmarks can now be removed through laser surgery.

The Innervation of the Skin

Figure 5-1

Nerve fibers in the skin control blood flow, adjust gland secretion rates, and monitor sensory receptors in the dermis and the deeper layers of the epidermis. We have already noted the presence of Merkel cells in the deeper layers of the epidermis. These cells are monitored by sensory terminals known as *Merkel's discs.* The epidermis also contains the processes of sensory neurons that provide sensations of pain and temperature. The dermis contains similar receptors, as well as other, more specialized receptors. Examples discussed in Chapter 17 and shown in Figure 5-1●, p. 150, include receptors sensitive to light touch (*Meissner's corpuscles,* located in dermal papillae), and deep pressure and vibration (*Pacinian corpuscles,* in the reticular layer).

DERMATITIS Because of the abundance of sensory receptors in the skin, regional infection or inflammation can be very painful. **Dermatitis** (der-muh-TĪ-tis) is an inflammation of the skin that primarily involves the papillary layer. In typical dermatitis, inflammation begins in a portion of the skin exposed to infection or irritated by chemicals, radiation, or mechanical stimuli. Dermatitis may cause no physical discomfort, or it may produce an annoying itch, as in poison ivy. Other forms of this condition can be quite painful, and the inflammation may spread rapidly across the entire integument.

There are many forms of dermatitis, some of them quite common:

- ***Contact dermatitis*** usually occurs in response to strong chemical irritants. It produces an itchy rash that may spread to other areas; poison ivy is an example.
- ***Eczema*** (EK-se-muh) is a dermatitis that can be triggered by temperature changes, fungus, chemical irritants, greases, detergents, or stress. Hereditary or environmental factors or both can encourage the development of eczema.
- ***Diaper rash*** is a localized dermatitis caused by a combination of moisture, irritating chemicals from fecal or urinary wastes, and flourishing microorganisms.
- ***Urticaria*** (ur-ti-KAR-ē-uh), also known as *hives,* is an extensive allergic response to a food, drugs, an insect bite, infection, stress, or other stimulus.

Where would you find the capillaries and sensory neurons that supply the epidermis?

What accounts for the ability of the dermis to undergo repeated stretching?

The Subcutaneous Layer

Figure 5-1

The connective tissue fibers of the reticular layer are extensively interwoven with those of the subcutaneous layer, or **hypodermis,** and the boundary between the two is usually indistinct (Figure 5-1●, p. 150). Although the subcutaneous layer is not a part of the integument, it is important in stabilizing the position of the skin in relation to underlying tissues, such as skeletal muscles or other organs, while permitting independent movement.

The subcutaneous layer consists of loose connective tissue with abundant fat cells. Infants and small children usually have extensive "baby fat," which helps reduce heat loss. Subcutaneous fat also serves as a substantial energy reserve and a shock absorber for the rough-and-tumble activities of our early years.

As we grow, the distribution of subcutaneous fat changes. Men accumulate subcutaneous fat at the neck, upper arms, along the lower back, and over the buttocks. In women, the breasts, buttocks, hips, and thighs are the primary sites of subcutaneous fat storage. In adults of either sex, the subcutaneous layer of the backs of the hands and the upper surfaces of the feet contain few fat cells, whereas distressing amounts of adipose tissue can accumulate in the abdominal region, producing a prominent "pot belly."

The subcutaneous layer is quite elastic. Only the superficial region contains large arteries and veins, and the rest contains a limited number of capillaries and no vital organs. This last characteristic makes **subcutaneous injection** a useful method for administering drugs. The familiar term **hypodermic needle** refers to the region targeted for injection.

Accessory Structures

The accessory structures of the integument include hair follicles, sebaceous glands, sweat glands, and nails. During embryological development these structures originate from the epidermis, and they are also known as *epidermal derivatives.* Although located in the dermis, they project through the epidermis to the integumentary surface.

Hair Follicles and Hair

Hairs project above the surface of the skin almost everywhere except over the sides and soles of the feet, the palms of the hands, the sides of the fingers and toes, the lips, and portions of the external genitalia.[1] There are about 5 million hairs on the human body, and 98 percent of them are on the general body surface, not on the head. Hairs originate in complex organs called **hair follicles,** and hair production is a complex process involving cooperation between the dermis and epidermis.

Hair Production

Figure 5-6

Hair follicles extend deep into the dermis, often projecting into the underlying subcutaneous layer (Figure 5-6●). The epithelium at the base of a hair follicle surrounds a small **hair papilla,** a peg of connective tissue containing capillaries and nerves. The **hair bulb** consists of epithelial cells that surround the papilla.

Hair production involves a specialization of the cornification process. The epithelial layer involved is called the **hair matrix.** Basal cells near the center of the matrix divide, producing daughter cells that are gradually pushed toward the surface. Those cells produced closest to the center of the matrix form the soft core, or **medulla,** of the hair, whereas cells closer to the edge of the developing hair form the relatively hard **cortex.** The medulla contains flexible **soft keratin. Hard keratin** in the cortex gives the hair its stiffness. Dead cells at the surface of the hair form the **cuticle,** a layer of hard keratin that coats the hair.

The **root** of the hair extends from the hair bulb to the point where the internal organization of the hair is complete, usually about halfway to the skin surface. The **shaft** extends from this point to the exposed tip of the hair. The size, shape, and color of the hair shaft are highly variable.

[1]The glans penis and prepuce of the male; the clitoris, labia minora, and inner surfaces of the labia majora in the female.

Follicle Structure

Figure 5-7

The cells of the follicle walls are organized into several concentric layers (Figure 5-7●). Beginning at the hair cuticle, these layers include:

- The **internal root sheath.** This layer surrounds the hair root and the deeper portion of the shaft. It is produced by the cells at the periphery of the hair matrix. The cells of the internal root sheath disintegrate relatively quickly, and this layer does not extend the entire length of the follicle.
- The **external root sheath.** This layer extends from the skin surface to the hair matrix. Over most of that distance it has all the cell layers found in the superficial epidermis. However, where the external root sheath joins the hair matrix, all the cells resemble those of the stratum germinativum.
- The **glassy membrane.** This layer, a thickened basement membrane, is wrapped in a dense connective tissue sheath.

Functions of Hair

Figure 5–6

The 5 million hairs on the human body have important functions. The roughly 100,000 hairs on the head protect the scalp from ultraviolet light, cushion a blow to the head, and provide insulating benefits for the skull. The hairs guarding the entrances to the nostrils and external ear canals help prevent the entry of foreign particles and insects, and eyelashes perform a similar function for the surface of the eye. A **root hair plexus** of sensory nerves surrounds the base of each hair follicle. As a result, the movement of the shaft of even a single hair can be felt at a conscious level. This sensitivity provides an early-warning system that may help to prevent injury. For example, you may be able to swat a mosquito before it reaches the skin surface.

Ribbons of smooth muscle, called **arrector pili** (a-REK-tōr PĪ-li) muscles (Figure 5-6b●), extend from the papillary dermis to the connective tissue sheath surrounding the hair follicle. When stimulated, the arrector pili pull on the follicles and erect the hairs. Contraction may be caused by emotional states, such as fear or rage, or as a response to cold, producing the characteristic "goose bumps" associated with shivering. In a furry mammal, this action

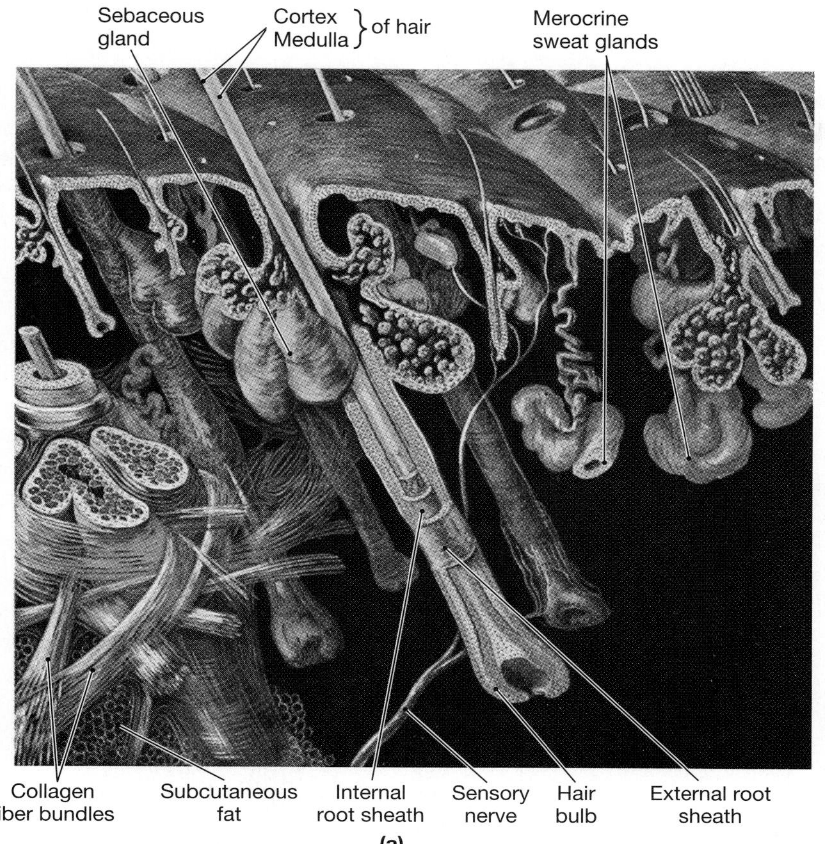
Sebaceous gland
Cortex
Medulla
of hair
Merocrine sweat glands
Collagen fiber bundles
Subcutaneous fat
Internal root sheath
Sensory nerve
Hair bulb
External root sheath
(a)

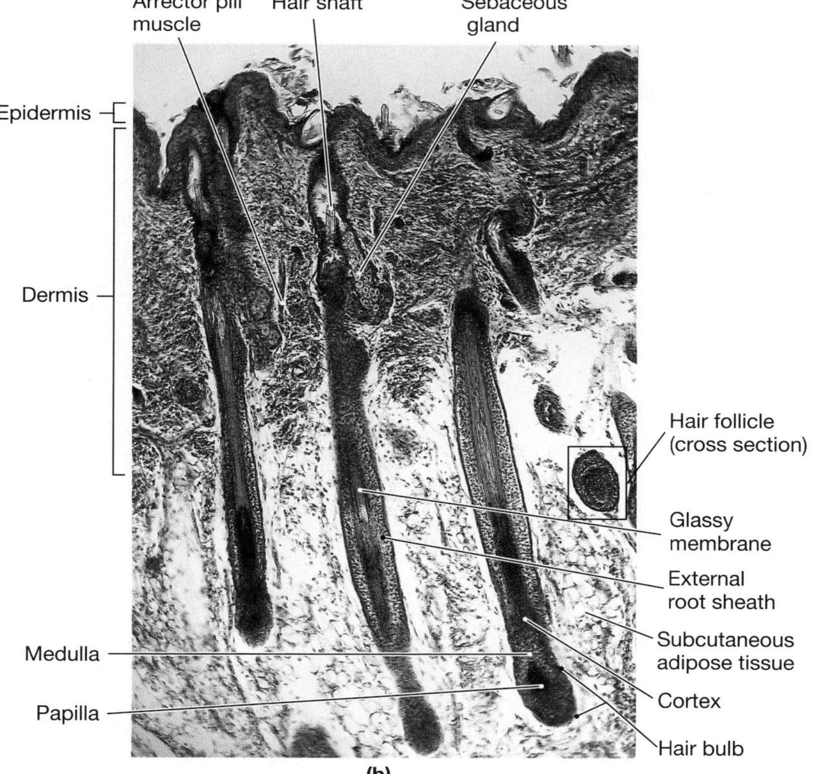
Arrector pili muscle
Hair shaft
Sebaceous gland
Epidermis
Dermis
Hair follicle (cross section)
Glassy membrane
External root sheath
Subcutaneous adipose tissue
Medulla
Papilla
Cortex
Hair bulb
(b)

● **FIGURE 5-6**
Accessory Structures of the Skin. (a) A three-dimensional view of the skin. **(b)** A light micrograph showing the sectional appearance of the skin of the scalp. (LM × 73)

increases the thickness of the insulating coat, rather like putting on an extra sweater. Although we do not receive any comparable insulating benefits, the reflex persists.

Types of Hairs

There are two major types of hairs in the adult integument, *vellus hairs* and *terminal hairs.*

- **Vellus hairs** are the fine "peach fuzz" hairs found over much of the body surface.
- **Terminal hairs** are heavy, more deeply pigmented, and sometimes curly. The hair on your head, including your eyebrows and eyelashes, are examples of terminal hairs.

Hair follicles may alter the structure of the hairs in response to circulating hormones, and this fact accounts for many of the changes in terminal hair distribution that begin at puberty.

Hair Color

Variations in hair color reflect differences in structure and variations in the pigment produced by melanocytes at the papilla. These characteristics are genetically determined, but the condition of your hair may be influenced by hormonal or environmental factors. As pigment production decreases with age, the hair color lightens toward gray. White hair results from the combination of a lack of pigment and the presence of air bubbles within the medulla of the hair shaft. Because the hair itself is dead and inert, changes in coloration are gradual.

Growth and Replacement of Hair

A hair in the scalp grows for 2–5 years, at a rate of around 0.33 mm/day. Variations in the hair growth rate and in the duration of the **hair growth cycle** account for individual differences in uncut hair length.

While hair growth is under way, the cells of the hair root are absorbing nutrients and incorporating them into their structure. Clipping or collecting hair for analysis can be helpful in diagnosing several disorders. For example, the hairs of individuals

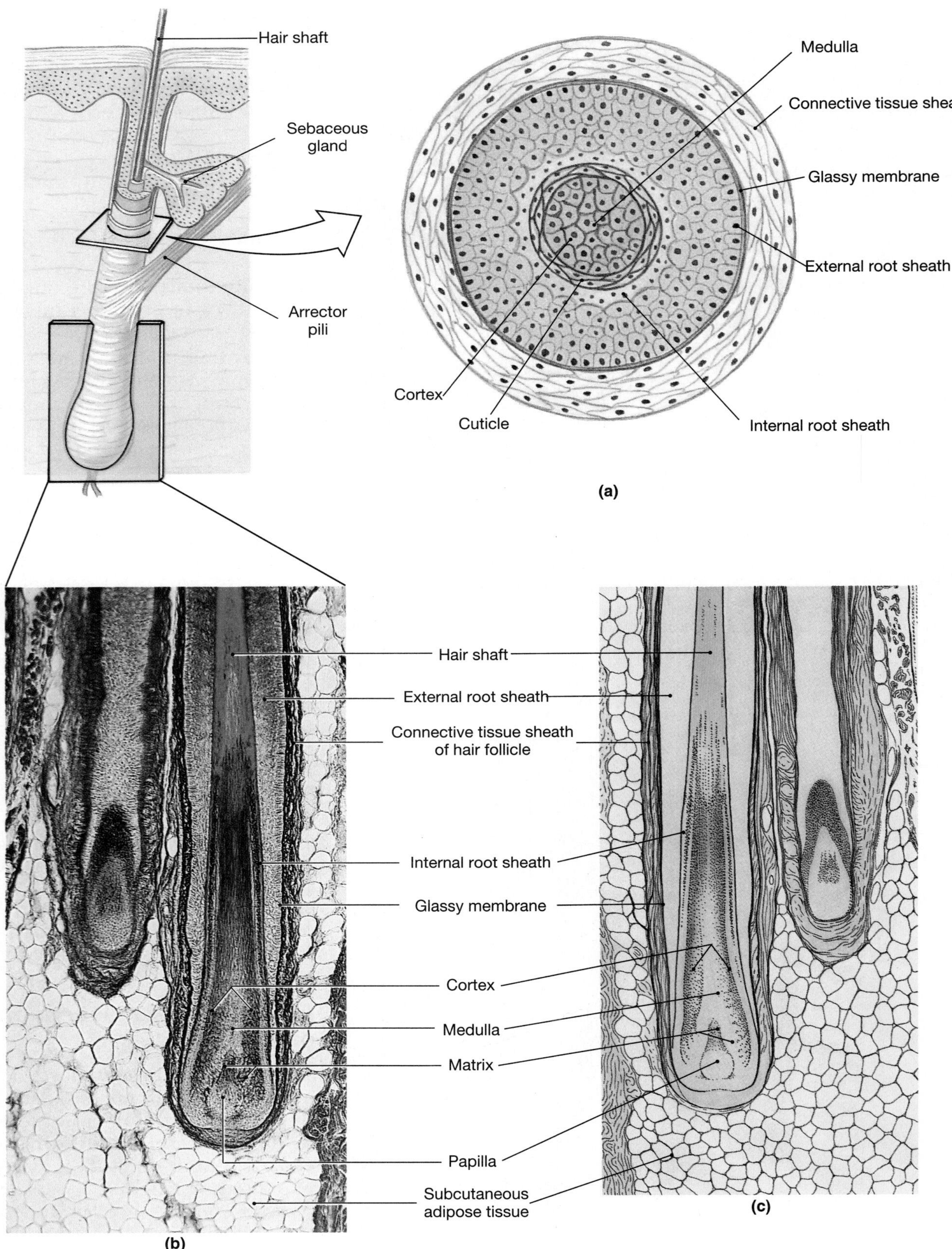

Hair shaft
Sebaceous gland
Arrector pili
Medulla
Connective tissue sheath
Glassy membrane
External root sheath
Internal root sheath
Cortex
Cuticle
(a)
Hair shaft
External root sheath
Connective tissue sheath of hair follicle
Internal root sheath
Glassy membrane
Cortex
Medulla
Matrix
Papilla
Subcutaneous adipose tissue
(b)
(c)

FIGURE 5-7

Hair Follicles. (a) A longitudinal section and a cross section through a hair follicle. **(b)** A light micrograph showing a closer view of the base of the follicle and hair shaft and the matrix and papilla at the hair root. (LM × 46) **(c)** A section along the longitudinal axis of a hair follicle.

suffering from lead or other heavy metal poisoning will contain high quantities of those ions. Hair samples can also be used for identification purposes, through the process of *DNA fingerprinting*, discussed in Chapter 3. [p. 94]

As it grows, the root of the hair is firmly attached to the matrix of the follicle. At the end of the growth cycle the follicle becomes inactive, and the hair is now termed a **club hair.** The follicle gets smaller, and over time the connections between the hair matrix and the root of the club hair break down. When another growth cycle begins, the follicle produces a new hair, and the old club hair gets pushed toward the surface.

HAIR LOSS In healthy adults about 50 hairs are lost from the head each day, but several factors may affect this rate. Sustained losses of more than 100 hairs per day usually indicate that something is wrong. Temporary increases in hair loss can result from drugs, dietary factors, radiation, vitamin A excess, high fever, stress, and hormonal factors related to pregnancy. In males, changes in the level of the sex hormones circulating in the blood can affect the scalp, causing a shift from terminal hair to vellus hair production, beginning at the temples and the crown of the head. This alteration is called **male pattern baldness.** Some cases of male pattern baldness will respond to drug therapies, such as topical application of *minoxidyl (Rogaine)*. For additional information, see the *Applications Manual.* AM Baldness and Hirsutism

What condition is produced by the contraction of the arrector pili muscles?

A person suffers a burn on the forearm that destroys the epidermis and extensive areas of the deep dermis. When the injury heals, would you expect to find hair growing again in the area of the injury?

Glands in the Skin

The skin contains two types of exocrine glands, *sebaceous glands* and *sweat glands.* Sebaceous glands produce an oily lipid that coats hair shafts and the epidermis. Sweat glands produce a watery solution and perform other special functions.

Sebaceous (Oil) Glands

Figure 5-8

Sebaceous (se-BĀ-shus) **glands,** or *oil glands,* are holocrine glands that discharge a waxy, oily secretion into hair follicles (Figure 5-8●). Several sebaceous glands communicate with a single follicle by sharing a single duct, and they are classified as *simple branched alveolar glands.* [p. 121] The gland cells manufacture large quantities of lipids as they mature, and the lipid product is released through *holocrine secretion,* a process that involves the rupture of the secretory cells. [p. 119]

The lipids released from these cells enter the open passageway, or **lumen,** of the gland. Contraction of the arrector pili muscle that erects the hair squeezes the sebaceous gland, forcing the oily secretions into the follicle and onto the surface of the skin. This secretion, called **sebum** (SĒ-bum), is a mixture of triglycerides, cholesterol, proteins, and electrolytes. Sebum provides lubrication and inhibits the growth of bacteria. Keratin is a tough protein, but dead, cornified cells become dry and brittle once exposed to the environment. Sebum lubricates and protects the keratin of the hair shaft and conditions the surrounding skin. Shampooing removes the natural oily coating, and excessive washing can make hairs stiff and brittle.

Sebaceous follicles are large sebaceous glands that communicate directly with the epidermis. These

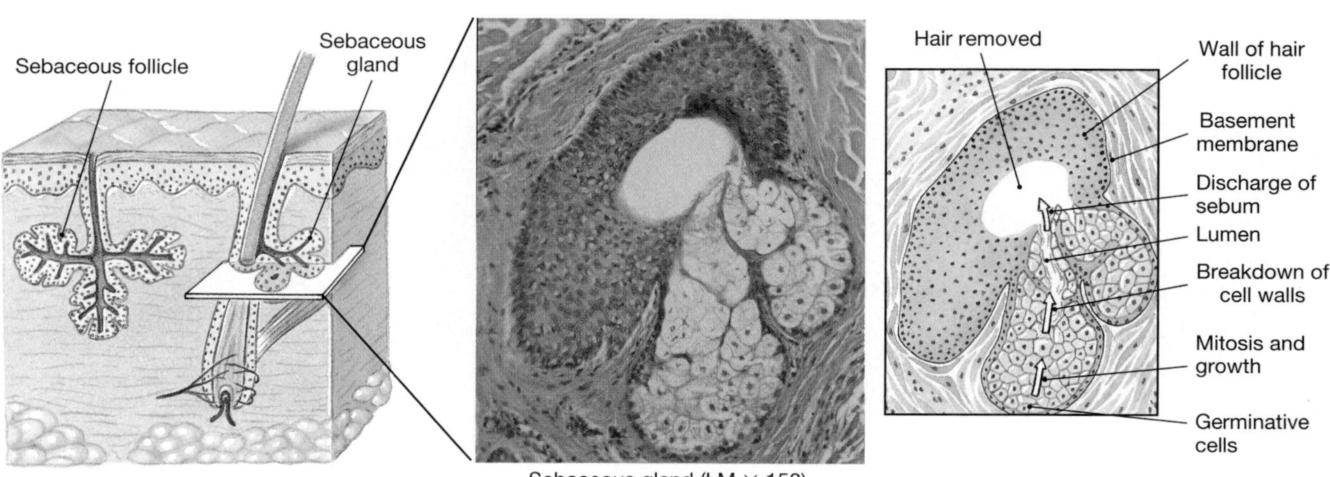
Sebaceous follicle
Sebaceous gland
Sebaceous gland (LM × 150)
Hair removed
Wall of hair follicle
Basement membrane
Discharge of sebum
Lumen
Breakdown of cell walls
Mitosis and growth
Germinative cells

● **FIGURE 5-8**
Sebaceous Glands and Follicles. The structure of sebaceous glands and sebaceous follicles in the skin.

follicles, which never produce hairs, are found on the integument covering the face, back, chest, nipples, and male sex organs. Although sebum has bactericidal (bacteria-killing) properties, under some conditions bacteria can invade sebaceous glands or follicles. The presence of bacteria in sebaceous glands or follicles can produce a local inflammation known as *folliculitis* (fo-lik-ū-LĪ-tis). If the duct of the gland becomes blocked, a distinctive abscess called a *furuncle* (FUR-ung-kl), or "boil," develops. The usual treatment for a furuncle is to cut it open, or "lance" it, so that normal drainage and healing can occur.

Sebaceous glands and sebaceous follicles are very sensitive to changes in the concentrations of sex hormones, and their secretory activities accelerate at puberty. For this reason an individual with large sebaceous glands may be especially prone to develop **acne** during adolescence. In acne, sebaceous ducts become blocked and secretions accumulate, causing inflammation and a raised "pimple." The trapped secretions provide a fertile environment for bacterial infection. AM Acne

Seborrheic dermatitis is an inflammation around abnormally active sebaceous glands. The affected area becomes red, and there is usually some epidermal scaling. Sebaceous glands of the scalp are most often involved. In infants, mild cases are called "cradle cap." This condition is one cause of "dandruff" in adults. Anxiety, stress, and food allergies can exaggerate the problem.

Sweat Glands

Figures 5-6, 5-9

The skin contains two different groups of sweat glands, *apocrine sweat glands* and *merocrine sweat glands* (Figure 5-6●, p. 159). These names refer to the mechanism of secretion, introduced in Chapter 4. [p. 119]

APOCRINE SWEAT GLANDS In the armpits, around the nipples, and in the groin, **apocrine sweat glands** communicate with hair follicles (Figure 5-9a●). These are coiled tubular glands that produce a sticky, cloudy, and potentially odorous secretion. Apocrine sweat glands begin secreting at puberty, and the sweat produced is a nutrient source for bacteria that intensify its odor. Special **myoepithelial cells** (*myo-*, muscle) surround the secretory cells. Myoepithelial cells are capable of contraction, and their contractions squeeze the gland and discharge the accumulated secretion into the hair follicles. The secretory activities of the gland cells and the contractions of myoepithelial cells are controlled by the nervous system and by circulating hormones.

MEROCRINE (ECCRINE) SWEAT GLANDS **Merocrine sweat glands,** also known as *eccrine* (EK-rin) *sweat glands*, *sudoriferous glands*, or *sudoriparous glands,* are far more numerous and widely distributed than apocrine glands (Figure

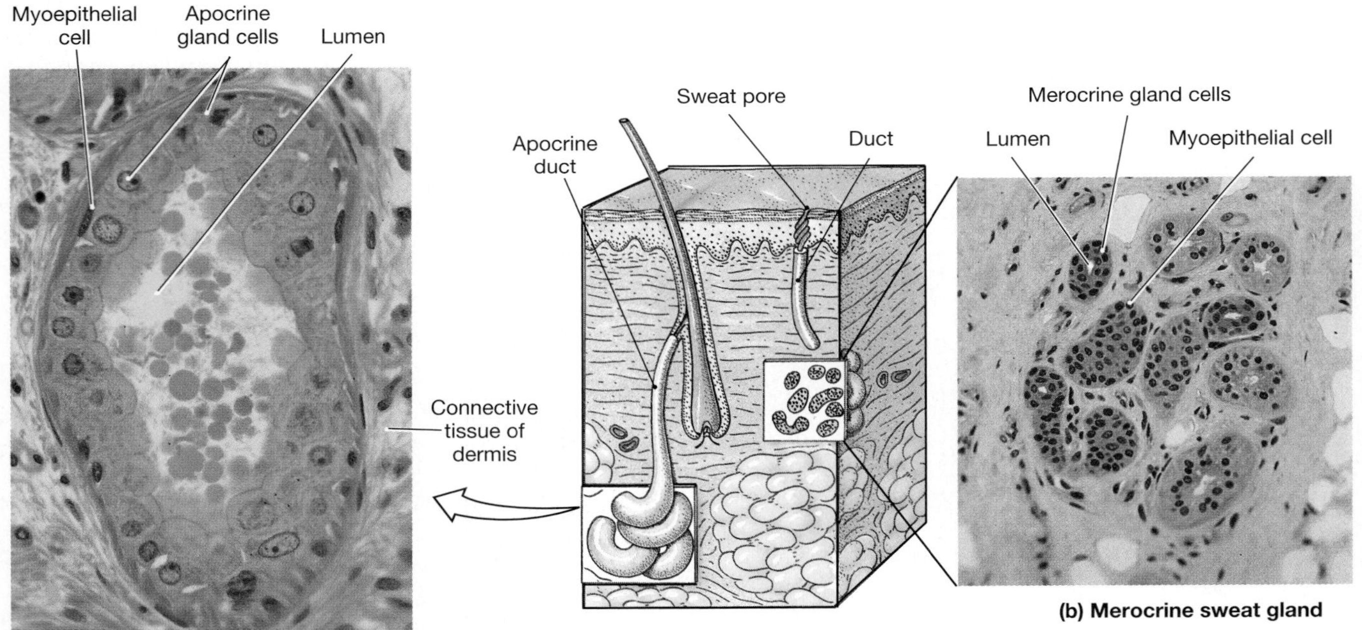
Myoepithelial cell
Apocrine gland cells
Lumen
Connective tissue of dermis
Apocrine duct
Sweat pore
Duct
Merocrine gland cells
Lumen
Myoepithelial cell
(a) Apocrine sweat gland
(b) Merocrine sweat gland

● **FIGURE 5-9**

Sweat Glands. (a) Apocrine sweat glands are found in the axillae, groin, and nipples. They produce a thick, odorous fluid by apocrine secretion. (LM × 390) **(b)** Merocrine sweat glands produce a watery fluid by merocrine secretion. (LM × 199)

5-9b●). The adult integument contains around 3 million merocrine sweat glands. They are smaller than apocrine sweat glands, and they do not extend as far into the dermis. Palms and soles have the highest numbers; estimates are that the palm of the hand has about 500 merocrine sweat glands per square centimeter (3000 per in.^{2}). These are coiled tubular glands that discharge their secretions directly onto the surface of the skin.

The clear secretion produced by merocrine sweat glands is termed **sweat,** or **sensible perspiration.** Sweat is chiefly water (99 percent), but it does contain some electrolytes (chiefly sodium chloride), metabolites, and waste products. It has a pH of 4–6.8, and the presence of sodium chloride gives sweat a salty taste. A complete analysis of the composition of normal sweat can be found in Appendix VII.

The functions of merocrine sweat gland activity include:

- ***Cooling the surface of the skin to reduce body temperature.*** This is the primary function of sensible perspiration, and the degree of secretory activity is regulated by neural and hormonal mechanisms. When all the merocrine sweat glands are working at maximum, the rate of perspiration may exceed a gallon per hour, and dangerous fluid and electrolyte losses can occur. For this reason athletes in endurance sports must pause at regular intervals to drink fluids.
- ***Excreting water and electrolytes.*** A number of ingested drugs are excreted as well.
- ***Providing protection from environmental hazards.*** The sweat dilutes harmful chemicals and discourages the growth of microorganisms.

Other Integumentary Glands

Sebaceous glands and merocrine sweat glands are found over most of the body surface. Apocrine sweat glands are found in relatively restricted areas. The skin also contains a variety of specialized glands that are restricted to specific locations. Many will be encountered in later chapters; two important examples will be noted here.

1. The **mammary glands** of the breasts are anatomically related to apocrine sweat glands. A complex interaction between sexual and pituitary hormones controls their development and secretion. Mammary gland structure and function will be discussed in Chapter 28.
2. **Ceruminous** (se-ROO-mi-nus) **glands** are modified sweat glands located in the external auditory canal. Their secretions combine with those of nearby sebaceous glands, forming a mixture called **cerumen,** or simply "ear wax." Ear wax, together with tiny hairs along the ear canal, probably helps trap foreign particles or small insects and keeps them from reaching the eardrum.

Control of Glandular Secretions

Sebaceous and apocrine glands can be collectively turned on or off by the nervous system, but no regional control is possible. When one sebaceous or apocrine gland is activated, so are all the other glands of that type in the body. Merocrine sweat glands are much more precisely controlled, and the amount of secretion and the area of the body involved can be varied independently. For example, when you are nervously awaiting an anatomy and physiology exam, your palms may begin to sweat.

The primary function of sensible perspiration is to cool the surface of the skin and to reduce body temperature. When a person sweats in the hot sun, all the eccrine glands are working together. The blood vessels beneath the epidermis are flushed with blood, and the skin assumes a reddish coloration. The skin surface is warm and wet, and as the moisture evaporates the skin cools. If body temperature falls below normal, sensible perspiration ceases, blood flow to the skin declines, and the cool dry surfaces release little heat into the environment. The negative feedback mechanisms involved in thermoregulation (temperature control) were introduced in Chapter 1, and additional details will be provided in Chapter 25. ∞ [pp. 14-15]

What are the functions of sebaceous secretions?

Deodorants are used to mask the effects of secretions from what type of skin gland?

Which type of skin gland is most affected by the hormonal changes that occur during puberty?

Nails

Figure 5-10

Nails form on the dorsal surfaces of the tips of the fingers and toes. The nails protect the exposed tips of the fingers and toes and help limit their distortion when they are subjected to mechanical stress—for example, in running or grasping objects. The structure of a nail can be seen in Figure 5-10●. The body of the nail covers the **nail bed,** but nail production occurs at the **nail root,** an epithelial fold not visible from the surface. The deepest portion of the nail root lies very close to the periosteum of the bone of the fingertip.

A portion of the stratum corneum of the fold extends over the exposed nail nearest the root,

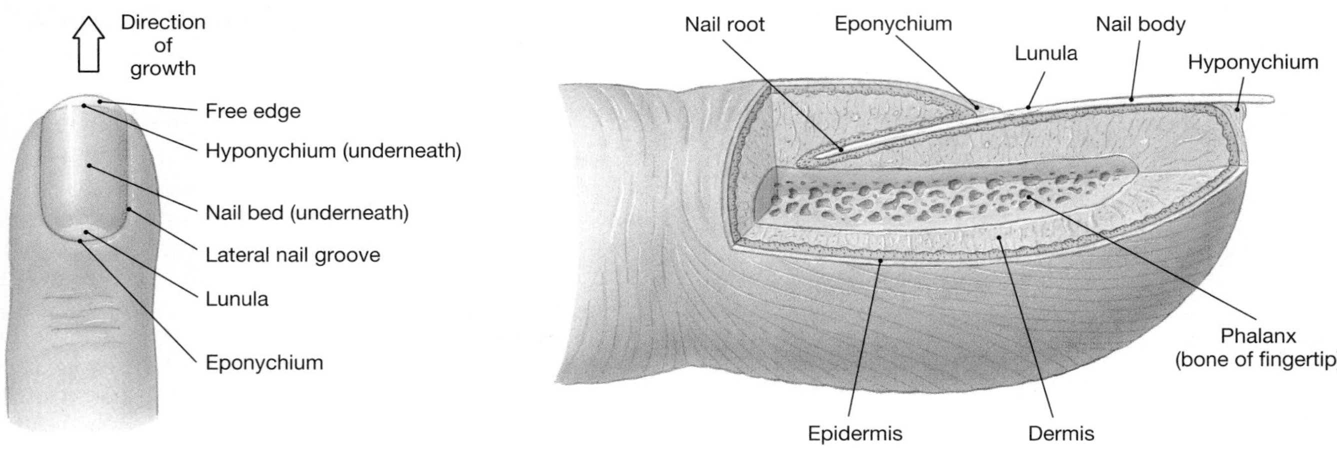
Direction of growth
Free edge
Hyponychium (underneath)
Nail bed (underneath)
Lateral nail groove
Lunula
Eponychium
Nail root
Eponychium
Lunula
Nail body
Hyponychium
Phalanx (bone of fingertip)
Epidermis
Dermis

FIGURE 5-10
Structure of a Nail. These drawings illustrate the prominent features of a typical fingernail as viewed from the surface and in section.

forming the **cuticle,** or **eponychium** (ep-ō-NIK-ē-um; *epi-*, over + *onyx,* nail). Underlying blood vessels give the nail its characteristic pink color, but near the root these vessels may be obscured, leaving a pale crescent known as the **lunula** (LOO-nu-la; *luna,* moon). The **nail body** is recessed beneath the level of the surrounding epithelium, and it is bounded by **nail grooves** and **nail folds.** The **free edge** of the nail extends over a thickened stratum corneum, the **hyponychium** (hi-pō-NIK-ē-um).

The body of the nail consists of dead, tightly compressed cells packed with keratin. The cells producing the nails can be affected by conditions that affect body metabolism, and changes in the shape, structure, or appearance of the nails can assist in diagnosis. For example, the nails may turn yellow in patients who have chronic respiratory disorders, thyroid gland disorders, or AIDS. They may become pitted and distorted in psoriasis, and concave in some blood disorders.

Local Control of Integumentary Function

The integumentary system displays a significant degree of functional independence. It often responds directly and automatically to local influences without the involvement of the nervous or endocrine system. For example, when the skin is subjected to mechanical stresses, stem cells in the stratum germinativum divide more rapidly, and the depth of the epithelium increases. That is why calluses form on your palms when you perform manual labor. A more dramatic display of local regulation can be seen after an injury to the skin.

Injury and Repair

Figure 5-11

The skin can regenerate effectively even after considerable damage has occurred, because stem cells persist in both the epithelial and connective tissue components. Germinative cell divisions replace epidermal cells, and mesenchymal cell divisions replace lost dermal cells. This process can be slow, and when large surface areas are involved problems of infection and fluid loss complicate the situation. The relative speed and effectiveness of skin repair vary depending on the type of wound involved. A slender, straight cut, or *incision,* may heal relatively quickly compared with a deep scrape, or *abrasion,* which involves a much greater area. AM *A Classification of Wounds*

Figure 5-11 shows stages in the regeneration of the skin after an injury. When damage extends through the epidermis and into the dermis, bleeding usually occurs. The blood clot, or **scab,** that forms at the surface temporarily restores the integrity of the epidermis and restricts the entry of additional microorganisms. Cells of the stratum germinativum undergo rapid divisions and begin to migrate along the sides of the wound in an attempt to replace the missing epidermal cells.

If the wound covers an extensive area, or involves a region covered by thin skin, dermal repairs must be under way before epithelial cells can cover the surface. Fibroblast and mesenchymal cell divisions produce mobile cells that invade the deeper areas of injury. Endothelial cells of damaged blood vessels also begin to divide, and capillaries follow the fibroblasts, providing a cir-

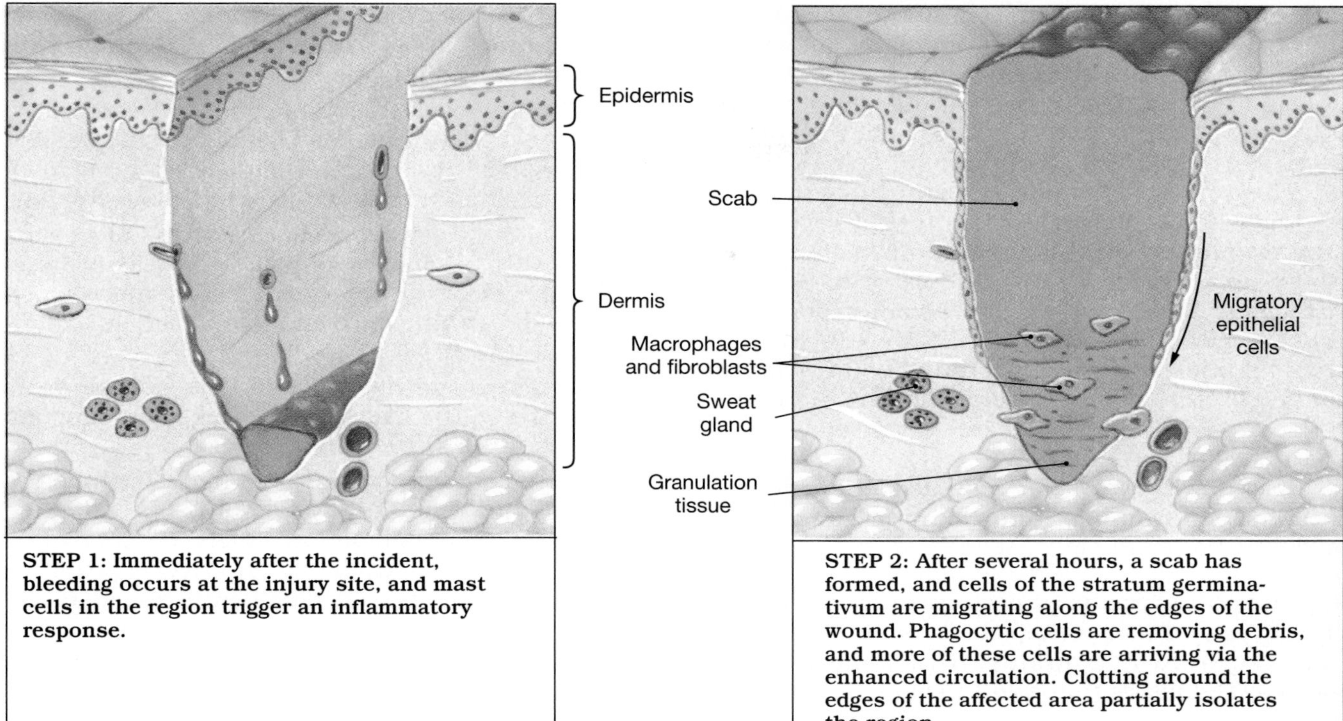
Epidermis
Dermis
Scab
Macrophages and fibroblasts
Sweat gland
Granulation tissue
Migratory epithelial cells

STEP 1: Immediately after the incident, bleeding occurs at the injury site, and mast cells in the region trigger an inflammatory response.

STEP 2: After several hours, a scab has formed, and cells of the stratum germinativum are migrating along the edges of the wound. Phagocytic cells are removing debris, and more of these cells are arriving via the enhanced circulation. Clotting around the edges of the affected area partially isolates the region.

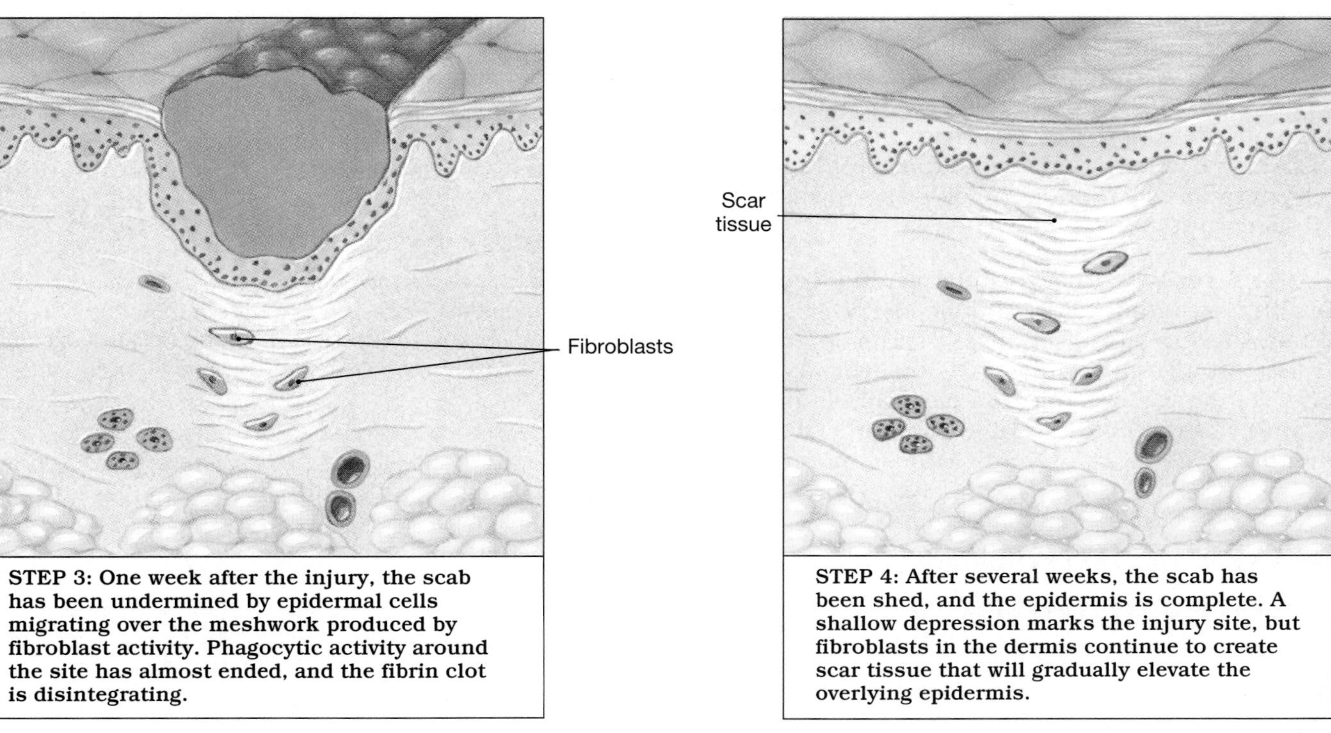
Fibroblasts
Scar tissue

STEP 3: One week after the injury, the scab has been undermined by epidermal cells migrating over the meshwork produced by fibroblast activity. Phagocytic activity around the site has almost ended, and the fibrin clot is disintegrating.

STEP 4: After several weeks, the scab has been shed, and the epidermis is complete. A shallow depression marks the injury site, but fibroblasts in the dermis continue to create scar tissue that will gradually elevate the overlying epidermis.

FIGURE 5-11
Integumentary Repair

culatory supply. The combination of blood clot, fibroblasts, and an extensive capillary network is called **granulation tissue.** Over time, the clot dissolves and the number of capillaries declines. Fibroblast activity leads to the appearance of collagen fibers and typical ground substance.

These repairs do not restore the integument to its original condition, however, for the region will contain an abnormally large number of collagen fibers and relatively few blood vessels. Severely damaged hair follicles, sebaceous or sweat glands, muscle cells, and nerves are seldom repaired, and they too are replaced by fibrous tissue. The formation of this rather inflexible, fibrous, noncellular scar tissue can be considered a practical limit to the healing process.

It is not known what regulates the extent of scar tissue formation, and the process is highly variable. For example:

- Surgical procedures performed on a fetus do not leave scars, perhaps because damaged fetal tissues do not produce the same types of growth factors that adult tissues do.
- In some adult individuals, most often those with dark skin, scar tissue formation may continue beyond the requirements of tissue repair. The result is a flattened mass of scar tissue that begins at the injury site and grows into the surrounding dermis. This thickened area of scar tissue, called a **keloid** (KĒ-loyd), is covered by a shiny, smooth epidermal surface. Keloids most often develop on the upper back, shoulders, anterior chest, and earlobes. They are harmless, and some aboriginal cultures intentionally produce keloids as a form of body decoration.

Skin repairs proceed most rapidly in young, healthy individuals. For example, barring infection, it takes 3–4 weeks to complete the repairs to a blister site in a young adult. The same repairs at age 65–75 take 6–8 weeks. However, this is just one example of the changes that occur in the integumentary system as a result of the aging process.

Aging and the Integumentary System

Aging affects all of the components of the integumentary system. These changes include:

1. The epidermis thins as germinative cell activity declines, making older people more prone to injury and skin infections.
2. The number of Langerhans cells decreases to around 50 percent of levels seen at maturity. This decrease may reduce the sensitivity of the immune system and further encourage skin damage and infection.
3. Vitamin D_3 production declines by around 75 percent. The result can be reduced calcium and phosphate absorption, eventually leading to muscle weakness and a reduction in bone strength.
4. Melanocyte activity declines, and in Caucasians the skin becomes very pale. With less melanin in the skin, the older persons are more sensitive to sun exposure and more likely to experience sunburn.
5. Glandular activity declines. The skin becomes dry and often scaly because sebum production is reduced; sweat glands are also less active.
6. The blood supply to the dermis is reduced at the same time that sweat glands become less active. This combination makes the elderly less able to lose body heat, and overexertion or overexposure to warm temperatures can cause dangerously high body temperatures.
7. Hair follicles stop functioning or produce thinner, finer hairs. With decreased melanocyte activity, these hairs are gray or white.
8. The dermis becomes thinner, and the elastic fiber network decreases in size. The integument therefore becomes weaker and less resilient; sagging and wrinkling occur. These effects are most pronounced in areas exposed to the sun.
9. With changes in levels of sex hormones, secondary sexual characteristics in hair and body fat distribution begin to fade. In consequence, people age 90–100 of both sexes and all races look very much alike.
10. Skin repairs proceed relatively slowly, and recurring infections may result.

✔ What is the name given to the combination of fibrin clot, fibroblasts, and the extensive network of capillaries found in healing tissue?

✔ Why can skin regenerate effectively even after considerable damage has occurred?

✔ Older individuals do not tolerate the summer heat as well as they did when they were young and are more prone to heat-related illness. What accounts for this change?

Integration with Other Systems

Figure 5-12

Although it can function independently, many activities of the integumentary system are integrated with those of other systems. Figure 5-12● diagrams the major functional relationships.

SKELETAL SYSTEM

Provides structural support.

Synthesizes vitamin D_3, essential for calcium and phosphorus absorption (bone maintenance and growth).

MUSCULAR SYSTEM

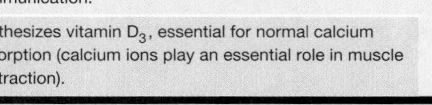

Contractions of skeletal muscles pull against skin of face, producing facial expressions important in communication.

Synthesizes vitamin D_3, essential for normal calcium absorption (calcium ions play an essential role in muscle contraction).

NERVOUS SYSTEM

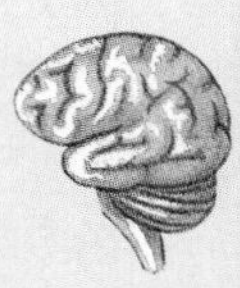

Controls blood flow and sweat gland activity for thermoregulation; stimulates contraction of arrector pili muscles to elevate hairs.

Receptors in dermis and deep epidermis provide sensations of touch, pressure, vibration, temperature, and pain.

THE INTEGUMENTARY SYSTEM

FOR ALL SYSTEMS

Provides mechanical protection against environmental hazards.

ENDOCRINE SYSTEM

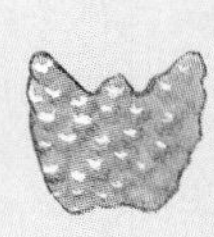

Sexual hormones stimulate sebaceous gland activity; male and female sex hormones influence hair growth, distribution of subcutaneous fat, and apocrine sweat gland activity; adrenal hormones alter dermal blood flow and help mobilize lipids from adipocytes.

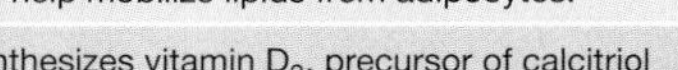

Synthesizes vitamin D_3, precursor of calcitriol

CARDIOVASCULAR SYSTEM

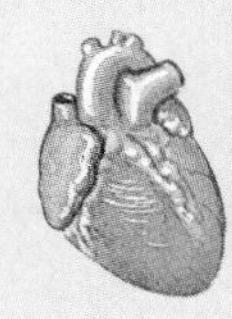

Provides O_2 and nutrients; delivers hormones and cells of immune system; carries away CO_2, waste products, and toxins; provides heat to maintain normal skin temperature.

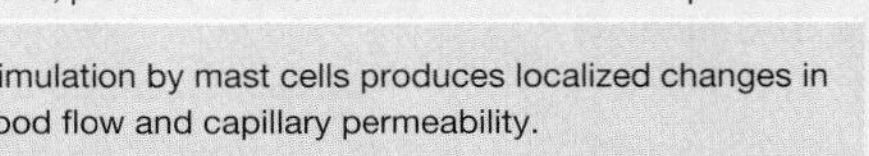

Stimulation by mast cells produces localized changes in blood flow and capillary permeability.

LYMPHATIC SYSTEM

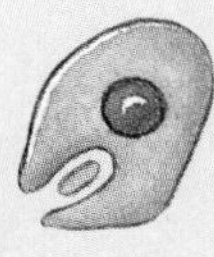

Assists in defending the integument by providing additional macrophages and mobilizing lymphocytes.

Provides physical barriers that prevent pathogen entry; Langerhans cells and macrophages resist infection; mast cells trigger inflammation and initiate the immune response.

RESPIRATORY SYSTEM

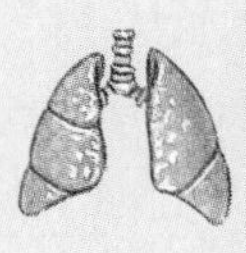

Provides oxygen and eliminates carbon dioxide.

Hairs guard entrance to nasal cavity

DIGESTIVE SYSTEM

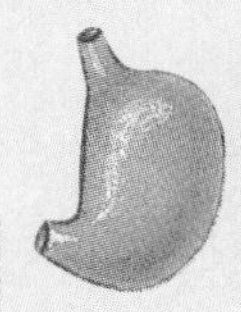

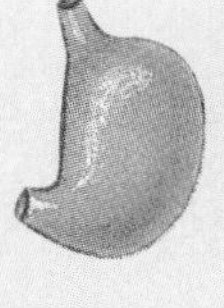

Provides nutrients for all cells and lipids for storage by adipocytes.

Synthesizes vitamin D_3, needed for absorption of calcium and phosphorus.

URINARY SYSTEM

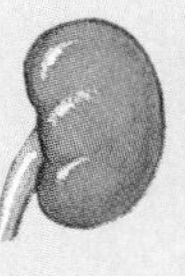

Excretes waste products, maintains normal body fluid pH and ion composition.

Assists in elimination of water and solutes; keratinized epidermis limits fluid loss through skin.

REPRODUCTIVE SYSTEM

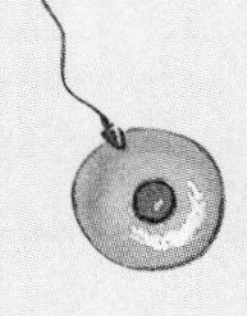

Covers external genitalia; provides sensations that stimulate sexual behaviors; mammary gland secretions provide nourishment for newborn infant.

FIGURE 5-12
Functional Relationships between the Integumentary System and Other Systems

EMBRYOLOGY SUMMARY Development of the Integumentary System

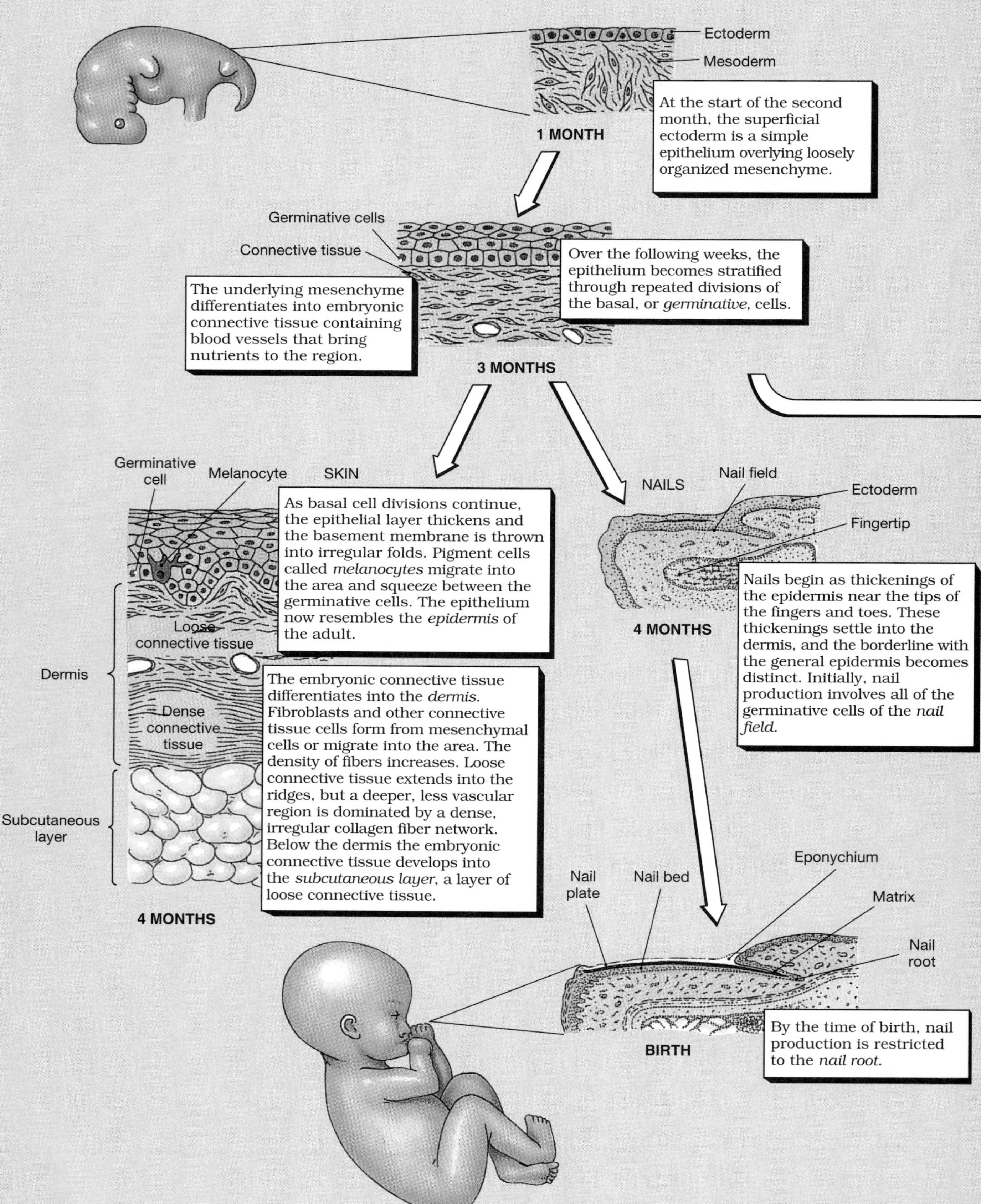

Ectoderm
Mesoderm
1 MONTH
At the start of the second month, the superficial ectoderm is a simple epithelium overlying loosely organized mesenchyme.
Germinative cells
Connective tissue
The underlying mesenchyme differentiates into embryonic connective tissue containing blood vessels that bring nutrients to the region.
Over the following weeks, the epithelium becomes stratified through repeated divisions of the basal, or *germinative*, cells.
3 MONTHS
Germinative cell
Melanocyte
SKIN
As basal cell divisions continue, the epithelial layer thickens and the basement membrane is thrown into irregular folds. Pigment cells called *melanocytes* migrate into the area and squeeze between the germinative cells. The epithelium now resembles the *epidermis* of the adult.
Loose connective tissue
Dermis
Dense connective tissue
Subcutaneous layer
The embryonic connective tissue differentiates into the *dermis*. Fibroblasts and other connective tissue cells form from mesenchymal cells or migrate into the area. The density of fibers increases. Loose connective tissue extends into the ridges, but a deeper, less vascular region is dominated by a dense, irregular collagen fiber network. Below the dermis the embryonic connective tissue develops into the *subcutaneous layer*, a layer of loose connective tissue.
4 MONTHS
NAILS
Nail field
Ectoderm
Fingertip
Nails begin as thickenings of the epidermis near the tips of the fingers and toes. These thickenings settle into the dermis, and the borderline with the general epidermis becomes distinct. Initially, nail production involves all of the germinative cells of the *nail field*.
4 MONTHS
Eponychium
Nail plate
Nail bed
Matrix
Nail root
By the time of birth, nail production is restricted to the *nail root*.
BIRTH

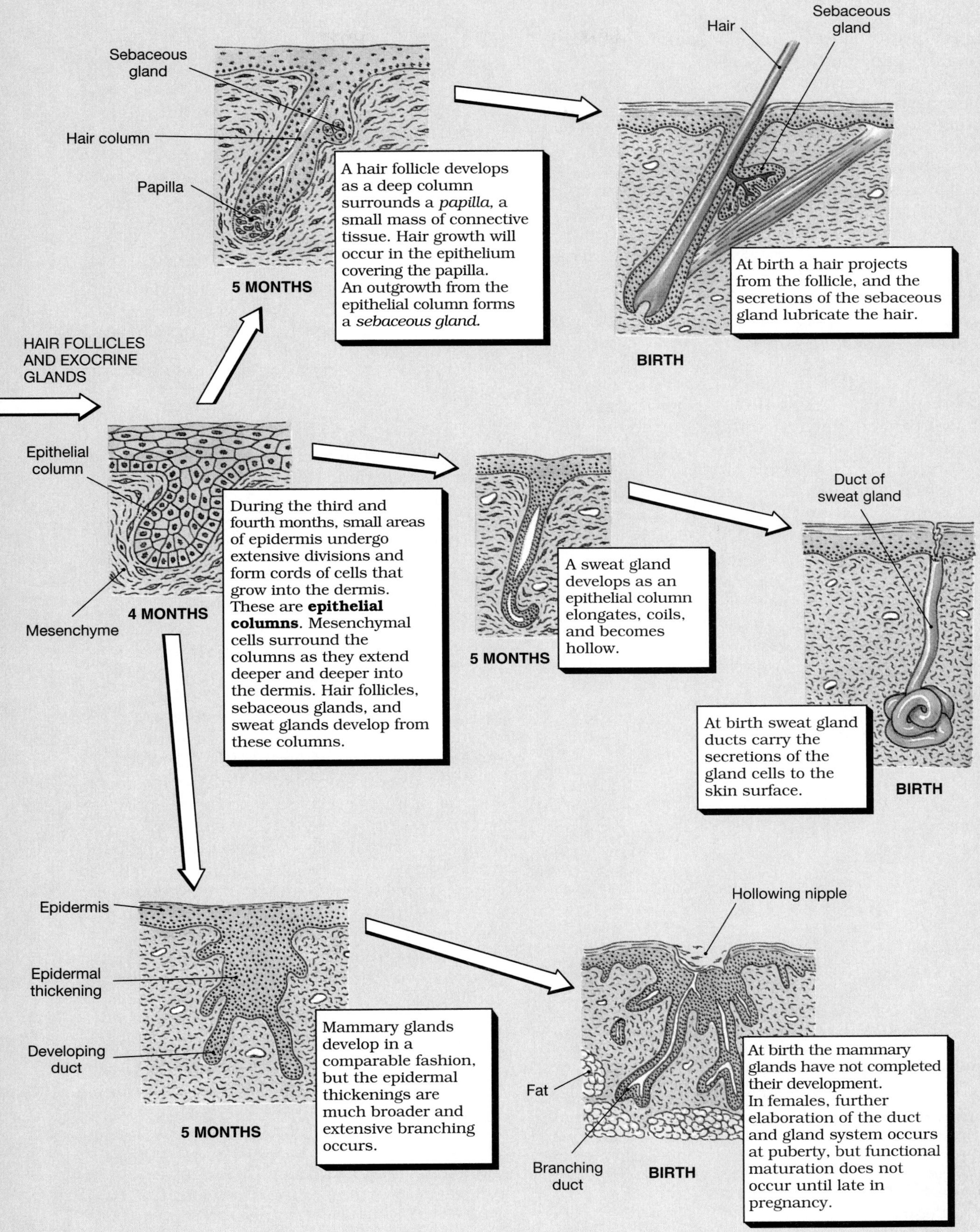
HAIR FOLLICLES AND EXOCRINE GLANDS
Epithelial column
Mesenchyme
4 MONTHS
During the third and fourth months, small areas of epidermis undergo extensive divisions and form cords of cells that grow into the dermis. These are **epithelial columns**. Mesenchymal cells surround the columns as they extend deeper and deeper into the dermis. Hair follicles, sebaceous glands, and sweat glands develop from these columns.
Sebaceous gland
Hair column
Papilla
5 MONTHS
A hair follicle develops as a deep column surrounds a *papilla,* a small mass of connective tissue. Hair growth will occur in the epithelium covering the papilla. An outgrowth from the epithelial column forms a *sebaceous gland.*
Hair
Sebaceous gland
BIRTH
At birth a hair projects from the follicle, and the secretions of the sebaceous gland lubricate the hair.
5 MONTHS
A sweat gland develops as an epithelial column elongates, coils, and becomes hollow.
Duct of sweat gland
BIRTH
At birth sweat gland ducts carry the secretions of the gland cells to the skin surface.
Epidermis
Epidermal thickening
Developing duct
5 MONTHS
Mammary glands develop in a comparable fashion, but the epidermal thickenings are much broader and extensive branching occurs.
Hollowing nipple
Fat
Branching duct
BIRTH
At birth the mammary glands have not completed their development. In females, further elaboration of the duct and gland system occurs at puberty, but functional maturation does not occur until late in pregnancy.

Burns and Grafts

Burns result from exposure of the skin to heat, radiation, electrical shock, or strong chemical agents. The severity of the burn reflects the depth of penetration and the total area affected.

First- and second-degree burns are also called **partial-thickness burns** because damage is restricted to the superficial layers of the skin. Accessory structures such as hair follicles and glands are usually unaffected. **Full-thickness burns,** or **third-degree burns,** destroy the epidermis and dermis, extending into subcutaneous tissues. These burns are actually less painful than second-degree burns, because sensory nerves are destroyed along with accessory structures, blood vessels, and other dermal components. Extensive third-degree burns cannot repair themselves, and the site remains exposed to potential infection.

Roughly 10,000 people die from burns each year in the United States. The larger the area burned, the more significant the effects on integumentary function. Figure 5-13● presents a standard reference for calculating the percentage of total surface area involved. Burns that cover more than 20 percent of the skin surface represent serious threats to life because they affect the following functions:

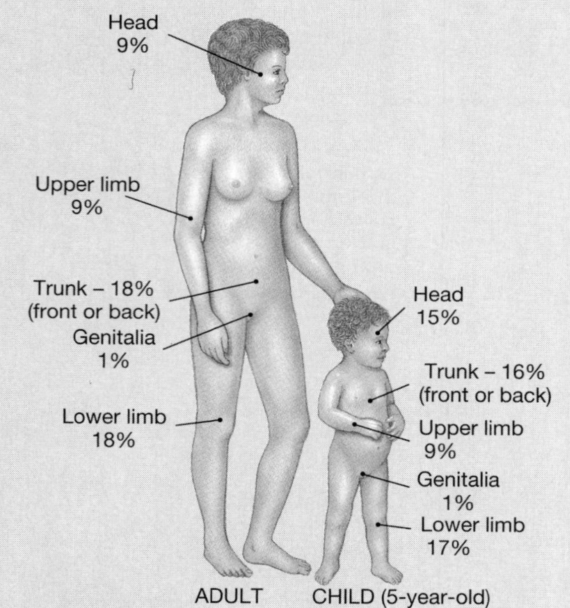
Head
9%
Upper limb
9%
Trunk – 18%
(front or back)
Genitalia
1%
Lower limb
18%
Head
15%
Trunk – 16%
(front or back)
Upper limb
9%
Genitalia
1%
Lower limb
17%
ADULT
CHILD (5-year-old)

● **FIGURE 5-13**

A Quick Method for Estimating the Percentage of Surface Area Affected by Burns in Adults and Small Children. The method of estimation is called the "rule of nines" because of the surface area proportion in the adult. This rule must be modified in children because their proportions are quite different.

- ***Fluid and electrolyte balance.*** Even areas with partial-thickness burns lose their effectiveness as barriers to fluid and electrolyte losses. In full-thickness burns, the rate of fluid loss through the skin may reach five times the normal level.
- ***Thermoregulation.*** Increased fluid loss means increased evaporative cooling. More energy must be expended to keep body temperature within acceptable limits.
- ***Protection from attack.*** The epidermal surface, damp from uncontrolled fluid losses, encourages bacterial growth. If the skin is broken at a blister or the site of a third-degree burn, infection is likely. Widespread bacterial infection, or **sepsis** (*septikos*, rotting), is the leading cause of death in burn victims.

Effective treatment of full-thickness burns focuses on these procedures:

1. Replacing lost fluids and electrolytes.
2. Providing sufficient nutrients to meet increased metabolic demands for thermoregulation and healing.
3. Preventing infection by cleaning and covering the burn while administering antibiotic drugs.
4. Assisting tissue repairs.

Because full-thickness burns cannot heal unaided, surgical procedures are necessary to encourage healing. In a **skin graft,** areas of intact skin are transplanted to cover the burn site. A **split-thickness graft** takes a shaving of the epidermis and superficial portions of the dermis. A **full-thickness graft** involves the epidermis and both layers of the dermis.

With the development of fluid replacement therapies, infection control methods, and grafting techniques, the recovery rate for severe burns has improved dramatically. At present, young patients with burns over 80 percent of the body have an approximately 50 percent chance of recovery.

Recent advances in cell culture techniques may improve survival rates further. It is now possible to remove a small section of undamaged epidermis and grow it under controlled laboratory conditions. Over time, the germinative cell divisions produce large sheets of epidermal cells that can then be used to cover the burn area. From initial samples the size of postage stamps, square yards of epidermis have been grown and transplanted onto body surfaces. Although questions remain concerning the strength and flexibility of the repairs, skin cultivation represents a substantial advance in the treatment of serious burns. AM *Synthetic Skin*

Selected Clinical Terminology

Terms Discussed in This Chapter

acne: A sebaceous gland inflammation caused by an accumulation of secretions. *(p. 162 and AM)*
contact dermatitis: A dermatitis usually caused by strong chemical irritants. It produces an itchy rash that may spread to other areas as scratching distributes the chemical agent; poison ivy is an example. *(p. 157)*
cyanosis (sī-uh-NŌ-sis): A bluish coloration of the skin resulting from reduced oxygenation of the blood in superficial vessels. *(p. 155)*
decubitis ulcers ("bedsores"): Ulcers that occur in areas subject to restricted circulation, especially common in hospitalized or bedridden persons. *(p. 155)*
dermatitis: An inflammation of the skin that primarily involves the papillary region of the dermis. *(p. 157)*
diaper rash: A localized dermatitis caused by a combination of moisture, irritating chemicals from fecal or urinary wastes, and flourishing microorganisms. *(p. 157)*
eczema (EK-se-muh): A dermatitis that can be triggered by temperature changes, fungus, chemical irritants, greases, detergents, or stress and that can be related to hereditary or environmental factors. *(p. 157)*
folliculitis (fo-lik-ū-LĪ-tis): A local inflammation caused by bacterial infection of a sebaceous gland or sebaceous follicle. *(p. 162)*
furuncle (FUR-ung-kl): An abscess or boil that develops when the duct of a sebaceous gland is blocked. *(p. 162)*
granulation tissue: A combination of fibrin, fibroblasts, and capillaries that forms during tissue repair following inflammation. *(p. 166)*
hypodermic needle: A needle used to administer drugs via subcutaneous injection. *(p. 158)*
keloid (KĒ-loyd): A thickened area of scar tissue covered by a shiny, smooth epidermal surface. *(p. 166)*
scab: A blood clot that forms at the surface of a wound to the skin. *(p. 164)*
seborrheic dermatitis: An inflammation around abnormally active sebaceous glands. *(p. 162)*
sepsis (*septikos*, rotting): A dangerous, widespread bacterial infection. Sepsis is the leading cause of death in burn patients. *(p. 170)*
skin graft: Transplantation of a section of skin (partial thickness or full thickness) to cover an extensive injury site, such as a third-degree burn. *(p. 170)*
ulcer: A localized shedding of an epithelium. *(p. 155)*
urticaria (ur-ti-KAR-ē-uh) or **hives:** An extensive dermatitis resulting from an allergic reaction to food, drugs, an insect bite, infection, stress, or other stimulus. *(p. 157)*

AM *Additional Terms Discussed in the Applications Manual*

basal cell carcinoma: A malignant cancer that originates in the stratum germinativum. This is the most common skin cancer, and roughly two-thirds of these cancers appear in areas subjected to chronic UV exposure. Metastasis seldom occurs.
capillary hemangioma: A birthmark caused by a tumor in the capillaries of the papillary layer of the dermis. It usually enlarges after birth, but subsequently fades and disappears.
cavernous hemangiomas ("port-wine stains"): Birthmarks caused by a tumor affecting larger vessels in the dermis. Such birthmarks usually last a lifetime.
erysipelas (er-i-SIP-e-lus): A widespread inflammation of the dermis caused by bacterial infection.
malignant melanoma (mel-uh-NŌ-muh): A skin cancer originating in malignant melanocytes.
pruritis (proo-RĪ-tis): An irritating itching sensation, common in skin conditions.
psoriasis (so-RĪ-uh-sis): A painless condition characterized by rapid stem cell divisions in the stratum germinativum of the scalp, elbows, palms, soles, groin, and nails. Affected areas appear dry and scaly.
skin signs: Characteristic abnormalities in the skin surface that can assist in diagnosing skin conditions. They include *flat macules, wheals, papules, nodules, vesicles (blisters), pustules, erosions (ulcers), crusts, scales, fissures.*
squamous cell carcinoma: A form of skin cancer less common than basal cell carcinoma, almost totally restricted to areas of sun-exposed skin. Metastasis seldom occurs.
xerosis (ze-RŌ-sis): "Dry skin," a common complaint of older persons and almost anyone living in an arid climate.

CHAPTER REVIEW

STUDY OUTLINE

INTRODUCTION, p. 150

1. The **integumentary system** consists of the **cutaneous membrane,** which includes the **epidermis** and **dermis,** and the **accessory structures.** Underneath lies the **subcutaneous layer.** *(Figure 5-1)*

THE EPIDERMIS, p. 151

1. **Thin skin** covers most of the body; heavily abraded body surfaces may be covered by **thick skin.** *(Figure 5-2)*

Layers of the Epidermis p. 151

2. Cell divisions in the **stratum germinativum** replace more superficial cells. *(Figure 5-3)*
3. As epidermal cells age they pass through the **stratum spinosum,** the **stratum granulosum,** the **stratum**

lucidum (if thick skin), and the **stratum corneum.** In the process they accumulate large amounts of **keratin.** Ultimately the cells are shed or lost. *(Figure 5-3)*

4. **Epidermal ridges,** such as those on the palms and soles, improve our gripping ability and increase the skin's sensitivity.
5. **Langherhans cells** are part of the immune system; **Merkel cells** provide information about objects touching the skin.

Skin Color, p. 153

6. The color of the epidermis depends on two factors: blood supply and pigment composition and concentration. **Melanocytes** protect us from **ultraviolet radiation.** *(Figure 5-4)*

The Epidermis and Vitamin D_3, p. 155

7. Epidermal cells synthesize vitamin D_3 when exposed to the UV radiation in sunlight.

Epidermal Growth Factor, p. 155

8. Epidermal growth factor stimulates maintenance and repair of the epidermis and the secretion of epithelial glands.

THE DERMIS, p. 155

Dermal Organization, p. 156

Dermal Circulation and Innervation, p. 157

1. The **dermis** consists of the **papillary layer** and the deeper **reticular layer.**
2. The papillary layer of the dermis contains blood vessels, lymphatics, and sensory nerves. This layer supports and nourishes the overlying epidermis. The reticular layer consists of a meshwork of collagen and elastic fibers oriented to resist tension in the skin. *(Figure 5-5)*

THE SUBCUTANEOUS LAYER, p. 157

1. The subcutaneous layer, or **hypodermis,** stabilizes the skin's position against underlying organs and tissues. *(Figure 5-1)*

ACCESSORY STRUCTURES, p. 158

Hair Follicles and Hair, p. 158

1. Hairs originate in complex organs called **hair follicles.** Each hair has a **root** and a **shaft.** Hair production involves cell specialization to form a soft core, or **medulla,** surrounded by a **cortex.** The **cuticle** is a hard layer that coats the hair. *(Figures 5-6, 5-7)*
2. The **arrector pili** muscles can erect the hairs. *(Figure 5-6)*
3. There are **vellus hairs** ("peach fuzz") and heavy **terminal hairs** on our bodies.
4. Our hairs grow and are shed according to the **hair growth cycle.** A single hair grows for 2–5 years, and is subsequently shed.

Glands in the Skin, p. 161

5. **Sebaceous glands** discharge the waxy **sebum** into hair follicles. **Sebaceous follicles** lack hairs but have large sebaceous glands. *(Figure 5-8)*
6. **Apocrine sweat glands** produce an odorous secretion; the more numerous **merocrine,** or **eccrine, sweat glands** produce a watery secretion, known as **sensible perspiration.** *(Figure 5-9)*
7. **Ceruminous glands** in the ear produce a waxy **cerumen.**

Nails, p. 163

8. Nail production occurs at the **nail root.** *(Figure 5-10)*

LOCAL CONTROL OF INTEGUMENTARY FUNCTION, p. 164

Injury and Repair, p. 164

1. The epidermis provides mechanical protection and keeps microorganisms outside of the body.
2. The skin can regenerate effectively even after considerable damage. The process includes formation of a **scab** and **granulation tissue.** *(Figure 5-11)*

AGING AND THE INTEGUMENTARY SYSTEM, p. 166

1. Aging affects all of the components of the integumentary system.

REVIEW QUESTIONS

LEVEL 1 Reviewing Facts and Terms

1. The two major components of the integumentary system are:
   (a) the cutaneous membrane and the accessory structures
   (b) the epidermis and the hypodermis
   (c) the hair and the nails
   (d) the dermis and the subcutaneous layer

2. Beginning at the basement membrane and traveling toward the free surface, the layers of the epidermis include the strata:
   (a) corneum, lucidum, granulosum, spinosum, germinativum
   (b) granulosum, lucidum, spinosum, germinativum, corneum
   (c) germinativum, spinosum, granulosum, lucidum, corneum
   (d) lucidum, granulosum, spinosum, germinativum, corneum

3. The fibrous protein that forms the basic structural component of hair and nails is:
   (a) collagen (b) melanin
   (c) elastin (d) keratin

4. The primary pigments contained in the epidermis are:
   (a) carotene and xanthophyll
   (b) carotene and melanin
   (c) melanin and chlorophyll
   (d) xanthophyll and melanin

5. The two major components of the dermis are the:
   (a) superficial fascia and cutaneous membrane
   (b) epidermis and hypodermis
   (c) papillary layer and reticular layer
   (d) stratum germinativum and stratum corneum

6. The cutaneous plexus and papillary plexus consist of:
   (a) a network of arteries providing the dermal blood supply
   (b) a network of nerves providing dermal sensations
   (c) specialized cells for cutaneous sensations
   (d) gland cells that release cutaneous secretions

7. The accessory structures of the integument include the:
   (a) blood vessels, glands, muscles, and nerves
   (b) Merkel's cells, Pacinian corpuscles, and Meissner's corpuscles
   (c) hair, skin, and nails
   (d) hair follicles, sebaceous glands, and sweat glands

8. The two major types of hairs in the integument include:
   (a) thick, thin
   (b) vellus, terminal
   (c) primary, secondary
   (d) short, long

9. The two types of exocrine glands found in the skin are:
   (a) merocrine and sweat glands
   (b) sebaceous and sweat glands
   (c) apocrine and sweat glands
   (d) eccrine and sweat glands

10. Sweat glands that communicate with hair follicles in the armpits and produce an odorous secretion are:
   (a) apocrine glands
   (b) merocrine glands
   (c) sebaceous glands
   (d) a, b, and c are correct

12. The primary function of sensible perspiration is to:
   (a) get rid of wastes
   (b) protect the skin from dryness
   (c) maintain electrolyte balance
   (d) reduce body temperature

13. The thickened stratum corneum over which the free edge of the nail extends is called the:
   (a) eponychium
   (b) hyponychium
   (c) cuticle
   (d) cerumen

14. Muscle weakness and a reduction in bone strength in the elderly results from decreased:
   (a) vitamin D_3 production
   (b) melanin production
   (c) sebum production
   (d) dermal blood supply

15. The reason older persons are more sensitive to sun exposure and more likely to experience sunburn is that with age:
   (a) melanocyte activity declines
   (b) vitamin D_3 production declines
   (c) glandular activity declines
   (d) skin thickness decreases

16. In what layer(s) of the epidermis does cell division occur?

17. What are the protein precursors of keratin, and in what layer(s) of the epidermis are these proteins produced?

18. What two skin pigments are found in the epidermis?

19. What widespread effects does epidermal growth factor (EGF) have on the integument?

20. What two major layers constitute the dermis, and what components are found in each layer?

21. Beginning at the hair cuticle, what three cell layers make up the walls of the hair follicle?

22. What two different groups of sweat glands are contained in the skin?

LEVEL 2 Reviewing Concepts

23. How does insensible perspiration differ from sensible perspiration?

24. During transdermal administration of drugs, why are fat-soluble drugs more desirable than those that are water-soluble?

25. In our society, a tan body is associated with good health. However, medical research constantly warns about the dangers of excessive exposure to the sun. What are the benefits of a tan?

26. Why is it important for a surgeon to choose an incision pattern according to the lines of cleavage of the skin?

27. Why is regional infection or inflammation of the skin usually very painful?

28. Why is a subcutaneous injection with a hypodermic needle a useful method for administering drugs?

29. How are changes in the shape, structure, or appearance of the nails clinically significant?

30. Why is the formation of a scab important in the healing process of the skin?

31. Why does skin sag and wrinkle as a person ages?

LEVEL 3 Critical Thinking and Clinical Applications

32. A new mother notices that her 6-month-old child has a yellow-orange complexion. Fearful that the child may have jaundice (a condition caused by a toxic yellow-orange pigment in the blood), she takes him to her pediatrician. After examining the child, the pediatrician declares him perfectly healthy and advises the mother to watch the child's diet. Why?

33. Vanessa remarks that her 80-year-old grandmother keeps her thermostat set at 80°F and wears a sweater on balmy spring days. When she asks her grandmother why, her grandmother tells her that she is cold. Vanessa can't understand this and asks you for an explanation. What would you tell Vanessa?

34. Exposure to optimum amounts of sunlight is necessary for proper bone maintenance and growth in children. (a) What does sunlight do to promote bone maintenance and growth? (b) If a child lives in an area where exposure to sunlight is rare because of pollution or overcast skies, what can be done to minimize impaired maintenance and growth of bone?

35. One of the factors that lie detectors respond to is an increase in skin conductivity caused by the presence of moisture. Explain the physiological basis for the use of this indicator.

CHAPTER

6

The Tin Man wore his skeleton on the outside. The armor plating was useful, but there were drawbacks, such as having to oil the joints. We avoid such problems by having internal skeletons with self-lubricating joints. Our skeletons are made of bone, a remarkable tissue that is as strong as steel-reinforced concrete, immune to rust (unlike the Tin Man's), and capable of repairing itself even after serious injuries. This chapter focuses on the nature of bone.

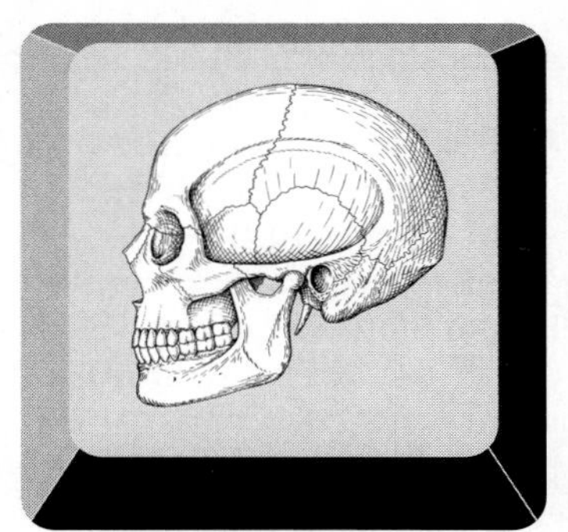

Osseous Tissue and Skeletal Structure

Chapter Outline and Objectives

Introduction p. 176

1 Describe the functions of the skeletal system.

Structure of Bone p. 176

HISTOLOGICAL ORGANIZATION p. 176
COMPACT AND SPONGY BONE p. 178
THE PERIOSTEUM AND ENDOSTEUM p. 180

2 Identify the cell types found in bone and list their major functions.

3 Compare the structures and functions of compact and spongy bone.

Bone Development and Growth p. 181

INTRAMEMBRANOUS OSSIFICATION p. 181
ENDOCHONDRAL OSSIFICATION p. 182
THE BLOOD AND NERVE SUPPLY p. 186

4 Compare the mechanisms of intramembranous and endochondral ossification.

5 Discuss the timing of bone development and growth and account for the differences in the internal structure of adult bones.

The Dynamic Nature of Bone p. 187

EXERCISE EFFECTS ON BONE p. 187
HORMONAL AND NUTRITIONAL EFFECTS ON BONE p. 187
THE SKELETON AS A CALCIUM RESERVE p. 189
FRACTURE REPAIR p. 190
AGING AND THE SKELETAL SYSTEM p. 191

6 Describe the remodeling and homeostatic mechanisms of the skeletal system.

7 Discuss the effects of nutrition, hormones, exercise, and aging on bone development and the skeletal system.

8 Describe the different types of fractures and explain how they heal.

Anatomy of Skeletal Elements p. 193

A CLASSIFICATION OF BONES p. 193
BONE MARKINGS (SURFACE FEATURES) p. 194

9 Classify bones according to their shapes and give examples of each type.

Integration with Other Systems p. 195

Selected Clinical Terminology p. 197

This chapter begins our examination of the skeletal system. This system includes the bones of the skeleton and the cartilages, ligaments, and other connective tissues that stabilize or connect them. Skeletal elements are more than just racks to hang muscles on; they have a great variety of vital functions. In addition to supporting the weight of the body, bones work together with muscles to maintain body position and to produce controlled, precise movements. Without the skeleton to pull against, contracting muscle fibers would be unable to make us sit, stand, walk, or run. Without something to hold onto, contracting muscles merely get shorter and fatter.

We begin our study of the skeletal system by identifying its primary functions:

1. ***Support.*** The skeletal system provides structural support for the entire body. Individual bones or groups of bones provide a framework for the attachment of soft tissues and organs.
2. ***Storage of minerals and lipids.*** The calcium salts of bone represent a valuable mineral reserve that maintains normal concentrations of calcium and phosphate ions in body fluids. Calcium is the most abundant mineral in the human body. In addition to acting as a mineral reserve, the bones of the skeleton store energy reserves as lipids in areas of *yellow marrow.*
3. ***Blood cell production.*** Red blood cells, white blood cells, and other blood elements are produced within the *red marrow* that fills the internal cavities of many bones. The role of the bone marrow in blood cell formation will be described in later chapters dealing with the cardiovascular and lymphatic systems (Chapters 19 and 22).
4. ***Protection.*** Delicate tissues and organs are often surrounded by skeletal elements. The ribs protect the heart and lungs, the skull encloses the brain, the vertebrae shield the spinal cord, and the pelvis cradles delicate digestive and reproductive organs.
5. ***Leverage.*** Many bones of the skeleton function as *levers* that can change the magnitude and direction of the forces generated by skeletal muscles. The movements produced range from the delicate motion of a fingertip to powerful changes in the position of the entire body.

The bones of the skeleton are actually complex, dynamic organs that contain osseous tissue, other connective tissues, smooth muscle tissue, and neural tissue. We will now consider the internal organization of a typical bone.

Structure of Bone

Bone tissue, or **osseous tissue,** is one of the supporting connective tissues. (You may wish to review the sections on dense connective tissues, cartilage, and bone in Chapter 4 at this time.) ∞ *[pp. 126–134]* Like other connective tissues, osseous tissue contains specialized cells and a matrix consisting of extracellular protein fibers and a ground substance. The matrix of bone tissue is solid and sturdy because of the deposition of calcium salts around the protein fibers.

Histological Organization

Figure 6-1

The basic organization of bone tissue was introduced in Chapter 4. ∞ *[pp. 132–134]* Four basic features of a representative sample of bone were discussed at that time:

1. The matrix of bone is very dense and contains deposits of calcium salts.
2. The matrix contains bone cells, or *osteocytes,* within pockets, or *lacunae.* The lacunae are often organized around blood vessels that branch through the bony matrix.
3. Narrow passageways through the matrix, called *canaliculi,* extend between the lacunae and nearby blood vessels, forming a branching network for the exchange of nutrients, waste products, and gases.
4. Except at joints, the outer surfaces of bones are covered by a *periosteum* that has outer fibrous and inner cellular layers.

With the aid of Figure 6-1•, we will now take a closer look at the organization of the matrix and cells of bone.

The Matrix of Bone

Calcium phosphate, $Ca_3(PO_4)_2$, accounts for almost two-thirds of the weight of bone. The calcium phosphate interacts with calcium hydroxide $[Ca(OH)_2]$ to form crystals of **hydroxyapatite,** $Ca_{10}(PO_4)_6(OH)_2$. As they form, these crystals also incorporate other calcium salts, such as calcium carbonate, and ions such as sodium, magnesium, and fluoride. Roughly one-third of the weight of bone is contributed by collagen fibers. Osteocytes and other cell types account for only 2 percent of the mass of a typical bone.

Calcium phosphate crystals are very hard, but relatively inflexible and quite brittle. They can withstand compression, but the crystals are likely to shatter when exposed to bending, twisting, or sud-

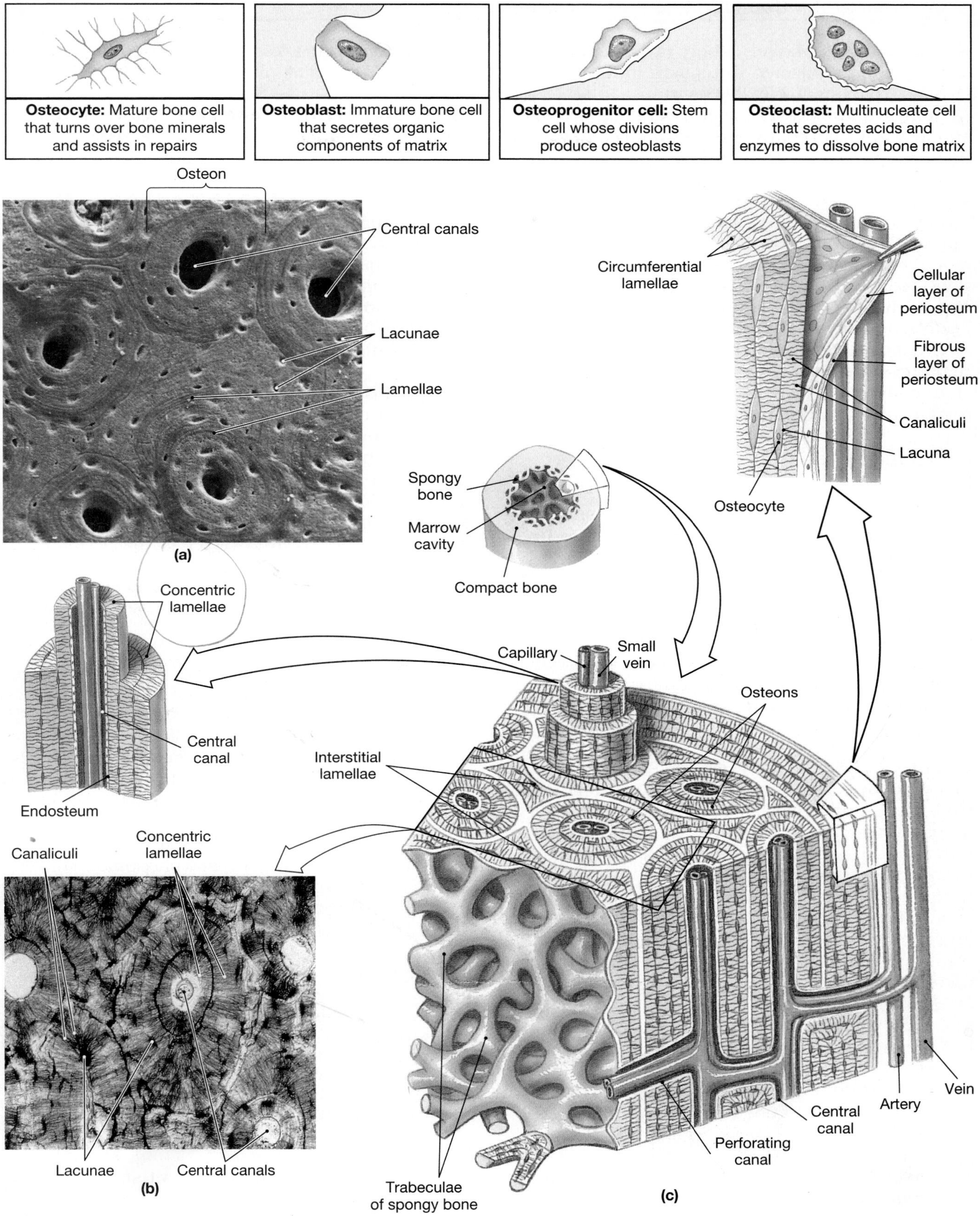

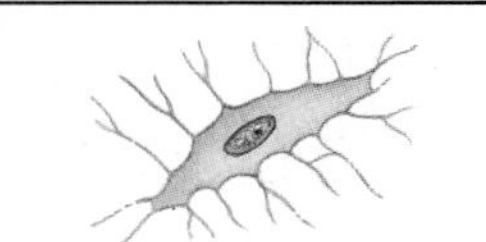

Osteocyte: Mature bone cell that turns over bone minerals and assists in repairs

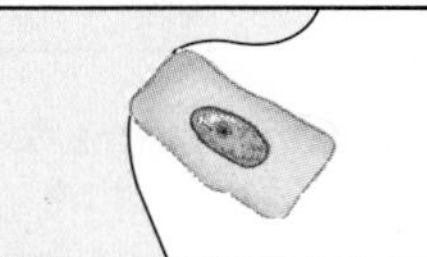

Osteoblast: Immature bone cell that secretes organic components of matrix

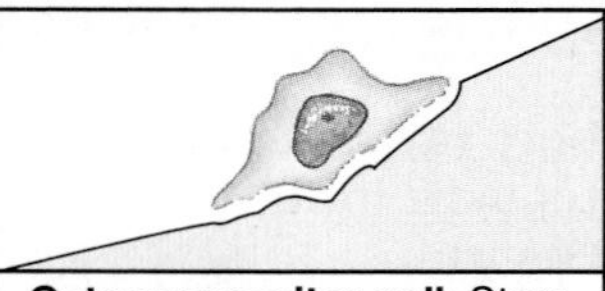

Osteoprogenitor cell: Stem cell whose divisions produce osteoblasts

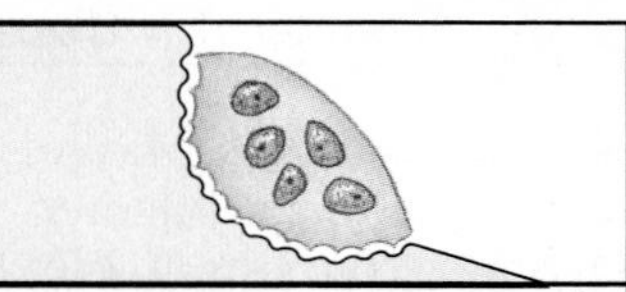

Osteoclast: Multinucleate cell that secretes acids and enzymes to dissolve bone matrix

Osteon
Central canals
Lacunae
Lamellae
(a)

Concentric lamellae
Central canal
Endosteum

Canaliculi
Concentric lamellae
Lacunae
Central canals
(b)

Spongy bone
Marrow cavity
Compact bone

Circumferential lamellae
Cellular layer of periosteum
Fibrous layer of periosteum
Canaliculi
Lacuna
Osteocyte

Capillary
Small vein
Osteons
Interstitial lamellae
Trabeculae of spongy bone
Perforating canal
Central canal
Artery
Vein
(c)

● **FIGURE 6-1**
Structure of a Typical Bone. **(a)** The cells of osseous tissue and a scanning electron micrograph of several osteons in compact bone. (SEM × 182) **(b)** A thin section through compact bone; in this procedure the intact matrix and central canals appear white, and the lacunae and canaliculi are shown in black. (LM × 272) **(c)** Diagrammatic view of the structure of a representative bone.

den impacts. Collagen fibers are remarkably strong; when subjected to tension (pull) they are stronger than steel. Flexible as well as tough, they can easily tolerate twisting and bending, but offer little resistance to compression—when compressed they simply bend out of the way.

In bone, the collagen fibers provide an organic framework for the formation of hydroxyapatite crystals. These crystals form small plates and rods that are locked into the collagen fibers. The result is a protein-crystal combination with properties intermediate between those of collagen and those of pure mineral crystals. Bone is strong, somewhat flexible, and very resistant to shattering. In its overall properties, bone is on a par with the best steel-reinforced concrete. In reality, bone is far superior to concrete because it can be remodeled on a regular basis and can repair itself after injury.

Cells in Bone

Figure 6-1a,b

Although osteocytes are most abundant, bone actually contains four different cell types (Figure 6-1a●):

1. ***Osteocytes,*** mature bone cells that account for most of the cell population.
2. ***Osteoblasts,*** cells that are actively depositing bone matrix.
3. ***Osteoprogenitor cells,*** stem cells whose divisions give rise to osteoblasts.
4. ***Osteoclasts,*** cells that actively remove bone matrix.

OSTEOCYTES Osteocytes (*osteon,* bone) are mature bone cells. Osteocytes are found in lacunae that are sandwiched between layers of calcified matrix. These layers of matrix are called **lamellae** (lah-MEL-lē; singular *lamella,* a thin plate). Osteocytes cannot divide, and there is never more than one osteocyte within each lacuna. Canaliculi penetrate the lamellae, radiating through the matrix and connecting lacunae with one another and with nutrient sources. The canaliculi contain cytoplasmic extensions of the osteocytes. Neighboring osteocytes are linked together by gap junctions that permit the exchange of ions and small molecules, including nutrients and hormones, between the cells. The interstitial fluid that surrounds the osteocytes and their extensions provides an additional route for the diffusion of nutrients and waste products.

Osteocytes have two major functions:

- ***They recycle the calcium salts in the matrix around them.*** Osteocytes release chemicals that dissolve the minerals in the adjacent matrix. These minerals enter the circulation, while others interact to form new hydroxyapatite crystals.
- ***They can participate in the repair of damaged bone.*** If released from their lacunae, these cells can convert to a less specialized type of cell, such as an osteoblast or osteoprogenitor cell.

OSTEOBLASTS Osteoblasts (OS-tē-ō-blasts; *blast,* precursor) are cuboidal cells responsible for the production of new bone, a process called **osteogenesis** (os-tē-ō-JEN-e-sis; *gennan,* to produce). These cells synthesize and release the proteins and other organic components of the bone matrix. The matrix prior to calcification is called **osteoid** (OS-tē-oyd). The osteoblasts also assist in elevating local concentrations of calcium phosphate and establishing conditions favorable to calcification. When an osteoblast becomes completely surrounded by calcified matrix, it differentiates into an osteocyte.

OSTEOPROGENITOR CELLS Bone contains small numbers of mesenchymal cells that can divide to produce daughter cells that differentiate into osteoblasts. These **osteoprogenitor** (os-tē-ō-prō-JEN-i-tor) **cells** (*progenitor,* ancestor) maintain populations of osteoblasts and play an important role in fracture repair. These cells are found (1) in the inner, cellular layer of the periosteum, (2) in an inner layer, or *endosteum,* that lines the marrow cavities of bone, and (3) accompanying blood vessels that travel through passageways in the matrix.

OSTEOCLASTS Osteoclasts (OS-tē-ō-klasts; *clast,* to break) are giant cells with 50 or more nuclei. They are not related to osteoprogenitor cells or their descendants. Osteoclasts are derived from circulating *monocytes,* phagocytic white blood cells.

Acids and proteolytic (protein-digesting) enzymes secreted by osteoclasts dissolve the matrix and release the stored minerals. This process, called **osteolysis** (os-tē-OL-i-sis), is important in the regulation of calcium and phosphate concentrations in body fluids. Osteoclasts are always removing matrix, and osteoblasts are always adding to it. The balance between the activities of osteoblasts and osteoclasts is very important. When osteoclasts remove calcium salts faster than osteoblasts deposit them, bones become weaker. When osteoblast activity predominates, bones become stronger and more massive.

❑ Compact and Spongy Bone

Figure 6-1c

There are two types of osseous tissue: *compact bone,* or *dense bone,* and *spongy bone.*

Compact bone is relatively dense and solid, whereas **spongy bone** forms an open network of struts and plates. Both compact and spongy bone are present in a typical bone of the skeleton, such as a humerus or vertebra. Compact

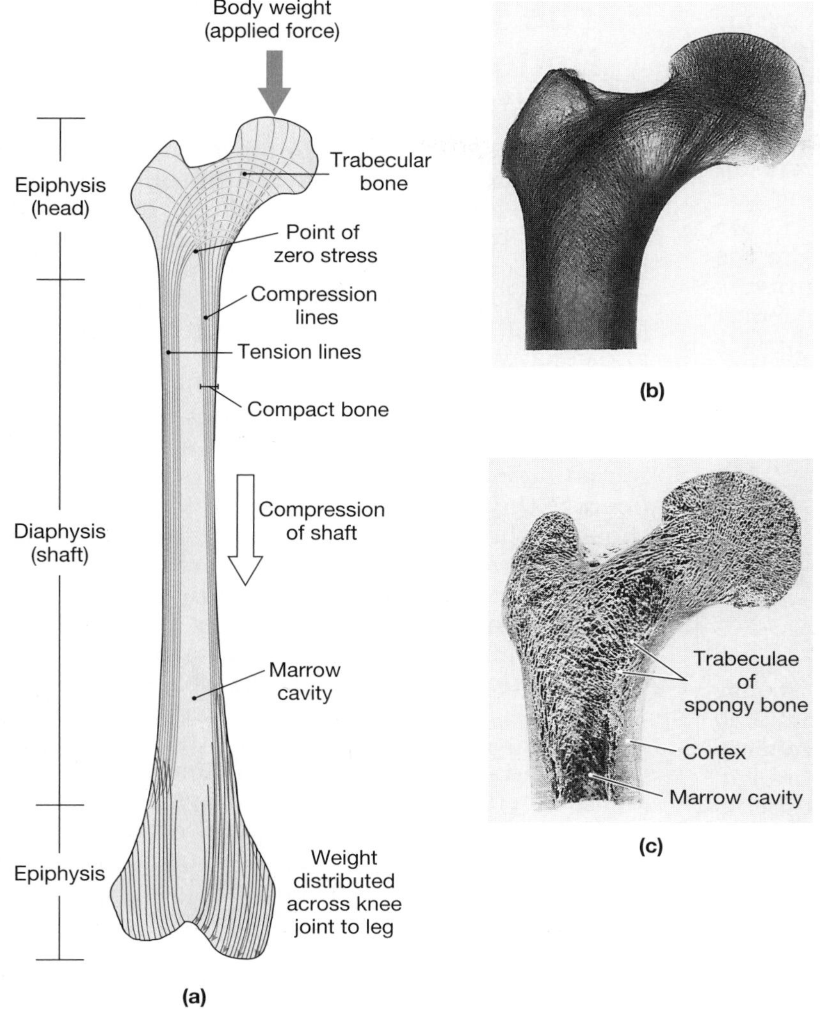
Body weight
(applied force)
Epiphysis
(head)
Trabecular
bone
Point of
zero stress
Compression
lines
Tension lines
Compact bone
Compression
of shaft
Diaphysis
(shaft)
Marrow
cavity
Epiphysis
Weight
distributed
across knee
joint to leg
(a)
(b)
Trabeculae
of
spongy bone
Cortex
Marrow cavity
(c)

● **FIGURE 6-2**
Lamellar Organization in a Long Bone. (a) The femur, or thigh bone, has a diaphysis (shaft) with walls of compact bone and epiphyses (heads) filled with spongy bone. The body weight is transferred to the femur at the hip joint. Because the hip joint is off-center relative to the axis of the shaft, the body weight is distributed along the bone so that the medial portion of the shaft is compressed and the lateral portion is stretched. **(b)** An X-ray showing the orientation of the trabeculae in the epiphysis. **(c)** A photograph showing the epiphysis after sectioning. Compare the orientation of trabeculae with the stress lines indicated in part (a).

bone tissue forms the walls, and an internal layer of spongy bone surrounds the **marrow cavity** (Figure 6-1c●). The marrow cavity contains **bone marrow,** a loose connective tissue that may be dominated by adipocytes **(yellow marrow)** or by a mixture of mature and immature red and white blood cells, and the stem cells that produce them **(red marrow).**

The matrix composition in compact bone is the same as that of spongy bone, but they differ in the three-dimensional arrangement of osteocytes, canaliculi, and lamellae.

Compact Bone

Figures 6-1, 6-2

The basic functional unit of mature compact bone is the **osteon** (OS-tē-on), or *Haversian system.* Within an osteon the osteocytes are arranged in concentric layers around a **central canal,** or *Haversian canal,* that contains one or more blood vessels that supply that osteon. Central canals usually run parallel to the surface of the bone. Other passageways, known as **perforating canals,** or the *canals of Volkmann,* extend roughly perpendicular to the surface. Blood vessels in the perforating canals deliver blood to osteons deeper in the bone and supply the interior marrow cavity.

The lamellae of each osteon are cylindrical and aligned parallel to the long axis of the central canal. These are known as *concentric lamellae.* Collectively the concentric lamellae resemble a bull's-eye around the central canal (Figure 6-1●). Canaliculi radiating through the lamellae interconnect the lacunae of the osteon with one another and with the central canal. *Interstitial lamellae* fill in the spaces between the osteons in compact bone, and *circumferential* (ser-kum-fer-EN-shul) *lamellae* (*circum,* around + *ferre,* to bear) lie beneath the periosteum and endosteum. Interstitial lamellae are remnants of osteons whose matrix components have been recycled by osteoclasts. Circumferential lamellae are produced during the growth of the bone.

DISTRIBUTION OF COMPACT BONE

A layer of compact bone covers the surfaces of bones; the thickness of that layer varies from region to region and from one bone to another. This superficial layer of compact bone is covered by the *periosteum,* a connective tissue component of the deep fascia, everywhere except inside joint capsules. Within joint cavities, hyaline *articular cartilages* cover opposing surfaces.

FUNCTION OF COMPACT BONE Compact bone is thickest where stresses arrive from a limited range of directions. Osteons in compact bone are all lined up the same way, making such bones very strong when stressed along that axis. You might envision a single osteon as a drinking straw with very thick walls. When you attempt to push the ends of a straw together or to pull them apart, it is quite strong. But if you hold the ends and push from the side, it will break easily.

Figure 6-2● shows the orientation of osteons and trabeculae in the femur, a typical long bone. The compact bone of the **cortex** surrounds the marrow cavity, also known as the *medullary cavity* (*medulla*, innermost part). Stresses are normally applied along the tubular **shaft,** or **diaphysis** (dī-AF-i-sis). The osteons are parallel to the long axis of the shaft, which does not bend when forces are applied to either end, although an impact to the side of the shaft can lead to a femoral fracture.

The hip joint consists of the head of the femur and a corresponding socket on the lateral surface of the hip bone. The femoral head projects medially, and body weight compresses the medial side of the diaphysis. Because the force is applied off-center, the bone has a tendency to bend into a lateral bow. The other side of the shaft, which resists this bending, is placed under a stretching load, or *tension* (Figure 6-2a●).

Spongy Bone

Figures 6-1c, 6-2

There are neither osteons nor blood vessels in spongy bone, or *cancellous* (KAN-sel-us) *bone.* The matrix in spongy bone forms struts or plates called **trabeculae** (tra-BEK-ū-lē). The thin trabeculae often branch, creating an open network (Figure 6-1c●). Nutrients reach the osteocytes by diffusion along canaliculi that open onto the surfaces of the trabeculae.

DISTRIBUTION OF SPONGY BONE Spongy bone is present at the expanded regions of long bones, where they articulate with other skeletal elements. These expanded regions are the ends, or **epiphyses** (ē-PIF-i-sēz), of the bones. Spongy bone may also be found extending into the marrow cavity.

FUNCTION OF SPONGY BONE Spongy bone is found where bones are not heavily stressed or where stresses arrive from many directions. Figure 6-2a● diagrams the trabecular alignment in the proximal epiphysis of the femur. The trabeculae are oriented along the stress lines, but with extensive cross-bracing. The actual alignment can be seen in Figure 6-2b,c●. At the proximal epiphysis the trabeculae transfer forces from the hip to the femoral shaft; at the distal epiphysis the trabeculae direct the forces across the knee joint to the leg.

In addition to being able to withstand stresses applied from many directions, spongy bone is much lighter than compact bone. Spongy bone reduces the weight of the skeleton and makes it easier for muscles to move the bones. Finally, the trabecular framework supports and protects the cells of the bone marrow. Yellow marrow, often found in the marrow cavity of the shaft, is an important energy reserve. Extensive areas of red marrow, such as that found in the spongy bone of the femoral epiphyses, are important sites of blood cell formation.

❑ The Periosteum and Endosteum

Figures 6-1, 6-3

The outer surface of a bone is covered by a **periosteum** that consists of a fibrous connective tissue outer layer and a cellular inner layer (Figure 6-1c●, p. 177). The periosteum (1) isolates the bone from surrounding tissues, (2) provides a route for circulatory and nervous supply, and (3) actively participates in bone growth and repair.

Near joints, the periosteum becomes continuous with the connective tissues that lock the bones together. At a synovial joint, the periosteum is continuous with the joint capsule. The fibers of the periosteum are also interwoven with those of the tendons attached to the bone. As the bone grows, these tendon fibers are cemented into the superficial lamellae by osteoblasts from the cellular layer of the periosteum. Collagen fibers incorporated into bone tissue from tendons and ligaments, as well as from the superficial periosteum, are called *perforating fibers* (*Sharpey's fibers*). This method of attachment bonds the tendons and ligaments into the general structure of the bone, providing a much stronger attachment than would otherwise be possible. An extremely powerful pull on a tendon or ligament will usually break a bone rather than snap the collagen fibers at the bone surface.

Inside the bone, a cellular **endosteum** lines the marrow cavity (Figure 6-3●). This layer covers the trabeculae of spongy bone and lines the inner surfaces of the central canals. The endosteum is active during the growth of bone and whenever repair or remodeling is under way. The endosteum consists of a simple flattened layer of osteoprogenitor cells that covers the bone matrix, usually without any intervening connective tissue fibers. The cellular layer is not a complete epithelium, and the matrix is occasionally exposed. At these exposed sites osteoclasts and osteoblasts can remove or deposit matrix components. The osteoclasts are usually found in shallow depressions (*Howship's lacunae*) that have eroded into the matrix.

How would the strength of a bone be affected if the ratio of collagen to hydroxyapatite increased?

A sample of bone shows concentric lamellae surrounding a central canal. Is the sample from the cortex or the marrow cavity of a long bone?

If the activity of osteoclasts exceeds the activity of osteoblasts in a bone, how will the mass of the bone be affected?

● **FIGURE 6-3**
The Periosteum and Endosteum

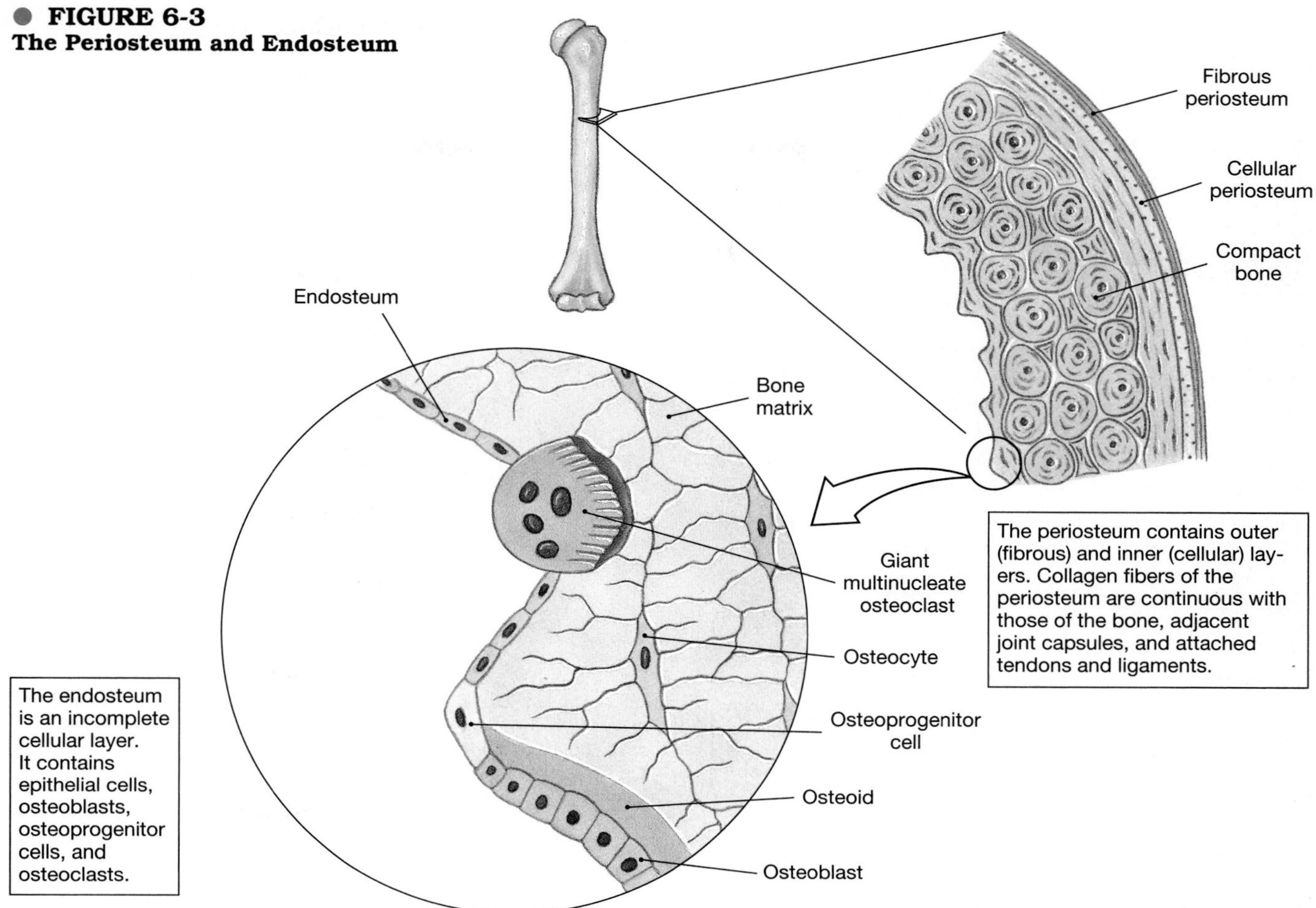
Fibrous periosteum
Cellular periosteum
Compact bone
Endosteum
Bone matrix
The periosteum contains outer (fibrous) and inner (cellular) layers. Collagen fibers of the periosteum are continuous with those of the bone, adjacent joint capsules, and attached tendons and ligaments.
Giant multinucleate osteoclast
Osteocyte
Osteoprogenitor cell
Osteoid
Osteoblast
The endosteum is an incomplete cellular layer. It contains epithelial cells, osteoblasts, osteoprogenitor cells, and osteoclasts.

Bone Development and Growth

The growth of the skeleton determines the size and proportions of our body. The bony skeleton begins to form about 6 weeks after fertilization, when the embryo is approximately 12 mm (0.5 in.) long. (At this stage the existing skeletal elements are cartilaginous.) During subsequent development, the bones undergo a tremendous increase in size. Bone growth continues through adolescence, and portions of the skeleton usually do not stop growing until age 25. The entire process is carefully regulated, and a breakdown in regulation will ultimately affect all of the body systems. In this section we will consider the physical process of osteogenesis (bone formation) and bone growth. The next section will examine the maintenance and replacement of mineral reserves in the adult skeleton. AM *Examination of the Skeletal System*

During development, mesenchyme or cartilage is replaced by bone. This process of replacing other tissues with bone is called **ossification.** Ossification refers specifically to the formation of bone. The process of **calcification** refers to the deposition of calcium salts within a tissue. There are two major forms of ossification. In *intramembranous ossification,* bone develops from mesenchyme or fibrous connective tissue. In *endochondral ossification,* bone replaces an existing cartilage model.

Intramembranous Ossification

Figures 6-4, 6-5

Intramembranous (in-tra-MEM-bra-nus) **ossification,** also called *dermal ossification,* begins when osteoblasts differentiate within a mesenchymal or fibrous connective tissue. This type of ossification normally occurs in the deeper layers of the dermis, and the bones that result are often called **dermal bones.** Examples of dermal bones include the roofing bones of the skull, the *mandible* (lower jaw), and the *clavicle* (collarbone). In response to abnormal stresses, bone may form in other dermal areas, or in connective tissues within tendons, around joints, in the kidneys, or in skeletal muscles. Intramembranous bones forming in abnormal locations are called *heterotopic bones* (*hetero,* different + *topos,* place), or *ectopic bones* (*ektos,* outside). AM *Heterotopic Bone Formation*

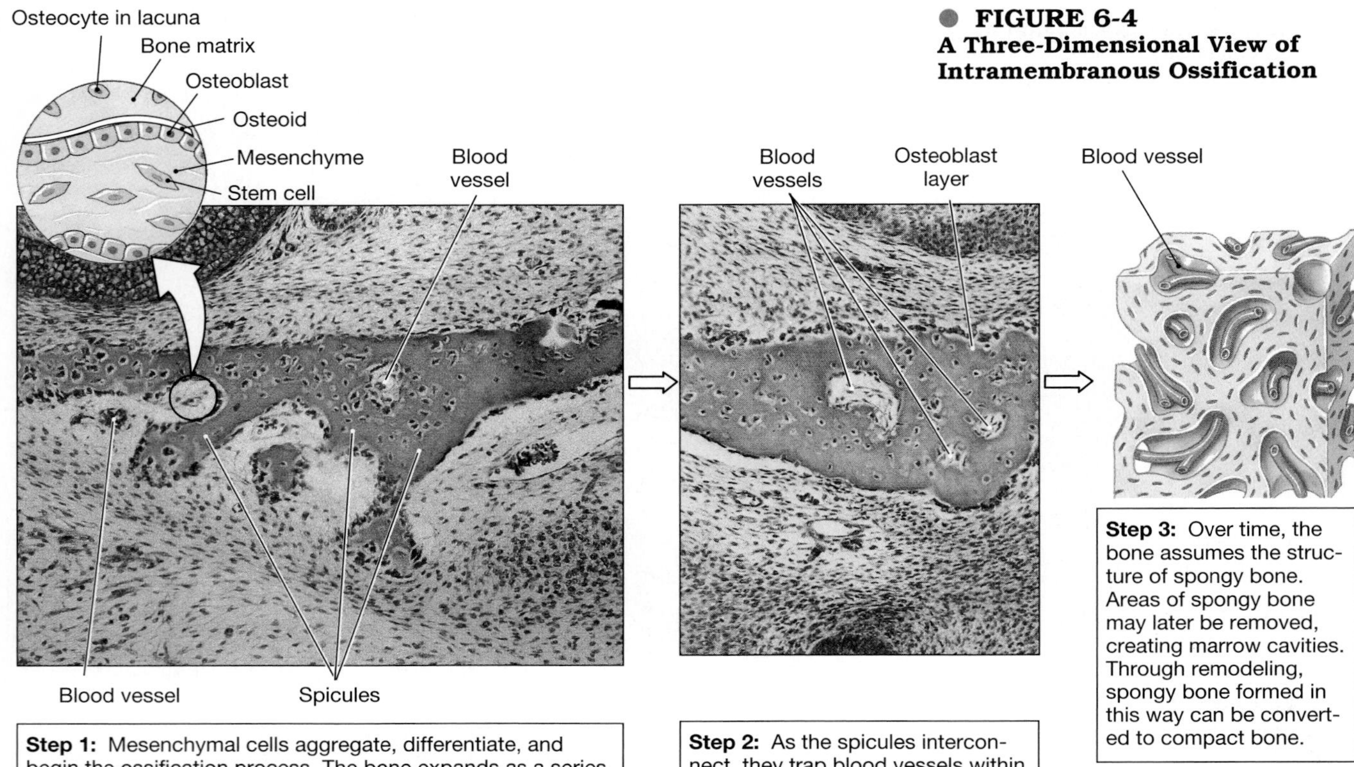

Osteocyte in lacuna
Bone matrix
Osteoblast
Osteoid
Mesenchyme
Stem cell
Blood vessel
Blood vessel
Spicules
Blood vessels
Osteoblast layer
Blood vessel

● **FIGURE 6-4**
A Three-Dimensional View of Intramembranous Ossification

Step 1: Mesenchymal cells aggregate, differentiate, and begin the ossification process. The bone expands as a series of spicules that spread into surrounding tissues. (LM × 24)

Step 2: As the spicules interconnect, they trap blood vessels within the bone. (LM × 24)

Step 3: Over time, the bone assumes the structure of spongy bone. Areas of spongy bone may later be removed, creating marrow cavities. Through remodeling, spongy bone formed in this way can be converted to compact bone.

The steps in the process of intramembranous ossification are illustrated in Figure 6-4● and may be summarized as:

Step 1. Osteoblasts first cluster together and start to secrete the organic components of the matrix. The resulting mixture of collagen fibers and osteoid then becomes mineralized through the crystallization of calcium salts. The location in a bone where ossification first occurs is called an **ossification center.** As ossification proceeds, it traps some osteoblasts inside bony pockets; these cells differentiate into osteocytes.

Step 2. The developing bone grows outward from the ossification center in small struts, called **spicules.** Although osteoblasts are still being trapped in the expanding bone, mesenchymal cell divisions continue to produce additional osteoblasts. Bone growth is an active process, and osteoblasts require oxygen and a reliable supply of nutrients. Blood vessels that branch between the spicules meet these demands. As a result, the rate of bone growth actually accelerates.

Step 3. Over time, the bone assumes the structure of spongy bone. Although initially the intramembranous bone resembles spongy bone, subsequent remodeling around the trapped blood vessels can produce compact bone. As growth slows, the connective tissue around the bone becomes organized into the fibrous layer of the periosteum. The osteoblasts closest to the bone surface become less active and form the inner cellular layer.

Figure 6-5● shows intramembranous ossification forming skull bones in the head of a fetus.

❑ Endochondral Ossification

Figures 6-5, 6-6

Endochondral (en-dō-KON-drul) **ossification** (*endo,* inside + *chondros,* cartilage) begins with the formation of a cartilaginous model. Most bones in the body form in this way, and limb bone development is a good example of this process. By the time an embryo is 6 weeks old, the proximal bone of the limb, either the humerus (arm) or femur (thigh), is present, but it is composed entirely of cartilage. This model continues to grow by expansion of the cartilage matrix (interstitial growth) and the production of new cartilage at the outer surface (appositional growth). These growth mechanisms were introduced in Chapter 4. ∞ *[p. 132]* Steps in the growth and ossification of one of the limb bones seen in Figure 6-5● are diagrammed in Figure 6-6●.

Step 1. As the cartilage enlarges, chondrocytes near the center of the shaft increase greatly in size. As these cells enlarge, their lacunae expand, and the matrix is reduced to a series of small struts that soon begin to calcify. The enlarged chondrocytes then die and disintegrate.

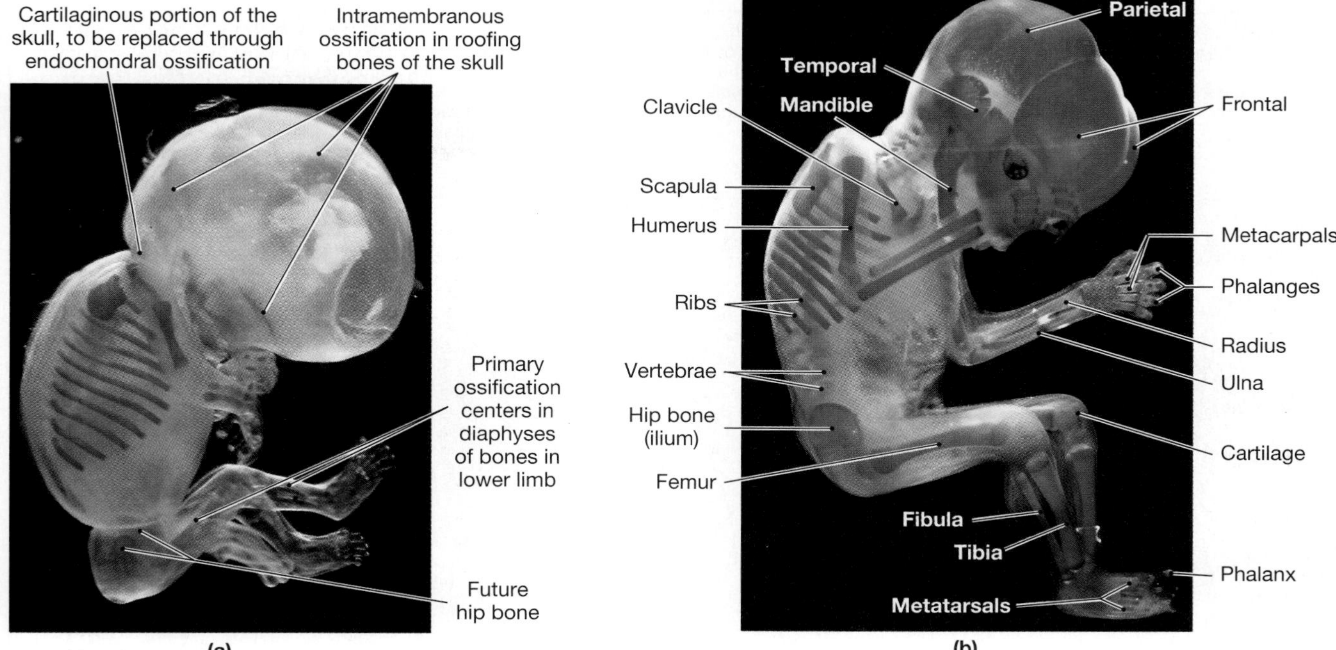
Cartilaginous portion of the skull, to be replaced through endochondral ossification
Intramembranous ossification in roofing bones of the skull
Primary ossification centers in diaphyses of bones in lower limb
Future hip bone
(a)
Parietal
Temporal
Mandible
Clavicle
Frontal
Scapula
Humerus
Metacarpals
Phalanges
Ribs
Radius
Vertebrae
Ulna
Hip bone (ilium)
Cartilage
Femur
Fibula
Tibia
Phalanx
Metatarsals
(b)

FIGURE 6-5
Intramembranous and Endochondral Ossification in a Fetus. **(a)** This 10-week human fetus has been specially stained to show developing skeletal elements. **(b)** At 16 weeks, intramembranous bones are completing the roof of the skull. Most of the appendicular skeleton at this stage consists of cartilage that will be replaced through endochondral ossification.

Enlarging chondrocytes within calcifying matrix

Epiphysis

Diaphysis

Bone formation

Hyaline cartilage

Marrow cavity

Center of ossification

Periosteal bone

Spongy bone

Blood vessel

See Figure 6-7

See Figure 6-8

Metaphysis

Step 1: As the cartilage enlarges, chondrocytes near the center of the shaft increase greatly in size. The matrix is reduced to a series of small struts that soon begin to calcify. The enlarged chondrocytes then die and disintegrate, leaving cavities within the cartilage.

Step 2: Blood vessels grow around the edges of the cartilage, and the perichondral cells convert to osteoblasts. The shaft of the cartilage then becomes ensheathed in a superficial layer of bone.

Step 3: Blood vessels penetrate the cartilage and invade the central region, growing toward the epiphyses at either end. Fibroblasts migrating with the blood vessels differentiate into osteoblasts and begin producing spongy bone.

Step 4: Remodeling occurs as growth continues, creating a marrow cavity. The bone of the shaft becomes thicker, and the cartilage near each epiphysis is replaced by shafts of bone. Further growth involves increases in length (Fig. 6-7) and diameter (Fig. 6-8).

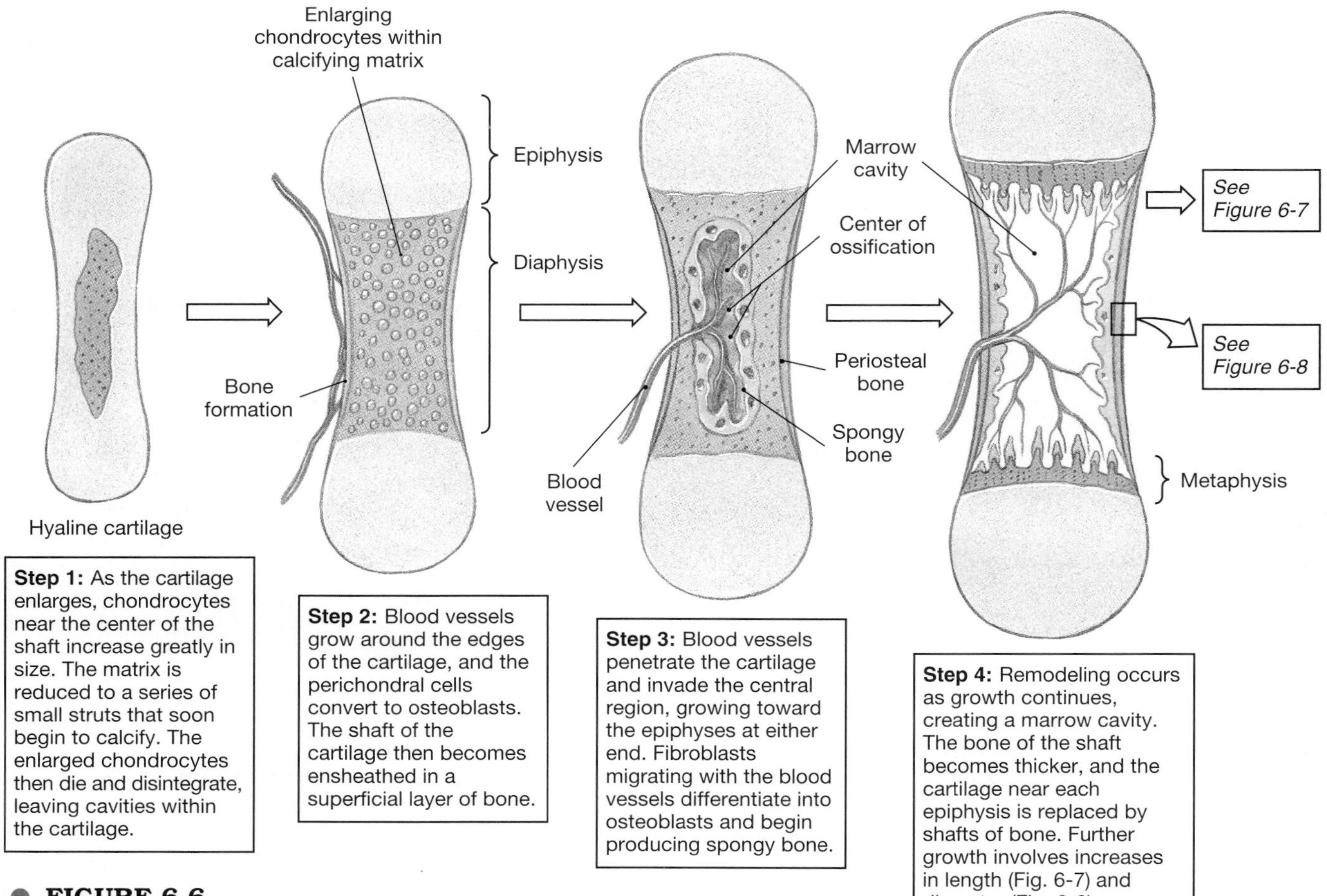

FIGURE 6-6
Initial Steps in Endochondral Ossification

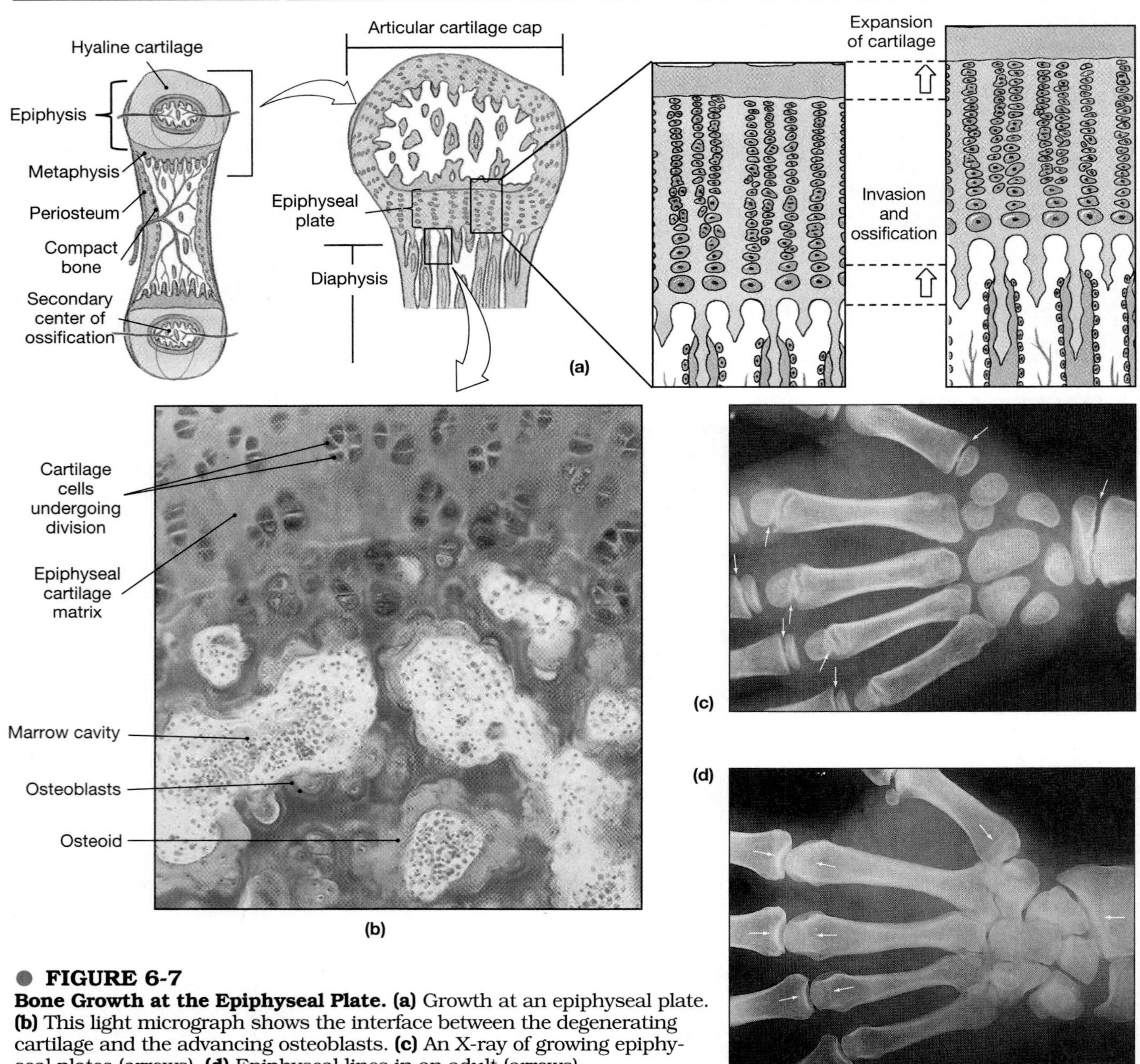
Capillaries and osteoblasts migrate into the epiphysis, creating a secondary ossification center.
Soon the epiphysis is filled with spongy bone. A thin cap of articular cartilage remains exposed to the joint cavity, and at the metaphysis an epiphyseal plate separates the epiphysis from the diaphysis.
Within the epiphyseal plate, chondrocytes closest to the epiphysis undergo division, thickening the plate and forcing the epiphysis farther from the shaft. As the daughter cells mature, they become enlarged, and the surrounding matrix becomes calcified. On the other side of the epiphyseal plate, osteoblasts and capillaries invade the lacunae and replace the dead cartilage with living bone. As long as the rate of cartilage growth keeps pace with the rate of osteoblast invasion, the shaft grows longer but the epiphyseal plate survives.
Hyaline cartilage
Epiphysis
Metaphysis
Periosteum
Compact bone
Secondary center of ossification
Articular cartilage cap
Epiphyseal plate
Diaphysis
(a)
Expansion of cartilage
Invasion and ossification
Cartilage cells undergoing division
Epiphyseal cartilage matrix
Marrow cavity
Osteoblasts
Osteoid
(b)
(c)
(d)

● **FIGURE 6-7**
Bone Growth at the Epiphyseal Plate. (a) Growth at an epiphyseal plate. **(b)** This light micrograph shows the interface between the degenerating cartilage and the advancing osteoblasts. **(c)** An X-ray of growing epiphyseal plates (arrows). **(d)** Epiphyseal lines in an adult (arrows).

Step 2. Blood vessels grow into the perichondrium of the cartilaginous shaft, and the cells of the perichondrium in this region develop into osteoblasts. The perichondrium has now been converted into a periosteum, and the inner **osteogenic layer** soon produces a thin layer of bone around the shaft of the cartilage.

Step 3. While these changes are under way, the blood supply to the periosteum increases, and capillaries and osteoblasts migrate into the heart of the cartilage, invading the spaces left by the disintegrating chondrocytes. The calcified cartilaginous matrix breaks down, and osteoblasts replace it with spongy bone. Bone development begins at this **primary center of ossification** and spreads toward both ends of the cartilaginous model.

Step 4. While the diameter is small, the entire diaphysis is filled with spongy bone, but as it enlarges osteoclasts appear and begin eroding the central portion to create a marrow cavity. Further growth involves two distinct processes: an increase in *length* and an enlargement in *diameter.*

Increasing the Length of a Developing Bone

Figure 6-7

Growth in bone length involves a process summarized in Figure 6-7•. During the initial stages of osteogenesis, osteoblasts move away from the center of ossification toward the epiphyses. But they do not manage to complete the ossification of the model immediately, because the epiphyseal cartilages continue to enlarge.

The region where the cartilage is being replaced by bone lies at the junction between the shaft (diaphysis) and each end (epiphysis) of the bone (Figure 6-7a•). This region is known as the **metaphysis** (me-TAF-i-sis). On the shaft side of the metaphysis, osteoblasts are continually invading the cartilage and converting it to bone. But on the epiphyseal side, new cartilage is continually being added. The osteoblasts are therefore moving toward the epiphysis, which is being pushed ahead by the growth of the epiphyseal cartilage. The situation is like a pair of joggers, one in front of the other. As long as they are running at the same speed, they can run for miles without colliding. The osteoblasts don't catch up to the epiphysis, but both the osteoblasts and the epiphysis "run away" from the primary ossification center. As they do so, the skeletal element grows longer and longer.

The next major change occurs when the centers of the epiphyses begins to calcify. Capillaries and osteoblasts then migrate into these areas, creating **secondary ossification centers.** The time of appearance of secondary ossification centers varies from one bone to another, and from individual to individual. Secondary ossification centers may be found at birth in the proximal and distal ends of the *humerus* (arm), *femur* (thigh), and *tibia* (leg), but the ends of some other bones remain cartilaginous until early adulthood.

The epiphyses eventually become filled with spongy bone. A thin cap of the original cartilage model will remain exposed to the joint cavity as the **articular cartilage.** This cartilage prevents damaging bone-to-bone contact within the joint. At the metaphysis a relatively narrow cartilaginous **epiphyseal plate** separates the epiphysis from the diaphysis.

Within the epiphyseal plate, the chondrocytes closest to the epiphysis continue to enlarge and divide, adding to the thickness of the plate. The light micrograph in Figure 6-7b• shows the interface between the degenerating cartilage and the advancing osteoblasts. The continual expansion of the epiphyseal cartilage forces the epiphysis farther from the shaft (Figure 6-7a•). As the daughter cells mature, they become enlarged, and the surrounding matrix becomes calcified. On the shaft side of the epiphyseal plate, osteoblasts and capillaries continue to invade these lacunae and replace the dead cartilage with living bone. As long as the rate of cartilage growth keeps pace with the rate of osteoblast invasion, the shaft grows longer but the epiphyseal plate survives.

At puberty, the combination of rising levels of sexual hormones, growth hormone, and thyroid hormones stimulates bone growth dramatically. Osteoblasts now begin producing bone faster than the rate of epiphyseal cartilage expansion. As a result, the epiphyseal plate gets narrower and narrower, until it ultimately disappears. The former location of the plate can still be detected in X-rays as a distinct **epiphyseal line** that remains after epiphyseal growth has ended (Figure 6-7c,d•).

Increasing the Diameter of a Developing Bone

Figure 6-8

The diameter of a bone enlarges through appositional growth at the outer surface. In this process, periosteal cells differentiate into osteoblasts and contribute to the growth of the bone matrix. Eventually they become surrounded by matrix and differentiate into osteocytes. As this process continues to add layers of bone tissue, blood vessels and collagen fibers of the periosteum become incorporated into the bony structure. Steps in appositional bone growth are described and illustrated in Figure 6-8•.

While bone matrix is being added to the outer surface of the growing bone, osteoclasts are removing bone matrix at the inner surface. As a result, the marrow cavity gradually enlarges as the bone gets larger in diameter.

 During the process of intramembranous ossification, what type of tissue is replaced by bone?

 In endochondral ossification, what is the original source of osteoblasts?

 How could X-rays of the femur be used to determine whether a person had reached full height?

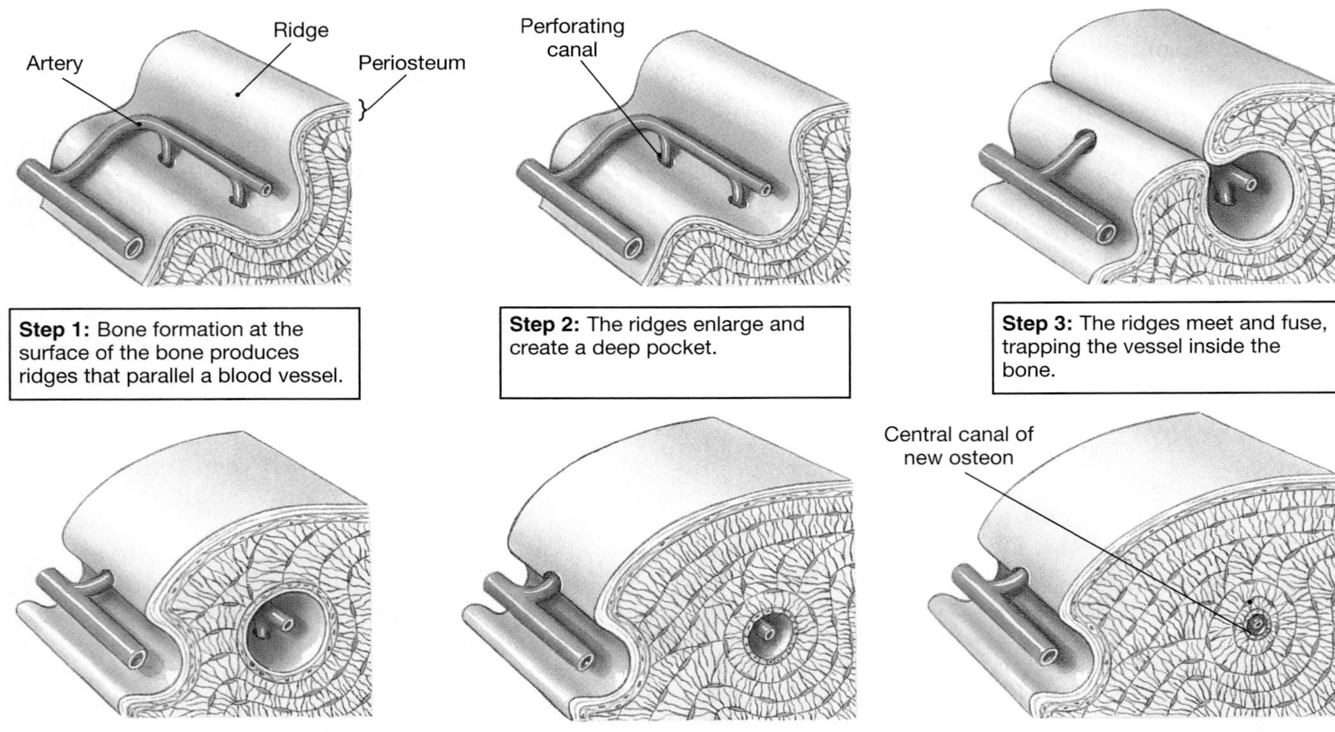
Artery
Ridge
Periosteum
Perforating canal
Step 1: Bone formation at the surface of the bone produces ridges that parallel a blood vessel.
Step 2: The ridges enlarge and create a deep pocket.
Step 3: The ridges meet and fuse, trapping the vessel inside the bone.
Central canal of new osteon
Steps 4–6: Bone deposition then proceeds toward the vessel, creating a typical osteon.

● **FIGURE 6-8**
Appositional Bone Growth. Three-dimensional diagrams illustrating the mechanism responsible for increasing the diameter of a growing bone.

❑ The Blood and Nerve Supply

Figures 6-8, 6-9

Osseous tissue is very vascular, and the bones of the skeleton have an extensive blood supply. In a typical bone such as the humerus, three major sets of blood vessels develop (Figure 6-9●):

1. ***The nutrient artery and vein.*** These vessels form as blood vessels invade the cartilage model at the start of endochondral ossification. There is usually only one **nutrient artery** and one **nutrient vein,** although a few bones, including the femur, have more than one pair. Branches of these vessels extend along the length of the shaft and into the osteons of the surrounding cortex.
2. ***Metaphyseal vessels.*** These vessels supply blood to the inner (diaphyseal) surface of each epiphyseal plate, where bone is replacing the cartilage.
3. ***Periosteal vessels.*** Blood vessels from the periosteum are incorporated into the developing bone surface as described previously (Figure 6-8●). These vessels provide blood to the superficial osteons of the shaft. During endochondral bone formation, branches of periosteal ves-

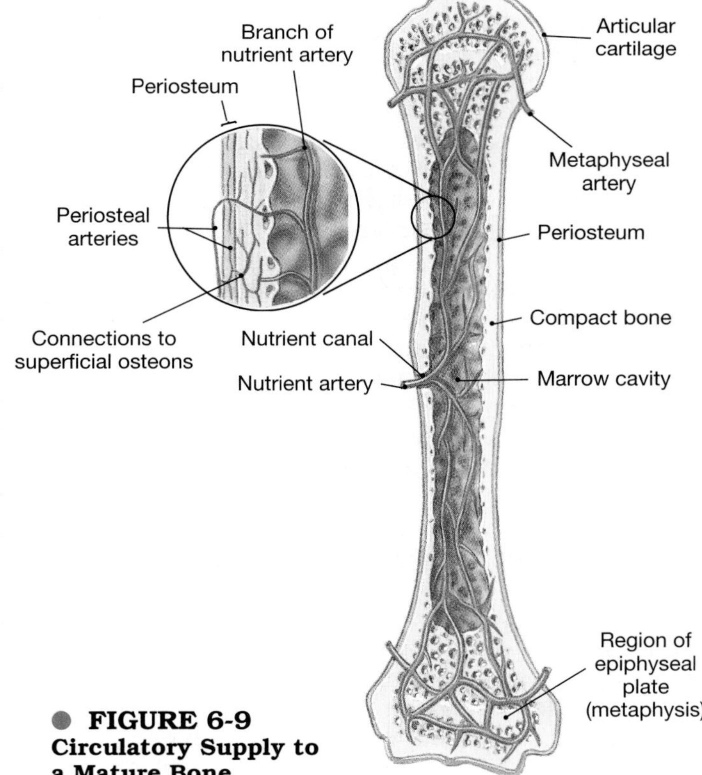
Branch of nutrient artery
Periosteum
Periosteal arteries
Connections to superficial osteons
Nutrient canal
Nutrient artery
Articular cartilage
Metaphyseal artery
Periosteum
Compact bone
Marrow cavity
Region of epiphyseal plate (metaphysis)

● **FIGURE 6-9**
Circulatory Supply to a Mature Bone

sels also enter the epiphyses, providing blood to the secondary ossification centers.

Following the closure of the epiphyses, all three sets of vessels become extensively interconnected.

The periosteum contains an extensive network of lymphatic vessels and sensory nerves. The lymphatics collect lymph from branches that enter the bone and reach individual osteons via the perforating canals. The sensory nerves penetrate the cortex with the nutrient artery to innervate the endosteum, marrow cavity, and epiphyses. Because of the rich sensory innervation, injuries to bones are usually very painful.

The Dynamic Nature of Bone

Many people believe that bone is static and inert, like a brick or a boulder. This idea is definitely incorrect, for bone is an extremely active tissue. The organic and mineral components of the bone matrix are continually being recycled and renewed through the process of **remodeling.** Bone remodeling is under way throughout life, as part of normal bone maintenance. Remodeling may replace the matrix but leave the bone as a whole unchanged, or there may be alterations in the shape, internal architecture, or mineral content of the bone.

Bone remodeling involves an interplay between the activities of osteocytes, osteoblasts, and osteoclasts. In the adult, osteocytes are continually removing and replacing the surrounding calcium salts. But osteoclasts and osteoblasts also remain active, even after the epiphyseal plates have closed. Normally their activities are balanced: As one osteon forms through the activity of osteoblasts, another is destroyed by osteoclasts. The turnover rate for bone is quite high. In a young adult, each year almost one-fifth of the adult skeleton is demolished and then rebuilt or replaced. Every part of every bone may not be affected, as there are regional and even local differences in the rate of turnover. For example, the spongy bone in the head of the femur may be replaced two or three times each year, whereas the compact bone along the shaft remains largely unchanged.

Exercise Effects on Bone

The turnover and recycling of minerals give each bone the ability to adapt to new stresses. Osteoblast sensitivity to electrical events has been theorized as the mechanism that controls the internal organization and structure of bone. Whenever a bone is stressed, the mineral crystals generate minute electrical fields. Osteoblasts are apparently attracted to these electrical fields, and once in the area they begin to produce bone. (Electrical fields may also be used to stimulate the repair of severe fractures.)

Because bones are adaptable, their shapes reflect the forces applied to them. For example, bumps and ridges on the surface of a bone mark the sites where tendons attach to the bone. If muscles become more powerful, the corresponding bumps and ridges enlarge to withstand the increased forces. Heavily stressed bones become thicker and stronger, whereas bones not subjected to ordinary stresses will become thin and brittle. Regular exercise is therefore important as a stimulus that maintains normal bone structure. Champion weight lifters have massive bones with thick, prominent ridges where muscles attach. In nonathletes, moderate amounts of physical activity and weight bearing are essential to stimulate normal bone maintenance and maintain adequate bone strength.

Degenerative changes in the skeleton occur after relatively brief periods of inactivity. For example, using a crutch while wearing a cast takes weight off the injured limb. After a few weeks, the unstressed bones will lose up to a third of their mass. The bones rebuild just as quickly when normal loading resumes. However, the removal of calcium salts can be a potentially serious health hazard for astronauts remaining in a weightless environment and for bedridden or paralyzed patients who spend months or years without stressing the skeleton.

Hormonal and Nutritional Effects on Bone

Normal bone growth and maintenance depend on a combination of nutritional and hormonal factors.

- Normal bone growth and maintenance cannot occur without a constant dietary source of calcium and phosphate salts. Lesser amounts of other minerals, such as magnesium, fluoride, iron, and manganese, are also required.
- The hormone *calcitriol*, synthesized in the kidneys, is essential for normal calcium and phosphate ion absorption at the digestive tract. Calcitriol synthesis is dependent on the availability of a related steroid, *cholecalficerol* (vitamin D_3), which may be synthesized in the skin or absorbed from the diet.
- Adequate levels of vitamin C must be present in the diet. This vitamin is required by certain key enzymatic reactions involved with collagen synthesis. It also stimulates osteoblast differ-

entiation. One of the signs of vitamin C deficiency, a condition called *scurvy*, is a loss of bone mass and strength.

- Three other vitamins have significant effects on bone structure. Vitamin A, which stimulates osteoblast activity, is particularly important for normal bone growth in children. Vitamins K and B_{12} are required for the synthesis of proteins in normal bone.
- *Growth hormone*, produced by the pituitary gland, and *thyroxine*, from the thyroid gland, stimulate bone growth. Growth hormone stimulates protein synthesis and cell growth throughout the body. Thyroxine stimulates cell metabolism and increases the rate of osteoblast activity. In proper balance, these hormones maintain normal activity at the epiphyseal plates until roughly the time of puberty.
- At puberty, rising levels of sex hormones (*estrogens* in the female and *androgens* in the male) stimulate osteoblasts to produce bone faster than the rate of epiphyseal cartilage expansion. Over time, the epiphyseal plates narrow and eventually ossify or "close." There are differences from bone to bone and individual to individual as to the timing of epiphyseal closure. The toes may complete their ossification by age 11, whereas portions of the pelvis or the wrist may continue to enlarge until age 25. Differences in the male and female sex hormones account for the variation between the sexes and for related variations in body size and proportions.

Two other hormones are important in the homeostatic control of calcium and phosphate levels in body fluids. The interactions between these hormones, *calcitonin* (kal-si-TŌ-nin) from the thyroid gland and *parathyroid hormone* from the parathyroid gland, will be considered in the next section. The major hormones affecting the growth and maintenance of the skeletal system are summarized in Table 6-1.

ABNORMAL BONE GROWTH AND DEVELOPMENT A variety of endocrine or metabolic problems can result in characteristic skeletal changes. **Gigantism** results from an overproduction of growth hormone before puberty. (The world record for height is 272 cm, or 8 ft 11 in., with a weight of 216 kg, or 475 lb.) The opposite extreme is **pituitary growth failure** (*pituitary dwarfism*), in which inadequate growth hormone production leads to reduced epiphyseal activity and abnormally short bones. This form of growth failure is becoming increasingly rare in the United States because children can be treated with human growth hormone.

Several inherited metabolic problems that affect many systems have effects on the growth and development of the skeletal system. These conditions are often recognizable because of characteristic variations in body proportions. For example, individuals with *Marfan's syndrome* are very tall, with long slender limbs due to excessive cartilage formation at the epiphyseal plates. Although this is an obvious physical distinction, the characteristic body proportions are not in themselves dangerous. However, the underlying metabolic defect often causes life-threatening problems with the cardiovascular system. AM *Inherited Abnormalities of Skeletal Development*

If growth hormone levels rise abnormally after epiphyseal closure, the skeleton does not grow larger, but cartilage growth and alterations in soft tissue structure lead to changes in bone density and in physical features, such as the contours of the face. AM *Hyperostosis and Acromegaly*

Why would you expect the arm bones of a weight lifter to be thicker and heavier than those of a jogger?

A child who enters puberty several years later than the average age is usually taller than average as an adult. Why?

A 7-year-old child has a pituitary tumor involving the cells that secrete growth hormone (GH), resulting in increased levels of GH. How will this condition affect the child's growth?

TABLE 6-1 Hormones Involved in the Regulation of Bone Growth

Hormone	*Primary Source*	*Effects on Skeletal System*
Calcitriol	Kidneys	Promotes calcium and phosphate ion absorption along the digestive tract.
Growth hormone	Pituitary gland	Stimulates osteoblast activity and the synthesis of bone matrix.
Thyroxine	Thyroid gland (follicle cells)	With growth hormone, stimulates osteoblast activity and synthesis of bone matrix.
Sex hormones	Ovaries (estrogens) Testes (androgens)	Stimulate osteoblast activity and synthesis of bone matrix.
Parathyroid hormone	Parathyroid glands	Stimulates osteoclast activity; elevates calcium ion concentrations in body fluids.
Calcitonin	Thyroid gland (C cells)	Inhibits osteoclast activity; reduces calcium ion concentrations in body fluids.

❑ The Skeleton as a Calcium Reserve

Figure 6-10

The bones of the skeleton are important mineral reservoirs, as indicated in Figure 6-10•. Other minerals will be considered in later chapters; for the moment we will focus on the homeostatic regulation of calcium ion concentrations in body fluids. Calcium is the most abundant mineral in the human body. A typical human body contains 1–2 kg (2.2–4.4 lb) of calcium, with roughly 99 percent of it deposited in the skeleton.

Calcium ion concentrations must be closely controlled to prevent damage to essential physiological systems. Even small variations from normal concentrations will have some effect on cellular operations. Larger changes can cause a clinical crisis. Calcium ions are particularly important to the membranes and intracellular activities of neurons and muscle cells, especially cardiac muscle cells. If the calcium concentration in body fluids increases by 30 percent, neurons and muscle cells become relatively unresponsive. If calcium levels decrease by 35 percent, neurons become so excitable that convulsions may occur. A 50 percent reduction in calcium concentrations usually causes death. Such gross disturbances in calcium metabolism are relatively rare, for the calcium ion concentrations are so closely regulated that daily fluctuations of more than 10 percent are very unusual.

Hormones and Calcium Balance

Figure 6-11

The calcium ion concentration in body fluids depends on activities under way in three organs: (1) the bones, (2) the intestinal tract, and (3) the kidneys. Interactions among these components are indicated in Figure 6-11•. Factors that promote an elevation in calcium levels are indicated by red arrows; blue arrows indicate factors that depress calcium ion concentrations.

Calcium ion homeostasis is maintained by a negative feedback system involving a pair of hormones with opposing effects. These hormones, *parathyroid hormone* and *calcitonin*, coordinate the storage, absorption, and excretion of calcium ions. The target organs are the bones (storage), the digestive tract (absorption), and the kidneys (excretion).

When calcium ion concentrations fall below normal, cells of the **parathyroid gland** release **parathyroid hormone** (**PTH**, also known as *parathormone*) into the bloodstream. Parathyroid hormone has three major effects:

1. ***Stimulating osteoclast activity*** and enhancing the recycling of minerals by osteocytes.
2. ***Increasing the rate of intestinal absorption of calcium ions*** by enhancing the action of calcitriol.
3. ***Decreasing the rate of excretion of calcium ions.***

Under these conditions more calcium ions enter body fluids, losses are restricted, and calcium ion concentrations increase to normal levels.

If the calcium ion concentration of the blood rises above normal, special cells within the thyroid gland secrete **calcitonin.** Calcitonin has two major functions:

1. ***Inhibiting osteoclast activity.***
2. ***Increasing the rate of excretion of calcium ions.***

Less calcium enters body fluids because osteoclasts leave the mineral matrix alone, and more calcium leaves body fluids because osteoblasts are stimulated to produce new bone matrix, and calcium ion excretion at the kidneys accelerates. The net result is a decline in the calcium ion concentration in body fluids, restoring homeostasis.

By providing a calcium reserve, the skeleton plays the primary role in the homeostatic maintenance of normal calcium ion concentrations in body fluids. This function can have a direct effect on the shape and strength of the bones in the skeleton. When large numbers of calcium ions are mobilized, the bones become weaker; when calcium salts are deposited, bones become denser and

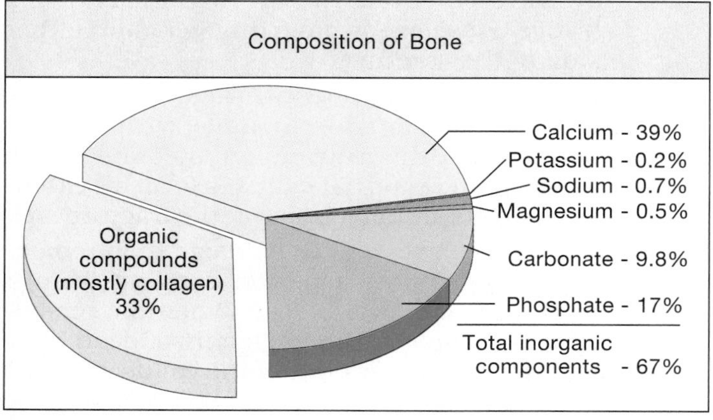
Composition of Bone
Calcium - 39%
Potassium - 0.2%
Sodium - 0.7%
Magnesium - 0.5%
Carbonate - 9.8%
Phosphate - 17%
Total inorganic components - 67%
Organic compounds (mostly collagen) 33%

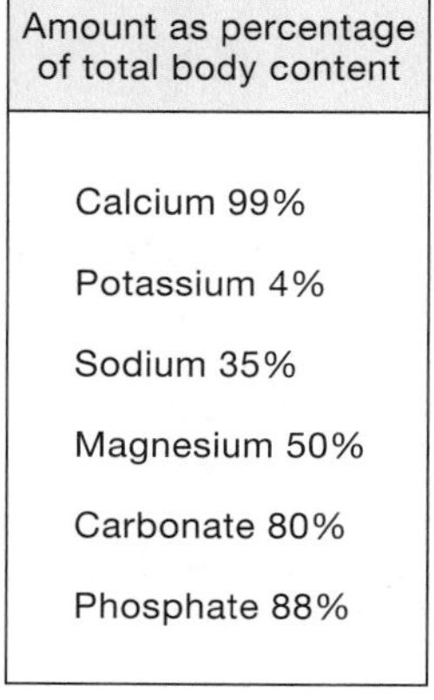
Amount as percentage of total body content
Calcium 99%
Potassium 4%
Sodium 35%
Magnesium 50%
Carbonate 80%
Phosphate 88%

• **FIGURE 6-10**
A Chemical Analysis of Bone

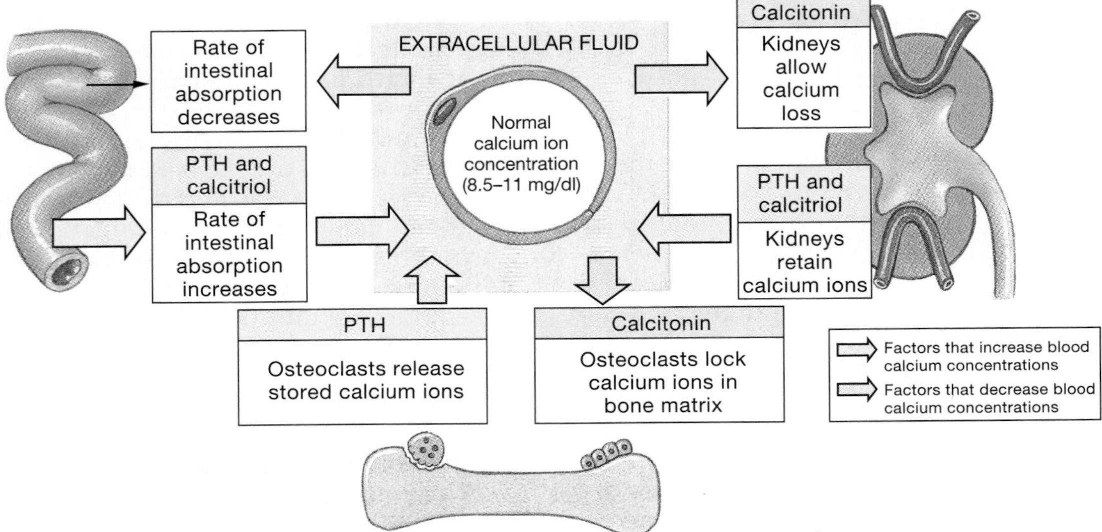
EXTRACELLULAR FLUID
Rate of intestinal absorption decreases
PTH and calcitriol
Rate of intestinal absorption increases
Normal calcium ion concentration (8.5–11 mg/dl)
Calcitonin
Kidneys allow calcium loss
PTH and calcitriol
Kidneys retain calcium ions
PTH
Osteoclasts release stored calcium ions
Calcitonin
Osteoclasts lock calcium ions in bone matrix
Factors that increase blood calcium concentrations
Factors that decrease blood calcium concentrations

● **FIGURE 6-11**
Factors That Alter the Concentration of Calcium Ions in Body Fluids

stronger. Because the bone matrix contains protein fibers, as well as mineral deposits, changes in mineral content do not necessarily affect the shape of the bone. In *osteomalacia* (os-tē-ō-ma-LĀ-shē-uh; *malakia*, softness) the bones appear normal although they are relatively weak and flexible because of poor mineralization. *Rickets*, a form of osteomalacia affecting children, usually results from a vitamin D_3 deficiency caused by inadequate sunlight exposure and inadequate dietary supply. In rickets the bones are so poorly mineralized that they become very flexible. Affected individuals develop a bowlegged appearance as the thigh and leg bones bend under the weight of the body.

 Why does a child with rickets have difficulty walking?

 What effect would increased PTH secretion have on blood calcium levels?

 How does the hormone calcitonin help to lower blood calcium levels?

❑ Fracture Repair

Figure 6-12

Despite its mineral strength, bone may crack or even break if subjected to extreme loads, sudden impacts, or stresses from unusual directions. The damage produced constitutes a **fracture.** (See the Focus, *A Classification of Fractures*, on p.192.) Healing of a fracture usually occurs even after severe damage, provided that the blood supply and the cellular components of the endosteum and periosteum survive. Steps in the repair process are illustrated in Figure 6-12●.

Step 1. In even a small fracture, many blood vessels are broken and extensive bleeding occurs. A large clot, or **fracture hematoma,** soon closes off the injured vessels and leaves a fibrous meshwork in the damaged area. The disruption of circulation kills osteocytes around the fracture, increasing the size of the area affected. Dead bone soon extends along the shaft in either direction from the break.

Step 2. In an adult, the cells of the periosteum and endosteum are relatively inactive. When a fracture occurs, the cells of the intact endosteum and periosteum undergo rapid mitoses, and the daughter cells migrate into the fracture zone. An enlarged collar, or **external callus** (*callum*, hard skin), forms, encircling the bone at the level of the fracture. An extensive **internal callus** organizes within the marrow cavity and between the broken ends of the shaft. At the center of the external callus, cells differentiate into chondrocytes and produce blocks of cartilage. At the edges of each callus the cells differentiate into osteoblasts and begin creating a bridge between the bone fragments on either side of the fracture.

Step 3. As the repair continues, osteoblasts replace the central cartilage with spongy bone. When this conversion is complete the external and internal calluses form an extensive and continuous brace at the fracture site.

Step 4. Osteoclasts and osteoblasts then remodel the region for a period ranging from 4 months to well over a year. When the remodeling is complete, the fragments of dead bone and the trabecular bone of the calluses will be gone, and only living compact bone will remain.

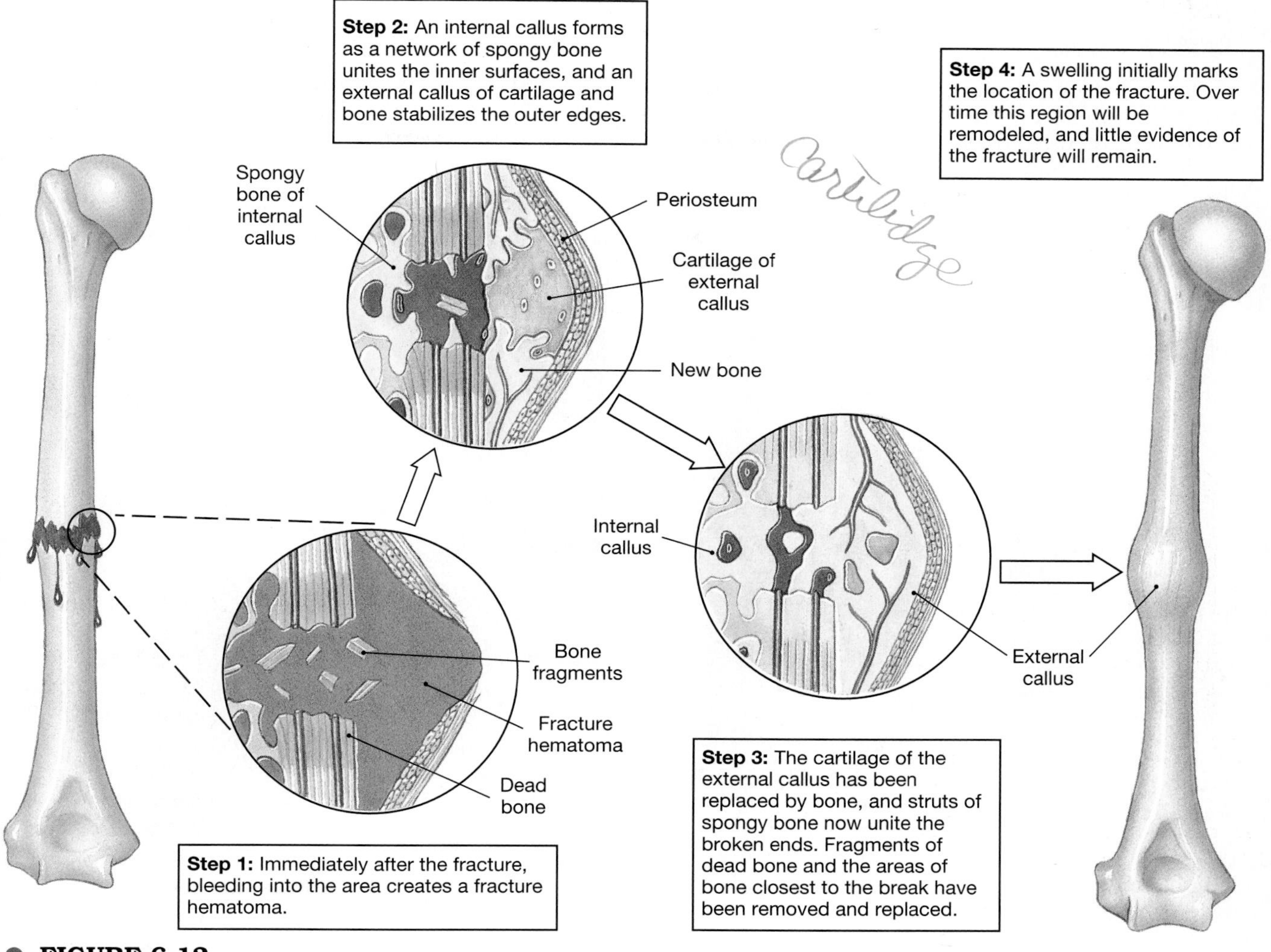
Step 2: An internal callus forms as a network of spongy bone unites the inner surfaces, and an external callus of cartilage and bone stabilizes the outer edges.
Step 4: A swelling initially marks the location of the fracture. Over time this region will be remodeled, and little evidence of the fracture will remain.
Spongy bone of internal callus
Periosteum
Cartilage of external callus
New bone
Internal callus
External callus
Bone fragments
Fracture hematoma
Dead bone
Step 3: The cartilage of the external callus has been replaced by bone, and struts of spongy bone now unite the broken ends. Fragments of dead bone and the areas of bone closest to the break have been removed and replaced.
Step 1: Immediately after the fracture, bleeding into the area creates a fracture hematoma.

FIGURE 6-12
Steps in the Repair of a Fracture

The repair may be "good as new," with no indications that a fracture ever occurred, but often the bone will be slightly thicker and stronger than normal at the fracture site. Under comparable stresses, a second fracture will usually occur at a different site.

Comparable events occur after more complex breaks or even after transplantation of a bone fragment. Each year in the United States, roughly 200,000 people receive *bone grafts* to stimulate bone repair. The bone fragments are often taken from another part of the body, such as the hip. However, because the inserted bone is ultimately destroyed and replaced, dead and sterilized bone fragments from other donors or even other species of animals can be used to establish a framework for the repair process. In situations where the fracture is so severe that normal repair processes cannot occur, even with bone grafts, surgeons may rely on synthetic bone or the use of strong electrical fields to stimulate osteoblast activity.
AM *Stimulation of Bone Growth and Repair*

Aging and the Skeletal System

The bones of the skeleton become thinner and relatively weaker as a normal part of the aging process. Inadequate ossification is called **osteopenia** (os-tē-ō-PĒ-nē-uh; *penia*, lacking), and all people become slightly osteopenic as they age. This reduction in bone mass begins between the ages of 30 and 40. Over this period osteoblast activity begins to decline while osteoclast activity continues at previous levels. Once the reduction begins, women lose roughly 8 percent of their skeletal mass every decade, whereas the skeletons of men deteriorate at the slower rate of about 3 percent per decade. All parts of the skeleton are not equally affected. Epiphyses, vertebrae, and the jaws lose more than their fair share, resulting in fragile limbs, a reduction in height, and the loss of teeth.

When reduction in bone mass is sufficient to compromise normal function, the condition is

FOCUS

A Classification of Fractures

Fractures are classified according to their external appearance, the site of the fracture, and the nature of the crack or break in the bone. Important fracture types are indicated below, with representative X-rays. Many fractures fall into more than one category. For example, a Colles' fracture is a transverse fracture, but depending on the injury, it may also be a comminuted fracture that can be either open or closed. **Closed,** or *simple*, fractures are completely internal; they do not involve a break in the skin. **Open,** or *compound*, fractures project through the skin; they are more dangerous because of the possibility of infection or uncontrolled bleeding.

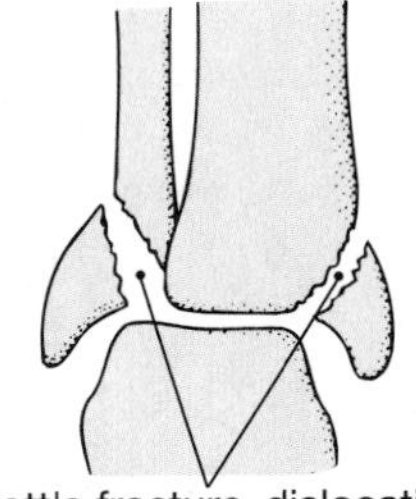
Pott's fracture–dislocation of the tibia and fibula

A **Pott's fracture** occurs at the ankle and affects both bones of the lower leg.

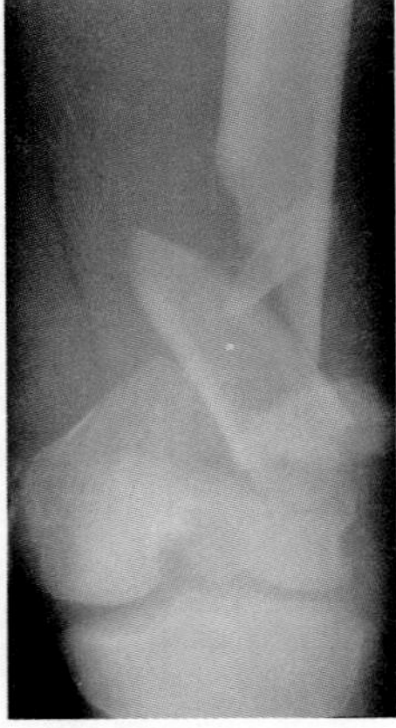
Comminuted fracture of distal femur

Comminuted fractures shatter the affected area into a multitude of bony fragments.

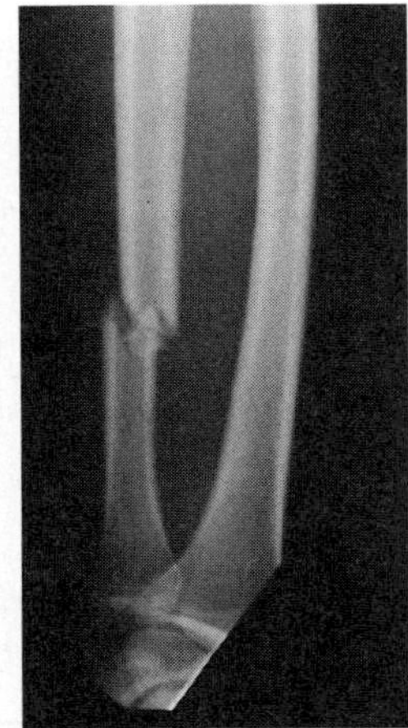

Transverse fractures break a shaft bone across its long axis.

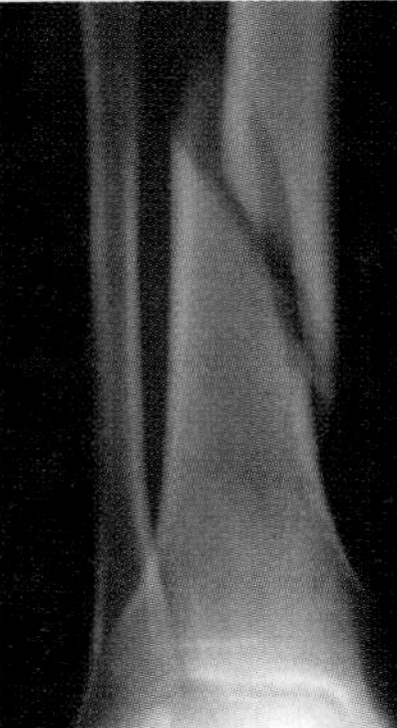
Spiral fracture of tibia

Spiral fractures, produced by twisting stresses, spread along the length of the bone.

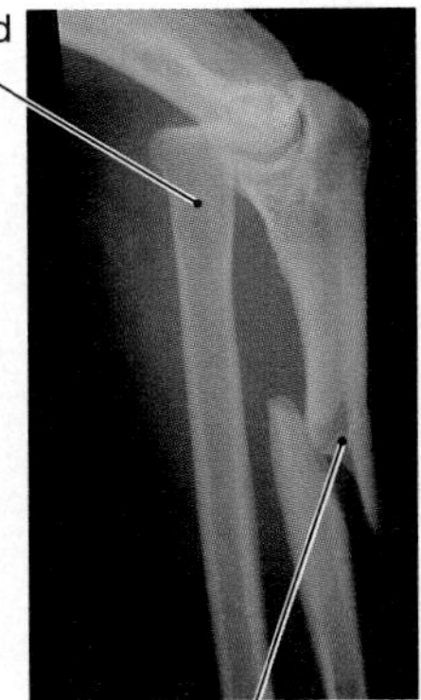
Dislocated radius
Displaced ulnar fracture

Displaced fractures produce new and abnormal arrangements of bony elements.

Nondisplaced fractures retain the normal alignment of the bone elements or fragments.

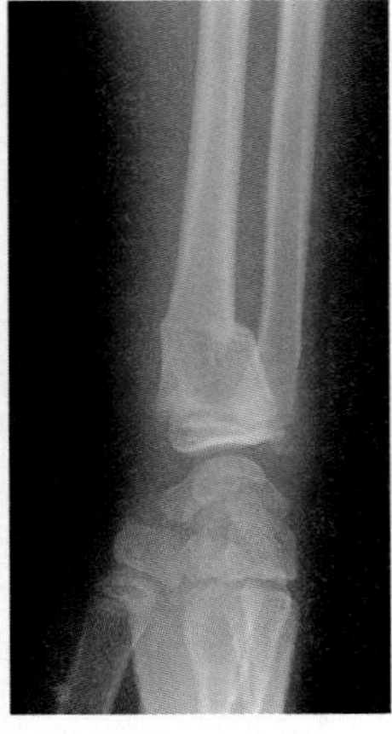

A **Colles' fracture** is a break in the distal portion of the radius, the slender bone of the forearm; it is often the result of reaching out to cushion a fall.

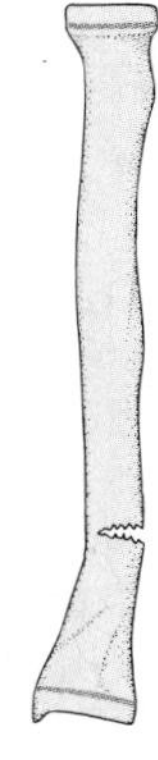

In a **greenstick fracture,** only one side of the shaft is broken, and the other is bent; this type usually occurs in children, whose long bones have yet to fully ossify.

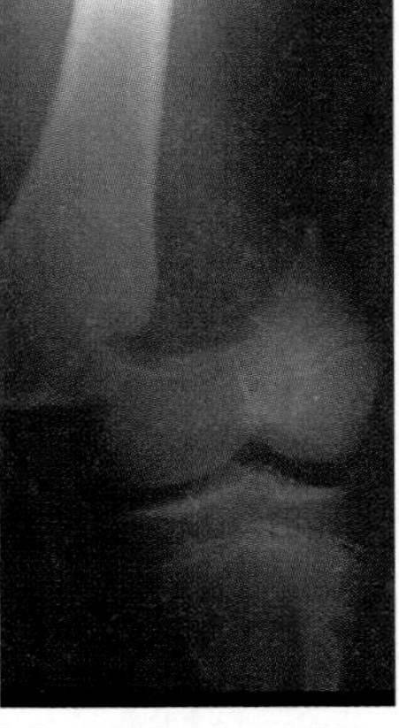
Epiphyseal fracture of femur

Epiphyseal fractures usually occur where the matrix is undergoing calcification and chondrocytes are dying. A clean transverse fracture along this line usually heals well. Fractures between the epiphysis and the epiphyseal plate can permanently halt further longitudinal growth unless carefully treated; often surgery is required.

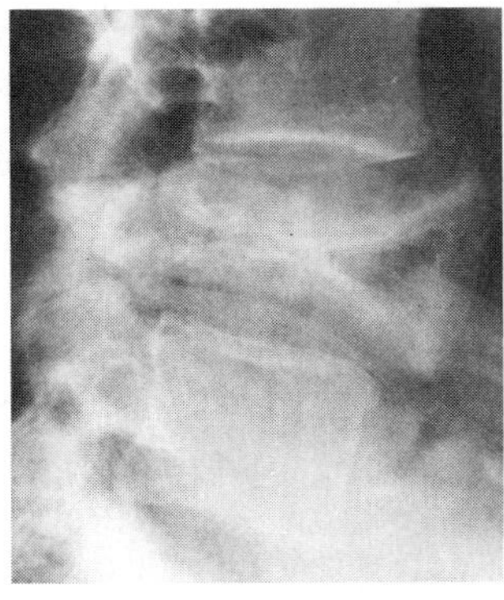
Compression fracture of vertebra

Compression fractures occur in vertebrae subjected to extreme stresses, as when landing on your seat after a fall.

known as **osteoporosis** (os-tē-ō-por-Ō-sis; *porosus*, porous). The fragile bones that result are likely to break when exposed to stresses that would be easily tolerated by younger individuals. For example, a hip fracture may occur when a woman in her nineties tries to stand up. Any fractures that do occur lead to a loss of independence and an immobility that further weakens the skeleton.

Sex hormones, primarily estrogens in females and androgens in males, are important in maintaining normal rates of bone deposition. A significant percentage of women over age 45 suffer from osteoporosis. The condition accelerates after menopause, owing to the decline in circulating estrogens after menopause. Because men continue to produce androgens until relatively late in life, severe osteoporosis is less common in males below age 60.

Osteoporosis can also develop as a secondary effect of many cancers. Cancers of the bone marrow, breast, or other tissues release a chemical known as **osteoclast-activating factor.** This compound increases both the number and activity of osteoclasts, and produces a severe osteoporosis. Osteoporosis is discussed further in the *Applications Manual.* AM *Osteoporosis and Age-Related Skeletal Abnormalities*

Anatomy of Skeletal Elements

The human skeleton contains 206 or more bones. We can divide these bones into six broad categories according to their individual shapes.

A Classification of Bones

Figure 6-13

Refer to Figure 6-13• as we describe the anatomical classification of bones.

1. **Long bones** are characterized by diaphysis and epiphysis portions and contain a marrow cavity (Figure 6-13a•). Long bones are found in the upper arm and forearm, thigh and lower leg, palms, soles, fingers, and toes.
2. **Short bones** are boxlike in appearance (Figure 6-13b•). Their external surfaces are covered by compact bone, but the interior contains spongy bone. Examples of short bones include the *carpals* (wrists) and *tarsals* (ankles).
3. **Flat bones** have thin, roughly parallel surfaces of compact bone. In structure a flat bone

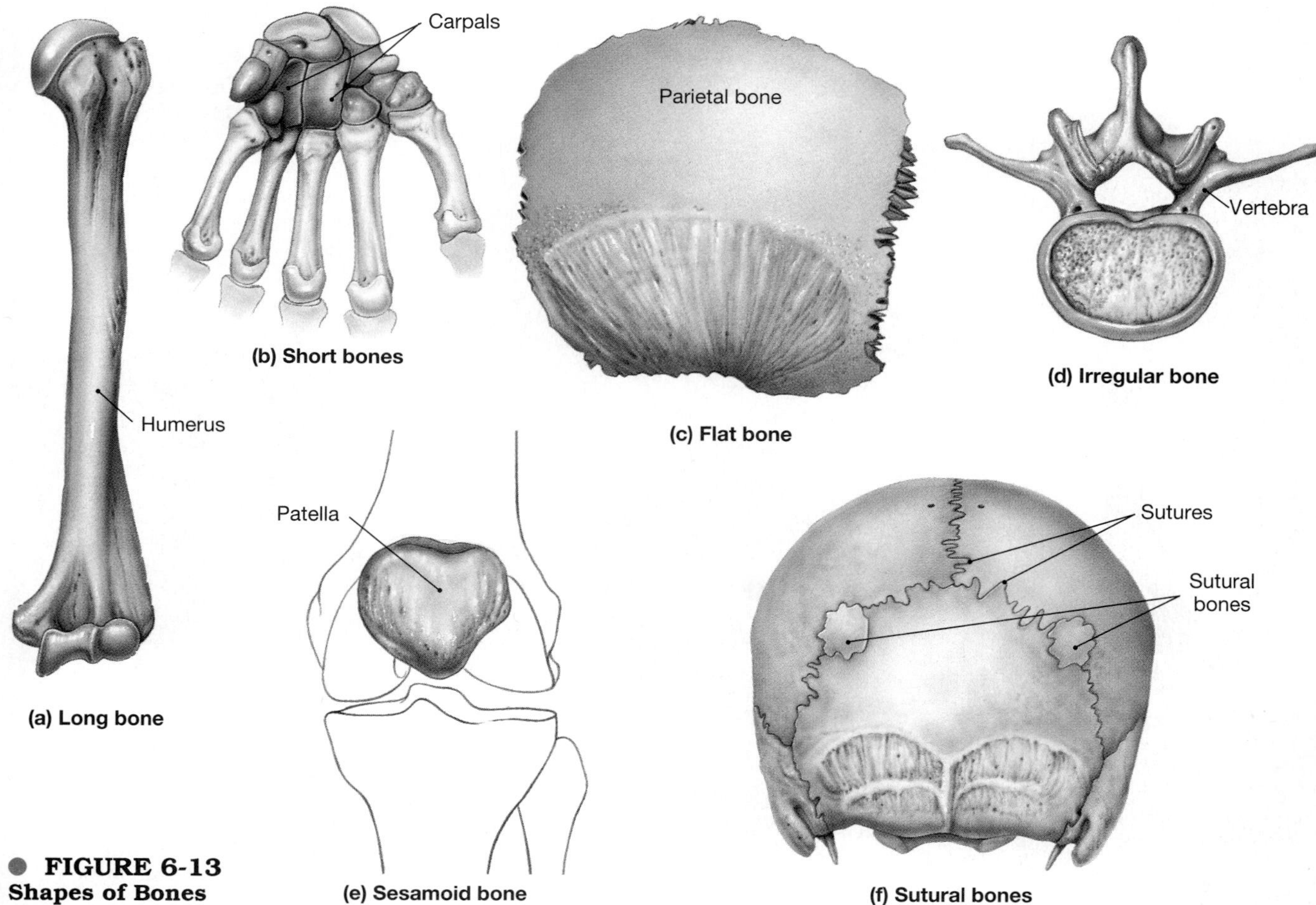

Carpals
(b) Short bones
Parietal bone
(c) Flat bone
Vertebra
(d) Irregular bone
Humerus
(a) Long bone
Patella
(e) Sesamoid bone
Sutures
Sutural bones
(f) Sutural bones

• **FIGURE 6-13**
Shapes of Bones

resembles a spongy bone sandwich; such bones are strong but relatively light. Flat bones form the roof of the skull (Figure 6-13c●), the sternum, the ribs, and the scapula. They provide protection for underlying soft tissues and offer an extensive surface area for the attachment of skeletal muscles. Special terms are used when describing the flat bones of the skull, such as the parietals. Their relatively thick layers of compact bone are called the *internal* and *external tables*, and the layer of spongy bone between the tables is called the *diploë* (DIP-lō-ē).

4. **Irregular bones** have complex shapes with short, flat, notched, or ridged surfaces (Figure 6-13d●). Their internal structure is equally varied. The spinal vertebrae and several bones in the skull are examples of irregular bones.
5. **Sesamoid bones** are usually small, round, and flat (Figure 6-13e●). They develop inside tendons and are most often encountered near joints at the knee, the hands, and the feet. Few individuals have sesamoid bones at every possible location, but everyone has sesamoid **patellae** (pa-TEL-ē), or kneecaps.
6. **Sutural bones** or *Wormian bones* are small, flat, irregularly shaped bones found between the flat bones of the skull in a suture line (Figure 6-13f●). There are individual variations in the number, shape, and position of the sutural bones. Their borders are like puzzle pieces and may range in size from a grain of sand to the size of a quarter.

❑ Bone Markings (Surface Features)

Figure 6-14

Each bone in the body has a distinctive shape and characteristic external and internal features. Elevations or projections form where tendons and ligaments attach and where adjacent bones articulate. Depressions, grooves, and tunnels in bone indicate sites where blood vessels and nerves lie alongside or penetrate the bone. Detailed examination of these **bone markings,** or *surface features,* can yield an abundance of anatomical information. For example, anthropologists, criminologists, and pathologists can often determine the size, weight,

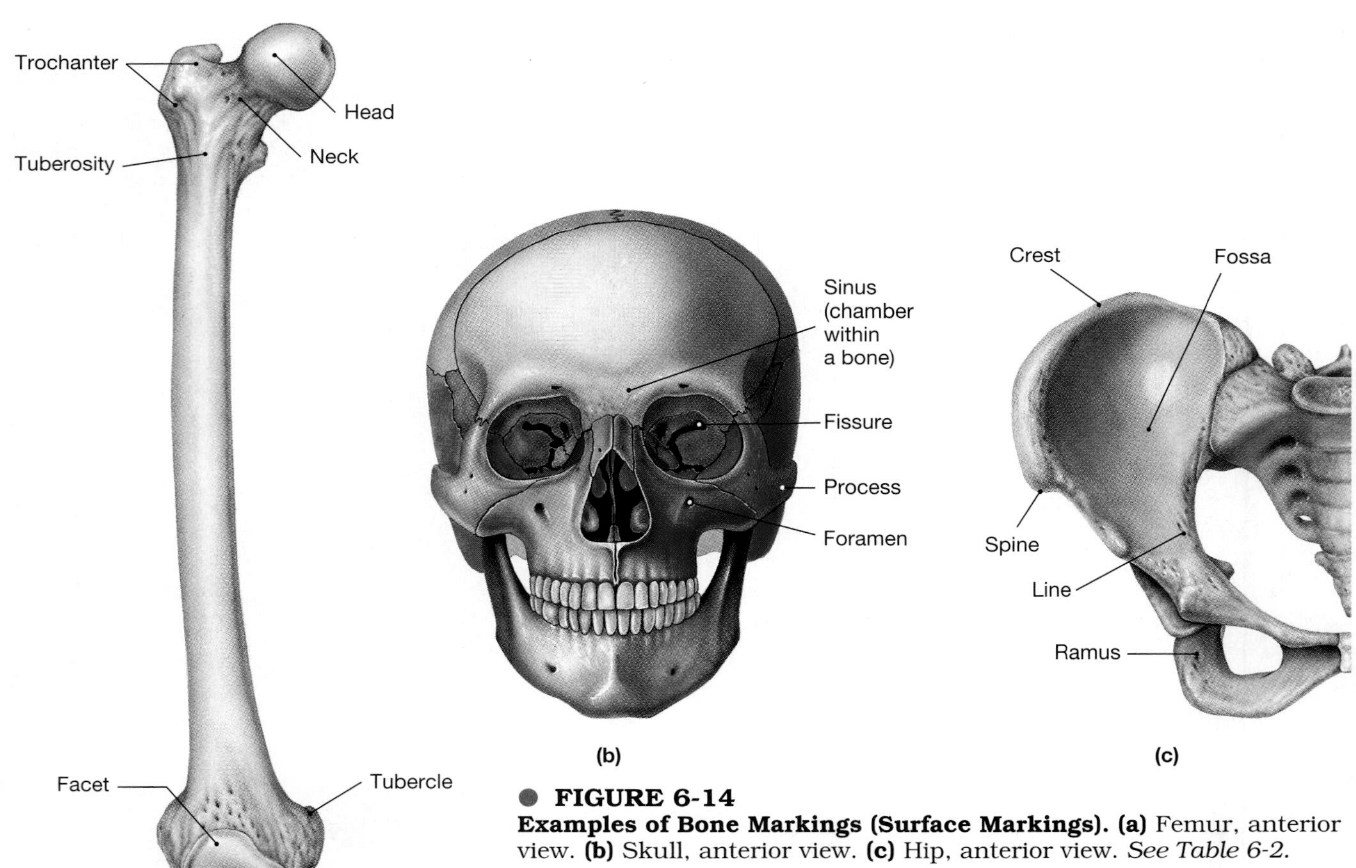

Trochanter
Head
Tuberosity
Neck
Facet
Tubercle
Condyle
(a)
Sinus (chamber within a bone)
Fissure
Process
Foramen
(b)
Crest
Fossa
Spine
Line
Ramus
(c)

● **FIGURE 6-14**
Examples of Bone Markings (Surface Markings). (a) Femur, anterior view. **(b)** Skull, anterior view. **(c)** Hip, anterior view. *See Table 6-2.*

TABLE 6-2 An Introduction to Skeletal Terminology

General Description	*Anatomical Term*	*Definition*
Elevations and projections (general)	Process	Any projection or bump
	Ramus	An extension of a bone making an angle to the rest of the structure
Processes formed where tendons or ligaments attach	Trochanter	A large, rough projection
	Tuberosity	A smaller, rough projection
	Tubercle	A small, rounded projection
	Crest	A prominent ridge
	Line	A low ridge
Processes formed for articulation with adjacent bones	Head	The expanded articular end of an epiphysis, separated from the shaft by a narrower neck
	Condyle	A smooth, rounded articular process
	Trochlea	A smooth, grooved articular process shaped like a pulley
	Facet	A small, flat articular surface
	Spine	A pointed process
Depressions	Fossa	A shallow depression
	Sulcus	A narrow groove
Openings	Foramen	A rounded passageway for blood vessels and/or nerves
	Fissure	An elongated cleft
	Meatus	A canal leading through the substance of a bone
	Sinus or antrum	A chamber within a bone, normally filled with air

sex, and general appearance of an individual on the basis of incomplete skeletal remains. (This topic will be discussed further in Chapter 8.)

We will ignore minor variations of individual bones and focus on prominent features that identify the bone. These markings are useful because they provide fixed landmarks that can help in determining the position of the soft tissue components of other systems. Specific anatomical terms are used to describe the various elevations and depressions. Bone marking terminology is presented in Table 6-2 and illustrated in Figure 6-14●.

Integration with Other Systems

Figure 6-15

Although the bones may seem inert, you should now realize that they are quite dynamic structures. The entire skeletal system is intimately associated with other systems. The bones of the skeleton are attached to the muscular system, extensively connected to the cardiovascular and lymphatic systems, and largely under the physiological control of the endocrine system. The digestive and excretory systems also play important roles in providing the calcium and phosphate minerals needed for bone growth. In return, the skeleton represents a reserve of calcium, phosphate, and other minerals that can compensate for changes in the dietary supply of those ions. The functional relationships between the skeletal system and other systems are diagrammed in Figure 6-15●.

At what point in fracture repair would you find an external callus?

Why is the condition known as osteoporosis more common in older women than older men?

X-rays of a child's skeleton indicate bones forming in the tendons of several joints. What kind of bones are these?

INTEGUMENTARY SYSTEM

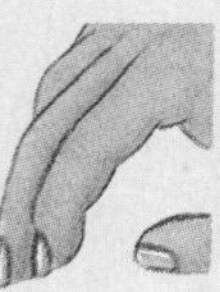

Synthesizes vitamin D_3, essential for calcium and phosphorus absorption (bone maintenance and growth)

Provides structural support

MUSCULAR SYSTEM

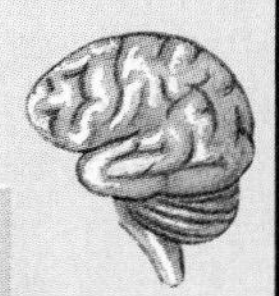

Stabilizes bone positions; tension in tendons stimulates bone growth and maintenance

Provides calcium needed for normal muscle contraction; bones act as levers to produce body movements

NERVOUS SYSTEM

Regulates bone position by controlling muscle contractions

Provides calcium for neural function; protects brain, spinal cord; receptors at joints provide information about body position

THE SKELETAL SYSTEM

ENDOCRINE SYSTEM

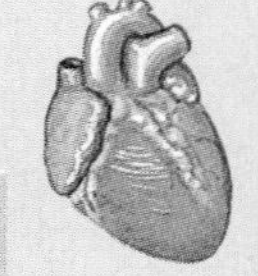

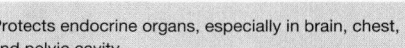

Skeletal growth regulated by growth hormone, thyroid hormones, and sex hormones; calcium mobilization regulated by parathyroid hormone and calcitonin

Protects endocrine organs, especially in brain, chest, and pelvic cavity

CARDIOVASCULAR SYSTEM

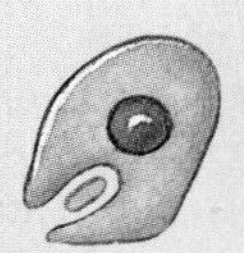

Provides oxygen, nutrients, hormones, blood cells; removes waste products and carbon dioxide

Provides calcium needed for cardiac muscle contraction; blood cells produced in bone marrow

LYMPHATIC SYSTEM

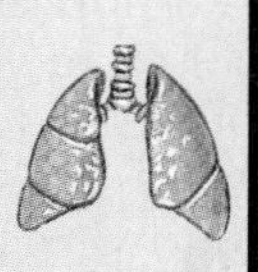

Lymphocytes assist in the defense and repair of bone following injuries

Lymphocytes and other cells of the immune response are produced and stored in bone marrow

RESPIRATORY SYSTEM

Provides oxygen and eliminates carbon dioxide

Movements of ribs important in breathing; axial skeleton surrounds and protects lungs

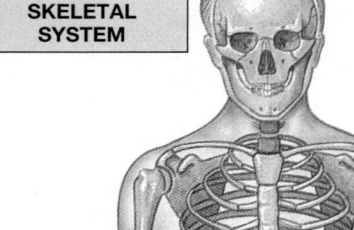

FOR ALL SYSTEMS

Provides mechanical support
Stores energy reserves
Stores calcium and phosphate reserves

DIGESTIVE SYSTEM

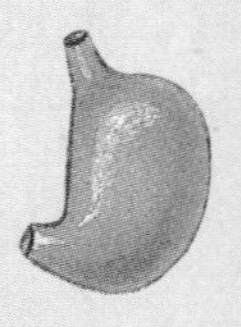

Provides nutrients, calcium, and phosphate

Ribs protect portions of liver, stomach, and intestines

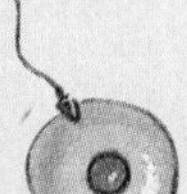

Sexual hormones stimulate growth and maintenance of bones; surge of sex hormones at puberty causes acceleration of growth and closure of epiphyseal plates

Pelvis protects reproductive organs of female, protects portion of ductus deferens and accessory glands in male

REPRODUCTIVE SYSTEM

URINARY SYSTEM

Conserves calcium and phosphate needed for bone growth, disposes of waste products

Axial skeleton provides some protection for kidneys and ureters; pelvis protects urinary bladder and proximal urethra

● **FIGURE 6-15**
Functional Relationships between the Skeletal System and Other Systems

Selected Clinical Terminology

Terms Discussed in This Chapter

acromegaly: A condition caused by excess secretion of growth hormone after puberty. Skeletal abnormalities develop affecting the cartilages and various small bones. *(p. 188* and *AM)*
external callus: A toughened layer of connective tissue that encircles and stabilizes a bone at a fracture site. *(p. 190)*
fracture: A crack or break in a bone. *(p. 190)*
fracture hematoma: A large blood clot that closes off the injured vessels and leaves a fibrous meshwork in the damaged area. *(p. 190)*
gigantism: A condition resulting from an overproduction of growth hormone before puberty. *(p. 188)*
internal callus: A bridgework of trabecular bone that unites the broken ends of a bone on the marrow side of the fracture. *(p. 190)*
Marfan's syndrome: An inherited condition linked to defective production of a connective tissue glycoprotein. Extreme height and long, slender limbs are the most obvious physical indications of this disorder. *(p. 188* and *AM)*
osteoclast-activating factor: A compound released by cancers of the bone marrow, breast, or other tissues. It produces a severe osteoporosis. *(p. 193)*
osteomalacia (os-tē-ō-ma-LĀ-shē-uh): A softening of bone due to a decrease in the mineral content. *(p. 190)*
osteopenia (os-tē-ō-PĒ-nē-uh): Inadequate ossification, leading to thinner, weaker bones. *(p. 191)*
osteoporosis (os-tē-ō-por-Ō-sis): A reduction in bone mass to a degree that compromises normal function. *(p. 193* and *AM)*
pituitary growth failure: A type of dwarfism caused by inadequate growth hormone production. *(p. 188)*
rickets: A childhood disorder that reduces the amount of calcium salts in the skeleton; often characterized by a bowlegged appearance because the leg bones bend. *(p. 190)*
scurvy: A condition involving weak, brittle bones; it results from a vitamin C deficiency. *(p. 188)*

AM *Additional Terms Discussed in the Applications Manual*

achondroplasia (ā-kon-drō-PLĀ-sē-uh): A condition resulting from abnormal epiphyseal activity; the epiphyseal plates grow unusually slowly, and the individual develops short, stocky limbs. The trunk is normal in size, and sexual and mental development remain unaffected.
hyperostosis (hī-per-os-TŌ-sis): The excessive formation of bone tissue.
osteogenesis imperfecta (im-per-FEK-tuh): An inherited condition affecting the organization of collagen fibers. Osteoblast function is impaired, growth is abnormal, and the bones are very fragile, leading to progressive skeletal deformation and repeated fractures.
osteomyelitis (os-tē-ō-mī-e-LĪ-tis): A painful infection in a bone, usually caused by bacteria.
osteopetrosis (os-tē-ō-pe-TRŌ-sis): A condition caused by a decrease in osteoclast activity, causing increased bone mass and various skeletal deformities.
Paget's disease *(osteitis deformans)* (os-tē-Ī-tis de-FŌR-manz): A condition characterized by gradual deformation of the skeleton.

CHAPTER REVIEW

STUDY OUTLINE

INTRODUCTION, p. 176

1. The skeletal system includes the bones of the skeleton and the cartilages, ligaments, and other connective tissues that stabilize or interconnect bones. Its functions include: structural support, storage, blood cell production, protection, and leverage.

STRUCTURE OF BONE, p. 176

1. **Osseous tissue** is a supporting connective tissue with a solid **matrix.**

Histological Organization, p. 176

2. There are two types of bone: **dense** *(compact)* **bone** and **spongy** *(cancellous)* **bone.** Both contain **osteocytes** in **lacunae.** Layers of calcified matrix are **lamellae,** and **canaliculi** within and between the lamellae interconnect the lacunae. *(Figure 6-1a,b)*

Compact and Spongy Bone, p. 178

3. The basic functional unit of compact bone is the **osteon,** containing osteocytes arranged around a **central canal.**
4. Spongy bone contains **trabeculae,** often in an open network. *(Figure 6-1c)*
5. Compact bone is found where stresses come from a limited range of directions; spongy bone is located where stresses are few or come from many different directions. *(Figure 6-2)*

The Periosteum and Endosteum, p. 180

6. A bone is covered by a **periosteum** and lined with an **endosteum. Osteoclasts** dissolve the bony matrix through **osteolysis.** Other cells called **osteoblasts** synthesize the matrix in the process of **osteogenesis.** (*Figures 6-1, 6-3*)

BONE DEVELOPMENT AND GROWTH, p. 181

1. **Ossification** is the process of converting other tissues to bone; **calcification** is the process of depositing calcium salts within a tissue.

Intramembranous Ossification, p. 181

2. **Intramembranous ossification** begins when osteoblasts differentiate within connective tissue and can produce spongy or compact bone. (*Figures 6-4, 6-5*)

Endochondral Ossification, p. 182

3. **Endochondral ossification** begins by forming a cartilaginous model that is gradually replaced by bone. (*Figures 6-5, 6-6*)
4. There are differences between bones and between individuals regarding the timing of epiphyseal closure. (*Figure 6-7*)
5. Bone diameter increases through appositional growth. (*Figure 6-8*)

The Blood and Nerve Supply, p. 186

6. An extensive supply of blood is supplied to bone through three major sets of blood vessels. (*Figure 6-9*)

THE DYNAMIC NATURE OF BONE, p. 187

1. The organic and mineral components of bone are continually recycled and renewed through the process of remodeling.

Exercise Effects on Bone, p. 187

2. The shapes and thicknesses of bones reflect the stresses applied to them.

Hormonal and Nutritional Effects on Bone, p. 187

3. Normal osteogenesis requires a reliable source of minerals, vitamins, and hormones.
4. **Growth hormone** and **thyroxine** stimulate bone growth. (*Table 6-1*)

The Skeleton as a Calcium Reserve, p. 189

5. Calcium is the most common mineral in the human body, with roughly 99 percent of it located in the skeleton. (*Figure 6-10*)
6. Interactions among the bones, intestinal tract, and kidneys affect calcium ion concentrations. (*Figure 6-11*)
7. Two hormones, **calcitonin** and **parathyroid hormone (PTH),** regulate calcium ion homeostasis. Calcitonin leads to a decline in calcium concentrations, whereas parathyroid hormone increases calcium concentrations.

Fracture Repair, p. 190

8. A **fracture** is a crack or break in a bone. Repair of a fracture involves the formation of a **fracture hematoma,** an **external callus,** and an **internal callus.** (*Figure 6-12*)
9. Various criteria are used to classify fractures. (*Focus: A Classification of Fractures*)

Aging and the Skeletal System, p. 191

10. Effects of aging on the skeleton can include **osteopenia** and **osteoporosis.**

ANATOMY OF SKELETAL ELEMENTS, p. 193

A Classification of Bones, p. 193

1. Categories of bones include: **long bones, short bones, flat bones, irregular bones, sesamoid bones,** and **sutural bones (Wormian bones).** (*Figure 6-13*)

Bone Markings (Surface Features), p. 194

2. **Bone markings** can be used to identify specific bones within each category. (*Figure 6-14; Table 6-2*)

INTEGRATION WITH OTHER SYSTEMS, p. 195

1. The skeletal system is dynamically associated with all the other systems of the body. (*Figure 6-15*)

REVIEW QUESTIONS

LEVEL 1 Reviewing Facts and Terms

1. The bones of the skeleton store energy reserves as lipids in areas of:
   (a) red marrow
   (b) yellow marrow
   (c) the matrix of bone tissue
   (d) the ground substance
2. Two-thirds of the weight of bone is accounted for by:
   (a) crystals of calcium phosphate
   (b) collagen fibers
   (c) osteocytes
   (d) calcium carbonate
3. The two types of osseous tissue are:
   (a) compact bone and spongy bone
   (b) dense bone and compact bone
   (c) spongy bone and cancellous bone
   (d) a, b, and c are correct
4. The basic functional units of mature compact bone are:
   (a) lacunae
   (b) osteocytes
   (c) osteons
   (d) canaliculi

5. Unlike dense bone, cancellous bone contains concentric lamellae that form struts or plates called:
(a) canaliculi (b) canals of Volkmann
(c) osteons (d) trabeculae

6 The vitamins essential for normal adult bone maintenance and repair are:
(a) A and E (b) C and D
(c) B and E (d) B complex and K

7. The hormones that coordinate the storage, absorption, and excretion of calcium ions are:
(a) growth hormone and thyroxine
(b) calcitonin and parathyroid hormone
(c) calcitriol and cholecalciferol
(d) estrogens and adrogens

8. Inadequate ossification due to the aging process is called:
(a) osteomalacia (b) osteolysis
(c) osteopenia (d) osteogenesis

9. The primary reason that osteoporosis accelerates after menopause in women is:
(a) reduced levels of circulating estrogens
(b) reduced levels of vitamin C
(c) diminished osteoclast activity
(d) increased osteoblast activity

10. A child suffering from rickets would have:
(a) oversized facial bones
(b) long arms and legs
(c) weak, brittle bones
(d) bowlegs

11. What are the five primary functions of the skeletal system?

12. List the four distinctive cell populations that are found in osseous tissue.

13. What is the functional difference between an osteoblast and an osteoclast?

14. What are the three primary functions of the periosteum?

15. What is the primary difference between intramembranous ossification and endochondral ossification?

16. What three major sets of blood vessels develop in typical bones in the skeleton?

17. What nutritional factors are essential for normal bone growth and maintenance? What hormonal factors are necessary for normal bone growth and maintenance?

18. What three organs and/or tissues interact to assist in the regulation of calcium ion concentration in body fluids?

19. What major effects of parathyroid hormone oppose those of calcitonin?

20. What are the major functions of the hormone calcitonin?

LEVEL 2 Reviewing Concepts

21. If there are no osteons in spongy bone, how do nutrients reach the osteocytes?

22. Why are stresses or impacts to the side of the shaft in a long bone more dangerous than stress applied to the long axis of the shaft?

23. Why do extended periods of inactivity cause degenerative changes in the skeleton?

24. What is the functional relationship between the skeleton and the digestive and excretory systems?

25. During the growth of a long bone, how is the epiphysis forced farther from the shaft?

26. What role do sex hormones in males and females play in bone growth and development?

27. How might damage to the thyroid gland influence calcium regulation in the body?

28. Why does a second fracture in the same bone usually occur at a site different from that of the first fracture?

29. Why are individuals in their eighties shorter than they were in their forties?

30. How might bone markings be useful in identifying the remains of a criminal who has been shot and killed?

31. What purpose do elevations or projections serve on bones?

LEVEL 3 Critical Thinking and Clinical Applications

32. While playing on her swing set, 10-year-old Sally falls and breaks her right leg. At the emergency room, the doctor tells her parents that the proximal end of the tibia where the epiphysis meets the diaphysis is fractured. The fracture is properly set and eventually heals. During a routine physical when she is 18, Sally learns that her right leg is 2 cm shorter than her left, probably because of her accident. What might account for this difference?

33. Sherry is a pregnant teenager. Her diet before she was pregnant consisted mostly of "junk" food and hasn't changed since she became pregnant. Approximately 8–10 weeks into her pregnancy, she falls and breaks her arm. She thinks that this is unusual, because it wasn't a hard fall. The attending physician in the emergency room informs her that test results reveal that a significant amount of bone demineralization is occurring. Explain what is happening to Sherry.

34. As a result of vitamin D_3 deficiency, would you expect to see changes in blood levels of the hormones calcitonin and PTH? Explain.

35. Why might a person suffering from kidney failure exhibit symptoms similar to those of osteoporosis?

CHAPTER 7

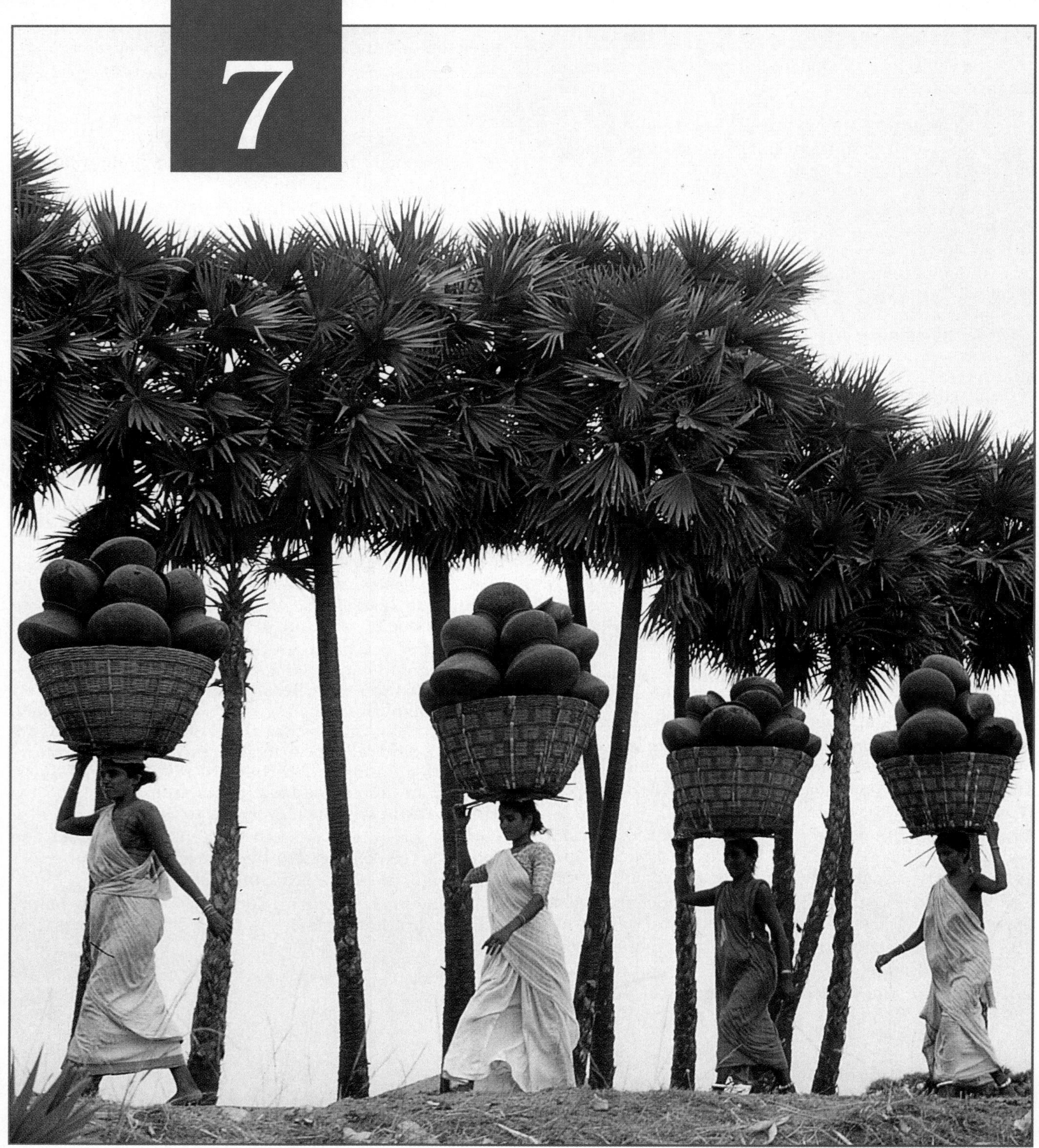

Many societies throughout the world have chosen this sensible method of carrying heavy loads. Rather than bearing them in their arms, and possibly straining their backs, people place bundles on their heads. The weight is thus distributed along the longitudinal axis of the body via the axial skeleton, and one or both hands are free to perform other tasks. Weight bearing is a major function of the axial skeleton, but as you will see, these bones have many other important roles.

The Axial Skeleton

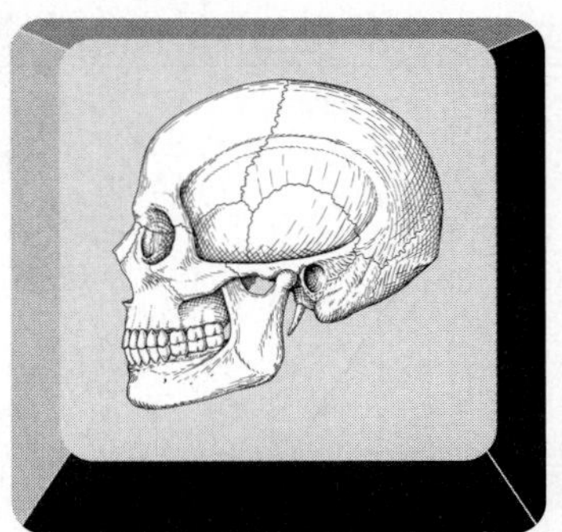

Chapter Outline and Objectives

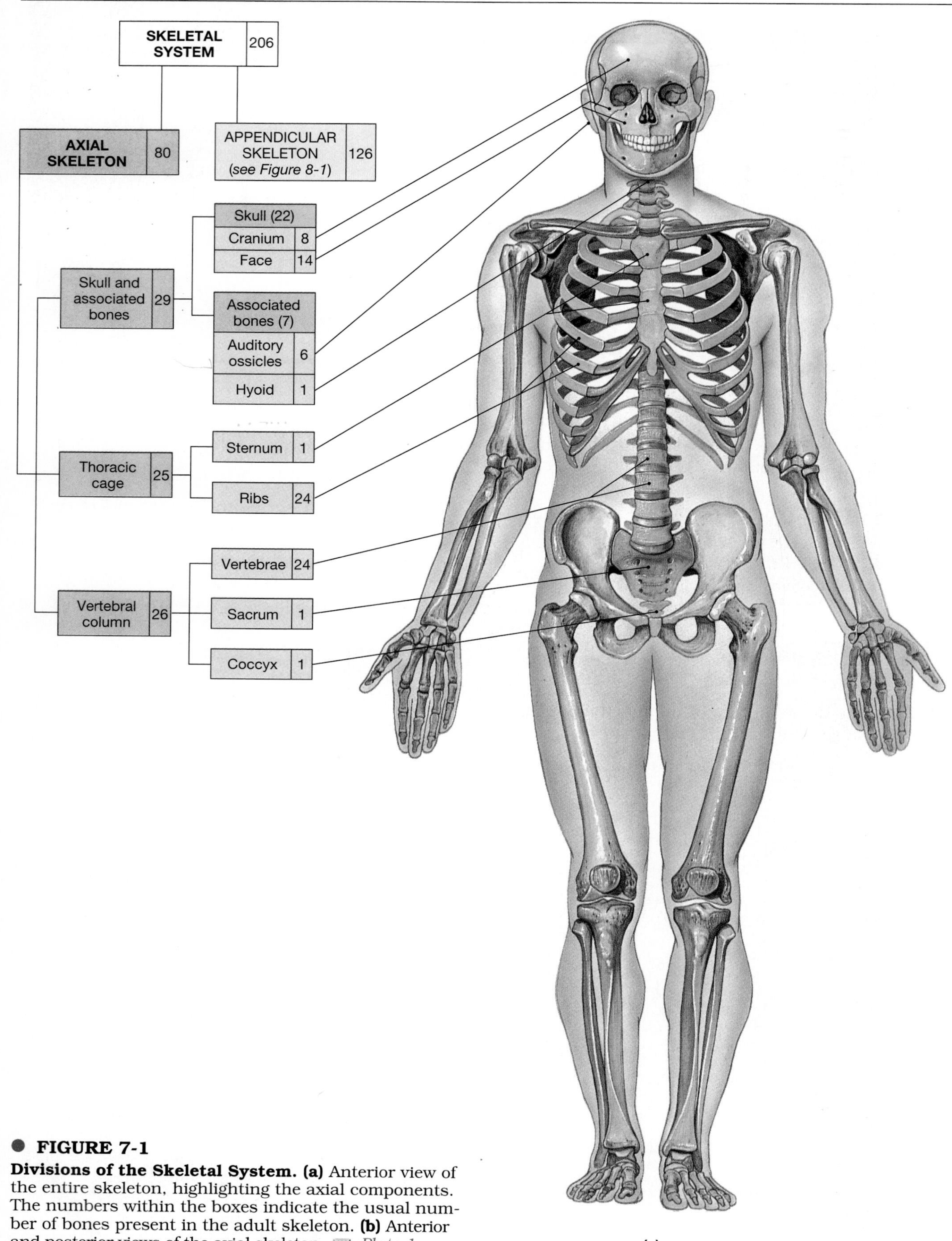
SKELETAL SYSTEM 206
AXIAL SKELETON 80
APPENDICULAR SKELETON (see Figure 8-1) 126
Skull and associated bones 29
Skull (22)
Cranium 8
Face 14
Associated bones (7)
Auditory ossicles 6
Hyoid 1
Thoracic cage 25
Sternum 1
Ribs 24
Vertebral column 26
Vertebrae 24
Sacrum 1
Coccyx 1
(a)

● **FIGURE 7-1**

Divisions of the Skeletal System. (a) Anterior view of the entire skeleton, highlighting the axial components. The numbers within the boxes indicate the usual number of bones present in the adult skeleton. **(b)** Anterior and posterior views of the axial skeleton. AM *Plate 1*

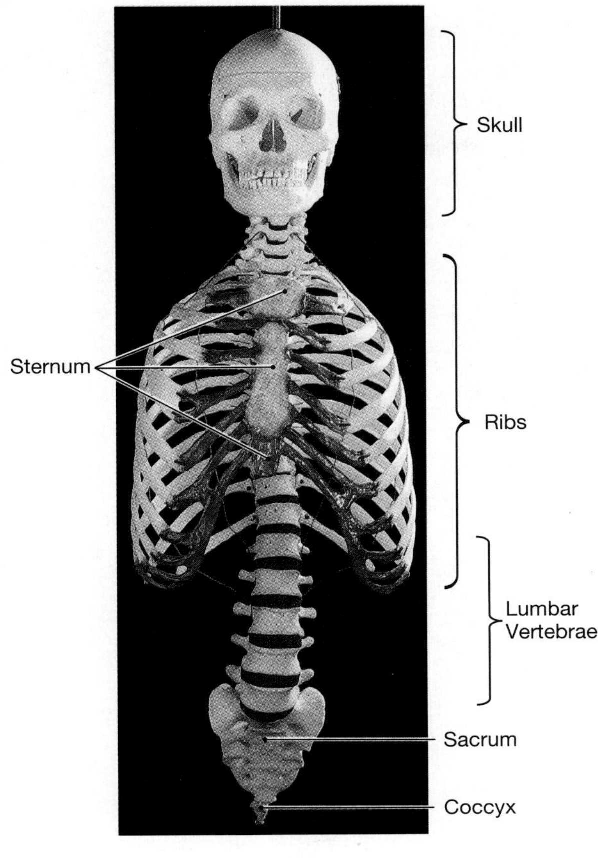
Skull
Sternum
Ribs
Lumbar
Vertebrae
Sacrum
Coccyx

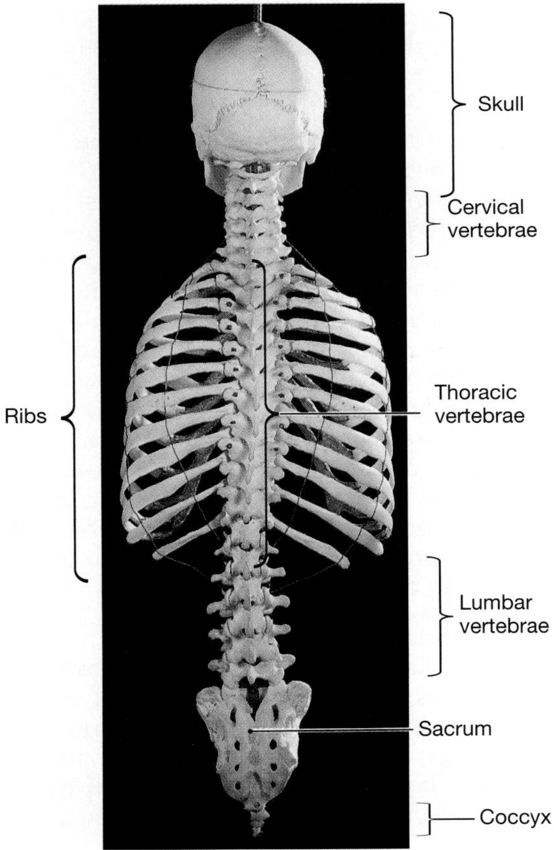
Skull
Cervical
vertebrae
Ribs
Thoracic
vertebrae
Lumbar
vertebrae
Sacrum
Coccyx

(b)

Bone first evolved in fishes roughly 400 million years ago. It was dermal bone, developing in the skin, and it provided a mineral reserve and a protective armor plating for defense against predators. The internal skeletons of these fishes were composed primarily of cartilage, but in the evolutionary line that led toward humans most of those cartilages were replaced by bones.

As you saw in Chapter 6, bones are remarkable structures. They are as strong as or stronger than reinforced concrete, but considerably lighter. [p. 178] Better yet, bones can be remodeled and reshaped to meet metabolic demands or to adapt to changing activity patterns. The basic features of the human skeleton have been shaped by evolution, but the detailed characteristics of each bone reflect the stresses placed upon it. As a result, the skeleton changes in the course of a lifetime. Examples discussed in the last chapter included the proportional changes at puberty and the gradual osteopenia of aging. This chapter will consider additional examples, such as the changes in the shape of the vertebral column during the transition from crawling to walking.

The skeletal system is divided into *axial* and *appendicular divisions;* the components of the axial skeleton are highlighted in Figure 7-1•. The skeletal system includes 206 separate bones and a number of associated cartilages. The **axial skeleton** consists of the bones of the skull, thorax, and vertebral column. These elements form the longitudinal axis of the body. There are 80 bones in the axial skeleton, roughly 40 percent of the bones in the human body. The axial components include:

- The ***skull*** (22 bones: 8 ***cranial bones*** and 14 ***facial bones***).
- Bones associated with the skull (6 ***auditory ossicles*** and the ***hyoid bone***).
- The ***vertebral column*** (24 ***vertebrae***, the ***sacrum,*** and the ***coccyx***).
- The ***thoracic cage*** (24 ***ribs*** and the ***sternum***).

The function of the axial skeleton is to create a framework that supports and protects organs in the dorsal and ventral body cavities. In addition, it provides an extensive surface area for the attachment of muscles that (1) adjust the positions of the head, neck, and trunk; (2) perform respiratory movements; and (3) stabilize or position structures of the appendicular skeleton. The joints of the axial skeleton permit limited movement, but they are very strong and heavily reinforced with ligaments.

The **appendicular skeleton** consists of 126 bones. This division includes the bones of the limbs and the pectoral and pelvic girdles that attach the limbs to the trunk. The appendicular skeleton will be examined in Chapter 8. This chapter describes the functional anatomy of the axial skeleton, and we will begin with the skull. Before proceeding, you may find it helpful to review the directional references included in Tables 1-1 and 1-2, and the terms introduced in Table 6-2. [pp. 18, 20, 195]

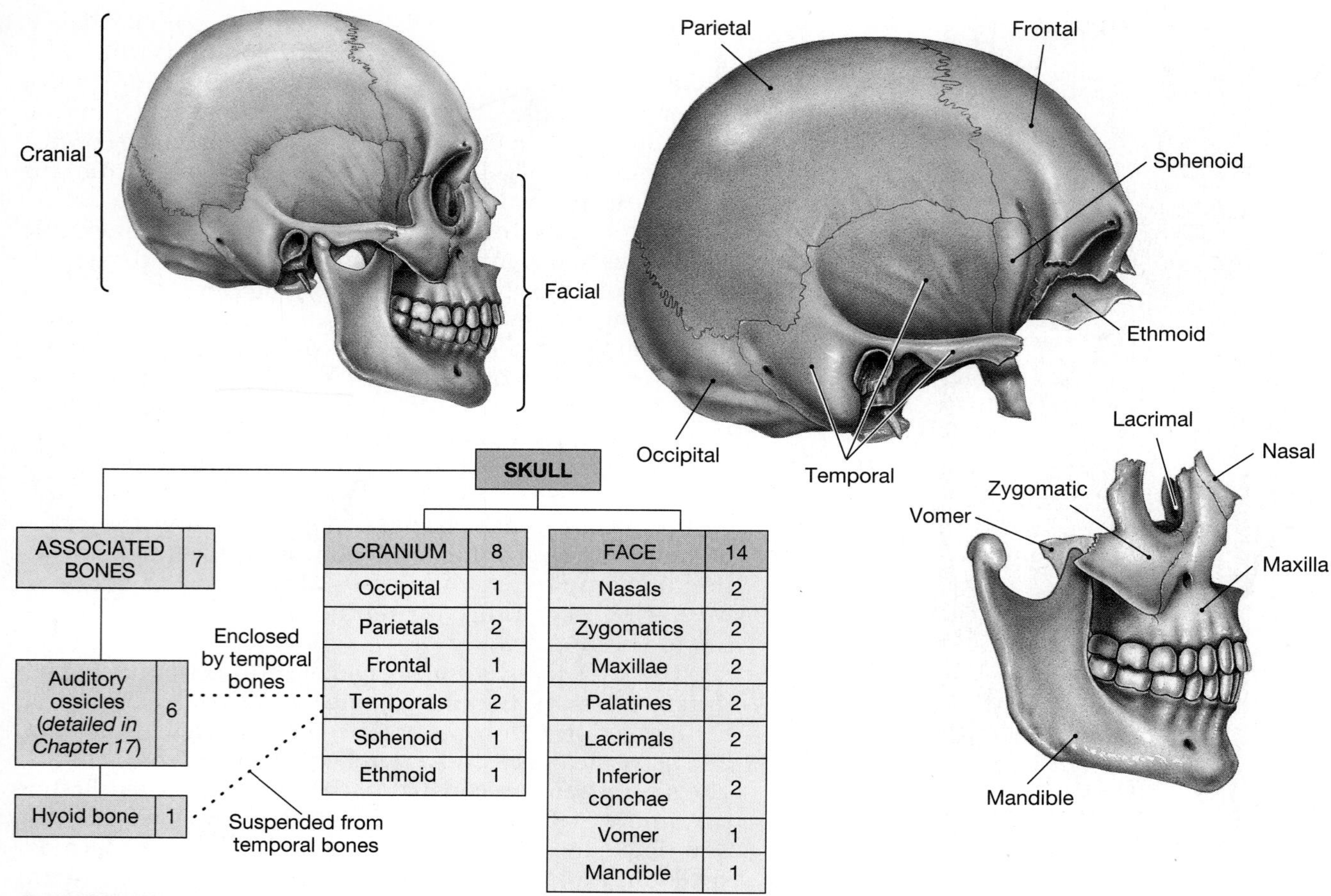

Cranial
Facial
Parietal
Frontal
Sphenoid
Ethmoid
Occipital
Temporal
Lacrimal
Nasal
Zygomatic
Vomer
Maxilla
Mandible
SKULL
ASSOCIATED BONES 7
Auditory ossicles (detailed in Chapter 17) 6
Hyoid bone 1
Enclosed by temporal bones
Suspended from temporal bones
CRANIUM 8
Occipital 1
Parietals 2
Frontal 1
Temporals 2
Sphenoid 1
Ethmoid 1
FACE 14
Nasals 2
Zygomatics 2
Maxillae 2
Palatines 2
Lacrimals 2
Inferior conchae 2
Vomer 1
Mandible 1

● **FIGURE 7-2**
Cranial and Facial Subdivisions of the Skull. The skull can be divided into the cranial and the facial divisions.

The Skull

Figures 7-2, 7-3

The bones of the **skull** protect the brain and guard the entrances to the digestive and respiratory systems. The skull itself contains 22 bones: 8 form the **cranium,** or *"braincase,"* and 14 are associated with the face (Figure 7-2●). Seven additional bones are associated with the skull. The 6 auditory ossicles are enclosed by the *temporal bones* of the cranium, and the hyoid bone is connected to the inferior surfaces of the temporal bones by a pair of ligaments.

The cranium consists of the *frontal, parietal, occipital, temporal, sphenoid,* and *ethmoid bones.* Together these **cranial bones** enclose the **cranial cavity,** a fluid-filled chamber that cushions and supports the brain. Blood vessels, nerves, and membranes that stabilize the position of the brain are attached to the inner surface of the cranium. Its outer surface provides an extensive area for the attachment of muscles that move the eyes, jaws, and head. A specialized joint between the occipital bone and the first spinal vertebra stabilizes the positions of the brain and spinal cord while permitting a considerable range of head movements.

If the cranium is the house where the brain resides, the *facial complex* is the front porch. **Facial bones** protect and support the entrances to the digestive and respiratory tracts. The superficial facial bones, the *lacrimal, nasal, maxilla, zygomatic,* and *mandible* (Figure 7-2●), provide areas for the attachment of muscles that control facial expressions and assist in the manipulation of food.

Except where the mandible articulates with the cranium, articulations between the skull bones of adults are immovable joints called **sutures.** At a suture the bones are tied firmly together with dense fibrous connective tissue. Each of the sutures of the skull has a name, but you need to know only four major sutures at this time: the *lambdoidal, coronal, sagittal,* and *squamosal* sutures.

- **Lambdoidal** (lam-DOYD-ul) **suture** (Greek *lambda,* Λ + *eidos,* shape). The lambdoidal suture arches across the posterior surface of the skull (Figure 7-3a●), separating the *occipital bone* from the *parietal bones.* One or more **sutural bones** (*Wormian bones*) may be found along this suture; they range from a bone the size of a grain of sand to one as large as a quarter.

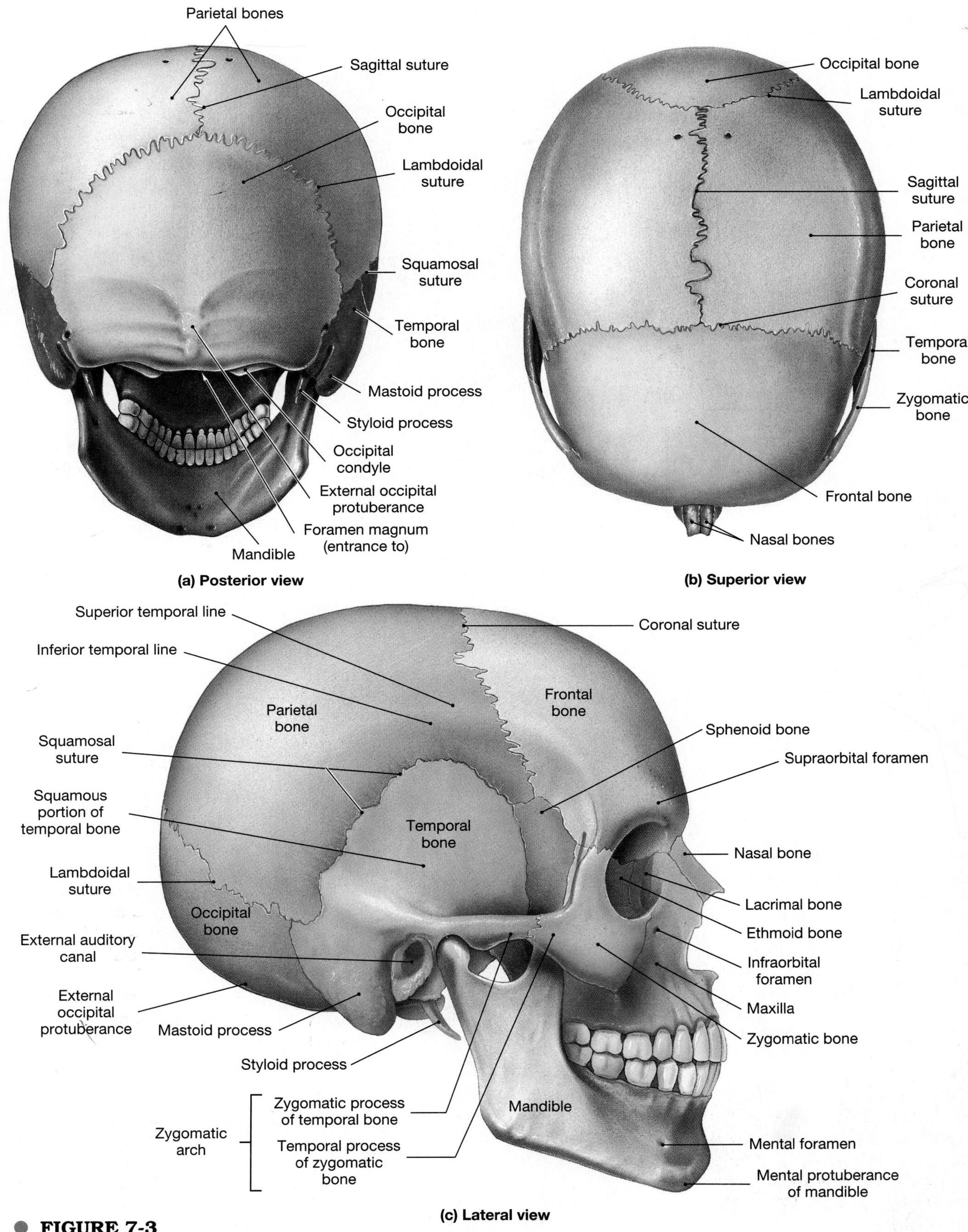
Parietal bones
Sagittal suture
Occipital bone
Lambdoidal suture
Squamosal suture
Temporal bone
Mastoid process
Styloid process
Occipital condyle
External occipital protuberance
Foramen magnum (entrance to)
Mandible
(a) Posterior view
Occipital bone
Lambdoidal suture
Sagittal suture
Parietal bone
Coronal suture
Temporal bone
Zygomatic bone
Frontal bone
Nasal bones
(b) Superior view
Superior temporal line
Inferior temporal line
Coronal suture
Frontal bone
Parietal bone
Sphenoid bone
Supraorbital foramen
Squamosal suture
Squamous portion of temporal bone
Temporal bone
Nasal bone
Lambdoidal suture
Lacrimal bone
Occipital bone
Ethmoid bone
External auditory canal
Infraorbital foramen
Maxilla
External occipital protuberance
Mastoid process
Zygomatic bone
Styloid process
Zygomatic process of temporal bone
Mandible
Zygomatic arch
Temporal process of zygomatic bone
Mental foramen
Mental protuberance of mandible
(c) Lateral view

● **FIGURE 7-3**
The Adult Skull. The adult skull is shown in **(a)** posterior view, **(b)** superior view, **(c)** lateral view. AM *Plates 2.1, 2.2*

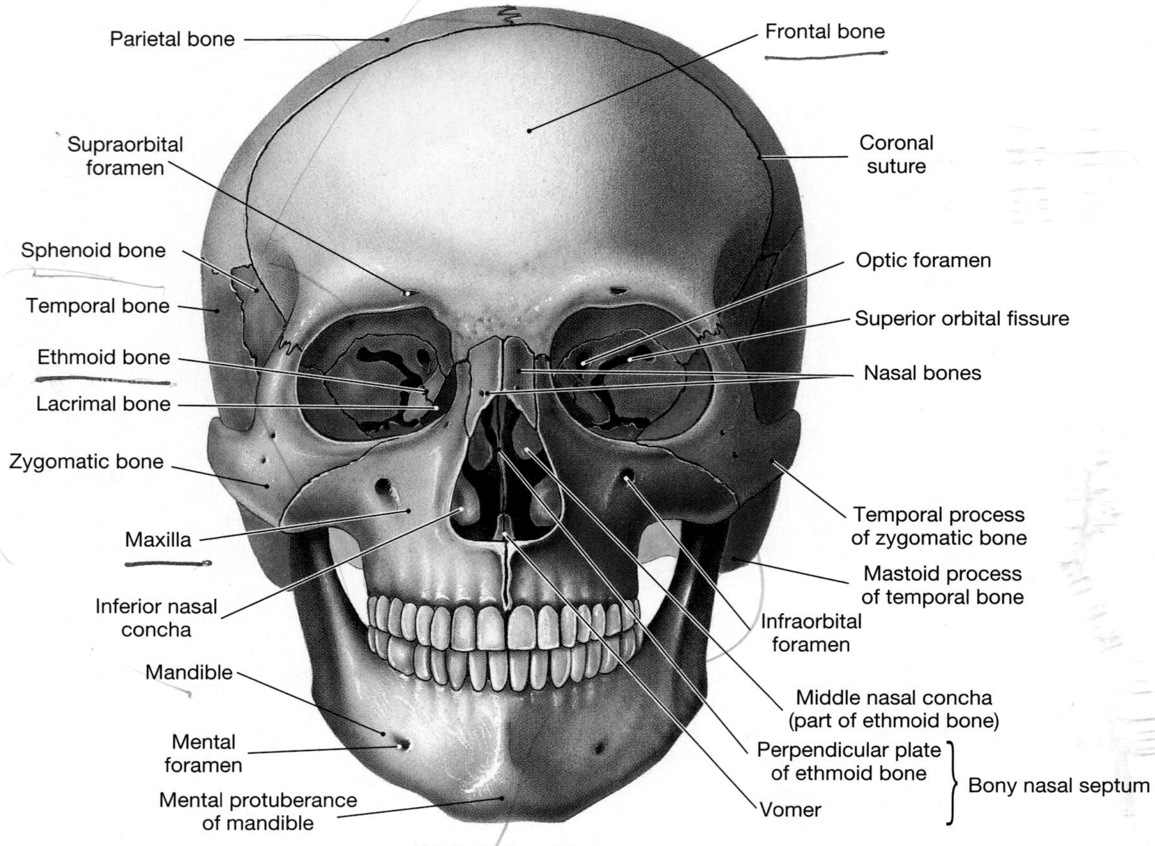

Parietal bone
Frontal bone
Supraorbital foramen
Coronal suture
Sphenoid bone
Temporal bone
Optic foramen
Superior orbital fissure
Ethmoid bone
Nasal bones
Lacrimal bone
Zygomatic bone
Temporal process of zygomatic bone
Maxilla
Mastoid process of temporal bone
Inferior nasal concha
Infraorbital foramen
Mandible
Middle nasal concha (part of ethmoid bone)
Mental foramen
Perpendicular plate of ethmoid bone
Bony nasal septum
Mental protuberance of mandible
Vomer

(d) Anterior view

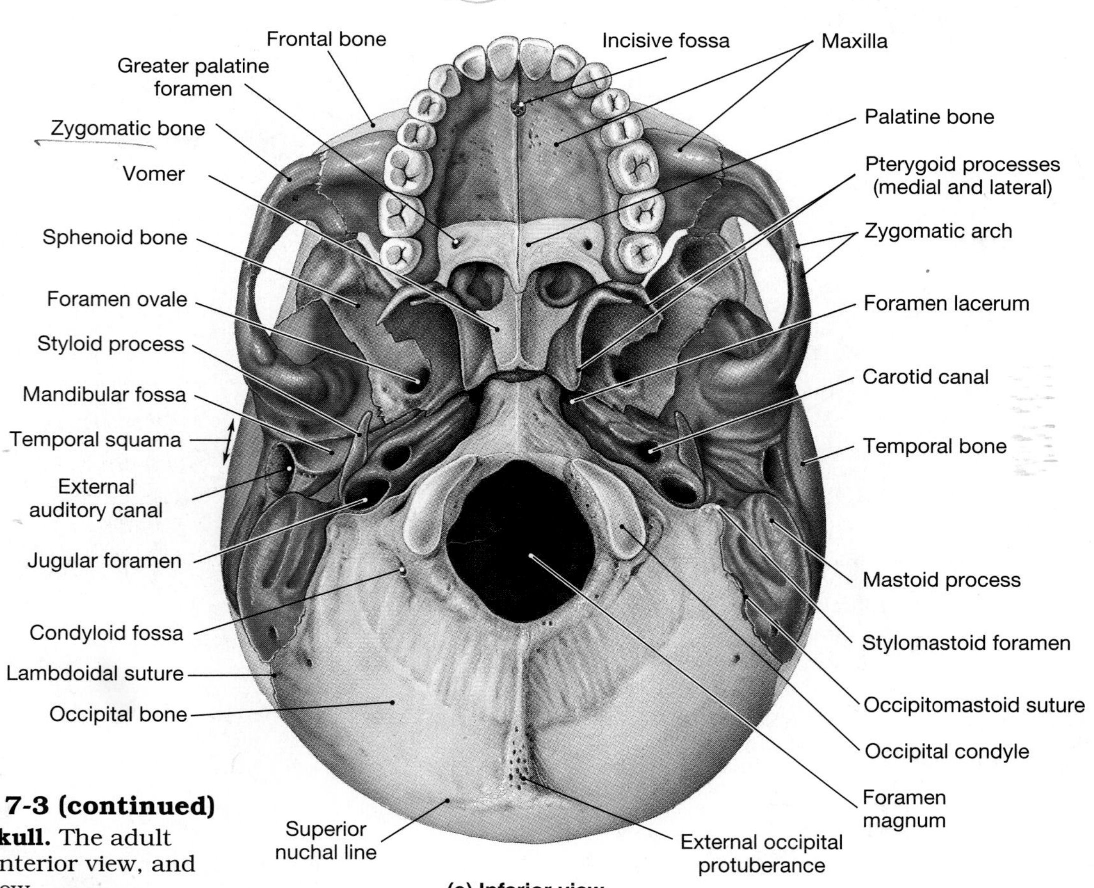
Frontal bone
Incisive fossa
Maxilla
Greater palatine foramen
Zygomatic bone
Palatine bone
Vomer
Pterygoid processes (medial and lateral)
Sphenoid bone
Zygomatic arch
Foramen ovale
Foramen lacerum
Styloid process
Carotid canal
Mandibular fossa
Temporal squama
Temporal bone
External auditory canal
Jugular foramen
Mastoid process
Condyloid fossa
Stylomastoid foramen
Lambdoidal suture
Occipitomastoid suture
Occipital bone
Occipital condyle
Foramen magnum
Superior nuchal line
External occipital protuberance

(e) Inferior view

● **FIGURE 7-3 (continued)**
The Adult Skull. The adult skull in **(d)** anterior view, and **(e)** inferior view.

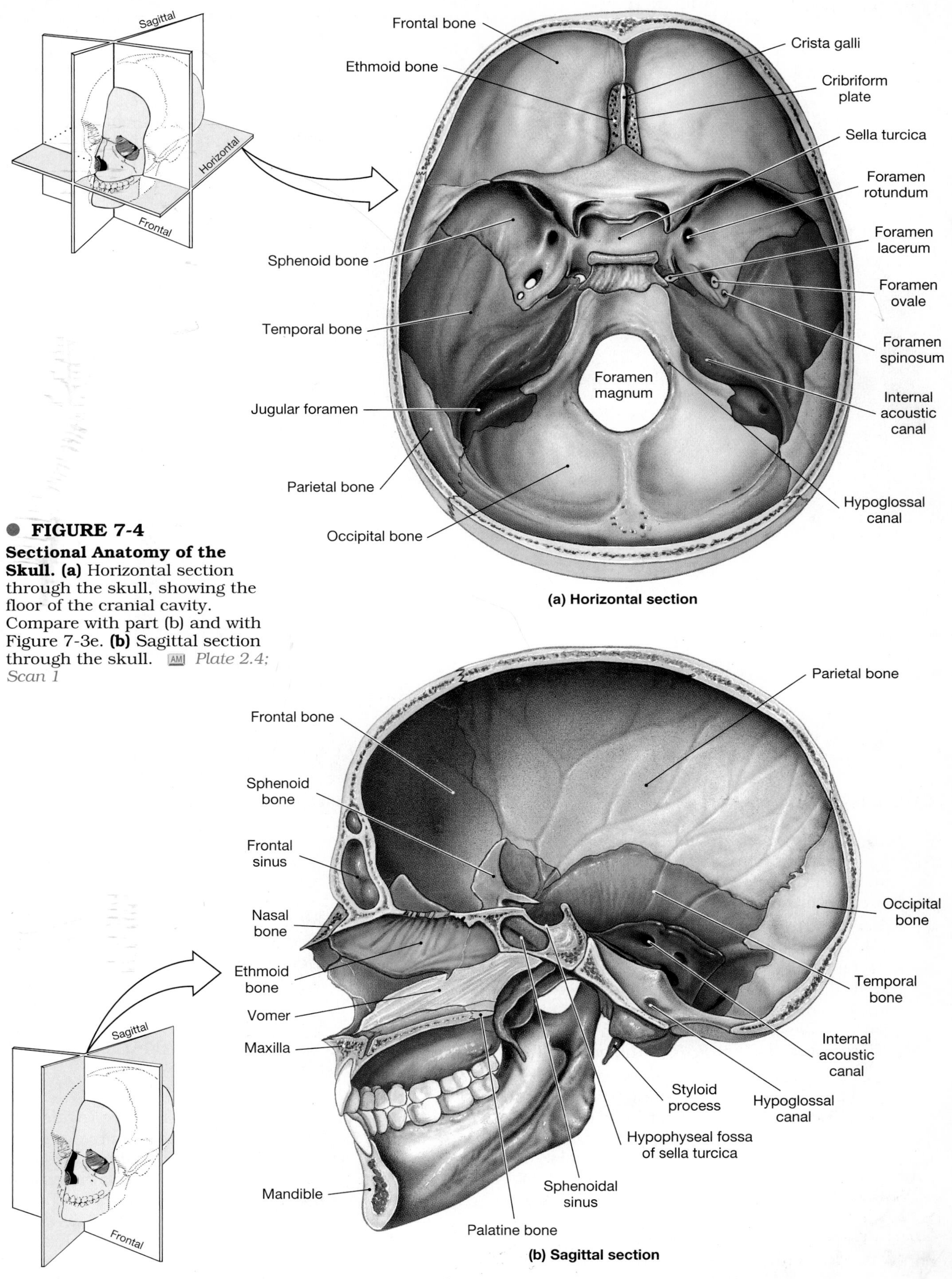
Sagittal
Horizontal
Frontal
Frontal bone
Crista galli
Ethmoid bone
Cribriform plate
Sella turcica
Foramen rotundum
Foramen lacerum
Sphenoid bone
Foramen ovale
Temporal bone
Foramen spinosum
Foramen magnum
Internal acoustic canal
Jugular foramen
Parietal bone
Hypoglossal canal
Occipital bone
(a) Horizontal section
Parietal bone
Frontal bone
Sphenoid bone
Frontal sinus
Nasal bone
Occipital bone
Ethmoid bone
Temporal bone
Vomer
Sagittal
Maxilla
Internal acoustic canal
Styloid process
Hypoglossal canal
Hypophyseal fossa of sella turcica
Mandible
Sphenoidal sinus
Palatine bone
Frontal
(b) Sagittal section

● **FIGURE 7-4**
Sectional Anatomy of the Skull. (a) Horizontal section through the skull, showing the floor of the cranial cavity. Compare with part (b) and with Figure 7-3e. **(b)** Sagittal section through the skull. AM *Plate 2.4; Scan 1*

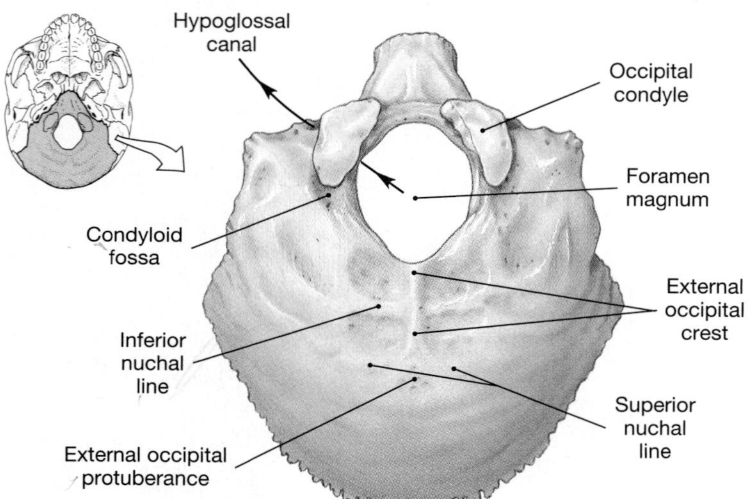

Hypoglossal canal
Occipital condyle
Foramen magnum
Condyloid fossa
External occipital crest
Inferior nuchal line
Superior nuchal line
External occipital protuberance

(a) Occipital bone, inferior (external) view

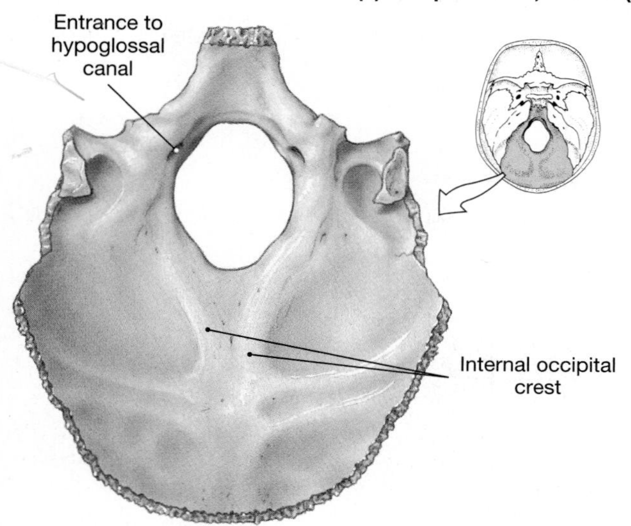

Entrance to hypoglossal canal
Internal occipital crest

(b) Occipital bone, superior (internal) view

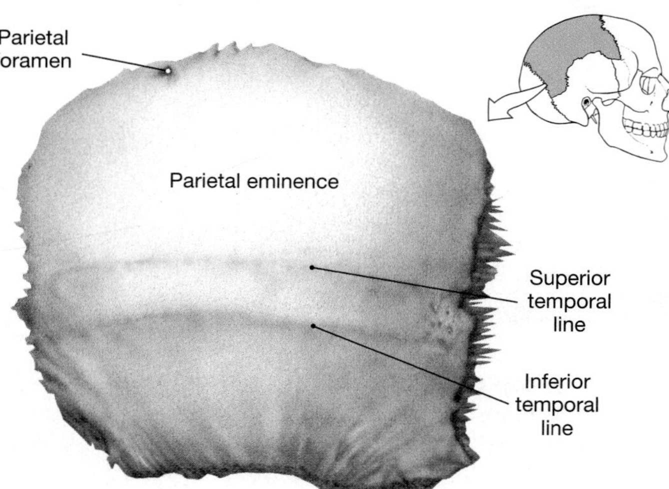

Parietal foramen
Parietal eminence
Superior temporal line
Inferior temporal line

(c) Parietal bone, right lateral view

● **FIGURE 7-5**
The Occipital and Parietal Bones. **(a)** The occipital bone, inferior (external) view. **(b)** The occipital bone, superior (internal) view. **(c)** Parietal bone, right lateral view.

- ***Sagittal suture.*** The sagittal suture begins at the superior midline of the lambdoidal suture and extends rostrally between the parietal bones on either side (Figure 7-3b●).
- ***Coronal suture.*** Anteriorly, the sagittal suture ends when it intersects the coronal suture. The coronal suture crosses the superior surface of the skull, separating the median *frontal bone* from the more posterior parietals (Figure 7-3b●). The occipital, parietal, and frontal bones form the **calvaria** (kal-VAR-ē-uh), or "skullcap."
- ***Squamosal suture.*** A squamosal suture on each side of the skull marks the boundary between the *temporal bone* and the parietal bone of that side. The squamosal sutures can be seen in Figure 7-3a●, where they intersect the lambdoidal suture. The path of the squamosal suture on the right side of the skull can be seen in Figure 7-3c●.

The frontal, temporal, sphenoid, and ethmoid bones of the cranium and the maxillary bone of the face contain internal, air-filled chambers called **sinuses.** Sinuses have two major functions: (1) The presence of a sinus makes a bone much lighter than it would be otherwise, and (2) the mucous membrane lining the sinuses produces mucus that helps moisten and clean the air in and adjacent to the sinus. The *paranasal sinuses*—all of the sinuses other than the ones in the temporal bones—empty into the nasal cavity.

❑ Bones of the Cranium

Figures 7-3, 7-4

We will now consider each of the bones of the cranium. As we proceed, use the figures provided to develop a three-dimensional perspective on the individual bones. Ridges and *foramina* (passageways) that are detailed here mark the attachment of muscles or passage of nerves and blood vessels that will be studied in later chapters. AM *Phrenology*

Figures 7-3 and 7-4● present the adult skull in superficial and sectional views.

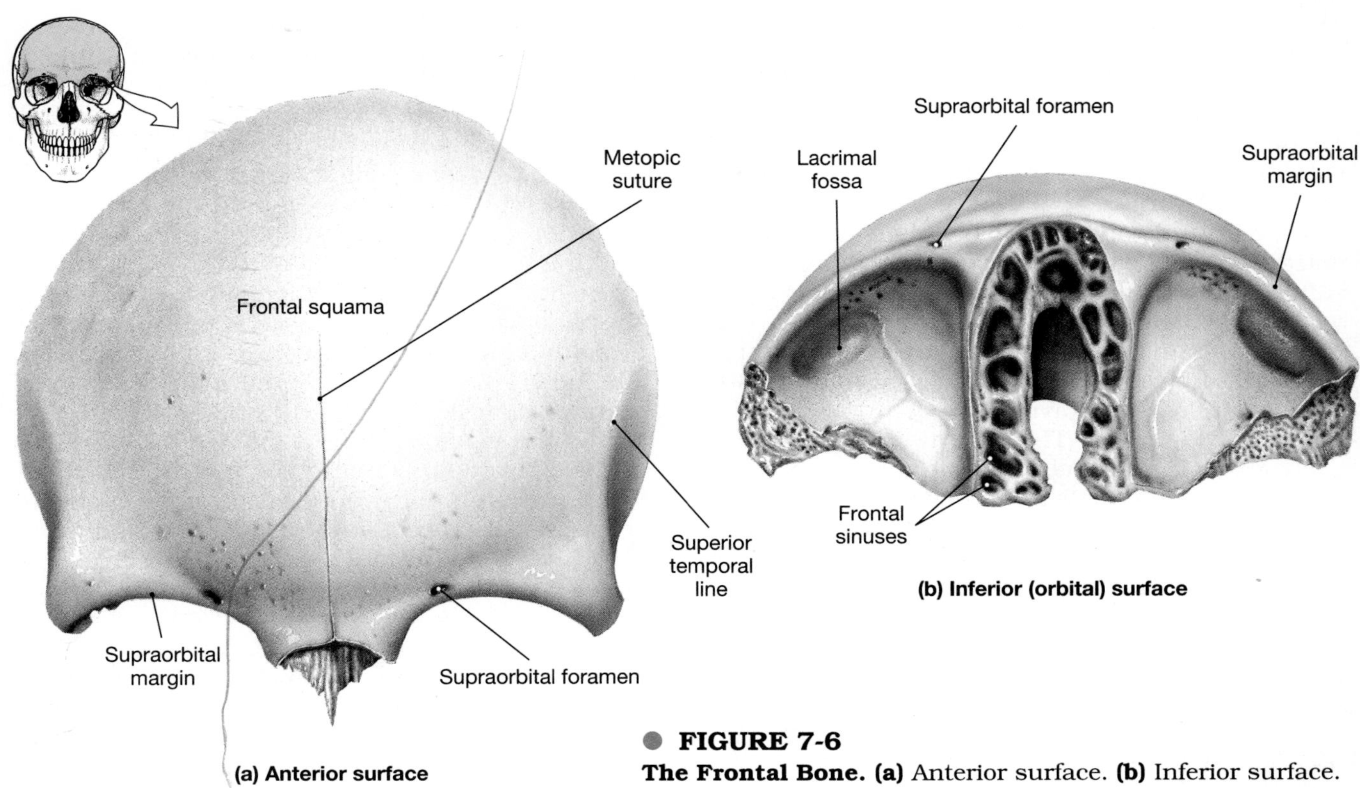
Metopic suture
Frontal squama
Superior temporal line
Supraorbital margin
Supraorbital foramen
(a) Anterior surface
Supraorbital foramen
Lacrimal fossa
Supraorbital margin
Frontal sinuses
(b) Inferior (orbital) surface

● **FIGURE 7-6**
The Frontal Bone. (a) Anterior surface. **(b)** Inferior surface.

The Occipital Bone

Figures 7-3, 7-5

The **occipital bone** contributes to the posterior and inferior surfaces of the cranium (Figure 7-3a,c,e●). The inferior surface of the occipital bone contains a large, circular opening, the **foramen magnum,** which connects the cranial cavity with the spinal cavity enclosed by the vertebral column. At the adjacent **occipital condyles** the skull articulates with the first cervical vertebra. The posterior, external surface of the occipital bone (Figure 7-5a●) bears a number of prominent ridges. The **occipital crest** extends posteriorly from the foramen magnum, ending in a small midline bump called the **external occipital protuberance.** Two horizontal ridges intersect the crest, the **inferior** and **superior nuchal** (NOO-kul) **lines.** These mark the attachment of muscles and ligaments that stabilize the articulation at the occipital condyles and balance the weight of the head over the vertebrae of the neck. The occipital bone forms part of the wall of the large **jugular foramen.** The *internal jugular vein* passes through this foramen to drain venous blood from the brain. The **hypoglossal canals** begin at the lateral base of each occipital condyle, just superior to the condyles. The *hypoglossal nerves,* cranial nerves that control the tongue muscles, pass through these canals.

Inside the skull, the hypoglossal canals begin on the inner surface of the occipital bone near the foramen magnum (Figure 7-5b●). Note the concave internal surface of the occipital bone, which closely follows the contours of the brain. The grooves follow the paths of major vessels, and the ridges mark the attachment site of membranes that stabilize the position of the brain.

The Parietal Bones

Figures 7-3b,c, 7-4a,b, 7-5c

The paired **parietal bones** contribute to the superior and lateral surfaces of the cranium (Figure 7-3b,c●). Grooves on the inner surface of the parietal bones (Figure 7-4a,b●) mark the path of cranial blood vessels. The lateral surface of each parietal bone (Figure 7-5c●) bears a pair of low ridges, the **superior** and **inferior temporal lines.** These lines mark the attachment of the *temporalis muscle,* a large muscle that closes the mouth. The smooth parietal surface above these lines is called the **parietal eminence.**

The Frontal Bone

Figures 7-3b,c,d, 7-4, 7-6

The dome-shaped **frontal bone** forms the forehead and roof of the orbits (Figure 7-3b,c,d●). During development, the bones of the cranium form through the fusion of separate centers of ossification, and at birth the fusions have not been completed. At this time there are two frontal bones that articulate along the **metopic suture.** Although the suture usually disappears by age 8 with the fusion of the bones, the adult skull often retains traces of the suture line. This suture, or what remains of it, runs down the center of the **frontal squama,** or forehead (Figure 7-6a●). To either side is the anterior continuation of the superior temporal line from the parietal surface.

The frontal squama ends at the **supraorbital margins.** Above each margin are raised ridges, the **supraciliary arches,** which are externally marked by the eyebrows. The center of each supraorbital margin contains either a single **supraorbital foramen** or a **supraorbital notch,** marking the path of blood vessels supplying the eyebrow, eyelids, and frontal sinuses. The orbital surface is relatively smooth, but it contains small openings for blood vessels and nerves heading to or from structures in the orbit. The shallow **lacrimal fossa** marks the location of the *lacrimal* (tear) *gland* that lubricates the surface of the eye (Figure 7-6b●).

The **frontal sinuses** (Figure 7-4b●, p. 207) are extremely variable in size and time of appearance. They usually appear after age 6, but some people never develop them at all. The frontal sinuses and other sinuses of the cranium and face will be described in a later section.

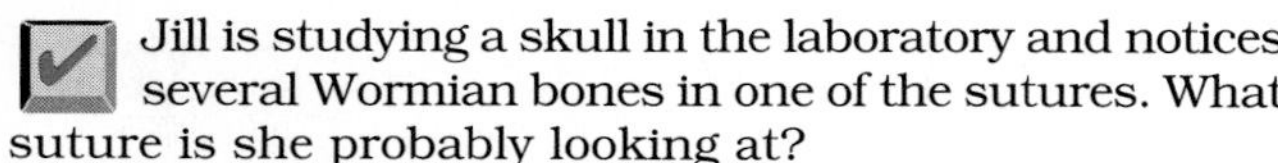
Jill is studying a skull in the laboratory and notices several Wormian bones in one of the sutures. What suture is she probably looking at?

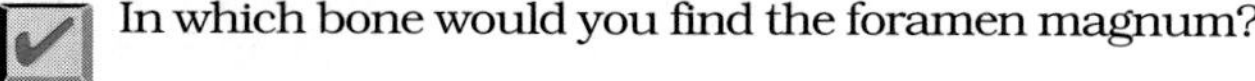
In which bone would you find the foramen magnum?

Tom suffers a blow to the skull that fractures the right superior lateral surface of his cranium. What bone is fractured?

The Temporal Bones

Figures 7-3c,e, 7-7

The **squama,** or **squamous portion,** of the **temporal bone** is the convex, irregular surface inferior to the squamosal suture (Figure 7-3c●, p. 205). Inferior to the squama, the temporal bone has a prominent **zygomatic process** that curves laterally and anteriorly to meet the **temporal process** of the *zygomatic bone.* Together these processes form the **zygomatic arch,** or *cheekbone.* Medially, each temporal bone surrounds the jugular foramen of that side, along with the adjacent portion of the occipital bone (Figure 7-3e●, p. 206). Anterior and slightly medial to the jugular foramen, the round **carotid foramen** is the entrance to the *carotid canal.* The *internal carotid artery,* a major artery that supplies blood to the brain, enters the skull through this foramen. Anterior and medial to this opening, an elongate, jagged slit, the **foramen lacerum** (LA-se-rum; *lacerare,* to tear) extends between the occipital and temporal bones. In life, this space contains hyaline cartilage and small arteries to the inner surface of the cranium.

Inferior to the base of the zygomatic process, the temporal articulates with the mandible. A depression called the **mandibular fossa** marks this site (Figure 7-7a,b●). The entrance to the *external auditory canal* is located immediately posterior and lateral to that mandibular fossa. In life, this canal ends at the delicate *tympanic membrane,* or *eardrum,* but this membrane disintegrates during the preparation of a dried skull. The prominent bulge just posterior and inferior to the meatus is the **mastoid process.** This process provides an attachment site for muscles that rotate or extend the head. Numerous interconnected mastoid sinuses, termed *mastoid air cells,* are contained within the mastoid process (Figure 7-7a●). These sinuses are connected to the middle ear cavity within the temporal bone. If pathogens invade these air cells, a condition called *mastoiditis* develops. The individual then experiences severe earaches, fever, and swelling behind the ear, among other symptoms.

Many of the important features of the temporal bone can be seen on its inferior surface (Figure 7-7b●). Near the base of the mastoid process, the **mastoid foramen** penetrates the temporal bone. Blood vessels travel through this passageway to reach the membranes surrounding the brain. Ligaments that support the hyoid bone attach to the sharp **styloid** (STĪ-loyd; *stylos,* pillar) **process,** as do some of the tongue muscles. The **stylomastoid foramen** lies posterior to the base of the styloid process. The *facial nerve* passes through this foramen to control the facial muscles.

Lateral and anterior to the carotid foramen, the temporal bone articulates with the sphenoid. A small canal begins at that articulation and ends inside the mass of the temporal bone. This canal has long been known as the *Eustachian* (ū-STĀ-kē-an) *tube,* although the terms **auditory tube** or **pharyngotympanic tube** are now used to designate this structure. In life, it contains an air-filled passageway that begins at the pharynx and ends at the *tympanic cavity,* a chamber inside the temporal bone (Figure 7-7c●).

The tympanic cavity, or *middle ear,* contains the **auditory ossicles**, or *ear bones.* These tiny bones, three on each side of the skull, transfer sound vibrations from the eardrum to the hearing receptors of the *inner ear.* (Details concerning the individual bones and their role in hearing will be discussed in Chapter 17.) The auditory ossicles and inner ear are completely enclosed by the temporal bone.

A medial view of the temporal bone (Figure 7-7d●) shows several additional features. The thick **petrous portion** of the temporal bone houses the inner ear structures that provide information about hearing and balance. The **internal acoustic canal** carries blood vessels and nerves to the inner ear and the facial nerve to the stylomastoid foramen.

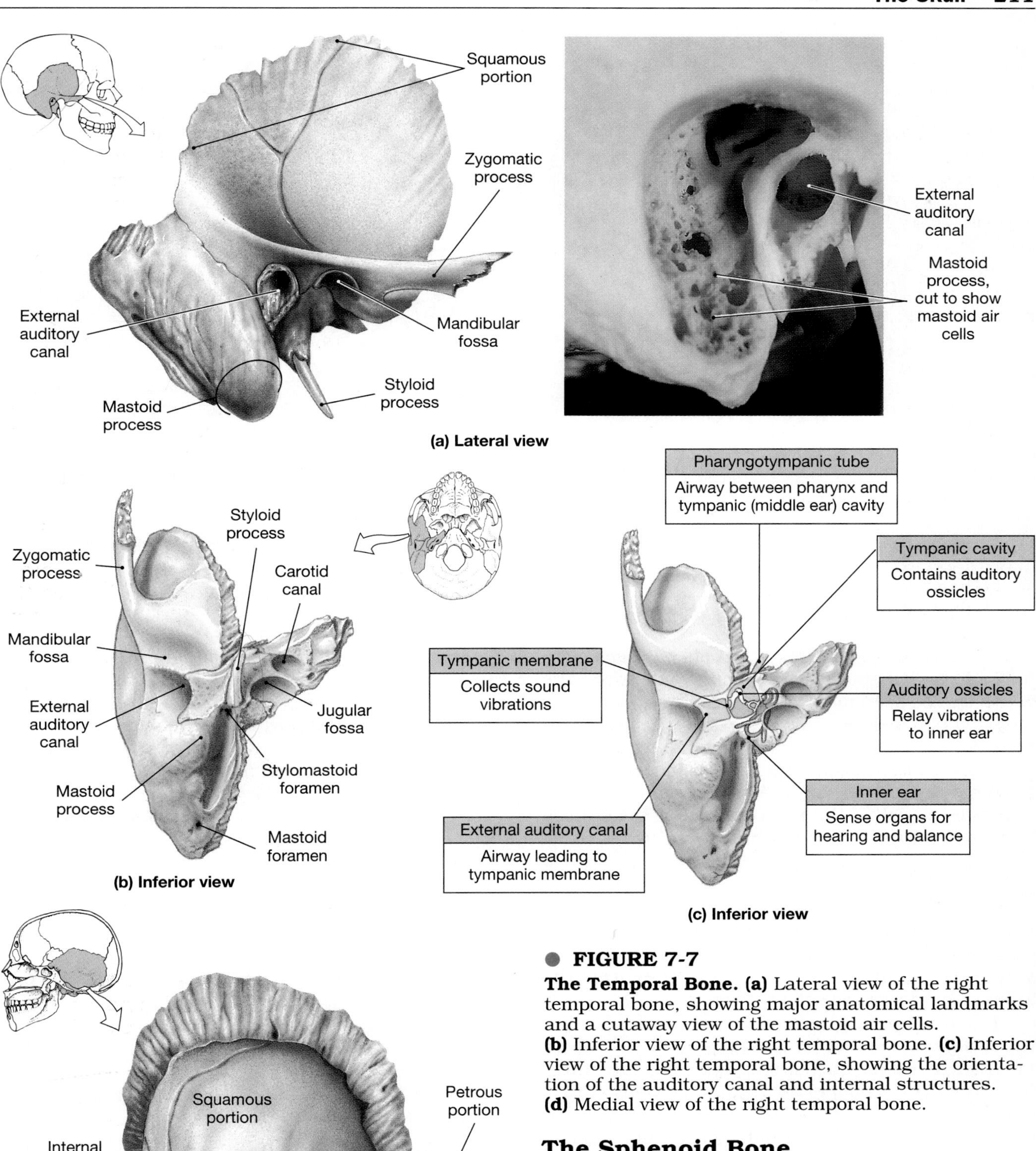
Squamous portion
Zygomatic process
External auditory canal
Mandibular fossa
Styloid process
Mastoid process
External auditory canal
Mastoid process, cut to show mastoid air cells
(a) Lateral view
Zygomatic process
Styloid process
Carotid canal
Mandibular fossa
External auditory canal
Jugular fossa
Stylomastoid foramen
Mastoid process
Mastoid foramen
(b) Inferior view
Pharyngotympanic tube
Airway between pharynx and tympanic (middle ear) cavity
Tympanic cavity
Contains auditory ossicles
Tympanic membrane
Collects sound vibrations
Auditory ossicles
Relay vibrations to inner ear
Inner ear
Sense organs for hearing and balance
External auditory canal
Airway leading to tympanic membrane
(c) Inferior view
Squamous portion
Petrous portion
Internal acoustic canal
Zygomatic process
Styloid process
Mastoid process
(d) Medial view

● **FIGURE 7-7**
The Temporal Bone. (a) Lateral view of the right temporal bone, showing major anatomical landmarks and a cutaway view of the mastoid air cells.
(b) Inferior view of the right temporal bone. **(c)** Inferior view of the right temporal bone, showing the orientation of the auditory canal and internal structures.
(d) Medial view of the right temporal bone.

The Sphenoid Bone

Figures 7-3c,d,e, 7-4, 7-8

The **sphenoid bone** contributes to the floor of the cranium. Although relatively large, much of it is hidden by more superficial bones. The sphenoid is a bridge uniting the cranial and facial bones; it articulates with the frontal, occipital, parietal, ethmoid, and temporal bones of the cranium and with the palatine, zygomatic, maxilla, and vomer of the facial complex (Figure 7-3c,d,e●, pp. 205–206). It also acts as a brace, strengthening the sides of the skull.

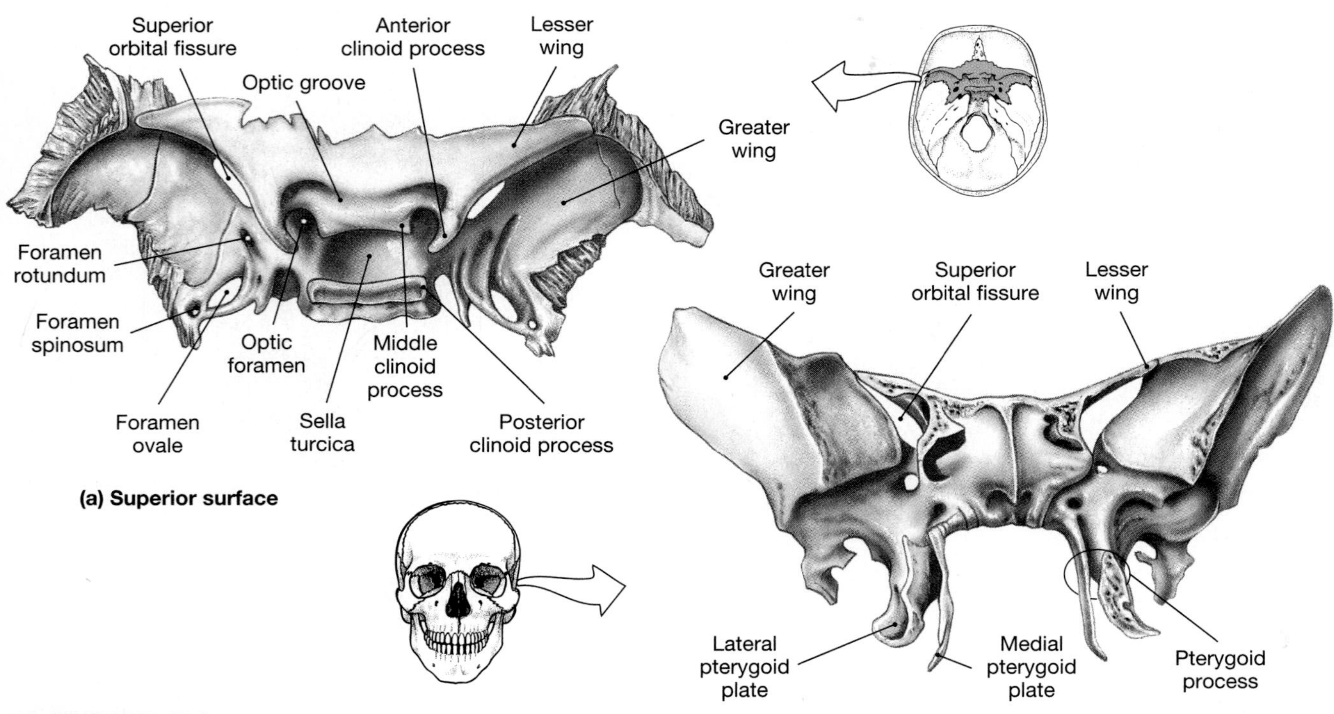
Superior
orbital fissure
Anterior
clinoid process
Lesser
wing
Optic groove
Greater
wing
Foramen
rotundum
Foramen
spinosum
Optic
foramen
Middle
clinoid
process
Foramen
ovale
Sella
turcica
Posterior
clinoid process
(a) Superior surface
Greater
wing
Superior
orbital fissure
Lesser
wing
Lateral
pterygoid
plate
Medial
pterygoid
plate
Pterygoid
process
(b) Anterior surface

FIGURE 7-8
The Sphenoid Bone. (a) Superior surface. **(b)** Anterior surface.

The general shape of the sphenoid has been compared to a giant bat, with its wings extended. The wings can be seen most clearly on the superior surface (Figures 7-4, p. 207, and 7-8a●). A prominent central depression between the wings cradles the pituitary gland below the brain. This recess is called the **hypophyseal** (hī-po-FIZ-ē-ul) **fossa,** and the bony enclosure is called the **sella turcica** (TUR-si-kuh) because it supposedly resembles a "Turkish saddle." If the Turk rode facing forward, his back would rest against the **posterior clinoid** (KLĪ-noyd) **process** and he could reach forward to grasp the **anterior clinoid processes** on either side. The anterior clinoid processes are posterior projections of the **lesser wings** of the sphenoid. The **middle clinoid process** forms the anterior border of the sella turcica.

The transverse groove that crosses the front of the saddle, above the level of the seat, is the **optic groove.** At either end of this groove is an **optic foramen.** The optic nerves that carry visual information from the eyes to the brain travel through these foramina. On either side of the sella turcica, the **superior orbital fissure,** the **foramen rotundum,** the **foramen ovale** (ō-VAH-le), and the **foramen spinosum** penetrate the sphenoid. These passages carry blood vessels and nerves to structures of the orbit, face, and jaws. These foramina penetrate the **greater wings** of the sphenoid. The optic foramina and the superior orbital fissures can also be seen in the anterior view of the sphenoid (Figure 7-8b●). On either side of the sphenoid, inferior to the sella turcica, the sphenoid contains a *sphenoidal sinus.* The sphenoid is a large bone, and the presence of the sinus makes it lighter than it would be otherwise. In addition, the epithelium lining the sphenoidal sinus produces mucus that helps flush clean the surfaces of the nasal cavity.

The **pterygoid** (TER-i-goyd; *pterygion,* wing) **processes** of the sphenoid are vertical projections that begin at the boundary between the greater and lesser wings. Each process forms a pair of *pterygoid plates* that are important sites for the attachment of muscles that move the lower jaw and soft palate.

The Ethmoid Bone

Figures 7-3d, 7-4, 7-9

The **ethmoid bone** is an irregularly shaped bone that forms part of the orbital wall (Figure 7-3d●, p. 206), the anteromedial floor of the cranium, and the roof of the nasal cavity and part of the nasal septum (Figure 7-4●, p. 207). The ethmoid has three parts: the *cribriform plate,* the paired *lateral masses,* and the *perpendicular plate* (Figure 7-9●).

1. The superior surface of the ethmoid (Figure 7-9a●) contains the **cribriform plate** (*cribrum,* sieve). The foramina in this plate permit passage of the *olfactory nerves,* which provide the sense of smell. A prominent ridge, the **crista galli** (*crista,* crest + *gallus,* chicken; "cock's comb") projects above the cribriform plate. The *falx cerebri,* a membrane that stabilizes the position of the brain, attaches to this bony ridge.
2. The lateral masses contain the **ethmoidal sinuses**, or *ethmoidal air cells,* which open into the

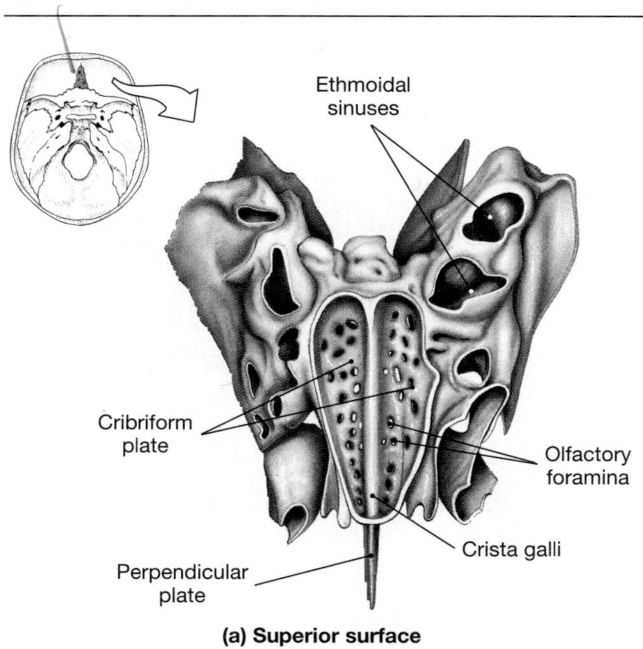
Ethmoidal sinuses
Cribriform plate
Olfactory foramina
Crista galli
Perpendicular plate
(a) Superior surface

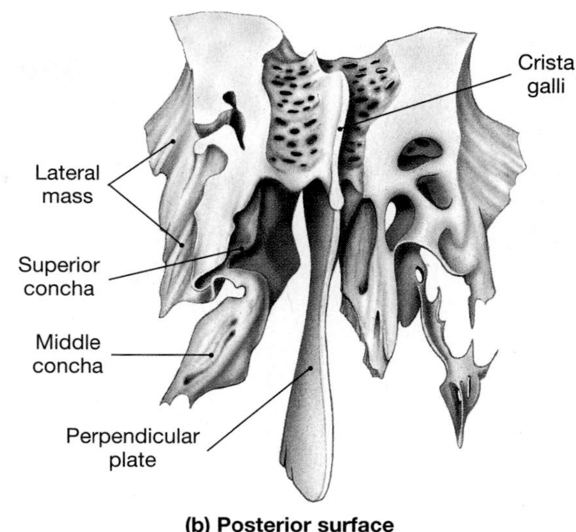
Crista galli
Lateral mass
Superior concha
Middle concha
Perpendicular plate
(b) Posterior surface

● **FIGURE 7-9**
The Ethmoid Bone. (a) Superior surface. **(b)** Posterior surface.

nasal cavity on each side. Mucous secretions from these sinuses flush the surfaces of the nasal cavities. Medial projections of the lateral masses, the **superior conchae** (KONG-kē; singular *concha*, a snail shell), and the **middle conchae** are best viewed from the posterior surface of the ethmoid (Figure 7-9b●). The conchae are thin scrolls of bone that break up the airflow in the nasal cavity, creating swirls and eddies that have three major functions: (1) The swirling throws any particles in the air against the sticky mucus that covers the walls of the nasal cavity; (2) the turbulence slows air movement, providing time for warming, humidification, and dust removal before the air reaches more delicate portions of the respiratory tract; and (3) the eddies direct air toward the superior portion of the nasal cavity, adjacent to the cribriform plate, where the olfactory (smell) receptors are located.

3. The **perpendicular plate** forms part of the nasal septum, along with the vomer and a piece of hyaline cartilage. Olfactory receptors are located in the epithelium covering the inferior surfaces of the cribriform plate, the medial surfaces of the superior conchae, and the superior portion of the perpendicular plate.

✔ The internal jugular veins are important blood vessels of the head. Between what bones do these blood vessels pass?

✔ What bone contains the depression called the sella turcica? What is located in the depression?

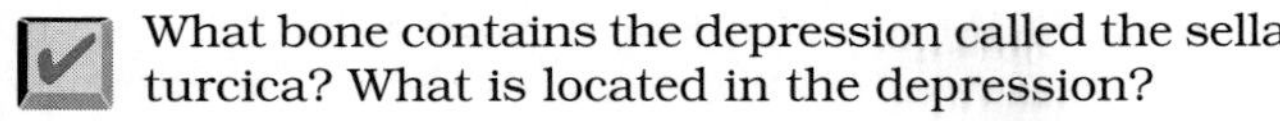
✔ Which of the five senses would be affected if the cribriform plate of the ethmoid failed to form?

❑ Bones of the Face

The facial bones are the *maxillae*, the *palatines*, the *nasals*, the *mandible*, the *zygomatics*, the *lacrimals*, the *inferior conchae*, and the *vomer*. Facial bones and cranial bones together form the *nasal complex* that surrounds the nasal cavities and the *orbital complex* surrounding each eye.

The Maxillae

Figure 7-3d, 7-10

The left and right **maxillae,** or *maxillary bones*, are the largest facial bones and form the upper jaw. They articulate with all other facial bones except the mandible (Figure 7-3d●, p. 206). The **orbital rim** (Figure 7-10a●) provides protection for the eye and other structures in the orbit. A large **infraorbital foramen** marks the entrance of a major sensory nerve from the face; it leaves the skull via the foramen rotundum of the sphenoid. An elongate **inferior orbital fissure** within each orbit lies between the maxilla and the sphenoid. The oral margins of the maxillae form the **alveolar processes** that contain the upper teeth.

The large **maxillary sinuses** are evident in horizontal section (Figure 7-10b●). These are the largest sinuses in the skull; they lighten the portion of the maxillae above the teeth and produce mucous secretions that flush the inferior surfaces of the nasal cavities. The sectional view also shows the extent of the **palatine processes** that form most of the bony roof, or **hard palate,** of the mouth.

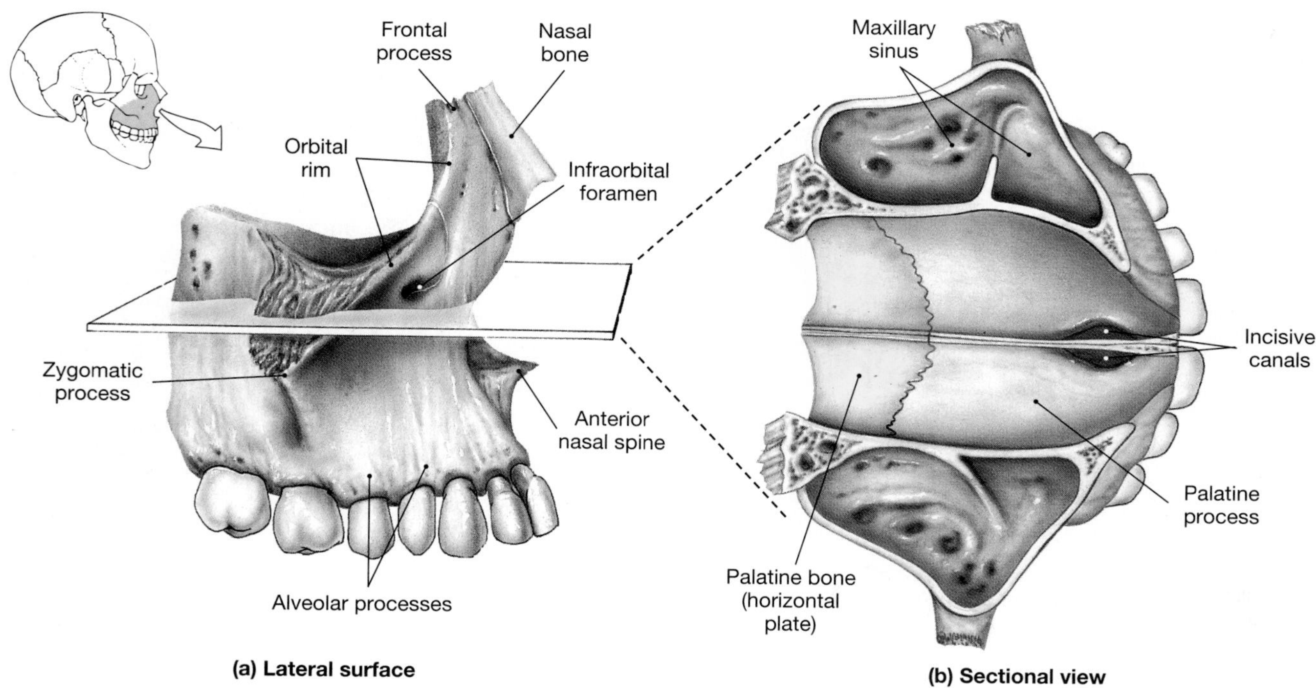
Frontal process
Nasal bone
Orbital rim
Infraorbital foramen
Zygomatic process
Anterior nasal spine
Alveolar processes
(a) Lateral surface
Maxillary sinus
Incisive canals
Palatine process
Palatine bone (horizontal plate)
(b) Sectional view

● **FIGURE 7-10**
The Maxillae and Palatine Bones. (a) Anterior and lateral surfaces of right maxilla, showing superficial landmarks and the sectional plane for part (b). **(b)** Horizontal section through both maxillary bones, showing the size and orientation of the maxillary sinuses.

The Palatine Bones

Figures 7-3e, 7-10b, 7-12

The **palatine bones** are small, L-shaped bones that articulate with the maxillae to form the posterior portions of the hard palate (Figures 7-3e, p. 206, and 7-10b●). The inferior surface of the palatine on each side has a prominent foramen, the *greater palatine foramen,* and usually one or more *lesser palatine foramina.* The vertical portion of the palatine bone extends superiorly (Figure 7-12●), where it forms a small portion of the floor of the orbit.

The Nasal Bones

Figures 7-3d, 7-10a

The paired **nasal bones** articulate with the frontal bone at the midline of the face (Figure 7-3d●, p. 206. The nasal bones support the superior portion of the bridge of the nose. They extend to the superior border of the **external nares** (NA-rēz), or nasal openings. Cartilage attached to the nasal bones forms the flexible portion of the nose. The lateral surfaces of the nasal bones articulate with the maxillae on either side (Figure 7-10a●).

The Vomer

Figure 7-4b

The **vomer** forms the inferior portion of the nasal septum (Figure 7-4b●, p. 207). It is based on the floor of the nasal cavity and articulates with both the maxillae and palatines along the midline. The vertical portion of the vomer is thin. Its curving superior surface articulates with the sphenoid and the perpendicular plate of the ethmoid, forming a bony **nasal septum** (*septum,* wall) that separates the right and left nasal cavities. Anteriorly, the vomer supports a cartilaginous extension of the nasal septum that continues into the fleshy portion of the nose and separates the external nares.

The Inferior Nasal Conchae

Figure 7-11a

The **inferior nasal conchae** are paired scroll-like bones that resemble the superior and middle conchae of the ethmoid. One inferior concha is located on each side of the nasal septum, attached to the lateral wall of the nasal cavity (Figure 7-11a●). They perform the same functions as the superior and middle conchae.

The Nasal Complex

Figure 7-11

The **nasal complex** includes the bones that enclose the nasal cavities and the *paranasal sinuses,* air spaces connected to the nasal cavities. The frontal, sphenoid, and ethmoid bones form the superior wall of the nasal cavities; the lateral walls are formed by the maxillae, the lacrimals, and the ethmoidal and inferior conchae (Figure 7-11●). Much of the anterior margin of the nasal cavity is formed by the soft tissues of the nose, but the bridge of the nose is supported by the maxillary and nasal bones.

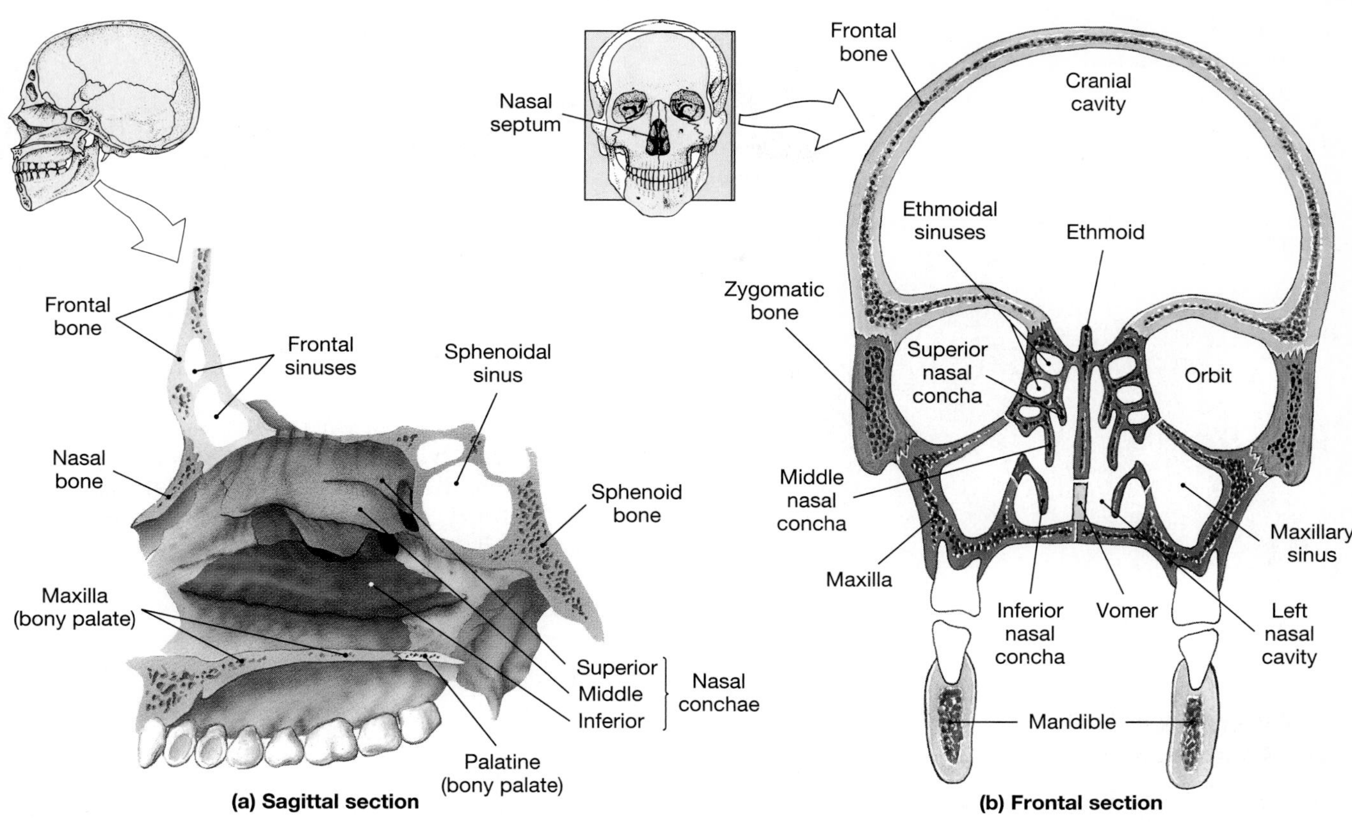
Nasal septum
Frontal bone
Cranial cavity
Ethmoidal sinuses
Ethmoid
Zygomatic bone
Superior nasal concha
Orbit
Middle nasal concha
Maxillary sinus
Maxilla
Inferior nasal concha
Vomer
Left nasal cavity
Mandible
(b) Frontal section
Frontal bone
Frontal sinuses
Sphenoidal sinus
Nasal bone
Sphenoid bone
Maxilla (bony palate)
Superior
Middle
Inferior
Nasal conchae
Palatine (bony palate)
(a) Sagittal section

● **FIGURE 7-11**

The Nasal Complex. (a) Sagittal section through the skull, with the nasal septum removed to show major features of the wall of the right nasal cavity. **(b)** A frontal section showing the positions of the paranasal sinuses. AM *Scans 1, 2*

PARANASAL SINUSES The frontal, sphenoid, ethmoid, and maxillary bones contain the **paranasal sinuses,** air-filled chambers that communicate with the nasal cavities. Figure 7-11a● shows the location of the **frontal** and **sphenoidal sinuses.** The **ethmoidal** and **maxillary sinuses** are shown in Figure 7-11b●. Sinuses make skull bones lighter and provide an extensive area of mucous epithelium. The mucous secretions are released into the nasal cavities, and the ciliated epithelium passes the mucus back toward the throat, where it is eventually swallowed. Incoming air is humidified and warmed as it flows across this thick carpet of mucus. Foreign particulate matter, such as dust or microorganisms, becomes trapped in this sticky mucus and is then swallowed. This mechanism helps protect more delicate portions of the respiratory tract.

SEPTAL DEFECTS AND SINUS PROBLEMS Flushing the nasal epithelium with mucus produced in the paranasal sinuses often succeeds in removing a mild irritant. But a viral or bacterial infection produces an inflammation of the mucous membrane of the nasal cavity. As swelling occurs, the communicating passageways narrow. Drainage of mucous slows, congestion increases, and the victim experiences headaches and a feeling of pressure within the facial bones. This condition of sinus inflammation and congestion is called **sinusitis.** The maxillary sinuses are often involved. Because gravity does little to assist mucus drainage from these sinuses, the effectiveness of the flushing action is reduced and pressure on the sinus walls typically increases.

Temporary sinus problems may accompany allergies or the exposure of the mucous epithelium to chemical irritants or invading microorganisms. Chronic sinusitis may occur as the result of a **deviated** (nasal) **septum.** In this condition the nasal septum has a bend in it, most often at the junction between the bony and cartilaginous regions. Septal deviation often blocks drainage of one or more sinuses, producing chronic cycles of infection and inflammation. A deviated septum can result from developmental abnormalities or injuries to the nose, and the condition can usually be corrected or improved by surgery.

The Zygomatic Bones

Figures 7-3c, 7-12

As noted previously, the temporal process of the **zygomatic bone** articulates with the zygomatic process of the temporal bone to form the zygomatic arch (Figure 7-3c●, p. 205. A **zygomaticofacial foramen** on the anterior surface of each zygomatic bone carries a sensory nerve innervating the cheek. The zygomatic bone also forms the lateral rim of the orbit (Figure 7-12●) and contributes to the interior orbital wall.

● **FIGURE 7-12**
The Orbital Complex. *Plate 2.3*

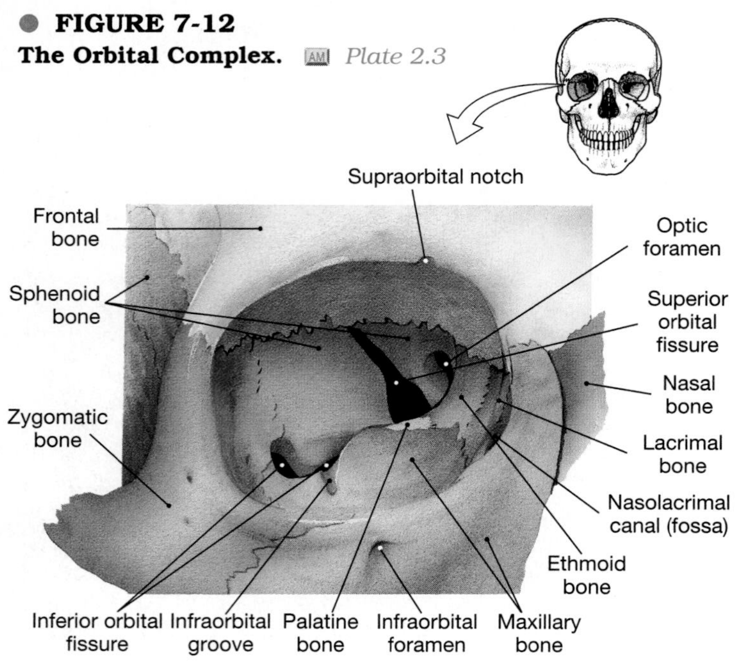
Supraorbital notch
Frontal bone
Sphenoid bone
Zygomatic bone
Optic foramen
Superior orbital fissure
Nasal bone
Lacrimal bone
Nasolacrimal canal (fossa)
Ethmoid bone
Inferior orbital fissure
Infraorbital groove
Palatine bone
Infraorbital foramen
Maxillary bone

The Orbital Complex

Figure 7-12

The **orbits** are the bony recesses that contain the eyes. Each orbit is made up of the seven bones of the **orbital complex** (Figure 7-12●). The frontal bone forms the roof, and the maxillary bone provides most of the orbital floor. Proceeding from medial to lateral, the orbital rim and the first portion of the wall are contributed by the maxilla, the lacrimal bone, and the lateral mass of the ethmoid, which articulates with the sphenoid and a small process of the palatine bone. Several prominent foramina and fissures penetrate the sphenoid or lie between the sphenoid and maxillary bone. Laterally, the sphenoid and maxillary articulate with the zygomatic bone, which forms the lateral wall and rim of the orbit.

The Lacrimal Bones

Figures 7-3d, 7-12

The paired **lacrimal bones** (*lacrima*, tear) are the smallest of the facial bones. One is situated in the medial portion of each orbit, where it articulates with the frontal, maxilla, and ethmoid (Figures 7-3d, p. 206, and 7-12●). A small passageway, the **lacrimal canal,** surrounds the tear duct as it passes toward the nasal cavity.

The Mandible

Figure 7-13

The **mandible** forms the entire lower jaw (Figure 7-13●). This bone can be subdivided into the horizontal **body** and the ascending **rami** (branches; singular *ramus*). The teeth are supported by the mandibular body. Each ramus meets the body at the mandibular **angle.** The **condylar processes** articulate with the mandibular fossae of the temporal bones at the *temporomandibular joint.* This joint is quite mobile, as evidenced by the jaw movements during chewing or talking. The disadvantage of such mobility is that the jaw can easily be dislocated by forceful protraction or lateral displacement.

At the **coronoid** (kō-RŌ-noyd) **processes** the *temporalis* muscle inserts onto the mandible. This is one of the most forceful muscles involved in closing the mouth. Anteriorly, the **mental foramina** (*mentalis*, chin) penetrate the

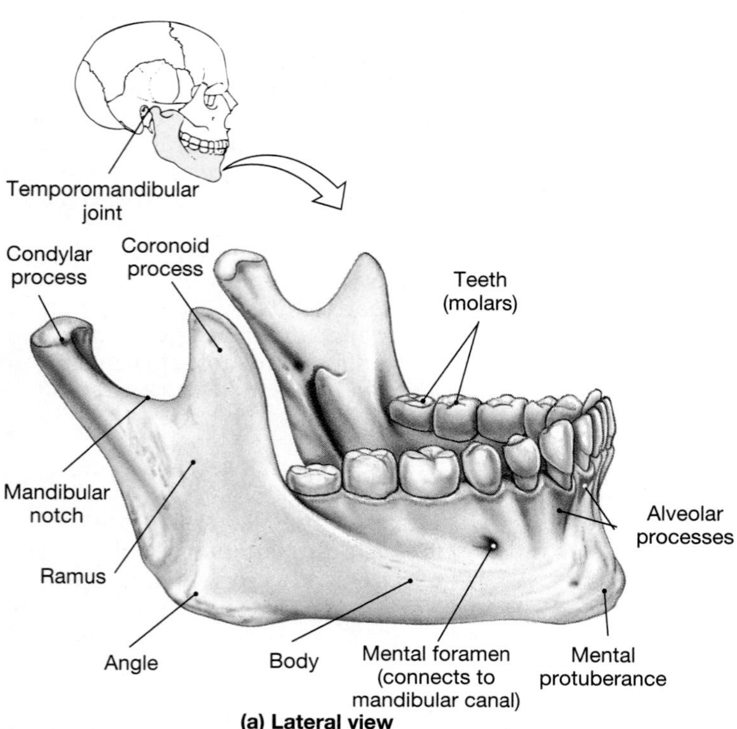
Temporomandibular joint
Condylar process
Coronoid process
Teeth (molars)
Mandibular notch
Ramus
Alveolar processes
Angle
Body
Mental foramen (connects to mandibular canal)
Mental protuberance

(a) Lateral view

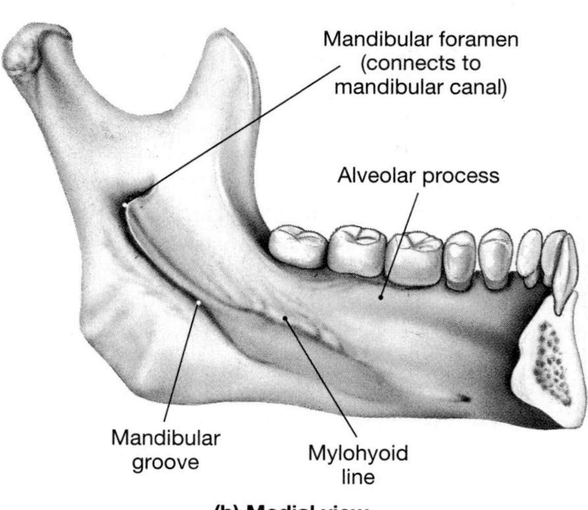
Mandibular foramen (connects to mandibular canal)
Alveolar process
Mandibular groove
Mylohyoid line

(b) Medial view

● **FIGURE 7-13**
The Mandible. (a) Lateral view. **(b)** Medial view of the left side of the mandible.

body on each side of the chin. Nerves pass through these foramina carrying sensory information from the lips and chin back to the brain. The **mandibular notch** is the depression that lies between the condylar and coronoid processes.

Medially (Figure 7-13b●), an **alveolar process** covers the alveoli and the roots of the teeth in the lower jaw. On the medial aspect of each ramus above the groove, a prominent **mandibular foramen** leads into the **mandibular canal.** This is a passageway for blood vessels to service the lower teeth and for nerves. The nerve that uses this passage carries sensory information from the teeth and gums; dentists typically anesthetize this nerve before working on the lower teeth.

TMJ SYNDROME The temporomandibular joint permits a considerable degree of mandibular movement. The connective tissue sheath, or *capsule,* that surrounds the joint is relatively loose, and the opposing bone surfaces are separated by a fibrocartilage pad. In **TMJ sydrome,** or *myofascial pain syndrome,* the mandible is pulled slightly out of alignment, usually by spasms in one of the jaw muscles. The individual experiences (1) facial pain that radiates around the ear on the affected side and (2) an inability to open the mouth fully.

TMJ syndrome is a repeating cycle of: muscle spasm → misalignment → pain → muscle spasm. It has been linked to unconscious behaviors, such as grinding of the teeth during sleep, and emotional stress. Treatment focuses on breaking the cycle of pain and muscle spasm and, when necessary, providing emotional support. The application of heat to the affected joint, coupled with the use of anti-inflammatory drugs or local anesthetics or both, may be helpful. If teeth grinding is suspected, special mouth guards may be worn at night.

The Hyoid Bone

Figure 7-14

The **hyoid bone** lies below the skull, suspended by the **stylohyoid ligaments** (Figure 7-14●). The U-shaped **body** of the hyoid serves as a base for several muscles concerned with movements of the tongue and larynx. Because muscles and ligaments form the only connections between the hyoid and other skeletal structures, the entire complex is quite mobile. The larger spinous processes on the hyoid are the **greater cornua** (horns; singular *cornu*), which help support the larynx and serve as the base for muscles that move the tongue. The **lesser cornua** are connected to the stylohyoid ligaments, and from these ligaments the hyoid and larynx hang beneath the skull like a child's swing from the limb of a tree.

Table 7-1 summarizes information concerning the foramina and fissures introduced thus far. This reference will be especially important in later chapters dealing with the nervous and cardiovascular systems.

❑ The Skulls of Infants and Children

Figure 7-15

Many different centers of ossification are involved in the formation of the skull, but as development proceeds, fusion of the centers produces a smaller number of composite bones. For example, the sphenoid begins as 14 separate ossification centers. At birth fusion has not been completed, and there are two frontal bones, four occipital bones, and a number of sphenoid and temporal elements.

The skull organizes around the developing brain, and as the time of birth approaches, the brain enlarges rapidly. Although the bones of the skull are also growing, they fail to keep pace, and at birth the cranial bones are connected by areas of fibrous connective tissue. These connections are quite flexible, and the skull can be distorted without damage. Such distortion normally occurs during delivery and eases the passage of the infant along the birth canal.

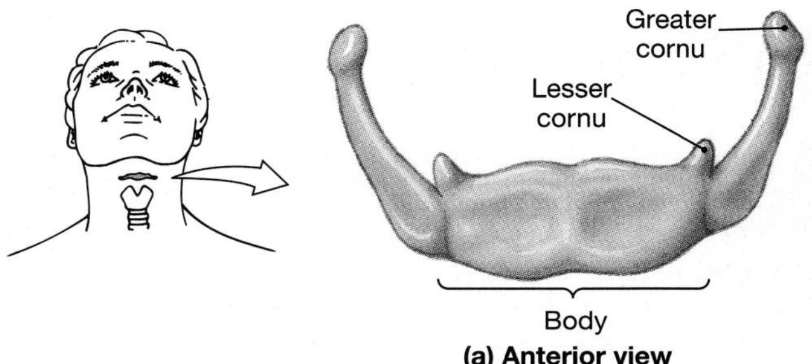
Greater cornu
Lesser cornu
Body
(a) Anterior view

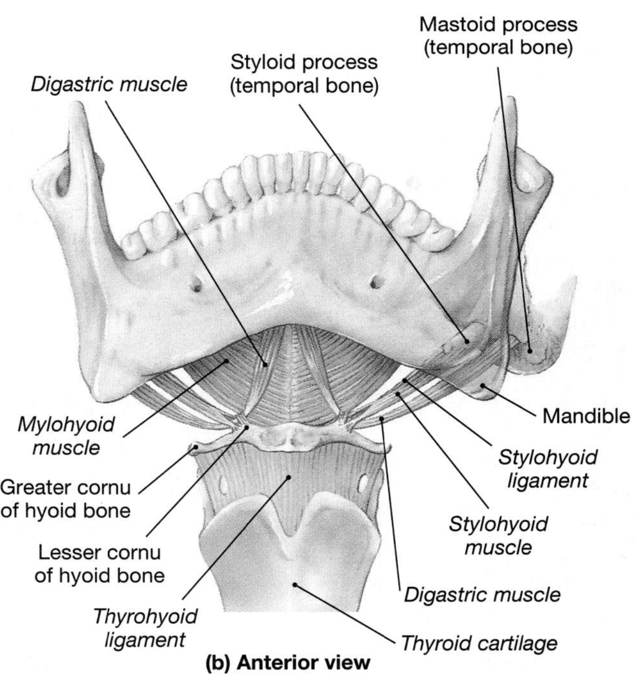
Digastric muscle
Styloid process (temporal bone)
Mastoid process (temporal bone)
Mandible
Mylohyoid muscle
Greater cornu of hyoid bone
Lesser cornu of hyoid bone
Thyrohyoid ligament
Stylohyoid ligament
Stylohyoid muscle
Digastric muscle
Thyroid cartilage
(b) Anterior view

● **FIGURE 7-14**

The Hyoid Bone. (a) The hyoid bone. **(b)** The hyoid bone shown in relation to the larynx and muscles of the mandible and tongue.

TABLE 7-1 A Key to the Foramina and Fissures of the Skull

		Major Structures Using Passageway	
Bone	*Foramen/ Fissure*	*Neural Tissue*	*Vessels and Other Structures*
OCCIPITAL	Foramen magnum	Medulla oblongata (last portion of brain) and accessory nerve (XI).	Vertebral arteries. Supporting membranes around central nervous system.
	Hypoglossal canal	Hypoglossal nerve (XII) provides motor control to muscles of the tongue.	
With temporal	Jugular foramen	Glossopharyngeal nerve (IX), vagus nerve (X), accessory nerve (XI). Nerve IX provides taste sensation; X is important for visceral functions; XI innervates important muscles of the back and neck.	Internal jugular vein, important vein returning blood from brain to heart.
FRONTAL	Supraorbital foramen (or notch)	Supraorbital nerve, sensory branch of the ophthalmic nerve, innervating the eyebrow, eyelid, and frontal sinus.	Supraorbital artery delivers blood to same region.
LACRIMAL	Lacrimal foramen		Tear duct, drains into the nasal chamber.
TEMPORAL	Mastoid foramen		Vessels to membranes around central nervous system.
	Stylomastoid foramen	Facial nerve (VII) provides motor control of facial muscles.	
	Carotid foramen		Internal carotid artery, major arterial supply to the brain.
	External auditory canal		Air in canal conducts sound to eardrum.
	Internal acoustic canal	Vestibulocochlear nerve (VIII) goes to sense organs for hearing and balance. Facial nerve (VII) enters here, exits at stylomastoid foramen.	Internal acoustic artery to inner ear.
SPHENOID	Optic foramen	Optic nerve (II) brings information from the eye to the brain.	Ophthalmic artery brings blood into orbit.
	Superior orbital fissure	Oculomotor nerve (III), trochlear nerve (IV), ophthalmic branch of trigeminal nerve (V), abducens nerve (VI). Ophthalmic nerve provides sensory information about eye and orbit; other nerves control muscles that move the eye.	Ophthalmic vein returns blood from orbit.
	Foramen rotundum	Maxillary branch of trigeminal nerve (V) provides sensation to the face.	
	Foramen ovale	Mandibular branch of trigeminal nerve (V) controls the muscles that move the lower jaw and provides sensory information from that area.	
	Foramen spinosum		Vessels to membranes around central nervous system.
With temporal and occipital	Foramen lacerum		Internal carotid artery leaves carotid canal, enters cranium via foramen lacerum.
With maxilla	Inferior orbital fissue	Maxillary branch of trigeminal nerve (V). See *Foramen rotundum.*	
ETHMOID	Cribriform plate	Olfactory nerve (I) provides sense of smell.	
MAXILLA	Infraorbital foramen	Infraorbital nerve, maxillary branch of trigeminal nerve (V) from the inferior orbital fissure to face.	Infraorbital artery with same distribution.
MANDIBLE	Mental foramen	Mental nerve, sensory branch of the mandibular nerve, provides sensation from the chin and lips.	Mental vessels to chin and lips.
	Mandibular foramen	Inferior alveolar nerve, sensory branch of the mandibular nerve, provides sensation from the gums, teeth.	Inferior alveolar vessels supply same region.
ZYGOMATIC	Zygomaticofacial foramen	Zygomaticofacial nerve, sensory branch of mandibular nerve to cheek.	

The fibrous areas between the cranial bones are known as **fontanels** (fon-tuh-NELZ; sometimes spelled *fontanelles*) (Figure 7-15●).

- The *anterior fontanel* is the largest fontanel. It lies at the intersection of the frontal, sagittal, and coronal sutures.
- The *posterior fontanel* is at the junction between the lambdoidal and sagittal sutures.
- The *sphenoidal fontanels* are at the junctions between the squamosal sutures and the coronal suture.
- The *mastoid fontanels* are at the junctions between the squamosal sutures and the lambdoidal suture.

The posterior, sphenoidal, and mastoid fontanels disappear within a month or two after birth. The anterior fontanel persists for a longer period, usually until a child is nearly 2 years old. Even after the fontanels disappear, the bones of the skull remain separated by fibrous connections.

The skulls of infants and adults differ in terms of the shape and structure of cranial elements, and this difference accounts for variations in proportions as well as in size. The most significant growth in the skull occurs before age 5, for at that time the brain stops growing and the cranial sutures develop. As a result, when compared with the skull as a whole, the cranium of a young child is relatively larger than that of an adult. The growth of the cranium is usually coordinated with the expansion of the brain. If one or more sutures form before the brain stops growing, the skull will be abnormal in shape or size or both.

CRANIOSTENOSIS AND MICROCEPHALY Unusual distortions of the skull result from the premature closure of one or more fontanels, a condition called *craniostenosis* (krā-nē-ō-sten-Ō-sis; *stenosis*, narrowing). As the brain continues to enlarge, the rest of the skull accommodates it. A long and narrow head will be produced by early closure of the sagittal suture, whereas a very broad skull results if the coronal suture forms prematurely. Closure of *all* of the cranial sutures restricts the development of the brain, and surgery must be performed to prevent brain damage. If the brain enlargement stops because of genetic or developmental abnormalities, however, skull growth ceases as well. This condition, which results in a very undersized head, is called *microcephaly* (mī-krō-SEF-uh-lē).

- What are the functions of the paranasal sinuses?
- Why would a fracture of the coronoid process of the mandible make it difficult to close the mouth?
- What symptoms would you expect to see in a person suffering from a fractured hyoid bone?

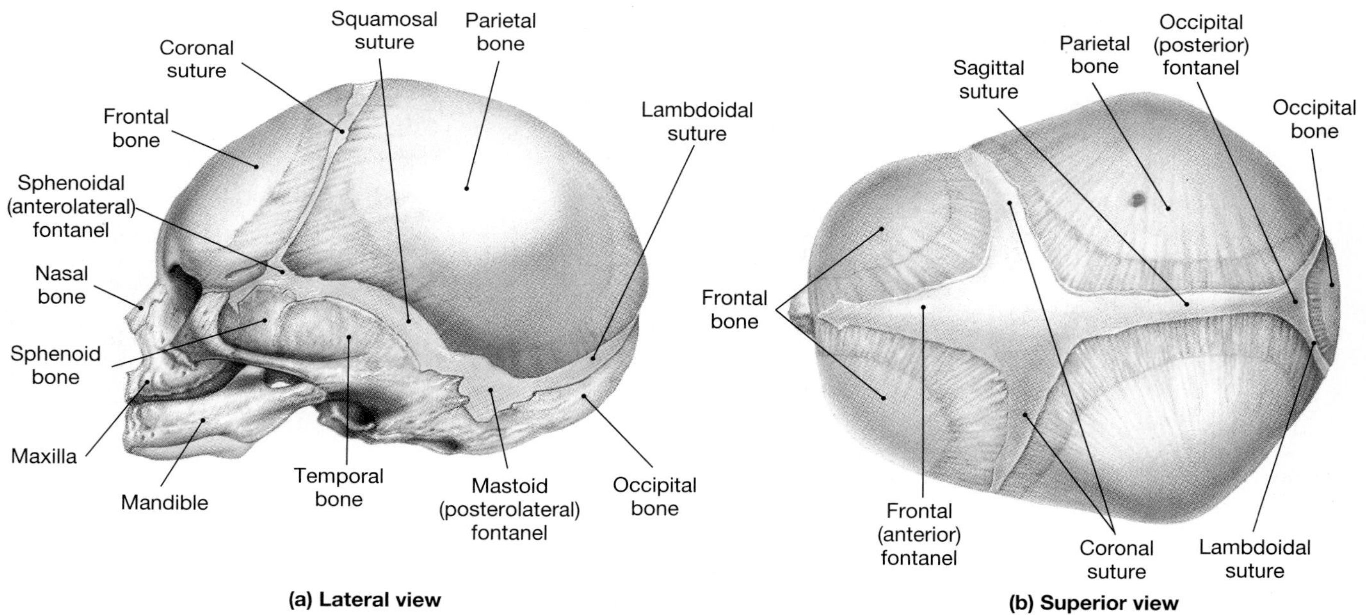
Squamosal suture
Parietal bone
Coronal suture
Frontal bone
Lambdoidal suture
Sphenoidal (anterolateral) fontanel
Nasal bone
Sphenoid bone
Maxilla
Mandible
Temporal bone
Mastoid (posterolateral) fontanel
Occipital bone
(a) Lateral view
Occipital (posterior) fontanel
Parietal bone
Sagittal suture
Occipital bone
Frontal bone
Frontal (anterior) fontanel
Coronal suture
Lambdoidal suture
(b) Superior view

● **FIGURE 7-15**

The Skull of an Infant. (a) Lateral view. The skull of an infant contains a greater number of individual bones than that of an adult. Many of the bones will eventually fuse; thus there will be fewer bones in the adult skull. The flat bones of the skull are separated by areas of fibrous connective tissue, allowing for cranial expansion and the distortion of the skull during birth. The large fibrous areas are called fontanels. By about age 4 these areas will disappear. **(b)** Superior view. AM *Plate 2.5*

EMBRYOLOGY SUMMARY Development of the Skull

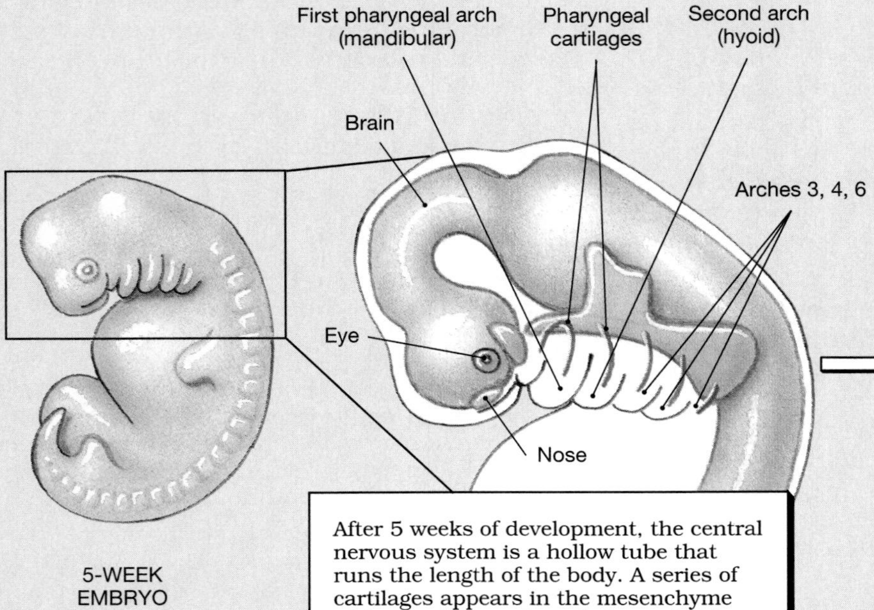

First pharyngeal arch (mandibular)
Pharyngeal cartilages
Second arch (hyoid)
Brain
Arches 3, 4, 6
Eye
Nose
5-WEEK EMBRYO

After 5 weeks of development, the central nervous system is a hollow tube that runs the length of the body. A series of cartilages appears in the mesenchyme of the head beneath and alongside of the expanding brain and around the developing nose, eyes, and ears. These cartilages are shown in light blue. Five additional pairs of cartilages develop in the walls of the pharynx. These cartilages, shown in dark blue, are located within the **pharyngeal,** or **branchial, arches**. (*Branchial* refers to gills—in fish the caudal arches develop into skeletal supports for the gills.) The first arch, or **mandibular arch**, is the largest.

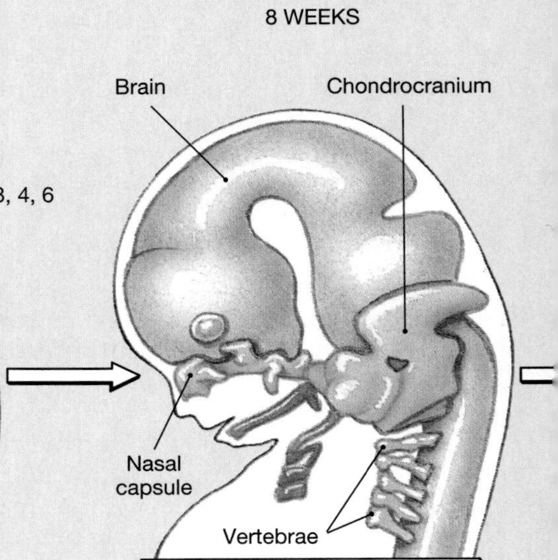

8 WEEKS
Brain
Chondrocranium
Nasal capsule
Vertebrae

The cartilages associated with the brain enlarge and fuse forming a cartilaginous **chondrocranium** (kon-drō-KRĀ-nē-yum; *chondros*, cartilage + *cranium*, skull) that cradles the brain and sense organs. At 8 weeks its walls and floor are incomplete, and there is no roof.

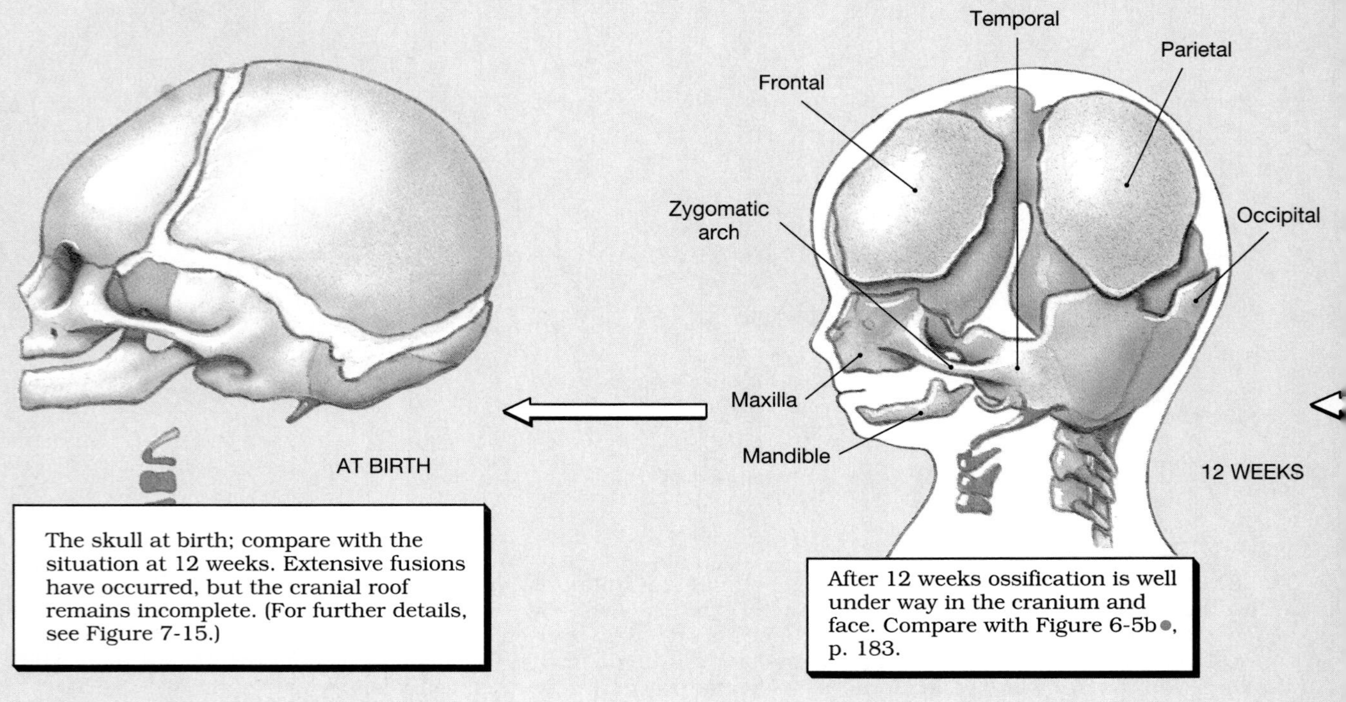

Temporal
Parietal
Frontal
Zygomatic arch
Occipital
Maxilla
Mandible
AT BIRTH
12 WEEKS

The skull at birth; compare with the situation at 12 weeks. Extensive fusions have occurred, but the cranial roof remains incomplete. (For further details, see Figure 7-15.)

After 12 weeks ossification is well under way in the cranium and face. Compare with Figure 6-5b, p. 183.

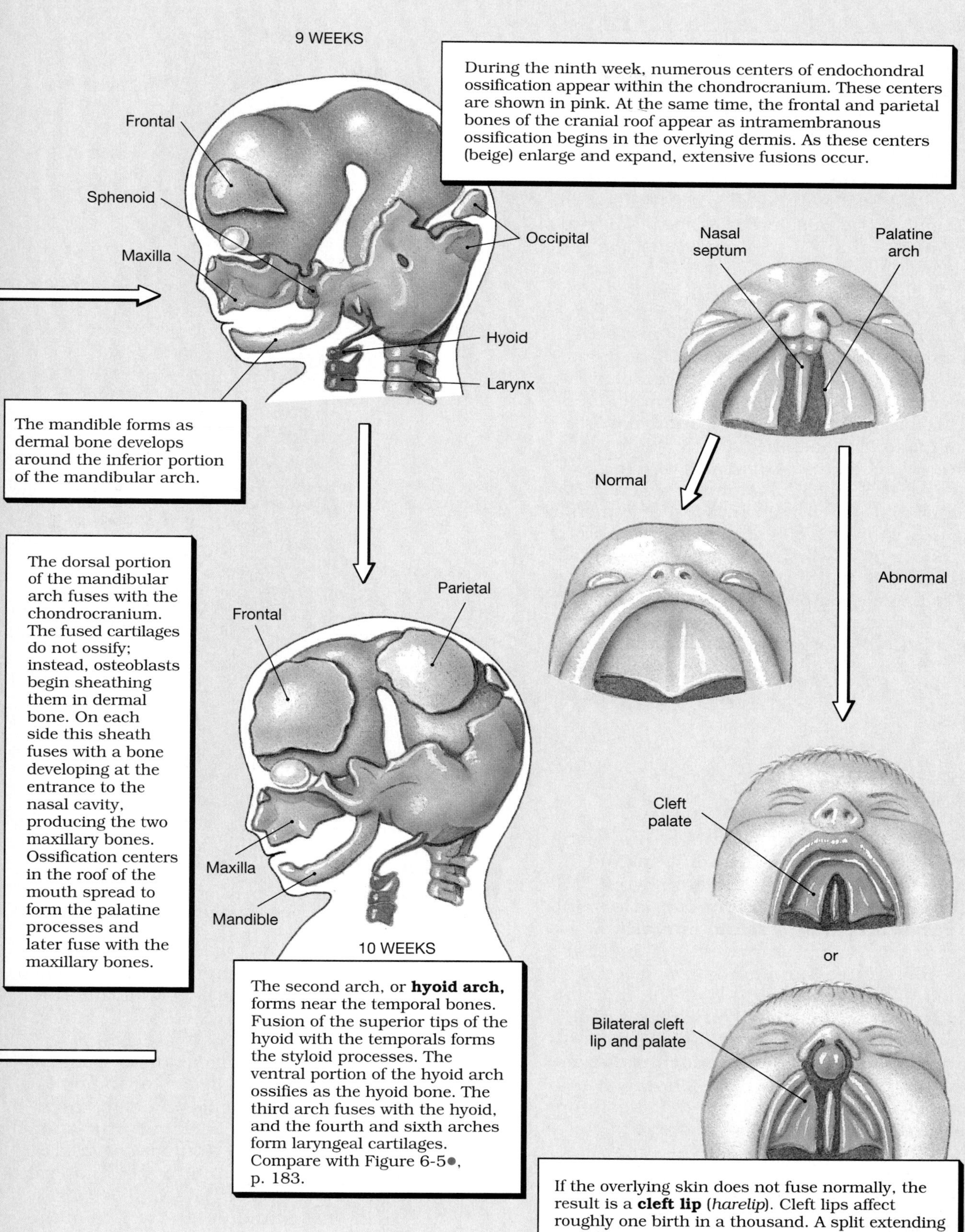
9 WEEKS
During the ninth week, numerous centers of endochondral ossification appear within the chondrocranium. These centers are shown in pink. At the same time, the frontal and parietal bones of the cranial roof appear as intramembranous ossification begins in the overlying dermis. As these centers (beige) enlarge and expand, extensive fusions occur.
Frontal
Sphenoid
Maxilla
Occipital
Hyoid
Larynx
The mandible forms as dermal bone develops around the inferior portion of the mandibular arch.
Nasal septum
Palatine arch
Normal
Abnormal
The dorsal portion of the mandibular arch fuses with the chondrocranium. The fused cartilages do not ossify; instead, osteoblasts begin sheathing them in dermal bone. On each side this sheath fuses with a bone developing at the entrance to the nasal cavity, producing the two maxillary bones. Ossification centers in the roof of the mouth spread to form the palatine processes and later fuse with the maxillary bones.
Frontal
Parietal
Maxilla
Mandible
10 WEEKS
The second arch, or **hyoid arch,** forms near the temporal bones. Fusion of the superior tips of the hyoid with the temporals forms the styloid processes. The ventral portion of the hyoid arch ossifies as the hyoid bone. The third arch fuses with the hyoid, and the fourth and sixth arches form laryngeal cartilages. Compare with Figure 6-5●, p. 183.
Cleft palate
or
Bilateral cleft lip and palate
If the overlying skin does not fuse normally, the result is a **cleft lip** (*harelip*). Cleft lips affect roughly one birth in a thousand. A split extending into the orbit and palate is called a **cleft palate**. Cleft palates are half as common as cleft lips. Both conditions can be corrected surgically.

The Vertebral Column

Figure 7-16

The rest of the axial skeleton is subdivided on the basis of vertebral structure. The adult **vertebral column** consists of 26 bones: the **vertebrae** (24), the **sacrum,** and the **coccyx.** The vertebrae provide a column of support, bearing the weight of the head, neck, trunk, and ultimately transferring the weight to the appendicular skeleton of the lower limbs. They also protect the spinal cord and help maintain an upright body position, as in sitting or standing.

The vertebral column is divided into regions. Beginning at the skull, the regions are: *cervical, thoracic, lumbar, sacral,* and *coccygeal* (Figure 7-16●). Seven **cervical vertebrae** constitute the neck and extend inferiorly to the trunk. Twelve **thoracic vertebrae** form the midback region; each articulates with one or more pairs of ribs. Five **lumbar vertebrae** form the lower back; the fifth articulates with the sacrum, which in turn articulates with the coccyx. The cervical, thoracic, and lumbar regions consist of individual vertebrae. During development, the sacrum originates as a group of five vertebrae, and the **coccyx** (KOK-siks), or "tailbone," begins as three to five very small vertebrae. The vertebrae of the sacrum usually complete their fusion by age 25. The distal coccygeal vertebrae do not complete their ossification before puberty, and thereafter fusion occurs at a variable pace. The total length of the vertebral column of an adult averages 71 cm (28 in.).

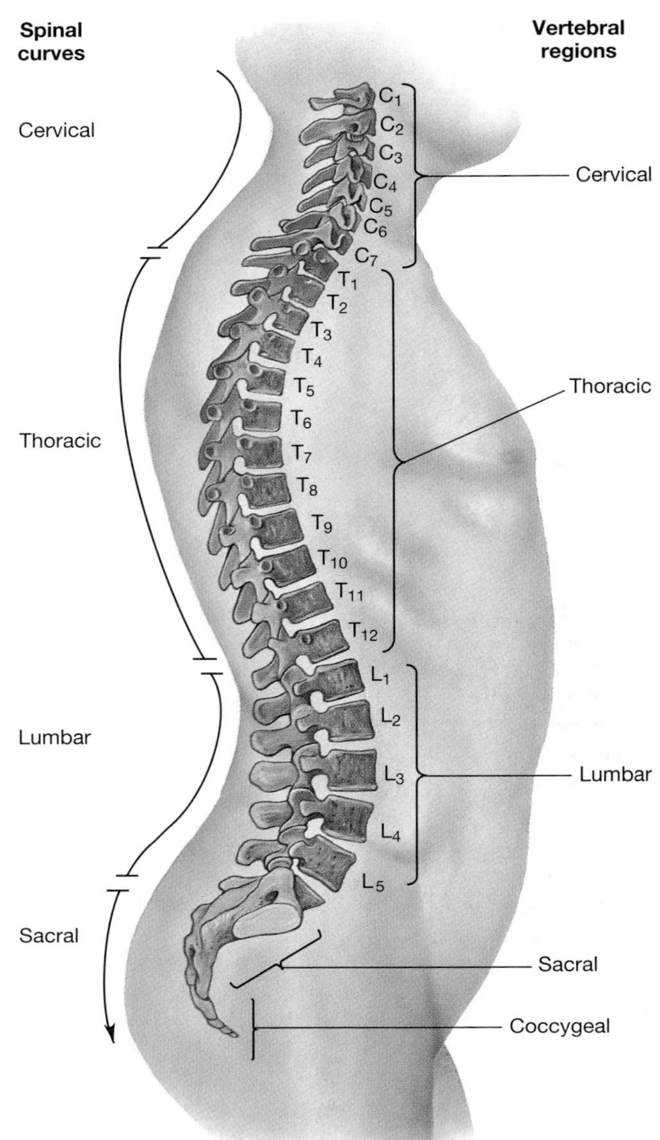

Spinal curves
Vertebral regions
Cervical
Thoracic
Lumbar
Sacral
C1
C2
C3
C4
C5
C6
C7
T1
T2
T3
T4
T5
T6
T7
T8
T9
T10
T11
T12
L1
L2
L3
L4
L5
Cervical
Thoracic
Lumbar
Sacral
Coccygeal

● **FIGURE 7-16**

The Vertebral Column. The major divisions of the vertebral column, showing the four spinal curves. AM *Scan 3*

Spinal Curvature

Figures 7-16, 7-17

The vertebrae do not form a straight and rigid structure. A lateral view of the adult spinal column shows four **spinal curves** (Figure 7-16●): (1) **cervical curvature,** (2) **thoracic curvature,** (3) **lumbar curvature,** and (4) **sacral curvature.**

The sequence of appearance of the spinal curvatures is illustrated from fetus to newborn, to child, and to adult in Figure 7-17●. The thoracic and sacral curves are called **primary curves** because they appear late in fetal development. These are also called **accommodation curves** because they accommodate the thoracic and abdominopelvic viscera. In the vertebral column of the newborn only the primary curves are present. The lumbar and cervical curves, known as **secondary curves,** do not appear until several months after birth. These are also called **compensation curves** because they help shift the trunk weight over the lower limbs. The cervical curve develops as the infant learns to balance the head upright, and the lumbar curve develops with the ability to stand. Both compensations become accentuated as the toddler learns to walk and run. All four curves are fully developed by the time a child is 10 years old.

There are several abnormal distortions of spinal curvature that may appear during childhood and adolescence. Examples include *kyphosis* (kī-FŌ-sis; exaggerated thoracic curvature), *lordosis* (lor-DŌ-sis; exaggerated lumbar curvature), and *scoliosis* (skō-lē-Ō-sis; an abnormal lateral curvature). AM *Kyphosis, Lordosis, and Scoliosis*

When a person is standing, the weight of the body must be transmitted through the vertebral column to the hips and ultimately to the lower extremities. Yet most of the body weight lies in

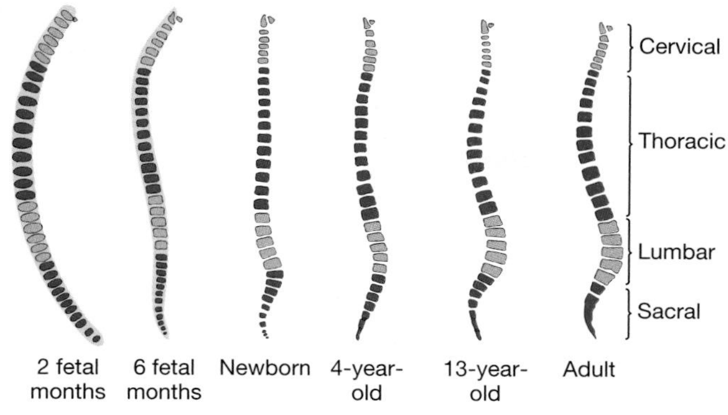
Cervical
Thoracic
Lumbar
Sacral
2 fetal months
6 fetal months
Newborn
4-year-old
13-year-old
Adult

FIGURE 7-17
The Development of Spinal Curvature. Primary curves are shown in blue. The secondary spinal curves (red) balance the body weight over the legs; these curves do not develop fully until the individual begins walking.

front of the vertebral column. The various curves bring that weight in line with the body axis. Consider what a person does automatically when standing with a heavy object hugged to the chest. The person avoids toppling forward by exaggerating the lumbar curvature and moving the weight back toward the body axis. This posture can lead to discomfort at the base of the spinal column. For example, women in the last three months of pregnancy often develop chronic back pain from the changes in lumbar curvature that must adjust for the increasing weight of the fetus. African and South American natives often carry heavy objects balanced on their heads, as seen in the photograph at the start of this chapter. This practice increases the load on the vertebral column, but the spinal curves are not affected because the weight is aligned with the axis of the spine.

Vertebral Anatomy

Figure 7-18

Each vertebra consists of three basic parts: (1) a *body* (*centrum*), (2) a *vertebral arch*, and (3) *articular processes* (Figure 7-18●).

The Vertebral Body

The **body,** or *centrum* (plural *centra*), is the part of a vertebra that transfers weight along the axis of the vertebral column. Each vertebra articulates with neighboring vertebrae; the bodies are interconnected by ligaments and separated by pads of fibrocartilage, the **intervertebral discs.**

The Vertebral Arch

The **vertebral arch,** also called the *neural arch*, surrounds each **vertebral foramen** that in life encloses a segment of the spinal cord. The vertebral arch has a floor (the posterior surface of the body); walls, called *pedicles* (PE-di-kulz); and a roof, formed by the *laminae* (LA-mi-nē; singular *lamina*, a thin plate). The **pedicles** arise along the posterior and lateral margins of the body. The **laminae** on either side extend dorsally and medially to complete the roof.

The vertebral foramina of successive vertebrae collectively form the **vertebral canal** that encloses the entire spinal cord. In the condition called *spina bifida* (SPĪ-nuh BI-fi-duh), the vertebral laminae fail to unite during development. The neural arch is incomplete, and the membranes, or *meninges*, that line the dorsal body cavity bulge outward. Mild cases involving the sacral and lumbar regions may pass unnoticed; in severe cases the entire spinal column and skull are affected. AM *Spina Bifida*

A **spinous process,** also known as a *spinal process*, projects posteriorly from the point where the vertebral laminae fuse to complete the vertebral arch. The spinous processes can be seen and felt through the skin of the back when the spine is flexed. **Transverse processes** project laterally or dorsolaterally on both sides from the point where the laminae join the pedicles. These processes are sites of muscle attachment, and they may also articulate with the ribs.

The Articular Processes

The **articular processes** also arise at the junction between the pedicles and laminae. There is a superior and inferior articular process on each side of the vertebra. The **superior articular processes** project superiorly; the **inferior articular processes** project inferiorly.

Vertebral Articulation

The inferior articular processes of one vertebra articulate with the superior articular processes of the vertebra below. Each articular process has a polished concave surface, called an **articular facet.** The superior processes have articular facets on their dorsal surfaces, whereas the inferior processes articulate along their ventral surfaces.

Adjacent vertebral bodies are separated by intervertebral discs, and there are gaps between the pedicles of successive vertebrae. These gaps, called **intervertebral foramina,** permit the passage of nerves running to or from the enclosed spinal cord.

✔ Why are there fewer vertebrae in the vertebral column of an adult than in the vertebral column of a newborn?

✔ What is the importance of the secondary curves of the spine?

✔ When you run your finger along a person's spine, what part of the vertebrae are you feeling just beneath the skin?

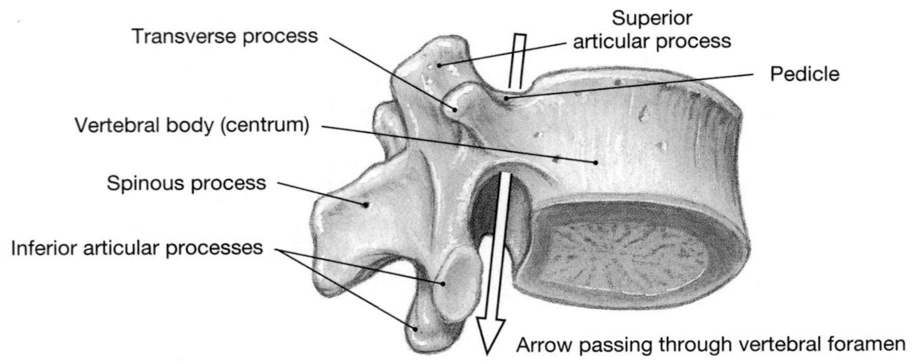
Transverse process
Superior articular process
Pedicle
Vertebral body (centrum)
Spinous process
Inferior articular processes
Arrow passing through vertebral foramen

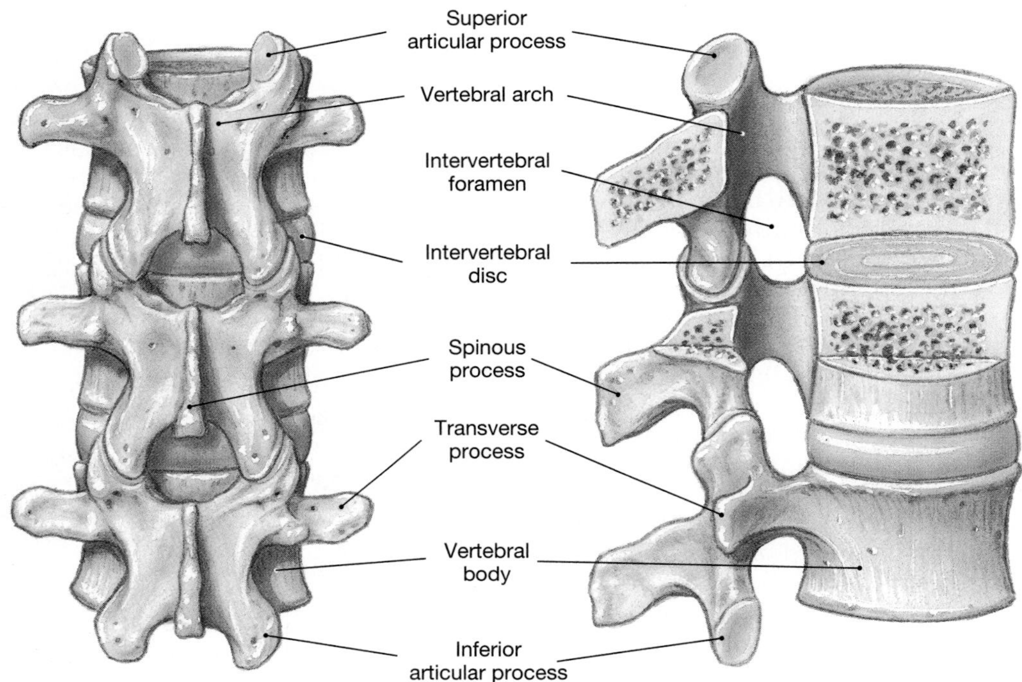
Superior articular process
Vertebral arch
Intervertebral foramen
Intervertebral disc
Spinous process
Transverse process
Vertebral body
Inferior articular process

● **FIGURE 7-18**
Vertebral Anatomy. The anatomy of a typical vertebra and the arrangement of articulations between vertebrae.

❑ Vertebral Regions

Figure 7-16

In references to the vertebrae, a capital letter indicates the vertebral region, and a number indicates the vertebra in question, starting with the cervical vertebra closest to the skull. For example, C_3 refers to the third cervical vertebra, with C_1 in contact with the skull; L_4 is the fourth lumbar vertebra, with L_1 in contact with the last thoracic vertebra (Figure 7-16●). This shorthand will be used throughout the text.

Although each vertebra bears characteristic markings and articulations, focus on the general characteristics of each region and how the regional variations determine the vertebral group's function. Table 7-2 compares typical vertebrae from the three regions of the vertebral column.

Cervical Vertebrae

Figures 7-19, 7-20

The seven cervical vertebrae are the smallest of the vertebrae (Figure 7-19●). They extend from the occipital bone of the skull to the thorax. Notice that the body of a cervical vertebra is small compared with the size of the vertebral foramen. At this level the spinal cord still contains most of the axons that connect the brain to the rest of the body. As you continue along the vertebral canal, the diameter of the spinal cord decreases, and so does the diameter of the vertebral arch. On the other hand, cervical vertebrae support only the weight of the head, so the vertebral bodies can be relatively small and light. As you continue inferiorly along the column, the loading increases and the bodies gradually enlarge.

TABLE 7-2 Regional Differences in Vertebral Structure and Function

	Type (Number)		
	Cervical Vertebrae (7)	*Thoracic Vertebrae (12)*	*Lumbar Vertebrae (5)*
Location	Neck	Chest	Lower back
Body	Small, oval, curved faces	Medium, heart-shaped, flat faces, facets for rib articulations	Massive, oval flat faces
Vertebral foramen	Large	Smaller	Smallest
Spinous process	Long, split tip, points inferiorly	Long, slender, not split, points inferiorly	Blunt, broad, points posteriorly
Transverse process	Has transverse foramen	All but two (T_{11}, T_{12}) have facets for rib articulations	Short, no articular facets or transverse foramen
Functions	Support skull, stabilize relative positions of brain and spinal cord, and allow controlled head movement	Support weight of head, neck, upper limbs, chest and organs of thoracic cavity; articulate with ribs to allow changes in volume of thoracic cage	Support weight of head, neck, upper limbs, and trunk

In a cervical vertebra, the superior surface of the body is concave from side to side, and it slopes, with the anterior edge inferior to the posterior edge. The spinous process is relatively stumpy, usually shorter than the diameter of the vertebral foramen, and the tip of each process, other than that of C_7, bears a prominent notch. A notched spinous process is described as **bifid** (BĪ-fid; *bi*, two + *findere*, to cut).

Laterally, the transverse processes are fused to the **costal processes** that originate near the ventrolateral portion of the body. The costal and transverse processes encircle prominent, round, **transverse foramina**. In life these passageways protect the *vertebral arteries* and *vertebral veins*, important blood vessels servicing the brain.

This description would be adequate to identify all but the first two cervical vertebrae. When cervical vertebrae C_3–C_7 articulate, their interlocking bodies permit a relatively greater degree of flexibility than do those of other regions. The first two cervical vertebrae are unique, and the seventh is modified. Table 7-2 summarizes the features of cervical vertebrae.

THE ATLAS (C_1) The **atlas** (C_1) holds up the head, articulating with the occipital condyles of the skull (Figure 7-19c●). It is named after Atlas, a figure in Greek mythology who held up the world. The articulation between the occipital condyles and the atlas is a joint that permits nodding (as when indicating "yes"). The atlas can be distinguished from the other vertebrae by the following features: (1) lack of a body or spinous process; (2) possession of semicircular **anterior** and **posterior vertebral arches,** each containing **anterior** and **posterior tubercles;** and (3) presence of oval **superior facets** and round **inferior articular facets.**

The atlas articulates with the second cervical vertebra, the *axis*. This articulation permits rotation (as when shaking the head to indicate "no").

THE AXIS (C_2) During development, the body of the atlas fuses to the body of the second cervical vertebra, called the **axis** (C_2) (Figure 7-19c●). This fusion creates the prominent **dens** (DENZ; *tooth*), or **odontoid** (ō-DON-toyd) **process** (*odontos*, tooth) of the axis. A transverse ligament binds the dens to the inner surface of the atlas, forming a pivot for rotation of the atlas and skull. Important muscles controlling the position of the head and neck attach to the especially robust spinous process of the axis.

In a child the fusion between the dens and axis is incomplete, and impacts or even severe shaking can cause dislocation of the dens and severe damage to the spinal cord. In an adult, a blow to the base of the skull can be equally dangerous because a dislocation of the axis-atlas joint can force the dens into the base of the brain, with fatal results.

THE VERTEBRA PROMINENS (C_7) The transition from one vertebral region to another is not abrupt, and the last vertebra of one region usually resembles the first vertebra of the next. The **vertebra prominens** (C_7) has a long, slender spinous process that ends in a broad tubercle that can be felt beneath the skin at the base of the neck. This vertebra, shown in Figure 7-20●, is the interface between the cervical curve, which arches forward, and the thoracic curve, which arches backward. The transverse processes are large, providing additional surface area for muscle attachment, and the transverse foramina are either reduced or absent. A large elastic ligament, the **ligamentum nuchae** (li-guh-MEN-tum NOO-kē; *nucha,* nape) begins at the vertebra prominens and extends cranially to an insertion along the external occipital crest. Along the way, it attaches to the spinous processes of the other cervical vertebrae. When the head is upright, this ligament acts like the string on a bow, maintaining the cervical curvature without muscular effort. If the neck has been bent forward, the elasticity in this ligament helps return the head to an upright position.

The head is relatively massive, and it sits atop the cervical vertebrae like a soup bowl on the tip of a finger. With this arrangement, small muscles can produce significant effects by tipping the balance one way or another. But if the body suddenly changes position, as in a fall or during rapid acceleration (a jet taking off) or deceleration (a car crash), the balancing muscles are not strong enough to stabilize the head. A dangerous partial or complete dislocation of the cervical vertebrae can result, with injury to muscles and ligaments and potential injury to the spinal cord. The term **whiplash** is used to describe such an injury, because the movement of the head resembles the cracking of a whip.

Thoracic Vertebrae

Figure 7-20

There are 12 thoracic vertebrae. A typical thoracic vertebra (Figure 7-20●) has a distinctive heart-shaped body that is more massive than that of a cervical vertebra. The vertebral foramen is relatively smaller, and the long, slender spinous process projects in a posterior and inferior direction. The spinous processes of T_{10}, T_{11}, and T_{12} increasingly resemble those of the lumbar series, as the transition between the thoracic and lumbar curvatures approaches. Because of the weight carried by the lower thoracic and lumbar vertebrae, it is difficult to stabilize the transition between the thoracic and lumbar curves. As a result, compression frac-

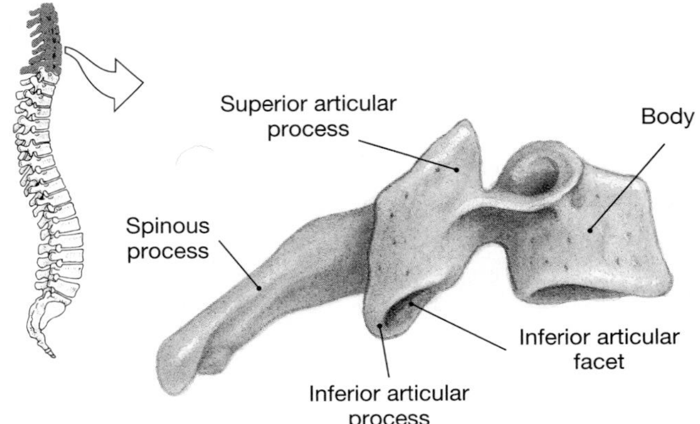
Superior articular process
Body
Spinous process
Inferior articular facet
Inferior articular process

(a) Typical cervical vertebra (lateral view)

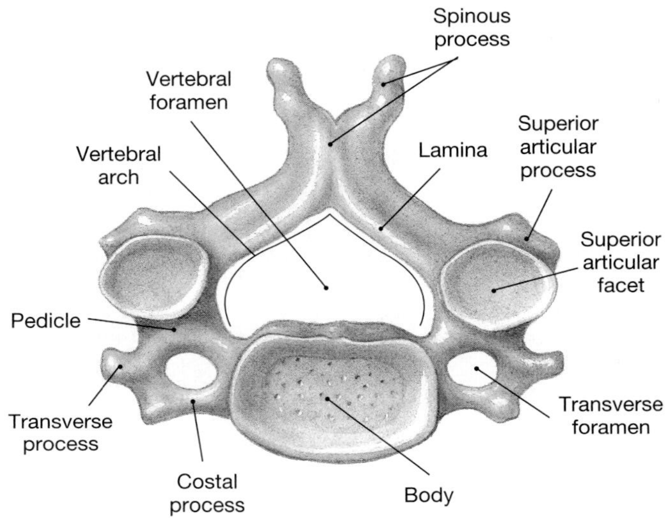
Spinous process
Vertebral foramen
Superior articular process
Vertebral arch
Lamina
Superior articular facet
Pedicle
Transverse foramen
Transverse process
Costal process
Body

(b) Typical cervical vertebra (superior view)

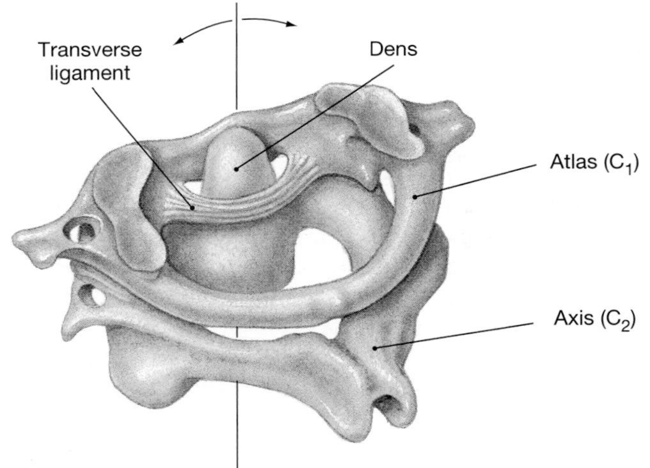
Transverse ligament
Dens
Atlas (C1)
Axis (C2)

(c) The atlas/axis complex

● **FIGURE 7-19**

The Cervical Vertebrae. (a) Lateral view of a typical cervical vertebra (C_3–C_6). **(b)** Superior view of the same vertebra. Note the characteristic features listed in Table 7-2. **(c)** The atlas (C_1) and axis (C_2). AM *Scan 3*

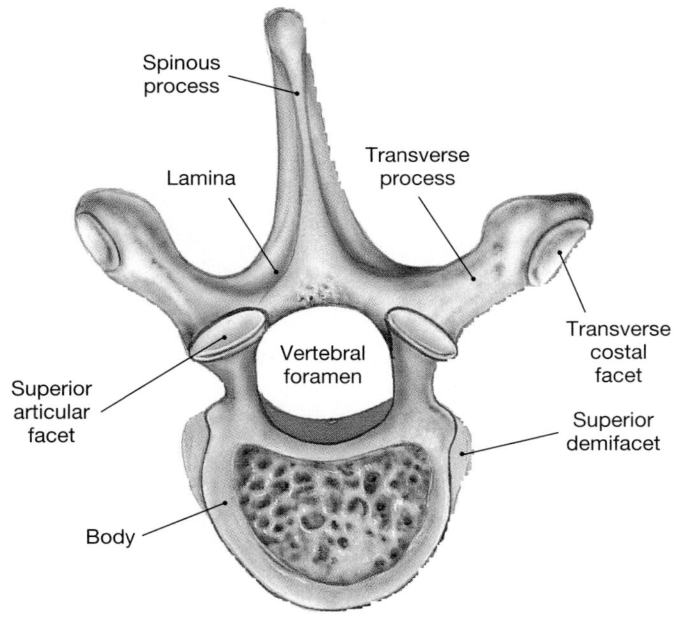
Spinous process
Lamina
Transverse process
Transverse costal facet
Superior articular facet
Vertebral foramen
Superior demifacet
Body

(a) Thoracic vertebra, superior view

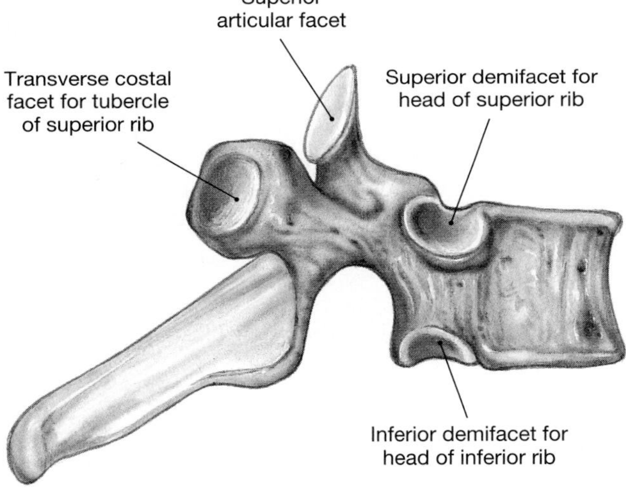
Superior articular facet
Transverse costal facet for tubercle of superior rib
Superior demifacet for head of superior rib
Inferior demifacet for head of inferior rib

(b) Thoracic vertebra, lateral view

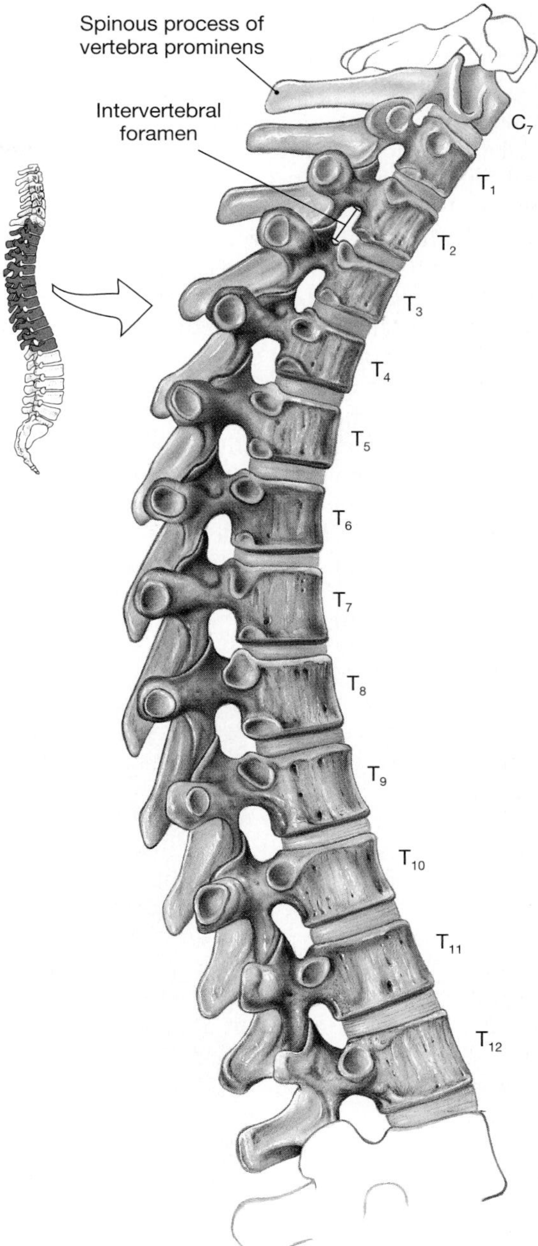
Spinous process of vertebra prominens
Intervertebral foramen
C7
T1
T2
T3
T4
T5
T6
T7
T8
T9
T10
T11
T12

(c) Thoracic vertebrae, lateral view

● **FIGURE 7-20**

The Thoracic Vertebrae. (a) Thoracic vertebra, superior view. **(b)** Thoracic vertebra, lateral view. Note the characteristic features listed in Table 7-2. **(c)** Lateral view of the thoracic region of the spinal column. The vertebra prominens (C_7) resembles T_1, but it lacks facets for rib articulation. Vertebra T_{12} resembles the first lumbar vertebra (L_1), but it has a facet for rib articulation.

tures or compression-dislocation fractures after a hard fall most often involve the last thoracic and first two lumbar vertebrae.

Each thoracic vertebra articulates with ribs along the dorsolateral surfaces of the body. The location and structure of the articulations vary somewhat from vertebra to vertebra (Figure 7-20b●). Rib pairs 2 through 8 originate between adjacent vertebrae, so vertebrae T_2–T_8 have **superior** and **inferior demifacets** on each side. The first rib originates at the body of T_1, so that vertebra has a *whole facet* and an inferior demifacet on each side. Vertebra T_9 has only a superior demifacet on each side, whereas T_{10}, T_{11}, and T_{12} have a single whole facet on each side.

The transverse processes of vertebrae T_1–T_{10} are relatively thick, and they contain **transverse costal facets** for rib articulation. Thus, rib pairs 1 through 10 contact their vertebrae at two points, at a whole facet or demifacet and at a transverse costal facet. Table 7-2 summarizes the features of the thoracic vertebrae.

Lumbar Vertebrae

Figure 7-21

The lumbar vertebrae are the largest of the vertebrae. The body of a typical lumbar vertebra (Figure 7-21●) is thicker than that of a thoracic vertebra, and the superior and inferior surfaces are oval rather than heart-shaped. Other noteworthy features include: (1) There are neither whole facets nor demifacets on the body; (2) the slender transverse processes, which lack costal facets, project dorsolaterally; (3) the vertebral foramen is triangular; (4) the stumpy spinous processes project dorsally; (5) the superior articular processes face medially ("up and in"); and (6) the inferior articular processes face laterally ("down and out").

The lumbar vertebrae bear the most weight, and their massive spinous processes provide surface area for the attachment of lower back muscles that reinforce or adjust the lumbar curvature. Table 7-2, p. 225, summarizes the characteristics of lumbar vertebrae.

The Sacrum

Figure 7-22

The sacrum consists of the fused components of five sacral vertebrae. These vertebrae begin fusing shortly after puberty and are usually completely fused between ages 25–30. The sacrum provides protection for reproductive, digestive, and excretory organs and, via paired articulations, attaches the axial skeleton to the pelvic girdle of the appendicular skeleton. The broad surface area of the sacrum provides an extensive area for the attachment of muscles, especially those responsible for movement of the thigh. Figure 7-22● shows the posterior, lateral, and anterior surfaces of the sacrum.

The sacrum is curved, with a convex dorsal surface (Figure 7-22a●). The narrow, inferior portion is the sacral **apex,** whereas the broad superior surface forms the **base.** The **articular processes** form synovial articulations with the last lumbar vertebra. The **sacral canal** begins between those processes and extends the length of the sacrum. Nerves and membranes that line the vertebral canal in the spinal cord continue into the sacral canal.

The spinous processes of the five fused sacral vertebrae form a series of elevations along the **median sacral crest.** The laminae of the fifth sacral vertebra fail to contact one another at the midline, and they form the **sacral cornua.** These ridges establish the margins of the **sacral hiatus** (hī-Ā-tus), the end of the sacral canal. In life, this

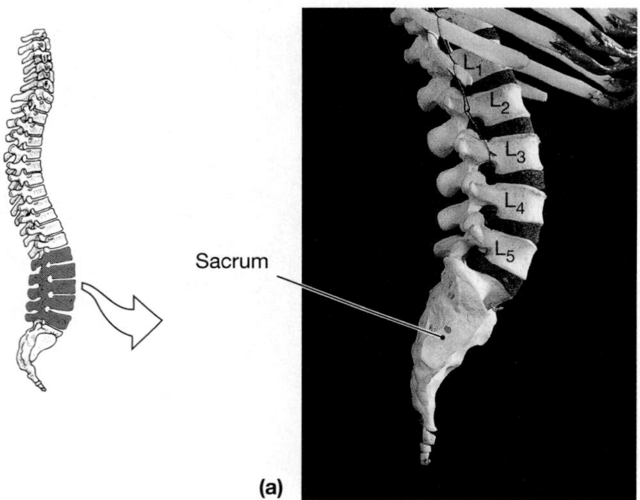

L_1
L_2
L_3
L_4
L_5
Sacrum
(a)

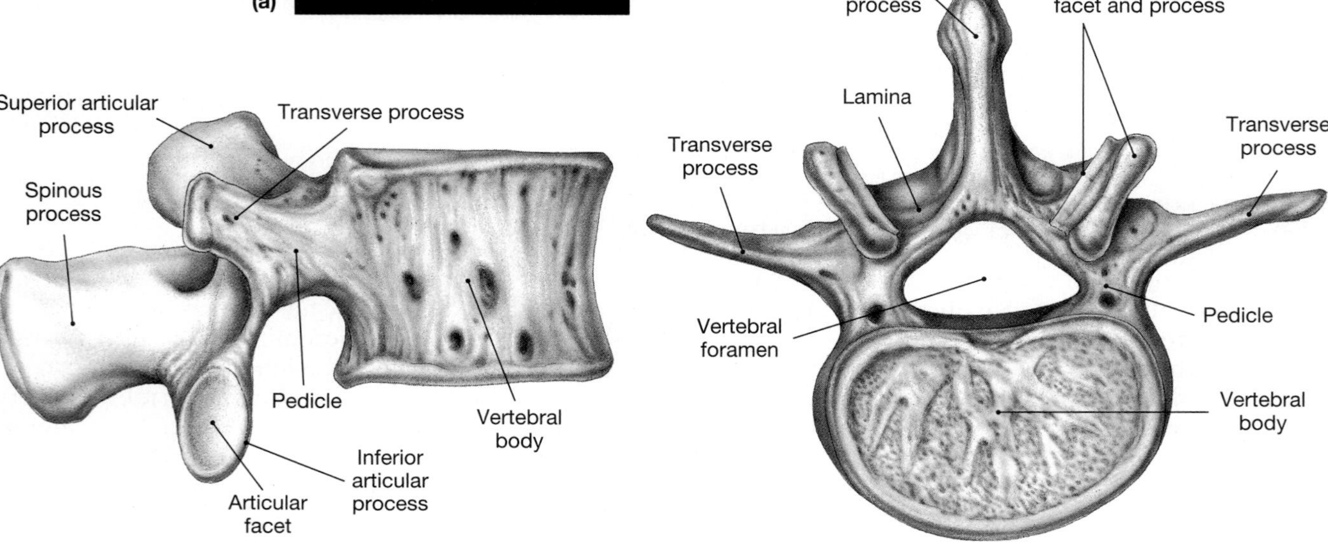

Superior articular process
Transverse process
Spinous process
Pedicle
Vertebral body
Inferior articular process
Articular facet
(b) Lateral view
Spinous process
Superior articular facet and process
Lamina
Transverse process
Transverse process
Vertebral foramen
Pedicle
Vertebral body
(c) Superior view

● **FIGURE 7-21**
The Lumbar Vertebrae. (a) The lumbar vertebrae, right lateral view. **(b)** Lateral view of a typical lumbar vertebra. **(c)** Superior view.

opening is covered by connective tissues. On either side of the median sacral crest, the **sacral foramina** represent the intervertebral foramina, now enclosed by the fused sacral bones. A broad sacral **ala,** or **wing,** extends laterally from each **lateral sacral crest.** The median and lateral sacral crests provide surface area for the attachment of muscles of the lower back and hip.

The sacral curvature is most apparent when viewed from the side (Figure 7-22b●); the curvature is more pronounced in males than in females. A thickened, flattened area, the **auricular surface,** marks the site of articulation with the pelvic girdle, the *sacroiliac joint.* Dorsal to the auricular surface is a roughened area, the **sacral tuberosity,** marking the attachment of a ligament that stabilizes this articulation. The anterior surface of the sacrum is concave (Figure 7-22c●), and, after fusion is completed, prominent *transverse lines* mark the former boundaries of individual vertebrae. At the apex, a flattened area marks the site of articulation with the coccyx.

The Coccyx

Figure 7-22

The small coccyx consists of 3–5 (most often 4) coccygeal vertebrae that have usually begun fusing by age 26 (Figure 7-22●). The coccyx provides an attachment site for a number of ligaments and for a muscle that constricts the anal opening. The first two coccygeal vertebrae have transverse processes and unfused neural arches. The prominent laminae of the first are known as the **coccygeal cornua,** and they curve to meet the cornua of the sacrum. The coccygeal vertebrae do not complete their fusion until late in adulthood. In very old people, the coccyx may fuse with the sacrum.

Joe suffered a hairline fracture at the base of the odontoid process. What bone is fractured, and where would you find it?

Examining a human vertebra, you notice that in addition to the large foramen for the spinal cord, there are two smaller foramina on either side of the bone in the region of the transverse processes. From which region of the spinal column did this vertebra come?

Why are the bodies of the lumbar vertebrae so large?

The Thoracic Cage

Figure 7-23

The skeleton of the chest, or **thoracic cage,** consists of the thoracic vertebrae, the ribs, and the sternum (Figure 7-23●). The ribs and the sternum form the *rib cage* and support the walls of the thoracic cavity. The thoracic cage serves two functions:

- It protects the heart, lungs, thymus, and other structures in the thoracic cavity.
- It serves as an attachment point for muscles involved with (1) respiration, (2) the position of the vertebral column, and (3) movements of the pectoral girdle and upper extremity.

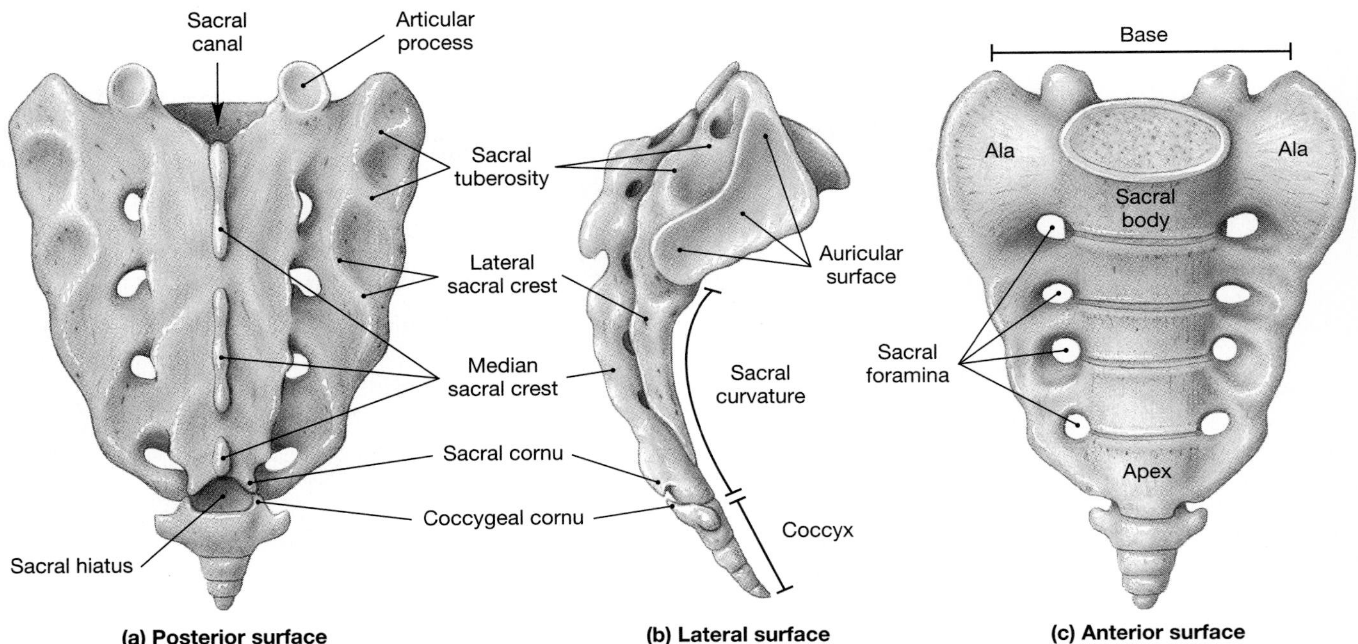

Sacral canal
Articular process
Sacral tuberosity
Lateral sacral crest
Median sacral crest
Sacral cornu
Coccygeal cornu
Sacral hiatus
Auricular surface
Sacral curvature
Coccyx
Base
Ala
Ala
Sacral body
Sacral foramina
Apex
(a) Posterior surface
(b) Lateral surface
(c) Anterior surface

● **FIGURE 7-22**
The Sacrum and Coccyx. (a) Posterior view. **(b)** Lateral view from the right side. **(c)** Anterior view.

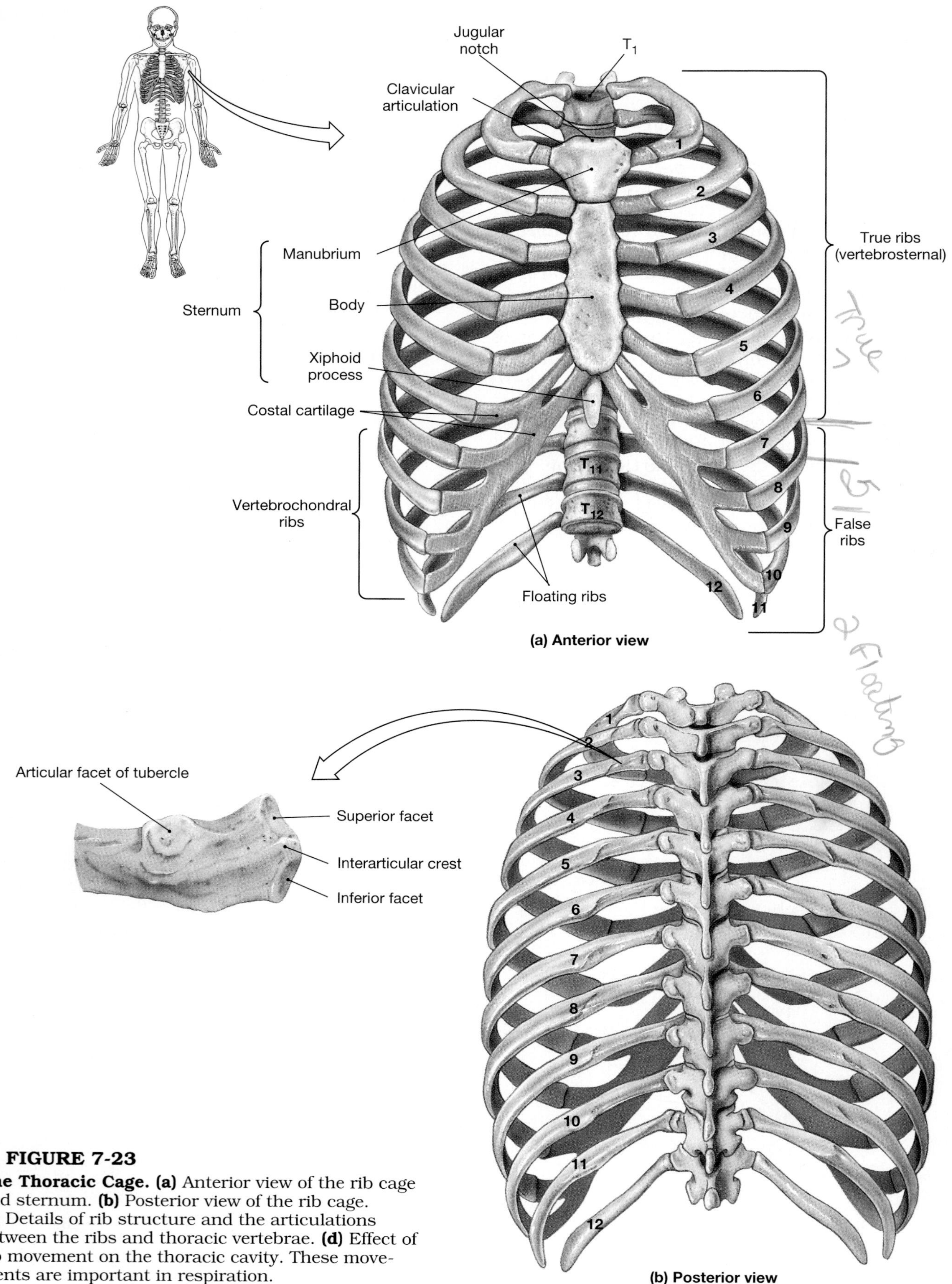

Jugular notch
T1
Clavicular articulation
1
2
3
4
5
6
7
8
9
10
11
12
Manubrium
Sternum
Body
Xiphoid process
Costal cartilage
True ribs (vertebrosternal)
T11
T12
Vertebrochondral ribs
False ribs
Floating ribs
(a) Anterior view
Articular facet of tubercle
Superior facet
Interarticular crest
Inferior facet
(b) Posterior view

● **FIGURE 7-23**

The Thoracic Cage. (a) Anterior view of the rib cage and sternum. **(b)** Posterior view of the rib cage. **(c)** Details of rib structure and the articulations between the ribs and thoracic vertebrae. **(d)** Effect of rib movement on the thoracic cavity. These movements are important in respiration.

❑ The Ribs

Figure 7-23

Ribs, or **costae,** are elongate, curved, flattened bones that (1) originate on or between the thoracic vertebrae and (2) end in the wall of the thoracic cavity. There are 12 pairs of ribs (Figure 7-23•). The first seven pairs are called **true,** or **vertebrosternal ribs.** They reach the anterior body wall and are connected to the sternum by separate cartilaginous extensions, the **costal cartilages.** Beginning with the first rib, the vertebrosternal ribs gradually increase in length and in the radius of curvature.

Ribs 8–12 are called **false ribs** because they do not attach directly to the sternum. The costal cartilages of ribs 8–10, the **vertebrochondral ribs,** fuse together before reaching the sternum (Figure 7-23a•). The last two pairs of ribs are called **floating ribs** because they have no connection with the sternum, or **vertebral ribs** because they are attached only to the vertebrae.

Figure 7-23b• shows the superior surface of a typical rib. The *vertebral end* of the rib articulates with the vertebral column at the **head,** or **capitulum** (ka-PIT-ū-lum). When articulating between adjacent vertebrae, the articular surface is divided into **superior** and **inferior articular facets** by the **interarticular crest.** After a short **neck,** the **tubercle,** or **tuberculum** (tū-BER-kū-lum), projects dorsally. The inferior portion of the tubercle contains an articular facet that contacts the transverse process of the thoracic vertebra.

Ribs 1 and 10 originate at whole facets on vertebrae T_1 and T_{10}, and their tubercular facets articulate with their respective vertebrae. Ribs 2–9 originate at demifacets, and their tubercular facets articulate with the inferior member of the vertebral pair. Ribs 11–12 originate at whole facets on T_{11} and T_{12}. These ribs do not have tubercular facets. The difference in rib orientation can be seen by comparing Figures 7-20, p. 227, and 7-23b•.

The bend, or **angle,** of the rib indicates the site where the tubular **body,** or **shaft,** begins curving toward the sternum. The internal rib surface is concave, and a prominent **costal groove** along its inferior border marks the path of nerves and blood vessels. The superficial surface is convex and provides an attachment site for muscles of the pectoral girdle and trunk. The *intercostal muscles* that move the ribs are attached to the superior and inferior surfaces.

With their complex musculature, dual articulations at the vertebrae, and flexible connection to the sternum, the ribs are quite mobile. Note how the ribs curve away from the vertebral column to angle downward (Figure 7-23b•). Functionally, a typical rib acts as if it were the handle on a bucket, lying just below the horizontal plane. Pushing it down forces it inward; pulling it up swings it outward (Figure 7-23d•). In addition, because of the curvature of the ribs, the same movements change the position of the sternum. Depressing the ribs moves the sternum posteriorly (inward), whereas elevation moves it anteriorly (outward). As a result, movements of the ribs affect both the width and the depth of the thoracic cage, increasing or decreasing its volume accordingly.

The ribs can bend and move to cushion shocks and absorb blows, but severe and sudden impacts can cause painful rib fractures. Because the ribs are tightly bound in connective tissues, a cracked rib can heal without a cast or splint; compound fractures of the ribs can send bone splinters or fragments into the thoracic cavity, with potential damage to internal organs.

Surgery on the heart, lungs, or other organs in the thorax often involves entering the thoracic cavity. The mobility of the ribs and the cartilaginous connections with the sternum allow the ribs to be temporarily moved out of the way. Special rib-spreaders are used, which push them apart in much the same way that a jack lifts a car off the ground for a tire change. If more extensive access is required, the sternal cartilages can be cut and the entire sternum can be folded out of the way. Once replaced, the cartilages are reunited by scar tissue, and the ribs heal fairly rapidly.

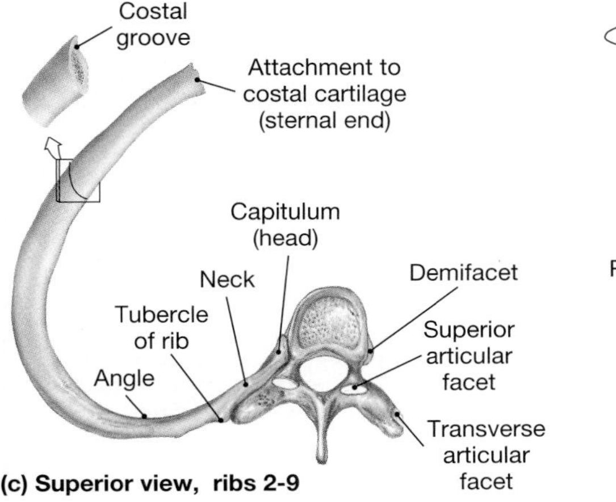
Costal groove
Attachment to costal cartilage (sternal end)
Capitulum (head)
Neck
Demifacet
Tubercle of rib
Superior articular facet
Angle
Transverse articular facet

(c) Superior view, ribs 2-9

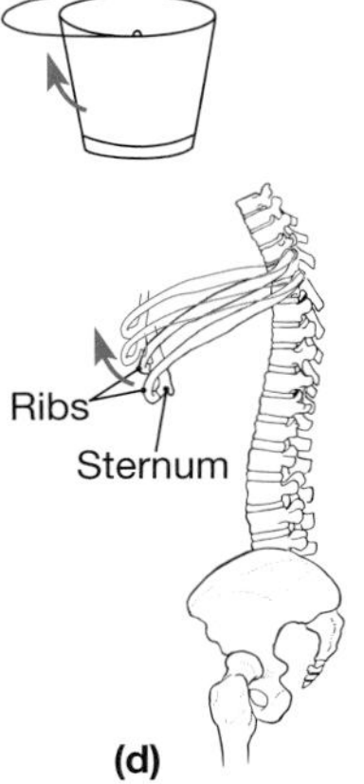
Ribs
Sternum

(d)

THORACENTESIS After thoracic surgery, *chest tubes* may be inserted through the thoracic wall to permit drainage of fluids. To install a chest tube or obtain a sample of pleural fluid, one must penetrate the wall of the thorax. This process, called *thoracentesis* (thō-ra-sen-TĒ-sis) or *thoracocentesis* (thō-ra-kō-sen-TĒ-sis; *kentesis,* perforating), involves the penetration of the thoracic wall along the superior border of one of the ribs. Penetration at this location avoids damage to vessels and nerves within the costal groove.

EMBRYOLOGY SUMMARY

Development of the Spinal Column

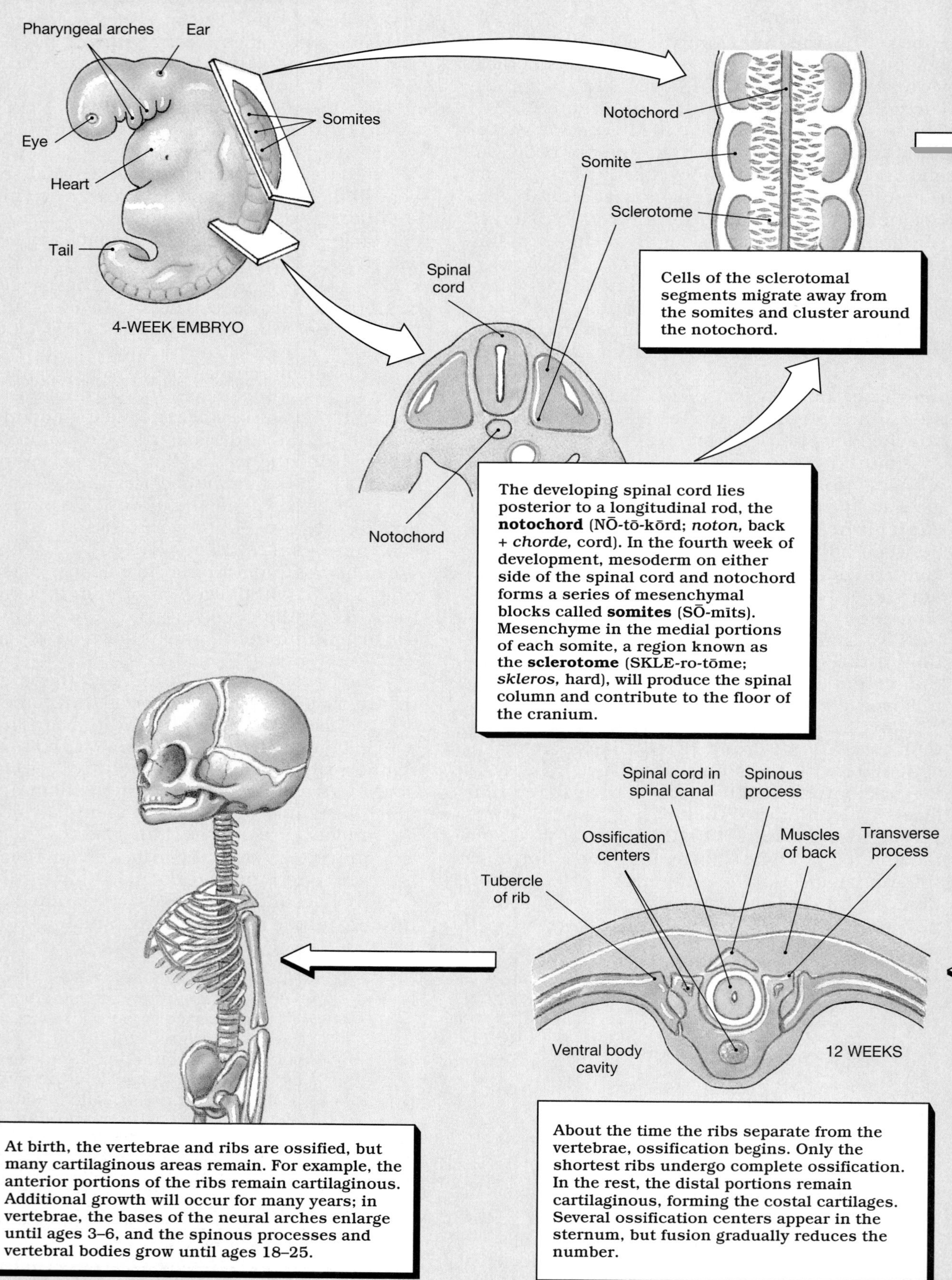
Pharyngeal arches
Ear
Eye
Heart
Tail
Somites
4-WEEK EMBRYO
Notochord
Somite
Sclerotome
Cells of the sclerotomal segments migrate away from the somites and cluster around the notochord.
Spinal cord
Notochord
The developing spinal cord lies posterior to a longitudinal rod, the **notochord** (NŌ-tō-kōrd; *noton*, back + *chorde*, cord). In the fourth week of development, mesoderm on either side of the spinal cord and notochord forms a series of mesenchymal blocks called **somites** (SŌ-mīts). Mesenchyme in the medial portions of each somite, a region known as the **sclerotome** (SKLE-ro-tōme; *skleros*, hard), will produce the spinal column and contribute to the floor of the cranium.
Spinal cord in spinal canal
Spinous process
Ossification centers
Muscles of back
Transverse process
Tubercle of rib
Ventral body cavity
12 WEEKS
About the time the ribs separate from the vertebrae, ossification begins. Only the shortest ribs undergo complete ossification. In the rest, the distal portions remain cartilaginous, forming the costal cartilages. Several ossification centers appear in the sternum, but fusion gradually reduces the number.
At birth, the vertebrae and ribs are ossified, but many cartilaginous areas remain. For example, the anterior portions of the ribs remain cartilaginous. Additional growth will occur for many years; in vertebrae, the bases of the neural arches enlarge until ages 3–6, and the spinous processes and vertebral bodies grow until ages 18–25.

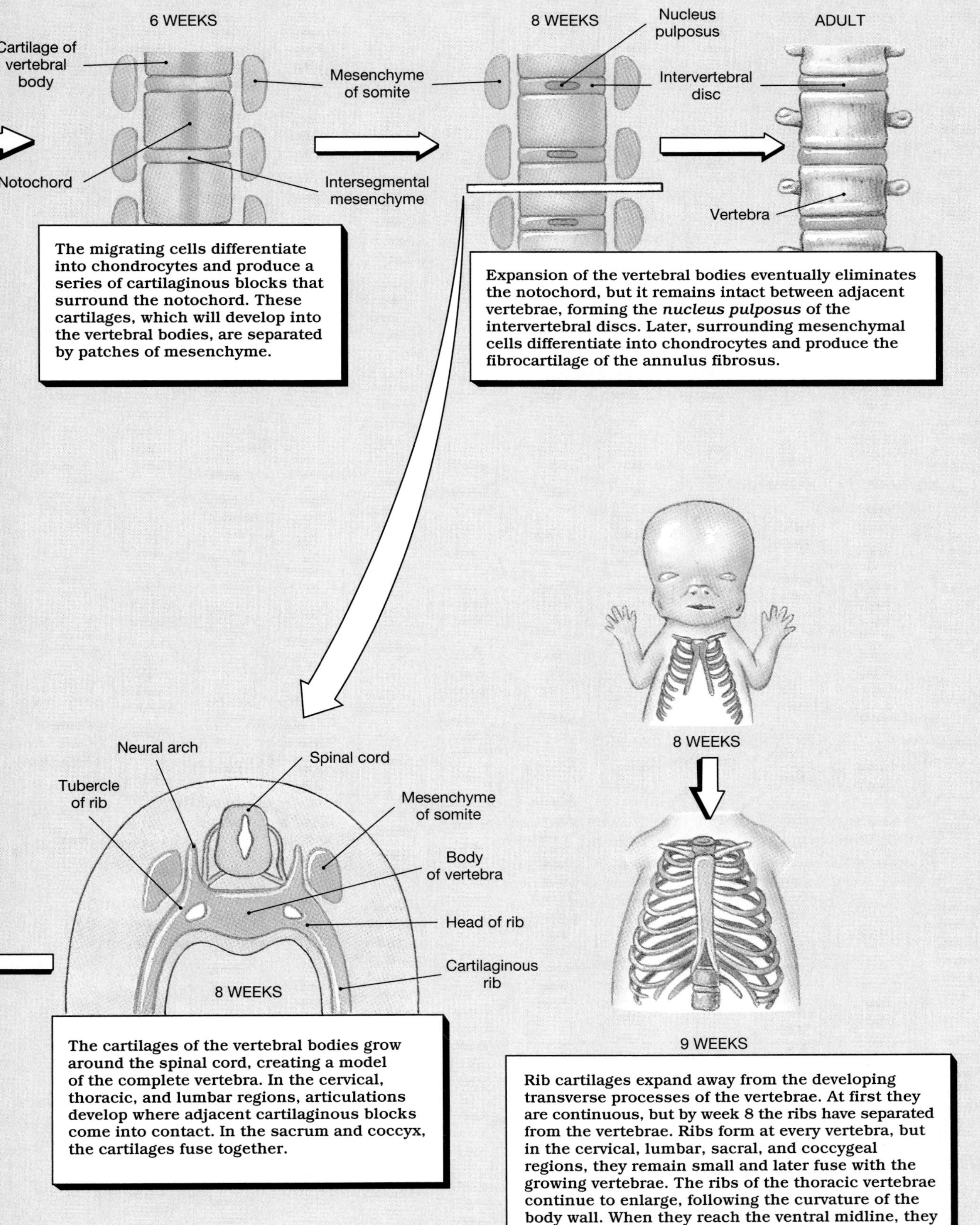
6 WEEKS
Cartilage of vertebral body
Notochord
Mesenchyme of somite
Intersegmental mesenchyme
8 WEEKS
Nucleus pulposus
Intervertebral disc
ADULT
Vertebra
Neural arch
Spinal cord
Tubercle of rib
Mesenchyme of somite
Body of vertebra
Head of rib
Cartilaginous rib
8 WEEKS
8 WEEKS
9 WEEKS

The migrating cells differentiate into chondrocytes and produce a series of cartilaginous blocks that surround the notochord. These cartilages, which will develop into the vertebral bodies, are separated by patches of mesenchyme.

Expansion of the vertebral bodies eventually eliminates the notochord, but it remains intact between adjacent vertebrae, forming the *nucleus pulposus* of the intervertebral discs. Later, surrounding mesenchymal cells differentiate into chondrocytes and produce the fibrocartilage of the annulus fibrosus.

The cartilages of the vertebral bodies grow around the spinal cord, creating a model of the complete vertebra. In the cervical, thoracic, and lumbar regions, articulations develop where adjacent cartilaginous blocks come into contact. In the sacrum and coccyx, the cartilages fuse together.

Rib cartilages expand away from the developing transverse processes of the vertebrae. At first they are continuous, but by week 8 the ribs have separated from the vertebrae. Ribs form at every vertebra, but in the cervical, lumbar, sacral, and coccygeal regions, they remain small and later fuse with the growing vertebrae. The ribs of the thoracic vertebrae continue to enlarge, following the curvature of the body wall. When they reach the ventral midline, they fuse with the cartilages of the sternum.

The Sternum

The adult **sternum,** or "breastbone," is a flat bone that forms in the anterior midline of the thoracic wall. The sternum has three components:

- The broad, triangular **manubrium** (ma-NŪ-brē-um) articulates with the clavicles and the cartilages of the first pair of ribs. This is the widest and most superior portion of the sternum. Only the first pair of ribs are attached by cartilage to this portion of the sternum. The **jugular notch** is the shallow indentation on the superior surface of the manubrium. It is located between the clavicular articulations.
- The tongue-shaped **body** attaches to the inferior surface of the manubrium and extends inferiorly along the midline. Individual costal cartilages from rib pairs 2–7 are attached to this portion of the sternum. The cartilages of rib pairs 8–10 are attached to the body by the same pair of cartilages that connect the seventh pair of ribs.
- The **xiphoid** (ZĪ-foyd) **process,** the smallest part of the sternum, is attached to the inferior surface of the body. The muscular *diaphragm* and *rectus abdominis muscles* attach to the xiphoid process.

Ossification of the sternum begins at 6 to 10 different centers, and fusion is not completed until at least age 25. Before age 25, the sternal body consists of four separate bones. Their boundaries can be detected as a series of transverse lines crossing the adult sternum. The xiphoid process is usually the last of the sternal components to undergo ossification and fusion. Its connection to the body of the sternum can be broken by an impact or strong pressure, creating a spear of bone that can severely damage the liver. Cardiopulmonary resuscitation (CPR) training strongly emphasizes the proper positioning of the hand to reduce the chances of breaking ribs or the xiphoid process.

How could you distinguish between true ribs and false ribs?

Improper administration of CPR (cardiopulmonary resuscitation) could result in a fracture of what bone?

Selected Clinical Terminology

Terms Discussed in This Chapter

chest tube: A drain installed after thoracic surgery to permit removal of blood and pleural fluid. *(p. 231)*
craniostenosis (krā-nē-ō-sten-Ō-sis): Premature closure of one or more fontanels, which can lead to unusual distortions of the skull. *(p. 219)*
deviated nasal septum: A bent nasal septum that slows or prevents sinus drainage. *(p. 215)*
kyphosis (kī-FŌ-sis): Abnormal exaggeration of the thoracic curvature that produes a "humpback" appearance. *(p. 222 and AM)*
lordosis (lōr-DŌ-sis): Abnormal lumbar curvature giving a "swayback" appearance. *(p. 222 and AM)*
microcephaly (mī-krō-SEF-uh-lē): An undersized head resulting from genetic or developmental abnormalities. *(p. 219)*
scoliosis (skō-lē-Ō-sis): Abnormal lateral curvature of the spine. *(p. 222 and AM)*
sinusitis: Inflammation and congestion of the sinuses. *(p. 215)*
spina bifida (SPĪ-nuh BI-fi-duh): A condition resulting from failure of the vertebral laminae to unite during development; often associated with developmental abnormalities of the brain and spinal cord. *(p. 223 and AM)*
thoracentesis (thō-ra-sen-TĒ-sis), or **thoracocentesis:** The penetration of the thoracic wall along the superior border of one of the ribs. *(p. 231)*
TMJ syndrome: A painful condition resulting from misalignment of the mandible at the temporomandibular joint. *(p. 217)*
whiplash: An injury resulting from a sudden change in the body position that can injure the cervical vertebrae. *(p. 226)*

CHAPTER REVIEW

STUDY OUTLINE

INTRODUCTION, p. 203

1. The skeletal system consists of the axial skeleton and the appendicular skeleton. The **axial skeleton** can be subdivided into the **skull,** the **auditory ossicles** and **hyoid,** the **vertebral column,** and the **thoracic cage,** composed of the **ribs** and **sternum.** (*Figure 7-1*)
2. The **appendicular skeleton** includes the pectoral and pelvic girdles that support the upper and lower limbs.

THE SKULL, p. 204

1. The **cranium** encloses the **cranial cavity,** a division of the dorsal body cavity. (*Figure 7-2*)

2. Prominent superficial landmarks on the skull include the **lambdoidal, sagittal, squamosal,** and **coronal sutures.** (*Figure 7-3*)

Bones of the Cranium, p. 208

3. The **occipital bone** surrounds the **foramen magnum** and articulates with the sphenoid, temporal, and parietal bones. (*Figures 7-3, 7-4, 7-5*)
4. The **parietal bones** articulate with the occipital bone, the **temporal bones,** and the frontal bone. (*Figures 7-3, 7-4, 7-5, 7-7*)
5. The **frontal bone** articulates with the parietal bones, the temporal bone, the sphenoid bone, the ethmoid bone, the nasal bones, the lacrimal bones, the zygomatic bones, and the maxillae. (*Figures 7-3, 7-4, 7-6*)
6. The **sphenoid bone** articulates with all the cranial bones as well as the maxillae, the zygomatic bones, and the palatine bones. (*Figures 7-3, 7-4, 7-8*)
7. The **cribriform plate** of the **ethmoid bone** contains foramina for olfactory nerves. The lateral masses support the **superior** and **middle conchae**. The **perpendicular plate** forms part of the nasal septum. (*Figures 7-3, 7-4, 7-9*)

Bones of the Face, p. 213

8. The left and right **maxillae,** or *maxillary bones,* are the largest facial bones and form the upper jaw. They articulate with all other facial bones except the mandible. (*Figures 7-3, 7-10*)
9. The **palatine bones** are small, L-shaped bones that articulate with the maxillae, forming the posterior portions of the hard palate, and contribute to the floor of the orbital cavities. (*Figures 7-3, 7-10*)
10. The paired **nasal bones** articulate with the frontal bone at the midline of the face and extend to the superior border of the **external nares.** (*Figures 7-3, 7-10*)
11. The **vomer** forms the inferior portion of the nasal septum. It is based on the floor of the nasal cavity and articulates with both the maxillae and palatine bones along the midline. (*Figure 7-4*)
12. One **inferior nasal concha** is located on each side of the nasal septum, attached to the lateral wall of the nasal cavity. (*Figure 7-11*)
13. The **paranasal sinuses** are hollow airways that connect with the nasal passages. They are found in the frontal, sphenoid, ethmoid, and maxillary bones. (*Figure 7-11*)
14. The **temporal process** of the **zygomatic bone** articulates with the **zygomatic process** of the temporal bone to form the **zygomatic arch.** (*Figures 7-3, 7-12*)
15. The paired **lacrimal bones,** the smallest bones of the face, are situated in the medial portion of each orbit. (*Figures 7-3, 7-12*)
16. Seven bones form each **orbital complex:** the frontal bone, the maxilla, the lacrimal bone, the ethmoid bone, the sphenoid bone, the palatine bone, and the zygomatic bone. (*Figure 7-12*)
17. The **mandible** is the bone of the lower jaw. (*Figure 7-13*)
18. The hyoid bone, suspended by **stylohyoid ligaments,** consists of a **body,** the **greater cornua,** and the **lesser cornua.** (*Figure 7-14*)

The Skulls of Infants and Children, p. 217

19. Fibrous connections at **fontanels** permit the skulls of infants and children to continue growing. (*Figure 7-15*)

THE VERTEBRAL COLUMN, p. 222

1. There are 7 **cervical vertebrae** (the first articulates with the skull), 12 **thoracic vertebrae** (which articulate with ribs), and 5 **lumbar vertebrae** (the last articulates with the sacrum). The **sacrum** and **coccyx** consist of fused vertebrae. (*Figure 7-16*)

Spinal Curvature, p. 222

2. The spinal column has four **spinal curves:** the **thoracic** and **sacral curvatures** are called **primary,** or **accommodation, curves;** the **lumbar** and **cervical curvatures** are known as **secondary,** or **compensation, curves.** (*Figures 7-16, 7-17*)

Vertebral Anatomy, p. 223

3. A typical vertebra has a **body** (*centrum*) and a **vertebral arch** (*neural arch*); it articulates with adjacent vertebrae at the **superior** and **inferior articular processes.** (*Figure 7-18*)
4. Adjacent vertebrae are separated by **intervertebral discs.** Spaces between successive **pedicels** form the **intervertebral foramina.** (*Figure 7-18*)

Vertebral Regions, p. 224

5. Cervical vertebrae are distinguished by the shape of the body, the relative size of the vertebral foramen, the presence of **costal processes** with **transverse foramina,** and notched **spinous processes.** (*Figure 7-19; Table 7-2*)
6. Thoracic vertebrae have distinctive heart-shaped bodies, long, slender spinous processes, and articulations for the ribs. (*Figures 7-20, 7-23; Table 7-2*)
7. The lumbar vertebrae are the most massive and least mobile; they are subjected to the greatest strains. (*Figure 7-21; Table 7-2*)
8. The sacrum protects reproductive, digestive, and excretory organs. It has **auricular surfaces** for articulation with the pelvic girdle. The sacrum articulates with the fused elements of the **coccyx.** (*Figure 7-22; Table 7-2*)

THE THORACIC CAGE, p. 229

1. The skeleton of the thoracic cage consists of the thoracic vertebrae, the ribs, and the sternum. The ribs and sternum form the **rib cage.** (*Figure 7-23*)

The Ribs, p. 231

2. Ribs 1–7 are **true,** or **vertebrosternal ribs.** Ribs 8–12 are called **false ribs;** they include the **vertebrochondral ribs** and two pairs of **floating (vertebral) ribs.** A typical rib has a **head,** or **capitulum; neck; tubercle,** or **tuberculum; angle;** and **body,** or **shaft.** An inferior **costal groove** marks the path of nerves and blood vessels. (*Figure 7-23*)

The Sternum, p. 234

3. The sternum consists of a **manubrium,** a **body,** and a **xiphoid process.** (*Figure 7-23*)

REVIEW QUESTIONS

LEVEL 1 Reviewing Facts and Terms

Match each item in Column A with the most closely related item in Column B. Use letters for answers in the spaces provided.

	Column A	Column B
___	1. foramina	a. skullcap
___	2. sinuses	b. tear ducts
___	3. sutures	c. atlas
___	4. calvaria	d. lumbar and cervical
___	5. auditory ossicles	e. air-filled chambers
___	6. hypophyseal fossa	f. axis
___	7. lacrimal bones	g. ear bones
___	8. accommodation	h. ribs
___	9. compensation	i. passageways
___	10. costae	j. sella turcica
___	11. C_1	k. thoracic and sacral
___	12. C_2	l. immovable joints

13. The axial skeleton consists of the bones of the:
    (a) pectoral and pelvic girdles
    (b) skull, thorax, and vertebral column
    (c) arms, legs, hands, and feet
    (d) limbs, pectoral girdle, and pelvic girdle

14. The appendicular skeleton consists of the bones of the:
    (a) pectoral and pelvic girdles
    (b) skull, thorax, and vertebral column
    (c) arms, legs, hands, and feet
    (d) limbs, pectoral girdle, and pelvic girdle

15. Which of the following contains *only* bones of the cranium?
    (a) frontal, parietal, occipital, sphenoid
    (b) frontal, occipital, zygomatic, parietal
    (c) occipital, sphenoid, temporal, lacrimal
    (d) mandible, maxilla, nasal, zygomatic

16. The facial bones include the:
    (a) lacrimal, nasal, maxilla, mandible
    (b) frontal, lacrimal, zygomatic, sphenoid
    (c) maxilla, mandible, ethmoid, sphenoid
    (d) ethmoid, sphenoid, temporal, parietal

17. The boundaries between skull bones are immovable joints called:
    (a) foramina (b) fontanels
    (c) lacunae (d) sutures

18. The major sutures of the skull include the:
    (a) frontal, parietal, occipital, sphenoid
    (b) frontal, lambdoidal, occipital, coronal
    (c) lambdoidal, coronal, sagittal, squamosal
    (d) coronal, sagittal, frontal, parietal

19. Of the following bones which one is unpaired?
    (a) vomer (b) maxilla
    (c) palatine (d) nasal

20. The cribriform plate, crista galli, and superior conchae are parts of the:
    (a) parietal bone (b) occipital bone
    (c) sphenoid bone (d) ethmoid bone

21. The bone that houses the inner ear structures associated with hearing and balance is the:
    (a) temporal (b) occipital
    (c) parietal (d) sphenoid

22. The bony enclosure that forms a prominent depression on the sphenoid bone and houses the pituitary gland is the:
    (a) clinoid process (b) sella turcica
    (c) acoustic meatus (d) Eustachian tube

23. The hyoid bone:
    (a) forms the inferior portion of the skull
    (b) is attached to the zygomatic bone
    (c) does not articulate with any other bone
    (d) is a cartilaginous structure

24. The membranous areas between the cranial bones of the fetal skull are called:
    (a) fontanels (b) sutures
    (c) Wormian bones (d) foramina

25. The part of the vertebra that transfers weight along the axis of the vertebral column is the:
    (a) neural arch (b) lamina
    (c) pedicles (d) body

26. Beginning superiorly and progressing inferiorly, the components of the sternum include:
    (a) xiphoid process, body, manubrium
    (b) body, manubrium, xiphoid process
    (c) manubrium, body, xiphoid process
    (d) manubrium, xiphoid process, body

27. What five bones make up the facial complex?

28. What four major sutures make up the boundaries between the skull bones?

29. What is the primary function of the vomer?

30. What bones contain the paranasal sinuses?

31. What unique characteristic of the hyoid bone makes it different from all the other bones in the body?

32. What purpose do the fontanels serve during delivery?

33. Beginning at the skull, cite the regions of the vertebral column and list the number of vertebrae in each region of the adult vertebral column.

34. What four spinal curves are apparent in the adult spinal column when viewed from the side?

35. What two primary functions are performed by the thoracic cage?

36. Why are the upper seven pairs of ribs called *true* ribs and the lower five pairs called *false* ribs?

LEVEL 2 Reviewing Concepts

37. The atlas (C_1) can be distinguished from the other vertebrae by:
    (a) the presence of anterior and posterior vertebral arches
    (b) the lack of a body
    (c) the presence of superior facets and inferior articular facets
    (d) a, b, and c are correct

38. What is the relationship between the temporal bone and the ear?

39. What is the relationship between the sphenoid bone and the pituitary gland?

40. Unlike that of other animals, the skull of humans is balanced above the vertebral column, allowing an upright posture. What anatomical adaptations are necessary in humans to maintain the balance of the skull?

41. The sternum is a part of the axial skeleton that is important in a variety of clinical situations because of its structural features and skeletal components. If you were teaching this information to prospective nursing students, what clinical applications would you cite?

42. Why are ruptured intervertebral discs more common in lumbar vertebrae and dislocations and fractures more common in cervical vertebrae?

43. We generally associate a "runny" nose with a cold. Does this nasal condition serve a purpose? What is the cause of this type of reaction?

44. Why is it important to keep the back straight when lifting a heavy object?

LEVEL 3 Critical Thinking and Clinical Applications

45. Tess is diagnosed with a disease that affects the membranes surrounding the brain. The doctor tells her family that the disease is caused by an airborne virus. Explain how this virus could have entered the cranium.

46. Joe is 40 years old and is 30 pounds overweight. Like many middle-aged men, Joe carries most of this weight in his abdomen and jokes with his friends about his "beer gut." During an annual physical, Joe's doctor advises him that his spine is developing an abnormal curvature. Why is the curvature of Joe's spine changing, and what is this condition called?

47. A babysitter is being tried for the death of a 10-month-old infant. The prosecutor contends that the child died as the result of being violently shaken. The defense claims that the child's head became stuck in the slats of his crib and, in trying to twist free, the child broke his neck. The medical examiner testifies that the child died as the result of a compression of the spinal cord between the sixth and seventh cervical vertebrae. The superior articular processes of the seventh cervical vertebra and the inferior articular processes of the sixth cervical vertebra were fractured, and the processes on the right side were laterally displaced, causing the sixth vertebra to slide laterally across the seventh, damaging the spinal cord. On the basis of this evidence, whom do you believe? Why?

48. While working at an excavation, an archaeologist finds several small skull bones. She examines the frontal, parietal, and occipital bones and concludes that the skulls are those of children not yet 1 year old. How can she tell their ages from examining the bones?

CHAPTER 8

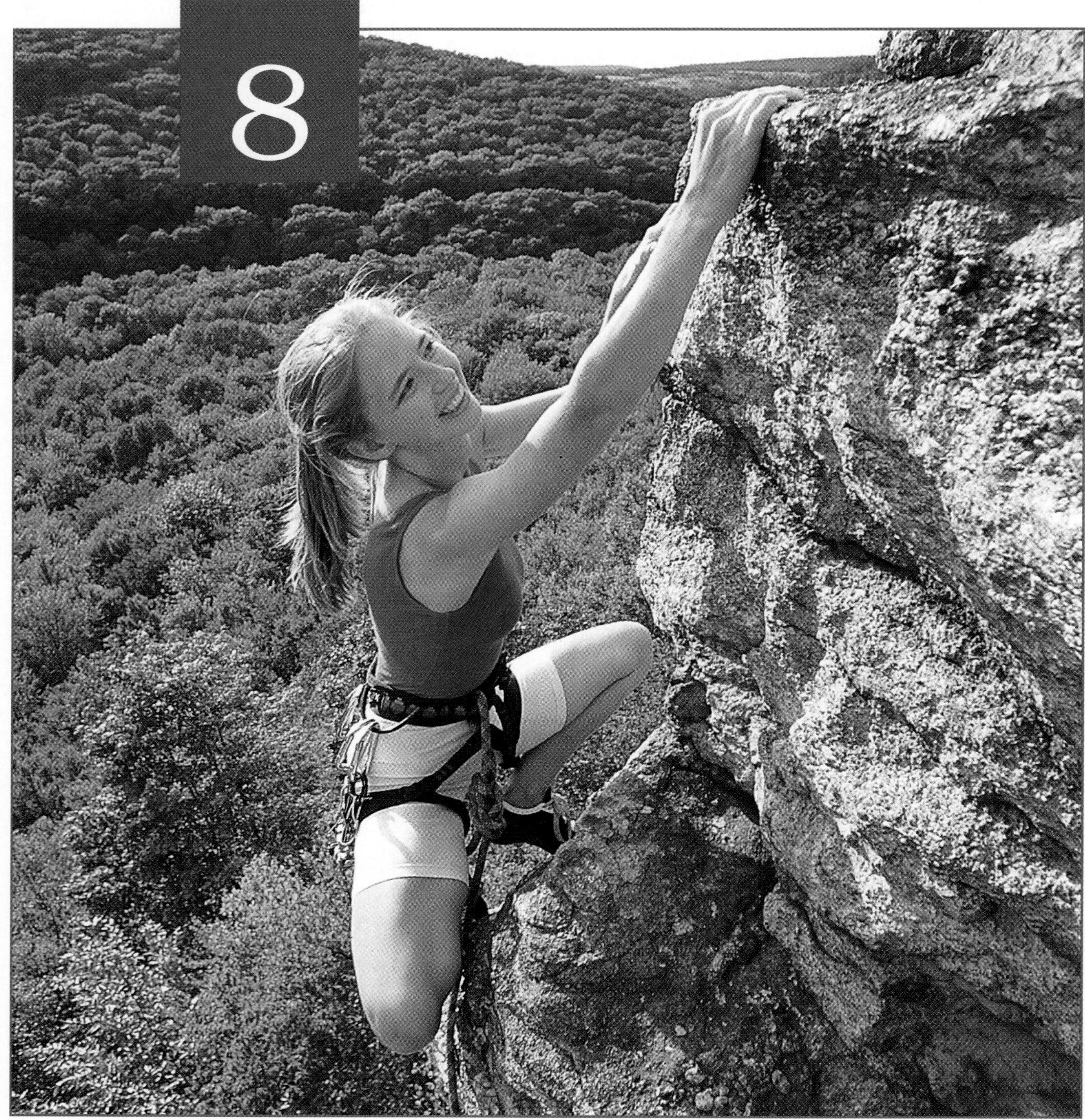

Clinging to a vertical cliff face by fingers and toes places unusually severe stresses on the appendicular skeleton and the associated muscles and supporting connective tissues. Although you may not climb mountains, your appendicular skeleton still plays a major role in daily life. Take a minute and make a mental list of all of the things you have done with your arms or legs in the last day. Standing, walking, writing, turning pages, eating, dressing, shaking hands, waving—the list quickly becomes so long that it is impractical to complete it. The axial skeleton protects and supports internal organs, and it participates in vital functions, such as respiration. But it is your appendicular skeleton that gives you control over your environment, changes your position in space, and provides mobility. This chapter examines the components of the appendicular skeleton; the next chapter will consider the ways those elements interact during normal movements.

The Appendicular Skeleton

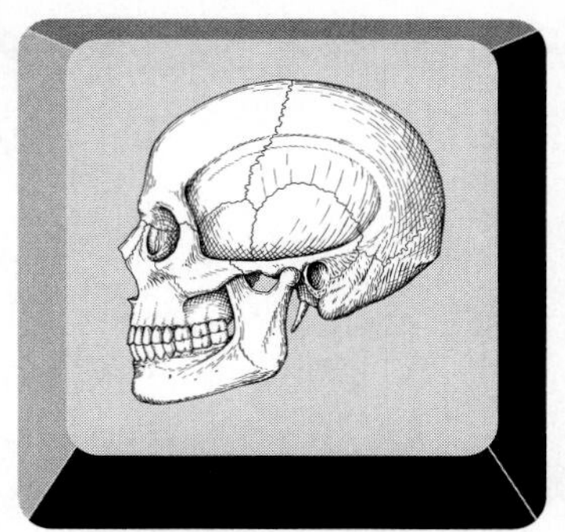

Chapter Outline and Objectives

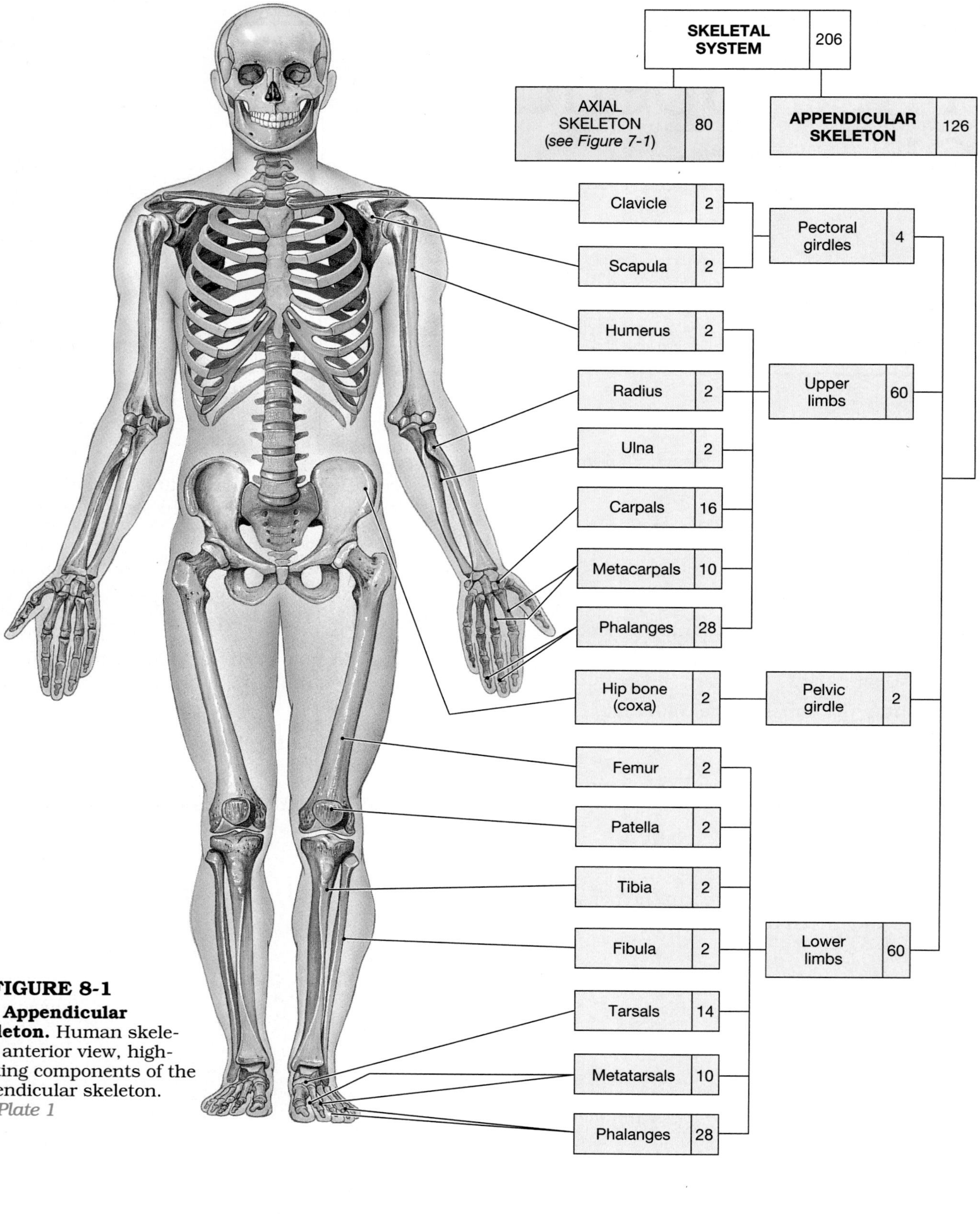

- SKELETAL SYSTEM 206
  - AXIAL SKELETON (*see Figure 7-1*) 80
  - APPENDICULAR SKELETON 126
    - Pectoral girdles 4
      - Clavicle 2
      - Scapula 2
    - Upper limbs 60
      - Humerus 2
      - Radius 2
      - Ulna 2
      - Carpals 16
      - Metacarpals 10
      - Phalanges 28
    - Pelvic girdle 2
      - Hip bone (coxa) 2
    - Lower limbs 60
      - Femur 2
      - Patella 2
      - Tibia 2
      - Fibula 2
      - Tarsals 14
      - Metatarsals 10
      - Phalanges 28

● **FIGURE 8-1**
The Appendicular Skeleton. Human skeleton, anterior view, highlighting components of the appendicular skeleton.
AM *Plate 1*

The **appendicular skeleton** includes the bones of the limbs and the supporting elements, or *girdles*, that connect them to the trunk (Figure 8-1●). The descriptions in this chapter emphasize surface features that have functional importance, such as the attachment sites for skeletal muscles or the paths of major nerves and blood vessels.

The Pectoral Girdles and Upper Limbs

Each arm articulates with the trunk at the **shoulder girdle,** or **pectoral girdle.** The shoulder girdle consists of the S-shaped *clavicle* (*collarbone*) and a broad, flat *scapula* (*shoulder blade*). The clavicle articulates with the manubrium of the sternum, and this is the *only* direct connection between the pectoral girdle and the axial skeleton. Skeletal muscles support and position the scapula, which has no direct bony or ligamentous connections to the thoracic cage. As a result, the shoulder is extremely mobile, but not very strong.

The Pectoral Girdle

Movements of the clavicle and scapula position the shoulder joint and provide a base for arm movement. Once the shoulder joint is in position, muscles that originate on the pectoral girdle help to move the upper limb. The surfaces of the scapula and clavicle are therefore extremely important as sites for muscle attachment. Where major muscles attach they leave their marks, creating bony ridges and flanges. Other bone markings, such as foramina, indicate the position of nerves or blood vessels that control the muscles and nourish the muscles and bones.

The Clavicle

Figure 8-2

The **clavicle** (KLAV-i-kul) is an S-shaped bone that originates at the superior and lateral border of the manubrium of the sternum, lateral to the jugular notch (Figure 8-2●). From the roughly triangular **sternal end** the clavicle curves laterally and dorsally until it articulates with the acromion of the scapula. The broad, flat, **acromial end** is larger than the sternal end.

The smooth superior surface of the clavicle lies just beneath the skin; the rough inferior surface of the acromial end is marked by prominent lines and tubercles. Two prominent landmarks on the inferior surface are the **conoid tubercle** at the acromial end and the **costal tuberosity** at the sternal end. These surface features are attachment sites for muscles and ligaments of the shoulder.

You can explore the interaction between scapulae and clavicles. With your fingers in the jugular notch, locate the clavicle on either side. When you move your shoulders you can feel the clavicles change their positions. Because the clavicles are so close to the skin, you can trace one laterally until it articulates with the scapula.

The clavicles are relatively small and fragile, and fractures of the clavicle are fairly common. For example, a clavicular fracture can result from a simple fall on the hand of an outstretched arm. Because it breaks rather easily, and causes intense pain, instructors teaching self-defense often advise their students to strike an opponent's clavicle in an emergency. Fortunately, in view of the clavicle's vulnerability, most clavicular fractures heal rapidly without a cast.

The Scapula

Figure 8-3

The anterior aspect of the **body** of the **scapula** (SCAP-ū-luh) forms a broad triangle (Figure 8-3a●).

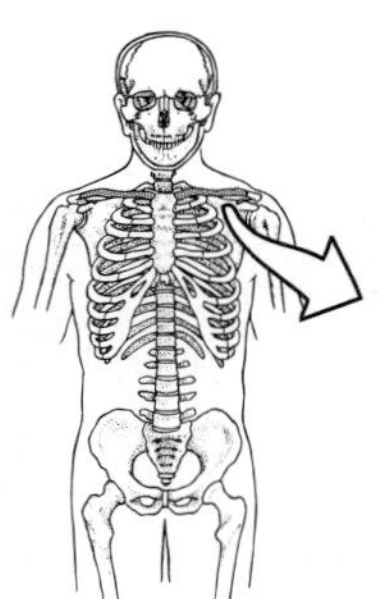

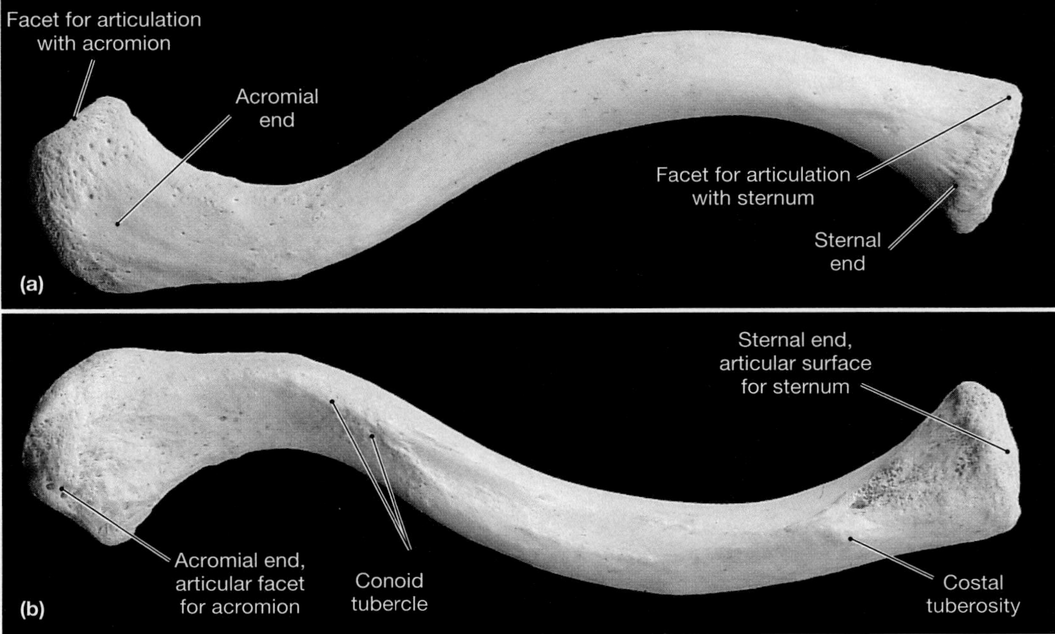

Facet for articulation with acromion
Acromial end
Facet for articulation with sternum
Sternal end
(a)
Sternal end, articular surface for sternum
Acromial end, articular facet for acromion
Conoid tubercle
Costal tuberosity
(b)

● **FIGURE 8-2**
The Clavicle.
(a) Superior and **(b)** inferior views of the left clavicle.

The three sides of that triangle are the **superior border;** the **medial border,** or *vertebral border;* and the **lateral border,** or *axillary border* (*axilla,* armpit). Muscles that position the scapula attach along these edges. The corners of the triangle are called the **superior angle,** the **inferior angle,** and the **lateral angle.** The region adjacent to the lateral angle is thickened and rounded, forming a **neck** that supports the cup-shaped **glenoid fossa.** At the glenoid fossa, the scapula articulates with the proximal end of the *humerus,* the bone of the arm. This articulation is the shoulder joint, also known as the *scapulohumeral joint.* The relatively smooth, concave **subscapular fossa** forms most of the anterior surface of the scapula.

The articular head of the humerus is smooth, round, and several times the diameter of the glenoid fossa (Figure 8-3b●). Two large scapular processes extend over the superior margin of the glenoid fossa, superior to the head of the humerus. The smaller, anterior projection is the **coracoid** (KŌR-uh-koyd) **process.** The **acromion** (a-KRŌ-mē-on) **process** is the larger posterior process. If you run your fingers along the superior surface of the shoulder joint, you will feel this process. The acromion articulates with the clavicle at the **acromioclavicular joint.** Both the acromion and the coracoid process are attached to ligaments and tendons associated with the shoulder joint, which will be considered further in Chapter 9.

The acromion process is continuous with the **scapular spine** (Figure 8-3c●). This ridge crosses the scapular body before ending at the medial border. The scapular spine divides the convex dorsal surface of the body into two regions. The area superior to the spine constitutes the **supraspinous fossa** (*supra,* above); the region below the spine is the **infraspinous fossa** (*infra,* beneath). The spine is an attachment site for two large muscles (the *supraspinatus* and the *infraspinatus*), and the entire dorsal surface is marked by small ridges and lines where smaller muscles attach to the scapula.

✓ Why would a broken clavicle affect the mobility of the scapula?

✓ What bone articulates with the scapula at the glenoid fossa?

● **FIGURE 8-3**
The Scapula. (a) Anterior, **(b)** lateral, and **(c)** posterior views of the right scapula.

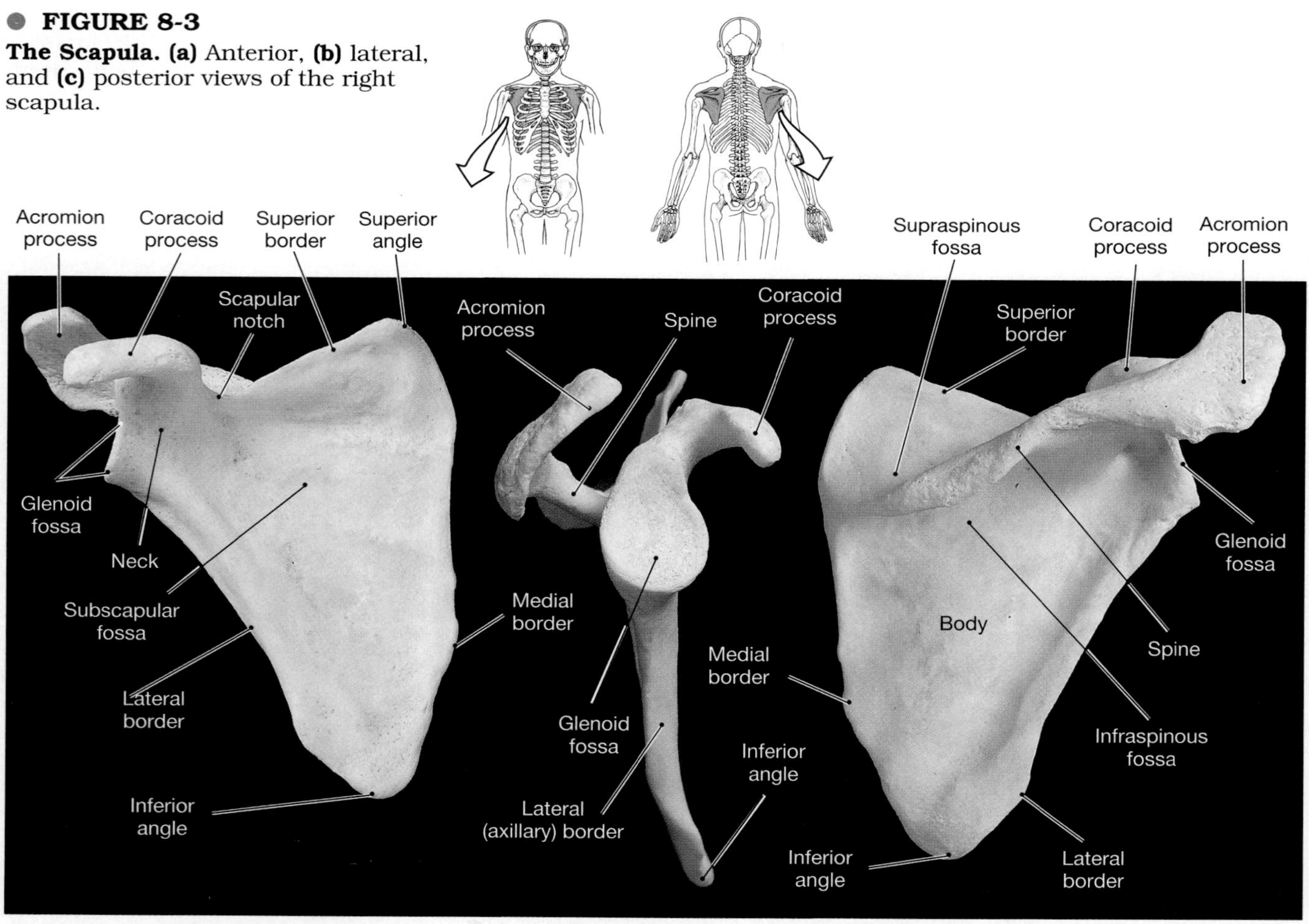
Acromion process
Coracoid process
Superior border
Superior angle
Scapular notch
Glenoid fossa
Neck
Subscapular fossa
Lateral border
Inferior angle
Acromion process
Spine
Coracoid process
Medial border
Glenoid fossa
Lateral (axillary) border
Inferior angle
Supraspinous fossa
Coracoid process
Acromion process
Superior border
Glenoid fossa
Body
Spine
Medial border
Infraspinous fossa
Inferior angle
Lateral border
(a) Anterior view
(b) Lateral view
(c) Posterior view

❑ The Upper Limb

The skeleton of the upper limb includes the bones of the arm, forearm, wrist, and hand. Note that in anatomical descriptions the term *arm* refers only to the proximal portion of the upper limb, not to the entire limb. The arm, or *brachium*, contains a single bone, the **humerus.** The humerus extends from the scapula to the elbow. At the proximal end of the humerus, the round **head** articulates with the scapula. At its distal end, the humerus articulates with the bones of the forearm, or *antebrachium*, the *radius* and *ulna*.

The Humerus

Figure 8-4

The prominent **greater tubercle** of the humerus is located near the head, on the lateral surface of the epiphysis (Figure 8-4●). The greater tubercle establishes the lateral contour of the shoulder. You can verify its position by feeling for a bump situated a few centimeters anterior and inferior to the tip of the acromion. The **lesser tubercle** lies on the anterior and medial surface of the epiphysis, separated from the greater tubercle by the **intertubercular groove,** or *intertubercular sulcus*. Both tubercles are important sites for muscle attachment; a large tendon runs along the groove. The **anatomical neck** marks the distal limit of the articular capsule. It lies between the tubercles and the articular surface of the head. Distal to the tubercles, the narrow **surgical neck** corresponds to the metaphysis of the growing bone. This name reflects the fact that fractures often occur at this site.

The proximal **shaft** of the humerus is round in section. The elevated **deltoid tuberosity** runs along the lateral border of the shaft, extending more than halfway down its length. It is named after the *deltoid muscle* that attaches to it. On the anterior surface of the shaft, the intertubercular groove continues alongside the deltoid tuberosity.

On the posterior surface (Figure 8-4b●), the deltoid tuberosity ends at the **radial groove.** This depression marks the path of the *radial nerve*, a large nerve that provides sensory information from the back of the hand and motor control over large muscles that extend

● **FIGURE 8-4**

The Humerus. Major landmarks on **(a)** the anterior and **(b)** the posterior surfaces of the right humerus.

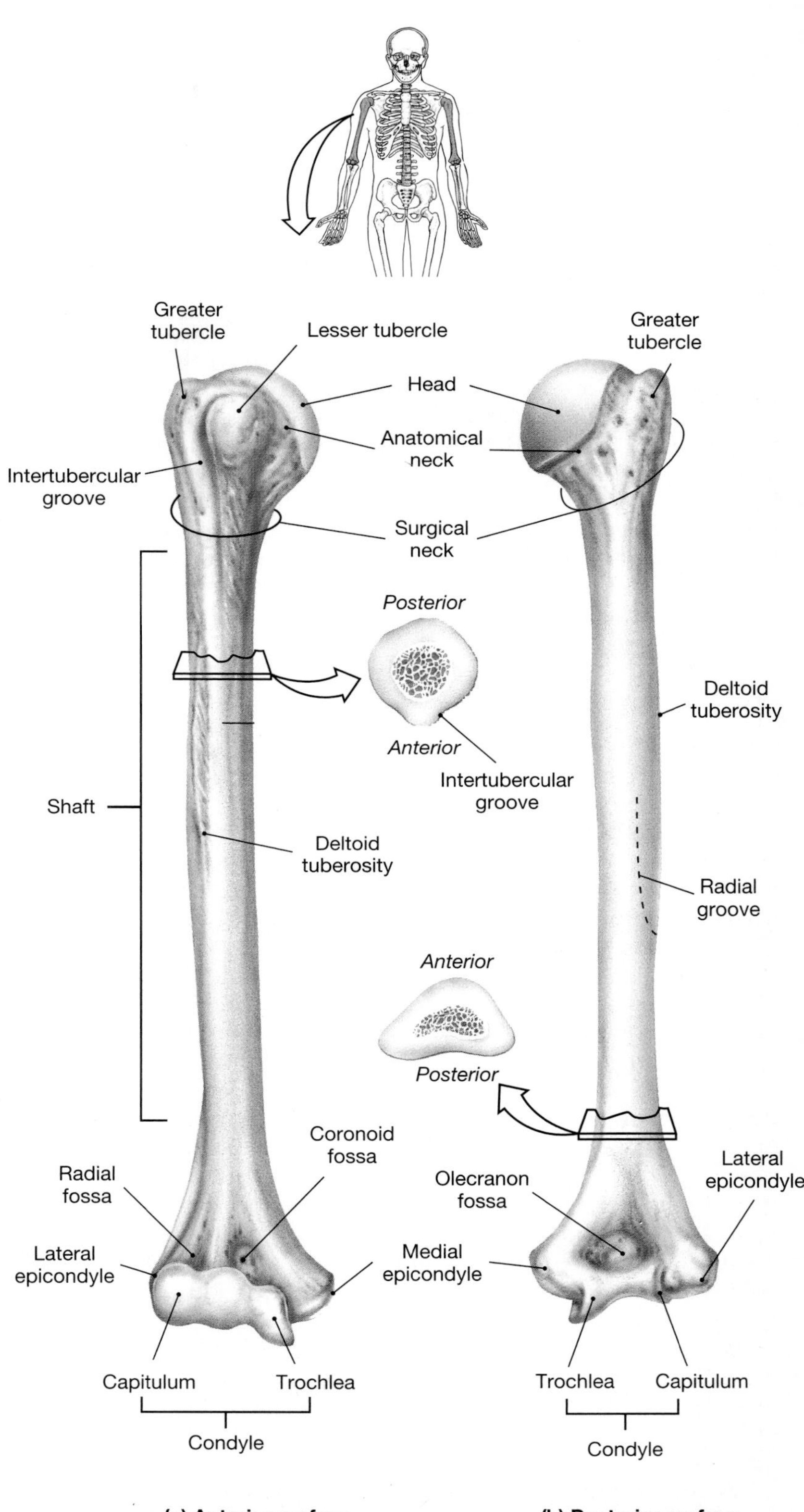
Greater tubercle
Lesser tubercle
Head
Anatomical neck
Intertubercular groove
Surgical neck
Posterior
Anterior
Intertubercular groove
Shaft
Deltoid tuberosity
Greater tubercle
Deltoid tuberosity
Radial groove
Anterior
Posterior
Coronoid fossa
Radial fossa
Olecranon fossa
Lateral epicondyle
Lateral epicondyle
Medial epicondyle
Capitulum
Trochlea
Condyle
Trochlea
Capitulum
Condyle
(a) Anterior surface
(b) Posterior surface

the forearm. Distal to the radial groove, the posterior surface of the humerus is relatively flat. Near the distal end, the shaft expands to either side, forming a broad triangle. The **medial** and **lateral epicondyles** that project to either side provide additional surface area for muscle attachment. The *ulnar nerve* passes across the posterior surface of the medial epicondyle. A blow at the humeral side of the elbow joint can strike this nerve and produce a temporary numbness and paralysis of muscles on the anterior surface of the forearm.

The articular **condyle** dominates the distal, inferior surface of the humerus. A low ridge crosses the condyle, dividing it into two distinct articular regions (Figure 8-4a,b•). The **trochlea** (*trochlea*, a pulley) is the spool-shaped medial portion that extends from the base of the **coronoid** (KOR-ō-noyd; *corona*, crown) **fossa** on the anterior surface to the **olecranon fossa** on the posterior surface (Figure 8-4b•). These depressions accept projections from the surface of the ulna as the elbow approaches full flexion or extension. The rounded **capitulum** forms the lateral surface of the condyle. A shallow **radial fossa** proximal to the capitulum accommodates a small projection on the radius.

The Ulna

Figure 8-5

The **ulna** and **radius** are parallel bones that support the forearm. In the anatomical position, the ulna lies medial to the radius.

The **olecranon** (ō-LEK-ruh-non), or *olecranon process*, is the point of the elbow. This process forms the superior and posterior portions of the proximal epiphysis of the ulna (Figure 8-5a•). On the anterior surface of the proximal epiphysis (Figure 8-5b•), the **trochlear notch** (or *semilunar notch*) articulates with the trochlea of the humerus at the elbow joint, also known as the *olecranal joint*, or *humeroulnar joint*. The olecranon forms the superior lip of the trochlear notch, and the **coronoid process** forms its inferior lip. During extreme extension, the olecranon swings into the olecranon fossa on the posterior surface of the humerus. In extreme flexion, the coronoid process projects into the coronoid fossa on the anterior humeral surface. Lateral to the coronoid process, a smooth **radial notch** accommodates the head of the radius at the proximal *radioulnar joint*.

When viewed in cross section, the shaft of the ulna is roughly triangular, with the smooth medial surface at the base of the triangle and the lateral margin at the apex. A fibrous sheet, the **interosseous membrane,** connects the lateral margin of the ulna to the radius (Figure 8-5a,b•). Distally, the ulnar shaft narrows before ending at a disc-shaped **head** whose posterior margin supports a short **styloid process** (*styloid*, long and pointed). A triangular **articular cartilage** attaches to the styloid process, isolating the ulnar head from the bones of the wrist. The distal radioulnar joint lies near the lateral border of the ulnar head.

The Radius

Figure 8-5

The radius is the lateral bone of the forearm. The disc-shaped radial **head** articulates with the capitulum of the humerus. A narrow **neck** extends from the radial head to a prominent **radial tuberosity** that marks the attachment site of a muscle that flexes the forearm. The shaft of the radius curves along its length, and the **distal extremity** is considerably larger than the distal portion of the ulna. Because the articular cartilage of the ulnar head separates it from the wrist, only the distal extremity of the radius participates in the wrist joint. The **styloid process** on the lateral surface of the distal extremity assists in the stabilization of the joint.

The medial surface of the distal extremity articulates with the ulnar head at the **ulnar notch.** The proximal radioulnar articulation permits rotation; when this movement occurs the ulnar notch rolls across the rounded surface of the ulnar head (Figure 8-5b•). This movement is called **pronation** (Figure 8-5c•). The reverse movement, which returns the radius and ulna to the anatomical position, is called **supination.**

The Carpals

Figure 8-6

The eight **carpal bones** of the wrist, or **carpus**, form two rows; there are four **proximal carpals** and four **distal carpals.** The proximal carpals are the *scaphoid, lunate, triangular,* and *pisiform* (PIS-i-form). The distal carpals are the *trapezium, trapezoid, capitate,* and *hamate* (Figure 8-6a•).

THE PROXIMAL CARPALS

- The **scaphoid** is the proximal carpal located on the lateral border of the wrist adjacent to the styloid process of the radius.
- The comma-shaped **lunate** (*luna*, moon) lies medial to the scaphoid. Like the scaphoid, the lunate articulates with the radius.
- The **triangular bone,** or *triquetral bone,* is medial to the lunate. It has the shape of a small pyramid. The triangular articulates with the cartilage that separates the ulnar head from the wrist.
- The small, pea-shaped **pisiform** lies anterior to the triangular bone and extends farther medially than any other carpal in the proximal or distal rows.

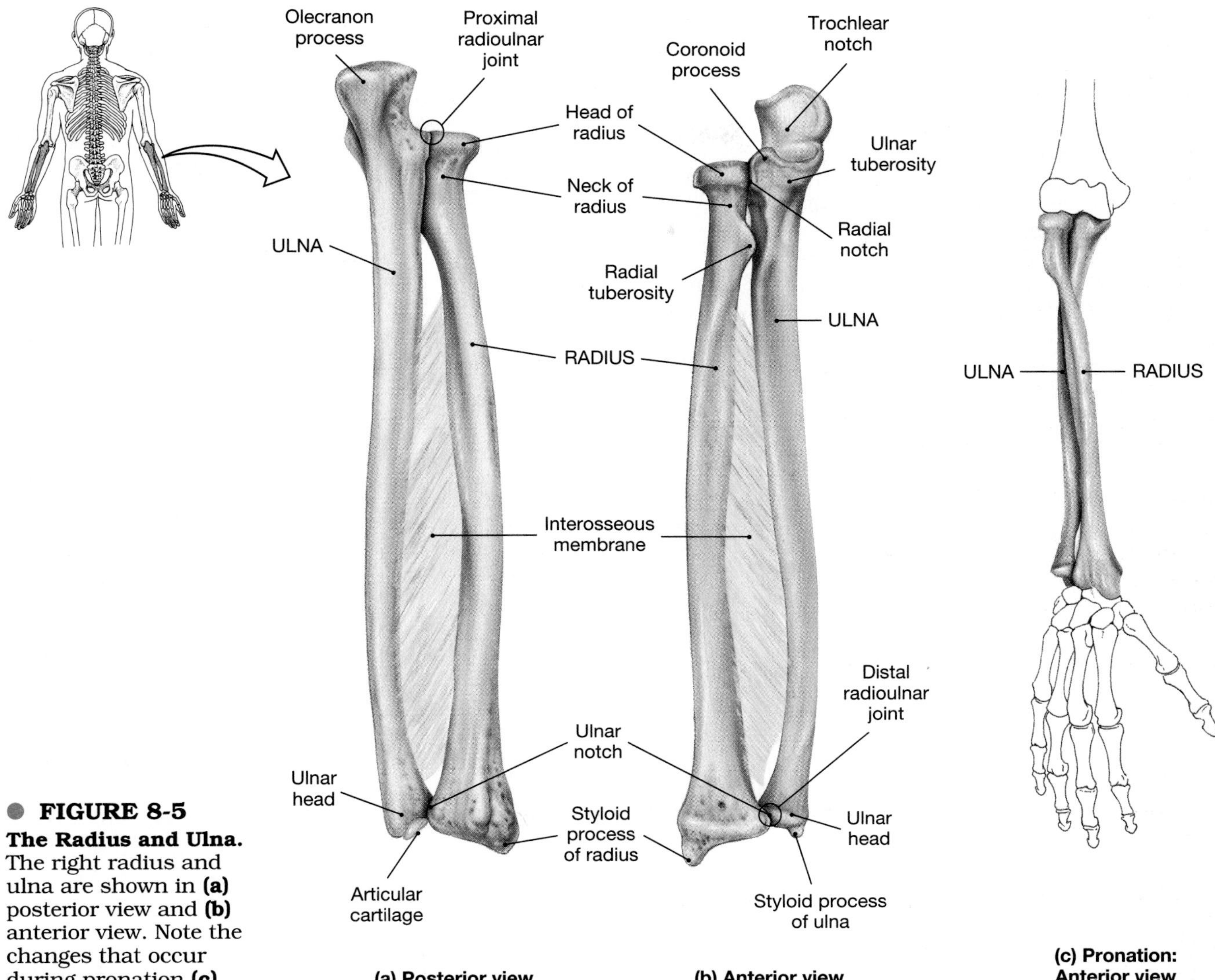
Olecranon process
Proximal radioulnar joint
Head of radius
Neck of radius
ULNA
RADIUS
Interosseous membrane
Ulnar head
Ulnar notch
Styloid process of radius
Articular cartilage
Coronoid process
Trochlear notch
Ulnar tuberosity
Radial notch
Radial tuberosity
ULNA
Distal radioulnar joint
Ulnar head
Styloid process of ulna
ULNA
RADIUS
(a) Posterior view
(b) Anterior view
(c) Pronation: Anterior view

● **FIGURE 8-5**
The Radius and Ulna. The right radius and ulna are shown in **(a)** posterior view and **(b)** anterior view. Note the changes that occur during pronation **(c)**.

THE DISTAL CARPALS

- The **trapezium** is the lateral bone of the distal row. It forms a proximal articulation with the scaphoid.
- The wedge-shaped **trapezoid** lies medial to the trapezium; it is the smallest distal carpal. Like the trapezium, it has a proximal articulation with the scaphoid.
- The **capitate** is the largest carpal bone. It sits between the trapezoid and the hamate.
- The **hamate** (*hamatum*, hooked) is a hook-shaped bone that is the medial distal carpal.

The carpals articulate with one another at joints that permit limited sliding and twisting. Ligaments interconnect the carpals and help stabilize the wrist joint. Tendons going to the fingers pass across the anterior surface of the wrist, sandwiched between these ligaments and a broad superficial transverse ligament, called the *flexor retinaculum*. Inflammation of the connective tissues between the flexor retinaculum and the carpals can squeeze the tendons and adjacent sensory nerves, producing pain and a loss of wrist mobility. This condition is called *carpal tunnel syndrome*.

The Hand

Figure 8-6

Five **metacarpals** (met-uh-KAR-pulz) articulate with the distal carpals and support the palm (Figure 8-6●). Roman numerals I–V are used to identify the metacarpals, beginning with the lateral metacarpal that articulates with the trapezium. Each metacarpal has a wide, concave proximal *base*, a small *body*, and a distal *head*. Distally, the metacarpals articulate with the finger bones, or **phalanges** (fa-LAN-jēz; singular *phalanx*). There are 14 phalangeal bones in each hand. The thumb, or **pollex** (POL-eks), has two phalanges (proximal and distal), and each of the other fingers has three phalanges (proximal, middle, and distal).

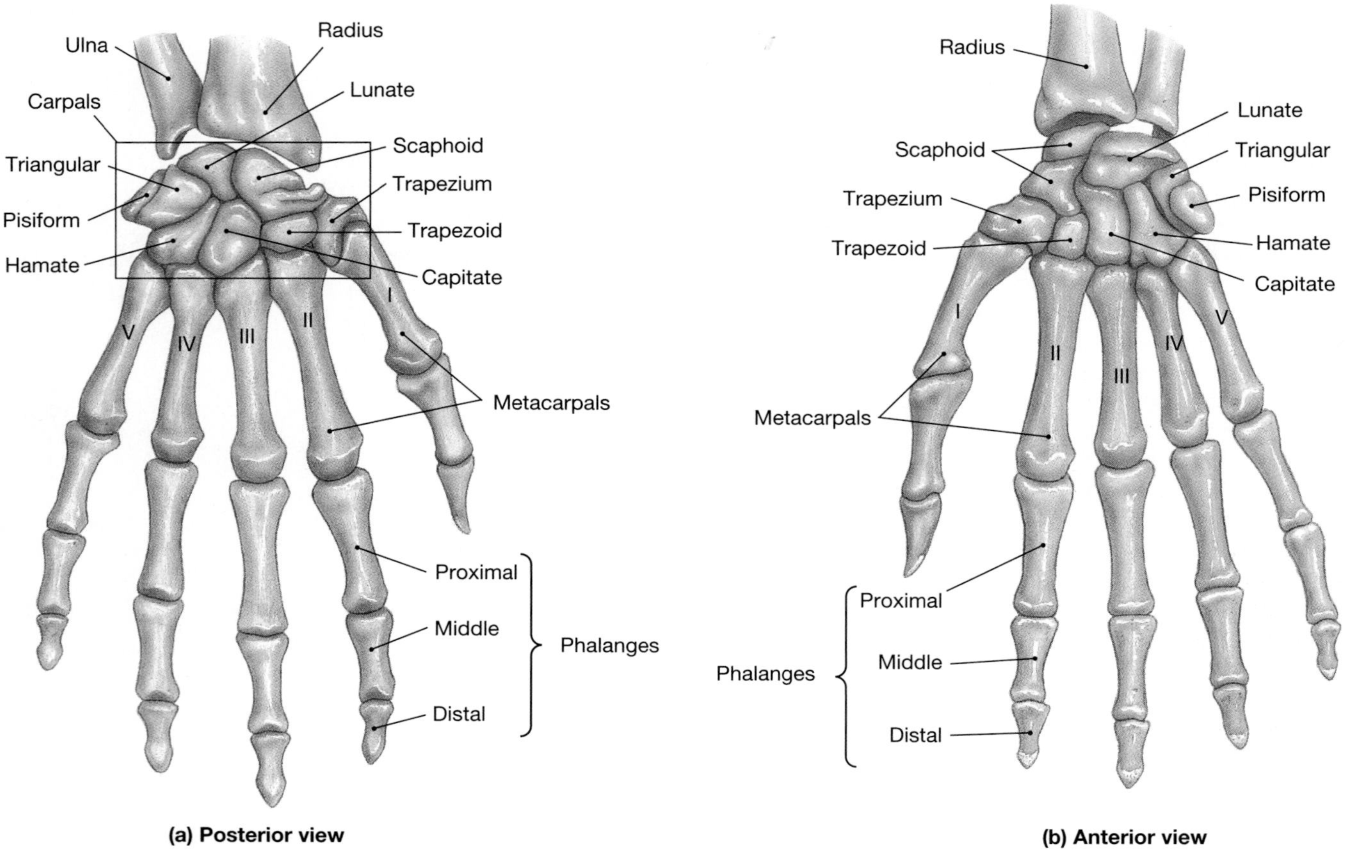

(a) Posterior view

(b) Anterior view

● **FIGURE 8-6**
Bones of the Wrist and Hand. (a) Posterior view of the right hand. **(b)** Anterior view of the right hand.

 The rounded projections on either side of the elbow are parts of what bone?

 Which antebrachial bone is lateral in the anatomical position?

 Bill accidentally strikes his first distal phalanx with a hammer and fractures it. Which finger is broken?

The Pelvic Girdle and Lower Limbs

The bones of the **pelvic girdle** are more massive than those of the pectoral girdle because of the stresses involved in weight bearing and locomotion. The bones of the lower extremities are more massive than those of the upper extremities, for similar reasons. The pelvic girdle consists of the two fused *coxae,* or *innominate bones.* The *pelvis* is a composite structure that includes the coxae of the appendicular skeleton and the sacrum and coccyx of the axial skeleton. Each lower limb consists of the *femur* (thigh), the *patella,* the *tibia* and *fibula* (leg), and the bones of the ankle (*tarsals*) and foot.

❑ The Pelvic Girdle

Figure 8-7

Each **coxa,** or "hip bone," forms through the fusion of three bones, an **ilium** (IL-ē-um), an **ischium** (IS-kē-um), and a **pubis** (PŪ-bis) (Figure 8-7●). The articulation between the coxa and the auricular surfaces of the sacrum occurs at the posterior and medial aspect of the ilium. Ventrally the coxae are connected by a pad of fibrocartilage at the *pubic symphysis.* On the lateral surface of each coxa the head of the femur articulates with the curved surface of the **acetabulum** (a-se-TAB-ū-lum; *acetabulum,* cup).

The acetabulum lies inferior and anterior to the center of the coxa (Figure 8-7a●). The space enclosed by the walls of the acetabulum is the **acetabular fossa,** which has a diameter of approximately 5 cm (2 in.). The acetabulum contains a smooth curved surface that forms the shape of the letter "C." This is the **lunate surface,** which articulates with the head of the femur.

The Coxa

Figure 8-7

The three coxal bones meet inside the acetabular fossa, as if it were a pie sliced into three pieces. The ilium, the largest coxal bone, provides the superior slice that includes around two-fifths of the acetabular surface. Above the acetabulum, the ilium forms a broad, curved surface that provides an extensive area for the attachment of muscles, tendons, and ligaments (Figure 8-7a●). The iliac expansion begins superior to the **arcuate line.** The anterior border includes the **anterior inferior iliac spine,** above the **inferior iliac notch,** and continues anteriorly to the **anterior superior iliac spine.** Curving posteriorly, the superior border supports the **iliac crest,** a ridge marking the attachments of both ligaments and muscles. The iliac crest ends at the **posterior superior iliac spine.** Below the spine, the ilial margin continues inferiorly to the rounded **posterior inferior iliac spine** that sits above the **greater sciatic** (sī-A-tik) **notch.**

Near the superior and posterior margin of the acetabulum, the ilium fuses with the ischium, which accounts for the posterior two-fifths of the acetabular surface. Posterior to the acetabulum, the prominent **ischial spine** projects above the **lesser sciatic notch.** The rest of the ischium forms a sturdy process that turns medially and inferiorly. A roughened projection, the **ischial tuberosity,** forms its posterolateral border. When a person is seated, the body weight is borne by the ischial tuberosities. The narrow **ramus** (branch) of the ischium continues toward its anterior fusion with the pubis.

At the point of fusion, the ramus of the ischium meets the **inferior ramus** of the pubis. Anteriorly, the inferior ramus ends at the **pubic tubercle,** where it meets the **superior ramus** of the pubis, which originates near the acetabular margin. The anterior, superior surface of the superior ramus bears a roughened ridge, the **pubic crest,** that ends at the **pubic tubercle.** The pubic and ischial rami encircle the **obturator** (OB-tū-rā-tor) **foramen.** In life, this space is closed by a sheet of collagen fibers whose inner and outer surfaces provide a firm base for the attachment of muscles and visceral structures. The superior ramus of the pubis originates at the anterior margin of the acetabulum. Inside the acetabulum, the pubis contacts the ilium and ischium.

● **FIGURE 8-7**
The Coxa

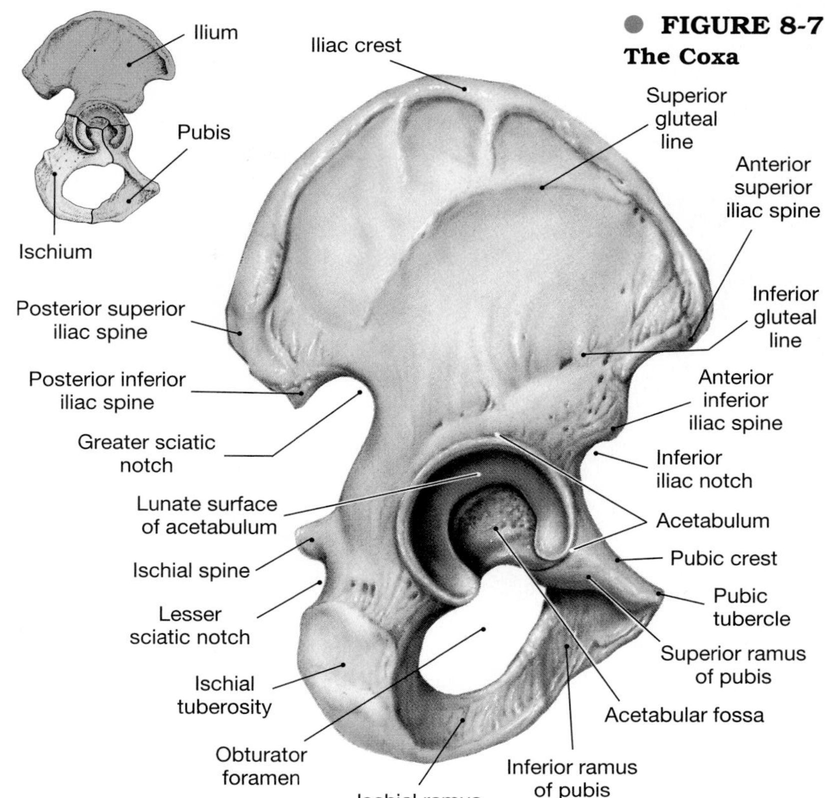
Ilium
Pubis
Ischium
Iliac crest
Superior gluteal line
Anterior superior iliac spine
Inferior gluteal line
Anterior inferior iliac spine
Inferior iliac notch
Acetabulum
Pubic crest
Pubic tubercle
Superior ramus of pubis
Acetabular fossa
Inferior ramus of pubis
Ischial ramus
Obturator foramen
Ischial tuberosity
Lesser sciatic notch
Ischial spine
Lunate surface of acetabulum
Greater sciatic notch
Posterior inferior iliac spine
Posterior superior iliac spine

(a) Right coxa, lateral view

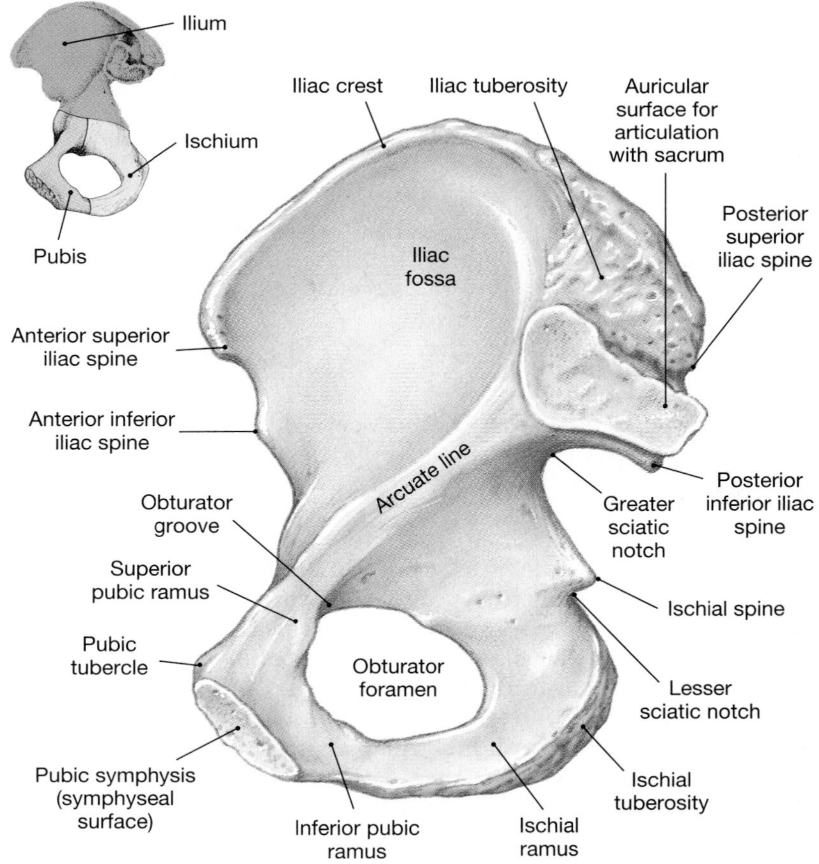
Ilium
Ischium
Pubis
Iliac crest
Iliac tuberosity
Auricular surface for articulation with sacrum
Posterior superior iliac spine
Iliac fossa
Anterior superior iliac spine
Anterior inferior iliac spine
Arcuate line
Posterior inferior iliac spine
Greater sciatic notch
Obturator groove
Superior pubic ramus
Ischial spine
Pubic tubercle
Obturator foramen
Lesser sciatic notch
Pubic symphysis (symphyseal surface)
Inferior pubic ramus
Ischial ramus
Ischial tuberosity

(b) Right coxa, medial view

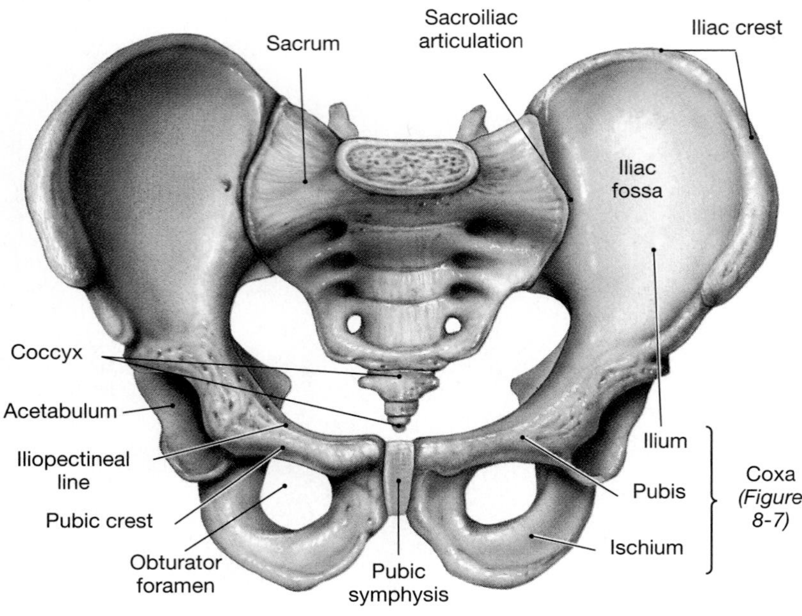
Sacroiliac articulation
Sacrum
Iliac crest
Iliac fossa
Coccyx
Acetabulum
Iliopectineal line
Pubic crest
Obturator foramen
Pubic symphysis
Ilium
Pubis
Ischium
Coxa (Figure 8-7)

(a) Anterior view

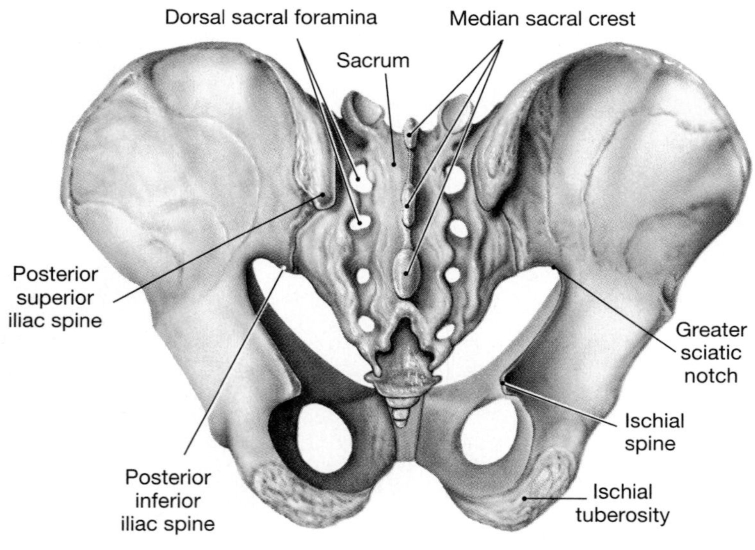
Dorsal sacral foramina
Median sacral crest
Sacrum
Posterior superior iliac spine
Greater sciatic notch
Ischial spine
Posterior inferior iliac spine
Ischial tuberosity

(b) Posterior view

● **FIGURE 8-8**
The Pelvis

The concave medial surface of the **iliac fossa** helps support the abdominal organs and provides additional surface area for muscle attachment (Figure 8-7b●). The arcuate line marks the inferior border of the iliac fossa. The anterior and medial surface of the pubis contains a roughened area that marks the site of articulation with the pubis of the opposite side. At this articulation, the **pubic symphysis,** the two pubic bones are attached to a median fibrocartilage pad. The **iliopectineal line** begins near the symphysis and extends diagonally across the pubis to merge with the arcuate line, which continues toward the **auricular surface** of the ilium. Ligaments arising at the **iliac tuberosity** stabilize the joint. On the medial surface of the superior ramus of the pubis lies the **obturator groove** for the obturator blood vessels and nerve.

The Pelvis

Figures 8-8, 8-9, 8-10

Figure 8-8● shows anterior and posterior views of the **pelvis,** which consists of the coxae, the sacrum, and the coccyx. An extensive network of ligaments connects the lateral borders of the sacrum with the iliac crest, the ischial tuberosity, the ischial spine, and the iliopectineal line. Other ligaments tie the ilia to the posterior lumbar vertebrae. These interconnections increase the stability of the pelvis.

The pelvis may be subdivided into the **false** (*greater*) **pelvis** and the **true** (*lesser*) **pelvis.** The boundaries of each are indicated in Figure 8-9●. The false pelvis consists of the expanded, blade-like portions of each ilium superior to the iliopectineal line. Structures inferior to that line, including the inferior portions of each ilium, both pubic bones, the ischia, the sacrum, and the coccyx, form the true pelvis.

The true pelvis encloses the *pelvic cavity,* a subdivision of the abdominopelvic cavity. *[p. 24]* In lateral view (Figure 8-9b●) the upper limit of the true pelvis is a line that extends from either side of the base of the sacrum, along the arcuate and iliopectineal lines, to the superior margin of the pubic symphysis. The bony edge of the true pelvis is called the **pelvic brim,** and the enclosed space is the **pelvic inlet.**

The **pelvic outlet** is the opening bounded by the inferior edges of the pelvis (Figure 8-9b,c●). In life this region, called the **perineum** (per-i-NĒ-um), is bounded by the coccyx, the ischial tuberosities, and the inferior border of the pubic symphysis. Perineal muscles form the floor of the pelvic cavity and support the enclosed organs.

The shape of the pelvis of a female is somewhat different from that of a male (Figure 8-10●). Some of these differences are the result of variations in body size and muscle mass. For example, in the female the pelvis is usually smoother, lighter, and has less prominent markings. Other differences are adaptations for childbearing, including:

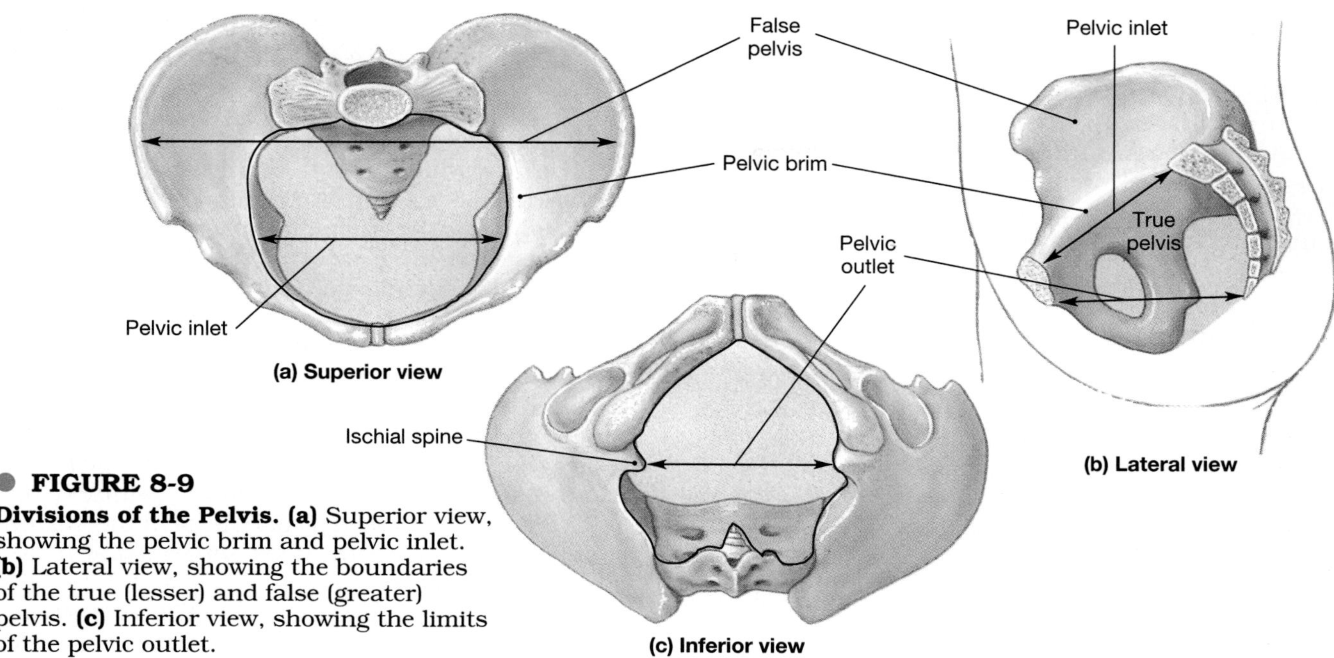
False pelvis
Pelvic inlet
Pelvic brim
Pelvic outlet
True pelvis
Pelvic inlet
(a) Superior view
Ischial spine
(b) Lateral view
(c) Inferior view

● **FIGURE 8-9**
Divisions of the Pelvis. (a) Superior view, showing the pelvic brim and pelvic inlet. **(b)** Lateral view, showing the boundaries of the true (lesser) and false (greater) pelvis. **(c)** Inferior view, showing the limits of the pelvic outlet.

- An enlarged pelvic outlet.
- Less curvature on the sacrum and coccyx, which in the male arc into the pelvic outlet.
- A wider, more circular pelvic inlet.
- A relatively broad, low pelvis.
- Ilia that project farther laterally, but do not extend as far above the sacrum.
- A broader pubic arch, with the inferior angle between the pubic bones greater than 100°.

These adaptations are related to (1) the support of the weight of the developing fetus and uterus and (2) passage of the newborn through the pelvic outlet at the time of delivery. In addition, a hormone produced during pregnancy loosens the pubic symphysis, allowing relative movement between the coxae that can further increase the size of the pelvic inlet and outlet.

 What three bones make up the coxa?

 How is the pelvis of the female adapted for childbearing?

 When seated, what part of the pelvis bears the body's weight?

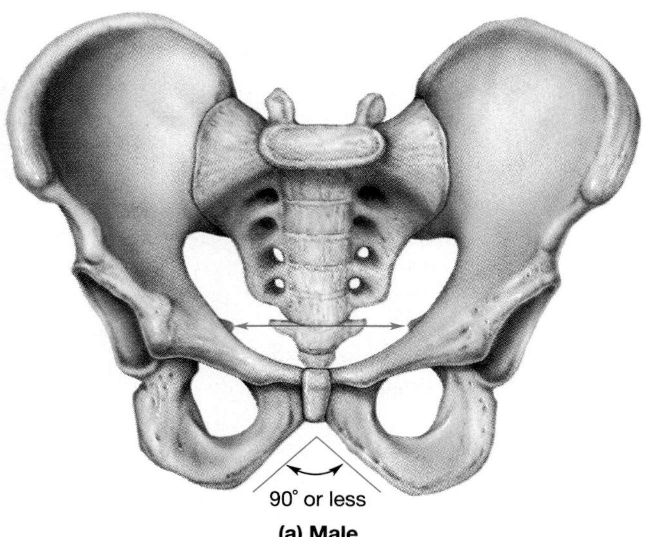
90° or less
(a) Male

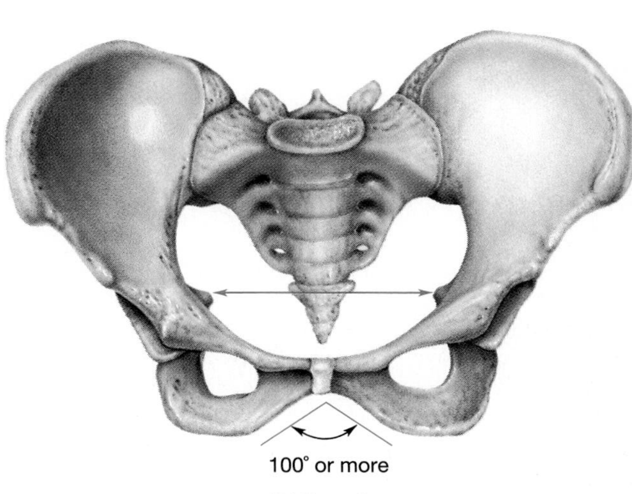
100° or more
(b) Female

● **FIGURE 8-10**
Anatomical Differences in the Pelvis of a Male and Female. Note the much sharper pubic angle in the pelvis of a male **(a)** than in that of a female **(b)**.

❑ The Lower Limb

The skeleton of each lower limb consists of a femur, a patella (kneecap), a tibia and fibula, and the bones of the ankle and foot. The functional anatomy of the lower limb is very different from that of the upper limb, primarily because the lower limb must transfer the body weight to the ground.

The Femur

Figure 8-11

The **femur** is the longest and heaviest bone in the body. Distally, the femur articulates with the tibia of the leg at the knee joint. The rounded epiphysis, or **head,** of the femur articulates with the pelvis at the acetabulum. Distally, the **neck** joins the **shaft** at an angle of about 125° (Figure 8-11•). The **greater trochanter** projects laterally from the junction of the neck and shaft. Nearby, the **lesser trochanter** originates on the posteromedial surface of the femur. Both trochanters develop where large tendons attach to the femur. On the anterior surface of the femur, the raised **intertrochanteric** (trō-kan-TER-ik) **line** marks the distal edge of the articular capsule. This line continues around to the posterior surface, passing inferior to the trochanters as the **intertrochanteric crest.**

The proximal portion of the femoral shaft is round in cross section. A prominent elevation, the **linea aspera** (*aspera*, rough), runs along the center of the posterior surface, marking the attachment site of powerful muscles that adduct the femur (Figure 8-11b•). Distally, the linea aspera divides into a **medial** and **lateral supracondylar ridge** to form a flattened triangular area, the **popliteal surface.** The medial supracondylar ridge terminates at the raised, roughened **medial epicondyle.** The lateral ridge ends at the **lateral epicondyle.** The smoothly rounded **medial** and **lateral condyles** are primarily distal and posterior to the epicondyles. On the posterior surface the two condyles are separated by a deep **intercondylar fossa.**

The condyles continue across the inferior surface of the femur, but the intercondylar fossa does not extend completely around to the anterior surface. As a result, the smooth articular faces merge,

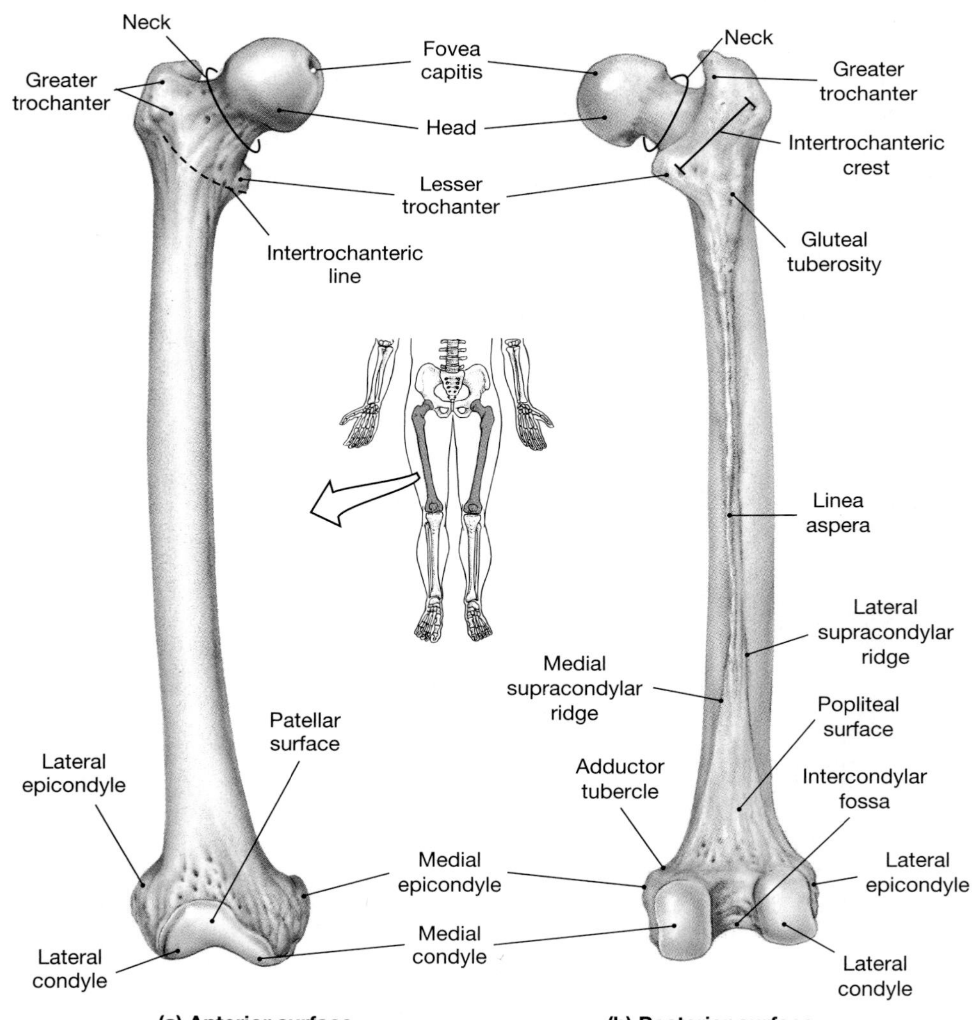
Neck
Greater trochanter
Fovea capitis
Head
Lesser trochanter
Intertrochanteric line
Lateral epicondyle
Patellar surface
Medial epicondyle
Medial condyle
Lateral condyle
(a) Anterior surface
Neck
Greater trochanter
Intertrochanteric crest
Gluteal tuberosity
Linea aspera
Lateral supracondylar ridge
Medial supracondylar ridge
Popliteal surface
Adductor tubercle
Intercondylar fossa
Lateral epicondyle
Lateral condyle
(b) Posterior surface

● **FIGURE 8-11**
The Femur. Bone markings on the right femur are presented as seen from the **(a)** anterior and **(b)** posterior surfaces.

producing an articular surface with elevated lateral borders (Figure 8-11a●). This is the **patellar surface** over which the patella glides.

The Patella

Figure 8-12a

The **patella** is a large sesamoid bone that forms within the tendon of the *quadriceps femoris*, a group of muscles that extend the leg. The patella has a rough, convex anterior surface (Figure 8-12a●). The roughened surface reflects the attachment of the quadriceps tendon (anterior and superior surfaces) and the *patellar ligament* (anterior and inferior surfaces). The patellar ligament connects the patella to the tibia. The posterior surface presents two concave **facets (medial** and **lateral)** for articulation with the medial and lateral condyles of the femur.

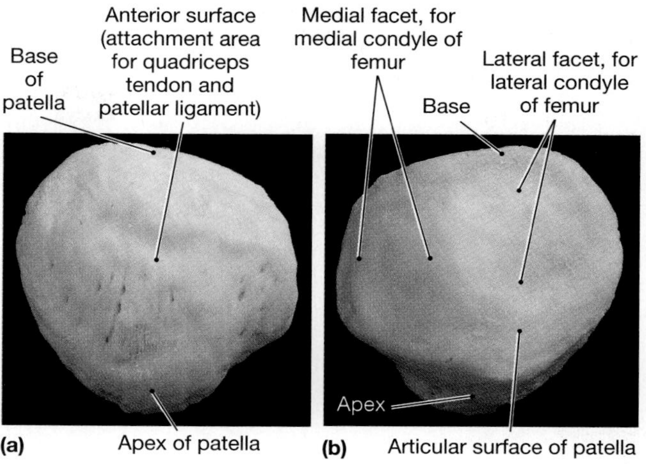
Base of patella
Anterior surface (attachment area for quadriceps tendon and patellar ligament)
Medial facet, for medial condyle of femur
Base
Lateral facet, for lateral condyle of femur
Apex
(a)
Apex of patella
(b)
Articular surface of patella

● **FIGURE 8-12**
The Right Patella. (a) Anterior surface. **(b)** Posterior surface.

The Tibia

Figure 8-13

The **tibia** (TI-bē-uh) is the large medial bone of the leg (Figure 8-13a●). The medial and lateral condyles of the femur articulate with the **medial** and **lateral tibial condyles** at the proximal end of the tibia. The anterior surface of the tibia near the condyles bears a prominent, rough **tibial tuberosity** that can easily be felt beneath the skin of the leg. This tuberosity marks the attachment of the stout patellar ligament.

The **anterior crest** is a ridge that begins at the tibial tuberosity and extends distally along the anterior tibial surface. The anterior crest of the tibia can also be easily felt through the skin. Distally, the tibia broadens, and the medial border ends in a large process, the **medial malleolus** (ma-LĒ-o-lus; *malleolus*, hammer). The inferior surface of the tibia forms a hinge joint with the proximal bone of the ankle; the medial malleolus provides medial support for the joint.

A ridge, the **intercondylar eminence,** separates the medial and lateral condyles of the tibia (Figure 8-13b●).

The Fibula

Figure 8-13

The slender **fibula** (FIB-yoo-luh) parallels the lateral border of the tibia (Figure 8-13a,b●). The fibular **head** articulates with the tibia at an articular facet located inferior and slightly posterior to the lateral tibial condyle. The medial border of the thin shaft is bound to the tibia by the **interosseous membrane,** which extends from the **interosseous crest** of the fibula to that of the tibia. A sectional view through the shafts of the tibia and fibula (Figure 8-13c●) shows the locations of the tibial and fibular interosseous crests, and the fibrous interosseous membrane that extends between them. This membrane helps stabilize the positions of the two bones and provides additional surface area for muscle attachment.

The fibula does not articulate with the femur, and it does not play a role in the transfer of weight to the ankle and foot. However, it is important as a site for muscle attachment. In addition, the distal tip of the fibula extends lateral to the ankle joint. This fibular process, the **lateral malleolus,** provides lateral stability to the ankle. The stability is somewhat limited, and forceful movement of the foot outward and backward can result in a dislocation of the ankle that breaks both the lateral malleolus of the fibula and the medial malleolus of the tibia. This injury is called a *Pott's fracture*, one of the fracture types introduced in Chapter 6. ∞ [p. 192]

The Ankle

Figure 8-14

The ankle, or **tarsus,** consists of seven **tarsals:** *talus, calcaneus, cuboid, navicular,* and three *cuneiforms* (Figure 8-14●).

- The **talus** transmits from the tibia the weight of the body anteriorly, from the tibia toward the toes. The talus is the second largest foot bone. The primary tibial articulation occurs across the smooth superior surface of the **trochlea.** The trochlea has lateral and medial extensions that articulate with the lateral malleolus (fibula) and medial malleolus (tibia). The lateral surfaces of the talus are roughened where ligaments connect it to the tibia and fibula, further stabilizing the ankle joint.
- The **calcaneus** (kal-KĀ-nē-us), or heel bone, is the largest of the tarsal bones. When you stand normally, most of your weight is trans-

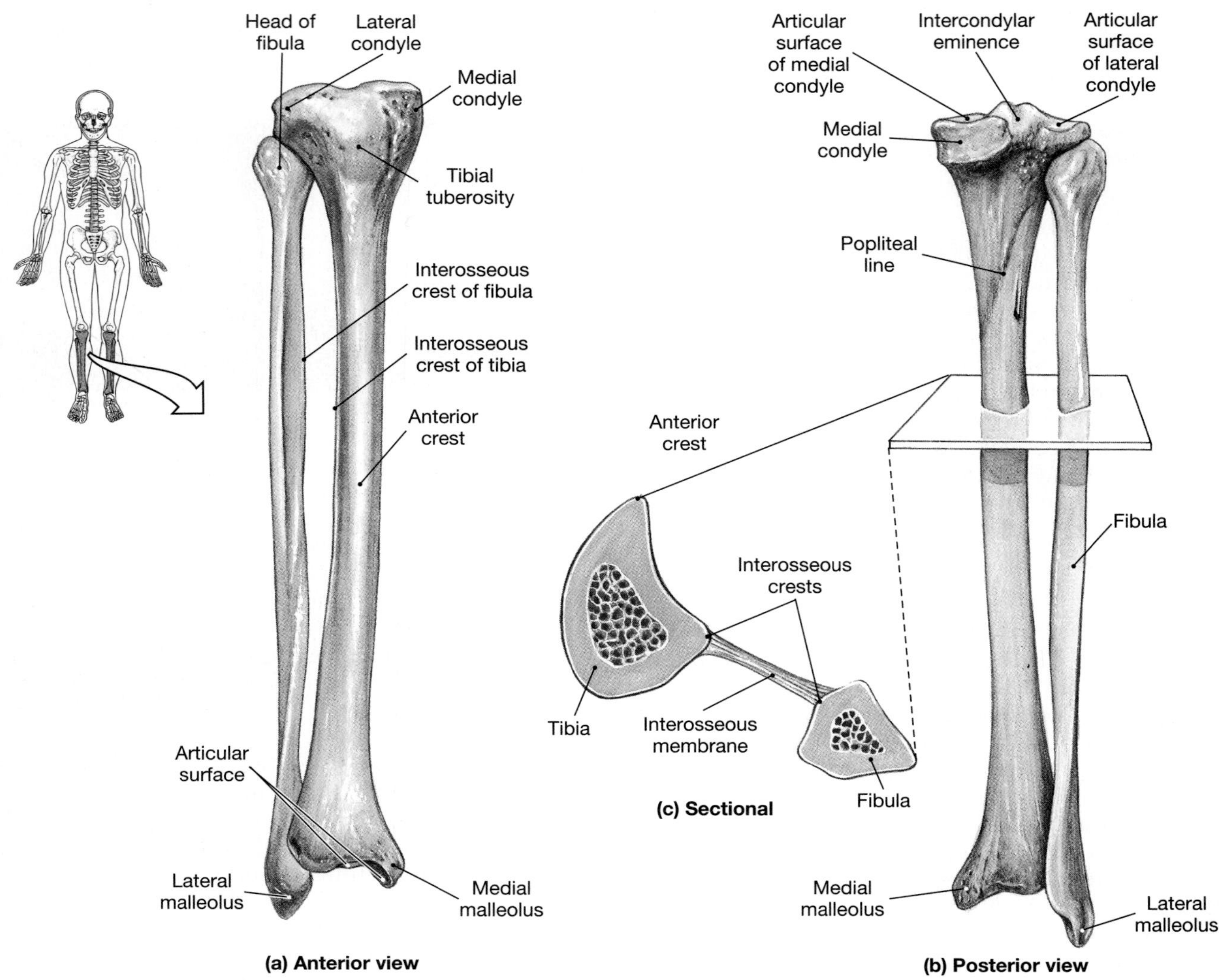
Head of fibula
Lateral condyle
Medial condyle
Tibial tuberosity
Interosseous crest of fibula
Interosseous crest of tibia
Anterior crest
Articular surface
Lateral malleolus
Medial malleolus
(a) Anterior view
Articular surface of medial condyle
Intercondylar eminence
Articular surface of lateral condyle
Medial condyle
Popliteal line
Anterior crest
Interosseous crests
Tibia
Interosseous membrane
Fibula
(c) Sectional
Fibula
Medial malleolus
Lateral malleolus
(b) Posterior view

● **FIGURE 8-13**
The Tibia and Fibula. (a) Anterior view of the right tibia and fibula. **(b)** Posterior view. **(c)** Sectional view at the level indicated in part (b).

mitted from the tibia to the talus to the calcaneus, and then to the ground. The posterior portion of the calcaneus is a rough, knob-shaped projection. This is the attachment site for the *calcaneal tendon* (*Achilles tendon*) that arises at strong calf muscles. These muscles raise the heel and depress the sole, as when standing on tiptoe. The superior and anterior surfaces of the calcaneus bear smooth facets for articulation with other tarsal bones.

- The **cuboid** articulates with the anterior, lateral surface of the calcaneus.
- The **navicular** is located anterior to the talus, on the medial side of the ankle. Its proximal surface articulates with the talus, its distal surface with the cuneiform bones.
- The three **cuneiforms** are wedge-shaped bones arranged in a row, with articulations between them. They are named according to their position: **medial cuneiform, intermediate cuneiform,** and **lateral cuneiform.** Proximally the cuneiforms articulate with the anterior surface of the navicular. The lateral cuneiform also articulates with the medial surface of the cuboid.

The distal surfaces of the cuboid and the cuneiforms articulate with the metatarsal bones of the foot.

The *ankle joint,* or **tibiotalar joint,** involves (1) the distal articular surface of the tibia, including the medial malleolus, (2) the lateral malleolus of the fibula, and (3) the trochlea and lateral articular facets of the talus. This joint works like a hinge that permits only dorsiflexion and plantar flexion. Articulations between the talus and other tarsals permit sliding and twisting of the foot. AM *Problems with the Ankle and Foot*

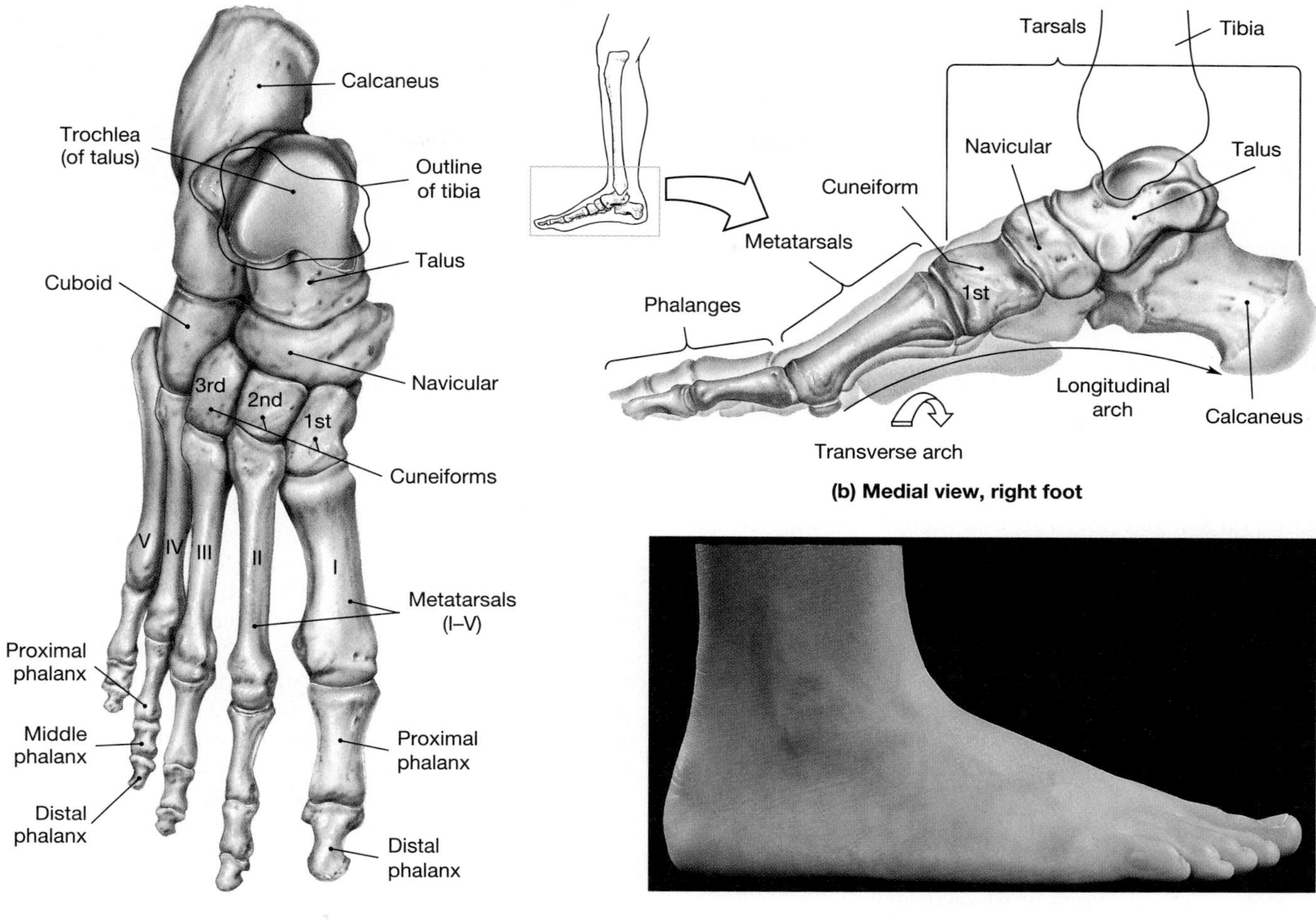
Calcaneus
Trochlea (of talus)
Outline of tibia
Talus
Cuboid
Navicular
3rd
2nd
1st
Cuneiforms
V
IV
III
II
I
Metatarsals (I–V)
Proximal phalanx
Middle phalanx
Distal phalanx
Proximal phalanx
Distal phalanx
Tarsals
Tibia
Navicular
Talus
Cuneiform
Metatarsals
Phalanges
1st
Longitudinal arch
Calcaneus
Transverse arch

(a) Superior view, right foot

(b) Medial view, right foot

(c) Lateral view, right foot

FIGURE 8-14

Bones of the Ankle and Foot. (a) Bones of the right foot as viewed from above. Note the orientation of the tarsals that convey the weight of the body to the heel and the plantar surfaces of the foot. **(b)** Medial view of bones of the foot. **(c)** Lateral view of the right foot. Note the firm contact with the surface along the lateral border of the sole. AM *Scan 8*

The Foot

Figure 8-14

The **metatarsal bones** are five long bones that form the sole of the foot (Figure 8-14•). The metatarsals are identified with Roman numerals I–V, proceeding from medial to lateral across the sole. Proximally, the first three metatarsals articulate with the three cuneiforms and the last two articulate with the cuboid. Distally, each metatarsal articulates with a proximal phalanx.

The **phalanges,** or toes (Figure 8-14•), have the same anatomical organization as the fingers. The toes contain 14 phalanges. The great toe, or **hallux,** has two phalanges (proximal and distal), and the other four toes have three phalanges apiece (proximal, middle, and distal).

ARCHES OF THE FOOT Weight transfer occurs along the **longitudinal arch** of the foot (Figure 8-14b•). Ligaments and tendons maintain this arch by tying the calcaneus to the distal portions of the metatarsals. The lateral, calcaneal side of the foot carries most of the weight of the body while standing normally. This portion of the arch has much less curvature than the medial, talar portion. The talar arch also has considerably more elasticity than the calcaneal arch. As a result, the medial, plantar surface remains elevated, and the muscles, nerves, and blood vessels that supply the inferior surface of the foot are not squeezed between the metatarsals and the ground. In the condition known as *flat feet* the normal arches do not form, and individuals with this condition cannot walk long distances without discomfort; for this reason they cannot enlist in the Army.

The elasticity of the talar arch absorbs the shocks that accompany sudden changes in weight loading. For example, the stresses involved with running or ballet dancing on the toes are cushioned by the elasticity of this portion of the arch. Because the degree of curvature changes from the medial to the lateral borders of the foot, a **transverse arch** also exists.

The amount of weight transferred forward depends on the position of the foot and the placement of body weight. During *dorsiflexion* of the foot, as when "digging in the heels," all of the body weight rests on the calcaneus. During *plantar flexion* and "standing on tiptoe," the talus and calcaneus transfer the weight to the metatarsals and phalanges through more anterior tarsal bones.

CONGENITAL TALIPES EQUINOVARUS
Congenital talipes equinovarus ("*clubfoot*") results from an inherited developmental abnormality that affects 2 in 1000 births. Boys are affected roughly twice as often as girls. One or both feet may be involved, and the condition may be mild, moderate, or severe. The underlying problem is abnormal muscle development that distorts growing bones and joints. Usually the tibia, ankle, and foot are affected, and the feet are turned medially and inverted. The longitudinal arch is exaggerated, and if both feet are involved, the soles face one another. Prompt treatment with casts or other supports in infancy helps to alleviate the problem, and fewer than half of the cases require surgery. Kristi Yamaguchi, Olympic Gold Medalist in figure skating, was born with this condition. AM *Problems with the Ankle and Foot*

- The fibula does not participate in the knee joint nor does it bear weight, but when it is fractured, walking becomes difficult. Why?
- While jumping off the back steps at his house, 10-year-old Mark lands on his right heel and breaks his foot. What foot bone is most likely broken?
- Which foot bone transmits the weight of the body from the tibia toward the toes?
- Ballet dancers strengthen their thigh muscles by placing a leg laterally on a waist-high bar, and contracting against the resistance. How would the increased strength in these muscles affect the shape of the femur?

Individual Variation in the Skeletal System

A comprehensive study of a human skeleton can reveal important information about the individual. For example, there are characteristic racial differences in portions of the skeleton, especially the skull and pelvis, and the development of various ridges and general bone mass can permit an estimation of muscular development and body weight. Details such as the condition of the teeth or the presence of healed fractures can provide information about the individual's medical history. Two important details, sex and age, can be determined or closely estimated on the basis of measurements indicated in Tables 8-1 and 8-2. Table 8-1 identifies characteristic differences between the skeletons of males and females, but not every skeleton shows every feature in classic detail. Many differences, including markings on the skull, cranial capacity, and general skeletal features, reflect differences in average body size, muscle mass, and muscular strength. The general changes in the skeletal system that take place with age are summarized in Table 8-2. Note how these changes begin at age 1 and continue throughout life. For example, fusion of the epiphyseal plates begins about age 3, and degenerative changes in the normal skeletal system, such as a reduction in mineral content in the bony matrix, typically do not begin until age 30–45.

An understanding of individual variation and of the normal timing of skeletal development is important in clinical diagnosis and treatment. Several professions focus on specific aspects of skeletal form and function. Each specialist has a different perspective, with its own techniques, traditions, and biases. For example, a person with back pain may consult an *orthopedist*, an *osteopath*, or a *chiropractor*. Information concerning these clinical specialties is included in the *Applications Manual*. AM *A Matter of Perspective*

Selected Clinical Terminology

Terms Discussed in This Chapter

carpal tunnel syndrome: Inflammation of the tissues beneath the flexor retinaculum, causing compression of the flexor tendons and adjacent sensory nerves. Symptoms are pain and a loss of wrist mobility. *(p. 245)*

congenital talipes equinovarus: A congenital deformity affecting one or both feet, commonly known as clubfoot. It develops secondary to abnormalities in muscular development. *(p. 254)*

flat feet: The loss or absence of a longitudinal arch. *(p. 253)*

AM *Additional Terms Discussed in the Applications Manual*

dancer's fracture: A fracture of the fifth metatarsal, usually near its proximal articulation.

TABLE 8-1 Sexual Differences in the Human Skeleton

Region/Feature	*Male (as compared to female)*	*Female (as compared to male)*
SKULL		
General appearance	Heavier, rougher	Lighter, smoother
Forehead	Sloping	More vertical
Sinuses	Larger	Smaller
Cranium	About 10% larger (average)	About 10% smaller
Mandible	Larger, robust	Lighter, smaller
Teeth	Larger	Smaller
PELVIS		
General appearance	Narrow, robust, heavy, rough	Broader, lighter, smoother
Pelvic inlet	Heart-shaped	Oval to round
Iliac fossa	Relatively deep	Relatively shallow
Ilium	Extends farther above sacral articulation	More vertical; less extension above sacroiliac joint
Angle inferior to pubic symphysis	Under 90°	100° or more
Acetabulum	Directed laterally	Faces slightly anteriorly as well as laterally
Obturator foramen	Oval	Triangular
Ischial spine	Points medially	Points posteriorly
Sacrum	Long, narrow triangle with pronounced sacral curvature	Broad, short triangle with less curvature
Coccyx	Points anteriorly	Points inferiorly
OTHER SKELETAL ELEMENTS		
Bone weight	Heavier	Lighter
Bone markings	More prominent	Less prominent

TABLE 8-2 Age-Related Changes in the Skeleton

Region/Feature	*Event(s)*	*Age (Years)*
GENERAL SKELETON		
Bony matrix	Reduction in mineral content	Values differ for males versus females between ages 45 and 65; similar reductions occur in both sexes after age 65.
Markings	Reduction in size, roughness	Gradual reduction with increasing age and decreasing muscular strength and mass.
SKULL		
Fontanels	Closure	Completed by age 2
Metopic suture	Fusion	2–8
Occipital bone	Fusion of ossification centers	1–4
Styloid process	Fusion with temporal bone	12–16
Hyoid bone	Complete ossification and fusion	25–30
Teeth	Loss of "baby teeth"; appearance of secondary dentition; eruption of posterior molars	Detailed in Chapter 24 (digestive system).
Mandible	Loss of teeth; reduction in bone mass; change in angle at mandibular notch	Accelerates in later years (60+).
VERTEBRAE		
Curvature	Appearance of major curves	Described in Figure 7-17.
Intervertebral discs	Reduction in size, percentage contribution to height	Accelerates in later years (60+).
LONG BONES		
Epiphyseal plates	Fusion	Ranges vary, but general analysis permits determination of approximate age.
PECTORAL AND PELVIC GIRDLES		
Epiphyses	Fusion	Overlapping ranges somewhat narrower than the above, including 14–16, 16–18, 22–25 years.

EMBRYOLOGY SUMMARY Development of the Appendicular Skeleton

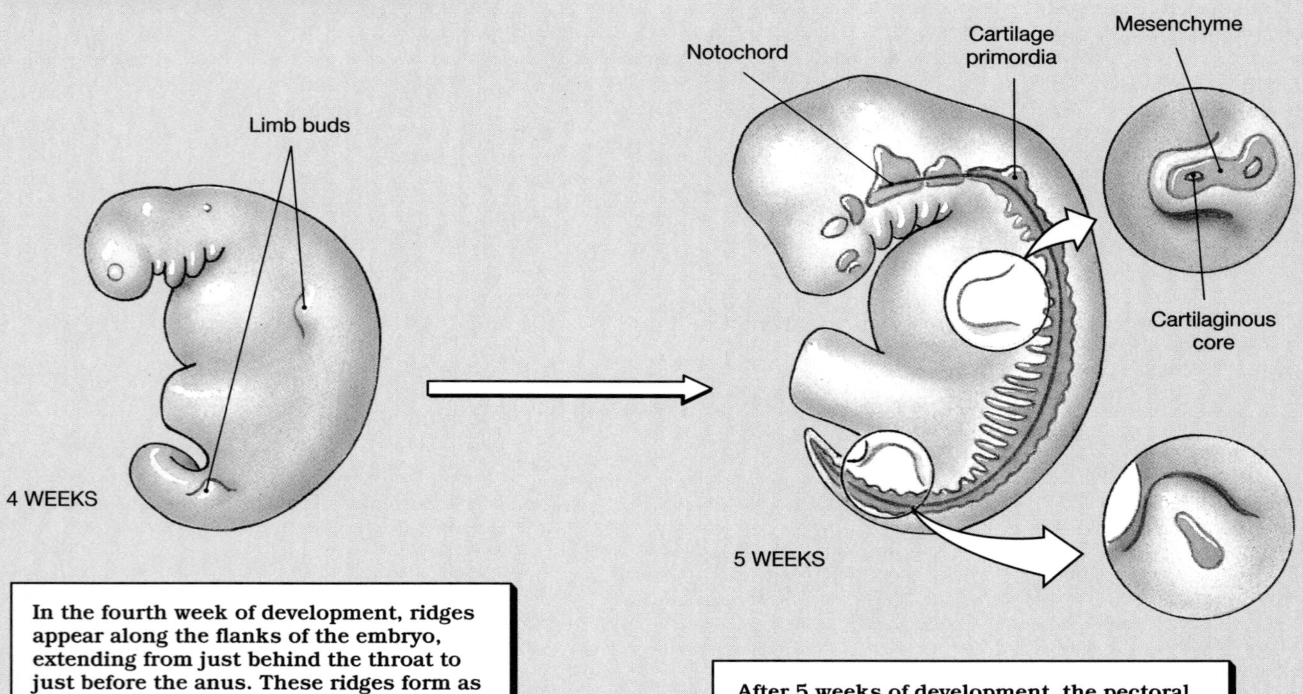
Limb buds
4 WEEKS
Notochord
Cartilage primordia
Mesenchyme
Cartilaginous core
5 WEEKS

In the fourth week of development, ridges appear along the flanks of the embryo, extending from just behind the throat to just before the anus. These ridges form as mesodermal cells congregate beneath the ectoderm of the flank. Mesoderm gradually accumulates at the end of each ridge, forming two pairs of limb buds.

After 5 weeks of development, the pectoral limb buds have a cartilaginous core and scapular cartilages are developing in the mesenchyme of the trunk.

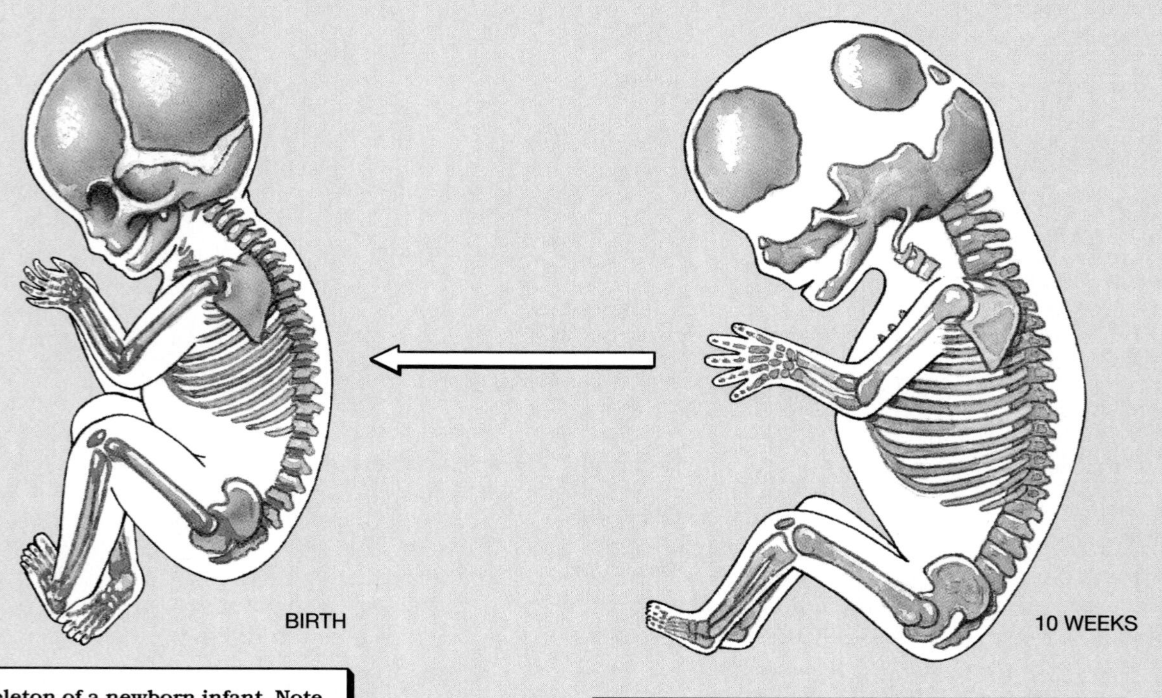
BIRTH
10 WEEKS

The skeleton of a newborn infant. Note the extensive areas of cartilage (blue) in the humeral head, in the wrist, between the bones of the palm and fingers, and in the coxae. Notice the appearance of the axial skeleton, with reference to the Embryology Summaries in Chapter 7.

Ossification in the embryonic skeleton after approximately 10 weeks of development. The shafts of the limb bones undergo rapid ossification, but the distal bones of the carpus and tarsus remain cartilaginous.

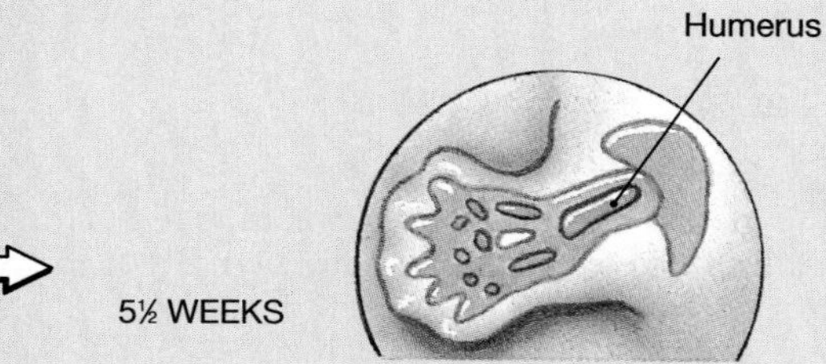
Humerus
5½ WEEKS

As the limb bud enlarges, bends develop at the future locations of the shoulder and elbow joints. Two cartilages form in the forearm, and a lateral rotation of the apical ridge places the elbow in its proper orientation.

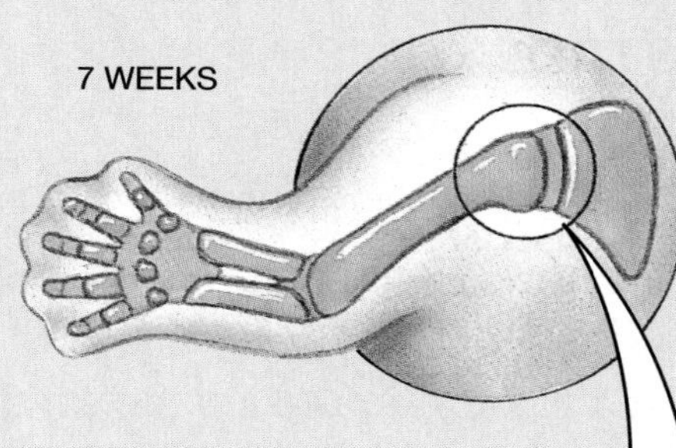
7 WEEKS

The hands originate as paddles, but the death of cells between the phalangeal cartilages produces individual fingers.

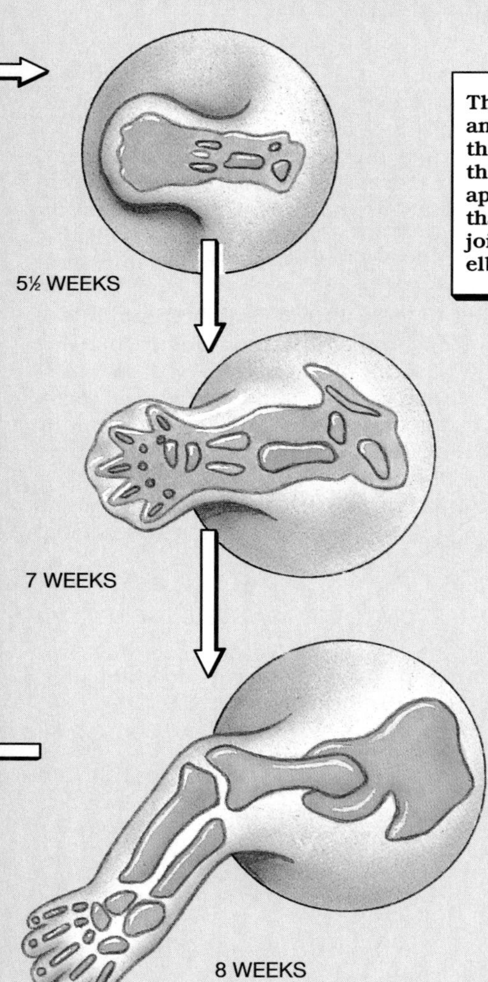
5½ WEEKS
7 WEEKS
8 WEEKS

The formation of the pelvic girdle and lower limbs closely parallels that of the pectoral complex. But as the pelvic limb bud enlarges, the apical ridge rotates medially rather than laterally. As a result, the knee joint faces posteriorly, while the elbow faces anteriorly.

By week 8, cartilaginous models of all of the major skeletal components are well formed, and endochondral ossification begins in the future limb bones. Ossification of the coxal bones begins at three separate centers that gradually enlarge.

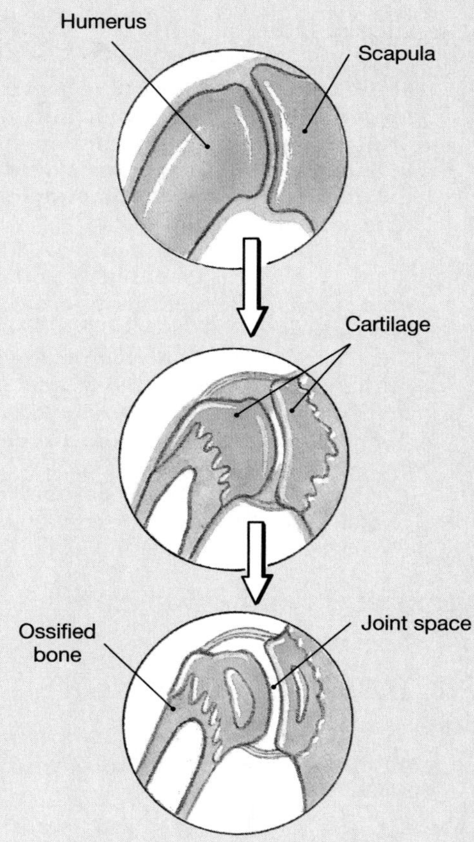
Humerus
Scapula
Cartilage
Ossified bone
Joint space

Joints form where two cartilages are in contact. The surfaces within the joint cavity remain cartilaginous, while the rest of the bones undergo ossification.

CHAPTER REVIEW

STUDY OUTLINE

INTRODUCTION, p. 240

1. The **appendicular division** includes the bones of the upper and lower extremities and the pectoral and pelvic girdles that connect the limbs to the trunk. (*Figure 8-1*)

THE PECTORAL GIRDLES AND UPPER LIMBS, p. 241

1. Each upper limb articulates with the trunk via the **shoulder,** or **pectoral girdle,** which consists of the **scapula** and **clavicle.**

The Pectoral Girdle, p. 241

2. The clavicle and scapula position the shoulder joint, help move the upper limb, and provide a base for muscle attachment. (*Figures 8-2, 8-3*)
3. The scapula articulates with the **humerus** at the shoulder (*scapulohumeral*) joint. Both the **coracoid process** and the **acromion** are attached to ligaments and tendons. The **scapular spine** crosses the scapular body. (*Figure 8-3*)

The Upper Limb, p. 243

4. The capsule of the shoulder attaches to the **humerus** at the **anatomical neck;** the **greater tubercle** and **lesser tubercle** are important sites for muscle attachment. Other prominent surface features include the **deltoid tuberosity,** the **radial groove,** the **medial** and **lateral epicondyles,** and the articular **condyle.** (*Figure 8-4*)
5. Distally the humerus articulates with the radius and ulna. The medial **trochlea** extends from the **coronoid fossa** to the **olecranon fossa.** (*Figure 8-4*)
6. The **radius** and **ulna** are the bones of the forearm. The olecranon fossa accommodates the **olecranon** during extension of the elbow joint (the *olecranal* or *humeroulnar* joint). The coronoid and radial fossae accommodate the **coronoid process** and a small projection of the radius. (*Figure 8-5*)
7. The bones of the wrist, or **carpus,** form two rows of **carpals.** Four of the fingers contain three **phalanges;** the **pollex** (thumb) has only two. (*Figure 8-6*)

THE PELVIC GIRDLE AND LOWER LIMBS, p. 246

The Pelvic Girdle, p. 246

1. The **pelvic girdle** consists of two **coxae;** each coxa forms through the fusion of an ilium, an ischium, and a pubis. (*Figure 8-7*)
2. The largest coxal bone is the **ilium.** Inside the **acetabulum** the ilium is fused to the **ischium** (posteriorly) and the **pubis** (anteriorly). The **pubic symphysis** limits movement between the pubic bones of the left and right coxae. (*Figures 8-7, 8-8*)
3. The **pelvis** consists of the coxae, sacrum, and coccyx. It may be subdivided into the greater, or **false, pelvis** and the lesser, or **true, pelvis.** (*Figures 8-8, 8-9, 8-10*)

The Lower Limb, p. 250

4. The **femur** is the longest bone in the body. It articulates with the **tibia** at the knee joint. The patellar ligament extends from the patella to the **tibial tuberosity.** (*Figures 8-11, 8-12, 8-13*)
5. Other tibial landmarks include the **anterior crest,** the **interosseous crest,** the **interosseous membrane,** which connects to the **fibula,** and the **medial malleolus.** The **fibular head** articulates with the tibia below the knee, and the **lateral malleolus** stabilizes the ankle. (*Figure 8-13*)
6. The **tarsus,** or ankle, includes seven bones; only the **talus** articulates with the tibia and fibula. When one stands normally, most of the body weight is transferred to the **calcaneus,** and the rest is passed on to the **metatarsals.** Weight transfer occurs along the **longitudinal arch;** there is also a **transverse arch.** (*Figure 8-14*)
7. The basic organizational pattern of the metatarsals and phalanges of the foot resembles that of the hand.

INDIVIDUAL VARIATION IN THE SKELETAL SYSTEM, p. 254

1. Studying a human skeleton can reveal important information such as race, medical history, weight, sex, body size, muscle mass, and age. (*Tables 8-1, 8-2*)
2. Age-related changes and events take place in the skeletal system. These changes begin about age 1 and continue throughout life. (*Table 8-2*)

REVIEW QUESTIONS

LEVEL 1 Reviewing Facts and Terms

1. The only direct connection between the pectoral girdle and the axial skeleton is where the:
   (a) clavicle articulates with the humerus
   (b) clavicle articulates with the manubrium of the sternum
   (c) coxa articulates with the femur
   (d) vertebral column articulates with the skull

2. The presence of foramina in bones indicates the position of:
   (a) tendons and ligaments
   (b) nerves and blood vessels
   (c) ridges and flanges
   (d) a, b, and c are correct

3. At the glenoid fossa, the scapula articulates with the proximal end of the:
   (a) humerus (b) radius
   (c) ulna (d) femur
4. In anatomical position, the ulna lies:
   (a) medial to the radius
   (b) lateral to the radius
   (c) inferior to the radius
   (d) superior to the radius
5. The proximal carpals include:
   (a) trapezium, trapezoid, capitate, hamate
   (b) scaphoid, capitate, lunate, hamate
   (c) trapezium, triangular, trapezoid, pisiform
   (d) scaphoid, lunate, triangular, pisiform
6. The distal carpals are the:
   (a) scaphoid, pisiform, capitate, hamate
   (b) trapezium, trapezoid, capitate, hamate
   (c) scaphoid, lunate, triangular, pisiform
   (d) trapezium, lunate, triangular, hamate
7. The bones of the hand articulate distally with the:
   (a) carpals (b) ulna and radius
   (c) metacarpals (d) phalanges
8. Each coxa of the pelvic girdle consists of three fused bones:
   (a) ulna, radius, humerus
   (b) ilium, ischium, pubis
   (c) femur, tibia, fibula
   (d) hamate, capitate, trapezium
9. The large foramen located between the pubic and ischial rami is the:
   (a) foramen magnum
   (b) suborbital foramen
   (c) acetabulum
   (d) obturator foramen
10. Which of the following represents an adaptation for childbearing in the female?
   (a) inferior angle between the pubic bones greater than 100°
   (b) a relatively broad, low pelvis
   (c) less curvature of the sacrum and coccyx
   (d) a, b, and c are correct
11. The epiphysis of the femur articulates with the pelvis at the:
   (a) pubic symphysis (b) acetabulum
   (c) sciatic notch (d) obturator foramen
12. The large medial bone of the leg is the:
   (a) tibia (b) femur
   (c) fibula (d) humerus
13. The selection that includes *only* tarsal bones is:
   (a) scaphoid, pisiform, capitate, hamate, talus
   (b) calcaneus, navicular, lunate, capitate, talus
   (c) talus, calcaneus, cuboid, navicular, cuneiforms
   (d) navicular, scaphoid, capitate, talus, calcaneus
14. The calcaneus is the attachment site for the:
   (a) tendon of Achilles
   (b) muscles of the calf
   (c) bones of the metatarsals
   (d) a, b, and c are correct
15. Name the skeletal components of the pectoral girdle.
16. What two large scapular processes are associated with the shoulder joint?
17. What bones make up the arm and forearm?
18. What two movements are associated with the proximal radioulnar articulation?
19. List the eight carpal bones of the wrist.
20. What bones constitute the lower limb?
21. Name the components of each coxa of the pelvic girdle.
22. How does the anatomist distinguish between the *false (greater) pelvis* and the *true (lesser) pelvis?*
23. What seven bones make up the ankle (tarsus)?
24. Distinguish between the *pollex* and the *hallux*.

LEVEL 2 Reviewing Concepts

25. Why are clavicular injuries common?
26. What is the skeletal structural difference between the *pelvic girdle* and the *pelvis*?
27. Why is the tibia involved in the transfer of weight to the ankle and foot but not the fibula?
28. What are the direct anatomical connections between the skeletal and muscular systems?
29. Why would an instructor teaching self-defense advise a student to strike an assailant's clavicle in an attack?
30. Why is it necessary for the bones of the pelvic girdle to be more massive than the bones of the pectoral girdle?

LEVEL 3 Critical Thinking and Clinical Applications

31. In order to settle a bet, you need to measure the length of your lower limb (femur and tibia). What landmarks would you use to make the measurement?
32. Fred, a fireman, is fighting a fire in a building when part of the ceiling collapses and a beam strikes him on his left shoulder. He is rescued by his friends, but he has a great deal of pain in his shoulder and cannot move his arm properly, especially in the anterior direction. His clavicle is not broken, and his humerus is intact. What is the probable nature of Fred's injury?
33. Cindy is anxiously awaiting the birth of her first child. As she gets closer to term, her physician orders an ultrasound. After seeing the results, he makes some calculations and informs Cindy that she will have to have a Cesarean section. Why can't Cindy deliver the baby by natural childbirth?

CHAPTER 9

Gymnasts make their movements look easy—that's part of the job. Yet every complex maneuver must be carefully planned so that the movements conform to the limitations of the human body. For example, it is not possible to bend the shaft of the humerus or femur—movements must be restricted to joints, where bones articulate. This gymnast is making use of almost every mobile joint in his body. Yet even at joints there are limits to permissible movement, and all of us (including gymnasts) must work within those limitations. Each joint will tolerate a specific range of motion, and a variety of bony surfaces, cartilages, ligaments, tendons, and muscles work together to keep movement within the normal range. The stronger the joint, the more restrictive those limits are likely to be. This chapter will examine representative joints and explore the relationship between structural stability and the range of motion.

Articulations

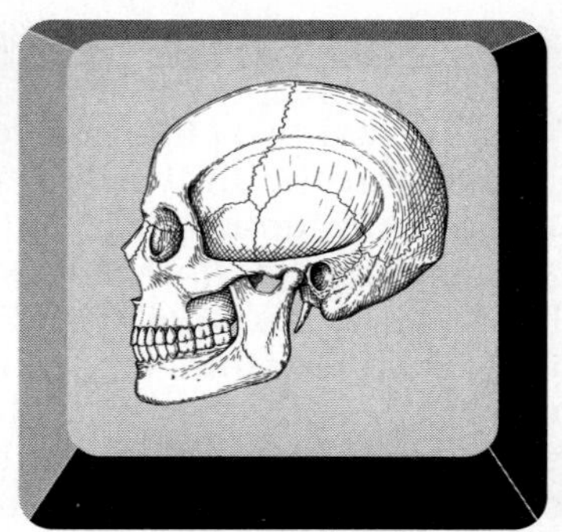

Chapter Outline and Objectives

1. Contrast the major categories of joints and explain the relationship between structure and function for each category.
2. Describe the basic structure of a synovial joint, identifying possible accessory structures and their functions.

3. Describe the dynamic movements of the skeleton.
4. List the different types of synovial joints and discuss how the characteristic motions of each type are related to its anatomical structure.

5. Describe the articulations between the vertebrae of the vertebral column.
6. Describe the structure and function of the shoulder, elbow, hip, and knee joints.
7. Explain the relationship between joint strength and mobility, using specific examples.

In the last two chapters you have become familiar with the individual bones of the skeleton. Your understanding of the skeletal system is still incomplete, however, because the skeleton in life is always in motion. Most of our daily activities, from breathing or speaking to writing or running, involve movements of the skeleton. The bones of the skeleton are solid, and movements can occur only at joints, or **articulations,** where two bones interconnect.

The characteristic structure of a joint determines the type of movement that may occur. Each joint reflects a workable compromise between the need for strength and the need for mobility.

This chapter compares the relationships between articular form and function, using several different examples that range from relatively immobile but very strong (the intervertebral articulations) to highly mobile and relatively weak (the shoulder).

A Classification of Joints

Three major categories of joints are based on the range of motion permitted (Table 9-1). These functional categories are then subdivided on the basis of the anatomical structure of the joint or the range of motion permitted. The basic categories are:

1. An *immovable joint* is a **synarthrosis** (sin-ar-THRŌ-sis; *syn,* together + *arthros,* joint). A synarthrosis may be *fibrous* or *cartilaginous,* depending on the nature of the connection, or the two elements may fuse together.
2. A *slightly movable joint* is an **amphiarthrosis** (am-fē-ar-THRŌ-sis; *amphi,* on both sides). An amphiarthrosis may be *fibrous* or *cartilaginous,* depending on the nature of the connection between the opposing bones.
3. A *freely movable joint* is a **diarthrosis** (dī-ar-THRŌ-sis; *dia,* through). Diarthroses are also called *synovial joints.* Their basic structure was introduced in Chapter 4 in the discussion of synovial membranes. [p. 134] Diarthroses are subdivided according to the degree of movement permitted.

Immovable Joints (Synarthroses)

At a synarthrosis the bony edges are quite close together and may even interlock. These extremely strong joints are found where movement between the bones must be prevented. There are four major types of synarthrotic joints:

- A **suture** (*sutura,* a sewing together) is a synarthrotic joint found only between the bones of the skull. The edges of the bones are

TABLE 9-1 A Functional Classification of Articulations

Functional Category	*Structural Category*	*Description*	*Example*
SYNARTHROSIS (no movement)	**Fibrous**		
	Suture	Fibrous connections plus interdigitation	Between the bones of the skull
	Gomphosis	Fibrous connections plus insertion in alveolus	Between the teeth and jaws
	Cartilaginous		
	Synchondrosis	Interposition of cartilage plate	Epiphyseal plates
	Bony fusion		
	Synostosis	Conversion of other articular form to solid mass of bone	Portions of the skull, epiphyseal lines
AMPHIARTHROSIS (little movement)	**Fibrous**		
	Syndesmosis	Ligamentous connection	Between the tibia and fibula
	Cartilaginous		
	Symphysis	Connection by a fibrocartilage pad	Between right and left halves of pelvis; between adjacent vertebrae of spinal column
DIARTHROSIS (free movement)	**Synovial**	Complex joint bounded by joint capsule and containing synovial fluid	Numerous; subdivided by range of movement *(see Figure 9-6)*
	Monaxial	Permits movement in one plane	Elbow, ankle
	Biaxial	Permits movement in two planes	Ribs, wrist
	Triaxial	Permits movement in all three planes	Shoulder, hip

interlocked and bound together at the suture by dense connective tissue.

- A **gomphosis** (gom-FŌ-sis; *gomphosis*, a bolting together) is a specialized synarthrosis that binds the teeth to bony sockets in the maxilla and mandible. The fibrous connection between a tooth and its socket is a **periodontal** (pe-rē-ō-DON-tal) **ligament** (*peri*, around + *odontos*, tooth).
- A **synchondrosis** (sin-kon-DRŌ-sis; *syn*, together + *chondros*, cartilage) is a rigid cartilaginous bridge between two articulating bones. The hyaline cartilage of an *epiphyseal plate* is a synchondrosis that connects the diaphysis with the epiphysis, even though the bones involved are part of the same skeletal element. [p. 185] Another example is the cartilaginous connection between the ends of the vertebrosternal ribs (ribs 1–7) and the sternum.
- A **synostosis** (sin-os-TŌ-sis) is a totally rigid, immovable joint that is created when two separate bones actually fuse together so that the boundary between them disappears. The *metopic suture* of the frontal bone and the epiphyseal lines of mature bones are examples of synostoses. [pp. 185, 209]

Slightly Movable Joints (Amphiarthroses)

An amphiarthrosis is another compromise between mobility and strength. It permits more movement than a synarthrosis, but is much stronger than a freely movable joint. The articulating bones may be connected by collagen fibers or cartilage. There are two major types of amphiarthrotic joints:

- At a **syndesmosis** (sin-dez-MŌ-sis; *desmos*, a band or ligament), bones are connected by a ligament. One example is the distal articulation between the tibia and fibula.
- At a **symphysis** the articulating bones are separated by a wedge or pad of fibrocartilage. The articulations between adjacent vertebral bodies (via the *intervertebral disc*) and the anterior connection between the two pubic bones (the *pubic symphysis*) are examples of this type of joint.

Freely Movable Joints (Diarthroses)

Figure 9-1

Diarthroses, or **synovial** (si-NŌ-vē-ul) **joints**, permit a wide range of motion. The basic structure of a synovial joint was introduced in Chapter 4 during the discussion of synovial membranes. [p. 134] Figure 9-1 • provides additional information about the structure of synovial joints. These joints are typically found at the ends of long bones, such as those of the upper and lower limbs. A synovial joint is surrounded by a fibrous **articular capsule**, and a *synovial membrane* lines the articular cavity.

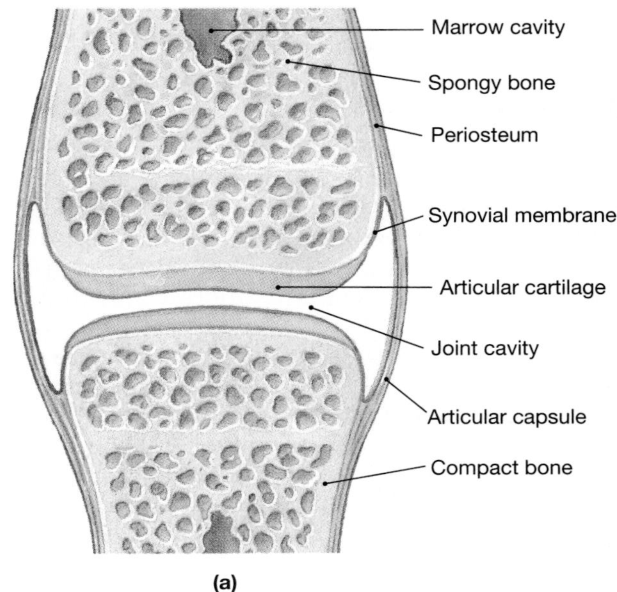

Marrow cavity
Spongy bone
Periosteum
Synovial membrane
Articular cartilage
Joint cavity
Articular capsule
Compact bone
(a)

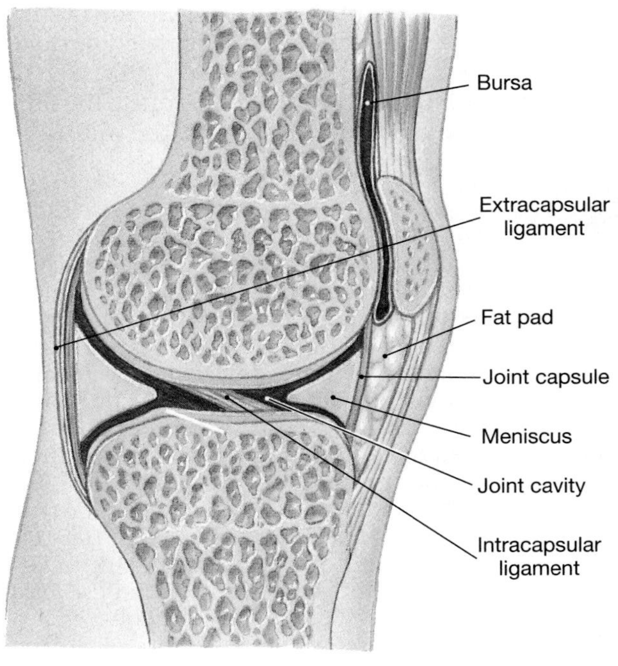

Bursa
Extracapsular ligament
Fat pad
Joint capsule
Meniscus
Joint cavity
Intracapsular ligament
(b)

● **FIGURE 9-1**
The Structure of a Synovial Joint. (a) Diagrammatic view of a simple articulation. **(b)** A simplified sectional view of the knee joint.

The Synovial Membrane

A synovial membrane consists of an area of loose connective tissue covered by a superficial layer of squamous or cuboidal cells that face the synovial cavity. Although usually called an epithelium, it is structurally distinct because there is no basement membrane and the cellular layer is incomplete. Some of the lining cells are phagocytic and others are secretory. The secretory cells regulate the composition of the *synovial fluid* within the joint cavity. A synovial membrane lines the inside of the joint cavity everywhere except at the *articular cartilages*.

Articular Cartilages

Under normal conditions the bony surfaces cannot contact one another because the articulating surfaces are covered by special **articular cartilages.** Articular cartilages resemble hyaline cartilages elsewhere in the body, but they have no perichondrium, and the matrix contains more water than do other cartilages.

The surfaces of the articular cartilages are slick and smooth, and this feature alone can reduce friction during movement at the joint. However, even when pressure is applied across a joint, the smooth articular cartilages do not actually touch one another because they are separated by a thin film of synovial fluid. This fluid acts as a lubricant, keeping friction at a minimum.

Proper synovial function can continue only if the articular cartilages retain their normal structure. If an articular cartilage is damaged, the matrix may begin to break down, and the exposed surface will change from a slick, smooth, gliding surface to a rough feltwork of bristly collagen fibers. This feltwork drastically increases friction at the joint.

RHEUMATISM AND ARTHRITIS *Rheumatism* (ROO-muh-tizm) is a general term that indicates pain and stiffness affecting the skeletal system, the muscular system, or both. There are several major forms of rheumatism. **Arthritis** (ar-THRĪ-tis) includes all the rheumatic diseases that affect synovial joints. Arthritis always involves damage to the articular cartilages, but the specific cause may vary. For example, arthritis can result from bacterial or viral infection, injury to the joint, metabolic problems, or severe physical stresses.

Osteoarthritis (os-tē-ō-ar-THRĪ-tis), also known as *degenerative arthritis,* or *degenerative joint disease (DJD),* usually affects individuals age 60 or older. DJD may result from cumulative wear and tear at the joint surfaces or from genetic factors affecting collagen formation. In the U.S. population 25 percent of women and 15 percent of men over 60 years of age show signs of this disease. *Rheumatoid arthritis* is an inflammatory condition that affects roughly 2.5 percent of the adult population. At least some cases result when the immune response mistakenly attacks the joint tissues. Allergies, bacteria, viruses, and genetic factors have all been proposed as contributing to or triggering the destructive inflammation.

Regular exercise, physical therapy, and drugs such as aspirin that reduce inflammation can slow the progress of the disease. Surgical procedures can realign or redesign the affected joint, and in extreme cases involving the hip, knee, elbow, or shoulder, the defective joint can be replaced by an artificial one. Additional information concerning the various forms of arthritis can be found in the *Applications Manual.*

AM *Rheumatism, Arthritis, and Synovial Function*

Degenerative changes comparable to those seen in arthritis may result from joint immobilization. When motion ceases, so does circulation of synovial fluid, and the cartilages begin to suffer. **Continuous passive motion (CPM)** of any injured joint appears to encourage the repair process by improving the circulation of synovial fluid. The movement is often performed by a physical therapist during the recovery process.

Synovial Fluid

Synovial fluid resembles interstitial fluid but contains a high concentration of hyaluronic acid, which is secreted by cells of the synovial membrane. It is a thick, viscous solution with the consistency of heavy molasses. The synovial fluid within a joint has three primary functions:

1. ***Lubrication.*** The articular cartilages are like sponges filled with synovial fluid. When part of an articular cartilage is compressed, some of the synovial fluid is squeezed out of the cartilage and into the space between the opposing surfaces. This thin layer of fluid markedly reduces friction between moving surfaces, just as a thin film of water reduces friction between a car's tires and a highway. When the compression stops, synovial fluid is sucked back into the articular cartilages.
2. ***Nutrient distribution.*** The total quantity of synovial fluid in a joint is normally less than 3 ml, even in a large joint such as the knee. This relatively small volume of fluid must be circulated continually to provide nutrients and a route for waste disposal for the chondrocytes of the articular cartilages. The synovial fluid circulates whenever the joint moves, and the compression and reexpansion of the articular cartilages pumps synovial fluid into and out of the cartilage matrix.
3. ***Shock absorption.*** Synovial fluid cushions shocks in joints that are subjected to compression. For example, the hip, knee, and ankle joints are compressed during walking, and more severely compressed during jogging or running. When the pressure across a joint suddenly increases, the synovial fluid lessens the shock by distributing it evenly across the articular surfaces.

Accessory Structures

Figure 9-1

Synovial joints may have a variety of accessory structures, including pads of cartilage or fat, ligaments, tendons, and bursae (Figure 9-1b●).

CARTILAGES AND FAT PADS In complex joints such as the knee (Figure 9-1●), accessory structures may lie between the opposing articular surfaces. These include *menisci* and *fat pads.*

- A **meniscus** (men-IS-kus; a crescent; plural *menisci*) is a pad of fibrocartilage that is placed between opposing bones within a synovial joint. Menisci, or *articular discs,* may subdivide a synovial cavity, channel the flow of synovial fluid, or allow for variations in the shapes of the articular surfaces.
- **Fat pads** are localized masses of adipose tissue that are often found around the edges of the joint, lightly covered by a layer of synovial membrane. Fat pads provide protection for the articular cartilages. They also act as packing material for the joint; when the bones move, the fat pads fill in the spaces created as the joint cavity changes shape.

LIGAMENTS The joint capsule that surrounds the entire joint is continuous with the periostea of the articulating bones. **Accessory ligaments** are localized thickenings of the capsule. These ligaments reinforce and strengthen the capsule, and they may also act to limit rotation at the joint. **Extracapsular ligaments** interconnect the articulating bones and pass across the outside of the capsule. These ligaments provide additional support to the wall of the joint. **Intracapsular ligaments,** found inside the capsule, help prevent extreme movements that might otherwise damage the joint. In a **sprain,** a ligament is stretched to the point where some of the collagen fibers are torn, but the ligament as a whole survives, and the joint is not damaged. Ligaments are very strong, and one of the attached bones often breaks before the ligament tears. In general, a broken bone heals much more quickly and effectively than does a torn ligament.

TENDONS While not part of the articulation itself, tendons passing across or around a joint may limit the range of motion and provide mechanical support. For example, tendons associated with the muscles of the arm provide much of the bracing for the shoulder joint.

BURSAE Bursae are small, fluid-filled pockets in connective tissue. They contain synovial fluid and are lined by a synovial membrane. Bursae may be connected to the joint cavity or may be completely separate from it. Bursae form where a tendon or ligament rubs against other tissues. Their function is to reduce friction and act as a shock absorber. Bursae are found around most synovial joints, such as the shoulder joint. **Synovial tendon sheaths** are tubular bursae that surround tendons where they pass across bony surfaces. Bursae may also appear beneath the skin covering a bone or within other connective tissues exposed to friction or pressure. Bursae that develop in abnormal locations, or because of abnormal stresses, are called *adventitious bursae.*

BURSITIS When bursae become inflamed, causing pain in the affected area whenever the tendon or ligament moves, a condition of **bursitis** exists. Inflammation can result from the friction associated with repetitive motion, pressure over the joint, irritation by chemical stimuli, infection, or trauma. Bursitis associated with repetitive motion often occurs at the shoulder; for example, golfers, pitchers, and tennis players may develop bursitis at this location. The most common pressure-related bursitis is a **bunion.** Bunions form over the base of the great toe as a result of the friction and distortion of the joint caused by tight shoes, especially those with pointed toes.

There are special names for bursitis at other locations; the names of these conditions indicate the occupations most often associated with them. In "housemaid's knee," which accompanies prolonged kneeling, the affected bursa lies between the patella (*kneecap*) and the skin. The condition of "student's elbow" is a form of bursitis that can result from propping your head above a desk while reading your A & P textbook.

Strength versus Mobility

A joint cannot be both highly mobile and very strong. The greater the range of motion at a joint, the weaker it becomes. A synarthrosis, the strongest type of joint, does not permit any movement at all. Any mobile diarthrosis may be damaged by movement beyond its normal range of motion. Several factors are responsible for limiting the range of motion at a diarthrosis and reducing the frequency of injury:

- The collagen fibers of the joint capsule and any accessory, extracapsular, or intracapsular ligaments.
- The shapes of the articulating surfaces, which often prevent movement in specific directions.
- The presence of other bones, skeletal muscles, or fat pads around the joint.
- Tension in tendons attached to the articulating bones. When a skeletal muscle contracts and pulls on a tendon, it may either encourage or oppose movement in a specific direction.

These reinforcing structures cannot always protect a joint from extreme stresses. A **dislocation,** or *luxation* (luks-Ā-shun), occurs when articulating surfaces are forced out of position. This displacement can damage the articular cartilages, tear ligaments, or distort the joint capsule. Although the *inside* of a joint has no pain receptors, nerves that monitor the capsule, ligaments, and tendons are quite sensitive, and dislocations are very painful. The damage accompanying a partial dislocation, or *subluxation* (sub-luks-Ā-shun), is less severe. People who are "double-jointed" have joints that are weakly stabilized. Although their joints permit a greater range of motion than those of other individuals, they are more likely to suffer partial or complete dislocations.

What common characteristics are found in typical synarthrotic and amphiarthrotic joints?

In a newborn infant, the large bones of the skull are joined by fibrous connective tissue. What type of joints are these? These bones later grow, interlock, and form immovable joints. What type of joints are these?

Why would improper circulation of synovial fluid lead to degeneration of articular cartilages in the affected joint?

Articular Form and Function

To *understand* human movement you must become aware of the relationship between structure and function at each articulation. To *describe* human movement you need a frame of reference that permits accurate and precise communication. The synovial joints can be classified according to their anatomical and functional properties. To demonstrate the basis for that classification, we will describe the movements that can occur at a typical synovial joint, using a simplified model.

Describing Dynamic Motion

Figure 9-2

Take a pencil (or pen) as your model and stand it upright on the surface of a desk or table, as shown in Figure 9-2a•. The pencil represents a bone, and the desk is an articular surface. A little imagination and a lot of twisting, pushing, and pulling will demonstrate that there are only three ways to move the model. Considering them one at a time will provide a frame of reference for analyzing any complex movement.

- ***Possible movement 1: The point can move.*** If you hold the pencil upright but do not secure the point, you can push the pencil across the surface. This kind of motion is called *gliding* (Figure 9-2b•), and it is an example of **linear motion.** You could slide the point forward or backward, from one side to the other, or diagonally. However you choose to move the pencil, the motion can be described using two lines of reference. One line represents forward/backward motion, and the other left/right movement. For example, a simple movement along one axis could be described as "forward 1 cm" or "left 2 cm." A diagonal movement could be described using both axes, as in "backward 1 cm and to the right 2.5 cm."
- ***Possible movement 2: The shaft can change its angle with the surface.*** With the tip held in position, you can still move the free (eraser) end forward and backward or from side to side. These movements, which change the angle between the shaft and the articular surface, are examples of **angular motion** (Figure 9-2c•).

Any angular movement can be described with reference to the same two axes (forward/backward, left/right) and the angular change (in degrees). In one instance, however, a special term is used to describe a complex angular movement. Grasp the free end of the pencil and move it in any direction until the shaft is no longer vertical. Now swing that end through a complete circle (Figure 9-2d•). This movement, which corresponds to the path of your arm when drawing a large circle on a blackboard, is very difficult to describe. Anatomists avoid the problem by using a special term, **circumduction** (sir-kum-DUK-shun; *circum*, around), for this type of angular motion.

- ***Possible movement 3: The shaft can rotate.*** If you prevent movement of the base and keep the shaft vertical, you can still spin the shaft around its longitudinal axis. This movement is called **rotation** (Figure 9-2e•). Several articulations will permit partial rotation, but none can rotate freely; such a movement would hopelessly tangle the blood vessels, nerves, and muscles that cross the joint.

If an articulation permits movement along only one axis, it is called **monaxial** (mon-AKS-ē-ul). Using the above model, if an articulation permits only angular movement in the forward/backward plane, or prevents any movement other than rotation around its longitudinal axis, it is monaxial. If movement can occur along two axes, the articulation is **biaxial** (bī-AKS-ē-ul). If the pencil could undergo angular motion in the forward/backward and left/right plane, but not rotation, it would be biaxial. The most mobile joints permit a combination of angular movement and rotation. These are called **triaxial** (trī-AKS-ē-ul).

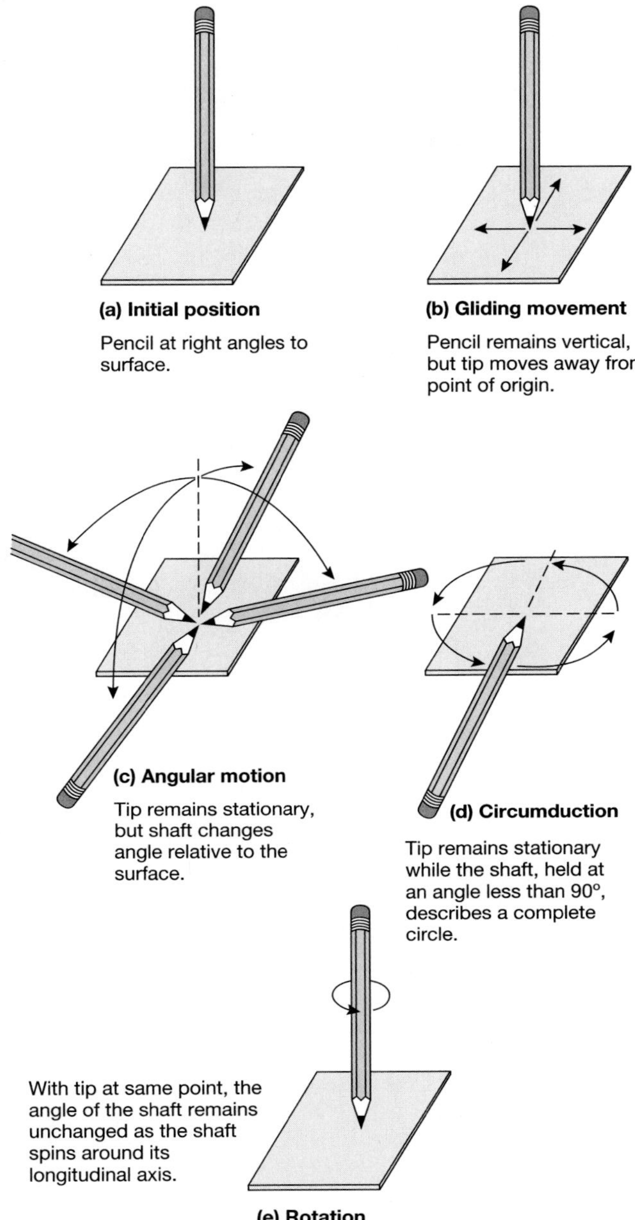
(a) Initial position
Pencil at right angles to surface.
(b) Gliding movement
Pencil remains vertical, but tip moves away from point of origin.
(c) Angular motion
Tip remains stationary, but shaft changes angle relative to the surface.
(d) Circumduction
Tip remains stationary while the shaft, held at an angle less than 90°, describes a complete circle.
With tip at same point, the angle of the shaft remains unchanged as the shaft spins around its longitudinal axis.
(e) Rotation

● **FIGURE 9-2**
A Simple Model of Articular Motion

❑ Types of Movements

In descriptions of motion at synovial joints, phrases such as "bend the leg" or "raise the arm" are not sufficiently precise. Anatomists use descriptive terms that have specific meanings. We will consider these motions with reference to the basic categories of movement discussed previously.

Gliding

In **gliding,** two opposing surfaces slide past one another, as in possible movement 1. Gliding occurs between the surfaces of articulating carpals and tarsals and between the clavicles and the sternum. The movement can occur in almost any direction, but the amount of movement is slight and rotation is usually prevented by the capsule and associated ligaments.

Angular Motion

Figure 9-3

Examples of angular motion include *flexion, extension, abduction,* and *adduction.* The descriptions of the movements are based on reference to an individual in the anatomical position.

- **Flexion** (FLEK-shun) can be defined as movement in the anterior/posterior plane that *reduces the angle between the articulating elements.* **Extension** occurs in the same plane, but it *increases the angle between articulating elements* (Figure 9-3●). When you bring your head toward your chest, you flex the head. When you bend down to touch your toes, you flex the spine. Extension reverses these movements.
- Flexion at the shoulder or hip moves the limbs forward, whereas extension moves them back. Flexion of the wrist moves the palm forward, and extension moves it back. In each of these examples, extension can be continued past the anatomical position, in which case **hyperextension** occurs. You can also hyperextend the head, a movement that allows you to gaze at the ceiling. Hyperextension of other joints is usually prevented by ligaments, bony processes, or soft tissues.
- **Abduction** (*ab,* from) is movement *away from the longitudinal axis of the body* in the frontal plane. For example, swinging the upper limb to the side is abduction of the limb. Moving it back constitutes **adduction** (*ad,* to). Adduction of the wrist moves the heel of the hand *toward* the body, whereas abduction moves it farther away. Spreading the fingers or toes apart abducts them, because they move *away* from a central digit (finger or toe). Bringing them together constitutes adduction. Abduction and adduction always refer to movements of the appendicular skeleton.
- A special type of angular motion, *circumduction,* was introduced in our model. A familiar example of circumduction is moving your arm in a loop, as when drawing a large circle on a chalkboard.

Rotation

Figure 9-4

Rotational movements are also described with reference to a figure in the anatomical position.

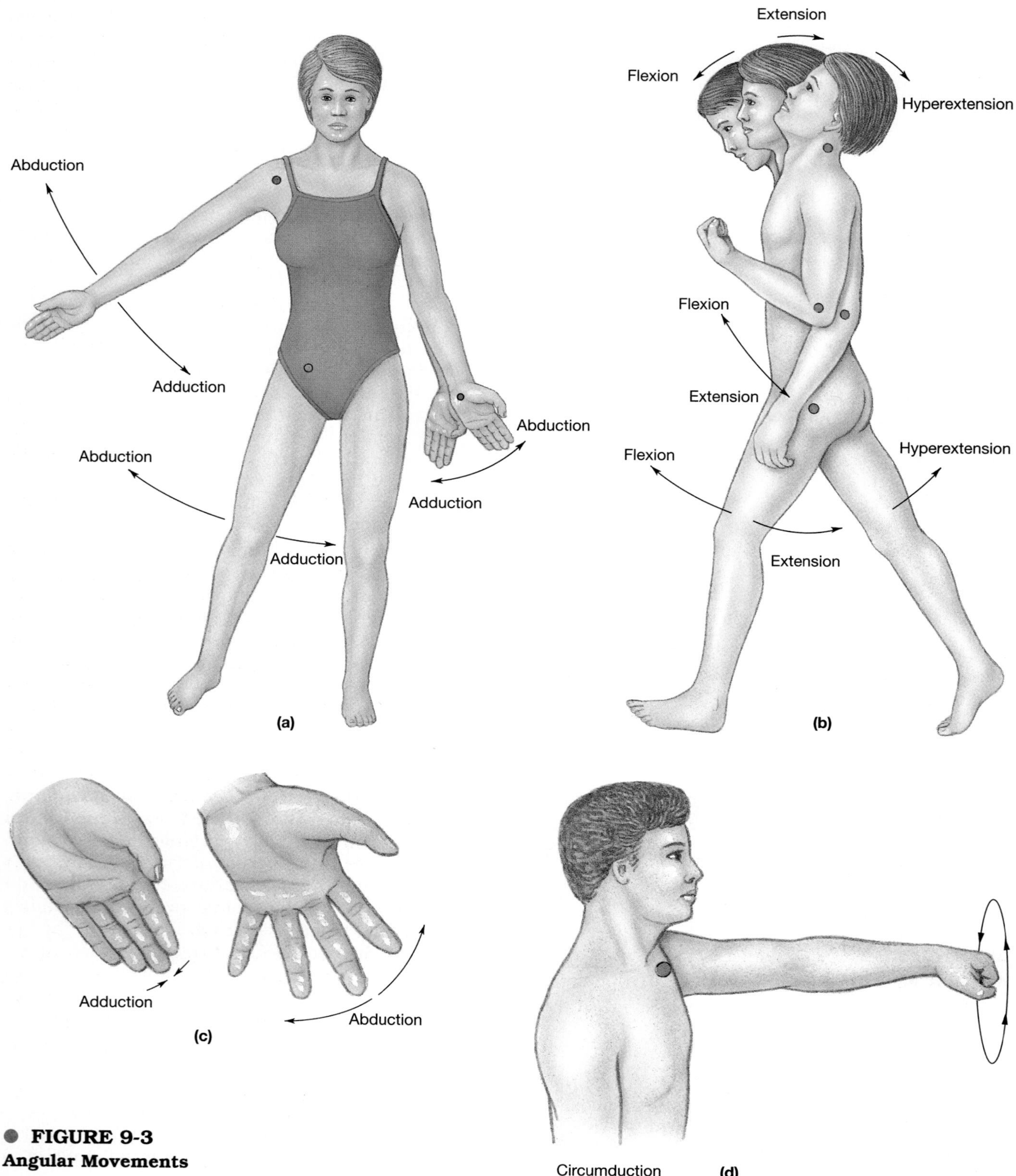
Abduction
Adduction
Abduction
Adduction
Abduction
Adduction
(a)
Extension
Flexion
Hyperextension
Flexion
Extension
Flexion
Hyperextension
Extension
(b)
Adduction
Abduction
(c)
Circumduction
(d)

● **FIGURE 9-3**
Angular Movements

Rotation of the head may involve **left rotation** or **right rotation.** In analyzing movements of the limbs, if the anterior aspect of the limb rotates *inward,* toward the ventral surface of the body, you have **internal rotation,** or **medial rotation.** If it turns *outward,* you have **external rotation,** or **lateral rotation.** These rotational movements are illustrated in Figure 9-4●.

The articulations between the radius and ulna permit the rotation of the distal end of the radius across the anterior surface of the ulna. This rotation moves the wrist and hand from palm-facing-

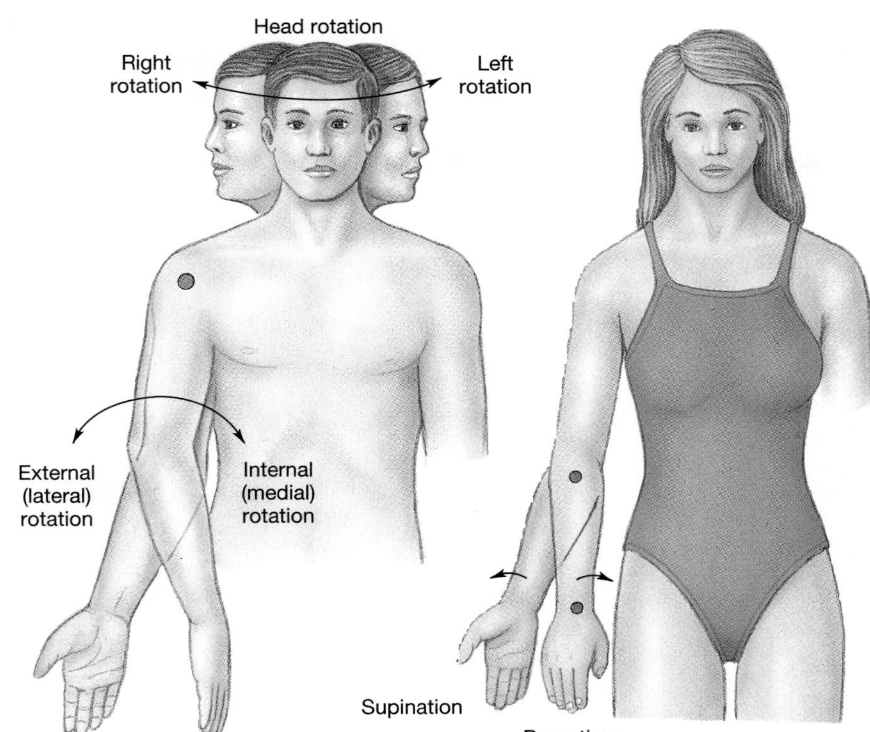
Head rotation
Right rotation
Left rotation
External (lateral) rotation
Internal (medial) rotation
Supination
Pronation

● **FIGURE 9-4**
Rotational Movements

front to palm-facing-back. This motion is called **pronation** (prō-NĀ-shun). The opposing movement, in which the palm is turned forward, is **supination** (sū-pi-NĀ-shun).

Special Movements

Figure 9-5

Several special terms apply to specific articulations or unusual types of movement (Figure 9-5●).

- **Inversion** (*in*, into + *vertere*, to turn) is a twisting motion of the foot that turns the sole inward. The opposite movement is called **eversion** (ē-VER-zhun; *e*, out).
- **Dorsiflexion** and **plantar flexion** (*planta*, sole) also refer to movements of the foot. Dorsiflexion is flexion of the ankle and elevation of the sole, as when "digging in the heels." Plantar flexion, the opposite movement, extends the ankle and elevates the heel, as when standing on tiptoes.
- **Opposition** is the special movement of the thumb that enables it to grasp and hold an object.
- **Protraction** entails moving a part of the body anteriorly in the horizontal plane. **Retraction** is the reverse movement. You protract your jaw when you grasp your upper lip with your lower teeth, and you protract your clavicles when you cross your arms.
- **Elevation** and **depression** occur when a structure moves in a superior or inferior direction. You depress your mandible when you open your mouth, and elevate it as you close it. Another familiar elevation occurs when you shrug your shoulders.

❑ A Structural Classification of Synovial Joints

Figure 9-6

Synovial joints can be described as *gliding, hinge, pivot, ellipsoidal, saddle,* or *ball-and-socket* joints on the basis of the shapes of the articulating surfaces. Each type of joint permits a different type and range of motion.

- **Gliding joints** (Figure 9-6a●), also called *planar joints,* have flattened or slightly curved faces. The relatively flat articular surfaces slide across one another, but the amount of movement is very slight. Although rotation is theoretically possible at such a joint, ligaments usually prevent or restrict such movement. Gliding joints are found at the ends of the clavicles, between the carpals, between the tarsals, and between the articular facets of adjacent spinal vertebrae. Gliding joints may be classified as *nonaxial,* because they permit only small sliding movements, or *multiaxial* because that sliding may occur in any direction.
- **Hinge joints** (Figure 9-6b●) permit angular movement in a single plane, like the opening and closing of a door. A hinge joint is a monaxial joint. Examples include the joint between the occipital bone and atlas in the axial skeleton, and the elbow, knee, ankle, and interphalangeal joints of the appendicular skeleton.
- **Pivot joints** (Figure 9-6c●) are also monaxial, but they permit only rotation. A pivot joint between the atlas and axis allows you to rotate your head to either side, and another between the head of the radius and the proximal shaft of the ulna permits pronation and supination of the palm.
- In an **ellipsoidal joint** (Figure 9-6d●) an oval articular face nestles within a depression on the opposing surface. With such an arrangement, angular motion occurs in two planes, along or across the length of the oval. It is thus an exam-

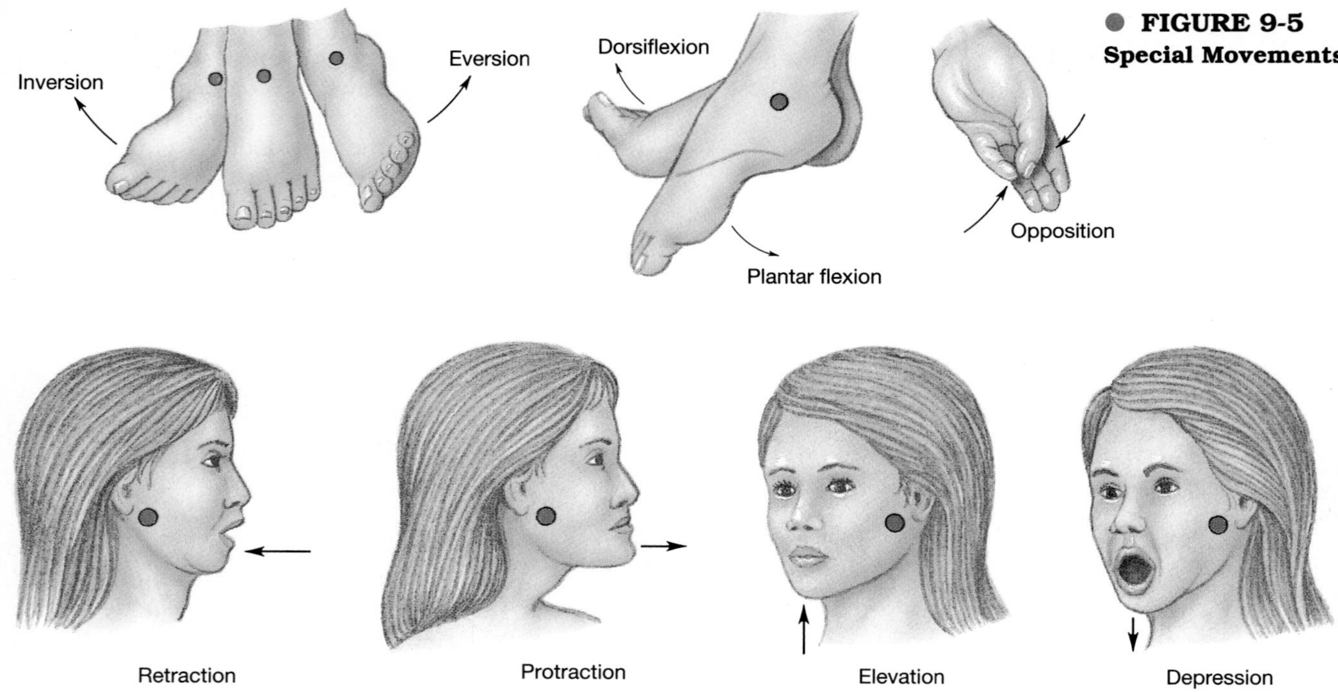
Inversion
Eversion
Dorsiflexion
Plantar flexion
Opposition
Retraction
Protraction
Elevation
Depression

● **FIGURE 9-5**
Special Movements

ple of a biaxial joint. Ellipsoidal joints connect the fingers and toes with the metacarpals and metatarsals.

- **Saddle joints** (Figure 9-6e●) have articular faces that resemble saddles. Each face is concave on one axis and convex on the other, and the opposing faces nest together. This arrangement permits angular motion, including circumduction, but prevents rotation. Saddle joints are usually considered to be biaxial. The carpometacarpal joint at the base of the thumb is the best example of a saddle joint, and "twiddling your thumbs" will demonstrate the possible movements.
- In a **ball-and-socket joint** (Figure 9-6f●) the round head of one bone rests within a cup-shaped depression in another. All combinations of movements, including circumduction and rotation, can be performed at ball-and-socket joints. These are triaxial joints, and examples include the shoulder and hip joints.

Give the proper term for each of the following types of motion: (a) moving the humerus away from the midline of the body, (b) turning the palms so that they face forward, (c) bending the elbow.

When a person does "jumping jacks," what lower limb movements are necessary?

What types of movement are usually associated with hinge joints?

Representative Articulations

This section considers examples of articulations that demonstrate important functional principles. We will first consider the *intervertebral articulations* of the axial skeleton. The articulations between adjacent vertebrae include gliding articulations and cartilaginous symphyses. We will then proceed to a discussion of the synovial articulations of the appendicular skeleton. The shoulder has great mobility, the elbow has great strength, and the wrist makes fine adjustments in the orientation of the palm and fingers. The functional requirements of the joints in the lower limb are very different from those of the upper limb. Articulations at the hip, knee, and ankle must transfer the body weight to the ground, and during movements such as running, jumping, or twisting, the applied forces are considerably greater than the weight of the body. Although this section considers representative articulations, Tables 9-2, 9-3, and 9-4 (pp. 274, 276, 280) summarize data concerning the majority of articulations in the body.

Intervertebral Articulations

Figure 9-7

The articulations between the superior and inferior articular processes of adjacent vertebrae are

● **FIGURE 9-6**
A Functional Classification of Synovial Joints

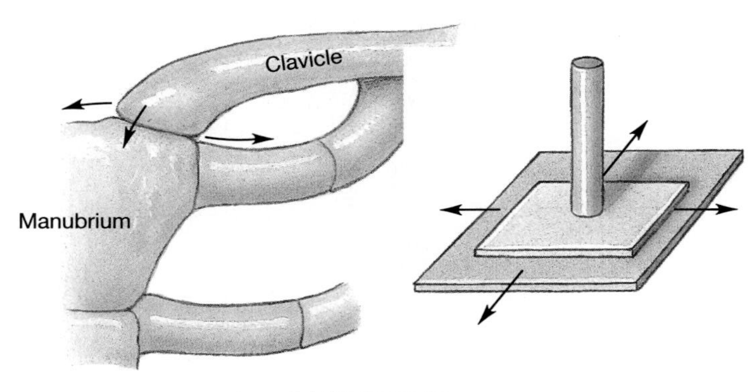
Clavicle
Manubrium

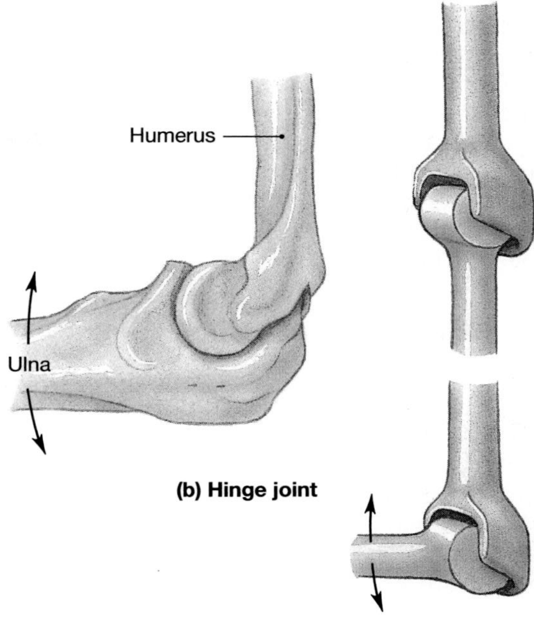
Humerus
Ulna

(a) Gliding joint

(b) Hinge joint

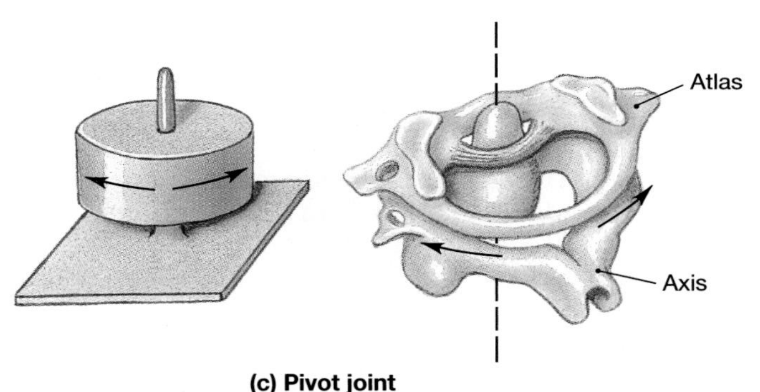
Atlas
Axis

(c) Pivot joint

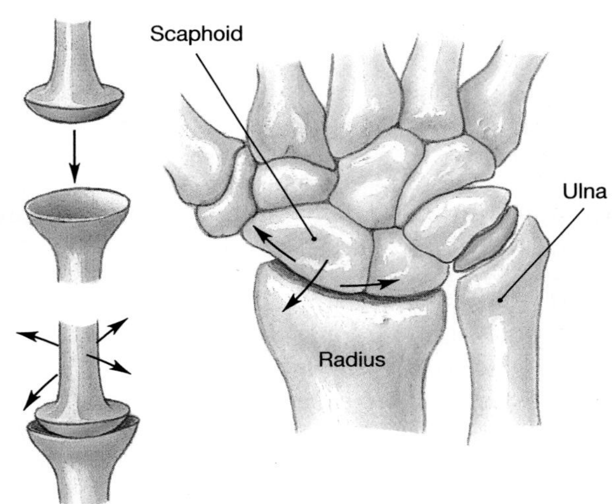
Scaphoid
Ulna
Radius

(d) Ellipsoidal joint

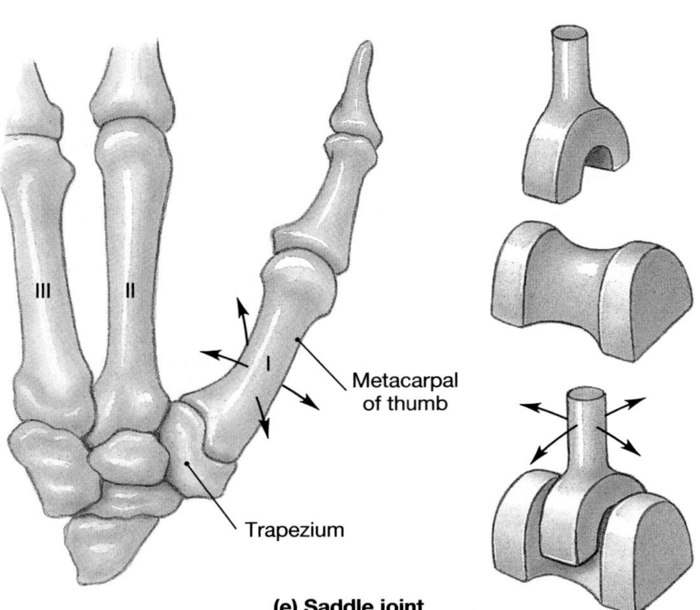
III
II
I
Metacarpal of thumb
Trapezium

(e) Saddle joint

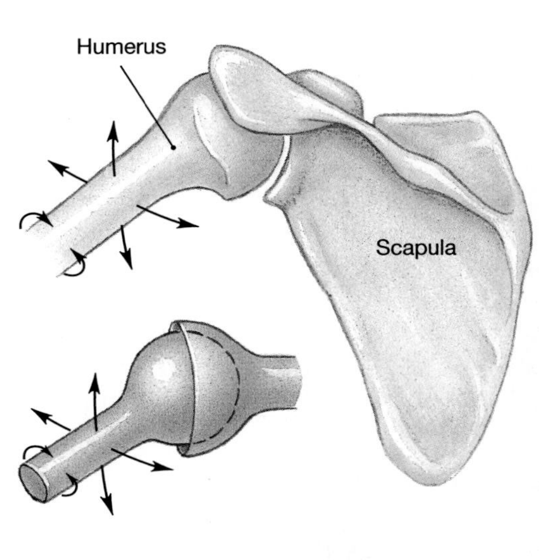
Humerus
Scapula

(f) Ball-and-socket joint

gliding joints that permit small movements associated with flexion and rotation of the vertebral column. Little gliding occurs between adjacent vertebral bodies. Figure 9-7● illustrates the structure of these joints. From axis to sacrum, the vertebrae are separated and cushioned by pads of fibrocartilage called **intervertebral discs.** Intervertebral discs are not found in the sacrum and coccyx, where vertebrae have fused, nor are they found between the first and second cervical vertebrae.

The Intervertebral Discs

Each intervertebral disc consists of a tough outer layer of fibrocartilage, the **annulus fibrosus** (AN-ū-lus fi-BRŌ-sus). The collagen fibers of the annulus fibrosis attach the disc to the bodies of adjacent vertebrae. The annulus fibrosis also surrounds a soft, elastic, and gelatinous core, the **nucleus pulposus** (pul-PŌ-sus). The nucleus pulposus is primarily composed of water (about 75 percent) with scattered reticular and elastic fibers. The nucleus pulposus gives the disc resiliency and enables it to act as a shock absorber.

Movement of the vertebral column compresses the nucleus pulposus and displaces it in the opposite direction. This displacement permits smooth gliding movements by each vertebra while still maintaining their alignment. The discs make a significant contribution to an individual's height; they account for roughly one-quarter of the length of the vertebral column above the sacrum. As we grow older, the water content of the nucleus pulposus within each disc decreases. The discs gradually become less effective as a cushion, and the chances for vertebral injury increase. Loss of water by the discs also causes shortening of the vertebral column, and this loss accounts for the characteristic decrease in height with advanced age.

Intervertebral Ligaments

Numerous ligaments are attached to the bodies and processes of all vertebrae to bind them together and stabilize the vertebral column. Ligaments interconnecting adjacent vertebrae include:

- The ***anterior longitudinal ligament,*** which connects the anterior surfaces of adjacent vertebral bodies.
- The ***posterior longitudinal ligament,*** which parallels the anterior longitudinal ligament, but connects the posterior surfaces of adjacent bodies.
- The ***ligamentum flavum*** (plural *ligamenta flava*), which connects the laminae of adjacent vertebrae.
- The ***interspinous ligament,*** which connects the spinous processes of adjacent vertebrae.
- The ***supraspinous ligament,*** which interconnects the tips of the spinous processes from C_7 to the sacrum. The *ligamentum nuchae*, discussed in Chapter 7, is a supraspinous ligament that extends from vertebra C_7 to the base of the skull. [p. 226]

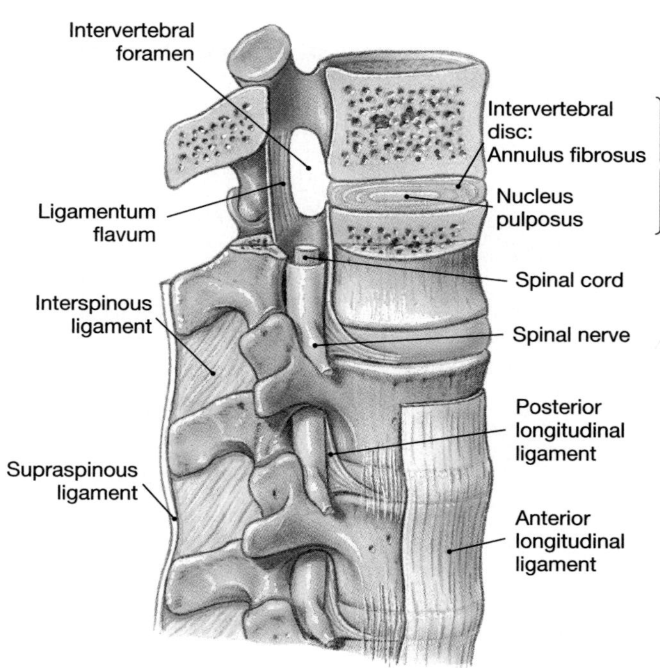
Intervertebral foramen
Intervertebral disc:
Annulus fibrosus
Nucleus pulposus
Ligamentum flavum
Spinal cord
Interspinous ligament
Spinal nerve
Posterior longitudinal ligament
Supraspinous ligament
Anterior longitudinal ligament

● **FIGURE 9-7**
Intervertebral Articulations

Vertebral Movements

The following movements of the vertebral column are possible: (1) *flexion*, bending forward; (2) *extension*, bending backward; (3) *lateral flexion*, bending to the side; and (4) *rotation*.

Table 9-2 summarizes information concerning other articulations of the axial skeleton.

- ☑ What regions of the vertebral column do not contain intervertebral discs? Why?
- ☑ What is the cause of the pain associated with a herniated disc?
- ☑ What vertebral movements are involved in (a) bending forward, (b) bending to the side, (c) moving the head to signify "no"?

❑ The Shoulder Joint

Figure 9-9

The shoulder joint, or *glenohumeral joint*, permits the greatest range of motion of any joint in the body. Because it is also the most frequently dislo-

Problems with the Intervertebral Discs

An intervertebral disc compressed beyond its normal limits may become temporarily or permanently damaged. If the posterior ligaments are weakened, as often occurs with advancing age, the compressed nucleus pulposus may distort the annulus fibrosus, forcing it partway into the vertebral canal. This condition is often called a *slipped disc* (Figure 9-8●), although the disc does not actually slip. The most common sites for disc problems are at C_5–C_6, L_4–L_5, and between L_5 and S_1. A disc problem can occur at any age as the result of an accidental injury, such as a hard fall or a "whiplash" injury to the neck, but with advanced age, the supporting ligaments may become so weak that the problem may occur without warning or apparent cause.

If the nucleus pulposus breaks through the annulus fibrosis, it often protrudes into the vertebral canal. This condition is called a **herniated disc.** When a disc herniates, sensory nerves are distorted, and the protruding mass can also compress the nerve roots passing through the adjacent intervertebral foramen. The result is severe backache, an abnormal posture (abnormal vertebral flexion), abnormal sensory function, often a burning or tingling sensation from the lower back and lower limbs, and in some cases a partial loss of control over skeletal muscles innervated by the compressed nerve fibers. The location of the injured disc can usually be determined by noting the distribution of abnormal sensations. For example, someone with a herniated disc at L_4–L_5 will experience pain in the hip, groin, the posterior and lateral surfaces of the thigh, the lateral surface of the calf, and the top of the foot; a herniation at L_5–S_1 produces pain in the buttocks, the posterior thigh, the posterior calf, and the sole of the foot.

Most lumbar disc problems can be successfully treated with some combination of rest, back braces, analgesic (pain-killing) drugs, and physical therapy. Surgery to relieve the symptoms is required in only about 10 percent of cases involving lumbar disc herniation. The primary method of treatment involves removing the offending disc, and, if necessary, fusing the vertebral bodies together to prevent relative movement. Accessing the disc requires removal of the nearest vertebral arch by shaving away the laminae. For this reason the procedure is known as a *laminectomy* (la-mi-NEK-to-mē).

In cases where the herniated portion of the disc does not extend well into the vertebral foramen, portions of the disc may be removed with a suction cutter guided to the site by radiological imaging. Although this procedure is faster and easier than a laminectomy, relatively few herniated discs fall within this category.

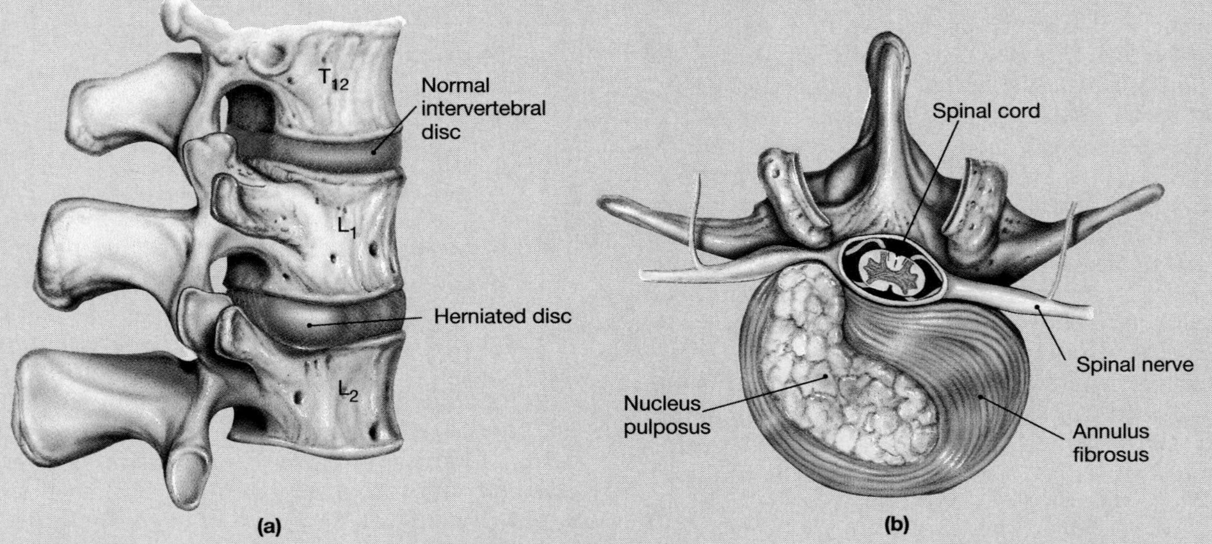
T12
Normal intervertebral disc
L1
Herniated disc
L2
(a)
Spinal cord
Spinal nerve
Nucleus pulposus
Annulus fibrosus
(b)

● **FIGURE 9-8**

Damage to the Intervertebral Discs. (a) Lateral view of the lumbar region of the spinal column, showing a herniated intervertebral disc. **(b)** Sectional view through a herniated disc, showing release of the nucleus pulposus and its effect on the spinal cord and adjacent nerves.

TABLE 9-2 Articulations of the Axial Skeleton

Element	Joint	Type of Articulation	Movements
Cranial and facial bones of skull	Various	Synarthroses (suture or synostosis)	None
Maxilla/teeth and mandible/teeth		Synarthrosis (gomphosis)	None
Temporal bone/ mandible	Temporomandibular	Hinge diarthrosis	Elevation and depression, lateral gliding
Occipital bone/ atlas	Atlanto-occipital	Ellipsoidal diarthrosis	Flexion/extension
Atlas/axis	Atlanto-axial	Pivot diarthrosis	Rotation
Other vertebral elements	Intervertebral (between vertebral bodies)	Amphiarthrosis (symphysis)	Slight movement
	Intervertebral (between articular processes)	Gliding diarthrosis	Slight rotation and flexion/extension
Thoracic vertebrae and ribs	Vertebrocostal	Gliding diarthrosis	Elevation/depression
Ribs and sternum	Sternocostal	Synarthrosis (synchondrosis)	No movement
Sternum and clavicle	Sternoclavicular	Gliding diarthrosis	Protraction/retraction, depression/elevation
L_5/sacrum	Between centrum and sacral body	Amphiarthrosis (symphysis)	Slight movement
	Between inferior articular processes of L_5 and articular processes of sacrum	Gliding diarthrosis	Slight flexion/extension
Sacrum/coxae	Sacroiliac	Gliding diarthrosis	Slight movement
Sacrum/coccyx	Sacrococcygeal	Gliding diarthrosis (may become fused)	Slight movement
Coccygeal bones		Synarthrosis (synostosis)	No movement

cated joint, it provides an excellent demonstration of the principle that strength and stability must be sacrificed to obtain mobility.

This joint is a ball-and-socket type, formed by the articulation of the head of the humerus with the glenoid fossa of the scapula (Figure 9-9●). In life, the surface of the glenoid fossa is covered by a fibrocartilaginous **glenoid labrum** (*labrum,* lip or edge) that encloses the glenoid cavity. The relatively loose articular capsule extends from the scapular neck to the humerus. It is a somewhat oversized capsule that permits an extensive range of motion. The bones of the pectoral girdle provide some stability to the superior surface, because the acromion and coracoid processes project laterally superior to the humeral head. However, most of the stability at this joint is provided by (1) ligaments and (2) surrounding skeletal muscles and their associated tendons.

Ligaments

Major ligaments involved with stabilizing the shoulder joint include those discussed below.

- The capsule surrounding the shoulder joint is relatively thin, but it thickens anteriorly in regions known as the **glenohumeral ligaments.** Because the capsular fibers are usually loose, these ligaments participate in joint stabilization only as the humerus approaches or exceeds the limits of normal motion.
- The large **coracohumeral ligament** originates at the base of the coracoid process and inserts on the head of the humerus.
- The **coracoacromial ligament** spans the gap between the coracoid process and the acromion, just above the capsule. This ligament provides additional support to the superior surface of the capsule.
- The **acromioclavicular ligament** reinforces the capsule of the acromioclavicular joint and provides support for the superior surface of the shoulder. A **shoulder separation** is a relatively common injury involving partial or complete dislocation of the acromioclavicular joint. This injury can result from a blow to the upper surface of the shoulder. The acromion is forcibly depressed, but the clavicle is held back by strong muscles.
- A large **coracoclavicular ligament** ties the clavicle to the coracoid process and helps to limit the relative motion between the clavicle and scapula.

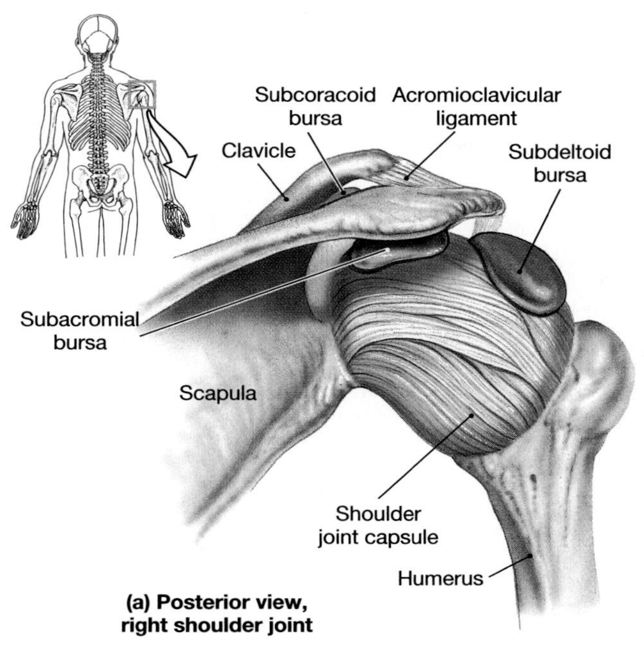
Subcoracoid bursa
Acromioclavicular ligament
Clavicle
Subdeltoid bursa
Subacromial bursa
Scapula
Shoulder joint capsule
Humerus
(a) Posterior view, right shoulder joint

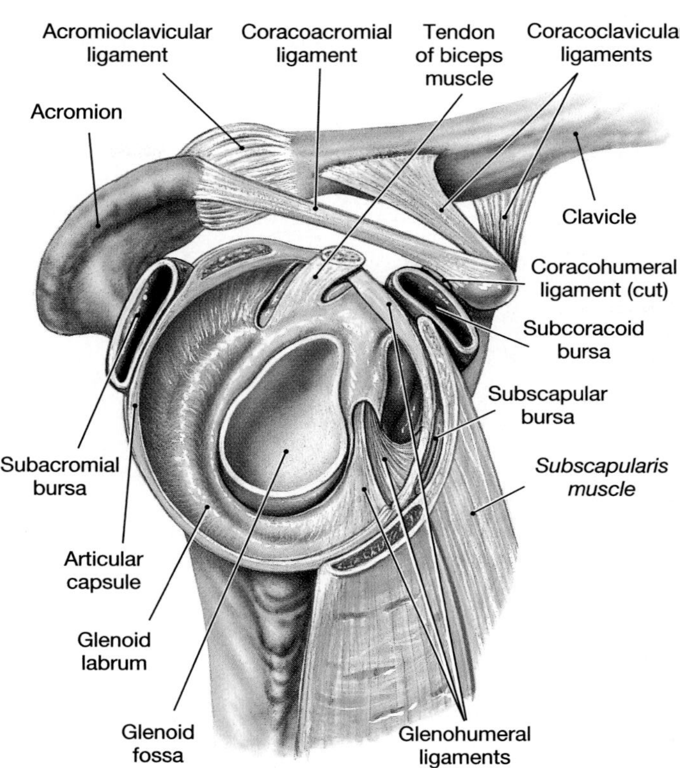
Acromioclavicular ligament
Coracoacromial ligament
Tendon of biceps muscle
Coracoclavicular ligaments
Acromion
Clavicle
Coracohumeral ligament (cut)
Subcoracoid bursa
Subscapular bursa
Subscapularis muscle
Subacromial bursa
Articular capsule
Glenoid labrum
Glenoid fossa
Glenohumeral ligaments
(b) Lateral view of shoulder joint, humerus removed

● **FIGURE 9-9**
The Shoulder Joint. (a) Posterior view showing the orientation and extent of the articular capsule and two of the associated bursae. **(b)** Lateral view of the shoulder joint with the humerus removed.

Skeletal Muscles and Tendons

Muscles that move the humerus do more to stabilize the shoulder joint than all the ligaments and capsular fibers combined. Muscles originating on the trunk, shoulder girdle, and humerus cover the anterior, superior, and posterior surfaces of the capsule. The tendons of these muscles, the *deltoid, supraspinatus, infraspinatus, subscapularis,* and *teres minor,* reinforce the joint capsule and limit the range of movement. These are known as the muscles of the *rotator cuff.* These muscles are the primary mechanism for supporting the joint and limiting the range of movement. Damage to the rotator cuff often occurs when individuals are engaged in sports that place severe strains on the shoulder. Whitewater kayakers, baseball pitchers, and quarterbacks are all at high risk for rotator cuff injuries.

The anterior, superior, and posterior surfaces of the shoulder joint are reinforced by ligaments, muscles, and tendons, but the inferior capsule is poorly reinforced. As a result, a dislocation caused by an impact or violent muscle contraction most often occurs at this site. Such a dislocation can tear the inferior capsular wall and the glenoid labrum. The healing process often leaves a weakness that increases the chances for future dislocations.

Bursae

Figure 9-9

As at other joints, *bursae* at the shoulder reduce friction where large muscles and tendons pass across the joint capsule. The shoulder has a relatively large number of important bursae. The **subacromial bursa** and the **subcoracoid bursa** (Figure 9-9a,b●) prevent contact between the acromial and coracoid processes and the capsule. The **subdeltoid bursa** and the **subscapular bursa** (Figure 9-9a,b●) lie between large muscles and the capsular wall. Inflammation of one or more of these bursae can restrict motion and produce the painful symptoms of *bursitis* (p. 265).

❑ The Elbow Joint

Figure 9-10

The elbow joint, or *olecranal joint* (Figure 9-10●) is a hinge joint. The trochlea of the humerus articulates with the trochlear notch of the ulna, and the capitulum of the humerus articulates with the head of the radius (Figure 9-10●).

Muscles that extend the elbow attach to the rough surface of the olecranon process. These muscles are primarily under the control of the radial nerve, which passes along the *radial groove* of the humerus. ∞ *[p. 243]* The large *biceps brachii* muscle covers the anterior surface of the arm. Its tendon is attached to the radius at the *radial tuberosity.* Contraction of this muscle produces supination of the forearm and flexion of the elbow.

The elbow joint is extremely stable because: (1) the bony surfaces of the humerus and ulna inter-

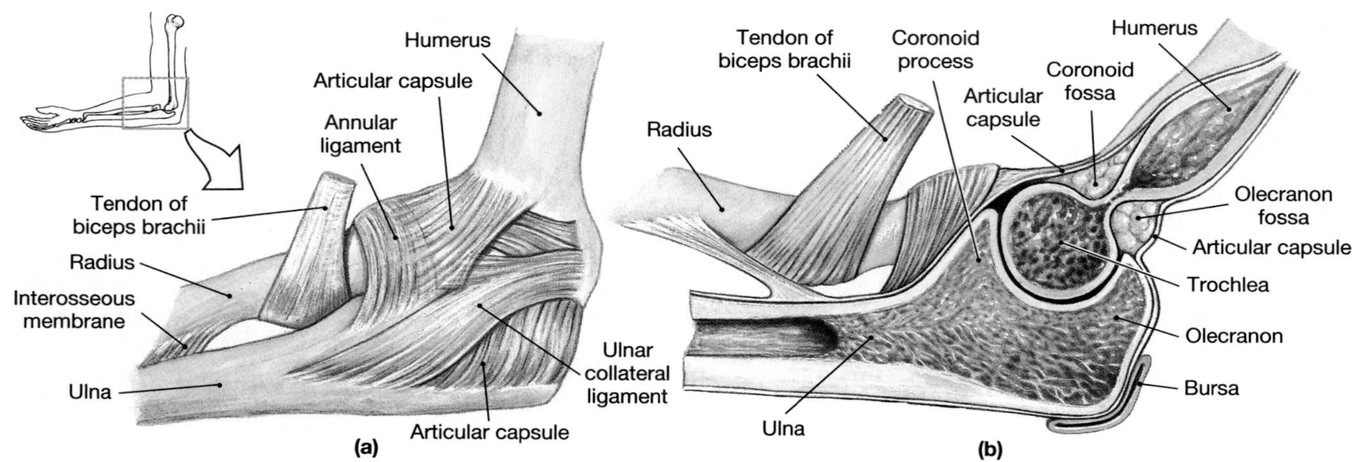
Humerus
Articular capsule
Annular ligament
Tendon of biceps brachii
Radius
Interosseous membrane
Ulna
Ulnar collateral ligament
Articular capsule
(a)
Tendon of biceps brachii
Coronoid process
Humerus
Coronoid fossa
Articular capsule
Radius
Olecranon fossa
Articular capsule
Trochlea
Olecranon
Bursa
Ulna
(b)

● **FIGURE 9-10**

The Elbow Joint. (a) Medial aspect of the right elbow joint, showing ligaments that stabilize the joint. **(b)** Longitudinal section through the right elbow.

lock; (2) the articular capsule is very thick; and (3) the capsule is reinforced by strong ligaments. The *radial collateral ligament* stabilizes the lateral surface of the joint. It extends between the lateral epicondyle of the humerus and the *annular ligament* that binds the proximal radial head to the ulna. The medial surface of the joint is stabilized by the *ulnar collateral ligament.* This ligament extends from the medial epicondyle of the humerus anteriorly to the coronoid processes of the ulna and posteriorly to the olecranon (Figure 9-10●).

Despite the strength of the capsule and ligaments, the elbow joint can be damaged by severe impacts or unusual stresses. For example, when a person falls on a hand with a partially flexed elbow, contractions of muscles that extend the elbow may break the ulna at the center of the trochlear notch. Less violent stresses can produce dislocations or other injuries to the elbow, especially if epiphyseal growth has not been completed. For example, parents in a hurry may drag a toddler along behind them, exerting an upward, twisting pull on the elbow joint that can result in a partial dislocation known as "nursemaid's elbow."

Table 9-3 summarizes the characteristics of the articulations of the upper limb.

TABLE 9-3 Articulations of the Pectoral Girdle and Upper Limb

Element	*Joint*	*Type of Articulation*	*Movement*
Clavicle/sternum	Sternoclavicular	Gliding diarthrosis	Gliding
Scapula/clavicle	Acromioclavicular	Gliding diarthrosis	Gliding
Scapula/humerus	Shoulder, or glenohumeral	Ball-and-socket diarthrosis	Flexion/extension, adduction/abduction, circumduction, rotation
Humerus/ulna and radius	Elbow, or olecranal	Hinge diarthrosis	Flexion/extension
Radius/ulna	Proximal radioulnar Distal radioulnar	Pivot diarthrosis Pivot diarthrosis	Pronation/supination Pronation/supination
Radius/carpals	Radiocarpal, or wrist	Ellipsoidal diarthrosis	Flexion/extension, adduction/abduction, circumduction
Carpal/carpal	Intercarpal	Gliding diarthrosis	Gliding
Carpal/metacarpal (first)	Carpometacarpal of thumb	Saddle diarthrosis	Flexion/extension, adduction/abduction, circumduction, opposition
Carpal/metacarpal (II-V)	Carpometacarpal	Gliding diarthrosis	Slight flexion/extension, adduction/abduction
Metacarpal/phalanx	Metacarpophalangeal	Ellipsoidal diarthrosis	Flexion/extension, adduction/abduction, circumduction
Phalanx/phalanx	Interphalangeal	Hinge diarthrosis	Flexion/extension

What tissues or structures provide most of the stability for the shoulder joint?

Would a tennis player or a jogger be more likely to develop inflammation of the subscapular bursa? Why?

A football player received a blow to the upper surface of his shoulder, causing a shoulder separation. What does this mean?

Terry suffers an injury to his forearm and elbow. After the injury, he notices an unusually large degree of motion between the radius and the ulna at the elbow. What ligament did Terry most likely damage?

❑ The Hip Joint

Figure 9-11

Figure 9-11● introduces the structure of the hip joint. A fibrocartilage pad covers the articular surface of the acetabulum and extends like a horseshoe along the sides of the **acetabular notch** (Figure 9-11a●). A fat pad covered by synovial membrane covers the central portion of the acetabulum. This pad acts as a shock absorber, and the adipose tissue stretches and distorts without damage.

The Articular Capsule

Figure 9-11b,c

The articular capsule of the hip joint is extremely dense and strong (Figure 9-11b,c●). It extends from the lateral and inferior surfaces of the pelvic girdle to the trochanteric line and trochanteric crest of the femur, enclosing both the femoral head and neck. This arrangement helps keep the head from moving away from the acetabulum.

Stabilization of the Hip

Figure 9-11b,c,d

Four broad ligaments reinforce the articular capsule (Figure 9-11b,c,d●). Three of them are regional thickenings of the capsule: the **iliofemoral, pubofemoral,** and **ischiofemoral ligaments.** The **transverse acetabular ligament** crosses the acetabular notch and completes the inferior border of the acetabular fossa. A fifth ligament, the **ligamentum teres** (*teres*, long and round) originates along the transverse acetabular ligament and attaches to the center of the femoral head. This ligament tenses only when the thigh is flexed and undergoing external rotation. Much more important stabilization is provided by the bulk of the surrounding muscles, aided by ligaments and capsular fibers.

The combination of an almost complete bony socket, a strong articular capsule, supporting ligaments, and muscular padding makes this an extremely stable joint. Fractures of the femoral neck or between the trochanters are actually more common than hip dislocations. As noted in Chapter 6, hip fractures are relatively common in elderly individuals with severe osteoporosis. ∞ [p. 193]

HIP FRACTURES Hip fractures are most often suffered by individuals over 60 years of age, when osteoporosis has weakened the thigh bones. These injuries may be accompanied by dislocation of the hip or pelvic fractures. For individuals with osteoporosis, healing proceeds very slowly. In addition, the powerful muscles that surround the joint can easily prevent proper alignment of the bone fragments. Trochanteric fractures usually heal well, if the joint can be stabilized; steel frames, pins, screws, or some combination of those devices may be needed to preserve alignment and permit healing to proceed normally.

Severe hip fractures are most common among those over age 60, but in recent years the frequency of hip fractures has increased dramatically among young, healthy professional athletes. Probably the best-known example is the case of Bo Jackson, discussed in the *Applications Manual.* AM *Hip Fractures, Aging, and Professional Athletes*

❑ The Knee Joint

The hip joint passes weight to the femur, and at the knee joint the femur transfers the weight to the tibia. The shoulder is mobile; the hip, stable; and the knee . . . ? If you had to choose one word, it would probably be "complicated." Although the knee functions as a hinge joint, the articulation is far more complex than that of the elbow or even the ankle. The rounded femoral condyles roll across the top of the tibia, so the points of contact are constantly changing.

Structurally the knee resembles three separate joints, two between the femur and tibia (medial condyle to medial condyle and lateral condyle to lateral condyle) and one between the patella and the patellar surface of the femur.

The Articular Capsule and Joint Cavity

Figure 9-12

There is no single unified capsule at the knee joint, nor is there a common synovial cavity (Figure 9-12a●). A pair of fibrocartilage pads, the **medial** and **lateral menisci,** lie between the femoral and tibial surfaces (Figure 9-12b,c●). The menisci (1) act as cushions, (2) conform to the shape of the articulating surfaces as the femur changes position, and (3) provide some lateral stability to the joint. Prominent fat pads provide padding around the margins of the joint and assist the bursae in reducing friction between the patella and other tissues.

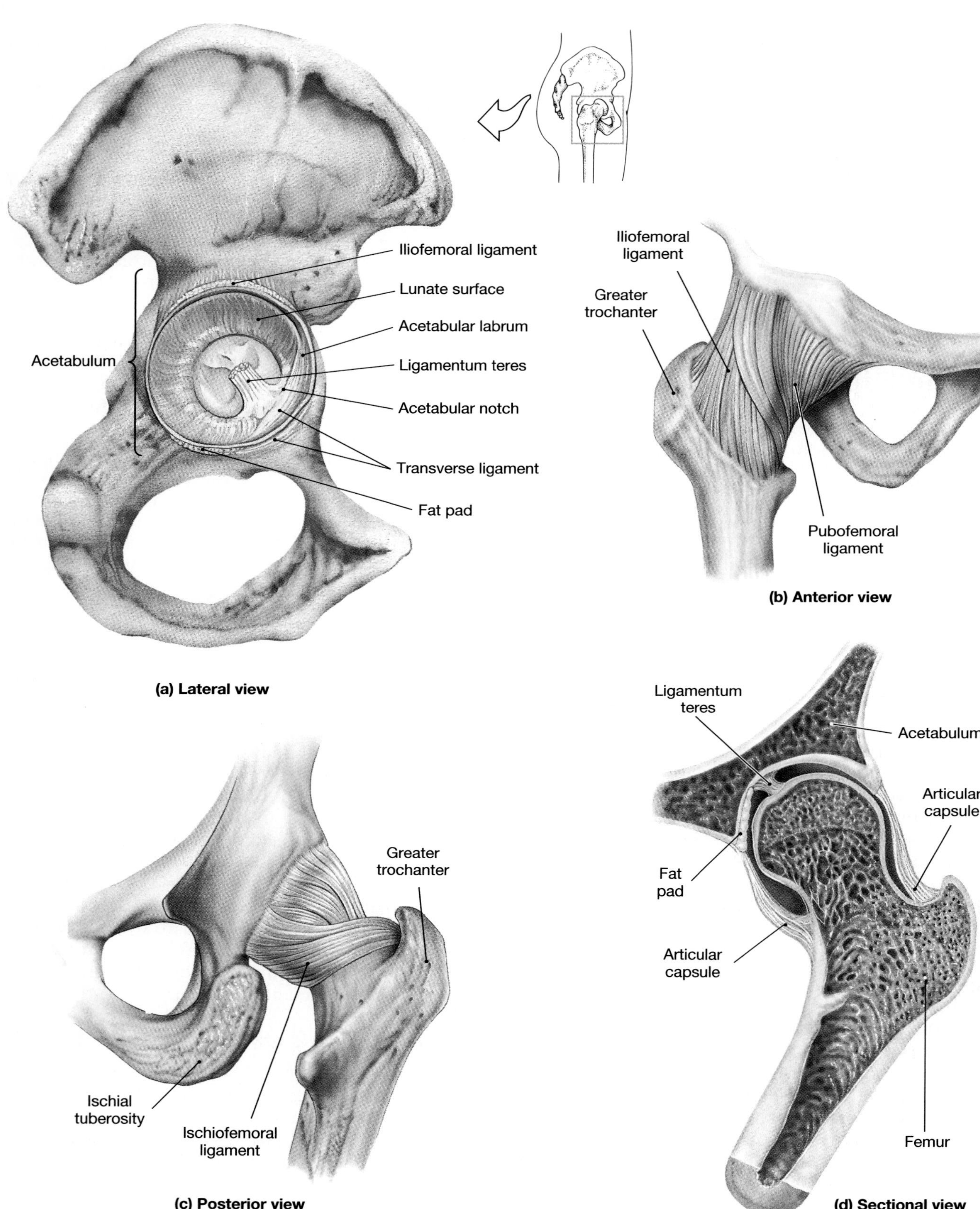
Iliofemoral ligament
Lunate surface
Acetabular labrum
Acetabulum
Ligamentum teres
Acetabular notch
Transverse ligament
Fat pad
(a) Lateral view
Iliofemoral ligament
Greater trochanter
Pubofemoral ligament
(b) Anterior view
Greater trochanter
Ischial tuberosity
Ischiofemoral ligament
(c) Posterior view
Ligamentum teres
Acetabulum
Articular capsule
Fat pad
Articular capsule
Femur
(d) Sectional view

● **FIGURE 9-11**

The Hip Joint. (a) Lateral view of the right hip joint with the femur removed. **(b)** Anterior view of the right hip joint. **(c)** Posterior view of the right hip joint, showing additional ligaments that add strength to the capsule. **(d)** Sectional view of the right hip joint. AM *Plate 7.3a; Scan 4*

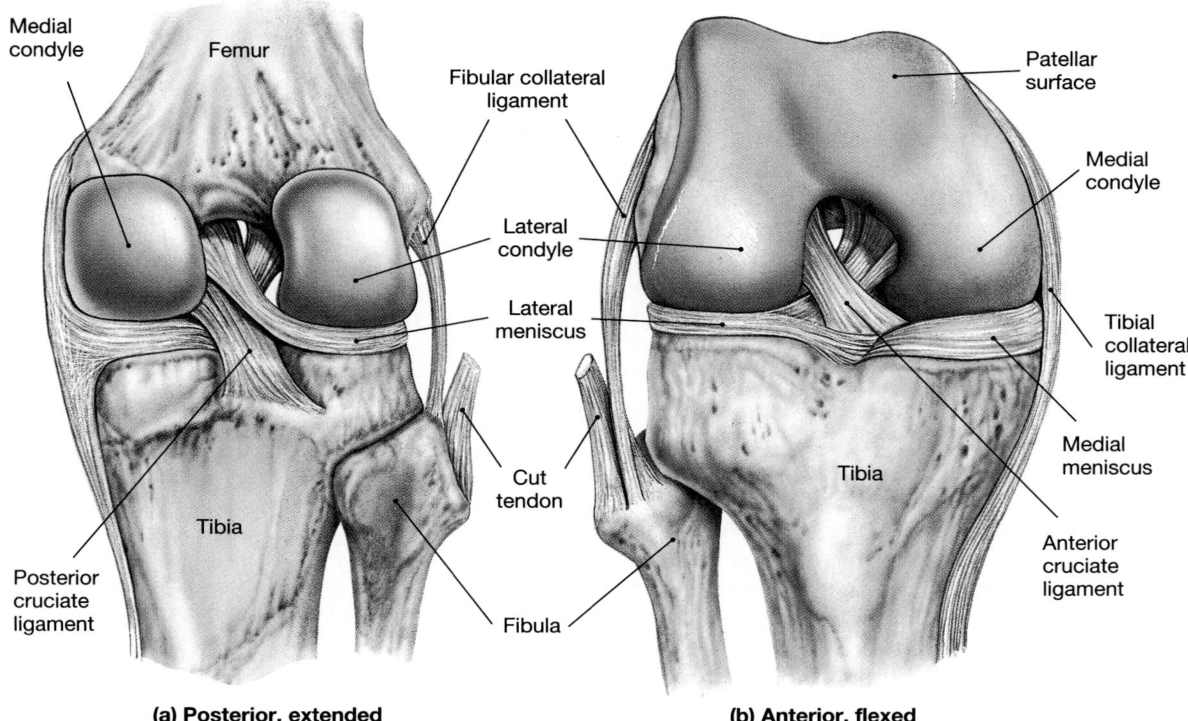
Medial condyle
Femur
Fibular collateral ligament
Lateral condyle
Lateral meniscus
Cut tendon
Tibia
Posterior cruciate ligament
Fibula
Patellar surface
Medial condyle
Tibial collateral ligament
Medial meniscus
Anterior cruciate ligament
(a) Posterior, extended
(b) Anterior, flexed

Supporting Ligaments

Figure 9-12

Seven major ligaments stabilize the knee joint, and a complete dislocation of the knee is very rare.

- The tendon from the muscles responsible for extending the knee passes over the anterior surface of the joint. The patella is embedded within this tendon, and the **patellar ligament** continues to its attachment on the anterior surface of the tibia. The patellar ligament provides support to the anterior surface of the knee joint (Figure 9-12c●).
- Two **popliteal ligaments** extend between the femur and the heads of the tibia and fibula. These ligaments reinforce the back of the knee joint.
- Inside the joint capsule, the **anterior cruciate** and **posterior cruciate ligaments** attach the intercondylar area of the tibia to the condyles of the femur. *Anterior* and *posterior* refer to their sites of origin on the tibia, and they cross one another as they proceed to their destinations on the femur (Figure 9-12b,c●). (The term *cruciate* is derived from the Latin word *crucialis*, meaning a cross.) These ligaments limit the anterior and posterior movement of the femur and maintain the alignment of the femoral and tibial condyles.
- The **tibial collateral ligament** reinforces the medial surface of the knee joint, and the

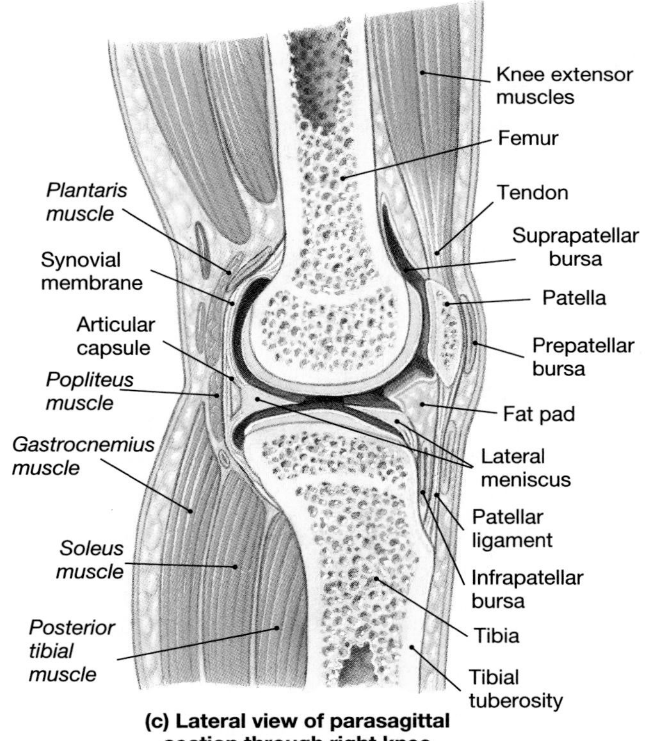
Knee extensor muscles
Femur
Tendon
Suprapatellar bursa
Patella
Prepatellar bursa
Fat pad
Lateral meniscus
Patellar ligament
Infrapatellar bursa
Tibia
Tibial tuberosity
Plantaris muscle
Synovial membrane
Articular capsule
Popliteus muscle
Gastrocnemius muscle
Soleus muscle
Posterior tibial muscle
(c) Lateral view of parasagittal section through right knee

● **FIGURE 9-12**

The Knee Joint. (a) Posterior view of the right knee at full extension. **(b)** Anterior view of the right knee at full flexion. **(c)** Lateral view of the extended right knee as seen in parasagittal section, showing major anatomical features. AM *Plate 7.3; Scans 5, 6, 7*

fibular collateral ligament reinforces the lateral surface (Figure 9-12b,c●). These ligaments tighten only at full extension, and in this position they act to stabilize the joint.

The knee is structurally complex and subjected to severe stresses in the course of normal activities. Painful knee injuries are all too familiar to both amateur and professional athletes. Treatment is often costly and prolonged, and repairs seldom make the joint "good as new."

Locking of the Knee

The knee joint can "lock" in the extended position. At full extension a slight external rotation of the tibia tightens the anterior cruciate ligament and jams the meniscus between the tibia and femur. This mechanism allows you to stand for prolonged periods without using (and tiring) the extensor muscles of the leg. Unlocking the joint requires muscular contractions that produce internal rotation of the tibia or external rotation of the femur. If the locked knee is struck from the side, the menisci can be torn and serious damage can occur to the supporting ligaments.

Table 9-4 summarizes information about the articulations of the lower limb.

Where would you find the following ligaments: iliofemoral ligament, pubofemoral ligament, and ischiofemoral ligament?

What symptoms would you expect to see in an individual who has damaged the menisci of the knee joint?

Why is the condition known as "clergyman's knee" (a type of bursitis), common in carpet layers and roofers?

KNEE INJURIES Athletes place tremendous stresses on their knees. Ordinarily, the medial and lateral menisci move as the femoral position changes. Placing a lot of weight on the knee while it is partially flexed can trap a meniscus between the tibia and femur, resulting in a break or tear in the cartilage. In the most common injury, the lateral surface of the lower leg is driven medially, tearing the medial meniscus. In addition to being quite painful, the torn cartilage may restrict movement at the joint. It can also lead to chronic problems and the development of a "trick knee," a knee that feels unstable. Sometimes the meniscus can be heard and felt popping in and out of position when the knee is extended.

Less common knee injuries involve tearing one or more stabilizing ligaments or damaging the patella. Torn ligaments are often difficult to correct surgically, and healing is slow. The patella can be injured in a number of ways. For example, if the leg is immobilized (as it might be in a football pileup) while you try to extend the knee, the muscles are powerful enough to pull the patella apart. Impacts to the anterior surface of the knee may also shatter the kneecap. (This was the intended target of the assault on Nancy Kerrigan prior to the 1994 Winter Olympics.) Treatment of a fractured patella is difficult and time-consuming. The fragments must be surgically removed, and the tendons and ligaments repaired, followed by immobilization of the joint. Total knee replacements are rarely performed on young people, but they are becoming increasingly common among elderly patients with severe arthritis. AM *Arthroscopic Surgery*

TABLE 9-4 Articulations of the Pelvic Girdle and Lower Limb

Element	*Joint*	*Type of Articulation*	*Movements*
Sacrum/ilium of coxa	Sacroiliac	Gliding diarthrosis	Gliding
Coxa/coxa	Symphysis pubis	Amphiarthrotic symphysis	Slight
Coxa/femur	Hip	Ball-and-socket diarthrosis	Flexion/extension, adduction/abduction, circumduction, rotation
Femur/tibia	Knee	Complex, functions as hinge	Flexion/extension, limited rotation
Tibia/fibula	Tibiofibular (proximal)	Gliding diarthrosis	Gliding
	Tibiofibular (distal)	Gliding diarthrosis and amphiarthrotic syndesmosis	Slight gliding
Tibia and fibula with talus	Ankle, or tibiotalar	Hinge diarthrosis	Dorsiflexion/plantar flexion
Tarsal/tarsal	Intertarsal	Gliding diarthrosis	Gliding
Tarsal/metatarsal	Tarsometatarsal	Gliding diarthrosis	Gliding
Metatarsal/ phalanx	Metatarsophalangeal	Ellipsoidal diarthrosis	Flexion/extension, adduction/abduction
Phalanx/ phalanx	Interphalangeal	Hinge diarthrosis	Flexion/extension

Bones and Muscles

The skeletal and muscular systems are structurally and functionally interdependent; their interactions are so extensive that they are often considered to be parts of a single *musculoskeletal system.* There are direct physical connections between the two systems. Many of the anatomical landmarks identified in this chapter are attachment sites of skeletal muscles, and the connective tissues that surround individual muscle fibers are continuous with those that form the organic framework of an attached bone. Muscles and bones are also physiologically linked through mutual dependence on stable calcium ion concentrations in body fluids. With most of the body's calcium tied up in the skeleton, abnormalities involving the bones can have a direct effect on the muscle.

The bones, in turn, are directly affected by muscular activity. When a muscle enlarges because of regular exercise, the increased forces exerted on the skeleton will make bones become stronger and more massive. The reverse holds true as well, for both muscle and bones decrease in size and strength after periods of inactivity.

The next two chapters will examine the structure and function of the muscular system and discuss how muscular contractions perform specific movements.

Selected Clinical Terminology

Terms Discussed in This Chapter

arthritis (ar-THRĪ-tis): Rheumatic diseases that affect synovial joints. Arthritis always involves damage to the articular cartilages, but the specific cause may vary. The diseases of arthritis are usually classified as either degenerative or inflammatory in nature. *(p. 264* and *AM)*

bunion: The most common pressure-related bursitis, involving a tender nodule formed around bursae over the base of the great toe. *(p. 265)*

bursitis: Inflammation of a bursa that causes pain whenever the associated tendon or ligament moves. *(p. 265)*

continuous passive motion (CPM): A therapeutic procedure involving passive movement of an injured joint to stimulate circulation of synovial fluid. The goal is to prevent degeneration of the articular cartilages. *(p. 264* and *AM)*

herniated disc: A condition caused by intervertebral compression severe enough to rupture the annulus fibrosus and release the nucleus pulposus, which may protrude beyond the intervertebral space. *(p. 273)*

laminectomy (la-mi-NEK-to-mē): Removal of vertebral laminae; may be performed to access the vertebral canal and relieve symptoms of a herniated disc. *(p. 273)*

luxation (luks-Ā-shun): A dislocation; a condition in which the articulating surfaces are forced out of position. *(p. 266)*

osteoarthritis (os-tē-ō-ar-THRĪ-tis) (*degenerative arthritis* or *degenerative joint disease,* DJD): An arthritic condition resulting from (1) cumulative wear and tear on joint surfaces or (2) genetic predisposition. In the U.S. population 25 percent of women and 15 percent of men over 60 years of age show signs of this disease *(p. 264* and *AM)*

rheumatism (ROO-muh-tizm): A general term that indicates pain and stiffness affecting the skeletal system, the muscular system, or both. *(p. 264* and *AM)*

rheumatoid arthritis: An inflammatory arthritis that affects roughly 2.5 percent of the adult population. The cause is uncertain, although allergies, bacteria, viruses, and genetic factors have all been proposed. The primary symptom is *synovitis,* swelling and inflammation of the synovial membrane. *(p. 264* and *AM)*

shoulder separation: The partial or complete dislocation of the acromioclavicular joint. *(p. 274)*

slipped disc: A common name for a condition caused by distortion of an intervertebral disc. The distortion applies pressure to spinal nerves, causing pain and limiting range of motion. *(p. 273)*

sprain: A condition caused when a ligament is stretched to the point where some of the collagen fibers are torn. The ligament remains functional, and the structure of the joint is not affected. *(p. 265)*

subluxation (sub-luks-Ā-shun): A partial dislocation; displacement of articulating surfaces sufficient to cause discomfort, but resulting in less physical damage to the joint than during a dislocation. *(p. 266)*

AM *Additional Terms Discussed in the Applications Manual*

ankylosis (an-kē-LŌ-sis): An abnormal fusion between articulating bones in response to trauma and friction within a joint.

arthroscope: A fiber-optic instrument used to view the inside of joint cavities.

arthroscopic surgery: The surgical modification of a joint using an arthroscope.

bony crepitus: A crackling or popping sound produced during movement of an abnormal joint.

meniscectomy: The surgical removal of a damaged meniscus.

monoarthritic: An arthritic condition affecting a single articulation.

polyarthritic: An arthritic condition affecting multiple articulations.

CHAPTER REVIEW

STUDY OUTLINE

INTRODUCTION, p. 262

A CLASSIFICATION OF JOINTS, p. 262

1. **Articulations** (joints) exist wherever two bones interact. Immovable joints are **synarthroses,** slightly movable joints are **amphiarthroses,** and those that are freely movable are called **diarthroses.** (*Table 9-1*)

Immovable Joints (Synarthroses), p. 262

2. Examples of synarthroses include a **suture,** a **gomphosis,** a **synchondrosis,** and a **synostosis.**

Slightly Movable Joints (Amphiarthroses), p. 263

3. Examples of amphiarthroses are a **syndesmosis** and a **symphysis.**

Freely Movable Joints (Diarthroses), p. 263

4. The bony surfaces at diarthroses are covered by **articular cartilages,** lubricated by **synovial fluid,** and enclosed within a **joint capsule.** Other synovial structures can include **menisci,** or **articular discs; fat pads;** and **accessory ligaments.** (*Figure 9-1*)

ARTICULAR FORM AND FUNCTION, p. 266

Describing Dynamic Motion, p. 266

1. Possible movements can be classified as **linear motion, angular motion,** or **rotation.** (*Figure 9-2*)
2. Joints are called **monaxial, biaxial,** or **triaxial** depending on the degree of movement they allow.

Types of Movements, p. 267

3. Important terms that describe dynamic motion are **flexion, extension, hyperextension, rotation, circumduction, abduction,** and **adduction.** (*Figures 9-3, 9-4*)
4. The bones in the forearm permit **pronation** and **supination.** (*Figure 9-4*)
5. The ankle undergoes **dorsiflexion** and **plantar flexion.** Movements of the foot include **inversion** and **eversion. Opposition** is the thumb movement that enables us to grasp objects. (*Figure 9-5*)
6. **Protraction** involves moving something anteriorly; **retraction** involves moving it posteriorly. **Depression** and **elevation** occur when we move a structure inferiorly or superiorly. (*Figure 9-5*)

A Structural Classification of Synovial Joints, p. 269

7. **Gliding joints** permit limited movement, usually in a single plane. (*Figure 9-6*)
8. **Hinge joints** and **pivot joints** are monaxial joints. (*Figure 9-6*)
9. Biaxial joints include **ellipsoidal joints** and **saddle joints.** (*Figure 9-6*)
10. Triaxial, or **ball-and-socket, joints,** permit rotation as well as other movements. (*Figure 9-6*)

REPRESENTATIVE ARTICULATIONS, p. 270

Intervertebral Articulations, p. 270

1. The articular processes form gliding joints with those of adjacent vertebrae. The bodies form symphyseal joints. They are separated by **intervertebral discs** containing an inner **nucleus pulposus** and an outer **annulus fibrosus.** (*Figures 9-7, 9-8; Table 9-2*)

The Shoulder Joint, p. 272

2. The shoulder joint, or *glenohumeral joint,* formed by the glenoid fossa and the head of the humerus, permits the greatest range of motion of any joint in the body. It is a ball-and-socket diarthrosis type of articulation. Strength and stability are sacrificed to obtain mobility. (*Figure 9-9; Table 9-3*)

The Elbow Joint, p. 275

3. The elbow joint, or *olecranal* joint, permits only flexion/extension. The articulation is a hinge diarthrosis. (*Figure 9-10; Table 9-3*)

The Hip Joint, p. 277

4. The hip joint is formed by the union of the acetabulum with the head of the femur. The joint permits flexion/extension, adduction/abduction, circumduction, and rotation. (*Figure 9-11*)

The Knee Joint, p. 277

5. The knee joint is a hinge joint formed by the union of the condyles of the femur with the superior condylar surfaces of the tibia. The joint permits flexion/extension and limited rotation. (*Figure 9-12; Table 9-4*)

Bones and Muscles, p. 281

REVIEW QUESTIONS

LEVEL 1 Reviewing Facts and Terms

1. A synarthrotic joint found only between the bones of the skull is a:
   (a) symphysis (b) syndesmosis
   (c) synchondrosis (d) suture

2. The distal articulation between the tibia and fibula is an example of a:
   (a) syndesmosis (b) symphysis
   (c) synchondrosis (d) synostosis

3. The anterior articulation between the two pubic bones is an example of a:
   (a) synchondrosis (b) synostosis
   (c) symphysis (d) synarthrosis

4. Joints typically found at the end of long bones are:
(a) synarthroses (b) amphiarthroses
(c) diarthroses (d) symphyses

5. The function of the synovial fluid is:
(a) to nourish chondrocytes
(b) to provide lubrication
(c) to absorb shock
(d) a, b, and c are correct

6. The structures responsible for channeling the flow of synovial fluid are:
(a) menisci (b) bursae
(c) carpal tunnels (d) articular cartilages

7. The structures that limit the range of motion and provide mechanical support across or around a joint are:
(a) bursae (b) tendons
(c) menisci (d) a, b, and c are correct

8. A partial dislocation of an articulating surface is called a:
(a) circumduction (b) hyperextension
(c) subluxation (d) supination

9. Abduction and adduction always refer to movements of the:
(a) axial skeleton (b) appendicular skeleton
(c) skull (d) vertebral column

10. Rotation of the forearm that makes the palm face posteriorly is called:
(a) supination (b) pronation
(c) proliferation (d) projection

11. Spreading the fingers or toes apart is:
(a) dorsiflexion (b) circumduction
(c) adduction (d) abduction

12. Standing on tiptoe is an example of a movement called:
(a) elevation (b) dorsiflexion
(c) plantar flexion (d) retraction

13. Examples of monoaxial joints that permit angular movement in a single plane are the:
(a) carpals and tarsals (b) shoulder and hip
(c) elbow and knee (d) a, b, and c are correct

14. Joints that connect the fingers and toes with the metacarpals and metatarsals are:
(a) ellipsoidal (b) saddle
(c) pivot (d) hinge

15. Movements that occur at the shoulder and the hip illustrate the action that occurs at a ________________ joint.
(a) hinge (b) ball-and-socket
(c) pivot (d) gliding

16. The annulus fibrosus and nucleus pulposus are structures associated with the:
(a) intervertebral discs (b) knee and elbow
(c) shoulder and hip (d) carpals and tarsals

17. Subacromial, subcoracoid, subdeltoid, and subscapular bursae reduce friction in the ______________ joint.
(a) hip (b) knee
(c) elbow (d) shoulder

18. The functional anatomy of the lower limbs is very different from that of the upper limbs primarily because:
(a) the upper limbs are used for grasping
(b) the upper limbs contain more bones than the lower limbs
(c) the lower limbs must transfer the body weight to the ground
(d) the lower limbs are used for bipedal mobility

LEVEL 2 Reviewing Concepts

19. The hip is an extremely stable joint because it has:
(a) a complete bony socket
(b) a strong articular capsule
(c) supporting ligaments
(d) a, b,and c are correct

20. Complete dislocation of the knee is an extremely rare event because:
(a) it is protected by the patella
(b) the femur articulates with the tibia at the knee
(c) it contains seven major ligaments
(d) it contains fat pads to absorb shocks

21. How does a meniscus (articular disc) function in a joint?

22. "The greater the range of motion at a joint, the weaker the joint becomes." Why?

23. How do articular cartilages differ from other cartilages in the body?

24. What is the significance of the fact that the pubic symphysis is a slightly movable joint?

25. How would you explain to your grandmother the characteristic decrease in height with advanced age?

26. When the biceps brachii muscle contracts, what movements does it produce?

27. When you stand for a long period of time, why should you "lock" your knee in extended position? How does the knee lock?

LEVEL 3 Critical Thinking and Clinical Applications

28. Dave "overturns" his ankle while playing tennis. He experiences swelling and pain, but after examination, he is told that there are no torn ligaments and the structure of the ankle is not affected. On the basis of the symptoms and the examination results, what do you think happened to Dave's ankle?

29. Bob injured his right knee during a basketball game when he jumped to retrieve the ball and then landed off-balance on his right leg. Since then he has pain and limited mobility of his right knee joint. What type of injury do you think Bob sustained?

30. A high-school student comes to the emergency room, complaining of persistent pain beneath his right shoulder blade. In talking with him, you discover that he has been spending many hours trying to improve his pitching skills for his school's softball team. What do you think is causing the pain?

CHAPTER 10

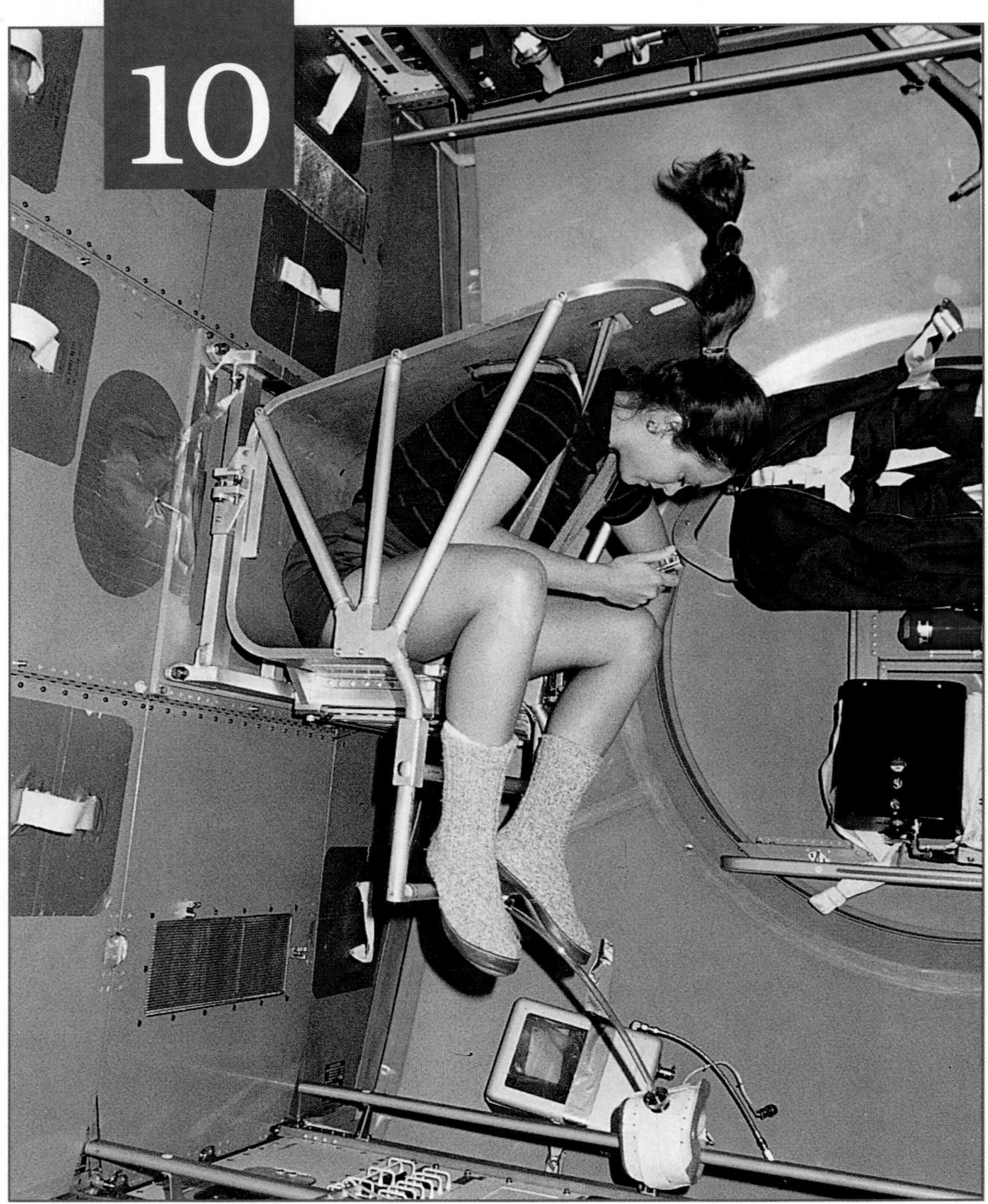

Bones and muscles work together, and our muscles need to work to remain strong. Prolonged weightlessness results in a loss of bone and muscle mass. Bones do not bear weight, and muscles do not need to work to oppose the force of gravity. The resulting reduction in bone and muscle mass can be potentially dangerous on return to normal gravity. This chair was part of the equipment carried aloft by the June 1991 mission of the space shuttle Columbia. *Astronaut Tamara Jernigan is engaged in experiments designed to assess—and reduce—the effects of weightlessness on muscle mass. How muscles work, and the effects of exercise on muscular strength and endurance, will be the focus of this chapter.*

Muscle Tissue

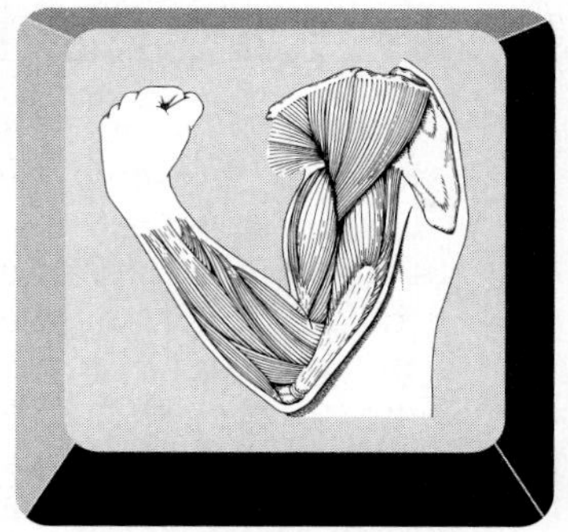

Chapter Outline and Objectives

Think for a moment what life would be like without muscle tissue. Imagine being unable to sit, stand, walk, speak, or grasp objects in your environment. Imagine, too, how your internal functions would be affected. Blood would not circulate, because there would be no heartbeat to propel it through the vessels. The lungs could not rhythmically empty and fill, nor could food move through the digestive tract. In fact, there would be practically no movement through any of the internal passageways of the body.

This is not to say that all life depends on muscle tissue. There are large organisms that get by very nicely without it—we call them plants. But life as we live it would be impossible, for many of our physiological processes, and virtually all our dynamic interactions with the environment, involve muscle tissue.

Skeletal Muscle Tissue and the Muscular System

Muscle tissue, one of the four primary tissue types, consists chiefly of muscle cells (usually called *muscle fibers*)—cells highly specialized for contraction. There are three types of muscle tissue: *skeletal muscle, cardiac muscle,* and *smooth muscle.* Without these muscle tissues, introduced in Chapter 4, nothing in the body would move, and no body movement could occur. [p. 136] Skeletal muscles move the body by pulling on bones of the skeleton, making it possible for us to walk, dance, bite into an apple, or play the ukulele. Cardiac muscle tissue pushes blood through the circulatory system. Smooth muscle tissue pushes fluids and solids along the digestive tract, regulates the diameters of small arteries, and performs a variety of other functions.

This chapter deals primarily with the structure and function of skeletal muscle tissue, in preparation for our discussion of the muscular system (Chapter 11). This chapter will also provide an overview of the differences among skeletal, cardiac, and smooth muscle tissue.

Skeletal muscle tissue is found within **skeletal muscles,** organs that also contain connective tissues, nerves, and blood vessels. As the name implies, skeletal muscles are directly or indirectly attached to the bones of the skeleton. Skeletal muscle tissue gives skeletal muscles the ability to perform the following functions.

1. ***Produce skeletal movement.*** Skeletal muscle contractions pull on tendons and move the bones of the skeleton. The effects range from simple motions such as extending the arm to the highly coordinated movements of swimming, skiing, or typing.
2. ***Maintain posture and body position.*** Tension in our skeletal muscles also maintains body posture—for example, holding the head in position when reading a book or balancing the weight of the body above the feet when walking. Without constant muscular activity we could not sit upright without collapsing into a heap or stand without toppling over.
3. ***Support soft tissues.*** The abdominal wall and the floor of the pelvic cavity consist of layers of skeletal muscle. These muscles support the weight of visceral organs and shield internal tissues from injury.
4. ***Guard entrances and exits.*** The openings of the digestive and urinary tracts are encircled by skeletal muscles. These muscles provide voluntary control over swallowing, defecation, and urination.
5. ***Maintain body temperature.*** Muscle contractions require energy, and whenever energy is used in the body, some of it is converted to heat. The heat released by working muscles keeps our body temperature in the range required for normal functioning.

We will begin our discussion with the gross anatomy of a typical skeletal muscle. We will then consider the structural features at the microscopic level that make contractions possible.

Anatomy of Skeletal Muscle

Figure 10-1

Figure 10-1● illustrates the appearance and organization of a representative skeletal muscle. A skeletal muscle contains connective tissues, blood vessels, nerves, and skeletal muscle tissue.

Connective Tissue Organization

Figure 10-1

Three layers of connective tissue are part of each muscle: an outer *epimysium,* a central *perimysium,* and an inner *endomysium.* These layers and their relationships are diagrammed in Figure 10-1●.

The entire muscle is surrounded by the **epimysium** (ep-i-MĪZ-ē-um; *epi-*, on + *mys*, muscle), a dense layer of collagen fibers. The epimysium separates the muscle from surrounding tissues and organs. It is one component of the deep fascia, the dense connective tissue layer described in Chapter 4 (see Figure 4-16●). [pp. 135–136] AM *Necrotizing Fasciitis*

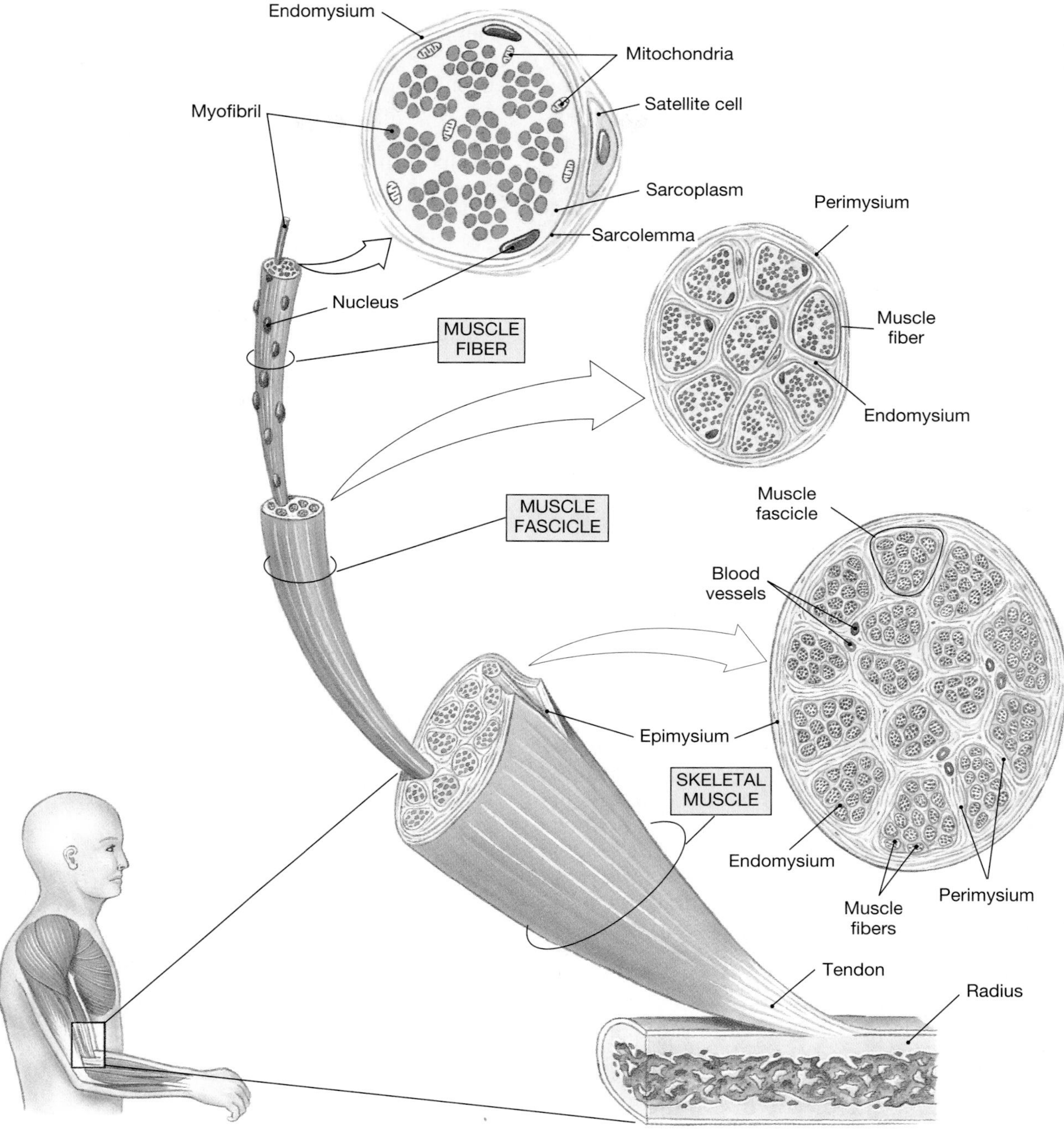
Endomysium
Mitochondria
Satellite cell
Myofibril
Sarcoplasm
Perimysium
Sarcolemma
Nucleus
MUSCLE FIBER
Muscle fiber
Endomysium
Muscle fascicle
MUSCLE FASCICLE
Blood vessels
Epimysium
SKELETAL MUSCLE
Endomysium
Muscle fibers
Perimysium
Tendon
Radius

● **FIGURE 10-1**

Organization of Skeletal Muscles. A skeletal muscle consists of fascicles (bundles of muscle fibers) enclosed by the epimysium. The bundles are separated by connective tissue fibers of the perimysium, and within each bundle the muscle fibers are surrounded by the endomysium. Each muscle fiber has many superficial nuclei, as well as mitochondria and other organelles detailed in Figure 10-2. D

The connective tissue fibers of the **perimysium** (per-i-MĪZ-ē-um; *peri-*, around) divide the skeletal muscle into a series of compartments, each containing a bundle of muscle fibers called a **fascicle** (FA-sik-ul; *fasciculus*, a bundle). In addition to collagen and elastic fibers, the perimysium contains blood vessels and nerves that maintain blood flow and innervate the fascicles. Each fascicle receives branches of the blood vessels and nerves in the perimysium.

Within the fascicle, the delicate connective tissue of the **endomysium** (en-do-MĪZ-ē-um; *endo-*, inside) surrounds the skeletal muscle fibers and ties adjacent muscle fibers together. Scattered **satellite cells** lie between the endomysium and the muscle fibers. These cells function in the repair of damaged muscle tissue. AM *Trichinosis; Fibromyalgia and Chronic Fatigue Syndrome*

The collagen fibers of the endomysium and perimysium are interwoven, and those of the perimysium blend into those of the epimysium. At each end of the muscle, the collagen fibers of the epimysium come together to form a bundle known as a **tendon** or a broad sheet called an **aponeurosis.** Tendons and aponeuroses, examples of dense regular connective tissue described in Chapter 4, attach skeletal muscles to bones. ∞ [p. 127] The tendon fibers are interwoven with the periosteum and extend into the matrix of the bone, providing a firm attachment. As a result, any contraction of the muscle will exert a pull on its tendon and thereby on the attached bone.

❑ Blood Vessels and Nerves

The connective tissues of the epimysium and perimysium contain the blood vessels and nerves that supply the muscle fibers. Muscle contraction requires tremendous quantities of energy. An extensive vascular network delivers the oxygen and nutrients and carries away the metabolic wastes generated by active skeletal muscles. The blood vessels usually enter the muscle in company with the nerve supply, and the vessels and nerves follow the same branching pattern through the perimysium. Once within the endomysium, arterioles supply blood to a capillary network that surrounds each individual muscle fiber.

Skeletal muscles contract only under stimulation from the central nervous system. Axons, or *nerve fibers,* penetrate the epimysium, branch through the perimysium, and enter the endomysium to innervate individual muscle fibers. Skeletal muscles are often called *voluntary muscles* because we have voluntary control over their contractions. Many of these skeletal muscles may also be controlled involuntarily. For example, skeletal muscles involved with breathing, such as the diaphragm, usually work under involuntary control.

Next we will examine the microscopic structure of a typical skeletal muscle fiber and then proceed to relate that microstructure to the physiology of the contraction process.

❑ Microanatomy of Skeletal Muscle Fibers

Figures 10-1, 10-2

Skeletal muscle fibers are quite different from the "typical" cells described in Chapter 3. One obvious difference is size, for skeletal muscle fibers are enormous. A muscle fiber from a thigh muscle could have a diameter of 100 μm and a length equal to that of the entire muscle (30–40 cm, or 10–16 in.). A second obvious difference is that skeletal muscle fibers are *multinucleate:* Each skeletal muscle fiber contains hundreds of nuclei just beneath the cell membrane. The genes contained in these nuclei direct the production of enzymes and structural proteins required for normal contraction, and the presence of multiple copies of these genes speeds up the process. This is particularly important in skeletal muscle fibers, where metabolic turnover tends to be very rapid. ∞ [p. 60]

The distinctive features of size and multiple nuclei are related. During development, groups of embryonic cells called **myoblasts** fuse together to create individual skeletal muscle fibers, as indicated in Figure 10-2•. Each nucleus in a skeletal muscle fiber reflects the contribution of a single myoblast. Some myoblasts do not fuse with developing muscle fibers. These unfused cells remain in adult skeletal muscle tissue as the satellite cells seen in Figure 10-1•. After an injury, satellite cells may enlarge, divide, and fuse with damaged muscle fibers, thereby assisting in the regeneration of the tissue.

The Sarcolemma and Transverse Tubules

The cell membrane, or **sarcolemma** (sar-cō-LEM-uh; *sarkos,* flesh + *lemma,* husk), of a muscle fiber surrounds the cytoplasm, or **sarcoplasm** (SAR-kō-plazm). Like other cell membranes, the sarcolemma has a characteristic transmembrane potential. In a skeletal muscle fiber, a sudden, conducted change in the transmembrane potential is the first step that leads to a contraction. Because a skeletal muscle fiber is very large, the signal to contract must be conveyed to the interior of the cell. This occurs at the **transverse tubules,** or **T tubules.** T tubules are narrow tubes that begin at the sarcolemma and extend into the sarcoplasm at right angles to the membrane surface. They are filled with extracellular fluid, and they form passageways through the muscle fiber, like a series of tunnels through a mountain. The T tubules open onto the sarcolemma, and the membrane that lines each tubule has the same general properties as the sarcolemma.

Myofibrils

Figure 10-2

Inside the muscle fiber, branches of the transverse tubules encircle cylindrical structures called *myofibrils.* A **myofibril** (Figure 10-2•) is a cylindrical structure 1–2 μm in diameter and as long as the entire cell. Each skeletal muscle fiber contains hundreds to thousands of myofibrils.

Myofibrils are bundles of **myofilaments,** protein filaments primarily composed of *actin* and *myosin.* Myofibrils can actively shorten; they are the organelles responsible for skeletal muscle fiber contraction. Chapter 3 noted that proteins, including those of the cytoskeleton, can be bound to the inner surface of the cell membrane. ∞ [p. 84] At each end of the skeletal muscle fiber, the myofibrils are attached to

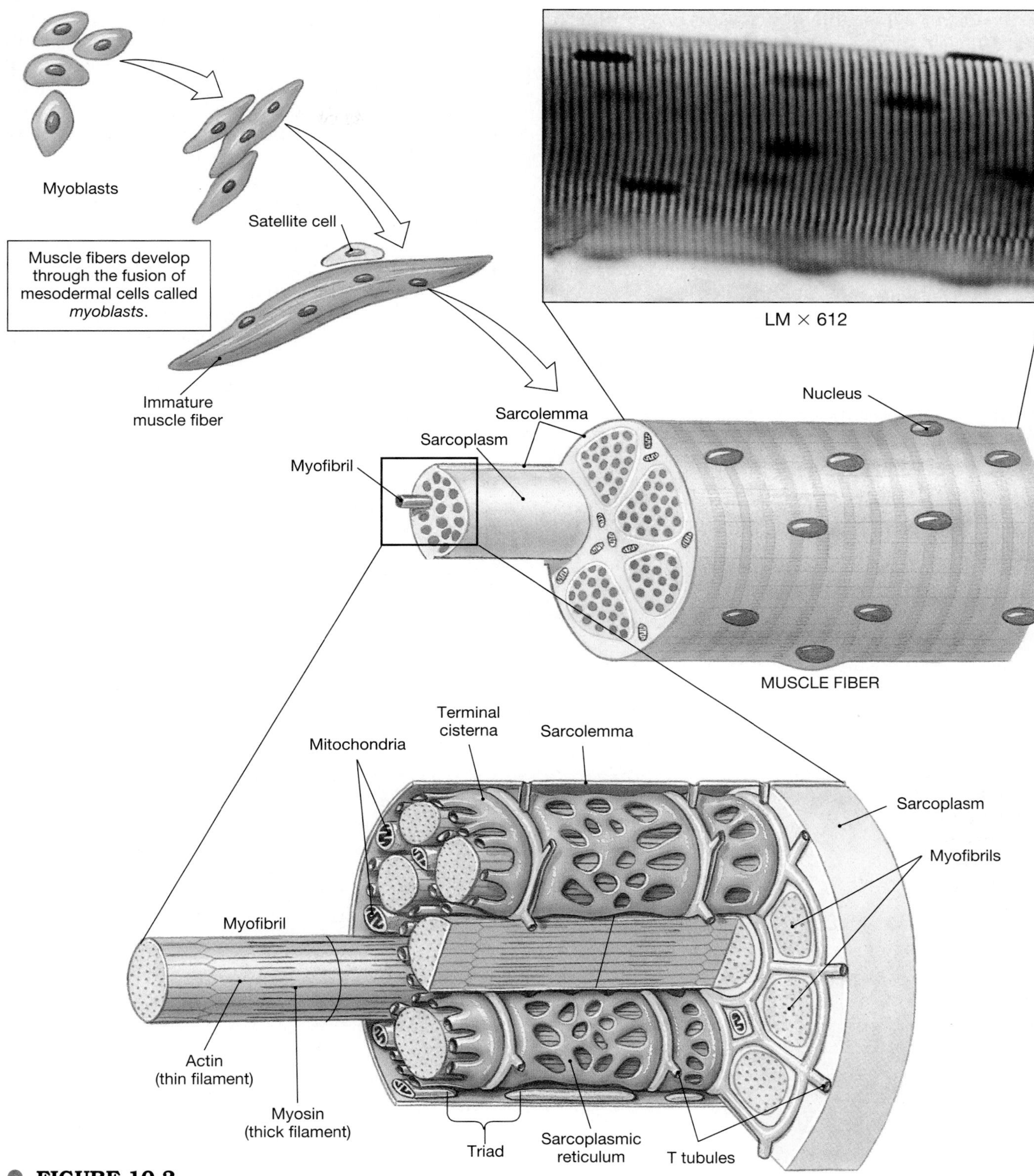
Myoblasts
Satellite cell
Muscle fibers develop through the fusion of mesodermal cells called *myoblasts*.
Immature muscle fiber
LM × 612
Nucleus
Sarcolemma
Sarcoplasm
Myofibril
MUSCLE FIBER
Terminal cisterna
Sarcolemma
Mitochondria
Sarcoplasm
Myofibrils
Myofibril
Actin (thin filament)
Myosin (thick filament)
Triad
Sarcoplasmic reticulum
T tubules

FIGURE 10-2
Formation and Structure of a Skeletal Muscle Fiber.

the sarcolemma. As a result, when the myofibrils contract the entire cell shortens. Scattered between the myofibrils are mitochondria and granules of glycogen, the storage form of glucose. Glucose breakdown via *glycolysis* and mitochondrial activity provide the ATP needed to power muscular contractions.

The Sarcoplasmic Reticulum

Figures 10-2, 10-3

Wherever a transverse tubule encircles a myofibril, it makes close contact with the membranes of the *sarcoplasmic reticulum*. The **sarcoplasmic**

reticulum (SR) is a membrane complex similar to the smooth endoplasmic reticulum of other cells. In skeletal muscle fibers the SR forms a tubular network that ensheaths each individual myofibril (Figures 10-2 and 10-3●). On either side of a transverse tubule, the tubules of the SR enlarge, fuse, and form expanded chambers called **terminal cisternae.** The combination of a pair of terminal cisternae plus a transverse tubule is known as a **triad.** Although the membranes of the triad are in close contact and tightly bound together, there is no direct connection between them, and their fluid contents remain separate and distinct.

In Chapter 3 we noted the existence of special ion pumps that keep the intracellular concentration of calcium ion very low. ∞ [p. 79] Most cells pump the calcium ions across their cell membranes and into the extracellular fluid. Although skeletal muscle fibers do pump calcium ions out of the cell in this way, they also remove them from the cytosol by actively transporting them into the cisternae of the sarcoplasmic reticulum. The sarcoplasm of a resting skeletal muscle fiber contains very low concentrations of calcium ions (Ca^{2+}), somewhere around 10^{-7} mmol/l. The free calcium ion concentration inside the terminal cisternae may be as much as 1000 times higher. In addition, the cisternae contain a protein, *calsequestrin*, that reversibly binds calcium ions. Between the free calcium and the bound calcium, the total concentration of calcium ions inside the cisternae may be 40,000 times that of the surrounding sarcoplasm.

A muscle contraction begins when stored calcium ions are released into the sarcoplasm. The calcium ions then diffuse into individual contractile units called *sarcomeres.*

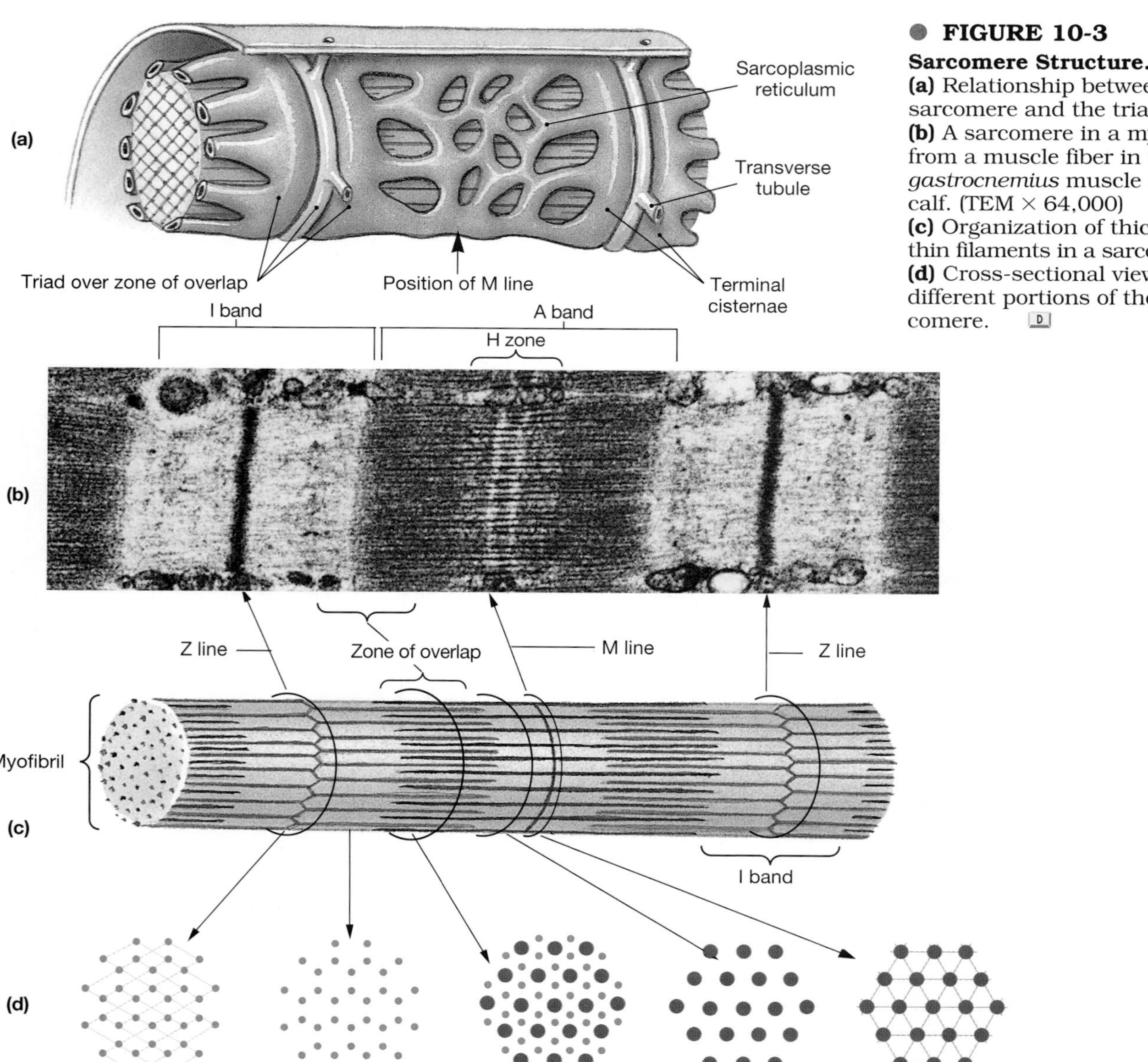
(a)
Sarcoplasmic reticulum
Transverse tubule
Triad over zone of overlap
Position of M line
Terminal cisternae
I band
A band
H zone
(b)
Z line
Zone of overlap
M line
Z line
Myofibril
(c)
I band
(d)
Z line
I band
Zone of overlap
H zone
M line

● **FIGURE 10-3**
Sarcomere Structure. **(a)** Relationship between the sarcomere and the triads. **(b)** A sarcomere in a myofibril from a muscle fiber in the *gastrocnemius* muscle of the calf. (TEM × 64,000) **(c)** Organization of thick and thin filaments in a sarcomere. **(d)** Cross-sectional views of different portions of the sarcomere. D

Sarcomeres

Figures 10-3, 10-4, 10-5

Myofibrils are bundles of actin and myosin myofilaments. The actin is found in **thin filaments,** and the myosin forms **thick filaments;** both types of filaments were introduced in Chapter 3. ∞ [p. 84] Myofilaments are organized in repeating functional units called **sarcomeres** (SAR-kō-mērz; *sarkos,* flesh + *meros,* part), detailed in Figure 10-3●.

SARCOMERE ORGANIZATION A myofibril consists of a linear series of approximately 10,000 sarcomeres, each with a resting length of 1.6–2.6 μm. *Sarcomeres are the smallest functional units of the muscle fiber—interactions between the thick and thin filaments of sarcomeres are responsible for muscle contraction.* A sarcomere contains (1) thick filaments, (2) thin filaments, (3) proteins that stabilize the positions of thick and thin filaments, and (4) proteins that regulate the interactions between thick and thin filaments.

Figure 10-3● indicates the structure of an individual sarcomere. Differences in the size, density, and distribution of thick filaments and thin filaments account for the banded appearance of the sarcomere. The dark areas, which contain the thick filaments and portions of the thin filaments, are called *A bands.* The lighter areas, which contain thin filaments only, are called *I bands.* (The terms *A band* and *I band* are derived from *anisotropic* and *isotropic,* which refer to the appearance of these bands when viewed under polarized light.) You may find it helpful to remember that the A bands are dArk and the I bands are lIght.

- ***A band.*** The thick filaments are located at the center of the sarcomere, within the **A band.** *The length of the A band is equal to the length of a typical thick filament.* The A band includes several subdivisions:
  1. ***M line.*** The central portion of each thick filament is connected to its neighbors by proteins of the **M line.** These densely staining proteins help stabilize the positions of the thick filaments.
  2. ***H zone.*** In a resting sarcomere the **H zone,** or *H band,* is a lighter region on either side of the M line. The H zone contains thick filaments but no thin filaments.
  3. ***Zone of overlap.*** In the **zone of overlap** thin filaments are found between the thick filaments. In this region each thin filament sits in a triangle formed by three thick filaments, and each thick filament is surrounded by six thin filaments.
- ***I Band.*** The **I band** extends from the A band of one sarcomere to the A band of the next sarcomere. **Z lines** mark the boundary lines between adjacent sarcomeres. The Z lines con-

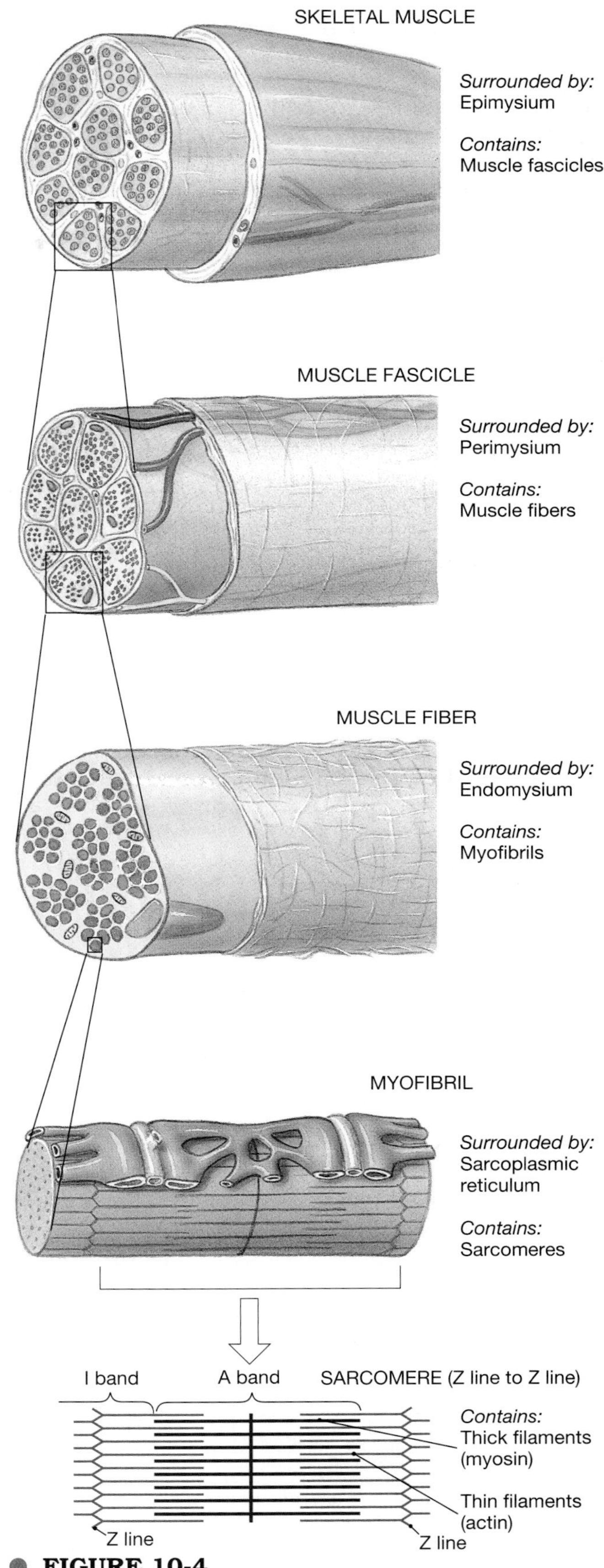
SKELETAL MUSCLE
Surrounded by:
Epimysium
Contains:
Muscle fascicles
MUSCLE FASCICLE
Surrounded by:
Perimysium
Contains:
Muscle fibers
MUSCLE FIBER
Surrounded by:
Endomysium
Contains:
Myofibrils
MYOFIBRIL
Surrounded by:
Sarcoplasmic reticulum
Contains:
Sarcomeres
I band
A band
SARCOMERE (Z line to Z line)
Contains:
Thick filaments (myosin)
Thin filaments (actin)
Z line
Z line

● **FIGURE 10-4**
Levels of Functional Organization in a Skeletal Muscle Fiber

sist of proteins, called *connectins,* that interconnect thin filaments associated with adjacent sarcomeres. From the Z lines at either end of the sarcomere, thin filaments extend toward the M line and into the zone of overlap.

Two transverse tubules encircle each sarcomere, and there are triads on each side of the M line in the region of the zone of overlap. As a result, calcium ions released by the SR enter the regions where thick and thin filaments can interact.

The entire muscle fiber has a banded appearance because the Z lines of adjacent myofibrils are aligned. Skeletal muscle tissue is also known as *striated muscle* because these bands, or *striations,* are visible with the light microscope. [p. 137]

Figure 10-4 reviews the levels of organization we have discussed so far. We will now proceed to the molecular level of organization and consider the structure of the myofilaments responsible for muscle contraction.

THIN FILAMENTS A typical thin filament is 5–6 nm in diameter and 1 μm in length. A single thin filament contains three different proteins: *F actin, tropomyosin,* and *troponin.*

- **F actin** is a twisted strand composed of 300–400 individual globular molecules of *G actin.* Each molecule of G actin contains an **active site** that can bind to a thick filament in much the same way that a substrate molecule binds to the active site of an enzyme. Under resting conditions myosin binding is prevented by the *troponin-tropomyosin complex.*
- Strands of **tropomyosin** (tro-po-MĪ-o-sin) cover the active sites and prevent actin-myosin interaction (Figure 10-5b). A single tropomyosin molecule is a protein strand that covers seven active sites. It is bound to one molecule of troponin midway along its length.
- A **troponin** (TRŌ-po-nin; *trope,* turning) molecule consists of three globular subunits:
  1. One subunit binds to tropomyosin, locking them together.
  2. A second subunit binds to G actin, holding the troponin-tropomyosin complex in position.
  3. The third subunit has a receptor that binds a calcium ion. In a resting muscle, intracellular calcium ion concentrations are very low, and that binding site is empty.

A contraction cannot occur unless there is a change in the position of the troponin-tropomyosin complex that exposes the active sites on F actin. The necessary change in position occurs when calcium ions bind to receptors on the troponin molecules.

At either end of the sarcomere, the thin filaments are attached to the Z line (Figure 10-5a). Although called a *line* because it looks like a dark line on the

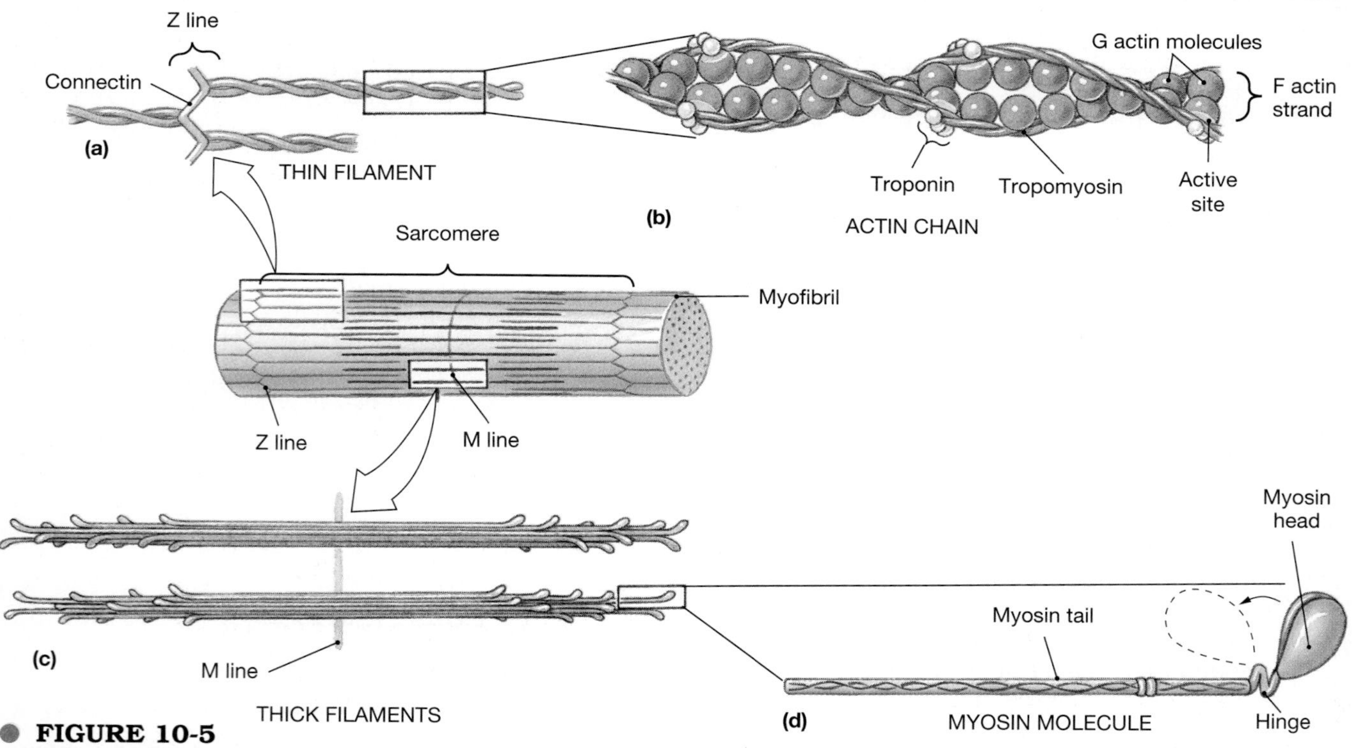
Z line
Connectin
(a)
THIN FILAMENT
G actin molecules
F actin strand
Troponin
Tropomyosin
Active site
(b)
ACTIN CHAIN
Sarcomere
Myofibril
Z line
M line
Myosin head
Myosin tail
(c)
M line
THICK FILAMENTS
(d)
MYOSIN MOLECULE
Hinge

FIGURE 10-5

Thick and Thin Filaments. (a) Gross structure of a thin filament, showing the attachment at the Z line. **(b)** The organization of G actin subunits in an F actin strand, and the position of the troponin-tropomyosin complex. **(c)** Structure of a thick filament, showing the orientation of the myosin molecules along the thick filaments. **(d)** Structure of a myosin molecule.

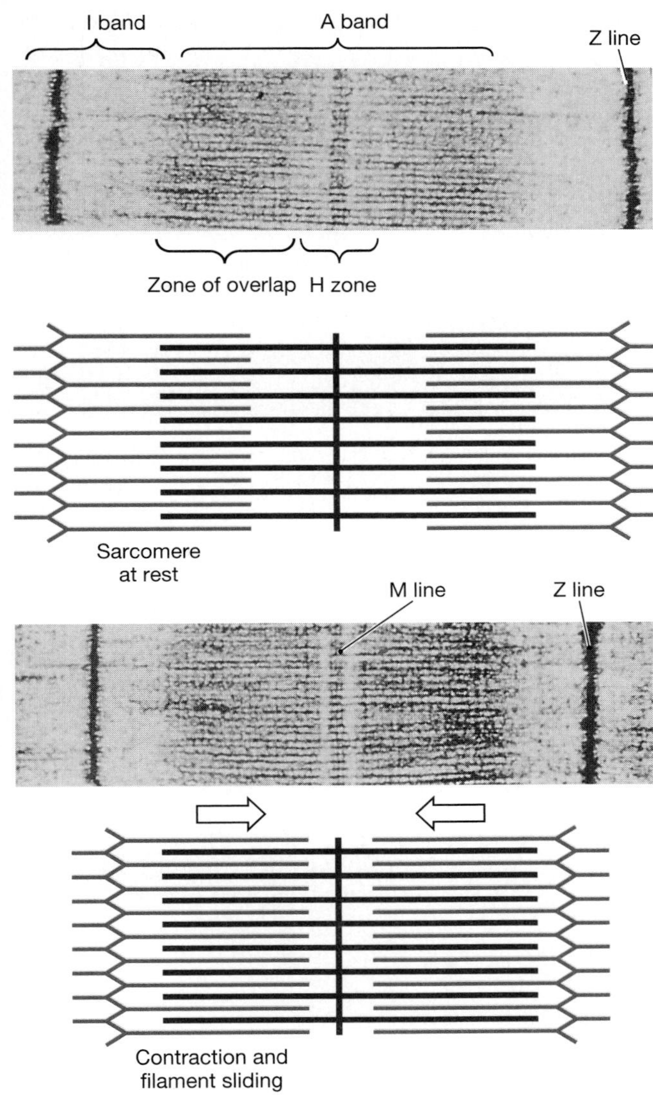
I band
A band
Z line
Zone of overlap
H zone
Sarcomere
at rest
M line
Z line
Contraction and
filament sliding

● **FIGURE 10-6**

Changes in the Appearance of a Sarcomere during Contraction of a Skeletal Muscle Fiber. During a contraction the A band stays the same width, but the Z lines move closer together and the I band gets smaller.

surface of the myofibril, in sectional view the Z line is more like an open meshwork (Figure 10-3d●). For this reason, the Z line is often called the *Z disc.*

THICK FILAMENTS Thick filaments (Figure 10-5c●) are 10–12 nm in diameter and 1.6 μm long. Each thick filament consists of roughly 500 myosin molecules. A single myosin molecule consists of a pair of myosin subunits twisted around one another (Figure 10-5d●). The long, attached **tail** is bound to other myosin molecules in the thick filament. The free globular **head** projects outward toward the nearest thin filament. Because myosin heads interact with thin filaments during a contraction, they are also known as **cross-bridges.** The connection between the head and the tail functions as a hinge that lets the head pivot at its base. When pivoting occurs, the head swings toward or away from the M line. As we will see in a later section, this pivoting is the key step in muscle contraction.

All of the myosin molecules are arranged with their tails pointing toward the center of the sarcomere (Figure 10-5c●). The H zone in the resting sarcomere corresponds to the region where there are no myosin heads. Outside the H zone, the myosin molecules are arranged in a spiral, with the myosin heads facing the G actin molecules on the surrounding thin filaments (Figures 10-3d and 10-5c●).

The Sliding Filament Theory

Figure 10-6

Several significant observations can be made while watching a skeletal muscle fiber contract. When contraction occurs, the H bands and I bands get smaller, the zones of overlap get larger, and the Z lines move closer together. The width of the A bands remains constant throughout the contraction. These observations make sense only if *the thin filaments are sliding toward the center of the sarcomere, alongside the thick filaments.* This sliding action is diagrammed in Figure 10-6●. This explanation for the physical changes that occur during the contraction process is called the **sliding filament theory.**

THE MUSCULAR DYSTROPHIES Abnormalities in the genes that code for structural and functional proteins in muscle fibers are responsible for a number of inherited diseases collectively known as the **muscular dystrophies** (DIS-tro-fēz). These conditions, which cause a progressive muscular weakness and deterioration, are the result of abnormalities in the sarcolemma or the structure of internal proteins. The best-known example is *Duchenne's muscular dystrophy,* which typically develops in male children from 3 to 7 years old. AM *The Muscular Dystrophies*

 How would severing the tendon attached to a muscle affect the ability of the muscle to move a body part?

 Why does skeletal muscle appear striated when viewed with a microscope?

 Where would you expect to find the greatest concentration of calcium ions in resting skeletal muscle?

Contraction of Skeletal Muscle

The sliding filament theory explains what happens to the sarcomere during a contraction. With that in mind, we will now follow the contraction sequence in greater detail.

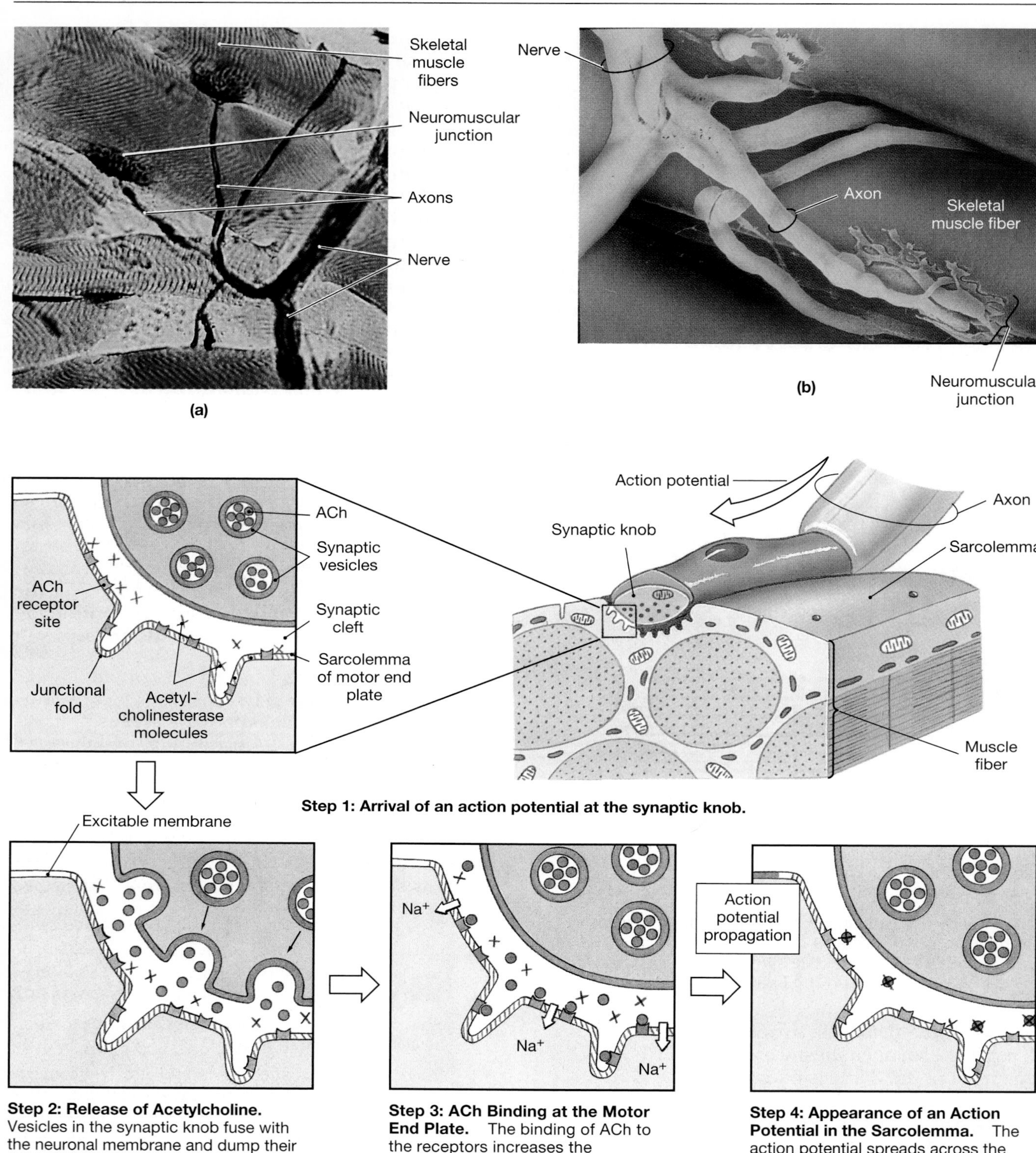

Skeletal muscle fibers
Neuromuscular junction
Axons
Nerve
(a)
Nerve
Axon
Skeletal muscle fiber
Neuromuscular junction
(b)
ACh
Synaptic vesicles
ACh receptor site
Synaptic cleft
Sarcolemma of motor end plate
Junctional fold
Acetylcholinesterase molecules
Action potential
Axon
Synaptic knob
Sarcolemma
Muscle fiber
Step 1: Arrival of an action potential at the synaptic knob.
Excitable membrane
Na+
Na+
Na+
Action potential propagation
Step 2: Release of Acetylcholine. Vesicles in the synaptic knob fuse with the neuronal membrane and dump their contents into the synaptic cleft.
Step 3: ACh Binding at the Motor End Plate. The binding of ACh to the receptors increases the membrane permeability to sodium ions. Sodium ions then enter the cell at an increased rate.
Step 4: Appearance of an Action Potential in the Sarcolemma. The action potential spreads across the membrane surface and travels down each of the transverse tubules, triggering the release of calcium ions at the terminal cisternae. While this occurs AChE removes the acetylcholine.
(c)

● **FIGURE 10-7**

Skeletal Muscle Innervation. (a) Several neuromuscular junctions are seen on the muscle fibers of this fascicle. (LM × 253) **(b)** An SEM showing additional details of the structure of the neuromuscular junction. **(c)** Diagrammatic view of the neuromuscular junction, showing major functional components, and steps in the transmission of action potentials across the neuromuscular junction.

❑ The Control of Skeletal Muscle Activity

Figure 10-7

Skeletal muscle fibers contract only under the control of the nervous system. Communication between the nervous system and skeletal muscle fibers occurs at specialized intercellular connections known as **neuromuscular junctions** (NMJs), or *myoneural junctions.* Several of these junctions are shown in Figure 10-7a,b●.

The Neuromuscular Junction

Figure 10-7

Each skeletal muscle fiber is controlled by a neuron at a single neuromuscular junction midway along its length. Figure 10-7c● summarizes key features of this structure. A single axon branches within the perimysium to form a number of fine branches. Each of these branches ends at an expanded **synaptic knob.** The cytoplasm of the synaptic knob contains mitochondria and vesicles filled with molecules of **acetylcholine** (as-ē-til-KŌ-lēn), usually abbreviated **ACh.** ACh is an example of a *neurotransmitter,* a chemical released by neurons to change the membrane properties of other cells. The release of ACh from the synaptic knob can result in changes in the sarcolemma that trigger the contraction of the muscle fiber.

A narrow space, the **synaptic cleft,** separates the synaptic knob from the opposing sarcolemmal surface. This surface, which contains membrane receptors that bind ACh, is known as the **motor end plate.** The motor end plate has deep creases, called *junctional folds,* that increase the membrane surface area, and thus the number of available ACh receptors. The synaptic cleft and sarcolemma also contain an enzyme, **acetylcholinesterase** (**AChE** or *cholinesterase*), that breaks down ACh.

When a neuron stimulates a muscle fiber, the process occurs in a series of steps.

Step 1: ***The Arrival of an Action Potential.*** The stimulus for ACh release is the arrival of an electrical impulse, or **action potential,** at the synaptic knob. An action potential is a sudden change in the transmembrane potential that is propagated along the length of the axon.

Step 2: ***The Release of ACh*** When that impulse reaches the synaptic knob, permeability changes in the membrane trigger the exocytosis of ACh into the synaptic cleft.

Step 3: ***ACh Binding at the Motor End Plate.*** ACh molecules diffuse across the synaptic cleft and bind to ACh receptors on the motor end plate's sarcolemmal surface. When ACh binding occurs, it changes the permeability of the sarcolemma to sodium ions. As you will recall from Chapter 3, the extracellular fluid contains a high concentration of sodium ions, whereas sodium ion concentrations inside the cell are very low. ∞ *[pp. 79, 82]* When the membrane permeability to sodium increases, sodium ions rush into the sarcoplasm. This influx continues until AChE removes the ACh from the receptors.

Step 4: ***Appearance of an Action Potential in the Sarcolemma.*** The sudden inrush of sodium ions results in the generation of an action potential in the sarcolemma at the edges of the motor end plate. This electrical impulse sweeps across the entire membrane surface and travels along each of the transverse tubules. The arrival of an action potential at the synaptic knob thus leads to the appearance of an action potential in the sarcolemma.

Excitation-Contraction Coupling

The link between the generation of an action potential in the sarcolemma and the start of a muscle contraction is called **excitation-contraction coupling.**

This coupling occurs at the triads. When an action potential reaches a triad, it triggers the release of calcium ions from the cisternae of the sarcoplasmic reticulum. The change in the permeability of the SR to calcium ions is temporary, lasting only around 0.03 second. Yet within a millisecond the calcium ion concentration in and around the sarcomere reaches 100 times resting levels. Because the cisternae are situated at the zones of overlap, where the thick and thin filaments interact, the effect of calcium release on the sarcomere is almost instantaneous. The rise in cytoplasmic calcium ion concentrations exposes the active sites along the thin filaments, initiating the contraction. The *contraction cycle* now begins.

INTERFERENCE WITH NEURAL CONTROL MECHANISMS Anything that interferes with either neural function or excitation-contraction coupling will cause muscular paralysis. Two examples are worth noting:

- Cases of *botulism* result from the consumption of contaminated canned or smoked food. The toxin, produced by bacteria, prevents the release of ACh at the synaptic terminals, leading to a potentially fatal muscular paralysis. AM *Botulism*
- The progressive muscular paralysis seen in *myasthenia gravis* results from the loss of ACh receptors at the junctional folds. The primary cause is a misguided attack on the ACh receptors by the immune system. Genetic factors play a role in predisposing individuals to develop this condition. AM *Myasthenia Gravis*

● **FIGURE 10-8**
Molecular Events of the Contraction Process

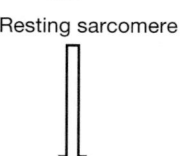

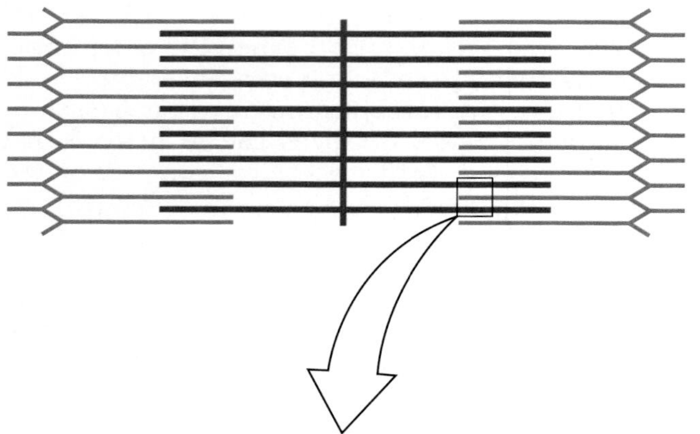

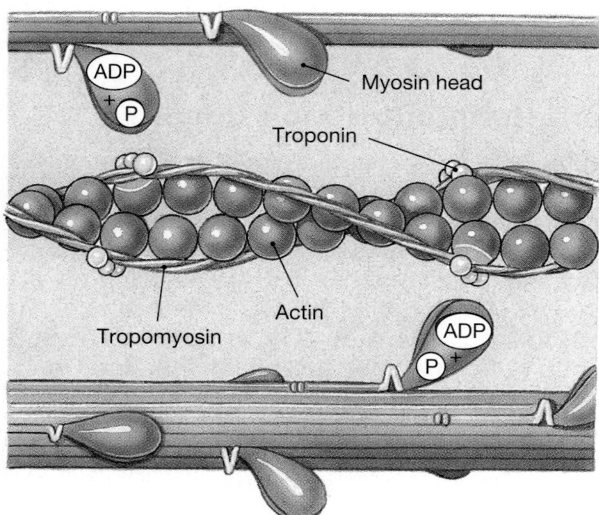
ADP
P
Myosin head
Troponin
Tropomyosin
Actin
ADP
P
Resting sarcomere

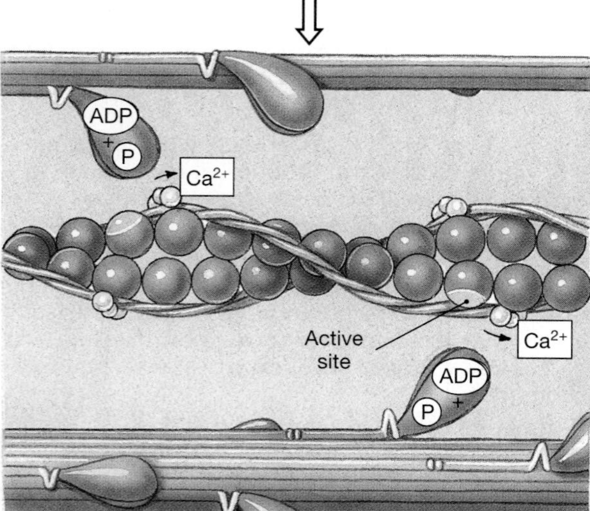
ADP
P
Ca2+
Active site
Ca2+
ADP
P
Step 1: Active-site exposure

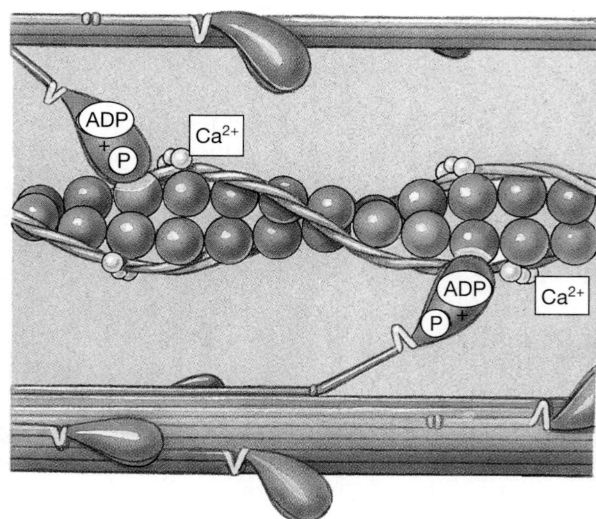
ADP
P
Ca2+
ADP
P
Ca2+
Step 2: Cross-bridge attachment

❑ The Contraction Cycle

Figure 10-8

Figure 10-8● details the molecular events that occur during the contraction process. In the resulting sarcomere, each cross-bridge is already "energized"—charged with the energy that will be used to power contraction. The cross-bridge incorporates the ATPase, an enzyme that can break down ATP. At the start of the contraction cycle, each cross-bridge has already split a molecule of ATP and stored the energy released in the process. The breakdown products, ADP and phosphate (PO_4^{3-}), remain bound to the cross-bridge.

The contraction process involves five interlocking steps:

Step 1: ***Active-site exposure.*** The calcium ions entering the sarcoplasm bind to troponin. This binding weakens the bond between the troponin complex and actin. The troponin molecule then changes position, pulling the tropomyosin molecule away from the active site (Figure 10-8●).

Step 2: ***Cross-bridge attachment.*** When the active sites are exposed, the myosin cross-bridges bind to them.

Step 3: ***Pivoting.*** In the resting sarcomere each cross-bridge points away from the M line. In this position the myosin head is "cocked" like the spring in a mousetrap. Cocking the myosin head required energy, and the energy was obtained by breaking down ATP into ADP and a phosphate group. In the cocked position, both the ADP and the phosphate are still bound to the myosin head. After cross-bridge attachment has occurred, the stored energy is released as the myosin head snaps toward the M line. When this occurs, the ADP and phosphate group are released.

Step 4: ***Cross-bridge detachment.*** When an ATP binds to the myosin head, it breaks the link between the active site on the actin molecule and the myosin head. The active site is now exposed and able to interact with another cross-bridge.

Step 5: ***Myosin activation.*** Myosin activation occurs when the free myosin head splits the ATP into ADP and a phosphate group. The energy released in this process is used to recock the myosin head. The entire cycle can now be repeated. If calcium ion concentrations remain elevated, and ATP reserves are sufficient, each myosin head will repeat this cycle about five times per second.

To appreciate the overall effect, imagine that you are pulling on a large rope. You are the myosin head, and the rope is a thin filament. You reach forward, grab the rope with both hands, and pull it toward you. This action corresponds to cross-bridge attachment and pivoting. You now release the rope, reach forward, and grab it once again. By repeating the cycle over and over, you can gradually pull in the rope.

Now consider the situation when several people are lined up, all pulling on the same rope. Each person reaches forward, grabs the rope, and pulls it. In the next cycle, each person reaches forward and grabs the rope where the person in front released it. This sequence corresponds to the way the myosin heads along the thick filament work together to pull a thin filament toward the center of the sarcomere.

❑ Relaxation

The duration of a contraction depends on the duration of stimulation at the neuromuscular junction. The ACh released after the arrival of a single action potential at the synaptic knob does not remain intact for long. ACh bound to the sarcolemma or free in the synaptic cleft is rapidly broken down and inactivated by *acetylcholinesterase*. Inside the muscle fiber, the permeability changes in the SR are also very brief. A contraction will therefore continue only if multiple additional action potentials arrive at the synaptic knob in rapid succession. When this occurs, the continual release of ACh into the synaptic cleft produces a series of action potentials in the sarcolemma that keeps the concentration of calcium ions in the sarcoplasm at a high level. Under these conditions the contraction cycle will be repeated over and over.

If just one action potential arrives at the NMJ, calcium ion concentrations in the sarcoplasm will quickly return to normal resting levels. Two mechanisms are involved in this process: (1) active

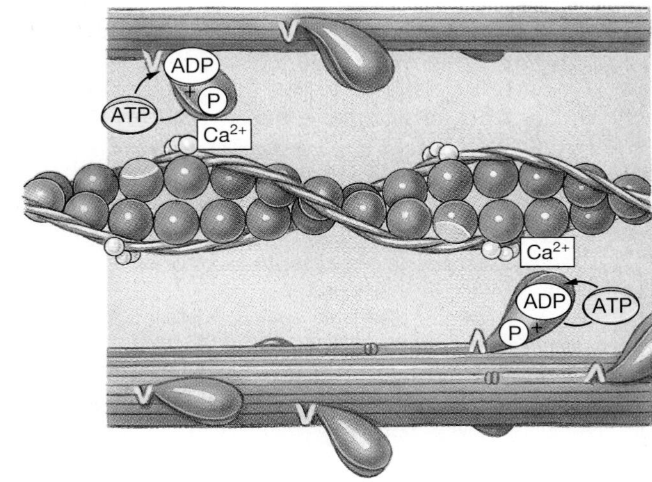
ADP
P
ATP
Ca^{2+}
Ca^{2+}
ADP
ATP
P

Step 5: Myosin reactivation

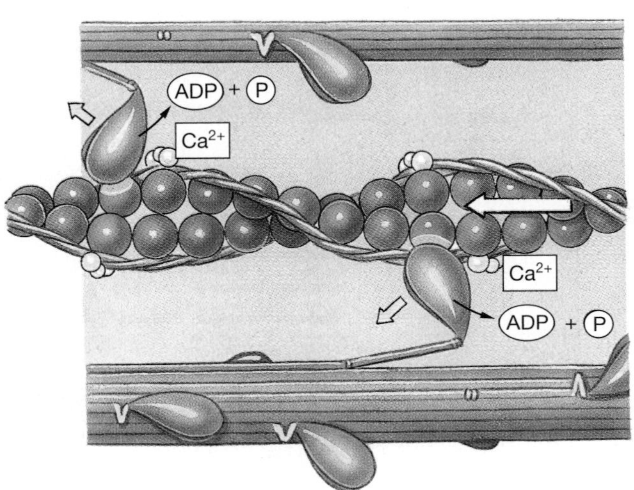
ADP + P
Ca^{2+}
Ca^{2+}
ADP + P

Step 3: Pivoting of myosin head

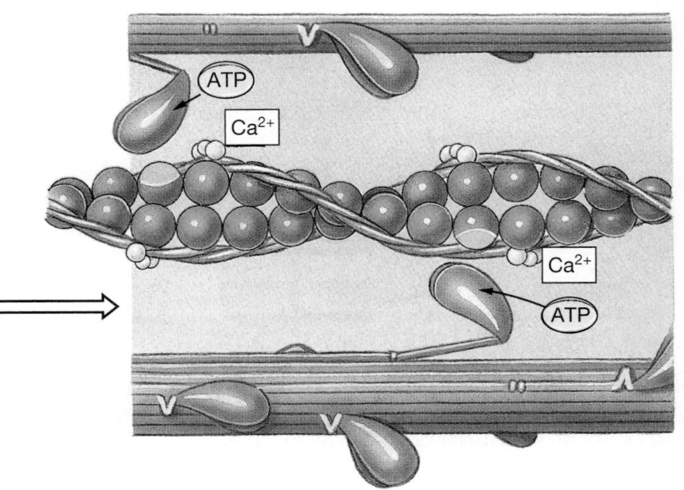
ATP
Ca^{2+}
Ca^{2+}
ATP

Step 4: Cross-bridge detachment

TABLE 10-1 A Summary of the Steps Involved in Skeletal Muscle Contraction

Key steps in the initiation of a contraction include:

1. At the neuromuscular junction, ACh released by the synaptic knob binds to receptors on the sarcolemma.
2. The resulting change in the transmembrane potential of the muscle fiber leads to the production of an action potential that spreads across its entire surface and reaches the triads via the transverse tubules.
3. The sarcoplasmic reticulum releases stored calcium ions, increasing the calcium concentration of the sarcoplasm in and around the sarcomeres.
4. Calcium ions bind to troponin, producing a change in the orientation of the troponin-tropomyosin complex that exposes active sites on the thin (actin) filaments.
5. Repeated cycles of cross-bridge binding, pivoting, and detachment occur, powered by the breakdown of ATP. These events produce filament sliding, and the muscle fiber shortens.

This process continues for a brief period, until:

6. Action potential generation ceases as ACh is removed by acetylcholinesterase.
7. The sarcoplasmic reticulum reabsorbs calcium ions, and the concentration of calcium ions in the sarcoplasm declines.
8. When calcium ion concentrations approach normal resting levels, the troponin-tropomyosin complex returns to its normal position. This change covers the active sites and prevents further cross-bridge interaction.
9. Without cross-bridge interactions, further sliding will not take place, and the contraction will end.

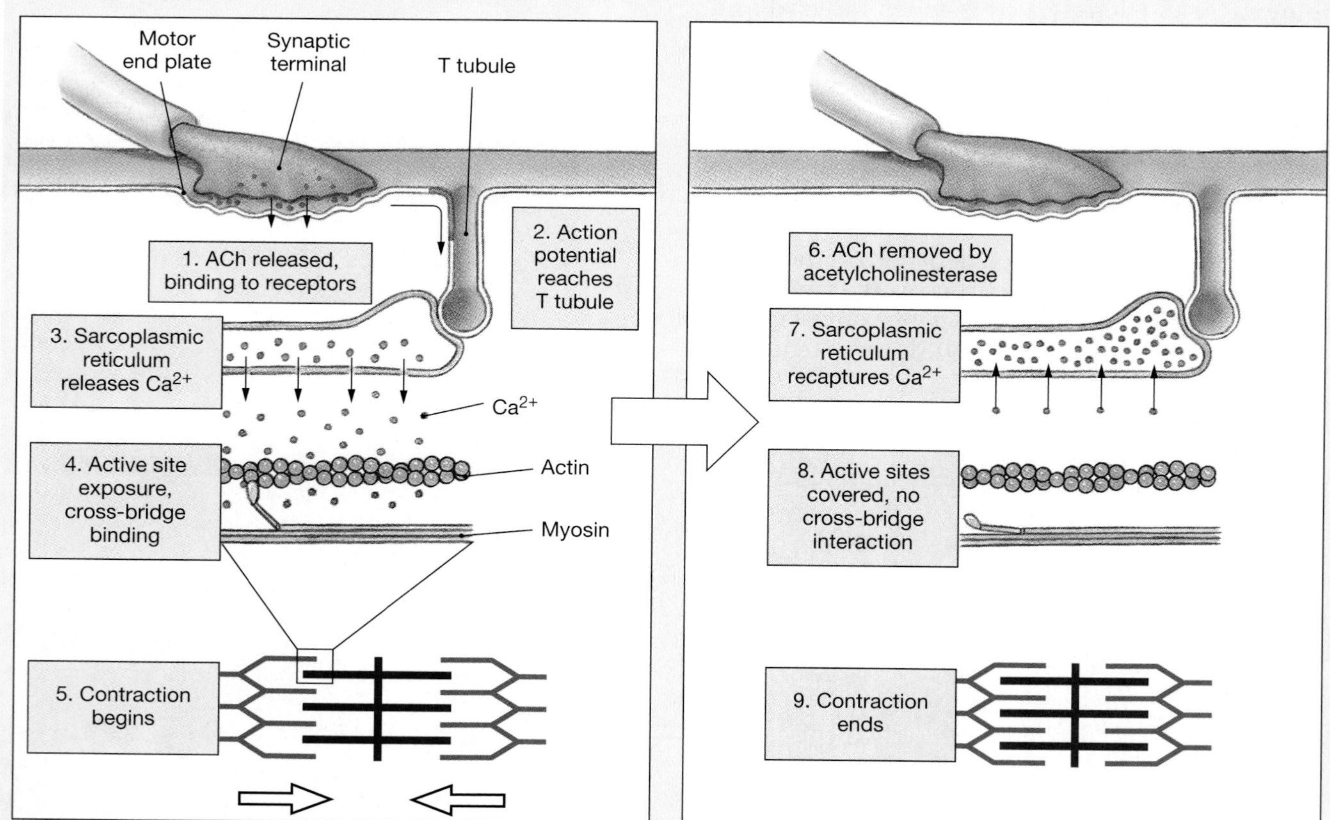
Motor end plate
Synaptic terminal
T tubule
1. ACh released, binding to receptors
2. Action potential reaches T tubule
3. Sarcoplasmic reticulum releases Ca2+
Ca2+
4. Active site exposure, cross-bridge binding
Actin
Myosin
5. Contraction begins
6. ACh removed by acetylcholinesterase
7. Sarcoplasmic reticulum recaptures Ca2+
8. Active sites covered, no cross-bridge interaction
9. Contraction ends

Steps in the initiation of a contraction.

Steps that end the contraction.

transport across the cell membrane into the extracellular fluid and (2) active transport into the sarcoplasmic reticulum. Of the two, transport into the sarcoplasmic reticulum is the more important. Virtually as soon as the calcium ions have been released, the SR returns to its normal permeability and begins actively absorbing calcium ions from the surrounding sarcoplasm. As calcium concentrations fall, (1) calcium ions detach from troponin, (2) troponin returns to its original position, and (3) the active sites are covered by tropomyosin. The contraction is now at an end.

Once the contraction has ended, the sarcomere does not automatically return to its original length. Sarcomeres shorten actively, but there is no mechanism for reversing the process. External forces must act on the contracted muscle fiber to stretch the myofibrils and sarcomeres to their original dimensions. These forces will be detailed in a later section.

Before proceeding further, let's step back and summarize the entire sequence of events from neural activation through excitation-contraction coupling to the completion of a contraction. Table 10-1 provides a summary of the contraction process, from ACh release to the end of the contraction.

RIGOR MORTIS When death occurs, circulation ceases and the skeletal muscles are deprived of nutrients and oxygen. Within a few hours, the skeletal muscle fibers have run out of ATP, and the sarcoplasmic reticulum becomes unable to pump calcium ions out of the sarcoplasm. Calcium ions diffusing into the sarcoplasm from the extracellular fluid or leaking out of the sarcoplasmic reticulum then trigger a sustained contraction. Without ATP, the cross-bridges cannot detach from the active sites, and skeletal muscles throughout the body become locked in the contracted position. Because all of the body's skeletal muscles are involved, the individual becomes "stiff as a board." This physical state, called **rigor mortis,** lasts until the lysosomal enzymes released by autolysis break down the myofilaments 15–25 hours later. The timing is dependent on environmental factors, such as temperature. Forensic pathologists can estimate the time of death on the basis of the degree of rigor mortis and the characteristics of the local environment.

❑ Length-Tension Relationships

Figure 10-9

When many people are pulling on a rope, the amount of tension produced is proportional to the number of people involved. In a muscle fiber, the amount of tension generated during a contraction depends on the number of cross-bridge interactions that occur in all of the sarcomeres along all of the myofibrils. The number of cross-bridge interactions is determined by the degree of overlap between thick and thin filaments. *When the muscle fiber is stimulated to contract, only myosin heads within the zone of overlap can bind to active sites and produce tension.* The tension produced by the intact muscle fiber can thus be related to the structure of an individual sarcomere (Figure 10-9●).

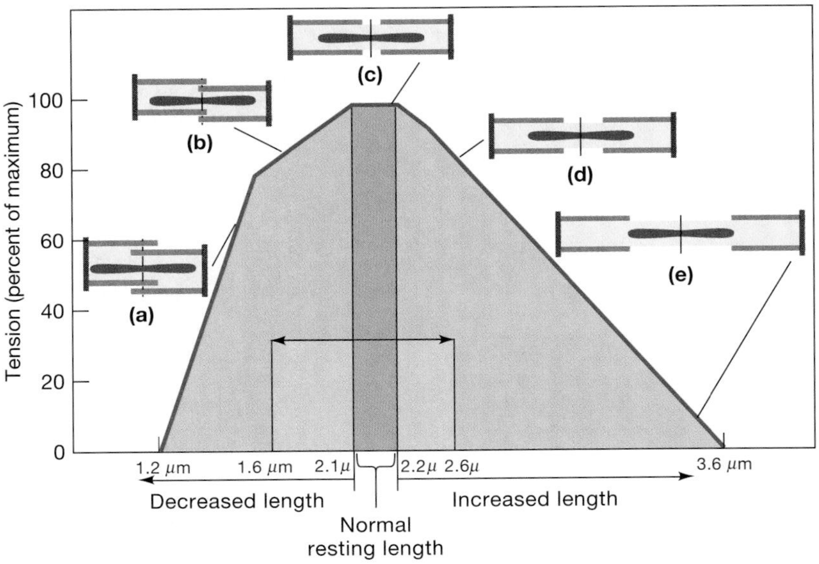

When the sarcomeres are as short as they can be, the thick filaments are jammed against the Z lines. Although cross-bridge binding can occur, the myosin heads cannot pivot, and no tension is produced (Figure 10-9a●). Even if the sarcomeres are somewhat longer, the thin filaments that extend across the center of the sarcomere collide with or overlap the thin filaments of the opposite side (Figure 10-9b●). This disruption of the normal arrangement of thick and thin filaments interferes with the binding of cross-bridges to active sites. The result is that little tension is produced when the muscle fiber is stimulated.

Within the optimal range of sarcomere lengths (Figure 10-9c●), the maximum number of cross-bridges can form, and the tension produced is highest. Any further increase in sarcomere length reduces the tension produced by reducing the size of the zone of overlap and the num-

Tension (percent of maximum)
100
80
60
40
20
0
(a)
(b)
(c)
(d)
(e)
1.2 μm
1.6 μm
2.1μ
2.2μ
2.6μ
3.6 μm
Decreased length
Increased length
Normal resting length

● **FIGURE 10-9**

The Effect of Sarcomere Length on Tension. When the sarcomeres are too short **(a,b),** contraction cannot occur because the thick filaments come in contact with the Z lines. If the sarcomeres are stretched too far, the zone of overlap is reduced **(d)** or disappears **(e),** and cross-bridge interactions are reduced or cannot occur. The tension produced reaches a maximum when the zone of overlap is large, but the thin filaments do not extend across the center of the sarcomere **(c).**

ber of potential cross-bridge interactions (Figure 10-9d●). When the zone of overlap disappears altogether, thin and thick filaments cannot interact (Figure 10-9e●). The muscle fiber cannot actively produce any tension, and a contraction cannot occur.

In summary, muscle fibers can contract most forcefully when stimulated over a relatively narrow range of resting lengths. The normal range of sarcomere lengths in the body, indicated in Figure 10-9●, is from 75 to 130 percent of the optimal length. The arrangement of skeletal muscles, connective tissues, and bones normally prevents extreme compression or excessive stretching that would correspond to Figure 10-9a or 10-9e●. For example, straightening the elbow stretches the biceps brachii muscle, but stabilizing ligaments and the bony structure of the elbow stop this movement before the muscle fibers stretch too far. During normal movements our muscle fibers perform over a broad range of intermediate lengths; the tension produced varies, depending on the initial length of the muscle fiber. During an activity such as walking, in which muscles contract and relax in a cyclical fashion, muscle fibers are stretched to a length very close to "ideal" before they are stimulated to contract.

- How would a drug that interferes with cross-bridge formation affect muscle contraction?
- What would you expect to happen to a resting skeletal muscle if the sarcolemma suddenly became very permeable to calcium ions?
- Predict what would happen to a muscle if the motor end plate failed to produce any acetylcholinesterase.

Muscle Mechanics

Now that you are familiar with the basic mechanisms of muscle contraction at the level of the individual muscle fiber, we can begin to examine the performance of *skeletal muscles*, organs of the muscular system. In this section we will consider the coordinated contractions of an entire population of muscle fibers.

The amount of tension produced by an individual muscle fiber depends solely on the number of cross-bridge interactions. If a muscle fiber at a given resting length is stimulated to contract, it will always produce the same amount of tension. There is no mechanism to regulate the amount of tension produced in that contraction: The muscle fiber is either "ON" (producing tension) or "OFF" (relaxed). This feature of muscle mechanics is known as the **all-or-none principle.** The amount of tension produced in the skeletal muscle *as a whole* is therefore determined by (1) the frequency of stimulation and (2) the number of muscle fibers stimulated.

The Frequency of Muscle Stimulation

Figures 10-10, 10-11

A **twitch** is a single stimulus-contraction-relaxation sequence in a muscle fiber. The duration of a single twitch is variable. Twitches in an eye muscle fiber can be as brief as 10 msec, but a twitch in a calf muscle fiber lasts around 100 msec. Figure 10-10a● is a graph, or **myogram,** of the development of tension in various muscles dur-

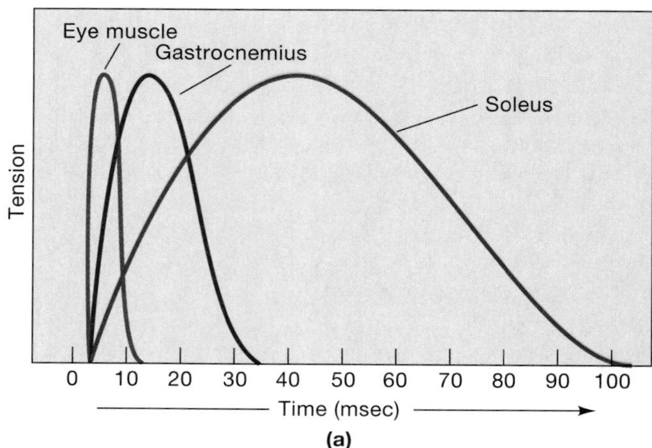
Eye muscle
Gastrocnemius
Soleus
Tension
0 10 20 30 40 50 60 70 80 90 100
Time (msec)
(a)

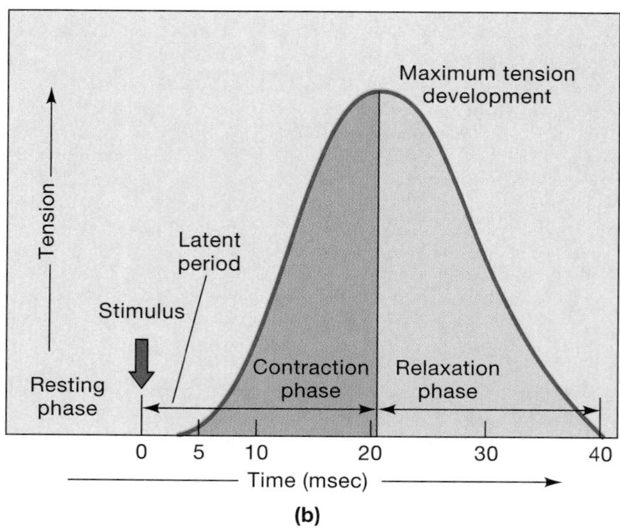
Maximum tension development
Tension
Latent period
Stimulus
Resting phase
Contraction phase
Relaxation phase
0 5 10 20 30 40
Time (msec)
(b)

● **FIGURE 10-10**
The Twitch and Development of Tension. (a) Myogram showing differences in the time course of a twitch contraction in different skeletal muscles in the body. **(b)** Details of the time course of a single isometric twitch contraction in the gastrocnemius muscle. Note the presence of a latent period, which corresponds to the time needed for the conduction of action potential and the subsequent release of calcium ions by the sarcoplasmic reticulum.

ing a twitch contraction. Twitch contractions may be as brief as 7.5 msec or as long as 100 msec. Figure 10-10b● details the phases of a 40-msec twitch in the *gastrocnemius* muscle, a prominent calf muscle.

A single twitch can be divided into a *latent period*, a *contraction phase*, and a *relaxation phase*.

- The **latent period** begins at stimulation and typically lasts about 2 msec. Over this period the action potential sweeps across the sarcolemma, and calcium ions are released by the sarcoplasmic reticulum. During the latent period the muscle fiber does not produce tension because the contraction cycle has yet to begin.
- In the **contraction phase** tension rises to a peak. Throughout this period the cross-bridges are interacting with the active sites on the actin filaments.
- The **relaxation phase** then continues for another 25 msec; during this period muscle tension falls to resting levels as the cross-bridges detach.

A single stimulation produces a single twitch, but twitches in a skeletal muscle do not accomplish anything useful. *All normal activities involve sustained muscle contractions.* The mechanism involved can most easily be understood by examining the responses of an isolated skeletal muscle when stimulated under laboratory conditions. Under these conditions the myogram of tension production changes dramatically as the rate of stimulation increases (Figure 10-11●).

WAVE SUMMATION AND INCOMPLETE TETANUS If a second stimulus arrives before the relaxation phase has ended, a second, more powerful contraction occurs (Figure 10-11a●). The addition of one twitch to another in this way constitutes the **summation of twitches,** or simply **wave summation.** The frequency of stimulation necessary to produce wave summation depends on the duration of a single twitch in that particular muscle fiber. For example, if a twitch lasts 20 msec, stimulation at a frequency of less than 50 per second will produce individual twitches. Stimulation at a higher frequency will result in wave summation.

If you continue to stimulate the muscle, never allowing it to relax completely, tension will rise to a peak (Figure 10-11b●). A muscle producing peak tension during rapid cycles of contraction and relaxation is said to be in **incomplete tetanus** (*tetanos*, convulsive tension).

COMPLETE TETANUS **Complete tetanus** can be obtained by increasing the rate of stimulation until the relaxation phase is completely eliminated

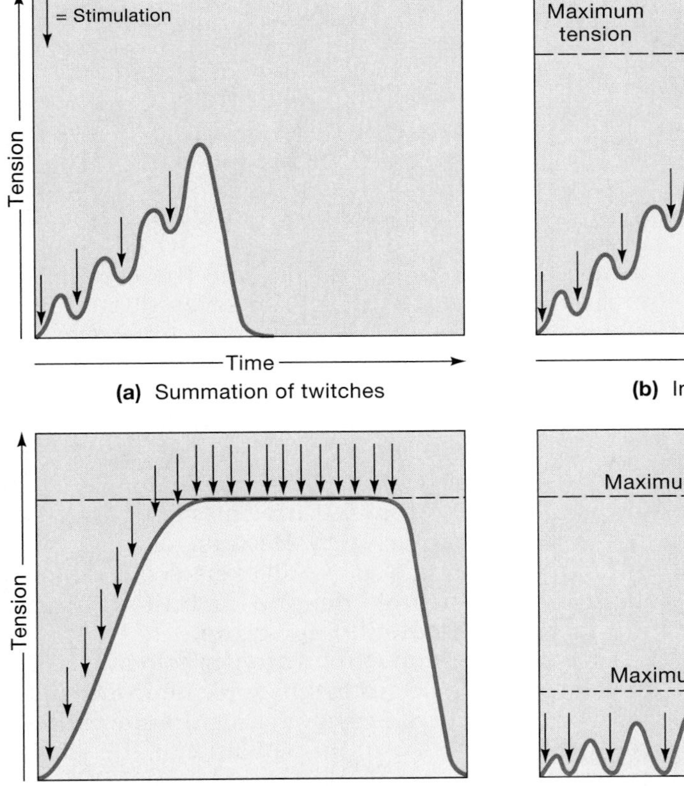
= Stimulation
Tension
Time
(a) Summation of twitches
Tension
Time
(c) Complete tetanus

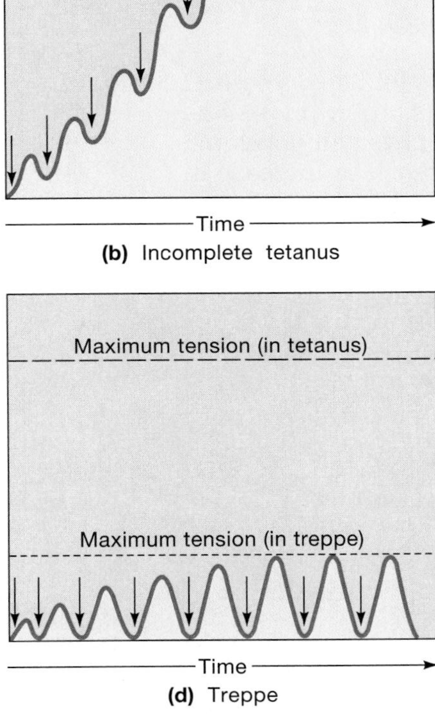
Maximum tension
Time
(b) Incomplete tetanus
Maximum tension (in tetanus)
Maximum tension (in treppe)
Time
(d) Treppe

● **FIGURE 10-11**
Effects of Repeated Stimulations. (a) Summation of twitches occurs when successive stimuli arrive before relaxation has been completed. **(b)** Incomplete tetanus occurs if the rate of stimulation increases further. Tension production will rise to a peak, and the periods of relaxation will be very brief. **(c)** In complete tetanus the frequency of stimulation is so high that the relaxation phase has been entirely eliminated, and tension plateaus at maximal levels. **(d)** Treppe is an increase in peak tension following repeated stimuli delivered shortly after the completion of the relaxation phase of each twitch.

Tetanus

Children are often told to "watch out" for rusty nails. Parents are not worrying about the rust or the nail, but about infection with a very common bacterium, *Clostridium tetani.* This bacterium can cause **tetanus,** a disease that has no relationship to the normal muscle response to neural stimulation. The *Clostridium* bacteria, although found virtually everywhere, can thrive only in tissues that contain abnormally low amounts of oxygen. For this reason a deep puncture wound, such as that from a nail, carries a much greater risk of producing tetanus than a shallow, open cut that bleeds freely.

When active in body tissues, these bacteria release a powerful toxin that affects the central nervous system. Motor neurons are particularly sensitive to it, and their stimulation produces a sustained, powerful contraction of skeletal muscles throughout the body. This toxin also has a direct effect on skeletal muscle fibers, producing contraction even in the absence of neural commands.

After exposure to the bacteria, the incubation period (the time before symptoms develop) is usually less than 2 weeks. The most common complaints are headache, muscle stiffness, and difficulty in swallowing. Because it soon becomes difficult to open the mouth, this disease is also called "lockjaw." Widespread muscle spasms usually develop within 2–3 days of the initial symptoms, and they often continue for a week before subsiding. After 2–4 weeks, symptoms in surviving patients disappear with no aftereffects.

Although severe tetanus has a 40–60 percent mortality rate, immunization is effective in preventing the disease. There are approximately 500,000 cases of tetanus worldwide each year, but only about 100 of them occur in the United States, thanks to an effective immunization program. (Most readers will have had "tetanus shots" or "tetanus boosters" sometime within the last 10 years.) Severe symptoms in unimmunized patients can be prevented by early administration of an antitoxin, usually *human tetanus immune globulin.* Such treatment does not reduce symptoms that have already appeared, however, and there is no generally effective treatment for this disease.

(Figure 10-11c●). In complete tetanus the action potentials are arriving so rapidly that the sarcoplasmic reticulum does not have time to reclaim the calcium ions. The high calcium ion concentration in the cytoplasm prolongs the state of contraction, making it continuous. *Virtually all normal muscular contractions involve the complete tetanus of the participating muscle fibers.*

TREPPE If a skeletal muscle is stimulated a second time immediately after the relaxation phase has ended, the contraction that occurs will develop a slightly higher maximum tension. The increase in peak tension indicated in Figure 10-11d● will continue over the first 30–50 stimulations. Thereafter the amount of tension produced will remain constant at roughly 25 percent of the maximal tension that would be produced in a tetanic contraction. Because the tension rises in stages, like the steps in a staircase, this phenomenon is called **treppe** (TREP-e), a German word meaning "stairs." The rise is accompanied by a gradual increase in the concentration of calcium ions in the cytoplasm, in part because the ion pumps in the sarcoplasmic reticulum are unable to recapture them in the time between stimulations.

Motor Units and Tension Production

Figure 10-12

We have a remarkable ability to control the amount of pull exerted by our skeletal muscles. During a normal contraction, tension rises smoothly, not jerkily, because activated muscle fibers are responding in complete tetanus. The *total force* exerted by the skeletal muscle depends on how many muscle fibers are activated.

A typical skeletal muscle contains thousands of muscle fibers. Although some individual motor neurons control a single muscle fiber, most control hundreds or thousands of muscle fibers. All of the muscle fibers controlled by a single motor neuron constitute a **motor unit.** The size of a motor unit is an indication of how fine the control of movement can be. In the muscles of the eye, where precise control is extremely important, a motor neuron may control two or three muscle fibers. We have much less precise control over our leg muscles, where more than 2000 muscle fibers respond to the call of a single motor neuron. As indicated in Figure 10-12●, the muscle fibers of each motor unit are intermingled with those of other motor units. Because of this intermingling, the direction of pull exerted on the tendon does not change despite variations in the numbers of activated motor units.

When a decision is made to perform a specific movement, specific groups of motor neurons within the central nervous system are stimulated. The contraction begins with the activation of the smallest motor units in the stimulated muscle. These motor units usually contain slow muscle fibers. Over time, larger motor units containing fast mus-

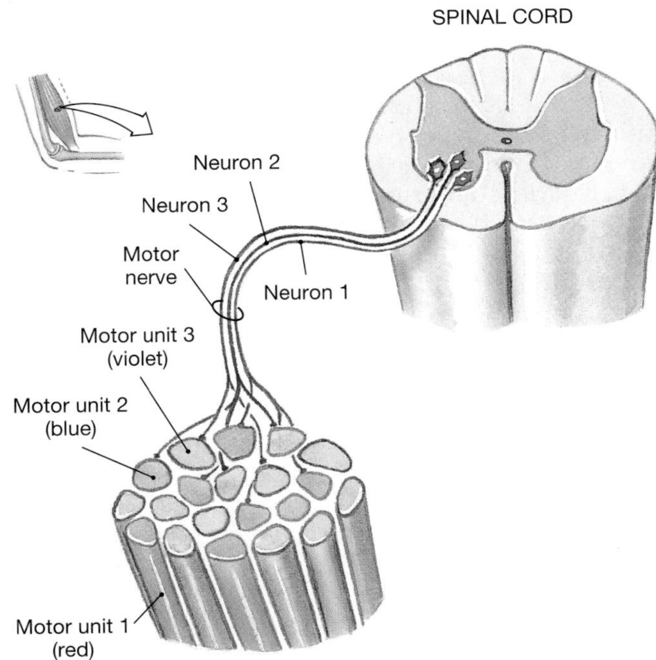
SPINAL CORD
Neuron 2
Neuron 3
Motor nerve
Neuron 1
Motor unit 3 (violet)
Motor unit 2 (blue)
Motor unit 1 (red)

● **FIGURE 10-12**
Arrangement of Motor Units within a Skeletal Muscle. Muscle fibers of different motor units are intermingled, so the net distribution of force applied to the tendon remains constant even when individual muscle groups cycle between contraction and relaxation.

cle fibers are activated, and tension production rises steeply. The smooth but steady increase in muscular tension produced by increasing the number of active motor units is called **recruitment,** or **multiple motor unit summation.**

Peak tension production occurs when all of the motor units in the muscle are contracting in complete tetanus. Such powerful contractions do not last long, however, because the individual muscle fibers soon use up their available energy reserves. During a sustained tetanic contraction motor units are activated on a rotating basis, so that some of them are resting and recovering while others are actively contracting. This "relay team" approach lets each motor unit recover somewhat before it is stimulated again. As a result our muscles usually contract for sustained periods at slightly reduced tension, rather than for brief periods at peak tension.

MUSCLE TONE Some of the motor units within any particular muscle are always active, even when the entire muscle is not contracting. Their contractions do not produce enough tension to cause movement, but they do tense and firm the muscle. This resting tension in a skeletal muscle is called **muscle tone.** A muscle with little muscle tone appears limp and flaccid, whereas one with moderate muscle tone is quite firm and solid. The identity of the stimulated motor units changes constantly, so that a constant tension in the attached tendon is maintained, but individual muscle fibers can relax.

Resting muscle tone stabilizes the position of bones and joints. For example, in muscles involved with balance and posture, enough motor units are stimulated to produce the tension needed to maintain body position. Muscle tone also helps prevent sudden, uncontrolled changes in the position of bones and joints. In addition to bracing the skeleton, the elastic nature of muscles and tendons lets skeletal muscles act as shock absorbers that cushion the impact of a sudden bump or shock.

Scattered within each skeletal muscle are clusters of specialized muscle fibers. These are sensory structures called **muscle spindles.** The individual *spindle fibers* are monitored by sensory neurons that deliver information about the length of the spindle fiber to the central nervous system. A *reflex* is an automatic, involuntary motor response to sensory stimulation. Reflexes triggered by the distortion of muscle spindles adjust the tone of skeletal muscles. Such reflexes play an important role in the automatic adjustment of body position and posture, a topic we will explore in more detail in Chapter 13.

The activity and sensitivity of muscle spindles increase with exercise training, producing an increase in background muscle tone. Heightened muscle tone accelerates the recruitment process during a voluntary contraction, because some of the motor units are already stimulated. Strong muscle tone also makes skeletal muscles appear firm and well-defined, even at rest.

Isotonic and Isometric Contractions

Figure 10-13

Muscle contractions may be classified as *isotonic* or *isometric* on the basis of the pattern of tension production.

ISOTONIC CONTRACTIONS In an **isotonic contraction** (*iso-*, equal + *tonos*, tension), tension rises to a plateau that extends until relaxation occurs. Consider Figure 10-13a●. A maximally stimulated skeletal muscle 1 cm^2 in cross-sectional area can develop roughly 4 kg of force. If we hang a 2-kg weight from that muscle and stimulate it, the muscle as a whole will shorten. Before the contraction can occur, the cross-bridges must produce enough tension to overcome the **resistance,** in this case a 2-kg weight. Tension in the muscle builds until it exceeds the amount of resistance, and then the muscle shortens. As it shortens, the tension in the muscle remains constant, at a value that just exceeds the applied resistance (Figure 10-13b●). Lifting an object off a desk, walking, running, and so forth involve isotonic contractions of this kind.

There are two different types of isotonic contractions: *concentric* and *eccentric.* In a **concentric contraction,** the muscle tension exceeds the resistance, and the muscle shortens, as in the examples given above. In an **eccentric contraction,** the tension developed is less than the resistance, and the muscle is stretched by the resistance. For example, you would use eccentric contractions of the biceps brachii when slowly lowering a book onto a desk.

ISOMETRIC CONTRACTIONS Now suppose we attach a 6-kg weight to the muscle and stimulate it, as in Figure 10-13c●. Although cross-bridges form, and tension rises to peak values, the muscle cannot overcome the resistance of the weight (Figure 10-13d●). This kind of contraction, in which tension rises but the resistance does not move, is called **isometric** (*metric,* measure). Examples of isometric contractions include holding a heavy weight above the ground, pushing against a locked door, or trying to pick up a car. These are rather unusual movements. However, many of the reflexive muscle contractions that keep our bodies upright when standing or sitting involve the isometric contractions of muscles that oppose gravity.

Normal daily activities therefore involve a combination of isotonic and isometric muscular contractions. As you sit reading this text, isometric contractions of postural muscles stabilize your vertebrae and maintain your upright position. When you next turn a page, the movements of your arm, forearm, hand, and fingers are produced by isotonic contractions.

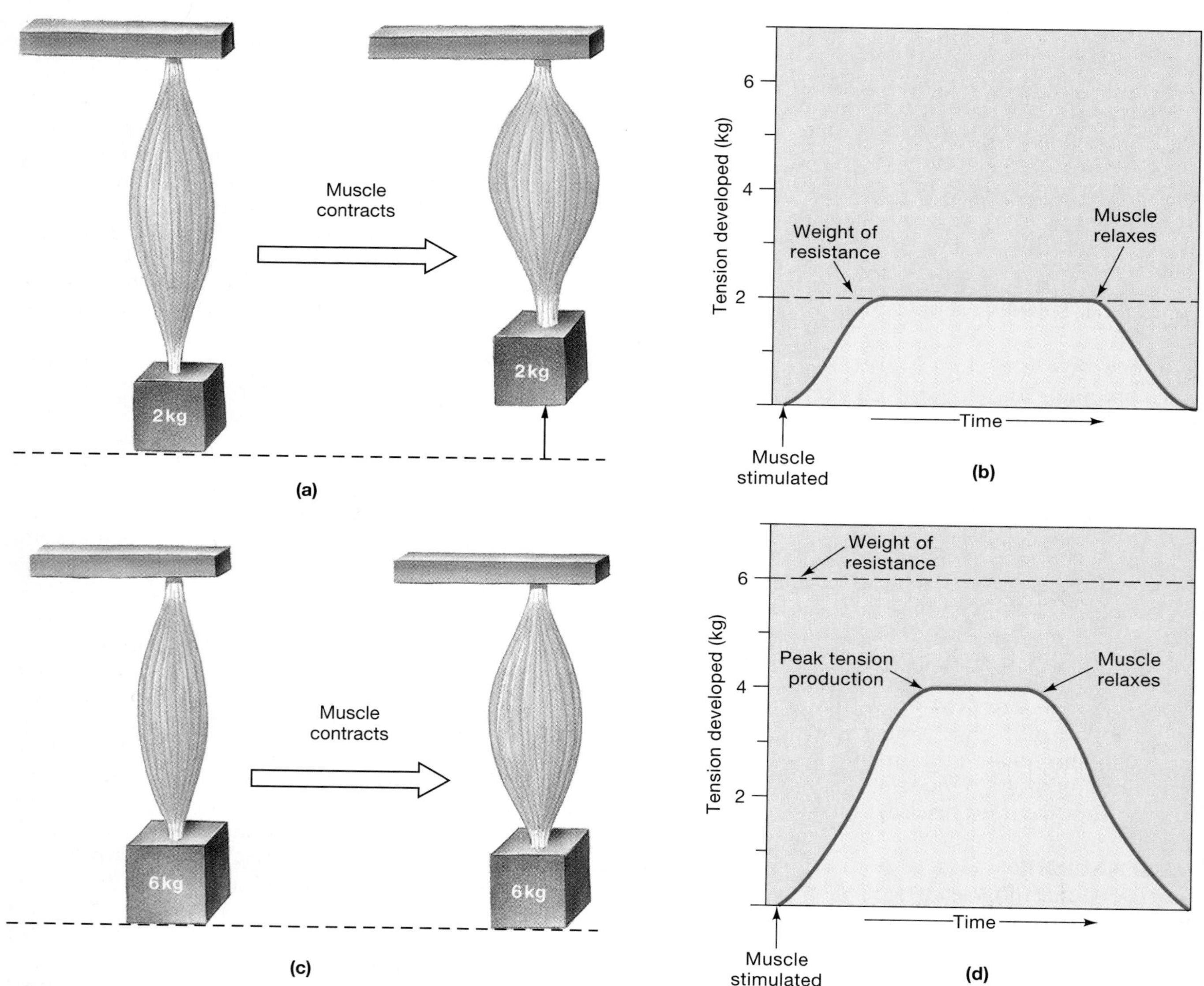
Muscle contracts
2kg
2kg
(a)
Tension developed (kg)
6
4
2
Weight of resistance
Muscle relaxes
Time
Muscle stimulated
(b)
Muscle contracts
6kg
6kg
(c)
Weight of resistance
Peak tension production
Muscle relaxes
Time
Muscle stimulated
(d)

● **FIGURE 10-13**

Isotonic and Isometric Contractions. (a,b) A muscle attached to a weight is stimulated and develops enough tension to lift the weight. Tension remains constant for the duration of the contraction, although the length of the muscle changes. This is an example of isotonic contraction. **(c,d)** If the muscle is attached to a larger weight, on stimulation, tension will rise to a peak, but the muscle will be unable to shorten. This is an example of isometric contraction.

Muscle Relaxation and the Return to Resting Length

As noted above, there is no active mechanism for muscle fiber elongation; contraction is active, but relaxation is passive. After a contraction, a muscle fiber returns to its original length through a combination of *elastic forces, opposing muscle contractions,* and *gravity.*

ELASTIC FORCES Intracellular structures and the extracellular fibers of the endomysium, perimysium, epimysium, and tendons are flexible and elastic. When a muscle fiber contracts, the extracellular fibers are stretched, and intracellular organelles are distorted. Some of the tension generated in a contraction is used to stretch these components, which are called *series elastic elements*. When the tension is relieved, these elements recoil to their original dimensions. Some of the energy initially "spent" in stretching the series elastic elements is recovered as they recoil. When a muscle contraction ends, the recoil of the series elastic elements helps to return the muscle fiber to its original resting length.

During an isotonic contraction, the skeletal muscle stretches the elastic components before the resistance begins to move. Once the resistance begins moving, the tension remains stable until the contraction ends. During an isometric contraction, the resistance does not move, but the elastic components are stretched, and the muscle fibers shorten to some degree. That is why a muscle undergoing isometric contraction bulges and becomes more prominent.

OPPOSING MUSCLE CONTRACTIONS Elastic forces gradually stretch relaxed muscle fibers toward their normal resting lengths. More rapid returns to resting length result from the contraction of opposing muscles, sometimes aided by the pull of gravity. For an example, consider the muscles of the arm that flex or extend the elbow. Contraction of the biceps brachii muscle on the anterior part of the arm flexes the elbow; contraction of the triceps brachii muscle on the posterior part of the arm extends the elbow. When the biceps brachii contracts, the triceps brachii is stretched. When the biceps brachii relaxes, contraction of the triceps brachii extends the elbow and stretches the muscle fibers of the biceps brachii to their original length.

GRAVITY Gravity may assist opposing muscle groups in returning a muscle to resting length after a contraction. For example, imagine the biceps brachii fully contracted with the elbow pointed at the ground. When the muscle relaxes, gravity will pull the forearm down and stretch the muscle. Although gravity can provide assistance in stretching muscles, some active muscle tension is needed to control the rate of movement and prevent damage to the joint. In the example above, eccentric contraction of the biceps brachii could control the movement.

Muscle Length, Tension, and Speed of Contraction

Figure 10-14

We have already noted that for any individual muscle fiber, there is a relationship between resting length and tension production. When we consider an entire skeletal muscle, the situation is complicated by the presence of series elastic elements. To lengthen a skeletal muscle, the series elastic elements must be stretched. Their resistance to stretching, called *passive tension,* increases as the length of the muscle increases. The tension measured when the muscle is stimulated therefore represents the sum of (1) the passive tension at that resting length and (2) the *active tension* generated by the myofibrils. Figure 10-14a● indicates the active and passive tension generated in the triceps brachii at various resting lengths. Under normal circumstances skeletal muscles are stimulated to contract when their length is within the range where the active tension is highest.

As you probably are aware, you can lift a light object more rapidly than a heavy one. That is because there is an inverse relationship between the amount of resistance and the speed of contraction (Figure 10-14b●). If the resistance is less than the peak tension production of the muscle, an isotonic, concentric contraction will occur, and the muscle will shorten. The heavier the resistance, the slower the contraction. An isometric contraction occurs if the resistance is equal to the tension produced during the contraction. An eccentric isotonic contraction occurs if the resistance is heavier than the tension produced by the muscle.

For each muscle there is an optimal combination of tension and speed for any given resistance. You are probably already aware of this fact, if you have ever ridden a 10-speed bicycle. When you are cruising along comfortably, your thigh and leg muscles are working at an optimal combination of speed and tension. When you come to a hill, the resistance increases. Your muscles must now develop more tension, and they move more slowly; they are no longer working at optimal efficiency. You then shift to a lower gear. The load on your muscles decreases, the speed increases, and the muscles are once again working efficiently.

Why is it difficult to contract a muscle that has been overstretched?

During treppe, why does tension in a muscle gradually increase even though the stimulus strength and frequency are constant?

Is it possible for a skeletal muscle to contract without shortening? Explain.

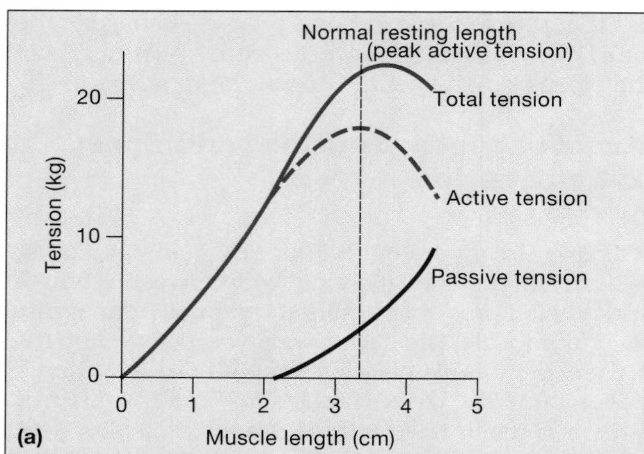
Normal resting length
(peak active tension)
Total tension
Active tension
Passive tension
Tension (kg)
20
10
0
0 1 2 3 4 5
(a)
Muscle length (cm)

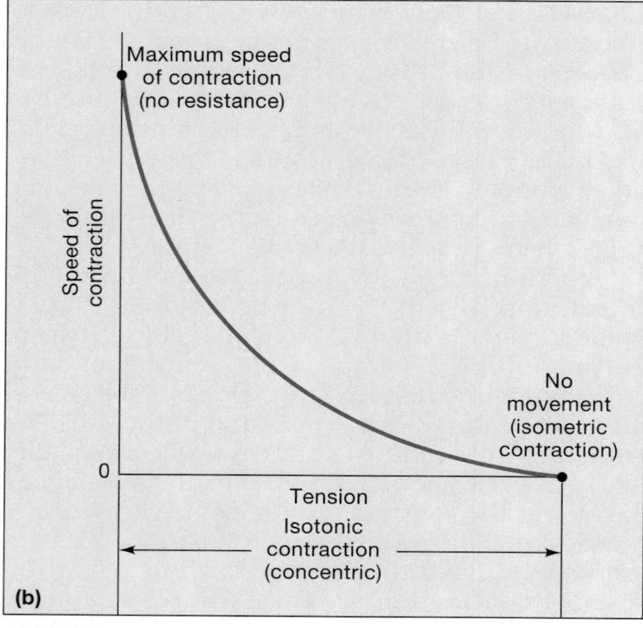
Maximum speed
of contraction
(no resistance)
Speed of contraction
No
movement
(isometric
contraction)
0
Tension
Isotonic
contraction
(concentric)
(b)

● **FIGURE 10-14**
Muscle Tension, Length, and Speed of Contraction. **(a)** A muscle produces a maximum amount of active tension over a relatively narrow range of lengths. Compare with Figure 10-9. Passive tension, which reflects primarily the presence of series elastic elements, increases as the muscle exceeds optimal lengths. **(b)** As the amount of tension increases, the speed of the contraction decreases. When the resistance equals the maximum tension produced by the muscle, speed is zero and the contraction is isometric.

Energetics of Muscular Activity

A single muscle fiber may contain 15 billion thick filaments and, when actively contracting, each thick filament breaks down roughly 2500 ATP molecules per second. Because even a small skeletal muscle contains thousands of muscle fibers, the ATP demands of a contracting skeletal muscle are enormous. In practical terms, the demand for ATP in a contracting muscle fiber is so high that it would be impossible to have all of the necessary energy available as ATP before the contraction begins. Instead, a resting muscle fiber contains only enough ATP and other high-energy compounds to sustain a contraction until additional ATP can be generated. Throughout the rest of the contraction the muscle fiber will generate ATP at roughly the same rate as it is used.

ATP and CP Reserves

The primary function of ATP is the transfer of energy from one location to another, rather than the long-term storage of energy. At rest a skeletal muscle fiber produces more ATP than it needs, and under these conditions ATP transfers energy to another high-energy compound, **creatine phosphate (CP),** or *phosphorylcreatine.* This reaction can be summarized as:

ATP + creatine → ADP + creatine phosphate

During a contraction each myosin cross-bridge breaks down ATP, producing ADP and a phosphate group. The energy stored in creatine phosphate is then used to "recharge" ADP, converting it back to ATP through the reverse reaction:

ADP + creatine phosphate → ATP + creatine

The enzyme that facilitates this reaction is called **creatine phosphokinase (CPK or CK)**. When muscle cells are damaged, CPK leaks across the cell membranes and into the circulation. Thus a high blood concentration of CPK usually indicates serious muscle damage.

The energy reserves of a representative muscle fiber are indicated in Table 10-2. A resting skeletal

TABLE 10-2 Sources of Energy Stored in a Typical Muscle Fiber

Energy Stored as	*Utilized through*	*Initial Quantity*	*Number of Twitches Supported by Each Energy Source Alone*	*Duration of Isometric Tetanic Contraction Supported by Each Energy Source Alone (in seconds)*
ATP	ATP → ADP + P	3 mmol	10	2
CP	CP + ADP → ATP + C	20 mmol	70	15
Glycogen	Glycolysis (anaerobic)	100 mmol	670	130
	Aerobic respiration		12,000	2400 (40 min)

muscle fiber contains about six times as much creatine phosphate as ATP. But when a muscle fiber is undergoing a sustained contraction, it takes only around 17 seconds to exhaust these energy reserves, and the muscle fiber must then rely on other mechanisms to convert ADP to ATP.

ATP Generation

As you may recall from Chapter 3, most cells in the body generate ATP through *aerobic metabolism* in mitochondria and through *glycolysis* in the cytoplasm. ∞ [p. 88]

AEROBIC METABOLISM **Aerobic metabolism** normally provides 95 percent of the ATP demands of a resting cell. In this process, mitochondria absorb oxygen, ADP, phosphate ions, and organic substrates from the surrounding cytoplasm. The substrates then enter the *TCA cycle*, an enzymatic pathway that breaks down organic molecules. The carbon atoms are oxidized to carbon dioxide. The hydrogen atoms are shuttled to *respiratory enzymes* of the mitochondrial cristae. After a series of intermediate steps they too are combined with oxygen to form water. Along the way, large amounts of energy are released and used to make ATP. The entire process is very efficient; for each molecule "fed" to the TCA cycle, the cell will gain 17 ATP.

Resting skeletal muscle fibers rely almost exclusively on the aerobic metabolism of fatty acids to generate ATP. When the muscle starts contracting, the mitochondria begin breaking down molecules of pyruvic acid instead of fatty acids. The pyruvic acid is provided by the enzymatic pathway of glycolysis.

GLYCOLYSIS **Glycolysis** is the breakdown of glucose to pyruvic acid in the cytoplasm of a cell. It is called an **anaerobic** process because it does not require oxygen. The reaction sequence of glycolysis provides a net gain of 2 ATP and generates 2 pyruvic acid molecules. The ATP produced by glycolysis is thus only a small fraction of that produced through aerobic metabolism, as the breakdown of the 2 pyruvic acid molecules in mitochondria would generate 34 ATP. *Yet because it can proceed in the absence of oxygen, glycolysis can be very important when the availability of oxygen limits the rate of mitochondrial ATP production.*

In most skeletal muscles, glycolysis is the primary source of ATP during peak periods of activity. The glucose broken down under these conditions is obtained primarily by mobilizing the reserves of glycogen present in the sarcoplasm. Glycogen, diagrammed in Figure 2-12●, p. 47, is a polysaccharide chain of glucose molecules. Typical skeletal muscle fibers contain large glycogen reserves that may account for 1.5 percent of their total weight. When the muscle fiber begins to run short of ATP and CP, enzymes split the glycogen molecules apart, releasing glucose that can be used to generate additional ATP.

As the level of muscular activity increases, and these reserves are mobilized, the pattern of energy production and use changes.

The Resting Skeletal Muscle

Figure 10-15a

In a resting skeletal muscle (Figure 10-15a●), the demand for ATP is low. More than enough oxygen is available for the mitochondria to meet that demand, and they produce a surplus of ATP. The extra ATP is used to build up reserves of CP and glycogen. Resting muscle fibers absorb glucose and fatty acids that are delivered by the bloodstream. The fatty acids are broken down in the mitochondria, and the ATP generated is used to convert creatine to creatine phosphate and glucose to glycogen.

Moderate Levels of Activity

Figure 10-15b

At moderate levels of activity (Figure 10-15b●) the demand for ATP increases, and the demand is met by the mitochondria. As the rate of mitochondrial ATP production rises, so does the rate of oxygen consumption. Oxygen availability is not a limiting factor, for oxygen can diffuse into the muscle fiber fast enough to meet mitochondrial needs. But all of the ATP produced is needed by the muscle fiber, and no surplus is available. The skeletal muscle now relies primarily on the aerobic metabolism of pyruvic acid to generate ATP. The pyruvic acid is provided by glycolysis, which breaks down glucose molecules obtained from glycogen inside the muscle fiber. If glycogen reserves are low, the muscle fiber can also break down other substrates, such as lipids or amino acids. As long as the demand for ATP can be met by mitochondrial activity, the ATP provided by glycolysis makes a relatively minor contribution to the total energy budget of the muscle fiber.

Peak Levels of Activity

Figure 10-15c

At peak levels of activity, the ATP demands are enormous, and mitochondrial ATP production rises to a maximum. This maximum rate is determined by the availability of oxygen, and oxygen cannot diffuse into the muscle fiber fast enough to enable the mitochondria to produce the required ATP. At peak levels of exertion, mitochondrial activity can provide only around one-third of the ATP needed. The rest is produced through glycolysis (Figure 10-15c●).

When glycolysis produces pyruvic acid faster than it can be utilized by the mitochondria, pyruvic acid levels rise in the sarcoplasm. Under these conditions the pyruvic acid is converted to **lactic acid,** a related three-carbon molecule. The enzyme responsible is called **lactate dehydrogenase,** or **LDH.**

The anaerobic process of glycolysis enables the cell to generate additional ATP when the mito-

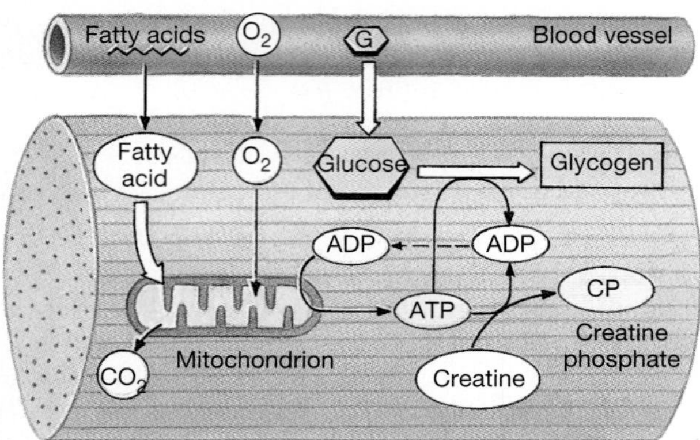
Fatty acids
O2
G
Blood vessel
Fatty acid
O2
Glucose
Glycogen
ADP
ADP
CP
ATP
Creatine phosphate
Mitochondrion
CO2
Creatine

(a) Resting muscle: Fatty acids are catabolized; the ATP produced is used to build energy reserves of ATP, CP, and glycogen.

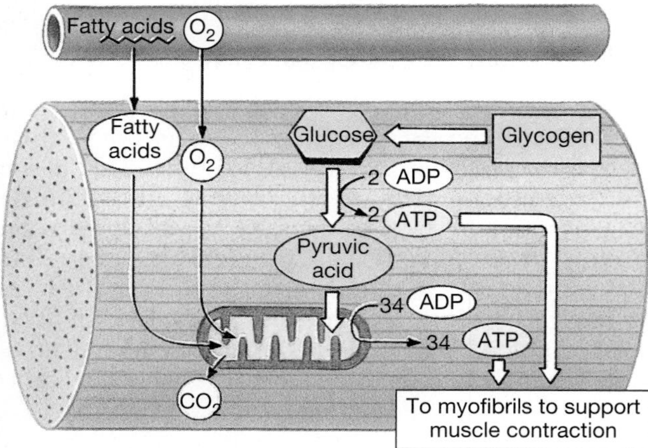
Fatty acids
O2
Fatty acids
O2
Glucose
Glycogen
2 ADP
2 ATP
Pyruvic acid
34 ADP
34 ATP
CO2
To myofibrils to support muscle contraction

(b) Moderate activity: Glucose and fatty acids are catabolized; the ATP produced is used to power contraction.

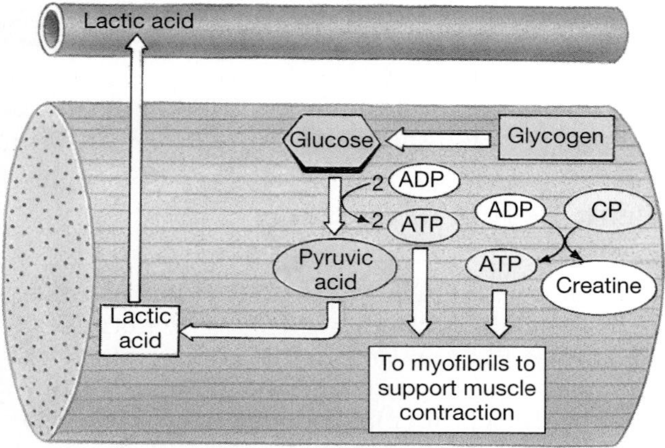
Lactic acid
Glucose
Glycogen
2 ADP
2 ATP
ADP
CP
Pyruvic acid
ATP
Creatine
Lactic acid
To myofibrils to support muscle contraction

(c) Peak activity: Most ATP is produced through glycolysis, with lactic acid as a byproduct.

FIGURE 10-15

Muscle Metabolism. (a) A resting muscle breaks down fatty acids via aerobic respiration to make ATP. Surplus ATP is used to build reserves of creatine phosphate and glycogen. **(b)** At modest-rate activity levels, mitochondria can meet the ATP demands through aerobic metabolism of fatty acids and glucose. **(c)** At peak levels of activity the mitochondria cannot get enough oxygen to meet ATP demands, and most of the ATP is provided by glycolysis, leading to the production of lactic acid.

chondria are unable to meet the current energy demands. However, anaerobic energy production has its drawbacks:

1. Lactic acid is an organic acid that in body fluids dissociates into a hydrogen ion and a negatively charged *lactate ion*. The production of lactic acid can therefore lower the intracellular pH. Buffers in the sarcoplasm can resist pH shifts, but these defenses are limited. Eventually changes in pH will alter the functional characteristics of key enzymes. The muscle fiber will then become unable to continue contracting.
2. Glycolysis is a relatively inefficient way to generate ATP. Under anaerobic conditions, each glucose molecule generates two pyruvic acid molecules that are converted to lactic acid. In return the cell gains 2 ATP through glycolysis. Had those two pyruvic acid molecules been catabolized aerobically in a mitochondrion, the cell would have gained *34 additional ATP.*

Muscle Fatigue

A skeletal muscle fiber is said to be **fatigued** when it can no longer contract despite continued neural stimulation. The cause of muscle fatigue varies, depending on the level of muscle activity. After peak levels of activity, such as a 100-meter dash, fatigue may result from the exhaustion of ATP and CP reserves or the fall in pH that accompanies the buildup of lactic acid. After prolonged exertion, such as running a marathon, fatigue may involve physical damage to the sarcoplasmic reticulum that interferes with the regulation of intracellular calcium ion concentrations. Muscle fatigue is cumulative, and the effects become more pronounced as more muscle fibers are affected. The result is a gradual reduction in the capabilities of the entire skeletal muscle.

If the muscle fiber is contracting at moderate levels, and ATP demands can be met through aerobic metabolism, fatigue will not occur until glycogen, lipid, and amino acid reserves are depleted. This type of fatigue affects the muscles of long-distance athletes, such as marathon runners, after hours of exertion.

When a muscle produces a sudden, intense burst of activity at peak levels, most of the ATP is provided by glycolysis. After a relatively short time (seconds to minutes), the rising lactic acid levels lower the tissue pH, and the muscle can no longer function normally. Athletes running sprints, such as the 100-meter dash, suffer from this type of muscle fatigue. We will return to the topics of fatigue, athletic training, and metabolic activity later in the chapter.

Normal muscle function requires (1) substantial intracellular energy reserves, (2) a normal circulatory supply, and (3) a normal blood oxygen concentration. Anything that interferes with one or more of these factors will promote premature muscle fatigue. For example, reduced blood flow from tight clothing, a circulatory disorder, or loss of blood slows the delivery of oxygen and nutrients, accelerates the buildup of lactic acid, and promotes fatigue.

The Recovery Period

Figure 10-16

When a muscle fiber contracts, the conditions in the sarcoplasm are changed. Energy reserves are consumed, heat is released, and, if the contraction was at peak levels, lactic acid is generated. In the **recovery period** conditions inside the muscle fibers are returned to normal, preexertion levels. It may take several hours for muscle fibers to recover from a period of moderate activity. After sustained activity at higher levels, complete recovery may take a week.

During the recovery period the muscle fibers must rebuild their energy reserves. Even when contracting at moderate levels, active skeletal muscles gradually deplete their glycogen reserves. When contracting at maximal levels, the depletion occurs much more rapidly, because glycolysis is relatively inefficient. Under these conditions, the energy reserves will be restored through the removal and recycling of lactic acid generated during the period of contraction. Rebuilding glycogen reserves is a synthetic activity that requires ATP. Because oxygen is readily available in the recovery period, the ATP is provided through aerobic metabolism in the mitochondria.

LACTIC ACID REMOVAL AND RECYCLING Glycolysis enables a skeletal muscle to continue contracting even when mitochondrial activity is limited by the availability of oxygen. As we have seen, however, lactic acid production is not an ideal way to generate ATP. It is wasteful because it squanders the glucose reserves of the muscle fibers, and it is potentially dangerous because lactic acid can alter the pH of the blood and tissues.

During the recovery period, when oxygen is available in abundance, that lactic acid can be recycled by converting it back to pyruvic acid. The pyruvic acid can then be used either by mitochondria to generate ATP, or as a substrate for enzyme pathways that synthesize glucose and rebuild glycogen reserves.

The reaction that converts pyruvic acid to lactic acid is freely reversible. When pyruvic acid concentrations rise, LDH generates lactic acid (Figure 10-16a●). During the recovery period, when lac-

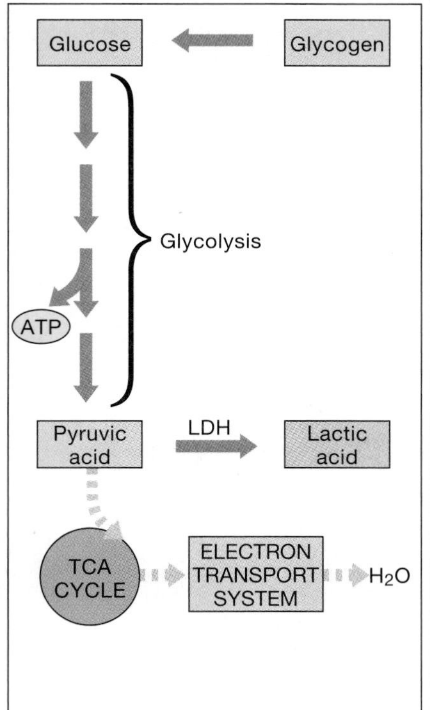
Glucose
Glycogen
Glycolysis
ATP
Pyruvic acid
LDH
Lactic acid
TCA CYCLE
ELECTRON TRANSPORT SYSTEM
H_2O

(a) Oxygen insufficient

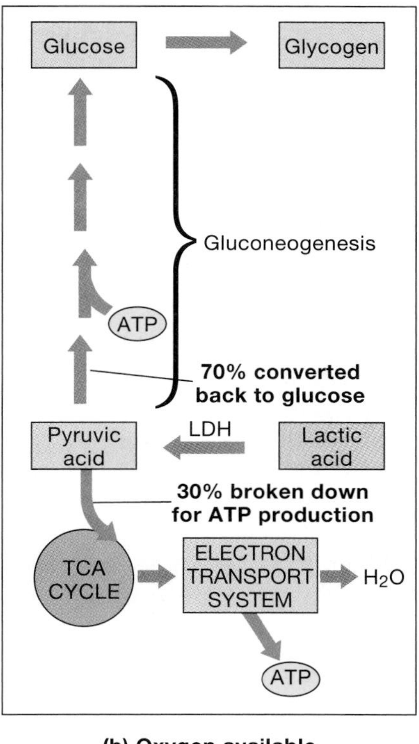
Glucose
Glycogen
Gluconeogenesis
ATP
70% converted back to glucose
Pyruvic acid
LDH
Lactic acid
30% broken down for ATP production
TCA CYCLE
ELECTRON TRANSPORT SYSTEM
H_2O
ATP

(b) Oxygen available

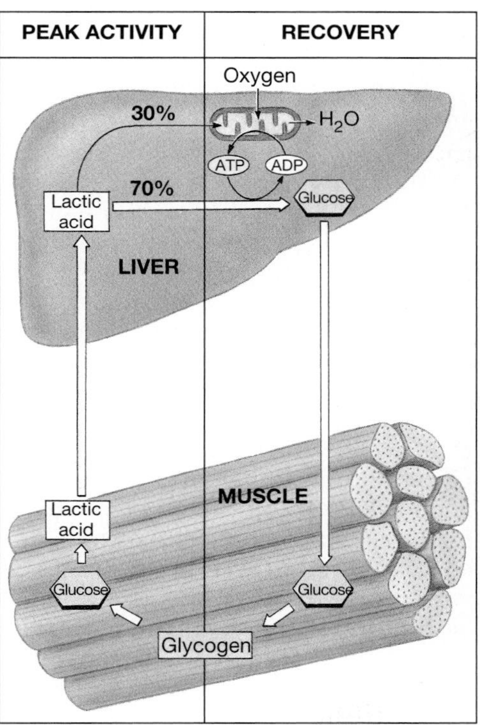
PEAK ACTIVITY
RECOVERY
Oxygen
30%
H_2O
ATP
ADP
70%
Lactic acid
Glucose
LIVER
MUSCLE
Lactic acid
Glucose
Glucose
Glycogen

(c)

● **FIGURE 10-16**

Lactic Acid Recycling and the Cori Cycle. (a) During strenuous activity skeletal muscles generate ATP through glycolysis under anaerobic conditions. LDH converts the pyruvate produced into lactic acid. **(b)** During the recovery period, when oxygen is plentiful, lactic acid is converted back to pyruvic acid. Some of the pyruvic acid is broken down aerobically, by the TCA cycle, to produce ATP. The rest is converted back to glucose. **(c)** The work of lactic acid recycling is divided between the liver and the muscles in the Cori cycle. The liver absorbs the circulating lactic acid and produces glucose for discharge into the bloodstream. Muscle fibers then use the glucose to rebuild their glycogen reserves. D

tic acid concentration is high and oxygen readily available, LDH converts the lactic acid back to pyruvic acid (Figure 10-16b•). Roughly 30 percent of the pyruvic acid will be used by mitochondria in the TCA cycle. Enough ATP will be generated to (1) restore normal ATP levels in the sarcoplasm, (2) convert creatine to creatine phosphate and restore normal CP levels, and (3) rebuild muscle glycogen reserves, using the remaining pyruvic acid molecules and additional glucose absorbed from the circulation.

During a period of exertion, lactic acid diffuses out of the muscle fibers and into the bloodstream. This process continues after the exertion has ended, because intracellular lactic acid concentrations are still relatively high. The liver absorbs the lactic acid and converts it to pyruvic acid. Roughly 30 percent of these molecules are broken down in the TCA cycle, providing the ATP needed to convert the other pyruvic acid molecules to glucose. The synthesis of glucose from other organic molecules is the process of **gluconeogenesis** (*gluco-*, sugar + *neo-*, new + *genesis*, production). The glucose molecules are then released into the circulation, where they are absorbed by skeletal muscle fibers and used to rebuild their glycogen reserves. This shuffling of lactate to the liver and glucose back to muscle cells is the **Cori cycle,** diagrammed in Figure 10-16c•.

THE OXYGEN DEBT During the recovery period, the body's oxygen demand goes up considerably. The recovery period is powered by the ATP generated through aerobic metabolism. The more ATP required, the more oxygen will be needed; on average, for each molecule of oxygen absorbed by a mitochondrion, the cell gains 4 ATP. The **oxygen debt** created during a period of exercise is the amount of oxygen needed to restore normal pre-exertion conditions. The major tissues involved in the additional oxygen consumption are skeletal muscle fibers, which must restore ATP, creatine phosphate, and glycogen concentrations to their former levels, and liver cells, which generate the ATP needed to convert excess lactic acid to glucose. Additional ATP is needed to restore normal conditions in other tissues. For example, sweat glands must increase their secretory activity to restore normal body temperature. While the oxygen debt is being repaid, the breathing rate and depth are increased; as a result, you continue to breathe heavily long after you stop exercising.

HEAT LOSS Muscular activity generates substantial amounts of heat. When a catabolic reaction occurs, such as the breakdown of glycogen or the reactions of glycolysis, the muscle fiber captures only a portion of the released energy. The rest is released as heat. A resting muscle fiber relying on aerobic metabolism captures about 42 percent of the energy released in catabolism. The other 58 percent warms the sarcoplasm, interstitial fluid, and circulating blood. Active skeletal muscles release roughly 85 percent of the heat needed to maintain normal body temperature.

When muscles become active, their consumption of energy skyrockets. As anaerobic energy production becomes the primary method of ATP generation, muscle fibers become less efficient at capturing energy. At peak levels of exertion, only about 30 percent of the released energy is captured as ATP, and the remaining 70 percent warms the muscle and surrounding tissues. Body temperature soon begins to climb, and heat loss at the skin accelerates through mechanisms introduced in Chapters 1 and 5. ∞ *[pp. 14, 163]*

Even after the exercise period ends and **initial heat** production stops, energy use remains high as the oxygen debt is repaid. The **recovery heat** produced keeps us perspiring for some time thereafter.

Hormones and Muscle Metabolism

Metabolic activities in skeletal muscle fibers are adjusted by hormones of the endocrine system.

- *Growth hormone* from the pituitary gland and *testosterone* (the primary sex hormone in males) stimulate the synthesis of contractile proteins and the enlargement of skeletal muscles.
- *Thyroid hormones* elevate the rate of energy consumption by resting and active skeletal muscles.
- During a sudden crisis, hormones of the adrenal gland, notably *epinephrine* (adrenaline), stimulate muscle metabolism and increase both the duration of stimulation and the force of contraction.

The effects of hormones on muscle and other tissues will be detailed in Chapter 18.

Muscle Performance

Muscle performance can be considered in terms of sheer **power,** the maximum amount of tension produced by a particular muscle or muscle group, and **endurance,** the amount of time for which the individual can perform a particular activity. Two major factors determine the capabilities of a particular skeletal muscle: (1) the types of muscle fibers within the muscle and (2) physical conditioning or training.

Types of Skeletal Muscle Fibers

Figure 10-17

There are three major types of skeletal muscle fibers in the human body: *fast fibers, slow fibers,* and *intermediate fibers.*

FAST FIBERS Most of the skeletal muscle fibers in the body are called **fast fibers** because they can contract in 0.01 second or less following stimulation. Fast fibers are large in diameter; they contain densely packed myofibrils, large glycogen reserves, and relatively few mitochondria. The tension produced by a muscle fiber is directly proportional to the number of sarcomeres, so muscles dominated by fast fibers produce powerful contractions. However, because these contractions use ATP in massive amounts, prolonged activity is supported primarily by anaerobic metabolism, and fast fibers fatigue rapidly.

SLOW FIBERS **Slow fibers** are only about half the diameter of fast fibers, and they take three times as long to contract after stimulation. Slow fibers are specialized to enable them to continue contracting for extended periods, long after a fast muscle would have become fatigued. The most important specializations improve mitochondrial performance. Slow muscle tissue contains a more extensive network of capillaries than is typical of fast muscle tissue, so the oxygen supply is dramatically increased. In addition, slow muscle fibers contain the red pigment **myoglobin** (MĪ-ō-glō-bin). This globular protein, detailed in Figure 2-18●, p. 54, is structurally related to hemoglobin, the oxygen-carrying pigment found in the blood. Both myoglobin and hemoglobin are red pigments that reversibly bind oxygen molecules. Although small amounts of myoglobin are found in other muscle fiber types, it is most abundant in slow muscle fibers. As a result, resting slow muscle fibers contain substantial oxygen reserves that can be mobilized during a contraction. Because slow muscle fibers have (1) an extensive capillary supply and (2) a high concentration of myoglobin, skeletal muscles dominated by slow muscle fibers are dark red.

With oxygen reserves and a more efficient blood supply, the mitochondria of slow muscle fibers are able to contribute a greater amount of ATP while contractions are under way, making slow muscle fibers less dependent on anaerobic metabolism than fast muscle fibers. Some of the mitochondrial energy production involves the breakdown of stored lipids rather than glycogen, so glycogen reserves are smaller than those of fast muscle cells. Slow muscles also contain a relatively larger number of mitochondria than do fast muscle fibers. Figure 10-17● compares the appearance of fast and slow skeletal muscle fibers.

INTERMEDIATE FIBERS **Intermediate fibers** have properties intermediate between those of fast fibers and slow fibers. In appearance intermediate fibers most closely resemble fast fibers, for they contain little myoglobin and are relatively pale. They have a more extensive capillary network around them, however, and are more resistant to fatigue than are fast fibers. The three types of muscle fibers are compared in Table 10-3. In muscles that contain a mixture of fast and intermediate muscle fibers, the proportion can change with physical conditioning. For example, if a muscle is used repeatedly for endurance events, some of the fast fibers will develop the appearance and functional capabilities of intermediate fibers. The muscle as a whole will thus become more resistant to fatigue.

Distribution of Muscle Fibers and Muscle Performance

The percentage of fast, intermediate, and slow muscle fibers in a particular skeletal muscle can be quite variable. Muscles dominated by fast fibers appear pale, and they are often called **white mus-**

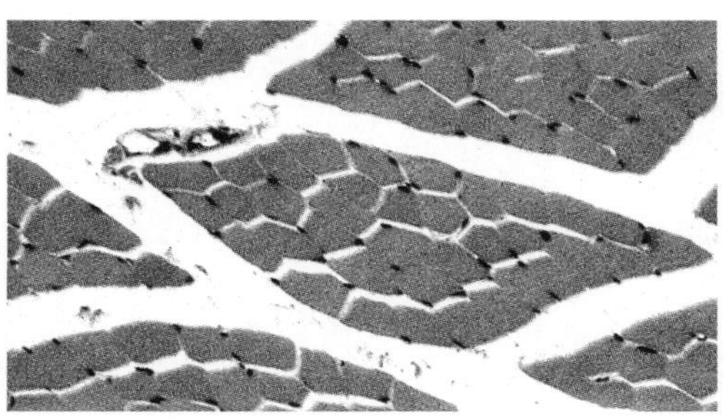

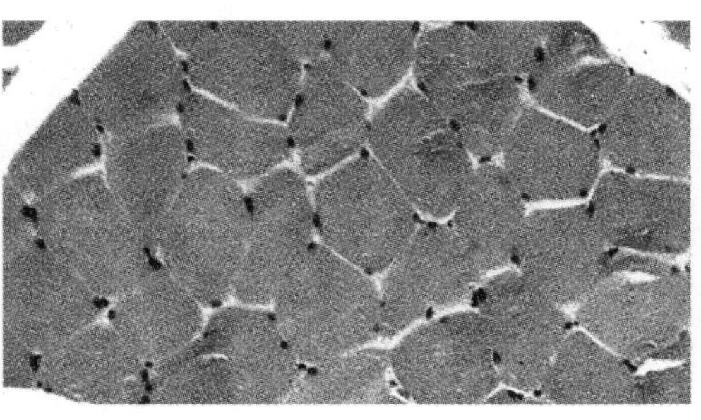

(a)

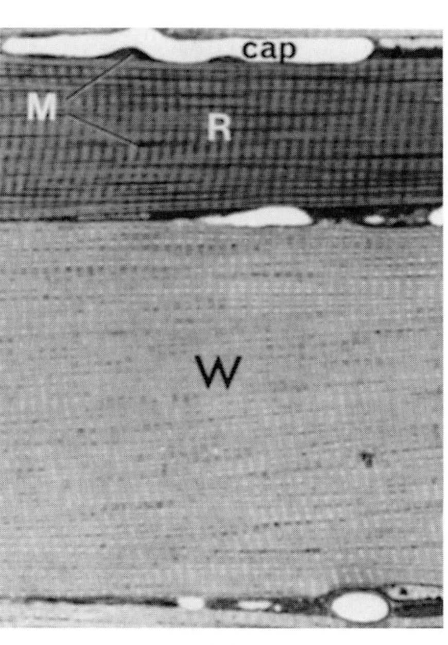
cap
M
R
W

(b)

● **FIGURE 10-17**
Fast and Slow Muscle Fibers. **(a)** Note the difference in the size of slow muscle fibers, above, and fast muscle fibers, below. (LM × 171) **(b)** The slender slow (red) muscle fiber (R) has more mitochondria (M) and a more extensive capillary supply (cap) than the fast (white) muscle fiber (W). (LM × 783)

cles. Chicken breasts contain "white meat" because chickens use their wings only for brief intervals, as when fleeing from a predator, and the power for flight comes from fast fibers in their breast muscles. The extensive blood vessels, mitochondria, and myoglobin in slow muscle fibers give them a reddish color, and muscles dominated by slow fibers are therefore known as **red muscles.** Chickens walk around all day, and the movements are performed by the slow muscle fibers in the "dark meat" of their legs.

Most human muscles contain a mixture of fiber types, and so appear pink. However, there are no slow fibers in muscles of the eye and hand, where swift but brief contractions are required. Many back and calf muscles are dominated by slow fibers; these muscles contract almost continually to maintain an upright posture. The percentage of fast versus slow fibers in each muscle is genetically determined. As noted above, the proportion of intermediate fibers to fast fibers can increase as the result of athletic training.

Muscle Hypertrophy

As a result of repeated, exhaustive stimulation, muscle fibers develop a larger number of mitochondria, a higher concentration of glycolytic enzymes, and larger glycogen reserves. These muscle fibers have more myofibrils, and each myofibril contains a larger number of thick and thin filaments. The net effect is an enlargement, or **hypertrophy,** of the stimulated muscle. The number of muscle fibers does not change, but the muscle as a whole enlarges because each muscle fiber increases in diameter. Hypertrophy occurs in muscles that have been repeatedly stimulated to produce near-maximal tension; the intracellular changes that occur increase the amount of tension produced when these muscles contract. A champion weight lifter or body builder is an excellent example of hypertrophied muscular development.

In general, the muscle mass of men exceeds that of women, because muscle fibers are stimulated by testosterone. In recent years amateur and professional athletes have used this steroid hormone, or related drugs, to stimulate development of hypertrophied muscles. The steroids appear to have the desired effect only when taken by individuals who are already engaged in an intensive weight-training program and are on a high-protein, high-calorie diet. It should be noted that this combination of training and diet would *by itself* produce muscular hypertrophy, even without the use of steroids. The benefits of steroids thus do not outweigh the known risks, which include liver failure and infertility. (The use of steroids and other hormones by athletes is discussed further in Chapter 18.)

Muscle Atrophy

When a skeletal muscle is not stimulated by a motor neuron on a regular basis, it loses muscle tone and mass. The muscle becomes flaccid, and the muscle fibers become smaller and weaker. This

TABLE 10-3 Properties of Skeletal Muscle Fiber Types

Property	*Slow*	*Intermediate*	*Fast*
Cross-sectional diameter	Small	Intermediate	Large
Tension	Low	Intermediate	High
Contraction speed	Slow	Fast	Fast
Fatigue resistance	High	Intermediate	Low
Color	Red	White	White
Myoglobin content	High	Low	Low
Capillary supply	Dense	Intermediate	Scarce
Mitochondria	Many	Intermediate	Few
Glycolytic enzyme concentration in sarcoplasm	Low	High	High
Substrates	Lipids, carbohydrates, amino acids (aerobic)	Primarily carbohydrates (anaerobic)	Carbohydrates (anaerobic)
Alternative names	Type I, S (slow), SO (slow oxidizing)	Type II, FR (fast resistant)	Type II, FF (fast fatigue)

reduction in muscle size, tone, and power is called **atrophy**. Individuals paralyzed by spinal injuries or other damage to the nervous system will gradually lose muscle tone and size in the areas affected. Even a temporary reduction in muscle use can lead to muscular atrophy; this effect may easily be seen by comparing limb muscles before and after a cast has been worn. Muscle atrophy is initially reversible, but dying muscle fibers are not replaced, and in extreme atrophy the functional losses are permanent. That is why physical therapy is crucial in cases in which people are temporarily unable to move normally.

Because skeletal muscles depend on their motor neurons for stimulation, disorders that affect the nervous system can have an indirect effect on the muscular system. In *polio*, a virus attacks motor neurons in the spinal cord and brain, producing muscular paralysis and atrophy. AM Polio

Physical Conditioning

Figure 10-18

Physical conditioning and training schedules enable athletes to improve both power and endurance. In practice, the training schedule varies depending on whether the activity is primarily supported by aerobic or anaerobic energy production.

ANAEROBIC ENDURANCE Anaerobic endurance is the length of time muscular contraction can continue supported by glycolysis and the existing energy reserves of ATP and CP. Anaerobic endurance is limited by (1) the amount of ATP and CP already at hand, (2) the amount of glycogen available for breakdown, and (3) the ability of the muscle to tolerate the lactic acid generated during the anaerobic period. Typically, the onset of muscle fatigue occurs within 2 minutes of the start of maximal activity. Examples of activities that require anaerobic endurance include a 50-meter dash or swim, a pole vault, or a weightlifting competition. Such activities are performed by fast muscle fibers. When these muscles begin contracting at maximal levels, the energy for the first 10–20 seconds comes from the ATP and CP (creatine phosphate) reserves of the cytoplasm. As these reserves dwindle, glycogen breakdown and glycolysis provide additional energy.

Athletes training to develop anaerobic endurance perform frequent, brief, intensive workouts that stimulate muscle hypertrophy.

AEROBIC ENDURANCE Aerobic endurance is the length of time that a muscle can continue to contract while supported by mitochondrial activities. Aerobic endurance is determined primarily by the availability of substrates for aerobic respiration, which the muscle fibers can obtain by breaking down carbohydrates, lipids, or amino acids. Initially, many of the nutrients catabolized by the muscle fiber are obtained from reserves within the muscle fibers themselves. Prolonged aerobic activity, however, must be supported by nutrients provided by the circulating blood. During exercise, blood vessels in the skeletal muscles dilate, and the increased blood flow brings oxygen and nutrients to the active muscle tissue. Warm-up periods are therefore important not only because of treppe, the increase in tension production noted on page 302, but also because they stimulate circulation in the muscles before the serious competition begins. Because glucose is a preferred energy source, aerobic athletes, such as marathon runners, often "load" or "bulk up" on carbohydrates for the last three days before an event. They may also consume glucose-rich "sports drinks" during a competition. The risks and benefits of these practices are considered in Chapter 25.

Training to improve aerobic endurance usually involves sustained low levels of muscular activity. Examples include jogging, distance swimming, and other exercises that do not require peak tension production. Improvements in aerobic endurance result from (1) altering the characteristics of muscle fibers and (2) improving the performance of the cardiovascular system.

1. ***Altering the characteristics of muscle fibers.*** The composition of fast and slow fibers in each muscle is genetically determined, and there are significant individual differences. These variations have an effect on aerobic endurance, for a person with more slow muscle fibers in a particular muscle will be better able to perform under aerobic conditions. However, skeletal muscle cells respond to changes in the pattern of neural stimulation. Fast muscle fibers trained for aerobic competition develop the characteristics of intermediate fibers, and this change improves aerobic endurance.
2. ***Improving cardiovascular performance.*** Cardiovascular activity affects muscular performance by delivering oxygen and nutrients to active muscles. Physical training alters cardiovascular function by accelerating blood flow, thus improving oxygen and nutrient availability. Factors involved in improving cardiovascular performance will be detailed in Chapter 21.

Aerobic activities do not promote hypertrophy, and many athletes prefer to train using a combination of aerobic and anaerobic exercises so that their muscles will enlarge and both anaerobic and aerobic endurance will improve. In this training scheme, an aerobic activity, such as swimming, is alternated with sprinting or weight lifting. The combination is known as *interval training.* Interval training is particularly useful for those engaged in popular racquet sports, such as tennis or squash. These sports are dominated by aerobic activities but are punctuated by brief periods of anaerobic effort.

POWER, ENDURANCE, AND ENERGY RESERVES Figure 10-18● compares the power/endurance curves for anaerobic and aerobic activities. The first 30 seconds of peak activity are totally supported by the mobilization of ATP and CP reserves (Figure 10-18a●). Thereafter roughly two-thirds of the energy requirements of skeletal muscles operating at peak activity levels are met via glycolysis, with associated lactic acid generation.

At modest activity levels, a skeletal muscle can rely on aerobic respiration to provide ATP. At peak levels of activity, the muscle relies primarily on anaerobic metabolism. The level of activity at which the muscle must begin relying upon anaerobic metabolism to meet its energy demands is called the **anaerobic threshold.** If energy demands are kept below the anaerobic threshold, muscular activity can be continued until nutrient sources are exhausted. In a trained athlete, muscle fatigue may not occur for several hours (Figure 10-18b●)

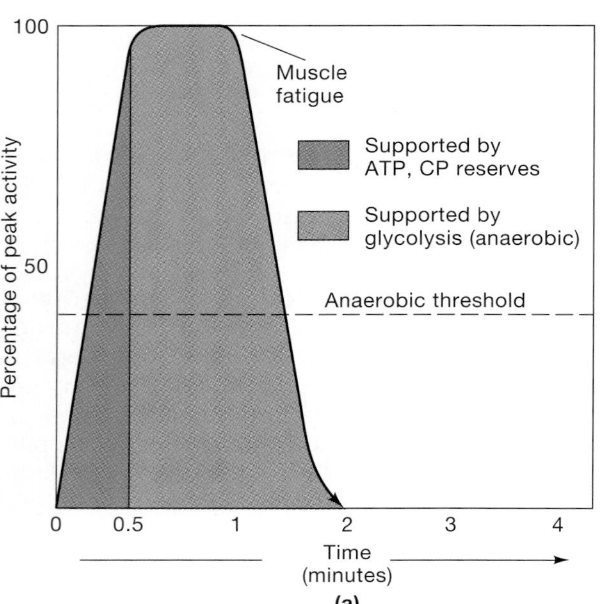
100
Muscle fatigue
Supported by ATP, CP reserves
Supported by glycolysis (anaerobic)
50
Anaerobic threshold
Percentage of peak activity
0
0.5
1
2
3
4
Time (minutes)
(a)

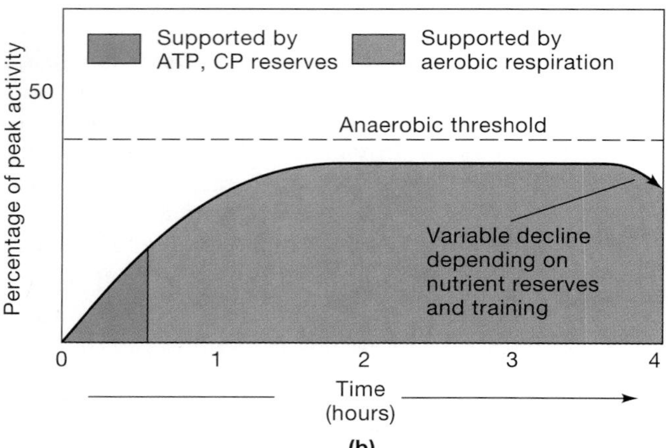
Supported by ATP, CP reserves
Supported by aerobic respiration
50
Anaerobic threshold
Variable decline depending on nutrient reserves and training
Percentage of peak activity
0
1
2
3
4
Time (hours)
(b)

● **FIGURE 10-18**

Muscular Performance and Endurance. (a) At peak levels of activity, skeletal muscles rely primarily on glycolysis for ATP production, with associated lactic acid production. Initial burst activity is supported by ATP and CP reserves. Muscles operating at peak levels fatigue rapidly. **(b)** Muscular activity can continue for extended periods when ATP demands are kept below the anaerobic threshold.

Aging and the Muscular System

As the body ages, there is a general reduction in the size and power of all muscle tissues. The effects on the muscular system can be summarized as follows:

1. ***Skeletal muscle fibers become smaller in diameter.*** This reduction in size primarily reflects a decrease in the number of myofibrils. In addition, the muscle fibers contain smaller ATP, CP, and glycogen reserves and less myoglobin. The overall effect is a reduction in muscle strength and endurance and a tendency to fatigue rapidly. Because cardiovascular performance also decreases with age, blood flow to active muscles does not increase with exercise as rapidly as it does in younger people. These factors interact to produce decreases in anaerobic and aerobic performance of 30–50 percent at age 65.
2. ***Skeletal muscles become smaller and less elastic.*** Aging skeletal muscles develop increasing amounts of fibrous connective tissue, a process called **fibrosis.** Fibrosis makes the muscle less flexible, and the collagen fibers can restrict movement and circulation.
3. ***Tolerance for exercise decreases.*** A lower tolerance for exercise results in part from the tendency for rapid fatigue and in part from the reduction in thermoregulatory ability described in Chapter 5. [p. 166] Because the elderly cannot eliminate the heat generated during muscular contraction as effectively as younger people, they are subject to overheating.
4. ***Ability to recover from muscular injuries decreases.*** The number of satellite cells steadily decreases with age, and the amount of fibrous tissue increases. As a result, when an injury occurs, repair capabilities are limited, and scar tissue formation is the usual result.

The rate of decline in muscular performance is the same in all individuals, regardless of their exercise patterns or lifestyle. Therefore to be in good shape late in life, one must be in *very* good shape early in life. Regular exercise helps control body weight, strengthens bones, and generally improves the quality of life at all ages. Extremely demanding exercise is not as important as regular exercise. In fact, extreme exercise in the elderly may lead to problems with tendons, bones, and joints. Although it has obvious effects on the quality of life, there is no clear evidence that exercise prolongs life expectancy.

Integration with Other Systems

Figure 10-19

To operate at maximum efficiency, the muscular system must be supported by many other systems. The changes that occur during exercise provide a good example of such interaction. As noted in earlier sections, active muscles consume oxygen and generate carbon dioxide and heat. Responses of other systems include:

1. ***Cardiovascular system:*** Dilation of blood vessels in the active muscles and the skin and an increase in the heart rate. These adjustments accelerate oxygen delivery and carbon dioxide removal at the muscle and bring heat to the skin for radiation into the environment.
2. ***Respiratory system:*** Increased respiratory rate and depth of respiration. Air moves into and out of the lungs more quickly, keeping pace with the increased rate of blood flow through the lungs.
3. ***Integumentary system:*** Dilation of blood vessels and increased sweat gland secretion. This combination helps promote evaporation at the skin surface and removes the excess heat generated by muscular activity.
4. ***Nervous and endocrine systems:*** Direct the responses of other systems by controlling heart rate, respiratory rate, and sweat gland activity.

The muscular system has extensive interactions with other systems even at rest. Figure 10-19• summarizes the range of interactions between the muscular system and other vital systems.

Why would a sprinter experience muscle fatigue before a marathon runner would?

Which activity would be more likely to create an oxygen debt, swimming laps or lifting weights?

What type of muscle fibers would you expect to predominate in the large leg muscles of someone who excels at endurance activities such as cycling or long-distance running?

Cardiac Muscle Tissue

Figure 10-20

Cardiac muscle tissue was introduced in Chapter 4, and its properties were briefly compared with those of other muscle types. Table 10-4 compares skeletal, cardiac, and smooth muscle tissue in greater detail. Cardiac muscle cells, also called **cardiocytes** or *cardiac myocytes*, are relatively small, averaging 10–20 μm in diameter and 50–100 μm in length. A typical cardiocyte (Figure 10-20•) has a single, centrally placed nucleus, although a few may have two or more. As the name implies, cardiac muscle tissue is found only in the heart.

TABLE 10-4 A Comparison of Skeletal, Cardiac, and Smooth Muscle Tissue

Property	*Skeletal Muscle Fiber*	*Cardiac Muscle Cell*	*Smooth Muscle Cell*
Fiber dimensions (diameter × length)	100 μm × up to 30 cm	15 μm × 100 μm	5–10 μm × 30–200 μm
Nuclei	Multiple, near sarcolemma	Usually single, centrally located	Single, centrally located
Filament organization	Sarcomeres along myofibrils	Sarcomeres along myofibrils	Scattered throughout sarcoplasm
SR	Terminal cisternae in triads at zones of overlap	SR tubules contact T tubules at Z lines	Dispersed throughout sarcoplasm, no T tubules
Control mechanism	Neural, at single neuromuscular junction	Automaticity (pacemaker cells)	Automaticity (pacesetter cells), neural or hormonal control
Ca^{2+} source	Release from SR	Across sarcolemma and release from SR	Across sarcolemma
Contraction	Rapid onset, may be tetanized, rapid fatigue	Slower onset, cannot be tetanized, resistant to fatigue	Slow onset, may be tetanized, resistant to fatigue
Energy source	Aerobic metabolism at moderate levels of activity; glycolysis (anaerobic) during peak activity	Aerobic metabolism, usually lipid or carbohydrate substrates	Primarily aerobic metabolism

INTEGUMENTARY SYSTEM

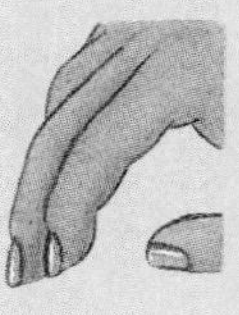

Removes excess body heat; synthesizes vitamin D_3 for Ca^{2+} and PO_4^{3-} absorption; protects underlying muscles

Skeletal muscles pulling on skin of face produce facial expressions

SKELETAL SYSTEM

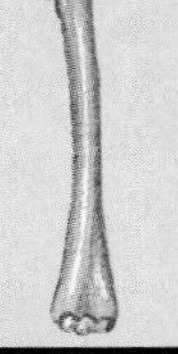

Maintains normal calcium and phosphate levels in body fluids; supports skeletal muscles; provides sites of attachment

Provides movement and support; stresses exerted by tendons maintain bone mass; stabilizes bones and joints

NERVOUS SYSTEM

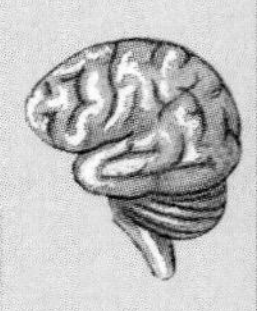

Controls skeletal muscle contractions; adjusts activities of respiratory and cardiovascular systems during periods of muscular activity

Muscle spindles monitor body position; facial muscles express emotions; intrinsic laryngeal muscles permit speech

THE MUSCULAR SYSTEM

FOR ALL SYSTEMS

Generates heat that maintains normal body temperature

ENDOCRINE SYSTEM

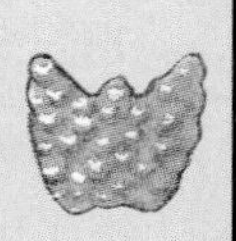

Hormones adjust muscle metabolism and growth; parathyroid hormone and calcitonin regulate calcium and phosphate ion concentrations

Skeletal muscles provide protection for some endocrine organs

CARDIOVASCULAR SYSTEM

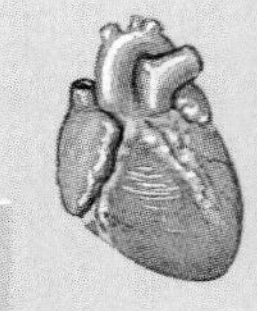

Delivers oxygen and nutrients; removes carbon dioxide, lactic acid, and heat

Skeletal muscle contractions assist in moving blood through veins; protects deep blood vessels

LYMPHATIC SYSTEM

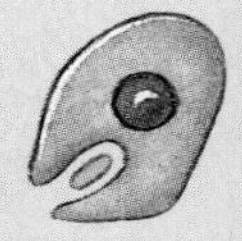

Defends skeletal muscles against infection and assists in tissue repairs after injury

Protects superficial lymph nodes and the lymphatic vessels in the abdominopelvic cavity

RESPIRATORY SYSTEM

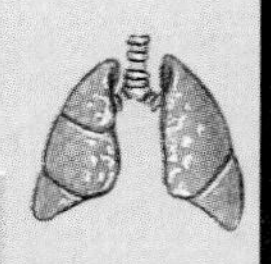

Provides oxygen and eliminates carbon dioxide

Muscles generate CO_2, control entrances to respiratory tract; fill and empty lungs, control airflow through larynx, and produce sounds

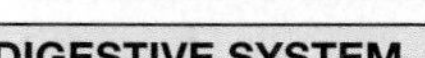

DIGESTIVE SYSTEM

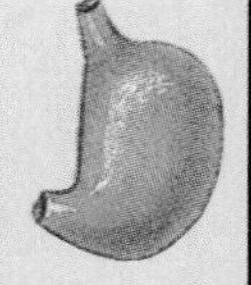

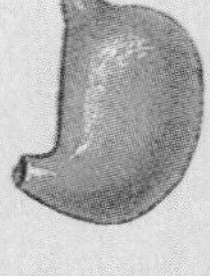

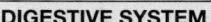

Provides nutrients; liver regulates blood glucose and fatty acid levels and removes lactic acid from circulation

Protects and supports soft tissues in abdominal cavity; controls entrances and exits to digestive tract

REPRODUCTIVE SYSTEM

Reproductive hormones accelerate skeletal muscle growth

Contractions of skeletal muscles eject semen from male reproductive tract; muscle contractions during sexual act produce pleasurable sensations

URINARY SYSTEM

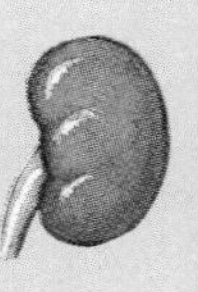

Removes waste products of protein metabolism; assists in regulation of calcium and phosphate concentrations

External sphincter controls urination by constricting urethra

● FIGURE 10-19
Functional Relationships between the Muscular System and Other Systems

❑ Differences between Cardiac and Skeletal Muscle

As in skeletal muscle fibers, each cardiac muscle cell contains organized myofibrils, and the presence of many aligned sarcomeres gives it a striated appearance. Cardiac muscle cells are much smaller than skeletal muscle fibers, and there are significant structural and functional differences between the two.

Structural Differences

Figure 10-20c

Important structural differences between skeletal muscle fibers and cardiac muscle cells include:

1. The T tubules in a cardiac muscle cell are short and broad, and there are no triads. The T tubules encircle the sarcomeres at the Z lines, rather than at the zone of overlap.
2. The SR lacks terminal cisternae, and its tubules contact the cell membrane as well as the T tubules. As in skeletal muscle fibers, the appearance of an action potential triggers calcium release from the SR and the contraction of sarcomeres; it also increases the permeability of the sarcolemma to extracellular calcium ions.
3. Cardiac muscle cells are almost totally dependent on aerobic metabolism to obtain the energy needed to continue contracting. The sarcoplasm of a cardiac muscle cell thus contains large numbers of mitochondria and abundant reserves of myoglobin (to store oxygen). Energy reserves are maintained in the form of glycogen and lipid inclusions.
4. Each cardiac muscle cell contacts several others at specialized sites known as **intercalated** (in-TER-ka-lā-ted) **discs.** Because intercalated discs play a vital role in the function of cardiac muscle, they will be discussed in greater detail below.

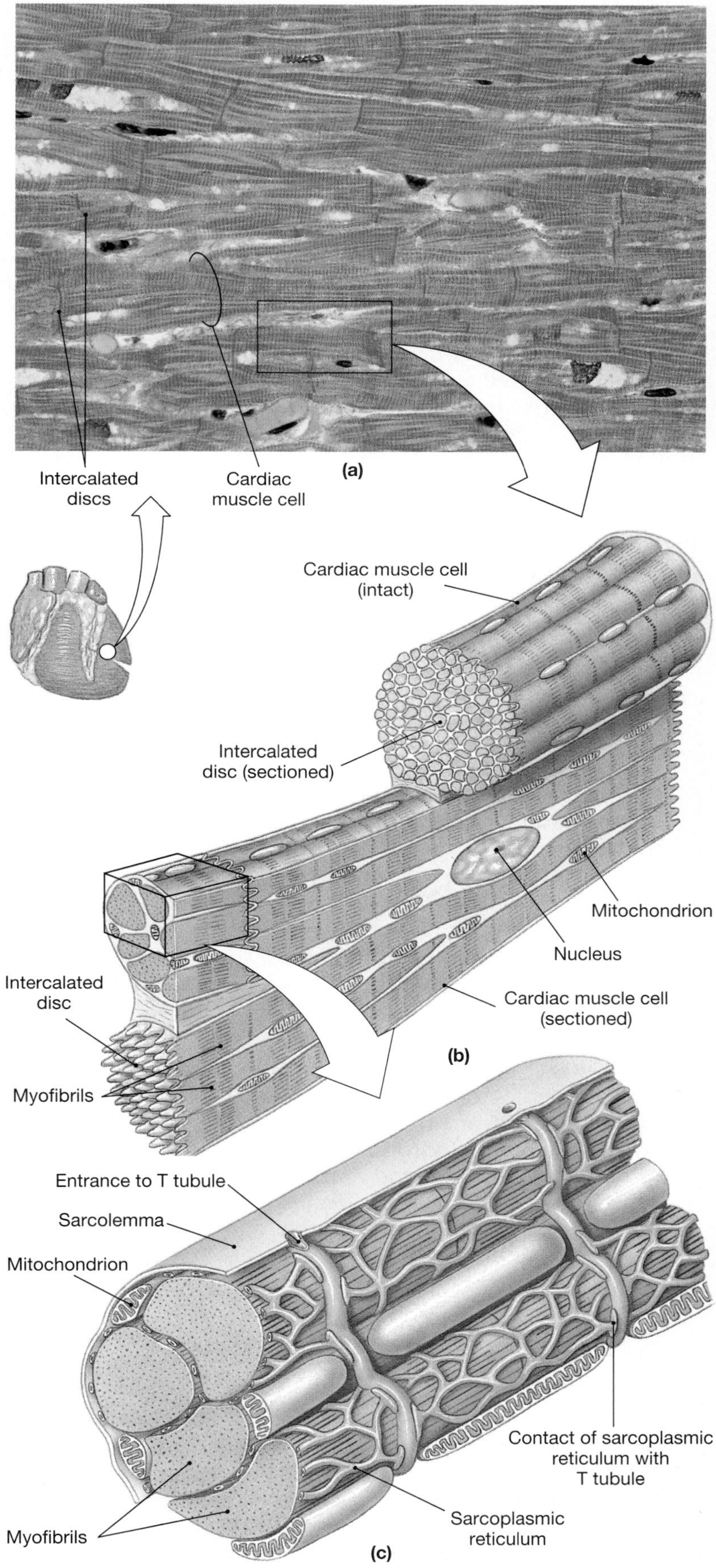
Intercalated discs
Cardiac muscle cell
(a)
Cardiac muscle cell (intact)
Intercalated disc (sectioned)
Mitochondrion
Nucleus
Intercalated disc
Cardiac muscle cell (sectioned)
(b)
Myofibrils
Entrance to T tubule
Sarcolemma
Mitochondrion
Contact of sarcoplasmic reticulum with T tubule
Sarcoplasmic reticulum
Myofibrils
(c)

● **FIGURE 10-20**
Cardiac Muscle Tissue. (a) Cardiac muscle tissue as seen with the light microscope. Note the striations and the intercalated discs. **(b, c)** Structure of a cardiac muscle cell; compare with Figure 10-3.

INTERCALATED DISCS Intercalated discs are diagrammed in Figure 10-20c●. At an intercalated disc the cell membranes of two cardiac muscle cells are extensively intertwined and bound together by gap junctions and desmosomes. These connections help to stabilize the relative positions of adjacent cells and maintain the three-dimensional structure of the tissue. The gap junctions allow ions and small molecules to move from one cell to another. This arrangement creates a direct electrical connection between the two muscle cells, and an action potential can travel across an intercalated disc, moving quickly from one cardiac muscle cell to another.

Myofibrils in the two interlocking muscle cells are firmly anchored to the membrane at the intercalated disc. Because their myofibrils are essentially locked together, the two muscle cells can "pull together" with maximum efficiency.

Because the cardiac muscle cells are mechanically, chemically, and electrically connected to one another, the entire tissue resembles a single, enormous muscle cell. For this reason cardiac muscle has been called a *functional syncytium* (sin-SISH-ē-um; a fused mass of cells).

Functional Differences

Chapter 20 will examine cardiac muscle physiology in detail; this section will provide only a brief overview.

1. Cardiac muscle tissue contracts without neural stimulation. This property is called **automaticity.** The timing of contractions is normally determined by specialized cardiac muscle cells called **pacemaker cells.**
2. Innervation by the autonomic nervous system can alter the pace established by the pacemaker cells and adjust the amount of tension produced during a contraction.
3. Cardiac muscle cell contractions last roughly 10 times longer than those of skeletal muscle fibers.
4. The properties of cardiac muscle cell membranes differ from those of skeletal muscle fibers. As a result, individual cardiocyte twitches cannot summate, and cardiac muscle tissue cannot undergo tetanic contractions. This difference is important because a heart in tetany could not pump blood.

Smooth Muscle Tissue

Figure 10-21

Smooth muscle cells range from 5–10 μm in diameter and from 30–200 μm in length. There is a single nucleus, centrally located within each spindle-shaped cell. Figure 10-21a● shows typical smooth muscle fibers as seen in light microscopy, and Figure 10-21b● shows diagrammatic views of relaxed and contracted smooth muscle cells.

Smooth muscle tissue is found within almost every organ, forming sheets, bundles, or sheaths around other tissues. In the skeletal, muscular, nervous, and endocrine systems smooth muscles around blood vessels regulate blood flow through vital organs. In the digestive and urinary systems, rings of smooth muscle, called *sphincters*, regulate movement along internal passageways. Smooth muscles in bundles, layers, or sheets play a variety of other roles.

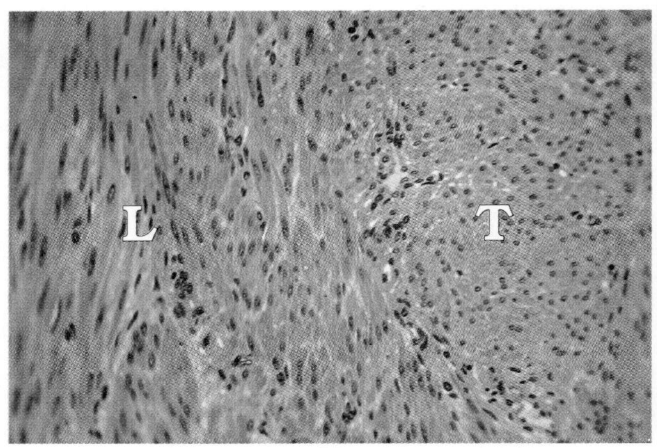
L
T

(a)

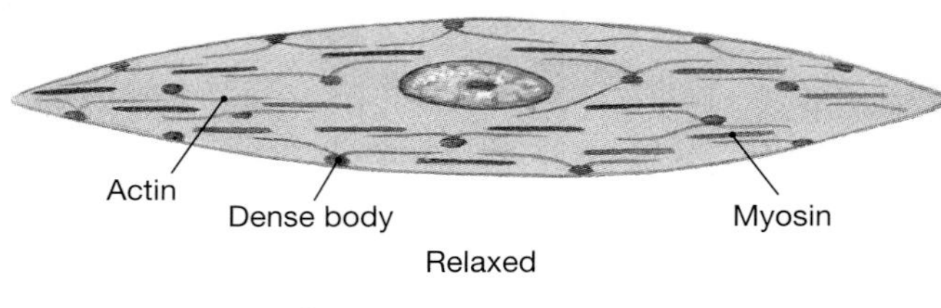
Actin
Dense body
Myosin
Relaxed

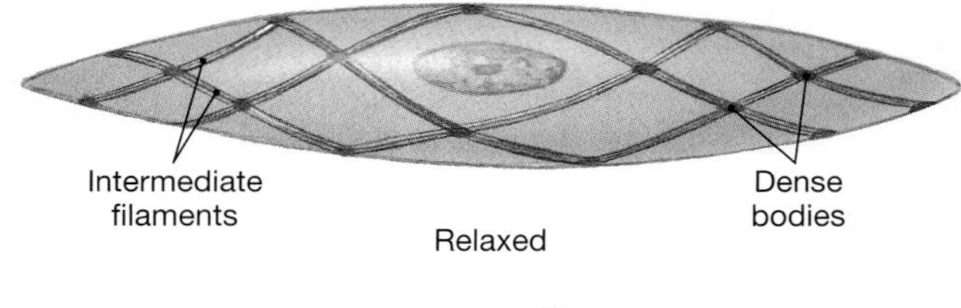
Intermediate filaments
Dense bodies
Relaxed

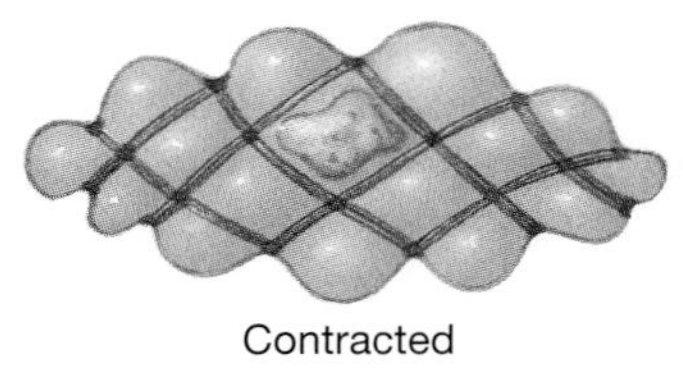
Contracted

(b)

● **FIGURE 10-21**

Smooth Muscle Tissue. (a) Many visceral organs contain several layers of smooth muscle tissue oriented in different ways. As a result, a single sectional view shows a smooth muscle cell in longitudinal (L) and transverse (T) section. **(b)** A single relaxed smooth muscle cell is spindle-shaped and has no striations. Note the changes in cell shape as contraction occurs.

- ***Integumentary system.*** Smooth muscles around blood vessels regulate the flow of blood to the superficial dermis; smooth muscles of the arrector pili elevate the hairs. [pp. 157, 158]
- ***Cardiovascular system.*** Smooth muscles encircling vessels of the circulatory system provide control over the peripheral distribution of blood and assist in the regulation of blood pressure.
- ***Respiratory system.*** Smooth muscle contraction or relaxation alters the diameters of the respiratory passageways and changes the resistance to airflow.
- ***Digestive system.*** Extensive layers of smooth muscle in the walls of the digestive tract play an essential role in mechanical processing and in moving materials along the tract. Smooth muscle in the walls of the gallbladder contract to eject bile into the digestive tracts.
- ***Urinary system.*** Smooth muscle tissue in the walls of small blood vessels alters the rate of filtration at the kidneys. Layers of smooth muscle in the walls of the ureters transport urine to the urinary bladder; contraction of the smooth muscle in the wall of the urinary bladder forces urine out of the body.
- ***Reproductive system.*** Layers of smooth muscle are important in the male for the movement of sperm along the reproductive tract and for the ejection of glandular secretions from the accessory glands into the reproductive tract. In the female, layers of smooth muscle are important in the movement of ova (and perhaps sperm) along the reproductive tract, and contraction of the smooth muscle in the walls of the uterus expels the fetus at delivery.

❑ Differences between Smooth Muscle Tissue and Other Muscle Tissues

The structure and organization of smooth muscle tissue differs from that of skeletal or cardiac muscle tissue.

Structural Differences

Figure 10-21b

Actin and myosin are present in all three muscle types. In skeletal and cardiac muscle cells, these proteins are organized in sarcomeres, with thin and thick filaments. The internal organization of a smooth muscle cell is very different.

1. There are no T tubules in a smooth muscle fiber, and the sarcoplasmic reticulum forms a loose network throughout the sarcoplasm. There are no myofibrils, sarcomeres, or striations in smooth muscle tissue.
2. Thick filaments are scattered throughout the sarcoplasm of a smooth muscle cell. The myosin proteins are organized differently than in skeletal or cardiac muscle cells, and there are more cross-bridges per thick filament.
3. The thin filaments in a smooth muscle cell are attached to **dense bodies** that are distributed throughout the sarcoplasm in a network of intermediate filaments (Figure 10-21b•). Some of the dense bodies are firmly attached to the sarcolemma. The dense bodies and intermediate filaments anchor the thin filaments so that, when sliding occurs between thin and thick filaments, the cell shortens. Because dense bodies are not arranged in straight lines, when a contraction occurs the muscle cell twists like a corkscrew.
4. Adjacent smooth muscle cells are bound together at dense bodies, transmitting the contractile forces from cell to cell throughout the tissue.
5. Although smooth muscle cells are surrounded by connective tissue, the collagen fibers never unite to form tendons or aponeuroses as they do in skeletal muscles.

Functional Differences

Smooth muscle tissue differs from other muscle types in terms of (1) excitation-contraction coupling, (2) length-tension relationships, and (3) control mechanisms.

EXCITATION-CONTRACTION COUPLING The trigger for smooth muscle contraction is the appearance of free calcium ions in the cytoplasm. Most of these calcium ions enter the cell from the extracellular fluid. Once in the sarcoplasm, the calcium ions interact with *calmodulin*, a calcium-binding protein. Calmodulin then activates an enzyme, **myosin light chain kinase,** that breaks down ATP and initiates the contraction. This situation is quite different from the one in skeletal or cardiac muscle, where the trigger for contraction is the binding of calcium ions to troponin.

LENGTH-TENSION RELATIONSHIPS Because the thick and thin filaments are not rigidly organized, there is no direct relationship between tension development and resting length in smooth muscle. A stretched smooth muscle soon adapts to its new length and retains the ability to contract on demand. This ability to function over a wide range of lengths is called **plasticity.** Smooth muscle can contract over a range of lengths four times greater than that of skeletal muscle. This ability is especially important for digestive organs that undergo great changes in volume, such as the stomach. Despite the lack of sarcomere organization, smooth muscle contractions can be just as powerful as those of skeletal muscles, and, like skeletal muscle fibers, smooth muscle cells often undergo sustained tetanic contractions.

CONTROL OF CONTRACTIONS Many smooth muscle fibers are not innervated by motor neurons, and the neurons that do innervate smooth muscles are not under voluntary control. The nature of the connection with the nervous system provides a means of categorizing smooth muscle cells.

Multiunit Smooth Muscle. **Multiunit smooth muscle cells** are innervated in motor units comparable to those of skeletal muscles, but each smooth muscle cell may be connected to several motor neurons. As in skeletal or cardiac muscles, an action potential is generated and propagated over the sarcolemma, but the contractions are more leisurely. Multiunit smooth muscle cells are not typical of the digestive tract. They are found in the iris of the eye, where they regulate the diameter of the pupil, along portions of the male reproductive tract, within the walls of large arteries, and in the arrector pili muscles of the skin.

Visceral Smooth Muscle. In contrast, many **visceral smooth muscle cells** lack a direct contact with any motor neuron. Visceral smooth muscle cells are found in the walls of the digestive tract, the gallbladder, the urinary bladder, and many other internal organs.

In visceral smooth muscle tissue the muscle cells are arranged in sheets or layers. Within each layer, adjacent muscle fibers are connected by gap junctions. Because they are connected in this way, whenever one muscle cell contracts, the electrical impulse that triggered the contraction can travel to adjacent smooth muscle cells. The contraction therefore spreads in a wave that soon involves every smooth muscle cell in the layer.

The initial stimulus may be the activation of a motor neuron that contacts one of the muscle cells in the region. But smooth muscle cells will contract or relax in response to chemicals, hormones, local concentrations of oxygen or carbon dioxide, or physical factors such as extreme stretching or irritation.

Many visceral smooth muscle networks show rhythmic cycles of activity in the absence of neural stimulation. This is especially true of the smooth muscle cells in the wall of the digestive tract, where **pacesetter cells** undergo spontaneous depolarization and trigger contraction of entire muscular sheets.

 What feature of cardiac muscle tissue allows the heart to act as a functional syncytium?

 Why are cardiac and smooth muscle contractions more affected by changes in extracellular calcium ions than are skeletal muscle contractions?

 Smooth muscle can contract over a wider range of resting lengths than skeletal muscle. Why?

Selected Clinical Terminology

Terms Discussed in This Chapter

botulism: A severe, potentially fatal paralysis of skeletal muscles, resulting from ingestion of the botulinus toxin. *(p. 295* and *AM)*

Duchenne's muscular dystrophy (DMD): One of the most common and best understood diseases in the muscular dystrophies category. *(p. 293* and *AM)*

fibrosis: A process in which increasing amounts of fibrous connective tissue develop, making muscles less flexible. *(p. 314)*

muscular dystrophies (DIS-trō-fēz): A varied collection of congenital diseases that produce progressive muscle weakness and deterioration. *(p. 293* and *AM)*

myasthenia gravis (mī-as-THĒ-nē-uh GRA-vis): A general muscular weakness resulting from fewer ACh receptors on the motor end plate. *(p. 295* and *AM)*

polio: A disease resulting from viral destruction of motor neurons and characterized by the paralysis and atrophy of motor units. *(p. 313* and *AM)*

rigor mortis: A state following death during which muscles are locked in the contracted position, making the body extremely stiff. *(p. 299)*

tetanus: A disease caused by a bacterial toxin that causes sustained, powerful contractions of skeletal muscles. *(p. 302)*

AM *Additional Terms Discussed in the Applications Manual*

cholinesterase inhibitor: A compound that "ties up" the active sites at which acetylcholinesterase normally binds ACh.

chronic fatigue syndrome: A condition resembling fibromyalgia that may develop suddenly shortly after a viral infection.

fibromyalgia syndrome: An inflammatory disorder characterized by a distinctive pattern of symptoms, including tender points on the body surface.

trichinosis (trik-i-NŌ-sis): A condition resulting from a parasitic infection by a nematode worm.

CHAPTER REVIEW

STUDY OUTLINE

SKELETAL MUSCLE TISSUE AND THE MUSCULAR SYSTEM, p. 286

1. There are three types of muscle tissue: skeletal muscle, cardiac muscle, and smooth muscle.
2. **Skeletal muscles** attach to bones directly or indirectly. Their functions are: (1) producing skeletal movement; (2) maintaining posture and body position; (3) supporting soft tissues; (4) guarding entrances and exits; and (5) maintaining body temperature.

ANATOMY OF SKELETAL MUSCLE, p. 286

Connective Tissue Organization, p. 286

Blood Vessels and Nerves, p. 288

1. Each muscle fiber is surrounded by an **endomysium.** Bundles of muscle fibers are sheathed by a **perimysium,** and the entire muscle is covered by an **epimysium.** At the ends of the muscle are **tendons** or **aponeuroses** that attach the muscle to other structures. (*Figure 10-1*)

Microanatomy of Skeletal Muscle Fibers, p. 288

2. A skeletal muscle fiber has a cell membrane, or **sarcolemma, sarcoplasm** (cytoplasm), and **sarcoplasmic reticulum (SR),** similar to the smooth endoplasmic reticulum of other cells. **Transverse tubules (T tubules)** and **myofibrils** aid in contraction. Filaments in a myofibril are organized into repeating functional units called **sarcomeres.** (*Figures 10-2, 10-3, 10-4*)
3. **Myofilaments** consist of **thin filaments** and **thick filaments.** (*Figures 10-2, 10-4*)
4. Thin filaments consist of **F actin, tropomyosin,** and **troponin.** Tropomyosin molecules cover **active sites** on the **G actin** subunits that form the F actin strand. Troponin binds to G actin and to tropomyosin and holds the tropomyosin in position. (*Figure 10-5*)
5. Thick filaments consist of a bundle of myosin molecules, each consisting of an elongate **tail** and a globular **head,** or **cross-bridge.** In a resting muscle cell the interactions between the active sites on G actin and myosin cross-bridges are prevented by tropomyosin. (*Figures 10-3, 10-5*)
6. The **sliding filament theory** explains how the relationship between thick and thin filaments changes as the muscle contracts and shortens. (*Figure 10-6*)

CONTRACTION OF SKELETAL MUSCLE, p. 293

The Control of Skeletal Muscle Activity, p. 295

1. Neural control of muscle function involves a link between electrical activity in the sarcolemma and the initiation of a contraction.
2. Each fiber is controlled by a neuron at a **neuromuscular** *(myoneural)* **junction.** (*Figure 10-7*)
3. When an **action potential** arrives at the **synaptic knob, acetylcholine (ACh)** is released into the **synaptic cleft.** ACh binding to receptors on the opposing **junctional folds** leads to the generation of an action potential in the sarcolemma. (*Figure 10-7*)
4. **Excitation-contraction coupling** occurs as the passage of an action potential along a T tubule triggers the release of calcium ions from the cisternae of the SR.

The Contraction Cycle, p. 296

5. A contraction involves a repeated cycle of "attach, pivot, detach, and return." It begins when calcium ions are released by the SR. The calcium ions bind to troponin, which changes position and moves tropomyosin away from the active sites of actin. Cross-bridge binding of myosin heads to actin can now occur. After binding, each myosin head pivots at its base, pulling the actin filament toward the center of the sarcomere. (*Figure 10-8*)

Relaxation, p. 297

6. **Acetylcholinesterase (AChE)** breaks down ACh and limits the duration of stimulation. (*Figure 10-9; Table 10-1*)
7. The contraction continues until either (1) calcium ion concentrations return to resting levels or (2) the muscle fiber runs out of ATP.

Length-Tension Relationships, p. 299

8. The number of cross-bridge interactions within a muscle fiber determines the amount of tension produced during a contraction. The fibers can contract most forcefully when stimulated over a relatively narrow range of resting lengths. (*Figure 10-9*)

MUSCLE MECHANICS, p. 300

1. A **twitch** is a cycle of contraction and relaxation produced by a single stimulus. (*Figure 10-10*)
2. Repeated stimulation before the relaxation phase ends may produce **summation of twitches (wave summation),** in which one twitch is added to another; **incomplete tetanus,** in which tension will peak because the muscle is never allowed to relax completely; or **complete tetanus,** in which the relaxation phase is completely eliminated. (*Figure 10-11*)
3. Repeated stimulation after the relaxation phase has been completed produces **treppe.** (*Figure 10-11*)
4. The number and size of a muscle's **motor units** indicate how precisely controlled its movements are. (*Figure 10-12*)
5. Resting **muscle tone** stabilizes bones and joints.
6. Normal activities usually include both **isotonic contractions** (in which the tension in a muscle remains constant but the length of the muscle changes) and **isometric contractions** (in which tension rises but the length of the muscle remains constant). (*Figure 10-13*)
7. The return to resting length after a contraction may involve **series elastic elements,** the contraction of opposing muscle groups, and/or gravity. (*Figure 10-14*)

ENERGETICS OF MUSCULAR ACTIVITY, p. 306

1. Muscle contractions require large amounts of energy. A resting muscle generates ATP through aerobic metabolism in mitochondria. (*Table 10-2*)

2. **Creatine phosphate (CP)** can release stored energy to convert ADP to ATP.
3. When a muscle fiber runs short of ATP and CP, enzymes can break down glycogen molecules to release glucose that can be broken down via glycolysis.
4. At rest or at moderate levels of activity, aerobic metabolism can provide most of the necessary ATP to support muscle contractions. (*Figure 10-15*)
5. At peak levels of activity the cell relies heavily on glycolysis to generate ATP, because the mitochondria cannot obtain enough oxygen to meet the existing ATP demands. (*Figure 10-15*)
6. A **fatigued** muscle can no longer contract, because of changes in pH, a lack of energy, or other problems.
7. The **recovery period** begins immediately after a period of muscle activity and continues until conditions inside the muscle have returned to preexertion levels. The **oxygen debt** created during exercise is the amount of oxygen used in the recovery period to restore normal conditions. (*Figure 10-16*)
8. Circulating hormones may alter metabolic activities in skeletal muscle fibers.

MUSCLE PERFORMANCE, p. 310

1. The three types of skeletal muscle fibers are **fast fibers, slow fibers,** and **intermediate fibers.** (*Table 10-3*)
2. Fast fibers, which are large in diameter, contain densely packed myofibrils, large glycogen reserves, and relatively few mitochondria. They produce rapid and powerful contractions of relatively brief duration. (*Figure 10-17*)
3. Slow fibers are only about half the diameter of fast fibers and take three times as long to contract after stimulation. Specializations such as abundant mitochondria, an extensive capillary supply, and high concentrations of **myoglobin** enable them to continue contracting for extended periods. (*Figure 10-17*)
4. Intermediate fibers are very similar to fast fibers, although they have a greater resistance to fatigue.
5. **Anaerobic endurance** is the time over which a muscle can support sustained, powerful muscle contractions through anaerobic mechanisms. Training to develop anaerobic endurance can lead to **hypertrophy** (enlargement) of the stimulated muscles.
6. **Aerobic endurance** is the time over which a muscle can continue to contract while supported by mitochondrial activities. For maximum endurance, energy demands should remain at or below the **anaerobic threshold.** (*Figure 10-18*)

AGING AND THE MUSCULAR SYSTEM, p. 314

1. The aging process reduces the size, elasticity, and power of all muscle tissues. Exercise tolerance and the ability to recover from muscular injuries both decrease.

INTEGRATION WITH OTHER SYSTEMS, p. 315

CARDIAC MUSCLE TISSUE, p. 315

Differences between Cardiac and Skeletal Muscle, p. 317

1. Cardiac muscle cells and skeletal muscle fibers differ structurally in terms of (1) size, (2) the number and location of nuclei, (3) the location of the T tubules at the sarcomeres, (4) their relative dependence on aerobic metabolism when contracting at peak levels, (5) the nature of the metabolic reserves, and (6) the presence or absence of **intercalated discs.** (*Figure 10-20; Table 10-4*)
2. Cardiac and skeletal muscle fibers differ functionally in terms of (1) the source of the stimulus for contraction, (2) the duration of contractions, (3) the ability to undergo tetany, and (4) the relationship with the autonomic nervous system.

SMOOTH MUSCLE TISSUE, p. 318

Differences between Smooth Muscle Tissue and Other Muscle Tissues, p. 319

1. Smooth muscle is nonstriated, involuntary muscle tissue that shows plasticity. (*Figure 10-21*)
2. Smooth muscle lacks sarcomeres and the resulting striations. The thin filaments are anchored to **dense bodies.** (*Table 10-4*)
3. **Visceral smooth muscle cells** are not always innervated by motor neurons. **Multiunit smooth muscle cells** are innervated by motor neurons. Neurons that innervate smooth muscle cells are not under voluntary control.

■ REVIEW QUESTIONS

LEVEL 1 Reviewing Facts and Terms

1. A skeletal muscle contains
   (a) connective tissues
   (b) blood vessels and nerves
   (c) skeletal muscle tissue
   (d) a, b, and c are correct

2. The contraction of a muscle will exert a pull on a bone because:
   (a) muscles are attached to bones by ligaments
   (b) muscles are directly attached to bones
   (c) muscles are attached to bones by tendons
   (d) a, b, and c are correct

3. Detachment of the cross-bridges is directly triggered by:
   (a) repolarization of T tubules
   (b) attachment of ATP to myosin heads
   (c) hydrolysis of ATP
   (d) calcium ions

4. A muscle producing peak tension during rapid cycles of contraction and relaxation is said to be in:
   (a) incomplete tetanus (b) treppe
   (c) tetany (d) twitch

5. The smooth steady increase in muscular tension produced by increasing the number of active motor units is called:
   (a) tetany (b) treppe
   (c) wave summation (d) recruitment

6. The type of contraction in which the tension rises but the resistance does not move is called:
(a) a wave summation
(b) a twitch
(c) an isotonic contraction
(d) an isometric contraction

7. Large diameter, densely packed myofibrils, large glycogen reserves, and few mitochondria are characteristics of:
(a) slow fibers (b) intermediate fibers
(c) fast fibers (d) red muscles

8. An action potential can travel quickly from one cardiac muscle cell to another because of the presence of:
(a) gap junctions (b) tight junctions
(c) intercalated discs (d) a and b are correct

9. List the three types of muscle tissue found in the body.

10. What are the five functions of skeletal muscle?

11. What three layers of connective tissue are a part of each muscle? What functional role does each layer play?

12. What is a motor unit?

13. What structural feature of a skeletal muscle fiber is responsible for conducting action potentials into the interior of the cell?

14. Under resting conditions what two proteins prevent interactions between cross-bridges and active sites?

15. What five interlocking steps are involved in the contraction process?

16. What two factors affect the amount of tension produced when a skeletal muscle contracts?

17. What specialized type of muscle fiber plays an important role in the reflex control of position and posture?

18. What forms of energy reserves are found in resting skeletal muscle fibers?

19. What two mechanisms are used to generate ATP from glucose in muscle cells?

20. What three factors are necessary to prevent premature muscle fatigue?

21. What is the calcium-binding protein found in smooth muscle tissue?

LEVEL 2 Reviewing Concepts

22. An activity that would require anaerobic endurance is:
(a) a 50-meter dash
(b) a pole vault
(c) a weight-lifting competition
(d) a, b, and c are correct

23. Areas of the body where you would not expect to find slow fibers include the:
(a) back and calf muscles
(b) eye and hand
(c) chest and abdomen
(d) a, b, and c are correct

24. When contraction occurs:
(a) the H and I bands get smaller
(b) the Z lines move closer together
(c) the width of the A band remains constant
(d) a, b, and c are correct

25. The amount of tension produced during a contraction is *not* affected by:
(a) the amount of ATP available to the muscle
(b) the degree of overlap between thick and thin filaments
(c) the interaction of troponin and tropomyosin
(d) the number of cross-bridge interactions within a muscle fiber

26. Describe the basic sequence of events that occurs at a neuromuscular junction.

27. Why is the multinucleate condition important in skeletal muscle fibers?

28. Describe the graphic events seen on a myogram as tension is developed in a stimulated calf muscle fiber during a twitch.

29. What three processes are involved in repaying the oxygen debt during the recovery period?

30. What are the effects of aging on the muscular system?

31. Why is cardiac muscle an example of a functional syncytium?

32. How does cardiac muscle tissue contract without neural stimulation?

33. Atracurium is a drug that blocks the binding of ACh. Give an example of a site where such binding normally occurs, and predict the physical effect of this drug.

34. Describe what happens to muscles that are not "used" on a regular basis. What can be done to offset this effect?

35. The time of a murder victim's death is often estimated by the flexibility of the body. Explain why this is possible.

36. Many visceral smooth muscle cells lack motor neuron innervation. How are their contractions coordinated and controlled?

37. A motor unit from a skeletal muscle contains 1500 muscle fibers. Would this muscle be involved in fine, delicate movements or powerful, gross movements? Explain.

LEVEL 3 Critical Thinking and Clinical Applications

38. Many potent insecticides contain toxins called organophosphates that interfere with the action of the enzyme acetylcholinesterase. Terry is using an insecticide containing organophosphates and is very careless. He does not use gloves or a dust mask and absorbs some of the chemical through his skin. He inhales a large amount as well. What symptoms would you expect to observe in Terry as a result of the organophosphate poisoning?

39. A rare (hypothetical) genetic disease causes the body to produce antibodies that compete with acetylcholine for receptors on the motor end plate. Patients suffering from this disease exhibit varying degrees of muscular weakness and flaccid paralysis in the affected muscles. If you were able to administer a drug that inhibits acetylcholinesterase or a drug that inhibits acetylcholine, which would you use to alleviate the symptoms of the disease described?

40. Mary has just completed a 10-km race. Thirty minutes later she begins to notice soreness and stiffness in her leg muscles. She wonders whether she may have damaged the muscle somehow during the race. She visits her doctor, who orders a blood test. How could the doctor tell from a blood test whether muscle damage had occurred?

CHAPTER 11

Skeletal muscles have a remarkable ability to adapt to the demands placed on them. Intensive weight training and physical conditioning will lead to an increase in the size and strength of the muscles being exercised. However, even the most dedicated body-builders cannot work on developing all of the 700 or so skeletal muscles in the human body. Instead, they concentrate on the largest, most prominent muscles responsible for powerful movements of the axial and appendicular skeleton. This chapter will introduce those skeletal muscles and the movements they perform.

The Muscular System

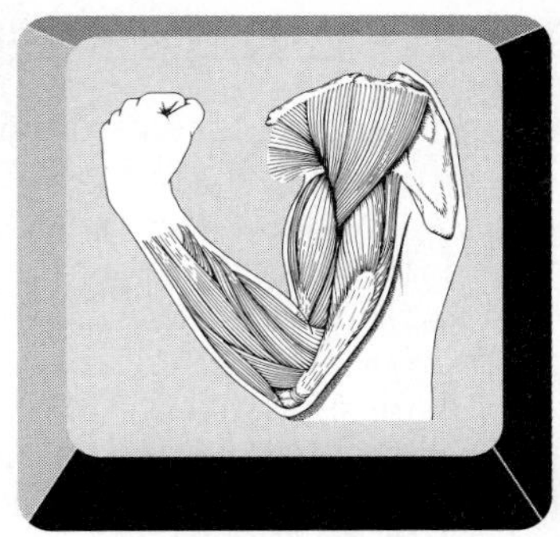

Chapter Outline and Objectives

1 Describe the arrangement of fascicles in the various types of muscles, and explain the resulting functional differences.

2 Describe the different classes of levers and how they make muscles more efficient.

3 Predict the actions of a muscle on the basis of the relative positions of its origin and insertion.

4 Explain how muscles interact to produce or oppose movements.

5 Explain how the name of a muscle can help identify its location, appearance, and/or function.

6 Identify the principal axial muscles of the body, together with their origins, insertions, actions, and innervation.

7 Identify the principal appendicular muscles of the body, together with their origins, insertions, actions, and innervation.

8 Compare the major muscle groups of the upper and lower limbs, and relate their differences to their functional roles.

The **muscular system** includes all of the skeletal muscles that can be controlled voluntarily. Most of the muscle tissue in the body is part of this system, and approximately 700 skeletal muscles have been identified. Some are attached to bony processes, others to broad sheets of connective tissue, but all are directly or indirectly associated with the skeletal system. Rather than attempt to survey all 700 skeletal muscles, we will focus on a relatively small but representative number of muscles, about 20 percent of the total. To ease the strain of memorization, these muscles have been organized into anatomical and functional groups.

The shape or appearance of each muscle provides clues to its primary function. Muscles involved with locomotion and posture work across joints, producing skeletal movement. Those that support soft tissue form slings or sheets between relatively stable bony elements, whereas those that guard an entrance or exit completely encircle the opening.

At the level of the individual skeletal muscle, two factors interact to determine the effects of its contraction: (1) the anatomical arrangement of the muscle fibers and (2) the way the muscle attaches to the bones of the skeletal system. The performance of muscles in the body can be understood in terms of basic mechanical laws. The analysis of biological systems in mechanical terms is the study of *biomechanics*. This chapter examines the biomechanics and gross anatomy of the muscular system.

Biomechanics and Muscle Anatomy

Although most skeletal muscle fibers contract at comparable rates and shorten to the same degree, variations in microscopic and macroscopic organization can dramatically affect the power, range, and speed of movement produced when a muscle contracts.

Organization of Skeletal Muscle Fibers

Muscle fibers within a skeletal muscle form bundles called *fascicles*. [p. 287] The muscle fibers within a single fascicle are arranged in parallel, but the organization of the fascicles in the skeletal muscle can vary, as can the relationship between the fascicles and the associated tendon. Four different patterns of fascicle organization produce *parallel muscles*, *convergent muscles*, *pennate muscles*, and *circular muscles*.

FIGURE 11-1
Different Arrangements of Skeletal Muscle Fibers

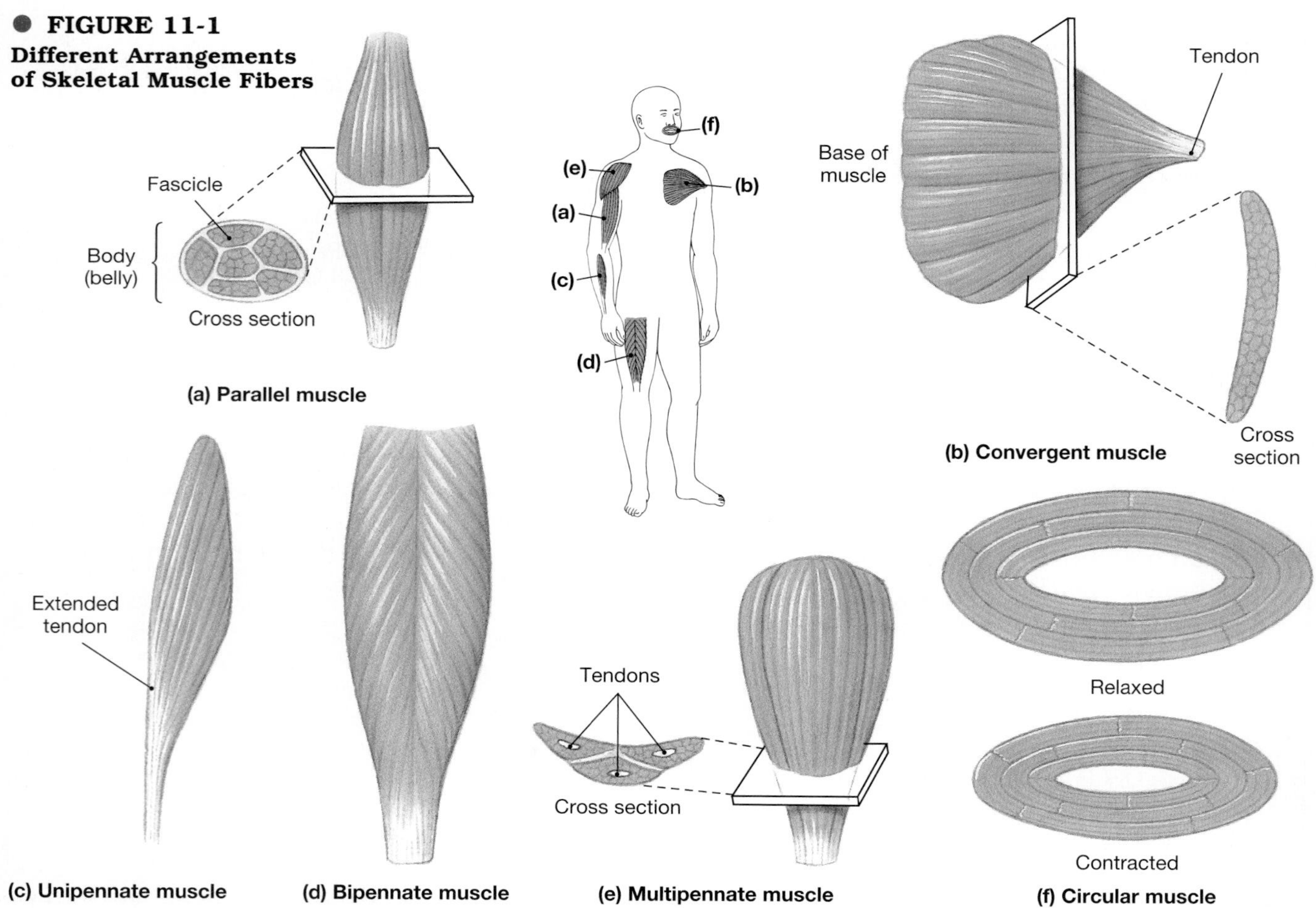
Fascicle
Body (belly)
Cross section
(a) Parallel muscle
(f)
(e)
(b)
(a)
(c)
(d)
Tendon
Base of muscle
Cross section
(b) Convergent muscle
Extended tendon
Tendons
Cross section
Relaxed
Contracted
(c) Unipennate muscle
(d) Bipennate muscle
(e) Multipennate muscle
(f) Circular muscle

Parallel Muscles

Figure 11-1a

In a **parallel muscle** the fascicles are parallel to the long axis of the muscle. The functional characteristics of a parallel muscle resemble those of an individual muscle fiber.

Most of the skeletal muscles in the body are parallel muscles. Some are flat bands, with broad attachments (*aponeuroses*) at each end; others are plump and cylindrical, with tendons at one or both ends. In the latter case the muscle is spindle-shaped (Figure 11-1a●), with a central **body,** also known as the *belly,* or *gaster* (GAS-ter; stomach). The *biceps brachii* muscle of the arm is an example of a parallel muscle with a central body. When it contracts, a parallel muscle gets shorter and larger in diameter; the bulge of the contracting biceps brachii can be seen on the anterior surface of the arm when the elbow is flexed.

A skeletal muscle cell can contract effectively until it has shortened by roughly 30 percent. Because the muscle fibers are parallel to the long axis of the muscle, when they contract together the entire muscle shortens by the same amount. For example, if the skeletal muscle is 10 cm long, the end of the tendon will move 3 cm when the muscle contracts. The tension developed by the muscle during this contraction depends on the total number of myofibrils it contains. Because the myofibrils are distributed evenly through the sarcoplasm of each cell, the tension can be estimated on the basis of the cross-sectional area of the resting muscle. A parallel skeletal muscle 6.25 cm^2 (1 $in.^2$) in cross-sectional area can develop approximately 23 kg (50 lb) of tension.

Convergent Muscles

Figure 11-1b

In a **convergent muscle** the muscle fibers are based over a broad area, but all the fibers come together at a common attachment site. They may pull on a tendon, a tendinous sheet, or a slender band of collagen fibers known as a **raphe** (RĀ-fē; a seam). The muscle fibers often spread out, like a fan or a broad triangle, with a tendon at the tip (Figure 11-1b●). The prominent chest muscles of the *pectoralis group* have this shape. Such a muscle has versatility, for the direction of pull can be changed by stimulating only one group of muscle cells at any one time. But when they all contract at once, they do not pull as hard on the tendon as a parallel muscle of the same size. The reason is that the muscle fibers on opposite sides of the tendon are pulling in different directions, rather than working together.

Pennate Muscles

Figures 11-1c,d,e, 11-14, 11-20, 11-23

In a **pennate muscle** (*penna,* feather) the fascicles form a common angle with the tendon. Because the muscle cells pull at an angle, contracting pennate muscles do not move their tendons as far as parallel muscles do. But a pennate muscle will contain more muscle fibers than a parallel muscle of the same size. A muscle that has more muscle fibers also has more myofibrils and sarcomeres, and as a result the contraction of the pennate muscle generates more tension than a parallel muscle of the same size.

If all of the muscle fibers are found on the same side of the tendon, the muscle is *unipennate* (Figure 11-1c●). A long muscle that extends the fingers, the *extensor digitorum* (Figure 11-20●, p. 361), is an example of a unipennate muscle. More commonly, there are muscle fibers on both sides of the tendon. A prominent muscle of the thigh, the *rectus femoris* (Figure 11-22c,d●, p. 368), is a *bipennate* muscle (Figure 11-1d●) that helps to extend the knee. If the tendon branches within the muscle, the muscle is *multipennate* (Figure 11-1e●). The triangular *deltoid* muscle that covers the superior surface of the shoulder joint (Figure 11-14●, p. 350) is an example of a multipennate muscle.

Circular Muscles

Figure 11-1f

In a **circular muscle,** or **sphincter** (SFINK-ter), the fibers are concentrically arranged around an opening or recess. When the muscle contracts, the diameter of the opening decreases. Circular muscles guard entrances and exits of internal passageways such as the digestive and urinary tracts. An example is the *orbicularis oris* of the mouth (Figure 11-1f●).

❑ Skeletal Muscle Length-Tension Relationships

Skeletal muscle fibers operate most efficiently, producing maximum tension, over a relatively narrow range of sarcomere lengths. ∞ *[p. 299]* Stretched too far, the muscle loses power because the overlap between myofilaments is reduced. Excessive overlap also reduces the production of tension by disrupting the normal relationship between thick and thin filaments.

Normally, other muscles, tendons, bones, and ligaments keep the degree of extension or compression of muscle fibers within tolerable limits. However, variations in resting sarcomere length still have an effect on the amount of tension that can be produced when the muscle is stimulated. For example, consider what happens when you try to lift a heavy bucket by flexing the elbow. This action is difficult if your forearm is extended, but it becomes easier as your elbow approaches a 90° angle. As you near full flexion, tension production declines, and movement becomes more difficult.

When complex movements occur, muscles often work in groups rather than as individuals. Their cooperation helps to improve the efficiency of a particular movement. For example, large arm and leg muscles produce flexion or extension over an extended range of motion. Although these muscles cannot develop much tension at full extension, they are usually paired with smaller muscles that provide assistance until the larger muscle can perform at maximum efficiency. The sarcomeres of the smaller muscle are at optimal length at full extension, so when the two contract together, the power of the smaller muscle will be at a maximum while that of the larger muscle is at a minimum. The importance of this smaller "assistant" decreases as the movement proceeds.

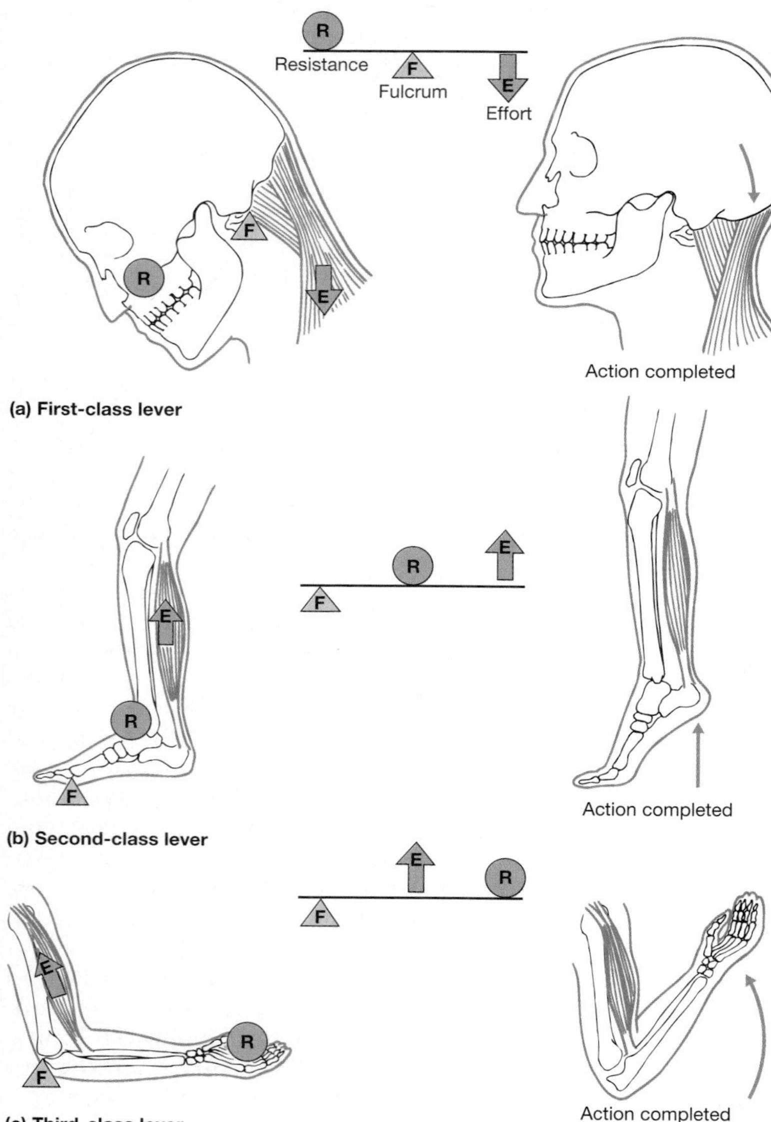
R
Resistance
F
Fulcrum
E
Effort
Action completed
(a) First-class lever
Action completed
(b) Second-class lever
Action completed
(c) Third-class lever

● **FIGURE 11-2**

The Three Classes of Levers. (a) In a first-class lever, the applied force and the resistance are on opposite sides of the fulcrum. First-class levers can change the amount of force transmitted to the resistance and alter the direction and speed of movement. **(b)** In a second-class lever, the resistance lies between the applied force and the fulcrum. This arrangement magnifies force at the expense of distance and speed; the direction of movement remains unchanged. **(c)** In a third-class lever, the force is applied between the resistance and the fulcrum. This arrangement increases speed and distance moved but requires a larger applied force. The example shows the force and distance relationships involved in the contraction of the biceps brachii muscle that flexes the forearm.

❑ Levers

Skeletal muscles do not work in isolation. When a muscle is attached to the skeleton, the nature and site of the connection will determine the force, speed, and range of the movement produced. These characteristics are interdependent, and the relationships can explain a great deal about the general organization of the muscular and skeletal systems.

The force, speed, or direction of movement produced by contraction of a muscle can be modified by attaching the muscle to a lever. A **lever** is a rigid structure—such as a board, a crowbar, or a bone—that moves on a fixed point, called the **fulcrum.** In the body, each bone is a lever and each joint a fulcrum. A child's teeter-totter, or seesaw, provides a more familiar example of lever action. Levers can change (1) the direction of an applied force, (2) the distance and speed of movement produced by a force, and (3) the effective strength of a force.

Classes of Levers

Figure 11-2

Three classes of levers are found in the human body. The seesaw is an example of a **first-class lever:** one in which the fulcrum lies between the applied force and the resistance. There are not many examples of first-class levers in the body. One, involving the muscles that extend the neck, is shown in Figure 11-2a●.

In a **second-class lever** (Figure 11-2b●) the resistance is located between the applied force and the fulcrum. A familiar example of such a lever is a loaded wheelbarrow. The weight of the load is the resistance, and the upward lift on the handle is the applied force. Because in this arrangement the force is always farther from the fulcrum than the resistance is, a small force can balance a larger weight. In other words, the effective force is increased. Notice, however, that when a force moves the handle, the resistance moves more slowly and covers a shorter distance. There are few examples of second-class levers in the body. In performing plantar flexion, the calf muscles act via a second-class lever (Figure 11-2b●).

Third-class levers are the most common levers in the body. In this lever system, a force is applied between the resistance and the fulcrum (Figure 11-2c●). The effect of this arrangement is just the reverse of that produced by a second-class lever: Speed and distance traveled are increased at the expense of effective force. In the example illustrated (the biceps brachii that flexes the forearm), the resistance is six times farther away from the fulcrum than is the applied force. The effective force is reduced accordingly. The muscle must generate 180 kg of tension at its attachment to the forearm to support 30 kg held in the hand. However, the distance traveled and the speed of movement are *increased* by the same ratio: The resistance will travel 45 cm when the insertion point moves 7.5 cm.

Although every muscle does not operate as part of a lever system, the presence of levers provides speed and versatility far in excess of what we would predict on the basis of muscle physiology alone. Skeletal muscle fibers resemble one another closely, and their abilities to contract and generate tension are quite similar. Consider a skeletal muscle that can contract in 500 msec and shorten 1 cm while exerting a 10-kg pull. Without using a lever, this muscle would be performing efficiently only when moving a 10-kg weight a distance of 1 cm. But by using a lever, the same muscle operating at the same efficiency could move 20 kg a distance of 0.5 cm, 5 kg a distance of 2 cm, or 1 kg a distance of 10 cm.

Why does a pennate muscle generate more tension than a parallel muscle of the same size?

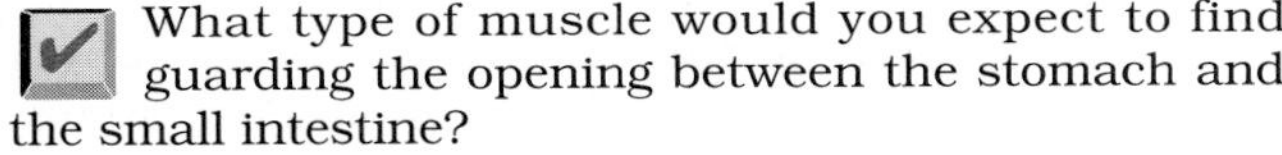

What type of muscle would you expect to find guarding the opening between the stomach and the small intestine?

The joint between the occipital bone of the skull and the first cervical vertebra (atlas) is an example of what type of lever system?

Muscle Terminology

This chapter focuses on the functional anatomy of skeletal muscles and muscle groups. Once again you will be faced with a number of new terms, and this section attempts to give you some assistance in understanding them.

Origins and Insertions

Each muscle begins at an **origin,** ends at an **insertion,** and contracts to produce a specific **action.** In general, the origin remains stationary while the insertion moves, or the origin is proximal to the insertion. For example, the triceps brachii inserts on the olecranon process and originates closer to the shoulder. Such determinations are made during normal movement. Part of the fun of studying the muscular system is that you can actually do the movements and think about the muscles involved, and laboratory discussions of the muscular system often resemble disorganized aerobics classes.

When the origins and insertions cannot be determined easily on the basis of movement, other rules are used. If a muscle extends between a broad aponeurosis and a narrow tendon, the aponeurosis is the origin and the tendon is the insertion. If there are several tendons at one end and just one at the other, there are multiple origins and a single insertion. These simple rules cannot cover every situation, and knowing which end is the origin and which is the insertion is ultimately less important that knowing where the two ends attach and what the muscle accomplishes when it contracts.

Actions

Almost all skeletal muscles either originate or insert upon the skeleton. When a muscle moves a portion of the skeleton, that movement may involve *flexion, extension, adduction, abduction, protraction, retraction, elevation, depression, rotation, circumduction, pronation, supination, inversion,* or *eversion.* Before proceeding, consider reviewing the discussions of planes of motion and Figures 9-2 to 9-5. [pp. 267–270]

Muscles can be grouped according to their **primary actions.**

- A **prime mover,** or **agonist,** is a muscle whose contraction is chiefly responsible for producing a particular movement. The *biceps brachii* is an example of a prime mover that produces flexion of the forearm.
- **Antagonists** are prime movers whose actions oppose that of the agonist under consideration. The *triceps brachii* is a prime mover that extends the forearm. It is therefore an antagonist of the biceps brachii, and the biceps brachii is an antagonist of the triceps brachii. Agonists and antagonists are functional opposites—if one produces flexion, the other will have extension as its primary action. When an agonist contracts to produce a particular movement the coresponding antagonist will be stretched, but it will usually not relax completely. Instead, its tension will be adjusted to control the speed of the movement and ensure its smoothness.
- When a **synergist** (*syn-*, together + *ergon*, work) contracts, it assists the prime mover in performing that action. Synergists may provide additional pull near the insertion or stabilize the point of origin. Their importance in assisting a particular movement may change as the movement progresses; in many cases they are most useful at the start, when the prime mover is stretched and unable to develop maximum tension. For example, the *latissimus dorsi* is a large trunk muscle that extends, adducts, and medially rotates the arm. A much smaller muscle, the *teres major,* assists in starting such movements with the arm fully flexed. Synergists may also assist an agonist by *preventing* movement at a joint and thereby stabilizing the origin of the agonist. These muscles are called *fixators.*

❑ Names of Skeletal Muscles

You will not need to learn every one of the approximately 700 muscles in the human body, but you will have to become familiar with the most important ones. Fortunately, anatomists of the past did not have memories any better than ours. Rather than writing the answers on their sleeves, they assigned names to the muscles that provided clues to their identification. If you can learn to recognize the clues, you will find it easier to remember the names and identify the muscles. The name of a muscle may include information concerning its *fascicle organization, location, relative position, structure, size, shape, origin and insertion,* or *action.*

Fascicle Organization

A muscle name may refer to the orientation of the muscle fibers within a particular skeletal muscle. **Rectus** means "straight," and rectus muscles are parallel muscles whose fibers generally run along the long axis of the body. Because there are several rectus muscles, the name usually includes a second term that refers to a precise region of the body. For example, the *rectus abdominis* is found on the abdomen, and the *rectus femoris* on the thigh. Other directional indicators include **transversus** and **obliquus** for muscles whose fibers run across or at an oblique angle to the longitudinal axis of the body.

Location

Table 11-1 includes a useful summary of terms that designate specific regions of the body. They are usually found as modifiers that help to identify individual muscles, as in the case of the rectus muscles noted previously. In a few cases, the muscle is such a prominent feature of the region that the regional name alone will identify it. Examples would include the *temporalis* of the head and the *brachialis* of the arm.

Relative Position

Muscles visible at the body surface are often called **externus** or **superficialis,** whereas those lying beneath are termed **internus** or **profundus.** Superficial muscles that position or stabilize an organ are called **extrinsic** muscles; those that operate within the organ are called **intrinsic** muscles.

Structure, Size, and Shape

Other muscles were named after specific and unusual structural features. The *biceps* muscle has two tendons of origin (*bi-*, two + *caput,* head), the *triceps* has three, and the *quadriceps* four. Shape is sometimes an important clue to the name of a muscle. For example, *trapezius* (tra-PĒ-zē-us), *deltoid, rhomboideus* (rom-BOY-dē-us), and *orbicularis* (or-bik-ū-LAR-is) refer to prominent muscles that look like a trapezoid, a triangle, a rhomboid, or a circle, respectively. Long muscles are called **longus** (long) or **longissimus** (longest), and **teres** muscles are both long and round. Short muscles are called **brevis;** large ones are called **magnus** (big), **major** (bigger), or **maximus** (biggest); and small ones are called **minor** (smaller) or **minimus** (smallest).

Origin and Insertion

Many names tell you the specific origin and insertion of each muscle. In such cases, the first part of the name indicates the origin and the second part the insertion. The *genioglossus,* for example, originates at the chin (*geneion*) and inserts in the tongue (*glossus*). Although the names may be long and occasionally difficult to pronounce, most can be figured out with the help of Table 11-1 and the anatomical terms introduced in Chapter 1. ∞ *[pp. 17–21]*

Action

Many muscles are named *flexor, extensor, retractor,* and so on. These are such common actions that the names almost always include other clues concerning the appearance or location of the muscle. For example, the *extensor carpi radialis longus* is a long muscle found along the radial (lateral) border of the forearm. When it contracts its primary function is extension of the carpus (wrist).

A few muscles are named after the specific movements associated with special occupations or habits. The *sartorius* (sar-TO-rē-us) muscle is active when crossing the legs. Before sewing machines were invented, a tailor would sit on the floor cross-legged, and the name of the muscle was derived from *sartor,* the Latin word for "tailor." On the face, the *buccinator* (BUK-si-nā-tor) muscle compresses the cheeks, as when pursing the lips and blowing forcefully. *Buccinator* translates as "trumpet player." Finally, another facial muscle, the *risorius* (ri-SO-rē-us), was supposedly named after the mood expressed. The Latin term *risor,* however, means "laugher"; a more appropriate description for the effect would be "grimace."

❑ Axial and Appendicular Muscles

Figure 11-3

The separation of the skeletal system into axial and appendicular divisions provides a useful guideline for subdividing the muscular system as well.

- The **axial musculature** arises on the axial skeleton. It positions the head and spinal column and also moves the rib cage, assisting in the movements that make breathing possible. It does not play a role in movement or support of the pectoral or pelvic girdles or appendages. This category encompasses roughly 60 percent of the skeletal muscles in the body.

TABLE 11-1 Muscle Terminology

Terms Indicating Direction Relative to Axes of the Body	*Terms Indicating Specific Regions of the Body**	*Terms Indicating Structural Characteristics of the Muscle*	*Terms Indicating Actions*
Anterior (front)	Abdominis (abdomen)	**Origin**	**General**
Externus (superficial)	Anconeus (elbow)	Biceps (two heads)	Abductor
Extrinsic (outside)	Auricularis (auricle of ear)	Triceps (three heads)	Adductor
Inferioris (inferior)	Brachialis (brachium)	Quadriceps (four heads)	Depressor
Internus (deep, internal)	Capitis (head)	**Shape**	Extensor
Intrinsic (inside)	Carpi (wrist)	Deltoid (triangle)	Flexor
Lateralis (lateral)	Cervicis (neck)	Orbicularis (circle)	Levator
Medialis/medius (medial, middle)	Cleido/clavius (clavicle)	Pectinate (comblike)	Pronator
Obliquus (oblique)	Coccygeus (coccyx)	Piriformis (pear-shaped)	Rotator
Posterior (back)	Costalis (ribs)	Platys- (flat)	Supinator
Profundus (deep)	Cutaneous (skin)	Pyramidal (pyramid)	Tensor
Rectus (straight, parallel)	Femoris (femur)	Rhomboideus (rhomboid)	**Specific**
Superficialis (superficial)	Genio- (chin)	Serratus (serrated)	Buccinator (trumpeter)
Superioris (superior)	Glosso/glossal (tongue)	Splenius (bandage)	Risorius (laugher)
Transversus (transverse)	Hallucis (great toe)	Teres (long and round)	Sartorius (like a tailor)
	Ilio- (ilium)	Trapezius (trapezoid)	
	Inguinal (groin)	**Other Striking Features**	
	Lumborum (lumbar region)	Alba (white)	
	Nasalis (nose)	Brevis (short)	
	Nuchal (back of neck)	Gracilis (slender)	
	Oculo- (eye)	Lata (wide)	
	Oris (mouth)	Latissimus (widest)	
	Palpebrae (eyelid)	Longissimus (longest)	
	Pollicis (thumb)	Longus (long)	
	Popliteus (behind knee)	Magnus (large)	
	Psoas (loin)	Major (larger)	
	Radialis (radius)	Maximus (largest)	
	Scapularis (scapula)	Minimus (smallest)	
	Temporalis (temples)	Minor (smaller)	
	Thoracis (thoracic region)	-tendinosus (tendinous)	
	Tibialis (tibia)	Vastus (great)	
	Ulnaris (ulna)		
	Uro- (urinary)		

*For other regional terms, refer to Figure 1-7, pp. 17–18, which deals with anatomical landmarks.

- The **appendicular musculature** stabilizes or moves components of the appendicular skeleton.

Figure 11-3• provides an overview of the axial and appendicular muscles of the human body. We will now examine representatives of these muscular divisions, paying attention to patterns of origin, insertion, and action. This discussion relies heavily on an understanding of skeletal anatomy, and you may find it helpful to review appropriate figures in Chapters 7, 8, and 9 as you proceed. For convenience, the relevant figures are indicated in the figure captions in this chapter.

✔ The gracilis muscle is attached to the anterior surface of the tibia at one end and to the pubis and ischium of the pelvis at the other. When the muscle contracts, the leg flexes. Which attachment point is the muscle's origin?

✔ Muscle A abducts the humerus, and muscle B adducts the humerus. On the basis of their actions what is the relationship between these two muscles?

✔ What does the name *flexor carpi radialis longus* tell you about this muscle?

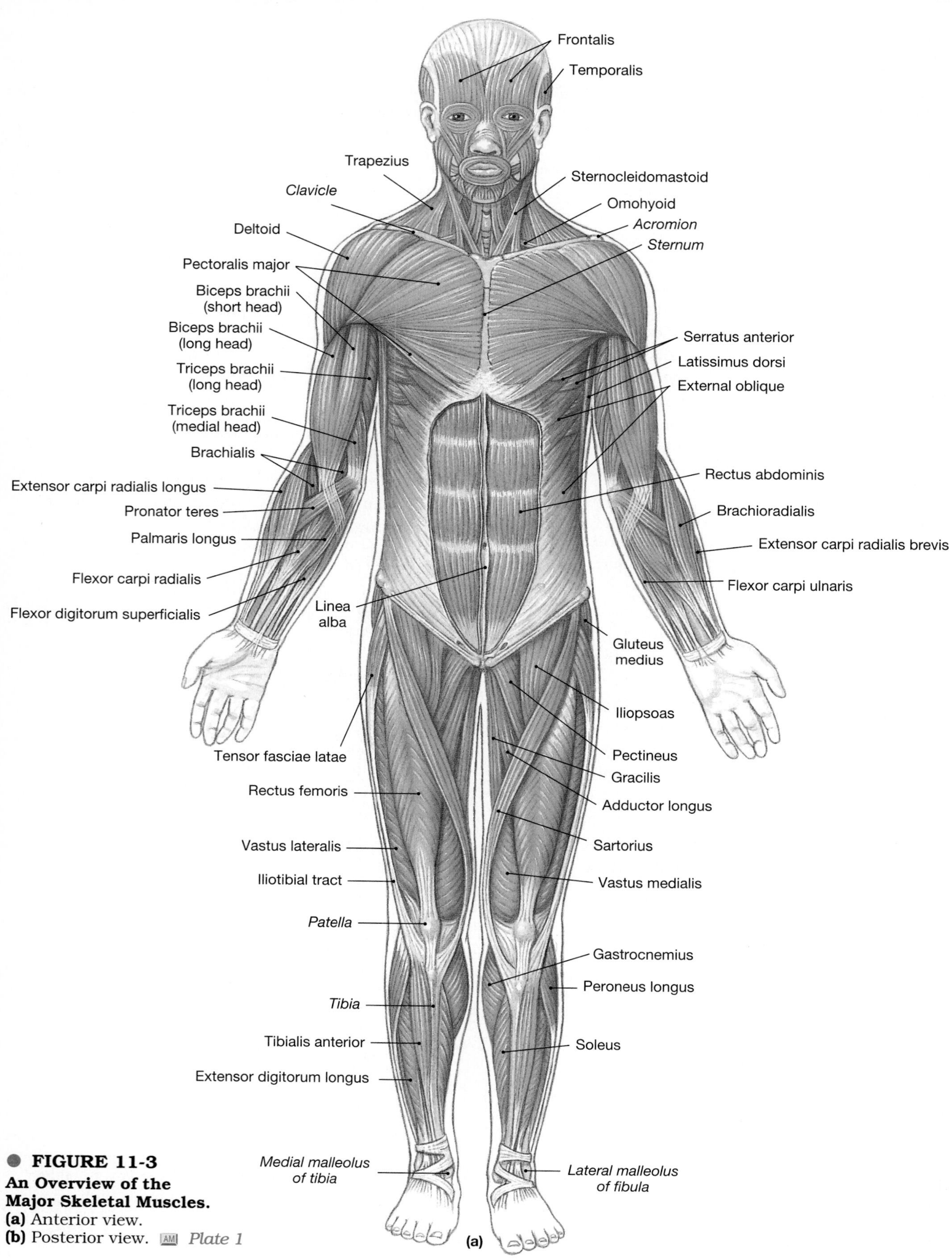
Frontalis
Temporalis
Trapezius
Clavicle
Sternocleidomastoid
Omohyoid
Acromion
Sternum
Deltoid
Pectoralis major
Biceps brachii (short head)
Biceps brachii (long head)
Triceps brachii (long head)
Triceps brachii (medial head)
Brachialis
Extensor carpi radialis longus
Pronator teres
Palmaris longus
Flexor carpi radialis
Flexor digitorum superficialis
Serratus anterior
Latissimus dorsi
External oblique
Rectus abdominis
Brachioradialis
Extensor carpi radialis brevis
Flexor carpi ulnaris
Linea alba
Gluteus medius
Iliopsoas
Tensor fasciae latae
Pectineus
Gracilis
Rectus femoris
Adductor longus
Vastus lateralis
Sartorius
Iliotibial tract
Vastus medialis
Patella
Gastrocnemius
Peroneus longus
Tibia
Tibialis anterior
Soleus
Extensor digitorum longus
Medial malleolus of tibia
Lateral malleolus of fibula
(a)

FIGURE 11-3
An Overview of the Major Skeletal Muscles.
(a) Anterior view.
(b) Posterior view. AM *Plate 1*

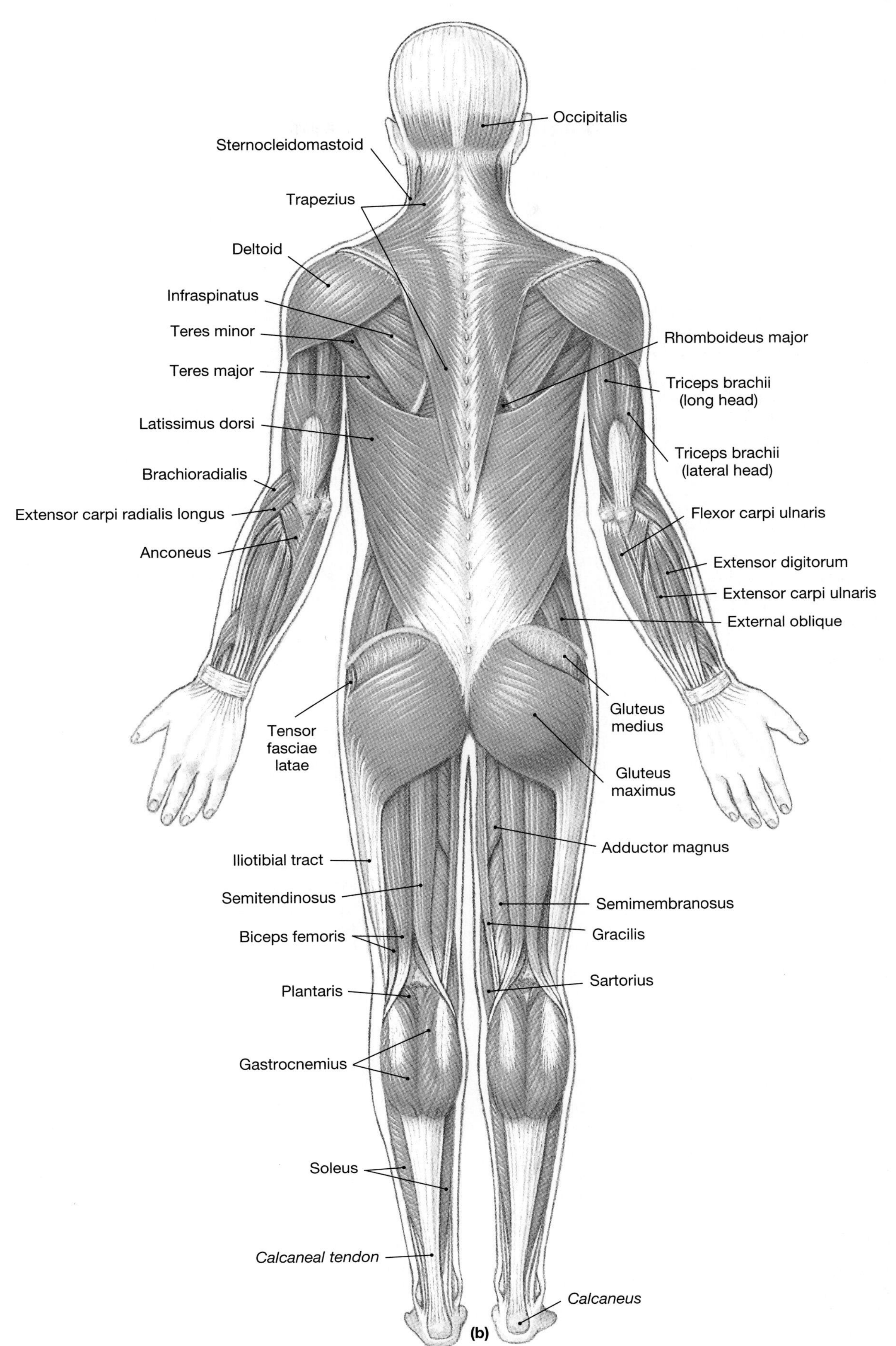
Occipitalis
Sternocleidomastoid
Trapezius
Deltoid
Infraspinatus
Teres minor
Teres major
Latissimus dorsi
Brachioradialis
Extensor carpi radialis longus
Anconeus
Rhomboideus major
Triceps brachii (long head)
Triceps brachii (lateral head)
Flexor carpi ulnaris
Extensor digitorum
Extensor carpi ulnaris
External oblique
Gluteus medius
Tensor fasciae latae
Gluteus maximus
Adductor magnus
Iliotibial tract
Semitendinosus
Semimembranosus
Gracilis
Biceps femoris
Sartorius
Plantaris
Gastrocnemius
Soleus
Calcaneal tendon
Calcaneus

(b)

The Axial Musculature

The axial muscles fall into logical groups based on location, function, or both. The groups do not always have distinct anatomical boundaries. For example, a function such as the extension of the spine involves muscles along the entire length of the spinal column.

1. The first group includes the *muscles of the head and neck* that are not associated with the spinal column. These muscles include those that move the face, tongue, and larynx. They are therefore responsible for verbal and nonverbal communication—laughing, talking, frowning, smiling, whistling. This group of muscles also performs feeding activities, such as sucking and chewing, as well as contractions of the eye muscles that help us look around for something to eat.
2. The second group, the *muscles of the spine,* includes numerous flexors and extensors of the head, neck, and spinal column.
3. The third group, the *oblique and rectus muscles,* form the muscular walls of the thoracic and abdominopelvic cavities between the first thoracic vertebra and the pelvis. In the thoracic area these muscles are partitioned by the ribs, but over the abdominal surface they form broad muscular sheets. There are also oblique and rectus muscles in the neck. Although they do not form a complete muscular wall, they are included in this group because they share a common developmental origin.
4. The fourth group, the *muscles of the pelvic floor,* extend between the sacrum and pelvic girdle, forming the muscular *perineum* that closes the pelvic outlet.

The tables that follow summarize information concerning the origin, insertion, and action of each muscle. You will also find information concerning the *innervation* of the individual muscles. **Innervation** refers to the identity of the nerve that controls the muscle. The names of the nerves provide information concerning the distribution of the nerve or the site at which the nerve leaves the cranial or spinal cavity. For example, the *facial nerve* innervates the facial musculature, and the various *spinal nerves* exit via the intervertebral foramina. You will learn more about the individual nerves involved when we discuss the peripheral nervous system in Chapters 14 and 15.

Muscles of the Head and Neck

The muscles of the head and neck can be divided into several groups. The *muscles of facial expression,* the *muscles of mastication,* the *muscles of the tongue,* and the *muscles of the pharynx* originate on the skull or hyoid bone. Muscles involved with sight and hearing are also based on the skull. The *extrinsic eye muscles*—those associated with movements of the eye—will be considered here. The intrinsic eye muscles, which control the diameter of the pupil and the shape of the lens, are discussed in Chapter 17, along with the muscles associated with the auditory ossicles. In the neck, the *extrinsic muscles of the larynx* adjust the position of the hyoid bone and larynx. The intrinsic laryngeal muscles, including those of the vocal cords, will be discussed in Chapter 23 (the respiratory system).

Muscles of Facial Expression

Figure 11-4

The muscles of facial expression originate on the surface of the skull (Figure 11-4• and Table 11-2). At their insertions the epimysial fibers are woven into those of the superficial fascia and the dermis of the skin; when they contract, the skin moves. These muscles are innervated by the seventh cranial nerve, the *facial nerve.*

The largest group of facial muscles is associated with the mouth. The **orbicularis oris** constricts the opening, while other muscles move the lips or the corners of the mouth. The **buccinator,** one of the muscles associated with the mouth, has two functions related to feeding (in addition to its importance to musicians). During chewing it cooperates with the masticatory muscles by moving food back across the teeth from the space inside the cheeks. In infants, the buccinator provides suction for suckling at the breast.

Smaller groups of muscles control movements of the eyebrows and eyelids, the scalp, the nose, and the external ear. The **epicranius** (ep-i-KRĀ-nē-us; *epi-*, on + *kranion,* skull) of the scalp consists of two muscles, the **frontalis** and the **occipitalis,** separated by a collagenous sheet, the **galea aponeurotica** (GĀ-lē-uh ap-ō-nū-RO-ti-kuh; *galea,* a helmet + aponeurosis). The **platysma** (pla-TIZ-muh; *platys,* flat) covers the ventral surface of the neck, extending from the base of the neck to the periosteum of the mandible and the fascia at the corner of the mouth.

FIGURE 11-4
Muscles of Facial Expression.
See also Figure 7-3, pp. 205–206.
(a) Anterolateral view. **(b)** Anterior view.
Plates 4.2, 4.3a, b, c

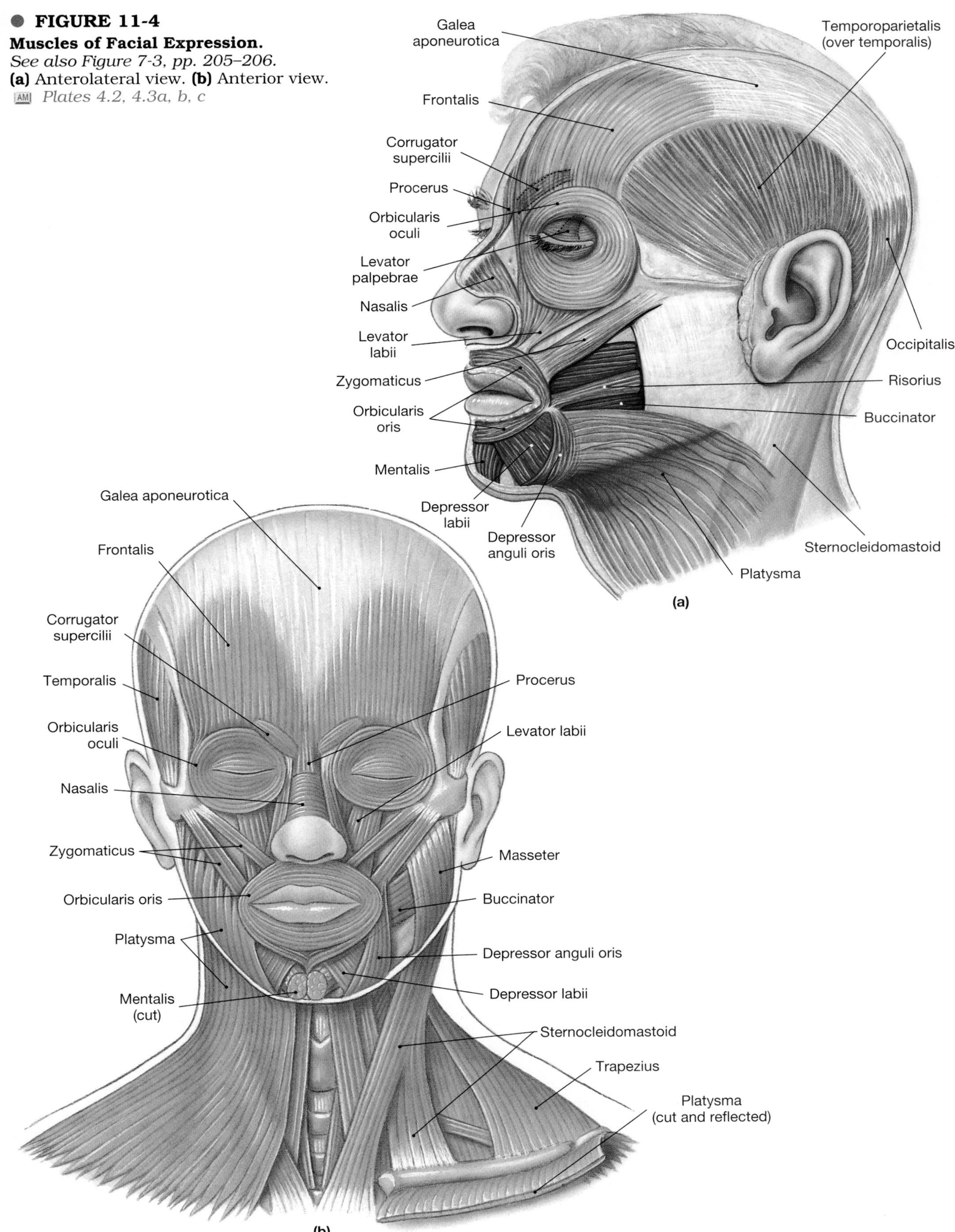
Galea aponeurotica
Temporoparietalis (over temporalis)
Frontalis
Corrugator supercilii
Procerus
Orbicularis oculi
Levator palpebrae
Nasalis
Levator labii
Zygomaticus
Orbicularis oris
Mentalis
Depressor labii
Depressor anguli oris
Occipitalis
Risorius
Buccinator
Sternocleidomastoid
Platysma
(a)
Galea aponeurotica
Frontalis
Corrugator supercilii
Temporalis
Orbicularis oculi
Nasalis
Zygomaticus
Orbicularis oris
Platysma
Mentalis (cut)
Procerus
Levator labii
Masseter
Buccinator
Depressor anguli oris
Depressor labii
Sternocleidomastoid
Trapezius
Platysma (cut and reflected)
(b)

TABLE 11-2 Muscles of Facial Expression

Region/Muscle	*Origin*	*Insertion*	*Action*	*Innervation*
Mouth				
Buccinator	Alveolar processes of maxillae and mandible	Blends into fibers of orbicularis oris	Compresses cheeks	Facial nerve (N VII)
Depressor labii	Mandible	Lower lip	Depresses lip	As above
Levator labii	Maxillae	Orbicularis oris	Raises upper lip	As above
Mentalis	Mandible	Skin of chin	Elevates and protrudes lower lip	As above
Orbicularis oris	Maxillae and mandible	Lips	Compresses, purses lips	As above
Risorius	Fascia surrounding parotid salivary gland	Angle of mouth	Draws corner of mouth to the side	As above
Depressor anguli oris	Anterolateral surface of mandibular body	Skin at angle of mouth	Depresses corner of mouth	Zygomatic branch of facial nerve (N VII)
Zygomaticus	Zygomatic bone near zygomaticomaxillary suture	Angle of mouth	Draws corner of mouth back and up	As above
Eye				
Corrugator supercilii	Frontal bone near nasal suture	Eyebrow	Pulls skin down and forward; wrinkles brow	As above
Levator palpebrae	Tendinous band around optic foramen	Upper eyelid	Raises upper eyelid	Oculomotor nerve (N III)*
Orbicularis oculi	Medial margin of orbit	Skin around eyelids	Closes eye	Facial nerve (N VII)
Nose				
Procerus	Nasal bone, lateral nasal cartilages	Skin between eyebrows	Moves nose, changes position and shape of nostrils	As above
Nasalis	Maxilla and alar cartilage of nose	Bridge of nose	Compresses bridge, depresses tip, elevates corners	As above
Ear (extrinsic)				
Temporoparietalis	Fascia around external ear	Galea aponeurotica	Tenses scalp, moves external ear	As above
Scalp				
Frontalis	Galea aponeurotica	Skin of eyebrow and bridge of nose	Raises eyebrows, wrinkles forehead	As above
Occipitalis	Superior nuchal line	Galea aponeurotica	Tenses, retracts scalp	As above
Neck				
Platysma	Upper thorax between cartilage of second rib and acromion of scapula	Mandible and skin of cheek	Tenses skin of neck, depresses mandible	As above

*This muscle originates in association with the extrinsic oculomotor muscles, so its innervation is unusual.

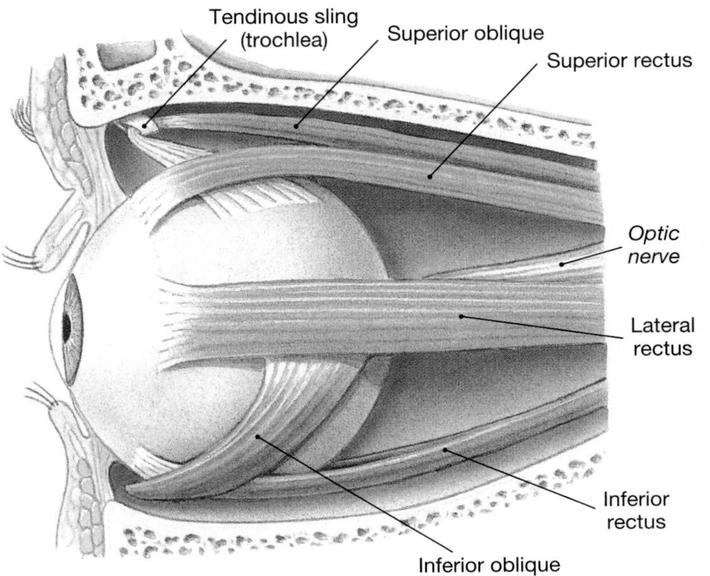
Tendinous sling (trochlea)
Superior oblique
Superior rectus
Optic nerve
Lateral rectus
Inferior rectus
Inferior oblique

(a) Lateral surface, left eye

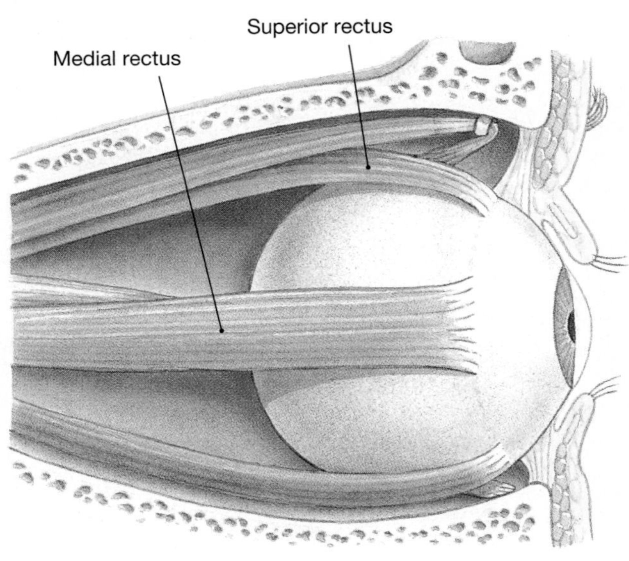
Superior rectus
Medial rectus

(b) Medial surface, left eye

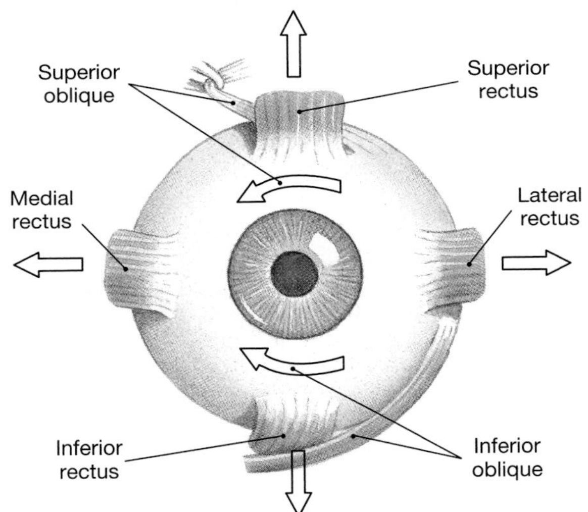
Superior oblique
Superior rectus
Medial rectus
Lateral rectus
Inferior rectus
Inferior oblique

(c) Anterior view. left eye

● **FIGURE 11-5**
Oculomotor Muscles. *See also Figure 7-12, p. 216.*
Plates 3.1, 3.3

Extrinsic Eye Muscles

Figure 11-5

Six **extrinsic eye muscles,** or **oculomotor** (ok-ū-lō-MŌ-ter) muscles, originating on the surface of the orbit control the position of the eye. These muscles, shown in Figure 11-5● and detailed in Table 11-3, are the **inferior rectus, lateral rectus, medial rectus, superior rectus, inferior oblique,** and **superior oblique.** The oculomotor muscles are innervated by the third (oculomotor), fourth (trochlear), and sixth (abducens) cranial nerves.

TABLE 11-3 Extrinsic Oculomotor Muscles

Muscle	*Origin*	*Insertion*	*Action*	*Innervation*
Inferior rectus	Sphenoid around optic foramen	Inferior, medial surface of eyeball	Eye looks down	Oculomotor nerve (N III)
Medial rectus	As above	Medial surface of eyeball	Eye rotates medially	As above
Superior rectus	As above	Superior, lateral surface of eyeball	Eye looks up	As above
Inferior oblique	Maxilla at front of orbit	Inferior, lateral surface of eyeball	Eye rolls, looks up and to the side	As above
Superior oblique	Sphenoid around optic foramen	Superior, medial surface of eyeball	Eye rolls, looks down and to the side	Trochlear nerve (N IV)
Lateral rectus	As above	Lateral surface of eyeball	Eye rotates laterally	Abducens nerve (N VI)

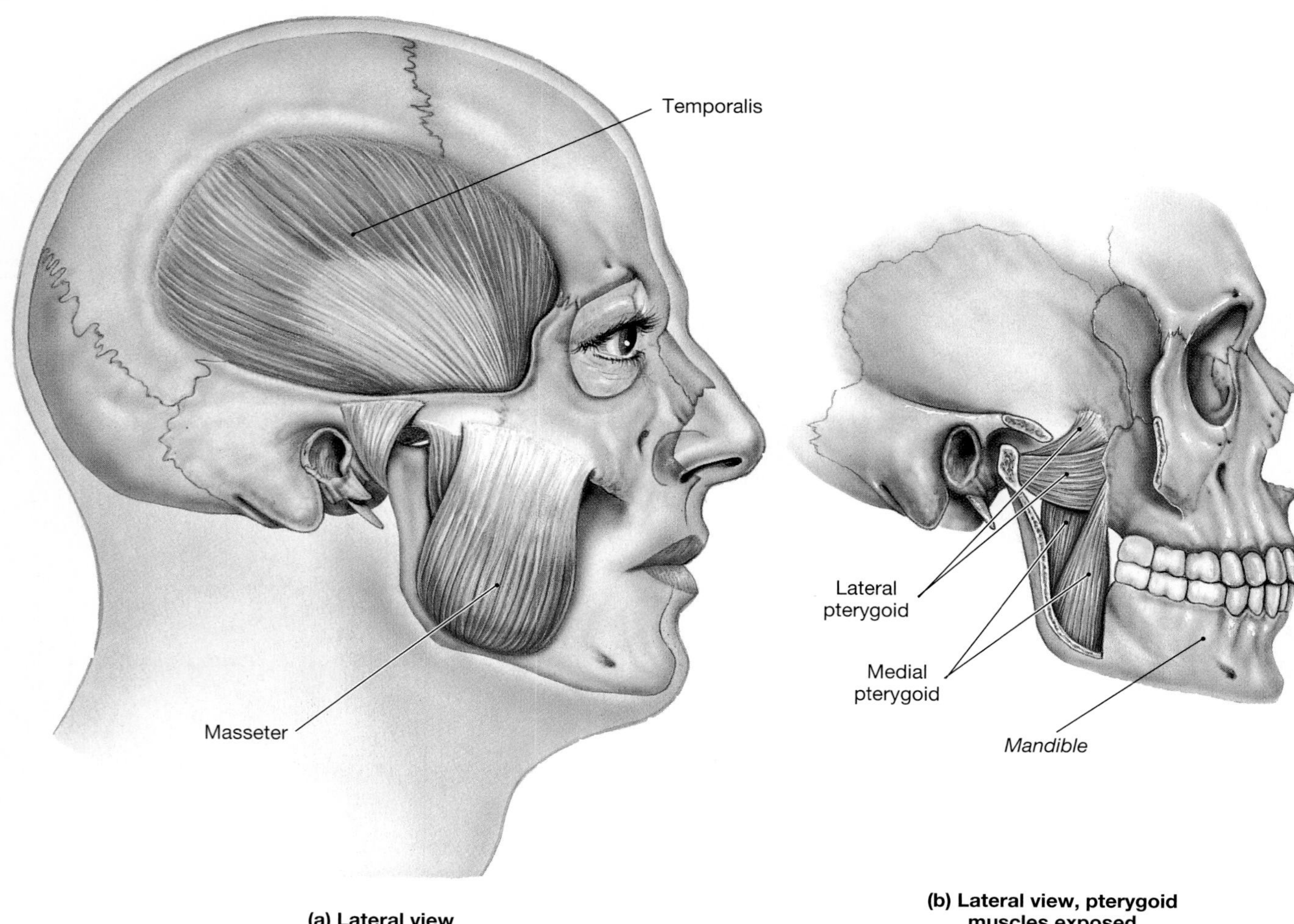
Temporalis
Masseter
Lateral pterygoid
Medial pterygoid
Mandible
(a) Lateral view
(b) Lateral view, pterygoid muscles exposed

● **FIGURE 11-6**
Muscles of Mastication. *See also Figures 7-3 and 7-13, pp. 205–206, 216.* AM *Plate 4.3*

Muscles of Mastication
Figure 11-6

The muscles of mastication (Figure 11-6● and Table 11-4) move the lower jaw at the temporomandibular joint. ∞ *[p. 216]* The large **masseter** is the strongest masticatory muscle. The **temporalis** assists in elevation of the mandible, whereas the **pterygoid** muscles, used in various combinations, can elevate, depress, or protract the mandible or slide it from side to side. These movements are important in maximizing the efficient use of the teeth while chewing foods of various consistencies. The muscles of mastication are innervated by the fifth cranial nerve, the *trigeminal nerve.*

TABLE 11-4 Muscles of Mastication

Muscle	*Origin*	*Insertion*	*Action*	*Innervation*
Masseter	Zygomatic arch	Lateral surface of mandibular ramus	Elevates mandible	Trigeminal nerve (N V), mandibular branch
Temporalis	Along temporal lines of skull	Coronoid process of mandible	Elevates mandible	As above
Pterygoideus (medial and lateral)	Medial pterygoid plate and processes	Medial surface of mandibular ramus	Elevates, protracts, and/or moves mandible laterally	As above

Muscles of the Tongue

Figure 11-7

The muscles of the tongue have names ending in *glossus*, the Greek word for "tongue." Once you can recall the structures referred to by *palato-*, *stylo-*, *genio-*, and *hyo-*, you shouldn't have much trouble with this group. The **palatoglossus** originates at the palate, the **styloglossus** at the styloid process, the **genioglossus** at the chin, and the **hyoglossus** at the hyoid bone (Figure 11-7●). These muscles, used in various combinations, move the tongue in the delicate and complex patterns necessary for speech, and they manipulate food within the mouth in preparation for swallowing. As indicated in Table 11-5, they are innervated by the *hypoglossal nerve* (N XII), a cranial nerve whose name indicates its function as well as its location.

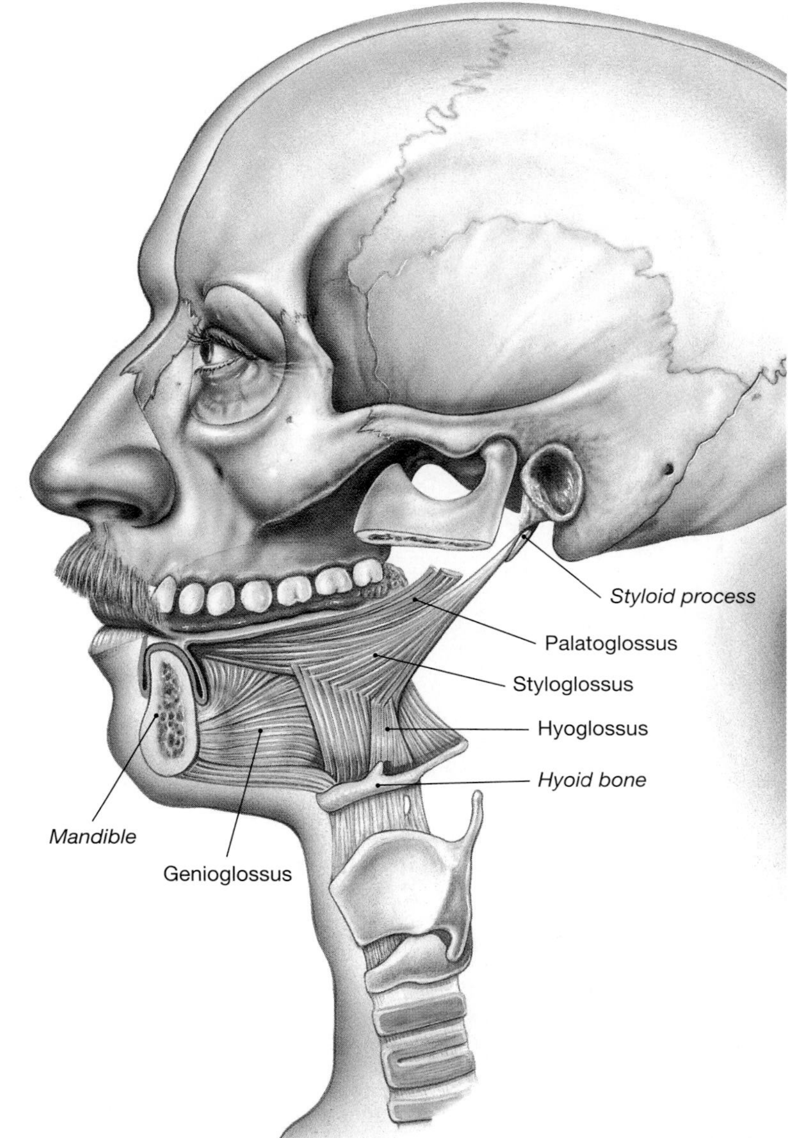
Styloid process
Palatoglossus
Styloglossus
Hyoglossus
Hyoid bone
Mandible
Genioglossus

● **FIGURE 11-7**
Muscles of the Tongue. *See also Figures 7-3 and 7-14, pp. 205–206, 217.*

TABLE 11-5 Muscles of the Tongue

Muscle	*Origin*	*Insertion*	*Action*	*Innervation*
Genioglossus	Medial surface of mandible around chin	Body of tongue, hyoid bone	Depresses and protracts tongue	Hypoglossal nerve (N XII)
Hyoglossus	Body and greater cornu of hyoid bone	Side of tongue	Depresses and retracts tongue	As above
Palatoglossus	Anterior surface of soft palate	As above	Elevates tongue, depresses soft palate	As above
Styloglossus	Styloid process of temporal bone	Via side to the tip and base of tongue	Retracts tongue, elevates side	As above

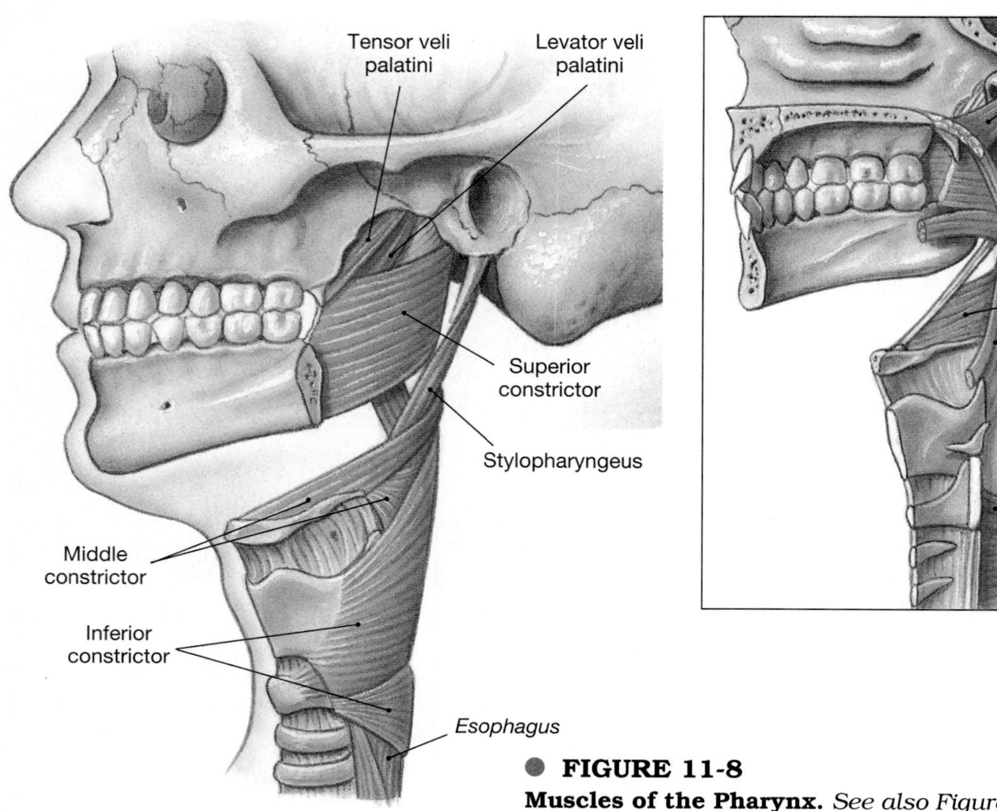

Tensor veli palatini
Levator veli palatini
Superior constrictor
Stylopharyngeus
Middle constrictor
Inferior constrictor
Esophagus
Levator veli palatini
Salpingopharyngeus
Superior constrictor
Palatopharyngeus
Middle constrictor
Stylopharyngeus
Inferior constrictor
Esophagus

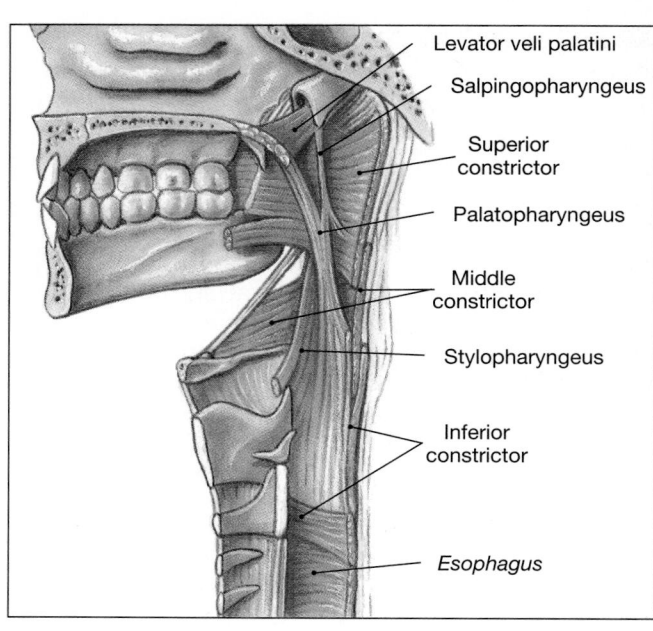

● **FIGURE 11-8**
Muscles of the Pharynx. *See also Figures 7-4 and 7-14, pp. 207, 217.*

TABLE 11-6 Muscles of the Pharynx

Muscle	*Origin*	*Insertion*	*Action*	*Innervation*
PHARYNGEAL CONSTRICTORS			Constrict pharynx to propel bolus into esophagus	Branches of pharyngeal plexus (N IX & X)
Superior constrictor	Pterygoid process of sphenoid, medial surfaces of mandible	Median raphe attached to occipital bone		
Middle constrictor	Cornu of hyoid	Median raphe		
Inferior constrictor	Cricoid and thyroid cartilages of larynx	Median raphe		
LARYNGEAL ELEVATORS*			Elevate larynx	Branches of pharyngeal plexus (N IX & X)
Palatopharyngeus	Soft palate	Thyroid cartilage		
Salpingopharyngeus	Cartilage around the inferior portion of the pharyngotympanic tube	Thyroid cartilage		
Stylopharyngeus	Styloid process of temporal	Thyroid cartilage		
PALATAL MUSCLES				
Levator veli palatini	Petrous portion of temporal and tissues around the pharyngotympanic tube	Soft palate	Elevates soft palate	N XI
Tensor veli palatini	Spine of sphenoid, tissues around the pharyngotympanic tube	Soft palate	As above	N V

*Assisted by the thyrohyoid, geniohyoid, stylohyoid, and hyoglossal muscles discussed in Tables 11-5 and 11-7.

Muscles of the Pharynx

Figure 11-8

The muscles of the pharynx (Figure 11-8● and Table 11-6) are responsible for initiating the swallowing process. The **pharyngeal constrictors** move materials into the esophagus. The **palatopharyngeus** (pal-a-tō-făr-IN-jē-us) and **stylopharyngeus** (stī-lō-făr-IN-jē-us) elevate the larynx, and the **palatal muscles** raise the soft palate and adjacent portions of the pharyngeal wall. The latter muscles also pull open the entrance to the *pharyngotympanic tube.* [p. 210] As a result, swallowing repeatedly can help one adjust to pressure changes when flying or diving.

Anterior Muscles of the Neck

Figures 11-4, 11-9

The anterior muscles of the neck (Figure 11-9● and Table 11-7) include (1) muscles that control the position of the larynx, (2) muscles that depress the mandible and tense the floor of the mouth, and (3) muscles that provide a stable foundation for muscles of the tongue and pharynx. The **digastricus** has two bellies, as the name implies (*di-*, two + *gaster*, stomach). One belly extends from the chin to the hyoid bone, and the other continues from the hyoid bone to the mastoid portion of the temporal bone. Depending on which of these bellies contracts, and whether or not fixator muscles are stabilizing the position of the hyoid, this muscle can open the mouth by depressing the mandible or elevate the larynx by raising the hyoid bone. The digastricus overlies the broad, flat **mylohyoid,** which provides a muscular floor to the mouth. The **stylohyoid** forms a muscular connection between the hyoid apparatus and the styloid process of the skull. The **sternocleidomastoid** (ster-nō-klī-dō-MAS-toid) extends from the clavicles and the sternum to the mastoid region of the skull (Figures 11-4, p. 335, and 11-9●). The other members of this group are straplike muscles that run between the sternum and the chin.

✓ If you were contracting and relaxing your masseter muscle, what would you probably be doing?

✓ What facial muscle would you expect to be well developed in a trumpet player?

✓ Why can swallowing help alleviate the pressure sensations at the eardrum when an airplane is changing altitude?

● **FIGURE 11-9**
Muscles of the Anterior Neck.
See also Figures 7-3 and 7-14, pp. 205–206, 217.

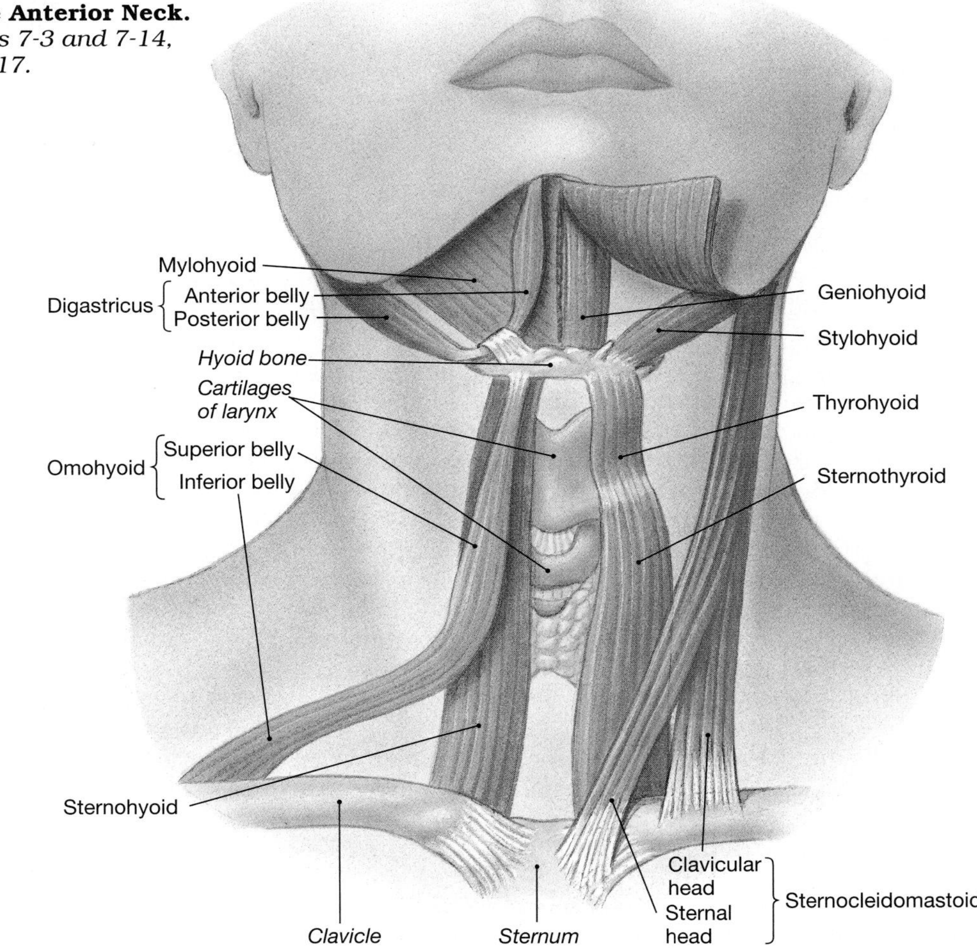
Mylohyoid
Digastricus
Anterior belly
Posterior belly
Hyoid bone
Cartilages of larynx
Omohyoid
Superior belly
Inferior belly
Sternohyoid
Clavicle
Sternum
Clavicular head
Sternal head
Sternocleidomastoid
Geniohyoid
Stylohyoid
Thyrohyoid
Sternothyroid

TABLE 11-7 Muscles of the Anterior Neck

Muscle	*Origin*	*Insertion*	*Action*	*Innervation*
Digastricus	Two bellies: *posterior* from mastoid region of temporal; *anterior* from inferior surface of mandible at chin	Hyoid bone	Depresses mandible and/or elevates larynx	Hypoglossal nerve (N XII) to posterior belly Trigeminal nerve (N V), mandibular branch, to anterior belly
Geniohyoid	Medial surface of mandible at chin	Hyoid bone	As above and pulls hyoid anteriorly	Cervical nerve C_1 via hypoglossal nerve (N XII)
Mylohyoid	Mylohyoid line of mandible	Medial connective tissue band (raphe) that runs to hyoid bone	Elevates floor of mouth, elevates hyoid, and/or depresses mandible	Trigeminal nerve (N V), mandibular branch
Omohyoid	Central tendon attaches to clavicle and 1st rib	Two bellies: *anterior* attaches to hyoid bone; *posterior* to superior margin of scapula	Depresses hyoid and larynx	Cervical spinal nerves
Sternohyoid	Clavicle and manubrium	Hyoid	As above	As above
Sternothyroid	Dorsal surface of manubrium and 1st rib	Thyroid cartilage of larynx	As above	As above
Stylohyoid	Styloid process of temporal bone	Hyoid	Elevates larynx	Facial nerve (N VII)
Thyrohyoid	Thyroid cartilage of larynx	Hyoid	Elevates thyroid, depresses hyoid	Hypoglossal nerve (N XII)
Sternocleido-mastoid	Superior margins of manubrium and clavicle	Mastoid region of skull	Together they flex the neck; alone one side bends head toward shoulder and turns face to opposite side	Accessory nerve (N XI) and cervical spinal nerves

❑ Muscles of the Spine

Figures 11-3, 11-10

The muscles of the spine are covered by more superficial back muscles, such as the trapezius and latissimus dorsi (see Figure 11-3●, pp. 332–333). The spinal extensors, or **erector spinae** muscles, include superficial and deep layers. The relatively superficial layer can be subdivided into **spinalis, longissimus,** and **iliocostalis** divisions (Figure 11-10● and Table 11-8). In the lower lumbar and sacral regions the boundary between the longissimus and iliocostalis muscles becomes indistinct, and they are sometimes known as the **sacrospinalis** muscles. When contracting together, the erector spinae extend the spinal column. When the muscles on only one side contract, the spine is bent laterally.

Beneath the spinalis muscles, deep muscles of the spine interconnect and stabilize the vertebrae. These muscles include the **multifidus, interspinales, intertransversarii,** and **rotatores** (Figure 11-10b●). In various combinations they produce slight extension or rotation of the spinal column. They are also important in making delicate adjustments in the positions of individual vertebrae, and they stabilize adjacent vertebrae. If injured, these muscles can start a cycle of pain → muscle stimulation → contraction → pain. This cycle can lead to pressure on adjacent spinal nerves, leading to sensory losses as well as limiting mobility. Many of the warm-up and stretching exercises recommended before athletic events are intended to prepare these small but very important muscles for their supporting role.

The muscles of the spine include many dorsal extensors, but few ventral flexors. The spinal column does not need a massive series of flexor muscles because (1) many of the large trunk muscles flex the spine when they contract, and (2) most of the body weight lies anterior to the spinal column, and gravity tends to flex the spine. However, there are a few spinal flexors associated with the anterior surface of the spinal column. In the neck (Figure 11-10c●), the **longus capitis** and the **longus cervicis** rotate or flex the neck, depending on whether the muscles of one or both sides are contracting. In the lumbar region (Figure 11-10a●), the large **quadratus lumborum** muscles flex the spine and depress the ribs.

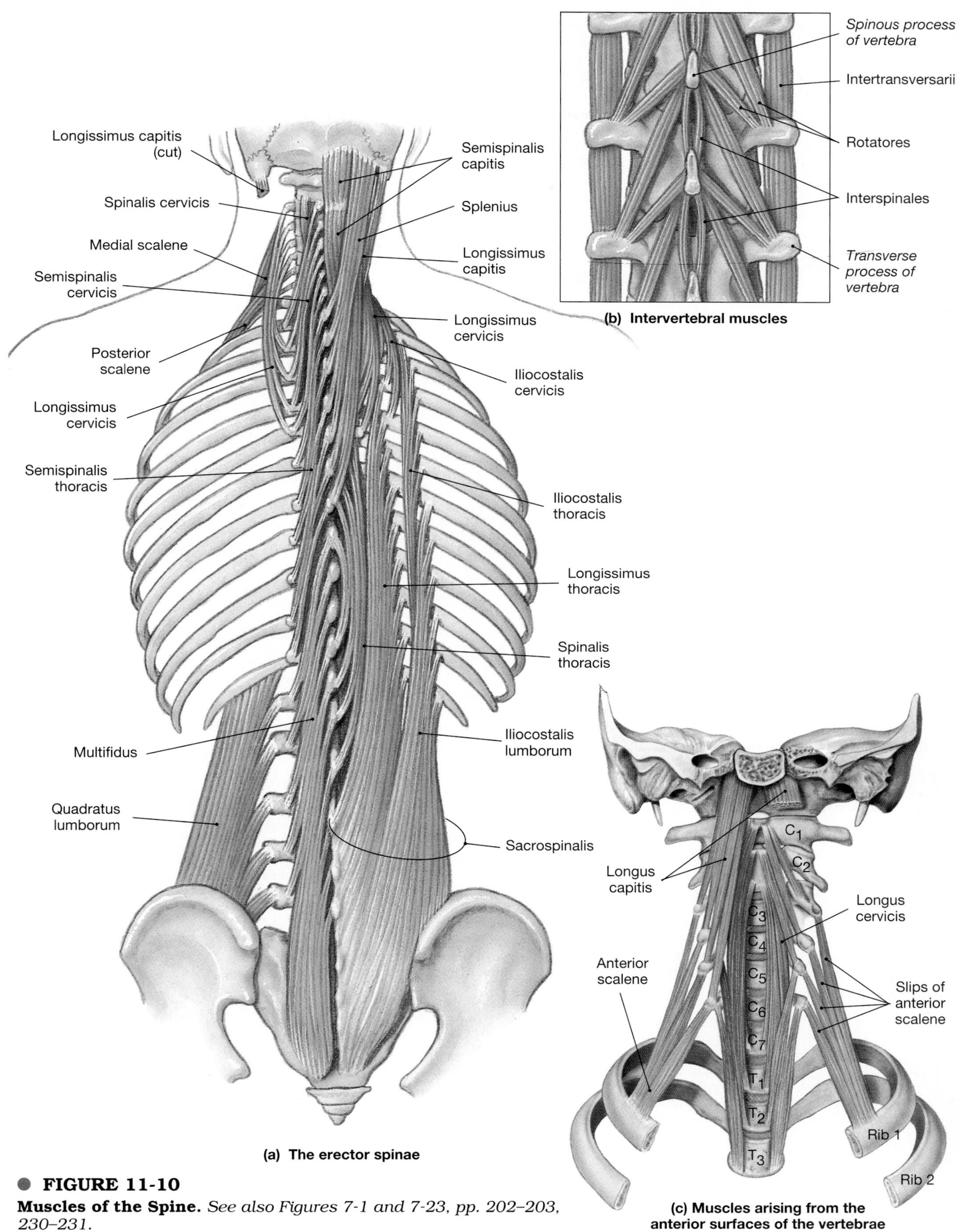
Longissimus capitis (cut)
Spinalis cervicis
Medial scalene
Semispinalis cervicis
Posterior scalene
Longissimus cervicis
Semispinalis thoracis
Multifidus
Quadratus lumborum
Semispinalis capitis
Splenius
Longissimus capitis
Longissimus cervicis
Iliocostalis cervicis
Iliocostalis thoracis
Longissimus thoracis
Spinalis thoracis
Iliocostalis lumborum
Sacrospinalis
(a) The erector spinae
Spinous process of vertebra
Intertransversarii
Rotatores
Interspinales
Transverse process of vertebra
(b) Intervertebral muscles
Longus capitis
Longus cervicis
Anterior scalene
Slips of anterior scalene
C_1
C_2
C_3
C_4
C_5
C_6
C_7
T_1
T_2
T_3
Rib 1
Rib 2
(c) Muscles arising from the anterior surfaces of the vertebrae

● **FIGURE 11-10**
Muscles of the Spine. *See also Figures 7-1 and 7-23, pp. 202–203, 230–231.*

TABLE 11-8 Muscles of the Spine

Group/Muscle	*Origin*	*Insertion*	*Action*	*Innervation*
SUPERFICIAL SPINAL EXTENSORS				
Spinalis group				
Semispinalis capitis	Processes of lower cervical and upper thoracic vertebrae	Occipital bone, between nuchal lines	The two sides act together to extend head; either alone extends and tilts head to that side	Cervical spinal nerves
Semispinalis cervicis	Transverse processes of T_1–T_5 or T_6	Spinous processes of C_2–C_5	Extends vertebral column and rotates toward opposite side	As above
Semispinalis thoracis	Transverse processes of T_6–T_{10}	Spinous processes of C_5–T_4	As above	Thoracic spinal nerves
Splenius (Splenius capitis, splenius cervicis)	Spinous processes and ligaments connecting upper cervical vertebrae	Mastoid process and occipital bone of skull	The two sides act together to extend head; either alone rotates and tilts head to that side	Cervical spinal nerves
Spinalis cervicis	Inferior portion of ligamentum nuchae and spinous process of C_7	Spinous process of axis	Extends neck	As above
Spinalis thoracis	Spinous processes of lower thoracic and upper lumbar vetebrae	Spinous processes of upper thoracic vertebrae	Extends spinal column	Thoracic and lumbar spinal nerves
Longissimus group				
Longissimus capitis	Processes of lower cervical and upper thoracic vertebrae	Mastoid process of temporal bone	The two sides act together to extend head; either alone rotates and tilts head to that side	Cervical and thoracic spinal nerves
Longissimus cervicis	Transverse processes of upper thoracic vertebrae	Transverse processes of middle and upper cervical vertebrae	As above	As above
Longissimus thoracis	Broad aponeurosis and at transverse processes of lower thoracic and upper lumbar vertebrae; joins iliocostalis to form "sacrospinalis"	Transverse processes of higher vertebrae and inferior surfaces of ribs	Extends and/or bends spine to the side	As above
Iliocostalis group				
Iliocostalis cervicis	Superior borders of vertebrosternal ribs near the angles	Transverse processes of middle and lower cervical vertebrae	Extends or bends neck, elevates ribs	As above
Iliocostalis thoracis	Superior borders of lower 7 ribs medial to the angles	Upper ribs and transverse process of last cervical vertebra	Stabilizes thoracic vertebrae in extension	Thoracic spinal nerves
Iliocostalis lumborum	Sacrospinal aponeurosis and iliac crest	Inferior surfaces of lower 7 ribs near their angles	Extends spine, depresses ribs	Lumbar spinal nerves
DEEP SPINAL EXTENSORS AND ROTATORS				
Multifidus	Sacrum and transverse processes of each vertebra	Spinous processes of the fourth or fifth more superior vertebrae	Extend vertebral column and rotate toward opposite side	Cervical, thoracic, and lumbar spinal nerves
Rotatores	Transverse processes of each vertebra	Spinous process of adjacent, more superior vertebra	Extend vertebral column and rotate toward opposite side	As above
Interspinales	Spinous processes of each vertebra	Spinous processes of preceding vertebra	Extend vertebral column	As above
Intertransversarii	Transverse processes of each vertebra	Transverse process of preceding vertebra	Bend the vertebral column laterally	As above
SPINAL FLEXORS				
Longus capitis	Transverse processes of cervical vertebrae	Base of the occipital bone	The two sides act together to bend head forward; either alone rotates head to that side	Cervical spinal nerves
Longus cervicis	Anterior surfaces of cervical and upper thoracic vertebrae	Transverse processes of upper cervical vertebrae	Flexes and/or rotates neck; limits hyperextension	As above
Quadratus lumborum	Iliac crest	Last rib and transverse processes of lumbar vertebrae	Together they depress ribs, flex spine; one side alone flexes spine laterally	Thoracic and lumbar spinal nerves

❑ Oblique and Rectus Muscles

Figures 11-3, 11-10, 11-11

The muscles of the oblique and rectus groups (Table 11-9) lie between the vertebral spines and the ventral midline (Figures 11-3, p. 332–333, and 11-11●). The oblique muscles can compress underlying structures or rotate the spinal column, depending on whether one or both sides are contracting. The rectus muscles are important flexors of the spinal column, acting in opposition to the erector spinae. The oblique

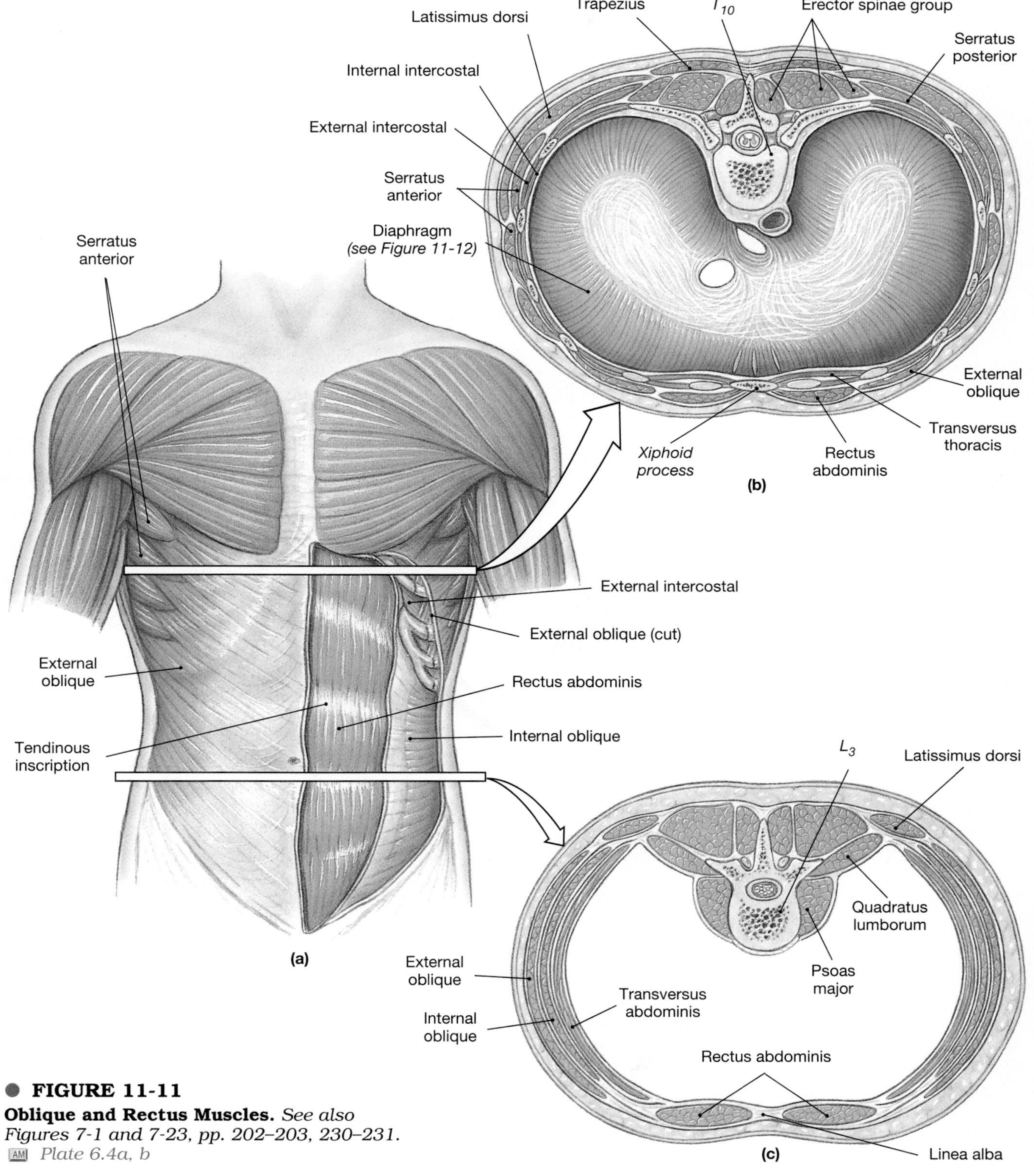

Latissimus dorsi
Trapezius
T_{10}
Erector spinae group
Serratus posterior
Internal intercostal
External intercostal
Serratus anterior
Diaphragm
(see Figure 11-12)
Serratus anterior
External oblique
Transversus thoracis
Xiphoid process
Rectus abdominis
(b)
External intercostal
External oblique (cut)
External oblique
Rectus abdominis
Internal oblique
Tendinous inscription
L_3
Latissimus dorsi
Quadratus lumborum
Psoas major
(a)
External oblique
Transversus abdominis
Internal oblique
Rectus abdominis
(c)
Linea alba

● **FIGURE 11-11**

Oblique and Rectus Muscles. *See also Figures 7-1 and 7-23, pp. 202–203, 230–231.*

AM *Plate 6.4a, b*

TABLE 11-9 Oblique and Rectus Muscles

Group/Muscle	*Origin*	*Insertion*	*Action*	*Innervation*
OBLIQUE GROUP				
Cervical region				
Scalenes	Transverse and costal processes of cervical vertebrae	Superior surfaces of first two ribs	Elevates ribs, and/or flexes neck	Cervical spinal nerves
Thoracic region				
External intercostals	Inferior border of each rib	Superior border of the next rib	Elevate ribs	Intercostal nerves (branches of thoracic spinal nerves)
Internal intercostals	Superior border of each rib	Inferior border of the previous rib	Depress ribs	As above
Transversus thoracis	Medial surface of sternum	Cartilages of ribs	As above	As above
Abdominal region				
External oblique	Lower eight ribs	Linea alba and iliac crest	Compresses abdomen, depresses ribs, flexes or bends spine	Intercostals and iliohypogastric nerves
Internal oblique	Lumbodorsal fascia and iliac crest	Lower ribs, xiphoid of sternum, and linea alba	As above	As above
Transversus abdominis	Cartilages of lower ribs, iliac crest, and lumbodorsal fascia	Linea alba and pubis	Compresses abdomen	Intercostals, iliohypogastric, and ilioinguinal nerves
RECTUS GROUP				
Cervical region	*See muscles in Table 11-6*			
Thoracic region				
Diaphragm	Xiphoid process, cartilages of ribs 4–10, and anterior surfaces of lumbar vertebrae	Central tendinous sheet	Contraction expands thoracic cavity, compresses abdominopelvic cavity	Phrenic nerves
Abdominal region				
Rectus abdominis	Superior surface of pubis around symphysis	Inferior surfaces of costal cartilages (ribs 5–7) and xiphoid process of sternum	Depresses ribs, flexes vertebral column	Thoracic spinal nerves (T_7–T_{12})

and rectus muscles are united by their common embryological origins. They can be divided into cervical, thoracic, and abdominal groups.

The oblique series includes the **scalenes** of the neck (Figure 11-10c●, p. 343) and the **intercostal** and **transversus** muscles of the thoracic region (Figure 11-11b●). In the thorax, the oblique muscles lie between the ribs, and the **external intercostals** cover the **internal intercostals.** Both muscles are important in respiratory movements of the ribs. A small **transversus thoracis** crosses the inner surface of the rib cage and is covered by the serous membrane that lines the pleural cavities. The sternum occupies the place where one might otherwise expect to find thoracic rectus muscles.

The same basic pattern of musculature extends unbroken across the abdominopelvic surface (Figure 11-11a,c●). Here the muscles are known as the **external obliques,** the **internal obliques,** the **transversus abdominis,** and the

rectus abdominis. The rectus abdominis begins at the xiphoid process and ends near the pubic symphysis. This muscle is longitudinally divided by a median collagenous partition, the **linea alba** (white line). Transverse **tendinous inscriptions** divide this muscle into segments.

When the abdominal muscles contract forcefully, pressures in the abdominopelvic cavity can skyrocket, and those pressures are applied to internal organs. If the individual exhales at the same time, the pressure is relieved, because the diaphragm can move upward as the lungs collapse. But during vigorous isometric exercises or when lifting a weight while holding one's breath, pressure in the abdominopelvic cavity can rise high enough to cause a variety of problems, among them the development of a *hernia.*

A **hernia** develops when an organ protrudes through an abnormal opening. The most common hernias are *inguinal hernias* and *diaphragmatic hernias.* Inguinal hernias typically occur in males, at the site where blood vessels, nerves, and reproductive ducts pass through the abdominal wall en route to the testes. Diaphragmatic hernias develop when visceral organs, such as a portion of the stomach, are forced into the left pleural cavity. If herniated structures become trapped or twisted, surgery may be required to prevent serious complications. AM *Hernias*

The Diaphragm
Figure 11-12

The term *diaphragm* refers to any muscular sheet that forms a wall. When used without a modifier, however, **diaphragm,** or *diaphragmatic muscle,* specifies the muscular partition that separates the abdominopelvic and thoracic cavities. This muscle, shown in Figure 11-12•, is included here because it is developmentally linked to the other muscles of the chest wall. The diaphragm is a major muscle of respiration.

❑ Muscles of the Pelvic Floor
Figure 11-13

The muscles of the pelvic floor extend from the sacrum and coccyx to the ischium and pubis. These muscles (Figure 11-13• and Table 11-10) (1) support the organs of the pelvic cavity, (2) flex the sacrum and coccyx, and (3) control the movement of materials through the urethra and anus.

The boundaries of the **perineum** are established by the inferior margins of the pelvis. If you draw a line between the ischial tuberosities you will divide the perineum into two triangles, an anterior, **urogenital triangle** and a posterior, **anal triangle.** The superficial muscles of the anterior triangle are the muscles of the external genitalia.

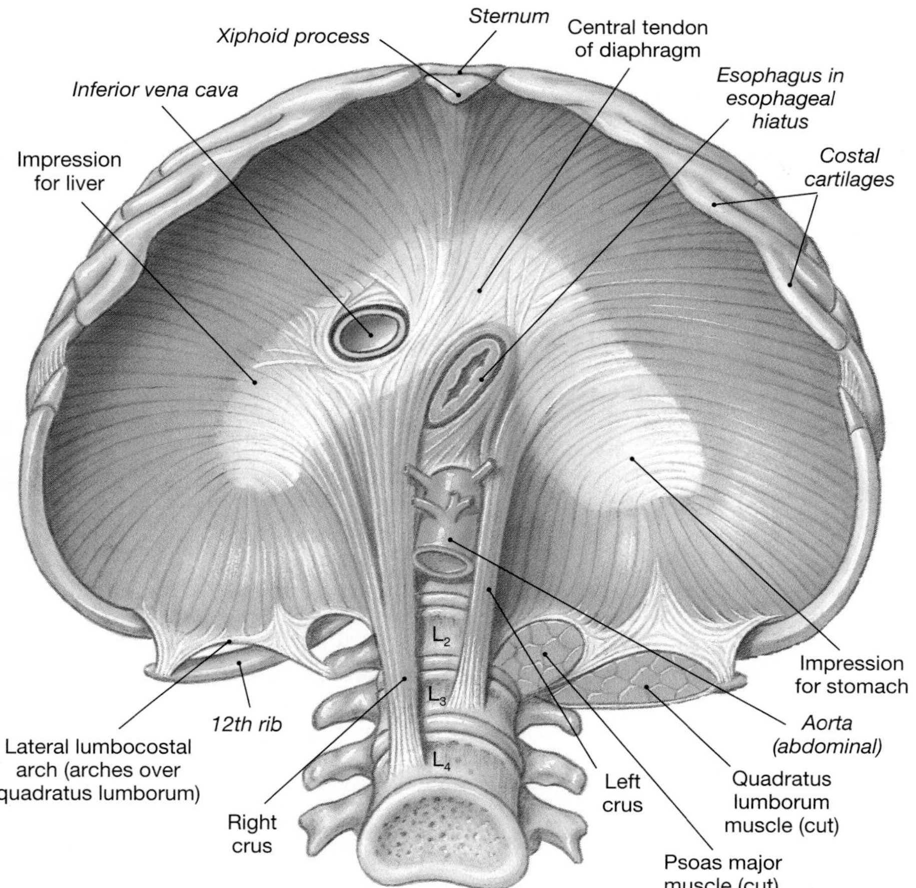
Sternum
Xiphoid process
Central tendon of diaphragm
Inferior vena cava
Esophagus in esophageal hiatus
Impression for liver
Costal cartilages
L_2
L_3
L_4
Impression for stomach
12th rib
Aorta (abdominal)
Lateral lumbocostal arch (arches over quadratus lumborum)
Left crus
Quadratus lumborum muscle (cut)
Right crus
Psoas major muscle (cut)

● **FIGURE 11-12**
The Diaphragm.
See also Figures 7-1 and 7-23, pp. 202–203, 230–231.

They overlie deeper muscles that strengthen the pelvic floor and encircle the urethra. These muscles constitute the **urogenital diaphragm,** a muscular layer that extends between the pubic bones.

An even more extensive muscular sheet, the **pelvic diaphragm,** forms the muscular foundation of the anal triangle. This layer extends anteriorly above the urogenital diaphragm as far as the pubic symphysis.

The urogenital and pelvic diaphragms do not completely close the pelvic outlet, for the urethra, vagina, and anus pass through them to open on the external surface. Muscular sphincters surround their openings and permit voluntary control of urination and defecation. Muscles, nerves, and blood vessels also pass through the pelvic outlet as they travel to or from the legs.

 Damage to the external intercostal muscles would interfere with what important process?

 If someone hit you in your rectus abdominis muscle, how would your body position change?

 After spending an afternoon carrying heavy boxes from his basement to his attic, Joe complains that the muscles in his back hurt. What muscle(s) are most likely sore?

TABLE 11-10 Muscles of the Pelvic Floor

Group/Muscle	*Origin*	*Insertion*	*Action*	*Innervation*
UROGENITAL TRIANGLE				
Superficial muscles				
Bulbocavernosus:				
male	Collagen sheath at base of penis; fibers cross over urethra	Median raphe and central tendon of perineum	Compresses base and stiffens penis, ejects urine or semen	Pudendal nerve, perineal branch
female	Collagen sheath at base of clitoris; fibers run on either side of urethral and vaginal openings	Central tendon of perineum	Compresses and stiffens clitoris, narrows vaginal opening	As above
Ischiocavernosus	Ramus and tuberosity of ischium	Symphysis pubis anterior to base of penis or clitoris	Compresses and stiffens penis or clitoris	As above
Superficial transverse perineus	Ischial ramus	Central tendon of perineum	Stabilizes central tendon of perineum	As above
Deep muscles: The urogenital diaphragm				
Deep transverse perineus	Ischial ramus	Median raphe of urogenital diaphragm	Stabilizes central tendon of perineum	Pudendal nerve, perineal branch
Urethral sphincter:				
male	Ischial and pubic rami	To median raphe at base of penis; inner fibers encircle urethra	Closes urethra, compresses prostate and bulbourethral glands	As above
female	Ischial and pubic rami	To median raphe; inner fibers encircle urethra	Closes urethra, compresses vagina and greater vestibular glands	As above
ANAL TRIANGLE				
Pelvic diaphragm				
Coccygeus	Ischial spine	Lateral, inferior borders of the sacrum	Flexes coccyx and coccygeal vertebrae	Pudendal nerve
External anal sphincter	Via tendon from coccyx	Encircles anal opening	Closes anal opening	Pudendal nerve, hemorrhoidal branch
Levator ani:				
Iliococcygeus	Ischial spine, pubis	Coccyx	Tenses floor of pelvis, flexes coccyx, elevates and retracts anus	Pudendal nerve
Pubococcygeus	Inner margins of pubis	Coccyx	As above	As above

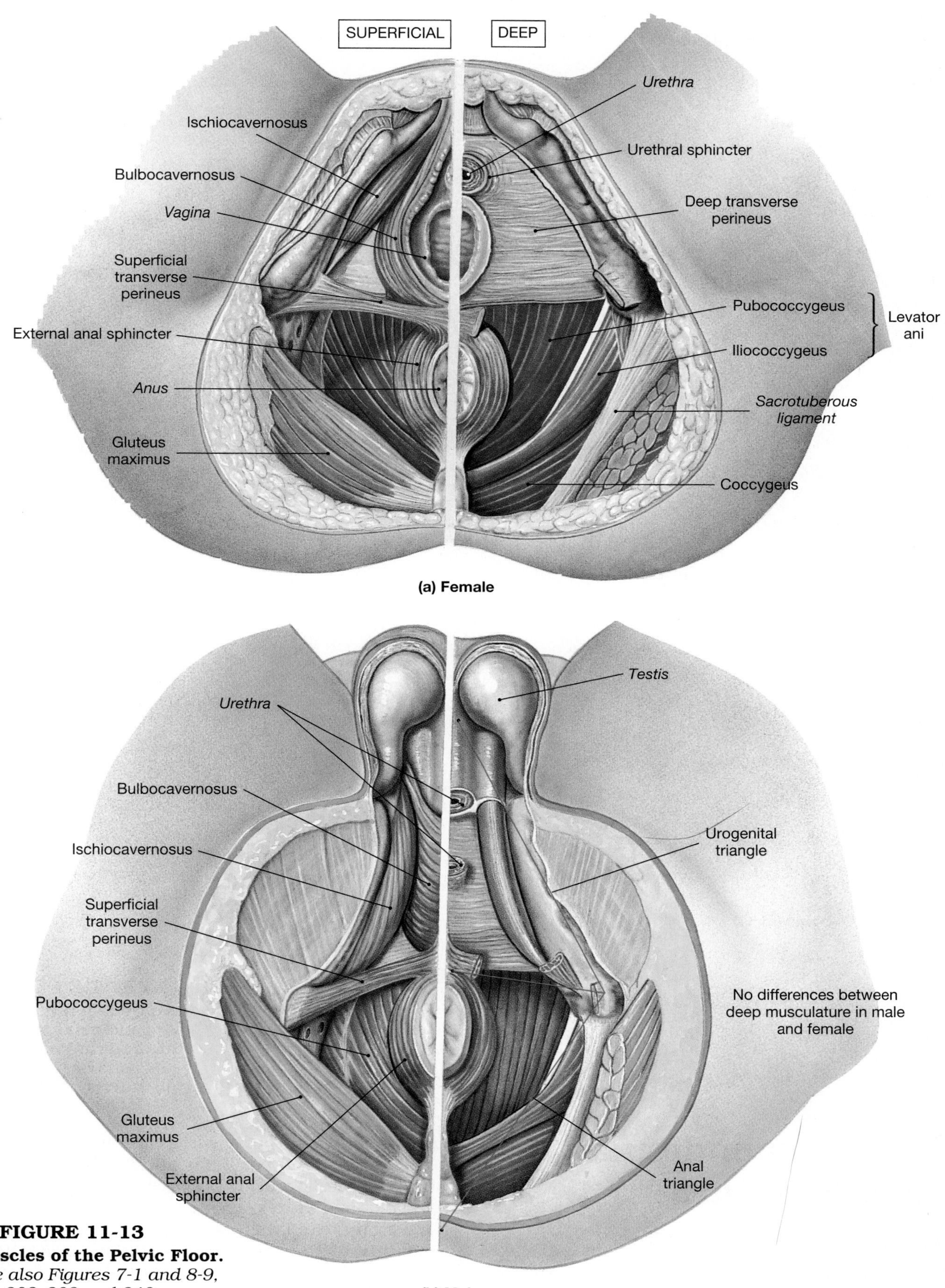
SUPERFICIAL
DEEP
Ischiocavernosus
Bulbocavernosus
Vagina
Superficial transverse perineus
External anal sphincter
Anus
Gluteus maximus
Urethra
Urethral sphincter
Deep transverse perineus
Pubococcygeus
Iliococcygeus
Levator ani
Sacrotuberous ligament
Coccygeus
(a) Female
Urethra
Bulbocavernosus
Ischiocavernosus
Superficial transverse perineus
Pubococcygeus
Gluteus maximus
External anal sphincter
Testis
Urogenital triangle
No differences between deep musculature in male and female
Anal triangle
(b) Male

FIGURE 11-13
Muscles of the Pelvic Floor. *See also Figures 7-1 and 8-9, pp. 202–203 and 249.*

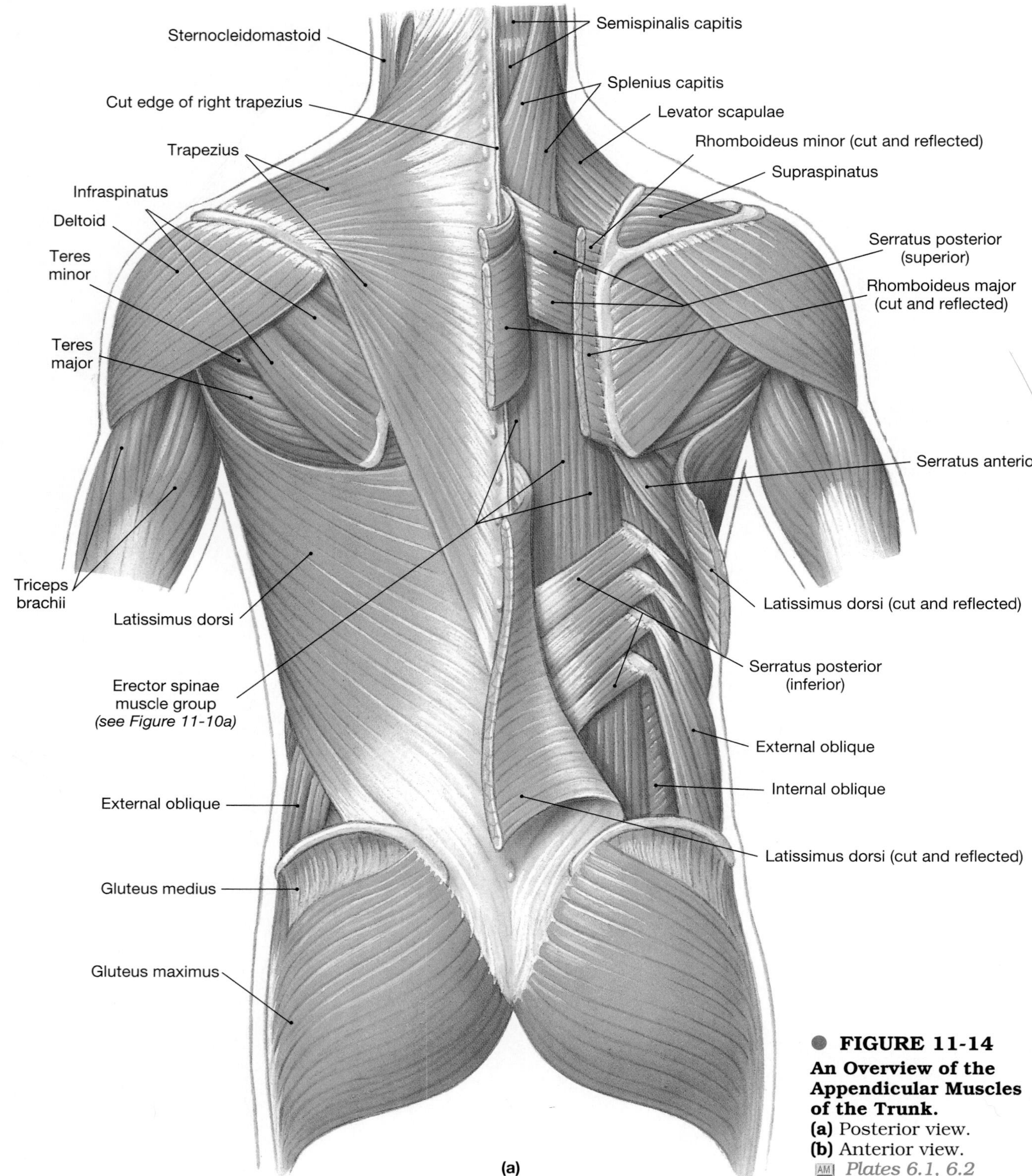
Sternocleidomastoid
Cut edge of right trapezius
Trapezius
Infraspinatus
Deltoid
Teres minor
Teres major
Triceps brachii
Latissimus dorsi
Erector spinae muscle group (see Figure 11-10a)
External oblique
Gluteus medius
Gluteus maximus
Semispinalis capitis
Splenius capitis
Levator scapulae
Rhomboideus minor (cut and reflected)
Supraspinatus
Serratus posterior (superior)
Rhomboideus major (cut and reflected)
Serratus anterior
Latissimus dorsi (cut and reflected)
Serratus posterior (inferior)
External oblique
Internal oblique
Latissimus dorsi (cut and reflected)
(a)

● **FIGURE 11-14**
An Overview of the Appendicular Muscles of the Trunk.
(a) Posterior view.
(b) Anterior view.
AM *Plates 6.1, 6.2*

The Appendicular Musculature

Figure 11-14

The appendicular musculature positions and stabilizes the pectoral and pelvic girdles and moves the upper and lower limbs. There are two major groups of appendicular muscles: (1) the muscles of the shoulders and upper limbs and (2) the muscles of the pelvic girdle and lower extremities. The functions and required ranges of motion are very different from one group to another. In addition to increasing the mobility of the arm, the muscular connections between the pectoral girdle and the axial skeleton must act as shock absorbers. For example, people

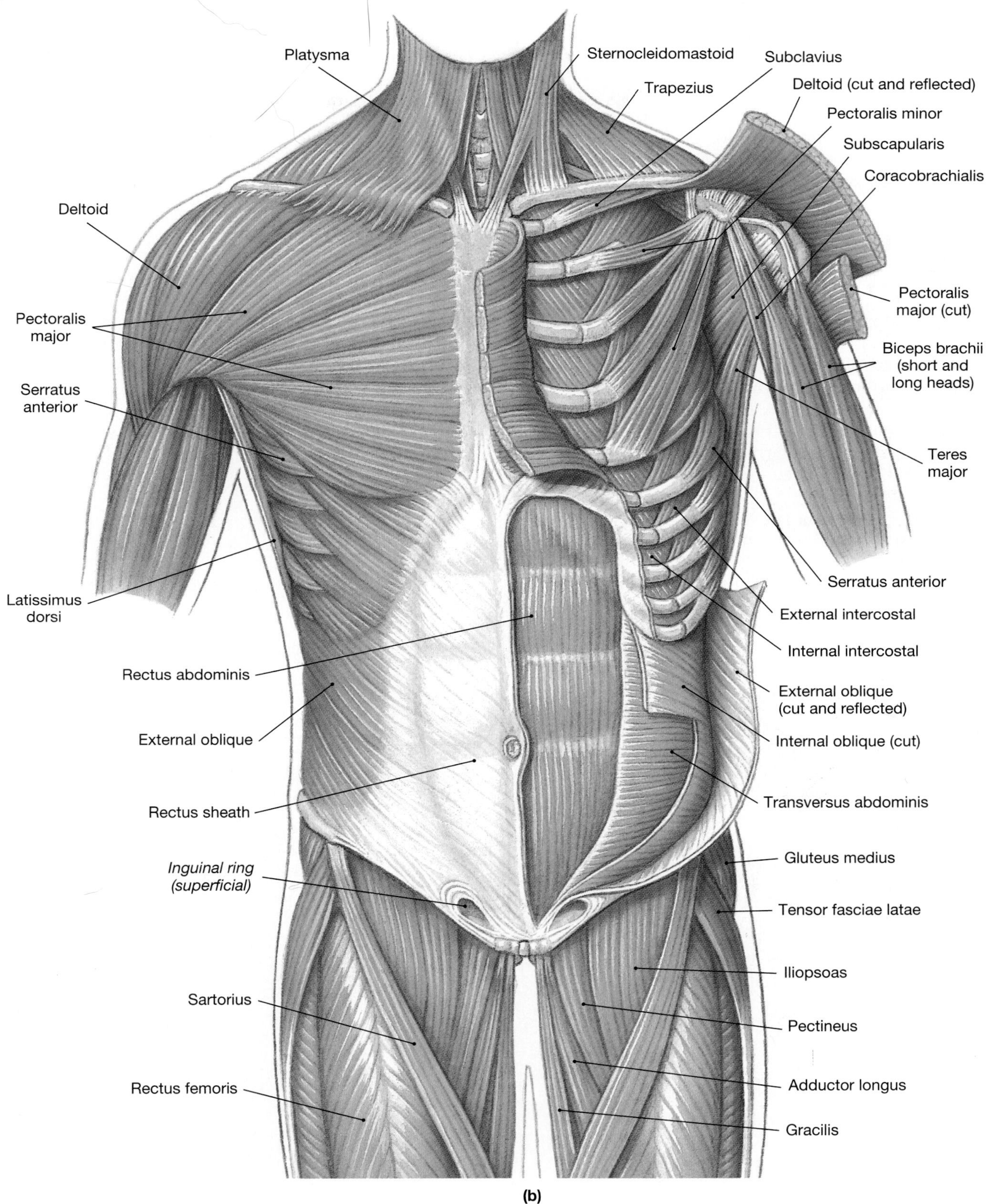
Platysma
Sternocleidomastoid
Subclavius
Trapezius
Deltoid (cut and reflected)
Pectoralis minor
Subscapularis
Coracobrachialis
Deltoid
Pectoralis major
Pectoralis major (cut)
Biceps brachii (short and long heads)
Serratus anterior
Teres major
Serratus anterior
Latissimus dorsi
External intercostal
Internal intercostal
Rectus abdominis
External oblique (cut and reflected)
External oblique
Internal oblique (cut)
Rectus sheath
Transversus abdominis
Gluteus medius
Inguinal ring (superficial)
Tensor fasciae latae
Iliopsoas
Sartorius
Pectineus
Adductor longus
Rectus femoris
Gracilis

(b)

who are jogging can still perform delicate hand movements because the muscular connections between the axial and appendicular skeleton smooth out the bounces in their stride. In contrast, the pelvic girdle has evolved to transfer weight from the axial to the appendicular skeleton. A muscular connection would reduce the efficiency of the transfer, and the emphasis is on strength rather than versatility. Figure 11-14• provides an introduction to the organization of the appendicular musculature.

Muscles of the Shoulders and Upper Limbs

Muscles associated with the shoulders and upper extremities can be divided into four groups: (1) muscles that position the shoulder girdle, (2) muscles that move the arm, (3) muscles that move the forearm and wrist, and (4) muscles that move the palm and fingers.

Muscles That Position the Shoulder Girdle

Figures 11-3, 11-14, 11-15

The large, superficial **trapezius** (tra-PĒ-zē-us) muscles cover the back and portions of the neck, reaching to the base of the skull. These muscles originate along the midline of the neck and back and insert upon the clavicles and the scapular spines. (Figures 11-14 and 11-15a●). The trapezius muscles are innervated by more than one nerve (Table 11-11), and specific regions can be made to contract independently. As a result, their actions are quite varied.

Removing the trapezius reveals the **rhomboideus** (rom-BOY-dē-us) muscles and the **levator scapulae** (Figure 11-15a●). These muscles are attached to the dorsal surfaces of the cervical and thoracic vertebrae. They insert along the vertebral border of each scapula, between the superior and inferior angles. Contraction of a rhomboideus muscle adducts the scapula on that side, and the levator scapulae, as the name implies, elevates the scapula.

On the chest, the **serratus anterior** originates along the anterior surfaces of several ribs (Figures 11-3, pp. 332–333, and 11-15b●). This fan-shaped muscle inserts along the anterior margin of the vertebral border of the scapula. When the serratus anterior contracts, it abducts the scapula and swings the shoulder forward.

Two other deep chest muscles arise along the ventral surfaces of the ribs on either side. A **subclavius** (sub-KLĀ-vē-us; *sub,* below + *clavius,* clav-

● **FIGURE 11-15**
Muscles That Position the Shoulder Girdle. (a) Posterior view. **(b)** Anterior view. *See also Figures 8-3, 8-4, and 9-9, pp. 242, 243, 275.* AM *Plates 5.1, 6.1*

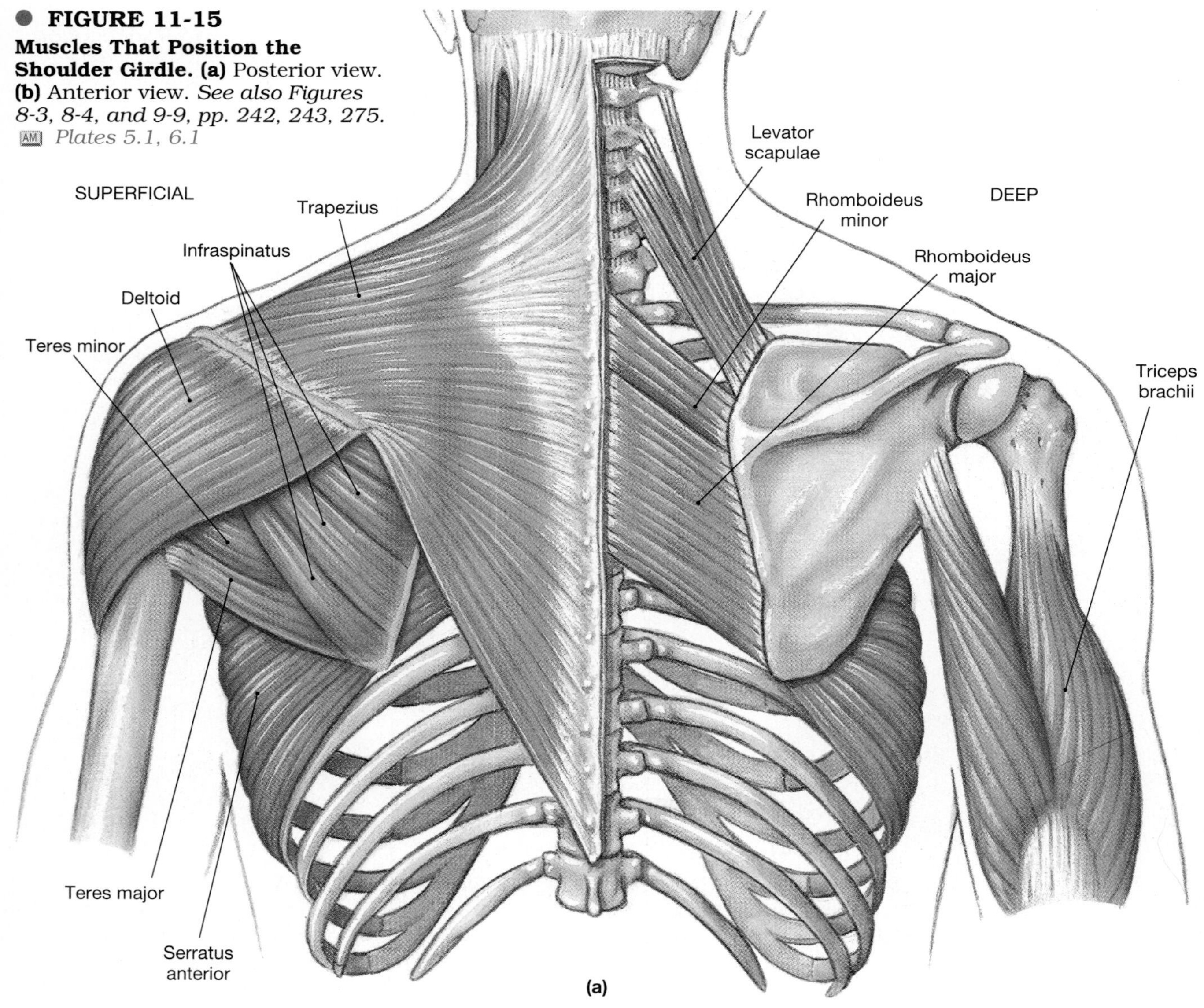
SUPERFICIAL
Trapezius
Infraspinatus
Deltoid
Teres minor
Teres major
Serratus anterior
Levator scapulae
Rhomboideus minor
DEEP
Rhomboideus major
Triceps brachii
(a)

TABLE 11-11 Muscles That Move the Shoulder Girdle

Muscle	*Origin*	*Insertion*	*Action*	*Innervation*
Levator scapulae	Dorsal surfaces of C_1–C_4	Vertebral border of scapula near superior angle	Elevates scapula	Dorsal scapular nerve
Pectoralis minor	Ventral surfaces of ribs 3–5	Coracoid process of scapula	Depresses and protracts shoulder; rotates scapula laterally; elevates ribs if scapula is stationary	Median pectoral nerve
Rhomboideus major	Spinal processes of upper thoracic vertebrae	Vertebral border of scapula from spine to inferior angle	Adducts and rotates scapula laterally	Dorsal scapular nerve
Rhomboideus minor	Spinal processes of vertebrae C_7–T_1	Vertebral border near spine	As above	As above
Serratus anterior	Ventral and superior margins of ribs 1–9	Ventral surface of vertebral border of scapula	Protracts shoulder, abducts and medially rotates scapula	Long thoracic nerve
Subclavius	First rib	Clavicle	Depresses and protracts shoulder	Subclavian nerve
Trapezius	Occipital bone, ligamentum nuchae, and spinal processes of thoracic vertebrae	Clavicle and scapula (acromion and scapular spine)	Depends on active region and state of other muscles; may elevate, adduct, depress, or rotate scapula and/or elevate clavicle; can also extend head and neck	Accessory nerve (N XI) and cervical spinal nerves

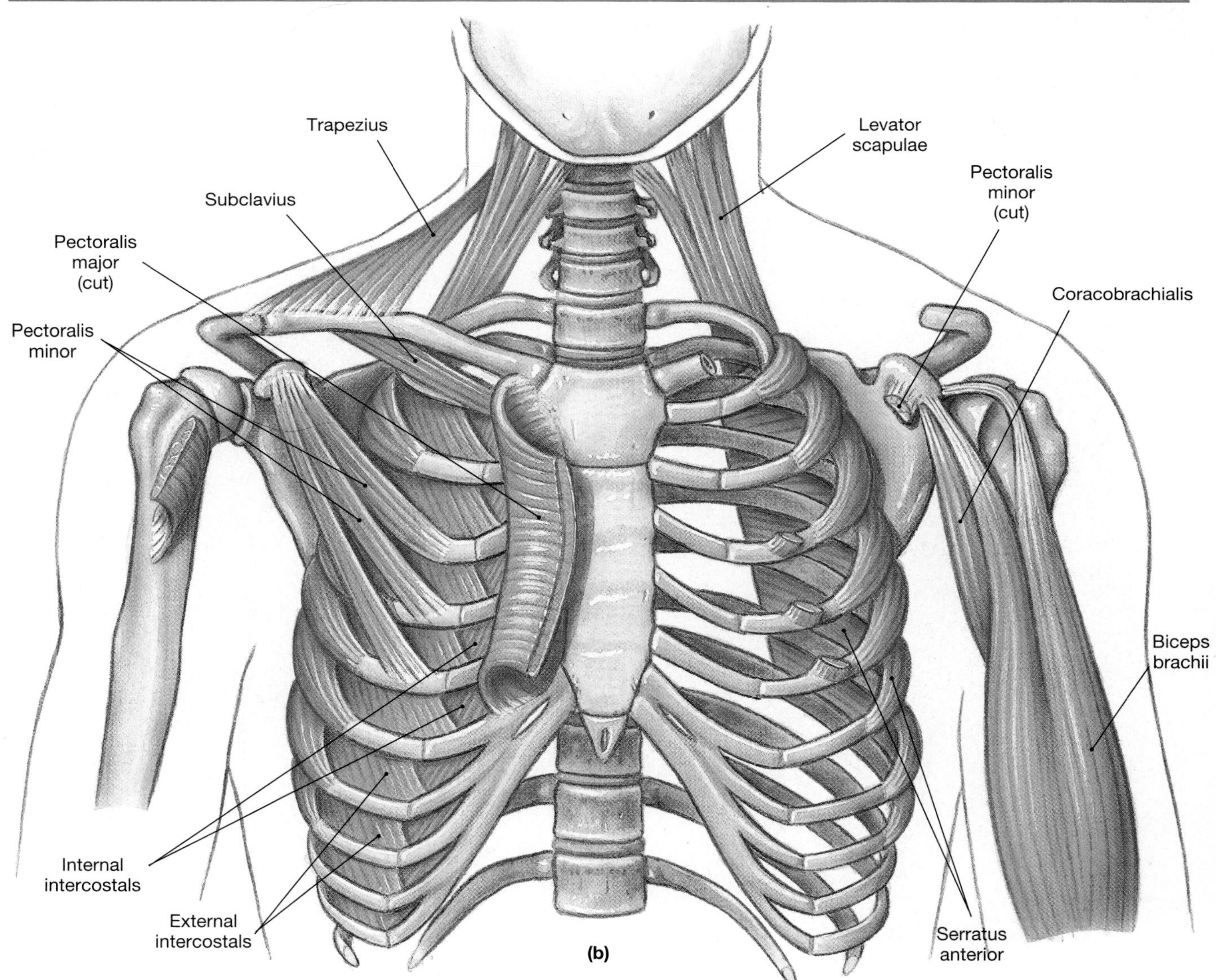
Trapezius
Levator scapulae
Subclavius
Pectoralis minor (cut)
Pectoralis major (cut)
Coracobrachialis
Pectoralis minor
Biceps brachii
Internal intercostals
External intercostals
Serratus anterior
(b)

icle) inserts upon the inferior border of the clavicle (Figure 11-15b●). When it contracts, it depresses and protracts the scapular end of the clavicle. Because ligaments connect this end to the shoulder joint and scapula, those structures move as well. The **pectoralis** (pek-tō-RA-lis) **minor** attaches to the coracoid process of the scapula. Its contraction usually complements that of the subclavius.

Muscles That Move the Arm

Figures 11-14, 11-16

The muscles that move the arm (Figures 11-14, pp. 350–351, and 11-16●) are easiest to remember when grouped by primary actions (Table 11-12). The **deltoid** is the major abductor of the arm, but the **supraspinatus** (su-pra-spī-NA-tus) assists at the start of this movement. The **subscapularis** and **teres** (TER-ēz) **major** rotate the arm medially, whereas the **infraspinatus** and the **teres minor** perform lateral rotation. All of these muscles originate on the scapula. The small **coracobrachialis** (KOR-uh-kō-brā-kē-A-lis) is the only muscle attached to the scapula that flexes and adducts the humerus. These muscles are detailed in Figure 11-16a●.

The **pectoralis major** extends between the chest and the greater tubercle of the humerus; the **latissimus dorsi** (la-TIS-i-mus DOR-sē) extends between the thoracic vertebrae and the lesser tubercle of the humerus (Figure 11-16b●). The pectoralis major flexes the arm, and the latissimus dorsi extends it. These two muscles can also work together to produce adduction and medial rotation of the humerus.

The supraspinatus, infraspinatus, subscapularis, and teres minor and associated tendons form the **rotator cuff,** a frequent site of sports injuries.

SPORTS INJURIES Exercise carries risks due to the stresses placed on muscles, joints, and connective tissues. Many Americans participate in exercise programs and sports on a regular basis; more than 30 million Americans go jogging, and millions more participate in various amateur and professional sports. As a result, *sports injuries* are very common, and sports medicine has become an active area of professional and academic research interest. For information on the incidence and classification of sports injuries, see the *Applications Manual.* AM Sports Injuries

Muscles That Move the Forearm and Wrist

Figures 11-16, 11-17, 11-18

Although most of the muscles that insert upon the forearm and wrist originate on the humerus, there are two noteworthy exceptions, the *biceps brachii* and the *triceps brachii.* These muscles are shown in their entirety in Figure 11-17●. The **biceps brachii**

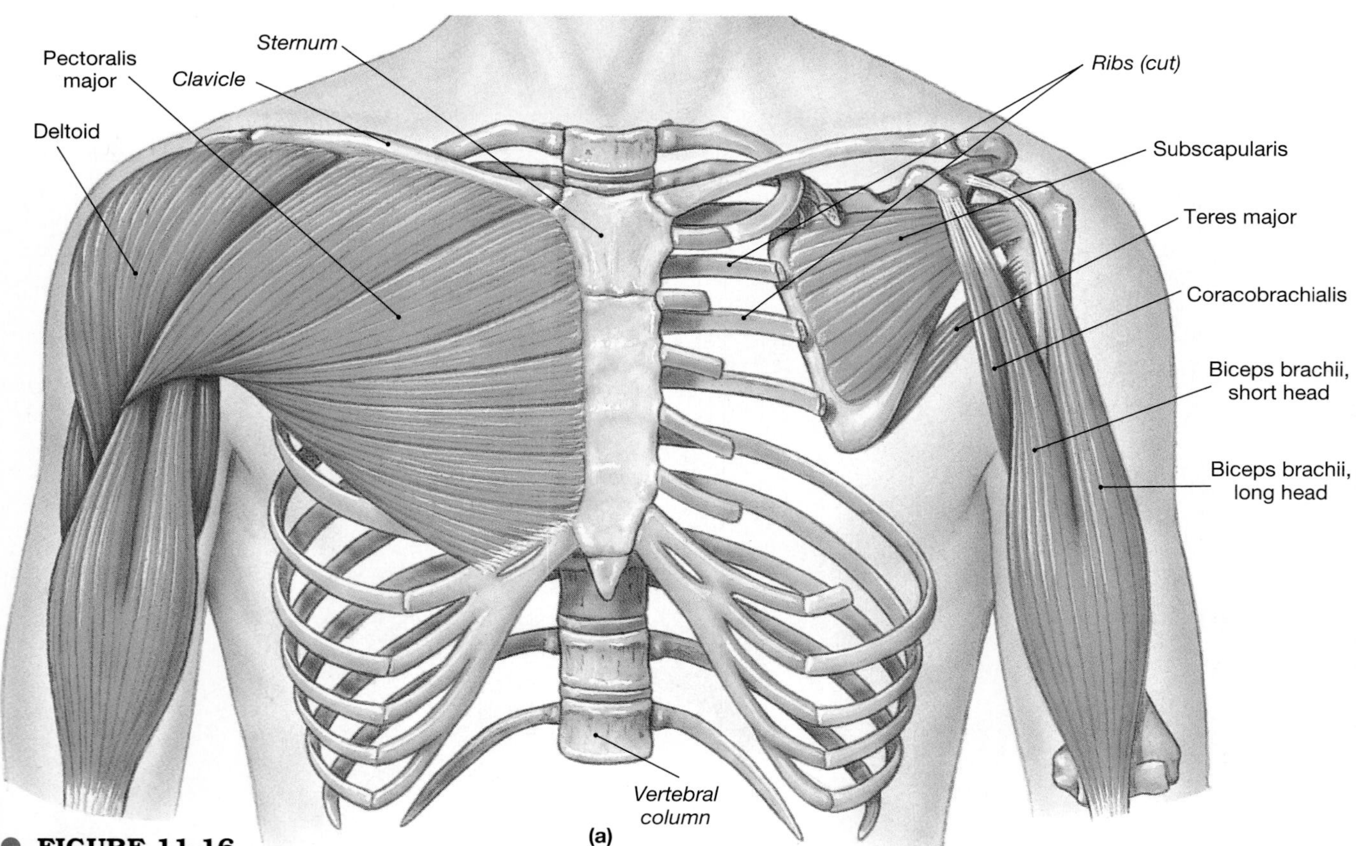

Sternum
Pectoralis major
Clavicle
Deltoid
Ribs (cut)
Subscapularis
Teres major
Coracobrachialis
Biceps brachii, short head
Biceps brachii, long head
Vertebral column
(a)

● **FIGURE 11-16**

Muscles That Move the Arm. (a) Anterior view. **(b)** Posterior view. *See also Figures 7-23, 8-3, and 9-9, pp. 230–231, 243, 275.* AM Plates 5.1, 6.1

TABLE 11-12 Muscles That Move the Arm

Muscle	*Origin*	*Insertion*	*Action*	*Innervation*
Coracobrachialis	Coracoid process	Medial margin of shaft of humerus	Adducts and flexes humerus	Musculocutaneous nerve
Deltoid	Clavicle and scapula (acromion and adjacent scapular spine)	Deltoid tuberosity of humerus	Abducts arm	Axillary nerve
Supraspinatus	Supraspinous fossa of scapula	Greater tubercle of humerus	Abducts arm	Suprascapular nerve
Infraspinatus	Infraspinous fossa of scapula	Greater tubercle of humerus	Lateral rotation of humerus	Suprascapular nerve
Subscapularis	Subscapular fossa of scapula	Lesser tubercle of humerus	Medial rotation of humerus	Subscapular nerve
Teres major	Inferior angle of scapula	Intertubercular groove of humerus	Adducts and medially rotates arm	Lower subscapular nerve
Teres minor	Axillary border of scapula	Greater tubercle of humerus	Lateral rotation of humerus	Axillary nerve
Triceps brachii (long head)	*See Table 11-13*			
Latissimus dorsi	Spinous processes of lower thoracic vertebrae, ribs 8–12, the spines of lumbar vetebrae, and the lumbodorsal fascia	Lesser tubercle, intertubercular groove of humerus	Extends, adducts, and medially rotates humerus	Thoracodorsal nerve
Pectoralis major	Cartilages of ribs 2–6, body of sternum, and inferior, medial portion of clavicle	Greater tubercle of humerus	Flexes, adducts, and medially rotates humerus	Pectoral nerves

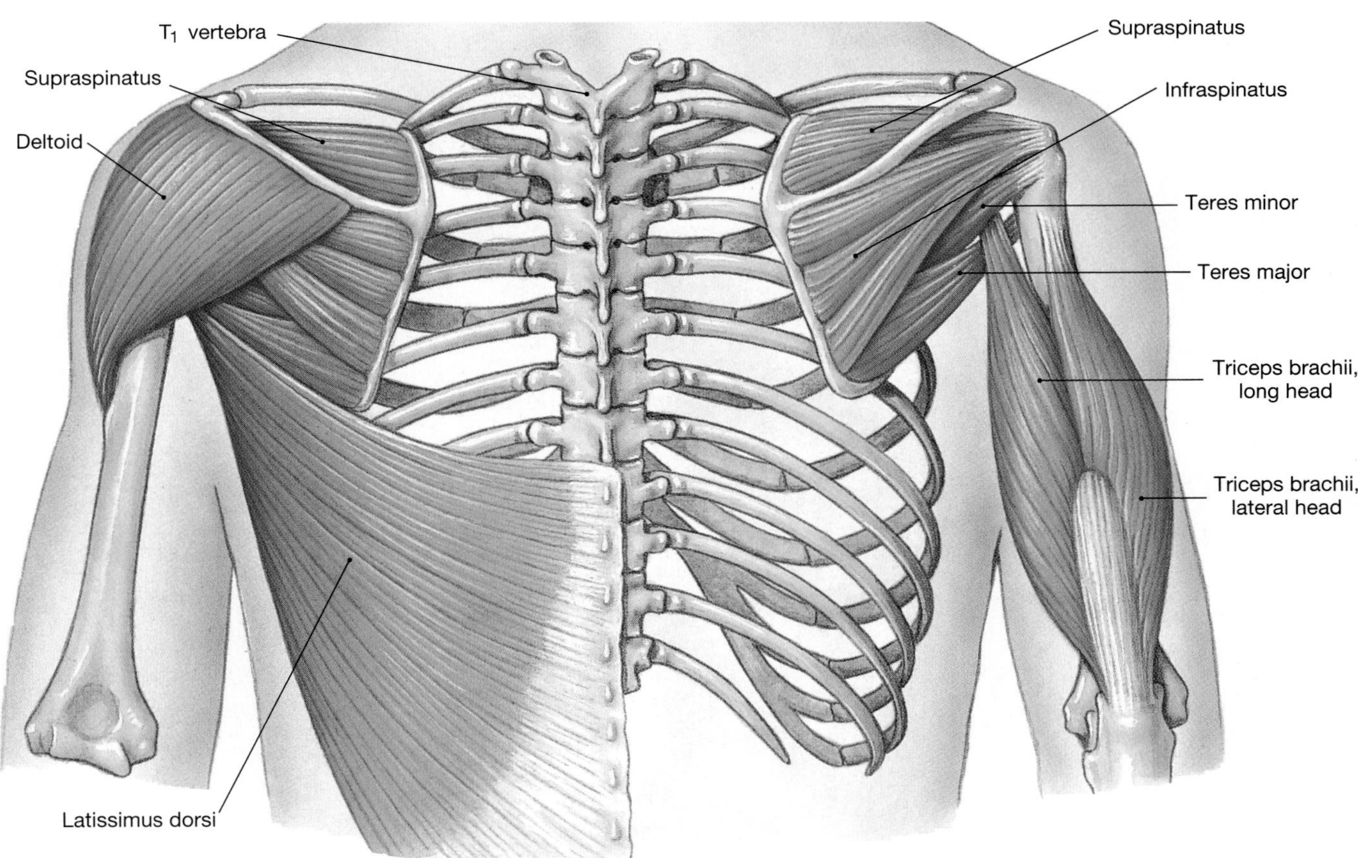
T_1 vertebra
Supraspinatus
Deltoid
Supraspinatus
Infraspinatus
Teres minor
Teres major
Triceps brachii, long head
Triceps brachii, lateral head
Latissimus dorsi

(b)

● **FIGURE 11-17**

Muscles That Move the Forearm and Wrist.

See also Figures 8-4, 8-5, and 9-10, pp. 243, 245, 276. AM *Plates 5.2, 5.3*

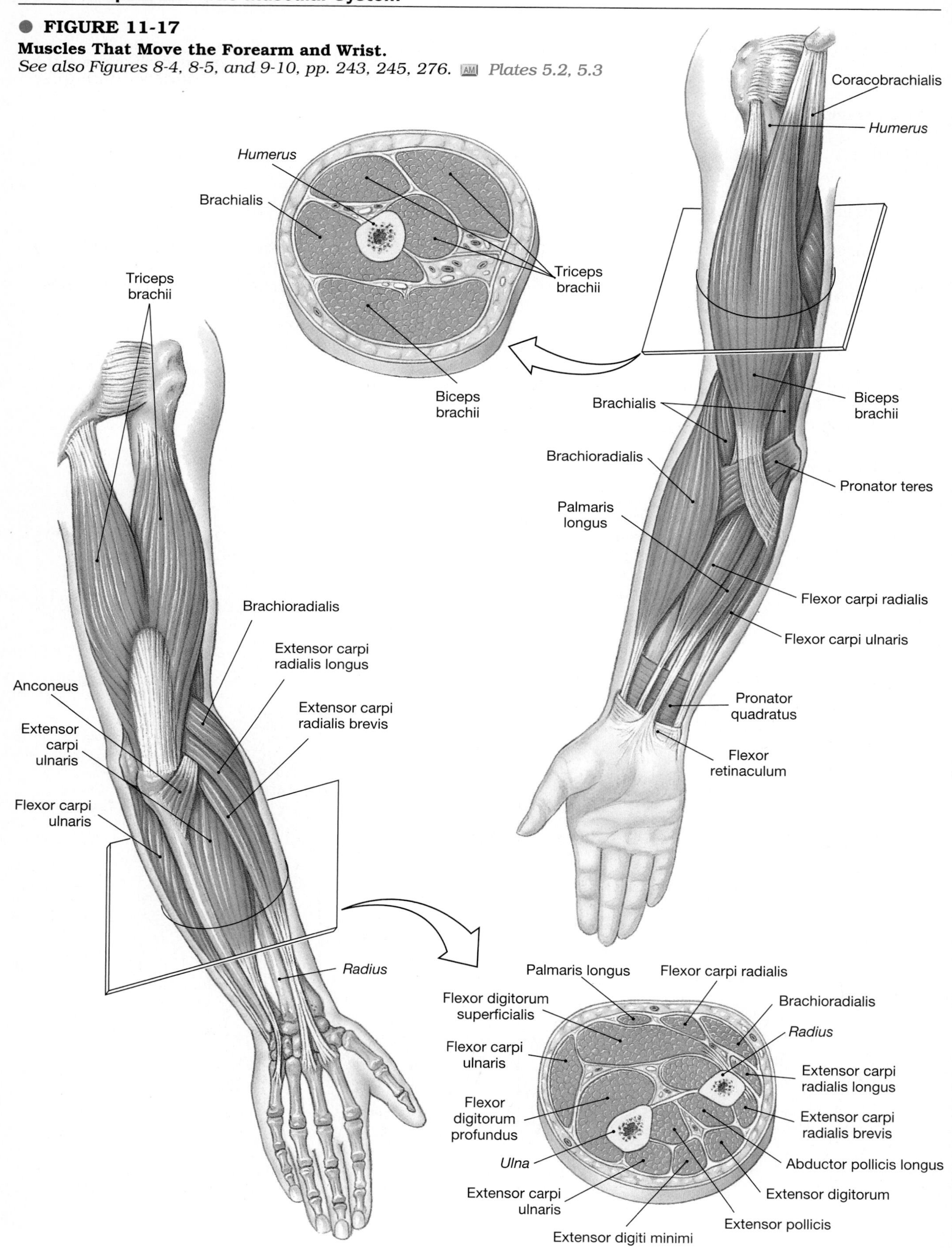
Coracobrachialis
Humerus
Humerus
Brachialis
Triceps brachii
Triceps brachii
Biceps brachii
Brachialis
Biceps brachii
Brachioradialis
Pronator teres
Palmaris longus
Flexor carpi radialis
Flexor carpi ulnaris
Pronator quadratus
Flexor retinaculum
Brachioradialis
Extensor carpi radialis longus
Anconeus
Extensor carpi radialis brevis
Extensor carpi ulnaris
Flexor carpi ulnaris
Radius
Palmaris longus
Flexor carpi radialis
Flexor digitorum superficialis
Brachioradialis
Radius
Flexor carpi ulnaris
Extensor carpi radialis longus
Flexor digitorum profundus
Extensor carpi radialis brevis
Ulna
Abductor pollicis longus
Extensor digitorum
Extensor carpi ulnaris
Extensor pollicis
Extensor digiti minimi

TABLE 11-13 Muscles That Move the Forearm and Wrist

Muscle	*Origin*	*Insertion*	*Action*	*Innervation*
PRIMARY ACTION AT THE ELBOW				
Flexors				
Biceps brachii	Short head from the coracoid process; long head from the supraglenoid tuberosity (both on the scapula)	Tuberosity of radius	Flexes and supinates forearm	Musculo-cutaneous nerve
Brachialis	Anterior, distal surface of humerus	Tuberosity of ulna	Flexes forearm	As above
Brachioradialis	Lateral epicondyle of humerus	Lateral aspect of styloid process of radius	As above	Radial nerve
Extensors				
Anconeus	Posterior surface of lateral epicondyle of humerus	Lateral margin of olecranon on ulna	Extends forearm, moves ulna laterally during pronation	As above
Triceps brachii				
lateral head	Superior, lateral margin of humerus	Olecranon process of ulna	Extends forearm	As above
long head	Infraglenoid tuberosity of scapula	As above	As above	As above
medial head	Posterior margin of humerus inferior to radial groove	As above	As above	As above
PRONATORS/ SUPINATORS				
Pronator quadratus	Medial surface of distal portion of ulna	Anterolateral surface of distal portion of radius	Pronates forearm	Median nerve
Pronator teres	Medial epicondyle of humerus and coronoid process of ulna	Distal lateral surface of radius	As above	As above
Supinator	Lateral epicondyle of humerus and ulna	Anterolateral surface of radius distal to the radial tuberosity	Supinates forearm	Radial nerve
PRIMARY ACTION AT THE WRIST				
Flexors				
Flexor carpi radialis	Medial epicondyle of humerus	Bases of 2nd and 3rd metacarpals	Flexes and abducts palm	Median nerve
Flexor carpi ulnaris	Medial epicondyle of humerus; adjacent medial surface of olecranon and anteromedial portion of ulna	Bases of 3rd and 4th metacarpals, pisiform, and hamate	Flexes and adducts palm	Ulnar nerve
Palmaris longus	Medial epicondyle of humerus	Palmar aponeurosis	Flexes palm	Median nerve
Extensors				
Extensor carpi radialis longus	Lateral supracondylar ridge of humerus	Base of 2nd metacarpal	Extends and abducts palm	Radial nerve
Extensor carpi radialis brevis	Lateral epicondyle of humerus	Base of 3rd metacarpal	As above	As above
Extensor carpi ulnaris	Lateral epicondyle of humerus; adjacent dorsal surface of ulna	Base of 5th metacarpal	Extends and adducts palm	Deep radial nerve

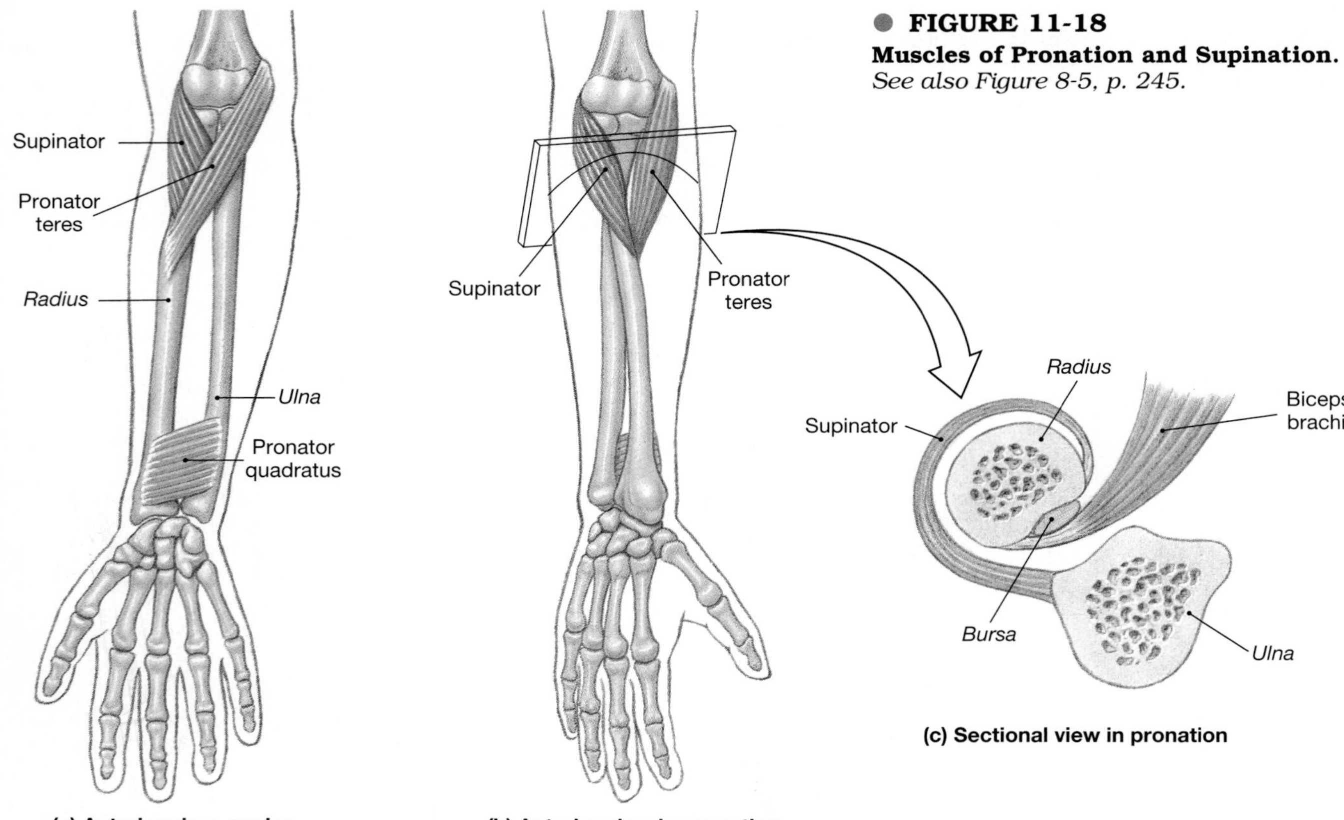
Supinator
Pronator teres
Radius
Ulna
Pronator quadratus
Supinator
Pronator teres
Radius
Supinator
Biceps brachii
Bursa
Ulna

(a) Anterior view, supine

(b) Anterior view in pronation

(c) Sectional view in pronation

● **FIGURE 11-18**
Muscles of Pronation and Supination.
See also Figure 8-5, p. 245.

and the long head of the **triceps brachii** originate on the scapula and insert on the bones of the forearm. The triceps brachii inserts on the olecranon process. Contraction of the triceps brachii extends the forearm, as in doing push-ups. The biceps brachii inserts on the radial tuberosity, a roughened area on the anterior surface of the radius. Contraction of the biceps flexes and supinates the forearm. With the forearm pronated (palm facing back) the biceps brachii cannot function effectively. As a result, we are strongest when flexing the supinated forearm; the biceps brachii then makes a prominent bulge.

Additional muscles are shown in Figure 11-17● and detailed in Table 11-13. The **brachialis** (brā-kē-A-lis) and **brachioradialis** (brā-kē-ō-rā-dē-A-lis) also flex the forearm, opposed by the **anconeus** and the triceps.

The **flexor carpi ulnaris,** the **flexor carpi radialis,** and the **palmaris longus** are superficial muscles that work together to produce flexion of the wrist. The flexor carpi radialis flexes and *ab*ducts, and the flexor carpi ulnaris flexes and *ad*ducts. *Pitcher's arm* is an inflammation at the origins of the flexor carpi muscles at the medial epicondyle. This condition results from forcibly flexing the wrist just before releasing a baseball.

The **extensor carpi radialis** muscles and the **extensor carpi ulnaris** have a similar relationship to that between the flexor carpi muscles. The extensor carpi radialis produces extension and *ab*duction, the extensor carpi ulnaris extension and *ad*duction.

The **pronator teres** and the **supinator** arise on both the humerus and forearm. They rotate the radius without producing either flexion or extension of the elbow. The **pronator quadratus** arises on the ulna and assists the pronator teres in opposing the actions of the supinator or biceps. The muscles involved in pronation and supination can be seen in Figure 11-18●. Note the changes in orientation that occur as the pronator teres and pronator quadratus contract. During pronation, the tendon of the biceps rolls under the radius, and a bursa prevents abrasion against the tendon (Figure 11-18c●).

As you study the muscles included in Table 11-13, note that in general the extensor muscles lie along the posterior and lateral surfaces of the arm, whereas the flexors are found on the anterior and medial surfaces.

Muscles That Move the Palm and Fingers

Figures 11-19, 11-20

Several superficial and deep muscles of the forearm (Table 11-14) perform flexion and extension of the digits. These muscles stop before reaching the wrist, and only their tendons cross the articulation. These are relatively large muscles (Figure 11-19●), and keeping them clear of the joints ensures maximum mobility at both the wrist and hand. The tendons that cross the dorsal and ventral surfaces of the wrist pass through **tendon sheaths,** elongate bursae that

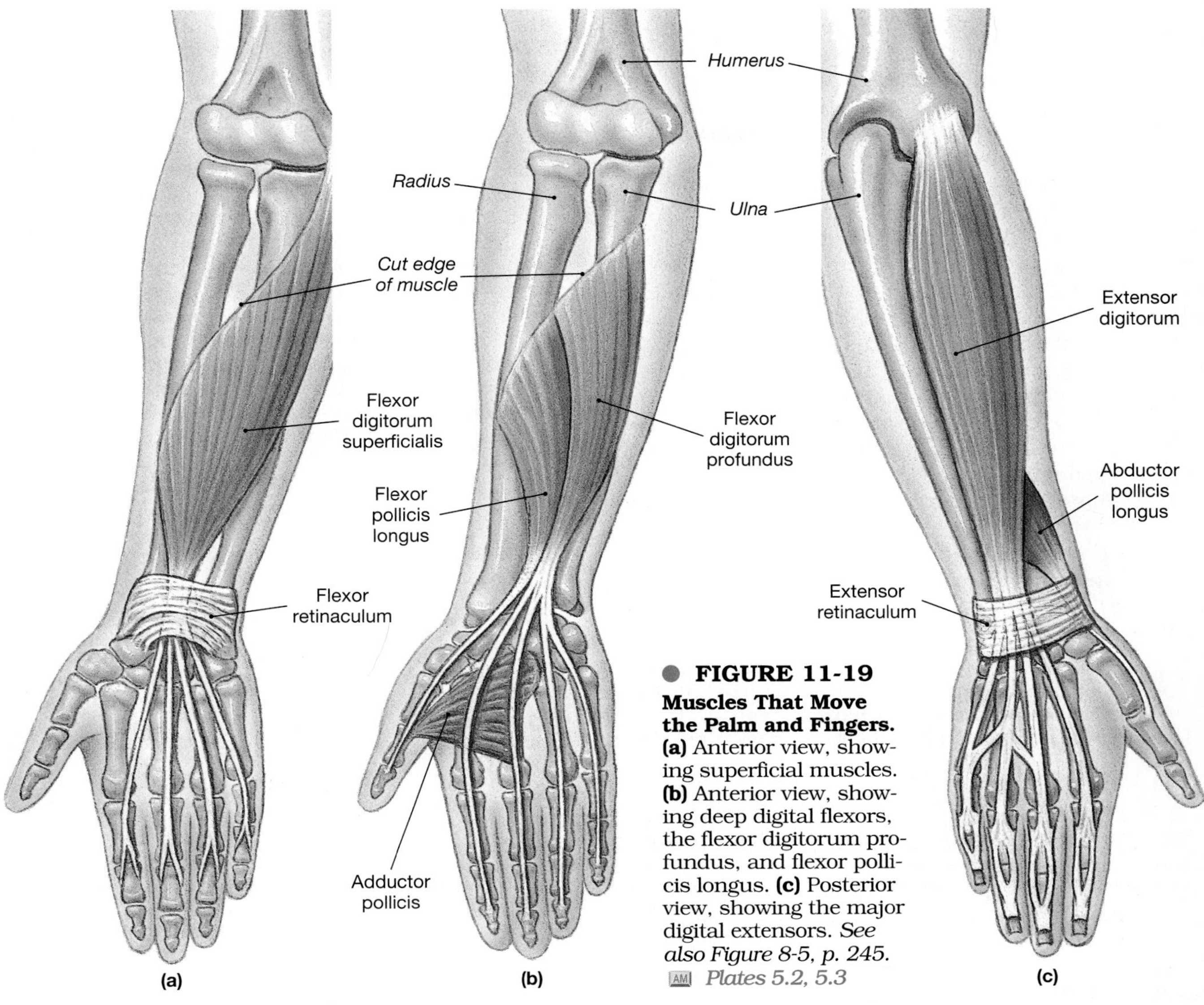
Humerus
Radius
Ulna
Cut edge
of muscle
Extensor
digitorum
Flexor
digitorum
superficialis
Flexor
digitorum
profundus
Flexor
pollicis
longus
Abductor
pollicis
longus
Flexor
retinaculum
Extensor
retinaculum
Adductor
pollicis
(a)
(b)
(c)

● **FIGURE 11-19 Muscles That Move the Palm and Fingers.** **(a)** Anterior view, showing superficial muscles. **(b)** Anterior view, showing deep digital flexors, the flexor digitorum profundus, and flexor pollicis longus. **(c)** Posterior view, showing the major digital extensors. *See also Figure 8-5, p. 245.* AM *Plates 5.2, 5.3*

TABLE 11-14 Muscles That Move the Palm and Fingers

Muscle	*Origin*	*Insertion*	*Action*	*Innervation*
Abductor pollicis longus	Proximal dorsal surfaces of ulna and radius	Lateral margin of 1st metacarpal	Abducts thumb	Deep radial nerve
Extensor digitorum	Lateral epicondyle of humerus	Dorsal surfaces of the phalanges	Extends fingers and palms	As above
Flexor digitorum profundus	Proximal anteromedial surface of ulna	Bases of distal phalanges	Flexes fingers, specifically 3rd phalanx on 2nd, and 2nd on 1st; flexes palm	Median nerve
Flexor digitorum superficialis	Medial epicondyle of humerus; adjacent anterior surfaces of ulna and radius	Midlateral surface of 2nd phalanx; connected by ligaments to others	Flexes fingers, specifically 2nd phalanx on 1st; flexes palm	As above
Flexor pollicis longus	Shaft of radius and interosseous membrane	Distal phalanx of thumb	Flexes thumb	As above

reduce friction. [p. 265] Figure 11-19● shows these muscles and their tendons in anterior and posterior views. The fascia of the forearm thickens on the posterior surface of the wrist to form a wide band of connective tissue, the **extensor retinaculum** (ret-i-NAK-ū-lum). The extensor retinaculum holds the tendons of the extensor muscles in place. On the anterior surface the fascia also thickens to form another wide band of connective tissue, the **flexor retinaculum,** which stabilizes the tendons of the flexor muscles. Inflammation of the retinacula and tendon sheaths can restrict movement and irritate the median nerve. This condition, known as *carpal tunnel syndrome,* causes chronic pain.

The muscles of the forearm provide strength and crude control of the palm and fingers. These are known as the *extrinsic muscles of the hand.* Fine control of the hand involves small *intrinsic muscles* that originate on the carpals and metacarpals. No muscles originate on the phalanges, and only tendons extend across the distal joints of the fingers. The intrinsic muscles of the hand are detailed in Figure 11-20● and Table 11-15.

CARPAL TUNNEL SYNDROME *Tenosynovitis* is the inflammation of a tendon sheath. **Carpal tunnel syndrome** results from tenosynovitis of the tendon sheath surrounding the flexor tendons of the palm. The inflammation leads to compression of the *median nerve,* a mixed (sensory and motor) nerve that innervates the palm. Symptoms include pain, especially on palmar flexion, a tingling sensation or numbness on the palm, and weakness in the abductor pollicis. This condition is fairly common and often strikes those engaged in repetitive hand movements, such as typing, working at a computer keyboard, or playing the piano. Treatment involves administration of anti-inflammatory drugs such as aspirin, injection of anti-inflammatory agents, such as *glucocorticoids* (steroid hormones produced by the adrenal cortex), and use of a splint to prevent wrist flexion and stabilize the region.

Carpal tunnel syndrome is an example of a *cumulative trauma disorder,* or *overuse syndrome.* These disorders are caused by repetitive movements of the arms, hands, and fingers. These musculoskeletal problems now account for over 50 percent of all work-related injuries in the United States.

What muscle are you using when you shrug your shoulders?

Baseball pitchers sometimes suffer from rotator cuff injuries. What muscles are involved in this type of injury?

Injury to the flexor carpi ulnaris would impair what two movements?

TABLE 11-15 Intrinsic Muscles of the Hand

Muscle	Origin	Insertion	Action	Innervation
Adductor pollicis	Metacarpals and carpals	Proximal phalanx of thumb	Adducts thumb	Ulnar nerve, deep branch
Opponens pollicis	Trapezium	Metacarpal of thumb	Opposition of thumb	Median nerve
Palmaris brevis	Palmar aponeurosis	Skin of medial border of hand	Moves skin on medial border toward midline of palm	Ulnar nerve, superficial branch
Abductor digiti minimi	Pisiform bone	Proximal phalanx of little finger	Abducts little finger and flexes its proximal phalanx	Ulnar nerve, deep branch
Flexor digiti minimi	Hamate bone	Proximal phalanx of little finger	Flexes little finger	Ulnar nerve, deep branch
Opponens digiti minimi	Hamate bone	Fifth metacarpal	Opposition of fifth metacarpal	Ulnar nerve, deep branch
Lumbricals (4)	Tendons of flexor digitorum profundus	Tendons of extensor digitorum	Flex metacarpophalangeal joints, extend middle and distal phalanges	#1 and #2 by median nerve, #3 and #4 by ulnar nerve, deep branch
Dorsal interossei (4)	Sides of metacarpals	Bases of proximal phalanges	Abduct fingers, flex metacarpophalangeal joints, extend fingertips	Ulnar nerve, deep palmar branch
Palmar interossei (3)	Sides of metacarpals I, IV, and V	Bases of proximal phalanges	Abduct fingers, flex metacarpophalangeal joints, extend fingertips	Ulnar nerve, deep palmar branch

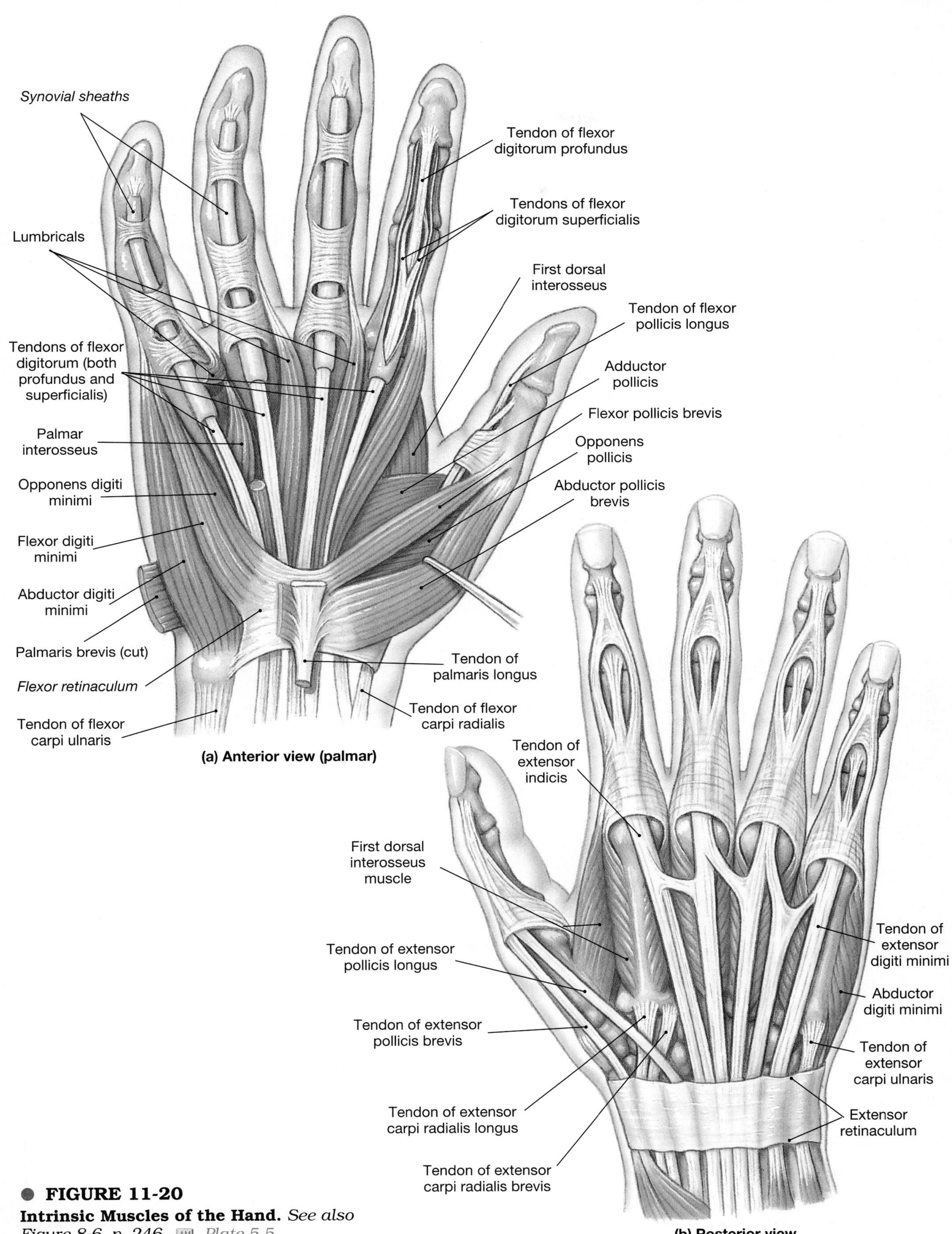
Synovial sheaths
Tendon of flexor digitorum profundus
Tendons of flexor digitorum superficialis
Lumbricals
First dorsal interosseus
Tendon of flexor pollicis longus
Tendons of flexor digitorum (both profundus and superficialis)
Adductor pollicis
Flexor pollicis brevis
Palmar interosseus
Opponens pollicis
Opponens digiti minimi
Abductor pollicis brevis
Flexor digiti minimi
Abductor digiti minimi
Palmaris brevis (cut)
Tendon of palmaris longus
Flexor retinaculum
Tendon of flexor carpi radialis
Tendon of flexor carpi ulnaris
(a) Anterior view (palmar)
Tendon of extensor indicis
First dorsal interosseus muscle
Tendon of extensor pollicis longus
Tendon of extensor digiti minimi
Abductor digiti minimi
Tendon of extensor pollicis brevis
Tendon of extensor carpi ulnaris
Extensor retinaculum
Tendon of extensor carpi radialis longus
Tendon of extensor carpi radialis brevis
(b) Posterior view

● **FIGURE 11-20**
Intrinsic Muscles of the Hand. *See also Figure 8-6, p. 246.* AM *Plate 5.5*

EMBRYOLOGY SUMMARY Development of the Muscular System

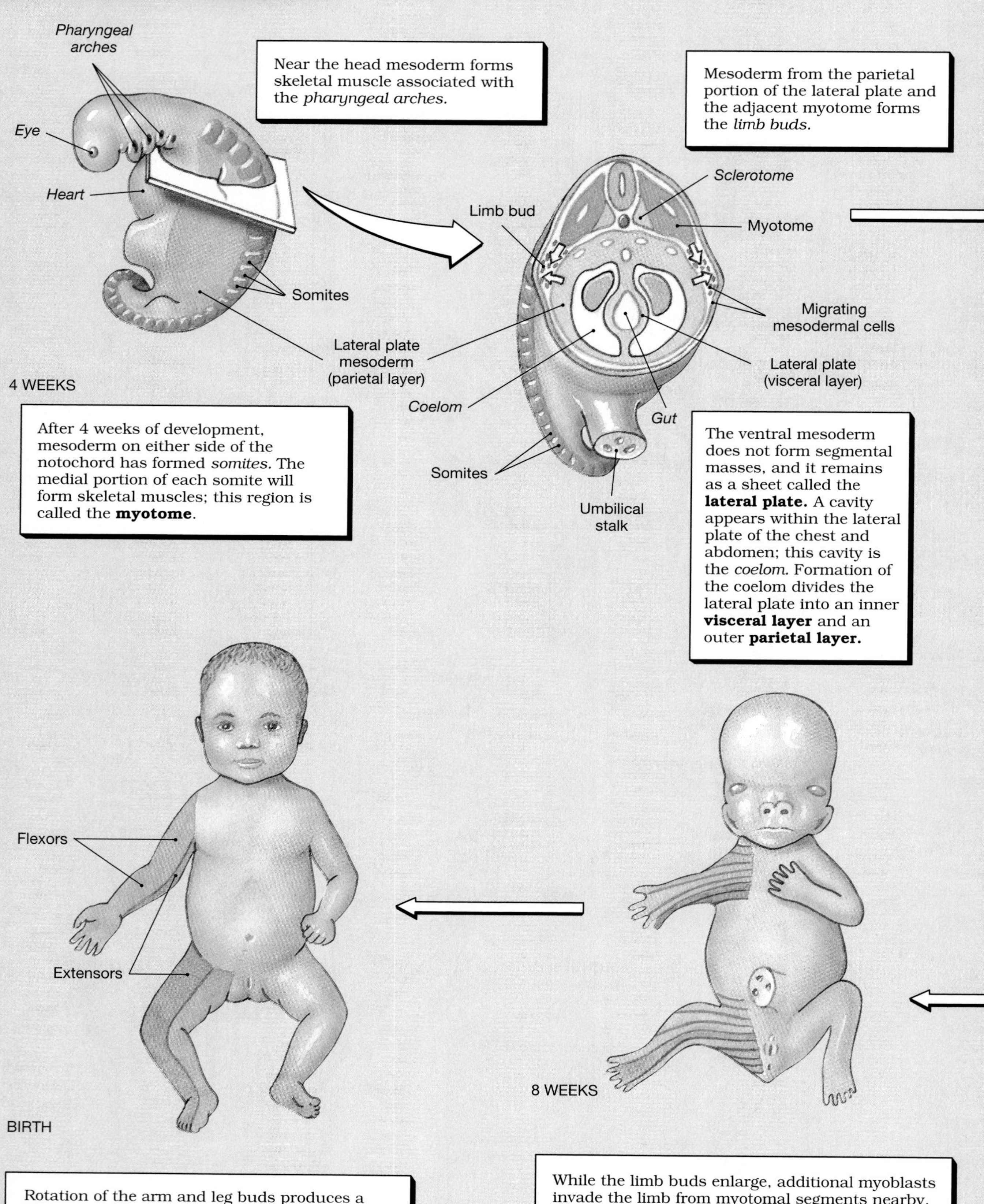

Pharyngeal arches
Eye
Heart
Somites
Lateral plate mesoderm (parietal layer)
4 WEEKS
Near the head mesoderm forms skeletal muscle associated with the *pharyngeal arches.*
Mesoderm from the parietal portion of the lateral plate and the adjacent myotome forms the *limb buds.*
Sclerotome
Limb bud
Myotome
Migrating mesodermal cells
Lateral plate (visceral layer)
Coelom
Gut
Somites
Umbilical stalk
After 4 weeks of development, mesoderm on either side of the notochord has formed *somites.* The medial portion of each somite will form skeletal muscles; this region is called the **myotome**.
The ventral mesoderm does not form segmental masses, and it remains as a sheet called the **lateral plate.** A cavity appears within the lateral plate of the chest and abdomen; this cavity is the *coelom.* Formation of the coelom divides the lateral plate into an inner **visceral layer** and an outer **parietal layer.**
Flexors
Extensors
BIRTH
8 WEEKS

Rotation of the arm and leg buds produces a change in the position of these masses relative to the body axis.

While the limb buds enlarge, additional myoblasts invade the limb from myotomal segments nearby. Lines indicate the boundaries between myotomes providing myoblasts to the limb.

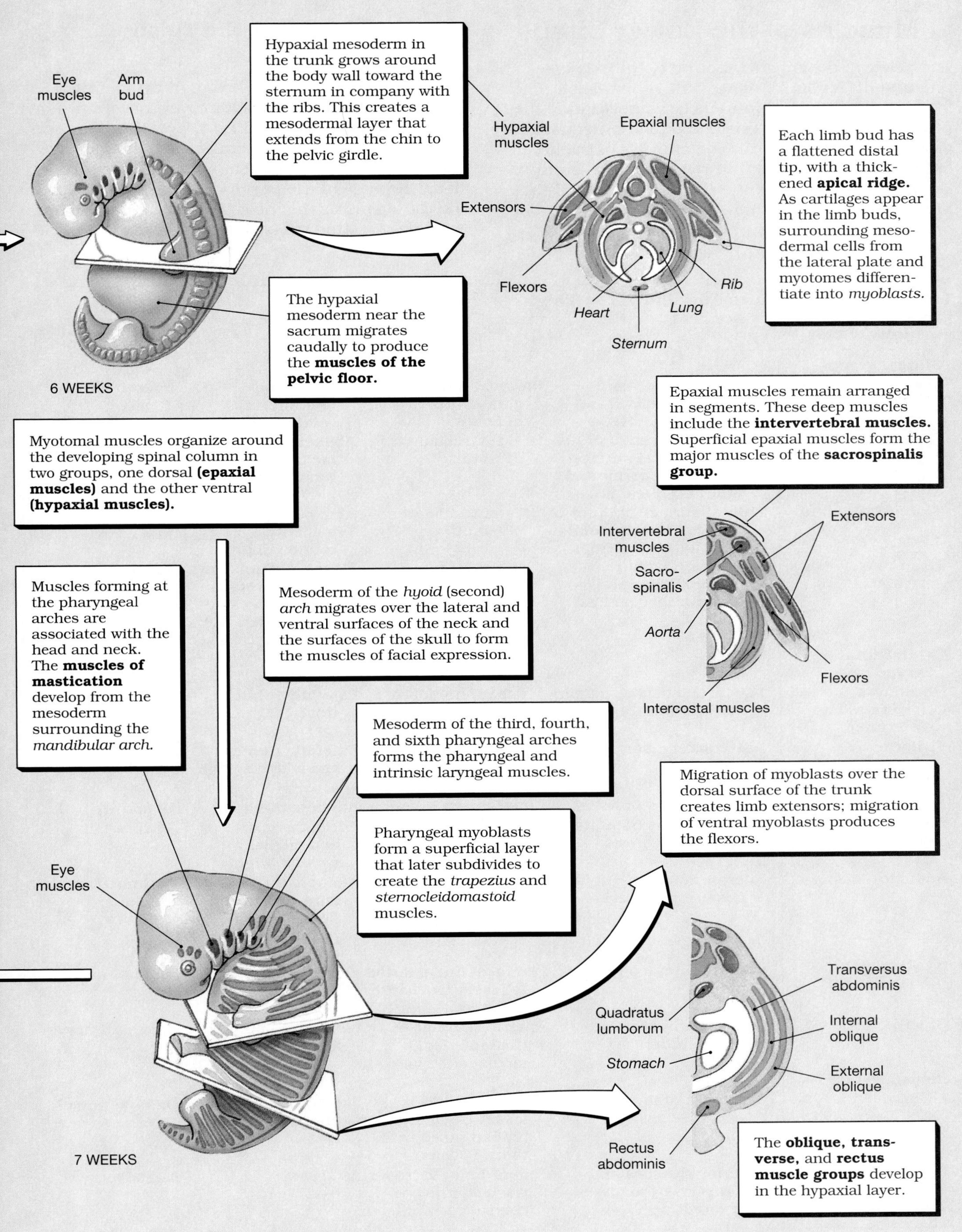
Eye muscles
Arm bud
Hypaxial mesoderm in the trunk grows around the body wall toward the sternum in company with the ribs. This creates a mesodermal layer that extends from the chin to the pelvic girdle.
The hypaxial mesoderm near the sacrum migrates caudally to produce the muscles of the pelvic floor.
6 WEEKS
Hypaxial muscles
Epaxial muscles
Extensors
Flexors
Heart
Sternum
Lung
Rib
Each limb bud has a flattened distal tip, with a thickened apical ridge. As cartilages appear in the limb buds, surrounding mesodermal cells from the lateral plate and myotomes differentiate into myoblasts.
Myotomal muscles organize around the developing spinal column in two groups, one dorsal (epaxial muscles) and the other ventral (hypaxial muscles).
Epaxial muscles remain arranged in segments. These deep muscles include the intervertebral muscles. Superficial epaxial muscles form the major muscles of the sacrospinalis group.
Intervertebral muscles
Sacro-spinalis
Aorta
Extensors
Flexors
Intercostal muscles
Muscles forming at the pharyngeal arches are associated with the head and neck. The muscles of mastication develop from the mesoderm surrounding the mandibular arch.
Mesoderm of the hyoid (second) arch migrates over the lateral and ventral surfaces of the neck and the surfaces of the skull to form the muscles of facial expression.
Mesoderm of the third, fourth, and sixth pharyngeal arches forms the pharyngeal and intrinsic laryngeal muscles.
Pharyngeal myoblasts form a superficial layer that later subdivides to create the trapezius and sternocleidomastoid muscles.
Migration of myoblasts over the dorsal surface of the trunk creates limb extensors; migration of ventral myoblasts produces the flexors.
Eye muscles
7 WEEKS
Transversus abdominis
Quadratus lumborum
Internal oblique
Stomach
External oblique
Rectus abdominis
The oblique, transverse, and rectus muscle groups develop in the hypaxial layer.

❑ Muscles of the Lower Limbs

The pelvic girdle is tightly bound to the axial skeleton, and little relative movement is permitted. The few muscles that can influence the position of the pelvis were therefore considered in our discussion of the axial musculature. The muscles of the lower limbs can be divided into three functional groups: (1) *muscles that move the thigh,* (2) *muscles that move the leg,* and (3) *muscles that move the foot and toes.*

Muscles That Move the Thigh

Figures 11-14, 11-21

The muscles that move the thigh are detailed in Table 11-16. **Gluteal muscles** cover the lateral surfaces of the ilia (Figures 11-14, pp. 350–351, and 11-21a●). The **gluteus maximus** is the largest and most posterior of the gluteal muscles. It originates along the edge of the posterior superior iliac spine and the ligaments that bind the sacrum to the ilium. Acting alone, this massive muscle extends and lat-

TABLE 11-16 Muscles That Move the Thigh

Group/Muscle	*Origin*	*Insertion*	*Action*	*Innervation*
Gluteal group				
Gluteus maximus	Iliac crest of ilium, sacrum, coccyx, and lumbodorsal fascia	Iliotibial tract and gluteal tuberosity of femur	Extends and laterally rotates thigh	Inferior gluteal nerve
Gluteus medius	Anterior iliac crest of ilium, lateral surface between superior and inferior gluteal lines	Greater trochanter of femur	Abducts and medially rotates thigh	Superior gluteal nerve
Gluteus minimus	Lateral surface of ilium between inferior and anterior gluteal lines	Greater trochanter of femur	Abducts and medially rotates thigh	As above
Tensor fasciae latae	Iliac crest and surface of ilium between anterior iliac spines	Iliotibial tract	Flexes, abducts, and medially rotates thigh; tenses fasciae latae, which laterally supports the knee	As above
Lateral rotator group				
Obturators (externus and internus)	Lateral and medial margins of obturator foramen	Trochanteric fossa of femur	Laterally rotates thigh	Obturator nerve
Piriformis	Anterolateral surface of sacrum	Greater trochanter of femur	Laterally rotates and adducts thigh	As above
Adductor group				
Adductor brevis	Inferior ramus of pubis	Linea aspera of femur	Adducts thigh	As above
Adductor longus	Inferior ramus of pubis anterior to brevis	As above	Adducts, flexes, and medially rotates thigh	As above
Adductor magnus	Inferior ramus of pubis posterior to brevis	As above	Adducts thigh; anterior portion flexes thigh, posterior portion extends thigh	Obturator and sciatic nerves
Pectineus	Superior surface of pubis	Pectineal line inferior to lesser trochanter of femur	Flexes and adducts thigh	Femoral nerve
Gracilis	Inferior rami of pubis and ischium	Anterior surface of tibia inferior to medial condyle	Flexes leg and adducts thigh	Obturator nerve
Iliopsoas group				
Iliacus	Iliac fossa of ilium	Femur distal to lesser trochanter; tendon fused with that of psoas	Flexes hip and/or lumbar spine	Femoral nerve
Psoas major	Anterior surfaces and transverse processes of vertebrae T_{12}–L_5	Femur distal to lesser trochanter in company with iliacus	As above	As above

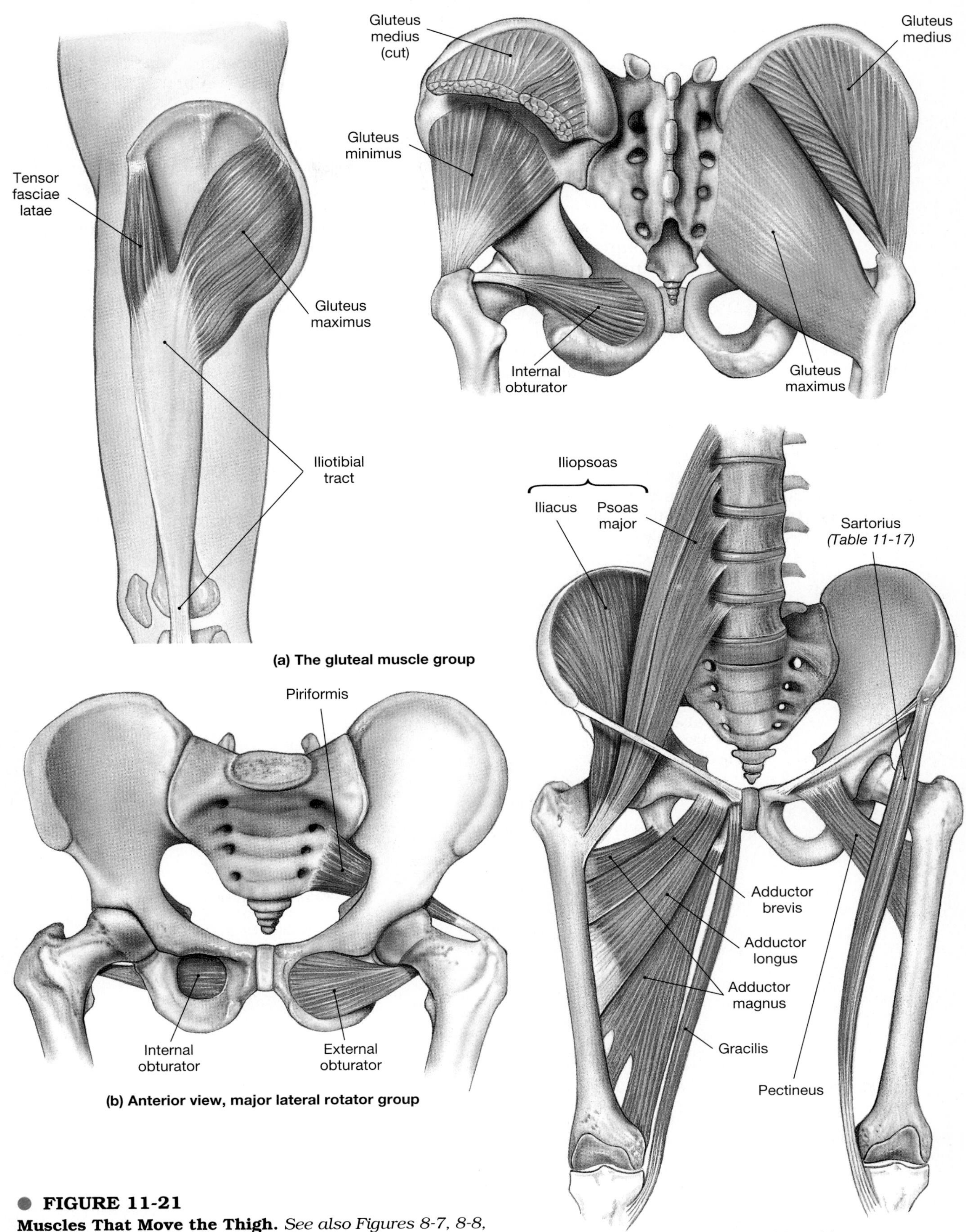
Tensor
fasciae
latae
Gluteus
maximus
Iliotibial
tract
Gluteus
medius
(cut)
Gluteus
minimus
Gluteus
medius
Internal
obturator
Gluteus
maximus
Piriformis
Internal
obturator
External
obturator
Iliopsoas
Iliacus
Psoas
major
Sartorius
(Table 11-17)
Adductor
brevis
Adductor
longus
Adductor
magnus
Gracilis
Pectineus

(a) The gluteal muscle group

(b) Anterior view, major lateral rotator group

(c) The iliopsoas muscle and the adductor group

● **FIGURE 11-21**

Muscles That Move the Thigh. *See also Figures 8-7, 8-8, 8-11, and 9-11, pp. 247, 248, 250, 278.* AM *Plates 7.2, 7.3*

erally rotates the thigh. The gluteus maximus shares an insertion with the **tensor fasciae latae** (TEN-sor FASH-ē-ē LĀ-tā), a muscle that originates on the anterior superior iliac spine. Together these muscles pull on the **iliotibial** (il-ē-ō-TIB-ē-ul) **tract,** a band of collagen fibers that extends along the lateral surface of the thigh and inserts upon the tibia. This tract provides a lateral brace for the knee that becomes particularly important when a person balances on one foot.

The **gluteus medius** and **gluteus minimus** (Figure 11-21a•) take their origins anterior to that of the gluteus maximus and insert upon the greater trochanter of the femur. The anterior and inferior gluteal lines on the lateral surface of the ilium mark the boundaries between these muscles.

The **lateral rotators** (Figure 11-21b•) arise close to or inferior to the horizontal axis of the acetabulum. There are six muscles in all, of which the **piriformis** (PIR-i-for-mis) and the **obturator** muscles are the most important.

The **adductors** (Figure 11-21c• and Table 11-16) are located inferior to the acetabular surface. This muscle group includes the **adductor magnus,** the **adductor brevis,** the **adductor longus,** the **pectineus** (pek-TIN-ē-us), and the **gracilis** (GRAS-i-lis). All but the magnus are found both anterior and inferior to the joint, so they perform flexion as well as adduction. The adductor magnus can produce either adduction and flexion or adduction and extension, depending on the region stimulated. The adductor magnus may also produce medial or lateral rotation. The other muscles produce medial rotation. These muscles insert upon low ridges along the posterior surface of the femur. When an athlete suffers a *pulled groin,* the problem is a strain in one of these adductor muscles.

The medial surface of the pelvis is dominated by a single pair of muscles. The large **psoas** (SŌ-us) **major** arises alongside the lower thoracic and lumbar vertebrae, and its insertion lies on the lesser trochanter of the femur. Before reaching this insertion, its tendon merges with that of the **iliacus** (il-Ē-uh-kus), which lies nestled within the iliac fossa. These two muscles are powerful flexors of the thigh and are often referred to as the **iliopsoas** (il-ē-ō-SŌ-us) muscle.

Muscles That Move the Leg

Figure 11-22

As in the upper limb, there is a pattern to muscle distribution in the lower limb. Extensor muscles are found along the anterior and lateral surfaces

TABLE 11-17 Muscles That Move the Leg

Muscle	*Origin*	*Insertion*	*Action*	*Innervation*
Flexors of the leg				
Biceps femoris	Tuberosity of ischium and linea aspera of femur	Head of fibula, lateral condyle of tibia	Flexes leg, extends and adducts thigh	Sciatic nerve, peroneal branch
Semimembranosus	Tuberosity of ischium	Posterior surface of medial condyle of tibia	Flexes leg, extends, adducts, and medially rotates thigh	Sciatic nerve
Semitendinosus	Tuberosity of ischium	Proximal, posteromedial surface of tibia near insertion of gracilis	As above	As above
Sartorius	Anterior superior spine of ilium	Medial surface of tibia near tibial tuberosity	Flexes leg, flexes and laterally rotates thigh	Femoral nerve
Popliteus	Proximal shaft of tibia	Lateral condyle of femur	Medially rotates tibia (or laterally rotates femur)	Sciatic nerve, tibial branch
Extensors of the leg				
Rectus femoris	Anterior inferior spine and superior acetabular rim of ilium	Tibial tuberosity via patellar ligament	Extends leg, flexes thigh	Femoral nerve
Vastus intermedius	Anterolateral surface of femur along linea aspera (distal half)	As above	Extends leg	As above
Vastus lateralis	Anterior and inferior to greater trochanter of femur and along linea aspera (proximal half)	As above	As above	As above
Vastus medialis	Entire length of linea aspera of femur	As above	As above	As above

● **FIGURE 11-22**

Muscles That Move the Leg. (a) Diagrammatic view of leg flexors. **(b)** Posterior view of thigh. **(c)** Anterior view of thigh. **(d)** Sectional view of thigh muscles. *See also Figures 8-11, 8-12, 8-13, and 9-12, pp. 250, 251, 252, 279.* AM Plates 7.2, 7.3; Scans 4, 5, 6, 7

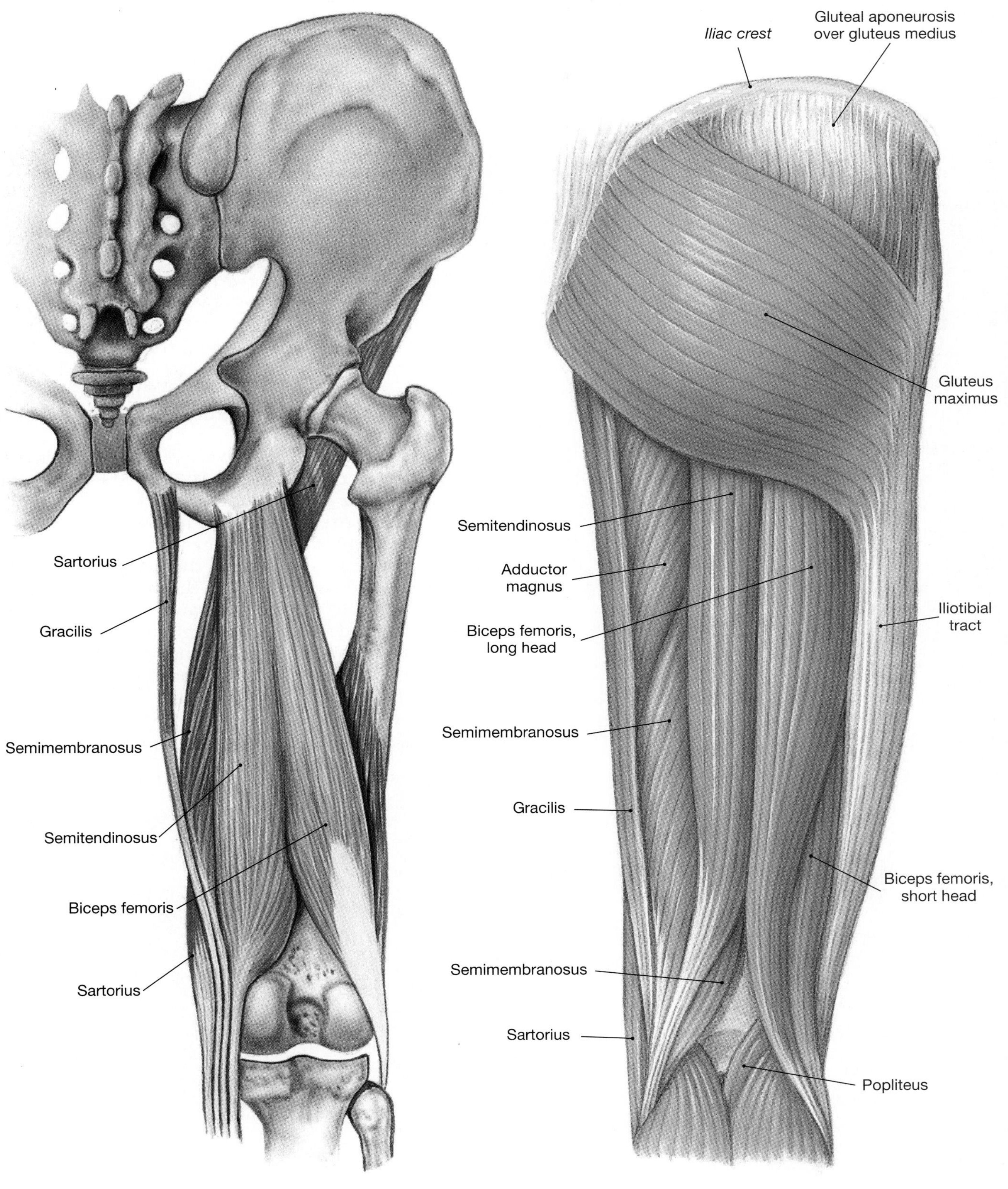
Sartorius
Gracilis
Semimembranosus
Semitendinosus
Biceps femoris
Sartorius
(a)
Iliac crest
Gluteal aponeurosis over gluteus medius
Gluteus maximus
Semitendinosus
Adductor magnus
Biceps femoris, long head
Iliotibial tract
Semimembranosus
Gracilis
Biceps femoris, short head
Semimembranosus
Sartorius
Popliteus
(b) Posterior view

of the leg, and flexors lie along the posterior and medial surfaces. Although the flexors and adductors originate on the pelvic girdle, most of the extensors originate on the femoral surface.

The *flexors of the leg* (Figure 11-22a,b,c●) include the **biceps femoris,** the **semimembranosus** (sem-ē-mem-bra-NŌ-sus), the **semitendinosus** (sem-ē-ten-di-NŌ-sus), and the **sartorius** (sar-TO-rē-us). These muscles arise along the edges of the pelvis and insert upon the tibia and fibula, and their contractions produce flexion of the knee. The sartorius is the only knee flexor that originates superior to the acetabulum, and its insertion lies along the medial aspect of the tibia. When it contracts, it flexes and laterally rotates the thigh, as when crossing the legs.

Because the biceps femoris, semimembranosus, and semitendinosus originate on the pelvic surface inferior and posterior to the acetabulum, their contractions also produce extension of the hip. These three muscles are often called the **hamstrings.** A *pulled hamstring* is a relatively common sports injury, caused by a strain affecting one of the hamstring muscles.

In Chapter 9 we noted that the knee joint can be locked at full extension by a slight lateral rotation of the tibia. [p. 279] The small **popliteus** (pop-LI-tē-us) muscle arises on the posterior tibial shaft and inserts on the femur near the lateral condyle. When flexion is initiated, this muscle contracts to produce a slight medial rotation of the tibia that unlocks the joint.

Collectively the *knee extensors* are known as the **quadriceps femoris** (Figure 11-22d●). The three **vastus** muscles originate along the body of the femur, and, with the assistance of the **rectus femoris,** they cradle the femur the way a bun surrounds a hot dog. All four muscles insert upon the patella and reach the tibial tuberosity by way of the patellar ligament. The rectus originates on the anterior inferior iliac spine, so in addition to producing extension of the knee it can assist in flexion of the thigh. The flexors and extensors of the leg are detailed in Table 11-17.

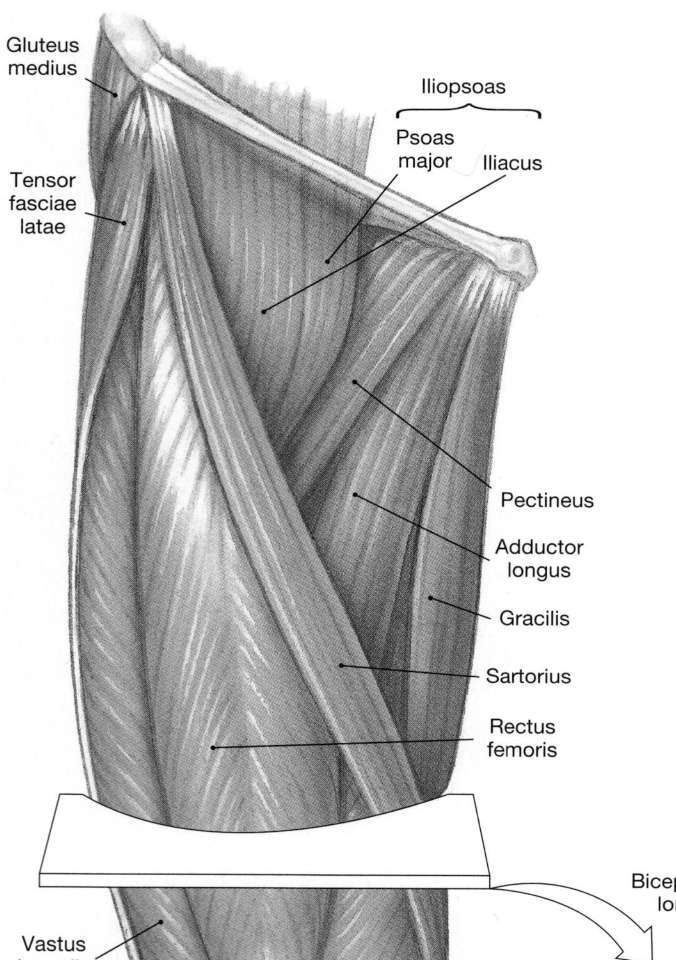
Gluteus medius
Iliopsoas
Psoas major
Iliacus
Tensor fasciae latae
Pectineus
Adductor longus
Gracilis
Sartorius
Rectus femoris
Vastus lateralis
Vastus medialis
Patella

(c) Anterior view

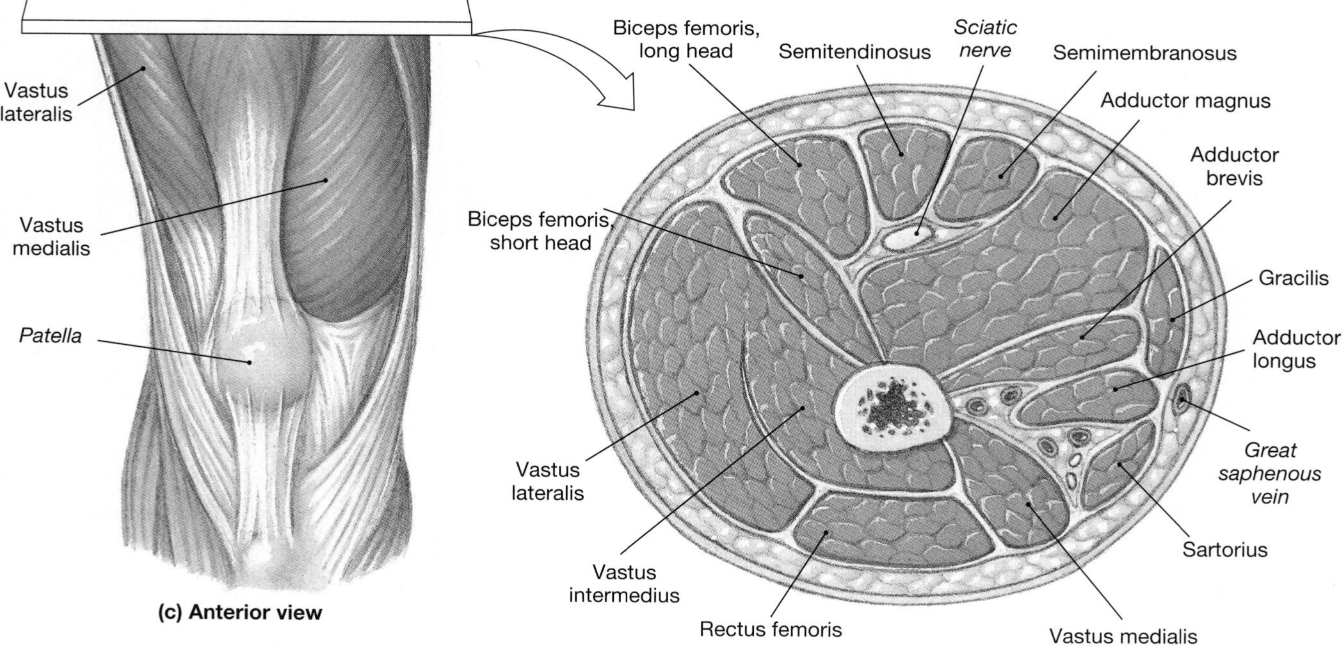
Biceps femoris, long head
Semitendinosus
Sciatic nerve
Semimembranosus
Adductor magnus
Adductor brevis
Biceps femoris, short head
Gracilis
Adductor longus
Great saphenous vein
Sartorius
Vastus lateralis
Vastus intermedius
Rectus femoris
Vastus medialis

(d)

● **FIGURE 11-22 continued**

INTRAMUSCULAR INJECTIONS Drugs are frequently injected into tissues rather than directly into the circulation. This method enables the physician to introduce a large amount of a drug at one time, yet have it enter the circulation gradually. In an **intramuscular (IM) injection** the drug is introduced into the mass of a large skeletal muscle. Uptake is usually faster and accompanied by less tissue irritation than when drugs are administered *intradermally* or *subcutaneously* (injected into the dermis or subcutaneous layers). [p. 158] Up to 5 ml of fluid may be injected at one time, and multiple injections are possible.

The most common complications involve accidental injection into a blood vessel or piercing of a nerve. The sudden entry of massive quantities of a drug into the bloodstream can have unpleasant or even fatal consequences, and damage to a nerve can cause motor paralysis or sensory loss. As a result, the site of injection must be selected with care. Bulky muscles that contain few large vessels or nerves make ideal targets, and the gluteus medius or the posterior, lateral, superior portion of the gluteus maximus is often selected. The deltoid muscle of the arm, about 2.5 cm (1 in.) distal to the acromion, is another popular site. Probably the most satisfactory from a technical point of view is the vastus lateralis of the thigh, for an injection into this thick muscle will not encounter vessels or nerves. This is the preferred injection site in infants and young children, whose gluteal and deltoid muscles are relatively small.

Muscles That Move the Foot and Toes

Figures 11-23, 11-24

The extrinsic muscles that move the foot and toes (Figure 11-23●) are detailed in Table 11-18. Most of the muscles that move the ankle produce the plantar flexion involved with walking and running movements. The large **gastrocnemius** (gas-trok-NĒ-mē-us; *gaster*, stomach + *kneme*, knee) of the calf is an important plantar flexor, but the slow muscle fibers of the underlying **soleus** (SŌ-lē-us) are more powerful. These muscles are best seen in posterior and lateral views (Figure11-23a,b●). The gastrocnemius arises from two heads located on the medial and lateral epicondyles of the femur just proximal to the knee. A sesamoid bone, the **fabella,** is usually found within the lateral head of the gastrocnemius. The gastrocnemius and soleus muscles share a common tendon, the **calcaneal tendon,** or *Achilles tendon.*

Deep to the gastrocnemius and soleus lie a pair of **peroneus** muscles (Figure 11-23b,c●). These muscles produce eversion as well as plantar flexion of the foot. Inversion is caused by contraction of the **tibialis** (tib-ē-A-lis) muscles; the large **tibialis anterior** (Figure 11-23c,d●) opposes the gastrocnemius and dorsiflexes the foot.

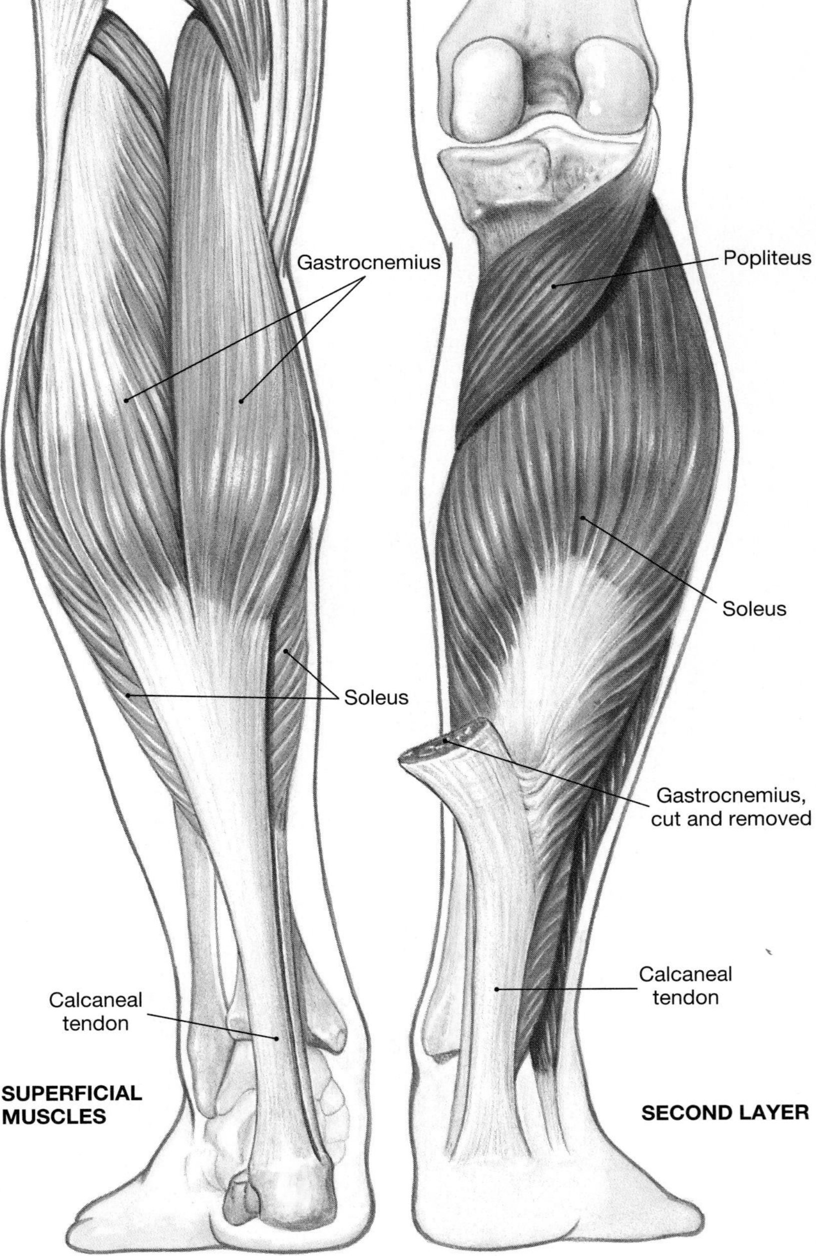

Gastrocnemius
Popliteus
Soleus
Soleus
Gastrocnemius, cut and removed
Calcaneal tendon
Calcaneal tendon
SUPERFICIAL MUSCLES
SECOND LAYER
(a) Posterior view

● **FIGURE 11-23**
Muscles That Move the Foot and Toes. *See also Figures 8-13 and 8-14, pp. 252, 253.* AM *Plate 7.4; Scans 5, 6, 7*

TABLE 11-18 Muscles That Move the Foot

Muscle	*Origin*	*Insertion*	*Action*	*Innervation*
Dorsiflexor				
Tibialis anterior	Lateral condyle and proximal shaft of tibia	Base of first metatarsal	Dorsiflexes foot	Sciatic nerve, deep peroneal branch
Plantar flexors				
Gastrocnemius	Above femoral condyles	Calcaneus via calcaneal tendon	Plantar flexes, inverts, and adducts foot; flexes leg	Sciatic nerve, tibial branch
Peroneus brevis	Midlateral margin of fibula	Base of 5th metatarsal	Everts foot	Sciatic nerve, superficial peroneal branch

● FIGURE 11-23 continued

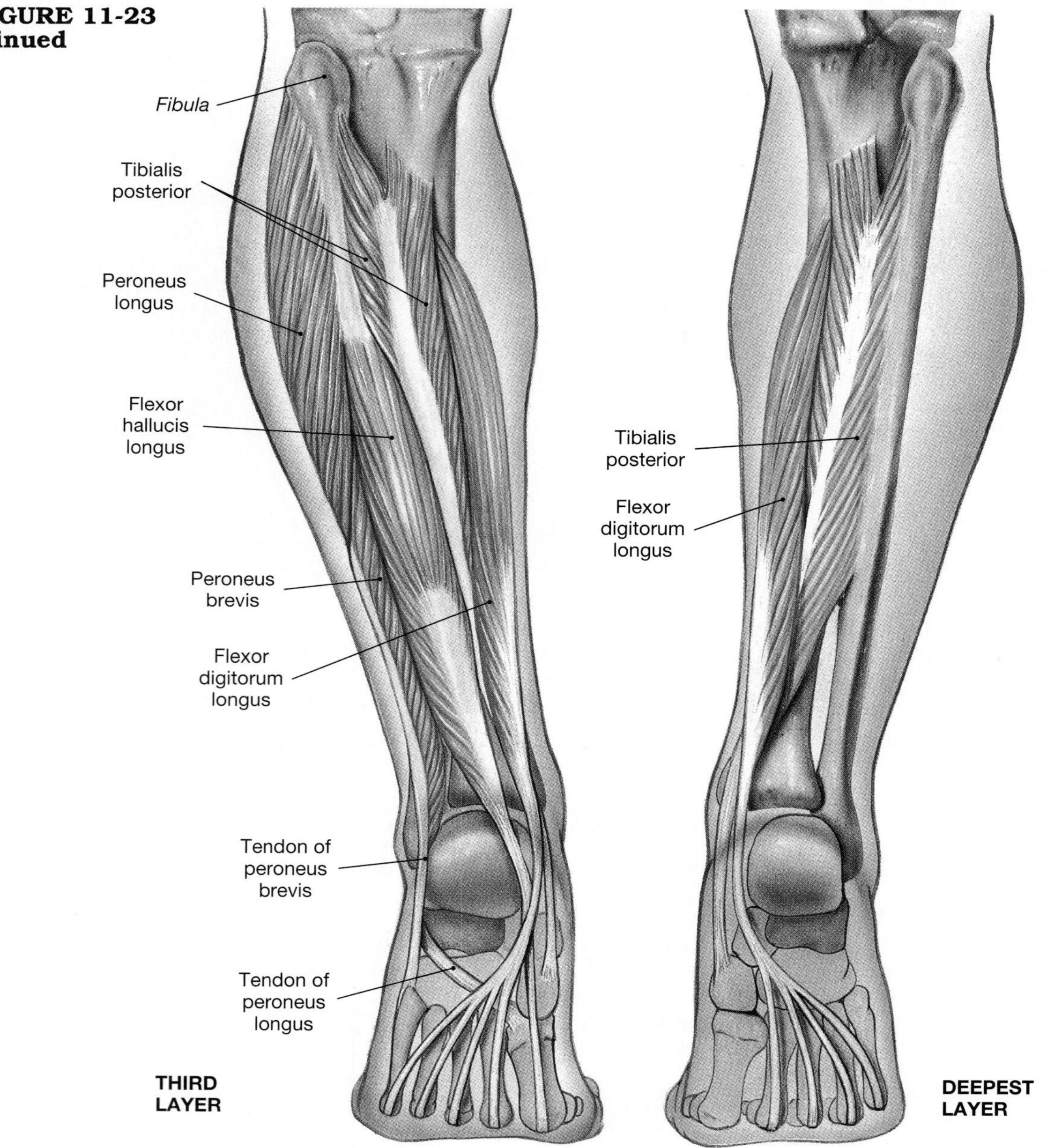
Fibula
Tibialis posterior
Peroneus longus
Flexor hallucis longus
Peroneus brevis
Flexor digitorum longus
Tendon of peroneus brevis
Tendon of peroneus longus
THIRD LAYER
Tibialis posterior
Flexor digitorum longus
DEEPEST LAYER

(b) Posterior view

TABLE 11-18 continued Muscles That Move the Foot

Muscle	*Origin*	*Insertion*	*Action*	*Innervation*
Plantar flexors continued				
Peroneus longus	Lateral condyle of tibia and head of fibula	Base of 1st metatarsal	Everts and plantar flexes foot; supports longitudinal arch	Sciatic nerve, superficial peroneal branch
Soleus	Head and proximal shaft of fibula, and adjacent posteromedial shaft of tibia	Calcaneus via calcaneal tendon (with gastrocnemius)	Plantar flexes, inverts, and adducts foot	Sciatic nerve, tibial branch
Tibialis posterior	Interosseous membrane and adjacent shafts of tibia and fibula	Tarsals and metatarsals	Adducts and inverts foot	As above

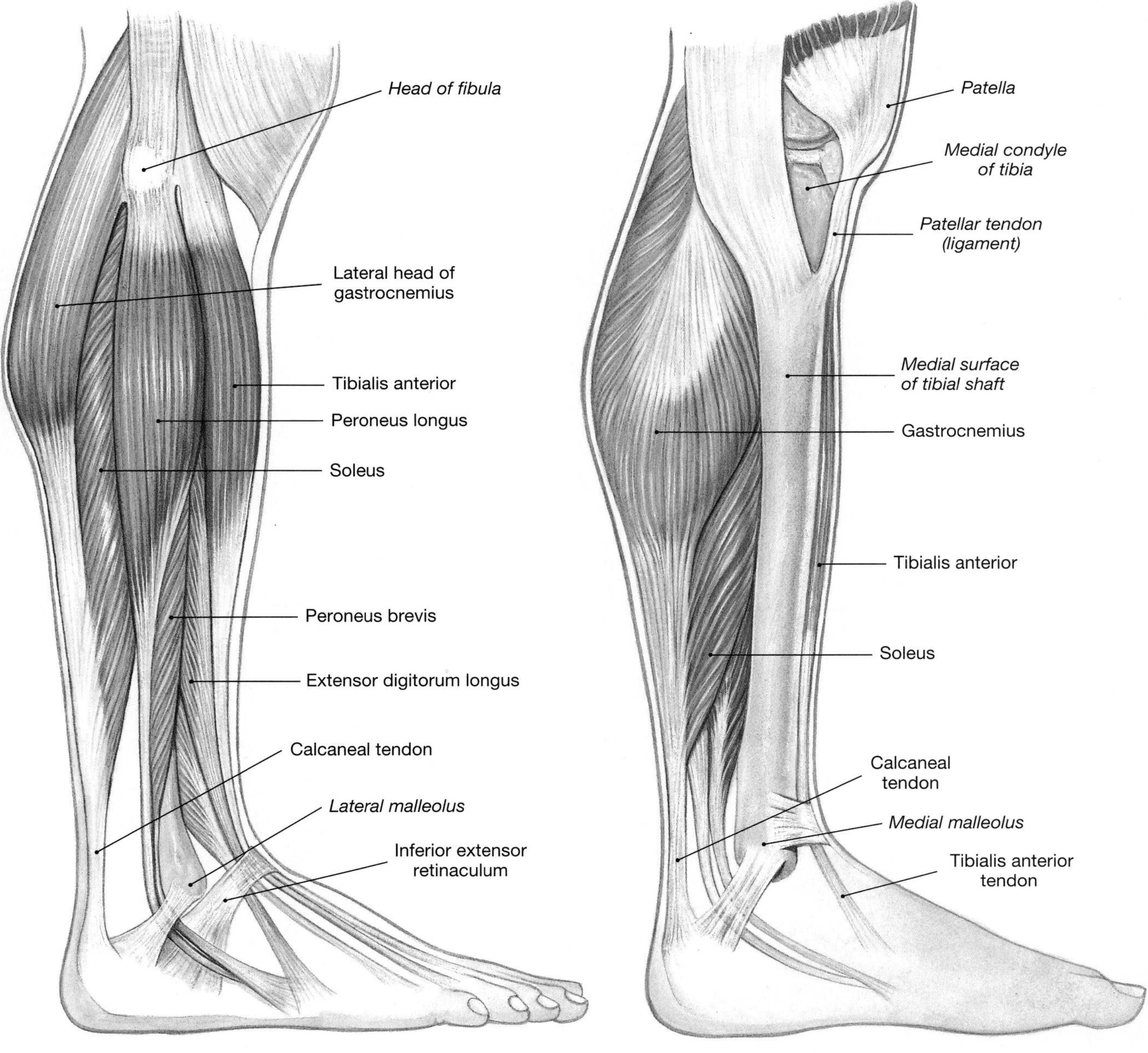
Head of fibula
Lateral head of gastrocnemius
Tibialis anterior
Peroneus longus
Soleus
Peroneus brevis
Extensor digitorum longus
Calcaneal tendon
Lateral malleolus
Inferior extensor retinaculum
Patella
Medial condyle of tibia
Patellar tendon (ligament)
Medial surface of tibial shaft
Gastrocnemius
Tibialis anterior
Soleus
Calcaneal tendon
Medial malleolus
Tibialis anterior tendon

Lateral view (c) **Medial view**

TABLE 11-19 Extrinsic Muscles of the Foot

Muscle	Origin	Insertion	Action	Innervation
Flexors				
Flexor digitorum longus	Posteromedial surface of tibia	Inferior surface of phalanges, toes 2–5	Flexes toes 2–5	Sciatic nerve, tibial branch
Flexor hallucis longus	Posterior surface of fibula	Inferior surface, terminal phalanx of great toe	Flexes great toe	As above
Extensors				
Extensor digitorum longus	Lateral condyle of tibia, anterior surface of fibula	Superior surfaces of phalanges, toes 2–5	Extends toes 2–5	Sciatic nerve, deep peroneal branch
Extensor hallucis longus	Anterior surface of fibula	Superior surface, terminal phalanx of great toe	Extends great toe	As above

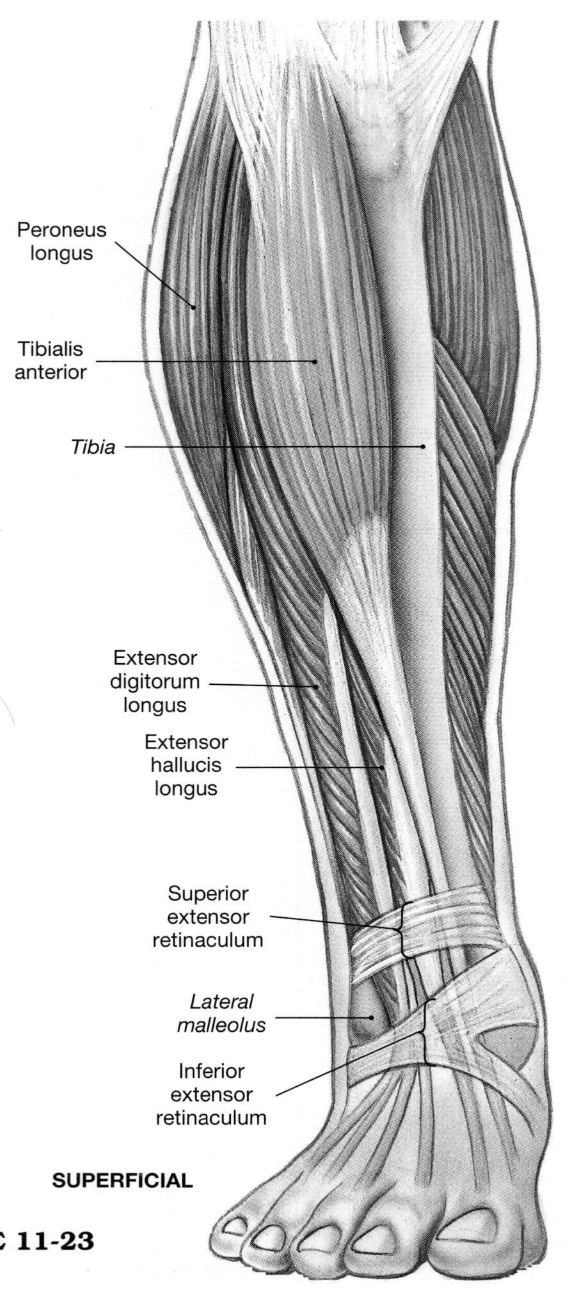
Peroneus longus
Tibialis anterior
Tibia
Extensor digitorum longus
Extensor hallucis longus
Superior extensor retinaculum
Lateral malleolus
Inferior extensor retinaculum
SUPERFICIAL

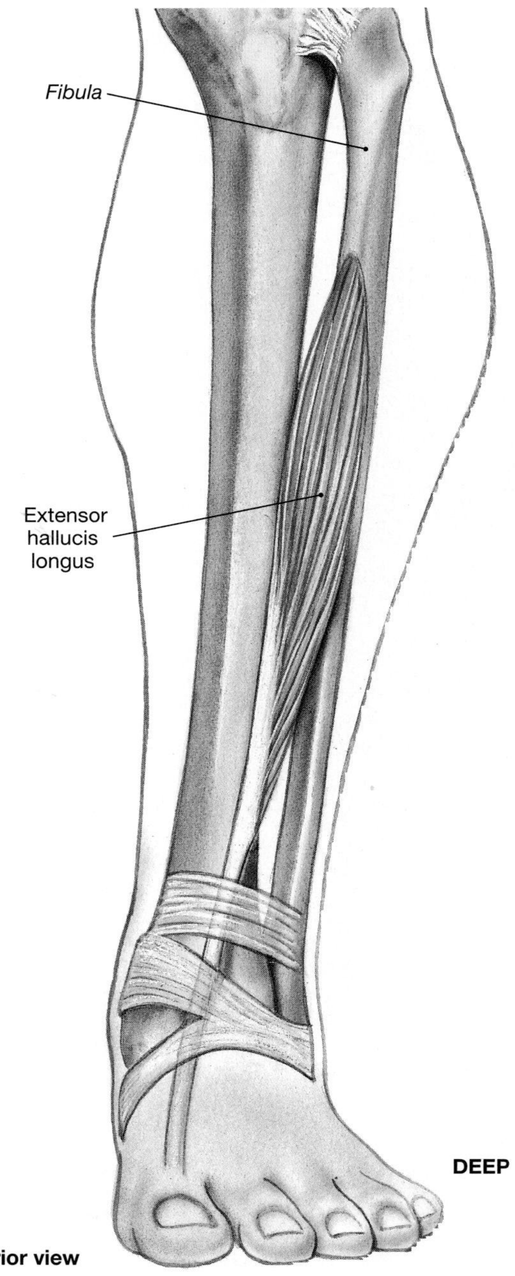
Fibula
Extensor hallucis longus
DEEP

(d) Anterior view

FIGURE 11-23 continued

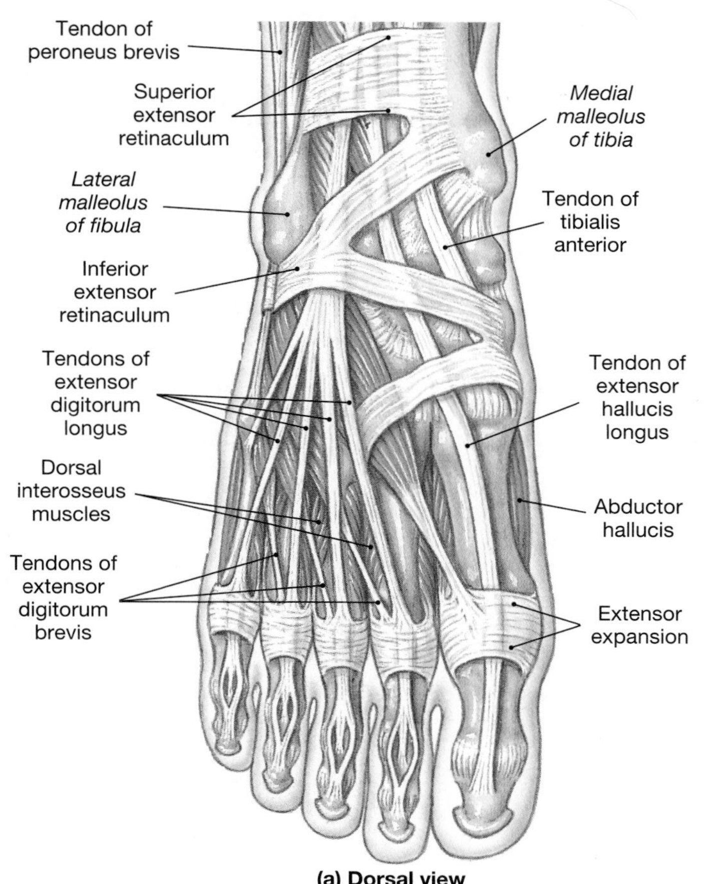
Tendon of peroneus brevis
Superior extensor retinaculum
Lateral malleolus of fibula
Inferior extensor retinaculum
Tendons of extensor digitorum longus
Dorsal interosseus muscles
Tendons of extensor digitorum brevis
Medial malleolus of tibia
Tendon of tibialis anterior
Tendon of extensor hallucis longus
Abductor hallucis
Extensor expansion

(a) Dorsal view

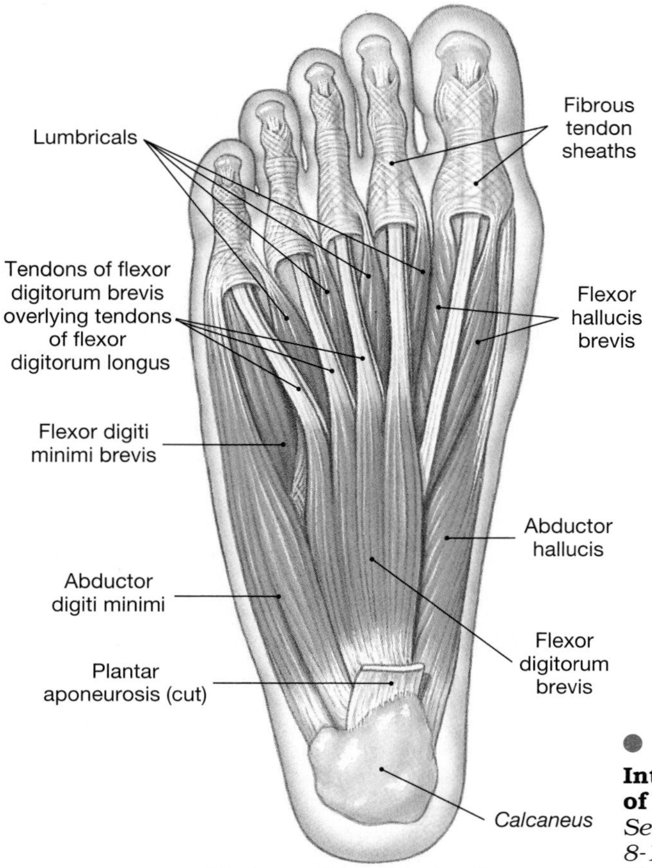

Important digital muscles originate on the surface of the tibia, the fibula, or both (Figure 11-23c,d● and Table 11-19). Large tendon sheaths surround the tendons of the tibialis anterior, **extensor digitorum longus,** and **extensor hallucis longus** where they cross the ankle joint. The positions of these sheaths are stabilized by superior and inferior **extensor retinacula** (Figure 11-23d●).

Several smaller muscles originate on the bones of the tarsus and foot, and their contractions move the toes. Some of the flexor muscles originate at the anterior border of the calcaneus; their muscle tone contributes to maintenance of the longitudinal arches of the foot. ∞ *[p. 253]* As in the hand, small **interosseus** muscles (plural *interossei*) originate on the lateral and medial surfaces of the metatarsals. Intrinsic muscles of the foot are detailed in Figure 11-24● and Table 11-20.

- ✔ What leg movement would be impaired by injury to the obturator muscle?
- ✔ You often hear of athletes who suffer a "pulled hamstring." To what does this phrase refer?
- ✔ How would you expect a torn calcaneal tendon to affect movement of the foot?

(b) Plantar view, superficial layer

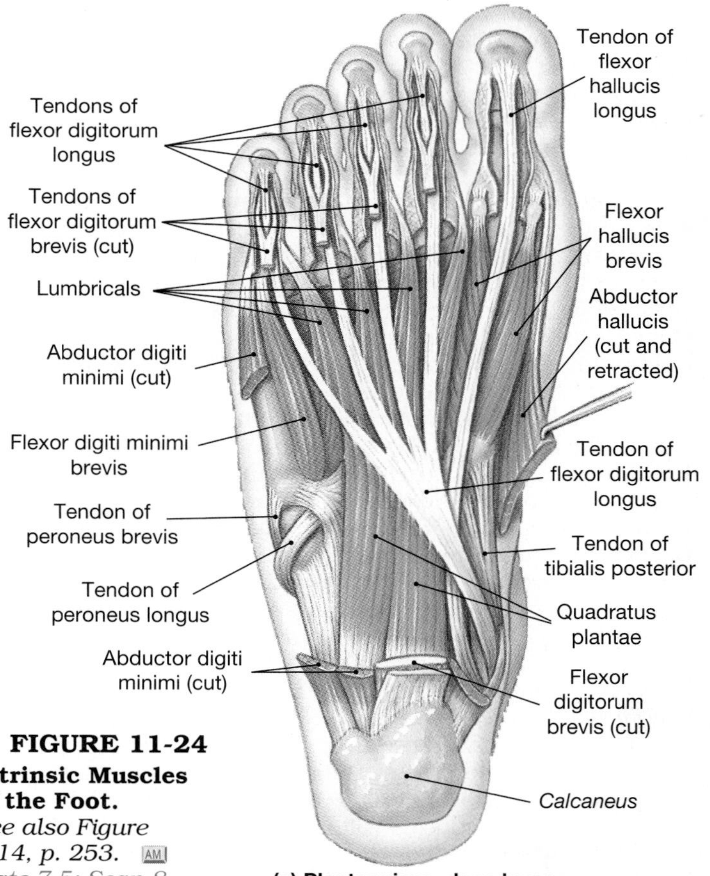

(c) Plantar view, deep layer

● **FIGURE 11-24 Intrinsic Muscles of the Foot.** *See also Figure 8-14, p. 253.* AM *Plate 7.5; Scan 8*

TABLE 11-20 Muscles of the Foot

Muscle	Origin	Insertion	Action	Innervation
Extensor digitorum brevis	Calcaneus	Dorsal surfaces of toes 1–4	Extends proximal phalanges of toes 1–4	Deep peroneal nerve
Abductor hallucis	Calcaneus	Proximal phalanx of great toe	Abducts great toe	Medial plantar nerve
Flexor digitorum brevis	Calcaneus	Sides of middle phalanges, toes 2–5	Flexes the middle phalanges of toes 2–5	As above
Abductor digiti minimi	Calcaneus	Lateral side of proximal phalanx of little toe	Abducts the little toe	Lateral plantar nerve
Quadratus plantae	Calcaneus	Tendon of flexor digitorum longus	Flexes toes 2–5	As above
Lumbricals (4)	Tendons of flexor digitorum longus	Insertions of extensor digitorum longus	Flexes proximal phalanges, extends middle phalanges, toes 2–5	Medial plantar nerve (toe 2), lateral plantar nerve (toes 2–5)
Flexor hallucis brevis	Cuboid and lateral cuneiform	Proximal phalanx of great toe	Flexes proximal phalanx of great toe	Medial plantar nerve
Adductor hallucis	Bases of metatarsals II–IV and plantar ligaments	Proximal phalanx of great toe	Adducts the great toe	Lateral plantar nerve
Interossei				
dorsal (4)	Sides of metatarsals	Sides of toes 2–4	Abducts the toes	As above
plantar (3)	Bases and medial sides of metatarsals	Sides of toes 3–5	Adducts the toes	As above

COMPARTMENT SYNDROMES In the arms and legs the interconnections between the superficial fascia, the deep fascia of the muscles, and the periostea of the appendicular skeleton are quite substantial. The muscles within a limb are effectively isolated in **compartments** formed by dense collagenous sheets, as shown in Figure 11-25●. Blood vessels and nerves traveling to specific muscles within the limb enter and branch within the appropriate compartments.

When a crushing injury, severe contusion, or muscle strain occurs, the blood vessels within one or more compartments may be damaged. These compartments then become swollen with blood and fluid leaked from damaged vessels. Because the connective tissue partitions are very strong, the accumulated fluid cannot escape, and pressures rise within the affected compartments. Eventually compartment pressures may become so high that they compress the regional blood vessels and eliminate the circulatory supply to the muscles and nerves of the compartment. This compression produces a condition of **ischemia** (is-KĒ-mē-uh), or "blood starvation," known as the **compartment syndrome.**

Slicing into the compartment along its longitudinal axis or implanting a drain are emergency measures used to relieve the pressure. If such steps are not taken, the contents of the compartment will suffer severe damage. Nerves in the affected compartment will be destroyed after 2–4 hours of ischemia, although they can regenerate to some degree if the circulation is restored. After 6 hours or more, the muscle tissue will also be destroyed, and no regeneration will occur. The muscles will be replaced by scar tissue, and shortening of the connective tissue fibers may result in *contracture.*

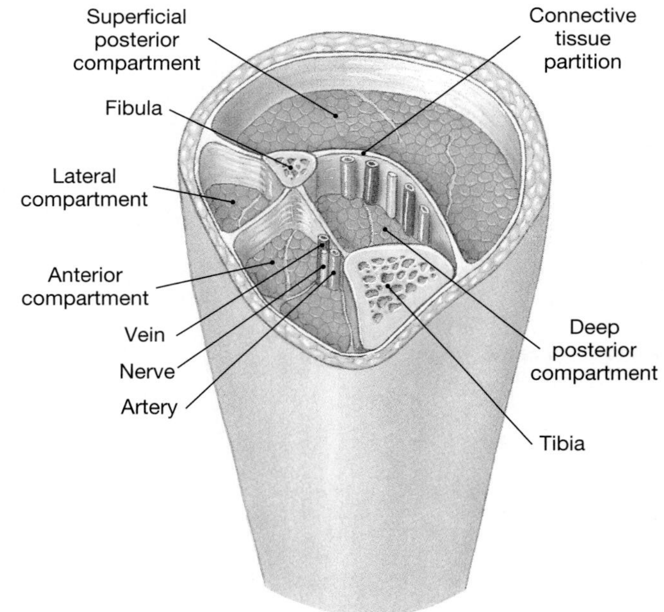
Superficial posterior compartment
Connective tissue partition
Fibula
Lateral compartment
Anterior compartment
Vein
Nerve
Artery
Deep posterior compartment
Tibia

● **FIGURE 11-25**
Musculoskeletal Compartments. A diagrammatic section through the leg, with the muscles removed to show the arrangement of the compartments. A section through the thigh or arm would show a comparable arrangement of dense connective tissue partitions. The anterior and lateral compartments of the leg contain muscles of the extensor/dorsiflexor series, and the posterior compartments ensheathe the flexor/plantar flexor muscles.

Selected Clinical Terminology

Terms Discussed in This Chapter

carpal tunnel syndrome: An inflammation of the sheath surrounding the flexor tendons of the palm. *(p. 360)*
compartment syndrome: Ischemia resulting from accumulated blood and fluid trapped within a musculoskeletal compartment. *(p. 374)*
diaphragmatic hernia (hiatal hernia): A hernia that occurs when abdominal organs slide into the thoracic cavity. *(p. 347)*
hernia: A condition involving an organ or body part that protrudes through an abnormal opening. *(p. 347)*
inguinal hernia: A condition in which the inguinal canal enlarges and abdominal contents are forced into it. *(p. 347)*
intramuscular (IM) injection: Administering a drug by injecting it into the mass of a large skeletal muscle. *(p. 369)*
ischemia (is-KĒ-mē-uh): A condition of "blood starvation" resulting from compression of regional blood vessels. *(p. 374)*
rotator cuff: The muscles that surround the shoulder joint; a frequent site of sports injuries. *(p. 354)*

AM *Additional Terms Discussed in The Application Manual*

bone bruise: Bleeding within the periosteum of a bone.
bursitis: Inflammation of the bursae around one or more joints.
muscle cramps: Prolonged, involuntary, painful muscular contractions.
sprains: Tears or breaks in ligaments or tendons.
strains: Tears or breaks in muscles.
stress fractures: Cracks or breaks in bones subjected to repeated stresses or trauma.
tendinitis: Inflammation of the connective tissue surrounding a tendon.

CHAPTER REVIEW

STUDY OUTLINE

INTRODUCTION, p. 326

1. The **muscular system** includes all the skeletal muscle tissue that can be controlled voluntarily.

BIOMECHANICS AND MUSCLE ANATOMY, p. 326

Organization of Skeletal Muscle Fibers, p. 326

1. A muscle can be classified according to the arrangement of fibers and fascicles as a **parallel muscle, convergent muscle, pennate muscle,** or **circular muscle (sphincter).** A pennate muscle may be **unipennate, bipennate,** or **multipennate.** (*Figure 11-1*)

Skeletal Muscle Length-Tension Relationships, p. 327

Levers, p. 328

2. A **lever** can change the direction, speed, or distance of muscle movements and modify the force applied to muscles.
3. Levers may be classified as **first-class, second-class,** or **third-class levers;** the last are the most common type of lever in the body. (*Figure 11-2*)

MUSCLE TERMINOLOGY, p. 329

Origins and Insertions, p. 329

Actions, p. 329

1. Each muscle may be identified by its **origin, insertion,** and **primary action.** A muscle may be classified as a **prime mover** or **agonist,** a **synergist,** an **antagonist,** or a **fixator.**

Names of Skeletal Muscles, p. 330

2. The names of muscles often provide clues to their location, orientation, or function. (*Table 11-1*)

Axial and Appendicular Muscles, p. 330

3. The **axial musculature** arises on the axial skeleton; it positions the head and spinal column and moves the rib cage. The **appendicular skeleton** stabilizes or moves components of the appendicular skeleton. (*Figure 11-3*)

THE AXIAL MUSCULATURE, p. 334

1. The axial muscles fall into logical groups based on location or function or both.
2. **Innervation** refers to the identity of the nerve that controls a muscle.

Muscles of the Head and Neck, p. 334

3. The muscles of facial expression are the **orbicularis oris, buccinator, epicranius (frontalis** and **occipitalis)**, and **platysma.** (*Figure 11-4; Table 11-2*)
4. Six **extrinsic eye muscles (oculomotor muscles)** control external eye movements: the **inferior** and **superior rectus, lateral** and **medial rectus,** and **superior** and **inferior obliques.** (*Figure 11-5; Table 11-3*)
5. The muscles of mastication (chewing) are the **masseter, temporalis,** and **pterygoid.** (*Figure 11-6; Table 11-4*)
6. The muscles of the tongue are necessary for speech and swallowing and assist in mastication. They are the **genioglossus, hyoglossus, palatoglossus,** and **styloglossus.** (*Figure 11-7; Table 11-5*)

7. The muscles of the pharynx constrict the pharyngeal walls **(pharyngeal constrictors)**, elevate the larynx **(laryngeal elevators)**, and raise the soft palate **(palatal muscles)**. *(Figure 11-8; Table 11-6)*
8. The muscles of the neck control the position of the larynx, depress the mandible, and provide a foundation for the muscles of the tongue and pharynx. The extrinsic muscles of the larynx are the **digastricus, geniohyoid, mylohyoid, omohyoid, sternohyoid, sternothyroid, stylohyoid,** and **thyrohyoid.** *(Figure 11-9; Table 11-7)*

Muscles of the Spine, p. 342

9. The superficial muscles of the spine can be classified into the **spinalis, longissimus,** and **iliocostalis** divisions. In the lower lumbar and sacral regions the longissimus and iliocostalis are sometimes called the **sacrospinalis** muscles. *(Figure 11-10; Table 11-8)*
10. Other muscles of the spine include the **longus capitis** and **longus cervicis** of the neck and the **quadratus lumborum** of the lumbar region. *(Figure 11-10; Table 11-8)*

Oblique and Rectus Muscles, p. 345

11. The oblique muscles include the **scalenes** and the **intercostal** and **transversus** muscles. The **external intercostals** and the **internal intercostals** are important in respiratory movements of the ribs. Also important to respiration is the **diaphragm.** *(Figures 11-10, 11-11, 11-12; Table 11-9)*

Muscles of the Pelvic Floor, p. 347

12. The **perineum** can be divided into an anterior **urogenital triangle** and a posterior **anal triangle.** The pelvic floor consists of the **urogenital diaphragm** and the **pelvic diaphragm.** *(Figure 11-3; Table 11-10)*

THE APPENDICULAR MUSCULATURE, p. 350

Muscles of the Shoulders and Upper Limbs, p. 352

1. The **trapezius** affects the positions of the shoulder girdle, head, and neck. Other muscles inserting on the scapula include the **rhomboideus,** the **levator scapulae,** the **serratus anterior,** the **subclavius,** and the **pectoralis minor.** *(Figures 11-14, 11-15; Table 11-11)*
2. The **deltoid** and the **supraspinatus** are important abductors. The **subscapularis** and the **teres major** rotate the arm medially; the **infraspinatus** and **teres minor** perform lateral rotation; and the **coracobrachialis** flexes and adducts the humerus. *(Figures 11-14, 11-16; Table 11-12)*
3. The **pectoralis major** flexes the arm, and the **latissimus dorsi** extends it. *(Figures 11-14, 11-16; Table 11-12)*
4. The primary actions of the **biceps brachii** and the **triceps brachii** (long head) affect the elbow joint. The **brachialis** and **brachioradialis** flex the forearm, opposed by the **anconeus.** The **flexor carpi ulnaris,** the **flexor carpi radialis,** and the **palmaris longus** cooperate to flex the wrist. They are opposed by the **extensor carpi radialis** and the **extensor carpi ulnaris.** The **pronator teres** and **pronator quadratus** pronate the forearm, opposed by the **supinator.** *(Figures 11-16–11-20; Tables 11-13–11-15)*

Muscles of the Lower Limbs, p. 364

5. **Gluteal muscles** cover the lateral surfaces of the ilia. The largest is the **gluteus maximus,** which shares an insertion with the **tensor fasciae latae;** together these muscles pull on the **iliotibial tract.** *(Figures 11-14, 11-21; Table 11-16)*
6. The **piriformis** and the **obturator muscles** are the most important **lateral rotators.** The **adductors** can produce a variety of movements. *(Figure 11-21; Table 11-16)*
7. The **psoas major** and the **iliacus** merge to form the **iliopsoas** muscle, a powerful flexor of the thigh. *(Figure 11-21; Table 11-16)*
8. The flexors of the leg include the **biceps femoris, semimembranosus,** and **semitendinosus** (the three **hamstrings**), and the **sartorius.** The **popliteus** unlocks the knee joint. *(Figure 11-22; Table 11-17)*
9. Collectively the knee extensors are known as the **quadriceps femoris.** This group includes the three **vastus** muscles and the **rectus femoris.** *(Figure 11-22; Table 11-17)*
10. The **gastrocnemius** and **soleus muscles** produce plantar flexion. A pair of **peroneus muscles** produce eversion as well as plantar flexion. *(Figure 11-23; Tables 11-18, 11-19)*
11. Smaller muscles of the calf and shin position the foot and move the toes. Precise control of the phalanges is provided by muscles originating at the tarsals and metatarsals. *(Figure 11-24; Table 11-20)*

REVIEW QUESTIONS

LEVEL 1 Reviewing Facts and Terms

1. Muscle fibers within a skeletal muscle form bundles called:
   (a) aponeuroses (b) fascicles
   (c) sarcomeres (d) myofibrils
2. Most muscles in the body are ______ muscles.
   (a) convergent (b) pennate
   (c) circular (d) parallel
3. Skeletal muscle fibers produce maximum tension over a relatively narrow range of:
   (a) complex movements (b) extensions
   (c) flexions (d) sarcomere lengths
4. The bones, which serve as levers in the body, change:
   (a) the direction of an applied force
   (b) the distance and speed of movement produced by a force
   (c) the effective strength of a force
   (d) a, b, and c are correct
5. A general rule to follow to understand the difference between an origin and insertion is:
   (a) the origin moves while the insertion remains stationary
   (b) each muscle begins and ends at the origin

(c) the origin remains stationary while the insertion moves
(d) each muscle begins and ends at the insertion

6. If an agonist produces flexion, the primary action of the antagonist will be:
(a) pronation (b) adduction
(c) abduction (d) extension

7. The muscles of facial expression are innervated by:
(a) cranial nerve VII (b) cranial nerve V
(c) cranial nerve IV (d) cranial nerve VI

8. The strongest masticatory muscle is the:
(a) pterygoid (b) masseter
(c) temporalis (d) mandible

9. The muscles of the tongue are innervated by the:
(a) hypoglossal nerve (N XII)
(b) trochlear nerve (N IV)
(c) abducens nerve (N VI)
(d) a, b, and c are correct

10. The muscle that opens the mouth by depressing the mandible is the:
(a) stylopharyngeus (b) digastricus
(c) sternocleidomastoid (d) hypoglossus

11. Important flexors of the spinal column that act in opposition to the erector spinae are the:
(a) rectus muscles (b) longus capitis
(c) longus cervices (d) scalenes

12. The linea alba is a median collagenous partition that longitudinally divides the:
(a) external obliques (b) rectus abdominis
(c) external intercostals (d) rectus femoris

13. The muscular partition that separates the abdominopelvic and thoracic cavities is the:
(a) masseter (b) perineum
(c) diaphragm (d) transversus abdominis

14. The major abductor of the arm is the:
(a) deltoid (b) biceps brachii
(c) triceps brachii (d) subscapularis

15. The muscles that rotate the radius without producing either flexion or extension of the elbow are:
(a) brachialis and brachioradialis
(b) pronator teres and supinator
(c) biceps and triceps brachii
(d) a, b, and c are correct

16. The powerful flexors of the thigh are referred to as the:
(a) piriformis (b) obturators
(c) pectineus (d) iliopsoas

17. Knee extensors known as the quadriceps include:
(a) three vastus muscles and rectus femoris
(b) biceps femoris, gracilis, sartorius
(c) popliteus, iliopsoas, gracilis
(d) gastrocnemius, tibialis, peroneus

18. What two factors interact to determine the effects of a muscle contraction?

19. List the four fascicle organizations that produce the different patterns of skeletal muscles.

20. What are the distinguishing characteristics of unipennate, bipennate, and multipennate muscles?

21. How do first-, second-, and third-class levers differ?

22. What three primary actions are used to identify muscle groups? Give a brief description of each action.

23. What is the functional difference between the axial musculature and the appendicular musculature?

24. What four muscle groups make up the axial musculature?

25. What axial muscles are used in the process of mastication?

26. What three functions are accomplished by the muscles of the pelvic floor?

27. What four groups of muscles are associated with the shoulders and upper extremities?

28. What three functional groups make up the muscles of the lower limbs?

29. What three muscular sites are most desirable for intramuscular injections? Why?

LEVEL 2 Reviewing Concepts

30. Of the following examples, the one that illustrates the action of a second-class lever is:
(a) leg extension
(b) plantar flexion
(c) flexion of the forearm
(d) a, b, and c are correct

31. An example of a prime mover that produces flexion of the forearm is the:
(a) brachioradialis (b) biceps brachii
(c) brachialis (d) biceps femoris

32. Removal of the trapezius muscle exposes the:
(a) serratus anterior and subclavian muscles
(b) infraspinatus and teres minor muscles
(c) deltoid and supraspinatus muscles
(d) rhomboideus and levator scapulae muscles

33. The muscles of the spine include many dorsal extensors, but few ventral flexors. Why?

34. What specific structural characteristic makes voluntary control of urination and defecation possible?

35. Why does a convergent muscle exhibit more versatility when contracting than does a parallel or pennate muscle?

36. Why can a pennate muscle generate more tension than a parallel muscle of the same size?

37. Why is it difficult to lift a heavy object when the forearm is at full flexion?

38. If the fifth cranial nerve (trigeminal) is damaged or severed, what function will be affected in the body?

39. What types of movements are affected when the hamstrings are injured?

LEVEL 3 Critical Thinking and Clinical Applications

40. Mary sees Jill coming toward her and immediately contracts her frontalis and procerus muscles. She also contracts her levator labii on the right side. Is Mary glad to see Jill? How can you tell?

41. Jeff is interested in building up his leg muscles, specifically the quadriceps group. What exercises would you recommend to help Jeff accomplish his goal?

42. Shelly gives her son an ice cream cone. The boy grasps the cone with his right hand, opens his mouth and begins to lick at the ice cream. What muscles are being used by the child to perform these actions?

CHAPTER 12

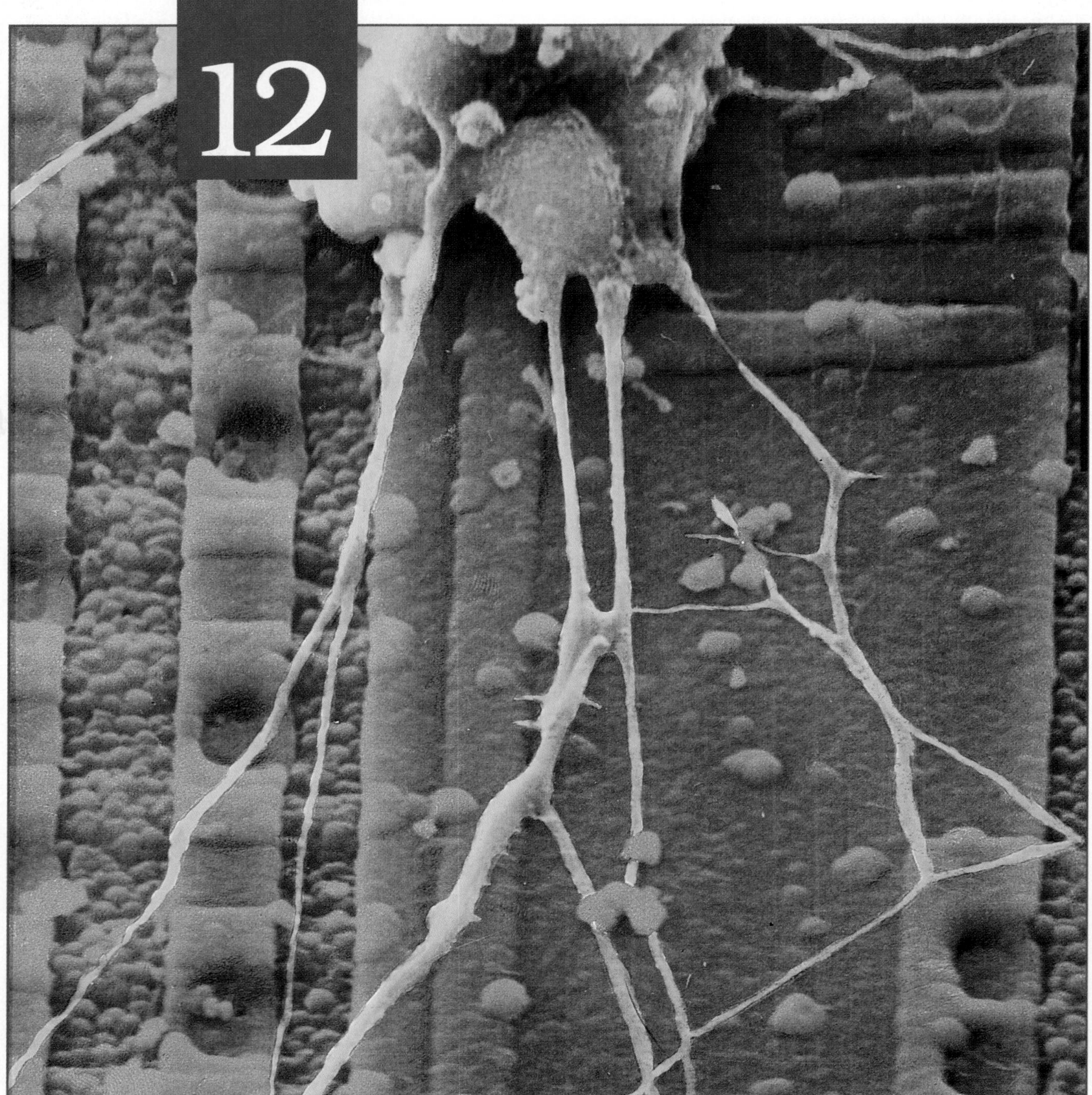

If you think of the nervous system as an organic computer, individual neurons are the "chips" that make it work. This scanning electron micrograph shows a human neuron growing on the surface of a silicon computer chip. The circuits visible on the chip are quite small, but the branches of the neuron—the organic circuits—are considerably smaller. Any comparison between the nervous system and a computer is actually rather misleading, because even the most sophisticated computer lacks the versatility and adaptability of a single neuron. A neuron may process information from 100,000 different sources, and there are more than 20 billion neurons in the nervous system. Comparing the intricacy of neural circuits, linkages, processing centers, and information pathways in the nervous system to today's computers is like comparing an eagle to a paper airplane. This chapter begins our examination of the nervous system by considering how neurons function and how groups of neurons interact.

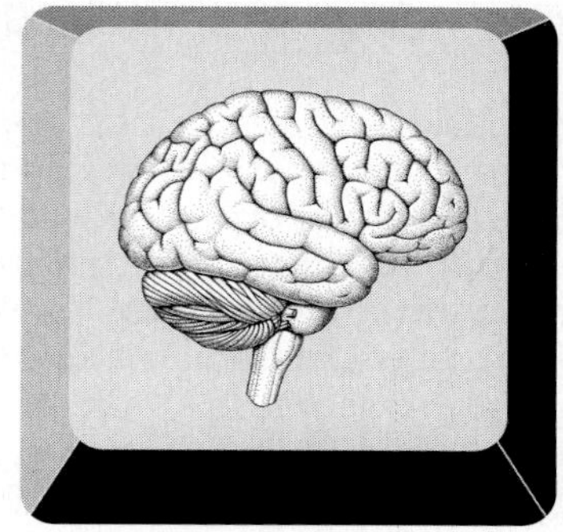

Neural Tissue

Chapter Outline and Objectives

In the next seven chapters our attention will shift to mechanisms that coordinate the activities of the body's organ systems. These activities are adjusted to meet changing situations and environmental conditions. You sit, stand, or walk by controlling muscular activities; your body temperature remains stable on a cold winter's day or in a warm kitchen because your rates of heat generation and heat loss are closely regulated.

Two organ systems, the *nervous system* and the *endocrine system*, provide the necessary regulation. These systems share several structural and functional characteristics, and they usually act in a complementary fashion. The endocrine system adjusts the metabolic operations of other systems in response to changes in the availability of nutrients and the demand for energy. It also directs activities that continue for extended periods, such as growth and maturation, sexual development, pregnancy, or responses to chronic environmental stresses. Endocrine responses are usually slow to develop, but often last much longer than those of the nervous system. The nervous system provides relatively swift but usually brief responses to stimuli, by temporarily modifying the activities of other organ systems. The modifications may appear almost immediately—in 1–2 msec—but the effects disappear soon after neural activity ceases.

The nervous system, which accounts for a mere 3 percent of the total body weight, is the most complex and important of our organ systems. It is vital not only to life, but to our experience and appreciation of life as well. This chapter details the structure and function of neural tissue and introduces principles of neurophysiology that are vital to an understanding of the nervous system's capabilities and limitations. Subsequent chapters will examine increasing levels of structural and functional complexity. AM *The Neurological Examination*

An Overview of the Nervous System

Figure 12-1

The nervous system includes all of the *neural tissue* in the body. Neural tissue was introduced in Chapter 4. *[p. 138]* The basic functional units of the nervous system are individual *neurons*. Supporting cells, or **neuroglia** (noo-RŌ-glē-ah; *glia*, glue), isolate the neurons, provide a supporting framework for neural tissue, act as phagocytes, and help regulate the composition of the interstitial fluid. Neuroglia, also called *glial cells*, far outnumber the neurons and account for roughly half the volume of the nervous system.

Neural tissue, with supporting blood vessels and connective tissues, forms the organs of the nervous system: the brain, the spinal cord, complex sense organs such as the eye and ear, and the *nerves* that interconnect those organs and link the nervous system with other systems. Two major *anatomical* subdivisions of the nervous system were introduced in Chapter 1: the *central nervous system* and the *peripheral nervous system*. *[p. 8]*

The **central nervous system (CNS)** consists of the brain and spinal cord. These are complex organs that include not only neural tissue but also blood vessels and the various connective tissues that provide physical protection and support. The CNS is responsible for integrating, processing, and coordinating sensory data and motor commands. Sensory data convey information about conditions inside or outside the body. Motor commands control or adjust the activities of peripheral organs, such as skeletal muscles. When you stumble, the CNS integrates information concerning balance and limb position and then coordinates your recovery by sending motor commands to appropriate skeletal muscles—all in a split second and without conscious thought. The CNS, specifically the brain, is also the seat of higher functions, such as intelligence, memory, learning, and emotion.

The **peripheral nervous system (PNS)** includes all of the neural tissue outside the CNS. The PNS delivers sensory information to the CNS and carries motor commands to peripheral tissues and systems. Bundles of *nerve fibers* (axons) carry sensory information and motor commands in the PNS. Such bundles, with associated blood vessels and connective tissues, are called *peripheral nerves*, or simply **nerves.** Nerves connected to the brain are called **cranial nerves;** those attached to the spinal cord are called **spinal nerves.**

Figure 12-1 • diagrams the *functional* divisions of the nervous system. The **afferent division** (*ad*, to + *ferre*, to carry) of the PNS brings sensory information to the CNS from **receptors** in peripheral tissues and organs. Receptors are sensory structures, ranging from the processes of single cells to complex organs, that either detect changes in the internal environment or respond to the presence of specific stimuli. Receptors may be neurons or specialized cells of other tissues, such as the *Merkel cells* of the epidermis.

The **efferent division** (*ex*, from) of the PNS carries motor commands from the CNS to muscles and glands. Because these target organs *do* something in response to these commands, they are called **effectors.** The efferent division has *somatic* and *visceral* components. The **somatic nervous system (SNS)** provides voluntary control over skeletal muscle contractions. The *visceral motor system*, or **autonomic nervous system (ANS),** provides automatic, involuntary regulation of

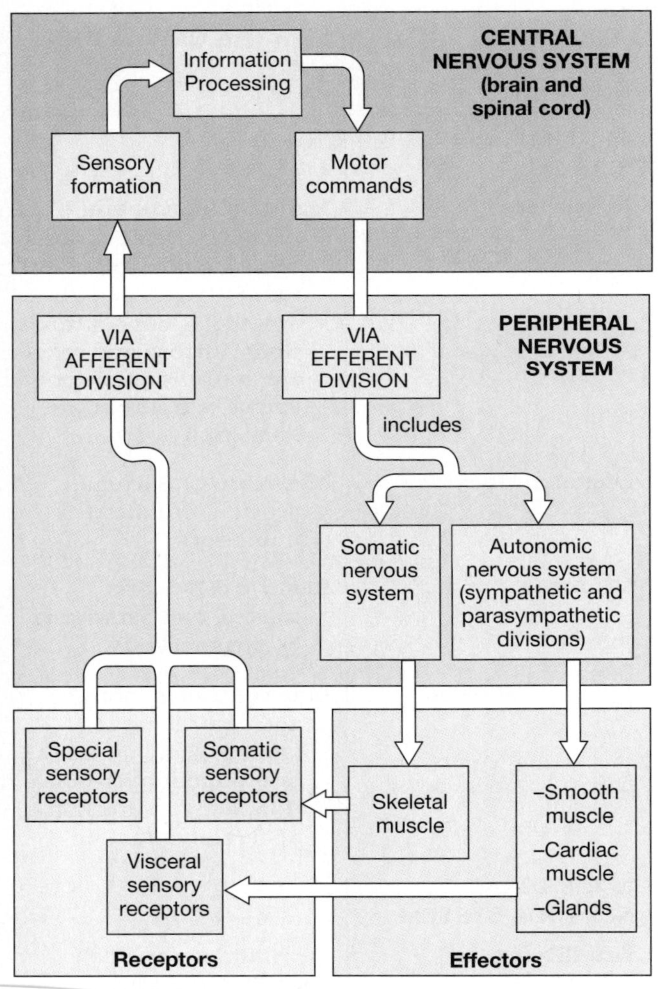
CENTRAL NERVOUS SYSTEM (brain and spinal cord)
Information Processing
Sensory formation
Motor commands
PERIPHERAL NERVOUS SYSTEM
VIA AFFERENT DIVISION
VIA EFFERENT DIVISION
includes
Somatic nervous system
Autonomic nervous system (sympathetic and parasympathetic divisions)
Special sensory receptors
Somatic sensory receptors
Skeletal muscle
–Smooth muscle
–Cardiac muscle
–Glands
Visceral sensory receptors
Receptors
Effectors

❑ Neuroglia of the Central Nervous System

There are four types of glial cells in the central nervous system: *ependymal cells, astrocytes, oligodendrocytes,* and *microglia.*

Ependymal Cells

Figure 12-2

The brain and spinal cord contain a fluid-filled central passageway. The thickness of the walls and the diameter of the passageway vary from one region to another. The narrow passageway within the spinal cord is termed the *central canal;* the expanded chambers found in portions of the brain are called *ventricles.* The ventricles and central canal are lined by a cellular layer of epithelial cells called the **ependyma** (e-PEN-di-mah) and filled with **cerebrospinal fluid (CSF).** This fluid, which also surrounds the brain and spinal cord, provides a protective cushion and distributes dissolved gases, nutrients, wastes, and other materials.

Ependymal cells (Figure 12-2a●) are cuboidal to columnar in form, and their free surfaces are usually covered with cilia or microvilli. In a few parts of the brain, specialized ependymal cells participate in the secretion of the CSF. Other regions of the ependyma assist in the circulation of the CSF and probably perform additional functions as well. Unlike typical epithelial cells, ependymal cells have slender processes that branch extensively and make direct contact with glial cells in the surrounding neural tissue. The functions of these connections are not known.

...tes

...ts; *astro-*, star + *cyte*, cell) are ... numerous glial cells. ...re 12-2b●, have a ... poorly under-...ized as:

processes form a complete cytoplasmic blanket around the capillaries, interrupted only where other glial cells contact the capillary walls. Chemicals secreted by astrocytes are somehow responsible for maintaining the special permeability characteristics of the endothelial cells. (The blood-brain barrier will be discussed further in Chapter 14.)

2. *Creating a three-dimensional framework for the CNS.* Astrocytes are packed with microfilaments that extend from foot to foot across the breadth of the cell. This reinforcement assists them in providing a structural framework for the neurons of the brain and spinal cord.
3. *Performing repairs in damaged neural tissue.* In the CNS, damaged neural tissue seldom regains normal function. However, astrocytes moving into the injury site can make structural repairs that stabilize the tissue and prevent further injury. Neural damage and subsequent repair will be detailed in a later section (p. 417).
4. *Guiding neuron development.* Astrocytes in the embryonic brain appear to be involved in directing the growth and interconnection of developing neurons.
5. *Controlling the interstitial environment.* Although much remains to be learned about astrocyte physiology, there is evidence that astrocytes adjust the interstitial fluid composition by: (1) regulating the concentration of sodium ions, potassium ions, and carbon dioxide; (2) providing a rapid-transit system for the transport of nutrients, ions, and dissolved gases between capillaries and neurons; (3) controlling the volume of blood flow through the capillaries; and (4) absorbing and recycling some of the neurotransmitters released by active neurons.

TABLE 12-1 Glial Cells in the CNS and PNS

Cell Type	*Functions*
CENTRAL NERVOUS SYSTEM	
Astrocytes	Maintain blood-brain barrier; provide structural support; regulate ion, nutrient, and dissolved gas concentrations; absorb and recycle neurotransmitters; assist in tissue repair after injury
Oligodendrocytes	Myelinate CNS axons; provide structural framework
Microglia	Remove cell debris, wastes, and pathogens by phagocytosis
Ependymal cells	Line ventricles (brain) and central canal (spinal cavity); assist in production, circulation, and monitoring of cerebrospinal fluid
PERIPHERAL NERVOUS SYSTEM	
Satellite cells	Surround neuron cell bodies in ganglia
Schwann cells	Enclose all axons in PNS; responsible for myelination of some peripheral axons; participate in repair process after injury

Oligodendrocytes

Figure 12-2b

Like astrocytes, **oligodendrocytes** (o-li-gō-DEN-drō-sīts; *oligo*, few) possess slender cytoplasmic extensions, but their cell bodies are smaller, and they have fewer processes (Figure 12-2b●). The processes usually contact the exposed surfaces of neurons. The

provided by the glial cell. Near the tip of each oligodendritic process, the cytoplasm becomes very thin, but the cell membrane expands to form an enormous membranous pad. This flattened pancake, diagrammed in Figure 12-2b●, somehow gets wound around the axon, creating a

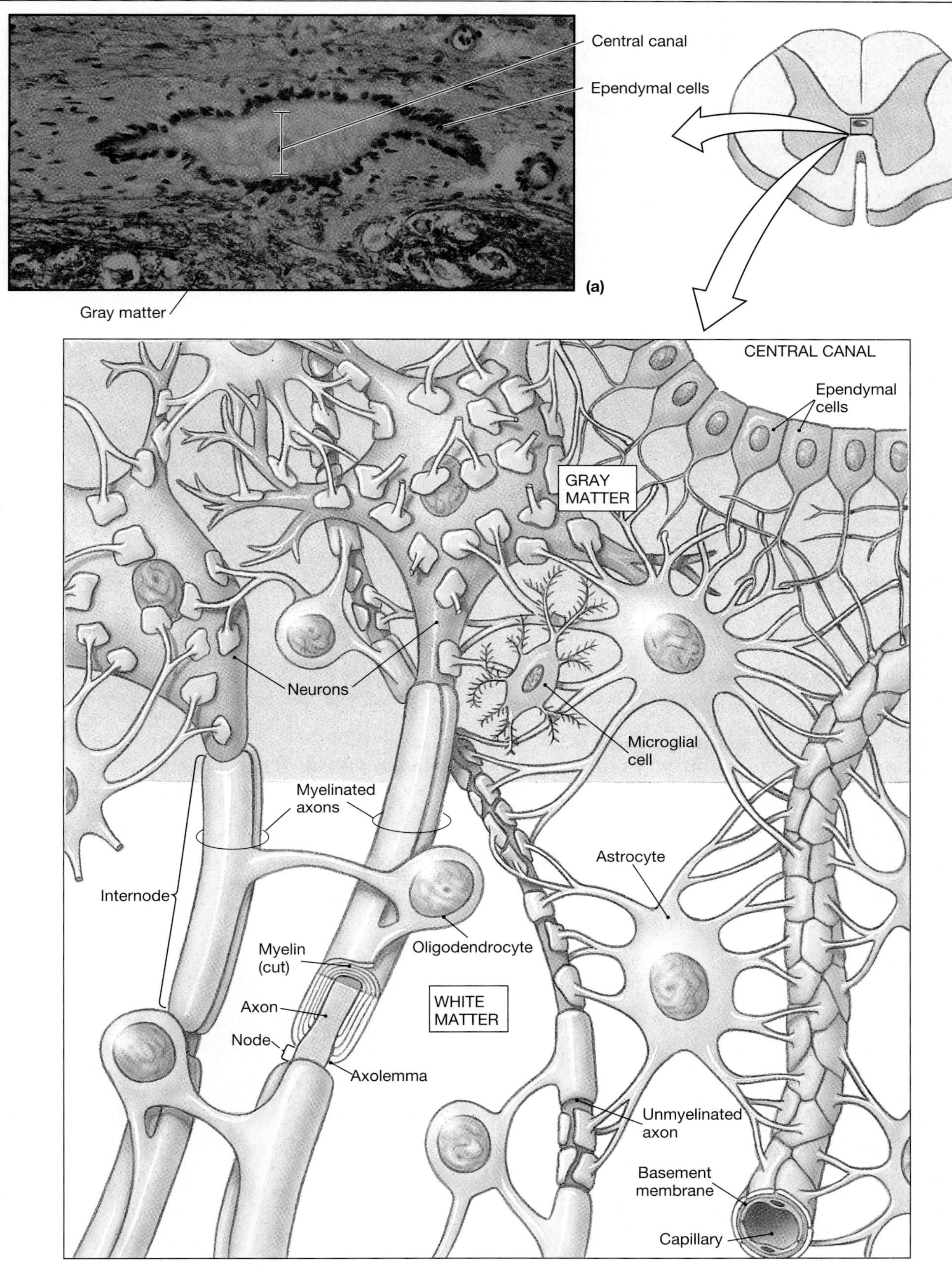
Central canal
Ependymal cells
(a)
Gray matter
CENTRAL CANAL
Ependymal cells
GRAY MATTER
Neurons
Microglial cell
Myelinated axons
Astrocyte
Internode
Oligodendrocyte
Myelin (cut)
Axon
Node
Axolemma
WHITE MATTER
Unmyelinated axon
Basement membrane
Capillary
(b)

● **FIGURE 12-2**

Neuroglia in the CNS. (a) Light micrograph, showing the ependymal lining of the central canal of the spinal cord. (LM × 236) **(b)** A diagrammatic view of neural tissue in the CNS, showing relationships between neuroglia and neurons.

These gaps are called **nodes,** or the *nodes of Ranvier* (RAHN-vē-ā), and the relatively large areas wrapped in myelin are called **internodes** (*inter*, between). In dissection, myelinated axons appear a glossy white, primarily because of the lipids present, and regions dominated by myelinated axons constitute the **white matter** of the CNS. In contrast, areas dominated by neuron cell bodies are called **gray matter** because of their dusky gray color. Not all axons in the CNS are myelinated, and **unmyelinated** axons may not be completely covered by glial cell processes.

In summary, oligodendrocytes play a role in structural organization by tying clusters of axons together, and they improve the functional performance of neurons by coating axons with myelin.

Microglia

Figure 12-2b

Roughly 5 percent of the CNS glial cells are **microglia** (mī-KRŌG-lē-uh). Microglial cells (Figure 12-2b●) are smaller than the other glial elements, and their slender cytoplasmic processes have many fine branches. They are also capable of migrating through neural tissue. Microglia do not develop in neural tissue; they are phagocytic white blood cells that have migrated across capillary walls in the neural tissue of the CNS. (See the Embryology Summary, p. 391). Microglia act as a wandering police force and janitorial service, engulfing cellular debris, waste products, and pathogens. Under ordinary circumstances the few microglia present are able to perform the necessary cleanup operations. In times of infection or injury their numbers increase dramatically, as other phagocytic cells are attracted to the damaged area.

CNS TUMORS AND NEUROGLIA Tumors of the brain, spinal cord, and associated membranes result in approximately 90,000 deaths in the United States each year. Tumors that originate in the CNS are called *primary CNS tumors*, to distinguish them from *secondary CNS tumors* that arise from the metastasis of cancer cells originating elsewhere. Roughly 75 percent of CNS tumors are primary tumors. Primary CNS tumors in adults result from the divisions of abnormal glial cells, since typical neurons in the adult are incapable of division. In childhood, however—usually before age 4, when neurons are increasing in number—primary CNS tumors may also result from the divisions of abnormal neurons or stem cells. Symptoms of CNS tumors vary depending on the location involved, and treatment may involve surgery, radiation, and/or chemotherapy.

❑ Neuroglia of the Peripheral Nervous System

Figures 12-2b, 12-3, 12-4

Neuron cell bodies in the PNS are clustered together in masses called **ganglia** (singular *ganglion*). Neuron cell bodies and axons in the PNS are completely insulated from their surroundings by the processes of glial cells. The two glial cell types involved are called *satellite cells* and *Schwann cells*.

Satellite cells, or *amphicytes* (AM-fi-sīts), surround the neuron cell bodies in peripheral ganglia, as indicated in Figure 12-3●. **Schwann cells,** or *neurolemmocytes*, form a sheath around every peripheral axon, whether unmyelinated or myelinated. Whereas an oligodendrocyte may myelinate portions of several adjacent axons, a Schwann cell can myelinate only one segment of a single axon (Figure 12-4a●). The presumed

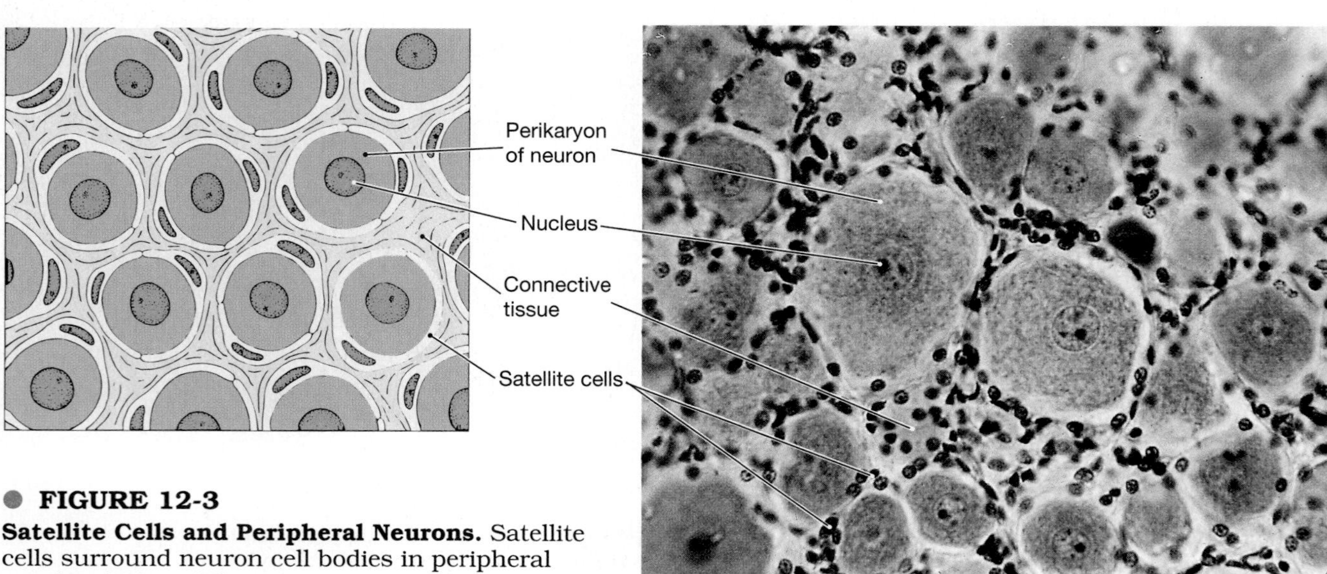
Perikaryon of neuron
Nucleus
Connective tissue
Satellite cells

● **FIGURE 12-3**
Satellite Cells and Peripheral Neurons. Satellite cells surround neuron cell bodies in peripheral ganglia. (LM × 128)

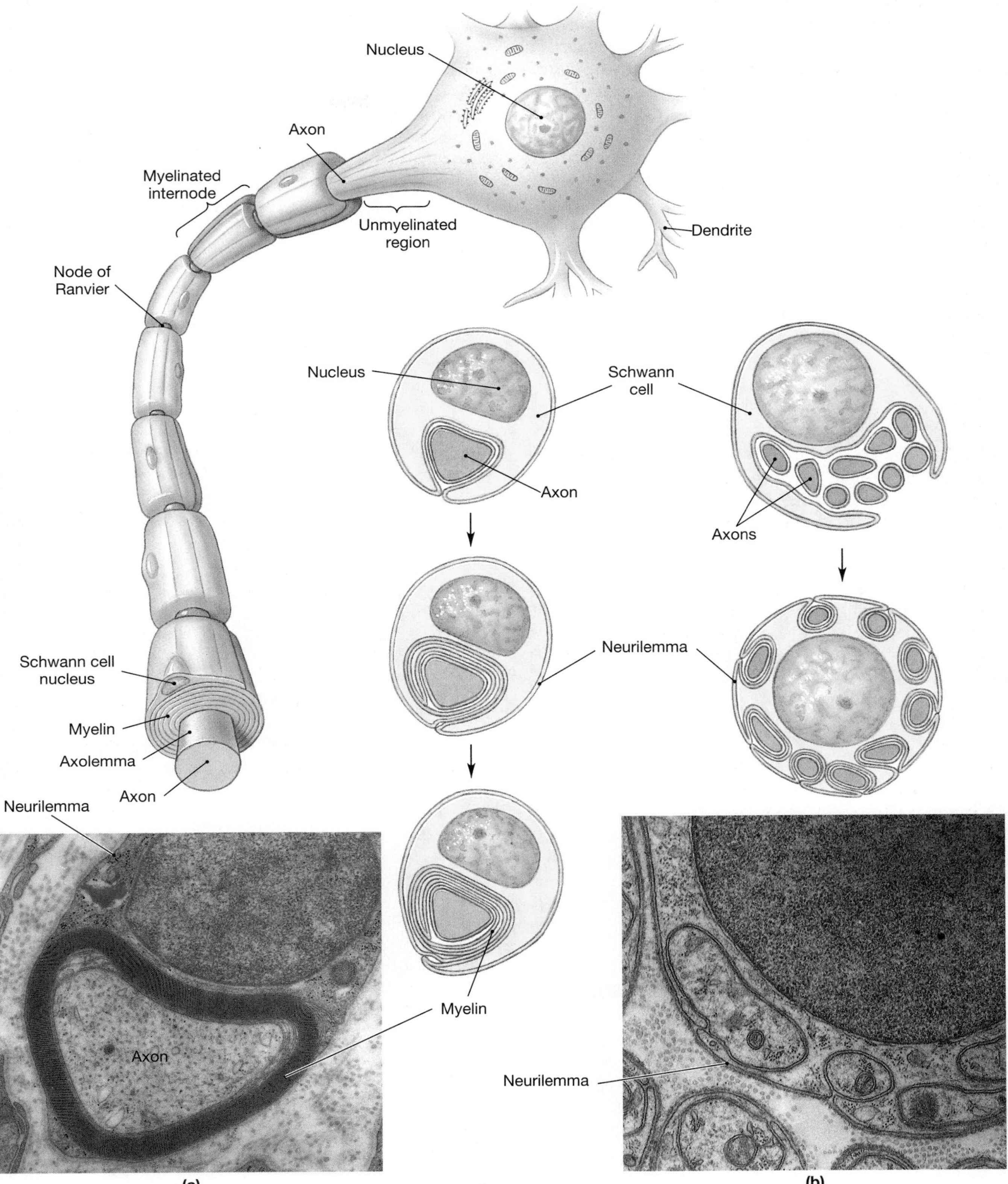
Nucleus
Axon
Myelinated internode
Unmyelinated region
Dendrite
Node of Ranvier
Nucleus
Schwann cell
Axon
Axons
Neurilemma
Schwann cell nucleus
Myelin
Axolemma
Axon
Neurilemma
Axon
Myelin
Neurilemma
(a)
(b)

● **FIGURE 12-4**

Schwann Cells and Peripheral Axons. (a) A single Schwann cell forms the myelin sheath around a portion of a single axon. This situation differs from the way myelin forms inside the CNS; compare with Figure 12-2b. (TEM × 20,603) **(b)** A single Schwann cell can encircle several unmyelinated axons. Unlike the situation inside the CNS (Figure 12-2b), every axon in the PNS has a complete neurilemmal sheath. (TEM × 27,627)

mode of myelin formation in the PNS is summarized in Figure 12-4a●, which should be compared with Figure 12-2b●, p. 383. However, a Schwann cell may *enclose* segments of several unmyelinated axons (Figure 12-4b●).

Demyelination is the progressive destruction of myelin sheaths in the CNS and PNS. The result is a gradual loss of sensation and motor control that leaves affected regions numb and paralyzed. Many unrelated conditions that result in myelin destruction can cause symptoms of demyelination. Several important examples of demyelinating disorders, including *heavy metal poisoning, diphtheria, multiple sclerosis* (MS), and *Guillain-Barré syndrome,* are detailed in the *Applications Manual.* AM Demyelination Disorders

Neurons

Before considering how neurons relay information, we need to take a closer look at the structure of an individual neuron. This section will consider (1) the structure of a "model neuron" and (2) the structural and functional classification of neurons.

Neuron Structure

Figure 12-5

Figure 12-5● shows the structure of a representative neuron in greater detail. Neurons can have a variety of shapes; this is a *multipolar neuron,* the most common type of neuron in the CNS.

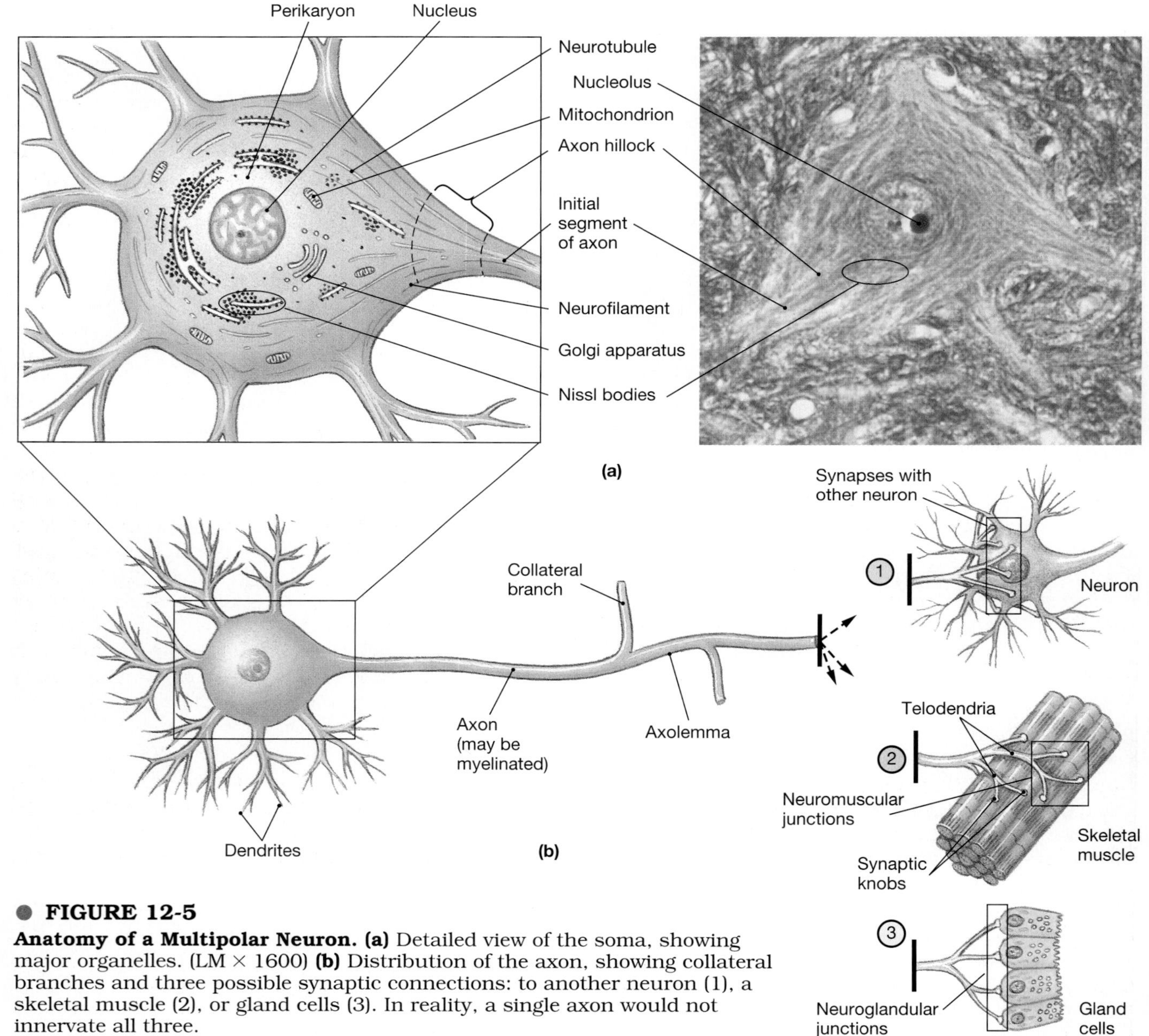

Perikaryon
Nucleus
Neurotubule
Nucleolus
Mitochondrion
Axon hillock
Initial segment of axon
Neurofilament
Golgi apparatus
Nissl bodies
(a)
Synapses with other neuron
1
Neuron
Collateral branch
Axon (may be myelinated)
Axolemma
Dendrites
(b)
Telodendria
2
Neuromuscular junctions
Skeletal muscle
Synaptic knobs
3
Neuroglandular junctions
Gland cells

● **FIGURE 12-5**
Anatomy of a Multipolar Neuron. (a) Detailed view of the soma, showing major organelles. (LM × 1600) **(b)** Distribution of the axon, showing collateral branches and three possible synaptic connections: to another neuron (1), a skeletal muscle (2), or gland cells (3). In reality, a single axon would not innervate all three.

The Soma

Figure 12-5a

The soma, or cell body (Figure 12-5a●), contains a relatively large, round nucleus with a prominent nucleolus. The surrounding cytoplasm constitutes the **perikaryon** (per-i-KAR-ē-on; *karyon*, nucleus). The cytoskeleton of the perikaryon contains **neurofilaments** and **neurotubules.** Bundles of neurofilaments, called **neurofibrils,** extend into the dendrites and axon, providing internal support for these relatively slender processes.

The perikaryon contains organelles that provide energy and synthesize organic materials, especially neurotransmitters. The numerous mitochondria, free and fixed ribosomes, and membranes of the rough endoplasmic reticulum (RER) give the perikaryon a coarse, grainy appearance. Mitochondria generate ATP to meet the high energy demands of an active neuron. The ribosomes and RER synthesize peptides and proteins. Some areas of the perikaryon contain clusters of RER and free ribosomes. These regions, which stain darkly, are called *Nissl bodies* because they were first described by the German microscopist Franz Nissl. Nissl bodies account for the gray color of areas containing neuron cell bodies—the *gray matter* seen in dissection.

Most neurons lack an important organelle complex, the *centrosome*, discussed in Chapter 3. [p. 85] In other cells, centrioles of the centrosome organize the cytoskeleton and the microtubules of the spindle fibers that move chromosomes during cell division. In some tissues, notably the olfactory (smell) epithelium of the nose, neural stem cells persist throughout life. Their divisions produce daughter cells that differentiate into highly specialized neurons, the olfactory receptors. However, stem cells are not present in the CNS, and CNS neurons usually lose their centrioles during differentiation. As a result, CNS neurons are incapable of dividing, and neurons lost to injury or disease cannot be replaced.

The Dendrites

Figure 12-5b

A variable number of dendrites extend out from the soma (Figure 12-5b●). Typical dendrites are highly branched, and each branch bears fine processes called *dendritic spines.* In the CNS, a neuron receives information from other neurons via synaptic connections at the dendritic spines. Although some synaptic connections occur on the soma, the majority of synapses occur on the dendrites, which represent 80–90 percent of the total surface area of the neuron. Chemicals released at these synapses cause localized changes in the transmembrane potential of the dendrites and soma.

The Axon

Figure 12-5a

An axon is a long cytoplasmic process capable of propagating an action potential. The **axoplasm** (AK-so-plazm), or cytoplasm of the axon, contains neurofibrils, neurotubules, small vesicles, lysosomes, mitochondria, and various enzymes. The *axolemma* is often covered by oligodendrocyte processes in the CNS and invariably is covered by Schwann cells in the PNS. The base, or **initial segment,** of the axon in a multipolar neuron is attached to the soma at a thickened region known as the **axon hillock** (Figure 12-5a●).

An axon may branch along its length, producing branches collectively known as **collaterals.** Collaterals enable a single neuron to communicate with several other cells. The main axon trunk and any collaterals end in a series of fine extensions, or **telodendria** (te-lō-DEN-drē-a; *telo-*, end + *dendron*, tree).

Expanded **synaptic knobs** at the tips of the telodendria form synaptic connections with other cells.[1] The communication between cells at a synapse most often involves the release of chemicals, called **neurotransmitters,** by the synaptic knob. As we saw in Chapter 10, this release is triggered by the arrival of an action potential at the synaptic knob. [p. 295]

AXOPLASMIC TRANSPORT Each synaptic knob contains mitochondria, portions of the endoplasmic reticulum, and thousands of vesicles filled with molecules of neurotransmitter. Breakdown products of neurotransmitter released at the synapse are reabsorbed and reassembled at the synaptic knob. The synaptic knob also receives a continual supply of neurotransmitter synthesized in the soma, along with enzymes and lysosomes. These products are exported to the synaptic knobs along the length of the axon. This movement, called **axoplasmic transport,** occurs along neurotubules and involves proteins, called *kinesins*, that require ATP. Some materials travel slowly, at rates of a few millimeters per day. This transport mechanism is known as the "slow stream." Vesicles containing neurotransmitter move much more rapidly, traveling in the "fast stream" at 5–10 mm per hour.

At the same time that some materials are traveling from the soma to the periphery of the neuron, other substances are being transported toward the soma. If debris or unusual chemicals appear in the synaptic knob, this **retrograde** (RET-rō-grād) **flow** (*retro*, backward) soon delivers them to the soma. The arriving materials may then alter

[1]The term *synaptic knob* is widely recognized and will be used throughout this text. However, the same structures may be called *synaptic terminals, axon terminals, terminal buttons, terminal boutons, end bulbs*, or *neuropodia.*

the activity of the cell by turning appropriate genes on or off. For example, Chapter 3 noted the role of various *growth factors* on cell development and differentiation. [p. 103] *Nerve growth factor* (NGF) targets neurons in the CNS and PNS. NGF absorbed at a synaptic knob in peripheral tissues is transported to the soma by retrograde flow. Once within the soma, NGF activates genes stimulating neuron growth and maintenance.

Rabies is perhaps the most dramatic example of a clinical condition directly related to retrograde flow. A bite on the hand from a rabid animal injects the rabies virus into peripheral tissues, where virus particles quickly enter synaptic knobs and peripheral axons. Retrograde flow then carries the virus into the central nervous system, with potentially fatal results. Many toxins, including heavy metals, some pathogenic bacteria, and other viruses also rely on some form of axoplasmic transport to bypass the blood-brain barrier and other CNS defenses. AM Axoplasmic Transport and Disease

The Synapse

Figures 12-5b, 12-6

A **synapse** is a specialized site of intercellular communication. A synapse may involve two neurons (Figure 12-6●) or a neuron and another cell type. When one neuron communicates with another, the synapse may occur on a dendrite, on the soma, or along the length of the axon of the receiving cell. A synapse may also permit communication between a neuron and another cell type. Such a synapse is called a **neuroeffector junction.** There are two major classes of neuroeffector junctions: *neuromuscular junctions* and *neuroglandular junctions* (Figure 12-5b●). At a neuromuscular junction the neuron communicates with a muscle cell; Chapter 10 examined the structure of a neuromuscular junction. [p. 295] At a **neuroglandular junction** a neuron controls or regulates the activity of a secretory cell. Neurons also innervate a variety of other cell types, such as fat cells, and examples will be encountered in later chapters.

❑ Neuron Classification

The billions of neurons in the nervous system are quite variable in form. Neurons are classified two ways: (1) on the basis of structure and (2) on the basis of function.

Structural Classification

Figure 12-7

The anatomical classification of neurons is based on the number of processes that project from the cell body.

- **Anaxonic** (an-ak-SON-ik) **neurons** are small, and there are no anatomical clues to distinguish dendrites from axons. Anaxonic neurons (Figure 12-7a●) are found in the CNS and in special sense organs, but their functions are poorly understood. Their processes are never myelinated.
- In a **unipolar neuron,** or *pseudounipolar neuron,* the dendritic and axonal processes are continuous, and the cell body lies off to one side (Figure 12-7b●). In a unipolar neuron the initial segment lies at the base of the dendritic branches, and the rest of the process is considered an axon on both structural and functional grounds. Sensory neurons of the peripheral nervous system are usually unipolar, and their axons may be myelinated.

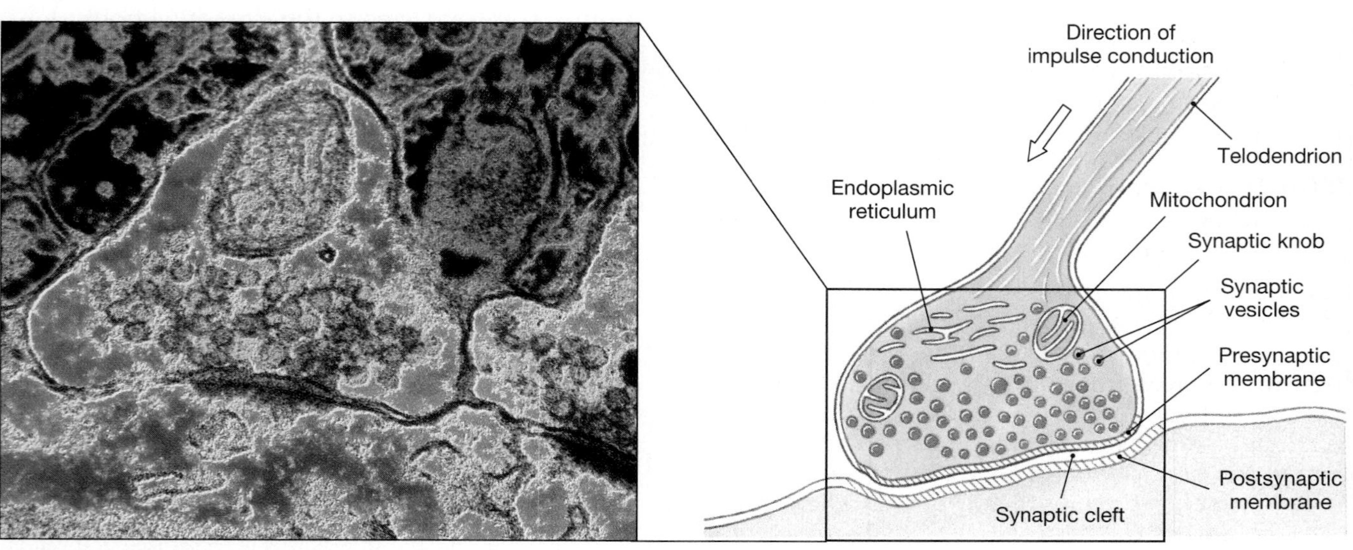
Direction of impulse conduction
Telodendrion
Endoplasmic reticulum
Mitochondrion
Synaptic knob
Synaptic vesicles
Presynaptic membrane
Postsynaptic membrane
Synaptic cleft

● **FIGURE 12-6**
Structure of a Chemical Synapse. (Color-enhanced TEM × 222,000) D

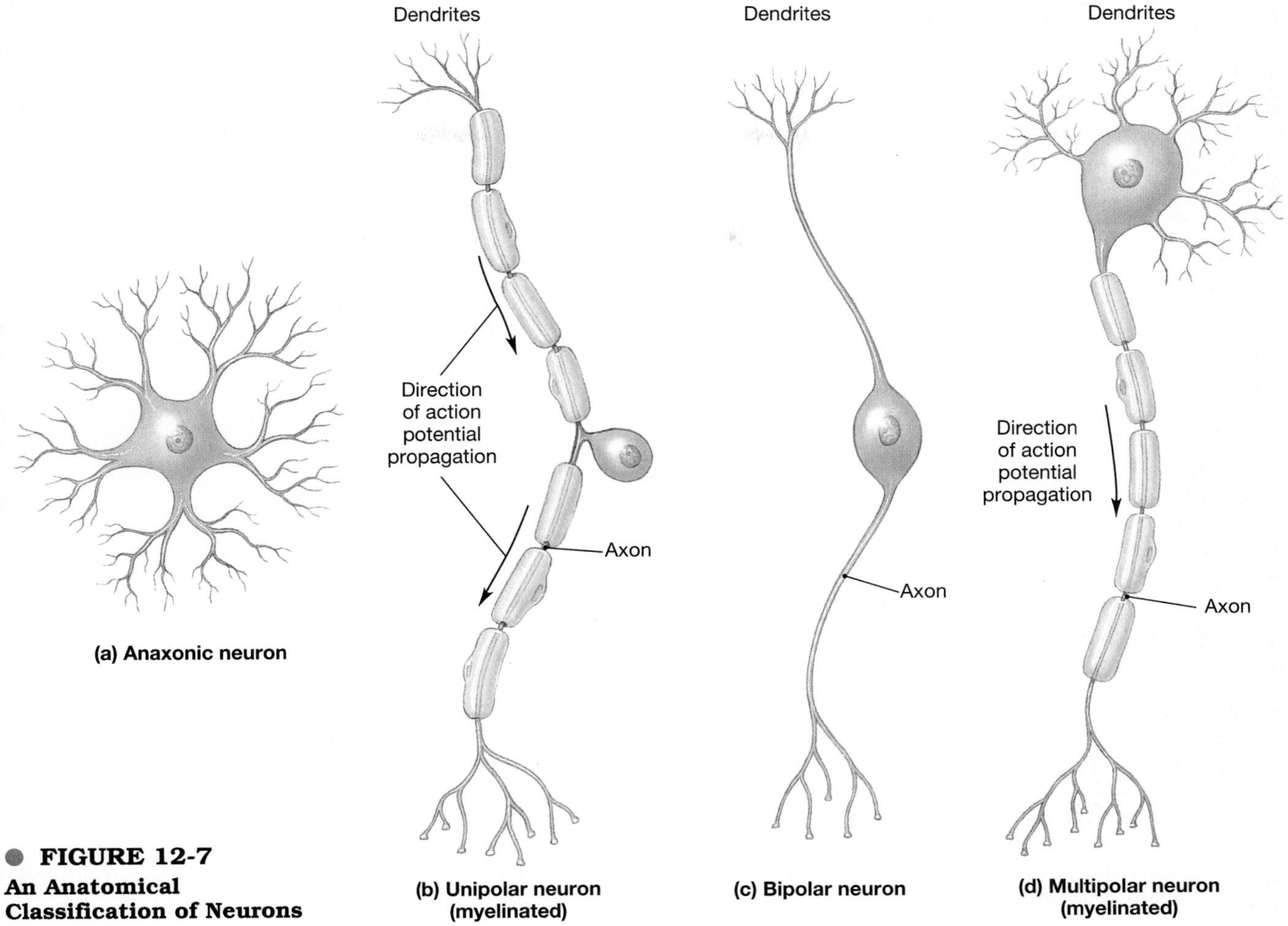

Dendrites
Dendrites
Dendrites
Direction of action potential propagation
Axon
Axon
Direction of action potential propagation
Axon
(a) Anaxonic neuron
(b) Unipolar neuron (myelinated)
(c) Bipolar neuron
(d) Multipolar neuron (myelinated)

● **FIGURE 12-7**
An Anatomical Classification of Neurons

- **Bipolar neurons** (Figure 12-7c●) have one dendrite and one axon, with the cell body between them. Bipolar neurons are relatively rare but play an important role in relaying information concerning sight, smell, and hearing. Their axons are not myelinated.
- **Multipolar neurons** have several dendrites and a single axon that may have one or more branches. Multipolar neurons (Figure 12-7d●) are the most common type of neuron in the CNS. For example, all of the motor neurons that control skeletal muscles are multipolar neurons with myelinated axons.

Functional Classification

Figure 12-8

Neurons can be categorized into three functional groups: (1) *sensory neurons,* (2) *motor neurons,* and (3) *interneurons,* or *association neurons.* Their relationships are diagrammed in Figure 12-8●.

SENSORY NEURONS **Sensory neurons** form the *afferent division* of the PNS. Their function is to deliver information to the CNS. The cell bodies of sensory neurons are found in peripheral *sensory ganglia.* These are unipolar neurons, and their processes, known as **afferent fibers,** extend between a sensory receptor and the spinal cord or brain. Sensory neurons collect information concerning the external or internal environment. There are about 10 million sensory neurons. **Somatic sensory neurons** monitor the outside world and our position within it. **Visceral sensory neurons** monitor internal conditions and the status of other organ systems.

Receptors may be the processes of specialized sensory neurons or cells monitored by sensory neurons. Receptors are broadly categorized as:

- **Exteroceptors** (*extero-,* outside) provide information about the external environment in the form of touch, temperature, and pressure sensations and the more complex senses of sight, smell, hearing, and taste.
- **Proprioceptors** (prō-prē-ō-SEP-torz) monitor the position and movement of skeletal muscles and joints.
- **Interoceptors** (*intero-,* inside) monitor the digestive, respiratory, cardiovascular, urinary, and reproductive systems and provide sensations of taste, deep pressure, and pain.

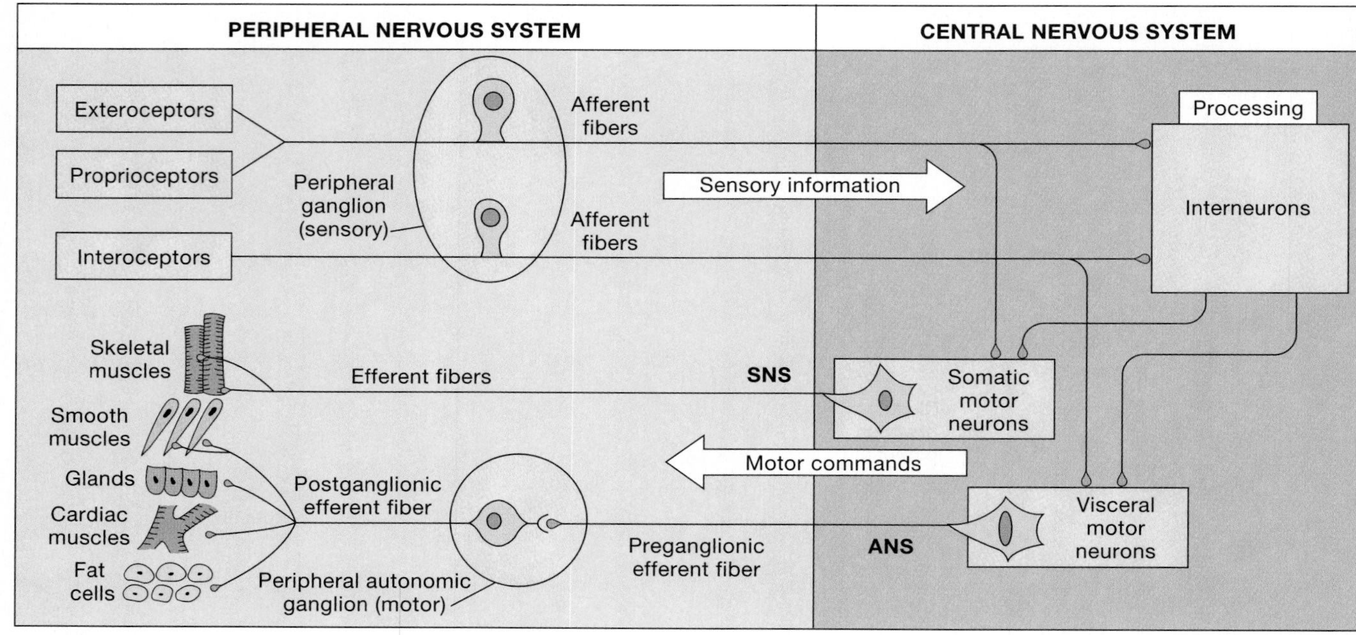
PERIPHERAL NERVOUS SYSTEM
CENTRAL NERVOUS SYSTEM
Exteroceptors
Proprioceptors
Interoceptors
Peripheral ganglion (sensory)
Afferent fibers
Afferent fibers
Sensory information
Processing
Interneurons
Skeletal muscles
Smooth muscles
Glands
Cardiac muscles
Fat cells
Efferent fibers
SNS
Somatic motor neurons
Motor commands
Postganglionic efferent fiber
Preganglionic efferent fiber
ANS
Visceral motor neurons
Peripheral autonomic ganglion (motor)

● **FIGURE 12-8**
A Functional Classification of Neurons

MOTOR NEURONS **Motor neurons** of the *efferent division* carry instructions from the CNS to peripheral effectors. A motor neuron stimulates or modifies the activity of a peripheral tissue, organ, or organ system. About a half a million motor neurons are found in the body. Axons traveling away from the CNS are called **efferent fibers.** There are two major efferent systems: the *somatic nervous system* and the *autonomic (visceral) nervous system.* The somatic nervous system includes all of the **somatic motor neurons** that innervate skeletal muscles. We have voluntary control over the activity of the somatic nervous system. The cell body of a somatic motor neuron lies inside the CNS, and its axon extends into the periphery to innervate one or more neuromuscular junctions.

The activities of the autonomic nervous system (ANS) are primarily under "automatic" involuntary control. **Visceral motor neurons** innervate all peripheral effectors other than skeletal muscles. The axons of visceral motor neurons inside the CNS synapse on neurons in peripheral *autonomic (motor) ganglia,* and the ganglionic neurons control peripheral effectors. Axons extending from the CNS to a ganglion are called **preganglionic fibers.** Axons connecting the ganglion cells with the peripheral effectors are known as **postganglionic fibers.** This arrangement clearly distinguishes the ANS from the SNS.

INTERNEURONS **Interneurons,** or **association neurons,** may be situated between sensory and motor neurons. Interneurons are located entirely within the brain and spinal cord. There are roughly 20 billion interneurons, outnumbering all other types of neurons combined. Interneurons are responsible for the distribution of sensory information and the coordination of motor activity. The more complex the response to a given stimulus, the greater the number of interneurons involved.

✓ What would damage to the afferent division of the nervous system affect?

✓ Examination of a tissue sample shows unipolar neurons. Are these more likely to be sensory neurons or motor neurons?

✓ What type of glial cell would you expect to find in larger than normal numbers in brain tissue from a person suffering from a CNS infection?

Neurophysiology

We will now begin to examine how neurons, aided by glial cells, process information and communicate with one another and with peripheral effectors. Chapter 3 introduced the concepts of *transmembrane potential* and *resting potential,* two characteristic features of living cells. [pp. 81–82] *All of the steps important to neural function involve changes in the transmembrane potentials of individual neurons.* Information is conveyed over relatively long distances in the form of *action potentials,* propagated changes in the transmembrane potential of axons. To understand these events, which occur at great speed, we must consider the origin and maintenance of the transmembrane potential in more detail.

EMBRYOLOGY SUMMARY Introduction to the Development of the Nervous System

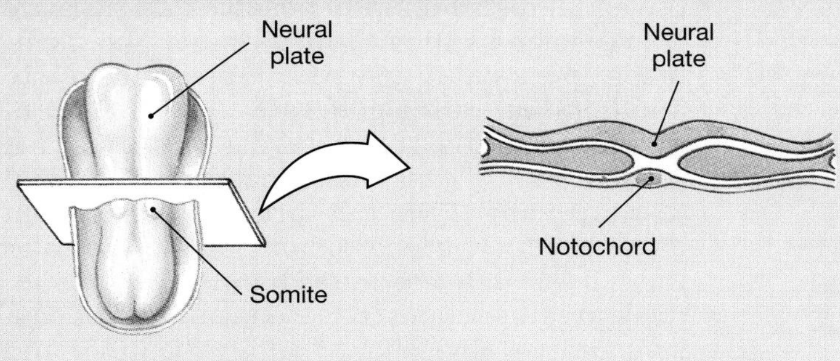

Neural plate
Somite
20 DAYS
Neural plate
Notochord

After 20 days of development *somites* begin appearing on either side of the *notochord* (page 366). The ectoderm near the midline thickens, forming an elevated **neural plate.** The neural plate is broadest near the future head of the developing embryo.

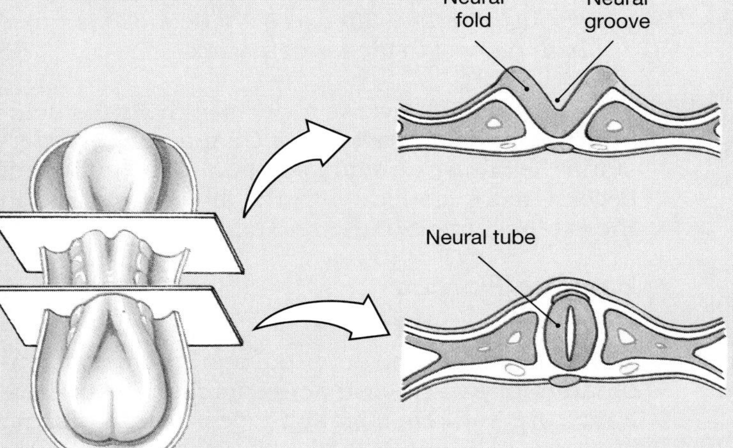

Neural fold
Neural groove
Neural tube
21 DAYS

The center of the plate sinks, creating the **neural groove.** The edges, or **neural folds,** gradually move together. They first contact one another midway along the axis of the neural plate, near the end of the third week.

Where the neural folds meet, they fuse to form a cylindrical **neural tube** that loses its connection with the superficial ectoderm. The process of neural tube formation is called **neurulation;** it is completed in less than a week. The formation of the axial skeleton and the musculature around the developing neural tube were described on pages 366 and 367.

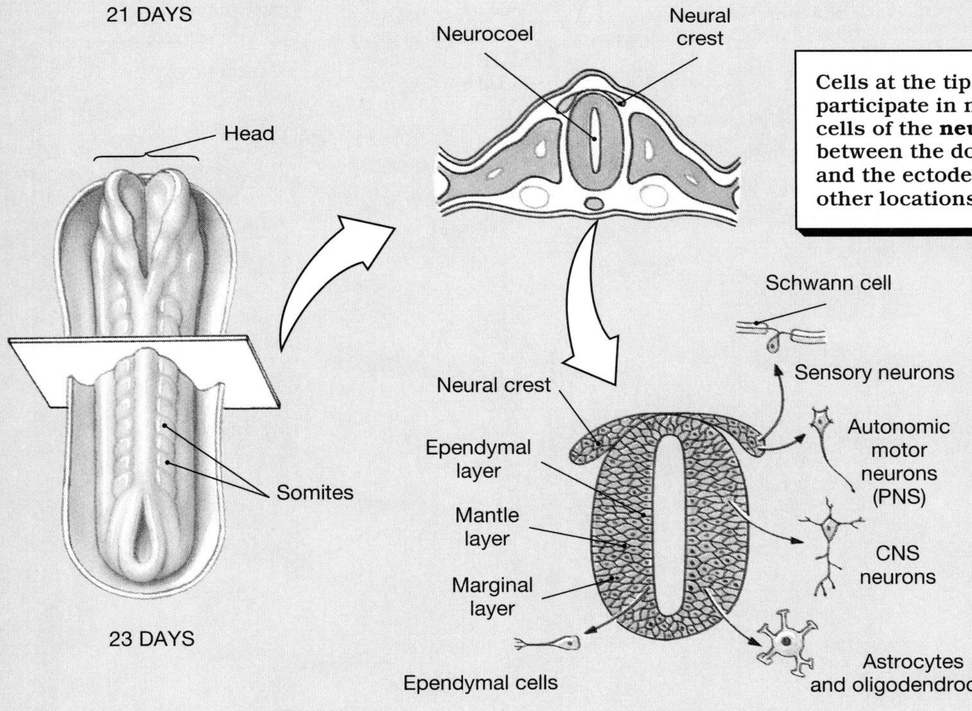

Head
Somites
23 DAYS
Neurocoel
Neural crest
Schwann cell
Sensory neurons
Neural crest
Ependymal layer
Mantle layer
Marginal layer
Autonomic motor neurons (PNS)
CNS neurons
Astrocytes and oligodendrocytes
Ependymal cells

Cells at the tips of the neural folds do not participate in neural tube formation. These cells of the **neural crest** at first remain between the dorsal surface of the **neural tube** and the ectoderm, but they later migrate to other locations.

The first cells to appear in the mantle differentiate into neurons, while the last cells to arrive become astrocytes and oligodendrocytes. Further development of the CNS and PNS will be found in the Embryology Summaries on pages 444–445 and 460–461.

The neural tube increases in thickness as its epithelial lining undergoes repeated mitoses. By the middle of the fifth developmental week, there are three distinct layers. The **ependymal layer** lines the enclosed cavity, or **neurocoel.** The ependymal cells continue their mitotic activities, and daughter cells create the surrounding **mantle layer.** Axons from developing neurons form a superficial **marginal layer.**

The Transmembrane Potential

Figure 12-9

The transmembrane potential was introduced in Chapter 3, but a brief review may prove useful.

1. The intracellular and extracellular fluids differ markedly in ionic composition. The extracellular fluid (ECF) contains relatively high concentrations of sodium ions (Na^+) and chloride ions (Cl^-), whereas the cytosol contains high concentrations of potassium ions (K^+) and negatively charged proteins (Pr^-).
2. If the cell membrane were freely permeable, diffusion would continue until these ions were evenly distributed across the membrane. This does not happen because living cells have selectively permeable membranes. Ions cannot freely cross the lipid portions of the cell membrane. They can enter or leave the cell only via membrane channels. There are many different kinds of membrane channels, each with its own properties. In addition, there are active transport mechanisms that move specific ions into or out of the cell.
3. These passive and active factors do not ensure the equal distribution of positive and negative charges across the cell membrane. The membrane permeability varies from one ion to another. For example, negatively charged proteins inside the cell cannot cross the membrane at all, and it is easier for a potassium ion (K^+) to diffuse out of the cell through a potassium channel than it is for a sodium ion (Na^+) to enter the cell through a sodium channel. As a result, there are differences in the distribution of positive and negative charges along the inner and outer surfaces of the cell membrane. *The inner surface contains an excess of negative charges with respect to the outer surface.*

Figure 12-9● reviews aspects of the transmembrane potential introduced in Chapter 3. To understand the nature of action potentials, you must first become more familiar with the factors involved in the establishment of the resting potential.

Passive Forces

Figure 12-10

The transmembrane potential results from a combination of passive and active forces. The passive forces are both chemical and electrical in nature.

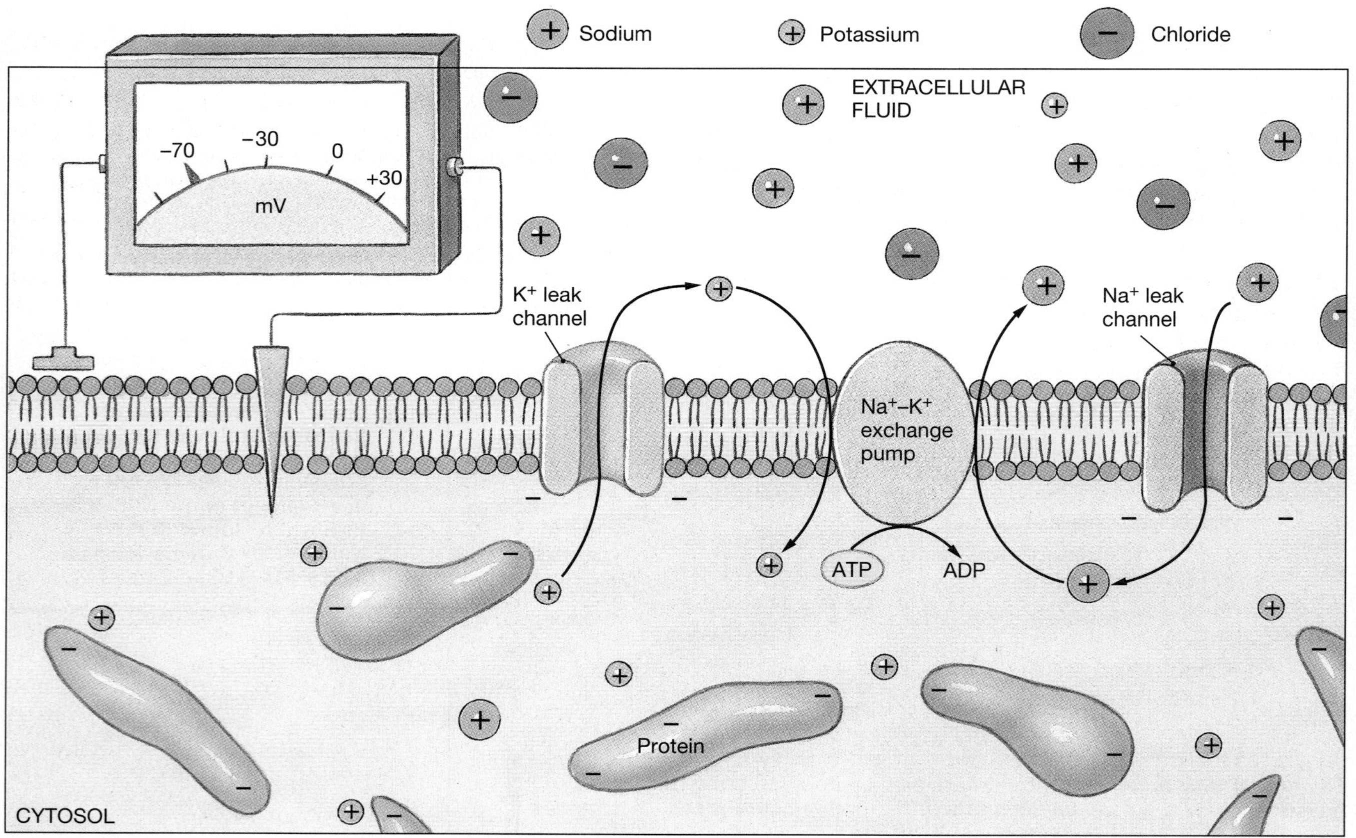
Sodium
Potassium
Chloride
EXTRACELLULAR FLUID
−70
−30
0
+30
mV
K⁺ leak channel
Na⁺ leak channel
Na⁺–K⁺ exchange pump
ATP
ADP
Protein
CYTOSOL

● **FIGURE 12-9**
The Cell Membrane at the Resting Potential

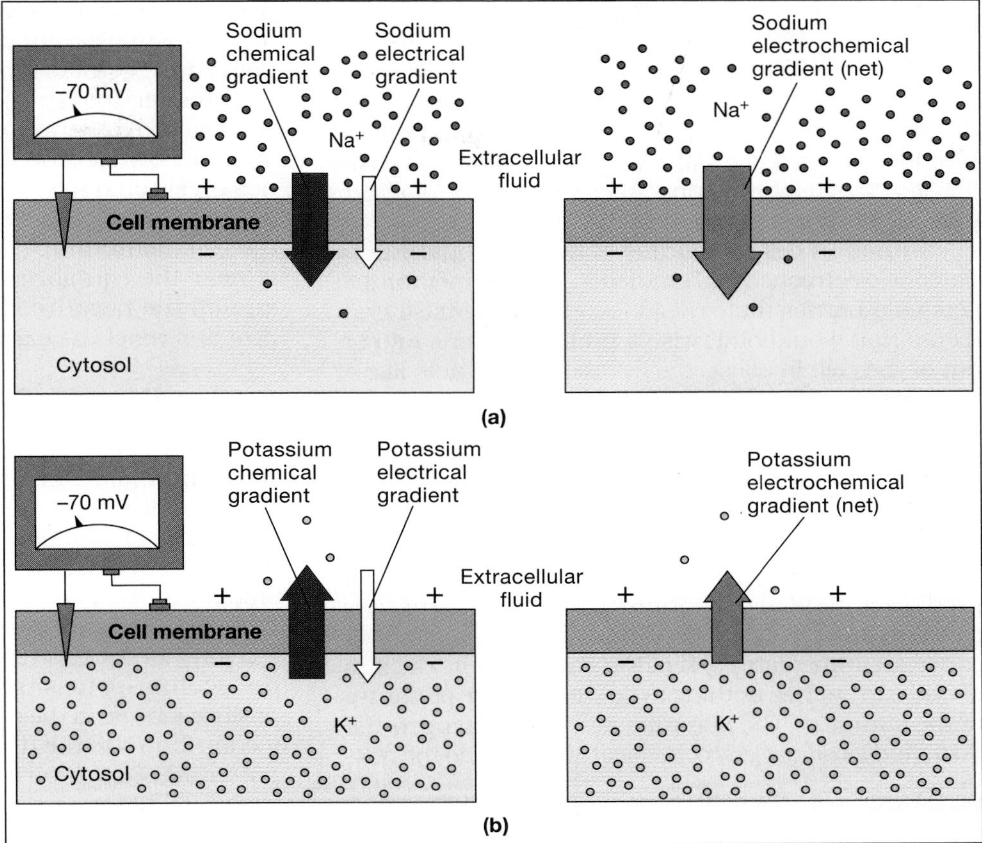

-70 mV
Sodium chemical gradient
Sodium electrical gradient
Na+
Extracellular fluid
Cell membrane
Cytosol
Sodium electrochemical gradient (net)
(a)
-70 mV
Potassium chemical gradient
Potassium electrical gradient
Extracellular fluid
Cell membrane
K+
Cytosol
Potassium electrochemical gradient (net)
(b)

● **FIGURE 12-10**
Electrochemical Gradients. **(a)** At the normal resting potential, large chemical and electrical gradients both tend to drive sodium ions into the cell. **(b)** At the normal resting potential, an electrical gradient opposes the chemical gradient for potassium ions. The net electrochemical gradient tends to force potassium ions out of the cell.

CHEMICAL GRADIENTS Because the intracellular concentration of potassium ions is relatively high, potassium ions tend to move out of the cell through open potassium channels. This movement is driven by a concentration gradient, or *chemical gradient.* Similarly, a chemical gradient for sodium ions tends to drive them into the cell.

ELECTRICAL GRADIENTS Because the membrane permeability to potassium is much greater than its permeability to sodium, potassium ions leave the cytoplasm more rapidly than sodium ions enter. As a result, the cytosol along the interior of the cell membrane experiences a net loss of positive charges, leaving an excess of negatively charged proteins. At the same time, the ECF near the exterior surface of the cell membrane experiences a net gain of positive charges. The positive and negative charges are separated by the cell membrane, which constitutes a **resistance** because it resists the free movement of ions. Whenever positive and negative ions are separated by a resistance, such as a cell membrane, a *potential difference* exists.

The size of a potential difference is measured in volts or millivolts. The *resting potential,* or transmembrane potential of an undisturbed cell, averages about 0.07 V for a neuron cell membrane. This value is usually expressed as –70 mV, with mV indicating millivolts (thousandths of a volt) and the minus sign signifying that the inner surface of the cell membrane is negatively charged with respect to the exterior.

Positive and negative charges attract one another, and if there is no resistance between them, oppositely charged ions move together and eliminate the potential difference. Such a movement of charges in response to voltage is called a **current.** The amount of current is inversely proportional to the resistance that separates the charges. If the resistance is high, the current is very small. If the resistance is low, the current is very large. The resistance of cell membranes can be changed by opening or closing ion channels, and these changes result in currents that bring ions into or out of the cytoplasm.

THE ELECTROCHEMICAL GRADIENT Electrical gradients can reinforce or oppose the chemical gradient for each ion. The **electrochemical gradient** for a specific ion is the sum of the chemical and electrical forces acting across the cell membrane.

- Sodium ions (Na^+) in the ECF are attracted by the excess of negative charges inside the cell membrane, so in this case both the chemical and the electrical forces push Na^+ into the cell (Figure 12-10a●).
- The chemical gradient for potassium ions tends to drive them out of the cell, but the movement is opposed by (1) attraction between K^+ and the negative charges on the inside of the cell mem-

brane and (2) repulsion between K^+ and the positive charges on the outside of the membrane (Figure 12-10b●).

An electrochemical gradient represents a form of *potential energy.* Potential energy is the energy of position, as in a stretched spring, a charged battery, or water behind a dam.

Without a cell membrane, diffusion would eliminate the electrochemical gradients. The cell membrane acts as a barrier that resists the electrochemical gradients that would otherwise tend to drive ions into or out of the cell. In effect, the cell membrane acts like a dam across a river. Without the dam, water would simply respond to gravity, flowing downstream, and its energy would be dissipated. With a dam in place, even a small opening will release water under tremendous pressure. Similarly, any stimulus that increases the membrane permeability to potassium, sodium, or other ions for which an electrochemical gradient exists will produce a sudden and dramatic ion movement. For example, an increase in sodium ion permeability will cause an immediate rush of Na^+ into the cell. The size of this current has nothing to do with the size or nature of the stimulus. The stimulus needs only to open the door, and existing electrochemical gradients do the rest.

POTASSIUM IONS AND THE RESTING POTENTIAL
If the cell membrane were freely permeable to potassium ions but impermeable to other positively charged ions, potassium ions would continue to leave the cell until the electrical gradient (holding K^+ inside the cell) was as large as the chemical gradient (pushing K^+ out of the cell). This equilibrium would occur at a transmembrane potential of around –90 mV, a value known as the *equilibrium potential* for potassium. Neuron membranes are very permeable to potassium, and this high permeability is the primary determinant of the resting potential. If other ion movements are prevented, the transmembrane potential of a neuron will reach –90 mV. Normally, however, the resting potential of a neuron is approximately –70 mV. The difference between the potassium equilibrium potential and the resting potential reflects primarily the small but significant membrane permeability to sodium ions.

SODIUM IONS AND THE RESTING POTENTIAL
If the cell membrane were freely permeable to sodium ions, they would continue to enter the cell until the interior of the cell membrane contained enough excess positive charges that the electrical gradient reversed itself. The interior would now be positively charged, and the repulsion between similar charges (+ and +) would prevent any further net movement of sodium ions into the cell. This event would occur at a transmembrane potential of around +66 mV. The resting potential is nowhere near this value, because the membrane permeability to sodium ions is very low. In a normal cell membrane, sodium permeability has only a modest effect on the resting potential, shifting it from –90 mV, the potassium equilibrium potential, to –70 mV.

The electrochemical gradients for potassium and sodium are the primary factors affecting the resting potential of a neuron. The contributions of the major negatively charged ions (Pr^- and Cl^-) can be ignored, because (1) proteins are too large to pass through the cell membrane, (2) the normal resting potential is near the equilibrium potential for chloride ions, and (3) the negative charges carried by the interior proteins repel the extracellular chloride ions.

Active Forces: Sodium-Potassium ATPase

At the normal resting potential, there is a slow leakage of sodium ions into the cell and a diffusion of potassium ions out of the cell. The cell must use ATP to eject those sodium ions and reclaim the lost potassium ions. As noted in Chapter 3, an ion pump, *sodium-potassium ATPase,* exchanges 3 intracellular sodium ions for 2 extracellular potassium ions. Because the sodium-potassium exchange pump removes 3 positive ions from the cytoplasm for every 2 reclaimed, it contributes to the negativity of the transmembrane potential. For this reason the sodium-potassium exchange pump is called an *electrogenic pump.* The effect is slight, however, and contributes only a few millivolts to the normal resting potential. The primary impact of this exchange pump is that it ejects sodium ions as quickly as they enter the cell. At –70 mV, the normal resting potential of a typical neuron, 3 sodium ions diffuse into the cell for every 2 potassium ions that diffuse out. Thus the activity of the exchange pump exactly balances the passive forces of diffusion, and the resting potential remains stable.

Summary

1. The resting potential reflects primarily the relatively high membrane permeability to potassium ions. It is therefore fairly close to –90 mV, the equilibrium potential for potassium ions.
2. Although the electrochemical gradient for sodium ions is very large, sodium permeability is very small. Consequently, sodium ions have only a small effect on the normal resting potential, making it somewhat less negative than it would otherwise be.
3. The sodium-potassium exchange pump stabilizes the resting potential when the ratio of passive sodium ion entry to passive potassium ion loss is 3:2.
4. The result of these passive and active factors is that a typical neuron has a resting potential of approximately –70 mV.

The transfer of information from neuron to neuron involves two integrated steps: (1) the propagation of one or more action potentials along an axon and (2) the chemical transmission of those signals across one or more synapses. In each instance, changes in membrane permeability, and resulting changes in the transmembrane potential, play a vital role.

❑ Ion Channels

Figures 12-7, 12-9, 12-11, 12-12

Ions cross the cell membrane via *membrane channels.* ∞ [p. 70] Our discussion will focus on the membrane permeability to sodium and potassium, as these are the primary determinants of the transmembrane potential of neurons. There are several types of sodium and potassium ion channels. **Passive channels,** or **leak channels,** are always open, although their permeability can vary from moment to moment as the proteins that make up the channel change shape in response to local conditions. Sodium and potassium leak channels are important in establishing the normal resting potential of the cell, as detailed above (Figure 12-9●).

Cell membranes also contain **active channels,** often called **gated channels.** Gated channels are ion channels that open or close in response to spe-

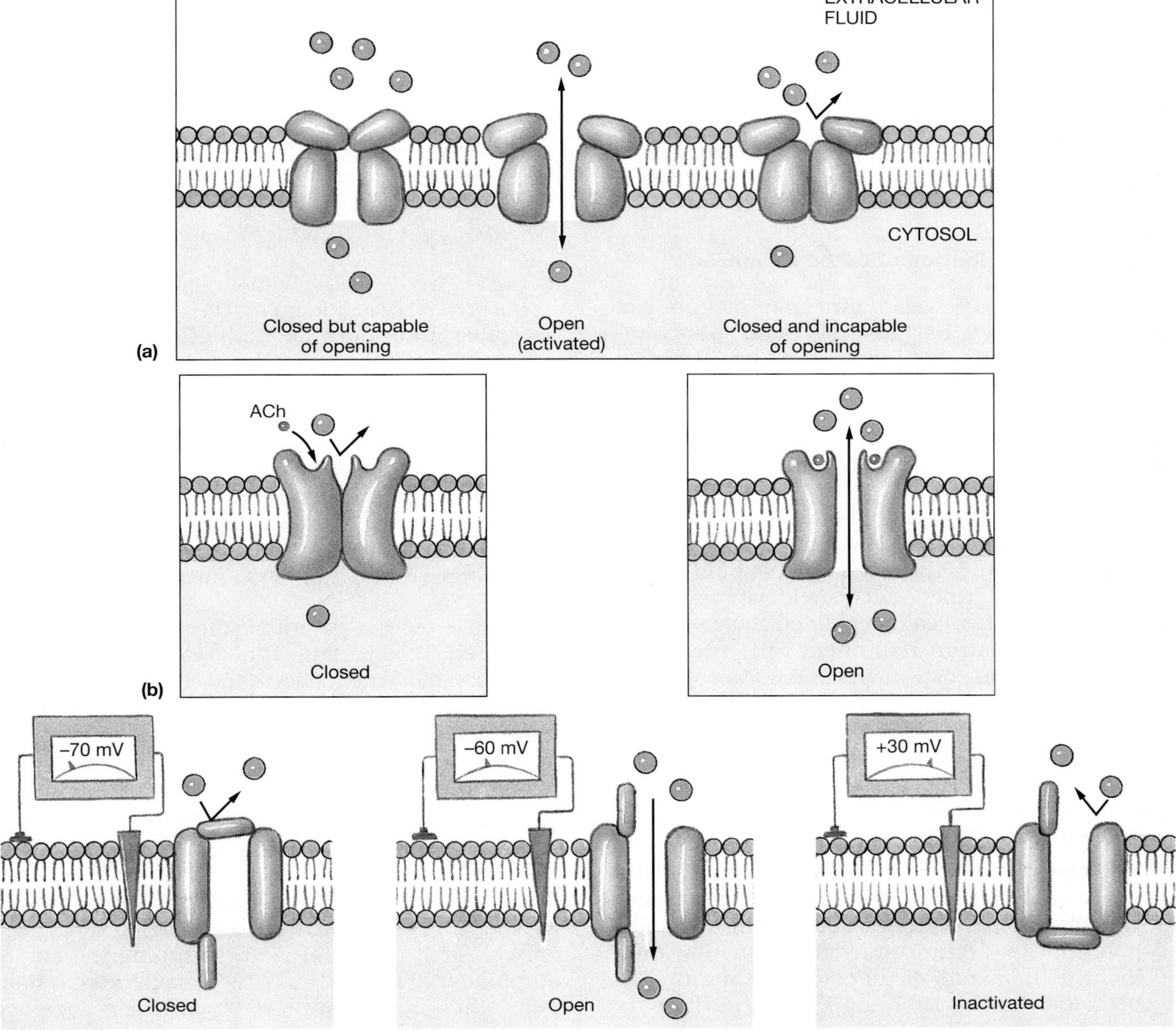
EXTRACELLULAR FLUID
CYTOSOL
Closed but capable of opening
Open (activated)
Closed and incapable of opening
(a)
ACh
Closed
Open
(b)
–70 mV
Closed
–60 mV
Open
+30 mV
Inactivated
(c)

● **FIGURE 12-11**

Gated Channels. (a) Possible configurations of a gated ion channel. **(b)** A chemically regulated channel for Na^+ that opens in response to the presence of ACh at a binding site. **(c)** A voltage-regulated sodium channel that responds to changes in the transmembrane potential.

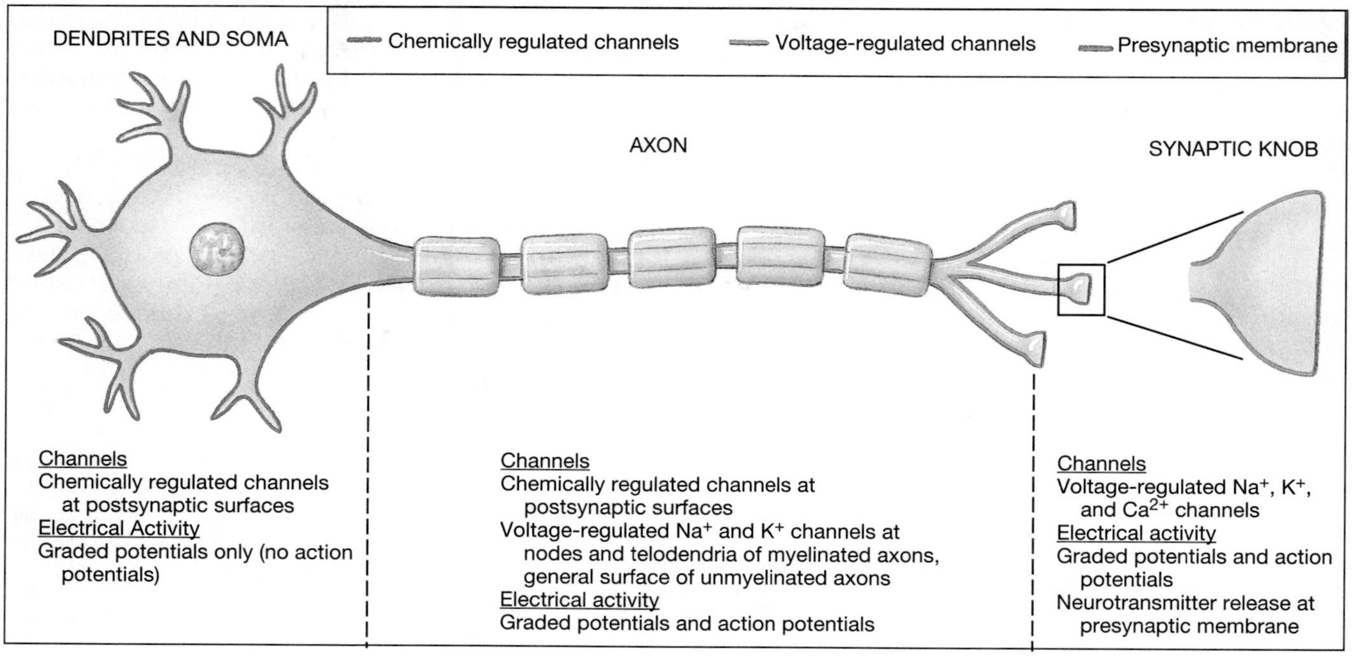

DENDRITES AND SOMA
Chemically regulated channels
Voltage-regulated channels
Presynaptic membrane
AXON
SYNAPTIC KNOB
Channels
Chemically regulated channels at postsynaptic surfaces
Electrical Activity
Graded potentials only (no action potentials)
Channels
Chemically regulated channels at postsynaptic surfaces
Voltage-regulated Na^+ and K^+ channels at nodes and telodendria of myelinated axons, general surface of unmyelinated axons
Electrical activity
Graded potentials and action potentials
Channels
Voltage-regulated Na^+, K^+, and Ca^{2+} channels
Electrical activity
Graded potentials and action potentials
Neurotransmitter release at presynaptic membrane

● **FIGURE 12-12**
The General Distribution of Gated Channels

cific stimuli. Each gated channel can be in one of three states (Figure 12-11a●): (1) closed but capable of opening, (2) open **(activated),** or (3) closed and incapable of opening **(inactivated).**

There are three classes of gated channels, *chemically regulated channels, voltage-regulated channels,* and *mechanically regulated channels.*

1. **Chemically regulated channels** open or close when they bind specific extracellular chemicals (Figure 12-11b●). The receptors that bind acetylcholine at the junctional folds (Chapter 10) are examples of chemically regulated channels. [p. 295] A neuron has chemically regulated channels on its postsynaptic membranes, where the receptor proteins are exposed to neurotransmitters released by other neurons. Chemically regulated channels are most abundant on the dendrites and soma of a neuron, as most synaptic communication occurs in these areas (Figure 12-12●).
2. **Voltage-regulated channels** are characteristic of areas of **excitable membrane,** a membrane capable of generating and conducting an action potential. Examples of excitable membranes include the axons of unipolar and multipolar neurons (Figure 12-7●, p. 389) and the sarcolemma (including T tubules) of skeletal and cardiac muscle fibers. Voltage-regulated channels open or close in response to changes in the transmembrane potential (Figure 12-11c●). The opening of voltage-regulated sodium channels is the first step in the generation of an action potential. These channels have two gates—an *activation gate* and an *inactivation gate.* The two gates function independently. In the resting membrane, the activation gate is closed. When the activation gate is open, sodium entry can be stopped by closing the inactivation gate.
3. **Mechanically regulated channels** open or close in response to physical distortion of the membrane surface. Such channels are important in sensory receptors that respond to touch, pressure, and vibration. These receptors will be discussed in detail in Chapter 17.

At the resting potential, most gated channels are closed. The opening of gated channels alters the rate of ion movement across the cell membrane and changes the transmembrane potential. The result may be a *graded potential* or an *action potential.*

❑ Graded Potentials

Figures 12-13, 12-14

Graded potentials, or *local potentials,* are changes in the transmembrane potential that cannot spread far from the area surrounding the site of stimulation. Figure 12-13a,b● considers what happens when a membrane is exposed to a chemical that opens chemically regulated sodium channels. Sodium ions enter the cell, and the arrival of additional positive charges shifts the transmembrane potential toward 0 mV. Any shift in the transmembrane potential away from resting levels toward 0 mV or above is called a **depolarization.** Thus it includes changes in potential from –70 mV

(a) Resting membrane with closed chemically regulated sodium channels

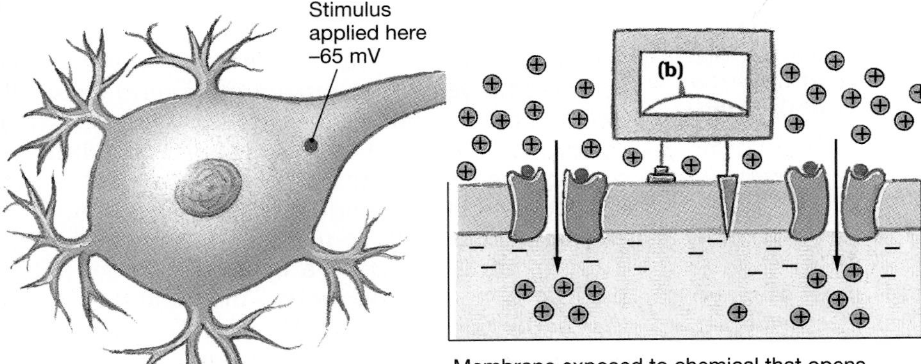
Stimulus
applied here
–65 mV
(b)

Membrane exposed to chemical that opens channels

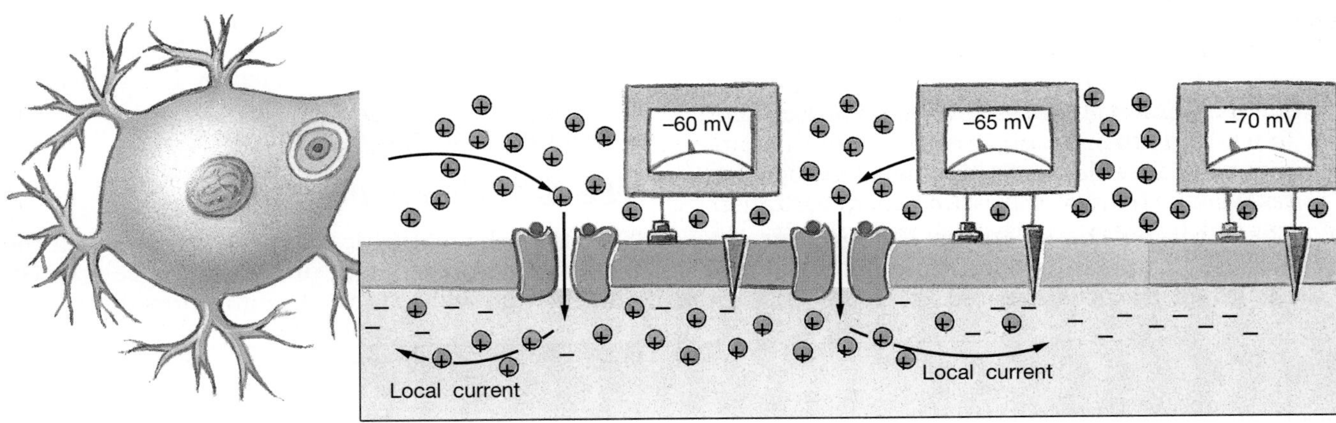
–60 mV
–65 mV
–70 mV
Local current
Local current

(c) Spread of sodium ions inside cell membrane produces a local current that depolarizes adjacent portions of the cell membrane

● **FIGURE 12-13**
Graded Potentials.
(a) Chemically regulated sodium channels are closed at the normal resting potential.
(b) Opening these channels leads to a sudden influx of sodium ions, which depolarizes the membrane. **(c)** Local currents develop as sodium ions on the outside of the membrane diffuse into the area, replacing those moving into the cell, and sodium ions inside the cell spread along the inner surface of the membrane. Note that the depolarizing effect radiates in all directions away from the source of stimulation. (For clarity, only gated channels are shown; leak channels as shown in Figure 12-9 are present but are not responsible for the production of graded potentials.)

to smaller negative values (–65 mV, –45 mV, –10 mV) as well as to membrane potentials past 0 mV (+10 mV, +30 mV).

The movement of sodium ions across the cell membrane in turn causes the depolarization of the surrounding membrane. Sodium ions entering the cell spread along the inner surface, attracted by the excess of negative charges (Figure 12-13c●). The arrival of these positive ions reduces the imbalance between negative and positive charges on the inside of the membrane. At the same time, extracellular sodium ions move toward the open channels, replacing those entering the cell. The ion movement parallel to the inner and outer surfaces of a membrane is called a **local current.**

In the area affected by the local current, there are fewer excess negative charges inside and fewer excess positive charges outside. In other words, the affected cell membrane depolarizes. The degree of depolarization decreases with distance, because the sodium ions entering at one point spread out in all directions along the inner surface of the membrane (Figure 12-13●). *The change in the transmembrane potential and the area affected by the local currents are directly related to the number of sodium channels opened by the chemical stimulus.* The more open channels, the more sodium ions enter and spread along the inner surface of the membrane, and the greater the depolarization.

Opening a chemically gated potassium channel would have the opposite effect: The rate of potassium outflow would increase, and the interior of the cell would lose additional positive ions. The loss of positive ions produces **hyperpolarization,** a shift in the resting potential away from 0 mV, perhaps to –80 mV or more. Again, a local current would distribute the effect to adjacent portions of the cell membrane.

Both depolarization and hyperpolarization move the transmembrane potential away from resting levels. When a chemical stimulus is removed and normal membrane permeability is restored, the transmembrane potential soon returns to resting levels (Figure 12-14●). The process of restoring the normal resting potential after depolarization is called **repolarization.**

Graded potentials, whether depolarizing or hyperpolarizing, share four basic characteristics:

1. The transmembrane potential is most affected at the site of stimulation, and the effect decreases with distance.
2. The effect spreads passively because of local currents.
3. The graded potential change may involve either depolarization or hyperpolarization. The nature of the change is determined by the properties of the membrane channels involved. For example, in a resting membrane, opening sodium channels will cause depolarization, whereas opening potassium channels will cause hyperpolarization.
4. The stronger the stimulus, the greater the change in the transmembrane potential and the larger the area affected.

Graded potentials occur in the membranes of many different cell types. For example, the depolarization of the junctional folds by ACh at a neuromuscular junction is a graded potential that may result in the appearance of an action potential in adjacent portions of the sarcolemma. Graded potentials also occur in epithelial cells, gland cells, fat cells, and a variety of sensory receptors. The dendrites, soma, and presynaptic surfaces of neurons support only graded potentials. Action potentials are restricted to areas of excitable membrane, which contain voltage-regulated channels.

❑ Action Potentials

Action potentials are propagated changes in the transmembrane potential that, once initiated, spread across an entire excitable membrane. In a representative neuron, an action potential usually begins at the initial segment of the axon. It is then conducted along the length of the axon, ultimately reaching the synaptic knobs. The stimulus that initiates an action potential is a depolarization large enough to open voltage-regulated sodium channels. That opening occurs at a transmembrane potential known as the **threshold.** Threshold for an axon is usually between –60 mV and –55 mV. A stimulus that lowers the resting membrane potential from –70 mV to –65 mV or –62 mV will not produce an action potential, only a graded depolarization. When such a stimulus is removed, the transmembrane potential returns to resting levels. Depolarization of the initial segment is caused by local currents resulting from the graded depolarization of the soma, especially the axon hillock.

The All-or-None Principle

The initial depolarization acts like pressure on the trigger of a gun. A slight pressure can be applied, and the gun will not fire. The gun fires only when

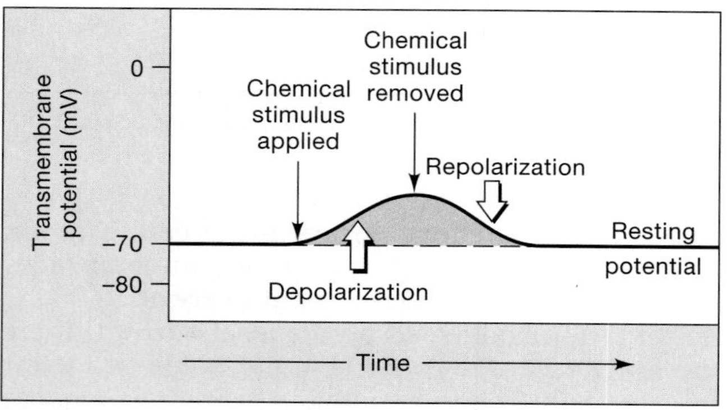
Transmembrane potential (mV)
0
-70
-80
Chemical stimulus applied
Chemical stimulus removed
Repolarization
Resting potential
Depolarization
Time

(a)

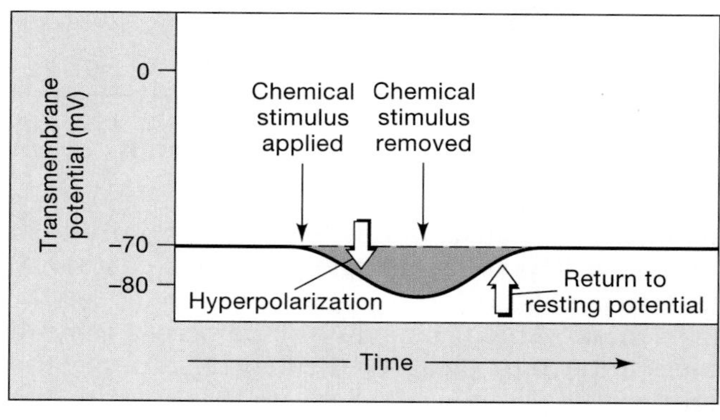
Transmembrane potential (mV)
0
-70
-80
Chemical stimulus applied
Chemical stimulus removed
Hyperpolarization
Return to resting potential
Time

(b)

● **FIGURE 12-14**

Depolarization and Hyperpolarization. (a) Depolarization and repolarization in response to the addition and removal of a stimulus that opens chemically regulated sodium channels. **(b)** Hyperpolarization in response to the addition of a stimulus that opens chemically regulated potassium channels. Upon removal of the stimulus, the membrane potential returns to the resting level.

a certain minimum pressure is applied to the trigger. Once the trigger pressure reaches this preset level, however, the firing pin drops and the gun discharges. At that point, it no longer matters whether the pressure was applied gradually or suddenly, or whether the shooter is a 4-year-old who can barely squeeze the trigger or a champion weight lifter with a crushing grip. The bullet that leaves the gun always has the same speed and range, regardless of the forces that were applied to the trigger.

In the case of an axon or another area of excitable membrane, a graded depolarization is the pressure on the trigger and the action potential is the firing of the gun. Any stimulus that brings the membrane to threshold will generate an identical action potential. In other words, *the properties of the action potential are independent of the relative strength of the depolarizing stimulus* as long as it exceeds threshold. This concept is called the **all-or-none principle,** because a given stimulus either triggers a typical action potential or does not produce one at all. If an action potential *is* produced, it will be propagated over the entire surface of the excitable membrane. The all-or-none principle applies to all excitable membranes. An action potential is *generated* at one site as the result of a localized depolarization. It is then *propagated* away from that site. We will consider each of these processes individually.

Action Potential Generation

Figures 12-15, 12-16

The steps involved in the generation of an action potential are shown in Figure 12-15•.

Step 1: ***Depolarization to Threshold*** (Figure 12-15b•)

Before an action potential can begin, an area of excitable membrane must be depolarized to threshold levels.

Step 2: ***Activation of Sodium Channels and Rapid Depolarization*** (Figure 12-15c•)

The next step is the activation of voltage-regulated sodium channels. At the normal resting potential the activation gate is all that prevents sodium entry. The activation gates open at threshold, and the cell membrane becomes much more permeable to sodium ions. Driven by the large electrochemical gradient, sodium ions rush into the cytoplasm, and rapid depolarization occurs at this site. In less than a millisecond the inner membrane surface contains more positive ions than negative ones, and the transmembrane potential has changed from –70 mV to +30 mV, a value much closer to the equilibrium potential for sodium.

Notice that the first two steps in action potential generation are an example of positive feedback, with a small depolarization acting as the trigger to produce a larger depolarization.

Step 3: ***Sodium Channel Inactivation and Potassium Channel Activation*** (Figure 12-15d•)

As the transmembrane potential approaches +30 mV the voltage-regulated sodium channels again change shape, and the inactivation gates begin closing. This step is known as **sodium channel inactivation.** At the same time that sodium channel inactivation is under way, voltage-regulated potassium channels are opening. At a transmembrane potential of +30 mV, the cytosol along the interior of the membrane contains an excess of positive charges. Thus, in contrast to the situation in the resting membrane (p. 403), both the electrical *and* chemical gradients favor potassium movement out of the cell. The result is the movement of potassium ions out of the cytoplasm and across the cell membrane. The sudden loss of positive charges then shifts the transmembrane potential back toward resting levels, and repolarization begins.

Step 4: ***The Return to Normal Permeability*** (Figure12-15e,f•)

The voltage-regulated sodium channels remain inactivated until the membrane has repolarized to threshold, around –60 mV. At this time they regain their normal status (closed but capable of opening). The voltage-regulated potassium channels begin closing as the membrane reaches the normal resting potential (around –70 mV), but the process takes about a millisecond. Over that period, potassium continues to move out of the cell at a faster rate than at rest, producing a brief hyperpolarization that brings the transmembrane potential very close to the equilibrium potential for potassium (–90 mV). As the voltage-regulated potassium channels close, the transmembrane potential returns to normal resting levels. The membrane is now in its prestimulation condition, and the action potential is over.

Table 12-2 and Figure 12-16a• summarize the membrane events during an action potential, and Figure 12-16b• shows the changes in membrane permeability that occur. Table 12-3, p. 402, summarizes important differences between graded potentials and action potentials.

THE REFRACTORY PERIOD From the time an action potential begins until the normal resting potential has stabilized, the membrane will not respond normally to additional depolarizing stimuli. This period is known as the **refractory period** of the membrane. From the moment that the voltage-regulated sodium channels open at threshold until the period of sodium channel inactivation ends, the membrane cannot respond to further stimulation because all of the voltage-regulated sodium channels are already either open or closed and incapable of opening. This is the **absolute refractory period.** The **relative refractory**

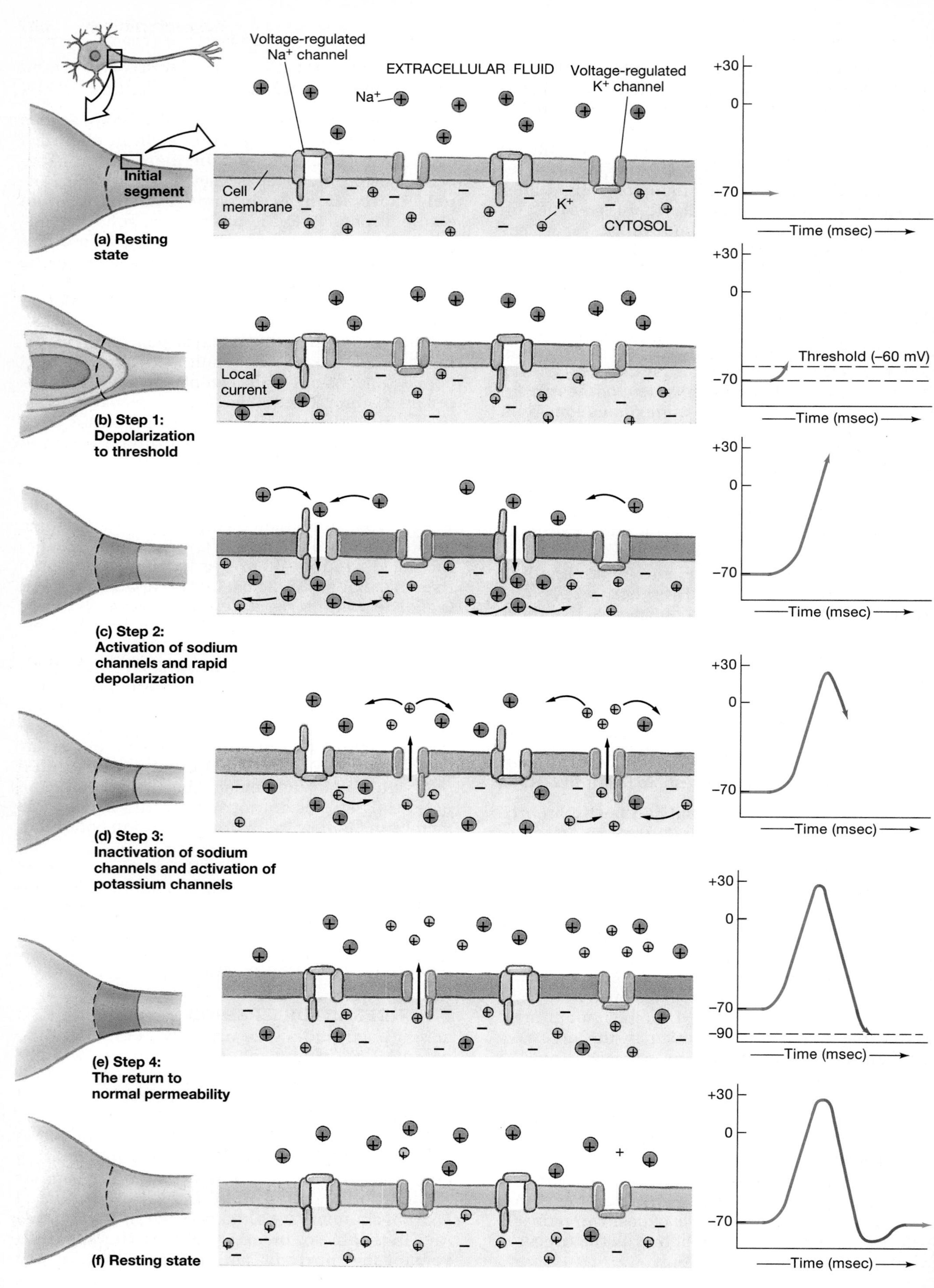

Voltage-regulated Na+ channel
EXTRACELLULAR FLUID
Voltage-regulated K+ channel
Na+
Initial segment
Cell membrane
K+
CYTOSOL
(a) Resting state
+30
0
–70
Time (msec)
(b) Step 1: Depolarization to threshold
Local current
Threshold (–60 mV)
(c) Step 2: Activation of sodium channels and rapid depolarization
(d) Step 3: Inactivation of sodium channels and activation of potassium channels
(e) Step 4: The return to normal permeability
–90
(f) Resting state

● **FIGURE 12-15 (Left)**

Generation of an Action Potential. (a) Resting condition. **(b)** ***Step 1:*** Depolarization to threshold. Local currents associated with a graded depolarization of the soma depolarize the proximal portion of the initial segment to –60 mV. **(c)** ***Step 2:*** Activation of sodium channels and rapid depolarization. The influx of sodium ions drives the membrane potential to +30 mV. **(d)** ***Step 3:*** Sodium channel inactivation and potassium channel activation. Potassium ions leave the cell, and repolarization begins. **(e)** ***Step 4:*** Return to normal permeability. At threshold, the sodium channels regain their resting condition, and the potassium channels begin closing. The transmembrane potential continues to drop until the voltage-regulated potassium channels have closed. The transmembrane potential then returns to resting levels **(f)**. For clarity, only gated channels are shown; leak channels as shown in Figure 12-9 are present but do not participate in the generation of an action potential. D

TABLE 12-2 Steps in the Generation of an Action Potential

Step 1:

- A graded depolarization brings an area of excitable membrane to threshold.

Step 2:

- The gates in voltage-regulated sodium channels open (sodium channel activation).
- Sodium ions, driven by charge attraction and the concentration gradient, flood into the cell.
- The transmembrane potential goes from –70 mV, the resting level, to +30 mV.

Step 3:

- The sodium channel gates close (sodium channel inactivation).
- Voltage-regulated potassium channels open, and potassium ions move out of the cell.
- Repolarization begins.

Step 4:

- Voltage-regulated sodium channels regain their normal properties when threshold is reached (–60 mV).
- The voltage-regulated potassium channels begin closing at –70 mV. Because they close slowly, a temporary hyperpolarization occurs. When all the voltage-regulated potassium channels have closed, the membrane regains its normal properties.

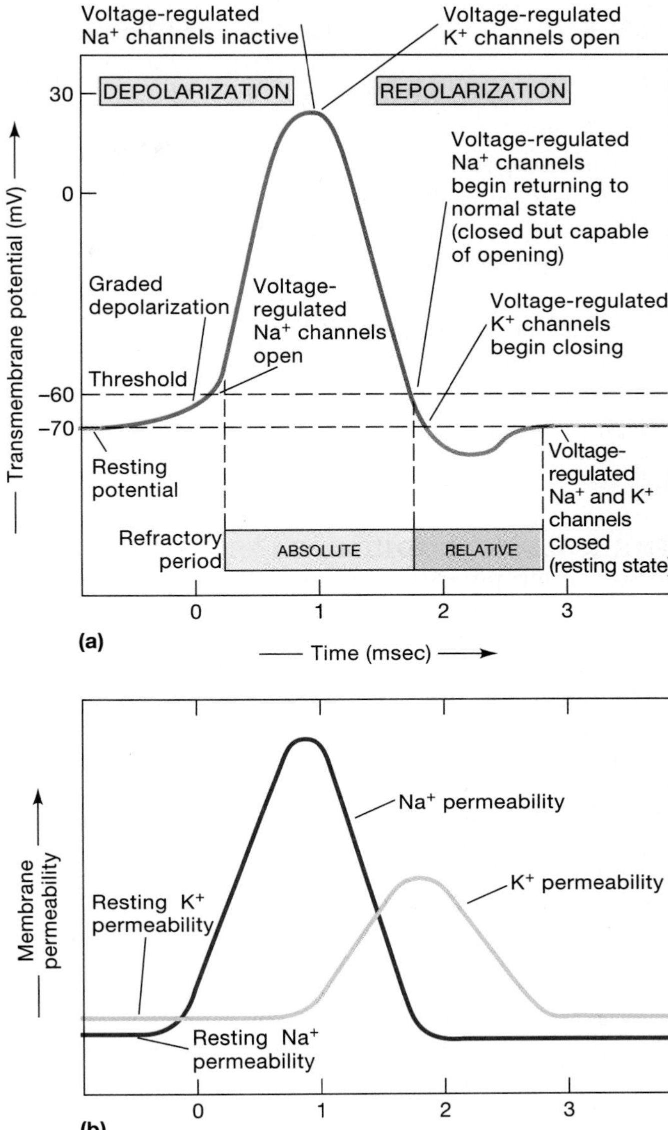
Voltage-regulated Na+ channels inactive
Voltage-regulated K+ channels open
DEPOLARIZATION
REPOLARIZATION
30
0
-60
-70
Transmembrane potential (mV)
Voltage-regulated Na+ channels begin returning to normal state (closed but capable of opening)
Graded depolarization
Voltage-regulated Na+ channels open
Voltage-regulated K+ channels begin closing
Threshold
Resting potential
Voltage-regulated Na+ and K+ channels closed (resting state)
Refractory period
ABSOLUTE
RELATIVE
0
1
2
3
(a)
Time (msec)
Membrane permeability
Na+ permeability
K+ permeability
Resting K+ permeability
Resting Na+ permeability
0
1
2
3
(b)
Time (msec)

● **FIGURE 12-16**

Changes in the Transmembrane Potential and Membrane Permeabilities during an Action Potential. (a) Changes in the transmembrane potential that occur during an action potential, and the factors involved. **(b)** Permeability changes in the membrane during an action potential.

period begins when the sodium channels regain their normal resting condition and continues until the transmembrane potential stabilizes at normal resting levels. Over this period voltage-regulated potassium channels are open. Another action potential can begin if the membrane is depolarized to threshold, but that depolarization requires a larger-than-normal depolarizing stimulus. The depolarizing stimulus must be larger than normal because (1) the local current must deliver enough sodium ions to counteract the loss of positively charged potassium ions, and (2) through most of the relative refractory period, the membrane is hyperpolarized to some degree.

NEUROTOXINS Potentially deadly forms of human poisoning result from eating seafood containing *neurotoxins*, poisons that affect primarily neurons. Several neurotoxins, such as *tetrodotoxin* (TTX), prevent sodium from entering voltage-regulated sodium channels. The result is an inability to generate action potentials. Motor neurons cannot function under these conditions, and death may result from paralysis of the respiratory muscles. AM *Neurotoxins in Seafood*

TABLE 12-3 Comparison of Graded Potentials and Action Potentials

Graded Potentials	*Action Potentials*
May be depolarizing or hyperpolarizing	Always depolarizing
No threshold value	Must depolarize to threshold before action potential begins
Amount of depolarization or hyperpolarization depends on intensity of stimulus	All-or-none phenomenon; any stimulus that exceeds threshold will produce an identical action potential
Passive spread from site of stimulation	Action potential at one site depolarizes adjacent sites to threshold
Effect on membrane potential decreases with distance from stimulation site	Propagated across entire membrane surface without decrease in strength
No refractory period	Refractory period
Occur in most cell membranes	Occur only in excitable membranes of neurons and muscle cells

THE ROLE OF SODIUM-POTASSIUM ATPase In the action potential detailed above, depolarization resulted from the influx of sodium ions and repolarization involved the loss of potassium ions. Over time, the sodium-potassium exchange pump, more formally known as *sodium-potassium ATPase,* returns intracellular and extracellular ion concentrations to prestimulation levels. Compared with the total number of ions inside and outside the cell, however, the number involved in a single action potential is relatively insignificant. *Tens of thousands of action potentials can occur before the intracellular ion concentrations change enough to disrupt the entire mechanism.* Thus the sodium-potassium exchange pump is not essential to any single action potential.

However, a maximally stimulated neuron may generate action potentials at a rate of 1000 per second. Under these circumstances the exchange pump is needed if ion concentrations are to remain within acceptable limits over a prolonged period. If sodium-potassium ATPase is inactivated by a metabolic poison, or if the cell runs out of ATP, a neuron will soon lose its ability to function.

Action Potential Propagation

Figures 12-17, 12-18

The sequence of events described above occurs in a relatively small portion of the total membrane surface. But we have already noted that unlike graded potentials, which diminish rapidly with distance, an action potential affects the entire excitable membrane. To understand the basic principle involved, imagine that you are standing by the doors of a movie theater at the start of a long line. Everyone is waiting for the doors to open. The manager steps outside and says to you, "Let everyone know that we are opening in 15 minutes." How would you spread the news? If you treated the line as an inexcitable membrane, you would shout, "The doors open in 15 minutes!" as loudly as you could. The closest people in the line would hear the news very clearly, but those farther away might not hear the entire message, and those at the end of the line probably wouldn't hear you at all. If you treated the crowd as an excitable membrane, you would give the message to the next person in line, with instructions to pass it on. In this way the message would travel along the line until everyone had heard the news. Such a message "moves" as each person repeats it to someone else. Distance is not a factor—the line can contain 50 people or 5000.

This situation is comparable to the way an action potential spreads across an excitable membrane. An action potential (message) is relayed from one location to another in a series of steps. At each step the message is repeated. Because the same events take place over and over, the term *propagation* is preferable to the term *conduction.*[1]

CONTINUOUS PROPAGATION The basic mechanism of action potential propagation along an unmyelinated axon is shown in Figure 12-17•. For convenience, we will consider the membrane as a series of adjacent segments. The action potential begins at the initial segment (Figure 12-17a•). For a brief moment at the peak of the action potential, the transmembrane potential becomes positive rather than negative. A local current then develops (Figure 12-17b•), and sodium ions begin moving in the cytoplasm and in the extracellular fluid. The local current spreads in all directions, depolarizing adjacent portions of the membrane. The axon hillock cannot respond with an action potential, but when the adjacent segment of the axon is depolarized to threshold, an action potential devel-

[1]In electrical conduction, there is a flow of electrons along a *conductor,* such as a copper wire. Axons are relatively poor conductors of electricity.

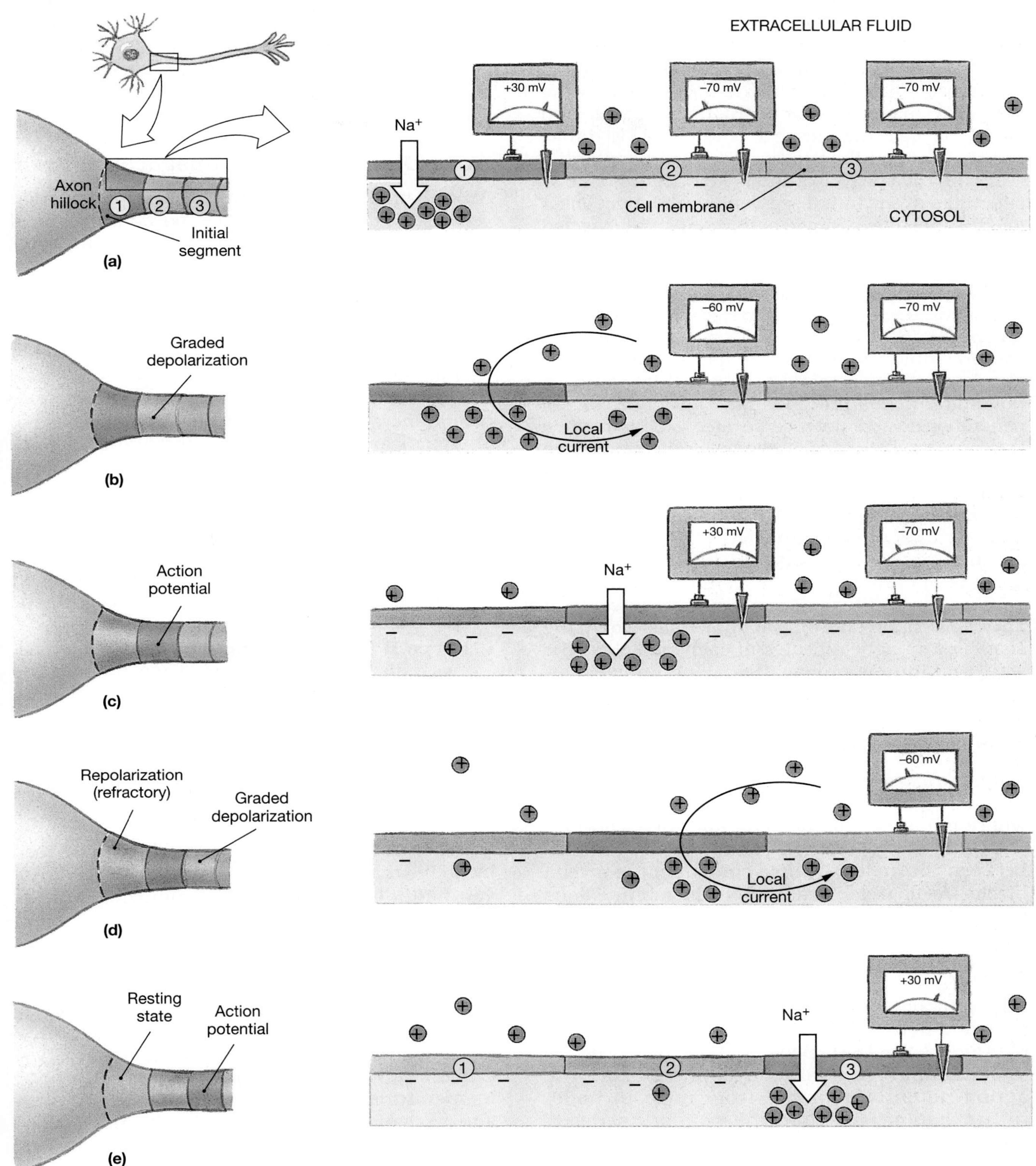
EXTRACELLULAR FLUID
+30 mV
−70 mV
−70 mV
Na+
1
2
3
Axon hillock
1 2 3
Initial segment
Cell membrane
CYTOSOL
(a)
−60 mV
−70 mV
Graded depolarization
Local current
(b)
+30 mV
−70 mV
Na+
Action potential
(c)
−60 mV
Repolarization (refractory)
Graded depolarization
Local current
(d)
+30 mV
Na+
Resting state
Action potential
(e)

● **FIGURE 12-17**

Action Potential Propagation along an Unmyelinated Axon. The axon can be considered as a series of adjacent segments. **(a)** As an action potential develops in the initial segment, the transmembrane potential depolarizes to +30 mV. **(b)** A local current depolarizes the adjacent portion of the membrane to threshold. **(c)** An action potential then develops in the affected membrane. **(d)** A local current then depolarizes the next segment of the axon, and an action potential develops. **(e)** The local current always drives the action potential forward because the previous segment is still in the absolute refractory period.

ops there. This process continues in a chain reaction (Figure 12-17c,d,e●), until the most distant portions of the cell membrane have been affected. As in our "crowd" model, the message is being relayed from one location to another. At each step along the way, the message is retold, so distance has no effect on the process. The action potential reaching the synaptic knob is identical to the one generated at the initial segment, and the net effect is the same as if a single action potential traveled across the membrane surface. This form of action potential propagation is known as **continuous propagation,** or **continuous conduction.**

Each time a local current develops, the action potential moves forward, not backward, because the previous segment of the axon is still in the absolute refractory period. As a result, *an action potential always proceeds away from the site of generation and cannot reverse direction.* By the time the refractory period in a segment of the membrane has ended, the action potential is too far away for local currents to have any effect. If there is to be a second action potential at the same site, a second stimulus must be applied.

In continuous propagation an action potential appears to move across the membrane surface in a series of tiny steps. Even though the events at any one location take only about a millisecond, the sequence must be repeated at each step along the way. Continuous propagation along unmyelinated axons occurs at speeds of about 1 meter per second (approximately 2 mph).

SALTATORY PROPAGATION In a myelinated axon, the axolemma is wrapped in layers of myelin. This wrapping is complete except at the nodes, where adjacent glial cells contact one another. Continuous propagation cannot occur, because between the nodes the lipid content of the myelin blocks the flow of ions across the membrane. Ions can cross the cell membrane only at the nodes. As a result, *only the nodes can respond to a depolarizing stimulus.*

When an action potential appears at the initial segment, the local current skips the internodes and depolarizes the closest node to threshold (Figure 12-18●). Because the nodes may be 1–2 mm apart in a large myelinated axon, the action potential "jumps" from node to node, rather than moving along the axon in a series of tiny steps. This process is called **saltatory propagation,** or **saltatory conduction** (Latin *saltare*, leaping). Saltatory propagation could be compared to relaying a message along a line of people spaced 5 meters apart. Each person shouts the message to the next person in line; by the time the message has been repeated four times, it has moved 20 meters. By comparison, in our model of continuous propagation, where people were closely packed, by the time the message had been repeated four times it would have moved only a few meters. Saltatory propagation in the CNS and PNS carries nerve impulses along an axon five to seven times more rapidly than does continuous propagation.

Axon Diameter and Propagation Speed

Axon diameter also has an effect on propagation velocity. To depolarize adjacent portions of the cell membrane, ions must move through the cytoplasm. Cytoplasm also offers resistance to ion movement, although the resistance is much less than that of the cell membrane. In this instance, an axon behaves like an electrical cable: the larger the diameter, the lower the resistance. That is why motors with large current demands, such as the starter on a car, an electric stove, or a big air conditioner, use such big wires.

Axons are classified according to the relationships among diameter, myelination, and propagation speed.

- **Type A fibers** are the largest axons, with diameters ranging from 4 to 20 μm. These are myelinated axons that carry action potentials at speeds of up to 140 meters per second, the equivalent of over 300 mph.
- **Type B fibers** are smaller myelinated axons, with diameters of 2–4 μm. Their propagation speeds average around 18 meters per second, or roughly 40 mph.
- **Type C fibers** are unmyelinated and under 2 μm in diameter. These axons propagate action potentials at the leisurely pace of 1 meter per second, a mere 2 mph.

The relative importance of myelin can be seen by noting that in going from Type C to Type A fibers you find a tenfold increase in diameter, but the propagation speed increases by 140 times!

Type A fibers carry sensory information to the CNS concerning position, balance, and delicate touch and pressure sensations from the surface of the skin. The motor neurons that control skeletal muscles also send their commands over large, myelinated Type A axons. Type B fibers and Type C fibers carry less urgent information concerning temperature, pain, and general touch and pressure sensations to the CNS and carry instructions to smooth muscle, cardiac muscle, glands, and other peripheral effectors.

When we need to tell a friend urgent news or receive an immediate response, we usually pick up the telephone. For general correspondence we usually send a first-class letter, which allows a swift but not immediate response. If we have to distribute an enormous volume of information to a huge mailing list, and there is no particular rush, bulk mail offers efficiency at a considerable savings. Instead of rep-

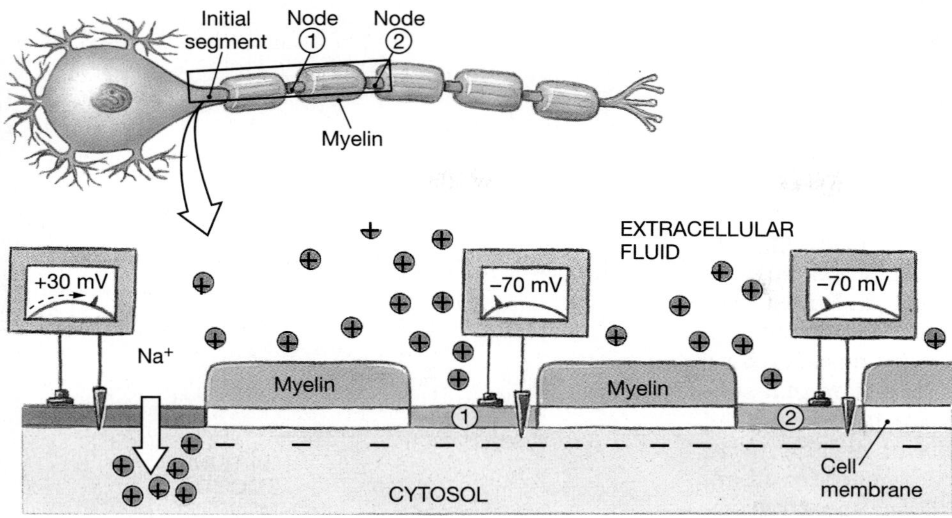
Initial segment
Node ①
Node ②
Myelin
+30 mV
EXTRACELLULAR FLUID
–70 mV
–70 mV
Na+
Myelin
Myelin
①
②
CYTOSOL
Cell membrane

(a) Action potential at initial segment

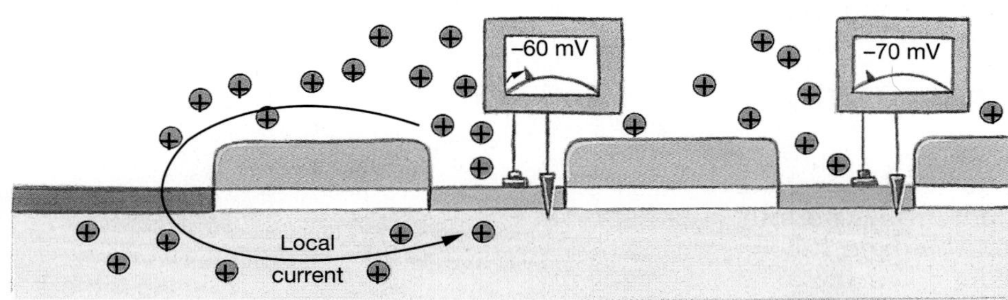
–60 mV
–70 mV
Local current

(b) Depolarization to threshold at node 1

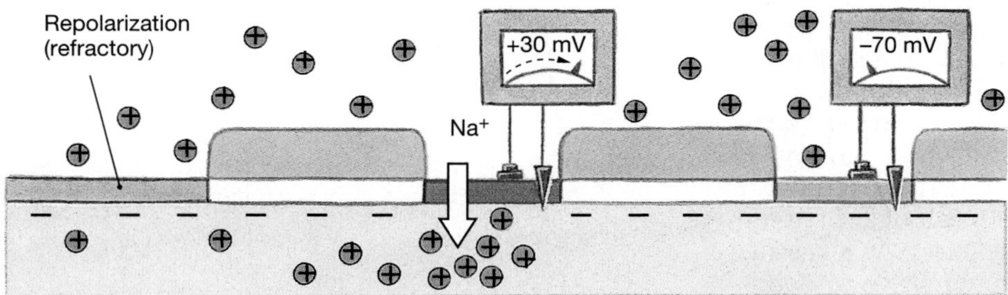
Repolarization (refractory)
+30 mV
–70 mV
Na+

(c) Action potential at node 1

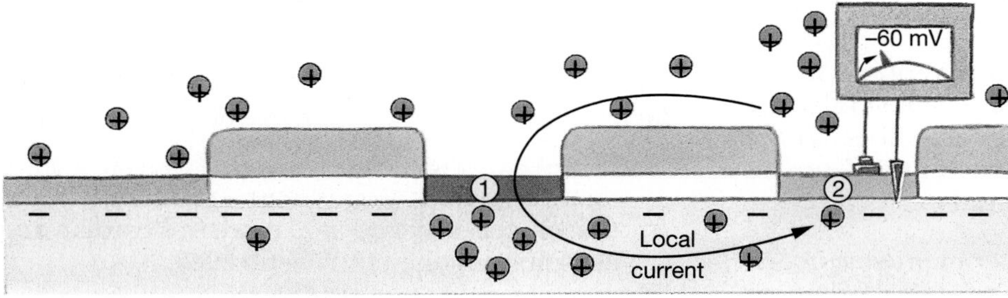
–60 mV
①
②
Local current

(d) Depolarization to threshold at node 2

FIGURE 12-18

Saltatory Propagation along a Myelinated Axon. (a) An action potential develops in the initial segment. **(b)** Local current depolarizes the next node to threshold. **(c)** An action potential develops at the node. **(d)** A local current depolarizes the membrane at the next node. This process will continue along the length of the axon.

resenting a compromise between time and money, information transfer within the nervous system reflects a compromise between time and space. Axons carry the information, and the larger the axon, the faster the rate of transmission. But if all sensory information were carried by Type A fibers, the peripheral nerves would be the size of garden hoses and the spinal cord would have the diameter of a garbage can. Instead, only around one-third of all axons carrying sensory information are myelinated, and most sensory information arrives over slender Type C fibers.

Myelination improves coordination and control by decreasing the time between reception of a sensation and initiation of an appropriate response. Myelination begins relatively late in development, and the myelination of sensory and motor fibers is not completed until early adolescence. In growing children, the pace of myelination and the pathways involved can be quite variable. This variability contributes to the observed range of physical capabilities within a given age group.

The Direction of Action Potential Propagation

Most synapses occur on the dendrites and soma of a multipolar neuron. Action potentials usually begin at the boundary between the axon hillock and the initial segment, traveling down the axon toward the synaptic knobs. Propagation in this case occurs in one direction only. However, synapses may also occur at a node or at a synaptic knob. An action potential originating at a node will be propagated along the axon in both directions; it will travel toward the soma as well as toward the telodendria. The action potential headed for the telodendria will depolarize the synaptic knob and cause the release of neurotransmitter. The one headed toward the soma will have no effect, as it is not propagated past the initial segment of the axon. As a result, the soma and dendrites remain sensitive to depolarizing or hyperpolarizing stimuli.

Action Potentials in Muscle Tissues

Figure 12-19

Figure 12-19● compares action potentials in the excitable membranes of a neuron, a skeletal muscle fiber, a multiunit smooth muscle fiber, and a cardiac muscle cell. Several important differences exist among these action potentials.

- ***Resting potentials are greater in muscle fibers.*** The resting potential of a skeletal, cardiac, or multiunit smooth muscle fiber is –85 mV to –90 mV, close to the potassium equilibrium potential. Threshold values are comparable to those of axons.
- ***Action potentials last longer in muscle fibers than in axons.*** An action potential in an axon may last 0.5 msec, versus 7.5 msec in a skeletal muscle fiber, 50 msec in a multiunit smooth muscle cell, and 250–300 msec in a cardiac muscle cell from one of the ventricles.
- ***Despite the large diameters of muscle fibers, action potentials are propagated at a relatively slow speed.*** Action potentials of muscle fibers travel via continuous propagation at speeds of 3–5 meters per second (6–10 mph).

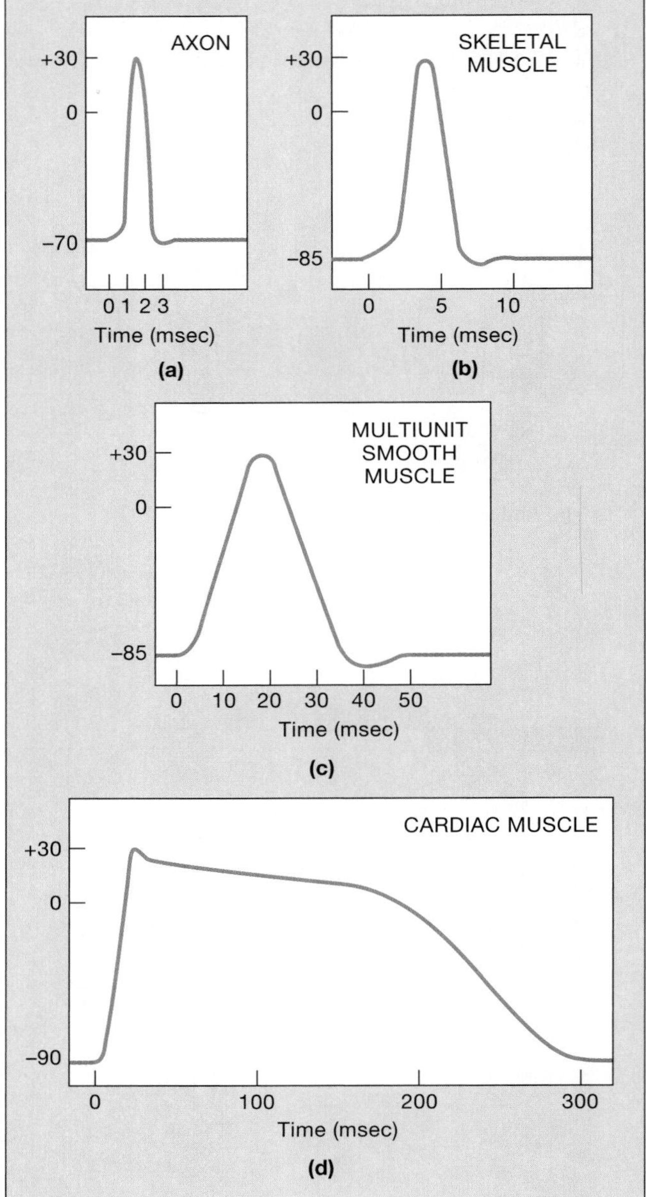
AXON
+30
0
–70
0 1 2 3
Time (msec)
(a)
SKELETAL MUSCLE
+30
0
–85
0
5
10
Time (msec)
(b)
MULTIUNIT SMOOTH MUSCLE
+30
0
–85
0 10 20 30 40 50
Time (msec)
(c)
CARDIAC MUSCLE
+30
0
–90
0
100
200
300
Time (msec)
(d)

● **FIGURE 12-19**
Action Potentials in Nerve and Muscle. An action potential as recorded in an axon **(a)** is compared with **(b)** an action potential in the sarcolemma of a skeletal muscle fiber, **(c)** a multiunit smooth muscle fiber, and **(d)** a cardiac muscle cell from one of the ventricles.

Visceral smooth muscle fibers do not have typical excitable membranes, and most have neither stable resting potentials nor action potentials comparable to those of other muscle types. When visceral smooth muscle fibers are stimulated to contract, changes in the transmembrane potential reflect primarily alterations in the membrane permeability to calcium ions (Ca^{2+}).

How would a chemical that blocks the sodium channels in neuron cell membranes affect a neuron's ability to depolarize?

What effect would decreasing the concentration of extracellular potassium ions have on the transmembrane potential of a neuron?

Two axons are tested for propagation velocities. One carries action potentials at 50 meters per second, the other at 1 meter per second. Which axon is myelinated?

Synaptic Transmission

In the nervous system, messages move from one location to another in the form of action potentials along axons. These electrical events are also known as **nerve impulses.** To be effective, a message must not only be propagated along an axon, it must also be transferred in some way to another cell. At a synapse involving two neurons, the impulse passes from the **presynaptic neuron** to the **postsynaptic neuron.** A synapse may also involve other types of postsynaptic cells. For example, the *neuromuscular junction* described in Chapter 10 is an example of a synapse where the postsynaptic cell is a skeletal muscle fiber. *[p. 295]*

At a synapse, a change in the transmembrane potential of the synaptic knob affects the activity of another cell. A synapse may be *electrical,* with direct physical contact between the cells, or *chemical,* involving a neurotransmitter.

Electrical Synapses

Electrical synapses are found in the CNS and PNS, but they are extremely rare.[1] At an electrical synapse, the presynaptic and postsynaptic membranes are locked together at *gap junctions* (Figure 4-2•, p. 112). The lipid portions of the opposing membranes, separated by only 2 nm, are held in position by binding between integral membrane proteins called *connexons.* These proteins contain pores that permit the passage of ions between the cells. Because the two cells are linked in this way, changes in the transmembrane potential of one cell will produce local currents that affect the other cell as if the two shared a common cell membrane. As a result, an electrical synapse quickly and efficiently conducts action potentials from one cell to another.

[1]They are present in some centers in the brain, including the *vestibular nuclei* of the medulla oblongata and in at least one pair of PNS ganglia (the *ciliary ganglia*).

Chemical Synapses

The situation at a **chemical synapse** is far more dynamic, because the cells are not directly coupled. For example, an action potential that reaches an electrical synapse will always be propagated to the next cell. But at a chemical synapse, an arriving action potential may or may not release enough neurotransmitter to bring the postsynaptic neuron to threshold. In addition, other factors may intervene and make the postsynaptic cell more or less sensitive to arriving stimuli. In essence, the postsynaptic cell at a chemical synapse is not a slave to the presynaptic neuron—its activity can be adjusted or "tuned" by a variety of factors.

Chemical synapses are by far the most abundant type of synapse. Most synapses between neurons and all communications between neurons and other cell types involve chemical synapses. Communication across a chemical synapse can normally occur in only one direction: from the presynaptic membrane to the postsynaptic membrane.

Although acetylcholine (ACh) is the neurotransmitter that has received the most attention, there are many other important chemical transmitters. Neurotransmitters are often classified as **excitatory** or **inhibitory** on the basis of their effects on postsynaptic membranes. *Excitatory neurotransmitters* cause depolarization and promote action potential generation, whereas *inhibitory neurotransmitters* cause hyperpolarization and depress action potential generation.

This classification is useful, but not always precise. For example, acetylcholine typically produces a depolarization in the postsynaptic membrane. But the ACh released at neuromuscular junctions in the heart has an inhibitory effect, producing a transient *hyperpolarization* of the postsynaptic membrane. This situation highlights an important aspect of neurotransmitter function. *The effect of a neurotransmitter on the postsynaptic membrane depends on the properties of the receptor, not on the nature of the neurotransmitter.*

Cholinergic Synapses

Figures 12-20, 12-21

We shall focus on chemical synapses releasing the neurotransmitter **acetylcholine,** or **ACh.** These are known as **cholinergic synapses.** The neuromuscular junction described in Chapter 10 is one example of a cholinergic synapse. *[p. 295]* ACh, the most widespread (and best-studied) neurotransmitter, is released (1) at all neuromuscular junctions involving skeletal muscle fibers, (2) at

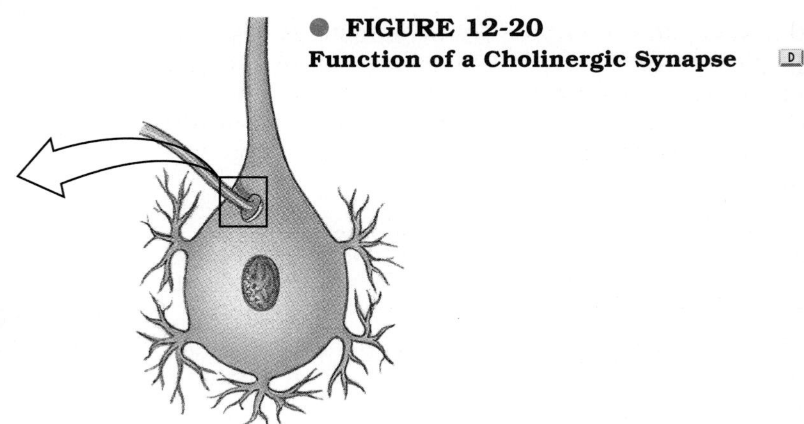

● **FIGURE 12-20**
Function of a Cholinergic Synapse

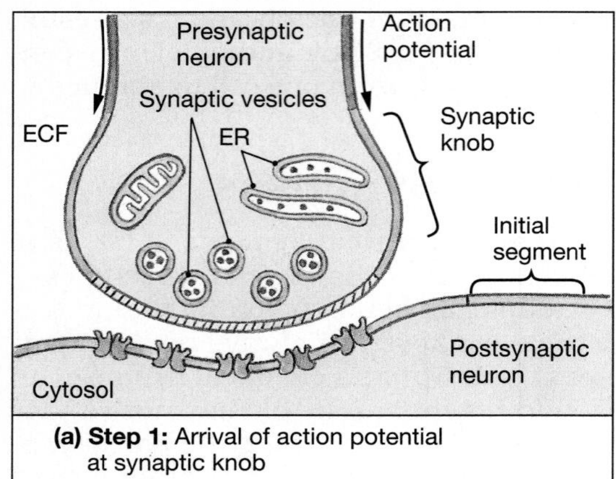

Presynaptic neuron
Action potential
Synaptic vesicles
ECF
ER
Synaptic knob
Initial segment
Postsynaptic neuron
Cytosol

(a) Step 1: Arrival of action potential at synaptic knob

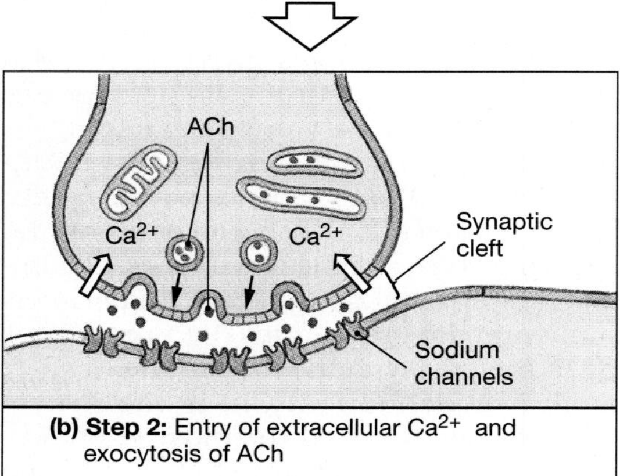

ACh
Ca2+
Ca2+
Synaptic cleft
Sodium channels

(b) Step 2: Entry of extracellular Ca^{2+} and exocytosis of ACh

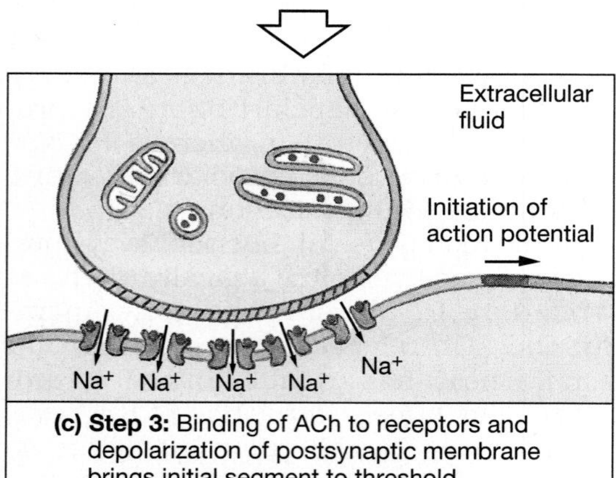

Extracellular fluid
Initiation of action potential
Na+
Na+
Na+
Na+
Na+

(c) Step 3: Binding of ACh to receptors and depolarization of postsynaptic membrane brings initial segment to threshold

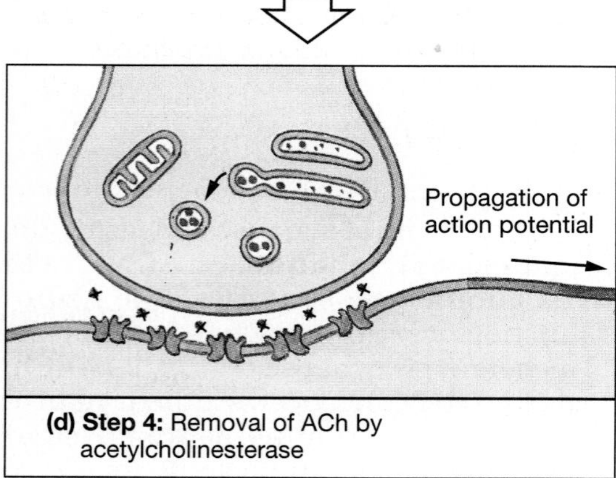

Propagation of action potential

(d) Step 4: Removal of ACh by acetylcholinesterase

many synapses inside the CNS, (3) at all neuron-to-neuron synapses in the PNS, and (4) at all neuroeffector junctions of the parasympathetic division of the ANS.

At a cholinergic synapse (Figure 12-20a●) the presynaptic membrane and the postsynaptic membrane are separated by a synaptic cleft that averages 20 nm (0.02 μm) in width. Most of the ACh in the synaptic knob is packaged within synaptic vesicles, each containing several thousand molecules of neurotransmitter. There may be a million such vesicles in a single synaptic knob at the neuromuscular junction.

Figure 12-20● diagrams the events that occur at a cholinergic synapse following the arrival of an action potential at the synaptic knob. The synapse under consideration is located at the boundary between the axon hillock and the initial segment of the axon.

Step 1: *Arrival of an Action Potential and Depolarization of the Synaptic Knob* (Figure 12-20a●)

The normal stimulus for neurotransmitter release is the depolarization of the presynaptic membrane following the arrival of an action potential.

Step 2: *Extracellular Calcium Ions Enter the Synaptic Knob and Trigger Exocytosis of ACh* (Figure 12-20b●)

The depolarization of the synaptic knob causes the opening of voltage-regulated calcium channels. During the brief period that these channels remain open, calcium ions flood into the axoplasm. Their arrival triggers exocytosis and the release of ACh into the synaptic cleft. The ACh is released in packets of roughly 3000 molecules. This number corresponds to the average number of ACh molecules within a single vesicle. The release of ACh stops very soon because the necessary calcium ions are rapidly removed from the cytoplasm. This is accomplished by means of active transport mechanisms that either pump them out of the cell or transfer them into mitochondria, vesicles, or the endoplasmic reticulum.

Step 3: *ACh Binding and Depolarization of the Postsynaptic Membrane* (Figure 12-20c●)

The ACh released through exocytosis diffuses across the synaptic cleft toward receptors on the postsynaptic membrane. These receptors are chemically regulated ion channels. The primary response to ACh binding is an increased permeability to sodium ions, producing a depolarization that lasts about 20 msec.[1] This depolarization is a graded potential; the more ACh released at the presynaptic membrane, the larger the depolarization. If the depolarization brings an adjacent area of excitable membrane to threshold, an action potential will appear in the postsynaptic neuron.

Step 4: ***Removal of ACh by AChE*** (Figure 12-20d●)

The effects on the postsynaptic membrane are temporary because the synaptic cleft and the postsynaptic membrane contain *acetylcholinesterase* (*AChE*, or *cholinesterase*). Roughly half of the ACh released at the presynaptic membrane is broken down before it reaches receptors on the postsynaptic membrane. ACh molecules that do succeed in binding to receptor sites are usually decomposed within 20 msec of their arrival.

AChE breaks down molecules of ACh into **acetate** and **choline.** The choline is actively absorbed by the synaptic knob and used to synthesize more ACh. Acetate diffusing away from the synapse can be absorbed and metabolized by the postsynaptic cell or by other cells and tissues.

Figure 12-21● shows the general pattern of ACh release and choline recycling, and Table 12-4 summarizes the events that occur at a cholinergic synapse.

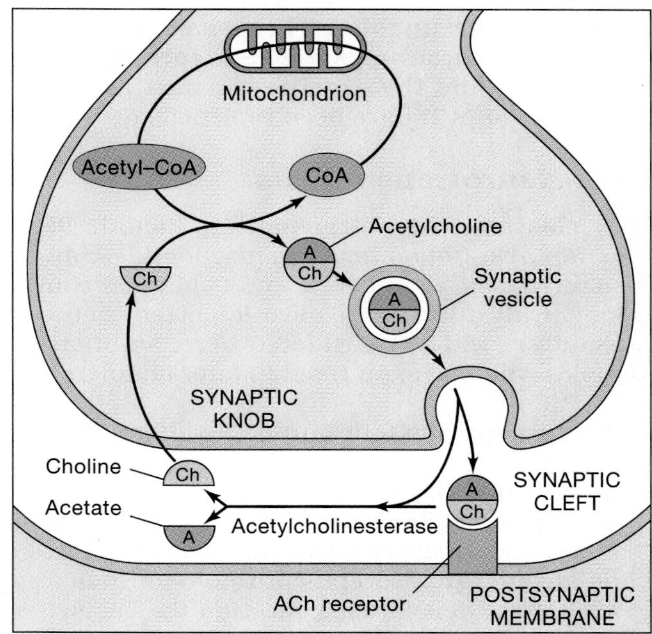
Mitochondrion
Acetyl–CoA
CoA
Acetylcholine
A
Ch
Ch
Synaptic vesicle
SYNAPTIC KNOB
Choline
Acetate
Acetylcholinesterase
SYNAPTIC CLEFT
ACh receptor
POSTSYNAPTIC MEMBRANE

● **FIGURE 12-21**

ACh Release and Recycling at a Cholinergic Synapse. ACh at the synaptic knob is stored in synaptic vesicles. After it is released into the synaptic cleft, ACh is broken down by hydrolysis into acetate and choline by the enzyme acetylcholinesterase. The choline is reabsorbed by the synaptic knob and used to synthesize ACh.

SYNAPTIC DELAY There is a 0.2–0.5 msec **synaptic delay** between the arrival of the action potential at the synaptic knob and the effect on the postsynaptic membrane. Most of that delay reflects the time involved in calcium influx and neurotransmitter release—the synaptic cleft is relatively narrow, and diffusion of neurotransmitters across it takes very little time. Although a delay of 0.5 msec is not very long, in that time an action potential may travel more than 7 cm (around 3 in.) along a myelinated axon. When information is being passed along a chain of interneurons in the CNS, the cumulative synaptic delay may exceed the propagation time. Reflexes, which provide rapid and automatic responses to stimuli, involve a small number of synapses. The fewer the synapses involved, the shorter the total synaptic delay and the faster the response. The fastest reflexes have just one synapse, with a sensory neuron directly controlling a motor neuron. The muscle spindle reflexes, introduced in Chapter 10 and detailed in Chapter 13, are an important example. ∞ [p. 303]

SYNAPTIC FATIGUE Because ACh molecules are recycled, the synaptic knob is not totally dependent on the ACh synthesized in the soma and delivered by axoplasmic transport. But under intensive stimulation, the resynthesis and transport mech-

[1]These channels also let potassium ions out of the cell, but because sodium ions are driven by a much stronger electrochemical gradient, the net effect is a slight depolarization of the postsynaptic membrane.

TABLE 12-4 Sequence of Events at a Typical Cholinergic Synapse

Step 1:
- **An arriving action potential depolarizes the synaptic knob and the presynaptic membrane.**

Step 2:
- **Calcium ions enter the cytoplasm of the synaptic knob.**
- **ACh release occurs through diffusion and exocytosis of neurotransmitter vesicles.**

Step 3:
- **ACh diffuses across the synaptic cleft and binds to receptors on the postsynaptic membrane.**
- **Chemically regulated sodium channels on the postsynaptic surface are activated, producing a graded depolarization.**
- **ACh release ceases because calcium ions are removed from the cytoplasm of the synaptic knob.**

Step 4:
- **The depolarization ends as ACh is broken down into acetate and choline by AChE.**
- **The synaptic knob reabsorbs choline from the synaptic cleft and uses it to resynthesize ACh.**

anisms may be unable to keep pace with the demand for neurotransmitter. **Synaptic fatigue** then occurs, and the synapse remains inactive until stores of ACh have been replenished.

Other Neurotransmitters

Major classes of neurotransmitters include *biogenic amines, amino acids, neuropeptides,* and *dissolved gases,* as well as a variety of other compounds. Only a few of the most important neurotransmitters will be considered here; additional examples will be encountered in later chapters.

- **Norepinephrine** (nor-ep-i-NEF-rin), or **NE,** is an important neurotransmitter in the brain and in portions of the autonomic nervous system. Norepinephrine is also called **noradrenaline,** and synapses releasing NE are described as **adrenergic.** Norepinephrine often has an excitatory, depolarizing effect on the postsynaptic membrane, but the mechanism is quite distinct from that of ACh. We will consider the mechanism in detail in Chapter 16, in our discussion of the ANS.
- **Dopamine** (DŌ-puh-mēn) and **serotonin** (ser-ō-TŌ-nin) are CNS neurotransmitters whose effects are usually inhibitory (hyperpolarizing). These neurotransmitters are released in a variety of centers in the brain. Inadequate dopamine production and release can lead to overstimulation of neurons controlling skeletal muscle tone. The result can be the characteristic rigidity and stiffness seen in *Parkinson's disease,* a condition detailed in Chapter 14. Serotonin is an important neurotransmitter involved in the regulation of endocrine activity. Inadequate serotonin production can have widespread effects on attention and emotional states and may be responsible for many cases of severe, chronic depression. Interactions among serotonin, norepinephrine, and other neurotransmitters are thought to be involved in the regulation of asleep/awake cycles.
- **Gamma aminobutyric** (GAM-ma a-MĒ-nō-bū-TIR-ik) **acid,** also known as **GABA,** usually has an inhibitory effect. GABA release appears to reduce anxiety, and some anti-anxiety drugs work by enhancing these effects.

There are many neurotransmitters whose functions are less well understood. In a clear demonstration of the principle "the more you look the more you see," at least 50 neurotransmitters have now been identified, including certain amino acids, peptides, polypeptides, prostaglandins, and ATP. In addition, two gases have recently been identified as important neurotransmitters. Nitric oxide (NO) is generated by synaptic knobs innervating smooth muscle in the walls of blood vessels in the PNS and at synapses in several regions of the brain. Carbon monoxide (CO), best known for its presence in automobile exhaust, can also be generated by specialized synaptic knobs in the brain, where it functions as a neurotransmitter. Our knowledge of the significance of these compounds and the mechanisms involved in their synthesis and release remains incomplete. Table 12-5 lists major neurotransmitters of the brain and spinal cord and their primary effects (if known).

NEUROMODULATORS Chemical synapses are always active to some degree, and small numbers of neurotransmitter molecules are continually leaking through the presynaptic membranes. Other chemicals may also be released by the synaptic knob, either via gradual diffusion or through exocytosis, in the company of neurotransmitter molecules. **Neuromodulators** (noo-rō-MOD-ū-lā-tōrz) are compounds that influence either the release of neurotransmitter by the presynaptic neuron or the postsynaptic cell's response to the neurotransmitter. Neuromodulators are often **neuropeptides,** small peptide chains synthesized and released by the synaptic knob. Many neuromodulators act by binding to receptors in the presynaptic or postsynaptic membranes and activating cytoplasmic enzymes.

Endorphins. Neuromodulators called **endorphins** (en-DOR-finz) are produced in the brain and spinal cord. The endorphins include several different **enkephalins** (en-KEF-a-linz), **beta-endorphin,** and **dynorphin** (dī-NOR-fin). Although their exact function is uncertain, their primary role is probably the relief of pain. These neuropeptides are structurally similar to morphine, and morphine binds to the same receptor sites as the enkephalins. Pain relief occurs through inhibition of the release of the neurotransmitter *Substance P* at synapses that relay pain sensations. Dynorphin has far more powerful analgesic (pain-relieving) effects than either morphine or the other endorphins.

Mechanisms of Neurotransmitter Action

Functionally, neurotransmitters and neuromodulators fall into three groups:

1. Compounds that have a direct effect on the membrane potential, by opening or closing chemically regulated channels. Examples include ACh and the amino acids *glutamate* and *aspartate.* In addition, some chemically regulated channels are sensitive to GABA and norepinephrine, although other membrane receptors for these neurotransmitters produce indirect effects.
2. Compounds that have an indirect effect on membrane potential, via *second messengers.* When epinephrine, norepinephrine, dopamine,

TABLE 12-5 Representative Neurotransmitters and Their Effects

Neurotransmitter	*Mechanism of Action*	*Distribution (Examples)*
ACETYLCHOLINE	Primarily direct, through binding to chemically regulated channels	*Widespread in CNS and PNS* *CNS:* Synapses throughout brain and spinal cord *PNS:* Neuromuscular junctions; preganglionic synapses of ANS; neuroeffector junctions of parasympathetic division and (rarely) sympathetic division of ANS
AMINES		
Norepinephrine (NE)	Indirect, via second messenger formation	*CNS and PNS* *CNS:* Cerebral cortex, hypothalamus, brain stem, cerebellum, spinal cord *PNS:* Most neuroeffector junctions of sympathetic division of ANS
Epinephrine	Indirect, via second messenger formation	*CNS:* Thalamus, hypothalamus, midbrain, spinal cord
Dopamine	Direct or indirect, depending on receptor type	*CNS:* Hypothalamus, limbic system
Serotonin	Direct or indirect, depending on receptor type	*CNS:* Limbic system, hypothalamus, cerebellum, spinal cord
Histamine	Indirect, via second messenger formation	*CNS:* Hypothalamus
EXCITATORY AMINO ACIDS		
Glutamate	Direct or indirect, depending on receptor type	*CNS:* Cerebral cortex, brain stem
Aspartate	Direct or indirect, depending on receptor type	*CNS:* Spinal cord
INHIBITORY AMINO ACIDS		
Gamma aminobutyric acid (GABA)	Direct or indirect, depending on receptor type	*CNS:* Cerebellum, cerebral cortex, inhibitory interneurons throughout brain and spinal cord
Glycine	Direct	*CNS:* Inhibitory interneurons throughout brain and spinal cord
NEUROPEPTIDES		
Substance P	Indirect	*CNS:* Synapses of sensory axons within spinal cord, hypothalamus, other areas of brain
Enkephalins and endorphins	Indirect, via second messenger formation	*CNS:* Hypothalamus, thalamus, brain stem
OTHER SYNAPTIC CHEMICALS		
High-energy compounds (ATP, GTP)	Indirect, via second messenger formation	*CNS:* Spinal cord
Hormones (ADH, oxytocin, insulin, glucagon, secretin, CCK, and many others)	Usually indirect	*CNS:* Brain
Prostaglandins	Indirect, via second messenger formation	*CNS:* Brain
DISSOLVED GASES		
Carbon monoxide (CO)	Indirect, via second messenger formation	*CNS:* Brain
Nitric oxide (NO)	Indirect, via second messenger formation	*CNS:* Brain *PNS:* Some sympathetic neuroeffector junctions (rare)

Drugs and Synaptic Function

Many drugs interfere with key steps in the process of synaptic transmission. These drugs may (1) interfere with transmitter synthesis, (2) alter the rate of transmitter release, (3) prevent transmitter inactivation, or (4) prevent transmitter binding to receptors. The discussion that follows is limited to clinically important compounds that exert their effects at cholinergic synapses. Their sites of activity are indicated in Figure 12-22●.

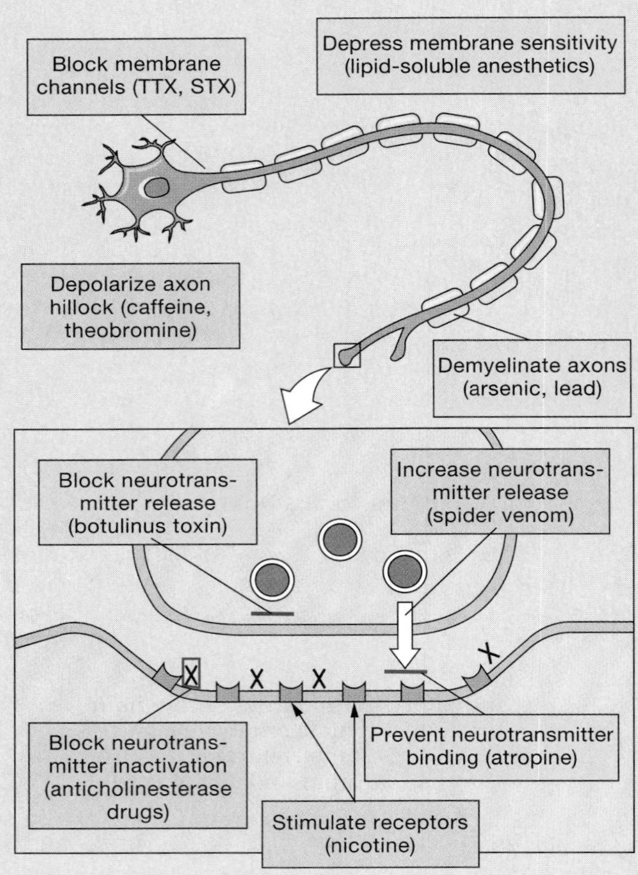
Block membrane channels (TTX, STX)
Depress membrane sensitivity (lipid-soluble anesthetics)
Depolarize axon hillock (caffeine, theobromine)
Demyelinate axons (arsenic, lead)
Block neurotrans-mitter release (botulinus toxin)
Increase neurotrans-mitter release (spider venom)
Block neurotrans-mitter inactivation (anticholinesterase drugs)
Prevent neurotransmitter binding (atropine)
Stimulate receptors (nicotine)

● **FIGURE 12-22**
Mechanism of Drug Action at a Cholinergic Synapse. Factors that facilitate neural function and make neurons more excitable are shown in violet. Factors that inhibit or depress neural function are shown in blue.

Botulinus toxin is responsible for the primary symptom of *botulism,* a widespread paralysis of skeletal muscles. [p. 295] Botulinus toxin blocks the release of ACh at the presynaptic membrane of cholinergic neurons. The venom of the black widow spider has the opposite effect. It causes a massive release of ACh that produces intense muscular cramps and spasms.

Anticholinesterase drugs, sometimes called *cholinesterase inhibitors,* block the breakdown of ACh by acetylcholinesterase. The result is an exaggerated and prolonged stimulation of the postsynaptic membrane. At the neuromuscular junctions, this abnormal stimulation produces an extended and extreme state of contraction. Military nerve gases block cholinesterase activity for weeks, although few persons exposed are likely to live long enough to regain normal synaptic function. Most animals utilize ACh as a neurotransmitter, and anticholinesterase drugs, such as *malathion,* are in widespread use in pest-control projects.

Drugs such as **atropine** or **d-tubocurarine** prevent ACh from binding to the postsynaptic receptors. The latter compound is a derivative of *curare,* a plant extract used by certain South American tribes to paralyze their prey. Curare and related compounds induce paralysis by preventing stimulation of the neuromuscular junction by ACh. Atropine can also be administered intentionally to counteract the effects of anticholinesterase poisoning. Other compounds, including **nicotine,** an active ingredient in cigarette smoke, bind to the receptor sites and stimulate the postsynaptic membrane. There are no enzymes to remove these compounds, and the effects are relatively prolonged. AM *Neuroactive Drugs*

serotonin, histamine, GABA, and any of the endorphins bind to membrane receptors, the receptors activate an intracellular enzyme, **adenylate cyclase,** also known as *adenylyl* (or *adenyl*) *cyclase.* This enzyme catalyzes the formation of **cyclic-AMP** from ATP at the inner surface of the cell membrane. Cyclic-AMP, or **cAMP,** is a **second messenger** that may open membrane channels and/or activate intracellular enzymes, depending on the nature of the postsynaptic cell. Other second messengers, including cGMP, calcium ions, and *inositol trisphosphate* will be encountered in later chapters.

3. Gases such as NO and CO are lipid-soluble so that they can diffuse across the cell membrane and bind to enzymes inside the cell. These enzymes then promote the appearance of second messengers that can have various effects on cellular activity.

It can be very difficult to distinguish between neurotransmitters and neuromodulators on either biochemical or functional grounds. In general, neuromodulators (1) have long-term effects that are relatively slow to appear, (2) trigger responses that involve second messengers, (3) may affect the presynaptic membrane, the postsynaptic membrane, or both, and (4) can be released alone or in the company of a neurotransmitter. However, the same compound may function as a neurotransmitter in one site and as a neuromodulator in another. For this reason, Table 12-5 does not draw distinctions between neurotransmitters and neuromodulators.

✔ What effect would blocking voltage-regulated calcium channels at a cholinergic synapse have on synaptic communication?

✔ While studying pathways in the central nervous system, you discover one pathway consisting of three neurons and another consisting of five neurons. If the neurons in both pathways are identical, which pathway will transmit impulses more rapidly?

✔ Norepinephrine causes hyperpolarization of smooth muscle in the blood vessels serving skeletal muscles, but causes smooth muscle in blood vessels serving the intestines to depolarize. How can this be?

Cellular Information Processing

At each synapse, an action potential arriving at a synaptic knob triggers chemical or electrical events that affect another cell. Some of the neurotransmitters arriving at the postsynaptic cell at any given moment may be excitatory, while others, from different presynaptic neurons, may be inhibitory. The net result may be a depolarization, a hyperpolarization, or no appreciable change in the transmembrane potential at the initial segment. The transmembrane potential at the initial segment therefore represents an integration of all the excitatory and inhibitory stimuli affecting the neuron at that moment. Excitatory and inhibitory stimuli are integrated through interactions between *postsynaptic potentials.* This interaction is the simplest level of **information processing** in the nervous system.

Postsynaptic Potentials

Postsynaptic potentials are graded potentials that develop in the postsynaptic membrane in response to a neurotransmitter. Two major types of postsynaptic potentials develop at neuron-to-neuron synapses: *excitatory postsynaptic potentials* and *inhibitory postsynaptic potentials.*

Excitatory Postsynaptic Potentials

Figure 12-13c

An **excitatory postsynaptic potential,** or **EPSP,** is a depolarization caused by the arrival of a neurotransmitter at the postsynaptic membrane. An EPSP results from the opening of chemically regulated ion channels; the depolarization produced by the binding of ACh (p. 408) is an example of an EPSP. Because it is a graded potential, an EPSP affects only the area immediately surrounding the synapse, as in Figure 12-13c●, p. 397.

Inhibitory Postsynaptic Potentials

Not all neurotransmitters have an excitatory (depolarizing) effect. An **inhibitory postsynaptic potential,** or **IPSP,** is a transient *hyperpolarization* of the postsynaptic membrane. An IPSP results from the opening of chemically regulated potassium or chloride channels. While the hyperpolarization continues, the neuron is **inhibited** because a larger-than-usual depolarizing stimulus must be provided to bring the membrane potential to threshold. For example, a stimulus sufficient to shift the transmembrane potential by 10 mV—from –70 mV to –60 mV—would normally produce an action potential. But if the transmembrane potential were reset at –85 mV by an IPSP, the same stimulus would depolarize it only to –75 mV—still well below threshold.

Summation

Figures 12-23, 12-24

Summation is the mechanism responsible for integration of EPSPs, IPSPs, or some combination of the two in the postsynaptic neuron. We will use the summation of EPSPs as an example of this process. An individual EPSP has a relatively small effect on the transmembrane potential; a typical EPSP produces a depolarization of around 0.5 mV at the postsynaptic membrane. Before an action potential will appear in the initial segment, local currents must depolarize that region by 15–20 mV. A single EPSP will therefore not result in an action potential, even if the synapse is on the axon hillock. But individual EPSPs can combine through the process of summation. There are two forms of summation: *temporal summation* and *spatial summation.*

TEMPORAL SUMMATION **Temporal summation** (*tempus,* time) is the addition of stimuli occurring in rapid succession. Temporal summation occurs at a *single synapse* that is active *repeatedly.* Consider this simple analogy: You cannot fill a bathtub with a single bucket of water, but you can if you keep using the bucket over and over. The water in each bucket corresponds to the sodium ions entering the cytoplasm during an EPSP. Consider what happens when a second EPSP arrives before the effects of the first have disappeared. A typical EPSP lasts about 20 msec, but under maximum stimulation one action potential may reach the synaptic knob each millisecond. Every time an action potential arrives, another group of vesicles discharges ACh into the synaptic cleft. Every time another batch of ACh molecules arrives at the postsynaptic membrane, additional receptor sites are occupied, more chemically regulated channels open, and the degree of depolarization increases. In this way, a series of small steps can eventually bring the initial segment to threshold (Figure 12-23a●).

SPATIAL SUMMATION **Spatial summation** occurs when simultaneous stimuli at different locations have a cumulative effect on the transmembrane potential. Spatial summation involves *multiple synapses* that are active *simultaneously*. In terms of our bucket analogy, you could fill the bathtub immediately if 50 friends with buckets, each standing at a different point around the tub, all poured at once.

The activity of a single synapse produces a graded potential with localized effects. If more than one synapse is active at the same time, all "pour" sodium ions across the postsynaptic membrane. At each active synapse, the sodium ions that produce the EPSP spread out along the inner surface of the membrane and mingle with those entering at other synapses. The effects on the initial segment are cumulative (Figure 12-23b●). The degree of depolarization will depend on how many synapses are active at any given moment, as well as their distance from the initial segment. As in the case of temporal summation, an action potential will appear when the transmembrane potential at the initial segment reaches threshold.

FACILITATION Spatial or temporal summation of EPSPs will not necessarily depolarize the initial segment to threshold, but every step closer to threshold makes it easier for the *next* stimulus to trigger an action potential. A neuron that has been brought closer to threshold is said to be **facilitated.** The larger the degree of facilitation, the smaller the additional stimulus needed to trigger an action potential. In a highly facilitated neuron, even a small depolarizing stimulus will produce an action potential.

Facilitation can also result from neuron exposure to certain drugs. For example, nicotine stimulates postsynaptic ACh receptors, producing prolonged EPSPs that facilitate CNS neurons. The active ingredients of coffee, cocoa, and tea (caffeine, theobromine, and theophylline) also cause facilitation, but they act in a different way. These compounds lower the threshold at the initial segment, so a smaller-than-usual depolarization will cause an action potential. Once an action potential reaches the synaptic knob, these compounds also increase the amount of ACh released. The entire nervous system becomes excitable, and you feel more alert. The phrase "coffee makes you jumpy" is not an exaggeration—after several cups you may jump at a sudden noise because your reflexes are primed to respond to the slightest stimulus. Because nicotine and caffeine work via different mechanisms, the combination of cigarettes and coffee has a very dramatic stimulatory effect on the CNS.

SUMMATION OF EPSPs AND IPSPs Like EPSPs, IPSPs can summate spatially and temporally. EPSPs

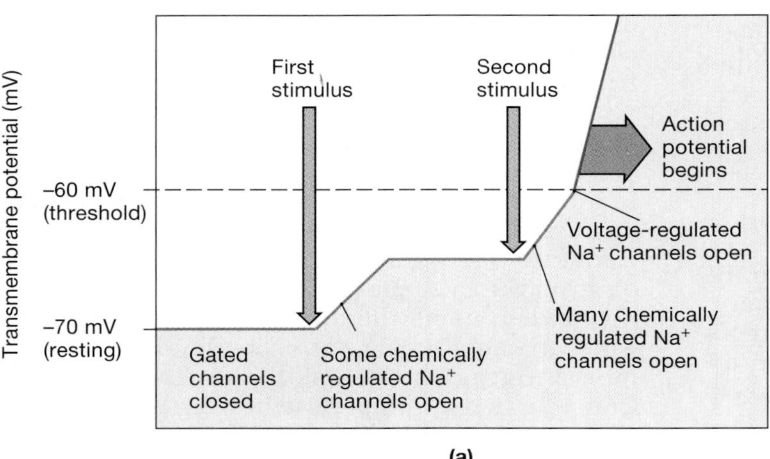
Transmembrane potential (mV)
First stimulus
Second stimulus
Action potential begins
−60 mV (threshold)
Voltage-regulated Na^+ channels open
−70 mV (resting)
Many chemically regulated Na^+ channels open
Gated channels closed
Some chemically regulated Na^+ channels open
(a)

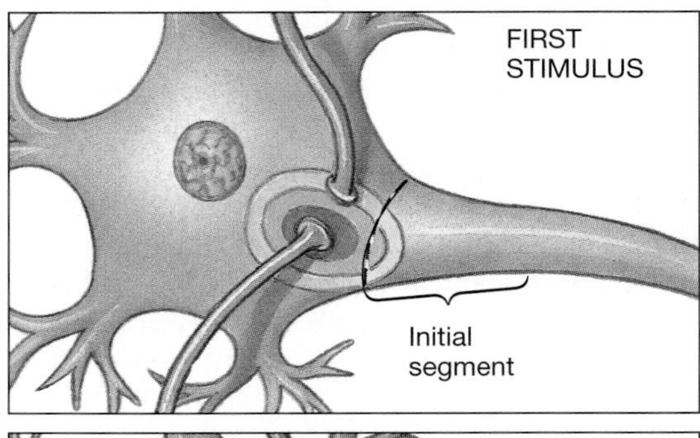
FIRST STIMULUS
Initial segment

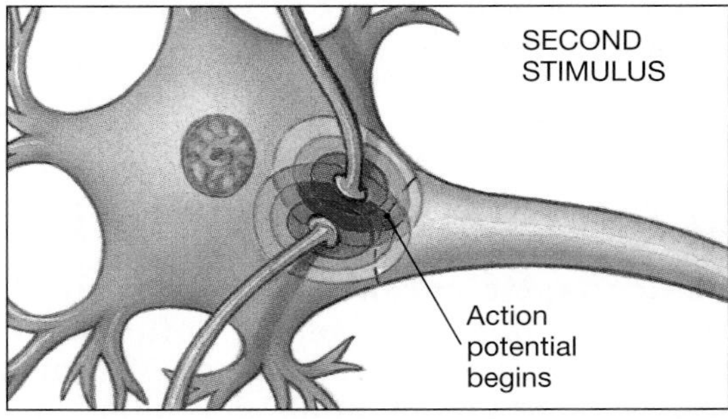
SECOND STIMULUS
Action potential begins

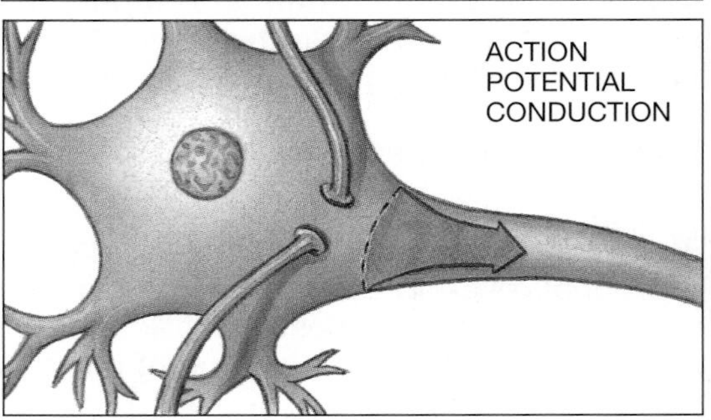
ACTION POTENTIAL CONDUCTION
(b)

● **FIGURE 12-23**

Temporal and Spatial Summation. (a) Temporal summation occurs on a membrane receiving two depolarizing stimuli separated in time. The effects of the second stimulus are added to those of the first. **(b)** Spatial summation occurs when sources of stimulation arrive simultaneously but at different locations. Local currents spread the depolarizing effects, and areas of overlap experience the combined effects.

and IPSPs reflect the activation of different types of chemically regulated channels, and they have opposing effects on the transmembrane potential. The antagonism between IPSPs and EPSPs is an important mechanism for cellular information processing. In terms of our bathtub analogy, EPSPs put water *into* the tub and IPSPs take water *out*. If there are more buckets adding water than there are buckets removing water, the water level in the tub will rise. If more buckets remove water, the water level will fall. If a bucket of water is removed every time another bucket is dumped into the tub, the water level will remain stable. Comparable interactions between EPSPs and IPSPs (Figure 12-24●) determine the transmembrane potential at the boundary between the axon hillock and the initial segment.

Neuromodulators or hormones or both can change the sensitivity of the postsynaptic membrane to excitatory or inhibitory neurotransmitters. By shifting the balance between EPSPs and IPSPs, these compounds can promote facilitation or inhibition of neurons in the CNS and PNS.

❑ The Rate of Action Potential Generation

Complex sensory information and motor commands are carried along axons, encoded in the frequency of action potentials. For example, action potentials arriving at a neuromuscular junction at the rate of 1 per second may produce a series of isolated twitches in the associated skeletal muscle fiber, whereas at the rate of 100 per second they will cause a sustained tetanic contraction. A few action potentials per second along a sensory fiber may be perceived as a feather-light touch, whereas hundreds of action potentials per second along that same axon may be perceived as unbearable pressure. This section will examine factors that vary the rate of action potential generation; the functional significance of these changes will be examined in later chapters.

If a graded potential briefly depolarizes the axon hillock to the point that the initial segment reaches threshold, an action potential will be propagated along the axon. Now consider what happens if the axon hillock *remains* depolarized for an extended period. As the membrane repolarizes after the first action potential, it becomes capable of generating a second one as soon as the absolute refractory period ends. Once the relative refractory period has begun, a larger-than-normal stimulus is required to produce an action potential. With the axon hillock already depolarized, there is a stimulus ready and waiting as soon as the relative refractory period begins. The larger the stimulus, the sooner the next action potential will occur. In other words, *the greater the degree of sustained depolarization at the axon hillock, the higher the frequency of action potential generation.* The rate of action potential generation reaches a maximum when the relative refractory period has been completely eliminated. The highest recorded action potential frequencies range between 500 and 1000 per second.

❑ Presynaptic Inhibition and Facilitation

Figure 12-25

Inhibitory or excitatory synapses may occur at an *axoaxonal synapse*, a synapse between the axon of one neuron and the axon of another. An axoaxonal synapse that occurs on the synaptic knob (Figure 12-25●) can modify the rate of neurotransmitter release at the presynaptic membrane. In one form of **presynaptic inhibition** (Figure 12-25a●), *GABA* release inhibits the opening of voltage-regulated calcium channels in the synaptic knob. This inhibition reduces the amount of neurotransmitter released when an action potential arrives at the synaptic knob and thus limits the effects on the postsynaptic membrane. In **presynaptic facilitation** (Figure 12-25b●), activity at an axoaxonal synapse increases the amount of neurotransmitter released when an action potential arrives at the synaptic knob. This increase enhances and prolongs the effects of the neurotransmitter on the postsynaptic membrane. *Serotonin* is a neurotransmitter involved in presynaptic facilitation. In the presence of serotonin released at an axoaxonal synapse, voltage-regulated calcium channels remain open for an extended period.

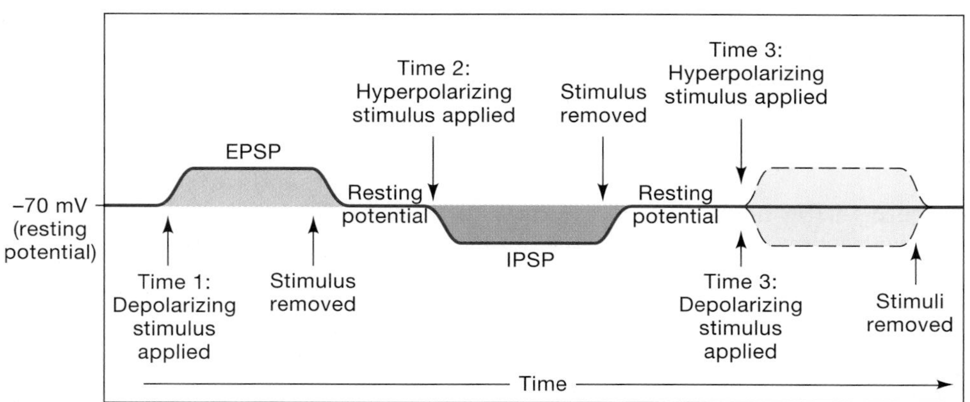
Time 2: Hyperpolarizing stimulus applied
Stimulus removed
Time 3: Hyperpolarizing stimulus applied
EPSP
Resting potential
Resting potential
–70 mV (resting potential)
IPSP
Time 1: Depolarizing stimulus applied
Stimulus removed
Time 3: Depolarizing stimulus applied
Stimuli removed
Time

● **FIGURE 12-24**
EPSP-IPSP Interactions. At time 1, a small depolarizing stimulus produces an EPSP. At time 2, a small hyperpolarizing stimulus produces an IPSP of comparable magnitude. If the two stimuli are applied simultaneously, as at time 3, summation occurs. Because the two are equal in size, the membrane potential remains at the resting level. If the EPSP were larger, a net depolarization would result; if the IPSP were larger, a net hyperpolarization would be seen.

General Factors That Affect Neural Function

We will consider briefly two factors that can alter the function of neurons in the CNS and PNS: (1) changes in the extracellular environment, and (2) the metabolic demands of active neurons.

Environmental Factors

Environmental factors, such as pH, ion, or temperature changes, can alter the resting membrane potential or disrupt the metabolic operations that support action potential generation.

- Changes in hydrogen ion concentration (pH) can have dramatic effects on neural activity. The normal extracellular pH averages 7.35–7.45. If the pH rises, neurons are facilitated; at a pH near 7.8 they begin to generate action potentials spontaneously, producing severe convulsions. If the pH declines, neurons are inhibited; at a pH around 7.0 the nervous system shuts down, and the individual becomes completely unresponsive. A variety of mechanisms exist to control the pH of the cerebrospinal fluid and other body fluids; these mechanisms will be discussed in later chapters.
- Changes in the ionic composition of the extracellular fluids have a direct effect on neural function. Fluctuations in sodium or potassium ion concentrations, such as those caused by dehydration or kidney disease, may facilitate or inhibit neural activity by depolarizing or hyperpolarizing the cell membrane. For example, an elevated extracellular potassium concentration, a condition called **hyperkalemia** (hī-per-ka-LĒ-mē-a; *kalium*, potassium + *haima*, blood) reduces the chemical gradient for potassium ions across the cell membrane. The rate of potassium loss decreases as a result, and the retention of positive charges gradually depolarizes the membrane. Hyperkalemia has damaging effects on all excitable membranes; it can lead to a general paralysis of skeletal muscles and death by cardiac arrest. Abnormally high or low extracellular calcium ion concentrations affect synaptic function by reducing or enhancing calcium entry into the synaptic knob, thereby modifying the amount of neurotransmitter released as well as changing the excitability of axonal membranes.

Metabolic Processes

The brain contributes just 2 percent to the body weight, but it accounts for 18 percent of resting energy consumption. Active neurons need ATP to support (1) the synthesis, release, and recycling of neurotransmitter molecules; (2) the movement of materials to and from the soma via axoplasmic flow; and (3) the recovery from action potentials through the activity of the Na^+/K^+ exchange pump. Each time an action potential occurs, sodium ions enter and potassium ions leave the cell, and over time ATP must be expended to maintain normal cytoplasmic ion concentrations. When impulses are generated at high frequencies, the energy demands are enormous.

Neurons normally derive ATP solely through aerobic mechanisms. Because their cytoplasm does not contain glycogen reserves, these cells are totally dependent on a continual and reliable supply of both oxygen and glucose from the blood. In severe malnutrition, neural function deteriorates as the body's energy reserves are exhausted. Neural function is

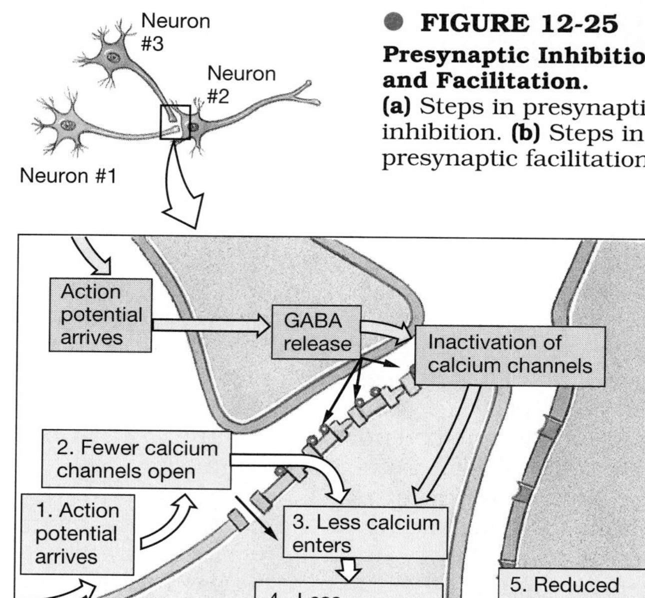
Neuron #3
Neuron #2
Neuron #1
Action potential arrives
GABA release
Inactivation of calcium channels
2. Fewer calcium channels open
1. Action potential arrives
3. Less calcium enters
4. Less neurotransmitter released
5. Reduced effect on postsynaptic membrane
(a) Presynaptic inhibition

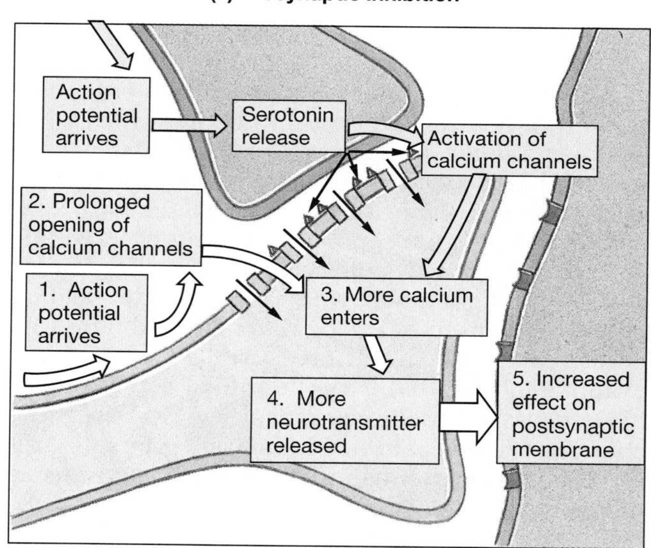
Action potential arrives
Serotonin release
Activation of calcium channels
2. Prolonged opening of calcium channels
1. Action potential arrives
3. More calcium enters
4. More neurotransmitter released
5. Increased effect on postsynaptic membrane
(b) Presynaptic facilitation

FIGURE 12-25 Presynaptic Inhibition and Facilitation. **(a)** Steps in presynaptic inhibition. **(b)** Steps in presynaptic facilitation.

FOCUS Neural Responses to Injuries

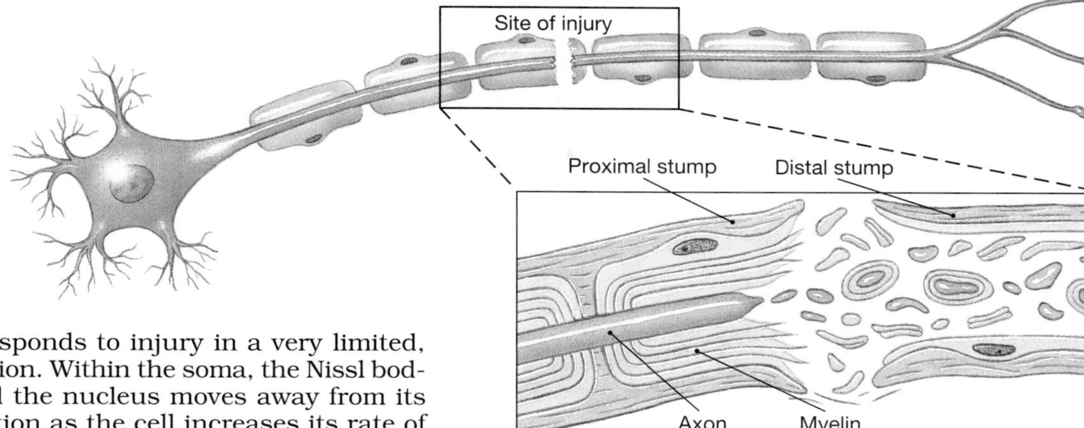

Site of injury
Proximal stump
Distal stump
Axon
Myelin

A neuron responds to injury in a very limited, stereotyped fashion. Within the soma, the Nissl bodies disperse and the nucleus moves away from its centralized location as the cell increases its rate of protein synthesis. If the neuron recovers its functional abilities, it will return to a normal appearance. The key to recovery seems to be the events under way in the axon. Consider the response of a neuron to mechanical stresses, such as the pressure applied during a crushing injury. The pressure produces a local decrease in blood flow and oxygen availability, and the affected membrane becomes inexcitable. If the pressure is released after an hour or two, the neuron will recover within a few weeks. More severe or prolonged pressure will produce effects similar to those caused by cutting off the distal portion of the axon.

In the peripheral nervous system, the Schwann cells participate in the repair of damaged nerves. In the process known as **Wallerian degeneration,** illustrated in Figure 12-26●, the axon distal to the injury site degenerates, and macrophages migrate into the area to phagocytize the debris. The Schwann cells do not degenerate; instead, they proliferate and form a solid cellular cord that follows the path of the original axon. As the neuron recovers, its axon grows into the injury site, and the Schwann cells wrap around it.

If the axon continues to grow into the periphery alongside the appropriate cord of Schwann cells, it may eventually reestablish its normal synaptic contacts. If it stops growing, or wanders off in some new direction, normal function will not return. The growing axon is most likely to arrive at its appropriate destination if the cut edges of the original nerve bundle remain in contact.

Limited regeneration can occur inside the central nervous system, but the situation is more complicated because (1) many more axons are likely to be involved, (2) astrocytes produce scar tissue that can prevent axon growth across the damaged area, and (3) astrocytes release chemicals that block the regrowth of axons.

STEP 1:
Fragmentation of axon and myelin occurs in distal stump.

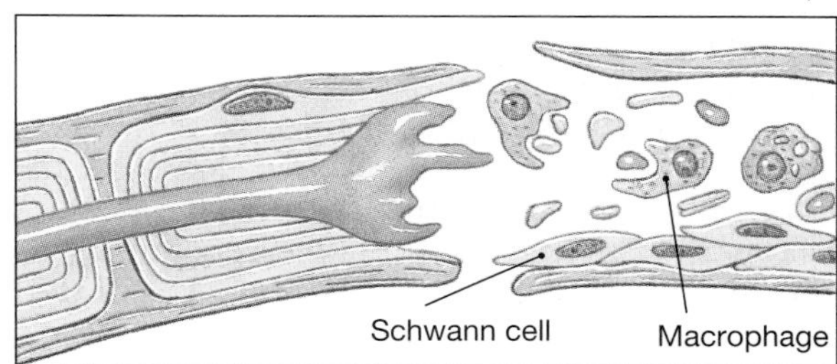

Schwann cell
Macrophage

STEP 2:
Schwann cells form cord, grow into cut, and unite stumps. Macrophages engulf degenerated axon and myelin.

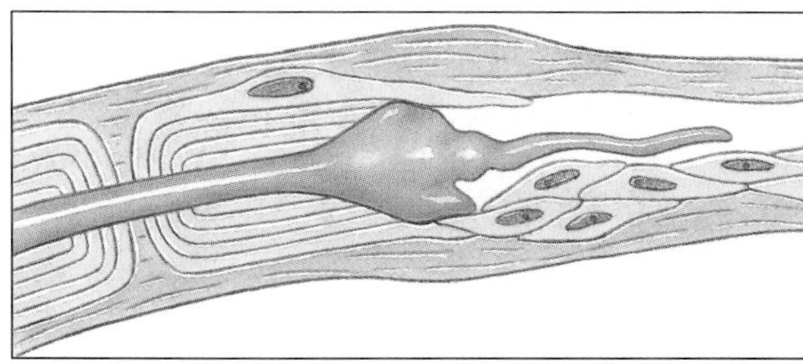

STEP 3:
Axon sends buds into network of Schwann cells and then starts growing along cord of Schwann cells.

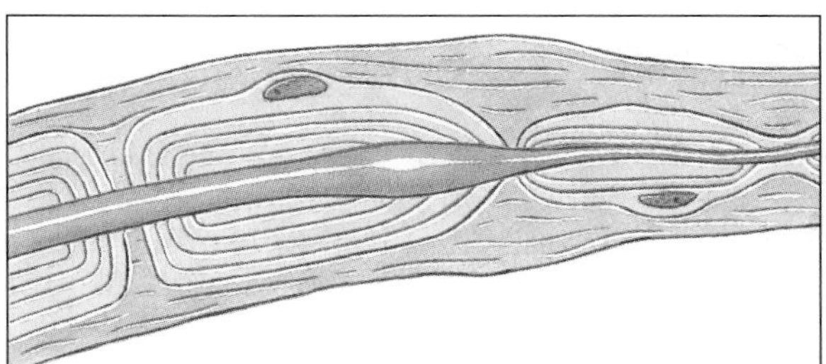

STEP 4:
Axon continues to grow into distal stump and is enfolded by Schwann cells.

● **FIGURE 12-26**
Peripheral Nerve Regeneration after Injury

also impaired if the local circulation is restricted or, worse yet, shut off. If the circulation to a region is interrupted for just a few seconds, neurons in that region will be injured; the longer the interruption, the more severe the injury will be. In a **stroke** the blood supply to the brain is interrupted by a circulatory blockage or vascular rupture. The degree of functional impairment after a stroke is determined by (1) the location and size of the region deprived of circulation and (2) the duration of the circulatory interruption. Subsequent chapters will consider the origins and treatment of strokes in greater detail.

There are several inherited abnormalities in neural function that are caused by metabolic problems within neurons. For example, *Tay-Sachs disease* is a genetic abnormality involving the metabolism of *gangliosides*, glycolipids that are important components of neuron cell membranes. AM *Tay-Sachs Disease*

❑ Summary

Let's briefly review the basic principles of information processing introduced in earlier discussions.

1. The neurotransmitters released at a synapse may have excitatory or inhibitory effects. The effect on the initial segment reflects an integration of the stimuli arriving at any given moment. The frequency of action potential generation is an indication of the degree of depolarization at the axon hillock.
2. Neuromodulators can alter the rate of neurotransmitter release or the response of a postsynaptic neuron to specific neurotransmitters.
3. Neurons may be facilitated or inhibited by extracellular chemicals other than neurotransmitters or neuromodulators.
4. The effect of a presynaptic neuron's activation on a postsynaptic neuron at a synapse may be altered by other neurons.

Higher Levels of Organization and Processing

❑ Functional Organization

Figure 12-27

There are around 10 million sensory neurons, 20 billion interneurons, and one-half million motor neurons in the body. The interneurons are organized into a smaller number of **neuronal pools.** A neuronal pool is a group of interconnected neurons with specific functions. Estimates concerning the actual number of pools vary, but they range between a few hundred and a few thousand. Each pool has a limited number of input sources and output destinations, and each may contain both excitatory and inhibitory neurons. The output of the entire pool may stimulate or depress the activity of other pools, or it may exert direct control over motor neurons or peripheral effectors.

The pattern of interaction among neurons provides clues to the functional characteristics of a neuronal pool. Five patterns may be distinguished.

1. **Divergence** is the spread of information from one neuron to several neurons, as in Figure 12-27a•, or from one pool to multiple pools. Divergence permits the broad distribution of a specific input. Considerable divergence occurs when sensory neurons bring information into the CNS, for the information is distributed to neuronal pools throughout the spinal cord and brain.
2. In **convergence,** several neurons synapse on the same postsynaptic neuron (Figure 12-27b•). Several different patterns of activity in the presynaptic neurons can therefore have the same effect on the postsynaptic neuron. Through convergence, the same motor neurons can be subject to both voluntary and involuntary control. For example, the movements of your diaphragm and ribs are now being involuntarily controlled by respiratory centers in the brain. But the same motor neurons can also be controlled voluntarily, as when you take a deep breath and hold it. Two different neuronal pools are involved, both synapsing on the same motor neurons.
3. Neuronal pools may function in sequence, with one pool activating a second, which in turn activates a third, and so forth. This pattern, called **serial processing,** is shown in Figure 12-27c•. Serial processing occurs as sensory information is relayed from one processing center to another in the brain.
4. **Parallel processing** occurs when several neuronal pools are processing the same information at one time (Figure 12-27d•). Thanks to parallel processing, many different responses occur simultaneously. For example, stepping on a sharp object stimulates sensory neurons that distribute the information to a number of neuronal pools. As a result of parallel processing you might withdraw your foot, shift your weight, move your arms, feel the pain, and shout "Ouch!" all at the same time.
5. Some neural circuits utilize positive feedback to produce **reverberation.** In this arrangement, collateral axons somewhere along the sequence extend back toward the source of an impulse and further stimulate the presynaptic neurons. Once a reverberating circuit has been activated it will continue to function until synaptic fatigue or inhibitory stimuli break the cycle. As with convergence or divergence, reverberation can occur within a single neuronal pool, or it may involve

a series of interconnected pools. A simple example of reverberation is shown in Figure 12-27e●. Much more complicated examples of reverberation between neuronal pools in the brain may be involved in the maintenance of consciousness, muscular coordination, and normal breathing.

❑ Anatomical Organization

The functions of the nervous system depend on the interactions between neurons in neuronal pools, and the most complex neural processing steps occur in the spinal cord and brain. Neurons and their axons are not randomly scattered in the CNS and PNS. Instead, they form masses or bundles with distinct anatomical boundaries, and they are identified by specific terms. These terms will be used in all later chapters, and a brief overview at this time will prove helpful.

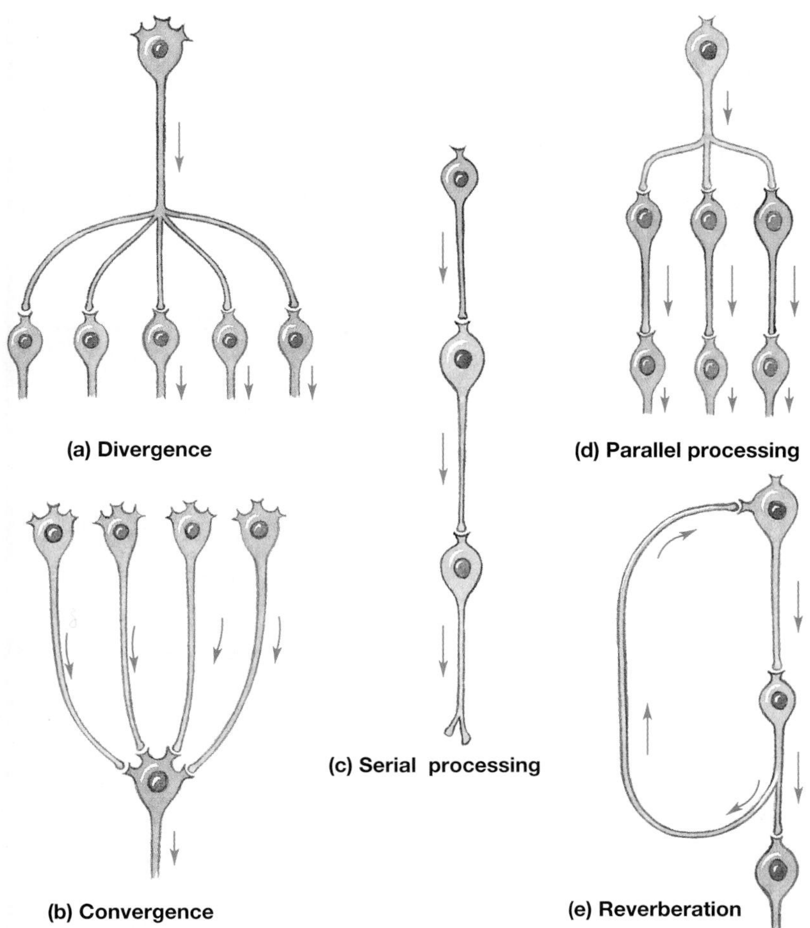
(a) Divergence
(b) Convergence
(c) Serial processing
(d) Parallel processing
(e) Reverberation

● **FIGURE 12-27**

Organization of Neuronal Pools. (a) Divergence, a mechanism for spreading stimulation to multiple neurons or neuronal pools in the CNS. **(b)** Convergence, a mechanism providing input to a single neuron from multiple sources. **(c)** Serial processing, in which neurons or pools work in a sequential manner. **(d)** Parallel processing, in which neurons or pools process information simultaneously. **(e)** Reverberation, a positive feedback mechanism.

In the PNS:

- Neuron cell bodies are found in *ganglia.*
- Axons are bundled together in *nerves*, with spinal nerves connected to the spinal cord, and cranial nerves connected to the brain.

In the CNS:

- A collection of neuron cell bodies with a common function is called a **center.** A center with a discrete anatomical boundary is called a **nucleus.** Portions of the brain surface are covered by a thick layer of gray matter, called the **neural cortex.** The term *higher centers* refers to the most complex integration centers, nuclei, and cortical areas of the brain.
- The white matter of the CNS contains bundles of axons that share common origins, destinations, and functions. These bundles are called **tracts.** Tracts in the spinal cord form larger groups, called **columns.**
- The centers and tracts that link the brain with the rest of the body are called **pathways.** For example, **sensory pathways** distribute information from peripheral receptors to processing centers in the brain, and **motor pathways** begin at CNS centers concerned with motor control and end at the effectors they control.

✔ A neuron from a cutaneous receptor synapses with a motor neuron located in the spinal cord. A motor neuron from the brain also synapses with the motor neuron in the spinal cord. What type of neuronal pool does this arrangement represent?

✔ In an injury involving a peripheral nerve, why is it important for the two ends of the damaged nerve to be closely aligned?

✔ How would damage to interneurons in the spinal cord affect nervous system function?

■ Integration with Other Systems

Figure 12-28

Figure 12-28● diagrams the relationships between the nervous system and other physiological systems. Many of these relationships will be explored in greater detail in subsequent chapters.

INTEGUMENTARY SYSTEM

Provides sensations of touch, pressure, pain, vibration, and temperature; hair provides some protection and insulation for skull and brain; protects peripheral nerves

Controls contraction of arrector pili muscles, secretion of sweat glands

SKELETAL SYSTEM

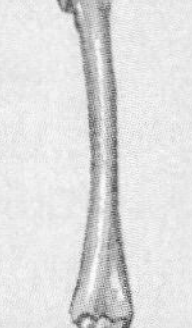

Provides calcium for neural function; protects brain and spinal cord

Controls skeletal muscle contractions that promote bone thickening and maintenance and determine bone position

MUSCULAR SYSTEM

Facial muscles express emotional state; intrinsic laryngeal muscles permit communication; muscle spindles provide proprioceptive sensations

Controls skeletal muscle contractions; coordinates respiratory and cardiovascular activities

THE NERVOUS SYSTEM

FOR ALL SYSTEMS

Monitors pressure, pain, and temperature; adjusts tissue blood flow patterns

ENDOCRINE SYSTEM

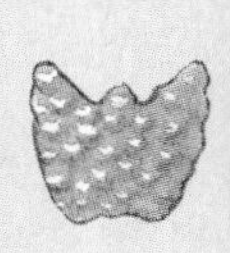

Many hormones affect CNS neural metabolism
Reproductive hormones and thyroid hormone influence CNS development

Controls pituitary gland and many other endocrine organs; secretes ADH and oxytocin

CARDIOVASCULAR SYSTEM

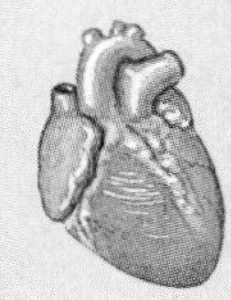

Endothelial cells maintain blood–brain barrier; blood vessels (with ependymal cells) produce CSF

Modifies heart rate and blood pressure

LYMPHATIC SYSTEM

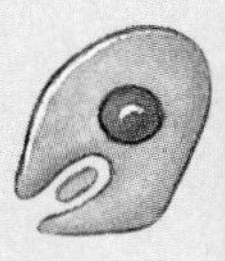

Defends against infection and assists in tissue repairs

Release of neurotransmitters and hormones affect sensitivity of immune response

RESPIRATORY SYSTEM

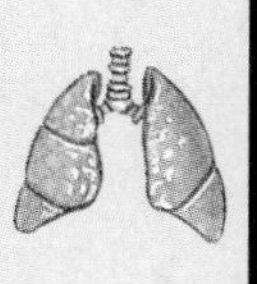

Provides oxygen and eliminates carbon dioxide

Controls pace and depth of respiration

DIGESTIVE SYSTEM

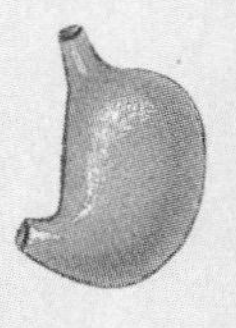

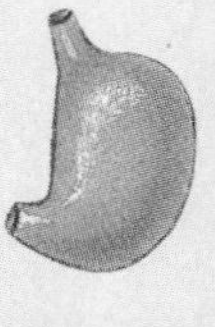

Provides nutrients for energy production and neurotransmitter synthesis

Regulates digestive tract (movement and secretion)

URINARY SYSTEM

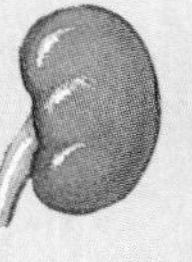

Eliminates metabolic wastes; regulates body fluid pH and electrolyte concentrations

Adjusts renal blood pressure; controls urination

REPRODUCTIVE SYSTEM

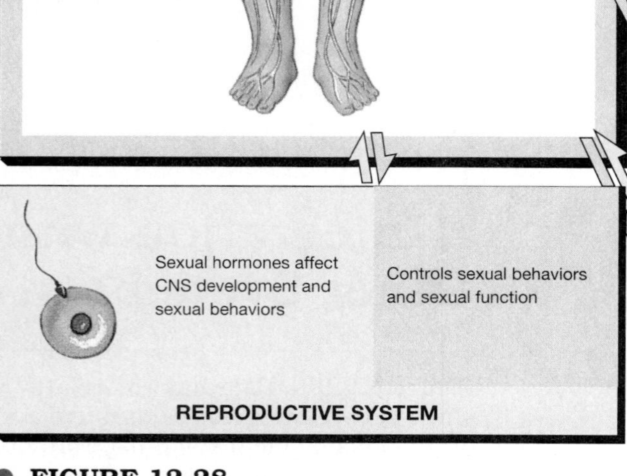

Sexual hormones affect CNS development and sexual behaviors

Controls sexual behaviors and sexual function

● **FIGURE 12-28**
Functional Relationships between the Nervous System and Other Systems

Selected Clinical Terminology

Terms Discussed in This Chapter

anticholinesterase drug: A drug that blocks the breakdown of ACh by acetylcholinesterase (AChE). *(p. 412)*
atropine: A drug that prevents ACh from binding to the postsynaptic membrane. *(p. 412)*
demyelination: Destruction of the myelin sheaths around axons in the CNS and PNS. *(p. 386* and *AM)*
d-tubocurarine: A drug, derived from curare, that prevents ACh from binding to the postsynaptic membrane. *(p. 412)*
endorphins (en-DOR-finz): Neuropeptides produced in the brain and spinal cord that appear to relieve pain and affect mood. *(p. 410)*
hyperkalemia (hī-per-ka-LĒ-mē-a): An abnormal physiological state resulting from a high extracellular concentration of potassium. *(p. 416)*
neurotoxin: A compound that disrupts normal nervous system function by interfering with the generation or propagation of action potentials. Examples include *tetrodotoxin (TTX), saxitoxin (STX), paralytic shellfish poisoning (PSP),* and *ciguatera (CTX). (p. 401* and *AM)*
nicotine: A compound found in tobacco that binds to specific ACh receptor sites and stimulates the postsynaptic membrane. *(p. 412)*
rabies: An acute disease of the central nervous system caused by a virus that reaches the CNS via retrograde flow along peripheral axons. *(p. 388* and *AM)*

AM *Additional Terms Discussed in the Applications Manual*

diphtheria (dif-THĒ-rē-a): A disease that results from a bacterial infection of the respiratory tract. Among other effects, the bacterial toxins damage Schwann cells and cause PNS demyelination.
Guillain-Barré syndrome: A progressive demyelination that appears to be linked to the immune response to a viral infection.
multiple sclerosis (skler-Ō-sis) **(MS):** A disease characterized by recurrent incidents of demyelination affecting axons in the optic nerve, brain, and/or spinal cord.
Tay-Sachs disease: A genetic abnormality involving the metabolism of gangliosides, important components of neuron cell membranes. The result is a gradual deterioration of neurons due to the buildup of metabolic byproducts and the release of lysosomal enzymes.

CHAPTER REVIEW

STUDY OUTLINE

INTRODUCTION, p. 380

1. Two organ systems, the nervous and endocrine systems, coordinate the activities of other organ systems. The nervous system provides swift but brief responses to stimuli; the endocrine system adjusts metabolic operations and directs long-term changes.

AN OVERVIEW OF THE NERVOUS SYSTEM, p. 380

1. The *nervous system* includes all the *neural tissue* in the body. Its anatomical divisions include the **central nervous system (CNS)** (the brain and spinal cord) and the **peripheral nervous system (PNS)** (all of the neural tissue outside the CNS).
2. Functionally the PNS can be divided into an **afferent division,** which brings sensory information to the CNS, and an **efferent division,** which carries motor commands to muscles and glands.
3. The efferent division includes the **somatic nervous system (SNS)** (voluntary control over skeletal muscle contractions) and the **autonomic nervous system (ANS)** (involuntary regulation of smooth muscle, cardiac muscle, and glandular activity). (*Figure 12-1*)

NEUROGLIA, p. 381

Neuroglia of the Central Nervous System, p. 381

1. There are four types of CNS neuroglia: (1) **astrocytes** (largest and most numerous); (2) **oligodendrocytes,** which form the **neurilemma** and are responsible for the **myelination** of CNS axons; (3) **microglia** (phagocytic derivatives of white blood cells); and (4) **ependymal cells,** with functions related to the **cerebrospinal fluid (CSF).** (*Figure 12-2; Table 12-1*)

Neuroglia of the Peripheral Nervous System, p. 384

2. Neuron cell bodies in the PNS are clustered into **ganglia** (singular *ganglion*), and their axons form **peripheral nerves.** (*Figures 12-3, 12-4; Table 12-1*)
3. **Satellite cells,** or *amphicytes,* surround neurons within ganglia. **Schwann cells** ensheath axons in the PNS. A single Schwann cell may myelinate one segment of an axon or enfold segments of several unmyelinated axons. (*Figure 12-4*)

NEURONS, p. 386

Neuron Structure, p. 386

1. The **perikaryon** of a multipolar neuron contains organelles, including **neurofilaments, neurotubules,** and **neurofibrils.** The **axon hillock** connects the **initial segment** of the axon to the cell body, and the **axoplasm** contains numerous organelles. (*Figure 12-5*)
2. **Collaterals** may branch from an axon, with **telodendria** branching from the axon's tip. (*Figure 12-5*)
3. A **synapse** is a site of intercellular communication. A synapse where neurons communicate with another cell type is a **neuroeffector junction.** (*Figures 12-5, 12-6*)

Neuron Classification, p. 388

4. Neurons may be described as **anaxonic, unipolar, bipolar,** or **multipolar.** (*Figure 12-7*)
5. There are three functional categories of neurons: *sensory neurons, motor neurons,* and *interneurons* (association neurons). (*Figure 12-8*)
6. **Sensory neurons** of the afferent division of the PNS deliver information to the CNS. (*Figure 12-8*)
7. **Motor neurons** form the efferent division of the PNS. These neurons stimulate or modify the activity of a peripheral tissue, organ, or organ system. (*Figure 12-8*)
8. **Interneurons (association neurons)** are always found in the CNS. They may be located between sensory and motor neurons; they analyze sensory inputs and coordinate motor outputs. (*Figure 12-8*)

NEUROPHYSIOLOGY, p. 390

The Transmembrane Potential, p. 392

1. The **electrochemical gradient** is the sum of all the chemical and electrical forces acting across the membrane. (*Figures 12-9, 12-10*)
2. The *resting potential* of a neuron, about –70 mV, is determined chiefly by the membrane permeability to potassium ions.

Ion Channels, p. 395

3. The cell membrane contains **passive (leak) channels** and **active (gated) channels.**
4. There are three types of gated channels: **chemically regulated channels, voltage-regulated channels,** and **mechanically regulated channels**. (*Figures 12-11, 12-12*)

Graded Potentials, p. 396

5. **Depolarization** or **hyperpolarization** can lead to a **graded potential** (a change in potential that results when the degree of depolarization decreases with distance). (*Figures 12-13, 12-14; Table 12-3*)

Action Potentials, p. 398

6. An **action potential** appears when a region of excitable membrane depolarizes to **threshold.** The steps involved include: membrane depolarization and activation of sodium channels; sodium channel inactivation; potassium channel activation; and return to normal permeability. (*Figures 12-15, 12-16; Tables 12-2, 12-3*)
7. The activity of the sodium-potassium exchange pump is necessary to maintain ion concentrations within acceptable limits over time.
8. In **continuous propagation (continuous conduction)** an action potential spreads across the entire excitable membrane surface in a series of small steps. (*Figure 12-17*)
9. During **saltatory propagation (saltatory conduction)** the action potential appears to leap from node to node, skipping the intervening membrane surface. Saltatory propagation carries nerve impulses five to seven times more rapidly than does continuous propagation. (*Figure 12-18*)
10. Axons can be classified as **Type A fibers, Type B fibers,** or **Type C fibers** on the basis of diameter, myelination, and propagation speed.
11. Action potentials in muscle tissue differ from those in neural tissue in that: (1) resting potentials are greater, (2) action potentials last longer, and (3) action potential propagation is slower along a sarcolemma than along an axon. (*Figure 12-19*)

SYNAPTIC TRANSMISSION, p. 407

1. An action potential traveling along an axon is called a **nerve impulse.** At a synapse involving two neurons, information passes from the **presynaptic neuron** to the **postsynaptic neuron.** A synapse may also involve other types of postsynaptic effector cells.
2. A synapse may be electrical (with direct physical contact between cells) or chemical (involving a neurotransmitter). **Chemical synapses** are more common than **electrical synapses.**

Electrical Synapses, p. 407

3. **Electrical synapses** are relatively rare in the CNS and PNS. At an electrical synapse, the presynaptic and postsynaptic cell membranes are bound together by interlocking membrane proteins at a gap junction. Pores within these proteins permit the passage of local currents, and the two neurons act as if they shared a common cell membrane.

Chemical Synapses, p. 407

4. Neurotransmitters are often classified as **excitatory** or **inhibitory** on the basis of their effects on postsynaptic membranes: *Excitatory neurotransmitters* cause depolarization and promote action potential generation, whereas *inhibitory neurotransmitters* cause hyperpolarization and depress action potential generation.
5. The effect of a neurotransmitter on the postsynaptic membrane depends on the properties of the receptor, not on the nature of the neurotransmitter.
6. **Cholinergic synapses** release the neurotransmitter **acetylcholine (ACh).** Communication moves from the presynaptic neuron to the postsynaptic neuron. The separation of the presynaptic membrane and postsynaptic membrane by a synaptic

cleft causes a **synaptic delay.** (*Figures 12-20, 12-21; Table 12-4*)

7. If stores of ACh are exhausted, **synaptic fatigue** can occur.
8. **Adrenergic** synapses release **norepinephrine (NE),** also called **noradrenaline.**
9. Other neurotransmitters include **dopamine, gamma aminobutyric acid (GABA),** and **serotonin.** (*Table 12-5*)
10. **Neuromodulators** influence the postsynaptic cell's response to neurotransmitters.

CELLULAR INFORMATION PROCESSING, p. 413

Postsynaptic Potentials, p. 413

1. Excitatory and inhibitory stimuli are integrated through interactions between **postsynaptic potentials.** This interaction is the simplest level of **information processing** in the nervous system.
2. A depolarization caused by a neurotransmitter is an **excitatory postsynaptic potential (EPSP).** Individual EPSPs can combine through **summation;** the two types of summation are **temporal summation** (which occurs at a single synapse when a second EPSP arrives before the effects of the first have disappeared) and **spatial summation** (which results from the cumulative effects of multiple synapses at various locations). (*Figures 12-13, 12-23*)
3. Hyperpolarization of the postsynaptic membrane is an **inhibitory postsynaptic potential (IPSP).**
4. EPSP-IPSP interactions are the most important determinants of neural activity. (*Figure 12-24*)

The Rate of Action Potential Generation, p. 415

5. The greater the degree of sustained depolarization at the axon hillock, the higher the frequency of action potential generation.

Presynaptic Inhibition and Facilitation, p. 415

6. In **presynaptic inhibition,** *GABA* release at an axoaxonal synapse inhibits the opening of voltage-regulated calcium channels in the synaptic knob. This inhibition reduces the amount of neurotransmitter released when an action potential arrives at the synaptic knob. (*Figure 12-25*)
7. In **presynaptic facilitation,** activity at an axoaxonal synapse increases the amount of neurotransmitter released when an action potential arrives at the synaptic knob. This increase enhances and prolongs the effects of the neurotransmitter on the postsynaptic membrane. (*Figure 12-25*)

General Factors That Affect Neural Function, p. 416

8. Environmental factors, such as pH, ion, or temperature changes, can alter the resting membrane potential or disrupt the metabolic operations that support action potential generation.
9. Active neurons consume a great deal of ATP. Nutritional factors may therefore play a role in neural activity. Neurons are also dependent on a reliable supply of oxygen.

Summary, p. 418

10. The neurotransmitters released at a synapse may have excitatory or inhibitory effects. The effect on the initial segment reflects an integration of the stimuli arriving at any given moment. The frequency of action potential generation is an indication of the degree of depolarization at the axon hillock.
11. Neuromodulators can alter the rate of neurotransmitter release or the response of a postsynaptic neuron to specific neurotransmitters.
12. Neurons may be facilitated or inhibited by extracellular chemicals other than neurotransmitters or neuromodulators.
13. The effect of a presynaptic neuron's activation on a postsynaptic neuron at a synapse may be altered by other neurons.

HIGHER LEVELS OF ORGANIZATION AND PROCESSING, p. 418

Functional Organization, p. 418

1. The roughly 20 billion interneurons can be classified into **neuronal pools** (groups of interconnected neurons with specific functions).
2. **Divergence** is the spread of information from one neuron to several, or from one pool to several pools. In **convergence,** several neurons synapse on the same postsynaptic neuron. Neuronal pools may also function in sequence **(serial processing),** or they may process the same information at one time **(parallel processing).** In **reverberation,** collateral axons establish a circuit that further stimulates presynaptic neurons. (*Figure 12-27*)

Anatomical Organization, p. 419

3. The functions of the nervous system as a whole depend on interactions between neurons in neuronal pools. In the PNS, **spinal nerves** communicate with the spinal cord, and **cranial nerves** are connected to the brain.
4. In the CNS, a collection of neuron cell bodies that share a particular function is called a **center.** A center with a discrete anatomical boundary is called a **nucleus.** Portions of the brain surface are covered by a thick layer of gray matter called the **neural cortex.** The term *higher centers* refers to the most complex integration centers, nuclei, and cortical areas of the brain.
5. The white matter of the CNS contains bundles of axons that share common origins, destinations, and functions. These bundles are called **tracts.** Tracts in the spinal cord form larger groups, called **columns. Sensory pathways** carry information from peripheral receptors to the brain; **motor pathways** extend from CNS centers concerned with motor control to the associated effectors

INTEGRATION WITH OTHER SYSTEMS, p. 419

REVIEW QUESTIONS

LEVEL 1 Reviewing Facts and Terms

1. Regulation by the nervous system provides:
   (a) relatively slow but long-lasting responses to stimuli
   (b) swift, long-lasting responses to stimuli
   (c) swift but brief responses to stimuli
   (d) relatively slow, short-lived responses to stimuli
2. The efferent division of the PNS:
   (a) brings sensory information to the CNS
   (b) carries motor commands to muscles and glands
   (c) processes and integrates sensory data
   (d) is the seat of higher functions in the body
3. The part of the nervous system that provides voluntary control over skeletal muscle contractions is the:
   (a) somatic nervous system
   (b) autonomic nervous system
   (c) visceral motor system
   (d) sympathetic division of the ANS
4. Smooth muscle, cardiac muscle, and glands are among the targets of the:
   (a) somatic nervous system
   (b) sensory neurons
   (c) afferent division of the PNS
   (d) autonomic nervous system
5. The cellular layer that lines the ventricles of the brain and the central canal of the spinal cord is the:
   (a) microglia (b) ganglia
   (c) ependyma (d) oligodendrocyte
6. Glial cells responsible for maintaining the blood-brain barrier are the:
   (a) microglia (b) ependymal cells
   (c) astrocytes (d) oligodendrocytes
7. Phagocytic cells found in neural tissue of the CNS are called:
   (a) astrocytes (b) ependymal cells
   (c) oligodendrocytes (d) microglia
8. Substances being transported from an axon terminal to the soma of the same neuron are delivered by:
   (a) axoplasmic transport
   (b) synaptic conduction
   (c) retrograde flow
   (d) active transport
9. All the motor neurons that control skeletal muscles are:
   (a) multipolar neurons
   (b) myelinated bipolar neurons
   (c) unipolar, unmyelinated sensory neurons
   (d) anaxonic neurons
10. The type of neural cells responsible for the analysis of sensory inputs and coordination of motor outputs are:
   (a) neuroglia (b) interneurons
   (c) sensory neurons (d) motor neurons
11. Depolarization of a neuron cell membrane will shift the membrane potential toward:
   (a) 0 mV (b) −70 mV
   (c) −90 mV (d) a, b, and c are correct
12. The primary determinant of the resting membrane potential is the:
   (a) membrane permeability to sodium
   (b) membrane permeability to potassium
   (c) intracellular negatively charged proteins
   (d) negatively charged chloride ions in the ECF
13. Receptors that bind acetylcholine at the motor end plate are:
   (a) chemically regulated channels
   (b) voltage-regulated channels
   (c) passive channels
   (d) mechanically regulated channels
14. Gated channels that open or close in response to a change in the transmembrane potential are:
   (a) mechanically regulated channels
   (b) voltage-regulated channels
   (c) chemically regulated channels
   (d) a, b, and c are correct
15. Changes in the transmembrane potential that are restricted to the area surrounding the site of stimulation are:
   (a) action potentials
   (b) graded potentials
   (c) inhibitory potentials
   (d) hyperpolarizing potentials
16. Neuromodulators are compounds that influence the:
   (a) postsynaptic cell's response to a neurotransmitter
   (b) synaptic vesicles in the synaptic knob
   (c) release of calcium ions into the axoplasm
   (d) a, b, and c are correct
17. A transient hyperpolarization of the postsynaptic membrane is referred to as:
   (a) a refractory period (b) an EPSP
   (c) an IPSP (d) threshold
18. When simultaneous stimuli have a cumulative effect on the transmembrane potential the process is called:
   (a) reverberation (b) spatial summation
   (c) divergence (d) temporal summation
19. The synapsing of several neurons on the same postsynaptic neuron is called:
   (a) serial processing (b) reverberation
   (c) divergence (d) convergence
20. Positive feedback in neural circuits produces:
   (a) reverberation (b) divergence
   (c) convergence (d) temporal summation
21. (a) What are the major components of the central nervous system?
   (b) What are the primary components of the peripheral nervous system?
22. What two major cell populations are found in the nervous system? What is the primary function of each cell type?
23. (a) What glial cells function to maintain the blood-brain barrier?
   (b) What glial cells serve a phagocytic function in the CNS?
24. What two types of glial cells insulate neuron cell bodies and axons in the PNS from their surroundings?
25. Classifying neurons on the basis of *structure*, what four types are found in the nervous system?
26. What three *functional* groups of neurons are found in the nervous system? What is the function of each type of neuron?

27. What two integrated steps are necessary for transfer of nerve impulses from neuron to neuron?

28. What is the *functional* difference between voltage-regulated, chemically regulated, and mechanically regulated channels?

29. What four basic characteristics are associated with graded potentials?

30. State the all-or-none principle regarding action potentials.

31. Describe the four steps involved in the generation of an action potential.

32. Describe the four steps that take place at a typical cholinergic synapse.

33. What environmental factors play a role in altering the function of neurons in the CNS and PNS?

LEVEL 2 Reviewing Concepts

34. If the resting membrane potential is –70 mV and the threshold is –55 mV, a membrane potential of –60 mV will:
   (a) produce an action potential
   (b) make it easier to produce an action potential
   (c) make it harder to produce an action potential
   (d) hyperpolarize the membrane

35. A graded potential:
   (a) decreases with distance from the point of stimulation
   (b) spreads passively because of local currents
   (c) may involve either depolarization or hyperpolarization
   (d) a, b, and c are correct

36. For an action potential to begin, an area of excitable membrane must:
   (a) have its voltage-regulated gates inactivated
   (b) be hyperpolarized
   (c) be depolarized to threshold level
   (d) not be in a relative refractory period

37. During an absolute refractory period, the membrane:
   (a) continues to hyperpolarize
   (b) cannot respond to further stimulation
   (c) can respond to a larger-than-normal depolarizing stimulus
   (d) will respond to summated stimulation

38. Action potentials are restricted to areas of excitable membranes that contain voltage-regulated channels.
   (a) true (b) false

39. The all-or-none principle states that:
   (a) the properties of an action potential are independent of the strength of the depolarizing stimulus
   (b) all stimuli will produce action potentials
   (c) all graded potentials will generate action potentials
   (d) any cell membrane can generate and propagate an action potential if stimulated to threshold

40. The loss of positive ions from the interior of a neuron produces:
   (a) depolarization (b) threshold
   (c) hyperpolarization (d) an action potential

41. Continuous propagation of an action potential cannot occur in:
   (a) myelinated axons (b) unmyelinated axons
   (c) Type A fibers (d) Type B fibers

42. Why can't neurons in the CNS be replaced when they are lost to injury or disease?

43. What purpose do collaterals serve in the nervous system?

44. What is the difference between axoplasmic transport and retrograde flow?

45. What is the difference between continuous propagation and saltatory propagation?

46. How does an action potential in a skeletal muscle fiber differ from one in a neuron?

47. How does the action of a neurotransmitter differ from that of a neuromodulator?

48. What is the difference between temporal summation and spatial summation?

49. How does the neuronal activity of divergence differ from that of convergence?

50. What functions of neurons necessitate the support of energy from ATP?

51. Why is an electrical synapse a more efficient carrier of nerve impulses from cell to cell than a chemical synapse?

52. When a runner experiences the "runner's high," why is the suppression of pain common?

53. Multiple sclerosis (MS) is a demyelination disorder. How does this condition produce muscular paralysis and sensory losses?

LEVEL 3 Critical Thinking and Clinical Applications

54. If neurons in the central nervous system lack centrioles and are unable to divide, how can a person develop brain cancer?

55. Harry suffers from a kidney condition that causes changes in his body's electrolyte levels (concentration of ions in the extracellular fluid). As a result of this problem, he is exhibiting tachycardia, an abnormally fast heart rate. What ion is involved, and how does a change in its concentration cause Harry's symptoms?

56. Twenty neurons synapse with a single receptor neuron. Fifteen of these neurons release neurotransmitters that produce EPSPs at the postsynaptic membrane, and the other five release neurotransmitters that produce IPSPs. Each time one of the neurons is stimulated, it releases enough neurotransmitter to produce a 2-mV change in potential at the postsynaptic membrane. If the threshold of the postsynaptic neuron is 10 mV, how many of the excitatory neurons must be stimulated to produce an action potential in the receptor neuron if all five inhibitory neurons are stimulated? (Assume that spatial summation occurs.)

57. Myelination of peripheral neurons occurs rapidly through the first year of life. How can this process explain the increased abilities of infants during their first year?

58. A drug that blocks ATPase is introduced into an experimental neuron preparation. The neuron is then repeatedly stimulated and recordings are made of the response. What effect would you expect to observe?

CHAPTER 13

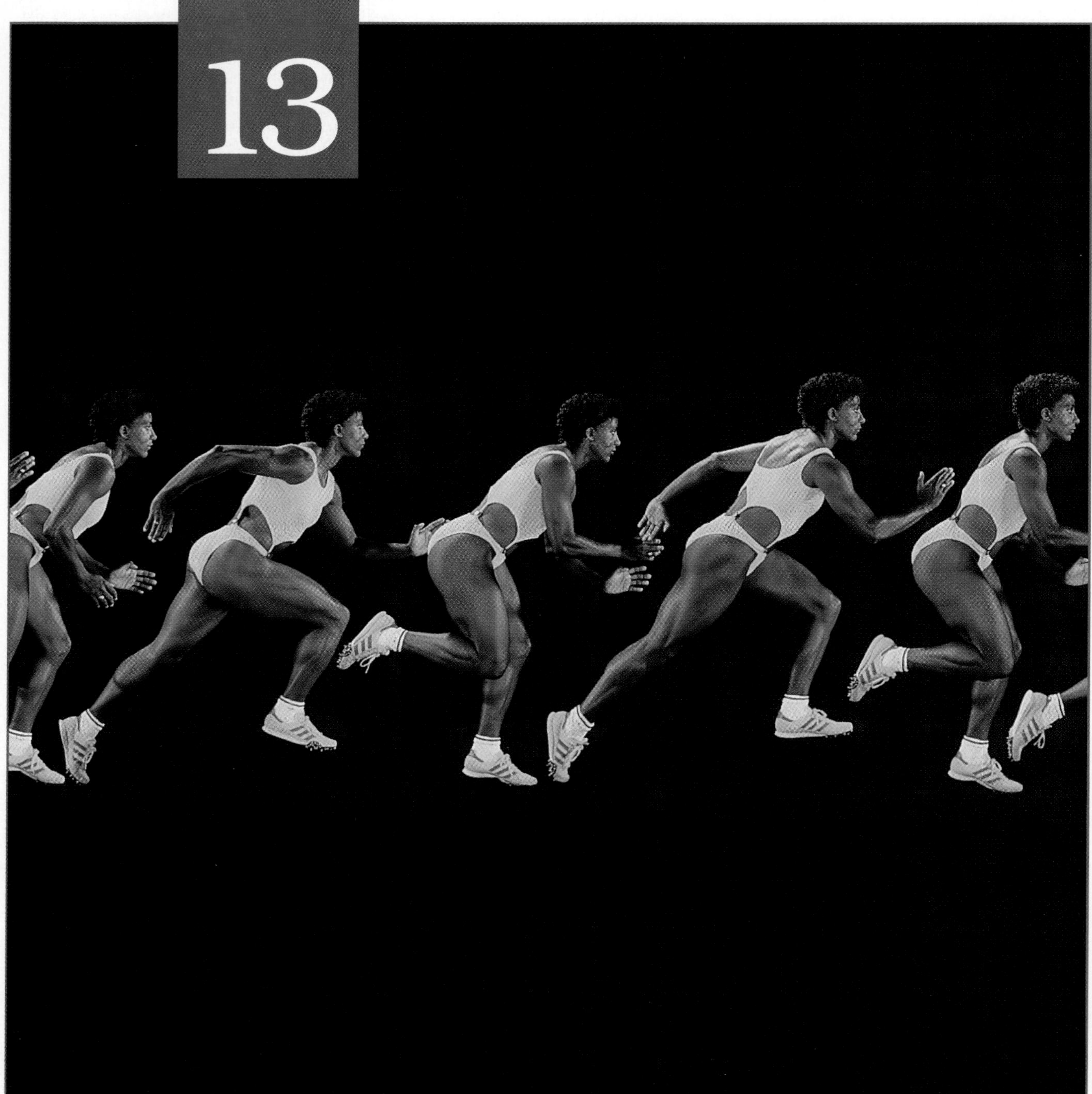

Running is physically demanding, but many people find it mentally relaxing. Imagine, though, what it would be like if you had to think about every detail of the process—if you had to direct the movements of each individual muscle group consciously, instant by instant. Running would be out of the question—it would probably take all your concentration just to take a single step.

Fortunately, you don't have to do so much mental work in order to run: there's no need for you to worry about which muscles will contract at any given moment. Although you make the conscious decision to start running, the actual movements are controlled not in the brain but in the spinal cord. This is possible because running involves a repetitive, stereotyped pattern of movements. (Notice in this photo series, for example, how the upper and lower limbs on each side are always out of phase - the right shoulder and left hip are flexed as the left shoulder and right hip are extended.) This chapter examines the spinal cord and its control over a variety of reflexes and motor patterns.

The Spinal Cord and Spinal Nerves

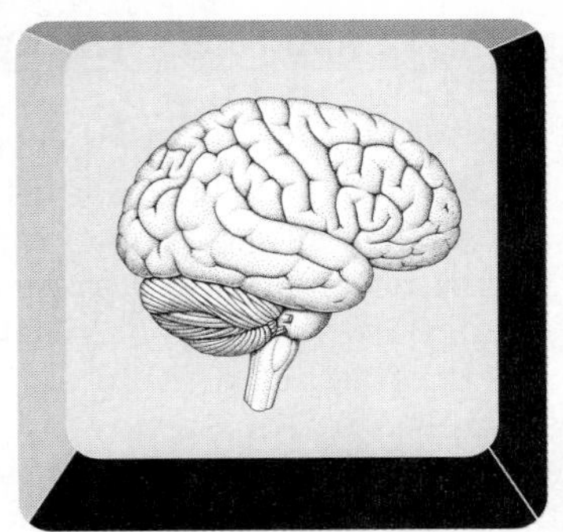

Chapter Outline and Objectives

As we saw in the previous chapter, the integration of thousands of EPSPs and IPSPs by a single neuron will determine (1) whether action potentials will appear and (2) at what frequency they will be generated. [p. 413] When a neuron is active, that activity will affect all of its synapses in the same way. For example, when a somatic motor neuron is active, an entire motor unit contracts—the neuron cannot stimulate one of the muscle fibers and not the rest.

Within a single neuronal pool, there may be thousands of neurons communicating with one another via *millions* of active synapses. The integration process is significantly more complex and time-consuming, but the output is more flexible and varied. For example, a neuronal pool may control hundreds of somatic motor neurons. The output of the pool may stimulate some neurons, facilitate others, and inhibit a third group.

Over the next four chapters we will examine the functional organization of the nervous system at increasing levels of complexity. We will begin with the spinal cord and spinal nerves and consider the interactions between neuronal pools that direct relatively simple *spinal reflexes*, stereotyped motor patterns that are controlled in the spinal cord. For example, information processing in the spinal cord, not in the brain, makes you drop a scalding-hot frying pan after accidentally grabbing it.

The spinal cord is structurally and functionally integrated with the brain. Chapter 14 provides an overview of the major components and functions of the brain and cranial nerves. It also discusses the *cranial reflexes* comparable to those of the spinal cord. We then proceed to consider the nervous system as an integrated unit. Chapter 15 discusses complex *higher-order functions*, such as memory and learning, that involve many different regions of the brain and affect all of the activities of the nervous system. That chapter also examines the interplay between centers in the brain and spinal cord that occurs in the processing of sensory information and in both the reflex control and the voluntary control of skeletal muscle activity. Chapter 16 completes the sequence with an examination of the autonomic nervous system. This system, which has processing centers in the brain, spinal cord, and PNS, is responsible for the control of visceral effectors, such as peripheral smooth muscles, cardiac muscle, and glands.

Gross Anatomy of the Spinal Cord

Figure 13-1

The adult spinal cord (Figure 13-1a,b●) measures approximately 45 cm (18 in.) in length and has a maximum width of roughly 14 mm (0.55 in.). The posterior (dorsal) surface of the spinal cord bears a shallow longitudinal groove, the **posterior median sulcus.** The **anterior median fissure** is a deeper groove along the anterior (ventral) surface. The diameter of the cord decreases as you proceed from cervical to sacral segments.

The amount of gray matter is substantially increased in segments of the spinal cord concerned with the sensory and motor control of the limbs. These areas are expanded, forming the **enlargements** of the spinal cord. The **cervical enlargement** supplies nerves to the shoulder girdles and upper limbs; the **lumbar enlargement** provides innervation to structures of the pelvis and lower limbs. Below the lumbar enlargement the spinal cord becomes tapered and conical; this region is known as the **conus medullaris.** A slender strand of fibrous tissue, the **filum terminale** ("terminal thread"), extends caudally from the inferior tip of the conus medullaris. This filamentous extension continues along the length of the vertebral canal as far as the second sacral vertebra. There it provides longitudinal support to the spinal cord as a component of the *coccygeal ligament.*

Figure 13-1b● provides a series of sectional views that demonstrate the variations in the relative mass of gray and white matter in the cervical, thoracic, lumbar, and sacral regions of the spinal cord. The entire spinal cord can be divided into 31 segments on the basis of the origins of the spinal nerve roots. Each segment is identified by a letter and number designation, as used in the identification of individual vertebrae. For example, C_3, the segment in the uppermost section in Figure 13-1b●, is the third cervical segment.

Every spinal segment is associated with a pair of **dorsal root ganglia** that contains the cell bodies of sensory neurons. The **dorsal roots,** which contain the axons of these neurons, bring sensory information into the spinal cord. A pair of **ventral roots** contains the axons of motor neurons that extend into the periphery to control somatic and visceral effectors. On either side, the dorsal and ventral roots of each segment pass between the vertebral canal and the periphery at the *intervertebral foramen* between successive vertebrae. The dorsal root ganglion lies between the pedicles of the adjacent vertebrae. (You may wish to review the description of vertebral anatomy in Chapter 7, Figure 7-18●, p. 224.)

Distal to each dorsal root ganglion the sensory and motor roots are bound together into a single **spinal nerve.** Spinal nerves are classified as **mixed nerves** because they contain both afferent (sensory) and efferent (motor) fibers. There are 31 pairs of spinal nerves, each identified by its association with adjacent vertebrae. For example, you may speak of "cervical spinal nerves" or even "cervical nerves" when making a general reference to

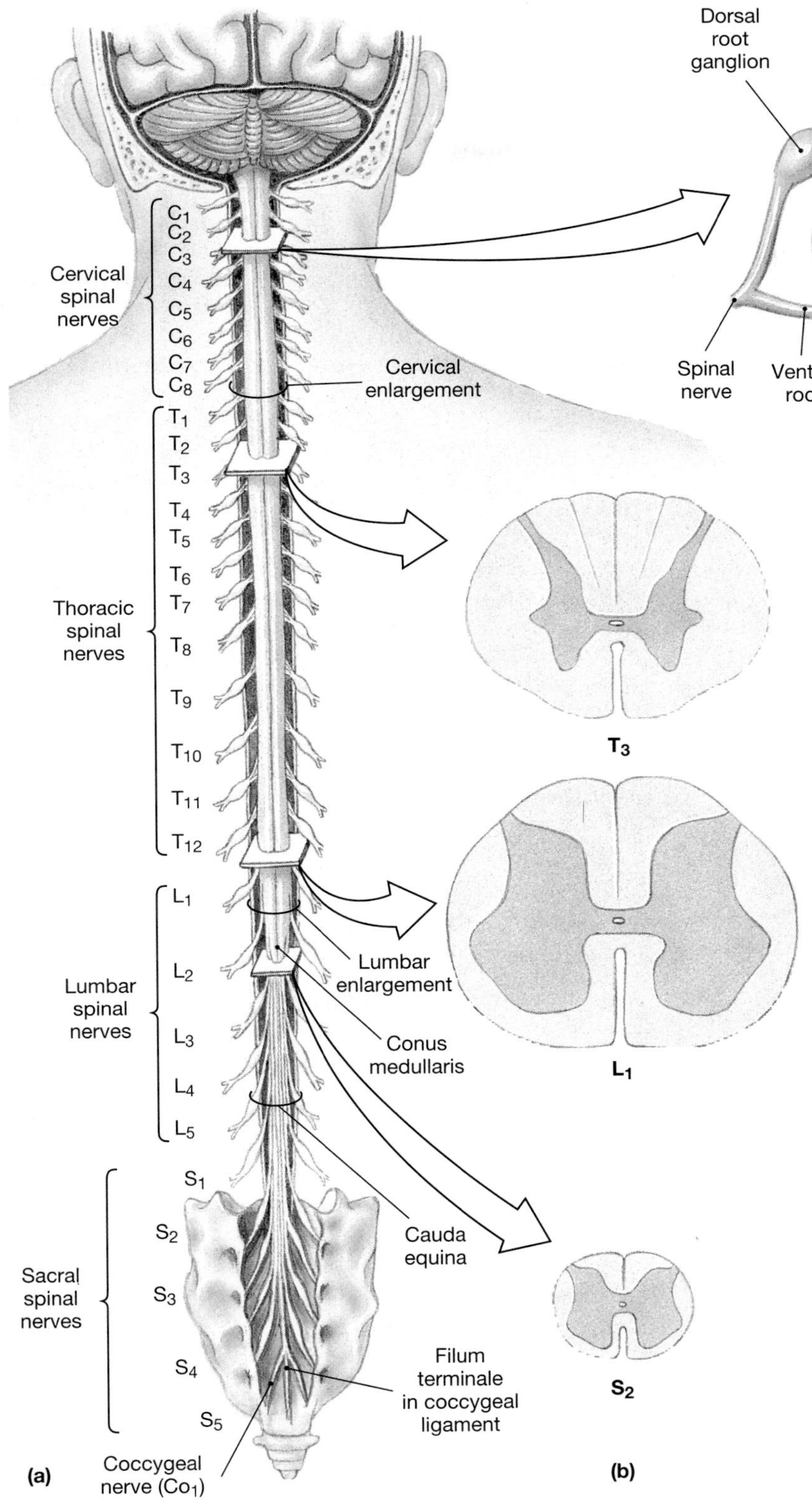

Dorsal root ganglion
Dorsal root
White matter
Posterior median sulcus
Central canal
Spinal nerve
Ventral root
Gray matter
C3
Anterior median fissure
Cervical spinal nerves
C1
C2
C3
C4
C5
C6
C7
C8
Cervical enlargement
T1
T2
T3
T4
T5
T6
T7
T8
T9
T10
T11
T12
Thoracic spinal nerves
T3
L1
L2
L3
L4
L5
Lumbar spinal nerves
Lumbar enlargement
Conus medullaris
L1
S1
S2
S3
S4
S5
Sacral spinal nerves
Cauda equina
Filum terminale in coccygeal ligament
S2
(a)
Coccygeal nerve (Co1)
(b)

FIGURE 13-1

Gross Anatomy of the Spinal Cord. (a) Superficial anatomy and orientation of the adult spinal cord. The numbers to the left identify the spinal nerves and indicate where the nerve roots leave the vertebral canal. Note, however, that the adult spinal cord extends only to the level of vertebrae L_1–L_2. **(b)** Inferior views of cross sections through representative segments of the spinal cord, showing the arrangement of gray and white matter.

spinal nerves of the neck. When indicating specific spinal nerves, it is customary to give them a regional number, as indicated in Figure 13-1a●. Each spinal nerve caudal to the first thoracic vertebra takes its name from the vertebra immediately preceding it. Thus spinal nerve T_1 emerges immediately caudal to vertebra T_1, spinal nerve T_2 follows vertebra T_2, and so forth.

The arrangement differs in the cervical region because the first pair of spinal nerves, C_1, emerges between the skull and the first cervical vertebra. For this reason, each cervical nerve takes its name from the vertebra immediately *following* it. In other words, cervical nerve C_2 *precedes* vertebra C_2, and the same system is used for the rest of the cervical series. The transition from one numbering system to another occurs between the last cervical and first thoracic vertebrae. Because the spinal nerve lying between them has been designated C_8, there are seven cervical vertebrae but *eight* cervical nerves.

The spinal cord continues to enlarge and elongate until an individual is approximately 4 years old. Up to that time enlargement of the spinal cord keeps pace with the growth of the vertebral column. Throughout this period the ventral and dorsal roots are very short, and they enter the intervertebral foramina immediately adjacent to their spinal segment. After age 4 the vertebral column continues to elongate, but the spinal cord does not. This vertebral growth carries the dorsal roots and spinal nerves farther and farther away from their original positions relative to the spinal cord. As a result, the dorsal and ventral roots gradually elongate, and the correspondence between spinal segment and the vertebral segment is lost. For example, in the adult the sacral segments of the spinal cord are at the level of vertebrae L_1–L_2 (Figure 13-1b●).

Because the adult spinal cord extends only to the level of the first or second lumbar vertebra, the dorsal and ventral roots of spinal segments L_2 to S_5 extend caudally, past the inferior tip of the conus medullaris. When seen in gross dissection, the filum terminale and the long ventral and dorsal roots reminded early anatomists of a horse's tail. As a result, they called the complex the **cauda equina** (KAW-duh ek-WĪ-nuh; *cauda*, tail + *equus*, horse).

Spinal Meninges

Figure 13-2

The vertebral column and its surrounding ligaments, tendons, and muscles isolate the spinal cord from the external environment. The delicate neural tissues must also be defended against damaging contacts with the surrounding bony walls of the vertebral canal. A series of specialized membranes, the **spinal meninges** (men-IN-jēz; *meninx*, membrane), provide the necessary physical stability and shock absorption. Blood vessels branching within these layers also deliver oxygen and nutrients to the spinal cord.

The relationships among the spinal meninges can be seen in Figure 13-2a●. There are three meningeal layers: the *dura mater*, the *arachnoid*, and the *pia mater*. At the foramen magnum of the skull, the spinal meninges are continuous with the **cranial meninges** that surround the brain. (The cranial meninges, which have the same three layers, are discussed in Chapter 14.)

Bacterial or viral infection can cause inflammation of the meningeal membranes, or **meningitis.** Meningitis is dangerous because it can disrupt the normal circulatory and cerebrospinal fluid supplies, damaging or killing neurons and glial cells in the affected areas. Although the initial diagnosis may specify the meninges of the spinal cord (*spinal meningitis*) or brain (*cerebral meningitis*), in later stages the entire meningeal system is usually affected. AM *Meningitis*

The Dura Mater

Figure 13-2

The tough, fibrous **dura mater** (DŪ-ruh MĀ-ter; *dura*, hard + *mater*, mother) forms the outermost covering of the spinal cord (Figure 13-2a●). The dense collagen fibers of the dura are oriented along the longitudinal axis of the cord. Between the dura mater and the walls of the vertebral canal lies the **epidural space,** which contains loose connective tissue, blood vessels, and a protective padding of adipose tissue (Figure 13-2b●).

The dura does not have extensive, firm connections to the surrounding vertebrae. Longitudinal stability is provided by localized attachment sites at either end of the vertebral canal. Cranially, the outer layer of the dura mater fuses with the periosteum of the occipital bone around the margins of the foramen magnum. Caudally the dura mater tapers from a sheath to a dense cord of collagen fibers that blend with components of the filum terminale to form the **coccygeal ligament.**

The coccygeal ligament continues along the sacral canal, ultimately blending into the periosteum of the coccyx. Lateral support for the dura mater is provided by loose connective tissue and adipose tissue within the epidural space. In addition, the dura mater extends between adjacent vertebrae at each intervertebral foramen, fusing with the connective tissues that surround the spinal nerves.

Injecting an anesthetic into the epidural space will affect only the spinal nerves in the immediate area of the injection. The result is a temporary sensory and motor paralysis known as an **epidural block.** Epidural blocks in the lower lumbar or sacral regions may be used to control pain during childbirth. AM *Spinal Anesthesia*

The Arachnoid

Figure 13-2

A narrow **subdural space** separates the inner surface of the dura mater from the second meningeal layer, the **arachnoid** (a-RAK-noyd; *arachne*, spider) (Figure 13-2b●). The inner surface of the dura and the outer surface of the arachnoid are lined by simple squamous epithelia. The intervening space contains a small quantity of serous fluid that reduces friction between the opposing surfaces. Beneath the arachnoid epithelium lies the **subarachnoid space,** which contains a delicate network of collagen and elastic fibers. The subarachnoid space is filled with **cerebrospinal fluid** that acts as a shock absorber as well as a diffusion medium for dissolved gases, nutrients, chemical messengers, and waste products.

The arachnoid membrane extends caudally as far as the filum terminale, and the dorsal and ventral roots of the cauda equina travel within the fluid-filled subarachnoid space. In adults, the withdrawal of cerebrospinal fluid, a procedure known as a **spinal tap,** involves inserting a needle into the subarachnoid space in the lower lumbar region. This placement avoids the possibility of damage to the spinal cord. Spinal taps are performed when CNS infection is suspected or when diagnosing severe back pain, headaches, disc problems, and some types of strokes. AM *Spinal Taps and Myelography*

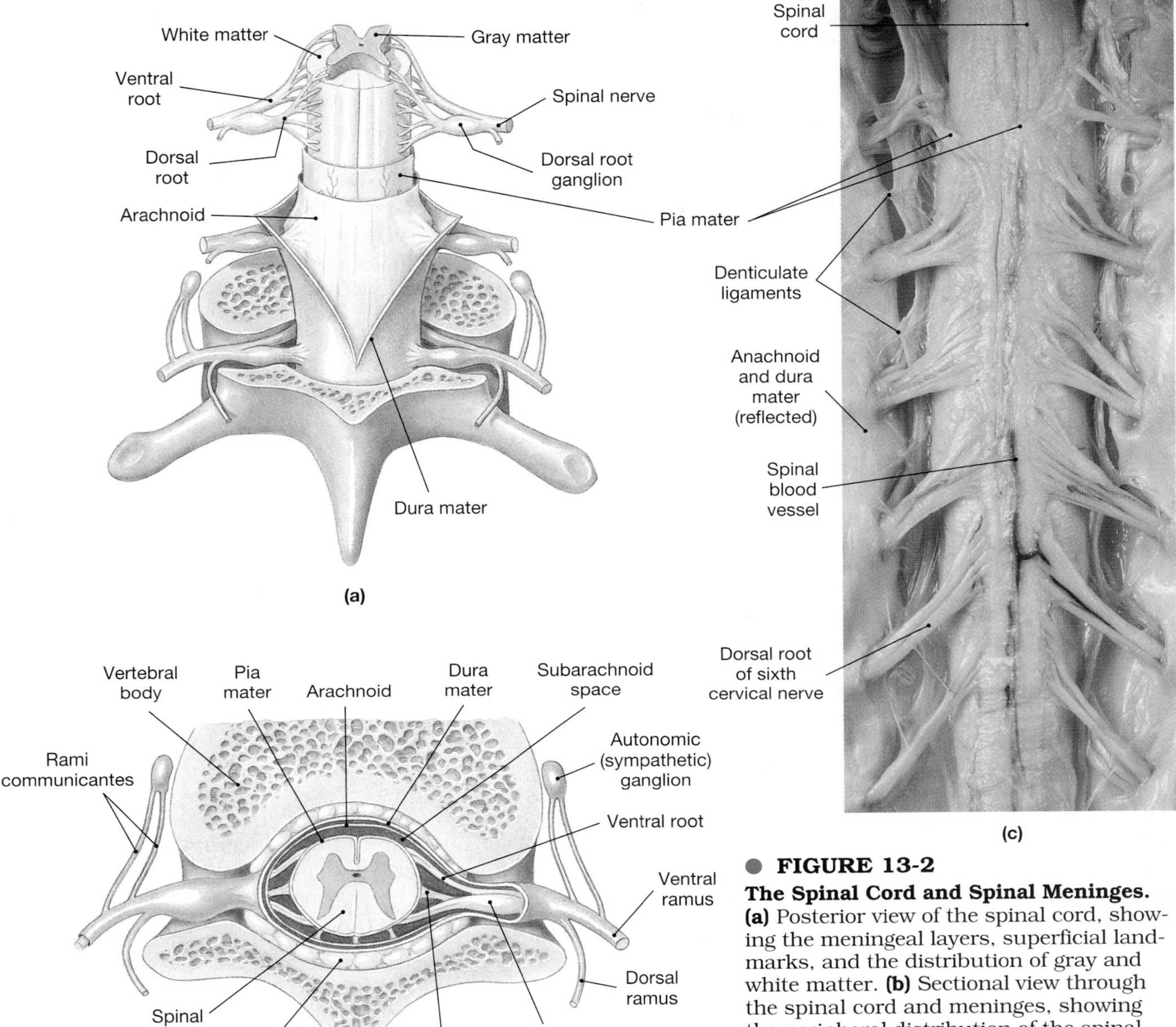
White matter
Gray matter
Ventral root
Spinal nerve
Dorsal root
Dorsal root ganglion
Arachnoid
Pia mater
Dura mater
(a)
Spinal cord
Denticulate ligaments
Anachnoid and dura mater (reflected)
Spinal blood vessel
Dorsal root of sixth cervical nerve
(c)
Vertebral body
Pia mater
Arachnoid
Dura mater
Subarachnoid space
Rami communicantes
Autonomic (sympathetic) ganglion
Ventral root
Ventral ramus
Dorsal ramus
Spinal cord
Dorsal root ganglion
Adipose tissue in epidural space
Denticulate ligament
(b)

● **FIGURE 13-2**
The Spinal Cord and Spinal Meninges. **(a)** Posterior view of the spinal cord, showing the meningeal layers, superficial landmarks, and the distribution of gray and white matter. **(b)** Sectional view through the spinal cord and meninges, showing the peripheral distribution of the spinal nerves. **(c)** Posterior view of the spinal cord and spinal nerve roots within the vertebral canal. The dura mater and arachnoid membrane have been removed; note the blood vessels in the delicate pia mater.

❑ The Pia Mater

Figure 13-2

The subarachnoid space bridges the gap between the arachnoid epithelium and the innermost meningeal layer, the **pia mater** (*pia*, delicate + *mater*, mother). The meshwork of elastic and collagen fibers of the pia are interwoven with those of the subarachnoid space. The blood vessels servicing the spinal cord are found here, and unlike more superficial meninges, the pia mater is firmly bound to the underlying neural tissue (Figure 13-2●).

Along the length of the spinal cord, paired **denticulate ligaments** extend from the pia mater through the arachnoid to the dura mater. Denticulate ligaments, which originate along either side of the spinal cord (Figure 13-2b●), prevent lateral (side-to-side) movement, while the dural connections at the foramen magnum and the sacrum, via the coccygeal ligament, prevent longitudinal (superior/inferior) movement.

The spinal meninges accompany the dorsal and ventral roots as they exit the vertebral cavity via the intervertebral foramina. As indicated in the sectional view of Figure 13-2b●, the meningeal membranes are continuous with the connective tissues surrounding the spinal nerves and their peripheral branches.

 Damage to which root of a spinal nerve would interfere with motor function?

Where is the cerebrospinal fluid that surrounds the spinal cord located?

Sectional Anatomy of the Spinal Cord

Figure 13-3

To understand the functional organization of the spinal cord, you must become familiar with its sectional organization. Consider Figure 13-3•, which presents a typical section through the spinal cord. The *anterior median fissure* and the *posterior median sulcus* mark the division between left and right sides of the spinal cord. The peripherally situated *white matter* contains large numbers of myelinated and unmyelinated axons. The *gray matter*, dominated by the cell bodies of neurons and glial cells, surrounds the narrow **central canal** and forms a rough H, or butterfly shape. The projections of gray matter toward the outer surface of the spinal cord are called **horns** (Figure 13-3•).

Organization of Gray Matter

Figure 13-3a

The cell bodies of neurons in the gray matter of the spinal cord are organized into functional groups, called *nuclei*. **Sensory nuclei** receive and relay sensory information from peripheral receptors. **Motor nuclei** issue motor commands to peripheral effectors. Although sensory and motor nuclei appear rather small in transverse section, they may extend for a considerable distance along the length of the spinal cord.

A frontal section along the length of the central canal of the spinal cord will separate the sensory (posterior, or dorsal) nuclei from the motor (anterior, or ventral) nuclei. The **posterior gray horns** contain somatic and visceral sensory nuclei, whereas the **anterior gray horns** are concerned with somatic motor control. The **lateral gray horns,** found in the thoracic and lumbar segments, are prominent expansions of the posterior and lateral portions of the anterior horns. Nuclei containing visceral motor neurons are found in this area. The **gray commissures** (*commissura*, a joining together) above and below the central canal contain axons crossing from one side of the cord to the other before reaching a destination within the gray matter.

Figure 13-3a• shows the relationship between the function of a particular nucleus (sensory or motor) and its relative position within the gray matter of the spinal cord. The nuclei within each gray horn are also highly organized.

Organization of White Matter

Figure 13-3a

The white matter on each side of the spinal cord can be divided into three regions called **columns,** or *funiculi* (Figure 13-3a•). The **posterior white column** extends between the posterior gray horns and the posterior median sulcus. The **anterior white column** lies between the anterior gray horns and the anterior median fissure. The anterior white columns on either side are interconnected by the **anterior white commissure.** The white matter between the anterior and posterior columns constitutes the **lateral white column.**

Each column contains *tracts* whose axons share functional and structural characteristics. A **tract,** or *fasciculus,* is a bundle of axons in the CNS that are relatively uniform with respect to diameter, myelination, and conduction speed. All of the axons within a tract relay the same type of information (sensory or motor) in the same direction. Short tracts carry sensory or motor signals between segments of the spinal cord, and larger tracts connect the spinal cord with the brain. **Ascending tracts** carry sensory information up toward the brain, and **descending tracts** convey motor commands down into the spinal cord.

DAMAGE TO SPINAL TRACTS All of the ascending sensory information and descending somatic motor commands associated with the trunk and limbs involve nuclei and tracts in the spinal cord. Multiple sclerosis (MS), a disorder mentioned in Chapter 12, produces muscular paralysis and sensory losses through demyelination. The initial symptoms appear as the result of myelin degeneration within the white matter of the lateral and posterior columns of the spinal cord or along tracts within the brain. During subsequent attacks the effects may become more widespread, and the cumulative sensory and motor losses can eventually lead to a generalized muscular paralysis. AM *Multiple Sclerosis*

Paralysis can also result from physical damage to the spinal cord in a severe auto crash or other accident, and the damaged tracts seldom undergo even partial repairs. Extensive damage to the spinal cord at the fourth or fifth cervical vertebra will eliminate sensation and motor control of the upper and lower limbs. The extensive paralysis produced is called *quadriplegia. Paraplegia*, the loss of motor control of the lower limbs, may follow damage to the thoracic spinal cord. AM *Spinal Cord Injuries and Experimental Treatments*

Less severe injuries affecting the spinal cord or cauda equina produce symptoms of sensory loss or motor paralysis that reflect the specific nuclei, tracts, or spinal nerves involved. One example, the loss of peripheral sensation along the distribution of a spinal nerve, is detailed in a later section.

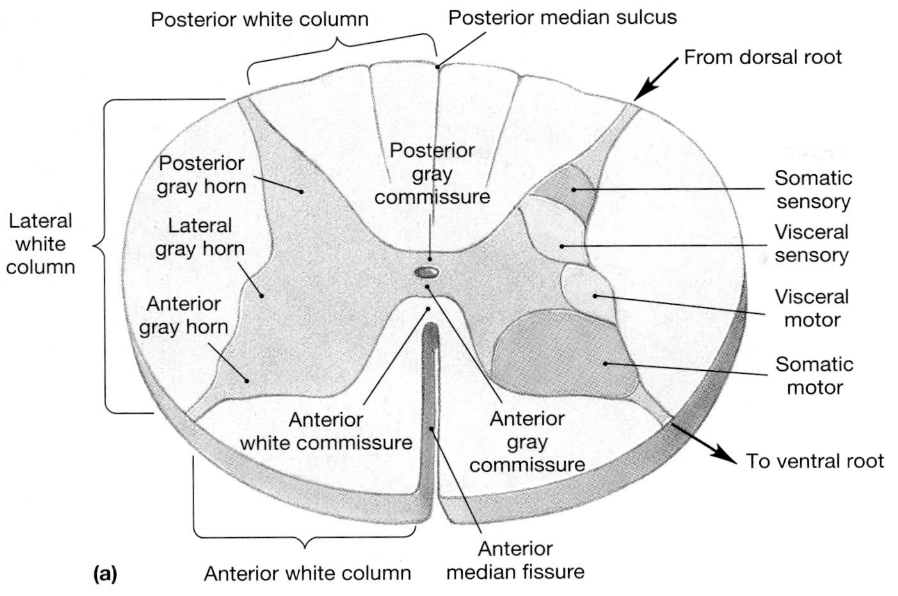
Posterior white column
Posterior median sulcus
From dorsal root
Posterior gray horn
Posterior gray commissure
Somatic sensory
Lateral white column
Lateral gray horn
Visceral sensory
Visceral motor
Anterior gray horn
Somatic motor
Anterior white commissure
Anterior gray commissure
To ventral root
Anterior median fissure
(a)
Anterior white column

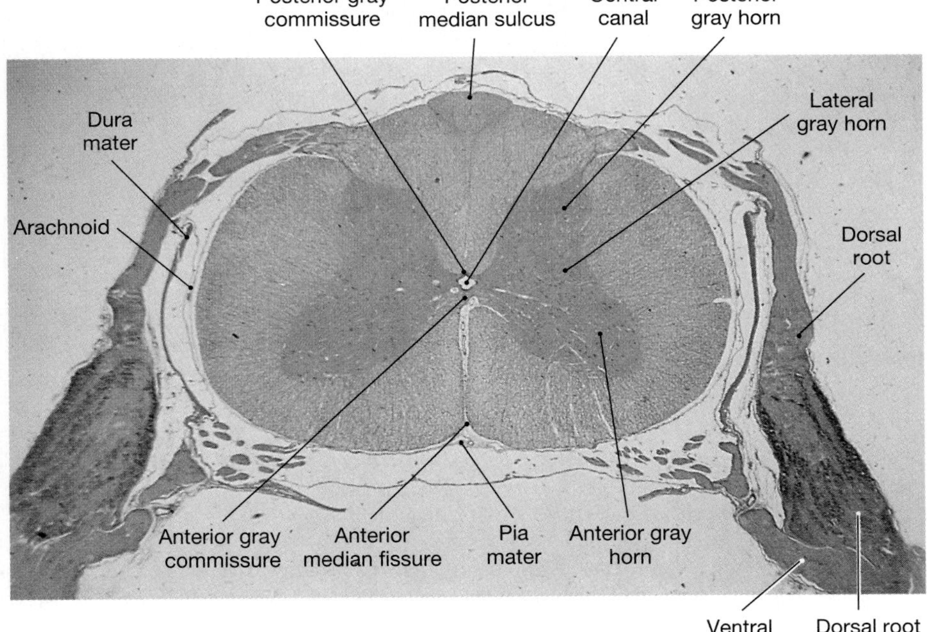
Posterior gray commissure
Posterior median sulcus
Central canal
Posterior gray horn
Lateral gray horn
Dura mater
Arachnoid
Dorsal root
Anterior gray commissure
Anterior median fissure
Pia mater
Anterior gray horn
Ventral root
Dorsal root ganglion
(b)

● **FIGURE 13-3**

Sectional Organization of the Spinal Cord. (a) The left half of this sectional view shows important anatomical landmarks and major regions of white matter in the posterior white column. The right half indicates the functional organization of the gray matter in the anterior, lateral, and posterior gray horns. **(b)** Micrograph of a section through the spinal cord, showing major landmarks; compare with part (a).

✔ A patient suffering from polio has lost the use of his leg muscles. In what area of the spinal cord would you expect to locate the virally infected motor neurons in this individual?

✔ What portion of the spinal cord would be affected by a disease that damages myelin sheaths?

Spinal Nerves

Figure 13-4

A series of connective tissue layers surrounds each spinal nerve and continues along all of its peripheral branches. These layers, best seen in sectional view (Figure 13-4●), are comparable to those associated with skeletal muscles (Figure 10-1●, p. 287). The outermost layer, or **epineurium,** consists of a dense network of collagen fibers. The fibers of the **perineurium** extend inward from the epineurium, dividing the nerve into a series of compartments that contain bundles of axons, or *fascicles.* Delicate connective tissue fibers of the **endoneurium** extend from the perineurium and surround individual axons.

Arteries and veins penetrate the epineurium and branch within the perineurium. Capillaries leaving the perineurium branch in the endoneurium and supply the axons and Schwann cells of the nerve and the fibroblasts of the connective tissues.

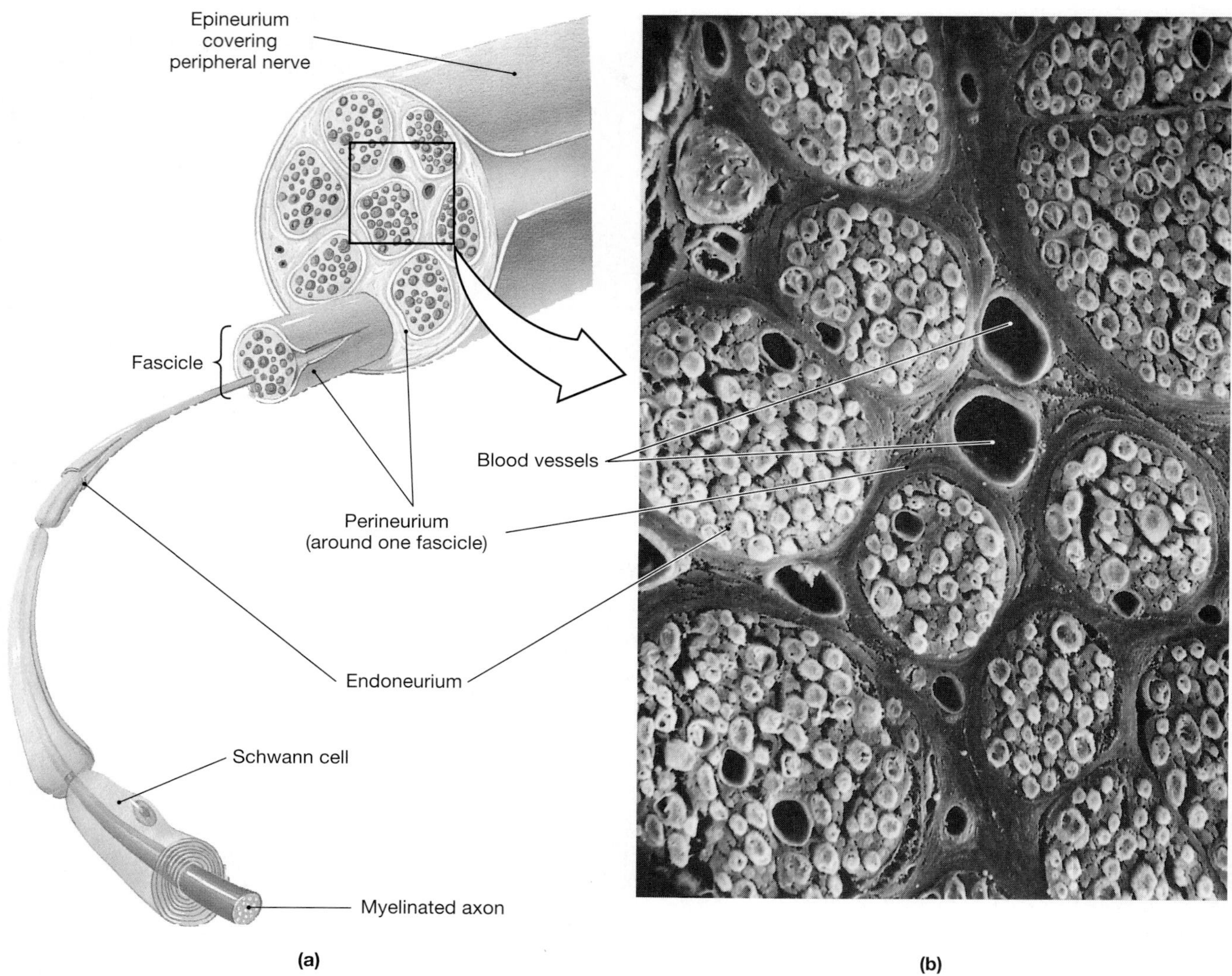

Epineurium covering peripheral nerve
Fascicle
Blood vessels
Perineurium (around one fascicle)
Endoneurium
Schwann cell
Myelinated axon
(a)
(b)

● **FIGURE 13-4**
Peripheral Nerves. (a) A typical peripheral nerve and its connective tissue wrappings. **(b)** A scanning electron micrograph showing the perineurium and endoneurium in great detail. (SEM × 340)

❑ Peripheral Distribution of Spinal Nerves

Figures 13-5, 13-6

Figure 13-5● diagrams the distribution of a typical spinal nerve. The spinal nerve forms just outside of the intervertebral foramen, where the dorsal and ventral roots unite. We will now follow the distribution of that nerve as it extends into the periphery.

In the thoracic and lumbar regions (segments T_1–L_2), the first branch from the spinal nerve carries visceral motor fibers to a nearby **autonomic ganglion** of the *sympathetic division* of the ANS. Because preganglionic axons are myelinated, this branch has a light color and is known as the **white ramus** ("branch"). The postganglionic fibers leaving the ganglion are unmyelinated. Those fibers headed for internal organs do not rejoin the spinal nerve, but form a series of separate **autonomic nerves.** Postganglionic fibers innervating glands and smooth muscles in the body wall or limbs form a **gray ramus** that rejoins the spinal nerve. The gray and white rami are collectively termed the **rami communicantes,** or "communicating branches."

The **dorsal ramus** of each spinal nerve provides sensory and motor innervation to the skin and muscles of the back. The relatively large **ventral ramus** supplies the ventrolateral body surface, structures in the body wall, and the limbs. The distribution of the dorsal and ventral rami provides a superb illustration of the segmental division of labor along the length of the spinal cord. Each pair of spinal nerves monitors a specific region of the body surface, an area known as a **dermatome.** Dermatomes, illustrated in Figure 13-6●, are clinically important because damage or infection of a spinal nerve or dorsal root ganglion will produce a characteristic loss of sensation in the skin. For example, in *shingles* a virus that infects dorsal root ganglia causes

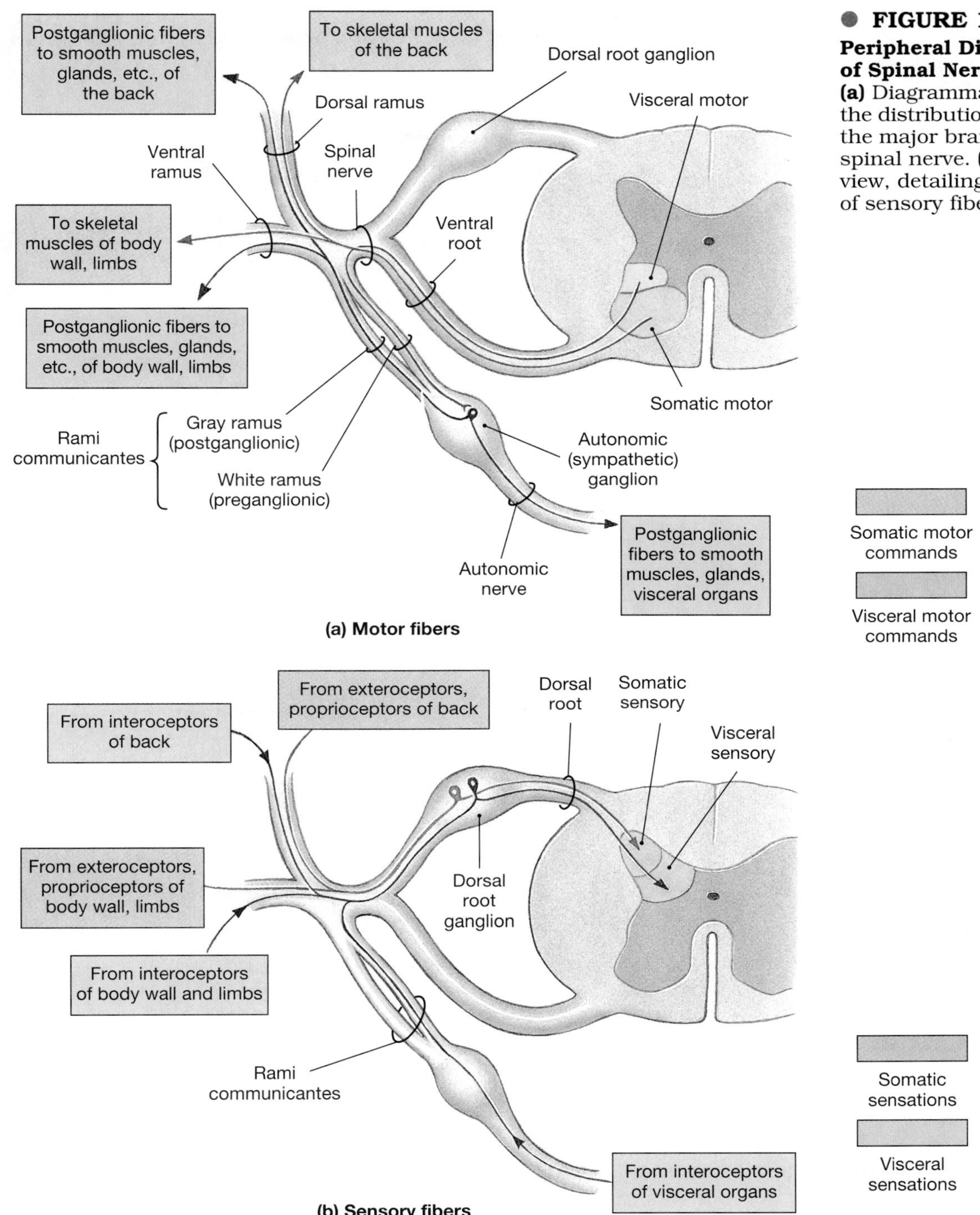
Postganglionic fibers to smooth muscles, glands, etc., of the back
To skeletal muscles of the back
Dorsal root ganglion
Dorsal ramus
Visceral motor
Ventral ramus
Spinal nerve
To skeletal muscles of body wall, limbs
Ventral root
Postganglionic fibers to smooth muscles, glands, etc., of body wall, limbs
Somatic motor
Rami communicantes
Gray ramus (postganglionic)
White ramus (preganglionic)
Autonomic (sympathetic) ganglion
Postganglionic fibers to smooth muscles, glands, visceral organs
Autonomic nerve
(a) Motor fibers
Somatic motor commands
Visceral motor commands
From interoceptors of back
From exteroceptors, proprioceptors of back
Dorsal root
Somatic sensory
Visceral sensory
From exteroceptors, proprioceptors of body wall, limbs
Dorsal root ganglion
From interoceptors of body wall and limbs
Rami communicantes
Somatic sensations
Visceral sensations
From interoceptors of visceral organs
(b) Sensory fibers

● **FIGURE 13-5**
Peripheral Distribution of Spinal Nerves.
(a) Diagrammatic view indicating the distribution of motor fibers in the major branches of a typical spinal nerve. **(b)** A comparable view, detailing the distribution of sensory fibers.

a painful rash whose distribution corresponds to that of the affected sensory nerves. AM *Shingles and Hansen's Disease*

❑ Nerve Plexuses

Figures 13-5, 13-7

The relatively simple distribution pattern of dorsal and ventral rami illustrated in Figure 13-5● applies to spinal nerves T_2–T_{12}. But in segments controlling the skeletal musculature of the neck, upper limbs, or lower limbs, the situation becomes more complicated. During development, small skeletal muscles that are innervated by different ventral rami often fuse together to form larger muscles with compound origins. Although the anatomical distinctions between the component muscles may disappear, separate ventral rami continue to provide sensory innervation and motor control to each portion of the compound muscle. As they converge, the ventral rami of adjacent spinal nerves blend their fibers to

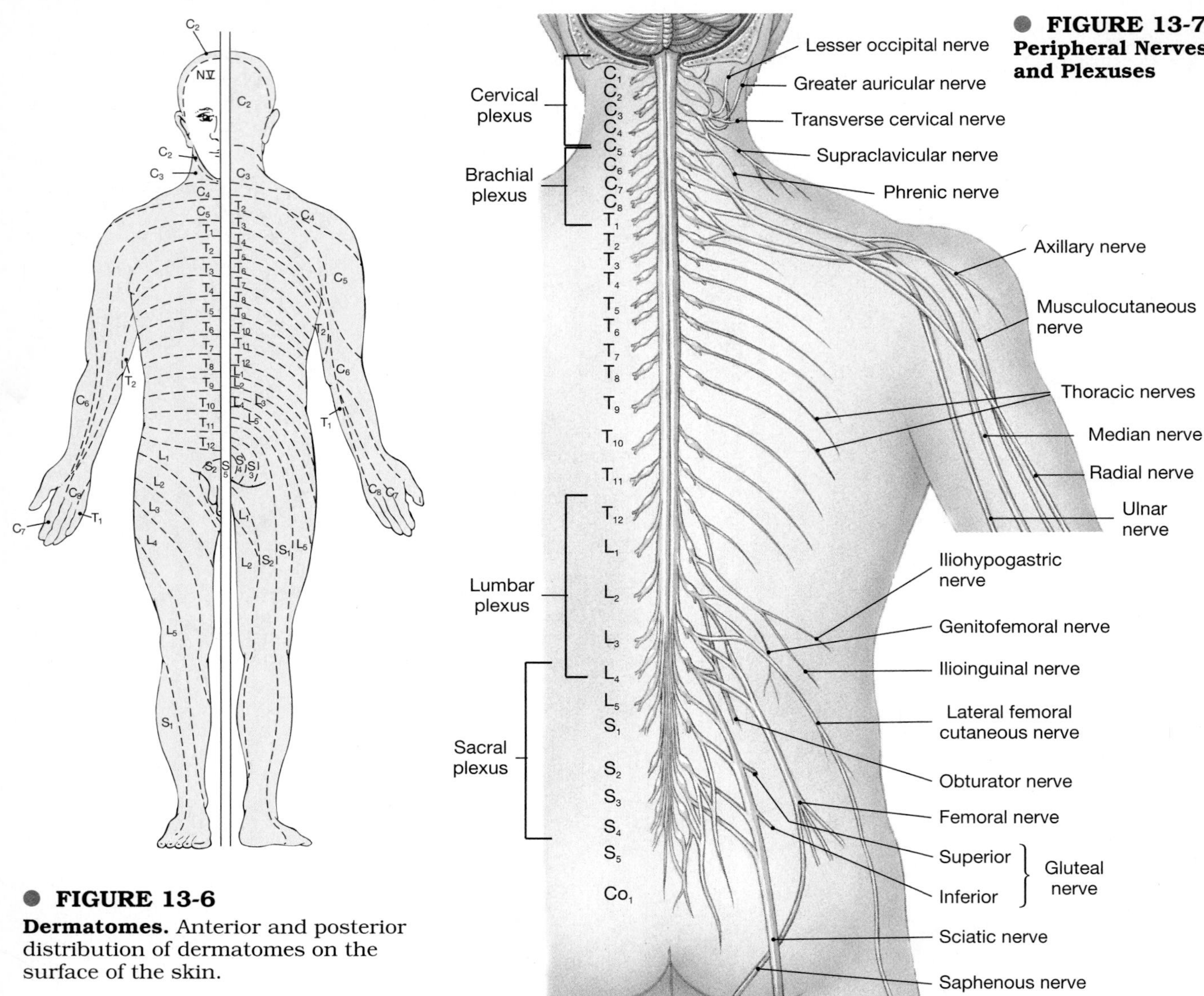
Lesser occipital nerve
Greater auricular nerve
Transverse cervical nerve
Supraclavicular nerve
Phrenic nerve
Cervical plexus
Brachial plexus
Lumbar plexus
Sacral plexus
Axillary nerve
Musculocutaneous nerve
Thoracic nerves
Median nerve
Radial nerve
Ulnar nerve
Iliohypogastric nerve
Genitofemoral nerve
Ilioinguinal nerve
Lateral femoral cutaneous nerve
Obturator nerve
Femoral nerve
Superior
Inferior
Gluteal nerve
Sciatic nerve
Saphenous nerve

● **FIGURE 13-6**
Dermatomes. Anterior and posterior distribution of dermatomes on the surface of the skin.

● **FIGURE 13-7**
Peripheral Nerves and Plexuses

produce a series of compound nerve trunks. Such a complex interwoven network of nerves is called a **nerve plexus** (PLEK-sus; *plexus*, a braid). There are three major plexuses: the *cervical plexus*, the *brachial plexus*, and the *lumbosacral plexus* (Figure 13-7●).

Chapter 11 introduced the peripheral nerves that control the major axial and appendicular muscles. At this point you should refer to the tables in that chapter to review the innervation of the skeletal muscle groups. [pp. 330–374]

Peripheral *nerve palsies*, also known as **peripheral neuropathies,** are characterized by regional losses of sensory and motor function as the result of nerve trauma or compression. You have experienced a mild, temporary palsy if your arm or leg has ever "fallen asleep" after leaning or sitting in an uncomfortable position. AM *Palsies* Although dermatomes can provide clues to the location of injuries along the spinal cord, the loss of sensation at the skin does not provide sufficiently precise information concerning the site of injury because the boundaries of dermatomes are not exact, clearly defined lines. More exact conclusions can be drawn from the loss of motor control, based on the origin and distribution of the peripheral nerves originating at nerve plexuses.

If a peripheral axon is damaged but not displaced, normal function may eventually return as the cut stump grows across the injury site away from the soma and along its former path. The mechanics of this process were detailed at the close of the previous chapter. [p. 417] Repairs made after damage to an entire peripheral *nerve* are generally incomplete, primarily because of problems with axon alignment and regrowth. A variety of technologically sophisticated procedures designed to improve nerve regeneration and repair are currently under evaluation. An entire family of *nerve growth factors* has been discovered in the last 3 years, and their use alone or in combination with other therapies may ultimately revolutionize the treatment of damaged neural tissue inside and outside of the CNS. AM *Damage and Repair of Peripheral Nerves*

The Cervical Plexus

Figures 13-7, 13-8

The **cervical plexus** (Figures 13-7, 13-8● and Table 13-1) consists of the ventral rami of spinal nerves C_1–C_5. Its branches innervate the muscles of the neck and extend into the thoracic cavity to control the diaphragmatic muscles. The **phrenic nerve,** the major nerve of this plexus, provides the entire nerve supply to the *diaphragm,* an important respi[ratory] muscle. Other branches are distributed to the [skin] of the neck and the upper portion of the chest.

The Brachial Plexus

Figures 13-7, 13-9

The **brachial plexus** (Table 13-2) innervates the shoulder girdle and upper limb, with contributions from the ventral rami of spinal nerves C_5–T_1 (Figures

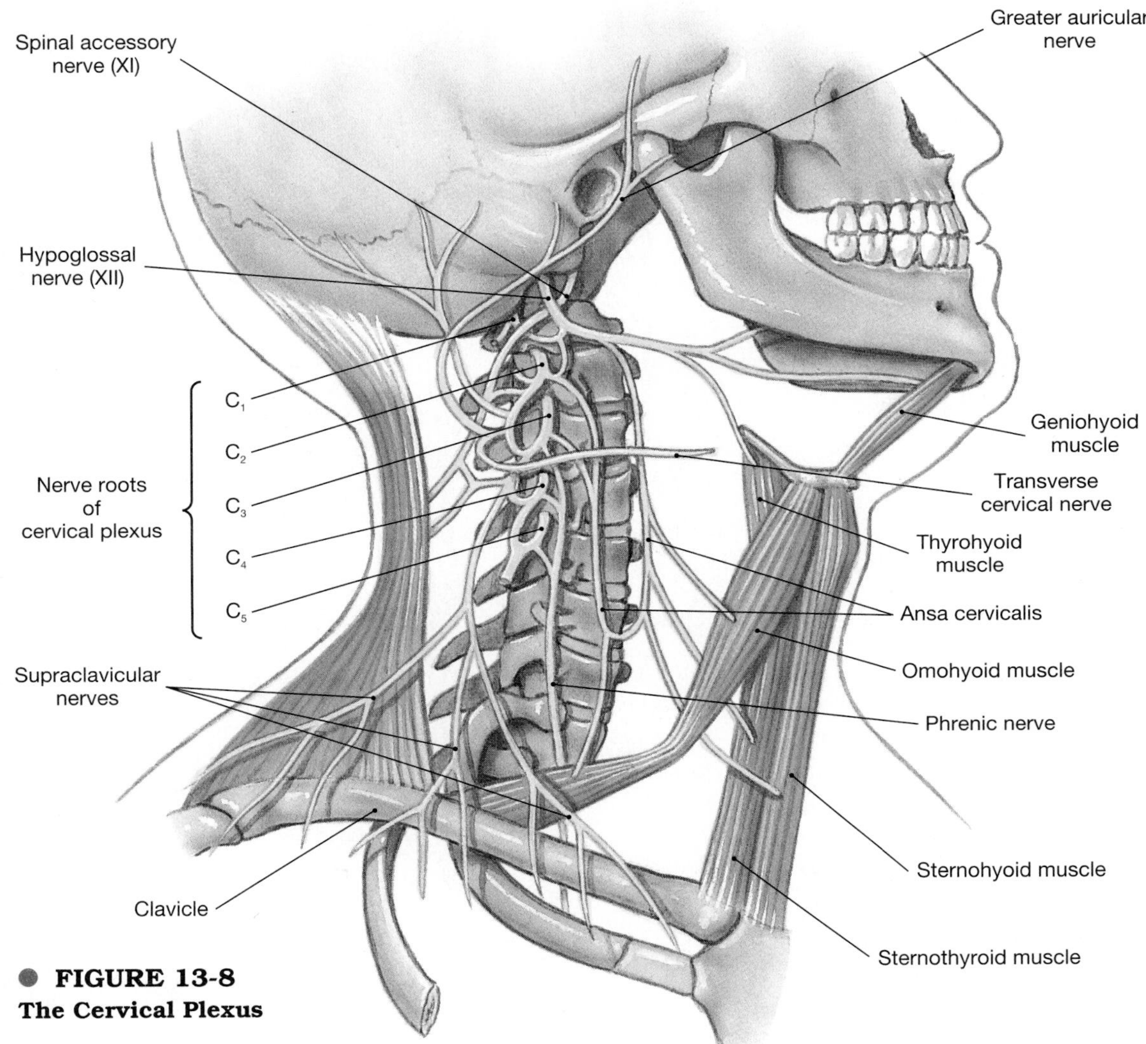

Spinal accessory nerve (XI)
Greater auricular nerve
Hypoglossal nerve (XII)
C_1
C_2
Nerve roots of cervical plexus
C_3
C_4
C_5
Geniohyoid muscle
Transverse cervical nerve
Thyrohyoid muscle
Ansa cervicalis
Omohyoid muscle
Supraclavicular nerves
Phrenic nerve
Sternohyoid muscle
Clavicle
Sternothyroid muscle

● **FIGURE 13-8**
The Cervical Plexus

TABLE 13-1 The Cervical Plexus

Spinal Segment	*Nerves*	*Distribution*
C_1–C_4	Ansa cervicalis (superior and inferior branches)	Five of the extrinsic laryngeal muscles: sternothyroid, sternohyoid, omohyoid, geniohyoid, and thyrohyoid (via N XII)
C_2–C_3	Lesser occipital, transverse cervical, supraclavicular, and greater auricular nerves	Skin of upper chest, shoulder, neck, and ear
C_3–C_5	Phrenic nerve	Diaphragm
C_1–C_5	Cervical nerves	Levator scapulae, scalenes, sternocleidomastoid, and trapezius (with N XI)

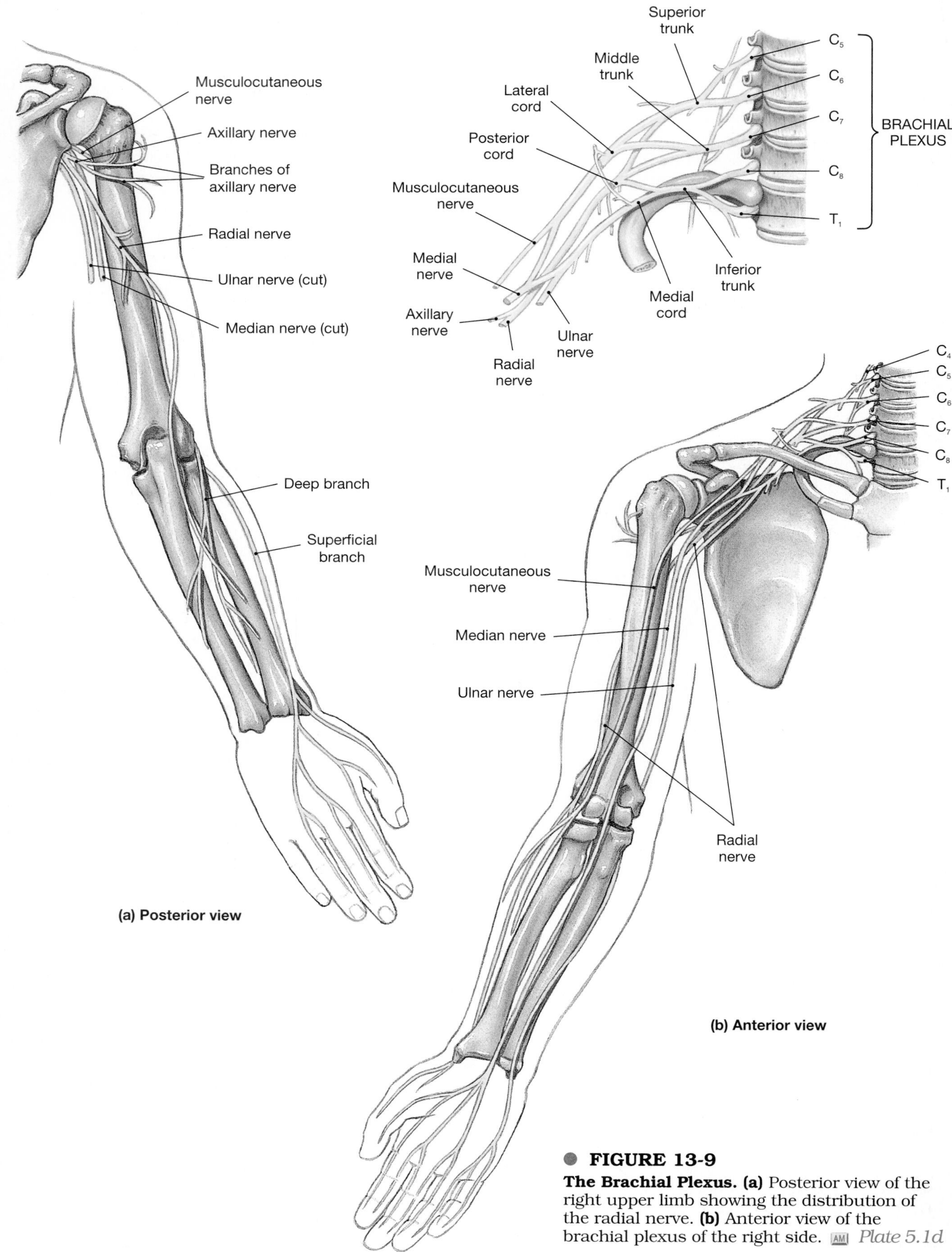

Musculocutaneous nerve
Axillary nerve
Branches of axillary nerve
Radial nerve
Ulnar nerve (cut)
Median nerve (cut)
Deep branch
Superficial branch
(a) Posterior view
Superior trunk
Middle trunk
Lateral cord
Posterior cord
Musculocutaneous nerve
Medial nerve
Axillary nerve
Radial nerve
Ulnar nerve
Medial cord
Inferior trunk
C_5
C_6
C_7
C_8
T_1
BRACHIAL PLEXUS
C_4
C_5
C_6
C_7
C_8
T_1
Musculocutaneous nerve
Median nerve
Ulnar nerve
Radial nerve
(b) Anterior view

FIGURE 13-9
The Brachial Plexus. (a) Posterior view of the right upper limb showing the distribution of the radial nerve. **(b)** Anterior view of the brachial plexus of the right side. AM *Plate 5.1d*

13-7 and 13-9●). The nerves that form this plexus originate from trunks and cords named according to their location. Axons from the spinal nerves pass through the *superior, middle,* and *inferior trunks* to reach the *lateral, medial,* and *posterior cords.* The major nerves of the lateral cord are the **musculocutaneous nerve** and the **median nerve.** The **ulnar nerve** is the major nerve of the medial cord. The **circumflex (axillary) nerve** and **radial nerve** are the major nerves of the posterior cord.

The Lumbosacral Plexus

Figures 13-7, 13-10

The **lumbosacral plexus** (Figures 13-7 and 13-10●) arises from the lumbar and sacral segments of the spinal cord, and the ventral rami of these nerves supply the pelvic girdle and lower limbs. This plexus can be subdivided into a **lumbar plexus** (T_{12}–L_4) and a **sacral plexus** (L_4–S_4). The individual nerves that form the lumbosacral plexus and their distributions are detailed in Table 13-3.

The major nerves of the lumbar plexus are the **genitofemoral nerve, lateral femoral cutaneous nerve,** and **femoral nerve.** The major nerves of the sacral plexus are the **sciatic nerve** and the **pudendal nerve.** The sciatic nerve passes posterior to the femur, deep to the long head of the biceps femoris muscle. As it approaches the popliteal fossa, the sciatic nerve divides into two branches, the **common peroneal nerve** and the **tibial nerve.**

An anesthetic blocks the function of the dorsal rami of the cervical spinal nerves. What areas of the body would be affected?

Injury to which of the nerve plexuses would interfere with the ability to breathe?

Compression of which nerve produces the sensation of your leg "falling asleep"?

An Introduction to Reflexes

Conditions inside or outside the body can change rapidly and unexpectedly. **Reflexes** are rapid, automatic responses to stimuli. Reflexes preserve homeostasis by making rapid adjustments in the function of organs or organ systems. The response shows little variability—activation of a particular reflex usually produces the same motor response. Chapter 1 introduced the basic functional components involved in all types of homeostatic regulation: a *receptor*, an *integration center*, and an *effector*. [p. 13] This discussion considers *neural reflexes*, in which sensory fibers deliver information to the CNS and motor fibers carry motor commands to peripheral effectors. Chapter 18 will examine *endocrine reflexes*, in which the com-

TABLE 13-2 The Brachial Plexus

Spinal Segment	*Nerves*	*Distribution*
C_5, C_6	Axillary nerve	Deltoid and teres minor muscles Skin of shoulder
C_5–T_1	Radial nerve	Extensor muscles on the arm and forearm (triceps brachii, brachioradialis, extensor carpi radialis, and extensor carpi ulnaris) Digital extensors and abductor pollicis Skin over the posterolateral surface of the arm
C_5–C_7	Musculocutaneous nerve	Flexor muscles on arm (biceps brachii, brachialis, and coracobrachialis) Skin over lateral surface of forearm
C_6–T_1	Median nerve	Flexor muscles on forearm (flexor carpi radialis and palmaris longus) Pronators (p. quadratus and p. teres) Digital flexors Skin over lateral surface of hand
C_8, T_1	Ulnar nerve	Flexor muscle on forearm (flexor carpi ulnaris) Adductor pollicis and small digital muscles Skin over medial surface of hand

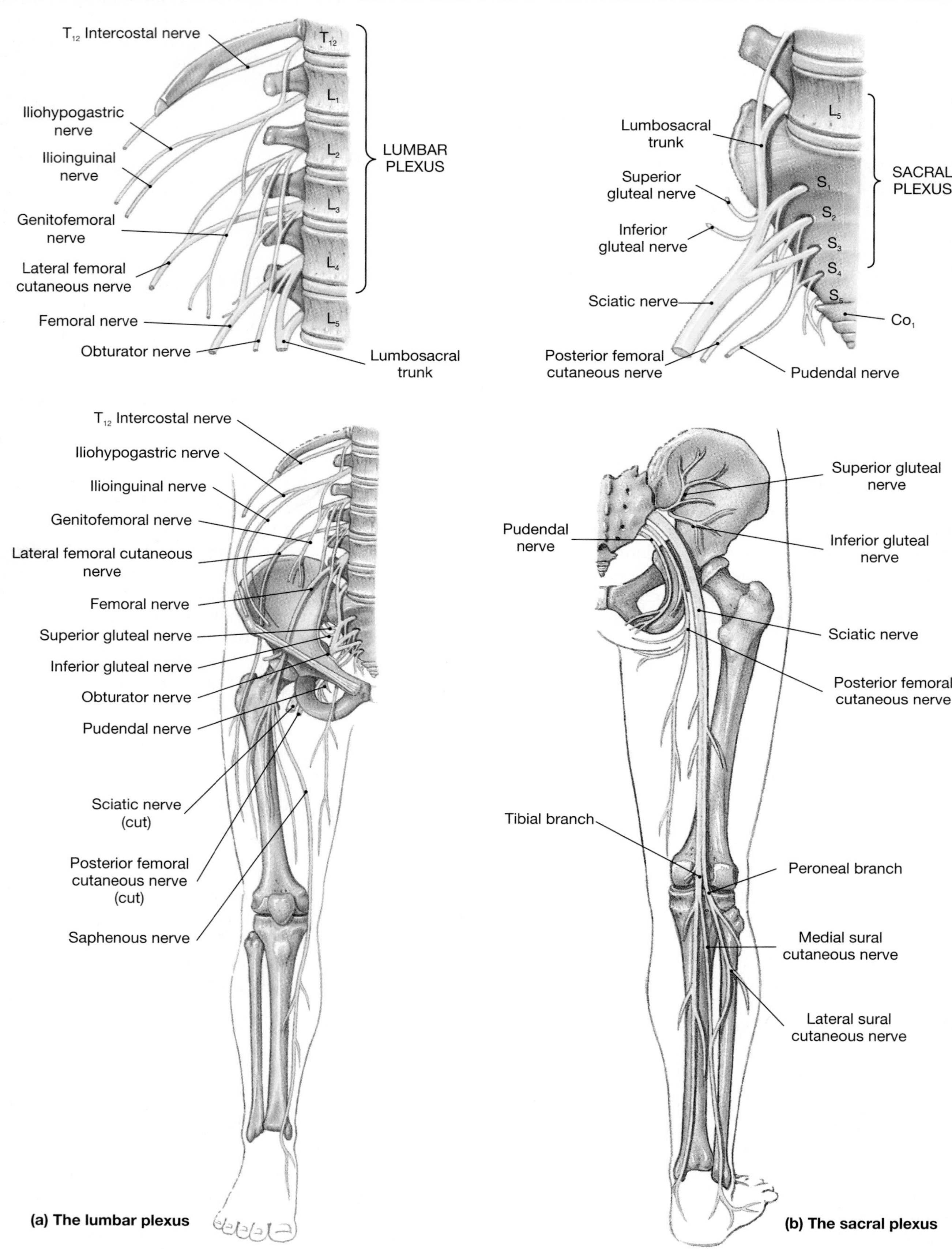

T12 Intercostal nerve
T12
L1
L2
L3
L4
L5
LUMBAR PLEXUS
Iliohypogastric nerve
Ilioinguinal nerve
Genitofemoral nerve
Lateral femoral cutaneous nerve
Femoral nerve
Obturator nerve
Lumbosacral trunk
L5
Lumbosacral trunk
Superior gluteal nerve
Inferior gluteal nerve
S1
S2
S3
S4
S5
SACRAL PLEXUS
Sciatic nerve
Co1
Posterior femoral cutaneous nerve
Pudendal nerve
T12 Intercostal nerve
Iliohypogastric nerve
Ilioinguinal nerve
Genitofemoral nerve
Lateral femoral cutaneous nerve
Femoral nerve
Superior gluteal nerve
Inferior gluteal nerve
Obturator nerve
Pudendal nerve
Sciatic nerve (cut)
Posterior femoral cutaneous nerve (cut)
Saphenous nerve
(a) The lumbar plexus
Superior gluteal nerve
Pudendal nerve
Inferior gluteal nerve
Sciatic nerve
Posterior femoral cutaneous nerve
Tibial branch
Peroneal branch
Medial sural cutaneous nerve
Lateral sural cutaneous nerve
(b) The sacral plexus

● **FIGURE 13-10**
The Lumbosacral Plexus. (a) Lumbar plexus. **(b)** Sacral plexus.

TABLE 13-3 The Lumbosacral Plexus

Spinal Segment	*Nerves*	*Distribution*
THE LUMBAR PLEXUS		
T_{12}, L_1	Iliohypogastric nerve	Abdominal muscles (external and internal obliques, transversus abdominis)
		Skin over lower abdomen and buttocks
L_1	Ilioinguinal nerve	Abdominal muscles (with iliohypogastric)
		Skin over medial upper thigh and portions of external genitalia
L_1, L_2	Genitofemoral nerve	Skin over anteromedial surface of thigh and portions of external genitalia
L_2, L_3	Lateral femoral cutaneous nerve	Skin over anterior, lateral, and posterior surfaces of thigh
L_2–L_4	Femoral nerve	Anterior muscles of thigh (sartorius and quadriceps)
		Adductors of thigh (pectineus and iliopsoas)
		Skin over anteromedial surface of thigh, medial surface of leg and foot
L_2–L_4	Obturator nerve	Adductors of thigh (adductor magnus, brevis, longus)
		Gracilis muscle
		Skin over medial surface of thigh
L_2–L_4	Saphenous nerve	Skin over medial surface of leg
THE SACRAL PLEXUS		
L_4–S_2	Gluteal nerves:	
	Superior	Abductors of thigh (gluteus minimus, gluteus medius, and tensor fasciae latae)
	Inferior	Extensor of thigh (gluteus maximus)
L_4–S_3	Sciatic nerve:	Two of the hamstrings (semimembranosus and semitendinosus)
		Adductor magnus (with obturator nerve)
	Tibial branch	Flexors of leg and plantar flexors of foot (popliteus, gastrocnemius, soleus, tibialis posterior)
		Flexors of toes
		Skin over posterior surface of leg, plantar surface of foot
	Peroneal branch	Biceps femoris of hamstrings
		Peroneus (brevis and longus) and tibialis anterior
		Extensors of toes
		Skin over anterior surface of leg and dorsal surface of foot
S_2–S_4	Pudendal nerve	Muscles of perineum
		Skin of external genitalia

mands to peripheral tissues and organs are delivered by hormones in the bloodstream.

❑ The Reflex Arc

Figure 13-11

The "wiring" of a single reflex is called a **reflex arc.** A reflex arc begins at a receptor and ends at a peripheral effector, such as a muscle fiber or gland cell. Figure 13-11• diagrams the five steps involved in a neural reflex: (1) *arrival of a stimulus and activation of a receptor,* (2) *activation of a sensory neuron,* (3) *information processing,* (4) *activation of a motor neuron,* and (5) *response by an effector* (muscle or gland).

Step 1: ***Arrival of a Stimulus and Activation of a Receptor.*** A receptor is a specialized cell or nerve ending sensitive to physical or chemical changes in the body or the external environment.

There are many types of sensory receptors; general categories were introduced in Chapter 12. [p. 389] In Figure 13-11●, leaning on a tack activates a pain receptor in the palm of the hand. (The linkage between receptor stimulation and sensory neuron activation will be discussed further in Chapter 17.)

Step 2: ***Activation of a Sensory Neuron.*** Stimulation of the pain receptor leads to activation of the sensory neuron and the propagation of action potentials to the CNS along an afferent fiber. The information in our example enters the spinal cord via one of the dorsal roots.

Step 3: ***Information Processing.*** Information processing begins when a neurotransmitter released by synaptic terminals of the sensory neuron reaches the postsynaptic membrane of a motor neuron or interneuron and produces an EPSP. [p. 413] This stimulus will be integrated with other stimuli arriving at the postsynaptic neuron at that moment. If that neuron is a motor neuron, the integration process will determine whether or not a specific motor response will occur. If the postsynaptic neuron is an interneuron, the next step may be the activation of a neuronal pool and subsequent activation of motor neurons that produce an appropriate motor response.

Step 4: ***Activation of a Motor Neuron.*** When activated, the axon of a motor neuron carries action potentials into the periphery, in this example over the ventral root of a spinal nerve.

Step 5: ***Response of a Peripheral Effector.*** The release of neurotransmitter by the motor neuron at synaptic knobs then leads to a response by a peripheral effector, in this case a skeletal muscle. A reflex response usually removes or opposes the original stimulus; in this case, the contracting muscle pulls the hand away from the painful stimulus. This reflex arc is therefore an example of a *negative feedback control mechanism.* [p. 14] By opposing potentially harmful changes in the internal or external environment, reflexes play an important role in homeostatic maintenance.

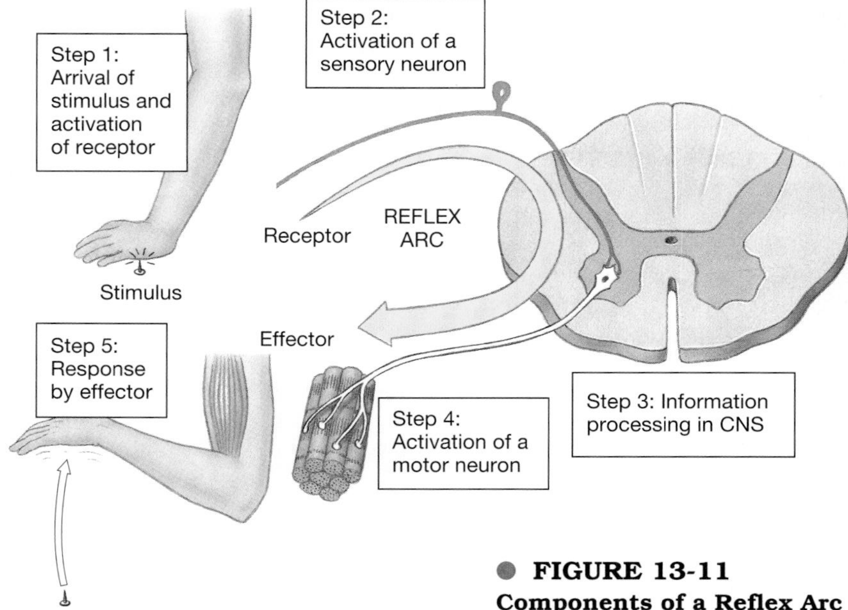
Step 1:
Arrival of stimulus and activation of receptor
Step 2:
Activation of a sensory neuron
Receptor
REFLEX ARC
Stimulus
Effector
Step 5:
Response by effector
Step 4:
Activation of a motor neuron
Step 3: Information processing in CNS

● **FIGURE 13-11**
Components of a Reflex Arc

❑ Classification of Reflexes

Figure 13-12

Reflexes can be classified according to (1) their development, (2) the site where information processing occurs, (3) the nature of the resulting motor response, or (4) the complexity of the neural circuit

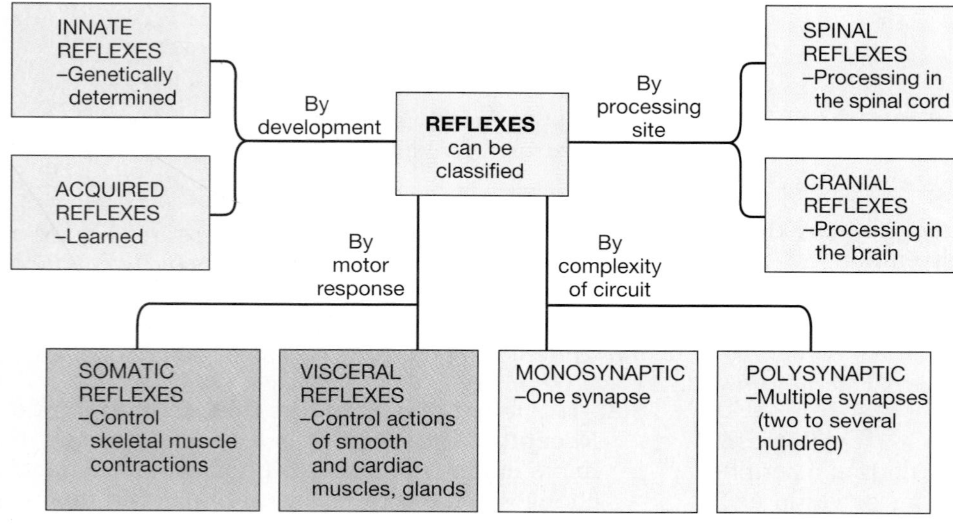
INNATE REFLEXES
–Genetically determined
ACQUIRED REFLEXES
–Learned
By development
REFLEXES can be classified
By processing site
SPINAL REFLEXES
–Processing in the spinal cord
CRANIAL REFLEXES
–Processing in the brain
By motor response
By complexity of circuit
SOMATIC REFLEXES
–Control skeletal muscle contractions
VISCERAL REFLEXES
–Control actions of smooth and cardiac muscles, glands
MONOSYNAPTIC
–One synapse
POLYSYNAPTIC
–Multiple synapses (two to several hundred)

● **FIGURE 13-12**
Methods of Classifying Reflexes

involved. These categories, diagrammed in Figure 13-12●, are not mutually exclusive; they represent different ways of describing a single reflex.

Development of Reflexes

Innate reflexes result from the connections that form between neurons during development. They usually appear in a predictable sequence, from the simplest reflex responses (withdrawal from pain) to more complex motor patterns (chewing, suckling, or tracking objects with the eyes). The neural connections responsible for basic motor patterns of an innate reflex are genetically or developmentally programmed. Examples include the reflexive removal of a hand from a hot stovetop and blinking when the eyelashes are touched.

More complex, learned motor patterns are sometimes called **acquired reflexes.** An experienced driver steps on the brake when trouble appears on the road ahead; a professional skier makes rapid, automatic adjustments in body position while racing. These motor responses are rapid and automatic, but they were learned rather than preestablished, and the reflexes enhanced by repetition. The distinction between innate and acquired reflexes is not absolute; some people can learn motor patterns more quickly than others, and the differences probably have a genetic basis.

Most reflexes, whether innate or acquired, can be modified over time or suppressed through conscious effort. For example, while walking a tightrope over the Grand Canyon you would probably ignore a bee sting, although under other circumstances it would probably cause you to abruptly withdraw the foot or hand involved.

Processing Sites

Reflexes processed in the brain are called **cranial reflexes.** Several examples of cranial reflexes will be found in Chapters 15 and 16. In a **spinal reflex** the important interconnections and processing events occur inside the spinal cord. These reflexes are discussed in detail in the next section.

Nature of the Motor Response

Somatic reflexes control activities of the muscular system, whereas **visceral reflexes,** or *autonomic reflexes,* control the activities of other systems. This chapter considers somatic reflexes in detail; visceral reflexes are examined in Chapter 16.

Complexity of the Neural Circuitry

Figure 13-13

In the simplest reflex arc a sensory neuron synapses directly on a motor neuron, which itself serves

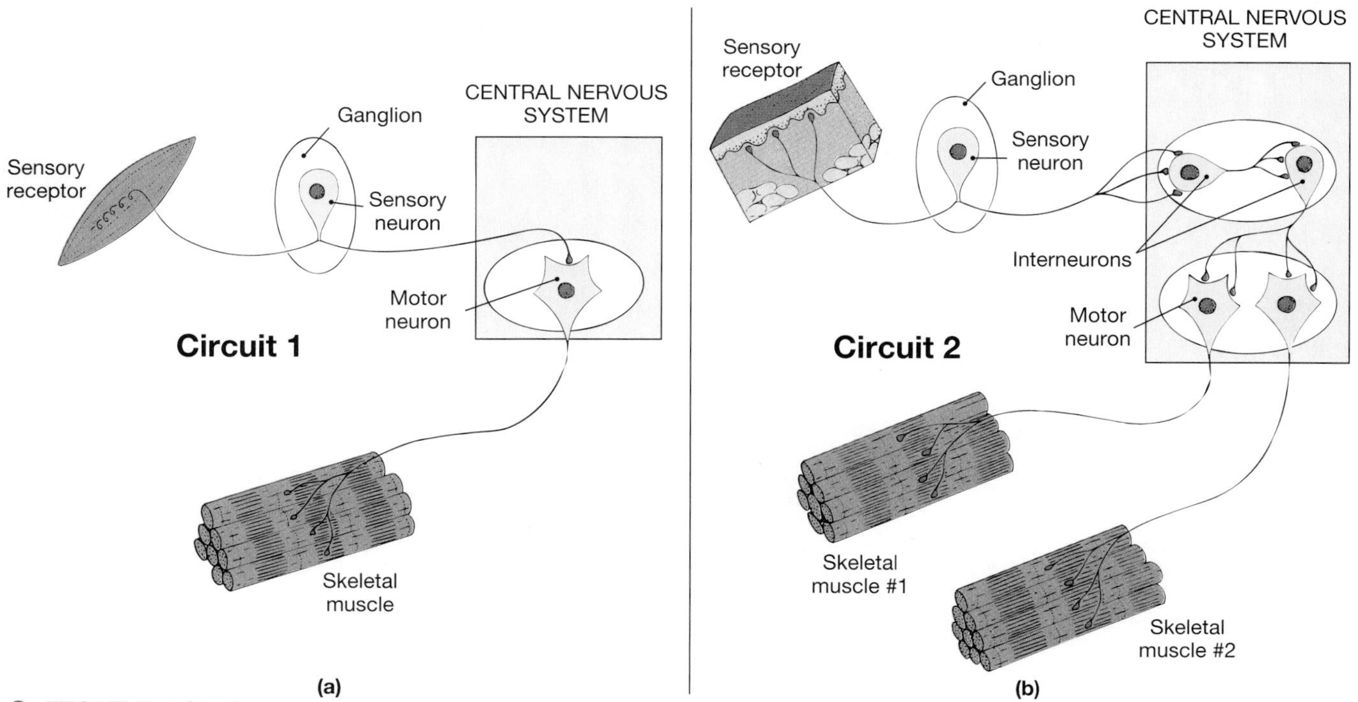
Sensory receptor
Ganglion
CENTRAL NERVOUS SYSTEM
Sensory neuron
Motor neuron
Circuit 1
Skeletal muscle
(a)
Sensory receptor
Ganglion
CENTRAL NERVOUS SYSTEM
Sensory neuron
Interneurons
Motor neuron
Circuit 2
Skeletal muscle #1
Skeletal muscle #2
(b)

● **FIGURE 13-13**

Neural Organization and Simple Reflexes. (a) A monosynaptic reflex involves a peripheral sensory neuron and a central motor neuron. In this example, stimulation of the receptor will lead to a reflexive contraction in a skeletal muscle. **(b)** A polysynaptic reflex involves a sensory neuron, interneurons, and motor neurons. In this example the stimulation of the receptor leads to the coordinated contractions of two different skeletal muscles.

EMBRYOLOGY SUMMARY Development of the Spinal Cord and Spinal Nerves

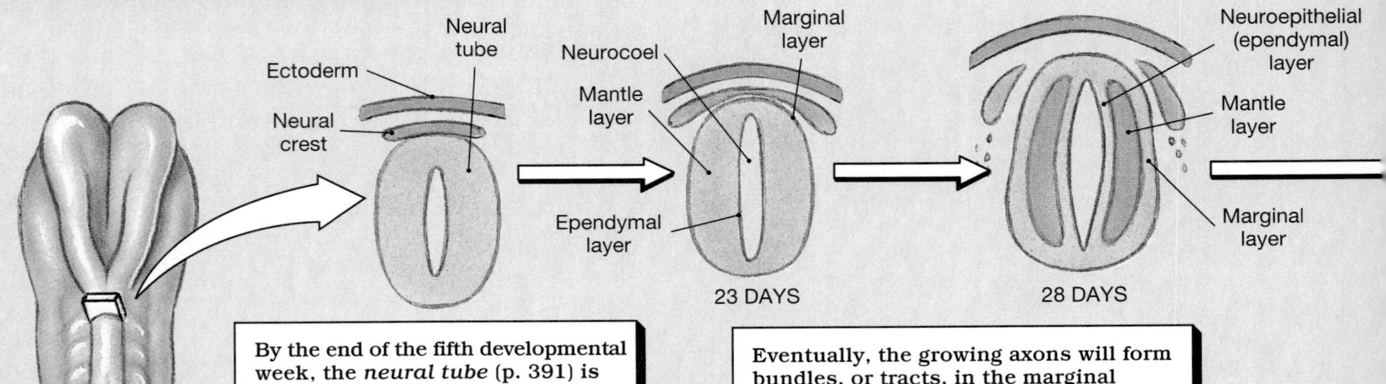
Neural tube
Ectoderm
Neural crest
Neurocoel
Marginal layer
Mantle layer
Ependymal layer
23 DAYS
Neuroepithelial (ependymal) layer
Mantle layer
Marginal layer
28 DAYS
22 DAYS

By the end of the fifth developmental week, the *neural tube* (p. 391) is almost completely closed. In the spinal cord the *mantle layer* that contains developing neurons and neuroglial cells will produce the gray matter that surrounds the *neurocoel*. As neurons develop in the mantle layer, their axons grow toward central or peripheral destinations. The axons leave the mantle layer and travel toward synaptic targets within a peripheral *marginal layer*.

Eventually, the growing axons will form bundles, or tracts, in the marginal layer, and these tracts will crowd together in the columns that form the white matter of the spinal cord.

DEVELOPMENTAL ABNORMALITIES

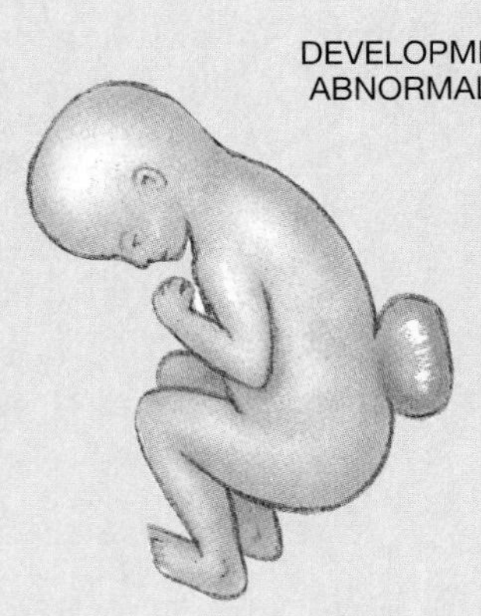

Spina bifida

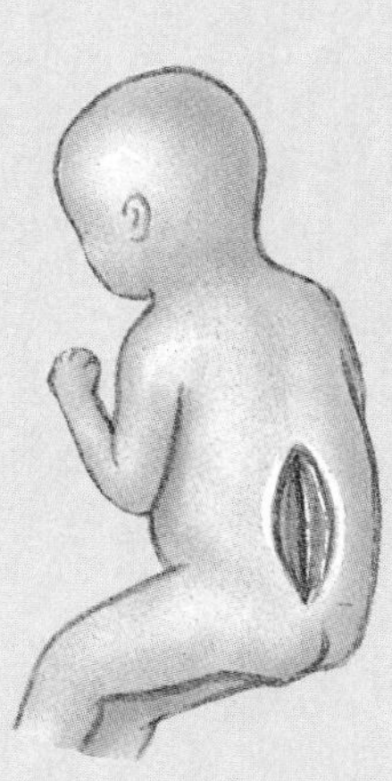

Neural tube defect

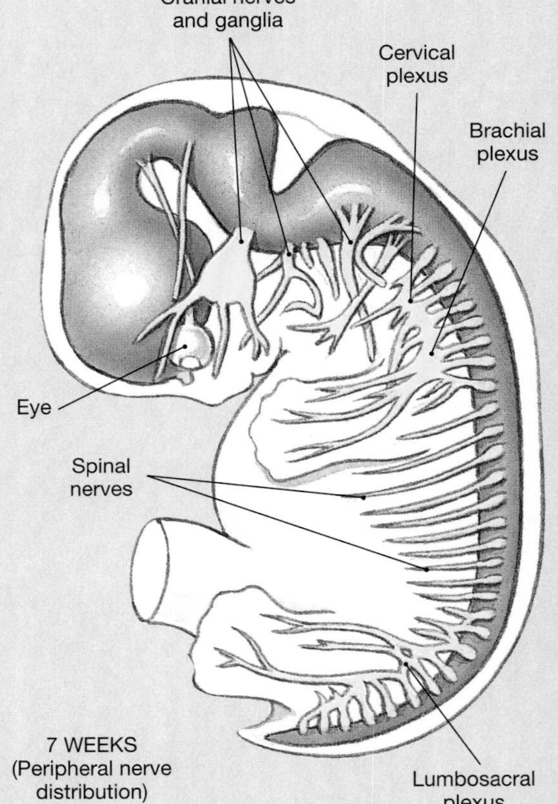
Cranial nerves and ganglia
Cervical plexus
Brachial plexus
Eye
Spinal nerves
Lumbosacral plexus

7 WEEKS (Peripheral nerve distribution)

Spina bifida (BI-fi-da) results when the developing vertebral laminae fail to unite. The neural arch is incomplete, and the meninges bulge outward beneath the skin of the back. The extent of the abnormality determines the severity of the defects. In mild cases the condition may pass unnoticed; extreme cases involve much of the length of the spinal column and are usually associated with abnormal nerve function.

A **neural tube defect (NTD)** is a condition that is secondary to a developmental error in the formation of the spinal cord. Instead of forming a hollow tube, a portion of the spinal cord develops as a broad plate. This is often associated with spina bifida. Neural tube defects affect roughly one individual in 1000; prenatal testing can detect the existence of these defects with an 80–85% success rate.

Several spinal nerves innervate each developing limb. When embryonic muscle cells migrate away from the myotome, the nerves grow right along with them. If a large muscle in the adult is derived from several myotomal blocks, connective tissue partitions will often mark the original boundaries, and the innervation will always involve more than one spinal nerve.

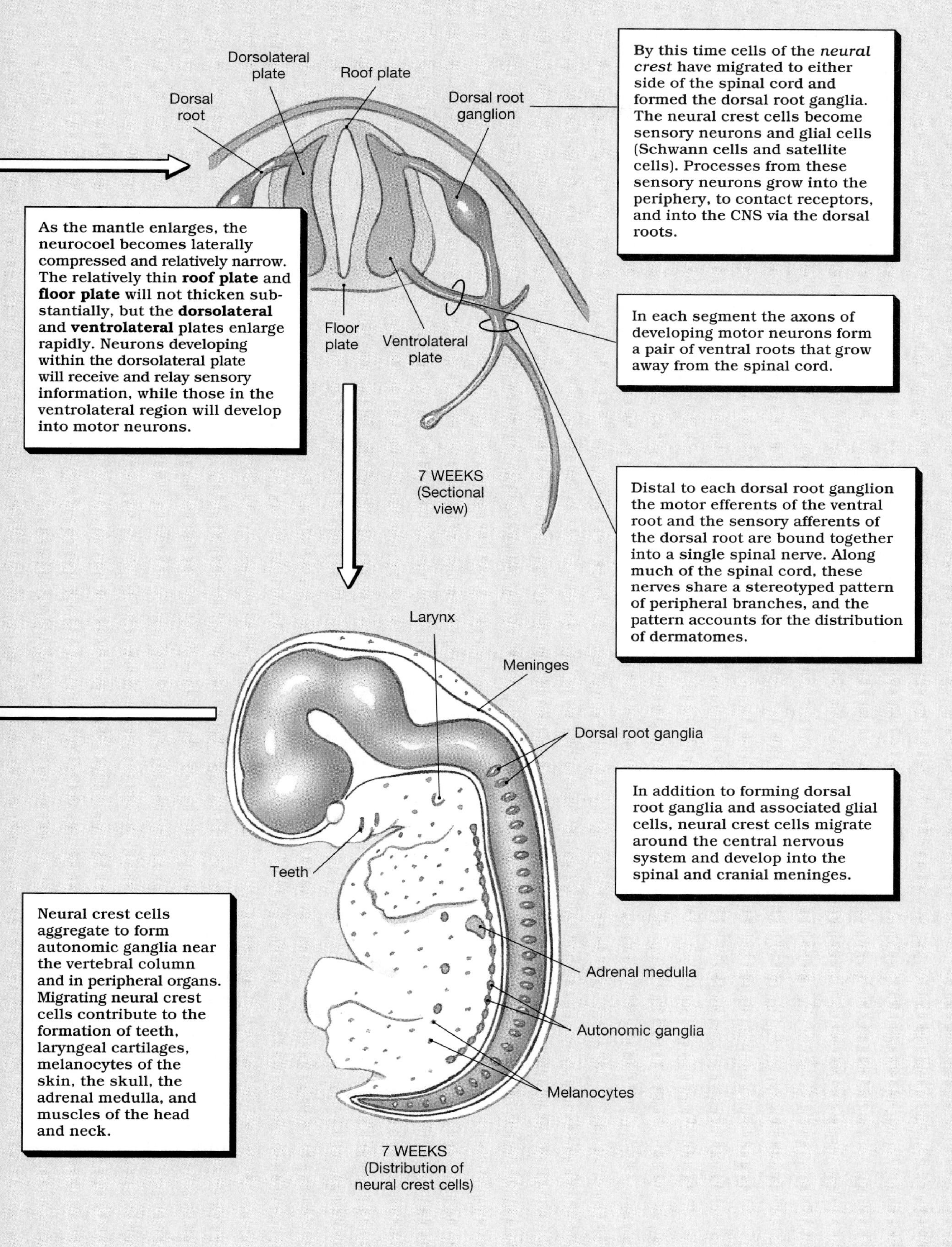
Dorsolateral plate
Roof plate
Dorsal root
Dorsal root ganglion
By this time cells of the *neural crest* have migrated to either side of the spinal cord and formed the dorsal root ganglia. The neural crest cells become sensory neurons and glial cells (Schwann cells and satellite cells). Processes from these sensory neurons grow into the periphery, to contact receptors, and into the CNS via the dorsal roots.
As the mantle enlarges, the neurocoel becomes laterally compressed and relatively narrow. The relatively thin **roof plate** and **floor plate** will not thicken substantially, but the **dorsolateral** and **ventrolateral** plates enlarge rapidly. Neurons developing within the dorsolateral plate will receive and relay sensory information, while those in the ventrolateral region will develop into motor neurons.
Floor plate
Ventrolateral plate
In each segment the axons of developing motor neurons form a pair of ventral roots that grow away from the spinal cord.
7 WEEKS (Sectional view)
Distal to each dorsal root ganglion the motor efferents of the ventral root and the sensory afferents of the dorsal root are bound together into a single spinal nerve. Along much of the spinal cord, these nerves share a stereotyped pattern of peripheral branches, and the pattern accounts for the distribution of dermatomes.
Larynx
Meninges
Dorsal root ganglia
In addition to forming dorsal root ganglia and associated glial cells, neural crest cells migrate around the central nervous system and develop into the spinal and cranial meninges.
Teeth
Neural crest cells aggregate to form autonomic ganglia near the vertebral column and in peripheral organs. Migrating neural crest cells contribute to the formation of teeth, laryngeal cartilages, melanocytes of the skin, the skull, the adrenal medulla, and muscles of the head and neck.
Adrenal medulla
Autonomic ganglia
Melanocytes
7 WEEKS (Distribution of neural crest cells)

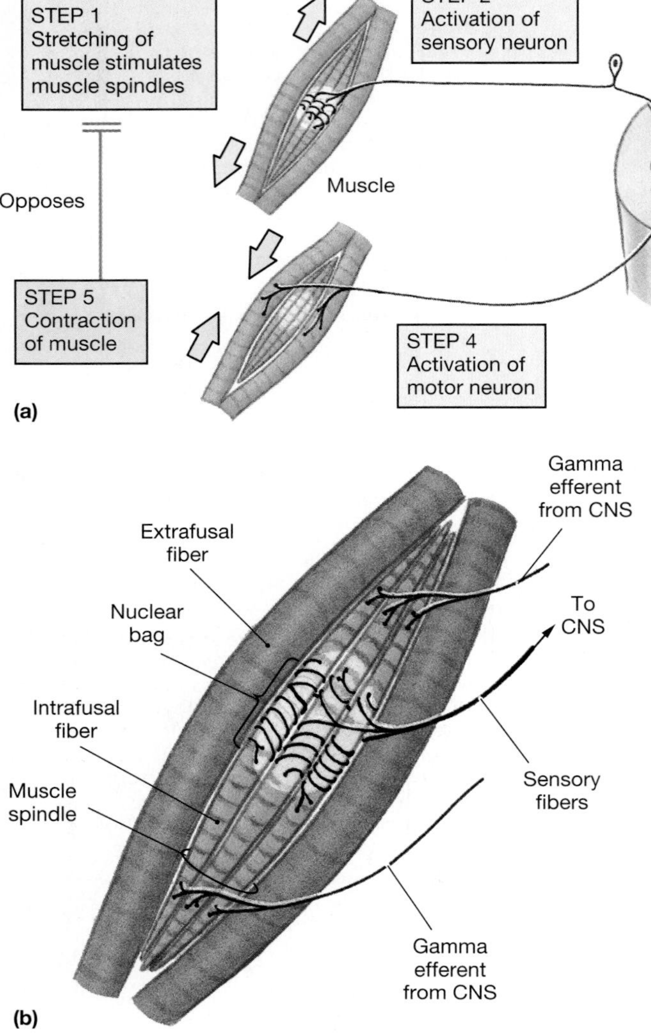
STEP 1
Stretching of muscle stimulates muscle spindles
STEP 2
Activation of sensory neuron
Muscle
Opposes
STEP 3
Information processing at motor neuron
STEP 5
Contraction of muscle
STEP 4
Activation of motor neuron
(a)
Gamma efferent from CNS
Extrafusal fiber
To CNS
Nuclear bag
Intrafusal fiber
Sensory fibers
Muscle spindle
Gamma efferent from CNS
(b)

● **FIGURE 13-14**
Components of the Stretch Reflex. **(a)** Diagram of the activities in a stretch reflex. **(b)** Structure of a muscle spindle.

as the processing center. Such a reflex is termed a **monosynaptic reflex.** Transmission across a chemical synapse always involves a synaptic delay, but with only one synapse the delay between stimulus and response is minimized. Circuit 1 in Figure 13-13●, p. 443, is a typical monosynaptic reflex.

Most other reflexes have at least one interneuron placed between the sensory afferent and the motor efferent, as indicated in Circuit 2. These **polysynaptic reflexes** have a longer delay between stimulus and response, the length of the delay being proportional to the number of synapses involved. Polysynaptic reflexes can produce far more complicated responses because the interneurons can control several different muscle groups.

Spinal Reflexes

Spinal reflexes range in complexity from simple monosynaptic reflexes involving a single segment of the spinal cord to polysynaptic reflexes that involve many different segments. The most complicated spinal reflexes are called **intersegmental reflex arcs** because many segments interact to produce a coordinated motor response. These polysynaptic reflexes include excitatory and inhibitory synapses and produce highly variable motor patterns.

Monosynaptic Reflexes

In a monosynaptic reflex there is little delay between sensory input and motor output. These reflexes control the most rapid, stereotyped motor responses of the nervous system to specific stimuli. The best known example of a monosynaptic reflex is the *stretch reflex.*

The Stretch Reflex

Figures 13-14, 13-15, 13-16

The **stretch reflex** (Figure 13-14a●) provides automatic regulation of skeletal muscle length. The stimulus (increasing muscle length) activates a sensory neuron that triggers an immediate motor response (contraction of the stretched muscle) that counteracts the stimulus. Action potentials traveling toward and away from the spinal cord are conducted along large myelinated Type A fibers (Chapter 12), and the entire reflex is completed within 20–40 msec. [p. 404]

MUSCLE SPINDLES The sensory receptors in the stretch reflex are the **muscle spindles.** Each muscle spindle consists of a bundle of small, specialized skeletal muscle fibers, called **intrafusal muscle fibers** (Figure 13-14b●). The muscle spindle is surrounded by larger extrafusal muscle fibers that are responsible for the resting muscle tone and, at greater levels of stimulation, for the contraction of the entire muscle.

Figure 13-15a● details the structure of a single intrafusal fiber. Each intrafusal fiber is innervated by both sensory and motor neurons. The dendrites of the sensory neuron spiral around the central portion of the intrafusal fiber, in a region known as the **nuclear bag.** Axons from a separate

● FIGURE 13-15

Intrafusal Fibers. (a) A single muscle spindle. The sensory processes wrap around the outside of the nuclear bags; myofibrils at either end of each intrafusal fiber are attached to the nuclear bag. **(b)** The effect of changes in the nuclear bag on muscle tone. **(c)** The nuclear bag will elongate when the entire skeletal muscle is stretched by an external force. This stretching increases the resistance to further stretching. **(d)** The nuclear bag will elongate when myofibrils contract within the intrafusal fiber. This contraction occurs under the command of gamma motor neurons. As a result, the muscle spindle remains sensitive to externally imposed changes in muscle length.

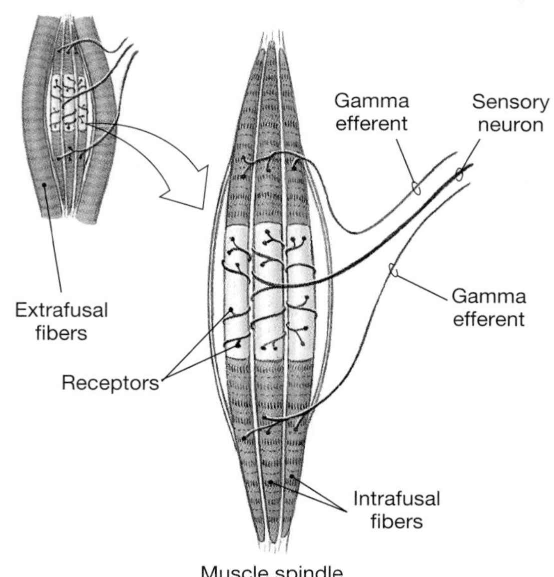

Gamma efferent
Sensory neuron
Extrafusal fibers
Gamma efferent
Receptors
Intrafusal fibers
Muscle spindle
(a)

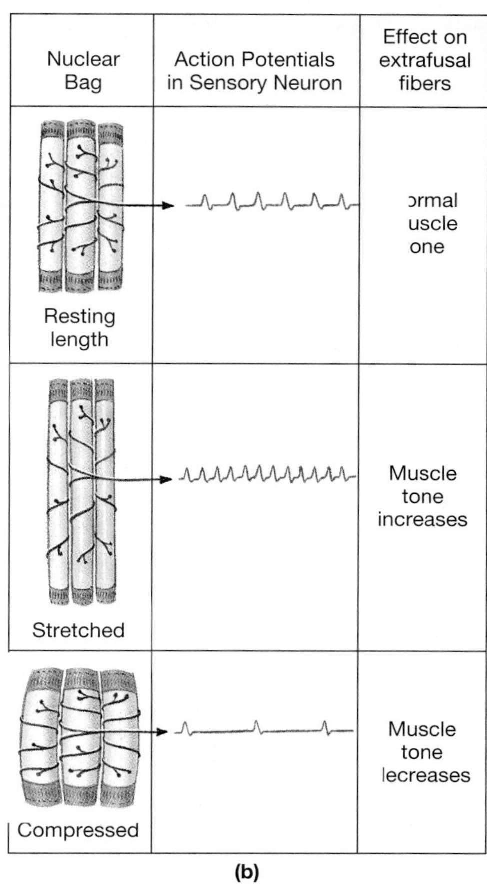

Nuclear Bag
Action Potentials in Sensory Neuron
Effect on extrafusal fibers
Resting length
ormal uscle one
Stretched
Muscle tone increases
Compressed
Muscle tone ecreases
(b)

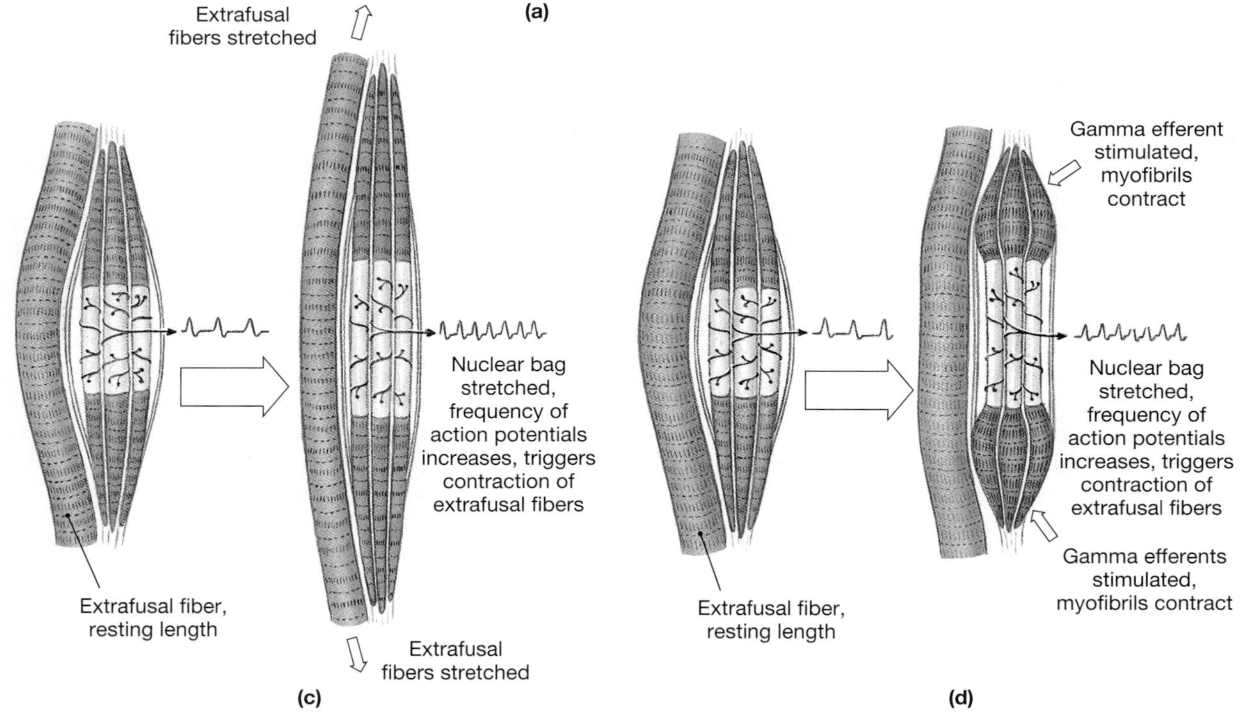

Extrafusal fibers stretched
Nuclear bag stretched, frequency of action potentials increases, triggers contraction of extrafusal fibers
Extrafusal fiber, resting length
Extrafusal fibers stretched
(c)
Gamma efferent stimulated, myofibrils contract
Nuclear bag stretched, frequency of action potentials increases, triggers contraction of extrafusal fibers
Gamma efferents stimulated, myofibrils contract
Extrafusal fiber, resting length
(d)

population of spinal motor neurons form neuromuscular junctions on either end of the intrafusal fiber. The motor neurons innervating intrafusal fibers are called **gamma** (γ) **motor neurons,** and their axons are called **gamma efferents.** There are two sets of myofibrils in the intrafusal fiber, one at each end, and each set is attached to the membrane of the nuclear bag.

The sensory neuron is always active, conducting impulses to the CNS. The axon enters the CNS in a dorsal root and synapses on motor neurons in the anterior gray horn of the spinal cord. Collaterals distribute the information to higher centers, providing proprioceptive information about the skeletal muscle.

Stretching the nuclear bag distorts the dendrites and stimulates the sensory neuron, increasing the frequency of action potentials. Compressing the nuclear bag inhibits the sensory neuron, decreasing the frequency of action potential generation (Figure 13-15b●).

The axon of the sensory neuron synapses on CNS motor neurons that control the extrafusal muscle fibers *of the same muscle.* An increase in stimulation of the sensory neuron, caused by stretching of the nuclear bag, will cause a rise in muscle tone. A decrease in stimulation of the sensory neuron, due to compression of the nuclear bag, will cause a decline in muscle tone.

There are two ways that the nuclear bag can be stretched: (1) passively, by stretching the entire skeletal muscle, including the intrafusal fibers (Figure 13-15c●), or (2) actively, by contraction of the myofibrils at the ends of the intrafusal fibers (Figure 13-15d●). The passive mechanism is important when the skeletal muscle is in the relaxed state; it adjusts muscle tone to stabilize the position of the muscle and to reduce resistance when agonistic muscles are contracting. The active mechanism is important when higher centers order the contraction of the skeletal muscle. We will consider these mechanisms separately.

STRETCH REFLEXES AND PASSIVE MUSCLE MOVEMENTS When a skeletal muscle is stretched, muscle spindles elongate and muscle tone increases (Figures 13-14a and 13-15b,c●). This increase provides automatic resistance that reduces the chance of muscle damage due to overextension. When a skeletal muscle is compressed, muscle spindles are also compressed, and muscle tone decreases (Figure 13-15b●). This helps decrease resistance to the movement under way. (In addition, the motor neurons innervating the extrafusal fibers are inhibited through *reciprocal inhibition,* discussed in a later section.)

The **knee jerk,** or **patellar, reflex,** detailed in Figure 13-16●, is an example of a stretch reflex triggered by passive muscle movement. This reflex, which responds to stretching of the quadriceps muscles of the thigh, can be demonstrated by tapping on the patellar tendon of a flexed knee, with the thigh and leg muscles relaxed. A tap on the patellar ligament stretches muscle spindles in the quadriceps group (Figure 13-16a●). This stretching produces a sudden burst of activity in the sensory neurons monitoring these muscle spindles, and muscle tone in the quadriceps group increases rapidly. The rise is so sudden that it usually produces a noticeable kick. As the extrafusal fibers contract, the muscle spindles return to their resting length, and the rate of action potential generation in the sensory neurons declines. This causes a drop in muscle tone, and the leg then swings back. If it swings far enough to stretch the muscle spindles a second time, there will be a small secondary kick (Figure 13-16b●).

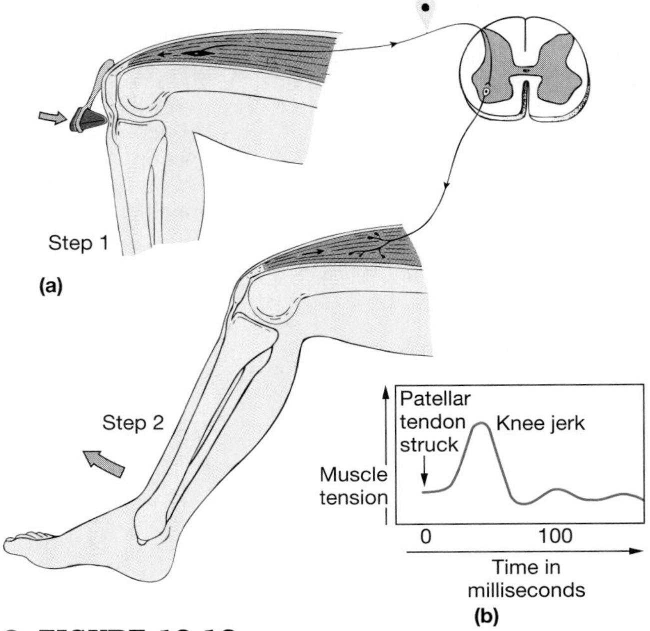
Step 1
(a)
Step 2
Patellar tendon struck
Knee jerk
Muscle tension
0
100
Time in milliseconds
(b)

● **FIGURE 13-16**
The Patellar Reflex. This stretch reflex is controlled by muscle spindles in the muscles that straighten the knee. **(a)** Step 1: A reflex hammer strikes the muscle tendon, stretching the spindle fibers. This stretching results in a sudden increase in the activity of the sensory neurons. These neurons synapse on spinal motor neurons. Step 2: The activation of extrafusal motor units produces an immediate increase in muscle tone and a reflexive kick. **(b)** The changes in muscle tension can be graphically recorded. Note the small secondary kick that may occur if the leg rebounds past its resting position.

Many stretch reflexes are examples of **postural reflexes,** reflexes that maintain normal upright posture. For example, standing involves a cooperative effort by many different muscle groups. Some of these muscles work in opposition to one another, exerting forces that keep the body's weight balanced over the feet. If the body leans forward, stretch receptors in the calf muscles are stimulated, and those muscles respond by contracting. This contraction returns the body to an upright position. If the muscles overcompensate, and the body begins to lean back, the calf muscles relax, but stretch receptors in muscles of the shins and thighs are stimulated, and the problem is corrected immediately.

Postural muscles usually have a firm muscle tone and extremely sensitive stretch receptors. As a result, very fine adjustments are continually being made, and you are not aware of the cycles of contraction and relaxation that occur. Stretch reflexes are only one type of postural reflex; there are many complex polysynaptic postural reflexes.

STRETCH REFLEXES AND ACTIVE MUSCLE MOVEMENTS The CNS adjusts the sensitivity of muscle spindles during a voluntary contraction. Throughout the contraction, impulses arriving over gamma efferents cause contraction of the myofibrils within the intrafusal fibers of that muscle. The myofibrils pull on the nuclear bag (Figure 13-15d●). This keeps the spindles sensitive to externally imposed changes in muscle length throughout the contraction. For example, if you hold your hand out palm up and someone drops an object onto it, the muscle spindles will automatically adjust muscle tone to compensate for the increased load.

 What is the minimum number of neurons needed for a reflex arc?

 One of the first somatic reflexes to develop is the suckling reflex. This is an example of what type of reflex?

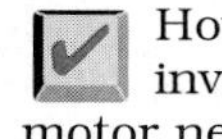 How would stimulation of the muscle spindles involved in the patellar (knee jerk) reflex by gamma motor neurons affect the speed of the reflex?

❑ Polysynaptic Reflexes

Polysynaptic reflexes can produce far more complicated responses than can monosynaptic reflexes. One reason is that the interneurons involved can control several different muscle groups. Moreover, these interneurons may produce either excitatory or inhibitory postsynaptic potentials (EPSPs or IPSPs) at CNS motor nuclei, so the response can involve the stimulation of some muscles and the inhibition of others.

The Tendon Reflex

The stretch reflex regulates the length of a skeletal muscle. The **tendon reflex** monitors the tension produced during a muscular contraction and prevents tearing or breaking of the tendons. The sensory receptors for this reflex are nerve endings called **Golgi tendon organs.** These receptors are located among the collagen fibers of the tendon, and the neuron is stimulated when the collagen fibers stretch. Within the spinal cord these neurons stimulate inhibitory interneurons that synapse on motor neurons controlling the skeletal muscles that pull on the tendon. The greater the tension in the tendon, the greater the inhibitory effect on the motor neurons. Before the tendon is strained to the point of breaking, the inhibitory effect may become so great that the entire muscle relaxes.

Withdrawal Reflexes

Figure 13-17a

Withdrawal reflexes move affected portions of the body away from a source of stimulation. The strongest withdrawal reflexes are triggered by painful stimuli, but these reflexes can sometimes be initiated by the stimulation of touch or pressure receptors.

There are several types of withdrawal reflexes. The **flexor reflex** (Figure 13-17a●) is a withdrawal reflex affecting the muscles of a limb. As you will recall from Chapters 9 and 11, *flexion* is a reduction in the angle between two articulating bones, and the contractions of *flexor muscles* perform this movement. ∞ [pp. 267, 330] Stepping on a tack produces a dramatic flexor reflex in the affected leg. When the pain receptors in the foot are stimulated, the sensory neurons activate interneurons in the spinal cord that stimulate motor neurons in the anterior gray horns. The result is a contraction of flexor muscles that yanks the foot off the ground.

When a specific muscle contracts, opposing muscles must relax to permit the movement. For example, the flexor muscles that bend the leg are opposed by extensor muscles that straighten it out. A potential conflict exists here: Contraction of a flexor muscle should theoretically trigger a stretch reflex in the extensors that would cause them to contract, opposing the movement that is under way. Interneurons in the spinal cord prevent such competition through **reciprocal inhibition.** When one set of motor neurons is stimulated, those controlling antagonistic muscles are inhibited. The term *reciprocal* refers to the fact that the system works both ways. When the flexors contract, the extensors relax; when the extensors contract, the flexors relax.

Withdrawal reflexes are more complex than the relatively stereotyped monosynaptic reflexes. They show tremendous versatility because the sensory neurons activate many pools of interneurons. If the stimuli are stong, interneurons will carry excitatory and inhibitory impulses up and down the spinal cord, affecting motor neurons in many different segments. The end result is always the same: a coordinated movement away from the source of stimulation. But the distribution of the effects and the strength and character of the motor responses depend on the intensity and location of the stimulus. When you step on something sharp, mild discomfort might provoke a brief contraction in muscles of the ankle and foot. More powerful stimuli would produce coordinated muscular contractions affecting the position of the ankle, foot, and leg. Severe pain would also stimulate contractions of shoulder, trunk, and arm muscles. These contractions could persist for several seconds, because of the activation of reverberating circuits. In contrast, monosynaptic reflexes are relatively invariable and brief; the entire knee jerk reflex is completed in roughly 20 msec.

Crossed Extensor Reflexes

Figure 13-17b

The stretch, tendon, and withdrawal reflexes are **ipsilateral reflex arcs** (*ipsi*, same + *lateral*, side), where the sensory stimulus and the motor response

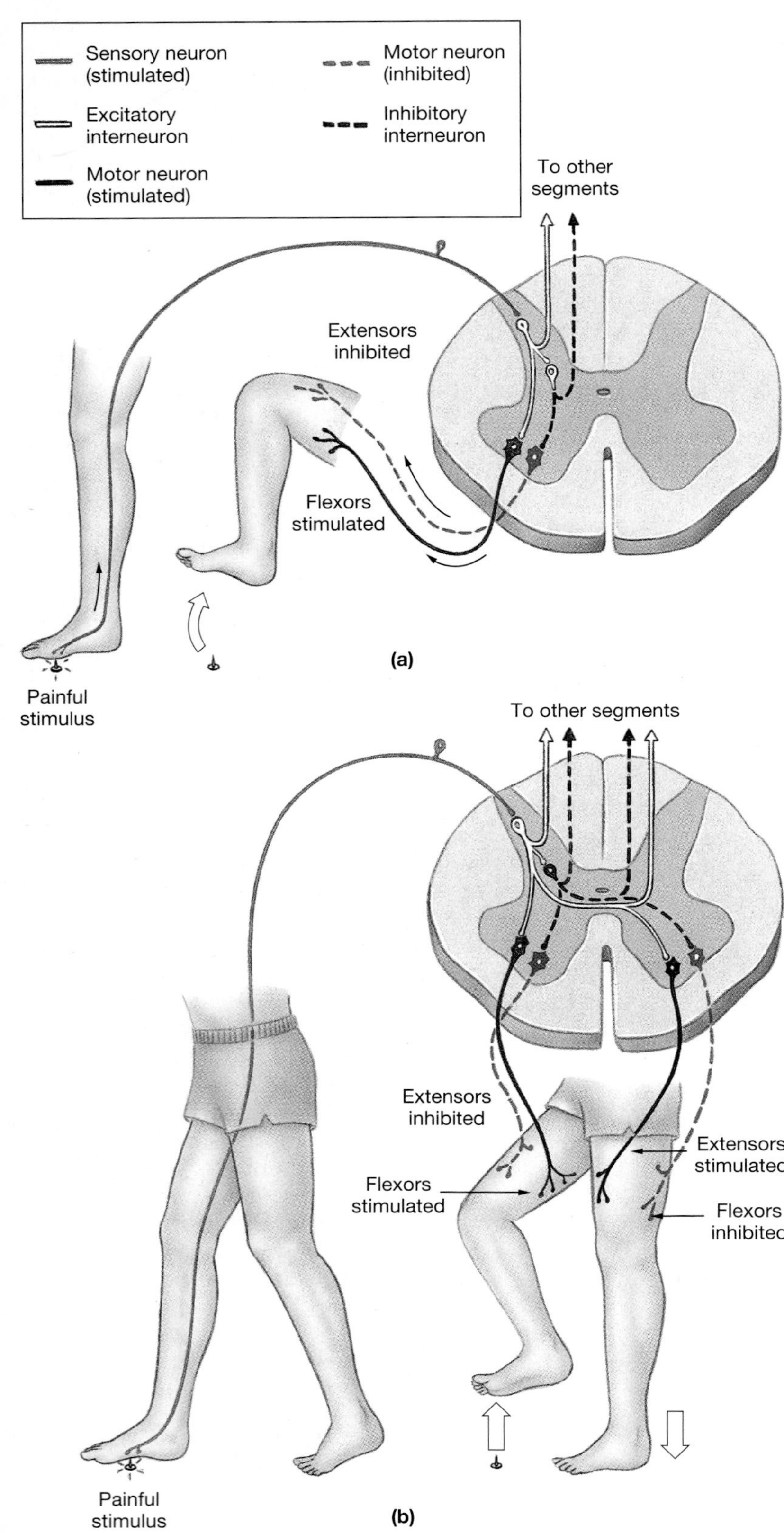
Sensory neuron (stimulated)
Excitatory interneuron
Motor neuron (stimulated)
Motor neuron (inhibited)
Inhibitory interneuron
To other segments
Extensors inhibited
Flexors stimulated
(a)
Painful stimulus
To other segments
Extensors inhibited
Flexors stimulated
Extensors stimulated
Flexors inhibited
Painful stimulus
(b)

● **FIGURE 13-17**
The Flexor and Crossed Extensor Reflexes. (a) The flexor reflex, an example of a withdrawal reflex. **(b)** The crossed extensor reflex.

occur on the same side of the body. The **crossed extensor reflex** (Figure 13-17b●) is a **contralateral reflex arc** (*contra*, opposite); the motor response occurs on the side opposite the stimulus.

The crossed extensor reflex complements the flexor reflex, and the two occur simultaneously. When you step on a thorn, the flexor reflex pulls your foot away from the ground as the crossed extensor reflex stiffens the other leg to support the body's weight. In the crossed extensor reflex, the axons of interneurons responding to the pain cross to the other side of the spinal cord and stimulate motor neurons controlling the extensor muscles of the uninjured leg. The opposite leg straightens to support the shifting weight. Reverberating circuits use positive feedback to ensure that the movement lasts long enough to be effective, despite the absence of motor commands from higher centers.

The flexor and crossed extensor reflexes are not the most complicated spinal reflexes. Postural and motor reflexes associated with standing, walking, and running are also spinal reflexes. But all polysynaptic reflexes share the same basic characteristics:

1. ***They involve pools of interneurons.*** Processing occurs in pools of interneurons before motor neurons are activated. The result may be excitation or inhibition; the tendon reflex produces inhibition of motor neurons, whereas the flexor and crossed extensor reflexes direct specific muscle contractions.
2. ***They are intersegmental in distribution.*** The interneuron pools extend across spinal segments, and they may activate muscle groups in many parts of the body.
3. ***They involve reciprocal innervation.*** Reciprocal innervation coordinates muscular contractions and reduces resistance to movement. In the flexor and crossed extensor reflexes the contraction of one muscle group is associated with the inhibition of opposing muscles.
4. ***They have reverberating circuits that prolong the reflexive motor***

response. Positive feedback between interneurons that innervate motor neurons and the processing pool maintains the stimulation even after the initial stimulus has faded.

5. ***Several reflexes may cooperate to produce a coordinated, controlled response.*** As a reflex movement is under way, any antagonistic reflexes are inhibited. There may also be interaction between monosynaptic and polysynaptic reflexes. For example, during the stretch reflex, antagonistic muscles are inhibited; during the tendon reflex, antagonistic muscles are stimulated.

Integration and Control of Spinal Reflexes

Reflex motor behaviors occur automatically, without instructions from higher centers. However, higher centers can have a profound effect on reflex performance. Processing centers in the brain can facilitate or inhibit reflex motor patterns based in the spinal cord. Descending tracts originating in the brain synapse on interneurons and motor neurons throughout the spinal cord. These synapses are continuously active, producing EPSPs or IPSPs at the postsynaptic membrane.

Reinforcement and Inhibition

Figure 13-18

Synapses producing EPSPs make postsynaptic neurons more sensitive to other excitatory stimuli. The resulting level of generalized facilitation (Chapter 12) rises and falls, depending on the activity in higher centers. [p. 415] For example, a voluntary effort to pull apart clasped hands elevates the general state of facilitation along the spinal cord. This elevated facilitation leads to an enhancement, or **reinforcement,** of spinal reflexes.

Other descending fibers have an inhibitory effect on spinal reflexes. Stroking an infant's foot on the side of the sole produces a fanning of the toes known as the **Babinski sign,** or *positive Babinski reflex* (Figure 13-18a•). This response disappears as descending motor pathways develop. In the adult, the same stimulus produces a curling of the toes, called a **plantar reflex,** or *negative Babinski reflex*, after about a 1-sec. delay (Figure 13-18b•). If either the higher centers or the descending tracts are damaged, the Babinski sign will reappear. As a result, this reflex is often tested if CNS injury is suspected.

Many reflexes can be assessed through careful observation and the use of simple tools. The procedures are easy to perform, and the results can provide valuable information about damage to the spinal cord or spinal nerves. By testing a series of spinal and cranial reflexes a physician can assess the function of sensory pathways and motor centers throughout the spinal cord and brain. AM *Reflexes and Diagnostic Testing; Abnormal Reflex Activity*

Voluntary Movements and Reflex Motor Patterns

Centers in the brain can interact with the stereotyped motor patterns programmed into the spinal cord. By making use of preexisting patterns, a relatively small number of descending fibers can control complex motor functions. For example, the motor patterns for walking, running, and jumping are primarily directed by neuronal pools in the spinal cord. The descending pathways provide appropriate facilitation, inhibition, or "fine-tuning" of the established patterns.

When complicated voluntary movements are under way, the neurons involved with spinal reflexes assist with muscular coordination and control. For example, the descending tracts that stimulate motor neurons controlling the biceps brachii,

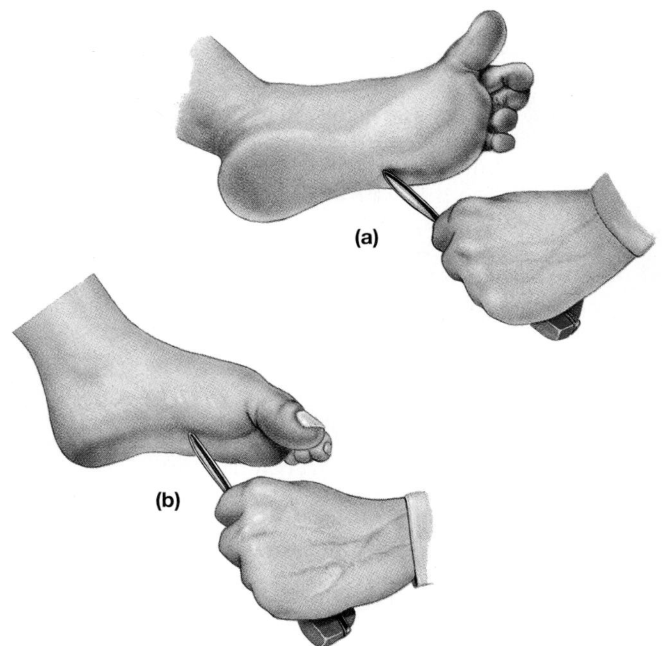
(a)
(b)

• **FIGURE 13-18**

The Babinski Reflex. (a) The Babinski sign (positive Babinski reflex) occurs in the absence of descending inhibition. It is normal in infants but pathological in adults. **(b)** The plantar reflex (negative Babinski reflex), a curling of the toes, is seen in normal adults.

brachialis, and other forearm flexors also (1) stimulate the gamma efferents to the intrafusal fibers in these muscles and (2) stimulate inhibitory interneurons to relax antagonistic muscle groups, such as the triceps brachii. The gamma efferents regulate the sensitivity of the stretch reflex, and the interneurons are the same as those activated in the withdrawal and crossed extensor reflexes. As the muscle contraction proceeds, tendon reflexes keep tension production within tolerable limits.

Motor control therefore involves a series of interacting levels. At the lowest level are monosynaptic reflexes that are rapid but stereotyped and relatively inflexible. At the highest level are centers in the brain that can modulate or build upon reflexive motor patterns.

✔ A weight lifter is straining to lift a 200-kg barbell. Shortly after he lifts it to chest height, his muscles appear to relax and he drops the barbell. What reflex does this response demonstrate?

✔ During a withdrawal reflex, what happens to the limb on the side opposite the stimulus? What is this response called?

✔ After suffering an injury to his back, Tom exhibits a positive Babinski reflex. What does this imply about Tom's injury?

Selected Clinical Terminology

Terms Discussed in This Chapter

Babinski sign (*positive Babinski reflex*): A spinal reflex in infants, consisting of a fanning of the toes, produced by stroking the foot on the side of the sole; in adults, a sign of CNS injury. *(p. 451)*

meningitis: Inflammation of the meninges involving the spinal cord (*spinal meningitis*) and/or brain (*cerebral meningitis*). Bacterial or viral pathogens are usually responsible. *(p. 430 and AM)*

paraplegia: Paralysis involving loss of motor control of the lower but not the upper limbs. *(p. 432)*

patellar reflex: The "knee jerk" reflex resulting from the stimulation of stretch receptors in the quadriceps muscles. *(p. 448)*

plantar reflex (*negative Babinski reflex*): A spinal reflex in adults, consisting of a curling of the toes, produced by stroking the foot on the side of the sole. *(p. 451)*

quadriplegia: Paralysis involving loss of sensation and motor control of the upper and lower limbs. *(p. 432 and AM)*

shingles: A viral condition caused by infection of neurons in dorsal root ganglia by the *Herpes varicella-zoster virus*. The primary symptom is a painful rash along the sensory distribution of the affected spinal nerves. *(p. 434 and AM)*

AM *Additional Terms Discussed in the Applications Manual*

abdominal reflex: A reflexive twitch in abdominal muscles that moves the navel toward a stimulus; often used to provide information about descending tracts in the spinal cord.

areflexia (ā-re-FLEK-sē-uh): A condition in which normal reflexes fail to appear, even with reinforcement.

caudal anesthesia: The introduction of anesthetics into the epidural space of the sacrum to paralyze lower abdominal and perineal structures.

clonus (KLŌ-nus): Oscillations that develop between opposing stretch reflexes in individuals with hyperreflexia.

epidural block: Injection of anesthetic into the epidural space to eliminate sensory and motor innervation via spinal nerves in the area of injection.

Hansen's disease (*leprosy*): A bacterial infection that begins in sensory nerves of the skin and gradually progresses to a motor paralysis of the same regions.

hyperreflexia: A condition resulting from a high degree of facilitation along the spinal cord; reflexes are easily triggered, and the responses may be grossly exaggerated.

hyporeflexia: A condition in which normal reflexes are weak, but demonstrable with reinforcement.

lumbar puncture: A spinal tap performed between adjacent lumbar vertebrae inferior to the level of the conus medullaris.

mass reflex: Sudden, abnormal, and prolonged activation of somatic and visceral reflexes throughout the spinal cord, resulting from an extreme state of motor neuron facilitation.

myelography: A diagnostic procedure in which a radiopaque dye is introduced into the cerebrospinal fluid in order to obtain an X-ray of the spinal cord and cauda equina.

nerve growth factor: A peptide factor that promotes the growth and maintenance of neurons. Other factors important to neuron growth and repair include *BDNF, NT-3, NT-4, GAP-43,* and a variety of other compounds.

palsies: Regional losses of sensory and motor function as a result of nerve trauma or compression; also called *peripheral neuropathies*. Common palsies include *radial nerve palsy, ulnar palsy, sciatica,* and *peroneal palsy*.

paresthesia: An abnormal tingling sensation, usually descibed as "pins and needles," that accompanies sensory return after a temporary palsy.

sciatica (sī-AT-i-kuh): The painful result of compression of the roots of the sciatic nerve.

spinal shock: A period of depressed sensory and motor function following any severe injury to the spinal cord. Spinal shock may accompany *spinal concussion, spinal contusion, spinal laceration, spinal compression,* or *spinal transection*.

spinal tap: A procedure in which cerebrospinal fluid is removed from the subarachnoid space through a needle, usually inserted between the lumbar vertebrae.

CHAPTER REVIEW

STUDY OUTLINE

INTRODUCTION, p. 428

1. In addition to relaying information to and from the brain, the spinal cord integrates and processes information on its own.

GROSS ANATOMY OF THE SPINAL CORD, p. 428

1. The adult spinal cord includes localized **enlargements** that provide innervation to the limbs. The spinal cord has 31 segments, each associated with a pair of **dorsal roots** and a pair of **ventral roots.** (*Figure 13-1*)
2. The **filum terminale** (a strand of fibrous tissue) that originates at the **conus medullaris** ultimately becomes part of the coccygeal ligament. (*Figure 13-1*)

SPINAL MENINGES, p. 430

1. The **spinal meninges** provide physical stability and shock absorption for neural tissues of the spinal cord; the **cranial meninges** surround the brain. (*Figure 13-2*)

The Dura Mater, p. 430

2. The **dura mater** covers the spinal cord; caudally it tapers into the **coccygeal ligament.** The **epidural space** separates the dura mater from the walls of the vertebral canal. (*Figure 13-2*)

The Arachnoid, p. 430

3. Beneath the inner surface of the dura mater are the **subdural space,** the **arachnoid** (the second meningeal layer), and the **subarachnoid space.** The latter space contains **cerebrospinal fluid,** which acts as a shock absorber and a diffusion medium for dissolved gases, nutrients, chemical messengers, and waste products. (*Figure 13-2*)

The Pia Mater, p. 431

4. The **pia mater,** a meshwork of elastin and collagen fibers, is the innermost meningeal layer. Unlike more superficial meninges, it is bound to the underlying neural tissue. (*Figure 13-2*)

SECTIONAL ANATOMY OF THE SPINAL CORD, p. 432

1. The *white matter* contains myelinated and unmyelinated axons, whereas the *gray matter* contains cell bodies of neurons and glial cells. The projections of gray matter toward the outer surface of the spinal cord are called **horns.** (*Figure 13-3*)

Organization of Gray Matter, p. 432

2. The **posterior gray horns** contain somatic and visceral sensory nuclei; nuclei in the **anterior gray horns** are concerned with somatic motor control. The **lateral gray horns** contain visceral motor neurons. The **gray commissures** contain axons that cross from one side of the cord to the other. (*Figure 13-3*)

Organization of White Matter, p. 432

3. The white matter can be divided into six **columns,** each of which contains **tracts** (*fasciculi*). **Ascending tracts** relay information from the spinal cord to the brain, and **descending tracts** carry information from the brain to the spinal cord. (*Figure 13-3*)

SPINAL NERVES, p. 433

1. There are 31 pairs of spinal nerves. Each has an **epineurium** (outermost layer), **perineurium,** and **endoneurium** (innermost layer). (*Figure 13-4*)

Peripheral Distribution of Spinal Nerves, p. 434

2. A typical spinal nerve has a **white ramus** (which contains myelinated axons), a **gray ramus** (containing unmyelinated fibers that innervate glands and smooth muscles in the body wall or limbs), a **dorsal ramus** (providing sensory and motor innervation to the skin and muscles of the back), and a **ventral ramus** (supplying the ventrolateral body surface, structures in the body wall, and the limbs). Each pair of nerves monitors a region of the body surface called a **dermatome.** (*Figures 13-5, 13-6*)

Nerve Plexuses, p. 435

3. A complex, interwoven network of nerves is called a **nerve plexus.** The three large plexuses are the **cervical plexus,** the **brachial plexus,** and the **lumbosacral plexus.** The latter can be divided into the **lumbar plexus** and the **sacral plexus.** (*Figures 13-7 to 13-10; Tables 13-1 to 13-3*)

AN INTRODUCTION TO REFLEXES, p. 439

1. A *neural reflex* is an automatic, involuntary motor response by the nervous system that helps preserve homeostasis by rapidly adjusting the functions of organs or organ systems.

The Reflex Arc, p. 441

2. A **reflex arc** is the neural "wiring" of a single reflex. (*Figure 13-11*)
3. There are five steps involved in a neural reflex: (1) arrival of a stimulus and activation of a receptor, (2) activation of a sensory neuron, (3) information processing, (4) activation of a motor neuron, and (5) response by an effector. (*Figure 13-11*)
4. A *receptor* is a specialized cell that monitors conditions in the body or external environment. Each receptor has a characteristic range of sensitivity.

Classification of Reflexes, p. 442

5. **Innate reflexes** result from the connections that form between neurons during development. **Acquired reflexes** are learned and often are more complex. (*Figure 13-12*)
6. Reflexes processed in the brain are **cranial reflexes.** In a **spinal reflex** the important interconnections and processing events occur inside the spinal cord. (*Figure 13-12*)

7. **Somatic reflexes** control skeletal muscles, and **visceral reflexes** (*autonomic reflexes*) control the activities of other systems.
8. A **monosynaptic reflex** is the simplest reflex arc, in which a sensory neuron synapses directly on a motor neuron, which acts as the processing center. **Polysynaptic reflexes,** which have at least one interneuron placed between the sensory afferent and the motor efferent, have a longer delay between stimulus and response. (*Figure 13-13*)

SPINAL REFLEXES, p. 446

1. Spinal reflexes range from simple monosynaptic reflexes to more complex polysynaptic reflexes in which many segments interact to produce a coordinated motor response.

Monosynaptic Reflexes, p. 446

2. The **stretch reflex** is a monosynaptic reflex that automatically regulates skeletal muscle length and muscle tone. The sensory receptors involved are **muscle spindles.** (*Figures 13-14 to 13-16*)
3. A **postural reflex** maintains normal upright posture.

Polysynaptic Reflexes, p. 449

4. Polysynaptic reflexes can produce more complicated responses. Examples include the **tendon reflex** (which monitors the tension produced during muscular contractions and prevents damage to tendons) and **withdrawal reflexes** (which move affected portions of the body away from a source of stimulation). The **flexor reflex** is a withdrawal reflex affecting the muscles of a limb, and the **crossed extensor reflex** complements the withdrawal reflex. (*Figure 13-17*)
5. In an **ipsilateral reflex arc** the sensory stimulus and motor response occur on the same side of the body. In a **contralateral reflex arc** the motor response occurs on the side opposite the stimulus. (*Figure 13-18*)
6. All polysynaptic reflexes share the following traits: (1) They involve pools of interneurons; (2) they are intersegmental in distribution; (3) they involve reciprocal innervation; (4) they have reverberating circuits that prolong the reflexive motor response; and (5) several reflexes may cooperate to produce a coordinated response.

INTEGRATION AND CONTROL OF SPINAL REFLEXES, p. 451

1. The brain can facilitate or inhibit reflex motor patterns based in the spinal cord.

Reinforcement and Inhibition, p. 451

2. The enhancement of spinal reflexes is called **reinforcement.** (*Figure 13-18*)

Voluntary Movements and Reflex Motor Patterns, p. 451

3. Motor control involves a series of interacting levels. Monosynaptic reflexes form the lowest level; at the highest level are the centers in the brain that can modulate or build upon reflexive motor patterns.

REVIEW QUESTIONS

LEVEL 1 Reviewing Facts and Terms

1. The expanded area of the spinal cord that supplies nerves to the shoulder girdle and arms is the:
   (a) conus medullaris
   (b) filum terminale
   (c) lumbar enlargement
   (d) cervical enlargement

2. The dorsal roots of each spinal segment:
   (a) bring sensory information to the spinal cord
   (b) control peripheral effectors
   (c) contain somatic and visceral motor neurons
   (d) a, b, and c are correct

3. Spinal nerves are called mixed nerves because:
   (a) they contain sensory and motor fibers
   (b) they exit at intervertebral foramina
   (c) they are associated with a pair of dorsal root ganglia
   (d) they are associated with dorsal and ventral roots

4. The spinal cord continues to enlarge and elongate until an individual is:
   (a) born (b) 1 year old
   (c) 4 years old (d) 10 years old

5. The tough, fibrous, outermost covering of the spinal cord is the:
   (a) arachnoid (b) pia mater
   (c) dura mater (d) epidural block

6. The gray matter of the spinal cord is dominated by:
   (a) large numbers of myelinated axons
   (b) large numbers of unmyelinated axons
   (c) Schwann cells and satellite cells
   (d) cell bodies of neurons and glial cells

7. The outermost layer of connective tissue that surrounds each spinal nerve is the:
   (a) perineurium (b) epineurium
   (c) endoneurium (d) epimysium

8. A sensory region monitored by the dorsal rami of a single spinal segment is:
   (a) a ganglion (b) a fascicle
   (c) a dermatome (d) a ramus

9. The major nerve of the cervical plexus that innervates the diaphragm is the:
   (a) median nerve (b) axillary nerve
   (c) phrenic nerve (d) peroneal nerve

10. The genitofemoral, femoral, and lateral femoral cutaneous nerves are major nerves of the:
    (a) lumbar plexus (b) sacral plexus
    (c) brachial plexus (d) cervical plexus

11. The final step in a neural reflex is:
    (a) activation of a receptor
    (b) response by an effector

(c) activation of a motor neuron
(d) information processing

12. The reflexes that control the most rapid, stereotyped motor responses of the nervous system to stimuli are:
(a) monosynaptic reflexes
(b) polysynaptic reflexes
(c) tendon reflexes

13. An example of a stretch reflex triggered by passive muscle movement is the:
(a) tendon jerk
(b) knee jerk
(c) flexor reflex
(d) ipsilateral reflex

14. The contraction of flexor muscles and the relaxation of extensor muscles illustrates the principle of:
(a) reverberating circuitry
(b) generalized facilitation
(c) reciprocal inhibition
(d) reinforcement

15. Where the sensory stimulus and the motor response occur on the same side of the body the reflex arcs are:
(a) contralateral (b) paraesthetic
(c) ipsilateral (d) monosynaptic

16. All polysynaptic reflexes:
(a) have reverberating circuits that prolong the reflexive motor response
(b) involve pools of interneurons
(c) involve reciprocal innervation
(d) a, b, and c are correct

17. What three meningeal layers provide physical stability and shock absorption for the spinal cord?

18. Beginning with the outermost layer, list the connective tissue layers that surround each spinal nerve.

LEVEL 2 Reviewing Concepts

19. Polysynaptic reflexes can produce far more complicated responses than can monosynaptic reflexes because:
(a) the response time is quicker
(b) the response is initiated by highly sensitive receptors
(c) motor neurons carry impulses at a faster rate than sensory neurons
(d) the interneurons involved can control several different muscle groups

20. Why do cervical nerves outnumber cervical vertebrae?

21. If the anterior gray horns of the spinal cord were damaged, what type of control would be affected?

22. Proceeding inward from the outermost layer, number the following in the correct sequence:
__ walls of vertebral canal __ subdural space
__ pia mater __ subarachnoid space
__ dura mater __ epidural space
__ arachnoid membrane __ spinal cord

23. In which of the above structures is the cerebrospinal fluid found? What is its function?

24. What five characteristics are common to all polysynaptic reflexes?

25. Which ramus would be involved if you felt something brushing against your:
(a) abdomen (b) left forearm
(c) upper back (d) right ankle

26. Which plexus has primary responsibility for innervating the following muscles?
(a) diaphragm (b) pelvic diaphragm
(c) deltoid (d) gastrocnemius

27. Why is response time in a monosynaptic reflex much faster than response time in a polysynaptic reflex?

28. List, in correct sequence, the five steps involved in a neural reflex.

29. What would happen if the ventral root of a spinal nerve was damaged or transected?

30. How does the stimulation of a sensory neuron that innervates an extrafusal muscle fiber affect muscle tone?

31. Why is it important that a spinal tap be done between the third and fourth lumbar vertebrae?

LEVEL 3 Critical Thinking and Clinical Applications

32. Mary complains that when she wakes up in the morning her thumb and forefinger are always "asleep." She mentions this condition to her physician, who asks if she sleeps with her arm above her head. She replies that she does. Her physician tells her that sleeping in that position compresses a portion of one of the nerves of the brachial plexus, producing her symptoms. Which nerve is involved?

33. Improper use of crutches can produce a condition known as "crutch paralysis," which is characterized by lack of response by the extensor muscles of the arm and a condition known as wrist drop. Which nerve is involved?

34. During childbirth, some women are given a local anesthetic that temporarily deadens sensory neurons in the region of the genitals. Which nerve does this anesthetic affect?

35. While playing football, Ramon is tackled hard and suffers an injury to his left leg. As he tries to get up, he finds that he cannot flex his left thigh or extend the leg. What nerve is damaged, and how would this damage affect sensory perception in the left leg?

CHAPTER 14

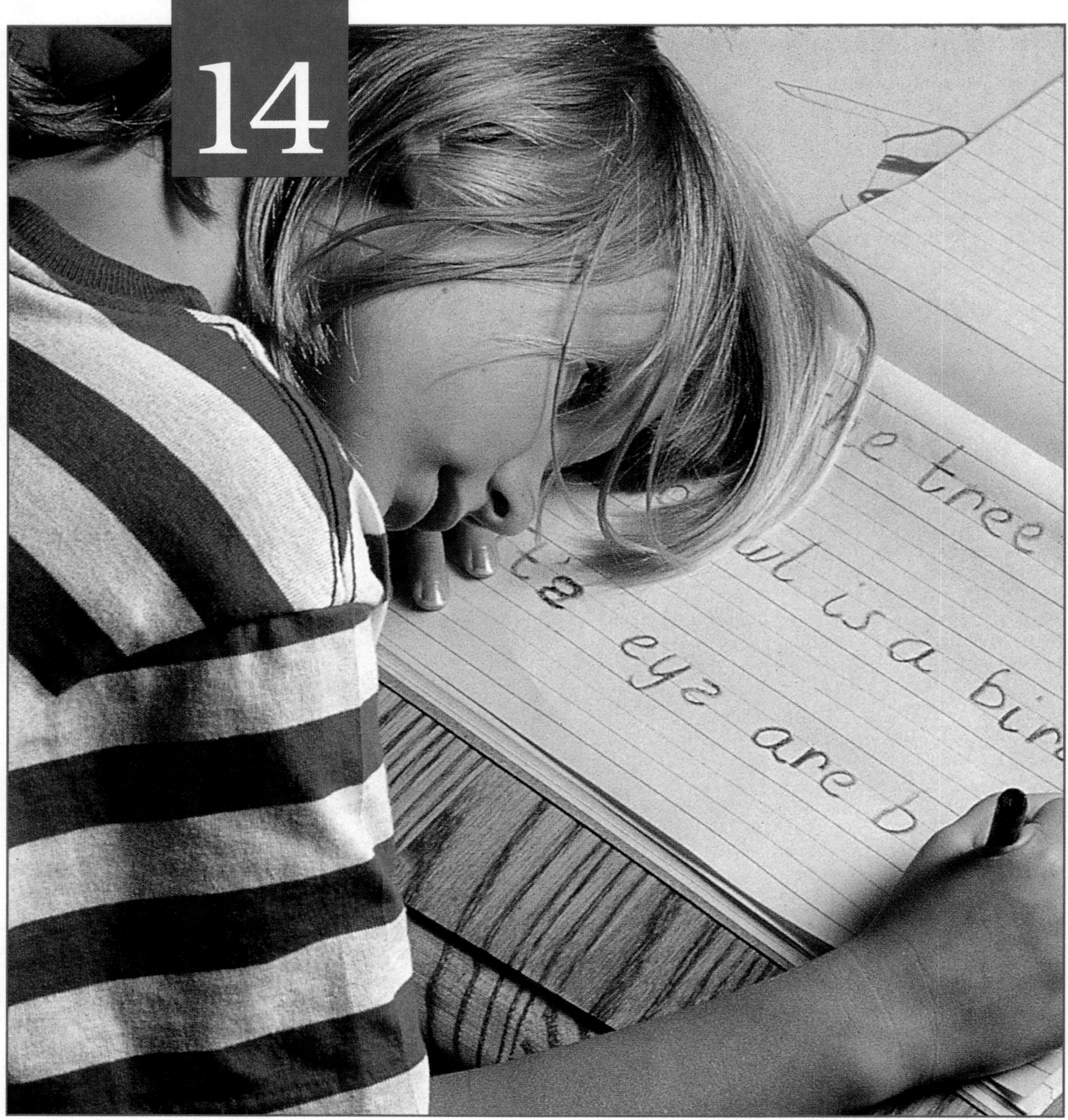

Our perceptions of the world around us depend on thousands of interactions among neurons within the central nervous system. We seldom realize how complex these processes are unless they go wrong in some way. This child has dyslexia, a condition characterized by difficulties with the recognition and use of words. Although the cause of dyslexia remains a mystery, there is general agreement that it results from problems with the integration and processing of visual or auditory information. There is much that we still do not understand about such activities, which ultimately create our consciousness and our unique personalities.

This chapter introduces the sensory and motor pathways involved in such activities as writing, and considers the neural basis for higher-order functions, such as memory and learning.

The Brain and Cranial Nerves

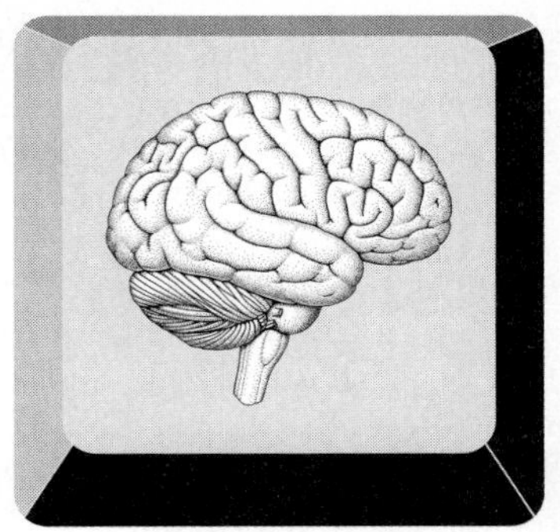

Chapter Outline and Objectives

1 Name the major regions of the brain and describe their functions.

2 Name the ventricles of the brain and describe their locations and the connections between them.

3 Explain how the brain is protected and supported.

4 Discuss the formation, circulation, and functions of the cerebrospinal fluid.

5 Locate the motor, sensory, and association areas of the cerebral cortex and discuss their functions.

6 Identify important structures within each region of the brain and explain their functions.

7 Identify the cranial nerves and relate each nerve to its principal destinations and functions.

8 Discuss important cranial reflexes.

The brain is probably the most fascinating organ in the body, and yet we know relatively little about its structural and functional complexities. What we *do* know is that all our dreams, passions, plans, and memories result from brain activity. The brain contains tens of billions of neurons organized into hundreds of neuronal pools; it has a complex three-dimensional structure and performs a bewildering array of functions. If a heart, liver, lung, or kidney stops working, heroic measures, including organ transplantation, can be taken to preserve the life of the individual. But if the brain permanently stops working, so that the person is classified as "brain dead," for all intents and purposes that individual ceases to exist.

The brain is far more complex than the spinal cord, and it can respond to stimuli with greater versatility. That versatility results from (1) the tremendous number of neurons and neuronal pools in the brain and (2) the complexity of the interconnections between those neurons and neuronal pools. Because the interconnections are complex and the pools large, the response can be varied to meet changing circumstances. But this versatility has a price: A response cannot be immediate, precise, and adaptable all at the same time. Adaptability requires multiple processing steps, and every synapse adds to the delay between stimulus and response. One of the major functions of spinal reflexes is to provide an *immediate* response that can be fine-tuned or elaborated on by more versatile—but slower—processing centers in the brain. For example, the flexor reflex will pull your hand from a hot stove before you become consciously aware that something is wrong. In effect, the spinal reflex prevents further injury and gives the brain time to decide what to do next.

This chapter considers the structure of the brain and the cranial nerves and indicates their primary functional roles. The organization of the chapter parallels that of Chapter 13 on the spinal cord and spinal nerves, and completes your introduction to the basic components of the central nervous system (CNS).

An Introduction to the Organization of the Brain

The adult human brain contains almost 98 percent of the neural tissue in the body. A "typical" brain weighs 1.4 kg (3 lb) and has a volume of 1200 cc (71 in.^{3}). There is considerable individual variation in brain size, and the brains of males average about 10 percent larger than those of females, owing to differences in average body size. No correlation exists between brain size and intelligence, and individuals with the smallest (750 cc) and largest (2100 cc) brains are functionally normal.

Embryology of the Brain

The organization of the adult brain is most easily introduced by considering its embryological origins. The development of the brain is detailed in the Embryology Summary on pages 460–461. The central nervous system begins as a hollow *neural tube*, with a fluid-filled internal cavity called the *neurocoel*. In the cephalic portion of the neural tube, three areas enlarge rapidly through expansion of the neurocoel. This enlargement creates three prominent **primary brain vesicles** named for their relative positions: the **prosencephalon** (prō-zen-SEF-a-lon; *proso*, forward + *enkephalos*, brain), or "forebrain;" the **mesencephalon** (*mesos*, middle), or "midbrain;" and the **rhombencephalon** (rom-ben-SEF-a-lon), or "hindbrain."

The fate of the three primary divisions of the brain is summarized in Table 14-1. The prosencephalon and rhombencephalon are subdivided further, forming **secondary brain vesicles.** The prosencephalon forms the **telencephalon** (tel-en-SEF-a-lon; *telos*, end), and the **diencephalon** (dī-en-SEF-a-lon; *dia*, through). The telencephalon will form the *cerebrum*, including the paired *cerebral hemispheres* that dominate the superior and lateral surfaces of the adult brain. The mesencephalon thickens, and the neurocoel becomes a relatively narrow passageway comparable to the central canal of the spinal cord. The portion of the rhombencephalon adjacent to the mesencephalon forms the **metencephalon** (met-en-SEF-a-lon; *meta*, after). The dorsal portion of the metencephalon will become the *cerebellum*, and the ventral portion will develop into the *pons*. The portion of the rhombencephalon closer to the spinal cord forms the **myelencephalon** (mī-el-en-SEF-a-lon; *myelon*, spinal cord), which will become the *medulla oblongata*.

Major Regions and Landmarks

The Cerebrum and Cerebellum

Figures 14-1, 14-2

Viewed from the superior surface (Figure 14-1a●, p. 462), the **cerebrum** (SER-e-brum) of the adult brain can be divided into large, paired **cerebral hemispheres.** Conscious thought processes, sen-

TABLE 14-1 Development of the Human Brain

Primary Brain Vesicles (3 weeks)	*Secondary Brain Vesicles (6 weeks)*	*Brain Regions at Birth*
Prosencephalon	Telencephalon	Cerebrum
	Diencephalon	Diencephalon
Mesencephalon	Mesencephalon	Mesencephalon
Rhombencephalon	Metencephalon	Cerebellum
		Pons
	Myelencephalon	Medulla oblongata

sations, intellectual functions, memory storage and retrieval, and complex motor patterns originate in the cerebrum. Immediately behind the cerebrum are the somewhat smaller hemispheres of the **cerebellum** (ser-e-BEL-um) (Figure 14-1b●, p. 462). The cerebellum adjusts voluntary and involuntary motor activities, comparing incoming sensory information with anticipated sensations during the execution of preestablished motor patterns. The surfaces of the cerebral hemispheres and cerebellum are highly folded and covered by **neural cortex** (*cortex*, rind or bark), a superficial layer of gray matter. The term *cerebral cortex* refers to the neural cortex of the cerebral hemispheres, as opposed to the *cerebellar cortex* of the cerebellar hemispheres.

The other major regions of the brain—the diencephalon, mesencephalon, pons, and medulla oblongata—can best be examined after removing the cerebral hemispheres and cerebellum. The diencephalon is a structural and functional link between the cerebral hemispheres and the components of the **brain stem.** The brain stem includes the mesencephalon, pons, and medulla oblongata.[1] It contains a variety of important processing centers and also provides relay stations for information headed to or from the cerebrum or cerebellum. Major landmarks and functions of the diencephalon and brain stem are indicated in Figure 14-2●, p. 463.

[1] Some sources consider that the brain stem includes the diencephalon. We will use the more restrictive definition here.

The Diencephalon

The walls of the diencephalon are composed of the **left** and **right thalamus** (THAL-a-mus). Each thalamus contains relay and processing centers for sensory information. A narrow stalk, the *infundibulum*, connects the floor of the diencephalon, or **hypothalamus** (*hypo-*, below), to the **pituitary gland,** a component of the endocrine system. The hypothalamus contains centers involved with emotions, autonomic function, and hormone production. As we shall see in Chapter 18, it is the primary link between the nervous and endocrine systems.

The Mesencephalon

Nuclei in the mesencephalon process visual and auditory information and generate involuntary somatic motor responses. For example, your immediate, reflexive responses to a loud, unexpected noise (eye movements and head turning) are directed by nuclei in the midbrain. This region also contains centers involved with the maintenance of consciousness.

The Pons

The term *pons* is Latin for "bridge," and the **pons** of the brain connects the cerebellum to the brain stem. In addition to tracts and relay centers, the pons also contains nuclei involved with somatic and visceral motor control.

EMBRYOLOGY SUMMARY

Development of the Brain and Cranial Nerves

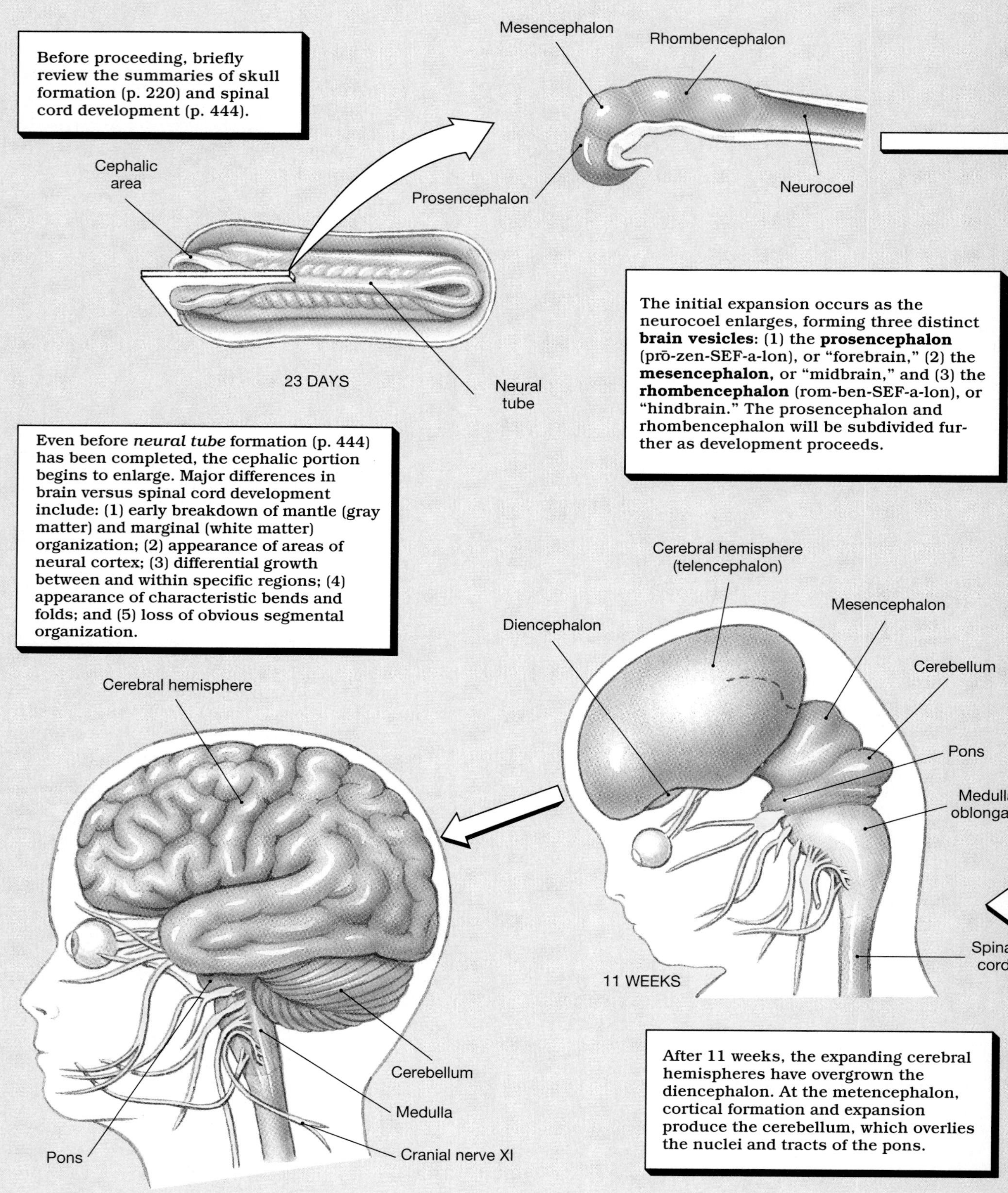

Mesencephalon
Rhombencephalon
Prosencephalon
Neurocoel
Cephalic area
Neural tube
23 DAYS
Cerebral hemisphere (telencephalon)
Mesencephalon
Diencephalon
Cerebellum
Pons
Medulla oblongat
Spinal cord
11 WEEKS
Cerebral hemisphere
Cerebellum
Medulla
Cranial nerve XI
Pons
ADULT

Before proceeding, briefly review the summaries of skull formation (p. 220) and spinal cord development (p. 444).

The initial expansion occurs as the neurocoel enlarges, forming three distinct **brain vesicles**: (1) the **prosencephalon** (prō-zen-SEF-a-lon), or "forebrain," (2) the **mesencephalon**, or "midbrain," and (3) the **rhombencephalon** (rom-ben-SEF-a-lon), or "hindbrain." The prosencephalon and rhombencephalon will be subdivided further as development proceeds.

Even before *neural tube* formation (p. 444) has been completed, the cephalic portion begins to enlarge. Major differences in brain versus spinal cord development include: (1) early breakdown of mantle (gray matter) and marginal (white matter) organization; (2) appearance of areas of neural cortex; (3) differential growth between and within specific regions; (4) appearance of characteristic bends and folds; and (5) loss of obvious segmental organization.

After 11 weeks, the expanding cerebral hemispheres have overgrown the diencephalon. At the metencephalon, cortical formation and expansion produce the cerebellum, which overlies the nuclei and tracts of the pons.

Metencephalon
Mesencephalon
Myelencephalon
Diencephalon
Telencephalon
4 WEEKS

The rhombencephalon first subdivides into the **metencephalon** (met-en-SEF-a-lon; *meta*, after) and the **myelencephalon** (mī-el-en-SEF-a-lon; *myelon*, spinal cord).

The prosencephalon forms the **telencephalon** (tel-en-SEF-a-lon; *telos*, end + *enkephalos*, brain) and the **diencephalon**. The telencephalon begins as a pair of swellings near the rostral, dorsolateral border of the prosencephalon.

Development of the **mesencephalon** produces a small mass of neural tissue with a constricted neurocoel, the mesencephalic (cerebral) aqueduct.

N III
N IV
N V
N VII
Myelencephalon
N XII
N IX
N X
N XI
5 WEEKS

Cranial nerves develop as sensory ganglia link peripheral receptors with the brain, and motor fibers grow out of developing cranial nuclei. Special sensory neurons of cranial nerves I, II, and VIII develop in association with the developing receptors (pp. 592–593). The somatic motor nerves III, IV, and VI grow to the eye muscles; the mixed nerves (IV, V, VII, IX,and X) innervate the *pharyngeal arches* (p. 220).

As differential growth proceeds and the position and orientation of the embryo change, a series of bends, or **flexures** (FLEK-sherz), appears along the axis of the developing brain.

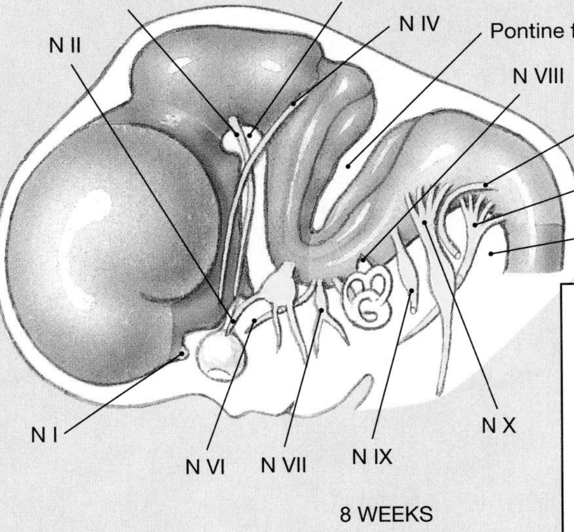
N III
Cephalic flexure
N II
N IV
Pontine flexure
N VIII
N XI
N XII
Cervical flexure
N I
N VI
N VII
N IX
N X
8 WEEKS

The roofs of the diencephalon and myelencephalon fail to develop, leaving a thin ependymal layer in contact with the developing meninges. Blood vessels invading these regions create areas of the choroid plexus.

As growth continues and the pontine flexure develops, the brain becomes more compact. The expanding cerebral hemispheres now dominate the superior and lateral surfaces of the brain. Migrating neuroblasts create the cerebral cortex, and underlying masses of gray matter develop into the cerebral nuclei (basal ganglia).

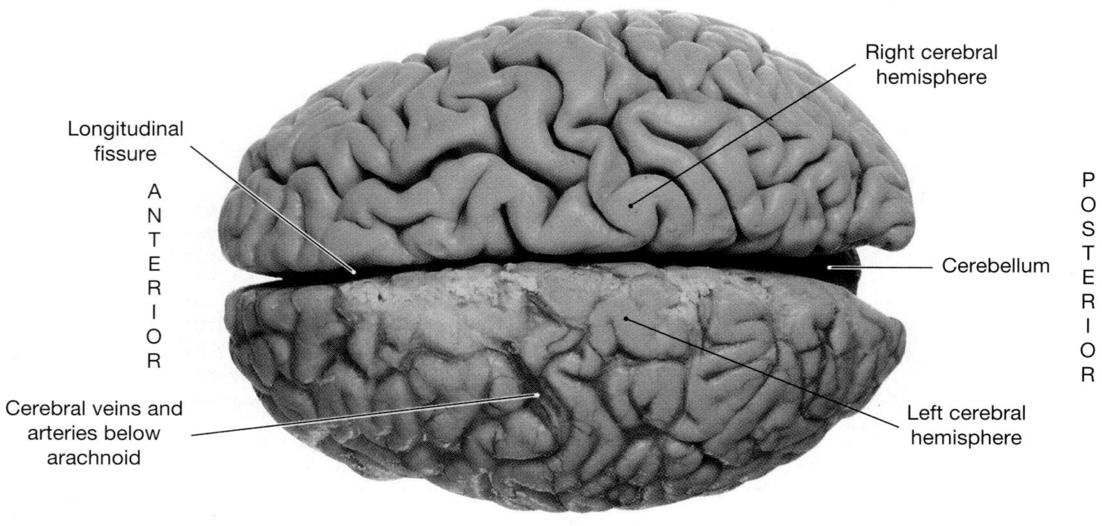

Longitudinal fissure
ANTERIOR
Right cerebral hemisphere
POSTERIOR
Cerebellum
Cerebral veins and arteries below arachnoid
Left cerebral hemisphere

(a) Superior view

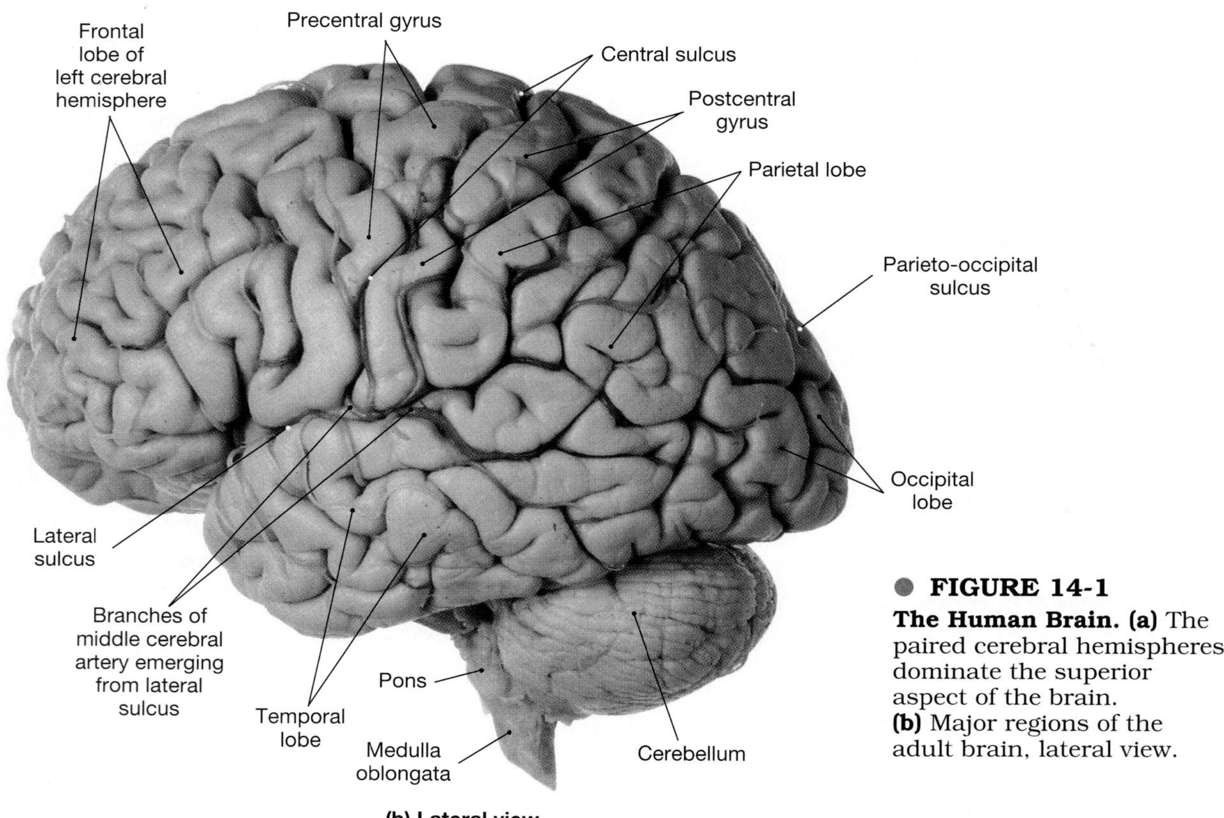

Frontal lobe of left cerebral hemisphere
Precentral gyrus
Central sulcus
Postcentral gyrus
Parietal lobe
Parieto-occipital sulcus
Occipital lobe
Lateral sulcus
Branches of middle cerebral artery emerging from lateral sulcus
Temporal lobe
Pons
Medulla oblongata
Cerebellum

(b) Lateral view

FIGURE 14-1
The Human Brain. (a) The paired cerebral hemispheres dominate the superior aspect of the brain.
(b) Major regions of the adult brain, lateral view.

The Medulla Oblongata

The spinal cord connects to the brain at the medulla oblongata. Near the pons, the roof of the medulla oblongata is thin and membranous. The caudal portion of the medulla oblongata resembles the spinal cord in having a narrow central canal. The medulla oblongata relays sensory information to the thalamus and to centers in portions of the brain stem. The medulla oblongata also contains major centers concerned with the regulation of autonomic function, such as heart rate, blood pressure, respiration, and digestion.

 What are the three primary brain vesicles, and what does each contribute to the structure of the adult brain?

 In response to a loud noise, your head automatically turns toward the source of the sound. What part of the brain directs this response?

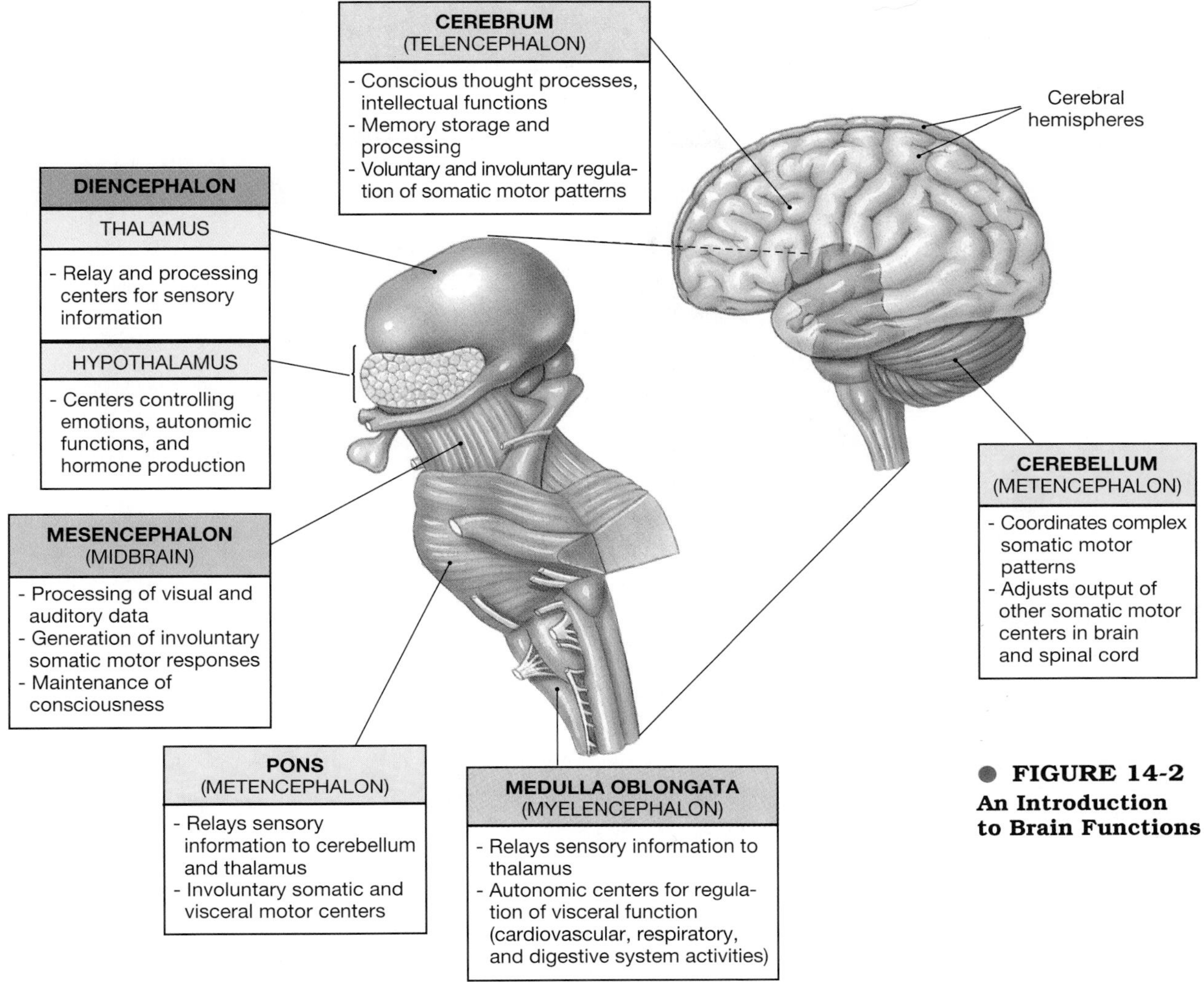
CEREBRUM
(TELENCEPHALON)
- Conscious thought processes, intellectual functions
- Memory storage and processing
- Voluntary and involuntary regulation of somatic motor patterns
Cerebral hemispheres
DIENCEPHALON
THALAMUS
- Relay and processing centers for sensory information
HYPOTHALAMUS
- Centers controlling emotions, autonomic functions, and hormone production
MESENCEPHALON
(MIDBRAIN)
- Processing of visual and auditory data
- Generation of involuntary somatic motor responses
- Maintenance of consciousness
CEREBELLUM
(METENCEPHALON)
- Coordinates complex somatic motor patterns
- Adjusts output of other somatic motor centers in brain and spinal cord
PONS
(METENCEPHALON)
- Relays sensory information to cerebellum and thalamus
- Involuntary somatic and visceral motor centers
MEDULLA OBLONGATA
(MYELENCEPHALON)
- Relays sensory information to thalamus
- Autonomic centers for regulation of visceral function (cardiovascular, respiratory, and digestive system activities)

● **FIGURE 14-2**
An Introduction to Brain Functions

❑ Ventricles of the Brain

Figure 14-3

During development the neurocoel within the cerebral hemispheres, diencephalon, metencephalon, and medulla oblongata expands to form chambers called **ventricles** (VEN-tri-kls). The ventricular cavities are lined by cells of the *ependyma.* ∞ [p. 381]

Each cerebral hemisphere contains an enlarged ventricular chamber. A thin partition, the **septum pellucidum,** separates this pair of **lateral ventricles.** There is no direct connection between the two lateral ventricles, but each communicates with the ventricle of the diencephalon through an **interventricular foramen** (*foramen of Monro*) (Figure 14-3a,b●). Because there are two lateral ventricles (first and second), the diencephalic chamber is called the **third ventricle.**

The mesencephalon has a slender canal known as the **mesencephalic aqueduct** (*aqueduct of Sylvius* or *cerebral aqueduct*). This passageway connects the third ventricle with the **fourth ventricle.** The superior portion of the fourth ventricle lies between the posterior surface of the pons and the anterior surface of the cerebellum. The fourth ventricle extends into the superior portion of the medulla oblongata (Figure 14-3a●). In the inferior portion of the medulla oblongata the fourth ventricle narrows and becomes continuous with the central canal of the spinal cord (Figure 14-3a,c●).

The ventricles are filled with cerebrospinal fluid. There is a continuous circulation of cerebrospinal fluid from the ventricles and central canal into the *subarachnoid space* of the meninges that surround the CNS via foramina in the roof of the fourth ventricle.

■ Protection and Support of the Brain

The delicate tissues of the brain are protected from mechanical forces by (1) the bones of the cranium, (2) the *cranial meninges,* and (3) cerebrospinal fluid. In addition, the neural tissue of the brain is biochemically isolated from the general circulation by the *blood-brain barrier.* You should refer to

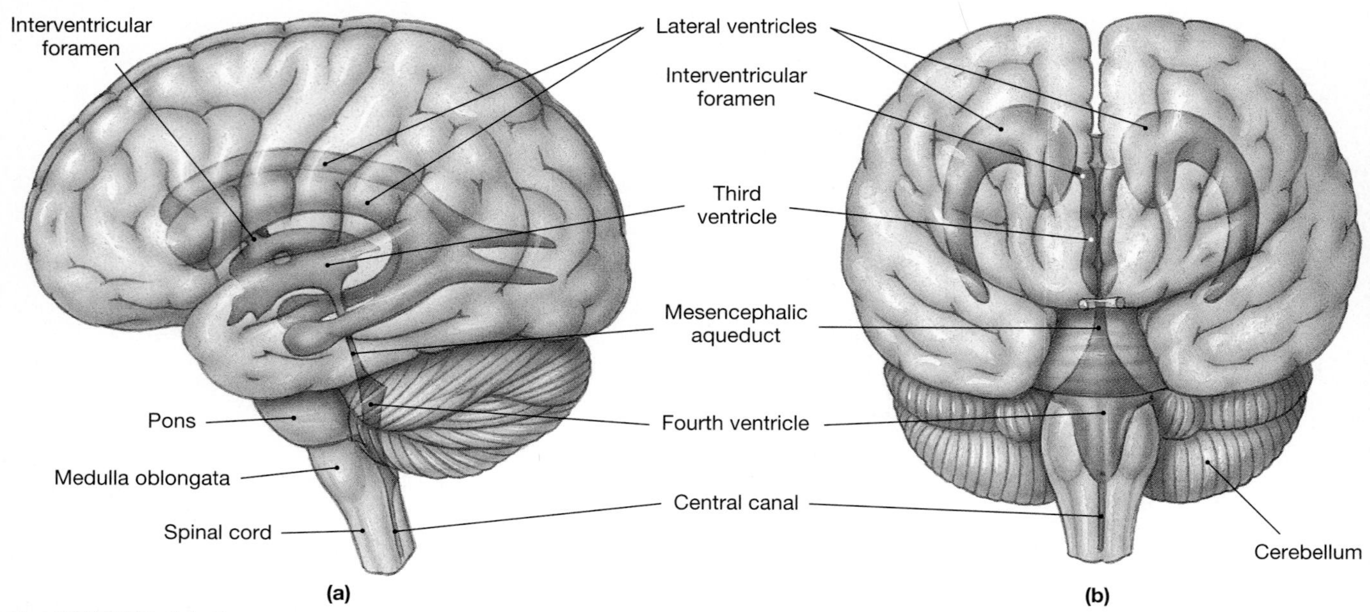
Interventricular foramen
Lateral ventricles
Interventricular foramen
Third ventricle
Mesencephalic aqueduct
Pons
Fourth ventricle
Medulla oblongata
Central canal
Spinal cord
Cerebellum
(a)
(b)

● **FIGURE 14-3**
Ventricles of the Brain. Orientation and extent of the ventricles as they would appear if the brain were transparent. **(a)** Lateral view. **(b)** Anterior view, showing the relationships between the lateral ventricles and the third ventricle. **(c)** A diagrammatic view of the CNS in frontal section, through the ventricles and central canal. AM *Plate 3.4; Scans 1,2*

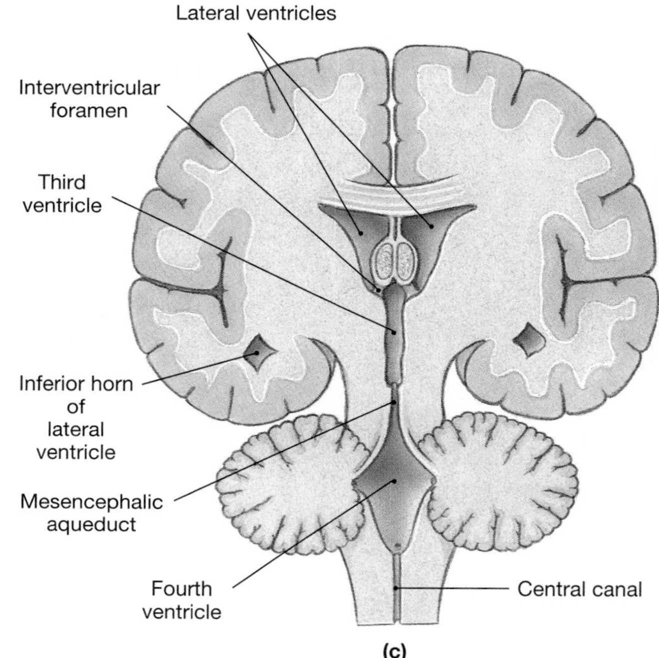
Lateral ventricles
Interventricular foramen
Third ventricle
Inferior horn of lateral ventricle
Mesencephalic aqueduct
Fourth ventricle
Central canal
(c)

Figures 7-3 and 7-4●, pp. 205–207, for a review of the bones of the cranium; we will discuss the other protective factors here.

❑ The Cranial Meninges

The layers that make up the cranial meninges—the *dura mater, arachnoid,* and *pia mater*—are continuous with those of the spinal cord. [p. 430] However, the cranial meninges have distinctive anatomical features and functional roles.

- The ***dura mater*** consists of outer and inner fibrous layers. The outer (*endosteal*) layer is fused to the periosteum of the cranial bones. The outer and inner layers are separated by a slender gap that contains tissue fluids and blood vessels, including large venous channels known as **dural sinuses.** The veins of the brain open into these sinuses, which deliver the venous blood to the *internal jugular veins* of the neck.
- The ***arachnoid*** covers the brain, providing a smooth surface that does not follow the underlying folds.
- The ***pia mater*** closely adheres to the surface contours of the brain, anchored by the processes of astrocytes. It extends into every fold and curve and accompanies the branches of cerebral blood vessels as they penetrate the surface of the brain to reach internal structures.

Functions of the Cranial Meninges

Figure 14-4

The brain lies cradled within the cranium of the skull, and there is an obvious correspondence between the shape of the brain and that of the cranial cavity (Figure 14-4a●). The massive cranial bones provide mechanical protection, but they also pose a threat. The brain is like a person driving a car. If the car hits a tree, the car protects the driver from contact with the tree, but serious injury will occur unless a seat belt or airbag protects the driver from contact with the car.

Cranial trauma is a head injury resulting from impact with another object. There are roughly 8

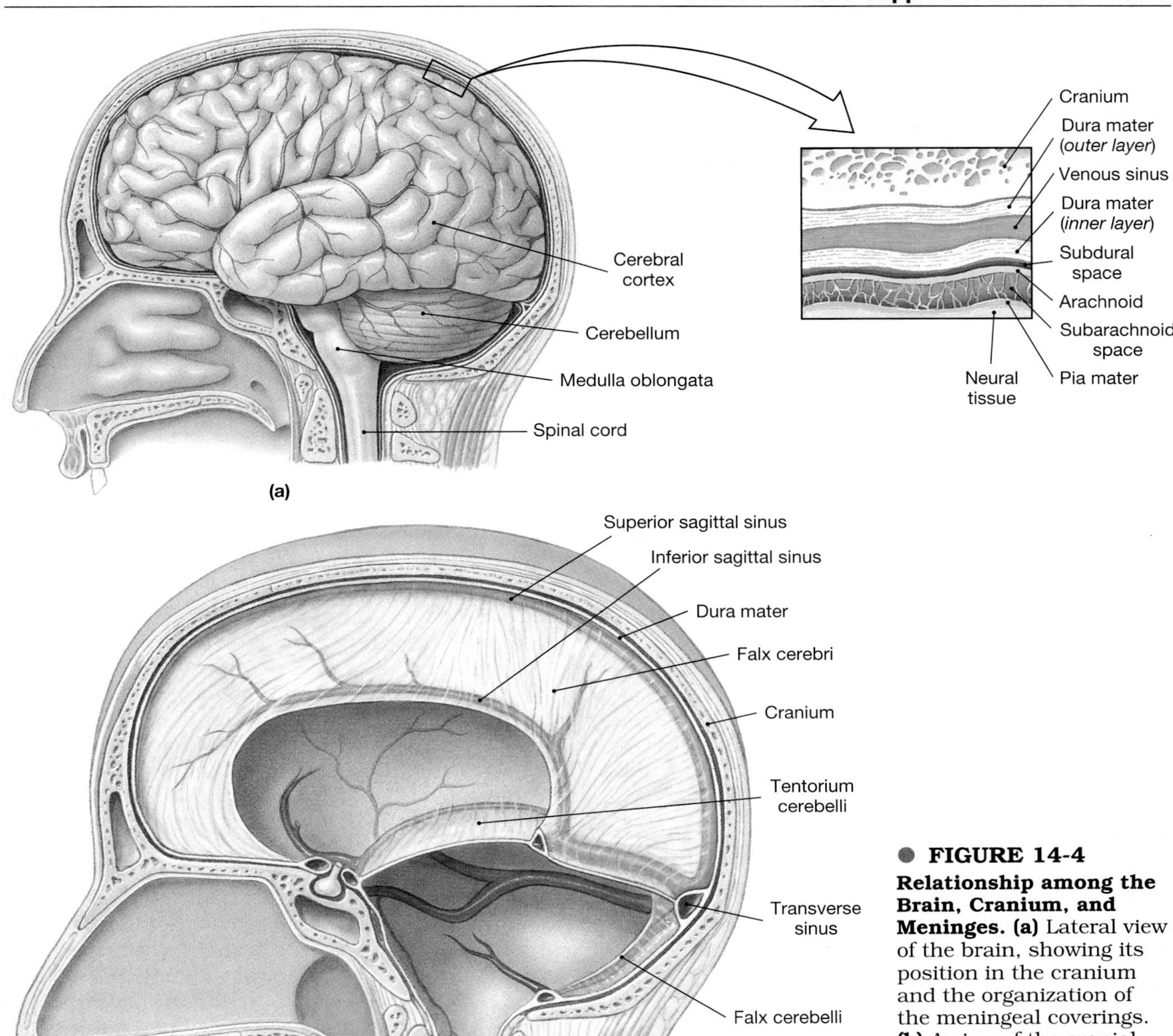
Cranium
Dura mater (outer layer)
Venous sinus
Dura mater (inner layer)
Subdural space
Arachnoid
Subarachnoid space
Pia mater
Neural tissue
Cerebral cortex
Cerebellum
Medulla oblongata
Spinal cord
(a)
Superior sagittal sinus
Inferior sagittal sinus
Dura mater
Falx cerebri
Cranium
Tentorium cerebelli
Transverse sinus
Falx cerebelli
(b)

● **FIGURE 14-4 Relationship among the Brain, Cranium, and Meninges. (a)** Lateral view of the brain, showing its position in the cranium and the organization of the meningeal coverings. **(b)** A view of the cranial cavity with the brain removed, showing the orientation and extent of the falx cerebri, tentorium cerebelli, and falx cerebelli.

million cases of cranial trauma each year in the United States, but only 1 case in 8 results in serious brain damage. The percentage is relatively low because the cranial meninges provide effective protection for the brain. Tough, fibrous *dural folds* act like safety belts, holding the brain in position. The cerebrospinal fluid contained in the subarachnoid space acts like an airbag that cushions sudden jolts and shocks. AM *Cranial Trauma*

DURAL FOLDS In several locations the innermost layer of the dura mater extends into the cranial cavity, forming a sheet that dips inward and then returns. These **dural folds** provide additional stabilization and support to the brain (Figure 14-4b●). Dural sinuses may be found between the two layers of a dural fold. The three largest dural folds are called the *falx cerebri*, the *tentorium cerebelli*, and the *falx cerebelli*.

- The **falx cerebri** (FALKS ser-Ē-brē; *falx*, curving, or sickle-shaped) is a fold of dura mater that projects between the cerebral hemispheres in the longitudinal fissure. Its inferior portions attach anteriorly to the crista galli (see Figure 7-4●, p. 207) and posteriorly to the internal occipital crest. Two large venous sinuses, the **superior sagittal sinus** and the **inferior sagittal sinus,** travel within this dural fold. The posterior margin of the falx cerebri intersects the *tentorium cerebelli.*

- The **tentorium cerebelli** (ten-TOR-ē-um ser-e-BEL-ē; *tentorium*, a covering) separates and protects the cerebellar hemispheres from those of the cerebrum. It extends across the cranium at right angles to the falx cerebri. The **transverse sinus** lies within the tentorium cerebelli.
- The **falx cerebelli** extends in the midsagittal line inferior to the tentorium cerebelli, dividing the two cerebellar hemispheres.

❑ Cerebrospinal Fluid

Cerebrospinal fluid (CSF) completely surrounds and bathes the exposed surfaces of the central nervous system. It has several important functions, including:

1. ***Cushioning delicate neural structures.***
2. ***Supporting the brain.*** In essence, the brain is suspended inside the cranium floating in the cerebrospinal fluid. A human brain weighs about 1400 g in air, but only about 50 g when supported by the cerebrospinal fluid.
3. ***Transporting nutrients, chemical messengers, and waste products.*** Except at the *choroid plexus*, where CSF is produced, the ependymal lining is freely permeable, and the CSF is in constant chemical communication with the interstitial fluid of the CNS.

Because free exchange occurs between the interstitial fluid and CSF, changes in CNS function may produce changes in the composition of the CSF. As noted in Chapter 13, a *spinal tap* can provide useful clinical information concerning CNS injury, infection, or disease. ∞ [p. 430]

The Formation of CSF

Figure 14-5

The **choroid plexus** (*choroid*, a vascular coat; *plexus*, a network) consists of a combination of specialized ependymal cells and permeable capillaries for the production of cerebrospinal fluid (CSF). Two extensive folds of the choroid plexus originate in the roof of the diencephalon and extend through the interventricular foramina. These folds cover the floors of the lateral ventricles (Figure 14-5a•). In the lower brain stem, a region of the choroid plexus in the roof of the fourth ventricle projects between the cerebellum and pons.

Large ependymal cells cover the capillaries of the choroid plexus and contact the CSF of the ventricles. These cells secrete cerebrospinal fluid into the ventricles. The regulation of CSF composition involves transport in both directions, and the choroid plexus removes waste products from the CSF and adjusts its composition over time. The differences in composition between cerebrospinal fluid and blood plasma (blood with the cellular elements removed) are quite pronounced. For example, the blood contains high concentrations of soluble proteins, but the CSF does not. There are also differences in the concentrations of individual ions and in the levels of amino acids, lipids, and waste products.

Circulation of CSF

Figure 14-5

The choroid plexus produces CSF at a rate of about 500 ml/day. The total volume of CSF at any given moment is approximately 150 ml; thus the entire volume of CSF is replaced roughly every 8 hours. Despite this rapid turnover, the composition of CSF is closely regulated, and the rate of removal normally keeps pace with the rate of production.

The CSF produced by the choroid plexus circulates through the ventricles and the central canal of the spinal cord. It reaches the subarachnoid space via two **lateral apertures** and a single **median aperture** in the roof of the fourth ventricle. CSF then flows through the subarachnoid space surrounding the brain, spinal cord, and cauda equina.

Along the axis of the superior sagittal sinus, fingerlike extensions of the arachnoid membrane penetrate the dura mater (Figure 14-5b•) and form **arachnoid villi** (singular *villus;* a tuft of hair). Cerebrospinal fluid is absorbed into the venous circulation at the arachnoid villi. If the normal circulation or reabsorption of CSF is interrupted, a variety of clinical problems may appear. For example, a problem with the reabsorption of CSF in infancy is responsible for symptoms of *hydrocephalus*, or "water on the brain." Infants with this condition have enormously expanded skulls because of the presence of an abnormally large volume of CSF. In an adult, failure of reabsorption or blockage of CSF circulation can cause distortion and damage to the brain. AM *Hydrocephalus*

❑ The Blood Supply to the Brain

The brain is an extremely active organ that requires a constant, reliable source of oxygen and nutrients. Arterial blood reaches the brain via the *internal carotid arteries* and the *vertebral arteries;* most of the venous blood from the brain leaves the cranium in the *internal jugular veins* that drain the dural sinuses. A head injury that damages cerebral blood vessels may cause bleeding into the subdural space or between the outer layer of the dura and the bones of the skull. These are serious conditions because the blood entering these spaces compresses and distorts the relatively soft tissues of the brain. AM *Epidural and Subdural Hemorrhages*

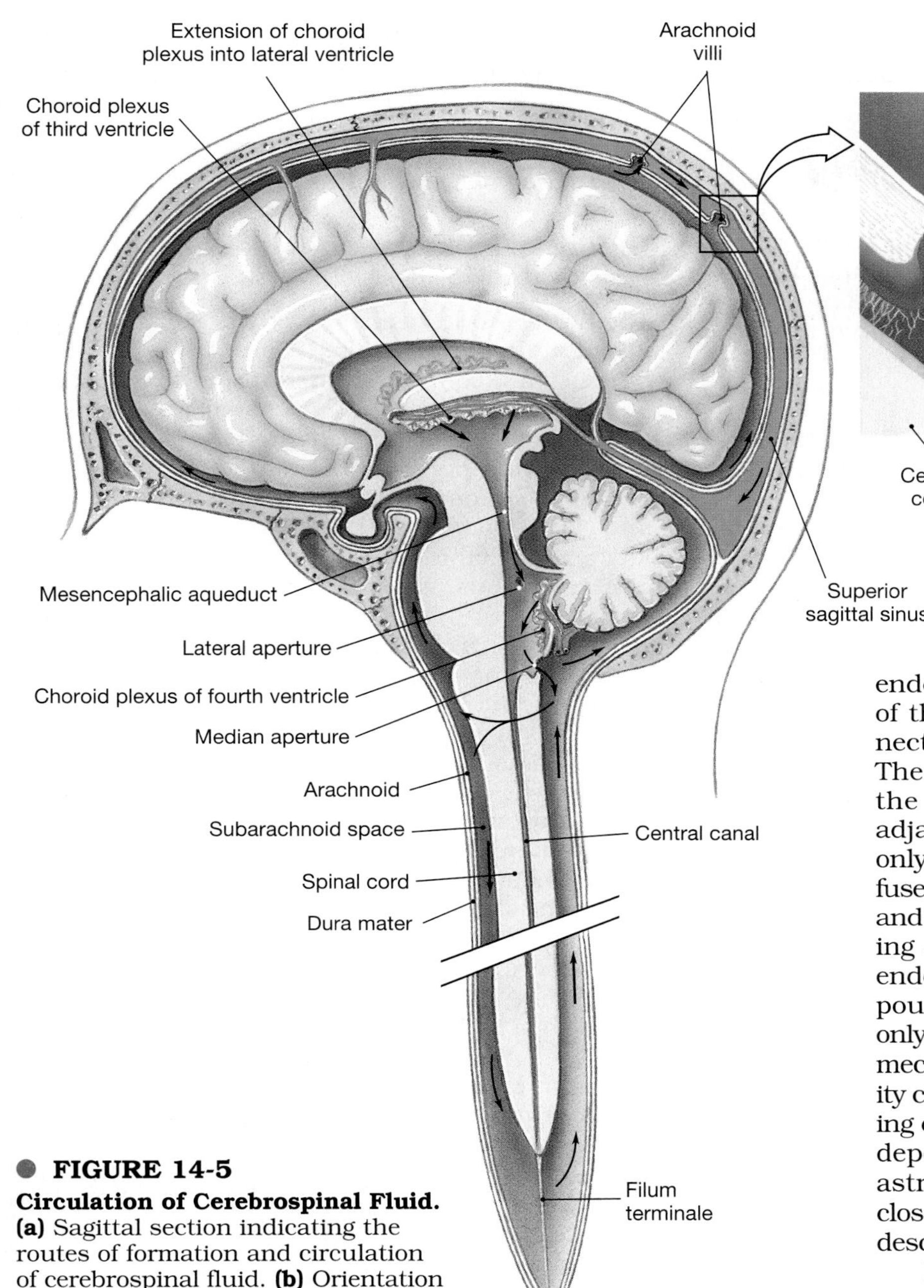
Extension of choroid plexus into lateral ventricle
Arachnoid villi
Choroid plexus of third ventricle
Mesencephalic aqueduct
Lateral aperture
Choroid plexus of fourth ventricle
Median aperture
Arachnoid
Subarachnoid space
Spinal cord
Dura mater
Central canal
Superior sagittal sinus
Filum terminale
(a)
Dura mater (outer layer)
Cranium
Endothelial lining
Superior sagittal sinus
Fluid movement
Arachnoid villus
Arachnoid
Dura mater (inner layer)
Cerebral cortex
Pia mater
Subdural space
(b)

FIGURE 14-5
Circulation of Cerebrospinal Fluid. **(a)** Sagittal section indicating the routes of formation and circulation of cerebrospinal fluid. **(b)** Orientation of the arachnoid villi. AM *Scan 2*

CEREBROVASCULAR DISEASES **Cerebrovascular diseases** are circulatory disorders that interfere with the normal circulatory supply to the brain. The particular distribution of the vessel involved will determine the symptoms, and the degree of oxygen or nutrient starvation will determine their severity. A stroke, or **cerebrovascular accident (CVA),** occurs when the blood supply to a portion of the brain is shut off. Affected neurons begin to die in a matter of minutes.

The symptoms of a stroke provide an indication of the vessel and region of the brain involved. Specific examples will be noted in Chapter 21, in our discussion of cranial circulation.

The Blood-Brain Barrier

Neural tissue in the CNS is isolated from the general circulation by the **blood-brain barrier.** This barrier exists because the endothelial cells lining the capillaries of the CNS are extensively interconnected by tight junctions. *[p. 113]* These junctional complexes prevent the diffusion of materials between adjacent endothelial cells. In general, only lipid-soluble compounds can diffuse into the interstitial fluid of the brain and spinal cord. They do so by diffusing across the cell membranes of the endothelial cells. Water-soluble compounds can cross the capillary walls only through active or passive transport mechanisms. The restricted permeability characteristics of the endothelial lining of brain capillaries are in some way dependent on chemicals secreted by astrocytes. These cells, which are in close contact with CNS capillaries, were described in Chapter 12. *[p. 381]*

Endothelial transport across the blood-brain barrier is selective and directional. Neurons have a constant need for glucose that must be met regardless of the relative concentrations in the blood and interstitial fluid. Even when circulating glucose levels are low, endothelial cells continue to transport glucose from the blood to the interstitial fluid of the brain.

The blood-brain barrier remains intact throughout the CNS, with three noteworthy exceptions:

1. In portions of the hypothalamus the capillary endothelium is extremely permeable. This permeability exposes hypothalamic nuclei to circulat-

ing hormones and permits the diffusion of hypothalamic hormones into the circulation.

2. Capillaries in the *pineal gland* are also very permeable. The pineal gland, an endocrine structure, is located in the posterior surface of the diencephalon. The capillary permeability allows pineal secretions into the general circulation.
3. Capillaries at the choroid plexus are extremely permeable. The secretion of CSF and the regulation of its composition depend entirely on the transport activities of specialized ependymal cells within the plexus.

What would happen if an interventricular foramen became blocked?

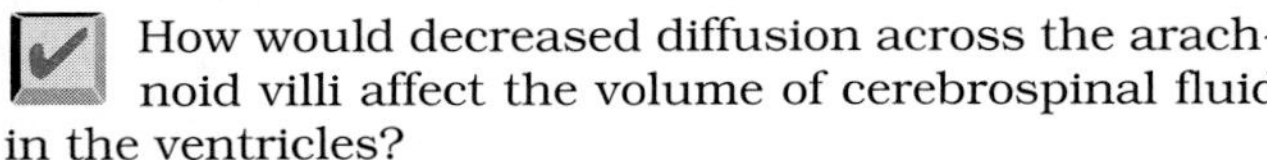

How would decreased diffusion across the arachnoid villi affect the volume of cerebrospinal fluid in the ventricles?

Many water-soluble molecules found in the blood in relatively large amounts are found in small amounts or not at all in the extracellular fluid of the brain. Why?

The Cerebrum

Figures 14-1, 14-6

The cerebrum is the largest region of the brain. Conscious thought processes and all intellectual functions originate in the cerebral hemispheres. Much of the cerebrum is involved in the processing of somatic sensory and motor information.

Figures 14-1, p. 462, and 14-6● provide perspective on the cerebrum and its relationships with other regions of the brain. Gray matter in the cerebrum is found in the *cerebral cortex* and in deeper *cerebral nuclei*. The *central white matter* lies beneath the neural cortex and around the cerebral nuclei.

The Cerebral Cortex

Figures 14-1, 14-6

A blanket of neural cortex covers the paired *cerebral hemispheres* that dominate the superior and lateral surfaces of the cerebrum (Figure 14-1, p. 462, and 14-6a●). The cortical surface forms a series of elevated ridges, or **gyri** (JĪ-rī), separated by shallow depressions, called **sulci** (SUL-kē), or deeper grooves, called **fissures.** Gyri increase the surface area of the cerebral hemispheres and the number of neurons in the cortical regions. The total surface area of the cerebral hemispheres is roughly equivalent to 2.25 square meters (25 sq. ft) of flat surface. An area that large can be packed into the skull only when folded, like a crumpled piece of paper. Complex analytical and integrative functions require large numbers of neurons. The entire brain has enlarged in the course of human evolution, but the cerebral hemispheres have enlarged at a much faster rate, even with extensive folding of the cerebral hemispheres.

Boundaries between Cerebral Lobes

Figures 14-1, 14-6

The two cerebral hemispheres are separated by a deep **longitudinal fissure** (Figure 14-1a●, p. 462), but they are connected inferiorly by a thick band of axons called the *corpus callosum*. Each cerebral hemisphere can be divided into **lobes** named after the overlying bones of the skull. There are individual differences in the appearance of the sulci and gyri of each brain, but the boundaries between lobes are reliable landmarks. A deep groove, the **central sulcus,** extends laterally from the longitudinal fissure. The area anterior to the central sulcus is the **frontal lobe,** and the **lateral sulcus** marks its inferior border (Figure 14-6a●). The region inferior to the lateral sulcus is the **temporal lobe.** Pushing this lobe to the side exposes the **insula** (IN-sū-luh; *insula*, island), an "island" of cortex that is otherwise invisible. The **parietal lobe** extends between the central sulcus and the **parieto-occipital sulcus.** What remains constitutes the **occipital lobe.**

Each lobe contains functional regions whose boundaries are less clearly defined. Some of these regions are concerned with sensory information and others with motor commands. Three points about the cerebral lobes should be kept in mind:

1. ***Each cerebral hemisphere receives sensory information and generates motor commands that concern the opposite side of the body.*** The left hemisphere controls the right side, and the right hemisphere controls the left side. This crossing over has no known functional significance.
2. ***The two hemispheres are functionally different even though anatomically they appear similar.*** These differences affect primarily higher-order functions, a topic that will be discussed in a later section.
3. ***The assignment of a specific function to a specific region of the cerebral cortex is imprecise.*** Because the boundaries are indistinct, with considerable overlap, any one region may have several different functions. Some aspects of cortical function, such as consciousness, cannot easily be assigned to any single region.

Our understanding of brain function is still incomplete, and we do not know the function of every anatomical feature. However, it is clear that all portions of the brain are used in a normal individual.

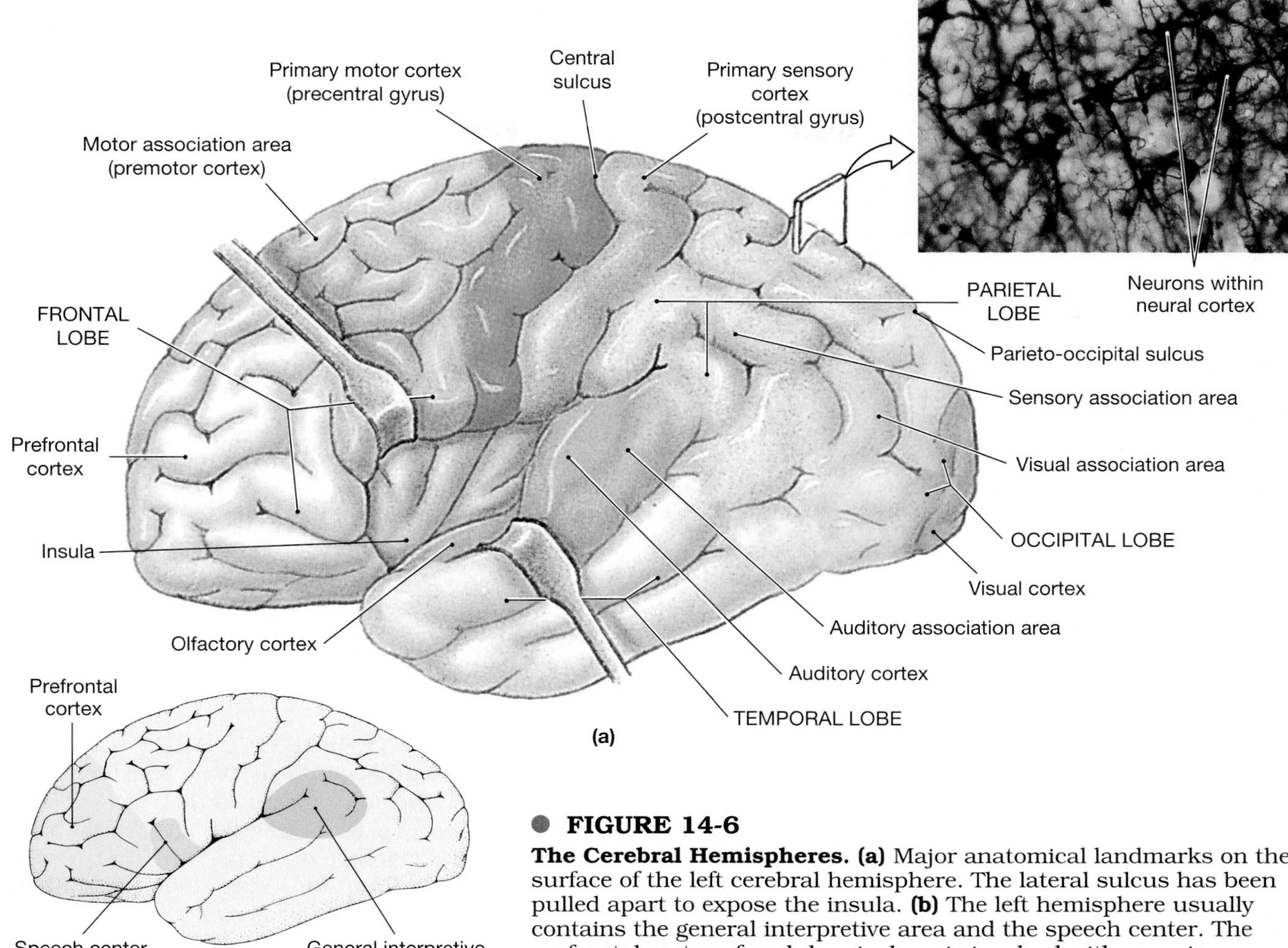
Primary motor cortex (precentral gyrus)
Central sulcus
Primary sensory cortex (postcentral gyrus)
Motor association area (premotor cortex)
FRONTAL LOBE
Prefrontal cortex
Insula
Olfactory cortex
PARIETAL LOBE
Neurons within neural cortex
Parieto-occipital sulcus
Sensory association area
Visual association area
OCCIPITAL LOBE
Visual cortex
Auditory association area
Auditory cortex
TEMPORAL LOBE
(a)
Prefrontal cortex
Speech center (Broca's area)
General interpretive area
(b)

● **FIGURE 14-6**

The Cerebral Hemispheres. (a) Major anatomical landmarks on the surface of the left cerebral hemisphere. The lateral sulcus has been pulled apart to expose the insula. **(b)** The left hemisphere usually contains the general interpretive area and the speech center. The prefrontal cortex of each hemisphere is involved with conscious intellectual functions.

The Central White Matter

The **central white matter** contains three major groups of axons: (1) *association fibers,* tracts that interconnect areas of neural cortex within a single cerebral hemisphere; (2) *commissural fibers,* tracts that connect the two cerebral hemispheres; and (3) *projection fibers,* tracts that link the cerebrum with other regions of the brain and the spinal cord.

- **Association fibers** interconnect portions of the cerebral cortex. The shortest association fibers are called **arcuate** (AR-kū-āt) **fibers** because they curve in an arc to pass from one gyrus to another. The longer association fibers are organized into discrete bundles, or *fasciculi.* The **longitudinal fasciculi** connect the frontal lobe to the other lobes of the same hemisphere.
- A dense band of **commissural** (kom-MIS-ū-rul) **fibers** (*commissura,* a crossing over) permits communication between the two hemispheres. Prominent commissural bundles linking the cerebral hemispheres include the **corpus callosum** and the **anterior commissure.** The corpus callosum alone contains over 200 million axons, carrying an estimated 4 billion impulses per second!
- **Projection fibers** link the cerebral cortex to the brain stem, cerebellum, and spinal cord. All must pass through the diencephalon, where the converging fibers heading to sensory areas of the cortex run past the diverging fibers leaving the motor cortex. In gross dissection the afferent fibers and efferent fibers look alike, and the entire collection of fibers is known as the **internal capsule.**

Motor and Sensory Areas of the Cortex

Figure 14-6

The major motor and sensory regions of the cerebral cortex are detailed in Figure 14-6● and Table 14-2. The central sulcus separates the motor and sensory portions of the cortex. The **precentral gyrus** of the frontal lobe forms the anterior margin of the central sulcus. The surface of this gyrus is the **primary motor cortex.** Neurons of the primary motor cortex direct voluntary movements by controlling somatic motor neurons in the brain stem and spinal

TABLE 14-2 The Cerebral Cortex

Region/Nucleus	*Functions*
FRONTAL LOBE	
Primary motor cortex	Voluntary control of skeletal muscles
PARIETAL LOBE	
Primary sensory cortex	Conscious perception of touch, pressure, vibration, pain, temperature, and taste
OCCIPITAL LOBE	
Visual cortex	Conscious perception of visual, auditory, and olfactory stimuli
TEMPORAL LOBE	
Auditory cortex and olfactory cortex	(see above)
ALL LOBES	
Association areas	Integration and processing of sensory data; processing and initiation of motor activities

cord. These cortical neurons are called **pyramidal cells,** and the pathway that provides voluntary motor control is known as the **pyramidal system.**

The **postcentral gyrus** of the parietal lobe forms the posterior margin of the central sulcus, and its surface contains the **primary sensory cortex.** Neurons in this region receive somatic sensory information from touch, pressure, pain, taste, and temperature receptors. We are consciously aware of these sensations because nuclei in the thalamus relay the information to the primary sensory cortex.

Sensory information concerning sensations of sight, sound, and smell arrive at other portions of the cerebral cortex (Figure 14-6a•). The **visual cortex** of the occipital lobe receives visual information, and the **auditory cortex** and **olfactory cortex** of the temporal lobe receive information concerned with hearing and smell, respectively.

Association Areas

Figure 14-6

The sensory and motor regions of the cortex are connected to nearby **association areas** that interpret incoming data or coordinate a motor response (Figure 14-6a•). The **somatic motor association area,** or **premotor cortex,** is responsible for the coordination of learned motor responses. The functional distinctions between the sensory and motor association areas are most evident after localized brain damage has occurred. For example, an individual with a damaged **visual association area** may see letters quite clearly but be unable to recognize or interpret them. This person would scan the lines of a printed page and see rows of clear symbols that convey no meaning. Someone with damage to the area of the premotor cortex concerned with coordination of eye movements can understand written letters and words but cannot read because his or her eyes cannot follow the lines on a printed page.

Integrative Centers

Figure 14-6b

Major integrative centers receive information from many different association areas. These centers direct extremely complex motor activities and perform complicated analytical functions. For example, the *prefrontal cortex* of the frontal lobe (Figure 14-6b•) integrates information from sensory association areas and performs abstract intellectual functions, such as predicting the consequences of possible responses.

Integrative centers are found on the lobes and cortical areas of both cerebral hemispheres. Integrative centers concerned with complex processes, such as speech, writing, mathematical computation, or understanding spatial relationships, are restricted to the left or right hemisphere. These centers include the *general interpretive area* and the *speech center.* The corresponding regions on the opposite hemisphere are also active, but their functions are less well defined.

THE GENERAL INTERPRETIVE AREA The **general interpretive area,** or *gnostic area,* receives information from all the sensory association areas. This analytical center is present in only one hemisphere (usually the left). Damage to the general interpretive area affects the ability to interpret what is read or heard, even though the words are understood as individual entities. For example, an individual might understand the meaning of the words *sit* and *here* but be totally bewildered by the request *sit here.*

Aphasia (*a-,* without + *phasia,* speech) is a disorder affecting the ability to speak or read. **Global aphasia** results from extensive damage to the general interpretive area or to the associated sensory tracts. Affected individuals are totally unable to speak, to read, or to understand the speech of others. Global aphasia often accompanies a severe stroke or tumor that affects a large area of cortex including the speech and language areas. Recovery is possible when the condition results from edema or hemorrhage, but the process often takes months or even years. AM *Aphasia*

Dyslexia (*lexis,* diction) is a disorder affecting the comprehension and use of words. **Developmental dyslexia** affects children; there are estimates that up to 15 percent of children in the United States suffer from some degree of dyslexia. These children have difficulty reading and writing, although their other intellectual functions may be normal or above normal. Their writing looks uneven and disorganized; letters are often reversed or written in the wrong order. Recent evidence suggests that at least some forms of dyslexia result from problems in processing, sorting, and integrating visual or auditory information.

THE SPEECH CENTER Efferents from the general interpretive area target the **speech center,** also called the *motor speech area* or *Broca's area.* This center lies along the edge of the premotor cortex in the same hemisphere as the general interpretive area. The speech center regulates the patterns of breathing and vocalization needed for normal speech. The motor commands are adjusted by feedback from the **auditory association area,** also called the *receptive speech area* or *Wernicke's area.* Damage to the speech center or related sensory areas can manifest itself in various ways. Some individuals have difficulty speaking, although they know exactly what words to use; others talk constantly but use all the wrong words.

THE PREFRONTAL CORTEX The **prefrontal cortex** of the frontal lobe coordinates information from the secondary and special association areas of the entire cortex. In doing so it performs such abstract intellectual functions as predicting the future consequences of events or actions. Damage to the prefrontal cortex leads to difficulties in estimating the temporal relationships between events; questions such as "How long ago did this happen?" or "What happened first?" become difficult to answer.

The prefrontal cortex has extensive connections with other cortical areas and with other portions of the brain. Feelings of frustration, tension, and anxiety are generated at the prefrontal cortex as it interprets ongoing events and makes predictions about future situations or consequences. If the connections between the prefrontal cortex and other brain regions are severed, the tensions, frustrations, and anxieties are removed. Earlier in this century this rather drastic procedure, called a **prefrontal lobotomy,** was used to "cure" a variety of mental illnesses, especially those associated with violent or antisocial behavior. After a lobotomy, the patient would no longer be concerned about what had previously been a major problem, whether psychological (hallucinations) or physical (severe pain). However, the individual was often equally unconcerned about tact, decorum, and toilet training. Now that drugs have been developed to target specific pathways and regions of the CNS, lobotomies are no longer used to change behavior.

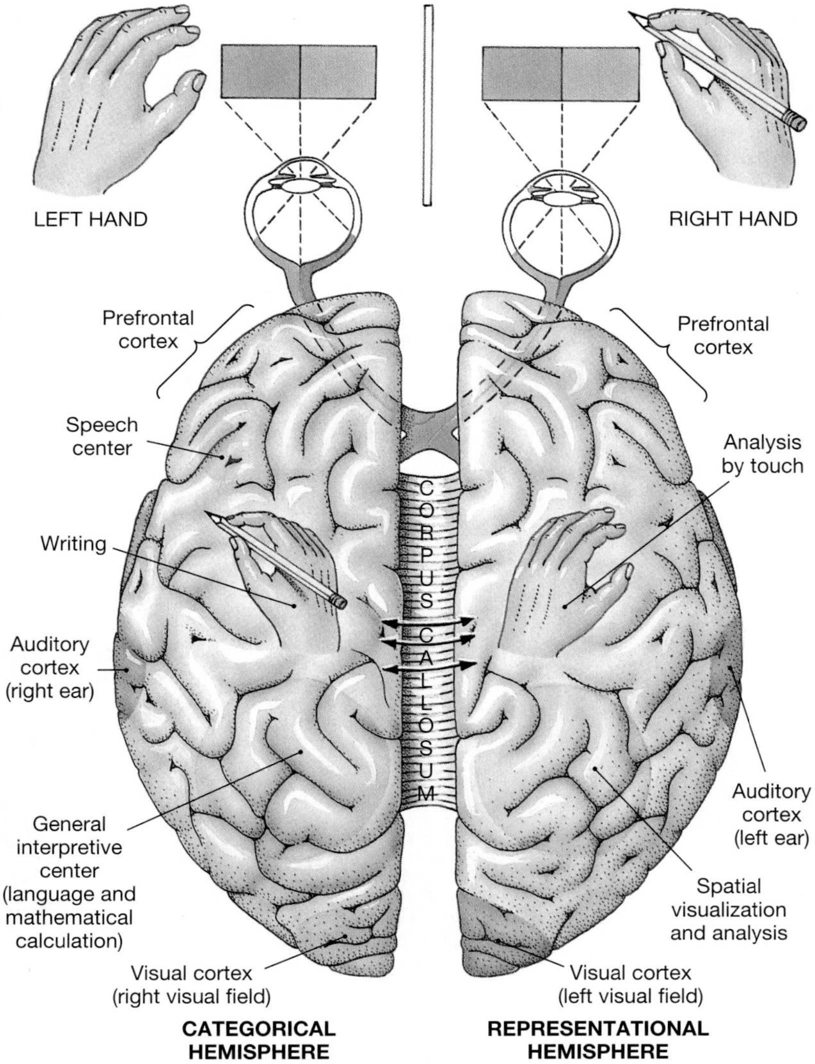

● **FIGURE 14-7**

Hemispheric Specialization. Functional differences between the left and right cerebral hemispheres.

Hemispheric Specialization

Figure 14-7

Figure 14-7● indicates the major functional differences between the hemispheres. In the majority of people in the United States, the left hemisphere contains the general interpretive and speech centers, and this is the hemisphere responsible for language-based skills. For example, reading, writing, and speaking are dependent on processing done in the left cerebral hemisphere. This hemisphere is also important in performing analytical tasks, such as mathematical calculations and logical decision making. Although the left hemisphere was formerly called the *dominant hemisphere,* a more appropriate term is the **categorical hemisphere,** because the right hemisphere has many important functions.

The right cerebral hemisphere analyzes sensory information and relates the body to the sensory environment. Interpretive centers in this hemisphere permit the identification of familiar objects by touch, smell, sight, taste, or feel. For example, the right hemisphere plays a dominant role in recognizing faces and in under-three-dimensional relationships. It is also important in analyzing the emotional context of a conversation—for example, determining whether the phrase "get lost" was intended as a threat or a question. Individuals with damaged right hemispheres may also be unable to add emotional inflections to their own words. Because it is concerned with spatial relationships and analyses, the term **representational hemisphere** is used to refer to the right hemisphere. Interestingly, an unusually high percentage of musicians and artists are left-handed. The complex motor activities performed by these individuals are directed by the primary motor cortex and association areas on the right (representational) hemisphere.

DISCONNECTION SYNDROME The functional differences between the hemispheres become apparent if the corpus callosum is cut, a procedure sometimes performed to treat epileptic seizures that cannot be controlled by other methods. This surgery produces symptoms of **disconnection syndrome.** In this condition the two hemispheres function independently, each remaining "unaware" of stimuli or motor commands involving their counterpart. The result is some rather interesting changes in the individual's abilities. For example, objects touched by the left hand can be recognized but not verbally identified, because the sensory information arrives at the right hemisphere and the speech center is on the left. The object can be verbally identified if felt with the right hand, but the person will not be able to say whether or not it is the same object previously touched with the left hand.

Shelly suffers a head injury that damages her primary motor cortex. Where is this area located?

What senses would be affected by damage to the temporal lobes of the cerebrum?

After suffering a stroke, Jake is unable to speak. He can understand what is said to him and he can understand written messages, but he cannot express himself verbally. What part of the brain has been affected by the stroke?

The Cerebral Nuclei

Figure 14-8

The **cerebral nuclei** (Figure 14-8•) lie within each hemisphere beneath the floor of the lateral ventricle. They are embedded within the central white

• FIGURE 14-8

The Cerebral Nuclei. (a) The relative positions of the cerebral nuclei in the intact brain. **(b)** The cerebral nuclei seen in frontal section. **(c)** The cerebral nuclei in horizontal section; compare this with the view in dissection **(d)**.

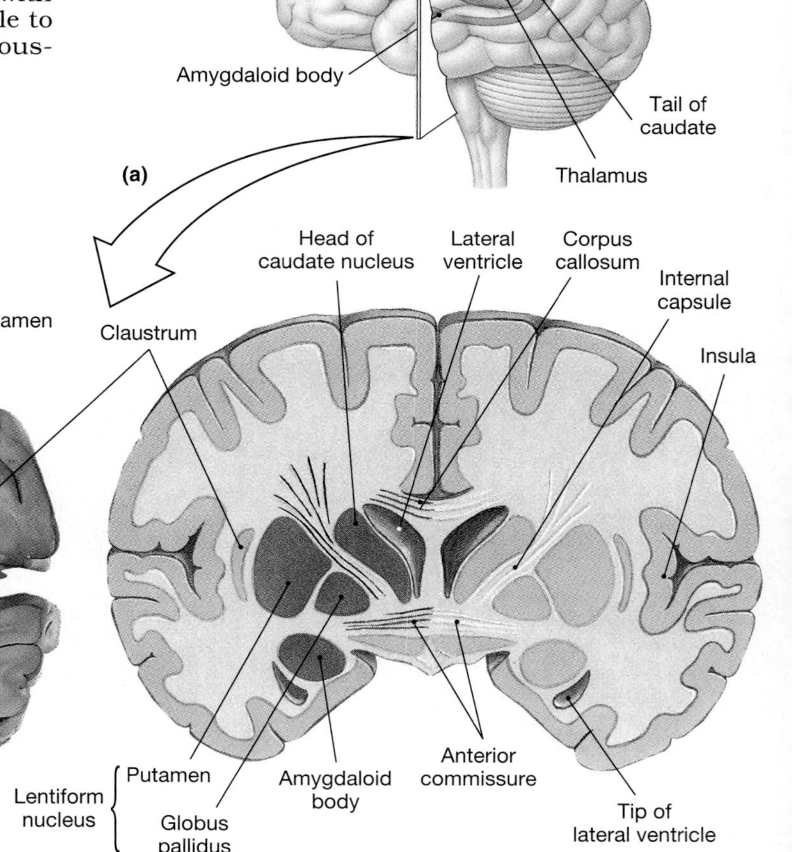
Head of caudate nucleus
Lentiform nucleus
Amygdaloid body
Tail of caudate
Thalamus
(a)
Corpus callosum
Longitudinal fissure
Caudate nucleus
Putamen
Lateral ventricles
Internal capsule
(b) Frontal section
Claustrum
Head of caudate nucleus
Lateral ventricle
Corpus callosum
Internal capsule
Insula
Lentiform nucleus
Putamen
Globus pallidus
Amygdaloid body
Anterior commissure
Tip of lateral ventricle

matter, and the radiating projection and commissural fibers travel around or between these nuclei. The cerebral nuclei are also known as the **basal nuclei,** or *basal ganglia.* The latter term has persisted despite the fact that ganglia are otherwise restricted to the peripheral nervous system.

The **caudate nucleus** has a massive head and a slender, curving tail that follows the curve of the lateral ventricle. The head of the caudate nucleus lies superior to the **lentiform** (lens-shaped) **nucleus,** which consists of a medial **globus pallidus** (GLŌ-bus PAL-i-dus; pale globe) and a lateral **putamen** (pū-TĀ-men). The term **corpus striatum** (striated body) has been used to refer to the caudate and lentiform nuclei, or to the caudate and putamen. The name refers to the striated (striped) appearance of the internal capsule as it passes among these nuclei. The **claustrum** (KLAWS-trum) is a small nucleus that lies between the lentiform nucleus and the surface of the insula; the **amygdaloid** (ah-MIG-da-loyd) **body** (*amygdale,* almond) is a separate nucleus located anterior to the tail of the caudate and inferior to the lentiform nucleus. The amygdaloid body is a component of the *limbic system* and will be considered further in a later section.

Functions of the Cerebral Nuclei

The cerebral nuclei are components of the **extrapyramidal system,** a motor system that controls skeletal muscle tone and coordinates learned movement patterns and other somatic motor activities. Under normal conditions these nuclei do not initiate particular movements, but once a movement is under way the cerebral nuclei provide the general pattern and rhythm, especially for movements of the trunk and proximal limb muscles.

Information arrives at the caudate nucleus and putamen from sensory, motor, and association areas of the cerebral cortex. Processing and integration occur in these nuclei and in the adjacent globus pallidus. Most of the output of the cerebral nuclei leaves the globus pallidus and synapses in the thalamus. The thalamic nuclei then project the information to appropriate areas of the cerebral cortex. The cerebral nuclei alter the motor commands issued by the cerebral cortex via this feedback loop. For example:

- When a person is walking, the cerebral nuclei control the cycles of arm and thigh movements that occur between the time the decision is made to "start walking" and the time the "stop" order is given.
- As a voluntary movement is starting, the cerebral nuclei control and adjust muscle tone, particularly in the appendicular muscles, to set body position. When you decide to pick up a pencil, you consciously reach and grasp with the forearm, wrist, and hand, while the cerebral nuclei position the shoulder and stabilize the arm.

The Limbic System

Figures 14-8d, 14-9, 14-10

The **limbic** (LIM-bik) **system** (*limbus,* a border) includes nuclei and tracts along the border between

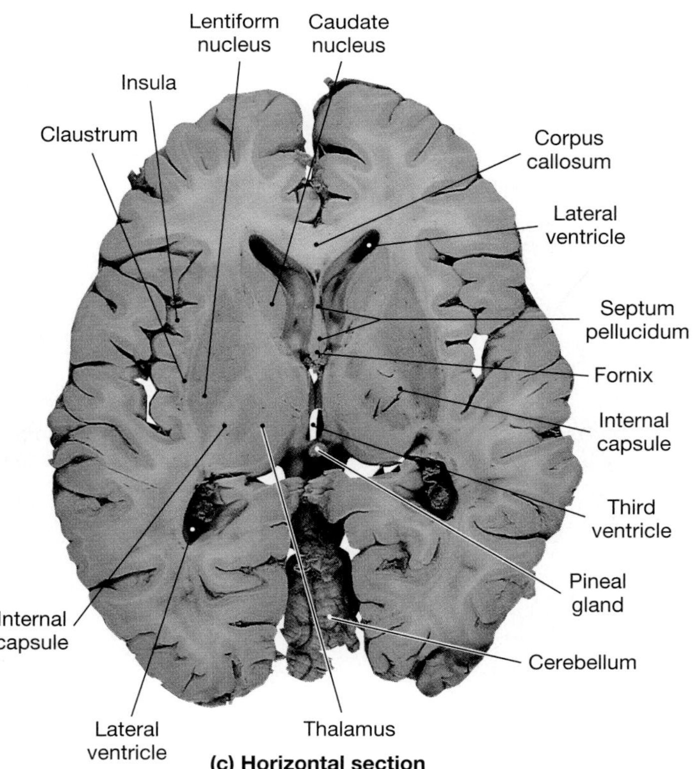
Lentiform nucleus
Caudate nucleus
Insula
Claustrum
Corpus callosum
Lateral ventricle
Septum pellucidum
Fornix
Internal capsule
Third ventricle
Pineal gland
Cerebellum
Internal capsule
Lateral ventricle
Thalamus

(c) Horizontal section

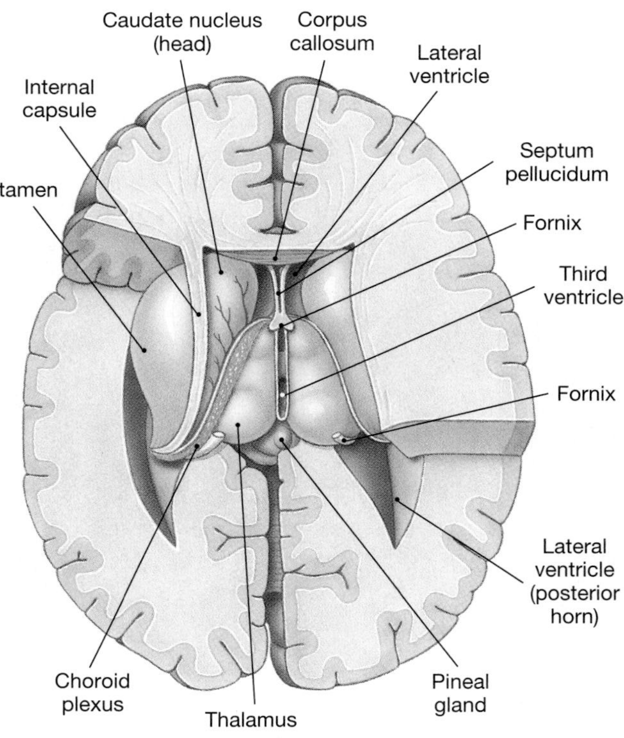
Caudate nucleus (head)
Corpus callosum
Lateral ventricle
Internal capsule
Septum pellucidum
Putamen
Fornix
Third ventricle
Fornix
Lateral ventricle (posterior horn)
Choroid plexus
Thalamus
Pineal gland

(d) Horizontal section, dissected

the cerebrum and diencephalon. The functions of the limbic system include: (1) establishing emotional states and related behavioral drives; (2) linking the conscious, intellectual functions of the cerebral cortex with the unconscious and autonomic functions of the brain stem; and (3) facilitating memory storage and retrieval.

Figures 14-8d and 14-9● indicate the relationships among some of the limbic system components and other structures of the cerebrum, diencephalon, and brain stem. Figure 14-10● focuses attention on major components of the limbic system. This system is a functional grouping, rather than an anatomical one, and the limbic system includes components of the cerebrum, diencephalon, and mesencephalon.

The amygdaloid body at the "tail end" of the caudate nucleus has already been described as a cerebral contribution to the limbic system (p. 473). It appears to act as an interface between the limbic system, the cerebrum, and various sensory systems. The **limbic lobe** of the cerebral hemisphere consists of two gyri that curve along the corpus callosum and onto the medial surface of the temporal lobe (Figure 14-10●). The **cingulate** (SIN-gū-lāt) **gyrus** (*cingulum*, a girdle or belt) sits above the corpus callosum. The **dentate gyrus** and the adjacent **parahippocampal** (pa-ra-hip-ō-KAM-pal) **gyrus** conceal the **hippocampus,** a folded area of cortex that lies beneath the floor of the lateral ventricle. Early anatomists thought this structure resembled a sea horse (*hippocampus*); it appears to be important in learning, especially in the storage and retrieval of new long-term memories.

The **fornix** (FŌR-niks) is a tract of white matter that connects the hippocampus with the hypothalamus. From the hippocampus the fornix curves medially to meet its counterpart from the opposing hemisphere and proceeds anteriorly, inferior to the corpus callosum, before arching downward to the hypothalamus. Many of the fibers end in the **mamillary** (MAM-i-lar-ē) **bodies** (*mamilla*, a breast or nipple), prominent nuclei in the floor of the hypothalamus (Figures 14-9 and 14-10●). The mamillary bodies process sensory information, including olfactory sensations, and they contain motor nuclei that control reflex movements associated with eating, such as chewing, licking, and swallowing.

Several other nuclei in the wall (thalamus) and floor (hypothalamus) of the diencephalon are components of the limbic system. The **anterior nucleus** of the thalamus relays information from the mamillary bodies (hypothalamus) to the cingulate gyrus. The boundaries between some of the hypothalamic nuclei are often poorly defined, but experimental stimulation has outlined a number of important hypothalamic centers responsible for the emotions of rage, fear, pain, sexual arousal, and pleasure.

Stimulation of specific regions of the hypothalamus can also produce heightened alertness and a generalized excitement or generalized lethargy and sleep. These responses are caused by stimulation or inhibition of the *reticular formation.* The **reticular formation,** an interconnected network of neurons and nuclei that extends the length of the brain stem, has its headquarters in the mesencephalon. Table 14-3 summarizes the organization and functions of the limbic system.

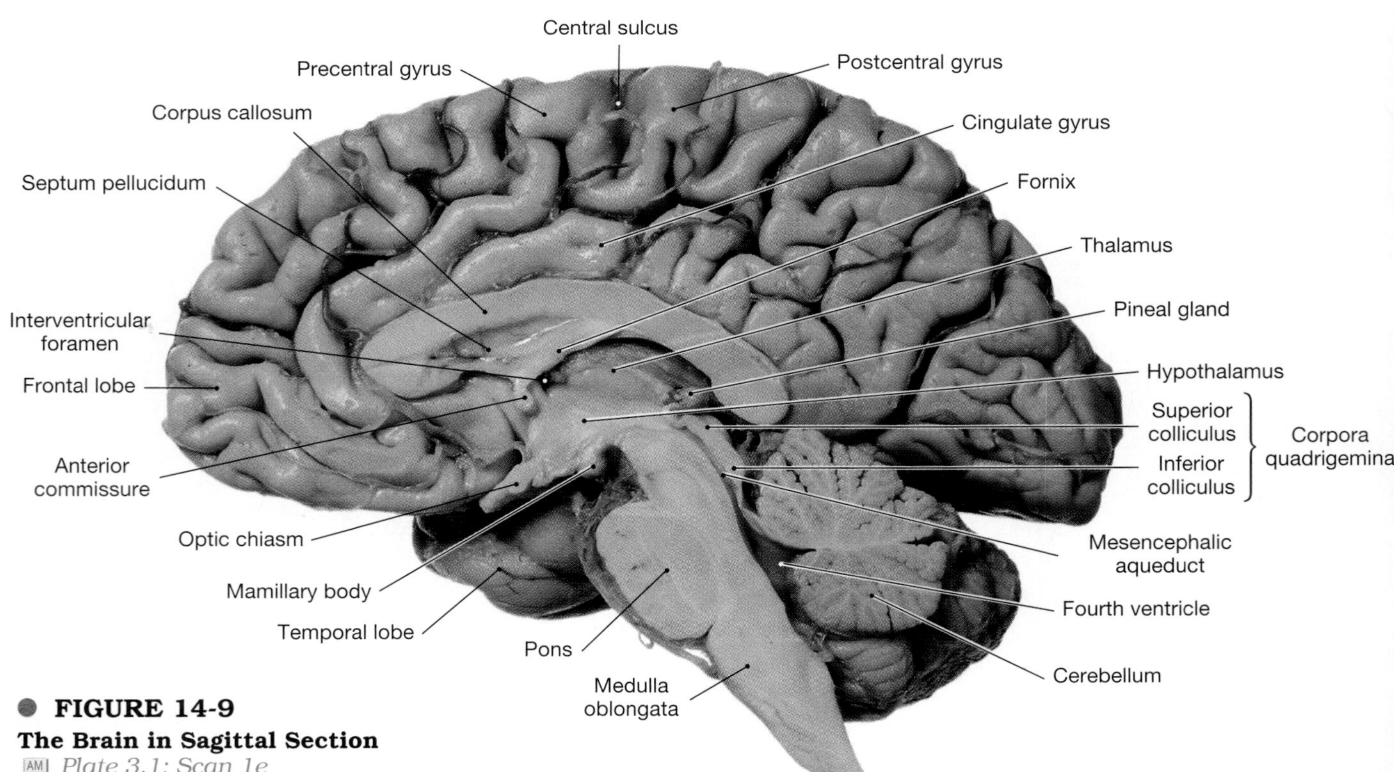
Central sulcus
Precentral gyrus
Postcentral gyrus
Corpus callosum
Cingulate gyrus
Septum pellucidum
Fornix
Thalamus
Pineal gland
Interventricular foramen
Frontal lobe
Hypothalamus
Superior colliculus
Inferior colliculus
Corpora quadrigemina
Anterior commissure
Optic chiasm
Mesencephalic aqueduct
Mamillary body
Fourth ventricle
Temporal lobe
Pons
Cerebellum
Medulla oblongata

● **FIGURE 14-9**
The Brain in Sagittal Section
AM *Plate 3.1; Scan 1e*

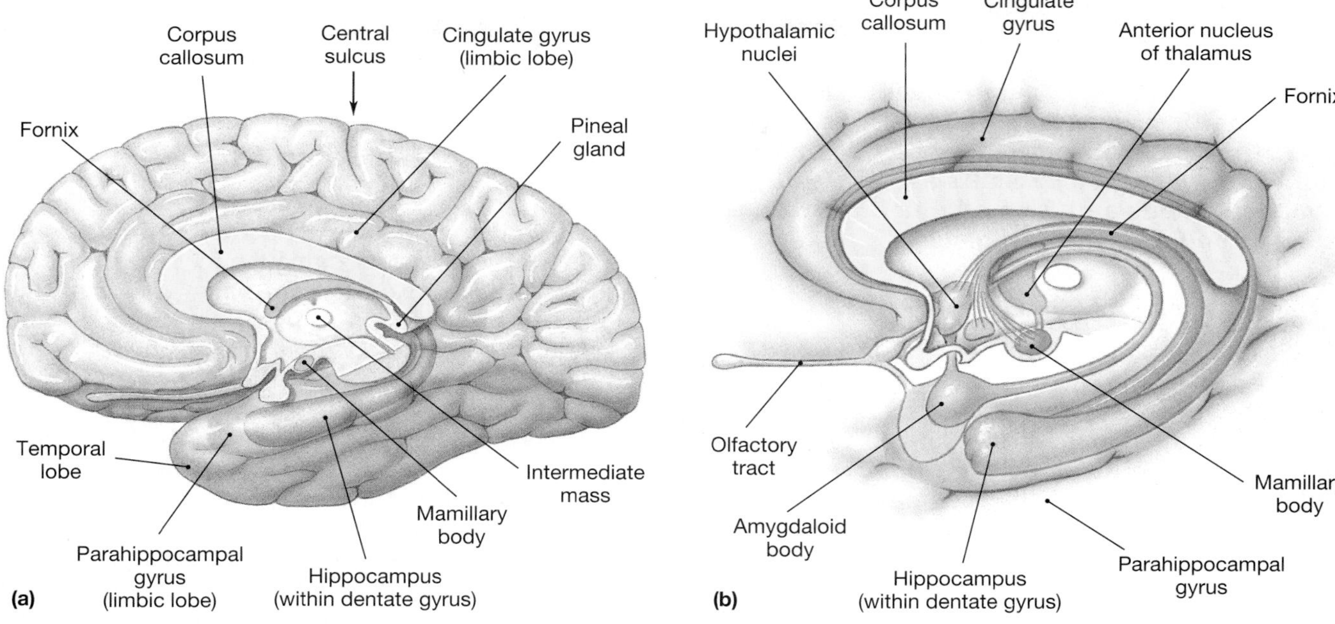
Corpus callosum
Central sulcus
Cingulate gyrus (limbic lobe)
Fornix
Pineal gland
Temporal lobe
Intermediate mass
Mamillary body
Parahippocampal gyrus (limbic lobe)
Hippocampus (within dentate gyrus)
(a)
Hypothalamic nuclei
Corpus callosum
Cingulate gyrus
Anterior nucleus of thalamus
Fornix
Olfactory tract
Mamillary body
Amygdaloid body
Hippocampus (within dentate gyrus)
Parahippocampal gyrus
(b)

● **FIGURE 14-10**

The Limbic System. (a) A diagrammatic sagittal section through the cerebrum, showing the cortical areas associated with the limbic system. The parahippocampal gyrus is shown as if transparent so that deeper limbic components can be seen. **(b)** Three-dimensional reconstruction of the limbic system, showing the relationships among the major components.

✔ What symptoms would you expect to observe in an individual who has damage to the extrapyramidal system?

✔ Paul is having a difficult time remembering facts and recalling long-term memories. What part of the cerebrum is probably involved?

✔ After suffering a head injury in an automobile accident, Terri finds she is having trouble chewing and swallowing. This difficulty suggests that Terry may have sustained damage to what part of her brain?

The Diencephalon

Figures 14-9, 14-10, 14-11

The diencephalon plays a vital role in integrating conscious and unconscious sensory information and motor commands. Figures 14-9, 14-10a, and 14-11● show the position of the diencephalon and its relationship to landmarks on the brain stem. Figure 14-11● includes the origins of 11 of the 12 pairs of cranial nerves. The individual cranial nerves are identified by Roman numerals; the full names of these nerves will be introduced in a later section (pp. 484–493).

The diencephalic roof, or *epithalamus,* contains an extensive area of choroid plexus that extends through the interventricular foramina into the lateral ventricles. The epithalamus also contains the **pineal gland** (Figure 14-11●), an endocrine structure that secretes the hormone **melatonin.** Melatonin may be important in the regulation of day/night cycles and of reproductive functions. (The role of melatonin will be described in Chapter 18.)

Most of the neural tissue in the diencephalon is concentrated in two structures, the *thalami* (walls) and *hypothalamus* (floor). Ascending sensory information from the spinal cord and cranial

TABLE 14-3 The Limbic System

FUNCTION

Processing of memories, creation of emotional states, drives, and associated behaviors

CEREBRAL COMPONENTS

Cortical areas: limbic lobe (cingulate gyrus, dentate gyrus, and parahippocampal gyrus)

Nuclei: hippocampus, amygdaloid body

Tracts: fornix

DIENCEPHALIC COMPONENTS

Thalamus: anterior nuclear group

Hypothalamus: centers concerned with emotions, appetites (thirst, hunger), and related behaviors (see Table 14-5)

OTHER COMPONENTS

Reticular formation: network of interconnected nuclei throughout brain stem

nerves (other than the olfactory tract) synapses in a nucleus in the left or right thalamus before reaching the cerebral cortex and our conscious awareness. The hypothalamus contains centers involved with emotions and visceral processes that affect the cerebrum as well as other components of the brain stem. It also controls a variety of autonomic functions and forms the link between the nervous and endocrine systems.

❑ The Thalamus

Figures 14-9, 14-11

The thalamus is the final relay point for ascending sensory information that will be projected to the primary sensory cortex. It acts as a filter, passing on only a small portion of the arriving sensory information. The thalamus also coordinates the activities of the pyramidal and extrapyramidal systems by relaying information between the cerebral nuclei and the cerebral cortex.

The left thalamus and right thalamus are separated by the third ventricle. Viewed in midsagittal section (Figure 14-9●), each thalamus extends from the anterior commissure to the inferior base of the pineal gland. A projection of gray matter, called an **intermediate mass,** extends into the ventricle from the thalamus on either side. In roughly 70 percent of the population, the two fuse in the midline.

The thalamus on each side bulges laterally, away from the third ventricle (Figure 14-11●), and anteriorly into the tissues of the cerebrum. The lateral border of the thalamus is established by the fibers of the *internal capsule* (p. 469). Each thalamus consists of a rounded mass of *thalamic nuclei.*

Functions of Thalamic Nuclei

Figure 14-12

The thalamic nuclei are concerned primarily with the relay of sensory information to the cerebral nuclei and cerebral cortex. The four major groups of thalamic nuclei, detailed in Figure 14-12● and Table 14-4, are (1) the *anterior group,* (2) the *medial group,* (3) the *ventral group,* and (4) the *posterior group.*

1. The **anterior nuclei** are part of the limbic system, discussed previously (p. 473).
2. The **medial nuclei** provide a conscious awareness of emotional states by connecting emotional centers in the hypothalamus with the frontal lobes of the cerebral hemispheres. The medial nuclei also receive and relay sensory information from other portions of the thalamus.
3. The **ventral nuclei** relay information from the cerebral nuclei and cerebellum to the primary motor cortex and related motor association areas. The ventral nuclei also relay sensory information concerning touch, pressure, pain, temperature, proprioception, and taste to the primary sensory cortex.
4. The **posterior nuclei** include the *pulvinar* and the *geniculates.* The **pulvinar** integrates sensory

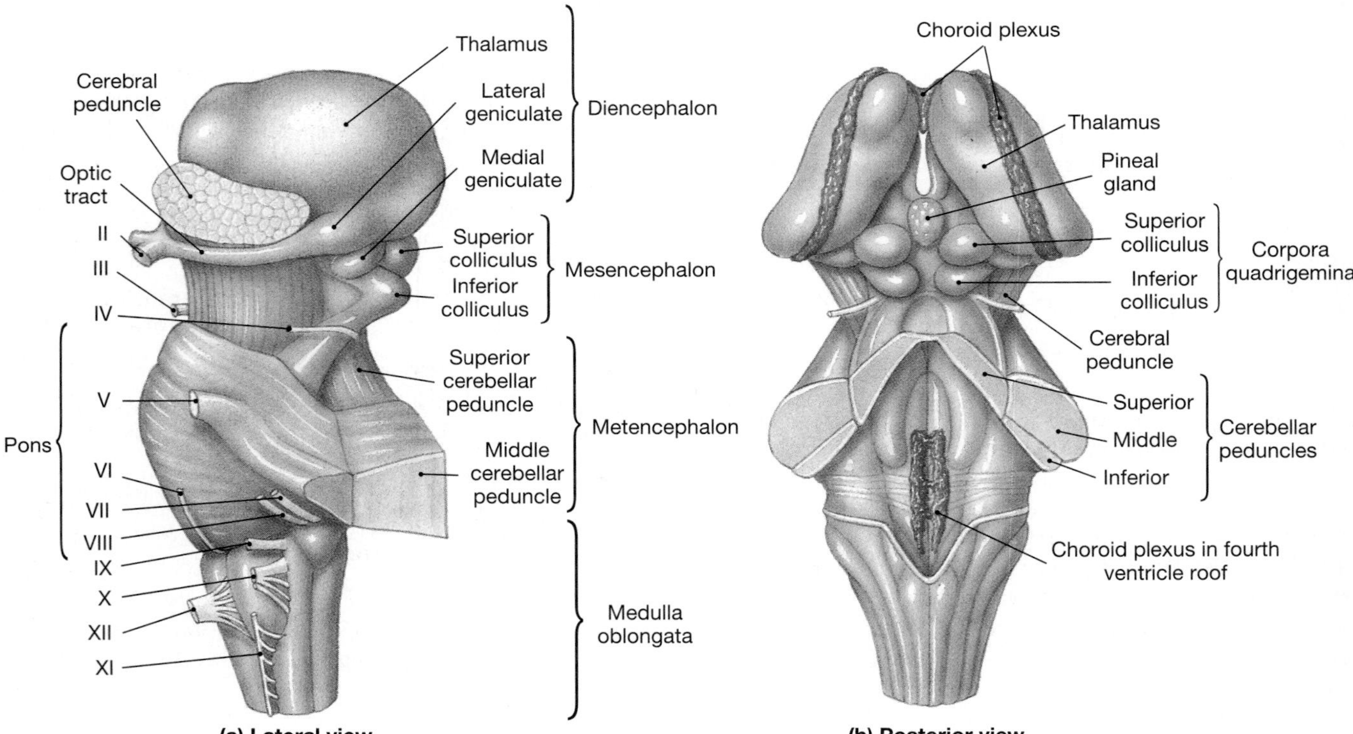
Thalamus
Lateral geniculate
Medial geniculate
Diencephalon
Cerebral peduncle
Optic tract
II
III
IV
Superior colliculus
Inferior colliculus
Mesencephalon
Pons
V
VI
VII
VIII
IX
X
XII
XI
Superior cerebellar peduncle
Middle cerebellar peduncle
Metencephalon
Medulla oblongata
(a) Lateral view
Choroid plexus
Thalamus
Pineal gland
Superior colliculus
Inferior colliculus
Corpora quadrigemina
Cerebral peduncle
Superior
Middle
Inferior
Cerebellar peduncles
Choroid plexus in fourth ventricle roof
(b) Posterior view

● **FIGURE 14-11**
The Diencephalon and Brain Stem. (a) Lateral view, as seen from the left side. **(b)** Posterior view.

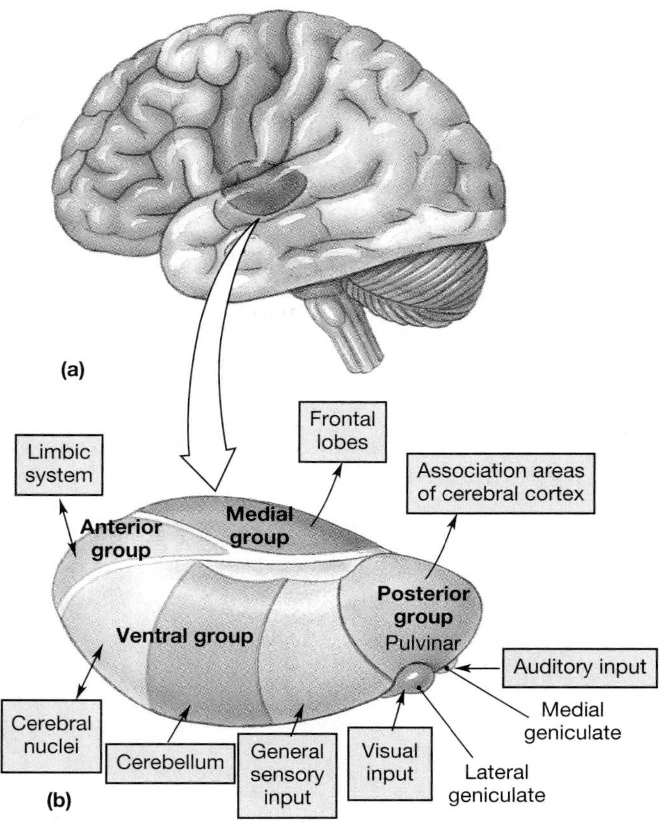
(a)
Limbic system
Frontal lobes
Association areas of cerebral cortex
Anterior group
Medial group
Posterior group
Pulvinar
Ventral group
Auditory input
Medial geniculate
Cerebral nuclei
Cerebellum
General sensory input
Visual input
Lateral geniculate
(b)

● **FIGURE 14-12**
The Thalamus. (a) Lateral view of the brain, showing the positions of the thalamic nuclei. The colored areas of the cerebral cortex are regions that receive input from the thalamus. **(b)** Enlarged view of the thalamic nuclei of the left side.

information for projection to the association areas of the cerebral cortex. The **lateral geniculate** (je-NIK-ū-lāt; *genicula*, a little knee) of each thalamus receives visual information from the eyes via the **optic tract.** The output of the lateral geniculate goes to the visual cortex and to the mesencephalon. The **medial geniculate** relays auditory information to the auditory cortex from the specialized receptors of the inner ear.

❑ The Hypothalamus
Figures 14-9, 14-13

The hypothalamus extends from the area superior to the *optic chiasm*, where the *optic tracts* from the eyes arrive at the brain, to the posterior margins of the mamillary bodies (Figure 14-9●). Posterior to the optic chiasm, the **infundibulum** (in-fun-DIB-ū-lum; *infundibulum*, a funnel) extends inferiorly, connecting the floor of the hypothalamus to the pituitary gland.

Viewed in sagittal section (Figure 14-13●) the floor of the hypothalamus between the infundibulum and the mamillary bodies is the **tuber cinereum** (sin-Ē-rē-um; *tuber*, a swelling + *cinereus*, ashen color). The tuber cinereum is a mass of gray matter whose neurons are involved with the control of pituitary gland function. The rostral portion of the tuber cinereum adjacent to the infundibulum is thickened and slightly elevated; this region is called the **median eminence.**

TABLE 14-4 The Thalamus

Structure/Nuclei	*Functions*
Anterior group	Part of the limbic system
Medial group	Integrates sensory information for projection to the frontal lobes
Ventral group	Projects sensory information to the primary sensory cortex; relays information from cerebellum and cerebral nuclei to motor area of cerebral cortex
Posterior group	
Pulvinar	Integrates sensory information for projection to association areas of cerebral cortex
Lateral geniculates	Project visual information to the visual cortex
Medial geniculates	Project auditory information to the auditory cortex

Functions of the Hypothalamus
Figure 14-13a

The hypothalamus contains a variety of important control and integrative centers, in addition to those associated with the limbic system. These centers and their functions are summarized in Figure 14-13a● and Table 14-5. Hypothalamic centers may be stimulated by (1) sensory information from the cerebrum, brain stem, and spinal cord; (2) changes in the CSF and interstitial fluid composition; or (3) chemical stimuli in the circulating blood that enter the hypothalamus across highly permeable capillaries.

Hypothalamic functions include:

1. ***Control of involuntary somatic motor activities.*** The hypothalamus directs somatic motor patterns associated with the emotions of rage, pleasure, pain, and sexual arousal by stimulating centers in other portions of the brain. For example, the changes in facial expression that accompany rage and the basic movements associated with sexual activity are controlled by hypothalamic centers.
2. ***Control of autonomic function.*** The hypothalamus adjusts and coordinates the activities of

● **FIGURE 14-13**
The Hypothalamus in Sagittal Section. (a) Diagrammatic view of the hypothalamus, showing the locations of major nuclei and centers. **(b)** Photograph of the hypothalamus and adjacent portions of the brain.

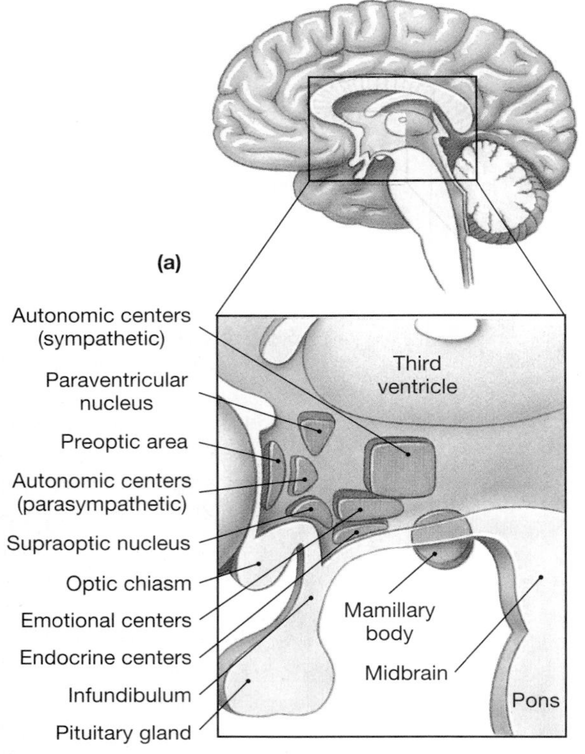
(a)
Autonomic centers (sympathetic)
Paraventricular nucleus
Preoptic area
Autonomic centers (parasympathetic)
Supraoptic nucleus
Optic chiasm
Emotional centers
Endocrine centers
Infundibulum
Pituitary gland
Third ventricle
Mamillary body
Midbrain
Pons

autonomic centers in the pons and medulla oblongata that regulate heart rate, blood pressure, respiration, and digestive functions.

3. ***Coordination of activities of the nervous and endocrine systems.*** The hypothalamus coordinates neural and endocrine activities by inhibiting or stimulating endocrine cells in the pituitary gland.
4. ***Secretion of hormones.*** The hypothalamus secretes two hormones: (1) *antidiuretic hormone* (ADH), produced by the **supraoptic nucleus,** restricts water loss at the kidneys, and (2) *oxytocin*, produced by the **paraventricular nucleus,**

TABLE 14-5 Components and Functions of the Hypothalamus

Region/Nucleus	*Functions*
Supraoptic nucleus	Secretes ADH, restricting water loss at the kidneys
Paraventricular nucleus	Secretes oxytocin
Preoptic area	Regulates body temperature
Tuberal area	Releases hormones that control endocrine cells of the anterior pituitary
Autonomic centers	Control medullary nuclei that regulate heart rate and blood pressure
Mamillary bodies	Control feeding reflexes (licking, swallowing, etc.)

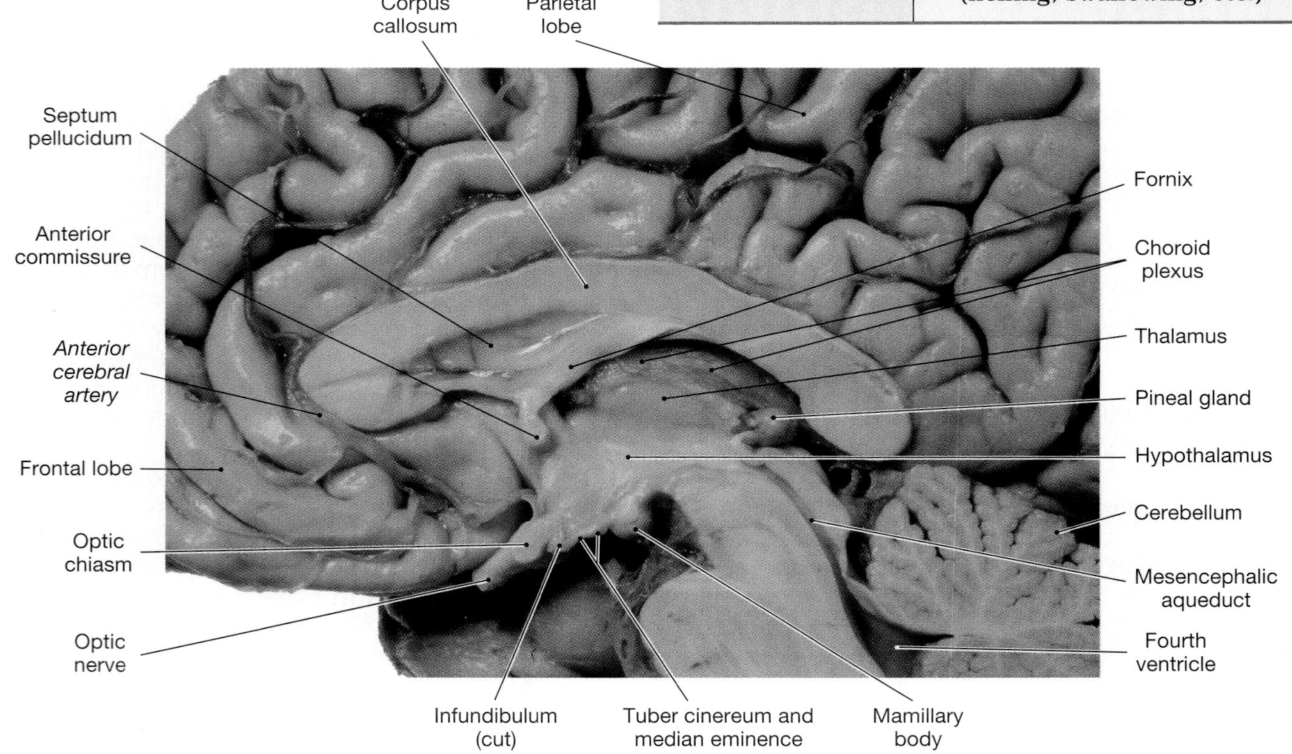
Corpus callosum
Parietal lobe
Septum pellucidum
Anterior commissure
Anterior cerebral artery
Frontal lobe
Optic chiasm
Optic nerve
Infundibulum (cut)
Tuber cinereum and median eminence
Mamillary body
Fornix
Choroid plexus
Thalamus
Pineal gland
Hypothalamus
Cerebellum
Mesencephalic aqueduct
Fourth ventricle
(b)

stimulates smooth muscle contractions in the uterus and mammary glands of females and the prostate gland of males.

5. ***Production of emotions and behavioral drives.*** Specific hypothalamic centers produce sensations that lead to voluntary or involuntary changes in behavior. These unfocused "impressions" originating in the hypothalamus are called **drives.** For example, stimulation of the **feeding center** produces the sensation of hunger, and stimulation of the **thirst center** produces the sensation of thirst. The conscious sensations are only part of the hypothalamic response. For example, the thirst center also orders the release of ADH by neurons in the supraoptic nucleus.
6. ***Coordination between voluntary and autonomic functions.*** When you think about a dangerous or stressful situation, your heart rate and respiratory rate go up and your body prepares for an emergency. These autonomic adjustments are made by the hypothalamus.
7. ***Regulation of body temperature.*** The **preoptic areas** of the hypothalamus coordinates the activities of other CNS centers and regulates other physiological systems to maintain normal body temperature. If body temperature falls, the preoptic area sends instructions to the *vasomotor center* in the medulla, an autonomic center that controls blood flow by regulating the diameter of peripheral blood vessels. In response, the vasomotor center decreases the blood supply to the skin, reducing the rate of heat loss.

✔ Damage to the lateral geniculate nuclei of the thalamus would interfere with the functions of which of the senses?

✔ What area of the diencephalon would be stimulated by changes in the body temperature?

The Mesencephalon

Figures 14-11, 14-14

The external anatomy of the mesencephalon can be seen in Figure 14-11●, and the major nuclei are detailed in Figure 14-14● and Table 14-6. The roof, or **tectum,** of the mesencephalon contains two pairs of sensory nuclei known collectively as the **corpora quadrigemina** (KŌR-pō-ra quad-ri-JEM-i-nuh). These nuclei, the *superior* and *inferior colliculi*, process visual and auditory sensations. Each **superior colliculus** (kol-IK-ū-lus; *colliculus*, a small hill) receives visual inputs from the lateral geniculate of the thalamus on that side. The **inferior colliculus** receives auditory data from nuclei in the medulla oblongata and pons. Some of this information may be forwarded to the medial geniculate on the same side. The superior and inferior colliculi control cranial reflexes triggered by "startling" stimuli such as a bright light or a loud noise.

On each side, the mesencephalon contains a *red nucleus* and the *substantia nigra.* The **red nucleus** contains numerous blood vessels that give

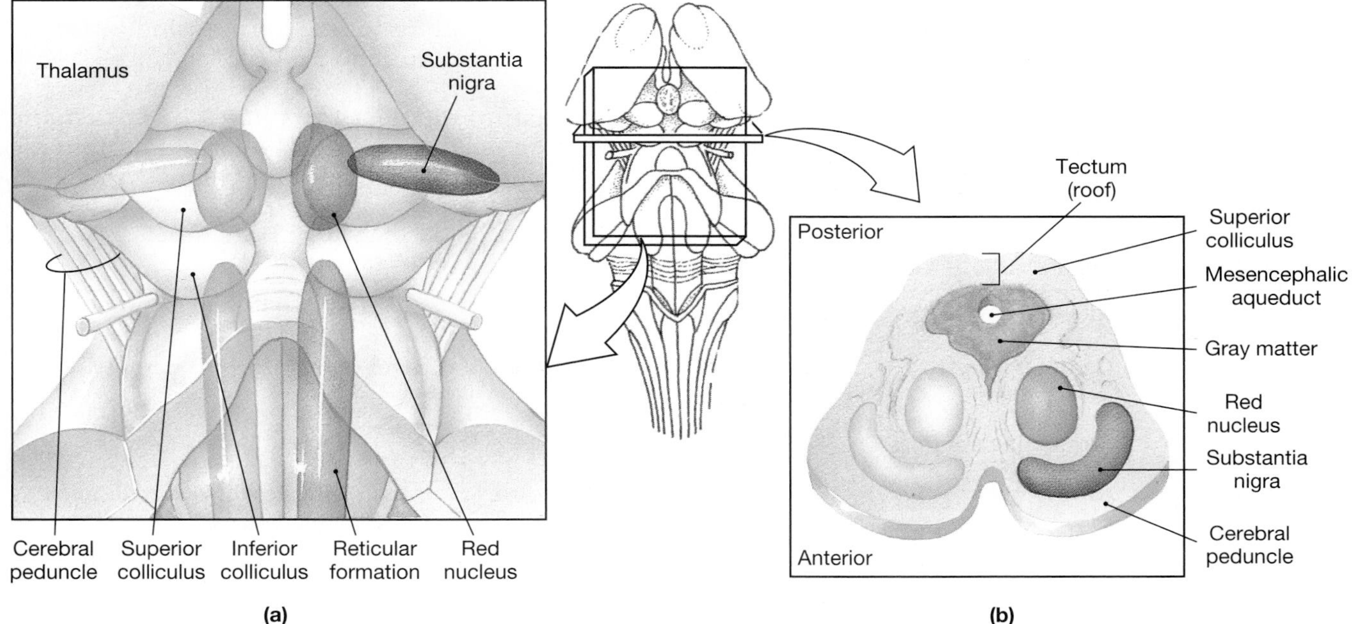

Thalamus
Substantia nigra
Cerebral peduncle
Superior colliculus
Inferior colliculus
Reticular formation
Red nucleus
(a)
Tectum (roof)
Posterior
Superior colliculus
Mesencephalic aqueduct
Gray matter
Red nucleus
Substantia nigra
Cerebral peduncle
Anterior
(b)

● **FIGURE 14-14**

The Mesencephalon. (a) Posterior view of the mesencephalon. Superficial structures are labeled on the left, and underlying nuclei are colored and labeled on the right. **(b)** Sectional view; nuclei are colored on the left side only, to match part (a).

TABLE 14-6 Components and Functions of the Mesencephalon

Subdivision	Region/Nucleus	Functions
GRAY MATTER		
Tectum (roof)	Superior colliculus	Integrates visual information with other sensory inputs; initiates involuntary motor responses
	Inferior colliculus	Relays auditory information to medial geniculate
Walls and floor	Red nucleus	Involuntary control of muscle tone and posture
	Substantia nigra	Regulates activity in the cerebral nuclei
	Reticular formation (headquarters)	Automatic processing of incoming sensations and outgoing motor commands; can initiate involuntary motor responses to stimuli; maintenance of consciousness (RAS)
	Other nuclei/centers	Nuclei associated with two cranial nerves (N III, IV)
WHITE MATTER	Cerebral peduncles	Connect primary motor cortex with motor neurons in brain and spinal cord; carry ascending sensory information to thalamus

it a rich red color. This nucleus receives information from the cerebrum and cerebellum and issues involuntary motor commands concerned with the maintenance of muscle tone and posture. The **substantia nigra** (NĪ-gruh; black) lies lateral to the red nucleus. The gray matter in this region contains darkly pigmented cells, giving it a black color.

Neurons in the substantia nigra inhibit the activity of the cerebral nuclei by releasing the neurotransmitter *dopamine.* [p. 410] If the substantia nigra is damaged, or the neurons reduce their secretion of dopamine, the cerebral nuclei become more active. The result is a gradual, generalized increase in muscle tone and the appearance of symptoms characteristic of **Parkinson's disease.** Individuals with Parkinson's disease have difficulty starting voluntary movements, because opposing muscle groups do not relax—they must be overpowered. Once a movement is under way, every aspect must be voluntarily controlled through intense effort and concentration. AM *The Cerebral Nuclei and Parkinson's Disease*

The nerve fiber bundles on the ventrolateral surfaces of the mesencephalon (Figures 14-11a, b, p. 476, and 14-14●) are the **cerebral peduncles** (*peduncles,* little feet). They contain (1) descending fibers that go to the cerebellum via the pons and (2) descending fibers carrying voluntary motor commands from the primary motor cortex of each cerebral hemisphere.

The mesencephalon also contains the headquarters of the **reticular activating system** (RAS), a specialized component of the reticular formation. Stimulation of the mesencephalic portion of the RAS makes an individual more alert and attentive. The role of the RAS in the maintenance of consciousness will be considered further in Chapter 15.

The Cerebellum

Figure 14-15

The cerebellum, detailed in Figure 14-15● and Table 14-7, is an automatic processing center. It has two primary functions:

1. *Adjusting the postural muscles of the body.* The cerebellum coordinates rapid, automatic adjustments that maintain balance and equi-

TABLE 14-7 Components of the Cerebellum

Subdivision	Region/Nucleus	Functions
Gray matter	Cerebellar cortex	Involuntary coordination and control of ongoing movements of body parts
	Cerebellar nuclei	As above
White matter	Arbor vitae	Connects cerebellar cortex and nuclei with cerebellar peduncles
	Cerebellar peduncles	
	Superior	Link the cerebellum with mesencephalon, diencephalon, and cerebrum
	Middle	Contain transverse fibers and carry communications between the cerebellum and pons
	Inferior	Link the cerebellum with the medulla oblongata and spinal cord
	Transverse fibers	Interconnect cerebellar hemispheres

● FIGURE 14-15
The Cerebellum. (a) Posterior and superior surfaces of the cerebellum, showing major anatomical landmarks and regions. **(b)** Sectional view of the cerebellum, showing the arrangement of gray matter and white matter. Purkinje cells, large neurons, are found in the cerebellar cortex. (LM × 112)

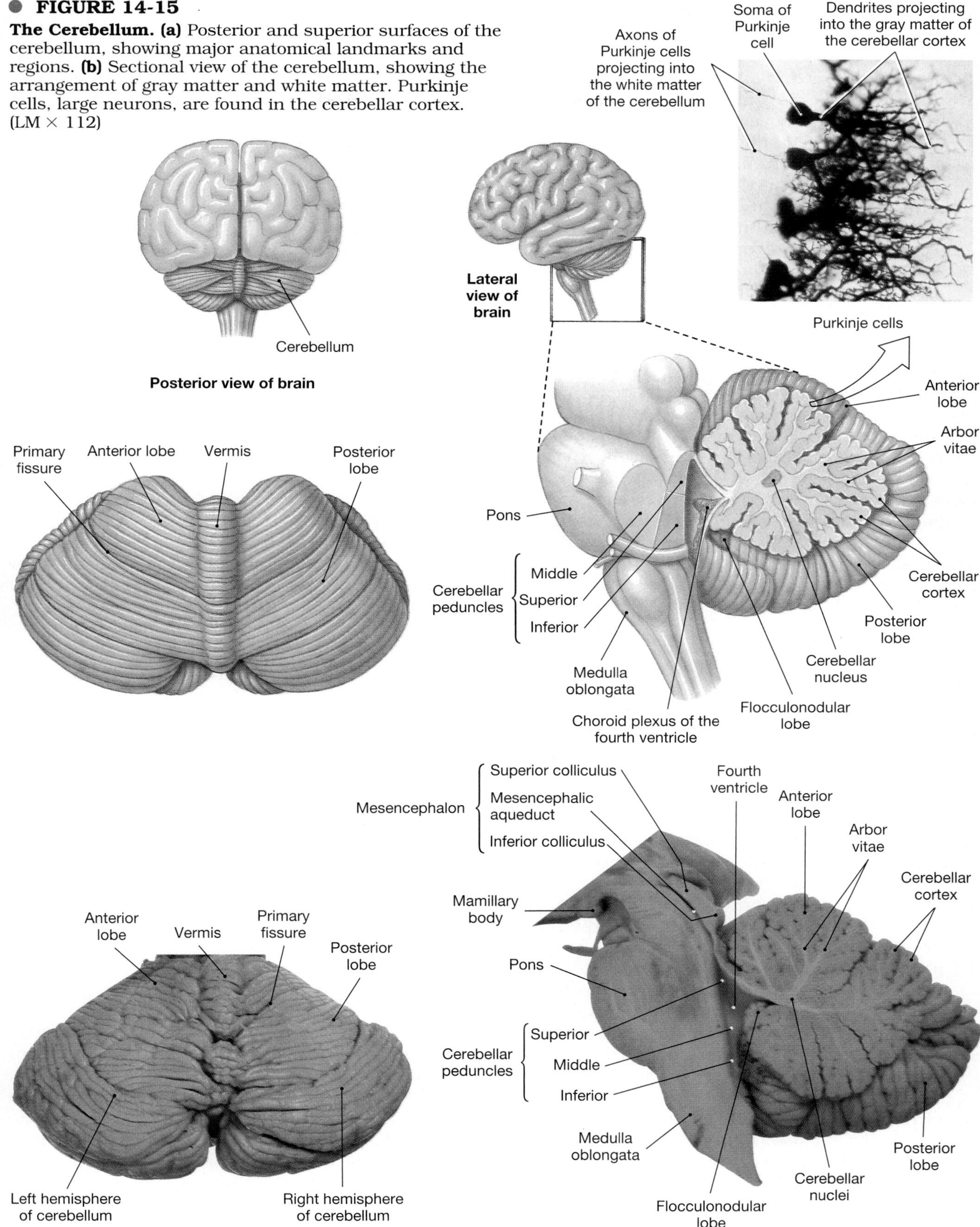

Cerebellum
Posterior view of brain
Lateral view of brain
Axons of Purkinje cells projecting into the white matter of the cerebellum
Soma of Purkinje cell
Dendrites projecting into the gray matter of the cerebellar cortex
Purkinje cells
Primary fissure
Anterior lobe
Vermis
Posterior lobe
Anterior lobe
Arbor vitae
Pons
Cerebellar peduncles
Middle
Superior
Inferior
Cerebellar cortex
Posterior lobe
Medulla oblongata
Cerebellar nucleus
Choroid plexus of the fourth ventricle
Flocculonodular lobe
Mesencephalon
Superior colliculus
Mesencephalic aqueduct
Inferior colliculus
Fourth ventricle
Anterior lobe
Arbor vitae
Cerebellar cortex
Mamillary body
Anterior lobe
Vermis
Primary fissure
Posterior lobe
Pons
Cerebellar peduncles
Superior
Middle
Inferior
Medulla oblongata
Left hemisphere of cerebellum
Right hemisphere of cerebellum
Flocculonodular lobe
Cerebellar nuclei
Posterior lobe

(a) Superior view of cerebellum

(b) Sagittal section of cerebellum

librium. These alterations in muscle tone and position are made by modifying the activity of the red nucleus.

2. ***Programming and fine-tuning voluntary and involuntary movements.*** The cerebellum refines learned movement patterns. This function is performed indirectly by regulating activity along both pyramidal and extrapyramidal motor pathways at the cerebral cortex, cerebral nuclei, and motor centers in the brain stem. The cerebellum compares the motor commands with proprioceptive information and performs any adjustments needed to make the movement smooth.

The cerebellum has a complex, highly convoluted surface composed of neural cortex. The folds, or **folia** (FŌ-lē-uh; leaves), of the surface are less prominent than the gyri of the cerebral hemispheres. The **anterior** and **posterior lobes** are separated by the **primary fissure** (Figure 14-15a●). Along the midline a narrow band of cortex known as the **vermis** (VER-mis; worm) separates the **cerebellar hemispheres** of the posterior lobe. The slender **flocculonodular** (flok-ū-lō-NOD-ū-ler) **lobe** lies between the roof of the fourth ventricle and the cerebellar hemispheres and vermis (Figure 14-15b●).

The cerebellar cortex contains huge, highly branched **Purkinje** (pur-KIN-jē) **cells,** or *basket cells.* Internally, the white matter of the cerebellum forms a branching array that in sectional view resembles a tree. Anatomists call it the **arbor vitae,** or "tree of life." The cerebellum receives proprioceptive information (position sense) from the spinal cord and monitors all proprioceptive, visual, tactile, balance, and auditory sensations received by the brain. Information concerning the voluntary and involuntary motor commands issued by higher centers reaches the cerebellum via tracts from the cerebral cortex and thalamus. These synapse within **cerebellar nuclei** before projecting to the cerebellar cortex. Tracts containing the axons of Purkinje cells then relay instructions to motor nuclei in the cerebrum and brain stem.

Tracts that link the cerebellum with the brain stem, cerebrum, and spinal cord leave the cerebellar hemispheres as the *superior, middle,* and *inferior cerebellar peduncles.* The **superior cerebellar peduncles** link the cerebellum with nuclei in the midbrain, diencephalon, and cerebrum. The **middle cerebellar peduncles** are connected to a broad band of fibers that cross the ventral surface of the pons at right angles to the axis of the brain stem. The middle cerebellar peduncles also connect the cerebellar hemispheres with sensory and motor nuclei in the pons. The **inferior cerebellar peduncles** permit communication between the cerebellum and nuclei in the medulla oblongata and carry ascending and descending cerebellar tracts from the spinal cord.

The cerebellum may be permanently damaged by trauma or stroke or temporarily affected by drugs such as alcohol. These alterations can produce **ataxia** (a-TAK-sē-uh; *ataxia,* a lack of order), a disturbance in balance. In severe ataxia the individual may be unable to sit or stand without assistance. AM *Cerebellar Dysfunction; Analysis of Gait and Balance*

The Pons

Figures 14-11, 14-16a

The pons links the cerebellar hemispheres with the mesencephalon, diencephalon, cerebrum, and spinal cord. Important features and regions are indicated in Figures 14-11, p. 476, and 14-16a●. The pons contains:

- ***Sensory and motor nuclei for four cranial nerves*** (N V, N VI, N VII, and N VIII). These cranial nerves innervate the jaw muscles, the anterior surface of the face, one of the extrinsic eye muscles (the lateral rectus), and the sense organs of the inner ear.
- ***Nuclei concerned with the involuntary control of respiration.*** On each side of the brain, the *reticular formation* in this region contains two respiratory centers, the *apneustic center* and the *pneumotaxic center.* These centers modify the activity of the *respiratory rhythmicity center* in the medulla oblongata.
- ***Nuclei that process and relay information heading to or from the cerebellum.***
- ***Ascending, descending, and transverse tracts.*** The longitudinal tracts interconnect other portions of the CNS. The middle cerebellar peduncles are connected to the **transverse fibers** of the pons that cross its anterior surface. These fibers are axons that link the pontine nuclei with the cerebellar hemisphere of the opposite side.

The Medulla Oblongata

Figures 14-11, 14-16b

The medulla oblongata is connected to the spinal cord. Figure 14-11●, p. 476, detailed the external surface of the medulla oblongata; Figure 14-16b● and Table 14-8 summarize its major components.

The medulla oblongata physically connects the brain with the spinal cord, and many of its functions are directly related to this fact. For example, all communication between the brain and spinal cord involves tracts that ascend or descend through the medulla oblongata.

Nuclei in the medulla oblongata include: (1) nuclei associated with the autonomic control of visceral activities, (2) sensory or motor nuclei associated with

TABLE 14-8 Components and Functions of the Medulla Oblongata

Subdivision	*Region/Nucleus*	*Functions*
Gray matter	Nucleus gracilis Nucleus cuneatus	Relay somatic sensory information to the thalamus
	Olivary nuclei	Relay information to the cerebellum
	Reflex centers:	
	Cardiac centers	Regulate heart rate and force of contraction
	Vasomotor center	Regulates distribution of blood flow
	Respiratory rhythmicity center	Sets the pace of respiratory movements
	Other nuclei/centers	Sensory and motor nuclei of five cranial nerves Nuclei relaying ascending sensory information from the spinal cord to higher centers
White matter	Ascending and descending tracts	Link the brain with the spinal cord

cranial nerves, or (3) relay stations along sensory or motor pathways. We will focus attention on nuclei that will be encountered in later chapters.

1. ***Autonomic nuclei.*** The reticular formation in the medulla oblongata contains nuclei and centers responsible for the regulation of vital autonomic functions. These **reflex centers** receive inputs from cranial nerves, the cerebral cortex, and the brain stem, and their output controls or adjusts the activities of one or more peripheral systems. Major centers regulating autonomic functions include:

   - The **cardiovascular centers** adjust heart rate, the strength of cardiac contractions, and the flow of blood through peripheral tissues. On functional grounds the cardiovascular centers may be subdivided into **cardiac** (*kardia*, heart) and **vasomotor** (*vas*, canal) **centers,** but their anatomical boundaries are difficult to determine.
   - The **respiratory rhythmicity centers** set the basic pace for respiratory movements, and their activity is regulated by inputs from the apneustic and pneumotaxic centers of the pons.

2. ***Nuclei of cranial nerves.*** The medulla oblongata contains sensory and motor nuclei associated with five of the cranial nerves (N VIII, N IX, N X, N XI, and N XII). These cranial nerves provide motor commands to muscles of the pharynx, neck, and back, as well as to the visceral organs of the thoracic and peritoneal cavities. They also carry sensory information from receptors in the inner ear to the pons and medulla.

3. ***Relay stations.*** The **nucleus gracilis** and the **nucleus cuneatus** pass somatic sensory information to the thalamus, and the **olivary nuclei** relay information from the spinal cord, the cerebral cortex, and brain stem to the cerebellar cortex. The bulk of these nuclei create the **olives,** prominent bulges along the ventrolateral surface of the medulla oblongata.

✔ In what part of the brain would you find a worm (vermis) and a tree (arbor vitae)?

✔ The medulla oblongata is one of the smallest sections of the brain, yet damage there can cause death, whereas similar damage in the cerebrum might go unnoticed. Why?

✔ What effect would decreased release of neurotransmitter from cells of the substantia nigra have on the activity of cerebral neurons?

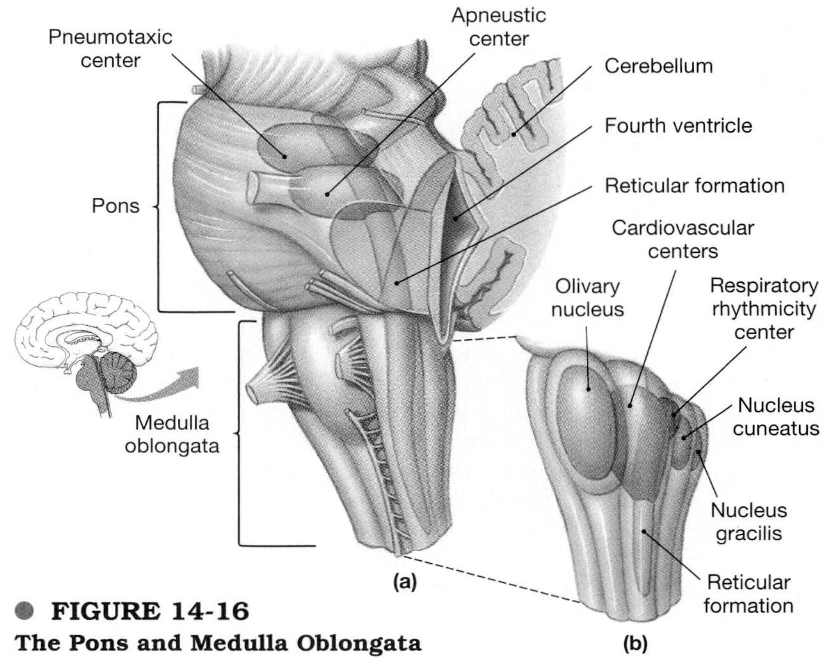
Pneumotaxic center
Apneustic center
Cerebellum
Fourth ventricle
Reticular formation
Pons
Cardiovascular centers
Olivary nucleus
Respiratory rhythmicity center
Medulla oblongata
Nucleus cuneatus
Nucleus gracilis
Reticular formation
(a)
(b)

● **FIGURE 14-16**
The Pons and Medulla Oblongata

FOCUS Cranial Nerves

Cranial nerves are components of the peripheral nervous system that connect to the brain rather than to the spinal cord. Twelve pairs of cranial nerves can be found on the ventrolateral surface of the brain (Figure 14-17●), each with a name related to its appearance or function.

The number assigned to a cranial nerve roughly corresponds to its position along the longitudinal axis of the brain, beginning at the cerebrum. Roman numerals are usually used, either alone or with the prefix N or CN. We will use the abbreviation N, which is generally preferred by neuroanatomists and clinical neurologists. CN, an equally valid abbreviation, is preferred by comparative anatomists.

Each cranial nerve attaches to the brain near the associated sensory or motor nuclei. The sensory nuclei act as switching centers, with the postsynaptic neurons relaying the information to other nuclei or to processing centers within the cerebral or cerebellar cortex. In a similar fashion the motor nuclei receive convergent inputs from higher centers or from other nuclei along the brain stem.

The next section classifies cranial nerves as primarily sensory, special sensory, motor, or mixed (sensory and motor). This is a useful method of classification, but it is based on the primary function, and a cranial nerve can have important secondary functions. Two examples are worth noting:

1. As elsewhere in the PNS, a nerve containing tens of thousands of motor fibers to a skeletal muscle will also contain sensory fibers from muscle spindles and tendon organs in that muscle. These sensory fibers are assumed to be present but are ignored in the classification of the nerve.
2. Regardless of their other functions, several cranial nerves (N III, VII, IX, and X) distribute autonomic fibers to peripheral ganglia, just as spinal nerves deliver them to ganglia along the spinal cord. The presence of small numbers of autonomic fibers will be noted (and discussed further in Chapter 16) but ignored in the classification of the nerve.

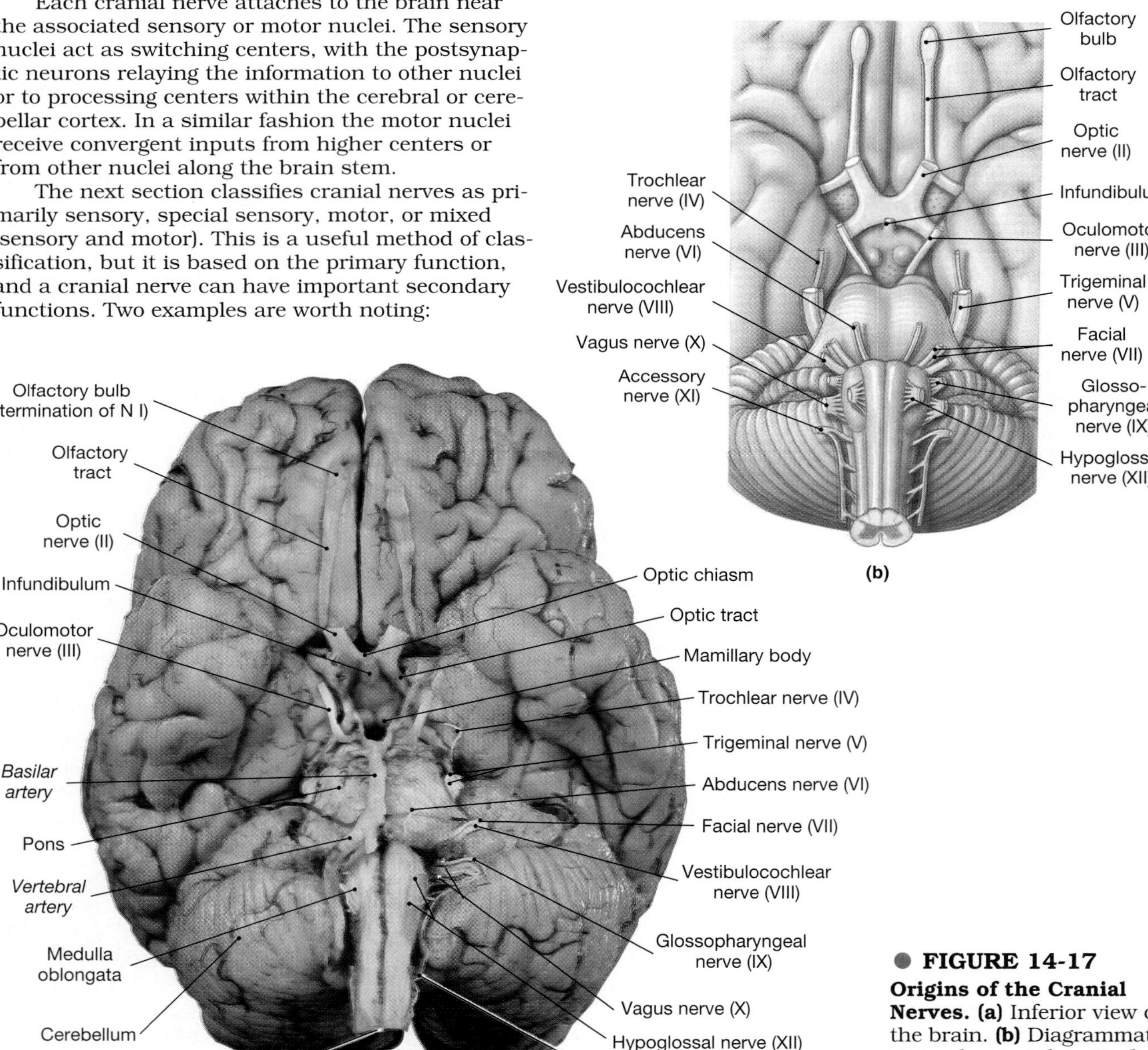

Olfactory bulb
Olfactory tract
Optic nerve (II)
Trochlear nerve (IV)
Infundibulum
Abducens nerve (VI)
Oculomotor nerve (III)
Trigeminal nerve (V)
Vestibulocochlear nerve (VIII)
Facial nerve (VII)
Vagus nerve (X)
Accessory nerve (XI)
Glosso-pharyngeal nerve (IX)
Hypoglossal nerve (XII)
(b)
Olfactory bulb (termination of N I)
Olfactory tract
Optic nerve (II)
Infundibulum
Oculomotor nerve (III)
Optic chiasm
Optic tract
Mamillary body
Trochlear nerve (IV)
Trigeminal nerve (V)
Basilar artery
Abducens nerve (VI)
Facial nerve (VII)
Pons
Vertebral artery
Vestibulocochlear nerve (VIII)
Medulla oblongata
Glossopharyngeal nerve (IX)
Vagus nerve (X)
Cerebellum
Hypoglossal nerve (XII)
Accessory nerve (XI)
Spinal cord
(a)

● **FIGURE 14-17**
Origins of the Cranial Nerves. (a) Inferior view of the brain. **(b)** Diagrammatic view, showing the attachment of the 12 pairs of cranial nerves.

The Olfactory Nerve (N I)

Figures 14-17, 14-18

Primary function: Special sensory (smell)
Origin: Receptors of olfactory epithelium
Passes through: Cribriform plate of ethmoid bone [p. 212]
Destination: Olfactory bulbs

The first pair of cranial nerves (Figure 14-18●) carries special sensory information responsible for the sense of smell. The olfactory receptors are specialized neurons in the epithelium covering the roof of the nasal cavity, the superior nasal conchae, and the superior portions of the nasal septum. Axons from these sensory neurons collect to form 20 or more bundles that penetrate the cribriform plate of the ethmoid bone. These bundles are components of the **olfactory nerve** (N I). Almost at once they enter the **olfactory bulbs,** neural masses that lie on either side of the crista galli. The olfactory afferents synapse within the olfactory bulbs, and the axons of the postsynaptic neurons proceed to the cerebrum along the slender **olfactory tracts** (Figures 14-17 and 14-18●).

Because the olfactory tracts looked like typical peripheral nerves, anatomists a hundred years ago misidentified the olfactory tracts as the first cranial nerve. Later studies demonstrated that the olfactory tracts and bulbs are actually part of the cerebrum, but by then the numbering system was already firmly established, and anatomists were left with a forest of tiny olfactory nerve bundles lumped together as N I.

The olfactory nerves are the only cranial nerves attached directly to the cerebrum. The rest originate or terminate within nuclei of the diencephalon or brain stem, and the ascending sensory information synapses in the thalamus before reaching the cerebrum.

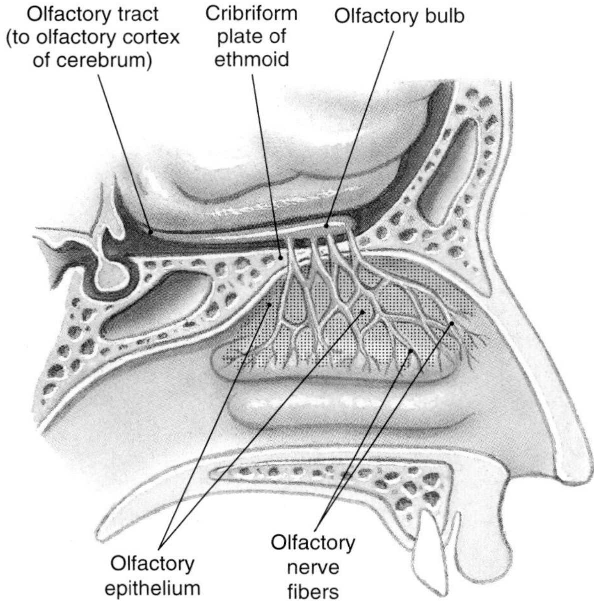
Olfactory tract (to olfactory cortex of cerebrum)
Cribriform plate of ethmoid
Olfactory bulb
Olfactory epithelium
Olfactory nerve fibers

● **FIGURE 14-18**
The Olfactory Nerve

The Optic Nerve (N II)

Figures 14-17, 14-19

Primary function: Special sensory (vision)
Origin: Retina of eye
Passes through: Optic foramen of sphenoid bone [p. 212]
Destination: Diencephalon via optic chiasm

The **optic nerves** (N II) carry visual information from special sensory ganglia in the eyes. These nerves, diagrammed in Figure 14-19●, contain about 1 million sensory nerve fibers. The optic nerves pass through the optic foramina of the sphenoid before converging at the ventral and anterior margin of the diencephalon, at the **optic chiasm** (*chiasma*, a crossing). At the optic chiasm, fibers from the nasal half of each retina cross over to the opposite side of the brain. The reorganized axons continue toward the lateral geniculates of the thalamus as the **optic tracts** (Figures 14-17 and 14-19●). After synapsing in the lateral geniculates, projection fibers deliver the information to the visual cortex of the occipital lobe. This arrangement results in each cerebral hemisphere's receiving visual information from the lateral half of the retina of the eye on that side and from the medial half of the retina of the eye of the opposite side. A relatively small number of axons in the optic tracts bypass the lateral geniculates and synapse in the superior colliculus of the midbrain. This pathway will be considered in Chapter 17.

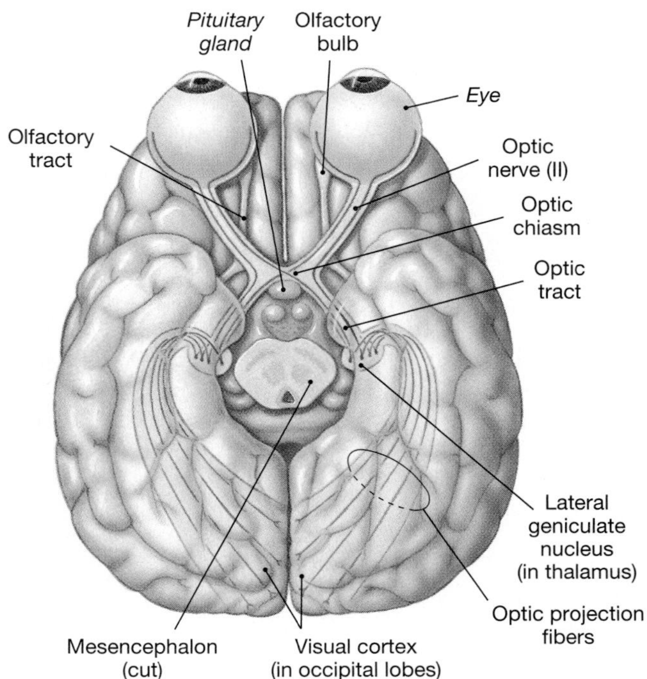
Pituitary gland
Olfactory bulb
Eye
Olfactory tract
Optic nerve (II)
Optic chiasm
Optic tract
Lateral geniculate nucleus (in thalamus)
Optic projection fibers
Mesencephalon (cut)
Visual cortex (in occipital lobes)

● **FIGURE 14-19**
The Optic Nerve

The Oculomotor Nerve (N III)

Figure 14-20

Primary function: Motor (eye movements)

Origin: Mesencephalon

Passes through: Superior orbital fissure of sphenoid bone [p. 212]

Destination: *Somatic motor:* superior, inferior, and medial rectus muscles; inferior oblique muscle; levator palpebrae superioris muscle. [p. 337]; *Visceral motor:* intrinsic eye muscles

The mesencephalon contains the motor nuclei controlling the third and fourth cranial nerves. The **oculomotor nerve** (N III) innervates four of the six extrinsic eye muscles and the levator palpebrae superioris, the muscle that raises the upper eyelid (Figure 14-20●). [p. 337] On each side of the brain, N III emerges from the ventral surface of the mesencephalon and penetrates the posterior orbital wall at the superior orbital fissure. [p. 212] Individuals with damage to this nerve often complain of pain over the eye and "double vision," because the movements of the left and right eyes cannot be coordinated properly.

The oculomotor nerve also delivers preganglionic autonomic fibers to neurons of the **ciliary ganglion.** The ganglionic neurons control intrinsic eye muscles. These muscles change the diameter of the pupil, adjusting the amount of light entering the eye, and change the shape of the lens to focus images on the retina.

The Trochlear Nerve (N IV)

Figure 14-20

Primary function: Motor (eye movements)

Origin: Mesencephalon

Passes through: Superior orbital fissure of sphenoid bone [p. 212]

Destination: Superior oblique muscle [p. 337]

The **trochlear** (TRŌK-lē-ar) **nerve** (*trochlea,* a pulley), smallest of the cranial nerves, innervates the superior oblique muscle of the eye (Figure 14-20●). The name "trochlear nerve" should remind you that the innervated muscle passes through a ligamentous pulley, or sling, on its way to its insertion on the surface of the eye. [p. 337] An individual with damage to this nerve or its nucleus will have difficulty looking upward.

The Abducens Nerve (N VI)

Figure 14-20

Primary function: Motor (eye movements)

Origin: Pons

Passes through: Superior orbital fissure of sphenoid bone [p. 212]

Destination: Lateral rectus muscle [p. 337]

The **abducens** (ab-DŪ-senz) **nerve** innervates the lateral rectus, the sixth of the extrinsic eye muscles. Innervation of this muscle makes lateral movements of the eyeball possible. The nerve emerges from the inferior surface of the brain stem at the border between the pons and the medulla oblongata. It reaches the orbit through the superior orbital fissure in company with the oculomotor and trochlear nerves (Figure 14-20●).

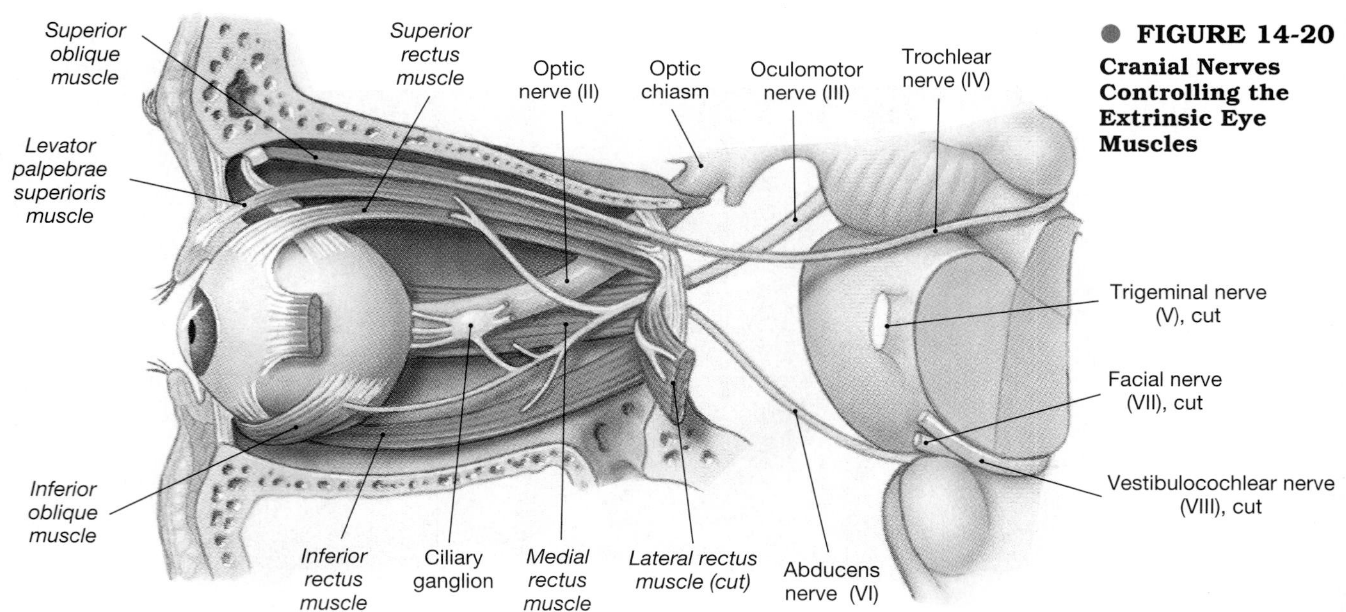
Superior oblique muscle
Superior rectus muscle
Optic nerve (II)
Optic chiasm
Oculomotor nerve (III)
Trochlear nerve (IV)
Levator palpebrae superioris muscle
Trigeminal nerve (V), cut
Facial nerve (VII), cut
Vestibulocochlear nerve (VIII), cut
Inferior oblique muscle
Inferior rectus muscle
Ciliary ganglion
Medial rectus muscle
Lateral rectus muscle (cut)
Abducens nerve (VI)

● **FIGURE 14-20**
Cranial Nerves Controlling the Extrinsic Eye Muscles

The Trigeminal Nerve (N V)

Figure 14-21

Primary function: Mixed (sensory and motor) to face; ophthalmic and maxillary branches (sensory), mandibular branch (mixed)

Origin: *Ophthalmic branch* (sensory): orbital structures, nasal cavity, skin of forehead, upper eyelid, eyebrow, nose (part). *Maxillary branch* (sensory): lower eyelid, upper lip, gums, and teeth; cheek; nose, palate, and pharynx (part). *Mandibular branch* (mixed): sensory from lower gums, teeth, and lips; palate and tongue (part); motor from motor nuclei of pons

Passes through: Ophthalmic branch via superior orbital fissure, maxillary branch via foramen rotundum, mandibular branch via foramen ovale [p. 212]

Destination: Ophthalmic and maxillary branches to sensory nuclei in pons; mandibular branch innervates muscles of mastication [p. 338]

The pons contains the nuclei associated with three cranial nerves (N V, VI, and VII) and contributes to a fourth (N VIII). The **trigeminal** (trī-JEM-i-nal) **nerve** (Figure 14-21●) is the largest cranial nerve. This mixed nerve provides somatic sensory information from the head and face and motor control over the muscles of mastication. Sensory (dorsal) and motor (ventral) roots originate on the lateral surface of the pons. The sensory branch is larger, and the enormous **semilunar ganglion** contains the cell bodies of the sensory neurons. As the name implies, the trigeminal has three major branches; the relatively small motor root contributes to only one of the three. **Tic douloureux** (doo-loo-ROO; *douloureux*, painful) is a painful condition affecting the area innervated by the maxillary and mandibular branches of the trigeminal nerve. Sufferers complain of severe, almost totally debilitating pain triggered by contact with the lip, tongue, or gums. The cause of the condition is unknown. AM Tic Douloureux

The trigeminal nerve branches are associated with the *ciliary, sphenopalatine, submandibular,* and *otic* ganglia. These are autonomic (parasympathetic) ganglia whose neurons innervate structures of the face. However, although its nerve fibers may pass around or through these ganglia, the trigeminal nerve does not contain visceral motor fibers. The ciliary ganglion was discussed previously (p. 486), and the other ganglia will be detailed below, with the branches of the *facial nerve* (N VII).

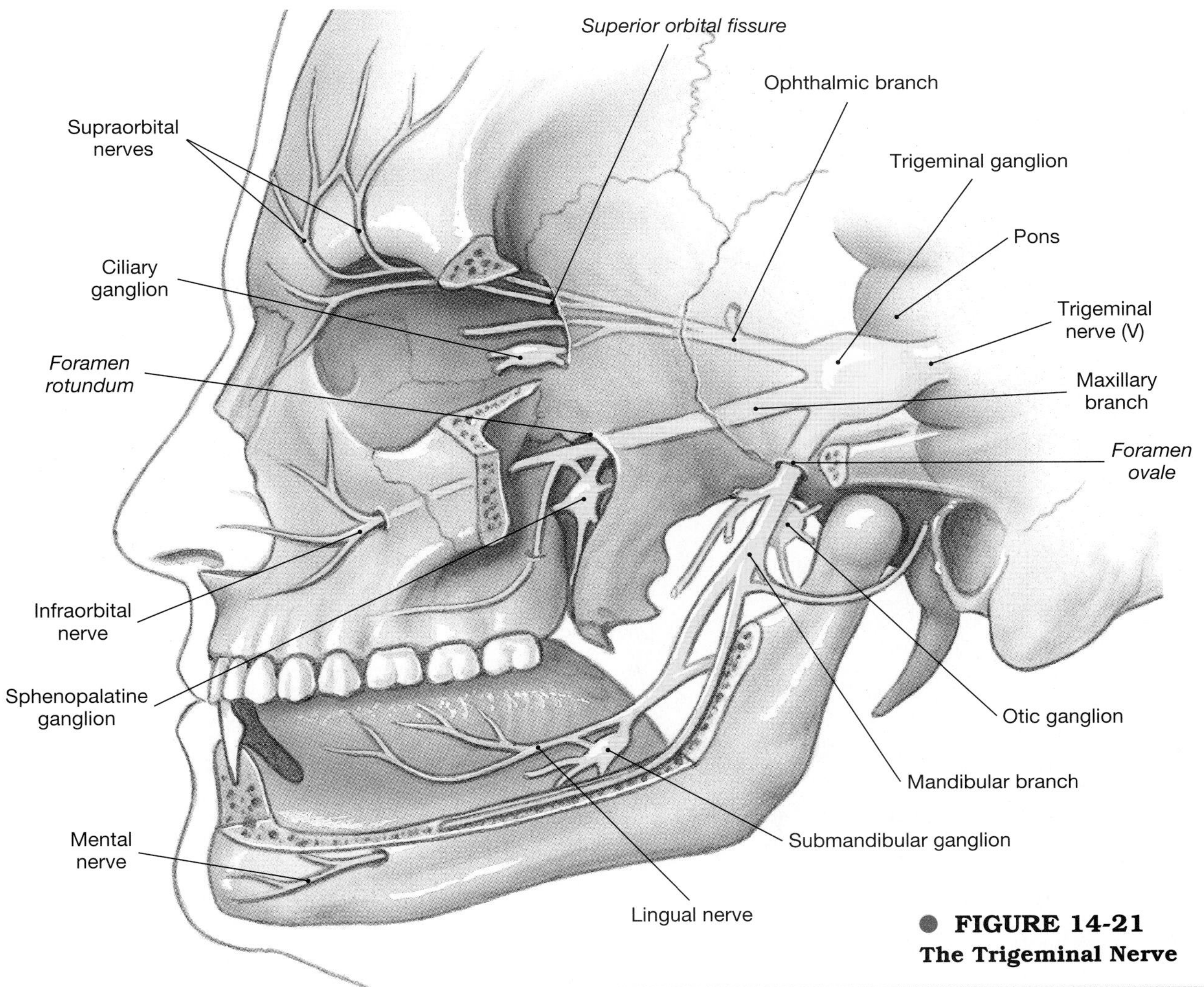
Superior orbital fissure
Ophthalmic branch
Supraorbital nerves
Trigeminal ganglion
Pons
Ciliary ganglion
Trigeminal nerve (V)
Foramen rotundum
Maxillary branch
Foramen ovale
Infraorbital nerve
Sphenopalatine ganglion
Otic ganglion
Mandibular branch
Submandibular ganglion
Mental nerve
Lingual nerve

● **FIGURE 14-21**
The Trigeminal Nerve

The Facial Nerve (N VII)

Figure 14-22

Primary function: Mixed (sensory and motor) to face

Origin: Sensory from taste receptors on anterior two-thirds of tongue; motor from motor nuclei of pons

Passes through: Internal acoustic canal of temporal bone, along internal acoustic canal and facial canal to reach stylomastoid foramen *[p. 210]*

Destination: Sensory to sensory nuclei of pons. *Somatic motor:* muscles of facial expression. *[p. 334]* *Visceral motor:* lacrimal (tear) gland and nasal mucous glands via sphenopalatine ganglion; submandibular and sublingual salivary glands via submandibular ganglion

The **facial nerve** is a mixed nerve. The cell bodies of the sensory neurons are located in the **geniculate ganglion,** and the motor nuclei are in the pons. The sensory and motor roots emerge from the side of the pons and enter the internal acoustic canal of the temporal bone (Figure 14-22●). The nerve then passes through the facial canal to reach the face via the stylomastoid foramen. *[p. 210]* The sensory neurons monitor proprioceptors in the facial muscles, provide deep pressure sensations over the face, and receive taste information from receptors along the anterior two-thirds of the tongue. Somatic motor fibers control the superficial muscles of the scalp and face and deep muscles near the ear.

The facial nerve carries preganglionic autonomic fibers to the *sphenopalatine* and *submandibular ganglia.* Postganglionic fibers from the **sphenopalatine ganglion** innervate the lacrimal gland and small glands of the nasal cavity and pharynx. The **submandibular ganglion** innervates the *submandibular* and *sublingual* (*sub,* under + *lingua,* tongue) salivary glands.

Bell's palsy is a cranial nerve disorder that results from an inflammation of the facial nerve. The condition probably results from a viral infection. Symptoms include paralysis of facial muscles on the affected side and loss of taste sensations from the anterior two-thirds of the tongue. The condition is usually painless, and in most cases, Bell's palsy "cures itself" after a few weeks or months.

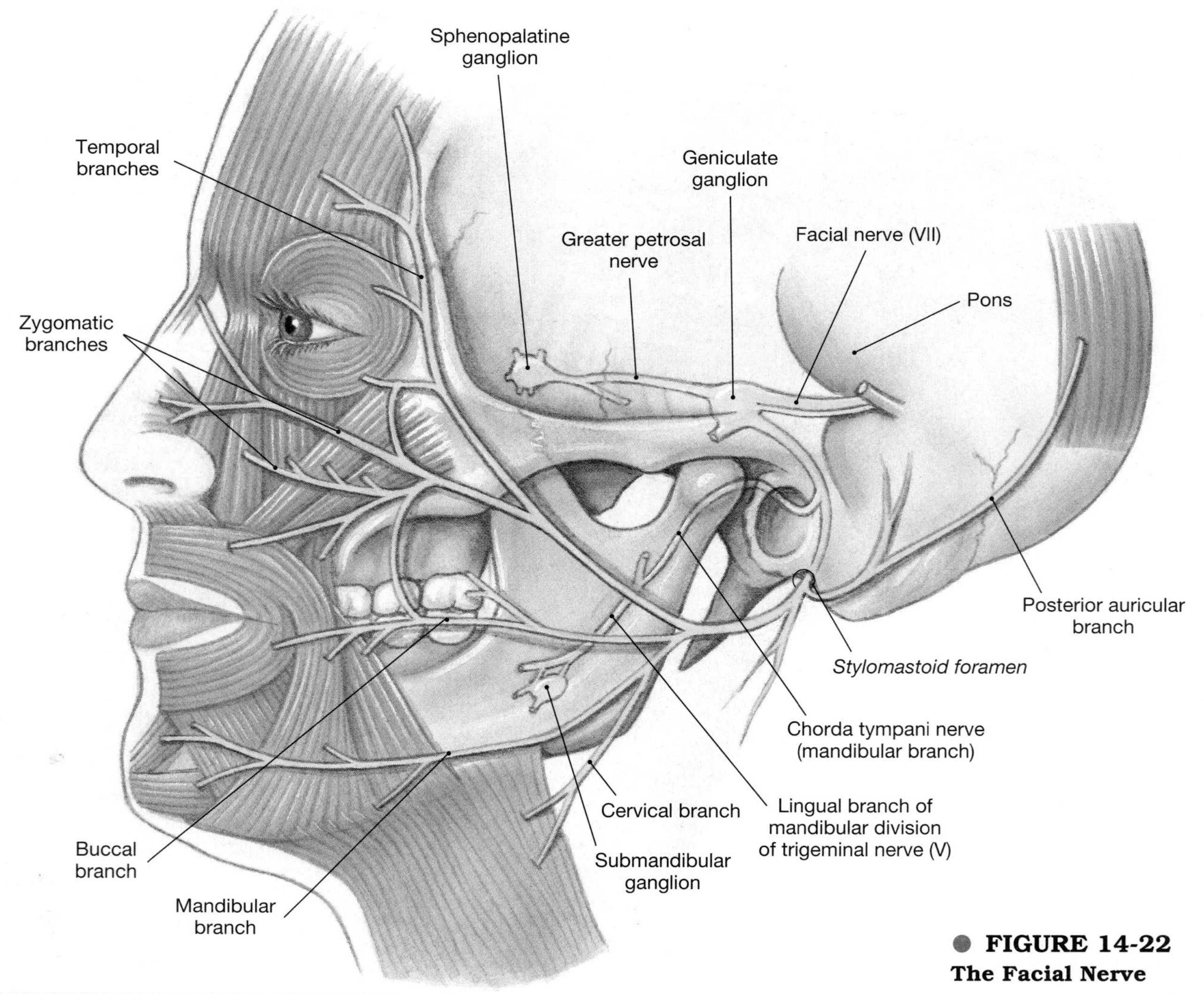
Sphenopalatine ganglion
Temporal branches
Geniculate ganglion
Greater petrosal nerve
Facial nerve (VII)
Pons
Zygomatic branches
Posterior auricular branch
Stylomastoid foramen
Chorda tympani nerve (mandibular branch)
Cervical branch
Lingual branch of mandibular division of trigeminal nerve (V)
Buccal branch
Submandibular ganglion
Mandibular branch

● **FIGURE 14-22**
The Facial Nerve

The Vestibulocochlear Nerve (N VIII)

Figure 14-23

Primary function: Special sensory: balance and equilibrium (vestibular branch) and hearing (cochlear branch)

Origin: Monitors receptors of the inner ear (vestibule and cochlea)

Passes through: Internal acoustic canal of temporal bone [p. 210]

Destination: Vestibular and cochlear nuclei of pons and medulla oblongata

The **vestibulocochlear nerve** is also known as the *acoustic nerve*, the *auditory nerve*, and the *statoacoustic nerve.* We will use the term *vestibulocochlear* because it indicates the names of its two major branches: the *vestibular branch* and the *cochlear branch.* The vestibulocochlear nerve lies posterior to the origin of the facial nerve, straddling the boundary between the pons and the medulla oblongata (Figure 14-23●). This nerve reaches the sensory receptors of the inner ear by entering the internal acoustic canal in company with the facial nerve. There are two distinct bundles of sensory fibers within the vestibulocochlear nerve. The **vestibular nerve** (*vestibulum*, a cavity) originates at the receptors of the *vestibule,* the portion of the inner ear concerned with balance sensations. The sensory neurons are located within an adjacent sensory ganglion, and their axons target the **vestibular nuclei** of the pons and medulla oblongata. These afferents convey information concerning the orientation and movement of the head. The **cochlear** (KOK-lē-ar) **nerve** (*cochlea*, snail shell) monitors the receptors in the cochlea that provide the sense of hearing. The neurons are located within a peripheral ganglion (the *spiral ganglion*), and their axons synapse within the **cochlear nuclei** of the pons and medulla oblongata. Axons leaving the vestibular and cochlear nuclei relay the sensory information to other centers or initiate reflexive motor responses. Balance and the sense of hearing will be discussed in Chapter 17.

The Glossopharyngeal Nerve (N IX)

Figure 14-24

Primary function: Mixed (sensory and motor) to head and neck

Origin: Sensory from posterior one-third of the tongue, part of the pharynx and palate, the carotid arteries of the neck; motor from motor nuclei of medulla oblongata

Passes through: Jugular foramen of temporal bone to foramen lacerum between occipital and temporal bones (superior surface) [p. 210]

Destination: Sensory to sensory nuclei of medulla oblongata; *Somatic motor:* pharyngeal muscles involved in swallowing; *Visceral motor:* parotid salivary gland, via otic ganglion

In addition to the vestibular nucleus of N VIII, the medulla oblongata contains the sensory and motor nuclei for the ninth, tenth, eleventh, and twelfth cranial nerves. The **glossopharyngeal** (glos-ō-fah-RIN-jē-al) **nerve** (*glossum*, tongue) innervates the tongue and pharynx. The glossopharyngeal nerve exits the cranium via the jugular foramen in company with N X and N XI (Figure 14-24●).

● FIGURE 14-23
The Vestibulocochlear Nerve

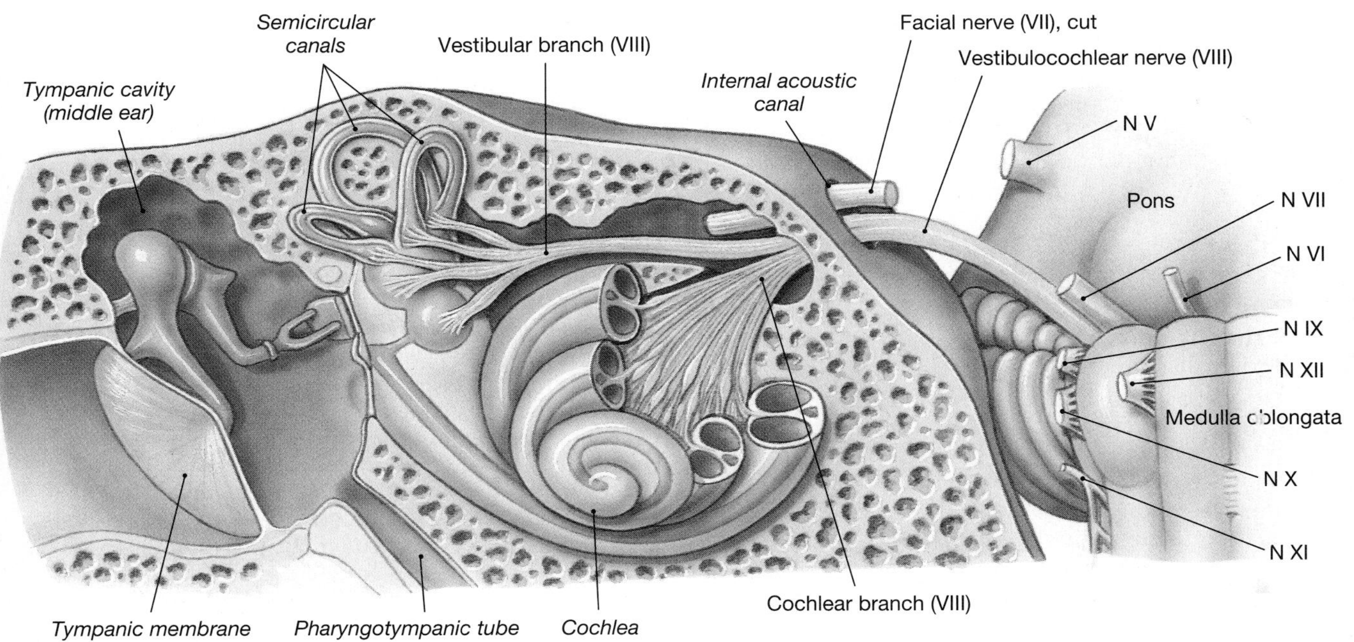
Semicircular canals
Vestibular branch (VIII)
Tympanic cavity (middle ear)
Internal acoustic canal
Facial nerve (VII), cut
Vestibulocochlear nerve (VIII)
N V
Pons
N VII
N VI
N IX
N XII
Medulla oblongata
N X
N XI
Cochlear branch (VIII)
Tympanic membrane
Pharyngotympanic tube
Cochlea

● **FIGURE 14-24**
The Glossopharyngeal Nerve

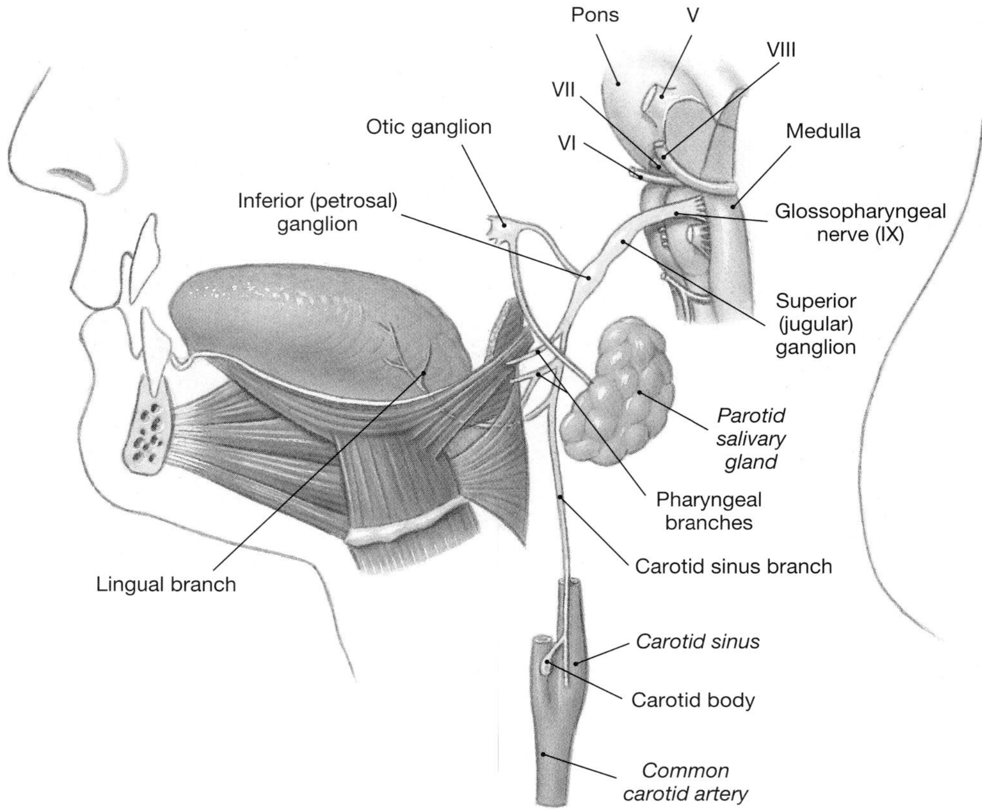
Pons
V
VIII
VII
VI
Otic ganglion
Medulla
Inferior (petrosal) ganglion
Glossopharyngeal nerve (IX)
Superior (jugular) ganglion
Parotid salivary gland
Pharyngeal branches
Carotid sinus branch
Lingual branch
Carotid sinus
Carotid body
Common carotid artery

The glossopharyngeal is a mixed nerve, but sensory fibers are most abundant. The sensory neurons are located in the **superior** *(jugular)* **ganglion** and **inferior** *(petrosal)* **ganglion.** The sensory fibers carry general sensory information from the lining of the pharynx and the soft palate to a nucleus in the medulla oblongata. The glossopharyngeal nerve also provides taste sensations from the posterior third of the tongue and has special receptors monitoring the blood pressure and dissolved gas concentrations within major blood vessels.

The somatic motor fibers control the pharyngeal muscles involved in swallowing. Visceral motor fibers synapse in the otic ganglion, and postganglionic fibers innervate the parotid salivary gland of the cheek.

The Vagus Nerve (N X)

Figure 14-25

Primary function: Mixed (sensory and motor), widely distributed

Origin: Sensory from pharynx (part), pinna and external auditory canal, diaphragm, and visceral organs in thoracic and abdominopelvic cavities; motor from motor nuclei in medulla oblongata

Passes through: Jugular foramen between occipital and temporal bones [p. 209]

Destination: Sensory fibers to sensory nuclei and autonomic centers of medulla oblongata; visceral motor fibers to muscles of the palate, pharynx, digestive, respiratory, and cardiovascular systems in the thoracic and abdominal cavities

The **vagus** (VĀ-gus) **nerve** arises immediately posterior to the glossopharyngeal. Many small rootlets contribute to its formation, and developmental studies indicate that this nerve probably represents the fusion of several smaller cranial nerves during our evolution. As its name suggests (*vagus*, wanderer) the vagus nerve branches and radiates extensively. Figure 14-25● shows only the general pattern of distribution.

Sensory neurons are located within the **jugular** (superior) and **nodose** (NŌ-dōs) (inferior) **ganglia** (*node*, knot). The vagus provides somatic sensory information concerning the external auditory canal, a portion of the ear, and the diaphragm and special sensory information from pharyngeal taste receptors. But most of the vagal afferents carry visceral sensory information from receptors along the esophagus, respiratory tract, and abdominal viscera as distant as the last portions of the large intestine. This visceral sensory information is vital to the autonomic control of visceral function.

The motor components of the vagus are equally diverse. The vagus nerve carries preganglionic autonomic fibers that affect the heart and control smooth muscles and glands within the areas monitored by its sensory fibers, including the stomach, intestines, and gallbladder. One of the most common signs of damage to either nerve IX or X is difficulty in swallowing, because this prevents coordination of the swallowing reflex.

● **FIGURE 14-25**
The Vagus Nerve

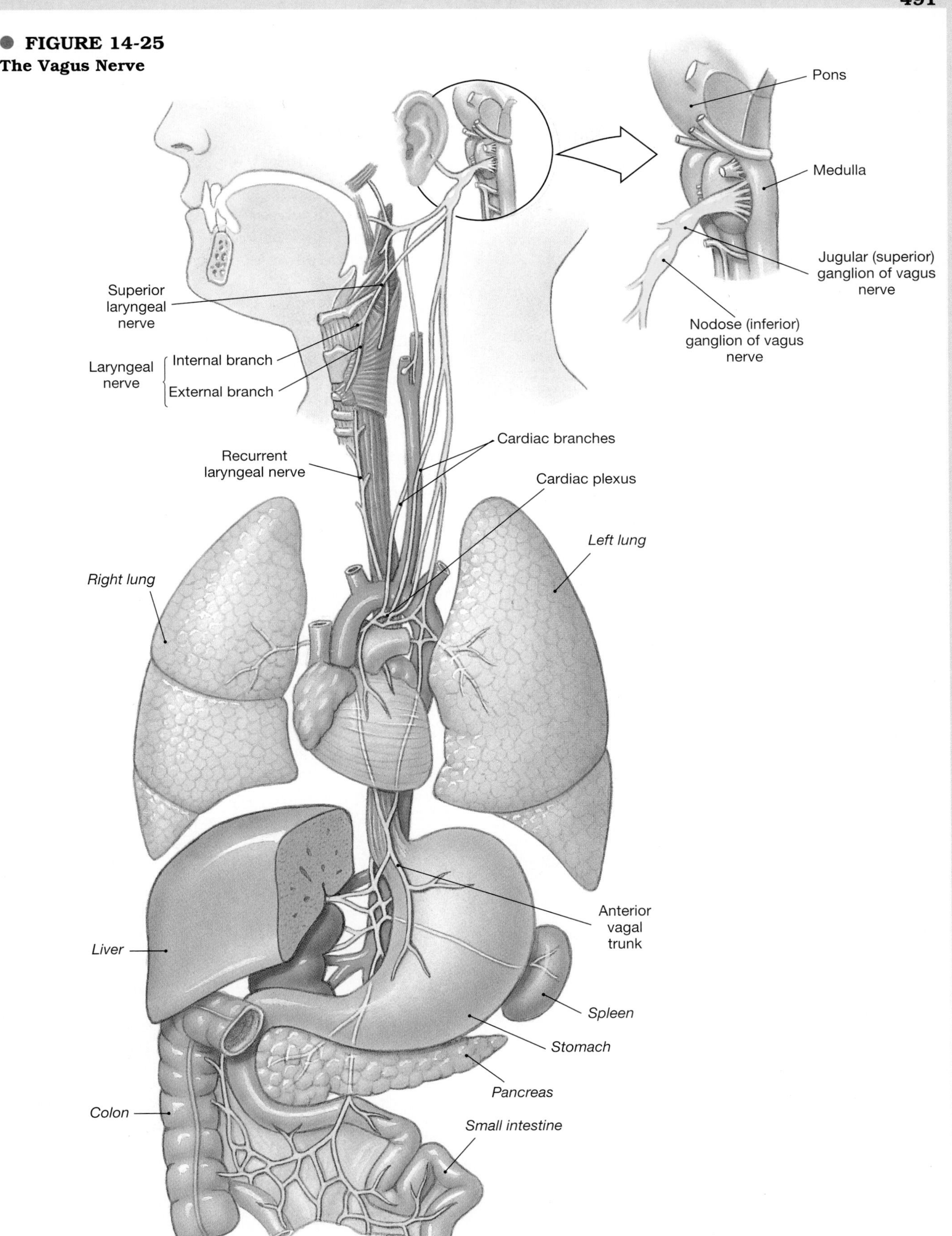
Pons
Medulla
Jugular (superior) ganglion of vagus nerve
Nodose (inferior) ganglion of vagus nerve
Superior laryngeal nerve
Laryngeal nerve
Internal branch
External branch
Cardiac branches
Recurrent laryngeal nerve
Cardiac plexus
Left lung
Right lung
Anterior vagal trunk
Liver
Spleen
Stomach
Pancreas
Colon
Small intestine

The Accessory Nerve (N XI)

Figure 14-26

Primary function: Motor to muscles of the neck and upper back

Origin: Motor nuclei of spinal cord and medulla oblongata

Passes through: Jugular foramen between occipital and temporal bones ⚭ *[p. 209]*

Destination: Medullary branch innervates voluntary muscles of palate, pharynx, and larynx; spinal branch controls sternocleidomastoid and trapezius muscles

The **accessory nerve,** also known as the *spinal accessory nerve* or the *spinoaccessory nerve,* differs from other cranial nerves in that some of its motor fibers originate in the lateral gray horns of the first five cervical vertebrae (Figure 14-26●). These somatic motor fibers enter the cranium through the foramen magnum, join the motor fibers from a nucleus in the medulla oblongata, and leave the cranium through the jugular foramen. The accessory nerve consists of two branches:

1. The **medullary branch** joins the vagus nerve and innervates the voluntary swallowing muscles of the soft palate and pharynx and the intrinsic muscles that control the vocal cords.
2. The **spinal branch** controls the sternocleidomastoid and trapezius muscles of the neck and back. ⚭ *[p. 341]* The motor fibers of this branch originate in the lateral gray horns of C_1 to C_5.

The Hypoglossal Nerve (N XII)

Figure 14-26

Primary function: Motor (tongue movements)

Origin: Motor nuclei of medulla oblongata

Passes through: Hypoglossal canal of occipital bone ⚭ *[p. 209]*

Destination: Muscles of the tongue ⚭ *[p. 339]*

The **hypoglossal** (hī-pō-GLOS-al) **nerve** leaves the cranium through the hypoglossal canal of the occipital bone (Figure 14-26●). ⚭ *[p. 209]* It then follows a curving path to reach the skeletal muscles of the tongue. This nerve provides voluntary motor control over movements of the tongue. Its condition is checked by "sticking out" the tongue; damage to the nerve or its nuclei will cause the tongue to veer toward the affected side.

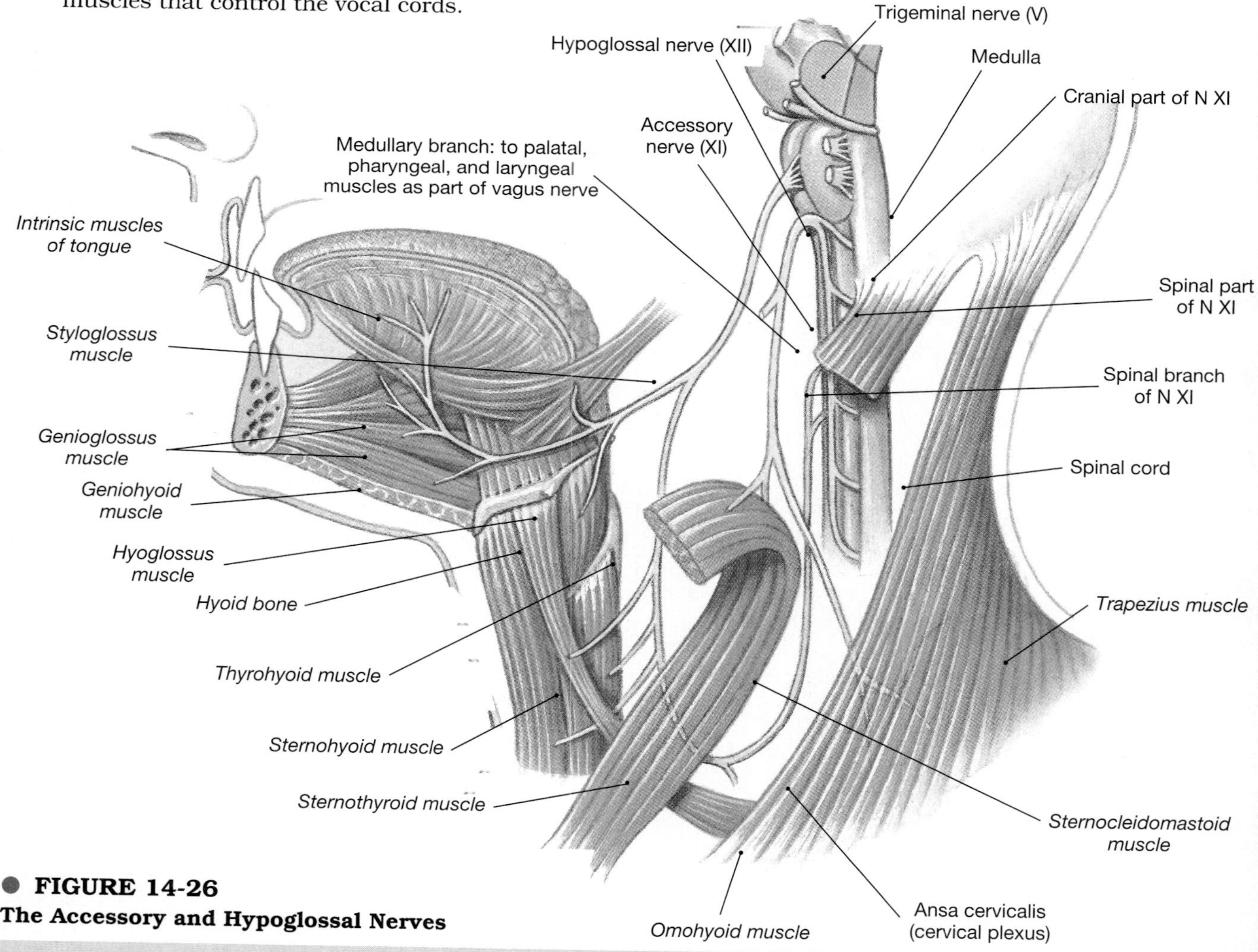

Trigeminal nerve (V)
Hypoglossal nerve (XII)
Medulla
Cranial part of N XI
Accessory nerve (XI)
Medullary branch: to palatal, pharyngeal, and laryngeal muscles as part of vagus nerve
Intrinsic muscles of tongue
Spinal part of N XI
Styloglossus muscle
Spinal branch of N XI
Genioglossus muscle
Spinal cord
Geniohyoid muscle
Hyoglossus muscle
Hyoid bone
Trapezius muscle
Thyrohyoid muscle
Sternohyoid muscle
Sternothyroid muscle
Sternocleidomastoid muscle
Omohyoid muscle
Ansa cervicalis (cervical plexus)

● **FIGURE 14-26**
The Accessory and Hypoglossal Nerves

A Summary of Cranial Nerve Branches and Functions

Few people are able to remember the names, numbers, and functions of the cranial nerves without a struggle. To aid memorization, mnemonic devices may prove useful. The most famous and oft-repeated is *On Old Olympus's Towering Top A Finn And German Viewed Some Hops.* (The *And* refers to the acoustic nerve, an alternative name for N VIII, and the *Some* is for spinal accessory (N XI).) A more modern one, *Oh, Once One Takes The Anatomy Final, Very Good Vacations Are Heavenly,* may be a bit easier to remember. A summary of the basic distribution and function of each cranial nerve is detailed in Table 14-9.

TABLE 14-9 The Cranial Nerves

Cranial Nerve (Number)	*Sensory Ganglion*	*Branch*	*Primary Function*	*Foramen*	*Innervation*
Olfactory (I)			Special sensory	Cribriform plate of ethmoid	Olfactory epithelium
Optic (II)			Special sensory	Optic foramen	Retina of eye
Oculomotor (III)			Motor	Superior orbital fissure	Inferior, medial, superior rectus, inferior oblique and levator palpebrae muscles; intrinsic muscles of eye
Trochlear (IV)			Motor	Superior orbital fissure	Superior oblique muscle
Trigeminal (V)	Semilunar		Mixed		Areas associated with the jaws
		Ophthalmic	Sensory	Superior orbital fissure	Orbital structures, nasal cavity, skin of forehead, upper eyelid, eyebrows, nose (part)
		Maxillary	Sensory	Foramen rotundum	Lower eyelid; upper lip, gums, and teeth; cheek, nose (part), palate and pharynx (part)
		Mandibular	Mixed	Foramen ovale	*Sensory* to lower gums, teeth, lips; palate (part) and tongue (part) *Motor* to muscles of mastication
Abducens (VI)			Motor	Superior orbital fissure	Lateral rectus muscle
Facial (VII)	Geniculate		Mixed	Internal acoustic canal to facial canal; exits at stylomastoid foramen	*Sensory* to taste receptors on anterior 2/3 of tongue *Motor* to muscles of facial expression, lacrimal gland, submandibular gland, sublingual salivary glands
Vestibulocochlear (Acoustic) (VIII)			Special sensory	Internal acoustic canal	
		Cochlear			Cochlea (receptors for hearing)
		Vestibular			Vestibule (receptors for motion and balance)
Glossopharyngeal (IX)	Superior, inferior		Mixed	Jugular foramen	*Sensory* from posterior 1/3 of tongue; pharynx and palate (part); blood pressure and composition *Motor* to pharyngeal muscles, parotid salivary gland
Vagus (X)	Jugular, nodose		Mixed	Jugular foramen	*Sensory* from pharynx; pinna and external canal; diaphragm; visceral organs in thoracic and abdominopelvic cavities *Motor* to palatal and pharyngeal muscles, and visceral organs in thoracic and abdominopelvic cavities
Accessory (XI)					
		Medullary	Motor	Jugular foramen	Voluntary muscles of palate, pharynx, and larynx (via vagus nerve)
		Spinal	Motor	Jugular foramen	Sternocleidomastoid and trapezius muscles
Hypoglossal (XII)			Motor	Hypoglossal canal	Tongue musculature

TABLE 14-10 Cranial Reflexes

Reflex	Stimulus	Afferents	Central Synapse	Efferents	Response
SOMATIC REFLEXES					
Corneal reflex	Contact with corneal surface	N V (trigeminal)	Motor nucleus for N VII (facial)	N VII	Blinking of eye
Tympanic reflex	Loud noise	N VIII (vestibulo-cochlear)	Inferior colliculus (midbrain)	N VII	Reduced movement of auditory ossicles
Auditory reflexes	Loud noise	N VIII	Motor nuclei of brain stem and spinal cord	N III, IV, VI, VII, X, cervical nerves	Eye and/or head movements triggered by sudden sounds
Vestibulo-ocular reflexes	Rotation of head	N VIII	Motor nuclei controlling eye muscles	N III, IV, VI	Opposite movement of eyes to stabilize field of vision
VISCERAL REFLEXES					
Direct light reflex	Light striking photoreceptors	N II (optic)	Superior colliculus (midbrain)	N III (oculomotor)	Constriction of ipsilateral pupil
Consensual light reflex	Light striking photoreceptors	N II	Superior colliculus	N III	Constriction of contralateral pupil

❑ Cranial Reflexes

Cranial reflexes are reflex arcs that involve the sensory and motor fibers of cranial nerves. Examples of cranial reflexes are discussed in later chapters, and this section will simply provide an overview and general introduction.

Table 14-10 lists representative examples of cranial reflexes and their functions. These reflexes are clinically important because they provide a quick and easy method for observing the condition of cranial nerves and specific nuclei and tracts in the brain, and thus localize the site of damage or disease.

Cranial somatic reflexes are seldom more complex than the somatic reflexes of the spinal cord. This table includes four somatic reflexes: the *corneal reflex*, the *tympanic reflex*, the *auditory reflexes*, and the *vestibulo-ocular reflexes*. The normal functions of these reflexes are indicated in Table 14-10. These reflexes are often used to check for damage to the cranial nerves or processing centers involved. The brain stem contains many reflex centers that control visceral motor activity. Many of these reflex centers are in the medulla oblongata, and they can direct very complex visceral motor responses to stimuli. These reflexes are essential to the control of respiratory, digestive, and cardiovascular functions, and Chapter 16 considers them in greater detail. AM *Cranial Nerve Tests*

Selected Clinical Terminology

Terms Discussed in This Chapter

ataxia: A disturbance of balance that in severe cases leaves the individual unable to stand without assistance. It is caused by problems affecting the cerebellum. *(p. 482)*

Bell's palsy: A condition resulting from an inflammation of the facial nerve; symptoms include paralysis of facial muscles on the affected side and loss of taste sensations from the anterior two-thirds of the tongue. *(p. 488)*

cerebrovascular accident (CVA): Also known as "stroke"; occurs when the blood supply to a portion of the brain is blocked off. *(p. 467)*

cerebrovascular diseases: Circulatory disorders that interfere with the normal circulatory supply to the brain. *(p. 467)*

cranial trauma: A head injury resulting from violent contact with another object. Cranial trauma may cause a **concussion,** a condition characterized by a temporary loss of consciousness and a variable period of amnesia. *(p. 464* and *AM)*

dyslexia: A disorder affecting the comprehension and use of words. *(p. 470)*

hydrocephalus: A condition resulting from interference with the normal circulation and/or reabsorption of cerebrospinal fluid. Also known as "water on the brain." *(p. 466* and *AM)*

Parkinson's disease *(paralysis agitans)*: A condition characterized by a pronounced increase in muscle tone, resulting from excitation of neurons in the cerebral nuclei. *(p. 480* and *AM)*
tic douloureux (doo-loo-ROO), or **trigeminal neuralgia:** A disorder of the maxillary and mandibular branches of N V characterized by severe, almost totally debilitating pain triggered by contact with the lip, tongue, or gums. *(p. 487* and *AM)*

AM *Additional Terms Discussed in the Applications Manual*

decerebrate rigidity: A generalized state of muscular contraction resulting from loss of CNS inhibitory control.
dysmetria (dis-MET-rē-a; *dys-*, bad + *metron*, measure): An inability to stop a movement at a precise, predetermined position; it often leads to an **intention tremor** in the affected individual. Dysmetria usually reflects cerebellar dysfunction.
epidural hemorrhage: A condition involving bleeding into the epidural spaces, usually resulting from cranial trauma.
subdural hemorrhage: A condition in which blood accumulates between the dura and the arachnoid membrane.
spasticity: A condition characterized by hesitant, jerky voluntary movements, increased muscle tone, and hyperactive stretch reflexes.
tremor: Cyclic oscillations in limb position resulting from a "tug of war" between antagonistic muscle groups.

CHAPTER REVIEW

STUDY OUTLINE

AN INTRODUCTION TO THE ORGANIZATION OF THE BRAIN, p. 458

1. The brain is far more complex and adaptable than the spinal cord.

Embryology of the Brain, p. 458

2. The brain forms from three swellings at the superior tip of the developing neural tube: the **prosencephalon,** the **mesencephalon,** and the **rhombencephalon.** (*Table 14-1*)

Major Regions and Landmarks, p. 458

3. There are six regions in the adult brain: **cerebrum, diencephalon, mesencephalon** (midbrain), **metencephalon, pons** and **cerebellum,** and **medulla oblongata.** (*Figures 14-1, 14-2; Table 14-1*)
4. Conscious thought, intellectual functions, memory, and complex involuntary motor patterns originate in the **cerebrum.** The **cerebellum** adjusts voluntary and involuntary motor activities on the basis of sensory data and stored memories. (*Figure 14-2*)
5. Each wall of the diencephalon forms a **thalamus** that contains relay and processing centers for sensory data. The **hypothalamus** contains centers involved with emotions, autonomic function, and hormone production. (*Figure 14-2*)
6. The **mesencephalon** processes visual and auditory information and generates involuntary somatic motor responses. (*Figure 14-2*)
7. The **pons** connects the cerebellum to the brain stem and is involved with somatic and visceral motor control. The spinal cord connects to the brain at the **medulla oblongata,** which relays sensory information and regulates autonomic functions. (*Figure 14-2*)
8. The brain contains extensive areas of **neural cortex,** a layer of gray matter on the surfaces of the cerebrum and cerebellum.

Ventricles of the Brain, p. 463

9. The central passageway of the brain expands to form chambers called **ventricles.** Cerebrospinal fluid continually circulates from the ventricles and central canal of the spinal cord into the subarachnoid space of the meninges that surround the CNS. (*Figure 14-3*)

PROTECTION AND SUPPORT OF THE BRAIN, p. 463

The Cranial Meninges, p. 464

1. The cranial meninges (the **dura mater, arachnoid,** and **pia mater**) are continuous with those of the spinal cord, but have anatomical and functional differences.
2. Folds of dura mater stabilize the position of the brain. The **falx cerebri, tentorium cerebelli,** and **falx cerebelli** are examples. (*Figure 14-4*)

Cerebrospinal Fluid, p. 466

3. The **choroid plexus** is the site of cerebrospinal fluid production.
4. Cerebrospinal fluid (CSF) (1) cushions delicate neural structures, (2) supports the brain, and (3) transports nutrients, chemical messengers, and waste products. Cerebrospinal fluid reaches the subarachnoid space via the **lateral apertures** and a **median aperture.** Diffusion across the **arachnoid villi** into the **superior sagittal sinus** returns CSF to the venous circulation. (*Figure 14-5*)

The Blood Supply to the Brain, p. 466

5. The **blood-brain barrier** isolates neural tissue from the general circulation.
6. The blood-brain barrier remains intact throughout the CNS except in portions of the hypothalamus, the pineal gland, and at the choroid plexus in the membranous roof of the diencephalon and medulla.

THE CEREBRUM, p. 468

The Cerebral Cortex, p. 468

1. The cortical surface contains **gyri** (elevated ridges) separated by **sulci** (shallow depressions) or **fissures** (deeper grooves). The **longitudinal fissure** separates the two

cerebral hemispheres. The **central sulcus** marks the boundary between the **frontal lobe** and the **parietal lobe**. Other sulci form the boundaries of the **temporal lobe** and the **occipital lobe.** (*Figures 14-1, 14-6*)

2. Each cerebral hemisphere receives sensory information and generates motor commands that concern the opposite side of the body. There are significant functional differences between the two hemispheres.
3. The **central white matter** contains **association fibers, commissural fibers,** and **projection fibers.**
4. The **primary motor cortex** of the **precentral gyrus** directs voluntary movements. The **primary sensory cortex** of the **postcentral gyrus** receives somatic sensory information from touch, pressure, pain, taste, and temperature receptors. (*Figure 14-6; Table 14-2*)
5. **Association areas,** such as the **visual association area** and **somatic motor association area** (**premotor cortex**), control our ability to understand sensory information and coordinate a motor response. (*Figure 14-6; Table 14-2*)
6. The **general interpretive area** receives information from all the sensory association areas. It is present in only one hemisphere, usually the left. (*Figure 14-6*)
7. The **speech center** regulates the patterns of breathing and vocalization needed for normal speech. (*Figure 14-6*)
8. The **prefrontal cortex** coordinates information from the secondary and special association areas of the entire cortex and performs abstract intellectual functions. (*Figure 14-6*)
9. The left hemisphere is usually the **categorical hemisphere;** it contains the general interpretive and speech centers and is responsible for language-based skills. The right hemisphere, or **representational hemisphere,** is concerned with spatial relationships and analyses. (*Figure 14-7*)

The Cerebral Nuclei, p. 472

10. The **cerebral nuclei** (*basal ganglia*) within the central white matter include the **caudate nucleus, amygdaloid body, claustrum,** and **putamen.** The cerebral nuclei are part of the **extrapyramidal system,** which controls muscle tone and coordinates learned movement patterns and other somatic motor activities. (*Figure 14-8*)

THE LIMBIC SYSTEM, p. 473

1. The **limbic system** includes the amygdaloid body, **cingulate gyrus, parahippocampal gyrus, hippocampus, mamillary bodies,** and the **fornix.** The functions of the limbic system involve emotional states and related behavioral drives. (*Figures 14-8 14-9, 14-10; Table 14-3*)

THE DIENCEPHALON, p. 475

1. The diencephalon provides the switching and relay centers necessary to integrate the conscious and unconscious sensory and motor pathways. (*Figures 14-9, 14-10, 14-11*)

The Thalamus, p. 476

2. The thalamus is the final relay point for ascending sensory information and coordinates the pyramidal and extrapyramidal systems. (*Figures 14-9, 14-11, 14-12; Table 14-4*)

The Hypothalamus, p. 477

3. The hypothalamus contains important control and integrative centers. It can (1) control involuntary somatic motor activities, (2) control autonomic function, (3) coordinate activities of the nervous and endocrine systems, (4) secrete hormones, (5) produce emotions and behavioral drives, (6) coordinate voluntary and autonomic functions, and (7) regulate body temperature. (*Figure 14-13; Table 14-5*)

THE MESENCEPHALON, p. 479

1. The **tectum** (roof) of the mesencephalon contains two pairs of sensory nuclei, the **corpora quadrigemina.** On each side the **superior colliculus** receives visual inputs from the thalamus, and the **inferior colliculus** receives auditory data from the medulla oblongata. The **red nucleus** integrates information from the cerebrum and issues involuntary motor commands related to muscle tone and posture. The **substantia nigra** regulates the motor output of the cerebral nuclei. The **cerebral peduncles** contain ascending fibers headed for thalamic nuclei and descending pyramidal fibers that carry voluntary motor commands from the primary motor cortex. (*Figures 14-11, 14-14; Table 14-6*)

THE CEREBELLUM, p. 480

1. The cerebellum oversees the body's postural muscles and programs and tunes voluntary and involuntary movements. The **cerebellar hemispheres** consist of neural cortex formed into folds, or **folia.** The surface can be divided into the **anterior** and **posterior lobes,** the **vermis,** and the **flocculonodular lobe.** (*Figure 14-15; Table 14-7*)

THE PONS, p. 482

1. The pons contains (1) sensory and motor nuclei for four cranial nerves; (2) nuclei concerned with involuntary control of respiration; (3) tracts linking the cerebellum with the brain stem, cerebrum, and spinal cord; and (4) ascending and descending tracts. (*Figures 14-11, 14-16*)

THE MEDULLA OBLONGATA, p. 482

1. The medulla oblongata connects the brain and spinal cord. It contains **olivary nuclei,** which relay information from the spinal cord, cerebral cortex, and brain stem to the cerebellar cortex. Its **reflex centers,** including the **cardiovascular centers** and the **respiratory rhythmicity centers,** control or adjust the activities of peripheral systems. (*Figures 14-11, 14-16; Table 14-8*)

CRANIAL NERVES, p. 484

1. There are 12 pairs of cranial nerves. Each nerve attaches to the brain near the associated sensory or motor nuclei on its ventrolateral surface. (*Figure 14-17*)
2. The **olfactory nerves** (N I) carry sensory information responsible for the sense of smell. The olfactory afferents synapse within the **olfactory bulbs.** (*Figures 14-17, 14-18*)
3. The **optic nerves** (N II) carry visual information from special sensory receptors in the eyes. (*Figures 14-17, 14-19*)

4. The **oculomotor nerves** (N III) are the primary source of innervation for four of the extrinsic oculomotor muscles. (*Figure 14-20*)
5. The **trochlear nerves** (N IV), the smallest cranial nerves, innervate the superior oblique muscles of the eyes. (*Figure 14-20*)
6. The **trigeminal nerves** (N V), the largest cranial nerves, are mixed nerves with **ophthalmic, maxillary,** and **mandibular branches.** (*Figure 14-21*)
7. The **abducens nerves** (N VI) innervate the lateral rectus muscles. (*Figure 14-21*)
8. The **facial nerves** (N VII) are mixed nerves that control muscles of the scalp and face. They provide pressure sensations over the face and receive taste information from the tongue. (*Figure 14-22*)
9. The **vestibulocochlear nerves** (N VIII) contain the **vestibular nerves,** which monitor sensations of balance, position, and movement, and **cochlear nerves,** which monitor hearing receptors. (*Figure 14-23*)
10. The **glossopharyngeal nerves** (N IX) are mixed nerves that innervate the tongue and pharynx and control the action of swallowing. (*Figure 14-24*)
11. The **vagus nerves** (N X) are mixed nerves that are vital to the autonomic control of visceral function. (*Figure 14-25*)
12. The **accessory nerves** (N XI) have **medullary branches,** which innervate voluntary swallowing muscles of the soft palate and pharynx, and **spinal branches,** which control muscles associated with the pectoral girdle. (*Figure 14-26*)
13. The **hypoglossal nerves** (N XII) provide voluntary motor control over tongue movements. (*Figure 14-26*)

A Summary of Cranial Nerve Branches and Functions, p. 493

14. The branches and functions of the cranial nerves are summarized in Table 14-9.

Cranial Reflexes, p. 494

15. **Cranial reflexes** are reflex arcs that involve the sensory and motor fibers of cranial nerves. (*Table 14-10*)

REVIEW QUESTIONS

LEVEL 1 Reviewing Facts and Terms

Match each item in Column A with the most closely related item in Column B. Use letters for answers in the spaces provided.

Set 1: The Brain

	Column A	Column B
___	1. prosencephalon	a. third ventricle
___	2. rhombencephalon	b. oxytocin
___	3. brain stem	c. auditory association area
___	4. diencephalic chamber	d. maintains muscle tone and posture
___	5. dural fold	e. melatonin
___	6. choroid plexus	f. white matter of cerebellum
___	7. Broca's area	g. mesencephalon, pons, medulla oblongata
___	8. Wernicke's area	h. link between nervous and endocrine systems
___	9. limbic system	i. releases neurotransmitter dopamine
___	10. pineal gland	j. forebrain
___	11. supraoptic nucleus	k. cerebellar cortex
___	12. paraventricular nucleus	l. falx cerebri
___	13. substantia nigra	m. a disturbance in balance
___	14. red nucleus	n. apneustic, pneumotaxic centers
___	15. Purkinje cells	o. hindbrain
___	16. arbor vitae	p. amygdaloid body
___	17. ataxia	q. medulla oblongata
___	18. reticular formation	r. production of CSF
___	19. nucleus gracilis	s. antidiuretic hormone
___	20. hypothalamus	t. speech center

Set 2: The Cranial Nerves

	Column A	Column B
___	21. facial nerve	a. sensory, smell
___	22. vagus nerve	b. N V
___	23. olfactory nerve	c. motor, tongue movements
___	24. trochlear nerve	d. sensory, vision
___	25. hypoglossal nerve	e. N X
___	26. accessory nerve	f. N III
___	27. trigeminal nerve	g. equilibrium, hearing
___	28. optic nerve	h. N VII
___	29. glossopharyngeal nerve	i. N VI
___	30. oculomotor nerve	j. N IX
___	31. vestibulocochlear nerve	k. motor, eye movments
___	32. abducens nerve	l. N XI

33. The structural and functional link between the cerebral hemispheres and the components of the brain stem is the:
    (a) neural cortex
    (b) medulla oblongata
    (c) mesencephalon
    (d) diencephalon
34. The primary link between the nervous and endocrine systems is the:
    (a) hypothalamus (b) mesencephalon
    (c) pons (d) medulla oblongata
35. Immediate reflexive responses to a loud, unexpected noise are directed by nuclei in the:
    (a) pons (b) diencephalon
    (c) mesencephalon (d) medulla oblongata
36. Regulation of autonomic function, such as heart rate and blood pressure, originates in the:
    (a) cerebrum (b) cerebellum
    (c) diencephalon (d) medulla oblongata

37. The ventricles in the brain are filled with:
(a) blood (b) cerebrospinal fluid
(c) air (d) neural tissue

38. The smooth surface that covers the brain but does not follow the underlying neural convolutions or sulci is the:
(a) neural cortex (b) dura mater
(c) pia mater (d) arachnoid membrane

39. The meningeal layer that adheres to the surface contour of the brain, extending into every fold and curve, is the:
(a) pia mater
(b) dura mater
(c) arachnoid membrane
(d) neural cortex

40. The dural fold that divides the two cerebellar hemispheres is the:
(a) transverse sinus (b) falx cerebri
(c) tentorium cerebelli (d) falx cerebelli

41. Production and secretion of cerebrospinal fluid occur in the:
(a) hypothalamus
(b) choroid plexus
(c) medulla oblongata
(d) crista galli

42. The primary purpose of the blood-brain barrier is to:
(a) provide the brain with oxygenated blood
(b) drain venous blood via the internal jugular veins
(c) isolate neural tissue in the CNS from the general circulation
(d) a, b, and c are correct

43. Conscious thought processes and all intellectual functions originate in the:
(a) cerebellum
(b) corpus callosum
(c) cerebral hemispheres
(d) medulla oblongata

44. The two cerebral hemispheres are functionally different even though anatomically they appear the same.
(a) true (b) false

45. The inability to speak, to read, or to understand the speech of others is called:
(a) dyslexia
(b) aphasia
(c) disconnection syndrome
(d) ataxia

46. The area of the brain that generates feelings of frustration, tension, and anxiety as it interprets ongoing events and makes predictions about future situations is the:
(a) Broca's area (b) postcentral gyrus
(c) prefrontal cortex (d) temporal lobes

47. Reading, writing, and speaking are dependent on processing in the:
(a) right cerebral hemisphere
(b) left cerebral hemisphere
(c) prefrontal cortex
(d) postcentral gyrus

48. Tracts in the central white matter that connect the two cerebral hemispheres are called:
(a) integrative fibers
(b) projection fibers
(c) association fibers
(d) commissural fibers

49. The motor system that controls skeletal muscle tone and coordinates learned movement patterns and other somatic activities is the:
(a) limbic system
(b) extrapyramidal system
(c) reticular system
(d) choroid plexus

50. Establishment of emotional states and related behavioral drives are functions of the:
(a) limbic system
(b) tectum
(c) mamillary bodies
(d) thalamus

51. The final relay point for ascending sensory information that will be projected to the primary sensory cortex is the:
(a) hypothalamus
(b) thalamus
(c) spinal cord
(d) medulla oblongata

52. The two hormones secreted by the hypothalamus are:
(a) epinephrine and norepinephrine
(b) antidiuretic hormone and oxytocin
(c) melatonin and serotonin
(d) FSH and ATP

53. Symptoms characteristic of Parkinson's disease appear if there is damage to the:
(a) red nucleus
(b) cerebral peduncles
(c) corpora quadrigemina
(d) substantia nigra

54. The part of the brain that coordinates rapid, automatic adjustments that maintain balance and equilibrium is the:
(a) cerebrum
(b) cerebellum
(c) pons
(d) medulla oblongata

55. The centers in the pons that modify the activity of the respiratory rhythmicity center in the medulla oblongata are the:
(a) apneustic and pneumotaxic centers
(b) inferior and superior peduncles
(c) cardiac and vasomotor centers
(d) nucleus gracilis and nucleus cuneatus

56. What are the primary functions of the cerebrum?

57. Briefly summarize the overall function of the cerebellum.

58. What role does the hypothalamus play in the body?

59. What three layers make up the cranial meninges?

60. What are the three important functions of the CSF?

61. What three areas in the brain are not isolated from the general circulation by the blood-brain barrier?

62. What are pyramidal cells, and what is their function?

63. What functional properties are associated with the neurons in the primary sensory cortex?

64. What is the extrapyramidal system, and what is its function?

65. Where is the limbic system located, and what are its three functions?

66. What three kinds of nuclei are found in the medulla oblongata?

67. Using the mnemonic device "Oh, Once One Takes The Anatomy Final, Very Good Vacations Are Heavenly," list the 12 pairs of cranial nerves with their functions.

LEVEL 2 Reviewing Concepts

68. A person with a damaged visual association area may:
   (a) be unable to scan the lines of a page or see rows of clear symbols
   (b) be declared legally blind
   (c) be unable to see letters but able to identify words and their meanings
   (d) be able to see letters quite clearly but unable to recognize or interpret them

69. If the corpus callosum is cut:
   (a) cross-referencing of sensory information is inhibited
   (b) the individual talks constantly but uses the wrong words
   (c) objects touched with the left hand cannot be recognized but can be verbally identified
   (d) objects touched with the right hand cannot be verbally identified

70. What structure in the brain would your A & P instructor be referring to if he talked about a nucleus that resembled a sea horse and that appeared to be important in the storage and retrieval of long-term memories? In what functional system of the brain would it be found?

71. Stimulation of what part of the brain would produce sensations of hunger and thirst?

72. Damage to the corpora quadrigemina would interfere with:
   (a) control of autonomic function
   (b) regulation of body temperature
   (c) processing of visual and auditory sensations
   (d) control of involuntary somatic motor activities

73. If symptoms characteristic of Parkinson's disease appear, what part of the mesencephalon is inhibited from secreting a neurotransmitter? Which neurotransmitter?

74. What are the principal functional differences between the categorical hemisphere and the representational hemisphere?

75. What major part of the brain is associated with identifying the clinical signs of life?

76. Why can the brain respond to stimuli with greater versatility than the spinal cord can?

77. What kinds of problems are associated with the presence of lesions in Wernicke's area and Broca's area?

LEVEL 3 Critical Thinking and Clinical Applications

78. Smelling salts can sometimes help restore consciousness after a person has fainted. The active ingredient of smelling salts is ammonia, and it acts by irritating the lining of the nasal cavity. Propose a mechanism by which smelling salts would raise a person to the conscious state from the unconscious state.

79. A police officer has just stopped Bill on suspicion of driving while intoxicated. The officer asks Bill to walk the yellow line on the road and then asks him to place the tip of his index finger on the tip of his nose. How would these activities indicate Bill's level of sobriety? What part of the brain is being tested by these activities?

80. While having some dental work performed, Tyler is given an injection of local anesthetic in his lower jaw. His dentist tells him not to eat until the anesthetic wears off, not because of his teeth but because of his tongue. Why is the dentist giving him this advice?

CHAPTER 15

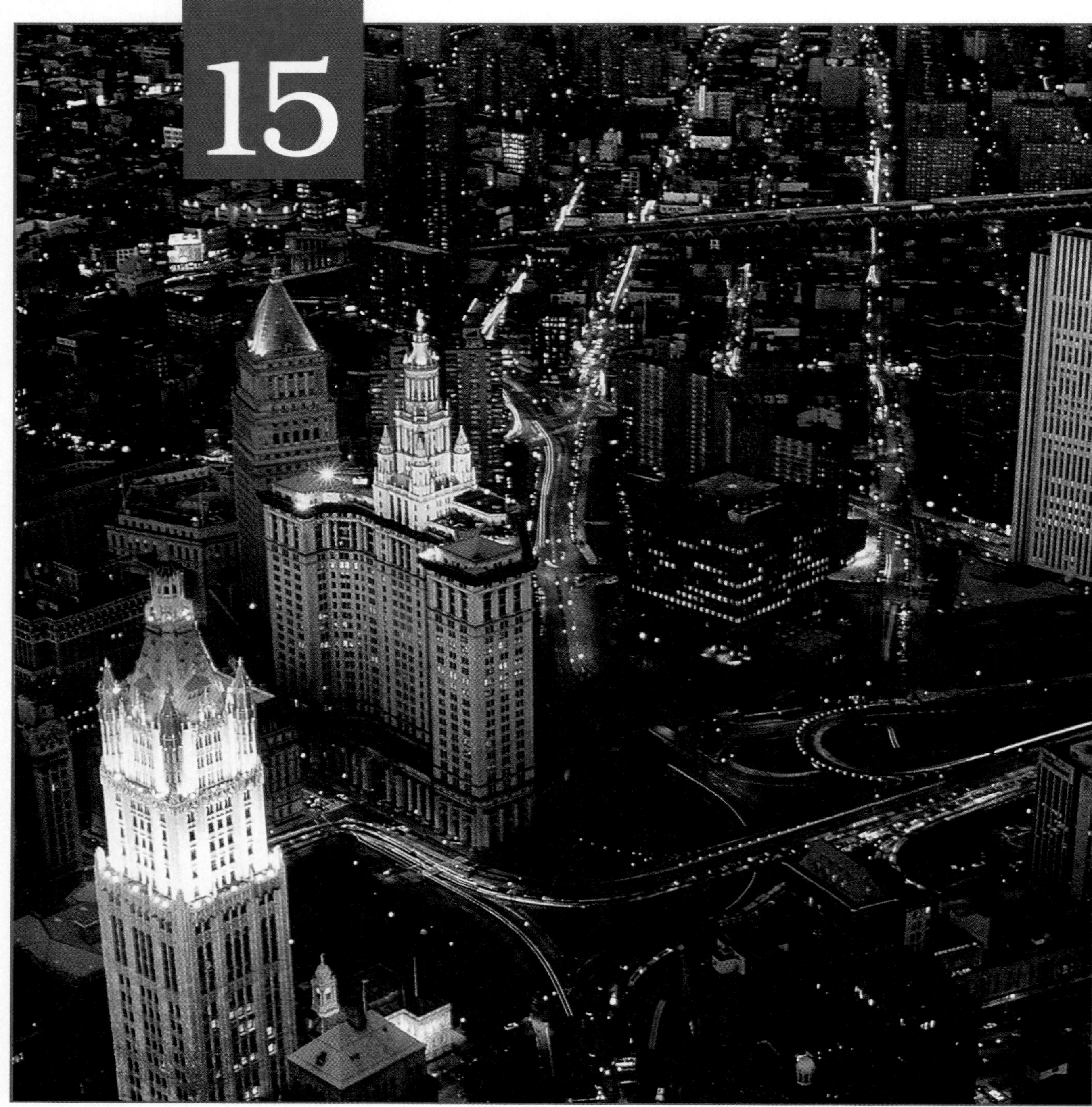

There is a saying that "big cities never sleep." If you visit New York or Los Angeles at 3:00 A.M., you will see shops open, deliveries being made, people on the street, and traffic moving briskly. The central nervous system (CNS) is much more complex than any city, and far busier. There is a continuous flow of information between the brain, spinal cord, and peripheral nerves. At any given moment, millions of sensory neurons are delivering information to processing centers in the CNS, and millions of motor neurons are controlling or adjusting the activities of peripheral effectors. This process continues 24 hours a day, whether we are awake or asleep. You may sleep soundly, but your nervous system does not; many brain stem centers are active throughout our lives, performing vital autonomic functions. This chapter examines the flow of sensory information and motor commands between the PNS and the CNS. It also examines complex aspects of human consciousness, such as memory and learning, that involve multiple regions of the brain.

Integrative Functions

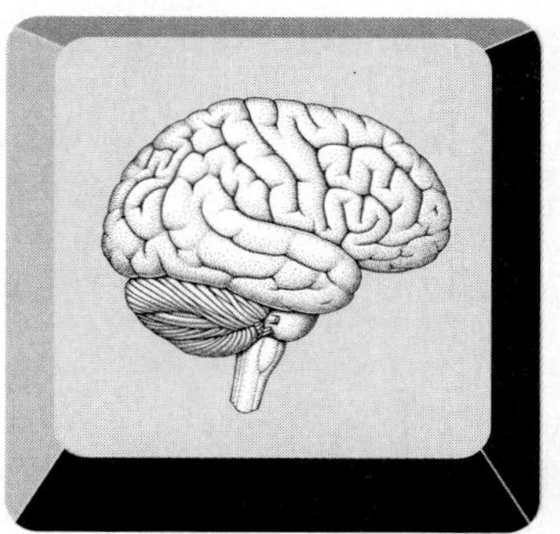

Chapter Outline and Objectives

1. Identify the principal sensory and motor pathways.
2. Compare the components, processes, and functions of the pyramidal and extrapyramidal systems.
3. Explain how we can distinguish among sensations that originate in different areas of the body.
4. Describe the levels of information processing involved in motor control.
5. Discuss how the brain integrates sensory information and coordinates responses.

6. Explain how memories are created, stored, and recalled.
7. Distinguish between the levels of consciousness and unconsciousness, and identify the characteristics of brain activity associated with the different levels of sleep.

8. Describe drug-related alterations in brain function.

9. Summarize the effects of aging on the nervous system.

Chapters 13 and 14 introduced the major components and functions of the CNS and PNS. Those chapters, however, could not tell the entire story, for the central nervous system is more than a collection of independent parts. You might think of the nervous system as a symphony orchestra. Thus far, you have studied the appearance and sound of the individual cellos, clarinets, and other instruments. While essential to an understanding of an orchestra, this information does not tell you how the orchestra performs, nor what a symphony sounds like. While you are reading this text and trying to memorize important information you are also breathing, balancing upright, controlling your body temperature, and possibly listening to the radio or TV. To accomplish these tasks, many centers in the CNS must work together in a coordinated and integrated fashion. This chapter considers the neural "orchestra" as it performs complex intellectual functions. In doing so it examines the integration of processing centers in the brain and spinal cord, as well as functional relationships between the CNS and PNS.

The essential communication between CNS and PNS occurs over tracts and nuclei that relay sensory information and motor commands. **Pathways** consist of the nuclei and tracts that link the brain with the rest of the body. Sensory and motor pathways involve a series of synapses, one after the other. For example, an ascending axon of a sensory neuron may synapse in the *nucleus gracilis* in the medulla oblongata. There it activates a neuron that synapses in the thalamus. Thalamic neurons relay the sensory information to the cerebral cortex. At each step there are opportunities for facilitation or inhibition that will magnify or suppress the sensations involved. There are also opportunities for divergence and the activation of additional neuronal pools that operate outside of our conscious awareness. For example, while the cerebral cortex considers a voluntary response to a stumble, collateral processing within the brain stem and cerebellum may already be issuing involuntary commands that prevent a fall.

Many subtle forms of interaction, feedback, and regulation link higher centers with the various components of the brain stem. Only a few are understood in any detail. The following sections focus on basic patterns and principles in the sensory, motor, and higher-order functions of the brain.

Sensory and Motor Pathways

We will begin by considering the pathways that utilize the major ascending (sensory) and descending (motor) tracts of the spinal cord. As we proceed with our discussion, it may help you to remember that the tract names indicate the origins and destinations of the axons:

- If the name of a tract begins with *spino-* it must *start* in the spinal cord and *end* in the brain. The rest of the name indicates the tract's destination. For example, a *spinothalamic tract* begins in the spinal cord and ends in the thalamus.
- If the name of a tract ends in *-spinal* its axons must *start* in the higher centers and *end* in the spinal cord. The rest of the name indicates the specific nucleus or cortical area of the brain where the tract originates. For example, a *corticospinal tract* carries motor commands from the cerebral cortex to the spinal cord.

Sensory Pathways

Sensory receptors represent the interface between the nervous system and the internal and external environments. A **sensory receptor** is a specialized cell that monitors conditions in the body or the external environment. When stimulated, a receptor passes information to the CNS in the form of action potentials along the axon of a sensory neuron. The arriving information is called a **sensation.**

The term **general senses** refers to sensations of temperature, pain, touch, pressure, vibration, and proprioception (body position). General sensory receptors are distributed throughout the body, and they are relatively simple in structure. Some of the information they send to the CNS reaches the primary sensory cortex and our conscious awareness. As noted in Chapter 12, sensory information is interpreted on the basis of the frequency of arriving action potentials. [p. 415] For example, when pressure sensations are arriving, the harder the pressure, the higher the frequency of action potentials. The conscious awareness of a sensation is called a **perception.**

Chapter 17 will consider the origins of sensations and the pathways involved in relaying the information to conscious and unconscious processing centers in the CNS. Most of the processing occurs in centers along the sensory pathways in the spinal cord or brain stem. Only about 1 percent of the information provided by afferent fibers reaches the cerebral cortex and our conscious awareness. For example, we usually do not feel the clothes we wear or hear the hum of the engine in our car.

Somatic Sensory Receptors

Receptors for the general senses are scattered throughout the body. They are relatively simple in structure. These receptors are classified according to the nature of the stimulus that excites them:

- **Nociceptors** (nō-sē-SEP-torz; *noceo,* hurt) respond to a variety of stimuli usually associated with tissue damage. Receptor activation causes the sensation of pain.
- **Thermoreceptors** respond to changes in temperature.
- **Mechanoreceptors** are stimulated or inhibited by physical distortion, contact, or pressure on their cell membranes. There are many different types of mechanoreceptors. Examples introduced in earlier chapters include *proprioceptors,* such as muscle spindles and Golgi tendon organs, and *tactile receptors,* such as touch receptors in the skin.
- **Chemoreceptors** monitor the chemical composition of body fluids and respond to the presence of specific molecules. Chemoreceptors are important components of several visceral reflexes.

This section discusses pathways that carry somatic sensory information to the primary sensory cortex of the cerebral hemispheres.

The Organization of Sensory Pathways

Figure 15-1

The sensory neuron that delivers the sensations to the CNS is often called a **first-order neuron.** The soma of a first-order sensory neuron is located in a dorsal root ganglion. Inside the CNS the axon of that sensory neuron synapses on an interneuron known as a **second-order neuron.** If the sensation is to reach our conscious awareness, the second-order neuron, which may be located in the spinal cord or brain stem, will synapse on a **third-order neuron** in the thalamus. Somewhere along its length, the axon of the second-order neuron crosses over to the opposite side of the body. As a result, the right side of the thalamus receives sensory information from the left side of the body, and vice versa.

The axons of the third-order neurons synapse on neurons of the primary sensory cortex of the cerebral hemisphere. Because the axons leaving the thalamus ascend on the same side of the brain, the right cerebral hemisphere receives sensory information from the left side of the body, and the left cerebral hemisphere receives sensations from the right side.

There are three major somatic sensory pathways, or **somatosensory pathways:** the *posterior column pathway,* the *spinothalamic pathway,* and the *spinocerebellar pathway.* There are two pairs of spinal tracts associated with each pathway; all the axons within a tract share a common origin and destination. Figure 15-1● indicates the relative positions of the spinal tracts involved, and Table 15-1, p. 506, summarizes information concerning these tracts and pathways.

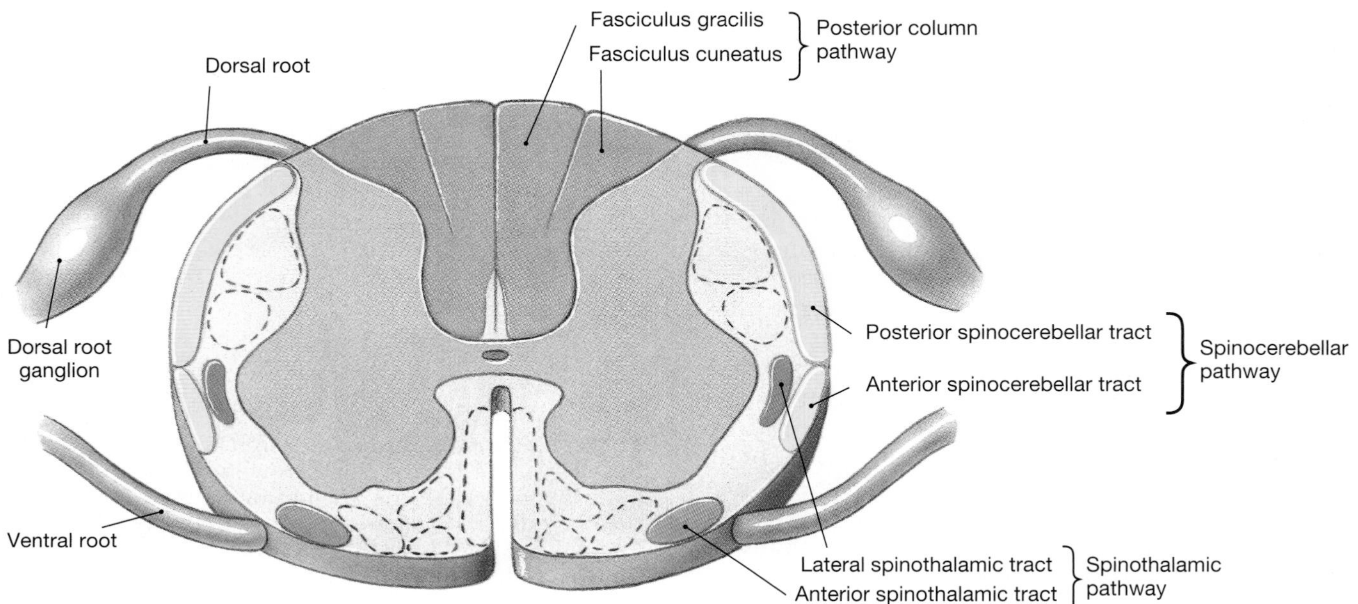
Fasciculus gracilis
Fasciculus cuneatus
Posterior column pathway
Dorsal root
Dorsal root ganglion
Posterior spinocerebellar tract
Anterior spinocerebellar tract
Spinocerebellar pathway
Ventral root
Lateral spinothalamic tract
Anterior spinothalamic tract
Spinothalamic pathway

● **FIGURE 15-1**

Ascending Tracts in the Spinal Cord. A cross-sectional view indicating the locations of the major ascending sensory tracts in the spinal cord. For information about these tracts, see Table 15-1. Descending motor tracts are shown in outline; these tracts are identified in Figure 15-5.

The Posterior Column Pathway

Figure 15-2

The **posterior column pathway** provides conscious sensations of highly localized ("fine") touch, pressure, vibration, and proprioception (position sense). This pathway begins at a peripheral receptor and ends at the primary sensory cortex of the cerebral hemispheres.

Figure 15-2a● traces sensations that originate on the right side of the body. These sensations will ultimately arrive at the primary sensory cortex of the left cerebral hemisphere. Sensations entering the posterior column pathway from the left side of the body will ultimately arrive at the right cerebral hemisphere. The spinal tracts involved are the left and right **fasciculus gracilis** (*gracilis*, delicate) and the left and right **fasciculus cuneatus** (*cuneus*, wedge-shaped). On each side of the posterior median sulcus, the fasciculus gracilis is medial to the fasciculus cuneatus.

The axons within the posterior column are organized according to the region innervated. Axons from the lower half of the body ascend within the fasciculus gracilis and synapse in the *nucleus gracilis* of the medulla oblongata. Axons from the upper half of the trunk, upper limbs, and neck ascend in the fasciculus cuneatus and synapse in the *nucleus cuneatus*. These nuclei were introduced in Chapter 14. ∞ [p. 483] Axons of the postsynaptic neurons then ascend to the thalamus. As the axons leave the nucleus, they cross over to the opposite side of the brain stem. Movement of an axon from the left side to the right side, or vice versa, is called **decussation** (*decussatio*, a crossing over). Once on the opposite side of the brain, the axons enter a tract called the **medial lemniscus** (*lemniskos*, ribbon). Before reaching the thalamus, the medial lemniscus receives sensory information from the face collected by sensory neurons of the trigeminal nerve (N V).

The axons in the medial lemniscus synapse on third-order neurons in the thalamus. Axons that carry sensory information from the entire body, excluding the face, synapse in the **ventral posterolateral nucleus** (VPL). Axons carring sensory information from the face synapse in the **ventral posteromedial nucleus** (VPM). These nuclei sort the arriving information according to (1) the nature of the stimulus and (2) the region of the body involved. Processing in the thalamus determines whether we perceive a given sensation as fine touch, rather than as pressure or vibration. Our ability to localize the sensation—to determine precisely where on the body a stimulus originated—depends on the projection of information from the ventral posterolateral nucleus to the primary sensory cortex. If the primary sensory cortex were damaged, or the projection fibers cut, one could detect a light touch but be unable to determine its source.

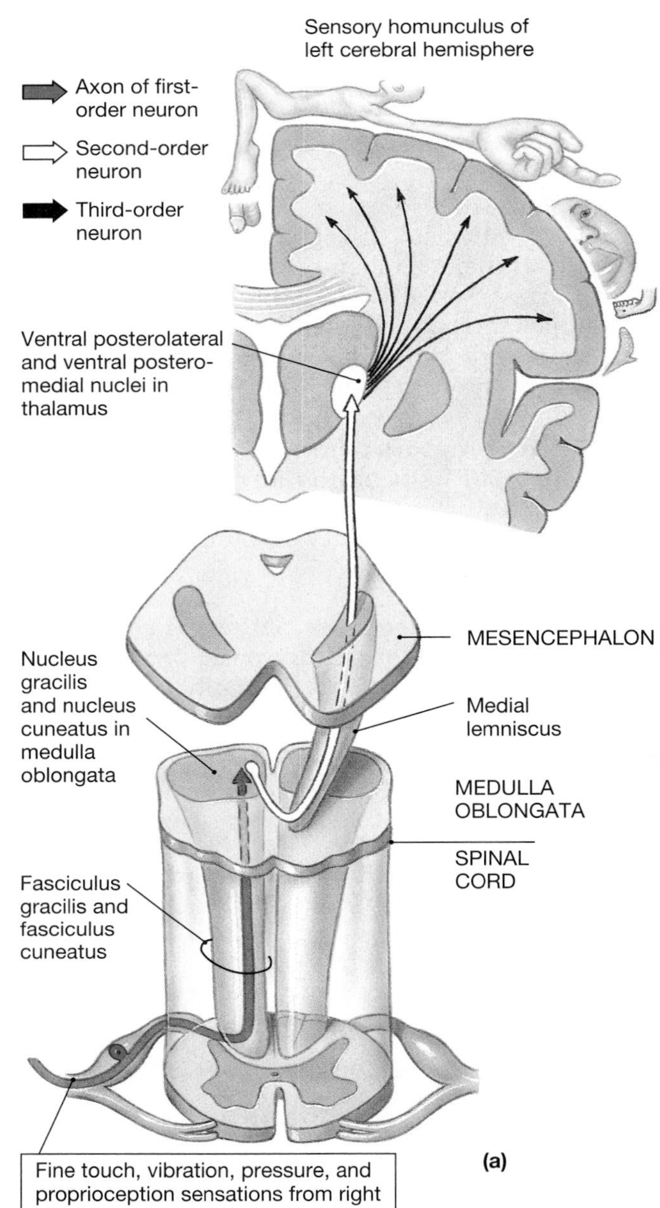

Sensory homunculus of left cerebral hemisphere
Axon of first-order neuron
Second-order neuron
Third-order neuron
Ventral posterolateral and ventral posteromedial nuclei in thalamus
MESENCEPHALON
Nucleus gracilis and nucleus cuneatus in medulla oblongata
Medial lemniscus
MEDULLA OBLONGATA
SPINAL CORD
Fasciculus gracilis and fasciculus cuneatus
Fine touch, vibration, pressure, and proprioception sensations from right side of body
(a)

● **FIGURE 15-2**
The Posterior Column and Spinothalamic Pathways. For clarity, this figure shows only the pathway for sensations originating on the right side of the body. **(a)** The posterior column pathway delivers fine touch, vibration, and proprioception information to the primary sensory cortex of the cerebral hemisphere on the opposite side of the body. **(b)** The anterior spinothalamic tracts carry sensations of crude touch and pressure to the primary sensory cortex of the opposite cerebral hemisphere. **(c)** The lateral spinothalamic tracts carry sensations of pain and temperature to the primary sensory cortex on the opposite side of the body.

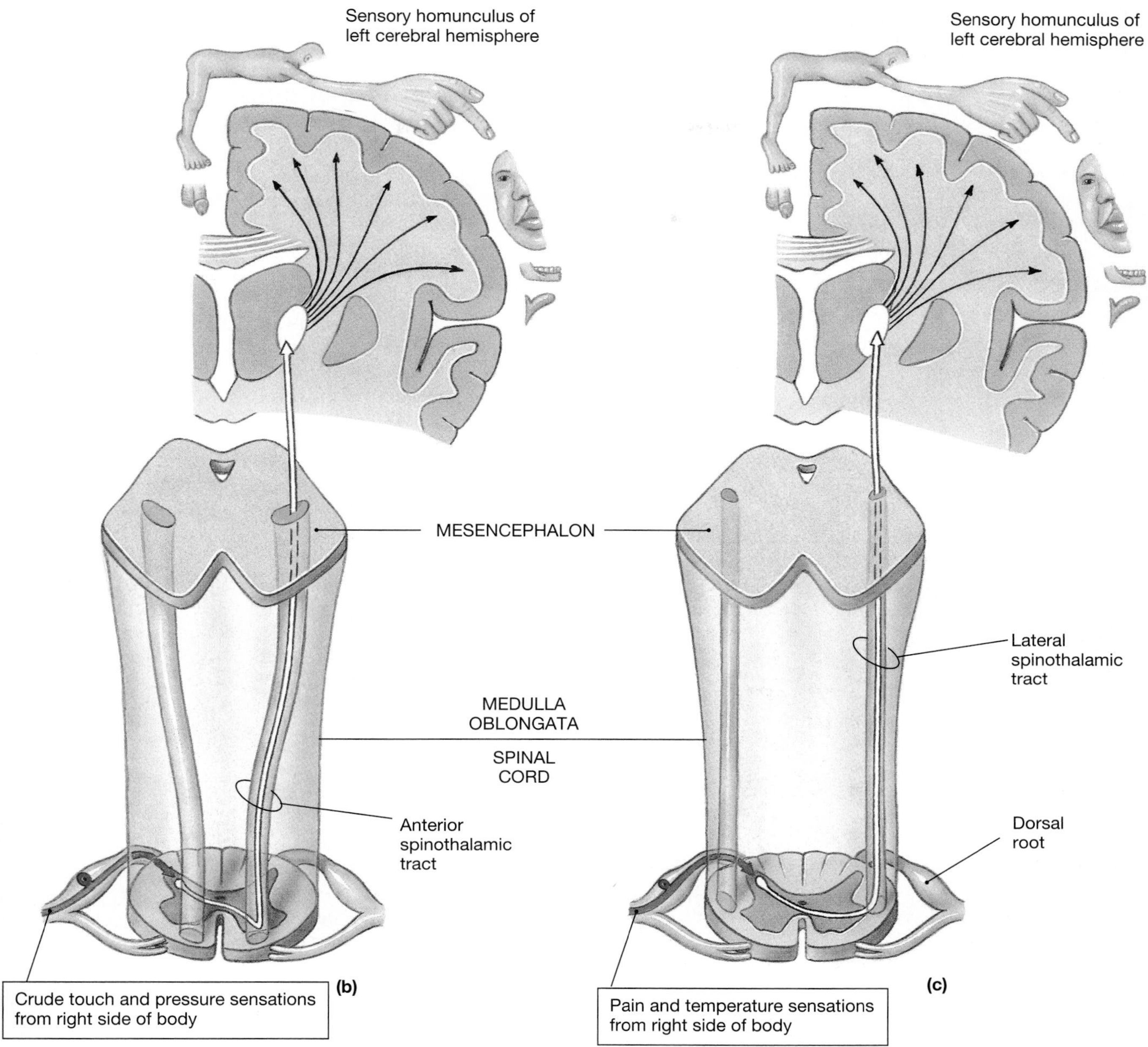
Sensory homunculus of left cerebral hemisphere
Sensory homunculus of left cerebral hemisphere
MESENCEPHALON
Lateral spinothalamic tract
MEDULLA OBLONGATA
SPINAL CORD
Anterior spinothalamic tract
Dorsal root
(b)
Crude touch and pressure sensations from right side of body
(c)
Pain and temperature sensations from right side of body

The ability to localize a specific stimulus depends on the organized distribution of sensory information to the primary sensory cortex. Sensory information from the toes arrives at one end of the primary sensory cortex, and information from the head arrives at the other. When neurons in one portion of the primary sensory cortex are stimulated, the individual becomes consciously aware of sensations originating at a specific location.

The same sensations are reported whether the cortical neurons are activated by axons ascending from the thalamus or by direct electrical stimulation. Researchers have electrically stimulated the primary sensory cortex in awake individuals during brain surgery and asked the subjects where they thought the stimulus originated. The results were used to create a functional map of the primary sensory cortex. This map, seen in Figure 15-2a●, is called a **sensory homunculus** ("little man").

The proportions of the sensory homunculus are obviously very different from those of the individual. For example, the face is huge and distorted, with enormous lips and tongue, whereas the back is relatively tiny. These distortions occur because the area of sensory cortex devoted to a particular region is proportional not to its absolute size, but rather to the *number of sensory receptors* it contains. In other words, it takes many more cortical neurons to process sensory information arriving from the tongue, which has tens of thousands of taste and touch receptors, than it does to analyze sensations originating on the back, where touch receptors are few and far between.

TABLE 15-1 Principal Ascending (Sensory) Tracts in the Spinal Cord and a Summary of the Functional Roles of the Associated Nuclei in the Brain

Pathway/ Tract	*Sensations*	*Location of Neurons*		
		First-Order	*Second-Order*	*Third-Order*
POSTERIOR COLUMN PATHWAY	Proprioception and fine touch, pressure, and vibration	Dorsal root ganglia of lower body; axons enter CNS via dorsal roots and join fasciculus gracilis	Nucleus gracilis of medulla oblongata; axons cross over before entering medial lemniscus	Ventral posterolateral nucleus of thalamus
		Dorsal root ganglia of upper body; axons enter CNS via dorsal roots and join fasciculus cuneatus	Nucleus cuneatus of medulla oblongata; axons cross over before entering medial lemniscus	Ventral posterolateral and ventral posteromedial nuclei of thalamus
SPINOTHALAMIC PATHWAY	Pain and temperature	Dorsal root ganglia; axons enter CNS via dorsal roots	Interneurons in posterior gray horn; axons enter lateral spinothalamic tract on opposite side	Ventral posterolateral nucleus of thalamus
	Crude touch and pressure	Dorsal root ganglia; axons enter CNS via dorsal roots	Interneurons in posterior gray horn; axons enter anterior spinothalamic tract on opposite side	Ventral posterolateral nucleus of thalamus
SPINOCEREBELLAR PATHWAY	Proprioception	Dorsal root ganglia; axons enter CNS via dorsal roots	Interneurons in posterior gray horn; axons enter posterior spinocerebellar tract on same side	
		Dorsal root ganglia; axons enter CNS via dorsal roots	Interneurons in same spinal segment; axons enter anterior spinocerebellar tract on same or opposite side	

The Spinothalamic Pathway

Figures 15-2, 15-3

The **spinothalamic pathway** provides conscious sensations of poorly localized ("crude") touch, pressure, pain, and temperature. This pathway begins at peripheral receptors and ends at the primary sensory cortex of the cerebral hemispheres.

The axons of first-order sensory neurons enter the spinal cord and synapse on second-order neurons within the posterior gray horns. The axons of these interneurons cross to the opposite side of the spinal cord and ascend within the **anterior** or **lateral spinothalamic tracts** (Figure 15-2b,c●). On each side of the brain, the anterior and lateral spinothalamic tracts end at third-order neurons in the ventral posterolateral nucleus of the thalamus. After they have been sorted and processed, the sensations are relayed to the primary sensory cortex.

As in the posterior column pathway, the determination that an arriving stimulus is painful, rather than cool, is determined by the projection of the sensation to different populations of second-order and third-order neurons. The ability to localize that stimulus depends on stimulation of an appropriate area of the primary sensory cortex. Any abnormality along the pathway can result in inappropriate sensations. Consider two examples:

- An individual can experience painful sensations that are not "real." For example, a person may continue to experience pain in an amputated limb. This condition, called *phantom limb pain,* is caused by activity in the sensory neurons or interneurons along the

Final Destination	*Site of Cross-Over*
Primary sensory cortex on side opposite stimulus	Axons of second-order neurons before joining medial lemniscus
Primary sensory cortex on side opposite stimulus	Axons of second-order neurons before joining medial lemniscus
Primary sensory cortex on side opposite stimulus	Axons of second-order neurons at level of entry
Primary sensory cortex on side opposite stimulus	Axons of second-order neurons at level of entry
Cerebellar cortex on side of stimulus	None
Cerebellar cortex on side opposite stimulus (and to side of stimulus)	Axons of most second-order neurons cross before entering tract

spinothalamic pathway; the neurons involved once monitored conditions in the intact limb.

- An individual can feel pain in an uninjured part of the body when the pain actually originates at another location. For example, strong visceral pain sensations arriving at a spinal segment can stimulate interneurons that are normally part of the spinothalamic pathway. Activity in these interneurons leads to stimulation of the primary sensory cortex, so that the individual feels pain in a specific part of the body surface. This phenomenon is called **referred pain.** Two familiar examples are (1) the pain of a heart attack is often felt in the left arm, and (2) the pain of appendicitis is usually felt in the right lower quadrant. Additional examples are illustrated in Figure 15-3●. We will consider the origins and distribution of pain sensations in greater detail in Chapter 17.

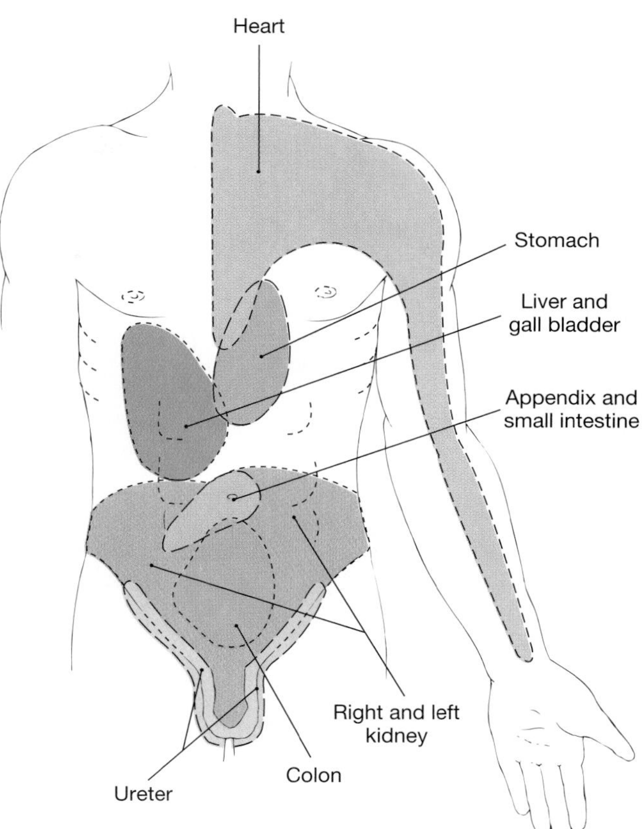
Heart
Stomach
Liver and
gall bladder
Appendix and
small intestine
Right and left
kidney
Colon
Ureter

● **FIGURE 15-3**

Referred Pain. Pain sensations originating in visceral organs are often perceived as involving specific regions of the body surface innervated by the same spinal segments.

The Spinocerebellar Pathway

Figure 15-4

The cerebellum receives proprioceptive information concerning the position of skeletal muscles, tendons, and joints via the **spinocerebellar pathway** (Figure 15-4●). This information does not reach our conscious awareness. The axons of first-order sensory neurons synapse on interneurons in the dorsal gray horns of the spinal cord. The axons of these second-order neurons ascend in one of the *spinocerebellar tracts.* Axons that enter the left or right **posterior spinocerebellar tract** do not cross over to the opposite side of the spinal cord. These axons reach the cerebellar cortex via the inferior cerebellar peduncle of that side. Axons that do cross over to the opposite side of the spinal cord enter the left or right **anterior spinocerebellar tract.**[1] These axons reach the cerebellar cortex via the superior cerebellar peduncle.

Information carried by the axons of the spinocerebellar tracts reaches the Purkinje cells of the

[1]Each anterior spinocerebellar tract also contains uncrossed axons from interneurons on the same side of the spinal cord as the stimulus.

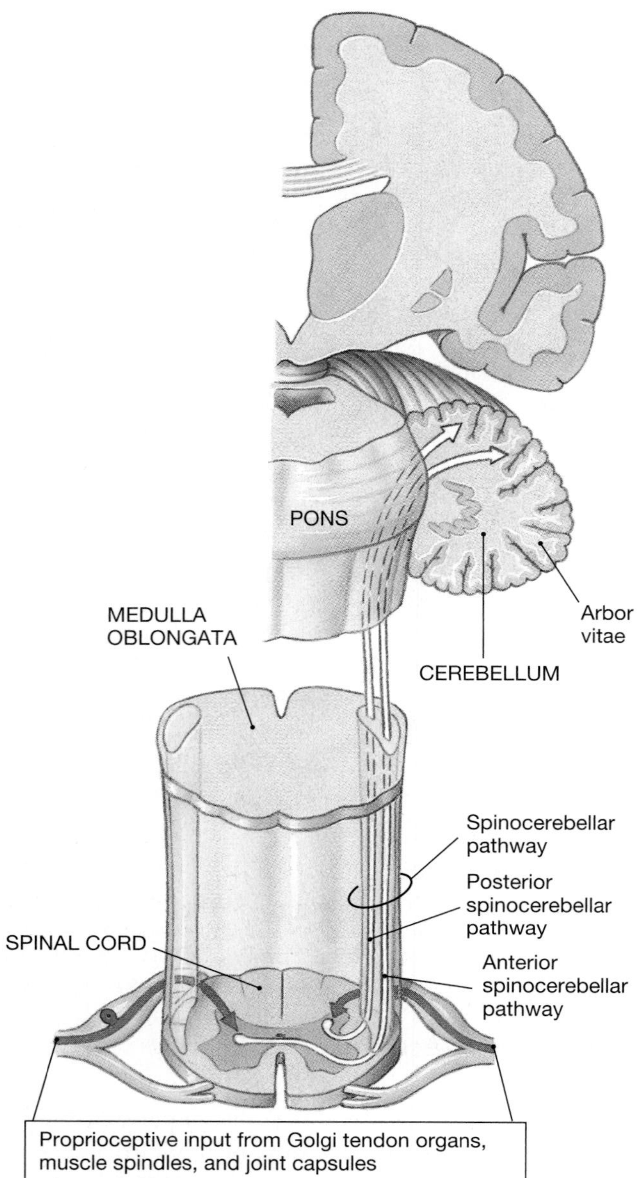
PONS
MEDULLA OBLONGATA
Arbor vitae
CEREBELLUM
Spinocerebellar pathway
Posterior spinocerebellar pathway
SPINAL CORD
Anterior spinocerebellar pathway
Proprioceptive input from Golgi tendon organs, muscle spindles, and joint capsules

● **FIGURE 15-4**
The Spinocerebellar Pathway

cerebellar cortex. These complex, highly branched cells were seen in Figure 14-15●, p. 481. Proprioceptive information from each part of the body is relayed to a specific portion of the cerebellar cortex. The integration of proprioceptive information and the role of the cerebellum in somatic motor control will be considered in a later section.

Table 15-1 reviews the three major sensory pathways discussed in this section.

 As a result of pressure on her spinal cord, Jill cannot feel touch or pressure on her lower limbs. What spinal tract is being compressed?

 What spinal tract carries action potentials generated by nociceptors?

 Which cerebral hemisphere receives impulses conducted by the right fasciculus gracilis?

❑ Motor Pathways

Figure 15-5

Motor commands issued by the CNS are distributed by the somatic nervous system (SNS) and the autonomic nervous system (ANS). The SNS, also called the *somatic motor system*, controls the contractions of skeletal muscles. The output of the somatic nervous system is under voluntary control. The autonomic nervous system, or *visceral motor system*, controls visceral effectors, such as smooth muscle, cardiac muscle, and glands. Chapter 16 will examine the organization of the ANS; our interest here will be the structure of the SNS. Throughout this discussion we will use the terms *motor neuron* and *motor control* to refer specifically to neurons and pathways of the SNS.

Somatic motor pathways always involve at least two motor neurons: an **upper motor neuron,** whose soma lies in a CNS processing center, and a **lower motor neuron** in a nucleus of the brain stem or spinal cord. The upper motor neuron synapses on the lower motor neuron, and the lower motor neuron innervates a single motor unit in a skeletal muscle. Activity in the upper motor neuron may facilitate or inhibit the lower motor neuron; activation of the lower motor neuron triggers a contraction in the innervated muscle. Only the axon of the lower motor neuron extends outside of the CNS. Destruction or damage to a lower motor neuron invariably produces a flaccid paralysis in the innervated motor unit.

Voluntary and involuntary somatic motor commands issued by the brain reach their peripheral targets by traveling over two integrated motor pathways, the *pyramidal system* and the *extrapyramidal system.* These systems were originally thought to be completely separate, with the pyramidal system providing voluntary control and the extrapyramidal system regulating involuntary motor activity. In fact, these systems are so extensively integrated that it is very difficult to distinguish between them on either anatomical or functional grounds. (Examples will be noted as the discussion proceeds.) Figure 15-5● indicates the positions of the major pyramidal and extrapyramidal tracts in the spinal cord.

The Pyramidal System

Figures 15-2, 15-6

The **pyramidal system** (Figure 15-6●) provides voluntary control over skeletal muscles. This system

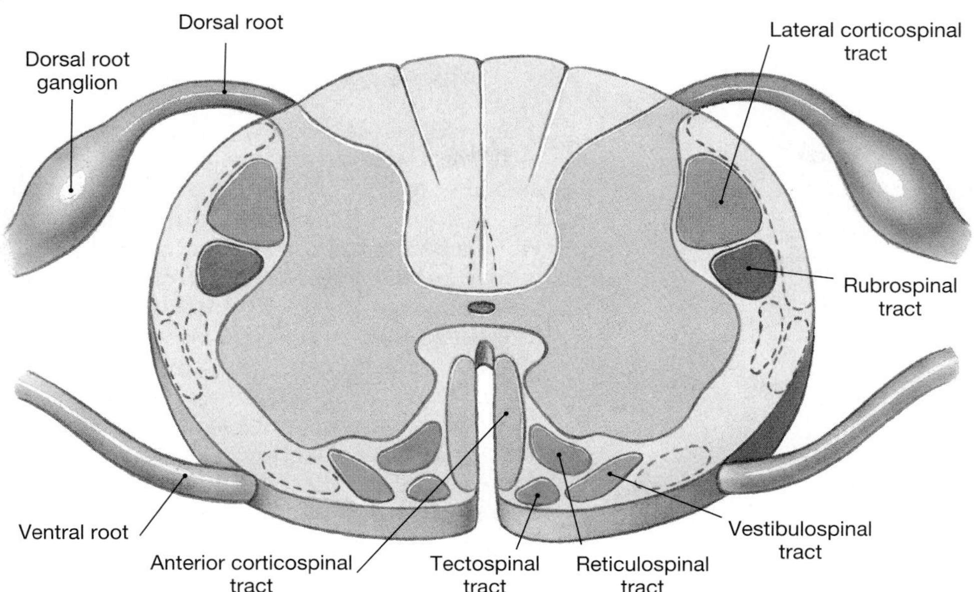
Dorsal root
Dorsal root ganglion
Lateral corticospinal tract
Rubrospinal tract
Ventral root
Anterior corticospinal tract
Tectospinal tract
Reticulospinal tract
Vestibulospinal tract

● **FIGURE 15-5 Descending Motor Pathways in the Spinal Cord.** A cross-sectional view indicating the locations of the major descending motor tracts that contain the axons of upper motor neurons. The origins and destinations of these tracts are detailed in Table 15-3. Sensory tracts, detailed in Figure 15-1, are outlined.

begins at the **pyramidal cells** of the primary motor cortex. These are upper motor neurons whose cell bodies are shaped like miniature pyramids. The axons of pyramidal cells descend into the brain stem and spinal cord to synapse on lower motor neurons.

TRACTS OF THE PYRAMIDAL SYSTEM The pyramidal system contains three pairs of descending tracts: (1) the *corticobulbar tracts,* (2) the *lateral corticospinal tracts,* and (3) the *anterior corticospinal tracts.* These tracts enter the white matter of the internal capsule, descend into the brain stem, and emerge on either side of the mesencephalon as the *cerebral peduncles* (Figures 15-6 and 14-11●, p. 476).

The Corticobulbar Tracts. Axons in the **corticobulbar** (kōr-ti-kō-BUL-bar) **tracts** (*bulbar,* brain stem) synapse on lower motor neurons in the motor nuclei of eight cranial nerves (N III, IV, V, VI, VII, IX, XI, and XII). They provide voluntary control over skeletal muscles that move the eye, jaw, and face and control some muscles of the neck and pharynx as well.

The Corticospinal Tracts. Axons in the **corticospinal tracts** synapse on lower motor neurons in the anterior gray horns of the spinal cord. As they descend, the corticospinal tracts are visible along the ventral surface of the medulla oblongata as a pair of thick bands, the **pyramids.** Along the length of the pyramids roughly 85 percent of the axons cross the midline *(decussate)* to enter the descending **lateral corticospinal tracts** on the opposite side of the spinal cord. The remaining 15 percent continue uncrossed along the spinal cord as the **anterior corticospinal tracts.** At the spinal segment it targets, an axon in the anterior corticospinal tract crosses over to the opposite side of the spinal cord in the anterior white commissure before synapsing on lower motor neurons in the anterior gray horns.

THE MOTOR HOMUNCULUS Activity of pyramidal cells in a specific portion of the primary motor cortex will result in the contraction of peripheral muscles. The identities of the stimulated muscles depend on the region of motor cortex that is active. As in the case of the primary sensory cortex, there is a fine point-to-point correspondence between the primary motor cortex and specific regions of the body. The cortical areas have been mapped out in diagrammatic form, creating a **motor homunculus.** A motor homunculus is seen in Figure 15-6●, which shows the pyramidal system components involved in controlling the right side of the body.

The proportions of the motor homunculus are quite different from those of the actual body, because the motor area devoted to a specific region of the cortex is proportional to the number of motor units involved in its control rather than its actual size. As a result, the homunculus provides an indication of the degree of fine motor control available. For example, the hands, face, and tongue, all of which are capable of varied and complex movements, are very large, whereas the trunk is relatively small. These proportions are similar to those of the sensory homunculus (Figure 15-2●, pp. 504–505). The sensory and motor homunculi differ in other respects because some highly sensitive regions, such as the sole of the foot, contain few motor units, and some areas with an abundance of motor units, such as the eye muscles, are not particularly sensitive.

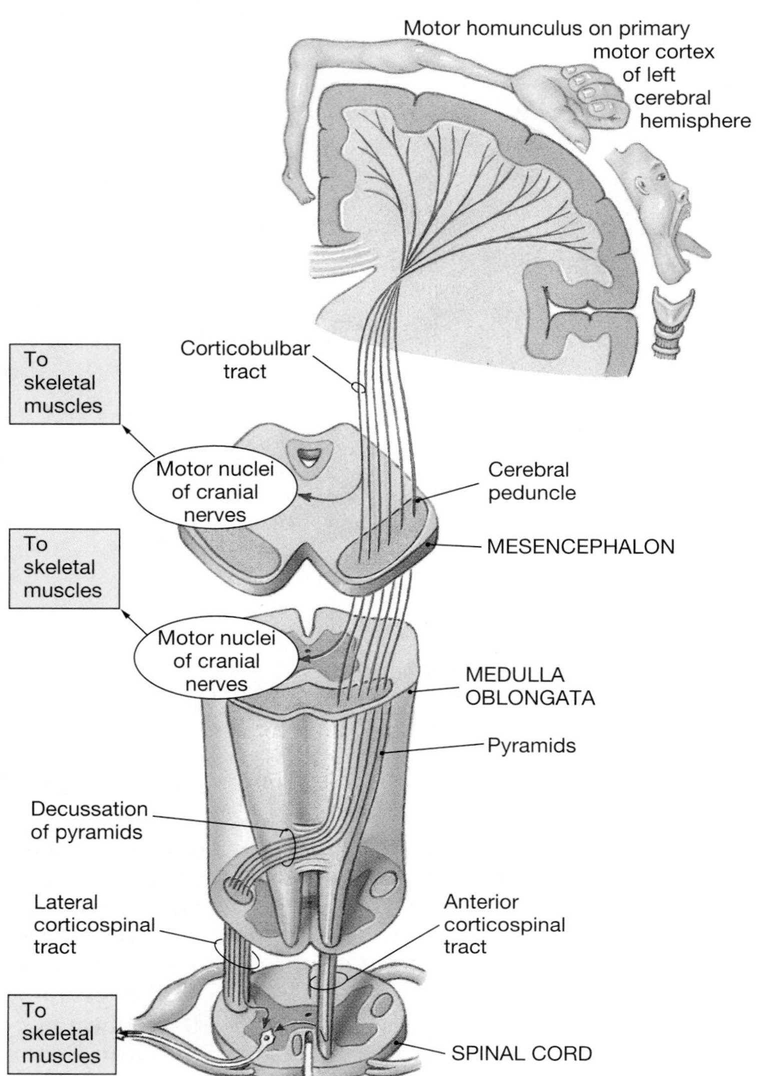
Motor homunculus on primary motor cortex of left cerebral hemisphere
To skeletal muscles
Corticobulbar tract
Motor nuclei of cranial nerves
Cerebral peduncle
To skeletal muscles
MESENCEPHALON
Motor nuclei of cranial nerves
MEDULLA OBLONGATA
Pyramids
Decussation of pyramids
Lateral corticospinal tract
Anterior corticospinal tract
To skeletal muscles
SPINAL CORD

● **FIGURE 15-6**
The Pyramidal System. The pyramidal system originates at the primary motor cortex. The corticobulbar tracts end at the motor nuclei of cranial nerves on the opposite side of brain. Most of the pyramidal fibers cross over in the medulla and enter the lateral corticospinal tracts; the rest descend in the anterior corticospinal tracts and cross over on reaching target segments in the spinal cord.

CEREBRAL PALSY The term **cerebral palsy** refers to a number of disorders affecting voluntary motor performance that appear during infancy or childhood and persist throughout the life of the affected individual. The cause may be trauma associated with premature or unusually stressful birth, maternal exposure to drugs, including alcohol, or a genetic defect that causes improper development of the motor pathways. Problems with labor and delivery result from compression or interruption of placental circulation or oxygen supplies. If the oxygen concentration in the fetal blood declines significantly for as little as 5–10 minutes, CNS function may be permanently impaired. The cerebral cortex, cerebellum, cerebral nuclei, hippocampus, and thalamus are likely targets, producing abnormalities in motor performance, involuntary control of posture and balance, memory, speech, and learning abilities.

The Extrapyramidal System

Figures 15-5, 15-7, 15-8

The axons of the pyramidal cells of the primary motor cortex descend to synapse on lower motor neurons in the brain stem and spinal cord. Because there are no intervening synapses, the pyramidal system provides a rapid and direct mechanism for the control of skeletal muscles. Several other centers may issue somatic motor commands as a result of processing performed at an unconscious, involuntary level. These centers and their associated tracts constitute the **extrapyramidal system (EPS).** The extrapyramidal system can modify or direct skeletal muscle contractions by stimulating, facilitating, or inhibiting lower motor neurons. In addition, feedback from the cerebral nuclei and cerebellum to the primary motor cortex can modify the commands issued by the upper motor neurons of the pyramidal system.

The components of the extrapyramidal system, introduced in Chapter 14, are spread throughout the brain (Table 15-2). The upper motor neurons of the extrapyramidal system are located in the *vestibular nuclei,* the *superior* and *inferior colliculi,* the *red nucleus,* and the *reticular formation.* The *cerebral nuclei* and *cerebellum* form a higher level of control. Their output either (1) stimulates or inhibits the activity of other extrapyramidal nuclei or (2) stimulates or inhibits the activity of pyramidal cells in the primary motor cortex.

The axons of upper motor neurons in the extrapyramidal system synapse on the same lower motor neurons innervated by the pyramidal system. In the brain stem they synapse in the motor nuclei of cranial nerves. Axons controlling lower motor neurons of the spinal cord descend within the *vestibulospinal, tectospinal, rubrospinal,* and *reticulospinal tracts* (Table 15-3, p.512). Figure 15-5● indicates the location of these tracts in the spinal cord.

THE VESTIBULAR NUCLEI AND VESTIBULOSPINAL TRACTS The vestibular nuclei receive information, via the eighth cranial nerve, from receptors in the inner ear that monitor the position and movement of the head. These nuclei respond to changes in the orientation of the head, sending motor commands that alter the muscle tone, extension, and position of the neck, eyes, head, and limbs. The primary goal is the maintenance of posture and balance. The descending fibers within the spinal cord constitute the **vestibulospinal tracts.**

TABLE 15-2 Components of the Extrapyramidal System

Center	*Location*	*Primary Function*
Cerebellar nuclei [p. 482]	Cerebellum	Coordination of movements, integration of sensory feedback
Cerebral nuclei [p. 472]	Cerebrum	Preparation and coordination of limb movements
Vestibular nuclei [p. 482]	Pons and medulla oblongata	Processing of equilibrium sensations
Superior colliculi [p. 479]	Mesencephalon	Processing of visual information
Inferior colliculi [p. 479]	Mesencephalon	Processing of auditory information
Red nucleus [p. 479]	Mesencephalon	Control of skeletal muscle tone
Reticular formation [p. 480]	Diencephalon and brain stem	Interconnected network of neurons and centers with varied functions
Thalamus [p. 476]	Diencephalon	Relay station for somatic sensory information and for feedback from cerebral nuclei and cerebellum

THE COLLICULI AND THE TECTOSPINAL TRACTS The superior and inferior colliculi are found in the roof, or *tectum*, of the mesencephalon (Figure 14-16●, p. 483). The colliculi receive visual (superior) and auditory (inferior) sensations. Axons of upper motor neurons in the superior and inferior colliculi descend in the **tectospinal tracts.** These axons cross to the opposite side immediately, before descending to synapse on lower motor neurons in the brain stem or spinal cord. The axons within the tectospinal tracts direct reflexive changes in the position of the head, neck, and upper limbs in response to bright lights, sudden movements, or loud noises.

THE RED NUCLEI AND THE RUBROSPINAL TRACTS The red nuclei of the mesencephalon control the level of muscle tone in the body's skeletal muscles. Axons of upper motor neurons in the red nuclei cross to the opposite side of the brain and descend into the spinal cord in the **rubrospinal tracts** (*ruber*, red). Motor neurons in the red nuclei may be stimulated or inhibited by input from the primary motor cortex, cerebral nuclei, cerebellum, or reticular formation. Stimulation increases muscle tone, and inhibition decreases it.

THE RETICULAR FORMATION AND THE RETICULOSPINAL TRACTS The reticular formation is a loosely organized network of neurons that extends throughout the brain stem. The reticular formation receives collateral input from almost every ascending and descending pathway. It also has extensive interconnections with the cerebrum, the cerebellum, and brain stem nuclei. Axons of upper motor neurons in the reticular formation descend into the **reticulospinal tracts** without crossing to the opposite side. The effects of reticular formation stimulation vary depending on the region stimulated. For example, stimulation of upper motor neurons in one portion of the reticular formation produces involuntary eye movements, whereas stimulation of another region activates respiratory muscles.

THE CEREBRAL NUCLEI The cerebral nuclei blur the distinctions between "conscious" and "unconscious" motor control. In effect, they provide the background patterns of movement involved in the performance of voluntary motor activities. These nuclei do not exert direct control over lower motor neurons; instead, they adjust the activities of upper motor neurons in both the extrapyramidal and pyramidal systems.

The cerebral nuclei use three major pathways to accomplish this adjustment.

1. One group of axons synapses with thalamic neurons, which then send their axons to the primary motor cortex. This arrangement creates a feedback loop that changes the sensitivity of the pyramidal cells and alters the pattern of instructions carried by the corticospinal tracts.
2. A second group of axons innervates the red nucleus and alters the activity in the rubrospinal tracts, thereby changing muscle tone and the degree of resistance to specific movements.

TABLE 15-3 Principal Descending (Motor) Tracts in the Spinal Cord, and a Summary of the Functional Roles of the Associated Nuclei in the Brain

Tract	*Location of Upper Motor Neurons*	*Destination*	*Site of Cross-Over*	*Actions*
PYRAMIDAL TRACTS				
Corticobulbar tracts	Primary motor cortex (cerebral hemispheres)	Lower motor neurons of cranial nerve nuclei in brain stem	Brain stem	Voluntary motor control of skeletal muscles
Lateral corticospinal tract	As above	Lower motor neurons of anterior gray horns of spinal cord	Pyramids of medulla oblongata	As above
Anterior corticospinal tract	As above	As above	At level of lower motor neuron	As above
EXTRAPYRAMIDAL TRACTS				
Rubrospinal tracts	Red nuclei of mesencephalon	Lower motor neurons of anterior gray horns	Brain stem (mesencephalon)	Involuntary regulation of posture and muscle tone
Reticulospinal tracts	Reticular formation (network of nuclei in brain stem)	As above	Uncrossed	Involuntary regulation of reflex activity
Vestibulospinal tracts	Vestibular nuclei (at border of pons and medulla oblongata)	As above	As above	Involuntary regulation of balance and muscle tone
Tectospinal tracts	Tectum (mesencephalon: superior and inferior colliculi)	Lower motor neurons of anterior gray horns of cervical spinal cord	Brain stem (mesencephalon)	Involuntary regulation of eye, head, neck and upper limb position in response to visual and auditory stimuli

3. A third group of axons synapses in the reticular formation, altering the output of the reticulospinal tracts.

Two distinct populations of neurons exist: one that stimulates motor neurons by releasing acetylcholine (ACh) and another that inhibits motor neurons by the release of gamma-aminobutyric acid (GABA). Under normal conditions the excitatory interneurons are kept inactive, and the tracts leaving the cerebral nuclei are normally responsible for inhibiting motor neuron activity. In *Parkinson's disease* (Chapter 14), the excitatory neurons become more active, leading to problems with the control of voluntary movements. ∞ [p. 480]

If the primary motor cortex is damaged, the individual will lose the ability to exert fine control over peripheral muscles. However, some voluntary movements could still occur under the direction of the cerebral nuclei. In effect, the extrapyramidal system functions as always, but the pyramidal system is not there to fine-tune the movements under way. For example, after damage to the primary motor cortex, the cerebral nuclei could still receive information about planned movements from the prefrontal cortex and perform preparatory movements of the trunk and limbs. But because the pyramidal system remains silent, precise movements of the forearms, wrists, and hands could not occur. An individual in this condition could stand, maintain balance, and even walk, but all movements would be hesitant, awkward, and poorly controlled.

THE CEREBELLUM The cerebellum integrates proprioceptive sensations, visual information from the eyes, and equilibrium-related sensations from the inner ear with the pattern of motor commands from higher centers. Axons relaying proprioceptive information reach the cerebellar cortex via the spinocerebellar tracts (p. 507). Visual information is relayed from the superior colliculi, and balance information from the vestibular nuclei. The output of the cerebellum affects upper motor neuron activity in both the extrapyramidal and pyramidal systems.

Both pyramidal and extrapyramidal centers send information to the cerebellum when motor commands are issued. As the movement proceeds, the cerebellum monitors proprioceptive (position) and vestibular (balance) information and compares the arriving sensations with the planned motor pattern. It then issues commands that adjust the activities of the voluntary and involuntary motor centers. In general, any voluntary movement begins with the activation of far more motor units than are actually required or even desirable. The cerebellum provides the necessary inhibition, reducing the number of motor commands to an efficient minimum. As the movement proceeds, the pattern and degree of inhibition change, producing the desired result.

The appropriate patterns of cerebellar activity are learned by trial and error over many repetitions. Many of the basic patterns are established early in life; examples would include the fine balancing adjustments made while standing and walking. Motor patterns involved in extremely complex behaviors require effort and practice until the movements become fluid and automatic. For example, consider the relaxed, smooth movements of acrobats, dancers, golfers, tennis players, and sushi chefs. These people perform the movements without consciously thinking about the details. This is important, because when concentrating on voluntary control, the rhythm and pattern of the movement usually fall apart because the pyramidal system begins overriding the commands of the cerebral nuclei and cerebellum.

Figure 15-7• provides an overview of the relationships between the components of the extrapyramidal system and between the extrapyramidal and pyramidal systems. The two systems control the same lower motor neurons. The individual motor nuclei of the extrapyramidal system are components of reflex arcs that produce motor responses in response to specific stimuli. The cerebellum and cerebral nuclei exert control over the upper motor neurons of this system. They also modify the output of the pyramidal system. Each time voluntary movement is under way, information is relayed to the cerebral nuclei and cerebellum. The output of these centers goes back to the cerebral cortex via the thalamus.

There are therefore two feedback loops that adjust the activity of the pyramidal system. The cerebellar loop provides precise control of complex motor skills (golfing, skating, and so forth). The loop involving the cerebral nuclei controls learned patterns of movement and the gross position of the trunk and limbs during voluntary movements.

AMYOTROPHIC LATERAL SCLEROSIS **Amyotrophic lateral sclerosis** (ALS) is a progressive, degenerative disorder that affects motor neurons in the spinal cord, brain stem, and cerebral hemispheres. Most people with ALS do not develop symptoms until after age 50, although there are inherited forms that affect individuals in their early teens. Affected individuals usually die within 5 years after symptoms appear.

The degeneration affects both upper and lower motor neurons of the pyramidal and extrapyramidal systems. Because of the intimate relationship between a motor neuron and its dependent muscle fibers, the destruction of CNS neurons causes atrophy of the associated skeletal muscles. The specificity of this disease is remarkable: Motor neurons are destroyed, but sensory neurons, sensory pathways, and intellectual functions remain unaffected. Roughly 90 percent of ALS cases have no known cause; 10 percent of cases can be linked to genetic factors, but the nature of the genetic abnormality is uncertain. There is no effective treatment. AM *Amyotrophic Lateral Sclerosis*

For what anatomical reason does the left side of the brain control motor function on the right side of the body?

An injury involving the superior portion of the motor cortex would affect what region of the body?

What effect would increased stimulation of the motor neurons of the red nucleus have on muscle tone?

Levels of Processing and Motor Control

Figure 15-8

When someone touches a hot stove, the hand is withdrawn before the individual is consciously aware of the injury. Voluntary motor responses, such as shaking the hand, stepping back, and crying out, occur somewhat later. This is a general pattern: Spinal and cranial reflexes provide rapid, automatic, preprogrammed responses that preserve homeostasis over the short term. Voluntary responses are more complex, but they require more time to prepare and execute.

Figure 15-8• reviews the primary sites of somatic motor control. Cranial and spinal reflexes control the most basic motor activities. Integrative centers in the brain perform more elaborate processing, and as one moves from the medulla oblongata to the cerebral cortex the motor patterns become increasingly complex and variable. The most complex and variable motor activities are directed by the primary motor cortex of the cerebral hemispheres.

During development, the spinal and cranial reflexes are the first to appear. More complex reflexes and motor patterns develop as CNS neurons multiply, enlarge, and interconnect. The process proceeds relatively slowly, as billions of neurons establish trillions of synaptic connections. At birth neither the cerebral nor the cerebellar cortex is fully functional, and the behavior of newborn infants is directed primarily by centers in the diencephalon and brain stem.

A number of anatomical factors contribute to the maturation of the CNS and the refinement of motor skills:

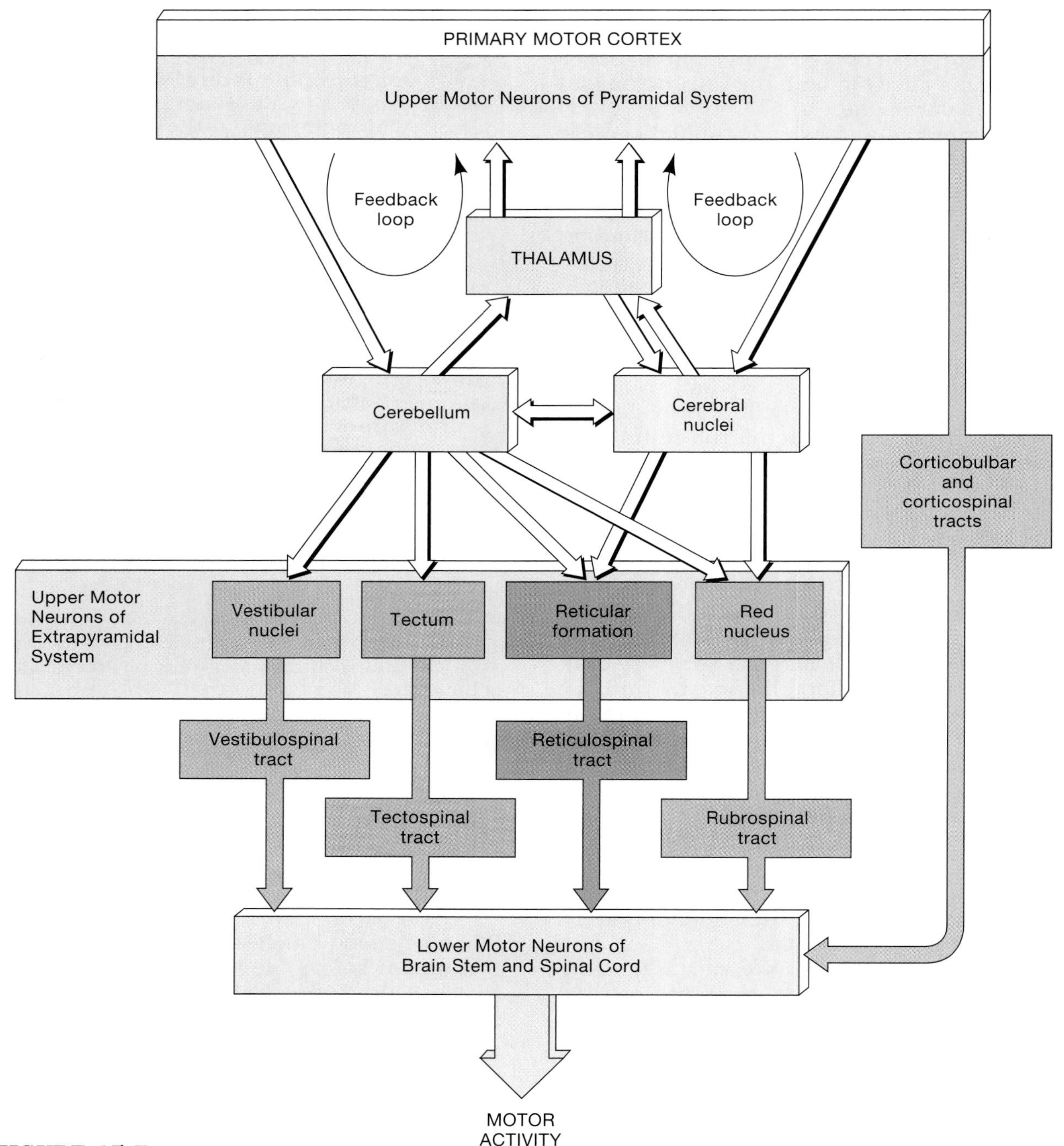
PRIMARY MOTOR CORTEX
Upper Motor Neurons of Pyramidal System
Feedback loop
Feedback loop
THALAMUS
Cerebellum
Cerebral nuclei
Corticobulbar and corticospinal tracts
Upper Motor Neurons of Extrapyramidal System
Vestibular nuclei
Tectum
Reticular formation
Red nucleus
Vestibulospinal tract
Reticulospinal tract
Tectospinal tract
Rubrospinal tract
Lower Motor Neurons of Brain Stem and Spinal Cord
MOTOR ACTIVITY

● **FIGURE 15-7**

The Extrapyramidal System.
This diagram indicates important interconnections and tracts of the pyramidal and extrapyramidal systems.

● **FIGURE 15-8**

Centers of Somatic Motor Control. Somatic motor control involves many portions of the brain; as one travels toward the cerebral cortex the motor patterns generated become increasingly complex and variable.

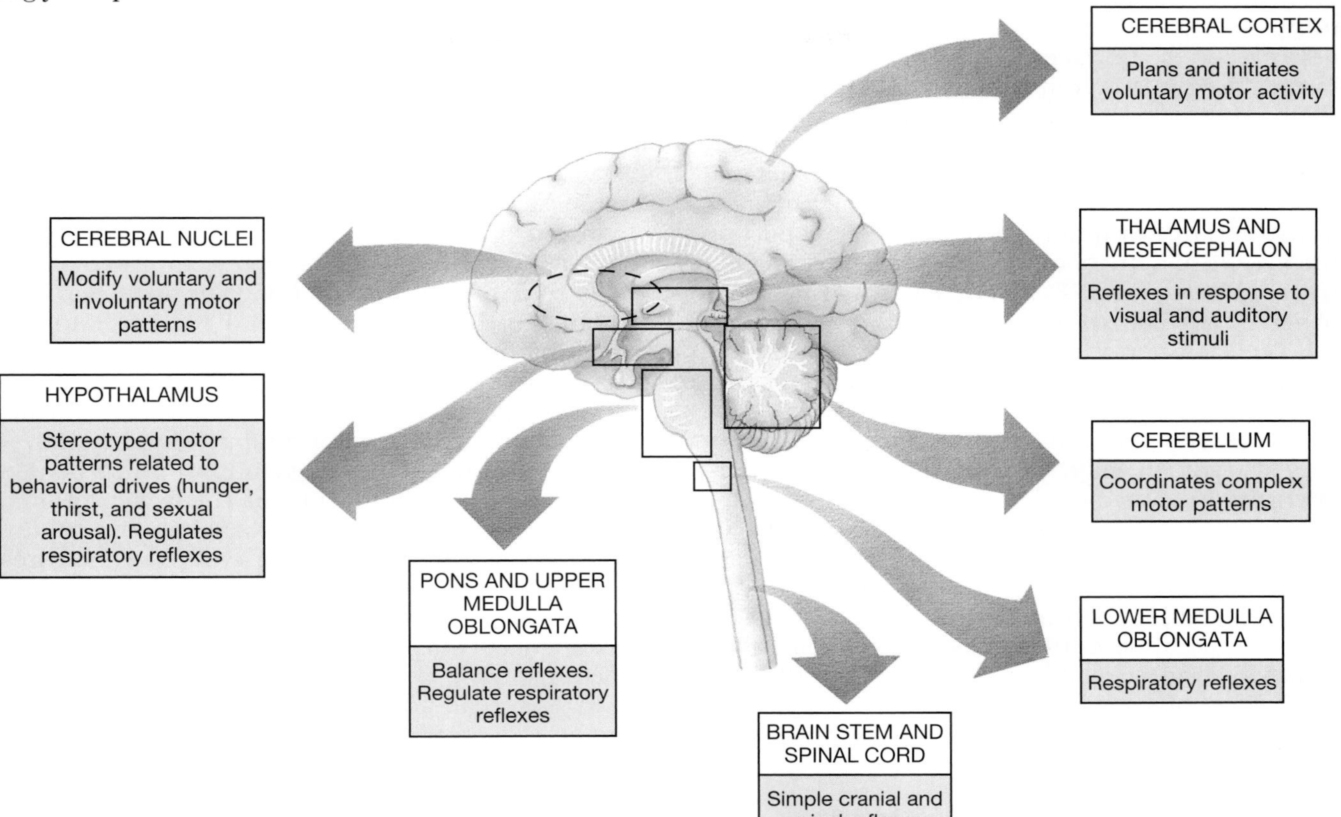
CEREBRAL CORTEX
Plans and initiates voluntary motor activity
CEREBRAL NUCLEI
Modify voluntary and involuntary motor patterns
THALAMUS AND MESENCEPHALON
Reflexes in response to visual and auditory stimuli
HYPOTHALAMUS
Stereotyped motor patterns related to behavioral drives (hunger, thirst, and sexual arousal). Regulates respiratory reflexes
CEREBELLUM
Coordinates complex motor patterns
PONS AND UPPER MEDULLA OBLONGATA
Balance reflexes. Regulate respiratory reflexes
LOWER MEDULLA OBLONGATA
Respiratory reflexes
BRAIN STEM AND SPINAL CORD
Simple cranial and spinal reflexes

1. The number of neurons in the cerebral cortex continues to increase until at least age 1.
2. The brain as a whole grows in size and complexity until at least age 4.
3. Myelination of CNS axons continues at least until age 1–2, and peripheral myelination may continue through puberty. Myelination reduces the delay between stimulus and response and improves motor control.

As these events occur, cortical neurons continue to establish new synaptic interconnections that will have a long-term effect on the functional capabilities of the individual.

ANENCEPHALY Although it may sound strange, physicians usually take a newborn infant into a dark room and shine a light against the skull. They are checking for **anencephaly** (an-en-SEF-uh-lē), a rare condition in which the brain fails to develop at levels above the mesencephalon or lower diencephalon. Usually the cranium also fails to develop, and diagnosis is easy, but in some cases, a normal skull forms. In such instances the cranial vault is empty and translucent enough to transmit light. Unless the condition is discovered right away, the parents may take the infant home, totally unaware of the problem. All the normal behavior patterns expected of a newborn are present, including suckling, stretching, yawning, crying, kicking, sticking fingers in the mouth, and tracking movements with the eyes. However, death will occur naturally over a period of days to months. This tragic condition provides a striking demonstration of the role of the brain stem in controlling complex involuntary motor patterns. During normal development these patterns become incorporated into variable and versatile behaviors as control and analytical centers appear in the cerebral cortex.

Monitoring Brain Activity: The Electroencephalogram

Figure 15-9

The primary sensory cortex and the primary motor cortex have been mapped by direct stimulation in patients undergoing brain surgery. The functions of other regions of the cerebrum can be inferred from behavioral changes that follow localized injuries or strokes, and the activities of specific regions can be examined using a PET scan or sequential MRI scans. AM *Scans of the Brain*

The electrical activity of the brain is often monitored to assess brain activity. Neural function depends on electrical events within the cell membrane. The brain contains billions of neurons, and their activity generates an electrical field that can be measured by placing electrodes on the brain or the outer surface of the skull. The electrical activity changes constantly, as nuclei and cortical areas are stimulated or quiet down. A printed report on the electrical activity of the brain is called an *electroencephalogram* (EEG). The electrical patterns observed are called *brain waves.*

Typical brain waves are shown in Figure 15-9●. **Alpha waves** are found in the brains of normal resting but awake adults when their eyes are closed. Alpha waves disappear during sleep, but they also vanish when the individual begins to concentrate on some specific task. During attention to stimuli or tasks, the alpha waves are replaced by higher-frequency **beta waves.** Beta waves are typical of individuals who are either concentrating on a task, under stress, or in a state of psychological tension. **Theta waves** may appear transiently during sleep in normal adults, but they are most often observed in children and in intensely frustrated adults. Their presence under other circumstances may indicate the presence of a brain disorder, such as a tumor. **Delta waves** are very large amplitude, low-frequency waves. They are normally seen during deep sleep in individuals of all ages. Delta waves are also seen in the brains of infants (in whom cortical development is still incomplete) and in the awake adult when a tumor, vascular blockage, or inflammation has damaged portions of the brain.

Electrical activity in the two hemispheres is usually synchronized; the thalamus appears to be involved in the "pacemaker" mechanism. Asynchrony between the hemispheres can therefore be used to detect localized damage or other cerebral abnormalities. For example, a tumor or injury affecting one hemisphere often changes the pattern in that hemisphere, and the two are no longer aligned. A **seizure** is a temporary cerebral disorder accompanied by abnormal involuntary movements, unusual sensations, inappropriate behavior, or some combination of these symptoms. Clinical conditions characterized by seizures are known as seizure disorders, or *epilepsies.* Seizures of all kinds are accompanied by a marked change in the pattern of electrical activity monitored in an electroencephalogram. The alteration begins in one portion of the cerebral cortex but may subsequently spread across the entire cortical surface, like a wave on the surface of a pond.

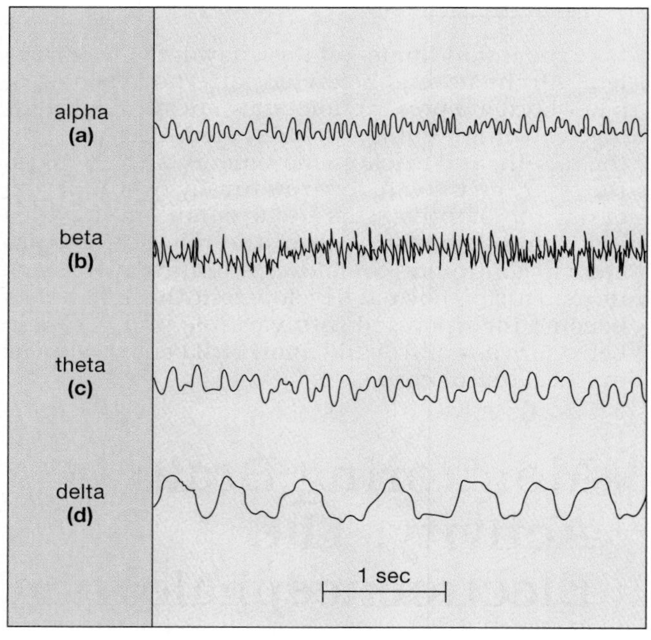
alpha
(a)
beta
(b)
theta
(c)
delta
(d)
1 sec

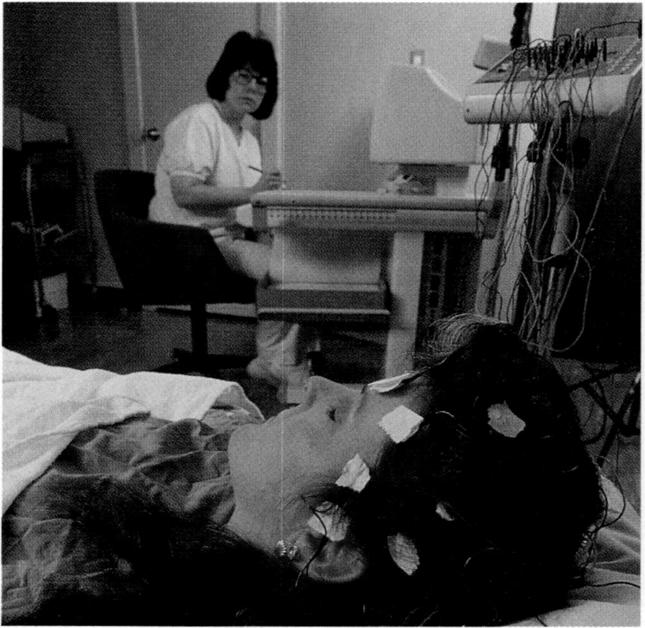

(e)

● **FIGURE 15-9**

Brain Waves. (a) Alpha waves are characteristic of normal resting adults. **(b)** Beta waves typically accompany intense concentration. **(c)** Theta waves are seen in children and in frustrated adults. **(d)** Delta waves occur during deep sleep and in certain pathological states. (Note: These wave patterns are not drawn to the same scale.) **(e)** A patient wired for EEG monitoring.

The nature of the symptoms produced depends on the region of the cortex involved. If the seizure affects the primary motor cortex, involuntary movements will occur; if it affects the auditory cortex, the individual will hear strange sounds. AM *Seizures and Epilepsy*

Higher-Order Functions

Higher-order functions share the following characteristics:

1. The cerebral cortex is required for their performance, and they involve complex interactions between areas of the cortex and between the cerebral cortex and other areas of the brain.
2. They involve both conscious and unconscious information processing.
3. They are not part of the programmed "wiring" of the brain; therefore, the functions are subject to modification and adjustment over time.

Chapter 14 considered functional areas of the cerebral cortex and the regional specializations of the left and right cerebral hemispheres. *[pp. 468–472]* In this section we consider the mechanisms of memory and learning and detail the neural interactions responsible for consciousness, sleep, and arousal. In the next section we provide an overview of brain chemistry and its effects on behavior and personality.

Memory

Figure 15-10

What was the topic of the last sentence you read? What do your parents look like? What is your Social Security number? When did Columbus discover America? What does a red traffic light mean? What does a hot dog taste like? Answering these questions involves accessing *memories,* stored bits of information gathered through prior experience. **Fact memories** are specific bits of information, such as the color of a stop sign or the smell of perfume. **Skill memories** are learned motor behaviors. For example, you can probably remember how to light a match or open a screw-top jar. With repetition, skill memories become incorporated at the unconscious level. Examples would include the complex motor patterns involved in skiing, playing the violin, and similar activities. Skill memories related to programmed behaviors, such as eating, are stored in appropriate portions of the brain stem. Complex skill memories involve integration of motor patterns in the cerebral nuclei, cerebral cortex, and cerebellum.

Two classes of memories appear to exist. **Short-term memories,** or *primary memories,* do not last long, but while they persist the information can be recalled immediately. Primary memories contain small bits of information, such as a person's name or a telephone number. Repeating a phone number or other bit of information reinforces the original short-term memory and helps ensure its conversion to a long-term memory. That conversion is called **memory consolidation. Long-term memories** remain for much longer periods, in some cases for an entire lifetime. There are two types of long-term memories. *Secondary memories* are long-term memories that fade with time and may require considerable effort to recall. *Tertiary memories* are long-term memories that seem to be part of consciousness, such as your name or the contours of your own body. Proposed relationships between these memory classes are diagrammed in Figure 15-10•.

Brain Regions Involved in Memory Consolidation and Access

Two components of the limbic system, the amygdaloid body and the hippocampus (Figure 14-10•, p. 475), are essential to memory consolidation. Damage to the hippocampus leads to an inability to convert short-term memories to new long-term memories, although existing long-term memories remain intact and accessible. Tracts leading from the amygdaloid body to the hypothalamus may link memories to specific emotions.

A cerebral nucleus near the diencephalon, the **nucleus basalis,** plays an uncertain role in memory storage and retrieval. Tracts connect this nucleus with the hippocampus, amygdaloid body, and all areas of the cerebral cortex. Damage to this nucleus is associated with changes in emotional states, memory, and intellectual function (see the discussion of Alzheimer's disease later in this chapter).

Most long-term memories are stored in the cerebral cortex. Conscious motor and sensory memories are referred to the appropriate association areas. For example, visual memories are stored in the visual association area, and memories of voluntary motor activity in the premotor cortex. Special portions of the occipital and temporal lobes are crucial to the memories of faces, voices, and words. In at least some cases, a specific memory probably depends on the activity of a single neuron. For example, in one portion of the temporal lobe an individual neuron responds to the sound of one word and ignores others. A specific neuron may also be activated by the proper combination of sensory stimuli associated with

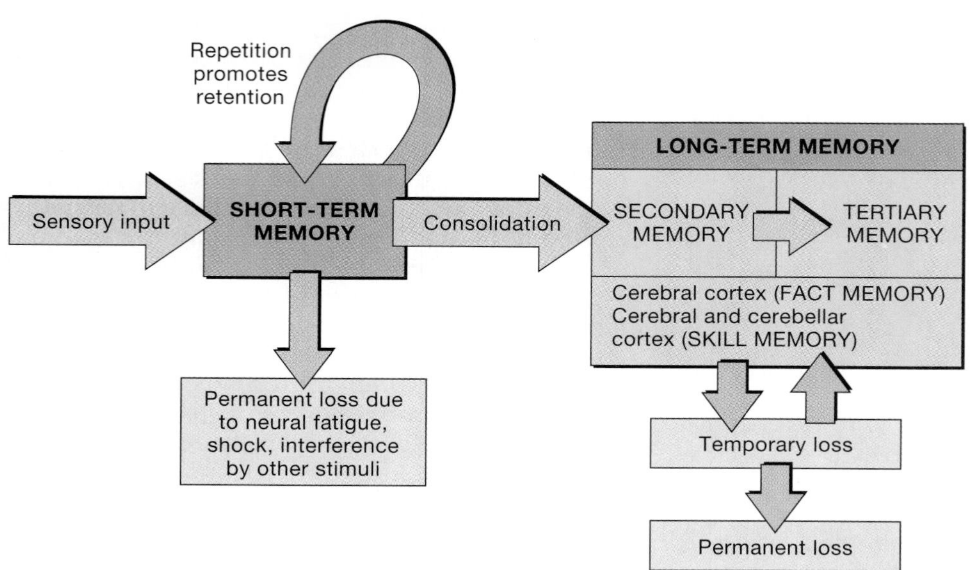
Repetition promotes retention
Sensory input
SHORT-TERM MEMORY
Consolidation
LONG-TERM MEMORY
SECONDARY MEMORY
TERTIARY MEMORY
Cerebral cortex (FACT MEMORY)
Cerebral and cerebellar cortex (SKILL MEMORY)
Permanent loss due to neural fatigue, shock, interference by other stimuli
Temporary loss
Permanent loss

● **FIGURE 15-10**
Memory Storage. Steps in the storage of memories and the conversion from short-term to long-term memories.

a particular individual, such as your grandmother. As a result these neurons are called "grandmother cells."

Information on one subject is parceled out to many different regions of the brain. Your memories of a cow are stored in the visual association area (what a cow looks like, how the letters C-O-W mean "cow"), the auditory association area (the "moo" sound and how the word *cow* sounds), the speech center (how to say the word *cow*), and the frontal lobes (how big cows are, what they eat). Related information, such as how you feel about cows and what milk tastes like, is stored in other locations. If one of these storage areas is damaged, the memory will be incomplete in some way. How these memories are accessed and assembled on demand remains a mystery.

Cellular Mechanisms of Memory Formation and Storage

Memory consolidation at the cellular level involves anatomical and physiological changes in neurons and synapses. For legal, ethical, and logistical reasons, it is not possible to perform much research on these mechanisms with human subjects. Research on other animals, often those with relatively simple nervous systems, has indicated that the following mechanisms may be involved:

1. ***Increased neurotransmitter release.*** A synapse that is frequently active increases the amount of neurotransmitter that it stores, and it releases more on each stimulation. The greater the amount of neurotransmitter released, the greater the effect on the postsynaptic neuron.
2. ***Facilitation at synapses.*** When a neural circuit is repeatedly activated, the synaptic terminals begin releasing neurotransmitter in small quantities on a continual basis. The neurotransmitter binds to receptors on the postsynaptic membrane, producing a graded depolarization that brings the membrane closer to threshold. The facilitation that results affects all the neurons in the circuit.
3. ***Formation of additional synaptic connections.*** There is evidence that when one neuron repeatedly communicates with another, the axon tip branches and forms additional synapses on the postsynaptic neuron. As a result, the presynaptic neuron will have a greater effect on the transmembrane potential of the postsynaptic neuron.

The facilitated circuit that results from these processes represents a **memory engram.** Efficient conversion of a short-term memory into a memory engram takes time, usually at least an hour and often longer. Whether or not that conversion occurs depends on several factors, including the nature, intensity, and frequency of the original stimulus. Very strong, repeated, or exceedingly pleasant (or unpleasant) events are excellent candidates for conversion to long-term memories. Drugs that stimulate the CNS, such as caffeine and nicotine, may enhance memory consolidation through facilitation; membrane effects of these drugs were discussed in Chapter 12. ∞ [p. 412]

Amnesia refers to the loss of memory from disease or trauma. The type of memory loss depends on the specific regions of the brain affected. For example, damage to the auditory association areas may make it difficult to remember sounds. Damage to thalamic and limbic structures, especially the hippocampus, will affect new memory storage and consolidation. AM *Amnesia*

Toward the end of her A & P test, Tina begins to feel a great deal of stress. What type of brain wave pattern would you expect her to exhibit?

After suffering a head injury in an automobile accident, David has difficulty comprehending what he hears or reads. This symptom might indicate damage to what portion of the brain?

As you recall facts while taking your A & P test, what type of memory are you using?

Consciousness

A conscious individual is alert and attentive; an unconscious individual is not. The difference is obvious, but there are many gradations of both the conscious and unconscious states. For example, a normal conscious person can be almost asleep, wide awake, or "high-strung" and jumpy; a sleeping person can be dozing lightly or so deeply asleep that he or she cannot easily be awakened. Normal individuals cycle between the alert, conscious state and the asleep state each day. The degree of wakefulness at any given moment is an indication of the level of ongoing CNS activity. When CNS function becomes abnormal or depressed, the state of wakefulness can be affected. As a result, clinicians are quick to note changes in the responsiveness of their patients. AM *Altered States*

Sleep

Figure 15-11

Conscious implies a state of awareness of and attention to external events and stimuli. *Unconscious* can imply a number of different conditions, ranging from the deep, unresponsive state induced by anesthesia before major surgery to the light drifting "nod" that occasionally plagues students reading anatomy and physiology textbooks. An individual is considered to be asleep when he or she is unconscious but can still be awakened by normal sensory stimuli.

There are several different levels of sleep, each typified by characteristic patterns of brain wave activity (Figure 15-11●).

- In **deep sleep,** also called *slow-wave* or *non-REM sleep,* the entire body relaxes, and activity at the cerebral cortex is at a minimum. Heart rate, blood pressure, respiratory rate, and energy utilization decline by up to 30 percent.
- During **rapid eye movement (REM) sleep,** active dreaming occurs, accompanied by alterations in blood pressure and respiratory rates. Although the EEG shows traces resembling those of the awake state, the individual becomes even less receptive to outside stimuli than in deep sleep, and muscle tone decreases markedly. Intense inhibition of somatic motor neurons probably prevents us from physically producing the responses we envision during the dream sequence. The oculomotor neurons escape this inhibitory influence, and the eyes move rapidly as the imaginary events unfold.

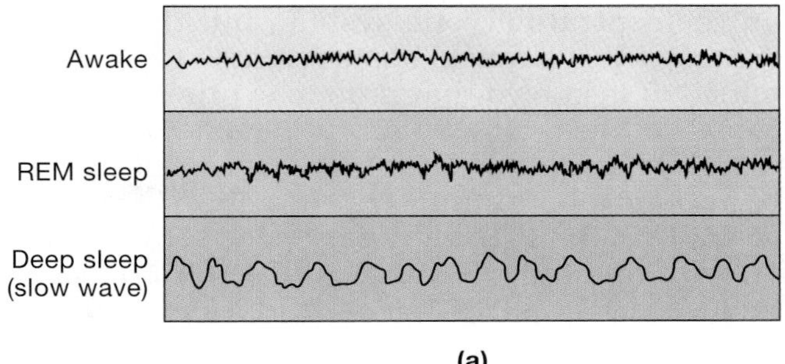

Awake
REM sleep
Deep sleep (slow wave)
(a)

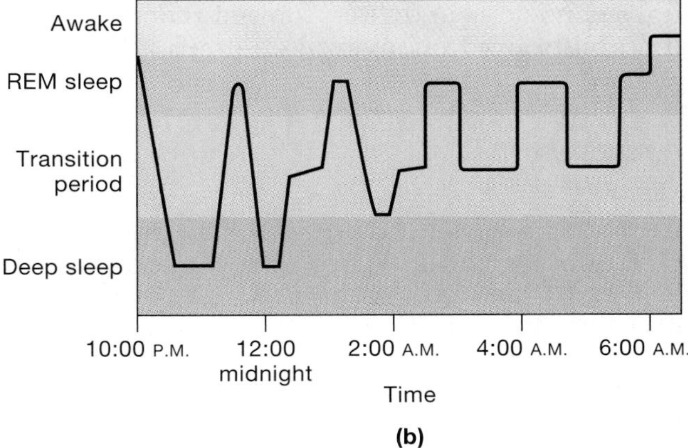

Awake
REM sleep
Transition period
Deep sleep
10:00 P.M.
12:00 midnight
2:00 A.M.
4:00 A.M.
6:00 A.M.
Time
(b)

● **FIGURE 15-11**
Levels of Sleep. (a) EEG from the awake, REM, and slow wave (deep) sleep states. The EEG pattern during REM sleep resembles the alpha waves typical of awake adults. **(b)** Typical pattern of oscillation between sleep stages during a single night's sleep for a healthy young adult.

Periods of REM and deep sleep alternate throughout the night, beginning with a period of deep sleep that lasts around an hour and a half. REM periods initially average about 5 minutes in length, but they gradually increase to about 20 minutes over an 8-hour night. Although each night we probably spend less than 2 hours dreaming, there are many sources of variation. For example, children devote more time to REM sleep than do adults, and extremely tired individuals have very short and infrequent REM periods.

Sleep produces only minor alterations in the physiological activities of other organs and systems, and none that appear to be essential to normal function. Its significance must lie in its impact on the central nervous system, but the physiological or biochemical basis remains to be determined. It is known that protein synthesis in neurons increases during sleep. Extended periods without sleep will lead to a variety of disturbances in mental function. Roughly 25 percent of the U.S. population experiences some form of a *sleep disorder.* Examples of such disorders include abnormal patterns or duration of REM sleep, or unusual behaviors, such as "sleep-walking," performed while sleeping. In some cases these problems begin to affect the individual's conscious activities; slowed reaction times, irritability, and behavioral changes may result. AM *Sleep Disorders*

Arousal

Figure 15-12

Arousal appears to be one of the functions of the reticular formation. The reticular formation is especially well suited for providing "watchdog" services, for it has extensive interconnections with the sensory, motor, and integrative nuclei and pathways all along the brain stem.

THE RETICULAR ACTIVATING SYSTEM The state of consciousness experienced by an individual is determined by complex interactions between the brain stem and the cerebral cortex. One of the most important brain stem components is a diffuse network in the reticular formation known as the **reticular activating system (RAS).** This network, diagrammed in Figure 15-12●, extends from the hypothalamus to the medulla oblongata. The output of the RAS projects to thalamic nuclei that influence large areas of the cerebral cortex. When the RAS is inactive, so is the cerebral cortex; stimulation of the RAS produces a widespread activation of the cerebral cortex.

The mesencephalic portion of the RAS appears to be the center of the system, and stimulation of this area produces the most pronounced and long-lasting effects on the cerebral cortex. Stimulating other portions of the RAS seems to have an effect only to the degree that it changes the activity of the mesencephalic region. The greater the stimulation to the mesencephalic region of the RAS, the more alert and attentive the individual will be to incoming sensory information. The thalamic nuclei associated with the RAS may also play an important role in focusing attention on specific mental processes.

Sleep may be ended by any stimulus sufficient to activate the reticular formation and RAS. Arousal occurs rapidly, but the effects of a single stimulation of the RAS last less than a minute. Thereafter consciousness can be maintained by positive feedback, since activity in the cerebral cortex, cerebral nuclei, and sensory and motor pathways will continue to stimulate the RAS.

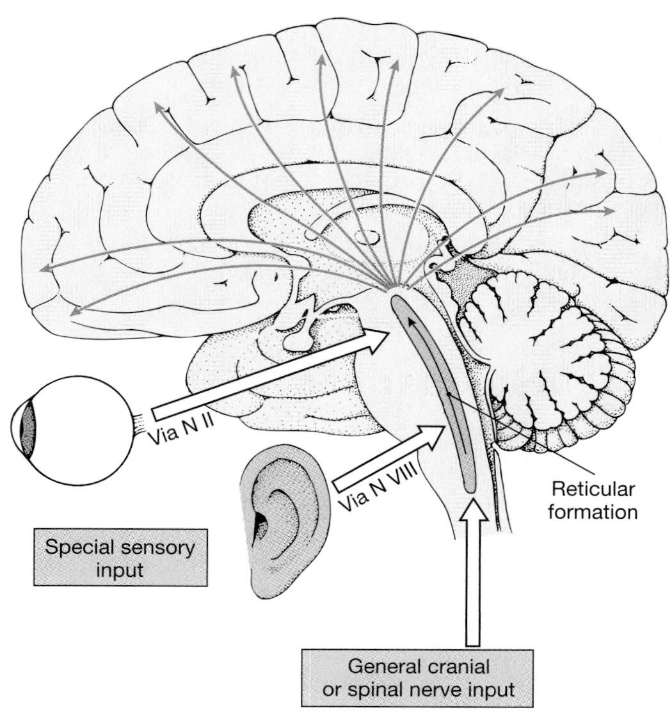
Via N II
Via N VIII
Reticular formation
Special sensory input
General cranial or spinal nerve input

● **FIGURE 15-12**
The Reticular Activating System. The mesencephalic headquarters of the reticular formation receives collateral inputs from a variety of sensory pathways. Stimulation of this region produces arousal and heightened states of attentiveness.

After many hours of activity, the reticular formation becomes less responsive to stimulation. The individual becomes less alert and more lethargic. The precise mechanism remains unknown, but neural fatigue probably plays a relatively minor role in the reduction of RAS activity. Evidence suggests that the regulation of awake/asleep cycles involves an interplay between brain stem nuclei using different neurotransmitters. One group of nuclei stimulates the RAS with norepinephrine and maintains the awake, alert state. The other group, which depresses RAS activity with serotonin, promotes deep sleep. These "dueling nuclei" are located in the brain stem.

Brain Chemistry and Behavior

The general distribution of neurotransmitters in the central nervous system was discussed in Chapter 12. ∞ *[p. 411]* Of the roughly 50 known neurotransmitters, the most important and widespread are acetylcholine (ACh), norepinephrine (NE), glutamate, aspartate, glycine, gamma aminobutyric acid (GABA), dopamine, serotonin, histamine, substance P, and the enkephalins and endorphins (see Table 12-5, p. 411). The distribution of neurons releasing each compound varies from region to region in the brain. However, tracts originating at each nucleus distribute these neurotransmitters throughout the CNS.

Changes in the normal balance between two or more neurotransmitters can also produce profound alterations in brain function. One example, the release of dopamine at the cerebral nuclei by neurons of the substantia nigra, was discussed in Chapter 14. [p. 480] A second example is the interplay between serotonin and norepinephrine that appears to be involved in the regulation of asleep/awake cycles. A third example is *Huntington's disease.* The primary problem in this inherited disease is the destruction of ACh-secreting and GABA-secreting neurons in the cerebral nuclei. The reason for this destruction is unknown. Symptoms appear as the cerebral nuclei and frontal lobes slowly degenerate. The individual has difficulties in performing voluntary and involuntary movements, and there is a gradual decline in intellectual abilities. AM *Huntington's Disease*

In many cases the importance of a specific neurotransmitter has been revealed in seeking a mechanism for the effects of administered drugs. Three examples demonstrate patterns that are emerging:

1. An extensive network of tracts delivers serotonin to nuclei and higher centers throughout the brain. Compounds that enhance the effects of serotonin produce hallucinations; for example, *LSD (lysergic acid diethylamide)* is a powerful hallucinogenic drug that activates serotonin receptors in the brain stem, hypothalamus, limbic system, and spinal cord. Compounds that inhibit serotonin production or block its action cause severe depression and anxiety. The most effective antidepressive drug now in use, *fluoxetine (Prozac),* slows the removal of serotonin at synapses, causing an increase in serotonin concentrations at the postsynaptic membrane.
2. Norepinephrine is another important neurotransmitter with pathways throughout the brain. Drugs that stimulate norepinephrine release cause exhilaration, and those that depress NE release cause depression. One inherited form of depression has been linked to a defective enzyme involved in norepinephrine synthesis.
3. Disturbances in dopamine transmission have been linked to several neurological disorders. Inadequate dopamine production causes the motor problems of Parkinson's disease. Excessive production of dopamine may be associated with **schizophrenia,** a psychological disorder marked by pronounced disturbances of mood, thought patterns, and behavior. Amphetamine, or "speed," stimulates dopamine secretion, and in large doses it can produce symptoms resembling those of schizophrenia. (It also affects other neurotransmitter systems, producing a variety of changes in CNS function.) The *Applications Manual* contains additional information concerning the effects and classification of CNS-active drugs. AM *Pharmacology and Drug Abuse*

Personality and Other Mysteries

The basic pathways and components of the CNS have been traced, and memory acquisition, learning, and associated mental processes can now be manipulated and explored in animal experiments. But the truly remarkable human characteristics of self-awareness and personality remain phenomena without discrete anatomical foundations. That is not to say that an anatomical basis does not exist, just that self-awareness and personality are more characteristic of the brain as an integrated system than a function of any one specific nucleus or region. As a result, the origins of human consciousness and personality may remain a mystery for the immediate future. But our knowledge of the anatomy and biochemistry of specific tracts, nuclei, and regions will continue to provide useful clinical information.

In recent years significant advances have been made by correlating anatomical or physiological deficits with observed behavioral disorders. In some cases, such as the use of L-dopa to treat Parkinson's disease, this procedure has led to new and effective forms of treatment.

Aging and the Nervous System

The aging process affects all bodily systems, and the nervous system is no exception. Anatomical and physiological changes begin shortly after maturity (probably by age 30) and accumulate over time. Although an estimated 85 percent of the elderly (above age 65) lead relatively normal lives, there are noticeable changes in mental performance and CNS functioning.

❑ Age-Related Anatomical and Functional Changes

Age-related anatomical changes in the nervous system that are commonly seen include:

1. A reduction in brain size and weight. This reduction results primarily from a decrease in the volume of the cerebral cortex. The brains of elderly individuals have narrower gyri and wider sulci than those of young persons, and the subarachnoid space is enlarged.
2. A reduction in the number of neurons. Brain shrinkage has been linked to a loss of cortical neurons, although evidence exists that neuronal loss does not occur (at least to the same degree) in brain stem nuclei.
3. A decrease in blood flow to the brain. With age, fatty deposits gradually accumulate in the walls of blood vessels. Like a kink in a garden hose or a clog in a drain, these deposits reduce the rate of blood flow through arteries. (This process, called *arteriosclerosis,* affects arteries throughout the body; it is discussed further in Chapter 21.) The reduction in blood flow may not cause an acute cerebral crisis, but it does increase the chances that the individual will suffer a stroke.
4. Changes in synaptic organization of the brain. In many areas, the number of dendritic branchings and interconnections appears to decrease. Synaptic connections are lost, and the rate of neurotransmitter production declines.
5. Intracellular and extracellular changes in CNS neurons. Many neurons in the brain begin accumulating abnormal intracellular deposits, including *lipofuscin, plaques,* and *neurofibrillary tangles.* **Lipofuscin** is a granular pigment that has no known function. **Plaques** are extracellular accumulations of an unusual fibrillar protein, **amyloid,** surrounded by abnormal dendrites and axons. **Neurofibrillary tangles** are masses of neurofibrils that form dense mats inside the soma and axon. The significance of these cellular and extracellular abnormalities remains to be determined. There is evidence that they appear in all aging brains (see below), but when present in excess they seem to be associated with clinical abnormalities.

These anatomical changes are linked to a series of functional alterations. In general, neural processing becomes less efficient. For example, memory consolidation often becomes more difficult, and secondary memories, especially those of the recent past, become harder to access. The sensory systems of the elderly, notably hearing, balance, vision, smell, and taste, become less acute. Light must be brighter, sounds louder, and smells stronger before they are perceived. Reaction times are slowed, and reflexes—even some withdrawal reflexes—become weaker or even disappear. There is a decrease in the precision of motor control, and it takes longer to perform a given motor pattern than it did 20 years earlier.

For roughly 85 percent of the elderly population, these changes do not interfere with their abilities to function in society. But for as yet unknown reasons, many elderly individuals become incapacitated by progressive CNS changes. By far the most common such incapacitating condition is *Alzheimer's disease.*

❑ Alzheimer's Disease

Alzheimer's disease is a progressive disorder characterized by the loss of higher cerebral functions. It is the most common cause of **senile dementia,** or "senility." The first symptoms usually appear at 50–60 years of age, although the disease occasionally affects younger individuals. Alzheimer's disease has widespread impact on the elderly; an estimated 2 million people in the United States, including roughly 15 percent of those over 65, suffer from some form of the condition, and it causes approximately 100,000 deaths each year.

Although the link remains uncertain, the areas containing concentrations of plaques and tangles are the same regions involved with memory, emotions, and intellectual function. It remains to be determined whether these deposits cause Alzheimer's or whether they are secondary signs of ongoing metabolic alterations that have an environmental, hereditary, or infectious basis.

Genetic factors certainly play a major role. The late-onset form of Alzheimer's disease has been traced to a gene on chromosome 19 that codes for proteins involved in cholesterol transport. Less than 5 percent of people with Alzheimer's disease have the early-onset form; these individuals develop the condition before age 50. The early-onset form of Alzheimer's disease has been linked to genes on chromosomes 21 and 14. Interestingly, the majority of individuals with *Down syndrome* develop Alzheimer's disease relatively early in life. (Down syndrome results from an extra copy of chromosome 21; this condition is discussed further in Chapter 29.) AM Alzheimer's Disease

✔ What would happen to a sleeping individual if his or her reticular activating system (RAS) were suddenly stimulated?

✔ What would you expect to be the effect of a drug that substantially increases the amount of serotonin released in the brain?

✔ One of the problems associated with aging is difficulty in recalling things and even a total loss of memory. What are some possible reasons for these changes?

Selected Clinical Terminology

Terms Discussed in This Chapter

Alzheimer's disease: A progressive disorder marked by the loss of higher cerebral functions. *(p. 522* and *AM)*
amnesia: A temporary or permanent loss of memory. *(p. 518* and *AM)*
amyotrophic lateral sclerosis (ALS): A progressive, degenerative disorder affecting motor neurons of the spinal cord, brain stem, and cerebral hemispheres. *(p. 513* and *AM)*
anencephaly (an-en-SEF-a-lē): A rare condition in which the brain fails to develop at levels above the mesencephalon or lower diencephalon. *(p. 515)*
cerebral palsy: A disorder affecting voluntary motor performance that arises in infancy or childhood as a result of prenatal trauma, drug exposure, or a congenital defect. *(p. 510)*
electroencephalogram (EEG): A printed record of the brain's electrical activity over time. *(p. 515)*
epilepsy: One of more than 40 different conditions characterized by a recurring pattern of seizures over extended periods. *(p. 516* and *AM)*
Huntington's disease: An inherited disease marked by a progressive deterioration of mental abilities and by motor disturbances. *(p. 521* and *AM)*
schizophrenia: A psychological disorder marked by pronounced disturbances of mood, thought patterns, and behavior. *(p. 521)*
seizure: A temporary disorder of cerebral function, accompanied by abnormal, involuntary movements, unusual sensations, and/or inappropriate behavior; includes *focal seizures, temporal lobe seizures, convulsive seizures, generalized seizures, grand mal seizures,* and *petit mal seizures. (p. 516* and *AM)*
senile dementia: A progressive loss of memory, spatial orientation, language, and personality as a consequence of aging. *(p. 522)*

AM *Additional Terms Discussed in the Applications Manual*

anterograde amnesia: Loss of the ability to store new memories after an injury, while earlier memories remain intact and accessible.
aphasia: A disorder affecting the ability to speak or read.
delirium: A state characterized by wild oscillations in the level of wakefulness, little or no grasp of reality, and possible hallucinations. Also called an *acute confusional state* (ACS).
dementia: A stable, chronic state of consciousness characterized by deficits in memory, spatial orientation, language, or personality.
hypersomnia: Extremely long periods of otherwise normal sleep.
insomnia: Shortened sleeping periods and difficulty in getting to sleep.
parasomnias: Abnormal behaviors performed during sleep.
post-traumatic amnesia (PTA): Amnesia caused by head injury, in which the individual can neither remember the past nor consolidate memories of the present.
retrograde amnesia (*retro-*, behind): Loss of memories of events that occurred prior to an injury.
sleep apnea: A condition in which a sleeping individual stops breathing for short periods of time throughout the night.

CHAPTER REVIEW

STUDY OUTLINE

INTRODUCTION, p. 502

1. There is continuous communication among the brain, spinal cord, and peripheral nerves.

SENSORY AND MOTOR PATHWAYS, p. 502

Sensory Pathways, p. 502

1. A **sensation** arrives in the form of action potentials in an afferent fiber. The **posterior column pathway** carries fine touch, pressure, and proprioceptive sensations. The axons ascend within the **fasciculus gracilis** and **fasciculus cuneatus,** to be relayed to the thalamus via the **medial lemniscus.** Before the axons enter the medial lemniscus, they *decussate* (cross over) to the opposite side of the brain stem. (*Figures 15-1, 15-2; Table 15-1*)
2. The **spinothalamic pathway** carries poorly localized sensations of touch, pressure, pain, and temperature. These decussate in the spinal cord and ascend via the **anterior** and **lateral spinothalamic tracts** to the **ventral posterolateral** and **ventral posteromedial nuclei** of the thalamus. (*Figures 15-2, 15-3; Table 15-1*)
3. The **spinocerebellar pathway,** including the **posterior** and **anterior spinocerebellar tracts,** carries sensations to the cerebellum concerning the position of muscles, tendons, and joints. (*Figure 15-4; Table 15-1*)

Motor Pathways, p. 508

4. Somatic motor pathways always involve an **upper motor neuron** (whose soma lies in a CNS processing center) and a **lower motor neuron** (located in a motor nucleus of the brain stem or spinal cord). (*Figure 15-5*)
5. The neurons of the primary motor cortex are **pyramidal cells;** the **pyramidal system** provides voluntary skeletal muscle control. The **corticobulbar tracts** terminate at the cranial nerve nuclei, while the **cor-**

ticospinal tracts synapse on motor neurons in the anterior gray horns of the spinal cord. The corticospinal tracts are visible along the medulla as a pair of thick bands, the **pyramids,** where most of the axons decussate to enter the descending **bilateral corticospinal tracts;** those that do not decussate enter the **anterior corticospinal tracts.** The pyramidal system provides a rapid, direct mechanism for controlling skeletal muscles. *(Figures 15-5, 15-6)*

6. The **extrapyramidal system (EPS)** consists of several other centers that may issue motor commands as a result of processing performed at an unconscious, involuntary level. Its outputs descend via the *vestibulospinal, tectospinal, rubrospinal,* and *reticulospinal tracts.* (*Figures 15-6, 15-7; Tables 15-2, 15-3*)
7. The **vestibulospinal tracts** carry information related to maintaining balance and posture. Commands carried by the **tectospinal tracts** change the position of the eyes, head, neck, and arms in response to bright lights, sudden movements, or loud noises. The **rubrospinal tracts** carry motor responses to spinal motor neurons. Motor commands carried by the **reticulospinal tracts** vary according to the region stimulated.
8. The cerebral nuclei (the most important and complex components of the extrapyramidal system) adjust the motor commands issued in other processing centers through synapses in the thalamus.
9. The **cerebellum** regulates the activity along both pyramidal and extrapyramidal motor pathways. The integrative activities performed by neurons in the cerebellar cortex and cerebellar nuclei are essential for precise control of voluntary and involuntary movements. (*Figures 15-7, 15-8*)

MONITORING BRAIN ACTIVITY: THE ELECTROENCEPHALOGRAM, p. 515

1. A printed report of the electrical activity of the brain over time is an *electroencephalogram.* EEGs can be used to check for a variety of clinical conditions. (*Figure 15-9*)

HIGHER-ORDER FUNCTIONS, p. 517

1. Higher-order functions (1) are performed by the cerebral cortex and involve complex interactions between areas of cerebral cortex and between the cortex and other areas of the brain, (2) involve conscious and unconscious information processing, and (3) are subject to modification and adjustment over time.

Memory, p. 517

2. Memories can be classified into **short-term,** or **primary, memories** (short-lived, but while they persist the information can be recalled immediately) and **long-term memories** (which remain for much longer periods).
3. Long-term memories in turn can be subdivided into **secondary memories** (which fade with time and may be difficult to recall) and **tertiary memories** (which seem to be part of consciousness). (*Figure 15-10*)
4. Conversion from short-term to long-term memory is **memory consolidation.** The amygdaloid body and the hippocampus are essential to memory consolidation.
5. Cellular mechanisms that seem to be involved in memory formation and storage include: (1) increased neurotransmitter release; (2) facilitation at synapses; and (3) formation of additional synaptic connections.

Consciousness, p. 519

6. In **deep sleep (slow-wave** or **non-REM sleep),** the body relaxes and cerebral cortex activity is low. During **rapid eye movement (REM) sleep**, active dreaming occurs. Periods of REM and deep sleep alternate throughout the night. (*Figure 15-11*)
7. Consciousness is determined by interactions between the brain stem and cerebral cortex; one of the most important brain stem components is a network in the reticular formation called the **reticular activating system (RAS).** (*Figure 15-12*)

BRAIN CHEMISTRY AND BEHAVIOR, p. 520

1. Changes in the normal balance between neurotransmitters can profoundly alter brain function. The most important and widespread neurotransmitters are acetylcholine (ACh), norepinephrine (NE), gamma-aminobutyric acid (GABA), dopamine, serotonin, histamine, substance P, and the enkephalins and endorphins.

Personality and Other Mysteries, p. 521

2. The human characteristics of self-awareness and personality remain a mystery; they appear to be characteristic of the brain as an integrated system, rather than a function of any specific component.

AGING AND THE NERVOUS SYSTEM, p. 521

Age-Related Anatomical and Functional Changes, p. 522

1. Age-related changes in the nervous system include: (1) reduction in brain size and weight, (2) reduction in number of neurons, (3) decrease in blood flow to the brain, (4) changes in synaptic organization of the brain, and (5) intracellular and extracellular changes in CNS neurons.

Alzheimer's Disease, p. 522

2. **Alzheimer's disease** is a progressive disorder characterized by the loss of higher cerebral functions; it is the most common cause of **senile dementia.**

REVIEW QUESTIONS

LEVEL 1 Reviewing Facts and Terms

1. The corticospinal tract:
   (a) carries motor commands from the cerebral cortex to the spinal cord
   (b) carries sensory information from the spinal cord to the brain
   (c) starts in the spinal cord and ends in the brain
   (d) a, b, and c are correct

2. The ability to localize a specific stimulus depends on the organized distribution of sensory information to the:
   (a) spinothalamic pathway
   (b) posterior column tract
   (c) spinocerebellar pathway
   (d) primary sensory cortex

3. Destruction or damage to the lower motor neuron in the somatic nervous system produces:
   (a) the inability to localize a stimulus
   (b) unconscious response to stimulation
   (c) a flaccid paralysis in the innervated motor unit
   (d) involuntary control of the innervated muscle

4. A progressive disorder characterized by the loss of higher cerebral functions is:
   (a) Parkinson's disease
   (b) parasomnia
   (c) Huntington's disease
   (d) Alzheimer's disease

5. What four kinds of sensory receptors for the general senses are found in the body? What is the nature of the stimuli that excite each one?

6. What are the three major somatic sensory pathways in the body, and what is the function of each pathway?

7. What three pairs of descending tracts make up the pyramidal system?

8. What four motor pathways make up the extrapyramidal system?

9. What are the two primary functional roles of the cerebellum?

10. What three anatomical factors contribute to the maturation of the CNS and the refinement of motor skills?

11. What three characteristics are shared by higher-order functions?

12. As a result of animal studies, what cellular mechanisms are thought to be involved in memory formation and storage?

13. What physiological activities distinguish non-REM sleep from REM sleep?

14. What anatomical and functional changes in the brain are linked to alterations that occur with aging?

LEVEL 2 Reviewing Concepts

15. Damage to the hippocampus, a component of the limbic system, leads to:
   (a) a loss of emotion due to forgetfulness
   (b) a loss of consciousness
   (c) a loss of long-term memory
   (d) an immediate loss of short-term memory

16. Describe the relationship among first-, second-, and third-order neurons.

17. How is the autonomic nervous system (ANS) *functionally* different from the somatic nervous system (SNS)?

18. What is a motor homunculus? Why does it differ from a sensory homunculus?

19. What effect does injury to the primary motor cortex have over peripheral muscles?

20. By what structures and in what part of the brain is the level of muscle tone in the body's skeletal muscles controlled? How is this control exerted?

21. One patient suffers a cerebrovascular accident (CVA) in the left hemisphere; another suffers a CVA in the right hemisphere. What functions could be affected in each case?

LEVEL 3 Critical Thinking and Clinical Applications

22. Kelly is having difficulty controlling movements of her eyes, and she notices that she has lost some control of her facial muscles. After an examination and testing, her physician tells her that her cranial nerves are perfectly normal, but she has a small tumor that is putting pressure on certain fiber tracts in her brain and this pressure is the cause of her symptoms. Where is the tumor most likely located?

23. Doris develops a clot that blocks the right branch of the middle cerebral artery, a blood vessel that serves the anterior portion of the right cerebral hemisphere. What symptoms would you expect to observe as a result of this blockage?

24. Researchers studying an illicit "street drug" find that it causes individuals to see strange shapes and hear sounds that aren't actually present. Perceptions of color are exaggerated, and individuals taking the drug report increased sexual appetites. What neurotransmitter is this drug mimicking and what part(s) of the brain is involved?

C H A P T E R

16

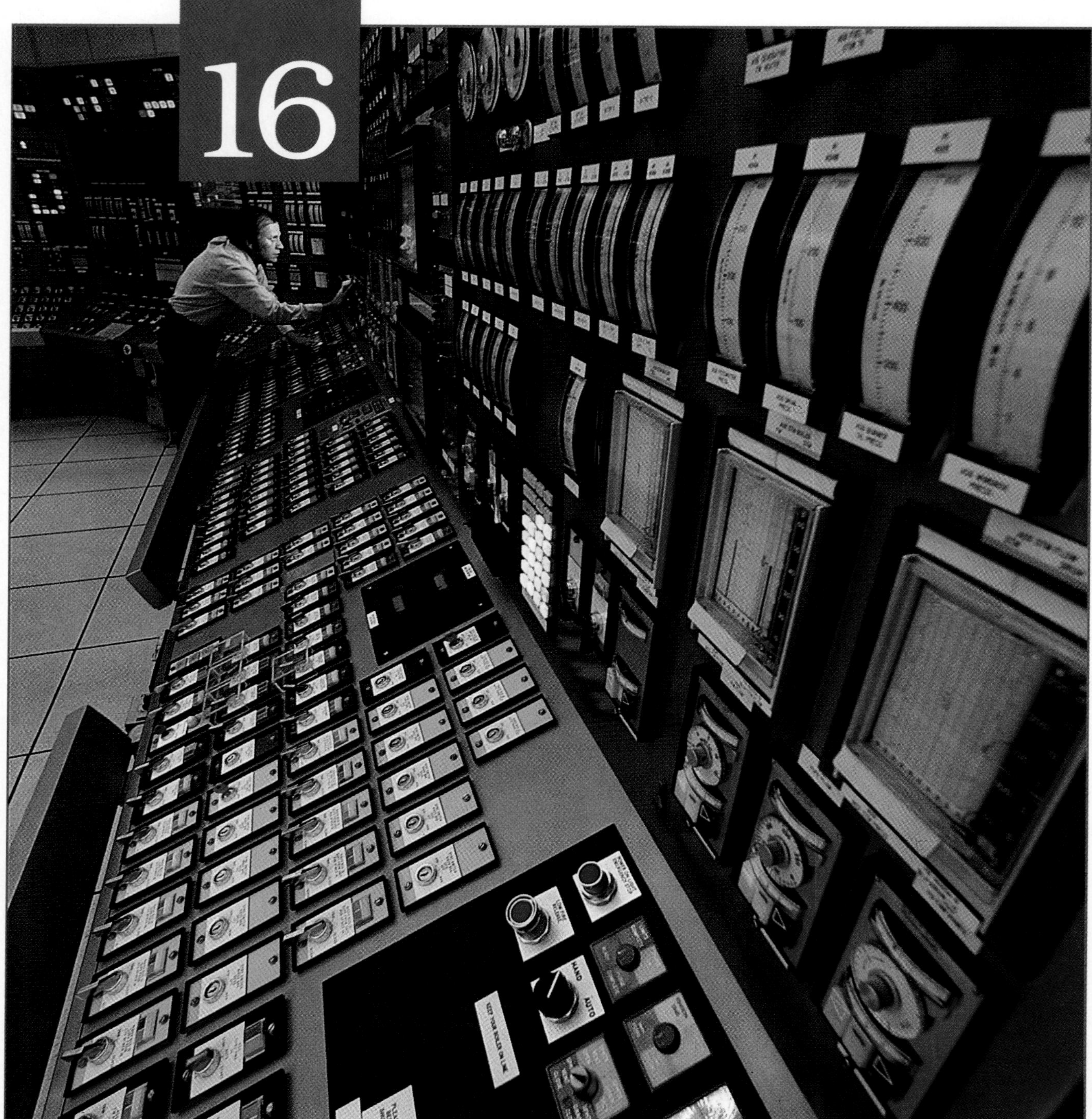

This is the control center of a nuclear power station. The generation and harnessing of nuclear energy are complex, potentially dangerous processes. This power station operates reliably because multiple systems—reactor, cooling, and electrical—are working smoothly and in balance. The systems involved are actually so complex, and the responses must be so swift and precise, that no one individual can manage a single system unaided, let alone the entire facility. Instead, computers are programmed to make adjustments in system operation, and the people look at the "big picture." The human body is an extremely complex biological mechanism. As in the power station, the individual systems are regulated automatically, and our conscious minds are aware of only the general patterns and results of those regulatory activities. This chapter will examine the regulatory mechanism and explore the structure and function of the autonomic nervous system.

The Autonomic Nervous System

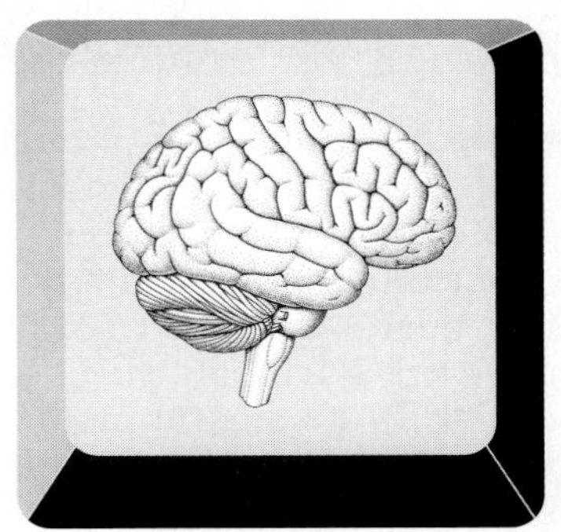

Chapter Outline and **Objectives**

1 Compare the autonomic nervous system with other divisions of the nervous system.

2 Contrast the structures and functions of the sympathetic and parasympathetic divisions of the autonomic nervous system.

3 Describe the mechanisms of neurotransmitter release in the autonomic nervous system.

4 Compare the effects of the various autonomic neurotransmitters on target organs and tissues.

5 Discuss the relationship between the divisions of the autonomic nervous system and the significance of dual innervation.

6 Explain the importance of autonomic tone.

7 Describe the hierarchy of interacting levels of control in the autonomic nervous system.

Your conscious thoughts, plans, and actions represent a tiny fraction of the activities of the nervous system. In practical terms, your conscious thoughts and the somatic motor system that operates under voluntary control seldom have a direct effect on your long-term survival. Of course, the somatic motor system can be important in moving you out of the way of a speeding bus or pulling a hand from a hot stove—but it was your conscious movements that put you in jeopardy in the first place. If all consciousness were eliminated, vital physiological processes would continue virtually unchanged; a night's sleep is not a life-threatening event. Longer, deeper states of unconsciousness are not necessarily more dangerous, as long as nourishment is provided in some way. People who have suffered severe brain injuries can survive in a coma for decades.

Survival under these conditions is possible because routine homeostatic adjustments in physiological systems are made by the autonomic nervous system (ANS). It is the ANS that coordinates cardiovascular, respiratory, digestive, excretory, and reproductive functions. In doing so, the ANS adjusts internal water, electrolyte, nutrient, and dissolved gas concentrations in body fluids—all without instructions or interference from the conscious mind.

This chapter examines the anatomical and physiological divisions of the autonomic nervous system. Our understanding of this system has had a profound effect on the practice of medicine. For example, in 1960, the 5-year survival rate for patients surviving their first heart attack was very low, primarily because it was difficult and sometimes impossible to control high blood pressure. Thirty years later, the survivors of heart attacks often lead normal lives. This dramatic change occurred as we learned to manipulate the autonomic nervous system with specific drugs and procedures.

An Overview of the ANS

Figure 16-1

It will be useful to compare the organization of the ANS with that of the somatic nervous system (SNS), which was considered in earlier chapters. We will focus on (1) the neural interactions that direct motor output and (2) the subdivisions of the ANS, based on structural and functional patterns of peripheral innervation.

Figure 16-1• compares the organization of the SNS and ANS. In the SNS, lower motor neurons exert direct control over skeletal muscles (Figure 16-1a•). In the ANS (Figure 16-1b•), *there is always a synapse between the CNS and the peripheral effector.* The visceral motor neurons in the CNS are known as **preganglionic neurons,** or *first-order neurons.* The axons of these neurons are called *preganglionic fibers.* The preganglionic fibers leave the CNS and synapse on **ganglionic neurons,** or *second-order neurons,* in peripheral **autonomic ganglia.** The axons of ganglionic neurons control peripheral effectors such as cardiac muscle, smooth muscles, glands, and adipose tissues. The axons of ganglionic neurons are called **postganglionic fibers** because they conduct impulses away from the ganglion.

Subdivisions of the ANS

Figure 16-2

The ANS contains two subdivisions, the *sympathetic division* and the *parasympathetic division.* Most often, the two divisions have opposing effects—if the sympathetic division causes excitation, the parasympathetic causes inhibition. However, this is not always the case, because (1) the two divisions may work independently, with some structures innervated only by one division or the other, and (2) the two divisions may work together, each controlling one stage of a complex process.

One way to recall the general functions of the sympathetic and parasympathetic divisions is to imagine a situation dominated by one system or the other. Although each of the responses described in the following scenarios will be considered in detail later in the chapter, this exercise will give you a sense of the overall *pattern* of response.

The sympathetic division prepares the body for heightened levels of somatic activity. When fully activated, it produces what is known as the "fight or flight" response, which prepares the body for a crisis that may require sudden, intense physical activity. To understand the nature of this response, imagine walking down a long, dark alley and hearing strange noises in the darkness ahead. Your body responds immediately, and you become more alert and aware of your surroundings. Your metabolic rate rises quickly, to as much as twice resting levels. Digestive and urinary activities are suspended temporarily, and blood flow to the skeletal muscles increases. You begin breathing more quickly and more deeply. Both heart rate and blood pressure increase, moving blood around the body more rapidly. You feel warm and begin to perspire. The general pattern can be summarized as follows: (1) heightened mental alertness, (2) increased metabolic rate, (3) reduced digestive and urinary function, (4) activation of energy reserves, (5) increased respiratory rate and dilation of respiratory passageways, (6) increased heart rate and blood pressure, and (7) activation of sweat glands.

The parasympathetic division stimulates visceral activity. For example, it is responsible for the state

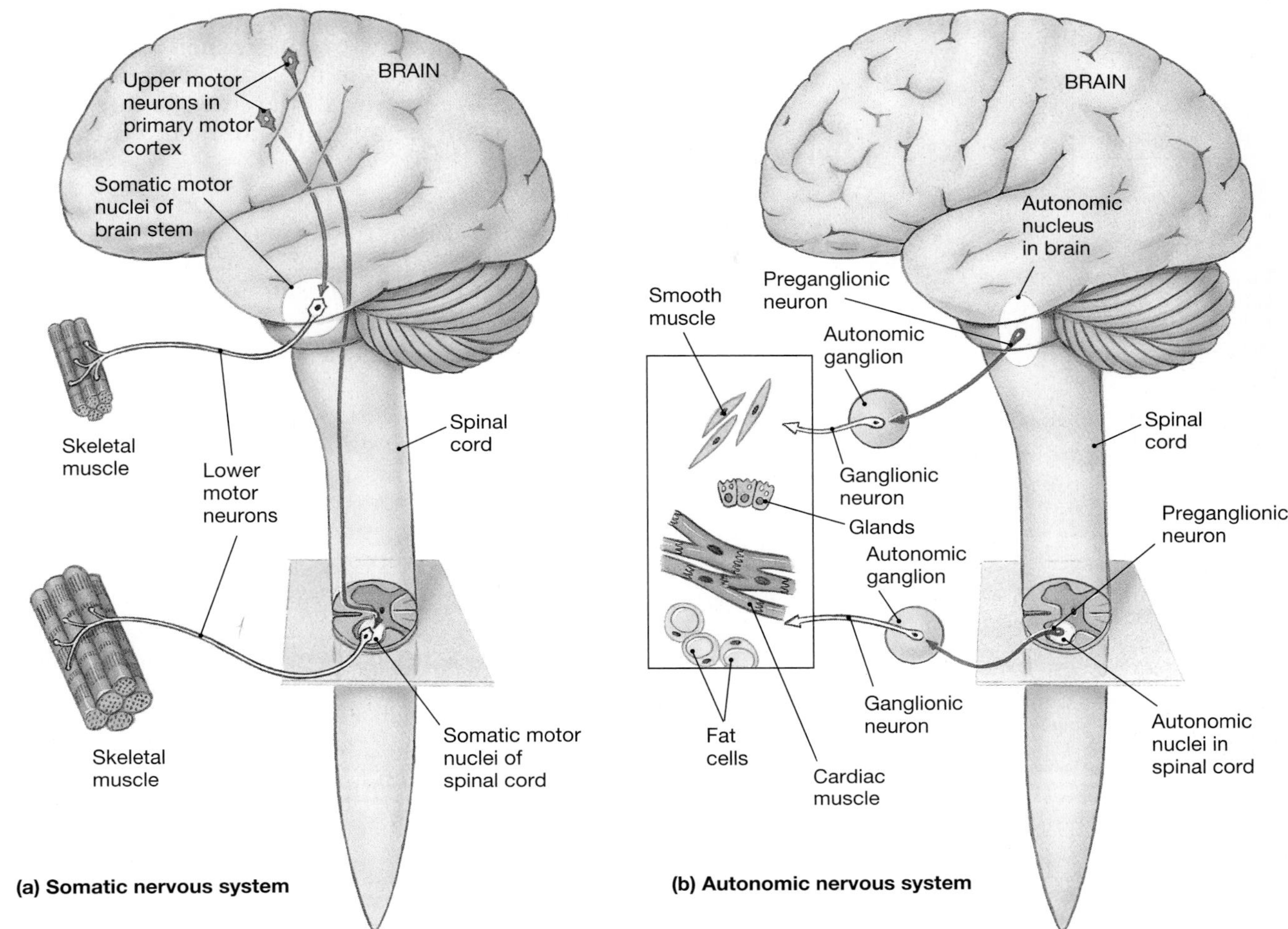
Upper motor neurons in primary motor cortex
BRAIN
Somatic motor nuclei of brain stem
Skeletal muscle
Lower motor neurons
Spinal cord
Skeletal muscle
Somatic motor nuclei of spinal cord
(a) Somatic nervous system
BRAIN
Autonomic nucleus in brain
Smooth muscle
Preganglionic neuron
Autonomic ganglion
Spinal cord
Ganglionic neuron
Glands
Preganglionic neuron
Autonomic ganglion
Ganglionic neuron
Fat cells
Cardiac muscle
Autonomic nuclei in spinal cord
(b) Autonomic nervous system

● **FIGURE 16-1**
Organization of the Somatic and Autonomic Nervous Systems. (a) In the SNS, an upper motor neuron in the CNS controls a lower motor neuron in the brain stem or spinal cord. The axon of the lower motor neuron has direct control over skeletal muscle fibers. Stimulation of the lower motor neuron always has an excitatory effect on the skeletal muscle fibers. **(b)** In the ANS, the axon of a preganglionic neuron in the CNS controls ganglionic neurons in the periphery. Stimulation of the ganglionic neurons may lead to excitation or inhibition of the visceral effector innervated.

of "rest and repose" that follows a big dinner. The body relaxes, energy demands are minimal, and both heart rate and blood pressure are relatively low. Meanwhile, the organs of the digestive tract are highly stimulated. Salivary glands and other secretory glands are active, the stomach is contracting, and smooth muscle contractions move materials along the digestive tract. This movement promotes defecation; at the same time, smooth muscle contractions along the urinary tract promote urination. The overall pattern is: (1) decreased metabolic rate, (2) decreased heart rate and blood pressure, (3) increased secretion by salivary and digestive glands, (4) increased motility and blood flow in the digestive tract, and (5) stimulation of urination and defecation.

The two systems are also anatomically distinct. Figure 16-2● provides an introduction to the major differences. We will now take a closer look at the structural and functional organization of each system.

Sympathetic (Thoracolumbar) Division

Figure 16-2

Preganglionic fibers from the thoracic and upper lumbar spinal segments synapse in ganglia near the spinal cord. These axons and ganglia are part of the **sympathetic division,** or *thoracolumbar* (thō-ra-kō-LUM-bar) *division,* of the ANS (Figure 16-2●). An increase in sympathetic activity generally stimulates tissue metabolism, increases alertness, and prepares the body to deal with emergencies.

Parasympathetic (Craniosacral) Division

Figure 16-2

Preganglionic fibers originating in the brain and the sacral spinal cord segments are part of the **parasympathetic division,** or *craniosacral* (krā-nē-ō-SĀ-krul) *division,* of the ANS (Figure 16-2●).

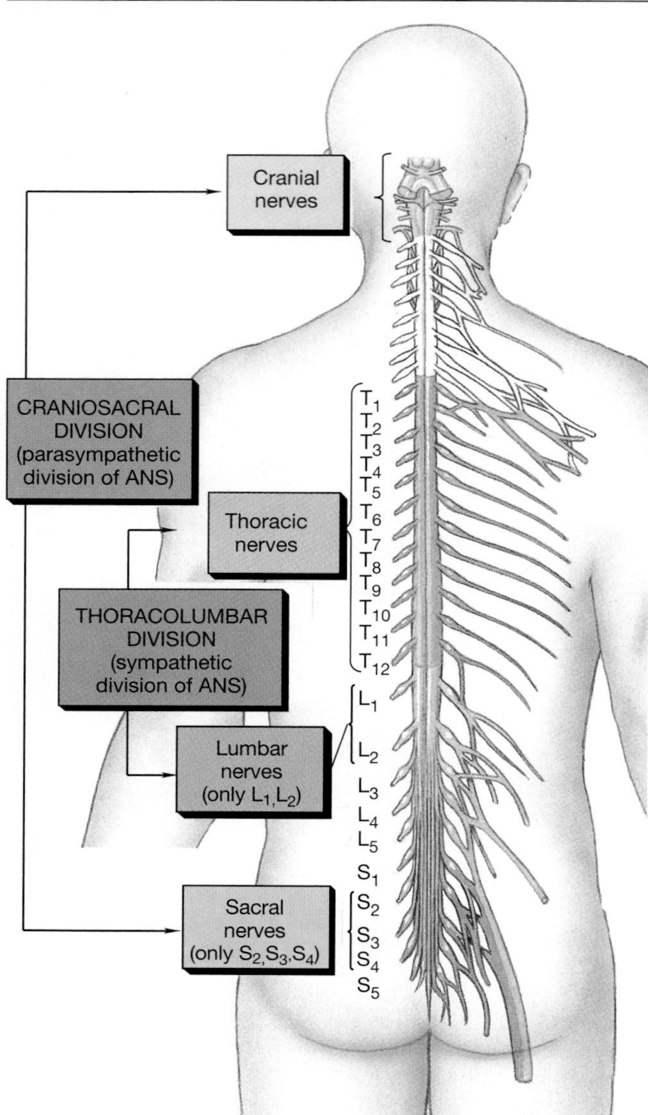
Cranial nerves
CRANIOSACRAL DIVISION (parasympathetic division of ANS)
Thoracic nerves
THORACOLUMBAR DIVISION (sympathetic division of ANS)
Lumbar nerves (only L_1,L_2)
Sacral nerves (only S_2,S_3,S_4)
T_1
T_2
T_3
T_4
T_5
T_6
T_7
T_8
T_9
T_{10}
T_{11}
T_{12}
L_1
L_2
L_3
L_4
L_5
S_1
S_2
S_3
S_4
S_5

● **FIGURE 16-2**

Divisions of the ANS. Anatomical divisions of the autonomic nervous system. At the cranial and sacral levels, the visceral efferent fibers from the CNS make up the parasympathetic, or craniosacral, division. At the thoracic and lumbar levels, the visceral efferent fibers that emerge form the sympathetic, or thoracolumbar, division.

The preganglionic fibers synapse on neurons of **terminal ganglia** or **intramural ganglia** (*murus*, wall) located near or within the tissues of visceral organs. The parasympathetic division is often called the "rest and repose" system because general parasympathetic activation conserves energy and promotes sedentary activities, such as digestion.

Innervation Patterns

The sympathetic and parasympathetic divisions affect target organs through the controlled release of neurotransmitters by postganglionic fibers. Whether the result is a stimulation or inhibition of activity depends on the response of the membrane receptor to the presence of the neurotransmitter. Three general statements can be made regarding the neurotransmitters and their effects:

1. All preganglionic autonomic fibers (sympathetic and parasympathetic) release *acetylcholine (ACh)* at their synaptic terminals. The effects are always excitatory.
2. Postganglionic parasympathetic fibers also release ACh, but the effects may be excitatory or inhibitory, depending on the nature of the receptor.
3. Most postganglionic sympathetic terminals release the neurotransmitter *norepinephrine (NE)*. The effects are usually excitatory, but the response depends on the nature of the postsynaptic receptor.

✔ How many motor neurons are required to conduct an action potential from the spinal cord to smooth muscles in the wall of the intestine?

✔ While out for a brisk walk, Jill is suddenly confronted by an angry dog. Which division of the autonomic nervous system is responsible for the physiological changes that occur as Jill turns and runs from the angry animal?

✔ How could you distinguish the sympathetic division from the parasympathetic division of the autonomic nervous system on the basis of anatomy?

The Sympathetic Division

Figures 16-3, 16-4

The sympathetic division, diagrammed in Figure 16-3●, consists of *preganglionic neurons located between segments T_1 and L_2 of the spinal cord,* and *ganglionic neurons located in ganglia near the vertebral column.* The preganglionic neurons are situated in the lateral gray horns, and their axons enter the ventral roots of these segments. The ganglionic neurons are found in three different locations:

1. ***Sympathetic chain ganglia.*** **Sympathetic chain ganglia,** also called *paravertebral ganglia* or *lateral ganglia,* lie on either side of the vertebral column. Neurons in these ganglia control effectors in the body wall and inside the thoracic cavity (Figure 16-4a●).
2. ***Collateral ganglia.*** **Collateral ganglia,** also known as *prevertebral ganglia,* are found anterior to the vertebral bodies (Figure 16-4b●).

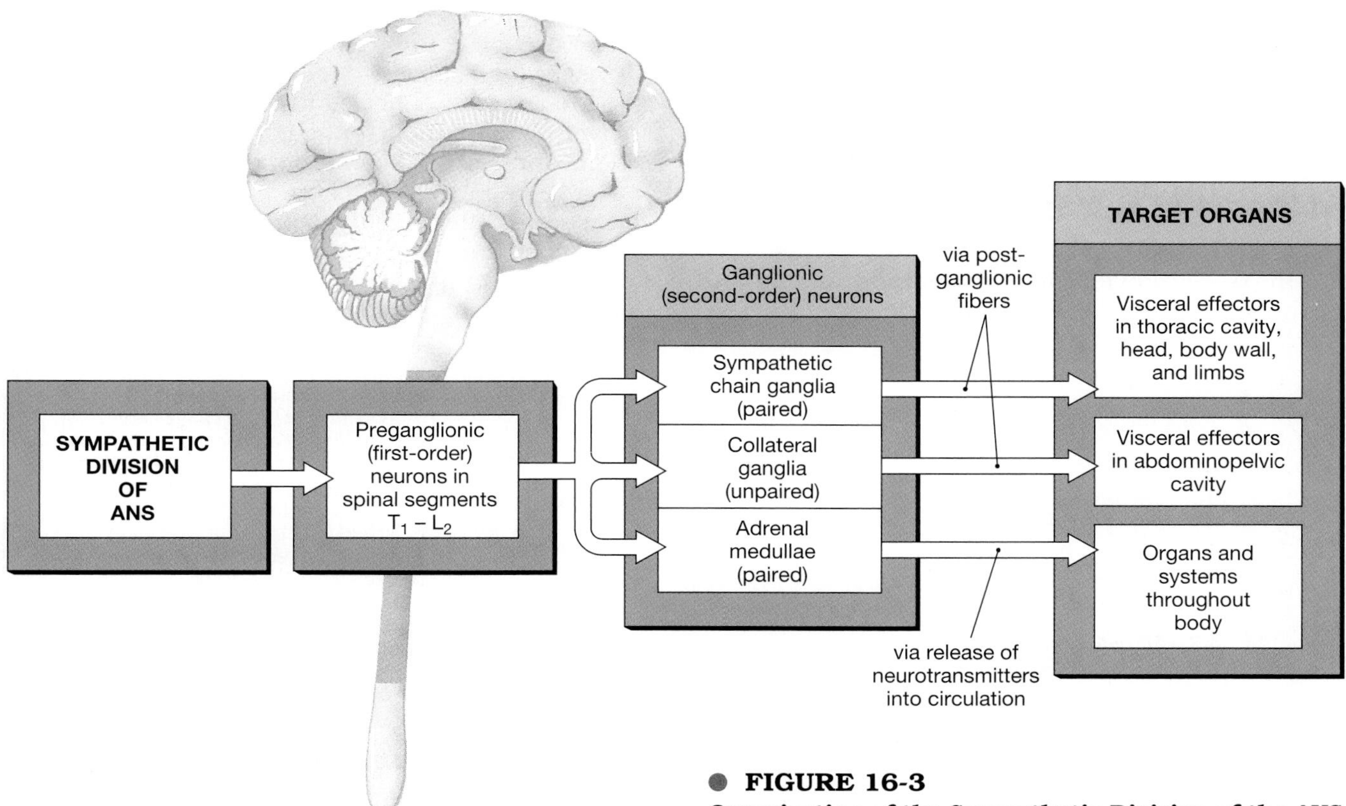
SYMPATHETIC DIVISION OF ANS
Preganglionic (first-order) neurons in spinal segments $T_1 - L_2$
Ganglionic (second-order) neurons
Sympathetic chain ganglia (paired)
Collateral ganglia (unpaired)
Adrenal medullae (paired)
via post-ganglionic fibers
via release of neurotransmitters into circulation
TARGET ORGANS
Visceral effectors in thoracic cavity, head, body wall, and limbs
Visceral effectors in abdominopelvic cavity
Organs and systems throughout body

FIGURE 16-3
Organization of the Sympathetic Division of the ANS

Collateral ganglia contain second-order neurons that innervate tissues and organs in the abdominopelvic cavity.

3. ***The adrenal medullae.*** The center of each adrenal gland, an area known as the *adrenal medulla,* is a modified sympathetic ganglion. The ganglionic neurons of the adrenal medulla have very short axons, and when stimulated they release neurotransmitters into the bloodstream (Figure 16-4c●).

❑ The Sympathetic Chain

Figure 16-4

The ventral roots of spinal segments T_1 to L_2 contain sympathetic preganglionic fibers. The basic pattern of sympathetic innervation in these regions was described in Figure 13-5a●, p. 435. After passing through the intervertebral foramen, each ventral root gives rise to a *white ramus,* or *white ramus communicans,* that carries preganglionic fibers into a nearby sympathetic chain ganglion. These fibers may synapse within the sympathetic chain ganglia (Figure 16-4a●), at one of the collateral ganglia (Figure 16-4b●), or in the adrenal medullae (Figure 16-4c●). Extensive divergence occurs, with one preganglionic fiber synapsing on two dozen or more ganglionic neurons. Preganglionic fibers running between the sympathetic chain ganglia interconnect them, making the chain resemble a string of beads. *Each ganglion in the sympathetic chain innervates a particular body segment or group of segments.*

If a preganglionic fiber carries motor commands that target structures in the body wall or the thoracic cavity, it will synapse in one or more of the sympathetic chain ganglia (Figure 16-4a●). The paths of the postganglionic fibers differ depending on whether their targets lie in the body wall or within the thoracic cavity. Postganglionic fibers that control visceral effectors in the body wall, such as the sweat glands of the skin or the smooth muscles in superficial blood vessels, enter the *gray ramus (gray ramus communicans)* and return to the spinal nerve for subsequent distribution. However, spinal nerves do not innervate structures in the ventral body cavities. Postganglionic fibers targeting structures in the thoracic cavity, such as the heart and lungs, form **autonomic nerves** that proceed directly to their peripheral targets. Although Figure 16-4a● shows autonomic nerves on the left side and spinal nerve distribution on the right, bear in mind that in reality *both* innervation patterns are found on *each* side of the body.

Functions of the Sympathetic Chain Ganglia

Figure 16-4a

Postganglionic fibers leaving the sympathetic chain reach their peripheral targets via spinal nerves and autonomic nerves. The primary results of increased activity in the postganglionic fibers leaving the sympathetic chain ganglia via spinal nerves and autonomic nerves are summarized in Figure 16-4a●. How these effects are brought about will be the focus of a later section.

Anatomy of the Sympathetic Chain

Figure 16-5

Figure 16-5● provides a more anatomical diagram of the structure of the sympathetic division. The left side represents the distribution to somatic structures; the right side depicts the innervation of visceral structures.

There are 3 cervical, 11 thoracic, 4 lumbar, and 4 sacral sympathetic ganglia in each sympathetic chain. First-order sympathetic neurons are limited to segments T_1 to L_2, and these spinal nerves have both white rami (preganglionic fibers) and gray rami (postganglionic fibers). The neurons in the cervical, lower lumbar, and sacral sympathetic chain ganglia are innervated by preganglionic fibers run-

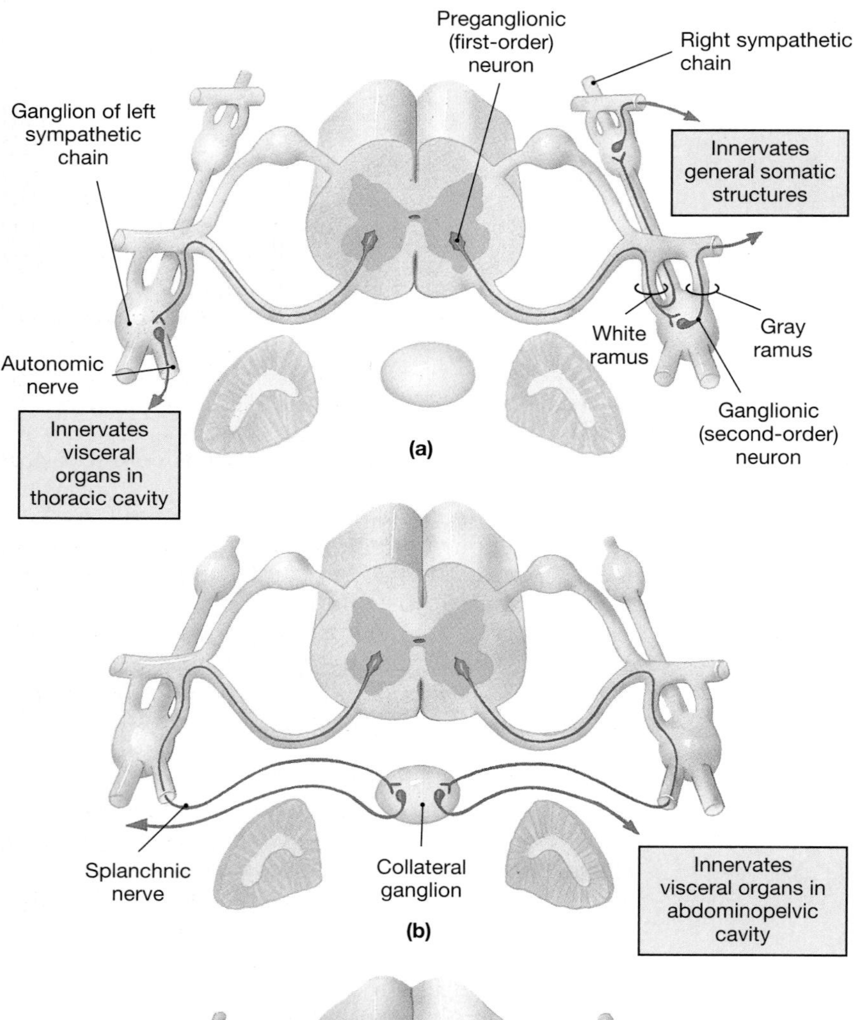
Preganglionic (first-order) neuron
Right sympathetic chain
Ganglion of left sympathetic chain
Innervates general somatic structures
White ramus
Gray ramus
Autonomic nerve
Ganglionic (second-order) neuron
Innervates visceral organs in thoracic cavity
(a)
Splanchnic nerve
Collateral ganglion
Innervates visceral organs in abdominopelvic cavity
(b)
Adrenal medullae
(c)

Major effects produced by sympathetic postganglionic fibers in spinal nerves:
- Constriction of cutaneous blood vessels, reduction in circulation to the skin and to most other organs in the body wall
- Acceleration of blood flow to skeletal muscles and brain
- Stimulation of energy production and use by skeletal muscle tissue
- Release of stored lipids from subcutaneous adipose tissue
- Stimulation of secretion by sweat glands
- Dilation of the pupils and focusing for distant objects

Major effects produced by postganglionic fibers entering the thoracic cavity in autonomic nerves:
- Accelerating the heart rate and increasing the strength of cardiac contractions
- Dilating the respiratory passageways

Major effects produced by preganglionic fibers innervating the collateral ganglia:
- Constricting small arteries and reducing the flow of blood to visceral organs
- Decreasing the activity of digestive glands and organs
- Stimulating the release of glucose from glycogen reserves in the liver
- Stimulating the release of lipids from adipose tissue
- Relaxing the smooth muscle in the wall of the urinary bladder
- Reducing the rate of urine formation at the kidneys
- Controlling some aspects of sexual function, such as ejaculation in the male

Major effect produced by preganglionic fibers innervating the adrenal medullae:
- Release of epinephrine and norepinephrine into the general circulation

● **FIGURE 16-4**
Sympathetic Pathways. (a) Sympathetic chain ganglia. **(b)** Collateral ganglia. **(c)** The adrenal medullae.

ning along the axis of the chain. In turn, these chain ganglia provide postganglionic fibers, via gray rami, to the cervical, lumbar, and sacral spinal nerves. *Every spinal nerve has a gray ramus that carries sympathetic postganglionic fibers.*

About 8 percent of the axons in each spinal nerve are sympathetic postganglionic fibers. As a result, the dorsal and ventral rami of the spinal nerves, which provide somatic motor innervation to skeletal muscles of the body wall and limbs, also distribute sympathetic postganglionic fibers. In the head, postganglionic sympathetic fibers leaving the cervical chain ganglia supply the regions and structures innervated by cranial nerves III, VII, IX, and X (Figure 16-5•).

In summary: (1) *Only the thoracic and upper lumbar ganglia receive preganglionic fibers via white rami,* (2) *the cervical, lower lumbar, and sacral chain ganglia receive preganglionic innervation via collateral fibers,* and (3) *every spinal nerve receives a gray ramus from a ganglion of the sympathetic chain.*

This anatomical arrangement means that if the ventral roots of thoracic spinal nerves are damaged, there will be no sympathetic motor function on the affected side of the head, neck, and trunk. Yet damage to the ventral roots of cervical spinal nerves will produce voluntary muscle paralysis on the affected side, *but leave sympathetic function intact* because the preganglionic fibers innervating the cervical ganglia originate in the white rami of thoracic segments, which are undamaged.

In contrast, damage to the cervical ganglia or thoracic segments can eliminate sympathetic innervation to the face, although sensation and muscle control remain unaffected. The affected side of the face becomes flushed, although sweating does not occur, and the pupil constricts. This combination of symptoms is known as *Horner's syndrome.* AM *Hypersensitivity and Sympathetic Function*

❑ Collateral Ganglia

Figure 16-4b

The abdominopelvic viscera receive sympathetic innervation via preganglionic fibers that pass through the sympathetic chain without synapsing. These fibers originate at first-order neurons in the lower thoracic and upper lumbar segments. They synapse within separate *collateral ganglia* (Figure 16-4b•). Preganglionic fibers that innervate the collateral ganglia form the **splanchnic** (SPLANK-nik) **nerves** in the dorsal wall of the abdominal cavity. Splanchnic nerves from both sides of the body converge on these ganglia. Although there are two sympathetic chains, one on each side of the vertebral column, most collateral ganglia are single rather than paired.

Functions of the Collateral Ganglia

Figure 16-4b

Postganglionic fibers leaving the collateral ganglia extend throughout the abdominopelvic cavity, innervating a variety of visceral tissues and organs. A summary of the effects of increased sympathetic activity along these postganglionic fibers is included in Figure 16-4b•. The general pattern is (1) a reduction of blood flow and energy use by visceral organs that are not important to short-term survival, such as the digestive tract, and (2) the release of stored energy reserves.

Anatomy of the Collateral Ganglia

Figures 16-5, 16-9

The splanchnic nerves innervate three collateral ganglia. Preganglionic fibers from the seven lower thoracic segments end at the **celiac** (SĒ-lē-ak) **ganglion** and the **superior mesenteric ganglion.** These ganglia are embedded in an extensive autonomic plexus that will be discussed in a later section. Preganglionic fibers from the lumbar segments form splanchnic nerves that end at the **inferior mesenteric ganglion.** These ganglia are diagrammed in Figure 16-5• and detailed in Figure 16-9•, p. 541. The ganglia are named by their association with adjacent arteries; the distribution of these arteries will be considered in Chapter 21.

THE CELIAC GANGLION The celiac ganglion is variable in appearance. It most often consists of a pair of interconnected masses of gray matter situated at the base of the *celiac artery.* The celiac ganglion may also form a single mass or many small, interwoven masses. Postganglionic fibers from the celiac ganglion innervate the stomach, liver, pancreas, and spleen.

THE SUPERIOR MESENTERIC GANGLION The superior mesenteric ganglion sits near the base of the *superior mesenteric artery.* Postganglionic fibers leaving the superior mesenteric ganglion innervate the small intestine and the initial segments of the large intestine.

THE INFERIOR MESENTERIC GANGLION The inferior mesenteric ganglion is located near the base of the *inferior mesenteric artery.* Postganglionic fibers from this ganglion provide sympathetic innervation to the terminal portions of the large intestine, the kidney and bladder, and the sex organs.

❑ The Adrenal Medullae

Figures 16-4c, 16-5

Preganglionic fibers entering an adrenal gland proceed to its center, a region called the **adrenal medulla** (Figures 16-4c and 16-5•). The adrenal medulla

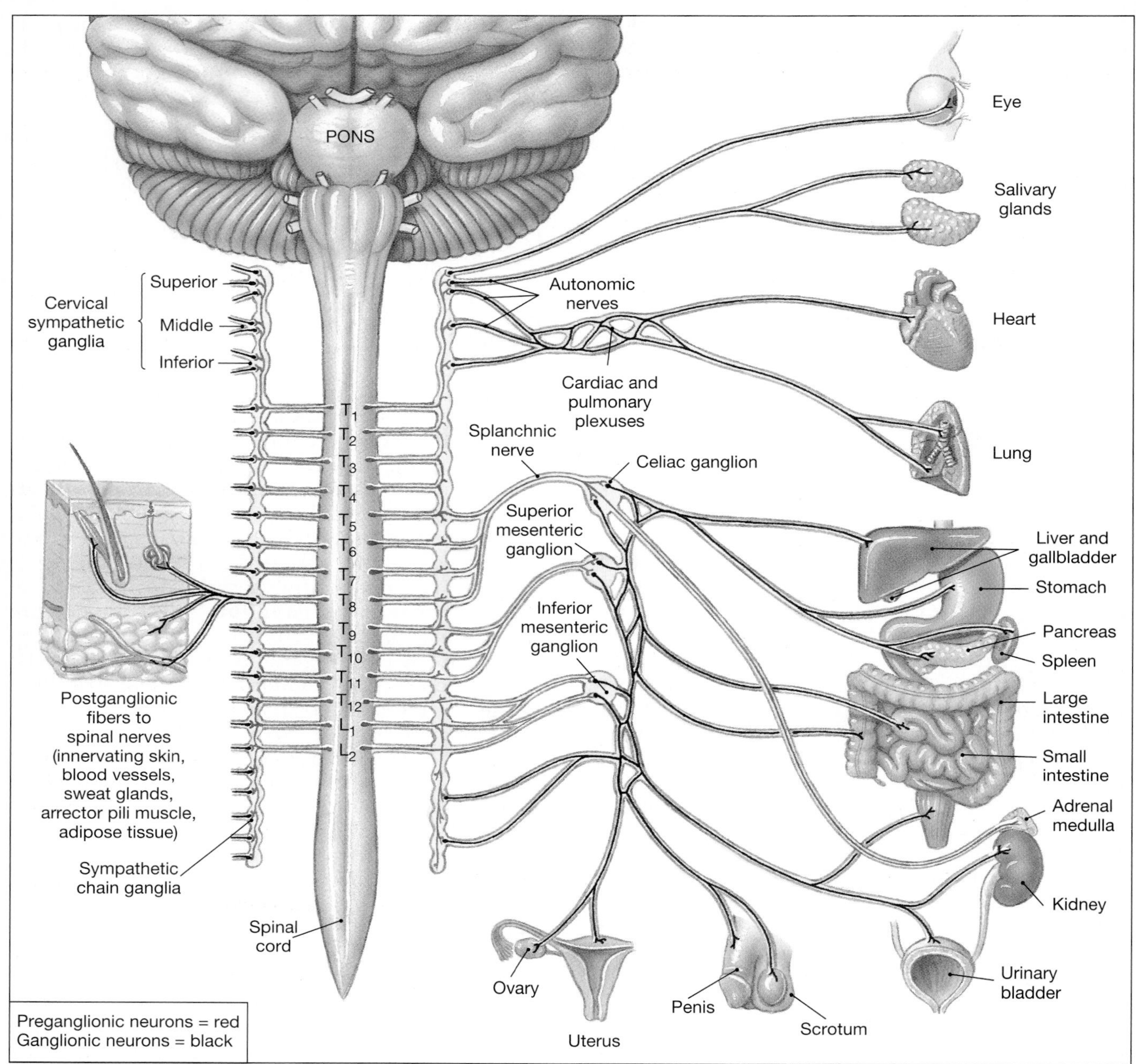
PONS
Eye
Salivary glands
Cervical sympathetic ganglia
Superior
Middle
Inferior
Autonomic nerves
Heart
Cardiac and pulmonary plexuses
Splanchnic nerve
Lung
Celiac ganglion
T1
T2
T3
T4
T5
T6
T7
T8
T9
T10
T11
T12
L1
L2
Superior mesenteric ganglion
Liver and gallbladder
Stomach
Inferior mesenteric ganglion
Pancreas
Spleen
Large intestine
Small intestine
Adrenal medulla
Postganglionic fibers to spinal nerves (innervating skin, blood vessels, sweat glands, arrector pili muscle, adipose tissue)
Sympathetic chain ganglia
Kidney
Spinal cord
Ovary
Uterus
Penis
Scrotum
Urinary bladder
Preganglionic neurons = red
Ganglionic neurons = black

● **FIGURE 16-5**
The Sympathetic Division. The distribution of sympathetic fibers is the same on both sides of the body. For clarity, the innervation of somatic structures is shown to the left and the innervation of visceral structures to the right.

is a modified sympathetic ganglion. Within the medulla, preganglionic fibers synapse on *neuroendocrine cells,* specialized neurons that release the neurotransmitters *norepinephrine* (NE) and *epinephrine* (E) into the general circulation.

The bloodstream then carries the neurotransmitters throughout the body, causing changes in the metabolic activities of many different cells. In general, these effects resemble those produced by the stimulation of sympathetic postganglionic fibers. They differ, however, in two respects: (1) Cells not innervated by sympathetic postganglionic fibers are affected as well; and (2) the effects last much longer than those produced by direct sympathetic innervation.

❑ Sympathetic Activation

The sympathetic division can change tissue and organ activities by releasing NE at peripheral synapses and distributing norepinephrine and epinephrine throughout the body in the bloodstream. The motor fibers that target specific effectors, such as smooth muscle fibers in blood vessels of the skin, can be activated in reflexes that do not involve other peripheral effectors. In a crisis, however, the entire division responds. This event is called **sympathetic activation.** Sympathetic activation is controlled by sympathetic centers in the hypothalamus. The effects are not limited to peripheral tissues—sympathetic activation also alters CNS activity. When sympathetic activation occurs, an individual experiences:

1. Increased alertness, via stimulation of the reticular activating system, causing the individual to feel "on edge."
2. A feeling of energy and euphoria, often associated with a disregard for danger and a temporary insensitivity to painful stimuli.
3. Increased activity in the cardiovascular and respiratory centers of the pons and medulla, leading to elevations in blood pressure, heart rate, breathing rate, and depth of respiration.
4. A general elevation in muscle tone through stimulation of the extrapyramidal system so that the person *looks* tense and may even begin to shiver.

These changes, coupled with the peripheral changes already noted, complete the preparations necessary for the individual to cope with stressful situations.

❑ Neurotransmitters and Sympathetic Function

We have examined the distribution of sympathetic impulses and the general effects of sympathetic activation. We will now consider the cellular basis for these effects on peripheral organs. On stimulation, sympathetic preganglionic fibers release ACh at synapses with ganglionic neurons. Synapses using ACh as a transmitter are called *cholinergic.* The effect on the ganglionic neurons is always excitatory.

Stimulation of the postganglionic neurons in the sympathetic division produces two widespread and distinctive results:

1. The release of norepinephrine at postganglionic neuroeffector junctions.
2. The secretion of epinephrine (and modest amounts of norepinephrine) into the general circulation.

Typical sympathetic postganglionic fibers release norepinephrine at neuroeffector junctions; these are examples of *adrenergic* synapses. The synaptic terminals do not resemble the neuroeffector junctions of the SNS. Instead, the terminal axon branches form a network or chain of enlarged terminal knobs that either contact the target cells or end in the adjacent connective tissues. Norepinephrine affects the postsynaptic membrane for several seconds before it is either reabsorbed by the terminal knobs, broken down by enzymes, or diffuses out of the region. The effect on the target cell varies, depending on the nature of the receptor on the postsynaptic membrane.

The specialized ganglionic neurons in the adrenal medullae release epinephrine and norepinephrine into the bloodstream. Epinephrine, also called *adrenaline,* accounts for 75–80 percent of the secretory output; the rest is norepinephrine.

The norepinephrine released by synaptic knobs affects its target for a few seconds before it is inactivated by enzymes. But the bloodstream does not contain the enzymes that break down epinephrine or norepinephrine, and most tissues contain relatively low concentrations of such enzymes. As a result, the effects of adrenal stimulation are widespread, and they continue for a relatively long time. For example, tissue concentrations of epinephrine may remain elevated for as long as 30 seconds, and the effects may persist for several minutes.

Membrane Receptors

Figure 16-6

The effects of sympathetic stimulation result primarily from interactions with membrane receptors sensitive to norepinephrine and epinephrine. There are two classes of sympathetic receptors, *alpha receptors* and *beta receptors.* In general, norepinephrine stimulates alpha receptors more than it does beta receptors; however, epinephrine stimulates both classes of receptors.

ALPHA AND BETA RECEPTORS Stimulation of alpha receptors activates enzymes on the inside of the cell membrane. There are two types of **alpha** (α) **receptors,** alpha-1 (α_1) and alpha-2 (α_2) (Figure 16-6a●). In the case of the most common type of alpha receptor, α_1, the result is the release of intracellular calcium ions from reserves in the endoplasmic reticulum (ER). This occurs following the release of *second messengers* inside the target cell. ∞ [p. 410] Stimulation of alpha-1 receptors on the surfaces of smooth muscle fibers is responsible for the constriction of peripheral blood vessels and the closure of sphincters along the digestive tract. Alpha-2 receptors are less common; their stimulation results in a lowering of cyclic-AMP levels in the cytoplasm. This usually has an inhibitory effect on the cell.

Beta receptors are found in many organs, including skeletal muscles, the lungs, the heart, and the liver. Stimulation of beta receptors at these sites triggers changes in the metabolic activity of the target cell. These alterations occur indirectly, as the beta receptor causes formation of a second messenger, cyclic-AMP (cAMP), that activates or inactivates key enzymes (Figure 16-6b●).

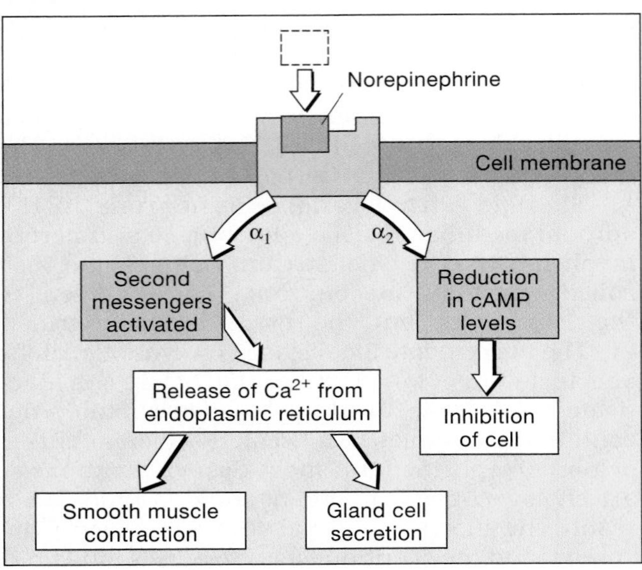
Norepinephrine
Cell membrane
α_1
α_2
Second messengers activated
Reduction in cAMP levels
Release of Ca^{2+} from endoplasmic reticulum
Inhibition of cell
Smooth muscle contraction
Gland cell secretion

(a) Alpha receptor stimulation

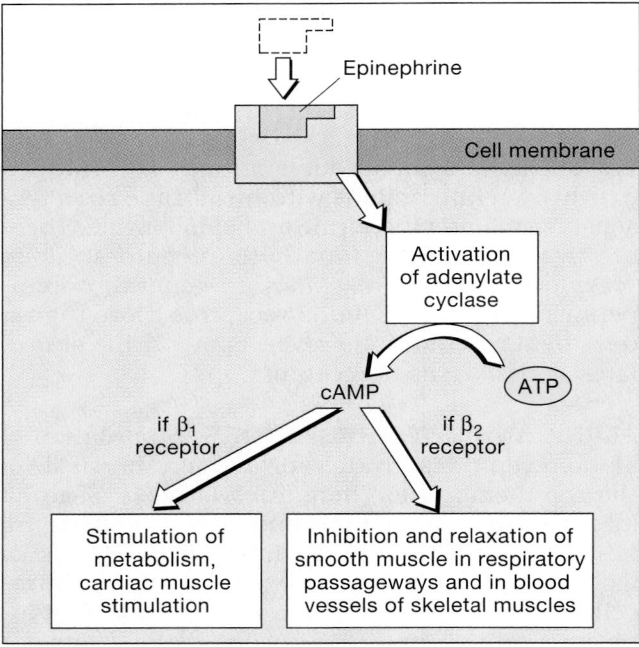
Epinephrine
Cell membrane
Activation of adenylate cyclase
ATP
cAMP
if β_1 receptor
if β_2 receptor
Stimulation of metabolism, cardiac muscle stimulation
Inhibition and relaxation of smooth muscle in respiratory passageways and in blood vessels of skeletal muscles

(b) Beta receptor stimulation

● **FIGURE 16-6**

Sympathetic Receptor Classification. (a) Stimulation of an alpha-1 receptor causes release of calcium ions into the cytoplasm. Stimulation of an alpha-2 receptor causes a reduction in the concentration of cyclic-AMP in the cytoplasm. **(b)** Beta receptor stimulation may lead to excitation or inhibition of the target cell.

There are two major types of **beta** (β) **receptors,** beta-1 (β_1) and beta-2 (β_2). Stimulation of β_1 receptors leads to an increase in metabolic activity. For example, stimulation of β_1 receptors on skeletal muscles accelerates their metabolic activities. Stimulation of β_1 receptors in the heart causes an increase in heart rate and force of contraction. Stimulation of β_2 receptors causes inhibition; when stimulated, these receptors trigger a relaxation of smooth muscles along the respiratory tract, increasing the diameter of the respiratory passageways and making it easier to breathe. This response accounts for the effectiveness of the inhalers used to treat asthma.

The effects of NE on the postsynaptic membrane last longer than those of ACh, because the NE is removed relatively slowly. From 50 to 80 percent of the NE is reabsorbed and recycled by the synaptic knob. The rest diffuses out of the area or is broken down by the enzymes *monoamine oxidase (MAO)* or *catechol-O-methyltransferase (COMT).*

Sympathetic Stimulation, ACh, and NO

Although the vast majority of sympathetic postganglionic fibers are adrenergic, releasing norepinephrine, a few postganglionic fibers are cholinergic. These postganglionic fibers innervate sweat glands of the skin and the blood vessels to skeletal muscles and the brain. Activation of these sympathetic fibers stimulates sweat gland secretion and dilates the blood vessels.

It may seem strange to find sympathetic terminals releasing ACh, as acetylcholine is the neurotransmitter used by the parasympathetic nervous system. However, (1) *ACh stimulates sweat gland secretion much more than NE does;* (2) *NE release causes constriction of most peripheral arteries;* and (3) *neither the body wall nor skeletal muscles are innervated by the parasympathetic division.* The distribution of cholinergic fibers via the sympathetic division provides a method of stimulating sweat gland secretion and selectively enhancing blood flow to skeletal muscles while the adrenergic terminals reduce the blood flow to other tissues in the body wall.

The sympathetic division also includes *nitroxidergic* synapses that release nitric oxide (NO) as a neurotransmitter. As mentioned in Chapter 12, such synapses occur where neurons innervate smooth muscles in the walls of blood vessels in many regions, notably in skeletal muscles and the brain. [p. 410] Activity of these synapses promotes immediate vasodilation and an increase in blood flow through the region.

❑ A Summary of the Sympathetic Division

1. The sympathetic division of the ANS includes two segmentally arranged sympathetic chains, one on each side of the spinal column, three collateral ganglia in front of the spinal column, and two adrenal medullae.
2. The preganglionic fibers are short, because the ganglia are close to the spinal cord. The postganglionic fibers are relatively long and

extend a considerable distance before reaching their target organs. (In the case of the adrenal medullae, very short axons end at capillaries that carry their secretions to the bloodstream.)

3. The sympathetic division shows extensive divergence, and a single preganglionic fiber may innervate as many as 32 second-order neurons in different ganglia. As a result, a single sympathetic motor neuron inside the CNS can control a variety of peripheral effectors and produce a complex and coordinated response.
4. All preganglionic neurons release ACh at their synapses with ganglionic neurons. Most of the postganglionic fibers release norepinephrine, but a few release ACh or NO.
5. The effector response depends on the nature of the channels or enzymes activated when NE or E binds to alpha or beta receptors.

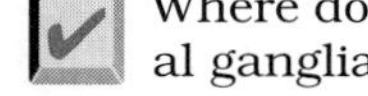
Where do the nerves that synapse in the collateral ganglia originate?

How would a drug that stimulates acetylcholine receptors affect the sympathetic nervous system?

Individuals with high blood pressure may be given a medication that blocks beta receptors. How would this medication help their condition?

The Parasympathetic Division

Figure 16-7

The parasympathetic division of the ANS (Figure 16-7●) includes:

1. ***Preganglionic (first-order) neurons in the brain stem and in sacral segments of the spinal cord.*** In the brain, the mesencephalon, pons, and medulla oblongata contain autonomic nuclei associated with cranial nerves III, VII, IX, and X. In the sacral segments of the spinal cord, the autonomic nuclei lie in the lateral gray horns of spinal segments S_2–S_4.
2. ***Ganglionic (second-order) neurons in peripheral ganglia located within or adjacent to the target organs.*** The preganglionic fibers of the parasympathetic division do not diverge as extensively as do those of the sympathetic division. A typical preganglionic fiber synapses on six to eight ganglionic neurons. In contrast to the pattern in the sympathetic division, these second-order neurons are all located in the same ganglion, and their postganglionic fibers influence the same target organ. As a result, *the effects of parasympathetic stimulation are more specific and localized than those of the sympathetic division.*

PARASYMPATHETIC DIVISION OF ANS

Preganglionic (first-order) neurons in brain stem

Preganglionic (first-order) neurons in spinal cord segments $S_2 - S_4$

Ganglionic (second-order) neurons

Target organs

N III → Ciliary ganglion → Intrinsic eye muscles (pupil and lens shape)

N VII → Sphenopalatine and submandibular ganglia → Nasal glands, tear glands, and salivary glands

N IX → Otic ganglion → Parotid salivary gland

N X → Intramural ganglia → Visceral organs of head, neck, thoracic cavity, and most of abdominal cavity

Pelvic splanchnic nerves → Intramural ganglia → Visceral organs in lower abdominopelvic cavity

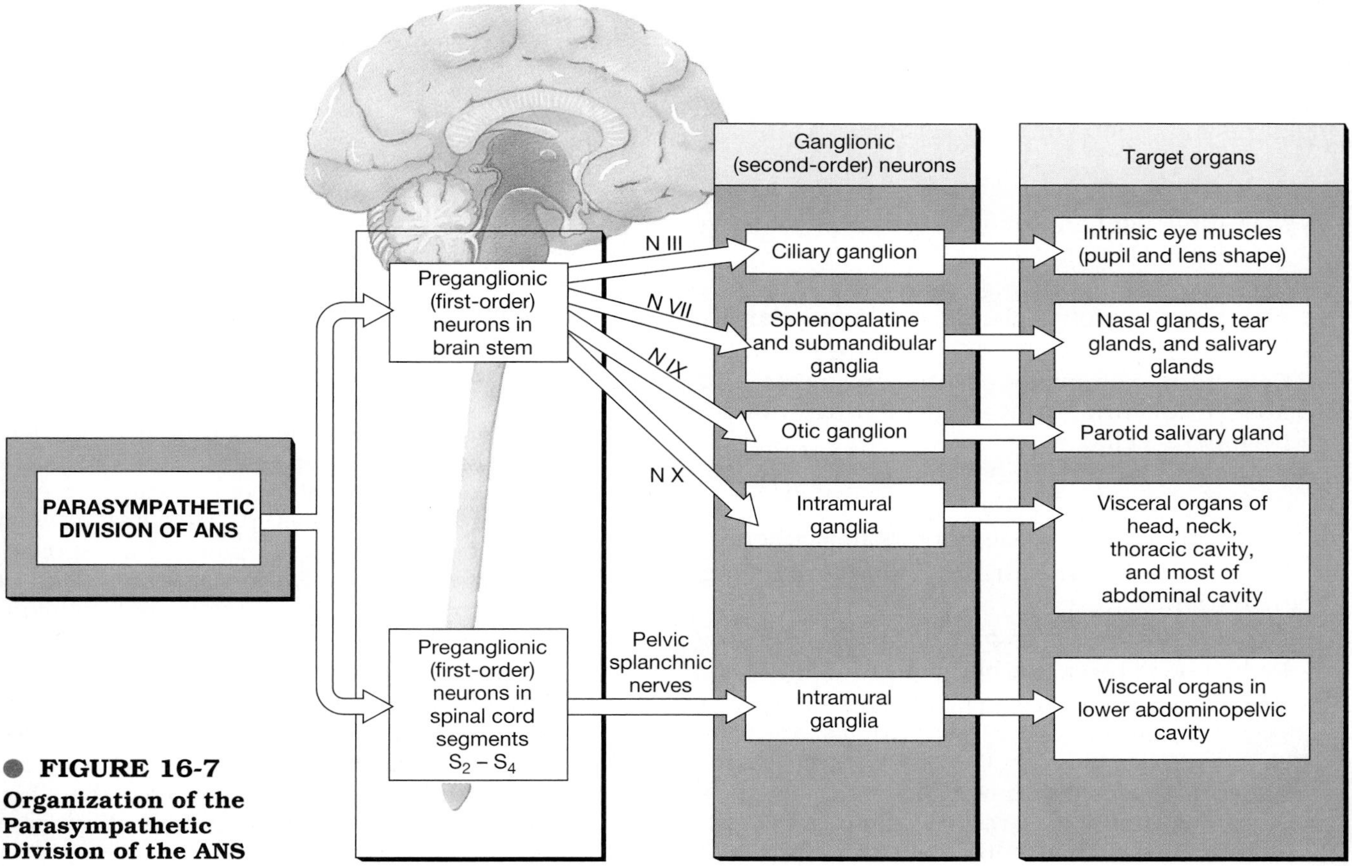

● **FIGURE 16-7**
Organization of the Parasympathetic Division of the ANS

❑ Organization and Anatomy of the Parasympathetic Division

Figure 16-8

Parasympathetic preganglionic fibers leave the brain as components of cranial nerves III (oculomotor), VII (facial), IX (glossopharyngeal), and X (vagus) (Figure 16-8●). Parasympathetic fibers in the oculomotor, facial, and glossopharyngeal nerves are concerned with the control of visceral structures in the head. These fibers synapse in the *ciliary, sphenopalatine, submandibular,* and *otic ganglia.* ∞ [pp. 486–492] Short postganglionic fibers then continue to their peripheral targets. The vagus nerve provides preganglionic parasympathetic innervation to structures in the thoracic and abdominopelvic cavity as distant as the last segments of the large intestine. The vagus nerve alone provides roughly 75 percent of all parasympathetic outflow.

The sacral parasympathetic outflow does not join the ventral roots of the spinal nerves. Instead, the preganglionic fibers form distinct **pelvic nerves** that innervate intramural ganglia in the kidney and bladder, the terminal portions of the large intestine, and the sex organs.

❑ General Functions of the Parasympathetic Division

A partial listing of the major effects produced by the parasympathetic division includes:

1. Constriction of the pupils to restrict the amount of light entering the eyes and focusing on nearby objects.
2. Secretion by digestive glands, including salivary glands, gastric glands, duodenal glands, intestinal glands, pancreas, and liver.
3. Secretion of hormones that promote the absorption and utilization of nutrients by peripheral cells.
4. Increased smooth muscle activity along the digestive tract.
5. Stimulation and coordination of defecation.
6. Contraction of the urinary bladder during urination.
7. Constriction of the respiratory passageways.
8. Reduction in heart rate and force of contraction.
9. Sexual arousal and stimulation of sexual glands in both sexes.

These functions center on relaxation, food processing, and energy absorption. The parasympathetic division has been called the *anabolic system* because stimulation leads to a general increase in the nutrient content of the blood. Cells throughout the body respond to this increase by absorbing nutrients and using them to support growth and other anabolic activities.

❑ Parasympathetic Activation and Neurotransmitter Release

All of the preganglionic and postganglionic fibers in the parasympathetic division release ACh at synapses and neuroeffector junctions. The neuroeffector junctions are small, with narrow synaptic clefts, as typified by the neuromuscular junctions of the muscular system. ∞ [p. 291] The effects of stimulation are short-lived, because most of the ACh released is inactivated by acetylcholinesterase within the synapse. Any ACh diffusing into the surrounding tissues will be inactivated by the enzyme *tissue cholinesterase.* As a result, the effects of parasympathetic stimulation are quite localized, and they last a few seconds at most.

Membrane Receptors and Responses

Although all the synapses (neuron to neuron) and neuroeffector junctions (neuron to effector) of the parasympathetic division use the same transmitter, acetylcholine, two different types of ACh receptors are found on the postsynaptic membranes.

1. **Nicotinic** (nik-ō-TIN-ik) **receptors** *are found on the surfaces of ganglion cells of both the parasympathetic and sympathetic divisions, as well as at neuromuscular junctions of the SNS.* Exposure to ACh always causes excitation of the ganglionic (second-order) neuron or muscle fiber via the opening of membrane ion channels.
2. **Muscarinic** (mus-kar-IN-ik) **receptors** *are found at cholinergic neuroeffector junctions in the parasympathetic division, as well as at the few cholinergic neuroeffector junctions in the sympathetic division.* Stimulation of muscarinic receptors produces longer-lasting effects than does stimulation of nicotinic receptors. The response, which reflects the activation or inactivation of specific enzymes, may be excitation or inhibition.

The names *nicotinic* and *muscarinic* indicate the chemical compounds that stimulate these receptor sites. Nicotinic receptors bind *nicotine,* a powerful toxin that can be obtained from a variety of sources, including tobacco leaves. Muscarinic receptors are stimulated by *muscarine,* a toxin produced by some poisonous mushrooms.

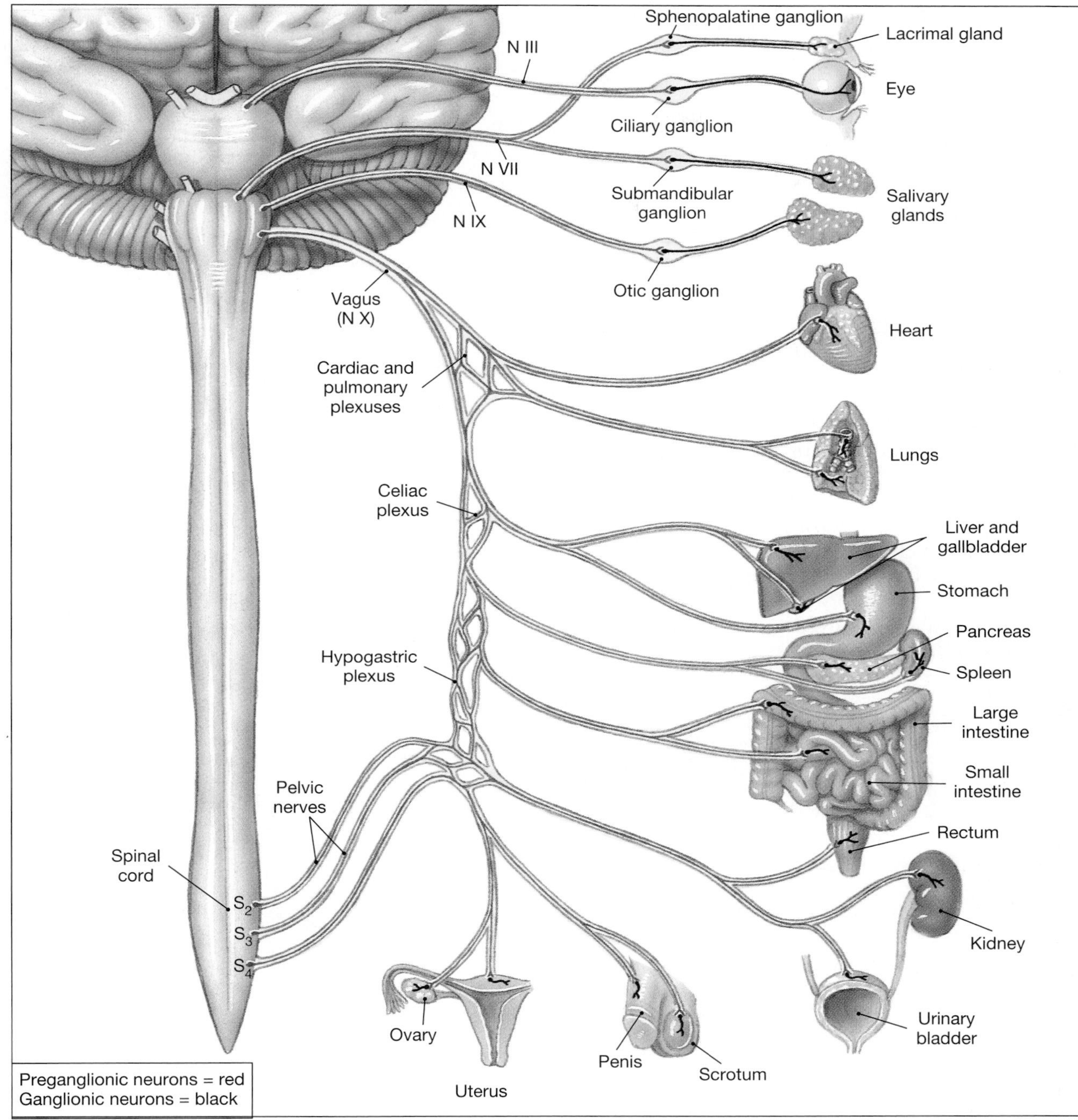
Sphenopalatine ganglion
Lacrimal gland
N III
Eye
Ciliary ganglion
N VII
Submandibular ganglion
Salivary glands
N IX
Otic ganglion
Vagus (N X)
Heart
Cardiac and pulmonary plexuses
Lungs
Celiac plexus
Liver and gallbladder
Stomach
Pancreas
Hypogastric plexus
Spleen
Large intestine
Small intestine
Pelvic nerves
Rectum
Spinal cord
S_2
S_3
S_4
Kidney
Ovary
Urinary bladder
Penis
Scrotum
Uterus
Preganglionic neurons = red
Ganglionic neurons = black

FIGURE 16-8
Distribution of Parasympathetic Innervation

These compounds have discrete actions, targeting either the autonomic ganglia and skeletal neuromuscular junctions (nicotine) or the parasympathetic neuroeffector junctions (muscarine). They produce dangerously exaggerated, uncontrolled responses that parallel those produced by normal receptor stimulation. For example, nicotine poisoning occurs if as little as 50 mg of the compound is ingested or absorbed through the skin. The symptoms reflect widespread autonomic activation—vomiting, diarrhea, high blood pressure, rapid heart rate, sweating, and profuse salivation. Because the neuromuscular junctions of the SNS are stimulated, convulsions occur. In severe cases, stimulation of nicotinic receptors inside the CNS may lead to coma and death within minutes. The symptoms of muscarine poisoning are almost entirely restricted to the parasympathetic division: salivation, nausea, vomiting, diarrhea, constricted respiratory passages, low blood pressure, and an abnormally slow heart rate.

Table 16-1 summarizes details concerning the adrenergic and cholinergic receptors in the ANS.

A Summary of the Parasympathetic Division

1. The parasympathetic division includes visceral motor nuclei associated with four cranial nerves (III, VII, IX, and X) and in sacral segments S_2–S_4.
2. The second-order neurons are situated in intramural ganglia or in ganglia closely associated with their target organs.
3. The parasympathetic division innervates areas serviced by the cranial nerves and organs in the thoracic and abdominopelvic cavities.
4. All parasympathetic neurons are cholinergic. Ganglionic neurons have nicotinic receptors that are excited by ACh. Muscarinic receptors present at neuroeffector junctions may produce either excitation or inhibition, depending on the nature of the enzymes activated when ACh binds to the receptor.
5. The effects of parasympathetic stimulation are usually brief and restricted to specific organs and sites.

- ✔ What nerve is responsible for parasympathetic innervation of the lungs, heart, stomach, liver, pancreas, and parts of the small and large intestine?
- ✔ What effect would stimulation of muscarinic receptors in cardiac muscle have on the heart?
- ✔ Why is the parasympathetic division of the ANS sometimes referred to as the anabolic system?

Interactions of the Sympathetic and Parasympathetic Divisions

The sympathetic division has widespread impact, reaching visceral organs and tissues throughout the body. The parasympathetic division innervates only visceral structures serviced by the cranial nerves or lying within the abdominopelvic cavity. Although some organs are innervated by one division or the other, most vital organs receive **dual innervation**—that is, they receive instructions from both sympathetic and parasympathetic divisions. Where dual

TABLE 16-1 Adrenergic and Cholinergic Receptors in the ANS

Receptor	*Location*	*Response*	*Mechanism*
ADRENERGIC RECEPTORS			
α_1	Widespread, found in most tissues; not in heart	Excitation, stimulation of metabolism	Activation of enzymes, release of intracellular calcium ions
α_2	Sympathetic neuroeffector junctions	Inhibition of effector cell	Reduction in cAMP concentrations
	Parasympathetic neuroeffector junctions	Inhibition of neurotransmitter release	Reduction in cAMP concentrations
β_1	Heart, kidneys, liver, adipose tissue*	Stimulation, increased energy consumption	Enzyme activation
β_2	Smooth muscle in vessels of heart and skeletal muscle, intestinal muscle, lungs, and bronchi	Inhibition, relaxation	Enzyme activation
CHOLINERGIC RECEPTORS			
Nicotinic	All autonomic synapses between preganglionic and ganglionic neurons; also neuromuscular junctions of the SNS	Stimulation, excitation	Opening of chemically regulated Na^+ channels
Muscarinic	All parasympathetic neuroeffector junctions; cholinergic sympathetic neuroeffector junctions	Variable	Enzyme activation causing changes in membrane permeability to K^+

*Adipocytes also contain an additional receptor type, β_3, not found in other tissues.

innervation exists, the two divisions often have opposing effects. Dual innervation is most prominent in the digestive tract, the heart, and the lungs. Secretory control of the salivary glands or the sexual functions of the male reproductive tract are also examples.

Anatomy of Dual Innervation

Figure 16-9

In the head, parasympathetic postganglionic fibers from the ciliary, sphenopalatine, submandibular, and otic ganglia accompany the cranial nerves to their peripheral destinations. The sympathetic innervation reaches the same structures by traveling directly from the superior cervical ganglia of the sympathetic chain.

In the thoracic and abdominopelvic cavities, the sympathetic postganglionic fibers mingle with parasympathetic preganglionic fibers at a series of plexuses (Figure 16-9●). These are the *cardiac plexus*, the *pulmonary plexus*, the *esophageal plexus*, the *celiac plexus*, the *inferior mesenteric plexus*, and the *hypogastric plexus*. Nerves leaving these plexuses travel with the blood vessels and lymphatics that supply visceral organs.

Autonomic fibers entering the thoracic cavity intersect at the **cardiac plexus** and the **pulmonary plexus.** These plexuses contain sympathetic and parasympathetic fibers bound for the heart and lungs, respectively, as well as the parasympathetic ganglia whose output affects those organs. The **esophageal plexus** contains descending branches of the vagus nerve and splanchnic nerves leaving the sympathetic chain on either side.

Parasympathetic preganglionic fibers of the vagus nerve enter the abdominopelvic cavity with the esophagus. There they join the network of the **celiac plexus,** also known as the *solar plexus.* The celiac plexus and associated smaller plexuses, such as the **inferior mesenteric plexus,** innervate viscera down to the initial segments of the large intestine. The **hypogastric plexus** contains the parasympathetic outflow of the pelvic nerves and sympathetic postganglionic fibers from the inferior mesenteric ganglion and splanchnic nerves from the sacral sympathetic chain. This plexus innervates the digestive, urinary, and reproductive organs of the pelvic cavity.

● **FIGURE 16-9**
The Peripheral Autonomic Plexuses

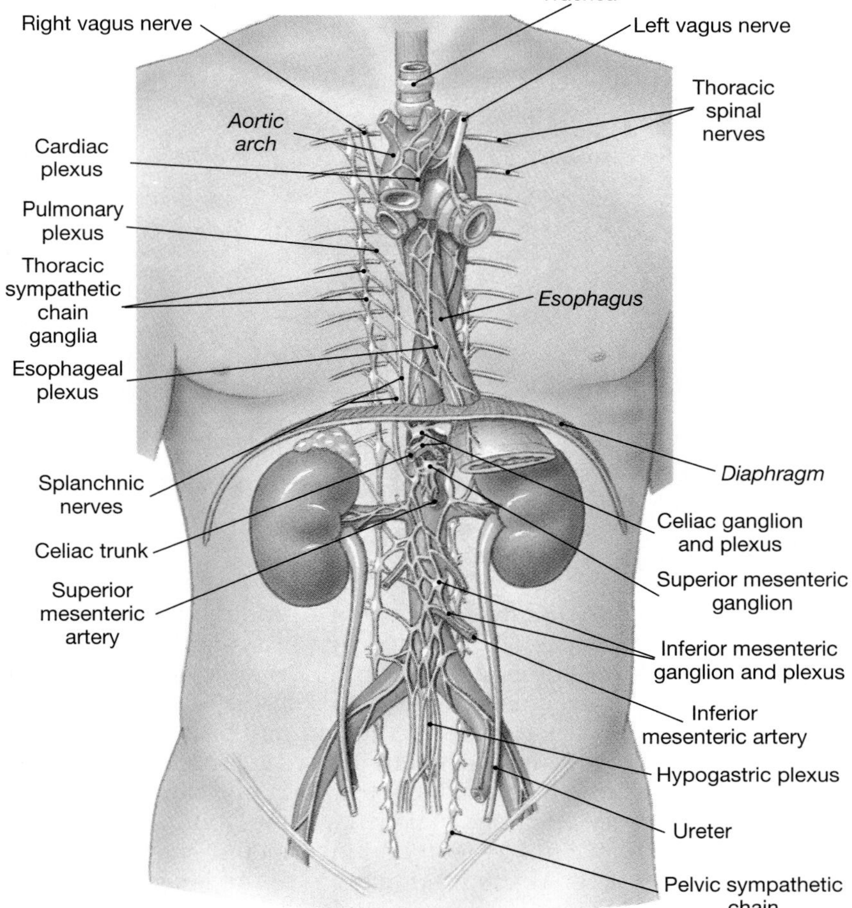
Trachea
Right vagus nerve
Left vagus nerve
Thoracic spinal nerves
Aortic arch
Cardiac plexus
Pulmonary plexus
Thoracic sympathetic chain ganglia
Esophagus
Esophageal plexus
Diaphragm
Splanchnic nerves
Celiac ganglion and plexus
Celiac trunk
Superior mesenteric ganglion
Superior mesenteric artery
Inferior mesenteric ganglion and plexus
Inferior mesenteric artery
Hypogastric plexus
Ureter
Pelvic sympathetic chain

A Comparison of the Sympathetic and Parasympathetic Divisions

Table 16-2 compares key features of the sympathetic and parasympathetic divisions of the ANS. The distinctions have physiological and functional correlates. Table 16-3 provides a more detailed comparison, considering the effect of sympathetic or parasympathetic activity on specific organs and systems.

Autonomic Tone

Even in the absence of stimuli, autonomic motor neurons show a resting level of spontaneous activity. The level of activation determines the **autonomic tone** of the individual. Autonomic tone is an important aspect of ANS function. If a nerve is absolutely silent under normal conditions, then all it can do is increase its activity on demand. But if the nerve maintains a background level of activity, it may either increase or decrease its activity, providing a range of control options.

Autonomic tone is significant where dual innervation occurs and the ANS

TABLE 16-2 A Comparison of the Sympathetic and Parasympathetic Divisions of the ANS

Characteristic	*Sympathetic Division*	*Parasympathetic Division*
Location of CNS visceral motor neuron	Lateral gray horns, spinal segments T_1–L_2	Brain stem and spinal segments S_2–S_4
Location of PNS ganglia	Near spinal column	Typically intramural
Preganglionic fibers		
Length	Relatively short	Relatively long
Neurotransmitter released	Acetylcholine	Acetylcholine
Postganglionic fibers		
Length	Relatively long	Relatively short
Neurotransmitter released	Usually norepinephrine	Always acetylcholine
Neuroeffector junction	Enlarged terminal knobs that release transmitter near target cells	Neuroeffector junctions releasing transmitter to special receptor surface
Degree of divergence from CNS to ganglion cells	Approximately 1:32	Approximately 1:6
General function	Stimulate metabolism, increase alertness, prepare for emergency ("fight or flight")	Promote relaxation, nutrient uptake, energy storage ("rest and repose")
CENTRAL NERVOUS SYSTEM		
PERIPHERAL NERVOUS SYSTEM		

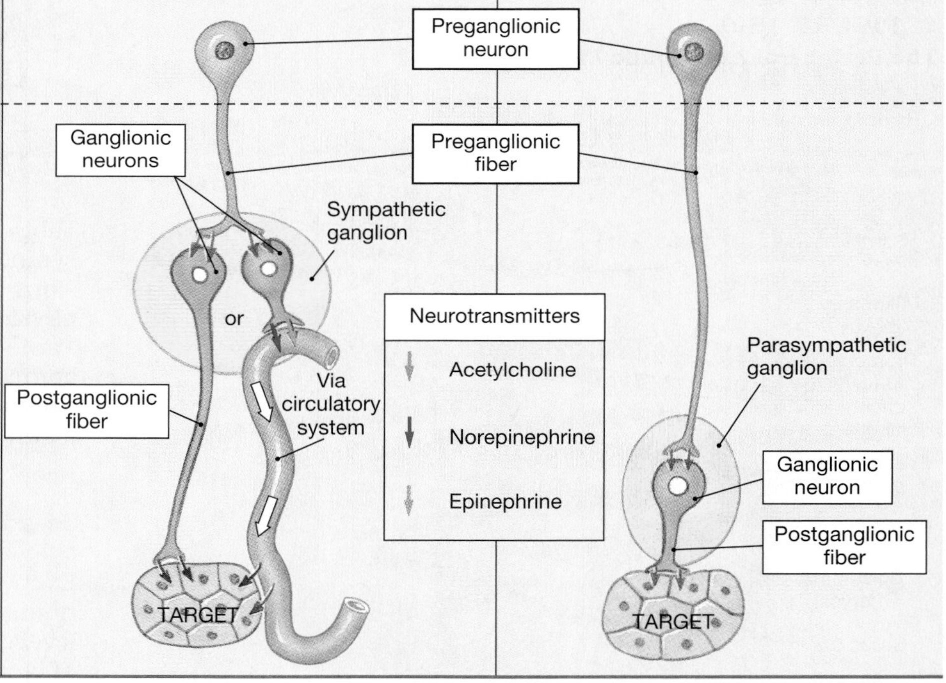

Preganglionic neuron
Ganglionic neurons
Preganglionic fiber
Sympathetic ganglion
or
Neurotransmitters
Acetylcholine
Norepinephrine
Epinephrine
Via circulatory system
Postganglionic fiber
Parasympathetic ganglion
Ganglionic neuron
Postganglionic fiber
TARGET
TARGET

divisions have opposing effects. It is even more important in situations where dual innervation does not occur. We will consider one example of each arrangement, to demonstrate how autonomic tone affects ANS function.

Autonomic Tone in Dual Innervation

The heart is an example of an organ that receives dual innervation, and the two autonomic divisions have opposing effects. ACh released by the parasympathetic division causes a reduction in heart rate, whereas NE released by the sympathetic division accelerates the heart rate. Because autonomic tone exists, small amounts of both these neurotransmitters are released on a continual basis. By means of small adjustments in parasympathetic versus sympathetic stimulation, the heart rate can be controlled very precisely. In a crisis, stimulation of the sympathetic innervation and inhibition of the parasympathetic innervation accelerate the heart rate to the maximum extent possible.

TABLE 16-3 A Functional Comparison of the Sympathetic and Parasympathetic Divisions of the ANS

Structure	*Sympathetic Receptor Type*	*Sympathetic Innervation Effect*	*Parasympathetic Innervation Effect (All muscarinic receptors)*
EYE	α_1	Dilation of pupil, accommodation for distance vision	Constriction of pupil, accommodation for close vision
SALIVARY GLANDS	α_1, β_1	Serous secretion stimulated	Mucous secretion stimulated
SKIN			
Sweat glands	α_1	Increased secretion	None (not innervated)
Arrector pili	α_1	Contraction, erection of hairs	None (not innervated)
TEAR GLANDS		None (not innervated)	Secretion
CARDIOVASCULAR SYSTEM			
Blood vessels			None (not innervated)
To integument	α_1	Vasoconstriction	
To skeletal muscles	β_2	Vasodilation	
To heart	β_2	Vasodilation	
To digestive viscera	α_1	Vasoconstriction	
Veins	α_1, β_1	Constriction	
Heart	β_1	Increased heart rate, force of contraction, and blood pressure	Decreased heart rate, force of contraction, and blood pressure
ADRENAL GLAND		Secretion of epinephrine, norepinephrine by medulla	None (not innervated)
POSTERIOR PITUITARY	β_1	Secretion of ADH	None (not innervated)
RESPIRATORY SYSTEM			
Airways	β_2	Increased diameter	Decreased diameter
Respiratory rate		Increased	Decreased
DIGESTIVE SYSTEM			
Sphincters	α_1	Constriction	Dilation
General level of activity	α_2, β_2	Decreased	Increased
Secretory glands	α_2	Inhibition	Stimulation
Liver	α_1, β_2	Glycogen breakdown, glucose synthesis and release	Glycogen synthesis
Pancreas	α_1	Decreased exocrine secretion	Increased exocrine secretion
	α_2	Decreased hormone (insulin) secretion	Increased hormone (insulin) secretion
SKELETAL MUSCLES	β_2	Increased force of contraction, glycogen breakdown	None (not innervated)
	α_2	Facilitation of ACh release at neuromuscular junction	None (not innervated)
ADIPOSE TISSUE	β_1, β_3	Lipolysis, fatty acid release	
URINARY SYSTEM			
Kidneys	β_2	Decreased urine production	Increased urine production
Bladder	α_1, β_2	Constriction of sphincter, relaxation of bladder	Tensing of bladder, relaxation of sphincter to eliminate urine
MALE REPRODUCTIVE SYSTEM	α_1	Increased glandular secretion and ejaculation	Erection
FEMALE REPRODUCTIVE SYSTEM	α_1	Increased glandular secretion; contraction of pregnant uterus	Variable (depending on hormones present)
	β_2	Relaxation of nonpregnant uterus	Variable (depending on hormones present)

Autonomic Tone in the Absence of Dual Innervation

Several organs are innervated by one division only. For example, most structures in the body wall, such as blood vessels, glands, adipose tissue, and skeletal muscles, do not receive parasympathetic innervation. On the other hand, the constrictor muscles of the pupil, the tear glands, and nasal glands are not innervated by the sympathetic division.

The sympathetic control of blood vessel diameter provides an example of how autonomic tone allows fine adjustment of peripheral activities when the target organ is not innervated by both ANS divisions. Sympathetic postganglionic fibers innervate the smooth muscle cells in the walls of peripheral vessels. The background sympathetic tone keeps these muscles partially contracted, so that the vessels are ordinarily at roughly half of their maximum diameter. Because the normal diameter is maintained by sympathetic tone, increasing or decreasing sympathetic stimulation provides precise control of vessel diameter over its entire range. When the vessel dilates, blood flow to the region increases; when the diameter decreases, blood flow is reduced.

Integration and Control of Autonomic Functions

Figure 15-7•, p. 514, diagrammed the relationships between centers involved with the control of somatic motor control. The lowest level of regulatory control consisted of the lower motor neurons involved in cranial and spinal reflex arcs. The highest level consisted of the pyramidal motor neurons of the primary motor cortex, operating with the assistance of extrapyramidal and cerebellar nuclei.

The ANS is also organized into a series of interacting levels. At the bottom are visceral motor neurons in the lower brain stem and spinal cord that are involved in cranial and spinal *visceral reflexes*. Visceral reflexes provide automatic motor responses that can be modified, facilitated, or inhibited by higher centers, especially those of the hypothalamus.

For example, shining a light in the eye triggers a visceral reflex that constricts the pupils of both eyes (the *consensual light reflex*, described in Chapter 14). [p. 494] The visceral motor commands are distributed by parasympathetic fibers. In darkness, the pupils dilate; this *pupillary reflex* is directed by sympathetic postganglionic fibers. However, the motor nuclei directing pupillary constriction or dilation are also controlled by hypothalamic centers concerned with emotional states. When you are queasy or nauseated, your pupils constrict; when you are sexually aroused, your pupils dilate.

MODIFYING AUTONOMIC NERVOUS SYSTEM FUNCTION Drugs may be administered to counteract or reduce symptoms either caused or aggravated by autonomic activities. These drugs are called **mimetic** if they mimic the activity of one of the normal autonomic transmitters. Drugs that reduce the effects of autonomic stimulation by keeping the neurotransmitter from affecting the postsynaptic membranes are known as **blocking agents.** Mimetic drugs often have advantages over natural neurotransmitters. For example, norepinephrine and epinephrine must be administered into the bloodstream by injection or infusion, and the side effects are short-lived. In contrast, **sympathomimetic drugs** may survive oral administration, produce longer-lasting effects, or have more specific actions than E or NE. Examples include drugs applied topically to reduce hemorrhaging, by spray to reduce nasal congestion, by inhalation to dilate the respiratory passageways, or in drops to dilate the pupils.

Sympathetic blocking agents bind to the receptor sites and prevent a normal response to the presence of neurotransmitters or sympathomimetic drugs. **Alpha-blockers** eliminate the peripheral vasoconstriction that accompanies sympathetic stimulation, and **beta-blockers** are effective and clinically useful for treating chronic high blood pressure and other forms of cardiovascular disease.

Parasympathomimetic drugs may also be used to increase the activity along the digestive tract and encourage defecation and urination. **Parasympathetic blocking agents** target the muscarinic receptors at the neuroeffector junctions. These drugs have diverse effects, but they are often used to control the diarrhea and cramps associated with various forms of food poisoning. For information on specific mimetic drugs and blocking agents and details on their modes of action, see the *Applications Manual.* AM *Pharmacology and the Autonomic Nervous System*

Visceral Reflexes

Figure 16-10

Each visceral reflex arc consists of a receptor, a sensory neuron, a processing center (interneuron or visceral motor neuron), and two visceral motor neurons (Figure 16-10•). All visceral reflexes are polysynaptic.

Visceral sensory neurons deliver information to the CNS along the dorsal roots of spinal nerves, within the sensory branches of cranial nerves, and within the autonomic nerves that innervate peripheral effectors. Figure 13-5•, p. 435, diagrammed the distribution of visceral sensory fibers in a representative spinal nerve.

As we examine other body systems in later chapters, we will encounter many examples of autonomic reflexes involved in respiration, cardiovascular function, and other visceral activities. Some

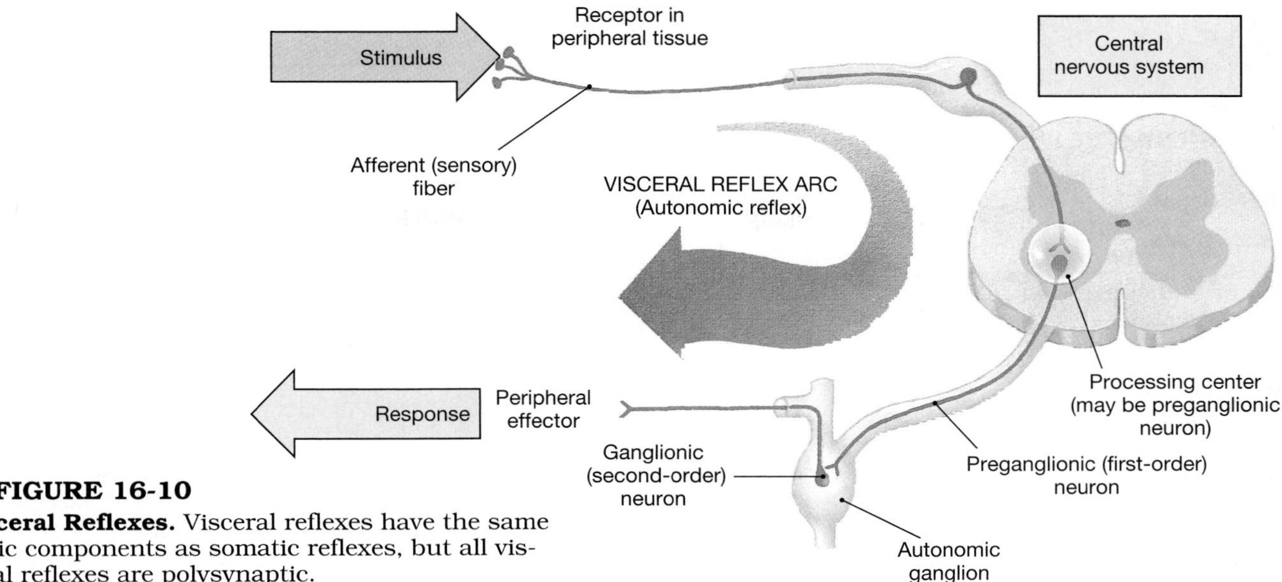
Stimulus
Receptor in peripheral tissue
Central nervous system
Afferent (sensory) fiber
VISCERAL REFLEX ARC (Autonomic reflex)
Response
Peripheral effector
Ganglionic (second-order) neuron
Processing center (may be preganglionic neuron)
Preganglionic (first-order) neuron
Autonomic ganglion

● **FIGURE 16-10**
Visceral Reflexes. Visceral reflexes have the same basic components as somatic reflexes, but all visceral reflexes are polysynaptic.

of the most important are previewed in Table 16-4. Note that the parasympathetic division participates in reflexes affecting individual organs and systems. This specialization reflects the relatively specific and restricted pattern of innervation. In contrast, there are fewer sympathetic reflexes. This division is typically activated as a whole, in part because of the degree of divergence and in part because the release of hormones by the adrenal medullae produces widespread peripheral effects.

❑ Higher Levels of Autonomic Control

Figure 16-11

The levels of activity in the sympathetic and parasympathetic divisions are controlled by centers in the brain stem concerned with specific visceral functions. Figure 16-11● diagrams the levels of autonomic control. As in the SNS, simple reflexes based in the spinal cord provide relatively rapid and automatic responses to stimuli. More complex sympathetic and parasympathetic reflexes are coordinated by processing centers in the medulla oblongata. In addition to the cardiovascular and respiratory centers, the medulla contains centers and nuclei involved with salivation, digestive secretions, peristalsis, and urinary function. These medullary centers are in turn subject to regulation by the hypothalamus. [pp. 477–479]

The term *autonomic* was originally applied to the visceral motor system because it was thought that the regulatory centers functioned without regard for other CNS activities. This view has been drastically revised in light of subsequent research. Because the hypothalamus interacts with all other

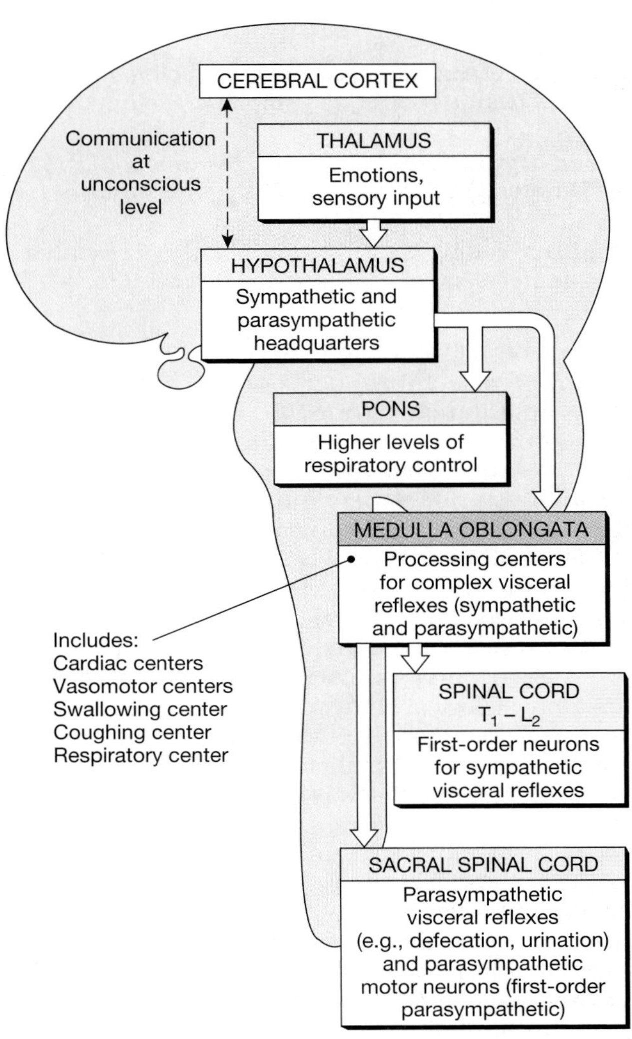
CEREBRAL CORTEX
Communication at unconscious level
THALAMUS
Emotions, sensory input
HYPOTHALAMUS
Sympathetic and parasympathetic headquarters
PONS
Higher levels of respiratory control
MEDULLA OBLONGATA
Processing centers for complex visceral reflexes (sympathetic and parasympathetic)
Includes:
Cardiac centers
Vasomotor centers
Swallowing center
Coughing center
Respiratory center
SPINAL CORD
$T_1 - L_2$
First-order neurons for sympathetic visceral reflexes
SACRAL SPINAL CORD
Parasympathetic visceral reflexes (e.g., defecation, urination) and parasympathetic motor neurons (first-order parasympathetic)

● **FIGURE 16-11**
Levels of Autonomic Control

TABLE 16-4 Representative Visceral Reflexes

Reflex	*Stimulus*	*Response*	*Comments*
PARASYMPATHETIC REFLEXES			
Urination *(Chapter 26)*	Distention of urinary bladder	Contraction of bladder walls	Requires voluntary relaxation of external sphincter
Light and consensual light reflexes *(Chapter 14, p. 494)*	Bright light shining in eye(s)	Constriction of pupils of both eyes	
Swallowing reflex *(Chapter 24)*	Movement of material into upper pharynx	Smooth muscle contractions moving material to stomach	Coordinated by medullary swallowing center
Coughing reflex *(Chapter 23)*	Irritation of respiratory tract	Sudden explosive ejection of air	Coordinated by medullary coughing center
Cardioinhibitory reflex *(Chapter 20)*	Sudden rise in carotid blood pressure	Reduction in heart rate and force of contraction	Coordinated in cardiac center in medulla oblongata
SYMPATHETIC REFLEXES			
Cardioacceleratory reflex *(Chapter 20)*	Sudden decline in carotid blood pressure	Increased heart rate and force of contraction	Coordinated in cardiac centers in medulla oblongata
Vasomotor reflexes *(Chapter 21)*	Changes in blood pressure in major arteries	Changes in diameter of peripheral vessels	Coordinated in vasomotor center in medulla oblongata
Pupillary reflex *(Chapter 17)*	Low light level reaching visual receptors	Dilation of pupil	

portions of the brain, activity in the limbic system, thalamus, or cerebral cortex can have dramatic effects on autonomic function. For example, when you become angry, your heart rate accelerates, your blood pressure rises, your respiratory rate increases; when you remember your last big dinner, your stomach "growls" and your mouth waters.

Although conscious thought processes have an effect on the autonomic nervous system, the normal individual does not perceive the effect because visceral sensory information does not reach the cerebral cortex. Even when a conscious mental process triggers a physiological shift, such as a change in blood pressure, sweat gland activity, skin temperature, or muscle tone, the information never arrives at the sensory cortex. **Biofeedback** is an attempt to bridge this gap. In this technique, a person's physiological processes are monitored, and a visual or auditory signal is used to alert the subject when a particular change takes place. These signals let the individual know when ongoing conscious thought processes have triggered a desirable change in autonomic function. For example, when biofeedback is used to regulate blood pressure, a light or tone informs the subject when blood pressure drops. With practice, some people can learn to recreate the proper mood or thought pattern that will lower their blood pressure.

Biofeedback techniques have been used to promote conscious control of blood pressure, heart rate, circulatory pattern, skin temperature, brain waves, and so forth. By reducing stress, lowering blood pressure, and improving circulation, these techniques can reduce the severity of clinical symptoms. In the process, they also lower the risks for serious complications, such as heart attacks or strokes, in patients with high blood pressure. Unfortunately, not everyone can learn to influence autonomic functions, and the combination of variable results and expensive equipment makes biofeedback unsuitable for widespread application.

What effect would loss of sympathetic tone have on blood flow to a tissue?

What physiological changes would you expect to observe in a patient who is about to have a root canal performed and who is quite anxious about the procedure?

Harry has a brain tumor that is interfering with the function of the hypothalamus. Would you expect this tumor to interfere with autonomic function? Why or why not?

Selected Clinical Terminology

Terms Discussed in This Chapter

alpha-blockers: Drugs that eliminate the peripheral vasoconstriction that accompanies sympathetic stimulation. *(p. 544* and *AM)*

beta-blockers: Drugs that decrease heart rate and force of contraction, lowering peripheral blood pressure. *(p. 544* and *AM)*

parasympathetic blocking agents: Drugs that target the muscarinic receptors at neuroeffector junctions. *(p. 544* and *AM)*

parasympathomimetic drugs: Drugs that mimic parasympathetic stimulation and increase the activity along the digestive tract. *(p. 544* and *AM)*

sympathetic blocking agents: Drugs that bind to receptor sites, preventing a normal response to neurotransmitters or sympathomimetic drugs. *(p. 544* and *AM)*

sympathomimetic drugs: Drugs that mimic the effects of sympathetic stimulation. *(p. 544* and *AM)*

CHAPTER REVIEW

STUDY OUTLINE

INTRODUCTION, p. 528

1. The autonomic nervous system (ANS) coordinates cardiovascular, respiratory, digestive, excretory, and reproductive functions.

AN OVERVIEW OF THE ANS, p. 528

1. **Preganglionic neurons,** or *first-order neurons,* in the CNS send axons to synapse on **ganglionic neurons,** or *second-order neurons,* in **autonomic ganglia** outside the CNS. (*Figure 16-1*)

Subdivisions of the ANS, p. 528

2. Visceral efferents from the thoracic and lumbar segments form the *thoracolumbar division,* or **sympathetic division** ("fight or flight" system) of the ANS. Visceral efferents leaving the brain and sacral segments form the *craniosacral division,* or **parasympathetic division** ("rest and repose" system). (*Figure 16-2*)

THE SYMPATHETIC DIVISION, p. 530

1. The sympathetic division consists of preganglionic (first-order) neurons between segments T_1 and L_2, ganglionic (second-order) neurons in ganglia near the vertebral column, and specialized neurons inside the adrenal gland. (*Figures 16-3, 16-4, 16-5*)
2. There are two types of sympathetic ganglia: **sympathetic chain ganglia** (paravertebral ganglia) and **collateral ganglia** (prevertebral ganglia). (*Figures 16-3, 16-4*)

The Sympathetic Chain, p. 531

3. Postganglionic fibers targeting structures in the body wall and limbs rejoin the spinal nerves and reach their destinations via the dorsal and ventral rami. (*Figure 16-4*)
4. Postganglionic fibers targeting structures in the thoracic cavity form **autonomic nerves** that go directly to their visceral destination. Preganglionic fibers run between the sympathetic chain ganglia and interconnect them. (*Figure 16-4*)

Collateral Ganglia, p. 531

5. The abdominopelvic viscera receive sympathetic innervation via preganglionic fibers that synapse within collateral ganglia. The preganglionic fibers that innervate the collateral ganglia form the **splanchnic nerves.** (*Figure 16-4*)
6. The **celiac ganglion** innervates the stomach, liver, pancreas, and spleen; the **superior mesenteric ganglion** innervates the small intestine and initial segments of the large intestine; and the **inferior mesenteric ganglion** innervates the kidney, bladder, sex organs, and terminal portions of the large intestine. (*Figures 16-5, 16-9*)

The Adrenal Medullae, p. 533

7. Preganglionic fibers entering an adrenal gland synapse within the **adrenal medulla.** (*Figures 16-4, 16-5*)

Sympathetic Activation, p. 534

8. In a crisis, the entire division responds, an event called **sympathetic activation.** Its effects include: increased alertness, a feeling of energy and euphoria, increased cardiovascular and respiratory activity, and general elevation in muscle tone.

Neurotransmitters and Sympathetic Function, p. 535

9. Stimulation of the sympathetic division has two distinctive results: the release of norepinephrine at specific locations and secretion of epinephrine and norepinephrine into the general circulation.
10. There are two types of sympathetic receptors: **alpha receptors** (which respond to NE or E by depolarizing the membrane) and **beta receptors** (which are particularly sensitive to E).
11. Most postganglionic fibers are adrenergic; a few are cholinergic or nitroxidergic. Postganglionic fibers innervating sweat glands and blood vessels to skeletal muscles release ACh. Postganglionic fibers innervating smooth muscle in the walls of small arteries in skeletal muscles and the brain release nitric oxide (NO), promoting vasodilation. (*Figure 16-6*)

A Summary of the Sympathetic Division, p. 536

12. Preganglionic sympathetic fibers are relatively short. Except for those of the adrenal medulla, postganglionic fibers are quite long. Each fiber typically synapses with many ganglionic neurons in different ganglia. A single effector cell may have more than one type of receptor,

so that the result of sympathetic stimulation depends on interaction among the receptors.

THE PARASYMPATHETIC DIVISION, p. 537

Organization and Anatomy of the Parasympathetic Division, p. 538

1. The parasympathetic division includes preganglionic (first-order) neurons in the brain stem and sacral segments of the spinal cord and ganglionic (second-order) neurons in peripheral ganglia located within or next to target organs. (*Figure 16-7*)
2. Preganglionic fibers leaving the sacral segments form **pelvic nerves.** (*Figure 16-8*)

General Functions of the Parasympathetic Division, p. 538

3. The effects produced by the parasympathetic division center on relaxation, food processing, and energy absorption.

Parasympathetic Activation and Neurotransmitter Release, p. 538

4. All parasympathetic preganglionic and postganglionic fibers release ACh at synapses and neuroeffector junctions. The effects are short-lived because of the actions of cholinesterase.
5. Two different ACh receptors are found in postsynaptic membranes. Stimulation of **muscarinic receptors** produces a longer-lasting effect than does stimulation of **nicotinic receptors.** (*Table 16-1*)

A Summary of the Parasympathetic Division, p. 540

6. The parasympathetic division innervates areas serviced by cranial nerves and organs in the thoracic and abdominopelvic cavities. All preganglionic and postganglionic parasympathetic neurons are cholinergic, and the effects of stimulation are usually brief and restricted to specific sites.

INTERACTIONS OF THE SYMPATHETIC AND PARASYMPATHETIC DIVISIONS, p. 540

1. The sympathetic division has widespread influence on visceral and somatic structures.
2. The parasympathetic division innervates only visceral structures serviced by cranial nerves or lying within the abdominopelvic cavity. Organs with **dual innervation** receive input from both divisions.

Anatomy of Dual Innervation, p. 541

3. In body cavities the parasympathetic and sympathetic nerves intermingle to form a series of characteristic nerve plexuses (nerve networks), which include the **cardiac, pulmonary, esophageal, celiac, inferior mesenteric,** and **hypogastric plexuses.** (*Figure 16-9*)

A Comparison of the Sympathetic and Parasympathetic Divisions, p. 541

4. There are important physiological and functional differences between the sympathetic and parasympathetic divisions. (*Tables 16-2, 16-3*)
5. Parasympathetic effects include decreased metabolic rate, heart rate, and blood pressure; salivary and digestive gland secretion; increased digestive tract activity; and urination and defecation.
6. Sympathetic effects include mental alertness; increased metabolic rate; suspended digestive and urinary tract function; activation of energy reserves; increased respiratory rate and efficiency; increased heart rate and blood pressure; and activation of sweat glands.

Autonomic Tone, p. 541

7. Even when stimuli are absent, autonomic motor neurons show a resting level of activation, the **autonomic tone.**

INTEGRATION AND CONTROL OF AUTONOMIC FUNCTIONS, p. 544

Visceral Reflexes, p. 544

1. *Visceral reflexes* are the simplest function of the ANS. (*Figure 16-10*)
2. Parasympathetic reflexes include: (1) urination; (2) light and consensual light reflexes; (3) swallowing, coughing, gagging, and sneezing; and (4) cardioinhibitory reflexes. Sympathetic reflexes include: (1) cardioacceleratory reflexes; (2) vasomotor reflexes; (3) pupillary dilation reflex. (*Table 16-4*)

Higher Levels of Autonomic Control, p. 545

3. The hypothalamus receives input from many other portions of the brain, including the cerebral cortex. Centers in the hypothalamus are concerned with the coordination of sympathetic and parasympathetic function. The regulatory activities occur outside conscious awareness and control. (*Figure 16-11*)

REVIEW QUESTIONS

LEVEL 1 Reviewing Facts and Terms

1. There is always a synapse between the CNS and the peripheral effector in the:
   (a) autonomic nervous system
   (b) somatic nervous system
   (c) reflex arc
   (d) a, b, and c are correct

2. The preganglionic fibers of the sympathetic division of the ANS originate in the:
   (a) cerebral cortex of the brain
   (b) medulla oblongata
   (c) craniosacral region of the brain stem and spinal cord
   (d) thoracolumbar region of the spinal cord

3. All preganglionic autonomic fibers release ______ at their synaptic terminals, and the effects are always ____.
(a) norepinephrine; inhibitory
(b) norepinephrine; excitatory
(c) acetylcholine; excitatory
(d) acetylcholine; inhibitory

4. Second-order neurons that innervate tissues and organs in the abdominopelvic cavity are located in:
(a) sympathetic chain ganglia
(b) collateral ganglia
(c) paravertebral ganglia
(d) lateral ganglia

5. In the dorsal wall of the abdominal cavity, preganglionic fibers that innervate the collateral ganglia form the:
(a) splanchnic nerves (b) phrenic nerve
(c) sciatic nerve (d) adrenal medulla

6. Postganglionic fibers from the celiac ganglion innervate the:
(a) heart, lungs, thymus, and diaphragm
(b) cerebrum, pons, medulla, and cerebellum
(c) stomach, liver, pancreas, and spleen
(d) large intestine, kidney, urinary bladder, and sex organs

7. The effect of the neurotransmitter on the target cell depends on the nature of the:
(a) neurotransmitter
(b) receptor on the presynaptic membrane
(c) receptor on the postsynaptic membrane
(d) a, b, and c are correct

8. Approximately 75 percent of parasympathetic outflow is provided by the:
(a) vagus nerve
(b) sciatic nerve
(c) glossopharyngeal nerves
(d) pelvic nerves

9. The neurotransmitter at all synapses and neuroeffector junctions in the parasympathetic division of the ANS is:
(a) epinephrine (b) norepinephrine
(c) cyclic-AMP (d) acetylcholine

10. How does the emergence of sympathetic fibers from the spinal cord differ from the emergence of parasympathetic fibers?

11. Starting in the spinal cord, trace an impulse through the sympathetic division of the ANS until it reaches a target organ in the abdominopelvic region.

12. What three collateral ganglia serve as origins for postganglionic neurons that innervate organs or tissues within the abdominopelvic region?

13. What two distinctive results are produced by stimulation of sympathetic postganglionic neurons?

14. What is the difference between a cholinergic synapse and an adrenergic synapse?

15. What four pairs of cranial nerves are associated with the cranial segment of the parasympathetic division of the ANS?

16. What four ganglia serve as origins for postganglionic fibers concerned with control of visceral structures in the head?

17. What six plexuses in the thorax and abdominopelvic cavities innervate visceral organs, and what are the effects of sympathetic versus parasympathetic stimulation?

18. What are the components of a visceral reflex arc?

19. What control centers in the brain stem influence the levels of activity in the sympathetic and parasympathetic divisions of the ANS?

LEVEL 2 Reviewing Concepts

20. Autonomic tone in autonomic motor neurons exists because:
(a) ANS neurons from both divisions often innervate the same organ
(b) ANS neurons rarely innervate the same organs as SNS neurons
(c) ANS neurons are inactive unless stimulated by higher centers
(d) ANS neurons are always active to some degree

21. Dual innervation refers to situations in which:
(a) vital organs receive instructions from sympathetic and parasympathetic fibers
(b) the atria and ventricles of the heart receive autonomic stimulation from the same nerves
(c) sympathetic and parasympathetic fibers have similar effects
(d) a, b, and c are correct

22. Compare the general effects of the sympathetic and parasympathetic divisions of the ANS.

23. Why does sympathetic function remain intact even when there is damage to the ventral roots of cervical spinal nerves?

24. Why is the adrenal medulla considered to be a modified sympathetic ganglion?

25. How does the result of stimulating alpha receptors differ from that of stimulating beta receptors?

26. Why are the effects of parasympathetic stimulation quite localized and of short duration?

27. Why is autonomic tone a significant part of ANS function?

28. You are alone in your home late at night when you hear what sounds like breaking glass. What physiological effects would this experience probably produce, and what would be their cause?

LEVEL 3 Critical Thinking and Clinical Applications

29. In some very severe cases of stomach ulcers, the branches of the vagus nerve (N X) that lead to the stomach are surgically severed. How would this procedure help control the ulcers?

30. Mr. Martin is suffering from a condition known as ventricular tachycardia, in which his heart beats too quickly. Would you suggest an alpha-blocker or a beta-blocker to help alleviate his problem? Why?

31. Little Billy is stung on his cheek by a wasp. He is allergic to wasp venom; his throat begins to swell, and his respiratory passages constrict. Would acetylcholine or epinephrine be more helpful in relieving his symptoms? Why?

32. While studying the activity of smooth muscle in blood vessels, Shelly discovers that a molecule chemically similar to a neurotransmitter triggers an increase in intracellular calcium ions when applied to muscle membrane. Which neurotransmitter is the molecule mimicking, and to what receptors is it binding?

CHAPTER 17

O*ur comprehension of the world around us is based on information provided by our senses. For example, if we can see something, it must be there, because seeing is believing...or is it? The more you study this image, the more confusing it becomes. We are so used to trusting what we see that when we see something that is physically impossible, it shocks us—some people actually become dizzy when they look at one of Escher's drawings. This chapter examines the way receptors provide sensory information, as well as the sensory pathways that distribute this information to provide us with our senses of smell, taste, vision, equilibrium, and hearing.*

Sensory Function

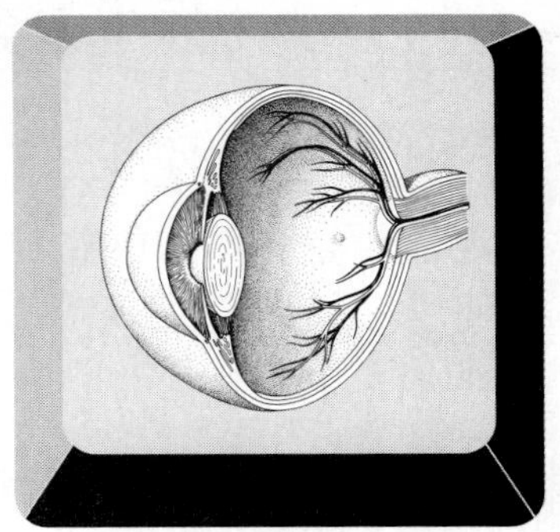

Chapter Outline and Objectives

1 Distinguish between the general and special senses.

2 Explain why receptors respond to specific stimuli and how the organization of a receptor affects its sensitivity.

3 Identify the receptors for the general senses and describe how they function.

4 Describe the sensory organ of smell and trace the olfactory pathways to their destinations in the brain.

5 Describe the sensory organ of taste and trace the gustatory pathways to their destinations in the brain.

6 Identify the accessory structures of the eye and explain their functions.

7 Describe the internal structures of the eye and explain their functions.

8 Explain how we are able to distinguish colors and perceive depth.

9 Explain how light stimulates production of nerve impulses, and trace the visual pathways to their destinations in the brain.

10 Describe the structures of the outer and middle ear and explain how they function.

11 Describe the parts of the inner ear and their roles in the process of equilibrium and hearing.

12 Trace the pathways for the sensations of equilibrium and hearing to their respective destinations in the brain.

Our knowledge of the world around us is limited to those characteristics that stimulate our sensory receptors. Although we may not realize it, our picture of the environment is incomplete. Colors invisible to us guide insects to flowers; sounds we cannot hear and smells we cannot detect provide dogs, cats, and dolphins with important information about their surroundings. What we *do* perceive varies considerably with the state of our nervous systems. For example, when sympathetic activation occurs we experience heightened awareness of sensory information and hear sounds that would normally escape our notice; yet when concentrating on a difficult problem, we may remain unaware of relatively loud noises. Finally, our perception of any stimulus reflects activity in the cerebral cortex, and that activity can be inappropriate. In phantom limb pain (Chapter 15) a person feels pain in a missing limb; during an epileptic seizure an individual may experience sights, sounds, or smells that have no physical basis. *[pp. 506, 516]*

This chapter begins by examining receptor function and basic concepts in sensory processing. We will then apply this information to each of the general and special senses.

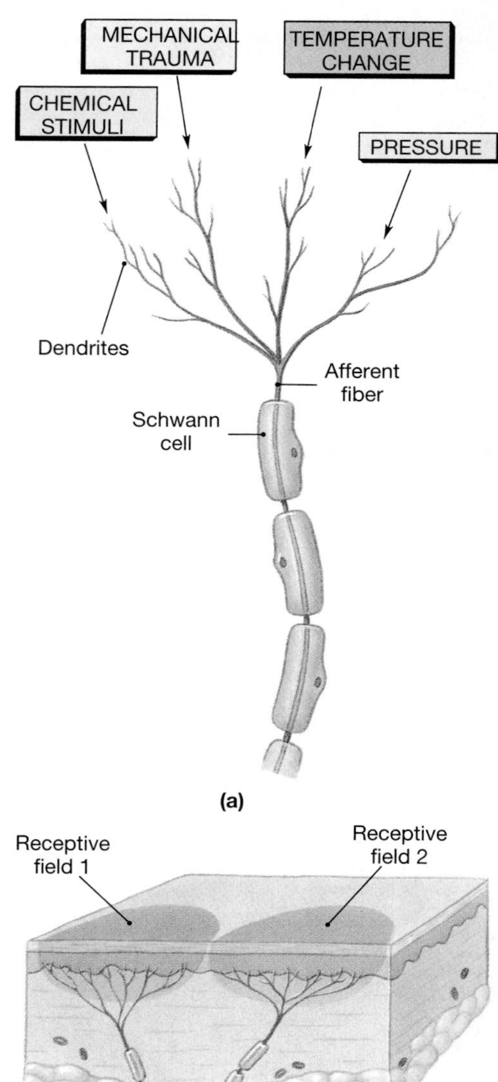
MECHANICAL TRAUMA
TEMPERATURE CHANGE
CHEMICAL STIMULI
PRESSURE
Dendrites
Afferent fiber
Schwann cell
(a)
Receptive field 1
Receptive field 2
(b)

FIGURE 17-1
Receptors and Receptive Fields. (a) A free nerve ending consists of sensory dendrites that may be stimulated by a variety of stimuli. **(b)** Each receptor monitors a specific area known as the receptive field.

Receptors

Sensory receptors are specialized cells that provide the central nervous system with information about conditions inside or outside the body. A sensory receptor detects an arriving stimulus and translates it into an action potential that can be conducted to the CNS. This translation process is called **transduction.**

Sensory receptors are the interface between the nervous system and the internal and external environments. The **general senses** of pain, temperature, touch, pressure, vibration, and proprioception (position sense) are distributed throughout the body. The ascending pathways providing conscious perception of those sensations were introduced in Chapter 15. *[pp. 502–508]* The **special senses** are smell **(olfaction)**, taste **(gustation)**, balance **(equilibrium)**, hearing, and sight **(vision).** These sensations are provided by receptors that are structurally more complex than those of the general senses. Special sensory receptors are found in **sense organs** such as the eye or ear, where they are protected by surrounding tissues. The information provided by these receptors is distributed to specific areas of the cerebral cortex (the auditory cortex, the visual cortex, and so forth) and to centers throughout the brain stem. AM *Analyzing Sensory Disorders*

Specificity

Figure 17-1

Each receptor has a characteristic sensitivity. For example, a touch receptor is very sensitive to pressure but relatively insensitive to chemical stimuli; a taste receptor is sensitive to dissolved chemicals, but insensitive to pressure. This concept is called **receptor specificity.**

Specificity may result from the structure of the receptor cell itself or from the presence of accessory cells or structures that shield it from other stimuli. The simplest receptors are the dendrites of sensory neurons. The dendritic processes, called **free nerve endings,** are not protected by accessory structures. They can be stimulated by many different stimuli (Figure 17-1a●). For example, free nerve endings that respond to tissue damage may be stimulated by chemical stimulation, pressure, temperature changes, or trauma. Complex receptors, such as the visual receptors of the eye, are protected by accessory cells and layers of connective tissue. These cells are seldom exposed to any stimulus other than light, and so provide very specific information.

The area monitored by a single receptor cell is its **receptive field** (Figure 17-1b●). Whenever a sufficiently strong stimulus arrives in the receptive field, the CNS receives the information "stimulus arriving at receptor X." The larger the receptive field, the poorer our ability to localize a stimulus. For example, a touch receptor on the general body surface may have a receptive field 7 cm (2.5 in.) in diameter. As a result, a light touch can be described as affecting only a general area, not an exact spot. On the tongue or fingertips, where the receptive fields are less than a millimeter in diameter, we can be very precise about the location of a stimulus.

An arriving stimulus can take many different forms; it may be a physical force (such as pressure), a dissolved chemical, a sound, or a beam of light. However, regardless of the nature of the stimulus, sensory information must be sent to the CNS in the form of action potentials, which are electrical events.

❑ Transduction

Figure 17-2

Transduction can be divided into three steps:

Step 1: ***An arriving stimulus alters the transmembrane potential of the receptor membrane.***

The nature of the interaction between stimulus and receptor determines the receptor's specificity. Regardless of the nature of the interaction, however, the result is always the same: The transmembrane potential of the receptor cell changes. The change, in the transmembrane potential that accompanies receptor stimulation is called a **receptor potential.**

The receptor potential may be a depolarization or a hyperpolarization. It is a graded potential change, and the stronger the stimulus, the larger the receptor potential.

Step 2: ***The receptor potential directly or indirectly affects a sensory neuron.***

A membrane depolarization that leads to an action potential in a sensory neuron is called a **generator potential.** The receptors for the general senses are themselves sensory neurons (Figure 17-2a●). Receptor potentials in the dendrites of such a cell can summate at the initial segment, producing a generator potential large enough to trigger an action potential in the afferent fiber. Alternatively, a single, strong stimulus can produce a receptor potential large enough to bring the initial segment to threshold. In this case, the terms *receptor potential* and *generator potential* can be used interchangeably.

Sensations of taste, hearing, equilibrium, and vision are provided by specialized receptor cells that communicate with sensory neurons across chemical synapses. The receptor cells develop graded receptor potentials in response to stimulation, and the change in membrane potential alters the rate of neurotransmitter release at the synapse. In the case of taste receptors (Figure 17-2b●), stimulation triggers the release of neurotransmitters that depolarize the membrane of the sensory neuron. The larger the receptor potential, the more neurotransmitter released, and the greater the depolarization. In this case the receptor potential and the generator potential are distinct in both location and timing. The receptor potential appears in the receptor cell, when the stimulus arrives. The generator potential develops later, in the sensory neuron, following the arrival of neurotransmitter at the postsynaptic membrane.

Step 3: ***Action potentials travel to the CNS along an afferent fiber.***

When a generator potential appears, action potentials develop in the afferent fiber. For reasons discussed in Chapter 12, the greater the amount of depolarization produced by the generator potential, the higher the frequency of action potentials in the afferent fiber. ∞ *[p. 415]* The arriving information is then processed and interpreted by the CNS at the conscious and unconscious levels.

❑ Interpretation of Sensory Information

When sensory information arrives at the CNS it is routed according to the location and nature of the stimulus. Previous chapters have emphasized the fact that axons in the CNS are organized in bundles with specific origins and destinations. Along sensory pathways, axons relay information from point A (the receptor) to point B (a neuron at a specific site in the cerebral cortex). For example, touch, pressure, pain, temperature, and taste sensations arrive at the primary sensory cortex; visual, auditory, and olfactory sensations reach the visual, auditory, and olfactory regions of the cortex.

The neural link between receptor and cortical neuron is called a **labeled line**. Each labeled line consists of axons carrying information concerning one type of stimulus (touch, pressure, light, sound, and so forth) from receptors in a specific part of the

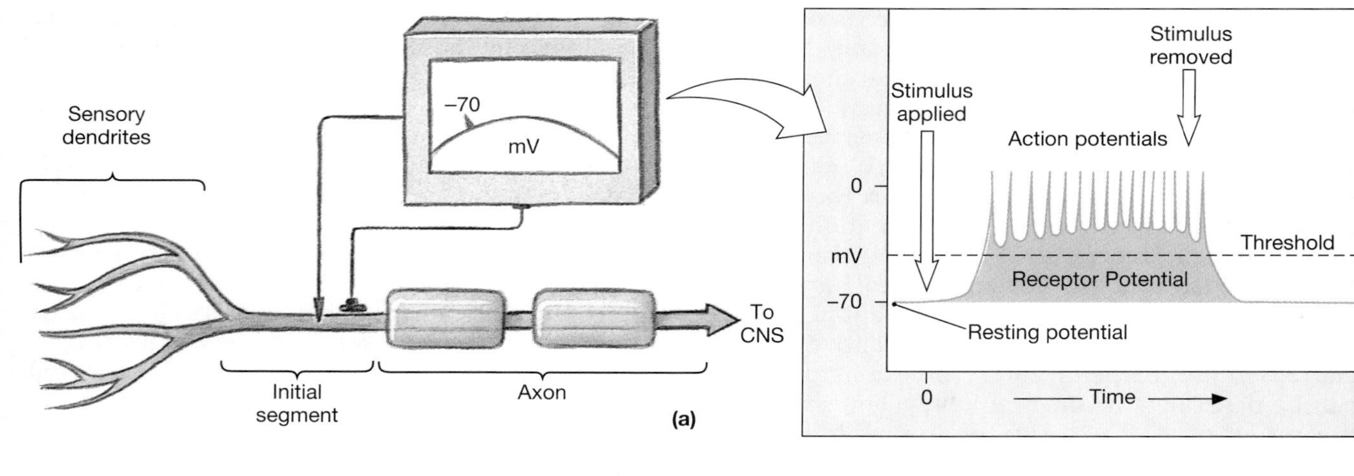
Sensory dendrites
−70
mV
Initial segment
Axon
To CNS
(a)
Stimulus applied
Stimulus removed
Action potentials
Threshold
Receptor Potential
Resting potential
0
mV
−70
0
Time

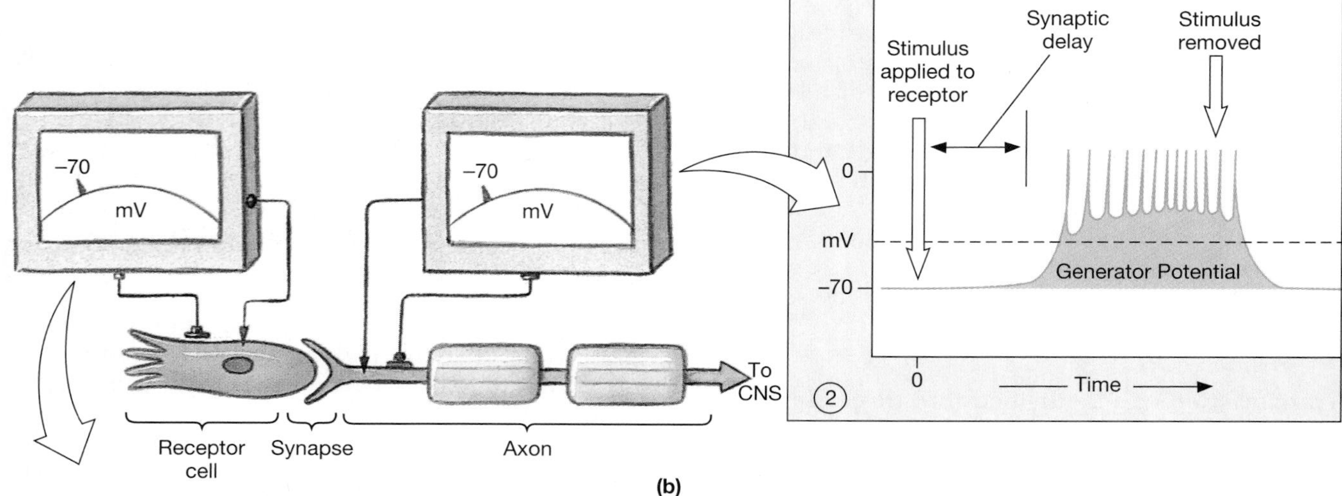
Axon
−70
mV
−70
mV
To CNS
Receptor cell
Synapse
Axon
(b)
Stimulus applied to receptor
Synaptic delay
Stimulus removed
0
mV
−70
Generator Potential
0
Time
2

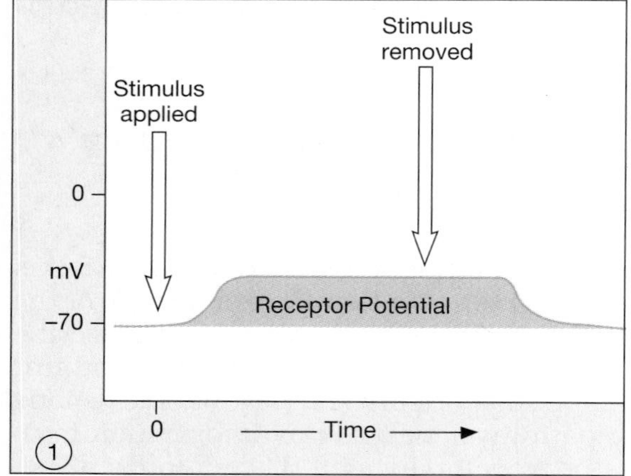
Receptor cell
Stimulus applied
Stimulus removed
0
mV
−70
Receptor Potential
0
Time
1

● **FIGURE 17-2**
Receptor and Generator Potentials. (a) When the sensory neuron acts as the receptor, a stimulus that depolarizes the dendrites may bring the initial segment of the axon to threshold. The receptor and the neuron are the same cell, and the receptor potential is a generator potential. **(b)** In the special senses of taste, equilibrium, hearing, and vision, the receptor cells are specialized cells that communicate with neurons across chemical synapses. The receptor cell shows a receptor potential in response to stimulation (1). In this example, the receptor potential is a depolarization that accelerates neurotransmitter release; the neurotransmitter produces a generator potential in the postsynaptic membrane (2).

body. *The CNS interprets sensory information entirely on the basis of the line over which it arrives.* As a result, you cannot tell the difference between a true sensation and a false one generated somewhere along the line. For example, when rubbing the eyes, one often sees flashes of light. Although the stimulus is mechanical rather than visual, any activity along the optic nerve is projected to the visual cortex and experienced as a visual perception.

The identity of the active labeled line indicates the location and nature of the stimulus. *All other characteristics* of the stimulus are conveyed by the frequency and pattern of action potentials in the afferent fibers. This **sensory coding** provides information about the strength, duration, variation, and movement of the stimulus.

Some sensory neurons, called **tonic receptors,** are always active. The frequency with which they generate action potentials indicates the background level of stimulation. When the stimulus increases or decreases, their rate of action potential generation changes accordingly. Other receptors are normally inactive, but become active for a short time whenever there is a change in the conditions they are monitoring. These **phasic receptors** provide information on the intensity and rate of change of a stimulus. Receptors that combine phasic and tonic coding can convey extremely complicated sensory information.

Central Processing and Adaptation

Adaptation is a reduction in sensitivity in the presence of a constant stimulus. **Peripheral adaptation** occurs when the receptors or sensory neurons alter their levels of activity. The receptor responds strongly at first, but thereafter the activity along the afferent fiber gradually declines, in part because of a gradual decrease in the size of the generator potential. This response is characteristic of phasic receptors, which are also called **fast-adapting receptors.** Tonic receptors show little peripheral adaptation, and so are called **slow-adapting receptors.**

Adaptation also occurs inside the CNS along the sensory pathways. For example, a few seconds after exposure to a new smell, conscious awareness of the stimulus virtually disappears, although the sensory neurons are still quite active. This process is known as **central adaptation.** Central adaptation usually involves the inhibition of nuclei along a sensory pathway.

Peripheral adaptation reduces the amount of information reaching the CNS. Central adaptation at the unconscious level further restricts the amount of detail arriving at the cerebral cortex. Most of the incoming sensory information is processed in centers along the spinal cord or brain stem, potentially triggering involuntary reflexes, such as the withdrawal reflex. Only about 1 percent of the information provided by afferent fibers reaches the cerebral cortex and our conscious awareness.

The output from higher centers can increase receptor sensitivity or facilitate transmission along a sensory pathway. The reticular activating system in the mesencephalon helps focus attention and thus heightens or reduces awareness of arriving sensations. This adjustment of sensitivity can occur under conscious or unconscious direction. When we "listen carefully" our sensitivity and awareness of auditory stimuli increase. Output from higher centers can also inhibit transmission along a sensory pathway. This occurs when we enter a noisy factory or walk along a crowded city street, as we automatically tune out the high level of background noise.

This discussion has introduced basic concepts of receptor function and sensory processing. We will next consider how these concepts apply to the general senses.

The General Senses

Receptors for the general senses are scattered throughout the body and are relatively simple in structure. A simple classification scheme divides them into *exteroceptors* and *interoceptors.* **Exteroceptors** provide information about the external environment; **interoceptors** monitor conditions inside the body.

A more detailed classification system divides the general sensory receptors into four types according to the nature of the stimulus that excites them: *nociceptors* (pain), *thermoreceptors* (temperature), *mechanoreceptors* (touch, pressure, proprioception), and *chemoreceptors* (chemical sense). Each class of receptors has distinct structural and functional characteristics.

Nociceptors

Figure 17-3f

Pain receptors, or **nociceptors,** are especially common in the superficial portions of the skin, in joint capsules, within the periostea of bones, and around the walls of blood vessels. There are few nociceptors in other deep tissues or in most visceral organs. Pain receptors are free nerve endings with large receptive fields (Figure 17-3f•). As a consequence it is often difficult to determine the exact origin of a painful sensation.

There are three different populations of nociceptors: (1) those sensitive to extremes of temperature, (2) those sensitive to mechanical damage, and (3) those sensitive to dissolved chemicals. Very strong stimuli, however, will excite all three receptor types. For this reason people describing very painful sensations—whether caused by acids, heat, or a deep cut—use similar descriptive terms, such as "burning."

Stimulation of the dendrites of a nociceptor causes depolarization, and when the initial segment of the axon reaches threshold, an action potential heads toward the CNS. The distribution of these sensations was detailed in Chapter 15: The axon synapses on an interneuron, whose axon crosses the spinal cord to ascend within the *lateral spinothalamic tract.* After a synapse in the thalamus, the information is relayed to the primary sensory cortex. [p. 506]

Two types of axons carry painful sensations. Myelinated Type A fibers carry sensations of **fast pain,** or *prickling pain.* An injection or a deep cut produces this type of pain. These sensations very quickly reach the CNS, where they often trigger somatic reflexes. They are also relayed to the primary sensory cortex, and so receive conscious attention. The arriving information usually permits localization of the stimulus to an area several inches in diameter.

Slower Type C fibers carry sensations of **slow pain,** or *burning and aching pain.* These sensations cause a generalized activation of the reticular formation and thalamus. The individual becomes aware of the pain but has only a general idea of the area affected.

Pain receptors exhibit a tonic response to stimulation. Significant peripheral adaptation does not occur and the receptors continue to respond as long as the painful stimulus remains. Painful sensations cease only after tissue damage has ended. However, central adaptation may reduce *perception* of the pain while the pain receptors are still stimulated. This effect involves the inhibition of centers in the thalamus, reticular formation, lower brain stem, and spinal cord.

Endorphins and enkephalins are neuromodulators whose release inhibits activity along pain pathways in the brain. These compounds, structurally similar to morphine, are found in the limbic system, hypothalamus, and reticular formation. Pain centers in these areas use the neuropeptide substance P as a neurotransmitter. Endorphins bind to the presynaptic membrane and prevent the release of substance P, thereby reducing the conscious perception of pain, although the painful stimulus remains. AM *The Control of Pain*

Thermoreceptors

Temperature receptors, or **thermoreceptors,** are found in the dermis of the skin, in skeletal muscles, in the liver, and in the hypothalamus. Cold receptors are three or four times more numerous than warm receptors. The receptors are free nerve endings, and there are no known structural differences between warm and cold thermoreceptors.

Temperature sensations are conducted along the same pathways that carry pain sensations. They are sent to the reticular formation, the thalamus, and (to a lesser extent) the primary sensory cortex. Thermoreceptors are phasic receptors: They are very active when the temperature is changing, but they quickly adapt to a stable temperature. When you enter an air-conditioned classroom on a hot summer day or a warm lecture hall on a brisk fall evening, the temperature seems unpleasant at first, but the discomfort fades as adaptation occurs.

Mechanoreceptors

Mechanoreceptors are sensitive to stimuli that distort their cell membranes. These membranes contain *mechanically regulated ion channels* whose gates open or close in response to stretching, compression, twisting, or other distortions of the membrane. There are three classes of mechanoreceptors:

1. **Tactile receptors** provide sensations of touch, pressure, and vibration.
2. **Baroreceptors** (bar-ō-rē-SEP-tōrz; *baro-*, pressure) detect pressure changes in the walls of blood vessels and in portions of the digestive, reproductive, and urinary tracts.
3. **Proprioceptors** monitor the positions of joints and muscles. The proprioceptors are the most structurally and functionally complex of the general sensory receptors.

Tactile Receptors

Figure 17-3

Fine touch and pressure receptors provide detailed information about a source of stimulation, including its exact location, shape, size, texture, and movement. These receptors are extremely sensitive and have relatively narrow receptive fields. The sensory information reaches our conscious awareness via the *posterior column pathway.* [p. 504] **Crude touch and pressure receptors** provide poor localization and, because of relatively large receptive fields, give little additional information about the stimulus. These sensations ascend within the *anterior spinothalamic tract,* and the thalamus relays the information to appropriate areas of the primary sensory cortex. [p. 506]

Tactile receptors range in complexity from free nerve endings to specialized sensory complexes with accessory cells and supporting structures. Figure 17-3● shows six different types of tactile receptors in the skin:

1. *Free nerve endings* are found between adjacent epidermal cells (Figure 17-3f●). They are sensitive to touch and pressure; there are no apparent structural differences between these receptors and the free nerve endings that provide temperature or pain sensations. These are the only sensory receptors found on the corneal surface of the eye, but in other portions of the body surface more specialized tactile receptors are probably more important.
2. Wherever hairs are found, the nerve endings of the **root hair plexus** (Figure 17-3a●) monitor distortions and movements across the body surface. When the hair is displaced, the movement of the follicle distorts the sensory den-

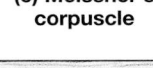

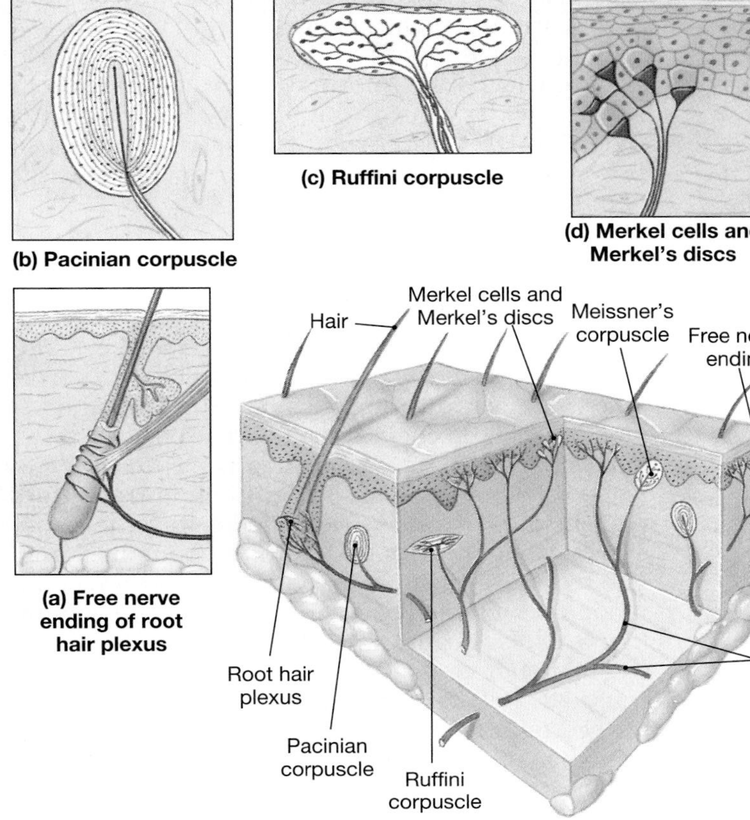
(c) Ruffini corpuscle
(d) Merkel cells and Merkel's discs
(b) Pacinian corpuscle
(e) Meissner's corpuscle
Hair
Merkel cells and Merkel's discs
Meissner's corpuscle
Free nerve ending
(a) Free nerve ending of root hair plexus
Root hair plexus
Pacinian corpuscle
Ruffini corpuscle
Sensory nerves

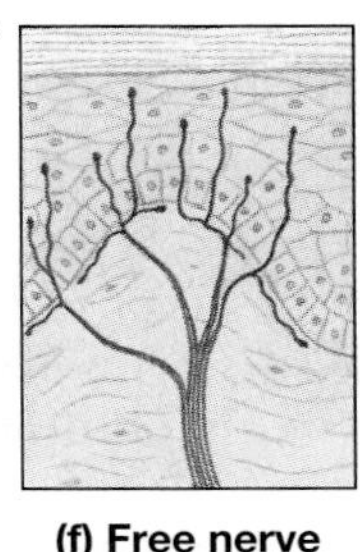
(f) Free nerve ending

● **FIGURE 17-3**
Tactile Receptors in the Skin. (a) Root hair plexus. **(b)** Pacinian corpuscle. **(c)** Ruffini corpuscle. **(d)** Merkel cells and Merkel's discs. **(e)** Meissner's corpuscle. **(f)** Free nerve endings.

drites and produces action potentials. These receptors adapt rapidly, so they are best at detecting initial contact and subsequent movements. For example, most people feel their clothing only when they move or when they consciously focus attention on their skin.

3. **Merkel's** (MER-kelz) **discs** are fine touch and pressure receptors (Figure 17-3d●). They are tonically active, extremely sensitive, and have narrow receptive fields. The dendritic processes of a single myelinated afferent fiber make close contact with unusually large epithelial cells in the stratum germinativum of the skin; these *Merkel cells* were described in Chapter 5. ∞ *[p. 152]*
4. **Meissner's** (MĪS-nerz) **corpuscles** (Figure 17-3e●) also provide fine touch and pressure, and low-frequency vibration sensations. They adapt to stimulation within a second after contact. Meissner's corpuscles are fairly large structures, measuring roughly 100 μm in length and 50 μm in width. These receptors are most abundant at the eyelids, lips, fingertips, nipples, and external genitalia. The dendrites are highly coiled and interwoven, and they are surrounded by modified Schwann cells. A fibrous capsule surrounds the entire complex and anchors it within the dermis.
5. **Pacinian** (pa-SIN-ē-an) **corpuscles,** or *lamellated corpuscles*, are most sensitive to pulsing or high-frequency vibrating stimuli. A single dendritic process lies within a series of concentric cellular layers, and the entire collection may reach 4 mm in length and 1 mm in diameter (Figure 17-3b●). The accessory structure shields the dendrite from virtually every source of stimulation other than direct pressure. Pacinian corpuscles adapt quickly, as distortion of the capsule relieves pressure on the sensory process. These receptors are scattered throughout the integument, notably in the fingers, breasts, and external genitalia. They are also encountered in the superficial and deep fasciae, in joint capsules, in mesenteries, and in the wall of the urinary bladder.
6. **Ruffini** (rū-FĒ-nē) **corpuscles** are also sensitive to pressure and distortion of the skin, but they are located in the reticular (deep) dermis. These receptors are tonically active and show little if any adaptation. The capsule surrounds a core of collagen fibers that are continuous with those of the surrounding dermis (Figure 17-3c●). Inside the capsule a network of dendrites is intertwined with the collagen fibers. Any tension or distortion of the dermis tugs or twists the capsular fibers, stretching or compressing the attached dendrites and altering the activity in the myelinated afferent fiber.

The central distribution of tactile sensations, via the posterior column and spinothalamic pathways, was considered in Chapter 15. ∞ *[pp. 504, 506]* Tactile sensitivities may be altered by peripheral infection, disease processes, and damage to sensory afferents or central pathways. As a result, mapping tactile responses can sometimes aid clinical assessment. Sensory losses with clear regional boundaries indicate trauma to spinal nerves. For example, sensory loss along a dermatomal boundary (described in Chapter 13) can permit a reasonably precise determination of the affected spinal nerve or nerves. [p. 434] Regional sensitivity to light touch can be checked by gentle contact with a fingertip or a slender wisp of cotton. Vibration receptors are tested by applying the base of a tuning fork to the skin. More detailed procedures, such as the *two-point discrimination test*, are considered in the *Applications Manual.* AM *Assessment of Tactile Sensitivities*

Baroreceptors

Baroreceptors monitor changes in pressure. The receptor itself consists of free nerve endings that branch within the elastic tissues in the wall of a distensible organ, such as a blood vessel or a portion of the respiratory, digestive, or urinary tract. When the pressure changes, the elastic walls of the tract recoil or expand. This movement distorts the dendritic branches and alters the rate of action potential generation. Baroreceptors respond immediately to a change in pressure, but they adapt rapidly and the output along the afferent fibers gradually returns to normal.

Baroreceptors monitor blood pressure in the walls of major vessels, including the carotid artery (at the *carotid sinus*) and the aorta (at the *aortic sinus*). The information plays a major role in regulating cardiac function and adjusting blood flow to vital tissues. Baroreceptors in the lungs monitor the degree of lung expansion. This information is relayed to the respiratory rhythmicity center, which sets the pace of respiration. Comparable stretch receptors in the digestive and urinary tracts trigger a variety of visceral reflexes, including those of urination and defecation. These baroreceptor reflexes will be detailed in chapters considering specific physiological systems.

Proprioceptors

Proprioceptors monitor the position of joints, the tension in tendons and ligaments, and the state of muscular contraction. Two representative examples were described in earlier chapters: *Golgi tendon organs*, which monitor the strain on a tendon, and *muscle spindles*, which monitor the length of a skeletal muscle.

In a Golgi tendon organ, dendrites branch repeatedly and wind around the densely packed collagen fibers in a tendon. Tension or distortion of the tendon stimulates the receptor, and excessive stimulation triggers a reduction in the strength of skeletal muscle contractions. This *tendon reflex* was discussed in Chapter 13. [p. 449] Chapters 10 and 13 described muscle spindles and considered their role in regulating skeletal muscle tone and maintaining normal posture and balance. [pp. 303, 446–449]

Proprioceptive information reaches our conscious awareness via the *posterior column pathway*. The *spinocerebellar pathway* delivers proprioceptive information to the cerebellum outside of our conscious awareness. [pp. 504, 507]

❑ Chemoreceptors

The free nerve endings of nociceptors may be stimulated by a variety of chemicals including ions, acids, histamine, various enzymes, and prostaglandins. More specialized chemoreceptive neurons can detect small changes in the concentration of specific chemicals or compounds. In general, **chemoreceptors** respond only to water-soluble and lipid-soluble substances that are dissolved in the surrounding fluid. The receptors show a pronounced peripheral adaptation over a period of seconds, and central adaptation may also occur.

The chemoreceptors included in the general senses do not send information to the primary sensory cortex, and we are not consciously aware of the sensations they provide. The arriving sensory information is routed to brain stem centers concerned with the autonomic control of respiratory and cardiovascular function. Neurons within the respiratory centers of the brain respond to the concentration of hydrogen ions and carbon dioxide molecules in the cerebrospinal fluid. Chemoreceptive neurons are also found within the **carotid bodies,** near the origin of the internal carotid arteries on each side of the neck, and in the **aortic bodies** between the major branches of the aortic arch. These receptors monitor the carbon dioxide and oxygen concentrations of arterial blood. The afferent fibers leaving the carotid and aortic bodies reach the respiratory centers by traveling along the ninth (glossopharyngeal) and tenth (vagus) cranial nerves.

✔ Receptor A has a circular receptive field with a diameter of 2.5 cm. Receptor B has a circular receptive field 7.0 cm in diameter. Which receptor will provide more precise sensory information?

✔ When the nociceptors in your hand are stimulated, what sensation do you perceive?

✔ What would happen to an individual if the information from proprioceptors in the legs were blocked from reaching the CNS?

We will next turn our attention to the five *special senses:* olfaction, gustation, vision, equilibrium, and hearing. Although the sense organs involved are structurally more complex than those of the general senses, the same basic principles of receptor function apply.

Olfaction

Figure 17-4

The sense of smell, more precisely called *olfaction*, is provided by paired **olfactory organs.** These organs are located in the nasal cavity on either side of the nasal septum. The olfactory organs (Figure 17-4●) consist of:

- An **olfactory epithelium,** which contains the **olfactory receptors,** supporting cells, and **basal cells** (*stem cells*).
- An underlying layer of loose connective tissue known as the *lamina propria.* This layer contains **olfactory glands,** also called *Bowman's glands,* which produce a thick, pigmented mucus. It also contains numerous blood vessels and nerves.

The olfactory epithelium covers the inferior surface of the cribriform plate, the superior portion of the nasal septum, and the superior nasal conchae.

When air is drawn in through the nose, the air swirls and eddies within the nasal cavity. This turbulence brings airborne compounds to the olfactory organs. A normal, relaxed inspiration provides a small sample (around 2 percent) of the inspired air to the olfactory organs. Sniffing repeatedly increases the flow of air across the olfactory epithelium, intensifying the stimulation of the receptors. Once compounds have reached the olfactory organs, water-soluble and lipid-soluble materials must diffuse into the mucus before they can stimulate the olfactory receptors.

Olfactory Receptors

Figure 17-4b

The olfactory receptors are highly modified neurons. The exposed tip of each receptor forms a prominent knob that projects above the epithelial surface (Figure 17-4b●). That projection provides a base for up to 20 cilia that extend into the surrounding mucus. These cilia lie parallel to the epithelial surface, exposing their considerable surface area to dissolved chemical compounds.

Olfactory reception occurs on the surfaces of the olfactory cilia, as dissolved chemicals interact with receptors on the membrane surface. When binding occurs, it leads to the activation of adenylate cyclase, the enzyme that converts ATP to cyclic-AMP. The cyclic-AMP then opens sodium channels in the membrane, which, as a result, begins to depolarize. If sufficient depolarization occurs, an action potential is triggered in the axon, and the information is relayed to the central nervous system (Figure 17-4b●).

Somewhere between 10 and 20 million olfactory receptors are packed into an area of roughly 5 cm^2.

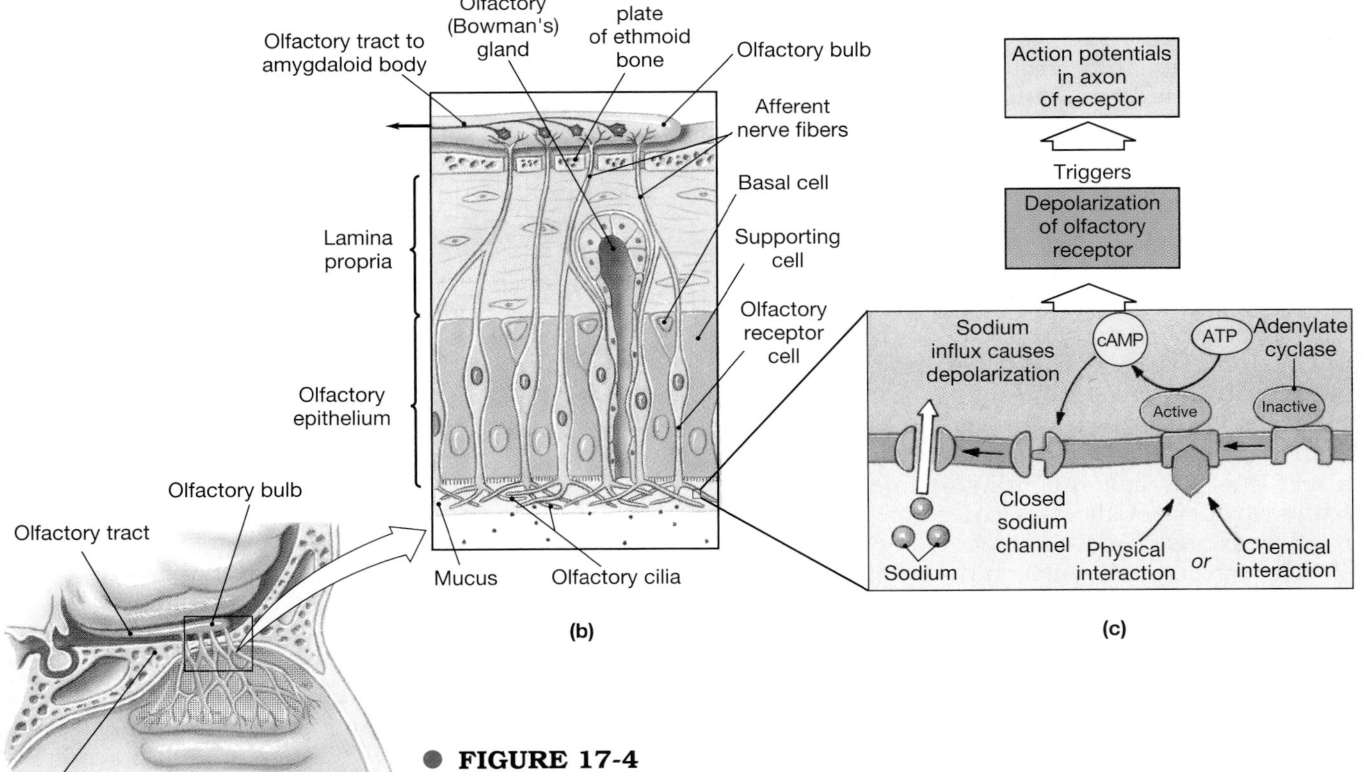
Olfactory tract to amygdaloid body
Olfactory (Bowman's) gland
Cribriform plate of ethmoid bone
Olfactory bulb
Afferent nerve fibers
Basal cell
Supporting cell
Olfactory receptor cell
Lamina propria
Olfactory epithelium
Mucus
Olfactory cilia
(b)
Olfactory bulb
Olfactory tract
Cribriform plate
(a)
Action potentials in axon of receptor
Triggers
Depolarization of olfactory receptor
Sodium influx causes depolarization
cAMP
ATP
Adenylate cyclase
Active
Inactive
Closed sodium channel
Sodium
Physical interaction
or
Chemical interaction
(c)

● **FIGURE 17-4**
The Olfactory Organs. (a) The structure of the olfactory organ on the left side of the nasal septum. **(b)** An olfactory receptor is a modified neuron with multiple cilia extending from its free surface. **(c)** Steps in the transduction process.

If we take the exposed ciliary surfaces into account, the actual sensory area probably approaches that of the entire body surface. Nevertheless, our olfactory sensitivities cannot compare with those of other vertebrates such as dogs, cats, or fishes. A German shepherd sniffing for smuggled drugs or explosives has an olfactory receptor surface 72 times greater than that of the nearby customs inspector.

❑ Olfactory Pathways

Figure 17-4a

The olfactory system is very sensitive. As few as four molecules of an odorous substance can activate an olfactory receptor. However, the activation of an afferent fiber does not guarantee a conscious awareness of the stimulus. Considerable convergence occurs along the olfactory pathway, and inhibition at the intervening synapses can prevent the sensations from reaching the cerebral cortex.

Axons leaving the olfactory epithelium collect into 20 or more bundles that penetrate the cribriform plate of the ethmoid bone to reach the *olfactory bulbs* of the cerebrum (Figure 17-4a●). Axons leaving the olfactory bulb travel along the olfactory tract to reach the olfactory cortex, the hypothalamus, and portions of the limbic system.

Olfactory stimulation is the only type of sensory information that reaches the cerebral cortex without first synapsing in the thalamus. The extensive limbic and hypothalamic connections help to explain the profound emotional and behavioral responses that can be produced by certain smells. The perfume industry understands the practical implications of these connections, expending considerable effort to develop odors that trigger sexual responses.

❑ Olfactory Discrimination

The olfactory system can make subtle distinctions among thousands of chemical stimuli. No apparent structural differences exist among the olfactory cells, but the epithelium as a whole contains receptor populations with distinctly different sensitivities. There are at least 50 different "primary smells," and our language does not allow us to describe these sensory impressions effectively. It appears likely that the CNS interprets each smell on the basis of the overall pattern of receptor activity.

❑ Aging and Olfactory Sensitivity

The olfactory receptor population shows considerable turnover, with new receptor cells being produced by division and differentiation of basal cells in the epithelium. This is the only proven example of neuronal replacement in the adult human. Despite this process, the total number of receptors declines with age, and the remaining receptors become less sensitive. As a result, elderly individuals have difficulty detecting odors in low concentrations. This decline in the number of receptors accounts for Grandmother's tendency to apply perfume in excessive quantities, and explains why Grandfather's aftershave seems so overdone; they must apply more to be able to smell it themselves.

Gustation

Figure 17-5

Gustation, or taste, provides information about the foods and liquids we consume. The **gustatory** (GUS-ta-tōr-ē), or **taste, receptors** are distributed over the superior surface of the tongue (Figure 17-5a●) and adjacent portions of the pharynx and larynx. By adulthood the taste receptors on the pharynx and larynx have decreased in importance, and the *taste buds* of the tongue are the primary gustatory receptors.

The superior surface of the tongue bears epithelial projections called **lingual papillae** (pa-PIL-lē; *papilla*, a nipple-shaped mound). There are three different types of lingual papillae on the human tongue: **filiform** (*filum*, thread) **papillae, fungiform** (*fungus*, mushroom) **papillae,** and **circumvallate** (sir-kum-VAL-āt) **papillae** (*circum-*, around + *vallum*, wall). There are regional differences in the distribution of the lingual papillae (Figure 17-5a●).The small fungiform papillae each contain about five taste buds; the large circumvallate papillae, which form a V near the posterior margin of the tongue, contain as many as 100 taste buds per papilla. An adult has approximately 3,000 taste buds.

❑ Gustatory Receptors

Figure 17-5

The taste receptors are clustered in individual **taste buds** (Figure 17-5b,c●). Each taste bud contains around 40 slender receptors, called **gustatory cells,** and a number of supporting cells. Taste buds are recessed into the surrounding epithelium and isolated from the relatively unprocessed oral contents. Each gustatory cell extends slender microvilli, sometimes called *taste hairs*, into the surrounding fluids through a narrow opening, the **taste pore.** Despite this relatively protected position, it's still a hard life, and a typical gustatory cell survives for only about 10 days before being replaced by the division of nearby epithelial cells.

Gustatory Pathways

Figure 17-6

Taste buds are monitored by cranial nerves VII (facial), IX (glossopharyngeal), and X (vagus) (Figure 17-6●). The sensory afferents synapse within the **nucleus solitarius** of the medulla oblongata, and the axons of the postsynaptic neurons enter the medial lemniscus, joining axons carrying somatic sensory information on touch, pressure, and proprioception. After another synapse in the thalamus, the information is projected to the appropriate portions of the primary sensory cortex.

A conscious perception of taste is produced as the information received from the taste buds is correlated with other sensory data. Information concerning the general texture of the food, together with taste-related sensations of "peppery" or "burning hot," is provided by sensory afferents in the trigeminal nerve (N V). In addition, the level of stimulation from the olfactory receptors plays an overwhelming role in taste perception. We are several thousand times more sensitive to "tastes" when our olfactory organs are fully functional. When you have a cold, airborne molecules cannot reach the olfactory receptors, and meals taste dull and unappealing. This reduction in taste perception will occur even though the taste buds may be responding normally.

Gustatory Discrimination

The mechanism behind gustatory reception seems to parallel that of olfaction. Dissolved chemicals contacting the taste hairs provide the stimulus that produces a change in the transmembrane potential of the taste cell. The dendrites of the sensory afferents are tightly wrapped by folds of the receptor cell membrane. Stimulation of the gustatory cell leads to neu-

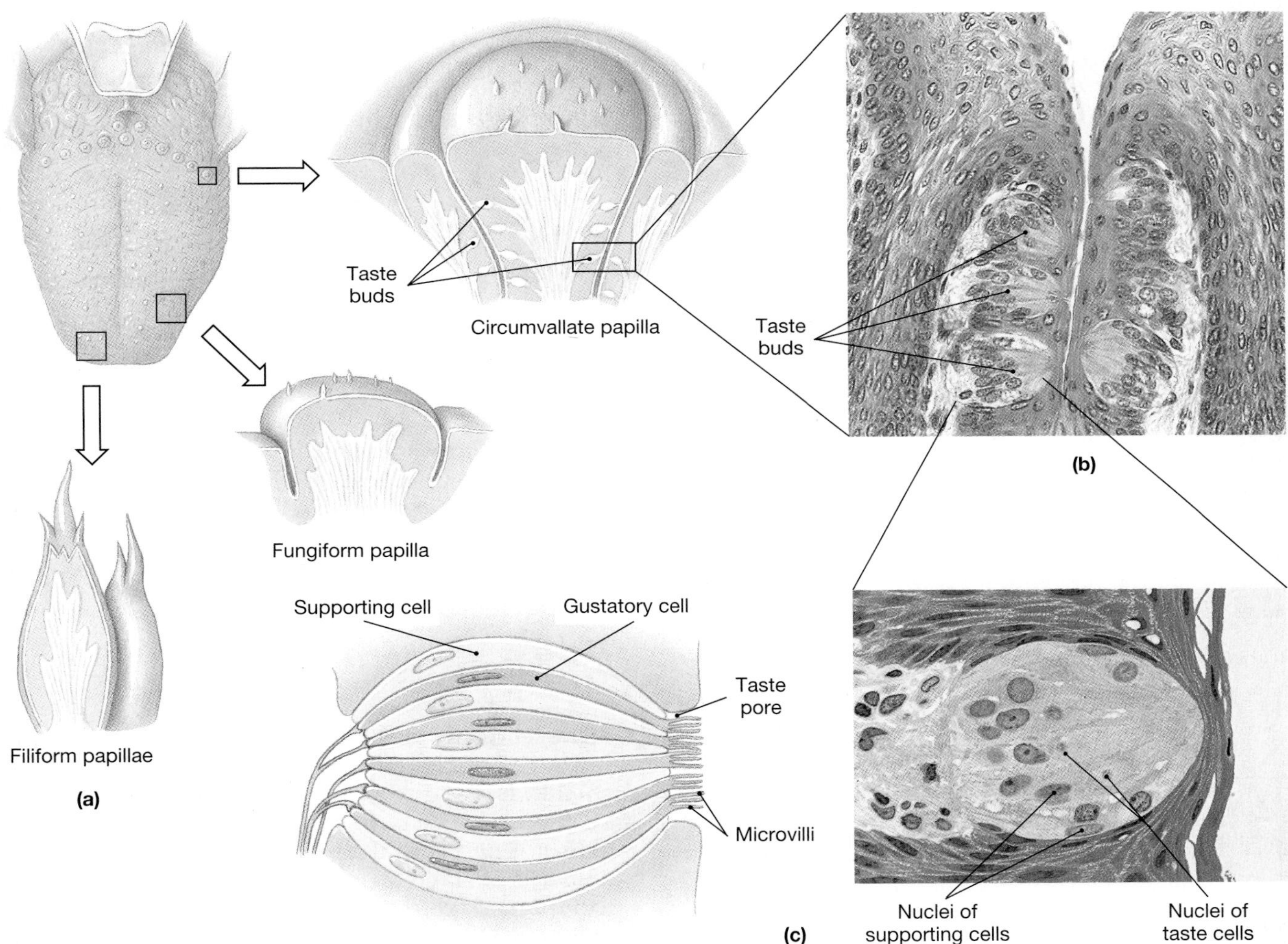

Taste buds
Circumvallate papilla
Taste buds
(b)
Fungiform papilla
Filiform papillae
(a)
Supporting cell
Gustatory cell
Taste pore
Microvilli
(c)
Nuclei of supporting cells
Nuclei of taste cells

● **FIGURE 17-5**

Gustatory Reception. (a) Gustatory receptors are found in taste buds that form pockets in the epithelium of the fungiform and circumvallate papillae. **(b)** Photomicrograph of taste buds in a circumvallate papilla. (LM × 280) **(c)** A taste bud, showing receptor cells and supporting cells. The diagrammatic view shows details of the taste pore not visible in the light micrograph. (LM × 650)

rotransmitter release and the generation of action potentials in the afferent fiber.

There are four **primary taste sensations:** sweet, salt, sour, and bitter. Each taste bud shows a particular sensitivity to one of these tastes, and a sensory map of the tongue indicates that there are regional differences in primary sensitivity (Figure 17-6●). The threshold for receptor stimulation varies for each of the primary taste sensations, and the taste receptors respond most readily to unpleasant rather than attractive stimuli. For example, we are almost a thousand times more sensitive to acids, which give a sour taste, than to either sweet or salty chemicals, and we are a hundred times more sensitive to bitter compounds than to acids. This sensitivity has survival value, for acids can damage the mucous membranes of the mouth and pharynx, and many potent biological toxins produce an extremely bitter taste.

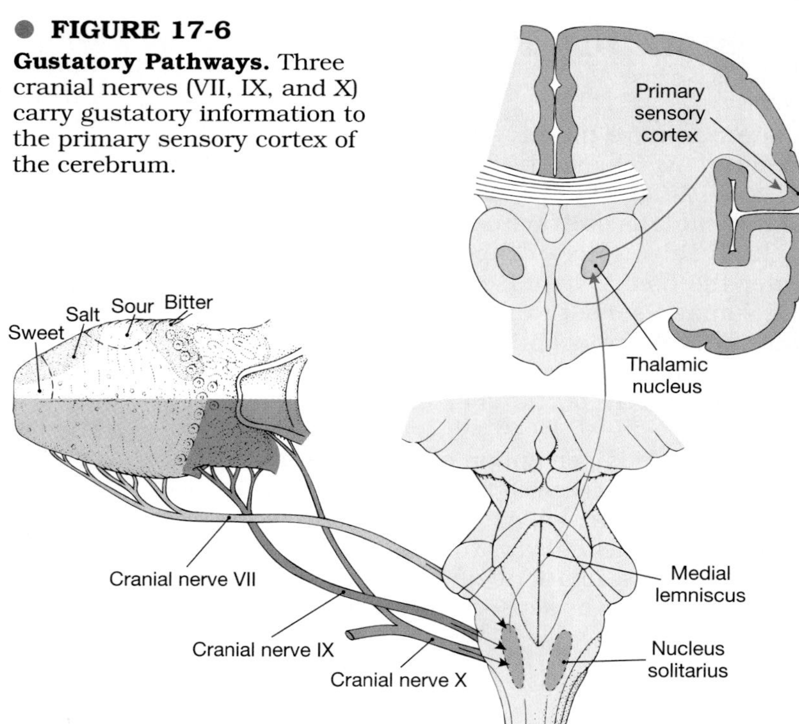

Primary sensory cortex
Thalamic nucleus
Sweet
Salt
Sour
Bitter
Cranial nerve VII
Medial lemniscus
Cranial nerve IX
Cranial nerve X
Nucleus solitarius

● **FIGURE 17-6**
Gustatory Pathways. Three cranial nerves (VII, IX, and X) carry gustatory information to the primary sensory cortex of the cerebrum.

There are significant individual differences in taste sensitivity. Many of these conditions are inherited. The best-known example involves sensitivity to the compound phenylthiourea, also known as phenylthiocarbamide, or **PTC.** In the Caucasian population roughly 70 percent can taste this substance; the rest are unable to detect it.

Our tasting abilities change with age. We begin life with over 10,000 taste buds, but the number begins declining dramatically by age 50. The sensory loss becomes especially significant because aging individuals also experience a decline in the population of olfactory receptors. As a result, many elderly people find that their food tastes bland and unappetizing, whereas children often find the same foods too spicy.

✔ When you first enter the A & P lab for dissection, you are very aware of the odor of preservatives, but by the end of the lab period the smell doesn't seem to be nearly as strong. Why?

✔ If you completely dry the surface of the tongue and then place salt or sugar crystals on it, they can't be tasted. Why not?

✔ Your grandfather tells you that he can't understand why foods he used to enjoy just don't taste the same anymore. What would you tell him?

Vision

We rely more on vision than on any other special sense. Our visual receptors are contained in elaborate structures, the eyes, which enable us not only to detect light but also to create detailed visual images. We will begin our discussion of these fascinating organs by considering the *accessory structures* of the eye that provide protection, lubrication, and support.

Accessory Structures of the Eye

Figure 17-7

The **accessory structures** of the eye include the eyelids, the superficial epithelium of the eye, and the structures associated with the production, secretion, and removal of tears. Figure 17-7● details the superficial anatomy of the eye and the accessory structures.

Eyelids

Figures 17-7, 17-8

The eyelids, or **palpebrae** (pal-PĒ-brē), are a continuation of the skin. The eyelids act like windshield wipers; their continual blinking movements keep the surface lubricated and free from dust and debris. They can also close firmly to protect the delicate surface of the eye. The free margins of the upper and lower eyelids are separated by the **palpebral fissure,** but the two are connected at the **medial canthus** (KAN-thus) and the **lateral canthus** (Figure 17-7a,c●). The **eyelashes** along the palpebral margins are very robust hairs. These help to prevent foreign matter and insects from reaching the surface of the eye.

The eyelashes are associated with large sebaceous glands, the *glands of Zeis* (ZĪS). Along the inner margin of the lid, **Meibomian** (mī-BŌ-mē-an) **glands** secrete a lipid-rich product that helps to keep the eyelids from sticking together. At the medial canthus the **lacrimal caruncle** (KAR-ung-kul) contains glands producing the thick secretions that contribute to the gritty deposits sometimes found after a good night's sleep. These various glands are subject to occasional invasion and infection by bacteria. A cyst, or **chalazion** (kah-LĀ-zē-on; small lump) usually results from the infection of a Meibomian gland. An infection in a sebaceous gland of one of the eyelashes, a Meibomian gland, or one of the many sweat glands that open to the surface between the follicles produces a painful localized swelling known as a **sty.**

The skin covering the visible surface of the eyelid is very thin. Beneath the skin lie the muscle fibers of the *orbicularis oculi* and the *levator palpebrae superioris* (Figure 11-4●, p. 339). These skeletal muscles are responsible for closing the eyelids (orbicularis oculi) and raising the upper eyelid (levator palpebrae superioris).

The epithelium covering the inner surfaces of the eyelids and the outer surface of the eye is called the **conjunctiva** (kon-junk-TĪ-va) (Figure 17-8b●). It is a mucous membrane covered by a specialized stratified squamous epithelium. The **palpebral conjunctiva** covers the inner surface of the eyelids, and the **ocular conjunctiva,** or *bulbar conjunctiva,* covers the anterior surface of the eye, extending to the edges of the transparent **cornea** (KŌR-nē-a). The cornea is covered by a very thin and delicate squamous *corneal epithelium* that is continuous with the ocular conjunctiva. A constant supply of fluid washes over the surface of the eyeball, keeping the ocular conjunctiva and cornea moist and clean. Goblet cells within the epithelium assist the various accessory glands in providing a superficial lubricant that prevents friction and drying of the opposing conjunctival surfaces.

Conjunctivitis, or "pink-eye," results from damage to and irritation of the conjunctival surface. The most obvious symptom results from dila-

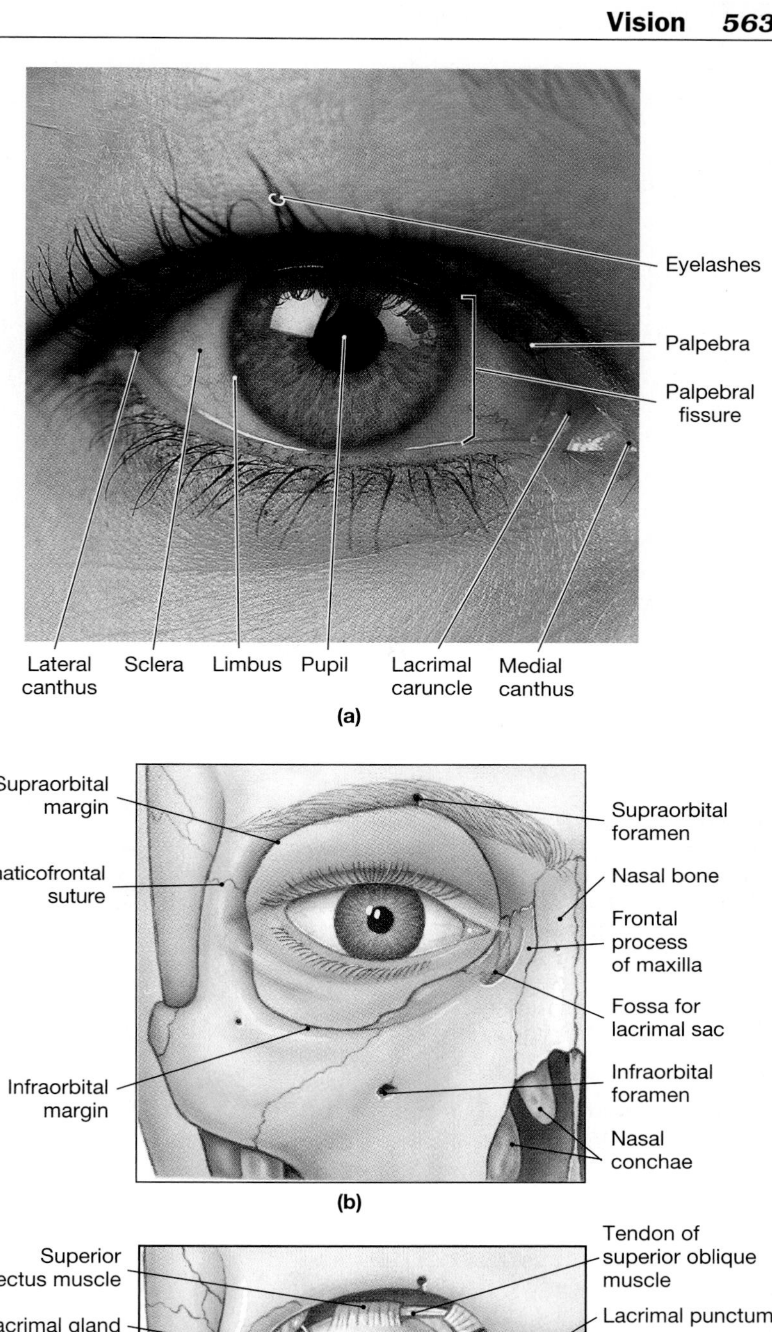
Eyelashes
Palpebra
Palpebral fissure
Lateral canthus
Sclera
Limbus
Pupil
Lacrimal caruncle
Medial canthus
(a)
Supraorbital margin
Zygomaticofrontal suture
Infraorbital margin
Supraorbital foramen
Nasal bone
Frontal process of maxilla
Fossa for lacrimal sac
Infraorbital foramen
Nasal conchae
(b)

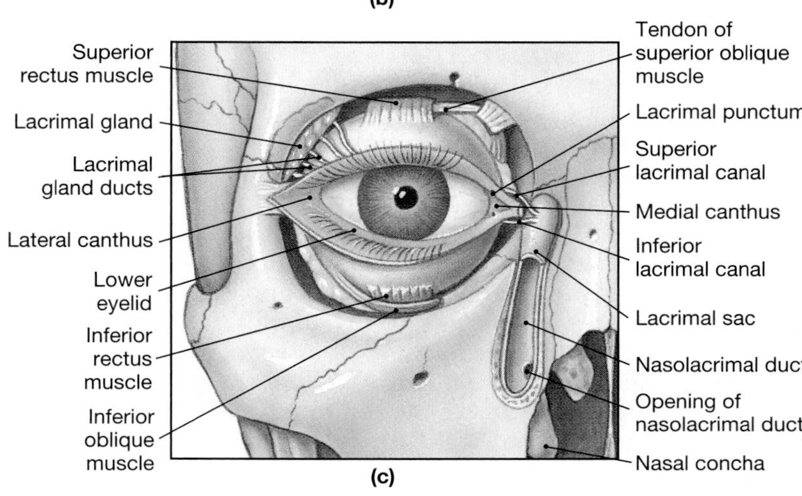
Superior rectus muscle
Lacrimal gland
Lacrimal gland ducts
Lateral canthus
Lower eyelid
Inferior rectus muscle
Inferior oblique muscle
Tendon of superior oblique muscle
Lacrimal punctum
Superior lacrimal canal
Medial canthus
Inferior lacrimal canal
Lacrimal sac
Nasolacrimal duct
Opening of nasolacrimal duct
Nasal concha
(c)

● **FIGURE 17-7**

External Features and Accessory Structures of the Eye. **(a)** Gross and superficial anatomy of the accessory structures. **(b)** Position of the eye within the orbit. **(c)** Details of the organization of the lacrimal apparatus.

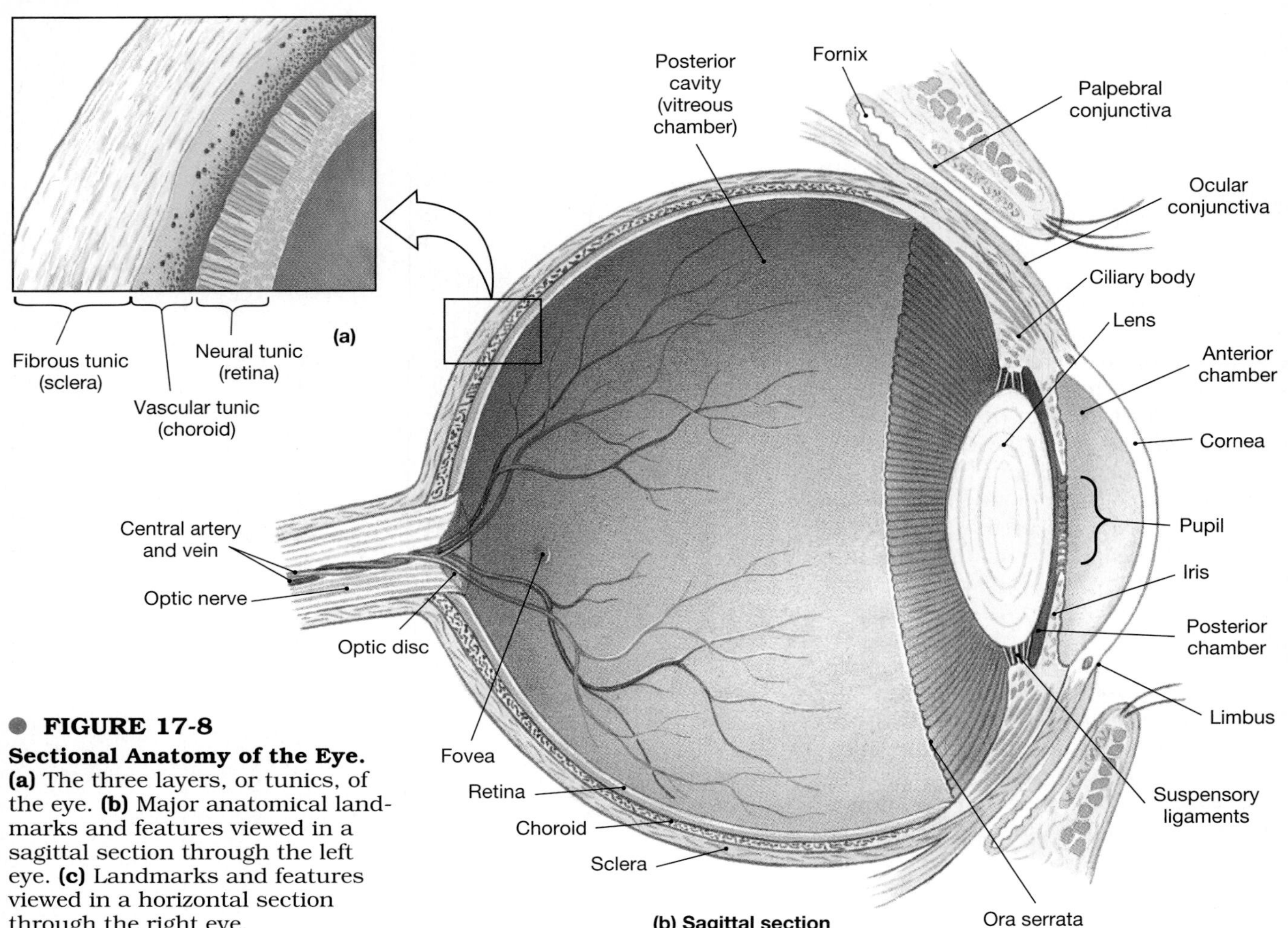

Fibrous tunic (sclera)
Vascular tunic (choroid)
Neural tunic (retina)
(a)
Posterior cavity (vitreous chamber)
Fornix
Palpebral conjunctiva
Ocular conjunctiva
Ciliary body
Lens
Anterior chamber
Cornea
Pupil
Iris
Posterior chamber
Limbus
Suspensory ligaments
Ora serrata
Central artery and vein
Optic nerve
Optic disc
Fovea
Retina
Choroid
Sclera
(b) Sagittal section

● **FIGURE 17-8**
Sectional Anatomy of the Eye. **(a)** The three layers, or tunics, of the eye. **(b)** Major anatomical landmarks and features viewed in a sagittal section through the left eye. **(c)** Landmarks and features viewed in a horizontal section through the right eye.

tion of the blood vessels beneath the conjunctival epithelium. This condition may be caused by pathogenic infection or by physical or chemical irritation of the conjunctival surface. AM *Conjunctivitis*

The Lacrimal Apparatus

Figures 17-7c, 17-8

A constant flow of tears keeps conjunctival surfaces moist and clean. Tears reduce friction, remove debris, prevent bacterial infection, and provide nutrients and oxygen to portions of the conjunctival epithelium. The **lacrimal apparatus** produces, distributes, and removes tears. The lacrimal apparatus of each eye consists of: (1) a *lacrimal gland* with associated ducts, (2) *lacrimal canals,* (3) a *lacrimal sac,* and (4) a *nasolacrimal duct* (Figure 17-7c●).

The pocket created where the conjunctiva of the eyelid connects with that of the eye is known as the **fornix** (FOR-niks) (Figure 17-8●). The lateral portion of the superior fornix receives 10–12 ducts from the **lacrimal gland,** or tear gland (Figure 17-17c●). The lacrimal gland is about the size and shape of an almond, measuring roughly 12–20 mm (0.5–0.75 in.). It nestles within a depression in the frontal bone, just inside the orbit and superior and lateral to the eyeball. ∞ *[p. 210]* The lacrimal gland normally provides the key ingredients and most of the volume of the tears that bathe the conjunctival surfaces. Its secretions are watery, slightly alkaline, and contain the enzyme **lysozyme,** which attacks microorganisms.

The lacrimal gland produces tears at a rate of around 1 ml/day. Once the lacrimal secretions have reached the ocular surface, they mix with the products of accessory glands and the oily secretions of the Meibomian and tarsal glands. The latter contributions produce a superficial "oil slick" that assists in lubrication and slows evaporation.

The blinking of the eye sweeps the tears across the ocular surface, and they accumulate at the medial canthus in an area known as the **lacus lacrimalis,** or "lake of tears." Two small pores, the **lacrimal puncta,** drain the lacrimal lake, emptying into the **lacrimal canals** that lead to the **lacrimal sac.** From the inferior portion of the lacrimal sac, the **nasolacrimal duct** passes through the nasolacrimal groove, formed by the lacrimal bone and the maxilla, and delivers the tears to the inferior meatus of the nasal cavity. ∞ *[p. 216]*

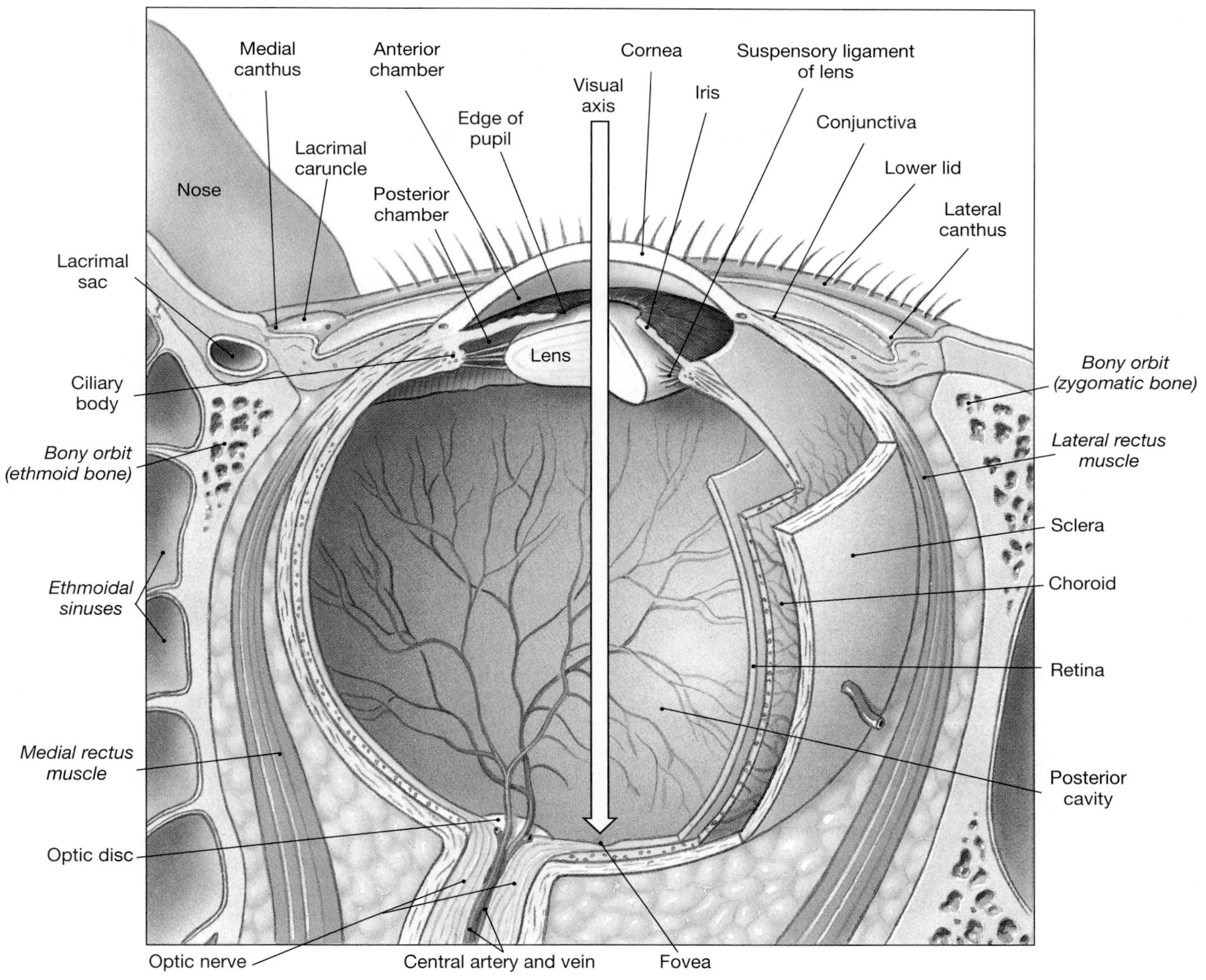
Medial canthus
Anterior chamber
Cornea
Suspensory ligament of lens
Visual axis
Iris
Edge of pupil
Conjunctiva
Lacrimal caruncle
Lower lid
Nose
Posterior chamber
Lateral canthus
Lacrimal sac
Lens
Ciliary body
Bony orbit (zygomatic bone)
Bony orbit (ethmoid bone)
Lateral rectus muscle
Sclera
Ethmoidal sinuses
Choroid
Retina
Medial rectus muscle
Posterior cavity
Optic disc
Optic nerve
Central artery and vein
Fovea

(c) Horizontal section

❑ The Eye

Figure 17-8

The eyes are extremely sophisticated visual organs, more versatile and adaptable than the most expensive cameras, yet light, compact, and durable. Each eye is a slightly irregular spheroid with an average diameter of 24 mm (almost 1 in.), a little smaller than a ping-pong ball, and a weight of around 8 g (0.28 oz). The eyeball shares space within the orbit with the extrinsic eye muscles, the lacrimal gland, and the cranial nerves and blood vessels that supply the eye and adjacent portions of the orbit and face. A mass of **orbital fat** provides padding and insulation.

The wall of the eye contains three distinct layers, or tunics (Figure 17-8a●): an outer *fibrous tunic*, an intermediate *vascular tunic*, and an inner *neural tunic*. The eyeball is hollow and the interior can be divided into two *cavities* (Figure 17-8b,c●). The large **posterior cavity** is also called the *vitreous chamber* because it contains the gelatinous *vitreous body*. The smaller **anterior cavity** is subdivided into two *chambers*, anterior and posterior. The shape of the eye is stabilized in part by the vitreous body and the clear *aqueous humor* that fills the anterior cavity.

The Fibrous Tunic

Figures 17-7, 17-8

The **fibrous tunic,** the outermost layer of the eye, consists of the *sclera* and the *cornea*. The fibrous tunic (1) provides mechanical support and some degree of physical protection, (2) serves as an attachment site for the extrinsic eye muscles, and (3) contains structures that assist in the focusing process.

Most of the ocular surface is covered by the **sclera** (SKLER-a) (Figure 17-8b,c●). The sclera, or "white of the eye," consists of a dense fibrous connective tissue containing both collagen and elastic fibers. This layer is thickest at the back of the eye, near the exit of the optic nerve, and thinnest over the anterior surface. The six extrinsic eye muscles insert upon the sclera, blending their collagenous fibers with those of the outer tunic. [p. 337]

The surface of the sclera contains small blood vessels and nerves that penetrate the sclera to reach internal structures. The network of small vessels that lie beneath the ocular conjunctiva usually does not carry enough blood to lend an obvious color to the sclera, but the vessels are visible, on close inspection, as red lines against the white background of collagen fibers.

The transparent cornea is structurally continuous with the sclera; the **limbus** (Figure 17-7a●, p. 563) is the border between the two. Beneath the corneal epithelium, the cornea consists primarily of a dense matrix containing multiple layers of collagen fibers. These fibers are organized into a series of layers that do not interfere with the passage of light. There are no blood vessels in the cornea, and the superficial epithelial cells must obtain oxygen and nutrients from the tears that flow across their free surfaces.

There are numerous free nerve endings in the cornea, and this is the most sensitive portion of the eye. Corneal damage will cause blindness even though the rest of the eye—photoreceptors included—is perfectly normal. Unfortunately, the cornea has a very restricted ability to repair itself, so corneal injuries must be treated immediately to prevent serious vision losses. To restore vision after corneal scarring it is usually necessary to replace the cornea through a *corneal transplant.* Corneal replacement is probably the most common form of transplant surgery. Such transplants can be performed between unrelated individuals because there are no blood vessels to carry white blood cells, which attack foreign tissues, into the area. Corneal grafts are obtained from the eyes of donors who have died from illness or accident; for best results the tissues must be removed within 5 hours after the donor's death.

The Vascular Tunic (Uvea)

Figure 17-8

The **vascular tunic,** or **uvea,** contains numerous blood vessels, lymphatics, and the intrinsic eye muscles. The functions of this layer include (1) providing a route for blood vessels and lymphatics that supply tissues of the eye, (2) regulating the amount of light entering the eye, (3) secreting and reabsorbing the *aqueous humor* that circulates within the eye, and (4) controlling the shape of the lens, an essential part of the focusing process. The vascular tunic includes the *iris,* the *ciliary body,* and the *choroid* (Figure 17-8a,b●).

THE IRIS The **iris,** which can be seen through the transparent corneal surface, contains blood vessels, pigment cells, and two layers of smooth muscle fibers. When these muscles contract they change the diameter of the central opening, or **pupil,** of the iris. One group of smooth muscle fibers forms a series of concentric circles around the pupil. When these **pupillary constrictor muscles** contract, the diameter of the pupil decreases. A second group of smooth muscles extends radially away from the edge of the pupil. Contraction of these **pupillary dilator muscles** enlarges the pupil. Both muscle groups are controlled by the autonomic nervous system. For example, parasympathetic activation causes pupillary constriction in response to bright light, and sympathetic activation causes pupillary dilation in response to dim light (the *pupillary reflex*). [p. 546]

The body of the iris consists of a connective tissue whose posterior surface is covered by an epithelium containing pigment cells. Pigment cells may also be present in the connective tissue of the iris and in the epithelium covering its anterior surface. Eye color is determined by the density and distribution of pigment cells. When there are no pigment cells in the connective tissue of the iris, light passes through it and bounces off the inner surface of the pigmented epithelium that covers the posterior surface of the iris. The eye then appears blue. Individuals with gray, brown, and black eyes have increasing numbers of pigment cells in the body and surface of the iris. The eyes of human albinos appear a very pale gray or blue-gray.

THE CILIARY BODY At the periphery the iris attaches to the anterior portion of the ciliary body. The **ciliary body** begins at the junction between the cornea and sclera. It extends posteriorly to the **ora serrata** (Ō-ra ser-RA-ta; serrated mouth) (Figure 17-8b,c●), the serrated anterior edge of the *neural retina,* which contains the visual receptors. The bulk of the ciliary body consists of the **ciliary muscle,** a muscular ring that projects into the interior of the eye. The epithelium is thrown into numerous folds, called **ciliary processes.** The **suspensory ligaments** of the lens attach to these processes. These connective tissue fibers hold the lens posterior to the iris and centered on the pupil. As a result, any light passing through the pupil and headed for the photoreceptors will pass through the lens.

THE CHOROID The **choroid** separates the fibrous and neural tunics posterior to the ora serrata (Figure 17-8●). It is covered by the sclera and attached to the outermost layer of the retina. The choroid contains an extensive capillary network that delivers oxygen and nutrients to the retina. It also contains scattered melanocytes, which are especially numerous near the sclera.

The Neural Tunic

Figures 17-8, 17-9, 17-10

The **neural tunic,** or **retina,** consists of a thin, outer **pigment layer** and a thick inner layer, the **neural**

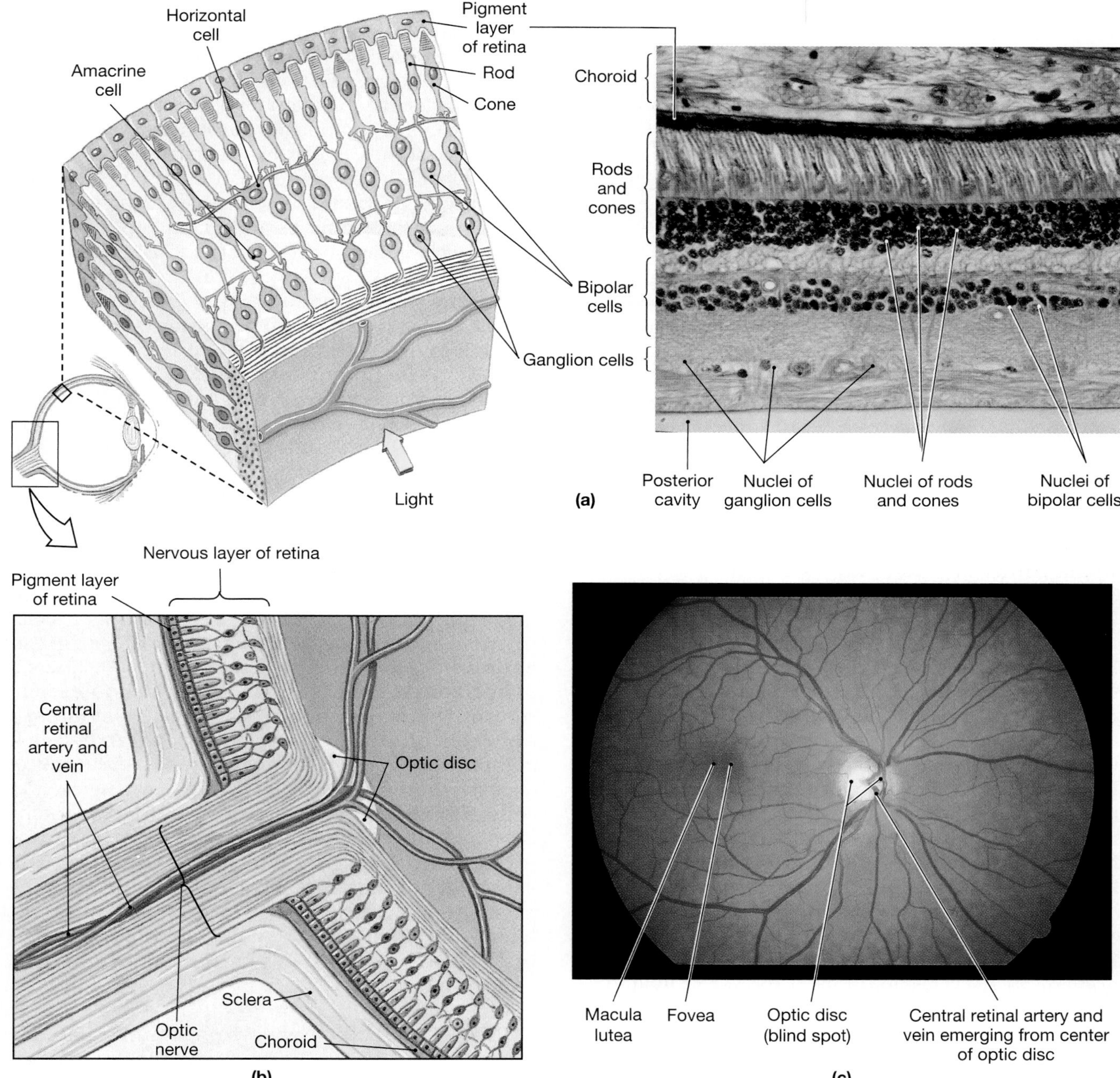
Horizontal cell
Amacrine cell
Pigment layer of retina
Rod
Cone
Bipolar cells
Ganglion cells
Light
Choroid
Rods and cones
Bipolar cells
Ganglion cells
Posterior cavity
Nuclei of ganglion cells
Nuclei of rods and cones
Nuclei of bipolar cells
(a)
Nervous layer of retina
Pigment layer of retina
Central retinal artery and vein
Optic disc
Sclera
Optic nerve
Choroid
(b)
Macula lutea
Fovea
Optic disc (blind spot)
Central retinal artery and vein emerging from center of optic disc
(c)

● **FIGURE 17-9**

Retinal Organization. (a) Cellular organization of the retina. Note that the photoreceptors are located closest to the choroid rather than near the vitreous chamber. (LM × 290) **(b)** The optic disc in diagrammatic horizontal section. **(c)** A photograph of the retina as seen through the pupil of the eye.

retina, that contains the visual receptors and associated neurons (Figure 17-9●). The pigment layer absorbs light after it passes through the neural retina, and the pigment cells have important biochemical interactions with retinal photoreceptors. The neural retina contains (1) the photoreceptors that respond to light, (2) supporting cells and neurons that perform preliminary processing and integration of visual information, and (3) blood vessels supplying tissues lining the posterior cavity.

The two retinal layers are normally very close together, but not tightly interconnected. The pigmented layer of the retina continues over the ciliary body and iris, although the neural retina extends anteriorly only as far as the ora serrata. The neural retina thus forms a cup that establishes the posterior and lateral boundaries of the posterior cavity (Figure 17-8b,c●).

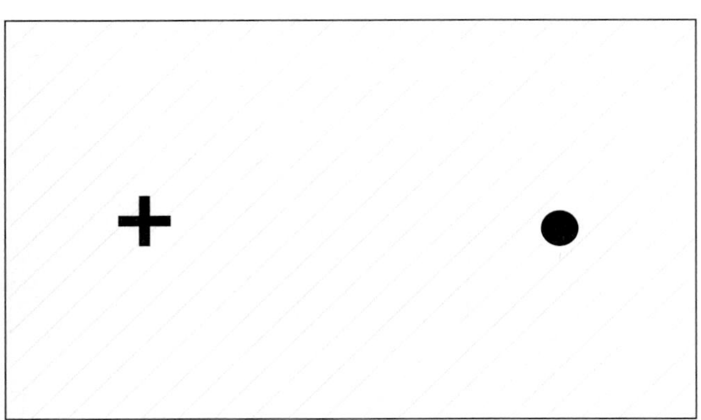

● **FIGURE 17-10**
The Optic Disc. Close your left eye and stare at the cross with your right eye, keeping it in the center of your field of vision. Begin with the page a few inches away, and gradually increase the distance. The dot will disappear when its image falls on the blind spot. To check the blind spot in the left eye, close your right eye and repeat this sequence staring at the dot.

RETINAL ORGANIZATION In sectional view, the retina contains several layers of cells (Figure 17-9a●). The outermost layer, closest to the pigment layer, contains the photoreceptors.

The photoreceptors are entirely dependent on the diffusion of oxygen and nutrients from blood vessels in the choroid. In a **detached retina** the neural retina becomes separated from the pigment layer. This condition can result from a sudden impact to the eye or from a variety of other factors. Unless the two layers of the neural tunic are reattached, the photoreceptors will degenerate and vision will be lost. The reattachment is usually performed by "welding" the two together using laser beams focused through the cornea.

There are two types of photoreceptors, **rods** and **cones.** Rods do not discriminate among different colors of light. They are very light-sensitive and enable us to see in dimly lit rooms, at twilight, or in pale moonlight. Cones provide us with color vision. There are three types of cones, and their stimulation in various combinations provides the perception of different colors. Cones give us sharper, clearer images, but they require more intense light than rods. If you sit outside at sunset you will probably be able to tell when your visual system shifts from cone-based vision (clear images in full color) to rod-based vision (relatively grainy images in black and white).

Rods and cones are not evenly distributed across the outer surface of the retina. Approximately 125 million rods form a broad band around the periphery of the retina. The posterior retinal surface is dominated by the presence of roughly 6 million cones. Most of these are concentrated in the area where a visual image arrives after passing through the cornea and lens. There are no rods in this region, which is known as the **macula lutea** (LOO-tē-a; yellow spot). The highest concentration of cones is found in the central portion of the macula lutea, an area called the **fovea** (FŌ-vē-a; shallow depression), or *fovea centralis* (Figure 17-9c●). The fovea is the site of sharpest vision; when you look directly at an object its image falls upon this portion of the retina (Figure 17-8c●, p. 565). A line drawn from the center of that object through the center of the lens to the fovea establishes the **visual axis** of the eye.

You are probably already aware of the visual consequences of this distribution. When you look directly at something, you are focusing the image on the fovea, the center of color vision. You see a very good image as long as there is enough light to stimulate the cones. In very dim light, cones simply cannot function. For example, when you try to stare at a dim star, you are unable to see it. But if you look a little to one side, rather than directly at the star, you will see it quite clearly. Shifting your gaze moves the image of the star from the fovea, where it does not provide enough light to stimulate the cones, to the periphery, where it can affect the more sensitive rods.

The rods and cones synapse with roughly 6 million **bipolar cells** (Figure 17-9●). Bipolar cells in turn synapse within the layer of **ganglion cells** adjacent to the posterior cavity. A network of **horizontal cells** extends across the outer portion of the retina at the level of the synapses between photoreceptors and bipolar cells. A comparable layer of **amacrine** (AM-a-krīn) **cells** occurs where bipolar cells synapse with ganglion cells. Horizontal and amacrine cells can facilitate or inhibit communication between photoreceptors and ganglion cells, adjusting the sensitivity of the retina. The effect could be compared to adjusting the "contrast" setting on a television. Their activities play an important role in the eye's adjustment to dim or brightly lit environments.

THE OPTIC DISC Axons from an estimated 1 million ganglion cells converge on the **optic disc,** a circular region just medial to the fovea. The optic disc is the origin of the optic nerve (N II). From this point, the axons turn, penetrate the wall of the eye, and proceed toward the diencephalon (Figure 17-9b●). The *central retinal artery* and *central retinal vein* that supply the retina pass through the center of the optic nerve and emerge on the surface of the optic disc (Figure 17-9b,c●). There are no photoreceptors or other retinal structures at the optic disc. Because light striking this area goes unnoticed, it is commonly called the **blind spot.** You do not "notice" a blank spot in the visual field primarily because involuntary eye movements keep the visual image moving and allow the brain to fill in the missing information. A simple experiment, shown in

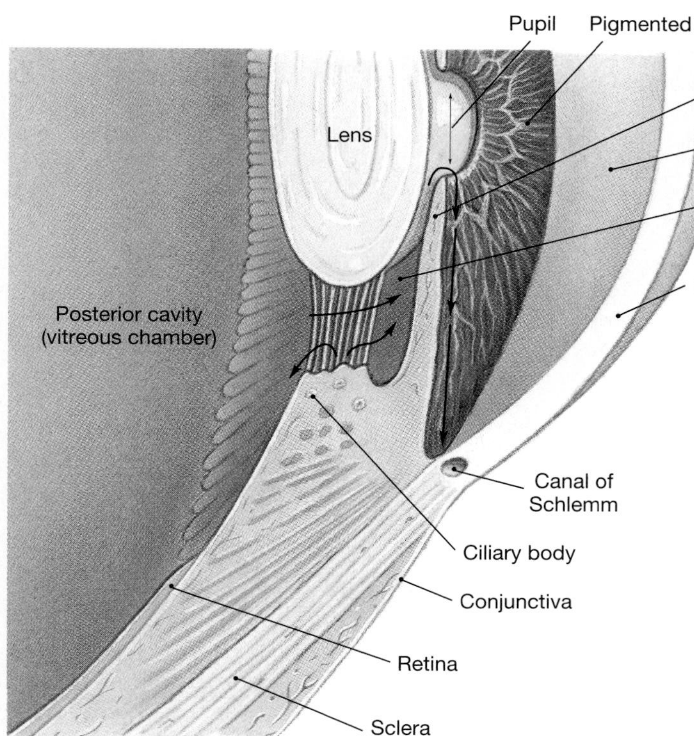
Pupil
Pigmented epithelium of iris
Iris
Lens
Anterior chamber
Posterior chamber
Anterior cavity
Cornea
Posterior cavity (vitreous chamber)
Canal of Schlemm
Ciliary body
Conjunctiva
Retina
Sclera

● **FIGURE 17-11**
The Circulation of Aqueous Humor. Aqueous humor secreted at the ciliary body circulates through the posterior and anterior chambers before being reabsorbed via the canal of Schlemm.

Figure 17-10●, will demonstrate the presence and location of the blind spot within the visual field.

DIABETIC RETINOPATHY *Diabetic retinopathy* develops in most individuals with *diabetes mellitus*, an endocrine disorder that affects primarily glucose metabolism. Many different systems are affected, but serious circulatory problems are very common. Diabetic retinopathy develops over a period of years because of degeneration and rupture of retinal blood vessels. Visual acuity is lost, and, over time, the photoreceptors are destroyed because of oxygen starvation.

The Chambers of the Eye

Figure 17-11

The ciliary body and lens divide the interior of the eye into a large *posterior cavity*, also called the *vitreous chamber*, and a smaller *anterior cavity*. The anterior cavity is further subdivided into the **anterior chamber,** which extends from the cornea to the iris, and a **posterior chamber** between the iris and the ciliary body and lens. The anterior and posterior chambers are filled with the fluid *aqueous humor*. Aqueous humor circulates within the anterior cavity, passing from the posterior to the anterior chamber via the pupil of the eye (Figure 17-11●). The posterior cavity is filled with a gelatinous substance known as the *vitreous body*, or *vitreous humor*.

AQUEOUS HUMOR **Aqueous humor** forms through active secretion by epithelial cells of the ciliary processes (Figure 17-11●). It is secreted to the posterior chamber at a rate of 1–2 μl per minute. The epithelial cells regulate its composition, which resembles that of cerebrospinal fluid. The aqueous humor circulates so that in addition to forming a fluid cushion it provides an important route for nutrient and waste transport.

The eye is filled with fluid, and fluid pressure in the aqueous humor helps retain its shape and stabilizes the position of the retina, pressing the photoreceptor layer against the pigment layer. In effect, the aqueous humor acts like the air inside a balloon. The **intraocular pressure** can be measured in the anterior chamber, where it pushes against the inner surface of the cornea. Intraocular pressure is most often checked by bouncing a tiny blast of air off the surface of the eye and measuring the deflection produced. Normal intraocular pressures range from 12 to 21 mm Hg.

Aqueous humor returns to the circulation from the anterior chamber. After filtering through a network of connective tissues located near the base of the iris, aqueous humor enters the **canal of Schlemm,** a passageway that extends completely around the eye at the level of the limbus. Collecting channels deliver the aqueous humor from the canal to veins in the sclera. The rate of removal normally keeps pace with the rate of generation at the ciliary processes, and aqueous humor is removed and recycled within a few hours of its formation. If aqueous humor cannot enter the canal of Schlemm, fluid pressure rises, distorting soft tissues within the eye. This condition, which may cause blindness, is called **glaucoma.** AM *Glaucoma*

THE VITREOUS BODY The posterior cavity of the eye contains the **vitreous body,** a gelatinous mass also known as the *vitreous humor*. The vitreous body helps to stabilize the shape of the eye and gives additional physical support to the retina. Specialized cells embedded within the vitreous body produce the collagen fibers and proteoglycans that account for its consistency. Aqueous humor produced in the posterior chamber freely diffuses through the vitreous body and across the retinal surface.

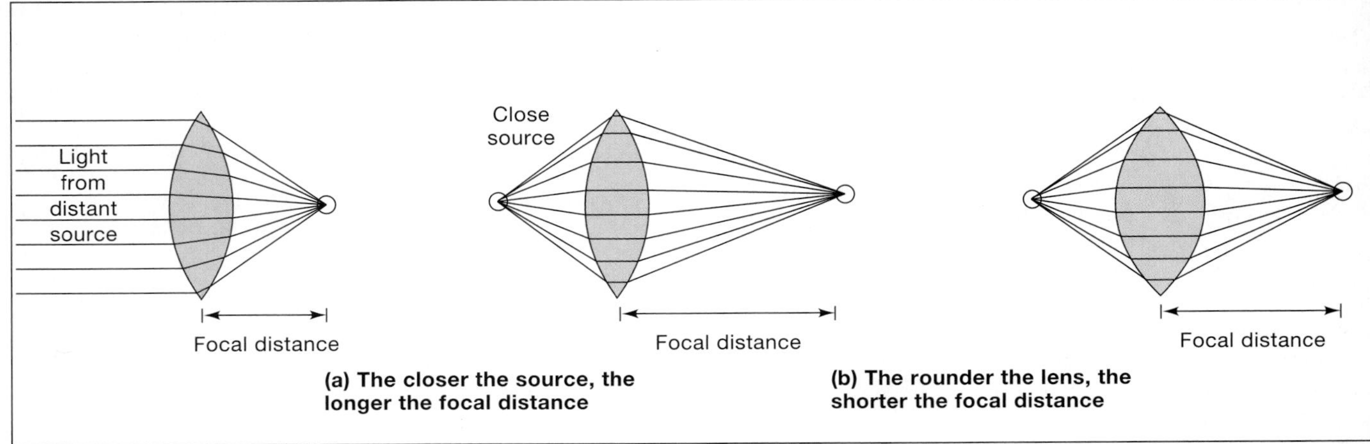
Light
from
distant
source
Focal distance
Close
source
Focal distance
Focal distance
(a) The closer the source, the longer the focal distance
(b) The rounder the lens, the shorter the focal distance

● **FIGURE 17-12**
Principles of Image Formation. (a) The closer the light source, the longer the focal distance. **(b)** The rounder the lens, the shorter the focal distance.

The Lens

Figures 17-8, 17-11, 17-12, 17-13, 17-14, 17-15

The **lens** lies behind the cornea, held in place by the suspensory ligaments that originate on the ciliary body of the choroid (Figures 17-8, p. 564, and 17-11●). The primary function of the lens is to focus the visual image on the retinal photoreceptors. It does so by changing its shape.

The lens consists of concentric layers of cells that are precisely organized. A dense fibrous capsule covers the entire lens. Many of the capsular fibers are elastic, and, unless an outside force is applied, they will contract and make the lens spherical. Around the edges of the lens, the capsular fibers intermingle with those of the suspensory ligaments.

CATARACTS The transparency of the lens depends on a precise combination of structural and biochemical characteristics. When that balance becomes disturbed the lens loses its transparency, and the abnormal lens is known as a **cataract.** Cataracts may result from drug reactions, injuries, or radiation, but **senile cataracts** are the most common form.

Over time, the lens takes on a yellowish hue, and eventually it begins to lose its transparency. As the lens becomes "cloudy," the individual needs brighter and brighter reading lights, and visual clarity begins to fade. If the lens becomes completely opaque, the person will be functionally blind, even though the retinal receptors are normal. Modern surgical procedures involve removing the lens, either intact or in pieces, after shattering it with high-frequency sound. The missing lens can be replaced by an artificial substitute, and vision can then be fine-tuned with glasses or contact lenses.

REFRACTION There are approximately 130 million photoreceptors in the retina, each monitoring a specific location. A visual image results from the processing of information provided by the entire receptor population. The eye has often been compared to a camera, and like a camera lens, the lens of the eye must focus the arriving image if it is to provide any useful information. "In focus" means that the rays of light arriving from the object strike the sensitive surface of the film or retina precisely ordered so as to form a miniature image of the original. If they are not perfectly focused, the image will be blurry. Focusing in the eye normally occurs in two steps, as light passes through the cornea and lens.

Light is bent, or **refracted,** when it passes between media of differing densities. You can see refraction clearly by sticking a pencil into a glass of water. Because refraction occurs as the light passes into the air from the much denser water, the pencil shaft appears to bend sharply at the air-water interface.

In the human eye, the greatest amount of refraction occurs when light passes from the air into the corneal tissues, which have a density close to that of water. When you open your eyes underwater, you cannot see clearly because the air-water interface has been eliminated, and light passes from one watery medium to another.

Additional refraction takes place when the light passes from the aqueous humor into the relatively dense lens. The lens provides the extra refraction needed to focus the light rays from an object toward a specific **focal point.** The distance between the center of the lens and the focal point is the **focal distance.** The focal distance is determined by:

1. ***The distance of the object from the lens.*** The focal distance increases as an object moves closer to a lens (Figure 17-12a●).
2. ***The shape of the lens.*** The rounder the lens, the more refraction occurs, so a very round lens has a shorter focal distance than a flatter one (Figure 17-12b●).

ACCOMMODATION In a camera an image can be focused by moving the lens toward or away from the film. This method of focusing cannot work in our eyes, because the distance from the lens to the macula lutea cannot change. We focus images on the retina by changing the shape of the lens so that the focal length remains constant. This process is called **accommodation.** During accommodation the lens becomes rounder to focus the image of a nearby object on the retina, and it flattens when focusing on a distant object.

The lens is held in place by the suspensory ligaments that originate at the ciliary body. Smooth muscle fibers in the ciliary body act like the sphincter muscles of the iris. When the ciliary muscle relaxes, the suspensory ligaments pull at the circumference of the lens, making the lens relatively flat (Figure 17-13a●). When the ciliary muscle contracts, the ciliary body moves toward the lens. This movement reduces the tension in the suspensory ligaments, and the elastic capsule pulls the lens into a more spherical shape. The rounder shape increases the refractive power of the lens, enabling it to bring light from nearby objects to a focus on the retina (Figure 17-13b●).

The greatest amount of refraction is required to view objects that are very close to the lens. The inner limit of clear vision, known as the *near point of vision,* is determined by the degree of elasticity in the lens. Children can usually focus on something 7–9 cm from the eye, but over time the lens tends to become stiffer and less responsive. A young adult can usually focus on objects 15–20 cm away, and as aging proceeds this distance gradually increases; the near point at age 60 is usually around 83 cm.

Changes in Pupillary Diameter. Most lenses have minor imperfections or irregularities in shape. Light passing through these regions fails to refract properly, disrupting the clarity of the visual image. This condition, called **astigmatism,** can be corrected by glasses or special contact lenses.

Minor astigmatism is very common, but many people are totally unaware of it. For any given lens shape the variation from "ideal" will usually be greatest around the edges of the lens. The effects of this variation, called *spherical aberration,* will be most noticeable when the lens rounds up to focus on nearby objects. The pupil constricts when focusing on nearby objects, however, and that constriction restricts the light path to central portions of the lens. Constriction or dilation of the pupil therefore not only limits the amount of light entering the eye but also reduces the effects of minor astigmatism and spherical aberration.

Image Reversal. Thus far we have considered light originating at a single source, either near or away from the viewer. An object in view is a complex light source that can be treated as a large number of individual points. Light from each source is focused on the retina as indicated in Figure 17-15a,b●. The result is the creation of a miniature image of the original (Figure 17-15c●). But the image arrives upside down and backward.

To understand why, consider Figure 17-15c● from two different perspectives. First imagine that this is a sagittal section through an eye that is looking at a telephone pole painted with colored bands. The image of the top of the pole lands at the bottom of the retina, and the image of the bottom is at the top of the retina. Now imagine that this is a horizontal section through an eye that is looking at a picket fence with each board painted a different color. The image of the left edge of the fence falls on the right side of the retina, and the image of the right edge falls on the left side of the retina. The brain compensates for this image reversal, and we are not consciously aware of any difference between the orientation of the image on the retina and the object.

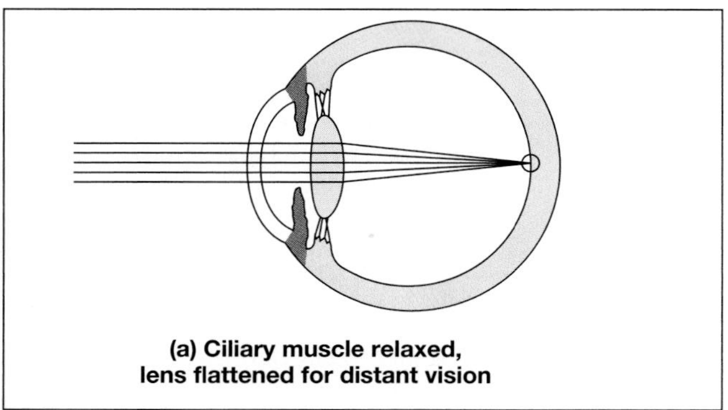

(a) Ciliary muscle relaxed, lens flattened for distant vision

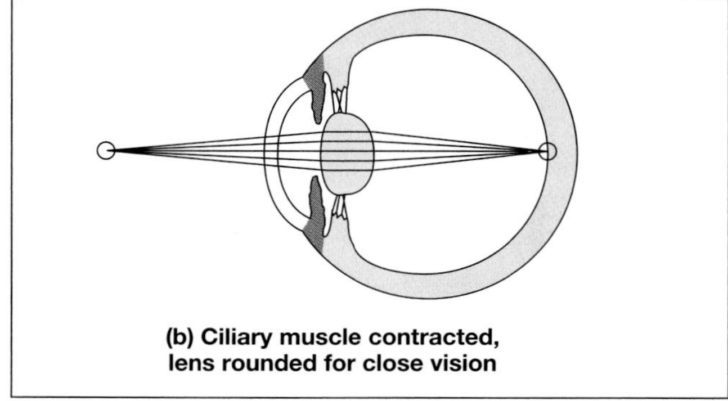

(b) Ciliary muscle contracted, lens rounded for close vision

● **FIGURE 17-13**

Accommodation. For the eye to form a sharp image, the focal distance must equal the distance between the center of the lens and the retina. **(a)** When the ciliary muscle relaxes, the ligaments pull against the margins of the lens and flatten it. **(b)** When the ciliary muscle contracts, the suspensory ligaments allow the lens to round up.

Accommodation Problems

In the normal eye, when the ciliary muscles are relaxed and the lens is flattened, a distant image will be focused on the retinal surface (Figure 17-14a●). This condition is called **emmetropia** (*emmetro-*, proper), or normal vision.

Figure 17-14b,d● diagrams two common problems with the accommodation mechanism. If the eyeball is too deep or the resting curvature of the lens too great, the image of a distant object will form in front of the retina (Figure 17-14b●). The individual will see distant objects as blurry and out of focus. Vision at close range will be normal, because the lens will be able to round up as needed to focus the image on the retina. As a result, such individuals are said to be "nearsighted." Their condition is more formally termed **myopia** (*myein*, to shut + *ops*, eye). Myopia can be treated by placing a diverging lens in front of the eye (Figure 17-14c●). This lens shape, typical of the lenses used in prescription corrective glasses, spreads the light rays apart as if the object were closer to the viewer.

If the eyeball is too shallow, or the lens too flat, **hyperopia** results (Figure 17-14d●). The ciliary muscles must contract to focus even a distant object on the retina, and at close range the lens cannot provide enough refraction to focus an image on the retina. These individuals are said to be "farsighted" because they can see distant objects most clearly. Older individuals become farsighted as their lenses lose elasticity; this form of hyperopia is called **presbyopia** (*presbys*, old man). Hyperopia can be corrected by placing a converging lens in front of the eye. This lens provides the additional refraction needed to bring nearby objects into focus (Figure 17-14e●).

Variable success at correcting myopia and hyperopia has been achieved by surgically reshaping the cornea to alter its refractive powers. This procedure, called **radial keratotomy,** remains controversial. Although roughly two-thirds of patients are satisfied with the results, corneal healing takes several years. Many ophthalmological surgeons have expressed concerns that the scarring that develops after radial keratotomy creates weak points in the cornea that may increase the chances for a dangerous corneal rupture.

Another controversial procedure is **photo-refractive keratectomy (PRK),** in which a computer-guided laser shapes the cornea to exact specifications. Tissue is removed only to a depth of 10–20 µm—no more than about 10 percent of the cornea's thickness. The entire procedure can be done in less than a minute. However, problems with persistent cloudiness of the cornea after PRK continue to limit its use.

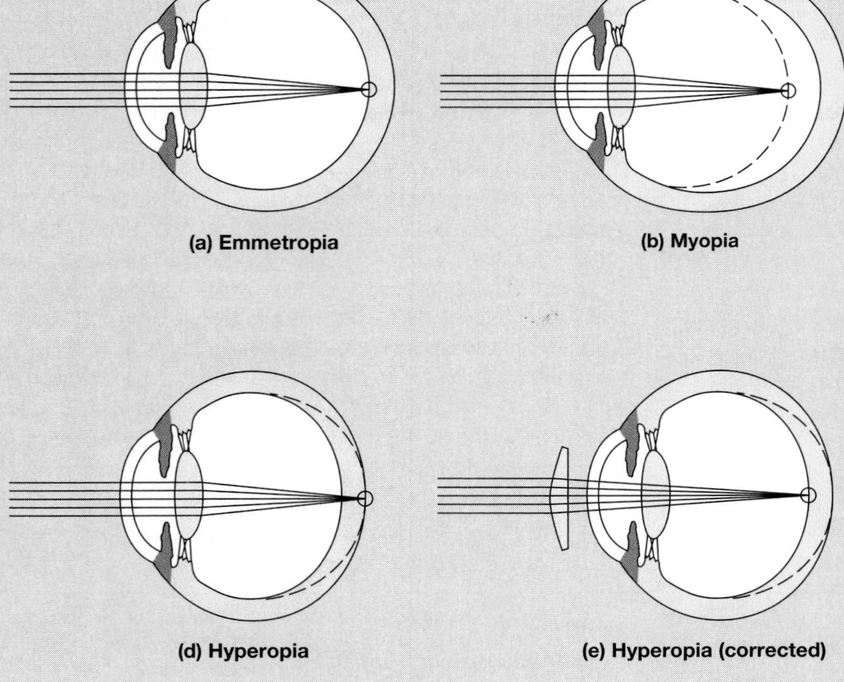
(a) Emmetropia
(b) Myopia
(c) Myopia (corrected)
(d) Hyperopia
(e) Hyperopia (corrected)

● **FIGURE 17-14**

Visual Abnormalities. In normal vision **(a)**, the lens focuses the visual image on the retina. Common problems with the accommodation mechanism involve an inability to lengthen the focal distance enough to focus the image of a distant object on the retina—myopia, **(b)**—and an inability to shorten the focal distance adequately for nearby objects—hyperopia, **(d)**. These conditions can be corrected by placing appropriately shaped lenses in front of the eyes. A diverging lens is used to correct myopia **(c)**, and a converging lens is used to correct hyperopia **(e)**.

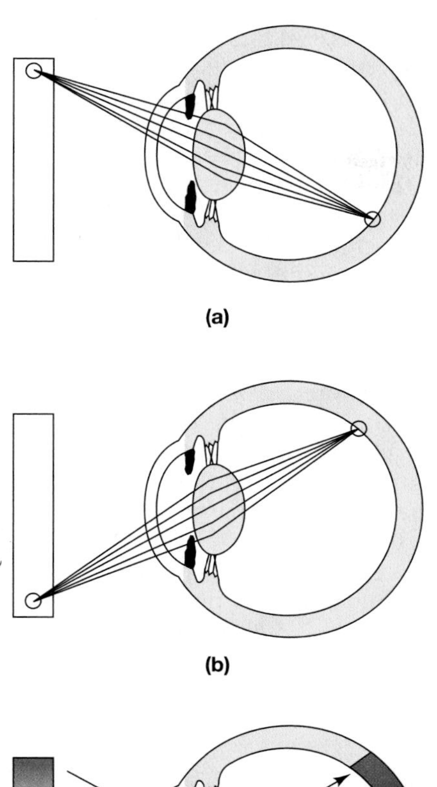
(a)
(b)

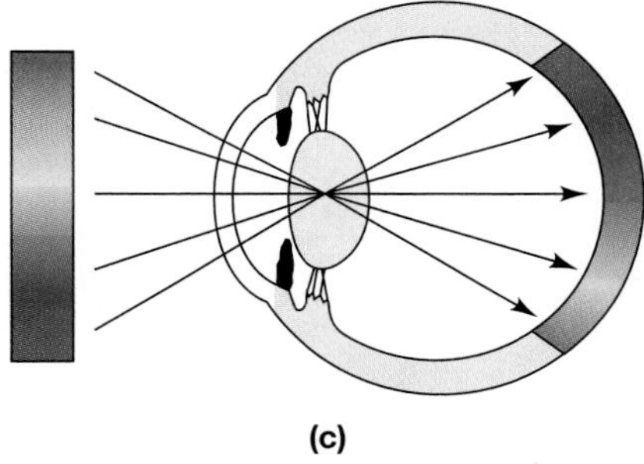
(c)

● **FIGURE 17-15**
Image Formation. (a,b) Light from each portion of an object is focused on a different part the retina. **(c)** The resulting image is reversed.

VISUAL ACUITY In rating the clarity of vision, or **visual acuity,** a person whose vision is rated 20/20 is seeing details at a distance of 20 feet as clearly as a "normal" individual would. Vision noted as 20/15 is better than average, for at 20 feet the person is able to see details that would be clear to a normal eye only at a distance of 15 feet. Conversely, a person with 20/30 vision must be 20 feet from an object to discern details that a person with normal vision could make out at a distance of 30 feet.

When visual acuity falls below 20/200, even with the help of glasses or contact lenses, the individual is considered to be legally blind. There are probably fewer than 400,000 legally blind people in the United States; more than half are over 65 years of age. Common causes of blindness include diabetes mellitus, cataracts, glaucoma, corneal scarring, retinal detachment, accidental injuries, and hereditary factors that are as yet poorly understood.

SCOTOMAS Abnormal blind spots, or **scotomas** (skō-TŌ-mahs), may appear elsewhere in the visual field. Scotomas are permanent abnormalities that are fixed in position. They may result from compression of the optic nerve, damage to photoreceptors, or central damage along the visual pathway. Most readers will probably be more familiar with *floaters,* small spots that drift across the field of vision. Floaters are usually temporary phenomena that result from blood cells or cellular debris within the vitreous body. They can be detected by staring at a blank wall or a white sheet of paper.

✓ What layer of the eye would be affected first by inadequate tear production?

✓ When the lens of the eye is very round, are you looking at an object that is close to you or distant from you?

✓ As Renee enters a dark room, most of the available light becomes focused on the fovea of her eye. Will she be able to see very clearly?

✓ How would blockage of the canal of Schlemm affect a person's vision?

❑ Visual Physiology

The rods and cones of the retina are called **photoreceptors** because they detect *photons,* basic units of visible light. Light, a form of *radiant energy,* is radiated in waves that have a characteristic *frequency* (cycles per second) and *wavelength* (distance between wave peaks). Visible light is one small part of the entire spectrum of electromagnetic radiation, which ranges from long-wavelength radio waves to short-wavelength gamma rays.

Our eyes are sensitive to wavelengths of 400–700 nm, the spectrum of visible light. This spectrum, seen in a rainbow, can be remembered by the acronym ROY G. BIV (red, orange, yellow, green, blue, indigo, violet). Photons of red light carry the least energy, and those from the violet portion of the spectrum contain the most. Rods provide the CNS with information on the presence or absence of photons, without regard to wavelength. Cones provide information on the wavelength, giving us a perception of color.

Anatomy of Rods and Cones

Figure 17-16

Figure 17-16● compares the structure of rods and cones. The elongate **outer segment** of a photoreceptor contains hundreds to thousands of flattened membranous plates, or **discs.** The names *rod* and *cone* refer to the shape of the outer segment. In a rod, each disc is an independent entity, and the outer segment forms an elongate cylinder. In cones, the discs are infoldings of the cell membrane, and the outer segment tapers to a blunt tip.

A narrow connecting stalk attaches the outer segment to the **inner segment,** a region that contains all of the usual cellular organelles. The inner segment makes synaptic contact with other cells, and it is here that neurotransmitters are released.

VISUAL PIGMENTS The discs of the outer segment in both rods and cones contain special organic compounds called **visual pigments.** Absorption of photons by visual pigments is the first key step in the process of photoreception. Visual pigments are derivatives of the compound **rhodopsin** (rō-DOP-sin). Rhodopsin consists of a protein, **opsin,** bound to a pigment, **retinal** (RET-i-nal), synthesized from **vitamin A.** Figure 17-16● diagrams the structure of a rhodopsin molecule.

The same retinal pigment is found in both rods and cones, but there are four different forms of opsin. The type of opsin present determines the wavelength of light that can be absorbed by the retinal. One form of opsin is characteristic of all rods. There are three different populations of cones, each with a different form of opsin, and each sensitive to a different range of wavelengths. Differential stimulation of these cone populations is the basis for color vision.

New discs containing visual pigment are continually assembled at the base of the outer segment. A completed disc then moves toward the tip of the outer segment. After about 10 days, it will be shed in a small droplet of cytoplasm. These droplets are absorbed by the pigment cells, which break down the membrane components and reconvert the retinal to vitamin A. The vitamin A is then stored within the pigment cells for subsequent transfer to the photoreceptors.

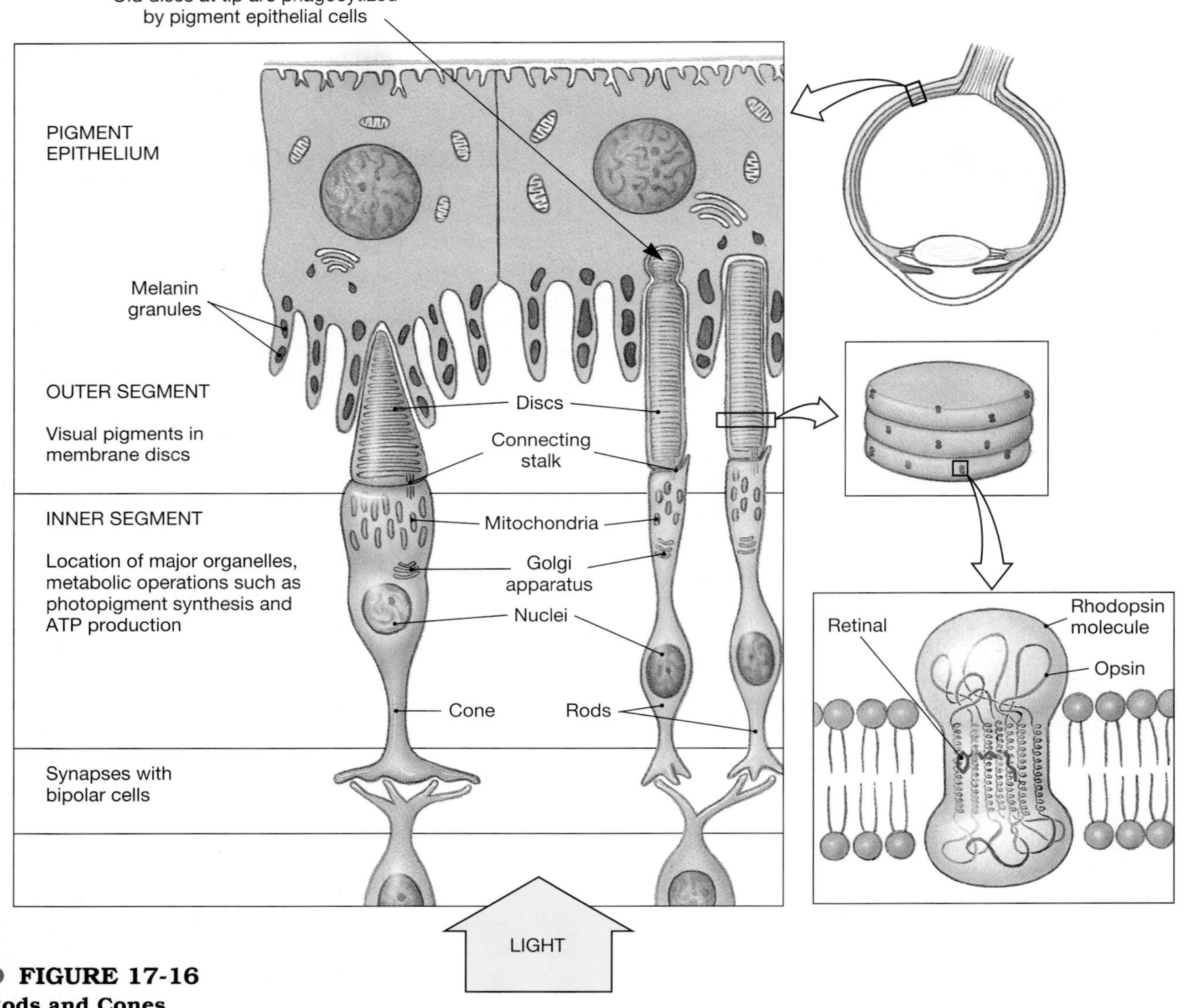
Old discs at tip are phagocytized by pigment epithelial cells
PIGMENT EPITHELIUM
Melanin granules
OUTER SEGMENT
Visual pigments in membrane discs
Discs
Connecting stalk
INNER SEGMENT
Location of major organelles, metabolic operations such as photopigment synthesis and ATP production
Mitochondria
Golgi apparatus
Nuclei
Cone
Rods
Synapses with bipolar cells
LIGHT
Retinal
Rhodopsin molecule
Opsin

● **FIGURE 17-16**
Rods and Cones

Photoreception

Figures 17-17, 17-18

We will now follow the steps in rhodopsin-based photoreception. In darkness, the cell membrane in the outer segment of the photoreceptor contains open, chemically regulated sodium ion channels. These gated channels are kept open in the presence of **cyclic-GMP** (cGMP), a derivative of the high-energy compound *guanosine triphosphate* (GTP). Because these channels are open, the transmembrane potential is approximately –40 mV, rather than the –70 mV typical of resting neurons. ∞ [pp. 394, 396] The inner segment continually pumps sodium ions out of the cytoplasm, and the movement of sodium ions into the outer segment, to the inner segment, and out of the cell is known as the *dark current* (Figure 17-17•).

At this transmembrane potential the photoreceptor is continually releasing neurotransmitters across synapses at the inner segment. The arrival of a photon reduces the dark current, alters the transmembrane potential, and changes the rate of neurotransmitter release.

The process begins when a photon strikes the retinal portion of a rhodopsin molecule. There are two possible configurations for the bound retinal molecule. It is normally in the **11-cis** form; on absorbing light, it adopts the more linear **11-trans** form. This change in shape triggers a chain of enzymatic steps, diagrammed in Figure 17-17•.

Step 1: ***Opsin activation occurs.***

Rhodopsin is a receptor protein embedded in the membrane of the disc. Opsin is an enzyme, and it is inactive in the dark. When light strikes the rhodopsin molecule, retinal switches from the 11-cis to the 11-trans form. This switch activates the opsin portion of the molecule.

Step 2: ***Opsin activates a second enzyme, called transducin, and transducin activates phosphodiesterase (PDE).***

Transducin is an example of a **G-protein,** a membrane-bound enzyme that is activated by interaction with receptor proteins in the cell membranes. In this case, transducin is activated by opsin, and transducin in turn activates **phosphodiesterase.** Phosphodiesterase (PDE) is an enzyme that breaks down cyclic-GMP (cGMP).

Step 3: ***Cyclic-GMP levels decline, and gated sodium ion channels close.***

The removal of cGMP from the gated sodium channels results in their inactivation. The rate of Na^+ entry into the cytoplasm then decreases.

Step 4: ***The rate of neurotransmitter release declines.***

Because active transport continues to remove sodium ions from the cytoplasm, when the sodium channels close the transmembrane potential drops toward –70 mV. As the transmembrane potential declines, so does the rate of neurotransmitter release. This decline indicates that this particular receptor has absorbed a photon.

RECOVERY AFTER STIMULATION Retinal does not spontaneously convert back to the 11-cis form after absorbing a photon. Instead, the entire rhodopsin molecule must be broken down and reassembled. Shortly after the conformational change occurs, the rhodopsin molecule begins to break down into retinal and opsin, a process known as *bleaching* (Figure 17-18•). The retinal must be enzymatically converted to the 11-cis form before it can recombine with opsin. This conversion requires energy in the form of ATP (adenosine triphosphate), and it takes time. Bleaching contributes to the lingering visual impression after a flashbulb goes off. After an intense exposure to light, a photoreceptor cannot respond to further stimulation until its rhodopsin molecules have been regenerated. As a result, a "ghost" image remains on the retina. Bleaching is seldom noticeable under ordinary circumstances because the eyes are constantly making small, involuntary changes in position that move the image across the retinal surface.

While the rhodopsin molecule is being reassembled, membrane permeability is returning to normal. Opsin is inactivated when bleaching occurs, and the breakdown of cGMP halts as a result. As other enzymes generate cGMP in the cytoplasm, the chemically gated sodium channels reopen.

RETINITIS PIGMENTOSA The term *retinitis pigmentosa* (RP) refers to a collection of inherited retinopathies. Together they are the most common inherited visual pathology, affecting approximately 1 individual in 3000. The visual receptors gradually deteriorate, and blindness eventually results. The mutations responsible affect the structure of the photoreceptors, specifically the visual pigments of the membrane discs. It is not known how the altered pigments lead to photoreceptor destruction.

NIGHT BLINDNESS The visual pigments of the photoreceptors are synthesized from vitamin A. The body contains vitamin A reserves sufficient for several months, and a significant amount is stored in the cells of the pigment layer of the neural tunic. If dietary sources are inadequate, these reserves are gradually exhausted, and the amount of visual pigment in the photoreceptors begins to decline. Daylight vision is affected, but during the day the light is usually bright enough to stimulate whatever visual pigments remain within the densely

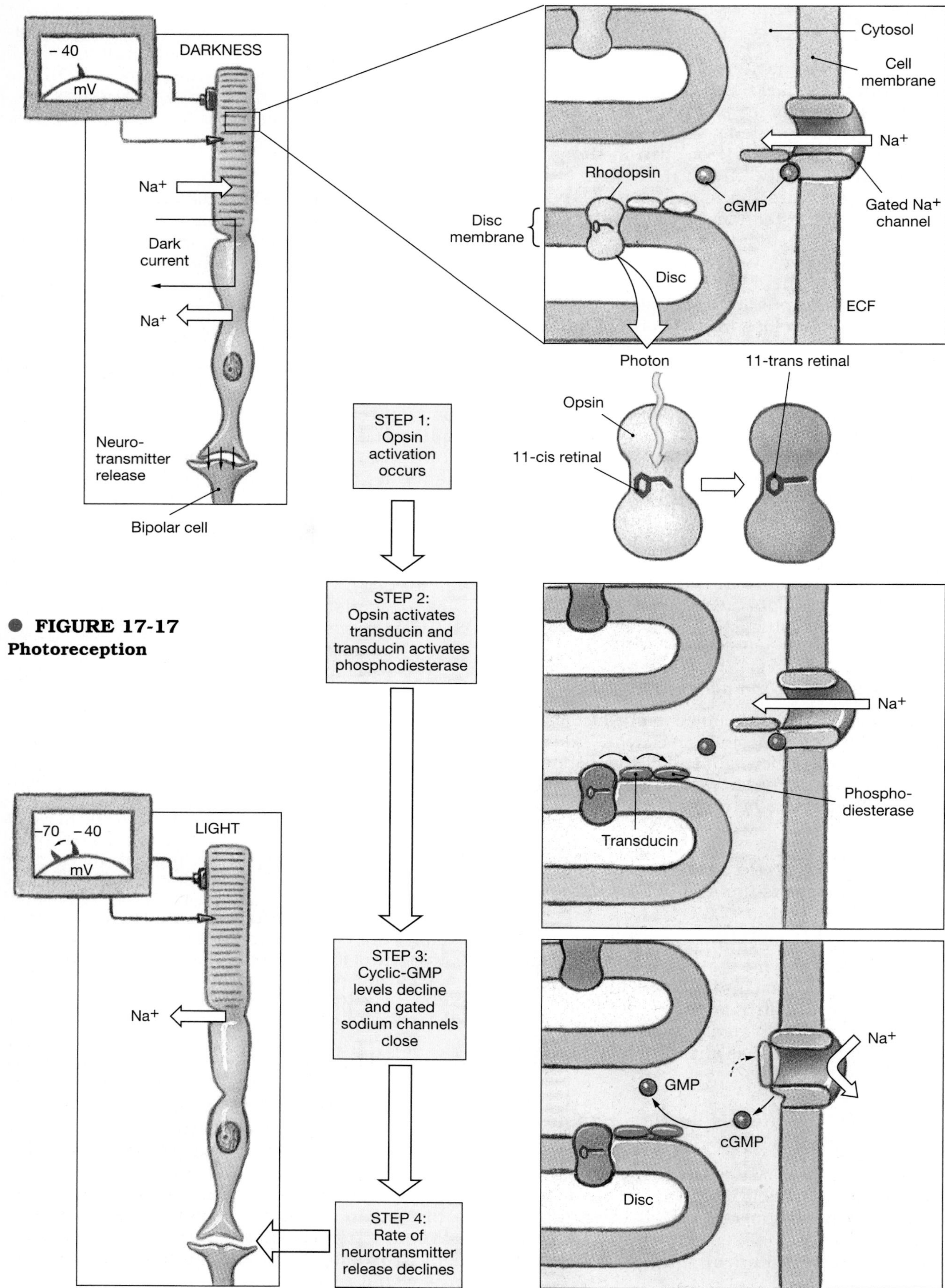
- 40
mV
DARKNESS
Na+
Dark current
Na+
Neuro-transmitter release
Bipolar cell
Cytosol
Cell membrane
Na+
Rhodopsin
cGMP
Gated Na+ channel
Disc membrane
Disc
ECF
Photon
11-trans retinal
Opsin
11-cis retinal
STEP 1: Opsin activation occurs
STEP 2: Opsin activates transducin and transducin activates phosphodiesterase
Na+
Phospho-diesterase
Transducin
-70 -40
mV
LIGHT
Na+
STEP 3: Cyclic-GMP levels decline and gated sodium channels close
Na+
GMP
cGMP
Disc
STEP 4: Rate of neurotransmitter release declines

● **FIGURE 17-17**
Photoreception

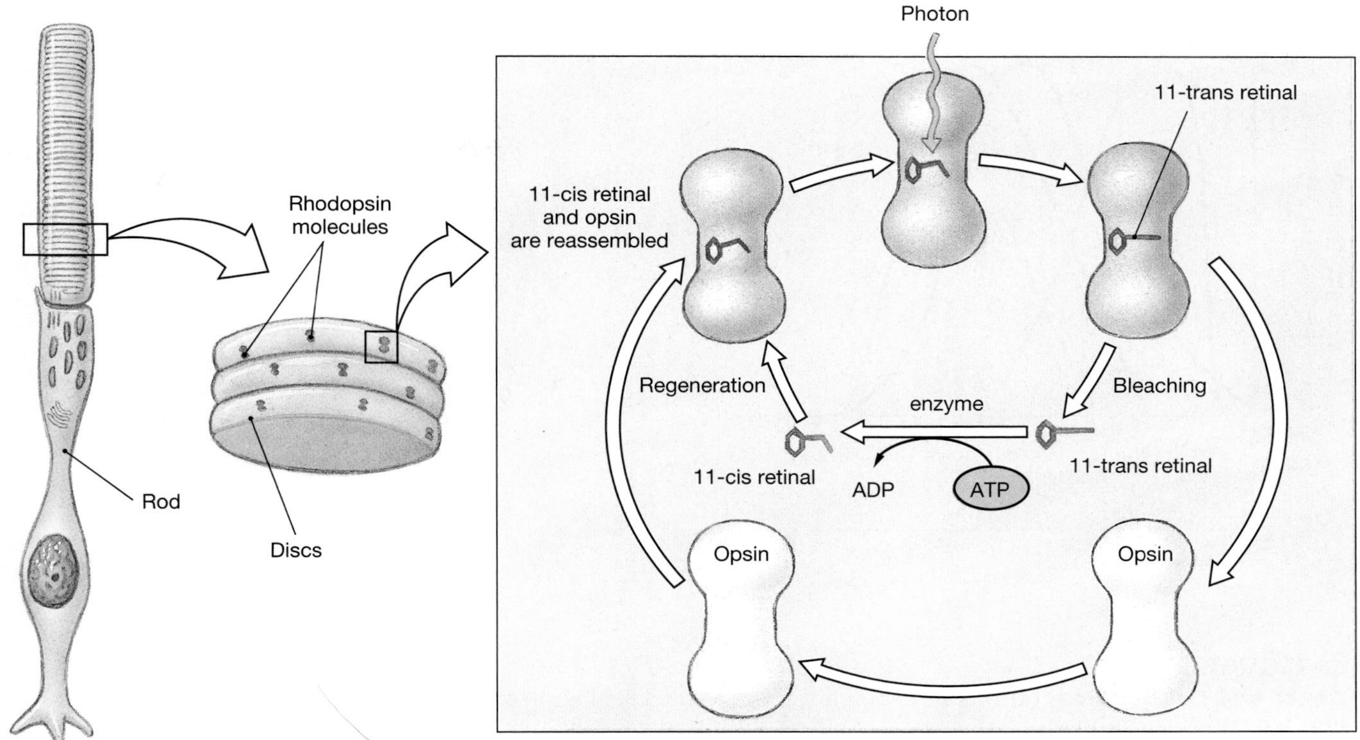
Photon
11-trans retinal
Rhodopsin molecules
11-cis retinal and opsin are reassembled
Regeneration
enzyme
Bleaching
11-cis retinal
ADP
ATP
11-trans retinal
Rod
Discs
Opsin
Opsin

● **FIGURE 17-18**
Bleaching and Recovery of Visual Pigments

packed cone population of the fovea. As a result, the problem first becomes apparent at night, when the dim light proves insufficient to activate the rods. This condition, known as **night blindness,** can be treated by administration of vitamin A. The carotene pigments found in many vegetables can be converted to vitamin A within the body. Carrots are a particularly good source of carotene, which explains the old saying that carrots are good for your eyes.

Color Vision

Figure 17-19

An ordinary light bulb or the sun emits photons of all wavelengths. These photons stimulate both rods and cones, producing a perception of "white" light. Our eyes also detect photons that reach the retina after bouncing off objects around us. If photons of all colors bounce off, the object appears white; if the photons are all absorbed, and none reach the retina, the object appears black. An object will appear to have a particular color if it reflects (or transmits) photons from one portion of the visible spectrum and absorbs the rest.

There are three types of cones: **blue cones, green cones,** and **red cones.** Each type has a different form of opsin and a different range of sensitivity. Their stimulation in various combinations accounts for the perception of colors. In a normal individual, the cone population consists of 16 percent blue cones, 10 percent green cones, and 74 percent red cones. Although their sensitivities overlap, each type is most sensitive to a specific portion of the visual spectrum (Figure 17-19a●).

Color discrimination occurs through the integration of information arriving from all three types of cones. For example, the perception of yellow results from a combination of inputs from green cones (highly stimulated), red cones (stimulated), and blue cones (relatively unaffected). If all three cone populations are stimulated, we perceive the color as white; we also perceive white if rods, rather than cones, are stimulated. For this reason everything appears to be black and white when we enter dimly lit surroundings or walk by starlight.

COLOR BLINDNESS Persons unable to distinguish certain colors have a form of **color blindness.** The standard tests for color vision involve picking numbers or letters out of a complex and colorful picture, such as those in Figure 17-19b●. Color blindness occurs because one or more classes of cones are nonfunctional. The cones may be absent entirely, or they may be present but unable to manufacture the necessary visual pigments. In the most common type of color blindness the red cones are missing, and the individual cannot distinguish red light from green light.

Blue cones
Rods
Green cones
Red cones
Light absorption (percent of maximum)
100
75
50
25
0
Violet Blue Green Yellow Orange Red
400 450 500 550 600 650 700
Wavelength (nm)

(a)

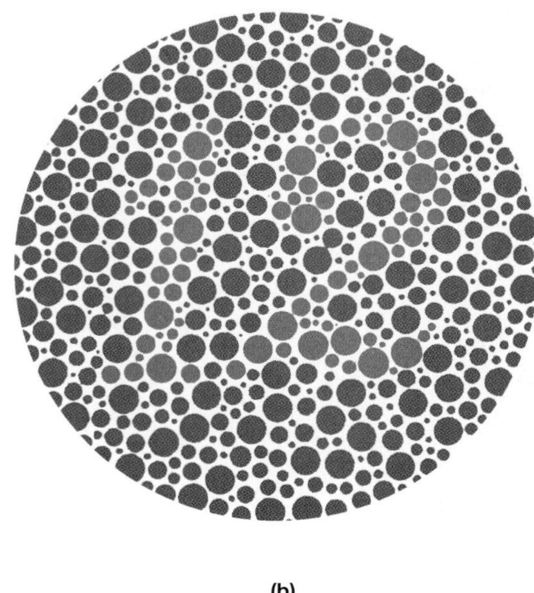

(b)

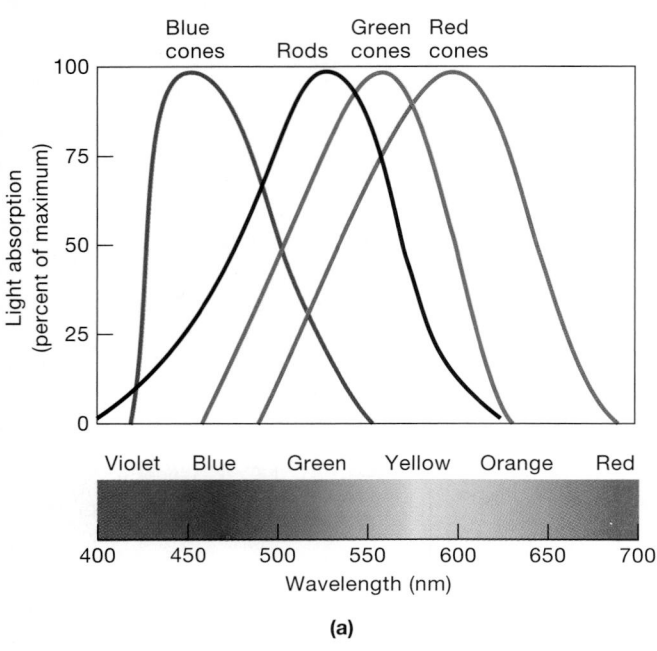

● **FIGURE 17-19**
Cones and Color Vision. (a) A graph comparing the absorptive characteristics of blue, green, and red cones with those of typical rods. Notice that the rod sensitivities overlap those of the cones, and the various cone types have overlapping sensitivity curves. **(b)** Part of a standard test for color vision. Lack of one or more populations of cones will produce an inability to distinguish the patterned images.

Inherited color blindness involving one or two cone pigments is not unusual. Ten percent of all men show some color blindness, while the incidence among women is only around 0.67 percent. Total color blindness is extremely rare; only 1 person in 300,000 fails to manufacture any cone pigments at all.

Light and Dark Adaptation

The sensitivity of the visual system varies, depending on the intensity of illumination. After 30 minutes or more in the dark, almost all visual pigments will be fully receptive to stimulation. This is the **dark-adapted state.** When dark-adapted, the visual system is extremely sensitive. For example, a single rod will hyperpolarize in response to a single photon of light. Even more remarkable, if as few as seven rods absorb photons at one time, the individual will see a flash of light.

When the lights come on, at first they seem almost unbearably bright, but over the next few minutes sensitivity decreases as bleaching occurs. Eventually the rate of visual pigment breakdown is balanced by the rate of re-formation. This condition is the **light-adapted state.** As a person moves from the depths of a cave to the full light of midday, the receptor sensitivity decreases by about 25,000 times.

A variety of central responses further adjust light sensitivity. Constriction of the pupil, via the pupillary constrictor reflex, reduces the amount of light entering the eye to 1/30 the maximum dark-adapted levels. Since dilating the pupil fully can produce a 30-fold increase in the amount of light entering the eye, and facilitating some of the synapses along the visual pathway can perhaps triple its sensitivity, the entire system may increase its efficiency by a factor of over 1 million.

❑ The Visual Pathway

The visual pathway begins at the photoreceptors and ends at the visual cortex of the cerebral hemispheres. In other sensory pathways we have examined, there is at most one synapse between a receptor and a sensory neuron that delivers information to the CNS. In the visual pathway, the message must cross two synapses (photoreceptor to bipolar cell and bipolar cell to ganglion cell) before it heads toward the brain. The extra synapse adds to the synaptic delay, but it provides an opportunity for the processing and integration of visual information before it leaves the retina.

Retinal Processing

Figure 17-20

Each photoreceptor in the retina monitors a specific receptive field. The retina contains approximately 130 million photoreceptors, 6 million bipolar cells, and 1 million ganglion cells. There is thus a considerable amount of convergence at the start of the visual path-

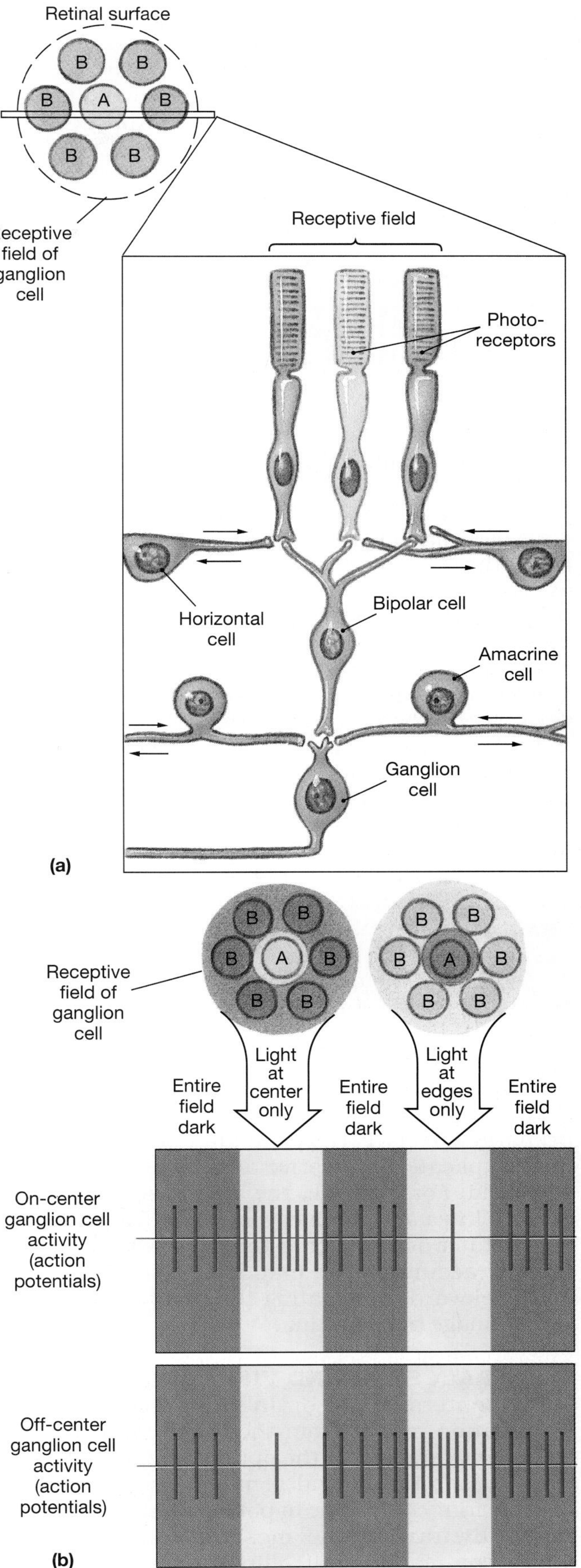
Retinal surface
B
B
B
A
B
B
B
Receptive field of ganglion cell
Receptive field
Photo-receptors
Horizontal cell
Bipolar cell
Amacrine cell
Ganglion cell
(a)
Receptive field of ganglion cell
Entire field dark
Light at center only
Entire field dark
Light at edges only
Entire field dark
On-center ganglion cell activity (action potentials)
Off-center ganglion cell activity (action potentials)
(b)

● **FIGURE 17-20**
Convergence and Ganglion Cell Function. (a) Ganglion cells monitor a well-defined portion of the visual field. **(b)** Some ganglion cells respond strongly to light arriving at the center of their receptive field (on-center neurons). Others respond most strongly to illumination of the edges of their receptive field (off-center neurons).

way. The degree of convergence differs, depending on whether you consider rods or cones. Regardless of the amount of convergence, each ganglion cell monitors a specific portion of the visual field.

As many as a thousand rods may pass information via their bipolar cells to a single ganglion cell. The ganglion cells that monitor rods are relatively large, and they are called **M cells** (*magnocells*). M cells provide information on general form, motion, and shadows under dim light conditions. Because there is so much convergence, when an M cell becomes active it indicates that light has arrived in a general *area*, rather than at a specific location.

The loss in specificity due to convergence is partially overcome by the fact that ganglion cells vary their activity depending on the pattern of activity in their sensory field. The sensory field is usually circular (Figure 17-20a●), and ganglion cells typically respond differently to stimuli arriving in the center and at the edges of their receptive fields. Some ganglion cells are excited by light arriving in the center of their sensory field **(on-center neurons)**, and inhibited when light strikes the edges of their receptive field. Others are inhibited by light in the central zone but are stimulated by illumination at the edges **(off-center neurons)**. The relationships are diagrammed in Figure 17-20b●. On-center and off-center neurons provide information about what portion of their sensory field is illuminated.

Cones typically show very little convergence; in the fovea the ratio of cones to ganglion cells is 1 to 1. The ganglion cells that monitor cones, called **P cells** (*parvo cells*), are smaller and more numerous than M cells. P cells are active in bright light, and they provide information on edges, fine detail, and color. Because little convergence occurs, the activation of a P cell means that light has arrived at one specific location. As a result, cones provide more precise information about a visual image. In photographic terms, pictures formed by rods have a coarse, grainy appearance that blurs the details, whereas those produced by cones are fine-grained, sharp, and clear.

Central Processing of Visual Information

Figure 17-21

Axons from the entire population of ganglion cells converge on the optic disc, penetrate the wall of the eye, and proceed toward the diencephalon as the optic nerve (N II). The two optic nerves, one from each eye, reach the diencephalon at the optic chiasm (Figure 17-21●). From this point approximately one-half of the fibers proceed toward the lateral geniculate of the same side of the brain, while the other half cross over to reach the lateral geniculate of the opposite side. The lateral geniculates act as switching and processing centers that relay visual information to reflex centers in the brain stem as well as to the cerebral cortex. For example, the pupillary reflexes and reflexes that control eye movement are triggered by information relayed to the superior colliculus by the lateral geniculates.

From each lateral geniculate, visual information travels to the occipital cortex of the cerebral hemisphere on that side. The bundle of projection fibers linking the lateral geniculate with the visual cortex is known as the **optic radiation.**

THE VISUAL FIELD The perception of a visual image reflects the integration of information arriving at the visual cortex of the occipital lobes. Each eye receives a slightly different visual image, because (1) the foveas are 5–7.5 cm apart, and (2) the nose and eye socket block the view of the opposite side. Depth perception, an interpretation of the three-dimensional relationships between objects in view, is obtained by comparing the relative positions of objects within the images received by the left and right eyes.

When you look ahead, the visual images from the left and right eyes overlap, as indicated in Figure 17-21●. The image received by the fovea of each eye lies in the center of the region of overlap. A vertical line drawn through the center of the region of overlap marks the division of visual information at the optic chiasm. Visual information from the left half of the combined visual field will reach the visual cortex of the right occipital lobe; visual information from the right half of the combined visual field will arrive at the left visual cortex.

The cerebral hemispheres thus contain a map of the entire field of vision. As in the case of the primary sensory cortex, the map does not faithfully duplicate the relative areas within the sensory field. For example, the area assigned to the macula lutea and fovea covers about 35 times the surface it would cover if the map were proportionally accurate. The map is also upside down and backward, duplicating the orientation of the visual image at the retina.

THE BRAIN STEM AND VISUAL PROCESSING Many centers in the brain stem receive visual information, either from the lateral geniculates or via collaterals from the optic tracts. Collaterals that bypass the lateral geniculates synapse in the superior colliculi or hypothalamus. The superior colliculus of the mesencephalon issues motor commands controlling unconscious eye,

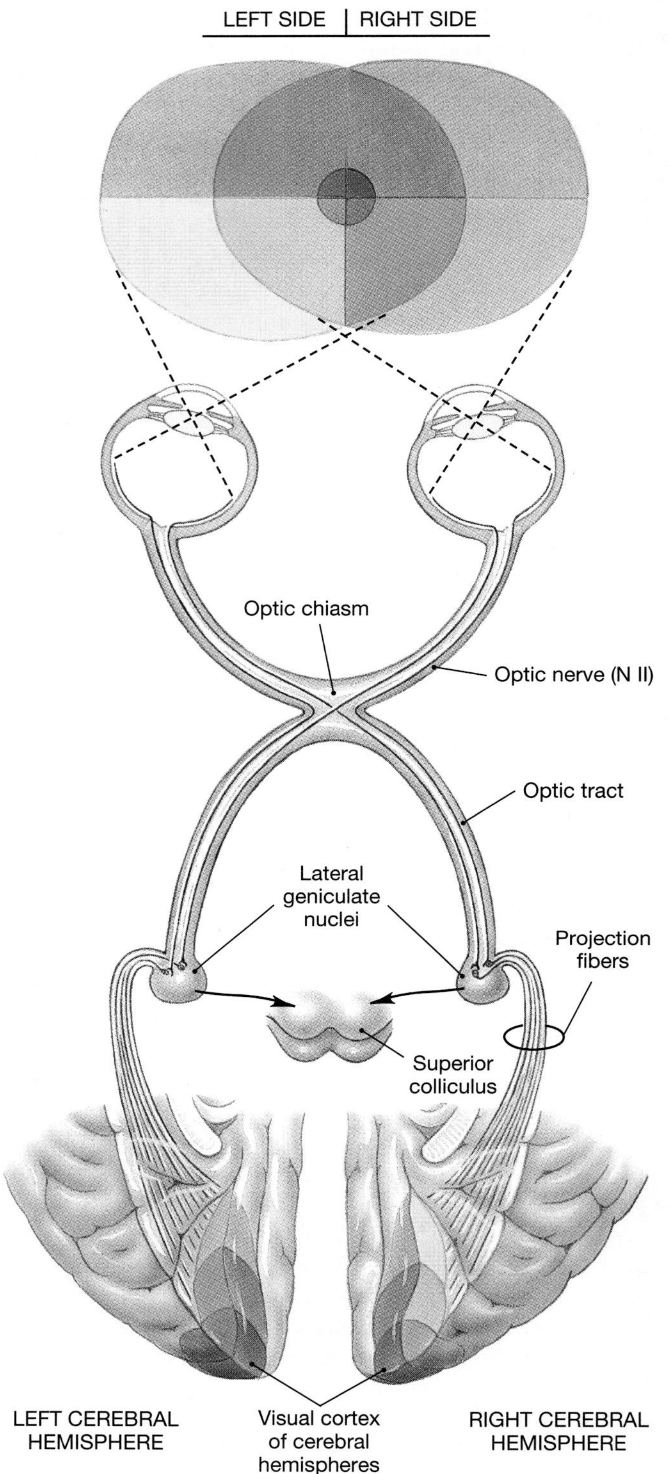
LEFT SIDE
RIGHT SIDE
Optic chiasm
Optic nerve (N II)
Optic tract
Lateral geniculate nuclei
Projection fibers
Superior colliculus
LEFT CEREBRAL HEMISPHERE
Visual cortex of cerebral hemispheres
RIGHT CEREBRAL HEMISPHERE

● **FIGURE 17-21**
The Visual Pathways. At the optic chiasm, a partial cross-over of nerve fibers occurs. As a result, each hemisphere receives visual information from the medial half of the visual field of the eye on that side and from the lateral half of the visual field of the eye on the opposite side. Visual association areas integrate this information to develop a composite picture of the entire visual field.

head, or neck movements in response to visual stimuli. Visual inputs to the **suprachiasmatic** (sū-pra-kī-az-MA-tik) **nucleus** of the hypothalamus affect the function of other brain stem nuclei. This nucleus and the *pineal gland* of the epithalamus establish a daily pattern of visceral activity that is tied to the day/night cycle. This **circadian rhythm** (*circa*, about + *dies*, day) affects metabolic rate, endocrine function, blood pressure, digestive activities, the awake/asleep cycle, and other physiological and behavioral processes discussed in Chapters 14–16.

 If a person was born without cones in her eyes, would she still be able to see? Explain.

 How could a diet that is deficient in vitamin A affect vision?

 What effect would a decrease in phosphodiesterase activity in photoreceptor cells have on vision?

Equilibrium and Hearing

The senses of equilibrium and hearing are provided by the *inner ear*, a receptor complex located in the petrous portion of the temporal bone of the skull. [p. 210] The basic receptor mechanism for these senses is the same. The receptors, or *hair cells*, are simple mechanoreceptors. The complex structure of the inner ear and the different arrangement of accessory structures account for the abilities of the hair cells to respond to different stimuli, and thus to provide the input for two different senses:

- *Equilibrium*, which informs us of the position of the body in space by monitoring gravity, linear acceleration, and rotation.
- *Hearing*, which enables us to detect and interpret sound waves.

Anatomy of the Ear

Figure 17-22

The ear is divided into three anatomical regions; the *external ear*, the *middle ear*, and the *inner ear* (Figure 17-22●). The external ear is the visible portion of the ear, which collects and directs sound waves to the *eardrum*. The *middle ear* is a chamber located within the petrous portion of the temporal bone. Structures within the middle ear collect sound waves and transmit them to an appropriate portion of the inner ear. The *inner ear* contains the sensory organs for equilibrium and hearing.

The External Ear

Figure 17-22

The **external ear** includes the fleshy flap and cartilaginous **pinna**, or *auricle*, that surrounds the **external auditory meatus.** The pinna protects the opening of the meatus and provides directional sensitivity to the ear. Sounds coming from behind the head are blocked by the pinna; sounds coming from the side or front are collected and channeled into the **external auditory canal** of the temporal bone. (When you "cup your ear" with your hand to hear a faint sound more clearly, you are exaggerating this effect.) The external auditory canal is a passageway that ends at the eardrum, also called the **tympanic membrane,** or **tympanum.** The tympanum is a thin, semitransparent sheet that separates the external ear from the middle ear (Figure 17-22●). The path of the external auditory canal and the location of the tympanum in the temporal bone were noted in Figure 7-7●, p. 211.

The tympanum is very delicate. The pinna and the narrow external auditory canal provide some protection from accidental injury. In addition, **ceruminous glands** along the external auditory canal secrete a waxy material, and many small, outwardly projecting hairs help deny access to foreign objects or insects. The waxy secretion of the ceruminous glands, called **cerumen,** also slows the growth of microorganisms in the external auditory canal and reduces the chances of infection.

The Middle Ear

Figures 17-22, 17-23

The **middle ear,** or **tympanic cavity,** is filled with air. It is separated from the external auditory canal by the tympanum, but it communicates with the superior portion of the pharynx, a region known as the *nasopharynx*, and with the mastoid air cells via a number of small and variable connections. [p. 211] The connection with the pharynx is the **pharyngotympanic tube,** also called the *auditory tube* and the *Eustachian tube* (Figure 17-22●). This tube, about 4 cm long, consists of two portions. The connection to the middle ear is relatively narrow and supported by cartilage (Figure 17-23●). The opening into the nasopharynx is relatively broad and funnel-shaped. The pharyngotympanic tube serves to equalize the pressure inside the eardrum with that outside the eardrum. Unfortunately, it can also allow microorganisms to travel from the nasopharynx into the tympanic cavity. Invasion by microorganisms can lead to an unpleasant middle ear infection known as *otitis media*. AM Otitis Media and Mastoiditis

THE AUDITORY OSSICLES The middle ear contains three tiny ear bones, collectively called **auditory ossicles.** The ear bones connect the tym-

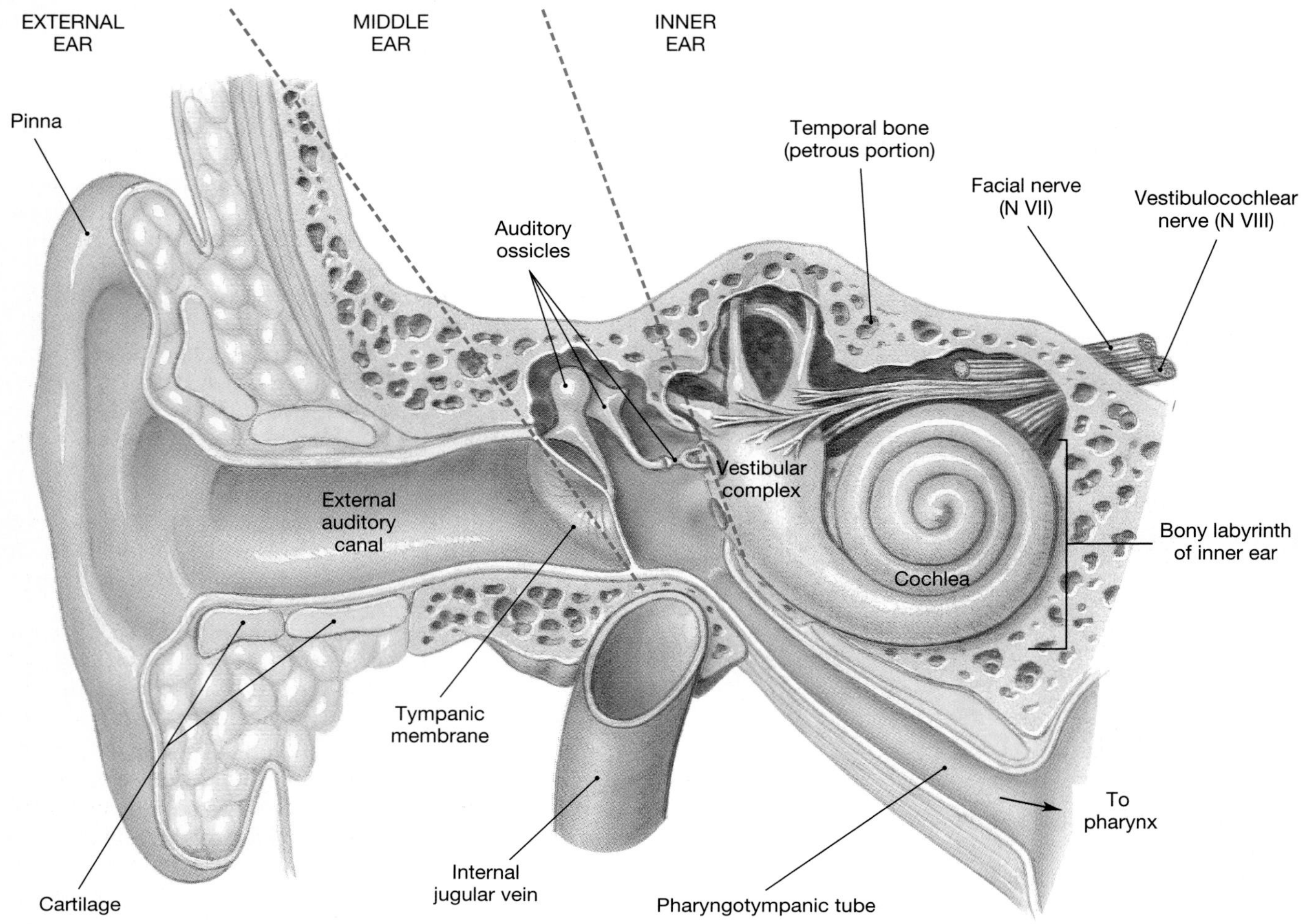
EXTERNAL EAR
MIDDLE EAR
INNER EAR
Pinna
Temporal bone (petrous portion)
Facial nerve (N VII)
Vestibulocochlear nerve (N VIII)
Auditory ossicles
Vestibular complex
External auditory canal
Bony labyrinth of inner ear
Cochlea
Tympanic membrane
To pharynx
Internal jugular vein
Cartilage
Pharyngotympanic tube

● **FIGURE 17-22**
Anatomy of the Ear. The orientation of the external, middle, and inner ear.

panic membrane with the receptor complex of the inner ear (Figures 17-22 and 17-23 ●). The three auditory ossicles are the *malleus*, the *incus*, and the *stapes*. The **malleus** (*malleus*, a hammer) attaches at three points to the interior surface of the tympanum. The middle bone, the **incus** (*incus*, an anvil), attaches the malleus to the inner **stapes** (*stapes*, a stirrup). The *base* of the stapes almost completely fills the *oval window* of the inner ear.

Vibration of the tympanum converts arriving sound waves into mechanical movements. The auditory ossicles act as levers that conduct those vibrations to the fluid-filled chamber of the inner ear. Because of the way these ossicles are connected, an in-out movement of the tympanum produces a rocking motion at the stapes. Because the tympanic membrane is 22 times larger and heavier than the oval window, a 1-μm movement of the tympanic membrane produces a 22-μm deflection of the base of the stapes. Thus, the amount of movement increases markedly from tympanic membrane to oval window.

Because this amplification occurs, we can hear very faint sounds. But that degree of magnification can be a problem when we are exposed to very loud noises. Within the tympanic cavity, two small muscles serve to protect the eardrum and ossicles from violent movements under very noisy conditions:

- The **tensor tympani** (TEN-sor tim-PAN-ē) **muscle** is a short ribbon of muscle whose origin is the petrous portion of the temporal bone and the pharyngotympanic tube and whose insertion is on the "handle" of the malleus. When the tensor tympani contracts, the malleus is pulled medially, stiffening the tympanum. This increased stiffness reduces the amount of possible movement. The tympani muscle is innervated by motor fibers of the mandibular branch of the trigeminal nerve (N V).
- The **stapedius** (sta-PĒ-dē-us) **muscle,** innervated by the facial nerve (N VII), originates from the posterior wall of the tympanic cavity and inserts on the stapes. Contraction of the stapedius pulls the stapes, reducing movement of the stapes at the oval window.

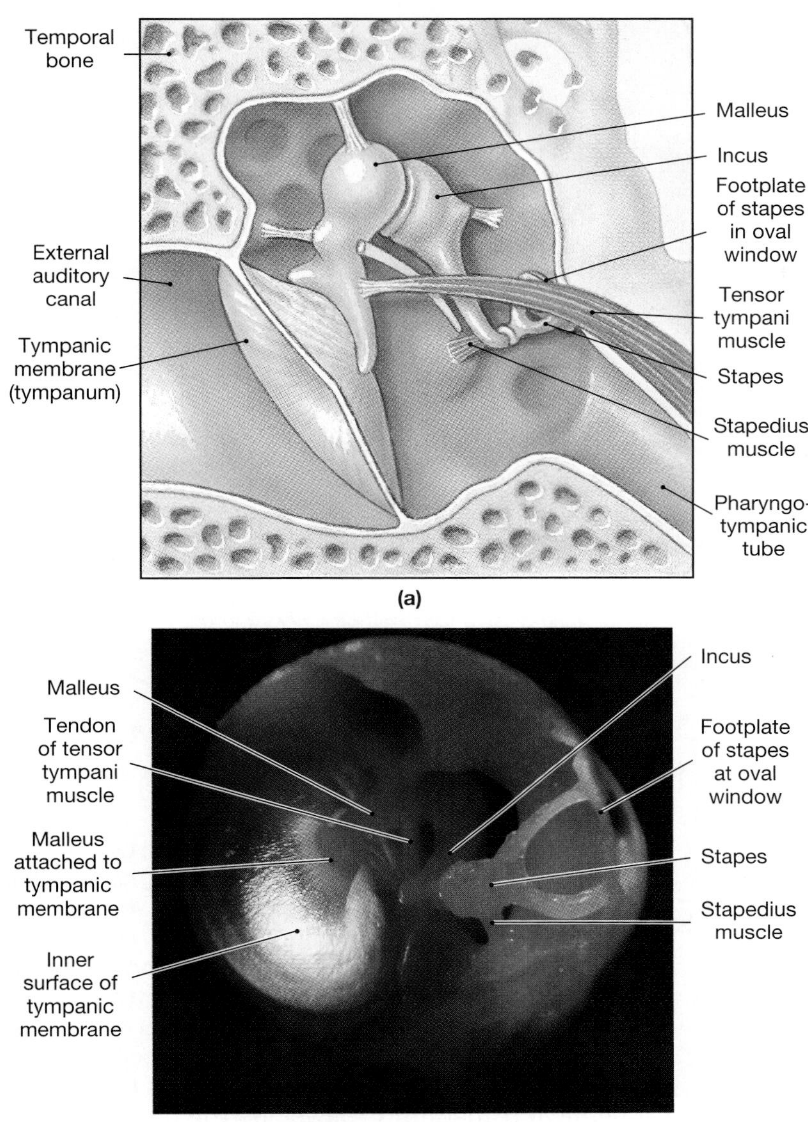
Temporal bone
External auditory canal
Tympanic membrane (tympanum)
Malleus
Incus
Footplate of stapes in oval window
Tensor tympani muscle
Stapes
Stapedius muscle
Pharyngo-tympanic tube
(a)
Malleus
Tendon of tensor tympani muscle
Malleus attached to tympanic membrane
Inner surface of tympanic membrane
Incus
Footplate of stapes at oval window
Stapes
Stapedius muscle
(b)

● **FIGURE 17-23**

The Middle Ear. (a) Detail of the structures associated with the middle ear. **(b)** Tympanic membrane and auditory ossicles.

The Inner Ear

Figures 17-22, 17-24a

The senses of equilibrium and hearing are provided by the receptors of the **inner ear** (Figures 17-22 and 17-24a●). The receptors lie within a collection of fluid-filled tubes and chambers known as the **membranous labyrinth** (*labyrinthos*, a network of canals). The membranous labyrinth contains a fluid, called **endolymph** (EN-dō-limf), with electrolyte concentrations different from those of other body fluids.

The **bony labyrinth** is a shell of dense bone that surrounds and protects the membranous labyrinth. Its inner contours closely follow the contours of the membranous labyrinth, and its outer walls are fused with the surrounding temporal bone. ∞ [p. 210] Between the bony and membranous labyrinths flows the **perilymph** (PER-i-limf), a liquid whose properties closely resemble those of cerebrospinal fluid.

The bony labyrinth can be subdivided into the **vestibule** (VES-ti-būl), the **semicircular canals,** and the **cochlea** (KOK-lē-a; *cochlea*, a snail shell) (Figure 17-24a●).

- The vestibule includes a pair of membranous sacs, the **saccule** (SAK-ūl) and the **utricle** (Ū-tre-kl), or the *sacculus* and *utriculus*. Receptors in the saccule and utricle provide sensations of gravity and linear acceleration.
- The semicircular canals enclose slender **semicircular ducts.** Receptors in the semicircular ducts are stimulated by rotation of the head. The combination of vestibule and semicircular canals is called the *vestibular complex* because the fluid-filled chambers within the vestibule are broadly continuous with those of the semicircular canals.
- The bony cochlea contains the **cochlear duct** of the membranous labyrinth. Receptors within the cochlear duct provide the sense of hearing. The cochlear duct sits sandwiched between a pair of perilymph-filled chambers, and the entire complex makes 2 1/2 turns around a central bony hub. In sectional view the spiral arrangement resembles that of a snail shell (*cochlea* in Latin).

The walls of the bony labyrinth consist of dense bone everywhere except at two small areas near the base of the cochlear spiral. The **round window** is a thin, membranous partition that separates the perilymph of the cochlear chambers from the air spaces of the middle ear. Collagen fibers connect the bony margins of the **oval window** to the base of the stapes. When a sound vibrates the tympanum, the movements are conducted by the auditory ossicles to the stapes. Movement of the stapes ultimately leads to the stimulation of receptors within the cochlear duct, and we hear the sound.

Receptor Function in the Inner Ear

Figure 17-24b

The sensory receptors of the inner ear are called **hair cells** (Figure 17-24b●). These receptor cells are surrounded by **supporting cells** and are mon-

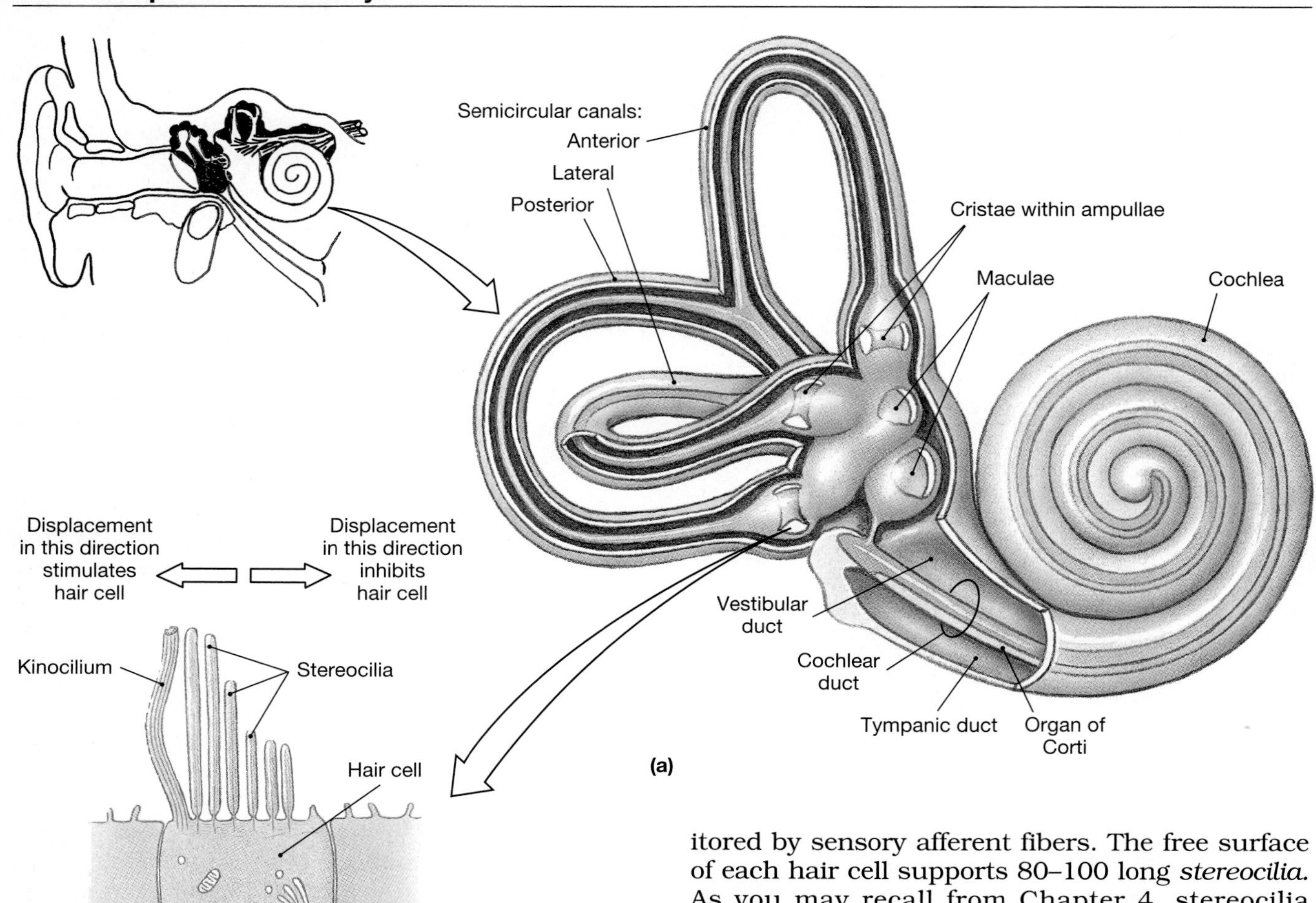
Semicircular canals:
Anterior
Lateral
Posterior
Cristae within ampullae
Maculae
Cochlea
Displacement in this direction stimulates hair cell
Displacement in this direction inhibits hair cell
Kinocilium
Stereocilia
Hair cell
Vestibular duct
Cochlear duct
Tympanic duct
Organ of Corti
(a)
Sensory nerve ending
(b) Hair cell

● **FIGURE 17-24**
The Inner Ear. (a) The bony and membranous labyrinths. Areas of the membranous labyrinth containing sensory receptors are highlighted. **(b)** A representative hair cell (receptor) from the vestibular complex. Bending the stereocilia toward the kinocilium depolarizes the cell and stimulates the sensory neuron. Displacement in the opposite direction inhibits the sensory neuron.

itored by sensory afferent fibers. The free surface of each hair cell supports 80–100 long *stereocilia.* As you may recall from Chapter 4, stereocilia resemble very long microvilli. *[p. 111]* Each hair cell in the vestibule also contains a **kinocilium,** a single large cilium. Hair cells do not actively move their kinocilia and stereocilia. However, when an external force pushes against these processes, the distortion of the cell membrane alters the rate of chemical transmitter release by the hair cell.

Hair cells provide information concerning the direction and strength of mechanical stimuli. The stimuli involved, however, are quite varied: gravity or acceleration in the vestibule, rotation in the semicircular canals, and sound in the cochlea. The sensitivities of the hair cells differ because in each of these regions there are different accessory structures that determine which stimulus will provide the force to deflect the kinocilia and stereocilia. The importance of these accessory structures will become apparent when we consider the way sensations of equilibrium and hearing are produced.

❑ Equilibrium

Equilibrium sensations are provided by receptors of the vestibular complex. The semicircular ducts provide information concerning rotational movements of the head. For example, when you turn your head to the left, stimulated receptors in the semicircular ducts tell you how rapid the movement is and in what direction. The saccule and the utricle provide information about your position with respect to gravity. If you stand with your head tilted to one side, these receptors will report the angle involved and whether your head tilts forward or backward. These receptors are also stimulated by sudden acceleration. When your car accelerates after stopping at a red light, the saccular and utricular receptors give you the impression of increasing speed.

The Semicircular Ducts

Figures 17-24a, 17-25

Receptors in the semicircular ducts respond to rotational movements of the head. They are active during a movement but quiet when the body is motionless. The **anterior, posterior,** and **lateral** semicircular ducts are continuous with the utricle (Figures 17-24 and 17-25a●). Each semicircular duct contains an expanded region, the **ampulla,** which contains the sensory receptors. Hair cells attached to the wall of the ampulla form a raised structure known as a **crista** (Figure 17-25b●). The kinocilia and stereocilia of the hair cells are embedded in a gelatinous structure, the **cupula** (KŪ-pū-la). Because the cupula has a density very close to that of the surrounding endolymph, it essentially floats above the receptor surface, nearly filling the ampulla. When the head rotates in the plane of the duct, movement of the endolymph along the canal axis pushes the cupula and distorts the receptor processes (Figure 17-25c●). Fluid movement in one direction stimulates the hair cells, and movement in the opposite direction inhibits them. When the endolymph stops moving, the elastic nature of the cupula makes it bounce back to its normal position.

Even the most complex movement can be analyzed in terms of motion in three rotational planes. Each semicircular duct responds to one of these rotational movements. A horizontal rotation, as in shaking the head "no," stimulates the hair cells of the lateral semicircular duct. Nodding "yes" excites the anterior duct, and tilting the head from side to side activates the receptors in the posterior duct.

The Utricle and Saccule

Figure 17-25

The utricle and saccule provide equilibrium sensations whether or not the body is moving. These chambers are connected by a slender passageway continuous with the narrow **endolymphatic duct** (Figure 17-25a●). The endolymphatic duct ends in a blind pouch, the **endolymphatic sac,** that projects through the dura mater lining the temporal bone into the subdural space. Portions of the cochlear duct continually secrete endolymph, and at the endolymphatic sac excess fluids return to the general circulation.

The hair cells of the utricle and saccule are clustered in oval **maculae** (MAK-ū-lē; *macula,* spot) (Figure 17-25a●). As in the ampullae, the hair cell processes are embedded in a gelatinous mass, but the macular receptors lie under a thin layer containing densely packed mineral crystals, called *otoconia* (*oto-*, ear + *conia,* dust), or **otoliths** ("ear stones").

The macula of the saccule is diagrammed in Figure 17-25a,d●. When the head is in the normal, upright position the otoliths sit atop the macula. Their weight presses down on the macular surfaces, pushing the sensory hairs down rather than to one side or another. When the head is tilted, the pull of gravity on the otoliths shifts the mass to the side. This shift distorts the sensory hairs, and the change in receptor activity tells the CNS that the head is no longer level (Figure 17-25e●).

A similar mechanism accounts for our perception of linear acceleration in a car that speeds up suddenly. The otoliths lag behind, and the effect on the hair cells is comparable to tilting the head back. Under normal circumstances the nervous system distinguishes between the sensations of tilting and linear acceleration by integrating vestibular sensations with visual information.

Pathways for Equilibrium Sensations

Figure 17-26

Hair cells of the vestibule and semicircular ducts are monitored by sensory neurons located in adjacent **vestibular ganglia.** Sensory fibers from each ganglion form the **vestibular branch** of the vestibulocochlear nerve (N VIII). These fibers synapse on neurons within the **vestibular nuclei** at the boundary between the pons and medulla. The two vestibular nuclei:

1. Integrate the sensory information arriving from each side of the head.
2. Relay information to the cerebellum.
3. Relay information to the cerebral cortex, providing a conscious sense of position and movement.
4. Send commands to motor nuclei in the brain stem and spinal cord.

The reflexive motor commands issued by the vestibular nuclei are distributed to the motor nuclei for cranial nerves involved with eye, head, and neck movements (N III, IV, VI, and XI). Instructions descending in the *vestibulospinal tracts* of the spinal cord adjust peripheral muscle tone and complement the reflexive movements of the head or neck. These pathways are indicated in Figure 17-26●; the role of the vestibulospinal tracts in the extrapyramidal system was considered in Chapter 15. ∞ [p. 510]

The automatic eye movements that occur in response to sensations of motion are directed by the *superior colliculus* of the mesencephalon. ∞ [p. 479] These movements attempt to keep the gaze focused on a specific point in space despite changes

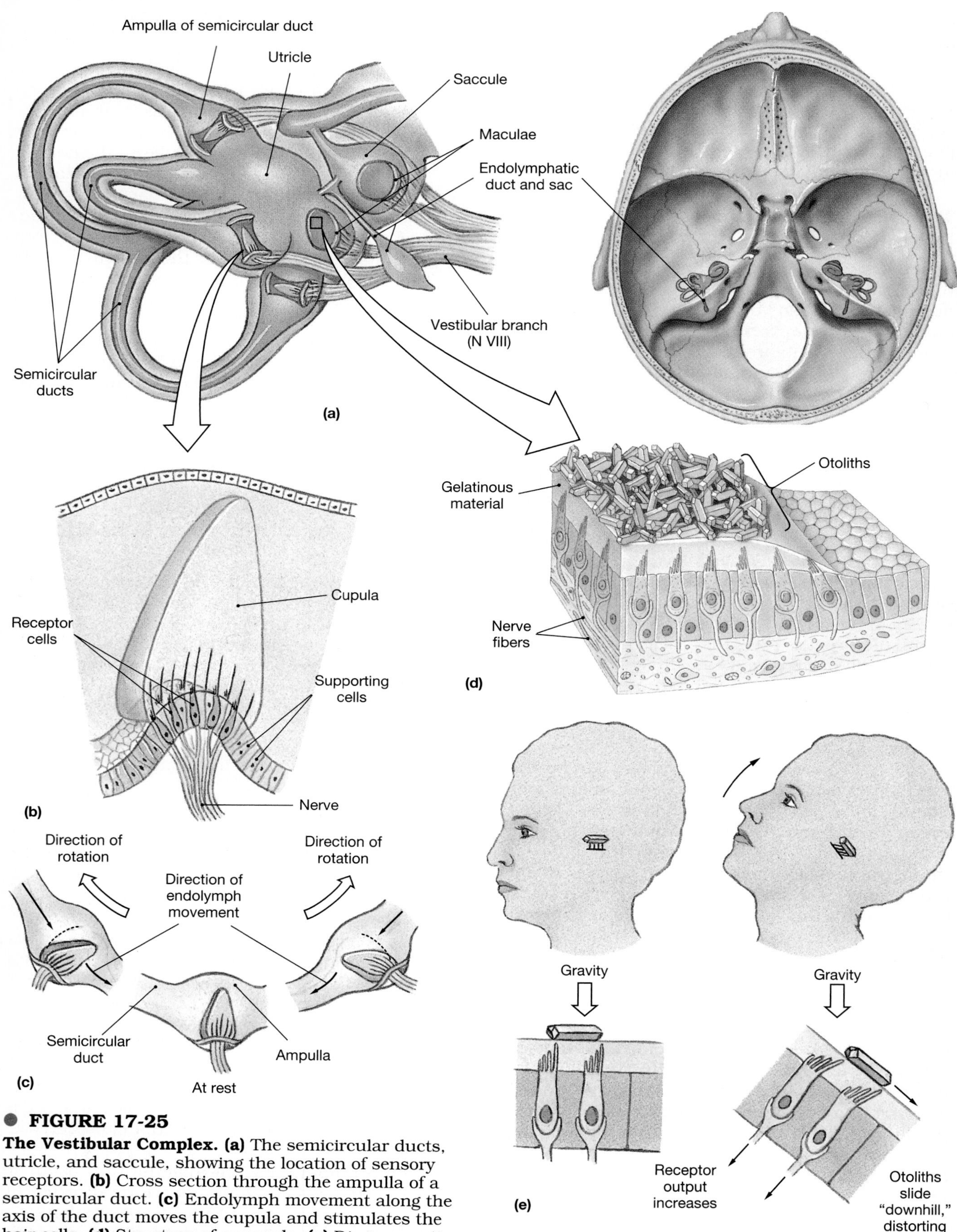

Ampulla of semicircular duct
Utricle
Saccule
Maculae
Endolymphatic duct and sac
Vestibular branch (N VIII)
Semicircular ducts
(a)
Gelatinous material
Otoliths
Nerve fibers
(d)
Cupula
Receptor cells
Supporting cells
Nerve
(b)
Direction of rotation
Direction of rotation
Direction of endolymph movement
Semicircular duct
Ampulla
At rest
(c)
Gravity
Gravity
Receptor output increases
Otoliths slide "downhill," distorting hair cell processes
(e)

● **FIGURE 17-25**

The Vestibular Complex. (a) The semicircular ducts, utricle, and saccule, showing the location of sensory receptors. **(b)** Cross section through the ampulla of a semicircular duct. **(c)** Endolymph movement along the axis of the duct moves the cupula and stimulates the hair cells. **(d)** Structure of a macula. **(e)** Diagrammatic view of macular function when the head is tilted back.

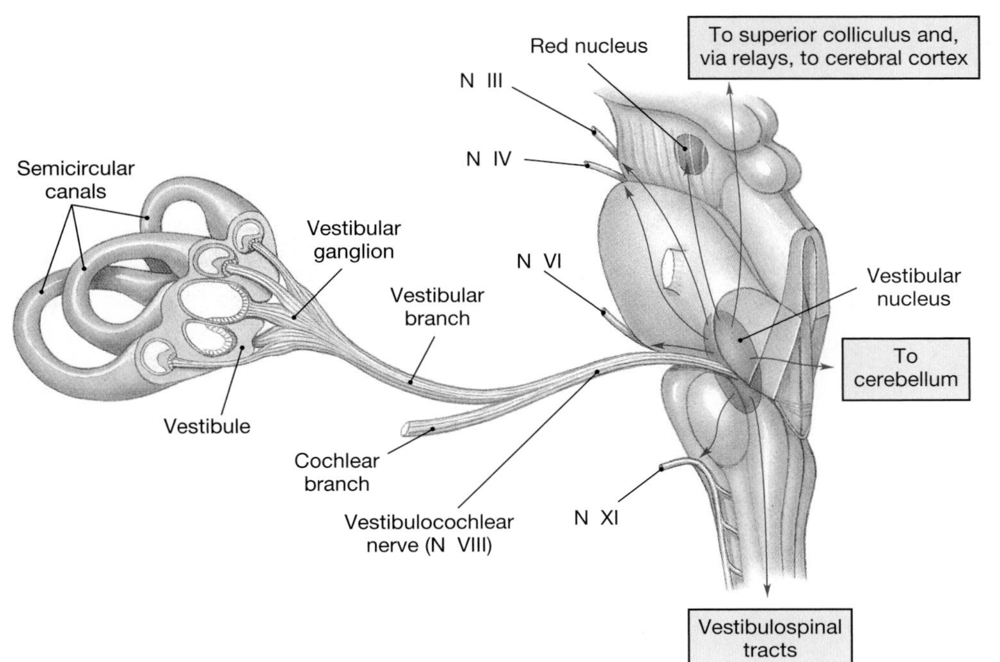

Red nucleus
To superior colliculus and, via relays, to cerebral cortex
N III
N IV
Semicircular canals
Vestibular ganglion
N VI
Vestibular branch
Vestibular nucleus
To cerebellum
Vestibule
Cochlear branch
Vestibulocochlear nerve (N VIII)
N XI
Vestibulospinal tracts

● **FIGURE 17-26**
Pathways for Equilibrium Sensations

in body position and orientation. If the body is turning or spinning rapidly, the eyes will fix on one point for a moment, then jump ahead to another in a series of short, jerky movements. This type of eye movement can occur even when the body is stationary if either the brain stem or the inner ear is damaged. An individual with this condition, called **nystagmus,** has trouble controlling eye movements. Physicians often check for nystagmus by asking the subject to watch a small penlight as it is moved across the field of vision. AM *Vertigo; Ménière's Disease and Motion Sickness*

❑ Hearing

The receptors of the cochlear duct provide us with a sense of hearing that enables us to detect the quietest whisper yet remain functional in a crowded, noisy room. The receptors responsible for auditory sensations are hair cells similar to those of the vestibular complex. However, their placement within the cochlear duct and the organization of the surrounding accessory structures shield them from stimuli other than sound. In conveying vibrations from the tympanic membrane to the oval window, the auditory ossicles convert pressure waves in air to pressure pulses in the perilymph of the cochlea. These pressure pulses stimulate hair cells along the cochlear spiral. The *frequency* of the perceived sound is determined by *which part* of the cochlear duct is stimulated. The *intensity* (volume) of the perceived sound is determined by *how many* of the hair cells at that location are stimulated. We will now consider the mechanics of this remarkably elegant process.

The Cochlear Duct

Figure 17-27

In sectional view (Figure 17-27●) the cochlear duct, or *scala media,* lies between a pair of perilymphatic chambers, the **vestibular duct** (*scala vestibuli*) and the **tympanic duct** (*scala tympani*). The vestibular and tympanic ducts are interconnected at the tip of the cochlear spiral. The outer surfaces of these ducts are encased by the bony labyrinth everywhere except at the oval window (base of the vestibular duct) and the round window (base of the tympanic duct).

THE ORGAN OF CORTI The hair cells of the cochlear duct are found in the **organ of Corti** (Figure 17-27b,c,d●). This sensory structure sits above the **basilar membrane** that separates the cochlear duct from the tympanic duct. The hair cells are arranged in a series of longitudinal rows. They lack kinocilia, and their stereocilia are in contact with the overlying **tectorial** (tek-TOR-ē-al) **membrane** (*tectum,* roof). This membrane is firmly attached to the inner wall of the cochlear duct. When a portion of the basilar membrane bounces up and down, the stereocilia of the hair cells are distorted. The basilar membrane moves in response to pressure waves within the perilymph. These waves are produced when sounds arrive at the tympanic membrane; to understand how these waves develop, we must consider the basic properties of sound.

An Introduction to Sound

Figure 17-28a

Hearing is the detection of sound, which consists of pressure waves conducted through air or water.

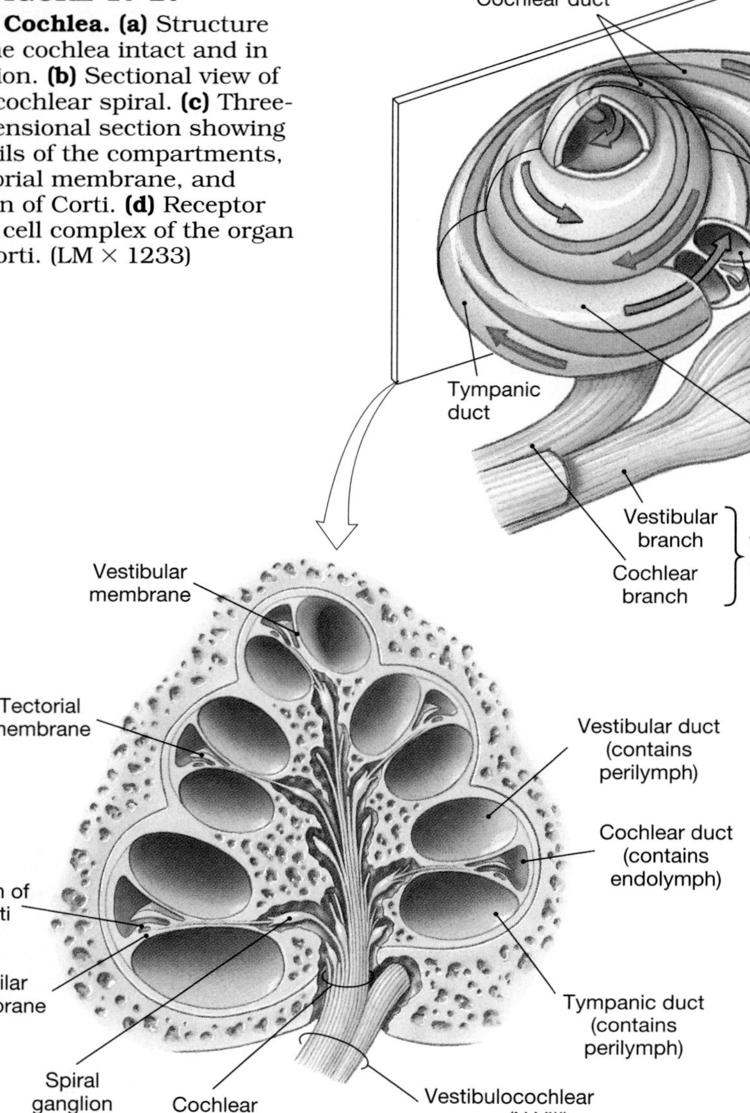
Cochlear duct
Round window
Stapes at oval window
Tympanic duct
Vestibular duct
Vestibular branch
Cochlear branch
Vestibulocochlear nerve (N VIII)
Semicircular canals
(a)
Vestibular membrane
Tectorial membrane
Organ of Corti
Basilar membrane
Spiral ganglion
Cochlear branch
(b)
Vestibular duct (contains perilymph)
Cochlear duct (contains endolymph)
Tympanic duct (contains perilymph)
Vestibulocochlear nerve (N VIII)

● **FIGURE 17-27**
The Cochlea. (a) Structure of the cochlea intact and in section. **(b)** Sectional view of the cochlear spiral. **(c)** Three-dimensional section showing details of the compartments, tectorial membrane, and organ of Corti. **(d)** Receptor hair cell complex of the organ of Corti. (LM × 1233)

In air, each pressure wave consists of a region where the air molecules are crowded together and an adjacent zone where they are relatively far apart (Figure 17-28a●). These waves travel through the air at approximately 1235 km/h (768 mph). Physicists use the term **cycles** rather than waves, and the number of cycles per second (cps), or **hertz (Hz),** represents the **frequency** of the sound. What we perceive as the **pitch** of a sound is our sensory response to its frequency. A *high-frequency* sound (high pitch) might have a frequency of 15,000 Hz or more; a very *low-frequency* sound (low pitch) could have a frequency of 100 Hz or less.

It takes energy to produce sound waves. When you strike a tuning fork it vibrates and pushes against the surrounding air, producing sound waves whose frequency depends on the frequency of vibration. The harder you strike it, the more energy you provide, and the louder the sound. The loudness increases because the sound waves carry energy away with them. The energy content, or power, of a sound determines its **intensity** (volume), which is reported in **decibels** (DES-i-belz).

When sound waves strike an object, their energy can be felt as a physical pressure. Given the right combination of frequencies and intensities, the object will begin to vibrate at the same frequency as the sound, a phenomenon called *resonance.* The greater the sound intensity, the greater the amount of movement produced.

The Hearing Process

Figures 17-27a, 17-28b

The process of hearing can be divided into six basic steps, diagrammed in Figure 17-28b● and summarized in Table 17-1.

Step 1: Sound waves arrive at the tympanic membrane.

Sound waves enter the external auditory canal and travel toward the tympanum. The orientation of the canal provides some directional sensitivity. Sound waves approaching the side of the head have direct access to the tympanic membrane on that side, whereas sounds arriving from another direction must bend around corners or pass through the pinna or other body tissues.

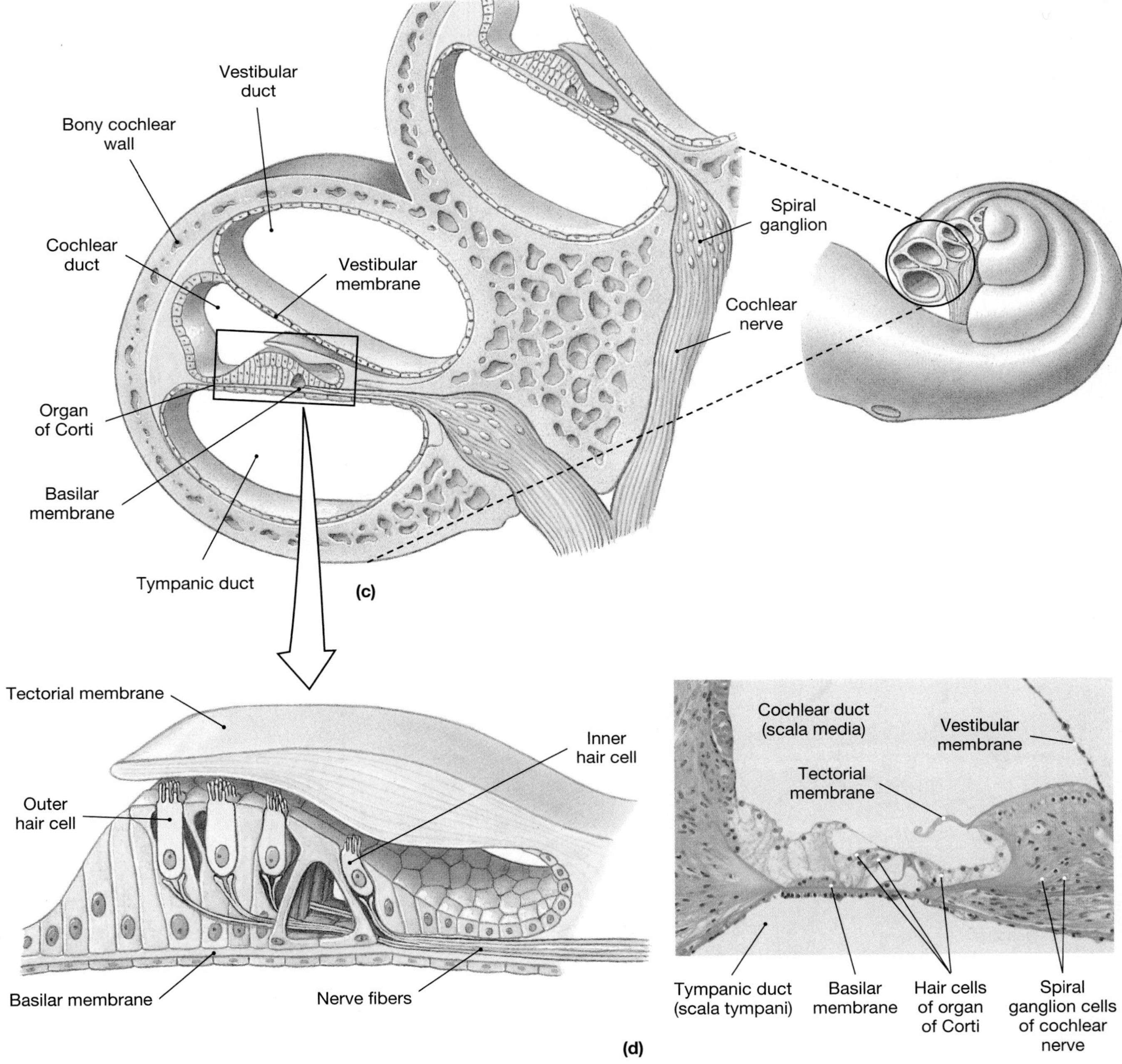
Vestibular duct
Bony cochlear wall
Cochlear duct
Vestibular membrane
Spiral ganglion
Cochlear nerve
Organ of Corti
Basilar membrane
Tympanic duct
(c)
Tectorial membrane
Inner hair cell
Outer hair cell
Basilar membrane
Nerve fibers
Cochlear duct (scala media)
Vestibular membrane
Tectorial membrane
Tympanic duct (scala tympani)
Basilar membrane
Hair cells of organ of Corti
Spiral ganglion cells of cochlear nerve
(d)

Step 2: ***Movement of the tympanic membrane causes displacement of the auditory ossicles.***

The tympanic membrane provides the surface for sound collection, and it vibrates in resonance to sound waves with frequencies between approximately 20 and 20,000 Hz (in a young child). When the tympanic membrane vibrates, so do the malleus and, via their articulations, the incus and stapes.

Step 3: ***Movement of the stapes at the oval window establishes pressure waves in the perilymph of the vestibular duct.***

Liquids are incompressible; when on a waterbed, if you push down *here* the waterbed bulges over *there.* Because the rest of the cochlea is sheathed in bone, pressure applied at the oval window can be relieved only at the round window. When the stapes moves inward, the round window bulges outward. As the stapes moves in and out, vibrating at the frequency of the sound arriving at the tympanic membrane, it creates pressure pulses, or *pressure waves,* within the perilymph.

Step 4: ***The pressure waves distort the basilar membrane on their way to the round window of the tympanic duct.***

The pressure waves established by movement of the stapes travel through the perilymph of the vestibular and tympanic ducts to reach the round

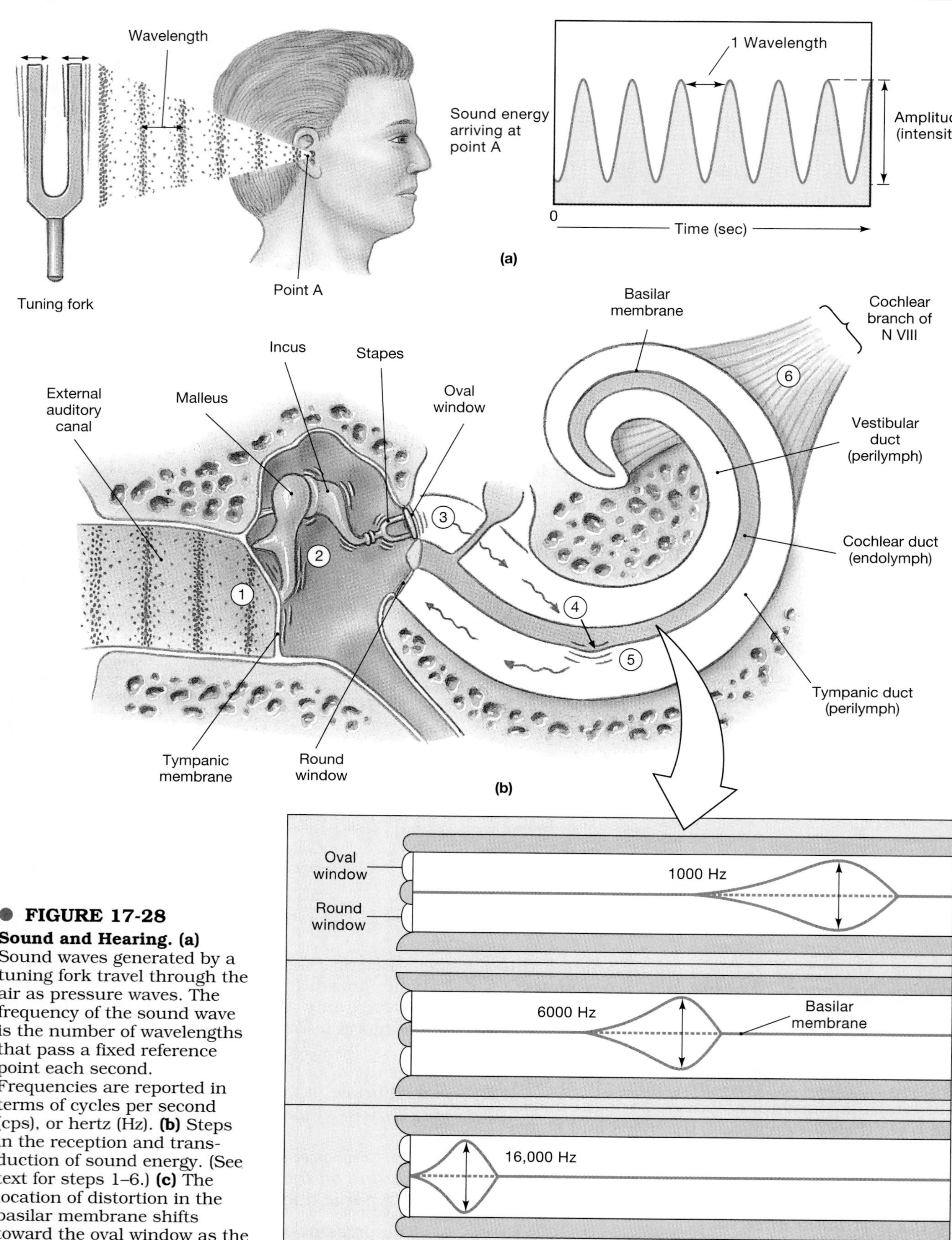
Wavelength
Tuning fork
Point A
Sound energy arriving at point A
1 Wavelength
Amplitude (intensity)
0
Time (sec)
(a)
Basilar membrane
Cochlear branch of N VIII
Incus
Stapes
External auditory canal
Malleus
Oval window
Vestibular duct (perilymph)
Cochlear duct (endolymph)
Tympanic duct (perilymph)
Tympanic membrane
Round window
(b)
Oval window
Round window
1000 Hz
6000 Hz
Basilar membrane
16,000 Hz
(c)

FIGURE 17-28
Sound and Hearing. (a) Sound waves generated by a tuning fork travel through the air as pressure waves. The frequency of the sound wave is the number of wavelengths that pass a fixed reference point each second. Frequencies are reported in terms of cycles per second (cps), or hertz (Hz). **(b)** Steps in the reception and transduction of sound energy. (See text for steps 1–6.) **(c)** The location of distortion in the basilar membrane shifts toward the oval window as the frequency of the sound increases.

window. In doing so, these pressure waves distort movement in the basilar membrane. The location of maximum distortion varies depending on the frequency of the sound. High-frequency sounds, which have a very short wavelength, vibrate the basilar membrane near the oval window. The lower the frequency of the sound, the longer the wavelength, and the farther away from the oval window the area of maximum distortion will be. Thus *frequency* information is translated into *position* information.

The actual *amount* of movement at a given location will depend on the amount of force applied by the stapes. This is a function of the intensity of the sound. The louder the sound, the greater the movement of the basilar membrane.

Step 5: ***Vibration of the basilar membrane causes vibration of hair cells against the tectorial membrane.***

Vibration of the affected region of the basilar membrane moves hair cells against the tectorial membrane. This movement leads to displacement of the stereocilia and stimulation of sensory neurons. The hair cells of the organ of Corti are arranged in several rows. A very soft sound may stimulate only a few hair cells in a portion of one row. As the volume of a sound increases, not only do these hair cells become more active, but additional hair cells—at first in the same row, and then in adjacent rows—are stimulated as well. The number of hair cells responding in a given region of the organ of Corti thus provides information on the volume of the sound.

Step 6: ***Information concerning the region and intensity of stimulation is relayed to the CNS over the cochlear branch of the vestibulocochlear nerve (N VIII).***

TABLE 17-1 Steps in the Production of an Auditory Sensation

1. **Sound waves arrive at the tympanic membrane.**
2. **Movement of the tympanic membrane causes displacement of the auditory ossicles.**
3. **Movement of the stapes at the oval window establishes pressure waves in the perilymph of the vestibular duct.**
4. **The pressure waves distort the basilar membrane on their way to the round window of the tympanic duct.**
5. **Vibration of the basilar membrane causes vibration of hair cells against the tectorial membrane.**
6. **Information concerning the region and intensity of stimulation is relayed to the CNS over the cochlear branch of N VIII.**

The cell bodies of the sensory neurons that monitor the cochlear hair cells are found at the center of the bony cochlea (Figure 17-27c,d●) in the **spiral ganglion.** This information is carried to the cochlear nuclei of the medulla oblongata for subsequent distribution to other centers in the brain.

Auditory Pathways

Figures 17-27, 17-29

Hair cell stimulation activates sensory neurons whose cell bodies are in the adjacent spiral ganglion (Figures 17-27 and 17-29●, p. 594). Their afferent fibers form the **cochlear branch** of the vestibulocochlear nerve (N VIII). These axons enter the medulla oblongata and synapse at the **cochlear nucleus.** From here the information crosses to the opposite side of the brain and ascends to the inferior colliculus of the mesencephalon. This processing center coordinates a number of responses to acoustic stimuli, including auditory reflexes involving skeletal muscles of the head, face, and trunk. These reflexes automatically change the position of the head in response to a sudden loud noise.

Before reaching the cerebral cortex and our conscious awareness, ascending auditory sensations synapse in the thalamus. Projection fibers then deliver the information to the auditory cortex of the temporal lobe. Information travels to the cortex over labeled lines: High-frequency sounds activate one portion of the cortex, and low-frequency sounds affect another. In effect, the auditory cortex contains a map of the organ of Corti. Thus, *frequency* information, translated into *position* information on the basilar membrane, is projected in this form onto the auditory cortex, where it is interpreted to produce our subjective sensation of pitch.

If the auditory cortex is damaged, the individual will respond to sounds and have normal acoustic reflexes, but sound interpretation and pattern recognition will be difficult or impossible. Damage to the adjacent association area leaves the ability to detect the tones and patterns, but produces an inability to comprehend their meaning.

Auditory Sensitivity

Our hearing abilities are remarkable, though it is difficult to assess the absolute sensitivity of the system. From the softest audible sound to the loudest tolerable blast represents a trillionfold increase in power. Theoretically, if we were to remove the stapes, the receptor mechanism is so sensitive that we could hear the sound of air molecules bouncing off the oval window, responding to displacements as small as one-tenth the diameter of a hydrogen atom. We never utilize the full

EMBRYOLOGY SUMMARY Development of Special Sense Organs

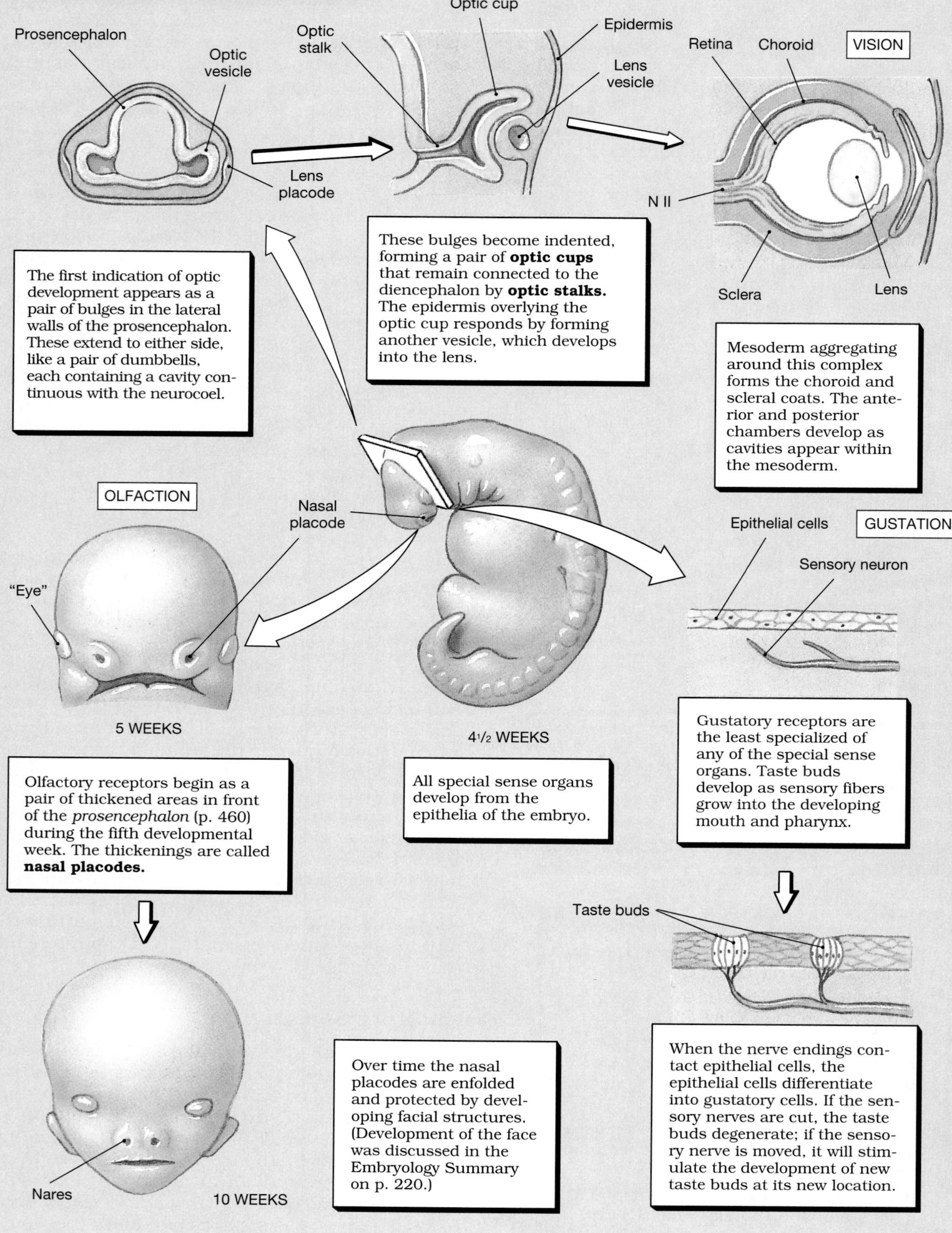
VISION
Prosencephalon
Optic vesicle
Lens placode
Optic stalk
Optic cup
Epidermis
Lens vesicle
Retina
Choroid
N II
Sclera
Lens
The first indication of optic development appears as a pair of bulges in the lateral walls of the prosencephalon. These extend to either side, like a pair of dumbbells, each containing a cavity continuous with the neurocoel.
These bulges become indented, forming a pair of **optic cups** that remain connected to the diencephalon by **optic stalks.** The epidermis overlying the optic cup responds by forming another vesicle, which develops into the lens.
Mesoderm aggregating around this complex forms the choroid and scleral coats. The anterior and posterior chambers develop as cavities appear within the mesoderm.
OLFACTION
Nasal placode
"Eye"
5 WEEKS
4 1/2 WEEKS
Olfactory receptors begin as a pair of thickened areas in front of the *prosencephalon* (p. 460) during the fifth developmental week. The thickenings are called **nasal placodes.**
All special sense organs develop from the epithelia of the embryo.
Nares
10 WEEKS
Over time the nasal placodes are enfolded and protected by developing facial structures. (Development of the face was discussed in the Embryology Summary on p. 220.)
GUSTATION
Epithelial cells
Sensory neuron
Gustatory receptors are the least specialized of any of the special sense organs. Taste buds develop as sensory fibers grow into the developing mouth and pharynx.
Taste buds
When the nerve endings contact epithelial cells, the epithelial cells differentiate into gustatory cells. If the sensory nerves are cut, the taste buds degenerate; if the sensory nerve is moved, it will stimulate the development of new taste buds at its new location.

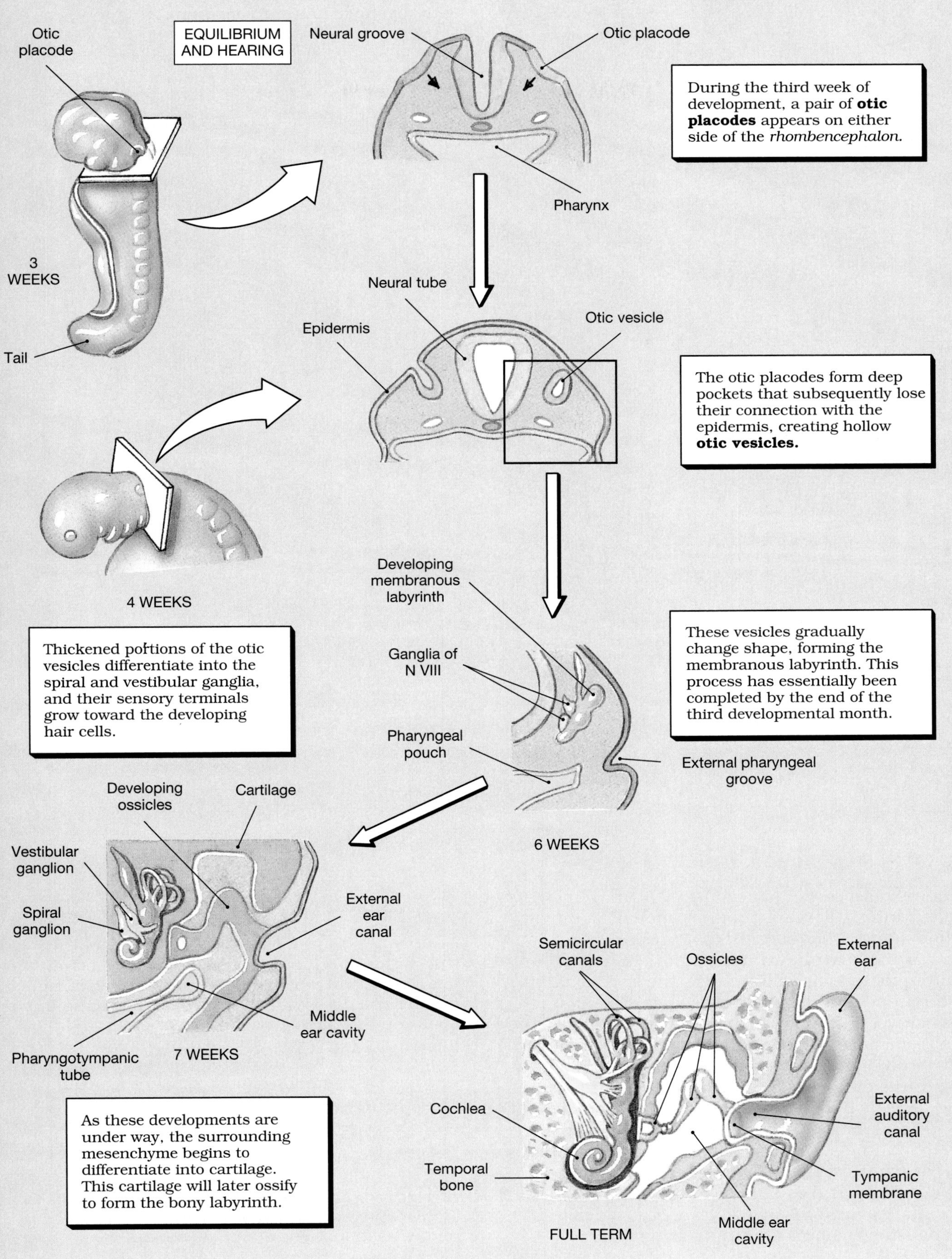
EQUILIBRIUM AND HEARING
Otic placode
3 WEEKS
Tail
Neural groove
Otic placode
Pharynx
During the third week of development, a pair of **otic placodes** appears on either side of the *rhombencephalon*.
Neural tube
Epidermis
Otic vesicle
4 WEEKS
The otic placodes form deep pockets that subsequently lose their connection with the epidermis, creating hollow **otic vesicles.**
Thickened portions of the otic vesicles differentiate into the spiral and vestibular ganglia, and their sensory terminals grow toward the developing hair cells.
Developing membranous labyrinth
Ganglia of N VIII
Pharyngeal pouch
External pharyngeal groove
6 WEEKS
These vesicles gradually change shape, forming the membranous labyrinth. This process has essentially been completed by the end of the third developmental month.
Developing ossicles
Cartilage
Vestibular ganglion
Spiral ganglion
External ear canal
Middle ear cavity
Pharyngotympanic tube
7 WEEKS
As these developments are under way, the surrounding mesenchyme begins to differentiate into cartilage. This cartilage will later ossify to form the bony labyrinth.
Semicircular canals
Ossicles
External ear
Cochlea
Temporal bone
External auditory canal
Tympanic membrane
Middle ear cavity
FULL TERM

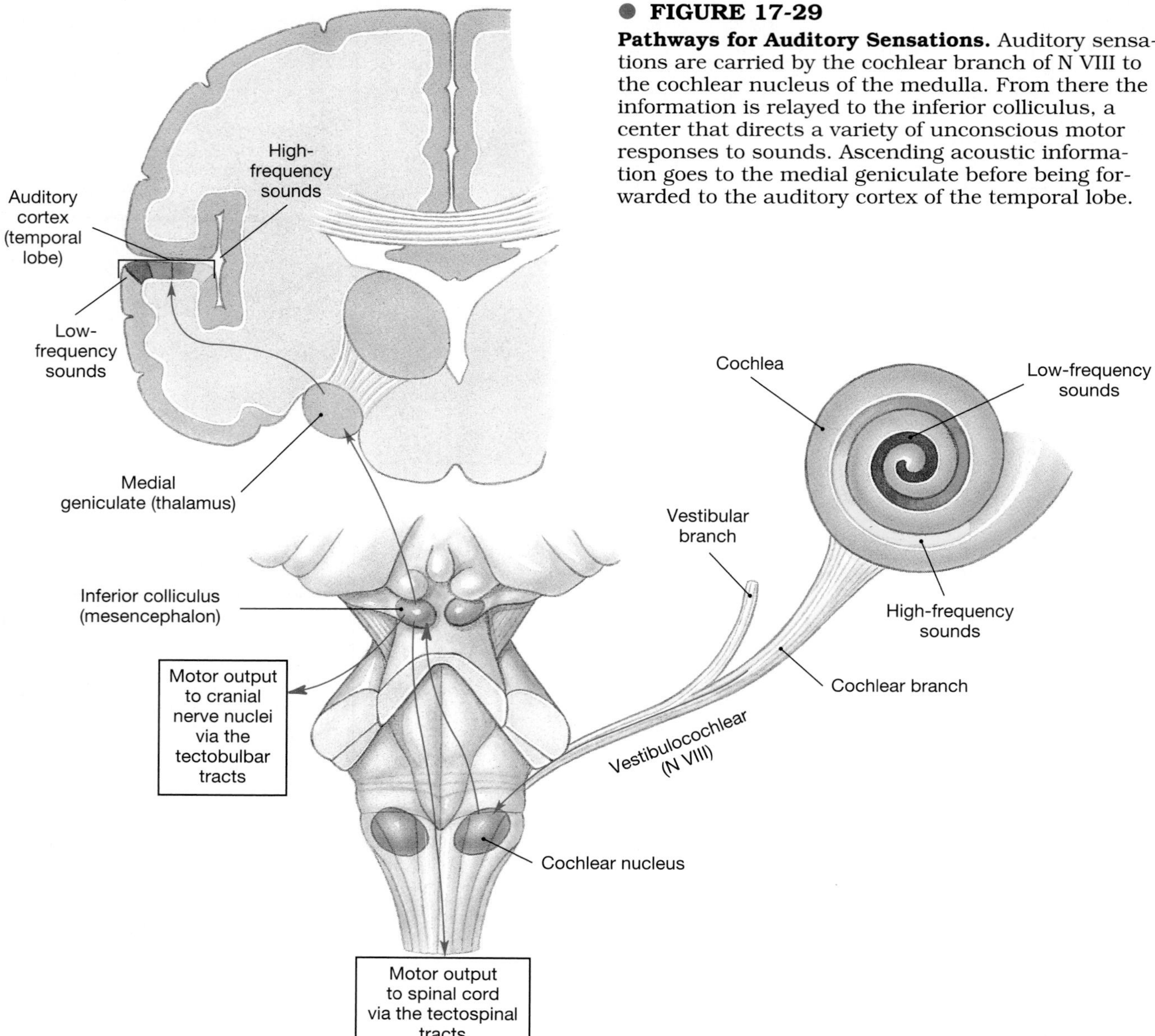
High-
frequency
sounds
Auditory
cortex
(temporal
lobe)
Low-
frequency
sounds
Medial
geniculate (thalamus)
Cochlea
Low-frequency
sounds
Vestibular
branch
High-frequency
sounds
Cochlear branch
Inferior colliculus
(mesencephalon)
Motor output
to cranial
nerve nuclei
via the
tectobulbar
tracts
Vestibulocochlear
(N VIII)
Cochlear nucleus
Motor output
to spinal cord
via the tectospinal
tracts

FIGURE 17-29
Pathways for Auditory Sensations. Auditory sensations are carried by the cochlear branch of N VIII to the cochlear nucleus of the medulla. From there the information is relayed to the inferior colliculus, a center that directs a variety of unconscious motor responses to sounds. Ascending acoustic information goes to the medial geniculate before being forwarded to the auditory cortex of the temporal lobe.

potential of this system because body movements and our internal organs produce squeaks, groans, thumps, and other sounds that are tuned out by central and peripheral adaptation. When other environmental noises fade away, the level of adaptation drops and the system becomes increasingly sensitive. If we relax in a quiet room, our heartbeat seems to get louder and louder as the auditory system adjusts to the level of background noise.

Young children have the greatest hearing range; they can detect sounds ranging from a 20-Hz buzz to a 20,000-Hz whine. With age, damage due to loud noises or other injuries accumulates; the eardrum gets less flexible, the articulations between the ossicles stiffen, and the round window may begin to ossify. As a result, older individuals show some degree of hearing loss.

HEARING DEFICITS There are probably over 6 million people in the United States alone who have at least a partial hearing deficit. **Conductive deafness** results from conditions in the outer or middle ear that block the normal transfer of vibration from the tympanic membrane to the oval window. An external auditory canal plugged with accumulated wax or trapped water may cause a temporary hearing loss. Scarring or perforation of the tympanum and immobilization of one or more of the auditory ossicles are more serious examples of conductive deafness.

In **nerve deafness** the problem lies within the cochlea or somewhere along the auditory pathway. The vibrations are reaching the oval window and entering the perilymph, but the receptors either cannot respond or their response cannot reach its central destinations. For example:

- Very loud (high-intensity) sounds can produce nerve deafness by breaking stereocilia off the surfaces of the hair cells. (The reflex contraction of the tensor tympani and stapedius in response to a dangerously loud noise occurs in less than 0.1 seconds, but this may not be fast enough.)
- Drugs such as the aminoglycoside antibiotics (*neomycin* or *gentamicin*) may diffuse into the endolymph and kill the hair cells. Because hair cells and sensory nerves can also be damaged by bacterial infection, the potential side effects must be balanced against the severity of infection.

There are many treatment options for conductive deafness; treatment options for nerve deafness are relatively limited. Because many of these problems become progressively worse, early diagnosis improves the chances for successful treatment. AM *Testing and Treating Hearing Deficits*

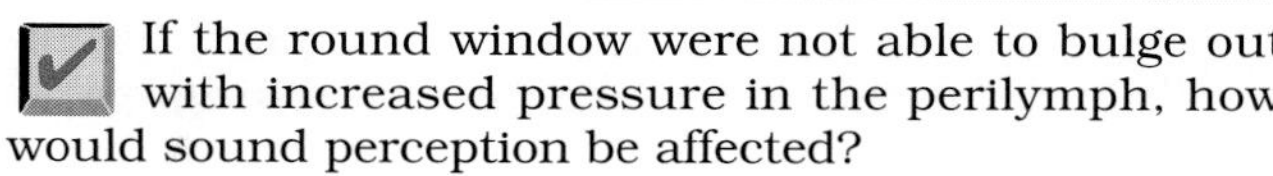
If the round window were not able to bulge out with increased pressure in the perilymph, how would sound perception be affected?

How would loss of stereocilia from the hair cells of the organ of Corti affect hearing?

Why does blockage of the pharyngotympanic tube produce an earache?

Selected Clinical Terminology

Terms Discussed in This Chapter

color blindness: A condition in which people are unable to distinguish certain colors. *(p. 577)*
conductive deafness: Deafness resulting from conditions in the outer or middle ear that block the transfer of vibrations from the tympanic membrane to the oval window. *(p. 594)*
hyperopia: A condition in which nearby objects are blurry but distant objects are clear; "farsightedness." *(p. 572)*
myopia: A condition in which vision at close range is normal but distant objects appear blurry; "nearsightedness." *(p. 572)*
nerve deafness: Deafness resulting from problems within the cochlea or along the auditory pathway. *(p. 594)*
nystagmus: Abnormal eye movements that may appear after damage to the brain stem or inner ear. *(p. 587)*
presbyopia: A type of hyperopia that develops with age as lenses become less elastic. *(p. 572)*

AM *Additional Terms Discussed in the Applications Manual*

anesthesia: A total loss of sensation.
cataract: An abnormal lens that has lost its transparency.
hypesthesia: A reduction in sensitivity.
paresthesia: Abnormal sensations, such as a "pins and needles" sensation.

CHAPTER REVIEW

STUDY OUTLINE

INTRODUCTION, p. 552

RECEPTORS, p. 552

1. The **general senses** are temperature, pain, touch, pressure, vibration, and proprioception; receptors for those sensations are distributed throughout the body. Receptors for the **special senses** (smell, taste, sight, balance, and hearing) are located in specialized areas, or in **sense organs.**
2. A *sensory receptor* is a specialized cell that, when stimulated, sends a *sensation* to the CNS.

Specificity, p. 552

3. **Receptor specificity** allows each receptor to respond to particular stimuli. The simplest receptors are **free nerve endings;** the area monitored by a single receptor cell is the **receptive field.** (*Figure 17-1*)

Transduction, p. 553

4. **Transduction** is the translation of a stimulus into an action potential.
5. Transduction involves the development of **receptor potentials** that can summate to produce a **generator potential** in an afferent fiber. The resulting action potential travels to the CNS. (*Figure 17-2*)

Interpretation of Sensory Information, p. 553

6. The identity of a sensation is indicated by the **labeled line** that carries the action potentials into the CNS. The intensity of the stimulus is indicated by the frequency or pattern of action potentials. This phenomenon is called **sensory coding.**
7. **Tonic receptors** are always sending signals to the CNS; **phasic receptors** become active only when the conditions that they monitor change.

Central Processing and Adaptation, p. 555

8. **Adaptation** (a reduction in sensitivity in the presence of a constant stimulus) may involve changes in receptor sensitivity **(peripheral,** or **sensory, adaptation)** or inhibition along the sensory pathways **(central**

adaptation). Fast-adapting receptors are phasic; **slow-adapting receptors** are tonic.

THE GENERAL SENSES, p. 555

Nociceptors, p. 555

1. **Nociceptors** respond to a variety of stimuli, usually ones associated with tissue damage. There are two types of these painful sensations: **fast (prickling) pain** and **slow (burning and aching) pain.** (*Figure 17-3*)

Thermoreceptors, p. 556

2. **Thermoreceptors** respond to changes in temperature.

Mechanoreceptors, p. 556

3. **Mechanoreceptors** respond to physical distortion, contact, or pressure on their cell membranes. **Tactile receptors** respond to touch, pressure, and vibration; **baroreceptors** respond to pressure changes in the walls of blood vessels and the digestive, reproductive, and urinary tracts. **Proprioceptors** monitor positions of joints and muscles.
4. **Fine touch and pressure receptors** provide detailed information about a source of stimulation; **crude touch and pressure receptors** are poorly localized. Important tactile receptors include the **root hair plexus, Merkel's discs, Meissner's corpuscles, Pacinian corpuscles,** and **Ruffini corpuscles.** (*Figure 17-3*)
5. Baroreceptors monitor changes in pressure; they respond immediately but adapt rapidly. Baroreceptors in the walls of major arteries and veins respond to changes in blood pressure. Receptors along the digestive tract help coordinate reflex activities of digestion.
6. Proprioceptors monitor the position of joints, tension in tendons and ligaments, and the state of muscular contraction. Proprioceptors include tendon organs and muscle spindles.

Chemoreceptors, p. 558

7. In general, **chemoreceptors** respond to water-soluble and lipid-soluble substances that are dissolved in the surrounding fluid. They monitor the chemical composition of body fluids.

OLFACTION, p. 559

Olfactory Receptors, p. 559

1. The **olfactory organs** contain the **olfactory epithelium** with **olfactory receptors** (neurons sensitive to chemicals dissolved in the overlying mucus), *supporting cells,* and **basal** (*stem*) **cells.** Their surfaces are coated with the secretions of the **olfactory glands.** (*Figure 17-4*)
2. The olfactory receptors are modified neurons.

Olfactory Pathways, p. 560

3. The highly sensitive olfactory system has extensive limbic and hypothalamic connections.

Olfactory Discrimination, p. 560

4. The olfactory system can distinguish thousands of chemical stimuli; the CNS interprets smells by the pattern of receptor activity.

Aging and Olfactory Sensitivity, p. 560

5. The olfactory receptor population shows considerable turnover. The total number of olfactory receptors declines with age.

GUSTATION, p. 560

Gustatory Receptors, p. 560

1. **Gustatory (taste) receptors** are clustered in **taste buds,** each of which contains **gustatory cells,** which extend taste hairs through a narrow **taste pore.** (*Figure 17-5*)
2. Taste buds are associated with epithelial projections on the dorsal surface of the tongue **(lingual papillae).** (*Figure 17-5*)

Gustatory Pathways, p. 561

3. The taste buds are monitored by cranial nerves that synapse within the **nucleus solitarius.** (*Figure 17-6*)

Gustatory Discrimination, p. 561

4. The **primary taste sensations** are sweet, salt, sour, and bitter. (*Figure 17-6*)
5. There are significant individual differences in taste sensitivity, some of which are inherited.
6. The number of taste buds declines with age.

VISION, p. 562

Accessory Structures of the Eye, p. 562

1. The **accessory structures** of the eye include the **palpebrae** (eyelids), which are separated by the **palpebral fissure.** The **eyelashes** line the palpebral margins. Along the inner margin of the lid are **Meibomian glands,** which secrete a lipid-rich product. Glands at the **lacrimal caruncle** produce other secretions. (*Figure 17-7*)
2. An epithelium called the **conjunctiva** covers most of the exposed surface of the eye; the **ocular** (*bulbar*) conjunctiva covers the anterior surface of the eye, and the **palpebral conjunctiva** lines the inner surface of the eyelids. The **cornea** is transparent. (*Figure 17-8*)
3. The secretions of the **lacrimal gland** bathe the conjunctiva; these secretions are slightly alkaline and contain **lysozymes** (enzymes that attack bacteria). Tears collect in the **lacus lacrimalis.** The tears reach the inferior meatus of the nose after passing through the **lacrimal puncta,** the **lacrimal canals,** the **lacrimal sac,** and the **nasolacrimal duct.** (*Figure 17-7*)

The Eye, p. 565

4. The eye has three layers: an outer **fibrous tunic,** a vascular tunic, and an inner neural tunic. Most of the ocular surface is covered by the **sclera** (a dense fibrous connective tissue); the **limbus** is the border between the sclera and the cornea. (*Figure 17-8*)
5. The **vascular tunic,** or **uvea,** includes the **iris,** the **ciliary body,** and the **choroid.** The iris forms the boundary between the anterior and posterior chambers. The ciliary body contains the **ciliary muscle** and the **ciliary processes,** which attach to the **suspensory ligaments** of the lens. (*Figure 17-8*)
6. The **neural tunic** consists of an outer **pigment layer** and an inner **neural retina;** the latter contains visual receptors and associated neurons. (*Figures 17-8, 17-9*)
7. The direct line to the CNS proceeds from the photoreceptors to **bipolar cells,** then to **ganglion cells,**

and to the brain via the optic nerve. **Horizontal cells** and **amacrine cells** modify the signals passed between other retinal components.

8. The ciliary body and lens divide the interior of the eye into a large **posterior cavity,** also called the *vitreous chamber,* and a smaller **anterior cavity.** The anterior cavity is further subdivided into the **anterior chamber,** which extends from the cornea to the iris, and a **posterior chamber** between the iris and the ciliary body and lens. (*Figure 17-11*)
9. The fluid **aqueous humor** circulates within the eye and reenters the circulation after diffusing through the walls of the anterior chamber and into the **canal of Schlemm.** (*Figure 17-11*)
10. The **lens,** held in place by the suspensory ligaments, lies behind the cornea and forms the anterior boundary of the posterior cavity (vitreous chamber). This cavity contains the **vitreous body,** a gelatinous mass that helps stabilize the shape of the eye and support the retina. (*Figures 17-8, 17-11*)
11. The lens focuses a visual image on the retinal receptors; a lens that has lost its transparency is a **cataract.** Light is **refracted** (bent) when it passes through the cornea and lens. During **accommodation** the shape of the lens changes to focus an image on the retina. Normal **visual acuity** is rated 20/20. (*Figures 17-12, 17-13, 17-14, and 17-15*)

Visual Physiology, p. 573

12. There are two types of photoreceptors (visual receptors of the retina): **Rods** respond to almost any photon, regardless of its energy content; **cones** have characteristic ranges of sensitivity. Many cones are densely packed within the **fovea** (the central portion of the **macula lutea**), the site of sharpest vision.
13. Each photoreceptor contains an **outer segment** with membranous **discs.** A narrow stalk attaches the outer to the **inner segment.** Light absorption occurs in the **visual pigments,** which are derivatives of **rhodopsin (opsin** plus a pigment, **retinal,** that is synthesized from **vitamin A).** The retinal molecule can be in either the normal **11-cis** or (after absorbing light) the **11-trans** form. (*Figures 17-16, 17-17, 17-18*)
14. **Color blindness** is the inability to detect certain colors. (*Figure 17-19*)
15. In the **dark-adapted state,** almost all visual pigments will be fully receptive to stimulation. The **light-adapted state** is characterized by constriction of the pupil and bleaching of the visual pigments.

The Visual Pathway, p. 578

16. Each photoreceptor monitors a specific receptive field. The ganglion cells that monitor rods, called **M cells** (*magnocells*), are relatively large. The ganglion cells that monitor cones, called **P cells** (*parvo cells*), are smaller and more numerous. (*Figure 17-20*)
17. Visual inputs to the **suprachiasmatic nucleus** affect the function of other brain stem nuclei. This nucleus establishes a visceral **circadian rhythm** that is tied to the day/night cycle and affects other metabolic processes. (*Figure 17-21*)

EQUILIBRIUM AND HEARING, p. 581

Anatomy of the Ear, p. 581

1. The senses of equilibrium and hearing are provided by the receptors of the **inner ear** (also known as the **membranous labyrinth**). Its chambers and canals contain the fluid **endolymph.** The **bony labyrinth** surrounds and protects the membranous labyrinth. The bony labyrinth can be subdivided into the **vestibule,** the **semicircular canals,** and the **cochlea.** The structures and airspaces of the **external ear** and **middle ear** help capture and transmit sound to the cochlea. (*Figures 17-22, 17-24*)
2. The external ear includes the **pinna** that surrounds the entrance to the **external auditory canal** that ends at the **tympanic membrane** (eardrum). (*Figure 17-22*)
3. The middle ear communicates with the nasopharynx via the **pharyngotympanic tube.** The middle ear encloses and protects the **auditory ossicles** that connect the tympanic membrane with the receptor complex of the inner ear. (*Figures 17-22, 17-23*)
4. The vestibule of the inner ear encloses a pair of membranous sacs, the **saccule** and **utricle,** whose receptors provide sensations of gravity and linear acceleration. The semicircular canals contain the **semicircular ducts,** whose receptors provide sensations of rotation. The cochlea contains the **cochlear duct,** an elongated portion of the membranous labyrinth. (*Figure 17-24*)
5. The basic receptors of the inner ear are **hair cells** whose surfaces support **stereocilia** and (except in the cochlear duct) a single **kinocilium** per hair cell. Hair cells provide information about the direction and strength of varied mechanical stimuli. (*Figure 17-24*)

Equilibrium, p. 584

6. The **anterior, posterior,** and **lateral** semicircular ducts are continuous with the utricle. Each contains an **ampulla** with sensory receptors. Here the cilia contact a gelatinous **cupula.** (*Figures 17-24, 17-25*)
7. The saccule and utricle are connected by a passageway continuous with the **endolymphatic duct,** which terminates in the **endolymphatic sac.** In the saccule and utricle hair cells cluster within **maculae,** where their cilia contact **otoliths** (densely packed mineral crystals). When the head tilts, the mass of otoliths shifts, and the resulting distortion in the sensory hairs signals the CNS. (*Figures 17-25, 17-26*)
8. The vestibular receptors activate sensory neurons of the **vestibular ganglia.** The axons form the **vestibular branch** of the vestibulocochlear nerve (N VIII), synapsing within the **vestibular nuclei.** (*Figure 17-26*)

Hearing, p. 587

9. The energy content of a sound determines its **intensity,** measured in **decibels.** Sound waves travel toward the tympanic membrane, which vibrates; the auditory ossicles conduct the vibrations to the inner ear. Movement at the oval window applies pressure to the perilymph of the **vestibular duct.** (*Table 17-1*
10. Pressure waves distort the **basilar membrane** and push the hair cells of the **organ of Corti** against the **tectorial membrane.** The **tensor tympani** and **stapedius muscles** contract to reduce the amount of motion when very loud sounds arrive. (*Figures 17-27, 17-28*)
11. The sensory neurons are located in the **spiral ganglion** of the cochlea. Their afferent fibers form the **cochlear branch** of the vestibulocochlear nerve (N VIII), synapsing at the **cochlear nucleus.** (*Figure 17-29*)

REVIEW QUESTIONS

LEVEL 1 Reviewing Facts and Terms

1. Regardless of the nature of a stimulus, sensory information must be sent to the CNS in the form of:
   (a) dendritic processes
   (b) action potentials
   (c) neurotransmitter molecules
   (d) generator potentials

2. A membrane depolarization that leads to an action potential in a sensory neuron is called:
   (a) a labeled line
   (b) a sensation
   (c) a neurotransmitter
   (d) a generator potential

3. A reduction in sensitivity in the presence of a constant stimulus is called:
   (a) transduction (b) sensory coding
   (c) line labeling (d) adaptation

4. Sensations produced by an injection or a deep cut would reach the CNS very quickly by:
   (a) Type A fibers (b) Type C fibers
   (c) tactile receptors (d) proprioceptors

5. Mechanoreceptors that detect pressure changes in the walls of blood vessels and in portions of the digestive, reproductive, and urinary tracts are:
   (a) tactile receptors (b) baroreceptors
   (c) proprioceptors (d) free nerve endings

6. Pacinian corpuscles in the skin detect:
   (a) deep pressure (b) fine touch
   (c) muscle stretch (d) temperature changes

7. Examples of proprioceptors that monitor the position of joints and the state of muscular contraction are:
   (a) Pacinian and Meissner's corpuscles
   (b) carotid and aortic sinuses
   (c) Merkel's discs and Ruffini corpuscles
   (d) Golgi tendon organs and muscle spindles

8. When chemicals dissolve in the nasal cavity, they stimulate:
   (a) gustatory cells (b) olfactory hairs
   (c) rod cells (d) tactile receptors

9. The taste sensation of sweetness is experienced on the:
   (a) posterior part of the tongue
   (b) anterior part of the tongue
   (c) right and left lateral sides of the tongue
   (d) the middle part of the tongue

10. The anterior, transparent part of the fibrous tunic is called the:
   (a) cornea (b) iris
   (c) sclera (d) retina

11. Meibomian glands secrete a lipid-rich product that helps:
   (a) produce and secrete tears in the lacrimal apparatus
   (b) keep the cornea moist and clean
   (c) dilate the blood vessels beneath the conjunctival epithelium
   (d) keep the eyelids from sticking together

12. The purpose of tears produced by the lacrimal apparatus is to:
   (a) keep conjunctival surfaces moist and clean
   (b) reduce friction and remove debris from the eye
   (c) provide nutrients and oxygen to the conjunctional epithelium
   (d) a, b, and c are correct

13. The thickened gel-like fluid that helps support the structure of the eyeball is known as the:
   (a) vitreous humor (b) aqueous humor
   (c) ora serrata (d) perilymph

14. Pupillary muscle groups are controlled by the ANS. Parasympathetic activation causes pupillary ________, and sympathetic activation causes ________.
   (a) dilation, constriction
   (b) dilation, dilation
   (c) constriction, dilation
   (d) constriction, constriction

15. The retina is considered to be a component of the:
   (a) vascular tunic (b) fibrous tunic
   (c) neural tunic (d) a, b, and c are correct

16. At sunset or sunrise your visual system adapts to:
   (a) fovea vision (b) rod-based vision
   (c) macular vision (d) cone-based vision

17. The primary function of the lens of the eye is to:
   (a) stabilize the shape of the eye
   (b) facilitate communication between photoreceptors and ganglion cells
   (c) focus the visual image on the retinal photoreceptors
   (d) a, b, and c are correct

18. In rating the clarity of vision, a better-than-average visual acuity rating is:
   (a) 20/20 (b) 20/30
   (c) 15/20 (d) 20/15

19. Ceruminous glands along the external auditory canal are responsible for:
   (a) slowing the growth of microorganisms in the exterior auditory canal
   (b) reducing chances of infection in the ear
   (c) helping to deny access to foreign objects
   (d) a, b, and c are correct

20. The malleus, incus, and stapes are the tiny ear bones located in the:
   (a) outer ear (b) middle ear
   (c) inner ear (d) membranous labyrinth

21. Receptors in the sacculus and utriculus provide sensations of:
   (a) balance and equilibrium
   (b) hearing
   (c) vibration
   (d) gravity and linear acceleration

22. The organ of Corti is located within the ________ of the inner ear.
   (a) utricle (b) bony labyrinth
   (c) vestibule (d) cochlea

23. Auditory information concerning the region and intensity of stimulation is relayed to the CNS over the cochlear branch of:
   (a) N IV (b) N VI
   (c) N VIII (d) N X

24. What three steps are necessary for the process of transduction to occur?

25. What three types of mechanoreceptors respond to stretching, compression, twisting, or other distortions of the cell membrane?

26. Identify six types of tactile receptors found in the skin, and their sensitivities.

27. Trace the olfactory pathway from the time an odor reaches the olfactory epithelium until it reaches its final destination in the brain.

28. What three different types of papillae are located on the human tongue?

29. (a) What structures make up the fibrous tunic of the eye? (b) What are the functions of the fibrous tunic?

30. What structures are included as parts of the vascular tunic of the eye?

31. Trace the pathway of a nerve impulse from the optic nerve to the visual cortex.

32. What three auditory ossicles are found in the middle ear and what are their functions?

33. What six basic steps are involved in the process of hearing?

LEVEL 2 Reviewing Concepts

34. The CNS interprets sensory information entirely on the basis of the:
   (a) strength of the action potential
   (b) number of generator potentials
   (c) line over which it arrives
   (d) a, b, and c are correct

35. The conscious perception of pain is reduced in the body by:
   (a) temperature receptors in the dermis of the skin
   (b) suppression of nociception in the epidermis
   (c) endorphins preventing the release of substance P
   (d) reduction of mechanoreceptor activity in the skin

36. If the auditory cortex is damaged, the individual will respond to sounds and have normal acoustic reflexes, but:
   (a) the sounds may produce nerve deafness
   (b) the auditory ossicle may be immobilized
   (c) sound interpretation and pattern recognition may be impossible
   (d) normal transfer of vibration to the oval window is inhibited

37. Distinguish between the general senses and the special senses in the human body.

38. In what form does the CNS receive a stimulus detected by a sensory receptor?

39. What type of information about a stimulus does sensory coding provide?

40. Why are olfactory sensations long-lasting and an important part of our memories and emotions?

41. What is the usual result if a sebaceous gland of an eyelash or a Meibomian gland becomes infected?

42. Jane makes an appointment with the optometrist for a vision test. Her test results are reported as 20/15. What does this test result mean? Is a rating of 20/20 better or worse?

LEVEL 3 Critical Thinking and Clinical Applications

43. You are at a park watching some deer 35 feet away from you when your friend taps you on the shoulder to ask a question. As you turn to look at your friend who is standing 2 feet away, what changes would occur regarding your eyes?

44. Sally's driver's license indicates that she must wear glasses when driving. She does not need glasses to read or to see objects that are close. What type of lenses are in Sally's glasses?

45. After attending a Fourth of July fireworks extravaganza, Millie finds it difficult to hear normal conversation, and her ears keep "ringing." What is causing her hearing problems?

46. After riding the express elevator from the twentieth floor to the ground floor, for a few seconds you still feel as if you are descending, even though you have obviously come to a stop. Why?

47. A person suffering from a disorder involving the saccule and the utricle is asked to stand with feet together and arms extended forward. As long as such people keep their eyes open they exhibit very little movement. When they are asked to close their eyes, the body begins to sway a great deal, and the arms tend to drift in the direction of the impaired vestibular receptors. Why does this occur?

CHAPTER 18

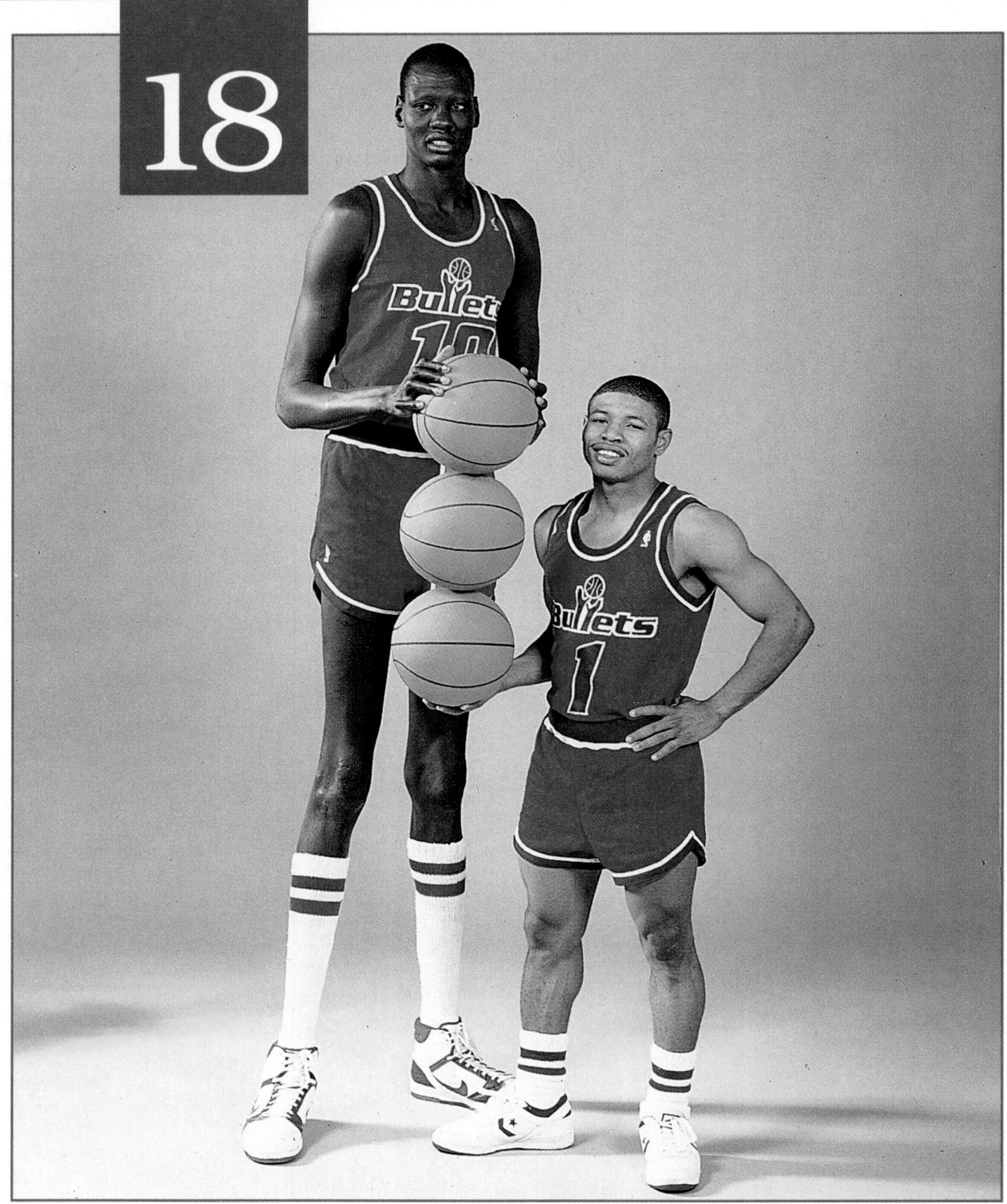

Bullets
1

These two basketball players, Manute Bol and Muggsy Bogues, differ in height by 0.7 m—more than 2 feet. Differences in genetically programmed levels of hormones, especially growth hormone, were probably responsible for this difference. Hormone levels during childhood and adolescence provided Manute with some physical advantages, but Manute and Muggsy both succeed in the NBA primarily because they have the right blend of coordination, balance, strength, speed, and intelligence. This is a common pattern: Hormones direct long-term changes (such as those that determine body form), while moment-to-moment adjustments (which make basketball games exciting) are controlled by the nervous system. This chapter introduces the major hormones and the endocrine organs that produce them.

The Endocrine System

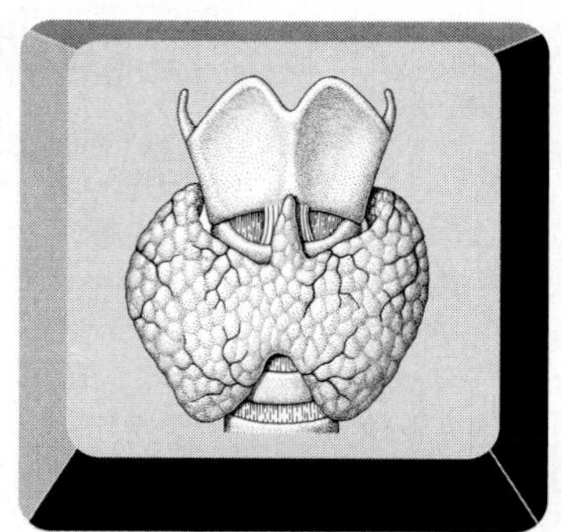

Chapter Outline and Objectives

Throughout life you communicate with other people. You talk with those around you, and you call and write to friends and relatives in distant locations. Roommates usually hear about everything that affects you, because you are in constant communication. Phone calls are direct and immediate, but they last only a short time, and your friends hear only the highlights of a few specific topics. In long letters, you can cover more topics in depth, but the information will not reach your correspondents for several days, and you will have to wait even longer for their replies.

To function effectively, every cell in the body must communicate with its neighbors and with cells and tissues in distant portions of the body. Most of the communication involves chemical messages. Each living cell is continually "talking" to its neighbors by releasing chemicals, often called *cytokines*, into the extracellular fluid. These chemicals let cells know what their neighbors are doing at any given moment, and the result is the coordination of tissue function at the local level. Examples of these "local hormones" include the various *growth factors* discussed in Chapter 3. [p. 103]

The nervous system acts like a telephone company, carrying high-speed "messages" from a receptor in one location to an effector in another. The source and the destination are quite specific, and the effects are generally short-lived. This form of communication is ideal for crisis management; if you are in danger of being hit by a speeding bus, the nervous system can coordinate and direct your leap to safety. Once the crisis is over and the neural circuit quiets down, things soon return to normal.

In cellular communication, hormones are long letters, and the circulatory system is the postal service. A hormone released into the circulation will be distributed throughout the body. Each hormone has specific *target cells* that will respond to its presence. These cells possess the receptors needed to bind and "read" the hormonal message. Although every cell in the body is exposed to the mixture of hormones in circulation at any given moment, a single cell will respond to only a few of the hormones present. The other hormones are treated like junk mail, and ignored, because the cell lacks the receptors to read the messages they contain.

Because the target cells can be anywhere in the body, a single hormone can alter the metabolic activities of multiple tissues and organs simultaneously. These effects may be slow to appear, but they often persist for days. Consequently, hormones are effective in coordinating cell, tissue, and organ activities on a sustained, long-term basis. For example, circulating hormones keep body water content and levels of electrolytes and organic nutrients within normal limits 24 hours a day throughout our entire lives.

Although the effects of a single hormone persist, a cell may receive additional instructions from other hormones. The result will be a further modification in cellular operations. Gradual changes in the quantities and identities of circulating hormones can produce complex changes in physical structure and physiological capabilities. Examples include the processes of embryological and fetal development, growth, and puberty.

When viewed from a general perspective, the differences between the nervous and endocrine systems seem relatively clear. In fact, these broad organizational and functional distinctions are the basis for treating them as two separate systems. Yet when considered in detail, the two systems are organized along parallel lines. For example:

- Both systems rely on the release of chemicals that bind to specific receptors on their target cells.
- The two systems use many of the same chemical messengers; for example, norepinephrine and epinephrine are called hormones when released into the circulation and neurotransmitters when released across synapses.
- Both systems are regulated primarily by negative feedback control mechanisms.
- Both systems share a common goal: to coordinate and regulate the activities of other cells, tissues, organs, and systems and preserve homeostasis.

This chapter introduces the components and functions of the endocrine system and explores the interactions between the nervous and endocrine systems. Subsequent chapters will consider specific endocrine organs, hormones, and functions in greater detail.

An Overview of the Endocrine System

Figure 18-1

The endocrine system includes all of the endocrine cells and tissues of the body. As noted in Chapter 4, *endocrine cells* are glandular secretory cells that release their secretions into the extracellular fluid. This characteristic distinguishes them from *exocrine cells*, which secrete onto epithelial surfaces. [p. 119]

The chemicals released by endocrine cells may affect only adjacent cells, as in the case of most prostaglandins, or they may affect cells throughout the body. **Paracrine factors,** or "local hormones," are chemicals that affect other cells only in the tissue of origin. This category includes most prostaglandins and related compounds. **Hormones** are chemical messengers that are released in one tissue and transported via the circulation to reach target cells in other tissues.

The components of the endocrine system are introduced in Figure 18-1●. This figure also lists the major hormones produced in each endocrine tissue and organ. Some of these organs, such as the pituitary gland, have endocrine secretion as a primary function; others, such as the pancreas, have many other functions in addition to endocrine secretion. The structure and functions of these organs will also be considered in chapters dealing with other systems. Examples include the hypothalamus (Chapter 14), the adrenal medulla (Chapter 16), the heart (Chapter 20), the thymus (Chapter 22), the pancreas and digestive tract (Chapter 24), the kidneys (Chapter 26), the reproductive organs (Chapter 28), and the placenta (Chapter 29).

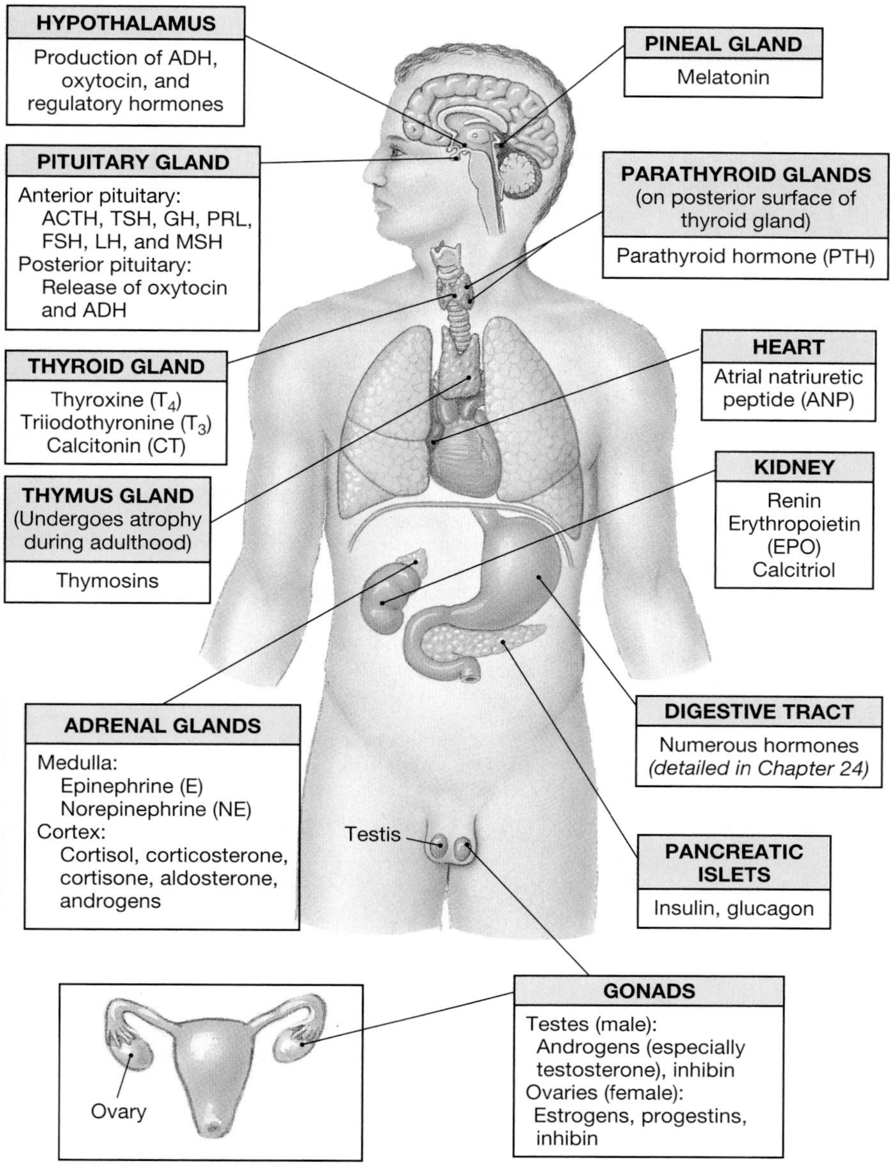

❑ Hormone Structure

Figure 18-2

Hormones can be divided into three different groups on the basis of chemical structure: *amino acid derivatives*, *peptide hormones*, and *lipid derivatives*. Examples of each group are included in Figure 18-2●.

Amino Acid Derivatives

Some hormones are relatively small molecules that are structurally similar to amino acids. (Amino acids, the building blocks of proteins, were introduced in Chapter 2.) ∞ *[p. 53]* This group includes *epinephrine, norepinephrine, dopamine,* the *thyroid hormones*, and the pineal hormone *melatonin*. Epinephrine, norepinephrine, and dopamine are structurally similar; these compounds are sometimes called **catecholamines** (kat-e-KŌ-la-mēnz). As we saw in Chapter 16, epinephrine and norepinephrine are secreted by the adrenal medulla during sympathetic activation. ∞ *[p. 534]* Dopamine, secreted by the hypothalamus, assists in the regulation of pituitary gland function. ∞ *[p. 410]* **Thyroid hormones** are released by the thyroid gland, and **melatonin** is produced by the pineal gland.

As indicated in Figure 18-2●, catecholamines and thyroid hormones are synthesized from molecules of the amino acid **tyrosine** (TĪ-rō-sēn). Melatonin is manufactured from molecules of the amino acid *tryptophan*.

Peptide Hormones

Figure 18-3

Peptide hormones are chains of amino acids. They range from short amino acid chains, such as ADH and oxytocin (9 amino acids apiece), to polypeptides, such as *growth hormone* (191 amino acids) and *prolactin* (198 amino acids). This is the largest class of hormones and includes all of the hormones secreted by the hypothalamus, pituitary gland, heart, kidneys, thymus, digestive tract, and pancreas.

Figure 18-3● diagrams stages in the synthesis of peptide hor-

HYPOTHALAMUS
Production of ADH, oxytocin, and regulatory hormones
PITUITARY GLAND
Anterior pituitary: ACTH, TSH, GH, PRL, FSH, LH, and MSH
Posterior pituitary: Release of oxytocin and ADH
THYROID GLAND
Thyroxine (T_4)
Triiodothyronine (T_3)
Calcitonin (CT)
THYMUS GLAND (Undergoes atrophy during adulthood)
Thymosins
ADRENAL GLANDS
Medulla: Epinephrine (E) Norepinephrine (NE)
Cortex: Cortisol, corticosterone, cortisone, aldosterone, androgens
Testis
Ovary
PINEAL GLAND
Melatonin
PARATHYROID GLANDS (on posterior surface of thyroid gland)
Parathyroid hormone (PTH)
HEART
Atrial natriuretic peptide (ANP)
KIDNEY
Renin
Erythropoietin (EPO)
Calcitriol
DIGESTIVE TRACT
Numerous hormones (detailed in Chapter 24)
PANCREATIC ISLETS
Insulin, glucagon
GONADS
Testes (male): Androgens (especially testosterone), inhibin
Ovaries (female): Estrogens, progestins, inhibin

● **FIGURE 18-1**
The Endocrine System

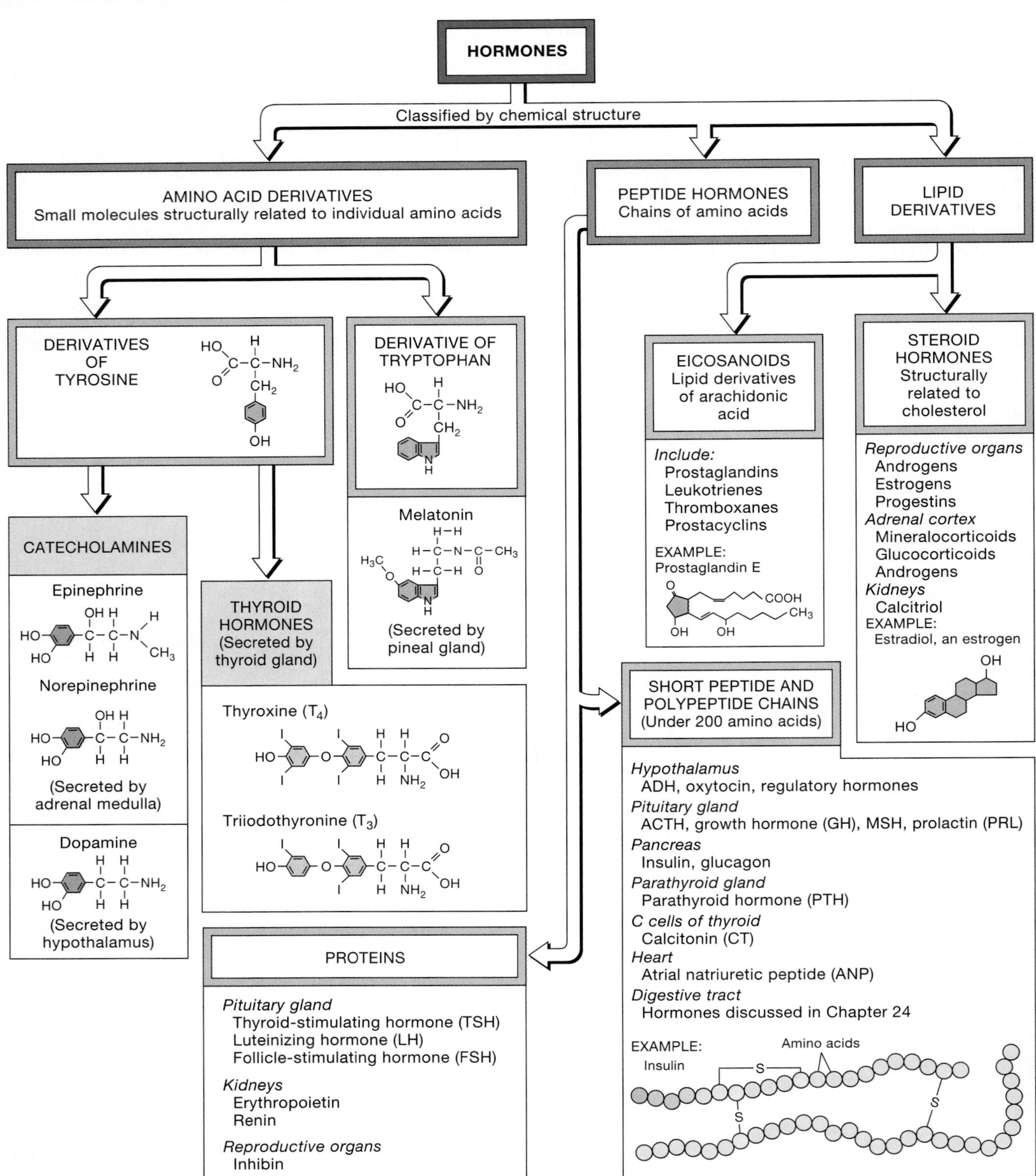
HORMONES
Classified by chemical structure
AMINO ACID DERIVATIVES
Small molecules structurally related to individual amino acids
PEPTIDE HORMONES
Chains of amino acids
LIPID DERIVATIVES
DERIVATIVES OF TYROSINE
DERIVATIVE OF TRYPTOPHAN
EICOSANOIDS
Lipid derivatives of arachidonic acid
STEROID HORMONES
Structurally related to cholesterol
CATECHOLAMINES
Epinephrine
Norepinephrine
(Secreted by adrenal medulla)
Dopamine
(Secreted by hypothalamus)
THYROID HORMONES
(Secreted by thyroid gland)
Thyroxine (T_4)
Triiodothyronine (T_3)
Melatonin
(Secreted by pineal gland)
Include:
Prostaglandins
Leukotrienes
Thromboxanes
Prostacyclins
EXAMPLE:
Prostaglandin E
COOH
CH_3
Reproductive organs
Androgens
Estrogens
Progestins
Adrenal cortex
Mineralocorticoids
Glucocorticoids
Androgens
Kidneys
Calcitriol
EXAMPLE:
Estradiol, an estrogen
SHORT PEPTIDE AND POLYPEPTIDE CHAINS
(Under 200 amino acids)
Hypothalamus
ADH, oxytocin, regulatory hormones
Pituitary gland
ACTH, growth hormone (GH), MSH, prolactin (PRL)
Pancreas
Insulin, glucagon
Parathyroid gland
Parathyroid hormone (PTH)
C cells of thyroid
Calcitonin (CT)
Heart
Atrial natriuretic peptide (ANP)
Digestive tract
Hormones discussed in Chapter 24
EXAMPLE:
Insulin
Amino acids
PROTEINS
Pituitary gland
Thyroid-stimulating hormone (TSH)
Luteinizing hormone (LH)
Follicle-stimulating hormone (FSH)
Kidneys
Erythropoietin
Renin
Reproductive organs
Inhibin

● **FIGURE 18-2**
A Structural Classification of Hormones

mones. In general, the process follows the pattern of protein synthesis and secretion discussed in Chapter 3. [p. 94] Synthesis begins at ribosomes of the rough endoplasmic reticulum (RER) of glandular cells. The peptide chains produced at the RER are called **prehormones** because they are structurally distinct from the final hormone product. For example, the prehormone may contain extra amino acids, or it may be missing important carbohydrate components.

Further modification in the endoplasmic reticulum converts the prehormone to a **prohormone,** an inactive form of the hormone. Final conversion to an active hormone may occur (1) prior to release, as the prohormone passes through the Golgi apparatus and secretory vesicles, or (2) after release, within the bloodstream. Once a functional peptide hormone appears, it remains in circulation for a relatively short period (typically less than 30 minutes). By that time the hormone will either (1) leave the circulation and bind to receptors in target tissues, (2) be absorbed and broken down by cells in the liver or kidneys, or (3) be broken down by enzymes in the plasma or interstitial fluids.

Lipid Derivatives

There are two classes of lipid-based hormones: (1) *steroid hormones,* derived from cholesterol, and (2) *eicosanoids* (ī-KŌ-sa-noyds), derived from *arachidonic acid,* a 20-carbon fatty acid.

STEROID HORMONES **Steroid hormones** are lipids structurally similar to cholesterol (Figure 2-15•, p. 50). Steroid hormones are released by male and female reproductive organs, the adrenal glands, and the kidneys. The individual hormones differ in the side chains attached to the central ring structure.

In the blood, steroid hormones are bound to specific transport proteins in the plasma. For this reason they remain in circulation longer than peptide hormones. The liver gradually absorbs these steroids and converts them to a soluble form that can be excreted in the bile or urine.

EICOSANOIDS **Eicosanoids** are small molecules with a five-carbon ring at one end. These compounds are important paracrine factors that coordinate cellular activities and affect enzymatic processes, such as blood clotting, that occur in extracellular fluids. Some of the eicosanoids also have secondary roles as hormones. Examples of important eicosanoids include the following:

- **Prostaglandins** are produced in most tissues of the body. Within each tissue, the prostaglandins released are involved primarily in coordinating local cellular activities.
- **Leukotrienes** are released by activated white blood cells, or *leukocytes.* Leukotrienes are important in coordinating tissue responses to injury or disease.
- **Thromboxanes** (throm-BOX-ānz) and **prostacyclins** (pros-ta-SĪ-klinz) are released by circulating *platelets* when blood clotting occurs. Platelets, normal components of the circulating blood, were introduced in Chapter 4. [p. 128]

Our focus in this chapter will be on circulating hormones whose primary functions are the coordination of activities in many different tissues and organs. Eicosanoids will be considered in chapters dealing with individual tissues and organs, including Chapters 19 (the blood), 22 (the lymphatic system), and 28 (the reproductive system).

• **FIGURE 18-3**
Peptide Hormone Production. Activated genes in the nucleus are transcribed, producing mRNA that travels to the rough endoplasmic reticulum. Translation at ribosomes of the RER then produces molecules of a prehormone. These molecules undergo further modification in the RER and Golgi apparatus. Final modifications of the prohormone are made within the secretory vesicles or after release via exocytosis.

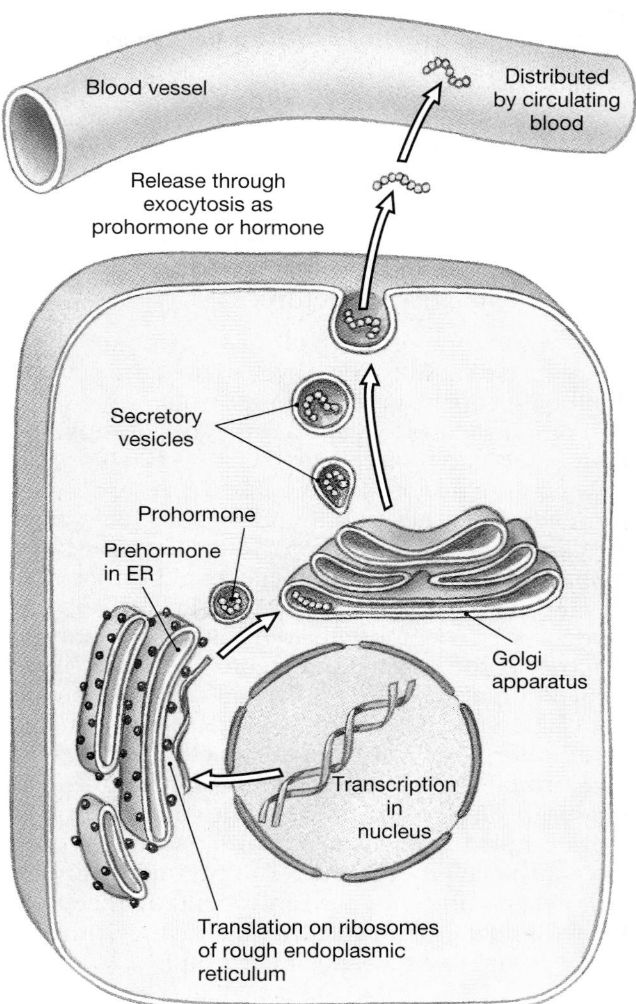
Blood vessel
Distributed by circulating blood
Release through exocytosis as prohormone or hormone
Secretory vesicles
Prohormone
Prehormone in ER
Golgi apparatus
Transcription in nucleus
Translation on ribosomes of rough endoplasmic reticulum

Hormone Function and Mechanisms of Action

All cellular structures and functions depend on proteins. Structural proteins determine the general shape and internal structure of a cell, and enzymes direct its metabolic activities. ∞ [p. 52] Hormones alter cellular operations by changing the *types, activities,* or *quantities* of important enzymes and structural proteins. In other words, a hormone may:

- Stimulate the synthesis of an enzyme or a structural protein not already present in the cytoplasm by activating appropriate genes in the nucleus.
- Increase or decrease the rate of synthesis of a particular enzyme or other protein by changing the rate of transcription or translation.
- Turn an existing enzyme "on" or "off" by changing its shape or structure.

Through one or more of these mechanisms, a hormone can modify the physical structure or biochemical properties of its target cells.

For a hormone to affect a target cell, it must first interact with an appropriate receptor. Each cell has the receptors needed to respond to several different hormones, but cells in different tissues have different combinations of receptors. This arrangement accounts for the differential effects of hormones on specific tissues. For every cell, the presence or absence of a receptor determines its hormonal sensitivities. If the cell has a receptor that will bind a particular hormone, that cell will respond to the hormone's presence; if the cell lacks the proper receptor, the hormone will have no effect.

Hormone receptors are found both on the cell membrane and inside the cell. The catecholamines, peptide hormones, and eicosanoids target receptors on the cell membrane. The thyroid hormones and steroids interact with intracellular receptors. We will now introduce the basic mechanisms involved, using a few specific examples.

Hormones and the Cell Membrane

Figure 18-4

The receptors for catecholamines (epinephrine, norepinephrine, dopamine), peptide hormones, and eicosanoids are found in the cell membranes of their respective target cells. These hormones do not exert their effects directly. For example, a hormone does not enter a cell and begin building a protein or catalyzing a specific reaction. The hormone acts as a **first messenger** that causes the appearance of a **second messenger** in the cytoplasm. The second messenger may function as an enzyme activator, inhibitor, or cofactor, but the net result is a change in the rates of various metabolic reactions. The most important second messengers are (1) *cyclic-AMP,* a derivative of ATP, (2) *cyclic-GMP,* a derivative of GTP, another high-energy compound, and (3) calcium ions.

The binding of a few hormone molecules to their membrane receptors may lead to the release of thousands of second messengers within a cell. This process, which magnifies the effect of the hormone on the cell, is called **amplification.** In addition, the arrival of a single hormone may promote the release of more than one type of second messenger. Thus the hormone can alter multiple aspects of cell function, all at the same time.

G-PROTEINS The link between the first messenger and the second messenger usually involves a **G-protein,** an integral membrane protein that interacts with the membrane receptor. There are several different types of G-proteins, but in each case the G-protein is activated when a hormone binds to its receptor at the membrane surface. Figure 18-4● diagrams important mechanisms that may be set in motion by activation of a G-protein. Each sequence affects levels of a second messenger in the cytoplasm.

1. Activation of adenylate cyclase. **Adenylate cyclase,** also called *adenyl cyclase* or *adenylyl cyclase,* converts ATP to a ring-shaped molecule of **cyclic-AMP** (cAMP). Cyclic-AMP then functions as the second messenger, opening ion channels or activating key enzymes in the cytoplasm. Many hormones encountered later in this chapter, including calcitonin, parathyroid hormone, ADH, ACTH, epinephrine, FSH, LH, TSH, and glucagon, produce their effects by increasing intracellular concentrations of cyclic-AMP. The increase is usually short-lived, because the cytoplasm contains an enzyme, **phosphodiesterase (PDE),** that inactivates cAMP by converting it to AMP (adenosine monophosphate).

2. Entry and/or release of calcium ions. An activated G-protein can trigger the opening of calcium ion channels in the membrane or the release of calcium ions from intracellular stores. The G-protein first activates **phospholipase C** (PLC), an enzyme that generates the second messengers **diacylglycerol** (DAG) and **inositol triphosphate** (IP_3) from membrane phospholipids known as *phosphatidylinositols* (fos-fa-ti-dil-i-NOS-i-tols) (PIP).

Step 1. Inositol triphosphate diffuses into the cytoplasm and triggers the release of calcium ions from intracellular reserves, such as those found in the smooth endoplasmic reticulum of many cells.

Step 2. The combination of diacylglycerol and intracellular calcium ions activates another membrane protein, **protein kinase C** (PKC). A *kinase* is an enzyme that performs *phosphorylation,* the attachment of a phosphate group (PO_4^{3-}) to another molecule. Activation of protein kinase C leads to the phosphorylation of calcium channel proteins. This opens the channels and permits the entry of extracellular calcium ions.

Step 3. The calcium ions then serve as second messengers, usually in combination with an intracellular protein called **calmodulin.** Once it has bound calcium ions, calmodulin can activate specific cytoplasmic enzymes. This chain of events is responsible for the stimulatory effects that follow activation of α_1 receptors by epinephrine or norepinephrine. *[p. 535]* Calmodulin activation is also involved in the responses to oxytocin and several regulatory hormones secreted by the hypothalamus.

3. Inhibition of cellular activities. There are also inhibitory G-proteins whose activation leads to a reduction in second messenger levels inside the cell. In the example shown, the activated G-protein inhibits adenylate cyclase activity. Levels of cAMP then decline because *phosphodiesterase*, an enzyme that breaks down cGMP and cAMP, is present in the cytoplasm. This decline has an inhibitory effect on the cell. This is the mechanism responsible for the inhibitory effects that follow stimulation of α_2 receptors, as discussed in Chapter 16. *[p. 535]* A different inhibitory effect was introduced in Chapter 17. *Transducin* is a G-protein that is activated by membrane events associated with photoreception. Transducin activates additional molecules of phosphodiesterase, which lowers cGMP levels and inhibits the cell. *[p. 575]*

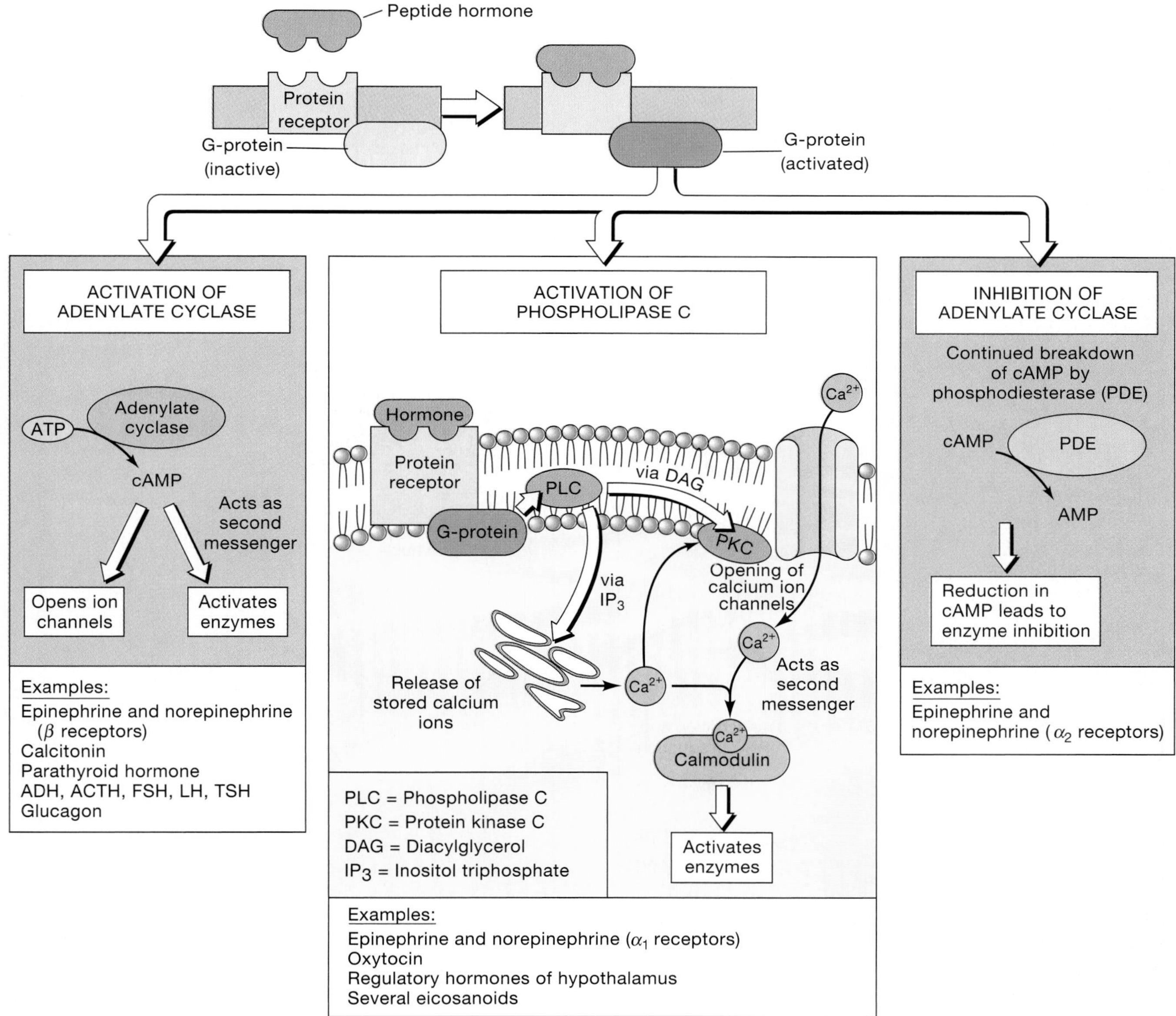
Peptide hormone
Protein receptor
G-protein (inactive)
G-protein (activated)
ACTIVATION OF ADENYLATE CYCLASE
ATP
Adenylate cyclase
cAMP
Acts as second messenger
Opens ion channels
Activates enzymes
Examples:
Epinephrine and norepinephrine (β receptors)
Calcitonin
Parathyroid hormone
ADH, ACTH, FSH, LH, TSH
Glucagon
ACTIVATION OF PHOSPHOLIPASE C
Hormone
Protein receptor
G-protein
PLC
via DAG
PKC
Ca^{2+}
via IP_3
Opening of calcium ion channels
Release of stored calcium ions
Acts as second messenger
Calmodulin
Activates enzymes
PLC = Phospholipase C
PKC = Protein kinase C
DAG = Diacylglycerol
IP_3 = Inositol triphosphate
Examples:
Epinephrine and norepinephrine (α_1 receptors)
Oxytocin
Regulatory hormones of hypothalamus
Several eicosanoids
INHIBITION OF ADENYLATE CYCLASE
Continued breakdown of cAMP by phosphodiesterase (PDE)
cAMP
PDE
AMP
Reduction in cAMP leads to enzyme inhibition
Examples:
Epinephrine and norepinephrine (α_2 receptors)

FIGURE 18-4

Mechanisms of Peptide Hormone Activity. Peptide hormones act by binding to membrane receptors and activating G-proteins. Three possible results are shown, each affecting the concentration of second messengers in the cytoplasm. G-protein activation may lead to (1) activation of adenylate cyclase and formation of cyclic-AMP, (2) opening of calcium ion channels and calcium release into the cytoplasm, or (3) inhibition of cellular activity by reduction of second messenger concentrations.

Hormones and Intracellular Receptors

Figure 18-5

Thyroid hormones and steroid hormones cross the cell membrane and bind to intracellular receptors. Steroid hormones diffuse across the lipid portion of the cell membrane and bind to receptors in the cytoplasm or nucleus (Figure 18-5a●). The hormone-receptor complex then binds to DNA segments called **hormone-responsive elements (HREs).** This event triggers the activation or inactivation of specific genes.

This mechanism enables steroid hormones to alter the rate of DNA transcription in the nucleus, thereby changing the pattern of protein synthesis. This may alter cell structure or, if enzymes are involved, the metabolic activity of the target cell. For example, the sex hormone *testosterone* stimulates the production of enzymes and proteins in skeletal muscle fibers, causing an increase in muscle size and strength.

Thyroid hormones cross the cell membrane via diffusion or by an unidentified carrier mechanism. Once within the cytosol, these hormones bind to receptors within the nucleus (Figure 18-5b●). The hormone-receptor complex then activates specific genes and alters transcription rates. This alteration affects the metabolic activities of the cell by changing the nature or number of enzyme molecules in the cytoplasm. Thyroid hormones also have a direct effect on mitochondria, increasing their rates of ATP production.

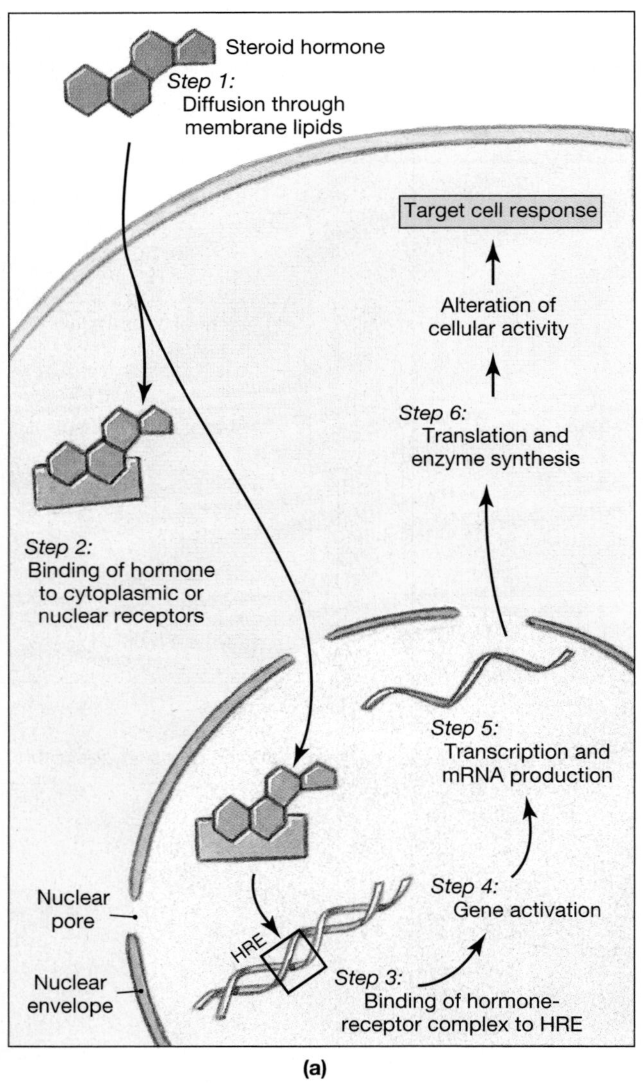
Steroid hormone
Step 1:
Diffusion through membrane lipids
Target cell response
Alteration of cellular activity
Step 6:
Translation and enzyme synthesis
Step 2:
Binding of hormone to cytoplasmic or nuclear receptors
Step 5:
Transcription and mRNA production
Step 4:
Gene activation
Nuclear pore
HRE
Nuclear envelope
Step 3:
Binding of hormone-receptor complex to HRE

(a)

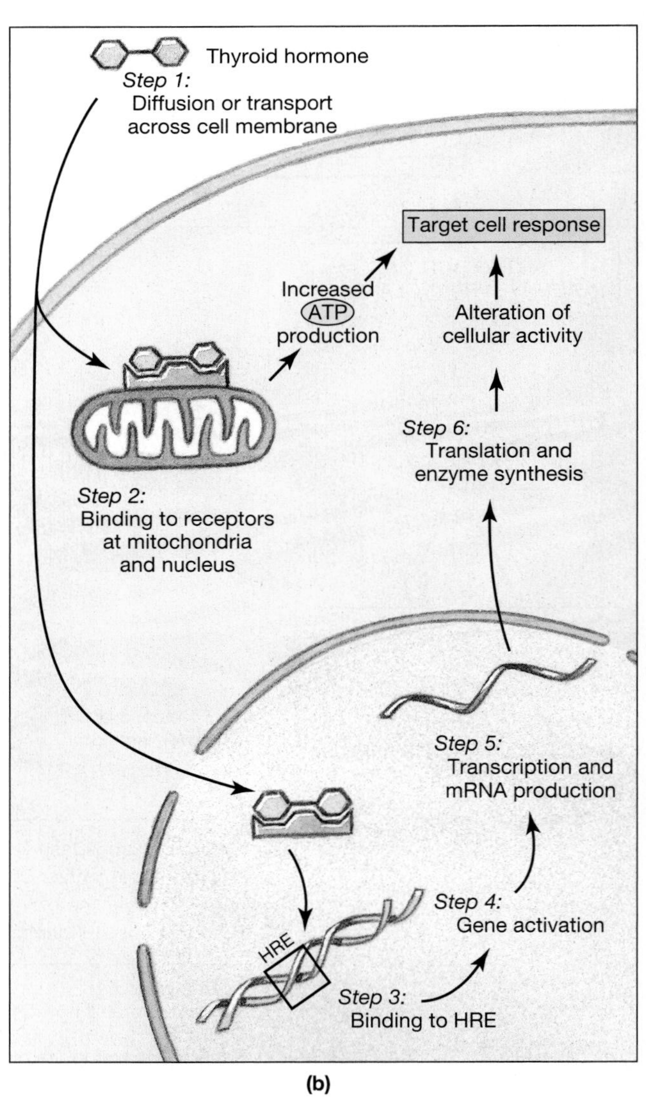
Thyroid hormone
Step 1:
Diffusion or transport across cell membrane
Target cell response
Increased ATP production
Alteration of cellular activity
Step 6:
Translation and enzyme synthesis
Step 2:
Binding to receptors at mitochondria and nucleus
Step 5:
Transcription and mRNA production
Step 4:
Gene activation
HRE
Step 3:
Binding to HRE

(b)

● **FIGURE 18-5**

Hormone Effects on Gene Activity. (a) Steroid hormones diffuse through the membrane lipids and bind to receptors in the cytoplasm or nucleus. The complex then binds to HREs in the nucleus and activates specific genes. **(b)** Thyroid hormones enter the cytoplasm and bind to receptors in the nucleus to activate specific genes. They also bind to receptors on mitochondria.

Control of Endocrine Activity

Figure 18-6a

There are many functional parallels between the organization of the nervous and endocrine systems. Chapter 13 considered the basic operation of neural reflex arcs, the simplest organizational units in the nervous system. [p. 441] The most direct arrangement was a monosynaptic reflex, such as the stretch reflex diagrammed in Figure 18-6a•. Polysynaptic reflexes provide more complex and variable responses to stimuli, and higher centers that integrate multiple inputs can facilitate or inhibit these reflexes as needed.

Endocrine Reflexes

Figure 18-6

Endocrine reflexes are the functional counterparts of neural reflexes. In most cases, these reflexes are controlled by negative feedback mechanisms: A stimulus triggers production of a hormone whose activities lead to a reduction in the magnitude of the stimulus.

Endocrine cells in a simple endocrine reflex (Figure 18-6b•) involve only one hormone. The endocrine cells involved respond directly to changes in the composition of the extracellular fluid. The hormone secreted adjusts the activities of target cells and restores homeostasis. Simple endocrine reflexes control hormone secretion by the heart, pancreas, parathyroid gland, and digestive tract.

More complex endocrine reflexes involve one or more intermediary steps and two or more hormones. The hypothalamus, the highest level of endocrine control, integrates activities of the nervous and endocrine systems. This integration involves three different mechanisms:

1. The hypothalamus secretes **regulatory hormones,** or *regulatory factors,* special hormones that control endocrine cells in the pituitary gland. The basic pattern of regulation is detailed in Figure 18-6c•. The hypothalamic regulatory factors control the secretory activities of endocrine cells in the pituitary gland. The hormones released by the pituitary, in turn, control the activities of endocrine cells in the thyroid, adrenal cortex, and reproductive organs. This arrangement closely parallels that of a complex neural reflex: Cells in the hypothalamus act as receptors, the endocrine organs are processing centers, and the target cells are effectors. The primary difference is that endocrine reflexes use hormones rather than axons to carry information and instructions from place to place.
2. The hypothalamus contains autonomic centers that exert direct neural control over the endocrine cells of the adrenal medullae. When the sympathetic division is activated, the adrenal medullae release hormones into the bloodstream (Figure 18-6d•).
3. The hypothalamus releases hormones into the circulation at the posterior pituitary. Two of these hormones, ADH and oxytocin, were introduced in Chapter 14. [p. 478]

The hypothalamus secretes ADH and regulatory hormones in response to changes in the composition of the circulating blood. The pathways leading to secretion of epinephrine, norepinephrine, and oxytocin involve both neural and hormonal components. For example, the commands issued to the adrenal medulla that trigger secretion of epinephrine and norepinephrine arrive in the form of action potentials along efferent fibers, rather than as circulating hormones. Pathways like this are called **neuroendocrine reflexes** (Figure 18-6d,e•) because they include both neural and endocrine components. We will consider these reflex patterns in more detail as we deal with specific endocrine tissues and organs.

✓ How could you distinguish between a neural response and an endocrine response on the basis of response time and response duration?

✓ How would the presence of a molecule that blocks adenylate cyclase affect the activity of a hormone that produces its cellular effects by way of the second messenger cAMP?

✓ What primary factor determines each cell's hormonal sensitivities?

The Pituitary Gland

Figure 18-7

Figure 18-7• details the anatomical organization of the pituitary gland, or **hypophysis** (hī-POF-i-sis). This small, oval gland lies nestled within the *sella turcica,* a depression in the sphenoid bone (Figure 7-8•, p. 212). The pituitary gland hangs beneath the hypothalamus, connected by a slender stalk, the **infundibulum** (in-fun-DIB-ū-lum; funnel). The base of the infundibulum lies between the optic chiasm and the mamillary bodies. The pituitary gland is cradled by the sella turcica and held in position by the *diaphragma sellae,* a dural sheet that encircles the infundibulum. The diaphragma sellae locks the pituitary in position and isolates it from the cranial cavity.

The pituitary gland can be divided into posterior and anterior divisions on anatomical and developmental grounds (see the Embryology Summary on pp. 632–633). Nine important peptide hormones are released by the pituitary gland, seven by the anterior pituitary and two at the posterior pituitary. All of these hormones bind to membrane receptors and use cyclic-AMP as a second messenger.

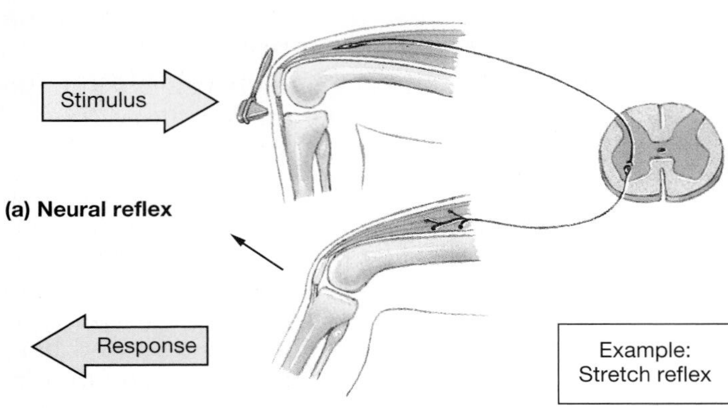
Stimulus
Response
Example:
Stretch reflex

(a) Neural reflex

FIGURE 18-6

Patterns of Endocrine Activity and Control. **(a)** A neural reflex, for comparison with endocrine control mechanisms. **(b)** A simple endocrine reflex. The endocrine cell responds to a change in the composition of the extracellular fluid, and the hormone released stimulates a target cell to respond in a way that restores homeostasis. **(c)** A more complex endocrine reflex, involving the hypothalamus. In this case the stimulus may be a chemical change in the blood or some change in CNS activity. In response, the hypothalamus produces a regulatory hormone that controls the hormonal output of the pituitary gland. In a neuroendocrine reflex a sensory or motor axon is part of the reflex arc, as in **(d)** the control of secretion by the adrenal medullae and **(e)** milk ejection at the mammary glands.

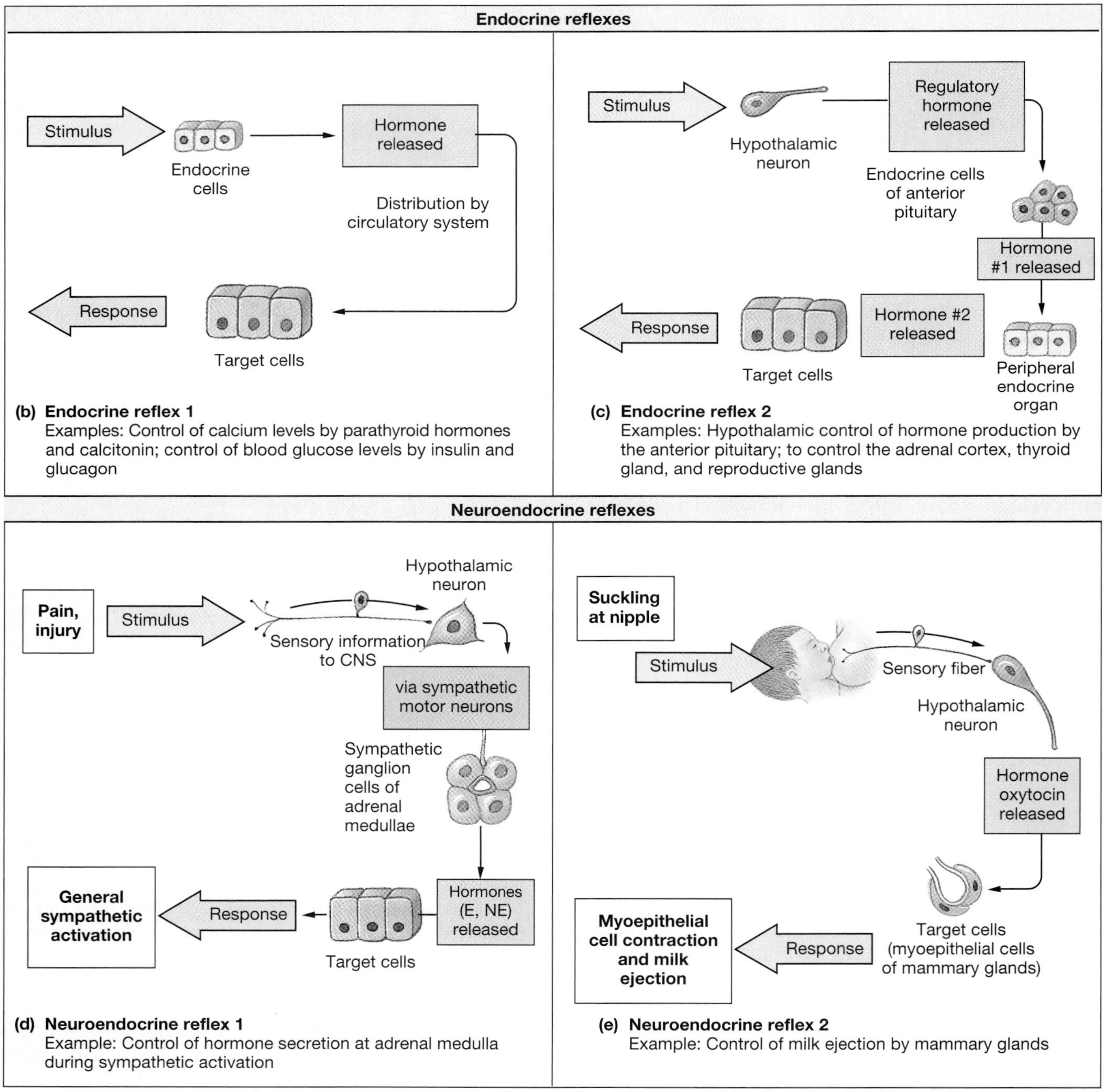

Endocrine reflexes

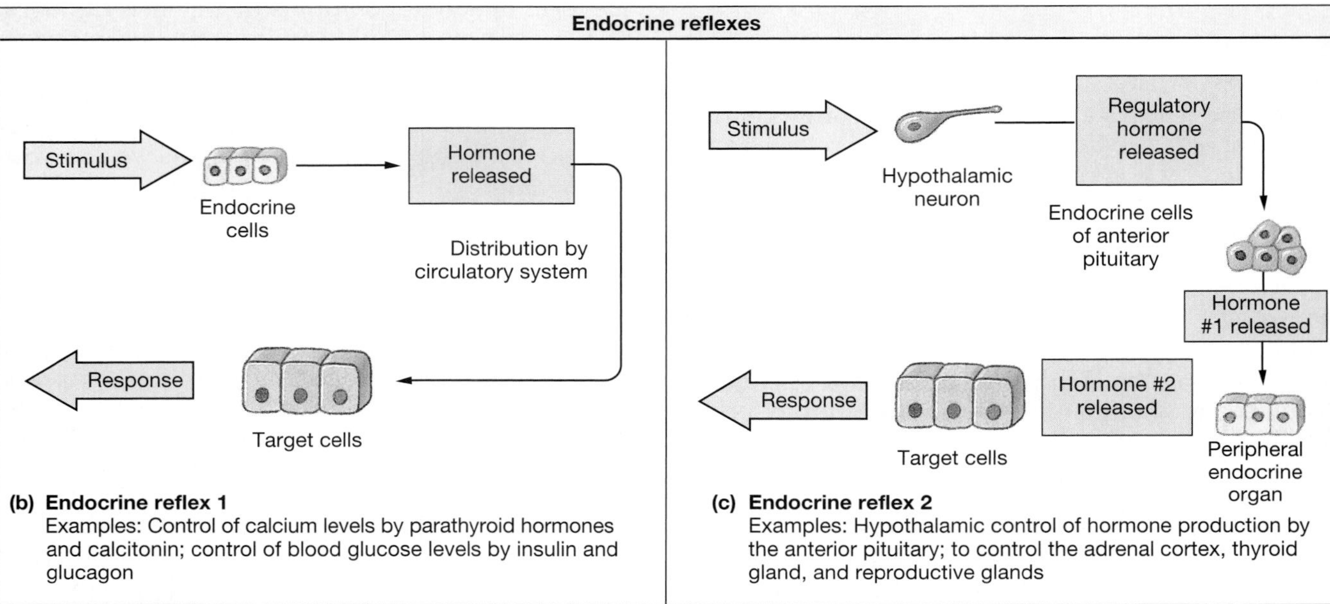
Stimulus
Endocrine
cells
Hormone
released
Distribution by
circulatory system
Response
Target cells
Stimulus
Hypothalamic
neuron
Regulatory
hormone
released
Endocrine cells
of anterior
pituitary
Hormone
#1 released
Hormone #2
released
Response
Target cells
Peripheral
endocrine
organ

(b) Endocrine reflex 1
Examples: Control of calcium levels by parathyroid hormones and calcitonin; control of blood glucose levels by insulin and glucagon

(c) Endocrine reflex 2
Examples: Hypothalamic control of hormone production by the anterior pituitary; to control the adrenal cortex, thyroid gland, and reproductive glands

Neuroendocrine reflexes

Pain, injury

Stimulus

Hypothalamic neuron

Sensory information to CNS

via sympathetic motor neurons

Sympathetic ganglion cells of adrenal medullae

Hormones (E, NE) released

General sympathetic activation

Response

Target cells

(d) Neuroendocrine reflex 1
Example: Control of hormone secretion at adrenal medulla during sympathetic activation

Suckling at nipple

Stimulus

Sensory fiber

Hypothalamic neuron

Hormone oxytocin released

Target cells (myoepithelial cells of mammary glands)

Myoepithelial cell contraction and milk ejection

Response

(e) Neuroendocrine reflex 2
Example: Control of milk ejection by mammary glands

The Anterior Pituitary

Figure 18-7

The **anterior pituitary,** or **adenohypophysis** (ad-ē-nō-hī-POF-i-sis), contains a variety of endocrine cell types. The anterior pituitary can be subdivided into two regions: (1) a large **pars distalis** (dis-TAL-is; distal part), which represents the major portion of the entire pituitary gland, and (2) a slender **pars intermedia** (in-ter-MĒ-dē-a; intermediate part), which forms a narrow band bordering the posterior pituitary (Figure 18-7●). Both regions of the anterior pituitary contain an extensive capillary network.

The Hypophyseal Portal System

Figure 18-8

The production of hormones in the anterior pituitary is controlled by the hypothalamus through the secretion of specific regulatory factors. At the *median eminence,* a swelling near the attachment of the infundibulum, hypothalamic neurons release regulatory factors into the surrounding interstitial fluids. Their secretions enter the circulation quite easily because the endothelial cells lining the capillaries in this region are unusually permeable. These capillaries, termed **fenestrated** (FEN-es-trā-ted) **capillaries** (*fenestra,* window), allow relatively large molecules to enter or leave the circulatory system. The capillary networks in the median eminence are supplied by the **superior hypophyseal artery.**

Before leaving the hypothalamus, the capillary network unites to form a series of larger vessels that spiral around the infundibulum to reach the pituitary gland. Once within the anterior pituitary, these vessels form a second capillary network that branches among the endocrine cells (Figure 18-8●).

This is an unusual vascular arrangement. A typical artery conducts blood from the heart to a capillary network, and a typical vein carries blood from a capillary network back to the heart. The vessels between the median eminence and the anterior pituitary, however, carry blood from one capillary network to another. Blood vessels that link two capillary networks are called **portal vessels,** and the entire complex is termed a **portal system.**

Portal systems provide an efficient means of chemical communication by ensuring that all of the blood entering the portal vessels will reach the intended target cells before returning to the general circulation. The communication is strictly one-way, however, because any chemicals released by the cells "downstream" must do a complete tour of the circulatory system before reaching the capillaries at the start of the portal system. Portal vessels are named after their destinations, so this particular network of vessels is the **hypophyseal portal system.**

Hypothalamic Control of the Anterior Pituitary

Figure 18-9

There are two classes of regulatory hormones. A **releasing hormone (RH)** stimulates synthesis and secretion of one or more hormones at the anterior pituitary, whereas an **inhibiting hormone (IH)** prevents the synthesis and secretion of hormones from the anterior pituitary. An endocrine cell in the anterior pituitary may be controlled by releasing hormones, inhibiting hormones, or some combination

● **FIGURE 18-7**
Anatomy and Orientation of the Pituitary Gland. (LM × 62)

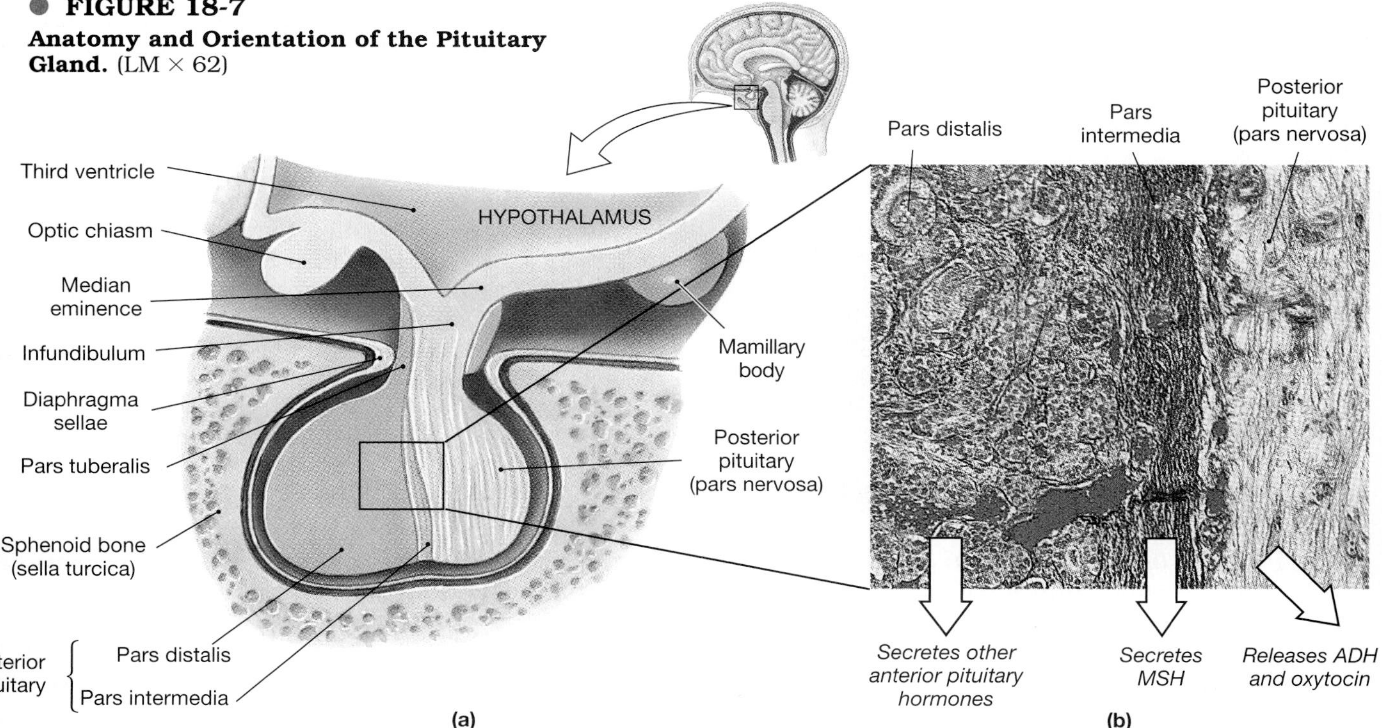

Third ventricle
Optic chiasm
Median eminence
Infundibulum
Diaphragma sellae
Pars tuberalis
Sphenoid bone (sella turcica)
Anterior pituitary
Pars distalis
Pars intermedia
HYPOTHALAMUS
Mamillary body
Posterior pituitary (pars nervosa)
(a)
Pars distalis
Pars intermedia
Posterior pituitary (pars nervosa)
Secretes other anterior pituitary hormones
Secretes MSH
Releases ADH and oxytocin
(b)

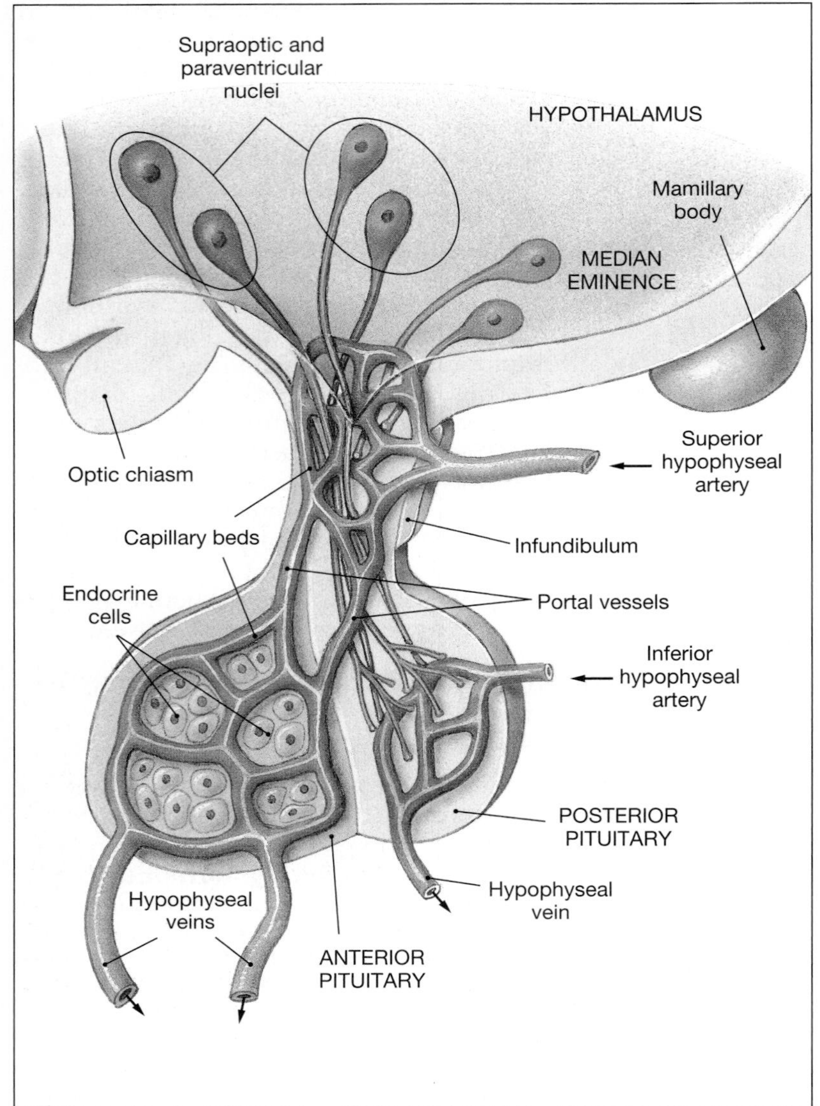

Supraoptic and paraventricular nuclei
HYPOTHALAMUS
Mamillary body
MEDIAN EMINENCE
Optic chiasm
Superior hypophyseal artery
Capillary beds
Infundibulum
Endocrine cells
Portal vessels
Inferior hypophyseal artery
POSTERIOR PITUITARY
Hypophyseal veins
Hypophyseal vein
ANTERIOR PITUITARY

● **FIGURE 18-8**
The Hypophyseal Portal System

of the two. The regulatory hormones released at the hypothalamus are transported directly to the anterior pituitary by the hypophyseal portal system.

The rate of regulatory hormone secretion by the hypothalamus is regulated through negative feedback mechanisms. The primary regulatory patterns are diagrammed in Figure 18-9●; these will be referenced as we examine specific pituitary hormones.

Hormones of the Anterior Pituitary

Figure 18-9

We will restrict our discussion to seven hormones whose functions and control mechanisms are reasonably well understood. Of the six hormones produced by the pars distalis, four regulate the production of hormones by other endocrine glands. The names of these hormones indicate their activities, but the phrases are often so long that abbreviations are used instead.

THYROID-STIMULATING HORMONE (TSH) Thyroid-stimulating hormone (TSH) targets the thyroid gland and triggers the release of thyroid hormones. TSH is released in response to *thyroid hormone–releasing hormone (TRH)* from the hypothalamus. As circulating concentrations of thyroid hormones rise, the rates of TRH and TSH production decline (Figure 18-9a●).

ADRENOCORTICOTROPIC HORMONE (ACTH) Adrenocorticotropic hormone (ACTH) (*trope*, a turning) stimulates the release of steroid hormones by the *adrenal cortex*, the outer portion of the adrenal gland. ACTH specifically targets cells producing hormones called *glucocorticoids* (gloo-kō-KOR-ti-koyds) that affect glucose metabolism. ACTH release occurs under the stimulation of **corticotropin-releasing hormone (CRH)** from the hypothalamus. As glucocorticoid levels increase, the rate of CRH and ACTH release declines (Figure 18-9a●).

THE GONADOTROPINS *FSH* and *LH* are called **gonadotropins** (gō-nad-ō-TRŌ-pinz) because they regulate the activities of the male and female sex organs (gonads). FSH and LH production occurs under stimulation by **gonadotropin-releasing hormone (GnRH)** from the hypothalamus.

Follicle-Stimulating Hormone (FSH). **Follicle-stimulating hormone (FSH)** promotes follicle development in women and, in combination with LH, stimulates the secretion of **estrogens** (female sex hormones) by ovarian cells. *Estradiol* is the most important estrogen. In men, FSH production supports the physical maturation of sperm in the testes by stimulating *sustentacular cells*. FSH production is inhibited by a hormone, *inhibin*, that is produced by cells in the testes and ovaries (Figure 18-9a●). (Disagreement exists as to whether inhibin also suppresses the release of GnRH as well as FSH.)

Luteinizing Hormone (LH). **Luteinizing** (LOO-tē-in-ī-zing) **hormone (LH)** induces ovulation in women and promotes the ovarian secretion of estrogens and the progestins (such as *progesterone*) that prepare the body for possible pregnancy. In men, LH was originally called *interstitial cell–stimulating hormone* (ICSH) because it stimulates the production of sex hormones by the *interstitial cells* of the

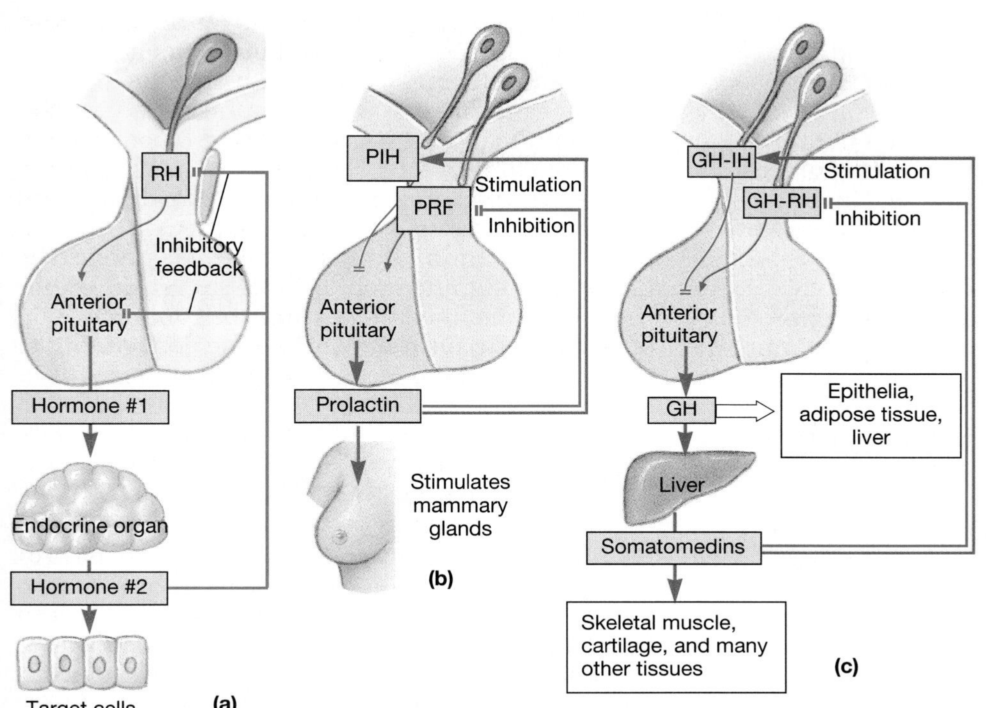
RH
Inhibitory feedback
Anterior pituitary
Hormone #1
Endocrine organ
Hormone #2
Target cells
(a)
PIH
PRF
Stimulation
Inhibition
Anterior pituitary
Prolactin
Stimulates mammary glands
(b)
GH-IH
GH-RH
Stimulation
Inhibition
Anterior pituitary
GH
Epithelia, adipose tissue, liver
Liver
Somatomedins
Skeletal muscle, cartilage, and many other tissues
(c)

● **FIGURE 18-9**
Feedback Control of Endocrine Secretion.
(a) Typical pattern of regulation when multiple endocrine organs are involved. The hypothalamus produces a releasing hormone to stimulate hormone production by other glands, and control is via negative feedback. **(b)** Regulation of prolactin production by the anterior pituitary. In this case the hypothalamus produces both a releasing hormone (PRH) and an inhibiting hormone (PIH); when one is stimulated, the other is inhibited. **(c)** Regulation of growth hormone production by the anterior pituitary; when GH-RH release is inhibited, GH-IH release is stimulated.

Regulatory Hormone	Pituitary Hormone #1	Endocrine Target	Hormone #2
TRH	TSH	Thyroid gland	Thyroid hormones
CRH	ACTH	Adrenal cortex	Gluco-corticoids
GnRH	FSH	Testes	Inhibin
		Ovaries	{ Inhibin, Estrogens
	LH	Ovaries	{ Progestins, Estrogens
		Testes	Androgens

testes. These hormones are called **androgens** (AN-drō-jenz; *andros*, man); the most important of them is *testosterone*. LH production, like FSH production, is stimulated by GnRH from the hypothalamus. GnRH production is inhibited by estrogens, progestins, and androgens (Figure 18-9a●).

PROLACTIN (PRL) Prolactin (prō-LAK-tin; *pro-*, before + *lac*, milk), or **PRL,** works with other hormones to stimulate mammary gland development. In pregnancy and during the nursing period that follows delivery, PRL stimulates milk production by the mammary glands. The functions of PRL in males are poorly understood, but there is evidence that PRL has an indirect role in the regulation of androgen production. (PRL appears to make interstitial cells more sensitive to LH.)

Prolactin production is inhibited by **prolactin-inhibiting hormone (PIH).** This hormone is identical to the neurotransmitter *dopamine*. The hypothalamus also secretes a prolactin-releasing hormone, but the identity of this *prolactin-releasing factor (PRF)* is a mystery. Circulating PRL stimulates PIH release and inhibits the secretion of prolactin-releasing factor (Figure 18-9b●).

Although PRL exerts the dominant effect on the glandular cells, normal development of the mammary glands is regulated by the interaction of a number of hormones. Prolactin, estrogens, progesterone, glucocorticoids, pancreatic hormones, and hormones produced by the placenta cooperate in preparing the mammary glands for secretion, and milk ejection occurs only in response to oxytocin release at the posterior pituitary. The functional development of the mammary glands will be covered in more detail in Chapter 29.

GROWTH HORMONE Growth hormone (GH), or **somatotropin** (*soma*, body), stimulates cell growth and replication by accelerating the rate of protein synthesis. Although virtually every tissue responds to some degree, skeletal muscle cells and chondrocytes (cartilage cells) are particularly sensitive to levels of growth hormone.

The stimulation of growth by GH involves two different mechanisms. The primary mechanism, which is indirect, is best understood. Liver cells respond to the presence of growth hormone by synthesizing and releasing **insulin-like growth factors (IGFs),** or **somatomedins,** hormones that bind to receptor sites on a variety of cell membranes (Figure 18-9c●). In skeletal muscle fibers, cartilage cells, and other target cells, somatomedins increase the rate of amino acid uptake and their incorporation into new proteins. These effects develop almost immediately after GH release occurs, and they are particularly important after a meal, when the blood contains high concentrations of glucose and amino acids. In functional terms, cells can now obtain ATP easily through the aerobic metabolism of glucose, and amino acids are readily available. Under these conditions GH, acting via the somatomedins, stimulates protein synthesis and cell growth.

The direct actions of GH are more selective and usually do not appear until after blood glucose and amino acid concentrations have returned to normal levels.

1. In epithelia and connective tissues, GH stimulates stem cell divisions and the differentiation of daughter cells. The subsequent growth of these daughter cells will be stimulated by somatomedins.
2. GH also has metabolic effects in adipose tissue and in the liver.
   - GH stimulates the breakdown of stored triglycerides by adipocytes (fat cells), which then release fatty acids into the blood. As circulating fatty acid levels rise, many tissues stop breaking down glucose and start breaking down fatty acids to generate ATP. This is termed a **glucose-sparing effect.**
   - GH stimulates the breakdown of glycogen reserves by liver cells. These cells then release glucose into the circulation. Because most other tissues are now metabolizing fatty acids rather than glucose, blood glucose concentrations begin climbing, perhaps to levels significantly higher than normal. The elevation of blood glucose levels by GH has been called a **diabetogenic effect,** because *diabetes mellitus,* an endocrine disorder considered later in the chapter, is characterized by abnormally high blood glucose concentrations.

Control of Growth Hormone Production. The production of growth hormone is regulated by **growth hormone–releasing hormone (GH-RH,** or *somatocrinin*) and **growth hormone–inhibiting hormone (GH-IH,** or *somatostatin*) from the hypothalamus. Somatomedins stimulate GH-IH and inhibit GH-RH (Figure 18-9c●).

The interactions between growth hormone and other hormones during normal development and maturation will be the focus of a later section.

MELANOCYTE-STIMULATING HORMONE The pars intermedia may secrete two forms of **melanocyte-stimulating hormone (MSH,** or *melanotropin*). As the name indicates, MSH stimulates the melanocytes of the skin, increasing their production of melanin, a brown, black, or yellow-brown pigment. MSH release is inhibited by *melanocyte-stimulating hormone–inhibiting hormone,* or *MSH-IH.* MSH is important in the control of skin pigmentation in fishes, amphibians, reptiles, and many mammals. The pars intermedia in adult humans is virtually nonfunctional, and the circulating blood usually does not contain MSH. However, MSH is secreted by the human pars intermedia (1) during fetal development, (2) in very young children, (3) in pregnant women, and (4) in some disease states. The functional significance of MSH secretion under these circumstances is not known. Administration of a synthetic form of MSH causes darkening of the skin, and it has been suggested as a means of obtaining a "sunless tan."

❑ The Posterior Pituitary

The **posterior pituitary** is also called the **neurohypophysis** (noo-rō-hī-POF-i-sis), or *pars nervosa* (nervous part), because it contains the axons of hypothalamic neurons. Neurons of the **supraoptic** and **paraventricular nuclei** manufacture antidiuretic hormone (ADH) and oxytocin, respectively. These products move via axoplasmic transport along the infundibulum to the basement membranes of capillaries in the posterior pituitary.

Antidiuretic Hormone

Antidiuretic hormone (ADH), or *vasopressin,* is released in response to a variety of stimuli, most notably a rise in the concentration of electrolytes in the blood or a fall in blood volume or pressure. The primary function of ADH is to decrease the amount of water lost at the kidneys. With losses minimized, any water absorbed from the digestive tract will be retained, reducing the concentration of electrolytes in the extracellular fluid. In high concentrations, ADH also causes the constriction of peripheral blood vessels, which helps to elevate blood pressure. ADH release is inhibited by alcohol, which explains the increased fluid excretion that follows the consumption of alcoholic beverages.

DISORDERS OF ADH PRODUCTION There are several different forms of **diabetes,** all characterized by excessive urine production **(polyuria).** Although diabetes can be caused by physical damage to the kidneys, most forms are the result of endocrine abnormalities. The two most important forms are *diabetes insipidus* and *diabetes mellitus.* Diabetes mellitus is described on p. 634. **Diabetes insipidus** usually develops because the posterior pituitary no longer releases adequate amounts of ADH. Water conservation at the kidneys is impaired, and excessive amounts of water are lost in the urine. As a result, the individual is constantly thirsty, but the fluids consumed are not retained by the body.

Mild cases of diabetes insipidus may not require treatment as long as fluid and electrolyte intake

Growth Hormone Abnormalities

Growth hormone stimulates muscular and skeletal development, and if it is administered before the epiphyseal plates have closed it will cause an increase in height, weight, and muscle mass. In **acromegaly** (*akron*, extremity + *megale*, great) an excessive amount of growth hormone is released after puberty, when most of the epiphyseal plates have already closed. Cartilages and small bones respond to the hormone, however, resulting in abnormal growth at the hands, feet, lower jaw, skull, and clavicle. Figure 18-22a●, p. 643, shows a typical acromegalic individual.

Children unable to produce adequate concentrations of growth hormone suffer from *pituitary growth failure*, sometimes called *pituitary dwarfism*, a condition introduced in Chapter 6. ∞ [p. 188] The steady growth and maturation that normally precede and accompany puberty do not occur in these individuals, who have short stature, slow epiphyseal growth, and larger-than-normal adipose tissue reserves.

Normal growth patterns can be restored by the administration of growth hormone. Before the advent of gene splicing and recombinant-DNA techniques, GH had to be carefully extracted and purified from the pituitaries of cadavers at considerable expense. It is now possible to use genetically manipulated bacteria to produce pure GH in commercial quantities.

The current availability of purified human growth hormone has led to its use under medically questionable circumstances. For example, it is now being praised as an "anti-aging" miracle cure. GH supplements do slow or even reverse the losses in bone and muscle mass that accompany aging; however, little is known about adverse side effects that may accompany long-term use in mature adults. GH is also being sought by some parents of short but otherwise normal children. These parents view short stature as a handicap that merits treatment by a physician. Whether considering GH treatment of adults or children, it is important to remember that GH and the somatomedins affect many different tissues and have widespread metabolic impacts. For example, children exposed to GH may grow faster, but their body fat content declines drastically. This decline is associated with metabolic changes in many organs. The range and significance of these metabolic side effects are now the subject of long-term studies.

keep pace with urinary losses. In severe cases the fluid losses can reach 10 liters per day, and fatal dehydration and electrolyte imbalances will occur unless treatment is provided. One innovative treatment method involves administering a synthetic form of ADH, *desmopressin acetate* (DDAVP), in a nasal spray. The drug enters the bloodstream after diffusing through the nasal epithelium.

Oxytocin

Figures 18-6e, 18-10

In women, **oxytocin** (*oxy-*, quick + *tokos*, childbirth), or **OT,** stimulates smooth muscle tissue in the wall of the uterus, promoting labor and delivery. After delivery, oxytocin stimulates the contraction of myoepithelial cells around the secretory alveoli and the ducts of the mammary glands.

The name oxytocin refers to this hormone's role in stimulating labor and delivery. Until the last stages of pregnancy the uterine smooth muscles are relatively insensitive to oxytocin, but sensitivity becomes more pronounced as the time of delivery approaches. The trigger for normal labor and delivery is probably a sudden rise in oxytocin levels at the uterus. There is good evidence, however, that the oxytocin released by the posterior pituitary plays only a *supporting* role and that most of the oxytocin involved is secreted by the uterus and fetus.

OT secretion and milk ejection are part of a neuroendocrine reflex diagrammed in Figure 18-6e●, p. 610. The stimulus is an infant suckling at the breast, and sensory nerves innervating the nipples relay the information to the hypothalamus. Oxytocin is then released into the circulation at the posterior pituitary, and the myoepithelial cells respond by squeezing milk from the secretory alveoli into large collecting ducts. This "milk let-down" reflex can be modified by any factor affecting the hypothalamus. For example, anxiety, stress, and other factors can prevent the flow of milk, even when the mammary glands are fully functional. By contrast, nursing mothers can become conditioned to associate a baby's crying with suckling. These women may begin milk let-down as soon as they hear a baby cry.

In the male, oxytocin stimulates smooth muscle contraction in the walls of the prostate gland. This action may be important in *emission*, the ejection of prostatic secretions, spermatozoa, and the secretions of other glands into the male reproductive tract prior to ejaculation.

Figure 18-10● and Table 18-1 summarize important information concerning the hormonal

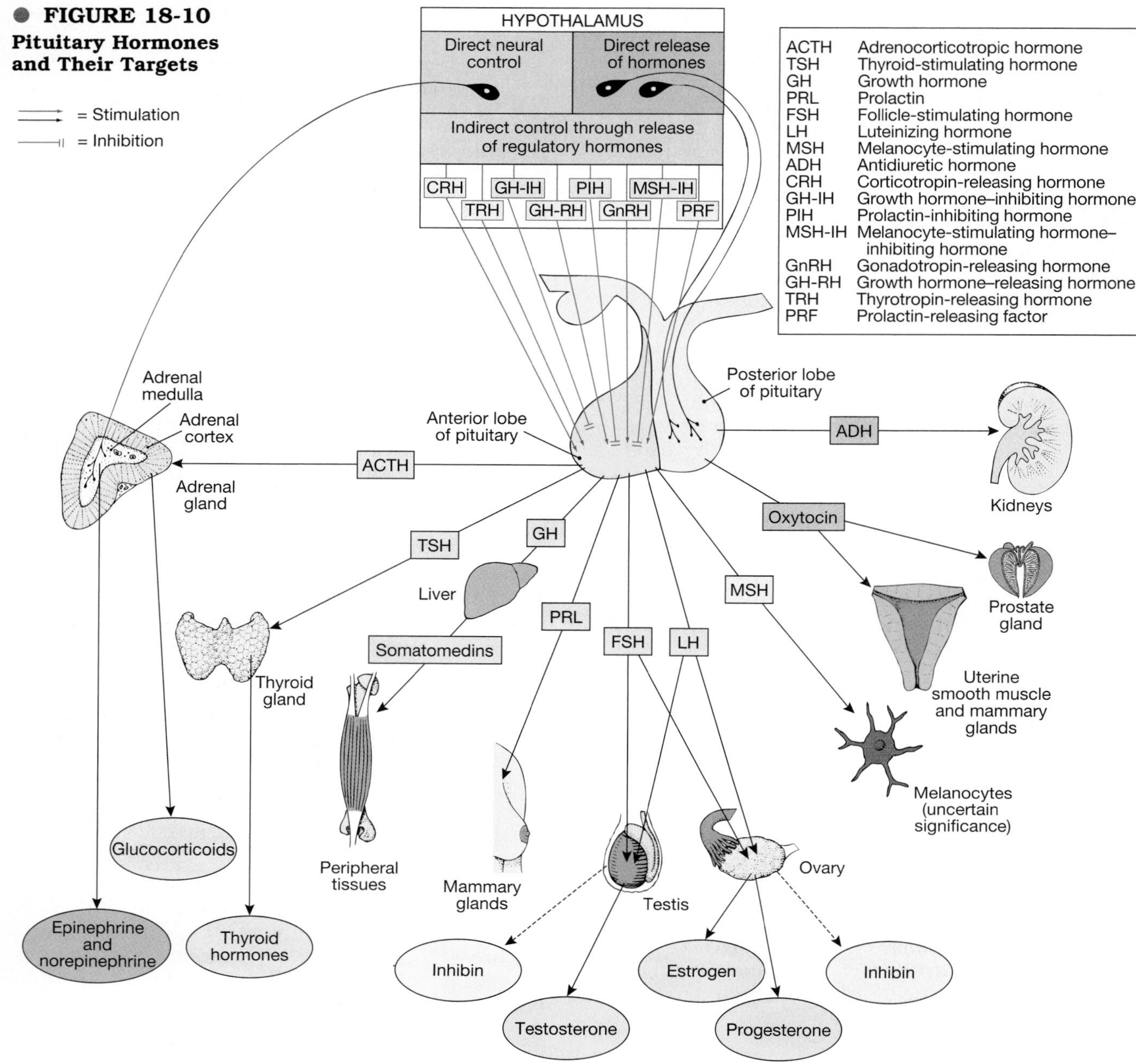

= Stimulation
= Inhibition
HYPOTHALAMUS
Direct neural control
Direct release of hormones
Indirect control through release of regulatory hormones
CRH
GH-IH
PIH
MSH-IH
TRH
GH-RH
GnRH
PRF
ACTH Adrenocorticotropic hormone
TSH Thyroid-stimulating hormone
GH Growth hormone
PRL Prolactin
FSH Follicle-stimulating hormone
LH Luteinizing hormone
MSH Melanocyte-stimulating hormone
ADH Antidiuretic hormone
CRH Corticotropin-releasing hormone
GH-IH Growth hormone–inhibiting hormone
PIH Prolactin-inhibiting hormone
MSH-IH Melanocyte-stimulating hormone–inhibiting hormone
GnRH Gonadotropin-releasing hormone
GH-RH Growth hormone–releasing hormone
TRH Thyrotropin-releasing hormone
PRF Prolactin-releasing factor
Adrenal medulla
Adrenal cortex
Adrenal gland
Anterior lobe of pituitary
Posterior lobe of pituitary
ACTH
ADH
Kidneys
Oxytocin
TSH
GH
Liver
MSH
Prostate gland
PRL
FSH
LH
Somatomedins
Thyroid gland
Uterine smooth muscle and mammary glands
Melanocytes (uncertain significance)
Glucocorticoids
Peripheral tissues
Mammary glands
Testis
Ovary
Epinephrine and norepinephrine
Thyroid hormones
Inhibin
Estrogen
Inhibin
Testosterone
Progesterone

● **FIGURE 18-10**
Pituitary Hormones and Their Targets

products of the pituitary gland. Review these carefully before considering the structure and function of other endocrine organs.

✔ If a person were suffering from dehydration, how would this condition affect the level of ADH released by the neurohypophysis?

✔ A blood sample shows elevated levels of somatomedins. What pituitary hormone would you expect to be elevated as well?

✔ What effect would elevated levels of cortisol, a steroid hormone from the adrenal cortex, have on the level of ACTH?

The Thyroid Gland

Figure 18-11

The thyroid gland curves across the anterior surface of the trachea just below the *thyroid* ("shield-shaped") *cartilage* that forms most of the anterior surface of the larynx (Figure 18-11●). The two **lobes** of the thyroid gland are united by a slender connection, the **isthmus** (IS-mus). Because of its location, the thyroid gland can be easily felt with the fingers; when something goes wrong with the gland, it may even become prominent. The size of the gland is quite variable, depending on heredity, environment, and nutritional factors, but the average

TABLE 18-1 The Pituitary Hormones

Region/Area	*Hormones*	*Targets*	*Hormonal Effects*	*Hypothalamic Regulatory Hormones*
ANTERIOR PITUITARY (ADENOHYPOPHYSIS)				
Pars distalis	Thyroid-stimulating hormone (TSH)	Thyroid gland	Secretion of thyroid hormones	Thyrotropin-releasing hormone (TRH)
	Adrenocorticotropic hormone (ACTH)	Adrenal cortex (fasciculata)	Glucocorticoid secretion	Corticotropin-releasing hormone (CRH)
	Gonadotropic hormones:			
	Follicle-stimulating hormone (FSH)	Follicle cells of ovaries	Estrogen secretion, follicle development	Gonadotropin-releasing hormone (GnRH)
		Sustentacular cells of testes	Sperm maturation	As above
	Luteinizing hormone (LH) or interstitial cell–stimulating hormone (ICSH)	Follicle cells of ovaries	Ovulation, formation of corpus luteum, progesterone secretion	As above
		Interstitial cells of testes	Testosterone secretion	As above
	Prolactin (PRL)	Mammary glands	Production of milk	Prolactin-inhibiting hormone (PIH) Prolactin-releasing factor (PRF)
	Growth hormone (GH)	All cells	Growth, protein synthesis, lipid mobilization and catabolism	Growth hormone–releasing hormone (GH-RH) Growth hormone–inhibiting hormone (GH-IH)
Pars intermedia (not active in normal adults)	Melanocyte-stimulating hormone (MSH)	Melanocytes	Increased melanin synthesis in epidermis	Melanocyte-stimulating hormone–inhibiting hormone (MSH-IH)
POSTERIOR PITUITARY (NEUROHYPOPHYSIS OR PARS NERVOSA)	Antidiuretic hormone (ADH)	Kidneys	Reabsorption of water, elevation of blood volume and pressure	Transported over axons from supraoptic nucleus
	Oxytocin (OT)	Uterus, mammary glands (female) Prostate gland (male)	Labor contractions, milk ejection Prostatic contractions	Transported over axons from paraventricular nucleus

weight is about 34 g (1.2 oz). An extensive blood supply gives the thyroid gland a deep red color.

❑ Thyroid Follicles and Thyroid Hormones

Figures 18-11, 18-12

The thyroid gland contains large numbers of **thyroid follicles.** Individual follicles are spherical, resembling miniature tennis balls that are lined by a simple cuboidal epithelium (Figure 18-11●). The follicle cells surround a **follicle cavity,** which contains a viscous *colloid,* a fluid containing large quantities of suspended proteins. A network of capillaries surrounds each follicle, delivering nutrients and regulatory hormones to the glandular cells and accepting their secretory products and metabolic wastes.

Follicular cells synthesize a globular protein called **thyroglobulin** (thī-rō-GLOB-ū-lin) and secrete it into the colloid of the thyroid follicles (Figure 18-12a●). Each thyroglobulin molecule contains amino acid derivatives of *tyrosine* that will be converted to thyroid hormones through the attachment of *iodide ions,* the ionic form of iodine (I_2). The follicle cells absorb iodide ions as I^- from the circulation. At the epithelial surface, these ions are converted to an activated form of iodide (I^+) by enzymes called *peroxidases.* This reaction sequence also attaches these iodide ions to the tyrosine molecules of thyroglobulin.

The hormone **thyroxine** (thī-ROKS-ēn) contains four iodide ions; it is also known as *tetraiodothyronine,* or simply $\mathbf{T_4}$. **Triiodothyronine,** or $\mathbf{T_3}$, is a related molecule containing three iodide ions. Eventually each molecule of thyroglobulin contains 4–8 molecules of T_3 or T_4 hormones or both.

Thyroid follicles are continually manufacturing and secreting thyroglobulin molecules. The major factor controlling the rate of thyroid hormone release is

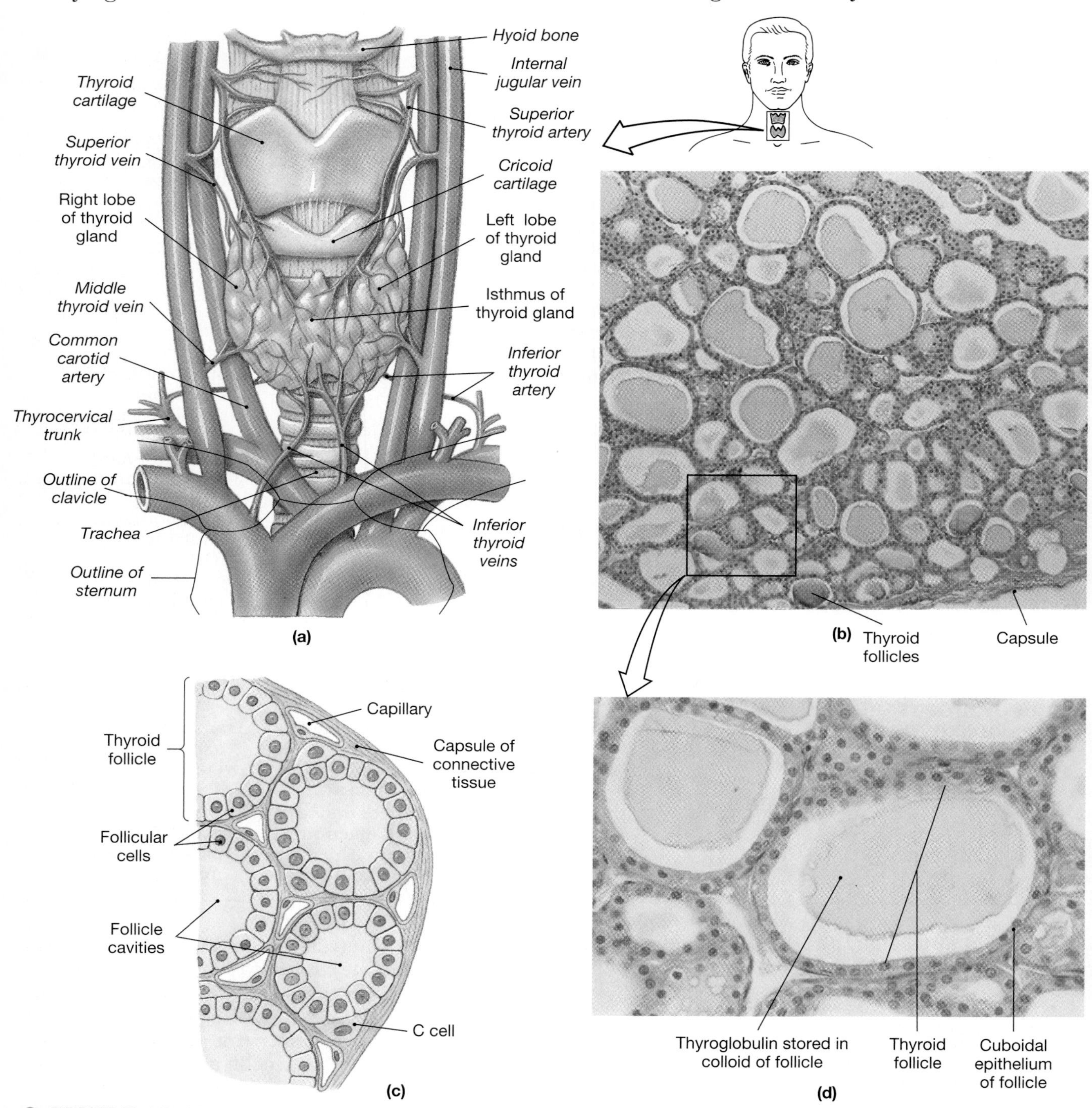
Hyoid bone
Internal jugular vein
Thyroid cartilage
Superior thyroid artery
Superior thyroid vein
Cricoid cartilage
Right lobe of thyroid gland
Left lobe of thyroid gland
Middle thyroid vein
Isthmus of thyroid gland
Common carotid artery
Inferior thyroid artery
Thyrocervical trunk
Outline of clavicle
Trachea
Inferior thyroid veins
Outline of sternum
(a)
(b) Thyroid follicles
Capsule
Thyroid follicle
Capillary
Capsule of connective tissue
Follicular cells
Follicle cavities
C cell
(c)
Thyroglobulin stored in colloid of follicle
Thyroid follicle
Cuboidal epithelium of follicle
(d)

● **FIGURE 18-11**

The Thyroid Gland. (a) Location and anatomy of the thyroid. **(b)** Histological organization of the thyroid. (LM × 99). **(c)** Diagrammatic view of a section through the wall of the thyroid gland. **(d)** Histological details of the thyroid gland, showing thyroid follicles. (LM × 211)

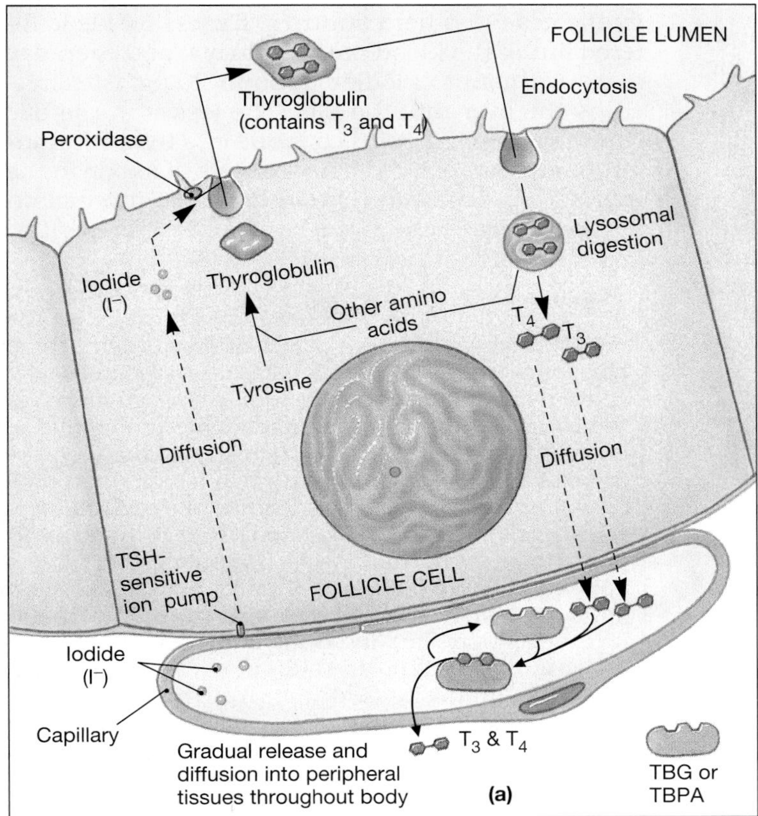
Follicle lumen
FOLLICLE LUMEN
Thyroglobulin (contains T_3 and T_4)
Endocytosis
Peroxidase
Lysosomal digestion
Iodide (I^-)
Thyroglobulin
Other amino acids
T_4
T_3
Tyrosine
Diffusion
Diffusion
TSH-sensitive ion pump
FOLLICLE CELL
Iodide (I^-)
Capillary
Gradual release and diffusion into peripheral tissues throughout body
T_3 & T_4
TBG or TBPA
(a)

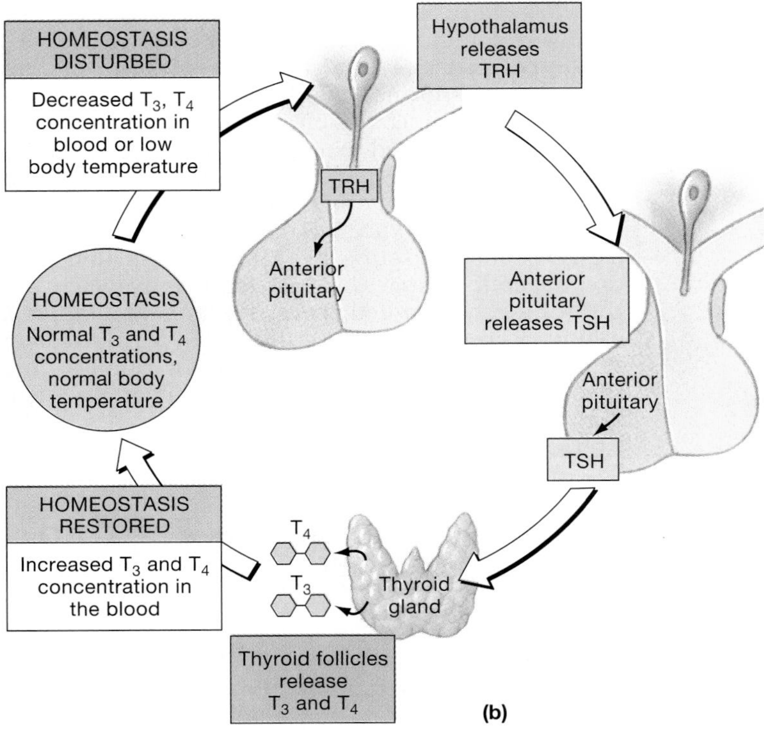
HOMEOSTASIS DISTURBED
Decreased T_3, T_4 concentration in blood or low body temperature
Hypothalamus releases TRH
TRH
Anterior pituitary
Anterior pituitary releases TSH
HOMEOSTASIS
Normal T_3 and T_4 concentrations, normal body temperature
Anterior pituitary
TSH
HOMEOSTASIS RESTORED
Increased T_3 and T_4 concentration in the blood
T_4
T_3
Thyroid gland
Thyroid follicles release T_3 and T_4
(b)

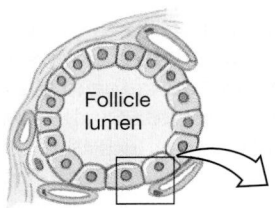

● **FIGURE 18-12**
The Functions of the Thyroid Follicles. **(a)** The synthesis, storage, and secretion of thyroid hormones. **(b)** The regulation of thyroid secretion.

the concentration of TSH in the circulating blood (Figure 18-12b●). Under the influence of TSH, follicle cells remove thyroglobulin from the follicles, break the protein down, and release molecules of T_3 and T_4, which then enter the circulation.

Thyroxine (T_4) accounts for roughly 90 percent of all thyroid secretions, and triiodothyronine (T_3) is secreted in small amounts. Roughly 75 percent of the T_4 and 70 percent of the T_3 molecules entering the circulation become attached to transport proteins called **thyroid-binding globulins,** or **TBGs.** Most of the rest of the T_4 and T_3 in the circulation is attached to **thyroid-binding prealbumin,** or **TBPA.** Only the relatively small quantities of thyroid hormones that remain unbound—roughly 0.3 percent of the circulating T_3 and 0.03 percent of the circulating T_4—can diffuse into peripheral tissues. As they leave the circulation, they are replaced by the release of bound thyroid hormones. The bound thyroid hormones represent a substantial reserve, and the bloodstream normally contains more than a week's supply of thyroid hormones.

Function of Thyroid Hormones

Thyroid hormones affect almost every cell in the body because they readily cross cell membranes. Inside a cell, they bind to (1) receptors in the nucleus, (2) receptors on the surfaces of mitochondria, and (3) receptors in the cytoplasm. Thyroid hormones bound to cytoplasmic receptors are essentially held in storage. If intracellular levels of thyroid hormones decline, the bound thyroid hormones are released into the cytoplasm.

The thyroid hormones binding to mitochondria increase the rates of mitochondrial ATP production. The binding to receptors in the nucleus activates genes that control the synthesis of enzymes involved with energy transformation and utilization. One specific effect is accelerated production of sodium-potassium ATPase, the membrane protein responsible for the ejection of intracellular sodium and the recovery of extracellular potassium. As noted in Chapter 3, this exchange pump consumes large amounts of ATP. ∞ *[p. 79]*

Thyroid hormones also activate genes coding for the synthesis of enzymes involved in glycolysis and ATP production. This effect, coupled with the direct effect of thyroid hormones on mitochondria, increases the metabolic rate of the cell. Because the cell consumes more energy, and energy use is measured in *calories,* this is called the **calorigenic effect** of thyroid hormones. When the metabolic rate increases, more heat is generated. In young children, TSH production increases in cold weather, and the calorigenic effect may help them to adapt to cold climates. (This response does not occur in adults.) In growing children, thyroid hormones are also essential to normal development of the skeletal, muscular, and nervous systems.

T_3 VERSUS T_4 The thyroid gland produces large amounts of T_4, but T_3 is primarily responsible for the observed effects of thyroid hormones: a strong, immediate, short-lived increase in the rate of cellular metabolism. There are actually two sources of T_3 in peripheral tissues:

- ***T_3 released by the thyroid gland.*** T_3 from the thyroid gland actually accounts for only around 10–15 percent of the T_3 in peripheral tissues at any given moment.
- ***The conversion of T_4 to T_3.*** T_4 can be converted to T_3 by enzymes in peripheral tissues. The circulation delivers much more T_4 than T_3 to peripheral tissues, and 85–90 percent of the T_3 reaching the target cells is produced by the conversion of T_4.

Table 18-2 summarizes the effects of thyroid hormones on major organs and systems.

IODINE AND THYROID HORMONES Iodine in the diet is absorbed at the digestive tract as iodide ions (I^-). The follicle cells in the thyroid gland absorb 120–150 μg of iodide ions per day, the minimum dietary amount needed to maintain normal thyroid function. The concentration of iodide inside thyroid follicle cells is usually around 30 times higher than that in the plasma. If plasma iodide levels rise, so do levels inside the follicle cells.

Iodide ions are actively transported into the thyroid follicle cells, and these follicles contain most of the iodide reserves in the body. The active transport mechanism for iodide is stimulated by TSH, and the increased movement of iodide into the cytoplasm accelerates the formation of thyroid hormones.

TABLE 18-2 Effects of Thyroid Hormones on Peripheral Tissues

1. Elevate oxygen consumption and rate of energy consumption; usually cause a rise in body temperature
2. Increase heart rate and force of contraction; usually cause a rise in blood pressure
3. Increase sensitivity to sympathetic stimulation
4. Maintain normal sensitivity of respiratory centers to changes in oxygen and carbon dioxide concentrations
5. Stimulate formation of red blood cells and thereby enhance oxygen delivery
6. Stimulate activity of other endocrine tissues
7. Accelerate turnover of minerals in bone

The typical American diet provides approximately 500 μg/day of iodide, roughly three times the minimum daily requirement. Much of the excess is due to the addition of iodide to the table salt sold in grocery stores ("iodized salt"). Thus iodide deficiency is seldom responsible for limiting the rate of thyroid hormone production. This is not necessarily the case in other countries. Excess iodide is filtered out of the blood at the kidneys, and each day a small amount of iodide (around 20 μg) is excreted by the liver into the bile. The losses in the bile continue even if the diet contains less than the minimum amount of iodide. As reserves of iodide in the thyroid are depleted, thyroid hormone production declines, regardless of the levels of TSH present.

THYROID GLAND DISORDERS Normal production of thyroid hormones establishes the background rates of cellular metabolism. These hormones exert their primary effects on metabolically active tissues and organs, including skeletal muscles, the liver, and the kidneys. Inadequate production of thyroid hormones is called **hypothyroidism.** Hypothyroidism in an infant produces a condition known as *cretinism,* marked by inadequate skeletal and nervous development and a metabolic rate as much as 40 percent below normal levels. This condition affects approximately 1 birth out of every 5000. Cretinism developing later in childhood will retard growth and mental development and delay puberty.

Adults with hypothyroidism are lethargic and unable to adjust to cold temperatures. The symptoms of adult hypothyroidism, collectively known as **myxedema** (miks-e-DĒ-ma), include subcutaneous swelling, dry skin, hair loss, low body temperature, muscular weakness, and slowed reflexes. Hypothyroidism and myxedema due to inadequate dietary iodide are very rare in the United States but may be relatively common in many poorer countries.

Thyrotoxicosis, or **hyperthyroidism,** occurs when thyroid hormones are produced in excessive quantities. The metabolic rate climbs, and the skin becomes flushed and moist with perspiration. Blood pressure and heart rate increase, and the heartbeat may become irregular as circulatory demands escalate. The effects on the CNS make the individual restless, excitable, and subject to shifts in mood and emotional states. Despite the drive for increased activity, the subject has limited energy reserves and fatigues easily. *Graves' disease* is a form of hyperthyroidism that afflicted President Bush and Mrs. Bush during their stay in the White House. Both hypothyroidism and hyperthyroidism are considered further in the *Applications Manual.* AM *Thyroid Gland Disorders*

❑ The C Cells of the Thyroid Gland: Calcitonin

Figure 18-6b

A second population of endocrine cells lies sandwiched between the cuboidal follicle cells and their basement membrane. These cells are larger than

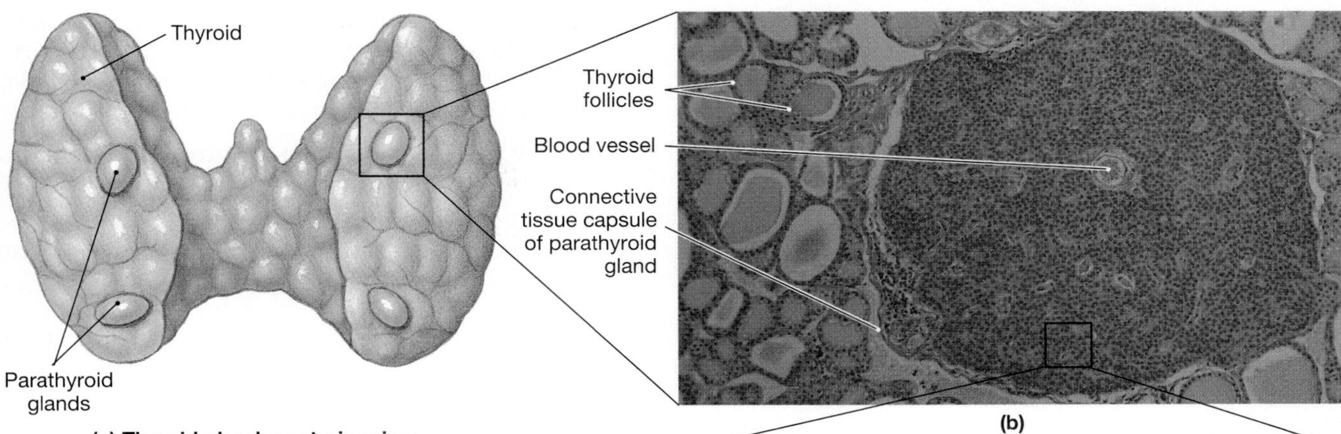
Thyroid
Parathyroid glands
Thyroid follicles
Blood vessel
Connective tissue capsule of parathyroid gland
(a) Thyroid gland, posterior view
(b)

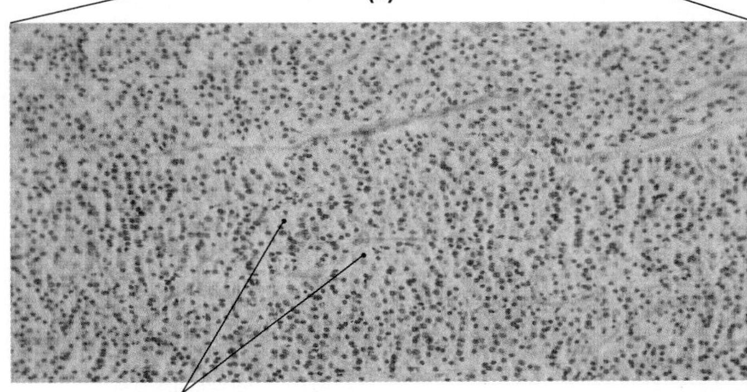
Principal (chief) cells
(c)

FIGURE 18-13

The Parathyroid Glands. (a) Location of parathyroids on posterior surface of thyroid lobes. **(b)** Photomicrograph showing both parathyroid and thyroid tissues. (LM × 94) **(c)** Photomicrograph of parathyroid cells. (LM × 221)

those of the follicular epithelium, and they do not stain as clearly. They are the **C cells,** or *parafollicular cells,* that produce the hormone **calcitonin (CT).** Calcitonin assists in the regulation of calcium ion concentrations in body fluids. The functions of this hormone were introduced in Chapter 6. *[p. 189]* The net effect of calcitonin release is a decrease in the concentration of calcium ions in body fluids. This reduction is accomplished through four mechanisms. Calcitonin

- stimulates osteoblasts, leading to increased bone deposition and the removal of calcium ions from body fluids.
- inhibits osteoclasts, slowing the rate of calcium ion release from bone.
- stimulates calcium ion excretion at the kidneys.
- inhibits secretion of the antagonistic hormone, *parathyroid hormone.*

The control of calcitonin secretion is an example of direct endocrine regulation (Figure 18-6b, p. 610) because the C cells respond directly to elevations in the calcium ion concentration of the blood. When those concentrations rise, calcitonin secretion increases. This lowers the calcium ion concentrations, eliminating the stimulus and "turning off" the C cells.

Calcitonin is most important during childhood, when it stimulates bone growth and mineral deposition in the skeleton. It is also important in reducing the loss of bone mass (1) during prolonged starvation and (2) in the late stages of pregnancy, when the maternal skeleton competes with the developing fetus for calcium ions absorbed by the digestive tract. The importance of calcitonin in the normal nonpregnant adult remains uncertain.

Several chapters have considered the importance of calcium ions in controlling muscle and nerve cell activities. Calcium ion concentrations also affect the sodium permeabilities of excitable membranes. At high calcium ion concentrations, sodium permeability decreases and membranes become less responsive. Such problems are relatively rare. Problems caused by lower-than-normal calcium ion concentrations are equally dangerous and much more common. When calcium ion concentrations decline, sodium permeabilities increase, and cells become extremely excitable. If calcium levels are allowed to fall too far, convulsions or muscular spasms may result. Maintenance of adequate calcium levels involves the parathyroid glands and parathyroid hormone.

The Parathyroid Glands

Figure 18-13

Two pairs of parathyroid glands are found embedded in the posterior surfaces of the thyroid gland. The gland cells are separated by the dense capsular fibers of the thyroid. All together the four parathyroid glands weigh a mere 1.6 g. The histological appearance of a parathyroid gland is shown in Figure 18-13. There are at least two different cell populations found in the parathyroid. The **chief cells** produce parathyroid hormone; the functions of the other cell types are unknown.

Parathyroid Hormone Secretion

Figure 18-14

Like the C cells of the thyroid, the chief cells monitor the circulating concentration of calcium ions. When the calcium concentration falls below normal, the chief cells secrete **parathyroid hormone (PTH),** or *parathormone.* The net result of parathyroid hormone secretion is an increase in the concentration of calcium ions in body fluids. Parathyroid hormone has five major effects. This hormone

- stimulates osteoclasts, accelerating mineral turnover and the release of calcium ions from bone.
- inhibits osteoblasts, reducing the rate of calcium deposition in bone.
- promotes the intestinal absorption of calcium and phosphate ions.
- reduces urinary excretion of calcium ions.
- stimulates the formation and secretion of *calcitriol* at the kidneys. In general, the effects of calcitriol complement or enhance those of PTH. Calcitriol will be considered in a later section (p. 627).

Figure 18-14• compares the functions of calcitonin and parathyroid hormone.

DISORDERS OF PARATHYROID FUNCTION When the parathyroid gland secretes inadequate or excessive amounts of parathyroid hormone, calcium concentrations move outside of normal homeostatic limits. Inadequate parathyroid hormone production, a condition called **hypoparathyroidism,** leads to low calcium concentrations in body fluids. The most obvious symptoms involve neural and muscle tissues: The nervous system becomes more excitable, and the affected individual may experience *hypocalcemic tetany,* prolonged muscle spasms affecting the limbs and face.

In **hyperparathyroidism,** calcium ion concentrations become abnormally high. Bones grow thin and brittle; skeletal muscles become weak, CNS function is depressed, nausea and vomiting occur, and in severe cases the patient may become comatose. AM *Disorders of Parathyroid Function*

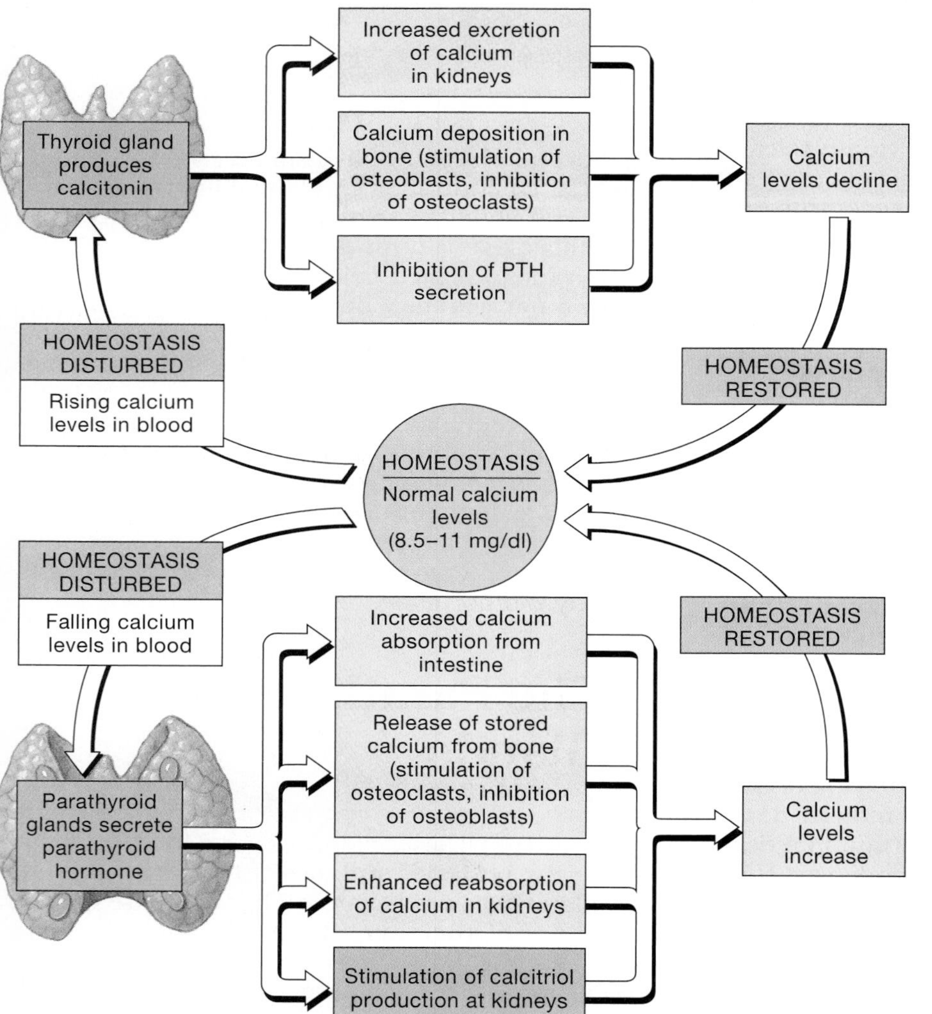
Thyroid gland produces calcitonin
Increased excretion of calcium in kidneys
Calcium deposition in bone (stimulation of osteoblasts, inhibition of osteoclasts)
Inhibition of PTH secretion
Calcium levels decline
HOMEOSTASIS RESTORED
HOMEOSTASIS DISTURBED
Rising calcium levels in blood
HOMEOSTASIS
Normal calcium levels (8.5–11 mg/dl)
HOMEOSTASIS DISTURBED
Falling calcium levels in blood
HOMEOSTASIS RESTORED
Increased calcium absorption from intestine
Release of stored calcium from bone (stimulation of osteoclasts, inhibition of osteoblasts)
Enhanced reabsorption of calcium in kidneys
Stimulation of calcitriol production at kidneys
Parathyroid glands secrete parathyroid hormone
Calcium levels increase

● **FIGURE 18-14**
The Homeostatic Regulation of Calcium Ion Concentrations

The Thymus

The **thymus** is located in the mediastinum, usually just behind the sternum. In a newborn infant and young child the thymus is relatively large, often extending from the base of the neck to the superior border of the heart. Although it continues to increase in size through childhood, the body as a whole grows even faster, so the size of the thymus relative to the other organs in the mediastinum gradually decreases. The thymus reaches its maximum absolute size just before puberty, at a weight of around 40 g. After puberty it gradually diminishes in size; by age 50 the thymus may weigh less than 12 g. It has been suggested that the gradual decrease in the size and secretory abilities of the thymus may make the elderly more susceptible to disease.

The thymus produces several hormones important to the

development and maintenance of normal immunological defenses (Table 18-3). **Thymosin** (THĪ-mō-sin) is the name originally given to a thymic extract that promotes the development and maturation of *lymphocytes,* the white blood cells responsible for immunity. The thymic extract actually contains a blend of several different, complementary hormones, including *thymosin-α, thymosin-β, thymosin V, thymopoietin, thymulin,* and others that will be considered in Chapter 22. The term *thymosins* is now used to refer to all of these thymic hormones.

Information concerning the hormones of the thyroid, parathyroid, and thymus glands is summarized in Table 18-3. The histological organization of the thymus and the functions of the thymosins will be further considered in Chapter 22.

The Adrenal Gland

Figure 18-15

A yellow, pyramid-shaped **adrenal gland,** or *suprarenal* (sū-pra-RĒ-nal) *gland* (*supra-,* above + *renes,* kidneys), sits on the superior border of each kidney. Each adrenal gland lies at the level of the twelfth thoracic rib and is firmly attached to the superior portion of each kidney by a dense fibrous capsule (Figure 18-15●). The adrenal gland on each side nestles between the kidney, the diaphragm, and the major arteries and veins running along the dorsal wall of the abdominopelvic cavity. The adrenal glands project into the peritoneal cavity, and their anterior surfaces are covered by a layer of parietal peritoneum. Like other endocrine glands, the adrenal glands are highly vascularized.

A typical adrenal gland weighs about 7.5 g, but adrenal size can vary greatly as secretory demands change. The adrenal gland is divided into two parts: a superficial **adrenal cortex** and an inner **adrenal medulla** (Figure 18-15●).

The Adrenal Cortex

Figure 18-15

The yellowish color of the adrenal cortex is due to the presence of stored lipids, especially cholesterol and various fatty acids. The adrenal cortex produces more than two dozen different steroid hormones, collectively called **adrenocortical steroids,** or simply **corticosteroids.** In the bloodstream, these hormones are bound to transport proteins called **transcortins.**

Corticosteroids are vital; if the adrenal glands are destroyed or removed, corticosteroids must be administered or the individual will not survive. Corticosteroids, like other steroid hormones, exert their effects by determining which genes are transcribed in the nuclei of their target cells and at what rates. The resulting changes in the nature and concentration of enzymes in the cytoplasm affect cellular metabolism.

Beneath the adrenal capsule there are three distinct regions, or *zones,* in the adrenal cortex: (1) an outer, *zona glomerulosa;* (2) a middle, *zona fasciculata;* and (3) an inner, *zona reticularis* (Figure 18-15●). Each zone synthesizes different steroid hormones, detailed in Table 18-4.

The Zona Glomerulosa

Figure 18-15

The **zona glomerulosa** (glo-mer-ū-LŌ-sa) produces **mineralocorticoids (MCs),** steroid hormones that

TABLE 18-3 Hormones of the Thyroid, Parathyroids, and Thymus

Gland/Cells	*Hormones*	*Targets*	*Effects*	*Regulatory Control*
THYROID				
Follicular epithelium	Thyroxine (T_4), triiodothyronine (T_3)	Most cells	Increases energy utilization, oxygen consumption, growth, and development (*see Table 18-2*)	Stimulated by TSH from anterior pituitary (*see Figure 18-9a*)
C cells	Calcitonin (CT)	Bone, kidneys, parathyroid glands	Decreases calcium ion concentrations in body fluids (*see Figure 18-14*)	Stimulated by elevated blood Ca^{2+} levels
PARATHYROIDS				
Chief cells	Parathyroid hormone (PTH)	Bone, kidneys, intestines	Increases calcium ion concentrations in body fluids (*see Figure 18-14*)	Stimulated by low blood Ca^{2+} levels; inhibited by calcitonin
THYMUS	Thymosins (thymosin-α, -β, thymosin V, thymopoietin, thymulin, thymolymphotropin, thymic-factor X)	Lymphocytes	Stimulates development and maturation of immune system	Unknown

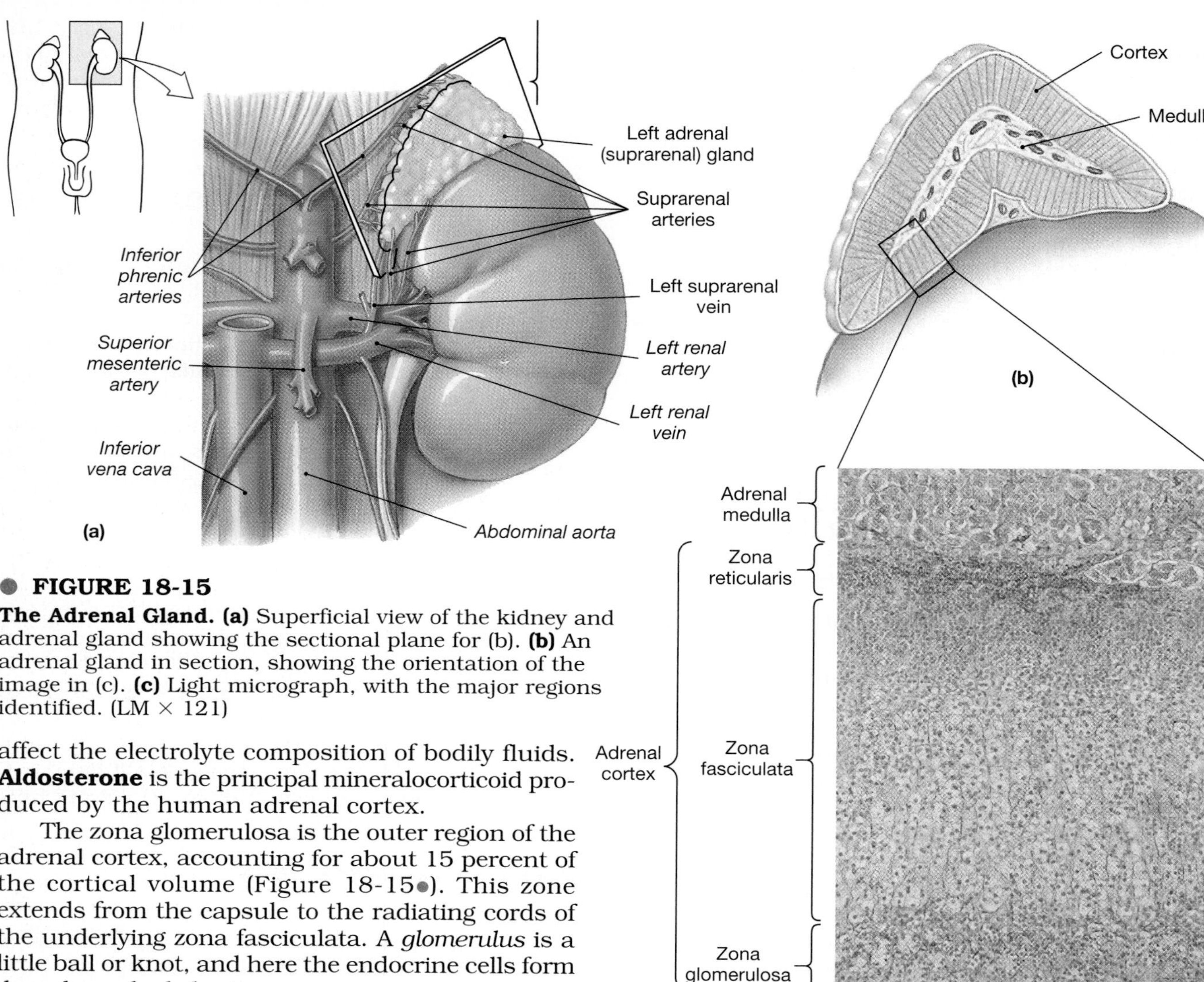

Left adrenal (suprarenal) gland
Suprarenal arteries
Inferior phrenic arteries
Left suprarenal vein
Superior mesenteric artery
Left renal artery
Left renal vein
Inferior vena cava
Abdominal aorta
(a)
Cortex
Medulla
(b)
Adrenal medulla
Zona reticularis
Adrenal cortex
Zona fasciculata
Zona glomerulosa
Capsule
(c)

FIGURE 18-15
The Adrenal Gland. (a) Superficial view of the kidney and adrenal gland showing the sectional plane for (b). **(b)** An adrenal gland in section, showing the orientation of the image in (c). **(c)** Light micrograph, with the major regions identified. (LM × 121)

affect the electrolyte composition of bodily fluids. **Aldosterone** is the principal mineralocorticoid produced by the human adrenal cortex.

The zona glomerulosa is the outer region of the adrenal cortex, accounting for about 15 percent of the cortical volume (Figure 18-15●). This zone extends from the capsule to the radiating cords of the underlying zona fasciculata. A *glomerulus* is a little ball or knot, and here the endocrine cells form densely packed clusters.

ALDOSTERONE Aldosterone targets kidney cells that regulate the ionic composition of the urine. It causes the retention of sodium ions at the kidneys, preventing their loss in the urine. As a secondary effect, the reabsorption of sodium ions will enhance water reabsorption if normal levels of ADH are present. Aldosterone also reduces sodium and water losses at the sweat glands, at the salivary glands, and along the digestive tract. The retention of sodium is accompanied by a loss of potassium ions, and elevated serum potassium levels will also trigger aldosterone secretion.

Aldosterone secretion occurs when the zona glomerulosa is stimulated by the hormone *angiotensin II* (*angeion*, vessel + *teinein*, to stretch). The appearance of this hormone in the circulation involves a series of steps that begins with the secretion of an enzyme, *renin*, by kidney cells. In addition to its other effects, angiotensin II stimulates the production of aldosterone and thereby assists in the restoration of normal blood volume and pressure. Renin and angiotensin will be discussed further in a later section.

DISORDERS OF ALDOSTERONE SECRETION In **hypoaldosteronism** the zona glomerulosa fails to produce enough aldosterone, usually because the kidneys are not releasing adequate amounts of renin. Affected individuals lose excessive amounts of water and sodium ions at the kidneys, and the water loss leads to low blood volume and low blood pressure. Changes in electrolyte concentrations affect transmembrane potentials, eventually disrupting neural and muscular tissues throughout the body.

Hypersecretion of aldosterone results in the condition of **aldosteronism.** Under continued aldosterone stimulation, the kidneys retain sodium ions very effectively, but potassium ions are lost in large quantities. A crisis eventually develops when low extracellular potassium concentrations disrupt normal cardiac, neural, and kidney cell functions. AM *Disorders of the Adrenal Cortex*

TABLE 18-4 The Adrenal Hormones

Region/Zone	*Hormone*	*Targets*	*Effects*	*Regulatory Control*
CORTEX				
Zona glomerulosa	Mineralocorticoids (MC), primarily aldosterone	Kidneys	Increases renal reabsorption of sodium ions and water; and accelerates urinary loss of potassium ions	Stimulated by angiotensin II; inhibited by ANP (*see Figure 18-17*)
Zona fasciculata	Glucocorticoids (GC): cortisol (hydrocortisone), corticosterone, cortisone	Most cells	Releases amino acids from skeletal muscles, lipids from adipose tissues; promotes liver glycogen and glucose formation; promotes peripheral utilization of lipids; anti-inflammatory effects	Stimulated by ACTH from anterior pituitary (*see Figure 18-9a*)
Zona reticularis	Androgens		Uncertain significance under normal conditions	Stimulated by ACTH; significance uncertain
MEDULLA	Epinephrine (adrenaline, E), norepinephrine (noradrenaline, NE)	Most cells	Increased cardiac activity, blood pressure, glycogen breakdown, blood glucose; release of lipids by adipose tissue (*see Chapter 16*)	Stimulated during sympathetic activation by sympathetic preganglionic fibers

The Zona Fasciculata

Figure 18-9a

The **zona fasciculata** (fa-sik-ū-LA-ta; *fasciculus*, little bundle) produces steroid hormones collectively known as **glucocorticoids (GC)** because of their effects on glucose metabolism. This zone begins at the inner border of the zona glomerulosa and extends toward the adrenal medulla. It contributes about 78 percent of the cortical volume. The endocrine cells are larger and contain more lipid than those of the zona glomerulosa, and the lipid droplets give the cytoplasm a pale, foamy appearance. The cells of the zona fasciculata form cords that radiate like a sunburst from the zona reticularis. The individual cords are stacks of cells, and adjacent cords are separated by flattened vessels with fenestrated walls.

THE GLUCOCORTICOIDS When stimulated by ACTH from the anterior pituitary, the zona fasciculata secretes primarily **cortisol** (KOR-ti-sol), also called *hydrocortisone*, along with smaller amounts of the related steroids **corticosterone** (kor-ti-KOS-te-rōn) and **cortisone.** Glucocorticoid secretion is regulated by negative feedback: The glucocorticoids released have an inhibitory effect on the production of corticotropin-releasing hormone (CRH) in the hypothalamus and ACTH in the anterior pituitary. This relationship was diagrammed in Figure 18-9a●, p. 613.

EFFECTS OF GLUCOCORTICOIDS Glucocorticoids accelerate the rates of glucose synthesis and glycogen formation, especially within the liver. Adipose tissue responds by releasing fatty acids into the blood, and other tissues begin to break down fatty acids and proteins instead of glucose. This process is another example of a *glucose-sparing effect.*

Glucocorticoids also show **anti-inflammatory** effects by inhibiting the activities of white blood cells and other components of the immune system. "Steroid creams" are often used to control irritating allergic rashes, such as those produced by poison ivy, and injections of glucocorticoids may be used to control more severe allergic reactions. Glucocorticoids slow the migration of phagocytic cells into an injury site and cause phagocytic cells already in the area to become less active. In addition, mast cells exposed to these steroids are less likely to release the chemicals that promote inflammation. As a result, swelling and further irritation are dramatically reduced. On the negative side, the rate of wound healing decreases, and the weakening of the region's defenses makes it an easy target for infecting organisms. For this reason, topical steroids are used to treat superficial rashes but are never applied to open wounds.

ABNORMAL GLUCOCORTICOID PRODUCTION **Addison's disease** is a condition caused primarily by inadequate glucocorticoid production (see Figure 18-22d•, p. 643). The usual cause is destruction of the zona fasciculata, either as the result of an *autoimmune response*, in which the body's defenses mistakenly attack normal tissues, or after infection by the bacteria responsible for *tuberculosis*. Affected individuals become weak and lose weight; they cannot effectively use their lipid reserves to generate ATP, and blood glucose concentrations fall sharply within hours after a meal. Because the zona glomerulosa is usually affected as well, symptoms similar to those of hypoaldosteronism may also appear.

Cushing's disease results from overproduction of glucocorticoids, usually because of hypersecretion of ACTH. The symptoms resemble those of a protracted and exaggerated response to stress (see Figure 18-22e•, p. 643). Glucose metabolism is suppressed, lipid reserves are mobilized, and peripheral proteins are broken down. Lipid reserves are redistributed, the distribution of body fat changes, and individuals develop a "moon-faced" appearance due to the deposition of lipids in the subcutaneous tissues of the face. Skeletal muscles become weak because of the breakdown of contractile proteins. AM *Disorders of the Adrenal Cortex*

The Zona Reticularis

Figure 18-15

The **zona reticularis** (re-tik-yū-LAR-is; *reticulum*, network) normally produces small quantities of the sex hormones called androgens. Androgens are produced in large quantities by the male testes, and the significance of the small adrenal contribution—in both males *and* females—remains uncertain. ACTH stimulates the zona reticularis to a slight degree, but the effects are usually insignificant.

The zona reticularis forms a narrow band bordering the adrenal medulla (Figure 18-15•). In total, the zona reticularis accounts for only around 7 percent of the total volume of the adrenal cortex. The endocrine cells of the zona reticularis form a folded, branching network, and fenestrated blood vessels wind between the cells.

ABNORMAL ACTIVITY IN THE ZONA RETICULARIS The zona reticularis ordinarily produces a negligible amount of androgens. If a tumor develops in this portion of the adrenal cortex, androgen secretion may increase dramatically, producing symptoms of the **adrenogenital syndrome.** In women, this condition leads to the gradual development of male secondary sexual characteristics, including body and facial hair patterns, adipose tissue distribution, and muscular development. Tumors affecting the zona reticularis of males may sometimes result in the production of large quantities of estrogens. In this condition the affected male develops enlarged breasts and, in some cases, other female secondary sexual characteristics. This array of symptoms is called **gynecomastia** (*gynaikos*, woman + *mastos*, breast).

❑ The Adrenal Medulla

The boundary between the adrenal cortex and the adrenal medulla does not form a straight line, and the supporting connective tissues and blood vessels are extensively interconnected. The adrenal medulla has a reddish brown coloration caused in part by the many blood vessels in this area. It contains large, rounded cells similar to those found in sympathetic autonomic ganglia, and these cells are innervated by preganglionic sympathetic fibers. The secretory activities of the adrenal medullae are controlled by the sympathetic division of the ANS, as detailed in Chapter 16. [p. 534]

The adrenal medullae contain two populations of secretory cells, one producing epinephrine (adrenaline) and the other norepinephrine (noradrenaline). There is evidence that the two cell types are distributed in different areas of the medulla and that their secretory activities can be independently controlled. The secretions are packaged in vesicles that form dense clusters just inside the cell membranes. The hormones within these vesicles are continually being released at low levels, through exocytosis. Sympathetic stimulation dramatically accelerates the rate of exocytosis and hormone release.

Epinephrine and Norepinephrine

Epinephrine makes up 75–80 percent of the secretions from the medullae, the rest being norepinephrine. The peripheral effects of these hormones, which result from interaction with alpha and beta receptors on cell membranes, were detailed in Chapter 16. [p. 535] Stimulation of α_1 and β_1 receptors, the most common types, accelerates cellular energy utilization and the mobilization of energy reserves.

When the adrenal medulla is activated:

- Epinephrine and norepinephrine trigger a mobilization of glycogen reserves in skeletal muscles and accelerate the breakdown of glucose to provide ATP. This combination increases muscular strength and endurance.
- In adipose tissue, stored fats are broken down to fatty acids and the fatty acids are released into the circulation for use by other tissues.
- In the liver, glycogen molecules are broken down and the resulting glucose molecules are released into the circulation, primarily for use by neural tissues, which cannot shift to fatty acid metabolism.
- In the heart, stimulation of β_1 receptors triggers an increase in the rate and force of cardiac muscle contraction.

The metabolic changes that follow catecholamine release are at their peak 30 seconds after adrenal stimulation and linger for several minutes thereafter. As a result, the effects produced by stimulation of the adrenal medullae outlast the other signs of sympathetic activation. AM *Disorders of the Adrenal Medulla*

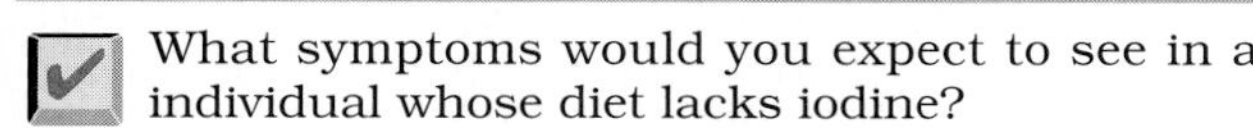

What symptoms would you expect to see in an individual whose diet lacks iodine?

When a person's thyroid gland is removed, signs of decreased thyroid hormone concentration do not appear until about one week later. Why?

Removal of the parathyroid glands would result in a decrease in the blood concentration of what important mineral?

What effect would elevated cortisol levels have on the level of glucose in the blood?

The Kidneys

The kidneys release three hormones: *calcitriol, erythropoietin,* and *renin.* Calcitriol is important for calcium ion homeostasis. Renin and erythropoietin are involved in the regulation of blood pressure and blood volume.

Calcitriol

Figure 18-16a

Calcitriol is a steroid hormone secreted by the kidney in response to the presence of parathyroid hormone (PTH) (Figure 18-16a●). Calcitriol synthesis is dependent on the availability of a related steroid, *cholecalciferol* (vitamin D_3), which may be synthe-

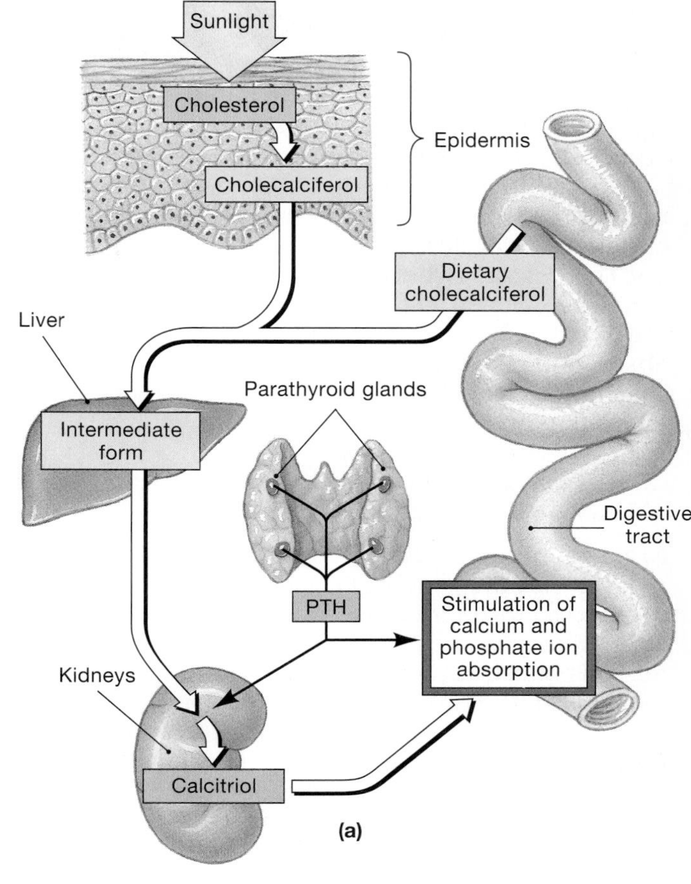
Sunlight
Cholesterol
Cholecalciferol
Epidermis
Dietary cholecalciferol
Liver
Intermediate form
Parathyroid glands
Digestive tract
PTH
Stimulation of calcium and phosphate ion absorption
Kidneys
Calcitriol
(a)

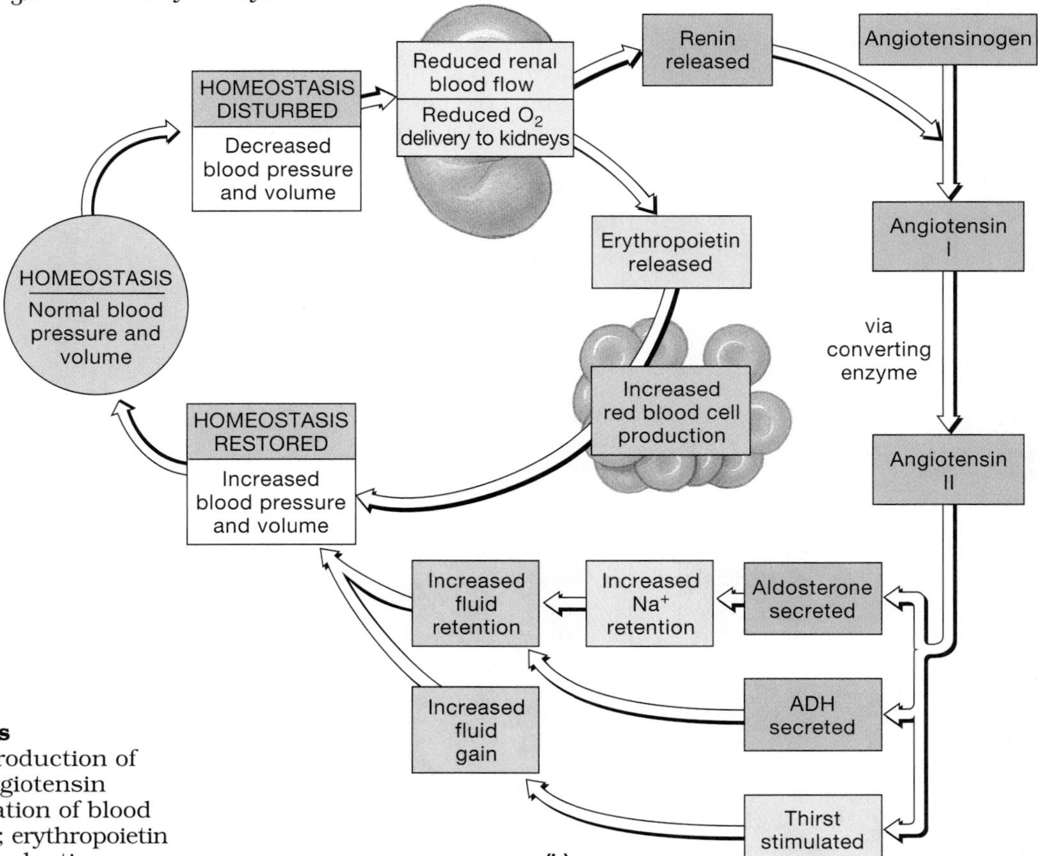
HOMEOSTASIS DISTURBED
Decreased blood pressure and volume
Reduced renal blood flow
Reduced O_2 delivery to kidneys
Renin released
Angiotensinogen
Angiotensin I
via converting enzyme
Angiotensin II
Erythropoietin released
Increased red blood cell production
HOMEOSTASIS
Normal blood pressure and volume
HOMEOSTASIS RESTORED
Increased blood pressure and volume
Increased fluid retention
Increased Na^+ retention
Aldosterone secreted
ADH secreted
Increased fluid gain
Thirst stimulated
(b)

● **FIGURE 18-16**

The Endocrine Functions of the Kidneys. (a) The production of calcitriol. **(b)** The renin-angiotensin system assists in the elevation of blood volume and blood pressure; erythropoietin enhances red blood cell production.

sized in the skin or absorbed from the diet. Cholecalciferol from either source is absorbed from the bloodstream by the liver and converted to an intermediary product. As mentioned in Chapter 5, this intermediary is released into the circulation and absorbed by the kidneys for conversion to calcitriol. [p. 155] The term *vitamin D* is used to indicate the entire group of related steroids, including calcitriol, cholecalciferol, and various intermediaries.

The best-known function of calcitriol is stimulation of calcium and phosphate ion absorption along the digestive tract. The effects of PTH on calcium ion absorption result at least partially via stimulation of calcitriol release. The effects of calcitriol on the skeletal system and kidney are not well understood.

❑ Erythropoietin

Figure 18-16b

Erythropoietin (e-rith-rō-poy-Ē-tin; *erythros*, red + *poiesis*, making), or **EPO,** is a peptide hormone released by the kidney in response to low oxygen levels in kidney tissues (Figure 18-16b●). The reduced oxygen concentrations may result from (1) reduced renal blood flow caused by blood loss, low blood pressure, or blockage of renal vessels; (2) a reduction in the number or the oxygen-carrying capacity of circulating red blood cells; (3) a reduction in the oxygen content of the air in the lungs, due to respiratory problems or increased altitude; or (4) problems at the respiratory surfaces of the lungs.

EPO stimulates the production of red blood cells by the bone marrow. The increase in the number of erythrocytes elevates blood volume to some degree, and, because these cells transport oxygen, the increase in their number improves oxygen delivery to peripheral tissues. EPO will be considered in greater detail when we discuss the formation of blood cells in Chapter 19.

❑ Renin

Figure 18-16b

Renin is released by kidney cells in response to (1) sympathetic stimulation or (2) a decline in renal blood flow. Reduced renal blood flow may be caused by a fall in blood volume, blood pressure, or both, resulting from hemorrhage, dehydration, or other factors. Once in the bloodstream, renin functions as an enzyme that starts an enzymatic chain reaction, diagrammed in Figure 18-16b●.

The initial step occurs when renin acts on **angiotensinogen,** a plasma protein produced by the liver. Renin converts angiotensinogen to **angiotensin I.** In the capillaries of the lungs, a **converting enzyme** then modifies angiotensin I to **angiotensin II,** an active hormone with diverse effects. Angiotensin II has three major effects. This hormone

- stimulates the adrenal production of aldosterone, causing sodium ion retention and enhancing water reabsorption at the kidneys. This effect prevents further reductions in blood volume due to urinary water losses.
- stimulates the secretion of ADH, stimulating water reabsorption at the kidneys and complementing the effects of aldosterone.
- stimulates thirst, resulting in increased fluid consumption and an elevation of blood volume.

We will examine the actions of renin and angiotensin II in more detail in Chapters 21 and 26.

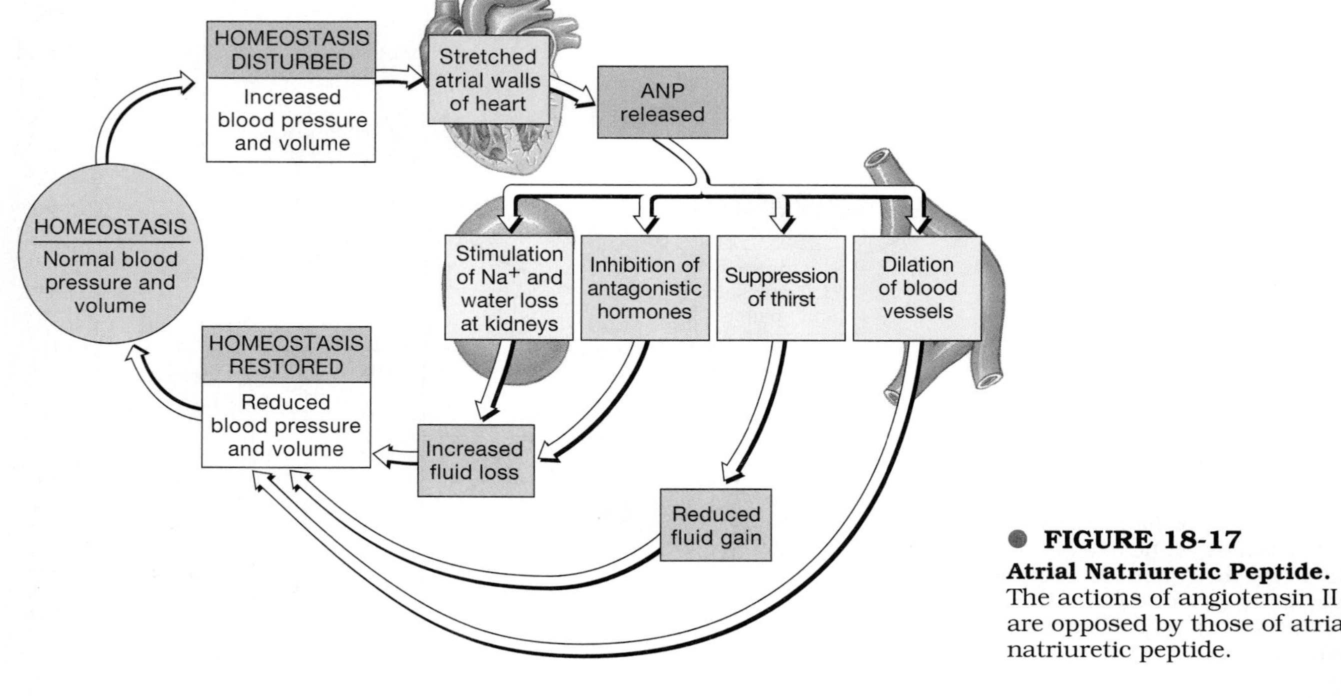
HOMEOSTASIS DISTURBED
Increased blood pressure and volume
Stretched atrial walls of heart
ANP released
HOMEOSTASIS
Normal blood pressure and volume
Stimulation of Na^+ and water loss at kidneys
Inhibition of antagonistic hormones
Suppression of thirst
Dilation of blood vessels
HOMEOSTASIS RESTORED
Reduced blood pressure and volume
Increased fluid loss
Reduced fluid gain

● **FIGURE 18-17**
Atrial Natriuretic Peptide. The actions of angiotensin II are opposed by those of atrial natriuretic peptide.

The Heart

Figure 18-17

The endocrine cells in the heart are cardiac muscle cells in the walls of the *atria*, chambers that receive venous blood. If the blood volume becomes too great, these cardiac muscle cells are excessively stretched. Under these conditions, they release **atrial natriuretic** (nā-trē-ū-RET-ik) **peptide (ANP)** (*natrium*, sodium + *ouresis*, making water). ANP has three effects that oppose those of angiotensin II (Figure 18-17●). This hormone

- promotes the loss of sodium ions and water at the kidneys.
- inhibits the secretion of water-conserving hormones, such as ADH and aldosterone.
- suppresses thirst.

This combination lowers blood volume and reduces the stretching of the cardiac muscle cells in the atrial walls. We will consider the actions of this hormone in greater detail when we discuss the control of blood pressure and volume in Chapter 21.

How would an increase in the amount of atrial natriuretic peptide affect the concentration of sodium ions in the urine?

Endocrine Tissues of the Digestive System

The linings of the digestive tract, the liver, and the pancreas produce a variety of exocrine secretions that are essential to the normal breakdown and absorption of food. Although the pace of digestive activities can be affected by the autonomic nervous system, most digestive processes are controlled locally. The various components of the digestive tract communicate with one another by means of hormones that will be considered in Chapter 24. One digestive organ, the pancreas, produces two hormones with widespread effects.

The Pancreas

Figures 18-18, 18-19

The **pancreas** lies within the abdominopelvic cavity in the J-shaped loop between the stomach and small intestine (Figure 18-18a●). It is a slender, usually pink organ with a nodular (lumpy) consistency. The adult pancreas ranges between 20 and 25 cm (8–10 in.) in length and weighs about 80 g (2.8 oz). The detailed anatomy of the pancreas is considered in Chapter 24 because it is primarily a digestive organ that manufactures digestive enzymes. The **exocrine pancreas,** roughly 99 percent of the pancreatic volume, produces large quantities

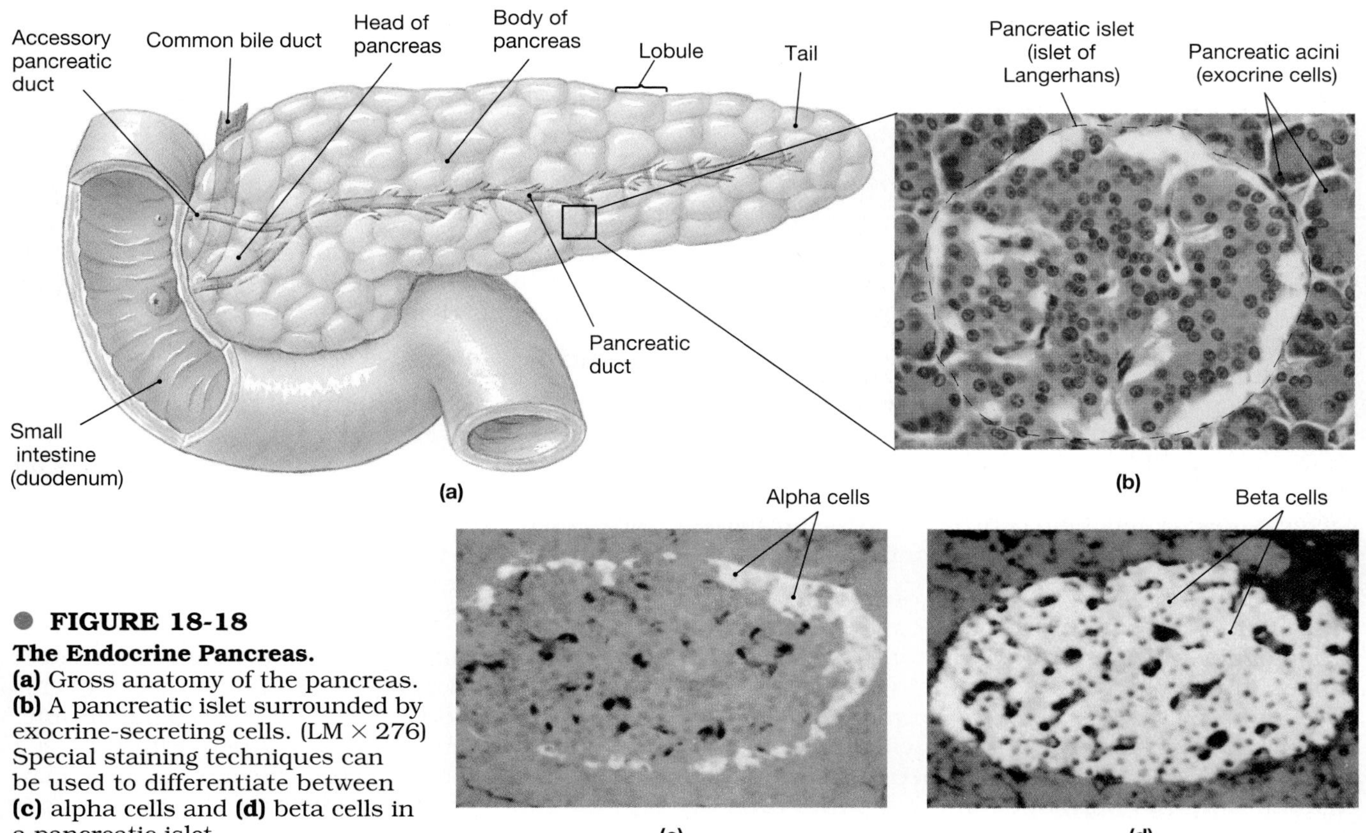
Accessory pancreatic duct
Common bile duct
Head of pancreas
Body of pancreas
Lobule
Tail
Pancreatic duct
Small intestine (duodenum)
(a)
Pancreatic islet (islet of Langerhans)
Pancreatic acini (exocrine cells)
(b)
Alpha cells
(c)
Beta cells
(d)

● **FIGURE 18-18**
The Endocrine Pancreas.
(a) Gross anatomy of the pancreas. **(b)** A pancreatic islet surrounded by exocrine-secreting cells. (LM × 276) Special staining techniques can be used to differentiate between **(c)** alpha cells and **(d)** beta cells in a pancreatic islet.

of an alkaline, enzyme-rich fluid. This secretion reaches the lumen of the digestive tract by traveling along a network of secretory ducts.

The **endocrine pancreas** consists of small groups of cells scattered among the exocrine cells. The endocrine clusters are known as **pancreatic islets,** or the *islets of Langerhans* (LAN-ger-hanz) (Figure 18-18b●). Pancreatic islets account for only about 1 percent of the pancreatic cell population. Nevertheless, the normal pancreas contains roughly 2 million pancreatic islets.

Like other endocrine tissue, the pancreatic islets are surrounded by an extensive, fenestrated capillary network that carries its hormones into the circulation. Each islet contains four different cell types:

1. **Alpha cells** (Figure 18-18c●) produce the hormone **glucagon** (GLOO-ka-gon). Glucagon raises blood glucose levels by increasing the rates of glycogen breakdown and glucose release by the liver.
2. **Beta cells** (Figure 18-18d●) produce **insulin** (IN-su-lin). Insulin lowers blood glucose by increasing the rate of glucose uptake and utilization by most body cells, and glycogen synthesis in skeletal muscles and the liver. Beta cells also secrete *amylin,* a recently discovered peptide hormone whose role is uncertain.
3. **Delta cells** produce a peptide hormone identical to **somatostatin** (GH-IH), a hypothalamic regulatory hormone. Somatostatin produced in the pancreas suppresses glucagon and insulin release by other islet cells and slows the rates of food absorption and enzyme secretion along the digestive tract.
4. **F-cells** produce the hormone **pancreatic polypeptide (PP).** Little is known about the specific functions of pancreatic polypeptide. It inhibits gallbladder contractions and it regulates the production of some pancreatic enzymes; it may help control the rate of nutrient absorption by the digestive tract.

We will focus our attention on insulin and glucagon, the hormones responsible for the regulation of blood glucose concentrations. These hormones interact to control blood glucose levels in the same way that parathyroid hormone and calcitonin interact to control blood calcium levels. When blood glucose levels rise, beta cells secrete insulin, and this hormone stimulates the transport of glucose across cell membranes. When blood glucose levels decline, alpha cells secrete glucagon, which stimulates glucose release by the liver. This relationship is diagrammed in Figure 18-19●.

● **FIGURE 18-19**
The Regulation of Blood Glucose Concentrations

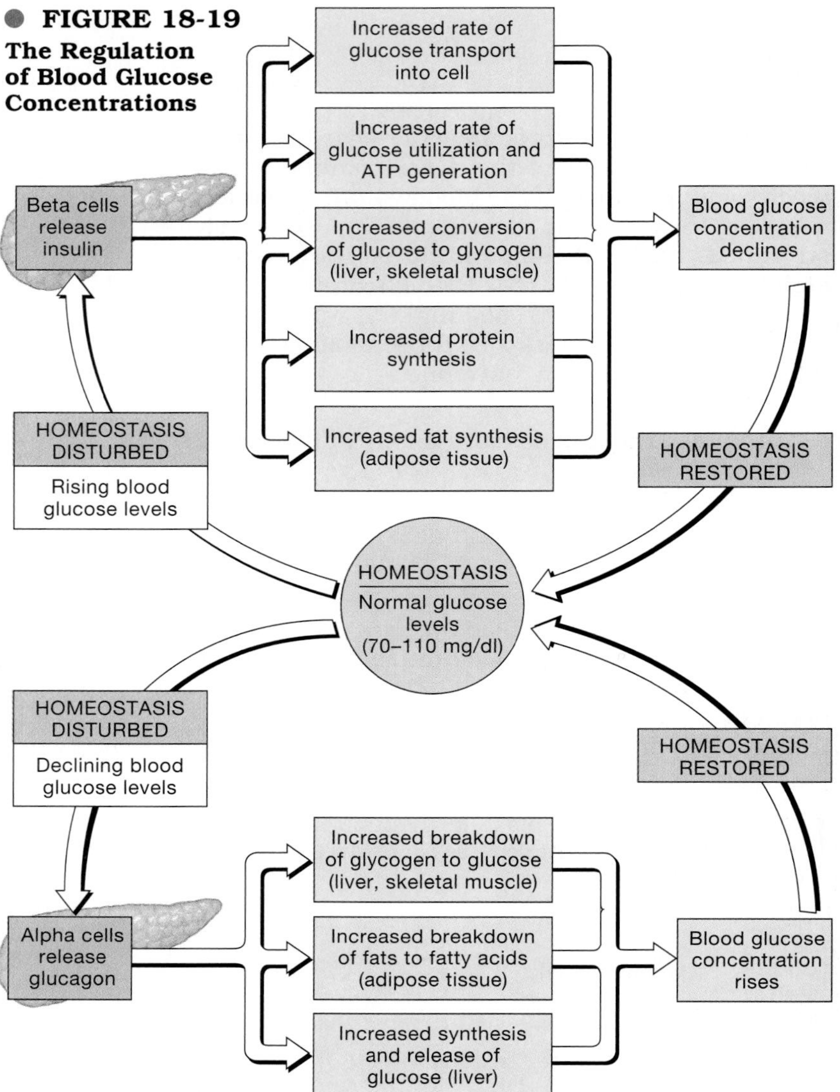
Increased rate of glucose transport into cell
Increased rate of glucose utilization and ATP generation
Beta cells release insulin
Increased conversion of glucose to glycogen (liver, skeletal muscle)
Blood glucose concentration declines
Increased protein synthesis
HOMEOSTASIS DISTURBED
Rising blood glucose levels
Increased fat synthesis (adipose tissue)
HOMEOSTASIS RESTORED
HOMEOSTASIS
Normal glucose levels (70–110 mg/dl)
HOMEOSTASIS DISTURBED
Declining blood glucose levels
HOMEOSTASIS RESTORED
Increased breakdown of glycogen to glucose (liver, skeletal muscle)
Alpha cells release glucagon
Increased breakdown of fats to fatty acids (adipose tissue)
Blood glucose concentration rises
Increased synthesis and release of glucose (liver)

Insulin

Insulin is a peptide hormone that is released by beta cells when glucose levels rise above normal levels (70–110 mg/dl). Insulin exerts its effects on cellular metabolism in a series of steps that begins when insulin binds to receptor proteins on the cell membrane. Second messengers in the cytoplasm then produce primary and secondary effects within the cell. Those effects include:

- ***Acceleration of glucose uptake.*** This effect results from an increase in the number of glucose transport proteins in the cell membrane. These proteins transport glucose into the cell via facilitated diffusion. This movement follows the concentration gradient for glucose, and ATP is not required.

- ***Acceleration of glucose utilization by the cell.*** This effect occurs because (1) the rate of glucose use is proportional to its availability—when more glucose enters the cell, more is used, and (2) second messengers activate a key enzyme involved in the initial steps of glycolysis.
- ***Stimulation of glycogen formation in skeletal muscles and liver cells.*** When excess glucose enters these cells, it is stored in the form of glycogen.
- ***Stimulation of triglyceride formation in adipose tissues.*** Insulin stimulates the absorption of fatty acids and glycerol by adipose cells. The cells then store these components as triglycerides.
- ***Stimulation of amino acid absorption and protein synthesis.***

In summary, insulin secretion occurs when glucose is abundant, and this hormone stimulates glucose utilization to support growth and the establishment of carbohydrate (glycogen) and lipid (triglyceride) reserves. The accelerated use of glucose soon brings circulating glucose levels within normal limits.

Insulin receptors are present in most cell membranes, and these cells are called *insulin-dependent.* However, cells in the brain and kidneys, the lining of the digestive tract, and red blood cells lack insulin receptors. These are called *insulin-independent* cells because they can absorb and utilize glucose without insulin stimulation.

Glucagon

When glucose concentrations fall below normal, alpha cells release glucagon, and energy reserves are mobilized. When glucagon binds to a receptor in the cell membrane, it activates adenylate cyclase, and cAMP acts as a second messenger that activates cytoplasmic enzymes. The primary effects of glucagon are:

- ***Stimulation of glycogen breakdown in skeletal muscles and liver cells.*** The glucose molecules released will either be metabolized for energy (skeletal muscle fibers) or released into the bloodstream (liver cells).
- ***Stimulation of triglyceride breakdown in adipose tissues.*** The adipose cells then release the fatty acids into the circulation for use by other tissues.
- ***Stimulation of glucose production at the liver.*** Liver cells absorb amino acids from the bloodstream, convert them to glucose, and release the glucose into the circulation. The process of glucose synthesis in the liver is called *gluconeogenesis* (glū-kō-nē-ō-JEN-e-sis).

The net result is a reduction in glucose use and the release of additional glucose into the bloodstream. As a result, blood glucose concentrations soon rise toward normal levels.

Pancreatic alpha and beta cells monitor blood glucose concentrations, and secretion of glucagon and insulin occur without endocrine or nervous instructions. Yet, because they are so sensitive to changes in blood glucose levels, any hormone that affects blood glucose concentrations will indirectly affect the production of insulin and glucagon. Insulin production is also influenced by autonomic activity. Parasympathetic stimulation enhances insulin release, and sympathetic stimulation inhibits it.

Information concerning insulin, glucagon, and other pancreatic hormones is summarized in Table 18-5.

TABLE 18-5 Hormones of the Pancreas

Structure/ Cells	*Hormone*	*Primary Targets*	*Effects*	*Regulatory Control*
PANCREATIC ISLETS				
Alpha cells	Glucagon	Liver, adipose tissues	Mobilizes lipid reserves; promotes glucose synthesis and glycogen breakdown in liver; elevates blood glucose concentrations	Stimulated by low blood glucose concentrations; inhibited by GH-IH from delta cells
Beta cells	Insulin	Most cells	Facilitates uptake of glucose by cells; stimulates lipid and glycogen formation and storage	Stimulated by high blood glucose concentrations and/or parasympathetic stimulation; inhibited by GH-IH from delta cells and by sympathetic activation
Delta cells	GH-IH (somatostatin)	Other islet cells, digestive epithelium	Inhibits insulin and glucagon secretion, slows rates of food absorption and enzyme secretion along digestive tract	Stimulated by protein-rich meal; mechanism uncertain
F-cells	Pancreatic polypeptide	Digestive organs	Inhibits gallbladder contraction, regulates production of pancreatic enzymes; influences rate of nutrient absorption by digestive tract	Stimulated by protein-rich meal and by parasympathetic stimulation

EMBRYOLOGY SUMMARY Development of the Endocrine System

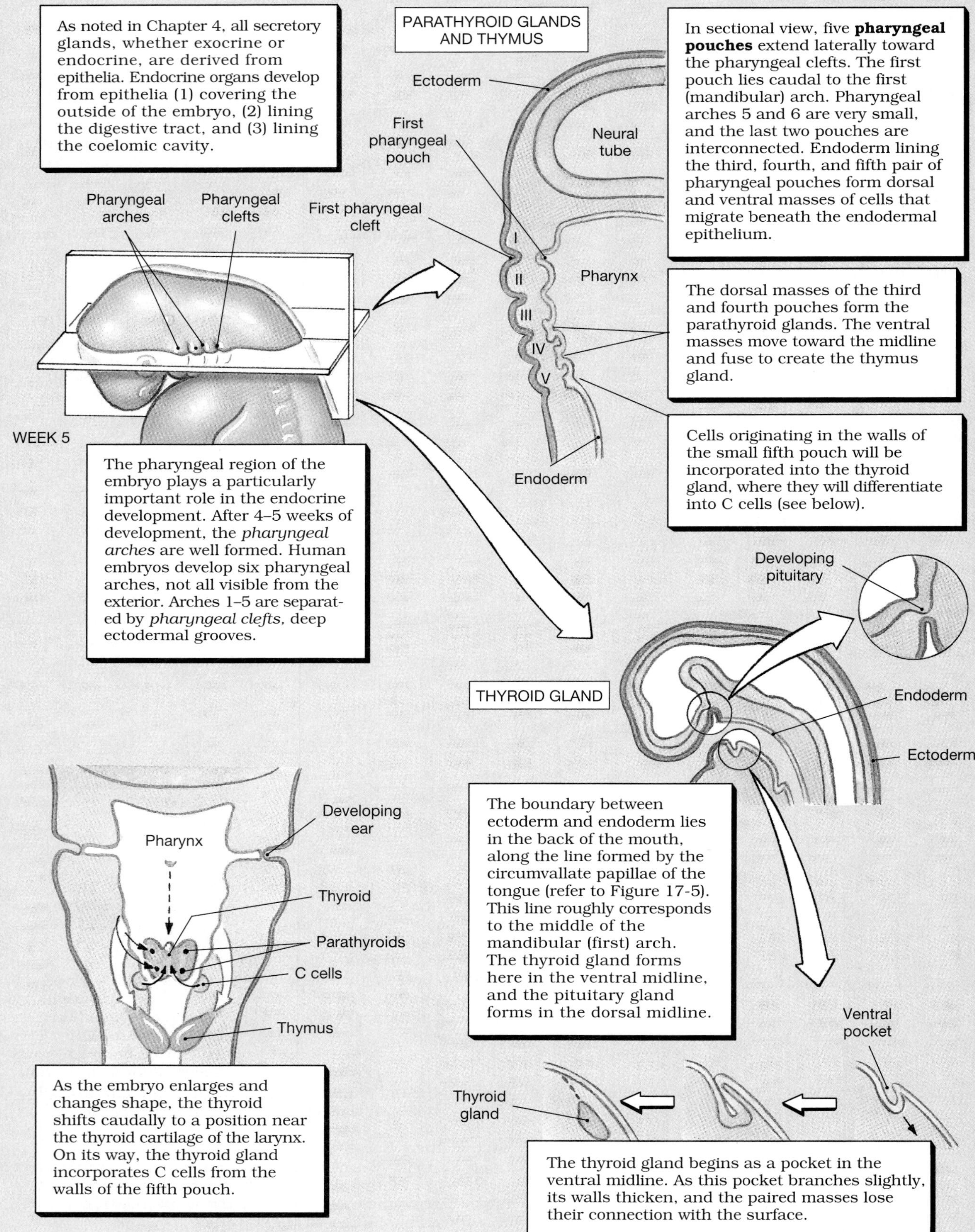
As noted in Chapter 4, all secretory glands, whether exocrine or endocrine, are derived from epithelia. Endocrine organs develop from epithelia (1) covering the outside of the embryo, (2) lining the digestive tract, and (3) lining the coelomic cavity.
PARATHYROID GLANDS AND THYMUS
Ectoderm
First pharyngeal pouch
Neural tube
In sectional view, five **pharyngeal pouches** extend laterally toward the pharyngeal clefts. The first pouch lies caudal to the first (mandibular) arch. Pharyngeal arches 5 and 6 are very small, and the last two pouches are interconnected. Endoderm lining the third, fourth, and fifth pair of pharyngeal pouches form dorsal and ventral masses of cells that migrate beneath the endodermal epithelium.
Pharyngeal arches
Pharyngeal clefts
First pharyngeal cleft
I
II
III
IV
V
Pharynx
The dorsal masses of the third and fourth pouches form the parathyroid glands. The ventral masses move toward the midline and fuse to create the thymus gland.
WEEK 5
Cells originating in the walls of the small fifth pouch will be incorporated into the thyroid gland, where they will differentiate into C cells (see below).
Endoderm
The pharyngeal region of the embryo plays a particularly important role in the endocrine development. After 4–5 weeks of development, the *pharyngeal arches* are well formed. Human embryos develop six pharyngeal arches, not all visible from the exterior. Arches 1–5 are separated by *pharyngeal clefts*, deep ectodermal grooves.
Developing pituitary
THYROID GLAND
Endoderm
Ectoderm
Developing ear
Pharynx
Thyroid
Parathyroids
C cells
Thymus
The boundary between ectoderm and endoderm lies in the back of the mouth, along the line formed by the circumvallate papillae of the tongue (refer to Figure 17-5). This line roughly corresponds to the middle of the mandibular (first) arch. The thyroid gland forms here in the ventral midline, and the pituitary gland forms in the dorsal midline.
Ventral pocket
Thyroid gland
As the embryo enlarges and changes shape, the thyroid shifts caudally to a position near the thyroid cartilage of the larynx. On its way, the thyroid gland incorporates C cells from the walls of the fifth pouch.
The thyroid gland begins as a pocket in the ventral midline. As this pocket branches slightly, its walls thicken, and the paired masses lose their connection with the surface.

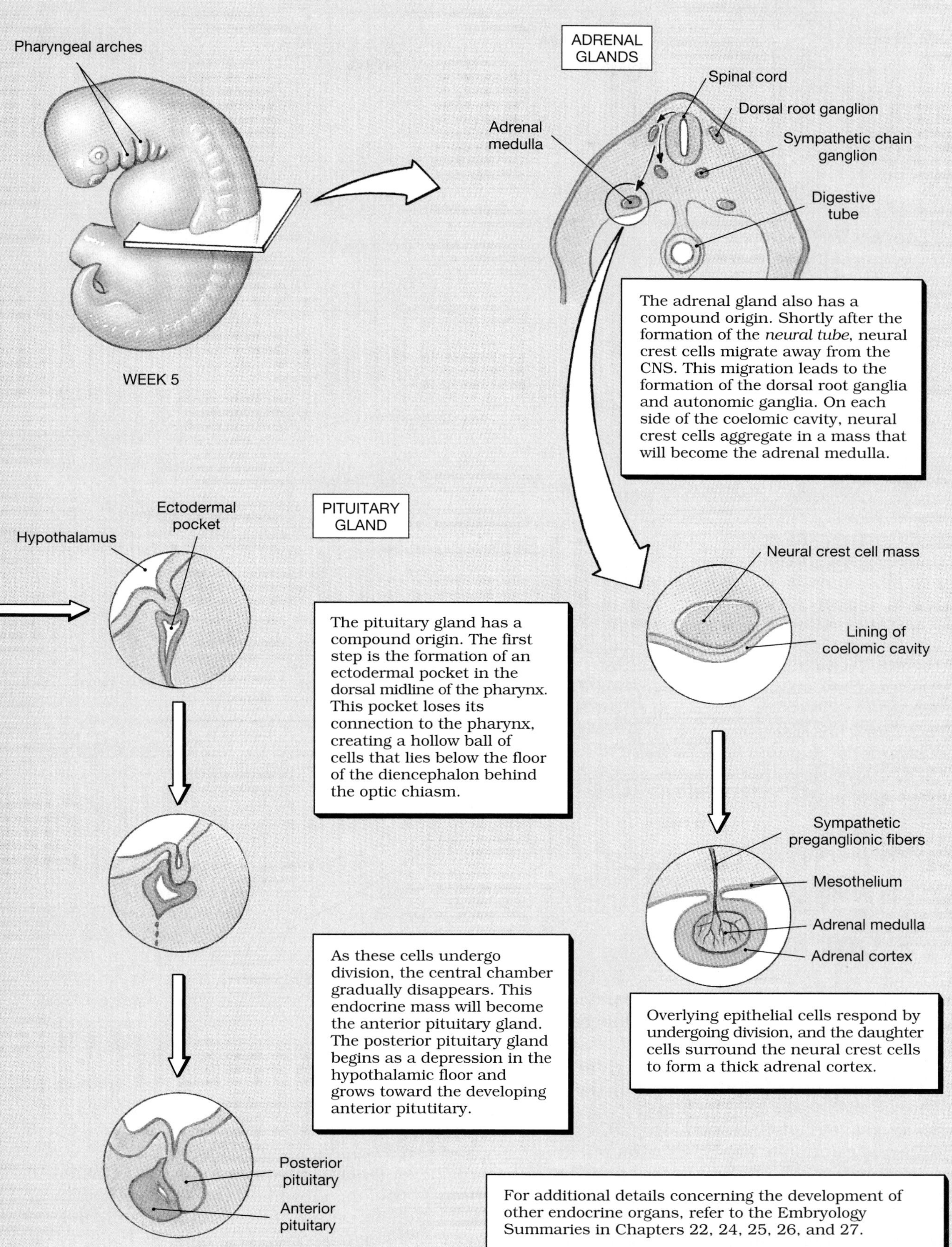

Pharyngeal arches
WEEK 5
ADRENAL GLANDS
Spinal cord
Dorsal root ganglion
Adrenal medulla
Sympathetic chain ganglion
Digestive tube
The adrenal gland also has a compound origin. Shortly after the formation of the *neural tube*, neural crest cells migrate away from the CNS. This migration leads to the formation of the dorsal root ganglia and autonomic ganglia. On each side of the coelomic cavity, neural crest cells aggregate in a mass that will become the adrenal medulla.
Ectodermal pocket
Hypothalamus
PITUITARY GLAND
Neural crest cell mass
Lining of coelomic cavity
The pituitary gland has a compound origin. The first step is the formation of an ectodermal pocket in the dorsal midline of the pharynx. This pocket loses its connection to the pharynx, creating a hollow ball of cells that lies below the floor of the diencephalon behind the optic chiasm.
Sympathetic preganglionic fibers
Mesothelium
Adrenal medulla
Adrenal cortex
As these cells undergo division, the central chamber gradually disappears. This endocrine mass will become the anterior pituitary gland. The posterior pituitary gland begins as a depression in the hypothalamic floor and grows toward the developing anterior pitutitary.
Overlying epithelial cells respond by undergoing division, and the daughter cells surround the neural crest cells to form a thick adrenal cortex.
Posterior pituitary
Anterior pituitary
For additional details concerning the development of other endocrine organs, refer to the Embryology Summaries in Chapters 22, 24, 25, 26, and 27.

Diabetes Mellitus

Blood glucose levels are usually very closely regulated by insulin and glucagon. Whether glucose is absorbed across the digestive tract or manufactured and released by the liver, very little leaves the body intact once it has entered the circulation. Glucose does get filtered out of the blood at the kidneys, but the cells lining the kidney tubules usually reabsorb virtually all of it, and urinary glucose losses are negligible.

Diabetes mellitus (mel-LI-tus; *mellitum,* honey) is characterized by glucose concentrations that are high enough to overwhelm the reabsorption capabilities of the kidneys. Glucose appears in the urine **(glycosuria),** and urine production usually becomes excessive (*polyuria*). Other metabolic products, such as fatty acids and other lipids, are also present in abnormal concentrations.

Diabetes mellitus may be caused by genetic abnormalities, and some of the genes responsible have been identified. For example, genes that result in inadequate insulin production, the synthesis of abnormal insulin molecules, or the production of defective receptor proteins will produce the same basic symptoms. Diabetes mellitus may also appear as the result of other pathological conditions, injuries, immune disorders, or hormonal imbalances. There are two major types of diabetes mellitus: *insulin-dependent (Type I) diabetes* and *non-insulin-dependent (Type II) diabetes.* Both are discussed in greater detail in the *Applications Manual.* AM Diabetes Mellitus

Probably because glucose levels cannot be stabilized adequately, even with treatment, patients with diabetes mellitus often develop chronic medical problems. In general, these problems are related to circulatory system abnormalities. The most common examples include the following:

1. Vascular changes at the retina, including proliferation of capillaries and hemorrhaging, which often cause partial or complete blindness. This condition, called *diabetic retinopathy,* was noted in Chapter 17. [p. 569]
2. Changes occur in the clarity of the lens, producing cataracts.
3. Small hemorrhages and inflammation at the kidneys cause degenerative changes that can lead to kidney failure. This condition is called **diabetic nephropathy.** Treatment with drugs that improve blood flow to the kidneys can slow the progression to kidney failure.
4. A variety of neural problems appear, including *nerve palsies* (Chapter 13) and abnormal autonomic function. [p. 436] These disorders, collectively termed **diabetic neuropathy,** are probably related to disturbances in the blood supply to neural tissues.
5. Degenerative changes in cardiac circulation can lead to early heart attacks. For a given age group, heart attacks are 3–5 times more likely in diabetic individuals.
6. Other peripheral changes in the vascular system can disrupt normal circulation to the extremities. For example, a reduction in blood flow to the feet can lead to tissue death, ulceration, infection, and loss of toes or a major portion of one or both feet.

Endocrine Tissues of the Reproductive System

The endocrine tissues of the reproductive system are restricted primarily to the male and female reproductive organs, the testes and ovaries. Details of the anatomy of the reproductive organs and the endocrinological control of reproductive function will be considered in Chapter 28. The primary regulatory hormones involved are FSH and LH from the anterior pituitary. Abnormally low production of either of these hormones will produce **hypogonadism.** Children with this condition will not undergo sexual maturation, and adults with hypogonadism will not produce and support functional sperm or eggs.

The Testes

In the male, the **interstitial cells** (*Leydig cells*) of the testis produce the male hormones known as androgens. **Testosterone** (tes-TOS-ter-ōn) is the most important androgen. This hormone promotes the production of functional sperm, maintains the secretory glands of the male reproductive tract, stimulates growth, and determines male secondary sexual characteristics such as the distribution of facial hair and body fat. Testosterone also affects metabolic operations throughout the body, notably stimulating protein synthesis and muscle growth, and it produces aggressive behavioral responses. During embryonic development, the production of testosterone affects the development of CNS structures, including hypothalamic nuclei that will later influence sexual behaviors.

Sustentacular cells (*Sertoli cells*) in the testes support the differentiation and physical maturation of spermatozoa. Under FSH stimulation these cells secrete **inhibin,** which inhibits the secretion of FSH at the anterior pituitary and perhaps suppresses GnRH release at the hypothalamus.

❑ The Ovaries

Female gametes are produced in specialized structures called **follicles.** Follicles develop under FSH stimulation, and the follicle cells surrounding each developing gamete, or *oocyte*, produce **estrogens** (ES-trō-jenz) under FSH and LH stimulation. Estrogens are steroid hormones that support the maturation of the oocytes and stimulate the growth of the uterine lining. **Estradiol** is the principal estrogen. Under FSH stimulation, follicle cells secrete inhibin, which suppresses FSH release through a feedback mechanism comparable to that in males.

A surge in LH secretion triggers ovulation, and LH then causes the follicular cells to reorganize into a **corpus luteum** (LOO-tē-um; "yellow body"). The luteal cells then begin to release a mixture of estrogens and **progestins. Progesterone** (prō-JES-ter-ōn), the principal progestin, has several important functions:

- It accelerates the movement of the oocyte or, if fertilization occurs, the developing embryo along the uterine tube.
- It prepares the uterus for the arrival of a developing embryo.
- It causes an enlargement of the mammary glands, working in combination with other hormones, such as estradiol, growth hormone, and prolactin.

Relaxin

During pregnancy the corpus luteum and uterus produce an additional hormone, **relaxin.** Relaxin levels are elevated throughout pregnancy. This hormone loosens the articulation at the pubic symphysis, permitting expansion of the lower portion of the uterus and the vagina during delivery. It also stimulates functional development of the mammary glands, in combination with many other hormones. Early in a pregnancy, relaxin production occurs under stimulation by LH from the anterior pituitary. Later in pregnancy, additional stimulation is provided by a placental hormone, *human chorionic gonadotropin,* or *hCG.* Relaxin secretion stops after delivery because of the expulsion of the placenta and the degeneration of the corpus luteum.

Information concerning the reproductive hormones of the testes and ovaries is summarized in Table 18-6.

During pregnancy, additional hormones produced by the placenta interact with those produced by the ovaries and the pituitary gland to

TABLE 18-6 Hormones of the Reproductive System

Structure/Cells	*Hormone*	*Primary Targets*	*Effects*	*Regulatory Control*
TESTES				
Interstitial cells	Androgens	Most cells	Support functional maturation of sperm, protein synthesis in skeletal muscles, male secondary sexual characteristics, and associated behaviors	Stimulated by LH from anterior pituitary (*see Figure 18-9a*)
Sustenacular cells	Inhibin	Anterior pituitary	Inhibits secretion of FSH	Stimulated by FSH from anterior pituitary (*see Figure 18-9a*)
OVARIES				
Follicular cells	Estrogens	Most cells	Support follicle maturation, female secondary sexual characteristics, and associated behaviors	Stimulated by FSH and LH from anterior pituitary (*see Figure 18-9a*)
	Inhibin	Anterior pituitary	Inhibits secretion of FSH	Stimulated by FSH from anterior pituitary (*see Figure 18-9a*)
Corpus luteum	Progestins	Uterus, mammary glands	Prepare uterus for implantation; prepare mammary glands for secretory functions	Stimulated by LH from anterior pituitary (*see Figure 18-9a*)
	Relaxin	Pubic symphysis, uterus, mammary glands	Loosens pubic symphysis, relaxes uterine (cervical) muscles, stimulates mammary gland development	Stimulated by LH from the anterior pituitary and by placental hCG (human chorionic gonadotropin)

promote normal fetal development and delivery. The endocrinological aspects of pregnancy will be considered in Chapter 29.

The Pineal Gland

The pineal gland is part of the epithalamus; it lies in the posterior portion of the roof of the third ventricle. The pineal gland contains neurons, glial cells, and special secretory cells called **pinealocytes** (pi-NĒ-al-ō-sīts). These cells synthesize the hormone **melatonin** (mel-a-TŌ-nin) from molecules of the neurotransmitter *serotonin*. Collaterals from the visual pathways enter the pineal gland and affect the rate of melatonin production. Melatonin production is lowest during daylight hours and highest at night.

Several functions have been suggested for melatonin in humans:

- In a variety of other mammals, melatonin slows the maturation of sperm, eggs, and reproductive organs by reducing the rate of GnRH secretion. The significance of this effect remains uncertain, but there is circumstantial evidence that melatonin may play a role in the timing of human sexual maturation. For example, melatonin levels in the blood decline at puberty, and pineal tumors that eliminate melatonin production will cause premature puberty in young children.
- Melatonin is a very effective *antioxidant* that may protect CNS neurons from free radicals, such as NO or H_2O_2, that may be generated in active neural tissue.
- Because of the cyclical nature of pineal activity, the pineal gland may also be involved with the establishment or maintenance of basic *circadian rhythms*, daily changes in physiological processes that follow a regular pattern. Increased melatonin secretion in darkness has been suggested as a primary cause for *seasonal affective disorder*. This condition, characterized by changes in mood, eating habits, and sleeping patterns, can develop during the winter in high latitudes, where sunshine is meager or lacking altogether. AM *Light and Behavior*

 Why does a person with Type I diabetes urinate frequently and have a pronounced thirst?

 What effect would increased levels of glucagon have on the amount of glycogen stored in the liver?

 Increased amounts of light would inhibit the production of which hormone?

Patterns of Hormonal Interaction

Although hormones are usually studied individually, the extracellular fluids contain a mixture of hormones whose concentrations change daily and even hourly. When a cell receives instructions from two different hormones at the same time, there are several possible results:

1. The two hormones may have opposing, or **antagonistic,** effects, as in the case of parathyroid hormone and calcitonin, or insulin and glucagon. The net result will depend on the balance between the two hormones. However, the presence of an antagonistic hormone means that the observed effects will be smaller than those produced by either hormone acting unopposed.
2. The two hormones may have additive effects. Sometimes the net result is not only greater than the effect that each would produce acting alone, but is actually greater than the *sum* of their individual effects. An example of such a **synergistic** (sin-er-JIS-tik; *synairesis*, a drawing together) effect is the glucose-sparing actions of growth hormone and glucocorticoids.
3. One hormone can have a **permissive** effect on another. In such cases the first hormone is needed for the second to produce its effect. For example, epinephrine has no apparent effect on energy consumption unless thyroid hormones are also present in normal concentrations.
4. Finally, hormones may produce different but complementary results in specific tissues and organs. These **integrative** effects are important in coordinating the activities of diverse physiological systems.

This section will present three examples of the ways hormones interact to produce complex, well-coordinated results. The patterns introduced will provide the background needed to understand the more detailed discussions found in chapters dealing with cardiovascular function, metabolism, excretion, and reproduction.

Hormones and Growth

Normal growth requires the cooperation of several endocrine organs. Six different hormones are especially important, although many others have secondary effects on growth. The circulating concentrations of these hormones are regulated independently, and every time the hormonal mixture changes, metabolic operations are modified to some degree. The alterations vary in duration and intensity, producing unique individual growth patterns.

- ***Growth hormone.*** Growth hormone assists in the maintenance of normal blood glucose concentrations and the mobilization of lipid reserves stored in adipose tissues. It is not the primary hormone involved, however, and an adult with a growth hormone deficiency but normal levels of thyroxine, insulin, and glucocorticoids will have no physiological problems.

  The effects of GH on protein synthesis and cellular growth are most apparent in children, where GH supports muscular and skeletal development. Undersecretion or oversecretion of GH can lead to dwarfism or gigantism, disorders considered in Chapter 6. [p. 188]
- ***Thyroid hormones.*** Normal growth also requires appropriate levels of thyroid hormones. If these hormones are absent for the first year after birth, the nervous system will fail to develop normally, and mental retardation and other signs of cretinism will result. If thyroxine concentrations decline later in life, but before puberty, normal skeletal development will not continue.
- ***Insulin.*** Growing cells need adequate supplies of energy and nutrients. Without insulin, produced by the pancreatic islets, the passage of glucose and amino acids across cell membranes will be drastically reduced or eliminated.
- ***Parathyroid hormone and calcitriol.*** Parathyroid hormone and calcitriol promote the absorption of calcium salts for subsequent deposition in bone. Without adequate levels of both hormones, bones can still enlarge, but they will be poorly mineralized, weak, and flexible. For example, in *rickets*, a condition caused by inadequate calcitriol production in growing children, the limb bones are so weak that they bend under the weight of the body. [p. 155]
- ***Gonadal hormones.*** The activity of osteoblasts in key locations and the growth of specific cell populations are affected by the presence or absence of sexual hormones (androgens in males, estrogens in females). Sexual hormones stimulate cell growth and differentiation in their target tissues. The targets differ for androgens and estrogens, and the differential growth induced by these hormones accounts for observed differences in skeletal proportions and secondary sexual characteristics.

Hormones and Stress

Any condition within the body that threatens homeostasis is a form of **stress.** Stresses are produced by the action of *stressors* that may be (1) physical, such as illness or injury; (2) emotional, such as depression or anxiety; (3) environmental, such as extreme heat or cold; or (4) metabolic, such as acute starvation. The stresses produced may be opposed by specific homeostatic adjustments. For example, a decline in body temperature will result in responses such as shivering or changes in the pattern of circulation in an attempt to restore normal body temperature.

In addition, the body has a *general* response to stress that can occur while other, more specific, responses are under way. *Exposure to a wide variety of stressors will produce the same general pattern of hormonal and physiological adjustments.* These responses are part of the **stress response,** also known as the **general adaptation syndrome,** or **GAS.** The GAS, first described by Hans Selye in 1936, can be divided into three phases: the *alarm phase*, the *resistance phase*, and the *exhaustion phase*.

The Alarm Phase

During the **alarm phase** there is an immediate response to the stress. This response is directed by the sympathetic division of the autonomic nervous system. In the alarm phase (1) energy reserves are mobilized, mainly in the form of glucose, and (2) the body prepares to deal with the stressor by "fight or flight."

Epinephrine is the dominant hormone of the alarm phase, and its secretion accompanies a generalized sympathetic activation. The characteristics of the alarm phase include:

- Increased mental alertness.
- Increased energy consumption by skeletal muscles and other peripheral tissues.
- The mobilization of energy reserves (glycogen and lipids).
- Increased blood flow to skeletal muscles.
- Decreased blood flow to the skin, kidneys, and digestive organs.
- A drastic reduction in digestion and urine production.
- Increased sweat gland secretion.
- Increased blood pressure, heart rate, and respiratory rate.

Although the effects of epinephrine are most apparent during the alarm phase, other hormones play supporting roles. Sympathetic activation also triggers the release of renin, leading to the activation of angiotensin II. The reduction of water losses by circulatory changes, ADH production, and aldosterone secretion may be very important if the stress(es) involve a loss of blood.

The Resistance Phase

The temporary adjustments of the alarm phase are often sufficient to remove or overcome the stress. But some stresses, such as starvation, acute illness, or severe anxiety, can persist for hours, days, or weeks. If a stress lasts longer than a few hours, the individual will enter the **resistance phase** of the GAS.

Glucocorticoids are the dominant hormones of the resistance phase. Epinephrine, growth hormone, and thyroid hormones play supporting roles. Energy demands in the resistance phase remain higher than normal because of the combined effects of these hormones.

Neural tissue has a high demand for energy, and neurons must have a reliable supply of glucose; if blood glucose concentrations fall too far, neural function will deteriorate. Glycogen reserves are adequate to maintain normal glucose concentrations during the alarm phase, but after several hours these reserves are nearly exhausted. The endocrine secretions of the resistance phase are coordinated to achieve three integrated results:

1. ***The mobilization of lipid and protein reserves.*** The hypothalamus produces GH-RH and CRH, stimulating the release of growth hormone and, via ACTH, the secretion of glucocorticoids. Adipose tissues respond to GH and glucocorticoids by releasing stored fatty acids. Skeletal muscles respond to glucocorticoids by breaking down proteins and releasing amino acids into the circulation.
2. ***The elevation and stabilization of blood glucose concentrations.*** As blood glucose levels decline, glucagon and glucocorticoids stimulate the liver to manufacture glucose from the fatty acids and amino acids provided by adipose tissue and skeletal muscles. The glucose molecules are then released into the circulation, and blood glucose concentrations return to normal levels.
3. ***The conservation of glucose for neural tissues.*** Glucocorticoids and growth hormone from the anterior pituitary stimulate lipid metabolism in peripheral tissues. These glucose-sparing effects maintain normal blood glucose levels even after long periods of starvation. Neural tissues do not alter their metabolic activities, however, and they continue to use glucose as an energy source.

The lipid reserves of the body are sufficient to maintain the resistance phase for a period of weeks or even months. (These reserves account for the ability to endure lengthy periods of starvation.) But the resistance phase cannot be sustained indefinitely. If starvation is the primary stress, the resistance phase ends as lipid reserves are exhausted, and structural proteins become the primary energy source. If starvation is not the problem, the resistance phase will end because of complications brought about by hormonal side effects. For example:

- Although the metabolic effects of glucocorticoids are essential to normal resistance, their anti-inflammatory action slows wound healing and increases susceptibility to infection.
- The continued conservation of fluids under the influence of ADH and aldosterone stresses the cardiovascular system by producing elevated blood volumes and higher-than-normal blood pressures.
- The adrenal cortex may become unable to continue producing glucocorticoids, and this failure quickly eliminates the ability to maintain acceptable blood glucose concentrations.

Poor nutrition, emotional or physical trauma, chronic illness, or damage to key organs such as the heart, liver, or kidneys will hasten the end of the resistance phase.

The Exhaustion Phase

Figure 18-20

When the resistance phase ends, homeostatic regulation breaks down and the **exhaustion phase** begins. Unless corrective actions are taken almost immediately, the failure of one or more organ systems will prove fatal.

Problems with mineral balance contribute to the existing problems with major systems. The production of aldosterone throughout the resistance phase results in a conservation of sodium ions at the expense of potassium ions. As the potassium content of the body declines, a variety of cells—notably neurons and muscle fibers—begin to malfunction. Although a single cause, such as heart failure, may be listed as the cause of death, the underlying problem is the inability to sustain the endocrine and metabolic adjustments of the resistance phase. The three phases of the GAS are summarized in Figure 18-20●.

❑ Hormones and Behavior

As we have seen, many endocrine functions are regulated by the hypothalamus, and hypothalamic neurons monitor the levels of many circulating hormones. Other portions of the central nervous system are also quite sensitive to hormonal stimulation.

The clearest demonstrations of the effects of specific hormones involve individuals whose endocrine glands are oversecreting or undersecreting. But even normal changes in circulating hormone levels can cause behavioral changes. In *precocious* (premature) *puberty*, sexual hormones are produced at an inappropriate time, perhaps as early as 5 or 6 years of age. The affected children not only begin to develop adult secondary sexual characteristics but also undergo significant behavioral changes. The "nice little kid" disappears, and the child becomes aggressive and assertive. These behavioral alterations represent the effects of sexual hormones on CNS function. Thus behaviors that in normal teenagers are usually attributed to environmental stimuli, such as peer pressure, actually have some physiological basis as well. In the adult, changes in the mixture of hormones reaching the CNS can have significant effects on intellectual capabilities, memory, learning, and emotional states.

ALARM PHASE

"Fight or Flight"
Immediate short-term responses to crises

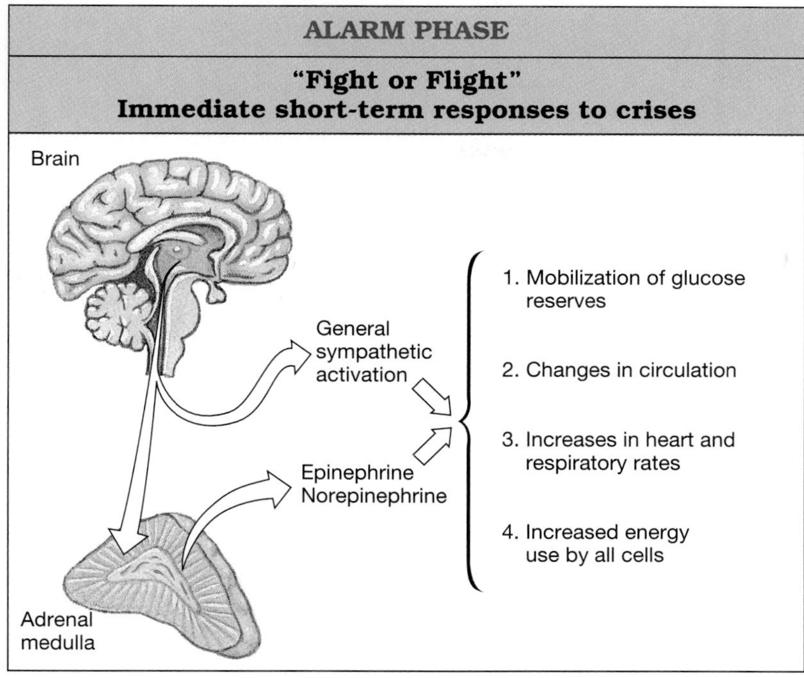
Brain
General sympathetic activation
Epinephrine
Norepinephrine
Adrenal medulla
1. Mobilization of glucose reserves
2. Changes in circulation
3. Increases in heart and respiratory rates
4. Increased energy use by all cells

RESISTANCE PHASE

Long-term metabolic adjustments occur

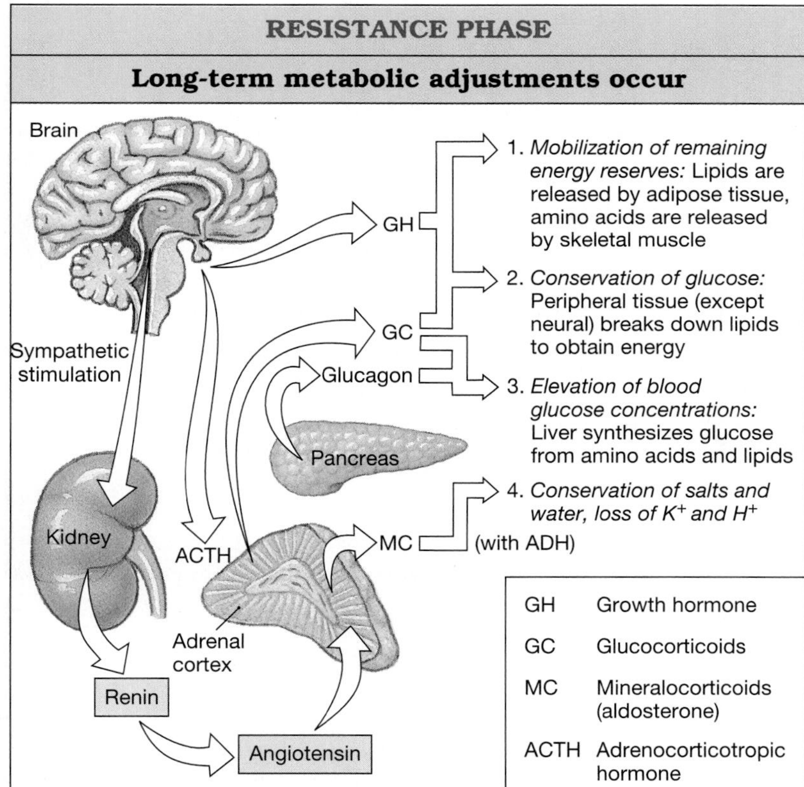
Brain
Sympathetic stimulation
GH
GC
Glucagon
Pancreas
Kidney
ACTH
Adrenal cortex
MC
Renin
Angiotensin
1. *Mobilization of remaining energy reserves:* Lipids are released by adipose tissue, amino acids are released by skeletal muscle
2. *Conservation of glucose:* Peripheral tissue (except neural) breaks down lipids to obtain energy
3. *Elevation of blood glucose concentrations:* Liver synthesizes glucose from amino acids and lipids
4. *Conservation of salts and water, loss of K+ and H+*
(with ADH)
GH Growth hormone
GC Glucocorticoids
MC Mineralocorticoids (aldosterone)
ACTH Adrenocorticotropic hormone

EXHAUSTION PHASE

Collapse of vital systems

Causes may include:
— Exhaustion of lipid reserves
— Inability to produce glucocorticoids
— Failure of electrolyte balance
— Cumulative structural or functional damage to vital organs

Hormones and Aging

The endocrine system shows relatively few functional changes with age. The most dramatic exception is the decline in the concentration of reproductive hormones. Effects of these hormonal changes on the skeletal system were noted in Chapter 6; further discussion will be found in Chapter 29.

Blood and tissue concentrations of many other hormones, including TSH, thyroid hormones, ADH, PTH, prolactin, and glucocorticoids, remain unchanged. Although circulating hormone levels may remain within normal limits, however, some endocrine tissues become less responsive to stimulation. For example, in elderly individuals less GH and insulin are secreted after a carbohydrate-rich meal or in a glucose tolerance test.

Finally, it should be noted that age-related changes in other tissues affect their abilities to respond to hormonal stimulation. As a result, peripheral tissues may become less responsive to some hormones. This loss of sensitivity has been documented for glucocorticoids and ADH.

✔ Insulin lowers the level of glucose in the blood and glucagon causes glucose levels to rise. What is this type of hormonal interaction called?

✔ The lack of which hormones would inhibit skeletal formation?

✔ Why do levels of GH-RH and CRH rise during the resistance phase of the general adaptation syndrome (GAS)?

Interactions with Other Systems

Figure 18-21

The relationships between the endocrine system and other systems are summarized in Figure 18-21●.

● **FIGURE 18-20**
The General Adaptation Syndrome

INTEGUMENTARY SYSTEM

Protects superficial endocrine organs; epidermis synthesizes cholecalciferol

Sexual hormones stimulate sebaceous gland activity, influence hair growth, fat distribution, and apocrine sweat gland activity; PRL stimulates development of mammary glands; adrenal hormones alter dermal blood flow, stimulate release of lipids from adipocytes; MSH stimulates melanocyte activity

SKELETAL SYSTEM

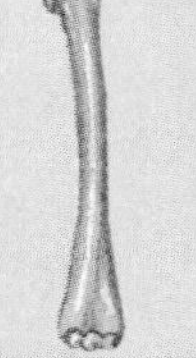

Protects endocrine organs, especially in brain, chest and pelvic cavity

Skeletal growth regulated by several hormones; calcium mobilization regulated by parathyroid hormone and calcitonin; sex hormones speed growth and closure of epiphyseal plates at puberty, and help maintain bone mass in adults

MUSCULAR SYSTEM

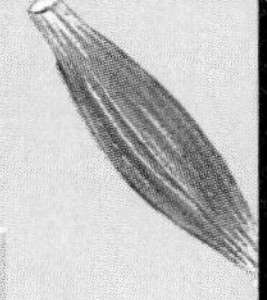

Skeletal muscles provide protection for some endocrine organs

Hormones adjust muscle metabolism, energy production, and growth; regulate Ca^{2+} and PO_4^{3-} levels in body fluids; speed skeletal muscle growth

THE ENDOCRINE SYSTEM

NERVOUS SYSTEM

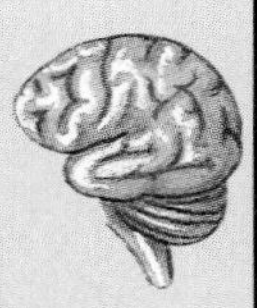

Hypothalamic hormones directly control pituitary and indirectly control secretions of other endocrine organs; controls adrenal medullae; secretes ADH and oxytocin

Several hormones affect neural metabolism; hormones help regulate fluid and electrolyte balance; reproductive hormones influence CNS development and behaviors

CARDIOVASCULAR SYSTEM

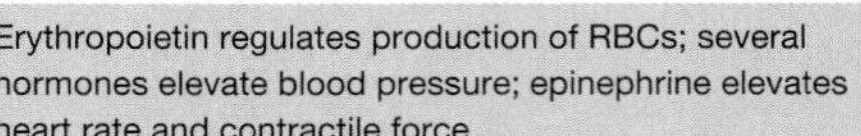

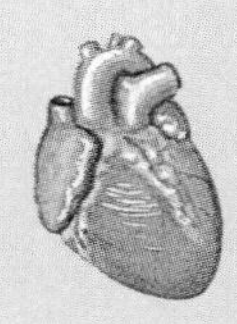

Circulatory system distributes hormones throughout the body; heart secretes ANP

Erythropoietin regulates production of RBCs; several hormones elevate blood pressure; epinephrine elevates heart rate and contractile force

LYMPHATIC SYSTEM

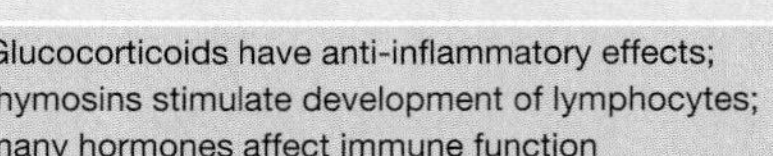

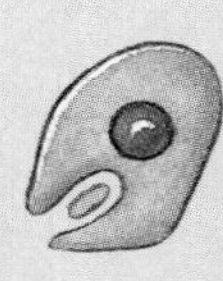

Lymphocytes provide defense against infection and, with other WBCs, assist in repair after injury

Glucocorticoids have anti-inflammatory effects; thymosins stimulate development of lymphocytes; many hormones affect immune function

FOR ALL SYSTEMS

Adjusts metabolic rates and substrate utilization; regulates growth and development

RESPIRATORY SYSTEM

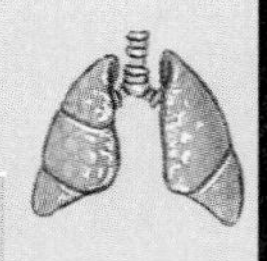

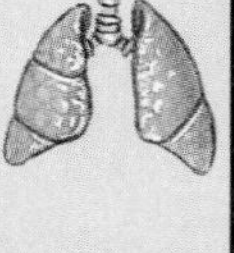

Provides O_2 and eliminates CO_2 generated by endocrine cells; converting enzyme in lung capillaries converts angiotensin I to angiotensin II

Epinephrine and norepinephrine stimulate respiratory activity and dilate respiratory passageways

DIGESTIVE SYSTEM

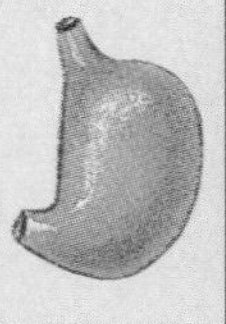

Provides nutrients and substrates to endocrine cells; endocrine cells of pancreas secrete insulin and glucagon; liver produces and releases angiotensinogen

E and NE stimulate constriction of sphincters and depress activity along digestive tract; digestive tract hormones coordinate secretory activities along tract

REPRODUCTIVE SYSTEM

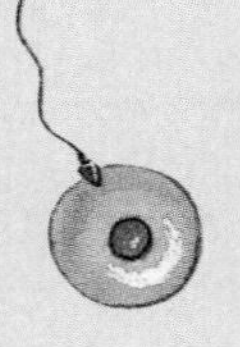

Steroid sex hormones and inhibin suppress secretory activities in hypothalamus and pituitary

Hypothalamic factors and pituitary hormones regulate sexual development and function; oxytocin stimulates uterine and mammary gland smooth muscle contractions

URINARY SYSTEM

Kidney cells (1) release renin and erythropoietin when local blood pressure declines; (2) produce calcitriol

Aldosterone, ADH, and ANP adjust rates of fluid and electrolyte reabsorption in kidneys

● **FIGURE 18-21**
Functional Relationships between the Endocrine System and Other Systems

Endocrinology and Athletic Performance

Medical research involving humans is generally subject to tight ethical constraints and meticulous scientific scrutiny. Yet a clandestine, unscientific, and potentially quite dangerous program of "experimentation" with hormones is today being pursued by athletes in many countries. Although the use of hormones to improve performance is banned by the International Olympic Committee, the U.S. Olympic Committee, the NCAA, and the NFL, and condemned by the American Medical Association and the American College of Sports Medicine, a significant number of amateur and professional athletes persist in this dangerous practice. Synthetic forms of testosterone are used most often, but athletes may use any combination of testosterone, growth hormone, EPO, and a variety of synthetic hormones.

Androgen Abuse. The use of androgens, or "anabolic steroids," has become popular with many athletes, both amateur and professional. The goal of steroid use is to increase muscle mass, endurance, and "competitive spirit." It has been suggested that as many as 30 percent of college and professional athletes and 10–20 percent of male high school athletes may be using anabolic steroids (with or without growth hormone) to improve their performance. Among body builders, the proportion using steroids in this country may be as high as 80 percent.

One supposed justification for this practice has been the unfounded opinion that compounds manufactured in the body are not only safe, but "good for you." In reality the administration of natural *or* synthetic androgens in abnormal amounts carries unacceptable health risks. Androgens affect many tissues in a variety of ways. Known complications include (1) premature epiphyseal closure, (2) various liver dysfunctions (including jaundice and hepatic tumors), (3) prostate enlargement and urinary tract obstruction, and (4) testicular atrophy and infertility. A link to heart attacks, impaired cardiac function, and strokes has also been suggested.

Moreover, the normal regulation of androgen production involves a feedback mechanism comparable to that described for adrenal steroids earlier in this chapter. GnRH stimulates the production of LH, and LH stimulates the secretion of testosterone and other androgens by the interstitial cells of the testes. Circulating androgens, in turn, inhibit the production of both GnRH and LH. Thus, when synthetic androgens are administered in high doses, they can (1) suppress the normal production of testosterone and (2) depress the manufacture of GnRH by the hypothalamus. *This suppression of GnRH release may be permanent.*

The use of androgenic "bulking agents" by female body builders may not only add muscle mass but may alter muscular proportions and secondary sexual characteristics as well. For example, women taking steroids can develop irregular menstrual periods and changes in body hair distribution (including baldness). Finally, androgen abuse may cause a generalized depression of the immune system.

EPO Abuse. Because it is now being synthesized using recombinant-DNA techniques, EPO is readily available. Athletes engaged in endurance sports, such as cycling or marathon running, may use it to boost the number of oxygen-carrying red blood cells in circulation. Although this effect improves the oxygen content of the blood, it also makes the blood denser, and the heart must work harder to push it around the circulatory system. Between 1991 and 1994 the deaths of 18 young, otherwise healthy European cyclists have been attributed to heart failure related to EPO abuse.

GHB and Clenbuterol. Androgens and EPO are known hormones with reasonably well-understood effects. Because drug testing is now widespread in amateur and professional sports, people interested in "getting an edge" are experimenting with drugs not easily detected by standard tests. The long-term and short-term effects of these drugs are difficult to predict. Two examples are the recent use of *GHB* and *clenbuterol* by amateur athletes.

Gamma-hydroxybutyrate, or **GHB,** was tested for use as an anesthetic 30 years ago and rejected, in part because it was linked to petit mal and grand mal seizures. In 1990 this drug appeared in health-food stores, where it was sold as an anabolic agent and diet aid. During a 5-month period in 1990, 16 cases of severe reaction to GHB were treated in San Francisco alone. Symptoms included confusion, hallucination, seizures, and coma at doses from 0.25 teaspoons to 4 tablespoons.

Clenbuterol abuse is reportedly widespread, although exact numbers are difficult to obtain. **Clenbuterol,** sometimes used to treat asthma, mimics epinephrine and stimulates β_2 receptors. This effect increases the diameter of the respiratory passageways and accelerates blood flow through active skeletal muscles. Although it is also rumored to have anabolic properties comparable to those of androgens, there is no evidence to support that notion. Heavy usage can cause severe headaches, tremors, insomnia, and potentially dangerous abnormal heartbeats. During the 1992 Olympics in Barcelona, Spain, two American athletes were disqualified because they tested positive for this drug.

TABLE 18-7 Clinical Implications of Endocrine Malfunctions

Hormone	*Underproduction Syndrome*	*Principal Symptoms*	*Overproduction Syndrome*	*Principal Symptoms*
Growth hormone (GH)	Pituitary growth failure (pp. 615, 188)	Retarded growth, abnormal fat distribution, low blood glucose hours after a meal	Gigantism, acromegaly (p. 615 and *AM*)	Excessive growth
Antidiuretic hormone (ADH)	Diabetes insipidus (p. 614)	Polyuria, dehydration, thirst	SIADH (syndrome of inappropriate ADH secretion) (*AM*)	Increased body weight and water content
Thyroxine (T_4), triiodothyronine (T_3)	Myxedema, cretinism (p. 620 and *AM*)	Low metabolic rate, body temperature; impaired physical and mental development	Hyperthyroidism, Graves' disease (p. 620 and *AM*)	High metabolic rate and body temperature
Parathyroid hormone (PTH)	Hypoparathyroidism (p. 622 and *AM*)	Muscular weakness, neurological problems, formation of dense bones, tetany due to low blood calcium concentrations	Hyperparathyroidism (p. 622 and *AM*)	Neurological, mental, muscular problems due to high blood calcium concentrations; weak and brittle bones
Insulin	Diabetes mellitus (Type I) (p. 634)	High blood glucose, impaired glucose utilization, dependence on lipids for energy, glycosuria	Excess insulin production or administration	Low blood glucose levels, possibly causing coma
Mineralocorticoids (MC)	Hypoaldosteronism (p. 624 and *AM*)	Polyuria, low blood volume, high blood potassium concentrations	Aldosteronism (p. 624 and *AM*)	Increased body weight due to water retention; low blood potassium concentration
Glucocorticoids (GC)	Addison's disease (p. 626 and *AM*)	Inability to tolerate stress, mobilize energy reserves, or maintain normal blood glucose concentrations	Cushing's disease (p. 626 and *AM*)	Excessive breakdown of tissue proteins and lipid reserves; impaired glucose metabolism
Epinephrine (E), norepinephrine (NE)	None identified		Pheochromocytoma (*AM*)	High metabolic rate, body temperature, and heart rate; elevated blood glucose levels
Estrogens (female)	Hypogonadism	Sterility, lack of secondary sexual characteristics	Adrenogenital syndrome (p. 626)	Overproduction of androgens by z. reticularis of adrenal leads to masculinization
	Menopause	Cessation of ovulation	Precocious puberty (p. 638)	Premature sexual maturation and related behavioral changes
Androgens (male)	Hypogonadism	Sterility, lack of secondary sexual characteristics	Adrenogenital syndrome (gynecomastia) (p. 626)	Abnormal production of estrogen, sometimes due to adrenal or interstitial cell tumors; leads to breast enlargement
			Precocious puberty (p. 638)	Premature sexual maturation and related behavioral changes

Endocrine Disorders

Endocrine disorders fall into two basic categories: They reflect either abnormal hormone production or abnormal cellular sensitivity. The symptoms are interesting because they highlight the significance of normally "silent" hormonal contributions. An abbreviated summary is presented in Table 18-7, and the characteristic features of many of these conditions are illustrated in Figure 18-22●.

● **FIGURE 18-22**
Endocrine Abnormalities

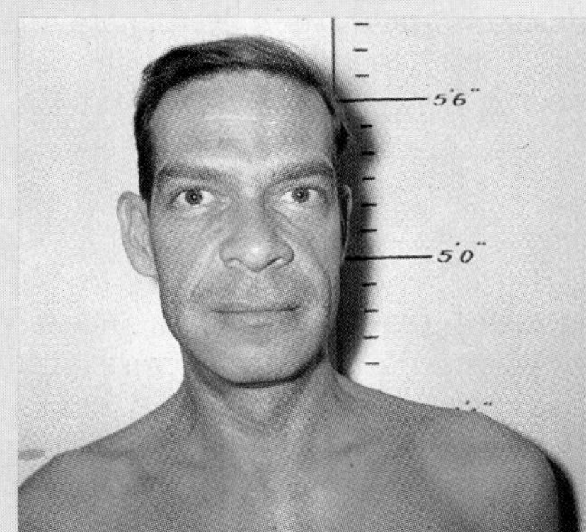
5'6"
5'0"

(a) *Acromegaly* results from the overproduction of growth hormone after the epiphyseal plates have fused. Bone shapes change and cartilaginous areas of the skeleton enlarge. Note the broad facial features and the enlarged lower jaw.

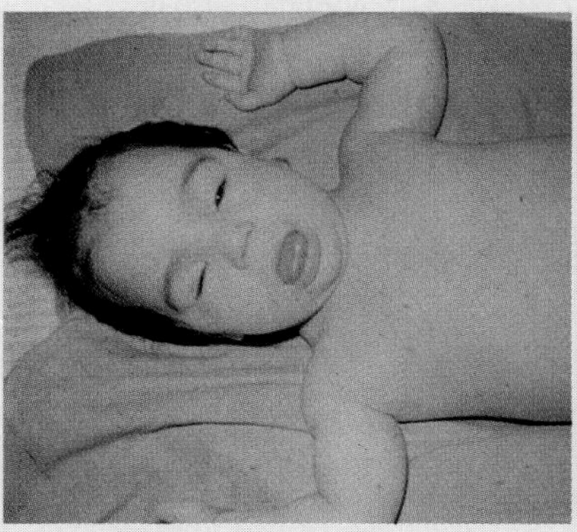

(b) *Cretinism*, which results from thyroid hormone insufficiency in infancy.

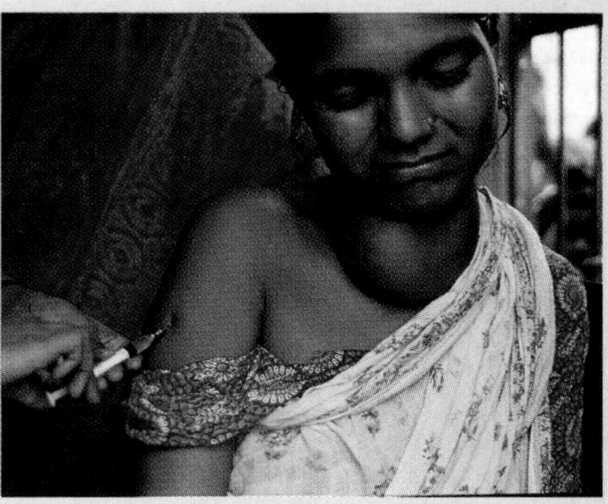

(c) An enlarged thyroid gland, or *goiter*, is usually associated with thyroid hyposecretion due to iodine insufficiency.

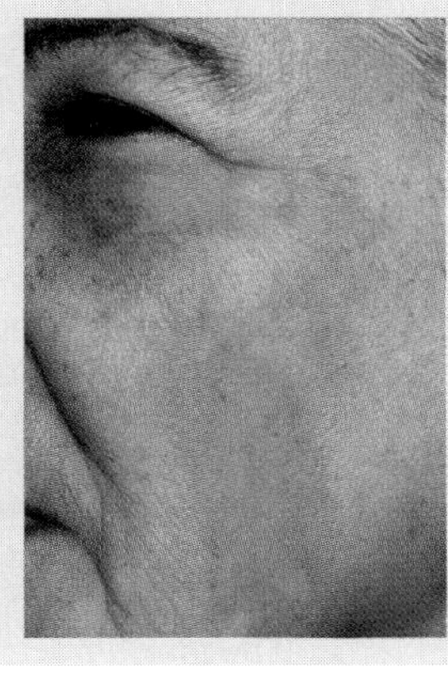

(d) *Addison's disease* is caused by hyposecretion of corticosteroids, especially glucocorticoids. Pigment changes result from stimulation of melanocytes by ACTH, which is structurally similar to MSH.

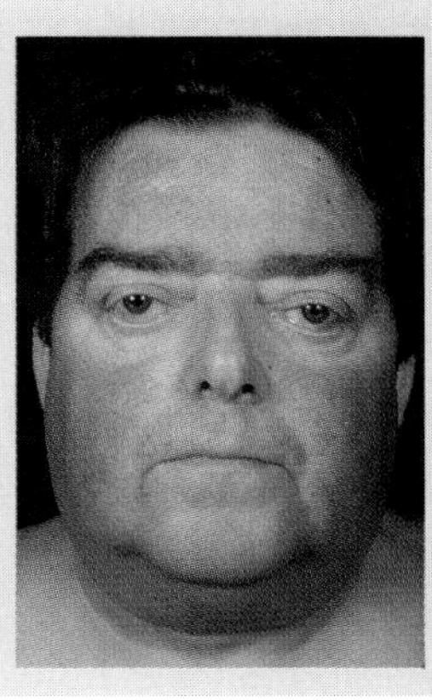

(e) *Cushing's disease* is caused by hypersecretion of glucocorticoids. Lipid reserves are mobilized, and adipose tissue accumulates in the cheeks and at the base of the neck.

Selected Clinical Terminology

Terms Discussed in This Chapter

Addison's disease: A condition caused by hyposecretion of glucocorticoids and mineralocorticoids; characterized by an inability to mobilize energy reserves and maintain normal blood glucose levels. *(p. 626* and *AM)*

cretinism: A condition, caused by hypothyroidism in infancy, marked by inadequate skeletal and nervous development, and a metabolic rate as much as 40 percent below normal levels. *(p. 620)*

Cushing's disease: A condition caused by hypersecretion of glucocorticoids; characterized by excessive breakdown and relocation of lipid reserves and proteins. *(p. 626* and *AM)*

diabetes insipidus: A disorder that develops when the posterior pituitary no longer releases adequate amounts of ADH. *(p. 614)*

diabetes mellitus (mel-LI-tus): A disorder characterized by glucose concentrations high enough to overwhelm the kidneys' reabsorption capabilities. *(p. 634* and *AM)*

diabetic retinopathy, nephropathy, neuropathy: Disorders of the retina, kidneys, and peripheral nerves related to diabetes mellitus; the conditions most often afflict middle-aged or older diabetics. *(p. 634)*

general adaptation syndrome (GAS): The pattern of hormonal and physiological adjustments with which the human body responds to all forms of stress. *(p. 637)*

glycosuria: The presence of glucose in the urine. *(p. 634)*
hypocalcemic tetany: Muscle spasms affecting the face and upper extremities; caused by low calcium ion concentrations in body fluids. *(p. 622* and *AM)*
insulin-dependent diabetes mellitus (IDDM); also known as **Type I diabetes** or **juvenile-onset diabetes:** A type of diabetes mellitus; the primary cause is inadequate insulin production by the beta cells of the pancreatic islets. *(p. 634* and *AM)*
myxedema: Symptoms of hyposecretion of thyroid hormones, including subcutaneous swelling, hair loss, dry skin, low body temperature, muscle weakness, and slowed reflexes. *(p. 620* and *AM)*
non-insulin-dependent diabetes mellitus (NIDDM); also known as **Type II diabetes** or **maturity-onset diabetes:** A type of diabetes mellitus in which insulin levels are normal or elevated, but peripheral tissues no longer respond normally. *(p. 634* and *AM)*
polyuria: Production of excessive amounts of urine; a symptom of diabetes. *(p. 614)*
seasonal affective disorder: A condition characterized by depression, lethargy, an inability to concentrate, and altered sleep and eating habits. It has been linked to enhanced melatonin levels in subjects living in areas or seasons that have short periods of daylight. *(p. 636* and *AM)*
thyrotoxicosis: A condition caused by oversecretion of thyroid hormones (hyperthyroidism). Symptoms include increased metabolic rate, blood pressure, and heart rate; excitability and emotional instability; and lowered energy reserves. *(p. 620* and *AM)*

AM *Additional Terms Discussed in the Applications Manual*

exophthalmos: Protrusion of the eyes from their sockets; often a symptom of hypersecretion of thyroid hormones.
goiter: An abnormal enlargement of the thyroid gland.
ketoacidosis: A condition in which large numbers of ketone bodies in the blood lead to a dangerously low blood pH.
ketone bodies: Organic acids produced during the breakdown of fatty acids.
thyrotoxic crisis: An acute state of hyperthyroidism, characterized by high fever, rapid heart rate, and the malfunctioning of multiple organs.

CHAPTER REVIEW

STUDY OUTLINE

INTRODUCTION, p. 602

1. In general, the nervous system performs short-term "crisis management," whereas the endocrine system regulates longer-term, ongoing metabolic processes. Endocrine cells release chemicals called **hormones** that alter the metabolic activities of many different tissues and organs simultaneously. (*Figure 18-1*)

AN OVERVIEW OF THE ENDOCRINE SYSTEM, p. 602

Hormone Structure, p. 603

1. Hormones can be divided into three groups on the basis of chemical structure: *amino acid derivatives* are structurally similar to amino acids; *peptide hormones* are chains of amino acids; and *lipid derivatives* include **steroid hormones** and **eicosanoids.** (*Figures 18-2, 18-3*)

Hormone Function and Mechanisms of Action, p. 606

2. Hormones exert their effects by modifying the activities of *target cells.* Receptors for catecholamine and peptide hormones are located on the cell membranes of target cells; the hormone acts as a **first messenger** that causes a **second messenger** to appear in the cytoplasm. Thyroid and steroid hormones cross the cell membrane and bind to receptors in the cytoplasm or nucleus. The hormone-receptor complex activates or inactivates specific genes. (*Figures 18-4, 18-5*)

Control of Endocrine Activity, p. 609

3. *Endocrine reflexes* are the functional counterparts of neural reflexes. (*Figure 18-6*)
4. The hypothalamus regulates the activities of the nervous and endocrine systems via three mechanisms: (1) It secretes **regulatory hormones** that control the activities of endocrine cells in the pituitary gland, (2) its autonomic centers exert direct neural control over the endocrine cells of the adrenal medulla, and (3) it acts as an endocrine organ itself by releasing hormones into the circulation at the posterior pituitary. (*Figure 18-6*)

THE PITUITARY GLAND, p. 609

1. The gland releases nine important peptide hormones; all bind to membrane receptors and use cyclic-AMP as a second messenger. (*Figure 18-7*)

The Anterior Pituitary, p. 611

2. The **anterior pituitary (adenohypophysis)** can be subdivided into the large **pars distalis** and the slender **pars intermedia.** (*Figure 18-7*)
3. At the median eminence, neurons release regulatory factors into the surrounding interstitial fluids. Their secretions enter the circulation easily because the capillaries in this area are **fenestrated.** (*Figure 18-8*)
4. The **hypophyseal portal system** ensures that all of the blood entering the **portal vessels** will reach the intended target cells before returning to the general circulation. (*Figure 18-8*)
5. **Thyroid-stimulating hormone (TSH)** triggers the release of thyroid hormones. **Thyroid hormone–releasing hormone (TRH)** promotes the secretion of TSH.

6. **Adrenocorticotropic hormone (ACTH)** stimulates the release of **glucocorticoids** by the adrenal gland. **Corticotropin-releasing hormone (CRH)** causes the secretion of ACTH.
7. **Follicle-stimulating hormone (FSH)** stimulates follicle development and estrogen secretion in women and sperm production in men. **Luteinizing hormone (LH)** causes ovulation and **progestin** production in women and **androgen** production in men. **Gonadotropin-releasing hormone (GnRH)** promotes the secretion of both FSH and LH.
8. **Prolactin (PRL)** with other hormones stimulates the development of the mammary glands, and milk production. Prolactin secretion is inhibited by **PIH** and stimulated by a prolactin-releasing factor. (*Figure 18-9*)
9. **Growth hormone (GH,** or **somatotropin)** stimulates cell growth and replication through the release of **somatomedins** from liver cells and has direct metabolic effects. The production of growth hormone is regulated by **growth hormone–releasing hormone (GH-RH)** and **growth hormone–inhibiting hormone (GH-IH).** (*Figure 18-9*)
10. **Melanocyte-stimulating hormone (MSH),** released by the pars intermedia, stimulates melanocytes to produce melanin in other species, but it is not normally secreted by the nonpregnant adult human.

The Posterior Pituitary, p. 614

11. The **posterior pituitary (neurohypophysis)** contains the axons of hypothalamic neurons. Neurons of the **supraoptic** and **paraventricular nuclei** manufacture antidiuretic hormone (ADH) and oxytocin, respectively. ADH decreases the amount of water lost at the kidneys. In women, **oxytocin** stimulates contractile cells in the mammary glands and has a stimulatory effect on uterine smooth muscles. (*Figure 18-10, Table 18-1*)

THE THYROID GLAND, p. 616

1. The thyroid gland lies near the thyroid cartilage of the larynx and consists of two **lobes** connected by a narrow **isthmus.** (*Figure 18-11*)

Thyroid Follicles and Thyroid Hormones, p. 617

2. The thyroid gland contains numerous **thyroid follicles.** Thyroid follicles release several hormones, including **thyroxine (T_4)** and **triiodothyronine (T_3).** (*Figures 18-11, 18-12; Table 18-3*)
3. The follicle cells manufacture **thyroglobulin** and store it as a colloid. (*Figure 18-12*)
4. Most of the thyroid hormones entering the bloodstream are attached to special **thyroid-binding globulins (TBGs)** or **thyroid-binding prealbumin (TBPA).** The unbound hormones affect peripheral tissues immediately; the binding proteins gradually release their hormones over a week or more. (*Figure 18-12; Table 18-2*)
5. Thyroid hormones exert a **calorigenic effect,** which enables us to adapt to cold temperatures.

The C Cells of the Thyroid Gland: Calcitonin, p. 620

6. The **C cells** of the follicles produce **calcitonin (CT),** which helps to regulate calcium ion concentrations in body fluids, especially during childhood and pregnancy. (*Figure 18-6; Table 18-3*)

THE PARATHYROID GLANDS, p. 621

Parathyroid Hormone Secretion, p. 622

1. Four parathyroid glands are embedded in the posterior surface of the thyroid gland. The **chief cells** of the parathyroid produce **parathyroid hormone (PTH)** in response to lower-than-normal concentrations of calcium ions. These and the C cells of the thyroid gland maintain calcium ion levels within relatively narrow limits. (*Figures 18-13, 18-14; Table 18-3*)

THE THYMUS, p. 622

1. The **thymus** produces several hormones, collectively known as **thymosins.** The thymosins play a role in developing and maintaining normal immunological defenses. (*Table 18-3*)

THE ADRENAL GLAND, p. 623

1. A single **adrenal** (*suprarenal*) **gland** lies along the superior border of each kidney. Each gland is surrounded by a fibrous *capsule.* The adrenal gland can be subdivided into the superficial **adrenal cortex** and the inner **adrenal medulla.** (*Figure 18-15*)

The Adrenal Cortex, p. 623

2. The adrenal cortex manufactures steroid hormones called **adrenocortical steroids (corticosteroids).** The cortex can be subdivided into three areas: (1) the **zona reticularis,** which produces androgens of uncertain significance; (2) the **zona fasciculata,** which produces **glucocorticoids (GCs),** notably **cortisol, corticosterone,** and **cortisone** (all of which exert a **glucose-sparing effect** on peripheral tissues); and (3) the **zona glomerulosa,** which releases **mineralocorticoids (MCs),** principally **aldosterone,** which restrict sodium losses at the kidneys, sweat glands, digestive tract, and salivary glands. The zona glomerulosa responds to the presence of **angiotensin II,** which appears after the secretion of the enzyme *renin* by kidney cells exposed to a decline in blood volume and/or blood pressure. (*Figure 18-15; Table 18-4*)

The Adrenal Medulla, p. 626

3. The adrenal medulla produces epinephrine (75–80 percent) and norepinephrine (20–25 percent.) (*Figure 18-15; Table 18-4*)

THE KIDNEYS, p. 627

1. Endocrine cells in the kidneys produce hormones important for the regulation of calcium metabolism, blood volume, and blood pressure.

Calcitriol, p. 627

2. **Calcitriol** stimulates calcium and phosphate ion absorption along the digestive tract. (*Figure 18-16*)

Erythropoietin, p. 628

3. **Erythropoietin (EPO)** stimulates red blood cell production by the bone marrow. (*Figure 18-16*)

Renin, p. 628

4. **Renin** converts **angiotensinogen** to **angiotensin I.** In the lungs this compound is converted to **angiotensin**

II, the hormone that stimulates the adrenal production of aldosterone. (*Figure 18-16*)

THE HEART, p. 629

1. Specialized muscle cells in the heart produce **atrial natriuretic peptide (ANP)** when blood volume becomes excessive. (*Figure 18-17*)

ENDOCRINE TISSUES OF THE DIGESTIVE SYSTEM, p. 629

1. The lining of the digestive tract, the liver, and the pancreas produce exocrine secretions that are essential to the normal breakdown and absorption of food.

The Pancreas, p. 629

2. The **pancreas** contains both exocrine and endocrine cells. The **exocrine pancreas** secretes an enzyme-rich fluid that travels to the lumen of the digestive tract. Cells of the **endocrine pancreas** form clusters called **pancreatic islets** (*islets of Langerhans*), containing **alpha cells** (which produce the hormone **glucagon**), **beta cells** (which secrete **insulin**), **delta cells** (which secrete **somatostatin**), and **F-cells** (which secrete **pancreatic polypeptide**). (*Figure 18-18; Table 18-5*)
3. Insulin lowers blood glucose by increasing the rate of glucose uptake and utilization; glucagon raises blood glucose by increasing the rates of glycogen breakdown and glucose manufacture in the liver. Somatostatin suppresses glucagon and insulin release and slows the rates of food absorption and enzyme secretion in the digestive tract. Pancreatic polypeptide inhibits gallbladder contractions and regulates the production of some pancreatic enzymes. (*Figure 18-19*)

ENDOCRINE TISSUES OF THE REPRODUCTIVE SYSTEM, p. 634

The Testes, p. 634

1. The **interstitial cells** of the male testis produce androgens. **Testosterone** is the most important sexual hormone in males.

The Ovaries, p. 635

2. In women, eggs develop in **follicles;** follicle cells produce **estrogens,** especially **estradiol.** After ovulation, the remaining follicle cells reorganize into a **corpus luteum** that releases a mixture of estrogens and **progestins,** especially **progesterone.** (*Table 18-6*)

THE PINEAL GLAND, p. 636

1. The pineal gland contains **pinealocytes** that synthesize **melatonin.** Although research and debate continue, melatonin appears to (1) slow the maturation of sperm, eggs, and reproductive organs by reducing the rate of GnRH secretion and (2) establish daily *circadian rhythms.*

PATTERNS OF HORMONAL INTERACTION, p. 636

1. The endocrine system functions as an integrated unit, and hormones often interact. These interactions may exert several effects: (1) **antagonistic** (opposing) effects; (2) **synergistic** (additive) effects; (3) **permissive** effects, in which one hormone is necessary for another to produce its effect; or (4) **integrative** effects, in which hormones produce different but complementary results.

Hormones and Growth, p. 636

2. Normal growth requires the cooperation of several endocrine organs. Five hormones are especially important: growth hormone, thyroid hormones, insulin, parathyroid hormone, and gonadal hormones.

Hormones and Stress, p. 637

3. Any condition that threatens homeostasis represents a **stress.** Our bodies respond to a variety of stressors with the **general adaptation syndrome (GAS).** The GAS can be divided into three phases: (1) the **alarm phase** (an immediate, "fight or flight" response to the stress, under the direction of the sympathetic division of the ANS); (2) the **resistance phase** dominated by glucocorticoids (mobilization of lipid and protein reserves, elevation and stabilization of blood glucose concentrations, and conservation of glucose for neural tissues); and (3) the **exhaustion phase** (the eventual breakdown of homeostatic regulation and failure of one or more organ systems). (*Figure 18-20*)

Hormones and Behavior, p. 638

4. Many hormones affect the nervous system, producing changes in mood, emotional states, and behavior.

Hormones and Aging, p. 639

5. The endocrine system shows few functional changes with advanced age, chiefly a decline in the concentration of reproductive hormones.

INTERACTIONS WITH OTHER SYSTEMS, p. 639

REVIEW QUESTIONS

LEVEL 1 Reviewing Facts and Terms

1. Peptide chains synthesized at ribosomes of the RER are called:
   (a) prehormones (b) prohormones
   (c) receptors (d) G-proteins
2. Cyclic-AMP functions as a second messenger to:
   (a) build proteins and catalyze specific reactions
   (b) activate adenylate cyclase
   (c) open ion channels and activate key enzymes in the cytoplasm
   (d) bind the hormone-receptor complex to DNA segments
3. Adrenocorticotropic hormone (ACTH) stimulates the release of:
   (a) thyroid hormones by the hypothalamus
   (b) gonadotropins by the adrenal glands

(c) somatotropins by the hypothalamus
(d) steroid hormones by the adrenal glands

4. FSH production in males supports:
(a) maturation of sperm by stimulating sustentacular cells
(b) development of muscles and strength
(c) production of male sex hormones
(d) increased desire for sexual activity

5. The hormone that induces ovulation in women and promotes the ovarian secretion of progesterone is:
(a) interstitial cell–stimulating hormone
(b) estradiol
(c) luteinizing hormone
(d) prolactin

6. The two hormones released by the posterior pituitary are:
(a) somatotropin and gonadotropin
(b) estrogen and progesterone
(c) growth hormone and prolactin
(d) antidiuretic hormone and oxytocin

7. The primary function of antidiuretic hormone (ADH) is to:
(a) increase the amount of water lost at the kidneys
(b) decrease the amount of water lost at the kidneys
(c) dilate peripheral blood vessels to decrease blood pressure
(d) increase absorption along the digestive tract

8. The element required for normal thyroid function is:
(a) magnesium (b) calcium
(c) potassium (d) iodine

9. Reduced fluid losses in the urine due to retention of sodium ions and water is a result of the action of:
(a) antidiuretic hormone
(b) calcitonin
(c) aldosterone
(d) cortisone

10. The adrenal medulla produces the hormones:
(a) cortisol and cortisone
(b) epinephrine and norepinephrine
(c) corticosterone and testosterone
(d) androgens and progesterone

11. Hormones released by the kidneys include:
(a) calcitriol and erythropoietin
(b) ADH and aldosterone
(c) epinephrine and norepinephrine
(d) cortisol and cortisone

12. What four events are set in motion by activation of a G-protein due to the binding of a hormone to its receptor site?

13. What three higher-level mechanisms are involved in integrating activities of the nervous and endocrine systems?

14. What seven hormones are released by the anterior pituitary gland?

15. What two hormones are released by the posterior pituitary gland?

16. What five primary effects result from the action of thyroid hormones?

17. What effects do calcitonin and parathyroid hormone have on blood calcium levels?

18. What three zones make up the adrenal cortex, and what kinds of hormones are produced by each zone?

19. What peripheral effects does activation of the adrenal medulla have on alpha and beta receptors?

20. What two hormones are released by the kidneys, and what is the importance of each one?

21. What are the three opposing effects of atrial natriuretic peptide and angiotensin II?

22. What four cell populations make up the *endocrine* pancreas? What hormone does each cell type produce?

23. (a) What three phases of the general adapation syndrome (GAS) constitute the body's response to stress? (b) What endocrine secretions play dominant roles in the alarm and resistance phases?

LEVEL 2 Reviewing Concepts

24. What is the primary difference in the way the nervous and endocrine systems communicate with their target cells?

25. How can a hormone modify the activities of its target cells?

26. What possible results occur when a cell receives instructions from two different hormones at the same time?

27. What is an endocrine reflex? Compare endocrine and neural reflexes.

28. How would blocking the activity of phosphodiesterase affect a cell that responds to hormonal stimulation by the cAMP second messenger system?

LEVEL 3 Critical Thinking and Clinical Applications

29. Roger M. has been suffering from extreme thirst; he drinks numerous glasses of water every day and urinates a great deal. Name two disorders that could produce these symptoms. What test could a clinician perform to determine which disorder is present?

30. Julie is pregnant and is not receiving any prenatal care. She has a poor diet consisting mostly of fast food. She drinks no milk, preferring colas instead. How would this situation affect Julie's level of parathyroid hormone?

31. Sherry tells her doctor that she has been restless and irritable lately. She has a hard time sleeping and complains of diarrhea and shortness of breath. During the examination, her physician notices an increased heart rate, decreased level of cholesterol, and slight exophthalmos. What tests would you suggest to help her physician make a positive diagnosis of Sherry's condition?

32. Patients receiving steroid hormone frequently retain large amounts of water. Why?

CHAPTER 19

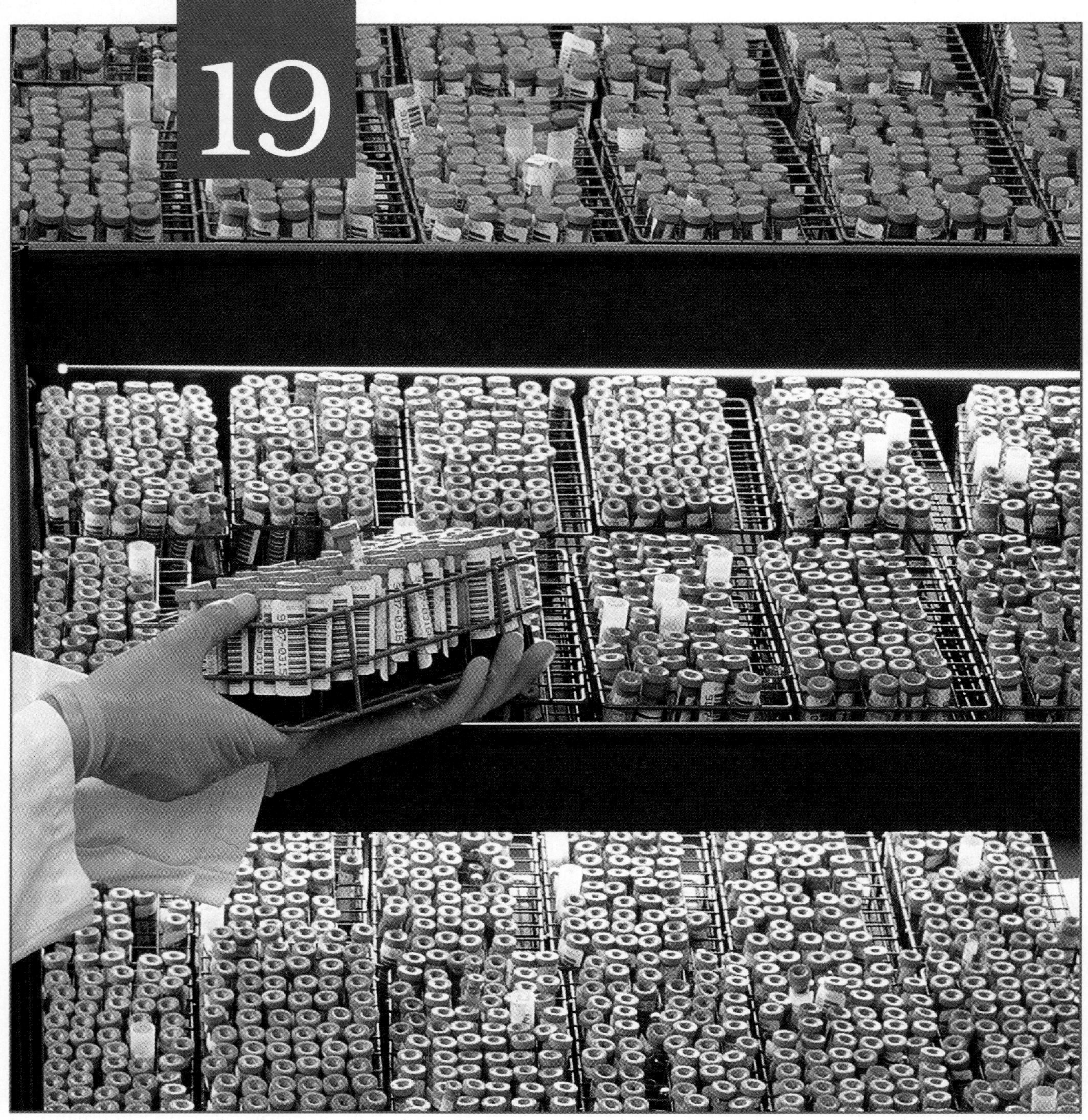

"Out of sight, out of mind." There's probably nothing this old saying applies to more fittingly than blood. As long as we don't see it, we don't think or worry much about it. But looking at vials or bags of it in a hospital--or, worse, watching even a drop of our own oozing from a cut or scrape—makes most of us feel distinctly uneasy; some people even faint at the sight. This anxiety probably comes from being reminded of how dependent we are on our relatively small supply of the precious fluid. If there are three half-gallon milk containers in your refrigerator at home, you probably have more milk than you do blood. You can easily spare a drop of blood, or even donate a pint, but if you suddenly lose more than about 15 to 20 percent of your supply, you will need to get more in a hurry. Fortunately, blood transfusions are now fairly routine matters—but as we will see in this chapter, not just anyone's blood will do. We'll also explore the many functions of this vital substance.

Blood

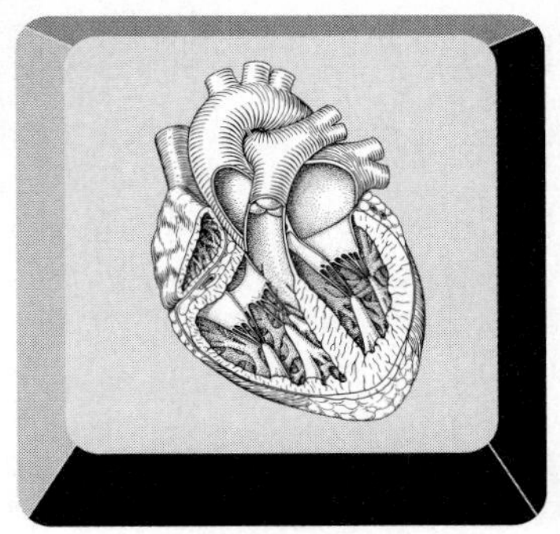

Chapter Outline and Objectives

1 Describe the important components and major functions of blood.

2 Identify locations on the body used for blood collection and list the basic physical characteristics of the blood samples drawn from these locations.

3 Discuss the composition and functions of plasma.

4 Describe the origin and production of the formed elements in blood.

5 List the characteristics and functions of red blood cells.

6 Describe the structure of hemoglobin and indicate its functions.

7 Describe the recycling system for aged or damaged erythrocytes.

8 Define erythropoiesis, identify the stages involved in erythrocyte maturation, and describe the homeostatic regulation of erythrocyte production.

9 List examples of important blood tests and cite the normal values for each test.

10 Explain the importance of blood typing on the basis of ABO and Rh incompatibilities.

11 Categorize the various white blood cells on the basis of their structures and functions and discuss the factors that regulate production of each class.

12 Describe the structure, function, and production of platelets.

13 Discuss mechanisms that control blood loss after an injury and describe the reaction sequences responsible for blood clotting.

Small embryos don't need circulatory systems because diffusion across their exposed surfaces provides adequate oxygen and removes waste products as rapidly as they are generated. By the time a human embryo reaches a few millimeters in length, however, developing tissues are consuming oxygen and nutrients faster than they can be provided by simple diffusion. At that stage the cardiovascular system must begin functioning to provide a rapid-transport system for oxygen, nutrients, and waste products. It is the first system to become fully operational; the heart begins beating by the end of the third week of embryonic life, when many other systems have barely begun their development.

Once blood begins circulating, the embryo can make more efficient use of the nutrients obtained from the maternal bloodstream, and it doubles its size in the next week. In the adult, circulating blood provides nutrients, oxygen, chemical instructions, and a mechanism for waste removal to each of the roughly 75 trillion individual cells in the human body. The blood also transports specialized cells that defend peripheral tissues from infection and disease. These services are absolutely essential, and a region completely deprived of circulation may die in a matter of minutes.

The cardiovascular system can be compared to the cooling system of a car. The basic components include a circulating fluid (blood), a pump (the heart), and an assortment of conducting pipes (the blood vessels). Although the cardiovascular system is far more complicated and versatile, both mechanical and biological systems can suffer from fluid losses, pump failures, or damaged pipes.

The next three chapters examine the individual components of the cardiovascular system. This chapter considers the nature of the circulating blood, Chapter 20 discusses the structure and function of the heart, and Chapter 21 examines the organization of blood vessels and the integrated functioning of the cardiovascular system.

The cardiovascular system is intimately bound up with the *lymphatic system.* The two systems are interconnected and interdependent: The vessels of the lymphatic system empty into the bloodstream, and the cells of the lymphatic system are carried throughout the body in the circulating blood. Chapter 22 details the components of the lymphatic system and considers the array of defenses that the body deploys to protect itself from internal and external hazards.

Functions of Blood

This chapter examines the structure and functions of blood, a specialized connective tissue that contains cells suspended in a fluid matrix. ∞ [p. 128] The functions of blood include:

1. ***The transportation of dissolved gases, nutrients, hormones, and metabolic wastes.*** Oxygen is carried from the lungs to the tissues, and carbon dioxide from the tissues to the lungs. Nutrients absorbed at the digestive tract or released from storage in adipose tissue or the liver are distributed throughout the body. Hormones are carried from endocrine glands toward their target tissues, and the wastes produced by tissue cells are absorbed by the blood and carried to the kidneys for excretion.
2. ***The regulation of the pH and electrolyte composition of interstitial fluids throughout the body.*** The blood absorbs and neutralizes the acids generated by active tissues, such as the lactic acid produced by skeletal muscles.
3. ***The restriction of fluid losses through damaged vessels or at other injury sites.*** Blood contains enzymes and factors that respond to breaks in the vessel walls by initiating the process of *blood clotting.* A blood clot acts as a temporary patch and prevents further blood loss.
4. ***Defense against toxins and pathogens.*** Blood transports *white blood cells,* specialized cells that migrate into peripheral tissues to fight infections or remove debris. It also delivers *antibodies,* special proteins that attack invading organisms or foreign compounds.
5. ***Stabilization of body temperature.*** Blood absorbs the heat generated by active skeletal muscles and redistributes it to other tissues. When body temperature is already high, that heat will be lost across the surface of the skin. If body temperature is too low, the warm blood is directed to the brain and to other temperature-sensitive organs.

Composition of Blood

Figure 19-1

Blood has a characteristic and unique composition (Figure 19-1•). **Plasma** (PLAZ-mah), the matrix of blood, has a density only slightly greater than that of water. It contains dissolved proteins rather than the network of insoluble fibers found in loose connective tissue or cartilage.

Formed elements are blood cells and cell fragments that are suspended in the plasma. **Red blood cells (RBCs),** or **erythrocytes** (e-RITH-rō-sīts; *erythros,* red + *cyte,* cell) are the most abundant blood cells. These specialized cells are essential for the transport of oxygen in the blood. The less numerous **white blood cells (WBCs),** or **leukocytes** (LOO-kō-sīts; *leukos,* white) are cells involved with the body's defense mechanisms. **Platelets** are small packets of cytoplasm that contain enzymes and factors important to the process of blood clotting.

Together the plasma and formed elements constitute **whole blood.** These components may be separated, or **fractionated,** for analytical or clinical purposes; examples of uses for fractionated blood will be encountered later in the chapter.

❑ Blood Collection and Analysis

Figure 19-1

Fresh whole blood is usually collected from a superficial vein, such as the *median cubital vein* on the anterior surface of the elbow (Figure 19-1•). This procedure is called a **venipuncture** (VEN-i-punk-chur; *vena*, vein + *punctura*, a piercing). Venipuncture is a common sampling technique because (1) superficial veins are easy to locate; (2) the walls of veins are thinner than those of comparably sized arteries; and (3) blood pressure in the venous system is relatively low, and the puncture wound seals quickly. The most common clinical procedures examine venous blood.

Blood from peripheral capillaries can be obtained by puncturing the tip of a finger, the lobe of an ear, or (in infants) the great toe or heel. A small drop of capillary blood can be used to prepare a *blood smear*, a thin film of blood on the surface of a microscope slide. The blood smear can then be stained with special dyes so that the different types of formed elements can be easily distinguished.

An **arterial puncture,** or "arterial stick," may be required when checking the efficiency of gas exchange at the lungs. Samples are usually drawn from the *radial artery* at the wrist or the *brachial artery* at the elbow.

Whole blood from any of these sources has the same basic physical characteristics.

- ***Temperature.*** Blood temperature is roughly 38° C (100.4° F), slightly higher than normal body temperature.
- ***Viscosity.*** Blood viscosity is five times that of water, because interactions among dissolved proteins, formed elements, and the surrounding water molecules make plasma relatively sticky, cohesive, and resistant to flow.
- ***pH.*** Blood pH ranges from 7.35 to 7.45, averaging 7.4. It is therefore slightly alkaline.
- ***Volume.*** There are 5–6 liters of whole blood in the cardiovascular system of an adult man and 4–5 liters in an adult woman. The differences in blood volume between the sexes primarily reflect differences in average body size. Blood volume in liters can be estimated for an individual of either sex by calculating 7 percent of the body weight in kilograms. For example, a 75-kg individual would have a blood volume of approximately 5.25 liters, or about 1.4 gallons. More precise determinations of blood volume can be obtained using special dyes or radioisotopes, such as phosphorus-32. Clinicians use the terms **hypovolemic** (hī-pō-vō-LĒ-mik), **normovolemic** (nor-mō-vō-LĒ-mik), and **hypervolemic** (hī-per-vō-LĒ-mik) to refer to low, normal, or excessive blood volumes, respectively. These conditions produce characteristic symptoms because variations in blood volume affect other components of the cardiovascular system. For example, an abnormally large blood volume can place a severe stress on the heart, which must push the extra fluid around the circulatory system.

Plasma

Figure 19-1

The composition of whole blood is summarized in Figure 19-1•. Plasma contributes approximately 55 percent of the volume of whole blood, with water accounting for 92 percent of the plasma volume. Together, plasma and interstitial fluid account for most of the volume of extracellular fluid (ECF) in the body.

❑ Differences between Plasma and Interstitial Fluid

In many respects, the composition of the plasma resembles that of interstitial fluid. The concentrations of the major plasma ions, for example, are similar to those of interstitial fluid and differ markedly from those found inside living cells. The primary differences between plasma and interstitial fluid involve (1) the concentrations of dissolved proteins and (2) the levels of respiratory gases (oxygen and carbon dioxide). The protein concentrations differ because plasma contains circulating *plasma proteins* that cannot cross the walls of blood vessels and thus cannot enter the interstitial fluids. We will consider the origins of the differences in oxygen and carbon dioxide levels in Chapter 23.

❑ Plasma Proteins

Plasma contains significant quantities of dissolved proteins. On average, there are 7.6 g of protein in each 100 ml of plasma, almost five times the concentration in interstitial fluid. The large size and globular shapes of most blood proteins prevent them from crossing capillary walls, and they remain trapped within the circulatory system. There are three primary classes of plasma proteins,

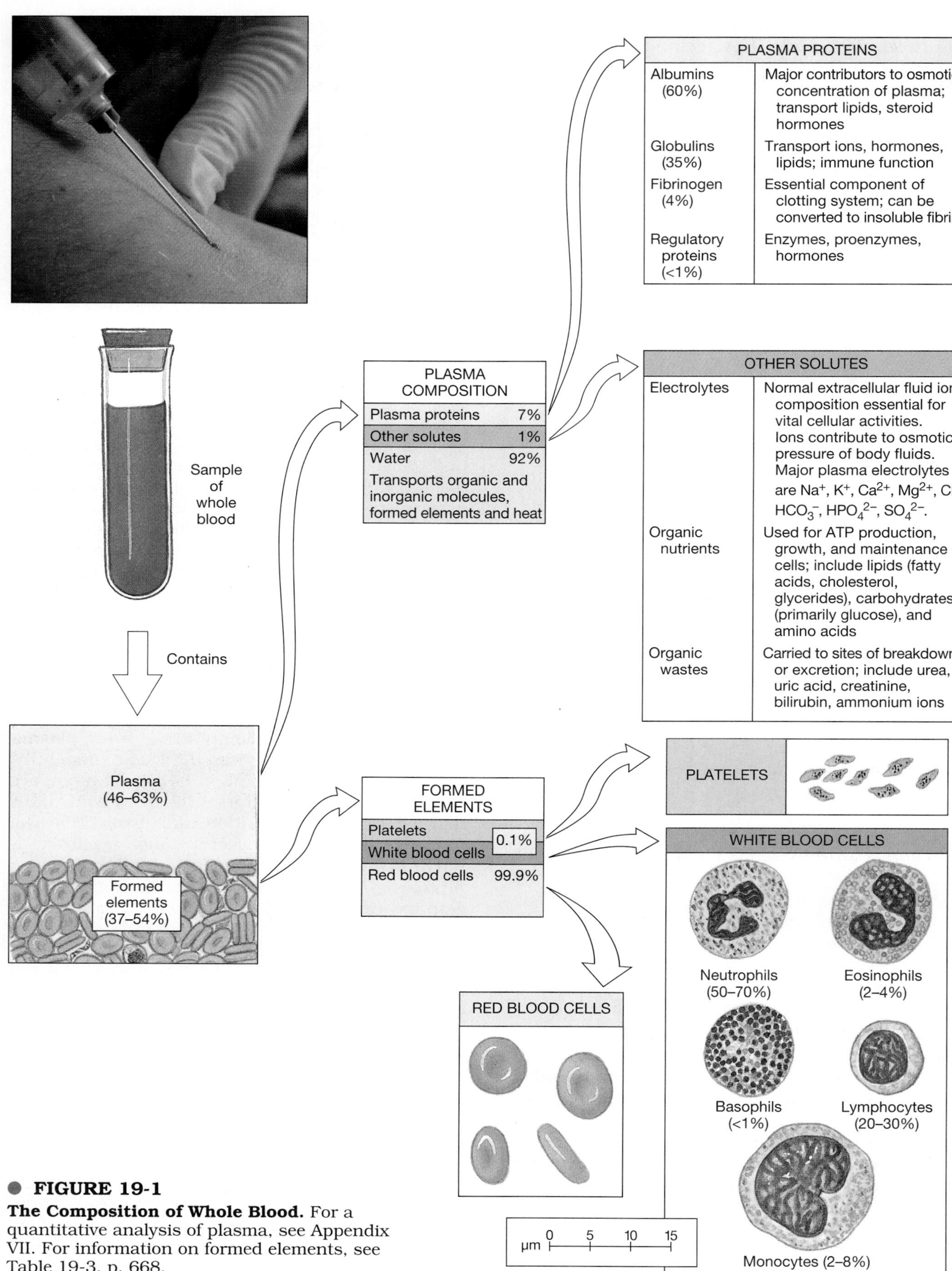

Sample of whole blood
Contains
Plasma (46–63%)
Formed elements (37–54%)
PLASMA COMPOSITION
Plasma proteins 7%
Other solutes 1%
Water 92%
Transports organic and inorganic molecules, formed elements and heat
PLASMA PROTEINS
Albumins (60%)
Major contributors to osmotic concentration of plasma; transport lipids, steroid hormones
Globulins (35%)
Transport ions, hormones, lipids; immune function
Fibrinogen (4%)
Essential component of clotting system; can be converted to insoluble fibrin
Regulatory proteins (<1%)
Enzymes, proenzymes, hormones
OTHER SOLUTES
Electrolytes
Normal extracellular fluid ion composition essential for vital cellular activities. Ions contribute to osmotic pressure of body fluids. Major plasma electrolytes are Na^+, K^+, Ca^{2+}, Mg^{2+}, Cl^-, HCO_3^-, HPO_4^{2-}, SO_4^{2-}.
Organic nutrients
Used for ATP production, growth, and maintenance of cells; include lipids (fatty acids, cholesterol, glycerides), carbohydrates (primarily glucose), and amino acids
Organic wastes
Carried to sites of breakdown or excretion; include urea, uric acid, creatinine, bilirubin, ammonium ions
FORMED ELEMENTS
Platelets
White blood cells
0.1%
Red blood cells 99.9%
PLATELETS
WHITE BLOOD CELLS
Neutrophils (50–70%)
Eosinophils (2–4%)
Basophils (<1%)
Lymphocytes (20–30%)
Monocytes (2–8%)
RED BLOOD CELLS
μm 0 5 10 15

● **FIGURE 19-1**

The Composition of Whole Blood. For a quantitative analysis of plasma, see Appendix VII. For information on formed elements, see Table 19-3, p. 668.

albumins (al-BŪ-minz), *globulins* (GLOB-ū-linz), and *fibrinogen* (fī-BRIN-ō-jen).

Albumins

Albumins constitute roughly 60 percent of the plasma proteins. As the most abundant proteins, they are major contributors to the colloid osmotic pressure of the plasma. Albumins are also important in the transport of fatty acids and several other substances useful in metabolism.

Globulins

Globulins account for approximately 35 percent of the protein population. Examples of important plasma globulins include:

1. **Transport globulins** bind small ions, hormones, or compounds that (1) might otherwise be filtered out of the blood at the kidneys or (2) have very low solubility in water. Transport globulins can be divided into *alpha globulins* and *beta globulins* on the basis of structure. Important examples of transport globulins include:
   - *Thyroid-binding globulin*, which binds and transports thyroxine, and *transcortin*, which transports ACTH.
   - **Metalloproteins** (me-tal-ō-PRŌ-tēnz), which transport metal ions. *Transferrin*, for example, is a metalloprotein that transports iron (Fe^{2+}).
   - *Apolipoproteins*, which carry triglycerides and other lipids in the blood. When bound to lipids, an apolipoprotein becomes a **lipoprotein** (lī-pō-PRŌ-tēn).
   - *Steroid-binding proteins*, which transport steroid hormones in the blood. For example, *testosterone-binding globulin* (TeBG) binds and transports testosterone in males.
2. **Immunoglobulins** (i-mū-nō-GLOB-ū-linz), also called **antibodies**, attack foreign proteins and pathogens. Several different classes of immunoglobulins will be described in Chapter 22.

Fibrinogen

The third type of plasma protein, **fibrinogen**, functions in blood clotting. Fibrinogen normally accounts for roughly 4 percent of plasma proteins. Under certain conditions fibrinogen molecules interact, forming large, insoluble strands of **fibrin** (FĪ-brin). These fibers provide the basic framework for a blood clot. If steps are not taken to prevent clotting, the conversion of fibrinogen to fibrin will occur in a sample of plasma. This conversion removes the clotting proteins, leaving a fluid known as **serum**. The clotting process also removes calcium ions and other materials from solution. As a result there are several significant differences between plasma and serum. Because the normal range of values may differ, the results of a blood test usually indicate the sample source, either plasma (P) or serum (S).

Other Plasma Proteins

Polypeptide hormones, including thyroid-stimulating hormone (TSH), follicle-stimulating hormone (FSH), luteinizing hormone (LH), insulin, and prolactin (PRL), are normally present in the circulating blood. Their plasma concentrations rise and fall from day to day and even hour to hour.

Origins of the Plasma Proteins

The liver synthesizes and releases more than 90 percent of the plasma proteins, including all of the albumins and fibrinogen and most of the globulins. However, the immunoglobulins and protein hormones are synthesized elsewhere. Immunoglobulins (antibodies) are produced by *plasma cells*. Plasma cells are derived from lymphocytes, the cells of the lymphatic system. Plasma hormones are produced in a variety of different endocrine organs.

Because the liver is the primary source of plasma proteins, liver disorders can alter the composition and functional properties of the blood. For example, some forms of liver disease can lead to uncontrolled bleeding due to inadequate synthesis of fibrinogen and other proteins involved in blood clotting.

Why is venipuncture a common technique for obtaining a blood sample?

What would be the effects of a decrease in the amount of plasma proteins?

Which plasma protein would you expect to be elevated during a viral infection?

Formed Elements

The major cellular components of blood are red blood cells and white blood cells. In addition, blood contains noncellular formed elements called *platelets*, small packets of cytoplasm that function in blood clotting.

Hemopoiesis

Formed elements are produced through the process of **hemopoiesis** (hēm-ō-poy-Ē-sis), or *hematopoiesis*. Blood cells appear in the circulation during the third week of embryonic development. These cells divide repeatedly, increasing their numbers. The vessels of the embryonic yolk

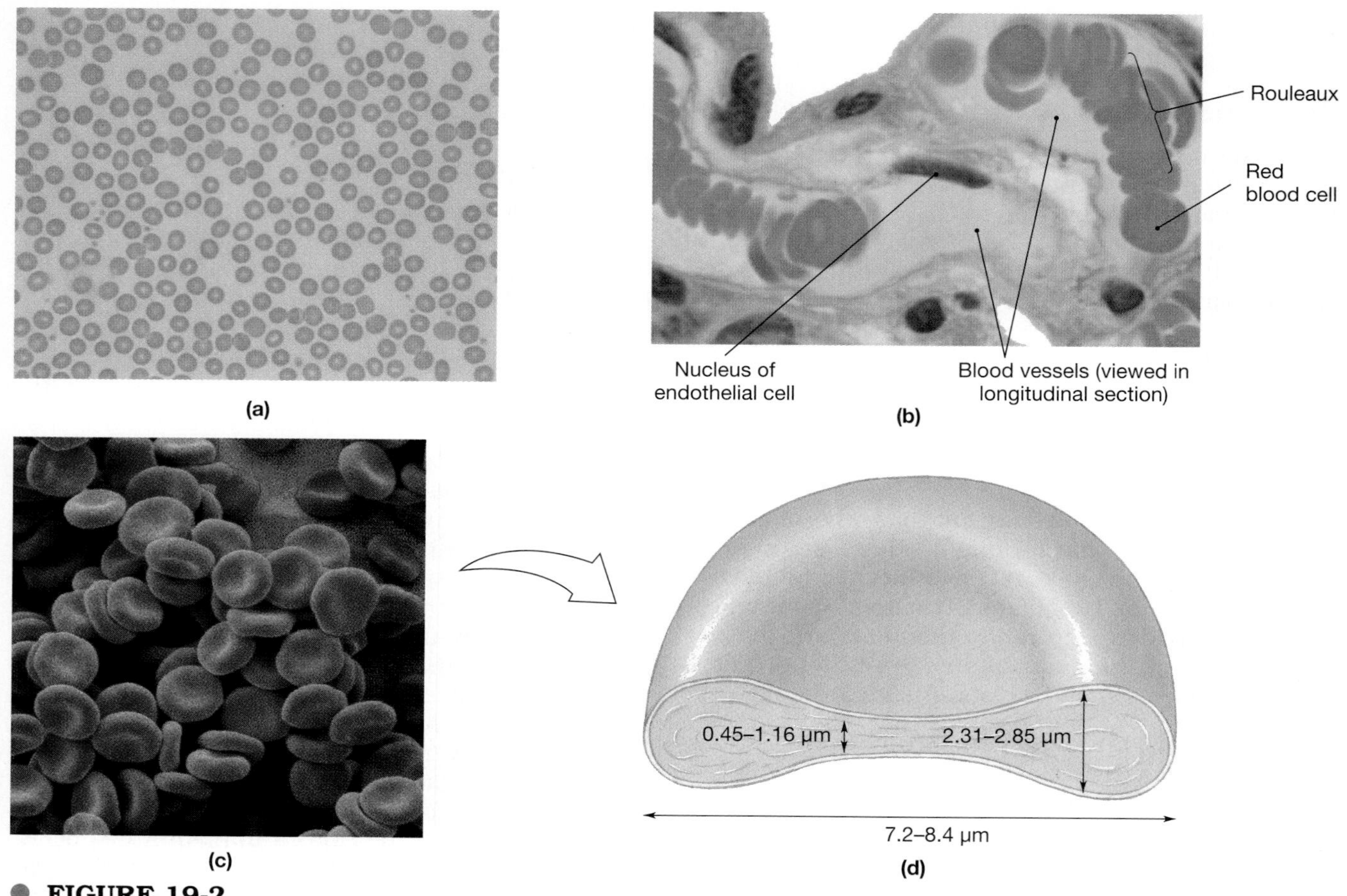
Rouleaux
Red blood cell
Nucleus of endothelial cell
Blood vessels (viewed in longitudinal section)
(a)
(b)
(c)
(d)
0.45–1.16 μm
2.31–2.85 μm
7.2–8.4 μm

● **FIGURE 19-2**
Anatomy of Red Blood Cells. (a) When viewed in a standard blood smear, red blood cells appear as two-dimensional objects because they are flattened against the surface of the slide. (LM × 320) **(b)** When traveling through relatively narrow capillaries, erythrocytes may stack like dinner plates, forming rouleaux. (LM × 1072) **(c)** A scanning electron micrograph of red blood cells reveals their three-dimensional structure quite clearly. (SEM × 1195) **(d)** A sectional view of a mature red blood cell, showing average dimensions.

sac are the primary site of blood formation for the first 8 weeks of development. As other organ systems appear, some of the embryonic blood cells move out of the circulation and into the liver, spleen, thymus, and bone marrow. These embryonic cells differentiate into **stem cells,** which produce blood cells by their divisions. The liver and spleen are the primary sites of hemopoiesis from the second to fifth months of development. As the skeleton enlarges, the bone marrow becomes increasingly important, and it predominates after the fifth developmental month. In the adult it is the only site of red blood cell production as well as the primary site of white blood cell formation.

Stem cells called **hemocytoblasts** produce all of the blood cells, but the process occurs in a series of steps. Hemocytoblast divisions produce at least four different types of stem cells with relatively restricted futures. These cells, called **progenitor cells,** remain capable of division, but their daughter cells will differentiate only into specific types of blood cells. For example, one type of progenitor cell produces daughter cells that mature into red blood cells; another gives rise to certain types of white blood cells. We will consider the fates of the different progenitor cells as we discuss the formation of each type of formed element.

❑ Red Blood Cells

Red blood cells (RBCs) contain the pigment *hemoglobin*, which binds and transports oxygen and carbon dioxide. RBCs are the most abundant blood cells, accounting for 99.9 percent of the formed elements, and they give whole blood its deep red color.

Abundance of RBCs

A standard blood test reports the number of RBCs per **microliter** (μl) of whole blood. One microliter, or *cubic millimeter* (mm^3), of whole blood from a man contains 4.5–6.3 million erythrocytes; a microliter from a woman contains 4.2–5.5 million. There are approximately 260 million red blood cells in a single drop of whole blood, and 25 trillion in the blood of an average adult. The number of erythrocytes thus accounts for roughly one-third of the cells in the human body.

The **hematocrit** (he-MA-tō-krit) is the percentage of whole blood occupied by cellular elements. The normal hematocrit in adult men averages 46 (range: 40–54); the average for adult women is 42 (range: 37–47). The difference in hematocrit between males and females reflects the fact that androgens stimulate red blood cell production, whereas estrogens have an inhibitory effect.

The hematocrit is determined by centrifuging a blood sample so that all the formed elements will come out of suspension. Whole blood contains roughly 1000 red blood cells for each white blood cell, and after centrifugation the white blood cells and platelets form a very thin *buffy coat* above a thick layer of RBCs. Because the hematocrit value is almost entirely due to the volume of erythrocytes, hematocrits are often reported as the **volume of packed red cells (VPRC)** or simply the **packed cell volume (PCV).**

Many different factors can alter the hematocrit. For example, the hematocrit increases in dehydration, because of a reduction in plasma volume, and after EPO (erythropoietin) stimulation. The hematocrit may decrease because of internal bleeding or from problems with RBC formation. As a result, the hematocrit alone does not provide specific diagnostic information. Yet a change in hematocrit is an indication that other, more specific tests are needed. (Some of those tests will be considered later in the chapter.) AM *Polycythemia*

Structure of RBCs

Figure 19-2

Red blood cells (Figure 19-2●) are among the most specialized cells of the body. A red blood cell is very different from the "typical cell" discussed in Chapter 3.

SHAPE Each red blood cell is a biconcave disc, with a thin central region and a thick outer margin. An average erythrocyte has a diameter of 7.8 μm and a maximum thickness of 2.6 μm, although the center narrows to about 0.8 μm. (Normal ranges of measurement are given in Figure 19-2d●.)

This unusual shape has three important effects on RBC function:

1. It gives each RBC a relatively large surface area. A large surface area is important because the RBC carries oxygen bound to intracellular proteins. That oxygen must be absorbed or released quickly as the RBC passes through the capillaries of the lungs or peripheral tissues. The larger the surface area, the faster the exchange between the interior of the cell and the surrounding plasma. The total surface area of the red blood cells in the blood of a typical adult is approximately 3800 square meters, 2000 times the total surface area of the body.
2. It enables them to form stacks, like dinner plates, that smooth the flow through narrow blood vessels. These stacks, called **rouleaux** (roo-LŌ; little rolls), form and dissociate repeatedly without affecting the cells involved. An entire rouleau can pass along a blood vessel little larger than the diameter of a single erythrocyte, whereas individual cells would bump the walls, bang together, and form log jams that could block the circulatory passageway.
3. It enables them to bend and flex when entering small capillaries and branches. RBCs are very flexible, and by changing shape, individual red blood cells can squeeze through capillaries as narrow as 4 μm.

ORGANELLES Red blood cells lack several organelles found in most other cells.

- Red blood cells lack mitochondria, so they can obtain energy only through anaerobic metabolism. They rely on glucose obtained from the surrounding plasma. This feature ensures that absorbed oxygen will be carried to peripheral tissues, not "stolen" by mitochondria in the cell.
- Mature RBCs in humans and other mammals lack both nuclei and ribosomes. (The RBCs of other vertebrates are nucleated cells.) Our RBCs can neither undergo cell division nor synthesize proteins.

Hemoglobin

Figure 19-3

In effect, a developing erythrocyte loses any intracellular component not directly associated with its primary function—the transport of respiratory gases. Molecules of **hemoglobin** (HĒ-mō-glō-bin) **(Hb)** account for over 95 percent of the intracellular proteins. The hemoglobin content of whole blood is reported in terms of grams of Hb per 100 ml of whole blood (g/dl). Normal ranges are 14–18 g/dl in males and 12–16 g/dl in females.

HEMOGLOBIN STRUCTURE A hemoglobin molecule has a complex quaternary shape (Figure 19-3●). There are two *alpha (α) chains* and two *beta (β) chains* of polypeptides in each hemoglobin molecule. Each individual chain is a globular protein subunit that resembles the *myoglobin* found in skeletal and cardiac muscle cells. Like myoglobin, each of the hemoglobin chains contains a single molecule of **heme,** a pigment complex diagrammed in Figure 19-3●. Heme is an example of a **porphyrin,** an organic compound usually associated with metal ions. Each heme unit

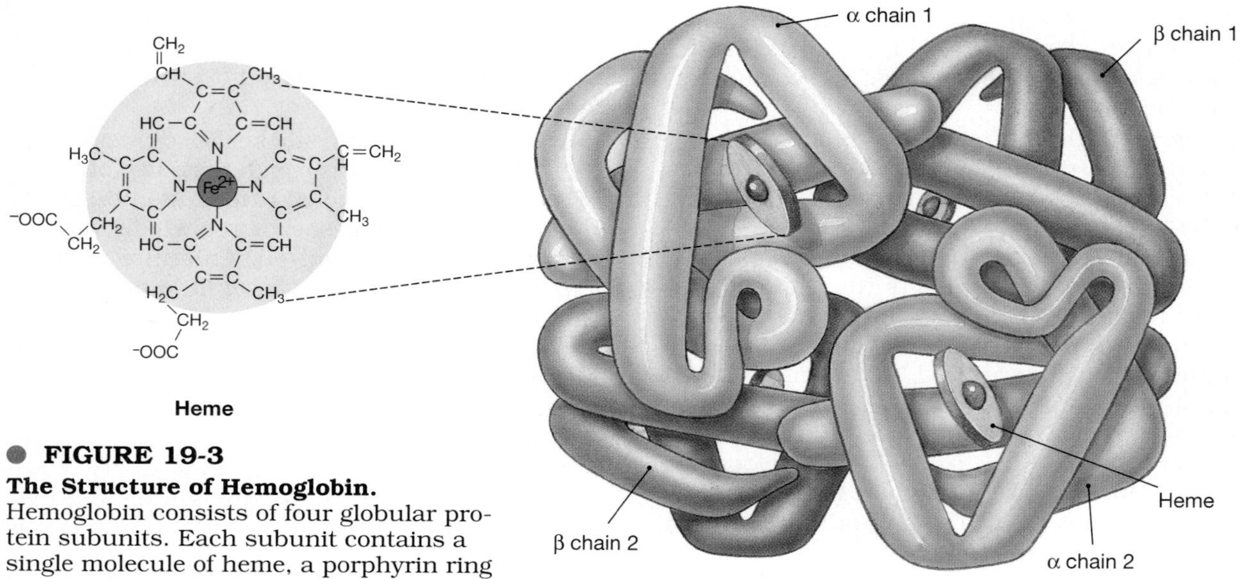
Heme
α chain 1
β chain 1
β chain 2
Heme
α chain 2
Hemoglobin molecule

FIGURE 19-3
The Structure of Hemoglobin.
Hemoglobin consists of four globular protein subunits. Each subunit contains a single molecule of heme, a porphyrin ring surrounding a single ion of iron.

holds an iron ion in such a way that it can interact with an oxygen molecule, forming *oxyhemoglobin*, HbO_2. The iron-oxygen interaction is very weak, and the two can be easily separated without damaging either the heme unit or the oxygen molecule.

The red blood cells of an embryo or fetus contain a different form of hemoglobin, known as *fetal hemoglobin*, which has a higher affinity for oxygen than does adult hemoglobin. This enables the developing fetus to "steal" oxygen from the maternal bloodstream at the placenta. The conversion from fetal hemoglobin to the adult form of hemoglobin begins shortly before birth and continues over the next year.

HEMOGLOBIN FUNCTION There are approximately 280 million molecules of hemoglobin in each red blood cell. Because a hemoglobin molecule contains four heme units, each erythrocyte can potentially carry more than a billion molecules of oxygen. Roughly 98.5 percent of the oxygen carried by the blood travels through the circulation bound to hemoglobin molecules inside red blood cells.

The amount of oxygen bound to hemoglobin depends primarily on the oxygen content of the plasma. When plasma oxygen levels are low, Hb releases oxygen. Under these conditions, typical of peripheral capillaries, plasma carbon dioxide levels are elevated. The α and β chains of hemoglobin then bind CO_2, forming **carbaminohemoglobin.** Carbaminohemoglobin carries approximately 23 percent of the carbon dioxide in the blood.

In the capillaries of the lungs, plasma oxygen levels are high and carbon dioxide levels are low. Upon reaching these capillaries, RBCs absorb and bind oxygen, and carbon dioxide is released.

Normal activity levels can be sustained only when tissue oxygen levels are kept within normal limits. If the hematocrit is low, or the hemoglobin content of the RBCs is reduced, the condition of **anemia** exists. Anemia produces clinical symptoms because it interferes with oxygen delivery to peripheral tissues. There are many different forms of anemia, and specific examples will be considered both in this chapter and in the *Applications Manual.*

ABNORMAL HEMOGLOBIN There are several inherited disorders characterized by the production of abnormal hemoglobin. Two of the best known are *thalassemia* and *sickle cell anemia* (SCA).

The various forms of **thalassemia** (thal-ah-SĒ-mē-ah) result from an inability to produce adequate amounts of alpha or beta chains. As a result, the rate of RBC production is slowed, and the mature RBCs are fragile and short-lived. The scarcity of healthy RBCs reduces the oxygen-carrying capacity of the blood and leads to problems with the development and growth of systems throughout the body. Individuals with severe thalassemia must undergo frequent transfusions. AM *Thalassemia*

Sickle cell anemia results from a mutation affecting the amino acid sequence of the β chains of the hemoglobin molecule. When the blood contains an abundance of oxygen, the hemoglobin molecules and the erythrocytes that carry them appear normal. But when the defective hemoglobin gives up enough of its stored oxygen, the adjacent hemoglobin molecules interact, and the cells change shape, becoming stiff and markedly curved (Figure 19-4●). This "sickling" does not affect the oxygen-carrying capabilities of the erythrocytes, but it causes them to become more fragile and easily damaged. Moreover, an RBC that has folded to squeeze into a narrow capillary may lose its oxygen, change shape, and become stuck. A circulatory blockage results, and nearby tissues become oxygen-starved. (More information on this condition and new treatment options are included in the *Applications Manual.*) AM *Sickle Cell Anemia*

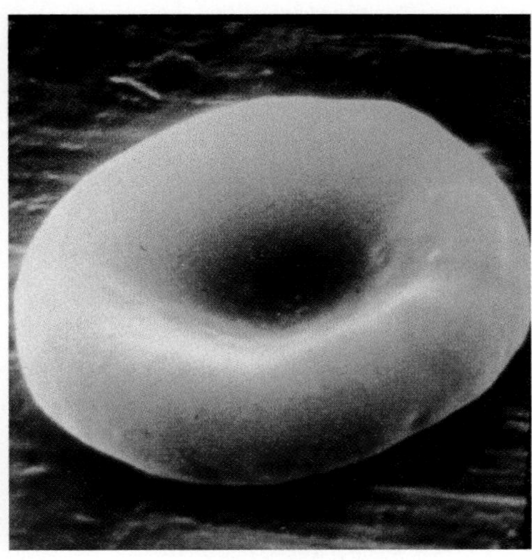

(a) Normal RBCs

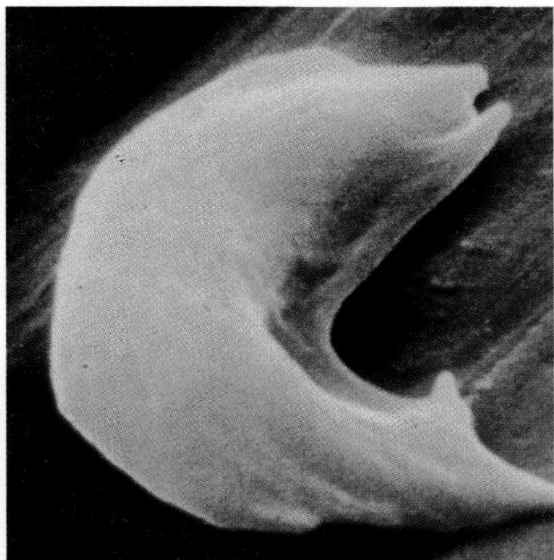

(b) Sickled RBCs

● **FIGURE 19-4**
Sickling in Red Blood Cells. (a) When fully oxygenated, the cells of an individual with the sickling trait appear relatively normal. **(b)** At lower oxygen concentrations the RBCs change shape, becoming relatively rigid and sharply curved. (SEM × 67,500)

RBC Life Span and Circulation

Figure 19-5

An erythrocyte is exposed to severe mechanical stresses. A single round-trip of the circulatory system usually takes less than 30 seconds. In that time a red blood cell gets pumped out of the heart and forced along vessels, where it bounces off the walls and collides with other red cells. It stacks in rouleaux, contorts and squeezes through tiny capillaries, and then joins its comrades in a headlong rush back to the heart for another round.

With all this wear and tear and no repair mechanisms, a typical red blood cell has a relatively short life span. After traveling about 700 miles in 120 days, it either breaks down or is destroyed by phagocytic cells. The continual elimination of red blood cells is usually unnoticed because new erythrocytes are entering the circulation at a comparable rate. About 1 percent of the circulating erythrocytes are replaced each day, and in the process approximately 3 million new erythrocytes enter the circulation *each second!*

HEMOGLOBIN CONSERVATION AND RECYCLING

Phagocytic cells of the liver, spleen, and bone marrow monitor the condition of circulating erythrocytes, and they usually recognize and engulf erythrocytes before they break down, or *hemolyze.* These phagocytes also remove hemoglobin and red blood cell fragments released by the small number (around 10 percent of the total recycled) that hemolyze in the bloodstream.

If the hemoglobin released by hemolysis is not phagocytized, it will be lost. Hemoglobin molecules remain intact only under the conditions found inside red blood cells. When hemolysis occurs, the hemoglobin molecule breaks down, and the α and β chains are filtered by the kidneys and excreted in the urine. When abnormally large numbers of erythrocytes are breaking down in the circulation, the urine may develop a reddish or brown coloration. This condition is called **hemoglobinuria**. **Hematuria** (hē-ma-TŪ-rē-uh), the presence of intact red blood cells in the urine, occurs only after kidney damage or damage to vessels along the urinary tract.

Once a red blood cell has been engulfed and broken down by a phagocytic cell, each component of the hemoglobin molecule has a different fate (Figure 19-5●).

Alpha and Beta Chains. The globular proteins are disassembled into their component amino acids. These amino acids are either metabolized by the cell or released into the circulation for use by other cells.

Heme Units. Each heme unit is stripped of its iron and converted to **biliverdin** (bil-ē-VER-din), a porphyrin derivative with a green color. (Bad bruises often take on a greenish coloration due to biliverdin formation in the blood-filled tissues.) Biliverdin is then converted to **bilirubin** (bil-ē-ROO-bin) and released into the circulation.

Liver cells absorb the bilirubin and use it to synthesize a complex molecule called **conjugated bilirubin,** or *bilirubin diglucuronide.* Most of the conjugated bilirubin is excreted in the bile. A small amount reenters the circulation; under normal circumstances it will be removed at the kidneys and eliminated in the urine. Urinary concentrations of conjugated bilirubin are very low, and unconjugated bilirubin does not normally enter the urine at all. If the bile ducts are blocked, however, or the liver cannot excrete bilirubin, circulating levels climb rapidly. Bilirubin then diffuses into peripheral tissues, giving them a yellow color that is most apparent in the skin and over the sclera of the eyes. This combination of signs (yellow skin and eyes) is called **jaundice** (JAWN-dis). AM *Bilirubin Tests and Jaundice*

Inside the large intestine, bacteria convert the conjugated bilirubin to *urobilinogens* and *stercobilinogens.* Some of the urobilinogens are absorbed into the circulation. These will be excreted in the urine. On exposure to oxygen some of the uro-

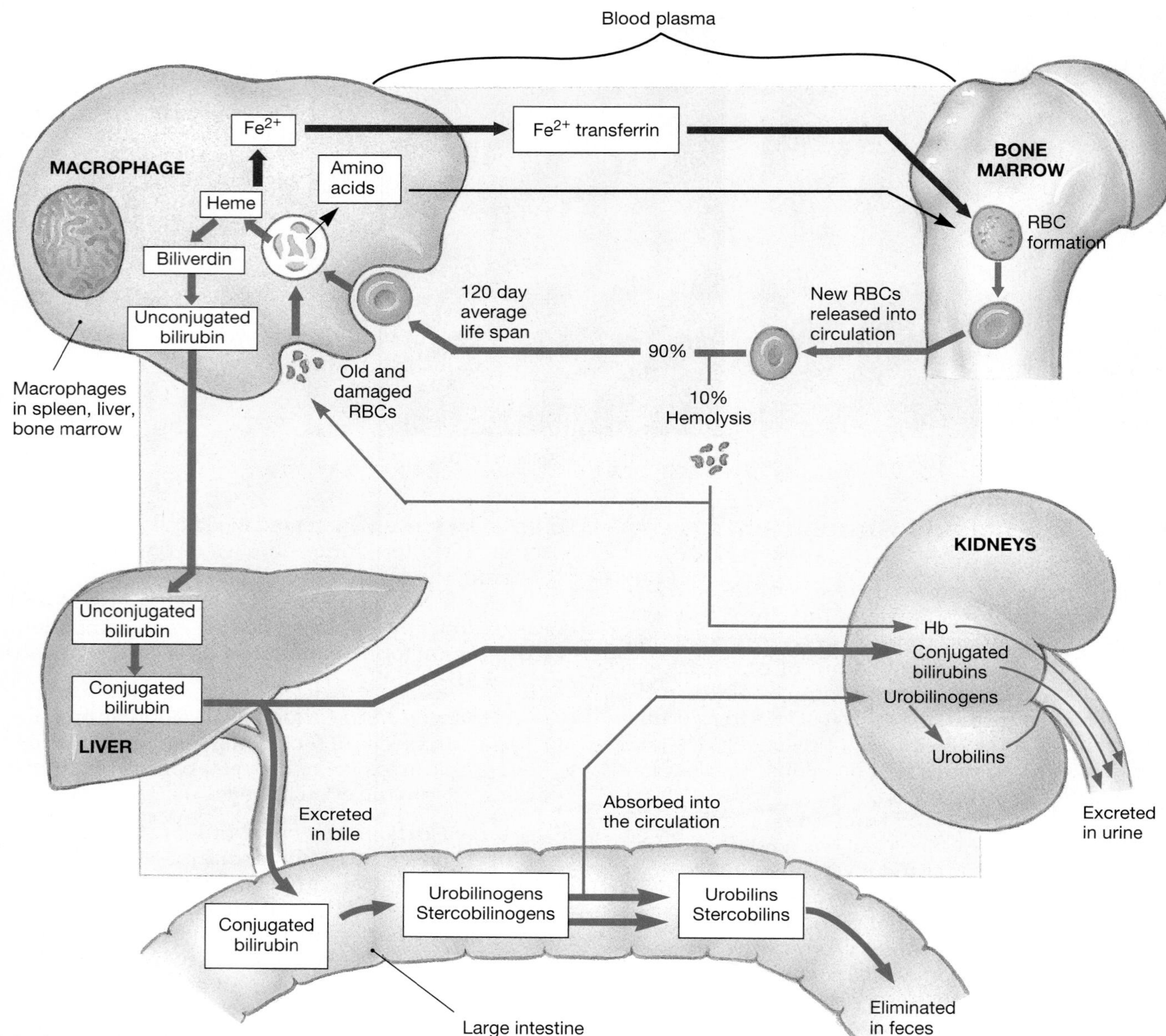
Blood plasma
MACROPHAGE
Fe^{2+}
Heme
Amino acids
Biliverdin
Unconjugated bilirubin
Fe^{2+} transferrin
BONE MARROW
RBC formation
120 day average life span
New RBCs released into circulation
90%
10% Hemolysis
Old and damaged RBCs
Macrophages in spleen, liver, bone marrow
KIDNEYS
Hb
Conjugated bilirubins
Urobilinogens
Urobilins
Unconjugated bilirubin
Conjugated bilirubin
LIVER
Excreted in bile
Absorbed into the circulation
Excreted in urine
Conjugated bilirubin
Urobilinogens Stercobilinogens
Urobilins Stercobilins
Eliminated in feces
Large intestine

FIGURE 19-5

Red Blood Cell Turnover. This diagram shows the normal pathways for recycling amino acids and iron from aging or damaged red blood cells. The amino acids are absorbed, especially by developing cells in the bone marrow. The iron is stored in many different sites. The porphyrins of the heme units are converted to bilirubin, absorbed by the liver, and subsequently excreted in the bile or urine; there is some recirculation of the breakdown products produced in the large intestine.

bilinogens and stercobilinogens convert to **urobilins** and **stercobilins.** Urobilins in the urine give it a yellow color; the brownish color of feces is due to urobilins and stercobilins.

Iron. Large quantities of free iron are toxic to cells, and iron within the body is usually bound to transport or storage proteins. Iron extracted from the heme molecules may be bound and stored within the phagocytic cell or released into the bloodstream, where it binds to a plasma protein, **transferrin** (tranz-FER-in). Red blood cells developing in the bone marrow absorb the amino acids and transferrins from the circulation and use them to synthesize new hemoglobin molecules. Excess transferrins are removed by the liver and bone marrow, and the iron is stored in two special protein-iron complexes, **ferritin** (FER-i-tin) and **hemosiderin** (hē-mō-SID-e-rin).

This recycling system is remarkably efficient; although roughly 26 mg of iron are incorporated into hemoglobin molecules each day, a dietary supply of 1–2 mg will keep pace with the incidental losses that occur at the kidney and the digestive tract.

Any impairment in iron uptake or metabolism can cause serious clinical problems due to impairment of red blood cell formation.

Too *much* iron can also cause problems because of excessive buildup in secondary storage sites, such as the liver and cardiac muscle tissue. Excessive iron deposition has recently been linked to heart disease. AM *Iron Deficiencies and Excesses*

Red Blood Cell Formation

Figure 19-6

Red blood cell formation, or **erythropoiesis** (e-rith-rō-poy-Ē-sis), occurs in the bone marrow, or **myeloid** (MĪ-e-loyd) **tissue** (*myelos,* marrow), of the adult. **Red marrow,** where active blood cell production occurs, is found in portions of the vertebrae, sternum, ribs, skull, scapulae, pelvis, and proximal limb bones. Other marrow areas contain a fatty tissue that is known as **yellow marrow.** Under extreme stimulation, as following a severe and sustained blood loss, areas of yellow marrow can convert to red marrow, increasing the rate of red blood cell formation.

STAGES IN RBC MATURATION During its maturation, a red blood cell passes through a series of stages. **Hematologists** (hē-ma-TOL-o-jists), specialists in blood formation and function, have given specific names to key stages. (See Figure 19-6•.) **Erythroblasts** are very immature red blood cells that are actively synthesizing hemoglobin. There are several types of erythroblasts, categorized on the basis of the amount of hemoglobin present, their total size, and the size of the nucleus.

After roughly 4 days of differentiation, the erythroblast, often called a *normoblast*, sheds its nucleus and becomes a **reticulocyte** (re-TIK-ū-lō-sīt). A reticulocyte contains 80 percent of the hemoglobin found in a mature red blood cell, and hemoglobin synthesis continues for 2–3 more days. Over this period, while the cells are synthesizing proteins, their cytoplasm still contains RNA, which can be visualized with appropriate stains. After 2 days in the bone marrow, reticulocytes enter the circulation. At this time they can still be detected by staining, and reticulocytes normally account for about 0.8 percent of the erythrocyte population in the blood. After 24 hours in circulation, the reticulocytes complete their maturation and become indistinguishable from other mature RBCs.

REGULATION OF ERYTHROPOIESIS For erythropoiesis to proceed normally, the myeloid tissues must receive adequate supplies of amino acids, iron, vitamin B_{12}, and other vitamins (B_6 and folic acid) required for protein synthesis. For example, if **vitamin B_{12}** is not obtained from the diet, normal stem cell divisions cannot occur, and *pernicious anemia* results. Vitamin B_{12} is obtained from dairy products and meat, and its absorption requires the presence of *intrinsic factor* produced in the stomach. Thus *pernicious anemia* may be caused by a diet deficient in vitamin B_{12}, a problem with gastric production of intrinsic factor, or a problem with B_{12} complex absorption.

Erythropoiesis is stimulated directly by erythropoietin and indirectly by several hormones, including thyroxine, androgens, and growth hormone. As noted above, estrogens have an inhibitory effect on erythropoiesis.

Erythropoietin, also called EPO or **erythropoiesis-stimulating hormone,** is a glycoprotein that appears in the

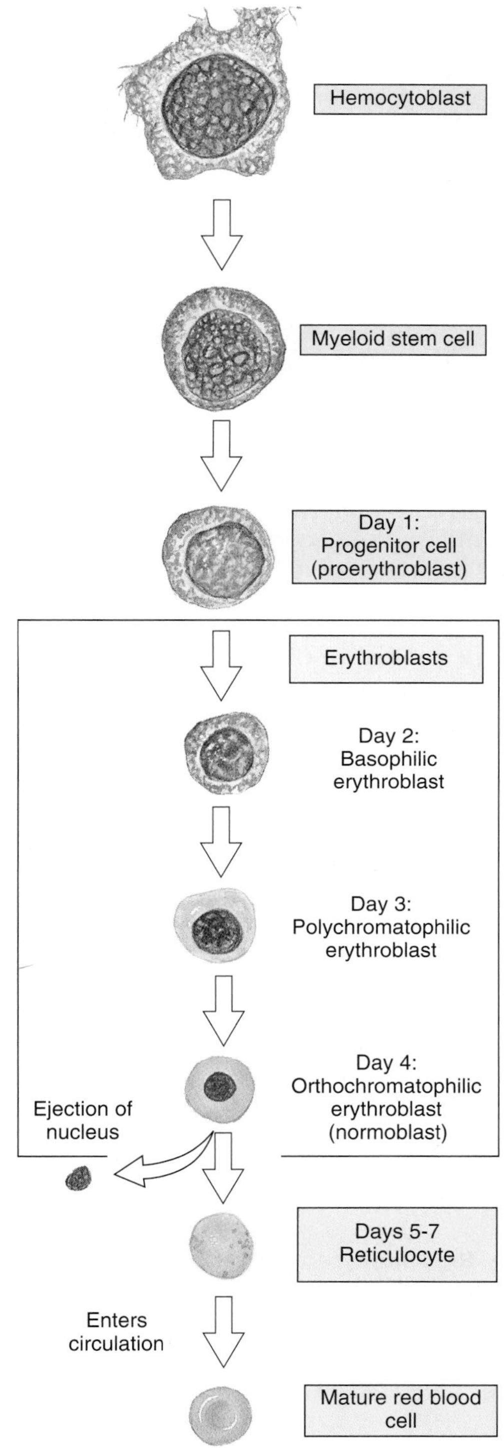
Hemocytoblast
Myeloid stem cell
Day 1: Progenitor cell (proerythroblast)
Erythroblasts
Day 2: Basophilic erythroblast
Day 3: Polychromatophilic erythroblast
Day 4: Orthochromatophilic erythroblast (normoblast)
Ejection of nucleus
Days 5-7 Reticulocyte
Enters circulation
Mature red blood cell

• **FIGURE 19-6**

Stages in RBC Maturation. RBC production occurs in the bone marrow. The color density indicates the abundance of hemoglobin in the cytoplasm. Note the reduction in the size of the cell and the size of the nucleus before a reticulocyte is formed.

plasma when peripheral tissues, especially the kidneys, are exposed to low oxygen concentrations (Figure 18-16●, p.627). The state of low tissue oxygen levels is called **hypoxia** (hī-POKS-ē-a; *hypo-*, below + *ox-*, presence of oxygen). For example, EPO release occurs (1) during anemia, (2) when blood flow to the kidneys declines, and (3) during adaptation to high altitudes. Once in the circulation, EPO travels to areas of red marrow, where it stimulates stem cells and developing erythrocytes.

Erythropoietin has two major effects: (1) It stimulates increased rates of cell divisions in erythroblasts and in the stem cells that produce erythroblasts; and (2) it speeds up the maturation of red blood cells, primarily by accelerating the rate of hemoglobin synthesis. Under maximum EPO stimulation the bone marrow can increase the rate of red blood cell formation tenfold, to around 30 million per second.

This reserve is important when recovering from a severe blood loss. However, if EPO is administered to a normal individual, as in the case of the cyclists mentioned in Chapter 18, the hematocrit may rise to 65 or more. *[p. 641]* This increase can place an intolerable strain on the heart. Comparable problems can occur following **blood doping.** In this practice, athletes reinfuse packed red blood cells removed and stored at an earlier date, with the goal of increasing hematocrit to improve performance. AM *Erythrocytosis and Blood Doping*

BLOOD TESTS AND RBCs Blood tests provide information about the general health of an individual, usually with a minimum of trouble and expense. Several common blood tests focus on red blood cells, the most abundant formed elements. These tests assess the number, size, shape, and maturity of circulating red blood cells. Such assessment provides an indication of the erythropoietic activities under way, and it can also be useful in detecting problems, such as internal bleeding, that may not produce other obvious symptoms. Table 19-1 lists examples of important blood tests and related terms. (A more detailed discussion, including sample calculations, can be found in the *Applications Manual.*) AM *Blood Tests and RBCs*

How would an individual's hematocrit change after suffering a hemorrhage?

Dave develops a blockage in his renal arteries that restricts blood flow to the kidneys. Will his hematocrit change?

How would the level of bilirubin in the blood be affected by diseases that cause damage to the liver?

Blood Types

Figure 19-7

An individual's **blood type** is determined by the presence or absence of specific components in the erythro-

TABLE 19-1 RBC Tests and Related Terminology

Test	*Determines*	*Terms Associated with Abnormal Values* *Elevated*	*Depressed*
Hematocrit (Hct)	Percentage of formed elements in whole blood Normal = 37–54	Polycythemia (may result from erythrocytosis or leukocytosis)	Anemia
Reticulocyte count (Retic.)	Circulating percentage of reticulocytes Normal = 0.8%	Reticulocytosis	
Hemoglobin concentration (Hb)	Concentration of hemoglobin in blood Normal = 12–18 g/dl		Anemia
RBC count	Number of RBCs per μl of whole blood Normal = 4.4–6.0 million/μl	Erythrocytosis	Anemia
Mean corpuscular volume (MCV)	Average volume of single RBC Normal = 82–101 μm^3 (normocytic)	Macrocytic	Microcytic
Mean corpuscular hemoglobin concentration (MCHC)	Average amount of Hb in one RBC Normal = 27–34 pg/μl (normochromic)	Hyperchromic	Hypochromic

Note: For additional details and sample calculations, see the *Applications Manual.*

cyte cell membranes. The cell membrane of a typical red blood cell contains integral glycoproteins or glycolipids whose characteristics are genetically determined. These molecules, called *surface antigens*, are detected by cells involved with immune defenses. The antigens on the surfaces of red blood cells are called **agglutinogens** (a-glu-TIN-o-jenz). There are at least 50 different kinds of agglutinogens on the surfaces of RBCs. Three of particular importance have been designated **agglutinogens A, B,** and **D (Rh).**

The red blood cells of a particular individual may have either, both, or neither agglutinogen A or B on their surfaces (Figure 19-7●). **Type A** blood has agglutinogen A only, **Type B** has agglutinogen B only, **Type AB** has both A and B, and **Type O** has neither A nor B. These blood types are not evenly distributed through the population. The average values for the U.S. population are: Type O 46 percent, Type A 40 percent, Type B 10 percent, and Type AB 4 percent. There are variations in these values due to racial and ethnic differences (Table 19-2).

The presence of the Rh agglutinogen, sometimes called the *Rh factor,* is indicated by the terms **Rh-positive** (present) or **Rh-negative** (absent). When the complete blood type is recorded, the term *Rh* is usually omitted, and the data are reported as O-negative, A-positive, and so forth.

TABLE 19-2 Differences in Blood Group Distribution

Population	*Percentage with Each Blood Type* O	A	B	AB	Rh$^+$
U.S. (average)	46	40	10	4	85
Caucasian	45	40	11	4	85
African-American	49	27	20	4	95
Chinese	42	27	25	6	100
Japanese	31	39	21	10	100
Korean	32	28	30	10	100
Filipino	44	22	29	6	100
Hawaiian	46	46	5	3	100
Native North American	79	16	4	<1	100
Native South American	100	0	0	0	100
Australian Aborigines	44	56	0	0	100

AGGLUTININS You are probably already aware that your blood type must be checked before you give or receive blood. The surface antigens on your own red blood cells are ignored by your immune system. (This ability to recognize normal body cells will be examined in Chapter 22.) However, your plasma contains antibodies called **agglutinins** (a-GLOO-ti-ninz) that will attack the antigens on "foreign" red blood cells. For example, if you have Type A blood, your plasma holds circulating anti-B agglutinins that will attack Type B erythrocytes. If you are Type B, your plasma contains anti-A agglutinins. The red blood cells of an individual with Type O blood lack A and B agglutinogens, and the plasma of such an individual contains both anti-A and anti-B agglutinins. At the other extreme, Type AB individuals lack agglutinins sensitive to either Type A or Type B blood cells.

In contrast to the situation with agglutinogens A and B, the plasma of an Rh-negative individual does not normally contain anti-Rh agglutinins. These agglutinins are present only if the individual has been **sensitized** by previous exposure to Rh-positive erythrocytes. Such exposure may occur accidentally, during a transfusion, but it may also accompany a seemingly normal pregnancy involving an Rh-negative mother and an Rh-positive fetus.

As in the distribution of A and B agglutinogens, there are racial and regional differences in Rh type. For example, the fraction of Rh-negative individu-

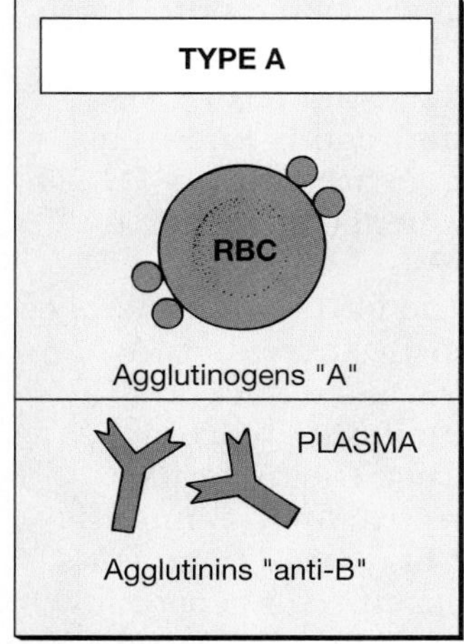
TYPE A
RBC
Agglutinogens "A"
PLASMA
Agglutinins "anti-B"

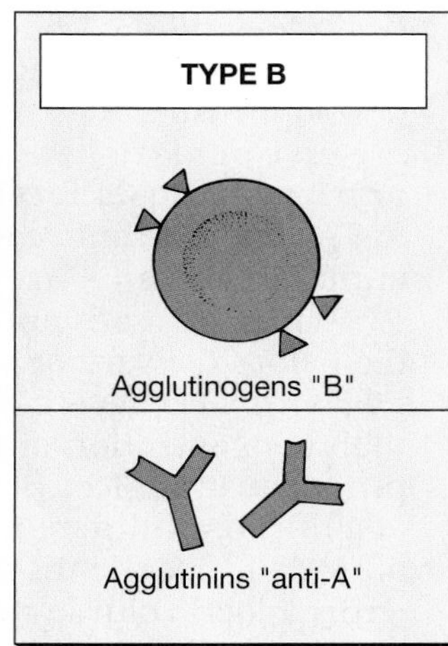
TYPE B
Agglutinogens "B"
Agglutinins "anti-A"

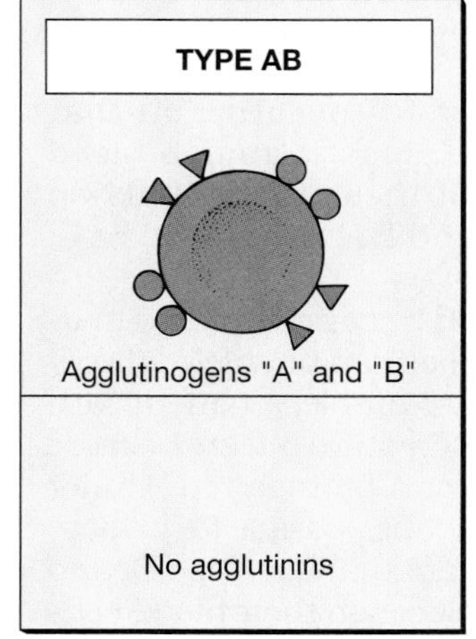
TYPE AB
Agglutinogens "A" and "B"
No agglutinins

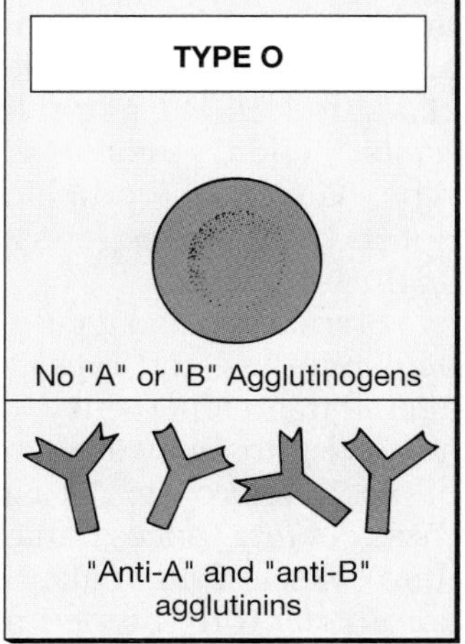
TYPE O
No "A" or "B" Agglutinogens
"Anti-A" and "anti-B" agglutinins

● **FIGURE 19-7**

Blood Typing. The blood type depends on the presence of agglutinogens on RBC surfaces. The plasma contains agglutinins, antibodies that will react with foreign agglutinogens. The relative frequencies of each blood type in the U.S. population are indicated in Table 19-2.

als in the U.S. Caucasian population is roughly 15 percent. That figure is 5 percent for the African-American population, essentially 0 for the black African population, and less than 1 percent for Asians everywhere.

CROSS-REACTIONS When an agglutinin meets its specific agglutinogen, a **cross-reaction** occurs. Initially the red blood cells clump together, a process called **agglutination** (a-glu-ti-NĀ-shun), and they may also **hemolyze,** or rupture. Clumps and fragments of red blood cells under attack form drifting masses that can plug small vessels in the kidneys, lungs, heart, or brain, damaging or destroying dependent tissues. Such reactions can be avoided by taking care to ensure that the blood types of the donor and the recipient are **compatible.** In practice, this involves choosing a donor whose blood cells will not undergo cross-reaction with the plasma of the recipient. Unless large volumes of whole blood or plasma are transferred, cross-reactions between the donor's plasma and the recipient's blood cells fail to produce significant agglutination because the donated plasma gets diluted through mixing with the plasma of the recipient.

Testing for Compatibility. Testing for compatibility normally involves two steps: a determination of blood type and a cross-match test. At least 50 agglutinogens have been identified on red cell surfaces, but the standard test for blood type categorizes a blood sample on the basis of the three most likely to produce dangerous cross-reactions. The test involves taking drops of blood and mixing them separately with solutions containing anti-A, anti-B, and anti-Rh (anti-D) agglutinins. Any cross-reactions are then recorded. For example, if the red blood cells clump together when exposed to anti-A *and* to anti-B, the individual has Type AB blood. If no reactions occur, the person must be Type O. The presence or absence of the Rh agglutinogen is also noted, and the individual is classified as Rh-positive or Rh-negative on that basis. Type O-positive is the most common blood type. The red blood cells of these individuals do not have agglutinogens A and B, but they do have agglutinogen D.

Standard blood typing can be completed in a matter of minutes, and Type O blood can be safely administered in an emergency. For example, a patient with a severe gunshot wound may require 5 *liters* or more of blood before the damage can be repaired. Under these circumstances it may not be possible to do more than collect blood, make certain that it is Type O, and administer it to the victim. Because their blood cells are unlikely to produce severe cross-reactions in a recipient, Type O individuals (specifically O⁻) are sometimes called *universal donors.* It is now possible to use enzymes to strip off the A or B agglutinogens

Hemolytic Disease of the Newb

Genes controlling the presence or absence of any agglutinogen in the erythrocyte membrane are provided by both parents, so a child may have a blood type different from that of either parent. During pregnancy, when fetal and maternal circulatory systems are closely intertwined, the mother's agglutinins may cross the placental barrier, attacking and destroying fetal red blood cells. The resulting condition is called **hemolytic disease of the newborn,** or **HDN.**

There are many forms of HDN, some so mild as to remain undetected, but those involving the Rh agglutinogen are potentially quite dangerous. Because HDN results from the passage of *maternal* agglutinins across the placental barrier, an Rh-positive mother (who lacks anti-Rh agglutinins) can carry an Rh-negative fetus without difficulty. Potential problems appear when an Rh-negative woman carries an Rh-positive fetus. Sensitization usually occurs at delivery, when bleeding occurs at the placenta and uterus. This event can trigger the maternal production of anti-Rh agglutinins. Within 6 months of delivery, roughly 20 percent of Rh-negative mothers who carried Rh-positive children have become sensitized.

Because the anti-Rh agglutinins are not produced in significant amounts until after delivery, this first infant will not be affected. But if a second pregnancy occurs involving an Rh-positive fetus, the mother will respond by producing massive amounts of anti-Rh agglutinins. These agglutinins then enter the fetal circulation, destroying fetal red blood cells and producing a dangerous anemia. The fetal demand for blood cells increases, and they begin leaving the bone marrow and entering the circulation before completing their development. Because these immature RBCs are erythroblasts, the condition is known as **erythroblastosis fetalis** (e-rith-rō-blas-TŌ-sis fē-TAL-is). The entire sequence of events is summarized in Figure 19-8●.

Without treatment, the fetus will probably die before delivery or shortly thereafter. A newborn with severe HDN is anemic, and the high concentration of circulating bilirubin produces jaundice. Because the maternal agglutinins will remain active for 1–2 months after delivery, the infant may need to have its entire blood volume replaced. This blood replacement removes most of the maternal agglutinins as well as the affected erythrocytes, reducing the complications and the chance that the infant will die.

When there is a danger that the fetus may not survive to full term, premature delivery may be induced after 7–8 months of development. In a severe case affecting a fetus at an earlier stage, one or more transfusions can be given while the fetus continues to develop within the uterus.

To avoid this problem completely, the maternal production of agglutinins can be prevented by administering anti-Rh antibodies (available under the name *RhoGam*) during and after delivery. These antibodies will destroy any fetal red blood cells that cross the placental barrier before they can stimulate an immune response in the mother. Because sensitization does not occur, anti-Rh agglutinins are not produced. This relatively simple procedure could almost entirely prevent HDN mortality caused by Rh incompatibilities.

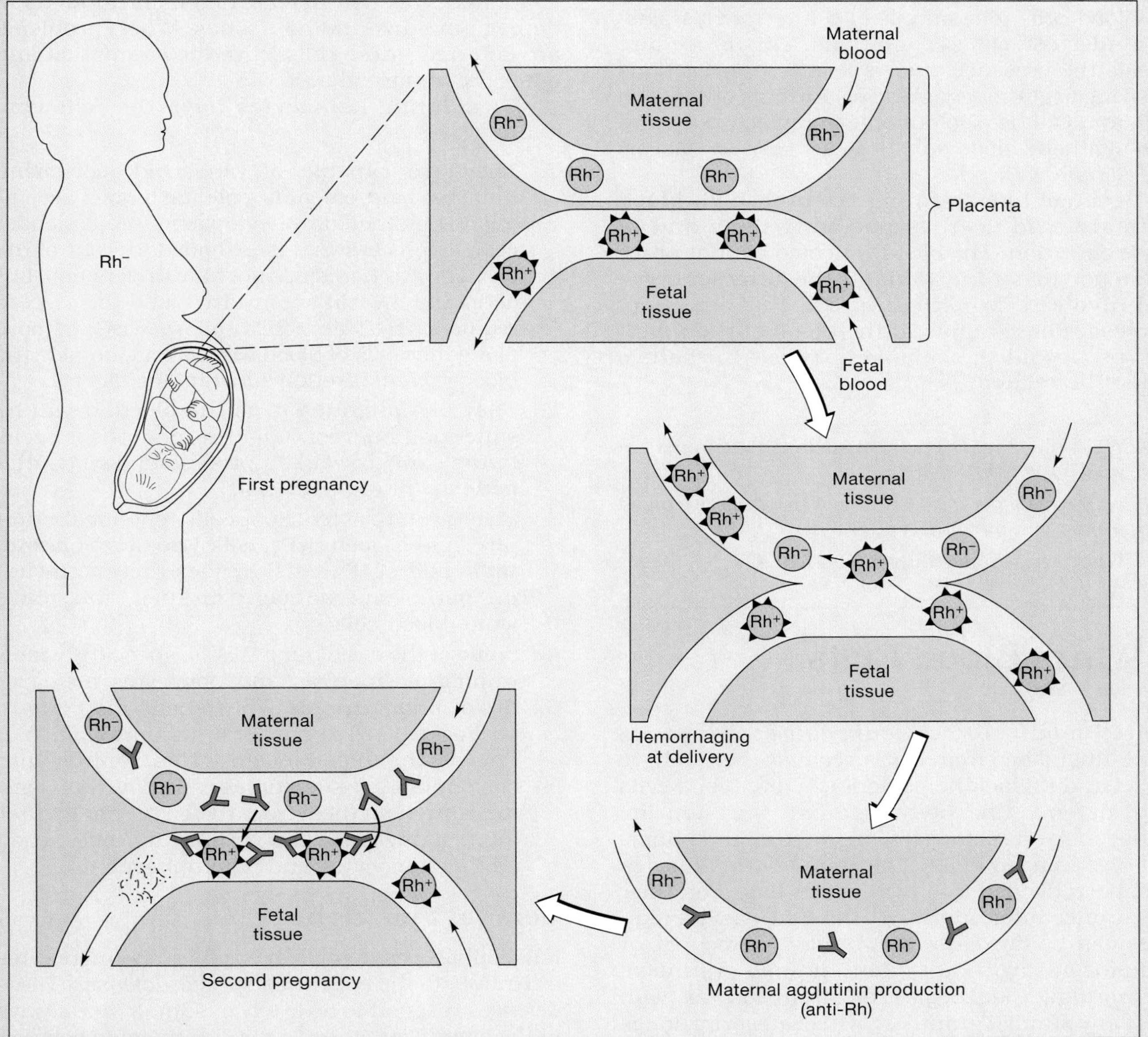
Maternal blood
Maternal tissue
Rh⁻
Rh⁺
Placenta
Fetal tissue
Fetal blood
First pregnancy
Hemorrhaging at delivery
Maternal agglutinin production (anti-Rh)
Second pregnancy

FIGURE 19-8

Rh Factors and Pregnancy. When an Rh-negative woman has her first Rh-positive child, mixing of fetal and maternal blood occurs at delivery when the placental connection breaks down. The appearance of Rh-positive blood cells in the maternal circulation sensitizes the mother, stimulating the production of anti-Rh agglutinins. If another pregnancy occurs with an Rh-positive fetus, maternal agglutinins can cross the placental barrier and attack fetal blood cells, producing symptoms of HDN (hemolytic disease of the newborn).

from the RBCs of a donor and create Type O blood. The procedure is expensive and time-consuming and has limited use in emergency treatment. In addition, it should be realized that with at least 48 other possible agglutinogens on the cell surface, cross-reactions can occur, even to Type O$^-$ blood. (Nevertheless, over 100,000 units of Type O$^-$ blood were transfused under emergency conditions during the war in Vietnam, with no deaths due to cross-reactions.)

Whenever time and facilities permit, further testing is performed to ensure complete compatibility. **Cross-match testing** involves exposing the donor's red blood cells to a sample of the recipient's plasma under controlled conditions. This procedure reveals the presence of significant cross-reactions involving agglutinogens and agglutinins other than A, B, and O. It is also possible to replace lost blood with synthetic blood substitutes. AM *Transfusions and Synthetic Blood*

Because blood groups are inherited, blood tests are also used as paternity tests and in crime detection. The blood collected cannot prove that a particular individual *is* the father or criminal involved, but it can prove that he is *not* involved. For example, if the blood of the infant is Type O, and the sample is Type AB, the two cannot be related.

 What blood types can be transfused into a person with Type AB blood?

 Why can't a person with Type A blood receive blood from a person with Type B blood?

 What are surface antigens on RBCs?

❑ White Blood Cells

Figure 19-9

White blood cells, or *leukocytes*, can easily be distinguished from RBCs because they (1) have nuclei and (2) lack hemoglobin. White blood cells help defend the body against invasion by pathogens and remove toxins, wastes, and abnormal or damaged cells. Traditionally, leukocytes have been divided into two groups based on their appearance after staining.[1] On that basis leukocytes can be divided into *granular leukocytes*, or *granulocytes* (with abundant stained granules), and *agranular leukocytes*, or *agranulocytes* (with few if any stained granules). This categorization is convenient but somewhat misleading, because the granules in granular leukocytes are secretory vesicles and lysosomes and the "agranular leukocytes" contain lysosomes as well—they are just smaller and difficult to see with the light microscope.

[1]The stains used in standard blood work are *Wright's stain* and *Giemsa stain*.

Typical leukocytes in the circulating blood are shown in Figure 19-9●. A typical microliter of blood contains 6000–9000 leukocytes. Most of the white blood cells in the body are found in peripheral tissues, and circulating leukocytes represent only a small fraction of the total population.

WBC Circulation and Movement

Unlike erythrocytes, leukocytes do not circulate for extended periods. They utilize the bloodstream for rapid transportation to areas of invasion or injury. As they travel along the miles of capillaries, leukocytes can detect the chemical signs of damage to surrounding tissues. When problems are detected, these cells leave the circulation and enter the abnormal area.

Circulating leukocytes have the following characteristics:

1. They are capable of *amoeboid movement*. Amoeboid movement is a gliding motion accomplished by the flow of cytoplasm into a slender cellular process that is extended in front of the cell. The mechanism is not fully understood, but actin and myosin are involved, and the process requires ATP. This mobility allows WBCs to move along the walls of blood vessels and, outside the bloodstream, through surrounding tissues.
2. They can migrate out of the bloodstream by squeezing between adjacent endothelial cells in the capillary wall, a process known as **diapedesis** (dī-a-pe-DĒ-sis).
3. They are attracted to specific chemical stimuli. This characteristic, called **positive chemotaxis** (kē-mō-TAK-sis), guides them to invading pathogens, damaged tissues, and active white blood cells.
4. Some of the circulating WBCs, specifically neutrophils, eosinophils, and monocytes, are capable of *phagocytosis*. These cells may engulf pathogens, cell debris, or other materials in or out of the bloodstream. Neutrophils and eosinophils are sometimes called *microphages*, to distinguish them from the larger phagocytes found in the blood and peripheral tissues, such as the monocytes and macrophages.

General Functions

Neutrophils, eosinophils, basophils, and monocytes contribute to the body's *nonspecific defenses*. These defenses respond to a variety of stimuli, but always in the same way—they do not discriminate between one threat and another. Lymphocytes, by contrast, are the cells responsible for *specific immunity:* the ability of the body to mount a counterattack against invading pathogens or foreign proteins *on an individual basis*. The interactions between white blood cells and the relationships between specific and nonspecific defenses will be discussed in Chapter 22.

Neutrophils

Figure 19-9a

Fifty to seventy percent of the circulating white blood cells are **neutrophils** (NOO-trō-filz). This name was selected because their granules are chemically neutral and thus difficult to stain with either acid or basic dyes. A mature neutrophil (Figure 19-9a●) has a very dense, segmented nucleus that forms two to five lobes resembling beads on a string. This characteristic has given neutrophils another name, **polymorphonuclear** (pol-ē-mor-fō-NŪ-klē-ar) **leukocytes** (*poly*, many + *morphe*, form), or **PMNs.** "Polymorphs," or "polys," as they are often called, are roughly 12 μm in diameter, and their cytoplasm is packed with pale granules containing lysosomal enzymes and bactericidal (bacteria-killing) compounds.

Neutrophils are highly mobile, and they are usually the first of the white blood cells to arrive at an injury site. They are very active cells that specialize in attacking and digesting bacteria that have been "marked" with antibodies or with *complement proteins*, plasma proteins involved in tissue defenses. (The complement system will be detailed in Chapter 22.)

When encountering a bacterium, the neutrophil quickly engulfs it. As it does, the metabolic rate of the neutrophil increases dramatically. This **respiratory burst** accompanies the production of destructive chemical agents, including hydrogen peroxide (H_2O_2) and *superoxide anions* (O_2^-). These highly reactive chemicals can kill bacteria.

Meanwhile, the endocytic vesicle fuses with lysosomes containing digestive enzymes and small peptides called **defensins.** This process, which reduces the number of granules in the cytoplasm, is called **degranulation.** Defensins kill a variety of pathogens, including bacteria, fungi, and some viruses, by combining to form large channels in their cell membranes. The digestive enzymes then break down the bacterial remains. While a neutrophil is actively engaged in attacking pathogens, it releases prostaglandins and *leukotrienes*. The prostaglandins contribute to local inflammation by increasing capillary permeability in the affected region. **Leukotrienes** are hormones of the immune system that attract other phagocytes and help coordinate the immune response. ∞ *[p. 605]*

Neutrophils usually have a short life span, surviving in the bloodstream for only about 10 hours. When actively engulfing debris or pathogens, they may last 30 minutes or less. After engulfing a couple of dozen bacteria a neutrophil dies, but its breakdown releases additional chemicals that attract other neutrophils to the site.

Eosinophils

Figure 19-9b

Eosinophils (ē-ō-SIN-ō-filz) were so named because their granules stained darkly with *eosin*, a red dye. The granules will also stain with other acid dyes, so the name **acidophils** (a-SID-ō-filz) applies as well. Eosinophils (Figure 19-9b●) usually represent 2–4 percent of the circulating white blood cells. These cells are similar in size to neutrophils, but the combination of deep red granules and a bilobed (two-lobed) nucleus makes an eosinophil easy to identify.

Eosinophils attack objects that have already been coated with antibodies. They are phagocytic cells and will engulf antibody-marked bacteria, protozoa, or cellular debris. However, their primary mode of attack involves the exocytosis of toxic compounds, including NO and cytotoxic enzymes, onto the surface of their targets. Eosinophils are important in the defense against large multicellular parasites, such as flukes or parasitic worms, and their numbers increase dramatically during a parasitic infection.[1] Because eosinophils are also sensitive to circulating *allergens* (materials that trigger allergies), their numbers increase during allergic reactions. Eosinophils are also attracted to sites of inflammation, where they release enzymes that reduce the degree of inflammation and control its spread to adjacent tissues.

Basophils

Figure 19-9c

Basophils (BĀ-sō-filz) have numerous granules that stain darkly with basic dyes. In a standard blood smear the inclusions are a deep purple or blue (Figure 19-9c●). Basophils are smaller than neutrophils or eosinophils, measuring 8–10 μm in diameter. They are also relatively rare, accounting for less than 1 percent of the circulating leukocyte population.

Basophils migrate to sites of injury and cross the capillary endothelium to accumulate within the damaged tissues, where they discharge their granules into the interstitial fluids. The granules contain histamine and *heparin*, a chemical that prevents blood clotting. Stimulated basophils release these chemicals into the interstitial fluids, and their arrival enhances the local inflammation initiated by mast cells. Although the same compounds are released by mast cells in damaged connective tissues, mast cells and basophils are distinct populations with separate origins. Other chemicals released by stimulated basophils attract eosinophils and other basophils to the area.

Monocytes

Figure 19-9d

Monocytes (MON-ō-sīts) in the blood are spherical cells that may exceed 15 μm in diameter, nearly twice the diameter of a typical erythrocyte (Figure 19-9d●). When flattened in a blood smear

[1] For a review of the major classes of pathogens, see *The Nature of Pathogens* in the *Applications Manual.*

they look even larger, so monocytes are relatively easy to identify. The nucleus is large and often has the shape of an oval or kidney bean. Monocytes normally account for 2–8 percent of the circulating leukocytes.

An individual monocyte uses the bloodstream as a highway, remaining in circulation for only around 24 hours before entering peripheral tissues. Monocytes outside the bloodstream are usually called *free macrophages*, to distinguish them from the immobile *fixed macrophages* found in many connective tissues. [p. 122]

Free macrophages are aggressive phagocytes, often attempting to engulf items as large or larger than themselves. When free macrophages encounter a foreign object too large for a single cell to engulf, several of them may fuse together to create a single **phagocytic giant cell** big enough to do the job. While doing so they release chemicals that attract and stimulate neutrophils, monocytes, and other phagocytic cells. Free macrophages also secrete substances that lure fibroblasts into the region. The fibroblasts then begin producing the scar tissue that will wall off the injured area.

Lymphocytes

Figure 19-9e

Typical **lymphocytes** (LIM-fō-sīts) are roughly the same size as red blood cells and lack abundant, deeply stained granules. In fact, when you see a lymphocyte in a blood smear of lymphocytes, you usually see just a thin halo of cytoplasm around a relatively large nucleus (Figure 19-9e●).

Lymphocytes account for 20–30 percent of the leukocyte population of the blood. Circulating lymphocytes represent a minute segment of the entire lymphocyte population, for lymphocytes are the primary cells of the lymphatic system, a network of lymphatic vessels and organs.

The circulating blood contains three different classes of lymphocytes, although they cannot be distinguished with a light microscope:

1. Lymphocytes called **T cells** are responsible for *cellular immunity*, a defense against foreign cells and tissues, and for the coordination of the immune response. Activated *cytotoxic T cells* enter peripheral tissues and destroy foreign cells directly, by physical and chemical attack. *Regulatory T cells*, including *helper T cells* and *suppressor T cells*, stimulate or inhibit the activities of other lymphocytes.
2. Lymphocytes called **B cells** are responsible for *humoral immunity*. Activated B cells differentiate into **plasma cells,** tissue cells that secrete antibodies that can attack alien cells or proteins in distant portions of the body.
3. **NK cells,** sometimes known as *large granular lymphocytes*, are responsible for *immune surveillance*, the destruction of abnormal tissue cells. These cells are important in preventing cancer.

The Differential Count and Changes in WBC Profiles

A variety of disorders, including pathogenic infection, inflammation, and allergic reactions, cause characteristic changes in circulating populations of WBCs. By examining a stained blood

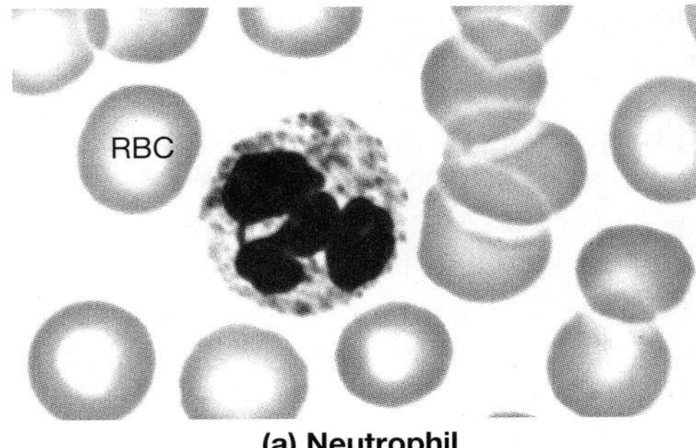
RBC

(a) Neutrophil

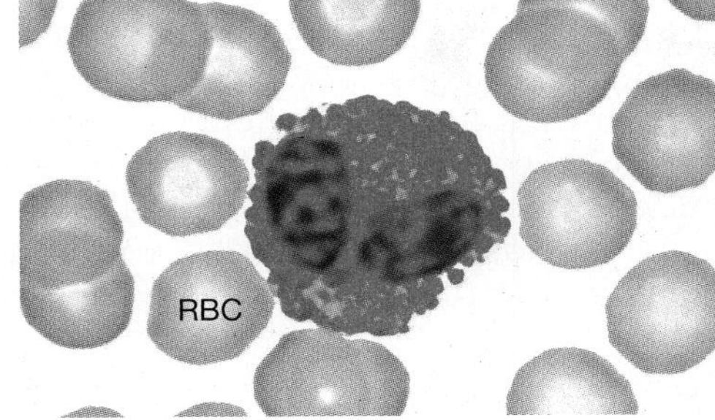
RBC

(b) Eosinophil

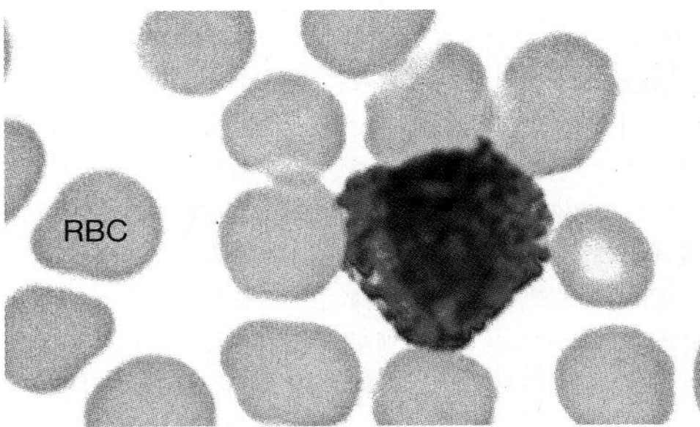
RBC

(c) Basophil

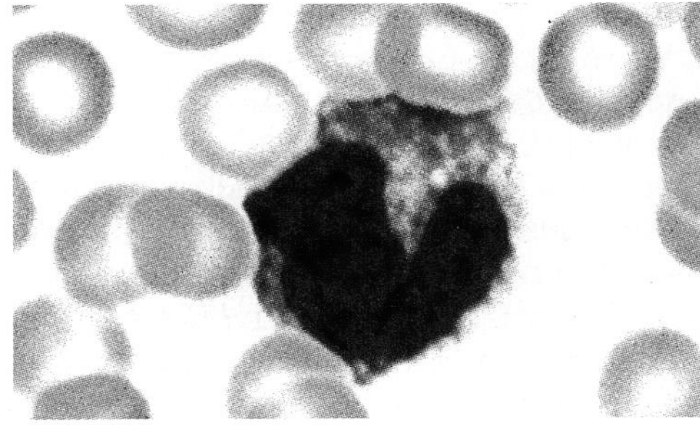

(d) Monocyte

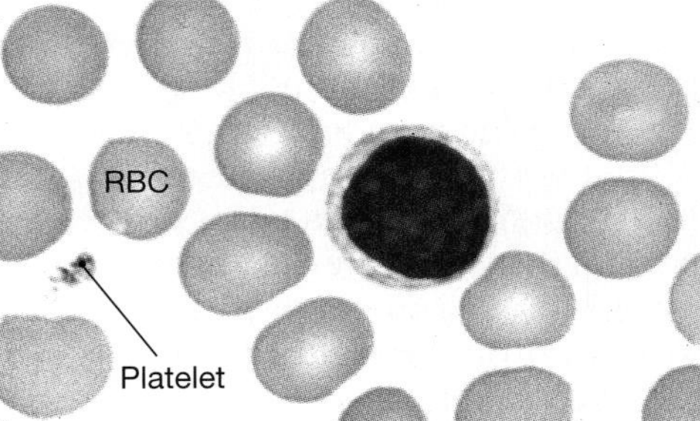
RBC
Platelet

(e) Lymphocyte

● **FIGURE 19-9**
White Blood Cells. (a) Neutrophil. **(b)** Eosinophil. **(c)** Basophil. **(d)** Monocyte. **(e)** Lymphocyte. (LMs × 1500)

smear it is possible to obtain a **differential count** of the white blood cell population. The values reported indicate the number of each type of cell encountered in a sample of 100 white blood cells.

The normal range for each cell type is indicated in Table 19-3. The term **leukopenia** (loo-kō-PĒ-nē-ah; *penia*, poverty) indicates inadequate numbers of white blood cells. **Leukocytosis** (loo-kō-sī-TŌ-sis) refers to excessive numbers. A modest leukocytosis is normal during an infection. Extreme leukocytosis (>100,000/μl or more) usually indicates the presence of some form of **leukemia** (loo-KĒ-mē-ah). There are many different types of leukemias, but unless treated all are fatal. *The Leukemias* The endings *-penia* and *-osis* can also be used to indicate low or high numbers of specific types of white blood cells. For example, *lymphopenia* means too few lymphocytes, whereas in *lymphocytosis* their numbers are unusually high.

White Blood Cell Production

Figure 19-10

Stem cells responsible for the production of white blood cells originate in the bone marrow. Neutrophils, eosinophils, and basophils complete their development in this *myeloid tissue*. Monocytes begin their differentiation in the marrow, enter the circulation, and complete their development when they become free macrophages in peripheral tissues. Each of these cell types goes through a characteristic series of maturational stages, proceeding from *blast cells* to *myelocytes* to *band cells* before becoming mature WBCs. For example, a cell differentiating into a neutrophil goes from a myeloblast to a *neutrophilic myelocyte* (*early*, *late*, and *meta* stages), and then becomes a *neutrophilic band cell*. Some band cells enter the circulation before completing their maturation—normally 3–5 percent of the circulating WBCs are band cells. Figure 19-10• summarizes the relationships between the various WBC populations and compares the formation of WBCs and RBCs.

Stem cells responsible for the production of lymphocytes, a process called **lymphopoiesis,** also originate in the bone marrow, but many of these subsequently migrate to peripheral **lymphoid tissues,** including the thymus, spleen, and lymph nodes. As a result, lymphocytes are produced in these organs as well as in the bone marrow.

REGULATION OF WBC PRODUCTION Factors that regulate lymphocyte maturation are as yet incompletely understood. Prior to maturity, the "thymosins" produced by the thymus gland promote the differentiation and maintenance of T cell populations. The importance of the thymus gland in adulthood, especially in aging, remains controversial. In the adult, production of B and T lymphocytes is regulated primarily by exposure to antigens (foreign proteins, cells, or toxins). When foreign antigens appear, lymphocyte production escalates; the control mechanisms will be described in Chapter 22.

Several hormones are involved in the regulation of other white blood cell populations. Figure 19-10• diagrams the targets of these hormones, called **colony-stimulating factors (CSF).** Four CSFs have been identified, each stimulating white blood cell or white *and* red blood cell formation. The abbreviation for each factor indicates its target:

- **M-CSF** stimulates activity in the monocyte/macrophage line.
- **G-CSF** stimulates production of granulocytes; a genetically engineered form of G-CSF, sold under the name of *filgrastim (Neupogen)*, is now used to stimulate production of neutrophils in patients undergoing cancer chemotherapy.
- **GM-CSF** stimulates production of both granulocytes and monocytes.
- **Multi-CSF** accelerates production of granulocytes, monocytes, megakaryocytes, and erythrocytes.

Chemical communication between lymphocytes and other leukocytes assists in the coordination of the immune response. For example, active macrophages release chemicals that make lymphocytes more sensitive to antigens and accelerate the development of specific immunity. In turn, active lymphocytes release multi-CSF and GM-CSF, reinforcing nonspecific defenses.

What type of white blood cell would you expect to find in the greatest numbers in an infected cut?

What cell type would you expect to find in elevated numbers in a person producing large amounts of circulating antibodies to combat a virus?

How do basophils respond during inflammation?

Platelets

Figure 19-9e

Platelets (PLĀT-lets) are flattened discs, round when viewed from above and spindle-shaped when seen in section or in a blood smear. They average about 4 μm in diameter and are roughly 1 μm thick; several platelets can be seen in Figure 19-9e•. Platelets in nonmammalian vertebrates are nucleated cells called **thrombocytes** (THROM-bō-sīts; *thrombos*, clot). Because in humans they are cell fragments, rather than individual cells, the term *platelet* is preferred when referring to human blood.

Platelets are continually replaced, and an individual platelet circulates for 9–12 days before being removed by phagocytes, primarily in the spleen.

TABLE 19-3 A Review of the Formed Elements of the Blood

Cell	*Abundance (Average per µl)*	*Appearance in Blood Smear*	*Functions*	*Remarks*
RED BLOOD CELLS	5.2 million (range: 4.4–6.0 million)	Flattened, circular cell; no nucleus, mitochondria, or ribosomes; red in color	Transport oxygen from lungs to tissues and carbon dioxide from tissues to lungs	Remain in circulation; 120-day life expectancy; amino acids and iron recycled; produced in bone marrow
WHITE BLOOD CELLS	7000 (range: 6000–9000)			
Neutrophils	4150 (range: 1800–7300) Differential count: 50–70%	Round cell; nucleus lobed and may resemble a string of beads; cytoplasm contains large, pale inclusions	Phagocytic: Engulf pathogens or debris in tissues, release cytotoxic enzymes and chemicals	Move into tissues after several hours; may survive minutes to days, depending on tissue activity; produced in bone marrow
Eosinophils	165 (range: 0–700) Differential count: 2–4%	Round cell; nucleus usually in two lobes; cytoplasm contains large granules that are usually stained bright red	Phagocytic: Engulf antibody-labeled materials, release cytotoxic enzymes, reduce inflammation	Move into tissues after several hours; survive minutes to days, depending on tissue activity; produced in bone marrow
Basophils	44 (range: 0–150) Differential count: <1%	Round cell; nucleus usually cannot be seen because of dense, blue-stained granules in cytoplasm	Enter damaged tissues and release histamine and other chemicals that promote inflammation	Survival time unknown; assist mast cells of tissues in producing inflammation; produced in bone marrow
Monocytes	456 (range: 200–950) Differential count: 2–8%	Very large cell; kidney bean–shaped nucleus; abundant pale cytoplasm	Enter tissues to become macrophages; engulf pathogens or debris	Move into tissues after 1–2 days; survive months or longer; primarily produced in bone marrow
Lymphocytes	2185 (range: 1500–4000) Differential count: 20–30%	Usually round cell, slightly larger than RBC; round nucleus; very little cytoplasm	Cells of lymphatic system, providing defense against specific pathogens or toxins	Survive for months to decades; circulate from blood to tissues and back; produced in bone marrow and lymphatic tissues
PLATELETS	350,000 (range: 150,000–500,000)	Spindle-shaped cytoplasmic fragment; contains enzymes and proenzymes; no nucleus	Hemostasis: Clump together and stick to vessel wall (platelet phase); activate intrinsic pathway of coagulation phase	Remain in circulation or in vascular organs; remain intact 7–12 days; produced by megakaryocytes in bone marrow

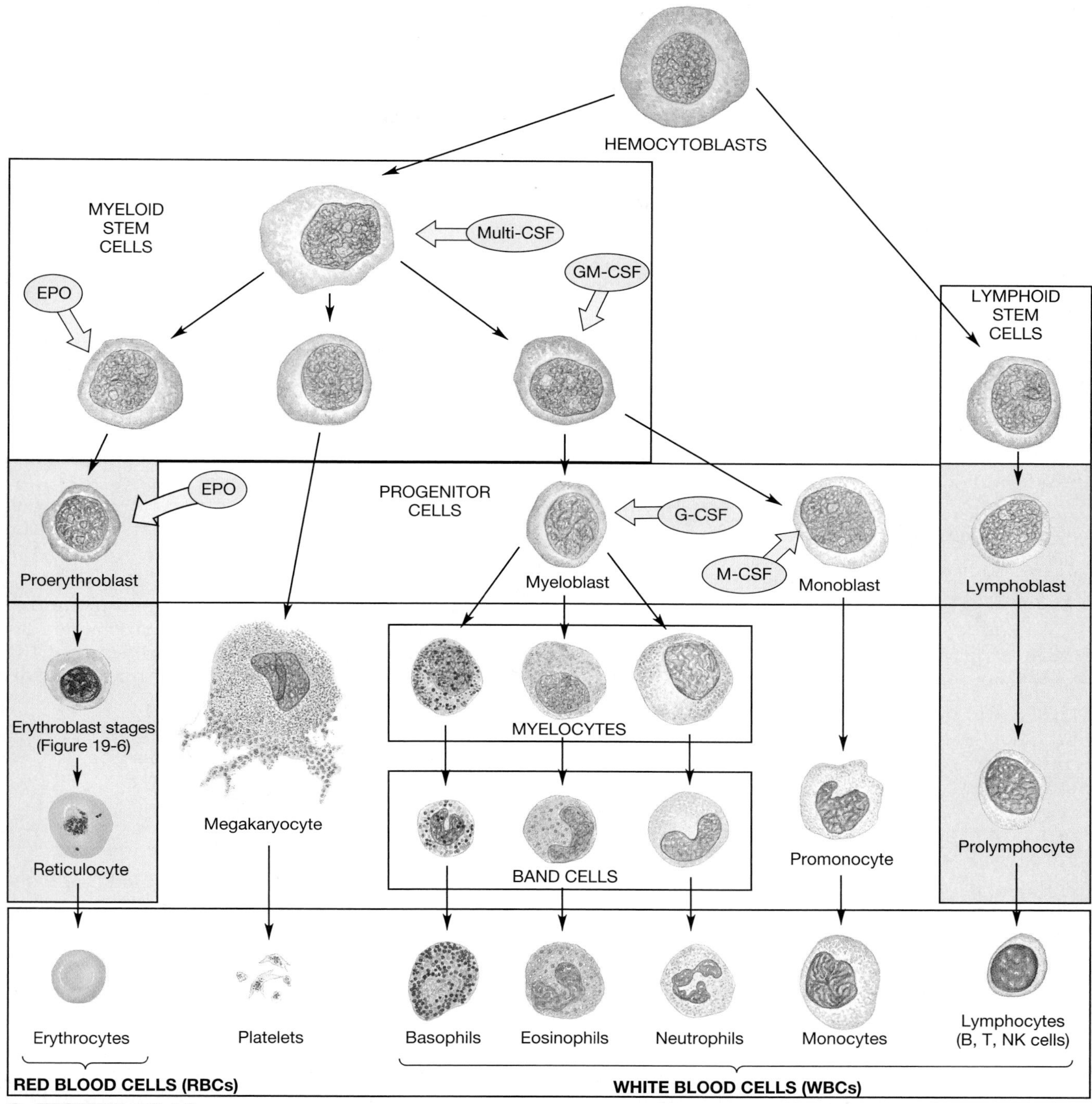
HEMOCYTOBLASTS
MYELOID STEM CELLS
Multi-CSF
GM-CSF
EPO
LYMPHOID STEM CELLS
EPO
PROGENITOR CELLS
G-CSF
M-CSF
Proerythroblast
Myeloblast
Monoblast
Lymphoblast
MYELOCYTES
Erythroblast stages (Figure 19-6)
Megakaryocyte
BAND CELLS
Promonocyte
Prolymphocyte
Reticulocyte
Erythrocytes
Platelets
Basophils
Eosinophils
Neutrophils
Monocytes
Lymphocytes (B, T, NK cells)
RED BLOOD CELLS (RBCs)
WHITE BLOOD CELLS (WBCs)

FIGURE 19-10

The Origins and Differentiation of Blood Cells. Hemocytoblast divisions give rise to myeloid and lymphoid stem cells. Lymphoid stem cells produce the various classes of lymphocytes. Myeloid stem cells produce progenitor cells with more restricted fates; these divide to produce the other classes of blood cells. The targets of the four colony-stimulating factors are indicated.

There are 150,000–500,000 platelets in each microliter of circulating blood; 350,000/µl represents the average concentration. Roughly one-third of the platelets in the body at any given moment are held in the spleen and other vascular organs, rather than in the circulation. These reserves are mobilized during a circulatory crisis, such as severe bleeding.

An abnormally low platelet count (80,000/µl or less) is known as **thrombocytopenia** (throm-bō-sī-tō-PĒ-nē-ah). Thrombocytopenia usually indicates excessive platelet destruction or inadequate platelet production. Symptoms include bleeding along the digestive tract, within the skin, and occasionally inside the CNS. Platelet counts in **thrombocytosis** (throm-bō-sī-TŌ-sis) may exceed 1,000,000/µl. Thrombocytosis usually results from accelerated platelet formation in response to infection, inflammation, or cancer.

Platelet Functions

Platelets are one participant in a vascular *clotting* system that also includes plasma proteins and the cells and tissues of the circulatory network. The functions of platelets include:

1. ***Transport of chemicals important to the clotting process.*** By releasing enzymes and other factors at the appropriate times, platelets help initiate and control the clotting process.
2. ***Formation of a temporary patch in the walls of damaged blood vessels.*** Platelets clump together at an injury site, forming a *platelet plug* that can slow the rate of blood loss while clotting occurs.
3. ***Active contraction after clot formation has occurred.*** Platelets contain filaments of actin and myosin. After a blood clot has formed, the contraction of platelets shrinks the clot and reduces the size of the break in the vessel wall.

Platelet Production

Figure 19-11

Platelet production, or **thrombocytopoiesis,** occurs in the bone marrow. Normal bone marrow contains a number of very unusual cells, called **megakaryocytes** (meg-a-KAR-ē-ō-sīts; *mega-*, big + *karyon*, nucleus + *-cyte*, cell). As the name suggests, these are enormous cells (up to 160 µm in diameter) with large nuclei. The dense nucleus may be lobed or ring-shaped, and the surrounding cytoplasm contains Golgi apparatus, ribosomes, and mitochondria in abundance. The cell membrane communicates with an extensive membrane network that radiates throughout the peripheral cytoplasm (Figure 19-11●).

During their development and growth, megakaryocytes manufacture structural proteins, enzymes, and membranes. They then begin shedding cytoplasm in small membrane-enclosed packets. These packets are the platelets that enter the circulation. A mature megakaryocyte gradually loses all of its cytoplasm, producing around 4000 platelets before the nucleus is engulfed by phagocytes and broken down for recycling.

The rate of megakaryocyte activity and platelet formation is regulated by several factors:

1. **Thrombopoietin** (TPO), or *thromobocyte-stimulating factor,* is a peptide hormone that accelerates platelet formation and stimulates the production of megakaryocytes. The result is a rapid rise in the platelet count. TPO is produced in the kidneys and perhaps other sites as well. Very little is known about this hormone, other than its existence. For example, we do not know its amino acid structure, the feedback control mechanism involved, or the kidney cells responsible for its production.
2. **Interleukin-6** (Il-6), a hormone of the immune system, has a stimulatory effect on platelet formation. Current evidence indicates that the effects of Il-6 are distinct from, and less pronounced than, the effects of TPO.
3. Multi-CSF stimulates the formation and growth of megakaryocytes, as noted above.

Hemostasis

The process of **hemostasis** (*haima*, blood + *stasis*, halt) prevents the loss of blood through the walls of damaged vessels. At the same time, it establishes a

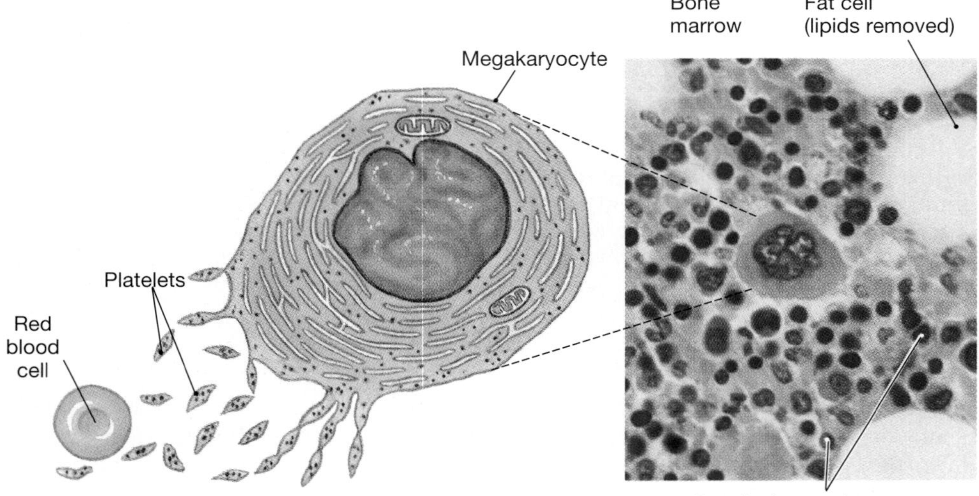
Megakaryocyte
Platelets
Red blood cell
Bone marrow
Fat cell (lipids removed)
Developing erythrocytes and granulocytes

● **FIGURE 19-11 Megakaryocytes and Platelet Formation.** Megakaryocytes stand out in bone marrow sections because of their relatively enormous size and the unusual shape of their nuclei. These cells are continually shedding chunks of cytoplasm that enter the circulation as platelets. (LM × 673)

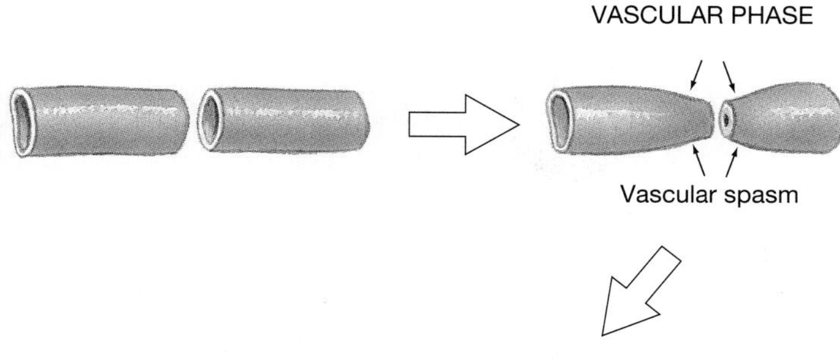
VASCULAR PHASE
Vascular spasm

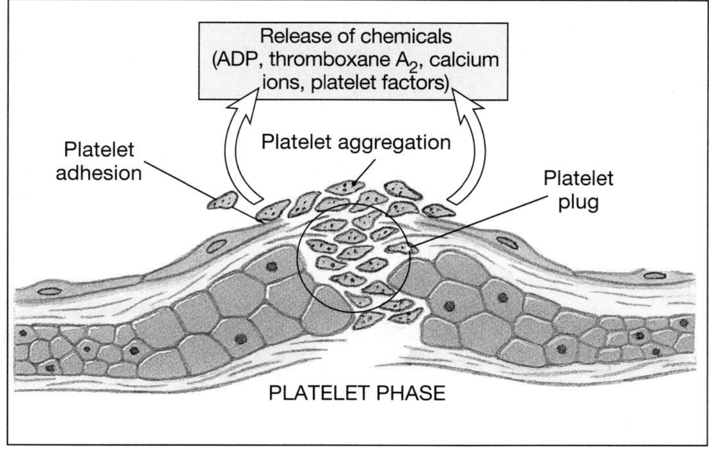
Release of chemicals
(ADP, thromboxane A2, calcium
ions, platelet factors)
Platelet aggregation
Platelet
adhesion
Platelet
plug
PLATELET PHASE

● **FIGURE 19-12**
The Vascular and Platelet Phases of Hemostasis

framework for tissue repairs. Although hemostasis can be analyzed as a series of steps, it is more like a chain reaction. In essence, each step modifies the events already under way. As a result, it is easier to say when a particular step begins than when it ends.

❑ The Vascular Phase

Figure 19-12

Cutting the wall of a blood vessel triggers a contraction in the smooth muscle fibers in the vessel wall (Figure 19-12●). This contraction produces a local **vascular spasm** that decreases the diameter of the vessel at the site of injury. Such a constriction can slow or even stop the loss of blood through the wall of a small vessel. The period of local vasoconstriction, called the **vascular phase** of hemostasis, lasts about 30 minutes.

During the vascular phase there are changes in the endothelium at the injury site:

- The endothelial cells contract and expose the underlying basement membrane to the bloodstream.
- The endothelial cells begin releasing chemical factors and local hormones. Several of these factors, including *ADP, tissue factor,* and *prostacyclin,* will be discussed in later sections. Endothelial cells also release **endothelins,** peptide hormones that (1) stimulate smooth muscle contraction and promote vascular spasms and (2) stimulate the division of endothelial cells, smooth muscle fibers, and fibroblasts to accelerate the repair process.
- The endothelial cell membranes become "sticky," and in small capillaries endothelial cells on opposite sides of the vessel may stick together and close off the passageway.

❑ The Platelet Phase

Figure 19-12

Platelets now begin to attach to sticky endothelial surfaces, to the basement membrane, and to exposed collagen fibers. This attachment marks the start of the **platelet phase** of hemostasis (Figure 19-12●). The attachment of platelets to exposed surfaces is called **platelet adhesion.** As more and more platelets arrive, they begin sticking to one another as well. This process, called **platelet aggregation,** forms a **platelet plug** that may close the break in the vascular lining if the damage is not severe or the vessel is relatively small. Platelet aggregation begins within 15 seconds after an injury occurs.

As they arrive, platelets become activated. The first sign of activation is that they change shape, becoming spherical and developing cytoplasmic processes that extend toward adjacent platelets. At this time the platelets begin synthesizing and releasing a wide variety of compounds, including ADP, *thromboxane* A_2, serotonin, calcium ions, and several *platelet factors.*

- ADP is the primary stimulus for platelet aggregation, shape changes, and platelet secretion. It is released by activated platelets and by endothelial cells at the injury site. ADP causes platelet aggregation by binding to **aggregin,** a receptor protein on the platelet membrane. This binding activates adenylate cyclase, and cAMP then activates a variety of cytoplasmic enzymes.
- *Thromboxanes* are structurally similar to prostaglandins, local messengers described in Chapter 18. [p. 605] **Thromboxane** $\mathbf{A_2}$ is released by activated platelets. It has two major effects: (1) It stimulates platelet aggregation and the secretory activities of individual platelets; and (2) it stimulates smooth muscle contractions in the vessel walls, enhancing the vascular spasms.
- Serotonin assists thromboxane A_2 in stimulating local vasoconstriction.
- Calcium ions are required for platelet aggregation and by several steps in the clotting process.

- Important platelet factors include four proteins, called *procoagulants*, that play a role in blood coagulation, and *platelet-derived growth factor* (PDGF), a peptide that promotes vessel repair by stimulating the division of endothelial cells, smooth muscle fibers, and fibroblasts.

Regulation of the Platelet Phase

The platelet phase proceeds rapidly because each arriving platelet releases ADP, thromboxane, and calcium ions that stimulate further aggregation. This is a positive feedback loop that ultimately produces a platelet plug that will be reinforced as clotting occurs. However, platelet aggregation must be controlled and restricted to the area of the injury. Several important factors limit the growth of the platelet plug:

1. **Prostacyclin,** a prostaglandin that inhibits platelet aggregation, is released by endothelial cells.
2. Inhibitory compounds are released by white blood cells entering the area, and circulating plasma enzymes break down ADP near the platelet plug.
3. Some of the compounds released by activated platelets have an inhibitory effect when present in high concentrations; for example, high serotonin levels block the action of ADP.
4. The development of a blood clot reinforces the platelet plug but separates it from the general circulation.

❑ The Coagulation Phase

Figures 19-13, 19-14

The vascular and platelet phases begin within a few seconds after the injury. The **coagulation** (cō-ag-ū-LĀ-shun) **phase** does not start until 30 seconds or more after vessel damage has occurred. Coagulation, or *blood clotting,* involves a complex sequence of steps leading to the conversion of circulating fibrinogen into the insoluble protein *fibrin.* As the fibrin network grows, it covers the surface of the platelet plug. Passing blood cells and additional platelets are trapped within the fibrous tangle, forming a **blood clot** that effectively seals off the damaged portion of the vessel. Figure 19-13● shows the structure of a blood clot viewed with a scanning electron microscope, and key stages of the coagulation phase are summarized in Figure 19-14●.

Clotting Factors

Figure 19-14

Normal coagulation cannot occur unless the plasma contains the necessary **clotting factors.** Important clotting factors include calcium ions and 11 different proteins, called **procoagulants.** Under the proper circumstances many of these procoagulants can be converted to active enzymes that direct essential reactions in the clotting response. These procoagulants are also called **proenzymes.**

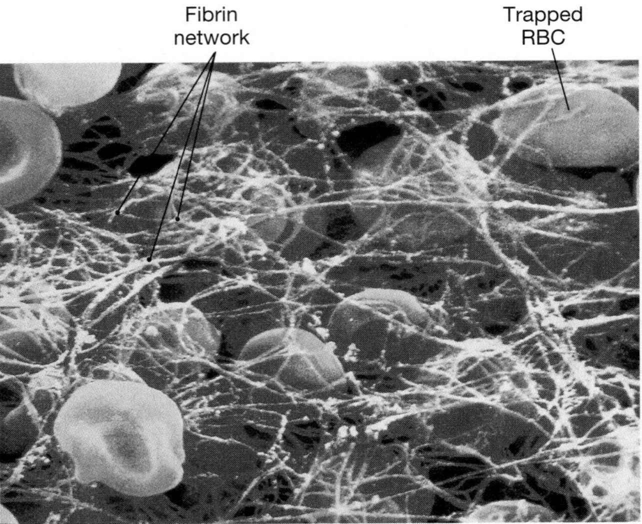
Fibrin network
Trapped RBC

● **FIGURE 19-13**
Structure of a Blood Clot. A scanning electron micrograph showing the network of fibrin that forms the framework of a clot. (SEM × 3561) Red blood cells trapped in those fibers add to the mass of the blood clot and give it a red color. Platelets that stick to the fibrin strands gradually contract, shrinking the clot and tightly packing the RBCs.

For reference, specific clotting factors are identified in Table 19-4. Many are identified by Roman numerals; calcium ions, for example, are also known as clotting Factor IV. All but three of the procoagulants (Factors III, IV, and VIII) are synthesized and released by the liver and are always present in the circulation. Activated platelets release five procoagulants (Factors III, IV, V, VIII, and XIII) during the platelet phase.

During the coagulation phase, proenzymes interact. The activation of one proenzyme often creates an enzyme that activates a second proenzyme, and so on, in a chain reaction, or *cascade.*

Figure 19-14● provides an overview of the cascades involved in the *extrinsic, intrinsic,* and *common pathways.* Readers interested in specific details concerning the steps in these pathways should consult Appendix V.

The Extrinsic Pathway

The **extrinsic pathway** begins with the release of **tissue factor** (TF) by damaged endothelial cells or peripheral tissues. The greater the damage, the more tissue factor is released and the faster clotting occurs. Tissue factor then combines with calcium ions and another procoagulant (Factor VII) to form an enzyme called **tissue thromboplastin** (Factor III).

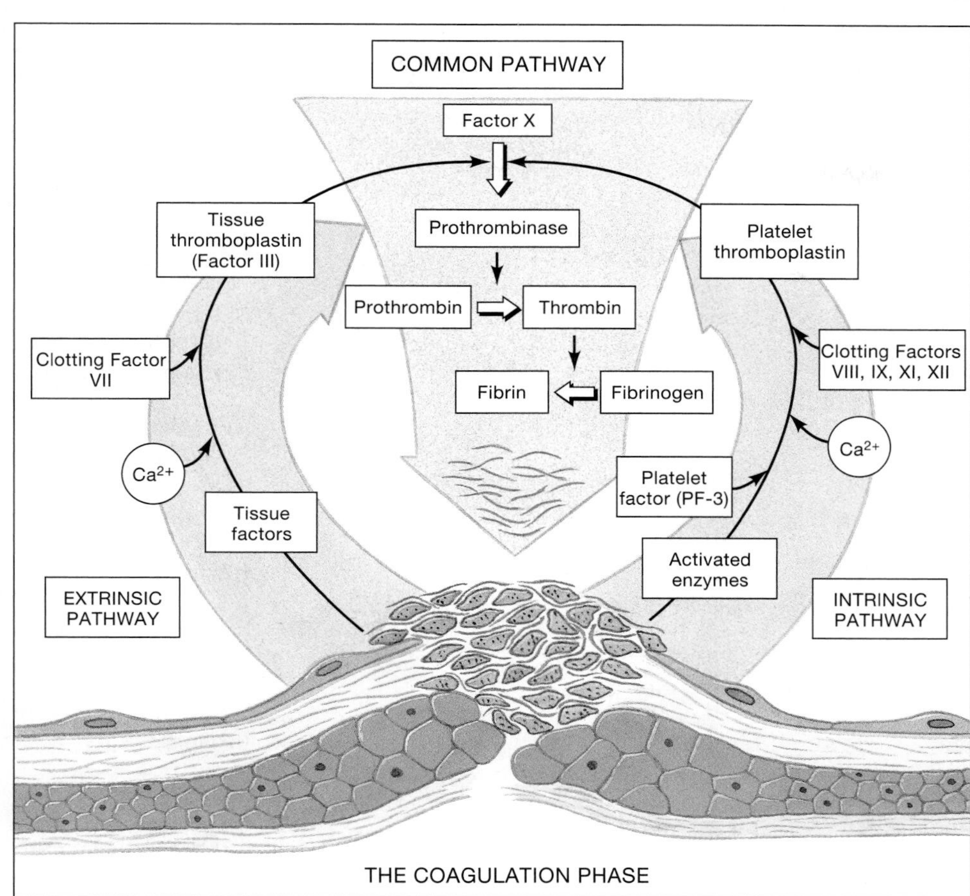
COMMON PATHWAY
Factor X
Tissue thromboplastin (Factor III)
Prothrombinase
Platelet thromboplastin
Prothrombin
Thrombin
Clotting Factor VII
Clotting Factors VIII, IX, XI, XII
Fibrin
Fibrinogen
Ca^{2+}
Ca^{2+}
Platelet factor (PF-3)
Tissue factors
Activated enzymes
EXTRINSIC PATHWAY
INTRINSIC PATHWAY
THE COAGULATION PHASE

FIGURE 19-14
The Coagulation Phase of Hemostasis

The Intrinsic Pathway

The **intrinsic pathway** begins with the activation of proenzymes exposed to collagen fibers at the injury site. This pathway proceeds with the assistance of a platelet factor **(PF-3)** released by aggregating platelets. Platelets also release a variety of other factors that accelerate the reactions of the intrinsic pathway. After a series of linked reactions, several enzymes, calcium ions, and PF-3 combine to form a complex called **platelet thromboplastin.**

The Common Pathway

The **common pathway** begins when thromboplastin from either the extrinsic or intrinsic pathways appears in the plasma. The first step involves the activation of a clotting factor (Factor X) and the formation of an enzyme, **prothrombinase.** Prothrombinase converts the proenzyme prothrombin into the enzyme **thrombin** (THROM-bin). Thrombin then completes the coagulation process by converting *fibrinogen*, a plasma protein, to insoluble strands of *fibrin.*

Interactions between the Pathways

When a blood vessel is damaged, both the extrinsic and intrinsic pathways respond, and clotting usually begins in about 15 seconds. The extrinsic pathway is shorter and faster than the intrinsic pathway, and it initiates the clotting process. In essence, the extrinsic pathway produces a small amount of thrombin very quickly. This quick patch is reinforced by the intrinsic pathway, which produces greater amounts of thrombin, but somewhat later. Once the intrinsic pathway begins generating platelet thromboplastin, it produces a larger, more effective clot than the extrinsic pathway alone.

The time required to complete clot formation varies depending on the site and the nature of the injury. In tests of the clotting system, blood held in fine glass tubes normally clots in 8–18 minutes (the *coagulation time*), and a small puncture wound typically stops bleeding in 1–4 minutes (the *bleeding time*).

Feedback Control of Coagulation

Thrombin generated in the common pathway stimulates the process of coagulation by (1) stimulating formation of tissue thromboplastin and (2) stimulating the release of PF-3 by platelets. Thus the activity of the common pathway has a stimulatory effect on both the intrinsic and extrinsic pathways. This positive feedback loop accelerates the clotting process, and speed may be very important in reducing blood loss after a severe injury.

TABLE 19-4 Procoagulants

Factor	*Structure*	*Name*	*Source*	*Concentration in Plasma (μg/ml)*	*Pathway*
I	Protein	Fibrinogen	Liver	2500–3500	Common
II	Protein	Prothrombin	Liver, requires vitamin K	100	Common
III	Lipoprotein	Tissue factor	Damaged tissue, activated platelets	0	Extrinsic and intrinsic
IV	Ion	Calcium ions	Bone, intestines, platelets	100	Entire process
V	Protein	Proaccelerin	Liver, platelets	10	Extrinsic and intrinsic
VI	(No longer used)				
VII	Protein	Proconvertin	Liver, requires vitamin K	0.5	Extrinsic
VIII	Protein	Antihemophilic factor (AHF)	Platelets, endothelial cells	15	Intrinsic
IX	Protein	Christmas factor	Liver, requires vitamin K	3	Intrinsic
X	Protein	Stuart-Prower factor	Liver, requires vitamin K	10	Extrinsic and intrinsic
XI	Protein	Plasma thromboplastin antecedent (PTA)	Liver	<5	Intrinsic
XII	Protein	Hageman factor	Liver	<5	Intrinsic; also activates plasmin
XIII	Protein	Fibrin-stabilizing factor (FSF)	Liver, platelets	20	Stabilizes fibrin, slows fibrinolysis

The coagulation process is restricted by factors that either inactivate or remove procoagulants and other stimulatory agents from the blood. Examples include:

- Normal plasma contains several enzymes, called **anticoagulants,** that inhibit coagulation. One of these, **antithrombin-III,** inhibits several different procoagulants, including thrombin.
- **Heparin,** a compound released by basophils and mast cells, is a cofactor that accelerates the activation of antithrombin-III. Heparin is used clinically to impede or prevent coagulation.
- **Thrombomodulin** is released by endothelial cells. This protein binds to thrombin and converts it to an enzyme that activates **protein C,** a plasma protein that inactivates several clotting factors and stimulates the formation of *plasmin,* an enzyme that breaks down fibrin strands.
- *Prostacyclin* released during the platelet phase inhibits platelet aggregation and opposes the stimulatory action of thrombin, ADP, and other factors.
- Other plasma proteins with anticoagulant properties include *alpha-2-macroglobulin,* which inhibits thrombin, and *C_1 inactivator,* which inhibits several procoagulants involved in the intrinsic pathway.

Calcium Ions, Vitamin K, and Blood Coagulation

Calcium ions and **vitamin K** affect almost every aspect of the clotting process. All three pathways (intrinsic, extrinsic, and common) require the presence of calcium ions, and a disorder that lowers plasma calcium concentrations will impair blood coagulation.

Adequate amounts of vitamin K must be present for the liver to be able to synthesize four of the clotting factors, including prothrombin. Vitamin K is a fat-soluble vitamin, present in green vegetables, grain, and organ meats, that is absorbed with dietary lipids. It is obtained from the diet, and a significant amount (roughly one-half of the daily requirement) is manufactured by bacteria within the large intestine. A diet inadequate in fats or vitamin K, or a disorder that affects fat digestion and absorption, such as problems with bile production, will lead to a vitamin K deficiency. This condition will cause the eventual breakdown of the common pathway, because of a lack of procoagulants, and the entire clotting system will be inactivated.

Abnormal Hemostasis

The coagulation process involves a complex chain of events, and a disorder that affects any individual clotting factor may disrupt the entire process. As a result, there are many different clinical conditions involving the clotting system. AM *Testing the Clotting System*

Excessive or Abnormal Coagulation

If the clotting response is inadequately controlled, clot formation will begin in the circulation, rather than at an injury site. These blood clots do not stick to the wall of the vessel but continue to drift around until either plasmin digests them or they become stuck in a small blood vessel. A drifting blood clot is a type of **embolus** (EM-bo-lus; *embolos,* plug), an abnormal mass within the bloodstream. When an embolus becomes stuck in a blood vessel, it blocks circulation to the area downstream, killing the affected tissues. The blockage is called an **embolism,** and the tissue damage caused by the circulatory interruption is an **infarct.**

An embolus in the arterial system may get stuck in capillaries in the brain, causing strokes. If an embolus forms in the venous system it will probably become lodged in one of the capillaries of the lungs, causing a condition known as a *pulmonary embolism.*

A **thrombus** (*thrombos,* clot) begins to form when platelets stick to the wall of an intact blood vessel. Often the platelets are attracted to areas called **plaques,** where endothelial and smooth muscle cells contain large quantities of lipids. The blood clot gradually enlarges, projecting into the lumen of the vessel and reducing its diameter. Eventually the vessel may be completely blocked, or a large chunk of the clot may break off, creating an equally dangerous embolus.

Treatment of these conditions must be prompt to prevent irreparable damage to the tissues whose vessels have been restricted or blocked by emboli or thrombi. The clots may be surgically removed or attacked by enzymes such as streptokinase or by plasmin stimulated by administered t-PA (tissue plasminogen activator). Controversy exists over the relative benefits of t-PA versus streptokinase or urokinase, when weighed against its relatively high cost.

Inadequate Coagulation

Hemophilia (hē-mō-FĒL-ē-a) is one of many inherited disorders characterized by inadequate production of clotting factors. The incidence of this condition in the general population is about 1 in 10,000, with males accounting for 80–90 percent of those affected. In hemophilia, production of a single clotting factor (most often Factor VIII) is reduced; the severity of the condition depends on the degree of reduction. In severe cases, extensive bleeding accompanies the slightest mechanical stresses, and hemorrhages occur spontaneously at joints and around muscles.

Transfusions of clotting factors can often reduce or control the symptoms of hemophilia, but plasma samples from many individuals must be pooled (combined) to obtain adequate amounts of clotting factors. This procedure makes the treatment very expensive and increases the risk of infection with blood-borne infections such as hepatitis or AIDS. Gene-splicing techniques have been used to manufacture clotting Factor VIII, an essential component of the intrinsic clotting pathway. Although supplies are now limited, this procedure should eventually provide a safer and cheaper method of treatment.

DISSEMINATED INTRAVASCULAR COAGULATION The clotting process is complex and normally is precisely regulated. In *disseminated intravascular coagulation* (DIC) bacterial toxins remove circulating fibrinogen by activating thrombin. The thrombin then converts fibrinogen to fibrin within the circulating blood. Much of the fibrin is removed by phagocytes or dissolved by plasmin, but small clots may block small vessels and damage peripheral tissues. If the liver cannot keep pace with the demand for fibrinogen, clotting abilities gradually decline, and uncontrolled bleeding may occur.

Clot Retraction

Once the fibrin meshwork has appeared, platelets and red blood cells stick to the fibrin strands. The platelets then contract, and the entire clot begins to undergo **clot retraction,** or **syneresis** (si-NER-e-sis; "a drawing together"). Clot retraction occurs over a period of 30–60 minutes. It is important because (1) it pulls the torn edges of the vessel closer together, reducing residual bleeding and stabilizing the injury site, and (2) it reduces the size of the damaged area, making it easier for fibroblasts, smooth muscle cells, and endothelial cells to complete repairs.

Fibrinolysis

As the repairs proceed, the clot gradually dissolves. This process, called **fibrinolysis** (fī-brin-OL-i-sis), begins with the activation of the proenzyme **plasminogen** (plaz-MIN-ō-jen) by two enzymes: thrombin, produced by the common pathway, and **tissue plasminogen activator (t-PA),** released by damaged tissues at the injury site. Activation of plasminogen produces the enzyme **plasmin** (PLAZ-min), which begins digesting the fibrin strands and eroding the foundation of the clot.

Manipulation of Hemostasis

Clinicians may attempt to prevent unwanted clotting by administering drugs that depress the clotting response or dissolve clots already present. Important anticoagulant drugs include:

- Heparin, which activates antithrombin-III.
- **Coumadin** (COO-ma-din), or *warfarin* (WAR-fa-rin), and **dicumarol,** which depress the synthesis of several clotting factors by blocking the action of vitamin K.
- t-PA synthesized by recombinant DNA techniques, which stimulates plasmin formation.
- **Streptokinase** (strep-tō-KĪ-nāz) or **urokinase** (ū-rō-KĪ-nāz), enzymes that convert plasminogen to plasmin.
- Aspirin, which inactivates platelet enzymes involved with the production of thromboxanes and prostaglandins and inhibits the endothelial cell production of prostacyclin. Daily ingestion of small quantities of aspirin reduces the sensitivity of the clotting process. This method has been proven to be effective in preventing heart attacks in people with significant heart disease.

These compounds have widespread use in the treatment of chronic or acute circulatory blockages. The use of the "clot-busting" agents t-PA, streptokinase, and urokinase in the treatment of heart attacks will be considered in Chapter 20.

Blood samples may be stabilized temporarily by adding either heparin or *EDTA (ethylenediaminetetroacetic acid)* to the sample. EDTA removes calcium ions from plasma, effectively preventing coagulation. In units of whole blood held for extended periods in a blood bank, *citratephosphate dextrose* (CPD) is often added. CPD, like EDTA, ties up plasma calcium ions.

☑ A sample of bone marrow has fewer than normal numbers of megakaryocytes. What body process would you expect to be impaired as a result?

☑ About half of our vitamin K is produced by bacteria in the large intestine. How would you expect extended use of broad-spectrum antibiotics to affect blood clotting?

☑ Unless chemically treated, blood will coagulate in a test tube. The process begins when Factor XII (Hageman factor) becomes activated. Which clotting pathway would be involved in this process?

To perform its vital functions, blood must be kept in motion; if the circulatory supply is cut off, dependent tissues may be destroyed in a matter of minutes. Individual RBCs complete two trips around the circulatory system each minute. The circulation of blood begins in the third week of embryonic development and continues throughout life. The next chapter examines the structure and function of the heart, the pump that maintains this vital blood flow.

Selected Clinical Terminology

Terms Discussed in This Chapter

anemia (a-NĒ-mē-ah): A condition in which the oxygen-carrying capacity of the blood is reduced because of low hematocrit or low blood hemoglobin concentrations. *(p. 656* and *AM)*

embolism: A condition in which a drifting blood clot becomes stuck in a blood vessel, blocking circulation to the area downstream. *(p. 675)*

hematocrit (he-MA-tō-krit): The value that indicates the percentage of whole blood occupied by cellular elements. *(p. 655)*

hematuria (hē-ma-TŪ-rē-uh): The presence of blood cells in the urine. *(p. 657)*

hemoglobinuria: The presence of hemoglobin in the urine. *(p. 657)*

hemolytic disease of the newborn (HDN): A condition in which fetal red blood cells have been destroyed by maternal agglutinins. *(p. 662)*

hemophilia (hē-mō-FĒL-ē-ah): Inherited disorders characterized by inadequate production of clotting factors. *(p. 675)*

hypovolemic (hī-pō-vō-LĒ-mik), **normovolemic** (nōr-mō-vō-LĒ-mik), and **hypervolemic** (hī-per-vō-LĒ-mik): Low, normal, or excessive blood volumes, respectively. *(p. 651)*

hypoxia (hī-POKS-ē-a): Low tissue oxygen levels. *(p. 660* and *AM)*

jaundice: Symptoms of yellow skin and eyes, caused by abnormally high levels of plasma bilirubin; examples include *hemolytic jaundice* and *obstructive jaundice*. *(p. 657* and *AM)*

leukemia: A condition characterized by extremely elevated levels of circulating WBCs; includes both *myeloid* and *lymphoid* forms. *(p. 667* and *AM)*

leukocytosis (loo-kō-sī-TŌ-sis): Excessive numbers of white blood cells in the circulation. *(p. 667)*

leukopenia (loo-kō-PĒ-nē-ah): Inadequate numbers of white blood cells in the circulation. *(p. 667)*

normochromic: The condition in which red blood cells contain normal amounts of hemoglobin. *(p. 660* and *AM)*

normocytic: A term referring to cells of normal size. *(p. 660* and *AM)*

plaque: An abnormal area within a blood vessel where large quantities of lipids accumulate. *(p. 675)*
sickle cell anemia: An anemia resulting from the production of hemoglobin S rather than normal hemoglobin; this hemoglobin form causes RBC sickling at low oxygen levels. *(p. 656* and *AM)*
tests of RBC status: These tests include a *reticulocyte count, hematocrit, hemoglobin concentration, RBC count, mean corpuscular volume,* and *mean corpuscular hemoglobin concentration.* *(p. 660* and *AM)*
thalassemia: A disorder resulting from production of an abnormal form of hemoglobin. *(p. 656* and *AM)*
thrombus: A blood clot. *(p. 675)*
venipuncture (VEN-i-punk-chur): The puncture of a vein for any purpose, including the withdrawal of blood or the administration of medication. *(p. 651)*

AM *Additional Terms Discussed in the Applications Manual*

autologous marrow transplant: Reinfusion of an individual's own bone marrow collected prior to chemotherapy or radiation treatment.
erythrocytosis (e-rith-rō-sī-TŌ-sis): A polycythemia affecting only red blood cells.
heterologous marrow transplant: Transplantation of bone marrow from one individual to another, to replace bone marrow destroyed during cancer therapy.
packed red cells: Red blood cells from which most of the plasma has been removed.
polycythemia (po-lē-sī-THĒ-mē-ah): A blood condition showing an elevated hematocrit with a normal blood volume.
tests of the clotting system: These tests include *bleeding time, coagulation time, partial thromboplastin time,* and *plasma prothrombin time.*
thrombocytopenic purpura: An immune system disorder that results in the production of antibodies that attack platelets. The platelet count is low, and the individual bruises easily because the remaining platelets are very fragile.
transfusion: A procedure in which blood components are given to someone whose blood volume has been reduced or whose blood is defective.

CHAPTER REVIEW

STUDY OUTLINE

INTRODUCTION, p. 650

1. The cardiovascular system provides a mechanism for the rapid transport of nutrients, waste products, respiratory gases, and cells within the body.

FUNCTIONS OF BLOOD, p. 650

1. Blood is a specialized connective tissue. Its functions include: (1) transporting dissolved gases, nutrients, metabolic wastes, and hormones; (2) regulating the pH and electrolyte composition of interstitial fluids; (3) restricting fluid losses through damaged vessels or injuries; (4) defending the body against toxins and pathogens; and (5) regulating body temperature by absorbing and redistributing heat.

COMPOSITION OF BLOOD, p. 650

1. Blood contains **plasma, red blood cells (RBCs), white blood cells (WBCs),** and **platelets.** The plasma and **formed elements** constitute **whole blood,** which can be **fractionated** for analytical or clinical purposes. (*Figure 19-1; Table 19-3*)

Blood Collection and Analysis, p. 651

PLASMA, p. 651

1. Plasma accounts for about 55 percent of the volume of blood; roughly 92 percent of plasma is water. (*Figure 19-1*)

Differences between Plasma and Interstitial Fluid, p. 651

2. Plasma differs from interstitial fluid because it has a higher dissolved oxygen concentration and large numbers of dissolved proteins. There are three primary classes of plasma proteins: *albumins, globulins,* and *fibrinogen.*

Plasma Proteins, p. 651

3. **Albumins** constitute about 60 percent of plasma proteins. **Globulins** constitute roughly 35 percent of plasma proteins: They include **immunoglobulins (antibodies),** which attack foreign proteins and pathogens, and **transport globulins,** which bind ions, hormones, and other compounds. **Fibrinogen** molecules function in the clotting reaction by interacting to form **fibrin:** Removing fibrinogen from plasma leaves a fluid called **serum.**

FORMED ELEMENTS, p. 653

Hemopoiesis, p. 653

1. **Hemopoiesis** is the process of blood cell formation. Circulating **stem cells** called **hemocytoblasts** divide to form all of the blood cells.

Red Blood Cells, p. 654

2. Red blood cells **(erythrocytes)** account for slightly less than half the blood volume and 99.9 percent of the formed elements. The **hematocrit** value indicates the percentage of whole blood occupied by formed elements.
3. RBCs transport oxygen and carbon dioxide within the bloodstream. They are highly specialized cells with large surface-to-volume ratios. Because RBCs

lack mitochondria, ribosomes, and nuclei they are unable to perform normal maintenance operations, so they usually degenerate after about 120 days in the circulation. (*Figure 19-2*)

4. Molecules of **hemoglobin (Hb)** account for over 95 percent of the RBCs proteins. Hemoglobin is a globular protein formed from four subunits. Each subunit contains a single molecule of **heme** (a **porphyrin**) and can reversibly bind an oxygen molecule. Damaged or dead RBCs are recycled by phagocytes. (*Figures 19-3, 19-5*)
5. **Erythropoiesis,** the formation of erythrocytes, occurs mainly within the **myeloid tissue** (bone marrow) in adults. RBC formation increases under **erythropoiesis-stimulating hormone** (erythropoietin, EPO) stimulation. Stages in RBC development include **erythroblasts** and **reticulocytes.** (*Figures 19-6, 19-10*)
6. One's **blood type** is determined by the presence or absence of specific **agglutinogens** in the RBC cell membranes: **agglutinogens A, B,** and **D (Rh). Agglutinins** within the plasma will react with RBCs bearing different agglutinogens. (*Figure 19-7; Table 19-2*)

White Blood Cells, p. 664

7. White blood cells **(leukocytes)** defend the body against pathogens and remove toxins, wastes, and abnormal or damaged cells. (*Figure 19-9*)
8. Leukocytes show **chemotaxis** (attraction to specific chemicals) and **diapedesis** (the ability to move through vessel walls).
9. Granular leukocytes are subdivided into **neutrophils, eosinophils,** and **basophils.** Fifty to seventy percent of circulating WBCs are neutrophils, which are highly mobile phagocytes. The much less common eosinophils are phagocytes that are attracted to foreign compounds that have reacted with circulating antibodies. The relatively rare basophils migrate to damaged tissues and release histamines, aiding the inflammation response. (*Figure 19-9*)
10. Agranular leukocytes are subdivided into **monocytes** and **lymphocytes.** Monocytes migrating into peripheral tissues become free macrophages. Lymphocytes, the primary cells of the lymphatic system, include **T cells** (which enter peripheral tissues and attack foreign cells directly), **B cells** (which produce antibodies), and **NK cells** (which destroy abnormal tissue cells). (*Figure 19-9; Table 19-3*)
11. Granulocytes and monocytes are produced by stem cells in the bone marrow. Stem cells responsible for **lymphopoiesis** (production of lymphocytes) also originate in the bone marrow, but many migrate to peripheral **lymphoid tissues.** (*Figure 19-10*)
12. Factors that regulate lymphocyte maturation are not completely understood. Several **colony-stimulating factors (CSFs)** are involved in regulating other WBC populations and coordinating RBC and WBC production. (*Figure 19-10*)

Platelets, p. 667

13. **Megakaryocytes** in the bone marrow release packets of cytoplasm **(platelets)** into the circulating blood. The functions of platelets include: (1) transporting chemicals important to the clotting process; (2) forming a temporary patch in the walls of damaged blood vessels; and (3) contracting after a clot has formed in order to reduce the size of the break in the vessel wall. (*Figure 19-11*)

HEMOSTASIS, p. 670

1. The five-step process of **hemostasis** prevents the loss of blood through the walls of damaged vessels.

The Vascular Phase, p. 671

2. The **vascular phase** is a period of local vasoconstriction resulting from **vascular spasm** at the injury site. (*Figure 19-12*)

The Platelet Phase, p. 671

3. The **platelet phase** follows as platelets are activated, aggregate at the site, and adhere to the damaged surfaces. (*Figure 19-12*)

The Coagulation Phase, p. 672

4. The **coagulation phase** occurs as factors released by platelets and endothelial cells interact with **clotting factors** to form a blood clot. In this reaction sequence, suspended **fibrinogen** is converted to large, insoluble fibers of **fibrin**. (*Figures 19-13, 19-14; Table 19-4*)

Clot Retraction, p. 675

5. During **clot retraction,** platelets contract and pull the torn edges of the damaged vessel closer together.

Fibrinolysis, p. 675

6. During the period of **fibrinolysis** the clot gradually dissolves through the action of plasmin, the activated form of circulating plasminogen.

Manipulation of Hemostasis, p. 676

7. Clotting may be prevented by administering drugs that depress the clotting response or dissolve clots already present. Important anticoagulant drugs include *heparin, coumadin, dicumarol, t-PA, streptokinase, urokinase,* and aspirin.

REVIEW QUESTIONS

LEVEL 1 Reviewing Facts and Terms

1. The formed elements of the blood include:
   (a) plasma, fibrin, serum
   (b) albumins, globulins, fibrinogen
   (c) WBCs, RBCs, platelets
   (d) a, b, and c are correct

2. Blood temperature is approximately ______, and the blood pH averages ______.
   (a) 98.6° F, 7.0 (b) 104° F, 7.8
   (c) 100.4° F, 7.4 (d) 96.8° F, 7.0

3. Plasma contributes approximately _______ percent of the volume of whole blood, and water accounts for _______ percent of the plasma volume.
(a) 55, 92 (b) 25, 55
(c) 92, 55 (d) 35, 72

4. When the clotting proteins are removed from plasma, _______ remains.
(a) fibrinogen (b) fibrin
(c) serum (d) heme

5. In an adult the only site of red blood cell production, and the primary site of white blood cell formation, is the:
(a) liver (b) spleen
(c) thymus (d) red bone marrow

6. The most numerous WBCs found in a differential count of a "normal" individual are:
(a) neutrophils (b) basophils
(c) lymphocytes (d) monocytes

7. The differential count of a person who has an allergy would show a high number of:
(a) neutrophils (b) eosinophils
(c) basophils (d) monocytes

8. Stem cells responsible for the process of lymphopoiesis are located in the:
(a) thymus and spleen (b) lymph nodes
(c) red bone marrow (d) a, b, and c are correct

9. The first step in the process of hemostasis is:
(a) coagulation (b) the platelet phase
(c) fibrinolysis (d) vascular spasm

10. The complex sequence of steps leading to the conversion of fibrinogen to fibrin is called:
(a) fibrinolysis (b) coagulation
(c) retraction (d) the platelet phase

11. What five major functions are performed by the blood?

12. What three primary classes of plasma proteins are found in the blood? What is the major function of each?

13. What type of agglutinins does the plasma contain for each of the following blood types?
(a) Type A (b) Type B
(c) Type AB (d) Type O

14. What four processes facilitate the movement of WBCs to areas of invasion or injury?

15. What kinds of WBCs contribute to the body's nonspecific defenses?

16. What three classes of lymphocytes are the primary cells of the lymphatic system? What are the functions of each class?

17. What kinds of WBCs are produced by each one of the four colony-stimulating factors (CSFs)?

18. What are the three functions of platelets during the clotting process?

19. What three factors regulate the rate of megakaryocyte activity and platelet formation?

20. What five steps are necessary for hemostasis to be completed after damage to the wall of a blood vessel?

21. What contribution from the intrinsic and extrinsic pathways is necessary for the common pathway to begin?

22. Distinguish between an embolus and a thrombus.

LEVEL 2 Reviewing Concepts

23. Dehydration would cause:
(a) an increase in the hematocrit
(b) a decrease in the hematocrit
(c) no effect in the hematocrit
(d) an increase in plasma volume

24. Erythropoietin directly stimulates RBC formation by:
(a) increasing rates of mitotic divisions in erythroblasts
(b) speeding up the maturation of red blood cells
(c) accelerating the rate of hemoglobin synthesis
(d) a, b, and c are correct

25. A person with Type A blood has:
(a) A agglutinogens in the plasma
(b) B agglutinins in the plasma
(c) A agglutinins on the red blood cells
(d) B agglutinogens on the red blood cells

26. Hemolytic disease of the newborn may result if:
(a) an Rh-positive woman marries an Rh-negative man
(b) both the man and the woman are Rh-negative
(c) both the man and the woman are Rh-positive
(d) an Rh-negative woman carries an Rh-positive fetus

27. How do red blood cells (RBCs) differ from typical cells found in the body?

28. How does the blood defend against toxins and pathogens in the body?

29. What is the role of blood in the stabilization and maintenance of body temperature?

30. If you are a respiratory therapist and you need blood to check the efficiency of gas exchange at the lungs, what type of "stick" will you use, and from what blood vessels might you draw the sample?

31. Linda S. was given RhoGam during and after the delivery of her baby. What does the administration of RhoGam imply, and what effect does it have on the mother?

32. Why is aspirin sometimes prescribed for the prevention of vascular problems?

LEVEL 3 Critical Thinking and Clinical Applications

33. A test for prothrombin time is used to determine deficiencies in the extrinsic clotting system and is prolonged if any of the factors are deficient. A test for activated partial thromboplastin time is used in a similar fashion to detect deficiencies in the intrinsic clotting system. What factor would be deficient if a person had a prolonged prothrombin time but a normal partial thromboplastin time?

34. Which of the formed elements would you expect to see increase after donating a pint of blood?

35. As a result of taking the antibiotic Cephalosporin, Mary's platelet count drops to 50,000/μl. What symptoms would you expect to observe?

36. Why do patients suffering from advanced kidney disease frequently become anemic?

37. After Randy was diagnosed with stomach cancer, nearly all of his stomach had to be removed. Postoperative treatment included weekly injections of vitamin B_{12}. Why was this vitamin prescribed, and why was this mode of administration specified?

38. How would you expect extended use of broad-spectrum antibiotics to affect blood clotting?

CHAPTER 20

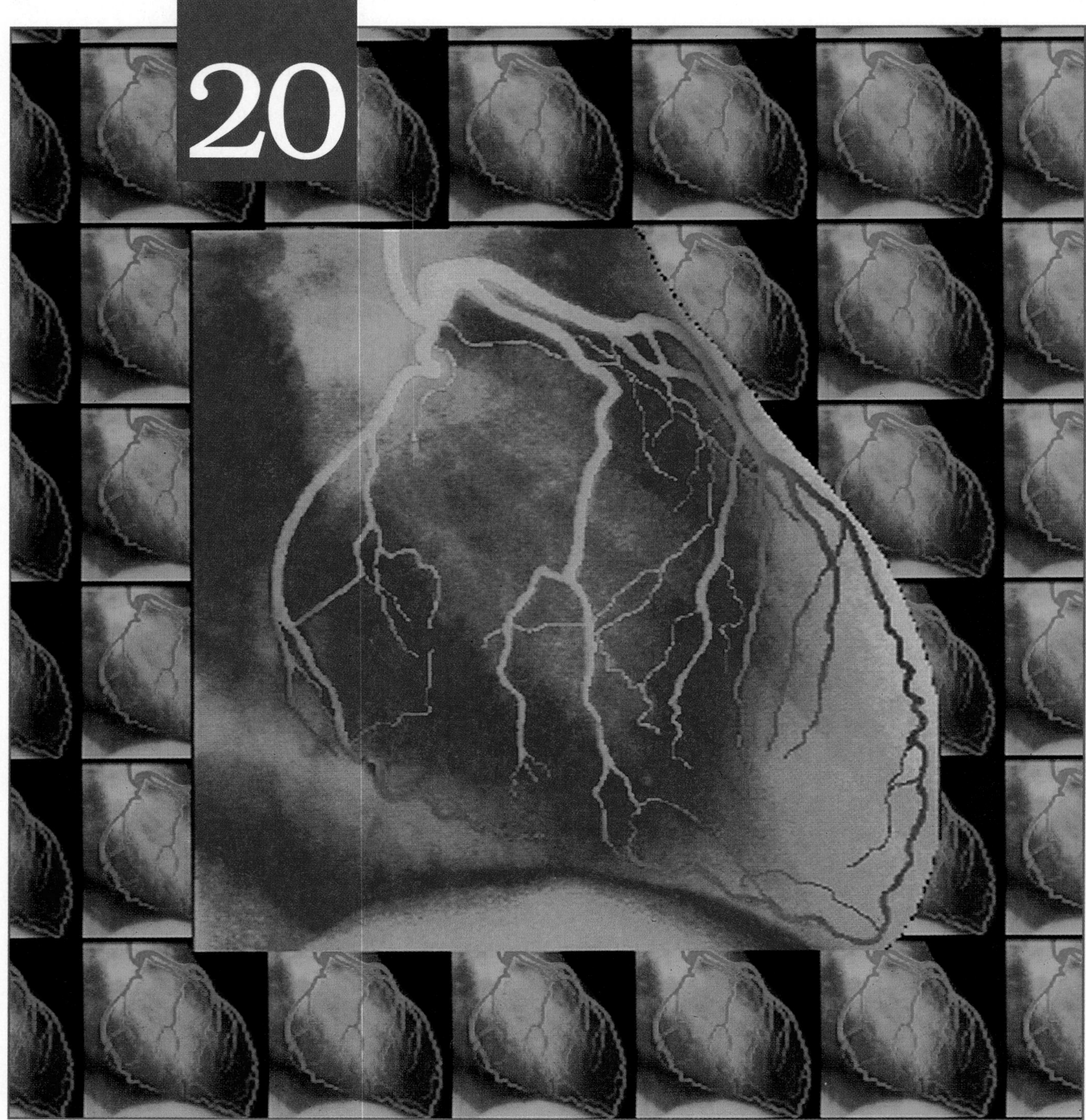

You know from previous chapters that hard-working muscles require a steady supply of blood to provide them with nutrients and oxygen. This is especially true for the hardest-working muscle of all: your heart. Unlike most other muscles, the heart never rests. Not surprisingly, then, any substantial interruption or reduction in the flow of blood to this organ has grave consequences: what we commonly call a heart attack. Such attacks typically occur when there is an obstruction in one of the arteries that supply the heart muscle. Physicians can check the status of these vessels using scanning techniques that produce computer-enhanced images like those shown here. In this chapter we'll see how the heart supplies blood, not only to itself, but to all other tissues and organs, and how its activities are adjusted so that the blood supply to active tissues--including the heart itself--can meet the body's constantly-changing demands.

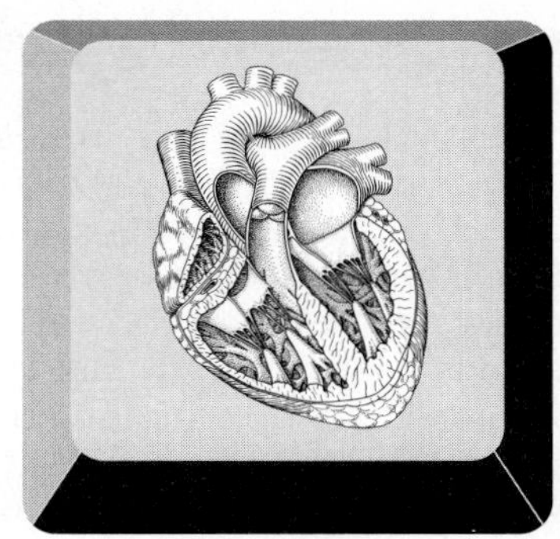

The Heart

Chapter Outline and Objectives

1 Describe the location and general features of the heart.

2 Describe the structure of the pericardium and explain its functions.

3 Trace the flow of blood through the heart, identifying the major blood vessels, chambers, and heart valves.

4 Identify the layers of the heart wall.

5 Describe the vascular supply and innervation of the heart.

6 Describe the events of an action potential in cardiac muscle and explain the importance of calcium ions to the contractile process.

7 Discuss the differences between nodal cells and conducting cells, and describe the components and functions of the conducting system of the heart.

8 Identify the electrical events associated with a normal electrocardiogram.

9 Explain the events of the cardiac cycle, including atrial and ventricular systole and diastole, and relate the heart sounds to specific events in this cycle.

10 Define stroke volume and cardiac output, and describe the factors that influence these values.

11 Explain how adjustments in stroke volume and cardiac output are coordinated at different levels of activity.

12 Describe the effects of autonomic activity on heart function.

13 Describe the effects of hormones, drugs, temperature, and changes in ion concentrations on the heart.

Every living cell relies on the surrounding interstitial fluid for oxygen, nutrients, and waste disposal. Conditions in the interstitial fluid are kept stable through continuous exchange between the peripheral tissues and the bloodstream. Yet the blood can help to maintain homeostasis only as long as it stays in motion. If blood remains stationary, its oxygen and nutrient supplies are quickly exhausted, its capacity to absorb wastes is soon saturated, and neither hormones nor white blood cells can reach their intended targets. Thus all the functions of the cardiovascular system ultimately depend on the heart. This muscular organ beats approximately 100,000 times *each day*, pumping roughly 8000 liters of blood—enough to fill 40 55-gallon drums, or 8800 quart-sized milk cartons.

This chapter begins by examining the structural features that enable the heart to perform so reliably. We will then consider the physiological mechanisms that regulate cardiac activity to meet changing circumstances.

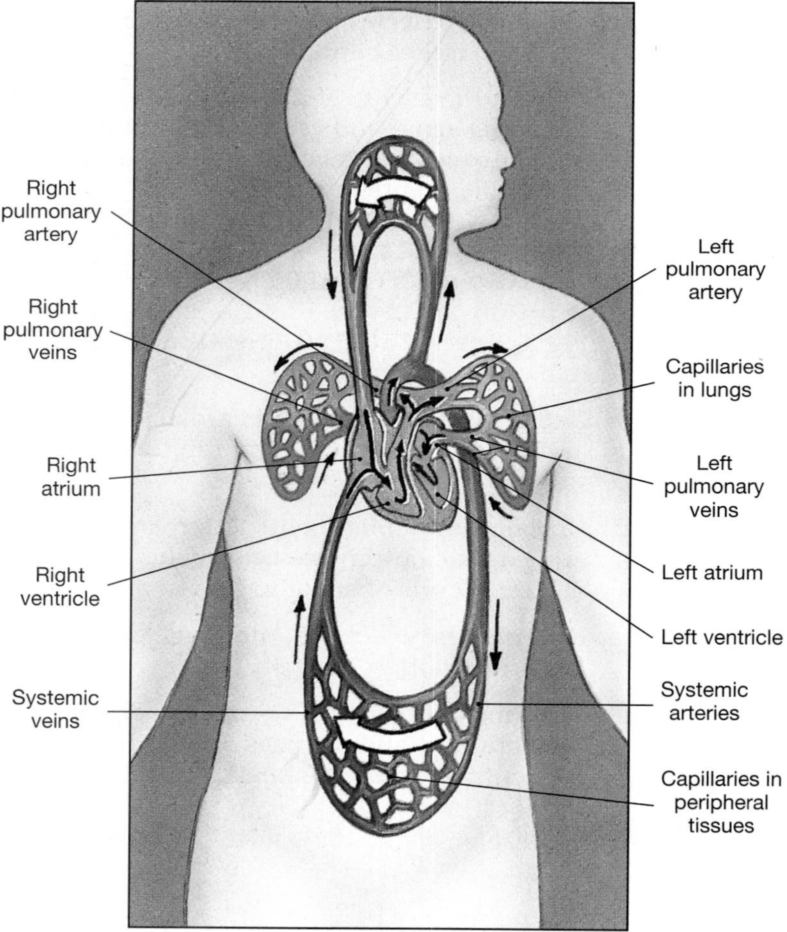
Right pulmonary artery
Right pulmonary veins
Right atrium
Right ventricle
Systemic veins
Left pulmonary artery
Capillaries in lungs
Left pulmonary veins
Left atrium
Left ventricle
Systemic arteries
Capillaries in peripheral tissues

● **FIGURE 20-1**
An Overview of the Cardiovascular System. Blood flows through separate pulmonary and systemic circuits, driven by the pumping of the heart. Each circuit begins and ends at the heart and contains arteries, capillaries, and veins.

An Overview of the Cardiovascular System

Figure 20-1

Blood flows through a network of blood vessels that extend between the heart and peripheral tissues. Those blood vessels can be subdivided into a **pulmonary circuit** that carries blood to and from the exchange surfaces of the lungs and a **systemic circuit** that transports blood to and from the rest of the body. Each circuit begins and ends at the heart (Figure 20-1●), and blood travels through these circuits in sequence. For example, blood returning to the heart from the systemic circuit must complete the pulmonary circuit before reentering the systemic circuit.

Arteries, or *efferent vessels,* carry blood away from the heart; **veins,** or *afferent vessels,* return blood to the heart. **Capillaries** are small, thin-walled vessels between the smallest arteries and veins. Capillaries are called **exchange vessels** because their thin walls permit exchange of nutrients, dissolved gases, and waste products between the blood and surrounding tissues.

Despite its impressive workload, the heart is a small organ, roughly the size of a clenched fist. The heart contains four muscular chambers, two associated with each circuit. The **right atrium** (Ā-trē-um; chamber; plural *atria*) receives blood from the systemic circuit and passes it to the **right ventricle** (VEN-tri-kul; little belly). The right ventricle discharges blood into the pulmonary circuit. The **left atrium** collects blood from the pulmonary circuit and empties it into the **left ventricle.** Contraction of the left ventricle ejects blood into the systemic circuit. When the heart beats, the two ventricles contract at the same time and eject equal volumes of blood.

Anatomy of the Heart

Figure 20-2

The heart is located near the anterior chest wall, directly behind the sternum (Figure 20-2a●). A midsagittal section through the trunk would not

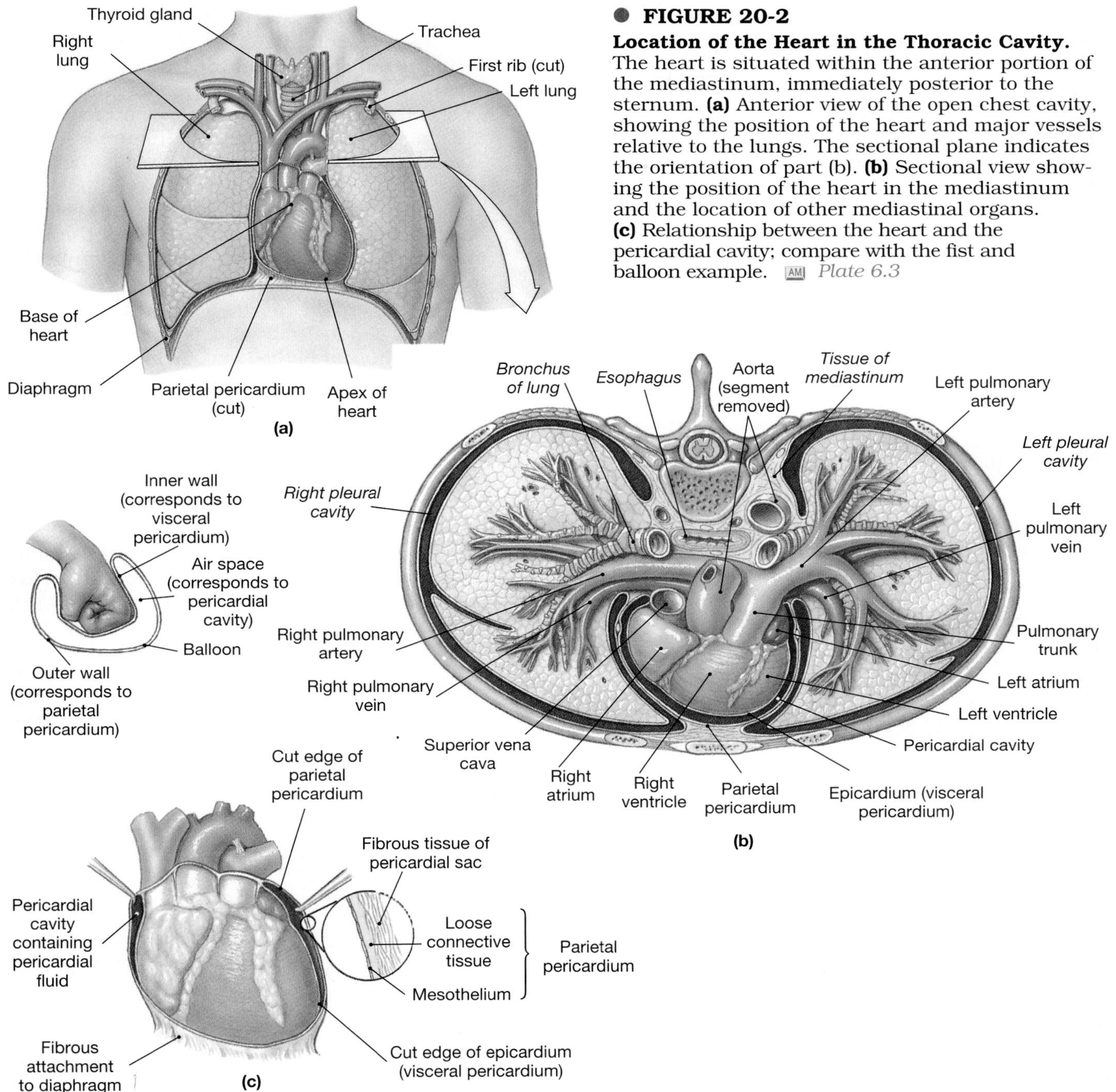

Thyroid gland
Right lung
Trachea
First rib (cut)
Left lung
Base of heart
Diaphragm
Parietal pericardium (cut)
Apex of heart
(a)
Inner wall (corresponds to visceral pericardium)
Air space (corresponds to pericardial cavity)
Balloon
Outer wall (corresponds to parietal pericardium)
Right pleural cavity
Right pulmonary artery
Right pulmonary vein
Bronchus of lung
Esophagus
Aorta (segment removed)
Tissue of mediastinum
Left pulmonary artery
Left pleural cavity
Left pulmonary vein
Pulmonary trunk
Left atrium
Left ventricle
Pericardial cavity
Superior vena cava
Right atrium
Right ventricle
Parietal pericardium
Epicardium (visceral pericardium)
(b)
Cut edge of parietal pericardium
Fibrous tissue of pericardial sac
Pericardial cavity containing pericardial fluid
Loose connective tissue
Mesothelium
Parietal pericardium
Fibrous attachment to diaphragm
Cut edge of epicardium (visceral pericardium)
(c)

● **FIGURE 20-2**
Location of the Heart in the Thoracic Cavity. The heart is situated within the anterior portion of the mediastinum, immediately posterior to the sternum. **(a)** Anterior view of the open chest cavity, showing the position of the heart and major vessels relative to the lungs. The sectional plane indicates the orientation of part (b). **(b)** Sectional view showing the position of the heart in the mediastinum and the location of other mediastinal organs. **(c)** Relationship between the heart and the pericardial cavity; compare with the fist and balloon example. AM *Plate 6.3*

divide the heart into two equal halves because the heart (1) lies slightly to the left of the midline, (2) sits at an angle to the longitudinal axis of the body, and (3) is rotated toward the left side. The heart is surrounded by the **pericardial** (per-i-KAR-dē-al) **cavity,** located in the anterior portion of the mediastinum. The mediastinum, which separates the two pleural cavities, also contains the thymus, esophagus, and trachea. *[pp. 23-24]* Figure 20-2b● is a sectional view that illustrates the position of the heart relative to other structures in the mediastinum.

❑ The Pericardium

Figure 20-2

The serous membrane lining the pericardial cavity is called the **pericardium.** To visualize the relationship between the heart and the pericardial cavity, imagine pushing your fist toward the center of a large balloon (Figure 20-2c●). The balloon represents the pericardium, and your fist is the heart. Your wrist, where the balloon folds back upon itself, corresponds to the base of the heart. The space inside the balloon is the pericardial cavity.

The pericardium can be subdivided into the *visceral pericardium* and the *parietal pericardium.* The **visceral pericardium,** or **epicardium,** covers the outer surface of the heart; the **parietal pericardium** lines the inner surface of the **pericardial sac** that surrounds the heart (Figure 20-2c●). The pericardial sac, which is reinforced by a dense network of collagen fibers, stabilizes the positions of the heart and associated vessels within the mediastinum.

The space between the opposing parietal and visceral surfaces is the pericardial cavity. This cavity normally contains 10–20 ml of **pericardial fluid** secreted by the pericardial membranes. Pericardial fluid acts as a lubricant, reducing friction between the opposing surfaces as the heart beats.

PERICARDITIS A variety of pathogens may infect the pericardium, producing a condition known as **pericarditis.** The inflamed pericardial surfaces rub against one another, producing a distinctive scratching sound. In addition, the pericardial irritation and inflammation often result in an increased production of pericardial fluid. Fluid then collects in the pericardial cavity, restricting the movement of the heart. This condition is called **cardiac tamponade** (tam-po-NĀD; *tampon,* plug). AM *Infection and Inflammation of the Heart*

Superficial Anatomy of the Heart

Figure 20-3

The four cardiac chambers can be identified easily in a superficial view of the heart (Figure 20-3●). The two atria have relatively thin muscular walls, and they are highly distensible. When not filled with blood, the outer portion of each atrium deflates and becomes a rather lumpy and wrinkled flap. This expandable extension of an atrium is called an **auricle** (AW-ri-k'l; *auris,* ear) because it reminded early anatomists of the external ear (Figure 20-3a●). A deep groove, the **coronary sulcus,** marks the border between the atria and the ventricles. Shallower depressions, the **anterior interventricular sulcus** and the **posterior interventricular sulcus,** mark the boundary line between the left and right ventricles (Figure 20-3a,b●).

The connective tissue of the epicardium at the coronary and interventricular sulci usually contains substantial amounts of fat, and in fresh or preserved hearts this fat must be stripped away to expose the underlying grooves. These sulci also contain the arteries and veins that supply blood to the cardiac muscle of the heart.

There are several superficial reference points that are useful when describing the heart or the locations of nerves, vessels, or other associated structures. The heart has an attached *base,* a free *apex,* four *borders,* and three *surfaces.*

- ***The base and apex.*** The great veins and arteries of the circulatory system are connected to the superior end of the heart at the **base.** The base sits posterior to the sternum at the level of the third costal cartilage, centered about 1.2 cm (0.5 in.) to the left side (Figure 20-3c●). The inferior, pointed tip of the heart is the **apex** (Ā-peks). A typical adult heart measures approximately 12.5 cm (5 in.) from the attached base to the apex. The apex reaches the fifth intercostal space approximately 7.5 cm (3 in.) to the left of the midline.
- ***The borders of the heart.*** The heart sits at an angle to the longitudinal axis of the body. The **superior border** of the heart includes the bases of the major vessels and the two atria. The right atrium forms the **right border** of the heart; the left ventricle and a small portion of the left atrium form the **left border.** The left border extends to the apex, where it meets the **inferior border.** The wall of the right ventricle forms most of the inferior border (Figure 20-3a,c●).
- ***The surfaces of the heart.*** The heart is rotated slightly toward the left, so the **anterior surface,** or **sternocostal** (ster-nō-KOS-tal) **surface,** consists primarily of the right atrium and right ventricle (Figure 20-3a,c●). The heart sits in a notch within the left lung, which comes in contact with the **pulmonary (left) surface of the heart.** The posterior wall of the left ventricle forms much of the sloping **diaphragmatic surface** between the base and the apex of the heart (Figure 20-3b,c●).

Internal Anatomy and Organization

Figure 20-4

The two atria are separated by the *interatrial septum* (*septum,* wall), and the two ventricles are separated by the *interventricular septum* (Figure 20-4a,c●). Each atrium communicates with the ventricle of the same side. **Atrioventricular (AV) valves,** folds of fibrous tissue, extend into the openings between the atria and ventricles. These valves prevent backflow and maintain a one-way flow of blood from the atria into the ventricles.

The Right Atrium

Figure 20-4

The right atrium receives blood from the systemic circuit via the two great veins, the **superior vena cava** (VĒ-na CĀ-va) and the **inferior vena cava.** The superior vena cava delivers blood to the right atrium from the head, neck, upper limbs, and chest. The superior vena cava opens into the posterior and superior portion of the right atrium. The inferior vena cava carries blood to the right atrium from the rest of the trunk, the viscera, and the lower limbs. The inferior vena cava opens into the posterior and inferior portion of the right atrium. The *coronary veins* of the heart return blood to the **coronary sinus,** which opens into the right atrium inferior to the connection with the inferior vena cava.

Prominent muscular ridges, the **pectinate muscles** (*pectin,* comb), or *musculi pectinati,* run along the inner surface of the auricle and across the adjacent anterior atrial wall (Figure 20-4a,c●). The **interatrial septum** divides the right atrium from the left atrium. From the fifth week of embryonic development until birth, an oval opening, the **foramen ovale,** penetrates the septum and connects the two atria. The foramen ovale permits blood flow from the right atrium to the left atrium while the lungs are devel-

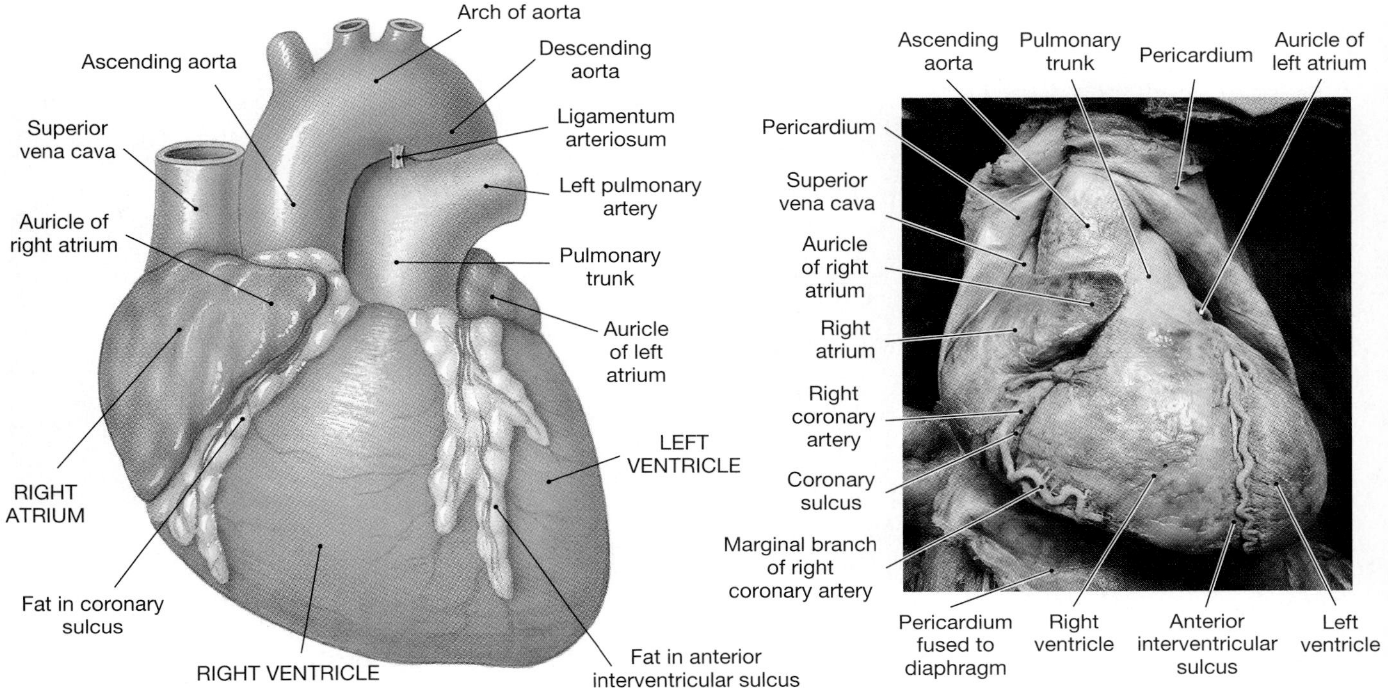
Arch of aorta
Ascending aorta
Descending aorta
Superior vena cava
Ligamentum arteriosum
Left pulmonary artery
Auricle of right atrium
Pulmonary trunk
Auricle of left atrium
LEFT VENTRICLE
RIGHT ATRIUM
Fat in coronary sulcus
RIGHT VENTRICLE
Fat in anterior interventricular sulcus
Ascending aorta
Pulmonary trunk
Pericardium
Auricle of left atrium
Pericardium
Superior vena cava
Auricle of right atrium
Right atrium
Right coronary artery
Coronary sulcus
Marginal branch of right coronary artery
Pericardium fused to diaphragm
Right ventricle
Anterior interventricular sulcus
Left ventricle

(a) Anterior (sternocostal) surface

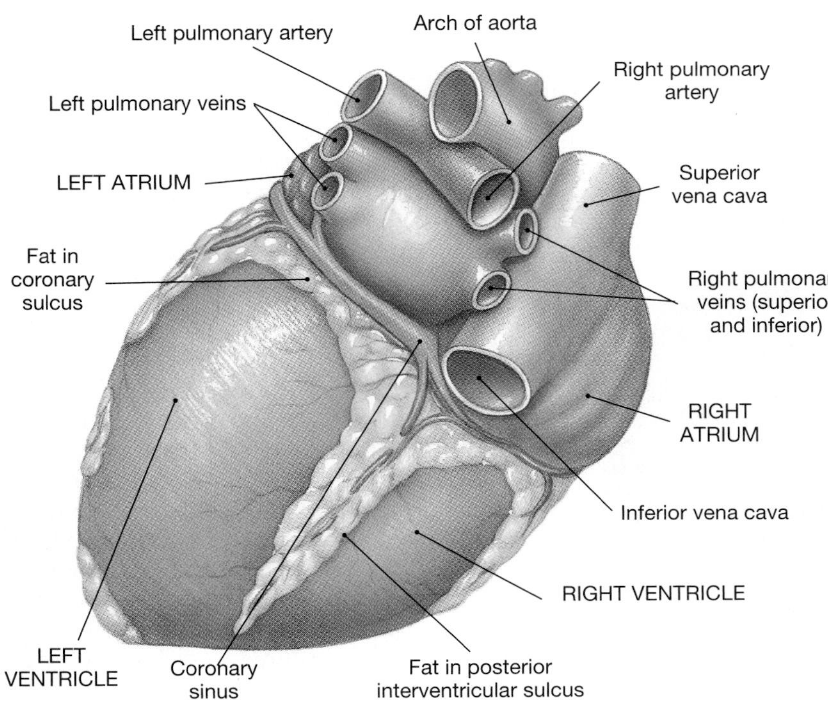
Left pulmonary artery
Arch of aorta
Right pulmonary artery
Left pulmonary veins
LEFT ATRIUM
Superior vena cava
Fat in coronary sulcus
Right pulmonary veins (superior and inferior)
RIGHT ATRIUM
Inferior vena cava
RIGHT VENTRICLE
LEFT VENTRICLE
Coronary sinus
Fat in posterior interventricular sulcus

(b) Posterior (diaphragmatic) surface

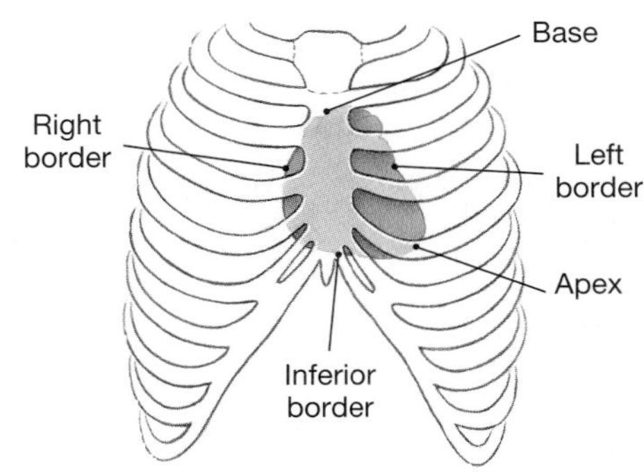
Base
Right border
Left border
Apex
Inferior border

(c) Position of the heart

● **FIGURE 20-3**
Superficial Anatomy of the Heart. **(a)** Anterior (sternocostal) view of the heart, showing major anatomical features. **(b)** The posterior (diaphragmatic) surface of the heart. (Coronary arteries are shown in red, coronary veins in blue.) **(c)** Anterior view of the chest, showing the position of the heart relative to the chest wall and the anatomical borders of the heart.

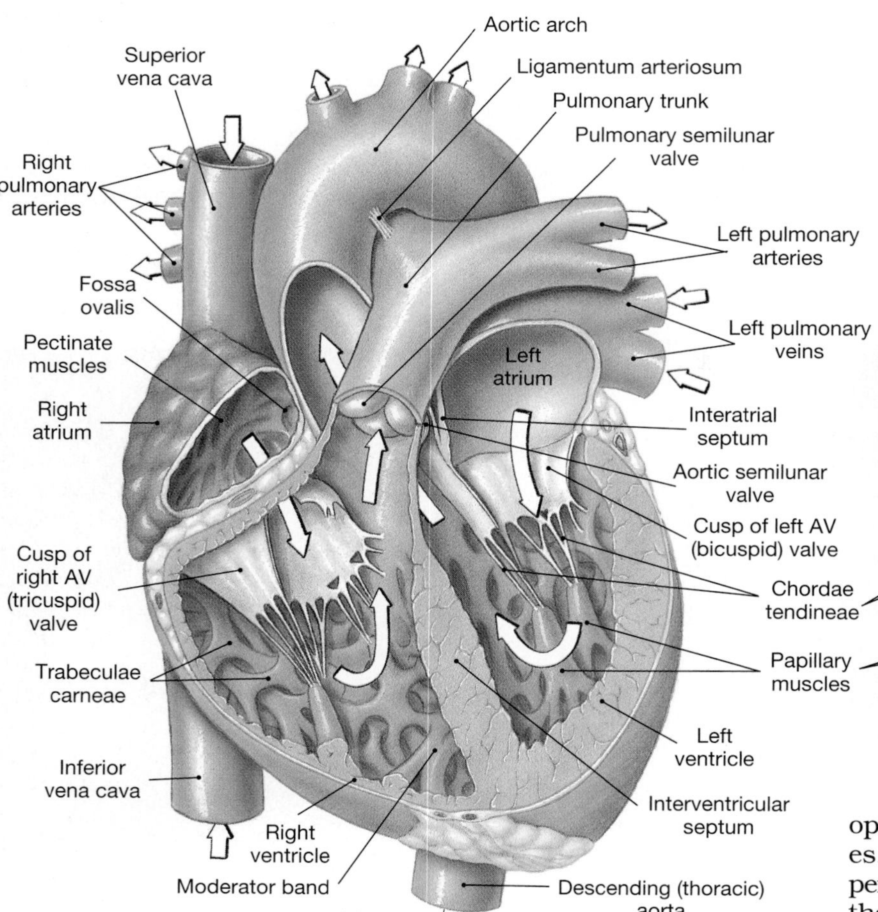

Aortic arch
Superior vena cava
Ligamentum arteriosum
Pulmonary trunk
Pulmonary semilunar valve
Right pulmonary arteries
Left pulmonary arteries
Fossa ovalis
Left pulmonary veins
Pectinate muscles
Left atrium
Right atrium
Interatrial septum
Aortic semilunar valve
Cusp of left AV (bicuspid) valve
Cusp of right AV (tricuspid) valve
Chordae tendineae
Trabeculae carneae
Papillary muscles
Left ventricle
Inferior vena cava
Interventricular septum
Right ventricle
Moderator band
Descending (thoracic) aorta
(a)

● FIGURE 20-4

Sectional Anatomy of the Heart. **(a)** A diagrammatic frontal section through the heart, showing major landmarks and the path of blood flow through the atria and ventricles. **(b)** Photograph of papillary muscles and chordae tendineae supporting the tricuspid (right AV) valve. The picture was taken inside the right ventricle, looking toward a light shining from the right atrium. **(c)** Sectional view of the heart.

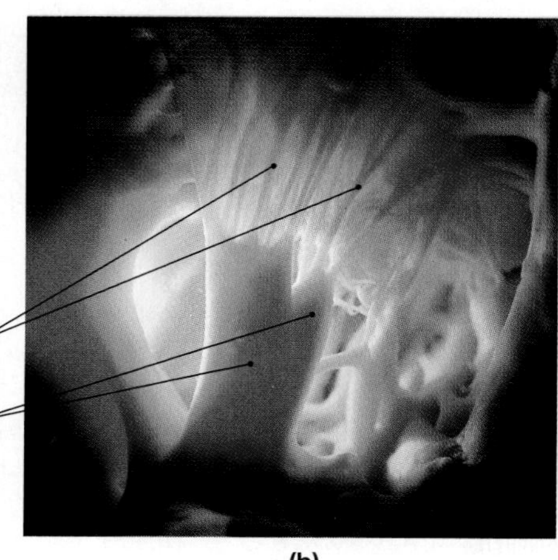

(b)

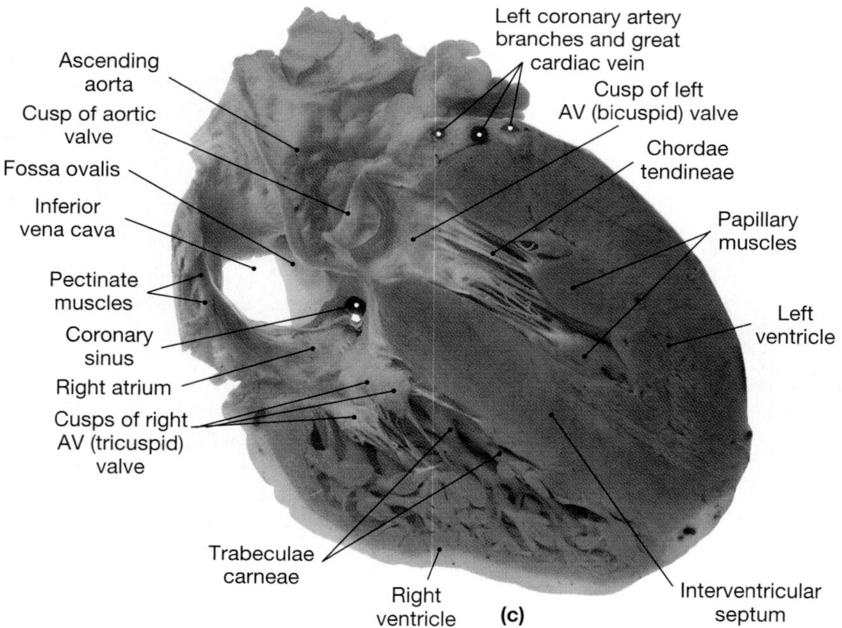

Ascending aorta
Left coronary artery branches and great cardiac vein
Cusp of aortic valve
Cusp of left AV (bicuspid) valve
Fossa ovalis
Chordae tendineae
Inferior vena cava
Papillary muscles
Pectinate muscles
Left ventricle
Coronary sinus
Right atrium
Cusps of right AV (tricuspid) valve
Trabeculae carneae
Right ventricle
(c)
Interventricular septum

oping. At birth the foramen ovale closes, and after 48 hours the opening is permanently sealed. A small depression, the **fossa ovalis,** persists at this site in the adult heart (Figure 20-4a,c●). Occasionally this sealing process does not occur, and the foramen ovale remains open, or *patent.* As a result, the circulation to the lungs is reduced, and blood oxygenation is low. The tissues of the newborn infant soon become starved for oxygen, and the *cyanosis* that develops makes the infant look blue. This is one factor that can produce a "blue baby." Other factors will be considered in Chapters 21 and 23.

The Right Ventricle

Figure 20-4b

Blood travels from the right atrium into the right ventricle through a broad opening bounded by three flaps of fibrous tissue. These flaps, or **cusps,** are part of the **right atrioventricular (AV) valve,** also known as the **tricuspid** (trī-KUS-pid; *tri,* three) **valve.** Each cusp is braced by tendinous connective tissue fibers, the **chordae tendineae** (KOR-dē TEN-di-nē-ē; tendinous cords), that originate at cone-shaped muscular projections, the **papillary** (PAP-i-ler-ē) **mus-**

cles, that arise from the inner surface of the right ventricle (Figure 20-4b●).

The internal surface of the ventricle also contains a series of muscular ridges, the **trabeculae carneae** (tra-BEK-ū-lē CAR-nē-ē; *carneus*, fleshy). The **moderator band** is a muscular ridge that extends horizontally from the inferior portion of the **interventricular septum,** a thick muscular partition that separates the right and left ventricles. It extends across the posterior wall of the right ventricle and connects to the anterior papillary muscle. The moderator band is variable in size in humans. It is noteworthy because it contains a portion of the *conducting system*, an internal network that coordinates the contractions of cardiac muscle fibers.

The superior end of the right ventricle tapers to a cone-shaped pouch, the **conus arteriosus,** which ends at the **pulmonary semilunar valve.** The pulmonary semilunar valve consists of three semilunar (half-moon–shaped) cusps of thick connective tissue. Blood flowing from the right ventricle passes through this valve to enter the **pulmonary trunk,** the start of the pulmonary circuit. Once within the pulmonary trunk, blood flows into the **left pulmonary arteries** and the **right pulmonary arteries.** These vessels branch repeatedly within the lungs before supplying the capillaries where gas exchange occurs.

The Left Atrium

Figure 20-4

From the respiratory capillaries blood collects into small veins that ultimately unite to form the four *pulmonary veins*. The posterior wall of the left atrium receives blood from two **left** and two **right pulmonary veins.** Like the right atrium, the left atrium has an auricle and a valve, the **left atrioventricular (AV) valve,** or **bicuspid** (bī-KUS-pid) **valve** (Figure 20-4a,c●). As the name *bicuspid* implies, the left AV valve contains a pair of cusps rather than a trio. Clinicians often use the term **mitral** (MĪ-tral; *mitre*, a bishop's hat) when referring to this valve. The left atrioventricular valve permits the flow of blood from the left atrium into the left ventricle.

The Left Ventricle

Figure 20-4

The left ventricle is the largest and thickest of the heart chambers. Its large size and thick myocardium enable it to pump blood to the entire body, whereas the right ventricle pumps blood only about 15 cm (6 in.) to reach the lungs. The internal organization of the left ventricle resembles that of the right ventricle (Figure 20-4a,c●). The trabeculae carneae are prominent, and a pair of large papillary muscles tense the chordae tendineae that brace the bicuspid valve.

Blood leaves the left ventricle by passing through the **aortic semilunar valve** into the **ascending aorta.** The aortic semilunar valve has the same construction as the pulmonary semilunar valve. Sac-like dilations of the base of the ascending aorta occur adjacent to each cusp. These sacs, called **aortic sinuses,** prevent the individual cusps from sticking to the wall of the aorta when the valve opens. The aortic semilunar valve prevents the backflow of blood into the left ventricle once it has been pumped out of the heart and into the systemic circuit. From the ascending aorta, blood flows on through the **aortic arch** and into the **descending aorta** (Figure 20-4a●).

Structural Differences between the Left and Right Ventricles

Figure 20-5

The function of an atrium is to collect blood returning to the heart and deliver it to the attached ventricle. The functional demands placed on the right and left atria are very similar, and the two chambers look almost identical. The demands placed on the right and left ventricles, however, are very different, and there are significant structural differences between the two.

Anatomical differences between the left and right ventricles are best seen in a three-dimensional view (Figure 20-5●). The lungs are close to the heart, and the pulmonary arteries and veins are relatively short and wide. Thus the right ventricle normally does not need to push very hard to propel blood through the pulmonary circuit. The wall of the right ventricle is relatively thin, and in sectional view it resembles a pouch attached to the massive wall of the left ventricle. When the right ventricle contracts it acts like a bellows pump, squeezing the blood against the mass of the left ventricle. This mechanism moves blood very efficiently with minimal effort, but it develops relatively low pressures.

A comparable pumping arrangement would not be suitable for the left ventricle, because six to seven times as much force must be exerted to push blood around the systemic circuit. The left ventricle has an extremely thick muscular wall, and it is round in cross section. When this ventricle contracts, two things happen: (1) The distance between the base and apex decreases, and (2) the diameter of the ventricular chamber decreases. If you imagine the effects of simultaneously squeezing and rolling up the end of a toothpaste tube, you will get the idea. The forces generated are quite powerful, more than enough to open the semilunar valve and eject blood into the ascending aorta. As the powerful left ventricle contracts, it also bulges into the right ventricular cavity. This dual action improves the efficiency of the right ventricle's efforts. Individuals whose right ventricular musculature has been severely damaged may continue to survive because of the extra push provided by the contraction of the left ventricle. AM *The Cardiomyopathies; Heart Transplants and Assist Devices*

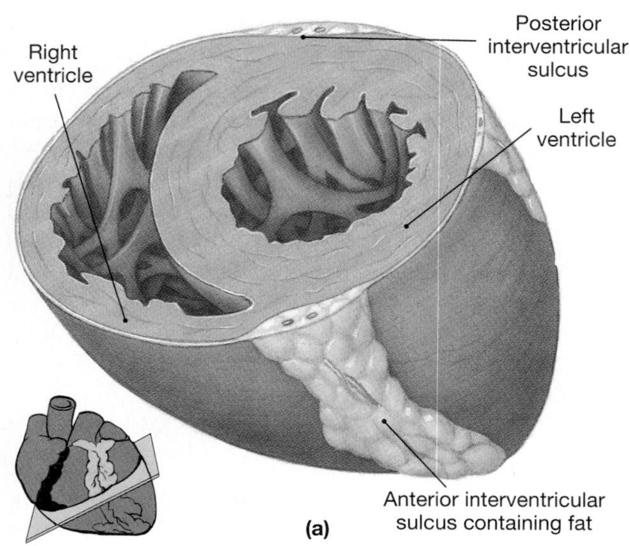
Right ventricle
Posterior interventricular sulcus
Left ventricle
Anterior interventricular sulcus containing fat
(a)

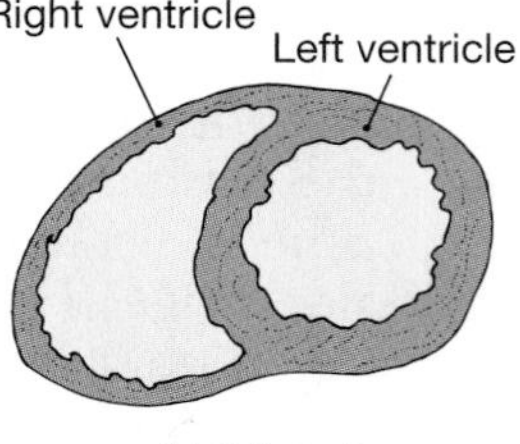
Right ventricle
Left ventricle
(b) Dilated (late diastole)

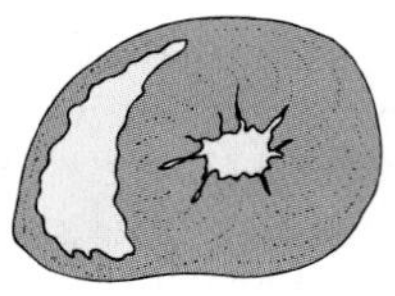
(c) Contracted (late systole)

● **FIGURE 20-5 Structural Differences between the Left and Right Ventricles. (a)** Detailed sectional view through the heart, showing the relative thicknesses of the two ventricles. Note the pouchlike shape of the right ventricle and the mass of the left ventricle. **(b)** Diagrammatic view of the relaxed ventricles just before a contraction, when they are filled with blood. **(c)** Comparable view of the contracted ventricles.

The Heart Valves

Figure 20-6

Details of the structure and function of the heart valves are shown in Figure 20-6a,b●.

THE ATRIOVENTRICULAR VALVES The atrioventricular valves prevent backflow of blood from the ventricles into the atria. The chordae tendineae and papillary muscles play an important role in the normal function of the AV valves. When a ventricle is filling with blood, the papillary muscles are relaxed and the AV valve offers no resistance to the flow of blood from atrium to ventricle (Figure 20-6a●). When the ventricle begins to contract, blood moving back toward the atrium swings the cusps together, closing the valve (Figure 20-6b●). During ventricular contraction, tension in the papillary muscles and chordae tendineae braces the cusps and keeps them from swinging into the atrium. This action prevents the backflow, or **regurgitation,** of blood into the atrium each time the ventricle contracts.

THE SEMILUNAR VALVES The pulmonary and aortic semilunar valves prevent backflow of blood from the pulmonary trunk and aorta into the right and left ventricles. The semilunar valves do not require muscular braces because the arterial walls do not contract, and the relative positions of the cusps are stable. When these valves close, the three symmetrical cusps support one another like the legs of a tripod (Figure 20-6a,c●).

VALVULAR HEART DISEASE Serious valve problems can interfere with cardiac function. If valve function deteriorates to the point that the heart cannot maintain adequate circulatory flow, symptoms of **valvular heart disease (VHD)** appear. Congenital malformations may be responsible, but often the condition develops after *carditis*, an inflammation of the heart. AM *Infection and Inflammation of the Heart*

One relatively common cause of carditis is *rheumatic* (roo-MA-tik) *fever*, an inflammatory condition that may develop after infection by streptococcal bacteria. Valve problems serious enough to affect cardiac function may not appear until 10–20 years after the initial infection, and the resulting clinical disorder is known as *rheumatic heart disease (RHD)*. AM *RHD and Valvular Stenosis*

❑ The Heart Wall

Figure 20-7

A section through the wall of the heart (Figure 20-7a●) reveals three distinct layers: (1) an outer *epicardium*, (2) a middle *myocardium*, and (3) an inner *endocardium*.

- The **epicardium** is the visceral pericardium that covers the outer surface of the heart. This serous membrane consists of an exposed mesothelium and an underlying layer of loose connective tissue that is attached to the myocardium.
- The **myocardium,** or muscular wall of the heart, forms both atria and ventricles. The myocardium contains cardiac muscle tissue and associated connective tissues, blood vessels, and nerves. The myocardium consists of concentric layers of cardiac muscle tissue. The atrial myocardium contains muscle bundles that wrap around the atria and form figure eights that pass through the interatrial septum. Superficial ventricular muscles wrap around both ventricles; deeper muscle layers spiral around and between the ventricles toward the apex (Figure 20-7b●).
- The inner surfaces of the heart, including the valves, are covered by a simple squamous epithelium, the **endocardium** (en-dō-KAR-dē-um; *endo-*, inside), that is continuous with the endothelium of the attached blood vessels.

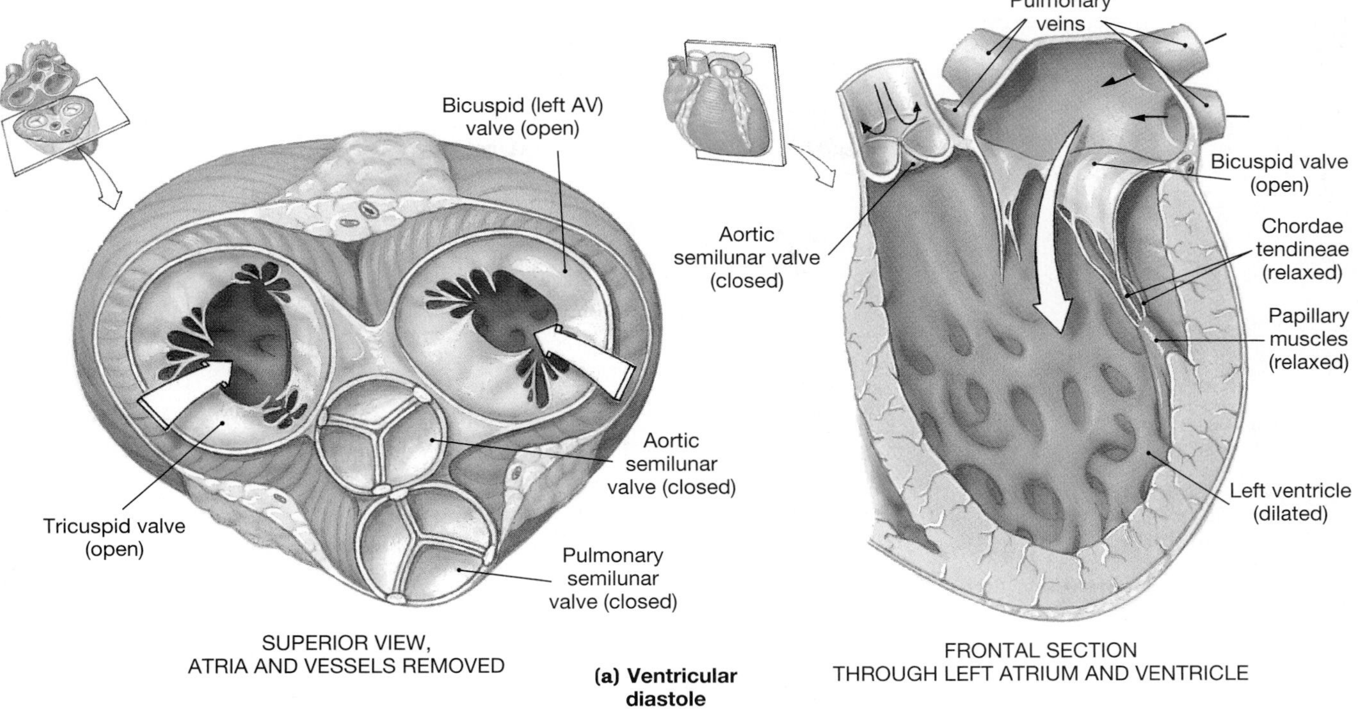
Bicuspid (left AV) valve (open)
Aortic semilunar valve (closed)
Tricuspid valve (open)
Pulmonary semilunar valve (closed)
SUPERIOR VIEW, ATRIA AND VESSELS REMOVED
Pulmonary veins
Aortic semilunar valve (closed)
Bicuspid valve (open)
Chordae tendineae (relaxed)
Papillary muscles (relaxed)
Left ventricle (dilated)
FRONTAL SECTION THROUGH LEFT ATRIUM AND VENTRICLE

(a) Ventricular diastole

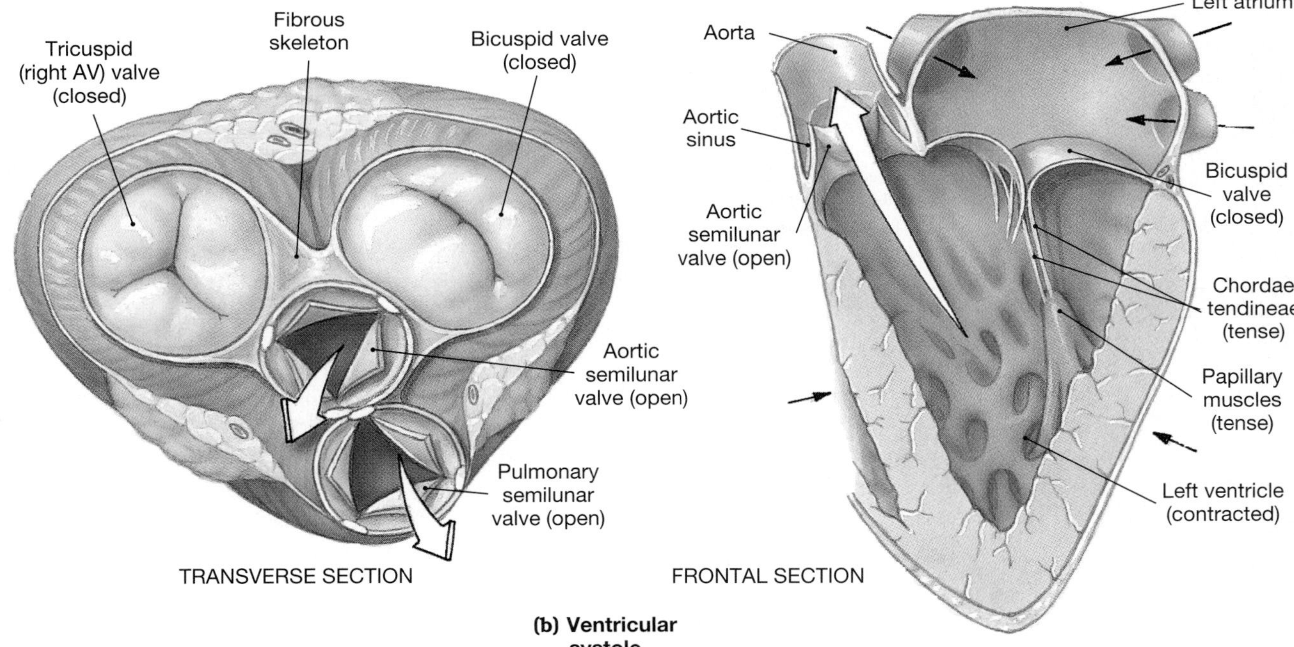
Tricuspid (right AV) valve (closed)
Fibrous skeleton
Bicuspid valve (closed)
Aortic semilunar valve (open)
Pulmonary semilunar valve (open)
TRANSVERSE SECTION
Aorta
Left atrium
Aortic sinus
Aortic semilunar valve (open)
Bicuspid valve (closed)
Chordae tendineae (tense)
Papillary muscles (tense)
Left ventricle (contracted)
FRONTAL SECTION

(b) Ventricular systole

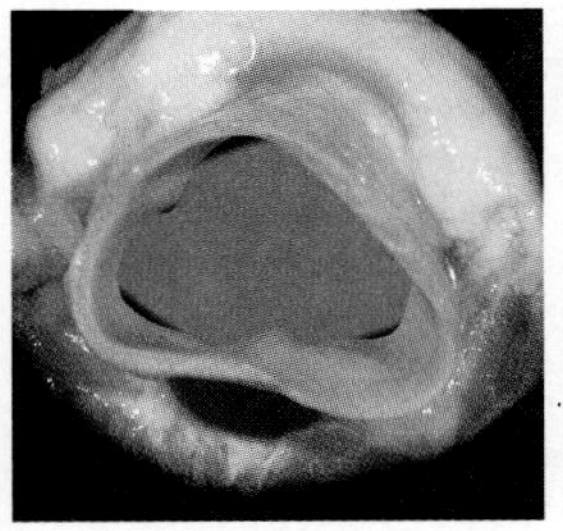

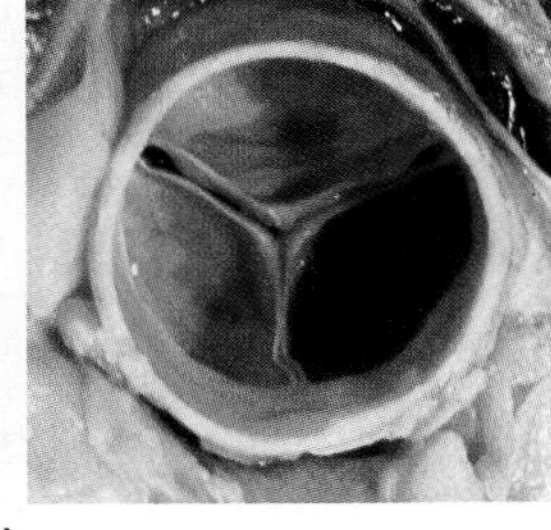

(c)

● **FIGURE 20-6**

Valves of the Heart. (a) Valve position during ventricular diastole (relaxation), when the AV valves are open and the semilunar valves are closed. Note that the chordae tendineae are slack and the papillary muscles are relaxed. **(b)** The appearance of the cardiac valves during ventricular systole (contraction), when the AV valves are closed and the semilunar valves are open. In the frontal section, note the attachment of the bicuspid valve to the chordae tendineae and papillary muscles. **(c)** The aortic semilunar valve in the open (left) and closed (right) positions. Note how the individual cusps brace one another in the closed position.

EMBRYOLOGY SUMMARY Development of the Heart

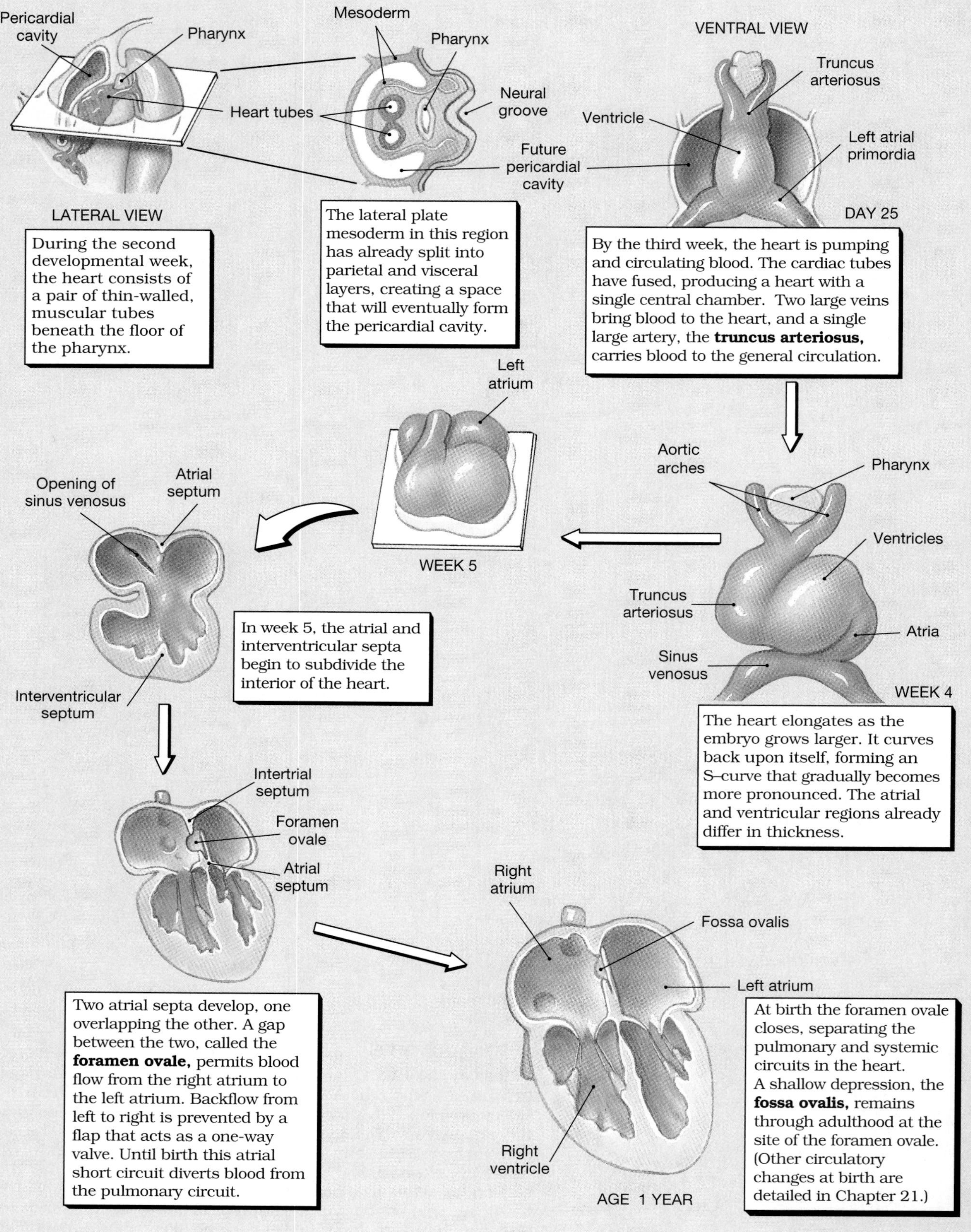
Pericardial cavity
Pharynx
Mesoderm
Pharynx
Heart tubes
Neural groove
Future pericardial cavity
LATERAL VIEW
VENTRAL VIEW
Truncus arteriosus
Ventricle
Left atrial primordia
DAY 25
Left atrium
Aortic arches
Pharynx
Opening of sinus venosus
Atrial septum
WEEK 5
Ventricles
Truncus arteriosus
Atria
Sinus venosus
WEEK 4
Interventricular septum
Intertrial septum
Foramen ovale
Atrial septum
Right atrium
Fossa ovalis
Left atrium
Right ventricle
AGE 1 YEAR

During the second developmental week, the heart consists of a pair of thin-walled, muscular tubes beneath the floor of the pharynx.

The lateral plate mesoderm in this region has already split into parietal and visceral layers, creating a space that will eventually form the pericardial cavity.

By the third week, the heart is pumping and circulating blood. The cardiac tubes have fused, producing a heart with a single central chamber. Two large veins bring blood to the heart, and a single large artery, the **truncus arteriosus,** carries blood to the general circulation.

The heart elongates as the embryo grows larger. It curves back upon itself, forming an S–curve that gradually becomes more pronounced. The atrial and ventricular regions already differ in thickness.

In week 5, the atrial and interventricular septa begin to subdivide the interior of the heart.

Two atrial septa develop, one overlapping the other. A gap between the two, called the **foramen ovale,** permits blood flow from the right atrium to the left atrium. Backflow from left to right is prevented by a flap that acts as a one-way valve. Until birth this atrial short circuit diverts blood from the pulmonary circuit.

At birth the foramen ovale closes, separating the pulmonary and systemic circuits in the heart.
A shallow depression, the **fossa ovalis,** remains through adulthood at the site of the foramen ovale. (Other circulatory changes at birth are detailed in Chapter 21.)

Cardiac Muscle Tissue

Figure 20-7

Cardiac muscle tissue was introduced in Chapter 4, and its properties were briefly compared with those of other muscle types in Chapter 10. [pp. 138, 315] Figure 20-7c,d and Table 20-1 provide a quick review of the structural and functional differences between cardiac muscle cells and skeletal muscle fibers that were noted in those discussions.

Connective Tissues and the Fibrous Skeleton

Figure 20-6

The connective tissues of the heart include large numbers of collagen and elastic fibers. Each cardiac muscle cell is wrapped in a strong but elastic sheath, and adjacent cells are tied together by fibrous cross-links, or "struts." These fibers are in turn interwo-

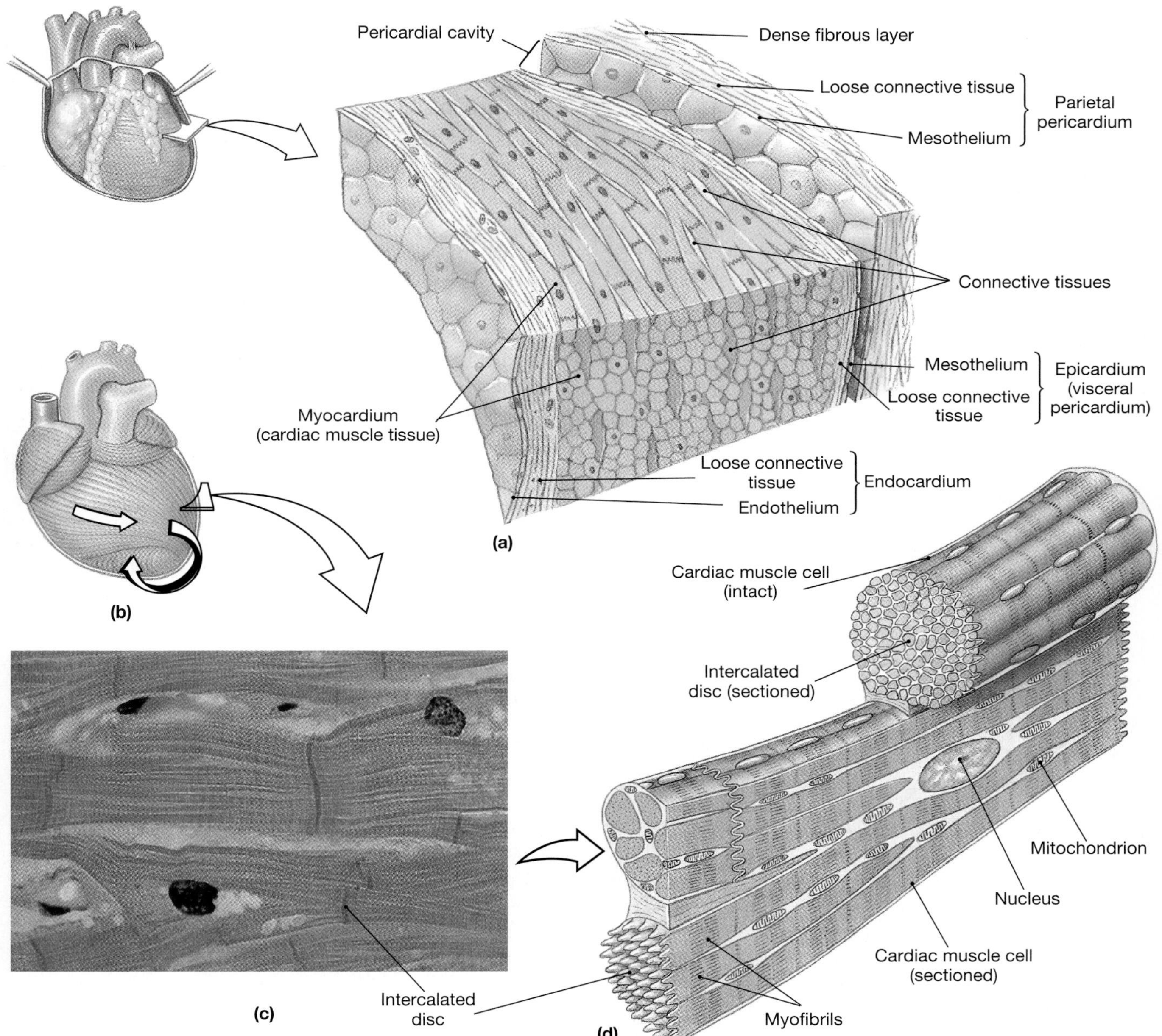
Pericardial cavity
Dense fibrous layer
Loose connective tissue
Mesothelium
Parietal pericardium
Connective tissues
Mesothelium
Loose connective tissue
Epicardium (visceral pericardium)
Myocardium (cardiac muscle tissue)
Loose connective tissue
Endothelium
Endocardium
(a)
(b)
Cardiac muscle cell (intact)
Intercalated disc (sectioned)
Mitochondrion
Nucleus
Cardiac muscle cell (sectioned)
Myofibrils
Intercalated disc
(c)
(d)

FIGURE 20-7

The Heart Wall. (a) A diagrammatic section through the heart wall showing the relative positions of the epicardium, myocardium, and endocardium. **(b)** Cardiac muscle tissue in the heart forms concentric layers that wrap around the atria and spiral within the walls of the ventricles. **(c,d)** Sectional and diagrammatic views of cardiac muscle tissue. Histological characteristics of cardiac muscle cells compared with those of skeletal muscle fibers include (1) small size, (2) single, centrally placed nucleus, (3) branching interconnections between cells, and (4) presence of intercalated discs. (LM × 575)

ven into sheets that separate the superficial and deep muscle layers. These connective tissue fibers (1) provide physical support for the cardiac muscle fibers, blood vessels, and nerves of the myocardium; (2) help distribute the forces of contraction; (3) add strength and prevent overexpansion of the heart; and (4) provide elasticity that helps return the heart to its original size and shape after a contraction.

The **fibrous skeleton** of the heart consists of four dense bands of fibroelastic tissue that encircle the bases of the pulmonary trunk and aorta and the valves of the heart (Figure 20-6•, p. 689). These bands stabilize the positions of the heart valves and ventricular muscle cells and physically isolate the ventricular cells from those in the atria.

❑ The Blood Supply to the Heart

Figure 20-8

The heart works continuously, and cardiac muscle cells require reliable supplies of oxygen and nutrients. The **coronary circulation** supplies blood to the muscles of the heart. During maximum exertion the oxygen demand rises considerably, and the blood flow to the heart may increase to nine times that of resting levels. The coronary circulation (Figure 20-8•) includes an extensive network of coronary blood vessels.

The Coronary Arteries

Figure 20-8

The left and right **coronary arteries** originate at the base of the ascending aorta (Figure 20-8a,b•). Blood pressure here is the highest found anywhere in the systemic circuit, and this pressure ensures a continuous flow of blood to meet the demands of active cardiac muscle tissue.

THE RIGHT CORONARY ARTERY The **right coronary artery,** which follows the coronary sulcus around the heart, supplies blood to (1) the right atrium; (2) portions of both ventricles; and (3) portions of the conducting system of the heart, including the *SA* (sinoatrial) and *AV nodes.* The cells of the SA node and AV node are essential to establishing the normal heart rate. Their functions and the regulation of the heart rate will be the focus of a later section. Near the right border of the heart the right coronary artery usually gives rise to one or more **marginal branches** that extend across the ventricular surface (Figure 20-8•). It then continues across the posterior surface of the heart, supplying the **posterior interventricular branch,** or *posterior descending artery,* which runs toward the apex within the posterior interventricular sulcus. The posterior interventricular branch supplies blood to the interventricular septum and adjacent portions of the ventricles.

THE LEFT CORONARY ARTERY The **left coronary artery** supplies blood to the left ventricle, left atrium, and the interventricular septum. As it reaches the anterior surface of the heart it gives rise to a *circumflex branch* and an *anterior interventricular branch* (Figure 20-8a,b•). The **circumflex branch** curves to the left around the coronary sulcus, eventually meeting and fusing with small branches of the right coronary artery. The much larger **anterior interventricular branch,** or *left anterior descending artery,* swings around the pulmonary trunk and runs along the anterior surface within the anterior interventricular sulcus. This branch supplies small tributaries continuous with those of the posterior interventricular branch of the right coronary artery. Such interconnections between arteries are called **anastomoses** (a-nas-tō-MŌ-sēz; *anastomosis,* outlet). Because the arteries are interconnected in this way, the blood supply to the cardiac muscle remains relatively constant, regardless of pressure fluctuations within the left and right coronary arteries.

THE CARDIAC VEINS The **great** and **middle cardiac veins** carry blood from the coronary capillaries to the **coronary sinus,** a large, thin-walled vein that lies in the posterior portion of the coronary sulcus (Figure 20-8c,d•). The coronary sinus communicates with the right atrium near the base of the inferior vena cava.

✔ Damage to the semilunar valves on the right side of the heart would interfere with blood flow to what vessel?

✔ What prevents the AV valves from opening back into the atria?

✔ Why are the left atrium and ventricle more muscular than the right atrium and ventricle?

❑ Innervation of the Heart

Figure 20-9

The sympathetic and parasympathetic divisions of the ANS provide innervation to the heart via the *cardiac plexus.* [p. 541] Postganglionic sympathetic neurons are found in the cervical and upper thoracic ganglia. The vagus nerve carries parasympathetic preganglionic fibers to small ganglia in the cardiac plexus. Both ANS divisions innervate the SA and AV nodes as well as the atrial and ventricular cardiac muscle cells, as indicated in Figure 20-9•.

The *cardiac centers* of the medulla oblongata contain the autonomic headquarters for cardiac control. (These centers were discussed in Chapter 14.) [p. 483] Stimulation of the **cardioacceleratory center** activates the necessary sympathetic neurons;

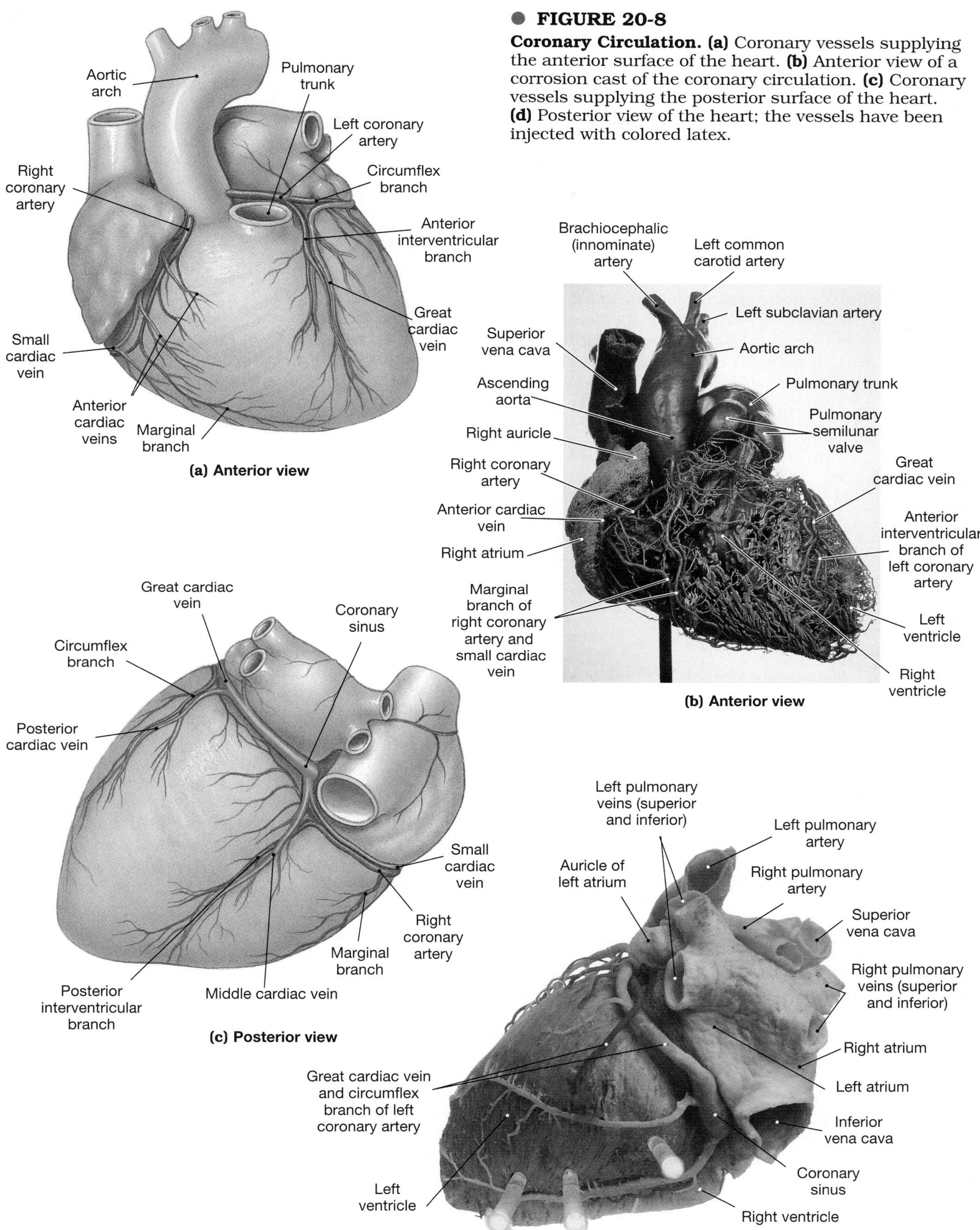

Aortic arch
Pulmonary trunk
Left coronary artery
Circumflex branch
Right coronary artery
Anterior interventricular branch
Great cardiac vein
Small cardiac vein
Anterior cardiac veins
Marginal branch
(a) Anterior view
Brachiocephalic (innominate) artery
Left common carotid artery
Left subclavian artery
Superior vena cava
Aortic arch
Ascending aorta
Pulmonary trunk
Pulmonary semilunar valve
Right auricle
Right coronary artery
Great cardiac vein
Anterior cardiac vein
Anterior interventricular branch of left coronary artery
Right atrium
Marginal branch of right coronary artery and small cardiac vein
Left ventricle
Right ventricle
(b) Anterior view
Great cardiac vein
Coronary sinus
Circumflex branch
Posterior cardiac vein
Small cardiac vein
Right coronary artery
Marginal branch
Posterior interventricular branch
Middle cardiac vein
(c) Posterior view
Left pulmonary veins (superior and inferior)
Left pulmonary artery
Auricle of left atrium
Right pulmonary artery
Superior vena cava
Right pulmonary veins (superior and inferior)
Right atrium
Left atrium
Great cardiac vein and circumflex branch of left coronary artery
Inferior vena cava
Coronary sinus
Left ventricle
Right ventricle
(d) Posterior view

FIGURE 20-8

Coronary Circulation. (a) Coronary vessels supplying the anterior surface of the heart. **(b)** Anterior view of a corrosion cast of the coronary circulation. **(c)** Coronary vessels supplying the posterior surface of the heart. **(d)** Posterior view of the heart; the vessels have been injected with colored latex.

the nearby **cardioinhibitory center** governs the activities of the parasympathetic neurons. The cardiac centers receive input from higher centers, especially from the parasympathetic and sympathetic headquarters in the hypothalamus. Information concerning the status of the cardiovascular system arrives via visceral sensory fibers accompanying the vagus nerve and the sympathetic nerves of the cardiac plexus.

The cardiac centers monitor baroreceptors and chemoreceptors innervated by the glossopharyngeal and vagus nerves (N IX and X). ∞ *[p. 488]* On the basis of this information, they adjust cardiac performance to maintain adequate circulation to vital organs, such as the brain. These centers respond to changes in blood pressure and in the arterial concentrations of dissolved oxygen and carbon dioxide. For example, a decline in blood pressure or oxygen concentrations or an increase in carbon dioxide levels usually indicates that the heart must work harder to meet the demands of peripheral tissues. The cardiac centers then call for an increase in cardiac activity. These reflexes and their effects on the heart and peripheral vessels will be detailed in Chapter 21.

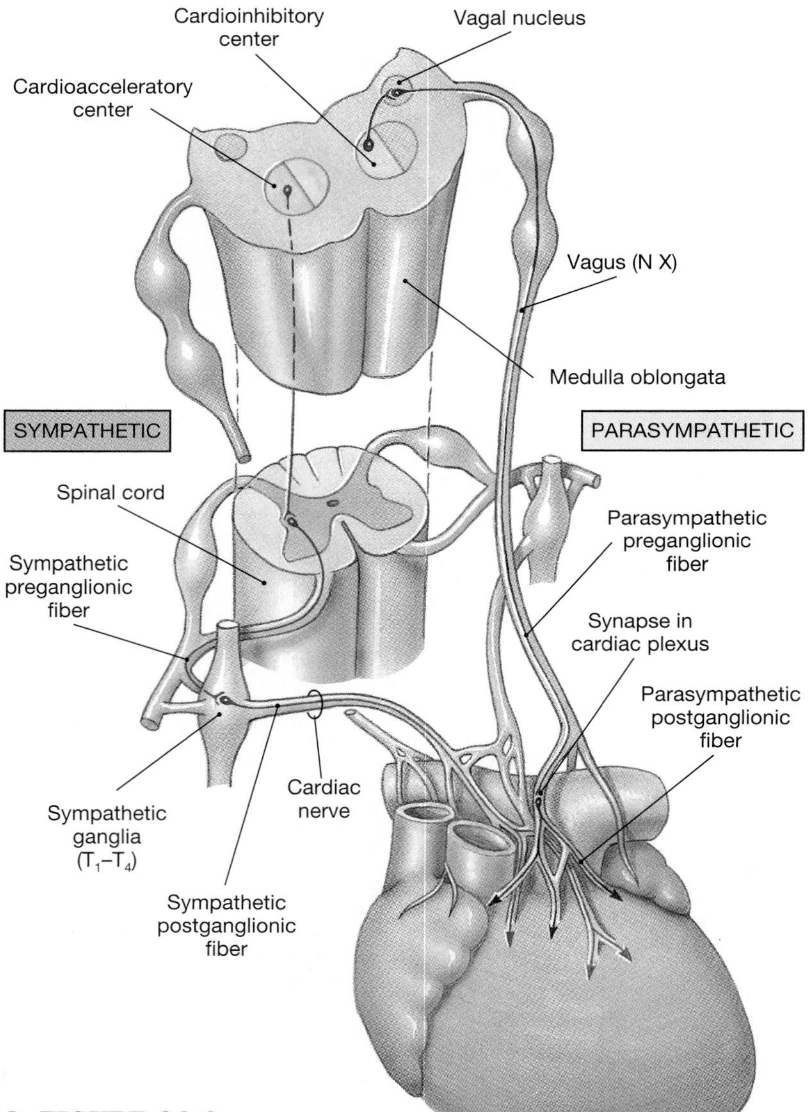

● **FIGURE 20-9**
Autonomic Innervation of the Heart

The Heartbeat

The myocardium consists of cardiac muscle tissue. Each time the heart beats, the contractions of individual cardiac muscle cells are coordinated and harnessed to ensure that blood flows in the right direction at the proper time. Two types of cardiac muscle cells are involved in a normal heartbeat. *Contractile cells* produce the powerful contractions that propel blood, while specialized muscle cells of the *conducting system* control and coordinate the activities of the contractile cells.

❑ Contractile Cells

Contractile cells form the bulk of the atrial and ventricular walls. Earlier discussions have considered the structure of contractile cells, which account for roughly 99 percent of the muscle cells in the heart. In both cardiac muscle cells and skeletal muscle fibers (1) an action potential leads to the appearance of calcium ions among the myofibrils, and (2) the binding of calcium ions to troponin on the thin filaments initiates the contraction (see Table 20-1). But skeletal and cardiac muscle cells differ in terms of the nature of the action potential, the source of the calcium ions, and the duration of the resulting contraction. The structural and functional differences between these muscle types were noted in Chapter 10. ∞ *[pp. 315-320]*

The Action Potential in Cardiac Muscle Cells

Figure 20-10

The basic appearance of action potentials in skeletal muscle fibers and cardiac muscle cells was compared in Chapter 12 and illustrated in Figure 12-19●, p. 406. We will now take a closer look at the origin and conduction of an action potential in a cardiac muscle fiber.

The resting potential of a cardiac muscle cell in one of the ventricles is approximately –90 mV, comparable to that of a resting skeletal muscle fiber. However, an action potential in a ventricular muscle cell (–85 mV), lasts 250–300 msec (Figure 20-10a,b●), 30 times as long as a typical action potential in a skeletal muscle fiber.

An action potential begins when the membrane of the ventricular muscle cell is brought to threshold, usually around –75 mV. This normally occurs in a portion of the membrane next to an intercalated disc. The usual stimulus is the excitation of an adjacent muscle fiber. Once threshold has been reached, the action potential proceeds in three basic steps.

Step 1: ***Rapid depolarization.*** The stage of rapid depolarization in a cardiac muscle cell resembles that of a skeletal muscle fiber. At threshold, voltage-regulated sodium channels open, and the membrane suddenly becomes permeable to sodium ions. The result is rapid depolarization of the sarcolemma. The channels involved are called **fast channels** because they open quickly and remain open for only a few milliseconds.

Step 2: ***The plateau.*** As the transmembrane potential approaches +30 mV, the voltage-regulated sodium channels close. They will remain closed and inactivated until the membrane potential reaches –60 mV. As the sodium channels are closing, voltage-regulated calcium channels are opening. The calcium channels are called **slow channels** because they open slowly and remain open for a relatively extended period—roughly 175 msec. As a result, the transmembrane potential remains near 0 mV for an extended period. This portion of the action potential curve is called the *plateau. The presence of a plateau is the major difference between action potentials in cardiac muscle cells and skeletal muscle fibers.* In a skeletal muscle fiber, rapid depolarization is immediately followed by a period of rapid repolarization.

Step 3: ***Repolarization.*** As the plateau continues, slow calcium channels begin closing and **slow potassium channels** begin opening. The result is a period of rapid repolarization that restores the resting potential.

THE REFRACTORY PERIOD After an action potential begins, the *refractory period* is the period during which the membrane will not respond normally to a second stimulus. In the *absolute refractory period,* the membrane cannot respond at all, because the sodium channels are either already

TABLE 20-1 Structural and Functional Differences between Cardiac Muscle Cells and Skeletal Muscle Fibers

Feature	*Cardiac Muscle Cells*	*Skeletal Muscle Fibers*
Size	10–20 μm × 50–100 μm	100 μm × up to 40 cm
Nuclei	Typically 1 (rarely 2–5)	Multiple (hundreds)
Contractile proteins	Sarcomeres along myofibrils	Sarcomeres along myofibrils
Internal membranes	Short T tubules; no triads formed with sarcoplasmic reticulum	Long T tubules form triads with cisternae of sarcoplasmic reticulum
Mitochondria	Abundant (25% of cell volume)	Relatively scarce
Inclusions	Myoglobin, lipids, glycogen	Extensive glycogen reserves
Circulatory supply	Extensive	Relatively sparse
Metabolism (resting)	Aerobic	Aerobic
Metabolism (active)	Aerobic, using any available substrate	Anaerobic, breakdown of glycogen reserves
Contractions	Twitches with brief relaxation periods; long refractory period prevents tetanic contractions	Usually sustained tetanic contractions
Stimulus for contraction	Autorhythmicity of pacemaker cells generates action potentials	Activity of somatic motor neuron generates action potentials
Trigger for contraction	Calcium entry from ECF and calcium release from sarcoplasmic reticulum	Calcium release from sarcoplasmic reticulum
Intercellular connections	Branching network with cell membranes locked together at intercalated discs	Adjacent fibers tied together by connective tissue fibers

open or closed and inactivated. In a ventricular muscle cell the absolute refractory period lasts approximately 200 msec, extending for the duration of the plateau and the initial period of rapid repolarization.

The absolute refractory period is followed by a shorter (50 msec) *relative refractory period.* During this period the voltage-regulated sodium channels are closed but capable of opening, and the membrane will respond to a stronger-than-normal stimulus by initiating another action potential.

Calcium Ions and Cardiac Contractions

Figure 20-10c

The appearance of an action potential in the cardiac muscle cell membrane produces a contraction by causing an increase in the concentration of calcium ions around the myofibrils. This process occurs in two steps:

1. Calcium ions entering the cell membrane during the plateau phase of the action potential provide roughly 20 percent of the calcium ions required for a contraction.
2. The arrival of extracellular calcium ions is the trigger for the release of additional calcium ions from reserves in the sarcoplasmic reticulum (SR).

Extracellular calcium ions thus have direct and indirect effects on cardiac muscle cell contraction. For this reason cardiac muscle tissue is very sensitive to changes in the calcium ion concentration of the extracellular fluid.

In a skeletal muscle fiber the action potential is relatively brief, and ends as the related twitch contraction begins (Figure 20-10c●). The twitch contraction is short and ends as the SR reclaims the calcium ions it released. In a cardiac muscle cell, as we have seen, the action potential is prolonged, and calcium ions continue to enter the cell throughout the plateau. As a result, the period of active muscle fiber contraction continues until the plateau ends. As the slow calcium channels close, the intracellular calcium ions are absorbed by the SR or pumped out of the cell, and the muscle cell relaxes.

In skeletal muscle fibers, the refractory period ends before the muscle fiber develops peak tension. As a result, twitches can summate, and tetanus can occur. In cardiac muscle cells the absolute refractory period continues until relaxation is under way. Because summation is not possible, tetanic contractions cannot occur in a normal cardiac muscle

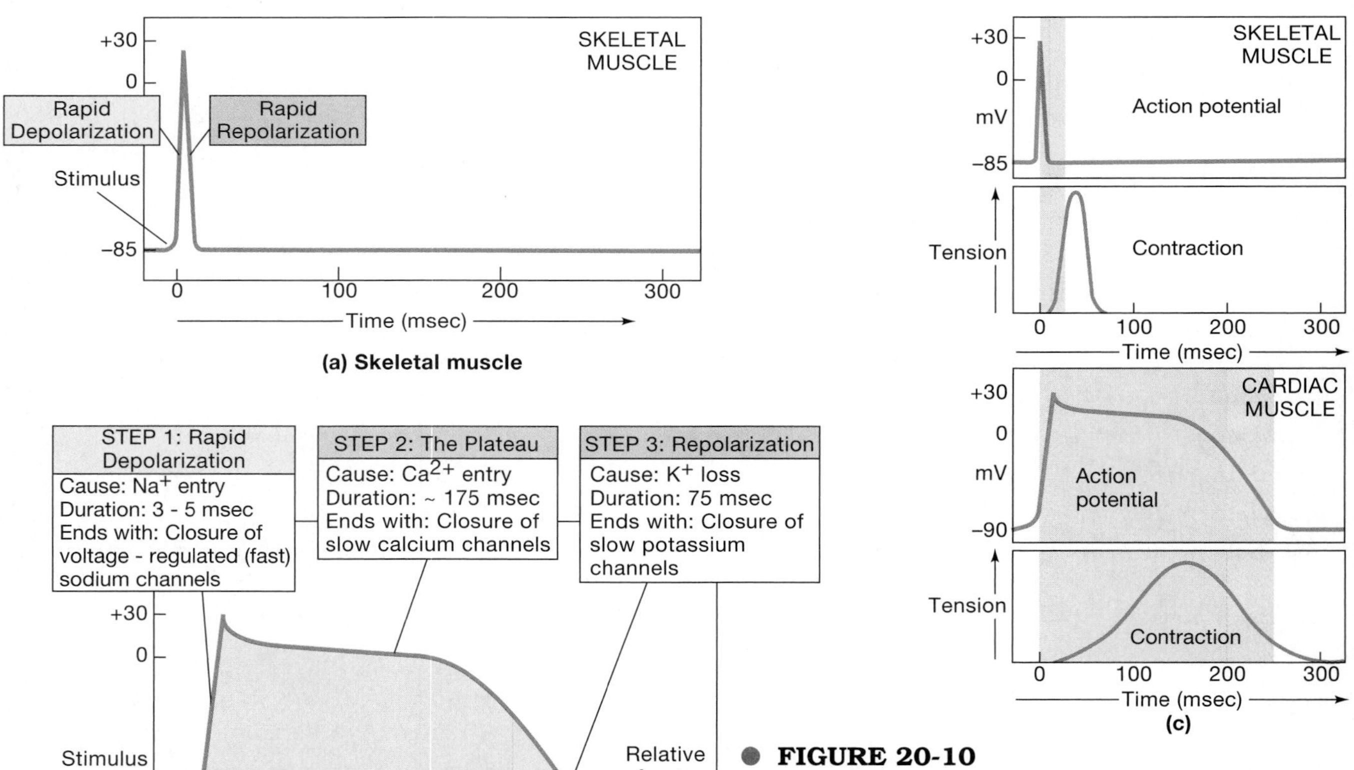
+30
0
-85
SKELETAL MUSCLE
Rapid Depolarization
Rapid Repolarization
Stimulus
0
100
200
300
Time (msec)
(a) Skeletal muscle
STEP 1: Rapid Depolarization
Cause: Na+ entry
Duration: 3 - 5 msec
Ends with: Closure of voltage - regulated (fast) sodium channels
STEP 2: The Plateau
Cause: Ca2+ entry
Duration: ~ 175 msec
Ends with: Closure of slow calcium channels
STEP 3: Repolarization
Cause: K+ loss
Duration: 75 msec
Ends with: Closure of slow potassium channels
+30
0
-90
Stimulus
Absolute refractory period
Relative refractory period
Time (msec)
(b) Cardiac muscle
SKELETAL MUSCLE
Action potential
mV
Tension
Contraction
CARDIAC MUSCLE
Action potential
Contraction
(c)

● **FIGURE 20-10**
The Action Potential in Skeletal and Cardiac Muscle. (a) Action potential in a representative skeletal muscle fiber; for more details see Figure 10-10, p. 300. **(b)** Action potential in a ventricular muscle cell. **(c)** The relationship between an action potential and a twitch contraction in skeletal and cardiac muscle.

cell, regardless of the frequency or intensity of stimulation. This feature is absolutely vital, since a heart in tetany could not pump blood. With a single twitch lasting 250 msec or more, a normal cardiac muscle cell can increase its rate of contraction to 300–400 per minute under maximum stimulation. This rate is not reached in a normal heart, because of limitations imposed by the conducting system.

Coronary Artery Disease

The term **coronary artery disease (CAD)** refers to degenerative changes in the coronary circulation. Cardiac muscle fibers need a constant supply of oxygen and nutrients, and any reduction in coronary circulation produces a corresponding reduction in cardiac performance. Such reduced circulatory supply, known as **coronary ischemia** (is-KĒ-mē-a), usually results from partial or complete blockage of the coronary arteries. The usual cause is the formation of a fatty deposit, or *plaque,* in the wall of a coronary vessel. The plaque, or an associated thrombus, then narrows the passageway and reduces blood flow. Spasms in the smooth muscles of the vessel wall can further decrease blood flow or even stop it altogether. Plaque development and growth will be considered in Chapter 21.

One of the first symptoms of CAD is often **angina pectoris** (an-JĪ-na PEK-tōr-is; *angina,* pain spasm + *pectoris,* of the chest). In the most common form of angina, temporary insufficiency and ischemia develop when the workload of the heart increases. Although the individual may feel comfortable at rest, any unusual exertion or emotional stress can produce a sensation of pressure, chest constriction, and pain that may radiate from the sternal area to the arms, back, and neck.

Angina can often be controlled by a combination of drug treatment and changes in lifestyle. Lifestyle changes to combat angina include (1) limiting activities known to trigger angina attacks, such as strenuous exercise, and avoiding stressful situations; (2) stopping smoking; and (3) modifying the diet to lower fat consumption. Medications useful for controlling angina include drugs that block sympathetic stimulation (*propranolol* or *metoprolol*); vasodilators such as *nitroglycerin* (nī-trō-GLIS-er-in) and *atrial natriuretic peptide* (*ANP*); and drugs that block calcium movement into the cardiac muscle cells (*calcium channel blockers*).

Angina can also be treated surgically. A single, soft plaque may be reduced with the aid of a slender, elongate **catheter** (KATH-e-ter). The catheter, a small-diameter tube, is inserted into a large artery and guided into a coronary artery to the plaque. A variety of surgical tools can be slid into the catheter, and the plaque can then be removed with laser beams or chewed to pieces by a miniature version of the Roto-Rooter machine. Debris created during plaque destruction is sucked up by the catheter, preventing blockage of smaller vessels.

In **balloon angioplasty** (AN-jē-ō-plas-tē; *angeion,* vessel) the catheter tip contains an inflatable balloon. Once in position, the balloon is inflated, pressing the plaque against the vessel walls. This procedure works best in small (under 10 mm) soft plaques. Several factors make this a highly attractive treatment: (1) The mortality rate during surgery is only around 1 percent; (2) the success rate is over 90 percent; and (3) it can be performed on an outpatient basis. Although in about 20 percent of patients the plaque deposit returns to its original size within 6 months, the process can be repeated as needed. Unfortunately, only about 10 percent of severe angina patients have isolated problems suitable for balloon angioplasty.

A **coronary artery bypass graft (CABG)** involves taking a small section from either a small artery (often the *internal thoracic artery*) or a peripheral vein (such as the *great saphenous vein* of the leg) and using it to create a detour around the obstructed portion of a coronary artery. As many as four coronary arteries can be rerouted this way during a single operation. The procedures are named according to the number of vessels repaired, so one speaks of *single, double, triple,* or *quadruple coronary bypass operations.* The mortality rate during surgery for operations performed before significant heart damage has occurred is relatively low (1–2 percent). Under these conditions the procedure completely eliminates the angina symptoms in 70 percent of the cases and provides partial relief in another 20 percent.

Although it does offer certain advantages, recent studies have shown that for mild angina, coronary bypass surgery does not yield significantly better results than drug therapy. Current recommendations are that coronary bypass surgery be reserved for cases of severe angina that do not respond to other treatment.

❑ The Conducting System

Figure 20-11

In contrast to skeletal muscle, cardiac muscle tissue contracts on its own, in the absence of neural or hormonal stimulation. This property is called *automaticity*, or *autorhythmicity*. The cells responsible for initiating and distributing the stimulus to contract are part of the **conducting system** of the heart. The conducting system is a network of specialized cardiac muscle cells that initiates and distributes electrical impulses. The actual contraction lags behind the passage of an electrical impulse, as excitation-contraction coupling occurs and cross-bridge interactions take place. The basic mechanical process of contraction in a cardiac muscle cell resembles that already described for skeletal muscle fibers in Chapter 10. ∞ [p. 296]

The conducting system (Figure 20-11a●) includes:

- The *sinoatrial (SA) node* in the wall of the right atrium.
- The *atrioventricular (AV) node* at the junction between the atria and ventricles.
- *Internodal pathways* that link the SA node to the AV node, and *conducting cells* that connect the AV node to the ventricular muscle cells. The conducting cells include those in the *AV bundle*, the *bundle branches*, and the *Purkinje fibers* that distribute the stimulus to the ventricular myocardium.

Most of the cells of the conducting system are smaller than the contractile cells of the myocardium and contain very few myofibrils. These cells share another important characteristic: They cannot maintain a stable resting potential. Each time repolarization occurs, the membrane again gradually drifts toward threshold. This gradual depolarization is called a **prepotential.**

The rate of spontaneous depolarization varies in different portions of the conducting system. It is fastest at the SA node, which in the absence of neural or hormonal stimulation will generate action potentials at a rate of 80–100 per minute (Figure 20-11b●). Isolated cells of the AV node depolarize more slowly, generating action potentials at a rate of 40–60 per minute. Cells of the AV bundle, the bundle branches, and Purkinje fibers depolarize even more slowly.

In the normal heart, the cells of the conducting system are interconnected, and an action potential appearing at any one location will be conducted across the entire network. Because the SA node reaches threshold first, it establishes the heart rate, so the SA node is known as the **cardiac pacemaker.**

We will now trace the path of an impulse from its initiation at the SA node and consider its effects on the surrounding myocardium.

The Sinoatrial (SA) Node

Figures 20-11, 20-12

The **sinoatrial** (sī-nō-Ā-trē-al) **node (SA node)** is embedded in the posterior wall of the right atrium,

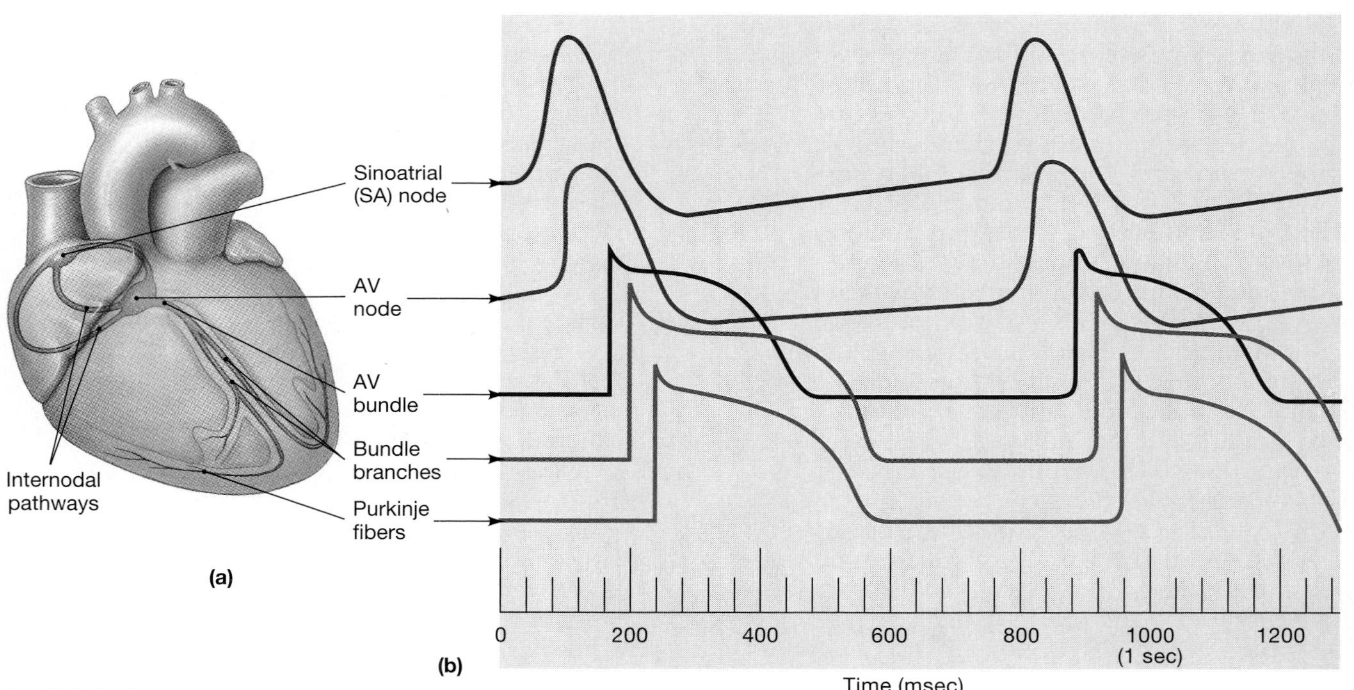
Sinoatrial (SA) node
AV node
AV bundle
Bundle branches
Purkinje fibers
Internodal pathways
(a)
(b)
0
200
400
600
800
1000 (1 sec)
1200
Time (msec)

● **FIGURE 20-11**
The Conducting System of the Heart. (a) Components of the conducting system. **(b)** This graph follows the spontaneous changes in the membrane potentials of nodal cells and the resulting action potentials in conducting cells. The rate of spontaneous depolarization is highest at the SA node; because these components are interconnected in the intact heart, the SA node is the cardiac pacemaker.

near the entrance of the superior vena cava (Figures 20-11a and 20-12[1]●). The SA node is also known as the *cardiac pacemaker* or the *natural pacemaker.* It is connected to the larger AV node via internodal pathways in the atrial walls. It takes roughly 50 msec for an action potential to travel from the SA node to the AV node over these pathways. Along the way, the conducting fibers pass the contractile stimulus to cardiac muscle cells of the right and left atria. The action potential then spreads across the atrial surfaces through cell-to-cell contact (Figure 20-12[2]●). The stimulus affects only the atria, because the fibrous skeleton separates the atrial myocardium from the ventricular myocardium.

The Atrioventricular (AV) Node

Figure 20-12

The relatively large **atrioventricular (AV) node** sits within the floor of the right atrium near the opening of the coronary sinus. The rate of conduction slows as the impulse travels from the internodal pathways to the AV node, because the nodal fibers are smaller in diameter. In addition, the connections between nodal cells are less efficient than those between conducting cells at conducting the impulse from one cell to another. As a result, it takes about 100 msec for the impulse to pass through the AV node and into the AV bundle (Figure 20-12(3)●).

The conduction delay at the AV node is important because the atria must contract before the ventricles. If all the chambers contracted at the same time, blood could not flow through the heart because the AV valves would prevent blood flow between the atria and ventricles. Because there is a delay at the AV node, the atrial myocardium completes its depolarization before the impulse reaches the ventricles.

The fibers of the AV node can conduct impulses at a maximum rate of 230 per minute. This value sets the upper limit for the normal heart rate, even though the SA node may generate impulses at a faster rate. The rate of ventricular contractions is therefore kept below the theoretical limit of 300–400 bpm (beats per minute). This limitation is important because mechanical factors, discussed below, decrease the pumping efficiency of the heart at rates above approximately 180 bpm.

The AV Bundle, Bundle Branches, and Purkinje Fibers

The connection between the AV node and the **AV bundle,** or *bundle of His,* is the only electrical connection between the atria and the ventricles. Once the impulse enters the AV bundle it travels to the interventricular septum and enters the **right** and **left bundle branches.** The left bundle branch, which supplies the relatively massive left ventricle, is much larger than the right bundle branch. Both bundle branches extend toward the apex of the heart and then turn and fan out beneath the

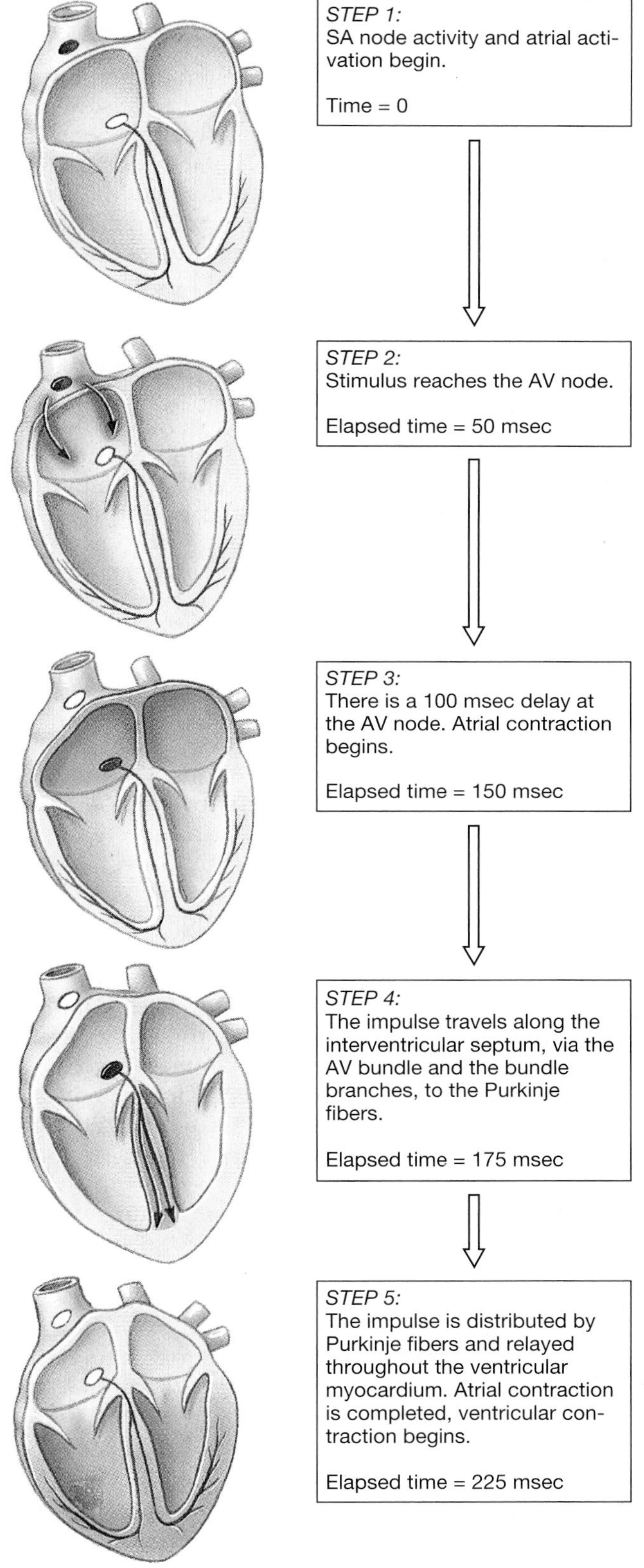
STEP 1:
SA node activity and atrial activation begin.
Time = 0
STEP 2:
Stimulus reaches the AV node.
Elapsed time = 50 msec
STEP 3:
There is a 100 msec delay at the AV node. Atrial contraction begins.
Elapsed time = 150 msec
STEP 4:
The impulse travels along the interventricular septum, via the AV bundle and the bundle branches, to the Purkinje fibers.
Elapsed time = 175 msec
STEP 5:
The impulse is distributed by Purkinje fibers and relayed throughout the ventricular myocardium. Atrial contraction is completed, ventricular contraction begins.
Elapsed time = 225 msec

● **FIGURE 20-12**
Impulse Conduction through the Heart

endocardial surface. As the branches diverge they conduct the impulse to **Purkinje** (pur-KIN-jē) **fibers** and to papillary muscles.

The Purkinje fibers conduct action potentials very rapidly, as fast as small myelinated axons. Within about 75 msec the signal to begin a contraction has reached all of the ventricular cardiac muscle fibers. The entire process, from generation of an impulse at the SA node to the complete depolarization of the ventricular myocardium, normally takes around 225 msec. The atria have now completed their contractions, and ventricular contraction can safely occur.

Because the bundle branches deliver the impulse to the papillary muscles directly, rather than via Purkinje fibers, the papillary muscles begin contracting before the rest of the ventricular musculature. This contraction applies tension to the chordae tendineae, bracing the AV valves. By limiting the movement of the cusps, tension in the papillary muscles prevents the backflow of blood into the atria when the ventricles contract. That contraction proceeds in a wave that begins at the apex and spreads toward the base. Because the ventricles contract in this way, blood is pushed toward the base of the heart, into the aortic and pulmonary trunks.

If the conducting pathways are damaged, the normal rhythm of the heart will be disturbed. The resulting problems are called *conduction deficits*. If the SA node or internodal tracts are damaged, the AV node will assume command, and the heart will continue beating normally, although at a slower rate. If an abnormal conducting cell or ventricular muscle cell begins generating action potentials at a more rapid rate, the impulses can override those of the SA or AV nodes. The origin of these abnormal signals is called an **ectopic** (ek-TOP-ik; out of place) **pacemaker.** The activity of an ectopic pacemaker partially or completely bypasses the conducting system, disrupting the timing of ventricular contraction. The result is a dangerous reduction in the efficiency of the heart. Such conditions are often diagnosed with the aid of an *electrocardiogram*. (For a more detailed discussion of conduction deficits and other problems, consult the *Applications Manual*.) AM *Diagnosing Abnormal Heartbeats*

If the cells of the SA node failed to function, what effect would this failure have on heart rate?

Why is it important for the impulses from the atria to be delayed at the AV node before passing into the ventricles?

If the cardioinhibitory center of the medulla oblongata were damaged, what part of the autonomic nervous system would be affected and what effect would this damage have on the heart?

How would an increase in the extracellular concentration of calcium ions affect the strength of a cardiac contraction?

The Electrocardiogram

Figures 20-12, 20-13

The electrical events occurring in the heart are powerful enough to be detected by electrodes on the body surface. A recording of these electrical activities constitutes an **electrocardiogram** (e-lek-trō-KAR-dē-ō-gram), also called an **ECG** or **EKG.** Each time the heart beats, a wave of depolarization radiates through the atria, reaches the AV node, travels down the interventricular septum to the apex, turns, and spreads through the ventricular myocardium toward the base (Figure 20-12•). This electrical activity can be monitored from the body surface.

By comparing the information obtained from electrodes placed at different locations, a clinician can monitor the performance of specific nodal, conducting, and contractile components. For example, when a portion of the heart has been damaged the affected muscle cells will no longer conduct action potentials, so an ECG will reveal an abnormal pattern of impulse conduction.

The appearance of the ECG tracing varies, depending on the placement of the monitoring electrodes, or *leads*. Figure 20-13• shows the important features of an electrocardiogram as analyzed with the leads in one of the standard configurations.

- The small **P wave** accompanies the depolarization of the atria. The atria begin contracting around 100 msec after the start of the P wave.
- The **QRS complex** appears as the ventricles depolarize. This is a relatively strong electrical signal because the mass of the ventricular muscle is much larger than that of the atria. The ventricles begin contracting shortly after the peak of the R wave.
- The smaller **T wave** indicates ventricular repolarization. You do not see a deflection corresponding to atrial repolarization because it occurs while the ventricles are depolarizing, and the electrical events are masked by the QRS complex.

ECG ANALYSIS Analyzing an ECG involves measuring the size of the voltage changes and determining the durations and temporal relationships of the various components. Attention usually focuses on the amount of depolarization occurring during the P wave and the QRS complex. For example, a smaller-than-normal electrical signal may mean that the mass of the heart muscle has decreased, and an excessively large QRS complex may indicate that the heart has become enlarged.

The size and shape of the T wave may also be affected by any condition that slows ventricular repolarization. For example, starvation and low cardiac energy reserves, coronary ischemia, or abnormal ion concentrations will reduce the size of the T wave.

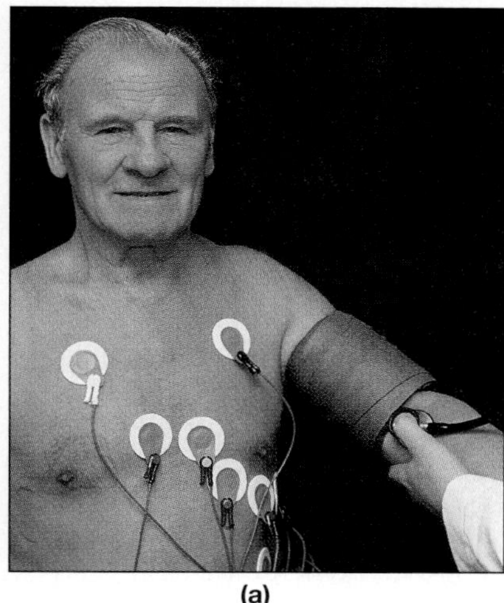

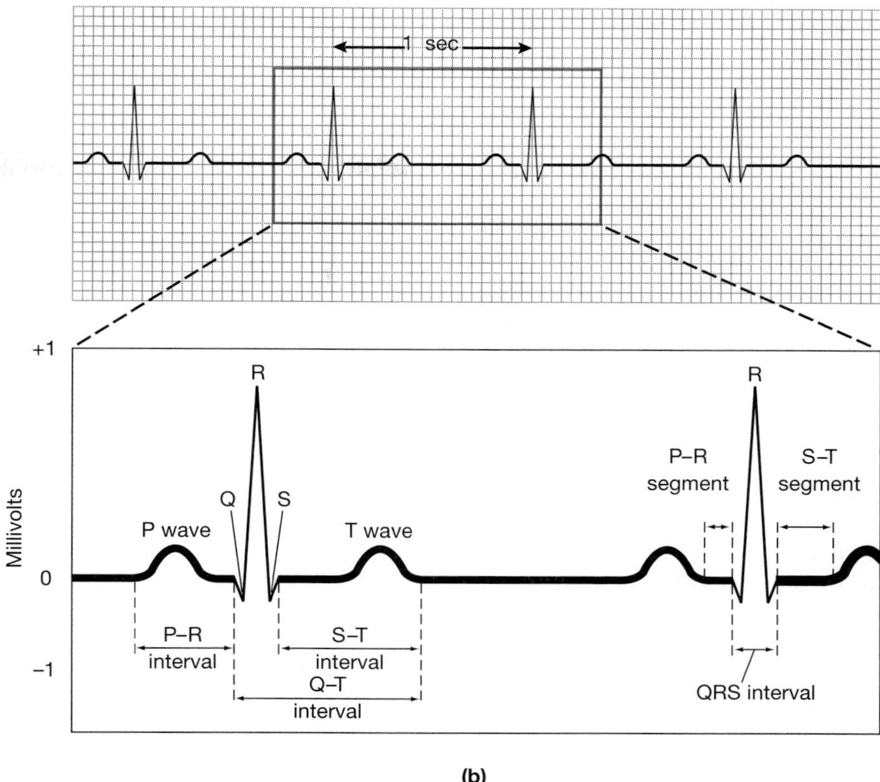
1 sec
+1
0
−1
Millivolts
R
R
P wave
Q
S
T wave
P–R segment
S–T segment
P–R interval
S–T interval
Q–T interval
QRS interval

(a) (b)

● **FIGURE 20-13**

An Electrocardiogram. (a) Electrode placement for recording a standard ECG. **(b)** An ECG printout is a strip of graph paper containing a record of the electrical events monitored by electrodes attached to the body surface. The placement of electrodes affects the size and shape of the waves recorded. This is an example of a normal ECG; the enlarged section indicates the major components of the ECG and the measurements most often taken during clinical analysis.

Measurements of the time between waves are also taken. The values are reported as *segments* or *intervals.* The terms used are indicated in Figure 20-13●, and you will find that the names do not always seem to fit. For example:

- The *P–R interval* extends from the start of atrial depolarization to the start of the QRS complex (ventricular depolarization), rather than to R. This is because in abnormal ECGs the peak can be difficult to determine. Extension of the P–R interval to more than 0.2 seconds can indicate damage to the conducting pathways or AV node.
- The *Q–T interval* indicates the time required for the ventricles to undergo a single cycle of depolarization and repolarization. It is usually measured from the end of the P–R interval rather than from the bottom of the Q wave. The Q–T interval, which roughly corresponds to the duration of a ventricular contraction, can be extended by conduction problems, coronary ischemia, or myocardial damage.

Despite the variety of sophisticated equipment available to assess or visualize cardiac function, in the majority of cases the electrocardiogram provides the most important diagnostic information. ECG analysis is especially useful in detecting and diagnosing **cardiac arrhythmias** (a-RITH-mē-az), abnormal patterns of cardiac electrical activity. Momentary arrhythmias are not inherently dangerous, and about 5 percent of the normal population experiences a few abnormal heartbeats each day. Clinical problems appear when the arrhythmias reduce the pumping efficiency of the heart. Serious arrhythmias may indicate damage to the myocardial musculature, injuries to the pacemakers or conduction pathways, exposure to drugs, or variations in the electrolyte composition of the extracellular fluids. (For a discussion of the most common types of arrhythmias detected with the ECG see the *Applications Manual.*) AM *Diagnosing Abnormal Heartbeats; Monitoring the Living Heart*

❑ The Cardiac Cycle

Figure 20-14

The period between the start of one heartbeat and the beginning of the next is a single **cardiac cycle.** The cardiac cycle therefore includes alternate periods of contraction and relaxation. For any one chamber in the heart, the cardiac cycle can be divided into two phases. During contraction, or **systole** (SIS-tō-lē), the chamber contracts and pushes blood into an adjacent chamber or into an arterial trunk. Systole is followed by the second phase, one of relaxation, or **diastole** (dī-AS-tō-lē). During diastole the chamber fills with blood and prepares for the start of the next cardiac cycle.

Fluids will move from an area of high pressure to one of relatively lower pressure. In the course of the cardiac cycle, the pressure within each chamber rises in systole and falls in diastole. Valves between adjacent chambers help to ensure that blood flows in the desired direction, but the mere presence of valves is not enough to accomplish this end. Blood will flow from one chamber to another only if the pressure in the first chamber exceeds that in the second. This basic principle governs the movement of blood between atria and ventricles, between ventricles and arterial trunks, and between the major veins and the atria.

The correct pressure relationships are dependent on the careful timing of contractions. For example, blood movement could not occur in the desired direction if an atrium and its attached ventricle contracted at precisely the same moment. The elaborate pacemaking and conduction systems normally provide the required spacing between atrial and ventricular systole. As a result, atrial systole and atrial diastole are slightly out of phase with ventricular systole and diastole. Figure 20-14• shows the duration and timing of systole and diastole for a heart rate of 75 bpm. At this heart rate a sequence of systole and diastole in either the atria or the ventricles lasts 800 msec. For convenience, we will consider that the cardiac cycle is determined by the atria, and that it includes one cycle of atrial systole and atrial diastole. This convention follows our description of the conducting system and the propagation of the stimulus for contraction.

Phases of the Cardiac Cycle

Figure 20-14

The phases of *atrial systole, ventricular systole,* and *ventricular diastole* are diagrammed in Figure 20-14a,b,c• for a heart rate of 75 bpm. The condition of the heart at the end of the previous cardiac cycle is indicated in Figure 20-14d•. The chambers are relaxed and the ventricles are partially filled with blood. Atrial systole lasts 100 msec, and at the end of this period the relaxed ventricles are filled with blood (Figure 20-14a•). The atria then enter a period of atrial diastole, which continues until the start of the next cardiac cycle. Atrial systole is followed by ventricular systole, which lasts 270 msec. During this period blood is pushed through the systemic and pulmonary circuits and begins flowing into the atria (Figure 20-14b•). The heart then enters a period of ventricular diastole, which lasts 530 msec. For the 430 msec remaining in this cardiac cycle, filling occurs passively, and both the atria and the ventricles are relaxed. Ventricular diastole then continues into the next cardiac cycle, lasting until the end of atrial systole (see Figure 20-14•).

When the heart rate increases, all these phases become shorter in duration. However, the great-

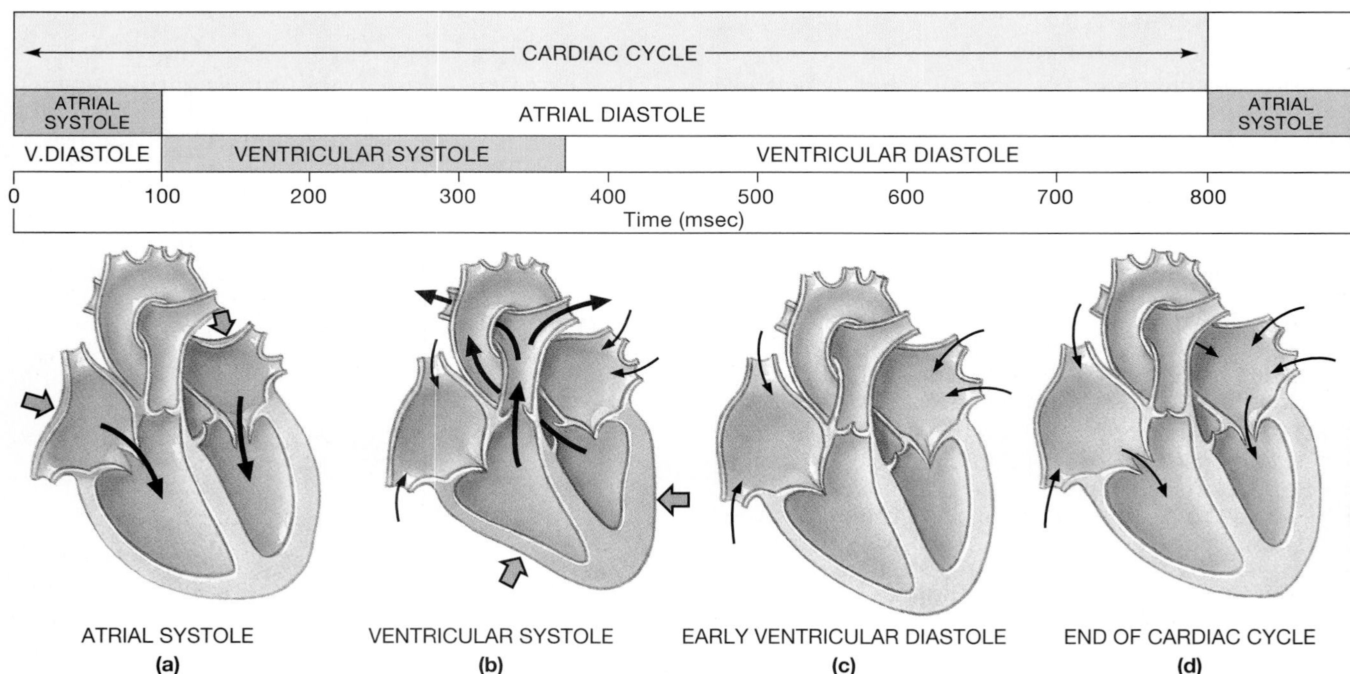
CARDIAC CYCLE
ATRIAL SYSTOLE
ATRIAL DIASTOLE
ATRIAL SYSTOLE
V.DIASTOLE
VENTRICULAR SYSTOLE
VENTRICULAR DIASTOLE
0
100
200
300
400
500
600
700
800
Time (msec)
ATRIAL SYSTOLE
(a)
VENTRICULAR SYSTOLE
(b)
EARLY VENTRICULAR DIASTOLE
(c)
END OF CARDIAC CYCLE
(d)

● **FIGURE 20-14**

Phases of the Cardiac Cycle. The atria and ventricles go through repeated cycles of systole and diastole. The timing of systole and diastole differs between the atria and ventricles. A cardiac cycle consists of one period of systole and diastole; we will consider a cardiac cycle as determined by the state of the atria. **(a)** Atrial systole. During this period the atria contract and the ventricles become filled with blood. **(b)** Ventricular systole. Blood is ejected into the pulmonary and aortic trunks. **(c)** Ventricular diastole. Once the AV valves open, passive filling of the ventricles occurs through the period of atrial systole in the next cardiac cycle. **(d)** Condition of the heart at the end of a cardiac cycle, with both the atria and ventricles in diastole.

est reduction occurs in the length of time spent in diastole. In going from a heart rate of 75 bpm to a rate of 200 bpm the time spent in systole drops by less than 40 percent, but the duration of diastole is reduced by almost 75 percent.

Pressure and Volume Changes in the Cardiac Cycle

Figure 20-15

We will consider the pressure and volume changes that occur during the cardiac cycle, using reference numbers that correspond to those in Figure 20-15●. The narrative description applies to both sides of the heart; although pressures are lower in the right atrium and right ventricle, both sides of the heart contract at the same time, and they eject equal volumes of blood.

ATRIAL SYSTOLE The cardiac cycle begins with atrial systole. This period lasts for approximately 100 msec in a resting adult.

1. As the atria contract, elevated atrial pressures push blood into the ventricles through the right and left AV valves.
2. At the start of atrial systole, the ventricles are already filled to around 70 percent of capacity, and atrial systole essentially "tops them off" by providing the additional 30 percent. (The ventricles filled during the previous cardiac cycle; the mechanism will be described below.)
3. At the end of atrial systole, each ventricle contains the maximum amount of blood that it will hold in this cardiac cycle. That quantity is called the **end-diastolic volume,** or **EDV.** The end-diastolic volume in a resting individual is typically around 130 ml.

VENTRICULAR SYSTOLE As atrial systole ends, ventricular systole begins. This period lasts for approximately 270 msec in a resting adult. As the pressures inside the ventricles rise above those in the atria, the AV valves swing shut.

4. During this stage of ventricular systole, the ventricles are contracting, but blood flow has yet to occur because ventricular pressures are not yet high enough to force open the semilunar valves and push blood into the pulmonary or aortic trunks. Over this period the ventricles contract *isometrically;* they generate tension, and ventricular pressures rise, but blood flow does not occur. The ventricle is now in the period of **isovolumetric contraction.** During isovolumetric contraction, all of the heart valves are closed, the volume of the ventricle remains constant, and ventricular pressure rises.
5. Once pressure in the ventricle exceeds that in the arterial trunks, the semilunar valves open and blood flows into the pulmonary and aortic trunks. This point marks the beginning of the period of **ventricular ejection.** The ventricles now contract *isotonically;* the muscle cells shorten, and tension production remains relatively constant. (For a review of isotonic versus isometric contractions, see Figure 10-13●, p. 304).

   After reaching a peak, ventricular pressures gradually decline near the end of ventricular systole. Figure 20-15● shows values for the left ventricle and aorta. Although pressures in the right ventricle and pulmonary trunk are much lower, the right ventricle also goes through periods of isovolumetric contraction and ventricular ejection. During this period the heart will eject approximately 80 ml of blood. This is the **stroke volume (SV)** of the heart. The stroke volume in this case is roughly 60 percent of the end-diastolic volume. This percentage, known as the *ejection fraction,* can vary in response to changing demands on the heart. (We will discuss the regulatory mechanisms in the next section.)
6. At the end of ventricular systole, ventricular pressures fall rapidly. Blood now starts to flow back toward the ventricles, and this movement closes the semilunar valves. As the backflow of blood begins, pressure decreases in the aorta. When the semilunar valves close, pressure rises once again as the elastic arterial walls recoil. This small, temporary rise produces a valley in the pressure tracing that is called a *dicrotic* (di-KRO-tik) *notch* (*dikrotos,* beating double). The amount of blood remaining in the ventricle when the semilunar valve closes is the **end-systolic volume (ESV).** The ESV in this example is 50 ml, about 40 percent of the end-diastolic volume.

VENTRICULAR DIASTOLE The period of ventricular diastole lasts for the 430 msec remaining in this cycle and continues through atrial systole in the next cardiac cycle.

7. The semilunar valves and AV valves are now closed, and the ventricular myocardium is relaxing. All of the heart valves are closed, and blood cannot flow into the ventricles. This is the period of **isovolumetric relaxation.** Ventricular pressures drop rapidly over this period because the elasticity of the connective tissues of the heart and the fibrous skeleton helps reexpand the ventricles toward their resting dimensions. When ventricular pressures fall below those of the atria, the AV valves open. Blood now flows from the atria into the ventricles.
8. Both the atria and the ventricles are now in diastole, but the ventricular pressures continue to

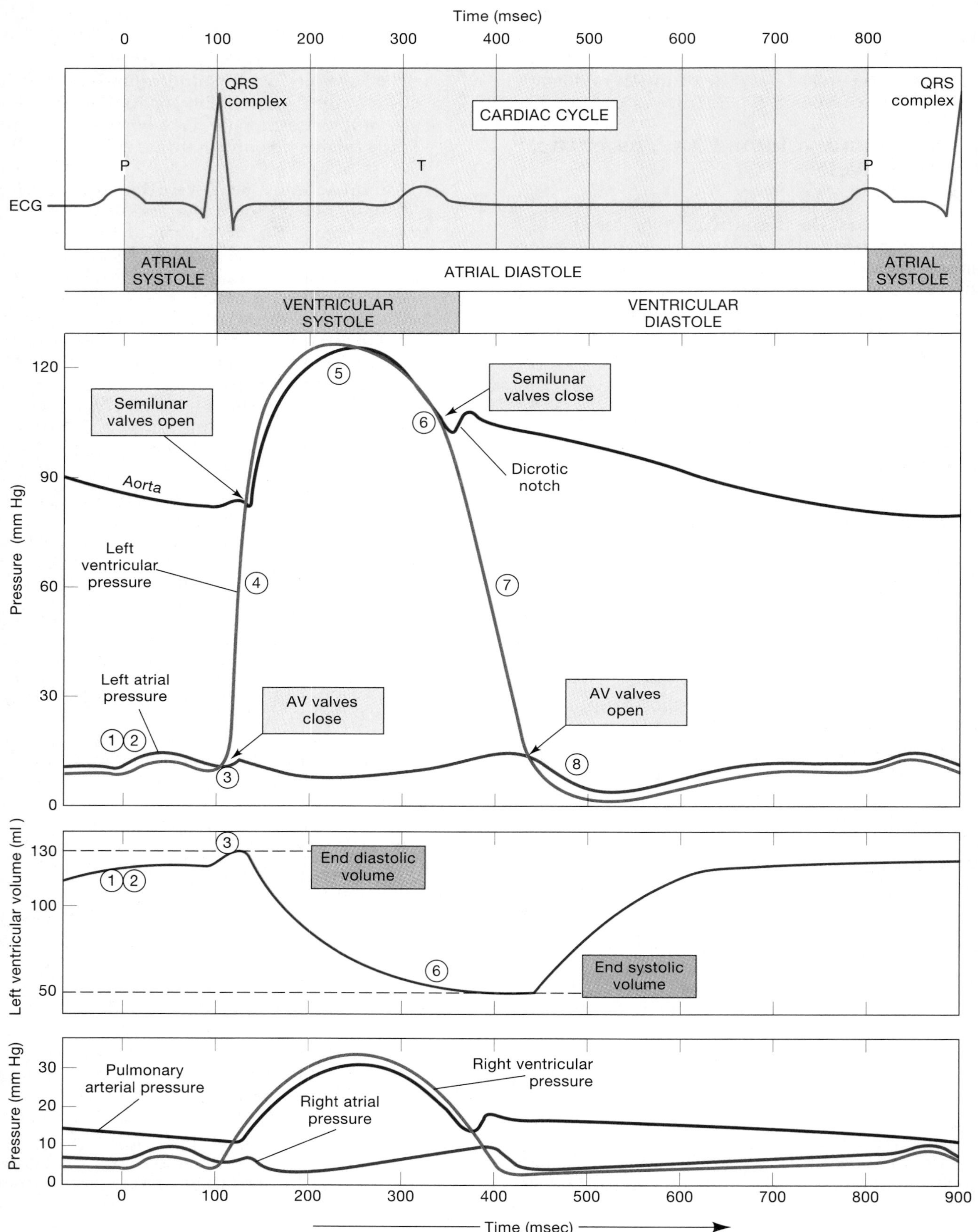

Time (msec)
0
100
200
300
400
500
600
700
800
QRS complex
CARDIAC CYCLE
QRS complex
P
T
P
ECG
ATRIAL SYSTOLE
ATRIAL DIASTOLE
ATRIAL SYSTOLE
VENTRICULAR SYSTOLE
VENTRICULAR DIASTOLE
120
90
60
30
0
Pressure (mm Hg)
Semilunar valves open
Semilunar valves close
Aorta
Dicrotic notch
Left ventricular pressure
Left atrial pressure
AV valves close
AV valves open
1 2 3 4 5 6 7 8
Left ventricular volume (ml)
130
100
50
End diastolic volume
End systolic volume
Pressure (mm Hg)
30
20
10
0
Pulmonary arterial pressure
Right ventricular pressure
Right atrial pressure
0
100
200
300
400
500
600
700
800
900
Time (msec)

● **FIGURE 20-15**

Pressure and Volume Relationships in the Cardiac Cycle. Major features of the cardiac cycle are shown for a heart rate of 75 bpm.

decline as the ventricular chambers expand. The expansion results from the elasticity of the myocardial connective tissues and the fibrous skeleton. Throughout this period pressures are low enough to pull blood from the major veins through the relaxed atria. This passive mechanism is the primary method of ventricular filling. By the time another cardiac cycle begins, the ventricles will be nearly three-quarters full.

The relatively minor contribution that atrial systole makes to ventricular volume explains why individuals can survive quite normally when their atria have been so severely damaged that they can no longer function. By contrast, damage to one or both ventricles can leave the heart unable to maintain adequate blood flow through peripheral tissues and organs. A condition of **heart failure** then exists. AM *Heart Failure*

Heart Sounds

Figure 20-16

Listening to the heart, a technique called *auscultation,* is a simple and effective method of cardiac diagnosis. Physicians use an instrument called a **stethoscope** (STETH-o-skōp) to listen to normal and abnormal heart sounds. The placement of the stethoscope varies, depending on the valve under examination (Figure 20-16●). Valve sounds must pass through the pericardium, surrounding tissues, and the chest wall, and some tissues muffle sounds more than others. As a result, the stethoscope placement does not always correspond to the position of the valve under review.

When you listen to your own heart, you usually hear the **first** and **second heart sounds.** These sounds accompany the action of the heart valves. The first heart sound, known as "lubb," lasts a little longer than the second. It marks the start of ventricular contraction, and it is produced as the AV valves close. The second heart sound, "dupp," occurs at the beginning of ventricular filling, when the semilunar valves close.

Third and **fourth heart sounds** may be audible as well, but they are usually very faint and seldom are detectable in healthy adults. These sounds are associated with atrial contraction (S_4) and blood flowing into the ventricles (S_3), rather than valve action.

MITRAL VALVE PROLAPSE Minor abnormalities in valve shape are relatively common. For example, an estimated 10 percent of normal individuals between 14 and 30 years of age have some degree of **mitral valve prolapse.** In this condition the mitral valve cusps do not close properly. The problem may involve abnormally long (or short) chordae tendineae or malfunctioning papillary muscles. Because the valve does not work perfectly, some regurgitation occurs during left ventricular systole. The surges, swirls, and eddies that occur during regurgitation create a rushing, gurgling sound known as a **heart murmur.** Most individuals with this condition are completely asymptomatic, and they live normal healthy lives unaware of any circulatory malfunction.

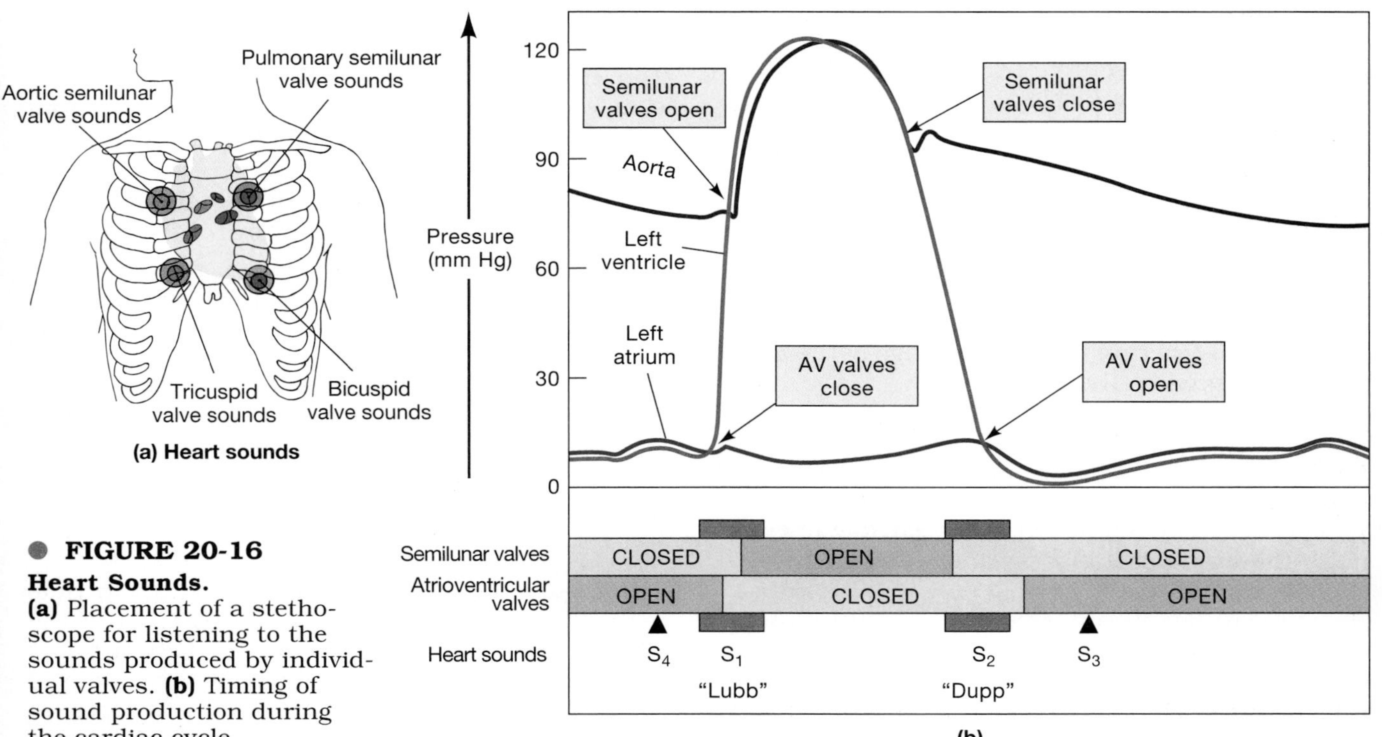
Pulmonary semilunar valve sounds
Aortic semilunar valve sounds
Tricuspid valve sounds
Bicuspid valve sounds
(a) Heart sounds
Pressure (mm Hg)
120
90
60
30
0
Semilunar valves open
Aorta
Left ventricle
Left atrium
Semilunar valves close
AV valves close
AV valves open
Semilunar valves
CLOSED
OPEN
CLOSED
Atrioventricular valves
OPEN
CLOSED
OPEN
Heart sounds
S_4
S_1
"Lubb"
S_2
"Dupp"
S_3
(b)

● **FIGURE 20-16**
Heart Sounds.
(a) Placement of a stethoscope for listening to the sounds produced by individual valves. **(b)** Timing of sound production during the cardiac cycle.

The Energy for Cardiac Contractions

When a normal heart is beating, the energy required is obtained by the mitochondrial breakdown of fatty acids (stored as lipid droplets) and glucose (stored as glycogen). These aerobic reactions can occur only when oxygen is readily available. In addition to obtaining oxygen from the coronary circulation, cardiac muscle cells maintain sizable reserves of oxygen bound to the heme units of myoglobin molecules. (The structure of this globular protein was detailed in Chapter 2, and its function in slow skeletal muscle fibers was discussed in Chapter 10.) ∞ *[pp. 54, 311]* Under normal circumstances the combination of circulatory supplies plus myoglobin reserves is enough to meet the oxygen demands of the heart, even when working at maximum capacity.

 Is the heart always pumping blood when pressure in the left ventricle is rising? Explain.

 What could cause an increase in the size of the QRS complex of an ECG tracing?

 Why is it a potential problem if the heart beats too rapidly?

Cardiodynamics

The term **cardiodynamics** refers to the movements and forces generated during cardiac contractions. Each time the heart beats, the two ventricles eject equal amounts of blood. The discussion above introduced the following important terms:

- ***End-diastolic volume (EDV).*** The amount of blood in each ventricle at the end of ventricular diastole (the start of ventricular systole).
- ***End-systolic volume (ESV).*** The amount of blood remaining in each ventricle at the end of ventricular systole (the start of ventricular diastole).
- ***Stroke volume (SV).*** The amount of blood pumped out of each ventricle during a single beat (EDV – ESV = SV).
- ***Ejection fraction.*** The percentage of the EDV represented by the SV.

In an examination of a single cardiac cycle, stroke volume is the most important factor. When considering cardiac function over time, physicians usually are more interested in the **cardiac output (CO),** the amount of blood pumped by each ventricle in 1 minute.

The cardiac output provides a useful indication of ventricular efficiency over time. It can be calculated by multiplying the average stroke volume by the heart rate (HR). This relationship can be summarized as:

CO	=	SV	×	HR
cardiac output (ml/min)		stroke volume (ml)		heart rate (bpm)

For example, if the stroke volume is 80 ml and the heart rate is 75 bpm, the cardiac output will be:

$$\text{CO} = 80 \text{ ml} \times 75 \text{ bpm} = 6000 \text{ ml/min (6 L/min)}$$

Cardiac output is precisely regulated so that peripheral tissues receive an adequate circulatory supply under a variety of conditions. When necessary, stroke volume in a normal heart can almost triple, and the heart rate can increase by 250 percent; increasing both the stroke volume *and* the heart rate to maximal levels can raise the cardiac output by 600–700 percent. The difference between resting and maximal cardiac output is the **cardiac reserve.**

For convenience, we will consider the regulation of stroke volume and heart rate separately, although alterations in cardiac output usually reflect changes in both aspects of cardiac function.

Factors Controlling Stroke Volume

The stroke volume is the difference between the end-diastolic volume and the end-systolic volume. Thus, changes in either the EDV or the ESV can alter the stroke volume and change cardiac output.

The EDV

The EDV is affected by the **filling time** (the duration of ventricular diastole) and the **venous return** (the rate of blood flow over this period).

- Filling time depends entirely on the heart rate. The faster the heart rate, the shorter the available filling time.
- Venous return changes in response to alterations in cardiac output, blood volume, patterns of peripheral circulation, skeletal muscle activity, or other factors that affect the rate of blood flow through the venae cavae. (These factors will be explored in Chapter 21.)

The ESV

Figure 20-18b

The EDV indicates how much blood a ventricle contains. After the ventricle contracts and the stroke volume has been ejected, the ESV remains in the ventricle. How much blood is forced out of the heart depends on (1) the *preload* (the degree of stretching at the end of ventricular diastole),

(2) the *contractility* of the ventricle (the forcefulness of its contraction), and (3) the *afterload* (the amount of pressure required to eject blood).

PRELOAD The degree of stretching experienced during ventricular diastole is called the **preload.** The amount of preload is directly proportional to the EDV; the greater the EDV, the larger the preload. Preload is significant because it affects the ability of muscle cells to produce tension. Chapter 10 considered the length-tension relationship for skeletal muscle fibers (Figure 10-8•, pp. 296–297), and the same principles apply to cardiac muscle fibers. If a muscle fiber is stimulated when the resting length is very short, a powerful contraction cannot occur because the sarcomeres are already at minimum size. The degree of shortening and the amount of tension produced during a contraction increase as the resting length increases. There is a relatively narrow range of optimal lengths; as a muscle fiber stretches *beyond* optimal length, its abilities to shorten and produce tension decline again. If stretched to the point that thick and thin filaments no longer overlap, a muscle fiber cannot contract at all.

The amount of preload varies depending on the demands placed on the heart. In a resting individual the end-diastolic volume is low, so the ventricular muscle is stretched very little and the sarcomeres are relatively short. During ventricular systole the cardiac muscle fibers contract only a short distance and develop little power. If the individual begins exercising, venous return increases, and more blood flows into the heart. The EDV increases, and the myocardium stretches further. As the sarcomeres approach optimal lengths the ventricular muscle fibers can contract to the greatest degree and produce the most forceful contractions. They also shorten more, and more blood is pumped out of the heart.

In the normal heart, the greater the EDV, the larger the stroke volume. Stretching *past* optimal length, which would reduce the force of contraction, does not normally occur because ventricular expansion is limited by myocardial connective tissues, the fibrous skeleton, and the pericardial sac.

The Frank-Starling Principle. The relationship between the amount of ventricular stretching and contractile force means that within normal physiological limits, *increasing the end-diastolic volume results in a corresponding increase in the stroke volume.* This general rule of "more in = more out" was first proposed by Starling, a physiologist, and was based on research performed by Frank. The relationship is therefore known as the **Frank-Starling principle,** or *Starling's law of the heart.*

Autonomic adjustments to cardiac output make it difficult to see the effects of the Frank-Starling principle. However, it can be demonstrated effectively in individuals who have received heart transplants, because the implanted hearts are not innervated by the ANS. The most obvious effect of the Frank-Starling principle in these hearts is that the outputs of the left and right ventricles remain balanced under a variety of conditions. For example, consider a situation in which an individual is at rest and the two ventricles are ejecting equal volumes of blood. The ventricles are in series: When the heart contracts and blood leaves the right ventricle, a comparable volume of blood arrives at the left atrium, to be ejected by the left ventricle at the *next* contraction. If the venous return decreases, the EDV of the right ventricle will decline. During ventricular systole it will then pump less blood through the pulmonary circuit to the left atrium. In the next cardiac cycle, the EDV of the left ventricle will be reduced, and it will eject a smaller volume of blood. The output of the two ventricles will again be in balance.

CONTRACTILITY Contractility is the amount of force produced during a contraction, independent of preload. Under normal circumstances contractility may be altered by autonomic innervation or circulating hormones. Under special circumstances, contractility can be altered by administered drugs or as the result of abnormal changes in ion concentrations in the extracellular fluid.

Factors that increase contractility are said to have a *positive inotropic action;* those that decrease contractility have a *negative inotropic action.* Positive inotropic agents often stimulate calcium entry into cardiac muscle cells, thus increasing the force and duration of ventricular contractions. Negative inotropic agents may block calcium ion movement or depress cardiac muscle metabolism in some way. Positive and negative inotropic factors include ANS activity, hormones, and changes in extracelluar ion concentrations.

Autonomic Activity. Autonomic activity alters the degree of contraction and changes the end-systolic volume.

- Sympathetic stimulation has a positive inotropic effect. It causes the release of norepinephrine by postganglionic fibers and the secretion of norepinephrine and epinephrine by the adrenal medullae. In addition to their effects on heart rate, discussed below, these compounds stimulate cardiac muscle cell metabolism and increase the force and degree of contraction. These effects are produced by stimulating alpha and beta receptors in cardiac muscle cell membranes. ∞ [p. 607] The net effect is that the ventricles contract more forcefully, increasing the ejection fraction and decreasing the ESV.

- Parasympathetic stimulation via the vagus nerve has a negative inotropic effect. The primary effect of ACh is at the membrane surface, where it produces hyperpolarization and inhibition. The result is a decrease in both the heart rate and force of cardiac contractions. Because parasympathetic innervation of the ventricles is relatively limited, the atria show the greatest changes in contractile force. However, under strong parasympathetic stimulation the ventricles contract less forcefully, the ejection fraction decreases, and the ESV enlarges.

Hormones. Several hormones have effects on the contractility of the heart.

- Epinephrine and norepinephrine have positive inotropic effects.
- Glucagon has a positive inotropic effect. Before synthetic inotropic agents were available, glucagon was widely used to stimulate cardiac function. It is still used in cardiac emergencies and to treat some forms of heart disease.
- Thyroid hormones have positive inotropic effects. Thyroid gland disorders that result in hypersecretion of thyroid hormones often produce excessive stimulation of cardiac muscle tissue, leading to cardiac arrhythmias. Former President George Bush experienced such problems during his term in the White House.

Changes in Ion Concentration. Changes in extracellular calcium ion concentrations affect primarily the strength and duration of cardiac contractions. If the extracellular concentration of calcium is elevated, a condition termed **hypercalcemia** (hī-per-kal-SĒ-mē-a), cardiac muscle cells become extremely excitable, and their contractions become powerful and prolonged. In extreme cases the heart goes into an extended state of contraction that is usually fatal. When the calcium ion concentration is abnormally low, a condition termed **hypocalcemia** (hī-pō-kal-SĒ-mē-a), the contractions become very weak and may cease altogether.

Changes in the extracellular potassium ion concentration affect primarily the transmembrane potential and the rates of depolarization and repolarization. When potassium ion concentrations are high, a condition termed **hyperkalemia** (hī-per-ka-LĒ-mē-a) exists. Under these circumstances, cardiac contractions become weak and irregular. When potassium ion concentrations are abnormally low, a condition termed **hypokalemia** (hī-pō-ka-LĒ-mē-a), the membranes of cardiac muscle cells hyperpolarize, and their rate of contraction declines. Severe hyperkalemia and hypokalemia are life-threatening conditions that require immediate corrective action.

EFFECTS OF DRUGS ON CONTRACTILITY The drugs *isoproterenol, dopamine,* and *dobutamine* have comparable effects because they mimic the action of epinephrine and norepinephrine and stimulate calcium entry through alpha-receptor stimulation. *Digitalis* and related drugs also stimulate calcium entry, but through a different mechanism.

Many of the drugs used to treat hypertension (high blood pressure) have a negative inotropic action. Drugs such as *propranolol, timolol, metoprolol, atenolol,* several *barbiturates,* and *labetalol* block beta or alpha receptors or both and prevent sympathetic stimulation of the heart.

AFTERLOAD The **afterload** is the amount of tension the contracting ventricle must produce to force open the semilunar valve and eject blood. The greater the afterload, the longer the period of isovolumetric contraction, the shorter the duration of ventricular ejection, and the larger the ESV. In other words, as the afterload increases, the stroke volume decreases.

Afterload is increased by any factor that restricts blood flow through the arterial system. For example, constriction of peripheral blood vessels or a circulatory blockage will elevate arterial blood pressure and increase afterload. If the afterload is too great, the ventricle will be unable to eject blood at all. Afterload that high is rare in the normal heart, but damage to the heart muscle can weaken the myocardium enough that even a modest rise in arterial blood pressure can reduce stroke volume to dangerously low levels, producing symptoms of *heart failure.*

Figure 20-18b• summarizes factors involved with the regulation of stroke volume.

What effect would stimulating the acetylcholine receptors of the heart have on cardiac output?

What effect would an increased venous return have on the stroke volume?

How would increased sympathetic stimulation of the heart affect the end-systolic volume (ESV)?

Joe's end-systolic volume is measured to be 40 ml, and his end-diastolic volume is measured to be 125 ml. What is Joe's stroke volume?

❑ Factors Affecting the Heart Rate

Autonomic activity, hormones and various drugs, changes in ion concentrations, and alterations in body temperature may alter the basic rhythm of contraction established by the SA node. Under normal circumstances, autonomic activity and circulating hormones are responsible for making delicate adjustments to the heart rate as circulatory demands change. Significant changes in ion concentration or body temperature are unusual and may indicate problems with homeostatic control mechanisms. All of these factors act by modifying the natural rhythm of the heart; even a heart removed for a heart transplant will continue to beat unless steps are taken to prevent it.

A number of clinical problems are the result of abnormal pacemaker function. **Bradycardia** (brā-dē-KAR-dē-a; *bradys*, slow) is the term used to indicate a heart rate that is slower than normal, whereas **tachycardia** (tak-ē-KAR-dē-a; *tachys*, swift) indicates a faster-than-normal heart rate. These are relative terms, and in clinical practice the definitions vary depending on the normal resting heart rate of the individual. AM *Problems with Pacemaker Function*

Autonomic Innervation

Figures 20-9, 20-17

Anatomical details concerning cardiac innervation were presented in Chapter 16 and diagrammed in Figure 20-9●, p. 694. *[p. 541]* As in other organs with dual innervation, there is a resting autonomic tone. Both autonomic divisions are normally active at a steady background level, releasing ACh and NE both at the nodes and into the myocardium. Thus, cutting the vagus nerves increases the heart rate, and sympathetic blocking agents slow the heart rate.

In a normal, resting individual, parasympathetic effects dominate. The heart rate in the absence of autonomic innervation is established by the pacemaker cells of the SA node. Such a heart beats at a rate of 80–100 bpm. At rest, a typical adult heart with intact innervation beats at 70–80 bpm, because of activity in the parasympathetic nerves innervating the SA node. If parasympathetic activity increases, the heart rate declines further; the heart rate will increase if either parasympathetic activity decreases or sympathetic activation occurs. Through dual innervation and adjustments in autonomic tone, the ANS can make very delicate adjustments in cardiovascular function to meet the demands of other systems.

EFFECTS ON THE SA NODE The sympathetic and parasympathetic divisions alter the heart rate by changing the permeabilities of cells in the conducting system. The effects are normally most pronounced at the SA node.

Consider the SA node of a resting individual, whose heart is beating at 75 bpm (Figure 20-17a●). Any factor that changes either the duration of repolarization or the time that it takes to reach threshold will alter the heart rate. Acetylcholine released by parasympathetic neurons extends the duration of repolarization and slows the rate of spontaneous depolarization (Figure 20-17b●). The result is a decline in heart rate.

When norepinephrine (NE) is released by sympathetic neurons, the repolarization period is shortened and nodal cells reach threshold more quickly. The result is an increase in the heart rate (Figure 20-17c●).

THE ATRIAL REFLEX The **atrial reflex,** or *Bainbridge reflex,* involves adjustments in heart rate in response to an increase in the venous return. When the walls of the right atrium are stretched, the cells of the SA node depolarize

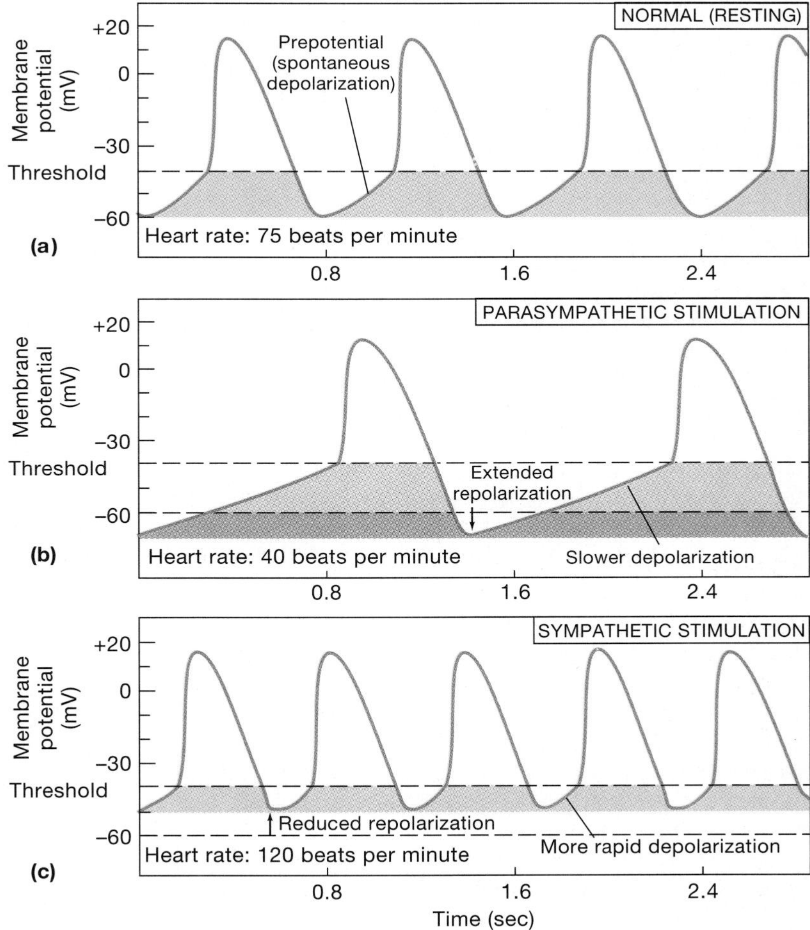
NORMAL (RESTING)
Membrane potential (mV)
+20
0
−30
Threshold
−60
Prepotential (spontaneous depolarization)
Heart rate: 75 beats per minute
(a)
0.8
1.6
2.4
PARASYMPATHETIC STIMULATION
Extended repolarization
Heart rate: 40 beats per minute
Slower depolarization
(b)
SYMPATHETIC STIMULATION
Reduced repolarization
Heart rate: 120 beats per minute
More rapid depolarization
(c)
Time (sec)

● **FIGURE 20-17**

Pacemaker Function. (a) Pacemaker cells have membrane potentials slightly lower than those of other cardiac muscle cells. Their cell membranes undergo spontaneous depolarization to threshold, producing action potentials at a frequency determined by (1) the resting membrane potential and (2) the slope of the prepotential. **(b)** Parasympathetic stimulation releases ACh, which extends repolarization and decreases the rate of spontaneous depolarization. The heart rate slows. **(c)** Sympathetic stimulation releases NE, which shortens repolarization and accelerates the rate of spontaneous depolarization. The heart rate increases as a result.

faster, and the heart rate accelerates. In addition, the stimulation of stretch receptors in the atrial walls triggers a reflexive increase in heart rate caused by increased sympathetic activity.

Hormones

Epinephrine, norepinephrine, and thyroid hormones cause an increase in the heart rate. The effects of epinephrine on the SA node are similar to those of norepinephrine. Epinephrine also stimulates other cells of the conducting system, and under massive sympathetic stimulation the myocardium may become so excitable that abnormal contractions occur.

EFFECTS OF DRUGS ON HEART RATE Chapter 12 noted several drugs, including caffeine and nicotine, that have a stimulatory effect on excitable membranes in the nervous system. ∞ *[p. 414]* These drugs also cause an increase in heart rate. Caffeine acts directly on the conducting system and increases the rate of depolarization at the SA node. Nicotine acts indirectly by stimulating the activity of sympathetic neurons innervating the heart.

Changes in Ion Concentration

The pace established by the SA node depends on (1) the resting membrane potential of the nodal cells and (2) the rate of spontaneous depolarization. Changes in the ionic composition of the extracellular fluid that affect either the resting potential or the rate of spontaneous depolarization will therefore alter the heart rate. For example, changes in extracellular potassium ion concentrations will alter the resting potential at the SA node and change the heart rate. When the extracellular concentration of potassium declines, the rate of potassium diffusion out of the muscle cell increases. As a result, a decrease in extracellular potassium concentration leads to membrane hyperpolarization and a reduction in heart rate. Changes in the concentration of other ions, such as calcium, affect the SA node and conducting system, but the effects on the contractility of the myocardium (noted in an earlier section) are more pronounced.

Changes in Body Temperature

Figure 20-18a

Temperature changes affect metabolic operations throughout the body. For example, lowering temperatures depresses cellular metabolism. At the heart, this depression slows the rate of depolarization at the SA node, lowers the heart rate, and reduces the strength of cardiac contractions. An elevated body temperature accelerates the heart rate and the contractile force, one reason why your heart seems to be racing and pounding whenever you have a fever.

Figure 20-18a● summarizes factors involved in the regulation of heart rate.

Cardiac Output and Heart Rate

Cardiac output cannot increase indefinitely, primarily because the available filling time becomes shorter and shorter as the heart rate increases. At heart rates of up to 160–180 bpm, the combination of an increased rate of venous return and increased contractility compensates for the reduction in filling time. Over this range, cardiac output and heart rate increase together. But if the heart rate continues to climb, the stroke volume begins to drop, and cardiac output first plateaus and then declines.

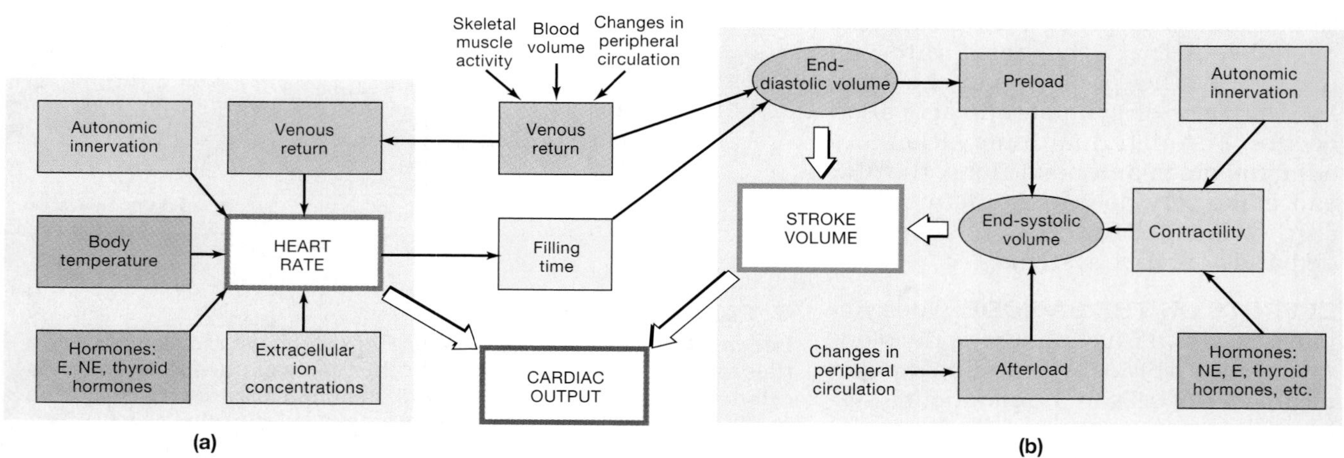

Skeletal muscle activity
Blood volume
Changes in peripheral circulation
Autonomic innervation
Venous return
Venous return
Body temperature
HEART RATE
Filling time
Hormones: E, NE, thyroid hormones
Extracellular ion concentrations
CARDIAC OUTPUT
(a)
End-diastolic volume
Preload
Autonomic innervation
STROKE VOLUME
End-systolic volume
Contractility
Changes in peripheral circulation
Afterload
Hormones: NE, E, thyroid hormones, etc.
(b)

● **FIGURE 20-18**
Factors Affecting Cardiac Output.
(a) Factors affecting heart rate.
(b) Factors affecting stroke volume.

Heart Attacks

In a **myocardial** (mī-ō-KAR-dē-al) **infarction (MI),** or *heart attack*, the coronary circulation becomes blocked and the cardiac muscle cells die from lack of oxygen. The affected tissue then degenerates, creating a nonfunctional area known as an *infarct*. Heart attacks most often result from severe coronary artery disease. The consequences depend on the site and nature of the circulatory blockage. If it occurs near the base of one of the coronary arteries, the damage will be widespread and the heart will probably stop beating. If the blockage involves one of the smaller arterial branches, the individual may survive the immediate crisis, but there are many potential complications, all unpleasant. As scar tissue forms in the damaged area, the heartbeat may become irregular, and other vessels can become constricted, creating additional circulatory problems.

Myocardial infarctions are most often associated with fixed blockages, such as those seen in CAD. When the crisis develops because of thrombus (clot) formation at a plaque, the condition is called **coronary thrombosis.** A vessel already narrowed by plaque formation may also become blocked by a sudden spasm in the smooth muscles of the vascular wall. The individual then may experience intense pain, similar to that of an angina attack but persisting even at rest. However, pain does not always accompany a heart attack. These *silent heart attacks* may be even more dangerous, because the condition may not be diagnosed and treated before a fatal MI occurs. Roughly 25 percent of heart attacks are not recognized when they occur.

The cytoplasm of a damaged cardiac muscle cell differs from that of a normal muscle cell. As the supply of oxygen decreases, the cells become more dependent on anaerobic metabolism to meet their energy needs. ∞ [p. 307] Over time the cytoplasm accumulates large numbers of enzymes involved with anaerobic energy production.

As the cardiac muscle cell membranes deteriorate, these enzymes enter the surrounding intercellular fluids. The appearance of such enzymes in the circulation thus indicates that an infarct has occurred. The enzymes tested for in a diagnostic blood test include **lactate dehydrogenase (LDH), serum glutamic oxaloacetic transaminase (SGOT,** also called *aspartate aminotransferase*), **creatine phosphokinase (CPK,** or **CK)**, and a special form of creatine phosphokinase found only in cardiac muscle **(CK-MB).**

Roughly 25 percent of MI patients die before obtaining medical assistance, and 65 percent of MI deaths among those under age 50 occur within an hour after the initial infarct. The goals of treatment are to limit the size of the infarct and prevent additional complications by preventing irregular contractions, improving circulation with vasodilators, providing additional oxygen, reducing the cardiac workload, and, if possible, eliminating the cause of the circulatory blockage. Anticoagulants may help prevent the formation of additional thrombi, and clot-dissolving enzymes may reduce the extent of the damage if they are administered within 6 hours after the MI has occurred. ∞ [p. 676] Controversy exists over the benefits of using t-PA, which is relatively expensive, versus other fibrinolytic agents, such as urokinase or streptokinase. A procedure that combines small amounts of t-PA with streptokinase has been proposed as a workable compromise that produces effects equivalent to high doses of t-PA alone. Follow-up treatment with heparin or aspirin or both is recommended; without further treatment the circulatory blockages will reappear in roughly 20 percent of patients.

There are roughly 1.3 million MIs in the United States each year, and half of the victims die within a year of the incident. A number of factors appear to increase the risk of a heart attack: smoking, high blood pressure, high blood cholesterol levels, high circulating levels of *low-density lipoproteins* (LDL), diabetes, male gender (below age 70), severe emotional stress, and obesity. The role of lipoproteins and cholesterol in plaque formation and heart disease will be considered in Chapter 21. Hereditary factors may also predispose an individual to coronary artery disease. Although the rate of heart attacks of women under age 70 is lower than that of men, their mortality rate is actually higher—perhaps because heart disease in women is neither diagnosed as early nor treated as aggressively as is heart disease in men.

The presence of two risk factors more than doubles the risk, so eliminating as many risk factors as possible will improve one's chances of preventing or surviving a heart attack. Changes in diet to limit cholesterol, exercise to lower weight, and seeking treatment for high blood pressure are steps in the right direction. It has been estimated that reduction of coronary risk factors could prevent 150,000 deaths each year in the United States alone.

The Heart and the Cardiovascular System

The goal of cardiovascular regulation is the maintenance of adequate tissue blood flow. The heart cannot accomplish this by itself, and it does not work in isolation. For example, when blood pressure changes, cardiovascular centers not only adjust the heart rate, but also alter the diameters of peripheral blood vessels. These adjustments work together to keep blood pressure within normal limits and to maintain circulation to vital tissues and organs. Chapter 21 begins by examining the structure of the circulatory network. It then completes our discussion of the cardiovascular system by considering its responses to changing activity patterns and circulatory emergencies.

What effect would drinking large amounts of caffeinated coffee have on the heart?

If a person suffers from bradycardia, is cardiac output likely to be increased or decreased? Explain.

How would a drug that increases the length of time required for repolarization of pacemaker cells affect a person's heart rate?

Selected Clinical Terminology

Terms Discussed in This Chapter

angina pectoris (an-JĪ-na PEK-tōr-is): A condition in which exertion or stress can produce severe chest pain, resulting from temporary insufficiency and ischemia when the heart's workload increases. *(p. 697)*

balloon angioplasty: A technique for reducing the size of a coronary plaque by compressing it against the arterial walls using a catheter with an inflatable collar. *(p. 697)*

bradycardia (brā-dē-KAR-dē-a): A heart rate that is slower than normal. *(p. 709* and *AM))*

calcium channel blockers: Drugs administered to reduce contractility of the heart by slowing the influx of calcium ions during the plateau phase of the cardiac muscle action potential. *(p. 697)*

cardiac arrhythmias (ā-RITH-mē-az): Abnormal patterns of cardiac electrical activity, indicating abnormal contractions. *(p. 701* and *AM)*

cardiac tamponade: A condition resulting from pericardial irritation and inflammation in which fluid collects in the pericardial sac and restricts cardiac output. *(p. 684* and *AM)*

carditis (kar-DĪ-tis): A general term indicating inflammation of the heart. *(p. 688* and *AM)*

conduction deficit: An abnormality in the conducting system of the heart that affects the timing and pacing of cardiac contractions. *(p. 700* and *AM)*

coronary artery disease (CAD): Degenerative changes in the coronary circulation. *(p. 697)*

coronary artery bypass graft: Replacement of a coronary artery or one of its branches by a vessel transplanted from another part of the body. *(p. 697)*

coronary ischemia: Restriction of the circulatory supply to the heart, potentially causing tissue damage and a reduction in cardiac efficiency. *(p. 697)*

coronary thrombosis: A blockage due to the formation of a clot (thrombus) at a plaque in a coronary artery. *(p. 711)*

electrocardiogram (ē-lek-trō-KAR-dē-ō-gram) **(ECG** or **EKG):** A recording of the electrical activities of the heart over time. *(p. 700)*

heart failure: A condition in which the heart weakens and peripheral tissues suffer from oxygen and nutrient deprivation. *(p. 705* and *AM)*

myocardial (mī-ō-KAR-dē-al) **infarction (MI):** A condition in which the coronary circulation becomes blocked and the cardiac muscle cells die from oxygen starvation; also called a *heart attack*. *(p. 711)*

pericarditis: Inflammation of the pericardium. *(p. 684)*

rheumatic heart disease (RHD): A disorder in which the heart valves become thickened and stiffen into a partially closed position, affecting the efficiency of the heart. *(p. 688* and *AM)*

tachycardia (tak-ē-KAR-dē-a): A heart rate that is faster than normal. *(p. 709* and *AM)*

valvular heart disease (VHD): A condition caused by abnormal functioning of one of the cardiac valves. The severity of the condition depends on the degree of damage and the identity of the valve involved. *(p. 688* and *AM)*

AM *Additional Terms Discussed in the Applications Manual*

atrial fibrillation: Rapid, uncontrolled, and uncoordinated spasms in the atrial walls.

atrial flutter: Rapid contractions of the atria that are ineffective in propelling blood.

cardiomyopathies (kar-dē-ō-mī-OP-a-thēz): A group of diseases characterized by the progressive, irreversible degeneration of the myocardium.

coronary arteriography: Production of an X-ray image of coronary circulation after the introduction of a radiopaque dye into one of the coronary arteries via a catheter; image is a *coronary angiogram*.

defibrillator: A device used to eliminate ventricular fibrillation and restore normal cardiac rhythm.

echocardiography: Ultrasound analysis of the heart and the blood flow through major vessels.

endocarditis: Inflammation of the endocardial lining.

heart block: A condition affecting the normal rhythm of the heart; characterized by impaired communication between the SA node, the AV node, and the ventricular myocardium resulting from damage to conduction pathways caused by mechanical distortion, ischemia, infection, or inflammation. There are three degrees of heart block.
myocarditis: Inflammation of the myocardium.
paroxysmal atrial tachycardia (PAT): Bursts of rapid atrial depolarization, leading to multiple atrial contractions with little pause between them.
premature atrial contractions (PACs): Atrial contractions that occur in advance of the rhythm established by the SA node.
premature ventricular contractions (PVCs): Premature ventricular beats that are directed by an ectopic pacemaker.
valvular stenosis (ste-NŌ-sis): A condition in which the opening between thickened heart valves becomes narrower.
ventricular escape: The initiation of ventricular contractions that are controlled by the AV node, Purkinje fibers, or other parts of the ventricular conducting system rather than the SA node.
ventricular fibrillation (VF): The most serious type of cardiac arrhythmia, in which the normal rhythm becomes disrupted because the cardiac muscle fibers are stimulating one another at such a rapid rate. Ventricular contractions are rapid, uncontrolled, and uncoordinated; the ventricles are incapable of pumping blood. Also known as *cardiac arrest.*
ventricular tachycardia (VT): Rapid ventricular beats that are not coordinated with atrial activity.

CHAPTER REVIEW

STUDY OUTLINE

AN OVERVIEW OF THE CARDIOVASCULAR SYSTEM, p. 682

1. The circulatory pathways can be subdivided into the **pulmonary circuit** (which carries blood to and from the lungs) and the **systemic circuit** (which transports blood to and from the rest of the body). **Arteries** carry blood away from the heart; **veins** return blood to the heart. **Capillaries** are narrow-diameter vessels that connect the smallest arteries and veins. (*Figure 20-1*)
2. The heart has four chambers: the **right atrium** and **right ventricle,** and the **left atrium** and **left ventricle.**

ANATOMY OF THE HEART, p. 682

1. The heart lies within the anterior portion of the mediastinum, which separates the two pleural cavities.

The Pericardium, p. 683

2. The heart is surrounded by the **pericardial cavity** (lined by the **pericardium**); the **visceral pericardium (epicardium)** covers the heart's outer surface, and the **parietal pericardium** lines the inner surface of the **pericardial sac** that surrounds the heart. (*Figure 20-2*)

Superficial Anatomy of the Heart, p. 684

3. A deep groove, the **coronary sulcus,** marks the boundary between the atria and the ventricles. Other surface markings also provide useful reference points in describing the heart and associated structures. (*Figure 20-3*)

Internal Anatomy and Organization, p. 684

4. The atria are separated by the **interatrial septum,** and the ventricles are divided by the **interventricular septum.** The right atrium receives blood from the systemic circuit via two large veins, the **superior vena cava** and **inferior vena cava.** The atrial walls contain prominent muscular ridges, the **pectinate muscles.** (*Figure 20-4*)
5. Blood flows from the right atrium into the right ventricle via the **right atrioventricular (AV) valve (tricuspid valve).** This opening is bounded by three **cusps** of fibrous tissue braced by the tendinous **chordae tendineae** that are connected to **papillary muscles.**
6. Blood leaving the right ventricle enters the **pulmonary trunk** after passing through the **pulmonary semilunar valve.** The pulmonary trunk divides to form the **left** and **right pulmonary arteries.** The **left** and **right pulmonary veins** return blood to the left atrium. Blood leaving the left atrium flows into the left ventricle via the **left atrioventricular (AV) valve (bicuspid valve** or **mitral valve).** Blood leaving the left ventricle passes through the **aortic semilunar valve** and into the systemic circuit via the **ascending aorta.**
7. Anatomical differences between the ventricles reflect the functional demands placed on them. The wall of the right ventricle is relatively thin, whereas the left ventricle has a massive muscular wall. (*Figure 20-5*)
8. Valves normally permit blood flow in only one direction, preventing the **regurgitation** (backflow) of blood. (*Figure 20-6*)

The Heart Wall, p. 688

9. The bulk of the heart consists of the muscular **myocardium.** The **endocardium** lines the inner surfaces of the heart. The **fibrous skeleton** supports the heart's contractile cells and valves. (*Figure 20-7*)
10. **Cardiac muscle cells** are interconnected by **intercalated discs** that convey the force of contraction from cell to cell and conduct action potentials. (*Table 20-1*)

Connective Tissues and the Fibrous Skeleton, p. 691

11. The connective tissues of the heart, mainly collagen and elastic fibers, are organized into a stabilizing **fibrous skeleton**.

The Blood Supply to the Heart, p. 692

12. The **coronary circulation** meets the high oxygen and nutrient demands of cardiac muscle cells. The **coronary arteries** originate at the base of the ascending aorta. Interconnections between arteries called **anastomoses** ensure a constant blood supply. The **great** and **middle cardiac veins** carry blood from the coronary capillaries to the **coronary sinus.** (*Figure 20-8*)

Innervation of the Heart, p. 692

13. The **cardioacceleratory center** in the medulla oblongata activates sympathetic neurons; the **cardioinhibitory center** governs the activities of the parasympathetic neurons. The cardiac centers receive inputs from higher centers and from receptors monitoring blood pressure and the concentrations of dissolved gases. (*Figure 20-9*)

THE HEARTBEAT, p. 694

1. The two general classes of cardiac muscle cells are involved in the normal heartbeat: **contractile cells** and cells of the *conducting system.* The conducting system includes **nodal cells,** or **pacemaker cells,** and **conducting cells.**

Contractile Cells, p. 694

2. Cardiac muscle cells have a long refractory period, so rapid stimulation produces isolated rather than tetanic contractions. (*Figure 20-10*)

The Conducting System, p. 698

3. The **conducting system,** composed of nodal and conducting cells, initiates and distributes electrical impulses within the heart. Nodal cells establish the rate of cardiac contraction, and conducting cells distribute the contractile stimulus to the general myocardium. (*Figure 20-11*)
4. Unlike skeletal muscle, cardiac muscle contracts without neural or hormonal stimulation. **Pacemaker cells** found in the **cardiac pacemaker (sinoatrial [SA] node)** normally establish the rate of contraction. From the SA node the stimulus travels to the **atrioventricular (AV) node,** then to the **AV bundle,** which divides into **bundle branches.** From here **Purkinje fibers** convey the impulses to the ventricular myocardium. (*Figure 20-12*)

The Electrocardiogram, p. 700

5. A recording of electrical activities in the heart is an **electrocardiogram (ECG** or **EKG).** Important landmarks of an ECG include the **P wave** (atrial depolarization), **QRS complex** (ventricular depolarization), and **T wave** (ventricular repolarization). (*Figure 20-13*)

The Cardiac Cycle, p. 701

6. The **cardiac cycle** consists of **systole** (contraction), followed by **diastole** (relaxation). (*Figure 20-14*)
7. Both sides of the heart contract at the same time and they eject equal volumes of blood. (*Figure 20-15*)
8. The closing of valves and rushing of blood through the heart cause characteristic heart sounds that can be heard during auscultation. (*Figure 20-16*)

CARDIODYNAMICS, p. 706

1. The amount of blood ejected by a ventricle during a single beat is the **stroke volume (SV);** the amount of blood pumped each minute is the **cardiac output (CO).** The difference between resting and maximal cardiac output is the **cardiac reserve.**

Factors Controlling Stroke Volume, p. 706

2. The stroke volume is the difference between the **end-diastolic volume (EDV)** and the **end-systolic volume (ESV).** The **filling time** and **venous return** interact to determine the end-diastolic volume. Normally, the greater the EDV, the more powerful the succeeding contraction **(Frank-Starling's law of the heart).**

Factors Affecting the Heart Rate, p. 708

3. The basic heart rate is established by the pacemaker cells, but it can be modified by the ANS. The **atrial reflex** accelerates the heart rate when the walls of the right atrium are stretched. (*Figure 20-17*)
4. Sympathetic activity produces more powerful contractions that reduce the ESV, and parasympathetic stimulation slows the heart rate, reduces the contractile strength, and raises the ESV.
5. Cardiac output is affected by various factors including reflexes, hormones, drugs, and temperature changes. (*Figure 20-18*)

THE HEART AND THE CARDIOVASCULAR SYSTEM, p. 712

REVIEW QUESTIONS

LEVEL 1 Reviewing Facts and Terms

1. Blood supply to the muscles of the heart is provided by the:
   (a) systemic circulation
   (b) pulmonary circulation
   (c) coronary circulation
   (d) coronary portal system
2. The autonomic centers for cardiac function are located in the:
   (a) myocardial tissue of the heart
   (b) cardiac centers of the medulla oblongata
   (c) cerebral cortex
   (d) a, b, and c are correct
3. The serous membrane covering the outer surface of the heart is the:
   (a) parietal pericardium (b) visceral pericardium
   (c) myocardium (d) endocardium
4. The squamous epithelium covering the valves of the heart constitutes the:
   (a) epicardium (b) endocardium
   (c) myocardium (d) fibrous skeleton
5. The structure that permits blood flow from the right atrium to the left atrium while the lungs are developing is the:
   (a) foramen ovale (b) interatrial septum
   (c) coronary sinus (d) fossa ovalis

6. Blood leaves the left ventricle by passing through the:
   (a) aortic semilunar valve
   (b) pulmonary semilunar valve
   (c) mitral valve
   (d) tricuspid valve

7. The QRS complex of the ECG appears as the:
   (a) atria depolarize (b) ventricles depolarize
   (c) ventricles repolarize (d) atria repolarize

8. In the cardiac cycle, during diastole the chambers of the heart:
   (a) relax and fill with blood
   (b) contract and push blood into an adjacent chamber
   (c) experience a sharp increase in pressure
   (d) reach a pressure of approximately 120 mm Hg

9. During the cardiac cycle the amount of blood remaining in the ventricle when the semilunar valve closes is the:
   (a) stroke volume (SV)
   (b) end-diastolic volume (EDV)
   (c) end-systolic volume (ESV)
   (d) cardiac output (CO)

10. What role do the chordae tendineae and papillary muscles have in the normal function of the AV valves?

11. What is the purpose of each of the three distinct layers that make up the wall of the heart?

12. What are the principal valves found in the heart and what is the function of each?

13. Trace the normal pathway of an electrical impulse through the conducting system of the heart.

14. (a) What is the cardiac cycle? (b) What phases and events are necessary to complete the cardiac cycle?

15. What three important factors regulate stroke volume to ensure that the left and right ventricles pump equal volumes of blood?

LEVEL 2 Reviewing Concepts

16. During the absolute refractory period, a cell membrane cannot respond at all because:
   (a) sodium channels are either open or closed and inactivated
   (b) potassium channels are temporarily closed
   (c) the membrane is hyperpolarized
   (d) a, b, and c are correct

17. Tetanic muscle contractions cannot occur in a normal cardiac muscle cell because:
   (a) cardiac muscle tissue contracts on its own
   (b) there is no neural or hormonal stimulation
   (c) the refractory period lasts until the muscle cell relaxes
   (d) the refractory period ends before the muscle cell reaches peak tension

18. The amount of blood forced out of the heart depends on:
   (a) the degree of stretching at the end of ventricular diastole
   (b) the contractility of the ventricle
   (c) the amount of pressure required to eject blood
   (d) a, b, and c are correct

19. The cardiac output cannot increase indefinitely because:
   (a) available filling time becomes shorter as the heart rate increases
   (b) cardiovascular centers adjust the heart rate
   (c) the rate of spontaneous depolarization decreases
   (d) the ion concentrations of pacemaker cell membranes decrease

20. Describe the association of the four muscular chambers of the heart with the pulmonary and systemic circuits.

21. What are the source and significance of the heart sounds?

22. (a) Differentiate between stroke volume and cardiac output. (b) How is cardiac output calculated?

23. What factors are used to calculate cardiac output?

24. What is the significance of preload to cardiac muscle cells?

25. (a) What effect does sympathetic stimulation have on the heart? (b) What effect does parasympathetic stimulation have on the heart?

26. Describe the effects that epinephrine, norepinephrine, glucagon, and thyroid hormones have on the contractility of the heart.

LEVEL 3 Critical Thinking and Clinical Applications

27. Most of the ATP produced in cardiac muscle is derived from the metabolism of fatty acids. During times of exertion, cardiac muscle cells can use lactic acid as an energy source. Why would this adaptation be advantageous to cardiac function?

28. A patient's ECG tracing shows a consistent pattern of two P waves followed by a normal QRS complex and T wave. What is the cause of this abnormal wave pattern?

29. The following measurements were made on two individuals. (The values recorded remained stable for 1 hour.)

Person 1: heart rate, 75 bpm; stroke volume, 60 ml
Person 2: heart rate, 90 bpm; stroke volume, 95 ml

Which person has the greater venous return? Which person has the longer ventricular filling time?

30. Karen is taking the medication *verapamil*, a drug that blocks the calcium channels in cardiac muscle cells. What effect would you expect this medication to have on Karen's stroke volume?

31. After a myocardial infarction, the cells surrounding the damaged tissue frequently become hyperexcitable and act as ectopic pacemakers. This condition can lead to abnormal ventricular rhythms, with fatal consequences. What would cause the excitability of the uninjured cells?

32. Preventricular contractions (PVCs) occur when a Purkinje cell or contractile cell in the ventricle depolarizes to threshold, triggering a premature contraction. A person experiencing a PVC may frequently feel that her heart has "skipped a beat." If the PVC causes an extra contraction, why does the person feel that a beat has been skipped?

CHAPTER 21

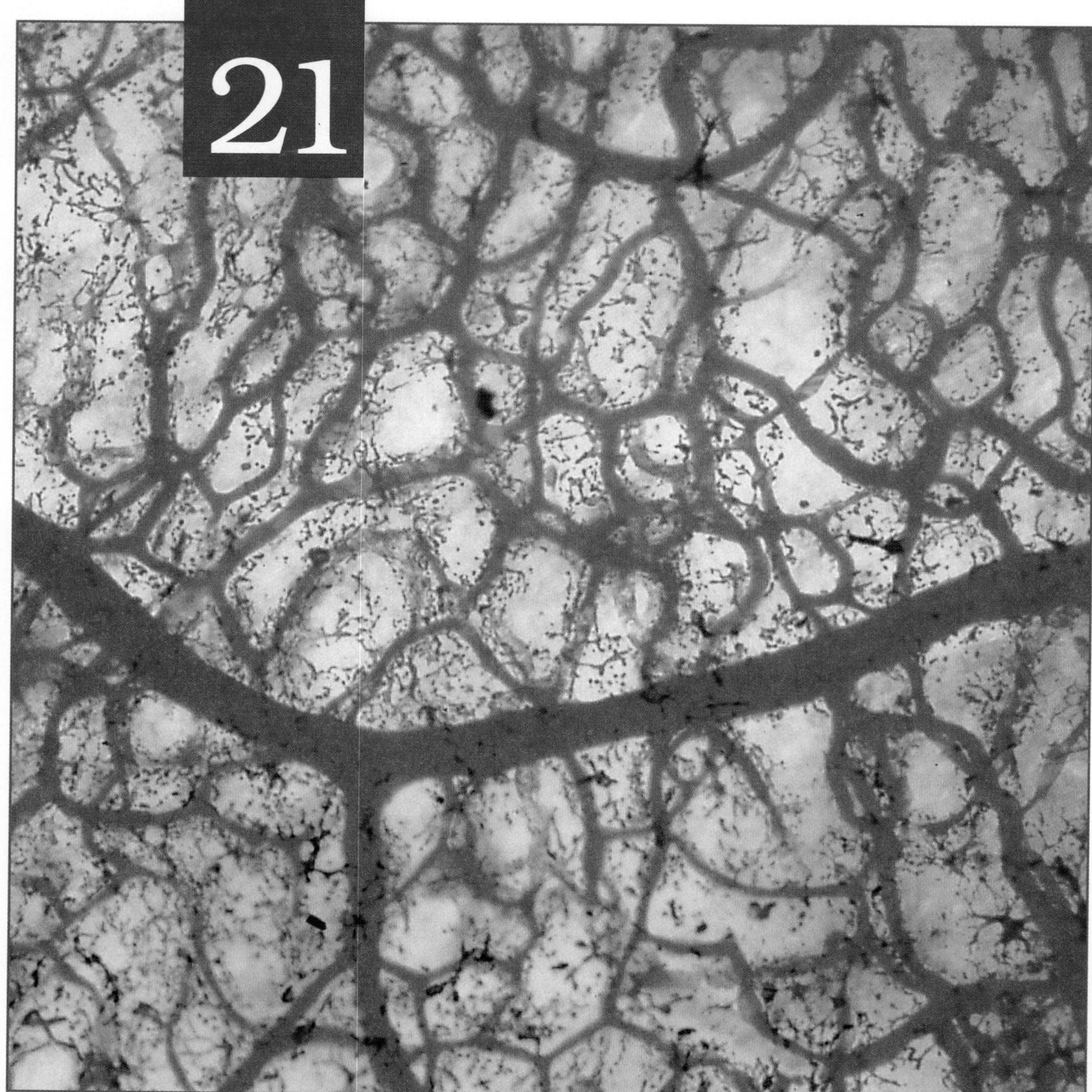

When we think of the cardiovascular system, we probably think first of the heart, or of the great blood vessels that leave it and return to it. But the real work of the cardiovascular system is done here, in microscopic vessels that permeate most tissues. This is a network of capillaries, delicate vessels that permit diffusion between the blood and the interstitial fluid. The cardiovascular system performs all of its major functions within capillary networks like this. Capillaries weave their way throughout active tissues, forming intricate networks that surround muscle fibers, radiate through connective tissues, and branch beneath the basement membranes of epithelia. These capillaries provide oxygen and nutrients and remove carbon dioxide and wastes. Homeostatic mechanisms operating at the local, regional, and systemic levels adjust blood flow through the capillaries to meet the demands of peripheral tissues. This chapter considers these homeostatic mechanisms in detail, and examines the structure and distribution of capillaries and other blood vessels.

Blood Vessels and Circulation

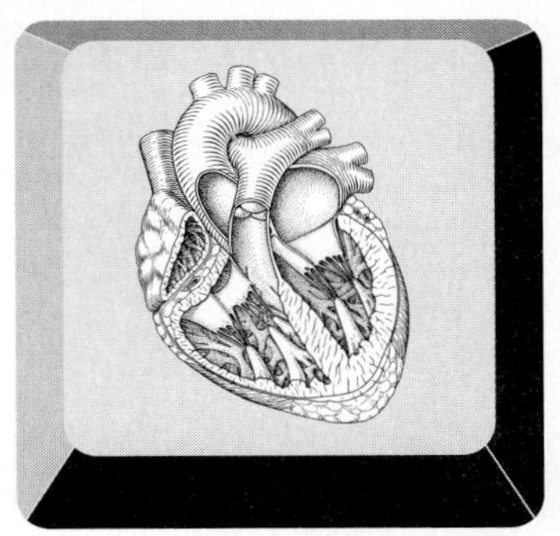

Chapter Outline and Objectives

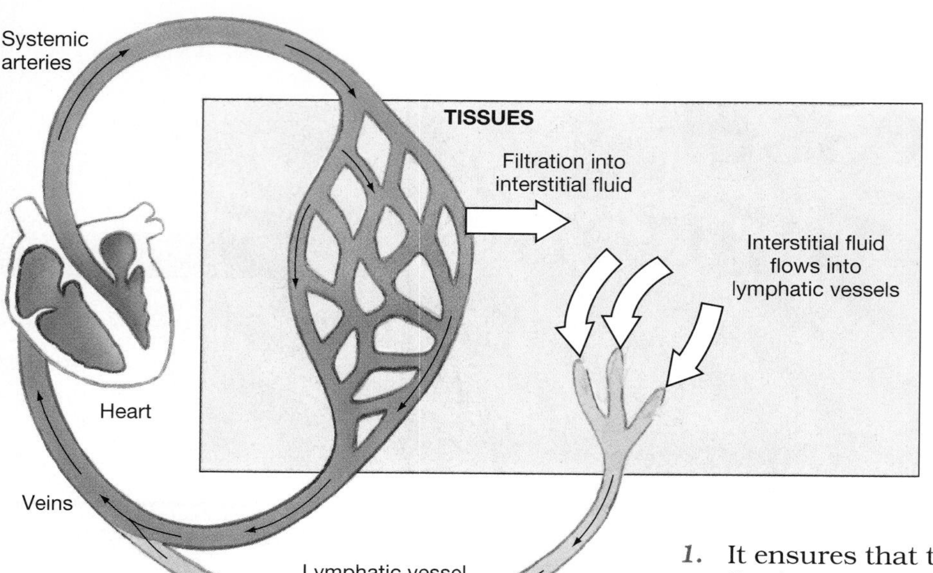
Systemic arteries
TISSUES
Filtration into interstitial fluid
Interstitial fluid flows into lymphatic vessels
Heart
Veins
Lymphatic vessel
Lymph flow

● **FIGURE 21-1**
The Circulation of Extracellular Fluid. There is continual movement of fluid from the plasma into the interstitial fluid at capillaries and back to the plasma via the lymphatic system.

In the last two chapters we examined the composition of the blood and the structure and function of the heart, whose pumping action keeps blood in motion. We will now consider the vessels that carry blood to peripheral tissues and the nature of the exchange that occurs between the blood and interstitial fluids of the body.

Blood leaves the heart in the pulmonary and aortic trunks, each with a diameter of around 2.5 cm (1 in.). These vessels branch repeatedly, forming the major arteries that distribute blood to body organs. Within these organs further branching occurs, creating several hundred million tiny arteries that provide blood to more than 10 billion capillaries barely the diameter of a single red blood cell. These capillaries form extensive, branching networks; if all the capillaries in the body were placed end to end they would circle the globe, with a combined length of over 25,000 miles.

The vital functions of the cardiovascular system depend entirely on events at the capillary level: *All chemical and gaseous exchange between the blood and interstitial fluid takes place across capillary walls.* Tissue cells rely on capillary diffusion to obtain nutrients and oxygen and to remove metabolic wastes, such as carbon dioxide and urea. Diffusion occurs very rapidly because the distances involved are very small—few living cells lie farther than 125 μm (0.005 in.) from a capillary.

As blood flows through peripheral tissues, blood pressure forces water and solutes across capillary walls. This fluid, about 3.6 liters per day, flows through peripheral tissues and enters the *lymphatic system,* which empties into the bloodstream (Figure 21-1●). The continual movement of water out of the capillaries, through peripheral tissues, then back to the bloodstream via the lymphatic system has several important functions:

1. It ensures that the plasma and the interstitial fluid, two major components of the extracellular fluid, are in constant communication.
2. It accelerates the distribution of nutrients, hormones, and dissolved gases throughout the tissue.
3. It assists in the transport of insoluble lipids and tissue proteins that cannot enter the circulation by crossing the capillary lining.
4. It has a flushing action that carries bacterial toxins and other chemical stimuli to lymphatic tissues and organs responsible for providing immunity from disease.

This chapter begins with a description of the histological organization of arteries, capillaries, and veins. We will then consider the functions of these vessels and basic principles of cardiovascular regulation. The final section of the chapter examines the distribution of major blood vessels in the body. We will then be ready to consider the organization and function of the lymphatic system, the focus of Chapter 22.

The Anatomy of Blood Vessels

Arteries carry blood away from the heart toward capillaries, and **veins** return blood from capillaries to the heart. As they approach the capillaries, arteries branch repeatedly, and the branches decrease in diameter. The smallest arterial branches are called **arterioles** (ar-TĒ-rē-ōlz). From the arterioles blood moves into the capillaries. By regulating smooth muscle tone in the walls of the arterioles, cardiovascular centers in the brain ensure that the volume, pressure, and speed of blood movement through peripheral capillaries remain within acceptable limits.

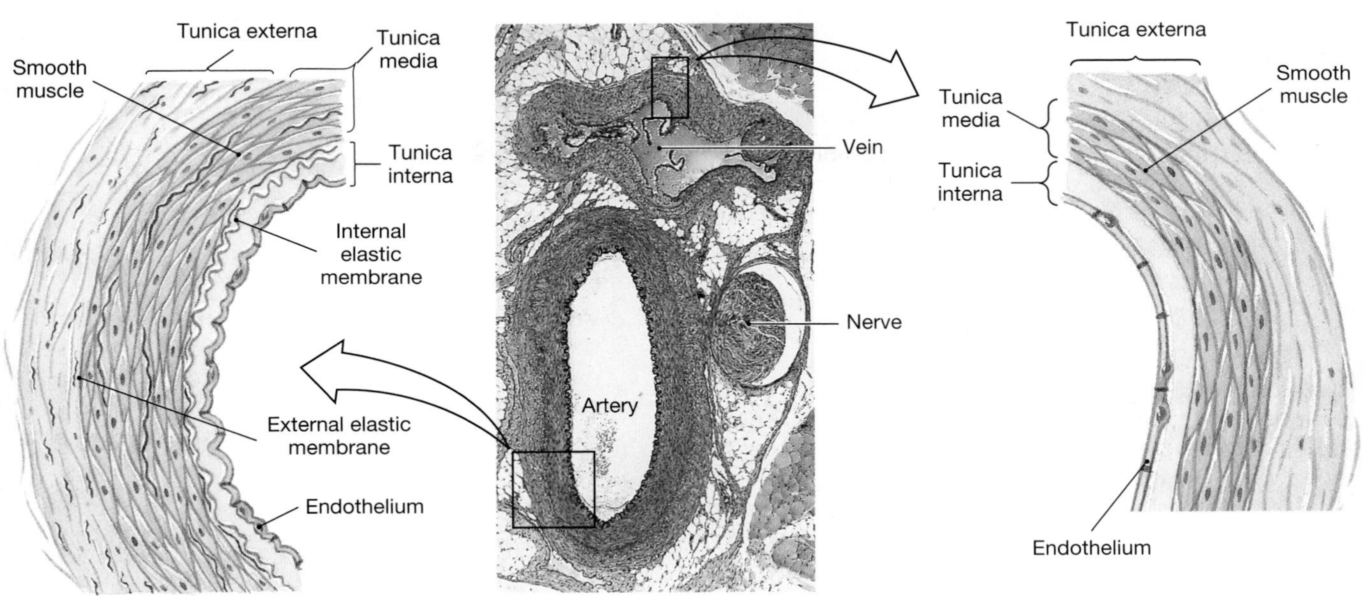
Tunica externa
Smooth muscle
Tunica media
Tunica interna
Internal elastic membrane
External elastic membrane
Endothelium
Vein
Nerve
Artery
Tunica externa
Tunica media
Tunica interna
Smooth muscle
Endothelium

Feature	*Typical Artery*	*Typical Vein*
General appearance in sectional view	Usually round, with relatively thick wall	Usually flattened or collapsed, with relatively thin wall
Tunica interna		
Endothelium	Usually rippled due to vessel constriction	Often smooth
Internal elastic membrane	Present	Absent
Tunica media	Thick, dominated by smooth muscle and elastic fibers	Thin, dominated by smooth muscle and collagen fibers
External elastic membrane	Present	Absent
Tunica externa	Collagen and elastic fibers	Collagen, elastic, and smooth muscle fibers

● **FIGURE 21-2**
A Comparison of a Typical Artery and Typical Vein. (LM × 74)

❑ Structure of Vessel Walls

Figure 21-2

The walls of arteries and veins contain three distinct layers: (1) an inner *tunica interna* (in-TER-na), or *tunica intima;* (2) a middle *tunica media;* and (3) an outer *tunica externa* (eks-TER-na), or *tunica adventitia* (ad-ven-TISH-ē-a) (Figure 21-2●).

- The **tunica interna,** or *tunica intima,* is the innermost layer of a blood vessel. This layer includes the endothelial lining and an underlying layer of connective tissue with abundant elastic fibers. In arteries the outer margin of the tunica interna contains a thick layer of elastic fibers called the **internal elastic membrane.**
- The **tunica media,** the middle layer, contains concentric sheets of smooth muscle tissue in a framework of loose connective tissue. The collagen and elastic fibers bind the tunica media to the tunica interna and tunica externa. The tunica media is often the thickest layer in the wall of a small artery. It is separated from the surrounding tunica externa by a thin band of elastic fibers, the **external elastic membrane.** The smooth muscle fibers of the tunica media encircle the endothelium lining the lumen of the blood vessel. When these smooth muscles contract, the vessel decreases in diameter, and when they relax, the diameter increases. Large arteries also contain layers of longitudinally arranged smooth muscle fibers.

- The outer **tunica externa,** or *tunica adventitia,* forms a connective tissue sheath around the vessel. In arteries this layer contains collagen fibers with scattered bands of elastic fibers. In veins this layer, which is usually thicker than the tunica media, contains networks of elastic fibers and bundles of smooth muscle fibers. The connective tissue fibers of the tunica externa typically blend into those of adjacent tissues, stabilizing and anchoring the blood vessel.

Their layered walls give arteries and veins considerable strength. The muscular and elastic components also permit controlled alterations in diameter as blood pressure or blood volume changes. However, the walls of arteries and veins are too thick to allow diffusion between the bloodstream and surrounding tissues, or even between the blood and the tissues of the vessel itself. For this reason the walls of large vessels contain small arteries and veins that supply the smooth muscle fibers and fibroblasts of the tunica media and tunica externa. These blood vessels are called the **vasa vasorum** ("vessels of vessels").

❑ Differences between Arteries and Veins

Figure 21-2

Arteries and veins supplying the same region typically lie side by side (Figure 21-2●). In sectional view, arteries and veins may be distinguished by the following characteristics:

1. In general, the walls of arteries are thicker than those of veins. The tunica media of an artery contains more smooth muscle and elastic fibers than does that of a vein. These contractile and elastic components resist the pressure generated by the heart as it forces blood into the circuit.
2. When not opposed by blood pressure, arterial walls contract. Thus, when seen on dissection or in sectional view (Figure 21-2●), the lumen of an artery looks smaller than that of the corresponding vein. Because the walls of arteries are relatively thick and strong, they retain their circular shape in section. Cut veins tend to collapse, and in section they often look flattened or grossly distorted.
3. The endothelial lining of an artery cannot contract, so when an artery constricts the endothelium is thrown into folds that give arterial sections a pleated appearance. The lining of a vein looks like a typical endothelial layer.

In gross dissection, arteries and veins are often seen running through a tissue side by side. They can usually be distinguished because:

1. The thicker walls of the arteries can be felt if the vessels are compressed.
2. Arteries usually retain their cylindrical shape, while veins often collapse.
3. Arteries are more resilient; when stretched, they keep their shape and elongate, and when released they snap back. A small vein cannot tolerate as much distortion without collapsing or tearing.
4. Veins often contain *valves,* internal structures that prevent the backflow of blood toward the capillaries. In an intact vein, the location of each valve is marked by a slight distension of the vessel wall. (Valve structure will be considered in a later section.)

❑ Arteries

Figure 21-3

Their relatively thick walls give arteries the properties of *elasticity* and *contractility.* Elasticity permits passive changes in vessel diameter in response to alterations in blood pressure. It allows arteries to absorb the pressure pulses that accompany the contractions of the ventricles.

The contractility of the arterial walls gives them the ability to change their diameters actively, primarily under the control of the sympathetic division of the ANS. When stimulated, arterial smooth muscles constrict, a process called **vasoconstriction.** Relaxation of the smooth muscles causes an increase in the diameter of the lumen, a process called **vasodilation.** Vasoconstriction and vasodilation affect (1) the afterload on the heart, (2) peripheral blood pressure, and (3) capillary blood flow. These effects will be explored in a later section. Contractility is also important during the *vascular phase* of hemostasis, when contraction of a damaged vessel wall helps reduce bleeding.

In traveling from the heart to peripheral capillaries, blood passes through *elastic arteries, muscular arteries,* and *arterioles* (Figure 21-3●).

Elastic Arteries

Figure 21-3

Elastic arteries, or *conducting arteries,* are large vessels with diameters of up to 2.5 cm (1 in.). These vessels transport large volumes of blood away from the heart. The pulmonary and aortic trunks and their major arterial branches (the *common carotid, subclavian,* and *common iliac arteries*) are examples of elastic arteries.

The walls of elastic arteries (Figure 21-3●) are not very thick relative to the vessel diameter, but they are extremely resilient. The tunica media of these vessels contains a high density of elastic fibers and relatively few smooth muscle fibers. As a result, elastic arteries are able to tolerate the pressure changes that occur during the cardiac cycle. When ventricular systole occurs, pressures rise rapidly and the elastic arteries are stretched. During ventricular diastole, blood pressure within the arterial system falls, and the elastic fibers recoil to their original dimensions. Their expansion cushions the sudden rise in pressure during ventricular systole, and their contraction slows the decline in pressure during ventricular diastole. This feature is important because blood pressure is the driving force behind blood flow, and the greater the pressure oscillations the greater the changes in blood flow. The elasticity of the arterial system dampens the pressure peaks and valleys that accompany the heartbeat. By the time blood reaches the arterioles, the pressure oscillations have completely disappeared, and blood flow is continuous.

Muscular Arteries

Figures 21-2, 21-3

Muscular arteries, also known as *medium-sized arteries* or *distribution arteries,* distribute blood to the body's skeletal muscle and internal organs. A typical muscular artery has a diameter of approx-

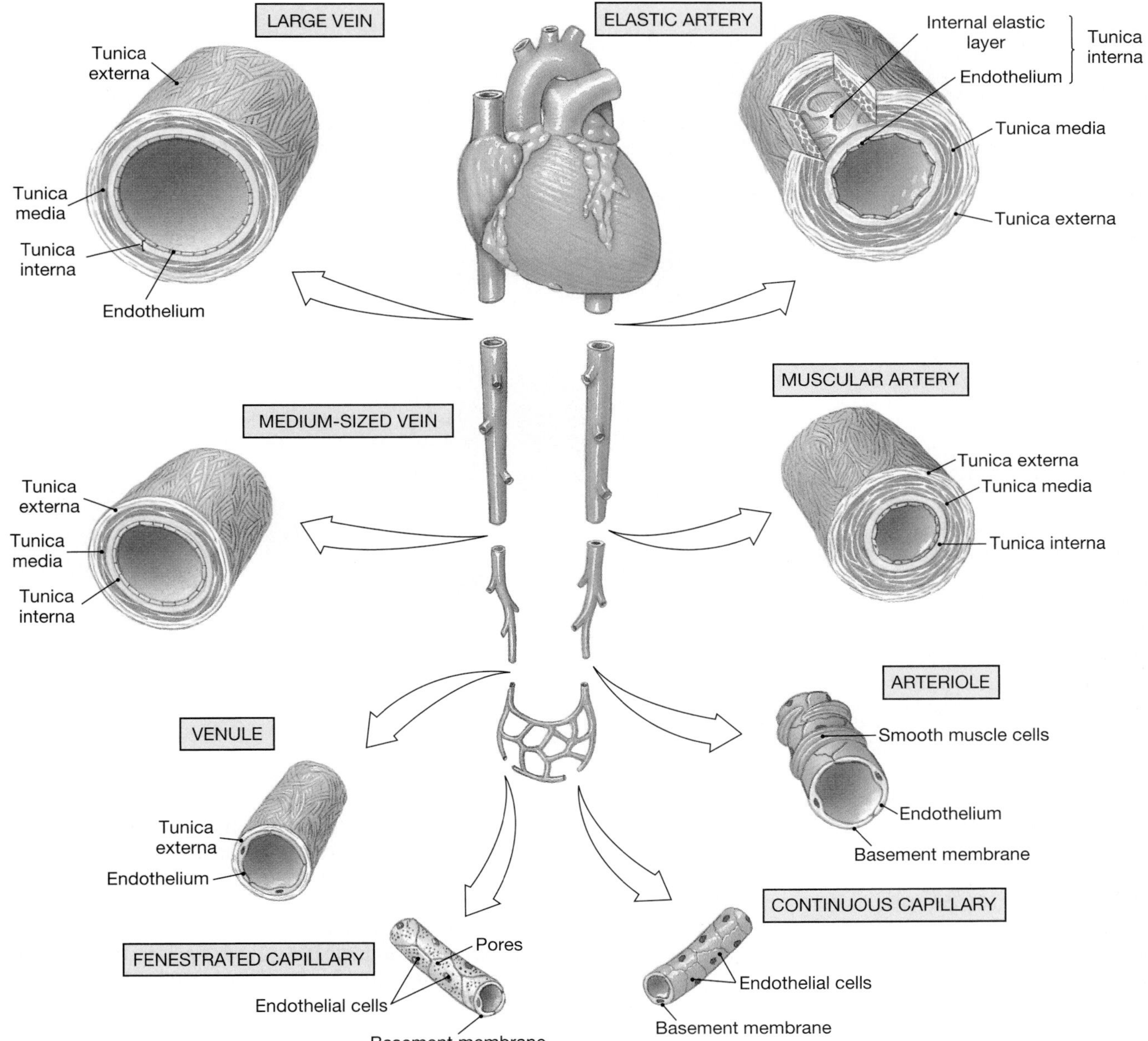
LARGE VEIN
Tunica externa
Tunica media
Tunica interna
Endothelium
ELASTIC ARTERY
Internal elastic layer
Endothelium
Tunica interna
Tunica media
Tunica externa
MEDIUM-SIZED VEIN
Tunica externa
Tunica media
Tunica interna
MUSCULAR ARTERY
Tunica externa
Tunica media
Tunica interna
ARTERIOLE
Smooth muscle cells
Endothelium
Basement membrane
VENULE
Tunica externa
Endothelium
FENESTRATED CAPILLARY
Pores
Endothelial cells
Basement membrane
CONTINUOUS CAPILLARY
Endothelial cells
Basement membrane

● **FIGURE 21-3**

Histological Structure of Blood Vessels. Representative cross-sectional views of the walls of arteries, veins, and capillaries. Note the differences in relative size of the layers in these vessels.

Arteriosclerosis

Arteriosclerosis (ar-tē-rē-ō-skle-RŌ-sis) is a thickening and toughening of arterial walls. Although this condition may not sound life-threatening, complications related to arteriosclerosis account for roughly one-half of all deaths in the United States. There are many different forms of arteriosclerosis; for example, arteriosclerosis of coronary vessels is responsible for *coronary artery disease* (CAD), and arteriosclerosis of arteries supplying the brain can lead to strokes. [p. 697]

There are two major forms of arteriosclerosis:

- **Focal calcification** is the gradual degeneration of smooth muscle in the tunica media and the subsequent deposition of calcium salts. This process typically involves arteries of the limbs and genital organs. Some focal calcification occurs as part of the aging process, and it may develop in association with atherosclerosis. Rapid and severe calcification may occur as a complication of diabetes mellitus, an endocrine disorder considered in Chapter 18. [p. 634]
- **Atherosclerosis** (ath-er-ō-skle-RŌ-sis) is associated with damage to the endothelial lining and the formation of lipid deposits in the tunica media. This is the most common form of arteriosclerosis.

Many factors may be involved in the development of atherosclerosis. One major factor is lipid levels in the blood. Atherosclerosis tends to develop in persons whose blood contains elevated levels of plasma lipids, specifically cholesterol. Circulating cholesterol is transported to peripheral tissues in lipoproteins, protein-lipid complexes. (The various types of lipoproteins and their interrelationships are discussed in Chapter 25.) Recent evidence indicates that many forms of atherosclerosis are associated with either (1) low levels of *apolipoprotein-E* (ApoE), a transport protein whose lipids are quickly removed by peripheral tissues, or (2) high levels of *lipoprotein(a)*, a *low-density lipoprotein* (LDL) that is removed at a much slower rate.

When ApoE levels are low, or lipoprotein(a) levels are high, cholesterol-rich lipoproteins remain in circulation for an extended period. Circulating monocytes then begin removing them from the bloodstream. Eventually the monocytes become filled with lipid droplets. Now called *foam cells*, they attach themselves to the endothelial walls of blood vessels, where they release growth factors. These cytokines stimulate the divisions of smooth muscle fibers near the tunica interna, thickening the vessel wall.

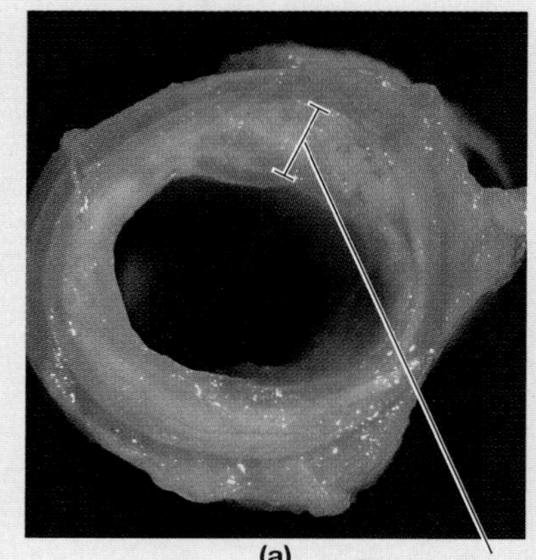
(a)

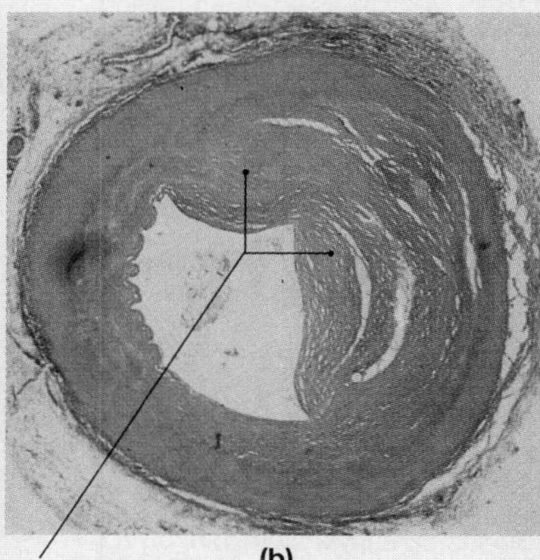
(b)
Plaque deposit in vessel wall

● **FIGURE 21-4**
A Plaque Blocking a Peripheral Artery. (a) A section of a coronary artery narrowed by plaque formation. **(b)** Sectional view of a large plaque. (LM × 18)

imately 0.4 cm (0.15 in.). Muscular arteries are characterized by a thick tunica media containing a greater amount of smooth muscle fibers than elastic arteries (Figures 21-2, p. 719, and 21-3●). The *external carotid arteries* of the neck, the *brachial arteries* of the arms, and the *femoral arteries* of the thighs are examples of muscular arteries.

Arterioles

Figure 21-3

Arterioles are considerably smaller than muscular arteries, with an internal diameter of 30 μm or less. Arterioles have a poorly defined tunica externa, and the tunica media in the larger arterioles consists of one or two layers of smooth muscle fibers. The tunica media of the smallest arterioles contains scattered smooth muscle fibers that do not form a complete layer (Figure 21-3●).

The smaller muscular arteries and arterioles change their diameter in response to local conditions or to sympathetic or endocrine stimulation. For example, arterioles in most tissues vasodilate when oxygen levels are low and, as we saw in Chapter 16, vasoconstrict under sympathetic stimulation. [p. 535] Because changes in their diameter affect the resistance to blood flow, arterioles are called **resistance vessels.**

Elastic and muscular arteries are interconnected, and vessel characteristics change gradu-

Other monocytes then invade the area, migrating between the endothelial cells. As these changes occur, the monocytes, smooth muscle fibers, and endothelial cells begin phagocytizing lipids as well. The result is a **plaque,** a fatty mass of tissue that projects into the lumen of the vessel. At this point the plaque has a relatively simple structure, and there is evidence that the process can be reversed if appropriate dietary adjustments are made.

If the conditions persist, the endothelial cells become swollen with lipids, and gaps appear in the endothelial lining. Platelets now begin sticking to the exposed collagen fibers, and the combination of platelet adhesion and aggregation leads to the formation of a localized blood clot that will further restrict blood flow through the artery. The structure of the plaque is now relatively complex. Plaque growth may be halted, but the structural changes are usually permanent.

Typical plaques can be seen in Figure 21-4●. Elderly individuals, especially elderly men, are most likely to develop atherosclerotic plaques. There is evidence that estrogens may slow plaque formation; this may account for the lower incidence of coronary artery disease, myocardial infarctions (MIs), and strokes in women. After menopause, when estrogen production declines, the risk of CAD, MIs, and strokes in women increases markedly.

In addition to advanced age and male sex, other important risk factors include high blood cholesterol levels, high blood pressure, and cigarette smoking. Roughly 20 percent of middle-aged men have all three of these risk factors; these individuals are four times more likely to experience an MI or cardiac arrest than are other men in their age group. Although fewer women develop this condition, elderly women smokers with high blood cholesterol and high blood pressure are at much greater risk than other women. Other factors that may promote development of atherosclerosis in both men and women include diabetes mellitus, obesity, and stress. There is also evidence that at least some forms of atherosclerosis may be linked to chronic infection with *Chlamydia pneumoniae,* a bacterium responsible for several types of respiratory infections, including some forms of pneumonia.

Potential treatments for atherosclerotic plaques, such as catheterization, balloon angioplasty, and bypass surgery, were discussed in Chapter 20. ∞ *[p. 697]* In cases where dietary modifications do not lower circulating LDL levels sufficiently, there are drug therapies that can bring them under control. Genetic engineering techniques have recently been used to treat an inherited form of *hypercholesterolemia* (high blood cholesterol) linked to extensive plaque formation. (The patients were unable to absorb and recycle cholesterol in the liver.) In this experimental procedure, circulating cholesterol levels declined after copies of appropriate genes were inserted into some of the individual's liver cells.

Without question, the best approach to atherosclerosis is to try and avoid it by eliminating or reducing associated risk factors. Suggestions include: (1) reducing the amount of dietary cholesterol and saturated fats by restricting consumption of fatty meats (such as beef, lamb, and pork), egg yolks, and cream; (2) giving up smoking (or never starting to begin with); (3) checking your blood pressure and taking steps to lower it if necessary; (4) having your blood cholesterol levels checked at annual physical examinations; (5) controlling your weight; and (6) exercising regularly.

ally as one travels away from the heart. For example, the largest of the muscular arteries contain a considerable amount of elastic tissue, while the smallest resemble heavily muscled arterioles.

ANEURYSMS An **aneurysm** (AN-ū-rizm) is a bulge in the weakened wall of an artery. This bulge resembles a bubble in the wall of a tire, and like a bad tire, the affected artery may suffer a catastrophic blowout. The most dangerous aneurysms are those involving arteries of the brain, where they cause strokes, and of the aorta, where a ruptured aneurysm will cause fatal bleeding in a matter of seconds.

Aneurysms most often occur in individuals with *arteriosclerosis.* Over time, arteriosclerosis causes vessel walls to become less elastic, and a weak spot may develop. Aneurysms may also be associated with other conditions that weaken arterial walls, such as arterial inflammation or infection, and with *Marfan's syndrome,* a connective tissue disorder introduced in Chapter 4. ∞ *[p. 125]* As you might expect, individuals with high blood pressure are most likely to develop dangerous aneurysms, because the elevated arterial pressures place greater stresses on the vessel walls. Unfortunately, because they are often painless, aneurysms are likely to go undetected. Treatment options for aneurysms are discussed in the *Applications Manual.* AM *Aneurysms*

❑ Capillaries

Figure 21-5

Capillaries are the only blood vessels whose walls permit exchange between the blood and the surrounding interstitial fluids. Because capillary walls are relatively thin, the diffusion distances are small, and exchange can occur quickly. In addition, blood flows through capillaries relatively slowly, allowing sufficient time for diffusion or active transport of materials across the capillary walls. Thus, the histological structure of capillaries permits a two-way exchange of substances between blood and interstitial fluid.

A typical capillary consists of an endothelial tube inside a delicate basement membrane. There is neither a tunica media nor a tunica externa. The average diameter of a capillary is a mere 8 µm, very close to that of a single red blood cell. There are two major types of capillaries: *continuous capillaries* and *fenestrated capillaries* (Figure 21-5●).

Continuous Capillaries

Figure 21-5a

In most regions of the body the capillary endothelium is a complete lining, with the endothelial cells connected by intermediate junctions. These vessels are called **continuous capillaries** (Figure 21-5a●). Continuous capillaries are found in connective tissues, muscle tissue, and neural tissue. In connective tissues, muscle tissues, and the PNS, continuous capillaries permit the diffusion of water, small solutes, and lipid-soluble materials into the surrounding interstitial fluid but prevent the loss of blood cells and plasma proteins. In addition, some transport may occur via vesicles that form at the inner endothelial surface. This form of vesicular transport, considered more closely in a later section, was introduced in Chapter 3. ∞ [p. 81]

The endothelial cells in specialized continuous capillaries in the CNS, thymus gland, and testes are bonded together by tight junctions. These capillaries have very restricted permeability characteristics. One example, the capillaries responsible for the *blood-brain barrier*, was detailed in Chapter 14. ∞ [p. 467]

Fenestrated Capillaries

Figure 21-5b

Fenestrated (FEN-es-trā-ted) **capillaries** (*fenestra*, window) are capillaries that contain "windows," or pores, that span the endothelial lining (Figure 21-5b●). The pores permit the rapid exchange of water and solutes as large as small peptides between the plasma and interstitial fluid. Examples of fenestrated capillaries noted in earlier chapters include the *choroid plexus* of the brain and the blood vessels in a variety of endocrine organs, including the hypothalamus, the pituitary, the pineal, the adrenal cortex, and the thyroid gland. Fenestrated capillaries are also found along absorptive areas of the intestinal tract and at filtration sites in the kidneys. Both the number of pores and their permeability characteristics may vary from one region of the capillary to another.

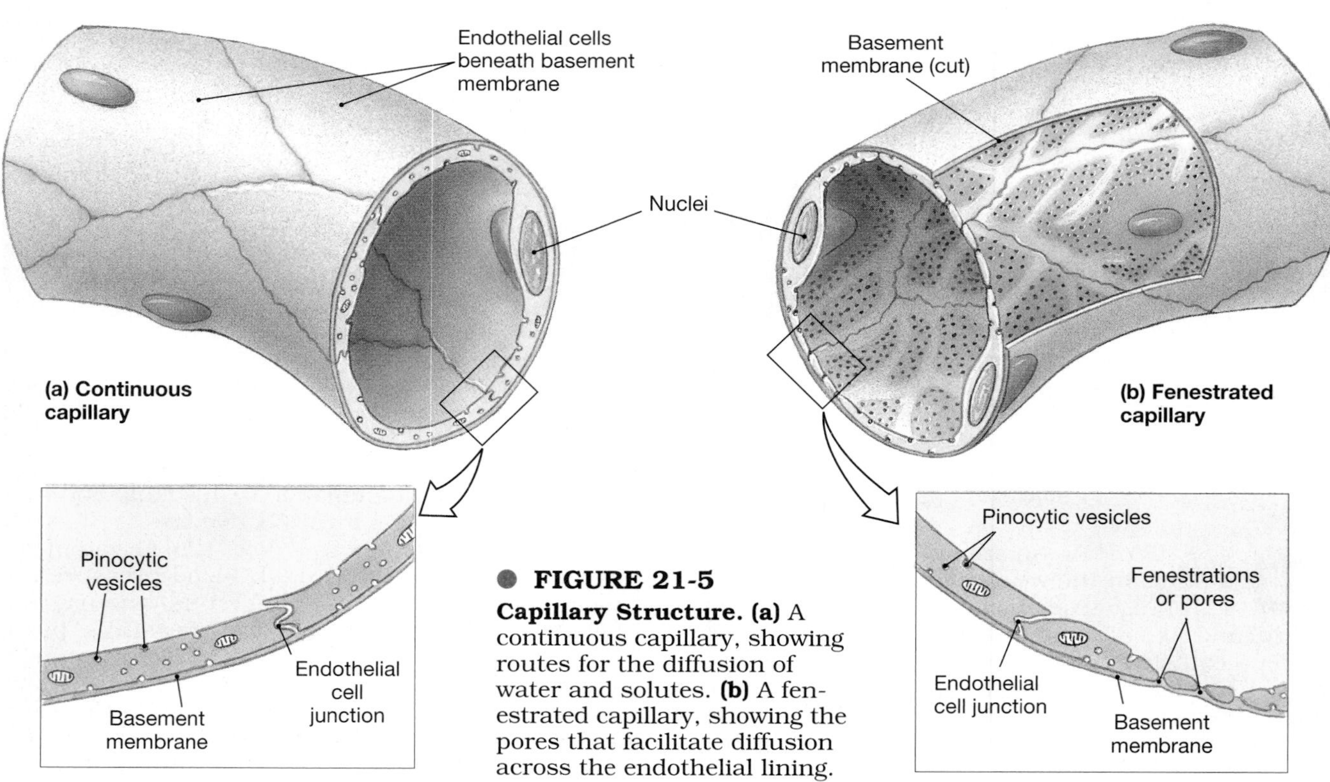
Endothelial cells beneath basement membrane
Nuclei
Basement membrane (cut)
(a) Continuous capillary
(b) Fenestrated capillary
Pinocytic vesicles
Endothelial cell junction
Basement membrane
Pinocytic vesicles
Fenestrations or pores
Endothelial cell junction
Basement membrane

● **FIGURE 21-5**
Capillary Structure. (a) A continuous capillary, showing routes for the diffusion of water and solutes. **(b)** A fenestrated capillary, showing the pores that facilitate diffusion across the endothelial lining.

SINUSOIDS **Sinusoids** (SĪ-nus-oydz) are specialized fenestrated capillaries that are flattened and irregular. In contrast to other fenestrated capillaries, sinusoids often have gaps between adjacent endothelial cells. As a result, sinusoids permit the free exchange of water and solutes as large as plasma proteins.

Blood moves through sinusoids relatively slowly, maximizing the time available for absorption and secretion across the sinusoidal walls. Sinusoids are found in the liver, bone marrow, and adrenal glands. At the liver sinusoids, plasma proteins secreted by the liver cells enter the circulation. Along sinusoids of the liver and bone marrow, phagocytic cells monitor the passing blood, engulfing damaged red blood cells, pathogens, and cellular debris.

Capillary Beds

Figure 21-6

Capillaries do not function as individual units, but as part of an interconnected network called a **capillary plexus,** or **capillary bed** (Figure 21-6•). A single arteriole usually gives rise to dozens of capillaries that empty into several *venules,* the smallest vessels of the venous system. The entrance to each capillary is guarded by a band of smooth muscle, called a **precapillary sphincter.** Contraction of the smooth muscle fibers constricts and narrows the diameter of the capillary entrance, thereby reducing the flow of blood. Relaxation of the sphincter dilates the opening, allowing blood to enter the capillary at a faster rate.

Within the capillary bed, **preferred channels,** or *central channels,* provide a relatively direct means of communication between arterioles and venules. The arteriolar segment of the channel contains smooth muscles capable of altering its diameter. This segment is often called a **metarteriole** (met-ar-TĒ-rē-ōl) (Figure 21-6a•). The rest of the central channel, which resembles a typical capillary in structure, is called a **thoroughfare channel.**

A single capillary bed may receive blood from more than one artery. The arteries, called **collaterals,** enter the region and fuse before giving rise to arterioles. The fusion of collaterals supplying one or more capillary beds is an example of an **arterial anastomosis.** The interconnections between the anterior and posterior interventricular arteries of the heart are arterial anastomoses. ∞ *[p. 692]* An arterial anastomosis acts like an insurance policy for the region, for if one artery is compressed or blocked, capillary circulation will continue.

Arteriovenous (ar-tē-rē-ō-VĒ-nus) **anastomoses** are direct connections between arterioles and venules. When an arteriovenous anastomosis is dilated, blood will bypass the capillary bed and flow directly into the venous circulation. The pattern of blood flow through an arteriovenous anastomosis is regulated primarily by sympathetic innervation under the control of the cardiovascular centers of the medulla oblongata.

Vasomotion

Figure 21-6b

Although blood normally flows from the arterioles to the venules at a constant rate, the flow within each capillary is quite variable. Each precapillary sphincter goes through cycles of alternately contracting and relaxing, perhaps a dozen times each minute. As a result, the blood flow within any one capillary occurs in a series of pulses rather than as a steady and constant stream. The net effect is that blood may reach the venules by one route now and by a quite different route later (Figure 21-6b•). The alteration of blood flow through capillary beds on a moment-to-moment basis is called **vasomotion.**

Vasomotion is controlled at the local level, by changes in the concentrations of chemicals and dissolved gases within the interstitial fluids. For example, when dissolved oxygen concentrations decline within a tissue, the capillary sphincters relax, and blood flow to the area increases. This process, an example of capillary *autoregulation,* will be the focus of a later section.

At rest, blood flows through roughly 25 percent of the vessels within a capillary bed. The circulatory system does not contain enough blood to maintain adequate blood flow to *all* of the capillaries in *all* of the capillary beds in the body at the same time. As a result, when many tissues become active, the blood flow through capillary beds must be coordinated. The coordination is performed by the cardiovascular centers via mechanisms detailed later in the chapter.

❑ Veins

Figures 21-2, 21-3

Veins collect blood from all tissues and organs and return it to the heart. The walls of veins are thinner than those of corresponding arteries because the blood pressure in veins is lower than that in the arteries. Veins are classified on the basis of their size, and in general veins are larger than their corresponding arteries. Review Figures 21-2, p. 719, and 21-3•, p. 721, to compare typical arteries and veins.

Venules

Venules, which collect blood from capillary beds, vary widely in size and character. An average venule has an internal diameter of roughly 20 μm. Venules smaller than 50 μm lack a tunica media altogether, and the smallest venules resemble expanded capillaries.

Medium-Sized Veins

Medium-sized veins range from 2 to 9 mm in diameter and correspond in general size to medium-sized arteries. In these veins the tunica media is thin and contains relatively few smooth muscle

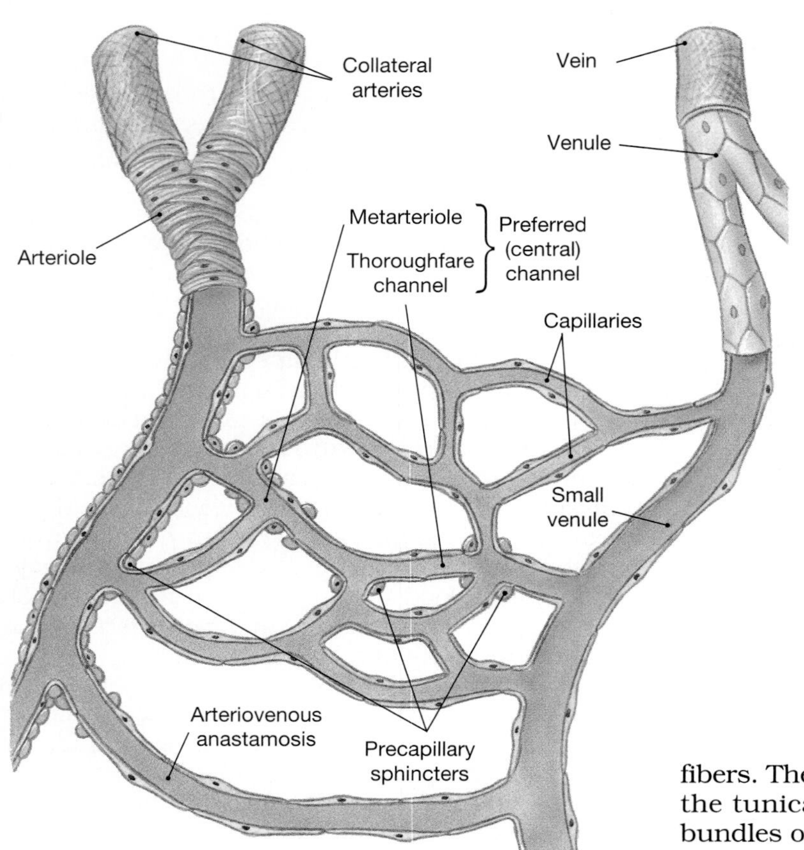
Collateral
arteries
Vein
Venule
Arteriole
Metarteriole
Thoroughfare
channel
Preferred
(central)
channel
Capillaries
Small
venule
Arteriovenous
anastamosis
Precapillary
sphincters
(a)

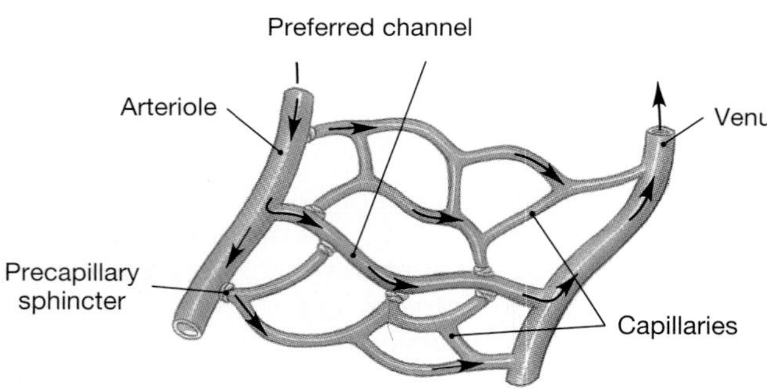
Preferred channel
Arteriole
Venule
Precapillary
sphincter
Capillaries

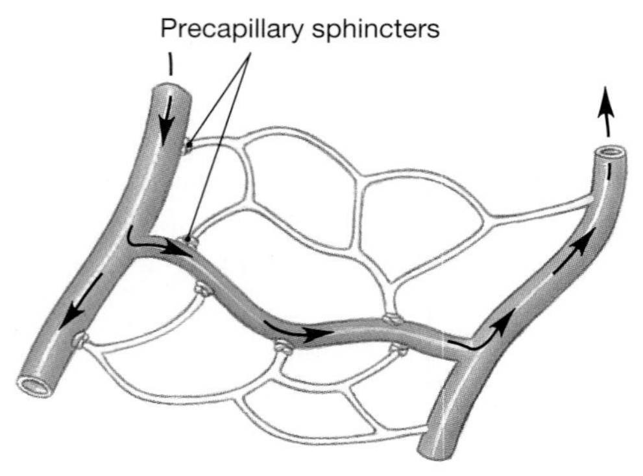
Precapillary sphincters
(c)

● **FIGURE 21-6**
Organization of a Capillary Bed. **(a)** Basic features of a typical capillary bed. **(b)** Micrograph of a capillary network. **(c)** Possible patterns of blood flow through the capillary network as vasomotion occurs. The pattern changes continually in response to local alterations in tissue oxygen demand.

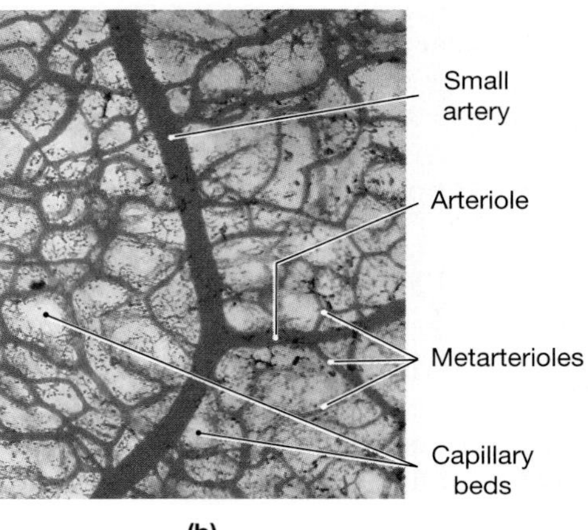
Small
artery
Arteriole
Metarterioles
Capillary
beds
(b)

fibers. The thickest layer of a medium-sized vein is the tunica externa, which contains longitudinal bundles of elastic and collagen fibers.

Large Veins

Large veins include the superior and inferior venae cavae and their tributaries within the abdominopelvic and thoracic cavities. All the tunica layers are present in large veins. The slender tunica media is surrounded by a thick tunica externa composed of a mixture of elastic and collagenous fibers.

Venous Valves

Figure 21-7

The arterial system is a high-pressure system, for it takes almost all of the force developed by the heart to push blood through the network of arteries and across miles of capillaries. Blood pressure within a peripheral venule is only about 10 percent of that in the ascending aorta, and pressures continue to fall along the venous system.

The blood pressure in venules and medium-sized veins is so low that it cannot oppose the force of gravity. In the limbs, veins of this size contain **valves** (Figure 21-7●). Valves are folds of the tunica interna that project from the vessel wall and point in the direction of blood flow. These valves act like the valves in the heart. They permit blood flow in one direction only and prevent the backflow of blood toward the capillaries.

As long as the valves function normally, any movement that distorts or compresses a vein will push blood toward the heart. For example, when a

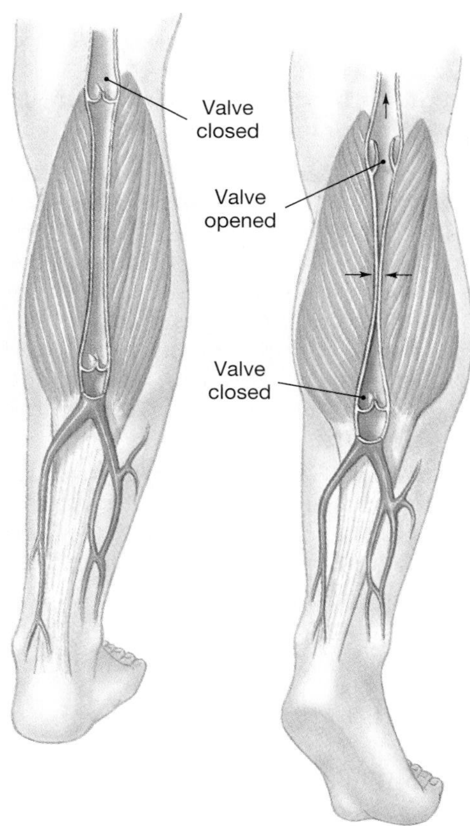
Valve
closed
Valve
opened
Valve
closed

● **FIGURE 21-7**
Function of Valves in the Venous System. Valves in the walls of medium-sized veins prevent the backflow of blood. Venous compression caused by the contraction of adjacent skeletal muscles assists in maintaining venous blood flow. Changes in body position and the thoracoabdominal pump may provide additional assistance.

person is standing, blood returning from the foot must overcome the pull of gravity to ascend to the heart. Valves compartmentalize the blood within the veins, thereby dividing the weight of the blood between the compartments. Any movement in the surrounding skeletal muscles squeezes the blood toward the heart. If the walls of the veins near the valves weaken, or if they become stretched and distorted, the valves may not work properly. Blood then pools in the veins, and the vessels become grossly distended. The effects range from mild discomfort and a cosmetic problem, as in superficial *varicose veins* in the thighs and legs, to painful distortion of adjacent tissues, as in *hemorrhoids*. These conditions and potential treatments are discussed in the *Applications Manual*.

AM *Problems with Venous Valve Function*

❑ The Distribution of Blood

Figure 21-8

The total blood volume is unevenly distributed among arteries, veins, and capillaries (Figure 21-8●). The heart, arteries, and capillaries normally contain 30–35 percent of the blood volume (roughly 1.5 liters of whole blood), and the venous system contains the rest (65–70 percent, or around 3.5 liters). Of the blood in the venous system, roughly one-third (about a liter) is held in **venous reservoirs** in the liver, bone marrow, and skin. These organs contain extensive venous networks through which large volumes of blood move very slowly.

Because their walls are thinner and contain a lower proportion of smooth muscle, veins are much more distensible than arteries. For a given rise in blood pressure, a typical vein will stretch about eight times as much as a corresponding artery. The *capacitance* of a blood vessel is the relationship between the volume of blood it contains and the blood pressure. The greater the distensibility of the vessel, the larger the capacitance. Veins are called **capacitance vessels** because they are easily distensible, and large changes in blood volume have little effect on blood pressure. If the blood volume rises or falls, the elastic walls stretch or recoil, changing the volume of blood in the venous system.

If serious hemorrhaging occurs, the *vasomotor center* of the medulla oblongata stimulates sympathetic nerves innervating smooth muscle fibers in the walls of medium-sized veins. This activity has two major effects:

1. Systemic veins contract, and this **venoconstriction** (vē-nō-kon-STRIK-shun) reduces the volume of the venous system. Reducing the amount of blood in the venous system

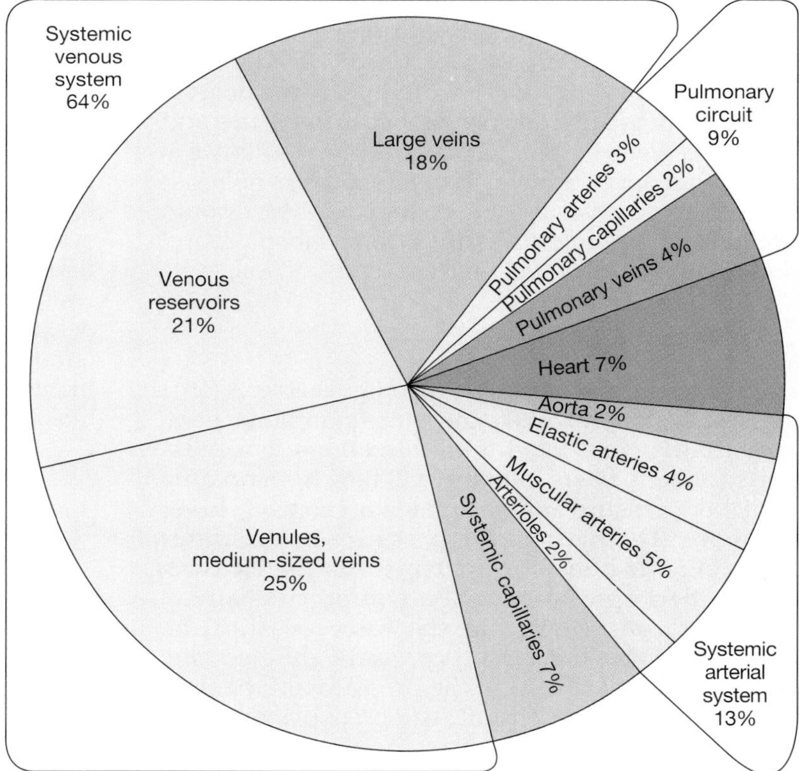
Systemic
venous
system
64%
Large veins
18%
Pulmonary
circuit
9%
Pulmonary arteries 3%
Pulmonary capillaries 2%
Pulmonary veins 4%
Venous
reservoirs
21%
Heart 7%
Aorta 2%
Elastic arteries 4%
Muscular arteries 5%
Arterioles 2%
Systemic capillaries 7%
Venules,
medium-sized veins
25%
Systemic
arterial
system
13%

● **FIGURE 21-8**
The Distribution of Blood in the Cardiovascular System

maintains the volume within the arterial system and capillaries at near-normal levels, despite a significant blood loss.

2. Blood normally held in the venous reservoirs is ejected into the circulation, increasing the volume of circulating blood. The change in volume constitutes the **venous reserve.** The venous reserve normally amounts to about 20 percent of the total blood volume.

Examination of a cross-section of tissue shows several small, thin-walled vessels with very little smooth muscle tissue in the tunica media. What type of vessels are these?

Why are valves found in veins but not in arteries?

Where in the body would you expect to find fenestrated capillaries?

Cardiovascular Physiology

The goal of cardiovascular regulation is the maintenance of adequate blood flow through peripheral tissues and organs. Under normal circumstances blood flow is equal to cardiac output. When cardiac output goes up, so does the blood flow through capillary beds; when cardiac output declines, capillary blood flow is reduced. The primary factors involved in the regulation of cardiac output were considered in Chapter 20. [p. 706] The afterload of the heart is determined by the interplay between *pressure* and *resistance* in the circulatory network. A resistance is a force that opposes blood flow. If there were no resistance in the cardiovascular system, the heart would not have to generate pressure to force blood around the pulmonary and systemic circuits.

Pressure

Liquids, including blood, are incompressible. A force exerted against a liquid generates *hydrostatic pressure* that is conducted in all directions. If a pressure gradient exists, a liquid will flow from an area of higher pressure toward an area of relatively lower pressure. The hydrostatic pressure in the water pipes of your house or apartment is higher than atmospheric pressure; when you open a faucet, water flows out. The greater the water pressure, the faster the water flow. In other words, the flow rate is directly proportional to the pressure gradient.

In the systemic circuit, the pressure gradient is the **circulatory pressure,** the pressure difference between the base of the ascending aorta and the entrance to the right atrium. Circulatory pressures average around 100 mm Hg. This relatively high pressure is needed primarily to force blood through the arterioles—*resistance vessels*—and into peripheral capillaries.

For convenience, the circulatory pressure is often divided into three components:

1. ***Blood pressure.*** When referring to arterial pressure, we will use the term **blood pressure (BP)** to distinguish it from the total circulatory pressure. Capillary blood flow is directly proportional to blood pressure, which is closely regulated by a combination of neural and hormonal mechanisms. Blood pressure in the systemic arterial system ranges from an average of 100 mm Hg to roughly 35 mm Hg at the start of a capillary network.
2. ***Capillary pressure.*** **Capillary pressure** is the pressure within capillary beds. Along the length of a typical capillary, pressures decline from roughly 35 mm Hg to about 18 mm Hg.
3. ***Venous pressure.*** **Venous pressure** is the pressure within the venous system. Venous pressure is quite low; the pressure gradient from the venules to the right atrium is only around 18 mm Hg.

Resistance

A *resistance* is a force that opposes movement. If you put a kink in a garden hose or place a piece of a screen across the mouth of the hose, you will increase the resistance and decrease the water flow. The resistance of the circulatory system opposes the movement of blood: the greater the resistance, the slower the blood flow.

For circulation to occur, the circulatory pressure must be great enough to overcome the **total peripheral resistance,** the resistance of the entire circulatory system. Because the resistance of the venous system is very low, for reasons detailed below, attention focuses on the **peripheral resistance (PR),** the resistance of the arterial system. For blood to flow into peripheral capillaries, blood pressure must be great enough to overcome the peripheral resistance; the higher the peripheral resistance, the lower the rate of blood flow. This relationship can be summarized as:

$$F \propto \frac{\Delta P}{R}$$

(Flow, F, is directly proportional to the pressure gradient, ΔP, and inversely proportional to resistance, R.)

Sources of peripheral resistance include *vascular resistance, viscosity,* and *turbulence.*

Vascular Resistance

Vascular resistance is the resistance of the blood vessels, and it is the largest component of peripheral resistance. *The most important factor in vascular resistance is friction between the blood and the* *vessel walls.* The amount of friction depends on the length of the vessel and its diameter.

VESSEL LENGTH Increasing the length of a blood vessel increases friction because the longer the vessel, the larger the surface area in contact with the blood. For example, you can easily blow water out of a snorkel that is 2.5 cm (1 in.) in diameter and 25 cm (10 in.) long, but you cannot blow the water out of a 15-m garden hose, because the total friction is too great. The most dramatic changes in vessel length occur between birth and maturity, as growth occurs. In an adult, vessel length can increase or decrease gradually when the individual gains or loses weight, but on a day-to-day basis this component of vascular resistance can be considered constant.

VESSEL DIAMETER Friction affects the blood primarily in a narrow zone closest to the vessel wall. In a small-diameter vessel, nearly all of the blood will be slowed down by friction with the walls. Resistance will therefore be relatively high. Blood near the center of a large-diameter vessel will not encounter any resistance from friction with the walls, and the resistance is therefore relatively low.

Differences in diameter have much more significant effects on resistance than differences in length. If there are two vessels of equal diameter, one twice as long as the other, the longer vessel will offer twice as much resistance to blood flow. But with two vessels of equal length, one twice the diameter of the other, the smaller one will offer 16 times as much resistance to blood flow. This relationship, expressed in terms of the vessel radius, r, and resistance, R, can be summarized as:

$$R \propto \frac{1}{r^4}$$

Even more significant, vessel length is constant, but vessel diameter can change. Most of the peripheral resistance occurs in the arterioles, the smallest vessels of the arterial system. As noted earlier in the chapter, arterioles are extremely muscular; the wall of an arteriole 30 μm in diameter may contain a 20-μm-thick layer of smooth muscle. When these smooth muscles contract or relax, peripheral resistance increases or decreases. Because a small change in diameter produces a large change in resistance, mechanisms that alter the diameters of arterioles provide control over peripheral resistance and blood flow.

Viscosity

Viscosity, introduced in Chapter 19, is resistance to flow caused by interactions among molecules and suspended materials in a liquid. ∞ [p. 651] Liquids of low viscosity, such as water (viscosity 1.0), flow at low pressures, whereas thick, syrupy fluids, such as molasses (viscosity 300), flow only under relatively high pressures. Whole blood has a viscosity about five times that of water, because of the presence of plasma proteins and blood cells. Under normal conditions the viscosity of the blood remains stable, but anemia, polycythemia, or other disorders that affect the hematocrit (Chapter 19) also change blood viscosity and alter peripheral resistance. ∞ [p. 660]

Turbulence

High flow rates, irregular surfaces, or sudden changes in vessel diameter upset the smooth flow of blood, creating eddies and swirls. This phenomenon, called **turbulence,** slows the rate of flow and increases resistance.

Turbulence normally occurs when blood flows between the atria and ventricles and between the ventricles and the aortic and pulmonary trunks. In addition to increasing resistance, this turbulence generates the *third* and *fourth heart sounds* that can often be heard through a stethoscope. Turbulence also develops in large arteries, such as the aorta, when cardiac output and arterial flow rates are very high. Turbulent blood flow across damaged or misaligned heart valves is responsible for the sound of *heart murmurs.* Heart sounds and heart murmurs were discussed in Chapter 20. ∞ [p. 705]

Turbulence seldom occurs in smaller vessels unless their walls are damaged. For example, scar tissue formation at an injury site or the development of an atherosclerotic plaque will create abnormal turbulence and restrict blood flow. Because of the sound, or *bruit* (broo-Ē), produced by this turbulence, the presence of large plaques can often be detected with a stethoscope.

Table 21-1 provides a quick review of the terms and relationships discussed in this section.

❑ An Overview of Circulatory Pressures

Figures 21-9, 21-10

The graphs in Figure 21-9• provide an overview of the pressures, vessel diameters, and velocity of blood flow in the systemic circuit. Notice that:

1. As you proceed from the aorta toward the capillaries, the arteries branch repeatedly, and each branch is smaller in diameter than the preceding one (Curve 1).
2. Although they are small in diameter, the number of arterioles and capillaries is very great (Curve 2). As the number of vessels increases, so does their total cross-sectional area. For example, the aorta has a cross-sectional area of 4.5 cm^2, but peripheral capillaries have a combined cross-sectional area of 5000 cm^2.
3. As branching proceeds, blood pressure falls rapidly. Most of the decline occurs in the small arteries and arterioles of the arterial system; venous pressures are relatively low (Curve 3).
4. Like a fast-flowing river delivering water to a floodplain, the blood flow decreases in veloci-

TABLE 21-1 A Mini-Glossary of Terms and Relationships Pertaining to Blood Circulation

Blood flow (F): The volume of blood flowing per unit of time through a vessel or group of vessels; may refer to circulation through a capillary, a tissue, an organ, or the entire vascular network. Total blood flow is equal to cardiac output.

Hydrostatic pressure: A pressure exerted by a liquid in response to an applied force.

Blood pressure (BP): The hydrostatic pressure in the arterial system that pushes blood through capillary beds.

Circulatory pressure: The pressure difference between the base of the ascending aorta and the entrance to the right atrium.

Venous pressure: The hydrostatic pressure in the venous system.

Resistance (R): A force that opposes movement.

Peripheral resistance (PR): The resistance of the arterial system. Factors that affect peripheral resistance include:

- ***Vascular resistance:*** Resistance due to friction between the blood and the walls of the blood vessels. Increases with increasing length or decreasing diameter; vessel length is constant, but vessel diameter may change.
- ***Viscosity:*** Resistance to flow due to interactions among molecules within a liquid.
- ***Turbulence:*** Resistance due to irregular, swirling movement of blood at high flow rates or on exposure to irregular surfaces.

Total peripheral resistance: The resistance of the entire circulatory system.

Relationships among these terms:

$F \propto \Delta P$ (flow is proportional to the pressure gradient)

$F \propto 1/R$ (flow is inversely proportional to resistance)

$F \propto \Delta P/R$ (flow is directly proportional to the pressure gradient and inversely proportional to resistance)

$F \propto BP/PR$ (flow is directly proportional to blood pressure and inversely proportional to peripheral resistance)

$R \propto 1/r^4$ (resistance is inversely proportional to the fourth power of the vessel radius)

ty as the total cross-sectional area of the vessels increases (Curve 4). On the venous side of the circuit, as venules combine to form small and medium-sized veins (Curve 1), the cross-sectional area decreases (Curve 2), blood pressure declines further (Curve 3), and the blood velocity gradually increases (Curve 4).

Figure 21-10● details the blood pressure throughout the cardiovascular system. Systemic pressures are highest in the aorta, peaking at around 120 mm Hg, and reach a minimum at the entrance to the right atrium. Pressures in the pulmonary circuit are much lower than those in the systemic circuit, primarily because pulmonary vessels are much shorter and more elastic than systemic vessels, and thus provide less resistance to blood flow.

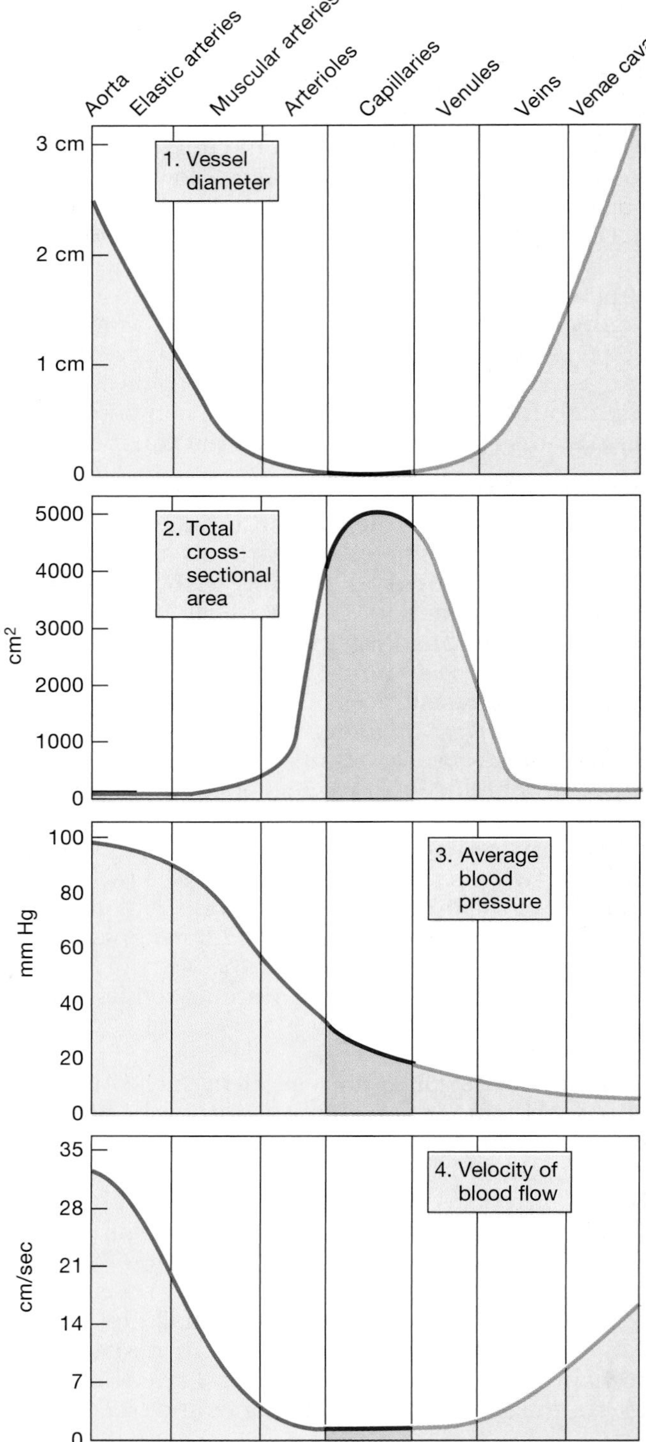
Aorta
Elastic arteries
Muscular arteries
Arterioles
Capillaries
Venules
Veins
Venae cavae
1. Vessel diameter
3 cm
2 cm
1 cm
0
2. Total cross-sectional area
cm²
5000
4000
3000
2000
1000
0
3. Average blood pressure
mm Hg
100
80
60
40
20
0
4. Velocity of blood flow
cm/sec
35
28
21
14
7
0

● **FIGURE 21-9**
The Relationships among Vessel Diameter, Blood Pressure, Cross-Sectional Area, and Blood Velocity

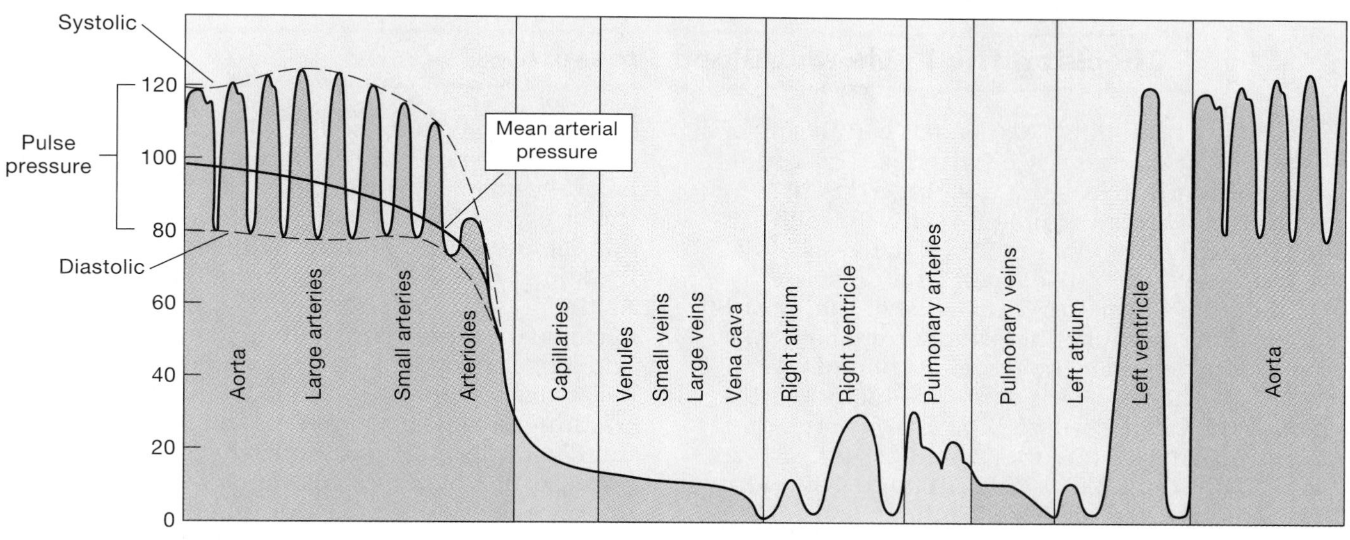
Systolic
Pulse pressure
Diastolic
Mean arterial pressure
120
100
80
60
40
20
0
Aorta
Large arteries
Small arteries
Arterioles
Capillaries
Venules
Small veins
Large veins
Vena cava
Right atrium
Right ventricle
Pulmonary arteries
Pulmonary veins
Left atrium
Left ventricle
Aorta

● **FIGURE 21-10**

Pressures within the Circulatory System. Notice the general reduction in circulatory pressures within the systemic circuit and the elimination of the pulse pressure within the arterioles. D

❑ Arterial Blood Pressure

Figure 21-10

Arterial pressures overcome peripheral resistance and maintain blood flow through capillary beds. Arterial pressure is not stable; it rises during ventricular systole and falls during ventricular diastole. The peak blood pressure measured during ventricular systole is called **systolic pressure,** and the minimum blood pressure at the end of ventricular diastole is called **diastolic pressure.** When the blood pressure is recorded, systolic and diastolic pressures are usually separated by a slashmark, as in "120/80" ("one twenty over eighty") or "110/75."

The difference between the systolic and diastolic pressures is the **pulse pressure** (Figure 21-10●). To report a single value for blood pressure, the **mean arterial pressure (MAP)** is used. The mean arterial pressure is calculated by adding one-third of the pulse pressure to the diastolic pressure. In other words:

$$\text{MAP} = P_{\text{dia}} + \frac{\text{Pulse pressure}}{3}$$

For example, if the systolic pressure is 120 mm Hg and the diastolic pressure is 90 mm Hg, the mean arterial pressure is 100 mm Hg:

$$90 + [(120 - 90)/3] = 90 + 10 = 100$$

Elastic Rebound

As systolic pressure climbs, the arterial walls stretch, just as an extra puff of air expands a partially inflated balloon. This expansion allows the arterial system to accommodate some of the blood provided by ventricular systole. When diastole begins and pressures fall, the arteries recoil to their original dimensions. Because the aortic semilunar valve prevents the return of blood to the heart, the arterial recoil pushes blood toward the capillaries. This phenomenon is called **elastic rebound.**

Pressures in Small Arteries and Arterioles

Figure 21-10

The mean arterial pressure and the pulse pressure become smaller as the distance from the heart increases (Figure 21-10●).

- The mean arterial pressure declines as the arterial branches become smaller and more numerous. In essence, the blood pressure decreases as it produces blood flow and overcomes friction with the arterial walls.
- The pulse pressure fades because of the cumulative effects of elastic rebound. Each arterial segment reduces the size of the pressure change experienced by its downstream neighbors. The effect is like a loud shout creating a series of ever-softer echoes. Each time an echo is produced, the reflecting surface absorbs some of the sound energy. Eventually, the echo disappears. The pressure pulse of ventricular ejection is the shout, and it is reflected by the wall of the aorta, echoing down the arterial system until it finally disappears completely at the level of the small arterioles. By the time blood reaches a precapillary sphincter, there are no pressure oscillations, and the blood pressure remains steady at about 35 mm Hg.

Checking the Pulse and Blood Pressure

The pulse can be felt within any of the large or medium-sized arteries. The usual procedure involves using the fingertips to squeeze an artery against a relatively solid mass, preferably a bone. When the vessel is compressed, the pulse is felt as a pressure against the fingertips. The inside of the wrist is often used, because the *radial artery* can easily be pressed against the distal portion of the radius. Other accessible arteries include the *temporal, facial, carotid, brachial, femoral,* and *popliteal arteries* (Figures 21-23, 21-24, 21-26, and 21-28•, pp. 752, 754, 756, and 759). Firm pressure exerted at these locations, called **pressure points,** can reduce or eliminate arterial bleeding distal to the site.

Blood pressure not only forces blood through the circulatory system, it also pushes outward against the walls of the containing vessels, just as air pushes against the walls of an inflated balloon. As a result, blood pressure can be measured indirectly by determining how forcefully the blood presses against the vascular walls.

The instrument used to measure blood pressure is called a **sphygmomanometer** (sfig-mō-ma-NOM-e-ter; *sphygmos,* pulse + *manometer,* device for measuring pressure). An inflatable cuff is placed around the arm in such a position that its inflation compresses the brachial artery (Figure 21-11•). A stethoscope is placed over the artery distal to the cuff, and the cuff is then inflated. A tube connects the cuff to a pressure gauge that reports the cuff pressure in millimeters of mercury (mm Hg). Inflation continues until cuff pressure is roughly 30 mm Hg above the pressure sufficient to collapse the brachial artery completely, stop the flow of blood, and eliminate the sound of the pulse.

The investigator then slowly lets the air out of the cuff. When the pressure in the cuff falls below systolic pressure, blood can again enter the artery. At first, blood enters only at peak systolic pressures, and the stethoscope picks up the sound of blood pulsing through the artery. As the pressure falls further, the sound changes because the vessel is remaining open for longer and longer periods. When the cuff pressure falls below diastolic pressure, blood flow becomes continuous and the sound of the pulse becomes muffled or disappears completely. Thus the pressure at which the pulse appears corresponds to the peak systolic pressure; when the pulse fades the pressure has reached diastolic levels. The distinctive sounds heard during this test are called *sounds of Korotkoff* (sometimes spelled *Korotkov* or *Korotkow*). These sounds are produced by turbulence as blood flows past the constricted portion of the artery.

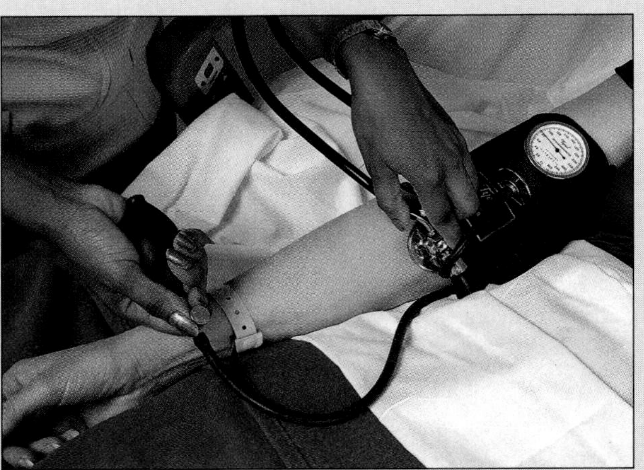

• **FIGURE 21-11**
Using a Sphygmomanometer to Measure Blood Pressure

HYPERTENSION AND HYPOTENSION **Hypertension** is the presence of abnormally high blood pressure. The usual criterion for hypertension in an adult is a blood pressure greater than 140/90. One study estimated that 20 percent of the U.S. Caucasian population has blood pressures greater than 160/95 and that another 25 percent is on the borderline, with pressures above 140/90. The figures for other racial groups vary; the percentages for African-Americans are roughly twice those for Caucasian Americans. The elevated incidence among African-Americans is associated with an increased risk of maturity-onset diabetes (Type II, NIDDM). ∞ [p. 634] It is not known to what extent the rate of either condition reflects genetic rather than environmental factors.

Hypertension significantly increases the workload on the heart, and the left ventricle gradually enlarges. More muscle mass means a greater oxygen demand. When the coronary circulation cannot keep pace, symptoms of coronary ischemia appear. Increased arterial pressures also place a physical stress on the walls of blood vessels throughout the body. This stress promotes or accelerates the development of arteriosclerosis and increases the risk of aneurysms, heart attacks, and strokes. **Hypotension,** or low blood pressure, is most often seen in patients who have received overly aggressive treatment for hypertension. AM *Hypertension and Hypotension*

❑ Capillary Exchange

Figure 21-3

Capillary walls are very thin and delicate (Figure 21-3●, p. 721); they consist of a single squamous endothelial cell, usually supported by a basement membrane. This arrangement minimizes the distance between the blood and the interstitial fluid, facilitating rapid diffusion or transport of materials into or out of the circulation. Chapter 3 introduced the forces responsible for the movement of water and solutes across membranes, and this section will assume some familiarity with basic concepts. If necessary, you should return to that chapter for a more detailed treatment of each mechanism.

The most important processes involved in the movement of materials across typical capillary walls are *diffusion, filtration,* and *reabsorption.* These mechanisms will be discussed in detail below. In specialized capillaries, such as the continuous capillaries of the CNS, *carrier-mediated transport* and *vesicular transport* are also important.

Diffusion

As we saw in Chapter 3, *diffusion* is the movement of ions and molecules as the result of random collisions. *[p. 73]* One result of diffusion is that there will be a net movement of material from an area where its concentration is relatively high to an area where its concentration is relatively low. The difference between the high and low concentrations represents a *concentration gradient,* and diffusion tends to eliminate that gradient. Diffusion occurs most rapidly when (1) the distances involved are small, (2) the concentration gradient is large, and (3) the ions or molecules involved are small.

Diffusion across capillary walls can occur via several different routes.

- Water, ions, and small organic molecules, such as glucose, amino acids, and urea, can usually enter or leave the circulation by diffusion between adjacent endothelial cells or through the pores of fenestrated capillaries. In the CNS, thymus, and testes, the endothelial cells of continuous capillaries are locked together, and there are no intervening gaps that permit passive diffusion of water, ions, and solutes. This arrangement is found at the *blood-brain barrier,* the *blood-thymus barrier* (Chapter 22), and the *blood-testis barrier* (Chapter 28). *[p. 467]*
- Many ions, including sodium, potassium, calcium, and chloride, can diffuse across the endothelial cells by passing through channels in the cell membranes.
- Large water-soluble compounds are unable to enter or leave the circulation except at fenestrated capillaries, such as those of the hypothalamus, the kidneys, many endocrine organs, and the intestinal tract.
- Lipids, such as fatty acids and steroids, and lipid-soluble materials, including soluble gases such as oxygen and carbon dioxide, can cross the capillary wall by diffusion through the endothelial cell membranes.
- Plasma proteins are normally unable to cross the endothelial lining anywhere except in sinusoids, such as those of the liver, where plasma proteins enter the circulation.

Filtration

Figure 21-12

The driving force for filtration is *hydrostatic pressure.* **Hydrostatic pressure (HP)** is a physical force that pushes water from an area of high pressure to an area of relatively lower pressure. Blood pressure is really the hydrostatic pressure of the blood (*BHP*).

In *capillary filtration,* water is forced across a capillary wall, and small molecules of solute travel with the water (Figure 21-12●). The solute molecules must be small enough to pass between adjacent endothelial cells or through the pores in a fenestrated capillary.

Along the length of a typical capillary, blood pressure gradually falls from about 35 mm Hg to roughly 18 mm Hg, the pressure at the start of the venous system. Filtration occurs primarily at the arterial end of a capillary, where BHP is highest.

Reabsorption

Reabsorption occurs as the result of osmosis. *Osmosis* is a special term used to refer to the diffusion of water across a selectively permeable membrane separating two solutions of differing solute concentrations. Water molecules will tend to diffuse across a membrane *toward the solution containing a higher solute concentration,* because in so doing they are moving down the concentration gradient for water (Figure 3-7●, p. 75).

The **osmotic pressure (OP)** of a solution is an indication of the force of water movement resulting from its solute concentration. The higher the solute concentration of a solution, the greater its osmotic pressure. The osmotic pressure of the blood is also called **blood colloid osmotic pressure (BCOP)** because much of the osmotic pressure results from the presence of suspended proteins.[1] *[p. 43]* Osmotic water movement will continue until either the solute concentrations are equalized or the movement is prevented by an opposing hydrostatic pressure.

[1]Clinicians often use the term *oncotic pressure* (*onkos,* a swelling) when referring to the colloid osmotic pressure of body fluids. The two terms are equivalent.

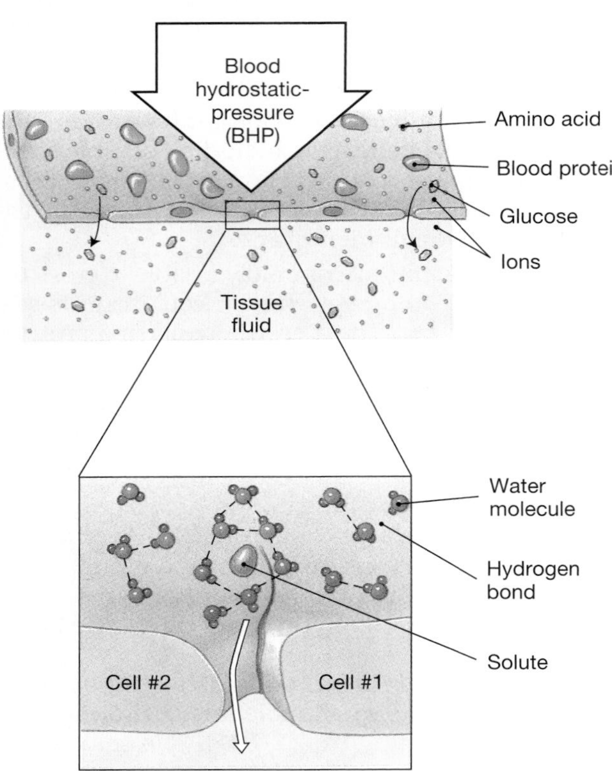
Blood hydrostatic-pressure (BHP)
Amino acid
Blood protein
Glucose
Ions
Tissue fluid
Water molecule
Hydrogen bond
Solute
Cell #2
Cell #1

● **FIGURE 21-12**
Capillary Filtration. Blood hydrostatic pressure forces water and solutes across capillary walls via the gaps between adjacent endothelial cells in continuous capillaries. Solute size is limited by the dimensions of the gaps.

We will now consider the interplay between filtration and reabsorption along the length of a typical capillary. As the discussion proceeds, remember that a hydrostatic pressure forces water *out* of a solution, whereas osmotic pressure draws water *in*.

The Interplay between Filtration and Reabsorption

Figure 21-13

The rates of filtration and reabsorption gradually change as blood passes along the length of a capillary. The factors involved are diagrammed in Figure 21-13●.

The *net hydrostatic pressure* tends to drive water and solutes into the interstitial fluid. The net hydrostatic pressure is the difference between:

- the **blood hydrostatic pressure (BHP),** which ranges from 35 mm Hg at the arterial end of a capillary to 18 mm Hg at the venous end, and
- the **hydrostatic pressure of the interstitial fluid (IHP).** Measurements of IHP have yielded very small values that differ from tissue to tissue. The range is from +6 mm Hg in the brain to –6 mm Hg in subcutaneous tissues. A positive IHP opposes BHP, and the tissue hydrostatic pressure must be overcome before fluid can move out of the capillary. A negative IHP assists BHP, and fluid will be pulled out of the capillary.

The *net colloid osmotic pressure* tends to pull water and solutes into the capillary. The net colloid osmotic pressure is the difference between:

- the **blood colloid osmotic pressure (BCOP),** which is roughly 25 mm Hg, and
- the **interstitial fluid colloid osmotic pressure (ICOP).** The ICOP is as variable and low as the IHP, because the interstitial fluid in most tissues contains negligible quantities of suspended proteins. Reported values of ICOP are all from 0 to 5 mm Hg, within the range of pressures recorded for the IHP.

Because (1) the colloid osmotic pressure and the hydrostatic pressure of the interstitial fluid oppose one another, (2) they are of comparable size, and (3) they are difficult to measure accurately, both values have been set at 0 mm Hg in our model. This value is acceptable because BHP and BOP are the primary forces acting across a capillary wall under normal circumstances. However, the method of calculation described here would still apply, regardless of the values selected for IHP and IOP.

The **net filtration pressure (NFP)** is the difference between the net hydrostatic pressure and the net osmotic pressure. In terms of the factors listed above, this means that:

$$NFP = (BHP - IHP) - (BCOP - ICOP)$$

At the arterial end of the capillary, the net filtration pressure is +10 mm Hg:

$$NFP = (35 - 0) - (25 - 0) = 35 - 25 = 10 \text{ mm Hg}$$

Because the value is positive, it indicates that fluid will tend to move *out of* the capillary and into the interstitial fluid. At the venous end of the capillary, the net filtration pressure is –7 mm Hg:

$$NFP = (18 - 0) - (25 - 0) = 18 - 25 = -7 \text{ mm Hg}$$

The minus sign indicates that there is a net movement of fluid *into* the capillary; that is, reabsorption is occurring.

The transition between filtration and reabsorption occurs where the blood hydrostatic pressure is 25 mm Hg, because at that point the hydrostatic and osmotic forces are equal, and

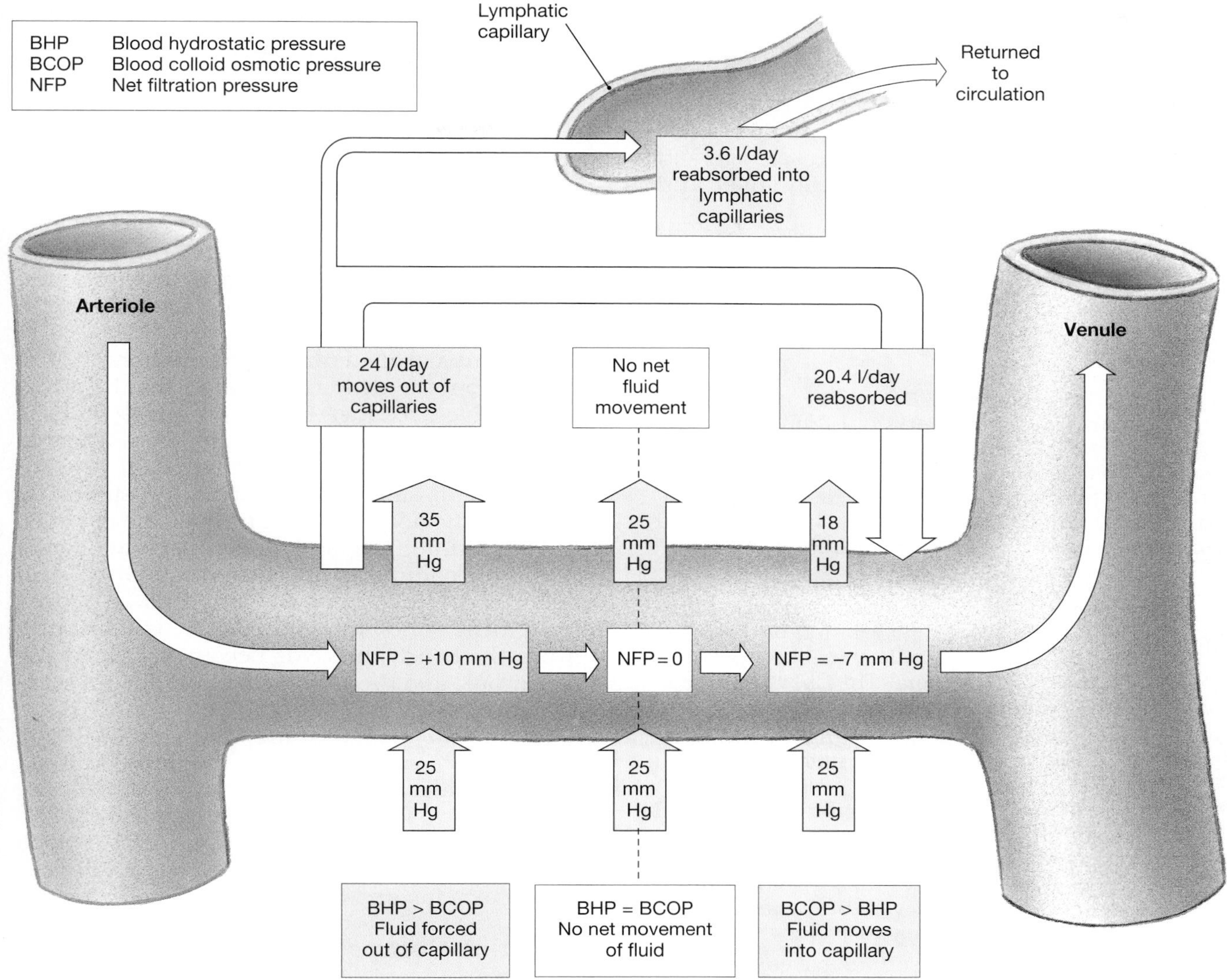
BHP Blood hydrostatic pressure
BCOP Blood colloid osmotic pressure
NFP Net filtration pressure
Lymphatic capillary
Returned to circulation
3.6 l/day reabsorbed into lymphatic capillaries
Arteriole
Venule
24 l/day moves out of capillaries
No net fluid movement
20.4 l/day reabsorbed
35 mm Hg
25 mm Hg
18 mm Hg
NFP = +10 mm Hg
NFP = 0
NFP = −7 mm Hg
25 mm Hg
25 mm Hg
25 mm Hg
BHP > BCOP Fluid forced out of capillary
BHP = BCOP No net movement of fluid
BCOP > BHP Fluid moves into capillary

● **FIGURE 21-13**

Forces Acting across Capillary Walls. At the arterial end of the capillary, blood hydrostatic pressure (BHP) is stronger than blood colloid osmotic pressure (BCOP), and fluid moves out of the capillary. Near the venule, BHP is lower than BCOP, and fluid moves into the capillary.

the net filtration pressure is 0 mm Hg. If the maximum filtration pressure at the arterial end of the capillary were equal to the maximum reabsorption pressure at the venous end, this transition point would lie midway along the length of the capillary. Under these circumstances, filtration would occur along the first half of the capillary, and an identical amount of reabsorption would occur along the second half. However, the maximum filtration pressure is higher than the maximum reabsorption pressure, so the transition point between filtration and reabsorption normally lies closer to the venous end of the capillary than to the arterial end. As a result, more filtration than reabsorption occurs along the length of the capillary. Of the roughly 24 liters of fluid that moves out of the plasma and into the interstitial fluid each day, 85 percent is reabsorbed. The remainder (3.6 liters) flows through the tissues and into lymphatic vessels, for eventual return to the venous system.

Any condition that affects hydrostatic or osmotic pressures in the blood or tissues will shift the balance between hydrostatic and osmotic forces. The effects can then be predicted on the basis of an understanding of capillary dynamics. For example:

- If hemorrhaging occurs, blood volume and blood pressure both decline. This reduction in BHP lowers the net filtration pressure and increases the amount of reabsorption. The result is a reduction in the volume of interstitial fluid and an increase in the circulating plasma volume. This process is known as a *recall of fluids.*

- If dehydration occurs, the plasma volume decreases owing to water loss, and the concentration of plasma proteins increases. The increase in BCOP accelerates reabsorption and a recall of fluids that delays the onset and severity of clinical symptoms.
- If BHP rises or BCOP declines, fluid will move out of the blood and build up within peripheral tissues, a condition called *edema.*

EDEMA **Edema** (e-DĒ-ma) is an abnormal accumulation of interstitial fluid. There are many different causes of edema, and specific examples will be encountered in later chapters. The underlying problem in all types of edema is a disturbance in the normal balance between hydrostatic and osmotic forces at the capillary level. For example:

- When a capillary is damaged, plasma proteins can cross the capillary wall and enter the interstitial fluid. The resulting elevation of the ICOP will reduce the rate of capillary reabsorption and produce a localized edema. This is why there is usually swelling at a bruise.
- In acute starvation, the liver cannot synthesize enough plasma proteins to maintain normal concentrations in the blood. BCOP declines, and fluids begin moving from the blood into peripheral tissues. In children, fluid accumulates in the abdominopelvic cavity, producing the swollen bellies typical of starvation victims. A reduction in BCOP is also seen after severe burns and in several types of liver or kidney diseases.
- In the U.S. population, serious cases of edema most often result from an increase in the blood pressure in the arterial system, the venous system, or both. This increase may result from heart problems, such as heart failure, venous blood clots that elevate venous pressures, or other circulatory abnormalities. The net result is an increase in BHP that accelerates the movement of fluid into the tissues.

Edema can also result from problems with other systems, such as the blockage of lymphatic vessels or impaired urine formation.

- If the lymphatic vessels in a region become blocked, the volume of interstitial fluid will continue to rise, and the IHP will gradually increase until capillary filtration ceases. In *filariasis,* a condition considered in Chapter 22, parasites can block lymphatic vessels and cause a massive regional edema known as *elephantiasis.*
- If the kidneys are unable to produce urine but the individual continues to drink liquids, the blood volume will rise. This situation ultimately leads to elevated blood hydrostatic pressure and enhances fluid movement into the peripheral tissues.

❑ Venous Pressure and Venous Return

Figure 21-9

Venous pressure, although low, determines the *venous return,* which has a direct impact on cardiac output and peripheral blood flow. Although pressure at the start of the venous system is only about one-tenth that at the start of the arterial system, the blood must still travel through a vascular network as complex as the arterial system before returning to the heart.

Pressures at the entrance to the right atrium fluctuate, but they average around 2 mm Hg. Thus the effective pressure in the venous system is roughly 16 mm Hg (18 mm Hg in the venules to 2 mm Hg in the venae cavae = 16 mm Hg), as compared with 65 mm Hg in the arterial system (100 mm Hg at the aorta to 35 mm Hg at the capillaries). Yet, although venous pressures are low, the veins offer comparatively little resistance, and pressure declines very slowly as blood moves through the venous system. As blood continues toward the heart, the veins become larger, resistance drops, and the velocity of blood flow increases (Figure 21-9•, p. 730).

When you are standing, the venous blood returning from your body inferior to the heart must overcome gravity as it ascends within the inferior vena cava. Two factors cooperate to assist the relatively low venous pressures in propelling blood toward the heart: (1) *muscular compression* of peripheral veins and (2) the *respiratory pump.*

Muscular Compression

Figure 21-7

The contractions of skeletal muscles near a vein compress it, helping to push blood toward the heart. The valves in small and medium-sized veins ensure that blood flow occurs in one direction only (Figure 21-7•, p. 727). During normal standing and walking, the cycles of contraction and relaxation that accompany normal movements assist venous return. If an individual stands at attention, with knees locked and leg muscles immobilized, this assistance is lost. The reduction in venous return then leads to a fall in cardiac output, which reduces the blood supply to the brain. This decline is sometimes enough to cause **fainting,** a temporary loss of consciousness. The person then collapses, and in the horizontal position both venous return and cardiac output return to normal.

The Respiratory Pump

On inhalation, the thoracic cavity expands, and pressures within the pleural cavities decline. This drop in pressure pulls air into the lungs. At the same time, blood is pulled into the inferior vena cava and right atrium from the smaller veins of the abdominal cavity and lower body. The effect on venous return from the superior vena cava is less pronounced, as blood in that vessel normally flows "downhill" and so has the assistance of gravity.

On exhalation, the thoracic cavity decreases in size. Internal pressures then rise, forcing air out of the lungs and pushing venous blood into the right atrium. This mechanism is called the **respiratory pump,** or *thoracoabdominal pump.* The importance of this pumping action increases during heavy exercise, when cardiac output is at maximal levels.

✔ In a normal individual, where would you expect the blood pressure to be greater, at the aorta or the inferior vena cava? Explain.

✔ While standing in the hot sun, Sally begins to feel light-headed and faints. Explain.

✔ Terry's blood pressure is 125/70. At what pressure did the nurse taking his blood pressure first hear the sounds of Korotkoff?

✔ Why do patients suffering from congestive heart failure frequently have swollen ankles and feet?

Cardiovascular Regulation

Figure 21-14

Homeostatic mechanisms regulate cardiovascular activity to ensure that tissue blood flow, also called **tissue perfusion,** meets the demand for oxygen and nutrients. The three variable factors are cardiac output, peripheral resistance, and blood pressure. Cardiac output was discussed in Chapter 20, and peripheral resistance and blood pressure were considered earlier in this chapter. ∞ *[p. 706]*

Most cells are relatively close to capillaries. When a group of cells becomes active, the circulation to that region must increase to deliver the necessary oxygen and nutrients and to carry away the waste products and carbon dioxide that they generate. The goal of cardiovascular regulation is to ensure that these blood flow changes occur (1) at an appropriate time, (2) in the right area, and (3) without drastically altering blood pressure and blood flow to any vital organs.

Factors involved in the regulation of cardiovascular function (Figure 21-14●) include:

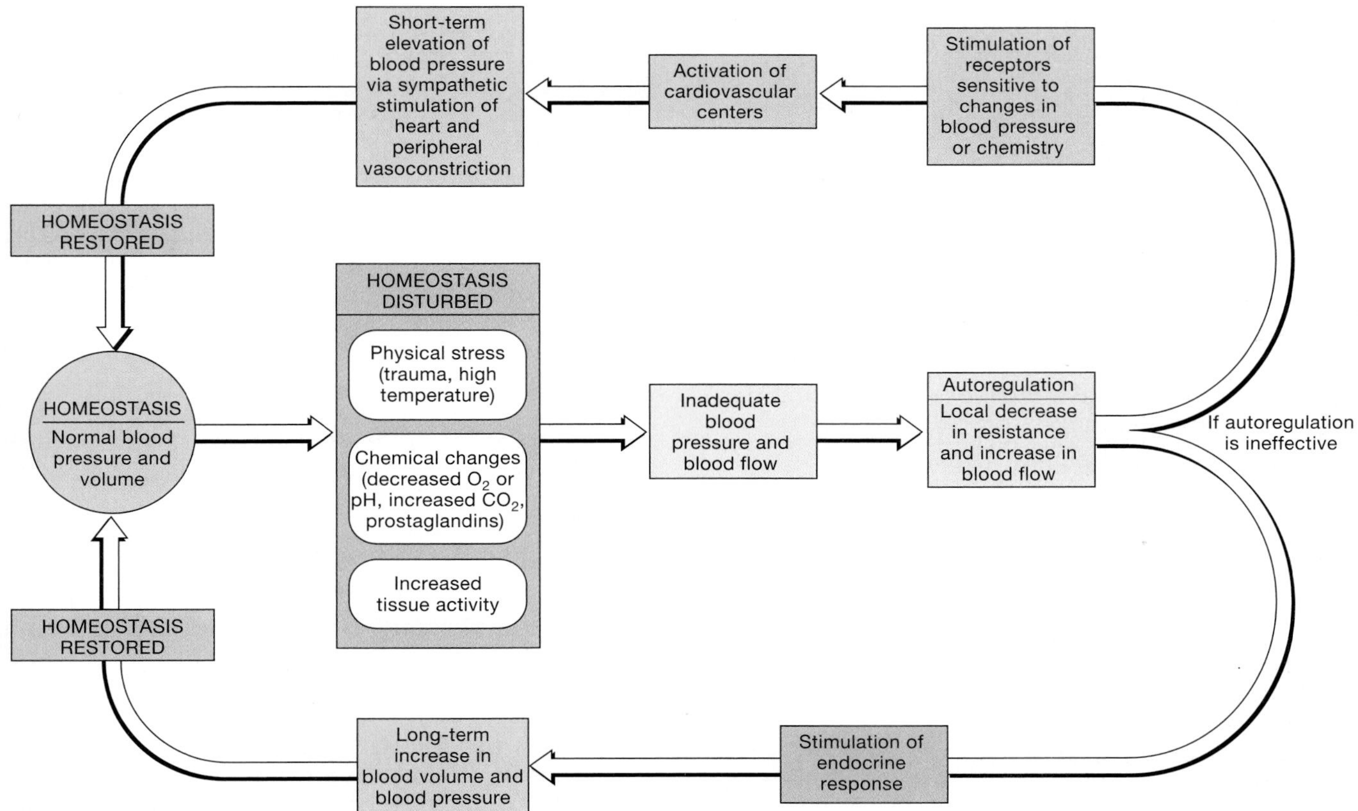
Short-term elevation of blood pressure via sympathetic stimulation of heart and peripheral vasoconstriction
Activation of cardiovascular centers
Stimulation of receptors sensitive to changes in blood pressure or chemistry
HOMEOSTASIS RESTORED
HOMEOSTASIS DISTURBED
Physical stress (trauma, high temperature)
Chemical changes (decreased O_2 or pH, increased CO_2, prostaglandins)
Increased tissue activity
HOMEOSTASIS
Normal blood pressure and volume
Inadequate blood pressure and blood flow
Autoregulation
Local decrease in resistance and increase in blood flow
If autoregulation is ineffective
HOMEOSTASIS RESTORED
Long-term increase in blood volume and blood pressure
Stimulation of endocrine response

● **FIGURE 21-14**
Homeostatic Adjustments That Maintain Blood Pressure and Blood Flow

1. ***Local factors.*** Local factors change the pattern of blood flow within capillary beds in response to chemical changes in the interstitial fluids. This is an example of *autoregulation* at the tissue level.
2. ***Central mechanisms.*** Central mechanisms respond to changes in arterial pressure or blood gas levels at specific sites. When those changes occur, the cardiovascular centers of the autonomic nervous system adjust cardiac output and peripheral resistance to maintain adequate blood flow.
3. ***Endocrine factors.*** The endocrine system releases hormones that enhance short-term adjustments and direct long-term changes in cardiovascular performance.

❑ Autoregulation of Blood Flow within Tissues

Under normal resting conditions, cardiac output remains stable, and peripheral resistance within individual tissues is adjusted to control local blood flow.

Local Vasodilators

Factors that promote dilation of precapillary sphincters are called **vasodilators.** *Local vasodilators* are produced at the tissue level and accelerate blood flow through the tissue of origin. Examples include:

- Decreased tissue oxygen levels or increased CO_2 levels
- The generation of lactic acid or other acids by tissue cells
- The release of nitric oxide, formerly known as *endothelium-derived relaxation factor (EDRF),* from endothelial cells
- Rising concentrations of potassium ions or hydrogen ions in the interstitial fluid
- Chemicals released during local inflammation, including histamine and nitric oxide
- Elevated local temperatures

These factors work by stimulating the relaxation of the smooth muscle fibers of the precapillary sphincters. All of these factors indicate that conditions in the tissue are abnormal in one way or another. An improvement in blood flow, which will bring oxygen, nutrients, and buffers, may be sufficient to restore homeostasis.

Local Vasoconstrictors

Figure 21-6

As noted in Chapter 19, aggregating platelets and damaged tissues produce compounds that stimulate constriction of precapillary sphincters. These compounds are **local vasoconstrictors.** Examples include prostaglandins and thromboxanes released by activated platelets and WBCs and the endothelins released by damaged endothelial cells. ∞ [p. 671]

Local vasodilators and vasoconstrictors control blood flow within a single capillary bed (Figure 21-6•, p. 726). When present in high concentrations, these factors also affect arterioles, increasing or decreasing blood flow to all of the capillary beds in a given area.

❑ The Neural Control of Blood Pressure and Blood Flow

The nervous system is responsible for adjusting cardiac output and peripheral resistance to maintain adequate blood flow to vital tissues and organs. Centers responsible for these regulatory activities include the *cardiac centers* and the *vasomotor centers* of the medulla oblongata. It is difficult to distinguish the cardiac and vasomotor centers anatomically, and they are often considered to be part of a complex **cardiovascular (CV) center.** In functional terms, however, the cardiac and vasomotor centers often act independently.

As noted in Chapter 20, the cardiac centers include a *cardioacceleratory center* that increases cardiac output via sympathetic innervation, and a *cardioinhibitory center* that reduces cardiac output via parasympathetic innervation. ∞ [pp. 692, 694]

The vasomotor centers contain two populations of neurons: (1) a very large group responsible for widespread vasoconstriction and (2) a relatively small group responsible for the vasodilation of arterioles in skeletal muscles and the brain. The vasomotor center exerts its effects by controlling the activity of sympathetic motor neurons.

1. ***Control of vasoconstriction.*** The neurons innervating peripheral blood vessels in most tissues are *adrenergic,* releasing norepinephrine. The response to NE release is the stimulation of smooth muscle in the walls of arterioles, producing vasoconstriction.
2. ***Control of vasodilation.*** Vasodilator neurons innervate blood vessels in skeletal muscles and in the brain. Stimulation of these neurons will relax smooth muscle fibers in the walls of arterioles, producing vasodilation. Relaxation of smooth muscle fibers is triggered by the appearance of nitric oxide in their surroundings. The vasomotor center may control NO release indirectly or directly.

- *Indirect control.* The most common vasodilator synapses are *cholinergic*, and their synaptic knobs release ACh. ACh stimulates the release of nitric oxide by endothelial cells in the area. The NO then causes local vasodilation.
- *Direct control.* Another population of vasodilator synapses are nitroxidergic, and release NO as a neurotransmitter. This compound has an immediate and direct relaxing effect on the vascular smooth muscle fibers in the area.

Vasomotor Tone

Chapter 16 discussed the significance of *autonomic tone* in setting a background level of neural activity that can increase or decrease on demand. [p. 541] The sympathetic vasoconstrictor nerves are chronically active, producing a significant **vasomotor tone.** Vasoconstrictor activity is normally sufficient to keep the arterioles partially constricted. Under maximal stimulation, arterioles constrict to around half their resting diameter, whereas a fully dilated arteriole increases its resting diameter by roughly 1.5 times. Constriction has a significant effect on resistance, because the resistance increases sharply as the diameter decreases. (The relationship was discussed on p. 729.) The resistance of a maximally contracted arteriole is roughly *80 times greater* than that of a fully dilated arteriole. Because blood pressure varies directly with peripheral resistance, the vasomotor center can provide very effective control over arterial blood pressure by making modest adjustments in vessel diameter. Extreme stimulation of the vasomotor center will also produce venoconstriction and a mobilization of the venous reserve.

Reflex Control of Cardiovascular Function

Figures 21-15, 21-16, 21-24

The cardiovascular centers detect changes in tissue demand by monitoring arterial blood, with particular attention to (1) blood pressure and (2) the pH and dissolved gas concentrations. The *baroreceptor reflexes* respond to changes in blood pressure, and the *chemoreceptor reflexes* monitor changes in the chemical composition of arterial blood. These reflexes are regulated through a negative feedback loop. Stimulation of a receptor by an abnormal condition leads to a response that counteracts the stimulus and restores normal conditions.

BARORECEPTOR REFLEXES *Baroreceptors* are specialized receptors that monitor the degree of stretch in the walls of distensible organs. [p. 558] The baroreceptors involved with cardiovascular regulation are located in the walls of (1) the **carotid sinuses,** expanded chambers near the bases of the *internal carotid arteries* of the neck (Figure 21-24•, p. 754), (2) the **aortic sinuses,** pockets in the walls of the ascending aorta adjacent to the heart (Figure 20-6•, p. 689), and (3) the wall of the right atrium. These receptors are components of the **baroreceptor reflexes** that adjust cardiac output and peripheral resistance to maintain normal arterial pressures.

Baroreceptor Reflexes of the Carotid and Aortic Sinuses. Aortic baroreceptors monitor blood pressure within the ascending aorta. The aortic reflex adjusts blood pressure in response to changes in pressure at this location. The goal is to maintain adequate blood pressure and blood flow through the systemic circuit. Carotid sinus baroreceptors respond to changes in blood pressure at the carotid sinus. These receptors trigger reflexes that maintain adequate blood flow to the brain. The blood flow to the brain must remain constant, and the carotid sinus receptors are extremely sensitive.

Figure 21-15• presents the basic organization of the baroreceptor reflexes triggered by changes in blood pressure at these locations.

When blood pressure climbs, the increased output from the baroreceptors alters activity in the CV centers and produces two major effects (Figure 21-15•):

1. ***A decrease in the cardiac output,*** due to parasympathetic stimulation and inhibition of sympathetic activity
2. ***Widespread peripheral vasodilation*** due to the inhibition of excitatory neurons in the vasomotor center

The decrease in cardiac output reflects primarily a reduction in heart rate due to ACh release at the SA node. [p. 709] The widespread vasodilation lowers peripheral resistance, and this effect, combined with a reduction in cardiac output, leads to a decline in blood pressure to normal levels.

When blood pressure falls, there is a corresponding reduction in baroreceptor output (Figure 21-15•). This change has two major effects:

1. ***An increase in cardiac output,*** caused by stimulation of the cardioacceleratory center, via the sympathetic innervation to the heart, and inhibition of the parasympathetic output of the cardioinhibitory center
2. ***Widespread peripheral vasoconstriction,*** caused by stimulation of sympathetic vasoconstrictor neurons by the vasomotor center

The effects on the heart result from the release of norepinephrine by sympathetic neurons innervating the SA node, the AV node, and the general myocardium. In a crisis situation, sympathetic acti-

● **FIGURE 21-15**
The Carotid and Aortic Sinus Baroreceptor Reflexes

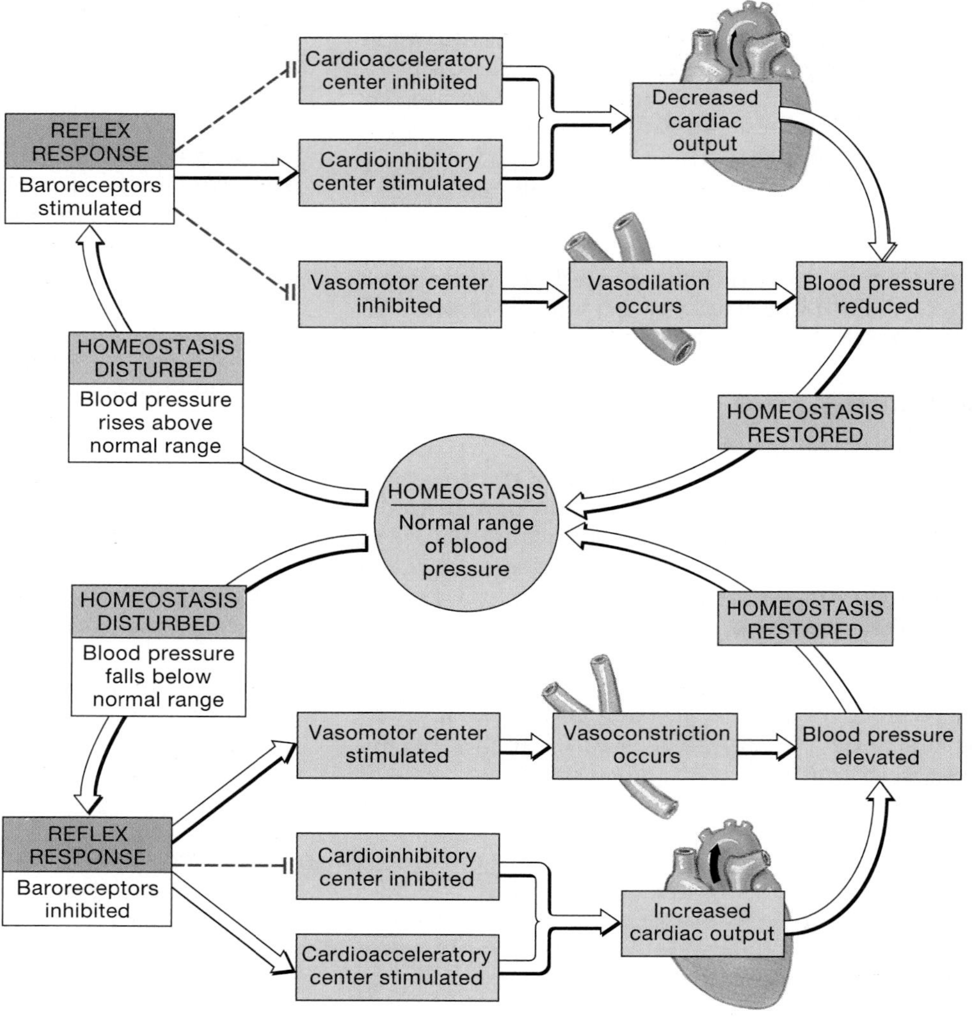

REFLEX RESPONSE
Baroreceptors stimulated
Cardioacceleratory center inhibited
Cardioinhibitory center stimulated
Decreased cardiac output
Vasomotor center inhibited
Vasodilation occurs
Blood pressure reduced
HOMEOSTASIS DISTURBED
Blood pressure rises above normal range
HOMEOSTASIS RESTORED
HOMEOSTASIS
Normal range of blood pressure
HOMEOSTASIS DISTURBED
Blood pressure falls below normal range
HOMEOSTASIS RESTORED
Vasomotor center stimulated
Vasoconstriction occurs
Blood pressure elevated
REFLEX RESPONSE
Baroreceptors inhibited
Cardioinhibitory center inhibited
Increased cardiac output
Cardioacceleratory center stimulated

vation occurs, and these effects will be enhanced by the release of epinephrine and norepinephrine from the adrenal medullae. The net effect is an immediate increase in heart rate and stroke volume and a corresponding rise in cardiac output. The vasoconstriction, which also results from NE release by sympathetic neurons, increases peripheral resistance. These adjustments, increased cardiac output and increased peripheral resistance, work together to elevate blood pressure.

BARORECEPTOR ACCOMMODATION *Receptor accommodation* occurs when a receptor becomes less sensitive to a sustained stimulus over time. Accommodation of baroreceptors is a significant complicating factor in the treatment of high blood pressure. If blood pressure rises very slowly or remains stable at an abnormally high level, the receptors gradually adjust their response to accept the new pressure as "normal." The baroreceptors will then respond to any drop in pressure with an increase in cardiac output and vasoconstriction. Initial treatment with beta-blockers and other drugs must therefore be strong enough to overcome the baroreceptor response.

Atrial Baroreceptors. Atrial baroreceptors monitor blood pressure at the end of the systemic circuit—at the venae cavae and the right atrium. The **atrial reflex,** introduced in Chapter 20, responds to stretching of the wall of the right atrium. [p. 709]

Under normal circumstances the heart pumps blood into the aorta at the same rate that it is arriving at the right atrium. When blood pressure rises at the right atrium it means that blood is arriving at the heart faster than it is being pumped out. The atrial baroreceptors solve the problem by stimulating the CV center and increasing cardiac output until the backlog of venous blood is removed. Atrial pressure then returns to normal.

CHEMORECEPTOR REFLEXES The **chemoreceptor reflexes** (Figure 21-16●) respond to changes in the carbon dioxide, oxygen, or pH levels in the blood and cerebrospinal fluid. The chemoreceptors involved are sensory neurons located in the **carotid bodies,** located in the neck near the carotid sinus, and the **aortic bodies,** situated near the arch of aorta. These receptors monitor the composition of the arterial blood. Additional chemoreceptors are found on the ventrolateral surfaces of the medulla oblongata. These receptors monitor the composition of the cerebrospinal fluid.

When chemoreceptors in the carotid bodies, the aortic bodies, or the medulla oblongata detect a rise in the carbon dioxide content or a fall in the pH of the arterial blood or cerebrospinal fluid, the cardiovascular centers are stimulated. The primary result is an elevation in arterial pressure via stimulation of the vasomotor center. A decline in the oxygen level at the aortic bodies will have the same effects. Strong chemoreceptor stimulation causes a more widespread sympathetic activation, increasing heart rate and cardiac output.

Arterial CO_2 levels can be reduced and O_2 levels increased most effectively by coordinating cardiovascular and respiratory activity. Chemoreceptor stimulation also affects the respiratory centers, and the rise in cardiac output and blood pressure is associated with an increased respiratory rate. Coordination of cardiovascular and respiratory activity is vital, because accelerating tissue blood flow is useful only if the circulating blood contains adequate oxygen. In addition, a rise in the respiratory rate accelerates venous return through the action of the respiratory pump. (Other aspects of chemoreceptor activity and respiratory control will be considered in Chapter 23.)

Special Circulation

The vasoconstriction that occurs in response to a fall in blood pressure or a rise in CO_2 levels affects multiple tissues and organs simultaneously. The term *special circulation* refers to the circulation through organs whose blood vessels are controlled by separate mechanisms. We will note three important examples: the circulation to the brain, the heart, and the lungs.

THE BRAIN Chapter 14 noted the existence of the blood-brain barrier that isolates most CNS tissue from the general circulation. ∞ [p. 467] The brain has a very high demand for oxygen, and it receives a substantial supply of blood. Under a variety of conditions, the blood flow to the brain remains steady at about 750 ml/min. This supply is roughly 12 percent of the cardiac output, delivered to an organ that represents less than 2 percent of the body weight. Neurons do not maintain significant energy reserves, and in functional terms most of the adjustments made by the cardiovascular system treat circulation to the brain as the number one priority. Even when a circulatory crisis is under way, blood flow through the brain remains as near normal as possible. While the cardiovascular centers are calling for widespread peripheral vasoconstriction, the cerebral vessels are told to dilate.

FIGURE 21-16
The Chemoreceptor Reflexes

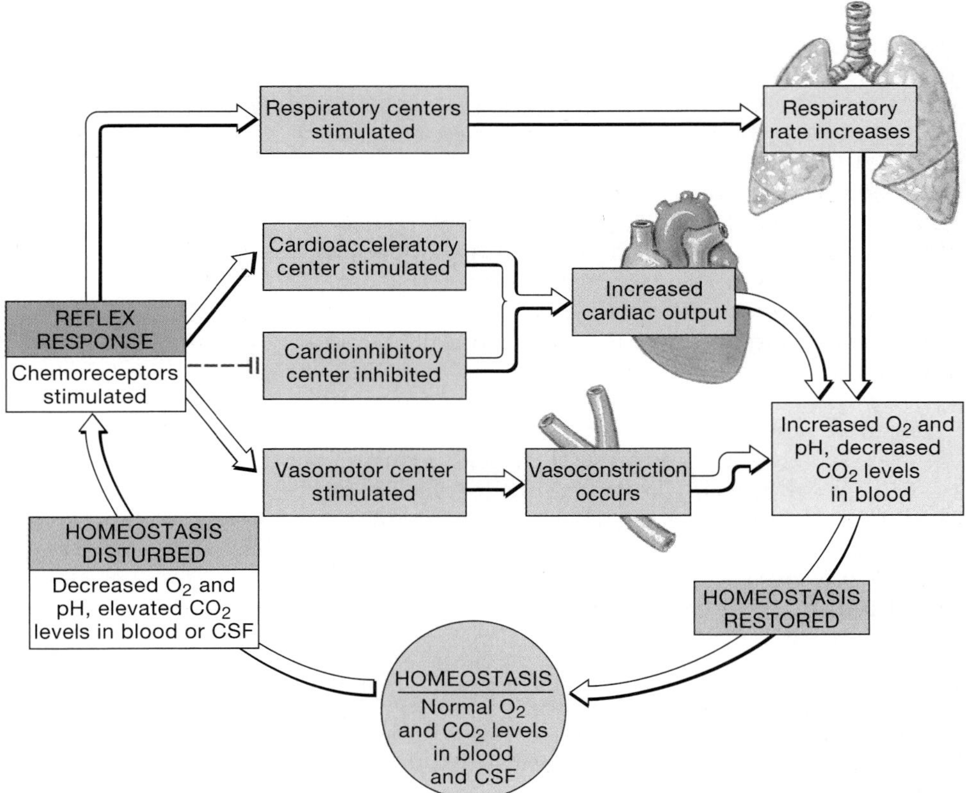
Respiratory centers stimulated
Respiratory rate increases
Cardioacceleratory center stimulated
Increased cardiac output
REFLEX RESPONSE
Chemoreceptors stimulated
Cardioinhibitory center inhibited
Increased O2 and pH, decreased CO2 levels in blood
Vasomotor center stimulated
Vasoconstriction occurs
HOMEOSTASIS DISTURBED
Decreased O2 and pH, elevated CO2 levels in blood or CSF
HOMEOSTASIS RESTORED
HOMEOSTASIS
Normal O2 and CO2 levels in blood and CSF

Although total blood flow to the brain remains relatively constant, circulation to specific regions within the brain changes on a moment-to-moment basis. These changes occur in response to local changes in the interstitial fluid composition that accompany neural activity. When you read, write, speak, or walk, specific regions of the brain become active. The circulation to those regions increases almost instantaneously, ensuring that the active neurons will continue to receive the oxygen and nutrients they require.

The brain receives arterial blood via four different arteries. Because these arteries form anastomoses inside the cranium, interruption of any one vessel will not compromise the circulatory supply to the brain. If a plaque or blood clot blocks an artery, or if an artery ruptures, dependent tissues will be injured or killed. Symptoms of a *stroke*, or *cerebrovascular accident* (CVA), then appear. The *Applications Manual* contains a discussion of current views on the causes and treatment of strokes. AM *The Causes and Treatment of Cerebrovascular Disease*

THE HEART The anatomy of the coronary circulation was described in Chapter 20. [p. 692] The coronary arteries arise at the base of the ascending aorta, where systemic pressures are highest. Each time the heart contracts, it squeezes the coronary vessels, and blood flow is reduced. In the left ventricle, systolic pressures are high enough that blood can flow into the myocardium only during diastole. Normal cardiac muscle cells can tolerate these brief circulatory interruptions because they have substantial oxygen reserves.

At rest the coronary blood flow is about 250 ml/min. When the workload on the heart increases, local factors, such as reduced O_2 levels and lactic acid production, dilate the coronary vessels and increase blood flow. Epinephrine released during sympathetic stimulation promotes vasodilation while increasing the heart rate and strength of cardiac contractions. As a result, when vasoconstriction is under way in other tissues, coronary blood flow increases.

For uncertain reasons, some individuals experience *coronary spasms* that can temporarily restrict coronary circulation and produce symptoms of angina. Permanent restriction or blockage of coronary vessels, as in coronary artery disease (CAD), and tissue damage, as caused by a myocardial infarction (MI), can limit the ability of the heart to increase cardiac output, even under maximal stimulation. These individuals will experience symptoms of *heart failure* when the cardiac workload increases much above resting levels. The causes and treatments of heart failure are detailed in the *Applications Manual.* AM *Heart Failure*

THE LUNGS The lungs contain roughly 300 million *alveoli*, delicate epithelial pockets where gas exchange occurs. Each alveolus is surrounded by an extensive capillary network. The primary mechanism of blood flow regulation through the lungs is via local responses to the levels of oxygen within individual alveoli. When an alveolus contains oxygen in abundance, the associated vessels dilate, and blood flow increases. This increase promotes the absorption of oxygen from the alveolar air. When the oxygen content of the air is very low, the vessels constrict, and blood is shunted to alveoli that still contain significant levels of oxygen. This mechanism maximizes the efficiency of the respiratory system, because there is no benefit to circulating blood through the capillaries of an alveolus unless it contains oxygen.

You will notice that this is precisely the opposite of the situation in other tissues, where a decline in oxygen levels causes local vasodilation, rather than vasoconstriction. The difference makes functional sense, but its physiological basis remains a mystery.

Blood pressure in the pulmonary capillaries is lower than the pressure in systemic capillaries, averaging 10 mm Hg. The blood colloid osmotic pressure is the same as elsewhere in the circulation (25 mm Hg). As a result, there is a continual movement of fluid into the pulmonary capillaries across the alveolar surfaces. This movement prevents fluid buildup in the alveoli that could interfere with the diffusion of respiratory gases. If the blood pressure in pulmonary capillaries rises above 25 mm Hg, fluid will enter the alveoli, causing *pulmonary edema*, a dangerous condition that is often seen in *congestive heart failure.*

CNS Activities and the Cardiovascular Centers

The output of the cardiovascular centers can also be influenced by activities in other areas of the brain. For example, activation of either division of the ANS will affect output from the cardiovascular centers:

- The cardioacceleratory and vasomotor centers are stimulated when a general sympathetic activation occurs. The result is an increase in cardiac output and blood pressure.
- When the parasympathetic division is activated, the cardioinhibitory center is stimulated, producing a reduction in cardiac output. Parasympathetic activity does not directly affect the vasomotor center, but vasodilation occurs as sympathetic activity declines.

The activities of higher brain centers can also affect blood pressure. Our thought processes or emotional states can produce significant changes in blood pressure by influencing cardiac output and vasomotor tone. For example, strong emotions of anxiety, fear, or rage are accompanied by an elevation in blood pressure, caused by cardiac stimulation and vasoconstriction.

Hormones and Cardiovascular Regulation

Figure 21-17

The endocrine system provides both short-term and long-term regulation of cardiovascular performance. Epinephrine and norepinephrine from the adrenal medullae stimulate cardiac output and peripheral vasoconstriction. Other hormones important in regulating cardiovascular function include (1) antidiuretic hormone (ADH), (2) angiotensin II, (3) erythropoietin (EPO), and (4) atrial natriuretic peptide (ANP). These hormones and their functions were described in Chapter 18, and we will provide only an overview here. [pp. 614, 628, 629]

Although ADH and angiotensin II affect blood pressure, all four hormones are concerned primarily with the long-term regulation of blood volume, as diagrammed in Figure 21-17•.

Antidiuretic Hormone

ADH is released at the posterior pituitary in response to a decrease in blood pressure or an increase in the osmotic concentration of the plasma. The immediate result is a peripheral vasoconstriction that elevates blood pressure. This hormone also stimulates the conservation of water at the kidneys, thus preventing a reduction in blood volume that would further reduce blood pressure.

Angiotensin II

Angiotensin II appears in the blood following the release of renin by specialized kidney cells in response to a fall in renal blood pressure. Renin starts a chain reaction that ultimately converts an inactive protein, *angiotensinogen*, to the hormone angiotensin II (Figure 18-16b•, p. 627).

Angiotensin II has two short-term effects: (1) an extremely powerful vasoconstriction that elevates blood pressure and (2) a *positive inotropic action* on the heart, which elevates cardiac output. The other effects of angiotension II have long-term significance:

- *It stimulates the secretion of ADH by the pituitary and aldosterone by the adrenal cortex.* Aldosterone stimulates the reabsorption of sodium ions at the kidneys, reducing sodium losses in the urine.
- *It stimulates thirst.* The presence of ADH and aldosterone ensures that the additional water consumed will be retained, elevating the plasma volume.

Erythropoietin

EPO is released at the kidneys if the blood pressure declines or if the oxygen content of the blood becomes abnormally low. This hormone stimulates red blood cell production and maturation, elevating the blood volume and improving the oxygen-carrying capacity of the blood (Figure 18-16b•, p. 627).

Atrial Natriuretic Peptide

ANP is produced by cardiac muscle fibers in the wall of the right atrium in response to excessive stretching during diastole. ANP reduces blood volume and blood pressure by (1) increasing water losses at the kidneys; (2) reducing thirst; (3) blocking the release of ADH, aldosterone, and catecholamines; and (4) stimulating peripheral vasodilation. As blood volume and blood pressure decline, the stress on the atrial walls is removed, and ANP production ceases (Figure 18-17•, p. 628).

Patterns of Cardiovascular Response

In this and the previous two chapters we have considered the blood, the heart, and the circulatory system as individual entities. Yet in our day-to-day lives the cardiovascular system operates as an integrated complex. The interactions are quite fascinating and of considerable importance when physical or physiological conditions are changing rapidly.

Two common stresses, exercise and blood loss, provide examples of the adaptability of this system and its ability to maintain homeostasis. The homeostatic responses involve an interplay between the cardiovascular system and other systems, and the central mechanisms are aided by automatic adjustments at the tissue level. We will also consider the physiological mechanisms involved in shock, an important homeostatic disorder.

Exercise and the Cardiovascular System

At rest the cardiac output averages around 5.6 liters per minute. That value changes dramatically during exercise. In addition, the pattern of blood distribution can change markedly, as detailed in Table 21-2.

Light Exercise

Before the exercise period begins, there is a slight increase in heart rate that accompanies a general rise in sympathetic activity as you think about the

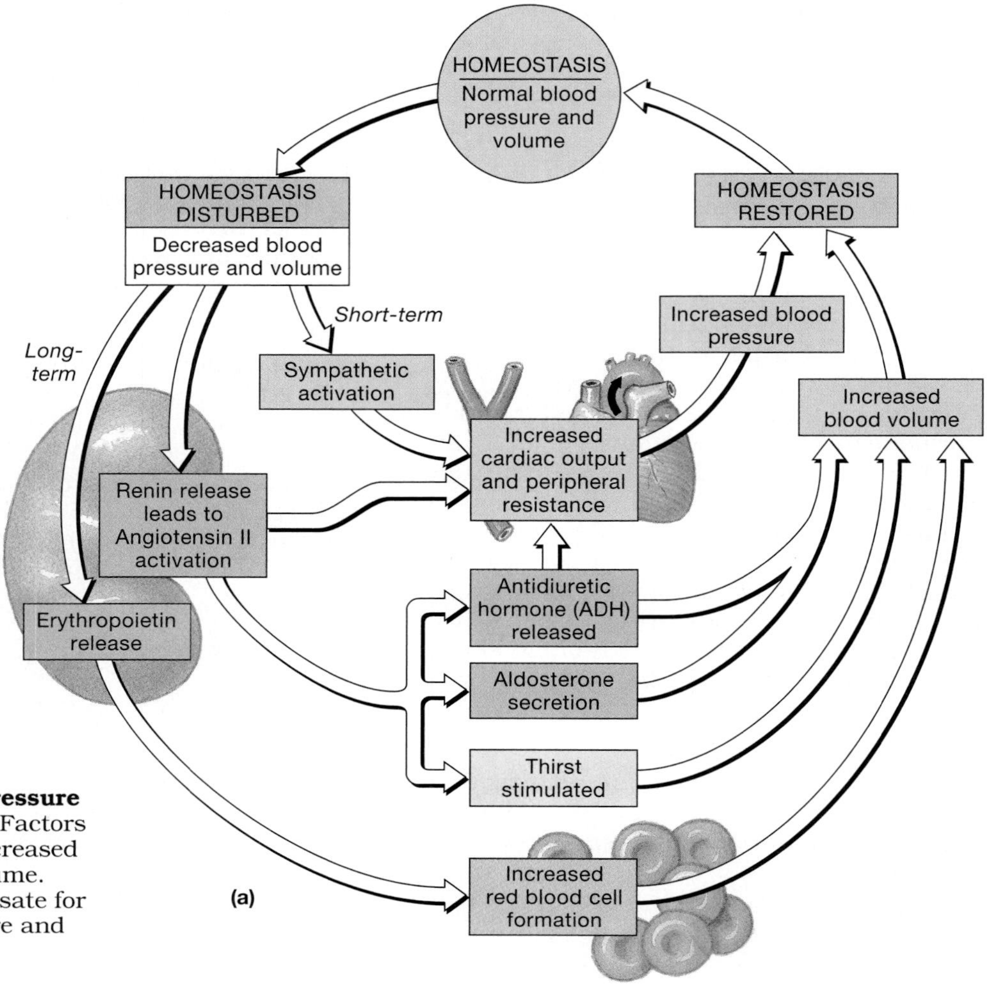

HOMEOSTASIS
Normal blood pressure and volume
HOMEOSTASIS DISTURBED
Decreased blood pressure and volume
Short-term
Long-term
Sympathetic activation
Renin release leads to Angiotensin II activation
Erythropoietin release
Increased cardiac output and peripheral resistance
Antidiuretic hormone (ADH) released
Aldosterone secretion
Thirst stimulated
Increased red blood cell formation
Increased blood pressure
Increased blood volume
HOMEOSTASIS RESTORED
(a)

● **FIGURE 21-17**
Regulation of Blood Pressure and Blood Volume. (a) Factors that compensate for decreased blood pressure and volume. **(b)** Factors that compensate for increased blood pressure and volume.

HOMEOSTASIS
Normal blood pressure and volume
HOMEOSTASIS DISTURBED
Increased blood pressure and volume
Atrial natriuretic peptide (ANP) released at right atrium
Blockage of ADH and aldosterone release
Reduction of thirst
Blockage of epinephrine and norepinephrine release
Peripheral vasodilation
Reduced blood volume
Reduced blood pressure
HOMEOSTASIS RESTORED
(b)

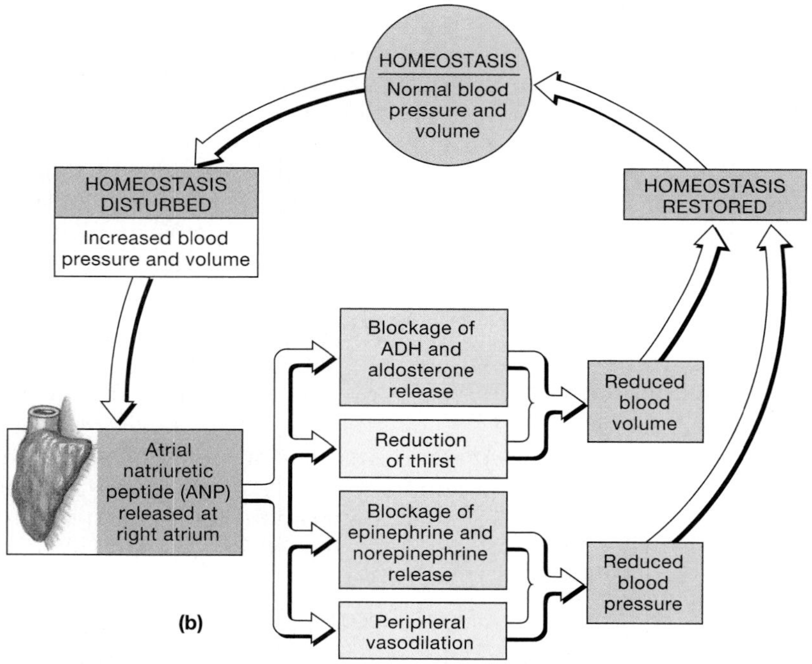

TABLE 21-2 Distribution of Blood During Exercise

	Blood Flow (ml/min)		
Tissue	*Rest*	*Light Exercise*	*Strenuous Exercise*
Muscle	1200	4500	12,500
Heart	250	350	750
Brain	750	750	750
Skin	500	1500	1900
Kidney	1100	900	600
Abdominal viscera	1400	1100	600
Miscellaneous	600	400	400
Total cardiac output	5800	9500	17,500

workout ahead. As light exercise actually begins, a number of interrelated changes occur.

1. ***Extensive vasodilation occurs*** as the rate of skeletal muscle oxygen consumption increases. Peripheral resistance drops, blood flow through the capillaries increases, and blood enters the venous system at an accelerated rate.
2. ***The venous return increases,*** as skeletal muscle contractions squeeze blood along the peripheral veins and an increased breathing rate pulls blood into the venae cavae, via the respiratory pump.
3. ***There is a rise in cardiac output*** caused by the increase in venous return. This increase occurs in direct response to ventricular stretching (the Frank-Starling law) and in a reflexive response to atrial stretching (the atrial reflex). The increased cardiac output keeps pace with the elevated demand, and arterial pressures are maintained despite the drop in peripheral resistance.

This regulation by venous feedback produces a gradual increase in cardiac output to about double resting levels. Over this range, typical of light exercise, the pattern of blood distribution remains relatively unchanged. However, there are increases in the blood flow to skeletal muscle, cardiac muscle, and the skin. The increased flow to the muscles reflects the dilation of precapillary sphincters in response to local factors; the increased blood flow to the skin occurs in response to the rise in body temperature.

Heavy Exercise

At higher levels of exertion, other physiological adjustments occur as the sympathetic nervous system stimulates the cardiac and vasomotor centers. Cardiac output increases toward maximal levels, and there are major changes in the peripheral distribution of blood, facilitating the blood flow to active skeletal muscles.

Under massive sympathetic stimulation the cardioacceleratory center can increase cardiac output to levels as high as 20–25 liters per minute. Even at these rates, the increased circulatory demands of the skeletal muscles can be met only if the vasomotor center severely restricts the blood flow to "nonessential" organs such as those of the digestive organs. During exercise at maximal levels, the blood essentially races between the skeletal muscles and the lungs and heart. Only the blood supply to the brain remains unaffected.

Exercise, Cardiovascular Fitness, and Health

Cardiovascular performance improves significantly with training. Table 21-3 compares the cardiac performance of athletes with that of nonathletes. Trained athletes have bigger hearts and larger stroke volumes than do nonathletes, and these are important functional differences.

Cardiac output is equal to the stroke volume times the heart rate, so for the same cardiac output an individual with a larger stroke volume will have a slower heart rate. A professional athlete at rest can maintain normal blood flow to peripheral tissues at a heart rate as low as 50 bpm (beats per minute), and when necessary the cardiac output can increase to levels 50 percent higher than those of nonathletes. Thus, a trained athlete can tolerate sustained levels of activity that are well outside the capabilities of nonathletes.

Exercise and Cardiovascular Disease

Regular exercise has several beneficial effects. Even a moderate exercise routine (jogging 5 miles per week, for example) can lower total blood cholesterol levels. A high cholesterol level is one of the major risk factors for atherosclerosis, leading to cardiovascular disease and strokes. In addition, a healthy lifestyle with regular exercise, a balanced diet, weight control, and no smoking, reduces stress, lowers blood pressure, and slows plaque formation.

Regular moderate exercise may cut the incidence of heart attacks almost in half. However, at present only an estimated 8 percent of adults in the United States exercise at recommended levels. Exercise is also beneficial in accelerating recovery after a heart attack. Regular light to moderate exercise, such as walking, jogging, or bicycling, coupled with a low-fat diet and low-stress lifestyle, not only reduces symp-

TABLE 21-3 Effects of Training on Cardiovascular Performance

Subject	Heart Weight (g)	Stroke Volume (ml)	Heart Rate (bpm)	Cardiac Output (l/min)	Blood Pressure (systolic/diastolic)
Nonathlete (rest)	300	60	83	5.0	120/80
Nonathlete (maximum)		104	192	19.9	187/75
Trained athlete (rest)	500	100	53	5.3	120/80
Trained athlete (maximum)		167	182	30.4	200/90[a]

[a]Diastolic pressures in athletes during maximal activity have not been accurately measured.

toms, such as angina, but also improves both mood and the overall quality of life. However, exercise does not remove the underlying medical problem, and atherosclerotic plaques do not disappear and seldom grow smaller with exercise.

There is no evidence that *intense* athletic training lowers the incidence of cardiovascular disease. On the contrary, the strains placed on all physiological systems, including the cardiovascular system, during an ultramarathon or other athletic extreme can be severe. Individuals with congenital aneurysms, cardiomyopathy, or cardiovascular disease risk fatal circulatory problems, such as an arrhythmia or heart attack, during severe exercise. Even normal individuals can develop acute physiological disorders, such as kidney failure, after extreme exercise. The effects of exercise on other systems will be detailed in later chapters.

❑ Cardiovascular Response to Hemorrhaging

Figure 21-18

In Chapter 19 we considered the local circulatory reaction to a break in the wall of a blood vessel. ∞ *[p. 670]* When hemostasis fails to prevent a significant blood loss, the entire cardiovascular system begins making adjustments to maintain blood pressure and restore blood volume (Figure 21-18•). The immediate problem is the maintenance of adequate blood pressure and peripheral blood flow. The long-term problem is the restoration of normal blood volume.

Short-Term Elevation of Blood Pressure

Almost as soon as the pressures start to decline, short-term adjustments begin. The steps include the following:

- The carotid and aortic reflexes increase cardiac output and cause peripheral vasoconstriction (p. 739). With the blood volume reduced, cardiac output is maintained by increasing the heart rate, often to rates of 180–200 bpm. ∞ *[p. 710]*
- Sympathetic activation provides assistance by increasing vasomotor tone, constricting the arterioles, and elevating blood pressure (p. 739). At the same time, the small muscular arteries, arterioles, and veins decrease in diameter.
- The venoconstriction demanded by the vasomotor center quickly improves venous return and mobilizes the venous reserve (p. 727).
- Hormonal adjustments occur. (1) Sympathetic activation causes the secretion of epinephrine and norepinephrine by the adrenal medulla, increasing cardiac output and extending peripheral vasoconstriction. (2) The release of ADH by the posterior pituitary and the activation of angiotensin II enhance the vasoconstriction while participating in the long-term response (p. 743).

This combination of adjustments elevates blood pressure and improves peripheral blood flow. After blood losses of up to 20 percent of the total blood volume, these short-term adjustments can restore normal arterial pressures and peripheral blood flow. For example, these adjustments are more than sufficient to compensate for the blood loss experienced when donating blood. (Most blood banks collect 500 ml of whole blood, roughly 10 percent of the total blood volume.)

Long-Term Restoration of Blood Volume

Short-term compensation involves adjusting to the loss in blood volume. Long-term compensation involves restoring normal blood volume. Following a serious hemorrhage, this process may take several days. The steps include:

- The decline in capillary blood pressure triggers the recall of fluids from the interstitial spaces (pp. 735–736).
- ADH and aldosterone promote fluid retention and reabsorption at the kidneys, preventing further reductions in blood volume.
- Thirst increases, and additional water is obtained by absorption across the digestive tract. This

● **FIGURE 21-18**
Cardiovascular Responses to Hemorrhaging and Blood Loss

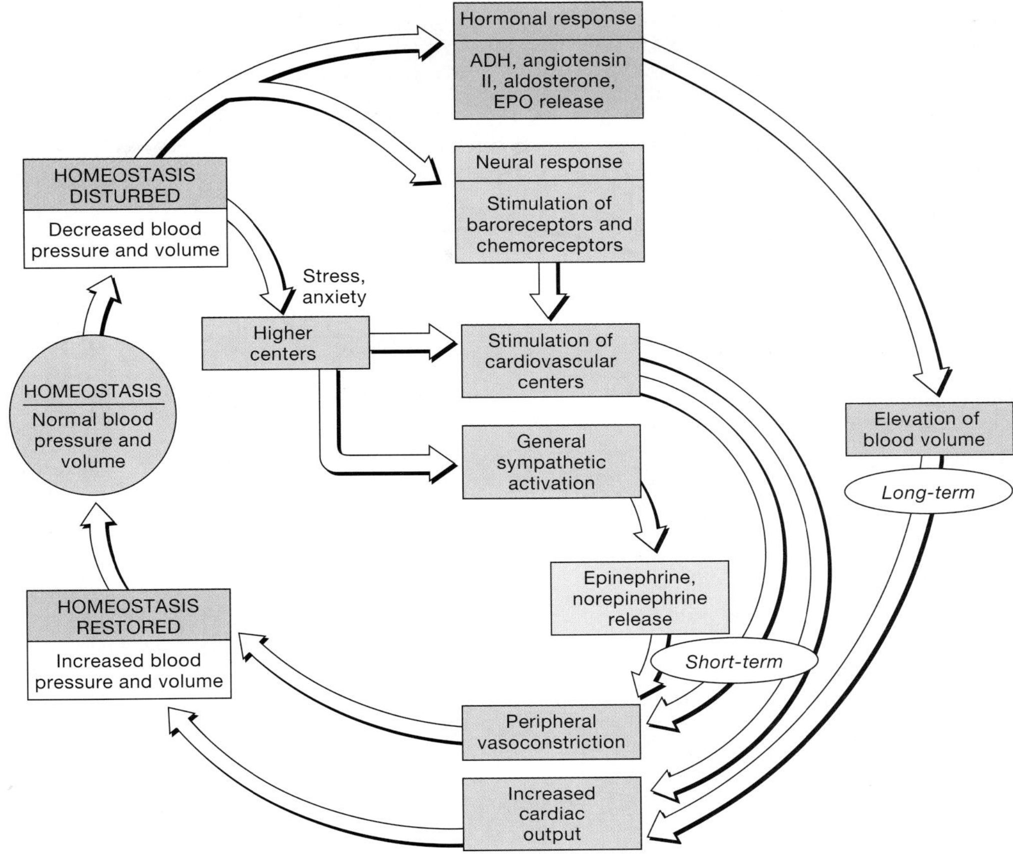
Hormonal response
ADH, angiotensin II, aldosterone, EPO release
Neural response
Stimulation of baroreceptors and chemoreceptors
HOMEOSTASIS DISTURBED
Decreased blood pressure and volume
Stress, anxiety
Higher centers
Stimulation of cardiovascular centers
HOMEOSTASIS
Normal blood pressure and volume
General sympathetic activation
Elevation of blood volume
Long-term
Epinephrine, norepinephrine release
HOMEOSTASIS RESTORED
Increased blood pressure and volume
Short-term
Peripheral vasoconstriction
Increased cardiac output

intake of fluid elevates plasma volume and ultimately replaces the interstitial fluids "borrowed" at the capillaries.

- Erythropoietin targets the bone marrow, stimulating the maturation of red blood cells, which increase the blood volume and improve oxygen delivery to peripheral tissues (p. 734).

Shock

Shock is an acute circulatory crisis marked by low blood pressure (hypotension) and inadequate peripheral blood flow. Severe and potentially fatal symptoms develop as vital tissues become starved for oxygen and nutrients. Common causes of shock are (1) a fall in cardiac output after hemorrhaging or other fluid losses, (2) damage to the heart, (3) external pressure on the heart, and (4) extensive peripheral vasodilation.

We will focus on the cause, symptoms, and treatment of circulatory shock. Readers interested in other types of shock should consult the *Applications Manual.* AM *Other Types of Shock*

Circulatory Shock

Figures 21-18, 21-19

A severe reduction in blood volume produces symptoms of **circulatory shock.** Symptoms of circulatory shock appear after fluid losses of about 30 percent of the total blood volume. The cause can be hemorrhaging or fluid losses to the environment, as in dehydration or after severe burns. All cases of circulatory shock share the same basic symptoms:

1. Hypotension, with systolic pressures below 90 mm Hg.
2. Pale, cool, and moist ("clammy") skin. The skin is pale and cool because of peripheral vasoconstriction; the moisture reflects sympathetic activation of the sweat glands.
3. Frequent confusion and disorientation, due to a fall in blood pressure at the brain.
4. Rapid, weak pulse.
5. Cessation of urination, because the reduced blood to the kidneys slows or stops urine production.
6. A drop in blood pH (*acidosis*), due to lactic acid generation in oxygen-deprived tissues.

Circulatory shock is often divided into *compensated, progressive,* and *irreversible* stages.

THE COMPENSATED STAGE (STAGE I) During the **compensated stage** homeostatic adjustments can cope with the situation; the short-term and long-term responses detailed in Figure 21-18● are part of the compensation process. During the period of compensation, peripheral blood flow is reduced but remains within tolerable limits.

● **FIGURE 21-19**

Shock. (a) The progressive stage is characterized by a gradual decline in systemic blood pressure, tissue blood flow, and cardiac output. **(b)** The irreversible stage involves a series of integrated chain reactions leading to a rapid decline in cardiac output, a dramatic and irreversible fall in blood pressure, circulatory collapse, and eventual death.

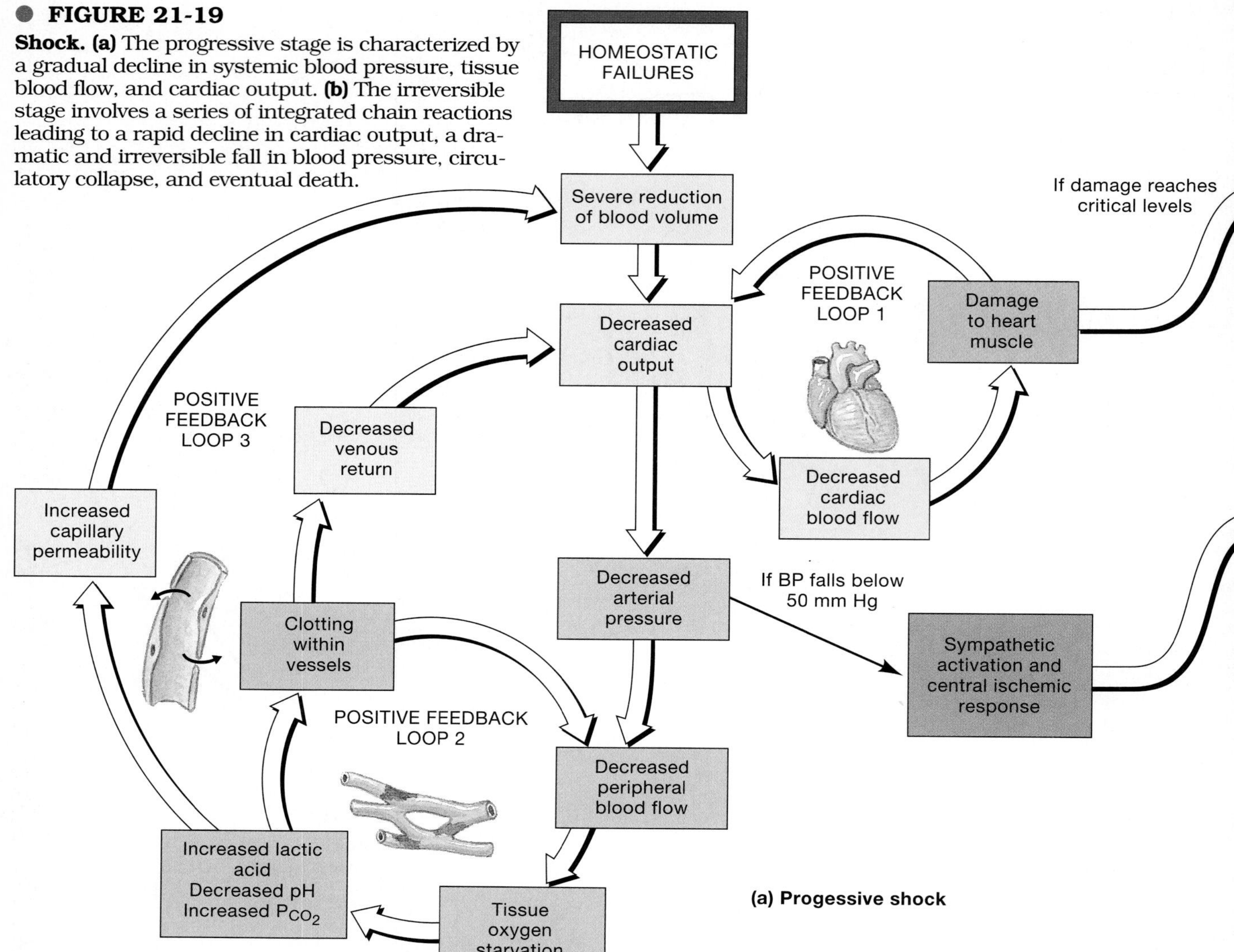

HOMEOSTATIC FAILURES
Severe reduction of blood volume
If damage reaches critical levels
POSITIVE FEEDBACK LOOP 1
Damage to heart muscle
Decreased cardiac output
POSITIVE FEEDBACK LOOP 3
Decreased venous return
Decreased cardiac blood flow
Increased capillary permeability
Decreased arterial pressure
If BP falls below 50 mm Hg
Clotting within vessels
Sympathetic activation and central ischemic response
POSITIVE FEEDBACK LOOP 2
Decreased peripheral blood flow
Increased lactic acid
Decreased pH
Increased P_{CO_2}
Tissue oxygen starvation
(a) Progessive shock

THE PROGRESSIVE STAGE (STAGE II) When blood volume declines by more than 35 percent, the individual enters the **progressive stage** of circulatory shock. Homeostatic mechanisms are now unable to cope with the situation. Despite sustained vasoconstriction and the mobilization of the venous reserve, blood pressure remains abnormally low, venous return is reduced, and cardiac output is inadequate. A vicious cycle begins when the low cardiac output causes myocardial damage. This damage leads to a further reduction in cardiac output and subsequent reductions in blood pressure and venous return. The sequence is diagrammed in Figure 21-19a●.

When the mean arterial blood pressure falls to about 50 mm Hg, carotid sinus baroreceptors trigger a massive activation of the vasomotor centers. In essence, the goal now is to preserve the circulation to the brain at any cost. Blood flow to cerebral vessels is not affected, and the blood pressure in the carotid arteries remains relatively high (70 mm Hg). But in other organs sympathetic output causes a sustained and maximal vasoconstriction. This reflex, called the **central ischemic response,** reduces peripheral circulation to an absolute minimum.

The central ischemic response is a last-ditch effort that maintains adequate blood flow to the brain at the expense of other tissues. Unless prompt treatment is provided, the condition will soon prove fatal. Treatment must concentrate on (1) preventing further fluid losses and (2) giving a transfusion of whole blood, plasma expanders, or blood substitutes.

THE IRREVERSIBLE STAGE (STAGE III) In the absence of treatment, progressive shock will soon turn into **irreversible shock** (Figure 21-19b●). At this point, conditions in the heart, liver, kidneys, and CNS are rapidly deteriorating to the point that death will occur, even *with* medical treatment.

Irreversible shock begins when conditions in the tissues become so abnormal that the arteriolar smooth muscles and precapillary sphincters become unable to contract, despite the commands

Irreversible damage to cardiac muscle and rapid drop in cardiac output

Decreased sympathetic activity

POSITIVE FEEDBACK LOOP 4

Falling arterial pressures

Irreversible CNS damage

Very low peripheral blood flow

Decreased blood flow to CNS

Drastic chemical changes in tissues

General vasodilation

Circulatory collapse

DEATH

(b) Irreversible shock

of the vasomotor center. The result is a widespread peripheral vasodilation and an immediate and fatal decline in blood pressure. This event is called **circulatory collapse.** The blood pressure in many tissues then falls so low that the capillaries collapse like deflating balloons. Blood flow through these capillary beds then stops completely, and the cells in the affected tissues die. The dying cells release more abnormal chemicals, and the effect quickly spreads throughout the body.

 Why does blood pressure increase during exercise?

 How would applying a small pressure to the common carotid artery affect your heart rate?

 What affect would vasoconstriction of the renal artery have on blood pressure and blood volume?

 Why does a person suffering from circulatory shock have a rapid and weak pulse?

The Blood Vessels

Figure 21-20

The circulatory system is divided into the *pulmonary circuit* and the *systemic circuit.* The pulmonary circuit is composed of arteries and veins that transport blood between the heart and the lungs. This circuit begins at the right ventricle and ends at the left atrium. From the left ventricle the arteries of the systemic circuit transport oxygenated blood and nutrients to all organs and tissues, ultimately returning deoxygenated blood to the right atrium. Figure 21-20● summarizes the primary circulatory routes within the pulmonary and systemic circuits.

In the pages that follow we will examine the vessels of the pulmonary and systemic circuits in detail. Two general functional patterns are worth noting at the outset:

1. The peripheral distributions of arteries and veins on the left and right sides are usually

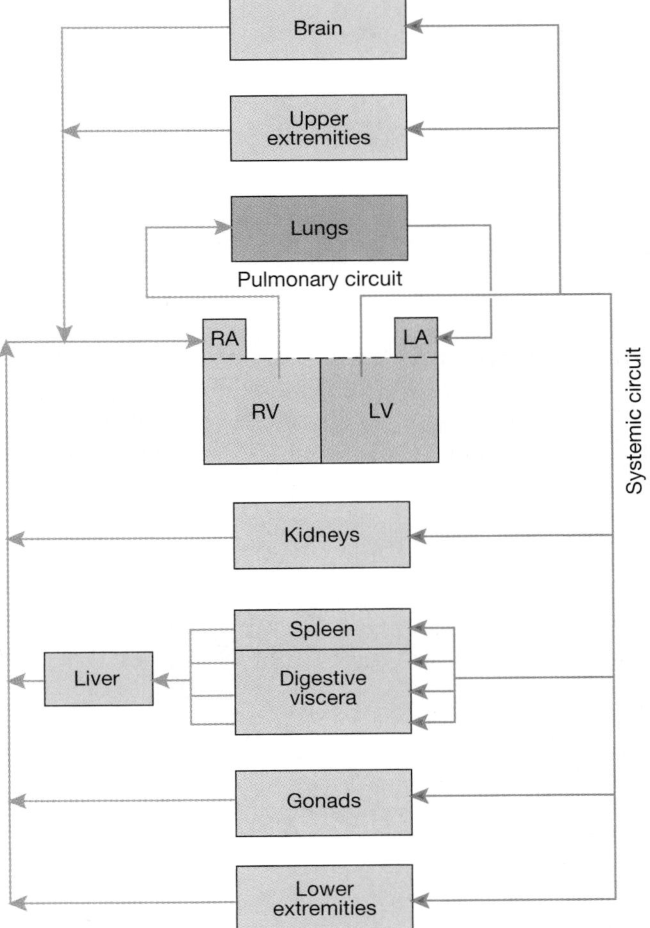
Brain
Upper extremities
Lungs
Pulmonary circuit
RA
LA
RV
LV
Systemic circuit
Kidneys
Spleen
Liver
Digestive viscera
Gonads
Lower extremities

● **FIGURE 21-20**
An Overview of the Pattern of Circulation

identical except near the heart, where the largest vessels connect to the atria or ventricles. For example, there are *left* and *right subclavian, axillary, brachial,* and *radial arteries* whose distribution parallels that of the *left* and *right subclavian, axillary, brachial,* and *radial veins.*

2. A single vessel may have several different names as it crosses specific anatomical boundaries, making accurate anatomical descriptions possible when the vessel extends far into the periphery. For example, the *external iliac artery* becomes the *femoral artery* as it leaves the trunk and enters the lower limb.

❑ The Pulmonary Circulation

Figure 21-21

Blood entering the right atrium has just returned from the peripheral capillary beds, where oxygen was released and carbon dioxide absorbed. After traveling through the right atrium and ventricle, blood enters the pulmonary trunk, the start of the pulmonary circuit. At the lungs (Figure 21-21●), oxygen will be replenished, carbon dioxide released, and the oxygenated blood returned to the heart for distribution via the systemic circuit. Compared with the systemic circuit, the pulmonary circuit is relatively short; the heart and lungs are only about 15 cm (6 in.) apart.

The arteries of the pulmonary circuit differ from those of the systemic circuit in that they carry deoxygenated blood. (For this reason, color-coded diagrams usually show the pulmonary arteries in blue, the same color as systemic veins.) As the pulmonary trunk curves over the superior border of the heart it gives rise to the **left** and **right pulmonary arteries.** These large arteries enter the lungs before branching repeatedly, giving rise to smaller and smaller arteries. The smallest branches, the *pulmonary arterioles,* provide blood to capillary networks that surround small air pockets, or *alveoli* (al-VĒ-ō-lī; *alveolus,* sac). The walls of alveoli are thin enough for gas exchange to occur between the capillary blood and inspired air. (Alveolar structure is described in Chapter 23.) As it leaves the

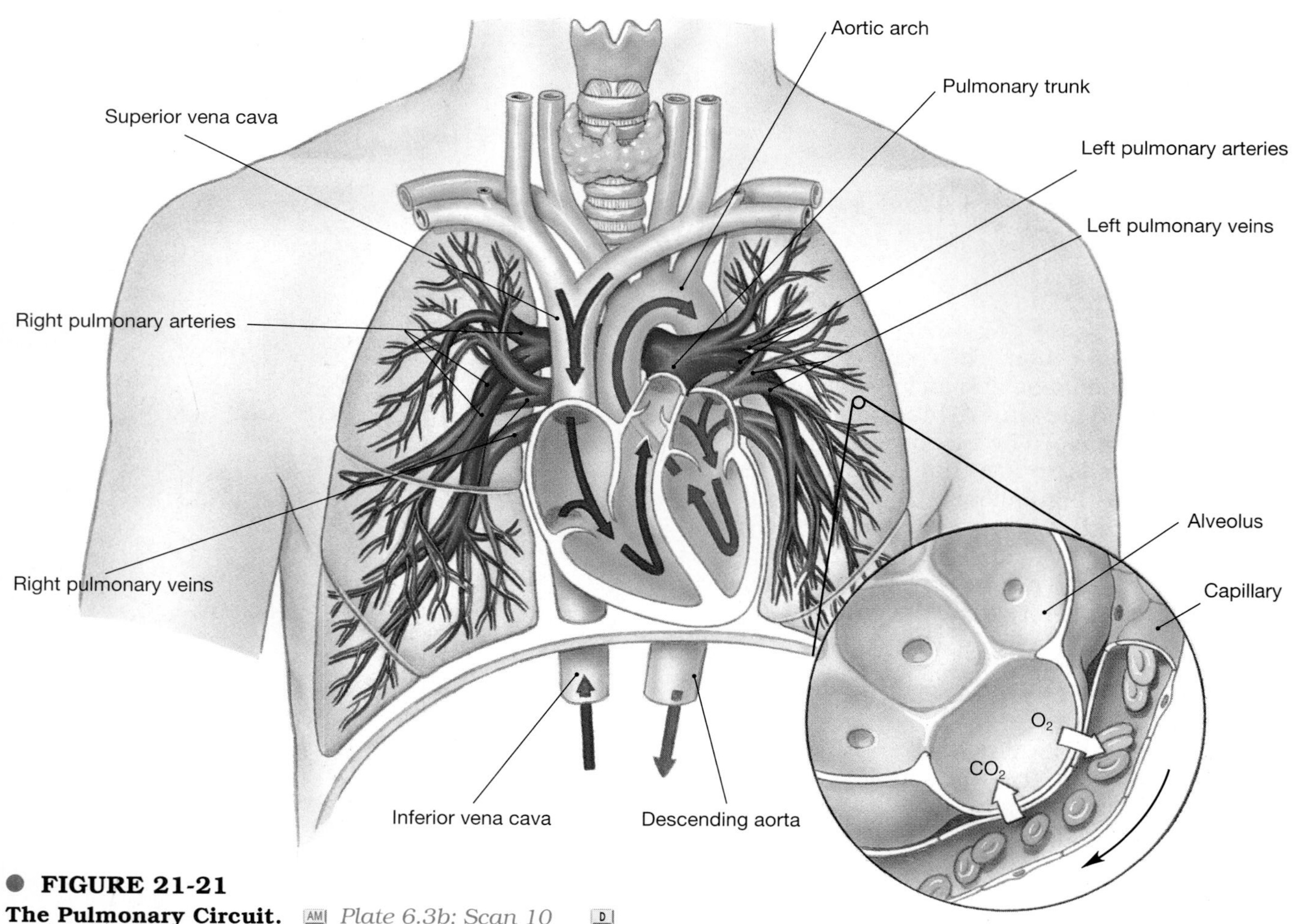
Aortic arch
Pulmonary trunk
Superior vena cava
Left pulmonary arteries
Left pulmonary veins
Right pulmonary arteries
Alveolus
Capillary
Right pulmonary veins
O_2
CO_2
Inferior vena cava
Descending aorta

● **FIGURE 21-21**
The Pulmonary Circuit. AM *Plate 6.3b; Scan 10* D

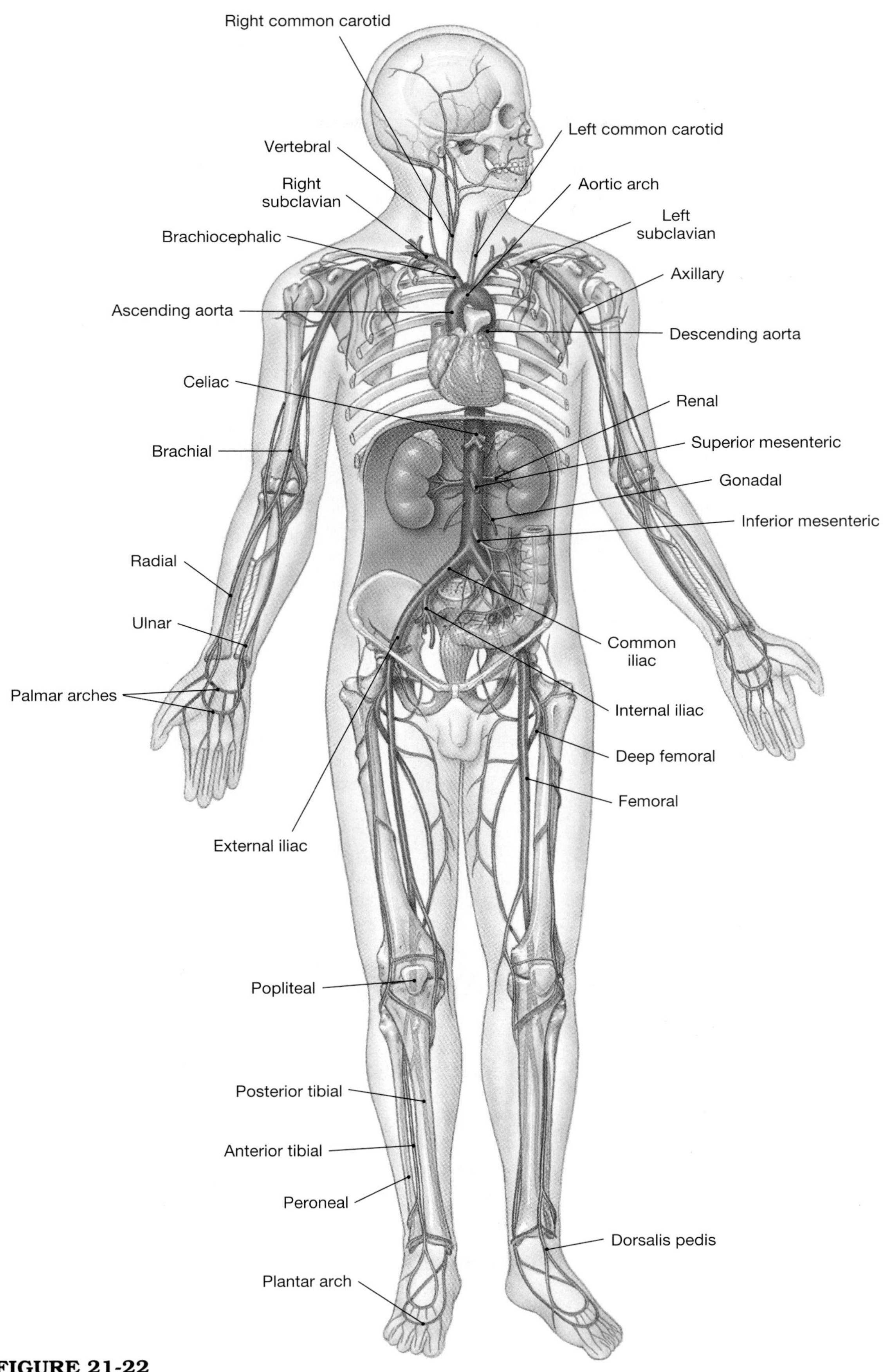
Right common carotid
Left common carotid
Vertebral
Aortic arch
Right
subclavian
Left
subclavian
Brachiocephalic
Axillary
Ascending aorta
Descending aorta
Celiac
Renal
Superior mesenteric
Brachial
Gonadal
Inferior mesenteric
Radial
Ulnar
Common
iliac
Palmar arches
Internal iliac
Deep femoral
Femoral
External iliac
Popliteal
Posterior tibial
Anterior tibial
Peroneal
Dorsalis pedis
Plantar arch

● **FIGURE 21-22**
An Overview of the Arterial System

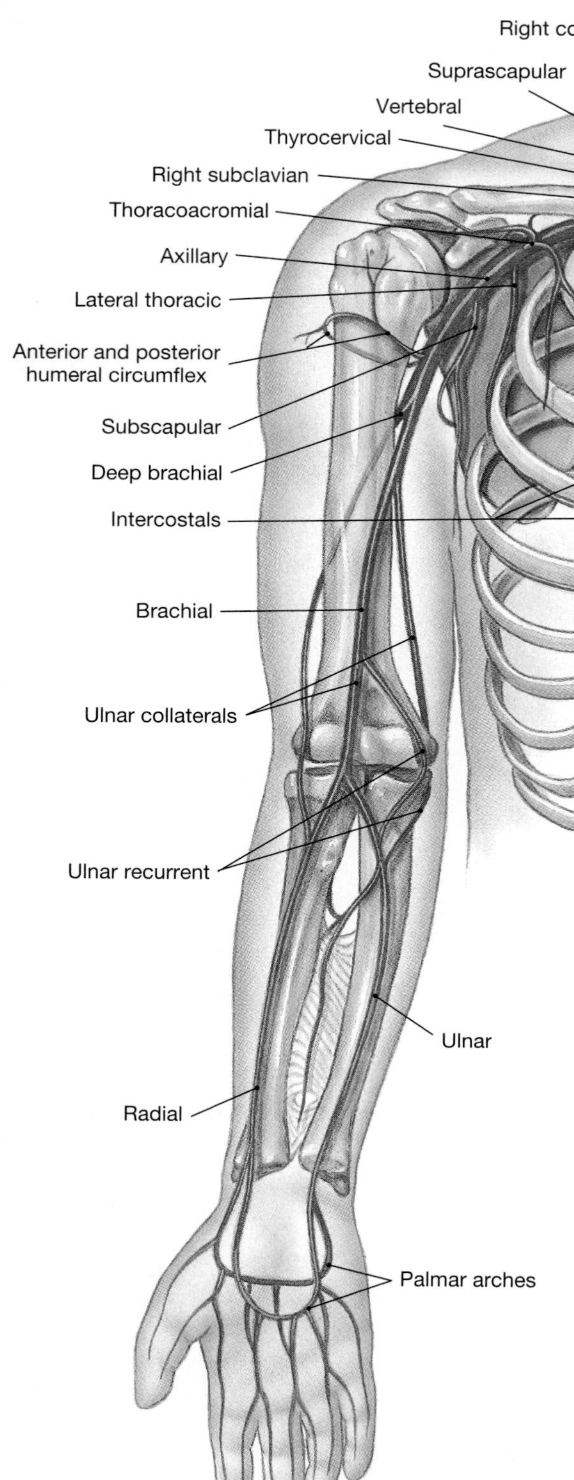

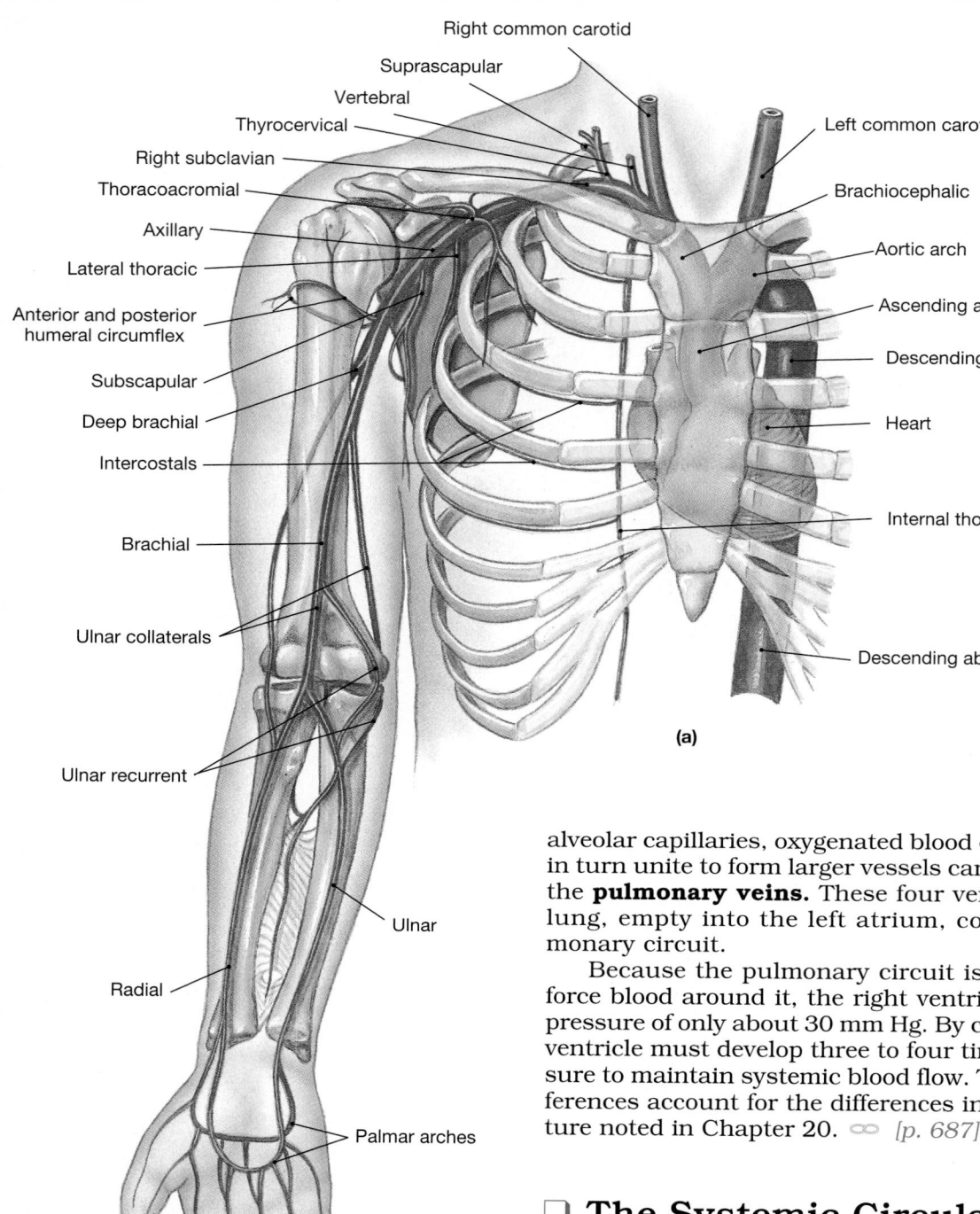

Right common carotid
Suprascapular
Vertebral
Thyrocervical
Right subclavian
Thoracoacromial
Axillary
Lateral thoracic
Anterior and posterior humeral circumflex
Subscapular
Deep brachial
Intercostals
Brachial
Ulnar collaterals
Ulnar recurrent
Ulnar
Radial
Palmar arches
Left common carotid
Brachiocephalic
Aortic arch
Ascending aorta
Descending thoracic aorta
Heart
Internal thoracic
Descending abdominal aorta
(a)

● **FIGURE 21-23**
Arteries of the Chest and Upper Limb. (a) Diagrammatic view. **(b)** Flow chart. AM *Plates 5.1, 5.2; Scan 11*

alveolar capillaries, oxygenated blood enters venules that in turn unite to form larger vessels carrying blood toward the **pulmonary veins.** These four veins, two from each lung, empty into the left atrium, completing the pulmonary circuit.

Because the pulmonary circuit is short, in order to force blood around it, the right ventricle need develop a pressure of only about 30 mm Hg. By comparison, the left ventricle must develop three to four times as much pressure to maintain systemic blood flow. These pressure differences account for the differences in ventricular structure noted in Chapter 20. *[p. 687]*

❑ The Systemic Circulation

The systemic circulation supplies the capillary beds in all other parts of the body. This circuit, which at any given moment contains about 84 percent of the total blood volume, begins at the left ventricle and ends at the right atrium.

Systemic Arteries

Figures 21-22, 21-23, 21-24, 21-25, 21-26, 21-27, 21-28, 21-29

Figure 21-22● is an overview of the arterial system. This figure indicates the relative locations of major systemic arteries. The detailed distribution of these vessels and their branches will be found in Figures 21-23 to 21-25●.

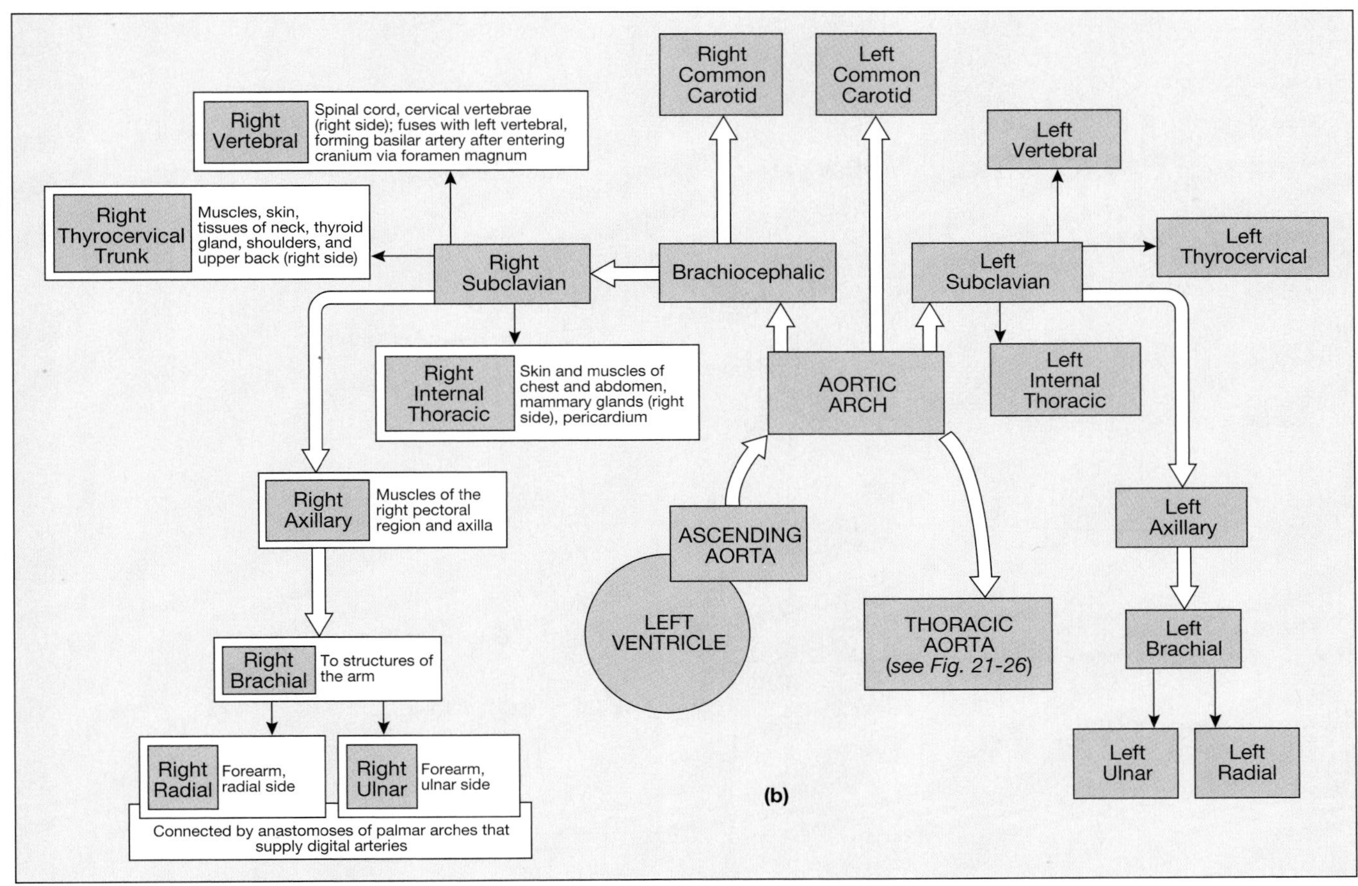

Right Common Carotid
Left Common Carotid
Right Vertebral
Spinal cord, cervical vertebrae (right side); fuses with left vertebral, forming basilar artery after entering cranium via foramen magnum
Left Vertebral
Right Thyrocervical Trunk
Muscles, skin, tissues of neck, thyroid gland, shoulders, and upper back (right side)
Right Subclavian
Brachiocephalic
Left Subclavian
Left Thyrocervical
Right Internal Thoracic
Skin and muscles of chest and abdomen, mammary glands (right side), pericardium
AORTIC ARCH
Left Internal Thoracic
Right Axillary
Muscles of the right pectoral region and axilla
ASCENDING AORTA
Left Axillary
LEFT VENTRICLE
THORACIC AORTA (see Fig. 21-26)
Left Brachial
Right Brachial
To structures of the arm
Left Ulnar
Left Radial
Right Radial
Forearm, radial side
Right Ulnar
Forearm, ulnar side
(b)
Connected by anastomoses of palmar arches that supply digital arteries

THE ASCENDING AORTA The **ascending aorta** begins at the aortic semilunar valve of the left ventricle (Figure 21-23●). The left and right coronary arteries originate at the base of the ascending aorta, just superior to the aortic semilunar valve. The distribution of coronary vessels was detailed in Chapter 20 and illustrated in Figure 20-8●, p. 693.

THE AORTIC ARCH The **aortic arch** curves like a cane handle across the superior surface of the heart, connecting the ascending aorta with the descending aorta. Three elastic arteries originate along the aortic arch (Figures 21-22, 21-23, and 21-24●). These arteries, the **brachiocephalic** (brā-kē-ō-se-FAL-ik), the **left common carotid,** and the **left subclavian** (sub-CLĀ-vē-an), deliver blood to the head, neck, shoulders, and upper limbs. The brachiocephalic artery, also called the *innominate artery* (i-NOM-i-nāt; unnamed), ascends for a short distance before branching to form the **right subclavian artery** and the **right common carotid artery.**

There is only one brachiocephalic artery, and the left common carotid and left subclavian arteries arise separately from the aortic arch. However, in terms of their peripheral distribution, the vessels on the left side are mirror images of those on the right side. Because the descriptions that follow focus on major branches found on both sides of the body, for clarity the terms *right* and *left* will not be used. Figures 21-23 and 21-24● illustrate the major branches of these arteries.

The Subclavian Arteries. The subclavian arteries supply blood to the arms, chest wall, shoulders, back, and central nervous system (Figures 21-22 and 21-23●). Three major branches arise before a subclavian artery leaves the thoracic cavity: (1) the **thyrocervical trunk,** which provides blood to muscles and other tissues of the neck, shoulder, and upper back; (2) an **internal thoracic artery,** supplying the pericardium and anterior wall of the chest; and (3) a **vertebral artery,** which provides blood to the brain and spinal cord.

After leaving the thoracic cavity over the outer border of the first rib, the subclavian is called the **axillary artery.** The axillary artery crosses the axilla (armpit) to enter the arm, where it becomes the **brachial artery.** The brachial artery supplies blood to the upper extremity. At the antecubital fossa, the brachial artery divides into the **radial artery,** which follows the radius, and an **ulnar artery,** which follows the ulna to the wrist. These arteries supply blood to the forearm. At the wrist these arteries

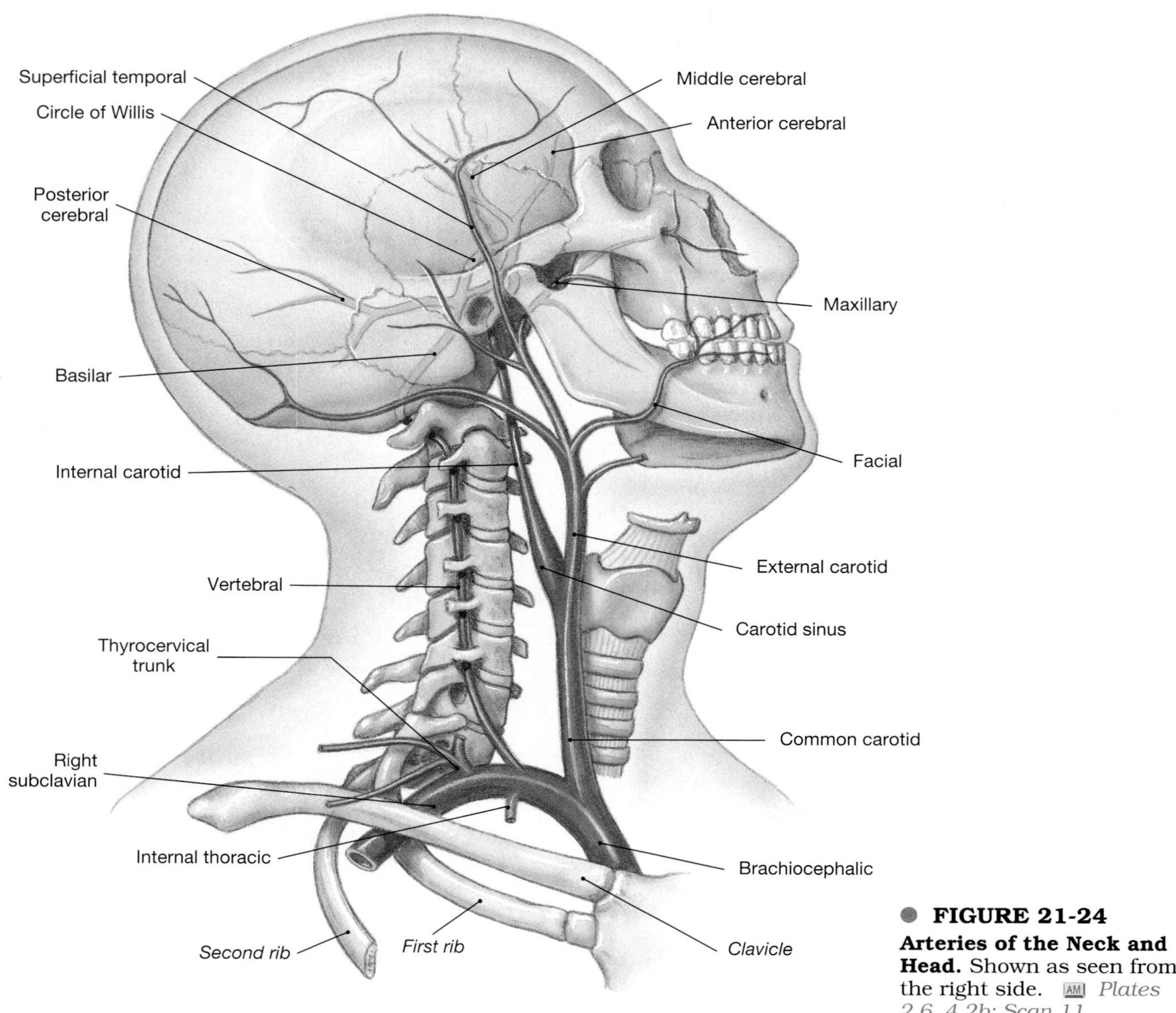

Superficial temporal
Circle of Willis
Posterior cerebral
Basilar
Internal carotid
Vertebral
Thyrocervical trunk
Right subclavian
Internal thoracic
Second rib
First rib
Middle cerebral
Anterior cerebral
Maxillary
Facial
External carotid
Carotid sinus
Common carotid
Brachiocephalic
Clavicle

● **FIGURE 21-24 Arteries of the Neck and Head.** Shown as seen from the right side. AM *Plates 2.6, 4.2b; Scan 11*

anastomose to form a **superficial palmar arch** and a **deep palmar arch** that supply blood to the palm and to the **digital arteries** of the thumb and fingers.

The Carotid Artery and the Blood Supply to the Brain. The common carotid arteries ascend deep in the tissues of the neck. A carotid artery can usually be located by pressing gently along either side of the windpipe (trachea) until a strong pulse is felt.

Each common carotid artery divides into an external carotid and an internal carotid artery at the carotid sinus (Figure 21-24●). The external carotids supply blood to the structures of the neck, pharynx, larynx, lower jaw, and face. The internal carotids enter the skull through the *carotid foramina* of the temporal bones, delivering blood to the brain (see Figures 7-3 and 7-4●, pp. 205–207).

The internal carotids ascend to the level of the optic nerves, where each divides into three branches: (1) an **ophthalmic artery,** which supplies the eyes; (2) an **anterior cerebral artery,** which supplies the frontal and parietal lobes of the brain; and (3) a **middle cerebral artery,** which supplies the mesencephalon and lateral surfaces of the cerebral hemispheres (Figure 21-25●).

The brain is extremely sensitive to changes in its circulatory supply. An interruption of circulation for several seconds will produce unconsciousness, and after 4 minutes there may be some permanent neural damage. Such circulatory crises are rare, because blood reaches the brain through the vertebral arteries as well as

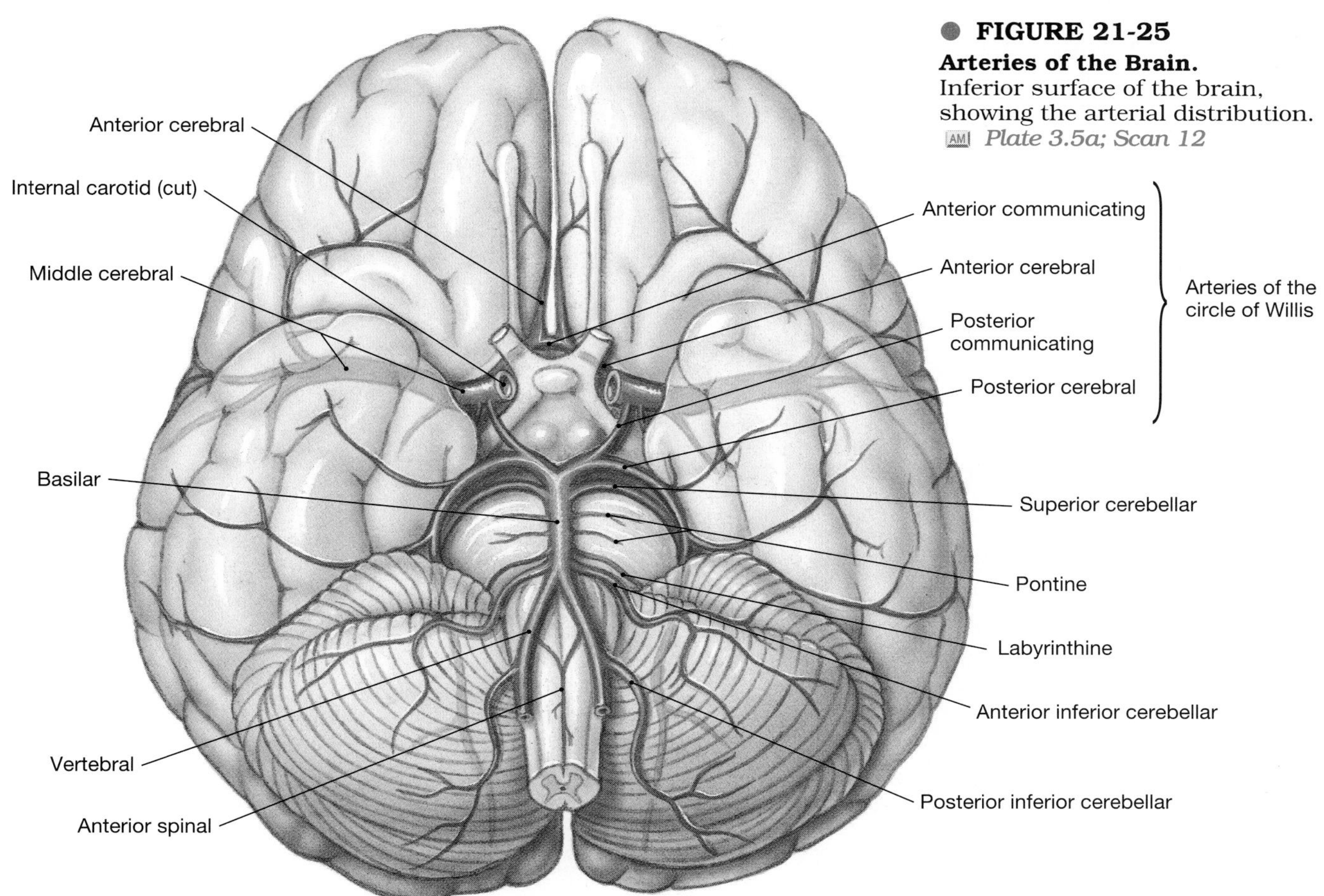
Anterior cerebral
Internal carotid (cut)
Middle cerebral
Basilar
Vertebral
Anterior spinal
Anterior communicating
Anterior cerebral
Posterior communicating
Posterior cerebral
Arteries of the circle of Willis
Superior cerebellar
Pontine
Labyrinthine
Anterior inferior cerebellar
Posterior inferior cerebellar

● **FIGURE 21-25**
Arteries of the Brain. Inferior surface of the brain, showing the arterial distribution. AM *Plate 3.5a; Scan 12*

by way of the internal carotids. The left and right vertebral arteries arise from the subclavian arteries and ascend within the *transverse foramina* of the cervical vertebrae (see Figure 7-19●, p. 226). The vertebral arteries enter the cranium at the foramen magnum, where they fuse along the ventral surface of the medulla oblongata to form the **basilar artery.** The basilar artery continues on the ventral surface along the pons, branching many times before dividing into the **posterior cerebral arteries.** The **posterior communicating arteries** branch off the posterior cerebral arteries (Figure 21-25●).

The internal carotids normally supply the arteries of the anterior half of the cerebrum, and the rest of the brain receives blood from the vertebral arteries. But this circulatory pattern can easily change, because the internal carotids and the basilar artery are interconnected in a ring-shaped anastomosis, the **cerebral arterial circle,** or *circle of Willis*, that encircles the infundibulum of the pituitary gland (Figure 21-25●). With this arrangement, the brain can receive blood from either the carotids or the vertebrals, and the chances for a serious interruption of circulation are reduced.

STROKES *Strokes*, or cerebrovascular accidents (CVAs), are interruptions of the vas cular supply to a portion of the brain. The *middle cerebral artery*, a major branch of the cerebral arterial circle, is the most common site of a stroke. Superficial branches deliver blood to the temporal lobe and large portions of the frontal and parietal lobes; deep branches supply the cerebral nuclei and portions of the thalamus. If a stroke blocks the middle cerebral artery on the left side of the brain, aphasia and a sensory and motor paralysis of the right side result. In a stroke affecting the middle cerebral on the right side, the individual experiences a loss of sensation and motor control over the left side and has difficulty drawing or interpreting spatial relationships. Strokes affecting vessels supplying the brain stem also produce distinctive symptoms; those affecting the lower brain stem are often fatal.

THE DESCENDING AORTA The descending aorta is continuous with the aortic arch. The diaphragm divides the descending aorta into a superior **thoracic aorta** and an inferior **abdominal aorta** (Figures 21-26, 21-27, p. 758, and 21-28●, p. 759).

The Thoracic Aorta. The thoracic aorta begins at the level of vertebra T_5 and penetrates the

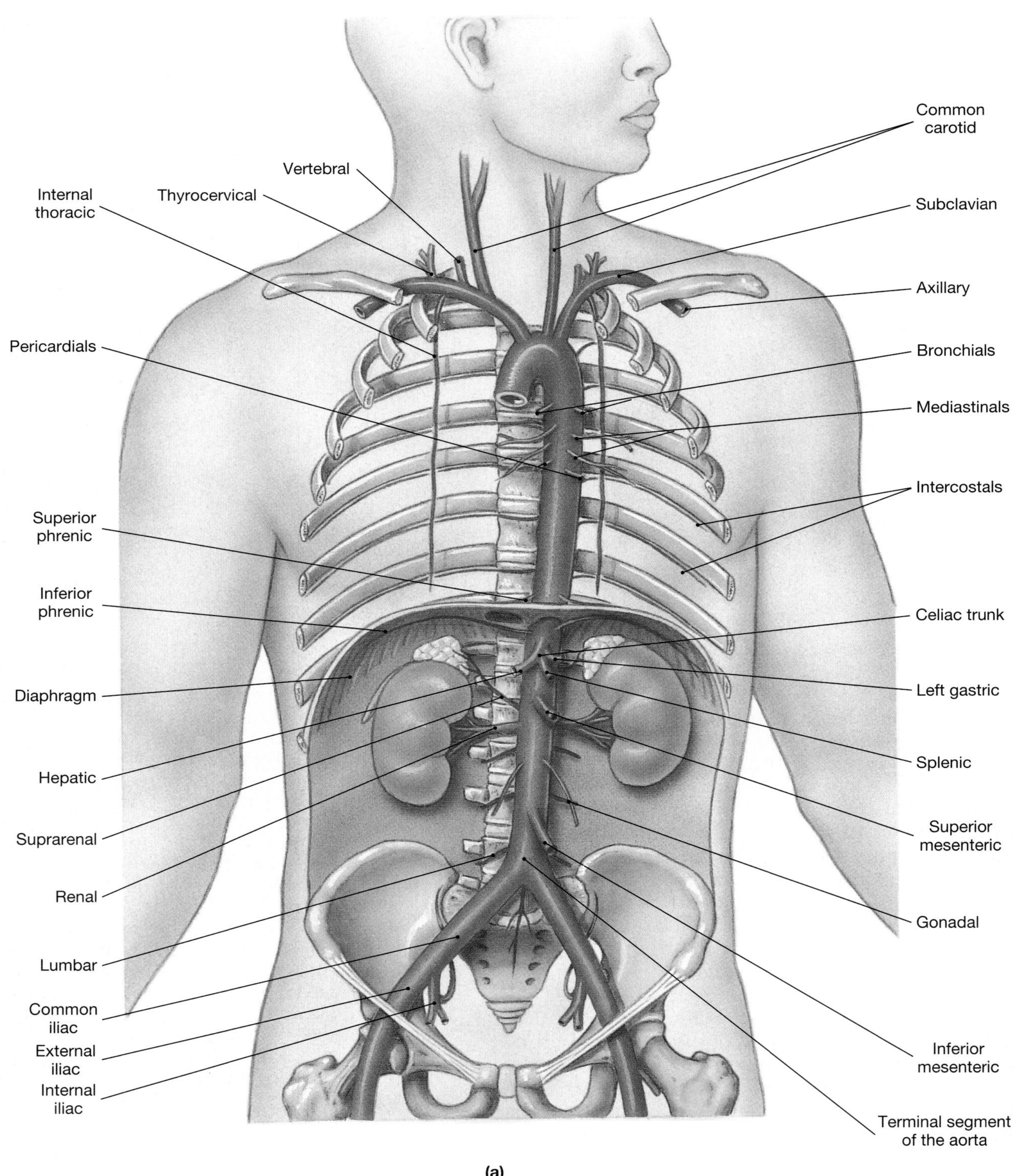
Common carotid
Vertebral
Internal thoracic
Thyrocervical
Subclavian
Axillary
Pericardials
Bronchials
Mediastinals
Intercostals
Superior phrenic
Inferior phrenic
Celiac trunk
Diaphragm
Left gastric
Hepatic
Splenic
Suprarenal
Superior mesenteric
Renal
Gonadal
Lumbar
Common iliac
External iliac
Internal iliac
Inferior mesenteric
Terminal segment of the aorta
(a)

● **FIGURE 21-26**
Major Arteries of the Trunk. (a) Diagrammatic view. **(b)** Flow chart. AM *Plate 6.8; Scan 13*

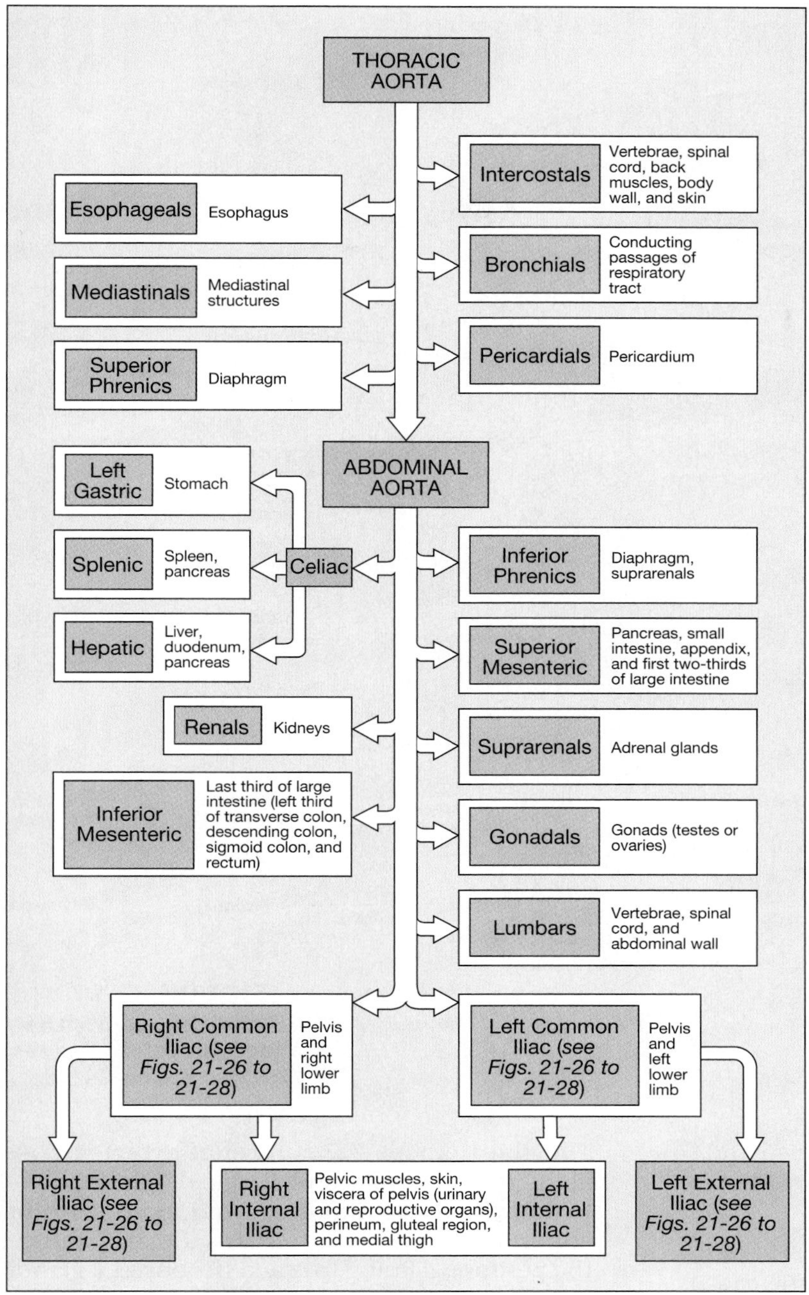
THORACIC AORTA
Intercostals
Vertebrae, spinal cord, back muscles, body wall, and skin
Esophageals
Esophagus
Bronchials
Conducting passages of respiratory tract
Mediastinals
Mediastinal structures
Pericardials
Pericardium
Superior Phrenics
Diaphragm
Left Gastric
Stomach
ABDOMINAL AORTA
Splenic
Spleen, pancreas
Celiac
Inferior Phrenics
Diaphragm, suprarenals
Hepatic
Liver, duodenum, pancreas
Superior Mesenteric
Pancreas, small intestine, appendix, and first two-thirds of large intestine
Renals
Kidneys
Suprarenals
Adrenal glands
Inferior Mesenteric
Last third of large intestine (left third of transverse colon, descending colon, sigmoid colon, and rectum)
Gonadals
Gonads (testes or ovaries)
Lumbars
Vertebrae, spinal cord, and abdominal wall
Right Common Iliac (see Figs. 21-26 to 21-28)
Pelvis and right lower limb
Left Common Iliac (see Figs. 21-26 to 21-28)
Pelvis and left lower limb
Right External Iliac (see Figs. 21-26 to 21-28)
Right Internal Iliac
Pelvic muscles, skin, viscera of pelvis (urinary and reproductive organs), perineum, gluteal region, and medial thigh
Left Internal Iliac
Left External Iliac (see Figs. 21-26 to 21-28)

(b)

diaphragm at the level of vertebra T_{12}. The thoracic aorta travels within the mediastinum, on the dorsal thoracic wall, just slightly to the left of the vertebral column. It supplies blood to branches servicing the viscera of the thorax, the muscles of the chest and the diaphragm, and the thoracic spinal cord.

The branches of the thoracic aorta are anatomically grouped as either *visceral branches* or *parietal branches.* Visceral branches supply the organs of the chest: the **bronchial arteries** supply the nonrespiratory tissues of the lungs, **pericardial arteries** supply the pericardium, and **esophageal arteries** supply the esophagus. The parietal branches supply the chest wall: The **intercostal arteries** supply the chest muscles and the vertebral column area, and the **superior phrenic** (FREN-ik) **arteries** deliver blood to the superior surface of the muscular diaphragm that separates the thoracic and abdominopelvic cavities. The branches of the thoracic aorta are detailed in Figure 21-26●.

The Abdominal Aorta. The abdominal aorta is a continuation of the thoracic aorta, immediately inferior to the diaphragm (Figure 21-26●). The abdominal aorta descends slightly to the left of the vertebral column but posterior to the peritoneal cavity and is often surrounded by a cushion of adipose tissue. At the level of vertebra L_4, the abdominal aorta splits into two major arteries that supply deep pelvic structures and the lower limbs. The region where the aorta splits is called the *terminal segment of the aorta.*

The abdominal aorta delivers blood to all of the abdominopelvic organs and structures. The major branches to visceral organs are unpaired, and they arise on the anterior surface of the abdominal aorta and extend into the mesenteries. Branches to the body wall, the kidneys, and other structures outside the abdominopelvic cavity are paired, and they originate along the lateral surfaces of the abdominal aorta. Figure 21-26● shows the major arteries and veins of the trunk after removal of the thoracic and abdominal organs. The distribution of those arteries to abdominopelvic organs is shown in Figure 21-27●.

The abdominal aorta gives rise to three unpaired arteries: (1) the *celiac trunk,* (2) the *superior mesenteric artery,* and (3) the *inferior mesenteric artery* (Figures 21-26 and 21-27●).

1. The **celiac** (SĒ-lē-ak) **trunk** delivers blood to the liver, stomach, and spleen. The celiac trunk divides into three branches:
   - The **left gastric artery** supplies the stomach and inferior portion of the esophagus.
   - The **common hepatic artery** supplies arteries to the liver (*hepatic*), stomach (*right gastric*), gallbladder (*cystic*), and duodenal area (*gastroduodenal* and *right gastroepiploic*).

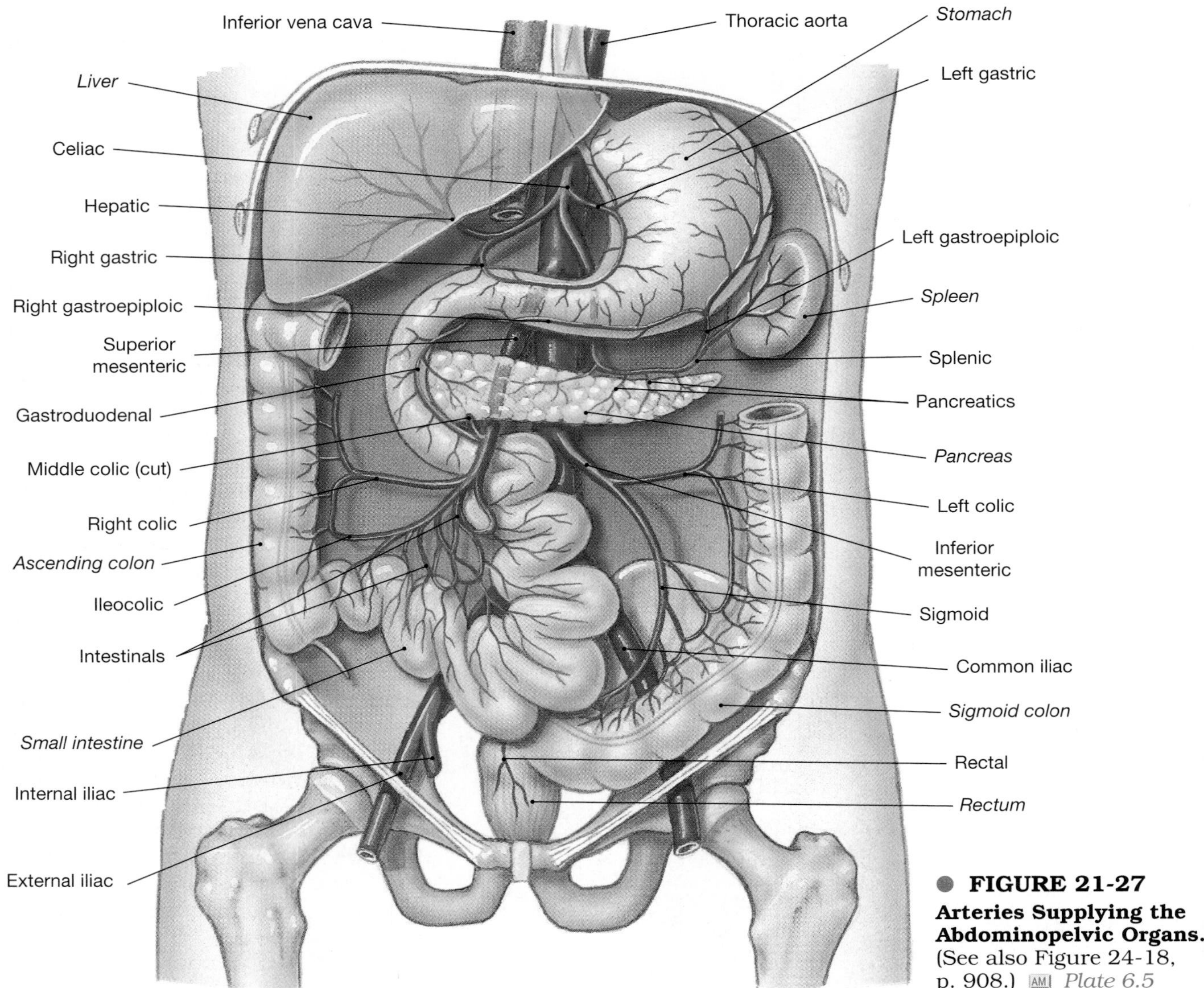
Inferior vena cava
Thoracic aorta
Stomach
Liver
Left gastric
Celiac
Hepatic
Right gastric
Left gastroepiploic
Right gastroepiploic
Spleen
Superior mesenteric
Splenic
Pancreatics
Gastroduodenal
Pancreas
Middle colic (cut)
Left colic
Right colic
Inferior mesenteric
Ascending colon
Ileocolic
Sigmoid
Intestinals
Common iliac
Sigmoid colon
Small intestine
Rectal
Internal iliac
Rectum
External iliac

● **FIGURE 21-27**
Arteries Supplying the Abdominopelvic Organs. (See also Figure 24-18, p. 908.) AM *Plate 6.5*

- The **splenic artery** supplies the spleen and arteries to the stomach *(left gastroepiploic)* and pancreas *(pancreatic).*

2. The **superior mesenteric** (mez-en-TER-ik) **artery** arises about 2.5 cm inferior to the celiac trunk to supply arteries to the pancreas and duodenum (*pancreaticoduodenal*), small intestine (*intestinal*), and most of the large intestine (*right* and *middle colic* and *ileocolic*).
3. The **inferior mesenteric artery** arises about 5 cm superior to the terminal aorta and delivers blood to the terminal portions of the colon (*left colic* and *sigmoid*) and the rectum (*rectal*).

There are four paired arteries: (1) the *suprarenals,* (2) the *renals,* (3) the *lumbars,* and (4) the *gonadals.*

1. The **suprarenal arteries** originate on either side of the aorta near the base of the superior mesenteric artery. Each suprarenal artery supplies one of the adrenal glands that cap the superior portion of the kidneys.
2. The short (about 7.5 cm) **renal arteries** arise along the posterolateral surface of the abdominal aorta, about 2.5 cm (1 in.) inferior to the superior mesenteric artery, and travel behind the peritoneal lining to reach the adrenal glands and kidneys. The branches of the renal arteries are considered in Chapter 26.
3. Small **lumbar arteries** arise on the posterior surface of the aorta and supply the spinal cord and the abdominal wall.
4. Paired **gonadal** (gō-NAD-al) **arteries** originate between the superior and inferior mesenteric arteries. In males, they are called *testicular arteries* and are long, thin arteries that supply blood to the testes and scrotum. In females, they are termed *ovarian arteries* and supply blood to the ovaries, uterine tubes, and uterus. The distribution of gonadal vessels (both arteries and veins) differs in males and females; the differences will be described in Chapter 28.

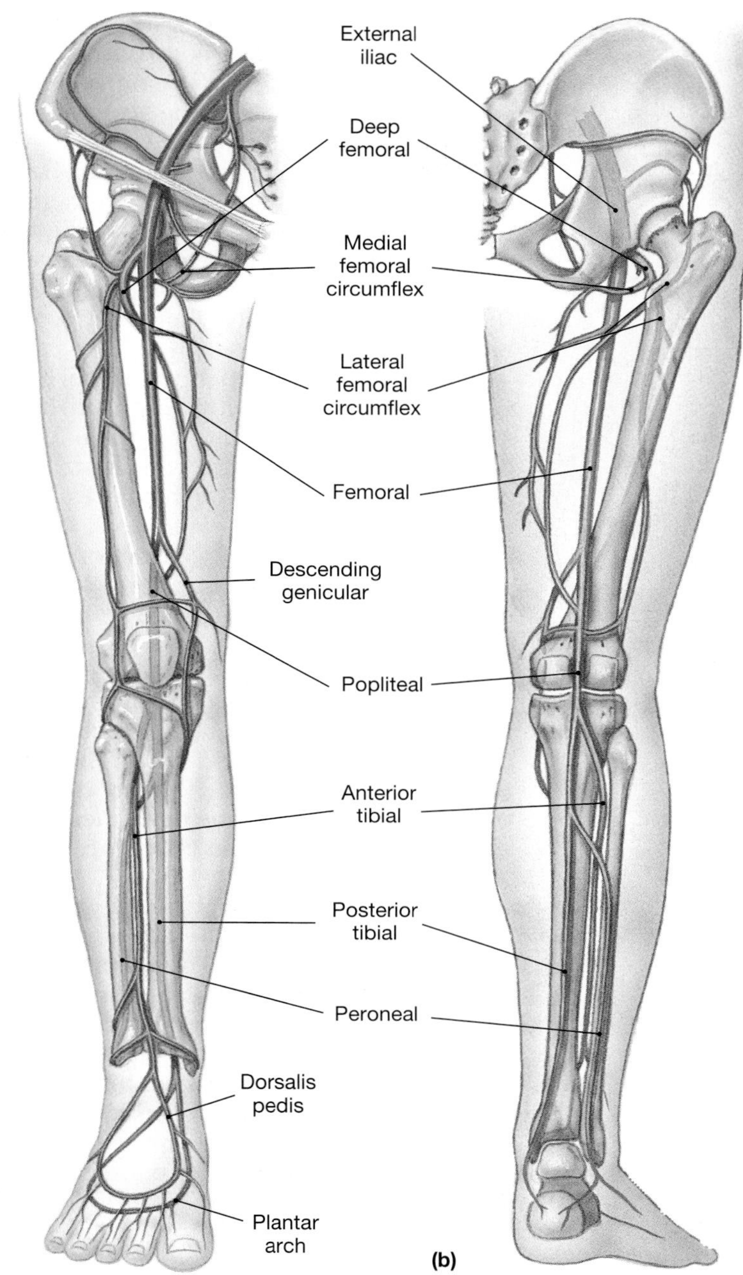
External iliac
Deep femoral
Medial femoral circumflex
Lateral femoral circumflex
Femoral
Descending genicular
Popliteal
Anterior tibial
Posterior tibial
Peroneal
Dorsalis pedis
Plantar arch
(a)
(b)

● **FIGURE 21-28**
Arteries of the Lower Limb. (a) Anterior view. **(b)** Posterior view.
AM · *Plate 7.2; Scans 5, 6*

ARTERIES OF THE PELVIS AND LOWER LIMBS
Near the level of vertebra L_4 the terminal segment of the abdominal aorta divides to form a pair of muscular arteries, the **right** and **left common iliac** (IL-ē-ak) **arteries.** These arteries carry blood to the pelvis and lower limbs (Figures 21-28 and 21-29●). As these arteries travel along the inner surface of the ilium, they descend behind the cecum and sigmoid colon, and, at the level of the lumbosacral joint, each common iliac divides to form an **internal iliac artery** and an **external iliac artery.** The internal iliac arteries enter the pelvic cavity to supply the urinary bladder, the internal and external walls of the pelvis, the external genitalia, the medial side of the thigh, and, in females, the uterus and vagina. The external iliac arteries supply blood to the lower limbs, and they are much larger in diameter than the internal iliac arteries.

Arteries of the Thigh and Leg. The external iliac artery crosses the surface of the iliopsoas muscle and penetrates the abdominal wall midway between the anterior superior iliac spine and the symphysis pubis. It emerges on the anteromedial surface of the thigh as the **femoral artery,** and roughly 5 cm distal to its emergence the **deep femoral artery** branches off its lateral surface (Figure 21-28a●). The deep femoral artery supplies blood to the ventral and lateral regions of the skin and deep muscles of the thigh.

The femoral artery continues inferiorly and posterior to the femur. As it reaches the *popliteal fossa* the femoral artery becomes the **popliteal** (pop-LIT-ē-al) **artery.** The popliteal artery crosses the popliteal fossa before branching to form the **posterior tibial artery** and the **anterior tibial artery.** The posterior tibial artery gives rise to the **peroneal artery** and continues inferiorly along the posterior surface of the tibia. The anterior tibial artery passes between the tibia and fibula, emerging on the anterior surface of the tibia. As it descends toward the foot, the anterior tibial provides blood to the skin and muscles of the anterior portion of the leg.

Arteries of the Foot. When the anterior tibial artery reaches the ankle, it becomes the **dorsalis pedis artery.** The dorsalis pedis branches repeatedly, supplying the ankle and dorsal portion of the foot (Figure 21-28a●).

As it reaches the ankle, the posterior tibial artery divides to form the **medial** and **lateral plantar arteries,** which supply blood to the plantar surface of the foot. The medial and lateral plantar arteries are connected to the dorsalis pedis artery via a pair of anastomoses. This arrangement produces a **dorsal arch** *(arcuate arch)* and a **plantar arch.** Small arteries branching off these arches supply the distal portions of the foot and the toes.

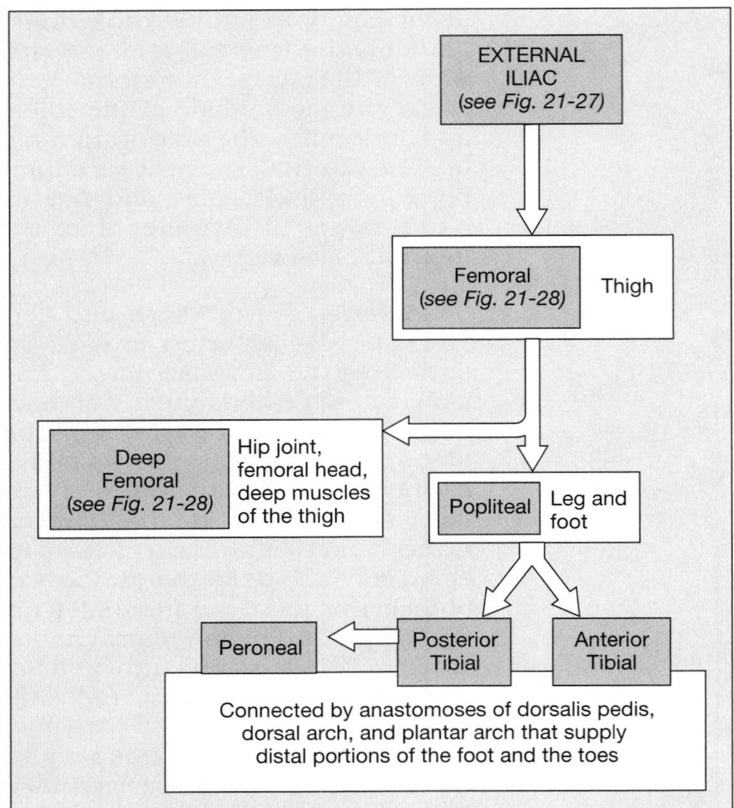
EXTERNAL ILIAC (see Fig. 21-27)
Femoral (see Fig. 21-28)
Thigh
Deep Femoral (see Fig. 21-28)
Hip joint, femoral head, deep muscles of the thigh
Popliteal
Leg and foot
Peroneal
Posterior Tibial
Anterior Tibial
Connected by anastomoses of dorsalis pedis, dorsal arch, and plantar arch that supply distal portions of the foot and the toes

● **FIGURE 21-29**
Flow Chart for Blood Flow to a Lower Limb

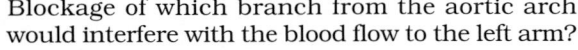

✔ Blockage of which branch from the aortic arch would interfere with the blood flow to the left arm?

✔ Why would compression of the common carotid artery cause a person to lose consciousness?

✔ Grace is in an automobile accident and ruptures her celiac trunk. What organs will be affected most directly by this injury?

Systemic Veins

Figures 21-22, 21-30, 21-31, 21-32, 21-33, 21-34, 21-35

Veins collect blood from the body's tissues and organs via an elaborate venous network that drains into the right atrium of the heart via the superior and inferior venae cavae (Figure 21-30●). A comparison between Figure 21-30 and Figure 21-22●, p. 751, reveals that complementary arteries and veins often run side by side, and in many cases they have comparable names. For example, the axillary arteries run alongside the axillary veins. In addition, arteries and veins often travel in the company of peripheral nerves that have the same names and innervate the same structures.

One significant difference between the arterial and venous systems concerns the distribution of major veins in the neck and limbs. Arteries in these areas are not found at the body surface; instead, they are deep beneath the skin, protected by bones and surrounding soft tissues. In contrast, the neck and limbs usually have two sets of peripheral veins, one superficial and the other deep. The superficial veins are so close to the surface that they can be seen quite easily. This location makes them easy targets for obtaining blood samples, and most blood tests are performed on venous blood collected from the superficial veins of the upper limb (usually where they cross the antecubital surface).

The branching pattern of peripheral veins is much more variable than that of arteries. Arterial pathways are usually direct, because developing arteries grow toward active tissues. By the time blood reaches the venous system, pressures are low, and routing variations make little functional difference. The discussion that follows is based on the most common arrangement of veins.

THE SUPERIOR VENA CAVA All of the body's systemic veins (except the cardiac veins) drain into either the *superior vena cava* or the *inferior vena cava.* The **superior vena cava** (SVC) receives blood from the tissues and organs of the head, neck, chest, shoulders, and upper limbs (Figure 21-30●).

Venous Return from the Cranium. Numerous *superficial cerebral veins* and *internal cerebral veins* drain the cerebral hemispheres. The **superficial cerebral veins** empty into a network of dural sinuses, including the *superior* and *inferior sagittal sinuses,* the *left* and *right transverse sinuses,* and the *straight sinus* (Figure 21-31●). The largest sinus, the **superior sagittal sinus,** is in the falx cerebri (see Figure 14-4●, p. 465). The majority of the **internal cerebral veins** collect inside the brain to form the **great cerebral vein,** which collects blood from the interior of the cerebral hemispheres and the choroid plexus and delivers it to the **straight sinus.** Other cerebral veins drain into the **cavernous sinus** in company with numerous small veins from the orbit.

The venous sinuses converge within the dura mater in the region of the lambdoidal suture. The left and right transverse sinuses converge at the base of the petrous portion of the temporal bone, forming the **sigmoid sinus,** which penetrates the jugular foramen and leaves the skull as the **internal jugular vein.** The internal jugular vein descends parallel to the common carotid artery in the neck (pp. 753–754).

Vertebral veins drain the cervical spinal cord and the posterior surface of the skull. These vessels descend within the transverse foramina of the cervical vertebrae, in company with the vertebral arteries. The vertebral veins empty into the *brachiocephalic veins* of the chest (p. 763).

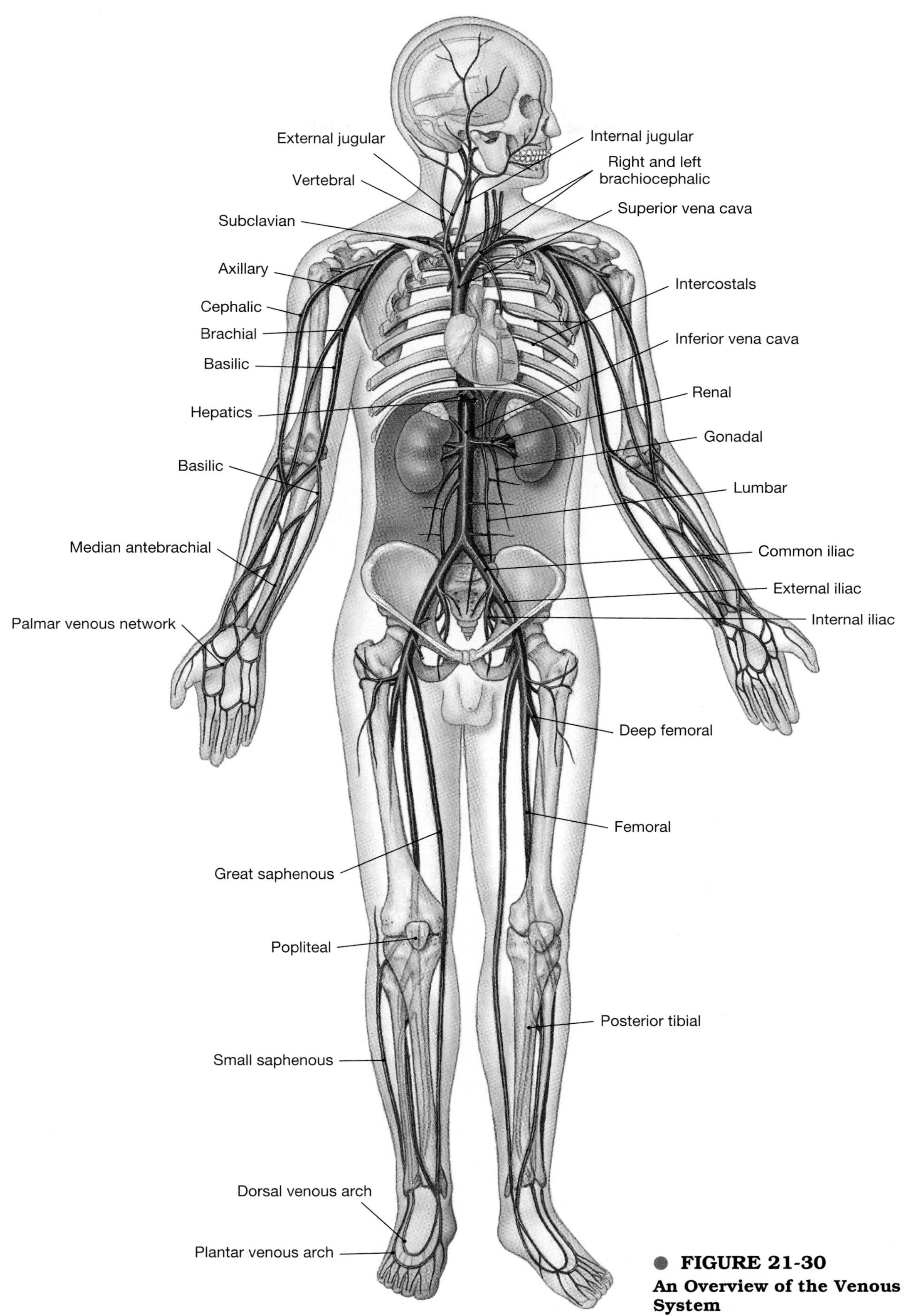
External jugular
Internal jugular
Vertebral
Right and left brachiocephalic
Subclavian
Superior vena cava
Axillary
Intercostals
Cephalic
Brachial
Inferior vena cava
Basilic
Renal
Hepatics
Gonadal
Basilic
Lumbar
Median antebrachial
Common iliac
External iliac
Palmar venous network
Internal iliac
Deep femoral
Femoral
Great saphenous
Popliteal
Posterior tibial
Small saphenous
Dorsal venous arch
Plantar venous arch

• FIGURE 21-30
An Overview of the Venous System

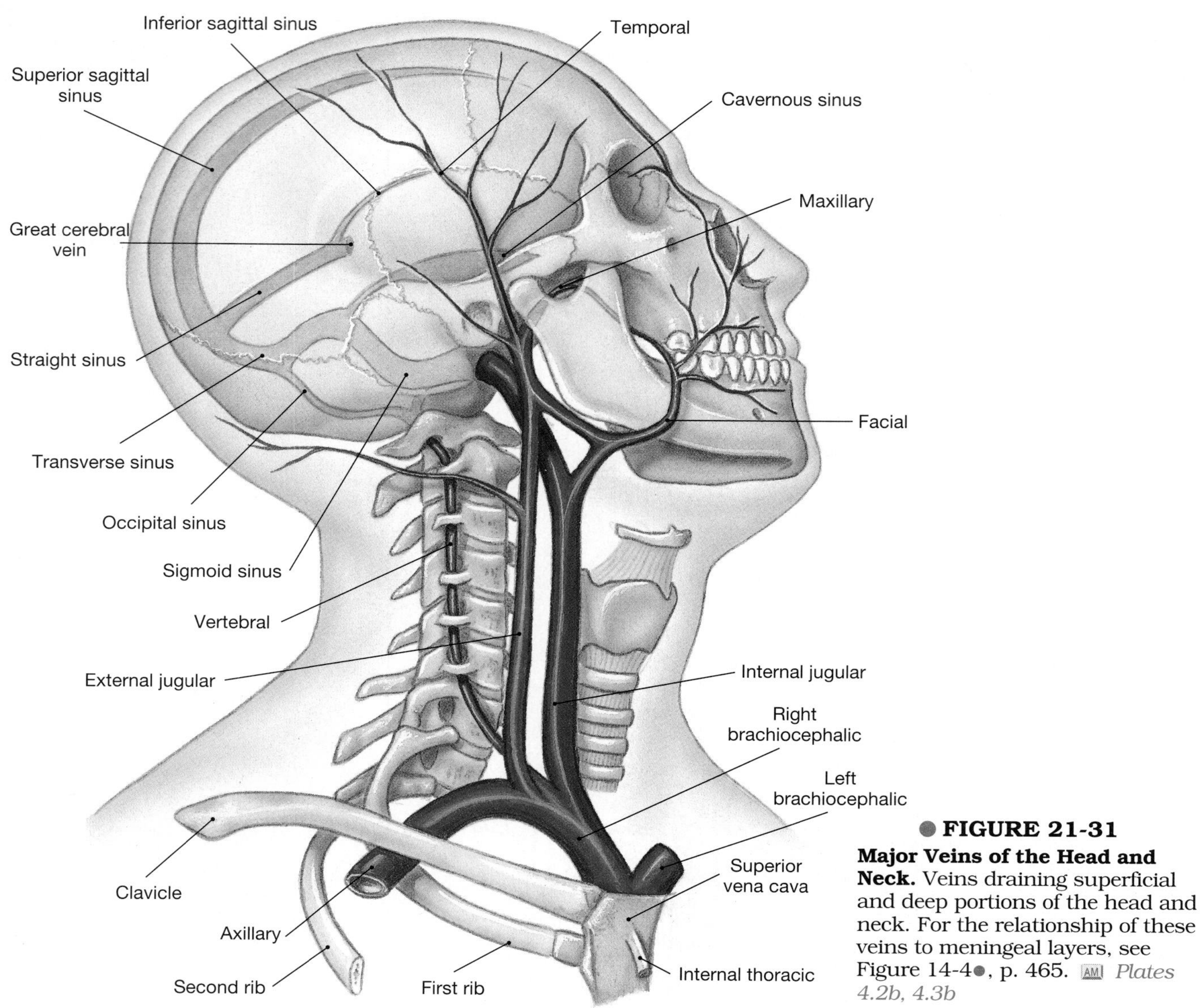
Inferior sagittal sinus
Temporal
Superior sagittal sinus
Cavernous sinus
Maxillary
Great cerebral vein
Straight sinus
Facial
Transverse sinus
Occipital sinus
Sigmoid sinus
Vertebral
External jugular
Internal jugular
Right brachiocephalic
Left brachiocephalic
Superior vena cava
Clavicle
Axillary
Second rib
First rib
Internal thoracic

● **FIGURE 21-31**
Major Veins of the Head and Neck. Veins draining superficial and deep portions of the head and neck. For the relationship of these veins to meningeal layers, see Figure 14-4●, p. 465. AM *Plates 4.2b, 4.3b*

Superficial Veins of the Head and Neck. Superficial veins of the head collect to form the **temporal, facial,** and **maxillary veins** (Figure 21-31●). The temporal and maxillary veins drain into the **external jugular vein.** The facial vein drains into the internal jugular vein; a broad anastomosis between the external and internal jugular veins at the angle of the mandible provides dual venous drainage of the face, scalp, and cranium. The external jugular vein descends toward the chest just beneath the skin on the outer surface of the sternocleidomastoid muscle. Posterior to the clavicle, the external jugular empties into the *subclavian vein.* In healthy individuals, the external jugular vein is easily palpable, and a *jugular venous pulse (JVP)* can sometimes be seen at the base of the neck.

Venous Return from the Upper Limbs. The **digital veins** empty into the **superficial** and **deep palmar veins** of the hand, which are interconnected to form the **palmar venous arches.** The superficial arch empties into the **cephalic vein,** which ascends along the radial side of the forearm, the **median antebrachial vein,** and the **basilic vein,** which ascends on the ulnar side. Anterior to the elbow is the superficial **median cubital vein.** This vein passes from the cephalic vein, medially and at an oblique angle, to connect to the basilic vein. Venous blood samples are typically collected from the median cubital vein.

From the elbow, the basilic vein passes superiorly along the median surface of the biceps brachii. As it approaches the axilla, the basilic vein joins the brachial vein to form the **axillary vein** (Figure 21-32●).

The deep palmar veins drain into the **radial vein** and the **ulnar vein.** After crossing the elbow, these veins fuse to form the **brachial vein.** The brachial vein lies parallel to the brachial artery. As the brachial vein continues toward the trunk, it receives blood from the basilic vein before entering the axilla as the axillary vein.

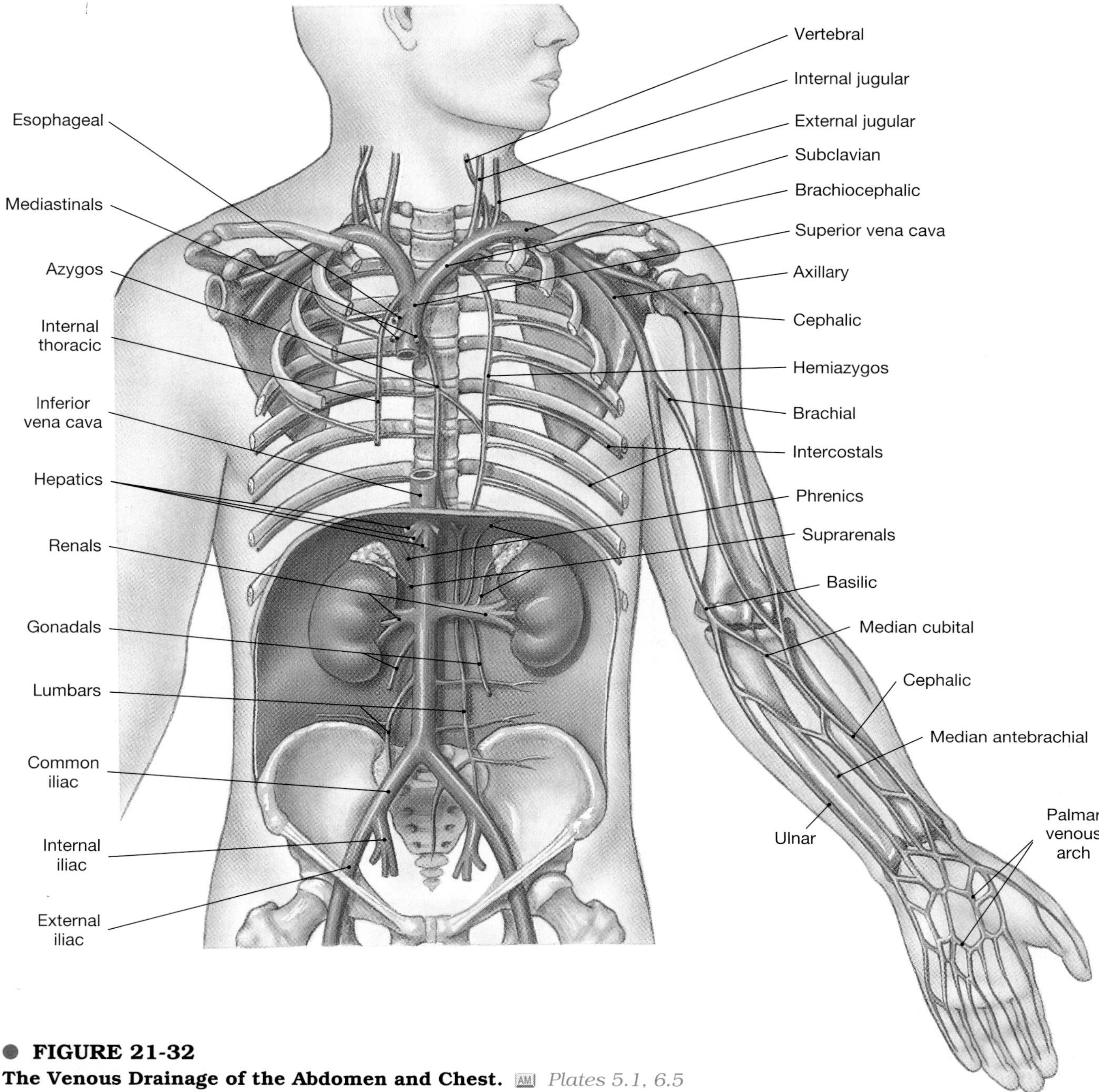
Vertebral
Internal jugular
External jugular
Subclavian
Brachiocephalic
Superior vena cava
Axillary
Cephalic
Hemiazygos
Brachial
Intercostals
Phrenics
Suprarenals
Basilic
Median cubital
Cephalic
Median antebrachial
Palmar venous arch
Ulnar
Esophageal
Mediastinals
Azygos
Internal thoracic
Inferior vena cava
Hepatics
Renals
Gonadals
Lumbars
Common iliac
Internal iliac
External iliac

FIGURE 21-32
The Venous Drainage of the Abdomen and Chest. AM *Plates 5.1, 6.5*

Formation of the Superior Vena Cava. The cephalic vein joins the axillary vein on the outer surface of the first rib, forming the **subclavian vein,** which continues into the chest. The subclavian vein passes over the surface of the first rib, along the clavicle, and into the thoracic cavity. After traveling a short distance inside the thoracic cavity, the subclavian meets and merges with the external and internal jugular veins of that side. This fusion creates the **brachiocephalic vein,** or *innominate vein.*

Each brachiocephalic vein receives blood from the **vertebral vein** of the same side, which drains the back of the skull and spinal cord. Near the heart, at the level of the first and second ribs, the left and right brachiocephalic veins combine, creating the **superior vena cava (SVC).** Close to the point of fusion, the **internal thoracic vein** empties into the SVC.

The **azygos** (AZ-i-gos) **vein** is the major tributary of the superior vena cava. This vessel ascends from the lumbar region over the right side of the vertebral column to invade the thoracic cavity through the diaphragm. The azygos joins the superior vena cava at the level of vertebra T_2. On the left side, the azygos receives blood from the smaller **hemiazygos vein.**

The azygos and hemiazygos veins are the chief collecting vessels of the thorax. They receive blood from (1) numerous **intercostal veins,** which receive blood from the chest muscles, (2) **esophageal veins,**

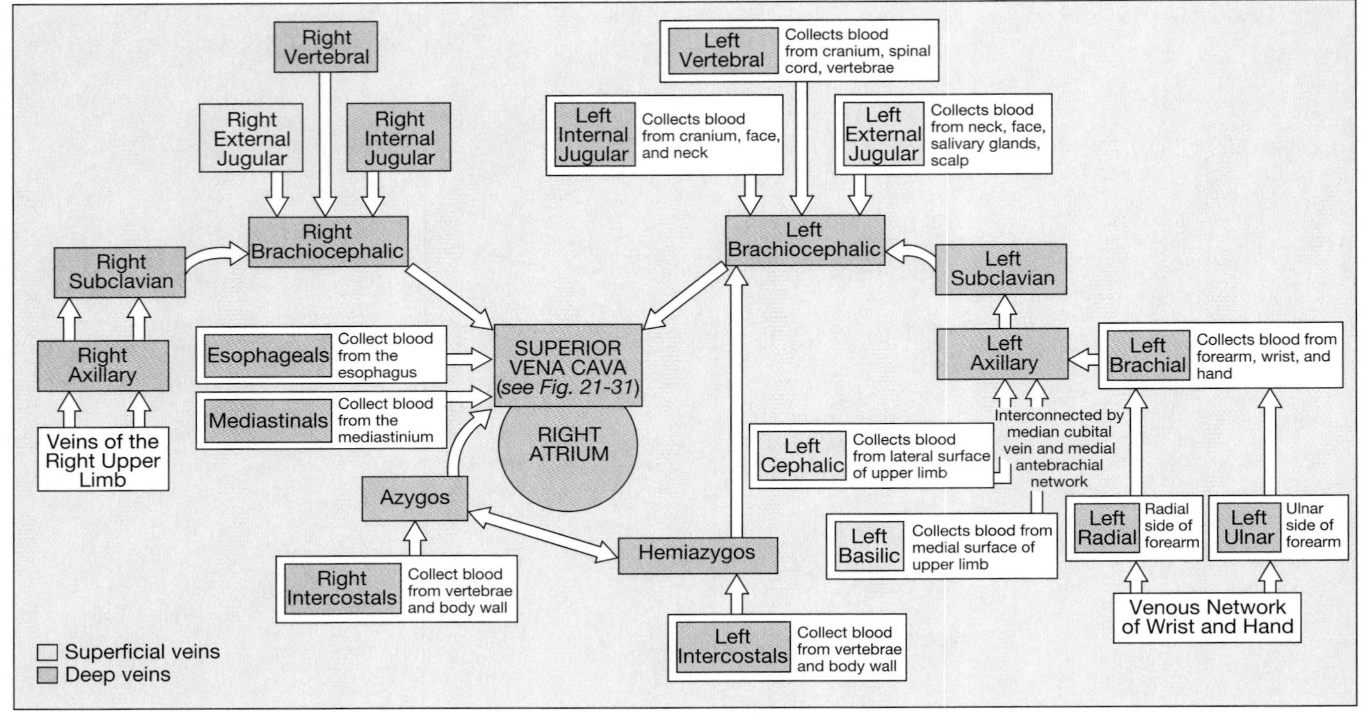
Right Vertebral
Right External Jugular
Right Internal Jugular
Left Vertebral
Collects blood from cranium, spinal cord, vertebrae
Left Internal Jugular
Collects blood from cranium, face, and neck
Left External Jugular
Collects blood from neck, face, salivary glands, scalp
Right Brachiocephalic
Left Brachiocephalic
Right Subclavian
Left Subclavian
Right Axillary
Esophageals
Collect blood from the esophagus
Mediastinals
Collect blood from the mediastinium
SUPERIOR VENA CAVA (see Fig. 21-31)
RIGHT ATRIUM
Left Axillary
Left Brachial
Collects blood from forearm, wrist, and hand
Interconnected by median cubital vein and medial antebrachial network
Veins of the Right Upper Limb
Left Cephalic
Collects blood from lateral surface of upper limb
Azygos
Hemiazygos
Left Basilic
Collects blood from medial surface of upper limb
Left Radial
Radial side of forearm
Left Ulnar
Ulnar side of forearm
Right Intercostals
Collect blood from vertebrae and body wall
Left Intercostals
Collect blood from vertebrae and body wall
Venous Network of Wrist and Hand
Superficial veins
Deep veins

(a)

RIGHT ATRIUM
INFERIOR VENA CAVA (see Fig. 21-32)
Phrenics
Collect blood from the diaphragm
Suprarenals
Collect blood from the adrenal glands
Gonadals
Collect blood from the gonads (testes or ovaries)
Hepatics
Collect blood from the liver
Renals
Collect blood from the kidneys
Lumbars
Collect blood from the spinal cord and body wall
Right Common Iliac
Left Common Iliac
Right External Iliac
Left External Iliac
Right Internal Iliac
Left Internal Iliac
Collect blood from the pelvic muscles, skin, urinary and reproductive organs of pelvic cavity
Blood from veins in right lower limb (see Fig. 21-34)
Blood from veins in left lower limb (see Fig. 21-34)

(b)

● **FIGURE 21-33**
Flow Chart of Circulation to the Superior and Inferior Venae Cavae. (a) Tributaries of the SVC. **(b)** Tributaries of the IVC.

which drain blood from the esophagus and (3) smaller veins draining other mediastinal structures.

Figure 21-33a● diagrams the venous tributaries of the superior vena cava.

THE INFERIOR VENA CAVA The *inferior vena cava (IVC)* collects most of the venous blood from organs below the diaphragm (a small amount reaches the superior vena cava via the azygos and hemiazygos veins). Figure 21-33b● diagrams the tributaries of the inferior vena cava.

Veins Draining the Lower Limb. Blood leaving capillaries in the sole of each foot collects into a network of **plantar veins.** The plantar network provides blood to the deep veins of the leg: the **anterior tibial vein,** the **posterior tibial vein,** and the **peroneal vein** (Figure 21-34a●). The **dorsal venous arch** collects blood from capillaries on the superior surface of the foot. There are extensive interconnections between the plantar arch and the dorsal arch, and the path of blood flow can easily shift from superficial to deep veins.

The dorsal venous arch is drained by two superficial veins, the **great saphenous** (sa-FĒ-nus) **vein** (*saphenes*, prominent) and the **small saphenous vein.** The great saphenous vein ascends along the medial aspect of the leg and thigh, draining into the *femoral vein* near the hip joint. The small saphenous vein arises from the dorsal venous arch and ascends along the posterior and lateral aspect of the calf. It then enters the popliteal fossa, where it meets the **popliteal vein** formed by the union of the tibial and peroneal veins. The popliteal vein may

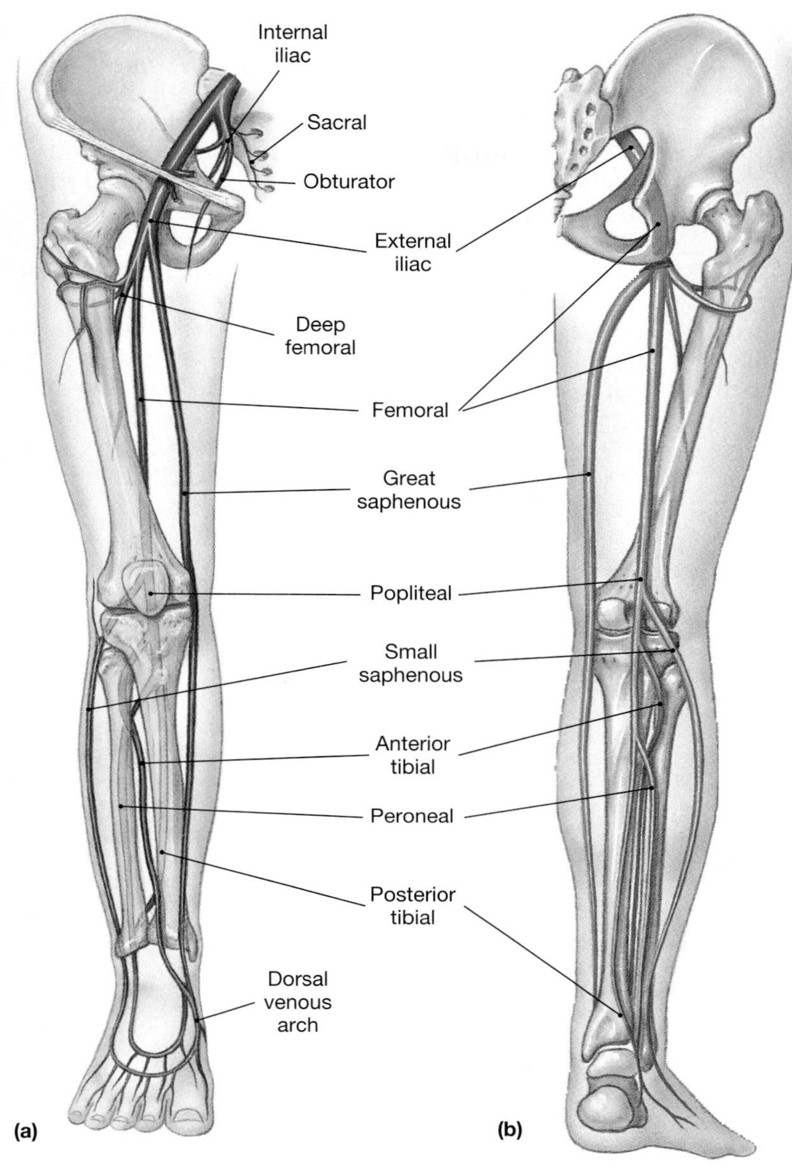
Internal iliac
Sacral
Obturator
External iliac
Deep femoral
Femoral
Great saphenous
Popliteal
Small saphenous
Anterior tibial
Peroneal
Posterior tibial
Dorsal venous arch
(a)
(b)

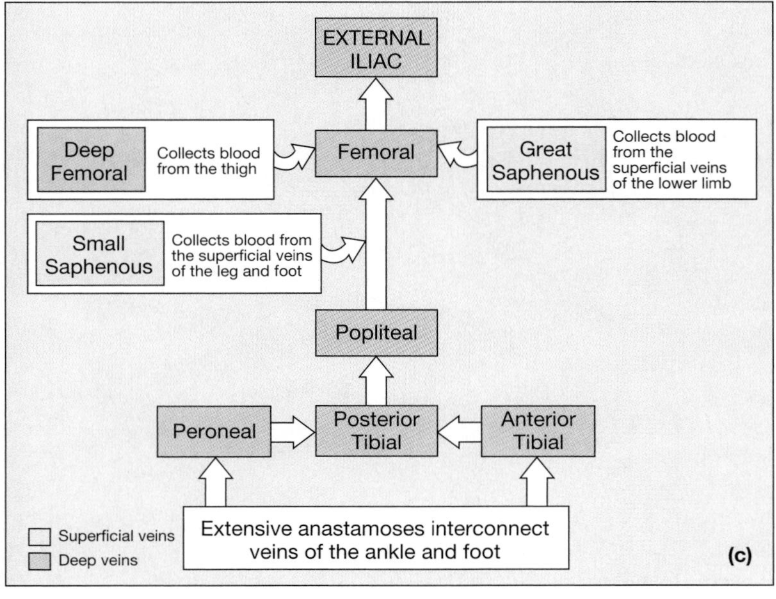
EXTERNAL ILIAC
Deep Femoral
Collects blood from the thigh
Femoral
Great Saphenous
Collects blood from the superficial veins of the lower limb
Small Saphenous
Collects blood from the superficial veins of the leg and foot
Popliteal
Peroneal
Posterior Tibial
Anterior Tibial
Extensive anastamoses interconnect veins of the ankle and foot
Superficial veins
Deep veins
(c)

● **FIGURE 21-34**

Venous Drainage from the Lower Limb. **(a)** Anterior view. **(b)** Posterior view. **(c)** Flow chart of venous circulation to a lower limb. AM *Plate 7.2; Scans 5, 6*

be easily palpated in the popliteal fossa adjacent to the adductor magnus muscle (Figure 21-34b●). Once it reaches the femur, the popliteal vein becomes the **femoral vein,** which ascends along the thigh next to the femoral artery. Immediately before penetrating the abdominal wall, the femoral vein receives blood from the great saphenous vein and the **deep femoral vein.** The femoral vein then penetrates the body wall and emerges in the pelvic cavity as the **external iliac vein.**

Veins Draining the Pelvis. The external iliac veins receive blood from the lower limbs, pelvis, and lower abdomen. As the left and right external iliacs travel across the inner surface of the ilium, they are joined by the **internal iliac veins,** which drain the pelvic organs. The union of external and internal iliac veins results in the **common iliac vein.** The right and left common iliacs ascend at an oblique angle, and anterior to vertebra L_5 they unite to form the **inferior vena cava (IVC)** (Figure 21-32●).

Veins Draining the Abdomen. The inferior vena cava ascends posterior to the abdominopelvic cavity, parallel to the aorta. Blood from both inferior and superior venae cavae flows into the right atrium to be pumped out the right ventricle to the lungs for oxygenation. The abdominal portion of the inferior vena cava collects blood from six major veins (Figures 21-32 and 21-33●):

1. **Lumbar veins** drain the lumbar portion of the abdomen.
2. **Gonadal** *(ovarian* or *testicular)* **veins** drain the ovaries or testes. The right gonadal vein empties into the inferior vena cava; the left gonadal usually drains into the left renal vein.
3. **Renal veins** collect blood from the kidneys. These are the largest tributaries of the inferior vena cava.
4. **Suprarenal veins** drain the adrenal glands. Usually only the right suprarenal vein drains into the

inferior vena cava, and the left drains into the left renal vein.

5. **Phrenic veins** drain the diaphragm. Only the *right phrenic vein* drains into the inferior vena cava; the *left* drains into the left renal vein.
6. **Hepatic veins** leave the liver and empty into the inferior vena cava at the level of vertebra T_{10}.

THE HEPATIC PORTAL SYSTEM The liver is the only digestive organ drained by the inferior vena cava. Instead of traveling directly to the inferior vena cava, blood leaving the capillaries supplied by the celiac, superior, and inferior mesenteric arteries flows into the **hepatic portal system.** As noted in Chapter 18, a blood vessel connecting two capillary beds is called a *portal vessel,* and the network is a *portal system.* [p. 611]

Blood flowing in the hepatic portal system is quite different from that in other systemic veins, because the hepatic portal vessels contain substances absorbed by the stomach and intestines. For example, levels of blood glucose and amino acids in the hepatic portal vein often exceed those found anywhere else in the circulatory system. The hepatic portal system delivers these and other absorbed compounds directly to the liver for storage, metabolic conversion, or excretion. After passing through the liver sinusoids, blood collects in the hepatic veins that empty into the inferior vena cava. Because blood from the digestive organs goes to the liver first, the composition of the blood in the general systemic circuit remains relatively stable, regardless of the digestive activities under way.

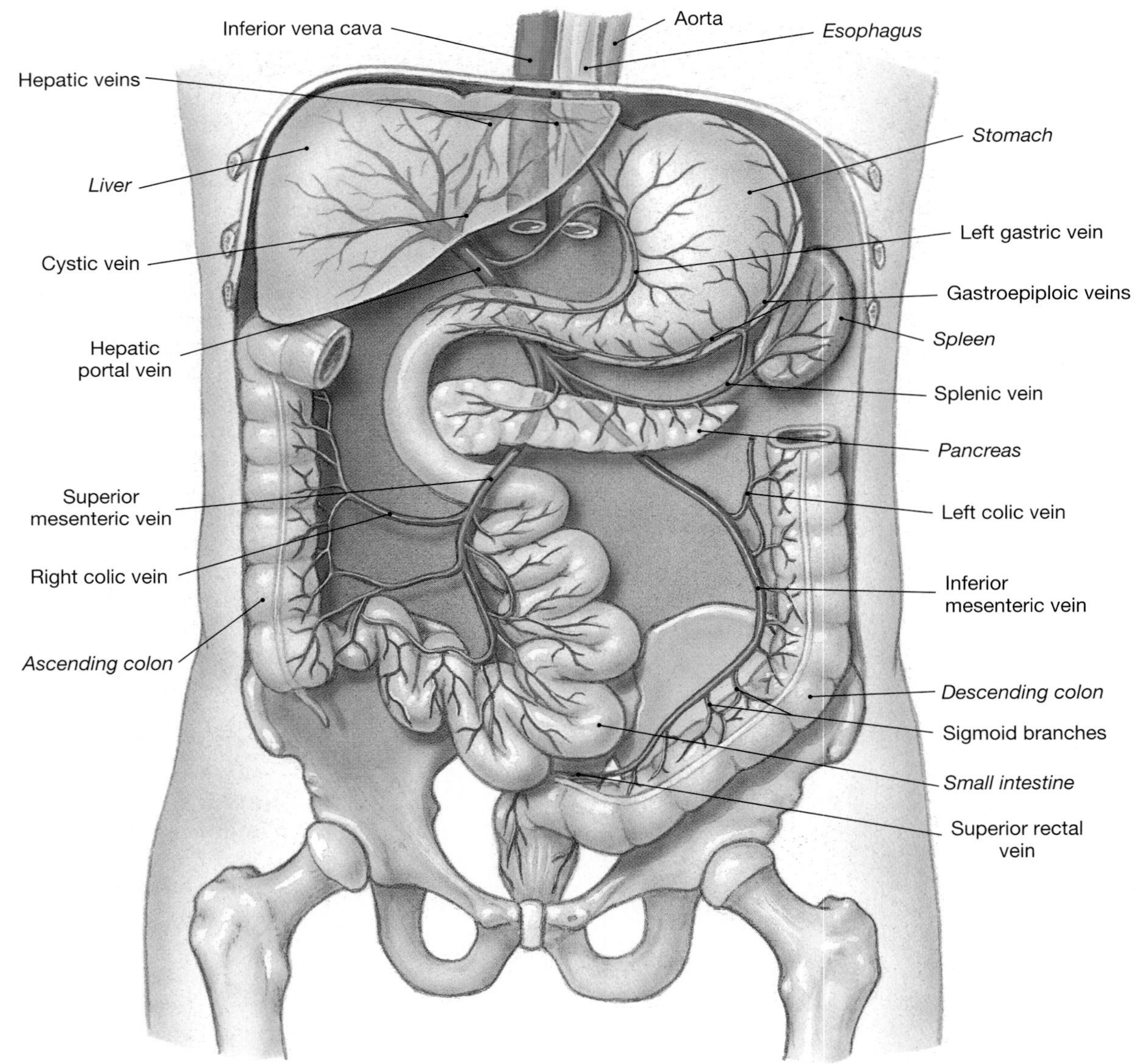
Inferior vena cava
Aorta
Esophagus
Hepatic veins
Stomach
Liver
Left gastric vein
Cystic vein
Gastroepiploic veins
Spleen
Hepatic portal vein
Splenic vein
Pancreas
Superior mesenteric vein
Left colic vein
Right colic vein
Inferior mesenteric vein
Ascending colon
Descending colon
Sigmoid branches
Small intestine
Superior rectal vein

FIGURE 21-35
The Hepatic Portal System. Vessels of the hepatic portal system. AM Plate 6.5

The hepatic portal system begins in the capillaries of the digestive organs and ends as the hepatic portal vein discharges blood into sinusoids in the liver. The tributaries of the hepatic portal vein (Figure 21-35●) include:

1. The **inferior mesenteric vein,** which collects blood from capillaries along the lower portion of the large intestine.
2. The **left colic vein** and the **superior rectal veins,** which drain the descending colon, sigmoid colon, and rectum.
3. The **splenic vein,** formed by the union of the inferior mesenteric vein and veins from the spleen, the lateral border of the stomach (*left gastroepiploic*), and the pancreas (*pancreatic*).
4. The **superior mesenteric vein,** which collects blood from veins draining the stomach (*right gastroepiploic*), the small intestine (*intestinal*), and two-thirds of the large intestine (*ileocolic, right colic,* and *middle colic*).

The hepatic portal vein forms through the fusion of the superior mesenteric and splenic veins. Of the two, the superior mesenteric normally contributes the greater volume of blood and most of the nutrients. As it proceeds toward the liver, the hepatic portal receives blood from the **gastric veins,** which drain the medial border of the stomach, and the **cystic vein** from the gallbladder.

❑ Fetal Circulation

There are significant differences between the fetal and adult circulatory systems that reflect differing sources of respiratory and nutritional support. The embryonic lungs are collapsed and nonfunctional, and the digestive tract has nothing to digest. All of the embryonic nutritional and respiratory needs are provided by diffusion across the placenta.

Placental Blood Supply

Figure 21-36

Fetal circulation is diagrammed in Figure 21-36●. Blood flow to the placenta is provided by a pair of **umbilical arteries** that arise from the internal iliac arteries and enter the umbilical cord. Blood returns

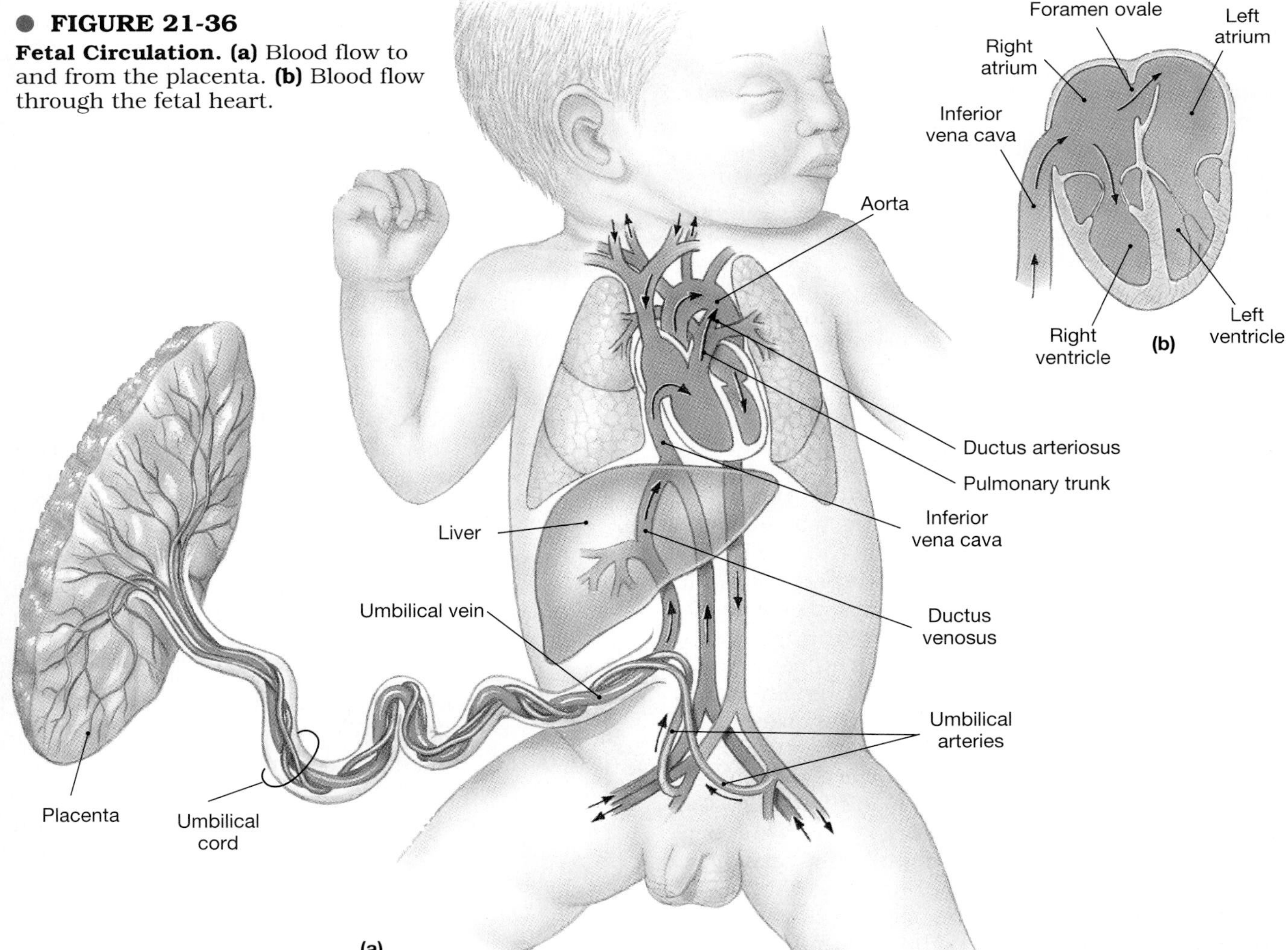
Foramen ovale
Left atrium
Right atrium
Inferior vena cava
Aorta
Right ventricle
(b)
Left ventricle
Ductus arteriosus
Pulmonary trunk
Inferior vena cava
Liver
Ductus venosus
Umbilical vein
Umbilical arteries
Placenta
Umbilical cord
(a)

● **FIGURE 21-36**
Fetal Circulation. (a) Blood flow to and from the placenta. **(b)** Blood flow through the fetal heart.

from the placenta in the **umbilical vein,** bringing oxygen and nutrients to the developing fetus. The umbilical vein delivers blood to capillaries within the developing liver and to the inferior vena cava via the **ductus venosus.** When the placental connection is broken at birth, blood flow ceases along the umbilical vessels, and they soon degenerate.

Circulation in the Heart and Great Vessels

One of the most interesting aspects of circulatory development reflects the differences between the life of an embryo or fetus and that of an infant. Throughout embryonic and fetal life, the lungs are collapsed; yet after delivery, the newborn infant must be able to extract oxygen from inspired air rather than across the placenta.

Although the interatrial and interventricular septa develop early in fetal life, the interatrial partition remains functionally incomplete up to the time of birth. The interatrial opening, or *foramen ovale,* is associated with an elongate flap that acts as a valve. Blood can flow freely from the right atrium to the left atrium, but any backflow will close the valve and isolate the two chambers. Thus blood can enter the heart at the right atrium and bypass the pulmonary circuit altogether. A second short-circuit exists between the pulmonary and aortic trunks. This connection, the **ductus arteriosus,** consists of a short, muscular vessel.

With the lungs collapsed, the capillaries are compressed, and little blood flows through the lungs. During diastole blood enters the right atrium and flows into the right ventricle, but it also passes into the left atrium via the foramen ovale. About 25 percent of the blood arriving at the right atrium bypasses the pulmonary circuit in this way. In addition, over 90 percent of the blood leaving the right ventricle passes through the ductus arteriosus and enters the systemic circuit, rather than continuing to the lungs.

Circulatory Changes at Birth

At birth dramatic changes occur. When the infant takes its first breath, the lungs expand, and so do the pulmonary vessels. The resistance in the pulmonary circuit declines suddenly, and blood rushes into the pulmonary vessels. Within a few seconds, rising O_2 levels stimulate constriction of the ductus arteriosus, isolating the pulmonary and aortic trunks. As pressures rise in the left atrium, the valvular flap closes the foramen ovale and completes the circulatory remodeling. In the adult, the interatrial septum bears a shallow depression, the *fossa ovalis,* that marks the site of the foramen ovale (see Figure 20-4•, p. 686). The remnants of the ductus arteriosus persist as a fibrous cord, the **ligamentum arteriosum.**

If the proper circulatory changes do not occur at birth or shortly thereafter, problems will eventually develop. The severity of the problem varies depending on which connection remains open, and the size of the opening. Treatment may involve surgical closure of the foramen ovale or the ductus arteriosus or both. Other forms of congenital heart defects result from abnormal cardiac development or inappropriate connections between the heart and major arteries and veins. (For further discussion of circulatory changes during development, see the Embryology Summary on p. 770.)

Aging and the Cardiovascular System

The capabilities of the cardiovascular system gradually decline with age. The major changes are listed and summarized in the same sequence as the cardiovascular chapters: blood, heart, and vessels.

- ***Age-Related Changes in the Blood.*** Age-related changes in the blood may include: (1) decreased hematocrit; (2) constriction or blockage of peripheral veins by a *thrombus* (stationary blood clot); the thrombus can become detached, pass through the heart, and become wedged in a small artery, most often in the lungs, causing a *pulmonary embolism;* (3) pooling of blood in the veins of the legs because valves are not working effectively.
- ***The Aging Heart.*** Age-related changes in the heart include: (1) a reduction in the maximum cardiac output; (2) changes in the activities of the nodal and conducting cells; (3) a reduction in the elasticity of the fibrous skeleton; (4) a progressive atherosclerosis that can restrict coronary circulation; (5) replacement of damaged cardiac muscle cells by scar tissue.
- ***Aging and Blood Vessels.*** Age-related changes in blood vessels are often related to arteriosclerosis. For example, (1) the inelastic walls of arteries become less tolerant of sudden pressure increases, which may lead to an *aneurysm* whose subsequent rupture may cause a stroke, infarct, or massive blood loss, depending on the vessel involved; (2) calcium salts can be deposited on weakened vascular walls, increasing the risk of a stroke or infarct; (3) thrombi can form at atherosclerotic plaques.

Whenever Tim gets mad, a large vein bulges in the lateral region of his neck. What vein is this?

A thrombus that blocks the popliteal vein would interfere with blood flow in which other veins?

A blood sample taken from the umbilical cord shows a high concentration of oxygen and nutrients and a low concentration of carbon dioxide and waste products. Did this blood sample come from an umbilical artery or the umbilical vein? Explain.

Congenital Circulatory Problems

Minor individual variations in the circulatory network are quite common. For example, very few individuals have identical patterns of venous distribution. Congenital circulatory problems serious enough to represent a threat to homeostasis are relatively rare. They usually reflect abnormal formation of the heart or problems with the interconnections between the heart and the great vessels. Several examples of congenital circulatory defects are illustrated in Figure 21-37●. All of these conditions can be surgically corrected, although multiple surgeries may be required.

The incomplete closure of the foramen ovale or ductus arteriosus (Figure 21-37a●) results in the bypassing of the lungs and the recirculation of blood into the pulmonary circuit. Because normal blood oxygenation does not occur, the circulating blood has a deep red color. The skin then develops the blue tones typical of *cyanosis*, a condition noted in Chapter 5, and the infant is known as a "blue baby."

Ventricular septal defects (Figure 21-37b●) are the most common congenital heart problems, affecting 0.12 percent of newborn infants. The opening between the left and right ventricles has the reverse effect of a connection between the atria: When it beats, the more powerful left ventricle ejects blood into the right ventricle and pulmonary circuit. Pulmonary hypertension, pulmonary edema, and cardiac enlargement are the results.

The *tetralogy of Fallot* (fa-LŌ) (Figure 21-37c●) is a complex group of heart and circulatory defects that affect 0.10 percent of newborn infants. In this condition (1) the pulmonary trunk is abnormally narrow, (2) the interventricular septum is incomplete, (3) the aorta originates where the interventricular septum normally ends, and (4) the right ventricle is enlarged.

In the *transposition of great vessels* (Figure 21-37d●) the aorta is connected to the right ventricle and the pulmonary artery is connected to the left ventricle. This malformation affects 0.05 percent of newborn infants.

In an *atrioventricular septal defect* (Figure 21-37e●) the atria and ventricles are incompletely separated. The results are quite variable, depending on the extent of the defect and the effects on the atrioventricular valves. This type of defect most often affects infants with *Down syndrome*, a disorder caused by the presence of an extra copy of chromosome 21.

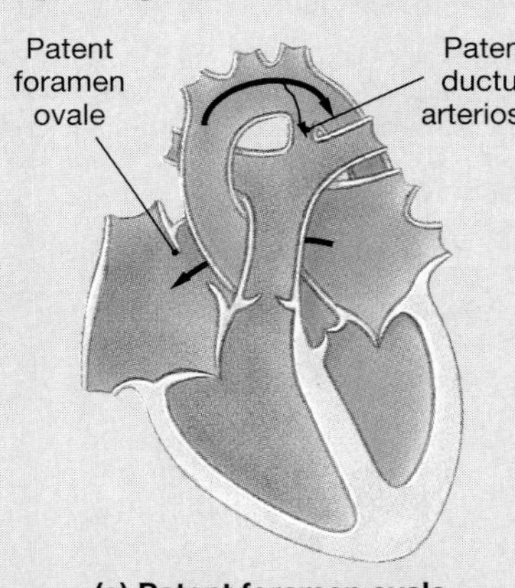
Patent foramen ovale
Patent ductus arteriosus
(a) Patent foramen ovale

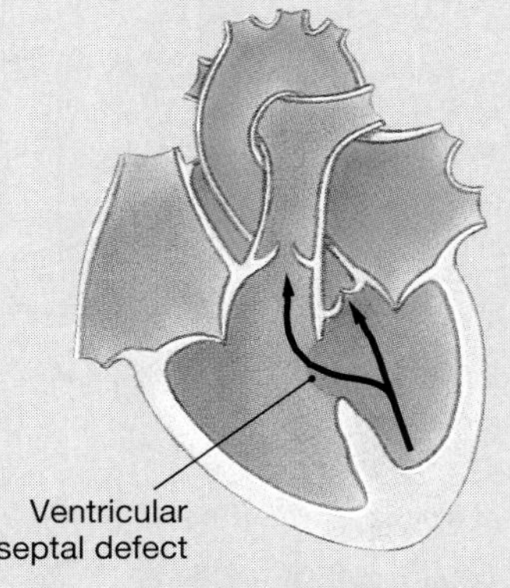
Ventricular septal defect
(b) Ventricular septal defect

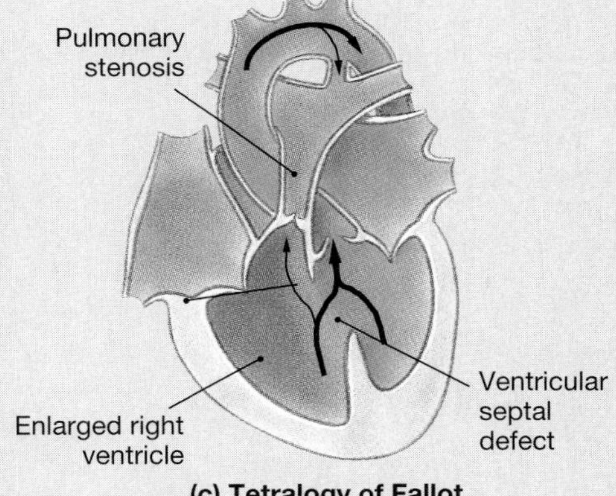
Pulmonary stenosis
Enlarged right ventricle
Ventricular septal defect
(c) Tetralogy of Fallot

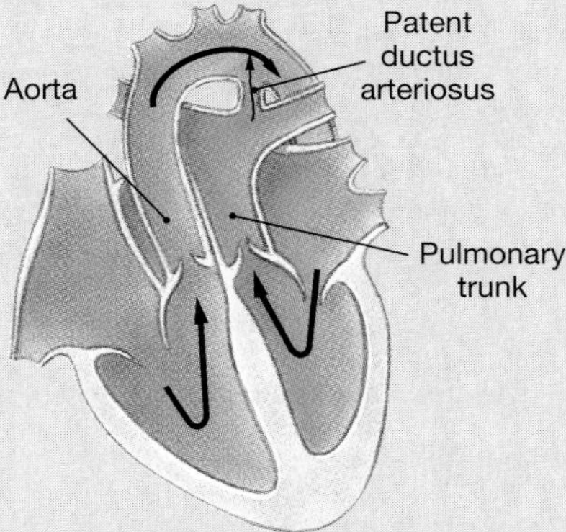
Aorta
Patent ductus arteriosus
Pulmonary trunk
(d) Transposition of great vessels

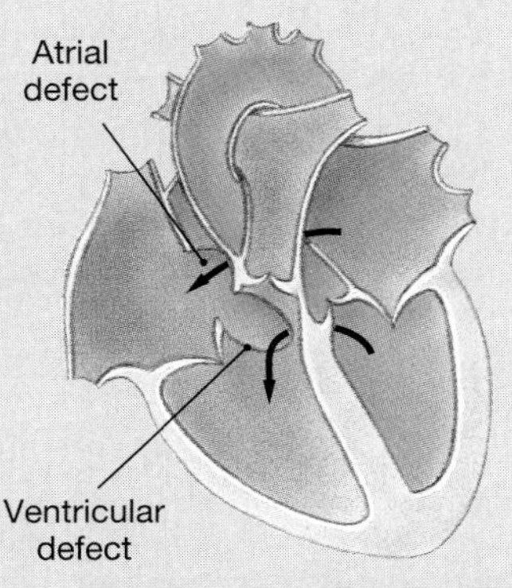
Atrial defect
Ventricular defect
(e) Atrioventricular septal defect

● **FIGURE 21-37**
Congenital Circulatory Problems

EMBRYOLOGY SUMMARY Development of the Circulatory System

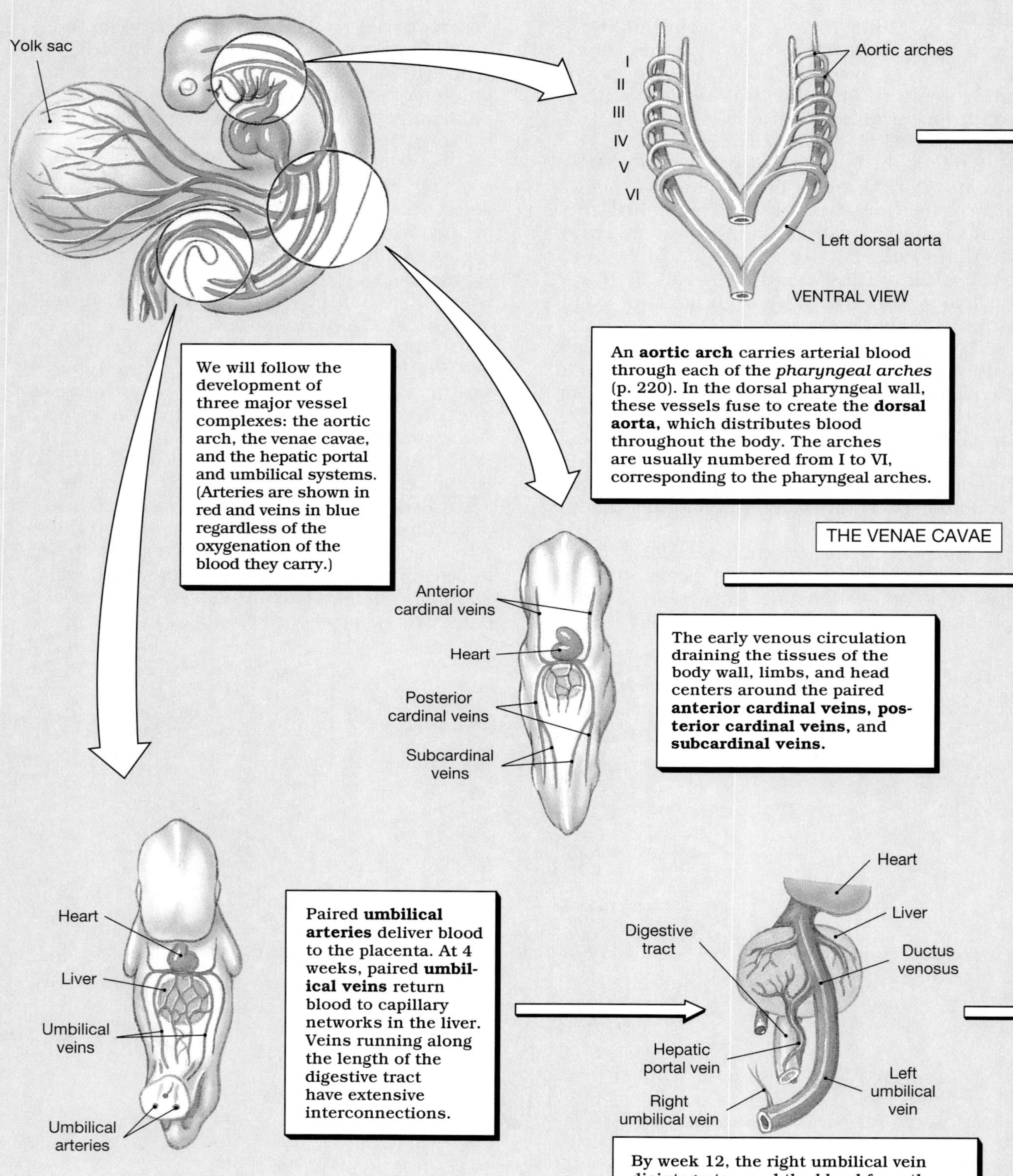

THE AORTIC ARCHES
Yolk sac
Aortic arches
I
II
III
IV
V
VI
Left dorsal aorta
VENTRAL VIEW
THE VENAE CAVAE
Anterior cardinal veins
Heart
Posterior cardinal veins
Subcardinal veins
Heart
Liver
Umbilical veins
Umbilical arteries
Heart
Liver
Digestive tract
Ductus venosus
Hepatic portal vein
Left umbilical vein
Right umbilical vein

We will follow the development of three major vessel complexes: the aortic arch, the venae cavae, and the hepatic portal and umbilical systems. (Arteries are shown in red and veins in blue regardless of the oxygenation of the blood they carry.)

An **aortic arch** carries arterial blood through each of the *pharyngeal arches* (p. 220). In the dorsal pharyngeal wall, these vessels fuse to create the **dorsal aorta,** which distributes blood throughout the body. The arches are usually numbered from I to VI, corresponding to the pharyngeal arches.

The early venous circulation draining the tissues of the body wall, limbs, and head centers around the paired **anterior cardinal veins, posterior cardinal veins,** and **subcardinal veins.**

Paired **umbilical arteries** deliver blood to the placenta. At 4 weeks, paired **umbilical veins** return blood to capillary networks in the liver. Veins running along the length of the digestive tract have extensive interconnections.

THE HEPATIC PORTAL AND UMBILICAL VESSELS

By week 12, the right umbilical vein disintegrates, and the blood from the placenta travels along a single umbilical vein. The **ductus venosus** allows some venous blood to bypass the liver. The veins draining the digestive tract have fused, forming the hepatic portal vein.

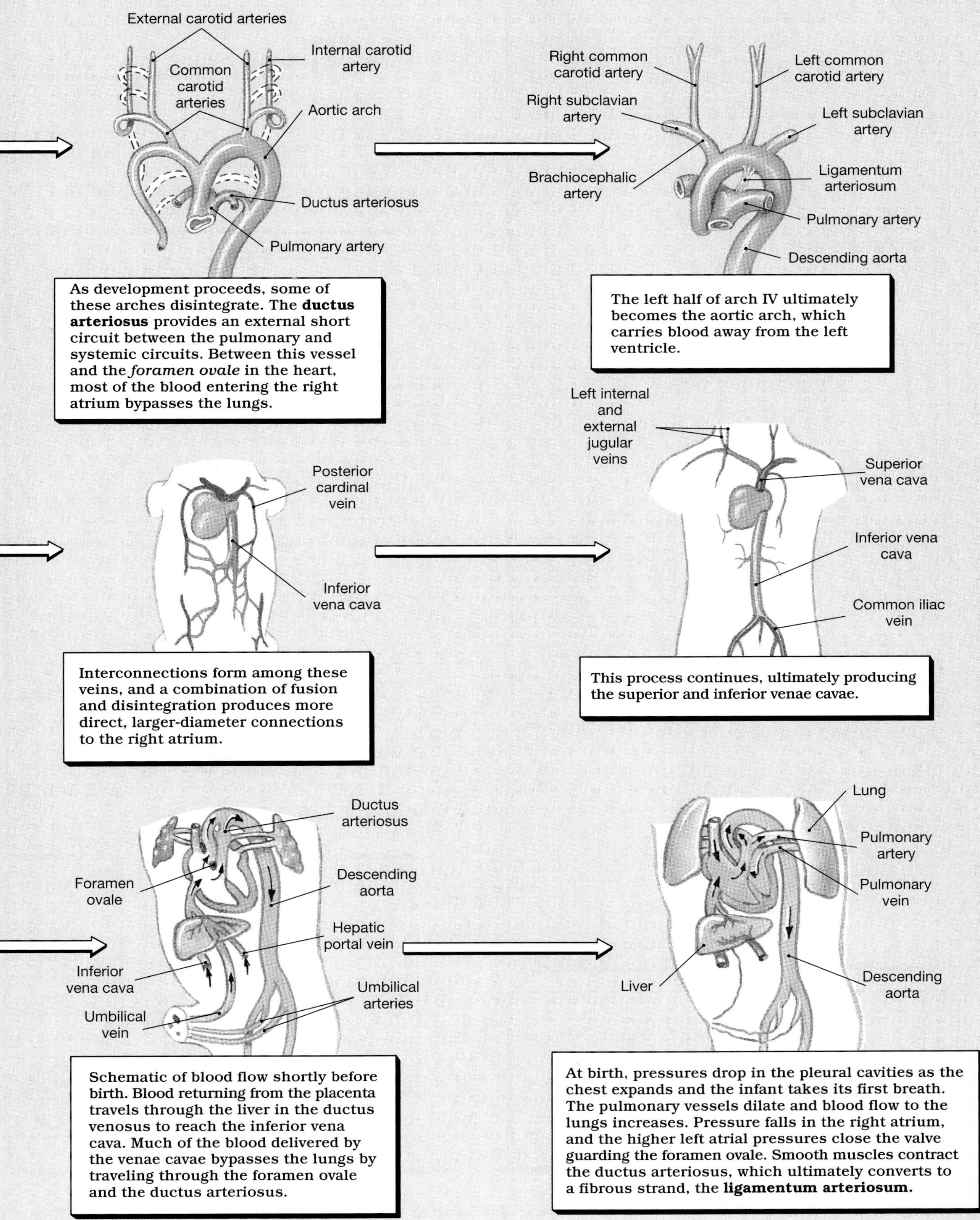
External carotid arteries
Internal carotid artery
Common carotid arteries
Aortic arch
Ductus arteriosus
Pulmonary artery
As development proceeds, some of these arches disintegrate. The **ductus arteriosus** provides an external short circuit between the pulmonary and systemic circuits. Between this vessel and the *foramen ovale* in the heart, most of the blood entering the right atrium bypasses the lungs.
Right common carotid artery
Left common carotid artery
Right subclavian artery
Left subclavian artery
Brachiocephalic artery
Ligamentum arteriosum
Pulmonary artery
Descending aorta
The left half of arch IV ultimately becomes the aortic arch, which carries blood away from the left ventricle.
Posterior cardinal vein
Inferior vena cava
Interconnections form among these veins, and a combination of fusion and disintegration produces more direct, larger-diameter connections to the right atrium.
Left internal and external jugular veins
Superior vena cava
Inferior vena cava
Common iliac vein
This process continues, ultimately producing the superior and inferior venae cavae.
Ductus arteriosus
Foramen ovale
Descending aorta
Hepatic portal vein
Inferior vena cava
Umbilical arteries
Umbilical vein
Schematic of blood flow shortly before birth. Blood returning from the placenta travels through the liver in the ductus venosus to reach the inferior vena cava. Much of the blood delivered by the venae cavae bypasses the lungs by traveling through the foramen ovale and the ductus arteriosus.
Lung
Pulmonary artery
Pulmonary vein
Liver
Descending aorta
At birth, pressures drop in the pleural cavities as the chest expands and the infant takes its first breath. The pulmonary vessels dilate and blood flow to the lungs increases. Pressure falls in the right atrium, and the higher left atrial pressures close the valve guarding the foramen ovale. Smooth muscles contract the ductus arteriosus, which ultimately converts to a fibrous strand, the **ligamentum arteriosum.**

INTEGUMENTARY SYSTEM

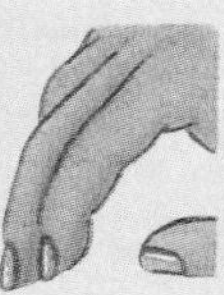

Mast cell stimulation produces localized changes in blood flow and capillary permeability

Delivers immune system cells to injury sites; clotting response seals breaks in skin surface; carries away toxins from sites of infection; provides heat

THE CARDIOVASCULAR SYSTEM

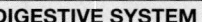

FOR ALL SYSTEMS

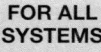

Delivers oxygen, hormones, nutrients, and white blood cells; removes carbon dioxide and metabolic wastes; transfers heat

SKELETAL SYSTEM

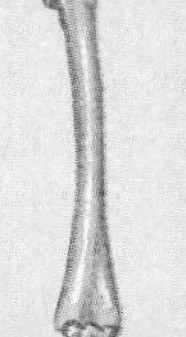

Provides calcium needed for normal cardiac muscle contraction; protects blood cells developing in bone marrow

Provides Ca^{2+} and PO_4^{3-} ions for bone deposition; delivers EPO to bone marrow, parathyroid hormone and calcitonin to osteoblasts and osteoclasts

MUSCULAR SYSTEM

Skeletal muscle contractions assist in moving blood through veins; protects superficial blood vessels, especially in neck and limbs

Delivers oxygen and nutrients, removes carbon dioxide, lactic acid, and heat during skeletal muscle activity

NERVOUS SYSTEM

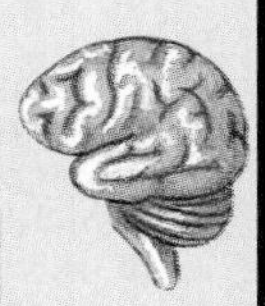

Controls patterns of circulation in peripheral tissues; modifies heart rate and regulates blood pressure; releases ADH

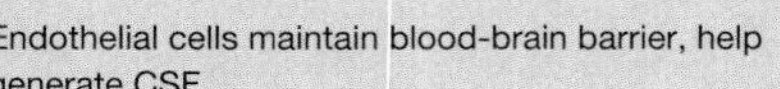

Endothelial cells maintain blood-brain barrier, help generate CSF

ENDOCRINE SYSTEM

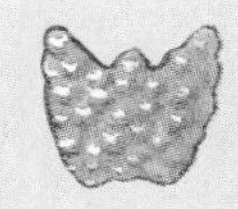

Erythropoietin regulates production of RBCs; several hormones elevate blood pressure; epinephrine stimulates cardiac muscle, elevating heart rate and contractile force

Distributes hormones throughout the body; heart secretes ANP

LYMPHATIC SYSTEM

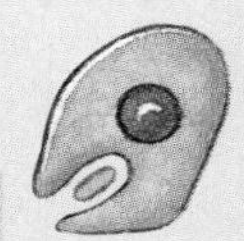

Defends against pathogens or toxins in blood; fights infections of cardiovascular organs; returns tissue fluid to circulation

Distributes WBCs; carries antibodies that attack pathogens; clotting response assists in restricting spread of pathogens; granulocytes and lymphocytes produced in bone marrow

RESPIRATORY SYSTEM

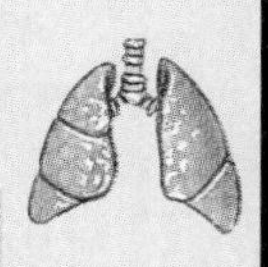

Provides oxygen to cardiovascular organs and removes carbon dioxide; enzyme in lung capillaries converts inactive angiotensin I to active angiotensin II

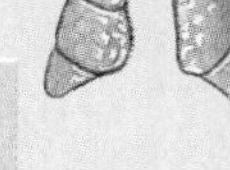

RBCs transport oxygen and carbon dioxide between lungs and peripheral tissues

DIGESTIVE SYSTEM

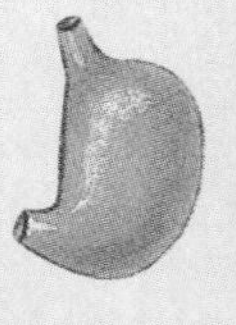

Provides nutrients to cardiovascular organs; absorbs water and ions essential to maintenance of normal blood volume

Distributes digestive tract hormones; carries nutrients, water, and ions away from sites of absorption; delivers nutrients and toxins to liver

URINARY SYSTEM

Releases renin to elevate blood pressure and erythropoietin to accelerate red blood cell production

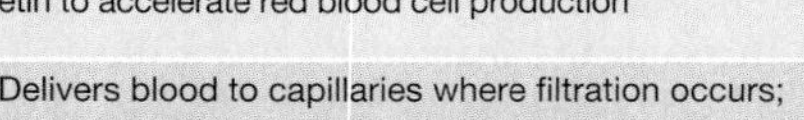

Delivers blood to capillaries where filtration occurs; accepts fluids and solutes reabsorbed during urine production

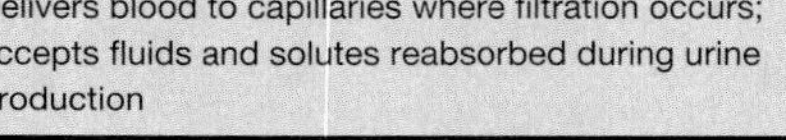

REPRODUCTIVE SYSTEM

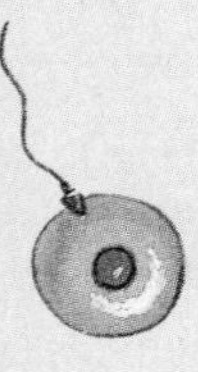

Estrogens may maintain healthy vessels and slow development of atherosclerosis with age

Distributes reproductive hormones; provides nutrients, oxygen, and waste removal for developing fetus; local blood pressure changes responsible for physical changes during sexual arousal

● **FIGURE 21-38**

Functional Relationships between the Cardiovascular System and Other Systems

Integration with Other Systems

Figure 21-38

The cardiovascular system is both anatomically and functionally linked to all other systems. The section on vessel distribution demonstrated the extent of the anatomical connections. Figure 21-38• summarizes the physiological relationships between the cardiovascular system and other organ systems.

The most extensive communication occurs between the cardiovascular and lymphatic systems. Not only are the two systems physically interconnected, but cell populations of the lymphatic system use the cardiovascular system as a highway to move from one part of the body to another. The next chapter examines the lymphatic system in detail and considers the role of the lymphatic system in the immune response.

Selected Clinical Terminology

Terms Discussed in This Chapter

aneurysm (AN-ū-rizm): A bulge in the weakened wall of a blood vessel, usually an artery. *(p. 723* and *AM)*
arteriosclerosis (ar-tē-rē-ō-skle-RŌ-sis): A thickening and toughening of arterial walls. *(p. 722)*
atherosclerosis (ath-er-ō-skle-RŌ-sis): A type of arteriosclerosis characterized by changes in the endothelial lining and the formation of a plaque. *(p. 722)*
edema (e-DĒ-ma): An abnormal accumulation of fluid in peripheral tissues. *(p. 736)*
hemorrhoids (HEM-ō-roydz): Varicose veins in the walls of the rectum and/or anus, often associated with frequent straining to force bowel movements. *(p. 727* and *AM)*
hypertension: Abnormally high blood pressure; usually defined in an adult as blood pressure greater than 150/90. *(p. 732* and *AM)*
hypotension: Blood pressure so low that circulation to vital organs may be impaired. *(p. 732* and *AM)*
pressure points: Locations where muscular arteries can be compressed against skeletal elements to restrict or prevent the flow of blood in an emergency. *(p. 732)*
pulmonary embolism: Circulatory blockage caused by the trapping of a freed thrombus in a pulmonary artery. *(p. 768)*
shock: An acute circulatory crisis marked by hypotension and inadequate peripheral blood flow. *(p. 747* and *AM)*
sounds of Korotkoff: Distinctive sounds caused by turbulent arterial blood flow heard while measuring a person's blood pressure. *(p. 732)*
sphygmomanometer: A device used to measure blood pressure using an inflatable cuff placed around one of the extremities. *(p. 732)*
thrombus: A stationary blood clot within a blood vessel. *(p. 768)*
varicose (VAR-i-kōs) **veins:** Sagging, swollen veins distorted by gravity and the failure of the venous valves. *(p. 727* and *AM)*

AM Additional Terms Discussed in the Applications Manual

angiotensin-converting enzyme (ACE) inhibitors: Drugs that block the conversion of angiotensin I to angiotensin II; sometimes used in the treatment of chronic hypertension and congestive heart failure
cardiogenic (kar-dē-ō-JEN-ik) **shock:** Progressive shock that develops when the heart becomes unable to maintain normal cardiac output.
cerebral embolism: Occlusion of one of the cerebral arteries by a blood clot, fatty mass, or air bubble that originated at another site; a cause of strokes.
cerebral hemorrhage: Rupture of a cerebral blood vessel, often following formation of an aneurysm; a cause of strokes.
cerebral thrombosis: Occlusion of one of the cerebral arteries by a clot formed at a plaque; a cause of strokes.
congestive heart failure: A condition that develops when the left ventricle can no longer keep pace with the right ventricle, and blood backs up into the pulmonary circuit. Elevated pulmonary pressures lead to *pulmonary edema.*
distributive shock: Progressive shock that results from a widespread, uncontrolled vasodilation, as in *neurogenic shock, septic shock,* or *anaphylactic shock.*
obstructive shock: Progressive shock that develops when ventricular output is reduced by tissues or fluids pressing against the heart.
orthostatic hypotension: Hypotension that develops when standing, due to an inability to make adjustments that increase blood pressure enough to maintain the blood supply to the brain.
phlebitis: Inflammation of a vein.
primary hypertension (essential hypertension): Hypertension without an obvious cause. Known risk factors include a hereditary history of hypertension, sex (males are at higher risk), high plasma cholesterol, obesity, chronic stresses, and cigarette smoking.
secondary hypertension: Hypertension that develops as the result of kidney problems or abnormal production of ADH, aldosterone, renin, epinephrine, or other hormones.
transient ischemic attack (TIA): A temporary loss of cerebral function due to a momentary blockage in a cerebral artery; an indication that cerebrovascular disease exists.

CHAPTER REVIEW

STUDY OUTLINE

INTRODUCTION, p. 718

1. Blood flows through a network of arteries, veins, and capillaries. All chemical and gaseous exchange between the blood and interstitial fluid takes place across capillary walls. *(Figure 21-1)*

THE ANATOMY OF BLOOD VESSELS, p. 718

1. Arteries and veins form an internal distribution system, propelled by the heart. Arteries branch repeatedly, decreasing in size until they become **arterioles;** from the arterioles blood enters the capillary networks. Blood flowing from the capillaries enters small **venules** before entering larger veins.

Structure of Vessel Walls, p. 719

2. The walls of arteries and veins contain three layers: the **tunica interna, tunica media,** and outermost **tunica externa.**

Differences between Arteries and Veins, p. 720

3. In general, the walls of arteries are thicker than those of veins. The endothelial lining of an artery cannot contract, so it is thrown into folds. Arteries constrict when blood pressure does not distend them; veins constrict very little. *(Figure 21-2)*

Arteries, p. 720

4. The arterial system includes the large **elastic arteries,** medium-sized **muscular arteries,** and smaller **arterioles.** As we proceed toward the capillaries the number of vessels increases, but the diameter of the individual vessels decreases and the walls become thinner. *(Figure 21-3)*
5. **Atherosclerosis,** a type of **arteriosclerosis,** is associated with changes in the endothelial lining of arteries. Fatty masses of tissue called **plaques** often develop during atherosclerosis. *(Figure 21-4)*

Capillaries, p. 724

6. Capillaries are the only blood vessels whose walls permit exchange between blood and interstitial fluid. Capillaries may be **continuous** or **fenestrated. Sinusoids** are specialized fenestrated capillaries found in certain tissues that allow very slow blood flow. *(Figure 21-5)*
7. Capillaries form interconnected networks called **capillary plexuses (capillary beds).** A **precapillary sphincter** (a band of smooth muscle) adjusts the blood flow into each capillary. Blood flow within a capillary changes as **vasomotion** occurs. The entire capillary plexus may be bypassed by blood flow through **arteriovenous anastomoses** or via **central (preferred) channels** within the capillary plexus. *(Figure 21-6)*

Veins, p. 725

8. Venules collect blood from the capillaries and merge into **medium-sized veins** and then **large veins.** The arterial system is a high-pressure system; blood pressure in veins is much lower. Valves in these vessels prevent the backflow of blood. *(Figures 21-2, 21-3, 21-7)*

The Distribution of Blood, p. 727

9. Peripheral **venoconstriction** helps maintain adequate blood volume in the arterial system after a hemorrhage. The **venous reserve** normally accounts for about 20 percent of the total blood volume. *(Figure 21-8)*

CARDIOVASCULAR PHYSIOLOGY, p. 728

Pressure, p. 728

1. Flow is proportional to pressure difference; blood will flow from an area of higher pressure to one of relatively lower pressure.

Resistance, p. 728

2. For circulation to occur, the **circulatory pressure** must be greater than the **total peripheral resistance** (the resistance of the entire circulatory system). For blood to flow into peripheral capillaries, blood pressure must be greater than the **peripheral resistance (PR)** (the resistance of the arterial system). Neural and hormonal control mechanisms regulate blood pressure.
3. The most important determinant of peripheral resistance is the diameter of arterioles.

An Overview of Circulatory Pressures, p. 729

4. The high arterial pressures overcome peripheral resistance and maintain blood flow through peripheral tissues. Capillary pressures are normally low; small changes in capillary pressure determine the rate of fluid movement into or out of the bloodstream. Venous pressure, normally low, determines venous return and affects cardiac output and peripheral blood flow. *(Table 21-1; Figure 21-9)*

Arterial Blood Pressure, p. 731

5. Arterial pressure rises during ventricular systole and falls during ventricular diastole. The difference between these two pressures is **pulse pressure.** *(Figure 21-10)*

Capillary Exchange, p. 733

6. At the capillaries, solute molecules diffuse across the capillary lining; water-soluble materials diffuse through small spaces between endothelial cells. Water will move when driven by either hydrostatic or osmotic pressure. The direction of water movement is determined by the balance between these two opposing pressures. Any change in hydrostatic or osmotic pressures in the blood or tissues will shift the point where the two forces are equal and determine whether there is a net loss or gain of fluid along the length of the capillary. *(Figures 21-11, 21-12, 21-13)*

Venous Pressure and Venous Return, p. 736

7. **Valves, muscular pumps,** and the **respiratory pump,** or *thoracoabdominal pump,* help the relatively low venuous pressures propel blood toward the heart. *(Figures 21-7, 21-9)*

CARDIOVASCULAR REGULATION, p. 737

1. Homeostatic mechanisms ensure that tissue blood flow **(perfusion)** delivers adequate oxygen and nutrients.
2. Local, autonomic, and endocrine factors influence the coordinated regulation of cardiovascular function. Local factors change the pattern of blood flow within capillary beds in response to chemical changes in interstitial fluids. Central mechanisms respond to changes in arterial pressure or blood gas levels. Hormones can assist in short-term adjustments (changes in cardiac output and peripheral resistance) and long-term adjustments (changes in blood volume that affect cardiac output and gas transport). *(Figure 21-14)*

Autoregulation of Blood Flow within Tissues, p. 738

3. Peripheral resistance is adjusted at the tissues by local factors that result in the dilation or constriction of precapillary sphincters.

The Neural Control of Blood Pressure and Blood Flow, p. 738

4. **Baroreceptor reflexes** are autonomic reflexes that adjust cardiac output and peripheral resistance to maintain normal arterial pressures. Baroreceptor populations include the aortic, carotid sinus, and atrial baroreceptors. *(Figure 21-15)*
5. **Chemoreceptor reflexes** respond to changes in the oxygen or carbon dioxide levels in the blood and cerebrospinal fluid. Sympathetic activation leads to stimulation of the cardioacceleratory and vasomotor centers; parasympathetic activation stimulates the cardioinhibitory center. Epinephrine and norepinephrine stimulate cardiac output and peripheral vasoconstriction. *(Figure 21-16)*

Hormones and Cardiovascular Regulation, p. 743

6. The endocrine system provides short-term regulation of cardiac output and peripheral resistance with epinephrine and norepinephrine from the adrenal medulla. Hormones involved in the long-term regulation of blood pressure and volume are antidiuretic hormone (ADH), angiotensin II, erythropoietin (EPO), and atrial natriuretic peptide (ANP). *(Figure 21-17)*
7. ADH and angiotensin II also promote peripheral vasoconstriction in addition to their other functions. ADH and aldosterone promote water and electrolyte retention and stimulate thirst. EPO stimulates red blood cell production. ANP encourages fluid loss, reduces blood pressure, inhibits thirst, and lowers peripheral resistance.

PATTERNS OF CARDIOVASCULAR RESPONSE, p. 743

Exercise and the Cardiovascular System, p. 743

1. During exercise, blood flow to skeletal muscles increases at the expense of circulation to nonessential organs, and cardiac output rises. Cardiovascular performance improves with training. Athletes have larger stroke volumes, slower resting heart rates, and increased cardiac reserves. *(Tables 21-2, 21-3)*

Cardiovascular Response to Hemorrhaging, p. 746

2. Blood loss causes an increase in cardiac output, mobilization of venous reserves, peripheral vasoconstriction, and the liberation of hormones that promote fluid retention and the manufacture of erythrocytes. *(Figure 21-18)*

Shock, p. 747

3. **Shock** is an acute circulatory crisis marked by hypotension and inadequate peripheral blood flow. A severe drop in blood volume produces symptoms of **circulatory shock.** Causes of fluid loss may include hemorrhaging, dehydration, and severe burns. *(Figure 21-19)*

THE BLOOD VESSELS, p. 749

1. The peripheral distributions of arteries and veins are usually identical on both sides of the body, except near the heart. *(Figure 21-20)*

The Pulmonary Circulation, p. 750

2. The pulmonary circuit includes the pulmonary trunk, the **left** and **right pulmonary arteries,** and the **pulmonary veins** that empty into the left atrium. *(Figure 21-21)*

The Systemic Circulation, p. 752

3. The **ascending aorta** gives rise to the coronary circulation. The **aortic arch** communicates with the **descending aorta.** *(Figures 21-22 to 21-29)*
4. Arteries in the neck and limbs are deep beneath the skin; in contrast, there are usually two sets of peripheral veins, one superficial and one deep. This dual-venous drainage is important for controlling body temperature.
5. The **superior vena cava** receives blood from the head, neck, chest, shoulders, and arms. *(Figures 21-30 to 21-33)* The **inferior vena cava** collects most of the venous blood from organs below the diaphragm. *(Figures 21-33, 21-34, 21-35)*

Fetal Circulation, p. 767

6. Major circulatory changes occur at birth as the lungs expand and the pulmonary circuit is activated. *(Figure 21-36)*
7. Congenital circulatory problems usually reflect abnormalities of the heart or of interconnections between the heart and great vessels. *(Figure 21-37)*

AGING AND THE CARDIOVASCULAR SYSTEM, p. 768

1. Age-related changes in the blood can include: (1) decreased hematocrit; (2) constriction or blockage of peripheral veins by a thrombus (stationary blood clot); (3) pooling of blood in the veins of the lower legs because valves are not working effectively.
2. Age-related changes in the heart include: (1) a reduction in the maximum cardiac output; (2) changes in the activities of the nodal and conducting fibers; (3) a reduction in the elasticity of the fibrous skeleton; (4) a progressive atherosclerosis that can restrict coronary circulation; (5) replacement of damaged cardiac muscle fibers by scar tissue.
3. Age-related changes in blood vessels, often related to arteriosclerosis, include (1) a weakening in the walls of arteries, potentially leading to *aneurysm* formation; (2) calcium salt deposition on weakened vascular walls, increasing the risk of a stroke or infarct; and (3) thrombus formation at atherosclerotic plaques.

REVIEW QUESTIONS

LEVEL 1 Reviewing Facts and Terms

Match each item in Column A with the most closely related item in Column B. Use letters for answers in the spaces provided

	Column A	Column B
___	1. veins	a. specialized fenestrated capillaries
___	2. vasa vasorum	b. minimum blood pressure
___	3. conducting arteries	c. vasomotor center
___	4. muscular arteries	d. drains the kidney
___	5. sinusoids	e. vasoconstrictor fibers
___	6. precapillary sphincter	f. ventricular stretching
___	7. medulla oblongata	g. elastic arteries
___	8. vascular resistance	h. blood supply to pelvis
___	9. systolic pressure	i. vasodilator fibers
___	10. diastolic pressure	j. largest superficial vein in body
___	11. blood pressure	k. capacitance vessels
___	12. arterioles	l. drains the liver
___	13. cholinergic	m. largest artery in body
___	14. adrenergic	n. stationary blood clot
___	15. baroreceptors	o. distribution arteries
___	16. Frank Starling law	p. foramen ovale
___	17. aorta	q. vasomotion
___	18. internal iliac artery	r. resistance vessels
___	19. renal vein	s. migrating blood clot
___	20. hepatic vein	t. carotid sinus
___	21. great saphenous	u. friction
___	22. interatrial opening	v. sounds of Korotkoff
___	23. thrombus	w. "vessels of vessels"
___	24. embolus	x. peak blood pressure

25. Blood vessels that carry blood away from the heart are called:
(a) veins (b) arterioles
(c) venules (d) arteries

26. The layer of the arteriole wall that provides the properties of contractility and elasticity is the:
(a) tunica adventitia (b) tunica media
(c) tunica intima (d) tunica externa

27. Blood vessels that supply the walls of arteries and veins with blood are called the:
(a) coronary vessels (b) capillaries
(c) vasa vasorum (d) metarterioles

28. Of the following arteries the *one* that is an elastic artery is the:
(a) subclavian artery
(b) external carotid artery
(c) brachial artery
(d) femoral artery

29. The two-way exchange of substances between blood and body cells occurs only through:
(a) arterioles (b) capillaries
(c) venules (d) a, b, and c are correct

30. Large molecules such as peptides and proteins move into and out of the circulation by way of:
(a) continuous capillaries
(b) fenestrated capillaries
(c) preferred channels
(d) metarterioles

31. The alteration of blood flow due to the action of precapillary sphincters is called:
(a) vasomotion (b) autoregulation
(c) selective resistance (d) turbulence

32. The blood vessels that collect blood from all tissues and organs and return it to the heart are the:
(a) veins (b) arteries
(c) capillaries (d) arterioles

33. Blood is compartmentalized within the veins because of the presence of:
(a) venous reservoirs (b) muscular walls
(c) clots (d) valves

34. The most important factor in vascular resistance is:
(a) the viscosity of the blood
(b) friction between the blood and the vessel walls
(c) turbulence due to irregular surfaces of blood vessels
(d) the length of the blood vessels

35. Plasma proteins are normally unable to cross the endothelial linings anywhere except in the:
(a) continuous capillaries
(b) fenestrated capillaries
(c) sinusoids
(d) arteriovenous anastomoses

36. Hydrostatic pressure forces water ________ a solution; osmotic pressure forces water _______ a solution.
(a) into, out of (b) out of, into
(c) out of, out of (d) a, b, and c are incorrect

37. The two arteries formed by the division of the brachiocephalic artery are the:
(a) aorta and internal carotid
(b) axillary and brachial
(c) external and internal carotid
(d) common carotid and subclavian

38. The unpaired arteries supplying blood to the visceral organs include the:
(a) suprarenal, renal, lumbar
(b) iliac, gonadal, femoral
(c) celiac, superior and inferior mesenteric
(d) a, b, and c are correct

39. The paired arteries supplying blood to the body wall and other structures outside the abdominopelvic cavity include the:
(a) left gastric, hepatic, splenic, phrenic
(b) suprarenals, renals, lumbars, gonadals
(c) iliacs, femorals, gonadals, ileocecals
(d) celiac, left gastric, superior and inferior mesenteric

40. The vein that drains the dural sinuses of the brain is the:
(a) cephalic (b) great saphenous
(c) internal jugular (d) superior vena cava

41. The vein that drains the thorax and empties into the superior vena cava is the:
(a) azygos (b) basilic
(c) cardiac (d) cephalic

42. The vein that collects most of the venous blood from below the diaphragm is the:
    (a) superior vena cava
    (b) great saphenous
    (c) inferior vena cava
    (d) azygos

43. The tributaries of the hepatic portal vein include the:
    (a) lumbar, gonadal, renal, and suprarenal veins
    (b) left colic, splenic, inferior and superior mesenteric veins
    (c) phrenic, hepatic, renal, and suprarenal veins
    (d) peroneal, iliac, saphenous, and femoral veins

44. Differentiate among the following types of pressure: (a) systolic (b) diastolic (c) pulse (d) mean arterial

45. (a) What are the primary forces that cause fluid to move out of a capillary and into the interstitial fluid at its arterial end? (b) What are the primary forces that cause fluid to move into a capillary from the interstitial fluid at its venous end?

46. What two factors assist relatively low venous pressures in propelling blood toward the heart?

47. Give at least five examples of local vasodilators produced at the tissue level that accelerate blood flow through the tissue.

48. What two effects occur when the baroreceptor response to elevated blood pressure is triggered?

49. What factors affect the activity of chemoreceptors in the carotid and aortic bodies?

50. List six hormones important in the regulation of cardiovascular function.

51. What interrelated cardiovascular changes occur before and during light exercise?

52. What criteria are used to identify the basic symptoms of all forms of shock?

53. What circulatory changes occur at birth?

54. What age-related changes take place in the blood, heart, and blood vessels?

LEVEL 2 Reviewing Concepts

55. When dehydration occurs there is:
    (a) accelerated reabsorption of water at the kidneys
    (b) a recall of fluids
    (c) an increase in the blood colloidal osmotic pressure
    (d) a, b, and c are correct

56. Increased CO_2 levels in tissues would promote:
    (a) constriction of precapillary sphincters
    (b) an increase in the pH of the blood
    (c) dilation of precapillary sphincters
    (d) decreased blood flow to tissues

57. Elevated levels of the hormones ADH and angiotensin II will produce:
    (a) increased peripheral vasodilation
    (b) increased peripheral vasoconstriction
    (c) increased peripheral blood flow
    (d) increased venous return

58. Secretion of ADH and aldosterone are typical of the body's long-term compensation following:
    (a) a heart attack
    (b) hypertension
    (c) a serious hemorrhage
    (d) heavy exercise

59. Relate the anatomical differences between arteries and veins to their functions.

60. What causes variations in blood flow in individual capillaries?

61. Why do capillaries permit the diffusion of materials whereas arteries and veins do not?

62. How is blood pressure maintained in veins to cope with the force of gravity?

63. How do pressure and resistance affect cardiac output and peripheral blood flow?

64. Why is blood flow to the brain relatively continuous and constant?

65. Compare the effects of the cardioacceleratory and cardioinhibitory centers on cardiac output and blood pressure.

66. A nurse practitioner tells Mrs. B. that her blood pressure is 150/90. Explain what these numbers represent and calculate Mrs. B's mean arterial pressure (MAP).

67. An accident victim displays the following symptoms: hypotension; pale, cool, moist skin; confusion and disorientation. Identify her condition and explain why these symptoms occur. If you took her pulse, what would you probably find?

68. Mr. Brown has been diagnosed with congestive heart failure. Because of this condition, his ankles and feet appear to be swollen. What is the relationship between congestive heart failure and accumulation of fluid in the feet and ankles?

LEVEL 3 Critical Thinking and Clinical Applications

69. Bob is sitting outside on a warm day and is sweating profusely. His friend Mary wants to practice taking blood pressures, and he agrees to play patient. Mary finds that Bob's blood pressure is elevated, even though he is resting and has lost fluid from sweating (she reasons that fluid loss should lower blood volume and thus blood pressure). Mary asks you why Bob's blood pressure is high instead of low. What will you tell her?

70. The most frequent site of varicose veins is the greater saphenous vein of the leg. Why do you think this is the case?

71. Would you expect a resting athlete or a resting person who never exercises and has a sedentary job to have a higher pulse pressure? Why?

72. People who suffer from allergies frequently take antihistamines and decongestants to relieve their symptoms. The medication's box warns that the medication should not be taken by individuals being treated for high blood pressure. Why not?

73. Jill awakens suddenly to the sound of her alarm clock. Realizing she is late for class, she jumps to her feet, feels light-headed, and falls back on her bed. What probably caused this to happen? Why doesn't this happen all the time?

CHAPTER 22

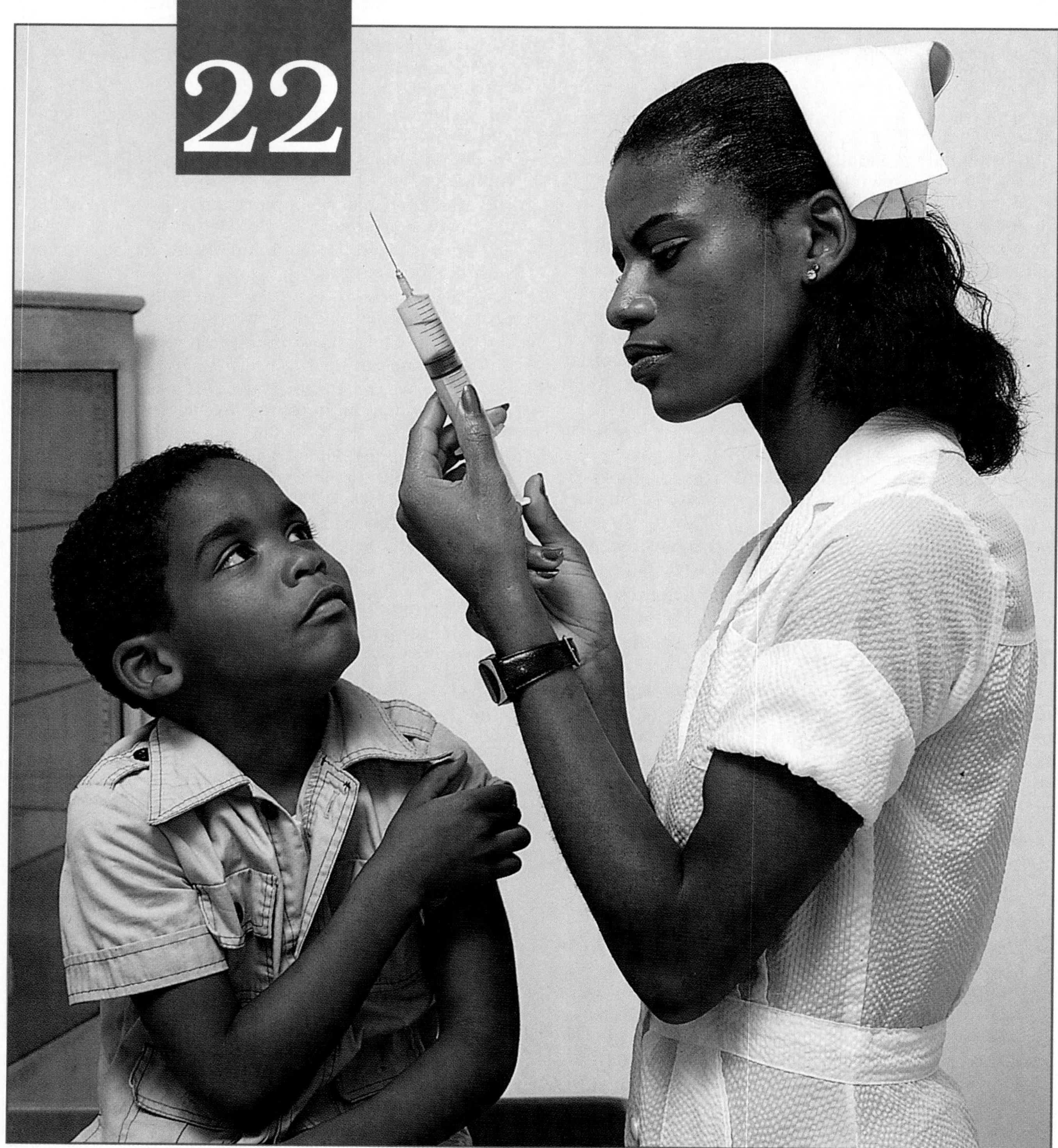

This is every child's nightmare: knowing that shot is for you! As children, all we notice is the sting of a vaccination. As adults, we realize that the shot hurts only for an instant (if at all), whereas the disease that it prevents can cripple or kill. We are constantly besieged by threats to our health from a variety of sources—viruses, bacteria, fungi, toxins, and renegade body cells that have turned malignant. In this chapter we will examine the defense mechanisms that the body deploys to fight off these dangers. We will also see how modern medicine can manipulate the immune system to enhance these defenses.

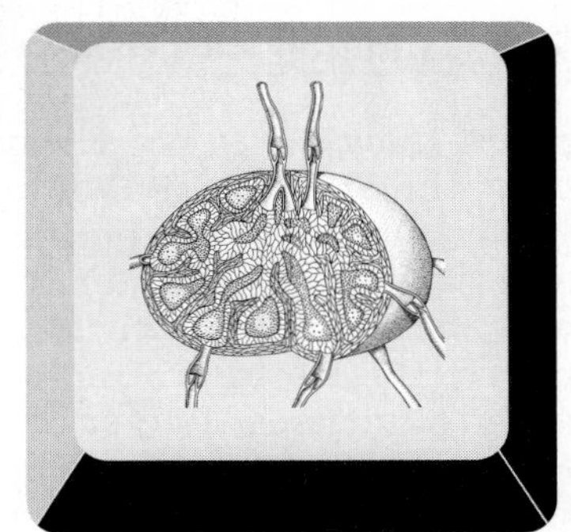

The Lymphatic System and Immunity

Chapter Outline and Objectives

The world is not always kind to the human body. Accidental bumps, cuts, and scrapes, chemical and thermal burns, extreme cold, and ultraviolet radiation are just a few of the hazards in the physical environment. Making matters worse, the world around us contains an assortment of viruses, bacteria, fungi, and parasites. Many of these organisms are perfectly capable of not only surviving, but thriving inside our bodies—and potentially causing us great harm in the process. These microorganisms, called **pathogens**, are responsible for many human diseases. Each has a different mode of life and attacks the body in a specific way. For example, viruses spend most of their time hiding within cells, whereas many bacteria multiply in the interstitial fluids, and the largest parasites burrow through internal organs. AM *The Nature of Pathogens*

Many different organs and systems work together in an effort to keep us alive and healthy. In this ongoing struggle, the **lymphatic system** plays a central role. The lymphatic system is an anatomical system consisting of (1) a fluid, *lymph;* (2) a network of *lymphatic vessels;* (3) specialized cells, called *lymphocytes;* and (4) an array of *lymphoid tissues* and *lymphoid organs* scattered throughout the body. Chapter 21 explored the relationship between lymph and plasma and the role of the lymphatic vessels in the circulation of extracellular fluid. [p. 718]

An Overview of the Lymphatic System

Lymphocytes, the primary cells of the lymphatic system, were introduced in Chapters 4 and 19. [pp. 124, 128, 666] These cells are vital to our ability to resist or overcome infection and disease. Lymphocytes respond to the presence of (1) invading pathogens, such as bacteria and viruses, (2) abnormal body cells, such as virus-infected cells and cancer cells, and (3) foreign proteins, such as the toxins released by some bacteria. They attempt to eliminate these threats or render them harmless by a combination of physical and chemical attack.

Lymphocytes respond to specific threats. If bacteria invade peripheral tissues, the lymphocytes organize a defense against that particular type of bacterium. For this reason, lymphocytes are said to provide a *specific defense,* known as the **immune response. Immunity** is the ability to resist infection and disease through the activation of specific defenses. All of the cells and tissues involved with the production of immunity are sometimes considered to be part of an *immune system,* a physiological system that includes not only the lymphatic system but also components of the integumentary, cardiovascular, respiratory, digestive, and other systems. For example, interactions between lymphocytes and Langerhans cells of the skin are important in mobilizing specific defenses against skin infections. This chapter begins by examining the organization of the lymphatic system. We will then consider how the lymphatic system interacts with cells and tissues of other systems to defend the body against infection and disease.

Organization of the Lymphatic System

Figure 22-1

The lymphatic system includes:

1. A network of **lymphatic vessels** that begin in peripheral tissues and end at connections to the venous system.
2. A fluid, called **lymph,** resembling plasma but containing a much lower concentration of suspended proteins.
3. **Lymphoid organs** connected to the lymphatic vessels and containing large numbers of lymphocytes.

Figure 22-1● provides a general overview of the components of this system.

Functions of the Lymphatic System

The primary functions of the lymphatic system are:

- ***The production, maintenance, and distribution of lymphocytes.*** Lymphocytes, the primary cells involved in the immune response, are produced and stored within (1) lymphoid tissues and organs, such as the spleen and thymus, and (2) areas of red bone marrow.
- ***The return of fluid and solutes from peripheral tissues to the blood.*** As we saw in the last chapter, capillaries normally deliver more fluid to the tissues than they carry away. [p. 735] The return of tissue fluids through lymphatic vessels maintains normal blood volume and eliminates local variations in the composition of the interstitial fluid.
- ***The distribution of hormones, nutrients, and waste products from their tissues of origin to the general circulation.*** Substances that originate in the tissues but are for some reason unable to enter the bloodstream directly may do so via the lymphatic vessels. For example, lipids absorbed by the digestive tract often fail to enter the circulation at the capillary level. However, they still reach the bloodstream via passage along lymphatic vessels (a process explored further in Chapter 24).

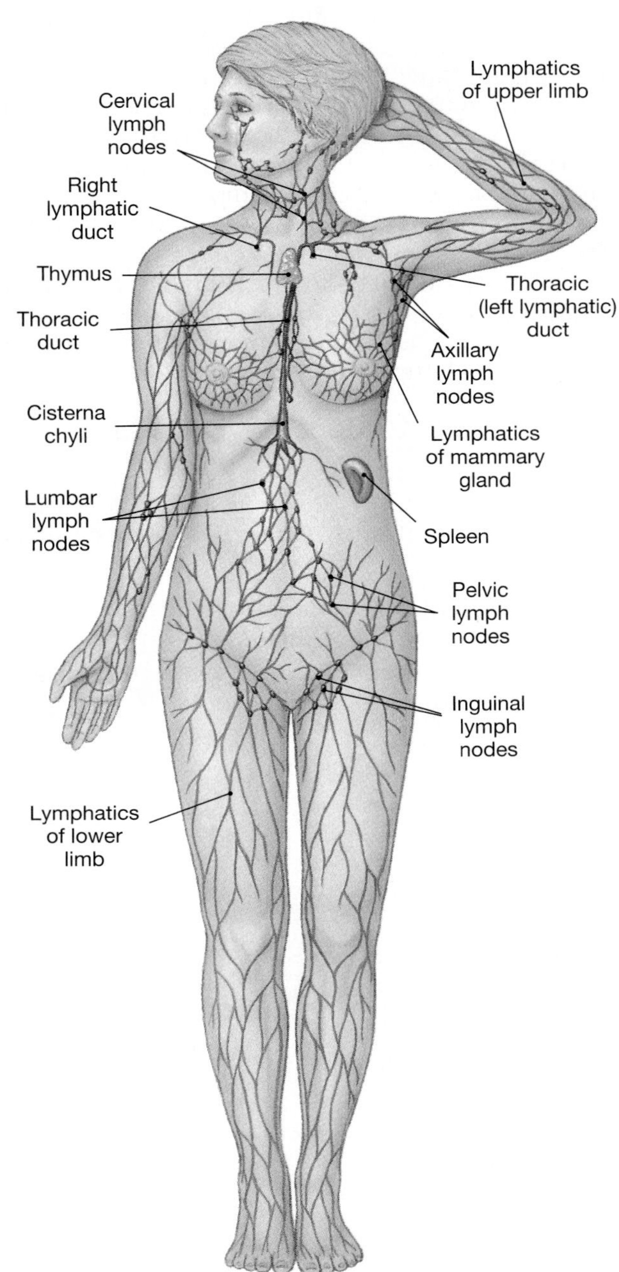
Cervical lymph nodes
Right lymphatic duct
Thymus
Thoracic duct
Cisterna chyli
Lumbar lymph nodes
Lymphatics of lower limb
Lymphatics of upper limb
Thoracic (left lymphatic) duct
Axillary lymph nodes
Lymphatics of mammary gland
Spleen
Pelvic lymph nodes
Inguinal lymph nodes

● **FIGURE 22-1**
Components of the Lymphatic System

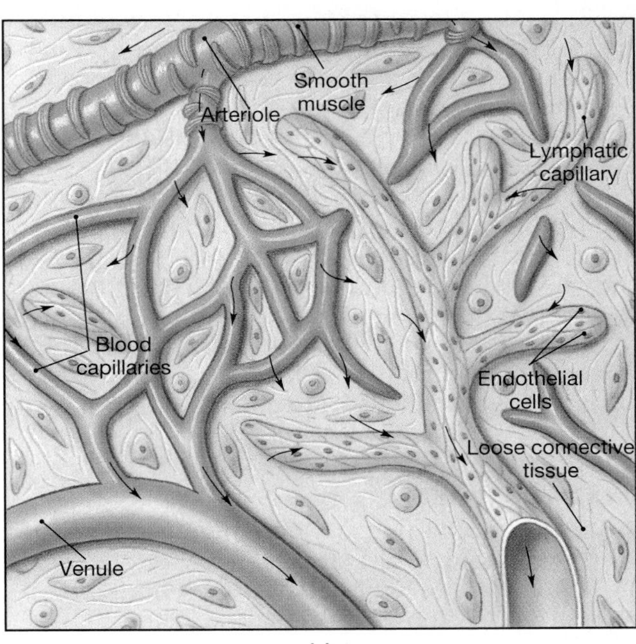
Smooth muscle
Arteriole
Lymphatic capillary
Blood capillaries
Endothelial cells
Loose connective tissue
Venule

(a)

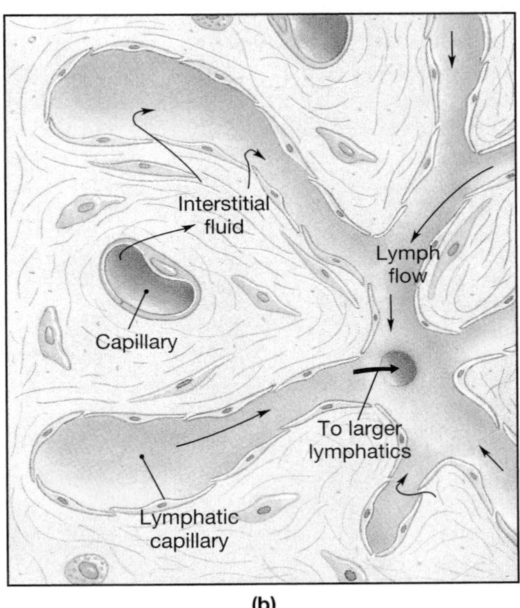
Interstitial fluid
Lymph flow
Capillary
To larger lymphatics
Lymphatic capillary

(b)

● **FIGURE 22-2**
Lymphatic Capillaries. (a) A diagrammatic view of the interwoven network formed by blood capillaries and lymphatic capillaries. Arrows show the net direction of interstitial fluid and lymph movement. **(b)** Sectional view of this association.

❑ Lymphatic Vessels

Lymphatic vessels, often called **lymphatics,** carry lymph from peripheral tissues to the venous system. The smallest lymphatic vessels are called *lymphatic capillaries.*

Lymphatic Capillaries

Figures 22-2, 22-3a

The lymphatic network begins with the **lymphatic capillaries**, or *terminal lymphatics*, that branch through peripheral tissues. They differ from blood capillaries in that lymphatic capillaries (1) originate as blind pockets, (2) are larger in diameter, (3) have thinner walls, and (4) in sectional view they often have a flattened or irregular outline (Figure 22-3a●). Although lined by endothelial cells, they have no underlying basement membrane, and no well-organized layers separate the lymphatic vessel from the surrounding tissue.

The endothelial cells of a lymphatic capillary are not tightly bound together, but they do overlap. The region of overlap acts as a one-way valve, permitting the entry of fluids and solutes, even those as large as proteins, but preventing its return to the intercellular spaces (Figure 22-2●).

Lymphatic capillaries are present in almost every tissue and organ in the body. They are absent in areas that lack a blood supply, such as the cornea of the eye, and there are no lymphatics in the central nervous system. Prominent lymphatic capillaries in the small intestine are called *lacteals;* these are important in the transport of lipids absorbed by the digestive tract.

Valves of Lymphatic Vessels

Figure 22-3

From the lymphatic capillaries, lymph flows into larger lymphatic vessels that lead toward the trunk. The walls of these lymphatic vessels contain layers comparable to those of veins, and, like veins, the larger lymphatic vessels contain valves. The valves are quite close together, and at each valve the lymphatic vessel bulges noticeably. As a result, large lymphatics have a beaded appearance (Figure 22-3a●). The valves prevent the backflow of lymph within lymph vessels, especially those of the limbs. Pressures within the lymphatic system are minimal, and the valves are essential to maintaining normal lymph flow toward the thoracic cavity.

Lymphatic vessels are often found in association with blood vessels (Figure 22-3a●). Note the differences in relative size, general appearance, and branching pattern that distinguish the lymphatic vessels from arteries and veins. There are also characteristic color differences that are apparent on examining living tissues. Arteries are usually a bright red, veins a dark red, and lymphatic vessels a pale golden color. In general, a tissue will contain many more lymphatics than veins, but the lymphatics are much smaller.

Major Lymph-Collecting Vessels

Figure 22-4

Two sets of lymphatic vessels collect lymph from the lymphatic capillaries. **Superficial lymphatics** are found:

- In the subcutaneous layer beneath the skin.
- In the loose connective tissues of the mucous membranes lining the digestive, respiratory, urinary, and reproductive tracts.
- In the loose connective tissues of the serous membranes lining the pleural, pericardial, and peritoneal cavities.

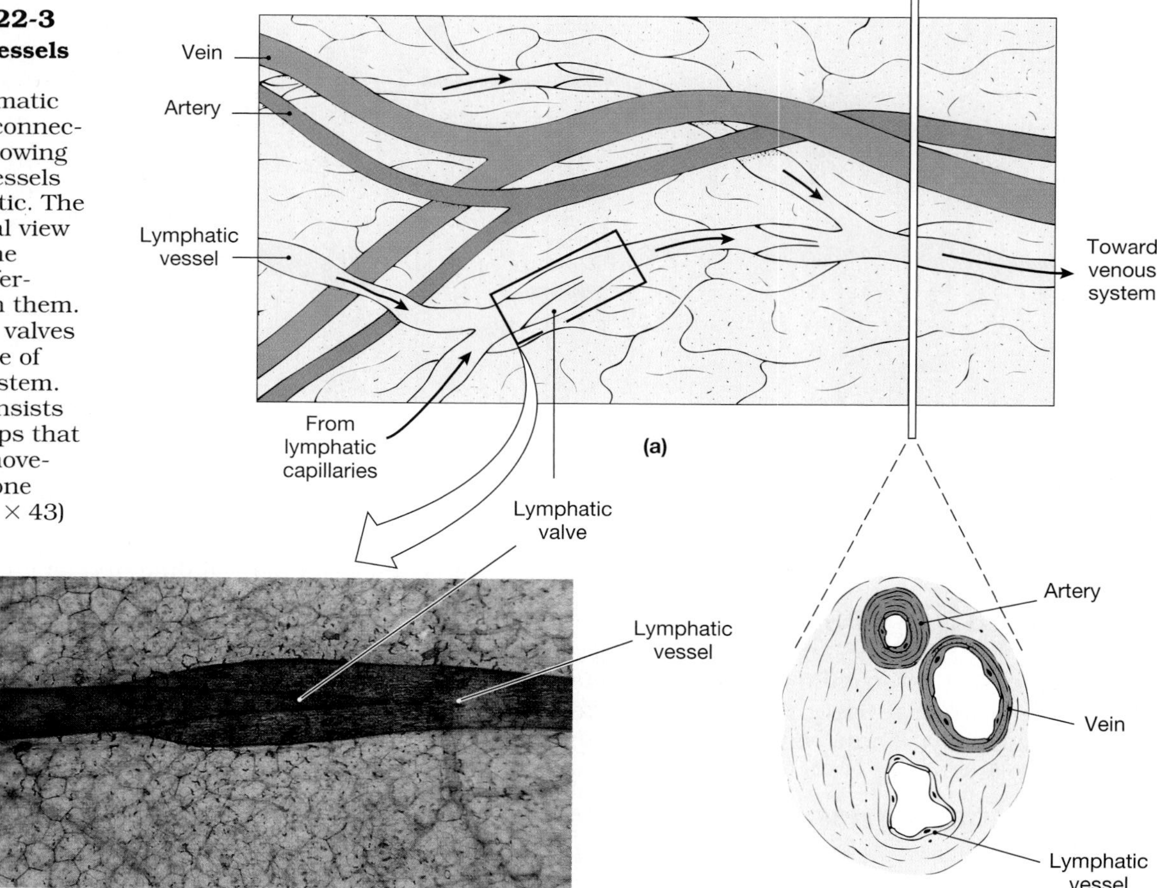
Vein
Artery
Lymphatic vessel
From lymphatic capillaries
Toward venous system
(a)
Lymphatic valve
Lymphatic vessel
Artery
Vein
Lymphatic vessel
(b)

● **FIGURE 22-3 Lymphatic Vessels and Valves.** **(a)** A diagrammatic view of loose connective tissue, showing small blood vessels and a lymphatic. The cross-sectional view emphasizes the structural differences between them. **(b)** Lymphatic valves resemble those of the venous system. Each valve consists of a pair of flaps that permit fluid movement in only one direction. (LM × 43)

Deep lymphatics are larger lymph vessels that accompany deep arteries and veins supplying skeletal muscles and other organs of the neck, limbs, and trunk, and the walls of visceral organs.

Superficial and deep lymphatics converge to form trunks that empty into two large collecting vessels, the *thoracic duct* and the *right lymphatic duct.* The **thoracic duct** collects lymph from the body below the diaphragm and from the left side of the body above the diaphragm. The relatively small **right lymphatic duct** collects lymph from the right side of the body above the diaphragm.

THE THORACIC DUCT The thoracic duct is formed inferior to the diaphragm at the level of vertebra L_2. The base of the thoracic duct is an expanded, sac-like chamber, the **cysterna chyli** (Figure 22-4●). The cysterna chyli receives lymph from the lower abdomen, pelvis, and lower limbs via the *right* and *left lumbar trunks* and the *intestinal trunk.*

The inferior segment of the thoracic duct lies anterior to the vertebral column. From the second lumbar vertebra, it penetrates the diaphragm in company with the aorta, at the *aortic hiatus,* and ascends along the left side of the vertebral column to the level of the left clavicle. After collecting lymph from the *left bronchomediastinal trunk,* the *left subclavian trunk,* and the *left jugular trunk,* it empties into the left subclavian vein near the left internal jugular vein (Figure 22-4●). Lymph collected from the left side of the head, neck, and thorax as well as lymph from the entire body inferior to the diaphragm reenter the venous system in this way.

THE RIGHT LYMPHATIC DUCT The right lymphatic duct is formed by the merging of the *right jugular, right subclavian,* and *right bronchomediastinal trunks* in the area near the right clavicle. This duct empties into the right subclavian vein, delivering lymph from the right side of the body above the diaphragm (Figure 22-4●).

LYMPHEDEMA Blockage of the lymphatic drainage from a limb produces **lymphedema** (lim-fe-DĒ-ma). In this painless condition, interstitial fluids accumulate, and the limb gradually becomes swollen and grossly distended. If the condition persists, the connective tissues lose their elasticity and the swelling becomes permanent. Lymphedema by itself does not pose a major threat to life. The danger comes from the constant risk that an uncontrolled infection will develop in the affected area. Because the interstitial fluids are essentially stagnant, toxins and pathogens can accumulate and overwhelm the local defenses without fully activating the immune system.

Temporary lymphedema may result from tight clothing. Chronic lymphedema can result from scar tissue formation or from parasitic infections. In **filariasis** (fil-a-RĪ-a-sis), larvae of a parasitic nematode (roundworm), usually *Wucheria bancrofti,* is transmitted by mosquitoes or blackflies. The adult worms form massive colonies within lymphatic vessels and lymph nodes. Repeated scarring of the passageways eventually blocks lymphatic drainage and produces extreme lymphedema with permanent distension of tissues. The limbs or external genitalia often become grossly distended, a condition known as **elephantiasis** (el-e-fan-TĪ-a-sis).

Therapy for chronic lymphedema consists of treating infections by the administration of antibiotics and (when possible) reducing the swelling. One possible treatment involves the application of elastic wrappings that squeeze the tissue. This external compression elevates the hydrostatic pressure of the interstitial fluids and opposes the entry of additional fluid from the capillaries.

Lymphocytes

Lymphocytes were introduced in Chapter 19 because they account for roughly 25 percent of the circulating white blood cell population. [p. 666] However, circulating lymphocytes are only a small fraction of the total lymphocyte population. The body contains around 10^{12} lymphocytes, with a combined weight of over a kilogram.

Types of Lymphocytes

There are three classes of lymphocytes in the blood: **T cells** (**t**hymus-dependent), **B cells** (**b**one marrow–derived), and **NK cells** (**n**atural **k**iller). Each type has distinctive biochemical and functional characteristics.

T CELLS Approximately 80 percent of circulating lymphocytes are classified as T cells. There are many different types of T cells, including:

- **Cytotoxic T cells,** which attack foreign cells or body cells infected by viruses. Their attack often involves direct contact. These lymphocytes are the primary cells involved in the production of *cell-mediated immunity,* or *cellular immunity.*
- **Helper T cells,** which stimulate the activation and function of both T cells and B cells.
- **Suppressor T cells,** which inhibit the activation and function of both T cells and B cells.

The interplay between suppressor and helper T cells helps establish and control the sensitivity of the immune response. For this reason, these cells are also known as *regulatory T cells.*

These are the T cells that we will examine in the course of this chapter. It is not a complete list, however, and there are other types of T cells that participate in the immune response. For example, *inflammatory T cells* stimulate regional inflammation and local defenses in an injured tissue, and *suppressor/inducer T cells* suppress B cell activity but stimulate other T cells.

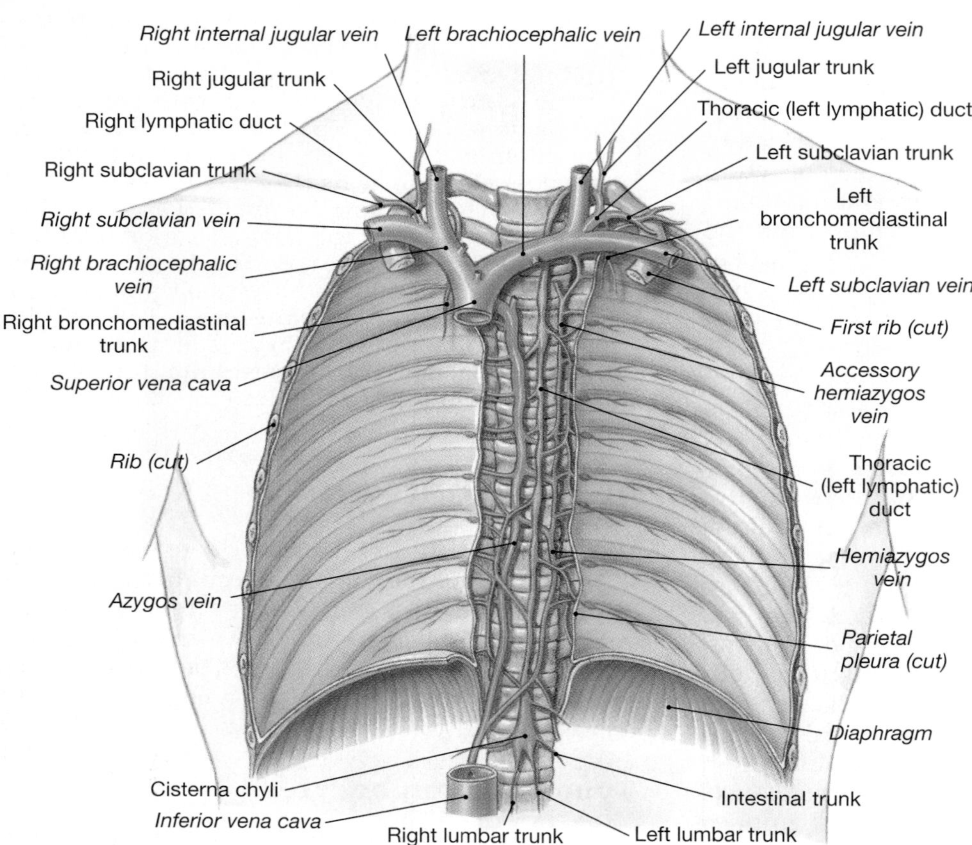
Right internal jugular vein
Left brachiocephalic vein
Left internal jugular vein
Right jugular trunk
Left jugular trunk
Right lymphatic duct
Thoracic (left lymphatic) duct
Right subclavian trunk
Left subclavian trunk
Right subclavian vein
Left bronchomediastinal trunk
Right brachiocephalic vein
Left subclavian vein
Right bronchomediastinal trunk
First rib (cut)
Superior vena cava
Accessory hemiazygos vein
Rib (cut)
Thoracic (left lymphatic) duct
Hemiazygos vein
Azygos vein
Parietal pleura (cut)
Diaphragm
Cisterna chyli
Intestinal trunk
Inferior vena cava
Right lumbar trunk
Left lumbar trunk

● **FIGURE 22-4 Relationship between the Lymphatic Ducts and the Venous System.** The thoracic duct carries lymph originating in tissues inferior to the diaphragm and from the left side of the upper body. It empties into the left subclavian vein. The right lymphatic duct drains the right half of the body superior to the diaphragm. It empties into the right subclavian vein. AM *Plate 6.3j; Scan 20*

B CELLS B cells account for 10–15 percent of circulating lymphocytes. When stimulated, B cells can differentiate into **plasma cells.** Plasma cells, introduced in Chapter 4, are responsible for the production and secretion of *antibodies*, soluble proteins that are also known as *immunoglobulins.* ∞ [p. 124] These proteins react with specific chemical targets, called **antigens.** Antigens are usually pathogens, parts or products of pathogens, or other foreign compounds. Most antigens are proteins, but some lipids, polysaccharides, and nucleic acids can also stimulate antibody production. When an antibody binds to its target antigen, it starts a chain of events leading to the destruction of the target compound or organism. B cells are responsible for *antibody-mediated immunity*, which is also known as *humoral* ("liquid") *immunity* because antibodies are found in body fluids.

NK CELLS The remaining 5–10 percent of circulating lymphocytes are NK cells, also known as **large granular lymphocytes.** These lymphocytes will attack foreign cells, normal cells infected with viruses, and cancer cells that appear in normal tissues. Their continual policing of peripheral tissues has been called *immunological surveillance.*

Life Span and Circulation of Lymphocytes

Lymphocytes are not evenly distributed in the blood, bone marrow, spleen, thymus, and peripheral lymphatic tissues. The ratio of B cells to T cells varies depending on the tissue or organ considered. For example, B cells are seldom found in the thymus, and in the blood T cells outnumber B cells by a ratio of 8:1. This ratio changes to 1:1 in the spleen and 1:3 in the bone marrow.

The lymphocytes within these organs are visitors, not residents. All types of lymphocytes move throughout the body; they wander through a tissue and then enter a blood vessel or lymphatic vessel for transport to another site.

T cells move relatively quickly. For example, a wandering T cell may spend about 30 minutes in the blood, 5–6 hours in the spleen, and 15–20 hours in a lymph node. B cells, which are responsible for antibody production, move more slowly; a typical B cell spends around 30 hours in a lymph node before moving to another location.

In general, lymphocytes have relatively long life spans. Roughly 80 percent survive for 4 years, and some last 20 years or more. Throughout life, normal lymphocyte populations are maintained through lymphopoiesis in the bone marrow and lymphatic tissues.

Lymphocyte Production

Figure 22-5

Chapter 19 discussed *hemopoiesis*, the formation of the cellular elements of blood. ∞ [p. 653] Erythropoiesis (red blood cell formation) in the adult

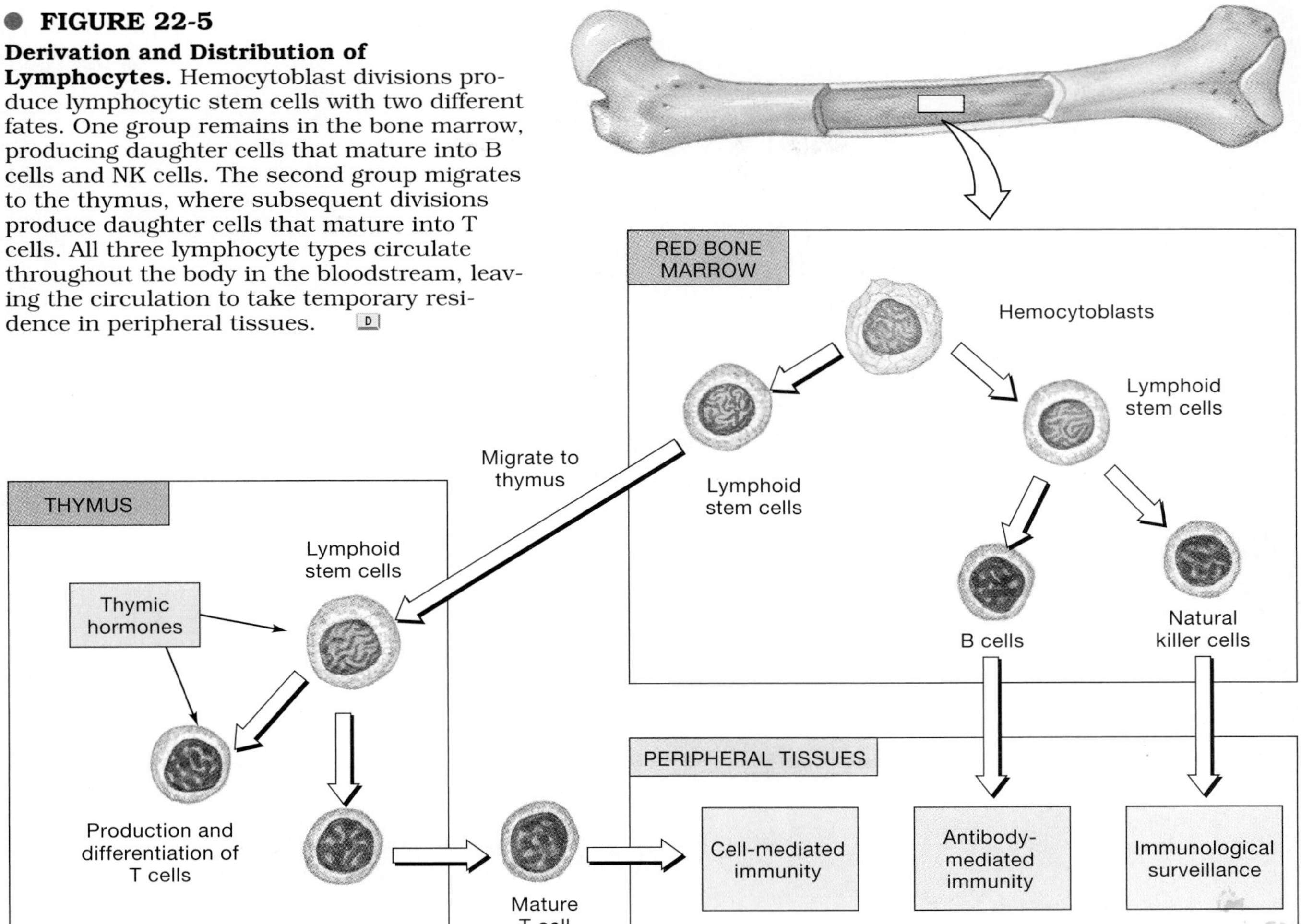
RED BONE MARROW
Hemocytoblasts
Lymphoid stem cells
Lymphoid stem cells
B cells
Natural killer cells
Migrate to thymus
THYMUS
Lymphoid stem cells
Thymic hormones
Production and differentiation of T cells
Mature T cell
PERIPHERAL TISSUES
Cell-mediated immunity
Antibody-mediated immunity
Immunological surveillance

FIGURE 22-5
Derivation and Distribution of Lymphocytes. Hemocytoblast divisions produce lymphocytic stem cells with two different fates. One group remains in the bone marrow, producing daughter cells that mature into B cells and NK cells. The second group migrates to the thymus, where subsequent divisions produce daughter cells that mature into T cells. All three lymphocyte types circulate throughout the body in the bloodstream, leaving the circulation to take temporary residence in peripheral tissues.

is normally confined to the bone marrow, but lymphocyte production, or **lymphopoiesis,** involves the bone marrow, thymus, and peripheral lymphatic tissues. The relationships are diagrammed in Figure 22-5●.

The bone marrow plays the primary role in the maintenance of normal lymphocyte populations. *Hemocytoblast divisions in the bone marrow of an adult generate the lymphocytic stem cells responsible for the production of all types of lymphocytes.* Two distinct populations of lymphocytic stem cells are produced in the bone marrow.

EVENTS IN THE BONE MARROW: B CELL AND NK CELL PRODUCTION One group of lymphocytic stem cells remains in the bone marrow. Divisions of these cells produce immature B cells and NK cells. B cell development involves intimate contact with large **stromal cells** (*stroma,* a bed) in the bone marrow. The cytoplasmic extensions of stromal cells contact or even wrap around the developing B cells. The stromal cells produce an immune system hormone, or *cytokine,* called *interleukin-7,* that promotes the differentiation of B cells. (Cytokines and their varied effects will be considered in a later section.)

As they mature, B cells and NK cells enter the circulation and migrate to peripheral tissues. Most of the B cells move into lymph nodes, the spleen, or other lymphoid tissues. The NK cells migrate throughout the body, moving through peripheral tissues in their search for abnormal cells.

EVENTS IN THE THYMUS: T CELL PRODUCTION
The second group of lymphocytic stem cells migrates to the thymus. While in the thymus, these cells and their descendants develop further in an environment that is isolated from the general circulation by the **blood-thymus barrier.** Under the influence of thymic hormones collectively known as **thymosins,** the lymphocytic stem cells divide repeatedly, producing the various kinds of T cells. At least seven different "thymosins" have been identified, including *thymosin-α, thymosin-β, thymosin V, thymopoietin, thymulin, thymolymphotropin,* and *thymic-factor X.* Their precise functions and interactions have yet to be determined.

When their development is nearing completion, T cells reenter the circulation and return to the bone marrow, as well as traveling to lymphoid tissues and organs, such as the spleen.

● FIGURE 22-6
Lymphoid Nodules.
(a) Appearance of a typical nodule as seen with the light microscope. Note the relatively pale germinal center, where lymphocyte cell divisions occur. (LM × 17) **(b)** The positions of the tonsils and the appearance of a tonsil in section.

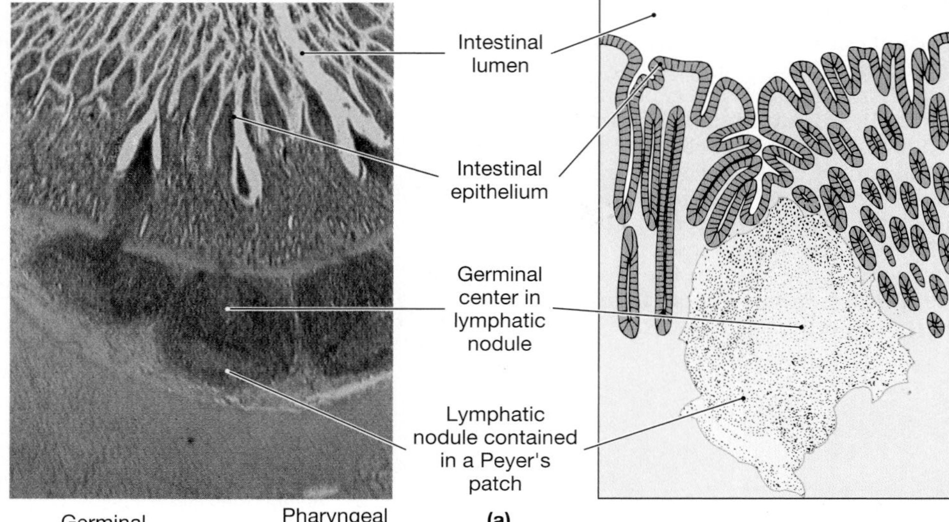
Intestinal lumen
Intestinal epithelium
Germinal center in lymphatic nodule
Lymphatic nodule contained in a Peyer's patch
(a)

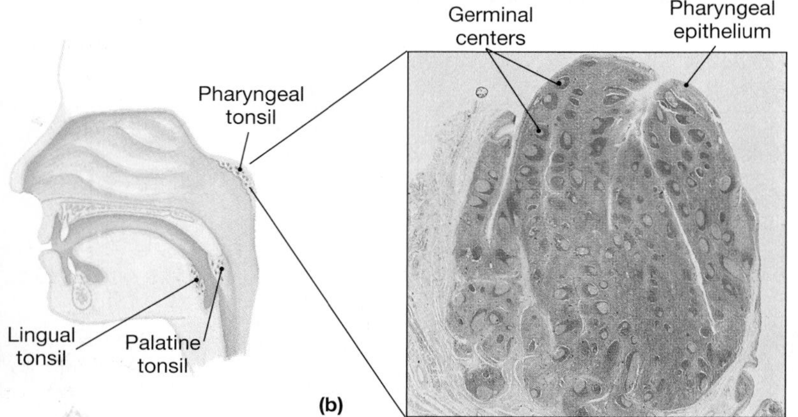
Germinal centers
Pharyngeal epithelium
Pharyngeal tonsil
Lingual tonsil
Palatine tonsil
(b)

LYMPHOCYTE PRODUCTION IN PERIPHERAL TISSUES The T cells and B cells that migrate from their sites of origin in the thymus and bone marrow retain the ability to divide. These divisions produce daughter cells of the same type; for example, a dividing B cell produces other B cells, not T cells or NK cells. As we shall see, the ability to increase the number of lymphocytes of a specific type is important to the success of the immune response.

❑ Lymphoid Tissues

Figure 22-6

Lymphoid tissues are connective tissues dominated by lymphocytes. In a **lymphoid nodule,** or *lymphatic nodule,* the lymphocytes are densely packed in an area of loose connective tissue (Figure 22-6●). Lymphoid nodules are found in the connective tissue beneath the epithelia lining the respiratory, digestive, and urinary tracts. Typical nodules average around a millimeter in diameter, but the boundaries are not distinct because no fibrous capsule surrounds them. They often have a central zone, called a **germinal center,** which contains dividing lymphocytes (Figure 22-6a●).

GALT

Figure 22-6a

The extensive collection of lymphoid tissues associated with the digestive system is called the **gut-associated lymphoid tissue (GALT).** Clusters of lymphoid nodules beneath the epithelial lining of the intestine are known as **aggregate lymphoid nodules,** or *Peyer's patches* (Figure 22-6a●). In addition, the walls of the appendix, a blind pouch that originates near the junction between the small and large intestines, contain a mass of fused lymphoid nodules.

Tonsils

Figure 22-6b

Large nodules in the walls of the pharynx are called **tonsils** (Figure 22-6b●). There are usually five tonsils:

- A single **pharyngeal tonsil,** often called the *adenoids,* located in the posterior superior wall of the pharynx.
- A pair of **palatine tonsils,** located at the back of the oral cavity, along the boundary with the pharynx.
- A pair of **lingual tonsils,** which are usually not visible because they are located under the attached base of the tongue.

INFECTED LYMPHOID NODULES The lymphocytes in a lymphoid nodule are not always able to destroy bacterial or viral invaders that have crossed the adjacent epithelium. If pathogens become established in a lymphoid nodule, an infection develops. Two examples are probably familiar to you: *tonsillitis,* an infection of one of the tonsils (usually the pharyngeal or palatine), and *appendicitis,* an infection of the appendix that begins in the lymphoid nodules. Treatment usually consists of antibiotic therapy, sometimes combined with surgical removal of the infected tissues. AM *Infected Lymphoid Nodules*

❑ Lymphoid Organs

Lymphoid organs are separated from surrounding tissues by a fibrous connective tissue capsule. Important lymphoid organs include the *lymph nodes*, the *thymus*, and the *spleen*.

Lymph Nodes

Figures 22-1, 22-4, 22-7

Lymph nodes are small, oval lymphoid organs ranging in diameter from 1 to 25 mm (up to around 1 in.). The general pattern of lymph node distribution in the body can be seen in Figure 22-1●, p. 781. Each lymph node is covered by a capsule of dense connective tissue. Bundles of collagen fibers extend from the capsule into the interior of the node. These fibrous partitions are called **trabeculae** (*trabecula*, a wall).

The shape of a typical lymph node resembles that of a kidney bean. Blood vessels and nerves attach to the lymph node at the indentation, or **hilus** (Figure 22-7●). Two sets of lymphatic vessels are connected to each lymph node: *efferent lymphatics* and *afferent lymphatics*.

- **Efferent lymphatics** are attached to the lymph node at the hilus. These lymphatic vessels carry lymph away from the lymph node and toward the venous system.
- **Afferent lymphatics** bring lymph to the lymph node from peripheral tissues. The afferent lymphatics penetrate the capsule of the lymph node on the side opposite the hilus.

LYMPH FLOW Lymph arriving via the afferent lymphatics flows through the lymph node within a network of sinuses, open passageways with incomplete walls. Lymph first enters a *subcapsular sinus* that contains a meshwork of branching reticular fibers, macrophages, and **dendritic cells.** Dendritic cells are involved in the initiation of the immune response, and their role will be considered in a later section. After passing through the subcapsular sinus, lymph flows through the **outer cortex** of the node. The outer cortex contains B cells within germinal centers resembling those of lymphoid nodules.

Lymph then continues through lymph sinuses in the **deep cortex** (*paracortical area*). Lymphocytes leave the circulation and enter the lymph node by crossing the walls of blood vessels within the deep cortex. The deep cortical area is dominated by T cells.

After flowing through the sinuses of the deep cortex, lymph continues into the core, or **medulla,** of the lymph node. The medulla contains B cells and plasma cells organized into elongate masses known as **medullary cords.** Lymph enters the efferent lymphatics at the hilus after passing through a network of sinuses in the medulla.

LYMPH NODE FUNCTION A lymph node functions like a kitchen water filter: It filters and purifies lymph before it reaches the venous system. As lymph flows through a lymph node, at least 99 percent of the antigens present in the arriving lymph will be removed. Fixed macrophages in the walls of the lymphatic sinuses engulf debris or pathogens in the lymph as it flows past. Antigens removed in

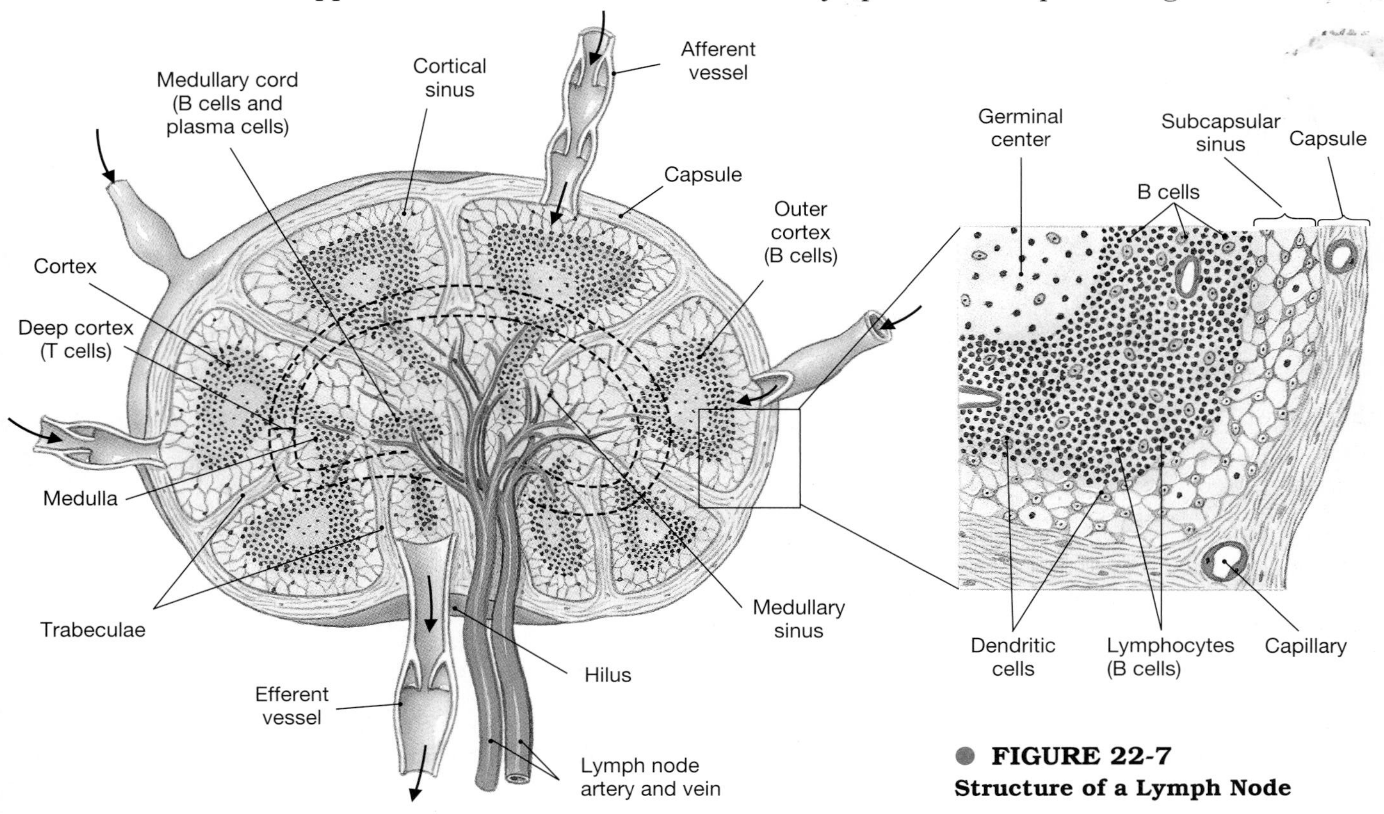
Medullary cord
(B cells and plasma cells)
Cortical sinus
Afferent vessel
Capsule
Outer cortex
(B cells)
Cortex
Deep cortex
(T cells)
Medulla
Trabeculae
Efferent vessel
Hilus
Lymph node artery and vein
Medullary sinus
Germinal center
Subcapsular sinus
Capsule
B cells
Dendritic cells
Lymphocytes
(B cells)
Capillary

● **FIGURE 22-7**
Structure of a Lymph Node

this way are then processed by the macrophages and "presented" to nearby lymphocytes. Other antigens stick to the surfaces of dendritic cells, where they can stimulate lymphocyte activity. This process, called *antigen presentation,* is usually the first step in the activation of the immune response.

In addition to serving as a filter, lymph nodes provide an early-warning system. Any infection or other abnormality in a peripheral tissue will introduce abnormal antigens into the interstitial fluid, and thus into the lymph leaving the area. These antigens will then stimulate macrophages and lymphocytes in nearby lymph nodes.

If we wanted to protect a house against intrusion we might guard all entrances and exits or place traps by the windows and doors. The distribution of lymphatic tissues and lymph nodes follows this pattern. The largest lymph nodes are found where peripheral lymphatics connect with the trunk, in regions such as the groin, the axillae, and the base of the neck. These nodes are often called *lymph glands.* Because lymph is monitored in the cervical, inguinal, or axillary lymph nodes, potential problems can be detected and dealt with before they affect the vital organs of the trunk. Aggregations of lymph nodes also exist within the mesenteries of the gut, near the trachea and passageways leading to the lungs, and in association with the thoracic duct (Figure 22-4•, p. 784). These lymph nodes provide protection against pathogens and other antigens within the digestive and respiratory systems.

A minor injury commonly produces a slight enlargement of the nodes along the lymphatics draining the region. This symptom, often called "swollen glands," typically indicates inflammation or infection of peripheral structures. The enlargement usually results from an increase in the number of lymphocytes and phagocytes in the node in response to a minor, localized infection. Chronic or excessive enlargement of lymph nodes constitutes **lymphadenopathy** (lim-fad-e-NOP-a-thē). This condition may occur in response to bacterial or viral infections, endocrine disorders, or cancer.

CANCER AND THE LYMPHATIC SYSTEM Lymphatics are found in all portions of the body except the central nervous system, and the lymphatic capillaries offer little resistance to the passage of cancer cells. As a result, metastasizing cancer cells often spread along the lymphatics. Under these circumstances the lymph nodes serve as way stations for migrating cancer cells. Thus an analysis of lymph nodes can provide information on the spread of the cancer cells, and such information has a direct influence on the selection of appropriate therapies. One example, the staging of breast cancer by the degree of nodal involvement, will be detailed in Chapter 29. *Lymphomas,* one group of cancers originating within the lymphatic system, are discussed in the *Applications Manual.* AM *Lymphomas*

The Thymus

Figure 22-8

The thymus lies behind the sternum, in the anterior portion of the mediastinum. It has a pinkish coloration and a grainy consistency. The thymus reaches its greatest size (relative to body size) in the first year or two after birth and its maximum absolute size during puberty, when it weighs between 30 and 40 g. Thereafter the thymus gradually decreases in size and becomes increasingly fibrous, a process called *involution.*

The capsule that covers the thymus divides it into two **thymic lobes** (Figure 22-8a,b•). Fibrous partitions, or **septae,** from the capsule divide the lobes into **lobules** averaging 2 mm in width (Figure 22-8b,c•). Each lobule consists of a densely packed outer **cortex** and a paler central **medulla.** Lymphocytes in the cortex are dividing, and as the T cells mature they migrate into the medulla. Eventually these T cells leave the thymus by entering one of the medullary blood vessels.

THE CORTEX Lymphocytes in the cortex are arranged in clusters that are completely surrounded by **reticular epithelial cells.** These cells, which developed from epithelial cells of the embryo, also encircle the blood vessels of the cortex. The reticular epithelial cells (1) maintain the blood-thymus barrier and (2) secrete the thymic hormones (thymosins) that stimulate stem cell divisions and T cell differentiation.

THE MEDULLA As they mature, T cells leave the cortex and enter the medulla of the thymus. There is no blood-thymus barrier in the medulla. The reticular epithelial cells in the medulla cluster together in concentric layers, forming distinctive structures known as **Hassall's corpuscles** (Figure 22-8d•). Despite their imposing appearance, the function of Hassall's corpuscles remains unknown. T cells within the medulla can enter or leave the circulation via the blood vessels in this region or within one of the efferent lymphatics that collect lymph from the thymus.

The Spleen

Figure 22-9

The adult spleen contains the largest collection of lymphoid tissue in the body. It is around 12 cm (5 in.) long and weighs on average nearly 160 g. The spleen lies along the curving lateral border of the stomach, extending between the ninth and eleventh ribs on the left side. It is attached to the lateral border of the stomach by a broad mesenterial band, the **gastrosplenic ligament** (Figure 22-9a•).

FUNCTIONS OF THE SPLEEN On gross dissection the spleen has a deep red color due to the blood it contains. In essence, the spleen performs the same

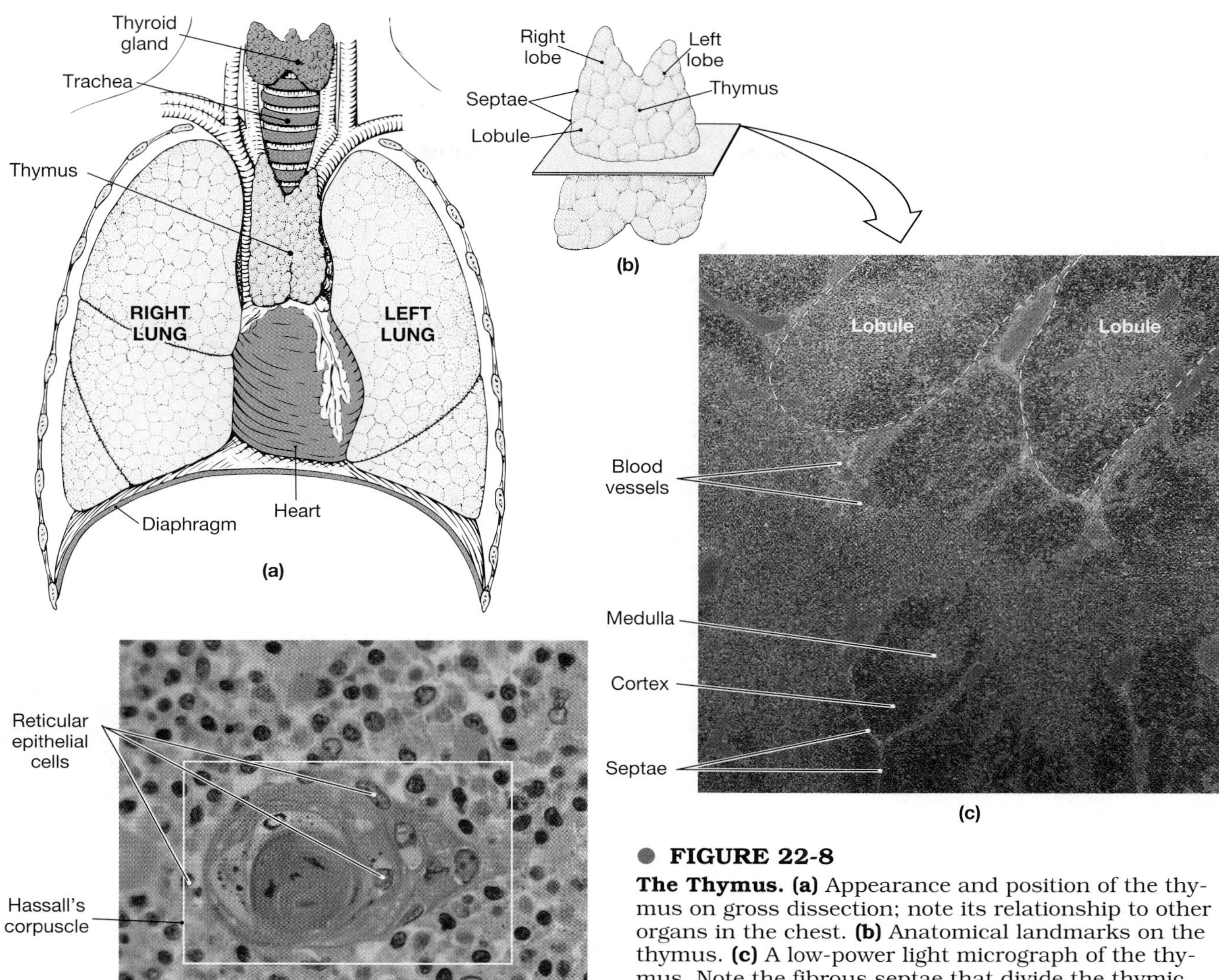
Thyroid gland
Trachea
Thymus
RIGHT LUNG
LEFT LUNG
Diaphragm
Heart
(a)
Right lobe
Left lobe
Thymus
Septae
Lobule
(b)
Lobule
Lobule
Blood vessels
Medulla
Cortex
Septae
(c)
Reticular epithelial cells
Hassall's corpuscle
T cells
(d)

● **FIGURE 22-8**

The Thymus. (a) Appearance and position of the thymus on gross dissection; note its relationship to other organs in the chest. **(b)** Anatomical landmarks on the thymus. **(c)** A low-power light micrograph of the thymus. Note the fibrous septae that divide the thymic tissue into lobes resembling interconnected lymphatic nodules. (LM × 40) **(d)** At higher magnification the unusual structure of Hassall's corpuscles can be examined. The small cells in view are lymphocytes in various stages of development. (LM × 532)

functions for the blood that lymph nodes perform for lymph. Splenic functions can be summarized as (1) the removal of abnormal blood cells and other blood components through phagocytosis, (2) the storage of iron from recycled red blood cells, and (3) the initiation of immune responses by B cells and T cells in response to antigens in the circulating blood.

SURFACES OF THE SPLEEN The spleen has a soft consistency, and its shape primarily reflects its association with the structures around it. It is in contact with the stomach, the left kidney, and the muscular diaphragm. The *diaphragmatic surface* is smooth and convex, conforming to the shape of the diaphragm and body wall. The *visceral surface* contains indentations that record the shape of the stomach (the *gastric area*) and of the kidney (the *renal area*) (Figure 22-9b●). Splenic blood vessels and lymphatics communicate with the spleen on the visceral surface at the **hilus,** a groove marking the border between the gastric and renal depressions. The **splenic artery, splenic vein,** and the lymphatics draining the spleen are attached at the hilus.

HISTOLOGY OF THE SPLEEN The spleen is surrounded by a capsule containing collagen and elastic fibers.[1] The cellular components within constitute the **pulp** of the spleen (Figure 22-9c●). Areas of **red pulp** contain large quantities of red blood cells, whereas areas of **white pulp** resemble lymphoid nodules.

[1]The spleens of dogs and cats have extensive layers of smooth muscle that can contract to eject blood into the circulation; the human spleen lacks those muscle layers and cannot contract.

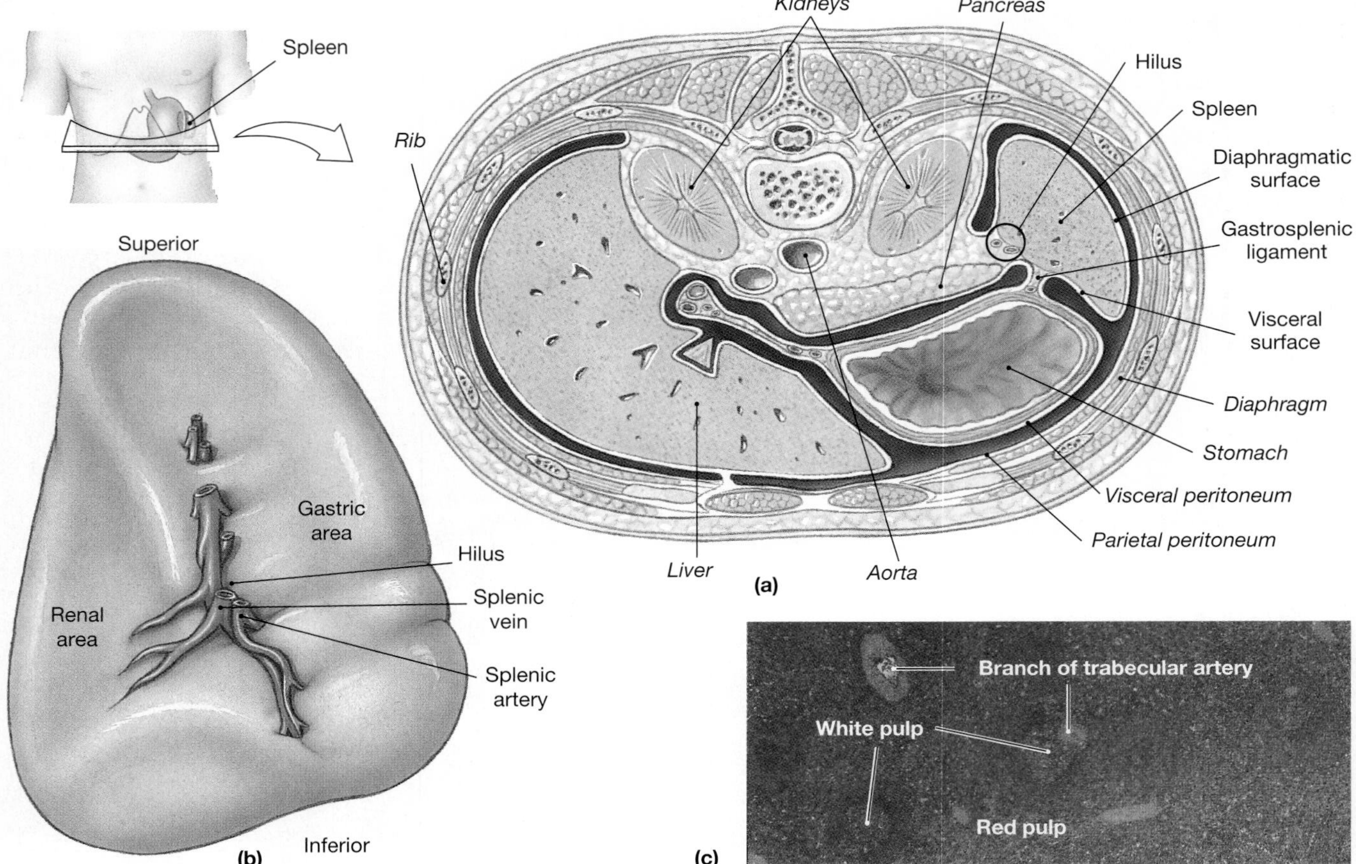
Spleen
Kidneys
Pancreas
Hilus
Spleen
Rib
Diaphragmatic surface
Gastrosplenic ligament
Visceral surface
Diaphragm
Stomach
Visceral peritoneum
Parietal peritoneum
Liver
(a)
Aorta
Superior
Gastric area
Hilus
Splenic vein
Renal area
Splenic artery
Inferior
(b)
(c)
Branch of trabecular artery
White pulp
Red pulp

● **FIGURE 22-9**

The Spleen. (a) The shape of the spleen roughly conforms to those of adjacent organs. This transverse section through the trunk shows the typical position of the spleen within the abdominopelvic cavity. **(b)** External appearance of the intact spleen, showing major anatomical landmarks. This view should be compared with that of part (a). **(c)** Histological appearance of the spleen. Areas of white pulp are dominated by lymphocytes; they appear blue because the nuclei of lymphocytes stain very darkly. Areas of red pulp contain a preponderance of red blood cells. (LM × 38). AM *Plate 6.5*

The splenic artery enters at the hilus and branches to produce a number of arteries that radiate outward toward the capsule. These **trabecular arteries** branch extensively, and their finer branches are surrounded by areas of white pulp. Capillaries then discharge the blood into the red pulp.

The cell population of the red pulp includes all of the normal components of the circulating blood, plus fixed and free macrophages. The structural framework of the red pulp consists of a network of reticular fibers. The blood passes through this meshwork and enters large sinusoids, also lined by fixed macrophages. The sinusoids empty into small veins, and these ultimately collect into **trabecular veins** that continue toward the hilus.

This circulatory arrangement gives the phagocytes of the spleen an opportunity to identify and engulf any damaged or infected cells in the circulating blood. Lymphocytes are scattered throughout the red pulp, and the *marginal zone* surrounding each area of white pulp has a high concentration of macrophages and dendritic cells. Thus any microorganism or other antigen in the blood will quickly come to the attention of the splenic lymphocytes.

INJURY TO THE SPLEEN An impact to the left side of the abdomen can distort or damage the spleen. Such injuries are known risks of contact sports, such as football and hockey, and more solitary athletic activities, such as skiing and sledding. However, the spleen tears so easily that a seemingly minor blow to the side may rupture the capsule. The result is serious internal bleeding and eventual circulatory shock.

Because the spleen is relatively fragile, it is very difficult to repair surgically. (Sutures usually tear out before they have been tensed enough to stop the bleeding.) Treatment for a severely ruptured spleen involves its complete removal, a process called a **splenectomy** (sple-NEK-tō-mē).

The spleen may also be damaged through infection, inflammation, or invasion by cancer cells. These conditions and related symptoms are considered in the *Applications Manual.* AM *Disorders of the Spleen*

 How would blockage of the thoracic duct affect the circulation of lymph?

 If the thymus gland failed to produce thymosins, what particular population of lymphocytes would be affected?

 Why do lymph nodes enlarge during some infections?

❑ The Lymphatic System and Body Defenses

The human body has multiple defense mechanisms, but these can be sorted into two general categories:

1. **Nonspecific defenses** do not discriminate between one threat and another. These defenses, which are present at birth, include *physical barriers, phagocytic cells, immunological surveillance, interferon, complement, inflammation,* and *fever.* They provide the body with a defensive capability known as *nonspecific resistance.*
2. **Specific defenses** provide protection against threats on an individual basis. For example, a specific defense may protect against infection by one type of bacteria but ignore other bacteria and viruses. Many specific defenses develop after birth, as a result of accidental or deliberate exposure to environmental hazards. *Specific defenses are dependent upon the activities of lymphocytes.* The body's specific defenses produce a state of protection known as **immunity** or **specific resistance.**

Nonspecific and specific resistance are complementary, and both must function normally to provide adequate resistance to infection and disease.

Nonspecific Defenses

Figure 22-10

Nonspecific defenses prevent the approach of, deny entrance to, or limit the spread of microorganisms or other environmental hazards. The major categories of nonspecific defenses are summarized in Figure 22-10●.

- ***Physical barriers*** keep hazardous organisms and materials outside the body. For example, a mosquito that lands on a full head of hair may be unable to reach the surface of the scalp.
- ***Phagocytes*** are cells that engulf pathogens and cell debris. Examples of phagocytes are the macrophages of peripheral tissues and the microphages of the blood.
- ***Immunological surveillance*** is the destruction of abnormal cells by NK cells in peripheral tissues.
- ***Interferons*** are chemical messengers that coordinate the defenses against viral infection.
- ***Complement*** is a system of circulating proteins that assist antibodies in the destruction of pathogens.
- ***Inflammation*** is a local response to injury or infection that is directed at the tissue level. Inflammation tends to restrict the spread of an injury, as well as combat an infection.
- ***Fever*** is an elevation in body temperature that accelerates tissue metabolism and defenses.

❑ Physical Barriers

To cause trouble, an antigenic compound or pathogen must enter the body tissues, and that means crossing an epithelium. The epithelial covering of the skin, described in Chapter 5, has multiple layers, a keratin coating, and a network of desmosomes that lock adjacent cells together. [p. 151] These barriers provide very effective protection for underlying tissues. Even along the more delicate internal passageways of the respiratory, digestive, and urinary tracts, the epithelial cells are tied together by tight junctions and usually are supported by a dense and fibrous basement membrane.

In addition to the barriers posed by the epithelial cells themselves, most epithelia are protected by specialized accessory structures and secretions. The hairs found in most areas of the body surface provide some protection against mechanical abrasion (especially on the scalp), and they often prevent hazardous materials or insects from contacting the skin's surface. The epidermal surface also receives the secretions of sebaceous and sweat glands. These secretions flush the surface, washing away microorganisms and chemical agents. The secretions also contain bactericidal chemicals, destructive enzymes (*lysozymes*), and antibodies.

The epithelia lining the digestive, respiratory, urinary, and reproductive tracts are more delicate, but they are equally well defended. Mucus bathes most surfaces of the digestive tract, and the stomach contains a powerful acid that can destroy many potential pathogens. Mucus moves across the lining of the respiratory tract, urine flushes the urinary passageways, and glandular secretions do the same for the reproductive tract. Special enzymes, antibodies, and an acidic pH may add to the effectiveness of these secretions.

❑ Phagocytes

Phagocytes perform janitorial and police services in peripheral tissues, removing cellular debris and responding to invasion by foreign compounds or

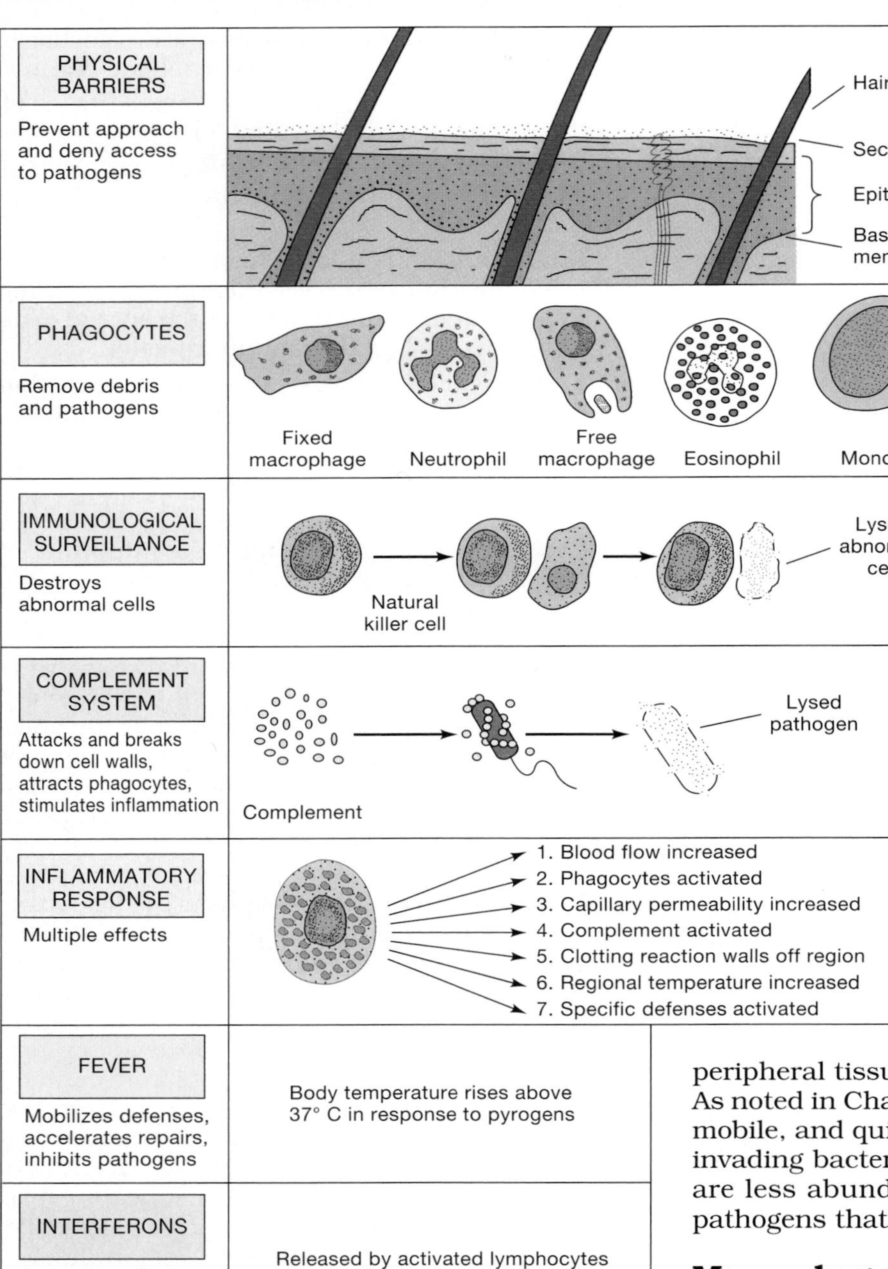
PHYSICAL BARRIERS
Prevent approach and deny access to pathogens
Hair
Secretions
Epithelium
Basement membrane
PHAGOCYTES
Remove debris and pathogens
Fixed macrophage
Neutrophil
Free macrophage
Eosinophil
Monocyte
IMMUNOLOGICAL SURVEILLANCE
Destroys abnormal cells
Natural killer cell
Lysed abnormal cell
COMPLEMENT SYSTEM
Attacks and breaks down cell walls, attracts phagocytes, stimulates inflammation
Complement
Lysed pathogen
INFLAMMATORY RESPONSE
Multiple effects
1. Blood flow increased
2. Phagocytes activated
3. Capillary permeability increased
4. Complement activated
5. Clotting reaction walls off region
6. Regional temperature increased
7. Specific defenses activated
FEVER
Mobilizes defenses, accelerates repairs, inhibits pathogens
Body temperature rises above 37° C in response to pyrogens
INTERFERONS
Increase resistance of cells to infection, slow the spread of disease
Released by activated lymphocytes and macrophages and by virally infected cells

FIGURE 22-10
Nonspecific Defenses

pathogenic organisms. These cells represent the "first line" of cellular defense against pathogenic invasion. Phagocytes often attack and remove the microorganisms before lymphocytes become aware of the incident. Two general classes of phagocytic cells are found in the human body: **microphages** and **macrophages.**

Microphages

Microphages are the neutrophils and eosinophils normally found in the circulating blood. These phagocytic cells leave the bloodstream and enter peripheral tissues subjected to injury or infection. As noted in Chapter 19, neutrophils are abundant, mobile, and quick to phagocytize cellular debris or invading bacteria. [p. 665] Eosinophils, which are less abundant, target foreign compounds or pathogens that have been coated with antibodies.

Macrophages

Macrophages are large, actively phagocytic cells. The body contains several different types of macrophages, and most are derived from the monocytes of the circulating blood. Although there are no purely phagocytic organs or tissues, almost every tissue in the body shelters resident or visiting macrophages. This relatively diffuse collection of phagocytic cells has been called the **monocyte-macrophage system.**

An activated macrophage may respond to a pathogen in several different ways. For example:

1. It may engulf a pathogen or other foreign object and destroy it using lysosomal enzymes.
2. It may bind or remove a pathogen from the interstitial fluid but be unable to destroy it until assisted by other cells.

3. It may destroy its target by releasing toxic chemicals, such as *tumor necrosis factor*, nitric oxide, or hydrogen peroxide, into the interstitial fluid.

We will consider these responses in more detail in a later section.

FIXED MACROPHAGES Fixed macrophages are permanent residents of specific tissues and organs. These cells are normally incapable of movement, so the objects of their phagocytic attention must diffuse or otherwise move through the surrounding tissue until they are within range. Fixed macrophages are scattered among connective tissues, usually in close association with collagen or reticular fibers. Their presence has been noted in the papillary and reticular layers of the dermis, in the pia-arachnoid layer of the meninges, and in bone marrow. In some organs the fixed macrophages have special names: **microglia** are macrophages inside the CNS, and **Kupffer cells** are found in and around the liver sinusoids.

FREE MACROPHAGES Free macrophages, or *mobile macrophages*, travel throughout the body, arriving at an injury site by migration through adjacent tissues or by recruitment from the circulating blood. In the skeletal system, the fusion of macrophages produces the osteoclasts that dissolve and recycle the mineral content of bone. Some tissues contain free macrophages with distinctive characteristics. For example, the exchange surfaces of the lungs are patrolled by **alveolar macrophages,** also known as *phagocytic dust cells.*

Both classes of macrophages are derived from the monocytes of the blood, and the primary difference is that fixed macrophages are always found within a given tissue, but the free macrophages come and go. During an infection this distinction often breaks down completely, for the fixed macrophages may lose their attachments and begin roaming around the damaged tissue.

Orientation and Phagocytosis

Mobile macrophages and microphages share a number of functional characteristics:

1. They can all move through capillary walls by squeezing between adjacent endothelial cells, a process known as **diapedesis.** The endothelial cells in an injured area develop membrane "markers" that let passing blood cells know that something is wrong. The cells then attach to the endothelial lining and migrate into the surrounding tissues.
2. They may be attracted or repelled by chemicals in the surrounding fluids, a phenomenon called **chemotaxis.** They are particularly sensitive to cytokines released by other body cells or by the chemicals released by pathogens.
3. Phagocytosis begins with **adhesion** of the phagocyte to its target. Adhesion depends on binding of the surface of the target to receptors on the cell membrane of the phagocyte. Adhesion is followed by the formation of a vesicle containing the bound target, as detailed in Figure 3-12•, p. 80. The contents of the vesicle are then digested when the vesicle fuses with lysosomes or peroxisomes.

All phagocytic cells function in much the same way, although the items selected for phagocytosis may differ from one cell type to another. The life span of an actively phagocytic cell can be rather brief; a neutrophil usually expires before it has engulfed more than 25 bacteria, and in an infection it may attack that many in an hour.

❑ Immunological Surveillance

The immune system generally ignores the body's own cells unless they become abnormal in some way. NK (natural killer) cells are responsible for recognizing and destroying abnormal cells when they appear in peripheral tissues. The constant monitoring of normal tissues by NK cells is called **immunological surveillance.**

The cell membrane of an abnormal cell usually contains antigens not found on the membranes of normal cells. NK cells recognize an abnormal cell by detecting the presence of these antigens. The recognition mechanism differs from that used by T cells or B cells. A T cell or B cell can be activated only by exposure to a *specific* antigen at a *specific* site on a cell membrane. An NK cell responds to a variety of abnormal antigens that may appear anywhere on a cell membrane. NK cells are therefore much less selective about their targets than other lymphocytes—if a membrane contains abnormal antigens, it will be attacked. As a result, NK cells are very versatile: A single NK cell can attack bacteria in the interstitial fluid, body cells infected with virus, or cancer cells.

NK cells also respond much more rapidly than T cells or B cells. The activation of T cells and B cells involves a relatively complex and time-consuming sequence of events. NK cells respond immediately upon contact with an abnormal cell.

NK Cell Activation

Figure 22-11

An activated NK cell reacts in a predictable way:

1. The Golgi apparatus moves around the nucleus until the maturing face points directly toward the abnormal cell. This process might be compared to the rotation of a tank turret to point the cannon toward the enemy.

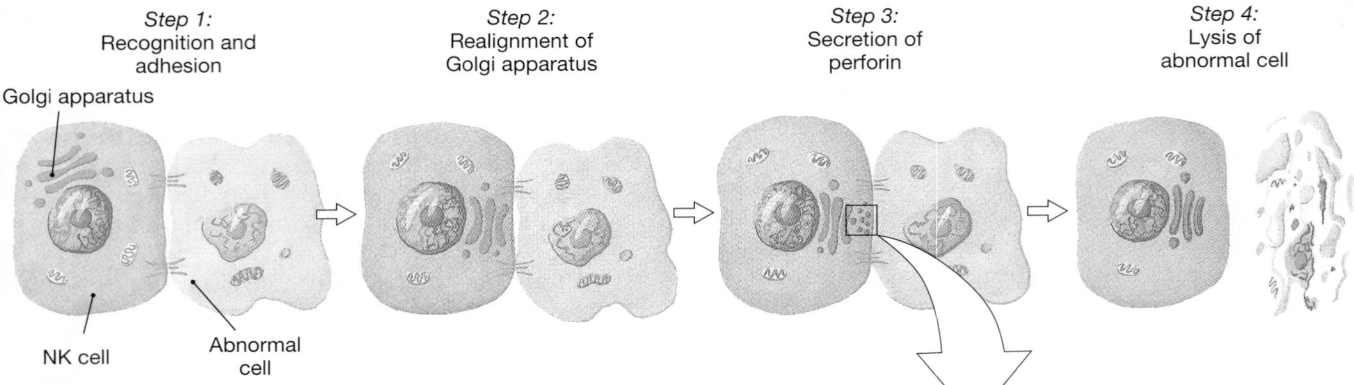

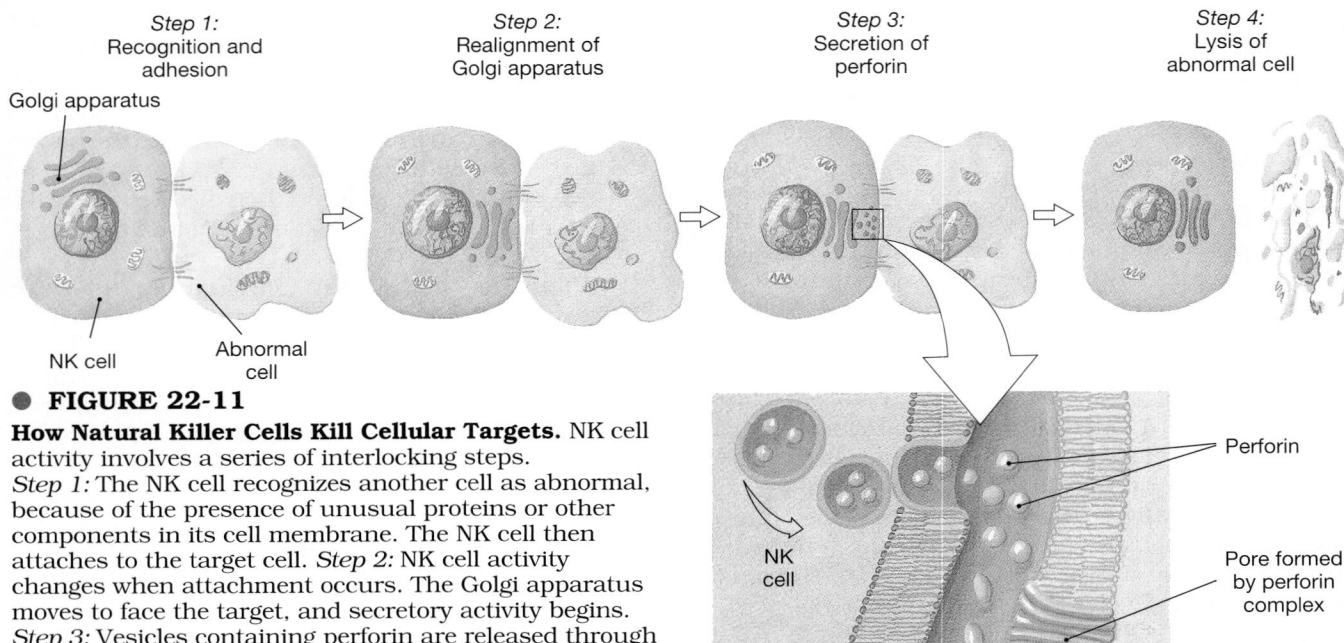

Step 1: Recognition and adhesion
Step 2: Realignment of Golgi apparatus
Step 3: Secretion of perforin
Step 4: Lysis of abnormal cell
Golgi apparatus
NK cell
Abnormal cell
NK cell
Perforin
Pore formed by perforin complex
Abnormal cell

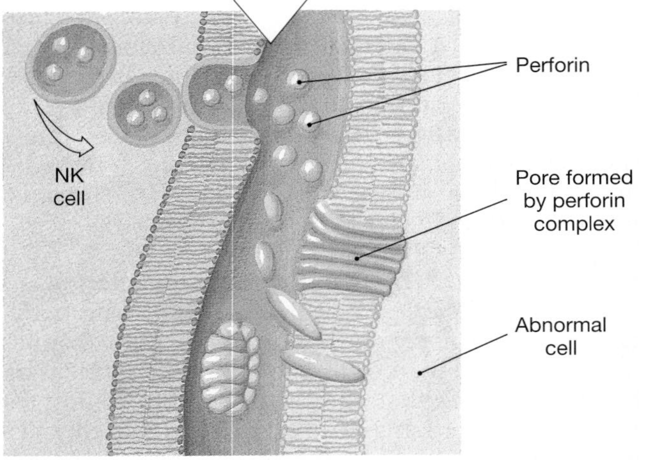

● **FIGURE 22-11**
How Natural Killer Cells Kill Cellular Targets. NK cell activity involves a series of interlocking steps. *Step 1:* The NK cell recognizes another cell as abnormal, because of the presence of unusual proteins or other components in its cell membrane. The NK cell then attaches to the target cell. *Step 2:* NK cell activity changes when attachment occurs. The Golgi apparatus moves to face the target, and secretory activity begins. *Step 3:* Vesicles containing perforin are released through exocytosis. *Step 4:* Perforin lyses the target cell by creating large pores through the cell membrane.

2. A flood of secretory vesicles is then produced at the Golgi apparatus. These vesicles travel through the cytoplasm, to be released at the cell surface through exocytosis. The secretory vesicles contain proteins called **perforins.**
3. Perforins then diffuse across the gap between the NK cell and its target.
4. On reaching the opposing cell membrane, perforin molecules interact with one another and with the membrane. The result, shown in Figure 22-11●, is the creation of a network of pores in the cell membrane.

The pores created by perforin are large enough to permit the free passage of ions, proteins, and other intracellular materials. The target cell can no longer maintain its internal environment, and it quickly disintegrates. It is not clear why perforin does not affect the membrane of the NK cell itself. It has been proposed that the NK cell membranes contain a second protein, called *protectin,* that binds and inactivates perforin.

NK CELLS AND CANCER NK cells attack cancer cells and cells infected with viruses. Cancer cells probably appear throughout life, but their cell membranes usually contain unusual proteins, called **tumor-specific antigens,** that NK cells recognize as abnormal. The affected cells are then destroyed, preserving tissue integrity. Unfortunately, some cancer cells avoid detection, perhaps because they lack tumor-specific antigens or because these antigens are covered in some way. This process of avoiding detection is called **immunological escape.** Once immunological escape has occurred, cancer cells can multiply and spread without interference by NK cells.

NK CELLS AND VIRAL INFECTIONS NK cells are also important in fighting viral infections. Viruses reproduce inside living cells, beyond the reach of circulating antibodies. However, infected cells incorporate viral antigens into their cell membranes, NK , because cells recognize these infected cells as abnormal. By destroying them, NK cells can slow or prevent the spread of a viral infection.

❑ Interferons

Interferons (in-ter-FĒR-onz) are small proteins released by activated lymphocytes and macrophages and by tissue cells infected with viruses. When an interferon reaches the membrane of a normal cell, it binds to surface receptors and, via second messengers, triggers the production of **antiviral proteins** in the cytoplasm. Antiviral proteins do not interfere with the entry of viruses, but they interfere with viral replication inside the cell. In addition to their role in slowing the spread of viral infections, interferons stimulate the activities of macrophages and NK cells.

There are at least three different interferons, each of which has additional specialized functions:

- *Alpha (α)-interferon,* produced by several different leukocytes, attracts and stimulates NK cells.

- *Beta (β)-interferon,* secreted by fibroblasts, slows inflammation in a damaged area.
- *Gamma (γ)-interferon,* secreted by T cells and NK cells, stimulates macrophage activity.

Most cell types other than lymphocytes and macrophages will respond to viral infection by secreting beta-interferon.

Interferons are examples of **cytokines** (SĪ-tō-kīnz), chemical messengers released by tissue cells to coordinate local activities. Cytokines are the hormones of the immune system; they are released to alter the activities of cells and tissues throughout the body. Their role in the regulation of specific defenses will be discussed below.

❑ Complement

Figure 22-12

The plasma contains 11 special **complement proteins** that form the **complement system.** The term *complement* refers to the fact that this system "complements," or supplements, the action of antibodies.

The complement proteins interact with one another in chain reactions comparable to those of the clotting system. Figure 22-12● provides an overview of the complement system. For readers interested in further details, Appendix V contains a detailed breakdown of the reactions involved in the complement system.

Known effects of complement activation include:

1. ***Destruction of target cell membranes.*** Five of the interacting complement proteins bind to the cell membrane. They form a group known as the **membrane attack complex** (MAC), which creates a pore in the membrane. The pores formed in this way are comparable to those produced by perforin and have the same effect; the target cell is soon destroyed.
2. ***Stimulation of inflammation.*** Activated complement proteins enhance the release of histamine by mast cells and basophils. These substances increase the degree of local inflammation and accelerate circulation to the region.
3. ***Attraction of phagocytes.*** Activated complement proteins attract neutrophils and macrophages to the area, improving the chances that phagocytic cells will be able to cope with the injury or infection.
4. ***Enhancement of phagocytosis.*** A coating of complement proteins and antibodies both attracts phagocytes and makes the target easier to engulf. Macrophage membranes contain receptors that can detect and bind to complement proteins and bound antibodies. After binding, the pathogens are easily engulfed. Because the antibodies involved are called **opsonins,** this effect is called **opsonization.**

Complement Activation

Figure 22-12

There are two routes of complement activation, the *classic pathway* and the *alternative pathway* (Figure 22-12●).

THE CLASSIC PATHWAY The most rapid and effective activation of the complement system occurs via the **classic pathway** (Figure 22-12●). The process begins when one of the complement proteins (C1) binds to an antibody molecule already attached to its specific antigen. The bound complement protein then acts as an enzyme, catalyzing a series of reactions involving other complement proteins. The classic pathway ends with the conversion of an inactive complement protein, **C3,** to an active form, **C3b.** C3b binds the surface of the antigen and stimulates phagocytosis. It also triggers additional reactions that end with the creation of the membrane attack complex.

THE ALTERNATIVE PATHWAY A less effective, slower activation of the complement system occurs in the absence of antibody molecules. This **alternative pathway,** or *properdin pathway,* is important in the defense against bacteria, some parasites, and virally infected cells. The pathway begins when several complement proteins, including **properdin,** or *factor P,* interact in the plasma (Figure 22-12●). This interaction can be triggered by exposure to foreign materials, such as the capsule of a bacterial cell. Like the classic pathway, the alternative pathway ends with the conversion of C3 to C3b, the stimulation of phagocytosis, and the subsequent formation of the membrane attack complex.

Some bacteria are completely unaffected by complement, because the complement proteins cannot interact with their cell membranes. These bacteria coat themselves in a protective capsule of carbohydrates. However, the classic pathway still provides a defense against these bacteria, once antibodies have bound to the capsule.

❑ Inflammation

Figure 22-13

Inflammation, a localized tissue response to injury introduced in Chapter 4, produces local sensations of swelling, redness, heat, and pain. ∞ [p. 139] Many stimuli can produce inflammation, including impact, abrasion, distortion, chemical irritation, infection by pathogenic organisms (such as bacteria or viruses), or extreme temperatures (hot or cold). The common factor is that each of these stimuli kills cells, damages connective tissue fibers, or injures the tissue in some other way. These changes alter the chemical composition of the interstitial fluid. Damaged cells release prostaglandins, proteins, and potassium ions, and the injury itself may have introduced foreign proteins or pathogens. The changes in the interstitial environment trigger a complex process called *inflammation*, or the *inflammatory response*.

Figure 22-13● contains a summary of the events that occur during inflammation of the skin, although comparable events will occur in almost any tissue subjected to physical damage or infection. The goals of inflammation are:

1. To perform a temporary repair at the injury site and prevent the access of additional pathogens.
2. To slow the spread of pathogens away from the injury site.
3. To mobilize local, regional, and systemic defenses that can overcome the pathogens and facilitate permanent repairs. The repair process is called *regeneration*.

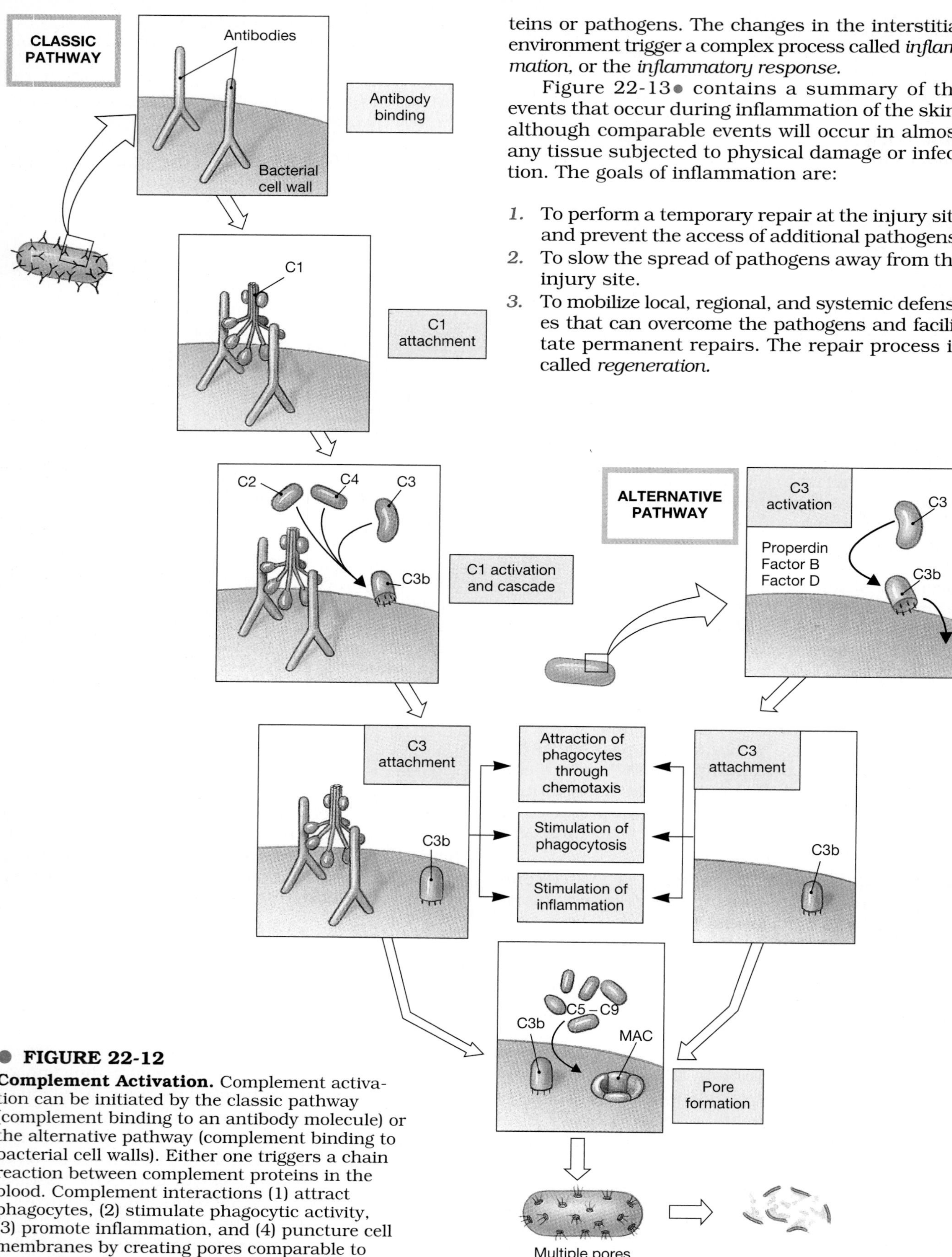
CLASSIC PATHWAY
Antibodies
Antibody binding
Bacterial cell wall
C1
C1 attachment
C2
C4
C3
C3b
C1 activation and cascade
ALTERNATIVE PATHWAY
C3 activation
C3
Properdin
Factor B
Factor D
C3b
C3 attachment
C3b
Attraction of phagocytes through chemotaxis
Stimulation of phagocytosis
Stimulation of inflammation
C3 attachment
C3b
C5 – C9
C3b
MAC
Pore formation
Multiple pores in bacterium
Cell lysis

● **FIGURE 22-12**
Complement Activation. Complement activation can be initiated by the classic pathway (complement binding to an antibody molecule) or the alternative pathway (complement binding to bacterial cell walls). Either one triggers a chain reaction between complement proteins in the blood. Complement interactions (1) attract phagocytes, (2) stimulate phagocytic activity, (3) promote inflammation, and (4) puncture cell membranes by creating pores comparable to those produced by perforin. D

The Response to Injury

Mast cells play a pivotal role in the inflammation process. When stimulated by mechanical stress or chemical changes in the local environment, these cells release histamine, heparin, prostaglandins, and other chemicals into the interstitial fluid. Events then proceed in a series of integrated steps:

- The released histamine increases capillary permeability and accelerates blood flow through the area. The increased circulation brings more cellular defenders to the site and carries away toxins and debris.
- Clotting factors and complement proteins leave the bloodstream and enter the injured or infected area. Clotting does not occur at the actual site of injury because of the presence of heparin. However, a clot soon forms around the damaged area, and this clot both isolates the region and slows the spread of the chemical or pathogen. Meanwhile, complement activation via the alternative pathway breaks down bacterial cell walls and attracts phagocytes.
- The increased blood flow elevates the local temperature, increasing the rate of enzymatic reactions and accelerating the activity of phagocytes. The rise in temperatures may also denature foreign proteins or vital enzymes of invading microorganisms.
- Debris and bacteria are attacked by neutrophils drawn to the area by chemotaxis. As they circulate through a blood vessel in an injured area, neutrophils undergo *activation.* In this process, (1) they stick to the side of the vessel and move into the tissue via diapedesis; (2) their metabolic rate goes up dramatically, and while this *respiratory burst* continues they generate reactive compounds, such as nitric oxide and hydrogen peroxide, that can destroy engulfed pathogens; and (3) they secrete cytokines that attract other neutrophils and macrophages to the area.
- Fixed and free macrophages are also actively engulfing pathogens and cell debris. At first these cells are outnumbered by neutrophils, but as the macrophages and neutrophils continue to secrete cytokines, the number of macrophages increases rapidly. Eosinophils may become involved if the foreign materials become coated with antibodies.
- Other cytokines released by active phagocytes stimulate fibroblasts to begin barricading the area with scar tissue, reinforcing the clot and further slowing the invasion of adjacent tissues.
- The combination of abnormal tissue conditions and chemicals released by mast cells stimulates local sensory neurons, producing sensations of pain. The individual is consciously aware of these sensations and may take steps to limit the damage, such as removing a splinter or cleaning a wound.
- As inflammation is under way, the foreign proteins, toxins, microorganisms, and active phagocytes in the area activate the body's specific defenses, via mechanisms described in a later section.

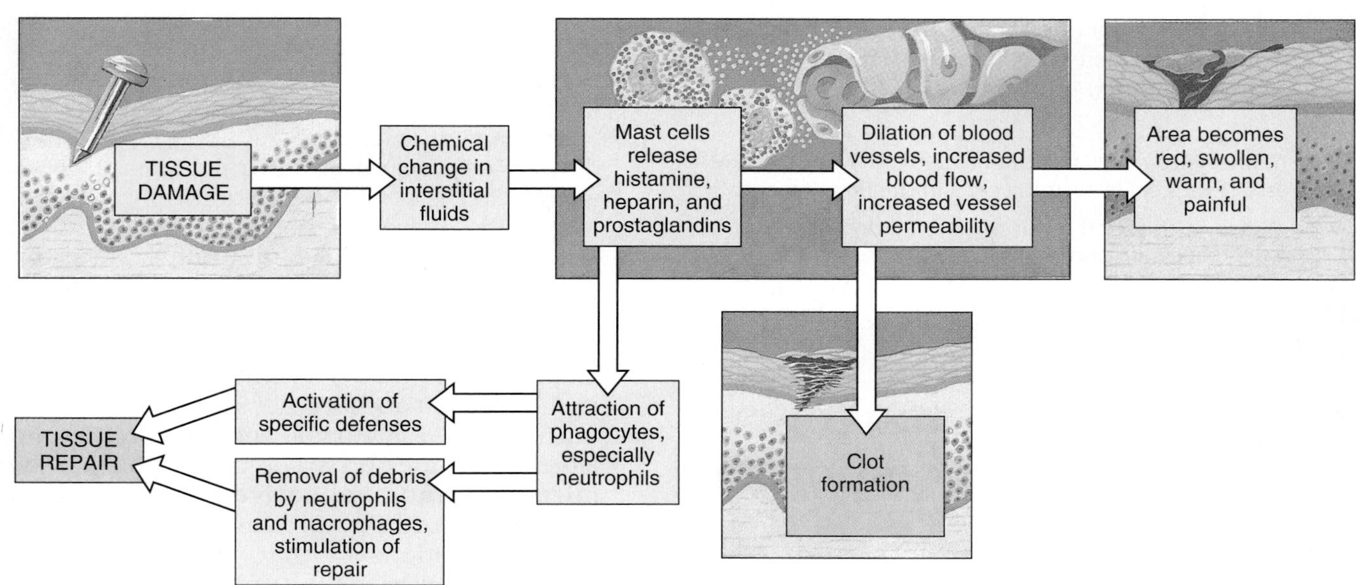
TISSUE DAMAGE
Chemical change in interstitial fluids
Mast cells release histamine, heparin, and prostaglandins
Dilation of blood vessels, increased blood flow, increased vessel permeability
Area becomes red, swollen, warm, and painful
Clot formation
Attraction of phagocytes, especially neutrophils
Activation of specific defenses
Removal of debris by neutrophils and macrophages, stimulation of repair
TISSUE REPAIR

● **FIGURE 22-13**
The Inflammation Process

NECROSIS In the period following an injury, tissue conditions usually become even more abnormal before they begin to improve. **Necrosis** (ne-KRŌ-sis) refers to the tissue degeneration that occurs after cells have been injured or destroyed. The process begins several hours after the initial event, and the damage is caused by lysosomal enzymes. Lysosomes, described in Chapter 3, break down through autolysis, releasing digestive enzymes that first destroy the injured cells and then attack surrounding tissues. [p. 91]

As local inflammation continues, debris, fluid, dead and dying cells, and necrotic tissue components accumulate at the injury site. This viscous fluid mixture is known as *pus*. An accumulation of pus in an enclosed tissue space is called an **abscess.** AM *Complications of Inflammation*

Fever

A **fever** is the maintenance of a body temperature greater than 37.2° C (99° F). Chapter 14 noted the presence of a temperature-regulating center in the preoptic area of the hypothalamus. [p. 479] Circulating proteins called **pyrogens** (PĪ-rō-jenz) can reset the thermostat and cause a rise in body temperature. A variety of stimuli, including pathogens, bacterial toxins, and antigen-antibody complexes, either act as pyrogens themselves or stimulate the release of pyrogens by macrophages. The pyrogen released by active macrophages is a cytokine called **endogenous pyrogen,** or **interleukin-1** (in-ter-LOO-kin), abbreviated **Il-1.**

Within limits, a fever may be beneficial. High body temperatures may inhibit some viruses and bacteria, but the most likely beneficial effect is on body metabolism. For each 1° C rise in temperature the metabolic rate jumps by 10 percent. Cells can move more quickly, and enzymatic reactions occur at a faster rate. The net results may be faster mobilization of tissue defenses and an accelerated repair process.

 What types of cells would be affected by a decrease in the monocyte-forming cells in the bone marrow?

 A rise in the level of interferon in the body would suggest what kind of infection?

 What effects do pyrogens have in the body?

Specific Resistance: The Immune Response

Specific resistance, or immunity, is provided by the coordinated activities of T cells and B cells, which respond to the presence of specific antigens. The basic functional relationship can be summarized as follows:

1. *T cells* are responsible for **cell-mediated immunity,** our defense against abnormal cells and pathogens inside living cells.
2. *B cells* provide **antibody-mediated immunity,** our defense against antigens and pathogenic organisms in body fluids.

Both mechanisms are important, because activated T cells do not respond to antigenic materials in solution, and the antibodies produced by activated B cells cannot cross cell membranes. Moreover, helper T cells play a crucial role in antibody-mediated immunity by stimulating the activity of B cells.

Our understanding of immunity has greatly improved in the past decade, and a comprehensive discussion would involve hundreds of pages and thousands of details. The discussion that follows emphasizes important patterns and introduces general principles that will provide a foundation for future courses in microbiology and immunology.

Forms of Immunity

Figure 22-14

Immunity may be either *innate* or *acquired* (Figure 22-14●).

- **Innate immunity** is genetically determined; it is present at birth and has no relationship to previous exposure to the antigen involved. For example, people are not subject to the same diseases as goldfish. Innate immunity breaks down only in the case of *AIDS* or other conditions that depress all aspects of specific resistance.
- **Active immunity** appears following exposure to an antigen, as a consequence of the immune response. The immune system has the *capability* of defending against a large number of antigens. However, the appropriate defenses are mobilized only after an individual encounters a particular antigen. Active immunity may develop as a result of natural exposure to an antigen in the environment (naturally acquired immunity) or from deliberate exposure to an antigen (induced active immunity). **Naturally acquired immunity** normally begins to develop after birth, and it is continually enhanced as the individual encounters "new" pathogens or other antigens. You might compare this process to vocabulary development—a child begins with a few basic common words and learns new ones on an as-needed basis. The purpose of **induced active immunity** is to stimulate antibody production under controlled conditions, so that the individual will be able to overcome natural exposure to the pathogen at some time in the future. This is the basic principle behind immunization to prevent disease.

FIGURE 22-14
Types of Immunity

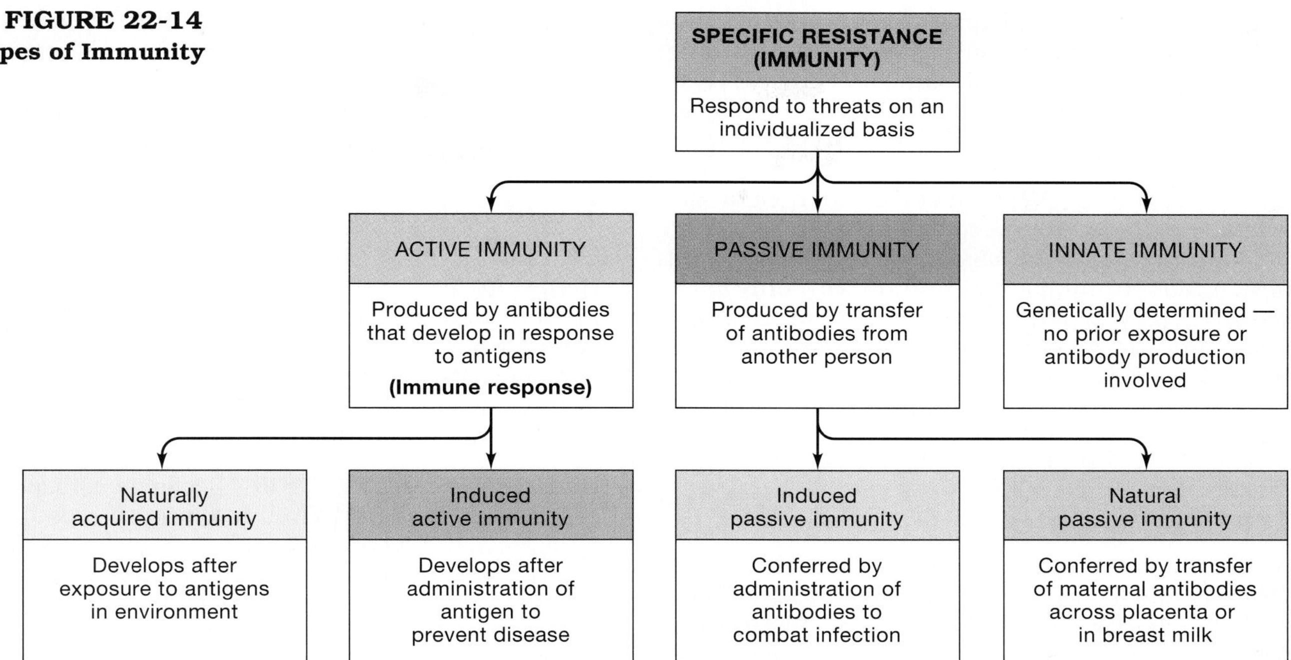
SPECIFIC RESISTANCE (IMMUNITY)
Respond to threats on an individualized basis
ACTIVE IMMUNITY
Produced by antibodies that develop in response to antigens
(Immune response)
PASSIVE IMMUNITY
Produced by transfer of antibodies from another person
INNATE IMMUNITY
Genetically determined — no prior exposure or antibody production involved
Naturally acquired immunity
Develops after exposure to antigens in environment
Induced active immunity
Develops after administration of antigen to prevent disease
Induced passive immunity
Conferred by administration of antibodies to combat infection
Natural passive immunity
Conferred by transfer of maternal antibodies across placenta or in breast milk

- **Passive immunity** is produced by the transfer of antibodies from another individual. **Natural passive immunity** results when antibodies produced by the mother provide protection against infections during development (across the placenta) or in early infancy (through breast milk). In **induced passive immunity** antibodies are administered to fight infection or prevent disease.

Properties of Immunity

We will begin our discussion of immunity by noting four general properties:

1. ***Specificity.*** A specific defense is activated by an antigen, and the response targets that particular antigen and not others. Specificity results from the activation of appropriate lymphocytes and the production of antibodies with targeted effects.
2. ***Versatility.*** In the course of a normal lifetime, an individual encounters tens of thousands of antigens. The immune system can differentiate among them, producing appropriate and specific responses to each. Versatility results in part from the large diversity of lymphocytes present in the body and in part from variability in the structure of synthesized antibodies.
3. ***Memory.*** The immune system "remembers" antigens that it encounters. As a result, the immune response that occurs following a second exposure to an antigen is stronger and lasts longer than the response to first exposure.
4. ***Tolerance.*** **Tolerance** is said to exist when the immune system does not respond to a particular antigen. For example, all cells and tissues in the body contain antigens that normally fail to stimulate an immune response.

Specificity

Specificity occurs because T cells and B cells respond to the molecular structure of an antigen. The antigen shape and size determine which lymphocytes will respond to its presence. Each T cell or B cell has receptors that will bind to one specific antigen, ignoring all others. The response of an activated T cell or B cell is equally specific. Either lymphocyte will destroy or inactivate that antigen without affecting other antigens or normal tissues.

Versatility

There are millions of different antigens in the environment that may pose a potential threat to health. In the lifetime of any single individual, only a relatively small fraction of those antigens will enter body tissues. Your immune system, however, has no way of anticipating which antigens it will actually encounter. It must be ready to confront *any* antigen at *any* time.

During development, differentiation of cells in the lymphatic system produces an enormous number of lymphocytes with varied antigen sensitivities. The trillion or more T cells and B cells in the human body include millions of different lymphocyte populations. Each population consists of several thousand cells with receptors in their membranes that differ from those of other lymphocyte populations. As a result, each group of lymphocytes will respond to the presence of a different antigen. Representatives of each population are distributed throughout the body.

A few thousand lymphocytes are not enough to overcome a pathogenic invasion. When activated in the presence of an appropriate antigen, however, a lymphocyte begins to divide, producing more lymphocytes with the same sensitivity. Thus whenever an antigen is encountered, more lymphocytes are produced. The term **clone** refers to all of the cells produced by the division of an activated lymphocyte. *All the members of a clone are sensitive to the same specific antigen.*

To understand the basic principle, you might think about running a snack shop with only samples on display. When a group comes in and makes a selection, the food is made on the spot. This is an efficient way to do business. You can offer a wide selection, because the samples don't take up much space, and you do not have to expend energy making food that won't be eaten.

Memory

During the initial response to an antigen, lymphocytes sensitive to its presence undergo repeated cycles of cell division. Two kinds of cells are produced: some that attack the invader and others that remain inactive unless they are exposed to the same antigen at a later date. These inactive *memory cells* enable the immune system to "remember" antigens that it has previously enountered and launch a faster, stronger counterattack if one of them ever appears again.

Tolerance

The immune response targets foreign cells and compounds, but it usually ignores normal tissues. During their differentiation in the bone marrow (B cells) and thymus (T cells), cells that react to antigens normally present in the body are destroyed. As a result, mature B and T cells will ignore normal, or *self,* antigens and attack foreign, or *nonself,* antigens. Tolerance can also develop over time in response to chronic exposure to an antigen in the environment. Such tolerance lasts only as long as the exposure continues.

We will now consider the origins and functions of T cells and B cells in greater detail.

❑ T Cells and Cell-Mediated Immunity

T cells play a key role in the initiation, maintenance, and control of the immune response. We have already introduced three major types of T cells:

1. ***Cytotoxic T cells*** (T_C) are responsible for cell-mediated immunity. These cells enter peripheral tissues and subject antigens to direct physical and chemical attack.
2. ***Helper T cells*** (T_H) stimulate the responses of both T cells and B cells. They are absolutely vital to the immune response because B cells must be activated by helper T cells before they will begin producing antibodies. The reduction in the helper T cell population that occurs in AIDS is largely responsible for the loss of immunity. (AIDS is discussed on p. 810.)
3. ***Suppressor T cells*** (T_S) inhibit T cell and B cell activities and moderate the immune response.

T Cell Activation

Figures 22-15, 22-16, 22-17

Before an immune response can begin, T cells must be activated by exposure to an antigen. This activation seldom happens by direct lymphocyte-antigen interaction, and foreign compounds or pathogens entering a tissue often fail to stimulate an immediate immune response.

ANTIGEN PRESENTATION T cells recognize antigens when they are bound to glycoproteins in cell membranes. Glycoproteins are integral membrane components introduced in Chapters 2, 3, and 19. ⚭ *[pp. 56, 71, 661]* The structure of these glycoproteins is genetically determined. The genes controlling their synthesis are found along one portion of chromosome 6, in a region called the **major histocompatibility complex (MHC).** These membrane glycoproteins are called **MHC proteins,** or *human leukocyte antigens (HLAs).*

The amino acid sequences and shapes of MHC proteins differ from individual to individual. Each MHC molecule has a distinct three-dimensional shape with a relatively narrow central groove. An antigen that fits into this groove can be held in position by hydrogen bonding.

Two major classes of MHC proteins are known: *Class I* and *Class II.*

MHC Class I. **Class I** MHC proteins are found in the membranes of all nucleated cells. These proteins are continually being synthesized and exported to the cell membrane in vesicles created at the Golgi apparatus. As they form, Class I proteins pick up small peptides from the surrounding cytoplasm and carry them to the cell surface. *Class I MHC proteins are always present in the cell membrane.* If the cell is healthy and the peptides are normal, T cells will ignore them. If the cytoplasm contains abnormal peptides, they will soon appear in the cell membrane, and T cells will be activated. Ultimately this will lead to the destruction of the abnormal cell.

Viruses entering body tissues enter their target cells and begin replicating. In this process, viral proteins appear in the cell membranes of infected cells (Figure 22-15a●), bound to Class I proteins. T cells activated by contact with these antigens play an important role in the defense against viral infection.

MHC Class II. **Class II** MHC proteins are found only in *antigen-presenting cells* and lymphocytes. **Antigen-presenting cells (APCs)** are specialized cells responsible for activating T cell defenses against foreign cells (including bacteria) and foreign proteins. Antigen-presenting cells include all of the phagocytic cells of the monocyte-macrophage group discussed in earlier chapters, including: (1) the free and fixed macrophages in connective tissues (Chapter 4); (2) the Kupffer cells of the liver, and (3) the microglia in the central nervous system (Chapter 12). [pp. 124, 384] The Langerhans cells of the skin and the dendritic cells of the lymph nodes and spleen are APCs that are not phagocytic. [p. 152]

Phagocytic APCs engulf and break down pathogens or foreign antigens. This **antigen processing** creates antigenic fragments, which are then bound to Class II MHC proteins and inserted into the cell membrane (Figure 22-15b●). *Class II MHC proteins appear in the cell membrane only when the cell is processing antigens.* Exposure to an APC membrane containing processed antigen can stimulate appropriate T cells.

The Langerhans cells of the skin and the dendritic cells of the lymph nodes and spleen remove antigenic materials from their surroundings via pinocytosis, rather than phagocytosis. The end result is the same, however, because antigens appear in the cell membrane bound to Class II MHC proteins.

ANTIGEN RECOGNITION Inactive T cells have receptors that recognize Class I or Class II MHC proteins. The receptors also have binding sites that will detect the presence of specific bound antigens. If an MHC protein contains any antigen other than its specific target, the T cell will remain inactive. If it contains the antigen that the T cell is programmed to detect, binding will occur. This process is called **antigen recognition,** because the T cell recognizes that it has found an appropriate target.

Some T cells can recognize antigens bound to Class I MHC proteins, whereas others can recognize antigens bound to Class II MHC proteins. Whether a T cell responds to antigens held by Class I or Class II proteins depends on the structure of the T cell membrane. The membrane proteins involved are members of a larger class of proteins called **CD** (*cluster designation*) **markers.**

CD Markers. CD markers are found only on lymphocytes, macrophages, and other related cells. There are over 50 different types of CD markers, each designated by an identifying number. We will consider only the two CD markers associated with the MHC receptor complex.

- **CD8** markers are found on cytotoxic T cells and suppressor T cells. These cells are often called

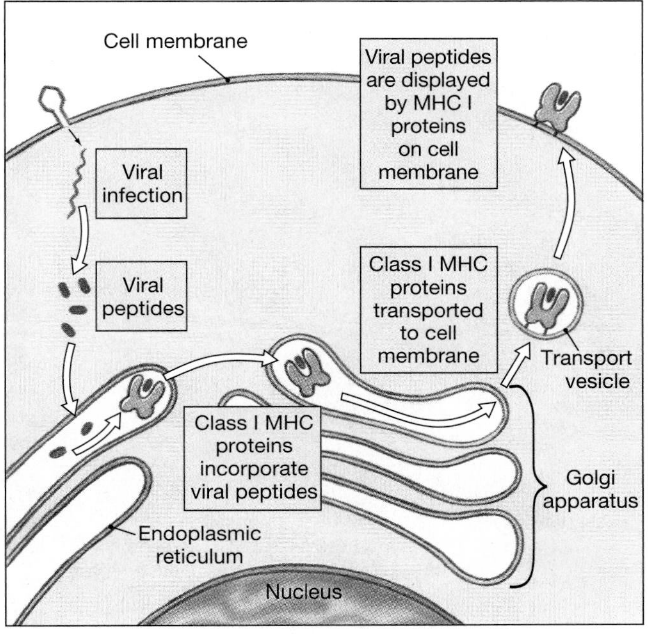
Cell membrane
Viral peptides are displayed by MHC I proteins on cell membrane
Viral infection
Viral peptides
Class I MHC proteins transported to cell membrane
Transport vesicle
Class I MHC proteins incorporate viral peptides
Golgi apparatus
Endoplasmic reticulum
Nucleus

(a) Virally infected cell

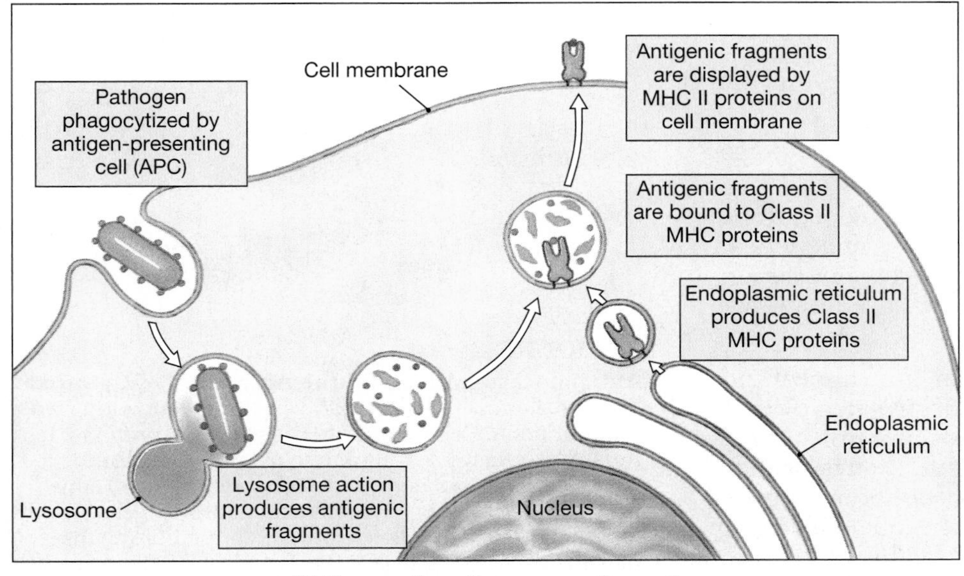
Cell membrane
Pathogen phagocytized by antigen-presenting cell (APC)
Antigenic fragments are displayed by MHC II proteins on cell membrane
Antigenic fragments are bound to Class II MHC proteins
Endoplasmic reticulum produces Class II MHC proteins
Endoplasmic reticulum
Lysosome
Lysosome action produces antigenic fragments
Nucleus

(b) Phagocytic antigen-presenting cell

● **FIGURE 22-15**
Antigens and MHC Proteins. **(a)** Viral or other foreign antigens appear in cell membranes bound to Class I MHC proteins. **(b)** Processed antigens appear on the surfaces of antigen-presenting cells bound to Class II MHC proteins.

CD8 T cells, or *CD8+ T cells.* CD8 T cells can be activated by exposure to antigens presented by Class I MHC proteins (Figure 22-16●).

- **CD4** markers are found on helper T cells. These cells are often called *CD4 T cells,* or *CD4+ T cells.* Helper T cells respond to antigens presented by Class II MHC proteins (Figure 22-17●).

COSTIMULATION A T cell does not become active immediately after recognizing an antigen. Antigen recognition simply prepares the cell for activation. Before activation can occur, a T cell must bind to the stimulating cell at a second site. This vital secondary binding process is called *costimulation.* Costimulation essentially confirms the "OK to activate" signal. Appropriate costimulation proteins appear in the presenting cell only if that cell either has engulfed antigens or is infected by viruses. Costimulation is like the safety on a gun: It helps prevent T cells from mistakenly attacking normal tissues. If a cell displays an unusual antigen, but does not display the "I am an active phagocyte" or "I am infected" signal, T cell activation will not occur. Costimulation is important only in determining whether a T cell will become activated. Once activation has occurred, the safety is off, and the T cell will attack any cells carrying the target antigens.

We will now consider the functions of CD8 and CD4 T cells in the immune response.

Cytotoxic T Cells

Figure 22-16

When activated, CD8 T cells undergo a series of divisions that generate *cytotoxic T cells,* or *T_C cells,* and **memory T_C cells.** The cytotoxic T cells, also called **killer T cells,** roam through the injured tissue. When a cytotoxic T cell encounters its target antigens bound to Class I MHC proteins of another cell, it immediately destroys that cell. A cytotoxic T cell may accomplish this destruction in several ways (Figure 22-16●):

- By rupturing the antigenic cell membrane through the release of perforin.
- By killing the target cell by secreting a poisonous **lymphotoxin** (lim-fō-TOK-sin).
- By activating genes within the nucleus of the cell that tell it to die. The process of genetically programmed cell death, called *apoptosis* (ap-op-TŌ-sis; *apo,* away + *ptosis,* a falling), was introduced in Chapter 3. *[p. 99]*

The entire sequence of events, from appearance of the antigen in a tissue to cell destruction by cytotoxic T cells, takes a significant amount of time. After first exposure to an antigen, 2 days or more may pass before the concentration of cytotoxic T cells reaches effective levels at the site of an injury or infection. Over this period, the damage or infection may spread, making it more difficult to control.

MEMORY T_C CELLS Memory T_C cells ensure that there will not be a delay in the response if the antigen appears in the future. These cells do not respond to the antigen on first exposure, although thousands of them are produced. Instead, they remain "in reserve." If the same antigen appears a second time, these cells will *immediately* differentiate into cytotoxic T cells, producing a swift and effective cellular response that can overwhelm an invading organism before it becomes well established in the tissues.

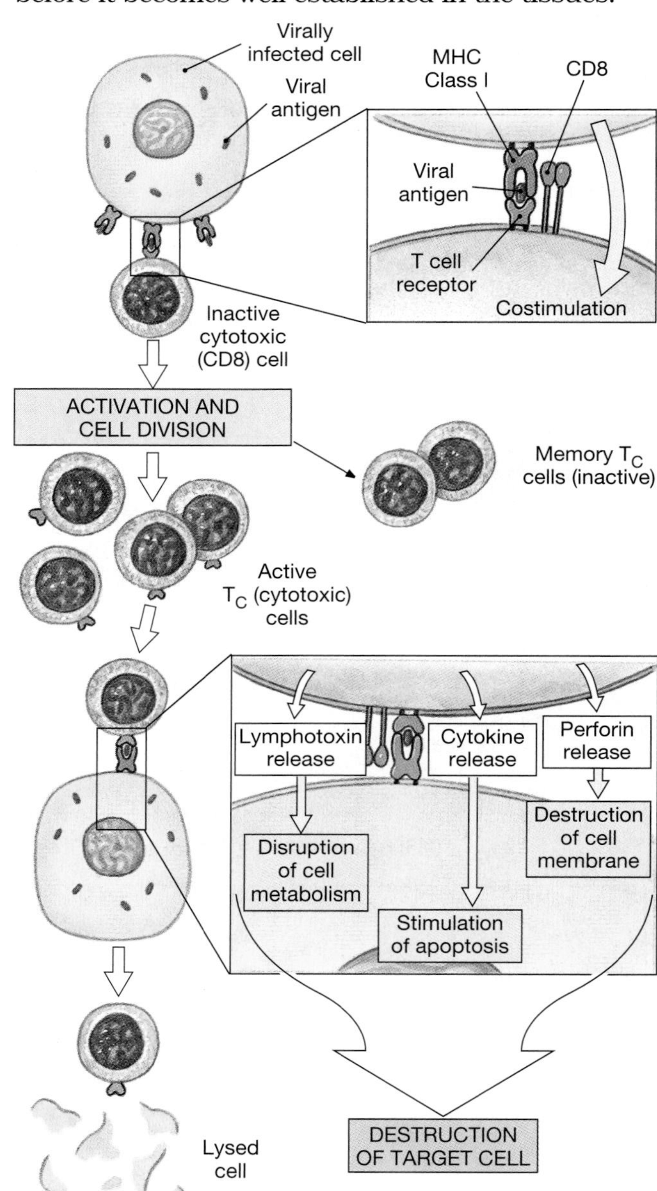
Virally infected cell
Viral antigen
MHC Class I
CD8
Viral antigen
T cell receptor
Costimulation
Inactive cytotoxic (CD8) cell
ACTIVATION AND CELL DIVISION
Memory T_C cells (inactive)
Active T_C (cytotoxic) cells
Lymphotoxin release
Cytokine release
Perforin release
Destruction of cell membrane
Disruption of cell metabolism
Stimulation of apoptosis
Lysed cell
DESTRUCTION OF TARGET CELL

● **FIGURE 22-16**

Antigen Recognition and the Activation of Cytotoxic T Cells. An inactive cytotoxic T cell must encounter an appropriate antigen bound to Class I MHC proteins. It must also receive costimulation from the membrane it contacts. It is then activated and undergoes divisions that produce memory T_C cells and active T_C cells. When one of these active cells encounters a membrane displaying the target antigen, the T_C cell will use one of several methods to destroy the cell.

Suppressor T Cells

Suppressor T cells, or T_S *cells,* depress the responses of other T cells and B cells. They produce these effects by secreting *suppression factors,* inhibitory cytokines of unknown structure. This suppression does not occur immediately, because suppressor T cell activation takes much longer than the activation of other types of T cells. In addition, most of the CD8 T cells in circulation will on activation produce cytotoxic T cells rather than suppressor T cells. As a result, suppressor T cells act *after* the initial immune response. In effect, these cells "put on the brakes" and limit the degree of immune system activation from a single stimulus.

Helper T Cells

Figure 22-17

Upon activation, CD4 T cells undergo a series of cell divisions that produce helper T cells, or T_H *cells,* and **memory T_H cells** (Figure 22-17●). The memory T_H cells remain in reserve, while the helper T cells secrete a variety of cytokines that (1) coordinate specific and nonspecific defenses and (2) stimulate cell-mediated and antibody-mediated immunity. For example, activated helper T cells secrete cytokines that have the following actions:

- Stimulate the T cell divisions that produce memory T cells and accelerate maturation of cytotoxic T cells.
- Enhance nonspecific defenses by (1) attracting macrophages to the area, (2) preventing their departure, and (3) stimulating their phagocytic activity and effectiveness.
- Attract and stimulate the activity of NK cells, providing another mechanism for the destruction of abnormal cells and pathogens.
- Promote B cell division, plasma cell maturation, and antibody production.

In addition, activated helper T cells participate in the direct activation of B cells sensitive to the antigen involved. We will consider this aspect of helper T cell function in the next section.

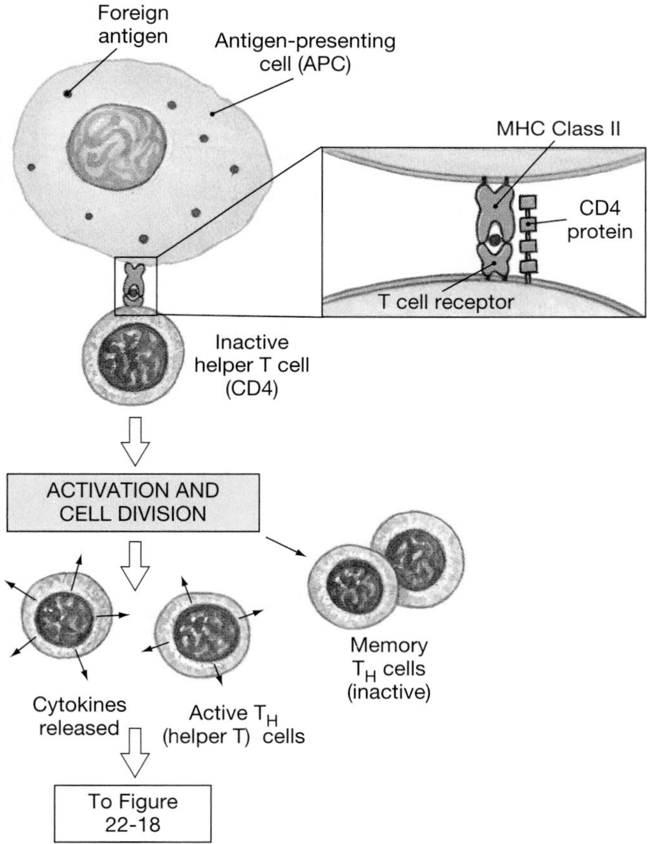
Foreign antigen
Antigen-presenting cell (APC)
MHC Class II
CD4 protein
T cell receptor
Inactive helper T cell (CD4)
ACTIVATION AND CELL DIVISION
Memory T_H cells (inactive)
Cytokines released
Active T_H (helper T) cells
To Figure 22-18

● **FIGURE 22-17**

Antigen Recognition and the Activation of Helper T Cells. Inactive helper T cells must be exposed to appropriate antigens bound to Class II MHC proteins. They then undergo activation, dividing to produce memory T_H cells and active helper T cells. The active T_H cells secrete cytokines that stimulate cell-mediated and antibody-mediated immunity. They also interact with sensitized B cells as shown in Figure 22-18.

GRAFT REJECTION AND IMMUNOSUPPRESSION Organ transplantation is a treatment option for patients with severe disorders of the kidneys, liver, heart, lungs, or pancreas. Finding a suitable donor is the first major problem. In the United States, each day 6 people die while awaiting an organ transplant and another 48 people are added to the transplant waiting list. After surgery has been performed, the major problem is **graft rejection.** In graft rejection, T cells are activated by contact with MHC proteins on cell membranes in the donated tissues. The cytotoxic T cells that develop then attack and destroy the foreign cells.

Significant improvements in transplant success can be made by reducing the sensitivity of the immune system. Until recently the drugs used to produce this **immunosuppression** did not selectively target the immune system. For example, **prednisone** (PRED-ni-sōn), a corticosteroid, was used because it has anti-inflammatory effects that reduce the number of circulating white blood cells and depress the immune response. However, corticosteroid use also caused undesirable changes in glucose metabolism. [p. 625]

An understanding of the communication among T cells, macrophages, and B cells has now led to the development of drugs with more selective effects. **Cyclosporin A (CsA),** a compound derived from a fungus, was the most important *immunosuppressive drug* developed in the 1980s. This compound depresses all aspects of the immune response, primarily by suppressing helper T cell activity while leaving suppressor T cells relatively unaffected. AM *Transplants and Immunosuppressive Drugs*

B Cells and Antibody-Mediated Immunity

Figure 22-18

B cells are responsible for launching a chemical attack on antigens through the production of appropriate *antibodies*. B cell activation proceeds in a series of steps diagrammed in Figure 22-18●.

B Cell Sensitization

As we noted earlier, there are millions of different populations of B cells. Each B cell carries its antibody molecules in its cell membrane. If antigens that will bind to those antibodies appear in the interstitial fluid, the B cell becomes **sensitized.**

During sensitization, antigenic materials appear on the surface of the B cell, bound to Class II MHC proteins. A sensitized B cell is now "ready to go," but it usually will not undergo activation unless it receives the OK from a helper T cell.[1] The need for activation by a helper T cell helps prevent inappropriate activation, the same way that costimulation acts as a "safety" for cell-mediated immunity.

B Cell Activation

Figure 22-18

A sensitized B cell next encounters a helper T cell already activated by antigen presentation. The helper T cell binds to the MHC complex, recognizes the presence of an antigen, and begins secreting cytokines that (1) promote B cell activation, (2) stimulate B cell division, (3) accelerate plasma cell formation, and (4) enhance antibody production.

Figure 22-18● diagrams the events that occur when a B cell is activated by a helper T cell. The activated B cell usually divides several times, producing daughter cells that differentiate into plasma cells and **memory B cells.** Plasma cells begin synthesizing and secreting large numbers of antibodies into the interstitial fluid. These antibodies have the same target as the antibodies on the surface of the sensitized B cell. When stimulated by cytokines from helper T cells, a plasma cell can secrete up to 100 million antibody molecules each hour.

MEMORY B CELLS Memory B cells perform the same role for antibody-mediated immunity that memory T cells perform for cell-mediated immunity. Memory B cells do not respond to a threat on first exposure. Instead, they remain in reserve to deal with subsequent injuries or infections that involve the same antigens. On subsequent exposure, the memory B cells respond by dividing and

[1]Under certain circumstances complex antigens can stimulate B cells directly. The B cell must bind multiple antigen molecules simultaneously, in a single, localized portion of its cell membrane. This process, called *capping*, is relatively rare.

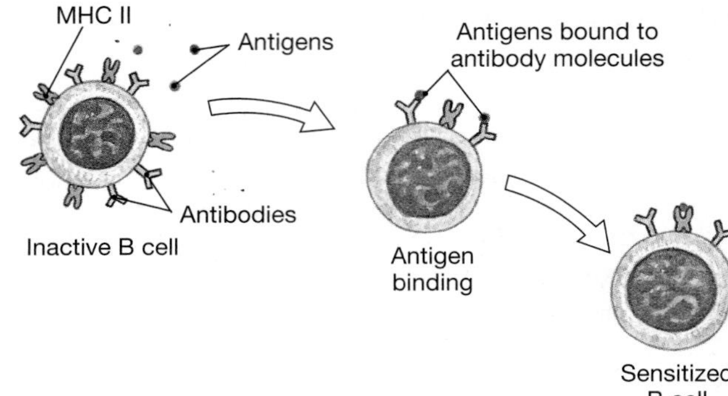
MHC II
Antigens
Antigens bound to antibody molecules
Antibodies
Inactive B cell
Antigen binding
Sensitized B cell

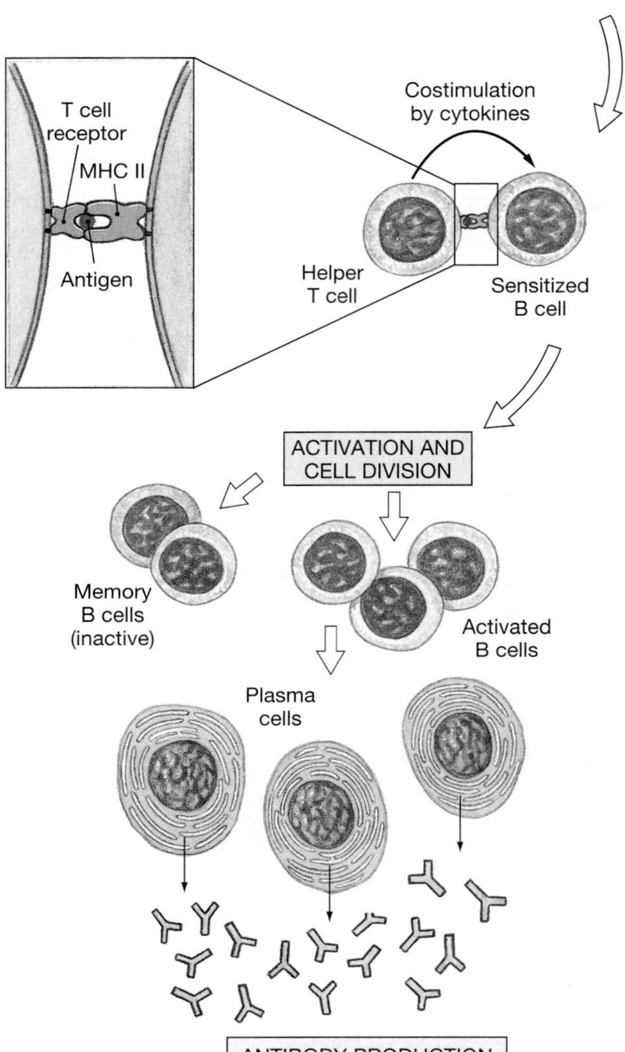
T cell receptor
MHC II
Antigen
Costimulation by cytokines
Helper T cell
Sensitized B cell
ACTIVATION AND CELL DIVISION
Memory B cells (inactive)
Activated B cells
Plasma cells
ANTIBODY PRODUCTION

● **FIGURE 22-18**
Sensitization and Activation of B Cells. A B cell is sensitized through exposure to antigens that bind to antibodies contained in the B cell membrane. The B cell then displays those antigens in its cell membrane. Activated helper T cells encountering those antigens release cytokines that trigger the activation of the B cell. The activated B cell then goes through cycles of division producing memory B cells and B cells that differentiate into plasma cells. The plasma cells secrete antibodies.

differentiating into plasma cells that secrete antibodies in massive quantities.

Antibody Structure

Figure 22-19

Figure 22-19● presents different views of a single antibody molecule. An antibody consists of two parallel pairs of polypeptide chains. Each chain contains *constant* and *variable segments.*

THE CONSTANT SEGMENTS The constant segments of the heavy chains form the base of the antibody molecule (Figure 22-19a,b●). B cells produce only five different types of constant segments. These form the basis of a classification scheme that identifies antibodies as *IgG, IgE, IgD, IgM,* or *IgA,* as discussed in the next section. The structure of the constant segments of the heavy chains determines the way the antibody is secreted and how it is distributed within the body. For example, antibodies in one class are found circulating in body fluids, whereas another type binds to the membranes of basophils and mast cells.

The heavy chain constant segments also contain binding sites that can activate the complement system. These binding sites are covered when the antibody is secreted, but they become exposed when the antibody binds to an antigen.

THE VARIABLE SEGMENTS The specificity of the antibody molecule depends on the structure of the variable segments of the light and heavy chains. The free tips of the two variable segments

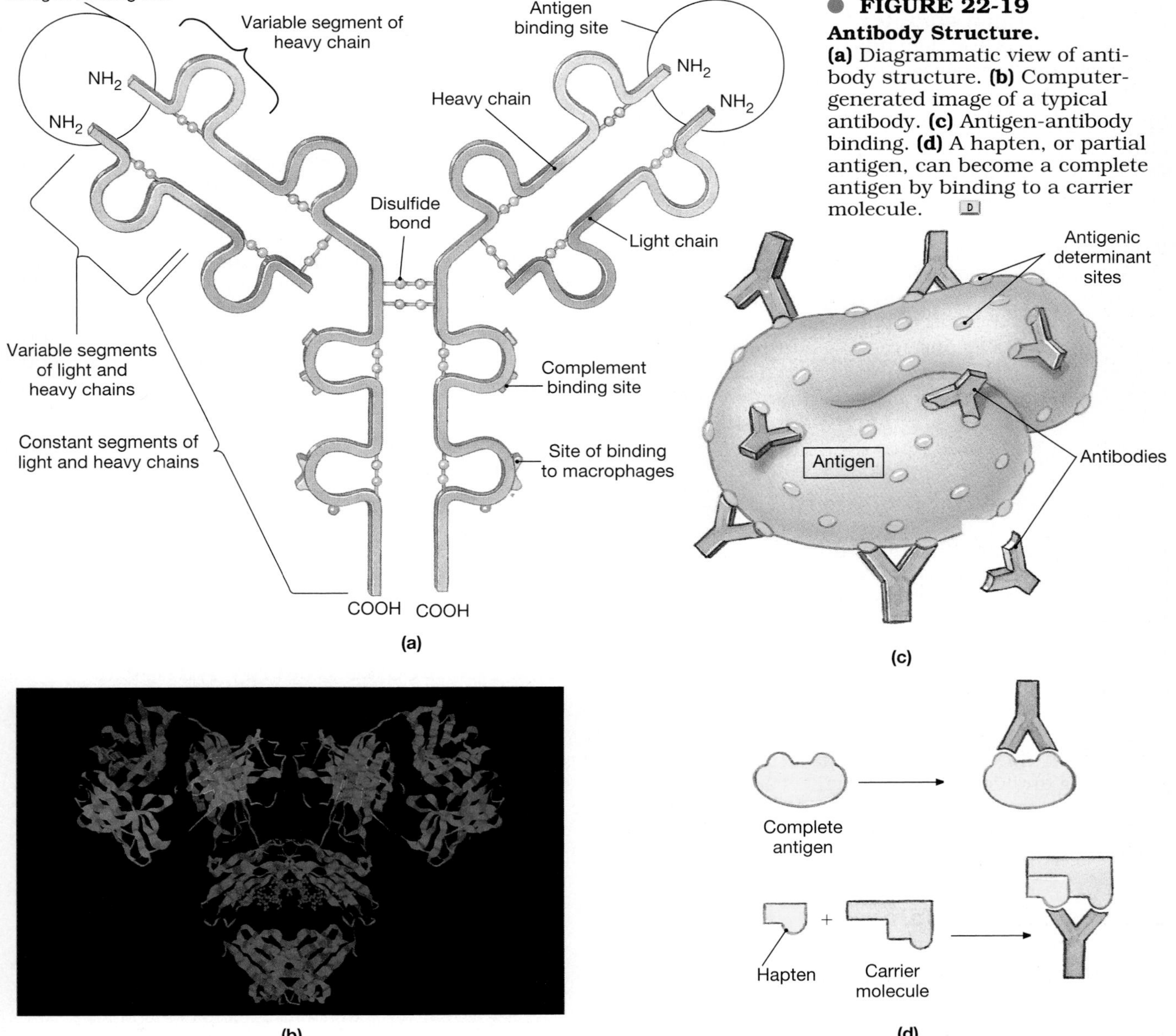
Antigen binding site
Variable segment of heavy chain
Antigen binding site
NH2
NH2
NH2
NH2
Heavy chain
Disulfide bond
Light chain
Variable segments of light and heavy chains
Constant segments of light and heavy chains
Complement binding site
Site of binding to macrophages
COOH COOH
(a)
Antigenic determinant sites
Antigen
Antibodies
(c)
(b)
Complete antigen
Hapten
Carrier molecule
(d)

● **FIGURE 22-19**
Antibody Structure.
(a) Diagrammatic view of antibody structure. **(b)** Computer-generated image of a typical antibody. **(c)** Antigen-antibody binding. **(d)** A hapten, or partial antigen, can become a complete antigen by binding to a carrier molecule.

contain the **antigen binding sites** of the antibody molecule (Figure 22-19b●). These sites can interact with an antigen in just the same way that the active site of an enzyme interacts with a substrate molecule. ∞ [p. 55]

Small differences in the amino acid sequence of the variable segments affect the precise shape of the antigen binding site. These differences account for differences in specificity between the antibodies produced by different B cells. The differences are the result of minor genetic variations that occur during the production, division, and differentiation of B cells. The normal adult body contains roughly 10 trillion B cells. It has been estimated that these cells can produce 100 million different types of antibodies, each with a different specificity.

The Antibody-Antigen Complex. When an antibody molecule binds to its proper antigen, an **antibody-antigen complex** is formed. The specificity of that binding depends initially on the three-dimensional "fit" between the variable segments of the antibody molecule and the corresponding antigenic determinant sites. Once the two molecules are in position, hydrogen bonding and other weak chemical forces lock them together.

Antibodies do not bind to the entire antigen. They bind to certain portions of its exposed surface, regions called **antigenic determinant sites** (Figure 22-19c●). A **complete antigen** is an antigen with at least two antigenic determinant sites, one for each of the antigen binding sites on an antibody molecule. Exposure to a complete antigen can lead to B cell sensitization and a subsequent immune response. Most environmental antigens have multiple antigenic determinant sites; entire microorganisms may have thousands.

DRUG REACTIONS **Partial antigens,** or **haptens,** do not ordinarily cause B cell activation and antibody production. Haptens include a variety of small compounds, among them several antibiotics, such as *penicillin.* Haptens may, however, become attached to carrier molecules, forming combinations that can function as complete antigens (Figure 22-19d). In some cases the carrier contributes an antigenic determinant site. The antibodies produced will attack both the hapten and the carrier molecule. If the carrier molecule is one normally present in the tissues, the antibodies may begin attacking and destroying normal cells. This process is the basis for several drug reactions, such as penicillin allergies. The mechanism involved will be considered in a later section.

Classes of Antibodies

Figure 22-20

There are five classes of antibodies, or **immunoglobulins (Ig)**, in body fluids: *IgG, IgE, IgD, IgM,* and *IgA.* The classes are determined by variations in the structure of the heavy chain constant segments, and so have no effect on the specificity of the antibody.

1. **Immunoglobulin G,** or **IgG,** is the largest class of antibodies. There are several different types of IgG, which together account for 80 percent of all immunoglobulins. The IgG antibodies are responsible for resistance against many viruses, bacteria, and bacterial toxins. IgG antibodies can cross the placenta, and during gestational development maternal IgG provides passive immunity to the developing fetus.
2. **IgE** attaches to the exposed surfaces of basophils and mast cells. When a suitable antigen appears and binds to the IgE molecules, the cell is stimulated to release histamine and other chemicals that accelerate inflammation in the immediate area. IgE is also important in the *allergic response* (p. 816).
3. **IgD** is found on the surfaces of B cells, where it may help to bind antigen molecules. IgG, IgE, and IgD consist of individual molecules, as shown in Figure 22-20●.
4. **IgM** is the first antibody type secreted after the arrival of an antigen. The concentration of IgM then declines as IgG production accelerates. Although plasma cells secrete individual IgM molecules, they circulate as a five-antibody starburst. This configuration makes them particularly effective in forming immune complexes. IgM antibodies, known as *agglutins,* are responsible for the agglutination of cross-matched blood, and so are used to determine an individual's blood type (Chapter 19). ∞ [p. 661] They may also attack bacteria that are insensitive to IgG.
5. **IgA** is found primarily in glandular secretions such as mucus, tears, and saliva. These antibodies attack pathogens *before* they gain access to internal tissues. IgA antibodies circulate in the blood as individual molecules or in pairs. Epithelial cells absorb them from the blood and attach a *secretory piece* before secreting them onto the epithelial surface.

How Antibodies Destroy Antigens

Figure 22-21

The formation of an antibody-antigen complex may cause the elimination of the antigen in several different ways.

1. ***Neutralization.*** Both viruses and bacterial toxins have specific sites that must bind to target regions on body cells in order to enter or injure them. Antibodies may bind to those sites, making the virus or toxin incapable of attaching itself to a cell. This mechanism is known as **neutralization.**
2. ***Agglutination and precipitation.*** Each antibody molecule has two antigen binding sites, and

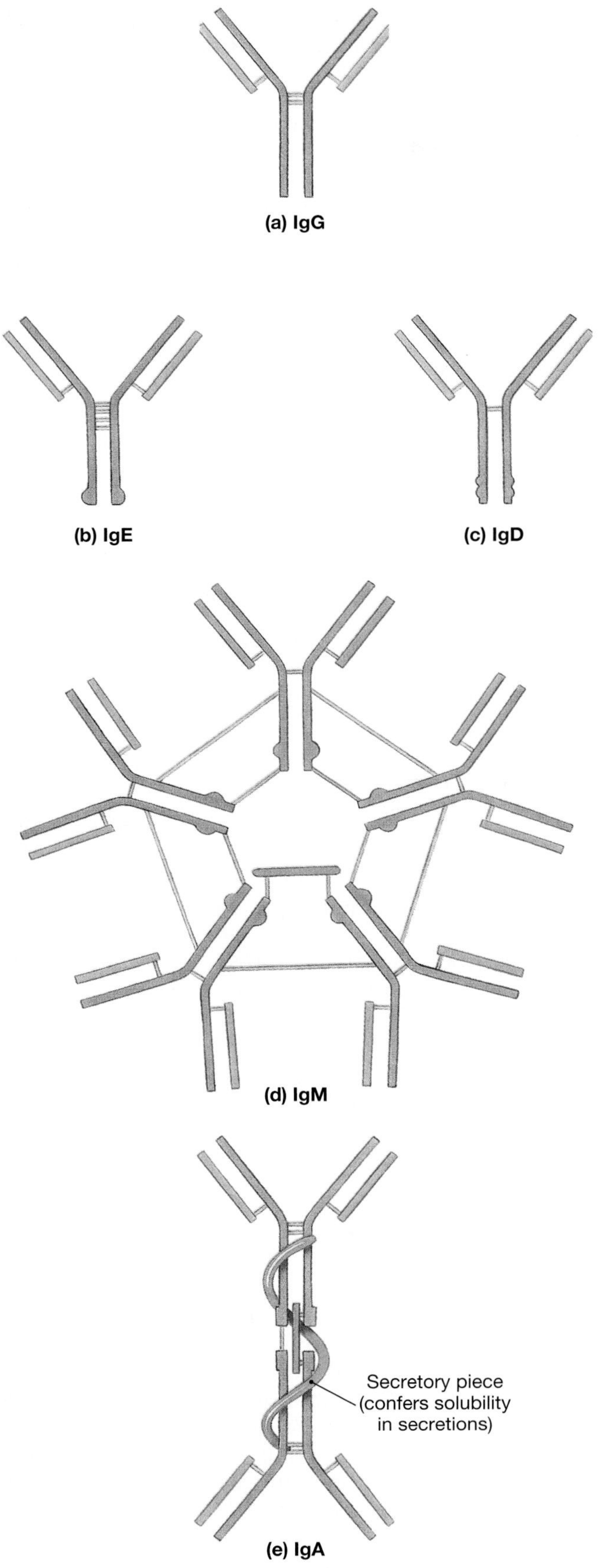

(a) IgG

(b) IgE

(c) IgD

(d) IgM

Secretory piece (confers solubility in secretions)

(e) IgA

● **FIGURE 22-20**
Antibody Classes

most antigens have many antigenic determinant sites. If separate antigens (such as macromolecules or bacterial cells) are far apart, an antibody molecule will necessarily bind to two antigenic sites on the same antigen. If antigens are close together, however, an antibody can bind to antigenic determinant sites on two different antigens. In this way antibodies can form "bridges" that tie large numbers of antigens together (Figure 22-21●). The three-dimensional structure created in this way is known as an **immune complex.** When the antigen is a soluble molecule, such as a toxin, this process may create complexes too large to remain in solution. The formation of insoluble immune complexes is called **precipitation.** When the antigenic target is on the surface of a cell or virus, the formation of large complexes is called **agglutination.** The clumping of red blood cells that occurs when incompatible blood types are mixed (Chapter 19) is an example of agglutination. ∞ [p. 661] AM *Immune Complex Disorders*

3. ***Activation of complement.*** Upon binding to an antigen, portions of the antibody molecule change shape, exposing areas that bind complement proteins. The bound complement molecules then activate the complement system, destroying the antigen as detailed previously.
4. ***Attraction of phagocytes.*** Antigens covered with antibodies attract eosinophils, neutrophils, and macrophages—cells that phagocytize pathogens and destroy foreign or abnormal cell membranes.
5. ***Opsonization.*** A coating of antibodies and complement proteins increases the effectiveness of phagocytosis. Microorganisms such as bacteria have slick cell membranes or capsules, and a

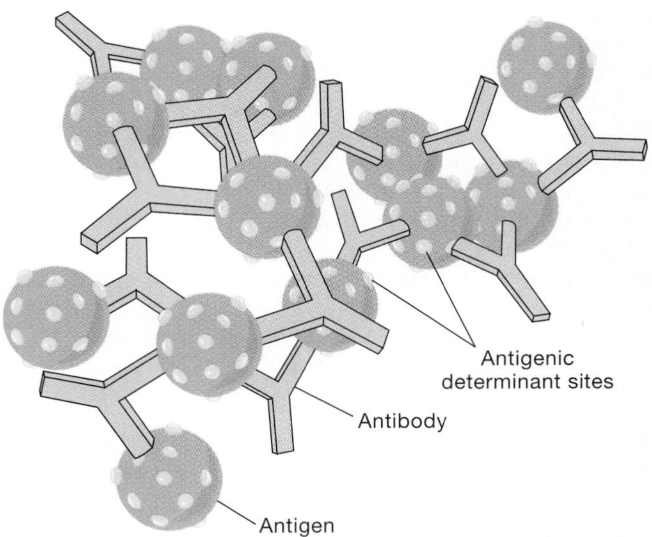
Antigenic determinant sites
Antibody
Antigen

● **FIGURE 22-21**
An Immune Complex. An immune complex forms when large numbers of antibodies and antigens interact. The antibodies bind to antigenic determinant sites on two different antigens, forming a complex structure. D

phagocyte must be able to hang onto its prey before it can engulf it. Phagocytes can bind more easily to antibodies and complement proteins on the surface of a pathogen than they can to the bare surface itself.

6. ***Stimulation of inflammation.*** Antibodies may promote inflammation through the stimulation of basophils and mast cells.
7. ***Prevention of bacterial and viral adhesion.*** Antibodies dissolved in saliva, mucus, and perspiration coat epithelia and provide an additional layer of defense. A covering of antibodies makes it difficult for pathogens to attach to body surfaces and penetrate their defenses.

 How can the presence of an abnormal peptide in the cytoplasm of a cell initiate an immune response?

 A decrease in the number of cytotoxic T cells would affect what type of immunity?

 How would a lack of helper T cells affect the humoral immune response?

 A sample of lymph contains an elevated number of plasma cells. On the basis of this observation would you expect the amount of antibodies in the blood to be increasing or decreasing? Why?

❑ Primary and Secondary Responses to Antigen Exposure

The initial response to antigen exposure is called the **primary response.** When an antigen appears a second time, it triggers a more extensive and prolonged **secondary response.** The secondary response reflects the presence of large numbers of memory cells that are already "primed" for the arrival of the antigen. Primary and secondary responses are characteristic of both cell-mediated and antibody-mediated immunity. The differences between these responses are most easily demonstrated by following the production of antibodies over time.

The Primary Response

Figure 22-22

Because the antigen must activate the appropriate B cells, and the B cells must then respond by differentiating into plasma cells, the primary response does not appear immediately (Figure 22-22•). As plasma cells differentiate and begin secreting, there is a gradual, sustained rise in the concentration of circulating antibodies.

During the primary response, the antibody activity, or **antibody titer** ("standard"), in the plasma does not peak until 1–2 weeks after the initial exposure. Thereafter the antibody concentration declines, assuming that the individual is no longer exposed to the antigen. This reduction in antibody production occurs because (1) plasma cells have very high metabolic rates and survive for only a short time, and (2) further production of plasma cells is inhibited by suppression factors released by suppressor T cells. However, suppressor T cell activity does not begin immediately after antigen exposure, and under normal conditions helper cells outnumber suppressors by more than 3 to 1. As a result, many B cells are activated before suppressor T cell activity has a noticeable effect.

Two different types of antibodies are involved in the primary response. Molecules of *immunoglobulin M,* or *IgM,* are the first to appear in the circulation. IgM is secreted by the plasma cells that form immediately after B cell activation. These lymphocytes do not pause to produce memory cells. Levels of *immunoglobulin G,* or *IgG,* rise more slowly, because the stimulated lymphocytes undergo repeated cell divisions and generate large numbers of memory cells as well as plasma cells. In effect, IgM provides an immediate but limited defense that fights the infection until massive quantities of IgG can be produced.

The Secondary Response

Memory B cells do not differentiate into plasma cells unless they are exposed to the same antigen a second time. If and when that exposure occurs, these cells respond immediately, dividing and differentiating into plasma cells that secrete antibodies in massive quantities. This represents the secondary response, or **anamnestic response** (an-am-NES-tik; *anamnesis,* a memory), to antigen exposure.

During the secondary response, antibody titers increase more rapidly and reach levels many times higher than they did in the primary response. The secondary response appears even if the second exposure occurs years after the first, for memory cells are long-lived, potentially surviving for 20 years or more.

The primary response to an antigen is not as effective a defense as the secondary response. The primary response develops slowly, and the antibodies are not produced in massive quantities. As a result the primary response may not prevent an infection, even though the secondary response would be more than adequate. From a practical standpoint this means that a person who survives the first infection will easily overcome a subsequent invasion by the same pathogen. Often the invading organisms will be destroyed before they have produced any symptoms of the disease. The effectiveness of the secondary response is one of the basic principles behind the use of immunization to prevent disease. AM *Immunization*

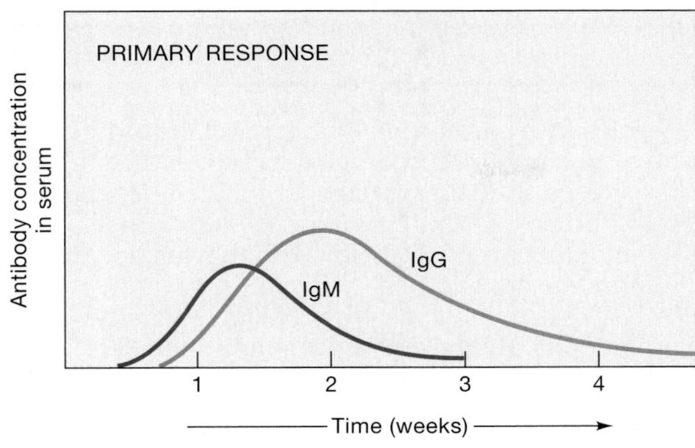

PRIMARY RESPONSE
Antibody concentration in serum
IgG
IgM
1
2
3
4
Time (weeks)

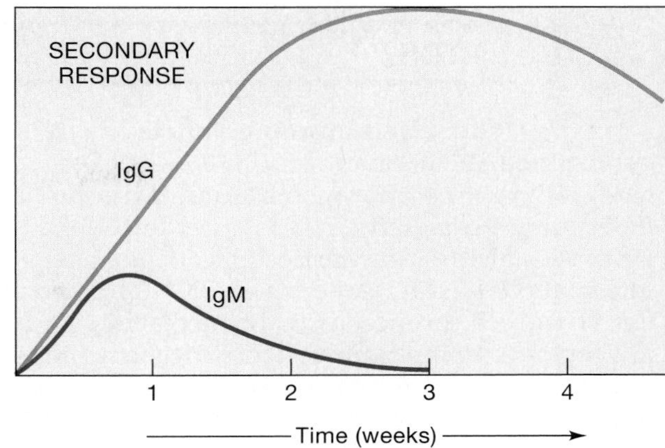

SECONDARY RESPONSE
IgG
IgM
1
2
3
4
Time (weeks)

FIGURE 22-22
The Primary and Secondary Immune Responses. The primary response takes about 2 weeks to develop peak antibody titers, and IgM and IgG antibody concentrations do not remain elevated. The secondary response is characterized by a very rapid increase in IgG antibody titer, to levels much higher than those of the primary response. Antibody activity remains elevated for an extended period following the second exposure to the antigen. D

Hormones of the Immune System

The specific and nonspecific defenses of the body are coordinated by physical interaction and by the release of chemical messengers. Examples of physical interaction include antigen presentation by activated macrophages and helper T cells. Examples of chemical messengers include the release of cytokines by many of the cells involved in the immune response. Cytokines are sometimes classified according to their origins: *Lymphokines* are secreted by lymphocytes, and *monokines* by active macrophages and other antigen-presenting cells. These terms are somewhat misleading, however, since lymphocytes and macrophages sometimes secrete the same cytokines, and cytokines can also be secreted by cells involved with nonspecific defenses and tissue repair.

Table 22-1, p. 811, contains an abbreviated summary of the cytokines identified to date. Five subgroups merit special attention: (1) *interleukins*, (2) *interferons*, (3) *tumor necrosis factors*, (4) chemicals that regulate phagocytic activities, and (5) *colony-stimulating factors*.

Interleukins

Interleukins are probably the most diverse and important chemical messengers in the immune system. Over a dozen different types of interleukins have now been identified (Table 22-1). Lymphocytes and macrophages are the primary sources of interleukins, but specific interleukins, such as interleukin-1 (Il-1), are also produced by endothelial cells, fibroblasts, and astrocytes.

Interleukins have the following general functions:

1. ***Increasing T cell sensitivity to antigens exposed on macrophage membranes.*** This heightened sensitivity accelerates the production of cytotoxic and regulatory T cells.
2. ***Stimulating B cell activity, plasma cell formation, and antibody production.*** These events promote the production of antibodies and the development of humoral immunity.
3. ***Enhancing nonspecific defenses.*** Known effects of interleukin production include: (1) the stimulation of inflammation, (2) the formation of scar tissue by fibroblasts, (3) the elevation of body temperature via the preoptic nucleus of the CNS, (4) the stimulation of mast cell formation, and (5) the promotion of ACTH secretion by the anterior pituitary.
4. ***Moderating the immune response.*** Some of the interleukins are involved with suppressing immune function and shortening the duration of an immune response.

Massive production of interleukins can cause problems at least as severe as those of the primary infection. For example, in *Lyme disease* the release of Il-1 by activated macrophages in response to a localized bacterial infection produces symptoms of fever, pain, skin rash, and arthritis that affect the entire body. AM *Lyme Disease*

You will notice that some interleukins enhance the immune response, whereas others suppress it. The relative quantities secreted at any given moment therefore have a significant effect on the nature and intensity of the response to an antigen. For example, in the course of a typical infection, the pattern of interleukin secretion is constantly changing.

AIDS

Acquired immune deficiency syndrome (AIDS), or *late-stage HIV disease,* is caused by a virus known as **human immunodeficiency virus (HIV).** HIV is an example of a *retrovirus,* a virus that carries its genetic information in RNA, rather than DNA. The virus enters the cell through receptor-mediated endocytosis. In the human body, the virus binds to CD4, the membrane protein characteristic of helper T cells. Several types of antigen-presenting cells, including those of the monocyte-macrophage line, are also infected by HIV, but it is the infection of helper T cells that leads to clinical problems.

Once inside a cell, a viral enzyme, *reverse transcriptase,* synthesizes a complementary strand of DNA that is then incorporated into the genetic complement of the cell. When these genes are activated, the infected cell begins synthesizing viral proteins. In effect, the introduced viral genes take over the synthetic machinery of the cell and force it to produce additional viruses. These new viruses are then shed at the cell surface. (For a more complete discussion refer to the *Applications Manual.*) AM *The Nature of Pathogens*

Cells infected with HIV are ultimately killed. Several mechanisms may be responsible for cell destruction, including (1) the formation of pores in the cell membrane as the viruses are shed, (2) cessation of cell maintenance, due to the continuing synthesis of viral components, (3) autolysis, and (4) stimulation of apoptosis.

The gradual destruction of helper T cells by itself impairs the immune response, because these cells play a central role in coordinating cellular and humoral responses to antigens. To make matters worse, suppressor T cells are relatively unaffected by the virus, and over time the excess of suppressing factors "turns off" the normal immune response. Circulating antibody levels decline, cellular immunity is reduced, and the body is left without defenses against a wide variety of microbial invaders. With immune function so reduced, ordinarily harmless pathogens can initiate lethal infections, known as *opportunistic infections.* Because immune surveillance is also depressed, the risk of cancer increases.

Infection with HIV occurs through intimate contact with the body fluids of infected individuals. Although all body fluids carry the virus, the major routes of transmission involve contact with blood, semen, or vaginal secretions. Most AIDS patients become infected through sexual contact with an HIV-infected person (who may *not* necessarily be suffering from the clinical symptoms of AIDS). The next largest group of patients consists of intravenous drug users who shared contaminated needles. A relatively small number of individuals have become infected with the virus after receiving a transfusion of contaminated blood or blood products. Finally, an increasing number of infants are born with the disease, having acquired it from infected mothers.

AIDS is a public health problem of massive proportions. By the end of 1994 an estimated 385,000 people had died from AIDS in the United States alone, and over 100,000 are living with AIDS. Estimates of the number of individuals infected with HIV in the United States range from 1 to 2 million. The virus has spread throughout the population. For example, a study performed in 1990 indicated that the incidence of infection at U.S. colleges and universities was 1 student in 500. The numbers worldwide are even more frightening: The World Health Organization estimates that over 20 million people are already infected, and the number is increasing rapidly in Africa, Asia, and South America. At the current rate of increase there may be 40 million to 110 million infected individuals by the year 2000. *Every 15 seconds another person becomes infected with the HIV virus.*

The best defense against AIDS consists of avoiding sexual contact with infected individuals. *All forms of sexual intercourse carry the potential risk of viral transmission.* The use of synthetic condoms greatly reduces the chance of infection (although it does not completely eliminate it). Condoms that are not made of synthetic materials are effective in preventing pregnancy but do not block the passage of viruses.

Clinical symptoms of AIDS may not appear for 5–10 years or more following infection, and when they do appear they are often mild, consisting of lymphadenopathy and chronic but nonfatal infections. So far as is known, however, AIDS is invariably fatal, and it appears likely that in the absence of a major clinical advance all those who carry the virus will eventually die of the disease.

Despite intensive efforts, a vaccine has yet to be developed that will provide immunity from HIV infection. While efforts to prevent the spread of HIV continue, the survival rate for AIDS patients has been steadily increasing because new drugs are available that slow the progress of the disease, and improved antibiotic therapies help combat secondary infections. This combination is extending the life span of patients while the search for more effective treatment continues. For more information on the distribution of HIV infection, current and future drug therapies, and additional details on AIDS, consult the *Applications Manual.* AM *AIDS*

Whether the individual succeeds in overcoming the infection may be determined in part by whether stimulatory or suppressive interleukins predominate. As a result, interleukins and their interactions are now the focus of an intensive research effort.

Interferons

Interferons make the synthesizing cell and its neighbors resistant to viral infection, thereby slowing the spread of the virus. These compounds may have other beneficial effects in addition to their antiviral activity. For example, **alpha-interferons** and **gamma-interferons** attract and stimulate NK cells, and **beta-interferons** slow the progress of inflammation associated with viral infection. Gamma-interferons also stimulate macrophages, making them more effective at killing bacterial or fungal pathogens.

Because they stimulate NK cell activity, interferons can be used to fight cancers. For example, alpha-interferons have been used in the treatment of malignant melanoma, bladder cancer, ovarian cancer, and two forms of leukemia. Alpha- or gamma-interferons may be used to treat Kaposi's sarcoma, a cancer that often develops in AIDS patients.

Tumor Necrosis Factors

Tumor necrosis factors (TNFs) slow tumor growth and kill sensitive tumor cells. Activated macrophages secrete one type of TNF and carry the molecules in their cell membranes. Cytotoxic T cells produce a different type of TNF. In addition to their effects on tumor cells, tumor necrosis factors stimulate granular leukocyte production, promote eosinophil activity, cause fever, and increase T cell sensitivity to interleukins.

Chemicals Regulating Phagocytic Activities

Several cytokines coordinate the specific and nonspecific defenses by adjusting the activities of phagocytic cells. These cytokines include factors that attract free macrophages and microphages to the area and prevent their premature departure.

Colony-Stimulating Factors

Colony-stimulating factors (CSFs) and their functions were introduced in Chapter 19. ∞ *[p. 667]* These factors are produced by active T cells, cells of the monocyte-macrophage group, endothelial cells, and fibroblasts. CSFs stimulate the production of blood cells in the bone marrow and lymphocytes in lymphoid tissues and organs.

TABLE 22-1 Chemical Mediators of the Immune Response

Compound	*Functions*
LEUKOTRIENES	Stimulate regional inflammation
TUMOR NECROSIS FACTOR (TNF)	Kills tumor cells, slows tumor growth, stimulates activities of T cells and eosinophils, inhibits parasites and viruses
PERFORIN	Destroys cell membranes by creating large pores
DEFENSINS	Kill cells by piercing cell membrane
INTERLEUKINS	
Interleukin-1 (Il-1)	Stimulates T cells to produce Il-2, promotes inflammation, causes fever
Interleukin-2 (Il-2)	Stimulates growth and activation of other T cells and NK cells
Interleukin-3 (Il-3)	Stimulates production of mast cells and other blood cells
Il-4 (B cell differentiating factor), Il-5 (B cell growth factor), Il-6, Il-7 (B-cell stimulating factors), Il-10, Il-11	Promote differentiation and growth of B cells and stimulate plasma cell formation and antibody production; each has somewhat different effects on macrophage and microphage activities
Il-8	Stimulates blood vessel formation (angiogenesis)
Il-9	Stimulates myeloid cell production (RBCs, platelets, granulocytes, monocytes)
Il-12	Stimulates T cell activity and cell-mediated immunity
LYMPHOTOXINS	Kill cells, damage tissue, promote inflammation
INTERFERONS (alpha, beta, gamma)	Activate other cells to prevent viral entry, inhibit viral replication, stimulate NK cells and macrophages
TRANSFER FACTOR	Sensitizes other T cells to same antigen
GROWTH-INHIBITORY FACTOR (GIF)	Reduces/inhibits replication of target cells
HEMOPOIESIS-STIMULATING FACTOR	Promotes blood cell production in bone marrow
MONOCYTE-CHEMOTACTIC FACTOR (MCF)	Attracts monocytes, activates them to macrophages
MACROPHAGE-INHIBITORY FACTOR (MIF)	Prevents macrophage migration from the area
MACROPHAGE-ACTIVATING FACTOR (MAF)	Makes macrophages more active and aggressive
MICROPHAGE-CHEMOTACTIC FACTOR	Attracts microphages from the blood
COLONY-STIMULATING FACTORS	Discussed in Chapter 19, p. 667
SUPPRESSION FACTORS	Inhibit T_C cell and B cell activity, depress immune response

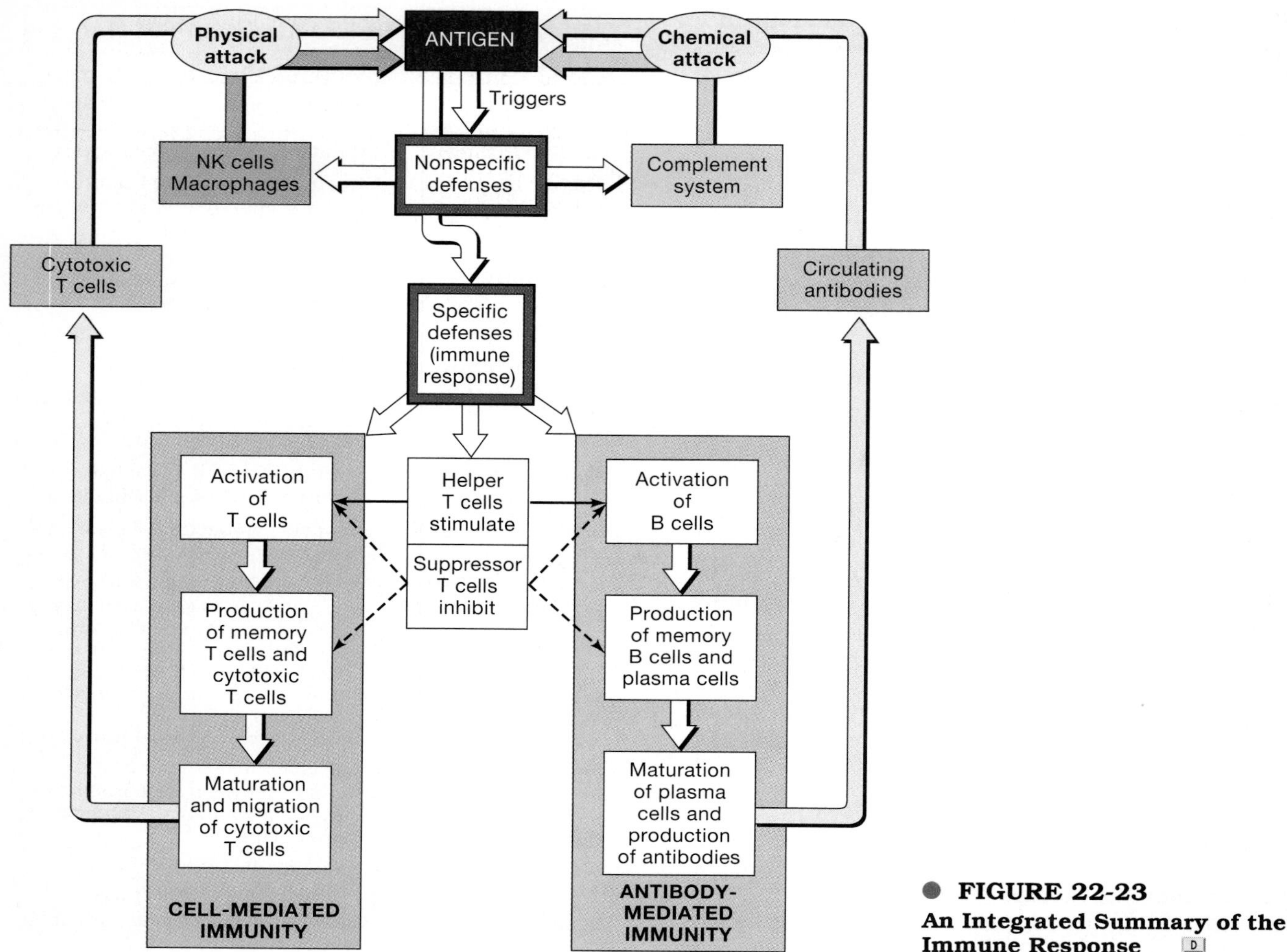

Physical attack
ANTIGEN
Chemical attack
Triggers
NK cells Macrophages
Nonspecific defenses
Complement system
Cytotoxic T cells
Circulating antibodies
Specific defenses (immune response)
Activation of T cells
Helper T cells stimulate
Activation of B cells
Suppressor T cells inhibit
Production of memory T cells and cytotoxic T cells
Production of memory B cells and plasma cells
Maturation and migration of cytotoxic T cells
Maturation of plasma cells and production of antibodies
CELL-MEDIATED IMMUNITY
ANTIBODY-MEDIATED IMMUNITY

● **FIGURE 22-23**
An Integrated Summary of the Immune Response

MANIPULATING THE IMMUNE RESPONSE As our understanding of the immune system grows, complex therapies involving combinations of cytokines, including interleukins, and *monoclonal antibodies*[1] are appearing. For example, in one procedure cytotoxic T cells were removed from patients with *malignant melanoma,* a particularly dangerous type of skin cancer. These lymphocytes were able to recognize tumor cells, and they had migrated to the tumor. However, for some reason they appeared to be unable to kill the tumor cells.

The extracted lymphocytes were cultured, and viruses were used to insert multiple copies of the genes responsible for the production of tumor necrosis factor. The patients were then given periodic infusions of these "supercharged" T cells. To enhance T cell activity further, the researchers also administered doses of interleukin-2. Initial results are promising. It is clear that the ability to manipulate the immune response will revolutionize the treatment of many serious diseases.

[1]Monoclonal antibodies are antibodies produced by a single clone of B cells under laboratory conditions. For a detailed discussion of the process, consult the *Applications Manual.* AM *Technology, Immunity, and Disease*

Patterns of Immune Response

Figures 22-23, 22-24, 22-25

The basic chemical and cellular interactions that follow the appearance of a foreign antigen have now been detailed. Figure 22-23● presents a broader, integrated view of the immune response and its relationship to nonspecific defenses. Figure 22-24● provides an overview of the time course of the events responsible for overcoming a bacterial infection.

In the early stages of infection, before antigen processing has occurred, neutrophils and NK cells migrate into the threatened area and destroy bacterial cells. Over time, cytokines draw increasing numbers of phagocytes to the region (Figure 22-24●). Cytotoxic T cells appear as arriving T cells are activated by antigen presentation. Last of all, the population of plasma cells rises as activated B cells differentiate. This rise is followed by a gradual, sustained increase in the level of circulating antibodies.

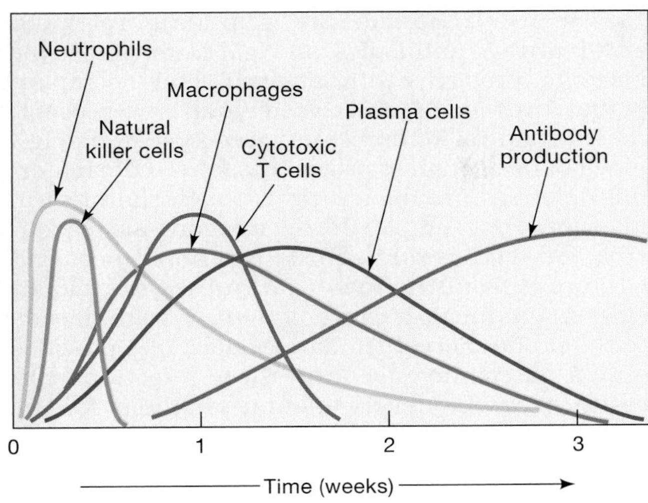
Neutrophils
Macrophages
Natural killer cells
Cytotoxic T cells
Plasma cells
Antibody production
0
1
2
3
Time (weeks)

● **FIGURE 22-24**
The Time Course of the Body's Response to Infection. This figure outlines the basic sequence of events that begins with the appearance of pathogens in peripheral tissues.

The basic sequence of events is very similar when a viral infection occurs. The initial steps are different, however, because cytotoxic T cells and NK cells can be activated by contact with infected cells. Figure 22-25● contrasts the steps involved in defense against bacteria with the defense against infection by viruses. Table 22-2 completes our summary by reviewing the cells involved.

The Development of Resistance

The ability to demonstrate an immune response upon exposure to an antigen is called **immunological competence.** Cell-mediated immunity can be demonstrated as early as the third month of fetal development, but active antibody-mediated immunity does not appear until somewhat later.

The first cells that leave the fetal thymus migrate to the skin and into the epithelia lining the

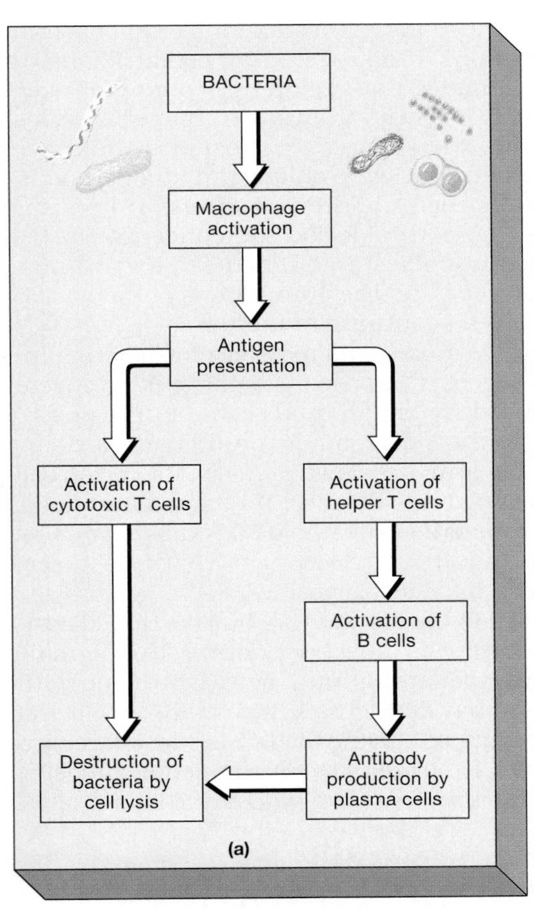
BACTERIA
Macrophage activation
Antigen presentation
Activation of cytotoxic T cells
Activation of helper T cells
Activation of B cells
Antibody production by plasma cells
Destruction of bacteria by cell lysis
(a)

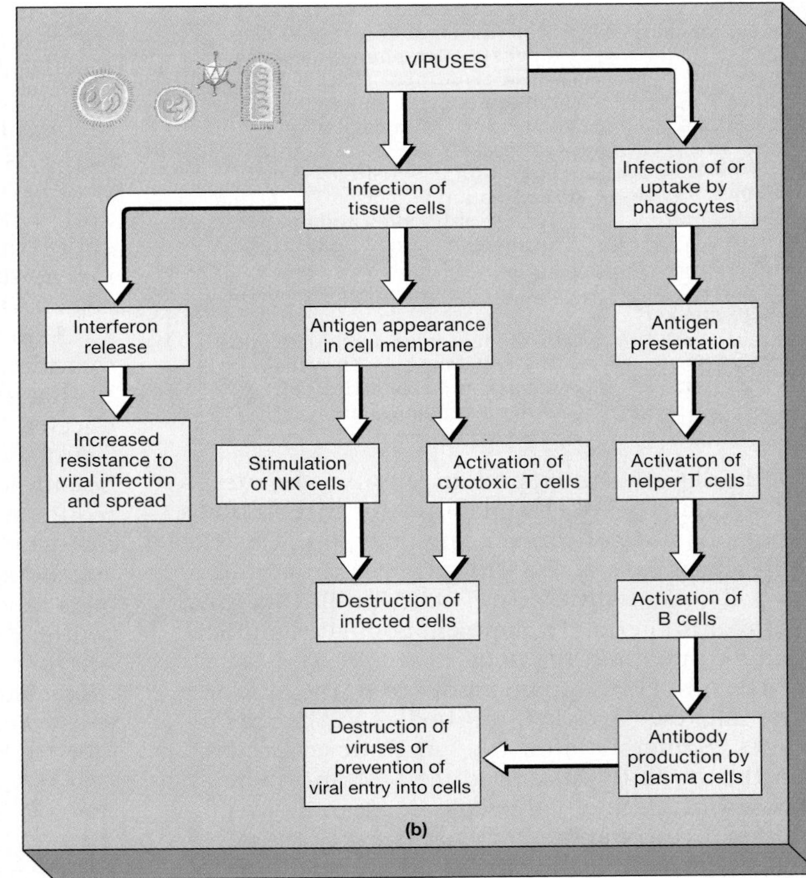
VIRUSES
Infection of tissue cells
Infection of or uptake by phagocytes
Interferon release
Antigen appearance in cell membrane
Antigen presentation
Increased resistance to viral infection and spread
Stimulation of NK cells
Activation of cytotoxic T cells
Activation of helper T cells
Destruction of infected cells
Activation of B cells
Destruction of viruses or prevention of viral entry into cells
Antibody production by plasma cells
(b)

● **FIGURE 22-25**
Defenses against Bacterial and Viral Pathogens. (a) Defenses against bacterial pathogens are usually initiated by active macrophages. **(b)** Defenses against viruses are usually activated following infection of normal cells.

TABLE 22-2 Cells That Participate in Tissue Defenses

Cell	Functions
Neutrophils	Phagocytosis; stimulation of inflammation
Eosinophils	Phagocytosis of antibody-antigen complexes; suppression of inflammation; involved in allergic response
Mast cells and basophils	Stimulation and coordination of inflammation via release of histamine, heparin, leukotrienes, prostaglandins
ANTIGEN-PRESENTING CELLS	
Macrophages (free and fixed macrophages, Kupffer cells, microglia, etc.)	Phagocytosis; antigen processing; antigen presentation with Class II MHC proteins; secretion of cytokines, especially interleukins and interferons
Dendritic cells, Langerhans cells	Antigen presentation bound to Class II MHC proteins
LYMPHOCYTES	
NK cells	Destruction of cell membranes containing abnormal antigens
Cytotoxic T cells (T_C, CD8 marker)	Lysis of cell membranes containing antigens bound to Class I MHC proteins; secretion of perforins, defensins, lymphotoxins, and other cytokines
Helper T cells (T_H, CD4 marker)	Secretion of cytokines that stimulate cell-mediated and antibody-mediated immunity; activation of sensitized B cells
B cells	Differentiation into plasma cells that secrete antibodies and provide antibody-mediated immunity
Suppressor T cells (T_S, CD8 marker)	Secretion of suppression factors that inhibit the immune response
Memory cells (T_C, T_H, B)	Cells produced during the activation of T cells and B cells; remain in tissues awaiting arrival of antigen at a later date

mouth, digestive tract, and, in females, the uterus and vagina. These cells take up residence in these tissues as antigen-presenting cells, such as the Langerhans cells of the skin, whose primary function will be the activation of T cells. T cells that leave the thymus later in development populate lymphoid organs throughout the body.

The first B cells to be produced in the liver and bone marrow carry IgM antibodies in their membranes. Sometime after the fourth developmental month, the fetus may produce IgM antibodies if exposed to specific pathogens. Fetal antibody production is uncommon, however, because the developing fetus has natural passive immunity due to the transfer of IgG antibodies from the maternal bloodstream. These are the only antibodies that cross the placenta, and they include the antibodies responsible for the clinical problems that accompany fetal-maternal Rh incompatibility, discussed in Chapter 19. ∞ [p. 662] Because the agglutinins anti-A and anti-B are IgM antibodies, and IgM cannot cross the placenta, problems with maternal-fetal incompatibilities involving the ABO blood groups rarely occur.

The natural immunity provided by maternal IgG may not be enough to protect the fetus if the maternal defenses are overrun by a bacterial or viral infection. For example, the microorganisms responsible for syphilis and rubella ("German measles") can cross from the maternal to the fetal bloodstream, producing a congenital infection that leads to the production of fetal antibodies. IgM provides only a partial defense, and these infections can result in severe developmental problems for the fetus. AM *Fetal Infections*

Delivery eliminates the maternal supply of IgG, and although the mother provides IgA antibodies in the breast milk, the infant gradually loses its passive immunity. The amount of maternal IgG in the infant's bloodstream declines rapidly over the first 2 months after birth. During this period the infant becomes vulnerable to infection by bacteria or viruses that were previously overcome by maternal antibodies. The infant also begins producing IgG on its own, as the immune system begins to respond to infections, environmental antigens, and the vaccinations administered by pediatricians. It has been estimated that children encounter a "new" antigen every 6 weeks from birth to age 12. (This fact explains why parents, exposed to the same antigens when they were children, usually remain healthy while their children develop runny noses and colds.) Over this period the concentration of circulating antibodies gradually rises toward normal adult levels, and the populations of memory B cells and T cells continue to increase.

Skin tests can be used to determine whether an individual has been exposed to a particular antigen. In this procedure, small quantities of antigen are injected into the skin, usually on the anterior surface of the forearm. If resistance exists, the region will become inflamed over the next 2–4 days. Many states require a tuberculosis test, called a *tuberculin skin test*, when someone applies for a job in the food-service industry. If the test is positive, the individual has been exposed to the disease and has developed antibodies. Further tests must be performed to determine whether an infection is under way. (Anyone with tuberculosis may accidentally transmit the bacteria when preparing or serving food.) Skin tests are also used to check for allergies to environmental antigens. AM *Delayed Hypersensitivity and Skin Tests*

Immune Disorders

Because the immune response is so complex, there are many opportunities for things to go wrong. A great variety of clinical conditions may result from disor-

TABLE 22-3 Autoimmune Disorders

Disorder	*Antibody Target*	*Discussion*
Psoriasis	Epidermis of skin	*AM*
Vitiligo	Melanocytes of skin	p. 154 and *AM*
Rheumatoid arthritis	Connective tissues at joints	p. 264 and *AM*
Myasthenia gravis	Synaptic ACh receptors	p. 295 and *AM*
Multiple sclerosis	Myelin sheaths of axons	p. 386 and *AM*
Addison's disease	Adrenal cortex	p. 626 and *AM*
Graves' disease	Thyroid follicles	p. 620 and *AM*
Hypoparathyroidism	Chief cells of parathyroid	p. 622 and *AM*
Thyroiditis	Thyroid-binding globulin	*AM*
Type I diabetes	Beta cells of pancreatic islets	p. 634 and *AM*
Rheumatic fever	Myocardium and heart valves	p. 688 and *AM*
Systemic lupus erythematosus	DNA, cytoskeletal proteins, other tissue components	*AM*
Ménière's disease	Collagen	*AM*
Thrombocytic purpura	Platelets	*AM*
Pernicious anemia	Parietal cells of stomach	*AM*
Chronic hepatitis	Hepatocytes of liver	p. 912 and *AM*

ders of immune function. **Autoimmune disorders** develop when the immune response mistakenly targets normal body cells and tissues. In an **immunodeficiency disease** either the immune system fails to develop normally or the immune response is blocked in some way. Autoimmune disorders and immunodeficiency diseases are relatively rare conditions—clear evidence of the effectiveness of the immune system's control mechanisms. A far more common, and usually far less dangerous, class of immune disorders is the **allergies.** We will consider examples of each type of disorder below. For a more extended discussion, including possible therapies, see the relevant sections of the *Applications Manual.*

❑ Autoimmune Disorders

Autoimmune disorders affect an estimated 5 percent of adults in North America and Europe. Many examples of autoimmune disorders were noted in other chapters, because of their effects on the function of major systems. Table 22-3 lists the major examples of autoimmune disorders considered in the text and *Applications Manual.*

The immune system usually recognizes and ignores the antigens normally found in the body. The recognition system can malfunction, however, and when it does the activated B cells begin to manufacture antibodies against other cells and tissues. The trigger may be a reduction in suppressor T cell activity, excessive stimulation of helper T cells, tissue damage that releases large quantities of antigenic fragments, haptens bound to compounds normally ignored, viral or bacterial toxins, or some combination of factors.

The symptoms produced depend on the identity of the antigen attacked by these misguided antibodies, called **autoantibodies.** For example:

- The inflammation of *thyroiditis* results from the release of autoantibodies against thyroglobulin.
- *Rheumatoid arthritis* occurs when autoantibodies form immune complexes within connective tissues, especially around the joints.
- *Insulin-dependent diabetes mellitus* (IDDM) is usually caused by autoantibodies that attack cells in the pancreatic islets.

Many autoimmune disorders appear to be cases of mistaken identity. For example, proteins associated with the measles, Epstein-Barr, influenza, and other viruses contain amino acid sequences that are similar to those of myelin proteins. As a result, antibodies that target these viruses may also attack myelin sheaths. This mechanism accounts for the neurological complications that sometimes follow a vaccination or viral infection. It is also the probable mechanism responsible for *multiple sclerosis.*

For unknown reasons, the risk of autoimmune problems increases if an individual has an unusual type of MHC proteins. At least 50 clinical conditions have been linked to specific variations in MHC structure. Many of these are disorders considered elsewhere in the text or in the *Applications Manual* (Table 22-3). AM *Systemic Lupus Erythematosus*

❑ Immunodeficiency Diseases

Immunodeficiency diseases are the result of (1) congenital problems with the development of immunity, (2) an infection with a virus, such as HIV, that depresses immune function, or (3) treatment with or exposure to immunosuppressive agents, such as radiation or drugs.

In **severe combined immunodeficiency disease (SCID)** the individual fails to develop either cellular or humoral immunity. Lymphocyte populations are reduced, and normal B and T cells are not present. Patients with SCID are unable to provide an immune defense, and even a mild infection can prove fatal. Total isolation offers protection at great cost and with severe restrictions on the individual's lifestyle. Bone marrow transplants from compatible donors, usually a close relative, have been used to colonize lymphoid tissues with functional lymphocytes. Gene-splicing techniques have led to therapies that can treat at least one form of SCID. AM *Genetic Engineering and Gene Therapy*

AIDS, an immunodeficiency disease considered on page 810, is the result of a viral infection that targets primarily helper T cells. As the number of T cells declines, the normal immune control mechanism breaks down. When an infection occurs, suppressor factors released by suppressor T cells inhibit an immune response before the few surviving helper T cells can stimulate the formation of cytotoxic T cells or plasma cells in adequate numbers.

Immunosuppressive agents may destroy stem cells and lymphocytes, leading to a complete immunological failure. This is one of the potentially fatal consequences of radiation exposure. Immunosuppressive drugs have been used for many years to prevent graft rejection following transplant surgery.

❑ Allergies

Allergies are inappropriate or excessive immune responses to antigens. The sudden increase in cellular activity or antibody titers can have a number of unpleasant side effects. For example, neutrophils or cytotoxic T cells may destroy normal cells while attacking the antigen, or the antibody-antigen complex may trigger a massive inflammatory response. Antigens that trigger allergic reactions are often called **allergens**.

There are several types of allergies. A complete classification recognizes four categories: *immediate hypersensitivity (Type I)*, *cytotoxic reactions (Type II)*, *immune complex disorders (Type III)*, and *delayed hypersensitivity (Type IV)*. We will consider immediate hypersensitivity at this time, as it is probably the most common form of allergy. One form, *allergic rhinitis*, includes "hay fever" and environmental allergies that may affect 15 percent of the U.S. population. An example of a cytotoxic (Type II) reaction, the cross-reactions that occur following the transfusion of an incompatible blood type, was discussed in Chapter 19. *[p. 661]* Other types of allergies are detailed in the *Applications Manual*. AM *Immune Complex Disorders; Delayed Hypersensitivity and Skin Tests*

Immediate Hypersensitivity

Figure 22-26

Immediate hypersensitivity begins with the process of *sensitization*. Sensitization is the initial exposure to an allergen that leads to the production of antibodies, specifically large quantities of IgE. The tendency to produce IgE antibodies in response to specific allergens may be genetically determined. In drug reactions, such as penicillin allergies, IgE antibodies are produced in response to a hapten (partial antigen) bound to a larger molecule inside the body.

Because of the lag time needed to activate B cells, produce plasma cells, and synthesize antibodies, the first exposure to an allergen does not produce symptoms. It merely sets the stage for the next encounter. After sensitization, the IgE antibodies become attached to the cell membranes of basophils and mast cells throughout the body. When exposed to the same allergen at a later date, the bound antibodies stimulate these cells to release histamine, heparin, several cytokines, prostaglandins, and other chemicals into the surrounding tissues. The result is a sudden, massive inflammation of the affected tissues.

Basophils, eosinophils, T cells, and macrophages are drawn to the area by the cytokines and other secretions of the mast cells. These cells release chemicals of their own, extending and exaggerating the responses initiated by mast cells.

The severity of the allergic reaction depends on the sensitivity of the individual and the location involved. If allergen exposure occurs at the body surface, the response may be restricted to that area. If the allergen enters the systemic circulation, the response may be more dramatic and occasionally lethal.

ANAPHYLAXIS In **anaphylaxis** (a-na-fi-LAK-sis; *ana-*, again + *phylaxis*, protection) a circulating allergen affects mast cells throughout the body (Figure 22-26•). The entire range of symptoms can develop within a matter of minutes. Changes in capillary permeabilities produce swelling and edema in the dermis, and raised welts, or *hives*, appear on the surface of the skin. Smooth muscles along the respiratory passageways contract, and the narrowed passages make breathing extremely difficult. In severe cases of anaphylaxis an extensive peripheral vasodilation occurs, producing a fall in blood pressure that may lead to a circulatory collapse. This response has been called **anaphylactic shock.**

TREATMENT OF ANAPHYLAXIS Many of the symptoms of immediate hypersensitivity can be prevented by the prompt administration of **antihistamines** (an-ti-HIS-ta-mēnz), drugs that block the action of histamine. *Benadryl* is a popular antihistamine available without a prescription. Treatment of severe anaphylaxis involves antihistamine, corticosteroid, and epinephrine injections.

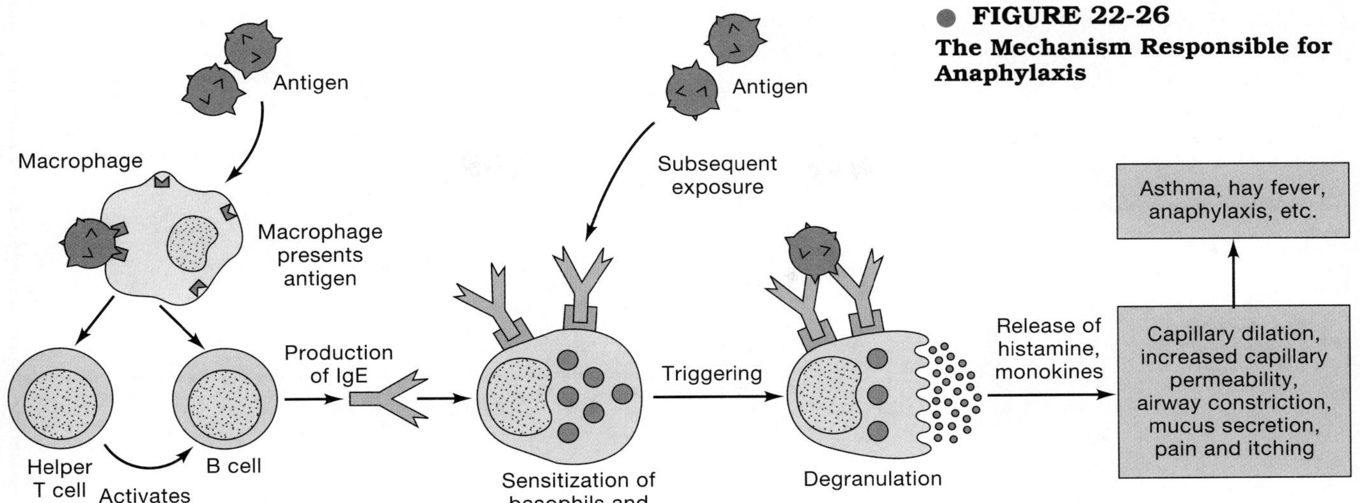
Antigen
Macrophage
Macrophage presents antigen
Helper T cell
Activates
B cell
Production of IgE
Antigen
Subsequent exposure
Sensitization of basophils and mast cells
Triggering
Degranulation
Release of histamine, monokines
Capillary dilation, increased capillary permeability, airway constriction, mucus secretion, pain and itching
Asthma, hay fever, anaphylaxis, etc.

● **FIGURE 22-26**
The Mechanism Responsible for Anaphylaxis

Stress and the Immune Response

One of the normal effects of interleukin-1 secretion is the stimulation of ACTH production by the anterior pituitary. ACTH production in turn leads to the secretion of glucocorticoids by the adrenal cortex. (The functions of ACTH and glucocorticoids were described in Chapter 18. *[p. 625]*) The anti-inflammatory effects of the glucocorticoids may help control the size of the immune response. However, the long-term secretion of glucocorticoids, as in the resistance phase of the *general adaptation syndrome* (Chapter 18), can inhibit the immune response and lower resistance to disease. *[p. 637]* Several of the tissue responses to glucocorticoids alter the effectiveness of specific and nonspecific defenses. Examples include the following:

1. ***Depression of the inflammatory response.*** Glucocorticoids inhibit mast cells and make capillaries less permeable. Inflammation is therefore less likely, and when it does occur the capillary effects slow the entry of fibrinogen, complement, and cellular defenders.
2. ***Reduction in the activities and numbers of phagocytes in peripheral tissues.*** This reduction further impairs nonspecific defense mechanisms and interferes with the processing and presentation of antigens to lymphocytes.
3. ***Inhibition of interleukin secretion.*** A reduction in interleukin production depresses the lymphocytic response, even to antigens bound to MHC proteins.

The mechanisms responsible for these changes are still under investigation, but it should be noted that immune system depression due to chronic stress may represent a serious threat to health.

Aging and the Immune Response

With advancing age the immune system becomes less effective at combating disease. T cells become less responsive to antigens; as a result, fewer cytotoxic T cells respond to an infection. Because the number of helper T cells is also reduced, B cells are less responsive, and antibody levels do not rise as quickly after antigen exposure. The net result is an increased susceptibility to viral and bacterial infection. For this reason, vaccinations for acute viral diseases, such as the flu (influenza), are strongly recommended for elderly individuals. The increased incidence of cancer in the elderly reflects the fact that immune surveillance declines, and tumor cells are not eliminated as effectively.

Integration with Other Systems

Figure 22-27

Figure 22-27● summarizes the interactions between the lymphatic system and other physiological systems. The relationships among the cells of the immune response and the nervous and endocrine systems are now the focus of intense research. It is now apparent that there is substantial interaction between these systems.

For example, the cells of the immune system can influence CNS and endocrine activity in the following ways:

- The thymus secretes oxytocin, ADH, and endorphins as well as the thymic hormones. The effects of these hormones on the CNS are not known, but removal or destruction of the thymus leads to a decrease in brain endorphin levels.

INTEGUMENTARY SYSTEM

Provides physical barriers to pathogen entry; Langerhans cells in epidermis and macrophages in dermis resist infection and present antigens to trigger immune response; mast cells trigger inflammation, mobilize cells of lymphatic system

Provides IgA for secretion onto integumentary surfaces

SKELETAL SYSTEM

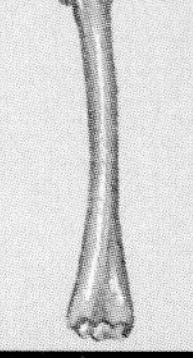

Lymphocytes and other cells involved in the immune response are produced and stored in bone marrow

MUSCULAR SYSTEM

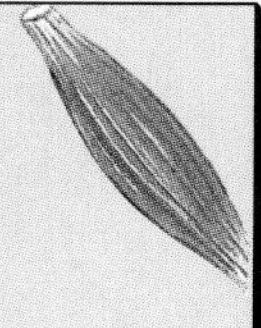

Protects superficial lymph nodes and the lymphatic vessels in the abdominopelvic cavity, muscle contractions help propel lymph along lymphatic vessels

THE LYMPHATIC SYSTEM

NERVOUS SYSTEM

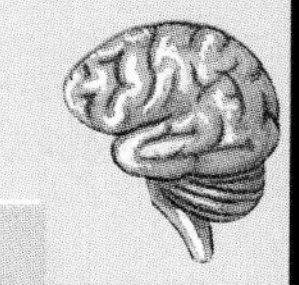

Microglia present antigens that stimulate specific defenses; glial cells secrete cytokines; innervation stimulates APCs

Cytokines affect hypothalmic production of CRH and TRH

ENDOCRINE SYSTEM

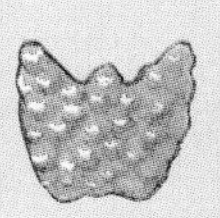

Glucocorticoids have antiinflammatory effects; thymosins stimulate development and maturation of lymphocytes; many hormones affect immune function

CARDIOVASCULAR SYSTEM

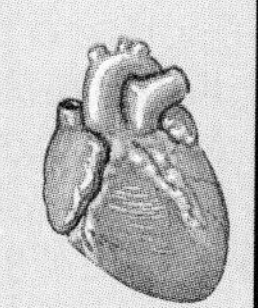

Distributes WBCs; carries antibodies that attack pathogens; clotting response helps restrict spread of pathogens; granulocytes and lymphocytes produced in bone marrow

Fights infections of cardiovascular organs; returns tissue fluid to circulation

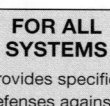

FOR ALL SYSTEMS

Provides specific defenses against infection; immune surveillance eliminates cancer cells

RESPIRATORY SYSTEM

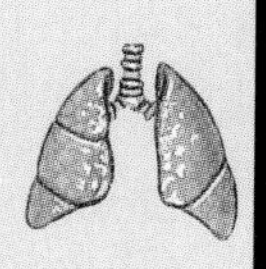

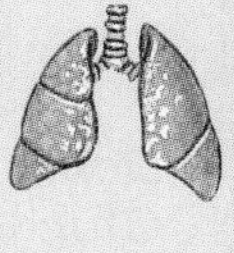

Alveolar phagocytes present antigens and trigger specific defenses; provides O_2 required by lymphocytes and eliminates CO_2 generated during their metabolic activities

Tonsils protect against infection at entrance to respiratory tract

DIGESTIVE SYSTEM

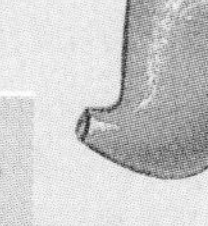

Provides nutrients required by lymphatic tissues, digestive acids, and enzymes; provides nonspecific defense against pathogens

Tonsils and GALT defend against infection and toxins absorbed from tract; lymphatics carry absorbed lipids to venous system

URINARY SYSTEM

Eliminates metabolic wastes generated by cellular activity; acid pH of urine provides nonspecific defense against urinary tract infection

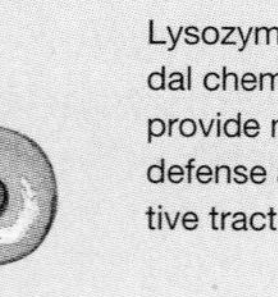

Lysozymes and bactericidal chemicals in secretions provide nonspecific defense against reproductive tract infections

Provides IgA for secretion by epithelial glands

REPRODUCTIVE SYSTEM

FIGURE 22-27
Functional Relationships between the Lymphatic System and Other Systems

- Thymic hormones as well as cytokines are involved in establishing the normal level of corticotropin-releasing hormone (CRH) produced by the hypothalamus. Removal or destruction of the thymus leads to a decrease in ACTH production and a decline in circulating adrenocortical hormones. Thymic hormones also stimulate the production of thyroid hormone–releasing hormone (TRH), leading to thyroid-stimulating hormone (TSH) secretion at the anterior pituitary. As a result, circulating thyroid hormone levels increase when an immune response is under way. The thyroid hormones further stimulate cell and tissue metabolism.
- Other thymic hormones affect the anterior pituitary directly, stimulating secretion of prolactin and growth hormone.

On the other hand, the nervous system can apparently adjust the sensitivity of the immune response. For example:

- The CNS innervates dendritic cells in the lymph nodes and spleen, Langerhans cells in the skin, and other antigen-presenting cells. The nerve endings release neurotransmitters that exaggerate and enhance local immune responses. For this reason, some skin conditions, such as *psoriasis*, become worse when an individual is under stress.
- Glial cells in the CNS produce cytokines that promote an immune response.
- A sudden decline in immune function can occur after even a brief period of emotional distress. Recent studies indicate that a heated argument can produce a steep reduction in circulating numbers of white blood cells over the next 24 hours.

The mechanisms involved and the functional significance of these interactions remain to be determined. The last decade saw an explosive growth in our understanding of the immune response; that growth will continue in the decades to come.

Would the primary response or the secondary response be more affected by a lack of memory B cells for a particular antigen?

What kind of immunity protects the developing fetus, and how is it produced?

How does increased stress decrease the effectiveness of the immune response?

Selected Clinical Terminology

Terms Discussed in This Chapter

acquired immune deficiency syndrome (AIDS): A disorder that develops following HIV infection, characterized by reduced circulating antibody levels and depressed cellular immunity. *(p. 810* and *AM)*

allergen: An antigen capable of triggering an allergic reaction. *(p. 816)*

allergy: An inappropriate or excessive immune response to antigens, triggered by stimulation of mast cells bound to IgE. *(p. 816* and *AM)*

anaphylactic shock: A drop in blood pressure that may lead to circulatory collapse, resulting from a severe case of anaphylaxis. *(p. 816)*

anaphylaxis (a-na-fi-LAK-sis): A type of allergy in which a circulating allergen affects mast cells throughout the body, producing numerous symptoms very quickly. *(p. 816)*

appendicitis: Infection and inflammation of the aggregate lymphoid nodules in the appendix. *(p. 786 and AM)*

autoimmune disorder: A disorder that develops when the immune response mistakenly targets normal body cells and tissues. *(p. 815)*

filariasis: An infection by larval stages of a parasitic nematode; the adult worms may scar and block lymphatic vessels, causing acute lymphadema, often affecting the external genitalia and lower limbs **(elephantiasis).** *(p. 783)*

human immunodeficiency virus (HIV): The virus responsible for AIDS and related immunodeficiency disorders. *(p. 810* and *AM)*

immune complex disorder: A disorder caused by the precipitation of immune complexes at sites such as the kidneys, where their presence disrupts normal tissue function. *(p. 816* and *AM)*

immunodeficiency disease: A disease in which either the immune system fails to develop normally or the immune response is somehow blocked. *(p. 815)*

immunosuppression: A reduction in the sensitivity of the immune system. *(p. 803)*

immunosuppressive drugs: Drugs administered to inhibit the immune response; examples include prednisone, cyclophosphamide, azathioprine, cyclosporin, and FK506. *(p. 803* and *AM)*

lymphadenopathy (lim-fad-e-NOP-a-thē): Chronic or excessive enlargement of lymph nodes. *(p. 788)*

lymphedema: A painless accumulation of lymph in a region whose lymphatic drainage has been blocked *(p. 783)*

lymphomas: Malignant cancers consisting of abnormal lymphocytes or lymphocytic stem cells; examples include *Hodgkin's disease* and *non-Hodgkin's lymphoma.* *(p. 788* and *AM)*

severe combined immunodeficiency disease (SCID): A congenital disorder resulting from the failure to develop cellular or humoral immunity. *(p. 815* and *AM)*

tonsillitis: Infection of one or more tonsils; symptoms include a sore throat, high fever, and leukocytosis (an abnormally high white blood cell count). *(p. 786* and *AM)*

AM Additional Terms Discussed in the Applications Manual

active immunization: An immunization that intentionally stimulates a primary response to a pathogen before it is encountered in the environment.
bacteria: Prokaryotic cells (cells lacking nuclei or other membranous organelles) that may be extracellular or intracellular pathogens.
bone marrow transplantation: Infusion of bone marrow from a compatible donor following the destruction of host marrow through radiation or chemotherapy; a treatment option for acute, late-stage lymphoma.
fungi (singular *fungus*): Eukaryotic organisms that absorb organic materials from the remains of dead cells; some fungi are pathogenic.
graft-versus-host disease (GVH): A condition that results when T cells in donor tissues, such as bone marrow, attack the tissues of the recipient.
hybridoma (hī-bri-DŌ-ma): A cell formed by fusion of a cancer cell with a plasma cell; used to produce large amounts of a single antibody.
hypersplenism: A condition caused by an overactive spleen; symptoms include anemia, leukopenia, and thrombocytopenia.
hyposplenism (hī-pō-SPLĒN-ism): A condition resulting from the absence or nonfunctional state of the spleen; there are few symptoms.
monoclonal (mo-nō-KLŌ-nal) **antibody:** An antibody produced in large quantities by a population of genetically identical cells.
mononucleosis: A condition, also known as the "kissing disease," resulting from chronic infection by the *Epstein-Barr virus (EBV);* symptoms include splenic enlargement, fever, sore throat, widespread swelling of lymph nodes, increased numbers of lymphocytes in the blood, and the presence of circulating antibodies to the virus.
passive immunization: An immunization in which the patient receives a dose of antibodies that will attack a pathogen immediately without involving the host's immune system.
protozoa: Unicellular eukaryotic organisms that are abundant in soil and water; some protozoa are pathogenic.
splenomegaly (splen-ō-MEG-a-lē): Enlargement of the spleen.
systemic lupus erythematosus (LOO-pus e-rith-ē-ma-TŌ-sis) **(SLE):** A condition resulting from a generalized breakdown in the antigen recognition mechanism, leading to the production of antibodies that destroy normal cells and tissues.
tonsillectomy: The removal of an infected tonsil.
vaccine (vak-SĒN): A preparation of antigens derived from a specific pathogen; administered during immunization.
viruses: Noncellular pathogens that replicate by directing the synthesis of virus-specific proteins and nucleic acids inside tissue cells.

CHAPTER REVIEW

STUDY OUTLINE

INTRODUCTION, p. 780

1. The cells, tissues, and organs of the **lymphatic** system play a central role in the body's defenses against a variety of **pathogens,** or disease-causing organisms.

AN OVERVIEW OF THE LYMPHATIC SYSTEM, p. 780

1. Lymphocytes, the primary cells of the lymphatic system, provide an **immune response** to specific threats to the body. **Immunity** is the ability to resist infection and disease through the activation of specific defenses.

ORGANIZATION OF THE LYMPHATIC SYSTEM, p. 780

1. The lymphatic system includes a network of **lymphatic vessels** that carry **lymph** (a fluid similar to plasma but with a lower concentration of proteins). A series of **lymphatic organs** is connected to the lymphatic vessels. (*Figure 22-1*)

Functions of the Lymphatic System, p. 780

2. The lymphatic system produces, maintains, and distributes lymphocytes (cells that attack invading organisms, abnormal cells, and foreign proteins). The system also helps maintain blood volume and eliminate local variations in the composition of the interstitial fluid.

Lymphatic Vessels, p. 781

3. Lymph flows along a network of **lymphatics** that originate in the **lymphatic capillaries** (*terminal* lymphatics). The lymphatic vessels empty into the **thoracic duct** and the **right lymphatic duct.** (*Figures 22-1 to 22-4*)

Lymphocytes, p. 783

4. There are three different classes of lymphocytes: **T cells** (**t**hymus-dependent), **B cells** (**b**one marrow–derived), and **NK cells** (**n**atural **k**iller).
5. **Cytotoxic T cells** attack foreign cells or body cells infected by viruses; they provide **cellular** immunity. **Regulatory T** cells regulate and coordinate the immune response.
6. B cells can differentiate into **plasma cells,** which produce and secrete *antibodies* that react with specific chemical targets, or **antigens.** Antibodies in body fluids are called **immunoglobulins.** B cells are responsible for **humoral immunity.**
7. NK cells (also called **large granular lymphocytes**) attack foreign cells, normal cells infected with viruses, and cancer cells. They provide *immunological surveillance.*
8. Lymphocytes continually migrate into and out of the blood through the lymphatic tissues and organs. **Lymphopoiesis** (lymphocyte production) involves the **bone marrow, thymus,** and **peripheral lymphatic tissues.** (*Figure 22-5*)

Lymphoid Tissues, p. 786

9. **Lymphoid tissues** are connective tissues dominated by lymphocytes. In a **lymphoid nodule** the lymphocytes are densely packed in an area of loose connective tissue. (*Figure 22-6*)

Lymphoid Organs, p. 787

10. Important **lymphoid organs** include the **lymph nodes**, the **thymus,** and the **spleen.** Lymphoid tis-

sues and organs are distributed in areas especially vulnerable to injury or invasion.

11. **Lymph nodes** are encapsulated masses of lymphatic tissue. The inner **cortex** is dominated by T cells; the outer cortex and **medulla** contain B cells. (*Figure 22-7*)
12. The thymus lies behind the sternum, in the anterior mediastinum. **Epithelial cells** scattered among the lymphocytes produce thymic hormones. (*Figure 22-8*)
13. The adult spleen contains the largest mass of lymphoid tissue in the body. The cellular components form the **pulp** of the spleen. **Red pulp** contains large numbers of red blood cells, and areas of **white pulp** resemble lymphoid nodules. (*Figure 22-9*)

The Lymphatic System and Body Defenses, p. 791

14. The lymphatic system is a major component of the body's defenses. These fall into two categories; (1) **nonspecific** defenses that do not discriminate between one threat and another and (2) **specific defenses** that protect against threats on an individual basis.

NONSPECIFIC DEFENSES, p. 791

1. Nonspecific defenses prevent the approach of, deny entrance to, or limit the spread of living or nonliving hazards. (*Figure 22-10*)

Physical Barriers, p. 791

2. Physical barriers include hair, epithelia, and various secretions of the integumentary and digestive systems.

Phagocytes, p. 791

3. There are two types of phagocytic cells: **microphages** and **macrophages** (cells of the **monocyte-macrophage system**). Microphages are the neutrophils and eosinophils in circulating blood.
4. Phagocytes move between cells through **diapedesis,** and they show **chemotaxis** (sensitivity and orientation to chemical stimuli).

Immunological Surveillance, p. 793

5. **Immunological surveillance** involves constant monitoring of normal tissues by NK cells sensitive to abnormal antigens on the surfaces of otherwise normal cells. Cancer cells with **tumor-specific antigens** on their surfaces are killed. (*Figure 22-11*)

Interferons, p. 794

6. **Interferons,** small proteins released by cells infected with viruses, trigger the production of **antiviral proteins** that interfere with viral replication inside the cell. Interferons are **cytokines,** chemical messengers released by tissue cells to coordinate local activities.

Complement, p. 795

7. At least 11 **complement proteins** make up the **complement system.** They interact with each other in chain reactions to destroy target cell membranes, stimulate inflammation, attract phagocytes, and/or enhance phagocytosis. (*Figure 22-12*)

Inflammation, p. 795

8. Inflammation represents a coordinated nonspecific response to tissue injury. (*Figure 22-13*)

Fever, p. 798

9. A **fever** (body temperature greater than 37.2° C or 99° F) can inhibit pathogens and accelerate metabolic processes.

SPECIFIC RESISTANCE: THE IMMUNE RESPONSE, p. 798

Forms of Immunity, p. 798

1. Specific immunity may involve **innate immunity** (genetically determined and present at birth) or **acquired immunity.** The two types of acquired immunity are **active immunity** (which appears after exposure to an antigen) and **passive immunity** (produced by the transfer of antibodies from another source). (*Figure 22-14*)

Properties of Immunity, p. 799

2. Lymphocytes provide specific immunity, which has four general characteristics: specificity, versatility, memory, and tolerance. **Memory cells** enable the immune system to "remember" previous target antigens. **Tolerance** refers to the ability of the immune system to ignore some antigens, such as those of body cells.

T Cells and Cell-Mediated Immunity, p. 800

3. Foreign antigens must usually be processed by **accessory cells** (most often macrophages) and incorporated into their membranes **(antigen presentation)** before they can activate lymphocytes.
4. All body cells have membrane glycoproteins. The genes controlling their synthesis make up a chromosomal region called the **major histocompatibility complex (MHC).** The membrane glycoproteins are called **MHC proteins.** Lymphocytes are not activated by lone antigens but will respond to an antigen bound to either a **Class I** or a **Class II** MHC protein.
5. Class I MHC proteins are found in all nucleated body cells. Class II MHC proteins are found only in **antigen-presenting cells (APCs)** and lymphocytes. (*Figure 22-15*)
6. Whether a T cell responds to antigens held in Class I or II MHC proteins depends on the structure of the T cell membrane. T cell membranes contain proteins called **CD (cluster designation) markers. CD8** markers are found on cytotoxic and suppressor T cells. **CD4** markers are found on all helper T cells.
7. **Cell-mediated immunity (cellular immunity)** results from the activation of CD8 T cells by antigens bound to Class I MHCs. When activated, most of these T cells divide to generate *cytotoxic T cells* and **memory T cells** that remain on reserve to guard against future such attacks. **Suppressor T cells** depress the responses of other T and B cells. (*Figure 22-16*)
8. **Helper,** or **CD4, T cells** respond to antigens presented by Class II MHC proteins. When activated, they secrete lymphokines that help coordinate specific and nonspecific defenses and regulate cellular and humoral immunity. (*Figure 22-17*)

B Cells and Antibody-Mediated Immunity, p. 804

9. B cells become **sensitized** when antigens bind to their membrane antibody molecules. The antigens are then displayed on the Class II MHC proteins of

the B cells and they become activated by helper T cells activated by the same antigen.
10. An activated B cell may differentiate into a **plasma cell** or produce daughter cells that differentiate into plasma cells and **memory B cells.** Antibodies are produced by the plasma cells. (*Figure 22-18*)
11. An antibody molecule consists of two parallel pairs of polypeptide chains containing constant and variable regions. (*Figure 22-19*)
12. When antibody molecules bind to an antigen they form an **antibody-antigen complex.** Effects that appear after binding include **neutralization; precipitation** (formation of an insoluble **immune complex**) and **agglutination** (formation of large complexes); **opsonization;** stimulation of **inflammation;** and prevention of bacterial or viral adhesion. (*Figure 22-21*)
13. There are five classes of antibodies in body fluids: (1) **immunoglobulin G (IgG),** responsible for resistance against many viruses, bacteria, and bacterial toxins; (2) **IgE,** which releases chemicals that accelerate local inflammation; (3) **IgD,** found on the surfaces of B cells; (4) **IgM,** the first antibody type secreted after an antigen arrives; and (5) **IgA,** found in glandular secretions. (*Figure 22-20*)

Primary and Secondary Responses to Antigen Exposure, p. 808

14. The antibodies first produced by plasma cells are the agents of the **primary response.** The maximum **antibody titer** appears during the **secondary (anamnestic) response** to antigen exposure. (*Figure 22-22*)

Hormones of the Immune System, p. 809

15. **Interleukins** increase T cell sensitivity to antigens exposed on macrophage membranes; stimulate B cell activity, plasma cell formation, and antibody production; and enhance nonspecific defenses.
16. **Interferons** slow the spread of a virus by making the synthesizing cell and its neighbors resistant to viral infections.
17. **Tumor necrosis factors** slow tumor growth and kill tumor cells.
18. Several lymphokines adjust the activities of phagocytic cells in order to coordinate specific and nonspecific defenses. (*Table 22-1*)

PATTERNS OF IMMUNE RESPONSE, p. 812

1. The initial steps in the immune response to viral and bacterial infections differ. (*Figures 22-23, 22-24, 22-25; Table 22-2*)

THE DEVELOPMENT OF RESISTANCE, p. 813

1. **Immunological competence** is the ability to demonstrate an immune response upon exposure to an antigen. The developing fetus receives passive immunity from the maternal bloodstream. After delivery the infant begins developing acquired immunity following exposure to environmental antigens.

IMMUNE DISORDERS, p. 814

Autoimmune Disorders, p. 815

1. **Autoimmune disorders** develop when the immune response mistakenly targets normal body cells and tissues. (*Table 22-3*)

Immunodeficiency Diseases, p. 815

2. In an **immunodeficiency disease** either the immune system does not develop normally or the immune response is somehow blocked.

Allergies, p. 816

3. **Allergies** are inappropriate or excessive immune responses to **allergens** (antigens that trigger allergic reactions). There are four types of allergies: *immediate hypersensitivity* (Type I), *cytotoxic reactions* (Type II), *immune complex disorders* (Type III), and *delayed hypersensitivity* (Type IV).
4. In **anaphylaxis** a circulating allergen affects mast cells throughout the body. (*Figure 22-26*)

STRESS AND THE IMMUNE RESPONSE, p. 817

1. Interleukin-1 released by active macrophages triggers the release of ACTH by the anterior pituitary. Glucocorticoids produced by the adrenal cortex moderate the immune response, but their long-term secretion can lower resistance to disease.

AGING AND THE IMMUNE RESPONSE, p. 817

1. With aging, the immune system becomes less effective at combating disease.

INTEGRATION WITH OTHER SYSTEMS, p. 817

1. There is substantial interaction between the cells and tissues involved with the immune response and the nervous and endocrine systems. (*Figure 22-27*)

REVIEW QUESTIONS

LEVEL 1 Reviewing Facts and Terms

1. Lymph from the lower abdomen, pelvis, and lower limbs is received by the:
 (a) right lymphatic duct (b) cysterna chyli
 (c) right thoracic duct (d) aorta

2. Lymphoid stem cells that can form all types of lymphocytes occur in the:
 (a) bloodstream (b) thymus
 (c) bone marrow (d) spleen

3. Lymphatics are found in all portions of the body except the:
 (a) lower limbs
 (b) central nervous system
 (c) integument
 (d) digestive tract

4. The largest collection of lymphoid tissue in the body is contained in the:

(a) adult spleen (b) adult thymus
(c) bone marrow (d) tonsils

5. Red blood cells that are damaged or defective are removed from the circulation by the:
(a) thymus (b) lymph nodes
(c) spleen (d) tonsils

6. Phagocytes move through capillary walls by squeezing between adjacent endothelial cells, a process known as:
(a) diapedesis (b) chemotaxis
(c) adhesion (d) perforation

7. Perforins and protectin are proteins associated with the activity of:
(a) T cells (b) B cells
(c) NK cells (d) plasma cells

8. Complement activation:
(a) stimulates inflammation
(b) attracts phagocytes
(c) enhances phagocytosis
(d) a, b, and c are correct

9. Inflammation:
(a) aids in temporary repair at an injury site
(b) slows the spread of pathogens
(c) facilitates permanent repair
(d) a, b, and c are correct

10. CD4 markers are associated with:
(a) cytotoxic T cells
(b) suppressor T cells
(c) helper T cells
(d) a, b, and c are correct

11. What two large collecting vessels are responsible for returning lymph to the veins of the circulatory system? What areas of the body does each serve?

12. Give a function for each of the following:
(a) cytotoxic T cells
(b) helper T cells
(c) suppressor (regulatory) T cells
(d) plasma cells
(e) NK cells
(f) stromal cells
(g) reticular epithelial cells
(h) interferons
(i) pyrogens
(j) B cells
(k) plasma cells
(l) interleukins
(m) tumor necrosis factor
(n) colony-stimulating factors

13. What are the three different classes of lymphocytes, and where does each class originate?

14. What seven defenses, present at birth, provide the body with the defensive capability known as nonspecific resistance?

LEVEL 2 Reviewing Concepts

15. Compared with nonspecific defenses, *specific defenses:*
(a) do not discriminate between one threat and another
(b) are always present at birth
(c) provide protection against threats on an individual basis
(d) deny entrance of pathogens to the body

16. T cells and B cells can be activated only by:
(a) pathogenic microorganisms
(b) interleukins, interferons, and colony-stimulating factors
(c) cells infected with viruses, bacterial cells, or cancer cells
(d) exposure to a specific antigen at a specific site on a cell membrane

17. Class II MHC proteins appear in the cell membrane only when:
(a) the plasma cells are releasing antibodies
(b) the cell is processing antigens
(c) cytotoxic T cells are inhibited
(d) NK cells are activated

18. List and explain the four general properties of immunity.

19. How does a cytotoxic T cell destroy another cell displaying antigens bound to Class I MHC proteins?

20. How does the formation of an antibody-antigen complex cause elimination of an antigen?

21. What effects follow activation of the complement system?

22. An anesthesia technician is advised that she should be vaccinated against hepatitis B, which is caused by a virus. She is given one injection and is told to come back for a second injection in a month and a third injection after 6 months. Why is this series of injections necessary?

LEVEL 3 Critical Thinking and Clinical Applications

23. An investigator at a crime scene discovers some body fluid on the victim's clothing. He carefully takes a sample and sends it to the crime lab for analysis. On the basis of analysis of immunoglobulins, could the crime lab determine whether the sample was blood plasma or semen? Explain.

24. Ted finds out that he has been exposed to the measles. He is concerned that he might have contracted the disease, so he goes to see his physician. The doctor takes a blood sample and sends it to a lab for antibody titers. The results show an elevated level of IgM antibodies to rubella (measles) virus, but very few IgG antibodies to the virus. Has Ted contracted the disease or not?

25. While walking along the street, you and your friend see an elderly lady whose left arm appears to be swollen to several times its normal size. Your friend remarks that she must have been in the tropics and contracted a filarial disease that produces elephantiasis. You disagree, saying that it is more likely that the woman had a radical mastectomy (removal of a breast because of cancer). Explain the rationale behind your answer.

26. Tilly has T cells that are capable of responding to antigen A. Tilly's friend Harry has been exposed to antigen A, and Tilly offers to have some of her T cells transfused so that Harry will not come down with an infection. You happen to overhear their conversation and observe that such a transfusion would not work. Why not?

27. You are a researcher interested in studying cells that can respond to the hormone FSH. How could you use antibodies to help you locate FSH responsive cells?

CHAPTER

23

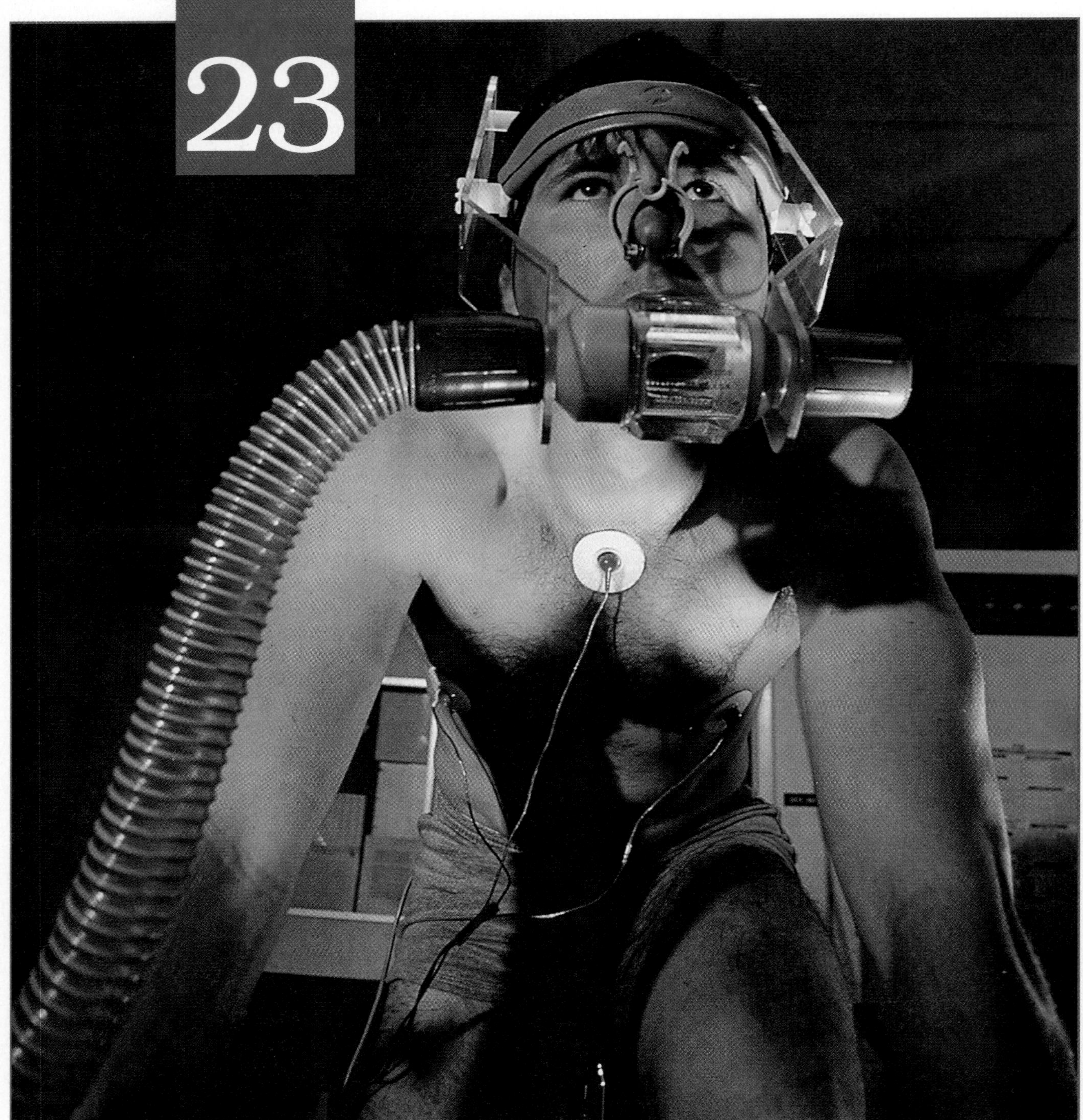

This is one way to go nowhere fast—pedaling on a treadmill. All the effort, however, is hardly wasted. The rather awkward-looking array of hoses and wires attached to this rider is designed to monitor respiratory and cardiovascular performance during exercise. Similar equipment can be used in a clinical setting to assess respiratory function in resting or active individuals.

The respiratory and cardiovascular systems are closely linked, both anatomically and functionally. In this chapter we will see how they cooperate to supply our tissues with vital oxygen and protect them from the carbon dioxide they generate. We will also examine the homeostatic mechanisms responsible for regulating respiratory activities as levels of physical activity change.

The Respiratory System

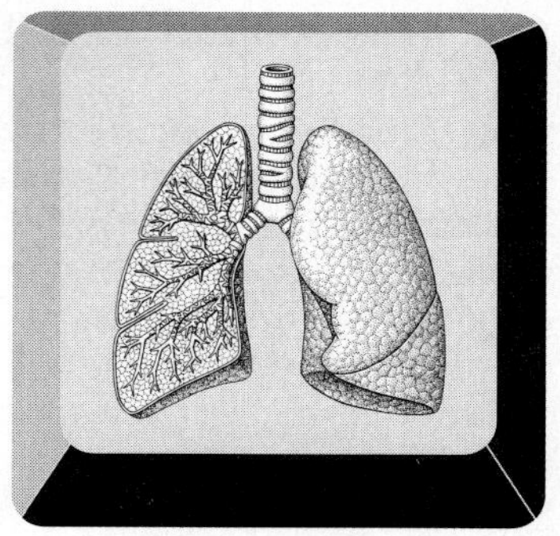

Chapter Outline and Objectives

Living cells need energy for maintenance, growth, defense, and replication. Our cells obtain that energy through aerobic mechanisms that require oxygen and produce carbon dioxide. [p. 88] Many aquatic organisms can obtain oxygen and excrete carbon dioxide by diffusion across the surface of the skin or in specialized structures, such as the gills of a fish. Such arrangements are poorly suited for life on land because the exchange surfaces must be very thin and relatively delicate to permit rapid diffusion. In air, the exposed membranes would collapse, evaporation and dehydration would reduce blood volume, and the delicate surfaces would be vulnerable to attack by pathogenic organisms. Our respiratory exchange surfaces are just as delicate as those of an aquatic organism, but they are confined to the inside of the lungs, in a warm, moist, and protected environment. Under these conditions diffusion can occur between the air and the blood.

The cardiovascular system provides the link between the interstitial fluids and the exchange surfaces of the lungs. The circulating blood carries oxygen from the lungs to peripheral tissues; it also accepts and transports the carbon dioxide generated by those tissues, delivering it to the lungs.

Diffusion between the blood and air occurs at the **alveoli** (al-VĒ-ō-lī) of the lungs. The distance between the blood in an alveolar capillary and the air inside an alveolus is usually less than 1 μm and sometimes as small as 0.1 μm. To meet the metabolic requirements of our peripheral tissues, the exchange surfaces of the lungs must be very large. In fact, the total surface area for gas exchange in the adult lungs is 20–25 times the surface area of the body.

Our discussion of the respiratory system begins by following the air as it travels toward the alveoli. We will then consider the mechanics of breathing and the physiology of respiration.

Functions of the Respiratory System

The functions of the respiratory system include:

1. Providing an extensive area for gas exchange between air and circulating blood.
2. Moving air to and from the exchange surfaces of the lungs.
3. Protecting respiratory surfaces from dehydration, temperature changes, or other environmental variations and defending the respiratory system and other tissues from invasion by pathogenic microorganisms.
4. Producing sounds involved in speaking, singing, and nonverbal communication.
5. Providing olfactory sensations to the CNS from the olfactory epithelium in the superior portions of the nasal cavity.
6. Assisting in the regulation of body fluid pH.

In addition, the capillaries of the lungs indirectly assist in the regulation of blood volume and blood pressure, through the conversion of angiotensin I to angiotensin II. [pp. 628, 743]

Organization of the Respiratory System

Figure 23-1

The **respiratory system** (Figure 23-1●) can be divided into an *upper respiratory system* and a *lower respiratory system.*

- The **upper respiratory system** consists of the nose, nasal cavity, paranasal sinuses, and pharynx. These passageways filter, warm, and humidify the air, protecting the more delicate surfaces of the lower respiratory system.
- The **lower respiratory system** includes the *larynx* (voicebox), *trachea* (windpipe), *bronchi, bronchioles,* and *alveoli* of the lungs.

The **respiratory tract** consists of the airways that carry air to and from the exchange surfaces of the lungs. The respiratory tract can be divided into a *conducting portion* and a *respiratory portion.* The conducting portion begins at the entrance to the nasal cavity and extends through the pharynx and larynx and along the trachea, bronchi, and bronchioles to the *terminal bronchioles.* The respiratory portion of the tract includes the delicate *respiratory bronchioles* and the alveoli that are the site of gas exchange.

Filtering, warming, and humidification of the inspired air begin at the entrance to the upper respiratory system and continue throughout the rest of the conducting system. By the time air reaches the alveoli, most foreign particles and pathogens have been removed, and the humidity and temperature are within acceptable limits. The success of this "conditioning process" is due primarily to the properties of the *respiratory mucosa.*

The Respiratory Mucosa

The **respiratory mucosa** (myū-KŌ-sa) lines the respiratory system. A *mucosa,* or mucous membrane, one of the four types of membranes introduced in Chapter 4, consists of an epithelium and an underlying layer of loose connective tissue. [p. 134]

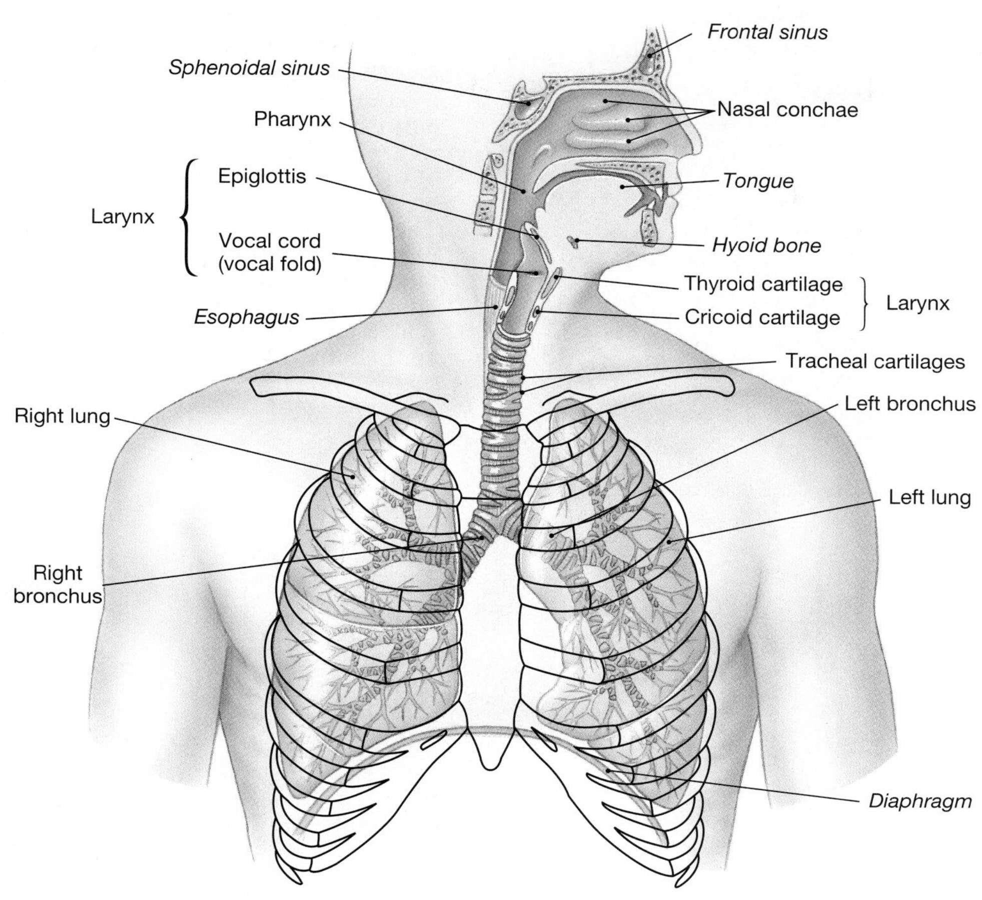
Frontal sinus
Sphenoidal sinus
Nasal conchae
Pharynx
Epiglottis
Tongue
Larynx
Vocal cord
(vocal fold)
Hyoid bone
Thyroid cartilage
Larynx
Esophagus
Cricoid cartilage
Tracheal cartilages
Left bronchus
Right lung
Left lung
Right
bronchus
Diaphragm

● **FIGURE 23-1**
Components of the Respiratory System.
Plates 3.1, 4.4, 6.3

The Respiratory Epithelium

Figure 23-2

The structure of the respiratory epithelium changes as you proceed along the respiratory tract.

- A pseudostratified, ciliated, columnar epithelium with numerous goblet cells (Figure 23-2●) lines the nasal cavity and the superior portion of the pharynx.
- The epithelium lining inferior portions of the pharynx is a stratified squamous epithelium similar to that of the oral cavity. The pharynx (1) conducts air to the lower respiratory tract and (2) conveys food to the esophagus. A stratified epithelium is essential to provide protection from mechanical and chemical damage as the relatively unprocessed food moves across the pharyngeal surfaces.
- A pseudostratified ciliated columnar epithelium similar to that lining the nasal cavity is found along the conducting portions of the lower respiratory system.
- In the smaller bronchioles the pseudostratified epithelium is replaced by a cuboidal epithelium with scattered cilia.
- A very delicate simple squamous epithelium forms the lining of the alveoli. Other, more specialized cells are scattered within the alveolar epithelium.

The Lamina Propria

The **lamina propria** (LA-mi-na PRŌ-prē-a) is the layer of loose connective tissue that supports the respiratory epithelium. In the upper respiratory system and in the trachea and bronchi, the lamina propria contains mucous glands that discharge their secretions onto the epithelial surface. The lamina propria in the conducting portions of the lower respiratory system contains bundles of smooth muscle fibers. At the level of the bronchioles the smooth muscles form relatively thick bands that encircle or spiral around the lumen.

The Respiratory Defense System

The delicate exchange surfaces of the respiratory system can be severely damaged if the inspired air becomes contaminated with debris or pathogens. Such contamination is prevented by a series of filtration mechanisms that together make up the **respiratory defense system.**

Along much of the length of the respiratory tract, goblet cells in the epithelium and mucous glands in the lamina propria produce a sticky

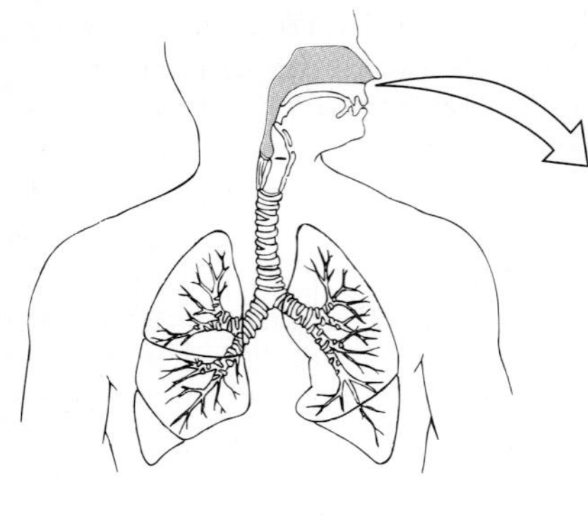

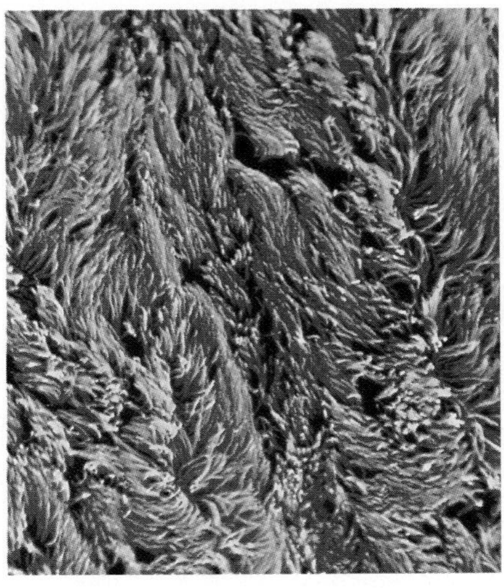

(a)

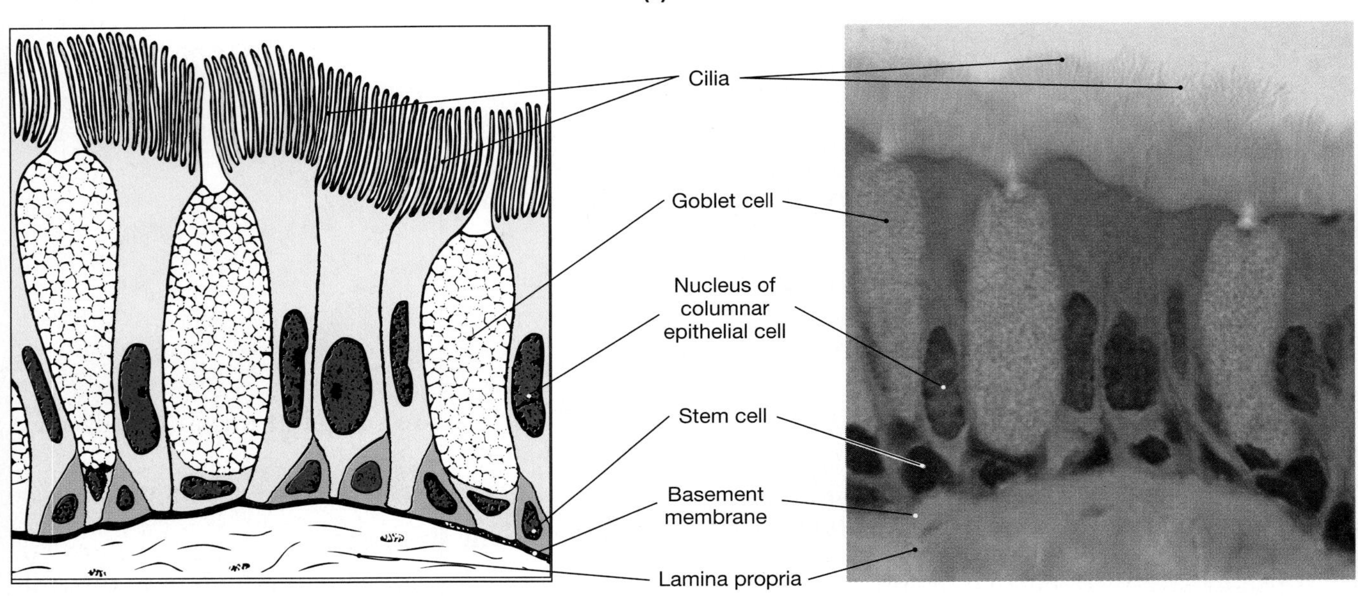
Cilia
Goblet cell
Nucleus of columnar epithelial cell
Stem cell
Basement membrane
Lamina propria

(b)

● **FIGURE 23-2**
The Respiratory Epithelium of the Nasal Cavity and Conducting System. **(a)** A surface view of the epithelium, as seen with the scanning electron microscope. The cilia of the epithelial cells form a dense layer that resembles a shag carpet. The movement of these cilia propels mucus across the epithelial surface. (SEM × 1614) **(b)** Sketch and light micrograph showing the sectional appearance of the respiratory epithelium. (LM × 1062)

mucus that bathes exposed surfaces. In the nasal cavity, cilia sweep that mucus and any trapped debris or microorganisms toward the pharynx, where it will be swallowed and exposed to the acids and enzymes of the stomach. In the lower respiratory system, the cilia also beat toward the pharynx, moving a carpet of mucus toward the pharynx and cleaning the respiratory surfaces. This process is often described as a *mucus escalator*.

Filtration in the nasal cavity removes virtually all particles larger than around 10 μm from the inspired air. Smaller particles may be trapped by the mucus of the nasopharynx or secretions of the pharynx before proceeding farther along the conducting system. Exposure to unpleasant stimuli, such as noxious vapors, large quantities of dust and debris, allergens, or pathogens, usually causes a rapid increase in the rate of mucus production in the nasal cavity and paranasal sinuses. (The familiar symptoms of the "common cold" result from the invasion of this respiratory epithelium by any of over 200 viruses.)

Particles 1–5 μm in diameter are usually trapped in the mucus coating the respiratory bronchioles or in the liquid covering the alveolar surfaces. These areas are outside the boundaries of the mucus escalator, but the foreign particles can be engulfed by alveolar macrophages. Particles smaller than about 0.5 μm usually remain suspended in the air.

Large quantities of airborne particles may overload the respiratory defenses and produce a variety of illnesses. For example, the presence of irritants in the lining of the conducting passageways may provoke the formation of abscesses, and damage to the epithelium in the affected area may allow the irritants to enter the surrounding tissues of the lung. There is also a strong link between airborne irritants and the development of lung cancer (p. 871). (Changes induced in the respiratory epithelium by smoking were noted in Chapter 4, Figure 4-19●, p. 140.) AM *Overloading the Respiratory Defenses*

CYSTIC FIBROSIS **Cystic fibrosis (CF)** is the most common lethal inherited disease affecting Caucasians of Northern European descent, occurring at a frequency of 1 birth in 2500. It occurs, with less frequency, in those with Southern European ancestry, in the Ashkenazi Jewish population, and in African-Americans. The condition results from a defective gene located on chromosome 7. Individuals with CF seldom survive past age 30; death is usually the result of a massive bacterial infection of the lungs and associated heart failure.

The most serious symptoms appear because the respiratory mucosa in these individuals produces a dense, viscous mucus that cannot be transported by the respiratory defense system. The mucus escalator stops working, and mucus blocks the smaller respiratory passageways. This blockage reduces the diameter of the airways, making breathing difficult, and the inactivation of the normal respiratory defenses leads to frequent bacterial infections. For information on the genetic basis of this disorder and current strategies for treatment, refer to the *Applications Manual.* AM *Cystic Fibrosis*

The Upper Respiratory System

Figures 23-1, 23-3

The upper respiratory system consists of the nose, nasal cavity, paranasal sinuses, and pharynx (Figures 23-1 and 23-3●).

The Nose and Nasal Cavity

Figure 23-3

The nose is the primary passageway for air entering the respiratory system. Air normally enters the respiratory system via the paired **external nares** (NĀR-ēz), or nostrils, that open into the **nasal cavity.** The **vestibule** (VES-ti-būl) is the portion of the nasal cavity contained within the flexible tissues of the external nose (Figure 23-3●). The epithelium of the vestibule contains coarse hairs that extend across the external nares. Large airborne particles, such as sand, sawdust, or even insects, are trapped in these hairs and are prevented from entering the nasal cavity.

The *nasal septum* divides the nasal cavity into left and right portions. The bony portion of the nasal septum is formed by the fusion of the perpendicular plate of the ethmoid and the plate of the vomer (Figure 7-3d●, p. 206). The anterior portion of the nasal septum is formed of hyaline cartilage. This cartilaginous plate supports the bridge, or *dorsum nasi* (DOR-sum NĀ-zī), and *apex* (tip) of the nose.

The maxillary, nasal, frontal, ethmoid, and sphenoid bones form the lateral and superior walls of the nasal cavity. The mucous secretions produced in the associated *paranasal sinuses*, aided by the tears draining through the nasolacrimal ducts, help keep the surfaces of the nasal cavity moist and clean. (The paranasal sinuses were discussed in Chapter 7—see Figure 7-11●, p. 215.) The superior portion, or *olfactory region*, of the nasal cavity includes the areas lined by olfactory epithelium: (1) the inferior surface of the cribriform plate, (2) the superior portion of the nasal septum, and (3) the superior nasal conchae. Receptors in the olfactory epithelium provide the sense of smell. [p. 559]

The *superior, middle,* and *inferior nasal conchae* project toward the nasal septum from the lateral walls of the nasal cavity. To pass from the vestibule to the internal nares, air tends to flow between adjacent conchae, through the **superior, middle,** and **inferior meatus** (mē-Ā-tus; *meatus,* a passage) (Figure 23-3●). These are narrow grooves rather than open passageways, and the incoming air bounces off the conchal surfaces and churns around like a stream flowing over rapids. This turbulence serves a purpose: As the air eddies and swirls, small airborne particles are likely to come in contact with the mucus that coats the lining of the nasal cavity. In addition to promoting filtration, the turbulence allows extra time for warming and humidifying the incoming air. It also creates eddy currents that bring olfactory stimuli to the olfactory receptors.

A bony **hard palate,** formed by portions of the maxillary and palatine bones, forms the floor of the nasal cavity and separates the oral and nasal cavities. A fleshy **soft palate** extends behind the hard palate, marking the boundary line between the superior **nasopharynx** (nā-zō-FAR-inks) and the rest of the pharynx. The nasal cavity opens into the nasopharynx at the **internal nares.**

The Nasal Mucosa

The mucosa of the nasal cavity prepares the air for arrival at the lower respiratory system. Throughout much of the nasal cavity the lamina propria contains an abundance of arteries, veins, and capillaries that bring nutrients and water to the secretory cells. The lamina propria of the nasal conchae also contains an extensive network of large and highly distensible veins. This extensive vascularization provides a mechanism for warming and humidifying the incoming air. As cool, dry air pass-

FIGURE 23-3
The Nose, Nasal Cavity, and Pharynx. (a) The nasal cartilages and external landmarks on the nose. **(b)** The meatuses and the maxillary and ethmoidal sinuses. **(c)** The nasal cavity and pharynx as seen in sagittal section, with the nasal septum removed. AM *Plates 3.1, 4.4; Scan 2*

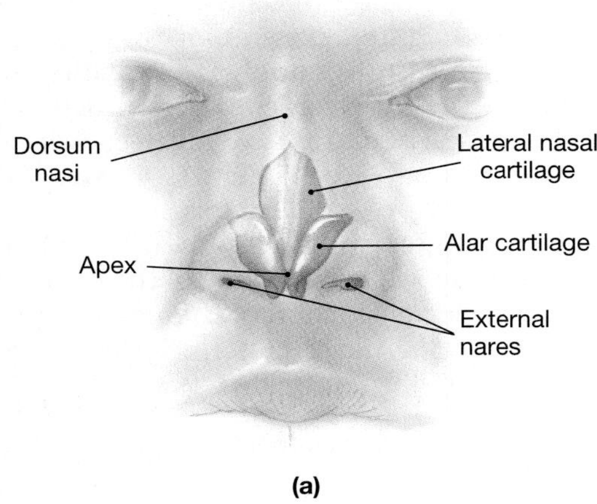
Dorsum nasi
Lateral nasal cartilage
Alar cartilage
Apex
External nares

(a)

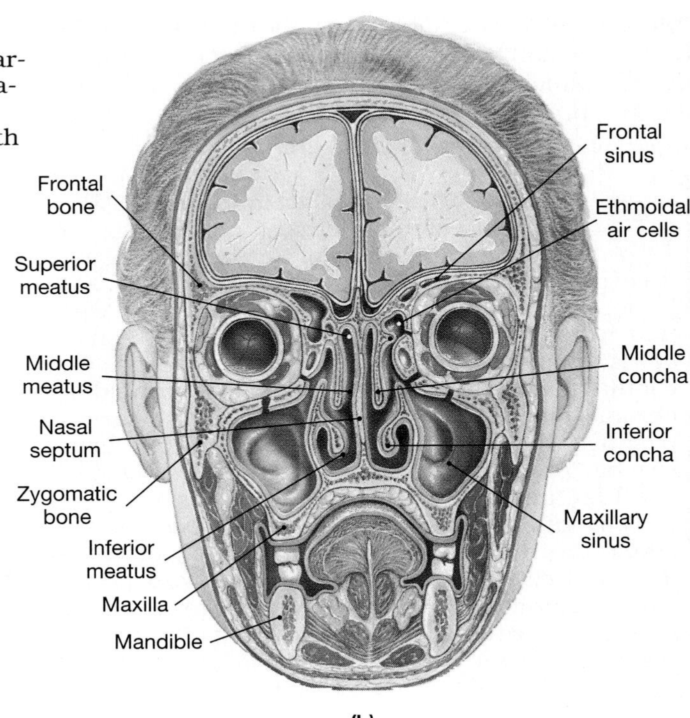
Frontal bone
Superior meatus
Middle meatus
Nasal septum
Zygomatic bone
Inferior meatus
Maxilla
Mandible
Frontal sinus
Ethmoidal air cells
Middle concha
Inferior concha
Maxillary sinus

(b)

Frontal sinus
Internal nares
Superior
Middle
Inferior
Nasal conchae
Nasopharynx
Pharyngeal tonsil
Nasal vestibule
Entrance to pharyngotympanic tube
External nares
Soft palate
Hard palate
C_2
Oral cavity
Palatine tonsil
Tongue
Oropharynx
Mandible
Epiglottis
Lingual tonsil
Glottis
Hyoid
Laryngopharynx
Thyroid cartilage
Vocal fold
Cricoid cartilage
Trachea
C_7
Esophagus
Thyroid gland

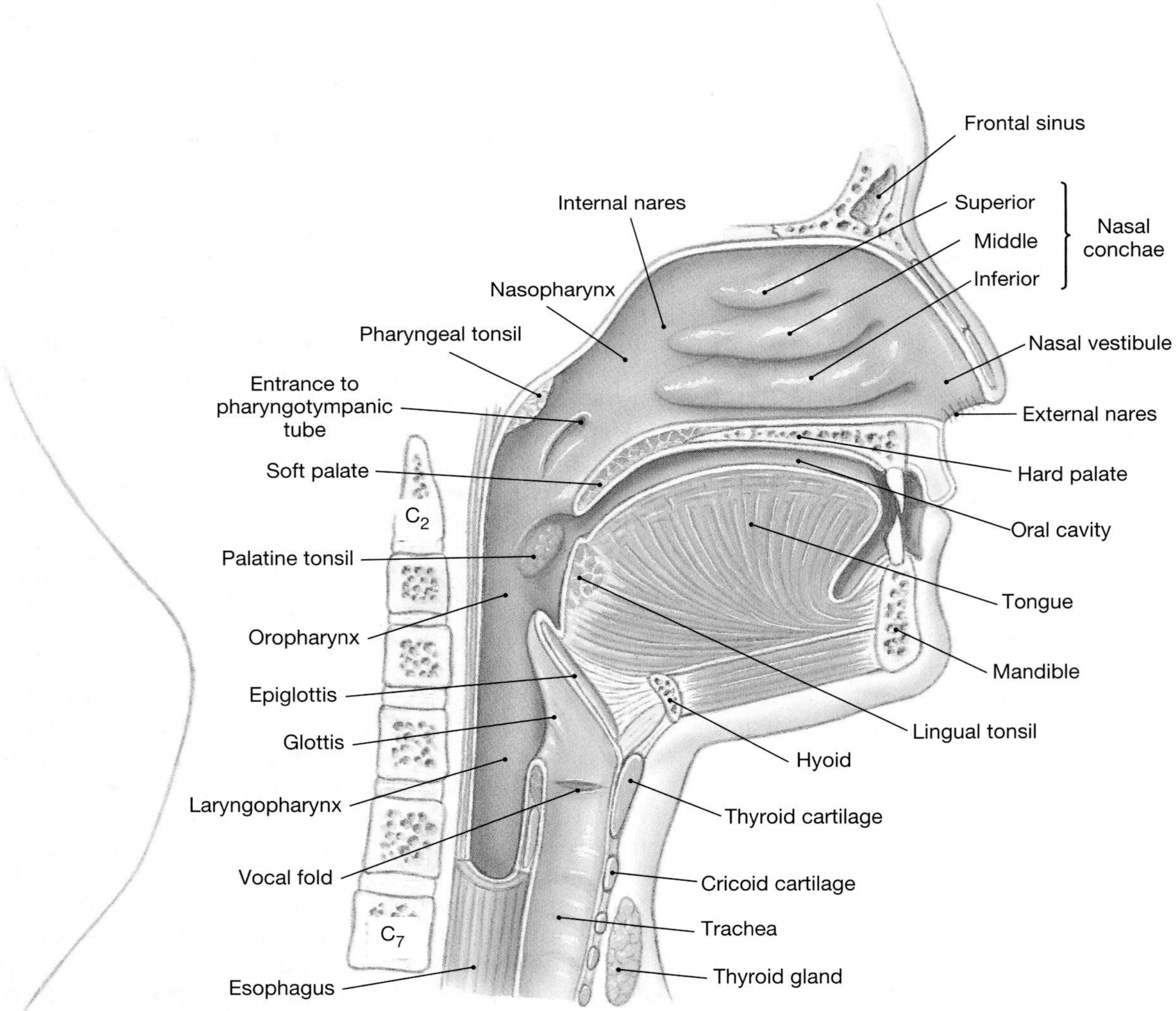

(c)

es over the exposed surfaces of the nasal cavity, the warm epithelium radiates heat and the water in the mucus evaporates. Air leaving the nasal cavity has been heated almost to body temperature, and it is nearly saturated with water vapor. This mechanism protects more delicate respiratory surfaces from chilling or drying out—two potentially disastrous events. Breathing through the mouth eliminates much of the preliminary filtration, heating, and humidifying of the inspired air. Patients breathing on a respirator, which utilizes a tube to provide air directly into the trachea, must receive air that has been externally filtered and humidified or risk alveolar damage.

NOSEBLEEDS The extensive vascularization of the nasal cavity and the relatively vulnerable position of the nose make a nosebleed, or **epistaxis** (ep-i-STAK-sis), a fairly common event. Bleeding usually involves vessels of the mucosa covering the cartilaginous portion of the septum. A variety of factors may be responsible. Examples include trauma, such as a punch in the nose, drying, infections, allergies, and clotting disorders. Hypertension may also provoke a nosebleed by rupturing small vessels of the lamina propria.

❑ The Pharynx

Figure 23-3

The nose, mouth, and throat are connected to the **pharynx** (FAR-inks). The pharynx is a chamber shared by the digestive and respiratory systems. It extends between the internal nares and the entrances to the larynx and esophagus. The curving superior and posterior walls are closely bound to the axial skeleton, but the lateral walls are flexible and muscular.

The pharynx is divided into three regions (Figure 23-3c•): the *nasopharynx*, the *oropharynx*, and the *laryngopharynx*.

1. The **nasopharynx** is the superior portion of the pharynx. It is connected to the posterior portion of the nasal cavity via the internal nares and is separated from the oral cavity by the soft palate (Figure 23-3c•).

   The nasopharynx is lined by a pseudostratified ciliated epithelium. The *pharyngeal tonsil* is located on the posterior wall of the nasopharynx, and, on each side, one of the *pharyngotympanic tubes* opens into the nasopharynx. ∞ *[p. 210, 786]*
2. The **oropharynx** (*oris*, mouth) extends between the soft palate and the base of the tongue at the level of the hyoid bone. The posterior portion of the oral cavity communicates directly with the oropharynx, as does the posterior inferior portion of the nasopharynx. At the boundary between the nasopharynx and oropharynx the epithelium changes from a pseudostratified columnar epithelium to a stratified squamous epithelium similar to that of the oral cavity.
3. The narrow **laryngopharynx** (la-rin-gō-FĀR-inks), the most inferior portion of the pharynx, includes that portion of the pharynx lying between the hyoid bone and the entrance to the larynx and esophagus (Figure 23-3•). Like the oropharynx it is lined by a stratified squamous epithelium that can resist mechanical abrasion, chemical attack, and pathogenic invasion.

The Larynx

Figure 23-4

Inspired (inhaled) air leaves the pharynx by passing through a narrow opening, the **glottis** (GLOT-is). The **larynx** (LAR-inks) surrounds and protects the glottis (Figure 23-4•). The larynx begins at the level of vertebra C_4 or C_5 and ends at the level of vertebra C_7. The larynx is essentially a cylinder whose cartilaginous walls are stabilized by ligaments and skeletal muscles.

❑ Cartilages of the Larynx

Figure 23-4

Three large cartilages form the body of the larynx (Figure 23-4•): the *thyroid cartilage*, the *cricoid cartilage*, and the *epiglottis*.

- The **thyroid cartilage** (*thyroid*, shield-shaped) is the largest laryngeal cartilage. Consisting of hyaline cartilage, it forms most of the anterior and lateral walls of the larynx. The thyroid cartilage in section has the form of a U; it is incomplete posteriorly. The prominent anterior surface of this cartilage, easily seen and felt, is commonly called the *Adam's apple*. The inferior surface of the thyroid cartilage articulates with the cricoid cartilage. The superior surface has ligamentous attachments to the epiglottis and smaller laryngeal cartilages.
- The thyroid cartilage sits atop the **cricoid** (KRĪ-koyd; ring-shaped) **cartilage,** another hyaline cartilage. The posterior portion of the cricoid is greatly expanded, providing support in the absence of the thyroid cartilage. The cricoid and thyroid cartilages protect the glottis and the entrance to the trachea, and their broad surfaces provide sites for the attachment of important laryngeal muscles and ligaments. Ligaments attach the inferior surface of the cricoid cartilage to the first cartilage of the trachea. The superior surface of the cricoid cartilage articulates with the small *arytenoid cartilages*.

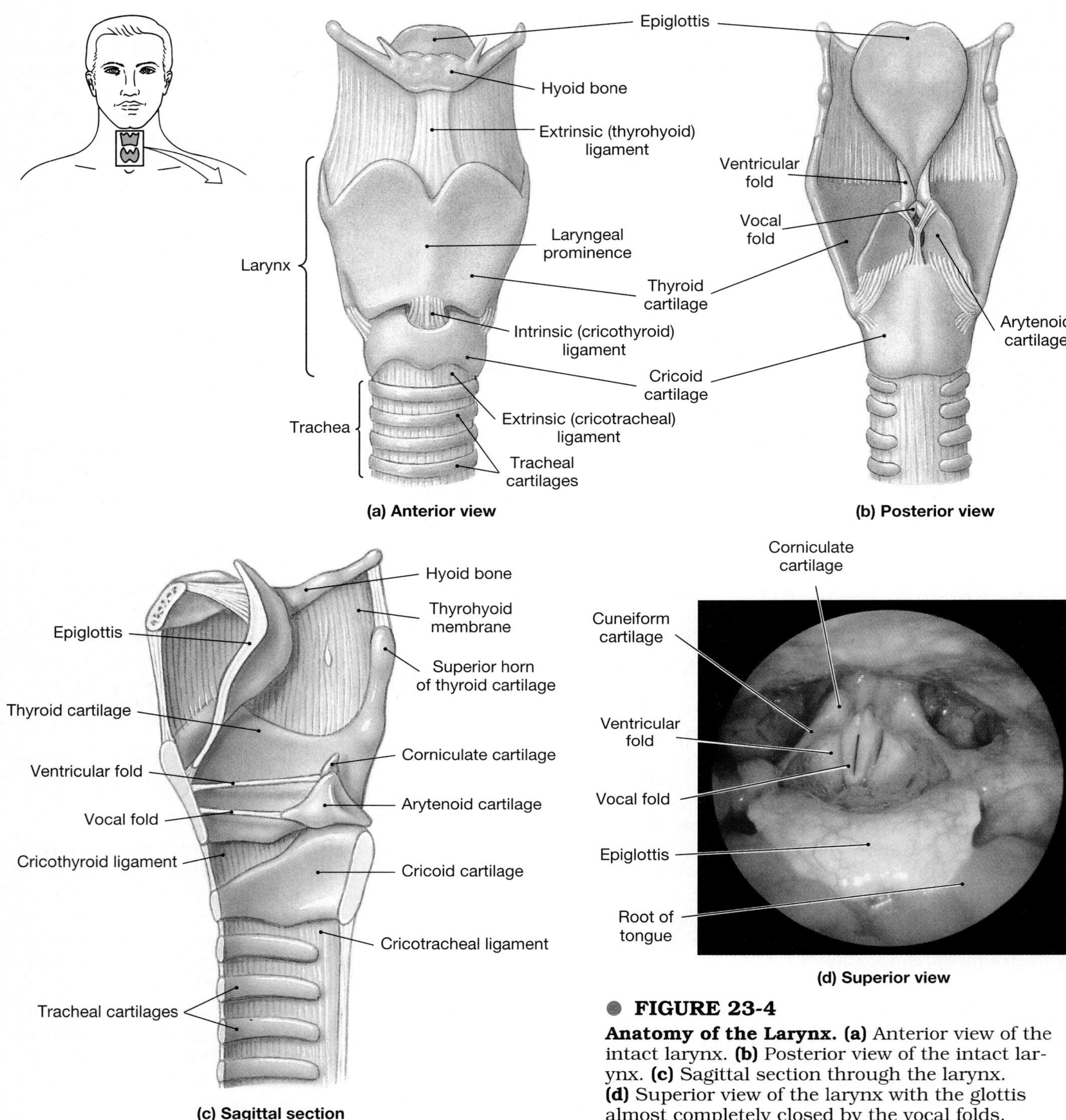
Epiglottis
Hyoid bone
Extrinsic (thyrohyoid) ligament
Larynx
Laryngeal prominence
Thyroid cartilage
Intrinsic (cricothyroid) ligament
Cricoid cartilage
Trachea
Extrinsic (cricotracheal) ligament
Tracheal cartilages
(a) Anterior view
Ventricular fold
Vocal fold
Arytenoid cartilage
(b) Posterior view
Hyoid bone
Thyrohyoid membrane
Epiglottis
Superior horn of thyroid cartilage
Thyroid cartilage
Ventricular fold
Corniculate cartilage
Arytenoid cartilage
Vocal fold
Cricothyroid ligament
Cricoid cartilage
Cricotracheal ligament
Tracheal cartilages
(c) Sagittal section
Corniculate cartilage
Cuneiform cartilage
Ventricular fold
Vocal fold
Epiglottis
Root of tongue
(d) Superior view

FIGURE 23-4
Anatomy of the Larynx. (a) Anterior view of the intact larynx. **(b)** Posterior view of the intact larynx. **(c)** Sagittal section through the larynx. **(d)** Superior view of the larynx with the glottis almost completely closed by the vocal folds.

- The shoehorn-shaped **epiglottis** (ep-i-GLOT-is) projects above the glottis. This structure, composed of elastic cartilage, has ligamentous attachments to the anterior and superior borders of the thyroid cartilage and the hyoid bone. During swallowing, the larynx is elevated, and the epiglottis folds back over the glottis, preventing the entry of liquids or solid food into the respiratory passageways.

The larynx also contains three pairs of smaller hyaline cartilages: the *arytenoid, corniculate,* and *cuneiform cartilages.*

- The paired **arytenoid** (ar-i-TĒ-noyd; ladle-shaped) **cartilages** articulate with the superior border of the enlarged portion of the cricoid cartilage.
- The **corniculate** (kor-NIK-ū-lāt; horn-shaped) **cartilages** articulate with the arytenoid carti-

lages. The corniculate and arytenoid cartilages are involved with the opening and closing of the glottis and the production of sound.

- Elongate, curving **cuneiform** (kyū-NĒ-i-form; wedge-shaped) **cartilages** lie within a fold of tissue that extends between the lateral aspect of each arytenoid cartilage and the epiglottis.

Laryngeal Ligaments

Figure 23-4

A series of **intrinsic ligaments** binds all nine cartilages together to form the larynx (Figure 23-4a,b●). **Extrinsic ligaments** attach the thyroid cartilage to the hyoid bone and the cricoid cartilage to the trachea. The **ventricular ligaments** and the **vocal ligaments** extend between the thyroid cartilage and the arytenoids.

The ventricular and vocal ligaments are covered by folds of laryngeal epithelium that project into the glottis. The ventricular ligaments lie within the superior pair of folds, known as the **ventricular folds** (Figure 23-4c,d●). The ventricular folds, which are relatively inelastic, help to prevent foreign objects from entering the glottis and provide protection for the more delicate **vocal folds.**

The vocal folds guard the entrance to the glottis. They are located inferior to the ventricular folds. The vocal folds are highly elastic, because they contain bands of elastic tissue called the *vocal ligaments.* The vocal folds are involved with the production of sounds, and for this reason they are known as the **true vocal cords.** Because the ventricular folds play no part in sound production, they are often called the **false vocal cords.**

Sound Production

Air passing through the glottis vibrates the vocal folds and produces sound waves. The pitch of the sound produced depends on the diameter, length, and tension in the vocal folds. The diameter and length are directly related to the size of the larynx. The tension is controlled by the contraction of voluntary muscles that change the relative positions of the thyroid and arytenoid cartilages. When the distance increases, the vocal folds tense and the pitch rises; when the distance decreases, the vocal folds relax and the pitch falls.

Anatomically, children of both sexes have slender, short vocal folds, and their voices tend to be high-pitched. At puberty the larynx of a male enlarges considerably more than that of a female. The true vocal cords of an adult male are thicker and longer and they produce lower tones than those of an adult female.

Sound production at the larynx is called **phonation** (fō-NĀ-shun; *phone,* voice). Phonation is one component of speech production, but clear speech also requires **articulation** (ar-tik-ū-LĀ-shun), the modification of those sounds by other structures. In a stringed instrument such as a guitar the quality of the sound produced does not depend solely on the nature of the vibrating string. The entire instrument becomes involved as the walls vibrate and the composite sound echoes within the hollow body. Similar amplification and resonance occur within the pharynx, the oral cavity, the nasal cavity, and the paranasal sinuses. The combination determines the particular and distinctive sound of each person's voice. The final production of distinct words further depends on voluntary movements of the tongue, lips, and cheeks.

❑ The Laryngeal Musculature

The larynx is associated with two groups of muscles, the *extrinsic laryngeal muscles* and the *intrinsic laryngeal muscles.* The **extrinsic laryngeal musculature,** part of the anterior muscles of the neck, positions and stabilizes the larynx. These muscles were considered in Chapter 11. ∞ *[p. 341]* The **intrinsic laryngeal muscles** have two major functions. One group regulates tension in the vocal folds, and a second set opens and closes the glottis. The muscles involved with the vocal folds insert upon the thyroid, arytenoid, and corniculate cartilages. Opening or closing the glottis involves rotational movements of the arytenoids that move the vocal folds apart or together.

During swallowing, both extrinsic and intrinsic muscles cooperate to prevent food or drink from entering the glottis. Before it is swallowed, the material is crushed and chewed into a pasty mass known as a *bolus.* Extrinsic muscles then elevate the larynx, bending the epiglottis over the entrance to the glottis, so that the bolus can glide across the epiglottis rather than falling into the larynx. While this movement is under way, intrinsic muscles close the glottis. Food particles or liquids that touch the surfaces of the ventricular or vocal folds will trigger the *coughing reflex.* The resulting blast of air from the trachea usually ejects the material and clears the entrance to the glottis.

LARYNGEAL PROBLEMS Infection or inflammation of the larynx is known as **laryngitis** (lar-in-JĪ-tis). This condition often affects the vibrational qualities of the vocal cords; hoarseness is the most familiar symptom. Mild cases are temporary and seldom serious. However, bacterial or viral infection of the epiglottis can be very dangerous because the resulting swelling may close the glottis and cause suffocation. This condition, acute **epiglottitis** (ep-i-glot-TĪ-tis), can develop relatively rapidly after a bacterial infection of the throat. Young children are most often affected.

Why is the vascularization of the nasal cavity important?

Why is the lining of the nasopharynx different from that of the oropharynx and laryngopharynx?

When the tension in the vocal cords increases, what happens to the pitch of the voice?

The Trachea

Figures 23-2, 23-5

The epithelium of the larynx is continuous with that of the **trachea** (TRĀ-kē-a), or "windpipe." The trachea is a tough, flexible tube with a diameter of around 2.5 cm (1 in.) and a length of approximately 11 cm (4.25 in.) (Figure 23-5●). The trachea begins anterior to vertebra C_6, in a ligamentous attachment to the cricoid cartilage; it ends in the mediastinum, at the level of vertebra T_5, where it branches to form the *right* and *left primary bronchi.*

The mucosa of the trachea resembles that of the nasal cavity and nasopharynx (Figure 23-2a●, p. 828). A thick layer of connective tissue, the **submucosa** (sub-myū-KŌ-sa), surrounds the mucosa. The submucosa contains mucous glands that communicate with the epithelial surface through a number of secretory ducts. The trachea contains 15–20 **tracheal cartilages** (Figure 23-5●). Each tracheal cartilage is bound to neighboring cartilages by elastic **annular ligaments.** The tracheal cartilages stiffen the tracheal walls and protect the airway. They also prevent its collapse or overexpansion as pressures change in the respiratory system.

Each tracheal cartilage is C-shaped. The closed portion of the C protects the anterior and lateral surfaces of the trachea. The open portions of the C-shaped tracheal cartilages face posteriorly, toward the esophagus. Because the cartilages do not continue around the trachea, the posterior tracheal wall can easily distort during swallowing, permitting the passage of large masses of food.

An elastic ligament and a band of smooth muscle, the **trachealis,** connect the ends of each tracheal cartilage (Figure 23-5b●). Contraction of the trachealis muscle alters the diameter of the trachea, changing the resistance to airflow. The normal diameter of the trachea changes from moment to moment, primarily under the control of the sympathetic division of the ANS. Sympathetic stimulation increases the diameter of the trachea and makes it easier to move large volumes of air along the respiratory passageways.

TRACHEAL BLOCKAGE Foreign objects that become lodged in the larynx or trachea are usually expelled by coughing. If the individual can speak or make a sound, the airway is still open, and no emergency measures should be taken. If the victim can neither breathe nor speak, an immediate threat to life exists. Unfortunately many victims become acutely embarrassed by this situation, and instead of seeking assistance, they run to the nearest restroom and die there.

In the **Heimlich** (HĪM-lik) **maneuver,** or *abdominal thrust,* a rescuer applies compression to the abdomen just beneath the diaphragm. This action elevates the diaphragm forcefully and may generate enough pressure to remove the blockage. The maneuver must be performed properly to avoid damage to internal organs. Many organizations such as the Red Cross, the local fire department, and other charitable groups periodically hold brief training sessions in the proper performance of the Heimlich maneuver.

If a tracheal blockage remains, professionally qualified rescuers may perform a **tracheostomy** (trā-kē-OS-to-mē). In this procedure, an incision is made through the anterior tracheal wall and a tube is inserted. The tube bypasses the larynx and permits air to flow directly into the trachea.

The Primary Bronchi

Figure 23-5

The trachea branches within the mediastinum, giving rise to the **right** and **left primary bronchi** (BRONG-kī). A ridge, the **carina** (ka-RĪ-na), marks the line of separation between the two bronchi (Figure 23-5a●). The histological organization of the primary bronchi is the same as that of the trachea, with cartilaginous C-shaped supporting rings. The right primary bronchus supplies the right lung, and the left supplies the left lung. The right primary bronchus has a larger diameter than the left, and it descends toward the lung at a steeper angle. Thus foreign objects that enter the trachea usually find their way into the right bronchus rather than the left.

Each primary bronchus travels to a groove along the medial surface of its lung before branching further. This groove, the **hilus** (HĪ-lus), also provides access for entry to pulmonary vessels and nerves. The entire array is firmly anchored in a meshwork of dense connective tissue. This complex, known as the **root** of the lung (Figure 23-5a●), attaches it to the mediastinum and fixes the positions of the major nerves, vessels, and lymphatics. The roots of the lungs are located anterior to vertebrae T_5 (right) and T_6 (left).

The Lungs

Figure 23-6

The left and right lungs (Figure 23-6●) are situated in the left and right pleural cavities. Each lung is a blunt cone, with the tip, or **apex,** pointing superi-

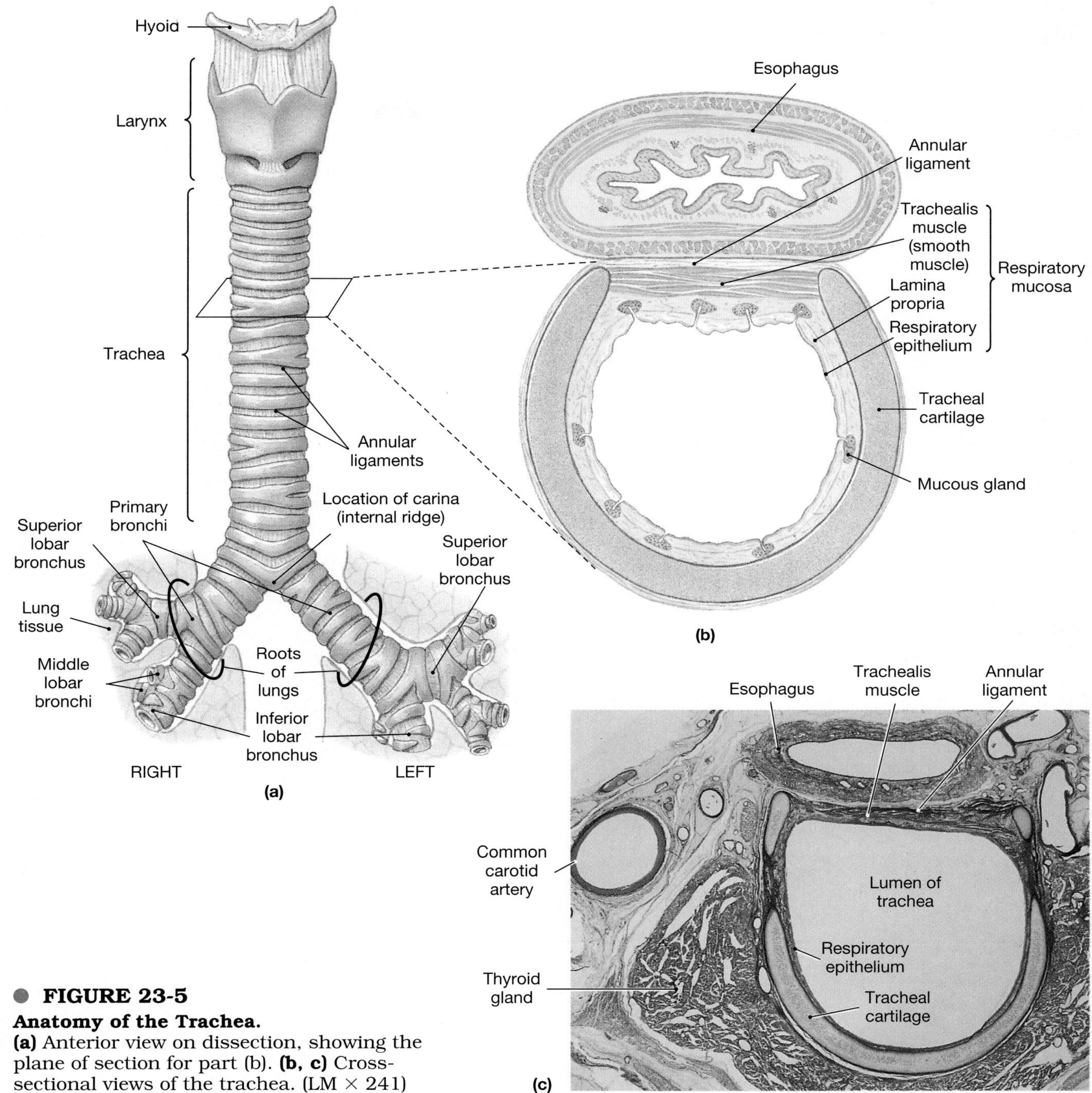
Hyoid
Larynx
Trachea
Annular ligaments
Primary bronchi
Location of carina (internal ridge)
Superior lobar bronchus
Superior lobar bronchus
Lung tissue
Middle lobar bronchi
Roots of lungs
Inferior lobar bronchus
RIGHT
LEFT
(a)
Esophagus
Annular ligament
Trachealis muscle (smooth muscle)
Lamina propria
Respiratory epithelium
Respiratory mucosa
Tracheal cartilage
Mucous gland
(b)
Esophagus
Trachealis muscle
Annular ligament
Common carotid artery
Lumen of trachea
Respiratory epithelium
Thyroid gland
Tracheal cartilage
(c)

● **FIGURE 23-5**
Anatomy of the Trachea.
(a) Anterior view on dissection, showing the plane of section for part (b). **(b, c)** Cross-sectional views of the trachea. (LM × 241)

orly. The apex on each side extends into the base of the neck above the first rib. The broad concave inferior portion, or **base,** of each lung rests on the superior surface of the diaphragm.

❑ Lobes of the Lungs

Figure 23-6

The lungs have distinct **lobes** separated by deep fissures (Figure 23-6●). The right lung has three lobes: *superior, middle,* and *inferior.* The left lung has only two lobes: *superior* and *inferior.* The right lung is broader than the left because most of the heart and great vessels project into the left thoracic cavity. However, the left lung is longer than the right lung, because the diaphragm rises on the right side to accommodate the mass of the liver.

❑ Lung Surfaces

Figures 23-6, 23-7

The curving anterior portion of the lung that follows the inner contours of the rib cage is the **costal surface.** The **mediastinal surface** containing the

● FIGURE 23-6
Gross Anatomy of the Lungs. AM *Plate 6.3*

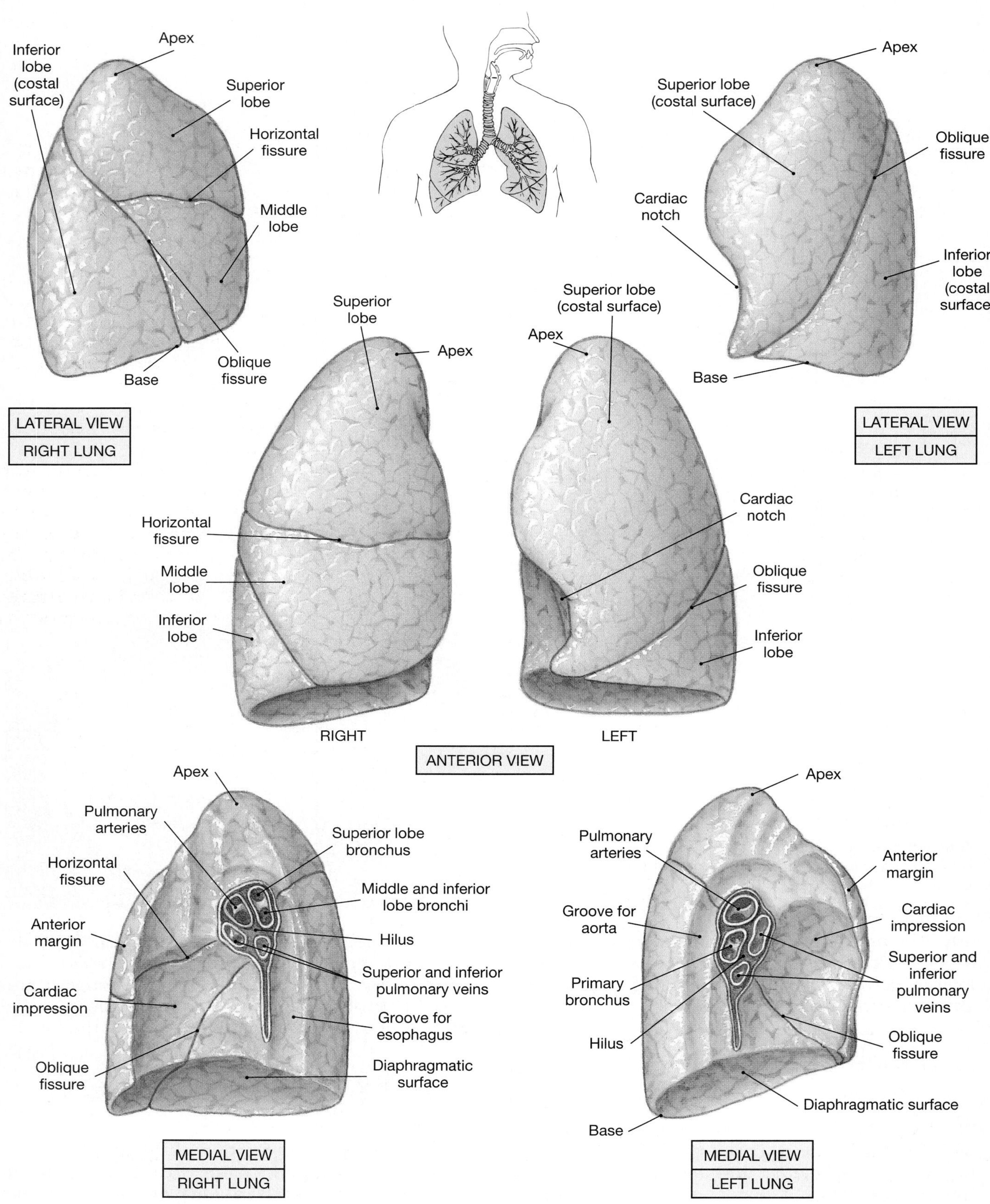
Apex
Inferior lobe (costal surface)
Superior lobe
Horizontal fissure
Middle lobe
Base
Oblique fissure
LATERAL VIEW
RIGHT LUNG
Apex
Superior lobe (costal surface)
Oblique fissure
Cardiac notch
Inferior lobe (costal surface)
Base
LATERAL VIEW
LEFT LUNG
Superior lobe
Apex
Horizontal fissure
Middle lobe
Inferior lobe
RIGHT
Superior lobe (costal surface)
Apex
Cardiac notch
Oblique fissure
Inferior lobe
LEFT
ANTERIOR VIEW
Apex
Pulmonary arteries
Horizontal fissure
Anterior margin
Cardiac impression
Oblique fissure
Superior lobe bronchus
Middle and inferior lobe bronchi
Hilus
Superior and inferior pulmonary veins
Groove for esophagus
Diaphragmatic surface
MEDIAL VIEW
RIGHT LUNG
Apex
Pulmonary arteries
Groove for aorta
Primary bronchus
Hilus
Anterior margin
Cardiac impression
Superior and inferior pulmonary veins
Oblique fissure
Diaphragmatic surface
Base
MEDIAL VIEW
LEFT LUNG

hilus has a more irregular shape (Figure 23-6●). The mediastinal surfaces of both lungs bear grooves that mark the passage of vessels traveling to and from the heart. The mediastinal surface of the left lung bears a concavity, the **cardiac notch,** that conforms to the shape of the pericardium (Figure 23-7●).

TUBERCULOSIS (tū-ber-kyū-LŌ-sis), or TB, results from a bacterial infection of the lungs, although other organs may be invaded as well. The bacteria, *Mycobacterium tuberculosis,* may colonize the respiratory passageways, the interstitial spaces, the alveoli, or a combination of the three. Symptoms are variable, but usually include coughing and chest pain, with fever, night sweats, fatigue, and weight loss. For information concerning the incidence and treatment of TB, see the *Applications Manual.* AM Tuberculosis

The Bronchi

Figures 23-5, 23-8, 23-9

The primary bronchi and their branches form the **bronchial tree** (Figure 23-8●). Because the left and right primary bronchi are outside of the lungs, they are also called **extrapulmonary bronchi.** As the primary bronchi enter the lungs, they divide to form smaller passageways (Figures 23-5, 23-8, and 23-9●). Those branches are collectively called the **intrapulmonary bronchi.**

Each primary bronchus divides to form **secondary bronchi,** also known as *lobar bronchi.*

- The right lung has three lobes, and the right primary bronchus divides into three secondary bronchi: a *superior lobar bronchus,* a *middle lobar bronchus,* and an *inferior lobar bronchus.*
- The left lung has two lobes, and the left primary bronchus divides into two secondary bronchi: a *superior lobar bronchus* and an *inferior lobar bronchus.*

Figure 23-9● follows the branching pattern of the left primary bronchus as it enters the lung. (The number of branches have been reduced for clarity.) Within each lung, the secondary bronchi branch to form **tertiary bronchi,** or *segmental bronchi.* The branching pattern differs depending on the lung considered, but each tertiary bronchus ultimately supplies air to a single **bronchopulmonary segment,** a specific region of one lung. There are 10 bronchopulmonary segments in the right lung. During development, the left lung also has 10 segments, but subsequent fusion of adjacent tertiary bronchi usually reduces that number to eight or nine.

The walls of the primary, secondary, and tertiary bronchi contain progressively lesser amounts of cartilage. In the secondary and tertiary bronchi, the cartilages form blocks arranged around the lumen. These cartilages serve the same purpose as the rings of cartilage in the trachea and primary bronchi. As the amount of cartilage decreases, the relative amount of smooth muscle increases. With less cartilaginous support, the amount of tension in those smooth muscles has a greater effect on bronchial diameter and the resistance to airflow.

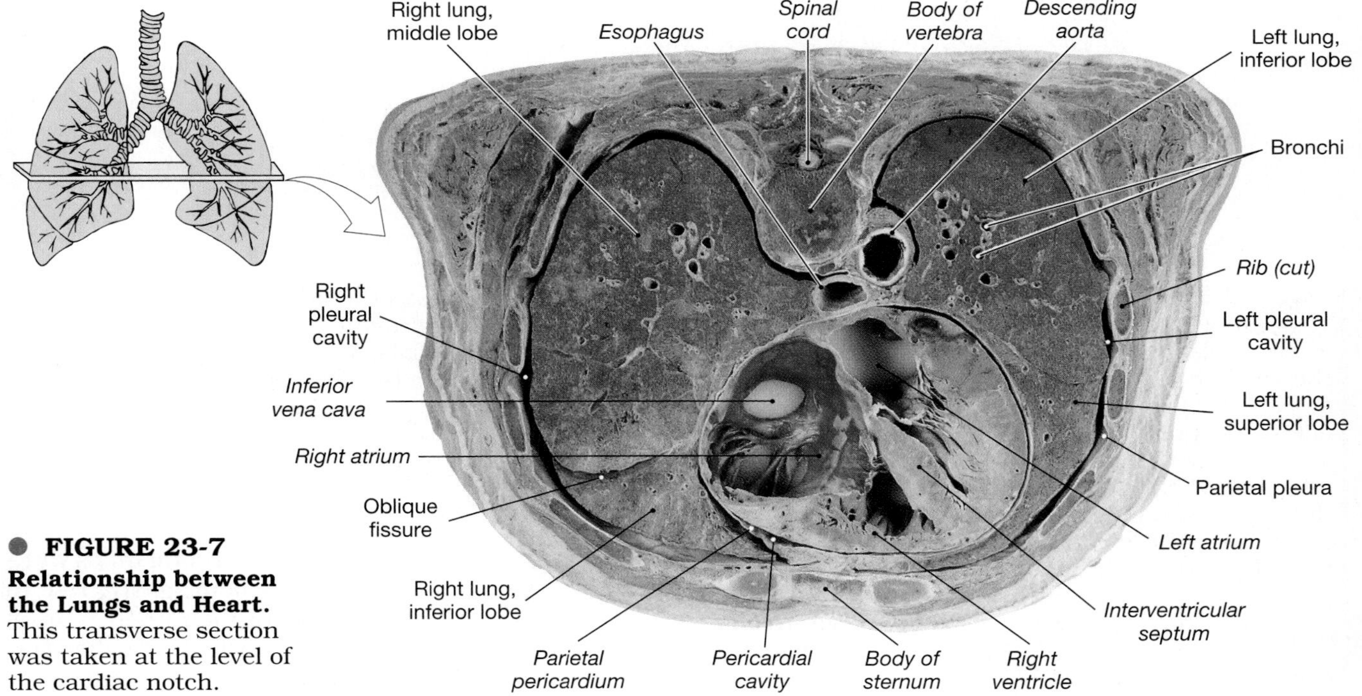

Right lung, middle lobe
Esophagus
Spinal cord
Body of vertebra
Descending aorta
Left lung, inferior lobe
Bronchi
Rib (cut)
Left pleural cavity
Left lung, superior lobe
Parietal pleura
Left atrium
Interventricular septum
Right ventricle
Body of sternum
Pericardial cavity
Parietal pericardium
Right lung, inferior lobe
Oblique fissure
Right atrium
Inferior vena cava
Right pleural cavity

● **FIGURE 23-7**
Relationship between the Lungs and Heart. This transverse section was taken at the level of the cardiac notch.

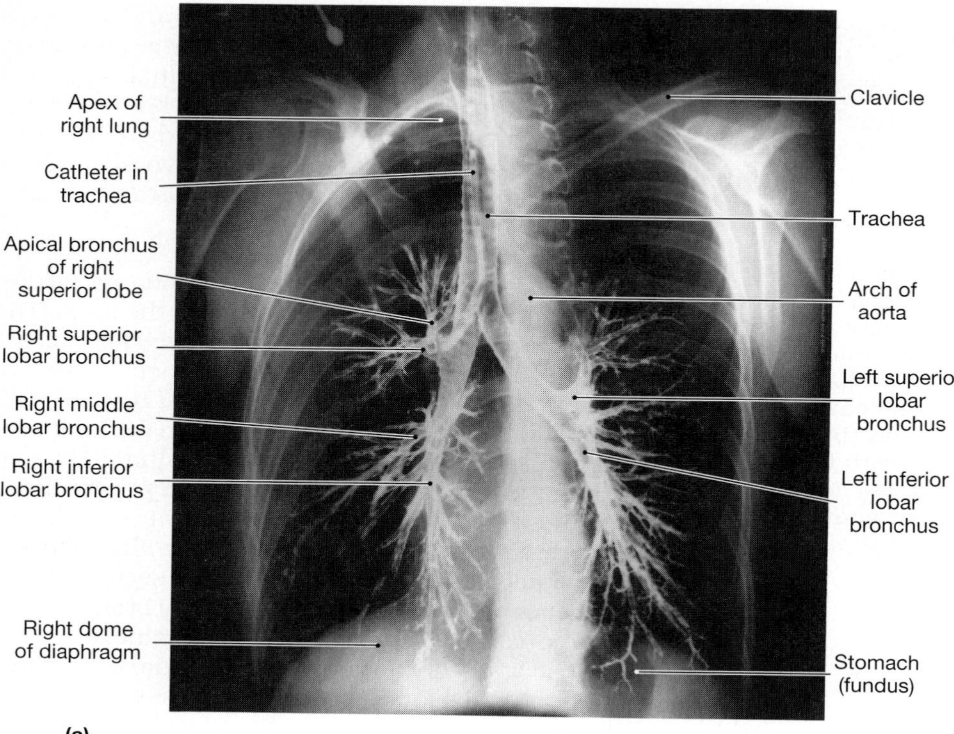
Apex of right lung
Catheter in trachea
Apical bronchus of right superior lobe
Right superior lobar bronchus
Right middle lobar bronchus
Right inferior lobar bronchus
Right dome of diaphragm
Clavicle
Trachea
Arch of aorta
Left superior lobar bronchus
Left inferior lobar bronchus
Stomach (fundus)
(a)

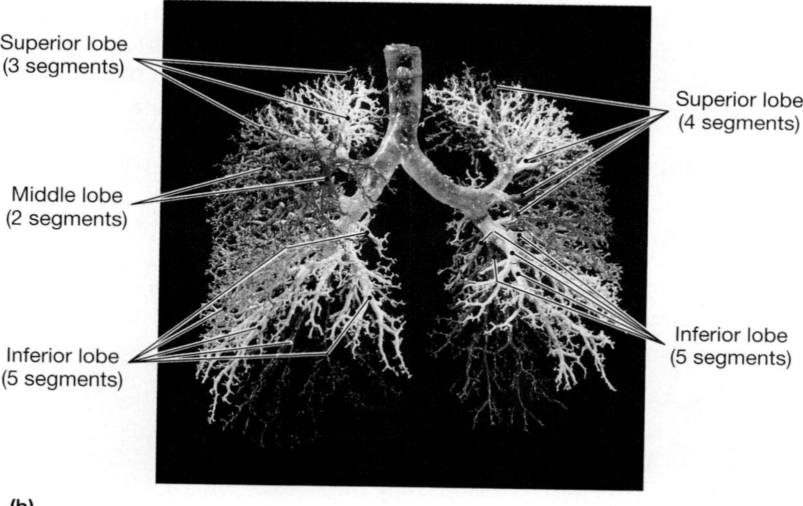
Superior lobe (3 segments)
Middle lobe (2 segments)
Inferior lobe (5 segments)
Superior lobe (4 segments)
Inferior lobe (5 segments)
(b)

● **FIGURE 23-8**
The Bronchial Tree. (a) An X-ray showing the major branches of the bronchial tree, including both extrapulmonary and intrapulmonary bronchi. **(b)** A cast of the bronchial tree. The finer branches have been painted to indicate the regions supplied by different bronchopulmonary segments. AM *Plate 6.3*

BRONCHOSCOPY During a respiratory infection, the bronchi and bronchioles may become inflamed. In this condition, called **bronchitis,** the smaller passageways may become greatly constricted, leading to difficulties in breathing. One method of investigating the status of the respiratory passageways is the use of a *bronchoscope.* A **bronchoscope** is a fiber-optic bundle small enough to be inserted into the trachea and steered along the conducting passageways to the level of the smaller bronchioles. This procedure is called **bronchoscopy** (brong-KOS-kō-pē). In addition to permitting direct visualization of bronchial structures, the bronchoscope can collect tissue or mucus samples from the respiratory tract. In **bronchography** (brong-KOG-ra-fē), a bronchoscope or catheter introduces a radiopaque material into the bronchi. This technique can permit detailed X-ray analysis of bronchial masses, such as tumors, or other obstructions along the bronchial tree (Figure 23-8●).

❑ The Bronchioles
Figure 23-9

Each tertiary bronchus branches several times within the bronchopulmonary segment, giving rise to multiple **bronchioles.** These branch further, until reaching the finest conducting branches, called **terminal bronchioles.** Roughly 6500 terminal bron-

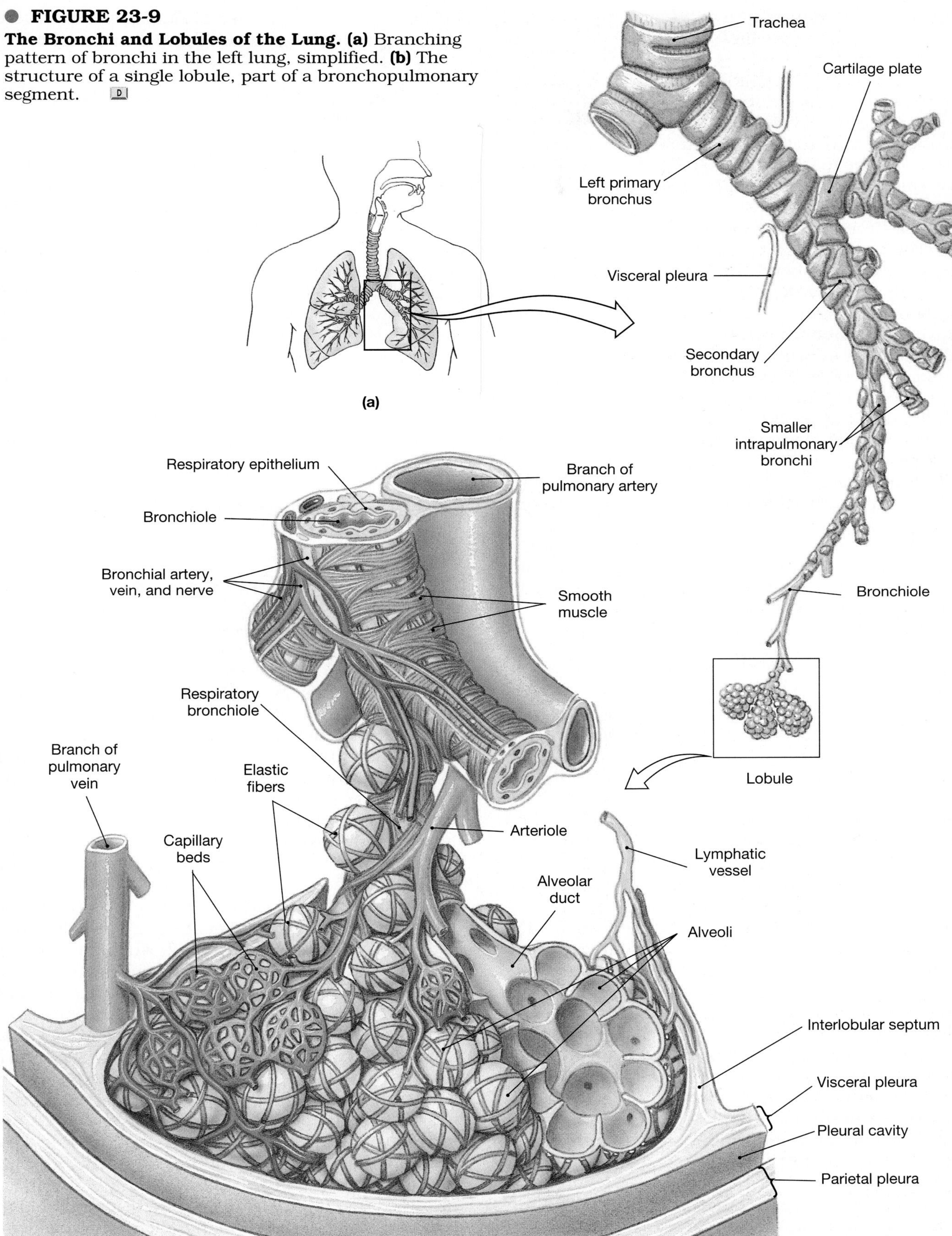
Trachea
Cartilage plate
Left primary bronchus
Visceral pleura
Secondary bronchus
Smaller intrapulmonary bronchi
Bronchiole
Lobule
(a)
Respiratory epithelium
Branch of pulmonary artery
Bronchiole
Bronchial artery, vein, and nerve
Smooth muscle
Respiratory bronchiole
Branch of pulmonary vein
Elastic fibers
Capillary beds
Arteriole
Lymphatic vessel
Alveolar duct
Alveoli
Interlobular septum
Visceral pleura
Pleural cavity
Parietal pleura
(b)

● **FIGURE 23-9**
The Bronchi and Lobules of the Lung. (a) Branching pattern of bronchi in the left lung, simplified. **(b)** The structure of a single lobule, part of a bronchopulmonary segment.

chioles are supplied by each tertiary bronchus. Terminal bronchioles have a lumenal diameter of 0.3–0.5 mm.

The walls of bronchioles, which lack cartilaginous supports, are dominated by smooth muscle tissue (Figure 23-9•). In functional terms, the bronchioles are to the respiratory system what the arterioles are to the cardiovascular system. Varying the diameter of the bronchioles provides control over the amount of resistance to airflow and the distribution of air within the lungs.

The autonomic nervous system regulates the activity in this smooth muscle layer and thereby controls the diameter of the bronchioles. Sympathetic activation leads to enlargement of the airways, or **bronchodilation.** Parasympathetic stimulation leads to **bronchoconstriction,** a reduction in the diameter of the airways. Bronchoconstriction also occurs during allergic reactions such as anaphylaxis (Chapter 22), in response to histamine release by activated mast cells and basophils. [p. 816]

Bronchodilation and bronchoconstriction alter the resistance to airflow toward or away from the respiratory exchange surfaces. Tension in the smooth muscles often throws the bronchiolar mucosa into a series of folds, and excessive stimulation, as in *asthma,* can almost completely prevent airflow along the terminal bronchioles. AM *Asthma*

Pulmonary Lobules

Figure 23-9b

The connective tissues of the root of each lung extend into its substance, or **parenchyma** (pa-REN-kī-muh). The fibrous partitions, or **trabeculae,** contain elastic fibers, smooth muscles, and lymphatics. The trabeculae branch repeatedly, dividing the lobes into smaller and smaller compartments, and the branches of the conducting passageways, pulmonary vessels, and nerves of the lungs follow these trabeculae to reach their peripheral destinations. The finest partitions, or **septa** (*septum,* a wall) divide the lung into **pulmonary lobules** (LOB-ūlz), each supplied by branches of the pulmonary arteries, pulmonary veins, and respiratory passageways (Figure 23-9b•). The connective tissues of the septa are in turn continuous with those of the visceral pleura.

Each terminal bronchiole delivers air to a single pulmonary lobule. Within the lobule, the terminal bronchiole branches to form several **respiratory bronchioles.** These are the thinnest and most delicate branches of the bronchial tree, and they deliver air to the exchange surfaces of the lungs.

The preliminary filtration and humidification of the incoming air are completed before air leaves the terminal bronchioles. The epithelial cells of the terminal bronchioles and respiratory bronchioles are cuboidal, with only scattered cilia, and there are no goblet cells or underlying mucous glands.

❑ Alveolar Ducts and Alveoli

Figures 23-9, 23-10, 23-11a

Respiratory bronchioles are connected to individual alveoli and to multiple alveoli along regions called **alveolar ducts.** These passageways end at **alveolar sacs,** common chambers connected to multiple individual alveoli (Figures 23-9 and 23-10•). Each lung contains approximately 150 million alveoli, and their abundance gives the lung an open, spongy appearance. An extensive network of capillaries is associated with each alveolus (Figure 23-11a•); the capillaries are surrounded by a network of elastic fibers. This elastic tissue helps maintain the relative positions of the alveoli and respiratory bronchioles. Recoil of these fibers during expiration reduces the size of the alveoli and helps push air out of the lungs.

The Alveolus and the Respiratory Membrane

Figure 23-11

The alveolar epithelium consists primarily of simple squamous epithelium (Figure 23-11b•). The squamous epithelial cells, called *Type I cells,* are unusually thin and delicate. Roaming **alveolar macrophages** (*dust cells*) patrol the epithelium, phagocytizing any particulate matter that has eluded the respiratory defenses and reached the alveolar surfaces. **Septal cells,** also called **surfactant** (sur-FAK-tant) **cells** or *Type II cells,* are scattered among the squamous cells. These large cells produce an oily secretion, or **surfactant,** containing a mixture of phospholipids and proteins. Surfactant is secreted onto the alveolar surfaces, where it forms a superficial coating over a thin layer of water.

Surfactant is important because it reduces surface tension in the liquid coating the alveolar surface. As we saw in Chapter 2, surface tension results from the attraction between water molecules at an air-water boundary. [p. 37] The alveolar walls are very delicate, and without surfactant the surface tension would be so high that the alveoli would collapse. The surfactant forms a thin surface layer that interacts with the water molecules, reducing the surface tension and keeping the alveoli open.

If surfactant cells produce inadequate amounts of surfactant, because of injury or genetic abnormalities, the alveoli will collapse and respiration will become difficult. On each breath the inhalation must be forceful enough to pop open the alveoli. An individual with this problem, called *respiratory distress syndrome,* is soon exhausted by the

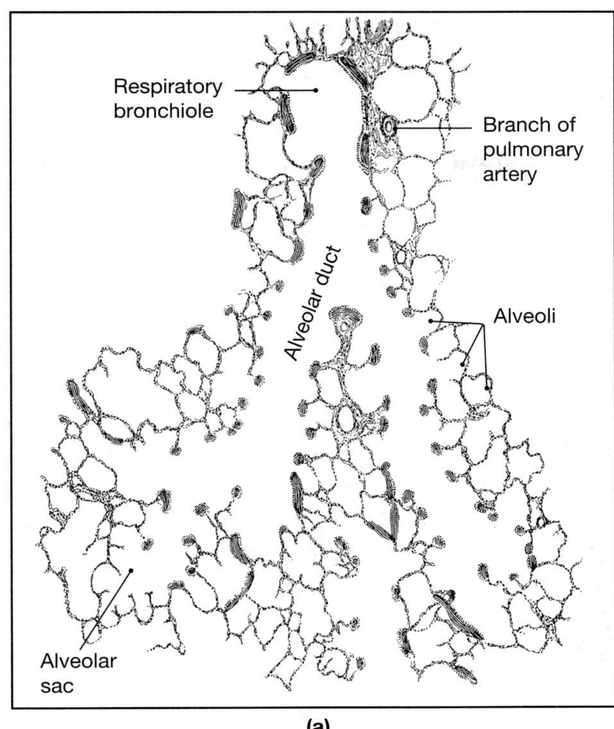

Respiratory bronchiole
Branch of pulmonary artery
Alveolar duct
Alveoli
Alveolar sac

(a)

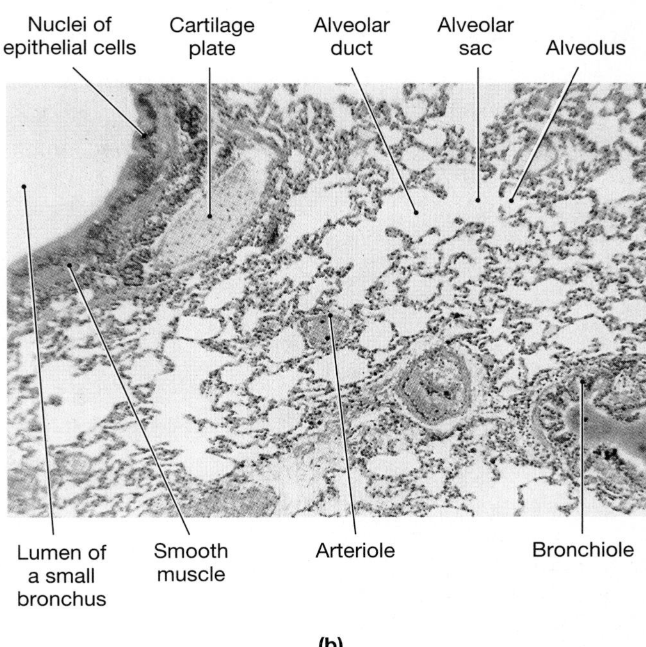

Nuclei of epithelial cells
Cartilage plate
Alveolar duct
Alveolar sac
Alveolus
Lumen of a small bronchus
Smooth muscle
Arteriole
Bronchiole

(b)

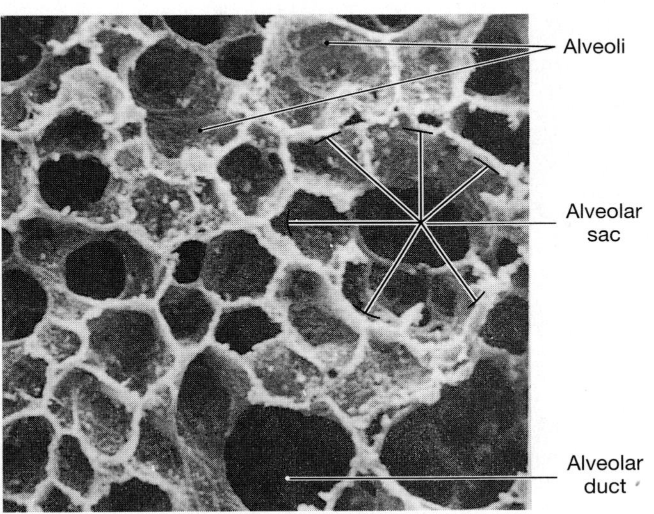

Alveoli
Alveolar sac
Alveolar duct

(c)

● **FIGURE 23-10**

The Bronchioles. (a) Distribution of a respiratory bronchiole supplying a portion of a lobule. **(b)** The alveolar sacs and alveoli. (LM × 42) **(c)** An SEM of the lung. Note the open, spongy appearance of the lung tissue. Compare with Figure 23-11b.

effort required to keep inflating and deflating the lungs. AM *Respiratory Distress Syndrome*

THE RESPIRATORY MEMBRANE Gas exchange occurs across the **respiratory membrane** of the alveoli. The respiratory membrane (Figure 23-11c●) is a composite structure consisting of:

1. The squamous epithelial cell lining the alveolus.
2. The endothelial cell lining an adjacent capillary.
3. The fused basement membranes that lie between the alveolar and endothelial cells.

At the respiratory membrane, the total distance separating the alveolar air and the blood can be as little as 0.1 μm. Diffusion across the respiratory membrane proceeds very rapidly, because (1) the distance is small and (2) both oxygen and carbon dioxide are lipid-soluble. The membranes of the epithelial and endothelial cells thus do not pose a barrier to the movement of oxygen and carbon dioxide between the blood and alveolar air spaces.

PNEUMONIA Pneumonia (nū-MŌ-nē-uh) develops from a pathogenic infection or any other stimulus causing inflammation of the lobules of the lung. As inflammation occurs, fluids leak into the alveoli and there is swelling and constriction of the respiratory bronchioles. Respiratory function deteriorates as a result. When bacteria are involved, they are usually species normally found in the mouth and pharynx that have somehow managed to evade the respiratory defenses. As a result, pneumonia becomes more likely when the respiratory defenses have been compromised by other factors, such as epithelial damage from smoking or the breakdown of the immune system in AIDS. The most common pneumonia that develops in AIDS patients results from infection by the fungus *Pneumocystis carinii*. These organisms are normally found in the alveoli, but in healthy individuals the respiratory defenses are able to prevent infection and tissue damage.

 Why are the cartilages that reinforce the trachea C-shaped?

 What would happen to the alveoli if surfactant were not produced?

Why do chronic smokers develop a hacking "smoker's cough"?

The Blood Supply to the Lungs

The respiratory exchange surfaces receive blood from arteries of the pulmonary circuit. The pulmonary arteries enter the lungs at the hilus and branch with the bronchi as they approach the lob-

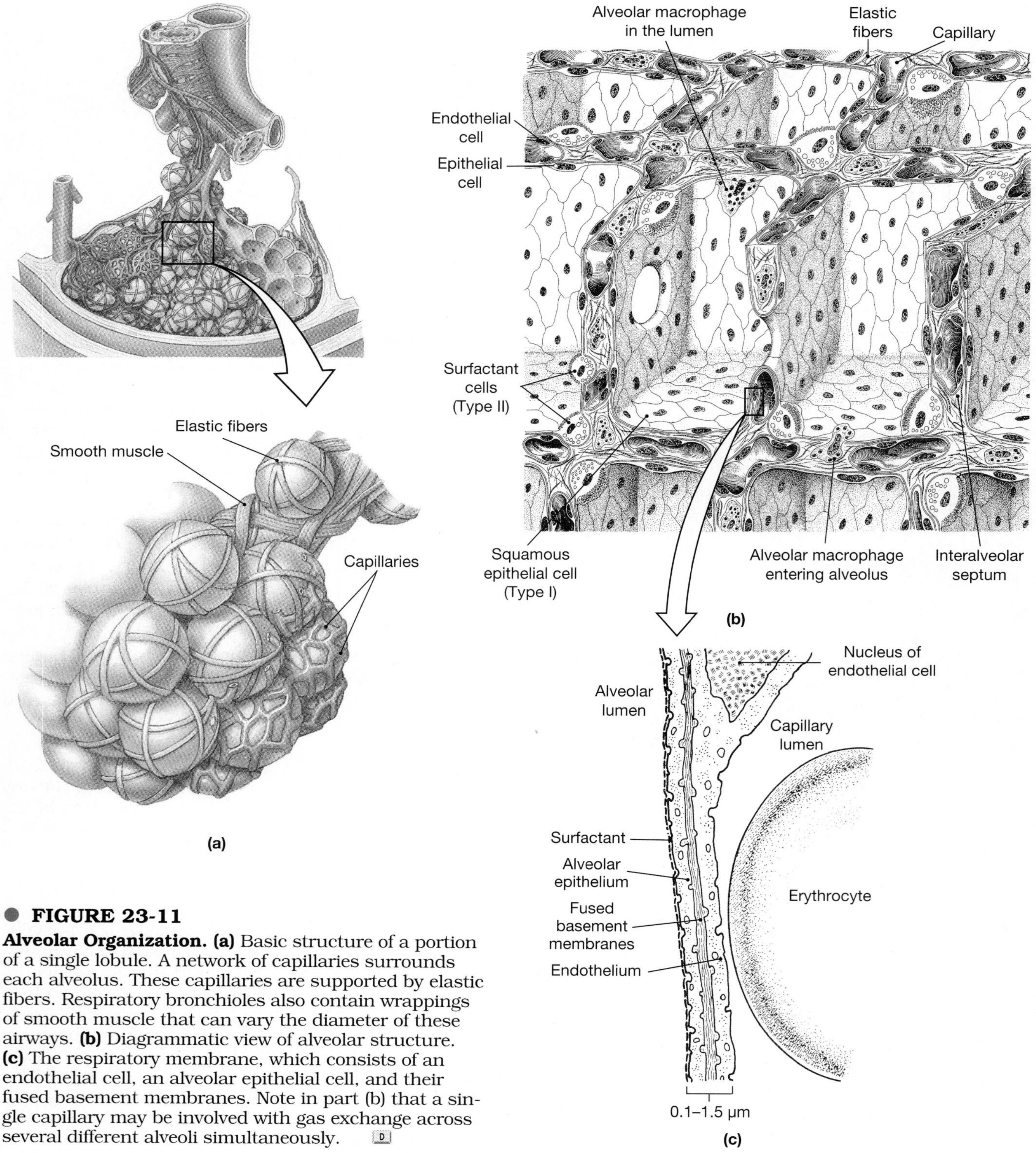

Elastic fibers
Smooth muscle
Capillaries
(a)
Alveolar macrophage in the lumen
Elastic fibers
Capillary
Endothelial cell
Epithelial cell
Surfactant cells (Type II)
Squamous epithelial cell (Type I)
Alveolar macrophage entering alveolus
Interalveolar septum
(b)
Nucleus of endothelial cell
Alveolar lumen
Capillary lumen
Surfactant
Alveolar epithelium
Fused basement membranes
Endothelium
Erythrocyte
0.1–1.5 μm
(c)

FIGURE 23-11
Alveolar Organization. (a) Basic structure of a portion of a single lobule. A network of capillaries surrounds each alveolus. These capillaries are supported by elastic fibers. Respiratory bronchioles also contain wrappings of smooth muscle that can vary the diameter of these airways. **(b)** Diagrammatic view of alveolar structure. **(c)** The respiratory membrane, which consists of an endothelial cell, an alveolar epithelial cell, and their fused basement membranes. Note in part (b) that a single capillary may be involved with gas exchange across several different alveoli simultaneously.

ules. Each lobule receives an arteriole and a venule, and a network of capillaries surrounds each alveolus directly beneath the respiratory membrane. In addition to providing a mechanism for gas exchange, the alveolar capillaries are the primary source of *converting enzyme*, and the endothelial cells convert circulating angiotensin I to angiotensin II, a hormone involved with the regulation of blood volume and blood pressure discussed in Chapters 18 and 21. [pp. 628, 743]

After passing through the pulmonary venules, blood from the alveolar capillaries enters the pulmonary veins, which deliver it to the left atrium. The conducting portions of the respiratory tract receive blood from the *external carotid arteries* (nasal passages and larynx), the *thyrocervical trunks* (the inferior larynx and trachea), and the *bronchial arteries* (the bronchi and bronchioles). (See Figures 21-23, 21-24, and 21-26, pp. 752, 754, 756.) The capillaries supplied by the bronchial arteries provide oxygen and nutrients to the conducting passageways of the lungs. The venous blood flows into the pulmonary veins, bypassing the rest of the systemic circuit and diluting the oxygenated blood leaving the alveoli.

PULMONARY EMBOLISM Blood pressure in the pulmonary circuit is usually relatively low, with systemic pressures of 30 mm Hg or less. With pressures that low, pulmonary vessels can easily become blocked by small blood clots, fat masses, or air bubbles in the pulmonary arteries. Because the lungs receive the entire cardiac output, any drifting masses in the blood are likely to cause problems almost at once. Blockage of a branch of a pulmonary artery will stop blood flow to a group of lobules or alveoli. This condition is called a **pulmonary embolism.** *Atherosclerosis*, described in Chapter 21, and *venous thrombosis* (the blockage of a vein by a blood clot) can promote development of a pulmonary embolism, because in each case there is a tendency for the formation of small blood clots that can break loose and drift in the circulation. [pp. 675, 722] If a pulmonary embolism remains in place for several hours, the alveoli will permanently collapse. If the blockage occurs in a major pulmonary vessel, rather than a minor tributary, pulmonary resistance increases. This resistance places extra strain on the right ventricle, which may be unable to maintain cardiac output. Congestive heart failure may be the result.

The Pleural Cavities and Pleural Membranes

Figure 23-12

The thoracic cavity has the shape of a broad cone. Its walls are the rib cage, and the muscular diaphragm forms the floor. The two pleural cavities are separated by the mediastinum (Figure 23-12●). Each lung occupies a single pleural cavity, which is lined by a serous membrane, or **pleura** (PLOO-ra). The **parietal pleura** covers the inner surface of the thoracic wall and extends over the diaphragm and mediastinum. The **visceral pleura** covers the outer surfaces of the lungs, extending into the fissures between the lobes. A small amount of **pleural fluid** is secreted by both pleural membranes. Pleural fluid gives a moist, slippery coating that provides lubrication, thereby reducing friction between the parietal and visceral surfaces during breathing. Samples of pleural fluid are sometimes obtained for diagnostic purposes, using a long needle inserted between the ribs. This sampling procedure is called *thoracentesis*. The fluid extracted is then examined for the presence of bacteria, blood cells, or other abnormal components.

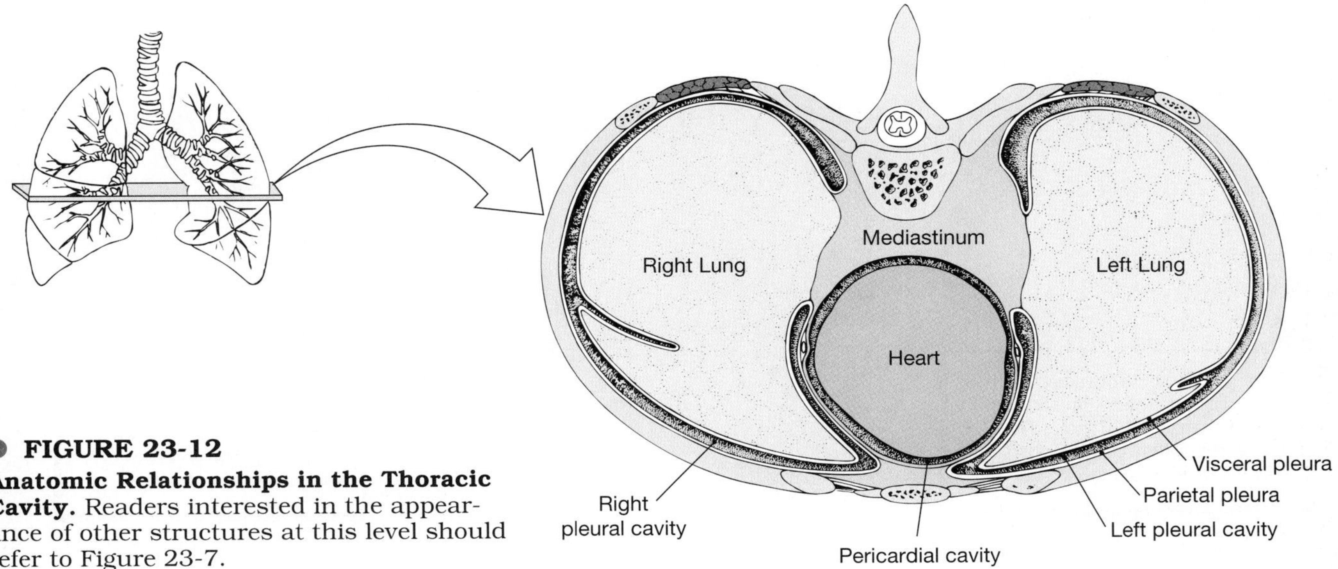
Mediastinum
Right Lung
Left Lung
Heart
Visceral pleura
Parietal pleura
Left pleural cavity
Right pleural cavity
Pericardial cavity

● **FIGURE 23-12**
Anatomic Relationships in the Thoracic Cavity. Readers interested in the appearance of other structures at this level should refer to Figure 23-7.

EMBRYOLOGY SUMMARY

Development of the Respiratory System

THE LUNGS

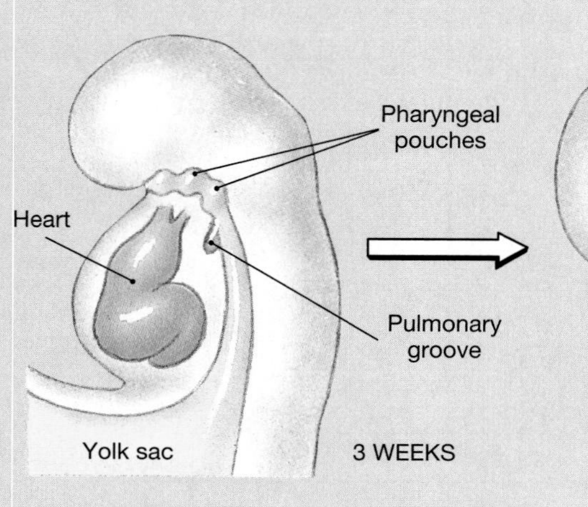
Pharyngeal pouches
Heart
Pulmonary groove
Yolk sac
3 WEEKS

A shallow groove appears in the midventral floor of the pharynx after roughly 3½ weeks of development. This groove, which lies near the level of the last pharyngeal arch, gradually deepens.

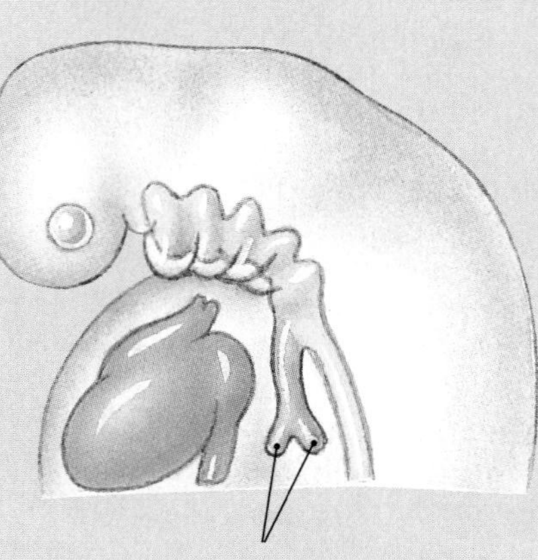
Lung buds

By week 4, the groove has become a tube that extends caudally, anterior to the esophagus. This tube will become the trachea. At its tip, the tube branches, forming a pair of **lung buds.**

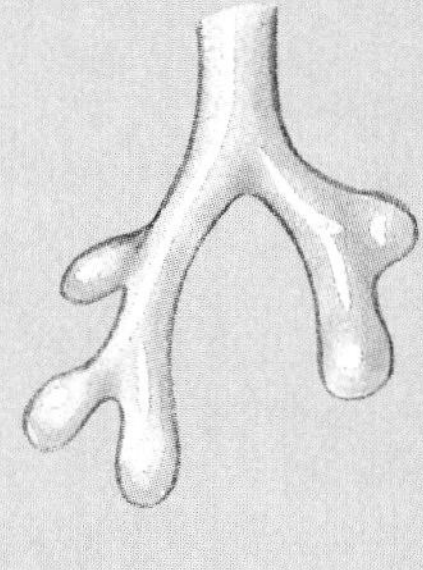

The lung buds continue to elongate and branch repeatedly.

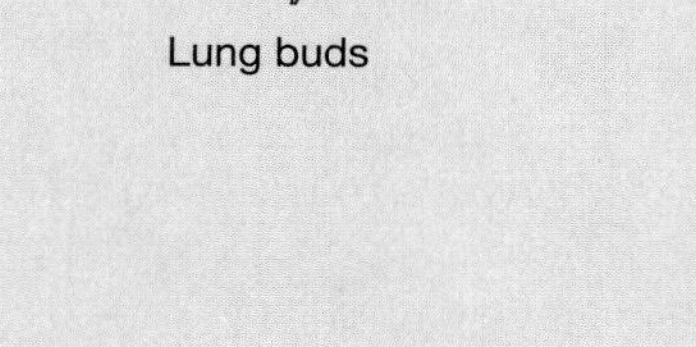

By the end of the sixth fetal month, there are around a million terminal branches, and the conducting passageways are complete to the level of the bronchioles.

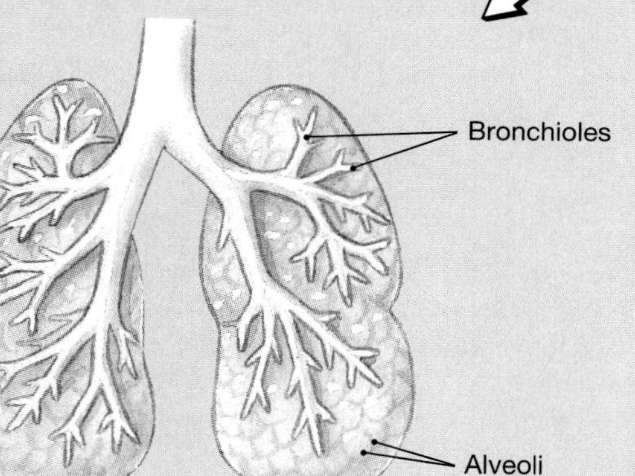
Bronchioles
Alveoli

Over the next 3 months, each of the bronchioles gives rise to several hundred alveoli. This process continues for a variable period after delivery.

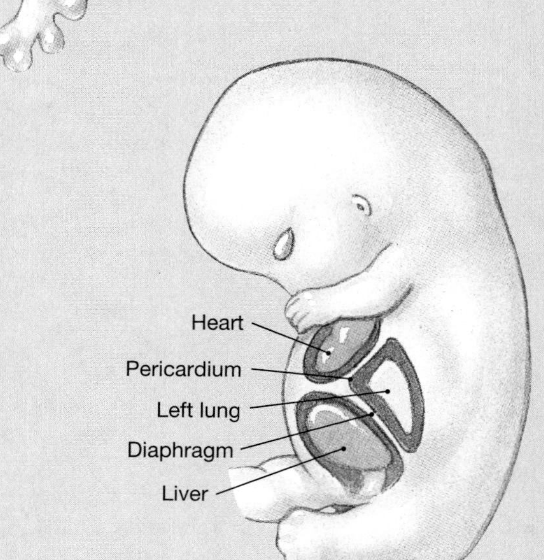
Heart
Pericardium
Left lung
Diaphragm
Liver

By week 9, the diaphragm completes its formation, forming a transverse sheet over the liver.

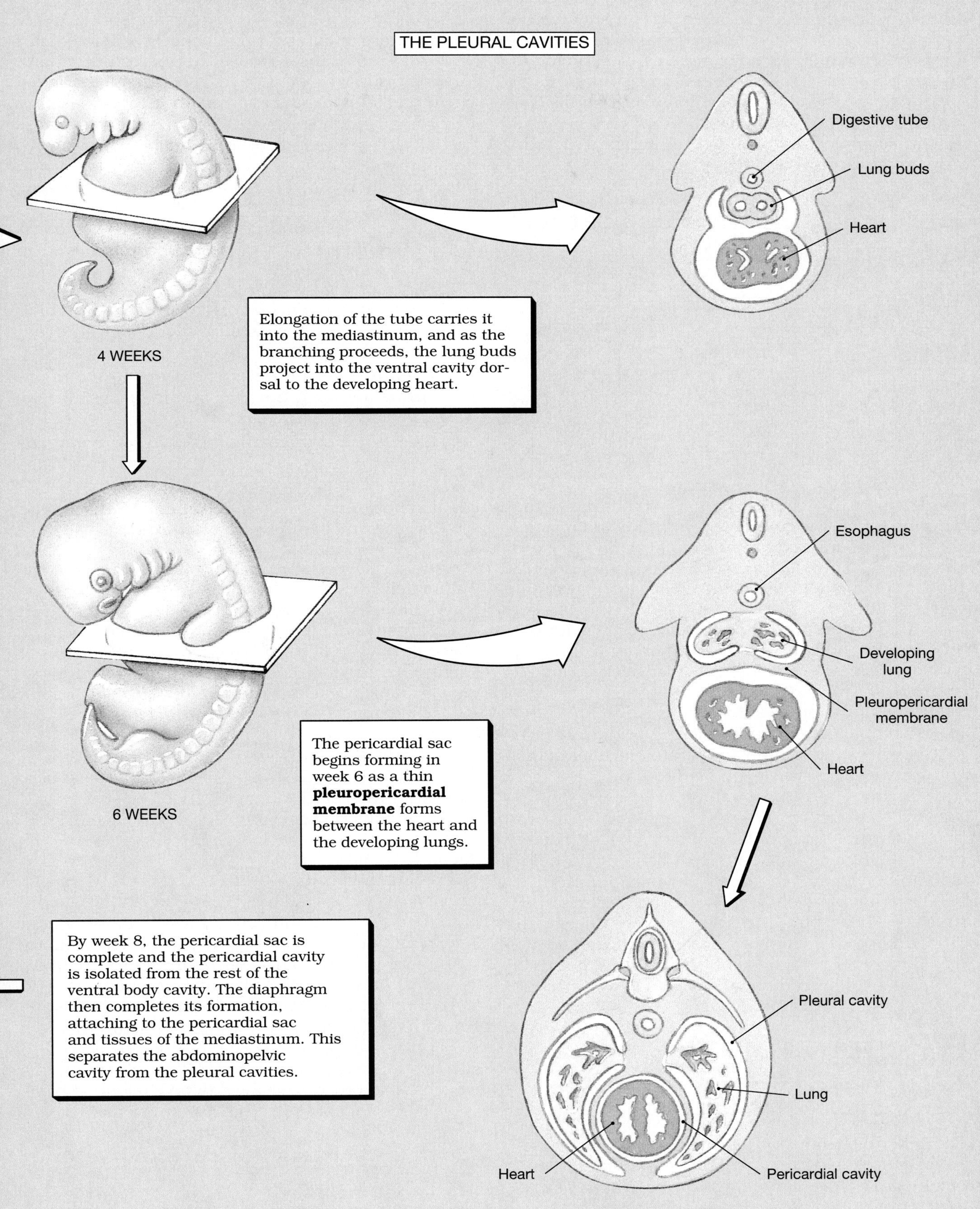
THE PLEURAL CAVITIES
Digestive tube
Lung buds
Heart
4 WEEKS
Elongation of the tube carries it into the mediastinum, and as the branching proceeds, the lung buds project into the ventral cavity dorsal to the developing heart.
Esophagus
Developing lung
Pleuropericardial membrane
Heart
6 WEEKS
The pericardial sac begins forming in week 6 as a thin **pleuropericardial membrane** forms between the heart and the developing lungs.
By week 8, the pericardial sac is complete and the pericardial cavity is isolated from the rest of the ventral body cavity. The diaphragm then completes its formation, attaching to the pericardial sac and tissues of the mediastinum. This separates the abdominopelvic cavity from the pleural cavities.
Pleural cavity
Lung
Heart
Pericardial cavity
8 WEEKS

The **pleural cavity** actually represents a potential space rather than an open chamber, for the parietal and visceral layers are usually in close contact. Inflammation of the pleurae causes the membranes to produce and secrete excess amounts of pleural fluid, a condition called **pleurisy.** In this disorder, breathing becomes difficult as the fluid buildup compresses the lungs, and prompt medical attention is required.

❑ Changes in the Respiratory System at Birth

There are several important differences between the respiratory systems of a fetus and a newborn infant. Prior to delivery, pulmonary arterial resistance is high, because the pulmonary vessels are collapsed. The rib cage is compressed, and the lungs and conducting passageways contain only small amounts of fluid and no air. At birth, the newborn infant takes a truly heroic first breath through powerful contractions of the diaphragmatic and external intercostal muscles. The inspired air enters the respiratory passageways with enough force to push the contained fluids out of the way and inflate the entire bronchial tree and most of the alveolar complexes. The same drop in pressure that pulls air into the lungs pulls blood into the pulmonary circulation; the changes in blood flow that occur lead to the closure of the *foramen ovale*, an interatrial connection, and the *ductus arteriosus*, the fetal connection between the pulmonary trunk and the aorta. (These events were detailed in Chapters 20 and 21.) ∞ [pp. 685, 768]

The exhalation that follows fails to empty the lungs completely, for the rib cage does not return to its former, fully compressed state. Cartilages and connective tissues keep the conducting passageways open, and the surfactant covering the alveolar surfaces prevents their collapse. Subsequent breaths complete the inflation of the alveoli. These physical changes are sometimes used by pathologists to determine whether a newborn infant died before delivery or shortly thereafter. Prior to the first breath, the lungs are completely filled with fluid, and they will sink if placed in water. After the first breath, even the collapsed lungs contain enough air to keep them afloat.

Respiratory Physiology

The general term *respiration* refers to two integrated processes, *external respiration* and *internal respiration.* The precise definitions of these terms varies from reference to reference. In this discussion, **external respiration** includes all of the processes involved in the exchange of oxygen and carbon dioxide between the interstitial fluids of the body and the external environment. The goal of external respiration, and the primary function of the respiratory system, is to meet the demands of living cells. **Internal respiration** is the absorption of oxygen and the release of carbon dioxide by those cells. We will consider the biochemical pathways responsible for oxygen consumption and carbon dioxide generation, often called *cellular respiration,* in Chapter 25.

Our discussion of respiratory physiology focuses on four integrated steps that are involved in external respiration:

1. ***Pulmonary ventilation,*** or breathing, which involves the physical movement of air into and out of the lungs.
2. ***Gas diffusion across the respiratory membrane,*** between the alveolar air spaces and the alveolar capillaries.
3. ***The storage and transport of oxygen and carbon dioxide*** between the alveolar capillaries and capillary beds in other tissues.
4. ***The exchange of dissolved gases between the blood and the interstitial fluids.***

Abnormalities affecting any one of these steps will ultimately affect the gas concentrations of the interstitial fluids, and thereby affect cellular activities. If the oxygen content declines, the affected tissues will become oxygen-starved. This *hypoxia* places severe limits on the metabolic activities of the tissue involved. For example, the effects of *coronary ischemia* (Chapter 20) result from chronic hypoxia affecting cardiac muscle cells. ∞ [p. 697] If the supply of oxygen is cut off completely, the condition of **anoxia** (a-NOKS-ē-a; *a-*, without + *ox-*, oxygen) results. Anoxia kills cells very quickly. Much of the damage caused by strokes and heart attacks is the result of localized anoxia.

❑ Pulmonary Ventilation

Pulmonary ventilation is the physical movement of air in and out of the respiratory tract. The primary function of pulmonary ventilation is to maintain adequate *alveolar ventilation,* air movement into and out of the alveoli. Alveolar ventilation prevents the buildup of carbon dioxide in the alveoli and ensures a continual supply of oxygen that keeps pace with absorption by the bloodstream. To understand this mechanical process we need to take a look at basic physical principles governing the movement of air.

Boyle's Law: Gas Pressure and Volume

Figure 23-13

The primary differences between liquids and gases such as air reflect the interactions between individual molecules. Although the molecules in a liquid are in constant motion, they are held closely together by weak interactions such as hydrogen bonding between the atoms of one molecule and those of another. As a result liquids are relatively dense and viscous. But because the electrons of adjacent atoms tend to repel one another, liquids tend to resist compression. If you squeeze a balloon filled with water, it will distort into a different shape, but the volume of the new shape will be the same as that of the original.

In a gas, the molecules are bouncing around as independent entities. At normal atmospheric pressures the molecules are fairly far apart, and the density of air is rather low. At such distances, the forces acting between the molecules are minimal, so an applied pressure can push them more closely together. For example, suppose that we have a sealed container of air at atmospheric pressure, such as the one depicted in Figure 23-13●. The pressure exerted by the enclosed gas results from the collision of gas molecules with the walls of the container. The greater the number of collisions, the higher the pressure.

The gas pressure within a container can be changed by altering the volume of the container, thereby giving the gas molecules more or less room in which to bounce around. Decreasing the volume of the container will result in more frequent collisions, and thus elevate the pressure of the gas (Figure 23-13a●). If the volume is increased, there will be fewer collisions per unit time, because it will take longer for a gas molecule to travel from one wall to another. As a result, the gas pressure inside the container will decline (Figure 23-13b●).

An inverse relationship thus exists between pressure and volume: *If you decrease the volume of a gas, its pressure rises; if you increase the volume of a gas, its pressure falls.* The relationship between pressure and volume is reciprocal: If you double the external pressure on a flexible container, its volume will decrease by one-half. If you reduce the external pressure by one-half, the volume of the container will double. This relationship, $P = 1/V$, first recognized by Robert Boyle in the 1600s, is called **Boyle's law.** AM *Boyle's Law and Air Overexpansion Syndrome*

Pressure and Airflow to the Lungs

Figure 23-14

Air will flow from an area of higher pressure to an area of relatively lower pressure. This tendency for directed airflow plus the pressure-volume relationships of Boyle's law provide the basis for pulmonary ventilation. A single respiratory cycle consists of an inhalation, or *inspiration*, and an exhalation, or *expiration*. Inhalation and exhalation involve changes in the volume of the lungs, and these changes create pressure gradients that move air into or out of the respiratory tract.

Each lung lies within a pleural cavity. The parietal and visceral pleurae are separated by only a thin film of pleural fluid, and, although the two membranes can slide across one another, they are held together by that fluid film. You encounter the same principle whenever you set a wet glass on a smooth surface. You can slide the glass quite easily, but when you try to lift it you encounter considerable resistance. As you pull the glass away from the tabletop you create a powerful suction, and the only way to defeat it is to tilt the glass so that air is pulled between the glass and the table, breaking the fluid bond.

A comparable fluid bond exists between the parietal pleura and the visceral pleura covering the lungs. As a result, the surface of each lung sticks to the inner wall of the chest and the superior surface of the diaphragm. Movements of the chest wall or diaphragm thus have a direct effect on the volume of the lungs. The basic principle is shown in Figure 23-14a●, p. 848. The volume of the thoracic cavity changes when (1) the diaphragm changes position or (2) the rib cage moves.

- The diaphragm forms the floor of the thoracic cavity. The relaxed diaphragm has the shape of a dome that projects upward into the thoracic cavity. When the diaphragm contracts, it tenses and moves inferiorly.

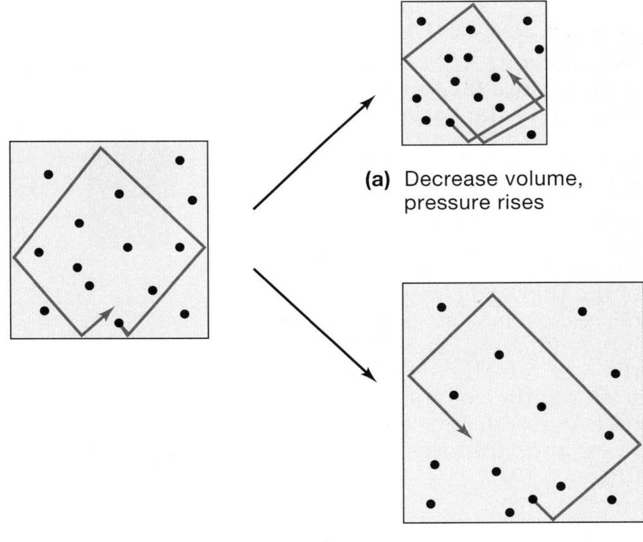
(a) Decrease volume, pressure rises
(b) Increase volume, pressure falls

● **FIGURE 23-13**

Pressure and Volume Relationships. Gas molecules within a container bounce off the walls and off one another, traveling the distance indicated in a given period of time. **(a)** If the volume decreases, each molecule will travel the same distance but will strike the walls more frequently. The pressure exerted by the gas increases. **(b)** Alternatively, if the volume of the container increases, the gas molecules will strike the walls less often, lowering the pressure in the container. D

This movement increases the volume of the thoracic cavity and exerts pressure on the contents of the abdominopelvic cavity. When the diaphragm relaxes, it returns to its original position, and the volume of the thoracic cavity decreases.

- Because of the nature of the articulations between the ribs and the vertebrae, superior movement of the ribs increases the depth and width of the thoracic cavity. Inferior movement of the rib cage reverses the process and reduces the volume of the thoracic cavity.

At the start of a breath, pressures inside and outside the thoracic cavity are identical, and there is no movement of air into or out of the lungs (Figure 23-14b●). When the thoracic cavity enlarges, the pleural cavities and lungs expand to fill the additional space (Figure 23-14c●). This expansion lowers the pressure inside the lungs. Air now enters the respiratory passageways because the pressure inside the lungs (P_i) is lower than atmospheric pressure (pressure outside, or P_o). Air continues to enter until the volume stops increasing and the internal pressure is the same as that outside. When the thoracic cavity decreases in volume, pressures rise inside the lungs, forcing air out of the respiratory tract (Figure 23-14d●).

COMPLIANCE The **compliance** of the lungs is an indication of their distensibility. The lower the compliance, the greater the force required to fill and empty the lungs. Factors affecting compliance include:

- The loss of supporting tissues resulting from damage to the alveoli, as in *emphysema*, increases compliance. (p. 863)
- The collapse of alveoli on expiration, due to inadequate surfactant, as in *respiratory distress syndrome*, reduces compliance. (p. 840)
- Arthritis or other skeletal disorders that affect the articulations of the ribs or spinal column will also reduce compliance.

At rest, the muscular activity involved in pulmonary ventilation accounts for 3–5 percent of the resting energy demand. If compliance is reduced, that figure climbs dramatically, and an individual may become exhausted simply trying to continue breathing.

Pressure Changes during Inhalation and Exhalation

Figure 23-15

To understand the mechanics of respiration and the principles of gas exchange we need to be familiar with the actual pressures recorded inside and outside of the respiratory tract. The pressure readings can be reported in several different ways (Table 23-1), but we will use millimeters of mercury (mm

● **FIGURE 23-14**
Mechanisms of Pulmonary Ventilation. (a) As the ribs are elevated or the diaphragm depressed, the thoracic cavity increases in volume. **(b)** Anterior view at rest, with no air movement. **(c)** *Inhalation*: Elevation of the rib cage and contraction of the diaphragm increase the size of the thoracic cavity. Pressure decreases, and air flows into the lungs. **(d)** *Exhalation*: When the rib cage returns to its original position, the volume of the thoracic cavity decreases. Pressure rises, and air moves out of the lungs.

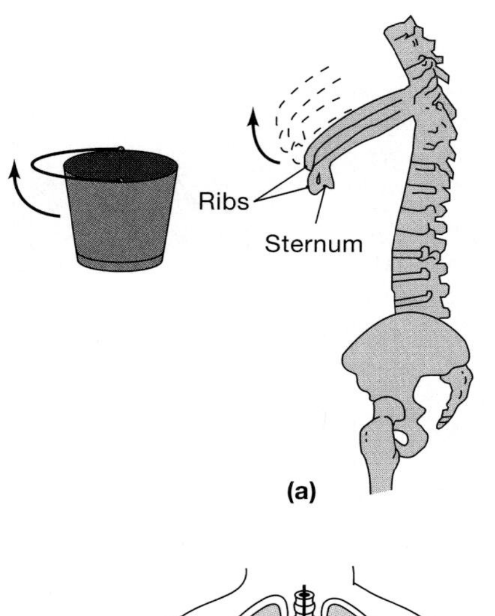
Ribs
Sternum
(a)

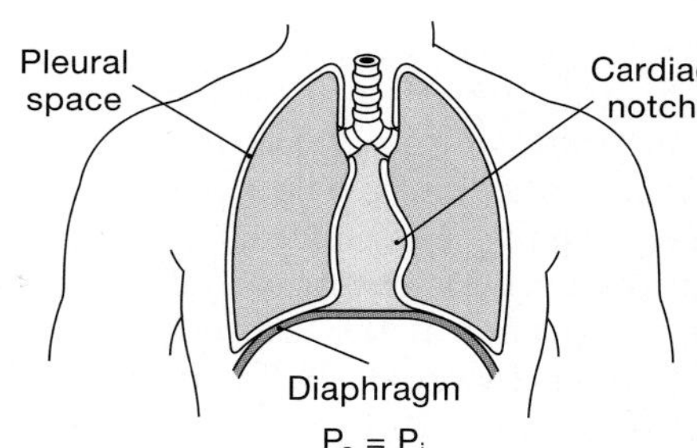
Pleural space
Cardiac notch
Diaphragm
$P_o = P_i$
Pressure outside and inside are equal, so no movement occurs
(b)

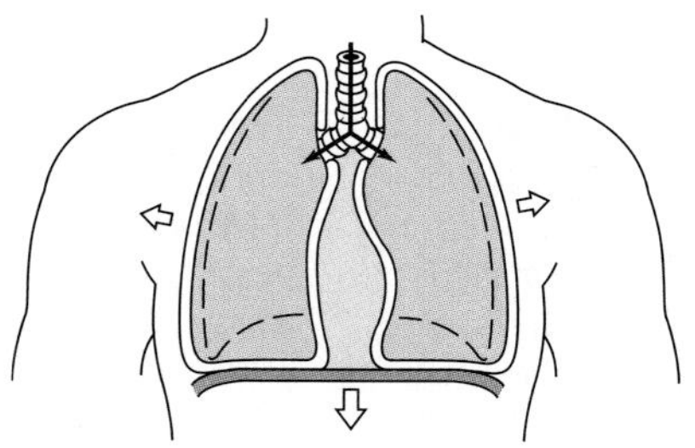
Volume increases
$P_o > P_i$
Pressure inside falls, and air flows in
(c)

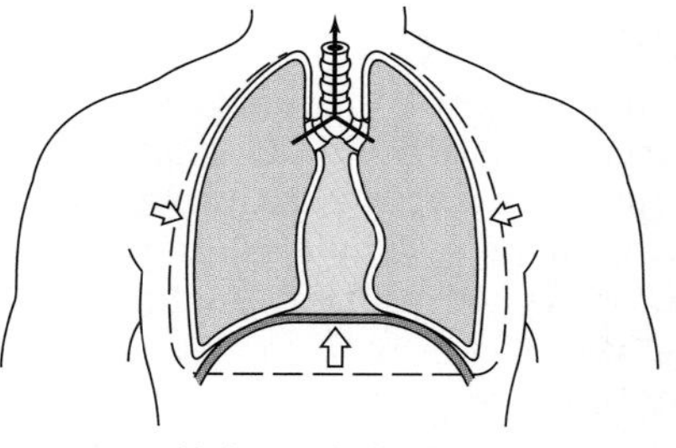
Volume decreases
$P_o < P_i$
Pressure inside rises, so air flows out
(d)

TABLE 23-1 The Four Most Common Methods of Reporting Gas Pressures

- ***millimeters of mercury*** (mm Hg): This is the most common method of reporting blood pressure and gas pressures. Normal atmospheric pressure is approximately 760 mm Hg.
- ***torr***: This unit of measurement is popular in Europe and in some technical journals. One torr is equivalent to 1 mm Hg; in other words, normal atmospheric pressure is equal to 760 torr.
- ***centimeters of water*** (cm H_2O): In a hospital setting, anesthetic gas pressures and oxygen pressures are often reported in terms of centimeters of water. One cm H_2O is equivalent to 0.735 mm Hg; normal atmospheric pressure is 1033.6 cm H_2O.
- ***pounds per square inch*** (psi): Pressures in compressed gas cylinders and other industrial applications are usually reported in terms of psi. Normal atmospheric pressure at sea level is approximately 15 psi.

Hg), as we did when considering blood pressure.

Although we are seldom reminded of the fact, the body is under considerable pressure due to the weight of the earth's atmosphere. This **atmospheric pressure** pushes against everything around us. One atmosphere of pressure **(1 atm)** is equivalent to 760 mm Hg.

THE INTRAPULMONARY PRESSURE The direction of airflow is determined by the relationship between atmospheric pressure and *intrapulmonary pressure*. The **intrapulmonary** (in-tra-PUL-mo-ner-ē) **pressure,** or **intra-alveolar** (in-tra-al-VĒ-ō-lar) **pressure,** is the pressure measured inside the respiratory tract, at the alveoli.

During quiet, relaxed breathing the pressure differences between atmospheric and intrapulmonary pressures are relatively small. On inhalation the lungs expand, and the intrapulmonary pressure drops to around 759 mm Hg. Because the intrapulmonary pressure is 1 mm Hg below atmospheric, it is usually reported as –1 mm Hg. On exhalation the lungs contract, and pressure rises to 761 mm Hg, or +1 mm Hg (Figure 23-15●).

The size of the pressure gradient increases during heavy breathing. When a trained athlete breathes at maximum capacity, the pressure differentials can reach –30 mm Hg during inspiration, and +100 mm Hg during forced expiration with the glot-

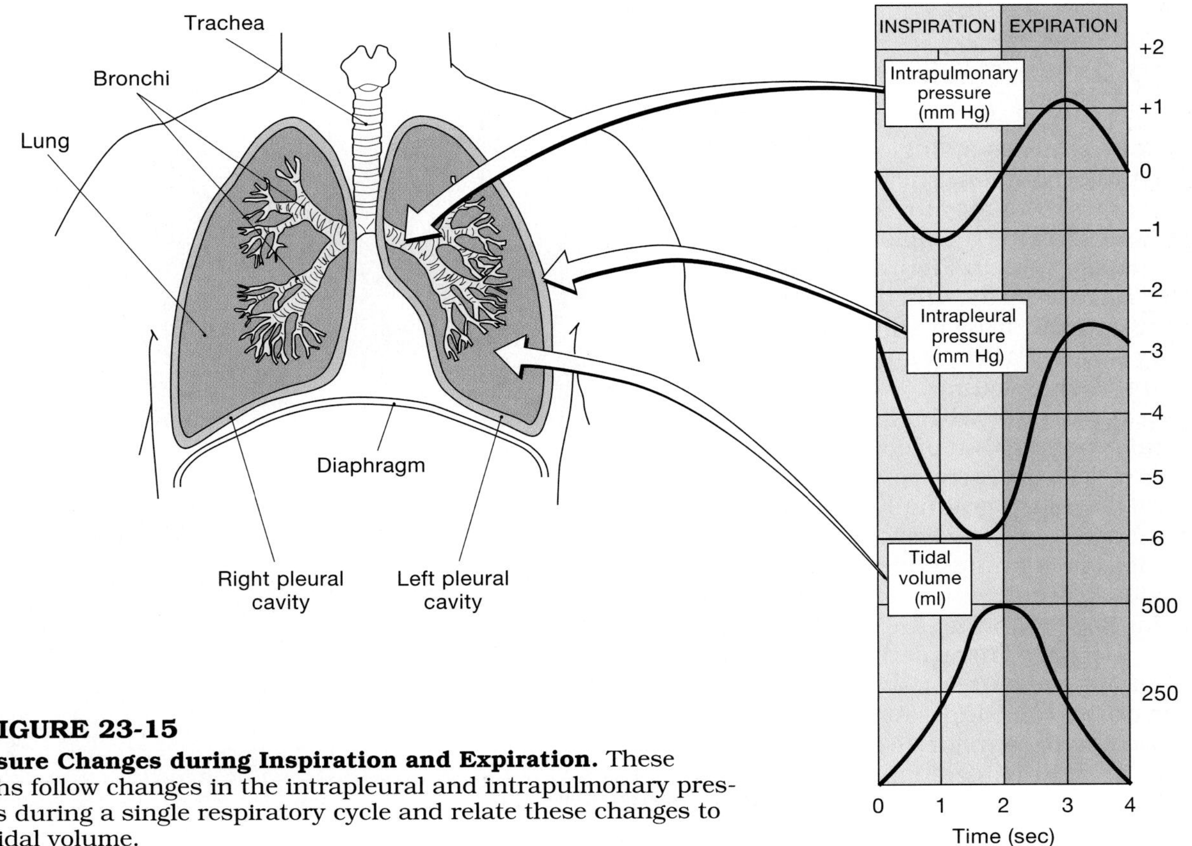
Trachea
Bronchi
Lung
Diaphragm
Right pleural cavity
Left pleural cavity
INSPIRATION
EXPIRATION
Intrapulmonary pressure (mm Hg)
Intrapleural pressure (mm Hg)
Tidal volume (ml)
+2
+1
0
–1
–2
–3
–4
–5
–6
500
250
0
1
2
3
4
Time (sec)

● **FIGURE 23-15**
Pressure Changes during Inspiration and Expiration. These graphs follow changes in the intrapleural and intrapulmonary pressures during a single respiratory cycle and relate these changes to the tidal volume.

tis closed. The latter procedure, known as the *Valsalva maneuver,* causes reflexive changes in blood pressure and cardiac output because of compression of the aorta and venae cavae. This maneuver is sometimes used to check the cardiovascular responses to changes in arterial pressure and venous return.

THE INTRAPLEURAL PRESSURE The **intrapleural pressure** is the pressure measured in the space between the parietal and visceral pleurae. The intrapleural pressure averages about –4 mm Hg and reaches –18 mm Hg during a powerful inspiration. The pressure is below atmospheric because of the relationship between the lungs and the body wall. We noted earlier that the lungs are highly elastic. In fact, they would collapse to around 5 percent of their normal resting volume if the elastic fibers were allowed to recoil completely. The elastic fibers cannot recoil significantly, however, because they are not strong enough to overcome the fluid bond between the parietal and visceral pleurae.

The elastic fibers are continually fighting that bond and pulling the lungs away from the chest wall and diaphragm. This pull lowers the intrapleural pressure slightly. Because the elastic fibers remain stretched, even after a full exhalation, the intrapleural pressure remains below atmospheric pressure during normal cycles of inhalation and exhalation. The cyclical changes in the intrapleural pressure are responsible for the *respiratory pump,* discussed in Chapter 21, that assists the venous return to the heart. [p. 736]

The Respiratory Cycle

Figure 23-15

A **respiratory cycle** is a single cycle of inhalation and exhalation. The **tidal volume** is the amount of air moved into or out of the lungs during a single respiratory cycle. The graphs in Figure 23-15• follow the intrapleural and intrapulmonary pressures during a single quiet respiratory cycle and relate those changes to the tidal volume.

Inhalation begins with the fall of intrapleural pressure that accompanies expansion of the thoracic cavity. The intrapleural pressure gradually falls to approximately –6 mm Hg. Over this period the intrapulmonary pressure declines to just under –1 mm Hg; it then begins to rise as air flows into the lungs. When exhalation begins, intrapleural and intrapulmonary pressures rise rapidly, forcing air out of the lungs. At the end of expiration, air movement again ceases when the pressure difference between intrapulmonary and atmospheric pressure has been eliminated.

PNEUMOTHORAX An injury to the chest wall that penetrates the parietal pleura or damages the alveoli and the visceral pleura can allow air into the pleural cavity. This **pneumothorax** (nū-mō-THŌ-raks) breaks the fluid bond between the pleurae and allows the elastic fibers to recoil. The result is called a collapsed lung, or **atelectasis** (at-e-LEK-ta-sis; *ateles,* imperfect + *ektasis,* expansion). Treatment for a collapsed lung involves removing as much of the air as possible from the affected pleural cavity before sealing the opening. This treatment lowers the intrapleural pressure and reinflates the lung.

Respiratory Muscles

Figure 23-16

The skeletal muscles involved in respiratory movements were introduced in Chapter 11. [pp. 345–347] Of these, the most important respiratory muscles are the *diaphragm* (Figure 11-12•, p. 347) and the *external and internal intercostals* (Figures 11-11, 11-14, and 11-15•, pp. 345, 350, and 352). These muscles are involved in normal quiet respiration. The **accessory muscles** become active when the depth and frequency of respiration must be increased markedly. The accessory muscles include the *sternocleidomastoid,* the *serratus anterior,* the *scalenes,* the *transversus thoracis,* the *transversus abdominis,* the *external and internal obliques,* and the *rectus abdominis.* These muscles are diagrammed in Figures 11-11, 11-12, and 23-16•.

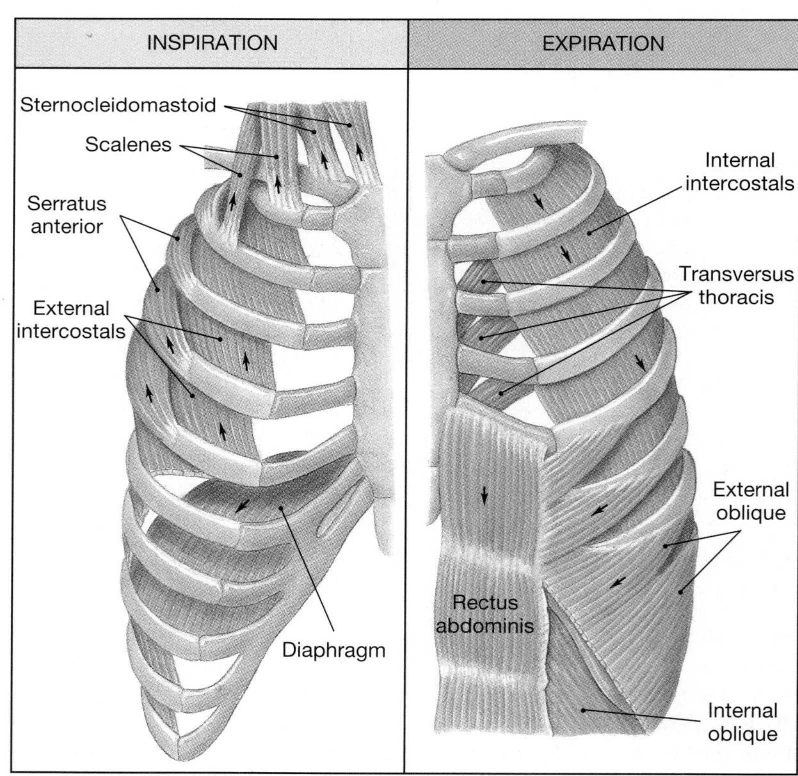
INSPIRATION
EXPIRATION
Sternocleidomastoid
Scalenes
Serratus anterior
External intercostals
Diaphragm
Internal intercostals
Transversus thoracis
External oblique
Rectus abdominis
Internal oblique

● **FIGURE 23-16**
The Respiratory Muscles

MUSCLES USED IN INHALATION Inhalation is an active process involving the contraction of one or more of these muscles:

- Contraction of the diaphragm increases the volume of the thoracic cavity by tensing and flattening its floor, and this increase draws air into the lungs. Diaphragmatic contraction is responsible for roughly 75 percent of the air movement in normal quiet breathing.
- The external intercostals assist in inhalation by elevating the ribs. This action contributes roughly 25 percent to the volume of air in the lungs.
- Accessory muscles, including the sternocleidomastoid, the serratus anterior, and the scalenes, can assist the external intercostals in elevating the ribs.

MUSCLES USED IN EXHALATION Exhalation may be passive or active, depending on the level of respiratory activity. When exhalation is active, it may involve one or more of the following muscles:

- The internal intercostals and transversus thoracis depress the ribs and reduce the width and depth of the thoracic cavity.
- The abdominal muscles, including the external and internal obliques, the transversus abdominis, and the rectus abdominis, can assist the internal intercostals in expiration by compressing the abdomen and forcing the diaphragm upward.

ARTIFICIAL RESPIRATION **Artificial respiration** is a technique to provide air to an individual whose respiratory muscles are no longer functioning. In **mouth-to-mouth resuscitation** a rescuer provides ventilation by exhaling into the mouth or mouth and nose of the victim. After each breath, contact is broken to permit passive exhalation by the victim. Air provided in this way contains adequate oxygen to meet the needs of the victim. If the victim's cardiovascular system is nonfunctional as well, a technique called *cardiopulmonary resuscitation (CPR)* is required to maintain adequate blood flow and tissue oxygenation. AM *CPR*

Modes of Breathing

Figure 23-17

The respiratory muscles may be used in various combinations depending on the volume of air that must be moved into or out of the system. Respiratory movements are usually classified as *quiet breathing* or *forced breathing,* depending on the pattern of muscle activity in the course of a single respiratory cycle. Relationships between these categories are indicated in Figure 23-17●.

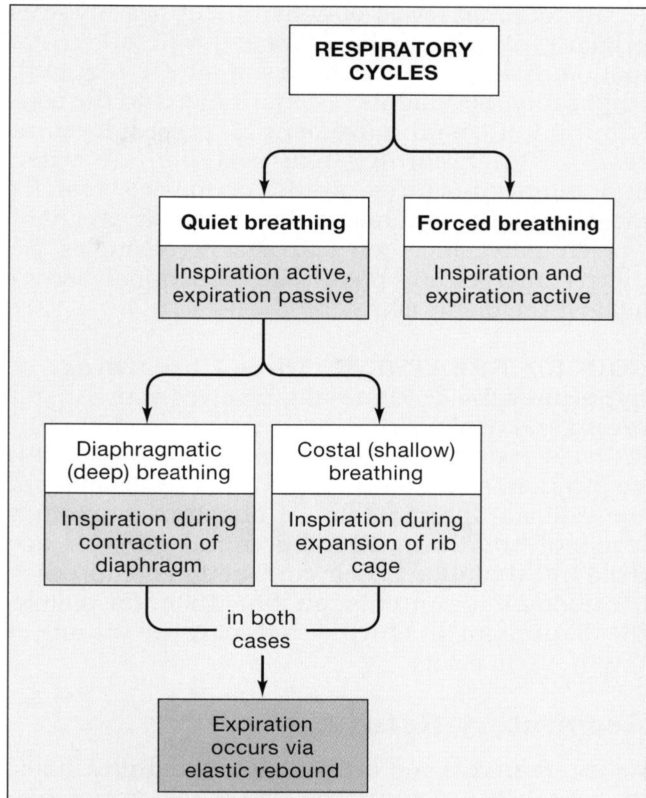
RESPIRATORY CYCLES
Quiet breathing
Inspiration active, expiration passive
Forced breathing
Inspiration and expiration active
Diaphragmatic (deep) breathing
Inspiration during contraction of diaphragm
Costal (shallow) breathing
Inspiration during expansion of rib cage
in both cases
Expiration occurs via elastic rebound

● **FIGURE 23-17**
A Classification of Respiratory Activity

QUIET BREATHING In **quiet breathing,** or **eupnea** (ŪP-nē-a), inhalation involves muscular contractions, but exhalation is a passive process. Inhalation usually involves the contraction of both the diaphragm and the external intercostals. The relative contributions of these muscles can vary, depending on the circumstances.

- During **diaphragmatic breathing,** or **deep breathing,** contraction of the diaphragm provides the necessary change in thoracic volume. Air is drawn into the lungs as the diaphragm contracts, and exhalation occurs when the diaphragm relaxes.
- In **costal breathing,** or **shallow breathing,** the thoracic volume changes because the rib cage changes shape. Inhalation occurs when contractions of the external intercostals elevate the ribs and enlarge the thoracic cavity. Exhalation occurs when these muscles relax.

During quiet breathing, expansion of the lungs stretches their elastic fibers. In addition, elevation of the rib cage stretches opposing skeletal muscles and elastic fibers in the connective tissues of the body wall. When the inspiratory muscles relax, these elastic components recoil, returning the diaphragm, the rib cage, or both to their original positions. This phenomenon is called **elastic rebound.**

At minimal levels of activity, eupnea involves primarily diaphragmatic movement with little costal motion. As increased volumes of air are required, inspiratory movements become larger and the contribution of costal movement increases. Even at rest, costal breathing can predominate when abdominal pressures, fluids, or masses restrict diaphragmatic movements. For example, pregnant women increasingly rely on costal breathing as the uterus enlarges and pushes the abdominal viscera against the diaphragm.

FORCED BREATHING **Forced breathing,** or **hyperpnea** (hī-PERP-nē-ah), involves active inspiratory and expiratory movements. Forced breathing calls upon the accessory muscles to assist with inspiration, and expiration involves contraction of the internal intercostals. At absolute maximum levels of breathing, the abdominal muscles are used in exhalation. Their contraction compresses the abdominal contents, pushing them up against the diaphragm and further reducing the volume of the thoracic cavity.

Respiratory Rates

As you read this you are probably breathing quietly, but when necessary the rate of respiration can change dramatically. The **respiratory rate** is the number of breaths per minute. The normal resting adult respiratory rate ranges from 12 to 18 breaths per minute. Children breathe more rapidly, at rates of around 18–20 breaths per minute.

The Respiratory Minute Volume

The amount of air moved each minute can be calculated by multiplying the respiratory rate (f) by the tidal volume (V_T). This value is called the **respiratory minute volume,** $\dot{V}_E$. The tidal volume at rest varies from individual to individual, but it averages around 500 ml. Therefore the respiratory minute volume at rest will be approximately 6 liters a minute:

$$\dot{V}_E \text{ (volume of air moved each minute)} = f \text{ (breaths per minute)} \times V_T \text{ (tidal volume)} = 12 \times 500 \text{ ml} = 6000 \text{ ml} = 6.0 \text{ liters}$$

Alveolar Ventilation

The respiratory minute volume ($\dot{V}_E$) measures pulmonary ventilation and provides an indication of how much air is moving into and out of the respiratory tract. Only a portion of the inspired air reaches the alveolar exchange surfaces. A typical tidal inspiration pulls around 500 ml of air into the respiratory system. The first 350 ml inspired travels along the conducting passageways and enters the alveolar spaces. The last 150 ml of inspired air never gets farther than the conducting passageways and does not participate in gas exchange with the blood. The volume of air in the conducting passages is known as the **anatomic dead space,** or V_D. **Alveolar ventilation,** or $\dot{V}_A$, is the amount of air reaching the alveoli each minute. Alveolar ventilation is less than the respiratory minute volume because some of the air never reaches the alveoli, but remains in the dead space of the lungs. Alveolar ventilation can be calculated by subtracting the dead space from the tidal volume, using the formula:

$$\dot{V}_A = f \times (V_T - V_D)$$

At rest, alveolar ventilation rates are approximately 4.2 liters per minute (12 × 350 ml). However, the gas arriving in the alveoli is significantly different from that of the surrounding atmosphere, because the air entering the respiratory tract on inhalation mixes with air in the anatomic dead space before reaching the exchange surfaces. The air in the alveoli thus contains less oxygen and more carbon dioxide than atmospheric air.

Relationships among V_T, $\dot{V}_E$, and $\dot{V}_A$

The respiratory minute volume can be increased by (1) increasing the tidal volume or (2) increasing the respiratory rate. Under maximum stimulation, the tidal volume can increase to roughly 4.8 liters, and at peak respiratory rates of 40–50 breaths per minute and maximum cycles of inspiration and expiration, the respiratory minute volume can approach 200 liters (about 55 gal) per minute.

In functional terms, the alveolar ventilation rate is more important than the respiratory minute volume, because it determines the rate of oxygen delivery to the alveoli.

- For a given respiratory rate, increasing the tidal volume will increase the alveolar ventilation rate.
- For a given tidal volume, increasing the respiratory rate will increase the alveolar ventilation rate.

It is important to note that *the alveolar ventilation rate can change independently of the respiratory minute volume.* In the above example, the respiratory minute volume at rest was 6 liters, and the alveolar ventilation rate was 4.2 l/min. If the respiratory rate increases to 20 breaths per minute, but the tidal volume drops to 300 ml, the respiratory minute volume remains the same (20 × 300 = 6000). However, the alveolar ventilation rate drops to only 3 l/min (20 × [300 – 150] = 3000). Thus, whenever the demand for oxygen increases, both tidal volume *and* respiratory rate must be regulated closely. The regulatory mechanisms involved will be the focus of a later section.

Respiratory Performance and Volume Relationships

Figure 23-18

Only a small proportion of the air in the lungs is exchanged during a single quiet respiratory cycle; the tidal volume can be increased by inhaling more vigorously and exhaling more completely. The total volume of the lungs (Figure 23-18●) can be divided into *volumes* and *capacities*. These values can be experimentally determined, and the values obtained are useful in diagnosing problems with pulmonary ventilation.

Pulmonary volumes include the following:

- The **resting tidal volume** is the amount of air moved into or out of the lungs during a single quiet respiratory cycle. The resting tidal volume averages about 500 ml in both males and females. Many other respiratory volumes and capacities differ between the sexes, because adult females on average have smaller bodies, and thus smaller lung volumes, than males. Representative values for both sexes are indicated in Figure 23-18●.
- The **expiratory reserve volume** (ERV) is the amount of air that could be voluntarily expelled after completing a normal quiet respiratory cycle. If, with maximum use of the accessory muscles, you could expel an additional 1000 ml of air, your expiratory reserve volume would be 1000 ml.
- The **residual volume** is the amount of air that remains in the lungs even after a maximal exhalation. The **minimal volume,** a component of the residual volume, is the amount of air remaining in the lungs if they are allowed to collapse. The minimal volume ranges from 30 to 120 ml, but unlike other volumes it cannot be measured in a normal subject. The minimal volume and the residual volume are very different because the fluid bond between the lungs and the chest wall prevents the recoil of the elastic fibers. Some air remains in the lungs, even at minimal volume, because the surfactant coating the alveolar surfaces prevents their collapse.
- The **inspiratory reserve volume** (IRV) is the amount of air that can be taken in over and above the tidal volume. There are significant differences between the inspiratory reserve volumes of males and females, because on average the lungs of males are larger. The inspiratory reserve volume of males averages 3300 ml versus 1900 ml in females.

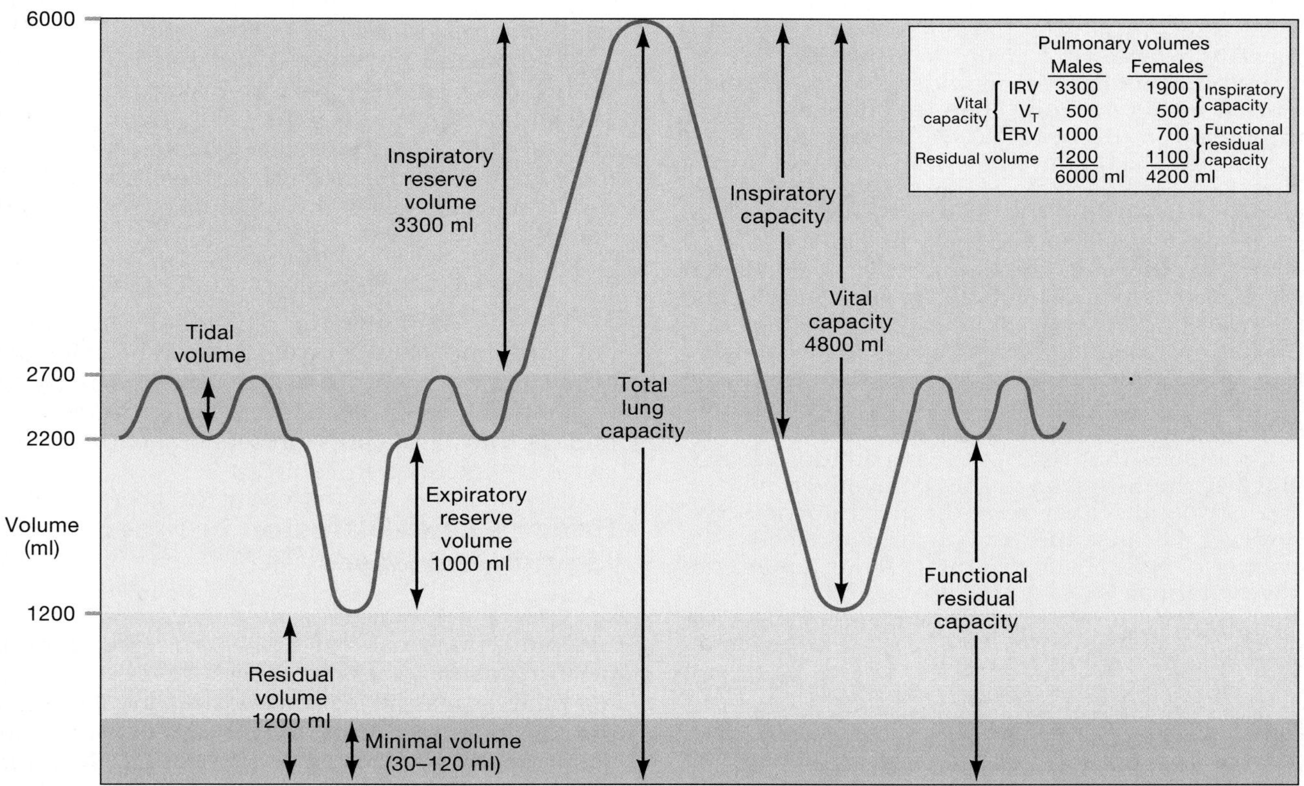

6000
2700
2200
1200
Volume (ml)
Tidal volume
Inspiratory reserve volume 3300 ml
Total lung capacity
Inspiratory capacity
Vital capacity 4800 ml
Expiratory reserve volume 1000 ml
Residual volume 1200 ml
Minimal volume (30–120 ml)
Functional residual capacity

		Pulmonary volumes Males	Females	
Vital capacity	IRV	3300	1900	Inspiratory capacity
	V_T	500	500	
	ERV	1000	700	Functional residual capacity
Residual volume		1200	1100	
		6000 ml	4200 ml	

● **FIGURE 23-18**

Respiratory Volumes and Capacities. The graph diagrams the relationships between the respiratory volumes and capacities of an average male.

Respiratory capacities are determined by adding the values of various volumes. Examples include:

- The **inspiratory capacity** is the amount of air that can be drawn into the lungs after completing a normal resting respiratory cycle. It is the sum of the tidal volume and the inspiratory reserve volume.
- The **functional residual capacity** is the amount of air remaining in the lungs after completing a quiet respiratory cycle. It is the sum of the expiratory reserve volume and the residual volume.
- The **vital capacity** is the maximum amount of air that can be moved into or out of the lungs in a single respiratory cycle. It is the sum of the expiratory reserve, the tidal volume, and the inspiratory reserve. The vital capacity averages around 4.8 liters in males and 3.1 liters in females.
- The **total lung capacity** is the total volume of the lungs. It is the sum of the vital capacity and the residual volume; it averages around 6 liters in males and 4.2 liters in females.

PULMONARY FUNCTION TESTS Pulmonary function tests monitor several aspects of respiratory function. A **spirometer** (spī-ROM-e-ter) measures parameters such as vital capacity, expiratory reserve, and inspiratory reserve. A **pneumotachometer** provides additional information by determining the rate of air movement throughout a respiratory cycle. A *peak flow meter* records the maximum rate of air movement during forced expiration. Although these tests are relatively simple to perform, they have considerable diagnostic significance. For example, in *asthma* the constricted airways tend to close before an exhalation is completed. As a result, both the vital capacity and the expiratory reserve volume are diminished, and the narrow respiratory passageways reduce the flow rate as well. Air whistling through the constricted airways produces the characteristic "wheezing" that accompanies an asthmatic attack. AM *Asthma*

Conditions that restrict the maximum distensibility of the lungs have the same effect on vital capacity because they lower the inspiratory reserve. However, because the airways are not affected the expiratory reserve and expiratory flow rates are relatively normal.

Mark breaks a rib that punctures the chest wall on his left side. What do you expect will happen to his left lung as a result?

In pneumonia, fluid accumulates in the alveoli of the lungs. How would this accumulation affect vital capacity?

Gas Exchange at the Respiratory Membrane

Pulmonary ventilation ensures that the alveoli are supplied with oxygen, and it removes the carbon dioxide arriving from the bloodstream. The actual process of gas exchange occurs between the blood and alveolar air across the respiratory membrane. To understand these events we consider (1) the *partial pressures* of the gases involved and (2) the diffusion of molecules from a gas into a liquid.

Dalton's Law and Partial Pressures

The air we breathe is not a single gas, but a mixture of gases. Nitrogen molecules (N_2) are the most abundant, accounting for about 78.6 percent of the atmospheric gas molecules. Oxygen molecules (O_2), the second most abundant, account for roughly 20.8 percent of the atmospheric population. Most of the remaining 0.5 percent consists of water molecules, with carbon dioxide (CO_2) contributing a mere 0.04 percent to the atmospheric total.

Atmospheric pressure, 760 mm Hg, represents the combined effects of collisions involving each type of molecule. At any given moment, 78.6 percent of those collisions will involve nitrogen molecules, 20.8 percent oxygen molecules, and so on. Thus each of the gases contributes to the total pressure in proportion to its relative abundance. This relationship is known as **Dalton's law.**

The **partial pressure** of a gas is the pressure contributed by a single gas within a mixture of gases. The partial pressure is abbreviated by the prefix *P*, or *p*. *All of the partial pressures added together equal the total pressure exerted by the gas mixture.* In the case of the atmosphere this relationship can be summarized as:

$$P_{N_2} + P_{O_2} + P_{H_2O} + P_{CO_2} = 760 \text{ mm Hg}$$

Because we know the individual percentages, the partial pressure for each gas can be calculated easily. For example, the partial pressure of oxygen, P_{O_2}, is 20.8 percent of 760 mm Hg, or roughly 159 mm Hg. The partial pressures for other atmospheric gases are included in Table 23-2.

Henry's Law: Diffusion between Liquids and Gases

Figure 23-19

Differences in pressure move gas molecules from one place to another. Pressure differences also affect the movement of gas molecules into and out of solution. At a given temperature, *the amount of a particular gas in solution is directly proportional to the partial pressure of that gas.* This principle is known as **Henry's law**.

When a gas under pressure contacts a liquid, the pressure tends to force gas molecules into solu-

TABLE 23-2 Partial Pressures (mm Hg) and Normal Gas Concentrations in Air

Source of Sample	*Nitrogen (N_2)*	*Oxygen (O_2)*	*Carbon Dioxide (CO_2)*	*Water Vapor (H_2O)*
Inspired air (dry)	597 (78.6 percent)	159 (20.8 percent)	0.3 (0.04 percent)	3.7 (0.5 percent)
Alveolar air (saturated)	573 (75.4 percent)	100 (13.2 percent)	40 (5.2 percent)	47 (6.2 percent)
Expired air (saturated)	569 (74.8 percent)	116 (15.3 percent)	28 (3.7 percent)	47 (6.2 percent)

tion. At a given pressure the number of dissolved gas molecules rises until an equilibrium becomes established. At equilibrium, gas molecules are diffusing out of the liquid as quickly as they are arriving, and the total number of gas molecules in solution remains constant. If the partial pressure goes up, more gas molecules go into solution; if the partial pressure goes down, gas molecules will come out of solution. These relationships are diagrammed in Figure 23-19●.

You see Henry's law in action whenever you grab a can of soda. The soda was put into the can under pressure, and the gas (carbon dioxide) is in solution. When you open the can, the pressure falls and the gas molecules begin coming out of solution. The process will theoretically continue until an equilibrium develops between the air and the gas in solution. In fact, the volume of the can is so small and the volume of the atmosphere is so great that over the course of a half-hour or so virtually all of the carbon dioxide comes out of solution. You are then left with a "flat" soda.

The actual *amount* of a gas in solution at a given partial pressure and temperature depends on the solubility of the gas in that particular liquid. Carbon dioxide is very soluble, oxygen somewhat less soluble, and nitrogen has very limited solubility in body fluids. The dissolved gas content is usually reported in terms of milliliters of gas per 100 ml of solution. For example, blood in a pulmonary vein, which has a P_{N_2} of 573 mm Hg, usually contains around 1.25 ml of nitrogen per 100 ml.

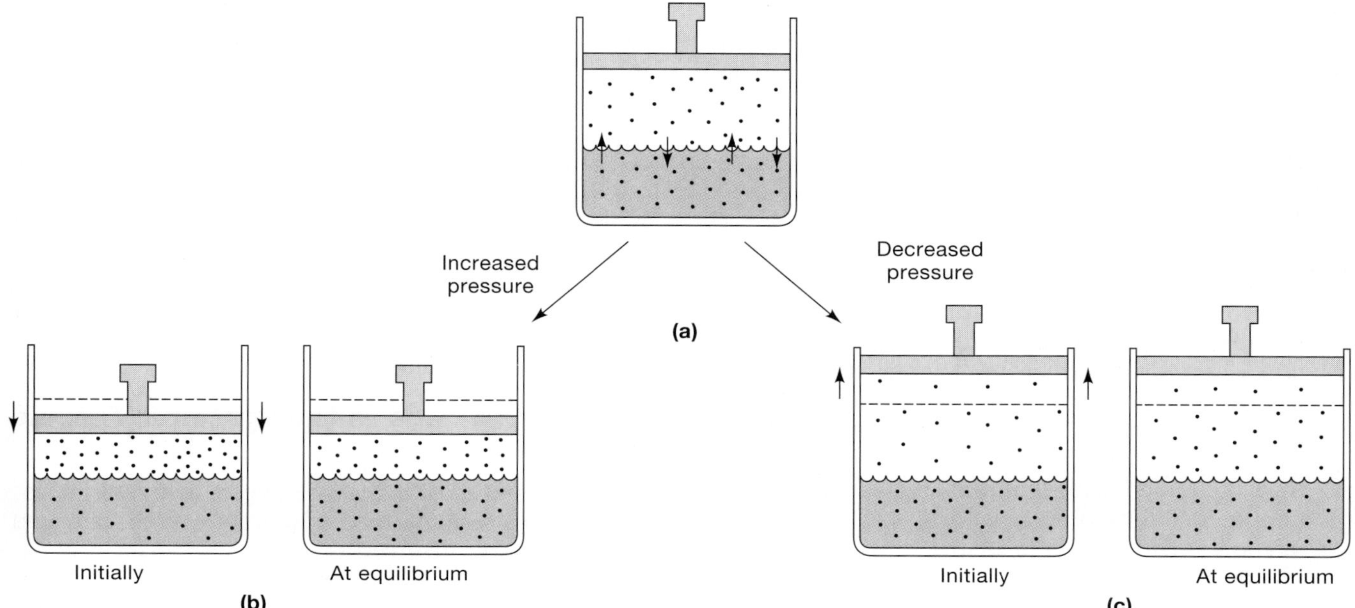
Increased pressure
Decreased pressure
(a)
Initially
At equilibrium
(b)
Initially
At equilibrium
(c)

● **FIGURE 23-19**
Henry's Law and the Relationship between Solubility and Pressure. (a) A solution containing dissolved gas molecules at equilibrium with air under a given pressure. **(b)** Increasing the pressure drives additional gas molecules into solution until a new equilibrium becomes established. **(c)** When the pressure decreases, some of the dissolved gas molecules leave the solution until equilibrium is restored.

DECOMPRESSION SICKNESS **Decompression sickness** is a painful condition that develops when a person is suddenly exposed to a drop in atmospheric pressure. Nitrogen is the gas responsible for the problems experienced, because of its high partial pressure. When the pressure drops, nitrogen comes out of solution, forming bubbles like those in a shaken can of soda. The bubbles may form in joint cavities, in the bloodstream, and in the cerebrospinal fluid. The victims often curl up because of the pain in affected joints, and this reaction accounts for the common name for the condition, *the bends.* Decompression sickness most often affects SCUBA divers, who are breathing air under greater-than-normal pressures while submerged. It also may affect people in airplanes subject to sudden losses of cabin pressure. AM *Decompression Sickness*

The Composition of Alveolar Air

As soon as air enters the respiratory tract its characteristics begin to change. In passing through the nasal cavity the air becomes warmer, and the amount of water vapor increases. Humidification and filtration continue as the air travels through the pharynx, trachea, and bronchial passageways. On reaching the alveoli, the incoming air mixes with air that remained in the alveoli after the previous respiratory cycle. The alveolar mixture thus contains more carbon dioxide and less oxygen than does atmospheric air.

The last 150 ml of inspired air never gets farther than the conducting passageways and remains in the anatomic dead space of the lungs. During the subsequent expiration, the departing alveolar air mixes with air in the dead space to produce yet another mixture that differs from both atmospheric and alveolar samples. The differences in composition between atmospheric (inspired) and alveolar air can be seen in Table 23-2.

Diffusion at the Respiratory Membrane

Gas exchange at the respiratory membrane is efficient because:

1. ***The differences in partial pressure across the respiratory membrane are substantial.*** This fact is important because the greater the difference in partial pressure, the faster the rate of gas diffusion. Conversely, if the P_{O_2} in the alveoli decreases, the rate of oxygen diffusion into the blood will decline. This is why many people feel light-headed at altitudes of 3000 m or more—the partial pressure of oxygen in the alveoli has dropped low enough that the rate of oxygen absorption is significantly reduced.
2. ***The distances involved are small.*** The fusion of capillary and alveolar basement membranes reduces the distance to an average of 0.5 μm. Inflammation of the lung tissue or fluid buildup inside the alveoli will increase the diffusion distance and impair alveolar gas exchange.
3. ***The gases are lipid-soluble.*** Both oxygen and carbon dioxide diffuse readily through the surfactant layer and the alveolar and endothelial cell membranes.
4. ***The total surface area is large.*** The combined alveolar surface area at peak inspiration is approximately 140 m^2. Damage to alveolar surfaces, as in *emphysema,* reduces the available surface area and the efficiency of gas transfer.
5. ***Blood flow and airflow are coordinated.*** This arrangement improves the efficiency of both pulmonary ventilation and pulmonary circulation. For example, blood flow is greatest around alveoli with the highest P_{O_2}s, where oxygen uptake can proceed with maximum efficiency. If the normal circulation is impaired, as in a *pulmonary embolism,* or the normal airflow is interrupted, as in various forms of *pulmonary obstruction,* this coordination is lost, and respiratory efficiency decreases. AM *Bronchitis, Emphysema, and COPD*

Partial Pressures in the Alveolar Air and Alveolar Capillaries

Figure 23-20a

Figure 23-20a• details the partial pressures of oxygen and carbon dioxide in the alveolar air and the alveolar capillaries. Blood arriving in the pulmonary arteries has a higher P_{CO_2} and a lower P_{O_2} than does alveolar air. Diffusion between the alveolar mixture and the pulmonary capillaries thus elevates the P_{O_2} of the blood while lowering its P_{CO_2}. By the time the blood enters the pulmonary venules it has reached equilibrium with the alveolar air, so it departs the alveoli with a P_{O_2} of about 100 mm Hg and a P_{CO_2} of roughly 40 mm Hg.

Diffusion between the blood in the pulmonary capillaries and the alveolar air occurs very rapidly. In a resting individual, a red blood cell moves through a pulmonary capillary in about 0.75 second, and during exercise the passage takes less than 0.3 second. This amount of time is usually sufficient to reach an equilibrium between the alveolar air and the blood.

Partial Pressures in the Systemic Circuit

Figure 23-20b

The oxygenated blood now leaves the alveolar capillaries and returns to the heart, to be discharged into the systemic circuit.

OXYGEN PRESSURES As it enters the pulmonary veins, the oxygenated blood from the alveolar capillaries mixes with blood that flowed through capillaries around conducting passageways rather than alveoli. Because gas exchange occurs only at alveoli, the blood leaving the conducting passageways carries relatively little oxygen. The partial pressure of oxygen in the pulmonary veins therefore drops to about 95 mm Hg. This is the P_{O_2} in the blood entering the systemic circuit, and no further changes in partial pressure occur until the blood reaches the peripheral capillaries (Figure 23-20b●).

Normal interstitial fluid has a P_{O_2} of 40 mm Hg. As a result, oxygen diffuses out of the capillaries and carbon dioxide diffuses in until the capillary partial pressures are the same as those in the adjacent tissues. At a normal tissue P_{O_2}, the blood entering the venous system still contains around 75 percent of its total oxygen content.

CARBON DIOXIDE PRESSURES Blood entering the systemic circuit normally has a P_{CO_2} of 40 mm Hg. Inactive peripheral tissues normally have a P_{CO_2} of around 45 mm Hg. As a result, carbon dioxide diffuses into the blood as oxygen diffuses out (Figure 23-20b●).

BLOOD GAS ANALYSIS Blood samples may be analyzed to determine the concentration of dissolved gases. The usual tests include determination of pH, P_{CO_2}, and P_{O_2} in an arterial sample. This analysis can be very helpful in monitoring patients after a heart attack or in chronic respiratory conditions such as obstructive disorders or asthma. A blood sample provides information about the degree of oxygenation in peripheral tissues. For example, if the P_{CO_2} is very high and the P_{O_2} very low, tissues are not receiving adequate oxygen. This condition may be corrected by providing a gas mixture that has a high P_{O_2} (or even pure oxygen, with a P_{O_2} of 760 mm Hg). In addition, blood gas measurements determine the efficiency of gas exchange at the lungs. If the arterial P_{O_2} remains low despite oxygen administration or the P_{CO_2} is very high, pulmonary exchange problems, such as pulmonary edema, emphysema, or pneumonia, must exist.

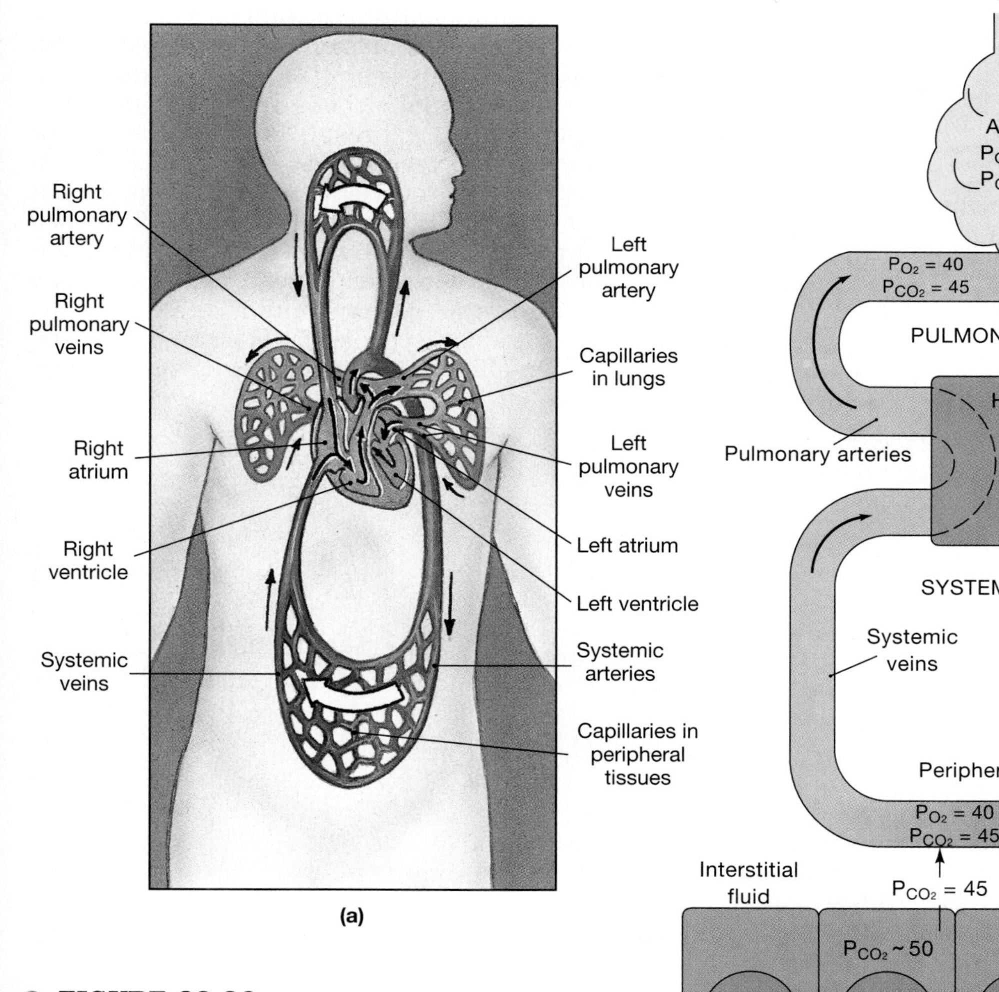
Right pulmonary artery
Right pulmonary veins
Right atrium
Right ventricle
Systemic veins
Left pulmonary artery
Capillaries in lungs
Left pulmonary veins
Left atrium
Left ventricle
Systemic arteries
Capillaries in peripheral tissues
(a)

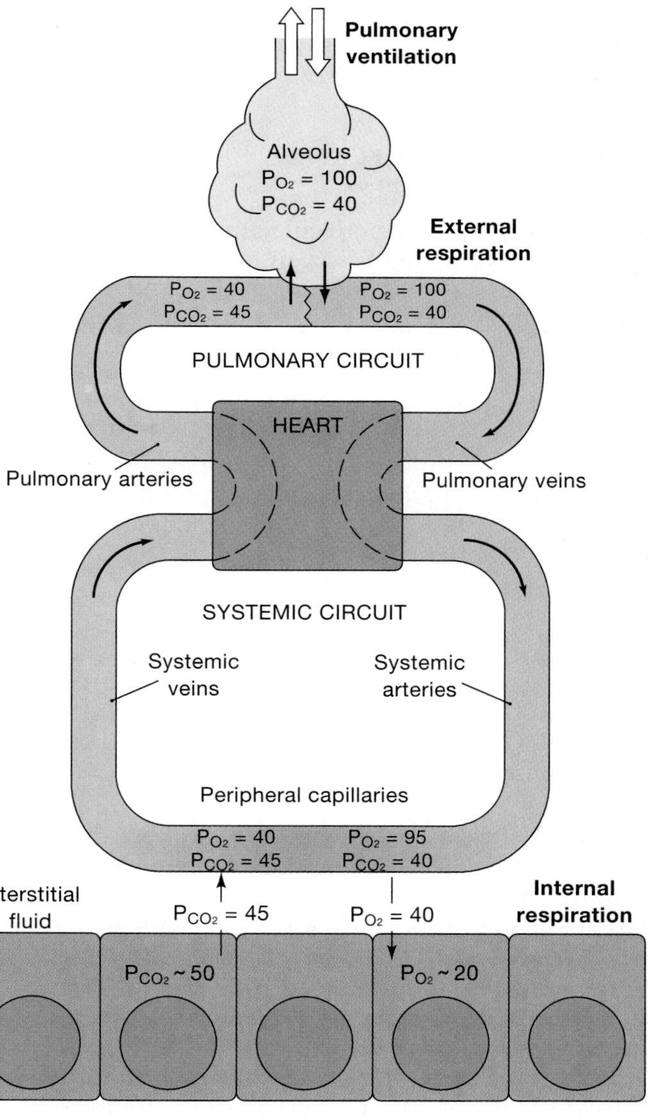
Pulmonary ventilation
Alveolus
P_{O_2} = 100
P_{CO_2} = 40
External respiration
P_{O_2} = 40
P_{CO_2} = 45
P_{O_2} = 100
P_{CO_2} = 40
PULMONARY CIRCUIT
HEART
Pulmonary arteries
Pulmonary veins
SYSTEMIC CIRCUIT
Systemic veins
Systemic arteries
Peripheral capillaries
P_{O_2} = 40
P_{CO_2} = 45
P_{O_2} = 95
P_{CO_2} = 40
Interstitial fluid
P_{CO_2} = 45
P_{O_2} = 40
Internal respiration
P_{CO_2} ~ 50
P_{O_2} ~ 20
(b)

● **FIGURE 23-20**

An Overview of Respiration and Respiratory Processes. (a) The basic pattern of circulation in the systemic and pulmonary circuits . **(b)** A diagrammatic view of respiratory processes.

Gas Pickup and Delivery

Oxygen and carbon dioxide have limited solubilities in blood plasma. For example, at the normal alveolar P_{O_2}, 100 ml of plasma will absorb only around 0.3 ml of oxygen. The limited solubilities of these gases are a problem because peripheral tissues need more oxygen and generate more carbon dioxide than the plasma can absorb and transport.

The problem is solved by the red blood cells, which remove dissolved oxygen and carbon dioxide molecules and either tie them up (in the case of oxygen) or use them to manufacture soluble compounds (in the case of carbon dioxide). Because these reactions remove dissolved gas from the plasma, gas will continue to diffuse into the blood but never reach equilibrium.

The important thing about these reactions is that they are (1) *temporary* and (2) *completely reversible.* When plasma oxygen or carbon dioxide concentrations are high, the excess molecules are removed by the red blood cells; when the plasma concentrations are falling, the red blood cells release their stored reserves.

Oxygen Transport

Figures 23-21, 23-22, 23-23

The blood leaving the alveolar capillaries carries away roughly 20 ml of oxygen per 100 ml. Of this amount, only around 1.5 percent (0.3 ml) consists of oxygen molecules in solution. The rest are bound to *hemoglobin (Hb) molecules,* specifically to the iron atoms in the center of heme units. The structure of hemoglobin was detailed in Chapter 19. *[p. 655]* A hemoglobin molecule consists of four globular protein subunits, each containing a heme unit. Thus each hemoglobin molecule can bind four molecules of oxygen, forming oxyhemoglobin. This is a reversible reaction that can be summarized as:

$$Hb + O_2 \rightleftharpoons HbO_2$$

There are approximately 280 million molecules of hemoglobin in each red blood cell, and because a hemoglobin molecule contains four heme units, each RBC potentially can carry more than a billion molecules of oxygen.

HEMOGLOBIN SATURATION Hemoglobin saturation is the percentage of heme units containing bound oxygen. At most, each hemoglobin molecule can carry four oxygen molecules. If all of the Hb molecules in the blood are fully loaded, there is 100 percent saturation. If, on average, each hemoglobin molecule carries two oxygen molecules, there is 50 percent saturation.

Chapter 2 noted that the shape and functional properties of a protein change in response to alterations in its environment. *[p. 55]* Hemoglobin is no exception, and any shape changes that occur can affect oxygen binding. Under normal conditions the most important environmental factor affecting hemoglobin are (1) the P_{O_2} of the blood, (2) blood pH, (3) temperature, and (4) ongoing metabolic activity within RBCs.

Hemoglobin and P_{O_2}. An **oxygen-hemoglobin saturation curve** (Figure 23-21●) is a graph that relates the saturation of hemoglobin to the partial pressure of oxygen. The graph is a curve, rather than a straight line, because the shape of a hemoglobin molecule changes slightly each time it binds an oxygen molecule, and the change affects its ability to bind *another* hemoglobin molecule. In other words, the attachment of the first oxygen molecule makes it easier to bind the second, binding the second promotes binding of the third, and binding of the third enhances binding of the fourth.

Because each arriving oxygen molecule increases the affinity of hemoglobin for the *next* oxygen molecule, the saturation curve takes the form of a ∫. The slope is gradual until the first oxygen molecule binds to the hemoglobin and then rises rapidly, with a prolonged plateau near 100 percent saturation. Over the steep initial slope, a very small decrease in the plasma P_{O_2} will result in a large change in the amount of oxygen bound to or released from HbO_2. Because the curve rises quickly, hemoglobin will be more than 90 percent saturated if exposed to an alveolar P_{O_2} above 60 mm Hg. This means that near-normal oxygen transport can continue despite a decrease in the oxygen content of alveolar air. Without this ability, people would be unable to survive at high altitudes, and conditions significantly

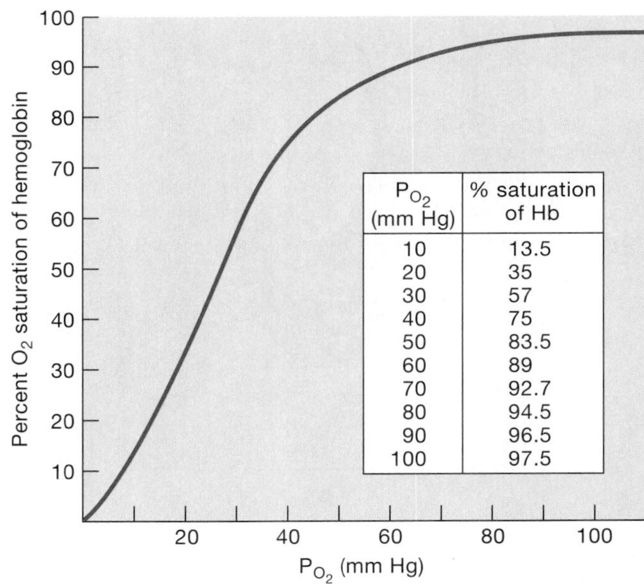
Percent O_2 saturation of hemoglobin
100
90
80
70
60
50
40
30
20
10
20
40
60
80
100
P_{O_2} (mm Hg)

P_{O_2} (mm Hg)	% saturation of Hb
10	13.5
20	35
30	57
40	75
50	83.5
60	89
70	92.7
80	94.5
90	96.5
100	97.5

● **FIGURE 23-21**

The Oxygen-Hemoglobin Saturation Curve. This curve indicates the normal saturation characteristics of hemoglobin at various partial pressures of oxygen.

reducing pulmonary ventilation would be immediately fatal.

At normal alveolar pressures (P_{O_2} = 100 mm Hg) the percent saturation is very high (97.5 percent), although complete saturation does not occur until the P_{O_2} reaches excessively high levels, around 250 mm Hg. In functional terms, the maximum percent saturation is not as important as the ability of hemoglobin to provide oxygen over the normal P_{O_2} range in body tissues. Over that range, from 100 mm Hg at the alveoli to perhaps 15 mm Hg in active tissues, the percent saturation drops from 97.5 percent to less than 20 percent, and a small change in P_{O_2} makes a big difference in terms of the amount of oxygen bound by hemoglobin.

Note that the relationship between P_{O_2} and percent saturation remains valid whether the P_{O_2} is rising or falling. *If the P_{O_2} increases, the percent saturation goes up, and hemoglobin stores oxygen. If the P_{O_2} decreases, hemoglobin releases oxygen into its surroundings.* When oxygenated blood arrives in the peripheral capillaries, the blood P_{O_2} declines rapidly as a result of gas exchange with the interstitial fluid. As the P_{O_2} falls, hemoglobin gives up its oxygen.

The relationship between P_{O_2} and hemoglobin saturation provides a mechanism for automatic regulation of oxygen delivery. Inactive tissues have little demand for oxygen, and the local P_{O_2} is usually around 40 mm Hg. Under these conditions hemoglobin will not release much oxygen. As it passes through the capillaries, it will go from 97 percent saturation (P_{O_2} = 95 mm Hg) to 75 percent saturation (P_{O_2} = 40 mm Hg). Because it still retains three-quarters of its oxygen, the venous blood has a relatively large oxygen reserve.

When tissue demands increase, more oxygen will be released. Active tissues consume oxygen at an accelerated rate, and the P_{O_2} may drop to 15–20 mm Hg. Hemoglobin passing through these capillaries will go from 97 percent saturation to around 20 percent saturation. In practical terms, this means that as blood circulates through peripheral capillaries, active tissues will receive 3.5 times as much oxygen as inactive tissues.

CARBON MONOXIDE POISONING Mystery writers often describe murder or suicide victims who died in their cars inside a locked garage. In real life, entire families are killed each winter by leaky furnaces or space heaters. The cause of death is carbon monoxide poisoning. The exhaust of automobiles, other petroleum-burning engines, oil lamps, and fuel-fired space heaters contains *carbon monoxide* (CO). Carbon monoxide competes with oxygen molecules for the binding sites on heme units. Unfortunately, the carbon monoxide usually wins, for it has a much stronger affinity for hemoglobin at very low partial pressures. The bond formed is extremely durable, and the attachment of a carbon monoxide molecule essentially inactivates that heme unit for respiratory purposes. Carbon monoxide will bind to hemoglobin at very low partial pressures. If carbon monoxide molecules make up just 0.1 percent of the components of inspired air, enough hemoglobin will be affected that survival will become impossible without medical assistance. Treatment may include (1) administration of pure oxygen, for at sufficiently high partial pressures the oxygen molecules will gradually replace carbon monoxide at the hemoglobin molecules, and, if necessary, (2) the transfusion of compatible red blood cells.

Hemoglobin and pH. The oxygen-hemoglobin saturation curve in Figure 23-21• was determined in normal blood, with a pH of 7.4 and a temperature of 37° C. In addition to consuming oxygen, active tissues generate acids that lower the pH of the interstitial fluid. When the pH declines, there is a change in the shape of hemoglobin molecules. As a result of this change, the slope of the hemoglobin saturation curve changes (Figure 23-22a•), and the hemoglobin molecules release their oxygen reserves more readily. Thus at a tissue P_{O_2} of 40 mm Hg, hemoglobin molecules will release 15 percent more oxygen at a pH of 7.2 than they would at a pH of 7.4. This effect of pH on the hemoglobin saturation curve is called the **Bohr effect.**

Carbon dioxide is the primary compound responsible for the Bohr effect. When CO_2 diffuses into the blood, an enzyme called **carbonic anhydrase** (an-HĪ-drās) catalyzes its reaction with water molecules:

$$CO_2 + H_2O \xrightleftharpoons{\text{carbonic anhydrase}} H_2CO_3$$

The product, H_2CO_3, is called *carbonic acid* because it dissociates into a bicarbonate ion (HCO_3^-) and a hydrogen ion (H^+). The rate of carbonic acid formation depends on the amount of carbon dioxide in solution, and, as we noted earlier, that amount depends on the P_{CO_2}. As a result, when the P_{CO_2} rises, the rate of carbonic acid accelerates and pH drops. When the P_{CO_2} declines, the rate of carbonic acid formation declines and pH rises.

Hemoglobin and Temperature. Temperature changes also affect the slope of the hemoglobin saturation curve, as indicated in Figure 23-22b•. As the temperature rises, hemoglobin releases more oxygen; as the temperature declines, hemoglobin holds oxygen more tightly. Temperature effects are significant only in active tissues where large amounts of heat are being generated. For example, active skeletal muscles generate heat, and the heat warms blood flowing through these organs. As the blood warms, the hemoglobin molecules release more oxygen than can be used by the active muscle fibers.

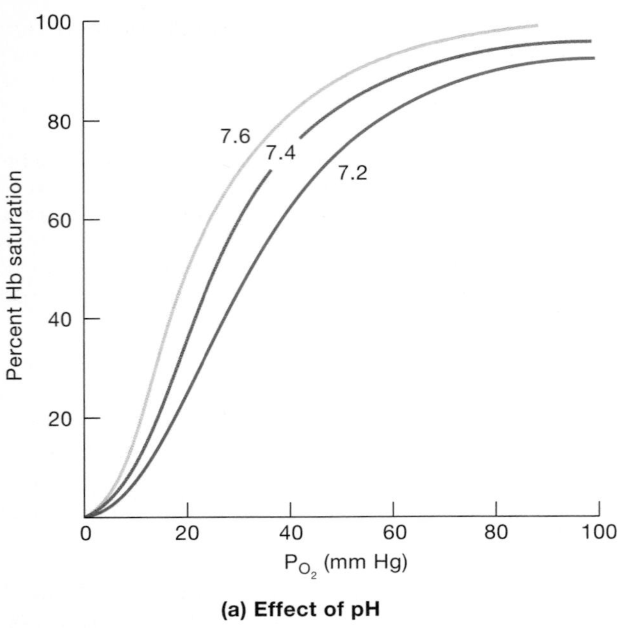
100
80
60
40
20
0
Percent Hb saturation
7.6
7.4
7.2
20
40
60
80
100
P_{O_2} (mm Hg)

(a) Effect of pH

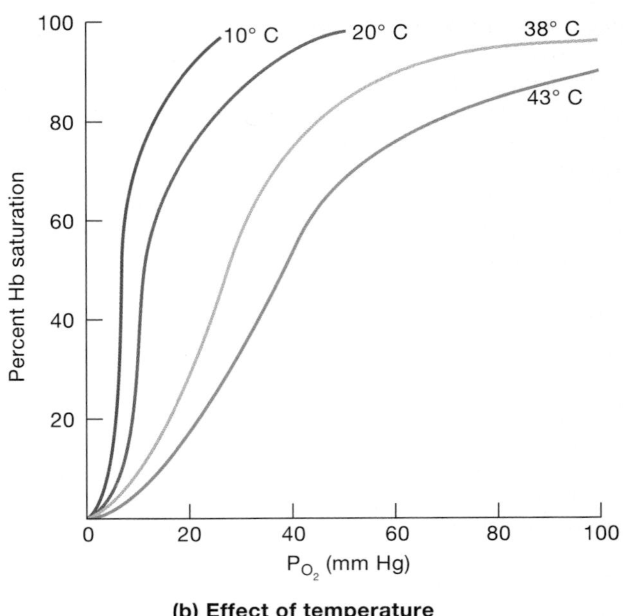
100
80
60
40
20
0
Percent Hb saturation
10° C
20° C
38° C
43° C
20
40
60
80
100
P_{O_2} (mm Hg)

(b) Effect of temperature

FIGURE 23-22
Effects of pH and Temperature on Hemoglobin Saturation. (a) When the pH declines below normal levels, more oxygen is released, and the hemoglobin saturation curve shifts to the right. If the pH increases, the curve shifts to the left as less oxygen is released. **(b)** When the temperature rises, the saturation curve shifts to the right.

Hemoglobin and BPG. Red blood cells can produce ATP only through glycolysis, which, as we saw in Chapter 10, involves the formation of lactic acid. [p. 307] The metabolic pathways involved in glycolysis in an RBC also generate a compound called **2,3-bisphosphoglycerate** (bis-fos-fō-GLIS-e-rāt), or **BPG,** also known as *2,3-diphosphoglycerate,* or *DPG.* BPG is always present in the normal red blood cell, and it has a direct effect on oxygen binding and release. For any partial pressure of oxygen, the higher the concentration of BPG, the more oxygen will be released by the hemoglobin molecules.

The concentration of BPG can be increased by thyroid hormones, growth hormone, epinephrine, androgens, and a high blood pH. These factors improve oxygen delivery to the tissues, because when BPG levels are elevated hemoglobin will release about 10 percent more oxygen at a given P_{O_2} than it would otherwise.

BPG production decreases as red blood cells age. As a result, the level of BPG can determine how long fresh whole blood can be held by a blood bank. When BPG levels get too low, hemoglobin becomes firmly bound to the available oxygen. The blood is then useless for transfusion purposes, because the red blood cells will no longer release oxygen to peripheral tissues, even at a disastrously low P_{O_2}.

FETAL HEMOGLOBIN **Fetal hemoglobin** is found in the red blood cells of a developing fetus. The structure of fetal hemoglobin, which differs from that of adult hemoglobin, gives it a much higher affinity for oxygen. Therefore, at the same P_{O_2}, fetal hemoglobin will bind more oxygen than adult hemoglobin (Figure 23-23●). This characteristic is important in transferring oxygen across the placenta.

A developing fetus obtains oxygen from the maternal bloodstream. The maternal blood at the placenta has a relatively low P_{O_2}, ranging from 35 to 50 mm Hg. If maternal blood arrives at the placenta with a P_{O_2} of 40 mm Hg, hemoglobin saturation will be roughly 75 percent. The fetal blood arriving at the placenta has a P_{O_2} close to 20 mm Hg. However, because fetal hemoglobin has a higher affinity for oxygen, it is still 58 percent saturated.

As diffusion occurs between fetal and maternal blood, oxygen enters the fetal circulation until the P_{O_2} equilibrates at 30 mm Hg. At this P_{O_2} the maternal hemoglobin will be less than 60 percent saturated, but the fetal hemoglobin will be over 80 percent saturated.

When the fetal red blood cells arrive in the peripheral tissues, the steep slope of the saturation curve for fetal hemoglobin means that it will release a large amount of oxygen in response to a very small change in P_{O_2}.

The functional difference between adult and fetal hemoglobin becomes apparent when you compare the uptake and delivery of oxygen in the adult and fetal circulations. Adult hemoglobin leaves the alveoli at a P_{O_2} of 100 mm Hg with 97.5 percent saturation and returns to the lungs with a P_{O_2} of 40 mm Hg and a 75 percent saturation. Thus it releases about 22 percent of its oxygen reserves over a 60 mm Hg drop in

P_{O_2}. Fetal hemoglobin leaves the placenta at a P_{O_2} of 30 mm Hg with an 80 percent saturation and returns to the placenta with a P_{O_2} of 20 mm Hg and 58 percent saturation. It has also released 22 percent of its oxygen reserves, but over a drop of just 10 mm Hg.

ADAPTATIONS TO HIGH ALTITUDE Atmospheric pressure declines with increasing altitude, and so do the partial pressures of the component gases, including oxygen. People living in Denver or Mexico City function normally with alveolar oxygen pressures in the 80–90 mm Hg range. At higher elevations, the alveolar P_{O_2} continues to decline. At 3300 m, an altitude familiar to many hikers and skiers, the alveolar P_{O_2} falls to around 60 mm Hg. Despite the low alveolar P_{O_2}, millions of people live and work at altitudes this high or higher. Important physiological adjustments include an increased respiratory rate, an increased heart rate, and an elevated hematocrit. Thus even though the hemoglobin is not fully saturated, there is more of it in circulation, and the round-trip between the lungs and the peripheral tissues takes less time. These responses represent an excellent example of the functional interplay between the respiratory and cardiovascular systems. However, most of these adaptations take time (days to weeks) to appear. As a result, athletes planning to compete in events held at high altitude (such as the Olympics in Mexico City) must begin training well in advance.

Not everyone is able to tolerate high-altitude conditions, and roughly 20 percent of people ascending to 2600 m or higher experience symptoms of *altitude sickness*, or *mountain sickness*. Symptoms may include headache, disorientation, and fatal pulmonary or cerebral edema. AM *Mountain Sickness*

Carbon Dioxide Transport

Figure 23-24

Carbon dioxide is generated by aerobic metabolism in peripheral tissues. After entering the bloodstream, a carbon dioxide molecule may be (1) converted to a molecule of carbonic acid, (2) bound to the hemoglobin of red blood cells, or (3) dissolved in the plasma. The mechanisms are diagrammed in Figure 23-24●. These are all completely reversible reactions. We will consider the events under way as blood enters peripheral tissues where the P_{CO_2} is 45 mm Hg.

CARBONIC ACID FORMATION Most of the carbon dioxide absorbed by the blood (roughly 70 percent of the total) will be transported as molecules of carbonic acid. Carbon dioxide is converted to carbonic acid through the activity of an enzyme, *carbonic anhydrase*, within red blood cells. The carbonic acid molecules immediately dissociate into a hydrogen ion and a bicarbonate ion. The entire reaction sequence can be summarized as:

$$CO_2 + H_2O \underset{}{\overset{\text{carbonic anhydrase}}{\rightleftharpoons}} H_2CO_3 \rightleftharpoons H^+ + HCO_3^-$$

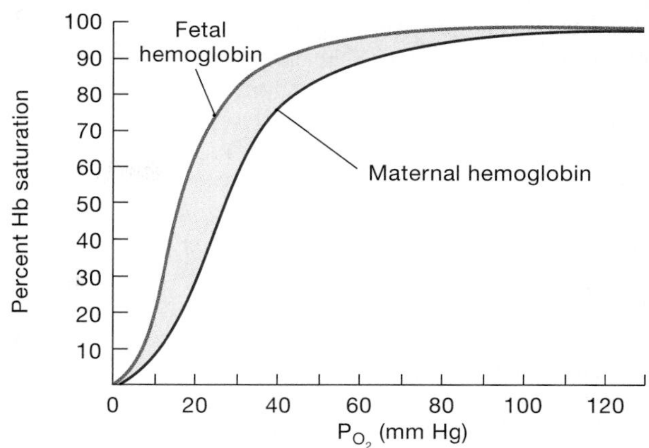
Fetal hemoglobin
Maternal hemoglobin
Percent Hb saturation
100
90
80
70
60
50
40
30
20
10
0
20
40
60
80
100
120
P_{O_2} (mm Hg)

● **FIGURE 23-23**
A Functional Comparison of Fetal and Adult Hemoglobin

The conversion of carbon dioxide to H^+ and HCO_3^- occurs very rapidly, and it is completely reversible. Because the carbonic acid dissociates into bicarbonate and hydrogen ions, we can ignore the intermediary step and summarize the reaction as:

$$CO_2 + H_2O \overset{\text{carbonic anhydrase}}{\rightleftharpoons} H^+ + HCO_3^-$$

In peripheral capillaries this reaction proceeds vigorously, tying up large numbers of carbon diox-

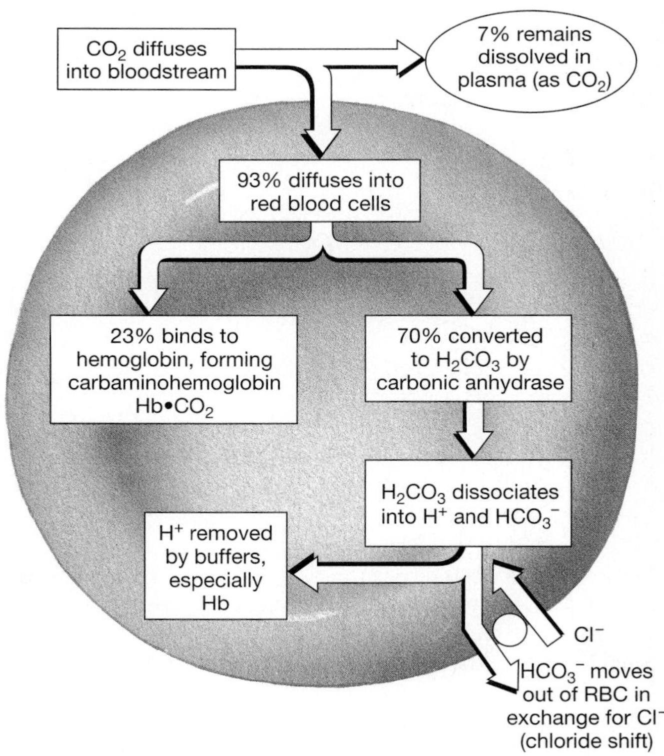
CO_2 diffuses into bloodstream
7% remains dissolved in plasma (as CO_2)
93% diffuses into red blood cells
23% binds to hemoglobin, forming carbaminohemoglobin Hb•CO_2
70% converted to H_2CO_3 by carbonic anhydrase
H_2CO_3 dissociates into H^+ and HCO_3^-
H^+ removed by buffers, especially Hb
Cl^-
HCO_3^- moves out of RBC in exchange for Cl^- (chloride shift)

● **FIGURE 23-24**
Carbon Dioxide Transport in the Blood D

ide molecules. The reaction continues as carbon dioxide diffuses out of the interstitial fluids.

The hydrogen ions and bicarbonate ions have different fates. Most of the hydrogen ions are tied up by hemoglobin molecules, forming H·Hb. The bicarbonate ions move into the surrounding plasma with the aid of a countertransport mechanism that exchanges intracellular bicarbonate ions for extracellular chloride ions. This exchange results in a mass movement of chloride ions into the red blood cells, an event known as the **chloride shift.**

HEMOGLOBIN BINDING Roughly 23 percent of the carbon dioxide carried by the blood will be bound to hemoglobin molecules inside red blood cells. These carbon dioxide molecules are attached to exposed amino groups ($-NH_2$) along the globin portions of hemoglobin molecules. The result is called **carbaminohemoglobin** (kar-ba-mē-nō-hē-mō-GLŌ-bin), $HbCO_2$. This is a reversible reaction that can be summarized as:

$$CO_2 + HbNH_2 \rightleftharpoons HbNHCOOH$$

This reaction is usually abbreviated as:

$$CO_2 + Hb \rightleftharpoons HbCO_2$$

PLASMA TRANSPORT The plasma becomes saturated with carbon dioxide quite rapidly, and only about 7 percent of the carbon dioxide absorbed by peripheral capillaries is transported in the form of dissolved gas molecules. The rest is absorbed by the red blood cells for conversion by carbonic anhydrase or storage as carbaminohemoglobin.

Gas Transport: A Summary

Figure 23-25

Figure 23-25● summarizes the transportation of oxygen and carbon dioxide. This system is obviously dynamic, capable of varying its responses to meet changing circumstances. Some of the responses are automatic and result from the basic chemistry of the transport mechanisms. Other responses require coordinated adjustments in the activities of the cardiovascular and respiratory systems. We will now consider these levels of control and regulation.

 Why does it take more energy to breathe on a hot, humid day than on a cool, dry day?

 During exercise, hemoglobin releases more oxygen to the active skeletal muscles than it does when the muscles are at rest. Why?

How would blockage of the trachea affect the blood pH?

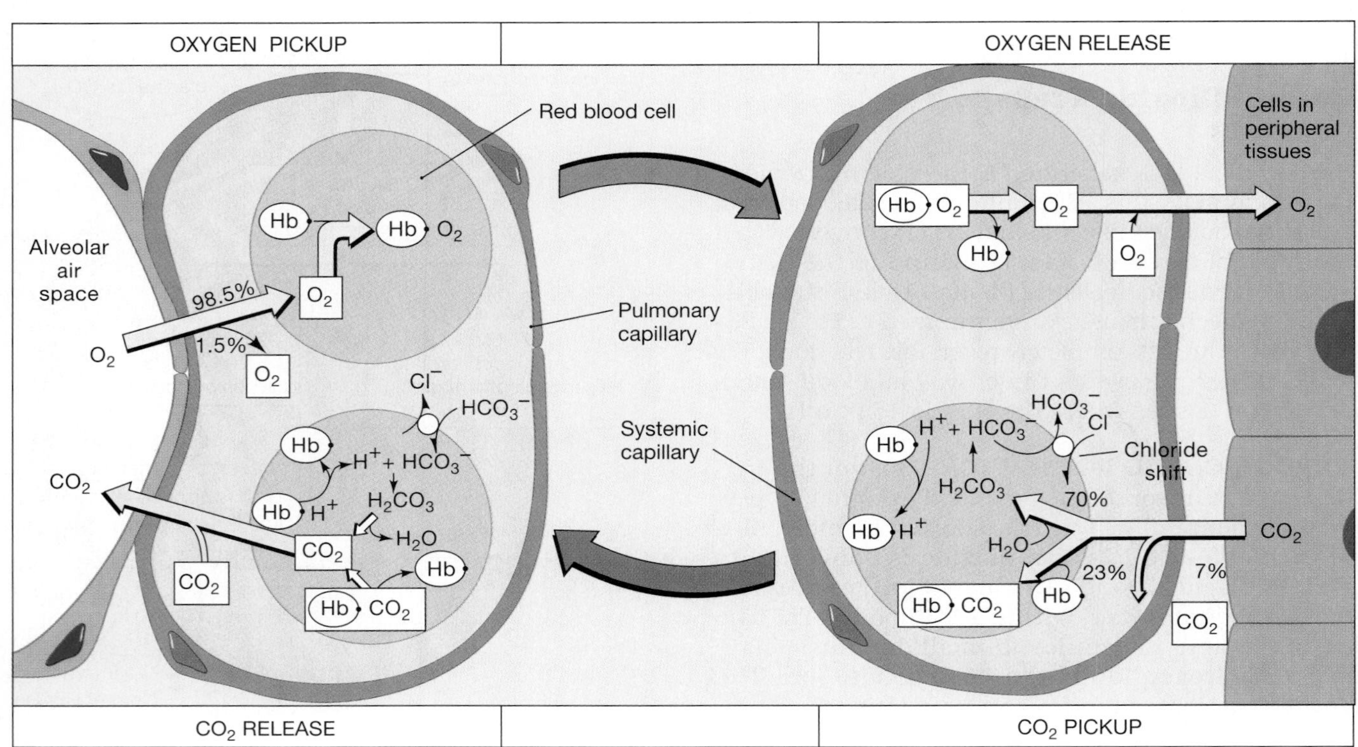
OXYGEN PICKUP
OXYGEN RELEASE
Red blood cell
Cells in peripheral tissues
Alveolar air space
Hb
Hb O2
98.5%
O2
1.5%
O2
O2
Pulmonary capillary
Systemic capillary
Cl⁻
HCO3⁻
H⁺ + HCO3⁻
H2CO3
H2O
Hb H⁺
CO2
Hb CO2
Hb O2
O2
Hb
Chloride shift
70%
23%
7%
CO2 RELEASE
CO2 PICKUP

● **FIGURE 23-25**
A Summary of the Primary Gas Transport Mechanisms

Control of Respiration

Peripheral cells are continually absorbing oxygen from the interstitial fluids and generating carbon dioxide. Under normal conditions the cellular rates of absorption and generation are matched by the capillary rates of delivery and removal. These rates are identical to those of oxygen absorption and carbon dioxide excretion at the lungs. If these rates become unbalanced, homeostatic mechanisms intervene to restore equilibrium. Those mechanisms involve (1) changes in blood flow and oxygen delivery that are regulated at the local level and (2) variations in the depth and rate of respiration under the control of the respiratory centers of the brain. The activities of the respiratory centers are coordinated with changes in cardiovascular function, such as variations in blood pressure and cardiac output.

Local Regulation of Gas Transport and Alveolar Function

The rate of oxygen delivery in each tissue and the efficiency of oxygen pickup at the lungs are largely regulated at the local level. For example, if a peripheral tissue becomes more active, the interstitial P_{O_2} falls and the P_{CO_2} rises. This change increases the difference between the partial pressures in the tissues and in the arriving blood, so more oxygen is delivered and more carbon dioxide is carried away. In addition, the rising P_{CO_2} has an inhibitory effect on arterioles and capillary smooth muscles, and the blood flow to the active region increases. [p. 738]

Local factors regulate blood flow, or *perfusion*, and airflow, or *ventilation*, over a wide range of conditions and activity levels.

- ***Changes in lung perfusion.*** As blood flows toward the alveolar capillaries, it is directed toward lobules where the P_{O_2} is relatively high. This movement occurs because alveolar capillaries constrict when the local P_{O_2} is low. (This response, the opposite of that seen in peripheral tissues, was noted in Chapter 21.) [p. 742] Such a shift in circulation tends to eliminate temporary problems due to oxygen depletion and carbon dioxide buildup in a single alveolus, an entire lobule, or a group of lobules.
- ***Changes in alveolar ventilation.*** Smooth muscles in the walls of bronchioles are sensitive to the P_{CO_2} of the air they contain. When the P_{CO_2} goes up, the bronchioles increase in diameter (bronchodilation). When the P_{CO_2} declines, the bronchioles constrict (bronchoconstriction). Airflow is therefore directed to lobules where the P_{CO_2} is high. Because the carbon dioxide is obtained from the blood, these lobules are actively engaged in gas exchange.

By directing blood flow to alveoli with low levels of carbon dioxide and improving airflow to alveoli with high levels of carbon dioxide, local adjustments improve the efficiency of gas transport. When activity levels increase and the demand for oxygen rises, the cardiac output and respiratory rates increase under neural control, but the adjustments in alveolar blood flow and bronchiolar diameter occur automatically.

EMPHYSEMA **Emphysema** (em-fi-SĒ-ma) is a chronic, progressive condition characterized by shortness of breath and an inability to tolerate physical exertion. The underlying problem is destruction of alveolar surfaces and inadequate surface area for oxygen and carbon dioxide exchange. In essence, respiratory bronchioles and alveoli are functionally eliminated. The alveoli gradually expand, their walls become incomplete, and adjacent alveoli merge to form larger, nonfunctional air spaces.

Emphysema has been linked to the inhalation of air containing fine particulate matter or toxic vapors, such as those found in cigarette smoke. There are also genetic factors that predispose individuals to this condition. Some degree of emphysema is a normal consequence of aging, and an estimated 66 percent of adult males and 25 percent of adult females have detectable areas of emphysema in their lungs. AM *Bronchitis, Emphysema, and COPD*

The Respiratory Centers of the Brain

Respiratory control has both involuntary and voluntary components. The involuntary centers regulate the activities of the respiratory muscles and control the respiratory minute volume by adjusting the frequency and depth of pulmonary ventilation. They do so in response to sensory information arriving from the lungs and other portions of the respiratory tract, as well as from a variety of other sites.

The voluntary control of respiration reflects activity in the cerebral cortex that affects either (1) the output of the involuntary respiratory centers in the pons and medulla oblongata or (2) motor neurons in the spinal cord that control respiratory muscles.

Respiratory Centers in the Medulla Oblongata

Figure 23-26

The *medullary rhythmicity center* of the medulla oblongata was introduced in Chapter 14. [p. 483] This center sets the basic pace for respiration.

It can be subdivided into a **dorsal respiratory group (DRG)** and a **ventral respiratory group (VRG).**

- The dorsal respiratory group, or *inspiratory center*, contains neurons that control lower motor neurons innervating the external intercostal muscles and the diaphragm. This group functions in every respiratory cycle, whether quiet or forced.
- The ventral respiratory group functions only during forced respiration. It includes neurons that innervate lower motor neurons controlling accessory muscles involved in active exhalation and maximal inhalation. The neurons involved with active exhalation are sometimes said to form an *expiratory center*.

There is reciprocal inhibition between the neurons involved with inhalation and exhalation. When the inspiratory neurons are active, the expiratory neurons are inhibited, and vice versa. The pattern of interaction between these groups varies depending on whether the respiratory activity is quiet or forced. During quiet respiration (Figure 23-26a●):

1. Activity in the dorsal respiratory group increases over a period of about 2 seconds, providing stimulation to the inspiratory muscles. Over this period inhalation occurs.
2. After 2 seconds, the DRG neurons become inactive, and they remain quiet for the next 3 seconds. Over this period passive exhalation occurs.

During forced respiration (Figure 23-26b●):

1. As the level of activity in the DRG increases, it in some way stimulates neurons of the VRG that activate the accessory muscles involved in inhalation.
2. At the end of each inhalation, active exhalation occurs as the neurons of the expiratory center stimulate the opposing set of accessory muscles.

The basic pattern of respiration thus reflects a cyclic interaction between the DRG and VRG. The pace of this interaction is thought to be established by pacemaker cells that spontaneously undergo rhythmic patterns of activity. Attempts to locate the pacemaker, however, have been unsuccessful.

Central nervous system stimulants, such as amphetamines or even caffeine, will increase the respiratory rate by facilitating the pacemaker neurons and respiratory centers. These actions can be opposed by CNS depressants, such as barbiturates and opiates.

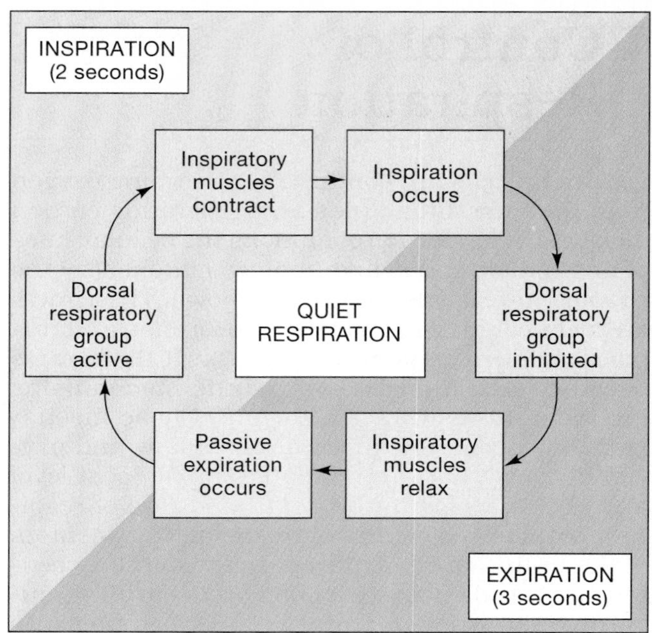
INSPIRATION (2 seconds)
Inspiratory muscles contract
Inspiration occurs
Dorsal respiratory group active
QUIET RESPIRATION
Dorsal respiratory group inhibited
Passive expiration occurs
Inspiratory muscles relax
EXPIRATION (3 seconds)

(a)

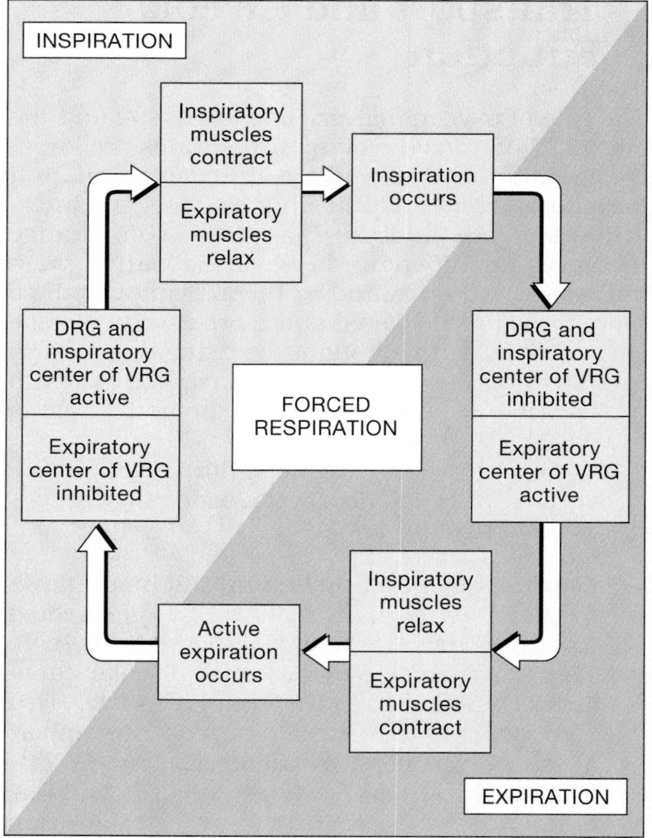
INSPIRATION
Inspiratory muscles contract
Expiratory muscles relax
Inspiration occurs
DRG and inspiratory center of VRG active
Expiratory center of VRG inhibited
FORCED RESPIRATION
DRG and inspiratory center of VRG inhibited
Expiratory center of VRG active
Inspiratory muscles relax
Expiratory muscles contract
Active expiration occurs
EXPIRATION

(b)

● **FIGURE 23-26**
Basic Regulatory Patterns. (a) Quiet respiration. **(b)** Forced respiration.

SIDS Sudden infant death syndrome (SIDS), also known as "crib death," kills an estimated 10,000 infants each year in the United States alone. Most crib deaths involve infants 2–4 months old, usually between midnight and 9:00 A.M., in the late fall or winter months. Eyewitness accounts indicate that the sleeping infant suddenly stops breathing, turns blue, and relaxes. There appear to be genetic factors involved, but controversy remains as to the relative importance of other factors, such as laryngeal spasms, cardiac arrhythmias, upper respiratory tract infections, viral infections, and CNS malfunctions. The age at the time of death corresponds with a period when the pacemaker complex and respiratory centers are establishing connections with other portions of the brain. It has recently been proposed that SIDS results from a problem in the interconnection process that disrupts the reflexive respiratory pattern.

The Apneustic and Pneumotaxic Centers

Figure 23-27

The **apneustic** (ap-NŪ-stik) **center** of the lower pons and the **pneumotaxic** (nū-mō-TAKS-ik) **center** of the upper pons adjust the output of the rhythmicity center. Their activities adjust the respiratory rate and the depth of respiration in response to sensory stimuli or input from other centers in the brain.

- The *apneustic center* provides continual stimulation to the dorsal respiratory group. During quiet respiration, stimulation from the apneustic center helps to increase the intensity of inspiration over the next 2 seconds. Under normal conditions, after 2 seconds the apneustic center is inhibited by signals from the pneumotaxic center. During forced respirations, the apneustic center also responds to sensory input from the vagus nerve concerning the amount of lung inflation. (This *inflation reflex* is detailed below.)
- The *pneumotaxic center* inhibits the apneustic center and promotes passive or active exhalation. Centers in the hypothalamus and cerebrum can alter the activity of the pneumotaxic center and change the respiratory rate and depth. However, essentially normal respiratory cycles continue even if the brain stem above the pons has been severely damaged.

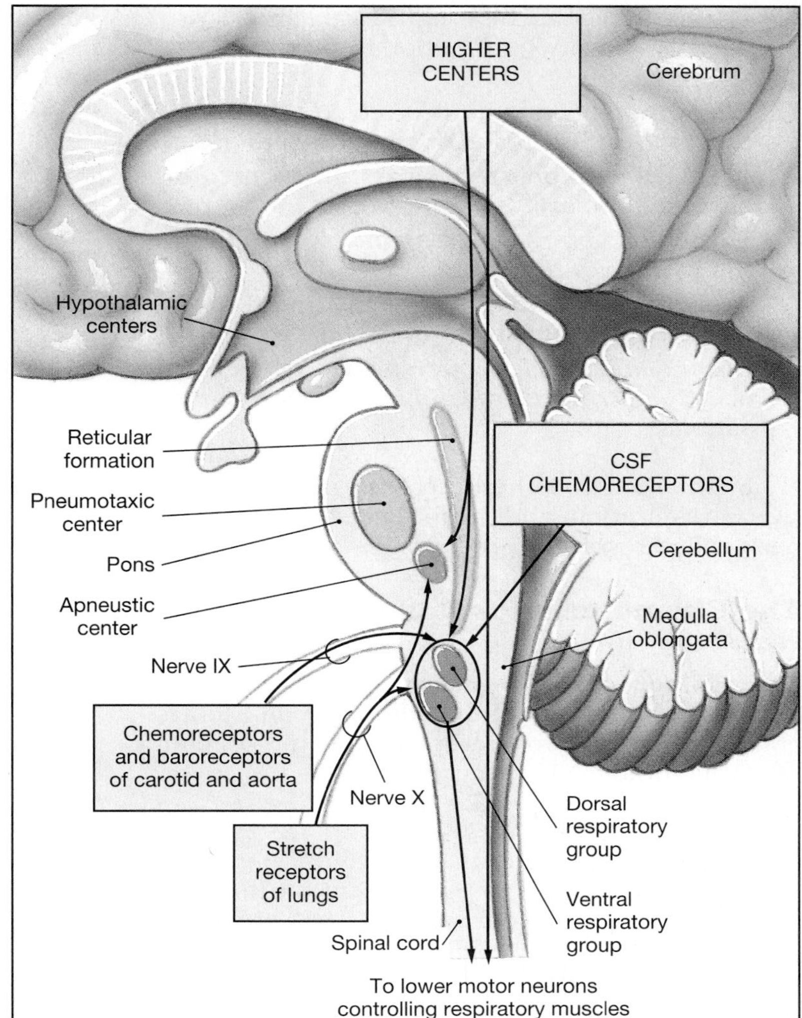
HIGHER CENTERS
Cerebrum
Hypothalamic centers
Reticular formation
Pneumotaxic center
Pons
Apneustic center
Nerve IX
CSF CHEMORECEPTORS
Cerebellum
Medulla oblongata
Chemoreceptors and baroreceptors of carotid and aorta
Nerve X
Stretch receptors of lungs
Dorsal respiratory group
Ventral respiratory group
Spinal cord
To lower motor neurons controlling respiratory muscles

FIGURE 23-27
Respiratory Centers and Reflex Controls. The positions of the major respiratory centers and other factors important to respiratory control.

If the inhibitory innervation from the pneumotaxic center is cut off by a stroke or other damage to the brain stem, and sensory innervation from the lungs is eliminated by cutting the vagus nerve, the individual inhales to maximum capacity and maintains that state for 10–20 seconds at a time. Intervening exhalations are brief, and little pulmonary ventilation occurs.

The CNS regions involved with respiratory control are diagrammed in Figure 23-27●. Interactions between the dorsal and ventral respiratory groups establish the basic pace and depth of respiration. The pneumotaxic center modifies that pace; an increase in pneumotaxic output quickens the pace of respiration by shortening the duration of each inhalation. A decrease in pneumotaxic output slows the respiratory pace but increases the depth of respiration, because the apneustic center is more active.

The activities of the respiratory centers are modified by sensory information from:

1. ***Chemoreceptors sensitive to the P_{CO_2}, pH, and/or P_{O_2} of the blood or cerebrospinal fluid.***
2. ***Changes in blood pressure in the aorta or carotid sinuses.***
3. ***Stretch receptors that respond to changes in the volume of the lungs.***
4. ***Irritating physical or chemical stimuli in the nasal cavity, larynx, or bronchial tree.***
5. ***Other sensations,*** including pain, changes in body temperature, and abnormal visceral sensations.

Information from these receptors alters the pattern of respiration, and the induced changes have been called *respiratory reflexes.*

The Chemoreceptor Reflexes

Figure 23-28

The respiratory centers are strongly influenced by chemoreceptor inputs from the ninth and tenth cranial nerves and from receptors monitoring cerebrospinal fluid (CSF) composition.

- The glossopharyngeal nerve (N IX) carries chemoreceptive information from the carotid bodies, adjacent to the carotid sinus. The carotid bodies are stimulated by a decrease in the pH or P_{O_2} of the blood. Because changes in P_{CO_2} affect pH, these receptors are also stimulated by a rise in the P_{CO_2}.
- The vagus nerve (N X) monitors chemoreceptors in the aortic bodies, near the aortic arch. These receptors are sensitive to the same stimuli as the carotid bodies. The receptors of the carotid and aortic bodies are often called *peripheral chemoreceptors.*
- Chemoreceptors are located on the ventrolateral surface of the medulla oblongata in a region known as the *chemosensitive area.* The neurons in that area respond only to the P_{CO_2} and pH of the cerebrospinal fluid. These neurons are often called *central chemoreceptors.*

Chemoreceptors and their effects on cardiovascular function were detailed in Chapters 17 and 21. ⚭ *[pp. 558,740]* Stimulation of these chemoreceptors leads to an increase in the depth and rate of respiration. Under normal conditions a reduction in the arterial P_{O_2} has relatively little effect on the respiratory centers until the arterial P_{O_2} drops by around 40 percent, to levels below 60 mm Hg. If the P_{O_2} of arterial blood drops to 40 mm Hg, the level in peripheral tissues, the respiratory rate will increase only by 50–70 percent. In contrast, a rise of just 10 percent in the arterial P_{CO_2} will cause a doubling of the respiratory rate, even if the P_{O_2} remains completely normal. Carbon dioxide levels are therefore responsible for regulating respiratory activity under normal conditions.

Although the receptors monitoring carbon dioxide levels are more sensitive, both oxygen and carbon dioxide receptors work together in a crisis. Carbon dioxide is generated during oxygen consumption, so when oxygen concentrations are falling rapidly carbon dioxide levels are usually increasing. This cooperation breaks down only under unusual circumstances. For example, many people know that if you prepare by taking deep, full breaths, you can hold your breath longer. Although most believe that this is due to "extra oxygen," it is really due to the loss of carbon dioxide. If the P_{CO_2} is driven down far enough, breath-holding ability may increase to the point that the individual becomes unconscious from oxygen starvation in the brain without ever feeling the urge to breathe. AM *Shallow Water Blackout*

The chemoreceptors are subject to adaptation if the P_{O_2} or P_{CO_2} remains abnormal for an extended period, and this adaption can complicate the treatment of chronic respiratory disorders. For example, if the P_{O_2} remains low for an extended period and the P_{CO_2} remains chronically elevated, the chemoreceptors will reset to those values. They will accept those values as normal and oppose any attempts to bring the partial pressures into the normal range. AM *Chemoreceptor Accommodation and Opposition*

Because the chemoreceptors monitoring carbon dioxide levels are also sensitive to pH, any condition altering the pH of the blood or CSF will affect respiratory performance. For example, the rise in lactic acid levels after exercise causes a drop in pH that helps stimulate respiratory activity.

HYPERCAPNIA AND HYPOCAPNIA Hypercapnia (hī-per-KAP-nē-a) is an increase in the P_{CO_2} of arterial blood. Figure 23-28● diagrams the central response to hypercapnia, which is triggered by stimulation of chemoreceptors in the carotid and aortic bodies. Carbon dioxide crosses the blood-brain barrier quite rapidly, so a rise in arterial P_{CO_2} almost immediately elevates CSF carbon dioxide levels, lowering the pH of the CSF and stimulating the chemoreceptive neurons of the medulla oblongata.

These receptors stimulate the respiratory centers to increase the rate and depth of respiration. Breathing becomes more rapid, and a greater amount of air moves into and out of the lungs with each breath. Because more air moves into and out of the alveoli each minute, alveolar concentrations of carbon dioxide decline, accelerating the diffusion of carbon dioxide from the alveolar capillaries. Thus homeostasis is restored.

If the rate and depth of respiration exceed the demands for oxygen delivery and carbon dioxide removal, the condition of **hyperventilation** exists.

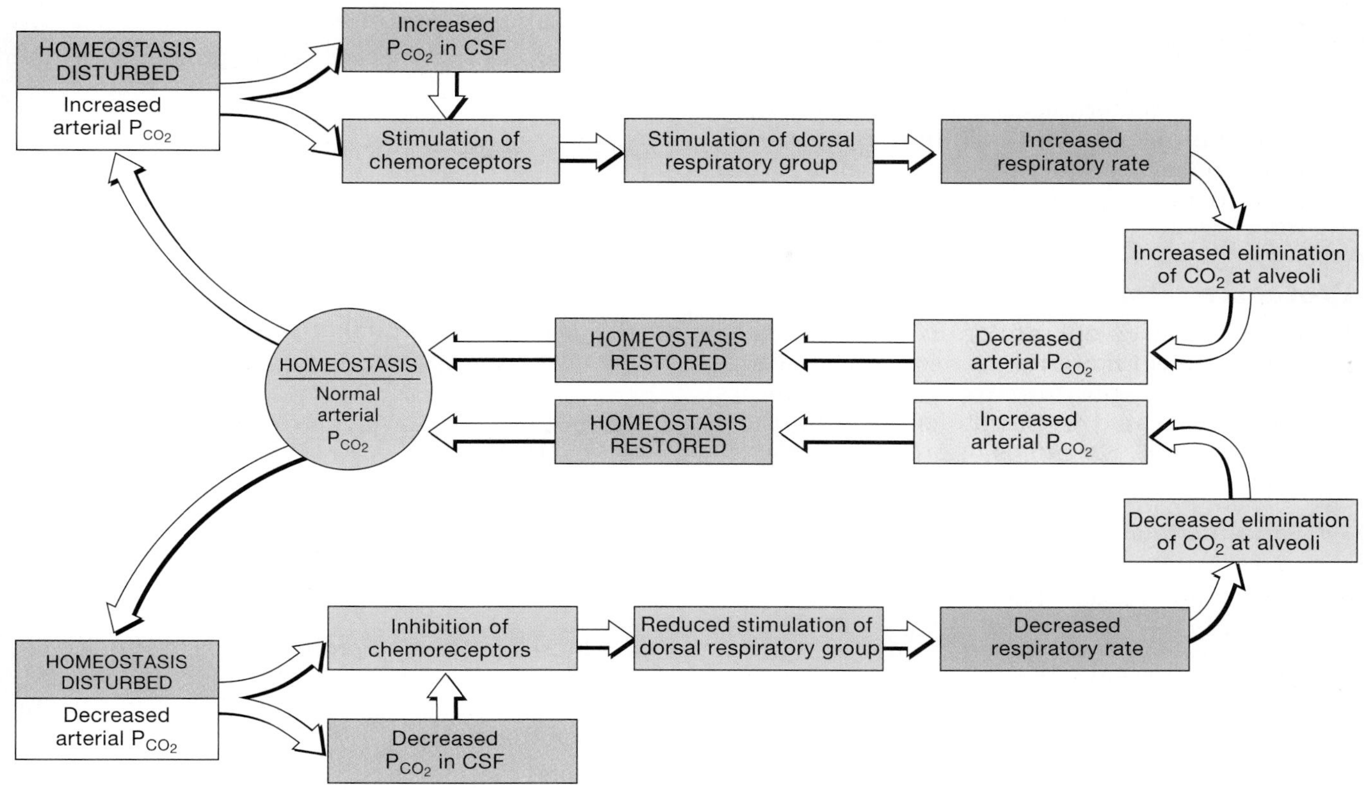
HOMEOSTASIS DISTURBED
Increased arterial P_{CO_2}
Increased P_{CO_2} in CSF
Stimulation of chemoreceptors
Stimulation of dorsal respiratory group
Increased respiratory rate
Increased elimination of CO_2 at alveoli
Decreased arterial P_{CO_2}
HOMEOSTASIS RESTORED
HOMEOSTASIS
Normal arterial P_{CO_2}
HOMEOSTASIS RESTORED
Increased arterial P_{CO_2}
Decreased elimination of CO_2 at alveoli
HOMEOSTASIS DISTURBED
Decreased arterial P_{CO_2}
Inhibition of chemoreceptors
Reduced stimulation of dorsal respiratory group
Decreased respiratory rate
Decreased P_{CO_2} in CSF

● **FIGURE 23-28**
Chemoreceptor Response to Changes in P_{CO_2}. A rise in arterial P_{CO_2} (top) stimulates chemoreceptors that accelerate breathing cycles at the inspiratory center. This change increases the respiratory rate, encourages CO_2 loss at the lungs, and lowers arterial P_{CO_2}. A decline in arterial P_{CO_2} (bottom) inhibits these chemoreceptors. In the absence of stimulation the rate of respiration decreases, slowing the rate of CO_2 loss and elevating arterial P_{CO_2}.

Hyperventilation will gradually lead to **hypocapnia,** an abnormally low P_{CO_2}. If the arterial P_{CO_2} declines below normal levels (Figure 23-28●), chemoreceptor activity declines and the respiratory rate falls. This situation continues until the P_{CO_2} returns to normal.

In **hypoventilation,** the respiratory rate remains abnormally low and is insufficient to meet the demands for normal oxygen delivery and carbon dioxide removal. Hypoventilation will ultimately produce hypercapnia, because of the retention of carbon dioxide.

The Baroreceptor Reflexes

The effects of carotid and aortic baroreceptor stimulation on systemic blood pressure were described in Chapter 21. ∞ [pp. 739–740] The output from these baroreceptors also affects the respiratory centers. When blood pressure falls, the respiratory rate increases; when blood pressure rises the respiratory rate declines. This adjustment results from stimulation or inhibition of the respiratory centers by sensory fibers in the glossopharyngeal (IX) and vagus (X) nerves.

The Hering-Breuer Reflexes

The Hering-Breuer reflexes are named after the physiologists who described them in 1865. There are two reflexes involved: the *inflation reflex* and the *deflation reflex.* The sensory information is distributed to the apneustic center and the ventral respiratory group. Neither reflex is involved in normal quiet breathing or in tidal volumes under 1000 ml.

- The **inflation reflex** prevents overexpansion of the lungs during forced breathing. The stretch receptors are located in the smooth muscle tissue around bronchioles and they are stimulated by lung expansion. Sensory fibers leaving the stretch receptors reach the respiratory centers over the vagus nerve. As lung volume increases, the dorsal respiratory group is gradually inhibited, and the expiratory center of the ventral respiratory group is stimulated. Inspiration then stops as the lungs near maximum volume, and active expiration then begins.
- The **deflation reflex** inhibits the expiratory center and stimulates the inspiratory center when the lungs are collapsing. The receptors, which

are distinct from those of inflation reflex, are located in the alveolar wall near the alveolar capillary network. The smaller the volume of the lungs, the greater the degree of inhibition, until expiration stops and inspiration begins. This reflex normally functions only during forced expiration, when both the inspiratory and expiratory centers are active.

Protective Reflexes

Protective reflexes operate on exposure to toxic vapors, chemical irritants, or mechanical stimulation of the respiratory tract. The receptors involved are located within the epithelium of the respiratory tract. Examples of protective reflexes include:

- ***Sneezing and coughing.*** Sneezing is triggered by irritation of the wall of the nasal cavity. Coughing is triggered by irritation of the larynx, trachea, or bronchi. Both of these reflexes involve a temporary period of **apnea** (AP-nē-a) when respiration is suspended. They are usually followed by a forceful expulsion of air intended to remove the offending stimulus. The glottis is forcibly closed while the lungs are still relatively full. The abdominal and internal intercostal muscles then contract suddenly, creating pressures that will blast air out of the respiratory passageways when the glottis reopens. Air leaving the larynx may travel at 160 kph, carrying mucus, foreign particles, and irritating gases out of the respiratory tract via the nose or mouth.
- **Laryngeal spasms** result from the entry of chemical irritants, foreign objects, or fluids into the area around the glottis. This reflex usually closes the airway temporarily, but a very strong stimulus, such as a toxic gas, may close the glottis so powerfully that the individual may lose consciousness and die without taking another breath. Fine chicken or fish bones that pierce the laryngeal walls may also stimulate laryngeal spasms, swelling, or both, restricting the airway.

Other Sensations That Affect Respiratory Function

Several other sensory stimuli can affect the activities of the respiratory centers. In some cases, the mechanism involved is not known. Examples include:

- Sudden pain or immersion in cold water can produce a temporary apnea. Chronic pain stimulates the sympathetic division of the ANS, leading to an increase in the respiratory rate.
- Vomiting and swallowing involve automatic adjustments in respiratory activity to prevent entrance of foreign objects into the trachea.
- Fever or an increase in body temperature due to exertion or overheating will cause an increase in the respiratory rate. A reduction in body temperature leads to a decrease in the respiratory rate.
- Curiously, stretching the anal sphincter stimulates the respiratory center and increases the rate of respiration. Although this reflex is occasionally used to stimulate respiration in an emergency situation, it is not clear what pathways are involved.

❑ Voluntary Control of Respiration

Activity of the cerebral cortex has an indirect effect on the respiratory centers. For example:

1. Conscious thought processes tied to strong emotions, such as rage or fear, affect the respiratory rate through stimulating centers in the hypothalamus.
2. Emotional states can also affect respiration via activation of the sympathetic or parasympathetic divisions of the ANS. Sympathetic activation causes bronchodilation and increases the respiratory rate, and parasympathetic stimulation has the opposite effects.
3. Anticipation of strenuous exercise can also trigger an automatic increase in the respiratory rate, along with increased cardiac output, via sympathetic stimulation.

We also have voluntary control over our respiratory activities. This control may bypass the respiratory centers completely, using pyramidal fibers that innervate the same lower motor neurons controlled by the dorsal and ventral respiratory groups. This control mechanism is an essential part of speaking, singing, and swimming, when respiratory activities must be precisely timed. Higher centers may also have an inhibitory effect on the apneustic center and the dorsal and ventral respiratory groups, and this is important in breath-holding.

There are limits to our abilities to override the respiratory centers, however. The chemoreceptor reflexes are extremely powerful respiratory stimulators, and they cannot be consciously suppressed. For example, you can hold your breath before diving into a swimming pool and thereby prevent the inhalation of water. But you cannot actually commit suicide by holding your breath "till you turn blue." Once the P_{CO_2} rises to critical levels, you will be forced to take a breath.

Aging and the Respiratory System

Figure 23-29

Many factors interact to reduce the efficiency of the respiratory system in elderly individuals. Three examples are particularly noteworthy:

1. With increasing age elastic tissue deteriorates throughout the body. This deterioration reduces the compliance of the lungs, lowering the vital capacity.
2. Movements of the chest cage are restricted by arthritic changes in the rib articulations and by decreased flexibility at the costal cartilages. In combination with the changes noted in (1), the stiffening and reduction in chest movement effectively limit the respiratory minute volume. This restriction contributes to the reduction in exercise performance and capabilities with increasing age.
3. Some degree of emphysema is normally found in individuals age 50 to 70. However, the extent varies widely depending on the lifetime exposure to cigarette smoke and other respiratory irritants. Figure 23-29● compares the respiratory performance of individuals who have never smoked with individuals who have smoked for varying periods of time. The message here is quite clear: Although some decrease in respiratory performance is inevitable, stopping smoking (or never starting) can prevent serious respiratory deterioration.

What effect would exciting the pneumotaxic center have on respiration?

Are peripheral chemoreceptors as sensitive to the levels of carbon dioxide as they are to the levels of oxygen?

Johnny is mad at his mother, so he tells her that he will just hold his breath until he turns blue and dies. Should Johnny's mother worry?

Integration with Other Systems

Figure 23-30

The respiratory system has extensive anatomical connections to the cardiovascular system, and examples have been noted throughout this chapter. The functional ties are just as extensive.

The goal of respiratory activity is the maintenance of homeostatic oxygen and carbon dioxide levels in peripheral tissues. Changes in respiratory activity alone are seldom sufficient to accomplish this. For example, during exercise, when tissue oxygen demands are high, alveolar ventilation increases through stimulation of the respiratory muscles. This response, which actually increases the demand for oxygen, serves no purpose unless cardiac output accelerates simultaneously, so that oxygen delivery to active tissues can be improved.

We will consider four examples of the integration between the respiratory and cardiovascular systems:

- At the local level, changes in lung perfusion in response to variations in alveolar P_{O_2} improve the efficiency of gas exchange within or among lobules.
- Chemoreceptor stimulation not only increases respiratory drive, it causes an elevation in blood pressure, through stimulation of the vasomotor center, and increased cardiac output. These responses were noted in Chapter 21. [pp. 738, 741]
- Stimulation of baroreceptors in the lungs has secondary effects on cardiovascular function. For example, airway stretch receptor stimulation not only triggers the inflation reflex, but also increases the heart rate. Thus as the lungs fill, cardiac output rises and more blood flows through the alveolar capillaries.
- Stimulation of baroreceptors in systemic blood vessels has secondary effects on respiratory function. When systemic blood pressure falls, stimulating carotid and aortic baroreceptors, the respiratory rate increases; when systemic blood pressure rises, the respiratory rate declines.

Of course, the respiratory system is functionally linked to all other systems, and Figure 23-30● details those interrelationships.

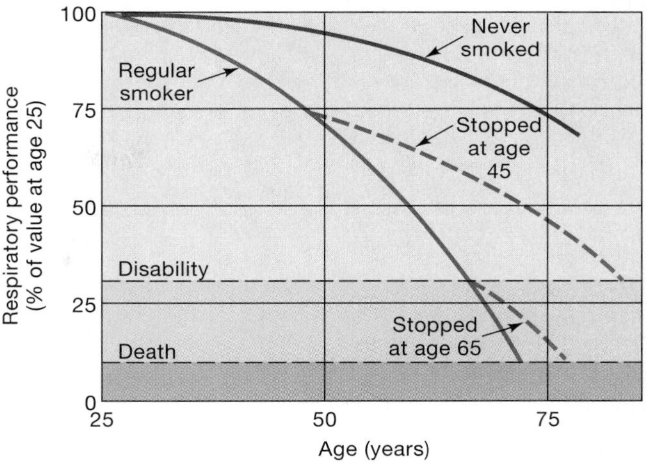
Respiratory performance (% of value at age 25)
100
75
50
25
0
Never smoked
Regular smoker
Stopped at age 45
Disability
Stopped at age 65
Death
25
50
75
Age (years)

● **FIGURE 23-29**

Aging and the Decline in Respiratory Performance. Graph comparing the relative respiratory performance of (1) individuals who have never smoked, (2) individuals who quit smoking at age 45, (3) individuals who quit smoking at age 65, and (4) lifelong smokers.

INTEGUMENTARY SYSTEM

Provides protection for portions of upper respiratory tract; hairs guard entry to external nares

SKELETAL SYSTEM

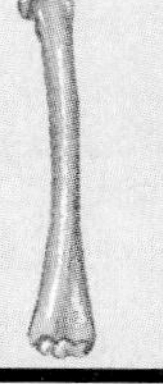

Movements of ribs important in breathing; axial skeleton surrounds and protects lungs

MUSCULAR SYSTEM

Muscle activity generates CO_2; respiratory muscles fill and empty lungs; other muscles control entrances to respiratory tract; intrinsic laryngeal muscles control airflow through larynx and produce sounds

RESPIRATORY SYSTEM

NERVOUS SYSTEM

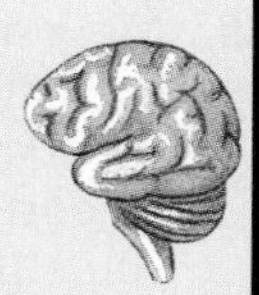

Monitors respiratory volume and blood gas levels; controls pace and depth of respiration

ENDOCRINE SYSTEM

Epinephrine and norepinephrine stimulate respiratory activity and dilate respiratory passageways

Converting enzyme along capillaries of lung converts angiotensin I to angiotensin II

CARDIOVASCULAR SYSTEM

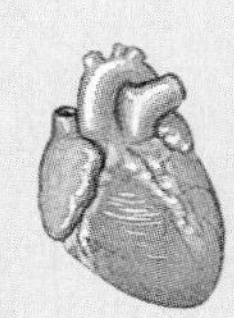

Red blood cells transport oxygen and carbon dioxide between the lungs and peripheral tissues

Activation of angiotensin II by converting enzyme important in regulation of blood pressure and volume; bicarbonate ions contribute to buffering capability of blood

FOR ALL SYSTEMS

Provides oxygen and eliminates carbon dioxide

LYMPHATIC SYSTEM

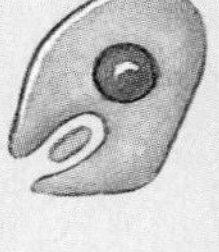

Tonsils protect against infection at entrance to respiratory tract; lymphatics monitor lymph drainage from lungs and mobilize specific defenses when infection occurs

Alveolar phagocytes present antigens to trigger specific defenses; respiratory defense system traps pathogens, protects deeper tissues

DIGESTIVE SYSTEM

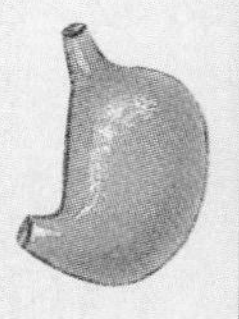

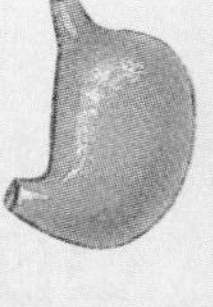

Provides substrates, vitamins, water, and ions that are necessary to all cells of the respiratory system

Increased thoracic and abdominal pressure through contraction of respiratory muscles can assist in defecation

REPRODUCTIVE SYSTEM

Changes in respiratory rate and depth occur during sexual arousal

URINARY SYSTEM

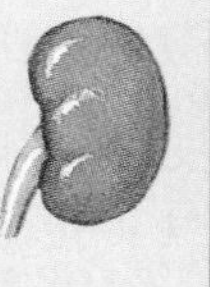

Eliminates organic wastes generated by cells of the respiratory system; maintains normal fluid and ion balance in the blood

Assists in the regulation of pH by eliminating carbon dioxide

FIGURE 23-30
Functional Relationships between the Respiratory System and Other Systems

LUNG CANCER **Lung cancer,** or *pleuropulmonary neoplasm,* is an aggressive class of malignancies originating in the bronchial passageways or alveoli. These cancers affect the epithelial cells lining conducting passageways, mucous glands, or alveoli. Symptoms usually do not appear until the condition has progressed to the point that the tumor masses are restricting airflow or compressing adjacent mediastinal structures. Chest pain, shortness of breath, a cough or wheeze, and weight loss are common symptoms. Treatment programs vary depending on the cellular organization of the tumor and whether or not metastasis (cancer cell migration) has occurred, but surgery, radiation exposure, or chemotherapy may be involved.

Deaths from lung cancer were rare at the turn of the century, but there were 29,000 in 1956, 105,000 in 1978, and 153,100 in 1994 in the United States. These figures continue to rise, with the number of diagnosed cases doubling every 15 years. Each year 22 percent of new cancers detected are lung cancers, and in 1994 100,000 men and 72,000 women were diagnosed with this condition. The rate is increasing markedly among women; the lung cancer rates in 1989 were 101,000 men and 54,000 women. For additional information concerning the diagnosis and treatment of lung cancer, see the *Applications Manual.*

AM *Lung Cancer*

Selected Clinical Terminology

Terms Discussed in This Chapter

anoxia (a-NOKS-ē-a): A condition of tissue oxygen starvation caused by (1) circulatory blockage, (2) respiratory problems, or (3) cardiovascular problems. *(p. 844)*

asthma (AZ-ma): An acute respiratory disorder characterized by unusually sensitive, irritable conducting passageways. *(p. 854* and *AM)*

atelectasis (at-e-LEK-ta-sis): A collapsed lung. *(p. 850)*

bronchitis (brong-KĪ-tis): An inflammation of the bronchial lining. *(p. 838* and *AM)*

bronchodilation (brong-kō-dī-LĀ-shun): Enlargement of the respiratory passageways. *(p. 840)*

bronchography (brong-KOG-ra-fē): A procedure in which radiopaque materials are introduced into the airways to improve X-ray imaging of the bronchial tree. *(p. 838)*

bronchoscope: A fiber-optic bundle small enough to be inserted into the trachea and finer airways; the procedure is called *bronchoscopy. (p. 838)*

cardiopulmonary resuscitation (CPR): The application of cycles of compression to the rib cage and mouth-to-mouth breathing to maintain circulatory and respiratory function. *(p. 851* and *AM)*

cystic fibrosis (CF): A relatively common lethal inherited disease caused by an abnormal membrane channel protein; a major symptom is that mucous secretions become too thick to be transported easily, leading to respiratory problems. *(p. 829* and *AM)*

decompression sickness: A condition caused by a drop in atmospheric pressure and the resulting formation of nitrogen gas bubbles in body fluids, tissues, and organs. *(p. 856* and *AM)*

emphysema (em-fi-SĒ-ma): A chronic, progressive condition characterized by shortness of breath and an inability to tolerate physical exertion. *(p. 863* and *AM)*

epistaxis (ep-i-STAK-sis): A nosebleed. *(p. 831)*

Heimlich (HĪM-lik) **maneuver,** or *abdominal thrust:* Compression applied to the abdomen just beneath the diaphragm, to force air out of the lungs and clear a blocked trachea or larynx. *(p. 834)*

hypercapnia (hī-per-KAP-nē-a): An increase in the P_{CO_2} of arterial blood. *(p. 866)*

hypocapnia: An abnormally low arterial P_{CO_2}. *(p. 867)*

hypoxia (hī-POKS-ē-a): A condition of reduced tissue P_{O_2}. *(p. 844)*

lung cancer (*pleuropulmonary neoplasm*): A class of aggressive malignancies originating in the bronchial passageways or alveoli. *(p. 871* and *AM)*

mountain sickness: Acute disorder resulting from CNS effects of the low gas partial pressures that occur at high altitudes. *(p. 861* and *AM)*

pleurisy: Inflammation of the pleurae and secretion of excess amounts of pleural fluid. *(p. 844)*

pneumonia (nū-MŌ-nē-a): A respiratory disorder characterized by fluid leakage into the alveoli and/or swelling and constriction of the respiratory bronchioles. *(p. 841)*

pneumothorax (nū-mō-THŌ-raks): The entry of air into the pleural cavity. *(p. 850)*

pulmonary embolism: Blockage of a branch of a pulmonary artery that stops blood flow to a group of lobules and/or alveoli. *(p. 843)*

respiratory distress syndrome: A condition resulting from inadequate surfactant production and associated alveolar collapse. *(p. 840* and *AM)*

sudden infant death syndrome (SIDS): Death of an infant due to respiratory arrest; the cause remains uncertain. *(p. 865)*

tracheostomy (trā-kē-OS-to-mē): Insertion of a tube directly into the trachea to bypass a blocked or damaged larynx. *(p. 834)*

tuberculosis: A respiratory disorder caused by infection of the lungs by the bacterium *mycobacterium tuberculosis. (p. 837* and *AM)*

AM *Additional Terms Discussed in the Applications Manual*

chronic obstructive pulmonary disease (COPD): A condition characterized by chronic bronchitis and chronic airways obstruction.

shallow water blackout: A potential risk for individuals who hyperventilate prior to swimming underwater; the CNS shuts down suddenly, because of hypoxia.

CHAPTER REVIEW

STUDY OUTLINE

INTRODUCTION, p. 826

1. Body cells must obtain oxygen and eliminate carbon dioxide. Gas exchange occurs at the **alveoli** of the lungs.

FUNCTIONS OF THE RESPIRATORY SYSTEM, p. 826

1. The functions of the respiratory system include: (1) providing an area for gas exchange between air and circulating blood; (2) moving air to and from exchange surfaces; (3) protecting respiratory surfaces; (4) defending the respiratory system and other tissues from pathogens; (5) permitting vocal communication; and (6) helping to regulate blood volume, blood pressure, and body fluid pH.

ORGANIZATION OF THE RESPIRATORY SYSTEM, p. 826

1. The **respiratory system** includes the nose, nasal cavity, sinuses, pharynx, larynx, trachea, and conducting passageways leading to the exchange surfaces of the lungs. (*Figure 23-1*)
2. The **respiratory tract** consists of the conducting passageways that carry air to and from the alveoli. The passageways of the **upper respiratory system** filter and humidify the incoming air. The **lower respiratory system** includes delicate conduction passages and the alveolar exchange surfaces.

The Respiratory Mucosa, p. 826

3. The **respiratory mucosa** (respiratory epithelium and underlying loose connective tissue) lines the upper and lower respiratory tracts.
4. The respiratory epithelium changes in structure along the respiratory tract. It is supported by a layer of loose connective tissue, the **lamina propria.** (*Figure 23-2*)

THE UPPER RESPIRATORY SYSTEM, p. 829

1. The components of the upper respiratory system include the nose, nasal cavity, and pharynx. (*Figures 23-1, 23-3*)

The Nose and Nasal Cavity, p. 829

2. Air normally enters the respiratory system via the **external nares** that open into the **nasal cavity.** The **vestibule** (entrance) is guarded by hairs that screen out large particles.
3. Incoming air flows through the **superior, middle,** and **inferior meatuses** (narrow grooves) and bounces off the conchal surfaces.
4. The **hard palate** separates the oral and nasal cavities. The **soft palate** separates the superior nasopharynx from the rest of the pharynx. The connections between the nasal cavity and nasopharynx represent the **internal nares.**
5. In addition to trapping particles, the nasal mucosa warms and humidifies the incoming air.

The Pharynx, p. 831

6. The **pharynx** is a chamber shared by the digestive and respiratory systems. The **oropharynx** is continuous with the oral cavity; the **laryngopharynx** includes the narrow zone between the hyoid and the entrance to the esophagus. (*Figure 23-3*)

THE LARYNX, p. 831

1. Inspired air passes through the **glottis** en route to the lungs; the **larynx** surrounds and protects the glottis. The **epiglottis** projects into the pharynx. (*Figure 23-4*)

Cartilages of the Larynx, p. 831

2. The cylindrical larynx is composed of three large cartilages (the **thyroid, cricoid,** and **epiglottis**) and three smaller pairs (the **arytenoid, corniculate,** and **cuneiform**).
3. Two pairs of folds span the glottal opening: the relatively inelastic **ventricular folds** and the more delicate **vocal folds.** Air passing through the glottis vibrates the vocal folds and produces sound.

The Laryngeal Musculature, p. 833

4. The **intrinsic laryngeal muscles** regulate tension in the vocal folds and open and close the glottis. The **extrinsic laryngeal muscles** position and stabilize the larynx. During swallowing, both sets of muscles help to prevent particles from entering the glottis.

THE TRACHEA, p. 834

1. The **trachea** ("windpipe") extends from the sixth cervical vertebra to the fifth thoracic vertebra. The **submucosa** contains C-shaped **tracheal cartilages** that stiffen the tracheal walls and protect the airway. The posterior tracheal wall can distort to permit large masses of food to pass. (*Figure 23-5*)

THE PRIMARY BRONCHI, p. 834

1. The trachea branches within the mediastinum to form the **right** and **left primary bronchi.** Each bronchus enters a lung at the **hilus** (a groove). The **root** of the lung is a connective tissue mass including the bronchus, pulmonary vessels, and nerves.

THE LUNGS, p. 834

Lobes of the Lungs, p. 835

1. The **lobes** of the lungs are separated by fissures; the right lung has three lobes and the left lung has two. The connective tissues of the root extend into the **parenchyma** of the lung as a series of **trabeculae** (partitions). These branch to form **septa** that divide the lung into **lobules.** (*Figure 23-6*)

Lung Surfaces, p. 835

2. The anterior **costal surfaces** follow the inner contours of the rib cage. The concavity of the **mediastinal surface** of the left lung is the **cardiac notch,** which conforms to the shape of the pericardium. (*Figure 23-7*)

The Bronchi, p. 837

3. The primary bronchi and their branches form the **bronchial tree.** The **secondary** and **tertiary bronchi** are called **intrapulmonary bronchi** because they are branches within the lungs. As they branch, the amount of cartilage in their walls decreases and the amount of smooth muscle increases. (*Figure 23-8*)
4. Each tertiary bronchus supplies air to a single **bronchopulmonary segment.** (*Figure 23-9*)

The Bronchioles, p. 838

5. Bronchioles within the bronchopulmonary segments ultimately branch into **terminal bronchioles.** Each terminal bronchiole delivers air to a single **pulmonary lobule.** Within the lobule, the terminal bronchiole branches into **respiratory bronchioles.** (*Figures 23-9, 23-10*)

Alveolar Ducts and Alveoli, p. 840

6. The respiratory bronchioles open into **alveolar ducts;** many alveoli are interconnected at each duct. The respiratory exchange surfaces are extensively connected to the circulatory system via the vessels of the pulmonary circuit.
7. The **respiratory membrane** consists of a simple squamous epithelium; **surfactant cells** scattered in it produce an oily secretion that keeps the alveoli from collapsing. **Alveolar macrophages** patrol the epithelium and engulf foreign particles. (*Figure 23-11*)

The Blood Supply to the Lungs, p. 842

8. The conducting portions of the respiratory tract receive blood from the external carotid arteries, the thyrocervical trunks, and the bronchial arteries. The venous blood flows into the pulmonary veins, bypassing the rest of the systemic circuit and diluting the oxygenated blood leaving the alveoli.

THE PLEURAL CAVITIES AND PLEURAL MEMBRANES, p. 843

1. Each lung occupies a single pleural cavity lined by a **pleura** (serous membrane). (*Figure 23-12*)

Changes in the Respiratory System at Birth, p. 844

2. Before delivery the fetal lungs are fluid-filled and collapsed. At the first breath, the lungs inflate and never collapse completely thereafter.

RESPIRATORY PHYSIOLOGY, p. 844

1. Respiratory physiology focuses on a series of integrated processes: **pulmonary ventilation** (movement of air into and out of the lungs); **external respiration** (the exchange of oxygen and carbon dioxide between the interstitial fluids of the body and the external environment); **internal respiration** (the exchange of oxygen and carbon dioxide between the interstitial fluid and living cells). If oxygen content declines, the affected tissues will suffer from **hypoxia;** if the oxygen supply is completely shut off, **anoxia** and tissue death result.

Pulmonary Ventilation, p. 844

2. As pressure on a gas decreases, its volume expands; as pressure increases, volume contracts; this inverse relationship is known as **Boyle's law.** Increased temperature elevates the volume of a gas, decreased temperature reduces it. (*Figure 23-13; Table 23-1*)
3. The relationship between **intrapulmonary pressure** (the pressure inside the respiratory tract) and **atmospheric pressure (atm)** determines the direction of airflow. The **intrapleural pressure** is the pressure in the space between the parietal and visceral pleurae. (*Figures 23-14, 23-15*)
4. The diaphragm and the external and internal intercostals are involved in normal **quiet breathing,** or **eupnea.** Accessory muscles become active during the active inspiratory and expiratory movements of **forced breathing,** or **hyperpnea.** (*Figures 23-16, 23-17*)
5. The **vital capacity** includes the **tidal volume** plus the **expiratory reserve** and the **inspiratory reserve.** The air left in the lungs at the end of maximum expiration is the **residual volume.** (*Figure 23-18*)

Gas Exchange at the Respiratory Membrane, p. 854

6. In a mixed gas the individual gases exert a pressure proportional to their abundance in the mixture **(Dalton's law).** The pressure contributed by a single gas is its **partial pressure.**
7. The relationship between pressure, solubility, and the number of gas molecules in solution is called **Henry's law.** (*Figure 23-19*)
8. **Alveolar ventilation** is the amount of air reaching the alveoli each minute. Alveolar and atmospheric air differ in their composition. (*Figure 23-20; Table 23-2*)

Gas Pickup and Delivery, p. 858

9. Blood entering peripheral capillaries delivers oxygen and absorbs carbon dioxide. The transport of oxygen and carbon dioxide in the blood involves reactions that are completely reversible.
10. Over the range of oxygen pressures normally present in the body, a small change in plasma P_{O_2} will mean a large change in the amount of oxygen bound or released. At alveolar P_{O_2} hemoglobin is almost fully saturated; at the P_{O_2} of peripheral tissues it retains a substantial oxygen reserve. When low plasma P_{O_2} continues for extended periods, red blood cells generate more **2,3-bisphosphoglycerate (BPG),** which reduces hemoglobin's affinity for oxygen. (*Figures 23-21, 23-22*)
11. Fetal hemoglobin has a stronger affinity for oxygen than does adult hemoglobin, aiding the removal of oxygen from the maternal blood. (*Figure 23-23*)
12. Aerobic metabolism in peripheral tissues generates carbon dioxide. About 7 percent of the CO_2 transported in the blood is dissolved in the plasma; 23 percent is bound as **carbaminohemoglobin;** the rest is converted to carbonic acid, which dissociates into H^+ and HCO_3^-. (*Figures 23-24, 23-25*)

CONTROL OF RESPIRATION, p. 863

1. Local regulation compensates for small oscillations affecting individual tissues and organs; large-scale or extended changes require the integration of cardiovascular and respiratory responses.

Local Regulation of Gas Transport and Alveolar Function, p. 863

2. Local factors regulate blood flow (perfusion) and airflow (ventilation). Alveolar capillaries constrict under conditions of low oxygen, and bronchioles dilate under conditions of low carbon dioxide.

The Respiratory Centers of the Brain, p. 863

3. The **respiratory centers** include three pairs of nuclei in the reticular formation of the pons and medulla. The **respiratory rhythmicity center** sets the pace for respiration. The **apneustic center** causes strong, sustained inspiratory movements, and the **pneumotaxic center** inhibits the apneustic center and the inspiratory center in the medulla. *(Figure 23-26)*
4. The **inflation reflex** prevents overexpansion of the lungs during forced breathing; the **deflation reflex** stimulates inspiration when the lungs are collapsing. Chemoreceptor reflexes respond to changes in the P_{O_2} and P_{CO_2} of the blood and cerebrospinal fluid. *(Figures 23-27, 23-28)*

Voluntary Control of Respiration, p. 868

5. Conscious and unconscious thought processes can affect respiration by affecting the respiratory centers.

AGING AND THE RESPIRATORY SYSTEM, p. 869

1. The respiratory system is generally less efficient in the elderly because (1) elastic tissue deteriorates, lowering the vital capacity of the lungs, (2) movements of the chest cage are restricted by arthritic changes and decreased flexibility of costal cartilages, and (3) some degree of emphysema is usually present. *(Figure 23-29)*

INTEGRATION WITH OTHER SYSTEMS, p. 869

1. The respiratory system has extensive anatomical connections to the cardiovascular system. *(Figure 23-30)*

REVIEW QUESTIONS

LEVEL 1 Reviewing Facts and Terms

1. The C shape of the tracheal cartilages is important because:
   (a) large masses of food can pass through the esophagus during swallowing
   (b) larger masses of air can pass through the trachea
   (c) it allows greater tracheal elasticity and flexibility
   (d) a, b, and c are correct
2. Control over the amount of resistance to airflow and the distribution of air within the lungs is provided by the:
   (a) diaphragm (b) trachea
   (c) bronchioles (d) alveoli
3. Pulmonary ventilation involves:
   (a) internal respiration (b) external respiration
   (c) cellular respiration (d) air movement
4. The presence of an abnormally low oxygen content in tissue fluids is called:
   (a) anoxia (b) edema
   (c) hypoxia (d) emphysema
5. During inhalation, the lungs expand and the intrapulmonary pressure:
   (a) rises to around 761 mm Hg
   (b) remains at 760 mm Hg
   (c) drops to around 759 mm Hg
   (d) does not change
6. During a normal inhalation, the intrapleural pressure is about:
   (a) –1 mm Hg (b) +6 mm Hg
   (c) +1 mm Hg (d) –6 mm Hg
7. According to Henry's law, if the partial pressure of a gas increases:
   (a) gas molecules will come out of solution
   (b) more gas molecules will enter solution
   (c) its solubility will decrease
   (d) its volume will decrease
8. At a normal tissue partial pressure of oxygen, the blood entering the venous system contains around ______ of its total oxygen content.
   (a) 25 percent (b) 50 percent
   (c) 75 percent (d) 90 percent
9. Approximately 70 percent of the carbon dioxide absorbed by the blood will be transported:
   (a) as bicarbonate ions
   (b) bound to hemoglobin
   (c) in the form of dissolved gas molecules
   (d) bound to oxygen molecules
10. The apneustic center of the lower pons provides:
   (a) inhibition of the pneumotaxic and inspiratory centers
   (b) continual stimulation to the inspiratory center
   (c) monitoring of blood gas levels
   (d) alterations in chemoreceptor sensitivity
11. All of the following provide chemoreceptor input to the medullary respiratory centers except the:
   (a) olfactory epithelium
   (b) medullary chemoreceptors
   (c) aortic body
   (d) carotid body
12. Sneezing and coughing are classic examples of:
   (a) inflation reflexes
   (b) deflation reflexes
   (c) protective reflexes
   (d) the Hering-Breuer reflexes
13. What are the six primary functions of the respiratory system?
14. Distinguish between the structures of the upper respiratory system and the lower respiratory system.
15. What defense mechanisms protect the respiratory system from becoming contaminated with debris or pathogens?
16. What are the three regions of the pharynx called, and where is each located?
17. What prevents the swallowing of food or liquids while breathing?
18. How does the parietal pleura differ from the visceral pleura?
19. What four integrated steps are involved in external respiration?

20. What important physiological adjustments are necessary to adapt to and tolerate high-altitude conditions?

21. By what three mechanisms is carbon dioxide transported in the bloodstream?

22. What effect does P_{CO_2} have on smooth muscles in the walls of bronchioles?

LEVEL 2 Reviewing Concepts

23. Parasympathetic stimulation to the smooth muscle tissue layer in the bronchioles causes:
(a) bronchoconstriction
(b) bronchodilation
(c) a relaxation of muscle tone
(d) an increase in tidal volume

24. If an individual has a respiration rate of 15 breaths per minute and a tidal volume of 500 ml of air, the respiratory minute volume is:
(a) 7.5 l/min (b) 75 l/min
(c) 750 l/min (d) 7500 l/min

25. If a tidal inspiration pulls in 1000 ml of air the amount reaching the alveolar spaces is about:
(a) 300 ml (b) 850 ml
(c) 150 ml (d) 700 ml

26. Gas exchange at the respiratory membrane is efficient because:
(a) the differences in partial pressure are substantial
(b) the gases are lipid-soluble
(c) the total surface area is large
(d) a, b, and c are correct

27. For any partial pressure of oxygen, if the concentration of 2,3-bisphosphoglycerate (BPG) increases:
(a) the amount of oxygen released by hemoglobin will decrease
(b) the oxygen levels in hemoglobin will be unaffected
(c) the amount of oxygen released by hemoglobin will increase
(d) the amount of carbon dioxide carried by hemoglobin will increase

28. The primary physiological adjustment(s) necessary for an athlete to compete at high altitudes is (are):
(a) increased respiratory rate
(b) increased heart rate
(c) elevated hematocrit
(d) a, b, and c are correct

29. An increase in the partial pressure of carbon dioxide in arterial blood causes the chemoreceptors to stimulate the respiratory centers, causing:
(a) hypoventilation (b) hyperventilation
(c) hypocapnia (d) hypercapnia

30. What is the functional significance of the decrease in the amount of cartilage and increase in the amount of smooth muscle in the lower respiratory passageways?

31. Why can't you swallow solid food or liquid properly while you are talking?

32. Why is breathing through the nasal cavity more desirable than breathing through the mouth?

33. How would you justify the statement, "the bronchioles are to the respiratory system what the arterioles are to the cardiovascular system"?

34. How are septal cells involved with keeping the alveoli from collapsing?

35. Why is diffusion across the respiratory membrane rapid and efficient?

36. How does pulmonary ventilation differ from alveolar ventilation, and what is the function of each type of ventilation?

37. What is the significance of Boyle's law, Dalton's law, and Henry's law to the process of respiration?

38. If the pleural cavity is penetrated because of an injury, why does the lung collapse?

39. What happens to the process of respiration when a person is sneezing or coughing?

40. What are the differences between pulmonary volumes and respiratory capacities? How are pulmonary volumes and respiratory capacities determined?

41. What are the primary effects of pH, temperature, and 2,3-bisphosphoglycerate (BPG) on hemoglobin saturation?

42. What is the functional difference between the dorsal respiratory group (DRG) and the ventral respiratory group (VRG) of the medulla oblongata?

43. What types of sensory information modify the activities of the respiratory centers?

LEVEL 3 Critical Thinking and Clinical Applications

44. Billy's normal alveolar ventilation rate (AVR) during mild exercise is 6.0 l/min. On a warm summer day while at the beach, he decides to go snorkeling. The snorkel has a volume of 50 ml. Assuming that the water is not too cold and that snorkeling for Billy is mild exercise, what would his respiratory rate have to be in order to maintain an AVR of 6.0 l/min while snorkeling? (Assume a constant tidal volume of 500 ml and an anatomic dead space of 150 ml.)

45. A decrease in blood pressure will trigger a baroreceptor reflex that leads to increased ventilation. What is the possible advantage of this reflex?

46. Mr. B. has suffered from chronic advanced emphysema for 15 years. While hospitalized with a respiratory infection, he goes into respiratory distress. Without thinking, his nurse immediately administers pure oxygen, which causes Mr. B. to stop breathing. Why did this occur?

47. You spend the night at a friend's house during the winter. Your friend's home is quite old, and the hot-air furnace lacks a humidifier. When you wake up in the morning you have a fair amount of nasal congestion and you decide you might be coming down with a cold. After a steamy shower and some juice for breakfast, the nasal congestion disappears. Explain.

48. Why would a person with kyphosis (see Chapter 7) exhibit a lower-than-normal vital capacity?

49. Why do patients suffering from anemia generally not exhibit an increase in respiratory rate or tidal volume even though their blood is not carrying enough oxygen?

50. Doris has an obstruction of her right primary bronchus. As a result, how would you expect the oxygen dissociation curve for her right lung to compare with the oxygen dissociation curve for her left?

CHAPTER 24

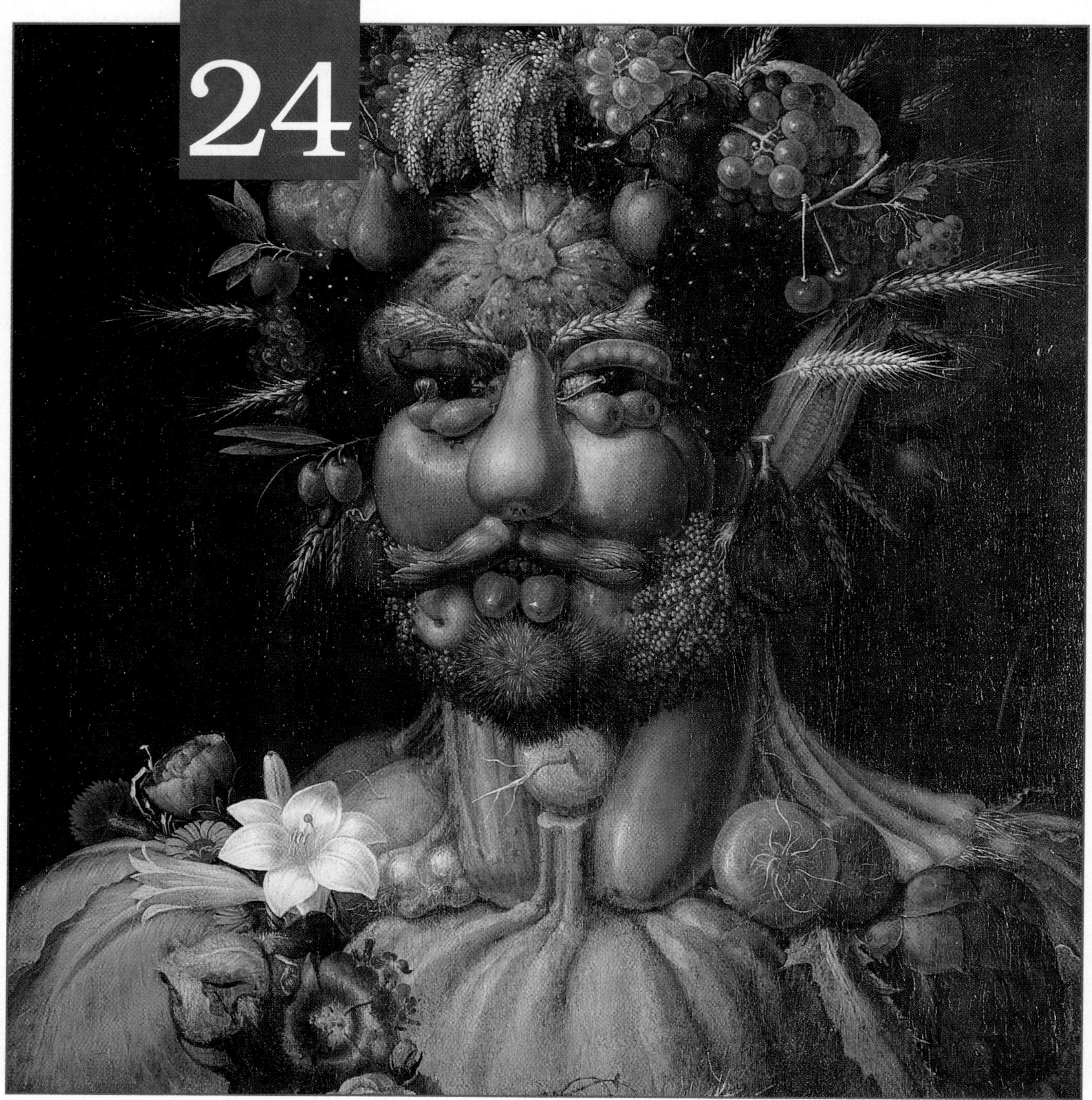

They say you are what you eat. Perhaps that's what Giuseppe Arcimboldo had in mind when he painted this whimsical picture in the sixteenth century. Though Arcimboldo may have taken the notion a bit too literally, modern physiology has confirmed that he had the right idea. Whether we are couch potatoes or marathoners, our bodies never stop using energy—and at least some of the body's components are always growing, being repaired, or being replaced. Thus, we all need a regular supply of nutrients in our diet. Unfortunately, the nutrients in food are not ready for use by our cells. The complex organic molecules must first be disassembled into smaller molecules that can be absorbed easily and distributed through the bloodstream. Without such processing by the digestive system, the food we eat would be of no more use to us than a lump of coal in the gas tank of a car. This chapter introduces the organs of the digestive system and their varied functions in the process of digestion.

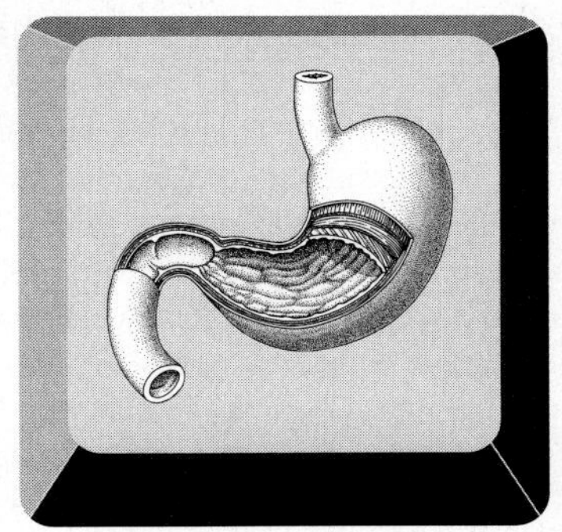

The Digestive System

Chapter Outline and Objectives

Few people give any serious thought to the digestive system unless it malfunctions. Yet we spend hours of conscious effort filling and emptying it. References to this system are part of our everyday language. Take a moment to think of how often people use expressions relating to the digestive system: They may "have a gut feeling," "want to chew on" something, or find your opinions "hard to swallow." (Other, less polite remarks may be heard almost as often.) When something does go wrong with the digestive system, even something minor, most people will seek treatment immediately. For this reason television advertisements promote toothpaste and mouthwash, diet supplements, antacids, and laxatives on an hourly basis.

Cells perform metabolic reactions that provide energy for the synthesis of ATP. These reactions require two essential ingredients: (1) oxygen and (2) organic molecules that can be broken down by intracellular enzymes. The respiratory system, working with the cardiovascular system, provides the necessary oxygen. The digestive system, working with the cardiovascular and lymphatic systems, provides the organic molecules. In effect, the digestive system provides the "fuel" that keeps all of the body's cells running and the building blocks needed for cell growth and repair.

The digestive system consists of a muscular tube, the **digestive tract,** and various **accessory organs.** Accessory digestive organs include the teeth and tongue, and various *glandular organs*, such as the salivary glands, liver, and pancreas, that secrete into ducts emptying into the digestive tract. Food enters the digestive tract and passes along its length. On the way, the secretions of the glandular organs are discharged into the digestive tract. These secretions, which contain water, enzymes, buffers, and other components, assist in preparing organic and inorganic nutrients for absorption across the epithelium of the digestive tract.

Digestive functions can be considered as a series of integrated steps:

1. ***Ingestion*** occurs when materials enter the digestive tract via the mouth. Ingestion is an active process involving conscious choice and decision making.
2. ***Mechanical processing*** is physical manipulation and distortion that makes materials easier to swallow and increases the surface area for enzymatic attack. Mechanical processing may or may not be required; liquids can be swallowed immediately, but solids must usually be processed first. Squashing with the tongue and tearing and crushing with the teeth are examples of mechanical processing that occurs before ingestion. Swirling, mixing, and churning motions of the digestive tract provide mechanical processing after ingestion.
3. ***Digestion*** refers to the chemical breakdown of food into small organic fragments suitable for absorption by the digestive epithelium. Simple molecules in food, such as glucose, can be absorbed intact, but epithelial cells have no way to deal with molecules the size and complexity of proteins, polysaccharides, and triglycerides. These molecules must be disassembled by digestive enzymes prior to absorption. For example, the starches in a potato are of no value until enzymes have broken them down to simple sugars that can be absorbed and distributed to our cells.
4. ***Secretion*** is the release of water, acids, enzymes, buffers, and salts by the epithelium of the digestive tract and by glandular organs.
5. ***Absorption*** is the movement of organic substrates, electrolytes, vitamins, and water across the digestive epithelium and into the interstitial fluid of the digestive tract.
6. ***Excretion*** is the elimination of waste products from the body. The digestive tract and glandular organs secrete waste products in secretions discharged into the lumen of the tract. Most of these waste products will leave the body through the process of **defecation** (def-e-KĀ-shun), which also eliminates the indigestible residue of the digestive process.

The lining of the digestive tract also plays a protective role by safeguarding surrounding tissues against (1) the corrosive effects of digestive acids and enzymes, (2) mechanical stresses, such as abrasion, and (3) bacteria that are either swallowed with food or reside inside the digestive tract. The digestive epithelium and its secretions provide a nonspecific defense against these bacteria. Bacteria reaching the underlying tissues are attacked by macrophages and other cells of the immune system.

An Overview of the Structure and Function of the Digestive Tract

Figure 24-1

The major components of the digestive system are shown in Figure 24-1●. The digestive tract begins at the oral cavity and continues through the pharynx, esophagus, stomach, small intestine, and large intestine before ending at the rectum and anus. Although these structures have overlapping functions, each has certain areas of specialization and shows distinctive histological characteristics.

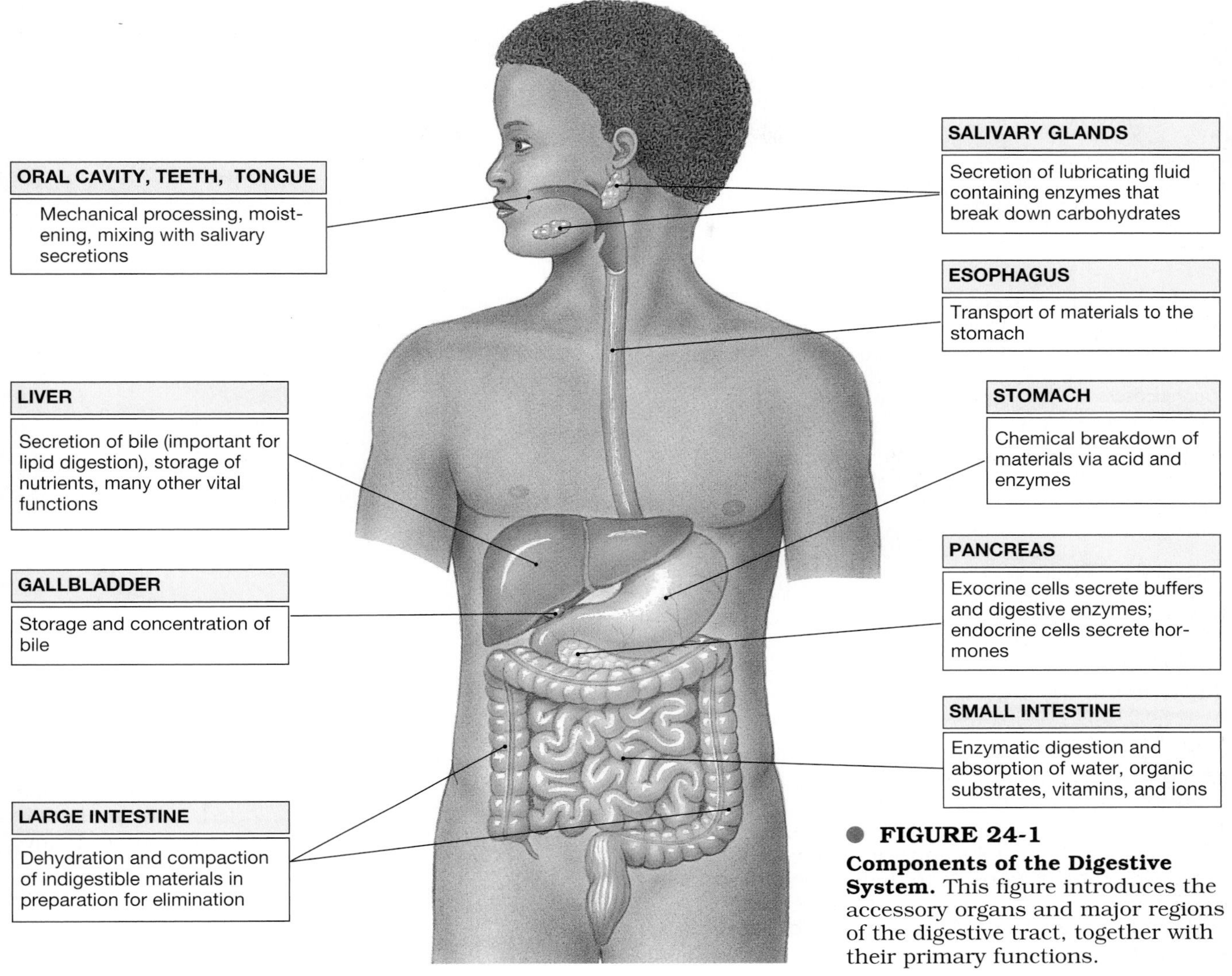
ORAL CAVITY, TEETH, TONGUE
Mechanical processing, moistening, mixing with salivary secretions
LIVER
Secretion of bile (important for lipid digestion), storage of nutrients, many other vital functions
GALLBLADDER
Storage and concentration of bile
LARGE INTESTINE
Dehydration and compaction of indigestible materials in preparation for elimination
SALIVARY GLANDS
Secretion of lubricating fluid containing enzymes that break down carbohydrates
ESOPHAGUS
Transport of materials to the stomach
STOMACH
Chemical breakdown of materials via acid and enzymes
PANCREAS
Exocrine cells secrete buffers and digestive enzymes; endocrine cells secrete hormones
SMALL INTESTINE
Enzymatic digestion and absorption of water, organic substrates, vitamins, and ions

● **FIGURE 24-1**
Components of the Digestive System. This figure introduces the accessory organs and major regions of the digestive tract, together with their primary functions.

❑ Histological Organization of the Digestive Tract

Figure 24-2

The major layers of the digestive tract include (1) the *mucosa*, (2) the *submucosa*, (3) the *muscularis externa*, and (4) the *serosa*. Sectional and diagrammatic views of these layers are presented in Figure 24-2●. There are regional variations in the structure of the layers; this figure is a composite view that most closely resembles the appearance of the small intestine, the longest segment of the digestive tract.

The Mucosa

Figure 24-2

The inner lining, or **mucosa,** of the digestive tract is an example of a *mucous membrane.* Mucous membranes, introduced in Chapter 4, consist of a layer of an epithelium moistened by glandular secretions and an underlying layer of loose connective tissue, the *lamina propria.* ∞ [p. 134]

THE DIGESTIVE EPITHELIUM The mucosal epithelium may be either simple or stratified, depending on the location and the stresses to which it is most often subjected. For example, the oral cavity and esophagus are lined by a stratified squamous epithelium, whereas the stomach, the small intestine, and almost the entire length of the large intestine have a simple columnar epithelium that contains goblet cells. Scattered among the columnar cells are **enteroendocrine cells.** These cells secrete hormones that coordinate the activities of the digestive tract and the accessory glands.

The lining of the digestive tract is often thrown into transverse or longitudinal folds (Figure 24-2●). The folding (1) dramatically increases the surface area available for absorption and (2) may permit expansion of the lumen to accommodate a large meal. Ducts opening onto the epithelial surfaces carry the secretions of gland cells located in the mucosa and submucosa or within accessory glandular organs.

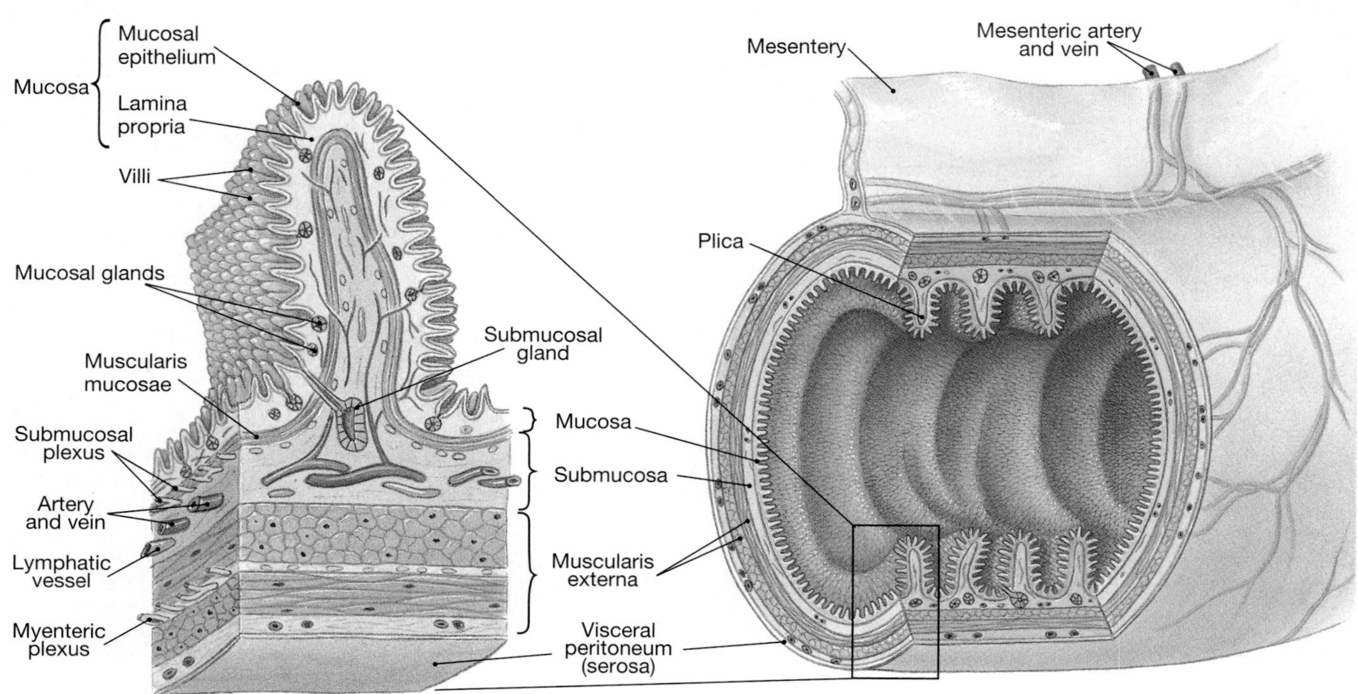

Mucosa
Mucosal epithelium
Lamina propria
Villi
Mucosal glands
Muscularis mucosae
Submucosal plexus
Artery and vein
Lymphatic vessel
Myenteric plexus
Submucosal gland
Mucosa
Submucosa
Muscularis externa
Visceral peritoneum (serosa)
Mesentery
Mesenteric artery and vein
Plica

FIGURE 24-2

Structure of the Digestive Tract. This is a diagrammatic view of a representative portion of the digestive tract. The features illustrated are those of the small intestine.

EPITHELIAL RENEWAL AND REPAIR The life span of a typical epithelial cell varies from 2 to 3 days in the esophagus to 6 days in the large intestine. The lining of the entire digestive tract is therefore continually being renewed through the divisions of epithelial stem cells, keeping pace with the rate of cell destruction and loss at epithelial surfaces. This high rate of cell division explains why radiation and anticancer drugs that inhibit mitosis have drastic effects on the digestive tract. Lost epithelial cells are no longer replaced, and the cumulative damage to the epithelial lining quickly leads to problems in nutrient absorption. In addition, the exposure of the lamina propria to digestive enzymes can cause internal bleeding and a variety of other serious problems.

THE LAMINA PROPRIA The lamina propria consists of a layer of loose connective tissue that also contains blood vessels, sensory nerve endings, lymphatic vessels, smooth muscle fibers, and scattered areas of lymphoid tissue. In many areas, such as the oral cavity, pharynx, esophagus, stomach, and *duodenum* (the proximal portion of the small intestine), the lamina propria also contains the secretory cells of mucous glands.

In most areas of the digestive tract the outer portion of the lamina propria consists of a narrow band of smooth muscle and elastic fibers. This band is called the **muscularis** (mus-kū-LAR-is) **mucosae.** The smooth muscle fibers in the muscularis mucosae are arranged in two concentric layers (Figure 24-2●). The inner layer encircles the lumen (the *circular muscle*), and the outer layer contains muscle fibers oriented parallel to the long axis of the tract (the *longitudinal layer*). Contractions in these layers alter the shape of the lumen and move the epithelial pleats and folds.

The Submucosa

Figure 24-2

The **submucosa** is a layer of dense connective tissue that surrounds the muscularis mucosae. Large blood vessels and lymphatics are found in this layer, and in some regions the submucosa also contains exocrine glands that secrete buffers and enzymes into the lumen of the digestive tract. Along its outer margin, the submucosa contains a network of nerve fibers and scattered neurons. This **submucosal plexus,** or *plexus of Meissner,* contains sensory neurons, parasympathetic ganglionic neurons, and sympathetic postganglionic fibers that innervate the mucosa and submucosa (Figure 24-2●).

The Muscularis Externa

The submucosal plexus lies along the inner border of the **muscularis externa,** a region dominated by smooth muscle fibers. As in the submucosa, the smooth muscle fibers are arranged in an inner, circular layer and an outer, longitudinal layer. These layers play an essential role in mechanical processing and in the movement of materials along the digestive tract. These movements are coordinated primarily by neurons of the **myenteric** (mī-en-TER-ik)

plexus (*mys*, muscle + *enteron*, intestine), or *plexus of Auerbach*. This network of parasympathetic ganglia and sympathetic postganglionic fibers lies sandwiched between the circular and longitudinal muscle layers. Parasympathetic stimulation increases muscular tone and activity, and sympathetic stimulation promotes muscular inhibition and relaxation.

The Serosa

Figure 24-2

Along most portions of the digestive tract inside the peritoneal cavity, the muscularis externa is covered by a *serous membrane* known as the **serosa** (Figure 24-2●). There is no serosa covering the muscularis externa of the oral cavity, pharynx, esophagus, and rectum, where a dense network of collagen fibers firmly attaches the digestive tract to adjacent structures. This fibrous sheath is called an *adventitia*.

❑ The Movement of Digestive Materials

The digestive tract contains layers of *visceral smooth muscle tissue*, a type of smooth muscle introduced in Chapter 10. [p. 318] The smooth muscle along the digestive tract shows rhythmic cycles of activity due to the presence of *pacesetter cells*. These smooth muscle fibers undergo spontaneous depolarization, and their contraction triggers a wave of contraction that spreads through the entire muscular sheet. Pacesetter cells are found in the muscularis mucosae and muscularis externa, whose layers surround the lumen of the digestive tract. The coordinated contractions of these layers play a vital role in the movement of materials along the tract, through *peristalsis*, and in mechanical processing, through *segmentation*.

Peristalsis

Figure 24-3

The muscularis externa propels materials from one portion of the digestive tract to another through the contractions of **peristalsis** (per-i-STAL-sis). Peristalsis consists of waves of muscular contractions that move along the length of the digestive tract. During a peristaltic movement, the circular muscles contract behind the digestive contents. Longitudinal muscles contract next, shortening adjacent segments. A wave of contraction in the circular muscles then forces the materials in the desired direction (Figure 24-3●).

Segmentation

Most areas of the small intestine and some portions of the large intestine undergo **segmentation.** These movements churn and fragment the digestive materials, mixing the contents with intestinal secretions. Because they do not follow a set pattern, segmentation movements do not produce directional movement of materials along the tract.

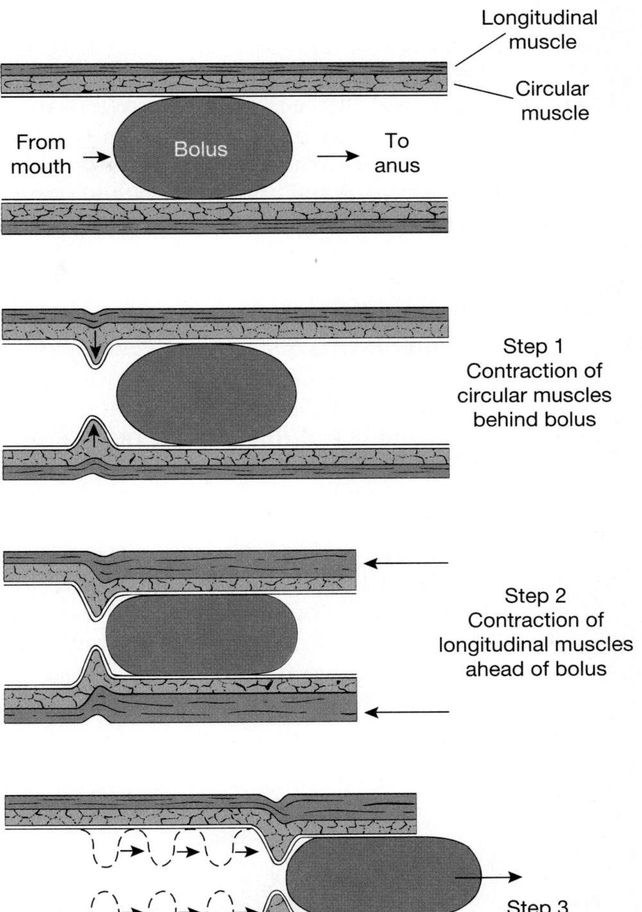
Longitudinal muscle
Circular muscle
From mouth
Bolus
To anus
Step 1
Contraction of circular muscles behind bolus
Step 2
Contraction of longitudinal muscles ahead of bolus
Step 3
Contraction in circular muscle layer forces bolus forward

● **FIGURE 24-3**
Peristalsis. Peristalsis propels materials along the length of the digestive tract.

❑ The Control of Digestive Function

Figure 24-4

The activities of the digestive system are regulated in three basic ways (Figure 24-4●): (1) by neural mechanisms, (2) by hormonal mechanisms, and (3) by local mechanisms.

- *Neural mechanisms.* The movement of materials along the digestive tract is controlled primarily by neural mechanisms. Local peristaltic movements limited to a few centimeters of the digestive tract are triggered by sensory receptors in the walls of the digestive tract. The afferent fibers synapse within the myenteric plexus on motor neurons that synapse in the muscularis externa. These **myenteric reflexes,** or *short reflexes*, operate independently of the CNS. The sensory information is also distributed to the CNS, where it can trigger *long reflexes*. Long reflexes provide a higher level of control that coordinates activities along the length of the digestive tract. These reflexes usually control large-scale peristaltic waves that move materials from one region of the digestive tract to another. Long reflexes may involve motor fibers in the glossopharyngeal, vagus, or pelvic nerves that synapse in the myenteric plexus. The sensitivity of the smooth muscle fibers to neural commands can be enhanced or inhibited by digestive hormones.

- ***Hormonal mechanisms.*** The digestive tract is a hormone factory that produces at least 18 different hormones. These hormones affect almost every aspect of digestive function, and some of them also affect the activities of other systems. The hormones, including *gastrin, secretin,* and others, are peptides produced by enteroendocrine cells in the digestive tract. They reach their target organs via the circulatory system. We will consider each of these hormones in more detail as we proceed down the length of the digestive tract.
- ***Local mechanisms.*** Prostaglandins, histamine, and other chemicals released into the interstitial fluid may affect cells within a small segment of the digestive tract. These local messengers are important in coordinating a response to conditions that affect only a portion of the digestive tract. For example, histamine release in the lamina propria of the stomach stimulates the secretion of acid by cells in the adjacent epithelium.

❑ The Peritoneum

The abdominopelvic cavity contains the *peritoneal cavity.* The peritoneal cavity is lined by a serous membrane consisting of a superficial mesothelium covering a layer of loose connective tissue. The serous membrane can be divided into the *visceral peritoneum,* or serosa that covers organs within the peritoneal cavity, and the *parietal peritoneum* that lines the inner surfaces of the body wall. [p. 24]

The serous membrane lining the peritoneal cavity continually produces peritoneal fluid that lubricates the peritoneal surfaces. About 7 liters of fluid are secreted and reabsorbed each day, although the volume within the peritoneal cavity at any one time is very small. Liver disease, kidney disease, or

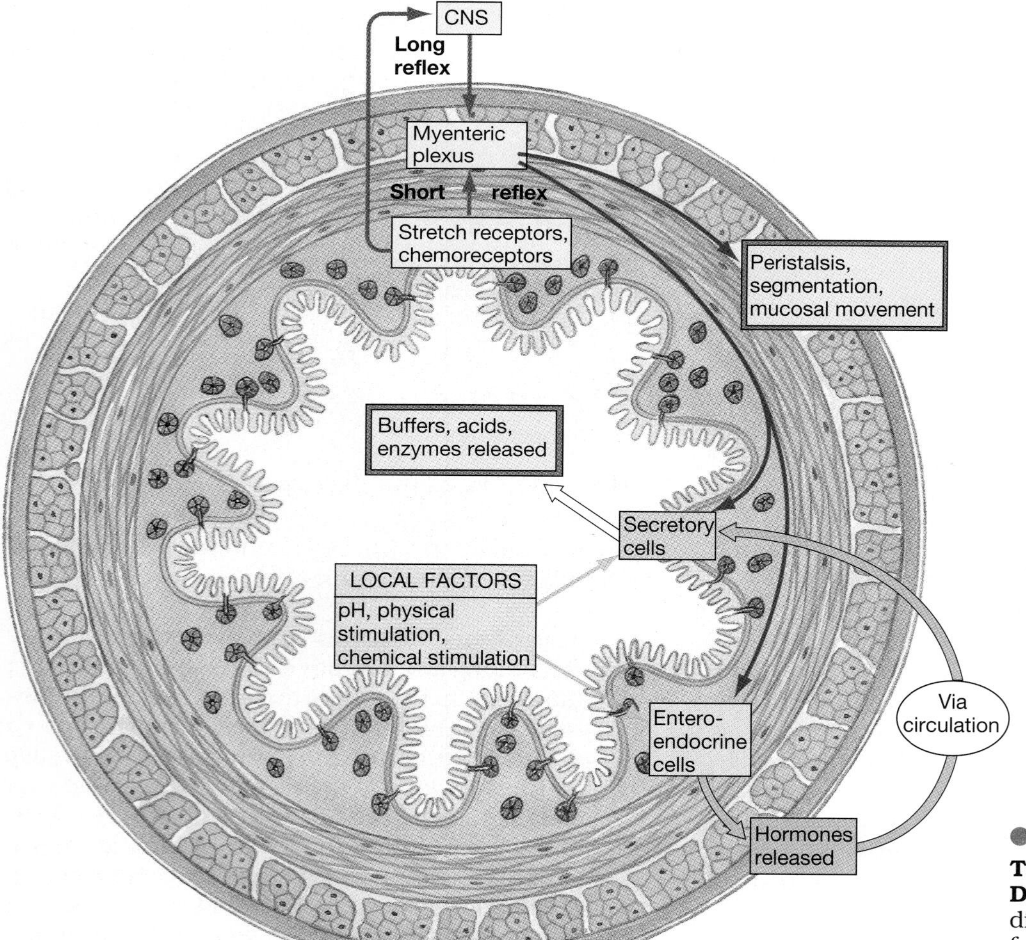
CNS
Long reflex
Myenteric plexus
Short reflex
Stretch receptors, chemoreceptors
Peristalsis, segmentation, mucosal movement
Buffers, acids, enzymes released
Secretory cells
LOCAL FACTORS
pH, physical stimulation, chemical stimulation
Via circulation
Entero-endocrine cells
Hormones released

● FIGURE 24-4
The Regulation of Digestive Activities. This diagram indicates the major factors responsible for regulating digestive activities.

heart failure can cause an increase in the rate of fluid movement through the peritoneal lining. The accumulation of fluid, called **ascites** (a-SĪ-tēz), creates a characteristic abdominal swelling. Distortion of internal organs by the contained fluid can result in a variety of symptoms; heartburn, indigestion, and low back pain are common complaints.

Mesenteries

Figures 24-2, 24-5, 24-13

Portions of the digestive tract are suspended within the peritoneal cavity by sheets of serous membrane that connect the parietal peritoneum with the visceral peritoneum. These **mesenteries** (MEZ-en-ter-ēz) are double sheets of peritoneal membrane (Figure 24-2●). The loose connective tissue between the mesothelial surfaces provides an access route for the passage of the blood vessels, nerves, and lymphatics to and from the digestive tract. Mesenteries also stabilize the positions of the attached organs and prevent the intestines from becoming entangled during digestive movements or sudden changes in body position.

During development, the digestive tract and accessory organs are suspended within the peritoneal cavity by dorsal and ventral mesenteries (Figure 24-5a●). The ventral mesentery later disappears along most of the digestive tract, persisting in the adult only (1) on the ventral surface of the stomach, between the stomach and liver (the *lesser omentum*), and (2) between the liver and the anterior abdominal wall (the *falciform ligament*). The lesser omentum stabilizes the position of the stomach and provides an access route for blood vessels and other structures entering or leaving the liver. The falciform ligament helps stabilize the position of the liver relative to the abdominal wall. (For additional information concerning the development of the digestive tract, accessory organs, and associated mesenteries see the Embryology Summary on pp. 886–887)

As the digestive tract elongates, it twists and turns within the relatively crowded peritoneal cavity. The dorsal mesentery of the stomach becomes greatly enlarged and forms an enormous pouch that extends inferiorly between the body wall and the anterior surface of the small intestine. This is the **greater omentum** (ō-MEN-tum; *omentum*, fat skin) (Figures 24-5c and 24-13●, p. 898), which hangs like an apron from the lateral and inferior borders of the stomach. Adipose tissue in the greater omentum conforms to the shapes of the surrounding organs, providing padding and protection across the anterior and lateral surfaces of the abdomen. The lipids in the adipose tissue represent an important energy reserve, and the greater omentum provides insulation that reduces heat loss across the anterior abdominal wall.

All but the first 25 cm of the small intestine is suspended by a thick **mesenterial sheet** that provides stability but permits a degree of independent movement. The mesentery associated with the initial portion of the small intestine (the *duodenum*) and the pancreas fuse with the posterior abdominal wall, locking these structures in position. Their anterior surfaces are covered by peritoneum, but the rest of the organs lie outside of the peritoneal cavity. Organs that lie behind, rather than within, the peritoneal cavity are called **retroperitoneal** (*retro*, behind).

A **mesocolon** is a mesentery associated with a portion of the large intestine. During development, the mesocolon of the *ascending colon*, the *descending colon*, and the *rectum* of the large intestine usually fuse to the body wall, and this fusion locks them in place. Thereafter these organs are retroperitoneal, with visceral peritoneum covering only their anterior surfaces and portions of their lateral surfaces (Figure 24-5b,c●).

PERITONITIS Inflammation of the peritoneal membrane produces symptoms of **peritonitis** (per-i-tō-NĪ-tis), a painful condition that interferes with the normal functioning of the affected organs. Physical damage, chemical irritation, or bacterial invasion of the peritoneum can lead to severe and even fatal cases of peritonitis. Peritonitis may be caused in untreated appendicitis by rupturing of the appendix and the release of bacteria into the peritoneal cavity. Peritonitis is always a potential complication of any surgery that involves opening the peritoneal cavity.

✓ What is the importance of the mesenteries?

✓ Would peristalsis or segmentation be more efficient in propelling intestinal contents from one place to another?

✓ What effect would a drug that blocks parasympathetic stimulation of the digestive tract have on peristalsis?

The Oral Cavity

Figure 24-6

Our exploration of the digestive tract will follow the path of food from the mouth to the anus. The mouth opens into the *oral cavity*. The functions of the oral cavity may be summarized as follows: (1) *analysis* of material before swallowing; (2) *mechanical processing* through the actions of the teeth, tongue, and palatal surfaces; (3) *lubrication* by mixing with mucus and salivary secretions; and (4) limited *digestion* of carbohydrates and lipids.

The oral cavity, or **buccal** (BUK-al) **cavity** (Figure 24-6●), is lined by the **oral mucosa,** which has a stratified squamous epithelium. Only the regions exposed to severe abrasion, such as the superior surface of the tongue and the opposing surfaces of

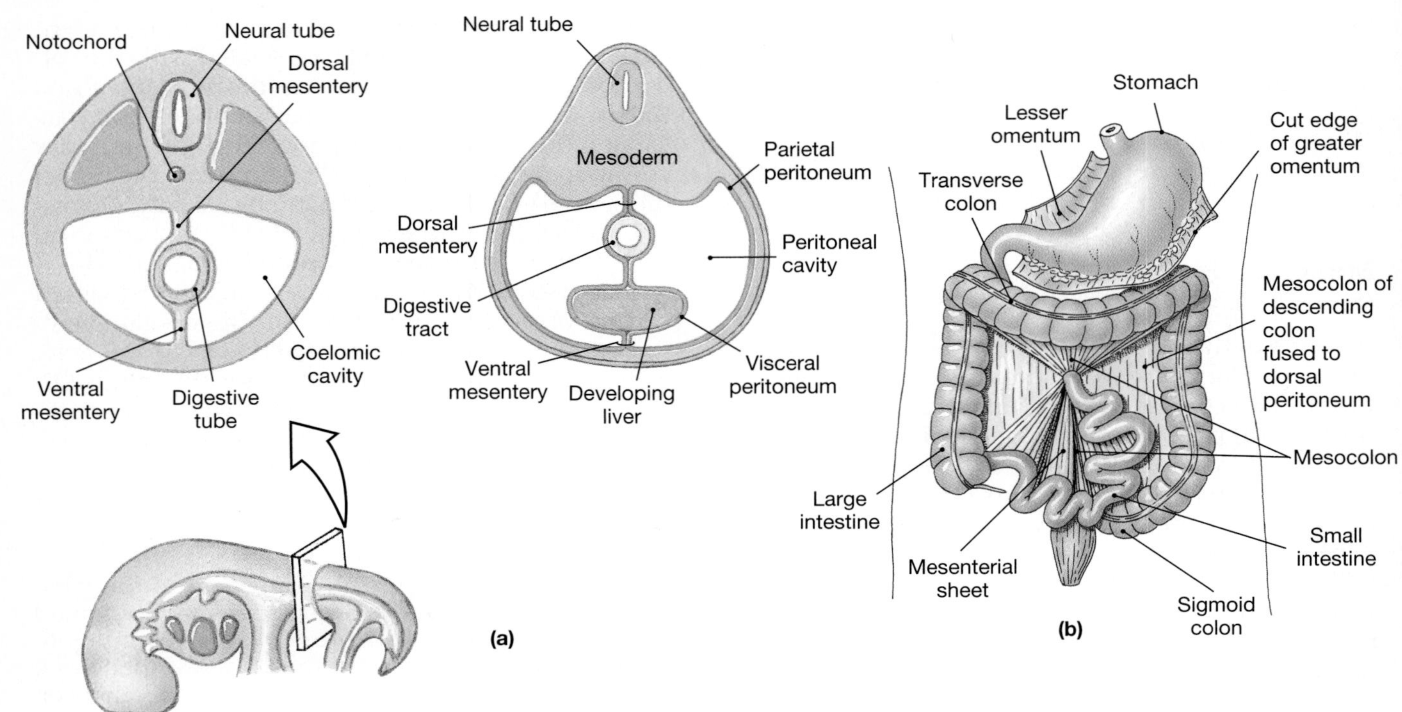
Notochord
Neural tube
Dorsal mesentery
Ventral mesentery
Digestive tube
Coelomic cavity
Neural tube
Mesoderm
Parietal peritoneum
Dorsal mesentery
Peritoneal cavity
Digestive tract
Ventral mesentery
Developing liver
Visceral peritoneum
(a)
Stomach
Lesser omentum
Cut edge of greater omentum
Transverse colon
Mesocolon of descending colon fused to dorsal peritoneum
Mesocolon
Large intestine
Small intestine
Mesenterial sheet
Sigmoid colon
(b)

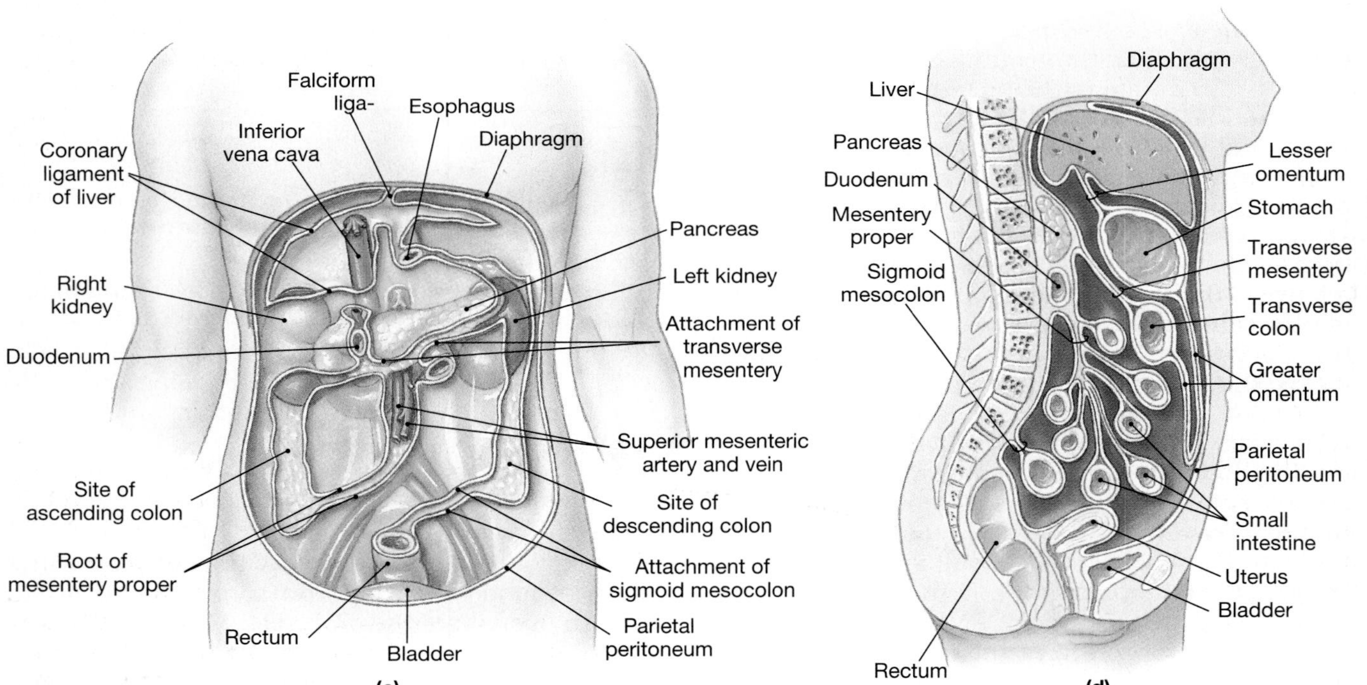
Falciform liga-
Esophagus
Inferior vena cava
Diaphragm
Coronary ligament of liver
Pancreas
Left kidney
Right kidney
Attachment of transverse mesentery
Duodenum
Superior mesenteric artery and vein
Site of ascending colon
Site of descending colon
Root of mesentery proper
Attachment of sigmoid mesocolon
Parietal peritoneum
Rectum
Bladder
(c)
Diaphragm
Liver
Pancreas
Lesser omentum
Duodenum
Stomach
Mesentery proper
Transverse mesentery
Sigmoid mesocolon
Transverse colon
Greater omentum
Parietal peritoneum
Small intestine
Uterus
Bladder
Rectum
(d)

● **FIGURE 24-5**

Mesenteries. (a) During development the digestive tube is initially suspended by dorsal and ventral mesenteries. The ventral mesentery in the adult is lost except where it connects the stomach to the liver *(lesser omentum)* and the liver to the anterior body wall and diaphragm *(falciform ligament)*. **(b)** As the digestive tract enlarges, mesenteries associated with the proximal portion of the small intestine, the pancreas, and the ascending and descending portions of the colon fuse to the body wall. **(c)** Anterior view of the empty peritoneal cavity, showing attachment sites where fusion occurs. **(d)** Mesenteries of the adult as seen in sagittal section. Note that the pancreas, duodenum, and rectum are retroperitoneal.

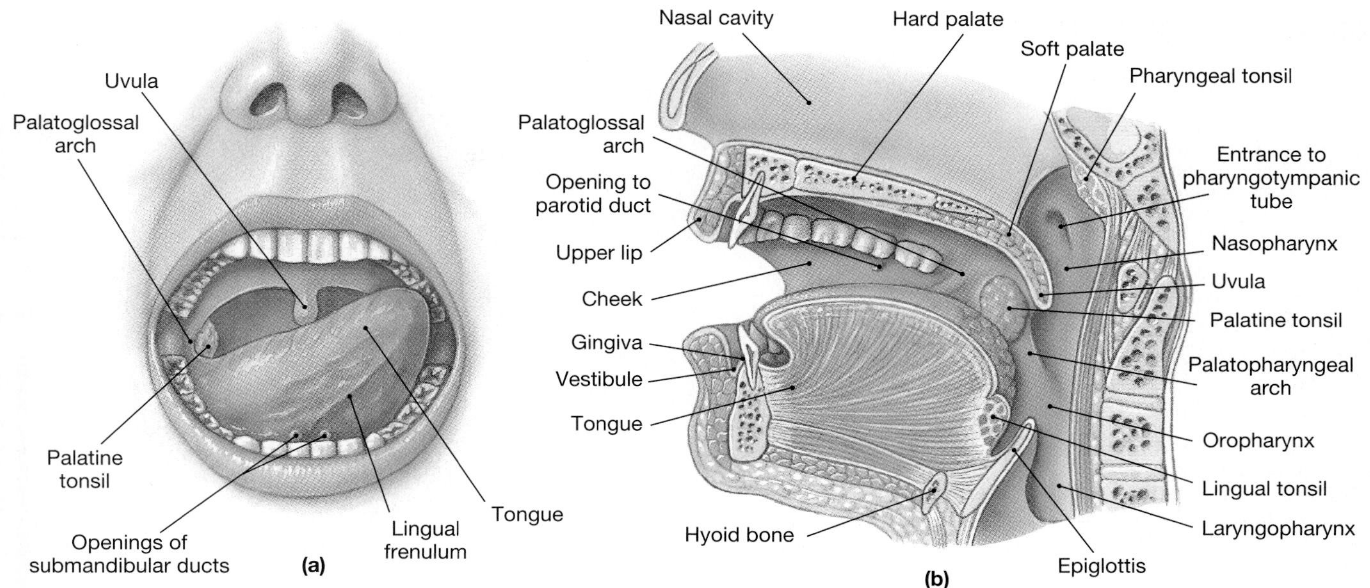
Uvula
Palatoglossal arch
Palatine tonsil
Openings of submandibular ducts
(a)
Lingual frenulum
Tongue
Nasal cavity
Hard palate
Soft palate
Pharyngeal tonsil
Palatoglossal arch
Opening to parotid duct
Upper lip
Cheek
Gingiva
Vestibule
Tongue
Hyoid bone
Entrance to pharyngotympanic tube
Nasopharynx
Uvula
Palatine tonsil
Palatopharyngeal arch
Oropharynx
Lingual tonsil
Laryngopharynx
Epiglottis
(b)

FIGURE 24-6
The Oral Cavity. (a) An anterior view of the oral cavity, as seen through the open mouth. **(b)** The oral cavity as seen in sagittal section. AM *Plate 4.4*

the hard palate, are covered by a layer of keratinized cells. The epithelial lining of the cheeks, lips, and undersurface of the tongue is relatively thin, nonkeratinized, and delicate. Although nutrient absorption does not occur in the oral cavity, the mucosa inferior to the tongue is thin enough and vascular enough to permit the rapid absorption of lipid-soluble drugs. *Nitroglycerin* is sometimes administered via this route to treat acute angina attacks.

The mucosae of the **cheeks,** or lateral walls of the oral cavity, are supported by pads of fat and the *buccinator muscles.* [p. 334] Anteriorly, the mucosa of each cheek is continuous with that of the lips, or **labia** (LĀ-bē-a; singular *labium*). The **vestibule** is the space between the cheeks (or lips) and the teeth. Ridges of oral mucosa, the gums, or **gingivae** (JIN-ji-vē), surround the base of each tooth on the alveolar surfaces of the maxilla and mandible. The gingivae in most regions are firmly bound to the periostea of the underlying bones.

The roof of the oral cavity is formed by the **hard** and **soft palates,** and the tongue dominates its floor (Figure 24-6b•). The floor of the mouth inferior to the tongue receives additional support from the *mylohyoid muscle.* [p. 341] The hard palate is formed by the palatine processes of the maxillary and palatine bones. A prominent central ridge, or *raphe,* extends along the midline of the hard palate. The mucosa lateral and anterior to the raphe is thick, with complex ridges. When the tongue compresses food against the hard palate, these ridges help provide traction. The soft palate lies posterior to the hard palate. A thinner and more delicate mucosa covers the posterior margin of the hard palate and extends onto the soft palate.

The posterior margin of the soft palate supports the dangling **uvula** (Ū-vū-lah) and two pairs of muscular **pharyngeal arches** (Figure 24-6•). On either side a palatine tonsil lies between an anterior **palatoglossal** (pal-a-tō-GLOS-al) **arch** and a posterior **palatopharyngeal** (pal-a-tō-fa-RIN-jē-al) **arch.** A curving line that connects the palatoglossal arches and uvula forms the boundaries of the **fauces** (FAW-sēz), the passageway between the oral cavity and the oropharynx.

The Tongue

Figure 24-6

The tongue (Figure 24-6•) manipulates materials inside the mouth and may occasionally be used to bring foods (such as ice cream) into the oral cavity. The primary functions of the tongue are: (1) mechanical processing by compression, abrasion, and distortion; (2) manipulation to assist in chewing and to prepare the material for swallowing; (3) sensory analysis by touch, temperature, and taste receptors, and (4) secretion of mucus and an enzyme, *lingual lipase.*

The tongue can be divided into an anterior **body,** or *oral portion,* and a posterior **root,** or *pharyngeal portion.* The superior surface, or **dorsum,** of the body contains a forest of fine *lingual papillae.* The structure of the lingual papillae was detailed in Chapter 17, in the discussion of taste buds and taste sensations. [p. 560] The thickened epithelium covering each papilla assists in the movement of materials by the tongue. A V-shaped line of circumvallate papillae roughly indicates the boundary between the body and the root of the tongue, which is situated in the pharynx.

EMBRYOLOGY SUMMARY

Development of the Digestive System

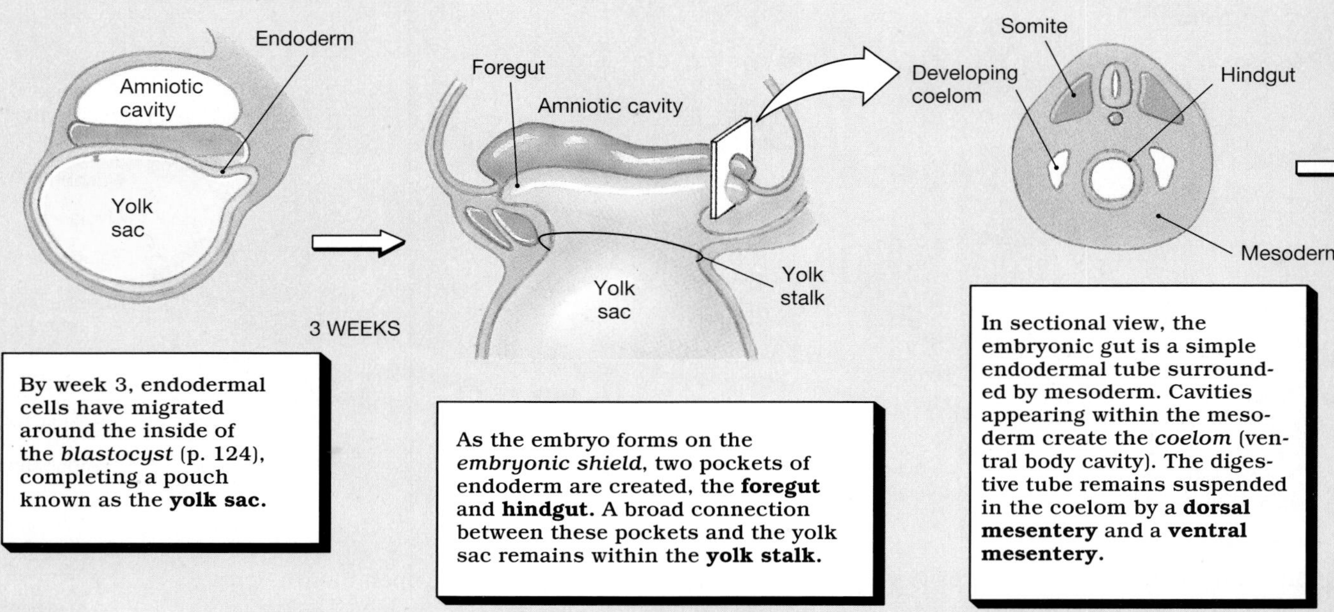

Endoderm
Amniotic cavity
Yolk sac
3 WEEKS
Foregut
Amniotic cavity
Yolk sac
Yolk stalk
Somite
Developing coelom
Hindgut
Mesoderm

By week 3, endodermal cells have migrated around the inside of the *blastocyst* (p. 124), completing a pouch known as the **yolk sac.**

As the embryo forms on the *embryonic shield*, two pockets of endoderm are created, the **foregut** and **hindgut.** A broad connection between these pockets and the yolk sac remains within the **yolk stalk.**

In sectional view, the embryonic gut is a simple endodermal tube surrounded by mesoderm. Cavities appearing within the mesoderm create the *coelom* (ventral body cavity). The digestive tube remains suspended in the coelom by a **dorsal mesentery** and a **ventral mesentery.**

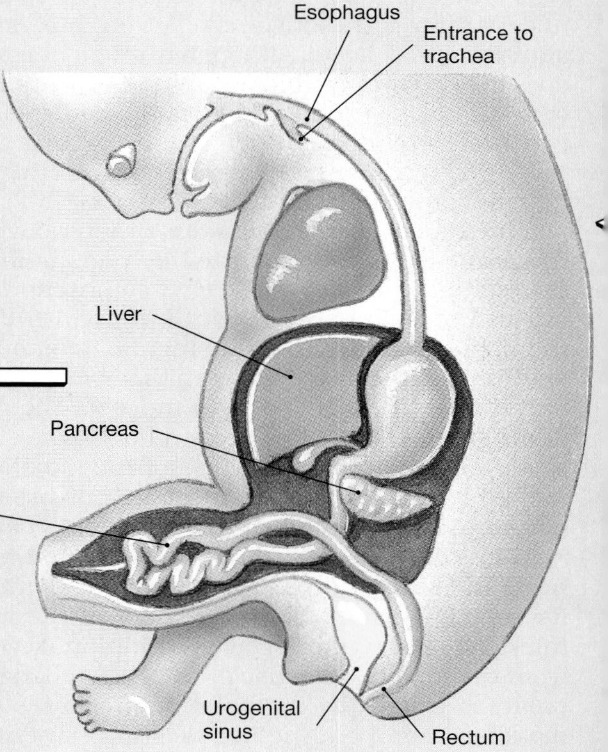

Esophagus
Entrance to trachea
Liver
Pancreas
Small intestine
Urogenital sinus
Rectum

8 WEEKS

A partition grows across the cloaca, dividing it into a posterior rectum and an anterior **urogenital sinus** that retains a connection to the allantois.

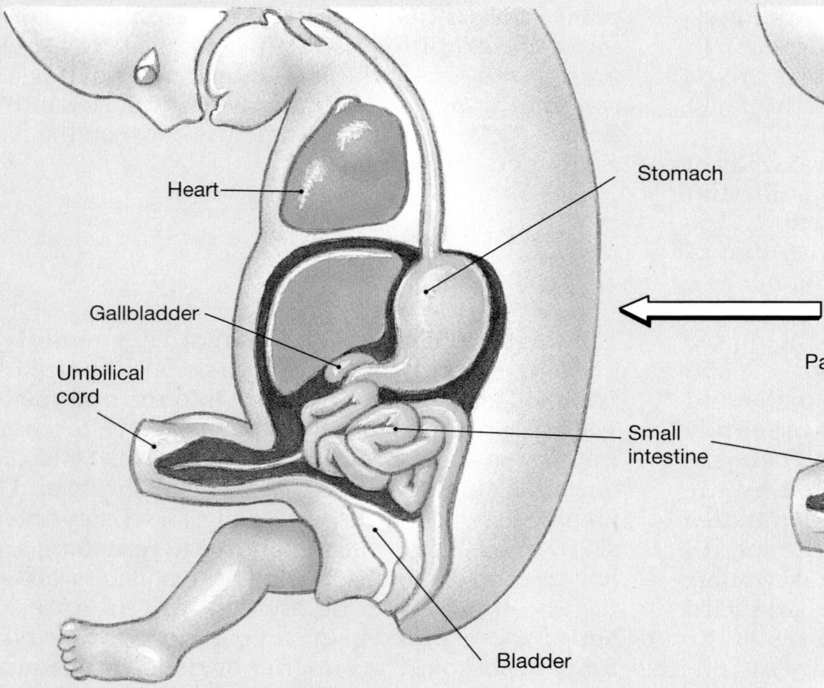

Heart
Stomach
Gallbladder
Umbilical cord
Small intestine
Bladder

10 WEEKS

By week 10, the intestines have begun moving back into the coelomic cavity, although they continue to grow longer.

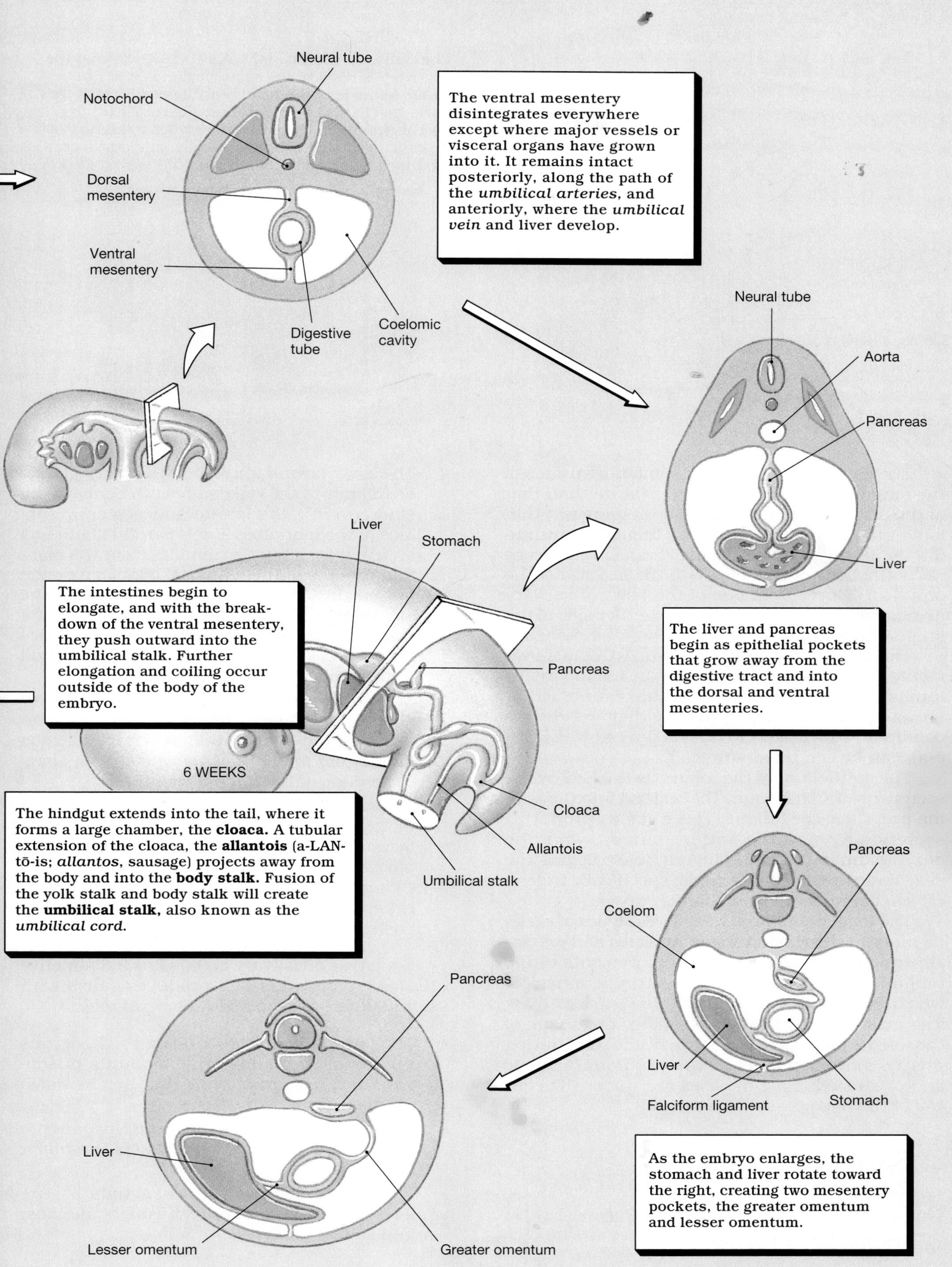

Neural tube
Notochord
Dorsal mesentery
Ventral mesentery
Digestive tube
Coelomic cavity
The ventral mesentery disintegrates everywhere except where major vessels or visceral organs have grown into it. It remains intact posteriorly, along the path of the *umbilical arteries,* and anteriorly, where the *umbilical vein* and liver develop.
Neural tube
Aorta
Pancreas
Liver
Liver
Stomach
The intestines begin to elongate, and with the breakdown of the ventral mesentery, they push outward into the umbilical stalk. Further elongation and coiling occur outside of the body of the embryo.
Pancreas
6 WEEKS
Cloaca
Allantois
Umbilical stalk
The liver and pancreas begin as epithelial pockets that grow away from the digestive tract and into the dorsal and ventral mesenteries.
The hindgut extends into the tail, where it forms a large chamber, the **cloaca.** A tubular extension of the cloaca, the **allantois** (a-LAN-tō-is; *allantos,* sausage) projects away from the body and into the **body stalk.** Fusion of the yolk stalk and body stalk will create the **umbilical stalk,** also known as the *umbilical cord.*
Pancreas
Coelom
Liver
Falciform ligament
Stomach
Pancreas
Liver
Lesser omentum
Greater omentum
As the embryo enlarges, the stomach and liver rotate toward the right, creating two mesentery pockets, the greater omentum and lesser omentum.

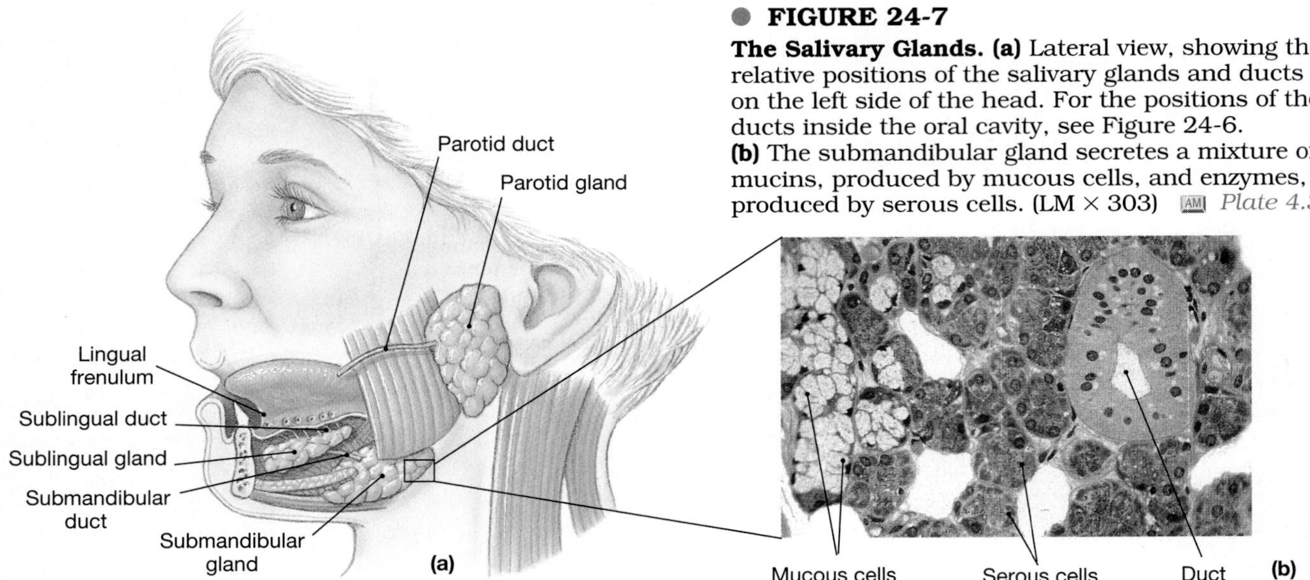

Parotid duct
Parotid gland
Lingual frenulum
Sublingual duct
Sublingual gland
Submandibular duct
Submandibular gland
(a)
Mucous cells
Serous cells
Duct
(b)

● **FIGURE 24-7**

The Salivary Glands. (a) Lateral view, showing the relative positions of the salivary glands and ducts on the left side of the head. For the positions of the ducts inside the oral cavity, see Figure 24-6.
(b) The submandibular gland secretes a mixture of mucins, produced by mucous cells, and enzymes, produced by serous cells. (LM × 303) AM *Plate 4.3*

The epithelium covering the inferior surface of the tongue is thinner and more delicate than that of the dorsum. Along the inferior midline is a thin fold of mucous membrane, the **lingual frenulum** (FREN-ū-lum; *frenulum*, a small bridle), which connects the body of the tongue to the mucosa of the oral floor. Ducts from one of the pairs of salivary glands are visible as they open on either side of the lingual frenulum.

The lingual frenulum prevents extreme movements of the tongue. However, if the lingual frenulum is *too* restrictive the individual cannot eat or speak normally. When properly diagnosed, this condition, called **ankyloglossia** (ang-ki-lō-GLOS-ē-a), can be corrected surgically.

The epithelium of the tongue is flushed by the secretions of small glands that extend into the lamina propria of the tongue. These glands produce a secretion containing water, mucus, and the enzyme **lingual lipase.** Lingual lipase begins the enzymatic breakdown of lipids, specifically triglycerides, before swallowing has occurred.

The tongue contains two different types of skeletal muscles, **intrinsic tongue muscles** and **extrinsic tongue muscles.** All gross movements of the tongue are performed by the extrinsic muscles, which were detailed in Chapter 11. ∞ *[p. 339]* The less massive intrinsic muscles alter the shape of the tongue and assist the extrinsic muscles during precise movements, as in speech. Both intrinsic and extrinsic tongue muscles are under the control of the hypoglossal nerve (N XII).

❑ Salivary Glands

Figures 24-6, 24-7

Three pairs of salivary glands (Figure 24-7●) secrete into the oral cavity:

1. The large **parotid** (pa-ROT-id) **salivary glands** lie inferior to the zygomatic arch beneath the skin covering the lateral and posterior surface of the mandible. Each parotid gland has an irregular shape, extending from the mastoid process of the temporal bone across the outer surface of the masseter muscle. The secretions of each gland are drained by a **parotid duct** (*Stensen's duct*), which empties into the vestibule at the level of the second upper molar.
2. The **sublingual** (sub-LING-gwal) **salivary glands** are located beneath the mucous membrane of the floor of the mouth. Numerous **sublingual ducts** *(Rivinus' ducts)* open along either side of the lingual frenulum.
3. The **submandibular salivary glands** are found in the floor of the mouth along the inner surfaces of the mandible within the *mandibular groove.* ∞ *[p. 216]* The **submandibular ducts** (*Wharton's ducts*) open into the mouth on either side of the lingual frenulum immediately posterior to the teeth (Figure 24-6a●, p. 885).

Each pair of salivary glands has a distinctive cellular organization and produces saliva with slightly different properties. For example:

- The parotid glands produce a thick, serous secretion containing large amounts of *salivary amylase,* an enzyme that breaks down complex carbohydrates.
- The sublingual salivary glands produce a watery, mucous secretion that acts as a buffer and lubricant.
- The submandibular salivary glands (Figure 24-7b●) secrete a mixture of buffers, mucins, and salivary amylase.

Saliva

The salivary glands produce 1.0–1.5 liters of saliva each day. Saliva is 99.4 percent water, and the remaining 0.6 percent includes an assortment of electrolytes (principally Na^+, Cl^-, and HCO_3^-), buffers, glycoproteins, antibodies, enzymes, and waste products. The glycoproteins, called **mucins,** are primarily responsible for the lubricating action of saliva. Saliva is a mixture of glandular secretions; about 70 percent of the saliva originates in the submandibular glands, 25 percent in the parotids, and 5 percent in the sublingual glands.

A continual background level of secretion flushes the oral surfaces, helping to keep them clean. Buffers within the saliva keep the pH of the mouth near 7.0 and prevent the buildup of acids produced through bacterial action. In addition, saliva contains immunoglobulins (IgA) and lysozymes that help to control populations of oral bacteria. A reduction or elimination of salivary secretions, caused by radiation exposure, emotional distress, or other factors, triggers a bacterial population explosion in the oral cavity. This proliferation rapidly leads to recurring infections and the progressive erosion of the teeth and gums.

The saliva produced at mealtimes has a variety of functions, including:

- Lubricating the mouth.
- Moistening and lubricating materials in the mouth.
- Dissolving chemicals that can stimulate the taste buds and provide sensory information about the material.
- Initiating the digestion of complex carbohydrates before the material is swallowed. The enzyme involved is **salivary amylase,** or *ptyalin.* Salivary amylase, also known as *alpha-amylase,* is secreted primarily by the parotid salivary glands, with the assistance of the submandibular salivary glands. Although the digestive process begins in the oral cavity, it is not completed, and there is no absorption of nutrients across the lining of the oral cavity.

MUMPS The mumps virus most often targets the salivary glands, especially the parotid salivary glands, although other organs may also become infected. Infection typically occurs at 5–9 years of age. The first exposure stimulates antibody production and usually confers permanent immunity; active immunity can be conferred by immunization. In postadolescent males the mumps virus may also infect the testes and cause sterility. Infection of the pancreas by the mumps virus may produce temporary or permanent diabetes; other organ systems, including the CNS, may be affected in severe cases. An effective mumps vaccine is available; widespread distribution has reduced the incidence of this disease in the United States.

Control of Salivary Secretion

Salivary secretions are usually controlled by the autonomic nervous system. Each gland receives parasympathetic and sympathetic innervation. The parasympathetic outflow originates in the **salivatory nuclei** of the medulla oblongata and synapses within the submandibular and otic ganglia. [p. 487] Any object placed within the mouth can trigger a salivary reflex by stimulating receptors monitored by the trigeminal nerve or by stimulating taste buds innervated by N VII, IX, or X. Parasympathetic stimulation accelerates secretion by all of the salivary glands, resulting in the production of large amounts of saliva. The role of the sympathetic innervation remains uncertain; some evidence exists that it provokes the secretion of small amounts of very thick saliva.

The salivatory nuclei are also influenced by other brain stem nuclei as well as by the activities of higher centers. For example, chewing with an empty mouth, the smell of food, or even thinking about food will initiate an increase in salivary secretion rates. The presence of irritating stimuli in the esophagus, stomach, or intestines will also accelerate saliva production, as will the sensation of nausea. In functional terms, increased saliva production in response to unpleasant stimuli helps to reduce the magnitude of the stimulus by dilution, a rinsing action, or by buffering strong acids or bases.

❑ The Teeth

Figure 24-8a

Movements of the tongue are important in passing food across the opposing surfaces of the teeth. The opposing, or **occlusal, surfaces** of the teeth perform chewing, or **mastication** (mas-ti-KĀ-shun), of food. Mastication breaks down tough connective tissues in meat and the plant fibers in vegetable matter, and it helps saturate the materials with salivary secretions and enzymes.

Figure 24-8a• is a sectional view through an adult tooth. The bulk of each tooth consists of a mineralized matrix similar to that of bone. This material, called **dentin** (DEN-tin), differs from bone in that it does not contain living cells. Instead, cytoplasmic processes extend into the dentin from cells in the central **pulp cavity.** The pulp cavity receives blood vessels and nerves via a narrow tunnel, the **root canal,** which is located at the base, or **root,** of the tooth. Blood vessels and nerves enter the root canal through the **apical foramen** to supply the pulp cavity.

The root of the tooth sits within a bony socket, or *alveolus.* Collagen fibers of the **periodontal ligament** extend from the dentin of the root to the alveolar bone, creating a strong articulation known as a *gomphosis.* [p. 263] A layer of **cementum** (se-MEN-tum) covers the dentin of the root, providing protection and firmly anchoring the periodontal ligament. Cementum is very similar in histological structure to bone, and it is less resistant to erosion than dentin.

The **neck** of the tooth marks the boundary between the root and the **crown.** The crown is the visible, exposed portion of the tooth. A shallow **gingival** (JIN-ji-val) **sulcus** surrounds the neck of each tooth. The mucosa of the gingival sulcus is very thin, and it is not tightly bound to the periosteum. The epithelium is bound to the tooth over an extensive area. This epithelial attachment prevents bacterial access to the lamina propria of the gingiva and the relatively soft cementum of the root. Brushing and gingival massage stimulate the epithelial cells and strengthen the attachment. If the epithelial attachment breaks down, bacterial infection of the gingivae can occur, a condition called **gingivitis.**

The dentin of the crown is covered by a layer of **enamel.** Enamel contains calcium phosphate in a crystalline form and is the hardest biologically manufactured substance. Adequate amounts of calcium, phosphates, and vitamin D during childhood are essential if the enamel coating is to be complete and resistant to decay.

Tooth decay usually results from the action of oral bacteria. Bacteria adhering to the surfaces of the teeth produce a sticky matrix that traps food particles and creates deposits of **plaque.** Over time this organic material can become calcified, forming a hard layer of *tartar,* or *dental calculus,* that can be difficult to remove. Tartar deposits most often develop at or near the gingival sulcus, where brushing cannot remove the relatively soft plaque deposits.

DENTAL PROBLEMS The mass of the plaque deposit protects the bacteria from salivary secretions, and as they digest nutrients the bacteria generate acids that erode the structure of the tooth. The results are *dental caries,* otherwise known as cavities. *Streptococcus mutans* is the most abundant bacteria at these sites, and vaccines are now being developed to promote resistance and prevent dental caries. If these bacteria reach the pulp and infect it, *pulpitis* (pul-PĪ-tis) results. Treatment usually involves the complete removal of the pulp tissue, especially the sensory innervation and all areas of decay, followed by packing the pulp cavity with appropriate materials. This procedure is called a *root canal.*

Brushing the exposed surfaces of the teeth after meals helps to prevent the settling of bacteria and the entrapment of food particles, but bacteria between the teeth, in the region known as the **interproximal space,** and within the gingival sulcus may elude the brush. Dentists therefore recommend the daily use of dental floss to clean these spaces and stimulate the gingival epithelium. If bacteria and plaque remain within the gingival sulcus for extended periods, the acids generated will begin eroding the connections between the neck of the tooth and the gingiva. The gums appear to recede from the teeth, and *periodontal disease* develops. As the disease progresses the bacteria attack the cementum, eventually destroying the periodontal ligament and eroding the alveolar bone. This deterioration loosens the tooth and it falls out or must be pulled. Periodontal disease is the most common cause for the loss of teeth.

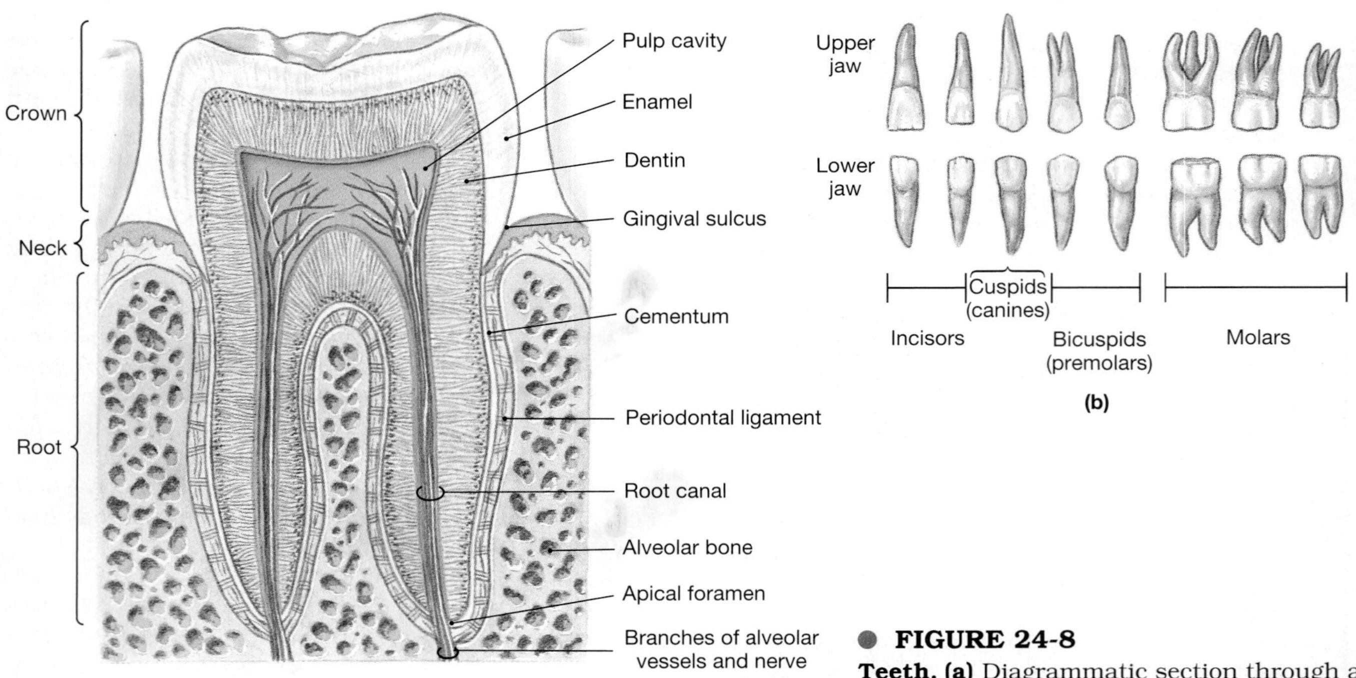
Crown
Neck
Root
Pulp cavity
Enamel
Dentin
Gingival sulcus
Cementum
Periodontal ligament
Root canal
Alveolar bone
Apical foramen
Branches of alveolar vessels and nerve
(a)
Upper jaw
Lower jaw
Cuspids (canines)
Incisors
Bicuspids (premolars)
Molars
(b)

FIGURE 24-8
Teeth. (a) Diagrammatic section through a typical adult tooth. **(b)** The adult teeth.

Types of Teeth

Figure 24-8b

There are four different types of teeth, each with specific functions (Figure 24-8b●):

1. **Incisors** (in-SĪ-zerz) are blade-shaped teeth found at the front of the mouth. Incisors are useful for clipping or cutting, as when nipping off the tip of a carrot stick.
2. The **cuspids** (KUS-pidz), or *canines*, are conical, with a sharp ridgeline and a pointed tip. They are used for tearing or slashing. A tough piece of celery might be weakened by the clipping action of the incisors, but then moved to one side to take advantage of the shearing action provided by the cuspids. Incisors and cuspids both have a single root.
3. **Bicuspids** (bī-KUS-pidz), or *premolars*, have flattened crowns with prominent ridges. They are used for crushing, mashing, and grinding. Bicuspids have one or two roots.
4. **Molars** have very large flattened crowns with prominent ridges. They excel at crushing and grinding. A tough nut or sparerib will usually be shifted to the bicuspids and molars for successful crunching. Molars typically have three or more roots.

Dental Succession

Figure 24-9

During development, two sets of teeth begin to form. The first to appear are the **deciduous teeth** (de-SID-ū-us; *deciduus*, falling off), also known as *primary teeth, milk teeth,* or *baby teeth.* There are usually 20 deciduous teeth, five on each side of the upper and lower jaws (Figure 24-9a●). These teeth will later be replaced by the adult **secondary dentition,** or *permanent dentition* (Figure 24-9b●). The larger adult jaws can accommodate more than 20 permanent teeth, and three additional molars appear on each side of the upper and lower jaws as the individual ages. These teeth extend the length of the tooth rows posteriorly and bring the permanent tooth count to 32.

On each side of the upper or lower jaw the primary dentition consists of two incisors, one cuspid, and a pair of deciduous molars. These are gradually replaced by the permanent dentition.

As replacement proceeds, the periodontal ligaments and roots of the primary teeth are eroded until eventually the teeth either fall out or are pushed aside by the emergence, or **eruption,** of the secondary teeth. The adult premolars take the place of the deciduous molars, and the definitive

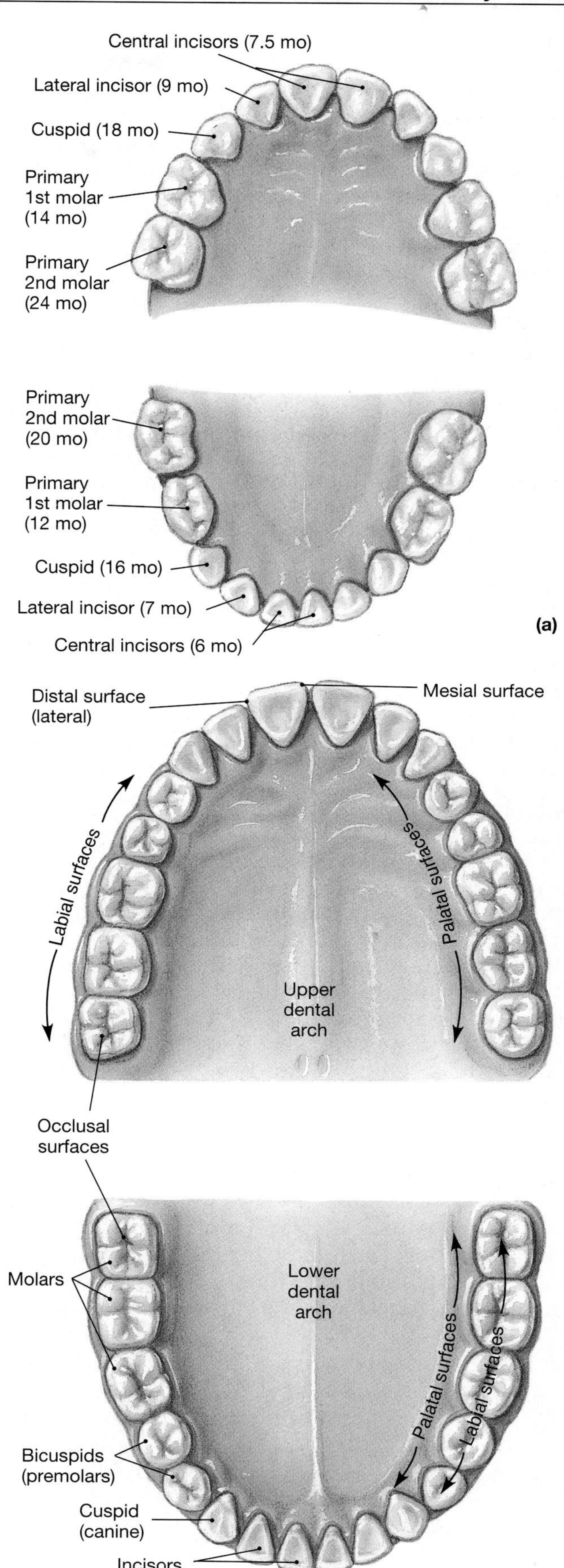

Central incisors (7.5 mo)
Lateral incisor (9 mo)
Cuspid (18 mo)
Primary 1st molar (14 mo)
Primary 2nd molar (24 mo)
Primary 2nd molar (20 mo)
Primary 1st molar (12 mo)
Cuspid (16 mo)
Lateral incisor (7 mo)
Central incisors (6 mo)
(a)
Distal surface (lateral)
Mesial surface
Labial surfaces
Palatal surfaces
Upper dental arch
Occlusal surfaces
Molars
Lower dental arch
Palatal surfaces
Labial surfaces
Bicuspids (premolars)
Cuspid (canine)
Incisors
(b)

● **FIGURE 24-9**

Primary and Secondary Teeth. (a) The primary teeth, with the age at eruption given in months. **(b)** The adult teeth, with the age at eruption given in years.

adult molars extend the tooth row as the jaw enlarges. The third molars, or *wisdom teeth*, may not erupt before age 21. Wisdom teeth fail to erupt because they either develop in inappropriate positions or because there is inadequate space on the dental arch. Any teeth that develop in locations that do not permit their eruption are called *impacted teeth*. Impacted wisdom teeth often will be surgically removed to prevent formation of abscesses in later life.

DENTAL IMPLANTS Lost or broken teeth are often replaced with "false teeth" attached to a plate or frame inserted into the mouth. Since the 1980s an alternative has been developed using dental implants. A ridged titanium cylinder is inserted into the alveolus, and osteoblasts lock the ridges into the surrounding bone. After 4–6 months, an artificial tooth is screwed into the cylinder. Dental implants are not suitable for everyone—for example, there must be enough alveolar bone present to provide a firm attachment—and there may be complications during or after surgery. Nevertheless, as the technique evolves, dental implants will become increasingly important. Roughly 42 percent of individuals over age 65 have lost all of their teeth; the rest have lost an average of 10 teeth.

Mastication

The *muscles of mastication* close the jaws and slide or rock the lower jaw from side to side. [p. 338] Chewing is not a simple process, as it can involve any combination of mandibular elevation, depression, protraction, retraction, and lateral excursion. (Try classifying the movements involved the next time you sit down to a meal.)

During mastication food is forced from the oral cavity to the vestibule and back, crossing and recrossing the occlusal surfaces. This movement results in part from the action of the masticatory muscles, but control would be impossible without the aid of the buccal, labial, and lingual muscles. Once the material has been shredded or torn to a satisfactory consistency and moistened with salivary secretions, the tongue begins compacting the debris into a small oval mass, or **bolus.** A compact, moist, cohesive bolus can be swallowed relatively easily.

What type of epithelium lines the oral cavity?

The digestion of which nutrient would be affected by damage to the parotid glands?

Which type of tooth is most useful for chopping off bits of relatively rigid foods?

The Pharynx

The pharynx serves as a common passageway for solid food, liquids, and air. The epithelial lining and divisions of the pharynx, the nasopharynx, the oropharynx, and the laryngopharynx were described in Chapter 23. [p. 831] Food normally passes through the oropharynx and laryngopharynx on its way to the esophagus. Both these divisions have a stratified squamous epithelium similar to that of the oral cavity, and the lamina propria contains scattered mucous glands and the lymphoid tissue of the pharyngeal, palatal, and lingual tonsils. Beneath the lamina propria lies a dense layer of elastic fibers, bound to the underlying skeletal muscles.

The specific pharyngeal muscles involved in swallowing were detailed in Chapter 11. [pp. 339–341] In summary:

- The *pharyngeal constrictors* push the bolus toward the esophagus.
- The *palatopharyngeus* and *stylopharyngeus* elevate the larynx.
- The *palatal muscles* raise the soft palate and adjacent portions of the pharyngeal wall.

These muscles cooperate with muscles of the oral cavity and esophagus to initiate the process of swallowing, or *deglutition*, which delivers the bolus along the esophagus and into the stomach.

The Esophagus

Figure 24-10

The esophagus (Figure 24-10•) is a hollow muscular tube with a length of approximately 25 cm (1 ft) and a diameter of about 2 cm (0.75 in.) at its widest point. The primary function of the esophagus is to carry solid food and liquids to the stomach.

The esophagus begins posterior to the cricoid cartilage, at the level of vertebra C_6. From this point, the narrowest portion of the esophagus, it descends toward the thoracic cavity posterior to the trachea, passes inferiorly along the dorsal wall of the mediastinum, and enters the abdominopelvic cavity through an opening in the diaphragm, the **esophageal hiatus** (hī-Ā-tus). It then empties into the stomach anterior to vertebra T_7.

The esophagus receives blood from the esophageal arteries and from branches of (1) the *thyrocervical trunk* and *external carotid arteries* of the neck, (2) the *bronchial arteries* of the mediastinum, and (3) the *celiac artery* and *inferior phrenic arteries* of the abdomen. Blood from the esophageal capillaries collects into the *inferior thyroid, azygos,* and *gastric veins.* [pp. 618, 763, 767]

The esophagus is innervated by the vagus and sympathetic trunks via the *esophageal plexus.*

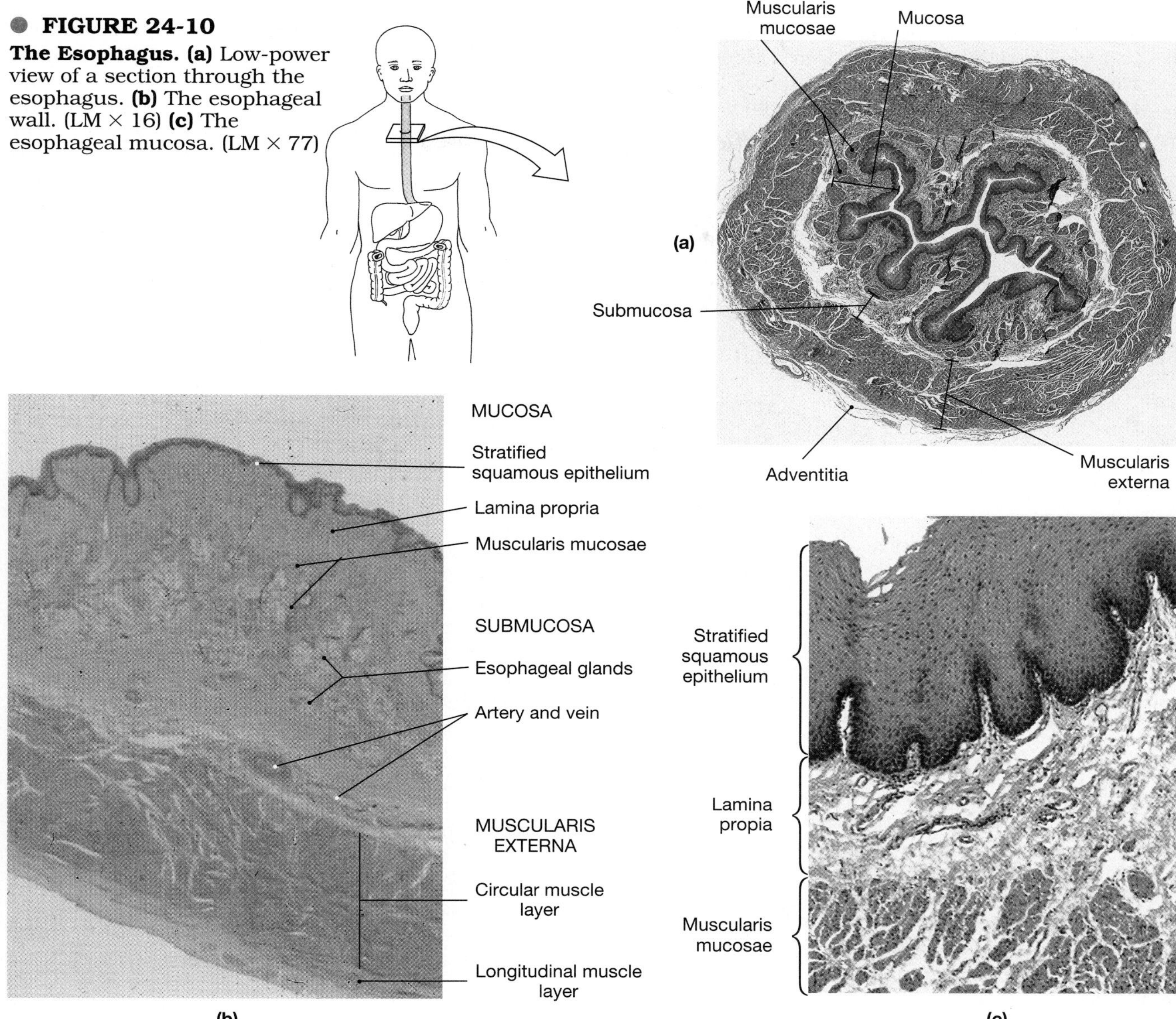
Muscularis mucosae
Mucosa
(a)
Submucosa
Adventitia
Muscularis externa
MUCOSA
Stratified squamous epithelium
Lamina propria
Muscularis mucosae
SUBMUCOSA
Esophageal glands
Artery and vein
MUSCULARIS EXTERNA
Circular muscle layer
Longitudinal muscle layer
(b)
Stratified squamous epithelium
Lamina propria
Muscularis mucosae
(c)

● **FIGURE 24-10**
The Esophagus. (a) Low-power view of a section through the esophagus. **(b)** The esophageal wall. (LM × 16) **(c)** The esophageal mucosa. (LM × 77)

[p. 541] Resting tension in the circular muscle layer in the upper 3 cm (1 in.) of the esophagus normally prevents the entry of air into the esophagus. A comparable zone at the inferior end of the esophagus normally remains in a state of active contraction. This condition prevents the backflow of materials from the stomach into the esophagus. Neither region contains a well-defined sphincter muscle comparable to those found elsewhere along the digestive tract. Nevertheless, the terms *upper esophageal sphincter* and *lower esophageal sphincter (cardiac sphincter)* are often used in recognition of the similarity in function.

Histology of the Esophagus

Figures 24-2, 24-10

The wall of the esophagus contains mucosal, submucosal, and muscularis layers comparable to those described in Figure 24-2●, p. 880. Distinctive features of the esophageal wall (Figure 24-10●) include:

- The mucosa of the esophagus contains a stratified squamous epithelium similar to that of the pharynx and oral cavity.
- The mucosa and submucosa are thrown into large folds that extend the length of the esophagus. These folds allow for expansion during the passage of a large bolus; except during swallowing, muscle tone in the walls keeps the lumen closed.
- The muscularis mucosae consists of an irregular layer of smooth muscle.
- The submucosa contains scattered *esophageal glands* that produce a mucous secretion that reduces friction between the bolus and the esophageal lining.
- The muscularis externa has inner circular and outer longitudinal layers. In the upper third of

the esophagus these layers contain skeletal muscle fibers; in the middle third there is a mixture of skeletal and smooth muscle tissue; along the lower third only smooth muscle is found.

- There is no serosa, and an adventitia of connective tissue outside of the muscularis externa anchors the esophagus in position against the dorsal body wall.

ESOPHAGEAL VARICES The veins draining the inferior portion of the esophagus empty into tributaries of the hepatic portal vein. If the venous pressure in the hepatic portal vein becomes abnormally high, because of liver damage or constriction of the vessels, blood will pool in the submucosal veins of the esophagus. The veins become grossly distended and create bulges in the esophageal wall. These distorted *esophageal varices* (VAR-i-sēz; *varices*, dilated veins) may constrict the esophageal passageway. They may also rupture, bleeding into the submucosal tissues or onto the epithelial surface. Esophageal varices often develop in individuals with advanced *cirrhosis*, a chronic liver disorder that restricts hepatic blood flow.

❑ Swallowing

Figure 24-11

Swallowing, or *deglutition*, is a complex process whose initiation can be voluntarily controlled, but it proceeds automatically once it begins. Although we are consciously aware of, and voluntarily control, swallowing when eating or drinking, swallowing can also occur unconsciously, as saliva collects at the back of the mouth. Each day you swallow approximately 2400 times.

Swallowing can be divided into *buccal, pharyngeal,* and *esophageal phases.*

1. The **buccal phase** begins with the compression of the bolus against the hard palate. Subsequent retraction of the tongue then forces the bolus into the pharynx and assists in the elevation of the soft palate, thereby isolating the nasopharynx (Figure 24-11a,b•). The buccal phase is strictly voluntary; once the bolus enters the oropharynx, involuntary reflexes are initiated, and the bolus is moved toward the stomach.
2. The **pharyngeal phase** begins as the bolus comes in contact with the palatal arches or the posterior pharyngeal wall or both (Figure 24-11c,d•). The **swallowing reflex** begins as tactile receptors on the palatal arches and uvula are stimulated by the passage of the bolus. The information is relayed to the **swallowing center** of the medulla oblongata over the trigeminal and glossopharyngeal nerves. Motor commands originating at this center then target the pharyngeal musculature, producing a coordinated and stereotyped pattern of muscle contraction. Elevation of the larynx and folding of the epiglottis direct the bolus past the closed glottis, while the uvula and soft palate block passage to the nasopharynx. It takes less than a second for the pharyngeal muscles to propel the bolus into the esophagus, and over this period the respiratory centers are inhibited, and breathing stops.
3. The **esophageal phase** of swallowing (Figure 24-11e–h•) begins as the contraction of pharyngeal muscles forces the bolus through the entrance to the esophagus. Once within the esophagus, the bolus is pushed toward the stomach by a peristaltic wave. The approach of the bolus triggers the opening of the lower esophageal sphincter, and the bolus then continues into the stomach.

Primary peristaltic contractions are coordinated by afferent and efferent fibers within the glossopharyngeal and vagus nerves. For a typical bolus the entire trip takes about 9 seconds to complete. Liquids may make the journey in a few seconds, arriving ahead of the peristaltic contractions with the assistance of gravity.

A relatively dry or poorly lubricated bolus travels much more slowly, and a series of **secondary peristaltic waves** may be required to push it all the way to the stomach. Secondary peristaltic waves are local reflexes triggered by stimulation of sensory receptors in the esophageal walls. These receptors relay information via the submucosal and myenteric plexuses, producing peristaltic contractions in the absence of CNS instructions. A completely dry bolus cannot be swallowed at all, for friction with the walls of the esophagus will make peristalsis ineffective. (For this reason it is not possible to swallow an entire slice of processed white bread without a drink.)

ACHALASIA AND ESOPHAGITIS In the condition known as *achalasia* (ak-a-LĀ-zē-a), a bolus descends relatively slowly, and its arrival does not cause the opening of the lower esophageal sphincter. Materials then accumulate at the base of the esophagus like cars at a stop light. Secondary peristaltic waves may occur repeatedly, adding to the individual's discomfort. The most successful treatment involves cutting the circular muscle layer at the base of the esophagus or expanding a balloon in the lower esophagus until the muscle layer tears.

A weakened or permanently relaxed sphincter can cause inflammation of the esophagus, or *esophagitis* (ē-sof-a-JĪ-tis), as powerful gastric acids enter the lower esophagus. The esophageal epithelium has few defenses against acid and enzyme attack, and inflammation, epithelial erosion, and intense discomfort are the result. Occasional incidents of reflux, or backflow, from the stomach are responsible for the symptoms of "heartburn." This relatively common problem supports a multimillion dollar industry devoted to producing and promoting antacids.

 What is unusual about the muscularis externa of the esophagus?

 Where would you find the fauces?

 What is occurring when the soft palate and larynx elevate and the glottis closes?

The Stomach

The stomach performs four major functions: (1) the bulk storage of ingested food, (2) the mechanical breakdown of ingested food, (3) the disruption of chemical bonds through the action of acids and enzymes, and (4) the production of *intrinsic factor*, a glycoprotein whose presence in the digestive tract is required for the absorption of vitamin B_{12}. The mixing of ingested substances with the gastric juices secreted by the glands of the stomach produces a viscous, highly acid, soupy mixture of partially digested food. This material is called **chyme** (kīm).

Anatomy of the Stomach

Figure 24-12

The stomach has the shape of an expanded J (Figure 24-12●). A short **lesser curvature** forms the **medial surface** of the organ and a long **greater curvature** forms the **lateral surface.** The **anterior** and **posterior surfaces** are smoothly rounded. The shape and size of the stomach are extremely variable from individual to individual and from one meal to the next. In an "average" stomach the lesser curvature has a length of approximately 10 cm (4 in.), and the greater curvature measures around 40 cm (16 in.). The stomach typically extends between the levels of vertebrae T_7 and L_3.

The stomach is divided into four regions (Figure 24-12●):

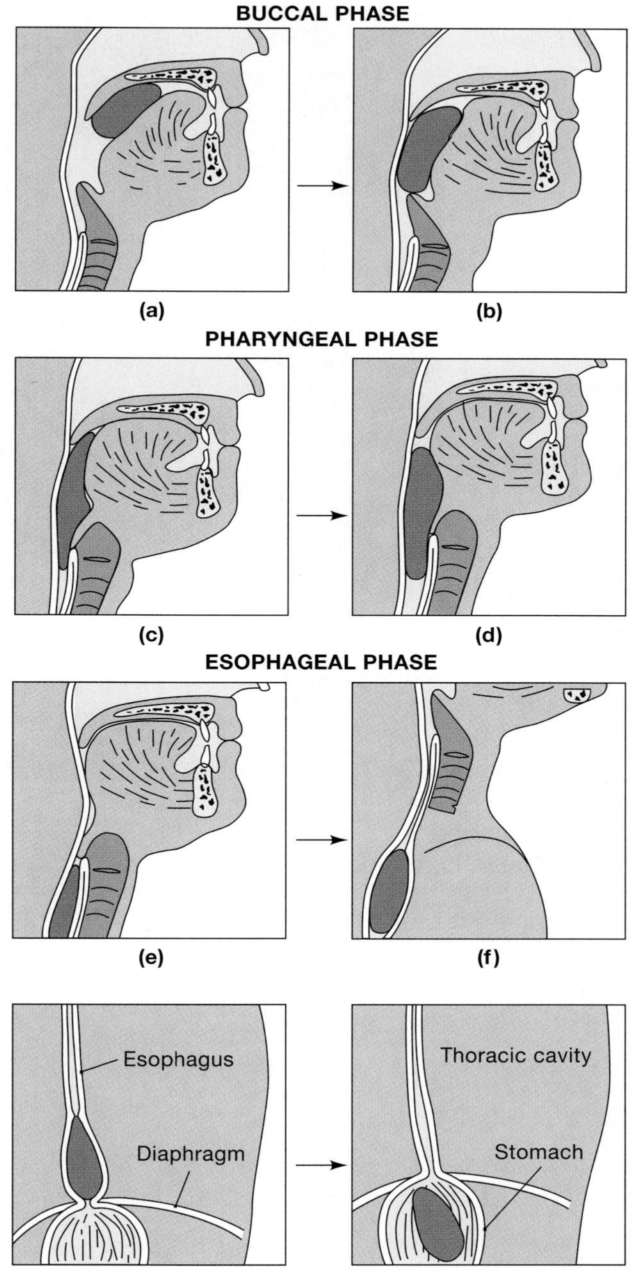

BUCCAL PHASE
(a)
(b)
PHARYNGEAL PHASE
(c)
(d)
ESOPHAGEAL PHASE
(e)
(f)
Esophagus
Diaphragm
Thoracic cavity
Stomach
(g)
(h)

● **FIGURE 24-11**
The Swallowing Process. This sequence, based on a series of X-rays, shows the stages of swallowing and the movement of materials from the mouth to the stomach.

1. ***The cardia.*** The **cardia** (KAR-dē-a) is the smallest part of the stomach. It consists of the superior and medial portion of the stomach within 3 cm (1.2 in.) of the junction with the esophagus. The cardia contains abundant mucous glands whose secretions coat the connection with the esophagus and help protect it from the acids and enzymes within the stomach.
2. ***The fundus.*** The portion of the stomach superior to the gastroesophageal junction is the **fundus** (FUN-dus). The fundus contacts the inferior and posterior surface of the diaphragm.
3. ***The body.*** The area between the fundus and the curve of the J is the **body** of the stomach. The body is the largest region of the stomach, and it functions as a mixing tank for ingested food and gastric secretions. *Gastric glands* in the fundus and body secrete most of the acids and enzymes involved in gastric digestion.
4. ***The pylorus.*** The **pylorus** (pī-LŌR-us) is the curve of the J. The pylorus is divided into a **pyloric antrum** (*antron*, cavity), which is connected to the body, and a **pyloric canal** that empties into the *duodenum*, the proximal segment of the small intestine. As mixing movements occur during digestion, the pylorus frequently changes shape. A muscular **pyloric sphincter** regulates the release of chyme into the duodenum. Glands in the pylorus secrete mucus and important

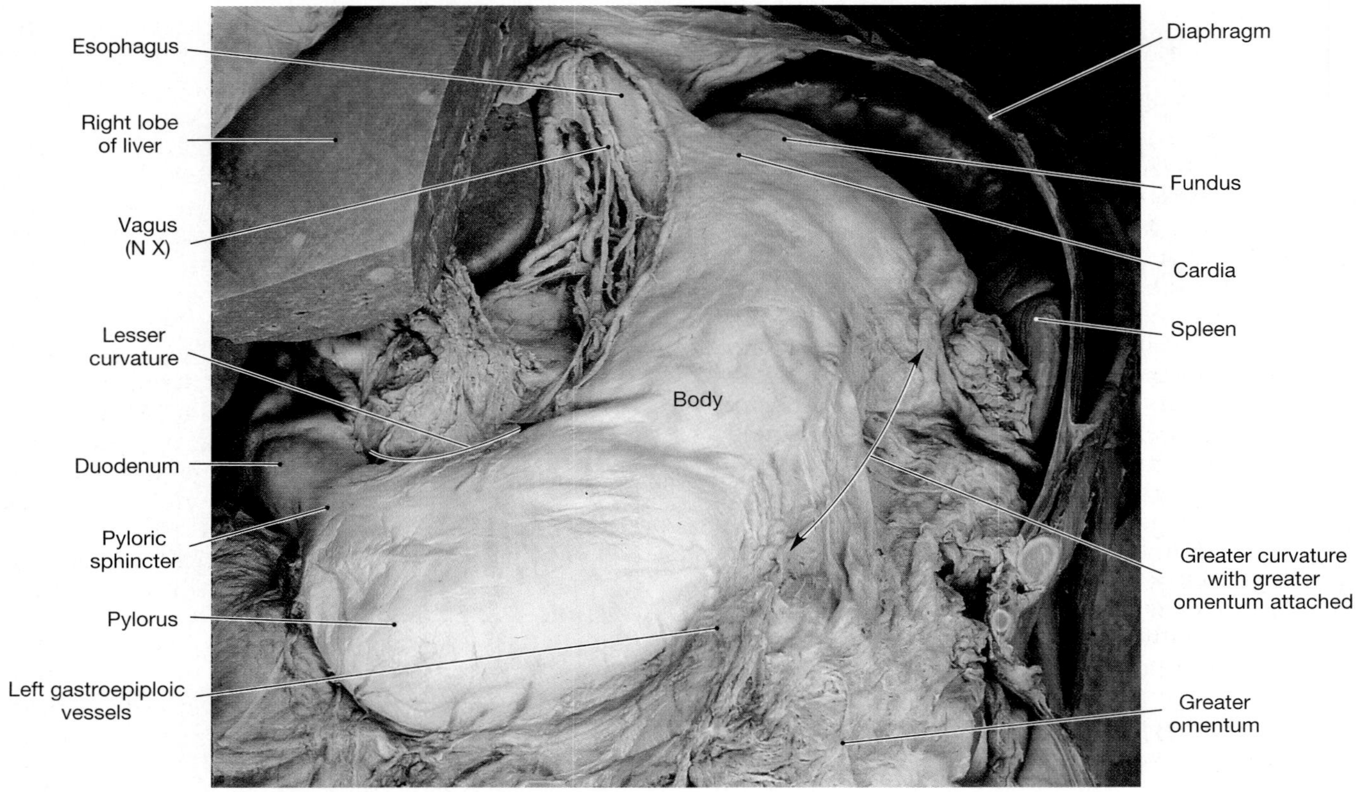

Esophagus
Right lobe of liver
Vagus (N X)
Lesser curvature
Duodenum
Pyloric sphincter
Pylorus
Left gastroepiploic vessels
Diaphragm
Fundus
Cardia
Spleen
Body
Greater curvature with greater omentum attached
Greater omentum
(a)
Esophagus
Fundus
Anterior surface
Cardia
Longitudinal muscle layer
Circular muscle layer
Pyloric sphincter
Lesser curvature (medial surface)
Duodenum
Body
Left gastroepiploic vessels
Oblique muscle layer overlying mucosa
Greater curvature (lateral surface)
Pylorus
Rugae
(b)

● **FIGURE 24-12**

The Stomach. (a) Position and external appearance of the stomach, showing superficial landmarks. **(b)** Diagrammatic view of the structure of the stomach wall. AM *Plate 6.5; Scans 17, 18*

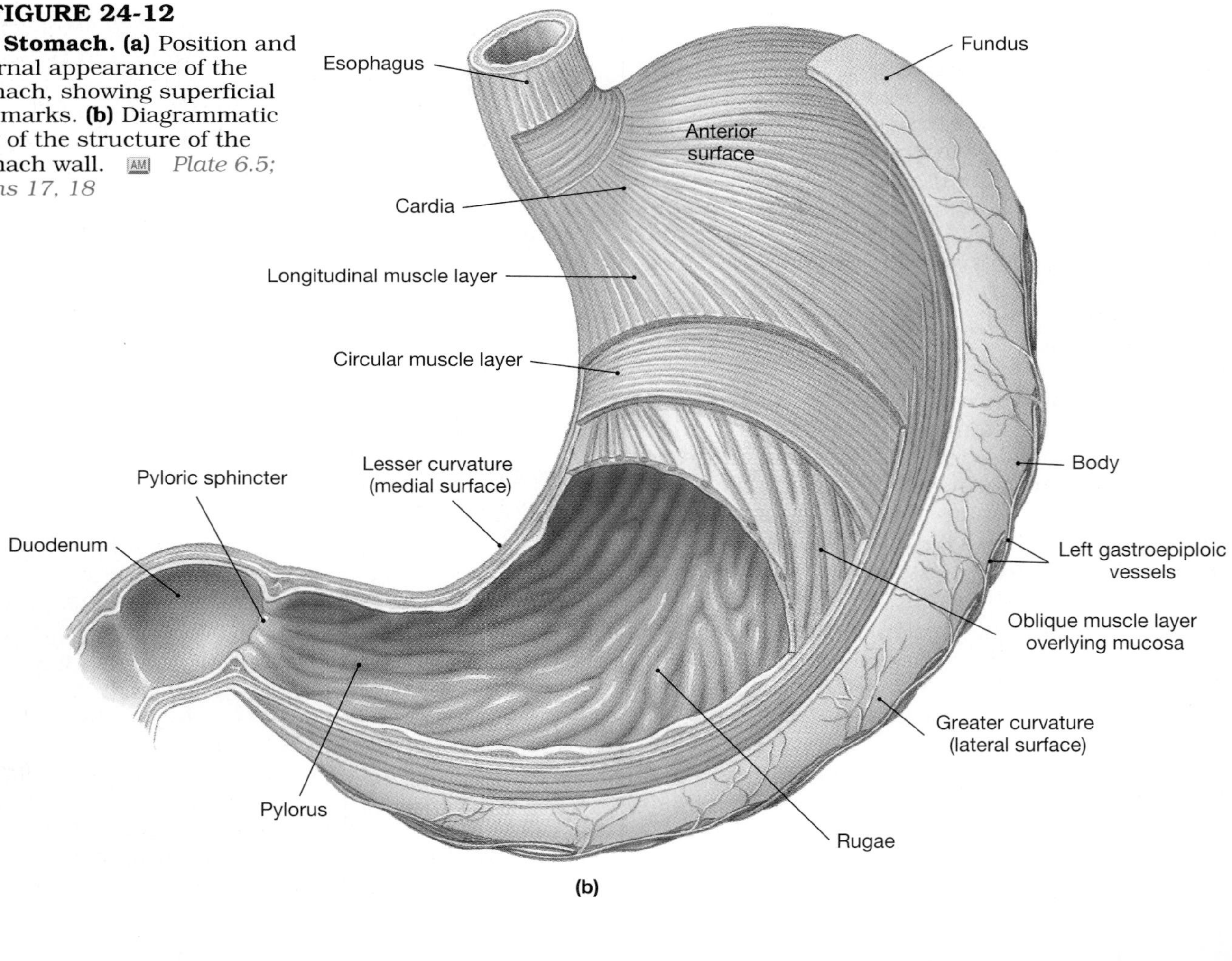

digestive hormones, including *gastrin,* a hormone that stimulates the activity of gastric glands.

The volume of the stomach increases at mealtimes and then decreases as chyme enters the small intestine. In the relaxed (empty) stomach the mucosa is thrown into a number of prominent longitudinal folds, called **rugae** (ROO-gē; wrinkles) (Figure 24-12b●). Rugae are temporary features of the mucosa that permit expansion of the gastric lumen. As the stomach fills, the rugae flatten out. In a full stomach, the rugae have almost disappeared. When empty, the stomach resembles a muscular tube with a narrow and constricted lumen. When full, it can expand to contain 1–1.5 liters of material.

Musculature of the Stomach

Figure 24-12b

The muscularis mucosae and muscularis externa of the stomach contain extra layers of smooth muscle fibers in addition to the usual circular and longitudinal layers. The muscularis mucosae usually contains an outer, circular layer of muscle fibers. The muscularis externa has an inner, **oblique layer** of smooth muscle (Figure 24-12b●). The extra layer of smooth muscle strengthens the stomach wall and assists in the mixing and churning activities essential to the formation of chyme.

Histology of the Stomach

Figures 24-13, 24-14

A simple columnar epithelium lines all portions of the stomach. The epithelium is a *secretory sheet* that produces a carpet of mucus that covers the interior surfaces of the stomach. The alkaline mucous layer provides protection against the acids and enzymes in the gastric lumen.

The stomach receives blood from (1) the *left gastric artery,* (2) the *splenic artery,* which supplies the *left gastroepiploic artery,* and (3) the *hepatic artery,* which supplies the *right gastroepiploic artery.* These vessels were detailed in Chapter 21. ∞ *[pp. 756–758]*

Shallow depressions, called **gastric pits,** open onto the gastric surface (Figure 24-13a,b●). The mucous cells at the base, or *neck,* of each gastric pit are actively dividing, replacing superficial cells that are shed into the chyme. The continual replacement of epithelial cells provides an additional defense against the gastric contents. A typical epithelial cell has a life span of 3–7 days, but exposure to strong alcohol or other chemicals can increase the rate of turnover.

GASTRIC GLANDS In the fundus and body of the stomach, each gastric pit communicates with several **gastric glands** that extend deep into the underlying lamina propria. Gastric glands (Figure 24-13b,c,d●) are dominated by two types of secretory cells: *parietal cells* and *chief cells.* Together they secrete about 1500 ml of **gastric juice** each day.

Parietal Cells. **Parietal cells** are especially common along the proximal portions of each gastric gland. These cells secrete *intrinsic factor* and *hydrochloric acid* (HCl). **Intrinsic factor** facilitates the absorption of **vitamin B_{12}** across the intestinal lining. This vitamin, essential for normal erythropoiesis, was introduced in Chapter 19. ∞ *[p. 659]*

The parietal cells do not produce HCl in the cytoplasm, because it is such a strong acid that it would erode a secretory vesicle and destroy the cell. Instead, the two ions are transported independently, by different mechanisms (Figure 24-14●). Hydrogen ions are generated inside the cell, as the enzyme carbonic anhydrase converts carbon dioxide and water to carbonic acid, which promptly dissociates into hydrogen ions and bicarbonate ions. The hydrogen ions are actively transported into the lumen of the gastric gland. The bicarbonate ions are ejected into the interstitial fluid by a countertransport mechanism that exchanges intracellular bicarbonate ions for extracellular chloride ions. The chloride ions then diffuse across the cell and through open chloride channels into the lumen of the gastric gland.

The bicarbonate ions released by the parietal cell diffuse through the interstitial fluid into the bloodstream. When gastric glands are actively secreting, enough bicarbonate ions enter the circulation to increase the pH of the blood significantly. This sudden influx of bicarbonate ions has been called the *alkaline tide.*

The secretory activities of the parietal cells can keep the stomach contents at a pH of 1.5–2.0. This highly acid environment does not by itself digest the chyme, but it has several important functions:

- The low pH of gastric juice kills most of the microorganisms ingested with food.
- The low pH denatures proteins and inactivates most of the enzymes in food.
- The acid helps break down plant cell walls and the connective tissues in meat.
- An acid environment is essential for the activation and function of *pepsin,* a protease (protein-digesting enzyme) secreted by the chief cells.

Chief Cells. **Chief cells** are most abundant near the base of a gastric gland. These cells secrete an inactive proenzyme, **pepsinogen** (pep-SIN-ō-jen), that is converted by the acid in the gastric lumen to an active proteolytic enzyme, **pepsin.** Pepsin functions most effectively at a strongly acid pH of 1.5–2.0. The stomachs of newborn infants also produce **rennin** and **gastric lipase,** enzymes important for the digestion of milk. Rennin coag-

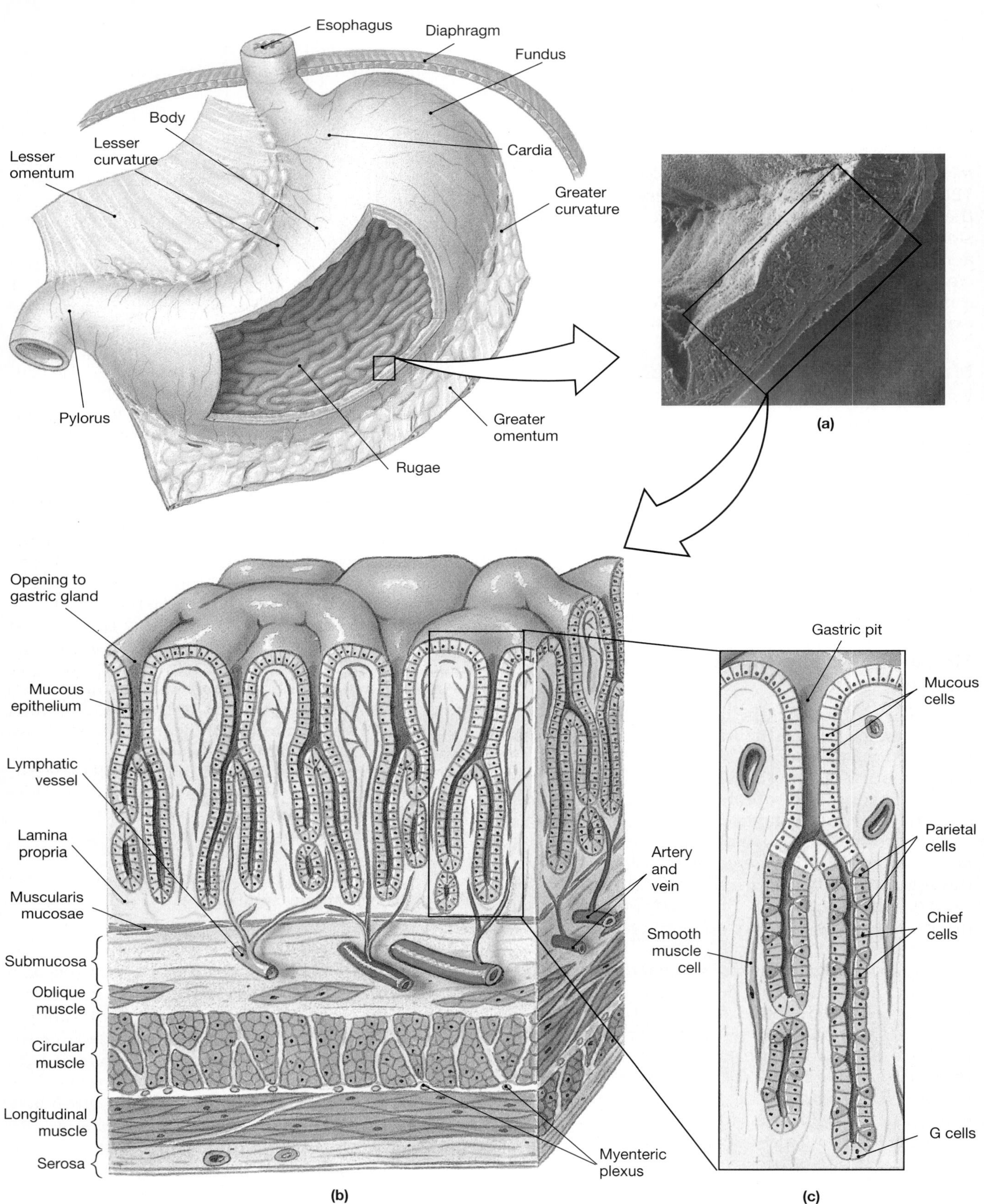

Esophagus
Diaphragm
Fundus
Body
Lesser curvature
Lesser omentum
Cardia
Greater curvature
Pylorus
Greater omentum
Rugae
(a)
Opening to gastric gland
Mucous epithelium
Lymphatic vessel
Lamina propria
Muscularis mucosae
Submucosa
Oblique muscle
Circular muscle
Longitudinal muscle
Serosa
Artery and vein
Myenteric plexus
(b)
Gastric pit
Mucous cells
Parietal cells
Chief cells
Smooth muscle cell
G cells
(c)

FIGURE 24-13

The Stomach Lining. (a) SEM of the empty stomach. (SEM × 35) **(b)** Diagrammatic view of the organization of the stomach wall. **(c)** A gastric gland. **(d)** The mucous epithelium lining the stomach cavity. (LM × 113)

ulates milk proteins, and gastric lipase initiates the digestion of milk fats.

PYLORIC GLANDS Glands in the pylorus produce primarily a mucous secretion, rather than enzymes and acid. In addition, several different types of enteroendocrine cells are scattered among the mucus-secreting cells. These cells produce at least seven different hormones, most notably the hormone **gastrin** (GAS-trin). Gastrin is produced by *G cells.* G cells are found in the gastric glands, but they are most abundant in the gastric pits of the pyloric antrum. Gastrin stimulates (1) the secretion of both parietal and chief cells and (2) contractions of the gastric wall that mix and stir the gastric contents. The pyloric glands also contain *D cells* that release **somatostatin,** a hormone that inhibits gastrin release. D cells are continually releasing their secretions into the interstitial fluid adjacent to the G cells. This inhibition of gastrin production can be overpowered by neural and hormonal stimuli when the stomach is preparing for digestion or actually is engaged in the digestion of food.

GASTRITIS AND PEPTIC ULCERS Inflammation of the gastric mucosa is called **gastritis** (gas-TRĪ-tis). This condition may develop after swallowing drugs, including alcohol and aspirin. Gastritis may also appear after severe emotional or physical stress, bacterial infection of the gastric wall, or the ingestion of strongly acid or alkaline chemicals.

A **peptic ulcer** develops when the digestive acids and enzymes manage to erode their way through the defenses of the stomach lining or proximal portions of the small intestine. The locations may be indicated by using the terms **gastric ulcer** (stomach) or **duodenal ulcer** (duodenum). Peptic ulcers result from the excessive production of acid or the inadequate production of the alkaline mucus that poses an epithelial defense. For the last decade, drugs such as *cimetidine (Tagamet)* have been used to inhibit acid production by parietal cells. Recent evidence suggests that infections involving the bacterium *Helicobacter pylori* are responsible for most peptic ulcers, and treatment for gastric ulcers today often involves the administration of antibiotic drugs. AM *Peptic Ulcers*

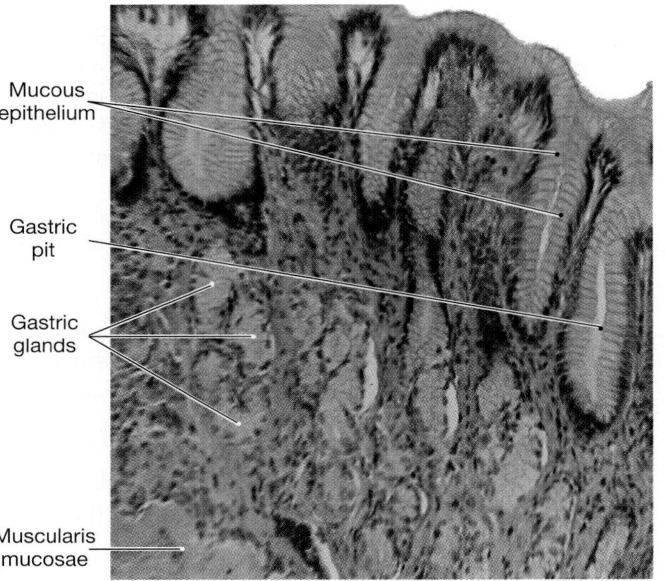
Mucous epithelium
Gastric pit
Gastric glands
Muscularis mucosae

(d)

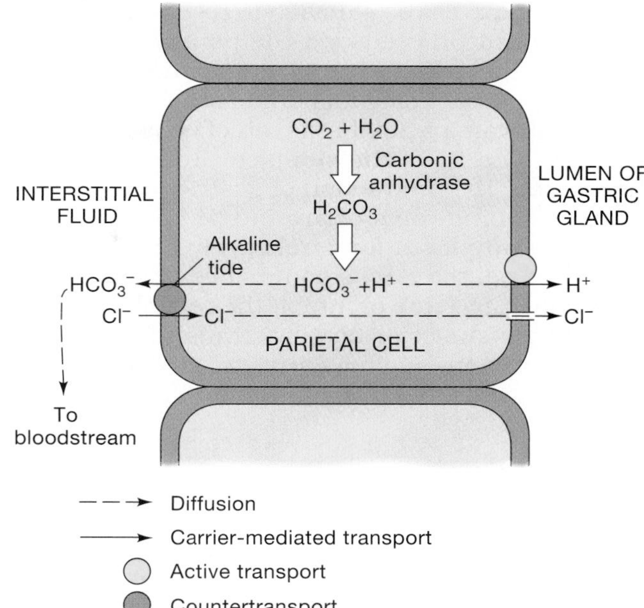
$CO_2 + H_2O$
Carbonic anhydrase
H_2CO_3
INTERSTITIAL FLUID
LUMEN OF GASTRIC GLAND
Alkaline tide
HCO_3^-
$HCO_3^- + H^+$
H^+
Cl^-
Cl^-
Cl^-
PARIETAL CELL
To bloodstream
Diffusion
Carrier-mediated transport
Active transport
Countertransport

FIGURE 24-14
Gastric Acid Secretion. An active parietal cell generates hydrogen ions through the dissociation of carbonic acid within the cell. The bicarbonate is exchanged for chloride ions in the interstitial fluid, and the chloride ions diffuse into the lumen of the gastric gland as the hydrogen ions are transported out of the cell.

Regulation of Gastric Activity

Figure 24-15

The production of acid and enzymes by the gastric mucosa can be (1) controlled by the central nervous system, (2) regulated by short reflexes coordinated in the wall of the stomach, and (3) regulated by digestive tract hormones. Several stages of gastric control can be identified, although considerable overlap exists among them. These stages are summarized in Figure 24-15●.

The Cephalic Phase

Figure 24-15a

The **cephalic phase** of gastric secretion (Figure 24-15a●) begins with the sight, smell, taste, or thought of food. This stage, which is directed by the CNS, prepares the stomach to receive food. The neural output proceeds via the parasympathetic division of the ANS and reaches the stomach over the vagus nerve. Postganglionic parasympathetic fibers innervate parietal cells, chief cells, mucous cells, and G cells of the stomach. In response to stimulation, the production of gastric juice accelerates, reaching rates of around 500 ml/h. This phase usually lasts for a relatively brief period before the *gastric phase* commences. Emotional states can exaggerate or inhibit the cephalic phase. For example, anger or hostility leads to excessive gastric secretion, whereas anxiety, stress, or fear decreases gastric secretion and motility.

The Gastric Phase

Figure 24-15b

The **gastric phase** (Figure 24-15b●) begins with the arrival of food in the stomach and builds on the stimulation provided during the cephalic phase. The stimuli that initiate the gastric phase are (1) distension of the stomach, (2) an increase in the pH of the gastric contents, and (3) the presence of undigested materials in the stomach, especially proteins and peptides.

- *Neural response.* Stimulation of stretch receptors in the stomach wall and chemoreceptors in the mucosa trigger short reflexes coordinated in the myenteric plexus. The motor neurons involved in these reflexes are parasympathetic ganglionic neurons, the same ones innervated by the vagus nerve. The postganglionic fibers innervate parietal cells and chief cells, and the release of ACh stimulates their secretion. Proteins, alcohol in small doses, and caffeine enhance gastric secretion markedly by stimulating chemoreceptors in the gastric lining.
- *Hormonal response.* Neural stimulation and the presence of peptides and amino acids in the chyme stimulate the secretion of gastrin, primarily by G cells of the pyloric antrum. Gastrin entering the interstitial fluid of the stomach must enter capillaries and complete a round-trip of the circulation before stimulating parietal and chief cells of the fundus and body. Both parietal and chief cells respond to the presence of gastrin by accelerating their rates of secretion. The effect on the parietal cells is the most pronounced, and the pH of the gastric juice declines as a result. Gastrin also stimulates gastric motility.
- *Local response.* Distortion of the gastric wall also stimulates the release of histamine in the lamina propria. The source of the histamine is thought to be mast cells within the connective tissue of this layer. Histamine binds to receptors on the parietal cells and stimulates acid secretion.

DURATION OF THE GASTRIC PHASE The gastric phase may continue for several hours while the ingested materials are processed by the acids and enzymes. Over this period, gastrin stimulates contractions in the muscularis externa of the stomach and intestinal tract. The effects are strongest in the stomach, where stretch receptors are stimulated as well. The initial contractions are weak pulsations in the gastric walls. These *mixing waves* occur several times per minute, and they gradually increase in intensity. After an hour, the material within the stomach is churning around like the clothing in a washing machine.

At the time the contractions begin, the pH of the gastric contents is high, and only the material in contact with the gastric epithelium is exposed to undiluted digestive acids and enzymes. As mixing occurs, the acid is diluted, and the pH remains elevated until a large volume of gastric juice has been secreted and the contents thoroughly mixed. This process usually takes several hours. As the pH throughout the chyme reaches 1.5–2.0, and the amount of undigested protein decreases, gastrin production declines, as do the rates of acid and enzyme secretion by parietal and chief cells.

The Intestinal Phase

Figure 24-15c

The **intestinal phase** of gastric secretion (Figure 24-15c●) begins when chyme starts to enter the small intestine. The intestinal phase usually starts after several hours of mixing contractions, when waves of contraction begin sweeping down the length of the stomach. Each time the pylorus contracts, a small quantity of chyme squirts through the pyloric sphincter. The purpose of the intestinal phase is to control the rate of gastric emptying and ensure that the secretory, digestive, and absorptive functions of the small intestine can proceed with reasonable efficiency. Although we will consider the intestinal phase now as it affects stomach activity, the arrival of chyme in the small intestine also triggers other neural and hormonal events that coordinate the activities of the intestinal tract and the pancreas, liver, and gallbladder.

The intestinal phase involves a combination of neural and hormonal responses.

- *Neural response.* Chyme leaving the stomach relieves some of the distension in the stomach wall, thus reducing the stimulation of stretch receptors. At the same time, distension of the duodenum by acidic chyme stimulates stretch receptors and chemoreceptors that trigger the **enterogastric reflex.** The enterogastric reflex is a short reflex that temporarily inhibits gastrin production and gastric contractions. The net result is that immediately after chyme enters the small intestine, gastric contractions decrease in

THE PHASES OF GASTRIC SECRETION

Function:
Preparation of stomach for arrival of food
Duration:
Short (minutes)
Mechanism:
Neural, via preganglionic fibers in vagus nerve and synapses in myenteric plexus
Actions:
Primary: increased volume of gastric juice by stimulating mucus, acid, and enzyme production
Secondary: stimulates release of gastrin by G cells

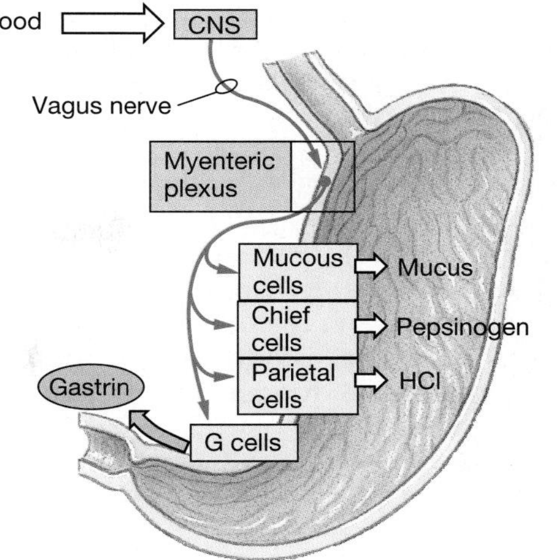
Sight, smell, thoughts of food
CNS
Vagus nerve
Myenteric plexus
Mucous cells
Mucus
Chief cells
Pepsinogen
Parietal cells
HCl
Gastrin
G cells

(a) The Cephalic Phase

Functions:
Enhance secretion started in cephalic stage; homogenization and acidification of chyme; initiate digestion of proteins by pepsin
Duration:
Long (3-4 hours)
Mechanisms:
Neural: short reflexes triggered by
(1) stimulation of stretch receptors as stomach fills
(2) stimulation of chemoreceptors as pH increases
Hormonal: stimulation of gastrin release by G cells by parasympathetic activity and presence of peptides and amino acids in chyme
Local: release of histamine by mast cells as stomach fills (not shown)
Actions:
Increased acid and pepsinogen production, increased motility and initiation of mixing waves

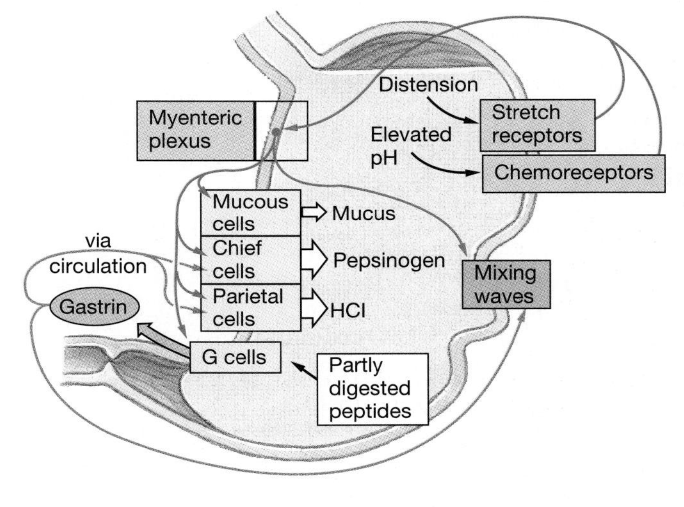
Distension
Myenteric plexus
Stretch receptors
Elevated pH
Chemoreceptors
Mucous cells
Mucus
via circulation
Chief cells
Pepsinogen
Mixing waves
Gastrin
Parietal cells
HCl
G cells
Partly digested peptides

(b) The Gastric Phase

Functions:
Control rate of chyme entry into duodenum
Duration:
Long (hours)
Mechanisms:
Neural: short reflexes (enterogastric reflex) triggered by extension of duodenum
Hormonal:
Primary: stimulation of GIP, secretin, CCK release by presence of acid, carbohydrates, and lipids
Secondary: release of gastrin stimulated by presence of undigested proteins and peptides (not shown)
Actions:
Feedback inhibition of gastric acid and pepsinogen production, reduction of gastric motility

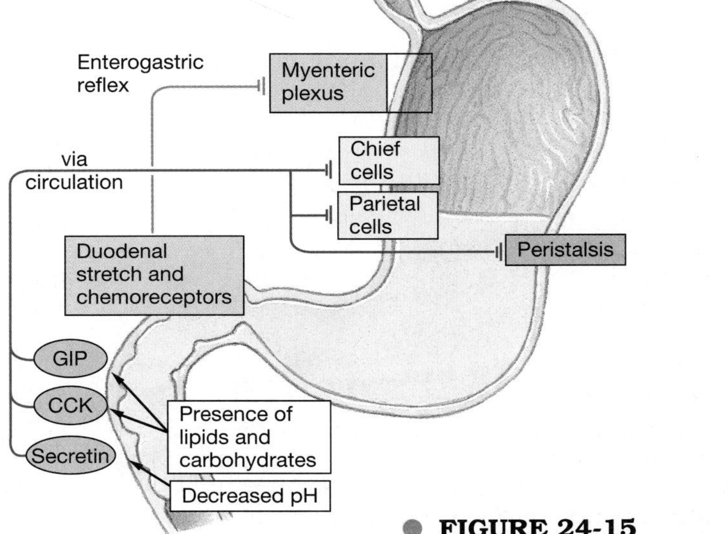
Enterogastric reflex
Myenteric plexus
via circulation
Chief cells
Parietal cells
Peristalsis
Duodenal stretch and chemoreceptors
GIP
CCK
Secretin
Presence of lipids and carbohydrates
Decreased pH

(c) The Intestinal Phase

FIGURE 24-15
The Phases of Gastric Secretion

strength and frequency. Thus the duodenum has time to deal with the arriving acids before the next wave of gastric contraction occurs.

- ***Hormonal response.*** Several hormonal responses are triggered by the arrival of acid chyme in the duodenum.

    1. A drop in pH below 4.5 stimulates secretion of *secretin* (se-KRĒ-tin) by enteroendocrine cells of the duodenum. Secretin inhibits parietal cell and chief cell activity in the stomach. It also targets the pancreas, where it stimulates the production of buffers that will protect the duodenum by neutralizing the acid in chyme.
    2. The arrival of lipids (especially triglycerides and fatty acids) and carbohydrates in the duodenum stimulates the secretion of the hormones *cholecystokinin* (kō-lē-sis-tō-KĪ-nin), or CCK, and *gastric inhibitory peptide* (GIP). CCK has multiple effects on the digestive system, including inhibition of gastric secretion of acids and enzymes. GIP, which also targets the pancreas, inhibits gastric secretion and reduces the rate and force of gastric contractions. As a result, a meal high in fats stays in the stomach longer and enters the duodenum at a more leisurely pace than does a meal low in fats. The extra time in the stomach allows more time for lipid digestion and absorption in the small intestine.
    3. The arrival of partially digested proteins in the duodenum stimulates G cells in the duodenal wall. These cells secrete gastrin, which circulates to the stomach and accelerates acid and enzyme production. In effect, this is a feedback mechanism that regulates the amount of gastric processing to meet the requirements of a specific meal.

In general, the rate of chyme movement into the small intestine is highest when the stomach is greatly distended and the meal contains relatively little protein. A large meal containing small amounts of protein, large amounts of carbohydrates (such as rice or pasta), wine (alcohol), and after-dinner coffee (caffeine) will leave the stomach extremely quickly because both alcohol and caffeine stimulate gastric secretion and motility.

❑ Digestion and Absorption in the Stomach

The stomach performs preliminary digestion of proteins by pepsin and, for a variable period, permits the digestion of carbohydrates and lipids by salivary amylase and lingual lipase. Until the pH throughout the material in the stomach falls below 4.5, the salivary amylase and lingual lipase continue to digest carbohydrates and lipids in the meal. These enzymes usually remain active 1–2 hours after a meal.

As the stomach contents become more fluid and the pH approaches 2.0, pepsin activity increases and protein disassembly begins. Protein digestion is not completed in the stomach, because (1) time is limited, and (2) pepsin attacks only specific types of peptide bonds, not all peptide bonds. However, there is usually enough time for pepsin to break down complex proteins into smaller peptide and polypeptide chains before the chyme enters the duodenum.

Although digestion does occur in the stomach, there is no nutrient absorption there because (1) the epithelial cells are covered by a blanket of alkaline mucus and are not directly exposed to the chyme, (2) the epithelial cells lack the specialized transport mechanisms found in cells lining the small intestine, (3) the gastric lining is impermeable to water, and (4) digestion has not proceeded to completion by the time chyme leaves the stomach. At this stage, most carbohydrates, lipids, and proteins are only partially broken down.

Some drugs can be absorbed in the stomach. For example, ethyl alcohol can diffuse through the mucous barrier and penetrate the lipid membranes of the epithelial cells. As a result, alcohol can be absorbed in the stomach before any nutrients in a meal reach the circulation. Meals containing large amounts of fat will slow the rate of alcohol absorption, because alcohol is lipid-soluble, and some of it will be dissolved in fat droplets within the chyme. Aspirin is another lipid-soluble drug that can enter the circulation across the gastric mucosa. Aspirin and related drugs alter the properties of the mucous layer and can promote epithelial damage by stomach acids and enzymes. Prolonged aspirin use can cause gastric bleeding, and aspirin is usually avoided by individuals with stomach ulcers.

STOMACH CANCER Stomach, or gastric, cancer is one of the most common lethal cancers, responsible for roughly 15,000 deaths in the United States each year. Because the symptoms may resemble those of gastric ulcers, the condition may not be reported in its early stages. Diagnosis usually involves X-rays of the stomach at various degrees of distension. The mucosa can also be visually inspected using a flexible instrument called a *gastroscope*. Attachments permit the collection of tissue samples for histological analysis. Treatment of gastric cancer involves the surgical removal of part or all of the stomach. Even a total *gastrectomy* (gas-TREK-to-mē) can be survived, because the only absolutely vital function of the stomach is the secretion of intrinsic factor. Protein breakdown can still be performed by the small intestine, although at reduced efficiency, and the loss of gastric functions such as food storage and acid production is not life-threatening.

How would a large meal affect the pH of the blood that leaves the stomach?

When a person suffers from chronic ulcers in the stomach, the branches of the vagus nerve that serve the stomach are sometimes severed. Why?

The Small Intestine and Associated Glandular Organs

The stomach is a holding tank where food is saturated with gastric juices and exposed to stomach acids and the digestive effects of pepsin. These are preliminary steps, for most of the important digestive and absorptive functions occur in the small intestine, where the products of digestion are absorbed. The mucosa of the small intestine produces only a few of the enzymes involved. The pancreas provides digestive enzymes as well as buffers that assist in the neutralization of acidic chyme. The liver and gallbladder provide *bile*, a solution that contains additional buffers and *bile salts*, compounds that facilitate the digestion and absorption of lipids.

The Small Intestine

Figure 24-5

The small intestine plays the primary role in the digestion and absorption of nutrients. The small intestine averages 6 meters (20 ft) in length (range: 4.5–7.5 m) and has a diameter ranging from 4 cm at the stomach to about 2.5 cm at the junction with the large intestine. It occupies all abdominal regions except the right and left hypochondriac and epigastric regions (Figure 1-8●, p. 19). Ninety percent of nutrient absorption occurs in the small intestine, and most of the rest in the large intestine.

The small intestine has three subdivisions: the *duodenum*, the *jejunum*, and the *ileum*.

- The **duodenum** (dū-A-de-num) is the 25 cm (10 in.) section closest to the stomach. This portion of the small intestine is a "mixing bowl" that receives chyme from the stomach and digestive secretions from the pancreas and liver. From its connection with the stomach, the duodenum curves in a C that encloses the pancreas. Except for the proximal 2.5 cm (1 in.), the duodenum is in a retroperitoneal position between vertebrae L_1 and L_4 (Figure 24-5●, p. 884).
- A rather abrupt bend marks the boundary between the duodenum and the **jejunum** (je-JOO-num). At this junction the small intestine reenters the peritoneal cavity, supported by a sheet of mesentery. The jejunum is about 2.5 meters (8 ft) long. The bulk of chemical digestion and nutrient absorption occurs in the jejunum.
- The **ileum** (IL-ē-um) is the third and last segment of the small intestine. It is also the longest, averaging 3.5 meters (12 ft) in length. The ileum ends at a sphincter, the *ileocecal valve*, which controls the flow of chyme from the ileum into the *cecum* of the large intestine.

The small intestine fills much of the peritoneal cavity, and its position is stabilized by mesenteries attached to the dorsal body wall (Figure 24-5●, p. 884). Movement of the small intestine during digestion is restricted by the stomach, the large intestine, the abdominal wall, and the pelvic girdle. Blood vessels, lymphatics, and nerves reach these segments of the small intestine via the connective tissue of the mesentery. The primary blood vessels involved are branches of the *superior mesenteric artery* and *superior mesenteric vein.* ∞ *[pp. 758, 767]*

Histology of the Small Intestine

Figures 24-16, 24-17

The intestinal lining bears a series of transverse folds called **plicae** (PLĪ-sē) **circulares** (Figure 24-16b●). Unlike the rugae in the stomach, each *plica* (PLĪ-ka) is a permanent feature that does not disappear when the small intestine fills with chyme. Roughly 800 plicae are found along the length of the small intestine, and their presence greatly increases the surface area available for absorption.

Figure 24-16c,d,e● presents a more detailed view of the intestinal wall.

INTESTINAL VILLI The mucosa of the small intestine is thrown into a series of fingerlike projections, the **intestinal villi.** The intestinal villi are covered by a simple columnar epithelium that is carpeted with microvilli. Because the microvilli project above the epithelium like the bristles on a brush, these cells are said to have a *brush border.*

If the small intestine were a simple tube with smooth walls, it would have a total absorptive area of roughly 3300 cm^2 (3.6 ft^2). Instead, the mucosa contains plicae circulares, each plica supports a forest of villi, and each villus is covered by epithelial cells whose exposed surfaces contain microvilli. This arrangement increases the total area for absorption by a factor of more than 600, to approximately 2 million square centimeters, or more than 2200 square feet.

The lamina propria of each villus contains an extensive network of capillaries. These capillaries originate in a vascular network within the submucosa. They transport respiratory gases and carry absorbed nutrients to the hepatic portal circulation for delivery to the liver. The liver adjusts the nutrient concentrations of the blood before it reaches the general systemic circulation.

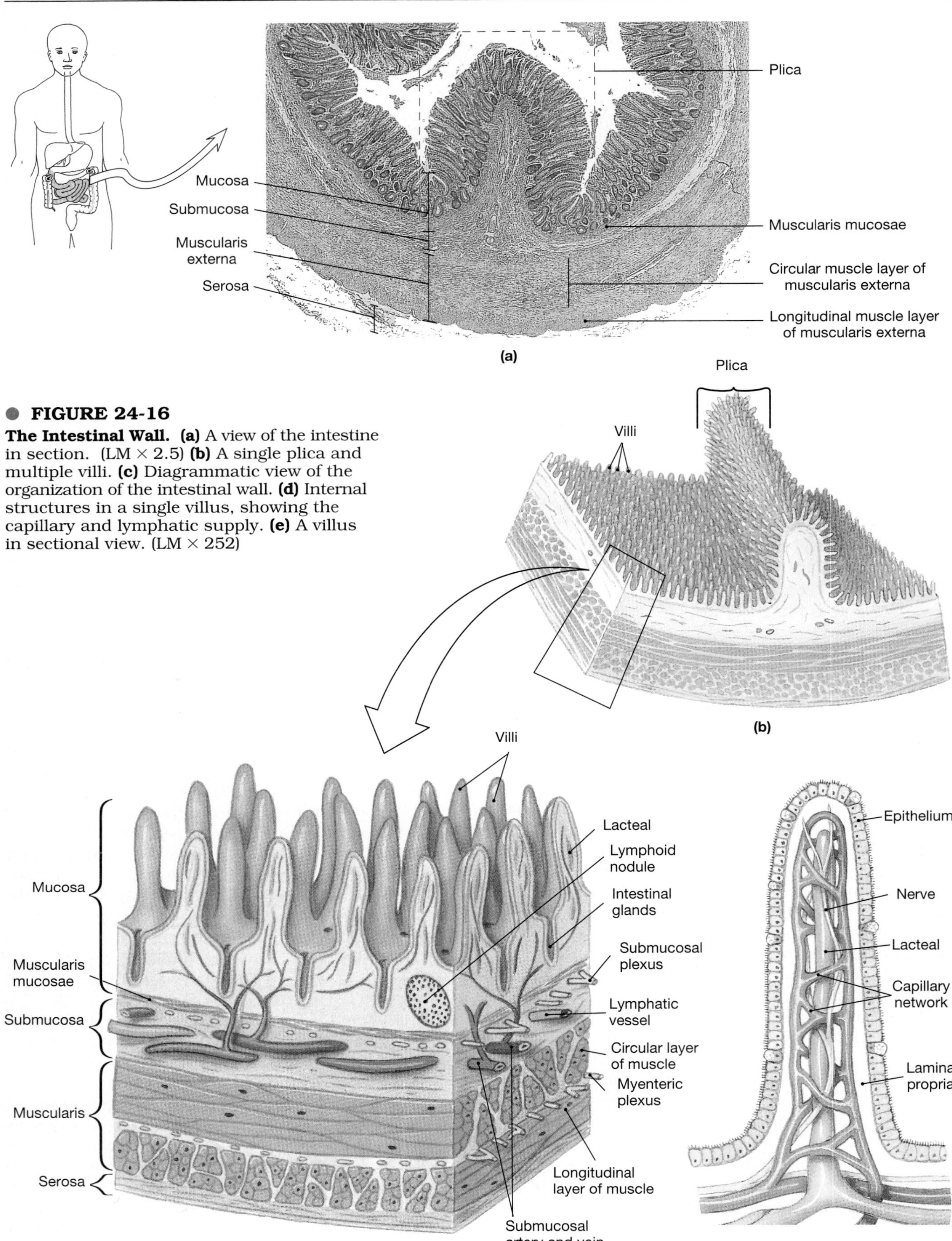
Plica
Mucosa
Submucosa
Muscularis externa
Serosa
Muscularis mucosae
Circular muscle layer of muscularis externa
Longitudinal muscle layer of muscularis externa
(a)
Plica
Villi
(b)
Villi
Mucosa
Muscularis mucosae
Submucosa
Muscularis
Serosa
Lacteal
Lymphoid nodule
Intestinal glands
Submucosal plexus
Lymphatic vessel
Circular layer of muscle
Myenteric plexus
Longitudinal layer of muscle
Submucosal artery and vein
(c)
Epithelium
Nerve
Lacteal
Capillary network
Lamina propria
(d)

FIGURE 24-16
The Intestinal Wall. (a) A view of the intestine in section. (LM × 2.5) **(b)** A single plica and multiple villi. **(c)** Diagrammatic view of the organization of the intestinal wall. **(d)** Internal structures in a single villus, showing the capillary and lymphatic supply. **(e)** A villus in sectional view. (LM × 252)

In addition to capillaries and nerve endings, each villus contains a lymphatic capillary called a **lacteal** (LAK-tē-al; *lacteus*, milky) (Figure 24-16d,e●). Lacteals transport materials that are unable to enter blood capillaries. For example, absorbed fatty acids are assembled into protein-lipid packages that are too large to diffuse into the bloodstream. These packets, called *chylomicrons*, reach the venous circulation via the thoracic duct, which delivers lymph into the left subclavian vein. [p. 763] The name *lacteal* refers to the pale, cloudy appearance of lymph containing large quantities of lipids.

Contractions of the muscularis mucosae and smooth muscles within the villi move villi back and forth, exposing the epithelial surfaces to the liquefied intestinal contents. This movement improves the efficiency of absorption, because local differences in the nutrient concentration of the chyme will be quickly eliminated. Movements of the villi also squeeze the lacteals, thus assisting in the movement of lymph out of the villi.

INTESTINAL CRYPTS Between the columnar epithelial cells, goblet cells eject mucus onto the intestinal surfaces. At the bases of the villi are found the entrances to the **intestinal glands,** or **intestinal crypts** (also known as *crypts of Lieberkuhn*). These glandular pockets extend deep into the underlying lamina propria (Figure 24-16c●). Near the base of each intestinal gland, stem cell divisions produce new generations of epithelial cells. These new cells are continually displaced toward the intestinal surface. Within a few days they will have reached the tip of a villus, where they are shed into the intestinal lumen. This ongoing process renews the epithelial surface and adds enzymes to the chyme. **Enterokinase** is one important enzyme that reaches the intestinal lumen in this way. Enterokinase does not directly participate in digestion, but it activates proenzymes secreted by the pancreas.

Intestinal crypts also contain enteroendocrine cells responsible for the production of several intestinal hormones, including gastrin, cholecystokinin, and secretin.

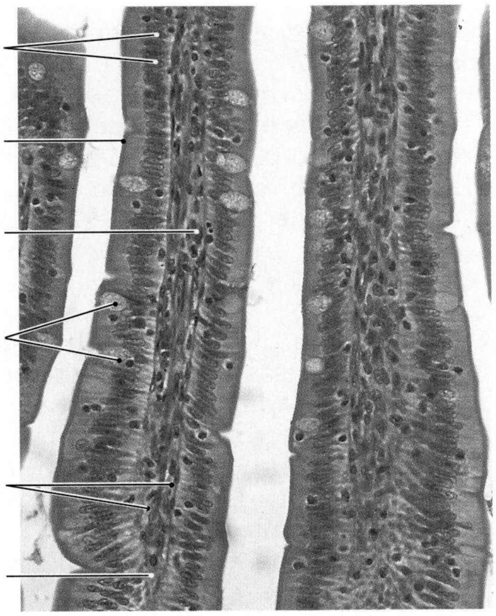

Nuclei of simple columnar epithelial cells
Brush border
Lacteal
Goblet cells
Capillary network
Lamina propria

(e)

REGIONAL SPECIALIZATIONS The regions of the small intestine have histological specializations related to their primary functions. The locations of these segments in the peritoneal cavity are indicated in Figure 24-17a●. Representative sections from each region of the small intestine are presented in Figure 24-17b●.

- ***Duodenum.*** The duodenum contains numerous mucous glands, both within the epithelium and beneath it. In addition to the intestinal crypts, the submucosa contains **submucosal glands,** also known as *Brunner's glands,* that produce copious quantities of mucus. Mucus produced by the submucosal glands protects the epithelium from the acid chyme arriving from the stomach. It also contains buffers that help elevate the pH of the chyme. Along the length of the duodenum, the pH of the chyme goes from 1–2 to 7–8. The submucosal glands also secrete a hormone, **urogastrone,** which inhibits gastric acid production. Urogastrone, also known as *epidermal growth factor* (EGF), stimulates the division of epithelial cells along the digestive tract, as well as stem cell activity in other areas.
- ***Jejunum.*** Plicae and villi are prominent over the proximal half of the jejunum. As chyme approaches the ileum, the plicae and villi become smaller and continue to diminish in size to the end of the ileum. This reduction parallels the reduction in absorptive activity; most nutrient absorption has occurred before chyme reaches the ileum. One rather drastic surgical method of promoting weight loss is the removal of a significant portion of the jejunum. The reduction in absorptive area causes a marked weight loss, but the side effects can be very troublesome. AM *Drastic Weight Loss Techniques*
- ***Ileum.*** The ileum adjacent to the large intestine lacks plicae altogether, and the scattered villi are stumpy and conical. The ileum also contains 20–30 masses of lymphoid tissue, called *aggregate lymphoid nodules,* or *Peyer's patches,* described in Chapter 22. [p. 786] The lymphocytes in these nodules protect the small intestine from bacteria that are normal inhabitants of the large intestine. Peyer's

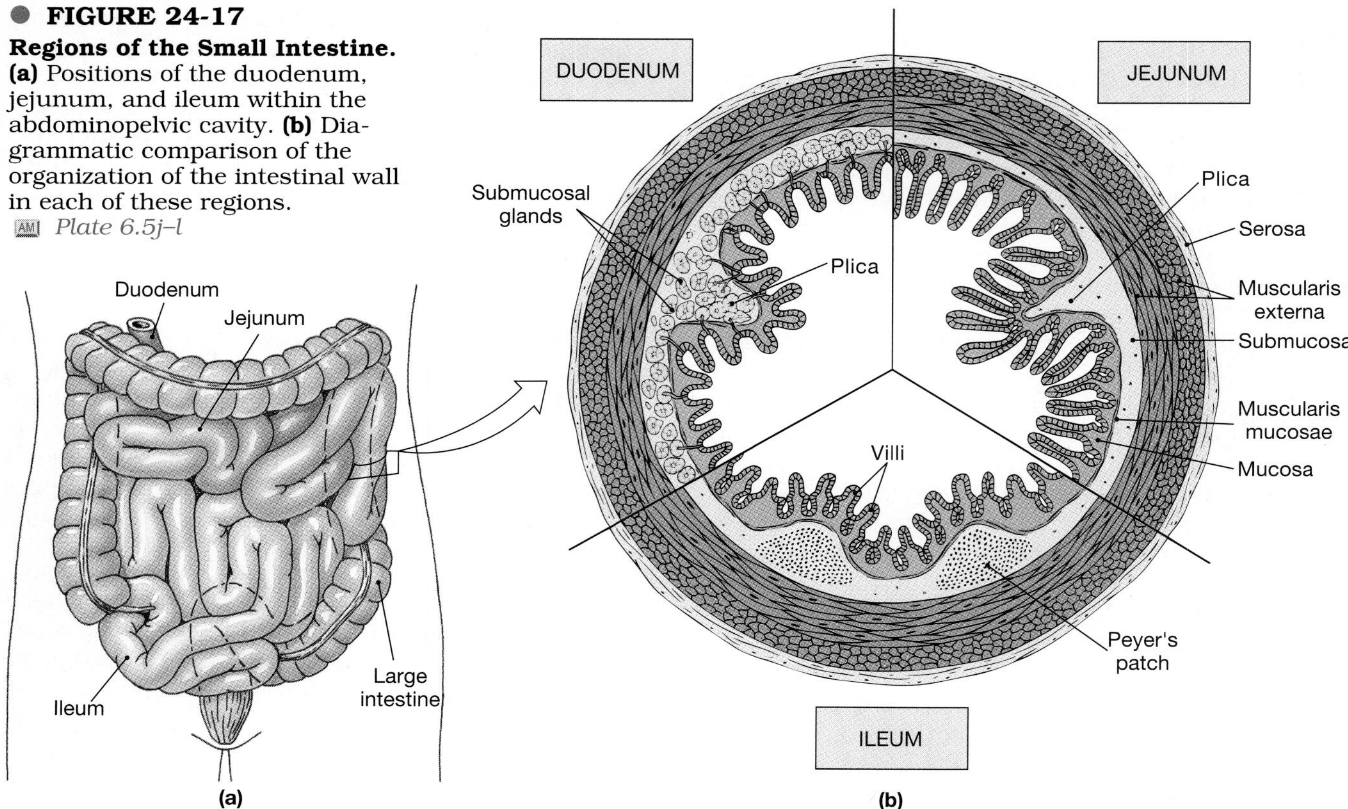

Duodenum
Jejunum
Ileum
Large intestine
(a)
DUODENUM
JEJUNUM
ILEUM
Submucosal glands
Plica
Plica
Serosa
Muscularis externa
Submucosa
Muscularis mucosae
Mucosa
Villi
Peyer's patch
(b)

● **FIGURE 24-17**
Regions of the Small Intestine. (a) Positions of the duodenum, jejunum, and ileum within the abdominopelvic cavity. **(b)** Diagrammatic comparison of the organization of the intestinal wall in each of these regions.
AM *Plate 6.5j–l*

patches are most abundant in the terminal portion of the ileum, near the entrance to the large intestine.

Intestinal Secretions

Roughly 1.8 liters of watery **intestinal juice** enter the intestinal lumen each day. Intestinal juice moistens the chyme, assists in buffering acids, and dissolves both the digestive enzymes provided by the pancreas and the products of digestion. Much of this fluid volume arrives through osmosis, as water flows out of the mucosa and into the relatively concentrated chyme. The rest is provided by intestinal glands, stimulated by activation of touch and stretch receptors in the intestinal walls.

The submucosal glands play an important role in protecting the duodenal epithelium from gastric acids and enzymes. These glands increase their secretory activities in response not only to local reflexes but also to parasympathetic stimulation via the vagus nerve. As a result, the duodenal glands begin secreting during the cephalic phase of gastric secretion, long before the chyme reaches the pyloric sphincter. Sympathetic stimulation will inhibit the activation of the submucosal glands, leaving the duodenal lining relatively unprepared for the arrival of acid chyme. This fact probably accounts for the common observation that duodenal ulcers can be caused by chronic stress or other factors that promote sympathetic activation.

❑ Intestinal Movements

After chyme has arrived in the duodenum, weak peristaltic contractions move it slowly toward the jejunum. These contractions are myenteric reflexes not under CNS control, and their effects are limited to within a few centimeters of the site of the original stimulus. These short reflexes are controlled by motor neurons in the submucosal and myenteric plexuses. In addition, some of the smooth muscle cells contract periodically, even without stimulation, establishing a basic contractile rhythm that then spreads from cell to cell.

Stimulation of the parasympathetic system increases the sensitivity of these reflexes and accelerates both local peristalsis and segmentation. More elaborate reflexes coordinate activities along the entire length of the small intestine. Two important examples are the *gastroenteric reflex* and the *gastroileal reflex.*

- Distension of the stomach initiates the **gastroenteric** (gas-trō-en-TER-ik) **reflex.** This reflex produces an immediate increase in the rates of glandular secretion and peristaltic activity in all segments of the small intestine. The increased peristalsis distributes the chyme along the length of the small intestine and empties the duodenum in preparation for the arrival of the next meal.
- The **gastroileal** (gas-trō-IL-ē-al) **reflex** is a combination of the gastroenteric reflex, direct-

ed by the CNS, and a hormonal response to circulating levels of gastrin. Gastrin, released in large quantities when food enters the stomach, triggers the relaxation of the ileocecal valve. Because the sphincter is relaxed, the peristalsis resulting from the gastroenteric reflex can push chyme into the large intestine.

VOMITING The responses of the digestive tract to chemical or mechanical irritation are rather predictable. Gastric secretion accelerates all along the digestive tract, and the intestinal contents are eliminated as quickly as possible. The *vomiting reflex* occurs in response to irritation of the fauces, pharynx, esophagus, stomach, or proximal portions of the small intestine. During the preparatory phase the pylorus relaxes, and the contents of the duodenum and proximal jejunum are discharged into the stomach by strong peristaltic waves that travel toward the stomach rather than toward the ileum. Vomiting, or *emesis* (EM-e-sis), then occurs as the stomach regurgitates its contents through the esophagus and pharynx. As regurgitation occurs, the uvula and soft palate block the entrance to the nasopharynx. Increased salivary secretion assists in buffering the stomach acids, thereby preventing erosion of the teeth. In conditions marked by repeated vomiting, severe tooth damage can occur; one example, the eating disorder *bulimia*, will be discussed in Chapter 25. Most of the force of vomiting comes from expiratory movements that elevate intra-abdominal pressures and force the stomach against the tensed diaphragm.

❑ The Pancreas

Figure 24-18a

The pancreas lies behind the stomach, extending laterally from the duodenum toward the spleen (Figure 24-18a●). The pancreas is an elongate, pinkish-gray organ with a length of approximately 15 cm (6 in.) and a weight of around 80 g (3 oz). The broad **head** of the pancreas lies within the loop formed by the duodenum as it leaves the pylorus. The slender **body** extends transversely toward the spleen, and the **tail** is short and bluntly rounded. The pancreas is retroperitoneal and is firmly bound to the posterior wall of the abdominal cavity.

The surface of the pancreas has a lumpy, nodular texture. A thin, transparent connective tissue capsule wraps the entire organ. The pancreatic lobules, associated blood vessels, and excretory ducts can be seen through the anterior capsule and the overlying layer of peritoneum. Arterial blood reaches the pancreas via branches of the *splenic, superior mesenteric,* and *hepatic arteries.* The *pancreatic arteries* and *posterior superior pancreaticoduodenal artery* are the major branches from these vessels. The *splenic vein* and its branches drain the pancreas.

The pancreas is primarily an exocrine organ, producing digestive enzymes and buffers. The large **pancreatic duct** *(duct of Wirsung)* delivers these secretions to the duodenum. A small **accessory duct,** or *duct of Santorini,* may drain into the pancreatic duct before it leaves the pancreas. The pancreatic duct extends within the attached mesentery to reach the duodenum, where it meets the *common bile duct* from the liver and gallbladder. The two ducts then empty into the *duodenal ampulla,* a chamber located roughly halfway along the length of the duodenum.

Histological Organization

Figure 24-18

Partitions of connective tissue divide the pancreatic tissue into distinct lobules (Figure 24-18●). The blood vessels and tributaries of the pancreatic ducts are found within these connective tissue septa. The pancreas is an example of a *compound tubuloacinar gland,* a gland structure described in Chapter 4. ∞ *[p. 121]* Within each lobule, the ducts branch repeatedly before ending in blind pockets, the **pancreatic acini** (AS-i-nī). Each pancreatic acinus is lined by a simple cuboidal epithelium. *Pancreatic islets,* the endocrine tissues of the pancreas, are scattered between the acini, but they account for only around 1 percent of the cellular population of the pancreas.

The pancreas has two distinct functions, one endocrine and the other exocrine. Endocrine cells of the pancreatic islets secrete insulin and glucagon into the bloodstream. These hormones and their actions were described in Chapter 18. ∞ *[p. 630]* The exocrine cells secrete an alkaline **pancreatic juice** into the small intestine. Pancreatic juice is a mixture of water, ions, and digestive enzymes. The fluids and ions assist in diluting and buffering the acids in the chyme. Pancreatic enzymes do most of the digestive work in the small intestine, breaking down ingested materials into small molecules suitable for absorption.

Physiology of the Pancreas

Each day the pancreas secretes around 1000 ml of pancreatic juice. The secretory activities are controlled primarily by hormones from the duodenum. When acid chyme arrives in the duodenum, *secretin* is released. This hormone triggers the pancreatic secretion of a watery buffer solution with a pH of 7.5–8.8. Among its other components, this secretion contains sodium bicarbonate, a buffer that helps elevate the pH of the chyme. A different duodenal hormone, *cholecystokinin,* stimulates the production and secretion of pancreatic enzymes. Pancreatic enzyme secretion also increases under stimulation by the vagus nerve. This stimulation occurs during the cephalic phase of gastric regulation, so the pancreas starts to synthesize enzymes before food even reaches the stomach. Such a head start is important because enzyme synthesis takes

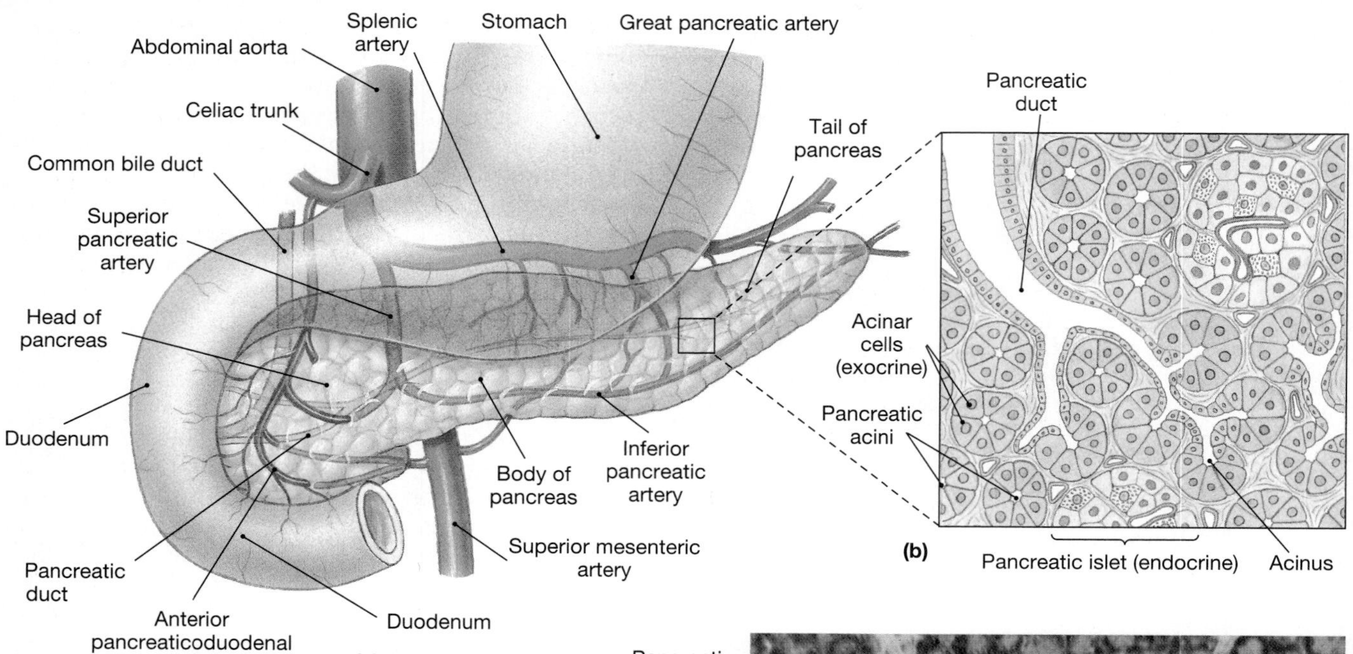
Splenic artery
Stomach
Great pancreatic artery
Abdominal aorta
Celiac trunk
Common bile duct
Superior pancreatic artery
Tail of pancreas
Pancreatic duct
Head of pancreas
Acinar cells (exocrine)
Duodenum
Pancreatic acini
Inferior pancreatic artery
Body of pancreas
Superior mesenteric artery
Pancreatic duct
(b)
Pancreatic islet (endocrine)
Acinus
Anterior pancreaticoduodenal artery
Duodenum
(a)

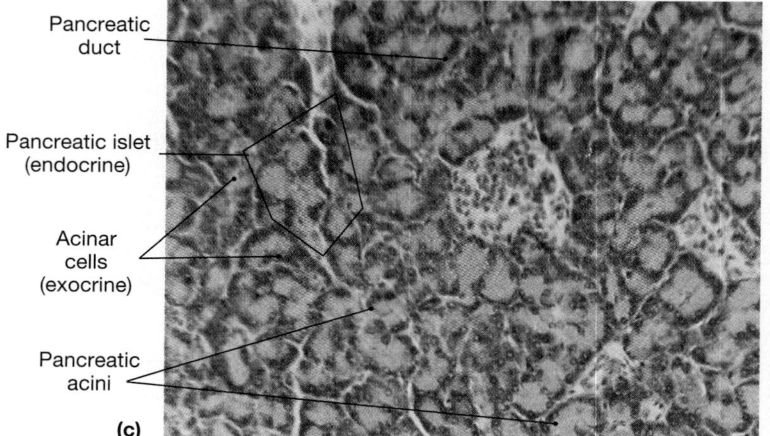
Pancreatic duct
Pancreatic islet (endocrine)
Acinar cells (exocrine)
Pancreatic acini
(c)

FIGURE 24-18
The Pancreas. (a) Gross anatomy of the pancreas. The head of the pancreas is tucked into a curve of the duodenum that begins at the pylorus of the stomach. **(b)** Diagrammatic view of the cellular organization of the pancreas, showing exocrine and endocrine regions. **(c)** Light micrograph of pancreatic tissue. (LM ×168) AM *Plate 6.5*

much longer than buffer production. By starting early, the pancreatic cells will be ready to meet the demand when chyme arrives in the duodenum.

The specific pancreatic enzymes involved include:

- **Pancreatic alpha-amylase** is a **carbohydrase** (kar-bō-HĪ-drās), an enzyme that breaks down starches. Pancreatic alpha-amylase is almost identical to salivary amylase.
- **Pancreatic lipase** breaks down complex lipids, releasing fatty acids and other products that can be easily absorbed.
- **Nucleases** break down nucleic acids.
- **Proteolytic enzymes** break proteins apart. The proteolytic enzymes of the pancreas include **proteases** and **peptidases;** proteases break apart large protein complexes, whereas peptidases break small peptide chains into individual amino acids.

Proteolytic enzymes account for around 70 percent of the total pancreatic enzyme production. The enzymes are secreted as inactive *proenzymes* that are activated only after reaching the small intestine. Examples of proenzymes discussed earlier in the text include pepsinogen, angiotensinogen, plasminogen, fibrinogen, and many of the clotting factors and enzymes of the complement system. As in the stomach, release of a proenzyme rather than an active enzyme protects the secretory cells from the destructive effects of their own products. Among the proenzymes secreted by the pancreas are **trypsinogen** (trip-SIN-ō-jen), **chymotrypsinogen** (kī-mō-trip-SIN-ō-jen), **procarboxypeptidase** (prō-kar-bok-sē-PEP-ti-dās), and **proelastase** (pro-ē-LAS-tās).

Once inside the duodenum, enterokinase produced in the small intestine triggers the conversion of trypsinogen to **trypsin,** an active protease. Trypsin then activates the other proenzymes, producing **chymotrypsin, carboxypeptidase,** and **elastase.** Each enzyme attacks peptide bonds with slightly different characteristics. Together, they break down complex proteins into a mixture of dipeptides, tripeptides, and amino acids.

PANCREATITIS **Pancreatitis** (pan-krē-a-TĪ-tis) is an inflammation of the pancreas. Blockage of the excretory ducts, bacterial or viral infections, ischemia, and drug reactions, especially those involving alcohol, are among the factors that may produce this condition. These stimuli provoke a crisis by injuring exocrine cells in at least a portion of the organ. Lysosomes within the damaged cells then activate the proenzymes, and autodigestion begins. The proteolytic enzymes digest the surrounding, undamaged cells, activating their enzymes and starting a chain reaction. In most cases only a portion of the pancreas will be affected, and the condition subsides in a few days. In 10–15 percent of pancreatitis cases the process does not subside, and the enzymes may ultimately destroy the pancreas.

The Liver

The liver, the largest visceral organ, is one of the most versatile organs in the body. Most of its mass lies within the right hypochondriac and epigastric regions, but it extends into the left hypochondriac and umbilical regions as well. The liver weighs about 1.5 kg (3.3 lb). This large, firm, reddish-brown organ provides essential metabolic and synthetic services that fall into three general categories: *metabolic regulation, hematological regulation,* and *bile production.* These general functions will be detailed following our examination of the anatomy and organization of the liver.

Anatomy of the Liver

Figure 24-19

The liver is wrapped in a tough fibrous capsule and covered by a layer of visceral peritoneum. On the anterior surface the **falciform** (FAL-si-fōrm) **ligament** marks the division between the **left lobe** and **right lobe** of the liver (Figure 24-19•). A thickening in the posterior margin of the falciform ligament is the **round ligament,** a fibrous band that marks the path of the fetal umbilical vein.

On the posterior surface of the liver, the impression left by the inferior vena cava marks the division between the right lobe and the small **caudate** (KAW-dāt) **lobe.** Below the caudate lobe lies the **quadrate lobe,** sandwiched between the left lobe and the gallbladder. Afferent blood vessels and other structures reach the liver by traveling within the connective tissue of the lesser omentum. They converge at the **hilus** of the liver, a region known as the *porta hepatis* ("doorway to the liver").

The *gallbladder* is a muscular sac that stores and concentrates bile prior to its excretion into the small intestine. The gallbladder is located in a recess, or fossa, in the posterior surface of the right lobe. The gallbladder and associated structures are described in a later section.

THE BLOOD SUPPLY TO THE LIVER The circulation to the liver was detailed in Chapter 21 and summarized in Figures 21-27 and 21-35•, pp. 756, 766. Roughly one-third of the blood supply to the liver is arterial blood from the hepatic artery. The remainder consists of venous blood from the hepatic portal vein, which begins in the capillaries of the esophagus, stomach, small intestine, and most of the large intestine. The distribution and major tributaries of the hepatic portal vein were detailed in Chapter 21. ∞ *[p. 766]* In the liver, the hepatocytes adjust circulating levels of nutrients by selective absorption and secretion. Blood leaving the liver returns to the systemic circuit via the hepatic veins that open into the inferior vena cava.

PORTAL HYPERTENSION Pressures in the hepatic portal system are usually low, averaging 10 mm Hg or less. This pressure can increase markedly if blood flow through the liver becomes restricted as a result of liver damage or a blood clot. A rise in portal pressure is called **portal hypertension.** As pressures rise, small peripheral veins and capillaries in the portal system become distended and are likely to rupture, and intestinal bleeding becomes a problem. Under these conditions, *esophageal varices* (p. 894) may develop, and there may also be leakage of fluid into the peritoneal cavity across the serosal surfaces of the liver and viscera, producing *ascites* (p. 883).

Histological Organization of the Liver

Figures 24-20, 24-21

Each lobe of the liver is divided by connective tissue into approximately 100,000 **liver lobules,** the basic functional units of the liver. The histological organization and structure of a typical liver lobule are shown in Figure 24-20•.

THE LIVER LOBULE Liver cells, or **hepatocytes** (he-PAT-ō-sīts), in a liver lobule form a series of irregular plates arranged like the spokes of a wheel (Figure 24-20•). The plates are only one cell thick, and exposed hepatocyte surfaces are covered with short microvilli. Sinusoids between adjacent plates empty into the **central vein.** The walls of the sinusoids contain large openings that allow substances to pass out of the circulation and into the spaces surrounding the hepatocytes.

In addition to typical endothelial cells, the sinusoidal lining includes a large number of **Kupffer** (KOOP-fer) **cells,** also known as *stellate reticuloendothelial cells.* These phagocytic cells, part of the monocyte-macrophage system, engulf pathogens, cell debris, and damaged blood cells. Kupffer cells are also responsible for storing (1) iron, (2) some lipids, and (3) heavy metals, such as tin or mercury, that are absorbed by the digestive tract.

FIGURE 24-19
Anatomy of the Liver. (a) Sectional view through the upper abdomen. **(b)** The anterior surface of the liver. **(c)** The posterior surface of the liver. AM *Plate 6.5*

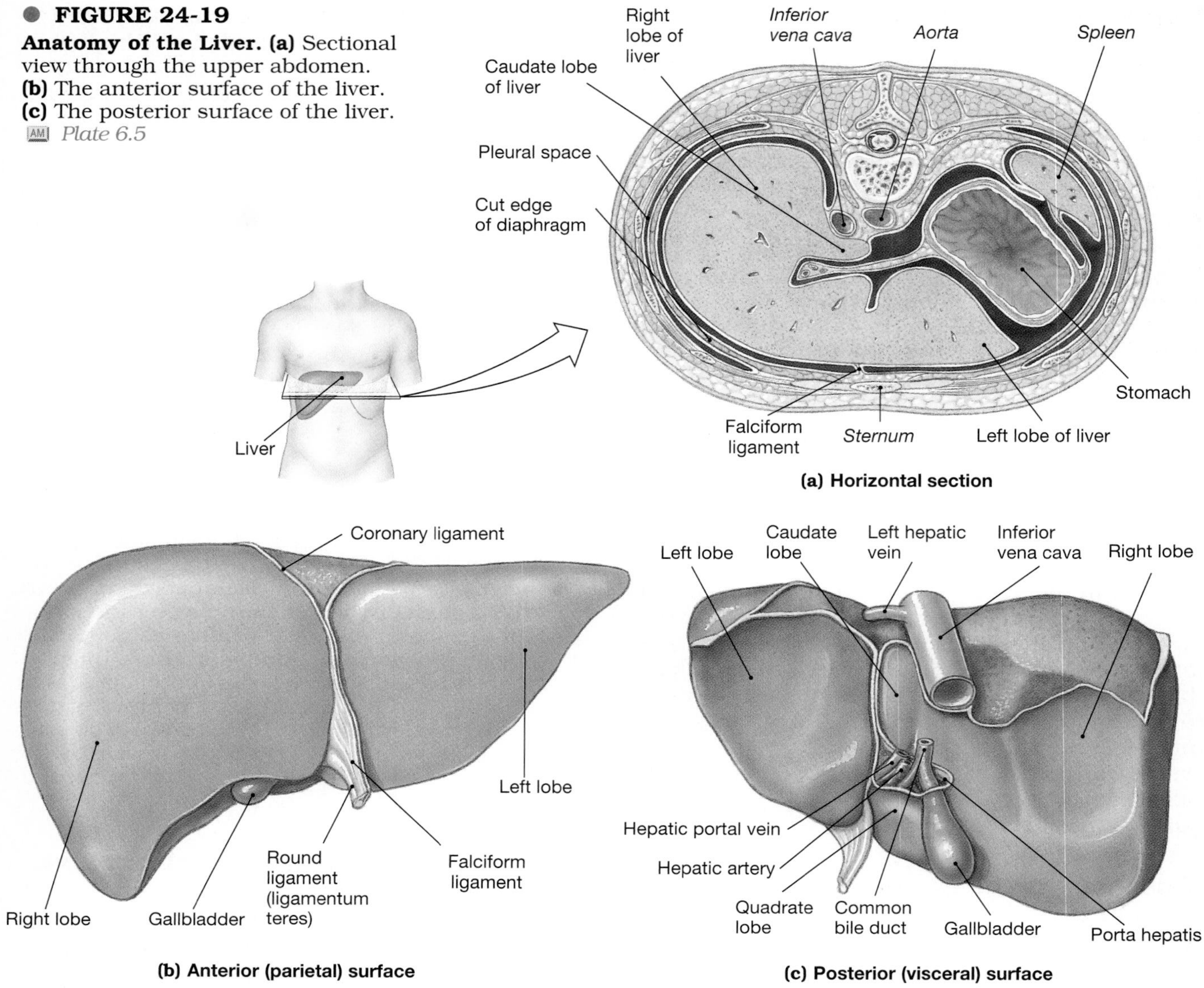

Right lobe of liver
Inferior vena cava
Aorta
Spleen
Caudate lobe of liver
Pleural space
Cut edge of diaphragm
Liver
Stomach
Falciform ligament
Sternum
Left lobe of liver

(a) Horizontal section

Coronary ligament
Left lobe
Right lobe
Gallbladder
Round ligament (ligamentum teres)
Falciform ligament

(b) Anterior (parietal) surface

Left lobe
Caudate lobe
Left hepatic vein
Inferior vena cava
Right lobe
Hepatic portal vein
Hepatic artery
Quadrate lobe
Common bile duct
Gallbladder
Porta hepatis

(c) Posterior (visceral) surface

Blood enters the liver sinusoids from small branches of the portal vein and hepatic artery. A typical lobule has a hexagonal shape in cross-section (Figure 24-20a,b●). There are six **portal areas,** or *hepatic triads,* one at each corner of the lobule. A portal area (Figure 24-20c●) contains three structures: (1) a branch of the hepatic portal vein, (2) a branch of the hepatic artery, and (3) a small branch of the bile duct.

Branches from the arteries and veins deliver blood to the sinusoids of adjacent lobules (Figure 24-20a,b●). As blood flows through the sinusoids, hepatocytes absorb solutes from the plasma and secrete materials, such as plasma proteins. Blood then leaves the sinusoids and enters the central vein of the lobule. The central veins ultimately merge to form the hepatic veins that empty into the inferior vena cava. Liver diseases, such as the various forms of *hepatitis,* and conditions such as alcoholism can lead to degenerative changes in the liver tissue and constriction of the circulatory supply. AM *Liver Disease*

Bile Secretion and Transport. Bile is secreted into a network of narrow channels between the opposing membranes of adjacent liver cells. These passageways, called **bile canaliculi,** extend outward, away from the central vein. Eventually they connect with fine **bile ductules** (DUK-tūlz) that carry bile to bile ducts in the nearest portal area. The **right** and **left hepatic ducts** collect bile from all the bile ducts of the liver lobes. These ducts unite to form the **common hepatic duct** that leaves the liver (Figure 24-21●). The bile within the common hepatic duct may either (1) flow into the *common bile duct* that empties into the duodenal ampulla or (2) enter the *cystic duct* that leads to the gallbladder.

The **common bile duct** is formed by the union of the **cystic duct** and the common hepatic duct. The common bile duct passes within the lesser omentum toward the stomach, turns, and penetrates the wall of the duodenum to meet the pancreatic duct at the duodenal ampulla.

The Physiology of the Liver

The liver is responsible for metabolic regulation, hematological regulation, and bile production. The liver has over 200 different functions; this discussion will provide only a general overview.

METABOLIC REGULATION The liver is the primary organ involved with regulating the composition of the circulating blood. All blood leaving the absorptive surfaces of the digestive tract enters the hepatic portal system and flows into the liver. This arrangement gives liver cells the opportunity to extract absorbed nutrients or toxins from the blood before it reaches the systemic circulation via the hepatic veins. Excess nutrients are removed and stored, and deficiencies are corrected by mobilizing stored reserves or performing appropriate synthetic activities. For example:

- ***Carbohydrate metabolism.*** The liver stabilizes blood glucose levels at around 90 mg/dl. If blood glucose levels decline, the hepatocytes break down glycogen reserves and release glucose into the circulation. They also synthesize glucose from other carbohydrates or from available amino acids. The synthesis of glucose from other compounds is a process called *gluconeo-*

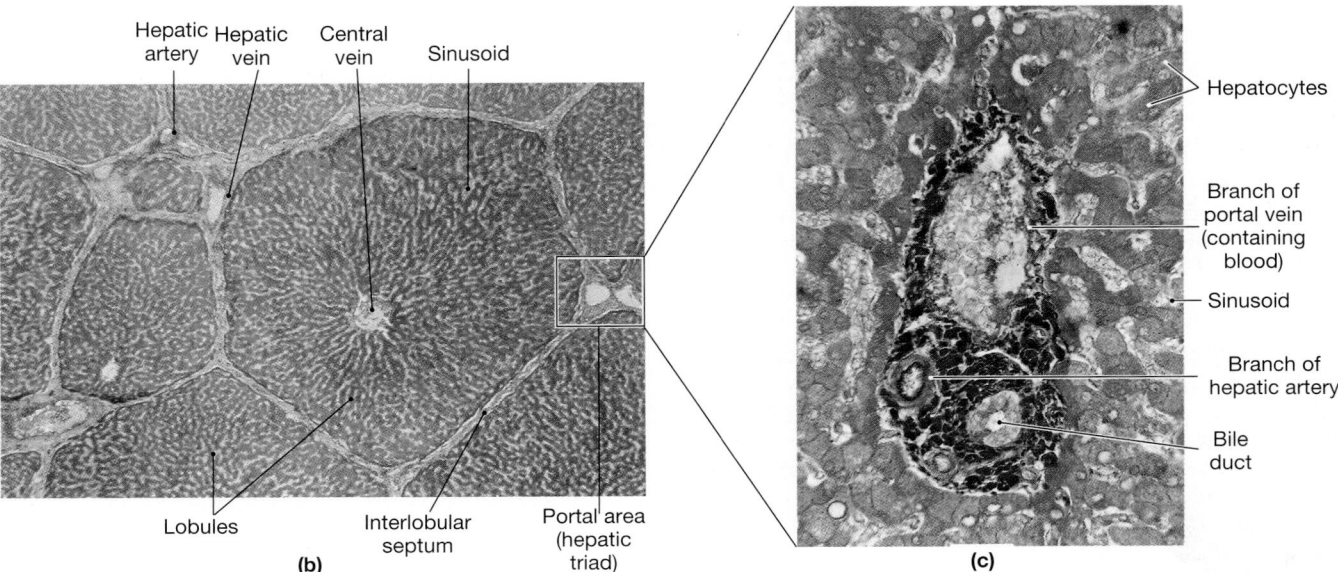

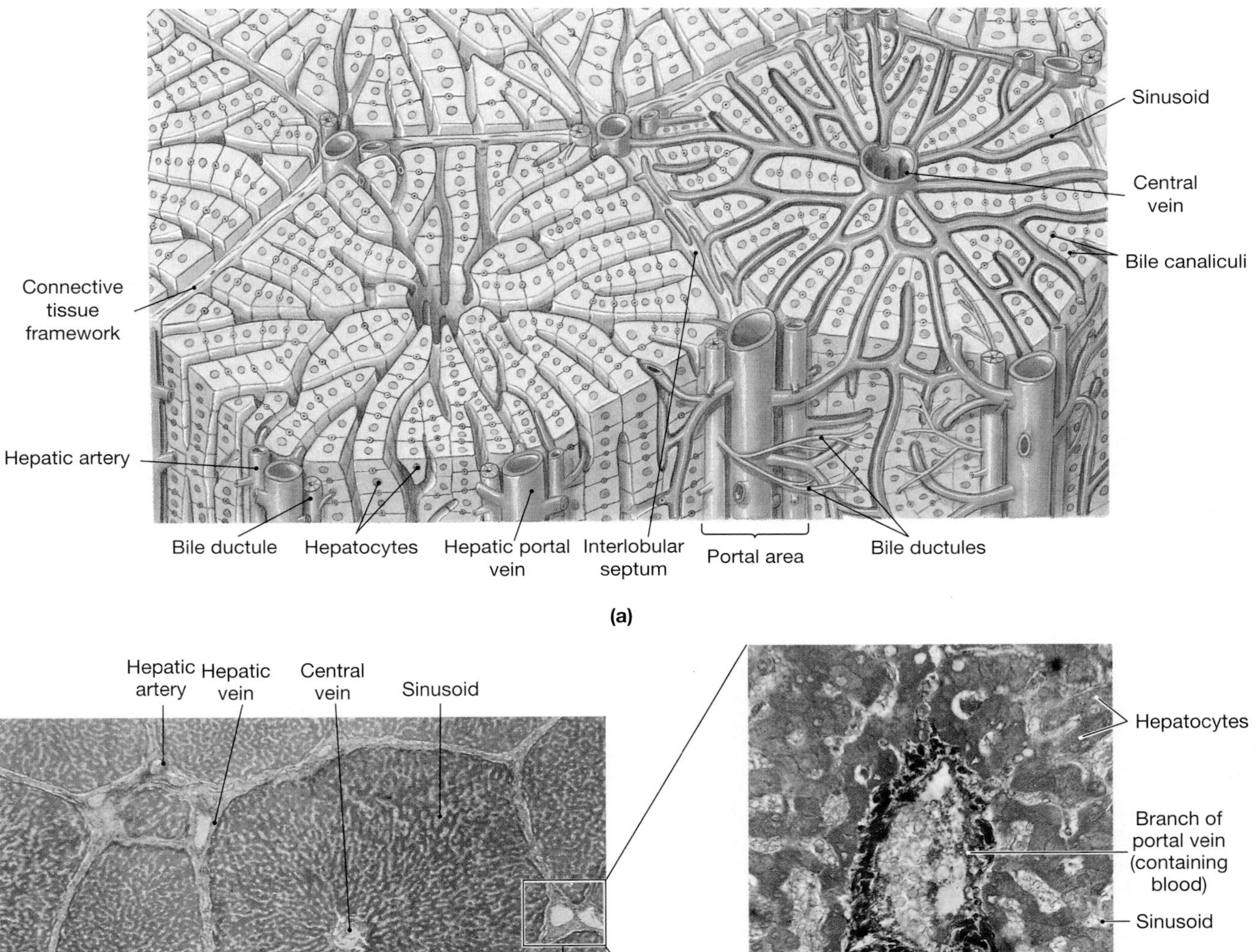
Sinusoid
Central vein
Bile canaliculi
Connective tissue framework
Hepatic artery
Bile ductule
Hepatocytes
Hepatic portal vein
Interlobular septum
Portal area
Bile ductules
(a)
Hepatic artery
Hepatic vein
Central vein
Sinusoid
Lobules
Interlobular septum
Portal area (hepatic triad)
(b)
Hepatocytes
Branch of portal vein (containing blood)
Sinusoid
Branch of hepatic artery
Bile duct
(c)

● **FIGURE 24-20**
Liver Histology. (a) Diagrammatic view of lobular organization. **(b)** Light micrograph showing a section through a liver lobule. (LM × 38) **(c)** A portal area. (LM × 31)

genesis. If blood glucose levels climb, liver cells remove glucose from the circulation and either store it as glycogen or use it to synthesize lipids that can be stored in the liver or other tissues. These metabolic activities are regulated by circulating hormones, such as insulin and glucagon, detailed in Chapter 18. [p. 630]

- ***Lipid metabolism.*** The liver regulates circulating levels of triglycerides, fatty acids, and cholesterol. When those levels decline, the liver breaks down its lipid reserves and releases them into the circulation. When the levels are high, the lipids are removed for storage.
- ***Amino acid metabolism.*** The liver removes excess amino acids from the circulation. These amino acids may be used to synthesize proteins, or they may be converted to lipids or to glucose for storage.
- ***Removal of waste products.*** When converting amino acids to lipids or carbohydrates, or when breaking down amino acids to obtain energy, the liver strips off the amino groups, a process called *deamination.* This process produces ammonia, a toxic waste product that the liver neutralizes through conversion to *urea,* a relatively harmless compound that is excreted at the kidneys. Other waste products, circulating toxins, and drugs are also removed from the blood for subsequent inactivation, storage, or excretion.
- ***Vitamin storage.*** Fat-soluble vitamins (A, D, E, and K) and vitamin B_{12} are absorbed from the blood and stored in the liver. These reserves are called upon when the diet contains inadequate amounts of these vitamins.
- ***Mineral storage.*** The liver contains important reserves of iron, stored as *ferritin.* (This function was discussed in Chapter 19.) [p. 658]
- ***Drug inactivation.*** The liver removes and breaks down circulating drugs, thereby limiting the duration of their effects. The rate at which the liver removes a particular drug must be taken into account when drugs are prescribed. For example, if the liver absorbs it relatively quickly, a drug must be administered every few hours to keep the plasma concentrations at therapeutic levels.

LIVER DISEASE Any condition that severely damages the liver represents a serious threat to life. The liver has a limited ability to regenerate after injury, but liver function will not fully recover unless the normal vascular pattern returns. Examples of important types of liver disease include *cirrhosis,* which is characterized by the replacement of lobules by fibrous tissue, and various forms of *hepatitis* that result from viral infections. Liver transplants are sometimes used to combat liver disease, but the supply of suitable donor tissue is limited, and the success rate is highest in young, otherwise healthy individuals. Clinical trials are now under way testing an artificial liver known as *ELAD* (Extracorporeal Liver Assist Device) that may prove suitable for the long-term support of persons with chronic liver disease. AM *Liver Disease*

HEMATOLOGICAL REGULATION The liver, the largest blood reservoir in the body, receives about 25 percent of the cardiac output. As blood passes by, the liver performs the following functions:

- ***Phagocytosis and antigen presentation.*** Kupffer cells in the liver sinusoids engulf old or damaged RBCs, cellular debris, and pathogens from the circulation. Kupffer cells are antigen-presenting cells that can stimulate an immune response.
- ***Plasma protein synthesis.*** The hepatocytes synthesize and release most of the plasma proteins. These include the albumins that contribute to the osmotic concentration of the blood, the various types of transport proteins, clotting proteins, and complement proteins.
- ***Removal of circulating hormones.*** The liver is the primary site for the absorption and recycling of epinephrine, norepinephrine, insulin, thyroid hormones, and steroid hormones such as the sex hormones (estrogens and androgens) and corticosteroids. The liver also absorbs cholecalciferol (vitamin D_3) from the blood. Liver cells then convert cholecalciferol, which may be synthesized in the skin or absorbed in the diet, into an intermediary product that is released back into the circulation. The intermediary is absorbed by the kidneys and used to generate calcitriol, a hormone important to calcium ion metabolism. [p. 627]
- ***Removal of antibodies.*** The liver absorbs and breaks down antibodies, releasing amino acids to be recycled.
- ***Removal or storage of toxins.*** Lipid-soluble toxins in the diet, such as DDT, are absorbed by the liver and stored in lipid deposits, where they do not disrupt cellular functions. Other toxins are removed from the circulation and either broken down or excreted in the bile.
- ***Synthesis and secretion of bile.*** **Bile** is synthesized in the liver and excreted into the lumen of the duodenum. Bile consists mostly of water, with minor amounts of ions, *bilirubin* (a pigment derived from hemoglobin), cholesterol, and an assortment of lipids collectively known as the **bile salts.** The water and ions assist in the dilution and buffering of acids in chyme as it enters the small intestine.

Bile salts are synthesized from cholesterol in the liver. Several related compounds are involved; the most abundant are derivatives of the steroids *cholate* and *chenodeoxycholate.*

THE FUNCTIONS OF BILE Most dietary lipids are not water-soluble. Mechanical processing in the stomach creates large drops containing a variety of lipids. Pancreatic lipase is not lipid-soluble, so the enzymes can interact with lipids only at the surface of a lipid drop. The larger the drop, the more lipids are found inside, isolated and protected from these enzymes. Bile salts break the drops apart, a process called **emulsification** (ē-mul-si-fi-KĀ-shun).

Emulsification creates tiny *emulsion droplets* with a superficial coating of bile salts. The formation of tiny droplets increases the surface area available for enzymatic attack. In addition, the layer of bile salts facilitates interaction between the lipids and lipid-digesting enzymes provided by the pancreas. After lipid digestion has been completed, bile salts promote absorption of lipids by the intestinal epithelium. Over 90 percent of the bile salts are themselves reabsorbed, primarily in the ileum, as lipid digestion is completed. The reabsorbed bile salts enter the hepatic portal circulation and are collected and recycled by the liver. The cycling of bile salts from the liver to the small intestine and back is called the **enterohepatic circulation of bile.**

❑ The Gallbladder

Figure 24-21

The **gallbladder** is a hollow, pear-shaped, muscular organ. It is divided into three regions: the **fundus,** the **body,** and the **neck** (Figure 24-21a●). The cystic duct leads from the gallbladder toward its union with the common hepatic duct to form the common bile duct. At the duodenum, the common bile duct meets the pancreatic duct before emptying into the **duodenal ampulla** (am-PUL-a) (Figure 24-21b●). The duodenal ampulla receives buffers and enzymes from the pancreas and bile from the liver and gallbladder. It opens into the duodenum at a small mound, the **duodenal papilla.**

A muscular sphincter, the **pancreaticohepatic sphincter** (*sphincter of Oddi*), encircles the lumen of the common bile duct, and usually the pancreatic duct and ampulla as well. This sphincter remains contracted unless stimulated by the intestinal hormone *cholecystokinin.*

Physiology of the Gallbladder

The gallbladder has two major functions, *bile storage* and *bile modification.* Liver cells produce roughly 1 liter of bile each day, but the pancreaticohepatic sphincter remains closed except at mealtimes. Over the interim, when bile cannot flow along the common bile duct, it enters the cystic duct for storage within the expandable gallbladder. When filled to capacity, the gallbladder contains 40–70 ml of bile. As bile remains in the gallbladder its composition gradually changes. Much of the water is absorbed, and the bile salts and other components of bile become increasingly concentrated.

● **FIGURE 24-21**

The Gallbladder. (a) A view of the inferior surface of the liver, showing the position of the gallbladder and ducts that transport bile from the liver to the gallbladder and duodenum. A portion of the lesser omentum has been cut away to make it easier to see the relationship of the common bile duct to the hepatic duct and cystic duct. **(b)** Interior view of the duodenum, showing the duodenal ampulla and related structures. AM *Plate 6.5; Scan 16*

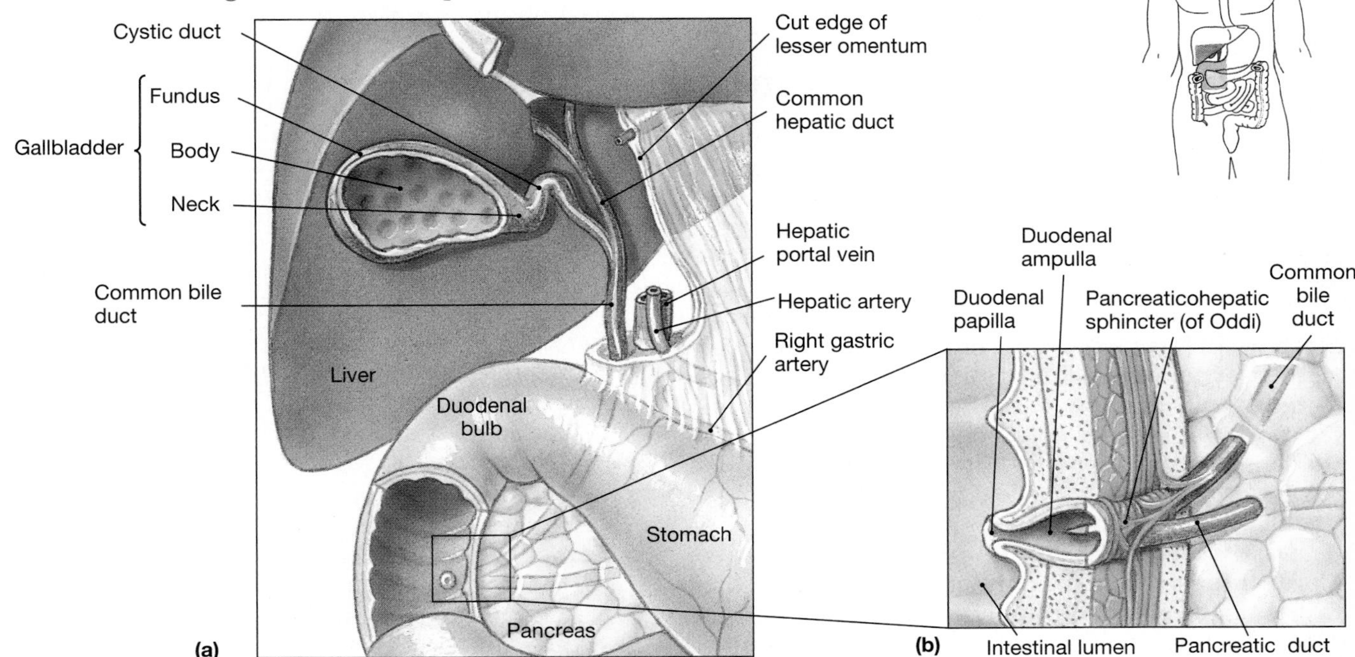
Cystic duct
Fundus
Gallbladder
Body
Neck
Common bile duct
Liver
Duodenal bulb
Stomach
Pancreas
(a)
Cut edge of lesser omentum
Common hepatic duct
Hepatic portal vein
Hepatic artery
Right gastric artery
Duodenal ampulla
Duodenal papilla
Pancreaticohepatic sphincter (of Oddi)
Common bile duct
(b)
Intestinal lumen
Pancreatic duct

Bile secretion occurs continually, but bile release into the duodenum occurs only under stimulation of cholecystokinin. In the absence of CCK, the pancreaticohepatic sphincter remains closed, and bile traveling out of the liver in the common hepatic duct goes to the gallbladder via the cystic duct. Cholecystokinin (1) relaxes the pancreaticohepatic sphincter and (2) stimulates contractions in the walls of the gallbladder. These contractions push bile into the small intestine. Cholecystokinin release occurs whenever acid chyme enters the duodenum, but the amount secreted increases markedly when the chyme contains large amounts of lipids.

PROBLEMS WITH BILE STORAGE AND EXCRETION If bile becomes too concentrated, crystals of insoluble minerals and salts begin to appear. These deposits are called **gallstones.** Merely having them, a condition termed **cholelithiasis** (kō-lē-li-THĪ-a-sis; *chole,* bile), does not represent a problem as long as the stones remain small. Small stones are normally flushed down the bile duct and excreted. In *cholecystitis,* the gallstones are so large that they can damage the wall of the gallbladder or block the cystic or common bile duct. A recent therapy for cholecystitis involves immersing the individual in water and shattering the stones with focused sound waves. The apparatus used is called a *lithotripter.* The particles produced are then small enough to pass through the bile duct without difficulty. In severe cases of cholecystitis, the gallbladder may become infected, inflamed, or perforated. Under these conditions it may be surgically removed, in a procedure known as a *cholecystectomy.* This loss does not seriously impair digestion, for bile production continues at normal levels. However, the bile is more dilute, and its entry into the small intestine is not as closely tied to the arrival of food in the duodenum. AM *Cholecystitis*

The Coordination of Secretion and Absorption

A combination of neural and hormonal mechanisms coordinates the activities of the digestive glands. These regulatory mechanisms are centered around the duodenum, for it is there that the acids must be neutralized and the appropriate enzymes added.

Neural mechanisms involving the CNS are concerned with (1) preparing the digestive tract for activity (parasympathetic innervation) or inhibiting gastrointestinal activity (sympathetic innervation) and (2) coordinating the movement of materials along the length of the digestive tract (the enterogastric, gastroenteric, and gastroileal reflexes).

In addition, motor neurons synapsing within the digestive tract release a variety of neurotransmitters. Many of these chemicals are also found inside the CNS, but in general their functions are poorly understood. Examples of potentially important neurotransmitters include substance P, enkephalins, and endorphins.

Hormones important to the regulation of intestinal and glandular function were introduced in the course of our discussion. Information on these hormones is summarized in Table 24-1.

Intestinal Hormones

Figure 24-22

The intestinal tract secretes a variety of hormones, but its hormonal secretions are poorly understood. It has proved very difficult to determine the primary effects of these hormones, largely because all are peptide hormones with similar chemical structures. Careful analyses have led to a marked increase in the number of identified intestinal hormones, but their specific functions have yet to be sorted out to everyone's satisfaction.

Duodenal enteroendocrine cells produce the following hormones known to coordinate digestive functions:

- **Enterocrinin,** a hormone released when acidic chyme enters the small intestine, stimulates the submucosal glands of the duodenum. The effects of enterocrinin are relatively straightforward, but other intestinal hormones have multiple effects that may target several regions of the digestive tract as well as affecting the accessory glandular organs.
- **Secretin** is released when acid chyme arrives in the duodenum. The primary effect of secretin is to cause an increase in the secretion of water and buffers by the pancreas and liver. Among its secondary effects, secretin further stimulates the duodenal submucosal glands and reduces gastric motility and secretory rates.
- **Cholecystokinin,** or **CCK,** is secreted when chyme arrives in the duodenum, especially when it contains lipids and partially digested proteins. In the pancreas, CCK accelerates the production and secretion of all types of digestive enzymes. It also causes relaxation of the pancreaticohepatic sphincter and contraction of the gallbladder, resulting in the ejection of bile and pancreatic juice into the duodenum. The net effects of CCK are thus (1) to increase the secretion of pancreatic enzymes and (2) to push pancreatic secretions and bile into the duodenum. The presence of CCK in high concentrations has two additional effects: (1) It inhibits gastric activity, and (2) it appears to have CNS effects that reduce the sensation of hunger.
- **Gastric inhibitory peptide,** or **GIP,** is secreted when fats and partially digested proteins enter the small intestine. The inhibition of gastric activity is accompanied by the stimulation of insulin release at the pancreatic islets. For this reason, GIP is also known as *glucose-*

TABLE 24-1 Important Gastrointestinal Hormones and Their Primary Effects

Hormone	Stimulus	Origin	Target	Effects
Gastrin	Vagal stimulation or arrival of food in the stomach	Stomach	Stomach	Stimulates production of acids and enzymes, increases motility
Enterocrinin	Arrival of acid chyme in the duodenum	Duodenum	Duodenal (Brunner's) glands	Stimulates production of alkaline mucus
Secretin	Arrival of acid chyme in the duodenum	Duodenum	Pancreas	Stimulates production of alkaline buffers
			Stomach	Inhibits gastric secretion and motility
Cholecystokinin (CCK)	Arrival of acid chyme containing lipids and partially digested proteins	Duodenum	Pancreas	Stimulates production of pancreatic enzymes
			Gallbladder	Stimulates contraction of gallbladder
			Duodenum	Causes relaxation of sphincter at base of bile duct
			Stomach	Inhibits gastric secretion and motion
Gastric-inhibitory peptide (GIP)	Arrival of chyme containing large quantities of glucose	Duodenum	Pancreas	Stimulates release of insulin by pancreatic islets
Vasoactive intestinal peptide (VIP)	Arrival of chyme in the duodenum	Duodenum	Duodenal glands, stomach	Stimulates buffer secretion, inhibits acid production, dilates intestinal capillaries
Gastrin	Arrival of chyme containing large quantities of undigested proteins	Duodenum	Stomach	Stimulates gastric secretion and motion (as above)

dependent insulinotropic peptide. Like secretin, GIP stimulates the activity of the duodenal submucosal glands. It also has effects in a variety of other tissues. For example, GIP works with insulin to stimulate lipid synthesis in adipose tissue and increases glucose use in skeletal muscle.

- **Vasoactive intestinal peptide (VIP)** stimulates the secretion of intestinal glands, dilates regional capillaries, and inhibits acid production in the stomach. By dilating capillaries in active areas of the intestinal tract, VIP provides an efficient mechanism for removing absorbed nutrients.
- **Gastrin** is secreted by G cells in the duodenum when they are exposed to large quantities of incompletely digested proteins.
- ***Other intestinal hormones.*** Several other hormones are produced in relatively small quantities. Examples include *motilin,* which stimulates intestinal contractions, *villikinin,* which promotes movement of villi and associated lymph flow, and *somatostatin,* which inhibits gastric secretion.

Functional interactions among gastrin, secretin, CCK, GIP, and VIP are diagrammed in Figure 24-22●.

Intestinal Absorption

On average it takes about 5 hours for chyme to pass from the duodenum to the end of the ileum, so the first of the materials to enter the duodenum after breakfast may leave the small intestine at lunch. Along the way, absorptive effectiveness is enhanced by the fact that so much of the mucosa is movable. The microvilli can be moved by their supporting microfilaments, the individual villi by smooth muscle cells, groups of villi by the muscularis mucosae, and the plicae by the muscularis mucosae and the muscularis externa. All of these movements tend to stir and mix the intestinal contents. As a result, the environment around each epithelial cell changes from moment to moment.

How is the small intestine adapted for the absorption of nutrients?

How would a meal that is high in fat affect the level of cholecystokinin (CCK) in the blood?

How would the pH of intestinal contents be affected if the small intestine did not secrete the hormone secretin?

The digestion of which nutrient would be most impaired by damage to the exocrine pancreas?

The Large Intestine

Figures 24-17a, 24-23

The horseshoe-shaped large intestine begins at the end of the ileum and ends at the anus. The large intestine lies inferior to the stomach and liver and almost completely frames the small intestine (Figure 24-17a●, p. 906). The major functions of the large intestine include: (1) the reabsorption of water and compaction of chyme into feces, (2) the absorption of important vitamins liberated by bacterial action, and (3) the storing of fecal material prior to defecation.

The large intestine, often called the **large bowel,** has an average length of about 1.5 meters (5 ft) and a width of 7.5 cm (3 in.). It can be divided into three parts (1) the pouchlike *cecum,* the first portion of the large intestine; (2) the *colon,* the largest portion of the large intestine; and (3) the *rectum,* the last 15 cm (6 in.) of the large intestine and the end of the digestive tract (Figure 24-23●).

The Cecum

Figure 24-23a

Material arriving from the ileum first enters an expanded pouch called the **cecum** (SĒ-kum). The ileum attaches to the medial surface of the cecum and opens into the cecum at the **ileocecal** (il-ē-ō-SĒ-kal) **valve** (Figure 24-23a●). The cecum collects and stores chyme and begins the process of compaction. The slender, hollow **vermiform appendix** (*vermis,* a worm) is attached to the posteromedial surface of the cecum. The appendix is usually about 9 cm (3.5 in.) long, but its size and shape are quite variable. A band of mesentery, called the **mesoappendix,** connects the appendix to the ileum and cecum. The mucosa and submucosa of the appendix are dominated by lymphoid nodules, and its primary function is as an organ of the lymphatic system. Inflammation of the appendix is known as *appendicitis,* a condition noted in Chapter 22. ∞ [p. 786]

The Colon

Figure 24-23

The colon has a larger diameter and a thinner wall than the small intestine. Distinctive features of the colon (Figure 24-23●) include:

1. The wall of the colon forms a series of pouches, or **haustra** (HAWS-truh), that permit considerable distension and elongation. Cutting into the intestinal lumen reveals that the creases between the haustra affect the mucosal lining as well, producing a series of internal folds.
2. Three separate longitudinal ribbons of smooth muscle, the **taenia coli** (TĒ-nē-a KŌ-lī), are visible on the outer surfaces of the colon just beneath the serosa. These bands correspond to the outer layer of the muscularis externa in other portions of the digestive tract. Muscle tone within these bands creates the haustra.
3. The serosa of the colon contains numerous teardrop-shaped sacs of fat, called **epiploic** (e-pip-LŌ-ik) **appendages.**

Regions of the Colon

Figures 24-5, 24-23

The colon can be subdivided into four regions: the *ascending colon,* the *transverse colon,* the *descending colon,* and the *sigmoid colon* (Figure 24-23●).

- The **ascending colon** begins at the superior border of the cecum and ascends along the right lateral and posterior wall of the peritoneal cavity to the inferior surface of the liver. At this point the colon makes a sharp bend to

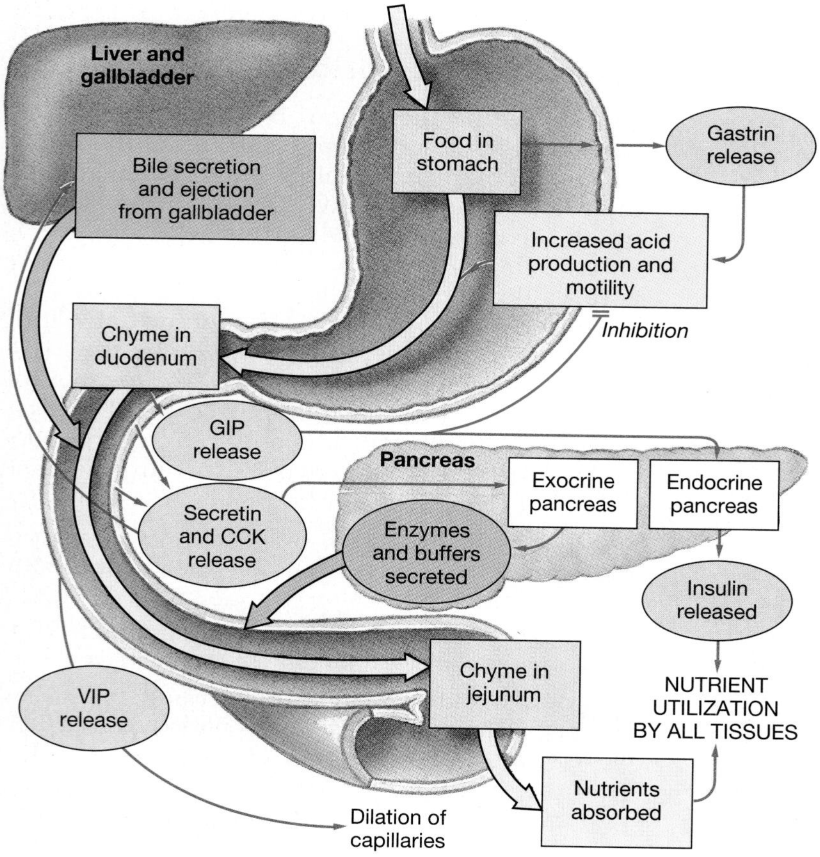

Liver and gallbladder
Food in stomach
Gastrin release
Bile secretion and ejection from gallbladder
Increased acid production and motility
Inhibition
Chyme in duodenum
GIP release
Pancreas
Exocrine pancreas
Endocrine pancreas
Secretin and CCK release
Enzymes and buffers secreted
Insulin released
Chyme in jejunum
NUTRIENT UTILIZATION BY ALL TISSUES
VIP release
Nutrients absorbed
Dilation of capillaries

● **FIGURE 24-22**
The Activities of Major Digestive Tract Hormones. This diagram follows the primary actions of gastrin, GIP, secretin, CCK, and VIP.

the left at the **right colic flexure,** or *hepatic flexure.* The right colic flexure marks the end of the ascending colon and the beginning of the *transverse colon.* The lateral and anterior surfaces of the ascending colon are covered by visceral peritoneum. There is no mesentery, and the ascending colon is retroperitoneal (Figure 24-5•, p. 884).

- The **transverse colon** begins at the right colic flexure. It curves anteriorly and crosses the abdomen from right to left. It is supported by the *transverse mesentery* and is separated from the anterior abdominal wall by the layers of the greater omentum. As the transverse colon reaches the left side, it passes below the greater curvature of the stomach. Near the spleen, the colon makes a 90° turn at the **left colic flexure,** or *splenic flexure,* and then proceeds inferiorly.
- The **descending colon** proceeds inferiorly along the left side until reaching the iliac fossa. The descending colon is retroperitoneal and firmly attached to the abdominal wall. At the iliac fossa the descending colon curves at the **sigmoid flexure** and enters an S-shaped segment, the *sigmoid colon.*
- The sigmoid flexure is the start of the **sigmoid** (SIG-moyd) **colon** (*sigmeidos,* the Greek letter S), an S-shaped segment that is only about 15 cm (6 in.) long. It lies posterior to the urinary bladder, suspended from the *sigmoid mesocolon* (Figure 24-5•, p. 884). The sigmoid colon empties into the *rectum.*

The large intestine receives blood from tributaries of the *superior mesenteric* and *inferior mesenteric arteries,* and venous blood is collected by the *superior mesenteric* and *inferior mesenteric veins.* These vessels were detailed in Chapter 21. [p. 767]

The Rectum

Figure 24-23b

The **rectum** (REK-tum) forms the last 15 cm (6 in.) of the digestive tract (Figure 24-23b•). The rectum is an expandable organ for the temporary storage of fecal material. Movement of fecal materials into the rectum triggers the urge to defecate.

The last portion of the rectum, the **anorectal** (ā-nō-REK-tal) **canal,** contains small longitudinal folds, the **rectal columns.** The distal margins of the rectal columns are joined by transverse folds that mark the boundary between the columnar epithelium of the proximal rectum and a stratified squamous epithelium similar to that found in the oral cavity. Very close to the anus the epidermis becomes keratinized and identical to the surface of the skin.

A network of veins in the lamina propria and submucosa of the anorectal canal occasionally becomes distended, producing *hemorrhoids.* The circular muscle layer of the muscularis externa in this region forms the **internal anal sphincter.** The smooth muscle fibers of the internal anal sphincter are not under voluntary control. The **external anal sphincter** guards the exit of the anorectal canal, termed **anal orifice.** This sphincter, which consists of a ring of skeletal muscle fibers, is under voluntary control.

COLON CANCER Colon cancers are relatively common. Approximately 149,000 cases are diagnosed in the United States each year, and in 1994 there were an estimated 58,000 deaths from colon and rectal cancers. The mortality rate for these cancers remains high, and the best defense appears to be early detection and prompt treatment. The standard screening test involves checking the feces for blood. This is a simple procedure that can easily be performed on a stool (fecal) sample in the course of a routine physical. AM *Colon Inspection and Cancer*

Histology of the Large Intestine

Figure 24-23c

Although the diameter of the colon is roughly three times that of the small intestine, its wall is much thinner. The major characteristics of the colon are the lack of villi, the abundance of goblet cells, and presence of distinctive intestinal glands (Figure 24-23c•). The glands in the large intestine are deeper than those of the small intestine, and they are dominated by goblet cells. The mucosa of the large intestine does not produce enzymes—any digestion that occurs in the feces results from enzymes introduced in the small intestine or from bacterial action. The mucus is important in providing lubrication as the fecal material becomes less moist and more compact. Mucous secretion occurs as local stimuli, such as friction or exposure to harsh chemicals, trigger short reflexes involving local nerve plexuses. Large lymphatic nodules are scattered throughout the lamina propria and submucosa.

The muscularis externa of the large intestine is unusual because the longitudinal layer has condensed to form the muscular bands of the taenia coli. However, the mixing and propulsive contractions of the colon resemble those of the small intestine.

DIVERTICULOSIS In *diverticulosis* (dī-ver-tik-ū-LŌ-sis), pockets (*diverticula*) form in the mucosa, usually in the sigmoid colon. These get forced outward, probably by the pressures generated during defecation. If the pockets push through weak points in the muscularis externa, they form semi-isolated chambers that are subject to recurrent infection and inflammation. The infections cause pain and occasional bleeding, a condition known as *diverticulitis* (dī-ver-tik-ū-LĪ-tis). Inflammation of other portions of the colon is called *colitis* (kō-LĪ-tis). AM *Inflammatory Bowel Disease*

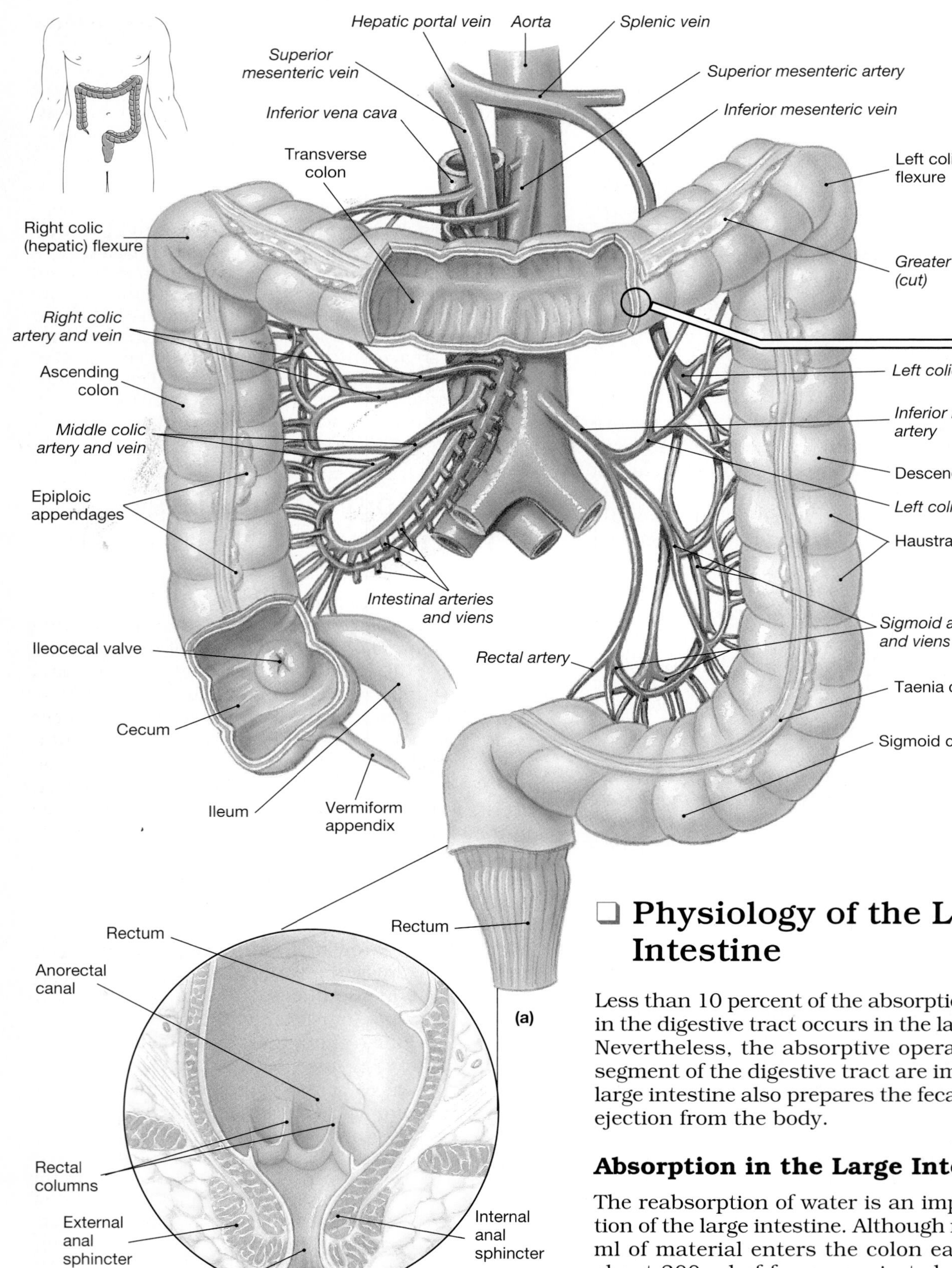
Hepatic portal vein
Aorta
Splenic vein
Superior mesenteric vein
Superior mesenteric artery
Inferior vena cava
Inferior mesenteric vein
Transverse colon
Left colic (splenic) flexure
Right colic (hepatic) flexure
Greater omentum (cut)
Right colic artery and vein
Ascending colon
Left colic vien
Middle colic artery and vein
Inferior mesenteric artery
Descending colon
Epiploic appendages
Left colic artery
Haustra
Intestinal arteries and viens
Sigmoid arteries and viens
Ileocecal valve
Rectal artery
Taenia coli
Cecum
Sigmoid colon
Ileum
Vermiform appendix
Rectum
Rectum
Anorectal canal
(a)
Rectal columns
External anal sphincter
Internal anal sphincter
Anus
(b)

● **FIGURE 24-23**
The Large Intestine. (a) Gross anatomy and regions of the large intestine. **(b)** Detailed anatomy of the rectum and anus. **(c)** The mucosa and glands of the colon. (LM × 114)
Scan 19

Physiology of the Large Intestine

Less than 10 percent of the absorption under way in the digestive tract occurs in the large intestine. Nevertheless, the absorptive operations in this segment of the digestive tract are important. The large intestine also prepares the fecal material for ejection from the body.

Absorption in the Large Intestine

The reabsorption of water is an important function of the large intestine. Although roughly 1500 ml of material enters the colon each day, only about 200 ml of feces are ejected. The remarkable efficiency of digestion can best be appreciated by considering the average composition of fecal wastes: 75 percent water, 5 percent bacteria, and the rest a mixture of indigestible materials, small quantities of inorganic matter, and the remains of epithelial cells.

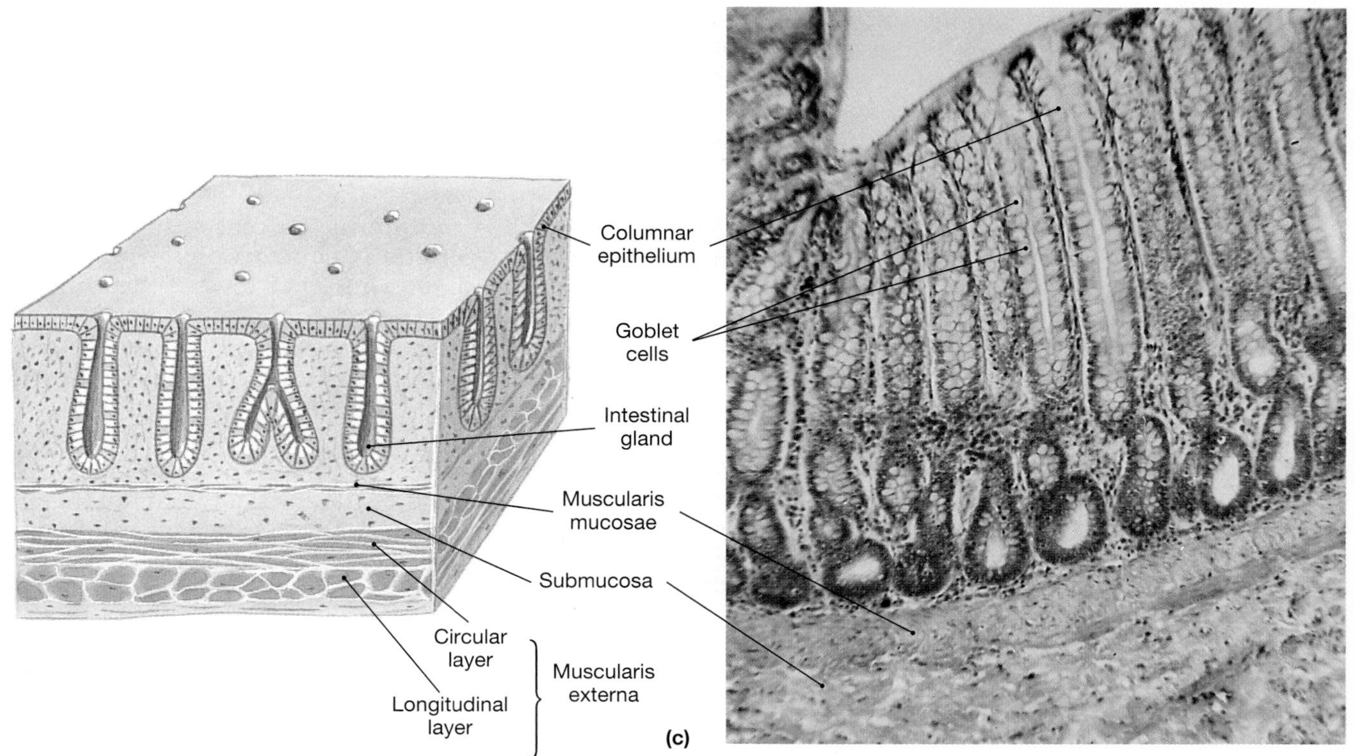
Columnar
epithelium
Goblet
cells
Intestinal
gland
Muscularis
mucosae
Submucosa
Circular
layer
Longitudinal
layer
Muscularis
externa
(c)

In addition to reabsorbing water, the large intestine absorbs a number of other substances that remain in the chyme or that were secreted into the chyme as it passed along the digestive tract.

- ***Vitamins.*** Vitamins are organic molecules that are important as cofactors or coenzymes in many metabolic pathways. Colonic bacteria generate three vitamins that supplement the dietary supply:

    1. *Vitamin K*, a fat-soluble vitamin that the liver needs to enable it to synthesize four clotting factors, including prothrombin. Intestinal bacteria produce roughly half of our daily vitamin K requirements.
    2. *Biotin*, a water-soluble vitamin important to a variety of reactions, notably those involved with glucose metabolism.
    3. *Vitamin B_5* (pantothenic acid), a water-soluble vitamin required in the manufacture of steroid hormones and some neurotransmitters.

Disorders resulting from deficiencies of biotin or vitamin B_5 are extremely rare after infancy because the intestinal bacteria produce sufficient amounts to supplement any shortage in the diet. Vitamin K deficiencies, which lead to impaired blood clotting, can result from (1) inadequate lipids in the diet, which impairs the absorption of all fat-soluble vitamins, or (2) problems affecting lipid processing and absorption, such as inadequate bile production or chronic diarrhea.

- ***Urobilinogen.*** Chapter 19 discussed the fate of conjugated bilirubin, a breakdown product of heme. ∞ [p. 657] Inside the large intestine, bacteria convert the conjugated bilirubin to *urobilinogens* and *stercobilinogens*. Some of the urobilinogens are absorbed into the circulation and excreted in the urine. On exposure to oxygen some of the urobilinogens and stercobilinogens are converted to **urobilins** and **stercobilins.** These pigments give feces a brown coloration.
- ***Bile salts.*** Some of the bile salts remaining in the feces will be reabsorbed in the cecum.
- ***Toxins.*** Bacterial action breaks down peptides remaining in the feces and generates (1) ammonia, in the form of soluble *ammonium ions* (NH_4^+), (2) *indole* and *skatole*, two nitrogen-containing compounds that are primarily responsible for the odor of feces, and (3) hydrogen sulfide (H_2S), a gas that produces a "rotten egg" odor. Significant amounts of ammonia and smaller amounts of other toxins cross the colonic epithelium and enter the hepatic portal circulation. These are removed by the liver and converted to relatively nontoxic compounds that can be released into the blood and excreted at the kidneys.

Indigestible carbohydrates are not altered by intestinal enzymes, and they arrive in the colon virtually intact. These complex polysaccharides provide a reliable nutrient source for colonic bacteria, whose metabolic activities are responsible for the small quantities of intestinal gas, or **flatus,** found in the large intestine. Meals containing large numbers of indigestible carbohydrates (such as "beans and franks") stimulate bacterial gas production, leading to colonic distension, cramps, and the frequent discharge of intestinal gases.

Movements of the Large Intestine

Figure 24-24

The gastroileal reflex moves chyme into the cecum at mealtimes. Movement from the cecum to the transverse colon occurs very slowly, allowing hours for water absorption that converts the already thick chyme into a sludgy paste. Movement from the transverse colon through the rest of the large intestine results from powerful peristaltic contractions, called **mass movements,** that occur a few times each day. The stimulus is distension of the stomach and duodenum, and the commands are relayed over the intestinal nerve plexuses. The contractions force fecal materials into the rectum and produce the conscious urge to defecate.

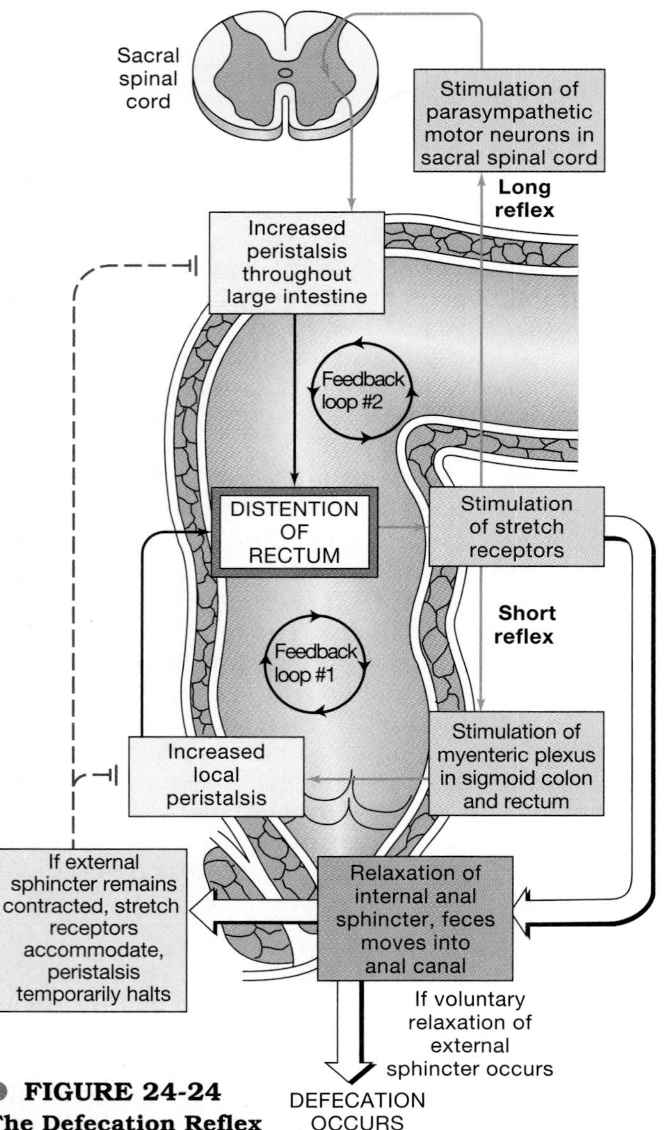

Sacral spinal cord
Stimulation of parasympathetic motor neurons in sacral spinal cord
Long reflex
Increased peristalsis throughout large intestine
Feedback loop #2
DISTENTION OF RECTUM
Stimulation of stretch receptors
Short reflex
Feedback loop #1
Stimulation of myenteric plexus in sigmoid colon and rectum
Increased local peristalsis
If external sphincter remains contracted, stretch receptors accommodate, peristalsis temporarily halts
Relaxation of internal anal sphincter, feces moves into anal canal
If voluntary relaxation of external sphincter occurs
DEFECATION OCCURS

● **FIGURE 24-24**
The Defecation Reflex

DEFECATION The rectal chamber is usually empty except when one of those powerful peristaltic contractions forces fecal materials out of the sigmoid colon. Distension of the rectal wall then triggers the defecation reflex. The **defecation reflex** (Figure 24-24●) involves two positive feedback loops:

1. Stretch receptors in the rectal walls order a series of peristaltic contractions in the colon and rectum, moving feces toward the anus.
2. The sacral parasympathetic system, also activated by the stretch receptors, stimulates peristalsis via motor commands distributed by the pelvic nerves.

Movement of feces through the anorectal canal requires relaxation of the internal anal sphincter, but when the internal sphincter relaxes, the external sphincter automatically clamps shut. Thus the actual release of feces requires conscious effort to open the external sphincter voluntarily. In addition to opening the external sphincter, consciously directed activities such as tensing the abdominal muscles or making expiratory movements while closing the glottis elevate intra-abdominal pressures and help to force fecal materials out of the rectum.

If the commands do not arrive, the peristaltic contractions cease until additional rectal expansion triggers the defecation reflex a second time. If pressure inside the rectum gets too high, the external sphincter will relax, and defecation will occur. This mechanism regulates defecation in infants and in adults with severe spinal cord injuries.

DIARRHEA AND CONSTIPATION **Diarrhea** (dī-a-RĒ-a) exists when an individual has frequent, watery bowel movements. Diarrhea results when the colonic mucosa becomes unable to maintain normal levels of absorption or the rate of fluid entry into the colon exceeds its maximum reabsorptive capacity. Bacterial, viral, or protozoan infection of the colon or small intestine may cause acute bouts of diarrhea lasting several days. Severe diarrhea can be life-threatening, because of cumulative fluid and ion losses. In *cholera* (KOL-e-ra), bacteria bound to the intestinal lining release toxins that stimulate a massive fluid secretion across the intestinal epithelium. Without treatment the victim may die of acute dehydration in a matter of hours. AM *Diarrhea: Cholera Epidemics*

Constipation is infrequent defecation, usually involving dry and hard feces. Constipation occurs when fecal materials are moving through the colon so slowly that excessive water reabsorption occurs. The feces then become extremely compact, difficult to move, and highly abrasive. Inadequate dietary fiber and fluids, coupled with a lack of exercise, are the usual causes. Constipation can usually be treated by oral administration of stool softeners, such as *Colace*, laxatives, or **cathartics** (ka-THAR-tiks) that promote defecation. These compounds promote water movement into the feces, increase fecal mass, or irritate the lining of the colon to stimulate peristalsis. For example, indigestible fiber adds bulk to the feces, retaining moisture and stimulating stretch receptors that promote peristalsis. Promotion of peristalsis is one of the benefits of "high-fiber" cereals. Active movement during exercise also assists in the movement of fecal materials through the colon.

Digestion and Absorption

A typical meal contains a mixture of carbohydrates, proteins, lipids, water, electrolytes, and vitamins. The digestive system handles each of these components differently. Large organic molecules must be broken down through digestion before absorption can occur. Water, electrolytes, and vitamins can be absorbed without preliminary processing, but special transport mechanisms are often involved.

The Processing and Absorption of Nutrients

Food contains large organic molecules, many of them insoluble. The digestive system first breaks down the physical structure of the ingested material and then proceeds to disassemble the component molecules into smaller fragments. This disassembly eliminates any antigenic properties, making the fragments suitable for absorption. The molecules released into the bloodstream will be absorbed by cells and either (1) broken down to provide energy for the synthesis of ATP or (2) used to synthesize carbohydrates, proteins, and lipids. This section will focus on the mechanics of digestion and absorption; the fate of the compounds inside cells will be the focus of Chapter 25.

Ingested organic materials are usually complex chains of simpler molecules. In a typical dietary carbohydrate the basic molecules are simple sugars. In a protein, the building blocks are amino acids, and in lipids they are usually fatty acids. Digestive enzymes break the bonds between the component molecules in a process called *hydrolysis.* (The hydrolysis of carbohydrates, lipids, and proteins was detailed in Chapter 2.)

The major classes of digestive enzymes differ with respect to their specific targets. *Carbohydrases* break the bonds between sugars, *proteases* split the linkages between amino acids, and *lipases* separate the fatty acids from glycerides. Specific enzymes within each class may be even more selective, breaking bonds involving specific molecular participants. For example, a particular carbohydrase might break bonds between glucose molecules but not those between glucose and another simple sugar.

The enzymes involved in digestion are localized in two different sites. The enzymes secreted by the salivary glands, tongue, stomach, and pancreas are mixed into the ingested material as it passes along the digestive tract. These enzymes break down large proteins, lipids, and carbohydrates into smaller fragments, which in turn must often be broken down further before absorption can occur. The final enzymatic steps involve enzymes attached to the exposed surfaces of microvilli, or *brush border enzymes.*

Figure 24-25 summarizes information concerning the digestive fates of carbohydrates, lipids, and proteins, and Table 24-2 reviews the major digestive enzymes and their functions. We will now take a closer look at the digestion and absorption of carbohydrates, lipids, and proteins.

Carbohydrate Digestion and Absorption

Carbohydrate digestion proceeds in two steps, one involving carbohydrases produced by the salivary glands and pancreas and the other involving brush border enzymes.

Salivary and Pancreatic Enzymes

Carbohydrate digestion involves two enzymes, salivary amylase and pancreatic amylase, that function effectively at a pH of 6.7–7.5. Carbohydrate digestion begins in the mouth during mastication, through the action of salivary amylase from the parotid and submandibular salivary glands. Salivary amylase breaks down complex carbohydrates into smaller fragments, producing a mixture composed primarily of *disaccharides* (two sugars) and *trisaccharides* (three sugars). Salivary amylase continues to digest the starches and glycogen in the meal for 1–2 hours before stomach acids render it inactive. Because the enzymatic content of saliva is not high, only a relatively small amount of digestion occurs over this period.

In the duodenum, the remaining complex carbohydrates are broken down through the action of pancreatic amylase. Any disaccharides or trisaccharides produced, as well as any already present in the food, are ignored by both salivary and pancreatic amylase. Additional hydrolysis does not occur until these molecules contact the intestinal mucosa.

Brush Border Enzymes

Prior to absorption, disaccharides and trisaccharides are fragmented into simple sugars by enzymes found on the surfaces of the intestinal microvilli.

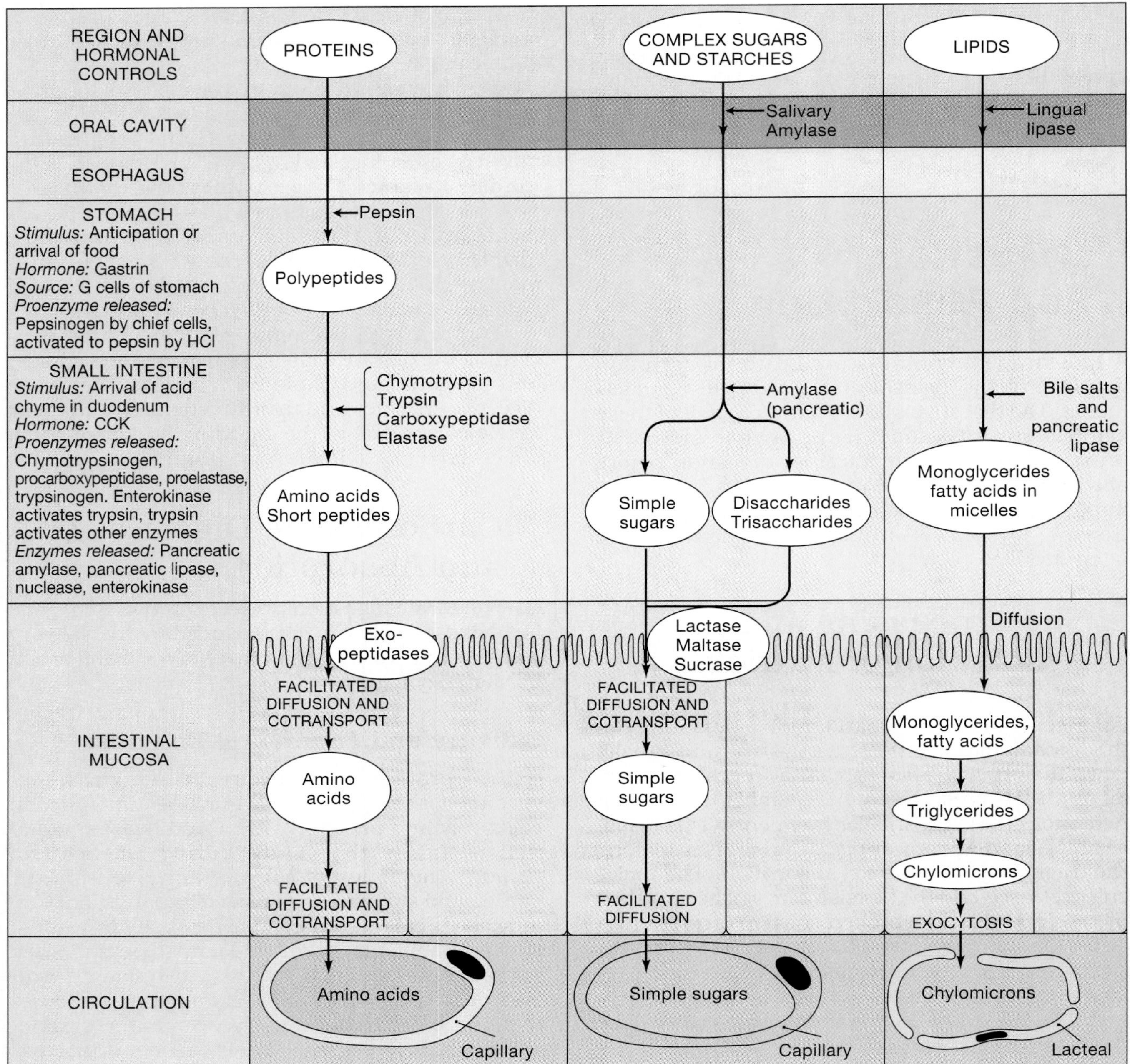
REGION AND HORMONAL CONTROLS
PROTEINS
COMPLEX SUGARS AND STARCHES
LIPIDS
ORAL CAVITY
Salivary Amylase
Lingual lipase
ESOPHAGUS
STOMACH
Stimulus: Anticipation or arrival of food
Hormone: Gastrin
Source: G cells of stomach
Proenzyme released: Pepsinogen by chief cells, activated to pepsin by HCl
Pepsin
Polypeptides
SMALL INTESTINE
Stimulus: Arrival of acid chyme in duodenum
Hormone: CCK
Proenzymes released: Chymotrypsinogen, procarboxypeptidase, proelastase, trypsinogen. Enterokinase activates trypsin, trypsin activates other enzymes
Enzymes released: Pancreatic amylase, pancreatic lipase, nuclease, enterokinase
Chymotrypsin
Trypsin
Carboxypeptidase
Elastase
Amylase (pancreatic)
Bile salts and pancreatic lipase
Amino acids Short peptides
Simple sugars
Disaccharides Trisaccharides
Monoglycerides fatty acids in micelles
INTESTINAL MUCOSA
Exo-peptidases
Lactase Maltase Sucrase
Diffusion
FACILITATED DIFFUSION AND COTRANSPORT
FACILITATED DIFFUSION AND COTRANSPORT
Amino acids
Simple sugars
Monoglycerides, fatty acids
Triglycerides
Chylomicrons
FACILITATED DIFFUSION AND COTRANSPORT
FACILITATED DIFFUSION
EXOCYTOSIS
CIRCULATION
Amino acids
Capillary
Simple sugars
Capillary
Chylomicrons
Lacteal

● **FIGURE 24-25**
A Summary of the Chemical Events in Digestion. For further details on the enzymes involved, see Table 24-2.

Maltase splits bonds between the two glucose molecules of the disaccharide *maltose.* **Sucrase** breaks the disaccharide **sucrose** into glucose and *fructose,* another six-carbon sugar. **Lactase** releases a molecule of glucose and one of *galactose* from the hydrolysis of the disaccharide **lactose.** Lactose is the primary carbohydrate in milk, so by breaking down lactose, lactase provides essential services throughout infancy and early childhood. The intestinal mucosa often stops producing lactase by adolescence, in which case the individual becomes *lactose intolerant.* Individuals who are lactose intolerant can have a variety of unpleasant digestive problems after eating a meal containing milk and other dairy products. AM *Lactose Intolerance*

Absorption of Monosaccharides

The intestinal epithelium then absorbs the simple sugars through *facilitated diffusion* and *cotransport mechanisms.* Facilitated diffusion, which involves a carrier protein, was detailed in Chapter

3 (see Figure 3-10●, p. 78). Cotransport also involves a carrier protein, but there are three major differences between cotransport and facilitated diffusion:

1. Cotransport moves more than one molecule or ion through the membrane at the same time. The transported materials move in the same direction, which is down the concentration gradient for at least one of the transported substances.
2. Cotransport *by itself* does not require ATP. However, the cell must often expend ATP to preserve homeostasis when cotransport is under way.
3. Cotransport can occur despite an opposing concentration gradient for one of the transported substances. For example, cells lining the small intestine will continue to absorb glucose when glucose concentrations inside the cells are much higher than they are in the intestinal contents.

The cotransport system responsible for the uptake of glucose also brings sodium ions into the cell. This is a passive process that resembles facilitated diffusion except that both a sodium ion and a glucose molecule must bind to the carrier protein before it will permit their movement into the cell. Glucose cotransport is an example of *sodium-linked cotransport.* Comparable cotransport mechanisms

TABLE 24-2 Digestive Enzymes and Their Functions

Enzyme (Proenzyme)	*Source*	*Optimal pH*	*Target and Action*	*Products*	*Remarks*
Alpha-amylase	Salivary glands, pancreas	6.7–7.5	Breaks bonds between carbohydrate molecules	Disaccharides and trisaccharides	
Lingual lipase	Glands of tongue	6.7–7.5	Triglycerides	Fatty acids and monoglycerides	
Pepsin (Pepsinogen)	Chief cells of stomach	1.5–2.0	Breaks bonds between amino acids in proteins	Short-chain polypeptides	Secreted as proenzyme, pepsinogen; activated by H^+ in stomach acid
Rennin			Coagulates milk proteins		Secreted only by stomachs of infants
Trypsin (Trypsinogen)	Pancreas	7–8	Proteins, polypeptides	Short-chain peptides	Proenzyme activated by enterokinase; activates other pancreatic proteases
Chymotrypsin (Chymotrypsinogen)	Pancreas	7–8	Proteins, polypeptides	Short-chain peptides	Activated by trypsin
Carboxypeptidase (Procarboxypeptidase)	Pancreas	7–8	Proteins, polypeptides	Short-chain peptides and amino acids	Activated by trypsin
Elastase (Proelastase)	Pancreas	7–8	Elastin	Short-chain peptides	Activated by trypsin
Pancreatic lipase	Pancreas	7–8	Triglycerides	Fatty acids and monoglycerides	Bile salts must be present for efficient action
Nuclease	Pancreas	7–8	Nucleic acids	Nitrogenous bases and simple sugars	Includes ribonuclease for RNA and deoxyribonuclease for DNA
Enterokinase	Intestinal glands	7–8	Trypsinogen	Trypsin	Reaches lumen through disintegration of shed epithelial cells
Maltase, sucrase, lactase	Brush border of small intestine	7–8	Maltose, sucrose, lactose	Monosaccharides	Found in membrane surface of microvilli
Exopeptidases	Brush border of small intestine	7–8	Dipeptides, tripeptides	Amino acids	Found in membrane surface of microvilli

exist for other simple sugars and for some amino acids. Although these mechanisms deliver valuable nutrients to the cytoplasm, they also bring sodium ions that must be ejected by the sodium-potassium exchange pump.

The simple sugars entering the cell diffuse through the cytoplasm and reach the interstitial fluid through facilitated diffusion at the base of the cell. Once in the interstitial fluid, the simple sugars diffuse into the capillaries of the villus for eventual transport to the liver in the hepatic portal vein.

❑ Lipid Digestion and Absorption

Lipid digestion involves lingual lipase from glands of the tongue and pancreatic lipase from the pancreas. Both enzymes are functional at a pH of 7–8. The most important and abundant dietary lipids are *triglycerides*, which consist of three fatty acids attached to a single molecule of glycerol (see Figure 2-14•, p. 49). The lingual and pancreatic lipases break off two of the fatty acids, leaving *monoglycerides*.

Lipases are water-soluble enzymes, and lipids tend to form large drops that exclude water molecules. As a result, lipases can attack only the exposed surfaces of the lipid drops. Lingual lipase begins breaking down triglycerides in the mouth and stomach, but the drops are so large and the time so brief that very little lipid digestion has occurred by the time the chyme enters the duodenum.

Bile salts improve the situation by emulsifying the lipid drops into tiny emulsion droplets, thus providing better access for pancreatic lipase. Pancreatic lipase then breaks apart the triglycerides to form a mixture of fatty acids and monoglycerides. As these molecules are released, they interact with bile salts in the surrounding chyme to form small lipid-bile salt complexes called **micelles** (mī-SELZ). A micelle is only about 2.5 nm (0.0025 μm) in diameter.

When a micelle contacts the intestinal epithelium, the lipids diffuse across the cell membrane and enter the cytoplasm. The intestinal cells take the monoglycerides and fatty acids and synthesize new triglycerides. These triglycerides, in company with absorbed steroids and phospholipids, are then coated with proteins, creating a complex known as a **chylomicron** (kī-lō-MĪ-kron; *chylos*, milky lymph + *mikros*, small).

The intestinal cells then secrete the chylomicrons into the interstitial fluids through exocytosis. Their superficial coating of proteins keeps the chylomicrons suspended in the interstitial fluids, but usually prevents them from diffusing into capillaries. Most of the chylomicrons released diffuse into the intestinal lacteals, which lack basement membranes and have large gaps between adjacent endothelial cells. From the lacteals they proceed along the lymphatics and through the thoracic duct, finally entering the circulation at the left subclavian vein.

❑ Protein Digestion and Absorption

Proteins have very complex structures, so protein digestion is both complex and time-consuming. The first problem is disrupting the three-dimensional organization of the food so that proteolytic enzymes can attack individual proteins. This step involves mechanical processing in the oral cavity, through mastication, and in the stomach, through the action of acidic gastric juices. Exposure of the bolus to a strongly acid environment breaks down plant cell walls and the connective tissues in animal products and has the extra benefit of killing most pathogenic microorganisms.

The acidic contents of the stomach also provide the proper environment for the activity of pepsin, the proteolytic enzyme secreted by chief cells of the stomach. Pepsin, which works effectively at a pH of 1.5–2.0, breaks peptide bonds within a polypeptide chain. A protease with this kind of activity is called an **endopeptidase.** For instance, pepsin might take a polypeptide 500 amino acids long and break it into two smaller polypeptides, one containing 200 amino acids and the other 300.

In the few hours that chyme spends in the stomach, pepsin has time to reduce the relatively huge proteins (10,000–100,000 amino acids in length) of the chyme into smaller polypeptide fragments.

When the chyme enters the duodenum, and the pH has been adjusted to 7–8, pancreatic proteases can begin working. Trypsin, chymotrypsin, and elastase are endopeptidases that, like pepsin, break peptide bonds within a polypeptide. Each enzyme has a somewhat different specialty. For example, trypsin breaks peptide bonds involving the amino acids *arginine* or *lysine*, whereas chymotrypsin targets peptide bonds involving *tyrosine* or *phenylalanine*. Carboxypeptidase is called an **exopeptidase** because it simply chops off the last amino acid of a polypeptide chain, ignoring the identities of the amino acids involved. Thus while the endopeptidases are generating peptide fragments of varying length, carboxypeptidase is producing free amino acids.

Absorption of Amino Acids

The epithelial surfaces of the small intestine contain exopeptidases and **dipeptidases** that break short peptide chains into individual amino acids. (Dipeptidases break apart *dipeptides*.) These amino acids, as well as those produced by the pancreatic enzymes, are absorbed through both facilitated diffusion and cotransport mechanisms.

After diffusing to the opposite end of the cell, the amino acids are released into the interstitial fluids by facilitated diffusion and cotransport. Once within the interstitial fluids, the amino acids diffuse into intestinal capillaries for transport to the liver via the hepatic portal vein.

❑ Water Absorption

Figure 24-26

Our cells are totally unable to absorb or secrete water actively. Thus all water movements across the lining of the digestive tract and the production of glandular secretions, such as saliva, involve passive water flow, following osmotic gradients. When two fluids are separated by a selectively permeable membrane, water will tend to flow into the solution containing the higher concentration of solutes. Osmotic movements are relatively rapid, so the interstitial fluids and those in the intestinal lumen always have the same osmolarity (concentration of solutes).

The epithelial cells are continually absorbing nutrients and ions from the intestinal contents, and these activities gradually lower the solute concentration in the chyme. As the solute concentration lowers, water moves into the surrounding tissues, maintaining osmotic equilibrium.

Each day 2–2.5 liters of water enters the digestive tract in the form of food or drink. The salivary, gastric, intestinal, pancreatic, and bile secretions provide another 6–7 liters. Out of that total, only about 150 ml is lost in the fecal wastes. The sites of secretion and absorption of water are indicated in Figure 24-26•.

❑ Electrolyte Absorption

Figure 24-27

For osmotic purposes, it really does not matter which ions are being transported. But for metabolic purposes it matters a great deal, and each ion is handled individually (Figure 24-27•). The regulatory mechanisms controlling the rates of absorption are often poorly understood.

Sodium ions, usually the most abundant cations in the chyme, may enter intestinal cells through diffusion or cotransport with another nutrient. These ions are then pumped into the interstitial fluid across the base of the cell. The rate of sodium uptake from the lumen is usually proportional to the concentration of sodium ions in the intestinal contents. As a result, eating heavily salted meals leads to increased sodium ion absorption and an associated gain of water through osmosis. The rate of sodium ion absorption by the digestive tract is increased by aldosterone, a steroid hormone from the adrenal cortex that was discussed in Chapter 18. ∞ [p. 624]

Calcium ion absorption involves active transport at the epithelial surface. The rate of transport is accelerated by parathyroid hormone and calcitriol and may be inhibited by calcitonin. The roles of these hormones in calcium ion homeostasis were considered in Chapters 6 and 18. ∞ [pp. 189, 621–622]

As other solutes move out of the chyme, the concentration of potassium ions increases. These ions can diffuse into the epithelial cells following

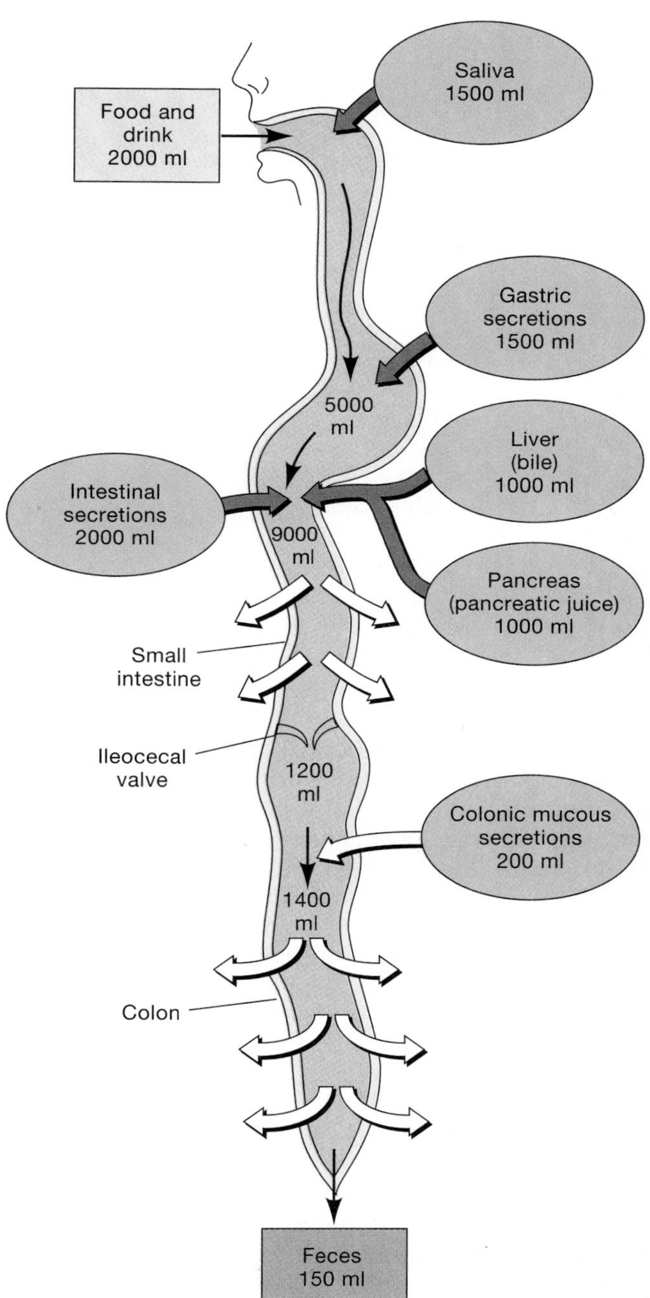

Saliva
1500 ml
Food and
drink
2000 ml
Gastric
secretions
1500 ml
5000
ml
Liver
(bile)
1000 ml
Intestinal
secretions
2000 ml
9000
ml
Pancreas
(pancreatic juice)
1000 ml
Small
intestine
Ileocecal
valve
1200
ml
Colonic mucous
secretions
200 ml
1400
ml
Colon
Feces
150 ml

● **FIGURE 24-26**

Secretion and Water Absorption along the Digestive Tract

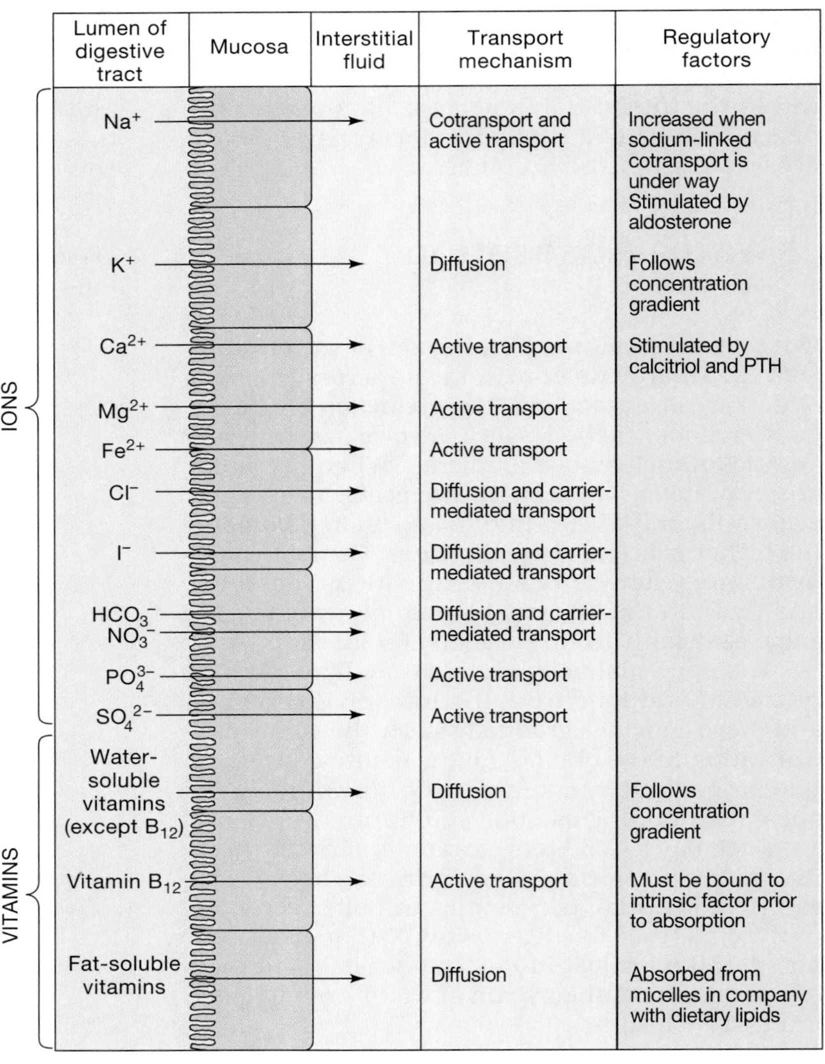
Lumen of digestive tract
Mucosa
Interstitial fluid
Transport mechanism
Regulatory factors
IONS
Na^+
Cotransport and active transport
Increased when sodium-linked cotransport is under way
Stimulated by aldosterone
K^+
Diffusion
Follows concentration gradient
Ca^{2+}
Active transport
Stimulated by calcitriol and PTH
Mg^{2+}
Active transport
Fe^{2+}
Active transport
Cl^-
Diffusion and carrier-mediated transport
I^-
Diffusion and carrier-mediated transport
HCO_3^-
NO_3^-
Diffusion and carrier-mediated transport
PO_4^{3-}
Active transport
SO_4^{2-}
Active transport
VITAMINS
Water-soluble vitamins (except B_{12})
Diffusion
Follows concentration gradient
Vitamin B_{12}
Active transport
Must be bound to intrinsic factor prior to absorption
Fat-soluble vitamins
Diffusion
Absorbed from micelles in company with dietary lipids

● **FIGURE 24-27**
Ion and Vitamin Absorption by the Digestive Tract

the concentration gradient. The absorption of magnesium, iron, and other cations involves specific carrier proteins, and the cell must use ATP to obtain and transport these ions to the interstitial fluid. Regulatory factors controlling the absorption of these cations are poorly understood.

The anions chloride, iodide, bicarbonate, and nitrate may be absorbed through diffusion or carrier-mediated transport. Phosphate and sulfate ions enter epithelial cells only through active transport.

❑ Vitamin Absorption

Figure 24-27

Vitamins are organic compounds required in very small quantities. There are two major groups of vitamins. Vitamins A, D, E, and K are called **fat-soluble vitamins** because their structure allows them to dissolve in lipids. The nine **water-soluble vitamins** include the B vitamins, common in milk and meats, and vitamin C, found in citrus fruits. Chapter 25 will consider the functions of vitamins and associated nutritional problems.

All but one of the water-soluble vitamins are easily absorbed by diffusion across the digestive epithelium. Vitamin B_{12} cannot be absorbed by the intestinal mucosa in normal amounts unless it has been bound to *intrinsic factor,* a protein secreted by the parietal cells of the stomach. The combination is then absorbed via active transport.

Fat-soluble vitamins in the diet enter the duodenum in fat droplets, mixed with triglycerides. These vitamins remain in association with these lipids as they form emulsion droplets and, after further digestion, micelles. The fat-soluble vitamins are then absorbed from the micelles in company with the fatty acids and monoglycerides. Vitamin K produced in the colon is absorbed with other lipids released through bacterial action. Taking supplements of fat-soluble vitamins on an empty stomach, while fasting, or on a low-fat diet will be relatively ineffective, as proper absorption requires the presence of other lipids.

The information contained in Figure 24-27● summarizes the digestive fates of electrolytes and vitamins. Before continuing, review this figure carefully to ensure that you are reasonably familiar with the outlined mechanisms and events. The next chapter examines the interplay between the respiratory and digestive systems as it considers the metabolism of the entire body.

MALABSORPTION SYNDROMES Difficulties in the absorption of all classes of compounds will result from damage to the accessory glands or the intestinal mucosa. If the accessory organs are functioning normally but their secretions cannot reach the duodenum, the condition is called *biliary obstruction* (bile duct blockage) or *pancreatic obstruction* (pancreatic duct blockage). Alternatively, the ducts may remain open but the glandular cells may be damaged and unable to continue normal secretory activities. Two examples, *pancreatitis* and *cirrhosis,* were noted earlier in the chapter.

Even with the normal enzymes present in the lumen, absorption will not occur if the mucosa cannot function properly. A genetic inability to manufacture specific enzymes will result in discrete patterns of malabsorption—*lactose intolerance* is a good example. Mucosal damage due to ischemia, radiation exposure, toxic compounds, or infection will affect absorption in general and will deplete nutrient and fluid reserves as a result.

Aging and the Digestive System

Essentially normal digestion and absorption occur in elderly individuals. However, there are many changes in the digestive system that parallel age-related changes already described for other systems.

1. ***The rate of epithelial stem cell division declines.*** The digestive epithelium becomes more susceptible to damage by abrasion, acids, or enzymes. Peptic ulcers therefore become more likely. In the mouth, esophagus, and anus the stratified epithelium becomes thinner and more fragile.
2. ***Smooth muscle tone decreases.*** General motility decreases, and peristaltic contractions are weaker. This change slows the rate of fecal movement and promotes constipation. Sagging and inflammation of the haustra in the colon can produce symptoms of *diverticulitis.* Straining to eliminate compacted fecal materials can stress the less-resilient walls of blood vessels, producing hemorrhoids. Problems are not restricted to the lower digestive tract. For example, weakening of muscular sphincters can lead to esophageal reflux and frequent bouts of "heartburn."
3. ***The effects of cumulative damage become apparent.*** A familiar example would be the gradual loss of teeth because of caries or gingivitis. Cumulative damage can involve internal organs as well. Toxins such as alcohol and other injurious chemicals that are absorbed by the digestive tract are transported to the liver for processing. The liver cells are not immune to these compounds, and chronic exposure can lead to *cirrhosis* or other types of liver disease.
4. ***Cancer rates increase.*** Not surprisingly, cancers are most common in organs where stem cells divide to maintain epithelial cell populations. ∞ *[pp. 102–103]* Rates of colon cancer and stomach cancer rise in the elderly; oral and pharyngeal cancers are particularly common in elderly smokers.
5. ***Changes in other systems have direct or indirect effects on the digestive system.*** For example, the reduction in bone mass and calcium content in the skeleton is associated with erosion of the tooth sockets and eventual tooth loss. The decline in olfactory and gustatory sensitivity with age can lead to dietary changes that affect the entire body.

What component of a meal would increase the number of chylomicrons in the lacteals?

The absorption of which vitamin would be impaired by removal of the stomach?

Why is it that diarrhea is potentially life-threatening but constipation is not?

Integration with Other Systems

Figure 24-28

The digestive system is functionally linked to all other systems, and it has extensive anatomical connections to the nervous, cardiovascular, endocrine, and lymphatic systems. Figure 24-28• summarizes the physiological relationships between the digestive system and other organ systems.

INTEGUMENTARY SYSTEM

Provides vitamin D_3 needed for the absorption of calcium and phosphorus

Provides lipids for storage by adipocytes in subcutaneous layer

SKELETAL SYSTEM

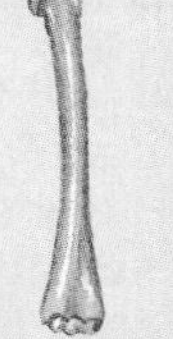

Skull, ribs, vertebrae, and pelvic girdle support and protect parts of digestive tract; teeth important in mechanical processing of food

Absorbs calcium and phosphate ions for incorporation into bone matrix; provides lipids for storage in yellow marrow

MUSCULAR SYSTEM

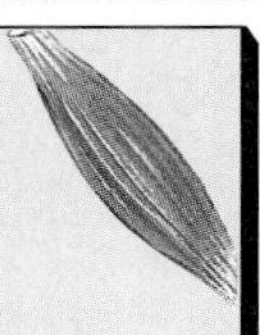

Protects and supports digestive organs in abdominal cavity; controls entrances and exits to digestive tract; liver metabolizes lactic acid from active muscles

Liver regulates blood glucose and fatty acid levels, removes lactic acid from circulation

DIGESTIVE SYSTEM

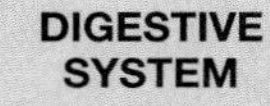

NERVOUS SYSTEM

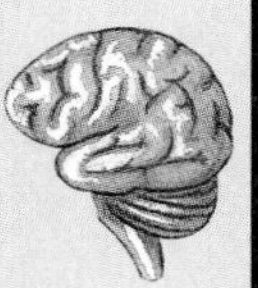

ANS regulates movement and secretion; reflexes coordinate passage of materials along tract; control over skeletal muscles regulates ingestion and defecation; hypothalamic centers control hunger, satiation, and feeding behaviors

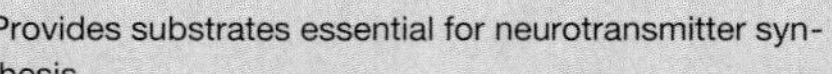

Provides substrates essential for neurotransmitter synthesis

ENDOCRINE SYSTEM

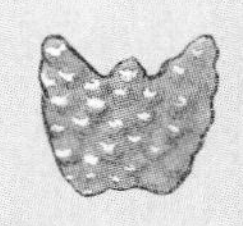

Epinephrine and norepinephrine stimulate constriction of sphincters and depress digestive activity; hormones coordinate activity along tract

Provides nutrients and substrates to endocrine cells; endocrine cells of pancreas secrete insulin and glucagon; liver produces angiotensinogen

CARDIOVASCULAR SYSTEM

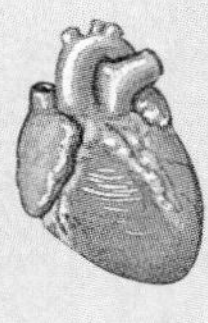

Distributes hormones of the digestive tract; carries nutrients, water, and ions from sites of absorption; delivers nutrients and toxins to liver

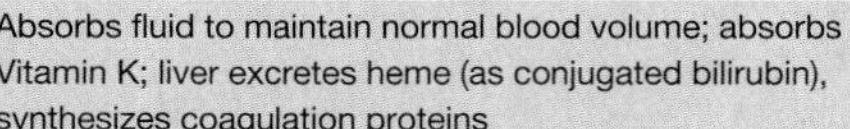

Absorbs fluid to maintain normal blood volume; absorbs Vitamin K; liver excretes heme (as conjugated bilirubin), synthesizes coagulation proteins

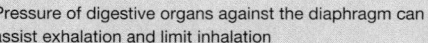

FOR ALL SYSTEMS

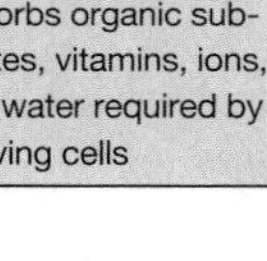

Absorbs organic substrates, vitamins, ions, and water required by all living cells

LYMPHATIC SYSTEM

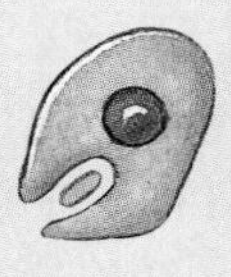

Tonsils and GALT defend against infection; GALT provides defense against toxins absorbed from the tract; lymphatics carry absorbed lipids to venous system

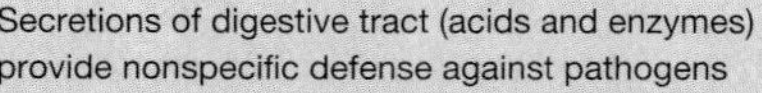

Secretions of digestive tract (acids and enzymes) provide nonspecific defense against pathogens

RESPIRATORY SYSTEM

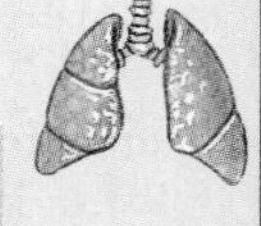

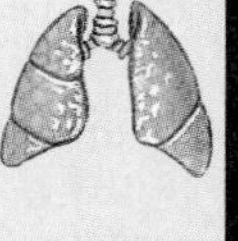

Increased thoracic and abdominal pressure through contraction of respiratory muscles can assist in defecation

Pressure of digestive organs against the diaphragm can assist exhalation and limit inhalation

REPRODUCTIVE SYSTEM

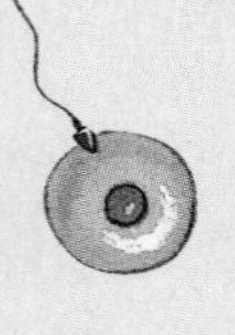

Provides additional nutrients required to support gamete production and (in pregnant women) embryonic and fetal development

URINARY SYSTEM

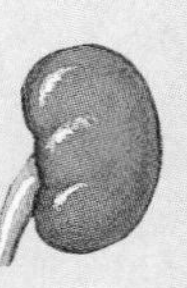

Excretes toxins absorbed by the digestive epithelium; excretes some conjugated bilirubin produced by liver

Absorbs water needed to excrete waste products at the kidneys; absorbs ions needed to maintain normal body fluid concentrations

FIGURE 24-28
Functional Relationships between the Digestive System and Other Systems

Selected Clinical Terminology

Terms Discussed in This Chapter

achalasia (ak-a-LĀ-zē-a): A condition that results when a bolus cannot reach the stomach because of constriction of the lower esophageal sphincter. *(p. 894)*

ascites: Fluid leakage into the peritoneal cavity across the serosal surfaces of the liver and viscera. *(p. 883)*

cathartics (ka-THAR-tiks): Drugs that promote defecation. *(p. 920)*

cholecystitis (kō-lē-sis-TĪ-tis): Inflammation of the gallbladder due to blockage of the cystic or common bile duct by gallstones. *(p. 914 and AM)*

cholelithiasis (kō-lē-li-THĪ-a-sis): The presence of gallstones within the gallbladder. *(p. 914 and AM)*

cholera: A bacterial infection of the digestive tract that causes massive fluid losses through diarrhea. *(p. 920 and AM)*

cirrhosis (sir-Ō-sis): A disease characterized by the widespread destruction of hepatocytes by exposure to drugs (especially alcohol), viral infection, ischemia, or blockage of the hepatic ducts. *(p. 912 and AM)*

colitis (kō-LĪ-tis): A general term for a condition characterized by inflammation of the colon. *(p. 917 and AM)*

constipation: Infrequent defecation, usually involving dry and hard feces. *(p. 920 and AM)*

diarrhea (dī-a-RĒ-a): Frequent, watery bowel movements. *(p. 920 and AM)*

diverticulitis (dī-ver-tik-ū-LĪ-tis): Infection and inflammation of diverticula. *(p. 917 and AM)*

diverticulosis (dī-ver-tik-ū-LŌ-sis): Formation of mucosal pockets (diverticula), usually along the sigmoid colon. *(p. 917)*

esophageal varices (VAR-i-sēz): Swollen and fragile esophageal veins that result from portal hypertension. *(p. 894)*

esophagitis (ē-sof-a-JĪ-tis): Inflammation of the esophagus. *(p. 894)*

gallstones: Deposits of minerals, bile salts, and cholesterol that form if bile becomes too concentrated. *(p. 914 and AM)*

gastrectomy (gas-TREK-to-mē): Surgical removal of the stomach, usually to treat advanced stomach cancer. *(p. 902)*

gastritis (gas-TRĪ-tis): Inflammation of the gastric mucosa. *(p. 899)*

gastroscope: A fiber-optic instrument inserted into the mouth and directed along the esophagus and into the stomach; used to examine the interior of the stomach and to perform minor surgical procedures. *(p. 902)*

hepatitis (hep-a-TĪ-tis): Virally induced disease of the liver; includes *hepatitis A, B, C,* and other forms. *(p. 912 and AM)*

lactose intolerance: A malabsorption syndrome that results from the lack of the enzyme *lactase* at the brush border of the intestinal epithelium. *(p. 922 and AM)*

mumps: A viral infection that in children usually focuses in the salivary glands, primarily the parotid. *(p. 889)*

pancreatitis (pan-krē-a-TĪ-tis): An inflammation of the pancreas. *(p. 909)*

peptic ulcer: Erosion of the gastric lining or duodenal lining by stomach acids and enzymes. *(p. 899 and AM)*

periodontal disease: A loosening of the teeth within the alveolar sockets caused by erosion of the periodontal ligaments by acids produced through bacterial action. *(p. 890)*

peritonitis: Inflammation of the peritoneal membrane. *(p. 883)*

portal hypertension: High venous pressures in the hepatic portal system. *(p. 909)*

pulpitis (pul-PĪ-tis): Infection of the pulp of a tooth; may be treated by a *root canal* procedure. *(p. 890)*

AM *Additional Terms Discussed in the Applications Manual*

colectomy (kō-LEK-tō-mē): The removal of all or a portion of the colon.

colonoscope (ko-LON-ō-skōp): A fiber-optic instrument inserted into the anus and moved into the colon to look for problems in the rectum, sigmoid colon, and ascending colon.

colostomy (kō-LOS-to-mē): Attachment of the cut end of the colon to an opening in the body wall after a colectomy.

gastric stapling: Surgical reduction in the size of the stomach by stapling closed a portion of the gastric lumen; an attempt to correct an overeating problem.

gastroenteritis (gas-trō-en-ter-Ī-tis): A condition characterized by vomiting and diarrhea, and resulting from bacterial toxins, viral infections, or various poisons.

giardiasis: An infection of the digestive tract by the protozoan *Giardia lamblia.* Some individuals do not develop symptoms, but others develop violent diarrhea, abdominal pains, cramps, nausea, and vomiting.

ileostomy (il-ē-OS-tō-mē): Attachment of the cut end of the ileum to an opening in the body wall after a colectomy.

inflammatory bowel disease *(ulcerative colitis):* Chronic inflammation of the digestive tract, most often affecting the colon.

irritable bowel syndrome: A disorder characterized by diarrhea, constipation, or an alternation between the two. When constipation is the primary problem, this condition may be called a *spastic colon,* or *spastic colitis.*

laparoscopy (lap-a-ROS-ko-pē): Use of a flexible fiber-optic instrument introduced through the abdominal wall to permit direct visualization of the viscera, tissue sampling, and limited surgical procedures.

liver biopsy: A sample of liver tissue usually taken through the anterior abdominal wall using a long needle.

liver scans: Radiological procedures that involve the injection of radioisotope-labeled compounds that will be selectively absorbed by either Kupffer cells, liver cells, or abnormal liver tissues.

perforated ulcer: A particularly dangerous ulcer in which the gastric acids erode their way through the wall of the digestive tract and enter the peritoneal cavity.

polyps: Small mucosal tumors that grow from the intestinal wall.

CHAPTER REVIEW

STUDY OUTLINE

INTRODUCTION, p. 878

1. The digestive system consists of the muscular **digestive tract** and various **accessory organs.**
2. Digestive functions include: ingestion, mechanical processing, digestion, secretion, absorption, compaction, and excretion.

AN OVERVIEW OF THE STRUCTURE AND FUNCTION OF THE DIGESTIVE TRACT, p. 878

1. The digestive tract includes the oral cavity, pharynx, esophagus, stomach, small intestine, large intestine, rectum, and anus. (*Figure 24-1*)

Histological Organization of the Digestive Tract, p. 879

2. The digestive tract is lined by a *mucous epithelium* moistened by glandular secretions of the epithelial and accessory organs.
3. The *lamina propria* and epithelium form the **mucosa** (mucous membrane) of the digestive tract. Proceeding outward one enters the **submucosa,** the **muscularis externa,** and a layer of loose connective tissue called the *adventitia.* Within the peritoneal cavity the muscularis externa is covered by a serous membrane called the **serosa.** (*Figure 24-2*)

The Movement of Digestive Materials, p. 881

4. Many smooth muscle cells are not innervated by motor neurons, and the neurons that do innervate smooth muscles are not under voluntary control.
5. The muscularis externa propels materials through the digestive tract by the contractions of **peristalsis. Segmentation** movements in areas of the small intestine churn digestive materials. (*Figure 24-3*)

The Control of Digestive Function, p. 881

6. Digestive tract activities are controlled by neural, hormonal, and local mechanisms. (*Figure 24-4*)

The Peritoneum, p. 882

7. Double sheets of peritoneal membrane called **mesenteries** suspend the digestive tract. The **greater omentum** lies in front of the abdominal viscera. Its adipose tissue provides padding, protection, insulation, and an energy reserve. (*Figure 24-5*)

THE ORAL CAVITY, p. 883

1. The functions of the oral cavity include (1) *analysis* of potential foods; (2) *mechanical processing* using the teeth, tongue, and palatal surfaces; (3) *lubrication* by mixing with mucus and salivary secretions; and (4) *digestion* by salivary and lingual enzymes.
2. The oral cavity, or **buccal cavity,** is lined by **oral mucosa.** The **hard** and **soft palates** form its roof, and the tongue forms its floor. (*Figure 24-6*)

The Tongue, p. 885

3. **Intrinsic** and **extrinsic** tongue muscles are controlled by the hypoglossal nerve. (*Figure 24-6*)

Salivary Glands, p. 888

4. The **parotid, sublingual,** and **submandibular glands** discharge their secretions into the oral cavity. (*Figure 24-7*)

The Teeth, p. 889

5. **Mastication** (chewing) occurs through the contact of the opposing **occlusal surfaces** of the teeth. The **periodontal ligament** anchors the tooth in an alveolar socket. **Dentin** forms the basic structure of a tooth. The **crown** is coated with **enamel,** and the **root** with **cementum.** (*Figure 24-8*)
6. The 20 primary teeth, or **deciduous teeth,** are replaced by the 32 teeth of the **secondary dentition** during development. (*Figure 24-9*)

THE PHARYNX, p. 892

1. Propulsion of the **bolus** results from the contractions of the *pharyngeal constrictors* and the *palatal muscles.*

THE ESOPHAGUS, p. 892

1. The esophagus carries solids and liquids from the pharynx to the stomach through an opening in the diaphragm, the **esophageal hiatus.**

Histology of the Esophagus, p. 893

2. The mucosa consists of a stratified epithelium. Mucous secretion by esophageal glands of the submucosa reduces friction during the passage of foods. The proportions of skeletal and smooth muscle of the muscularis externa change from the pharynx to the stomach. (*Figure 24-10*)

Swallowing, p. 894

3. Swallowing can be divided into **buccal, pharyngeal,** and **esophageal phases.** Swallowing begins with the compaction of a bolus and its movement into the pharynx, followed by the elevation of the larynx, reflection of the epiglottis, and closure of the glottis. After opening of the *upper esophageal sphincter,* peristalsis moves the bolus down the esophagus to the *lower esophageal sphincter.* (*Figure 24-11*)

THE STOMACH, p. 895

1. The stomach has four major functions: (1) bulk storage of ingested matter, (2) mechanical breakdown of resistant materials, (3) disruption of chemical bonds using acids and enzymes, and (4) production of intrinsic factor.

Anatomy of the Stomach, p. 895

2. The four regions of the stomach are the **cardia, fundus, body,** and **pylorus.** The **pyloric sphincter** guards the exit from the stomach. In a relaxed state the stomach lining contains numerous **rugae** (ridges and folds). (*Figure 24-12*)
3. Within the **gastric glands, parietal cells** secrete **intrinsic factor** and hydrochloric acid. **Chief cells** secrete **pepsinogen,** which acids in the gastric lumen convert to the enzyme **pepsin. Enteroendocrine** cells of the stomach secrete several compounds, notably the hormone **gastrin.** (*Figures 24-13, 24-14*)

Regulation of Gastric Activity, p. 899

4. Gastric secretion includes: (1) the **cephalic phase,** which prepares the stomach to receive ingested materials; (2) the **gastric phase,** which begins with the arrival of food in the stomach; and (3) the **intestinal phase,** which controls the rate of gastric emptying. (*Figure 24-15*)

Digestion and Absorption in the Stomach, p. 902

5. The preliminary digestion of proteins by pepsin begins in the stomach. Very little nutrient absorption occurs in the stomach.

THE SMALL INTESTINE AND ASSOCIATED GLANDULAR ORGANS, p. 903

1. Most of the important digestive and absorptive functions occur in the small intestine. Digestive secretions and buffers are provided by the pancreas, liver, and gallbladder.

The Small Intestine, p. 903

2. The small intestine includes the **duodenum,** the **jejunum,** and the **ileum.** A sphincter, the **ileocecal valve,** marks the transition between the small and large intestines.
3. The intestinal mucosa bears transverse folds called **plicae,** and small projections called **intestinal villi.** These increase the surface area for absorption. Each villus contains a terminal lymphatic called a **lacteal.** Pockets called **intestinal crypts** are lined by enteroendocrine, goblet, and stem cells. (*Figure 24-16*)
4. The **submucosal** (*Brunner's*) **glands** of the duodenum produce mucus, buffers, and a hormone, **urogastrone.** The absorptive plicae and villi of the jejunum are reduced in size near the ileum. The ileum contains masses of lymphoid tissue called *Peyer's patches* near the entrance to the large intestine. (*Figure 24-17*)
5. **Intestinal juice** moistens the chyme, helps to buffer acids, and holds digestive enzymes and the products of digestion in solution.

Intestinal Movements, p. 906

6. Some of the smooth muscle cells contract periodically, without stimulation, to produce a pattern of peristaltic contractions that move the chyme.

The Pancreas, p. 907

7. The **pancreatic duct** penetrates the wall of the duodenum. Within each lobule, ducts branch repeatedly before ending in the **pancreatic acini** (blind pockets). (*Figure 24-18*)
8. The pancreas has two functions: endocrine (secreting insulin and glucagon into the blood) and exocrine (secreting water, ions, and digestive enzymes into the small intestine). Pancreatic enzymes include **lipases, carbohydrases,** and **proteolytic enzymes.**

The Liver, p. 909

9. The liver performs metabolic and hematological regulation and produces **bile.** The bile ducts from all the lobules unite to form the **common hepatic duct,** which meets the **cystic duct** to form the **common bile duct** that empties into the duodenum. (*Figure 24-19*)
10. The **liver lobule** is the organ's basic functional unit. **Bile canaliculi** carry bile to the **bile ductules,** which lead to portal areas. (*Figure 24-20*)

The Gallbladder, p. 913

11. The **gallbladder** stores and concentrates bile. During **emulsification** bile salts break apart large drops of lipids and make them accessible to lipases secreted by the pancreas. (*Figure 24-21*)

The Coordination of Secretion and Absorption, p. 914

12. Neural and hormonal mechanisms coordinate the activities of the digestive glands. Gastrointestinal activity is stimulated by parasympathetic innervation and inhibited by sympathetic innervation. The enterogastric, gastroenteric, and gastroileal reflexes coordinate movement.
13. Intestinal hormones include **enterocrinin, secretin, cholecystokinin (CCK), gastric inhibitory peptide (GIP), vasoactive intestinal peptide (VIP), gastrin,** and others. (*Table 24-1; Figure 24-22*)

THE LARGE INTESTINE, p. 916

1. The main functions of the large intestine are to (1) reabsorb water and compact the feces, (2) absorb vitamins produced by bacteria, and (3) store fecal material prior to defecation. (*Figure 24-23a*)

The Cecum, p. 916

2. The cecum collects and stores material from the ileum and begins the process of compaction. The **vermiform appendix** is attached to the cecum.

The Colon, p. 916

3. The colon has a larger diameter and a thinner wall than the small intestine. It bears **haustra** (pouches) and the **taenia coli** (longitudinal bands of muscle).

The Rectum, p. 917

4. The **rectum** terminates in the **anorectal canal** leading to the anus. (*Figure 24-23b*)

Histology of the Large Intestine, p. 917

5. Characteristics of the colon include an absence of villi and the presence of goblet cells and deep intestinal glands. (*Figure 24-23c*)

Physiology of the Large Intestine, p. 918

6. The large intestine reabsorbs water and other substances such as vitamins, urobilinogen, bile salts, and toxins. Bacteria are responsible for intestinal gas, or **flatus.**
7. The gastroileal reflex moves chyme into the cecum at mealtimes. Distension of the stomach and duodenum stimulates peristalsis, or **mass movements,** of chyme from the transverse colon through the rest of the large intestine and into the rectum. Muscular sphincters control the passage of fecal materal to the anus. Distension of the rectal wall triggers the defecation reflex. (*Figure 24-24*)

DIGESTION AND ABSORPTION, p. 921

The Processing and Absorption of Nutrients, p. 921

1. The digestive system breaks down the physical structure of the ingested material and then disassembles the component molecules into smaller fragments through hydrolysis. (*Figure 24-25; Table 24-2*)

Carbohydrate Digestion and Absorption, p. 921

2. Salivary and pancreatic amylases break down complex carbohydrates into disaccharides and trisaccharides. These are broken down into monosaccharides by enzymes at the epithelial surface and absorbed by the intestinal epithelium through facilitated diffusion and cotransport.

Lipid Digestion and Absorption, p. 924

3. Triglycerides are emulsified into large lipid drops. The resulting fatty acids and monoglycerides interact with bile salts to form **micelles,** from which they diffuse across the intestinal epithelium.

Protein Digestion and Absorption, p. 924

4. Protein digestion involves the gastric enzyme pepsin and the various pancreatic proteases. Peptidases liberate amino acids that are absorbed and exported to the interstitial fluids.

Water Absorption, p. 925

5. About 2–2.5 liters of water are ingested each day, and digestive secretions provide 6–7 liters. Nearly all is reabsorbed by osmosis. (*Figure 24-26*)

Electrolyte Absorption, p. 925

6. Various processes such as diffusion, cotransport, and carrier-mediated and active transport are responsible for the movements of cations (sodium, calcium, potassium, and so on) and anions (chloride, iodide, bicarbonate, and so on) into epithelial cells. (*Figure 24-27*)

Vitamin Absorption, p. 926

7. The nine **water-soluble vitamins** are important as cofactors in enzymatic reactions. **Fat-soluble vitamins** are enclosed within fat droplets and are absorbed with the products of lipid digestion. (*Figure 24-27*)

AGING AND THE DIGESTIVE SYSTEM, p. 927

1. Age-related changes include a thinner and more fragile epithelium due to a reduction in epithelial stem cell division and weaker peristaltic contractions as smooth muscle tone decreases.

INTEGRATION WITH OTHER SYSTEMS, p. 927

1. The digestive system has extensive anatomic connections to the nervous, cardiovascular, endocrine, and lymphatic systems. (*Figure 24-28*)

REVIEW QUESTIONS

LEVEL 1 Reviewing Facts and Terms

1. The enzymatic breakdown of large molecules into their basic building blocks is called:
   (a) absorption
   (b) secretion
   (c) mechanical digestion
   (d) chemical digestion
2. The inner layer of the gastrointestinal tract is the:
   (a) mucosa (b) submucosa
   (c) serosa (d) muscularis
3. The muscularis externa propels materials from one portion of the digestive tract to another through the contractions of:
   (a) segmentation (b) propulsion
   (c) mass movements (d) peristalsis
4. The activities of the digestive system are regulated by:
   (a) hormonal mechanisms
   (b) local mechanisms
   (c) neural mechanisms
   (d) a, b, and c are correct
5. Double sheets of peritoneum that provide support and stability for the organs of the peritoneal cavity are called:
   (a) mediastinums (b) mucous membranes
   (c) omenta (d) mesenteries
6. The layer of the peritoneum that lines the inner surfaces of the body wall is the:
   (a) visceral peritoneum (b) parietal peritoneum
   (c) greater omentum (d) lesser omentum
7. The peritoneal fold that supports the small intestine is the:
   (a) serosa (b) lesser omentum
   (c) mesentery (d) parietal peritoneum
8. Intrinsic factor and hydrochloric acid are secreted by cells in the stomach wall called:
   (a) parietal cells (b) chief cells
   (c) acinar cells (d) G cells
9. Protein digestion in the stomach results primarily from secretions released by:
   (a) G cells (b) parietal cells
   (c) chief cells (d) D cells
10. The part of the gastrointestinal tract that plays the primary role in the digestion and absorption of nutrients is the:
    (a) large intestine (b) small intestine
    (c) stomach (d) cecum and colon
11. The duodenal hormone that stimulates the production and secretion of pancreatic enzymes is:
    (a) enterokinase (b) gastrin
    (c) secretin (d) cholecystokinin
12. The essential metabolic and synthetic service(s) provided by the liver is (are):
    (a) metabolic regulation
    (b) hematological regulation
    (c) bile production
    (d) a, b, and c are correct
13. Bile release from the gallbladder into the duodenum occurs only under the stimulation of:
    (a) cholecystokinin (b) secretin
    (c) gastrin (d) enterokinase

14. The major function(s) of the large intestine is (are):
    (a) reabsorption of water and compaction of chyme into feces
    (b) absorption of vitamins liberated by bacterial action
    (c) storage of fecal material prior to defecation
    (d) a, b, and c are correct
15. The part of the colon that empties into the rectum is the:
    (a) ascending colon (b) descending colon
    (c) transverse colon (d) sigmoid colon
16. Three vitamins generated by colonic bacteria are:
    (a) vitamins A, D, and E
    (b) B complex vitamins and vitamin C
    (c) vitamin K, biotin, and pantothenic acid
    (d) niacin, thiamine, and riboflavin
17. The final enzymatic steps in the digestive process are accomplished by:
    (a) brush border enzymes of the microvilli
    (b) enzymes secreted by the stomach
    (c) enzymes secreted by the pancreas
    (d) the action of bile from the gallbladder
18. What six steps are involved in digestion?
19. What is the purpose of the transverse or longitudinal folds of the epithelium in the digestive tract?
20. Name and describe the layers of the digestive tract, proceeding from the innermost to the outermost layer.
21. What three basic mechanisms regulate the activities of the digestive tract?
22. What is the relationship between the parietal peritoneum, visceral peritoneum, and mesenteries?
23. What are the four primary functions of the oral (buccal) cavity?
24. List the three pairs of salivary glands and identify the ducts used by their glandular secretions.
25. What are the functions of saliva, and how are salivary secretions controlled?
26. What specific function does each of the four types of teeth perform in the oral cavity?
27. What role do the pharyngeal muscles play in swallowing?
28. What are the three phases of swallowing, and how are they controlled?
29. What contributions do the gastric and pyloric glands of the stomach make to the digestive process?
30. What three subdivisions of the small intestine are involved in the digestion and absorption of food?
31. What are the primary functions of the pancreas, liver, and gallbladder in the digestive process?
32. What hormones produced by duodenal enteroendocrine cells effectively coordinate digestive functions?
33. What are the three major functions of the large intestine?
34. What two positive feedback loops are involved in the defecation reflex?
35. What five age-related changes occur in the digestive system?

LEVEL 2 Reviewing Concepts

36. If the lingual frenulum is too restrictive, an individual:
    (a) has difficulty tasting food
    (b) cannot swallow properly
    (c) cannot control movements of the tongue
    (d) cannot eat or speak normally
37. Increased secretion by all the salivary glands results from:
    (a) sympathetic stimulation
    (b) hormonal stimulation
    (c) parasympathetic stimulation
    (d) myenteric reflexes
38. The production of acid and enzymes by the gastric mucosa is controlled and regulated by:
    (a) the central nervous system
    (b) short reflexes coordinated in the stomach wall
    (c) digestive tract hormones
    (d) a, b, and c are correct
39. The gastric phase of secretion is initiated by:
    (a) distension of the stomach
    (b) an increase in the pH of the gastric contents
    (c) the presence of undigested materials in the stomach
    (d) a, b, and c are correct
40. Chyme reaches the small intestine most quickly when:
    (a) the stomach is greatly distended
    (b) a large amount of gastrin is released
    (c) the meal contains a relatively small quantity of proteins
    (d) a and c are correct
41. A drop in pH below 4.5 in the duodenum stimulates secretion of:
    (a) secretin (b) cholecystokinin
    (c) gastrin (d) a, b, and c are correct
42. If you wanted to nip the tip off a carrot stick, which teeth would you most likely use?
43. Differentiate between the action and outcome of peristalsis and segmentation.
44. How does the stomach facilitate and assist in the digestive process?
45. How do the three phases of gastric secretion promote and facilitate gastric control?
46. Why should aspirin be avoided if a person has stomach ulcers?
47. Nutritionists have found that after a heavy meal there is a slight increase in the pH of the blood, especially in the veins that carry blood away from the stomach. What causes this "postenteric alkaline tide" to occur?

LEVEL 3 Critical Thinking and Clinical Applications

48. Sometimes patients suffering from gallstones will develop pancreatitis. How could this occur?
49. Barb suffers from Crohn's disease, a regional inflammation of the intestine that is thought to have some genetic basis, although the actual cause is as yet unknown. When the disease flares up she experiences abdominal pain, weight loss, and anemia. What part(s) of the intestine is (are) probably involved, and what is the cause of her symptoms?
50. What symptoms would you expect to observe in a patient who suffers from a blockage of the small intestine at the level of the jejunum?

CHAPTER 25

TROL

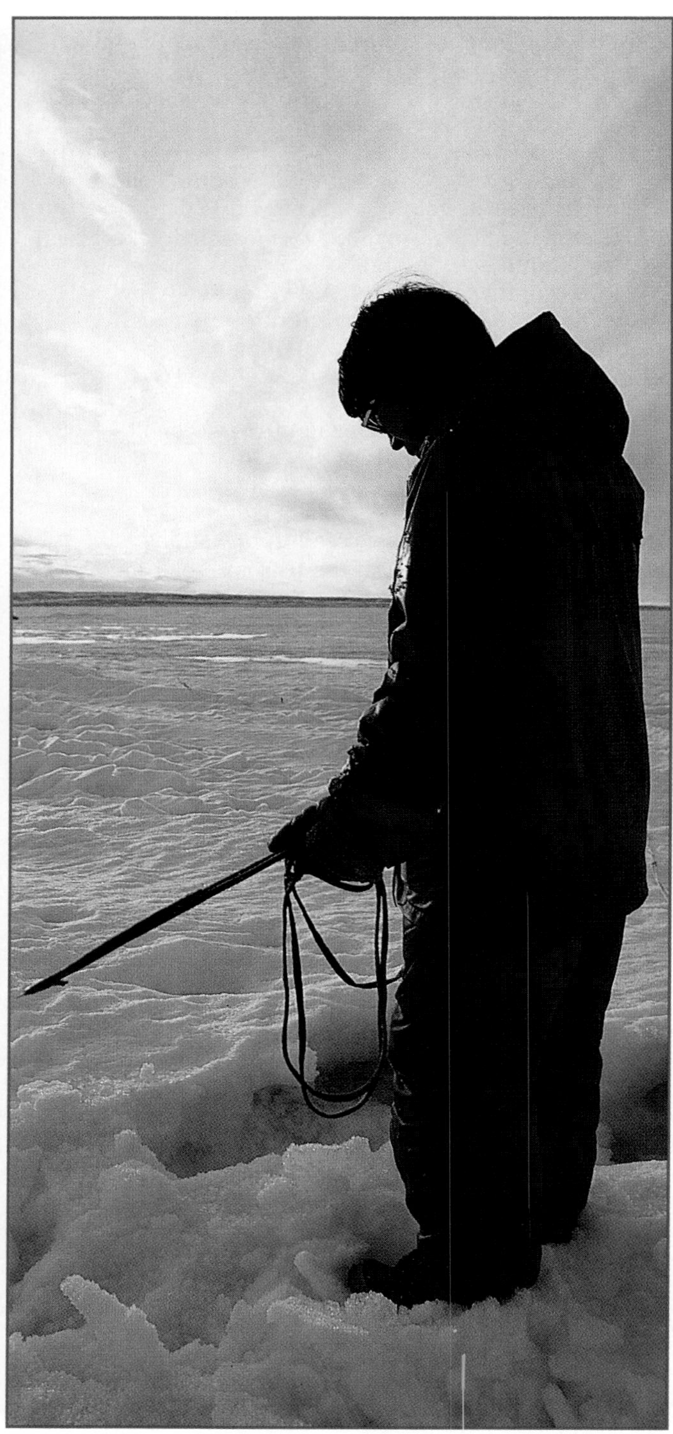

Human beings are remarkably adaptable animals— they can maintain homeostasis near the equator, where the year-round temperature averages 27°C, and near the poles, where the temperature remains below 0°C most of the time. The key to survival under such varied environmental conditions is the preservation of normal body temperature. In part this can be accomplished by shedding or adding layers of clothing, but hormonal and metabolic adjustments are also crucially important. Thus, the metabolic rate rises in cold climates, generating additional heat that warms the body. This chapter examines several aspects of metabolism, including nutrition, energy use, heat production, and the regulation of body temperature under normal and abnormal conditions.

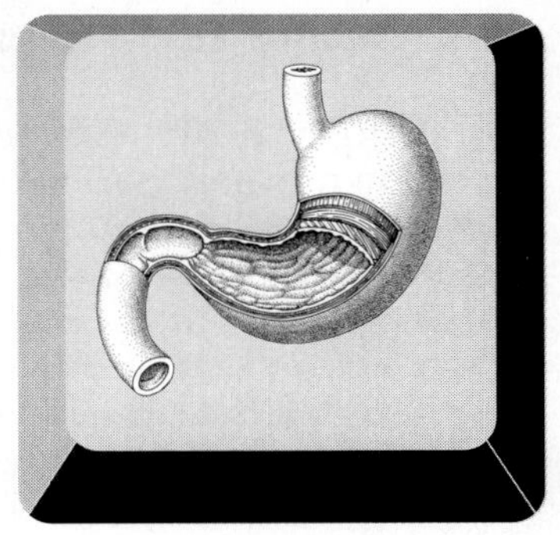

Metabolism and Energetics

Chapter Outline and Objectives

1 Define metabolism and explain why cells need to synthesize new organic components.

2 Describe the basic steps in glycolysis, the TCA cycle, and the electron transport chain.

3 Summarize the energy yield of glycolysis and cellular respiration.

4 Describe the pathways involved in lipid metabolism and the mechanisms necessary for lipid transport and distribution.

5 Discuss protein metabolism and the use of proteins as an energy source.

6 Discuss nucleic acid metabolism.

7 Differentiate between the absorptive and postabsorptive metabolic states and summarize the characteristics of each.

8 Explain what constitutes a balanced diet, and why such a diet is important.

9 Define metabolic rate and discuss the factors involved in determining an individual's BMR.

10 Discuss the homeostatic mechanisms that maintain a constant body temperature.

Living cells are chemical factories that break down organic molecules to obtain energy that can be used to generate ATP. As noted in Chapter 3, most of the reactions that release energy within cells occur inside mitochondria. [p. 88] To carry out these metabolic reactions, our cells must have a reliable supply of oxygen and nutrients, including water, organic substrates, vitamins, and mineral ions. Oxygen is absorbed at the lungs; the other substances are obtained by absorption at the digestive tract, and the cardiovascular system ensures prompt distribution to interstitial fluids throughout the body, where they become accessible to individual cells.

The energy released inside the cell supports growth, cell division, contraction, secretion, and a variety of other special functions that vary from cell to cell and tissue to tissue. Because different tissue types contain different populations of cells, the energy and nutrient requirements of any two tissues, such as loose connective tissue and cardiac muscle, are quite different. When cells, tissues, and organs vary their patterns of activity, the metabolic needs of the body change. There are short-term and long-term changes. For example, nutrient requirements vary from moment to moment (resting versus active), hour to hour (asleep versus awake), and year to year (growing child versus adult).

Differentiation creates tissues and organs that are specialized to store excess nutrients; the storage of lipids in adipose tissue is one familiar example. When nutrients are abundant, energy reserves are built up. These reserves, usually glycogen deposits or lipid inclusions, can then be called upon whenever the digestive tract cannot provide the right quantity or quality of nutrients. The endocrine system, with the assistance of the nervous system, adjusts and coordinates the metabolic activities of the body's tissues and controls the storage and mobilization of nutrient reserves.

An Overview of Cellular Metabolism

Figures 25-1, 25-2

Figure 25-1 • provides an overview of the ways cells use organic nutrients absorbed from the interstitial fluid. Simple sugars, amino acids, and lipids cross the cell membrane to join nutrients already in the cytoplasm. All of the cell's organic building blocks collectively form a *nutrient pool.*

Catabolism breaks down organic substrates, releasing energy that can be used to synthesize ATP or other high-energy compounds. Catabolism proceeds in a series of steps. In general, prelimi-

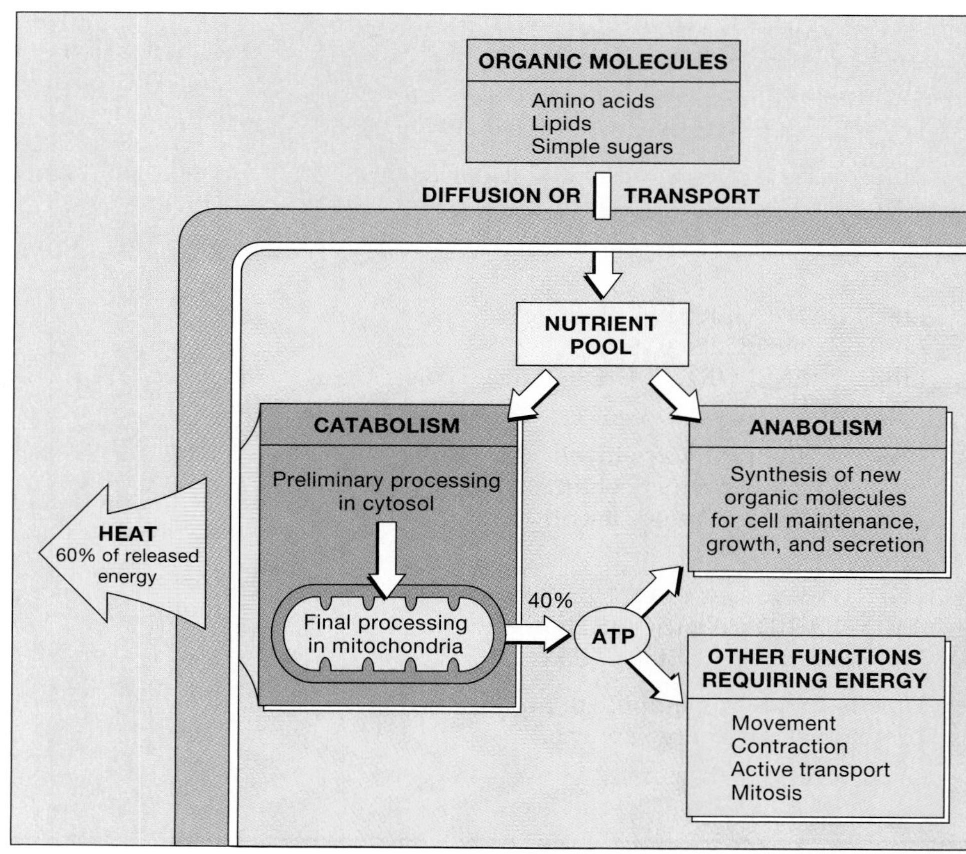
ORGANIC MOLECULES
Amino acids
Lipids
Simple sugars
DIFFUSION OR TRANSPORT
NUTRIENT POOL
CATABOLISM
Preliminary processing in cytosol
Final processing in mitochondria
HEAT
60% of released energy
40%
ATP
ANABOLISM
Synthesis of new organic molecules for cell maintenance, growth, and secretion
OTHER FUNCTIONS REQUIRING ENERGY
Movement
Contraction
Active transport
Mitosis

• **FIGURE 25-1**
Cellular Metabolism. The cell obtains organic molecules from the extracellular fluid and breaks them down to obtain ATP. Only about 40 percent of the energy released through catabolism is captured in ATP; the rest is radiated as heat. The ATP generated through catabolism provides energy for all vital cellular activities, including anabolism.

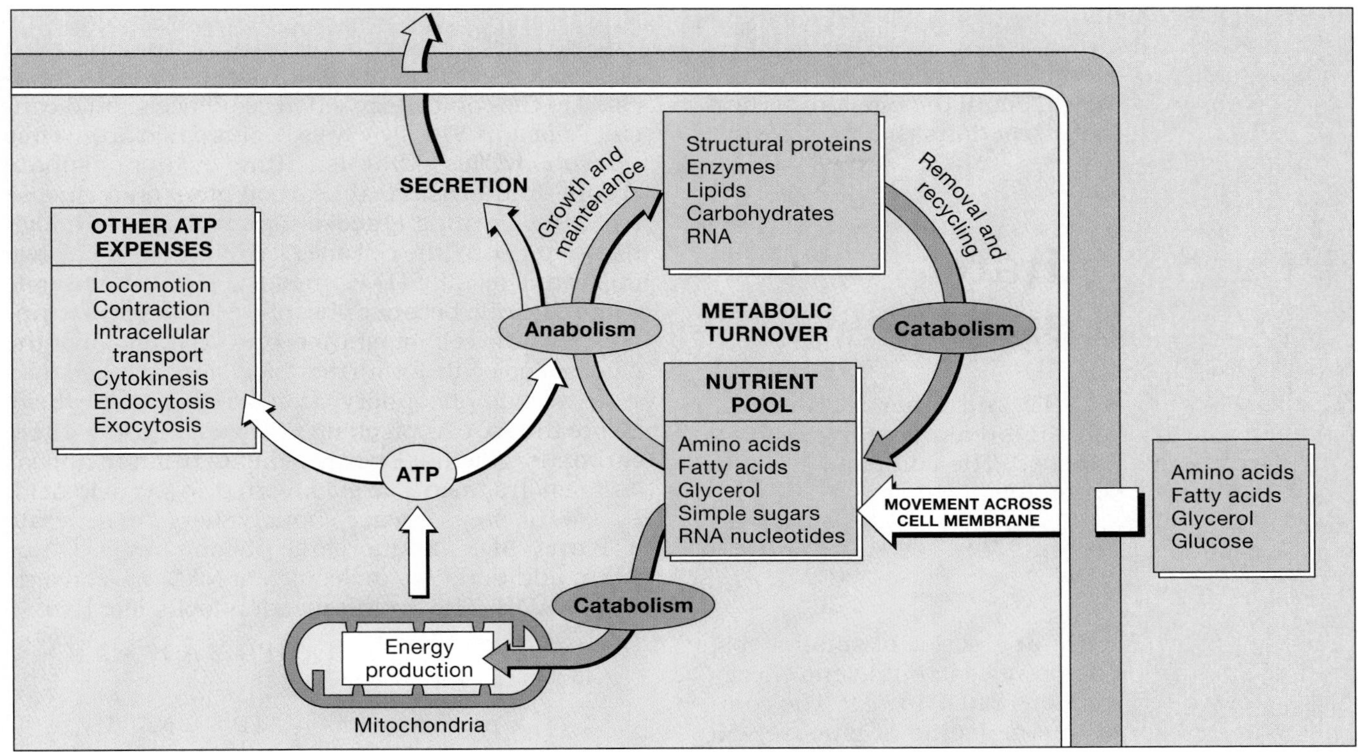
SECRETION
Growth and maintenance
Structural proteins
Enzymes
Lipids
Carbohydrates
RNA
Removal and recycling
OTHER ATP EXPENSES
Locomotion
Contraction
Intracellular transport
Cytokinesis
Endocytosis
Exocytosis
Anabolism
METABOLIC TURNOVER
Catabolism
NUTRIENT POOL
Amino acids
Fatty acids
Glycerol
Simple sugars
RNA nucleotides
ATP
MOVEMENT ACROSS CELL MEMBRANE
Amino acids
Fatty acids
Glycerol
Glucose
Catabolism
Energy production
Mitochondria

● **FIGURE 25-2**

Metabolic Turnover and Cellular ATP Production. Metabolic turnover is integrated into other cellular functions. Organic molecules released during metabolic turnover may be recycled via anabolic pathways and used to build other large molecules. They can also be broken down for energy production. The cell must also provide ATP for both anabolism and other cellular functions.

nary processing occurs in the cytosol, where enzymes break down large organic molecules into smaller fragments. For example, carbohydrates are broken down to short carbon chains, triglycerides are split into fatty acids and glycerol, and proteins are broken down to individual amino acids.

Relatively little ATP is produced during these preparatory steps. However, the simple molecules formed can be absorbed and processed by mitochondria, and the mitochondrial steps release significant amounts of energy. As mitochondrial enzymes break the covalent bonds that hold these molecules together, they capture roughly 40 percent of the energy released. The captured energy is used to convert ADP to ATP, and the rest escapes as heat that warms the interior of the cell and the surrounding tissues.

The ATP produced by mitochondria provides energy to support *anabolism*, the synthesis of new organic molecules, as well as other cell functions. Those additional functions, such as contraction, ciliary or cell movement, active transport, or cell division, vary from one cell to another. For example, muscle fibers need ATP to provide energy for contraction, and gland cells need ATP to synthesize and transport their secretions. Such specialized functions have been considered in other chapters, and we will restrict our focus here to anabolic processes.

Anabolism is an "uphill" process that involves the formation of new chemical bonds. Cells synthesize new organic components for three basic reasons (Figure 25-2●).

1. ***To perform structural maintenance or repairs.*** All cells must expend energy to perform ongoing maintenance and repairs, because most structures in the cell are temporary rather than permanent. Their removal and replacement are part of the process of *metabolic turnover* described in Chapter 2. ∞ [p. 60]
2. ***To support growth.*** Cells preparing for division increase in size and synthesize additional proteins and organelles.
3. ***To produce secretions.*** Secretory cells must synthesize their products and deliver them to the interstitial fluid.

The nutrient pool is the source for the substrates for both catabolism and anabolism. As you might expect, the cell tends to conserve the materials needed to build new compounds and breaks down the rest. The cell is continually replacing membranes, organelles, enzymes, and structural proteins. These anabolic activities require more amino acids than lipids and relatively few carbohydrates. *In general, a cell with excess carbohy-*

drates, lipids, and amino acids will break down carbohydrates first. Lipids are a second choice, and amino acids are seldom broken down.

The next section will detail the most important catabolic and anabolic reactions that occur within our cells.

Carbohydrate Metabolism

Most cells generate ATP and other high-energy compounds through the breakdown of carbohydrates, especially glucose. The complete reaction sequence can be summarized as:

$$\underset{\text{glucose}}{C_6H_{12}O_6} + \underset{\text{oxygen}}{6\ O_2} \rightarrow \underset{\text{carbon dioxide}}{6\ CO_2} + \underset{\text{water}}{6\ H_2O}$$

The breakdown occurs in a series of small steps, and several of the steps release sufficient energy to support the conversion of ADP to ATP. The complete catabolism of one molecule of glucose will provide most cells in the body with a net gain of 36 molecules of ATP.

Although most of the actual ATP production occurs inside mitochondria, the first steps take place in the cytosol. The steps involved were outlined in Chapter 10. [p. 307] In this reaction sequence, called *glycolysis*, six-carbon glucose molecules are broken down into a pair of three-carbon molecules of *pyruvic acid*. The steps are said to be *anaerobic* because oxygen is not needed. The pyruvic acid can be absorbed by mitochondria and used for ATP production. Mitochondrial activity, which requires oxygen, is called *cellular respiration*, or *aerobic metabolism*.

Glycolysis

Figure 25-3

Glycolysis (glī-KOL-i-sis; *glykus*, sweet + *lysis*, dissolution) is the breakdown of glucose to **pyruvic acid.** In this process, a series of enzymatic steps breaks the six-carbon glucose molecule into two three-carbon molecules of pyruvic acid (CH_3-CO-COOH). At the normal pH inside the cell, the pyruvic acid molecule loses a hydrogen ion and exists as the negatively charged ion CH_3-CO-COO^-. This ionized form is usually called *pyruvate* rather than pyruvic acid.

Glycolysis requires (1) glucose molecules, (2) appropriate cytoplasmic enzymes, (3) ATP and ADP, (4) inorganic phosphates, and (5) **NAD** (**n**icotinamide **a**denine **d**inucleotide), a coenzyme that removes hydrogen atoms during one of the enzymatic reactions. If any of these participants is missing, glycolysis cannot take place.

The steps in glycolysis are diagrammed in Figure 25-3•. This is an overview rather than a detailed description of the pathway involved. (Readers interested in the complete reaction sequence should consult Appendix VI.) Glycolysis begins when an enzyme *phosphorylates*—that is, attaches a phosphate group—to the last (sixth) carbon atom on a glucose molecule, creating **glucose-6-phosphate.** Although this step "costs" the cell one ATP molecule, it has two important results: (1) It traps the glucose molecule within the cell, because phosphorylated glucose cannot cross the cell membrane; and (2) it prepares the glucose molecule for further biochemical reactions.

A second phosphorylation occurs in the cytosol before the six-carbon chain is broken into two three-carbon fragments. Energy benefits begin to appear as these fragments are converted to pyruvic acid. Two of the steps release enough energy to generate ATP from ADP and inorganic phosphate (PO_4^{3-}, or P_i). In addition, two molecules of NAD are converted to NADH. The overall reaction looks like this:

$$\underset{\text{(6-carbon)}}{\text{Glucose}} + 2\ \text{NAD} + 2\ \text{ATP} + 2\ \text{ADP} + 2\ P_i \rightarrow$$

$$2\ \underset{\text{(3-carbon)}}{\text{pyruvic acid}} + 4\ \text{ATP} + 2\ \text{NADH}$$

This anaerobic reaction sequence provides a net gain of two molecules of ATP for the cell for each glucose molecule converted to two pyruvic acid molecules. A few highly specialized cells, such as red blood cells, lack mitochondria and derive all of their ATP through glycolysis. Skeletal muscle fibers rely on glycolysis for ATP production during periods of active contraction, and most cells can survive for brief periods using the ATP provided by glycolysis. However, when oxygen is readily available, mitochondrial activity provides most of the ATP required by our cells.

Mitochondrial ATP Production

Glycolysis yields an immediate net gain of two ATP molecules for the cell. However, a great deal of additional energy is still locked in the chemical bonds of pyruvic acid. The ability to capture that energy depends on the availability of oxygen. If oxygen supplies are adequate, mitochondria absorb the pyruvic acid molecules and break them down. The hydrogen atoms are removed by coenzymes, and they will ultimately be the source of most of the energy gain for the cell. The carbon and oxygen atoms are removed and released as carbon dioxide, a process called **decarboxylation** (dē-kar-boks-i-LĀ-shun).

Each mitochondrion has a double layer of membrane around it. The *outer mitochondrial membrane* contains large-diameter pores that are permeable to ions and small organic molecules such as pyruvic acid. The ions and molecules then enter

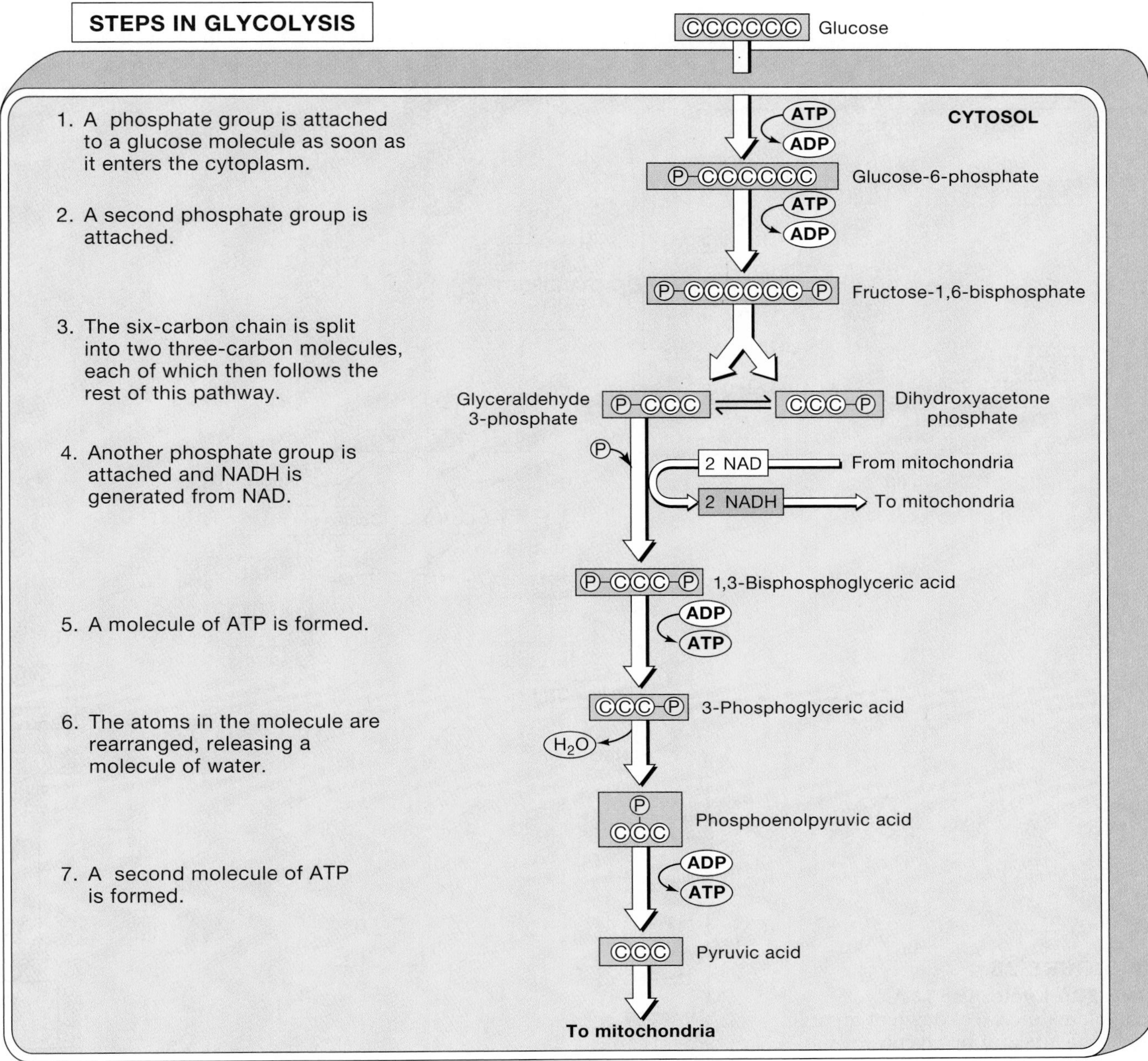

STEPS IN GLYCOLYSIS
Glucose
CYTOSOL
1. A phosphate group is attached to a glucose molecule as soon as it enters the cytoplasm.
ATP
ADP
Glucose-6-phosphate
2. A second phosphate group is attached.
ATP
ADP
Fructose-1,6-bisphosphate
3. The six-carbon chain is split into two three-carbon molecules, each of which then follows the rest of this pathway.
Glyceraldehyde 3-phosphate
Dihydroxyacetone phosphate
4. Another phosphate group is attached and NADH is generated from NAD.
2 NAD
From mitochondria
2 NADH
To mitochondria
1,3-Bisphosphoglyceric acid
5. A molecule of ATP is formed.
ADP
ATP
3-Phosphoglyceric acid
6. The atoms in the molecule are rearranged, releasing a molecule of water.
H_2O
Phosphoenolpyruvic acid
7. A second molecule of ATP is formed.
ADP
ATP
Pyruvic acid
To mitochondria

● **FIGURE 25-3**

Glycolysis. Glycolysis breaks down a six-carbon glucose molecule into two three-carbon molecules of pyruvic acid. This process involves a series of enzymatic steps. This diagram follows the fate of the carbon chain (additional biochemical details will be found in Appendix VI). There is a net gain of two ATP molecules for each glucose molecule converted to pyruvic acid. In addition, two molecules of the coenzyme NAD are converted to NADH. Both the pyruvic acid and the NADH can still yield a great deal more energy via mitochondrial respiration. The further catabolism of pyruvic acid begins with its entry into a mitochondrion (*see Figure 25-4*).

the *intermembrane space* that separates the outer mitochondrial membrane from the *inner mitochondrial membrane.* The inner mitochondrial membrane contains a carrier protein that moves pyruvic acid into the mitochondrial matrix.

Once inside the mitochondrion, a pyruvic acid molecule first loses one carbon atom in a reaction involving NAD and an additional coenzyme called **coenzyme A,** commonly abbreviated as **CoA.** This reaction yields one molecule of carbon dioxide, one molecule of NADH, and one molecule of **acetyl-CoA** (as-e-til-CŌ-ā). Acetyl-CoA consists of a two-carbon **acetyl group** (CH_3CO) bound to coenzyme A. Next, the acetyl group is transferred from CoA to a four-carbon molecule of *oxaloacetic acid,* producing *citric acid.*

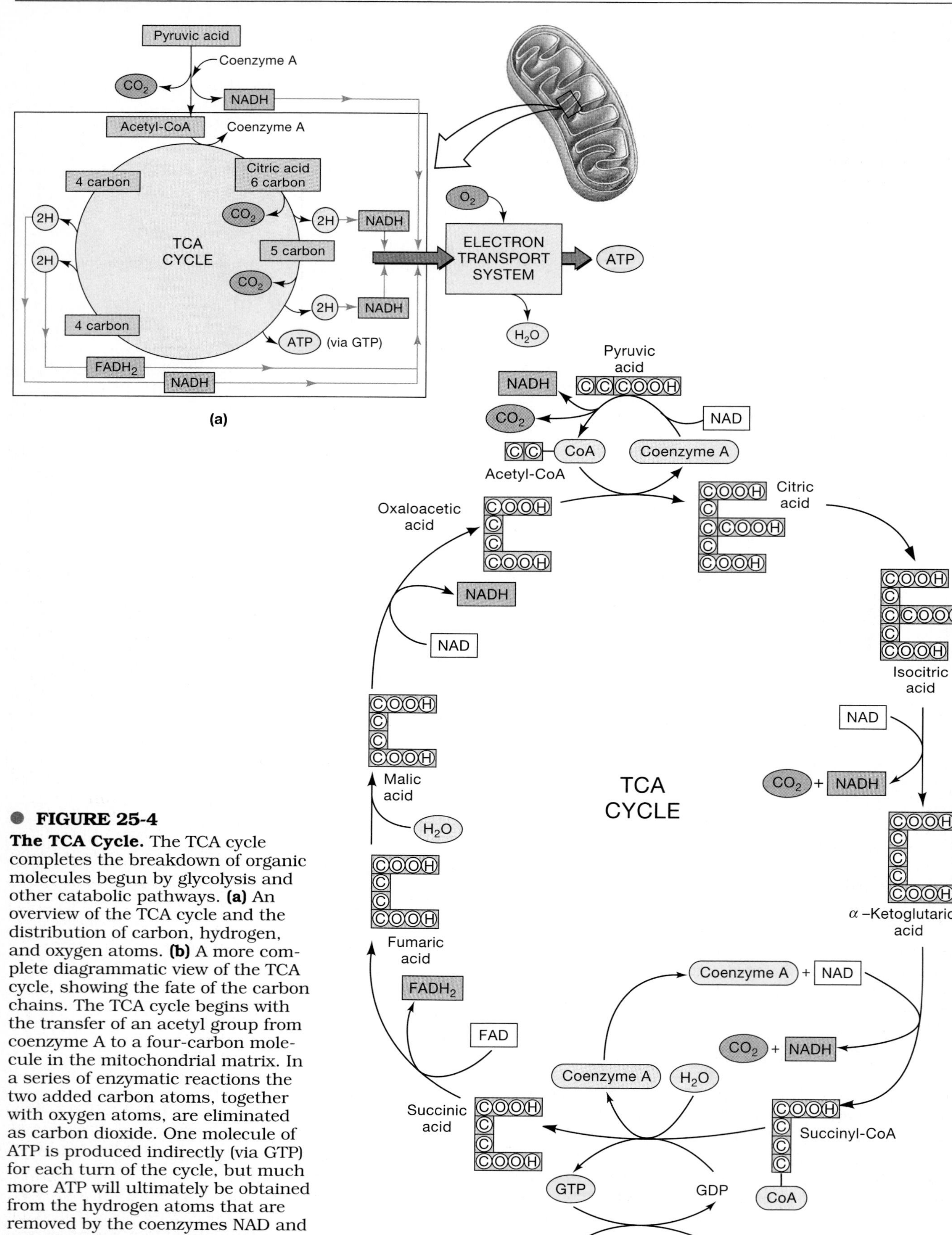
Pyruvic acid
Coenzyme A
CO_2
NADH
Acetyl-CoA
Coenzyme A
Citric acid 6 carbon
4 carbon
2H
CO_2
2H
NADH
TCA CYCLE
5 carbon
2H
CO_2
2H
NADH
4 carbon
ATP (via GTP)
$FADH_2$
NADH
(a)
O_2
ELECTRON TRANSPORT SYSTEM
ATP
H_2O
Pyruvic acid
NADH
CO_2
NAD
CoA
Coenzyme A
Acetyl-CoA
Citric acid
Oxaloacetic acid
NADH
NAD
Isocitric acid
NAD
CO_2 + NADH
TCA CYCLE
Malic acid
H_2O
α –Ketoglutaric acid
Fumaric acid
Coenzyme A + NAD
$FADH_2$
FAD
CO_2 + NADH
Coenzyme A
H_2O
Succinic acid
Succinyl-CoA
GTP
GDP
CoA
ADP
ATP
(b)

FIGURE 25-4
The TCA Cycle. The TCA cycle completes the breakdown of organic molecules begun by glycolysis and other catabolic pathways. **(a)** An overview of the TCA cycle and the distribution of carbon, hydrogen, and oxygen atoms. **(b)** A more complete diagrammatic view of the TCA cycle, showing the fate of the carbon chains. The TCA cycle begins with the transfer of an acetyl group from coenzyme A to a four-carbon molecule in the mitochondrial matrix. In a series of enzymatic reactions the two added carbon atoms, together with oxygen atoms, are eliminated as carbon dioxide. One molecule of ATP is produced indirectly (via GTP) for each turn of the cycle, but much more ATP will ultimately be obtained from the hydrogen atoms that are removed by the coenzymes NAD and FAD. (Additional biochemical details will be found in Appendix VI.)

The TCA Cycle

Figure 25-4

The formation of citric acid is the first step in a sequence of enzymatic reactions called the **tricarboxylic** (trī-kar-bok-SIL-ik) **acid (TCA) cycle,** or **citric acid cycle.** The function of the cycle is to remove hydrogen atoms from organic molecules and transfer them to coenzymes. The overall pattern of the cycle, sometimes known as the *Krebs cycle* in honor of the biochemist who described these reactions in 1937, is shown in Figure 25-4•. (Readers interested in a more complete reaction sequence should consult Appendix VI.)

At the start of the TCA cycle, the two-carbon acetyl group carried by CoA is attached to a four-carbon molecule to make the six-carbon compound citric acid. CoA is released intact to bind another acetyl group. A complete revolution of the TCA cycle removes the two added carbon atoms, regenerating the four-carbon chain. (This is why the reaction sequence is called a *cycle*.) The fate of the atoms in the acetyl group can be summarized as follows:

- The two carbon atoms are removed in enzymatic reactions that generate two molecules of carbon dioxide (CO_2), a metabolic waste product.
- The hydrogen atoms are removed by the coenzyme NAD or a related coenzyme, called **FAD** (**f**lavin **a**denine **d**inucleotide).

Several of the steps involved in a revolution of the TCA cycle involve more than one reaction and require more than one enzyme. Water molecules are tied up in two of those steps. The entire sequence can be summarized as:

$$CH_3CO\text{—}CoA + 3\ NAD + FAD + GDP + P_i + 2\ H_2O \rightarrow CoA + 2\ CO_2 + 3\ NADH + FADH_2 + GTP$$

The only immediate energy benefit of one revolution of the TCA cycle is the formation of a single molecule of GTP. In practical terms, GTP is the equivalent to ATP, because GTP readily transfers a phosphate group to ADP, producing ATP: GTP + ADP → GDP + ATP.

The formation of GTP from GDP in the TCA cycle is an example of **substrate-level phosphorylation.** In this process, an enzyme uses the energy released by a chemical reaction to transfer a phosphate group to a suitable acceptor molecule. Although GTP is formed in the TCA cycle, many reaction pathways in the cytosol phosphorylate ADP and form ATP directly. For example, the ATP produced during glycolysis is generated through substrate-level phosphorylation. Normally, however, substrate-level phosphorylation provides a relatively small amount of energy compared with *oxidative phosphorylation.*

Oxidative Phosphorylation

Figure 25-5

Oxidative phosphorylation is the generation of ATP within mitochondria through a reaction sequence that requires coenzymes and consumes oxygen. This process produces over 90 percent of the ATP used by our cells. The foundation of this reaction sequence is very simple:

$$2\ H_2 + O_2 \rightarrow 2\ H_2O$$

Living cells can easily obtain the ingredients for this reaction. Hydrogen is a component of all organic molecules, and oxygen is an atmospheric gas. The only problem is that the reaction releases a tremendous amount of energy all at once. In fact, this reaction releases so much energy that it is used to launch the space shuttle into orbit. Cells cannot handle energy explosions; energy release must be gradual, and this is what occurs in oxidative phosphorylation. During oxidation phosphorylation, this powerful reaction proceeds in a series of small, enzymatically controlled steps. Under these conditions energy can be captured and ATP generated.

OXIDATION, REDUCTION, AND ENERGY TRANSFER The enzymatic steps of oxidative phosphorylation involve **oxidation** and **reduction.** There are different types of oxidation and reduction reactions, but the most important for our purposes are those involving the transfer of electrons. The loss of electrons is a form of oxidation; the acceptance of electrons is a form of reduction. When electrons pass from one molecule to another, the donor is oxidized and the recipient reduced. Oxidation and reduction are important because electrons carry chemical energy. In a typical oxidation-reduction reaction, *the reduced molecule gains energy at the expense of the oxidized molecule.*

In such an exchange, the reduced molecule does not acquire all of the energy released by the oxidized molecule. Some energy is always released as heat, and additional energy may be used to perform physical or chemical work, such as the formation of ATP. By leading the electrons through a series of oxidation-reduction reactions before they ultimately combine with oxygen atoms, cells can capture and use much of the energy released in the formation of water.

Coenzymes play a key role in this process. A coenzyme acts as an intermediary that accepts electrons from one molecule and transfers them to another. In the TCA cycle, NAD and FAD remove hydrogen atoms from organic substrates. Each hydrogen atom consists of an electron (e^-) and a proton (a hydrogen ion, H^+), so when it accepts hydrogen atoms a coenzyme is reduced and gains energy. The donor molecule loses electrons and energy as it gives up its hydrogen atoms.

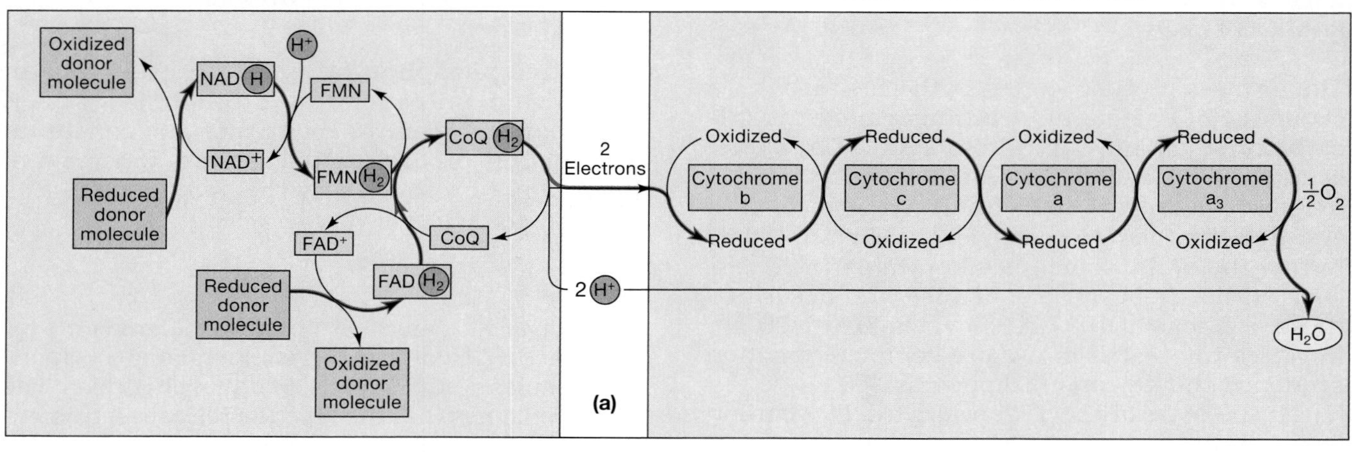
COENZYMES
ELECTRON TRANSPORT SYSTEM
Oxidized donor molecule
H+
NAD H
FMN
NAD+
FMN H2
CoQ H2
Reduced donor molecule
FAD+
CoQ
FAD H2
Reduced donor molecule
Oxidized donor molecule
2 Electrons
2 H+
Oxidized
Reduced
Cytochrome b
Cytochrome c
Cytochrome a
Cytochrome a3
Reduced
Oxidized
Reduced
Oxidized
1/2 O2
H2O
(a)

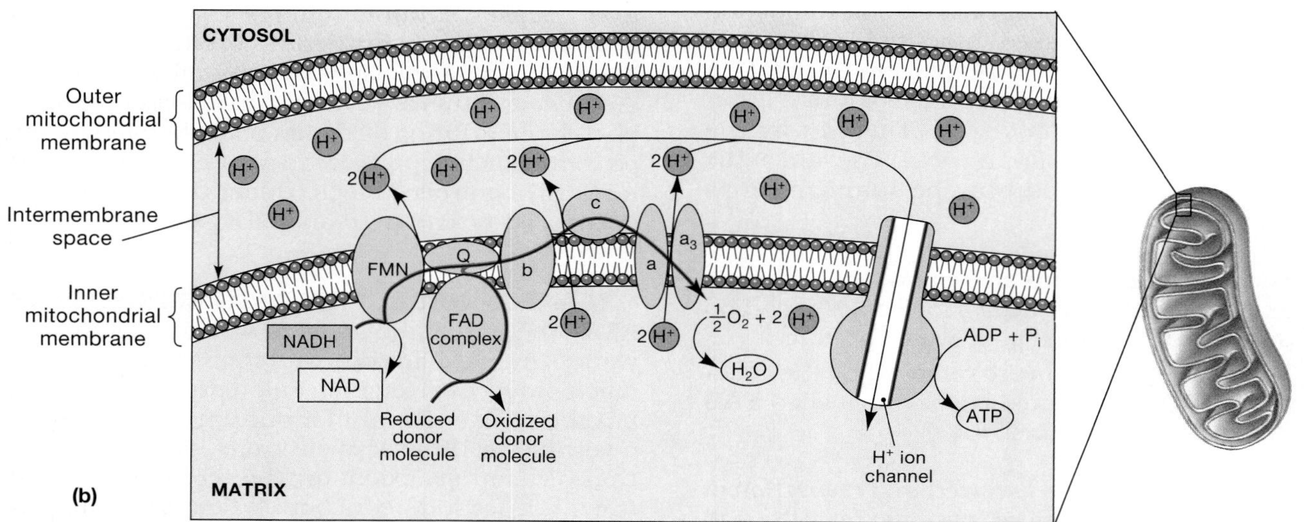
CYTOSOL
Outer mitochondrial membrane
Intermembrane space
Inner mitochondrial membrane
H+
2 H+
FMN
Q
b
c
a
a3
FAD complex
NADH
NAD
Reduced donor molecule
Oxidized donor molecule
1/2 O2 + 2 H+
H2O
ADP + Pi
ATP
H+ ion channel
MATRIX
(b)

FIGURE 25-5

Oxidative Phosphorylation. (a) The sequence of oxidation-reduction reactions involved. **(b)** Location of the coenzymes and the electron transport system. Note the sites where hydrogen ions are pumped into the intermembrane space.

NADH and $FADH_2$ then transfer their hydrogen atoms to other coenzymes. The protons are subsequently released, and the electrons, which carry the chemical energy, enter a sequence of oxidation-reduction reactions that ends with their transfer to oxygen and the formation of a water molecule. At several steps along the oxidation-reduction sequence, enough energy is released to support the synthesis of ATP from ADP. We will now consider this reaction sequence in greater detail.

FAD accepts two hydrogen atoms from the TCA cycle, and in doing so gains two electrons. NAD also gains two electrons as two hydrogen atoms are removed from the donor molecule. However, NAD binds one electron in the form of a hydrogen atom, forming NADH; it removes the electron from the other hydrogen atom and releases the proton (H^+). For this reason the reduced form of NAD is often described as "NADH + H^+."

Other coenzymes involved in the initial steps of oxidative phosphorylation are **FMN** (**f**lavin **m**ono**n**ucleotide) and **coenzyme Q** (*ubiquinone*). These coenzymes are either free in the mitochondrial matrix (NAD) or bound to the inner mitochondrial membrane (FAD, FMN, coenzyme Q). Coenzyme Q passes electrons to the *electron transport system.*

THE ELECTRON TRANSPORT SYSTEM The **electron transport system (ETS),** or *respiratory chain,* is a sequence of metalloproteins called **cytochromes** (SĪ-tō-krōmz; *cyto-,* cell + *chroma,* color). Each cytochrome has two components: a protein and a pigment. The protein, embedded in the inner mitochondrial membrane, surrounds the pigment complex, which contains a metal ion, either iron (Fe^{3+}) or copper (Cu^{2+}). There are four different cytochromes: **b, c, a,** and $\mathbf{a_3}$.

Figure 25-5• summarizes the major steps in oxidative phosphorylation. These steps are:

Step 1: ***A coenzyme strips a pair of hydrogen atoms from a substrate molecule.***

As we have seen, different coenzymes are used for different substrate molecules. During glycolysis, which occurs in the cytoplasm, NAD is reduced, producing NADH. Within mitochondria both NAD and FAD are reduced through reactions of the TCA cycle, producing NADH and $FADH_2$.

Step 2: ***NADH and $FADH_2$ deliver hydrogen atoms to coenzymes embedded in the inner mitochondrial membrane.***

It is the electrons that carry the energy, and the protons (H^+) that accompany them will be released before the electrons are transferred to the electron transport system. As indicated in Figure 25-5a●, the path taken to the ETS differs depending on whether the donor is NADH or $FADH_2$.

Step 3: ***Electrons are passed along the electron transport system, losing energy in a series of small steps.***

The sequence is cytochrome b to c to a to a_3.

Step 4: ***At the end of the electron transport system an oxygen atom accepts the electrons, creating an oxygen ion (O^{2-}).***

This ion has a very strong affinity for hydrogen ions (H^+); it quickly combines with two hydrogen ions in the mitochondrial matrix, forming a molecule of water.

Notice that this reaction sequence started with the removal of two hydrogen atoms from a substrate molecule, and it ended with the formation of water from two hydrogen ions and one oxygen ion. This is the reaction that we described initially as releasing too much energy if performed in a single step. Because the reaction has occurred in a series of small steps, however, the combination of hydrogen and oxygen can take place quietly rather than explosively. At several steps along the way, enough energy is released to establish conditions that will result in the formation of ATP.

ATP GENERATION AND THE ETS The cytochromes of the electron transport system do not produce ATP directly. Instead, they create the conditions necessary for ATP production. The electrons release energy as they pass from coenzyme to cytochrome and from cytochrome to cytochrome. At three points in this process enough energy is released to power ion pumps that transport hydrogen ions from the mitochondrial matrix into the space between the inner and outer mitochondrial membranes. The pumping activity occurs (1) when $FMNH_2$ reduces CoQ, (2) when cytochrome b reduces cytochrome c, and (3) when electrons are passed from cytochrome a to cytochrome a_3.

The hydrogen ion pumping by the respiratory enzymes creates a large concentration gradient for hydrogen ions across the inner mitochondrial membrane. This concentration gradient is a form of potential energy, and it is used to provide the energy to convert ADP to ATP.

Despite the concentration gradient, the hydrogen ions cannot diffuse into the matrix because they are not lipid-soluble. However, there are hydrogen ion channels in the inner mitochondrial membrane that will permit the diffusion of H^+ into the matrix. The entire complex of an ion channel and associated proteins looks like a miniature Tootsie-roll pop. These ion channels and their attached *coupling factors* in some way use the kinetic energy of passing hydrogen ions to generate ATP. This process is called **chemiosmosis** (kem-ē-oz-MŌ-sis).

The amount of ATP produced depends on how many hydrogen ions move across the membrane, which in turn depends on how many are pumped into the intermembranous space by the electron transport chain.

- For each pair of electrons removed by NAD from a substrate in the TCA cycle, six hydrogen ions are pumped across the inner mitochondrial membrane and into the intermembranous space (Figure 25-5●). Their reentry into the matrix provides the energy to generate three molecules of ATP.
- For each pair of electrons removed by FAD from a substrate in the TCA cycle, four hydrogen ions are pumped across the inner mitochondrial membrane and into the intermembranous space. Their reentry into the matrix provides the energy to generate two molecules of ATP.

Oxidative phosphorylation represents the single most important mechanism for the generation of ATP. A chronic suspension or even a significant reduction in the rate of oxidative phosphorylation will usually kill the cell. If many cells are affected, the individual may die. Oxidative phosphorylation requires both oxygen and electrons; the rate of ATP generation can thus be limited by the availability of either oxygen or electrons.

Cells obtain oxygen by diffusion from the extracellular fluid. If the supply of oxygen is cut off, mitochondrial ATP production will cease because reduced cytochrome a_3 becomes unable to get rid of its electrons. With the last reaction stopped, the entire ETS comes to a halt, like cars at a washed-out bridge. Oxidative phosphorylation can no longer take place, and cells quickly succumb to energy starvation. Because neurons have a high demand for energy, the brain is one of the first organs to be affected. Hydrogen cyanide gas is sometimes used as a pesticide to kill rats or mice; in some states where capital punishment is legal, it is used to execute criminals. The cyanide ion (CN^-) binds to cytochrome a_3 and prevents the transfer of electrons to oxygen. As a result, cells die from energy starvation.

❑ Energy Yield of Glycolysis and Cellular Respiration

Figure 25-6

The complete reaction pathway that begins with glucose and ends with carbon dioxide and water is for most cells the primary method of generating ATP. Figure 25-6• reviews the entire process from an energy standpoint.

- ***Glycolysis.*** During glycolysis, the cell gains two molecules of ATP directly for each glucose molecule broken down to pyruvic acid. Two molecules of NADH are also produced. In most cells, electrons are passed from NADH to FAD via an intermediary in the intermembranous space and thence to CoQ and the electron transport system. This sequence of events will ultimately provide an additional four ATP molecules.
- ***The TCA cycle.*** The two pyruvic acid molecules derived from each glucose molecule are fully broken down in mitochondria. Two revolutions of the TCA cycle, each yielding a molecule of ATP, provide a total gain of two more molecules of ATP.
- ***The electron transport system.*** Both the TCA cycle itself and the step leading into it (formation of acetyl-CoA from pyruvic acid) produce additional molecules of reduced coenzymes

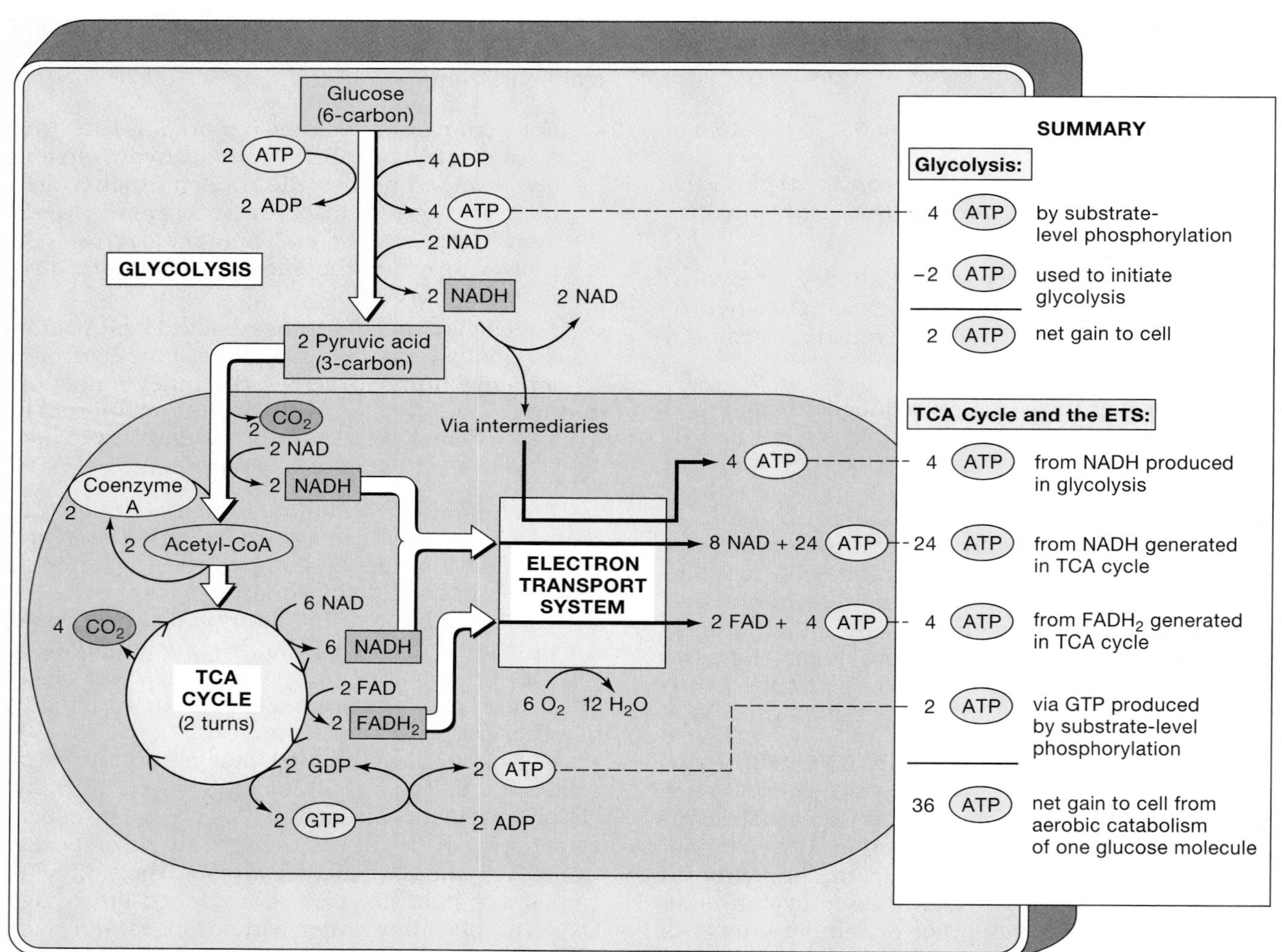
Glucose (6-carbon)
2 ATP
2 ADP
4 ADP
4 ATP
2 NAD
2 NADH
2 NAD
GLYCOLYSIS
2 Pyruvic acid (3-carbon)
2 CO2
2 NAD
2 NADH
Via intermediaries
Coenzyme A
2
2 Acetyl-CoA
4 CO2
6 NAD
6 NADH
TCA CYCLE (2 turns)
2 FAD
2 FADH2
2 GDP
2 GTP
2 ATP
2 ADP
ELECTRON TRANSPORT SYSTEM
4 ATP
8 NAD + 24 ATP
2 FAD + 4 ATP
6 O2 12 H2O
SUMMARY
Glycolysis:
4 ATP by substrate-level phosphorylation
−2 ATP used to initiate glycolysis
2 ATP net gain to cell
TCA Cycle and the ETS:
4 ATP from NADH produced in glycolysis
24 ATP from NADH generated in TCA cycle
4 ATP from FADH2 generated in TCA cycle
2 ATP via GTP produced by substrate-level phosphorylation
36 ATP net gain to cell from aerobic catabolism of one glucose molecule

● **FIGURE 25-6**

A Summary of the Energy Yield of Aerobic Respiration. For each glucose molecule broken down via glycolysis, only two molecules of ATP (net) are produced. The TCA cycle generates an additional two ATP molecules in GTP. However, glycolysis, the formation of acetyl-CoA, and the TCA cycle all yield molecules of reduced coenzymes (NADH and $FADH_2$). Many additional ATP molecules are produced when electrons from these coenzymes pass down the electron transport chain. Each NADH produced inside the mitochondria yields three ATP molecules. In most cells, each NADH molecule produced in glycolysis provides only two ATP molecules, the same amount gained from each molecule of $FADH_2$.

($NADH$ and $FADH_2$). The formation of acetyl-CoA generates a molecule of NAD, while each revolution of the TCA cycle generates three molecules of NADH and one molecule of $FADH_2$. These coenzymes feed electrons into the electron transport chain. Through the ETS, each molecule of NADH yields three molecules of ATP and one of water; the $FADH_2$ yields two molecules of ATP and one of water. The total yield per glucose molecule is an additional 28 molecules of ATP (see Figure 25-6●).

Summing up, for each glucose molecule processed, the cell gains 36 molecules of ATP: two from glycolysis, two from the TCA cycle, and 32 from the ETS (four from the NADH generated in glycolysis, and the rest from coenzymes generated within the mitochondrion).

Heart muscle cells and liver cells are able to gain an additional 2 ATP molecules for each glucose molecule broken down. This gain is accomplished by increasing the energy yield from the NADH generated during glycolysis from 4 ATP to 6 ATP molecules. In these cells, each NADH passes electrons to an intermediary that crosses the inner mitochondrial membrane and generates NADH in the mitochondrial matrix. The subsequent transfer of electrons to FMN, CoQ, and the ETS results in the production of 6 ATP molecules, just as if the two NADH molecules had been generated in the TCA cycle.

CARBOHYDRATE LOADING Although other nutrients can be broken down to provide substrates for the TCA cycle, carbohydrates require the least processing and preparation. It is not surprising, therefore, that athletes have tried to devise ways of exploiting these compounds as ready sources of energy.

Eating carbohydrates *just before* exercising does not improve performance and may actually decrease endurance by slowing the mobilization of existing energy reserves. Runners or swimmers preparing for lengthy endurance contests, such as a marathon or 5-km swim, do not eat immediately before competing, and for 2 hours before the race their intake is limited to drinking water. However, these athletes often eat carbohydrate-rich meals for 3 days before the event. This **carbohydrate loading** increases the carbohydrate reserves of muscle tissue that will be called upon during the competition.

Maximum effects can be obtained by exercising to exhaustion for 3 days before starting the high-carbohydrate diet; this practice is called *carbohydrate depletion/loading.* There are a number of potentially unpleasant side effects to carbohydrate depletion/loading, including muscle and kidney damage, and sports physiologists recommend that athletes use this routine fewer than three times per year.

What is the primary role of the TCA cycle in the production of ATP?

How is it that the NADH produced by glycolysis in skeletal muscles leads to the production of 2 ATP molecules in the mitochondria, but the NADH produced by glycolysis in cardiac muscle leads to the production of 3 ATP molecules?

How would a decrease in the level of cytoplasmic NAD affect ATP production in the mitochondria?

Other Catabolic Pathways

Figure 25-7

Aerobic metabolism is relatively efficient and capable of generating large amounts of ATP. It is the cornerstone for normal cellular metabolism, but it has one obvious limitation: The cell must have adequate supplies of both oxygen and glucose. Cells can survive only for brief periods without adequate oxygen, but low glucose concentrations have a much smaller effect on most cells because they can break down other nutrients to provide substrates for the TCA cycle (Figure 25-7●). Many cells can switch from one nutrient source to another as the need arises. For example, many cells can shift from glucose-based to lipid-based ATP production when necessary. Skeletal muscles catabolize glucose when actively contracting, but they utilize fatty acids when at rest.

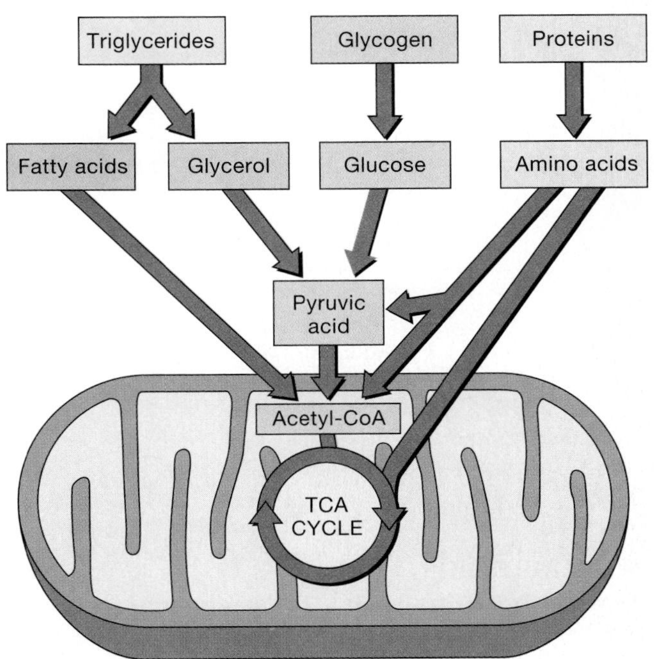
Triglycerides
Glycogen
Proteins
Fatty acids
Glycerol
Glucose
Amino acids
Pyruvic acid
Acetyl-CoA
TCA CYCLE

● **FIGURE 25-7**
Alternative Catabolic Pathways

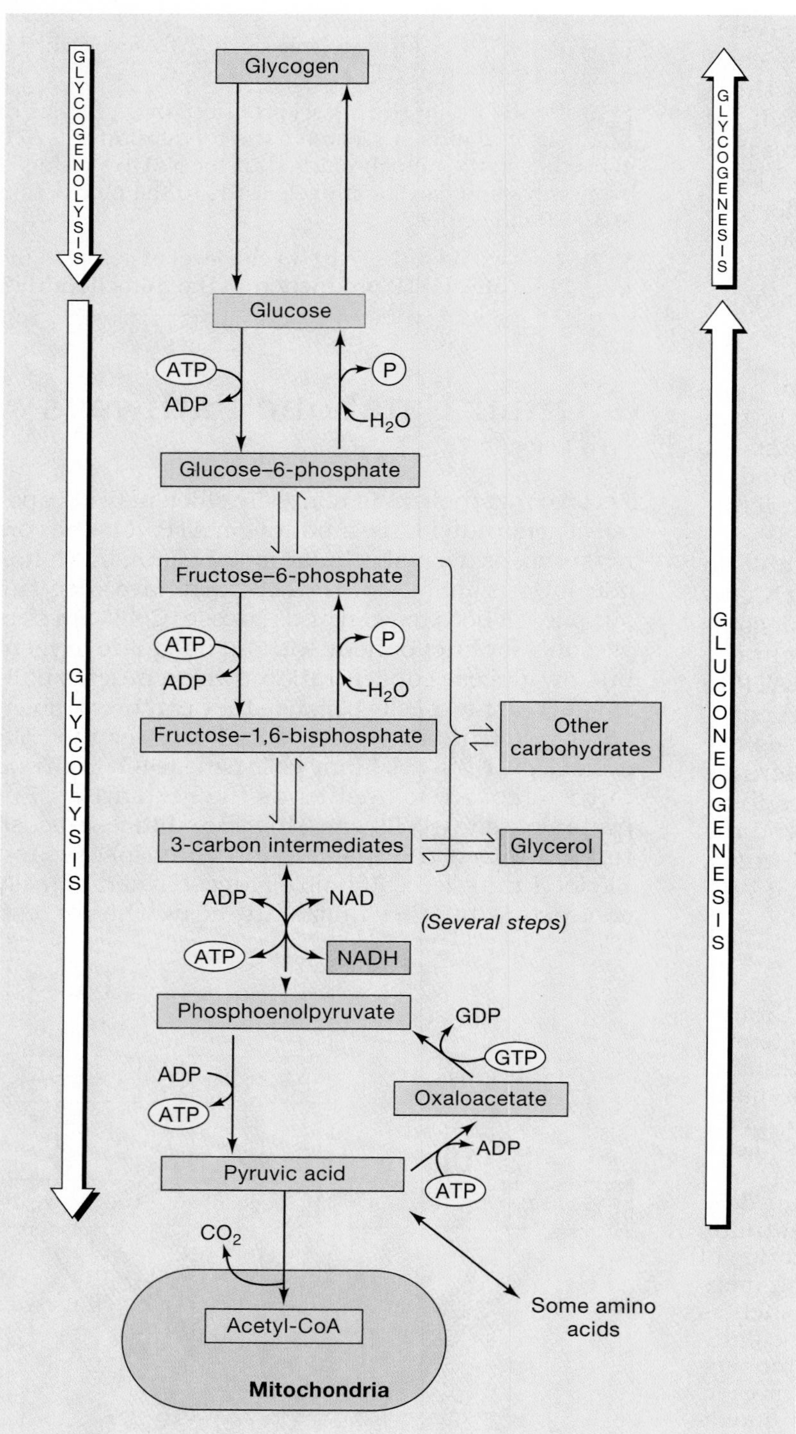
GLYCOGENOLYSIS
Glycogen
GLYCOGENESIS
Glucose
ATP
ADP
P
H_2O
Glucose–6-phosphate
Fructose–6-phosphate
ATP
ADP
P
H_2O
GLYCOLYSIS
GLUCONEOGENESIS
Fructose–1,6-bisphosphate
Other carbohydrates
3-carbon intermediates
Glycerol
ADP
NAD
(Several steps)
ATP
NADH
Phosphoenolpyruvate
GDP
GTP
ADP
ATP
Oxaloacetate
ADP
Pyruvic acid
ATP
CO_2
Some amino acids
Acetyl-CoA
Mitochondria

● **FIGURE 25-8**
Carbohydrate Synthesis. This flow chart diagrams the major pathways for glycolysis and gluconeogenesis. Many of the reactions are freely reversible, but separate regulatory enzymes control key steps. Some amino acids, other carbohydrates, and glycerol can be converted to glucose. Notice that the enzymatic reaction that converts pyruvic acid to acetyl-CoA cannot be reversed.

Cells break down proteins for energy only when lipids or carbohydrates are unavailable, primarily because proteins make up the enzymes and organelles that the cell needs to survive. Nucleic acids are present only in small amounts, and they are seldom catabolized for energy, even when the cell is dying of acute starvation. This restraint makes sense, because it is the DNA in the nucleus that determines all of the structural and functional characteristics of the cell. We will consider the catabolism of other compounds in later sections as we discuss lipid, protein, and nucleic acid metabolism.

❑ Gluconeogenesis

Figure 25-8

The synthesis of glucose is called **gluconeogenesis** (gloo-kō-nē-ō-JEN-e-sis). Because some of the steps in glycolysis are essentially irreversible, gluconeogenesis involves a different set of regulatory enzymes, and glucose breakdown and synthesis are independently regulated. Pyruvic acid or other three-carbon molecules can be used as starting materials. As indicated in Figure 25-8●, this reaction sequence enables a cell to create glucose molecules from other carbohydrates, glycerol, or some amino acids. However, acetyl-CoA cannot be used to make glucose because the decarboxylation step between pyruvic acid and acetyl-CoA cannot be reversed. Fatty acids and many amino acids cannot be converted to glucose because their catabolic pathways, described in later sections, produce acetyl-CoA.

Glucose molecules created by gluconeogenesis can be used to manufacture other simple sugars, complex carbohydrates, proteoglycans, or nucleic acids. In the liver and in skeletal muscle, glucose molecules are stored as glycogen. The process of glycogen formation is known as **glycogenesis.** Glycogen is an important energy reserve that can be broken down when the cell cannot obtain enough glucose from the interstitial fluid. Although glycogen molecules are large, glycogen reserves take up very little space because they form compact, insoluble granules.

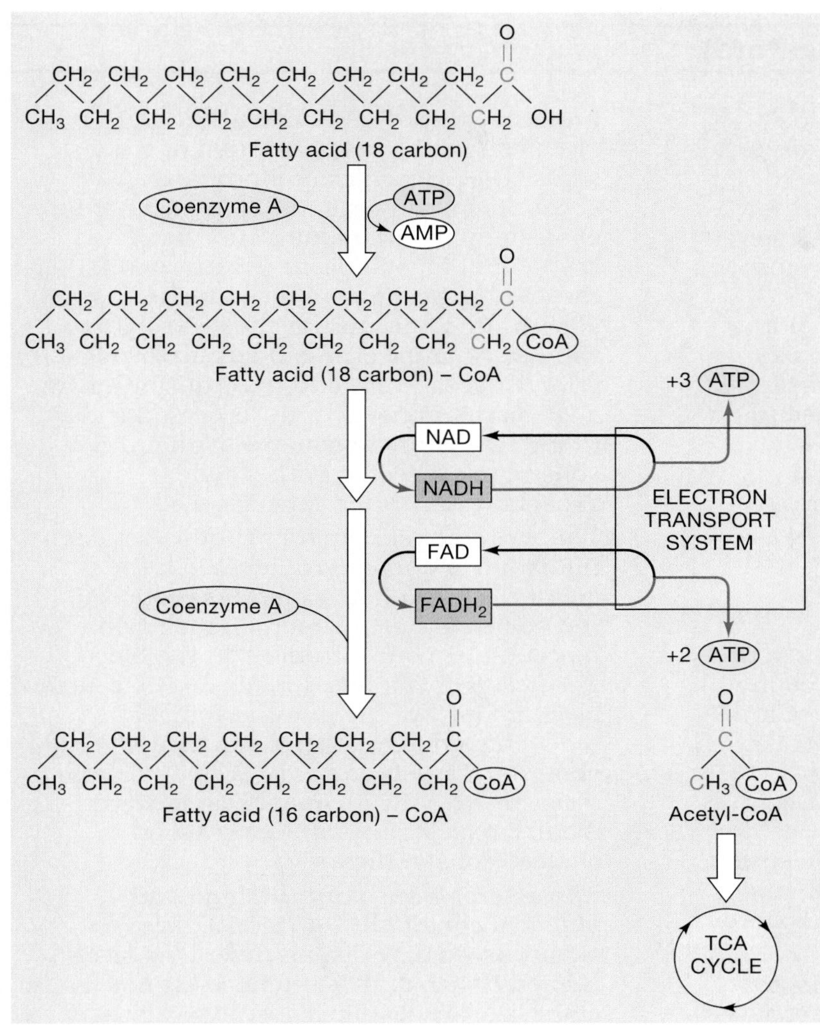
Fatty acid (18 carbon)
Coenzyme A
ATP
AMP
Fatty acid (18 carbon) – CoA
+3 ATP
NAD
NADH
ELECTRON TRANSPORT SYSTEM
FAD
FADH2
Coenzyme A
+2 ATP
Fatty acid (16 carbon) – CoA
Acetyl-CoA
TCA CYCLE

● **FIGURE 25-9**
Beta-Oxidation. During beta-oxidation the carbon chains of fatty acids are broken down to yield molecules of acetyl-CoA that can be used in the TCA cycle. The reaction also donates hydrogen atoms to coenzymes that deliver them to the electron transport system.

Lipid Metabolism

Lipid molecules, like carbohydrates, contain carbon, hydrogen, and oxygen, but these atoms are present in different proportions. Because triglycerides are the most abundant lipid in the body, our discussion will focus on pathways for triglyceride breakdown and synthesis.

Lipid Catabolism

During lipid catabolism, or **lipolysis,** lipids are broken down into pieces that can be converted to pyruvic acid or channeled directly into the TCA cycle.

A triglyceride is first split into its component parts through hydrolysis. This step yields one molecule of glycerol and three fatty acid molecules. Glycerol enters the TCA cycle after enzymes in the cytosol convert it to pyruvic acid. The catabolism of fatty acids involves a completely different set of enzymes.

Beta-Oxidation

Figure 25-9

Fatty acid molecules are broken down into two-carbon fragments by means of a sequence of reactions known as **beta-oxidation.** This process occurs inside mitochondria, so the carbon chains can enter the TCA cycle immediately. Figure 25-9● diagrams one step in the process of beta-oxidation. Each step generates molecules of acetyl-CoA, NADH, and $FADH_2$, leaving a shorter carbon chain bound to coenzyme A.

Beta-oxidation provides substantial energy benefits. For each two-carbon fragment removed from the fatty acid, the cell gains 12 ATP molecules from the processing of acetyl-CoA in the TCA cycle, plus 5 ATP molecules from the NADH and $FADH_2$. The cell can therefore gain 144 ATP molecules from the breakdown of an 18-carbon fatty acid molecule. This result is almost 1.5 times the energy obtained by the complete breakdown of three six-carbon glucose molecules. The catabolism of other lipids follows similar patterns, usually ending with the formation of acetyl-CoA.

Lipids and Energy Production

Lipids are important as an energy reserve because they can provide large amounts of ATP. Because they are insoluble, lipids can be stored in compact droplets in the cytosol. This storage method saves space, but when the lipid droplets are large it is difficult for water-soluble enzymes to get at them. Lipid reserves are therefore more difficult to access than carbohydrate reserves. In addition, most lipids are processed inside mitochondria, and mitochondrial activity is limited by the availability of oxygen. The net result is that lipids cannot provide large amounts of ATP in a short amount of time. However, cells with modest energy demands can shift over to lipid-based energy production when glucose supplies are limited. Skeletal muscle fibers normally cycle between lipid and carbohydrate metabolism. At rest, when energy demands are low, they break down fatty acids.

Dietary Fats and Cholesterol

Elevated cholesterol levels are associated with the development of atherosclerosis (Chapter 21) and coronary artery disease (CAD, Chapter 20). [pp. 722, 697] Current nutritional advice suggests reducing cholesterol intake to under 300 mg per day. This amount represents a 40 percent reduction for the average American adult. As a result of rising concerns about cholesterol, such phrases as "low in cholesterol," "contains no cholesterol," and "cholesterol-free" are now widely used in the advertising and packaging of foods. What do they really mean in terms of individual health and diet planning? Before answering that question we must consider some basic information about cholesterol and about lipid metabolism in general.

1. ***Cholesterol has many vital functions in the human body.*** It serves as a waterproofing for the epidermis, a lipid component of all cell membranes, a key constituent of bile, and the precursor of several steroid hormones and one vitamin (vitamin D_3). Because cholesterol is so important, dietary restrictions should have the goal of keeping cholesterol levels within acceptable limits. The goal is *not* the elimination of cholesterol in the diet or the circulating blood.
2. ***The cholesterol content of the diet is not the only source of circulating cholesterol.*** The human body can manufacture cholesterol from the acetyl-CoA obtained through glycolysis or the beta-oxidation of other lipids. Dietary cholesterol probably accounts for only around 20 percent of the cholesterol in circulation, and the rest is the result of metabolism of saturated fats in the diet. If the diet contains an abundance of saturated fats, serum cholesterol levels will rise because excess lipids are broken down to acetyl-CoA and used to synthesize cholesterol. Consequently, a person trying to lower serum cholesterol by dietary control must restrict other lipids—especially saturated fats—as well.
3. ***There are genetic factors that affect each individual's cholesterol level.*** If the dietary supply of cholesterol is reduced, the body synthesizes more to maintain "acceptable" concentrations in the blood. The acceptable level depends on the genetic programming of the individual. Because individuals differ in genetic makeup, their cholesterol levels can vary even on similar diets. In virtually all instances, however, dietary restrictions can lower blood cholesterol significantly.
4. ***Cholesterol levels vary with age and physical condition.*** At age 19, three out of four males have fasting cholesterol levels below 170 mg/dl. Cholesterol levels in females of this age are slightly higher, typically at or below 175 mg/dl. With increasing age, the

When active, and energy demands are both large and immediate, skeletal muscle fibers shift over to metabolizing glucose.

Lipid Synthesis

Figure 25-10

The synthesis of lipids is known as **lipogenesis** (li-pō-JEN-e-sis). Figure 25-10• shows the major pathways of lipogenesis. Glycerol synthesis reverses the steps of glycolysis; the synthesis of most other lipids begins with acetyl-CoA. Lipogenesis can use almost any organic substrate because lipids, amino acids, and carbohydrates can be converted to acetyl-CoA.

Fatty acid synthesis involves a reaction sequence quite distinct from that of beta-oxidation, and our cells cannot *build* every fatty acid they can break down. **Linoleic acid** and **linolenic acid,** both 18-carbon, unsaturated fatty acids, cannot be synthesized at all. These are called **essential fatty acids** because they must be included in the diet. They are synthesized by plants, and essential fatty acid deficiency usually occurs only among hospitalized individuals receiving nutrients in an intravenous solution. A diet poor in linoleic acid slows growth and alters the appearance of the skin. These fatty acids are also needed to synthesize prostaglandins and some of the phospholipids incorporated in cell membranes throughout the body.

Lipid Transport and Distribution

Like glucose, lipids are needed throughout the body. All cells need lipids to maintain their cell membranes, and there are important steroid hormones that must reach target cells in many different tissues. Because most lipids are not soluble in water, special transport mechanisms are required to carry them from one region of the body to another. Most lipids circulate through the bloodstream as *free fatty acids* and *lipoproteins.*

cholesterol values gradually climb, and over age 70 the values are 230 mg/dl (males) and 250 mg/dl (females). Cholesterol levels are considered unhealthy if they are higher than those of 90 percent of the population in that age group. For males, this value ranges from 185 mg/dl at age 19, to 250 mg/dl at age 70. For females the comparable values are 190 mg/dl and 275 mg/dl.

To determine whether you need to do anything about your cholesterol level without performing calculations regarding age, weight, and sex, just remember three simple rules:

1. Individuals of any age with total cholesterol values below 200 mg/dl probably do not need to change their lifestyles unless they have a family history of coronary artery disease and atherosclerosis.
2. Those with cholesterol levels between 200 and 239 mg/dl should modify their diets, lose weight (if overweight), and have annual checkups.
3. Cholesterol levels over 240 mg/dl warrant drastic changes in dietary lipid consumption, perhaps coupled with drug treatment. Drug therapies are always recommended in cases where the serum cholesterol level exceeds 350 mg/dl. Examples of drugs used to lower cholesterol levels include *cholestyramine*, *colestipol*, and *lovastatin*.

Most physicians, when ordering a blood test for cholesterol, also request information on circulating triglycerides and the total lipid content of the plasma. In fasting individuals, triglycerides are usually present at levels of 40–150 mg/dl, and the total lipid content of the blood may be from 400 to 1000 mg/dl. (After fatty meals, triglyceride and total lipid levels may be temporarily elevated.)

When cholesterol levels are high, or when an individual has a family history of atherosclerosis or CAD, further tests may be performed to determine the relative amounts of cholesterol circulating in LDLs and HDLs. A high total cholesterol value linked to a high LDL level spells trouble. In effect, an unusually large amount of cholesterol is being exported to peripheral tissues. Problems can also exist if the individual has high total cholesterol—or even normal total cholesterol—and low HDL levels (below 35 mg/dl). In this case, excess cholesterol delivered to the tissues cannot easily be returned to the liver for excretion. In either event the amount of cholesterol in peripheral tissues, and especially in arterial walls, is likely to increase. A standard guideline is that the ratio of total cholesterol to HDL cholesterol should be less than 4.5 to 1. If it is 4.5 or more, the individual is at risk for developing atherosclerosis.

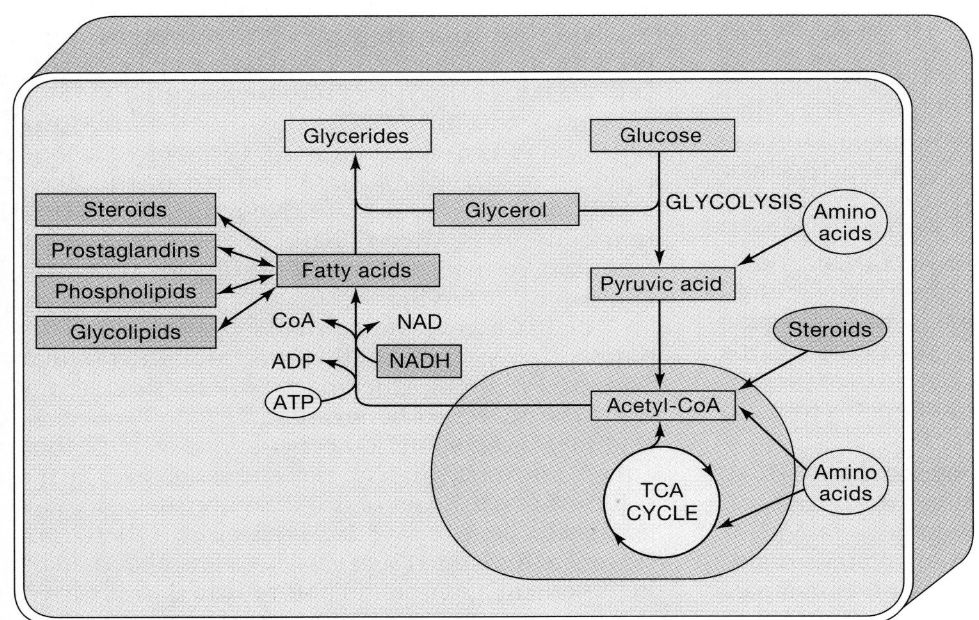
Glycerides
Glucose
Steroids
Glycerol
GLYCOLYSIS
Amino acids
Prostaglandins
Fatty acids
Phospholipids
Pyruvic acid
Glycolipids
CoA
NAD
Steroids
ADP
NADH
ATP
Acetyl-CoA
TCA CYCLE
Amino acids

FIGURE 25-10
Lipid Synthesis. Pathways of lipid synthesis begin with acetyl-CoA. Molecules of acetyl-CoA can be strung together in the cytosol, yielding fatty acids. Those fatty acids can be used to synthesize glycerides or other lipid molecules. Lipids can be synthesized from amino acids or carbohydrates via acetyl-CoA.

Free Fatty Acids

Free fatty acids (FFAs) are lipids that can diffuse easily across cell membranes. Free fatty acids in the blood are usually bound to albumin, the most abundant plasma protein.

Sources of free fatty acids in the blood include:

- Fatty acids not used in the synthesis of triglycerides, which diffuse out of the intestinal epithelium and into the blood.
- Fatty acids that diffuse out of lipid stores, such as the liver and adipose tissue, when triglycerides are broken down.

Liver cells, cardiac muscle cells, skeletal muscle fibers, and many other body cells can metabolize free fatty acids. They are an important energy source during periods of starvation, when glucose supplies are limited.

Lipoproteins

Figure 25-11

Lipoproteins are lipid-protein complexes that contain large insoluble glycerides and cholesterol with a superficial coating of phospholipids and proteins. The proteins and phospholipids make the entire complex soluble, and the proteins play a role in the regulation of lipid absorption by cells.

Lipoproteins are usually classified according to size and the relative proportions of lipid versus protein. The relationships among the different classes of lipoproteins will be examined after we introduce the major groups recognized at present:

1. **Chylomicrons.** Roughly 95 percent of the weight of a chylomicron consists of triglycerides. Chylomicrons are the largest lipoproteins, ranging in diameter from 0.03 to 0.5 μm. They are produced by intestinal epithelial cells, as described in Chapter 24. [p. 924] Chylomicrons carry absorbed lipids from the intestinal tract to the circulation. The other lipoproteins shuttle lipids among various tissues in the body, such as between the liver and adipose tissue. The liver is the primary source for all other types of lipoproteins.
2. **Very low-density lipoproteins (VLDLs).** These lipid masses contain triglycerides manufactured by the liver plus small amounts of phospholipids and cholesterol. The primary function of VLDLs is the transport of these triglycerides to peripheral tissues. VLDLs range in diameter from 25 to 75 nm (0.025–0.075 μm).
3. **Intermediate-density lipoproteins (IDLs).** Intermediate-density lipoproteins are intermediate in size and lipid composition between VLDLs and LDLs. They contain smaller amounts of triglycerides and relatively more phospholipids and cholesterol.
4. **Low-density lipoproteins (LDLs).** Low-density lipoproteins contain cholesterol, lesser amounts of phospholipids, and very few triglycerides. These lipoproteins, which are around 25 nm in diameter, deliver cholesterol to peripheral tissues. Because this cholesterol may wind up in arterial plaques, LDL cholesterol is often called "bad cholesterol."
5. **High-density lipoproteins (HDLs).** High-density lipoproteins, around 10 nm in diameter, have roughly equal amounts of lipid and protein. The lipids are primarily cholesterol and phospholipids. The primary function of HDLs is transporting excess cholesterol from peripheral tissues back to the liver for storage or excretion in the bile. Because HDL cholesterol is returning from peripheral tissues and will not cause circulatory problems, it is sometimes called "good cholesterol." Actually, applying the terms *good* and *bad* to cholesterol can be misleading, for cholesterol metabolism is complex and variable. (For more details, see the discussion of dietary fats and cholesterol on pp. 948–949.)

Figure 25-11• diagrams the probable relationships among these lipoproteins. Chylomicrons produced in the intestinal tract reach the venous circulation by entering lymphatic capillaries and traveling through the thoracic duct (Figure 25-11a•). Although chylomicrons are too large to diffuse across capillary walls, the endothelial lining of capillaries in adipose tissue, skeletal muscle, cardiac muscle, and the liver contains an enzyme, **lipoprotein lipase,** that breaks down complex lipids. When lipid complexes contact these endothelial walls, enzymatic activity releases fatty acids and monoglycerides that can diffuse across the endothelium and into the interstitial fluid.

The liver controls the distribution of other lipoproteins (Figure 25-11b•). Liver cells synthesize VLDLs for discharge into the circulation (Step 1). On arrival in peripheral capillaries, lipoprotein lipase removes many of the triglycerides, leaving an IDL (Step 2). On return to the liver, additional triglycerides are removed, and the protein content is altered. This process creates an LDL that returns to peripheral tissues to deliver cholesterol (Step 3).

The LDLs leave the circulation through capillary pores or cross the endothelium through vesicular transport (Step 4). Once in peripheral tissues, the LDLs are absorbed by *receptor-mediated endocytosis* (Step 5). The amino acids and cholesterol then enter the cytoplasm. The cholesterol not used by the cell in the synthesis of lipid membranes or other products diffuses out of the cell (Step 6). It then reenters the circulation (Step 7), where it is absorbed by high-density lipoproteins and returned to the liver. On arrival in the liver, the HDLs are absorbed, and

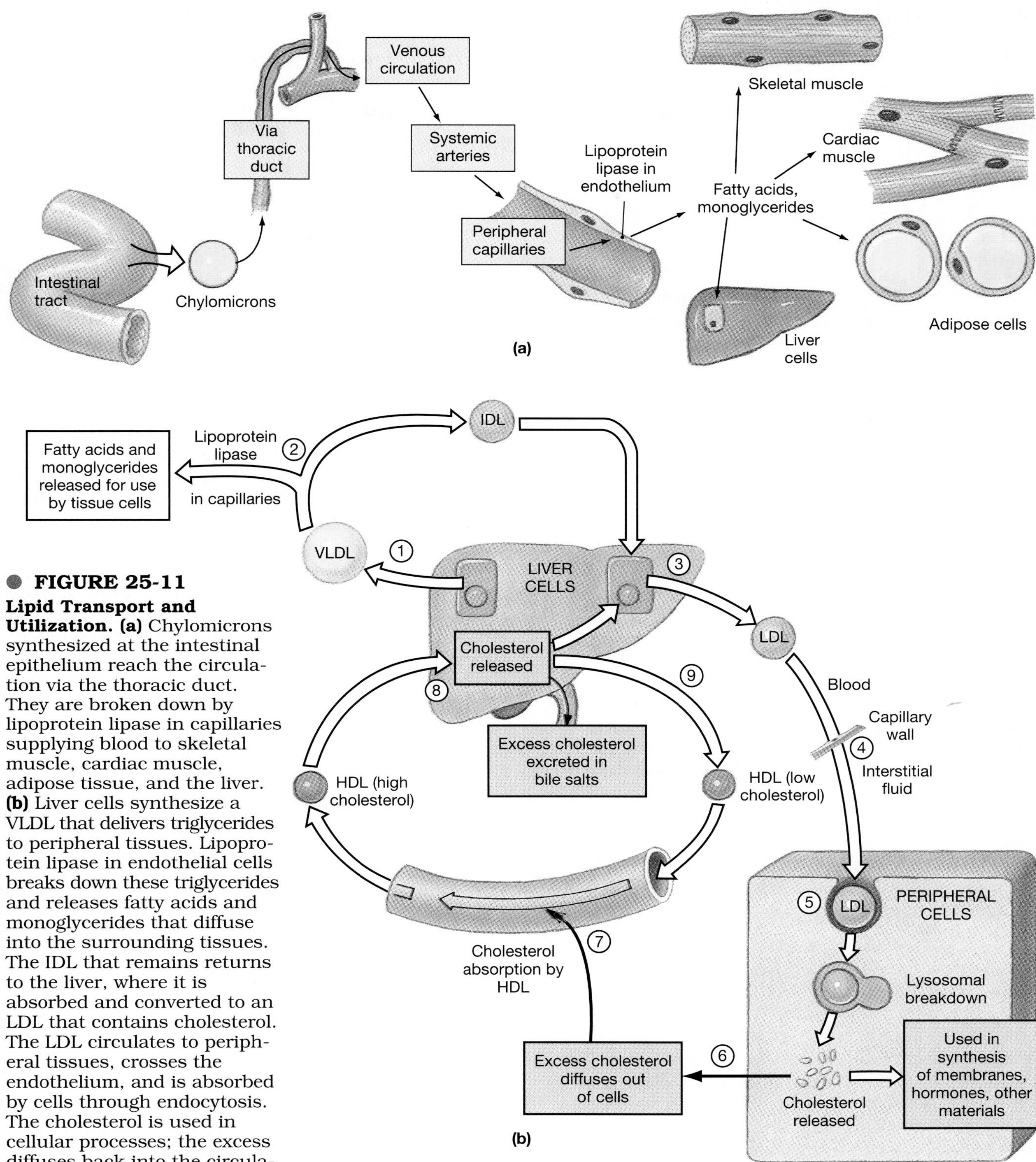

Venous circulation
Via thoracic duct
Systemic arteries
Skeletal muscle
Cardiac muscle
Lipoprotein lipase in endothelium
Fatty acids, monoglycerides
Peripheral capillaries
Intestinal tract
Chylomicrons
Adipose cells
Liver cells
(a)
IDL
Fatty acids and monoglycerides released for use by tissue cells
Lipoprotein lipase
in capillaries
2
VLDL
1
LIVER CELLS
3
LDL
Cholesterol released
8
9
Blood
Capillary wall
4
Interstitial fluid
Excess cholesterol excreted in bile salts
HDL (high cholesterol)
HDL (low cholesterol)
5
LDL
PERIPHERAL CELLS
Lysosomal breakdown
7
Cholesterol absorption by HDL
Excess cholesterol diffuses out of cells
6
Cholesterol released
Used in synthesis of membranes, hormones, other materials
(b)

● FIGURE 25-11
Lipid Transport and Utilization. (a) Chylomicrons synthesized at the intestinal epithelium reach the circulation via the thoracic duct. They are broken down by lipoprotein lipase in capillaries supplying blood to skeletal muscle, cardiac muscle, adipose tissue, and the liver. **(b)** Liver cells synthesize a VLDL that delivers triglycerides to peripheral tissues. Lipoprotein lipase in endothelial cells breaks down these triglycerides and releases fatty acids and monoglycerides that diffuse into the surrounding tissues. The IDL that remains returns to the liver, where it is absorbed and converted to an LDL that contains cholesterol. The LDL circulates to peripheral tissues, crosses the endothelium, and is absorbed by cells through endocytosis. The cholesterol is used in cellular processes; the excess diffuses back into the circulation. In the plasma, the cholesterol is absorbed by an HDL produced by the liver. On returning to the liver, the HDL is absorbed and the cholesterol extracted. Some of the cholesterol will be exported once again, in an LDL. Excess cholesterol will be excreted in bile salts.

their cholesterol extracted (Step 8). Some of the cholesterol recovered will be used in the synthesis of LDL, and the rest will be excreted in bile salts. The HDLs stripped of their cholesterol are released into the circulation, to travel into peripheral tissues and absorb additional cholesterol (Step 9).

Protein Metabolism

There are roughly 100,000 different proteins in the human body, with various forms, functions, and structures. All contain varying combinations of the same 20+ amino acids. (Appendix IV details the structures of these amino acids.) Under normal conditions, there is a continual turnover of cellular proteins. Peptide bonds are broken, and the free amino acids are used to manufacture new proteins. This recycling occurs in the cytosol.

If other energy sources are inadequate, mitochondria can break down amino acids in the TCA cycle to generate ATP. Not all amino acids enter the TCA cycle at the same point, so the ATP benefits vary. However, the average yield is comparable to that of carbohydrate catabolism.

Amino Acid Catabolism

Figure 25-12

The first step in amino acid catabolism is the removal of the amino group. This process requires a coenzyme derivative of **vitamin B_6** (*pyridoxine*). The amino group may be removed by **transamination** (trans-am-i-NĀ-shun) or **deamination** (dē-am-i-NĀ-shun). Figure 25-12● illustrates transamination and deamination reactions. Other details of amino acid catabolism will be considered in a later section.

Transamination

Figure 25-12a

Transamination (Figure 25-12a●) attaches the amino group of an amino acid to a **keto acid.** A keto acid resembles an amino acid except that the second carbon binds an oxygen atom rather than an amino group. Transamination produces a "new" amino acid that can enter the cytosol and be used for protein synthesis. It also converts the original amino acid to a keto acid that can enter the TCA cycle. Many different tissues perform transaminations. These reactions enable a cell to synthesize many of the amino acids needed for protein synthesis. Cells of the liver, skeletal muscles, heart, lung, kidney, and brain, which are particularly active in protein synthesis, perform large numbers of transaminations.

There are several inherited metabolic disorders that result from an inability to produce specific enzymes involved with amino acid metabolism. *Phenylketonuria* (fen-il-kē-tō-NU-rē-a), or PKU, is an example. Individuals with PKU cannot convert phenylalanine to tyrosine because of a defect in the enzyme *phenylalanine hydroxylase.* This reaction is an essential step in the synthesis of norepinephrine, epinephrine, dopamine, and melanin. If PKU is not detected in infancy, CNS development is inhibited, and severe brain damage results. AM *Phenylketonuria*

Deamination

Figure 25-12b,c

Deamination (Figure 25-12b●) is performed in preparing an amino acid for breakdown in the TCA cycle. Deamination is the removal of an amino group in a reaction that generates an ammonia molecule. Ammonia molecules are highly toxic, even in low concentrations. The liver, the primary site of deamination, has the enzymes needed to deal with the problem of ammonia generation. Liver cells convert the ammonia to **urea,** a relatively harmless, water-soluble compound that is excreted in the urine. The **urea cycle** is the reaction sequence involved (Figure 25-12c●).

When glucose supplies are low and lipid reserves are inadequate, liver cells break down internal proteins and absorb additional amino acids from the blood. The amino acids are deaminated, and the carbon chains are broken down to provide ATP.

Proteins and ATP Production

Several factors make protein catabolism an impractical source of quick energy:

1. Proteins are more difficult to break apart than are complex carbohydrates or lipids.
2. One of the byproducts, ammonia, is a toxin that can damage cells.
3. Proteins form the most important structural and functional components of any cell. Extensive protein catabolism therefore threatens homeostasis at the cellular and systems levels.

Protein Synthesis

Figure 25-13

The basic mechanism for protein synthesis was detailed in Chapter 3 (Figures 3-24 and 3-25●, pp. 96, 97). The human body can synthesize roughly half of the different amino acids needed to build proteins. There are 10 **essential amino acids.** Eight of them (*isoleucine, leucine, lysine, threonine, tryptophan, phenylalanine, valine,* and *methionine*) cannot be synthesized at all; the other two (*arginine* and *histidine*) can be synthesized in amounts that are insufficient for growing children. The carbon frameworks of the amino acids not in the essential group can be synthesized readily, and a nitrogen group

(a)

(b)

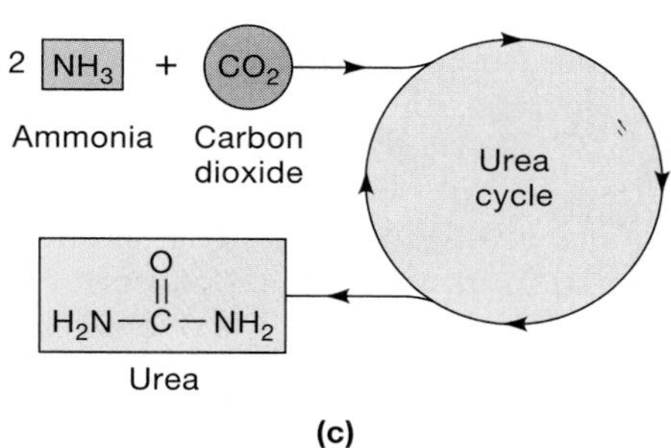
2 NH_3 + CO_2
Ammonia
Carbon dioxide
Urea cycle
O
||
$H_2N-C-NH_2$
Urea

(c)

● **FIGURE 25-12**

Amino Acid Catabolism. (a) During transamination an enzyme removes the amino group from one molecule and attaches it to a keto acid. **(b)** During deamination an enzyme strips the amino group from an amino acid and produces a keto acid and ammonia. **(c)** The urea cycle takes two metabolic waste products, carbon dioxide and ammonia, and produces a molecule of urea. Urea is a relatively harmless, soluble compound that is excreted in the urine.

can be attached through transamination or *amination*. **Amination** is the attachment of an amino group, as indicated in Figure 25-13●. Because the body can make these amino acids on demand, they are called the **nonessential amino acids.**

Protein deficiency diseases develop when an individual does not consume adequate amounts of all essential amino acids. All amino acids must be available if protein synthesis is to occur. Every tRNA must appear at the proper location bearing its individual amino acid; as soon as the amino acid called for by a particular codon is missing, the entire process comes to a halt. Regardless of the energy content of the diet, if it is deficient in essential amino acids the individual will be malnourished to some degree. Examples of protein deficiency diseases include *marasmus* and *kwashiorkor*. Although over 100 million children worldwide have symptoms of these disorders, neither condition is common in the United States today. AM *Protein Deficiency Diseases*

● **FIGURE 25-13**

Amination. Amination attaches an amino group to a keto acid. This is an important step in the synthesis of nonessential amino acids. Amino groups can also be attached through transamination (*see Figure 25-12*).

Nucleic Acid Metabolism

Living cells contain both DNA and RNA. Chapter 3 considered the replication of DNA, the mechanics of cell division, and the importance of DNA in regulating the structural and functional characteristics of the cell. *[pp. 93–104]* The genetic information contained in the DNA of the nucleus is absolutely essential to the long-term survival of the cell. As a result, the DNA in the nucleus is never catabolized for energy, even if the cell is dying of starvation. The RNA in the cell is involved with protein synthesis, and RNA molecules are broken down and replaced on a regular basis.

RNA Catabolism

In breaking down a strand of RNA, the bonds between nucleotides are broken, and the molecule is disassembled into individual nucleotides. These nucleotides are usually recycled into new nucleic acids. However, they can be catabolized to simple sugars and nitrogenous bases. The sugars can enter the glycolytic pathways. Pyrimidines (cytosine and uracil) are converted to acetyl-CoA and metabolized via the TCA cycle. The purines, which cannot be catabolized, are excreted.

RNA and Energy Production

RNA catabolism makes a relatively insignificant contribution to the total energy budget of the cell. Proteins account for 30 percent of the weight of the cell, and much more energy can be provided through the catabolism of nonessential proteins. Even when RNA is broken down, only the sugars and pyrimidines provide energy. The purines (adenine and guanine) cannot be catabolized at all. Instead they are deaminated and excreted as **uric acid.** Uric acid is another relatively nontoxic waste product, but it differs from urea in that it is far less soluble. Urea and uric acid are called **nitrogenous wastes** because they are waste products that contain nitrogen atoms.

Normal plasma uric acid concentrations average 2.7–7.4 mg/dl, depending on sex and age. When plasma concentrations exceed 7.4 mg/dl, *hyperuricemia* (hī-per-ū-ri-SĒ-mē-a) exists. This condition may affect 18 percent of the U.S. population. At concentrations over 7.4 mg/dl, body fluids are saturated with uric acid. Although symptoms may not appear at once, uric acid crystals may begin to form in body fluids. The condition that then develops is called *gout.* AM *Gout* Most cases of hyperuricemia and gout are associated with problems with the renal excretion of uric acid.

Nucleic Acid Synthesis

Most cells synthesize RNA, but DNA synthesis occurs only in cells that are preparing for mitosis and cell division or meiosis (nuclear events involved in gamete production). The process of DNA replication was described in Chapter 3. *[p. 100]* Messenger RNA, tRNA, and rRNA are transcribed by different forms of RNA polymerase.

- Messenger RNA is manufactured only when specific genes are activated, and a strand of mRNA has a life span measured in minutes or hours.
- Ribosomal RNA and transfer RNA are more durable than mRNA strands. For example, the half-life of ribosomal RNA is just over 5 days. However, because a typical cell contains roughly 100,000 ribosomes and many times that number of tRNA molecules, their replacement involves a considerable amount of synthetic activity.

How would a diet that is deficient in pyridoxine (vitamin B_6) affect protein metabolism?

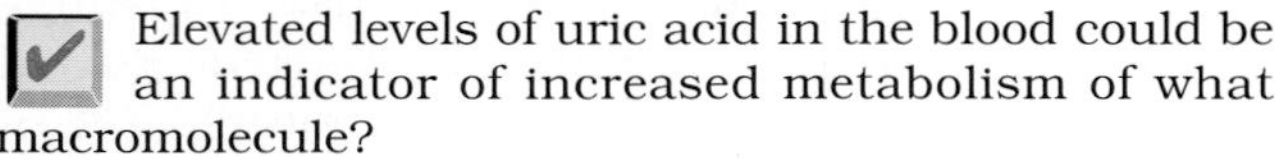
Elevated levels of uric acid in the blood could be an indicator of increased metabolism of what macromolecule?

Why are high-density lipoproteins (HDLs) considered to be beneficial?

Metabolic Interactions

Figure 25-14

Figure 25-14• summarizes the major pathways of cellular metabolism as described thus far. This diagram follows the reactions in a "typical" cell. Yet no one cell can perform all of the anabolic and catabolic operations and interconversions required by the body as a whole. As differentiation proceeds, each cell type develops its own complement of enzymes, and this enzyme complement determines the cell's metabolic capabilities. In the presence of such cellular diversity, homeostasis can be preserved only when the metabolic activities of tissues, organs, and organ systems are coordinated.

The nutrient requirements of each tissue vary depending on the types and quantities of enzymes present in the cytoplasm. From a metabolic standpoint, the body can be considered in terms of five distinctive components: the *liver, adipose tissue, skeletal muscle, neural tissue,* and *other peripheral tissues.*

- ***The liver.*** The liver represents the focal point for metabolic regulation and control. Liver cells contain a great diversity of enzymes, and they can break down or synthesize most of the carbohydrates, lipids, and amino acids needed by other cells in the body. Because of their extensive circulatory supply, liver cells are in an excellent position to monitor and adjust the nutrient composition of the circulating blood. The liver also contains significant energy reserves in the form of glycogen deposits.
- ***Adipose tissue.*** Adipose tissue stores lipids, primarily as triglycerides. Adipocytes are found in many areas of the body; previous chapters have noted the presence of fat cells in loose connective tissue, in mesenteries, within red and yellow marrow, beneath the epicardium, and behind the eyes.
- ***Skeletal muscle.*** Skeletal muscle accounts for almost half of an individual's body weight, and these cells maintain substantial glycogen reserves. In addition, their contractile proteins can be broken down and the amino acids used as an energy source if other nutrients are unavailable.
- ***Neural tissue.*** Neural tissue has a high demand for energy, but the cells do not maintain reserves of carbohydrates, lipids, or proteins. Neurons must be provided with a reliable supply of glucose, because they are usually unable to metabolize other molecules. If blood glucose becomes too low, neural tissue in the CNS cannot continue to function, and the individual becomes unconscious.

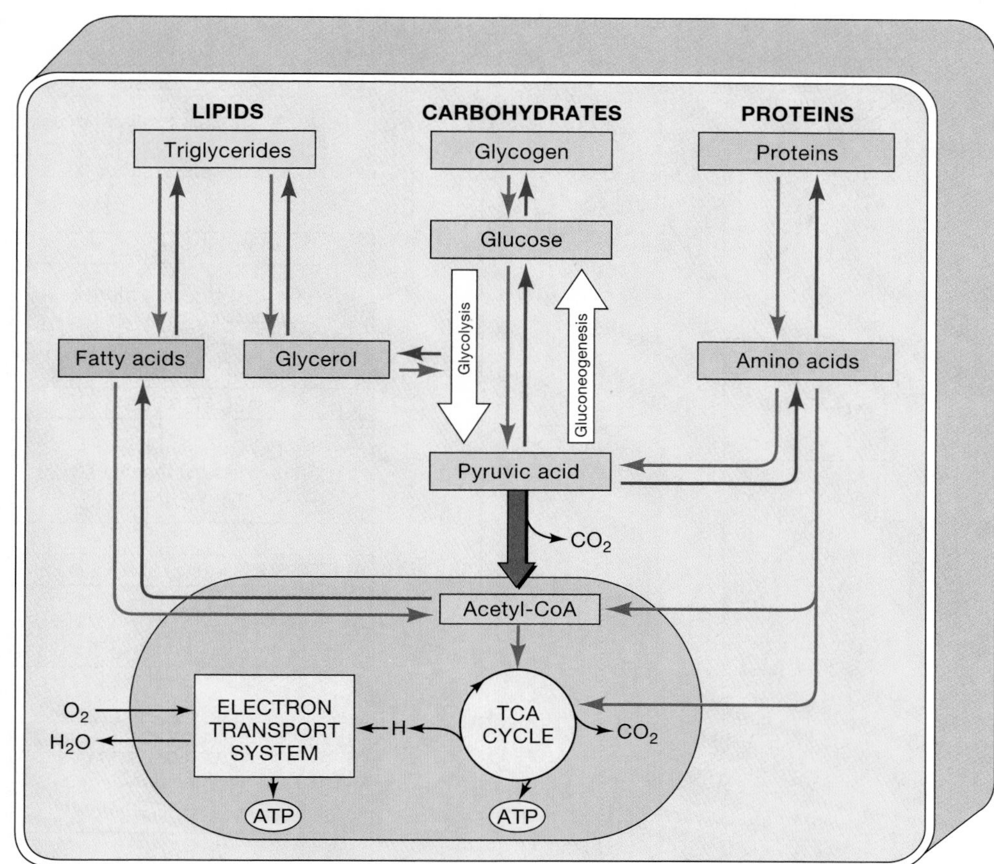
LIPIDS
CARBOHYDRATES
PROTEINS
Triglycerides
Glycogen
Proteins
Glucose
Glycolysis
Gluconeogenesis
Fatty acids
Glycerol
Amino acids
Pyruvic acid
CO_2
Acetyl-CoA
O_2
H_2O
ELECTRON TRANSPORT SYSTEM
H
TCA CYCLE
CO_2
ATP
ATP

● **FIGURE 25-14**
A Summary of the Pathways of Catabolism and Anabolism. This diagram provides an overview of major catabolic (red) and anabolic (blue) pathways.

- ***Other peripheral tissues.*** Other peripheral tissues do not maintain large metabolic reserves, but they are able to metabolize glucose, fatty acids, or other substrates. Their preferred source of energy varies, depending on the instructions provided by the endocrine system.

The interrelationships among these five components can best be understood by considering the events that occur over a 24-hour period.

❑ The Absorptive State

Figure 25-15

The **absorptive state** is the period following a meal, when nutrient absorption is under way. After a typical meal the absorptive state continues for around 4 hours. If you are fortunate enough to eat three meals a day you spend 12 hours out of every 24 in the absorptive state.

A typical meal contains proteins, lipids, and carbohydrates in varying proportions, and while in the absorptive state the intestinal mucosa busily absorbs these nutrients. Glucose and amino acids enter the circulation, and the hepatic portal vein carries them to the liver. Most of the absorbed fatty acids enter the lacteals packaged in chylomicrons.

Some of the carbohydrates, lipids, and amino acids will be broken down at once to provide the energy needed to support cellular operations. The remainder will be stored, lessening the impact of future shortages. Insulin is the primary hormone of the absorptive state, although various other hormones stimulate amino acid uptake (growth hormone) and protein synthesis (growth hormone, androgens, and estrogens). We will now consider the activities under way in specific sites, with particular reference to Figure 25-15● and Table 25-1.

The Liver

The liver regulates the levels of glucose and amino acids in the hepatic portal vein before that blood reaches the inferior vena cava.

GLUCOSE REGULATION Despite the continual absorption of glucose at the intestinal mucosa, blood glucose levels do not skyrocket because the liver cells, under insulin stimulation, remove glucose from the hepatic portal circulation. Blood glucose levels do rise, but only from about 90 mg/dl to perhaps 150 mg/dl, even after a meal rich in carbohydrates. The liver uses some of the absorbed glucose to generate the ATP required to perform synthetic operations, such as glycogen formation,

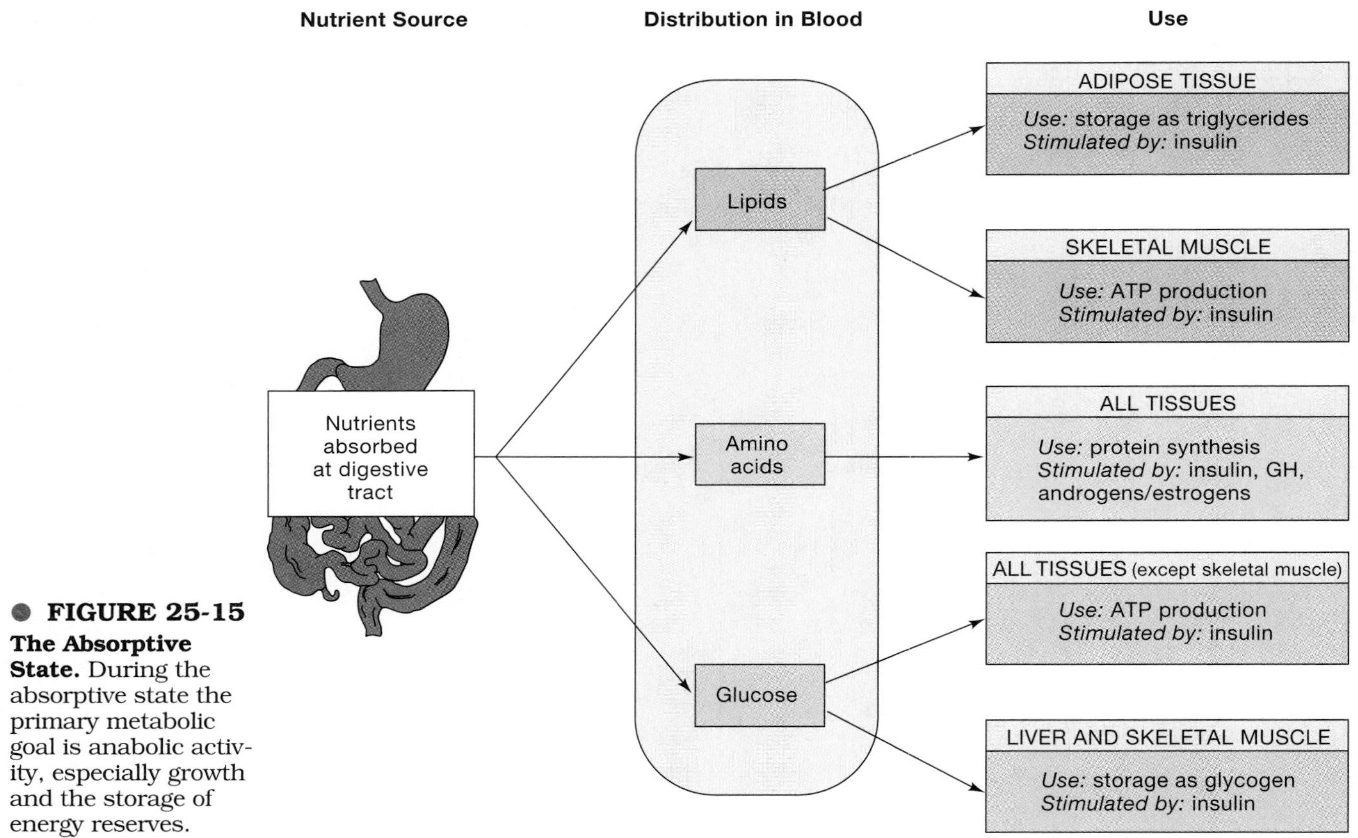
Nutrient Source
Distribution in Blood
Use
Nutrients absorbed at digestive tract
Lipids
Amino acids
Glucose
ADIPOSE TISSUE
Use: storage as triglycerides
Stimulated by: insulin
SKELETAL MUSCLE
Use: ATP production
Stimulated by: insulin
ALL TISSUES
Use: protein synthesis
Stimulated by: insulin, GH, androgens/estrogens
ALL TISSUES (except skeletal muscle)
Use: ATP production
Stimulated by: insulin
LIVER AND SKELETAL MUSCLE
Use: storage as glycogen
Stimulated by: insulin

FIGURE 25-15 The Absorptive State. During the absorptive state the primary metabolic goal is anabolic activity, especially growth and the storage of energy reserves.

plasma protein synthesis, or the manufacture of proenzymes. Glycogenesis (glycogen formation) continues until glycogen accounts for about 5 percent of the total liver weight. If excess glucose still remains in the circulation, the hepatocytes use glucose to synthesize triglycerides. Although small quantities of lipids are normally stored in the liver, most of the synthesized triglycerides are bound to transport proteins and released into the bloodstream as VLDLs. Peripheral tissues, primarily adipose tissues, then absorb these lipids for storage.

AMINO ACID REGULATION The liver does not control circulating levels of amino acids as precisely as it does glucose concentrations. Plasma amino acid levels normally range between 35 and 65 mg/dl, but they may become elevated after a protein-rich meal. The absorbed amino acids are used to support the synthesis of proteins, including plasma proteins and the proenzymes of the clotting system. Liver cells are also capable of synthesizing many amino acids, and an amino acid present in abundance may be converted to another, less common type and released into the circulation.

THE LIVER AND CIRCULATING LIPIDS Most of the lipids absorbed by the digestive tract do not reach the liver. Triglycerides, cholesterol, and large fatty acids reach the general venous circulation in chylomicrons that are transported in the thoracic duct. Most of these lipids will be absorbed by other tissues, and relatively few will reach the liver.

Adipose Tissue

During the absorptive state adipocytes remove fatty acids and glycerol from the circulation. Removal of lipids from the blood continues for 4–6 hours after a fatty meal. Over this period the presence of chylomicrons may give the plasma a milky appearance, a characteristic called **lipemia** (lip-Ē-mē-a).

Adipocytes are particularly active in absorbing these lipids and synthesizing new triglycerides for storage. At normal blood glucose concentrations any glucose entering these cells will be catabolized to provide the energy needed for lipogenesis (lipid synthesis). Adipocytes also absorb amino acids as required for protein synthesis. Although these cells can use glucose or amino acids to manufacture triglycerides, they do so only if circulating concentrations are unusually high.

If on a daily basis an individual takes in more nutrients during the absorptive state than he or she catabolizes during the postabsorptive state, the fat deposits in adipose tissue will enlarge. Most of the increase represents an increase in the size of individual adipocytes. An increase in the total number of adipocytes does not ordinarily occur, except in children before puberty and in extremely obese adults.

TABLE 25-1 Regulatory Hormones and Their Effects on Peripheral Metabolism

State/Hormone	*Effect on General Peripheral Tissues*	*Selective Effects on Target Tissues*
ABSORPTIVE		
Insulin	Increased glucose uptake and utilization	*Liver:* Glycogenesis *Adipose tissue:* Lipogenesis *Skeletal muscle:* Glycogenesis
Insulin and growth hormone	Increased amino acid uptake and protein synthesis	*Skeletal muscle:* Fatty acid catabolism
Androgens, estrogens	Increased amino acid use in protein synthesis	
POSTABSORPTIVE		
Glucagon		*Liver:* Glycogenolysis
Epinephrine		*Liver:* Glycogenolysis *Adipose tissue:* Lipolysis
Glucocorticoids	Decreased use of glucose; increased reliance on ketone bodies and fatty acids	*Liver:* Glycogenolysis *Adipose tissue:* Lipolysis, gluconeogenesis *Skeletal muscle:* Glycogenolysis, protein breakdown, amino acid release
Growth hormone	Complements effects of glucocorticoids	

OBESITY A useful definition of *obesity* is 20 percent over ideal weight, because this is the point at which serious health risks appear. Using that criterion, 20–30 percent of men and 30–40 percent of women in the United States can be considered obese. Simply stated, obese individuals are taking in more food energy than they are using. Unfortunately, there is very little agreement as to the underlying cause for this situation. The two major categories of obesity, *regulatory obesity* and *metabolic obesity*, are detailed in the *Applications Manual*. AM *Obesity; Liposuction*

Skeletal Muscle, Neural Tissue, and Other Peripheral Tissues

When blood glucose and amino acid concentrations are elevated, insulin is released from the pancreatic islets, and all of the body's tissues increase their rates of absorption and utilization. Glucose molecules are catabolized for energy, and the amino acids are used to build proteins. Glucose is normally retained in the body because the kidneys reabsorb any glucose molecules entering the urine. This ability to conserve glucose breaks down only when blood glucose concentrations are extraordinarily high, somewhere in excess of 180 mg/dl. Amino acids are not reclaimed as readily, and amino acids often appear in the urine after a protein-rich meal.

When blood glucose levels are elevated, most cells ignore the circulating lipids, and the adipocytes have little competition. In resting skeletal muscles, a significant portion of the metabolic demand is met through the catabolism of fatty acids, and glucose molecules are used to build glycogen reserves that may account for 0.5–1 percent of the weight of each muscle fiber.

❑ The Postabsorptive State

Figure 25-16

The **postabsorptive state** is the period when nutrient absorption is not under way and the body must rely on internal energy reserves to continue meeting its energy demands. Roughly 12 hours each day are spent in the postabsorptive state, although a person who is skipping meals can extend it considerably. Metabolic activity in the postabsorptive state is focused on the mobilization of energy reserves and the maintenance of normal blood glucose levels. These activities are coordinated by several hormones, including glucagon, epinephrine, the glucocorticoids, and growth hormone (Table 25-1).

The metabolic reserves of a typical 70-kg individual include carbohydrates, lipids, and proteins (Figure 25-16a●). Because of its high energy content, the adipose tissue represents a disproportionate percentage of the total reserve in the form of triglycerides. Most of the available protein reserve is found in the contractile proteins of skeletal muscle. Carbohydrate reserves are relatively small and sufficient for only a few hours or, at most, overnight.

We will now examine the events under way in specific tissues during this period, as diagrammed in Figure 25-16b●.

Carbohydrates
0.59%
Liver glycogen
Muscle glycogen
Glucose (body fluids)
Proteins
14.46%
Triglycerides
84.95%
(a)

The Liver

Figure 25-17

As the absorptive state ends, the intestinal cells stop providing glucose to the portal circulation. At first the peripheral tissues continue to remove glucose from the blood, and blood glucose levels begin to decline. The liver responds by reducing its synthetic activities, and when plasma concentrations fall below 80 mg/dl, liver cells begin breaking down glycogen reserves and releasing glucose into the circulation. This *glycogenolysis* occurs in response to a rise in circulating levels of glucagon and epinephrine. The liver contains 75–100 g of glycogen that is readily available, and this reserve would be adequate to maintain blood glucose levels for about 4 hours.

As glycogen reserves decline and plasma glucose levels fall to around 70 mg/dl, liver cells begin to manufacture glucose in an attempt to stabilize blood glucose concentrations. The shift from glycogenolysis to gluconeogenesis occurs under stimulation by *glucocorticoids*, steroid hormones from the adrenal cortex.

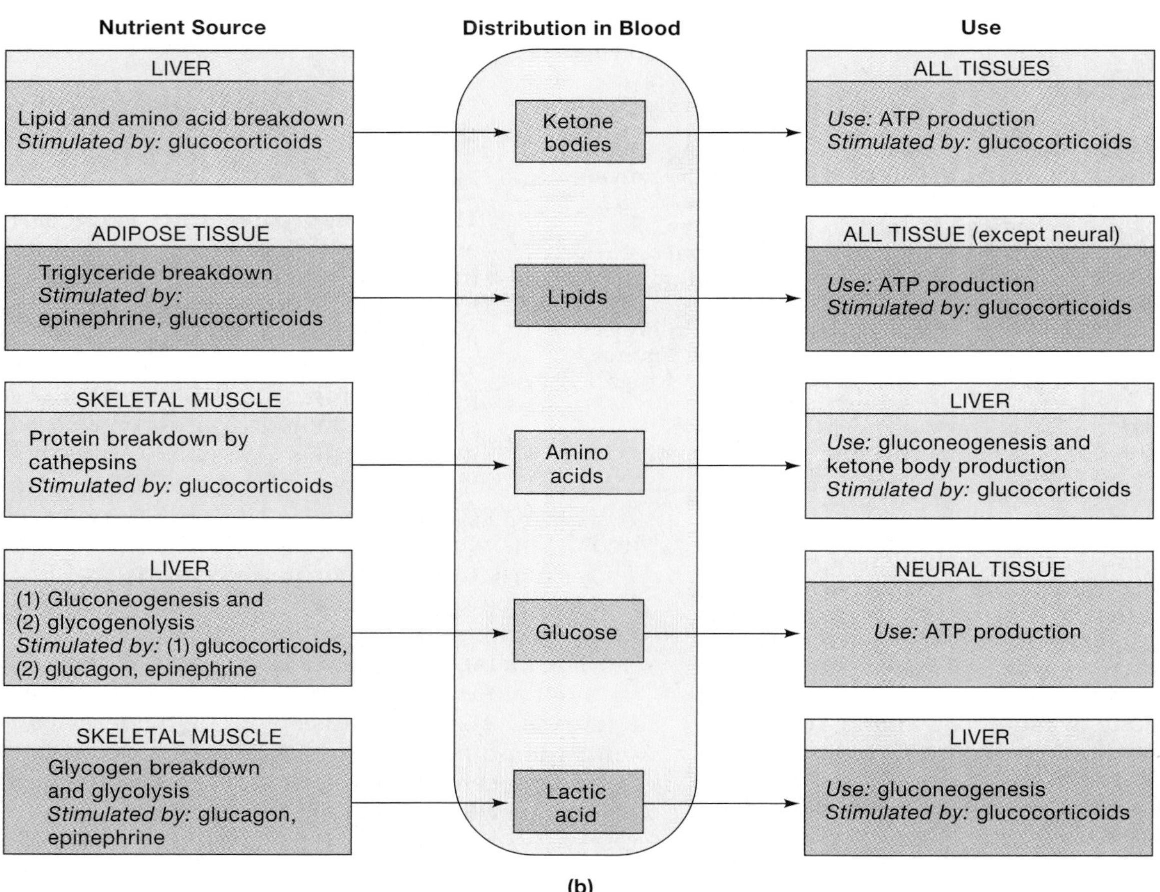
Nutrient Source
Distribution in Blood
Use
LIVER
Lipid and amino acid breakdown
Stimulated by: glucocorticoids
Ketone bodies
ALL TISSUES
Use: ATP production
Stimulated by: glucocorticoids
ADIPOSE TISSUE
Triglyceride breakdown
Stimulated by: epinephrine, glucocorticoids
Lipids
ALL TISSUE (except neural)
Use: ATP production
Stimulated by: glucocorticoids
SKELETAL MUSCLE
Protein breakdown by cathepsins
Stimulated by: glucocorticoids
Amino acids
LIVER
Use: gluconeogenesis and ketone body production
Stimulated by: glucocorticoids
LIVER
(1) Gluconeogenesis and (2) glycogenolysis
Stimulated by: (1) glucocorticoids, (2) glucagon, epinephrine
Glucose
NEURAL TISSUE
Use: ATP production
SKELETAL MUSCLE
Glycogen breakdown and glycolysis
Stimulated by: glucagon, epinephrine
Lactic acid
LIVER
Use: gluconeogenesis
Stimulated by: glucocorticoids
(b)

FIGURE 25-16

The Postabsorptive State. (a) The distribution of the estimated metabolic reserves of a 70-kg individual. **(b)** In the postabsorptive state, energy reserves are mobilized, and peripheral tissues (except neural) shift from glucose catabolism to fatty acid or ketone body catabolism to obtain energy.

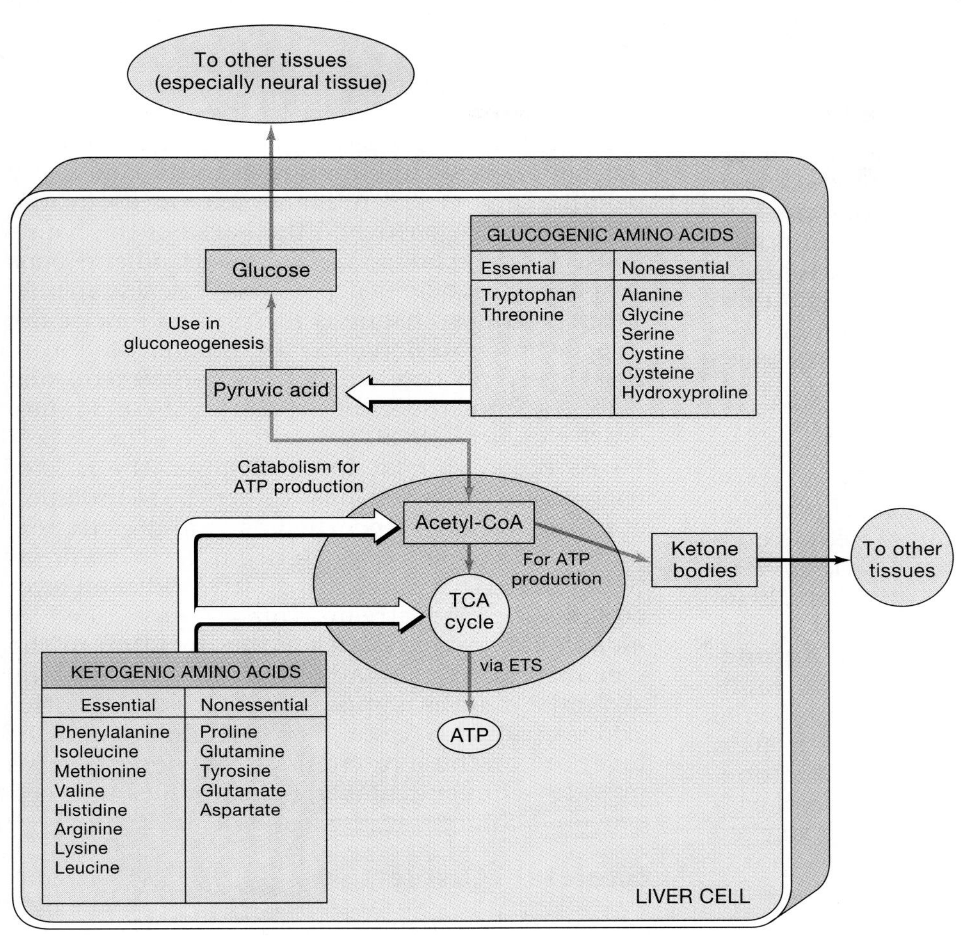
To other tissues (especially neural tissue)
Glucose
Use in gluconeogenesis
Pyruvic acid
GLUCOGENIC AMINO ACIDS
Essential
Tryptophan
Threonine
Nonessential
Alanine
Glycine
Serine
Cystine
Cysteine
Hydroxyproline
Catabolism for ATP production
Acetyl-CoA
For ATP production
TCA cycle
via ETS
ATP
Ketone bodies
To other tissues
KETOGENIC AMINO ACIDS
Essential
Phenylalanine
Isoleucine
Methionine
Valine
Histidine
Arginine
Lysine
Leucine
Nonessential
Proline
Glutamine
Tyrosine
Glutamate
Aspartate
LIVER CELL

● **FIGURE 25-17**
Routes of Amino Acid Catabolism

GLUCONEOGENESIS Through gluconeogenesis, liver cells synthesize glucose molecules from smaller carbon fragments. In effect, any carbon fragment that can be converted to pyruvic acid or one of the three-carbon compounds involved in the cytoplasmic reactions of glycolysis can be used to synthesize glucose. (The conversion of lactic acid to glucose in the liver was discussed in Chapter 10 and diagrammed in Figure 10-16●, p. 309.) With glucose already in short supply, lipids and amino acids must be catabolized to provide the ATP molecules needed for these synthetic operations.

UTILIZATION OF LIPIDS During the postabsorptive state the liver absorbs fatty acids and glycerol from the blood. The three-carbon glycerol molecules are converted to glucose. Fatty acids are broken down through beta-oxidation to produce large quantities of acetyl-CoA. However, because the enzymatic reaction that converts pyruvic acid to acetyl-CoA cannot be reversed, acetyl-CoA cannot be used to synthesize glucose. Instead,

- Some of the acetyl-CoA molecules deliver their two-carbon acetyl fragments to the TCA cycle, where they are broken down as described previously. The ATP generated can then be used to support gluconeogenesis, the synthesis of glucose from other carbohydrates or from amino acids.
- In addition, some of the molecules of acetyl-CoA are converted to special compounds, known as *ketone bodies*, that can be utilized by peripheral tissues. Ketone bodies are organic acids that are also produced during the catabolism of amino acids.

UTILIZATION OF AMINO ACIDS Before an amino acid can be used for either gluconeogenesis or energy production via the TCA cycle, the amino group (—NH_2) must be removed. The structure of the remaining carbon chain determines its subsequent fate. After deamination, some amino acids can be converted to molecules of pyruvic acid and then used for gluconeogenesis. These are known as the **glucogenic amino acids.** Other amino acids can be converted only to acetyl-CoA and must be either broken down further or converted to ketone bodies. These are called the **ketogenic amino acids.** *Most of the essential amino acids are ketogenic.* The metabolic fates of glucogenic and ketogenic amino acids are diagrammed in Figure 25-17●.

In the liver, glucogenic and ketogenic amino acids are broken down, and the ammonia generated by the deamination reactions is converted to urea. This relatively harmless, water-soluble compound is later excreted in the urine. The urea concentration in the blood rises during the postabsorptive period, because of an increase in the rate of amino acid catabolism.

KETONE BODIES During the postabsorptive state, liver cells conserve glucose and break down lipids and amino acids. Both lipid catabolism and amino acid catabolism generate acetyl-CoA. As the concentration of acetyl-CoA rises, ketone bodies begin to form. There are three different ketone bodies: **acetoacetate** (as-ē-tō-AS-e-tāt), **acetone** (AS-e-tōn), and **betahydroxybutyrate** (bā-ta-hī-droks-ē-BŪ-te-rāt). Liver cells do not metabolize any of the ketone bodies, and these compounds diffuse through the cytoplasm and into the general circulation. Cells in peripheral tissues then absorb the ketone bodies and reconvert them to acetyl-CoA for introduction into the TCA cycle.

The normal concentration of ketone bodies in the blood is around 30 mg/dl, and very few appear in the urine. During even a brief period of fasting, the increased production of ketone bodies results in **ketosis** (kē-TŌ-sis), a high concentration of ketone bodies in body fluids. In ketosis there is an elevated concentration of ketone bodies in the blood, a condition called **ketonemia** (kē-tō-NĒ-mē-a), and in the urine, a condition called **ketonuria** (kē-tō-NU-rē-a). Ketonemia and ketonuria are clear indications that protein and lipid catabolism are under way. Acetone, which diffuses out of the pulmonary capillaries and into the alveoli very readily, may be smelled on the breath.

KETOACIDOSIS A ketone body is also called a "keto acid" because it dissociates in solution, releasing a hydrogen ion. As a result, the appearance of ketone bodies in the circulation presents a threat to the plasma pH that must be controlled by buffers. During prolonged starvation, ketone levels continue to rise. Eventually the buffering capacities are exceeded, and a dangerous drop in pH occurs. This acidification of the blood is called **ketoacidosis** (kē-tō-as-i-DŌ-sis). In severe ketoacidosis the circulating concentration of ketone bodies can reach 200 mg/dl, and the pH may fall below 7.05. A pH this low can disrupt normal tissue activities and cause coma, cardiac arrhythmias, and death.

In *diabetes mellitus* most peripheral tissues cannot utilize glucose, because of a lack of insulin. *[p. 634]* Under these circumstances cells survive by catabolizing lipids and proteins. The result is the production of large numbers of ketone bodies, and this condition leads to *diabetic ketoacidosis,* the most common and life-threatening form of ketoacidosis.

SUMMARY In summary, during the postabsorptive state the liver attempts to stabilize blood glucose concentrations, first by the breakdown of glycogen reserves and later by gluconeogenesis. Over the remainder of the postabsorptive state, the combination of lipid and amino acid catabolism provides the necessary ATP and generates large quantities of ketone bodies that diffuse into the circulation.

Adipose Tissue

Adipose tissue contains a tremendous storehouse of energy in the form of triglycerides. The average individual owes approximately 15 percent of his or her body weight to fat, enough to provide a 1–2 month reserve of ATP. Although some areas, including the eyelids, nose, and the backs of the hands and feet, rarely contain adipose tissue, other regions are preferential sites of deposition. Typically, an individual's adipose tissue is distributed among the hypodermis (50 percent), the greater omentum (10–15 percent), between muscles (5–8 percent), and packed around the kidneys (12 percent) and reproductive organs (15–20 percent).

As blood glucose levels decline, the rate of triglyceride synthesis falls. Under the stimulation of epinephrine, glucocorticoids, and growth hormone, the adipocytes soon begin breaking down their lipid reserves, releasing fatty acids and glycerol into the bloodstream. This process, called *fat mobilization,* continues for the duration of the postabsorptive state. A normal individual retains around 2 months' supply of energy in the triglycerides of adipose tissue. The evolutionary advantages are obvious; retention of an energy reserve provides a buffer against daily, monthly, and even seasonal changes in the available food supply.

Skeletal Muscle

At the start of the postabsorptive state, skeletal muscles obtain energy by breaking down their glycogen reserves and catabolizing the glucose released. As the concentrations of fatty acids and ketone bodies in the circulation increase, these substrates become increasingly important as an energy source.

Skeletal muscle as a whole contains twice as much glycogen as the liver, but it is distributed throughout the muscular system. These glucose reserves are not directly available to other tissues because the lack of a key enzyme makes skeletal muscle cells unable to release glucose into the circulation. Once skeletal muscle fibers are metabolizing fatty acids as an energy source, they continue to break down their glycogen reserves and convert the pyruvic acid molecules to lactic acid. Lactic acid then diffuses out of the muscle fibers and into the circulation. However, even if all of the available glycogen reserves in the body were mobilized as glucose or as lactic acid, the energy provided would be only enough to get a person through a good night's sleep. If the postabsorptive state continues for an unusually long period, long enough that lipid reserves are being depleted, muscle proteins will be broken down by special enzymes called **cathepsins** (ka-THEP-sinz). The amino acids released by cathepsins diffuse into the blood for use by the liver in gluconeogenesis.

Other Peripheral Tissues

With rising plasma concentrations of lipids and ketone bodies and falling blood glucose levels, peripheral tissues gradually decrease their reliance upon glucose. Circulating ketone bodies and fatty acids are absorbed and converted to acetyl-CoA for entry into the TCA cycle.

Neural Tissue

Figures 25-15, 25-16

Neurons are unusual in that they continue "business as usual" during the postabsorptive state. Neurons are dependent on a reliable supply of glucose, and the alterations in the activity of liver, adipose tissue, skeletal muscle, and other peripheral tissues are intended to ensure that the supply of glucose to the nervous system continues unaffected, despite daily or even weekly changes in the availability of nutrients. Only after a prolonged period of starvation will neural tissue begin to metabolize ketone bodies and lactic acid molecules as well as glucose.

You should now take the time to compare Figures 25-15, p. 956, and 25-16, p. 958,• until the functional relationships in each state are clearly understood. The adaptations to starvation are exaggerations of the basic themes seen each day in the postabsorptive state. For a discussion of the physiological adaptations to prolonged starvation, consult the *Applications Manual.* AM *Adaptations to Starvation*

What process in the liver would you expect to increase after a meal that is high in carbohydrates?

Why is there an increase in the amount of urea in the blood during the postabsorptive state?

If a cell accumulates more acetyl-CoA than it can metabolize by way of the TCA cycle, what products are likely to be formed?

Diet and Nutrition

The postabsorptive state can be maintained for a considerable period, but homeostasis can be maintained indefinitely only if the digestive tract absorbs fluids, organic substrates, minerals, and vitamins on a regular basis, keeping pace with cellular demands. The absorption of nutrients from food is called **nutrition.**

The individual requirement for each nutrient varies from day to day and from person to person. *Nutritionists* attempt to analyze a diet in terms of its ability to meet the needs of a specific individual. A **balanced diet** contains all of the ingredients necessary to maintain homeostasis, including adequate substrates for energy generation, essential amino acids and fatty acids, minerals, and vitamins. In addition, the diet must include enough water to replace losses in urine, feces, and evaporation. A balanced diet prevents **malnutrition,** an unhealthy state resulting from inadequate or excessive intake of one or more nutrients. AM *Nutrition and Nutritionists*

❑ Food Groups and Food Pyramids

Figure 25-18

For several decades the traditional American method of avoiding malnutrition was to include members of each of the four **basic food groups** in the diet. These were the *milk and dairy group,* the *meat group,* the *vegetable and fruit group,* and the *bread and cereal group.* Each group differs from the others in the typical balance of proteins, carbohydrates, and lipids contained as well as in the amount and identity of vitamins and minerals.

In an attempt to provide more guidance as to how much one should rely on each food source, the four groups have recently been increased to six by separating the fruit group and establishing a *fats, oils, and sweets group.* The six groups are now arranged in a *food pyramid* with the bread and cereal group at the bottom (Figure 25-18•). The aim of this display is to emphasize the need to restrict dietary fats, oils, and sugar and to increase consumption of breads and cereals, which are rich in complex carbohydrates (polysaccharides such as starch).

It should be realized that these are rather artificial groupings at best, and downright misleading at worst. What is important is to obtain nutrients in sufficient *quantity* (adequate to meet energy needs) and *quality* (including essential amino acids, fatty acids, vitamins, and minerals). How these are packaged is a secondary concern. There is nothing magical about the number six—since 1940 there have been 11, 7, 4, and 6 food groups advocated by the U.S. government at various times. The key is making intelligent choices about what you eat. The wrong choices can lead to malnutrition even if all six groups are represented.

For example, consider the case of the essential amino acids. The liver cannot synthesize any of these ketogenic amino acids, and they must be obtained from the diet. Some members of the meat and milk groups, such as beef, fish, poultry, eggs, and milk, contain all of the essential amino acids in sufficient quantities. They are said to contain **complete proteins.** Many plants contain adequate *amounts* of protein, but they contain **incomplete proteins** that are deficient in one or more of the essential amino acids. True vegetarians, who restrict themselves to the fruit and vegetable groups (with or without the bread and cereal group), must become adept at juggling the constituents of their meals to include a combination of ingredients that will meet all of their amino acid

● **FIGURE 25-18**
The Food Pyramid

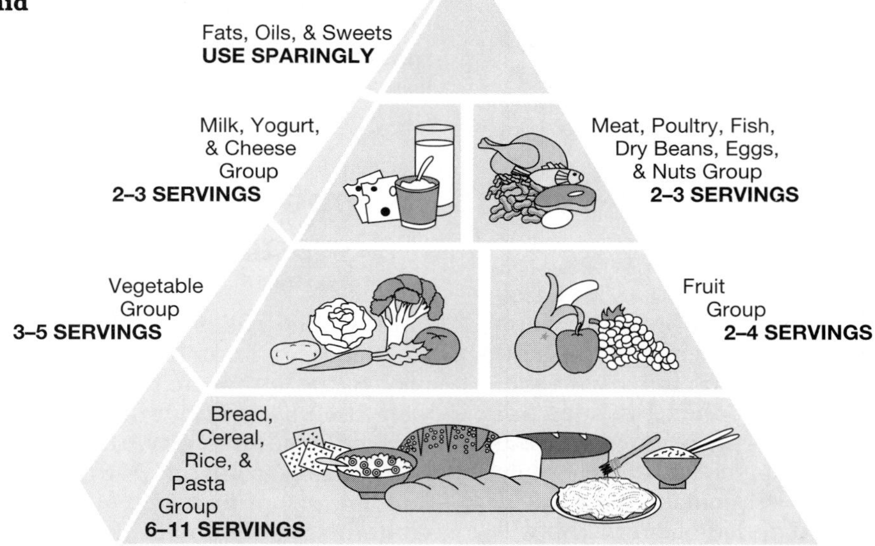
Fats, Oils, & Sweets
USE SPARINGLY
Milk, Yogurt, & Cheese Group
2–3 SERVINGS
Meat, Poultry, Fish, Dry Beans, Eggs, & Nuts Group
2–3 SERVINGS
Vegetable Group
3–5 SERVINGS
Fruit Group
2–4 SERVINGS
Bread, Cereal, Rice, & Pasta Group
6–11 SERVINGS

A Guide to Daily Food Choices

Nutrient Group	*Provides*	*Deficiencies*
Fats, oils, sweets	Calories	Usually deficient in most minerals and vitamins
Milk, yogurt, cheese	Complete proteins, fats, carbohydrates, calcium, potassium, magnesium, sodium, phosphorus, vitamins A, B_{12}, pantothenic acid, thiamine, riboflavin	Dietary fiber, vitamin C
Meat, poultry, fish, dry beans, eggs, nuts	Complete proteins, fats, potassium, phosphorus, iron, zinc, vitamins E, thiamine, B_6	Carbohydrates, dietary fiber, several vitamins
Vegetables and fruits	Carbohydrates, vitamins A, C, E, folacin, dietary fiber, potassium	Often low in fats, calories, and protein
Bread, cereal, rice, pasta	Carbohydrates, vitamins E, thiamine, niacin, folacin, calcium, phosphorus, iron, sodium, dietary fiber	Fats

requirements. Even with a proper balance of amino acids, the vegetarian faces a significant problem, because vitamin B_{12} is obtained only from animal products. (Although some health-food products, such as *Spirulina*, are marketed as sources of this vitamin, the B_{12} present is in a form that cannot be utilized by human beings.)

❑ Nitrogen Balance

A variety of important compounds in the body contain nitrogen atoms. These **N compounds** include:

1. *Amino acids* that are part of the framework of all proteins and protein derivatives, such as glycoproteins and lipoproteins.
2. *Purines* and *pyrimidines*, the nitrogenous bases of RNA and DNA.
3. *Creatine*, important in energy storage in muscle tissue (as creatine phosphate).
4. *Porphyrins*, complex ring-shaped molecules that bind metal ions and are essential to the function of hemoglobin, myoglobin, and the cytochromes.

N compounds are essential components of living systems, for they play key roles in determining the direction and rates of intracellular processes, and they form structural proteins. When a person consumes an adequate diet of fats, carbohydrates, and proteins, the fats and carbohydrates are catabolized, and the amino acids are incorporated into proteins or converted to other N compounds. As a result, dietary fats and carbohydrates are often called **protein sparers.**

Despite the importance of nitrogen to these compounds, the body neither stores nitrogen nor maintains reserves of N compounds as it does carbohydrates (glycogen) and lipids (triglycerides). The carbon chains of the N compounds can be synthesized in the body, but the nitrogen atoms must be obtained either by recycling N compounds already in the body or by absorbing nitrogen from the diet.

The individual is in **nitrogen balance** if the amount of nitrogen absorbed from the diet balances the amount lost in the urine and feces. This is the normal condition, and it means that the rates of N compound synthesis and breakdown are equivalent.

Growing children, athletes, persons recovering from an illness or injury, and pregnant or lactating women are actively synthesizing N compounds, so they must absorb more nitrogen than they excrete. Such individuals are in a state of **positive nitrogen balance.** When excretion exceeds ingestion, **negative nitrogen balance** exists. This is an extremely unsatisfactory situation; the body contains only around a kilogram of nitrogen tied up in N compounds, and a decrease of one-third can be fatal. Even when energy reserves are mobilized, as during starvation, carbohydrates and lipid reserves are broken down first, and N compounds are conserved.

Such conservation is relative, not absolute. Proteins are the most abundant organic constituents of living cells, accounting for roughly 20 percent of total body weight. If periods of energy shortage are prolonged, protein catabolism becomes increasingly important as other reserves are exhausted. For example, during the first week of starvation a person may catabolize 1–1.5 kg of proteins, but in the eighth week the same person may be catabolizing that amount each *day*.

❑ Minerals and Vitamins

Minerals and vitamins are essential components of the diet. The body cannot synthesize minerals, and our cells can generate only a small quantity of very few vitamins.

Minerals

Minerals are inorganic ions released through the dissociation of electrolytes. Minerals are important because:

1. Ions such as sodium and chloride determine the osmolarities of body fluids. Potassium is important in maintaining the osmolarity of the cytoplasm inside body cells.
2. Ions in various combinations play major roles in important physiological processes discussed, including:
   - The maintenance of transmembrane potentials (Chapters 3, 10, and 12).
   - Action potential generation (Chapter 12).
   - Neurotransmitter release (Chapters 10 and 12).
   - Muscle contraction (Chapters 10 and 20).
   - The construction and maintenance of the skeleton (Chapter 6).
   - The transport of respiratory gases (Chapter 23).
   - Buffer systems (Chapters 2 and 27).
   - Fluid absorption (Chapter 24).
   - Waste removal (Chapters 26 and 27).
3. Ions are essential *cofactors* in a variety of enzymatic reactions. For example, the calcium-dependent ATPase in skeletal muscle also requires the presence of magnesium ions, and another ATPase required for the conversion of glucose to pyruvic acid needs both potassium and magnesium ions. Carbonic anhydrase, important in CO_2 transport, buffering systems, and gastric acid secretion, requires the presence of zinc ions. Finally, the components of the electron transport system each require an iron atom, and the terminal cytochrome must bind a copper ion as well.

The major minerals and a summary of their functional roles are presented in Table 25-2. The body contains significant reserves of several important minerals; these reserves help to reduce the effects of variations in dietary supply. The reserves are often relatively small, however, and chronic dietary reductions can lead to a variety of clinical problems. Alternatively, because storage capabilities are limited, a dietary excess of mineral ions can prove equally dangerous.

Problems involving iron are particularly common. The body of a normal man contains around 3.5 g of iron in the ionic form Fe^{2+}. Of that amount, 2.5 g are bound to the hemoglobin of circulating red blood cells, and the rest is stored in the liver and bone marrow. In women, the total body iron content averages 2.4 g, with roughly 1.9 g incorporated into red blood cells. Thus a woman's iron reserves consist of only 0.5 g, half that of a typical man. If the diet contains inadequate amounts of iron, women are therefore more likely to develop signs of iron deficiency than men.

Vitamins

Vitamins can be assigned to either of two groups, depending on their chemical structure and characteristics: *fat-soluble vitamins* and *water-soluble vitamins.*

FAT-SOLUBLE VITAMINS Vitamins A, D, E, and K are the **fat-soluble vitamins.** These vitamins are absorbed primarily from the digestive tract along with the lipid contents of micelles, but the skin can synthesize small amounts of vitamin D when exposed to sunlight. There is considerable uncertainty over the mode of action of these vitamins. Vitamin D is ultimately converted to calcitriol, which binds to cytoplasmic receptors within the intestinal epithelium and promotes an increase in the rate of intestinal calcium and phosphorus absorption. Vitamin A has long been recognized as a structural component of the visual pigment retinal, but its more general metabolic effects are not well understood. Vitamin K is a necessary participant in a reac-

TABLE 25-2 Minerals and Mineral Reserves*

Mineral	*Significance*	*Total Body Content*	*Primary Route of Excretion*	*Recommended Daily Intake*
BULK MINERALS				
Sodium	Major cation in body fluids; essential for normal membrane function	110 g, primarily in body fluids	Urine, sweat, feces	0.5–1.0 g
Potassium	Major cation in cytoplasm; essential for normal membrane function	140 g, primarily in cytoplasm	Urine	1.9–5.6 g
Chloride	Major anion in body fluids	89 g, primarily in body fluids	Urine, sweat	0.7–1.4 g
Calcium	Essential for normal muscle and neuron function, bone structure	1.36 kg, primarily in skeleton	Urine, feces	0.8–1.2 g
Phosphorus	As phosphate in high-energy compounds, nucleic acids, and bone matrix	744 g, primarily in skeleton	Urine, feces	0.8–1.2 g
Magnesium	Cofactor of enzymes, required for normal membrane functions	29 g (skeleton, 17 g; cytoplasm and body fluids, 12 g)	Urine	0.3–0.4 g
TRACE MINERALS				
Iron	Component of hemoglobin, myoglobin, cytochromes	3.9 g, 1.6 stored (ferritin or hemosiderin)	Urine (traces)	10–18 mg
Zinc	Cofactor of enzyme systems, notably carbonic anhydrase	2 g	Urine, hair (traces)	15 mg
Copper	Required as cofactor for hemoglobin synthesis	127 mg	Urine, feces (traces)	2–3 mg
Manganese	Cofactor for some enzymes	11 mg	Feces, urine (traces)	2.5–5 mg

*For information on the effects of deficiencies and excesses, see Table 27-2, p. 1039.

tion essential to the synthesis of several proteins, including at least three of the clotting factors. Vitamin E probably stabilizes intracellular membranes. Current information concerning the fat-soluble vitamins is summarized in Table 25-3.

Because they dissolve in lipids, fat-soluble vitamins normally diffuse into cell membranes and other lipids in the body, including the lipid inclusions in the liver and adipose tissue. The body therefore contains a significant reserve of these vitamins, and normal metabolic operations can continue for several months after dietary sources have been cut off. As Table 25-3 points out, *too much* of a vitamin may produce effects just as unpleasant as *too little.* **Hypervitaminosis** (hī-per-vī-ta-min-Ō-sis) occurs when the dietary intake exceeds the abilities to store, utilize, or excrete a particular vitamin. This condition most often involves one of the fat-soluble vitamins because the excess is retained and stored in body lipids.

HYPERVITAMINOSIS "If a little is good, a lot must be better" is a common but dangerously incorrect attitude about vitamins. When the dietary supply of fat-soluble vitamins is excessive, the tissue lipids absorb the additional vitamins. Because these vitamins will later diffuse back into circulation, once the symptoms of hypervitaminosis appear they are likely to persist.

When absorbed in massive amounts (from ten to thousands of times the recommended daily allowance), fat-soluble vitamins can produce acute symptoms of *vitamin toxicity*. Vitamin A toxicity is the most common condition; it sometimes afflicts children whose parents are overanxious about proper nutrition and vitamins. A single enormous overdose can produce nausea, vomiting, headache, dizziness, lethargy, and even death. Chronic overdose can lead to hair loss, joint pain, hypertension, weight loss, and liver enlargement.

At least 19 cases of vitamin D toxicity were reported in the Boston area during 1992. Symptoms included fatigue, weight loss, and potentially severe damage to the kidneys and cardiovascular system. The problems resulted from drinking milk fortified with vitamin D. Because of problems at the dairy, instead of containing 400 units of vitamin D per quart, some of the milk sold had over 230,000 units per quart. The incident highlighted the need for quality control and care when taking vitamin supplements.

WATER-SOLUBLE VITAMINS Most of the **water-soluble vitamins** (Table 25-4) are components of coenzymes. For example, NAD is derived from niacin, FAD from riboflavin, and coenzyme A from vitamin B_5 (pantothenic acid).

Water-soluble vitamins are rapidly exchanged between the fluid compartments and the circulating blood, and excessive amounts are readily excreted in the urine. For this reason, hypervitaminosis involving water-soluble vitamins is relatively uncommon, except among individuals taking large doses of vitamin supplements. Only vitamins B_{12} and C are stored in significant quantities, and insufficient intake of other water-soluble vitamins may lead to initial symptoms of vitamin deficiency within a period of days to weeks. The condition that results is termed a **deficiency disease,** or **avitaminosis** (ā-vī-ta-min-Ō-sis). Avitaminosis involving either fat-soluble or water-soluble vitamins can be caused by a variety of factors other than dietary deficiencies. An inability to absorb a vitamin from the digestive tract, inadequate storage, or excessive demand may each produce the same result.

The bacterial inhabitants of our intestines help prevent deficiency diseases by producing small amounts of five of the nine water-soluble vitamins, in addition to fat-soluble vitamin K. All of the water-soluble vitamins except B_{12} can be easily absorbed by the intestinal epithelium. The B_{12} molecule is large, and, as you will recall from Chapter 24, it must be bound to the *intrinsic factor* from the gastric mucosa before absorption can occur. ∞ *[p. 926]*

❑ Diet and Disease

Diet has a profound influence on general health. We have already considered the effects of too many or too few nutrients, hypervitaminosis or avitaminosis, and above or below normal concentra-

TABLE 25-3 The Fat-Soluble Vitamins

Vitamin	*Significance*	*Sources*	*Daily Requirement*	*Effects of Deficiency*	*Effects of Excess*
A	Maintains epithelia; required for synthesis of visual pigments	Leafy green and yellow vegetables	1 mg	Retarded growth, night blindness, deterioration of epithelial membranes	Liver damage, skin peeling, CNS effects (nausea, anorexia)
D (steroids including cholecalciferol, or D_3)	Required for normal bone growth, calcium and phosphorus absorption at gut and retention at kidneys	Synthesized in skin exposed to sunlight	None*	Rickets, skeletal deterioration	Calcium deposits in many tissues, disrupting functions
E (tocopherols)	Prevents breakdown of vitamin A and fatty acids	Meat, milk, vegetables	12 mg	Anemia, other problems suspected	None reported
K	Essential for liver synthesis of prothrombin and other clotting factors	Vegetables; production by intestinal bacteria	0.7–0.14 mg	Bleeding disorders	Liver dysfunction, jaundice

*Unless sunlight exposure is inadequate for extended periods and alternative sources (fortified milk products) are unavailable.

TABLE 25-4 The Water-Soluble Vitamins

Vitamin	*Significance*	*Sources*	*Daily Requirement*	*Effects of Deficiency*	*Effects of Excess*
B_1 (thiamine)	Coenzyme in decarboxylation reactions	Milk, meat, bread	1.9 mg	Muscle weakness, CNS and cardiovascular problems including heart disease; called *beriberi*	Hypotension
B_2 (riboflavin)	Part of FMN and FAD	Milk, meat	1.5 mg	Epithelial and mucosal deterioration	Itching, tingling sensations
Niacin (nicotinic acid)	Part of NAD	Meat, bread, potatoes	14.6 mg	CNS, GI, epithelial, and mucosal deterioration; called *pellagra*	Itching, burning sensations, vasodilation, death after large dose
B_5 (pantothenic acid)	Part of acetyl-CoA	Milk, meat	4.7 mg	Retarded growth, CNS disturbances	None reported
B_6 (pyridoxine)	Coenzyme in amino acid and lipid metabolism	Meat	1.42 mg	Retarded growth, anemia, convulsions, epithelial changes	CNS alterations, perhaps fatal
Folacin (folic acid)	Coenzyme in amino acid and nucleic acid metabolism	Vegetables, cereal, bread	0.1 mg	Retarded growth, anemia, gastrointestinal disorders	Few noted except at massive doses
B_{12} (cobalamin)	Coenzyme in nucleic acid metabolism	Milk, meat	4.5 μg	Impaired RBC production causing *pernicious anemia*	Polycythemia
Biotin	Coenzyme in decarboxylation reactions	Eggs, meat, vegetables	0.1–0.2 mg	Fatigue, muscular pain, nausea, dermatitis	None reported
C (ascorbic acid)	Coenzyme; delivers hydrogen ions, antioxidant	Citrus fruits	60 mg	Epithelial and mucosal deterioration; called *scurvy*	Kidney stones

tions of minerals. More subtle, long-term problems may be encountered when the diet includes the wrong proportions or combinations of nutrients. The average American diet contains too many calories, and too great a proportion of those calories is provided by lipids. This diet increases the incidence of obesity, heart disease, atherosclerosis, hypertension, and diabetes in the U.S. population. AM *Perspectives on Dieting*

 Would an athlete in intensive training try to maintain positive or negative nitrogen balance?

 How would a decrease in the amount of bile salts in the bile affect the amount of vitamin A in the body?

Bioenergetics

When chemical bonds are broken, energy is released. Inside cells, a significant amount of energy may be used to synthesize ATP, but much of it is lost to the environment as heat. The process of **calorimetry** (kal-o-RIM-e-trē) determines the total amount of energy released when the bonds of organic molecules are broken. The unit of measurement is the **calorie** (KAL-o-rē), defined as the amount of energy required to raise the temperature of 1 g of water one degree centigrade. One gram of water is not a very practical measure when you are interested in the metabolic operations that keep a 70-kg human alive, so the **kilocalorie** (KIL-

Alcohol: A Risky Diversion

Alcohol production and sales are big business throughout the Western world. Beer commercials on television, billboards advertising various brands of liquors, characters on screen enjoying a drink—all demonstrate the significance of alcohol in our society. Most people are unaware of the medical consequences of this cultural fondness for alcohol. Problems with alcohol are usually divided into those stemming from alcohol abuse and those involving alcoholism. The boundary between these conditions is rather hazy. Alcohol abuse is the general term for overuse and the resulting behavioral and physical effects of overindulgence. Alcoholism is chronic alcohol abuse with the physiological changes associated with addiction to other CNS-active drugs. Alcoholism has received the most attention in recent years, although alcohol abuse—especially when combined with driving an automobile—is now in the limelight.

Consider the following:

- Alcoholism affects more than 10 million people in the United States alone.
- Alcoholism is probably the most expensive health problem today, with an annual estimated direct cost of more than $110 billion. Indirect costs, in terms of damage to automobiles, property, and innocent pedestrians, are unknown.
- An estimated 25–40 percent of U.S. hospital patients are undergoing treatment related to alcohol consumption. There are approximately 200,000 deaths annually due to alcohol-related medical conditions. Some major clinical conditions are caused almost entirely by alcohol consumption. For example, alcohol is responsible for 60–90 percent of all liver disease in the United States.
- Alcohol affects all physiological systems. Major clinical symptoms of alcoholism include: (1) disorientation and confusion (nervous system); (2) ulcers, diarrhea, and cirrhosis (digestive system); (3) cardiac arrhythmias, cardiomyopathy, anemia (cardiovascular system); (4) depressed sexual drive and testosterone levels (reproductive system); and (5) itching and angiomas (integumentary system).
- The toll on newborn infants has risen steadily since the 1960s as the number of women drinkers has increased. Women consuming 1 ounce of alcohol per day during pregnancy have a higher rate of spontaneous abortion and produce children with lower birth weights than do those who consume no alcohol. Heavier drinking causes *fetal alcohol syndrome (FAS)*. This condition is marked by characteristic facial abnormalities, a small head, slow growth, and mental retardation.
- Perhaps most disturbing of all, the problem of alcohol abuse is considerably more widespread than alcoholism. Although the medical effects are less well documented, they are certainly significant.

Several factors interact to produce alcoholism. The primary risk factors are gender (males are more likely to become alcoholics than are females) and a family history of alcoholism. There does appear to be a genetic component, but the relative importance of genes versus social environment has been difficult to assess. It is likely that alcohol abuse and alcoholism can result from a variety of factors.

Treatment may consist of counseling and behavior modification. To be successful, treatment must involve a total avoidance of alcohol. Supporting groups, such as Alcoholics Anonymous, can be very helpful in providing a social framework for abstinence. Use of the drug *disulfiram* (*Antabuse*) has not proved to be as successful as originally anticipated. Antabuse sensitizes the individual to alcohol so that a drink produces intense nausea, and it was anticipated that this would be an effective deterrent. Clinical tests indicated that it could increase the time between drinks, but not prevent drinking altogether.

o-kal-o-rē) (kc) or **Calorie** (with a capital *C*), also known as "large calorie," is used instead. Each Calorie represents the amount of energy needed to raise the temperature of 1 *kilo*gram of water one degree centigrade. The numbers at the back of a dieting guide that give the caloric value of various foods, indicate Calories, not calories.

❑ Food and Energy

In living cells, organic molecules are oxidized to carbon dioxide and water. Oxidation also occurs when something burns, and this process can be experimentally controlled. A known amount of material is placed in a chamber, called a **calorimeter** (kal-o-

RIM-e-ter), that is filled with oxygen and surrounded by a known volume of water. Once the material is inside, the chamber is sealed and the substance is electrically ignited. When it has completely oxidized and only ash remains in the chamber, the number of Calories released can be determined by comparing the water temperatures before and after the test. When reporting the energy potential of food, the results are usually expressed in Calories per gram (C/g). The catabolism of lipids entails the release of a considerable amount of energy, roughly 9.46 C/g. The catabolism of carbohydrates and proteins is not as rewarding, since many of the carbon and hydrogen atoms are already bound to oxygen. Their average yields are comparable: 4.18 C/g for carbohydrate and 4.32 C/g for protein. Most foods are mixtures of fats, proteins, and carbohydrates, and the values in a "Calorie counter" vary as a result.

❑ Metabolic Rate

It is possible to examine the metabolic state of an individual and determine how many Calories are being utilized. The result can be expressed as Calories per hour, Calories per day, or Calories per unit of body weight per day, but what is actually measured is the sum of all the varied anabolic and catabolic processes occurring in the body. This value represents the **metabolic rate** of the individual at that time. The metabolic rate will change according to the activity under way—sprinting and sleeping measurements are quite different. In an attempt to reduce the variations, the testing conditions are standardized so as to determine the **basal metabolic rate (BMR).** Ideally, the BMR would represent the minimum, resting energy expenditures of an awake, alert person.

A direct method of determining the BMR simply monitors respiratory activity, for in resting subjects energy utilization is proportional to oxygen consumption. Assuming that average amounts of carbohydrates, lipids, and proteins are being catabolized, the ratio works out to 4.825 Calories per liter of oxygen consumed.

An average individual has a BMR of 70 C per hour or about 1680 C per day. Although the test conditions are standardized, there are many uncontrollable factors that can influence the BMR. These include age, sex, physical condition, body weight, and genetic differences such as variations among ethnic groups.

Because measuring the BMR is technically difficult, and circulating thyroid hormone levels have a profound effect on the BMR, clinicians usually monitor the concentration of thyroid hormones rather than the actual metabolic rate. The results are then compared with normal values to obtain an index of metabolic activity. One such test, the **T_4 assay,** measures the amount of thyroxine in the blood.

Daily energy expenditures for a given individual vary widely depending on the activities undertaken. For example, a person leading a sedentary life may have near-basal energy demands, but a single hour of swimming can increase the daily caloric requirements by 500 C or more. If the daily energy intake exceeds the total energy demands of the individual, the excess will be stored, primarily as triglycerides in adipose tissue. If the daily caloric expenditures exceed the dietary supply, there will be a net reduction in the body's energy reserves and a corresponding loss in weight. This relationship accounts for the significance of calorie-counting and exercise in a weight-control program.

The control of appetite is poorly understood. Stretch receptors along the digestive tract, especially in the stomach, do play a role, but other factors are probably more important. Social factors, psychological pressures, and dietary habits are all important. There is also evidence that complex hormonal stimuli interact to affect appetite. For example, the hormones *cholecystokinin* and *ACTH* will suppress appetite.

❑ Thermoregulation

The BMR estimates the rate of energy use by the body. The energy not captured and harnessed by living cells is released as heat. This heat loss serves an important homeostatic purpose. Humans are subject to vast changes in environmental temperatures, but our complex biochemical systems have a major limitation: The enzyme systems will operate over only a relatively narrow range of temperatures. Our bodies have anatomical and physiological mechanisms that keep body temperatures within acceptable limits, regardless of the environmental conditions. This homeostatic process is called **thermoregulation.** Failure to control body temperature can result in a series of physiological changes. For example, a body temperature below 36°C or above 40°C can cause disorientation, and a temperature above 42°C can cause convulsions and permanent cell damage.

We are continually producing heat as a byproduct of metabolism, and that heat must be lost to the environment at the same rate if body temperature is to remain constant. When the environmental conditions vary from "ideal," becoming too warm or too cold, the gains or losses must be controlled to maintain homeostasis.

Mechanisms of Heat Transfer

Figure 25-19

Heat exchange with the environment involves four basic processes: *radiation, conduction, convection,* and *evaporation.* These mechanisms are illustrated in Figure 25-19●.

- ***Radiation.*** Warm objects lose heat energy as infrared radiation. When we feel the heat from the sun, we are experiencing radiant heat. Our

Eating Disorders

Eating disorders are psychological problems that result in either inadequate or excessive food consumption. The most common conditions are *anorexia nervosa*, characterized by self-induced starvation, and *bulimia*, characterized by feeding binges followed by vomiting, laxative use, or both. Adolescent females account for most cases. These conditions are less common in males, who account for only 5–10 percent of anorexia or bulimia cases. A common thread in the two conditions is an obsessive concern about food and body weight.

Current estimates are that the incidence of **anorexia nervosa** in the United States ranges from 0.5 to 1.0 percent of young women age 12–18. A typical person with this condition is an adolescent Caucasian woman whose weight is roughly 30 percent below normal levels. Although underweight, she is usually convinced that she is still too fat and refuses to eat normal amounts of food.

The psychological factors responsible for anorexia are complex. Young women with this condition tend to be high achievers who are attempting to reach an "ideal" weight that will be envied and admired, and thereby achieve a sense of security and accomplishment. The factors thus tend to be a combination of their view of society ("thin is desirable or demanded"), their view of themselves ("I am not yet thin enough"), and a desire to be able to control their fate ("I can decide when to eat"). Female models and theater and arts majors at any age may feel forced to drop weight to remain competitive. The few male anorexics diagnosed often face comparable stresses. They tend to be athletes, such as jockeys or wrestlers, who need to maintain a minimal weight to maintain their careers.

Young anorexic women will often continue to starve themselves down to a weight of 30–35 kg. Dry skin, peripheral edema, an abnormally low heart rate and blood pressure, a reduction in bone and muscle mass, and a cessation of menstrual cycles are relatively common symptoms. Some of the changes, especially in bone mass, may be permanent. Treatment is difficult, and only 50–60 percent of patients who regain normal weight stay there for 5 years or more. Death rates from severe anorexia nervosa range from 10 to 15 percent.

Bulimia is more common than anorexia. In this condition the individual goes on an "eating binge" that may involve a meal that lasts 1–2 hours and may include 20,000 or more calories. The meal is followed by induced vomiting, often accompanied by the use of laxatives (to promote movement of the material through the digestive tract) and diuretics (drugs that promote fluid loss in the urine). These often expensive binges may occur several times each week, separated by periods of either normal eating or fasting.

Bulimia most often involves women of the same age class as anorexia, and recent estimates of the incidence of bulimia in young women range from 5 to 20 percent or more. However, many bulimics are not diagnosed until they are age 30–40. Because of the amount of food ingested, bulimics may have normal body weight, and therefore the condition is harder to diagnose than anorexia nervosa.

The health risks of bulimia result from (1) cumulative damage to the stomach, esophagus, oral cavity, and teeth by repeated exposure to stomach acids; (2) electrolyte imbalances resulting from the loss of sodium and potassium ions in the gastric juices, diarrhea, and urine; (3) edema; and (4) cardiac arrhythmias.

The underlying cause of bulimia remains uncertain. Societal factors are certainly involved, but bulimia has also been strongly correlated with depression and with elevated CSF levels of ADH.

bodies lose heat the same way, but in proportionately smaller amounts. Over half of our heat loss is attributable to radiation; the exact percentage varies, depending on both body temperature and skin temperature.

- ***Conduction.*** Conduction refers to the direct transfer of energy through physical contact. When you arrive in an air-conditioned classroom and sit down on a cold plastic chair, you are immediately aware of this process. Conduction is usually not an effective mechanism for gaining or losing heat.
- ***Convection.*** Convection is the result of conductive heat loss to the air that overlies the surface of the body. Warm air is lighter than cooler air, and so it rises. As the body conducts heat to air next to the skin, that air warms and rises, moving away from the surface. Cooler air replaces it, and as it in turn becomes warmed, the cycle repeats. Convection accounts for roughly 15 percent of our heat loss; convection is insignificant as a mechanism for heat gain.
- ***Evaporation.*** When water evaporates, it changes from a liquid to a vapor. This process absorbs

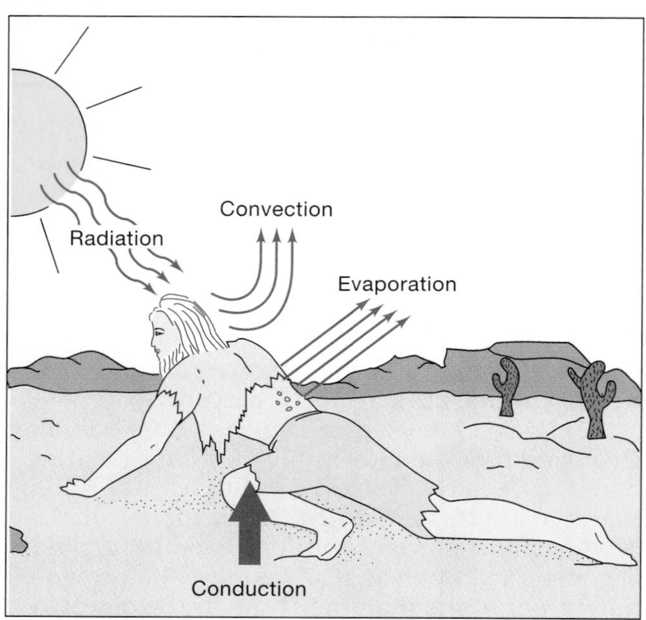
Radiation
Convection
Evaporation
Conduction

FIGURE 25-19
Routes of Heat Gain and Loss

energy, roughly 0.58 C per gram of water evaporated. The rate of evaporation occurring at the skin is highly variable. Each hour, 20–25 ml of water crosses epithelia to be evaporated from the alveolar surfaces and the surface of the skin. This **insensible perspiration** remains relatively constant; at rest it accounts for roughly one-fifth of the average heat loss. The sweat glands responsible for **sensible perspiration** have a tremendous scope of activity, ranging from virtual inactivity to secretory rates of 2–4 liters per hour.

Mechanisms of Heat Distribution

Biochemical reactions produce heat, and that heat is retained in the water that accounts for nearly 66 percent of the body weight. Water is an excellent conductor of heat, so the heat produced in one region of the body is rapidly distributed by diffusion, as well as via the circulation. The greater the number of biochemical reactions under way, the more heat will be produced. As a result, heat production can be quite variable, and the maintenance of a constant body temperature requires a finely tuned and adaptable homeostatic mechanism.

Heat Gain and Heat Loss

Heat loss and heat gain involve the coordinated activity of many different systems coordinated by the **heat-loss center** and **heat-gain center** in the preoptic nucleus of the anterior hypothalamus, discussed in Chapter 14. [p. 479] These centers modify the activities of other hypothalamic nuclei. The overall effect is to control temperature by influencing two events—the rate of heat production and the rate of heat loss to the environment. These may be further supported by behavioral modifications.

Mechanisms for Increasing Heat Loss

When the temperature at the preoptic nucleus exceeds its thermostat setting, the heat-loss center is stimulated. Stimulation of this center has three major effects:

1. Inhibition of the vasomotor center causes peripheral vasodilation, and warm blood flows to the surface of the body. The skin takes on a reddish color, skin temperatures rise, and radiational and convective losses increase.
2. As integumentary blood flow increases, sweat glands are stimulated to increase their secretory output. The perspiration flows across the body surface, and evaporative losses accelerate. A maximal secretion rate would, if it were completely evaporated, remove 2320 C per hour.
3. The respiratory centers are stimulated, and the depth of respiration increases. Often the individual begins respiring through an open mouth, rather than through the nasal passageways, increasing evaporative losses through the lungs.

HEAT EXHAUSTION AND HEAT STROKE *Heat exhaustion* and *heat stroke* represent malfunctions of the thermoregulatory system. In **heat exhaustion,** also known as heat prostration, the individual experiences difficulties with the maintenance of blood volume. The heat-loss center is stimulating sweat glands whose secretions moisten the surface of the skin to provide evaporative cooling. As fluid losses mount, blood volume decreases. The resultant decline in blood pressure is not countered by peripheral vasoconstriction, because the heat-loss center is actively stimulating peripheral vasodilation. As blood flow to the brain declines, headache, nausea, and eventual collapse follow. Treatment is straightforward: Provide fluids, salts, and a cooler environment.

Heat stroke is more serious and may occur following an untreated case of heat exhaustion. Predisposing factors include any preexisting condition, such as heart disease or diabetes, that affects peripheral circulation. The thermoregulatory center ceases to function, the sweat glands are inactive, and the skin becomes hot and dry. Unless the situation is recognized in time, body temperature may climb to 41°–45°C (106°–113°F). Temperatures in this range will quickly disrupt a variety of vital physiological systems and destroy brain, liver, skeletal muscle, and kidney cells. Proper treatment involves lowering body temperature as rapidly as possible.

Mechanisms for Promoting Heat Gain

Figure 25-20

The function of the heat-gain center of the brain is to prevent **hypothermia** (hī-pō-THER-mē-uh),

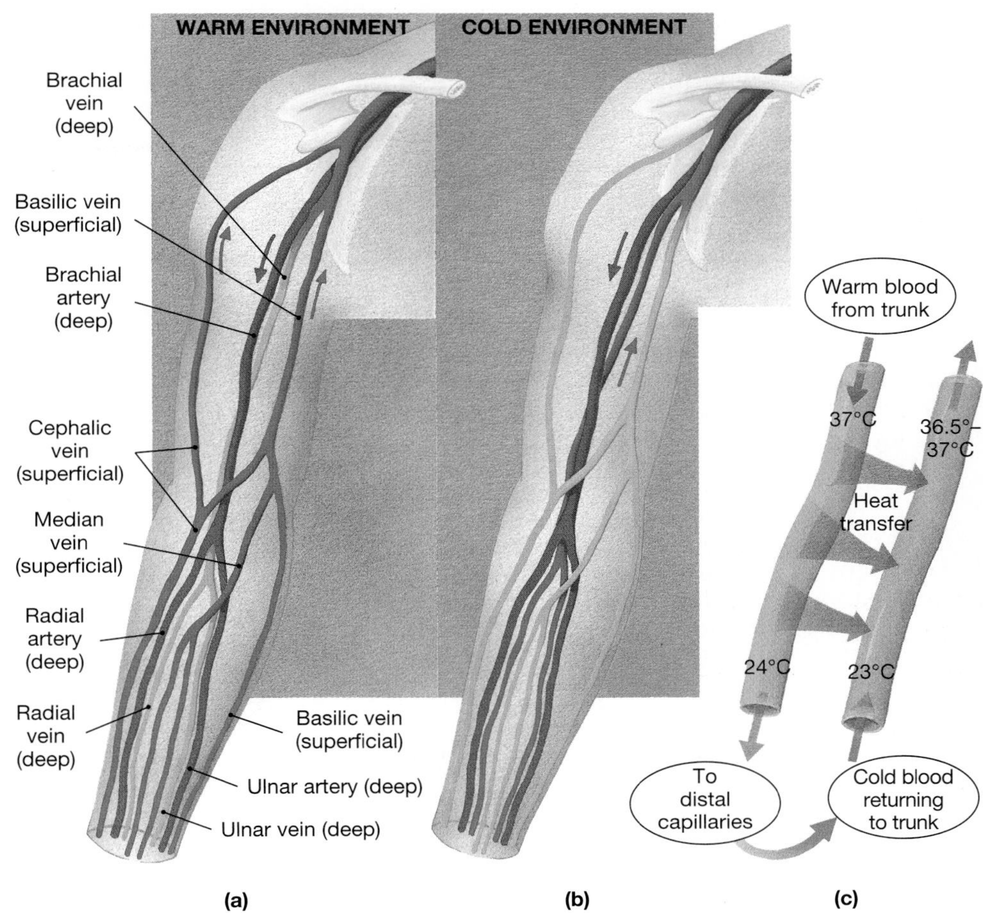
WARM ENVIRONMENT
COLD ENVIRONMENT
Brachial vein (deep)
Basilic vein (superficial)
Brachial artery (deep)
Cephalic vein (superficial)
Median vein (superficial)
Radial artery (deep)
Radial vein (deep)
Basilic vein (superficial)
Ulnar artery (deep)
Ulnar vein (deep)
Warm blood from trunk
37°C
36.5°–37°C
Heat transfer
24°C
23°C
To distal capillaries
Cold blood returning to trunk
(a)
(b)
(c)

● **FIGURE 25-20 Countercurrent Heat Exchange.** **(a)** Circulation through the blood vessels of the forearm in a warm environment. Blood enters the limb in a deep artery and returns to the trunk in a network of superficial veins. These veins radiate heat into the environment through the overlying skin. **(b)** Circulation through the blood vessels of the forearm in a cold environment. Blood now returns to the trunk via a network of deep veins that flow around the artery. The amount of heat loss is reduced, as indicated in part (c). **(c)** Countercurrent heat exchange occurs as heat radiates from the warm arterial blood into the cooler venous blood flowing in the opposite direction. By the time the arterial blood reaches distal capillaries, where most of the heat loss to the environment occurs, it is already 13°C cooler than it was when it left the trunk. This mechanism reduces the rate of heat loss while conserving body heat. In effect, the countercurrent exchange traps heat near the trunk.

or below-normal body temperature. When the temperature at the preoptic nucleus drops below acceptable levels, the heat-loss center is inhibited, and the heat-gain center is activated.

HEAT CONSERVATION The sympathetic vasomotor center decreases blood flow to the dermis of the skin, thus reducing radiational, convective, and conductive losses. The skin cools, and with the circulation restricted it may take on a bluish or pale coloration. The epithelial cells are not damaged, because they can tolerate extended periods at temperatures as low as 25°C (87°F) or as high as 49°C (110°F).

In addition, blood returning from the extremities is shunted into a network of deep veins described in Chapter 21. ∞ *[pp. 764–765]* Figure 25-20● shows the changes in circulation that occur. In warm weather, blood flows in a superficial venous network (Figure 25-20a●). In cold weather, blood is diverted to a network of deep veins that lie beneath an insulating layer of subcutaneous fat. This venous network wraps around the deep arteries (Figure 25-20b●). Heat diffuses from the warm blood flowing outward to the limbs into the cooler blood returning from the periphery. This arrangement traps the heat close to the body core and restricts heat loss by reducing the temperature gradient between the arterial blood and the outside world. Diffusion between fluids moving in opposite directions is called *countercurrent exchange* (Figure 25-20c●). (We will return to this topic in Chapter 26.)

HEAT GENERATION The mechanisms available to generate heat can be divided into two broad categories. In **shivering thermogenesis** (ther-mō-JEN-e-sis), a gradual increase in muscle tone increases the energy consumption of skeletal muscle tissue throughout the body. Both agonists and antagonists are involved, and the degree of stimulation varies, depending on the demand.

If the heat-gain center is extremely active, muscle tone increases to the point where stretch receptor stimulation will produce brief, oscillatory contractions of antagonistic muscles. In other words, the person begins to **shiver.** Shivering increases the workload of the muscles and further elevates oxygen and energy consumption. The heat that is produced warms the deep vessels, to which blood has been shunted by the sympathetic vasomotor center. Shivering can elevate body temperature quite effectively because it increases the rate of heat generation by as much as 400 percent.

In **nonshivering thermogenesis,** hormones are released that increase the metabolic activity of all tissues.

- The heat-gain center stimulates the adrenal medulla, via the sympathetic division of the ANS, and epinephrine is released. Epinephrine increases the rates of glycogenolysis in liver and skeletal muscle and the rate of cellular metabolism in most tissues. These effects are immediate.
- The preoptic nucleus itself directs the activity of the thyroid gland. When temperatures are below normal the thyroid gland releases the hormone thyroxine into the blood. Thyroxine increases not only the rate of carbohydrate catabolism but also the rate of catabolism of all other nutrients. These effects develop gradually, over a period of days to weeks.

INDUCED HYPOTHERMIA Hypothermia may be intentionally produced during surgery to reduce the metabolic rate of a particular organ or of the entire body. In controlled hypothermia, the individual is first anesthetized to prevent the shivering that would otherwise fight the process.

During open-heart surgery the body is often cooled to 25°–32°C (79°–89°F). This cooling reduces the metabolic demands of the body, which will be receiving blood from an external pump or oxygenator. The heart must be stopped completely during the operation, and it cannot be well supplied with blood over this period. So the heart is exposed to an *arresting solution* at 0°–4°C (32°–39°F) and maintained at a temperature below 15°C (60°F) for the duration of the operation. At these temperatures the cardiac muscle can tolerate several hours of ischemia without damage.

When cardiac surgery is performed on infants, a deep hypothermia may be produced by cooling the entire body to temperatures as low as 11°C (52°F) for an hour or more. In effect this procedure duplicates the conditions experienced by the accidental drowning victims discussed in the *Applications Manual.* AM *Accidental Hypothermia*

Sources of Individual Variation

The timing of thermoregulatory responses may differ from individual to individual. A person may undergo **acclimatization** (a-klī-ma-ti-ZĀ-shun), making physiological adjustment to a particular environment over time. For example, natives of Tierra del Fuego once lived naked in the snow, but Hawaii residents unpack their sweaters when the temperature drops below 22°C.

Another interesting source of variation is body size. Although heat *production* occurs within the mass of the body, heat *loss* must occur across a body surface. The relationship between heat production and heat loss is thus linked to what is called the *surface-to-volume ratio.* As an object (or person) gets larger, the surface area increases at a much slower rate than does the total volume.

For example, consider a cube that measures 1 meter in width, depth, and height. A cube has six sides, and each side has 1 m^2 of surface area (1×1), so it has a total surface area of 6 m^2. Its volume is 1 m^3 ($1 \times 1 \times 1$), and it has a surface-to-volume ratio of 6 to 1. Now consider a cube that is twice as big, measuring 2 meters in each dimension. Each side has 4 m^2 of surface area (2×2), so it has a total surface area of 24 m^2. Its volume is 8 m^3 ($2 \times 2 \times 2$), and it has a surface-to-volume ratio of 3 to 1. In doubling the volume, the surface-to-volume ratio was reduced by one half.

The smaller the object (or person) the larger the surface-to-volume ratio. Because heat is lost across the body surface, small individuals lose heat more readily than do large individuals.

THERMOREGULATORY PROBLEMS OF INFANTS Infants have problems with thermoregulation because of their relatively high surface-to-volume ratios. During embryonic development, temperature regulation is no concern of theirs, as the maternal surroundings are at normal body temperature. At birth, the temperature-regulating mechanisms are not fully functional. With such high surface-to-volume ratios, newborns must be dried quickly and kept bundled up; for those born prematurely, a thermally regulated incubator is required. Infants' body temperatures are more unstable than those of adults, and their metabolic rates decline when they are sleeping, then rise after arousal.

Infants cannot shiver, but they have a different mechanism for raising body temperature rapidly. The adipose tissue between the shoulder blades, around the neck, and possibly elsewhere in the upper body is histologically and functionally different from most of the adipose tissue in the adult. The tissue is highly vascularized, and the individual adipocytes contain numerous mitochondria. Together these characteristics give the tissue a deep, rich color responsible for the name **brown fat.** The individual adipocytes are innervated by sympathetic autonomic fibers. When these nerves are stimulated, lipolysis accelerates in the adipocytes. The cells do not capture the energy released through fatty acid catabolism, and it radiates into the surrounding tissues as heat. This heat quickly warms the blood passing through the surrounding network of vessels, and it is then distributed throughout the body. In this way an infant can accelerate metabolic heat generation by 100 percent very quickly, while nonshivering thermogenesis in the adult will raise heat production by only 10–15 percent after a period of weeks.

With increasing age and size, body temperature becomes more stable, and the importance of this thermoregulatory mechanism declines. There is little if any brown fat in the adult; with increased body size, skeletal muscle mass, and insulation, shivering thermogenesis is significantly more effective in elevating body temperature.

THERMOREGULATORY VARIATIONS AMONG ADULTS Adults of the same body weight may differ in their thermal responses if their weight is distributed differently such that they have different surface-to-volume ratios—just consider two 70-kg individuals, one 2 meters tall, and the other 1.5 meters tall. Which tissues account for the weight is also a factor. Adipose tissue is an excellent insulator, conducting heat at only about one-third the rate of other tissues. As a result, individuals with a more substantial layer of subcutaneous fat may not begin to shiver until long after their more slender companions.

In addition to hormone levels, environmental acclimatization, body weight, age, tissue distribution, and surface-to-volume ratios, our hypothalamic thermostats also affect our thermoregulatory responses. Two otherwise similar individuals may differ in their response to temperature changes because their hypothalamic thermostats are at different settings. There are daily oscillations in body temperature, with temperatures falling 1–2° C at night and peaking sometime during the day or early evening. Individuals vary in terms of their time of maximum temperature setting, and some have a series of peaks, with an afternoon low. The origin of these patterns is uncertain, since it is not the result of daily activity regimens—people who work at night still show their temperature peaks over the same range of times as the rest of the population.

Fevers

An elevated body temperature, or **pyrexia** (pī-REK-sē-a), may occur for a variety of reasons, not all of them pathological. In young children, transient fevers may result from exercise in warm weather, with no ill effects. Similar exercise-related elevations were rarely encountered in adults until running marathons became a national pastime. Temperatures ranging from 39° to 41°C (103° to 106°F) may result, and it is for this reason that competitions are usually held when the air temperature is below 28°C (82°F).

Fevers were discussed when we examined nonspecific defenses in Chapter 22. [p. 798] A fever is the maintenance of a body temperature greater than 37.2°C (99°F). Fevers may result from:

- Abnormalities affecting the entire thermoregulatory mechanism, such as heat exhaustion or heat stroke.
- Clinical problems that restrict circulation, such as congestive heart failure.
- Conditions that impair sweat gland activity, such as drug reactions and some skin conditions.
- The resetting of the hypothalamic thermostat by circulating *pyrogens*, most notably interleukin-1.

Fevers may be classifed as *chronic* or *acute;* the classification and treatment of fevers are discussed in the *Applications Manual.* AM Fevers

---

✔ How would the BMR (basal metabolic rate) of a pregnant woman compare with her BMR in the nonpregnant state?

✔ What effect would vasoconstriction of peripheral blood vessels have on body temperature on a hot day?

✔ Why do infants have greater problems with thermoregulation than adults?

Selected Clinical Terminology

Terms Discussed in This Chapter

avitaminosis (ā-vī-ta-min-Ō-sis): A vitamin deficiency disease. *(p. 965)*
carbohydrate loading: Eating large quantities of carbohydrates in the days preceding an athletic competition to increase endurance. *(p. 945)*
eating disorders: Psychological problems that result in inadequate or excessive food consumption. Examples include *anorexia nervosa* and *bulimia. (p. 969)*
heat exhaustion: A malfunction of the thermoregulatory system caused by excessive fluid loss in perspiration. *(p. 970)*
heat stroke: A condition in which the thermoregulatory center stops functioning and body temperature rises uncontrollably. *(p. 970)*
hyperuricemia (hī-per-ū-ri-SĒ-mē-a): Levels of plasma uric acid above 7.4 mg/dl; may result in the condition called *gout. (p. 954)*
hypervitaminosis (hī-per-vī-tah-min-Ō-sis): A disorder caused by ingestion of excessive quantities of one or more vitamins. *(p. 964)*
hypothermia (hī-pō-THER-mē-uh): Below-normal body temperature. *(p. 970 and AM)*
ketoacidosis (kē-tō-as-i-DŌ-sis): Acidification of the blood due to the presence of ketone bodies. *(p. 960 and AM)*
ketonemia (kē-tō-NĒ-mē-a): Elevated levels of ketone bodies in the blood. *(p. 960)*
ketonuria (kē-tō-NŪ-rē-a): The presence of ketone bodies in the urine. *(p. 960)*
ketosis (kē-TŌ-sis): Abnormally high concentration of ketone bodies in body fluids. *(p. 960)*
obesity: A body weight more than 20 percent above the ideal weight for a given individual. *(p. 957 and AM)*
phenylketonuria (fen-il-kē-tō-NU-rē-a): An inherited metabolic disorder resulting from an inability to convert phenylalanine to tyrosine. *(p. 952 and AM)*
protein deficiency diseases: Nutritional disorders resulting from a lack of essential amino acids. *(p. 953 and AM)*

CHAPTER REVIEW

STUDY OUTLINE

INTRODUCTION, p. 936

1. Cells in the human body are chemical factories that break down organic substrates to obtain energy.

AN OVERVIEW OF CELLULAR METABOLISM, p. 936

1. In general, cells will break down excess carbohydrates first, then lipids, while conserving amino acids. Only about 40 percent of the energy released through catabolism is captured in ATP; the rest is released as heat *(Figure 25-1)*
2. Cells synthesize new compounds: (1) to perform structural maintenance or repair, (2) to support growth, and (3) to produce secretions. *(Figure 25-2)*

CARBOHYDRATE METABOLISM, p. 938

1. Most cells generate ATP and other high-energy compounds through the breakdown of carbohydrates.

Glycolysis, p. 938

2. Glycolysis and aerobic respiration provide most of the ATP used by typical cells. Glycogen can be broken down to glucose molecules. In glycolysis, each molecule of glucose yields two molecules of pyruvic acid (as pyruvate ions), 2 NADH, and two molecules of ATP. *(Figure 25-3)*

Mitochondrial ATP Production, p. 938

3. In the presence of oxygen the pyruvic acid molecules enter the mitochondria, where they are broken down completely in the TCA cycle. The carbon and oxygen atoms are lost as carbon dioxide, and the hydrogen atoms are passed to coenzymes, which initiate the oxygen-consuming and ATP-generating reaction **oxidative phosphorylation.** *(Figure 25-4)*
4. **Cytochromes** pass electrons along the respiratory chain of the **electron transport system** to eventually generate ATP and water as the electrons and hydrogen ions combine with oxygen to form water. *(Figure 25-5)*

Energy Yield of Glycolysis and Cellular Respiration, p. 944

5. For each glucose molecule processed through aerobic respiration, most cells gain 36 molecules of ATP. *(Figure 25-6)*

Other Catabolic Pathways, p. 945

6. Cells can break down other nutrients to provide substrates for the TCA cycle if supplies of glucose are limited. *(Figure 25-7)*

Gluconeogenesis, p. 946

7. **Gluconeogenesis,** the synthesis of glucose, enables a liver cell to create glucose molecules from other carbohydrates, glycerol, or some amino acids. **Glycogenesis** is the process of glycogen formation (glycogen is an important energy reserve when the cell cannot obtain enough glucose from the extracellular fluid). *(Figure 25-8)*

LIPID METABOLISM, p. 947

Lipid Catabolism, p. 947

1. During **lipolysis** (lipid catabolism), lipids are broken down into pieces that can be converted into pyruvic acid or channeled into the TCA cycle.
2. Triglycerides are the most abundant lipids in the body. Triglycerides are split into glycerol and fatty acids. The glycerol enters the glycolytic pathways, and the fatty acids enter the mitochondria.
3. **Beta-oxidation** is the breakdown of a fatty acid molecule into two-carbon fragments that can be used in the TCA cycle. The steps of beta-oxidation cannot be reversed, and the body cannot manufacture all of the fatty acids needed for normal metabolic operations. *(Figure 25-9)*
4. Lipids cannot provide large amounts of ATP in a short amount of time. However, cells can shift to lipid-based energy production when glucose reserves are limited.

Lipid Synthesis, p. 948

5. In **lipogenesis,** the synthesis of lipids, almost any organic substrate can be used to form glycerol. **Essential fatty acids** cannot be synthesized and must be included in the diet. *(Figure 25-10)*

Lipid Transport and Distribution, p. 948

6. Lipids circulate as **free fatty acids (FFA)** (water-soluble lipids that can diffuse easily across cell membranes) and as **lipoproteins** (lipid-protein complexes that contain large glycerides and cholesterol). The largest lipoproteins, chylomicrons, carry absorbed lipids from the intestinal tract to the circulation. All other lipoproteins are derived from the liver and carry lipids to and from various tissues of the body. *(Figure 25-11a)*
7. Capillary walls of adipose tissue, skeletal muscle, cardiac muscle, and the liver contain **lipoprotein lipase,** an enzyme that breaks down compex lipids, releasing a mixture of fatty acids and monoglycerides. *(Figure 25-11b)*

PROTEIN METABOLISM, p. 952

Amino Acid Catabolism, p. 952

1. If other energy sources are inadequate, mitochondria can break down amino acids in the TCA cycle to generate ATP. In the mitochondria the amino group may be removed by **transamination** or **deamination,** and the carbon skeleton is converted to one of the compounds involved in glycolysis or oxidative respiration. *(Figure 25-12)*
2. Protein catabolism is an impractical source for quick energy.

Protein Synthesis, p. 952

3. Roughly half of the amino acids needed to build proteins can be synthesized. There are 10 **essential amino acids** that need to be acquired through the diet. **Amination,** the attachment of an amino acid group to a carbon framework, is an important step in the synthesis of **nonessential amino acids.** *(Figure 25-13)*

NUCLEIC ACID METABOLISM, p. 953

1. DNA in the nucleus is never catabolized for energy.

RNA Catabolism, p. 953

2. RNA molecules are broken down and replaced regularly; usually they are recycled as new nucleic acids, but the nucleotides can be catabolized to simple sugars and nitrogenous bases. In general, nucleic acids do not contribute significantly to the cell's energy reserves.

Nucleic Acid Synthesis, p. 954

3. Most cells synthesize RNA, but DNA synthesis occurs only in cells preparing for mitosis or meiosis.

METABOLIC INTERACTIONS, p. 954

1. No one cell of a human can perform all the anabolic and catabolic operations necessary to support life. Homeostasis can be preserved only when metabolic activities of different tissues are coordinated. *(Figure 25-14)*
2. The body has five metabolic components: the liver, adipose tissue, skeletal muscle, neural tissue, and other peripheral tissues.
3. The liver represents the focal point for metabolic regulation and control. Adipose tissue stores lipids, primarily as triglycerides. Skeletal muscle contains substantial glycogen reserves, and the contractile proteins can be mobilized and the amino acids used as an energy source. Neural tissue, which does not contain energy reserves, depends on aerobic respiration for energy production. Other peripheral tissues are able to metabolize glucose, fatty acids, or other substrates under the direction of the endocrine system.

The Absorptive State, p. 955

4. For around 4 hours after a meal, nutrients enter the blood as intestinal absorption proceeds. *(Figure 25-15; Table 25-1)*
5. The liver closely regulates the glucose content of blood and the circulating levels of amino acids.
6. **Lipemia** (milky appearance of the plasma due to the presence of lipids) often marks the absorptive state. Adipocytes remove fatty acids and glycerol from the circulation and synthesize new triglycerides for storage.
7. During the absorptive state glucose molecules are catabolized and amino acids are used to build proteins. Skeletal muscles may also catabolize circulating fatty acids, and the energy is used to increase glycogen reserves.

The Postabsorptive State, p. 957

8. The postabsorptive state extends from the end of the absorptive state to the next meal. *(Figure 25-16)*
9. When blood glucose falls, the liver begins breaking down glycogen reserves and releasing the glucose into the circulation. As the duration of the fast increases, liver cells synthesize glucose molecules from smaller carbon fragments and from glycerol molecules. Fatty acids undergo beta-oxidation; the fragments enter the TCA cycle or combine to form ketone bodies. *(Table 25-1)*
10. **Glucogenic amino acids** can be converted to pyruvic acid and used for gluconeogenesis. **Ketogenic amino acids** can be converted to acetyl-CoA and catabolized or converted to ketone bodies. *(Figure 25-17)*
11. The average individual carries a 1–2 month energy reserve in adipose tissue. During the postabsorptive state lipolysis increases and the fatty acids are released into the circulation for catabolism.
12. Skeletal muscles metabolize ketone bodies and fatty acids. Their glycogen reserves are broken down to yield lactic acid, which diffuses into the bloodstream. After a prolonged fast, **cathepsins** (proteolytic enzymes) begin breaking down contractile proteins.
13. Neural tissue continues to be supplied with glucose as an energy source until blood glucose levels become extremely low.

DIET AND NUTRITION, p. 961

1. **Nutrition** is the absorption of nutrients from food. A **balanced diet** contains all the ingredients necessary to maintain homeostasis; it prevents **malnutrition.**

Food Groups and Food Pyramids, p. 961

2. The six **basic food groups** are the milk; meat; vegetable; fruit; fats, oils, and sweets; and bread and cereal groups. These are arranged in a food pyramid with the bread and cereal group forming the base. *(Figure 25-18)*

Nitrogen Balance, p. 962

3. Amino acids, purines, pyrimidines, creatine, and porphyrins are **N compounds** that contain nitrogen atoms. An adequate dietary supply of nitrogen is essential, because the body does not maintain large nitrogen reserves.

Minerals and Vitamins, p. 963

4. **Minerals** act as cofactors in a variety of enzymatic reactions. They also contribute to the osmolarity of body fluids, and they play a role in transmembrane potentials, action potentials, construction and maintenance of the skeleton, transport of gases, buffer systems, fluid absorption, and waste removal. *(Table 25-2)*
5. Vitamins are needed in very small amounts. Vitamins A, D, E, and K are the **fat-soluble vitamins;** taken in excess, they can lead to **hypervitaminosis. Water-soluble vitamins** are not stored in the body; lack of adequate dietary supplies may lead to **deficiency disease (avitaminosis).** *(Tables 25-3, 25-4)*

Diet and Disease, p. 965

6. A balanced diet can improve general health.

BIOENERGETICS, p. 966

1. The energy content of food is usually expressed as **Calories** per gram (C/g). Less than half of the energy content of glucose or any other nutrient can be captured by our cells.

Food and Energy, p. 967

2. The catabolism of lipids releases 9.46 C/g, about twice the amount as equivalent weights of carbohydrates and proteins.

Metabolic Rate, p. 968

3. The total of all the anabolic and catabolic processes under way in the body represents the **metabolic rate**

of an individual. The **basal metabolic rate (BMR)** is the rate of energy utilization at rest.

Thermoregulation, p. 968

4. The homeostatic regulation of body temperature is **thermoregulation.** Heat exchange with the environment involves four processes: radiation, conduction, convection, and evaporation. *(Figure 25-19)*
5. The **preoptic area** of the hypothalamus acts as the body's thermostat, affecting the **heat-loss center** and the **heat-gain center.**
6. Mechanisms for increasing heat loss include both physiological mechanisms (peripheral vasodilation, increased perspiration, and increased respiration) and behavioral modifications.
7. Responses that conserve heat include decreased blood flow to the dermis and countercurrent exchange. *(Figure 25-20)*
8. Heat may be generated by **shivering thermogenesis** and **nonshivering thermogenesis.**
9. Thermoregulatory responses differ between individuals. One important source of variation is **acclimatization** (adjusting physiologically to an environment over time).
10. **Pyrexia** (fever), a body temperature above 37.2°C (99°F), can result from problems with the thermoregulatory mechanism, circulation, or sweat gland activity, or from the resetting of the hypothalamic thermostat by circulating pyrogens.

REVIEW QUESTIONS

LEVEL 1 Reviewing Facts and Terms

1. Cells synthesize new organic components to:
   (a) perform structural maintenance and repairs
   (b) support growth
   (c) produce secretions
   (d) a, b, and c are correct

2. During the complete catabolism of one molecule of glucose, a typical cell gains:
   (a) 4 ATP (b) 18 ATP
   (c) 36 ATP (d) 144 ATP

3. The breakdown of glucose to pyruvic acid is:
   (a) glycolysis
   (b) gluconeogenesis
   (c) cellular respiration
   (d) oxidative phosphorylation

4. Glycolysis yields an *immediate* net gain of ________ for the cell.
   (a) 1 ATP (b) 2 ATP
   (c) 4 ATP (d) 36 ATP

5. The only *immediate* energy benefit of one turn of the TCA cycle is the formation of a single molecule of:
   (a) GTP (b) NAD
   (c) FAD (d) CoA

6. The process that produces over 90 percent of the ATP used by our cells is:
   (a) glycolysis
   (b) the TCA cycle
   (c) substrate-level phosphorylation
   (d) oxidative phosphorylation

7. The electron transport chain and phosphorylation associated with it usually yield a total of ________ molecules of ATP in the complete catabolism of one glucose molecule.
   (a) 2 (b) 4
   (c) 32 (d) 36

8. The synthesis of glucose from nonglucose precursors is called:
   (a) glycolysis
   (b) glycogenesis
   (c) gluconeogenesis
   (d) beta-oxidation

9. The sequence of reactions responsible for the breakdown of fatty acid molecules is called:
   (a) beta-oxidation (b) the TCA cycle
   (c) lipogenesis (d) a, b, and c are correct

10. The essential fatty acids that cannot be synthesized by the body but must be included in the diet are:
    (a) linoleic and linolenic
    (b) leucine and lysine
    (c) cholesterol and glycerol
    (d) HDLs and LDLs

11. The lipoproteins that transport excess cholesterol from peripheral tissues back to the liver for storage or excretion in the bile are the:
    (a) chylomicrons (b) VLDLs
    (c) LDLs (d) HDLs

12. The removal of an amino group in a reaction that generates an ammonia molecule is called:
    (a) ketoacidosis (b) transamination
    (c) deamination (d) denaturation

13. The part of the RNA molecule that *cannot* be catabolized to provide energy is the:
    (a) phosphate (b) sugar
    (c) pyrimidine (d) purine

14. The focal point for metabolic regulation and control is the:
    (a) brain (b) liver
    (c) heart (d) kidneys

15. When the body is relying on internal energy reserves to continue meeting its energy demands it is in the:
    (a) postabsorptive state
    (b) absorptive state
    (c) starvation state
    (d) deprivation state

16. A complete protein contains:
    (a) the proper balance of amino acids
    (b) all the essential amino acids in sufficient quantities
    (c) a combination of nutrients selected from the food pyramid
    (d) N compounds produced by the body

17. All minerals and most vitamins:
    (a) are fat-soluble
    (b) cannot be stored by the body
    (c) cannot be synthesized by the body
    (d) must be synthesized by the body because they are not present in adequate amounts in the diet

18. The vitamins generally associated with vitamin toxicity are:
    (a) fat-soluble vitamins
    (b) water-soluble vitamins
    (c) the B complex vitamins
    (d) vitamins C and B_{12}

19. The basal metabolic rate (BMR) represents the:
    (a) maximum energy expenditure when exercising
    (b) minimum, resting energy expenditure of an awake, alert person
    (c) minimum amount of energy expenditure during light exercise
    (d) muscular energy expenditure added to the resting energy expenditure

20. The abuse of laxatives and diuretics is often associated with:
    (a) avitaminosis
    (b) kwashiorkor
    (c) bulimia
    (d) marasmus

21. Over half of the heat loss from our bodies is attributable to:
    (a) radiation
    (b) conduction
    (c) convection
    (d) evaporation

22. The most effective mechanism for elevating body temperature in an adult is:
    (a) lipolysis in brown fat
    (b) carbohydrate loading
    (c) fatty acid catabolism
    (d) shivering thermogenesis

23. The resetting of the hypothalamic thermostat by circulating pyrogens such as interleukin-1 produces:
    (a) hypothermia
    (b) fever
    (c) maintenance of normal body temperature
    (d) shivering

24. Define the terms *metabolism, anabolism,* and *catabolism.*

25. Write the complete reaction sequence for carbohydrate metabolism.

26. What is a lipoprotein? What are the major groups of lipoproteins, and how do they differ?

27. Why are vitamins and minerals essential components of the diet?

28. What energy yields in Calories per gram are associated with the catabolism of carbohydrates, lipids, and proteins?

29. What is the basal metabolic rate (BMR)?

30. What four basic mechanisms are involved in the body's thermoregulation and the environment?

31. What causative factors or conditions are associated with the production of fever?

LEVEL 2 Reviewing Concepts

32. The function of the TCA cycle is to:
    (a) produce energy during periods of active muscle contraction
    (b) break six-carbon chains into three-carbon fragments
    (c) prepare the glucose molecule for further reactions
    (d) remove hydrogen atoms from organic molecules and transfer them to coenzymes

33. During periods of fasting or starvation, the presence of ketone bodies in the circulation causes:
    (a) an increase in the pH
    (b) a decrease in the pH
    (c) lipemia
    (d) diabetes insipidus

34. What happens during the process of glycolysis? What conditions are necessary for this process to take place?

35. Why is the TCA cycle called a cycle? What substance(s) enter(s) the cycle, and what substances leave it?

36. What is oxidative phosphorylation? Explain how the electron transport system is involved in this process.

37. How are lipids catabolized in the body? How is beta-oxidation involved with lipid catabolism?

38. How is RNA catabolized to produce energy?

39. How do the absorptive and postabsorptive states serve to maintain normal blood glucose levels?

40. Why is the liver the focal point for metabolic regulation and control?

41. How can the food pyramid be used as a tool to obtain nutrients in sufficient quantity and quality? Why are the dietary fats, oils, and sugars at the top of the pyramid and the bread and cereal at the bottom?

42. How is the brain involved in body temperature regulation?

43. Articles in popular magazines sometimes refer to "good cholesterol" and "bad cholesterol." To what types and functions of cholesterol might these terms refer? Explain your answer.

LEVEL 3 Critical Thinking and Clinical Applications

44. While resting and alert, Mary has a respiratory rate of 10 breaths per minute and a tidal volume of 300 ml. The air she is breathing contains 18 percent oxygen. Estimate Mary's BMR (basal metabolic rate) for a 24-hour period.

45. Why is a starving person more susceptible to infectious disease than one who is well nourished?

46. Individuals suffering from anorexia nervosa frequently exhibit bradycardia, hypotension, and de-creased heart size. These problems can eventually lead to death from heart failure. How does anorexia cause these symptoms?

47. The drug *colestipol* binds bile salts in the intestine, forming complexes that cannot be absorbed. How would this drug affect cholesterol levels in the blood?

CHAPTER 26

Urinating in public is considered bad form these days. But this whimsical fountain in Brussels, Belgium, depicts what is, after all, a very natural and necessary process. After our digestive system assimilates a meal and our body harvests and utilizes the nutrients it needs, we must somehow discharge the resulting waste products. The urinary system performs this service, while efficiently conserving water and other valuable substances. It also, as we shall see, carries out a variety of other vital but less obvious functions (none of which has been immortalized in sculpture, so far as we know).

The Urinary System

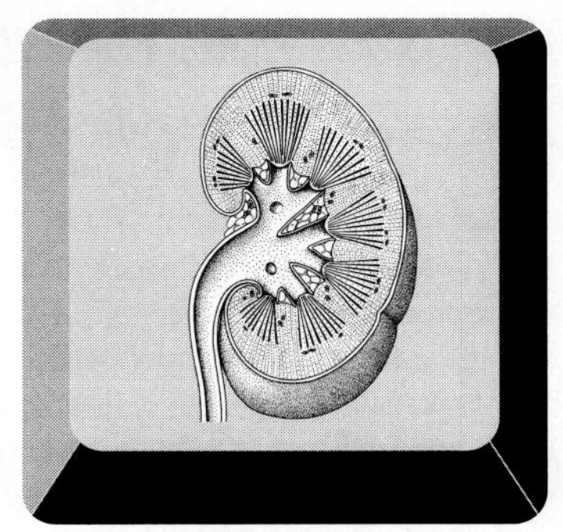

Chapter Outline and Objectives

The human body contains trillions of cells bathed in extracellular fluid. Previous chapters compared these cells to factories that burn nutrients to obtain energy. Imagine what would happen if *real* factories were built as close together as cells in the body. What a mess they would make! Each would generate piles of garbage, and the smoke they produced, together with the depletion of oxygen, would drastically reduce air quality. In short, there would be a serious pollution problem.

The coordinated activities of the digestive, cardiovascular, respiratory, and urinary systems prevent the development of similar pollution problems inside the body. The digestive tract absorbs nutrients from food, and the liver adjusts the nutrient concentration of the circulating blood. The cardiovascular system delivers these nutrients and oxygen from the respiratory system to peripheral tissues. As blood leaves these tissues it carries the carbon dioxide and waste products to sites of excretion. The carbon dioxide is eliminated at the lungs, as described in Chapter 23. Most of the organic waste products are removed and excreted by the urinary system, the focus of this chapter.

The urinary system performs vital excretory functions and eliminates the organic waste products generated by cells throughout the body. But it also has other essential functions that are often overlooked. A more complete list of urinary system functions includes the following:

1. Regulating blood volume and blood pressure by (a) adjusting the volume of water lost in the urine, (b) releasing erythropoietin, and (c) releasing renin. ∞ [p. 628]
2. Regulating plasma concentrations of sodium, potassium, chloride, and other ions by controlling the quantities lost in the urine and controlling calcium ion levels by the synthesis of calcitriol. ∞ [p. 627]
3. Contributing to the stabilization of blood pH by controlling the loss of hydrogen ions (H^+) and bicarbonate ions (HCO_3^-) in the urine.
4. Conserving valuable nutrients by preventing their excretion in the urine, while eliminating organic waste products, especially nitrogenous wastes such as *urea* and *uric acid.*
5. Assisting the liver in detoxifying poisons and deaminating amino acids so that they can be broken down by other tissues.

These activities are carefully regulated to keep the composition of the blood within acceptable limits. A disruption of any one of these functions will have immediate and potentially fatal consequences.

This chapter considers the functional organization of the urinary system and describes the major regulatory mechanisms that control urine production and concentration.

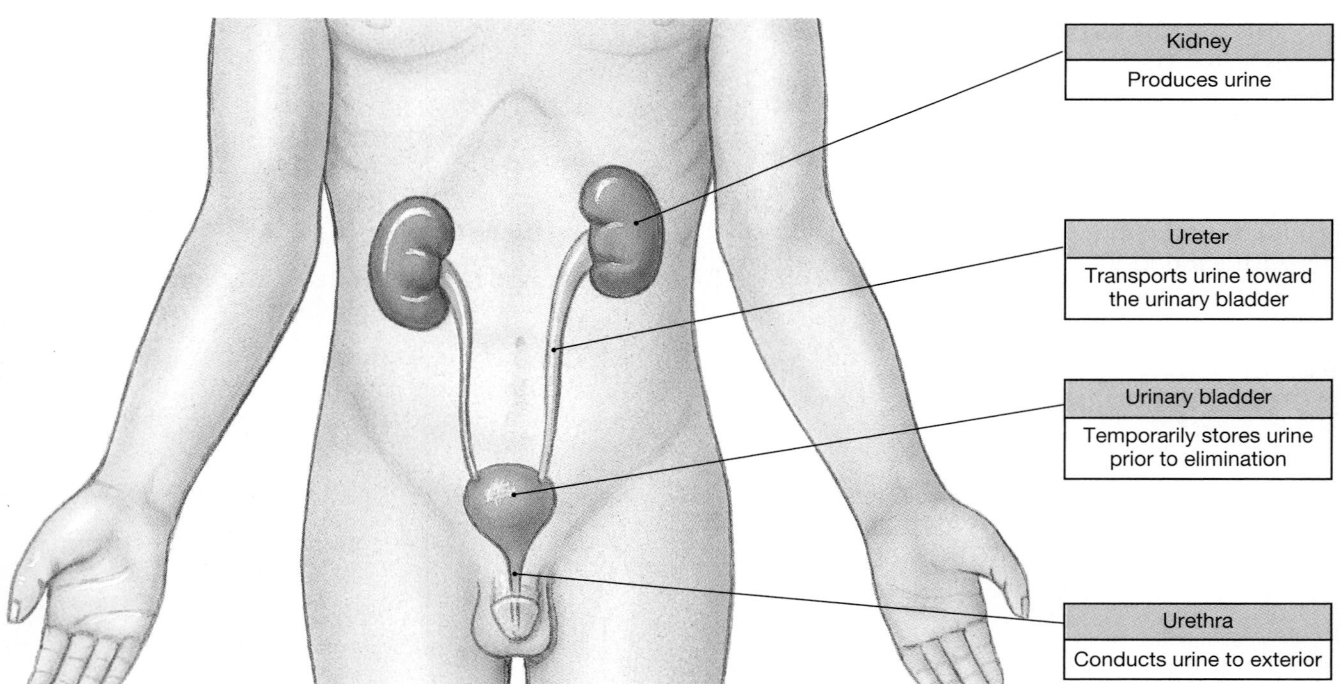
Kidney
Produces urine
Ureter
Transports urine toward the urinary bladder
Urinary bladder
Temporarily stores urine prior to elimination
Urethra
Conducts urine to exterior

● FIGURE 26-1
Components of the Urinary System

Organization of the Urinary System

Figure 26-1

The urinary system (Figure 26-1●) includes the *kidneys, ureters, urinary bladder,* and *urethra.* The excretory functions of the urinary system are performed by the two **kidneys.** These organs produce **urine,** a fluid containing water, ions, and small soluble compounds. Urine leaving the kidneys travels along the paired **ureters** (ū-RE-terz) to the **urinary bladder** for temporary storage. When **urination,** or **micturition** (mik-tu-RI-shun), occurs, contraction of the muscular urinary bladder forces urine through the **urethra** and out of the body.

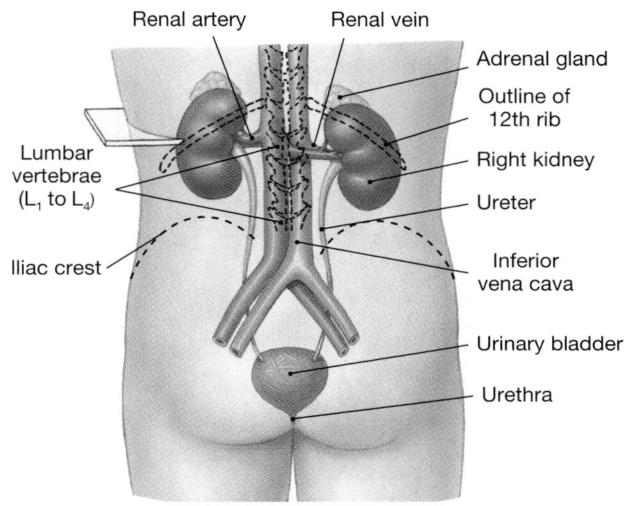
Renal artery
Renal vein
Adrenal gland
Outline of 12th rib
Right kidney
Ureter
Lumbar vertebrae (L_1 to L_4)
Iliac crest
Inferior vena cava
Urinary bladder
Urethra

(a) Posterior view

The Kidneys

Figures 26-2, 26-3b

The kidneys are located on either side of the vertebral column between the last thoracic and third lumbar vertebrae. The left kidney lies slightly higher than the right kidney (Figure 26-2a●).

On gross dissection, the anterior surface of the right kidney is covered by the liver, the hepatic flexure of the colon, and the duodenum. The anterior surface of the left kidney is covered by the stomach, pancreas, jejunum, and splenic flexure of the colon. The superior surface of each kidney is capped by an adrenal gland (Figure 26-2a●). The kidneys and adrenal glands lie between the muscles of the dorsal body wall and the parietal peritoneum in a retroperitoneal position (Figures 26-2b and 26-3b●).

The position of the kidneys in the abdominal cavity is maintained by (1) the overlying peritoneum, (2) contact with adjacent visceral organs, and (3) supporting connective tissues. Each kidney is protected and stabilized by three concentric layers of connective tissue (Figure 26-2b●):

1. A layer of collagen fibers covers the outer surface of the entire organ. This layer, the **renal capsule,** is also known as the *fibrous tunic* of the kidney.
2. A layer of adipose tissue, the **adipose capsule,** surrounds the renal capsule. This layer can be quite thick, and on dissection the adipose capsule usually obscures the outline of the kidney.
3. Collagen fibers extend outward from the renal capsule through the adipose capsule to a dense outer layer known as the **renal fascia.** The renal fascia anchors the kidney to surrounding structures. Posteriorly the renal fascia fuses with the deep fascia surrounding the muscles of the body wall. Anteriorly the renal fascia forms a thick fibrous layer that fuses with the peritoneum.

External oblique
Colon
Ureter
Stomach
Renal vein
Renal artery
Jejunum
Peritoneum
Pancreas
Aorta
Adipose capsule
Kidney
Vertebra L_1
Spleen
Psoas
Renal capsule
Renal fascia
Quadratus lumborum

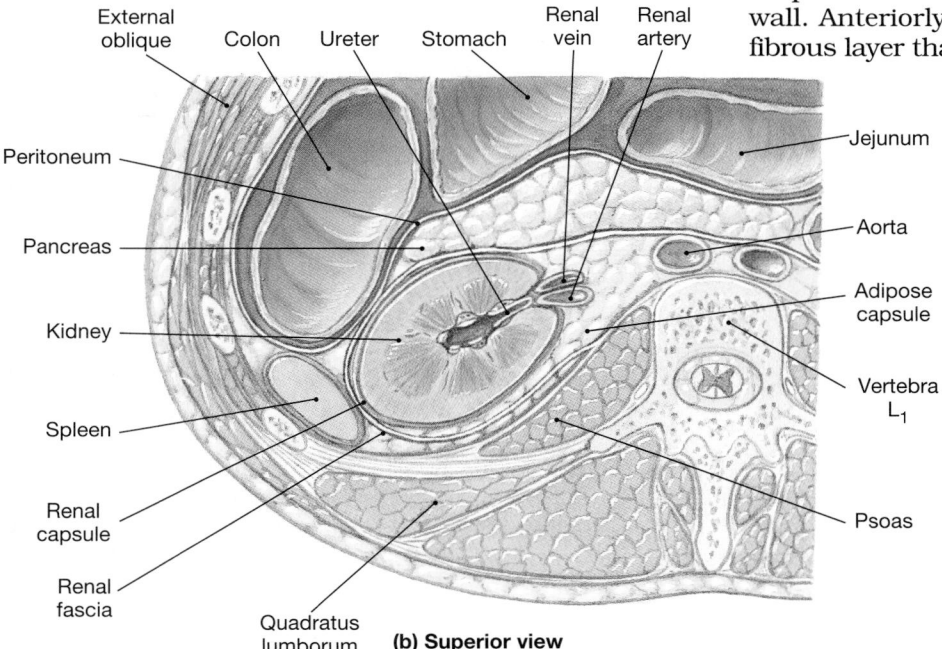

(b) Superior view

● **FIGURE 26-2**
An Introduction to Renal Anatomy. (a) Posterior view of the trunk, showing the positions of the kidneys and other components of the urinary system.
(b) Sectional view at the level indicated in part (a).

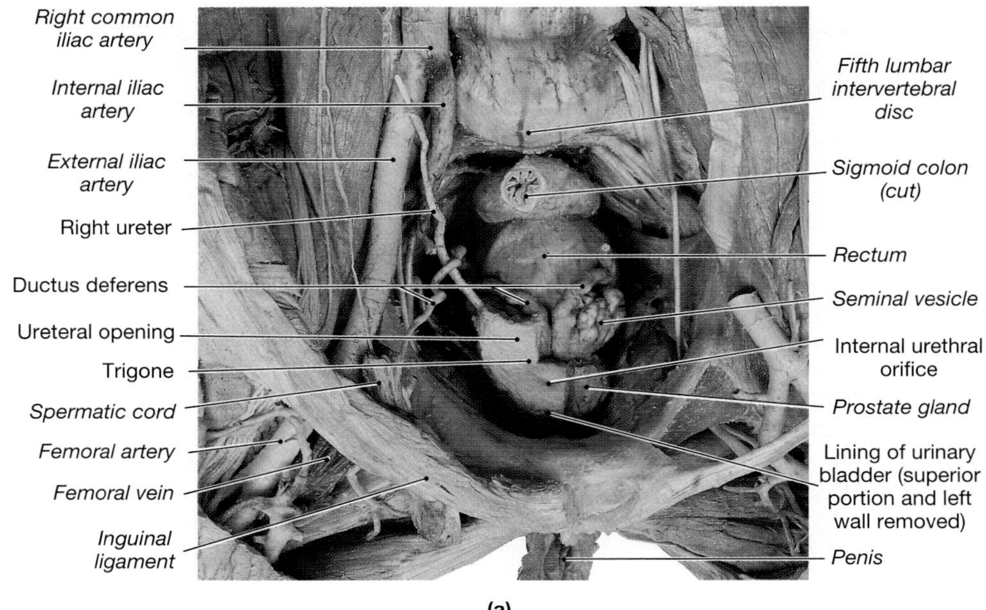
Right common iliac artery
Internal iliac artery
External iliac artery
Right ureter
Ductus deferens
Ureteral opening
Trigone
Spermatic cord
Femoral artery
Femoral vein
Inguinal ligament
Fifth lumbar intervertebral disc
Sigmoid colon (cut)
Rectum
Seminal vesicle
Internal urethral orifice
Prostate gland
Lining of urinary bladder (superior portion and left wall removed)
Penis
(a)

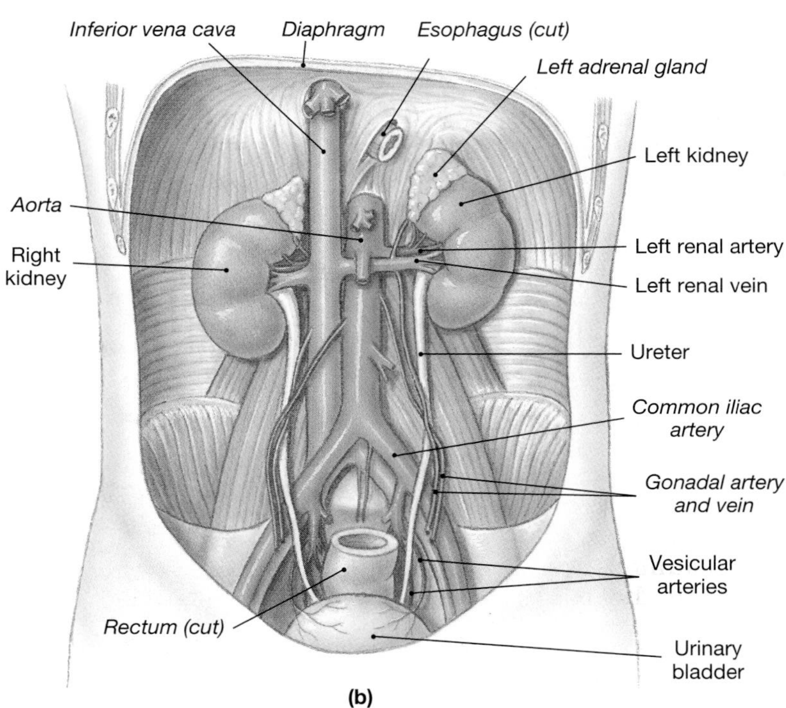
Inferior vena cava
Diaphragm
Esophagus (cut)
Left adrenal gland
Left kidney
Aorta
Right kidney
Left renal artery
Left renal vein
Ureter
Common iliac artery
Gonadal artery and vein
Vesicular arteries
Rectum (cut)
Urinary bladder
(b)

● **FIGURE 26-3**
The Urinary System in Gross Dissection.
(a) Anterior view of the male urogenital system with abdominal organs removed. **(b)** Diagrammatic view of the abdominopelvic cavity showing the kidneys, ureters, urinary bladder, and the blood supply to the urinary structures.

In effect, the kidney hangs suspended by collagen fibers from the renal fascia, packed in a soft cushion of adipose tissue. This arrangement prevents the jolts and shocks of day-to-day existence from disturbing normal kidney function. If the suspensory fibers break or become detached, a slight bump or jar may displace the kidney and stress the attached vessels and ureter. This condition, called a **floating kidney,** can be especially dangerous because the ureters or renal blood vessels may become twisted or kinked during movement.

❑ Superficial Anatomy of the Kidney

Figures 26-3, 26-4

Each brownish-red kidney has the shape of a kidney bean. A typical adult kidney (Figures 26-3 and 26-4●) measures approximately 10 cm (4 in.) in length, 5.5 cm (2.2 in.) in width, and 3 cm (1.2 in.) in thickness. A single kidney averages around 150 g (5.25 oz). A prominent medial indentation, the **hilus,** is the point of entry for the *renal artery* and exit for the *renal vein* and *ureter.*

❑ Sectional Anatomy of the Kidney

Figures 26-4, 26-5

The fibrous renal capsule has inner and outer layers. In sectional view (Figure 26-4a●) the inner layer folds inward at the hilus and lines an internal cavity, the **renal sinus.** Renal blood vessels and the ureter draining the kidney pass through the hilus and branch within the renal sinus. A thickened, external layer of the capsule extends across the hilus and stabilizes the position of these structures.

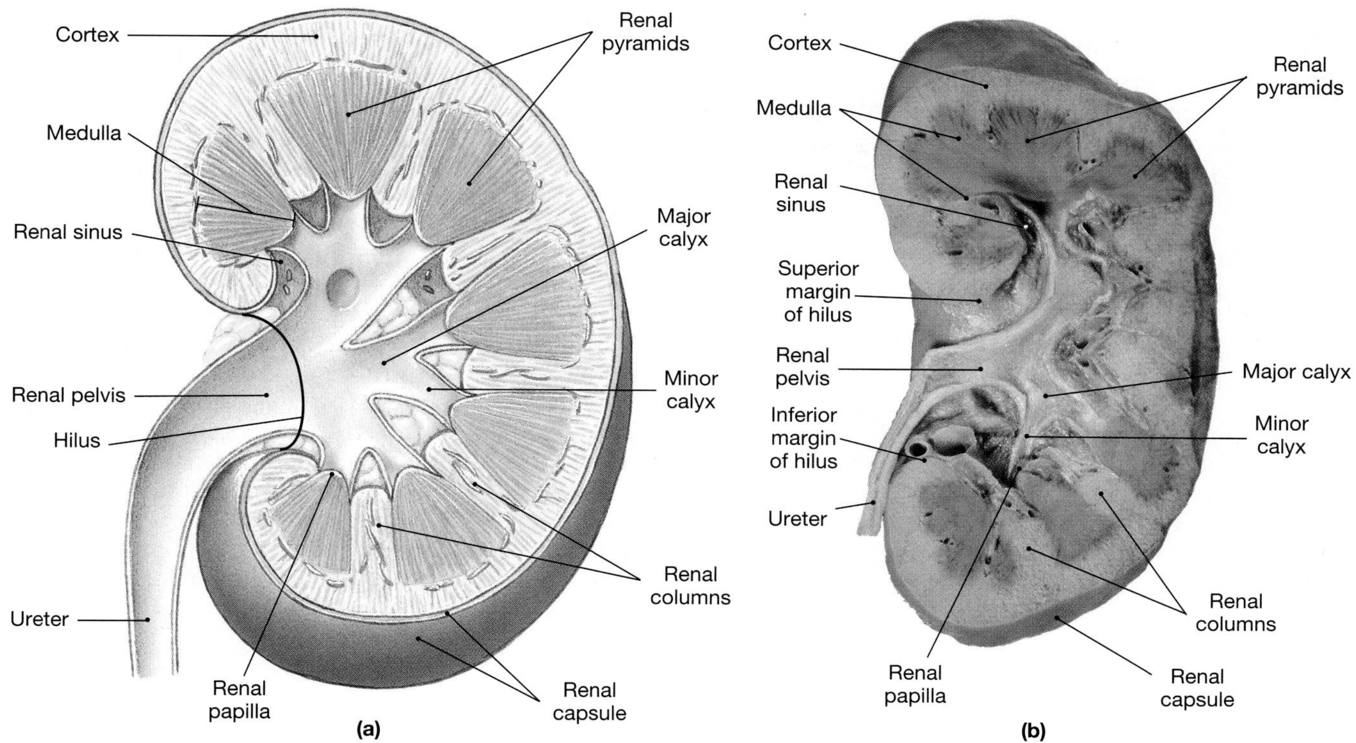
Cortex
Medulla
Renal sinus
Renal pelvis
Hilus
Ureter
Renal papilla
Renal pyramids
Major calyx
Minor calyx
Renal columns
Renal capsule
(a)
Cortex
Medulla
Renal sinus
Superior margin of hilus
Renal pelvis
Inferior margin of hilus
Ureter
Renal papilla
Renal pyramids
Major calyx
Minor calyx
Renal columns
Renal capsule
(b)

● **FIGURE 26-4**
Structure of the Kidney. (a) Frontal section through the left kidney, showing major structures. **(b)** Frontal section of kidney showing internal structures.

The renal **cortex** is the outer layer of the kidney in contact with the capsule. The cortex is reddish-brown and granular in texture. The **medulla** consists of 6 to 18 distinct conical or triangular structures, called **renal pyramids.** The base of each pyramid faces the cortex, and the tip, or **renal papilla,** projects into the renal sinus. Each pyramid has a series of fine grooves that converge at the papilla. Adjacent renal pyramids are separated by bands of cortical tissue, called **renal columns.** The columns have a distinctly granular texture, similar to that of the cortex. A **renal lobe** contains a renal pyramid, the overlying area of renal cortex, and adjacent tissues of the renal columns.

Urine production occurs in the renal lobes, and ducts within each renal papilla discharge urine into a cup-shaped drain, called a **minor calyx** (KĀ-liks). Four or five minor calyces (KĀL-i-sēz) merge to form a **major calyx,** and two of the major calyces combine to form a large, funnel-shaped chamber, the **renal pelvis.** The renal pelvis, which fills most of the renal sinus, is connected to the ureter at the hilus of the kidney.

Urine production begins in the cortex of each renal lobe, at microscopic structures called **nephrons** (NEF-ronz). There are roughly 1.25 million nephrons in each kidney, with a combined length of around 145 km (85 miles). Each nephron consists of a *renal tubule* that is roughly 50 mm long. The tubule has two *convoluted* (coiled or twisted) segments separated by a simple U-shaped tube. The convoluted segments are in the cortex, and the tube extends partially or completely into the medulla. For clarity, the nephrons diagrammed in Figure 26-5 ● have been shortened and straightened out.

❑ The Nephron

Figure 26-5

The renal tubule begins at an expanded chamber, called a *renal corpuscle* (KŌR-pus-ul), that contains a capillary knot, or *glomerulus* (glo-MER-ū-lus) (Figure 26-5 ●). A typical glomerulus consists of around 50 intertwining capillaries. Blood arrives at the glomerulus via the *afferent arteriole* and departs in the *efferent arteriole.* Filtration occurs in the renal corpuscle as blood pressure forces fluid and dissolved solutes out of the glomerular capillaries and into the capsular space. This produces an essentially protein-free solution known as a **filtrate** that is otherwise very similar to blood plasma.

From the renal corpuscle the filtrate enters a long tubular passageway that is subdivided into regions with varied structural and functional characteristics. Major subdivisions include the *proximal convoluted tubule,* the *loop of Henle* (HEN-lē), and the *distal convoluted tubule.* As it travels along the tubule, the filtrate, now called *tubular fluid,* gradually changes its composition. The changes that occur and the characteristics of the urine that results depend on the specialized activities under way in each segment of the nephron.

Each nephron empties into the **collecting system.** A *collecting tubule* connected to the distal convoluted tubule carries the tubular fluid toward a nearby *collecting duct.* The collecting duct, which receives tubular fluid from many different nephrons, leaves the cortex and descends into the medulla, carrying fluid toward a *papillary duct* that drains into a minor calyx.

The urine arriving at the renal pelvis is very different from the filtrate produced at the renal corpuscle. Filtration is a passive process that permits or prevents movement across a barrier solely on the basis of solute size. A filter with pores large enough to permit the passage of organic waste products is unable to prevent the passage of water, ions, and other organic molecules, such as glucose, fatty acids, and amino acids. These useful substances must be reclaimed and the waste products excreted. The segments of the nephron distal to the renal corpuscle are responsible for:

- Reabsorbing all of the useful organic substrates from the filtrate.
- Reabsorbing over 90 percent of the water in the filtrate.
- Secreting into the tubular fluid any waste products that were missed by the filtration process.

Additional water and salts will be removed in the collecting system before the urine is released into the renal sinus. You will find an overview of important information concerning the regions of the nephron and collecting system in Table 26-1.

Nephrons differ slightly in structure, depending on their location. Roughly 85 percent of the nephrons are called **cortical nephrons** because they are found in the superficial cortex of the kid-

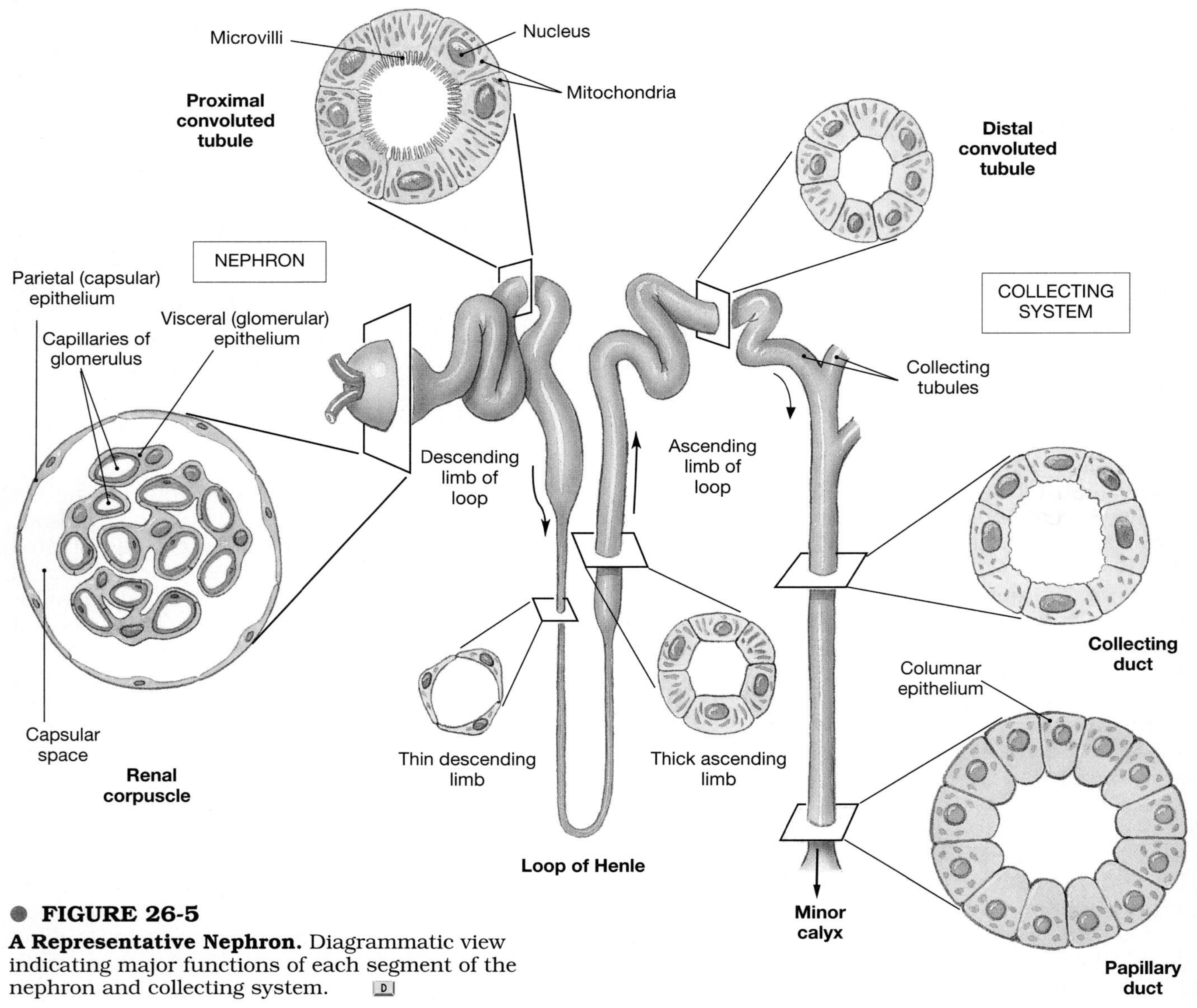

Microvilli
Nucleus
Mitochondria
Proximal convoluted tubule
Distal convoluted tubule
NEPHRON
COLLECTING SYSTEM
Parietal (capsular) epithelium
Visceral (glomerular) epithelium
Capillaries of glomerulus
Collecting tubules
Descending limb of loop
Ascending limb of loop
Collecting duct
Columnar epithelium
Capsular space
Renal corpuscle
Thin descending limb
Thick ascending limb
Loop of Henle
Minor calyx
Papillary duct

● **FIGURE 26-5**
A Representative Nephron. Diagrammatic view indicating major functions of each segment of the nephron and collecting system.

ney. The loops of Henle of cortical nephrons do not extend deeply into the medulla. The remaining 15 percent of nephrons, termed **juxtamedullary** (juks-ta-MED-ū-lar-ē) **nephrons** (*juxta*, near), are located closer to the medulla, and their loops of Henle extend deep into the renal pyramids. Although there are relatively few juxtamedullary nephrons, they play a vital role in the formation and concentration of urine. We will now examine each of the segments of a juxtamedullary nephron in greater detail.

The Renal Corpuscle

Figure 26-6

The renal corpuscle (Figure 26-6a,b,c•) has a diameter averaging 150–250 µm. It includes (1) the capillary knot of the glomerulus and (2) the expanded initial segment of the renal tubule, a region known as **Bowman's capsule.**

THE GLOMERULAR CAPILLARIES The glomerular capillaries are fenestrated capillaries with large-diameter pores (Figure 26-6d•). The openings are small enough to prevent the passage of blood cells, but too large to restrict the diffusion of dissolved or suspended compounds, even those the size of plasma proteins.

The capillaries are surrounded by an elaborate basement membrane known as the **lamina densa.** During filtration, the lamina densa restricts the passage of large plasma proteins, but permits the movement of smaller molecules, including albumin, many organic nutrients, and ions. Unlike basement membranes elsewhere, the lamina densa sometimes encircles two or more capillaries. When it does, **mesangial cells** are found between the capillaries. Mesangial cells have several important functions. They:

- Provide physical support to the capillaries.
- Engulf organic materials that might otherwise clog the filter at the lamina densa.
- Contract or relax to change the diameter of the glomerular capillaries and change the rate of filtration.

BOWMAN'S CAPSULE The glomerulus projects into Bowman's capsule much as the heart projects into the pericardial cavity (Figure 26-6c•). The outer wall of the capsule is lined by a simple squamous **parietal epithelium** (*capsular epithelium*). This layer is continuous with the **visceral epithelium** (*glomerular epithelium*) that covers the glomerular capillaries. The visceral epithelium consists of large cells with complex processes, or "feet," that wrap

TABLE 26-1 The Organization of the Nephron and Collecting System in the Kidney

Region	Length (mm)	Diameter (µm)	Primary Function	Histological Characteristics
Renal corpuscle	Spherical	150–250	Filtration of plasma to initiate urine formation	Glomerulus (capillary knot), mesangial cells, visceral epithelium (podocytes) with capsular space enclosed by parietal epithelium
Proximal convoluted tubule (PCT)	14	60	Reabsorption of ions, organic molecules, vitamins, water	Cuboidal cells with microvilli
Loop of Henle	30	15	Descending limb: reabsorption of water from filtrate	Low cuboidal or squamous
		30	Ascending limb: reabsorption of ions; assists in creation of the medullary concentration gradient	
Distal convoluted tubule (DCT)	5	30–50	Reabsorption of sodium ions; secretion of acids, ammonia, drugs	Cuboidal cells without microvilli
Collecting tubule	Variable	50	Reabsorption of water, sodium ions	Cuboidal cells without microvilli; pale compared with DCT
Collecting duct	15	50–100	Reabsorption of water, sodium ions, bicarbonate	Cuboidal to columnar cells
Papillary duct	5	100–200	Conduction of urine to minor calyx	

around the glomerular capillaries (Figure 26-6c,d,e●). These unusual cells are called **podocytes** (PŌ-do-sīts; *podos*, foot + *-cyte*, cell). Materials filtered out of the blood at the glomerulus must be small enough to pass between the narrow gaps, or *filtration slits*, between adjacent podocyte feet. These slits are small enough to prevent the loss of all but the smallest plasma proteins.

The **capsular space** separates the visceral and parietal epithelia. The connection between the two epithelial layers lies at the **vascular pole** of the renal corpuscle. At the vascular pole blood flows into and out of the glomerular capillaries. Blood arrives via an **afferent arteriole** and departs in an **efferent arteriole.**

The diffusion barriers within the renal corpuscle primarily limit the upper size of the materials gaining access to the capsular space. This kind of size-selection process has functional limitations. In addition to metabolic wastes and excess ions, compounds such as glucose, free fatty acids, amino acids, vitamins, and other solutes enter the capsular space. These potentially useful materials are recaptured before the filtrate leaves the kidneys, with much of the reabsorption occurring in the proximal convoluted tubule.

GLOMERULONEPHRITIS *Glomerulonephritis* (glo-mer-ū-lō-nef-RĪ-tis) is an inflammation of the renal cortex that affects the filtration mechanism. This condition, which may develop after an infection involving *Streptococcus* bacteria, is an example of an *immune complex disorder*, a class of diseases introduced in Chapter 22. [p. 816] The primary infection may not occur in or near the kidneys. However, as the immune system responds to the infection, the number of circulating antigen-antibody complexes skyrockets. These complexes are small enough to pass through the lamina densa but too large to fit between the slit pores. As a result, the filtration mechanism clogs up, and filtrate production declines. Any condition that leads to a massive immune response can cause glomerulonephritis, including viral infections and autoimmune disorders.

The Proximal Convoluted Tubule

Figures 26-5, 26-6, 26-7

The entrance to the **proximal convoluted tubule (PCT)** lies almost directly opposite the vascular pole of the glomerulus, at the **tubular pole** of the renal corpuscle (Figure 26-6c●). The lining of the PCT consists of a simple cuboidal epithelium whose exposed surfaces are blanketed with microvilli (Figures 26-5 and 26-7a●). The cuboidal tubular cells actively absorb organic nutrients, ions, and plasma proteins (if any) from the tubular fluid and release them into the **peritubular fluid,** the interstitial fluid surrounding the renal tubule. As these solutes are absorbed and transported, osmotic forces pull water across the wall of the PCT and into the peritubular fluid. Although reabsorption represents the primary function of the PCT, the epithelial cells can also secrete substances into the lumen.

The Loop of Henle

Figures 26-6a, 26-7c

The proximal convoluted tubule ends at an acute bend that turns the renal tubule toward the medulla. This bend marks the start of the loop of Henle (Figure 26-6a●). The loop of Henle can be divided into a **descending limb** and an **ascending limb.** The descending limb travels in the medulla toward the renal pelvis, and the ascending limb returns toward the cortex. Each limb contains a **thick segment** and a **thin segment.** (The terms *thick* and *thin* refer to the height of the epithelium, not to the diameter of the lumen.)

Thick segments are found closest to the cortex, whereas a thin squamous epithelium lines the deeper medullary portions (Figure 26-7c●). The thick ascending limb, which begins deep in the medulla, contains active transport mechanisms that pump sodium and chloride ions out of the tubular fluid. As a result of these transport activities, the interstitial fluid of the medulla contains an unusually high concentration of solutes.

The Distal Convoluted Tubule

Figures 26-6c, 26-7a

The ascending limb of the loop of Henle ends where it forms a sharp angle that places the tubular wall in close contact with the glomerulus and its accompanying vessels. The **distal convoluted tubule (DCT)** begins at that bend. The initial portion of the DCT crosses the vascular pole of the glomerulus, passing between the afferent and efferent arterioles (Figure 26-6c●).

In sectional view (Figure 26-7a●) the DCT differs from the PCT in that (1) the DCT has a smaller diameter, (2) the epithelial cells of the DCT lack microvilli, and (3) the boundaries between the epithelial cells in the DCT are distinct.

The distal convoluted tubule is an important site for:

- The active secretion of ions, acids, and other materials.
- The selective reabsorption of sodium ions from the tubular fluid.

In the distal portions of the DCT an osmotic flow of water may assist in concentrating the tubular fluid.

THE JUXTAGLOMERULAR APPARATUS The epithelial cells adjacent to the vascular pole are taller than those seen elsewhere along the DCT, and their nuclei are clustered together. This region, detailed in Figure 26-6c●, is called the **macula densa** (MAK-ū-la DEN-sa). The cells of the macu-

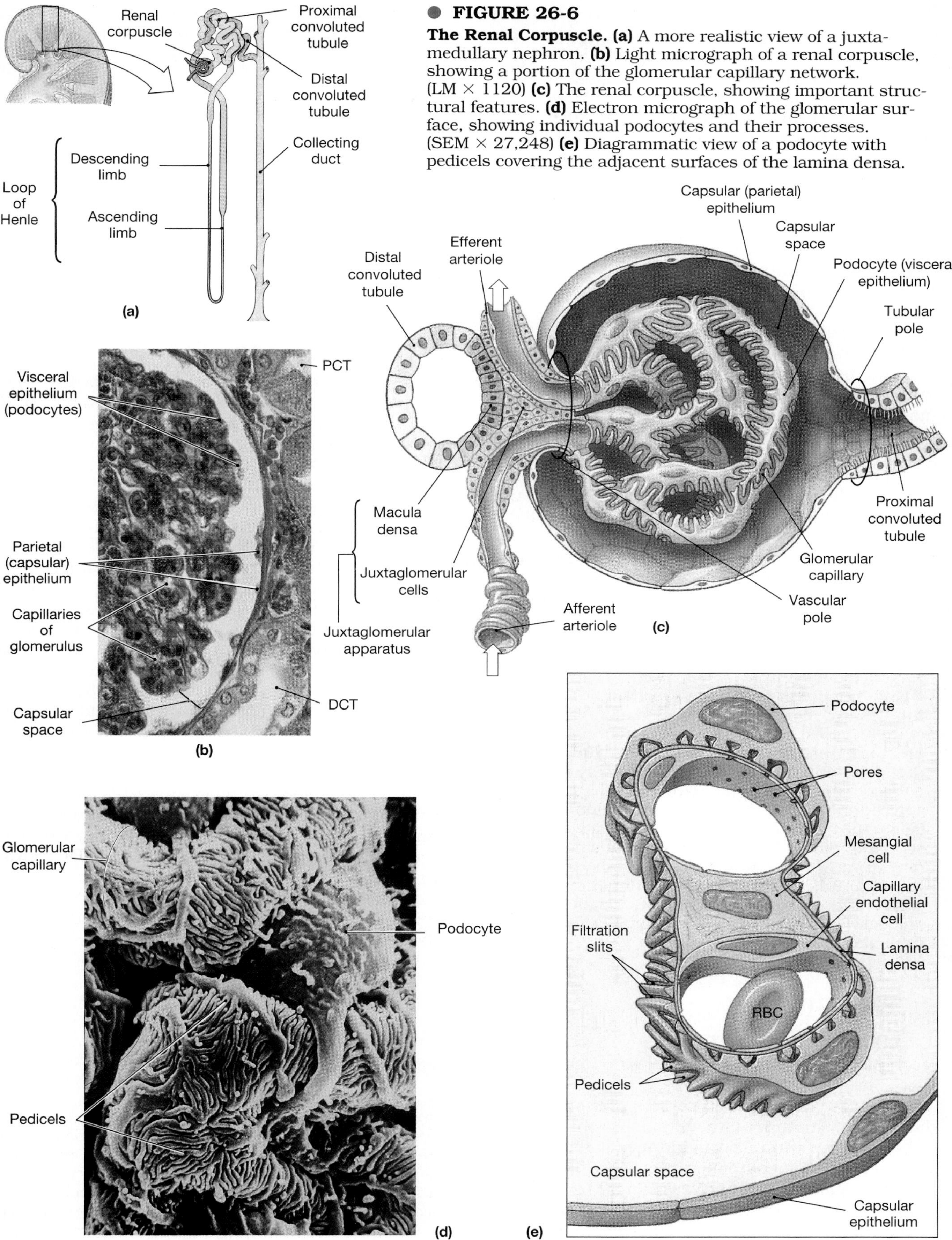
Renal corpuscle
Proximal convoluted tubule
Distal convoluted tubule
Collecting duct
Loop of Henle
Descending limb
Ascending limb
(a)
Visceral epithelium (podocytes)
PCT
Parietal (capsular) epithelium
Capillaries of glomerulus
Capsular space
DCT
(b)
Capsular (parietal) epithelium
Capsular space
Efferent arteriole
Distal convoluted tubule
Podocyte (visceral epithelium)
Tubular pole
Proximal convoluted tubule
Macula densa
Juxtaglomerular cells
Juxtaglomerular apparatus
Glomerular capillary
Vascular pole
Afferent arteriole
(c)
Glomerular capillary
Podocyte
Pedicels
(d)
Podocyte
Pores
Mesangial cell
Capillary endothelial cell
Filtration slits
Lamina densa
RBC
Pedicels
Capsular space
Capsular epithelium
(e)

● **FIGURE 26-6**

The Renal Corpuscle. (a) A more realistic view of a juxtamedullary nephron. **(b)** Light micrograph of a renal corpuscle, showing a portion of the glomerular capillary network. (LM × 1120) **(c)** The renal corpuscle, showing important structural features. **(d)** Electron micrograph of the glomerular surface, showing individual podocytes and their processes. (SEM × 27,248) **(e)** Diagrammatic view of a podocyte with pedicels covering the adjacent surfaces of the lamina densa.

la densa are closely associated with unusual smooth muscle fibers in the wall of the afferent arteriole. These muscle fibers are known as **juxtaglomerular cells** (*juxta*, near). Together the macula densa and juxtaglomerular cells form the **juxtaglomerular apparatus (JGA).** The juxtaglomerular apparatus is an endocrine structure that secretes two hormones, *renin* and *erythropoietin*, that were described in Chapter 18. ∞ [p. 628]

The Collecting System

Figures 26-5, 26-7c

The distal convoluted tubule, the last segment of the nephron, opens into the collecting system. The collecting system consists of *collecting tubules, collecting ducts,* and *papillary ducts* (Figure 26-5•, p. 984). Individual **collecting tubules** connect each nephron to a nearby **collecting duct** (Figure 26-7c,d•). Each collecting duct receives urine from many collecting tubules, and several collecting ducts converge to empty into the larger **papillary duct** that empties into a minor calyx. The epithelium lining the collecting system begins with simple cuboidal cells in the collecting tubules and changes to a columnar epithelium in the collecting and papillary ducts.

In addition to transporting urine from the nephron to the renal pelvis, the collecting system makes final adjustments to its osmotic concentration and volume.

POLYCYSTIC KIDNEY DISEASE *Polycystic* (po-lē-SIS-tik) *kidney disease* is an inherited condition affecting the structure of kidney tubules. Swellings (cysts) develop along the length of the tubules, some growing large enough to compress adjacent nephrons and vessels. Kidney function deteriorates, and the nephrons may eventually become nonfunctional. However, the process is so gradual that serious problems seldom appear before the individual is 30–40 years of age. Common symptoms include sharp pain in the sides, recurrent urinary infections, and the presence of blood in the urine. Treatment is symptomatic, focusing on prevention of infection and reduction of pain with analgesics. In severe cases, hemodialysis or kidney transplantation may be required.

❑ The Blood Supply to the Kidneys

Figures 26-6c, 26-8, 26-9

The kidneys receive 20–25 percent of the total cardiac output. In normal individuals, about 1200 ml of blood flows through the kidneys each minute. That is a phenomenal amount of blood for organs with a combined weight of less than 300 g (10.5 oz)!

Each kidney receives blood from a *renal artery* that originates along the lateral surface of the abdominal aorta near the level of the superior mesenteric artery (Figure 21-26•, p. 756). As the renal artery enters the renal sinus, it provides blood to the **segmental arteries** (Figure 26-8a•). Segmental arteries further divide into **lobar arteries** and a series of **interlobar arteries** that radiate outward through the renal columns between the renal pyramids. The interlobar arteries supply blood to the **arcuate** (AR-kū-āt) **arteries** that arch along the boundary between the cortex and medulla of the kidney. Each arcuate artery gives rise to a number of **interlobular arteries** supplying portions of the adjacent renal lobe. Branching from each interlobular artery are numerous afferent arterioles (Figure 26-8b•).

Blood reaches the vascular pole of each glomerulus through an *afferent arteriole* and leaves in an *efferent arteriole* (Figure 26-6c•). Blood travels from the efferent arteriole to form a capillary plexus, consisting of **peritubular capillaries,** that supplies the proximal and distal convoluted tubules. Each efferent arteriole supplies blood to the peritubular capillaries that form a network around the PCT and DCT (Figure 26-8c•). The peritubular capillaries provide a route for the pickup or delivery of substances that are reabsorbed or secreted by these portions of the nephron.

In juxtamedullary nephrons, the efferent arterioles and peritubular capillaries are connected to a series of long, slender capillaries that accompany the loops of Henle into the medulla (Figure 26-8d•). These capillaries, known as the **vasa recta** (*rectus*, straight), absorb and transport solutes and water reabsorbed into the medulla from tubular fluid in the loops of Henle and collecting ducts. Under normal conditions, the removal of solutes and water by the vasa recta precisely balances the rates of solute and water reabsorption in the medulla.

From the peritubular capillaries and vasa recta, blood enters a network of venules and small veins that converge on the **interlobular veins.** In a mirror image of the arterial distribution, the interlobular veins deliver blood to **arcuate veins** that empty into **interlobar veins.** The interlobar veins drain into the **segmental veins** that merge to form the renal vein. Many of the blood vessels just described are visible in the corrosion casts of the human kidneys shown in Figure 26-9•.

ANALYSIS OF RENAL BLOOD FLOW The rate of blood flow through the kidneys can be monitored by administering a compound, *para-aminohippuric acid,* or *PAH,* that is removed by filtration at the glomeruli and by active secretion along the proximal and distal convoluted tubules. Virtually all of the PAH contained in the blood arriving at the kidneys will be removed before the blood departs in the renal veins. By comparing plasma concentrations of PAH with the amount secreted in the urine, the clinician can calculate the renal blood flow. The *Applications Manual* describes the calculations involved. AM *PAH and the Calculation of Renal Blood Flow*

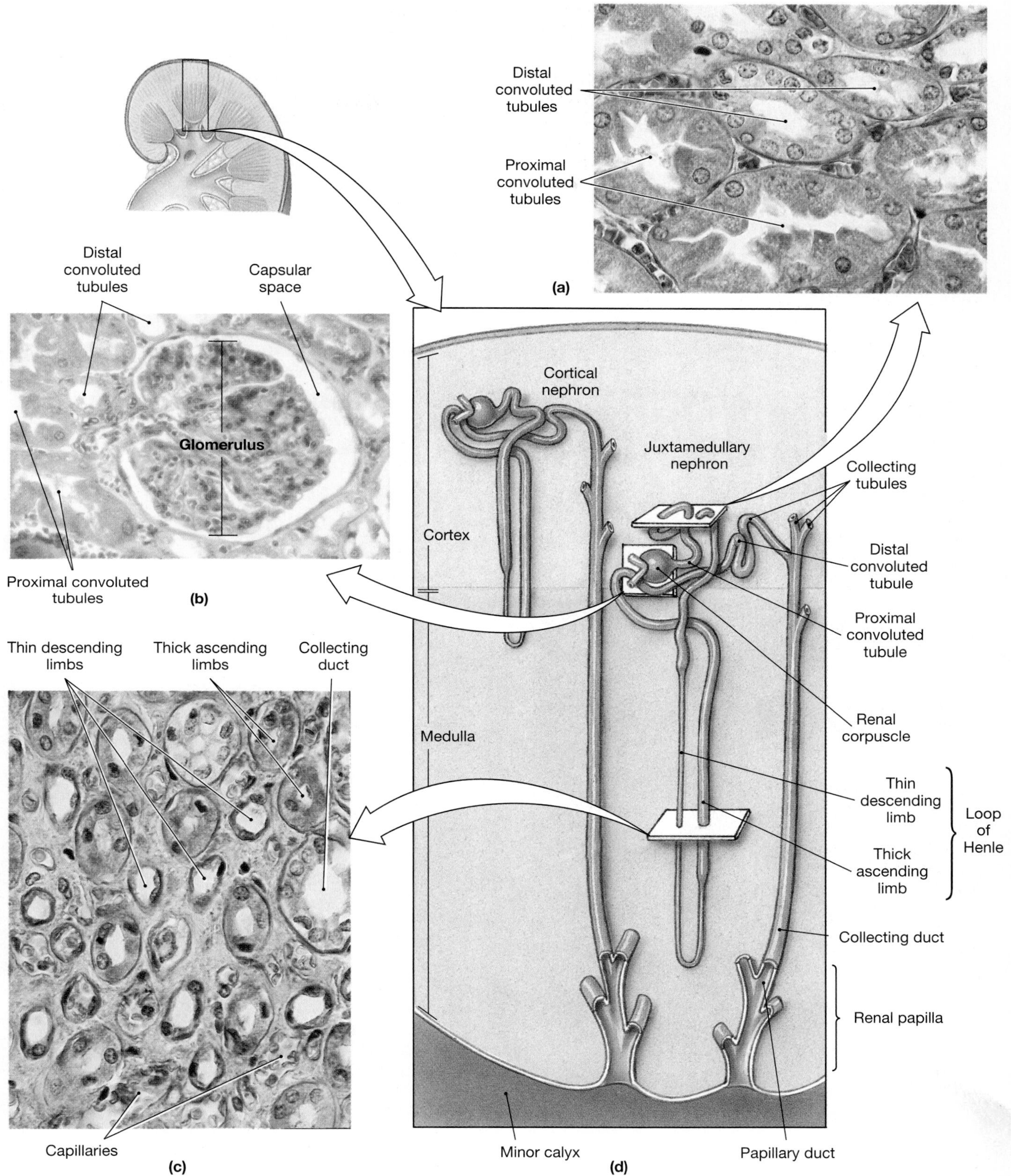
Distal
convoluted
tubules
Proximal
convoluted
tubules
(a)
Distal
convoluted
tubules
Capsular
space
Glomerulus
Proximal convoluted
tubules
(b)
Thin descending
limbs
Thick ascending
limbs
Collecting
duct
Capillaries
(c)
Cortical
nephron
Juxtamedullary
nephron
Cortex
Medulla
Collecting
tubules
Distal
convoluted
tubule
Proximal
convoluted
tubule
Renal
corpuscle
Thin
descending
limb
Thick
ascending
limb
Loop
of
Henle
Collecting duct
Renal papilla
Minor calyx
Papillary duct
(d)

FIGURE 26-7

Sectional Views of the Nephron. (a) Proximal and distal convoluted tubules. (LM × 560) **(b)** The association of proximal and distal convoluted tubules with a glomerulus. (LM × 370) **(c)** Descending and ascending limbs of the loop of Henle and a collecting duct. (LM × 500) **(d)** Approximate locations of the sections in parts (a), (b), and (c).

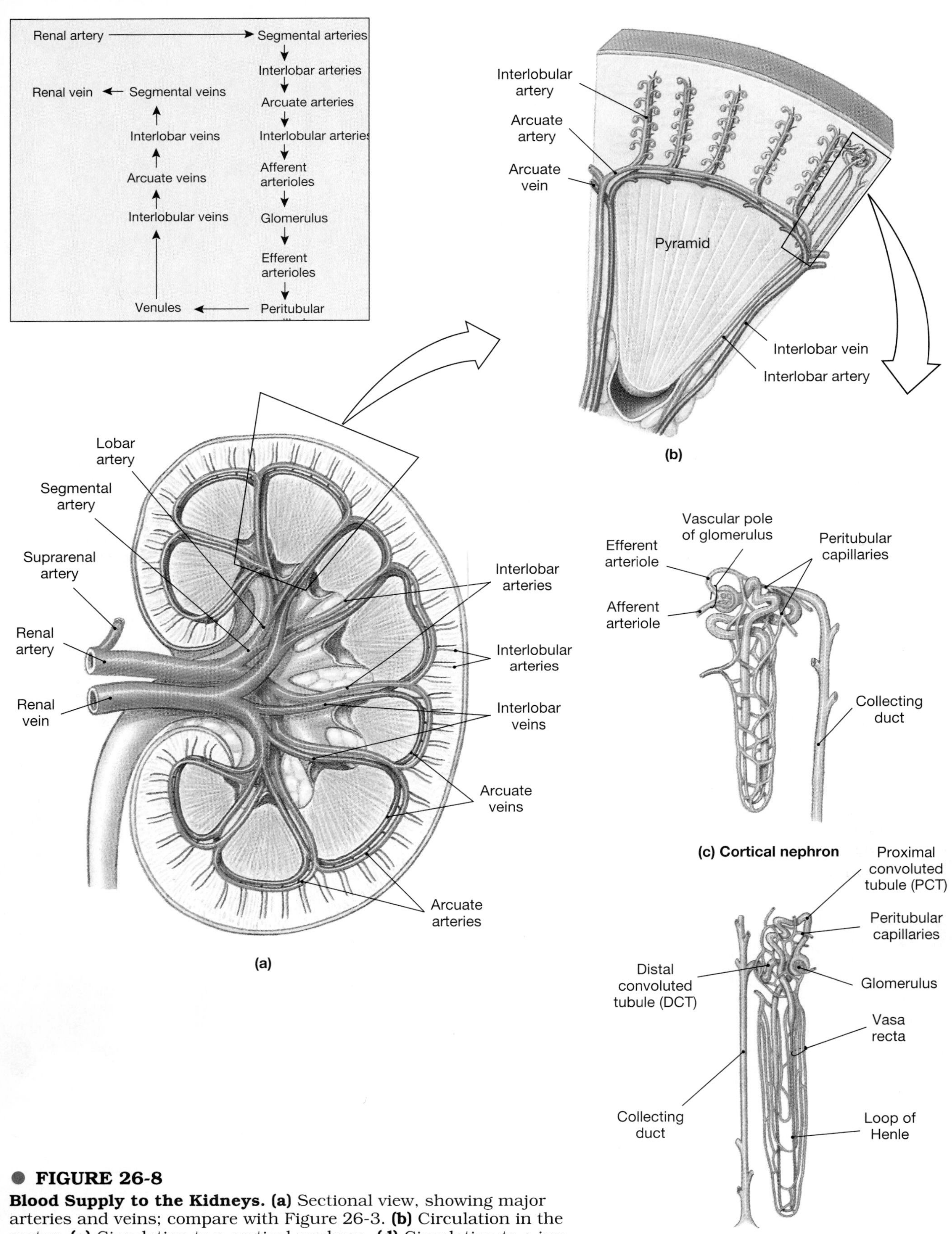
Renal artery
Segmental arteries
Interlobar arteries
Arcuate arteries
Interlobular arteries
Afferent arterioles
Glomerulus
Efferent arterioles
Peritubular
Venules
Interlobular veins
Arcuate veins
Interlobar veins
Segmental veins
Renal vein
Interlobular artery
Arcuate artery
Arcuate vein
Pyramid
Interlobar vein
Interlobar artery
(b)
Lobar artery
Segmental artery
Suprarenal artery
Renal artery
Renal vein
Interlobar arteries
Interlobular arteries
Interlobar veins
Arcuate veins
Arcuate arteries
(a)
Vascular pole of glomerulus
Efferent arteriole
Afferent arteriole
Peritubular capillaries
Collecting duct
(c) Cortical nephron
Proximal convoluted tubule (PCT)
Peritubular capillaries
Distal convoluted tubule (DCT)
Glomerulus
Vasa recta
Collecting duct
Loop of Henle
(d) Juxtamedullary nephron

● **FIGURE 26-8**

Blood Supply to the Kidneys. (a) Sectional view, showing major arteries and veins; compare with Figure 26-3. **(b)** Circulation in the cortex. **(c)** Circulation to a cortical nephron. **(d)** Circulation to a juxtamedullary nephron.

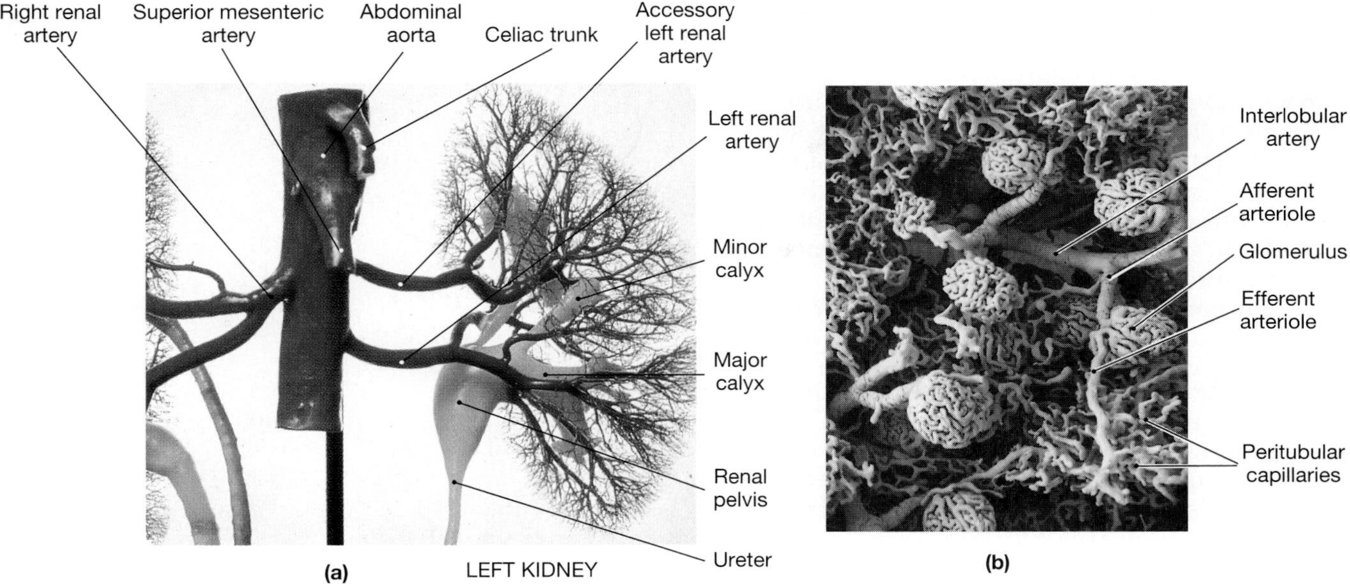
Right renal artery
Superior mesenteric artery
Abdominal aorta
Celiac trunk
Accessory left renal artery
Left renal artery
Minor calyx
Major calyx
Renal pelvis
Ureter
(a)
LEFT KIDNEY
Interlobular artery
Afferent arteriole
Glomerulus
Efferent arteriole
Peritubular capillaries
(b)

● **FIGURE 26-9**

Corrosion Cast of the Blood Supply to the Kidneys. (a) A corrosion cast of the circulation and conducting passageways of the kidneys. This individual had an accessory renal artery delivering blood to the left kidney. Accessory renal arteries are relatively common; they may originate at the aorta or from the common or external iliac arteries. **(b)** The arteriolar and glomerular vessels. (SEM × 75) AM *Scans 13, 14, 15*

Innervation of the Kidneys

The kidneys and ureters are innervated by **renal nerves.** Most of the nerve fibers involved are sympathetic postganglionic fibers from the superior mesenteric ganglion. A renal nerve enters each kidney at the hilus and follows the tributaries of the renal arteries to reach individual nephrons. The sympathetic innervation targets (1) the juxtaglomerular apparatus, (2) the smooth muscles in the walls of the afferent and efferent arterioles, and (3) mesangial cells. Known functions of sympathetic innervation include:

- Regulation of glomerular blood flow and pressure, through control of the diameters of the afferent and efferent arterioles and the glomerular capillaries.
- Stimulation of renin release from the juxtaglomerular apparatus.
- Direct stimulation of water and sodium ion reabsorption.

 What portions of the nephron are located in the renal cortex?

 Why don't plasma proteins pass into the capsular space under normal circumstances?

 Damage to what part of the nephron would interfere with the control of blood pressure?

Renal Physiology

Basic Principles of Urine Formation

The goal in urine production is the maintenance of homeostasis by regulating the volume and composition of the blood. This process involves the excretion and elimination of solutes, specifically metabolic waste products. The most noteworthy organic waste products are:

- ***Urea.*** Urea is the most abundant organic waste, and roughly 21 g of urea are generated each day. Most of it is produced during the breakdown of amino acids.
- ***Creatinine.*** Creatinine is generated in skeletal muscle tissue through the breakdown of *creatine phosphate,* a high-energy compound that plays an important role in muscle contraction. The body generates roughly 1.8 g of creatinine each day, and virtually all of it is excreted in the urine.
- ***Uric acid.*** Approximately 480 mg of uric acid are produced each day during the recycling of nitrogenous bases from RNA molecules.

These waste products must be excreted in solution, so their elimination is accompanied by an unavoidable water loss. The normal kidneys are capable of producing a concentrated urine with an osmotic concentration of 1200–1400 mOsm, more than four times that of plasma.[1] *Without this ability to concentrate the filtrate produced by glomerular filtration, fluid losses would lead to fatal dehydration in hours.* At the same time, the kidneys ensure that the fluid that *is* lost does not contain potentially useful organic substrates, such as sugars or amino acids, that are found in blood plasma. These valuable materials must be reabsorbed and retained for use by other tissues.

To accomplish these goals the kidneys rely on three distinct processes:

1. In **filtration,** blood pressure forces water across a filtration membrane. At the kidneys, the filtration membrane includes the glomerular endothelium, the lamina densa, and the filtration slits. Solute molecules small enough to pass through this filtration complex are carried along by the surrounding water molecules.
2. **Reabsorption** is the removal of water and solute molecules from the filtrate after it enters the renal tubules. Materials reabsorbed from the filtrate are usually nutrients that can be used by the body. Whereas filtration occurs solely on the basis of size, reabsorption is a selective process. Solute reabsorption may involve simple diffusion or the activity of carrier proteins in the tubular epithelium. The reabsorbed substances pass into the peritubular fluid and eventually reenter the blood. Water reabsorption occurs passively, through osmosis.
3. **Secretion** is the transport of solutes from the peritubular fluid, across the tubular epithelium, and into the tubular fluid. Secretion is necessary because filtration does not force all of the dissolved materials out of the plasma. Tubular secretion provides a backup for the filtration process that can further lower the plasma concentration of undesirable materials. Secretion can represent the primary method of excretion for some compounds, including many drugs.

Together these processes create a fluid that is very different from other body fluids. Table 26-2 provides an indication of the efficiency of the renal system by comparing the composition of the urine and plasma and the concentrations of the substances present. The osmotic concentration, or *osmolarity,* of a solution is the total number of solutes in each liter. Osmolarity is usually expressed in terms of **osmoles** or **milliosmoles** per liter. If each liter of a fluid contains a mole of dissolved particles, it is a 1 **osmolar (Osm)** or 1000 **milliosmolar (mOsm)** solution. Our body fluids have an osmotic concentration of about 300 mOsm. By comparison, seawater has an osmolarity of about 1000 mOsm, and

[1]For a review of the methods of reporting solute concentrations, see Chapter 2. [p. 43]

TABLE 26-2 Significant Differences between Urinary and Plasma Solute Concentrations

Component	*Urine*	*Plasma*
Ions (mEq/l)		
Sodium (Na^+)	147.5	138.4
Potassium (K^+)	47.5	4.4
Chloride (Cl^-)	153.3	106
Bicarbonate (HCO_3^-)	1.9	27
Metabolites and Nutrients (mg/dl)		
Glucose	0.009	90
Lipids	0.002	600
Amino acids	0.188	4.2
Proteins	0.000	7.5 g/dl
Nitrogenous Wastes (mg/dl)		
Urea	1800	305
Creatinine	150	8.6
Ammonia	60	0.2
Uric acid	40	3

fresh water about 5 mOsm. (For a more detailed discussion of methods of reporting solute concentrations, consult the *Applications Manual.*) AM *Solutions and Concentrations*

All segments of the nephron and collecting system participate in the process of urine formation. Most regions of the nephron perform a combination of reabsorption and secretion, but the balance between the two shifts from one region to another.

1. Filtration occurs exclusively in the renal corpuscle, across the glomerular walls.
2. Nutrient reabsorption occurs primarily at the proximal convoluted tubule.
3. Active secretion occurs primarily at the distal convoluted tubule.
4. The loop of Henle and the collecting system interact to regulate the amount of water and the number of sodium and potassium ions lost in the urine.

Normal kidney function can continue only as long as the processes of filtration, reabsorption, and secretion function within relatively narrow limits. A disruption in kidney function has immediate effects on the composition of the circulating blood. If both kidneys are affected, death will occur within a few days unless medical assistance is provided.

❑ Filtration

Figure 26-6e

Filtration occurs within the renal corpuscle, as fluids move across the wall of the glomerulus and into the capsular space. The process of **glomeru-**

lar filtration involves passage across three physical barriers (Figure 26-6e●, p. 987):

1. ***The capillary endothelium.*** The glomerular capillaries are *fenestrated capillaries* with pores ranging from 60 to 100 nm (0.06–0.1 μm) in diameter. These openings are small enough to prevent the passage of blood cells, but they are too large to restrict the diffusion of solutes, even those the size of plasma proteins.
2. ***The basement membrane.*** The lamina densa that surrounds the capillary endothelium has several times the density and thickness of a typical basement membrane. This layer restricts the passage of the larger plasma proteins but permits the movement of smaller plasma proteins, nutrients, and ions.
3. ***The filtration slits.*** The filtration slits between the processes of podocytes are only 6–9 nm wide, small enough to block the passage of most of the smaller protein molecules. As a result, under normal circumstances none of the larger plasma proteins and very few albumin molecules (average diameter 7 nm) enter the capsular space. The filtrate contains dissolved ions and small organic molecules in roughly the same concentrations as seen in the plasma.

Filtration Pressures

Figure 26-10

The major forces acting across capillary walls were detailed in Chapters 21 and 22. (You may find it helpful to review Figures 21-12 and 21-13● before proceeding.) *[pp. 734, 735]* The primary factor involved in glomerular filtration is basically the same as that governing fluid and solute movement across capillaries throughout the body: the balance between hydrostatic and colloid osmotic pressures on either side of the capillary walls. These forces are diagrammed in Figure 26-10●.

HYDROSTATIC PRESSURE The *net hydrostatic pressure* tends to drive water and solutes out of the plasma and into the intercapsular space. The net hydrostatic pressure is made up of two opposing components: the *glomerular hydrostatic pressure* and the *capsular hydrostatic pressure.*

- The **glomerular hydrostatic pressure (GHP)** is the blood pressure in the glomerular capillaries. It tends to push water and solute molecules out of the plasma and into the filtrate. GHP is significantly higher than capillary pressures elsewhere in the systemic circuit because of the arrangement of vessels at the glomerulus.

  Blood pressure is low in typical systemic capillaries because the capillary blood flows into the venous system, where resistance is relatively low. However, at the glomerulus

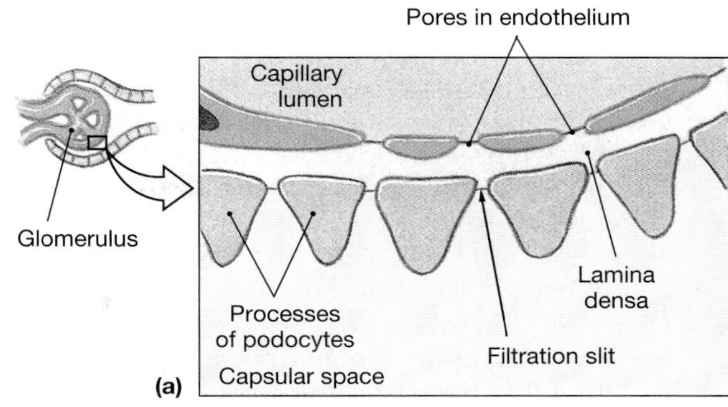

Pores in endothelium
Capillary lumen
Glomerulus
Lamina densa
Processes of podocytes
Capsular space
Filtration slit
(a)

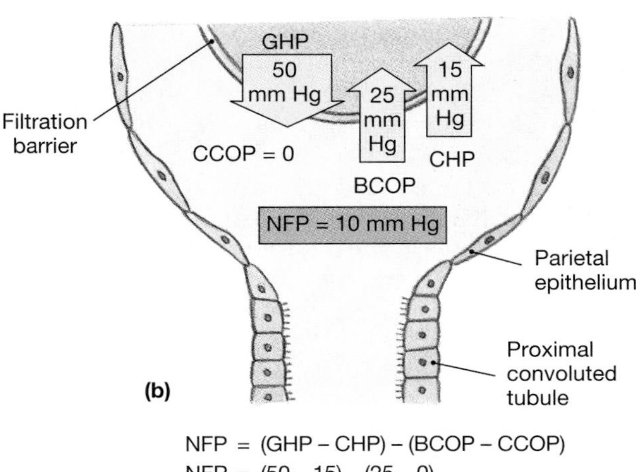

GHP
50 mm Hg
25 mm Hg
15 mm Hg
Filtration barrier
CCOP = 0
CHP
BCOP
NFP = 10 mm Hg
Parietal epithelium
Proximal convoluted tubule
(b)

NFP = (GHP – CHP) – (BCOP – CCOP)
NFP = (50 – 15) – (25 – 0)
NFP = 10 mm Hg

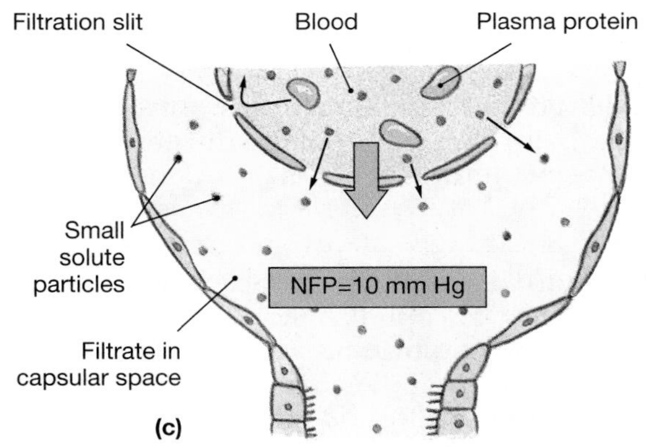

Filtration slit
Blood
Plasma protein
Small solute particles
NFP=10 mm Hg
Filtrate in capsular space
(c)

GHP = Glomerular (blood) hydrostatic pressure
BCOP = Blood colloid osmotic pressure
CHP = Capsular hydrostatic pressure
NFP = Net filtration pressure
CCOP = Capsular colloid osmotic pressure

● **FIGURE 26-10**

Glomerular Filtration. (a) The filtration membrane. **(b)** Forces acting across the glomerular wall. **(c)** The net filtration pressure.

blood leaving the glomerular capillaries flows into an efferent arteriole whose diameter is *smaller* than that of the afferent arteriole. This vessel offers considerable resistance, and relatively high pressures are needed to force blood into it. As a result, glomerular pressures are comparable to those of small arteries, averaging 45–55 mm Hg instead of the 35 mm Hg typical of peripheral capillaries.

- Glomerular hydrostatic pressure is opposed by the **capsular hydrostatic pressure, CHP.** The capsular hydrostatic pressure tends to push water and solutes out of the filtrate and into the plasma. The CHP results from the resistance to flow along the nephron and conducting system. (Before additional filtrate can enter the capsule, some of the filtrate already present must be forced into the proximal convoluted tubule.) The capsular hydrostatic pressure averages around 15 mm Hg.

The net hydrostatic pressure (N_{hp}) is the difference between the glomerular hydrostatic pressure and the capsular hydrostatic pressure:

$$N_{hp} = \text{GHP} - \text{CHP} = 50 \text{ mm Hg} - 15 \text{ mm Hg}$$
$$N_{hp} = 35 \text{ mm Hg}$$

COLLOID OSMOTIC PRESSURE As you may recall from Chapter 21, the colloid osmotic pressure of a solution is the osmotic pressure resulting from the presence of suspended proteins. [p. 734] The *net colloid osmotic pressure* also has two components: the *blood colloid osmotic pressure* and the *capsular colloid osmotic pressure.*

- The **blood colloid osmotic pressure, BCOP,** tends to draw water out of the filtrate and into the plasma. It thus opposes filtration. Over the entire length of the glomerular capillary bed the BCOP averages about 25 mm Hg.
- Under normal conditions, very few plasma proteins enter the capsular space, and the **capsular colloid osmotic pressure (CCOP)** is negligible. However, if the glomeruli are damaged by disease or injury and blood proteins begin passing into the capsular space, the CCOP increases. When present, the CCOP tends to draw water out of the plasma and into the filtrate. It thus promotes filtration.

The net colloid osmotic pressure (N_{cop}) is the difference between the blood colloid osmotic pressure and the capsular colloid osmotic pressure:

$$N_{cop} = \text{BCOP} - \text{CCOP} = 25 \text{ mm Hg} - 0 \text{ mm Hg}$$
$$N_{cop} = 25 \text{ mm Hg}$$

NET FILTRATION PRESSURE The **net filtration pressure** (NFP) at the glomerulus is the difference between the net hydrostatic pressure and the net colloid osmotic pressure. Under normal circumstances, this relationship can be summarized as:

$$\text{NFP} = N_{hp} - N_{cop} = 35 \text{ mm Hg} - 25 \text{ mm Hg}$$
$$\text{NFP} = 10 \text{ mm Hg}$$

This value represents the average pressure forcing water and dissolved materials out of the glomerular capillaries and into the capsular spaces.

Abnormal changes in the net filtration pressure can result in significant alterations in kidney function. For example:

- Damage to the glomerulus that increases the CCOP will enhance the movement of water and solutes into the capsular space.
- Interference with normal fluid movement along the tubules or urine flow along the ureters will elevate CHP and reduce or eliminate the net filtration pressure. AM *Conditions Affecting Filtration*

Glomerular Filtration Rate

The **glomerular filtration rate (GFR)** is the amount of filtrate produced in the kidneys each minute. Each kidney contains around 6 m^2 of filtration surface, and the GFR averages an astounding *125 ml per minute.* This means that almost 20 percent of the fluid delivered to the kidneys by the renal arteries leaves the bloodstream and enters the capsular spaces.

A *creatinine clearance test* is often used to estimate the GFR. Creatinine results from the breakdown of creatine phosphate in muscle tissue, and it is normally excreted in the urine. Creatinine enters the filtrate at the glomerulus, and it is not reabsorbed in significant amounts. By monitoring the concentrations in the blood and amount excreted in the urine, one can easily estimate the GFR. Sample calculations are provided in the *Applications Manual.* AM *Clinical Determination of the GFR*

In the course of a single day the glomeruli generate about 180 liters (50 gal) of filtrate, roughly 70 times the total plasma volume. But as the filtrate passes through the renal tubules, about 99 percent of it is reabsorbed. The significance of tubular reabsorption should now be apparent!

BLOOD PRESSURE AND GLOMERULAR FILTRATION The glomerular filtration rate depends on the filtration pressure acting across the glomerular capillaries. Any factor that alters the filtration pressure will therefore alter the GFR and affect kidney function. One of the most significant factors is a decline in renal blood pressure.

Despite the large volume of filtrate generated, the filtration pressure is relatively low. Whenever the mean arterial pressure falls by 10 percent (from a normal value of about 100 mm Hg to 90 mm Hg), the GFR is severely restricted. If blood pressure at the glomeruli drops by 20 percent, from 50 mm Hg to 40 mm Hg, kidney filtration will cease altogether, because the net filtration pressure will be 0 mm Hg. The kidneys are therefore sensitive to changes in blood pressure that have little or no effect on other organs. Hemorrhaging, shock, and dehydration are relatively common clinical conditions that can cause a dangerous decline in the GFR.

Controlling the GFR

Figure 26-11

Glomerular filtration is the vital first step essential to all other kidney functions. If filtration does not occur, waste products are not excreted, pH control is jeopardized, and an important mechanism for blood volume regulation is eliminated. It should be no surprise to find that a variety of regulatory mechanisms exist to ensure that the GFR remains within normal limits.

Filtration depends on adequate circulation to the glomerulus and the maintenance of normal filtration pressures. There are three interacting levels of control that act to stabilize the GFR: (1) *autoregulation* at the local level, (2) *hormonal regulation* initiated by the kidneys, and (3) *autonomic regulation*, primarily by the sympathetic division of the ANS.

AUTOREGULATION OF THE GFR The nephron is capable of a considerable amount of autoregulation through changes in the diameter of the afferent arterioles, the efferent arterioles, and the glomerular capillaries. These adjustments occur in response to changes in local blood pressure and blood flow. For example, a reduction in blood flow and a decline in glomerular pressure trigger (1) the dilation of the afferent arteriole, (2) the relaxation of mesangial cells and dilation of the glomerular capillaries, and (3) the constriction of the efferent arteriole. This combination increases blood flow and elevates glomerular blood pressure to normal levels. As a result, filtration rates remain relatively constant despite a general decline in systemic arterial pressures.

HORMONAL REGULATION OF THE GFR The GFR is regulated by the hormones *renin* and *ANP*. These hormones and their actions were introduced in Chapters 18 and 21. ∞ *[pp. 628–629, 743]*

The JGA and Renin Secretion. Renin release occurs when (1) glomerular blood pressure declines or (2) the osmolarity of the tubular fluid reaching the DCT decreases.

Renin is released by the juxtaglomerular apparatus (JGA) when renal blood flow declines because of a decrease in blood volume, a fall in systemic pressures, or a blockage in the renal artery or its tributaries. The functions of renin and the renin-angiotensin system were introduced in Chapter 18. ∞ *[p. 628]* Renin converts the inactive protein angiotensinogen to angiotensin I, and a converting enzyme in the capillaries of the lungs subsequently converts this to the active form, angiotensin II. Figure 26-11• diagrams the primary effects of this hormone. These include:

1. ***In peripheral capillary beds,*** angiotensin II causes a brief but powerful vasoconstriction of arterioles and precapillary sphincters, elevating pressures in the renal arteries and their tributaries.
2. ***At the nephron,*** angiotensin II causes the constriction of the efferent arteriole, further elevating glomerular pressures and filtration rates. It also (1) directly stimulates the reabsorption of sodium ions and water at the proximal convoluted tubule and (2) indirectly stimulates the reabsorption of sodium by the DCT and collecting system by triggering aldosterone secretion.
3. ***In the CNS,*** angiotensin II triggers the release of ADH, stimulating the reabsorption of water in the distal portion of the DCT and the collecting system, and causing the sensation of thirst.
4. ***At the adrenal gland,*** angiotensin II stimulates the secretion of aldosterone by the cortex and epinephrine by the medulla. The aldosterone accelerates sodium reabsorption in the DCT and cortical portion of the collecting system. The heart responds to the epinephrine by increasing its rate and force of contraction, elevating systemic (and thus renal) blood pressure.

Renin is also released in response to a decline in the osmolarity of the tubular fluid reaching the JGA. When the rate of glomerular filtration declines, so does the rate of fluid movement along the nephron. Because the tubular fluid then spends more time in the ascending limb of the loop of Henle, the concentration of sodium and chloride ions becomes abnormally low. When the osmolarity of the tubular fluid at the JGA decreases, the cells of the macula densa release renin. The final results are an increase in systemic blood volume and blood pressure and the restoration of normal glomerular filtration rates.

ANP and the GFR. Atrial natriuretic peptide (ANP) is released in response to stretching of the atrial walls of the heart by increased blood volume or blood pressure. Among its other effects, ANP triggers dilation of the afferent arteriole and constriction of the efferent arteriole. This effect elevates glomerular pressures and increases the GFR.

AUTONOMIC REGULATION OF THE GFR Most of the autonomic innervation of the kidneys consists of sympathetic postganglionic fibers. (The role of the few parasympathetic fibers in regulating kidney function is not known.) Sympathetic activation has one direct

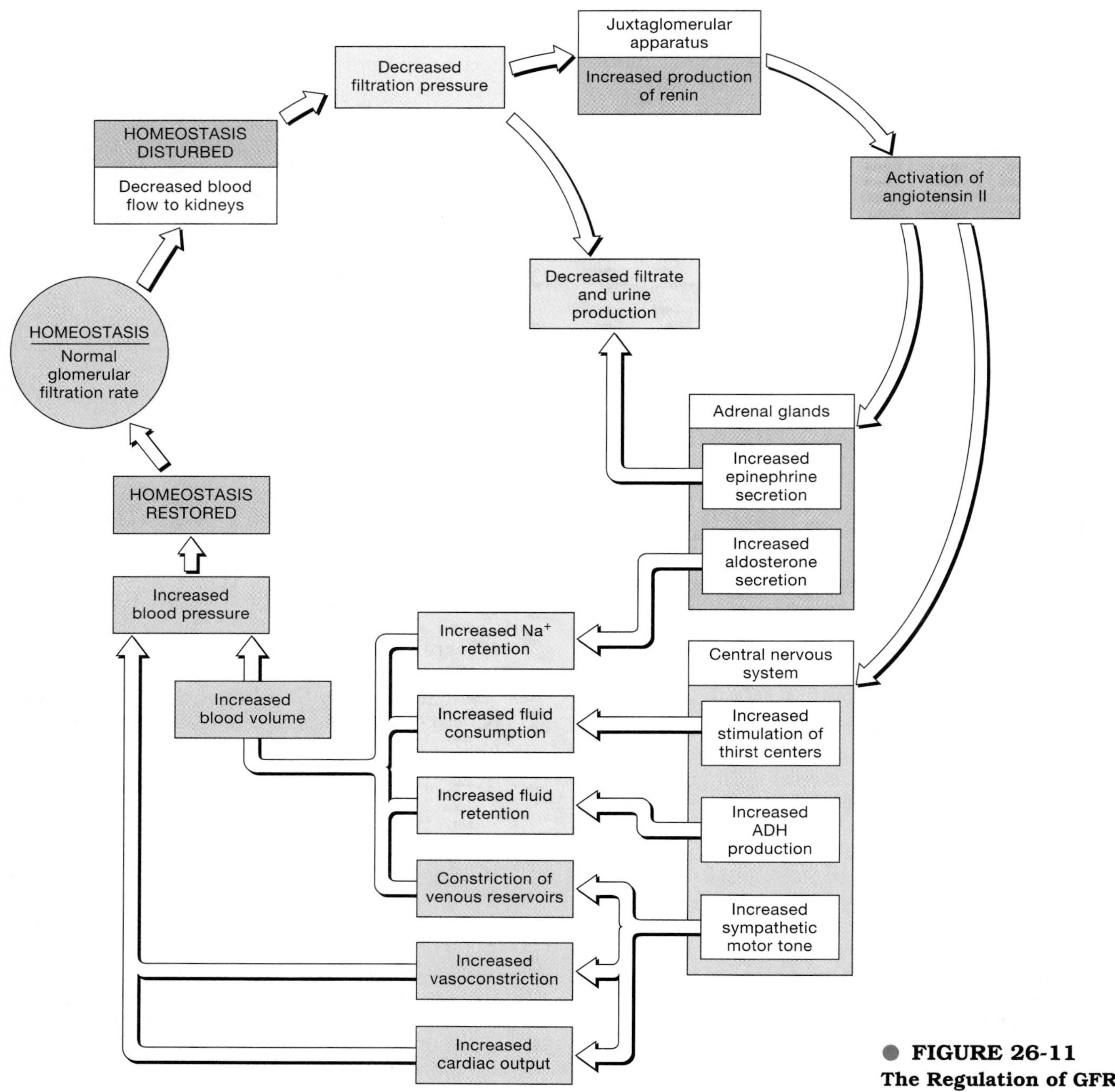
Juxtaglomerular apparatus
Increased production of renin
Decreased filtration pressure
HOMEOSTASIS DISTURBED
Decreased blood flow to kidneys
Activation of angiotensin II
Decreased filtrate and urine production
HOMEOSTASIS
Normal glomerular filtration rate
Adrenal glands
Increased epinephrine secretion
Increased aldosterone secretion
HOMEOSTASIS RESTORED
Increased blood pressure
Increased Na+ retention
Central nervous system
Increased blood volume
Increased fluid consumption
Increased stimulation of thirst centers
Increased fluid retention
Increased ADH production
Constriction of venous reservoirs
Increased sympathetic motor tone
Increased vasoconstriction
Increased cardiac output

FIGURE 26-11
The Regulation of GFR

effect on the GFR: It produces a powerful vasoconstriction of the afferent arterioles, decreasing the GFR and slowing the production of filtrate. The sympathetic activation triggered by an acute fall in blood pressure or a heart attack will override the local regulatory mechanisms that act to stabilize the GFR. As the crisis passes and sympathetic tone decreases, the filtration rate gradually returns to normal levels. Sustained sympathetic stimulation at high levels may lead to renal ischemia and kidney damage.

Widespread sympathetic activation also has an indirect effect on GFR because the epinephrine released by the adrenal medulla stimulates the secretion of renin by the juxtaglomerular apparatus. Renin release leads to angiotensin II activation, and angiotensin II causes constriction of the efferent arteriole and elevation of the GFR.

The normal activities of the sympathetic division also have an indirect effect on GFR. When the regional pattern of blood circulation changes, it alters the blood flow to the kidneys. For example, the dilation of superficial vessels in warm weather shunts blood away from the kidneys, and glomerular filtration declines temporarily. The effect becomes especially pronounced during periods of strenuous exercise. As the blood flow to the skin and skeletal muscles increases, kidney perfusion gradually declines. These changes may be opposed, with variable success, by autoregulation at the local level.

At maximal levels of exertion, renal blood flow may be less than 25 percent of normal resting levels. This reduction can create problems for distance swimmers and marathon runners because of the buildup of metabolic wastes over the course of a long competition. *Proteinuria* (protein loss in the urine) often occurs after such events, because the glomerular cells have been injured by prolonged hypoxia (low oxygen levels). If the damage is substantial, *hematuria* (blood loss in the urine) will occur. Hematuria develops in roughly 18 percent of marathon runners. Proteinuria and hematuria usually disappear within 48 hours, as the glomerular tissues are repaired. However, a small number of marathon and ultramarathon runners experience *renal failure*, with permanent impairment of kidney function.

RENAL FAILURE **Renal failure** occurs when the kidneys become unable to perform the excretory functions needed to maintain homeostasis. When kidney filtration slows for any reason, urine production declines. As the decline continues, symptoms of renal failure appear because of the retention of water, ions, and metabolic wastes. Virtually all systems in the body are affected. For example, there are disturbances in fluid balance, pH, muscular contraction, metabolism, and digestive function. The individual usually becomes hypertensive, anemia develops because of a decline in erythropoietin production, and central nervous system problems may lead to sleeplessness, seizures, delirium, and even coma.

Acute renal failure occurs when filtration suddenly slows or stops altogether because of exposure to toxic drugs, renal ischemia, urinary obstruction, or trauma. Sensitized individuals may also develop acute renal failure after exposure to antibiotics or anesthetics. In these cases recovery may occur if the individual survives the incident. In *chronic renal failure* kidney function deteriorates gradually, and the associated problems accumulate over time. The condition usually cannot be reversed, only delayed, and symptoms of acute renal failure eventually develop. Chronic and acute renal failure may be treated by kidney transplantation or the use of dialysis equipment, as discussed later in the chapter.

❑ Reabsorption and Secretion

Reabsorption and secretion at the kidneys involve a combination of diffusion, osmosis, and carrier-mediated transport. Diffusion and osmosis have been considered in many different chapters, so we will pause at this time only for a brief review of carrier-mediated transport mechanisms.

Carrier-Mediated Transport

Previous chapters have considered four major types of carrier-mediated transport: *facilitated diffusion, active transport, cotransport,* and *countertransport.*

1. ***Facilitated Diffusion.*** In facilitated diffusion (Figure 3-10•, p. 78) a carrier protein transports a molecule across the cell membrane without an expenditure of energy. The transport always follows the concentration gradient for the ion or molecule transported. Facilitated diffusion is important in the reabsorption of glucose and amino acids when their concentrations in the tubular fluid are relatively high.
2. ***Active Transport.*** Active transport is driven by the hydrolysis of ATP to ADP on the inner membrane surface. The sodium-potassium exchange pump (Figure 3-11•, p. 79), which ejects three intracellular sodium ions in exchange for two extracellular potassium ions, is a familiar example. Exchange pumps are just one of several types of ion pumps active along the kidney tubules. Many other carrier proteins transport only one type of ion across the membrane. Examples include the active transport mechanisms for calcium, magnesium, chloride, iodide, and iron. Active transport mechanisms can operate despite an opposing concentration gradient.
3. ***Cotransport.*** In cotransport, carrier protein activity is not directly linked to the hydrolysis of ATP. In this process, two substrates (ions, molecules, or both) cross the membrane while bound to the carrier protein. The movement always occurs following the concentration gradient *for at least one of the transported substances.* In most instances, a sodium ion is one of the substrates transported. This fact is significant, because there is always a concentration gradient for sodium ions across the cell membrane. This concentration gradient then "drives" the transport of some other ion or molecule. Cotransport mechanisms are responsible for the reabsorption of glucose and other simple sugars, amino acids, lactic acid, phosphate, chloride ions, and hydrogen ions from the tubular fluid.
4. ***Countertransport.*** Countertransport resembles cotransport in all respects except that the two transported ions move in opposite directions.

- Most cells in the body use sodium-calcium countertransport to keep intracellular calcium concentrations low. Along the kidney tubules, carriers on the epithelial surface bring calcium ions *into* the cell, while those on the surface attached to the basement membrane pump them into the peritubular fluid. Parathyroid hormone and calcitriol stimulate these carrier proteins and reduce the urinary loss of calcium ions.

- The DCT and collecting system contain countertransport mechanisms that exchange chloride ions, Cl^-, for bicarbonate ions, HCO_3^-. Bicarbonate ions are important buffers, and their reabsorption can help prevent a decline in blood pH.
- Hydrogen ion secretion along the DCT and collecting system occurs via a countertransport mechanism that reabsorbs sodium ions from the tubular fluid.

CHARACTERISTICS OF CARRIER-MEDIATED TRANSPORT All four carrier-mediated processes share features that are important for an understanding of kidney function:

1. ***In carrier-mediated transport, a specific substrate binds to a carrier protein that facilitates its movement across the membrane.***
2. ***A given carrier protein normally works in one direction only.*** In facilitated diffusion, that direction is determined by the concentration gradient of the substance being transported. In active transport, cotransport, and countertransport, the location and orientation of the carrier proteins determine whether a particular substance will be reabsorbed or secreted. For example, the carrier protein that transports amino acids from the tubular fluid to the cytoplasm will not carry amino acids back into the tubular fluid.
3. ***The distribution of carrier proteins can vary from one portion of the cell surface to another.*** Transport between the tubular fluid and the interstitial fluid involves two steps, because the material must cross both the outer (lumenal) and inner (attached) surfaces of the tubular cells. For example, in the case of amino acids, glucose, or other nutrients, carrier proteins facing the basement membrane accept substrates from the cytoplasm and deliver them to the interstitial fluid.
4. ***The membrane of a single tubular cell contains more than one type of carrier protein.*** Each cell can have multiple functions, and the same cell that reabsorbs one compound can secrete another.
5. ***Carrier proteins, like enzymes, have saturation limits.*** For any substance, the concentration at saturation is called the **transport maximum** (T_m), or *tubular maximum*. Although both carriers involved with reabsorption and those responsible for secretion have saturation limits, the concentration is most often applied to tubular reabsorption.

THE T_m AND RENAL THRESHOLD Normally any plasma proteins and nutrients, such as amino acids and glucose, are removed via active transport, cotransport, or facilitated diffusion. If the concentrations of these nutrients in the tubular fluid rise, the rates of reabsorption increase until the carrier saturation limits are reached. *A concentration higher than the tubular maximum will exceed the reabsorptive abilities of the nephron, and some of the material will remain in the tubular fluid and appear in the urine.* The transport maximum thus determines the **renal threshold,** the plasma concentration at which a specific compound or ion will begin appearing in the urine.

The renal threshold varies, depending on the substance involved. The renal threshold for glucose is approximately 180 mg/dl. When plasma glucose concentrations remain higher than this level, glucose concentrations in the tubular fluid will exceed the T_m of the tubular cells, and glucose will appear in the urine. The presence of glucose in the urine is a condition called *glycosuria.* After a meal rich in carbohydrates, plasma glucose levels may exceed 180 mg/dl, but the liver quickly lowers circulating glucose levels, and very little glucose is lost in the urine. Chronically elevated plasma and urinary glucose concentrations are very unusual. (Glycosuria is one of the key symptoms of *diabetes mellitus,* a condition detailed in Chapter 18. [p. 634])

The renal threshold for amino acids is lower than that for glucose, and amino acids appear in the urine when plasma concentrations exceed 65 mg/dl. Plasma levels often reach these levels after a protein-rich meal, causing some amino acids to appear in the urine. This condition is termed *aminoaciduria.*

The T_m values for water-soluble vitamins are relatively low, and as a result excess quantities of these vitamins are excreted in the urine. (This is often the fate of daily water-soluble vitamin supplements.) A number of other compounds in the tubular fluid are completely ignored by the tubular cells of the nephron. As water and other compounds are removed, the concentration of those materials gradually rises. Table 26-3 contains a partial listing of substances actively reabsorbed, secreted, or ignored by the renal tubules.

We will now proceed along the nephron and consider the changes in the composition and concentration of the filtrate produced at the glomerulus.

The Proximal Convoluted Tubule

Figure 26-12

The cells of the proximal convoluted tubule (PCT) normally reabsorb 60–70 percent of the volume of filtrate produced in the renal corpuscle. The reabsorbed materials enter the peritubular fluid and diffuse into the peritubular capillaries.

TABLE 26-3 Tubular Absorption and Secretion

Reabsorbed	*Secreted*	*No Active Transport*
Ions	**Ions**	Urea
Na^+, Cl^-, K^+, Ca^{2+}, Mg^{2+}, SO_4^{2-}, HCO_3^-	K^+, H^+, Ca^{2+}, PO_4^{3-}	Water
	Wastes	Urobilinogen
	Creatinine	Bilirubin
	Ammonia	
Metabolites	Organic acids and bases	
Glucose		
Amino acids		
Proteins	**Miscellaneous**	
Vitamins	Neurotransmitters (ACh, NE, E, dopamine)	
	Histamine	
	Drugs (penicillin, atropine, morphine, numerous others)	

The proximal convoluted tubule has five major functions:

1. ***Organic nutrient reabsorption.*** Under normal circumstances virtually all (more than 99 percent) of the glucose, amino acids, and other organic nutrients are reabsorbed before the tubular fluid leaves the PCT. This reabsorption involves a combination of facilitated transport and cotransport mechanisms.
2. ***Active ion reabsorption.*** The PCT actively transports ions, including sodium, potassium, calcium, magnesium, bicarbonate, phosphate, and sulfate ions. The ion pumps involved are individually regulated and may be influenced by circulating ion or hormone levels. For example, the presence of parathyroid hormone stimulates calcium ion reabsorption, and angiotensin II stimulates sodium ion reabsorption. The PCT recaptures roughly 90 percent of the bicarbonate ions in the tubular fluid indirectly, by absorbing carbon dioxide (Figure 26-12•). Bicarbonate is important in stabilizing blood pH, a process examined further in Chapter 27.
3. ***Water reabsorption.*** The reabsorptive processes have a direct effect on the solute concentrations inside and outside of the tubules. The filtrate entering the PCT has the same osmotic concentration as the peritubular fluid. As transport activities proceed, the solute concentration of the tubular fluid decreases and that of the peritubular fluid and adjacent capillaries increases. Osmotic forces then draw water out of the tubular fluid and into the peritubular fluid. In this way the PCT reabsorbs around 108 liters of water a day.
4. ***Passive ion reabsorption.*** As active ion reabsorption occurs and water leaves the tubular fluid by osmosis, the concentration of other solutes in the tubular fluid increases above that found in the peritubular fluid. If the tubular cells are permeable to these solutes, they will move through passive diffusion across the tubular cells and into the peritubular fluid. Urea, chloride ions, and lipid-soluble materials may diffuse out of the PCT in this way. As this diffusion occurs, it further reduces the solute concentration of the tubular fluid and promotes additional water reabsorption.
5. ***Secretion.*** A relatively small amount of active secretion also occurs along the PCT. The mechanisms involved resemble those of the distal convoluted tubule, which are discussed below.

Sodium ion reabsorption (Figure 26-12•) plays an important role in these processes. These ions may enter the tubular cells (1) by diffusion through sodium leak channels, (2) in the sodium-linked cotransport of glucose, amino acids, or other organic molecules, and (3) through countertransport for hydrogen ions. Once inside the tubular cell, the sodium ions diffuse toward the basement membrane. The cell membrane in this region contains sodium-potassium exchange pumps that eject sodium ions in exchange for extracellular potassium ions. The ejected sodium ions then diffuse into the adjacent peritubular capillaries. Most of the potassium ions reclaimed by the cell will diffuse back out of the cell and into the peritubular fluid through potassium leak channels.

The absorption of ions and compounds along the PCT involves many different carrier proteins. Some individuals have an inherited inability to manufacture one or more of these carrier proteins and are therefore unable to recover specific solutes from the tubular fluid. For example, in *renal glycosuria* (glī-cō-SŪ-rē-a) a defective carrier protein makes it impossible for the PCT to reabsorb glucose from the tubular fluid. AM *Inherited Problems with Tubular Function*

How does the reabsorption of calcium ions by the kidney affect the sodium ion concentration of the filtrate?

What occurs when the plasma concentration of a substance exceeds its tubular maximum?

How would a decrease in blood pressure affect the glomerular filtration rate?

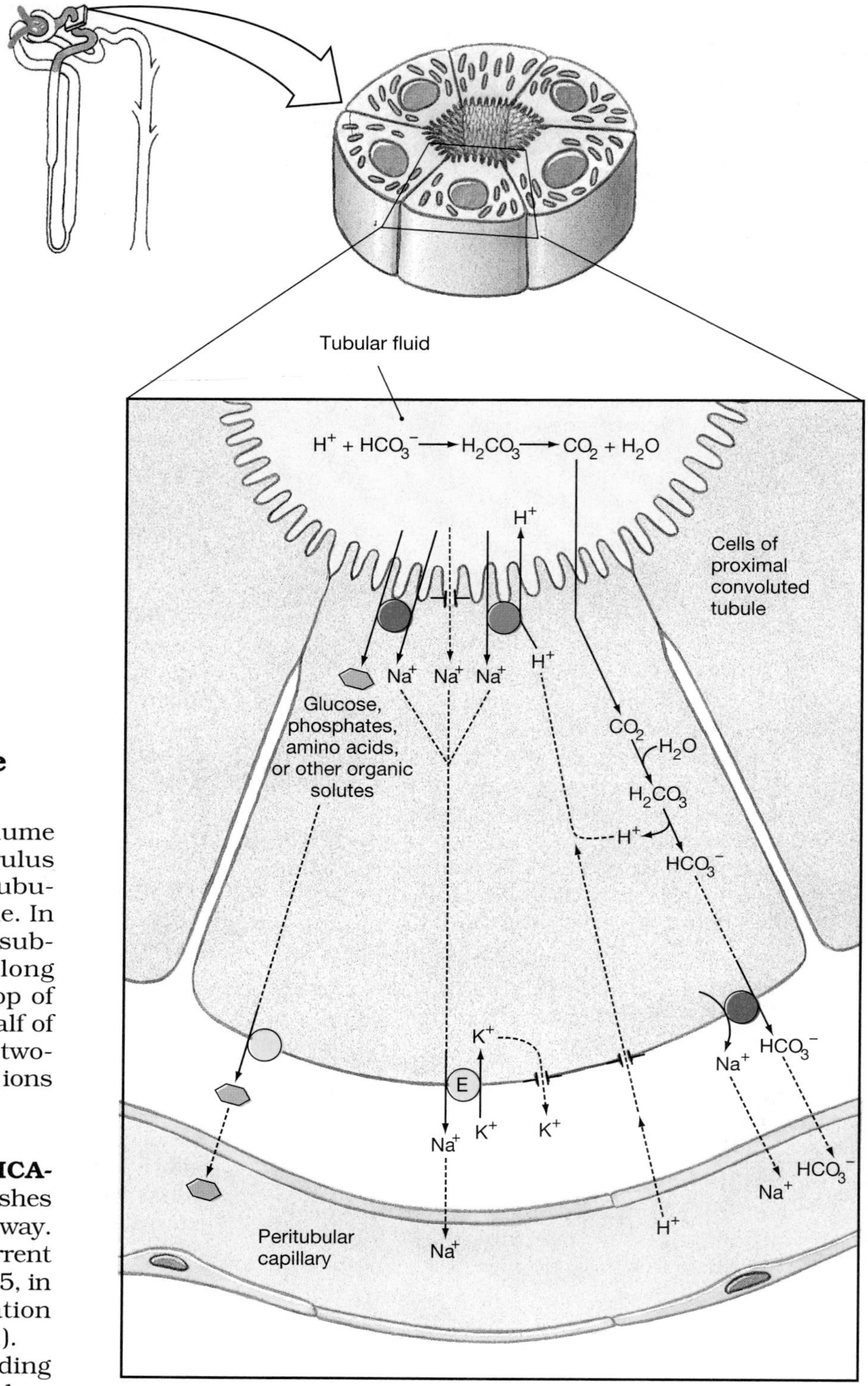

● **FIGURE 26-12**
Transport Activities at the Proximal Convoluted Tubule. Sodium ions may enter the tubular cell from the filtrate by diffusion, cotransport, or countertransport mechanisms. The sodium ions are then pumped into the peritubular fluid using the sodium-potassium exchange pump. Other reabsorbed solutes may be ejected into the peritubular fluid by separate active transport mechanisms.

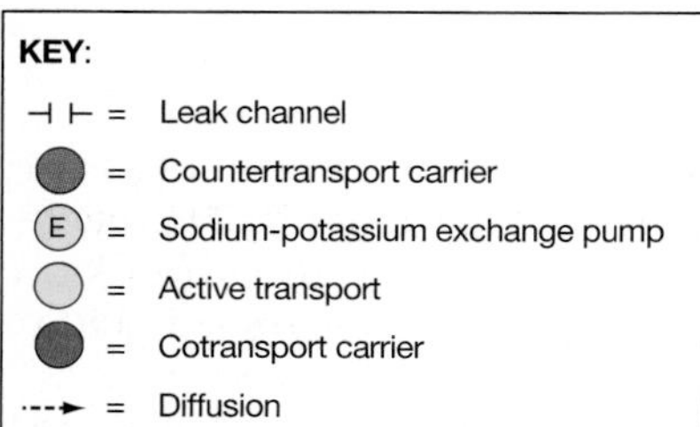
KEY:
= Leak channel
= Countertransport carrier
E = Sodium-potassium exchange pump
= Active transport
= Cotransport carrier
= Diffusion

The Loop of Henle and Countercurrent Exchange

Figures 26-13, 26-14

Roughly 60–70 percent of the volume of filtrate produced at the glomerulus has been reabsorbed before the tubular fluid reaches the loop of Henle. In the process, the useful organic substrates have been reclaimed, along with many mineral ions. The loop of Henle will reabsorb more than half of the remaining water, as well as two-thirds of the sodium and chloride ions in the tubular fluid.

COUNTERCURRENT MULTIPLICATION The loop of Henle accomplishes this reabsorption in a very efficient way. It uses the principle of countercurrent exchange introduced in Chapter 25, in our discussion of heat conservation techniques (Figure 25-20●, p. 971).

The descending and ascending limbs of the loop of Henle are very close together, separated by peritubular fluid. The exchange that occurs between these parallel lengths of the loop of Henle has been called **countercurrent multiplication.** *Countercurrent* refers to the fact that the exchange occurs between fluids moving in opposite directions—the tubular fluid in the descending limb is moving toward the renal pelvis, whereas the tubular fluid in the ascending limb is moving toward the cortex. *Multiplication* refers to the fact that the effect of exchange between the limbs increases as fluid movement continues.

The two parallel segments of the loop of Henle have very different permeability characteristics. The descending limb is permeable to water but relatively impermeable to solutes. The thick ascending limb is relatively impermeable to both water and solutes. The basic concept involved in countercurrent multiplication is:

- ***Sodium and chloride are pumped out of the ascending limb and into the peritubular fluid.***
- ***This pumping elevates the osmotic concentration in the peritubular fluid around the descending limb.***
- ***The result is an osmotic flow of water out of the descending limb and into the peritubular fluid, increasing the solute concentration in the descending limb.***
- ***The arrival of the highly concentrated solution in the ascending limb accelerates the transport of sodium and chloride into the peritubular fluid.***

Note that this arrangement is a simple positive feedback loop: Solute pumping at the ascending limb leads to higher solute concentrations in the descending limb, which then results in accelerated solute pumping in the ascending limb.

The countercurrent exchange mechanism is diagrammed in Figure 26-13•. For demonstration purposes, part (a) shows the loop of Henle containing a fluid isotonic with other body fluids (roughly 300 mOsm). Parts (b) through (e) follow the changes in osmotic concentration that occur as active transport begins in the ascending limb and tubular fluid moves through the loop of Henle. Part (e) approximates the actual situation in the normal kidney.

Active transport along the thick ascending limb moves sodium, potassium, and chloride ions out of the tubular fluid. The carrier is called a **Na^+-K^+/2 Cl^- transporter** because each cycle of the pump carries a sodium ion, a potassium ion, and two chloride ions into the tubular cell. The fates of these ions are indicated in Figure 26-13c•. Potassium and chloride ions are pumped across the basal membrane of the cell, but the potassium ions are recovered as the sodium-potassium exchange pump ejects sodium ions from the tubular cell. The reclaimed potassium ions then diffuse back into the lumen. The net result is that sodium ions and chloride ions enter the peritubular fluid of the medulla.

The removal of sodium and chloride ions from the tubular fluid in the ascending limb elevates the osmotic concentration of the peritubular fluid around the descending limb (Figure 26-13b•). The descending limb is permeable to water. Thus, as the tubular fluid travels along the descending limb, water moves out of the tubule by osmosis. Solutes remain behind, so the tubular fluid reaching the turn of the loop has a higher osmolarity than it did when it entered the descending limb of the loop of Henle (Figure 26-13d•).

The pumping mechanism of the thick ascending limb is very effective: Almost two-thirds of the sodium and chloride ions are pumped out of the tubular fluid before it reaches the DCT. In other tissues, solute concentration differences are quickly eliminated through osmosis. Osmosis cannot occur across the wall of the thick ascending limb, however, because it is impermeable to water. Thus as sodium and chloride ions are removed, the solute concentration in the tubular fluid declines.

The rate of ion transport by the thick ascending limb is proportional to the concentration in the tubular fluid. As a result, more sodium and chloride enter the medulla at the start of the thick ascending limb, where NaCl concentrations are highest, than near the cortex, where the concentrations are relatively low (Figure 26-13e•). The tubular fluid arrives at the DCT with an osmotic concentration of only around 100 mOsm, one-third of the concentration found in the surrounding peritubular fluid of the renal cortex.

THE CONCENTRATION GRADIENT OF THE MEDULLA Countercurrent multiplication creates a concentration gradient in the medulla, with the highest solute concentrations near the turn of the loop. As fluid movement continues, the concentration gradient gets larger. This increase triggers additional water losses from the tubular fluid in the descending limb. The feedback between the ascending thick limb and the descending thin limb continues until the sodium and chloride concentrations in the peritubular fluid reach approximately 750 mOsm. The concentration gradient does not increase further because the vasa recta is continually absorbing solutes as blood flows through the medulla. (We will consider the vasa recta in more detail in a separate section.)

In the normal kidney the maximum solute concentration near the turn of the loop is around 1200 mOsm. Sodium and chloride ions pumped out of the ascending limb of the loop of Henle account for roughly two-thirds of the solutes involved. The rest of the concentration gradient results from the presence of urea. To understand how the urea arrived in the medulla we must look ahead to events in the last segments of the collecting system.

Figure 26-14• diagrams the primary factors involved in the formation and maintenance of the osmotic gradient in the medulla, as detailed in Figure 26-13•. Figure 26-14a• summarizes the countercurrent mechanism responsible for elevated sodium and chloride concentrations in the medulla. A separate mechanism, diagrammed in Figure 26-14b•, accounts for the elevated urea concentrations in the medulla. The thick ascending limb of the loop of Henle and the DCT, collecting tubules, and collecting ducts are all impermeable to urea. As water reabsorption occurs, the concentration of urea gradually rises, and the urine reaching the papillary duct and minor calyx contains urea at a concentration of around 450 mOsm. Because the papillary ducts are permeable to urea, the urea concentration in the deepest portions of the medulla also averages 450 mOsm.

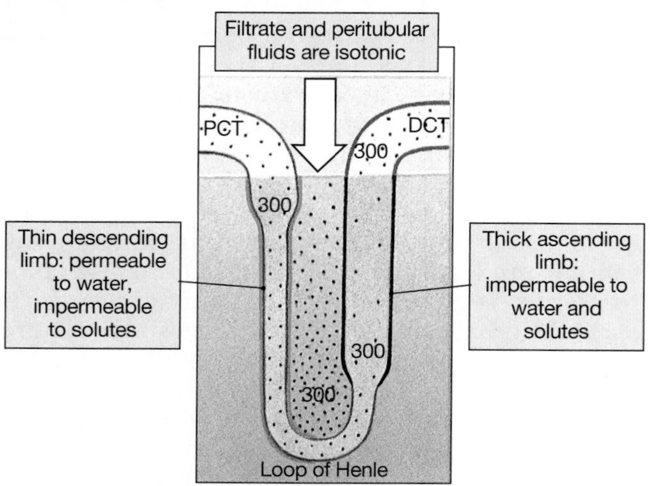
Filtrate and peritubular fluids are isotonic
PCT
DCT
300
300
300
300
Thin descending limb: permeable to water, impermeable to solutes
Thick ascending limb: impermeable to water and solutes
Loop of Henle

(a) The tubule contains a fluid with the same osmolarity as other body fluids. Note the differences in permeability between the descending and ascending limbs.

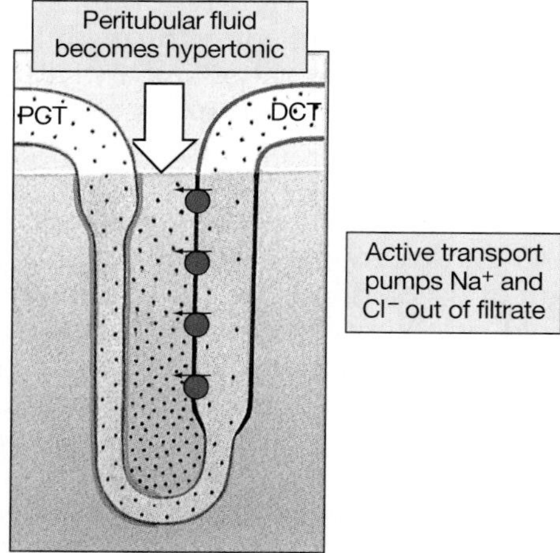
Peritubular fluid becomes hypertonic
PCT
DCT
Active transport pumps Na^+ and Cl^- out of filtrate

(b) The thick ascending limb actively transports sodium and chloride ions into the peritubular fluid, elevating the osmotic concentration around the descending limb.

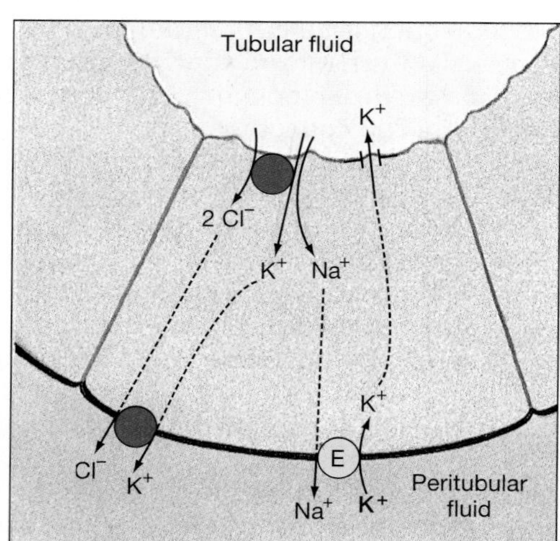
Tubular fluid
K^+
2 Cl^-
K^+
Na^+
K^+
Cl^-
K^+
E
Na^+
K^+
Peritubular fluid

(c) The mechanism of sodium and chloride transport involves (1) a Na^+-K^+/2 Cl^- carrier at the lumenal surface and (2) two carriers, a potassium-chloride cotransport pump and a sodium-potassium exchange pump, at the basal surface of the tubular cell. Because the potassium ions removed from the lumen of the tubule ultimately diffuse back in, through leak channels, the net result is the transport of sodium and chloride ions into the peritubular fluid.

KEY:

Membrane Permeabilities:

- = Water- and solute-impermeable. Example: thick ascending limb
- = Variable permeability to water; impermeable to urea. Example: DCT
- = Water-permeable, solute-impermeable. Example: thin and thick descending limbs
- = Water- and solute-permeable. Example: thin ascending limb, PCT
- ⊣ ⊢ = Leak channel

Transport mechanisms:

- (E) = $Na^+ - K^+$ exchange pump
- ● = Cotransport pump

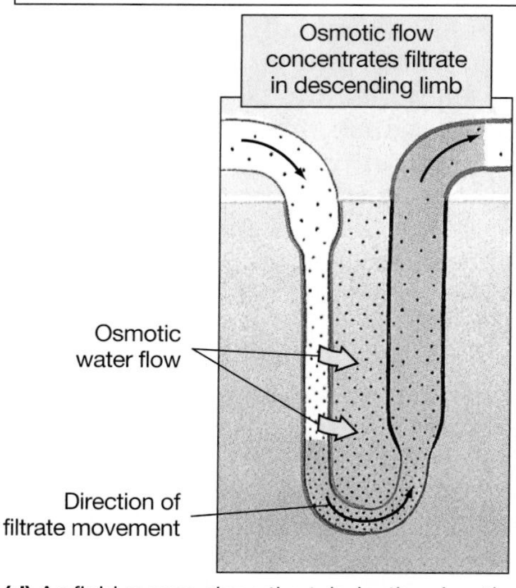
Osmotic flow concentrates filtrate in descending limb
Osmotic water flow
Direction of filtrate movement

(d) As fluid moves along the tubule, the elevation of the solute concentration in the peritubular fluid causes an osmotic flow of water out of the fluid in the descending limb. Water loss concentrates the tubular fluid, and when the concentrated fluid reaches the thick ascending limb additional Na^+ and Cl^- ions are pumped into the peritubular fluid. This leads to increased water loss, and the cycle continues.

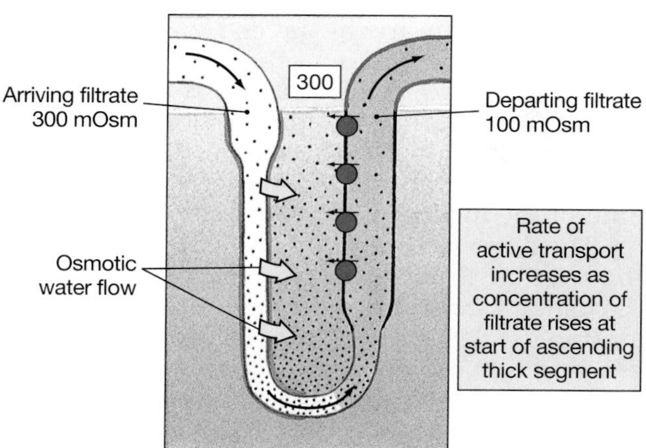
300
Arriving filtrate 300 mOsm
Departing filtrate 100 mOsm
Osmotic water flow
Rate of active transport increases as concentration of filtrate rises at start of ascending thick segment

(e) The fluid arriving in the DCT has a lower osmolarity than the fluid entering the loop of Henle. The missing ions have been trapped in the medulla.

● **FIGURE 26-13**

Countercurrent Multiplication. A simple diagram showing the mechanism of solute concentration in the medulla by countercurrent exchange at the loop of Henle. The color tint indicates a single volume of tubular fluid as it moves through the system. D

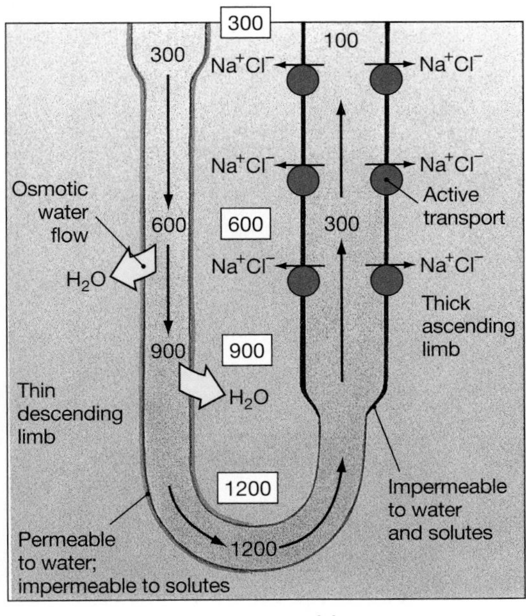
300
300
100
Na^+Cl^-
Na^+Cl^-
Na^+Cl^-
Na^+Cl^-
Osmotic water flow
Active transport
600
600
300
H_2O
Na^+Cl^-
Na^+Cl^-
Thick ascending limb
900
900
H_2O
Thin descending limb
1200
Impermeable to water and solutes
Permeable to water; impermeable to solutes
1200

(a)

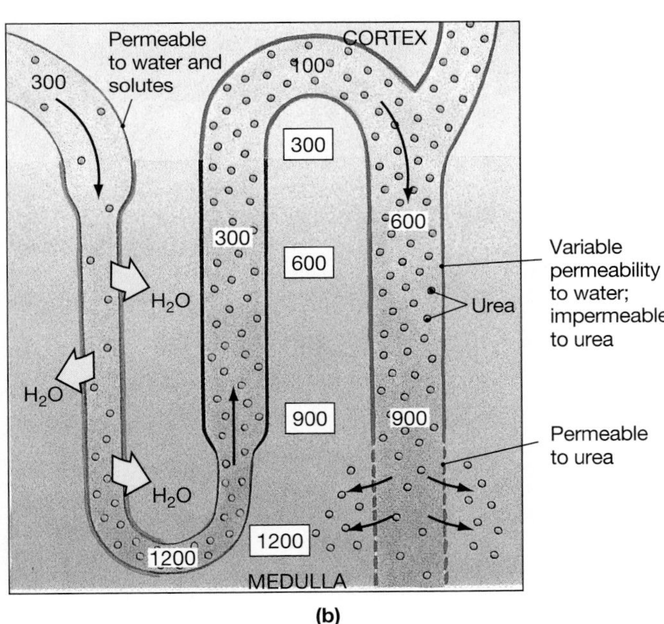
Permeable to water and solutes
CORTEX
300
100
300
600
300
600
Variable permeability to water; impermeable to urea
H_2O
Urea
H_2O
900
900
Permeable to urea
H_2O
1200
1200
MEDULLA

(b)

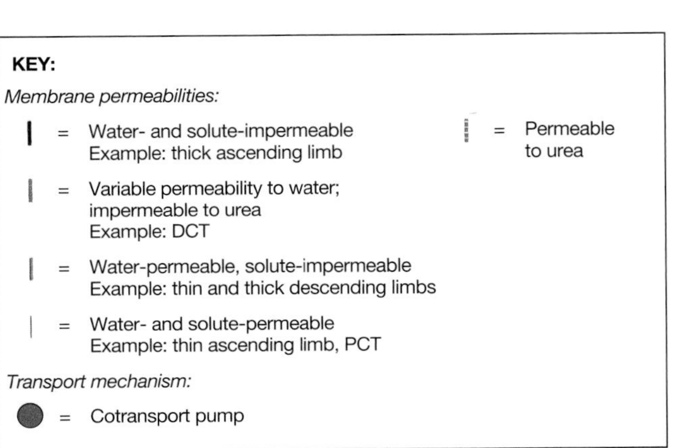
KEY:
Membrane permeabilities:
= Water- and solute-impermeable
Example: thick ascending limb
= Permeable to urea
= Variable permeability to water; impermeable to urea
Example: DCT
= Water-permeable, solute-impermeable
Example: thin and thick descending limbs
= Water- and solute-permeable
Example: thin ascending limb, PCT
Transport mechanism:
= Cotransport pump

● **FIGURE 26-14**

Fluid and Solute Movements along the Loop of Henle. (a) Active transport of NaCl along the ascending thick limb results in the movement of water from the descending limb (see Figure 26-13). **(b)** The permeability characteristics of the loop and the collecting duct tend to concentrate urea in the tubular fluid and in the medulla. The ascending thick limb, DCT, and collecting duct are impermeable to urea. As water reabsorption occurs, the urea concentration rises. The papillary ducts are permeable to urea, and urea accounts for roughly one-third of the solutes in the deepest portions of the medulla. D

BENEFITS OF COUNTERCURRENT MULTIPLICATION The countercurrent mechanism performs two services for the kidneys:

1. It is an efficient way to reabsorb solutes and water before the tubular fluid reaches the DCT and collecting system.
2. It establishes a concentration gradient that will permit the passive reabsorption of water from the urine in the collecting system. This reabsorption is regulated by circulating levels of ADH (antidiuretic hormone).

The tubular fluid arriving at the descending limb of the loop of Henle has an osmolarity of roughly 300 mOsm, most of that due to the presence of ions such as sodium and chloride. The concentration of organic wastes, such as urea, is relatively low. Roughly half that volume reaches the DCT, and it arrives with an osmolarity of 100 mOsm. Urea and other organic wastes, which were not pumped out of the ascending limb, are now a significant percentage of the dissolved solutes.

The Distal Convoluted Tubule

Figures 26-15, 26-16

The composition and volume of the tubular fluid change dramatically as it travels from the capsular space to the distal convoluted tubule. Roughly 60 percent of the water and 65 percent of the solutes are reabsorbed in the PCT, and another 20 percent of the water and 25 percent of the dissolved materials, specifically sodium and chloride ions, enter the peritubular fluid of the medulla along the loop of Henle. Selective reabsorption or secretion, primarily along the distal convoluted tubule, then performs the final adjustments in the composition of the tubular fluid.

SODIUM AND CHLORIDE ION REABSORPTION Throughout most of the DCT the tubular cells actively transport sodium and chloride ions out of the tubular fluid (Figure 26-15●). Tubular cells near the end of the DCT also contain aldosterone-sensitive pumps that exchange tubular sodium

● **FIGURE 26-15**
Tubular Secretion and Solute Reabsorption at the DCT. (Left) Basic pattern for the absorption of sodium and chloride ions and the secretion of potassium ions. (Right) Aldosterone-regulated absorption of sodium ions, linked to the passive loss of potassium ions. D

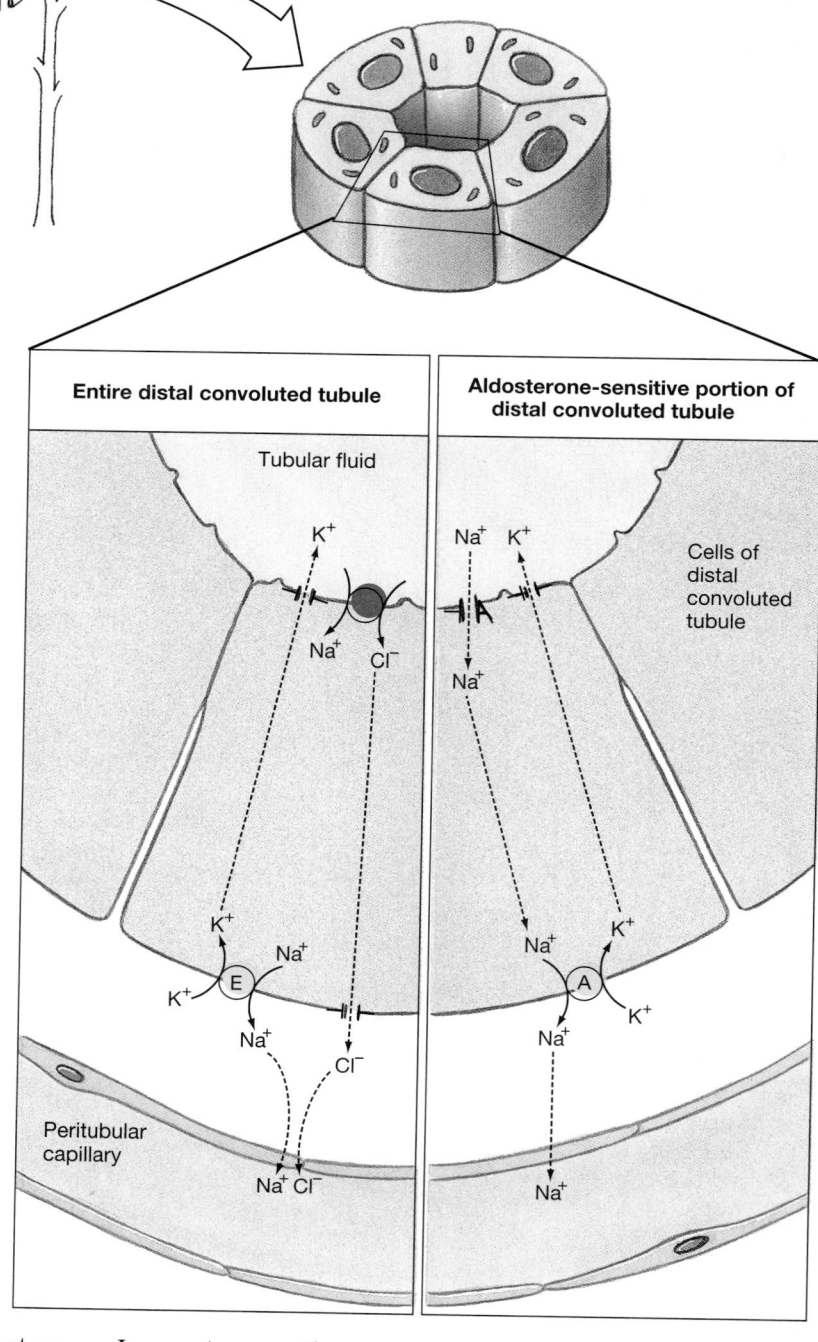

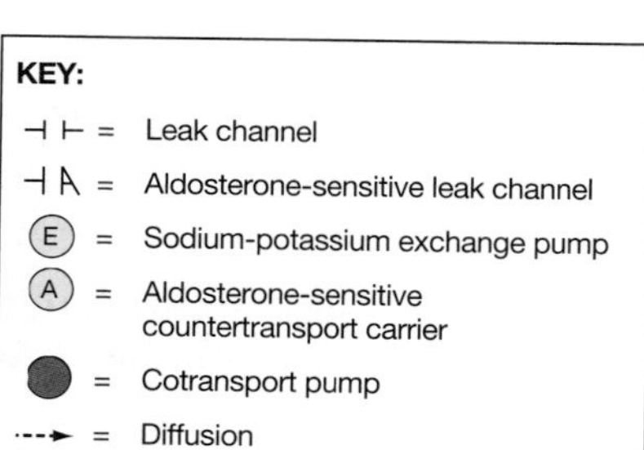
KEY:
⊣ ⊢ = Leak channel
⊣ ⩘ = Aldosterone-sensitive leak channel
E = Sodium-potassium exchange pump
A = Aldosterone-sensitive countertransport carrier
● = Cotransport pump
---➤ = Diffusion

Entire distal convoluted tubule

Aldosterone-sensitive portion of distal convoluted tubule

Tubular fluid

Cells of distal convoluted tubule

Peritubular capillary

K^+ Na^+ Cl^- E A

ions for peritubular potassium ions. Their activities will be considered below.

REABSORPTION OF WATER Water reabsorption by osmosis occurs in the proximal convoluted tubule and the descending limb of the loop of Henle. The water permeabilities of these regions cannot be adjusted, and water reabsorption occurs whenever the osmotic concentration of the peritubular fluid exceeds that of the tubular fluid. Because this water movement cannot be prevented, it is known as **obligatory water reabsorption.**

The distal convoluted tubule is impermeable to water except in the presence of ADH from the posterior pituitary. The DCT normally reabsorbs roughly 9 liters per day, or around 25 percent of the water arriving from the loop of Henle. This is 5 percent of the original volume of filtrate produced at the glomerulus. When ADH levels rise, the DCT becomes more permeable to water, and the amount of water reabsorbed increases. Because the flow of water out of the tubular fluid can be controlled, it is called **facultative water reabsorption.**

SECRETION Filtration does not force all of the dissolved materials out of the plasma, and blood entering the peritubular capillaries still contains a number of potentially undesirable substances. In most cases their presence is not significant, for the remaining concentrations are too low to cause physiological problems. However, any ions or compounds present in the peritubular capillaries will diffuse into the peritubular fluid of the DCT. If those concentrations become too high, the tubular cells may absorb these materials from the peritubular fluid and secrete them into the tubular fluid. A partial listing of substances secreted into the tubular fluid by the DCT can be found in Table 26-3, p. 999.

The rate of potassium ion and hydrogen ion secretion rises or falls in response to changes in their concentration in the peritubular capillaries. The higher the concentration outside of the tubules, the more they secrete. Potassium and hydrogen

● **FIGURE 26-16**
Hydrogen Ion Secretion. Hydrogen ion secretion and the acidification of the urine occur via two different routes. The central theme is the exchange of hydrogen ions in the cytoplasm for sodium ions in the tubular fluid and the reabsorption of the bicarbonate ions generated in the process.

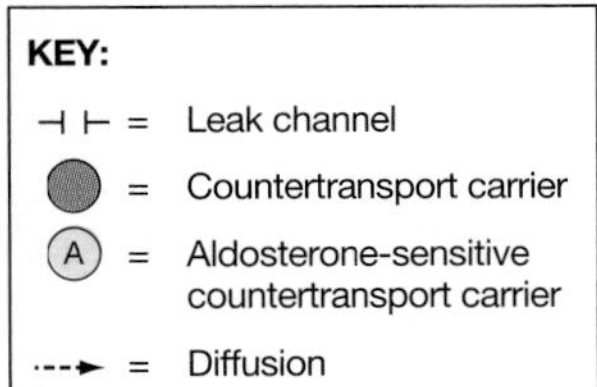
KEY:
= Leak channel
= Countertransport carrier
= Aldosterone-sensitive countertransport carrier
= Diffusion

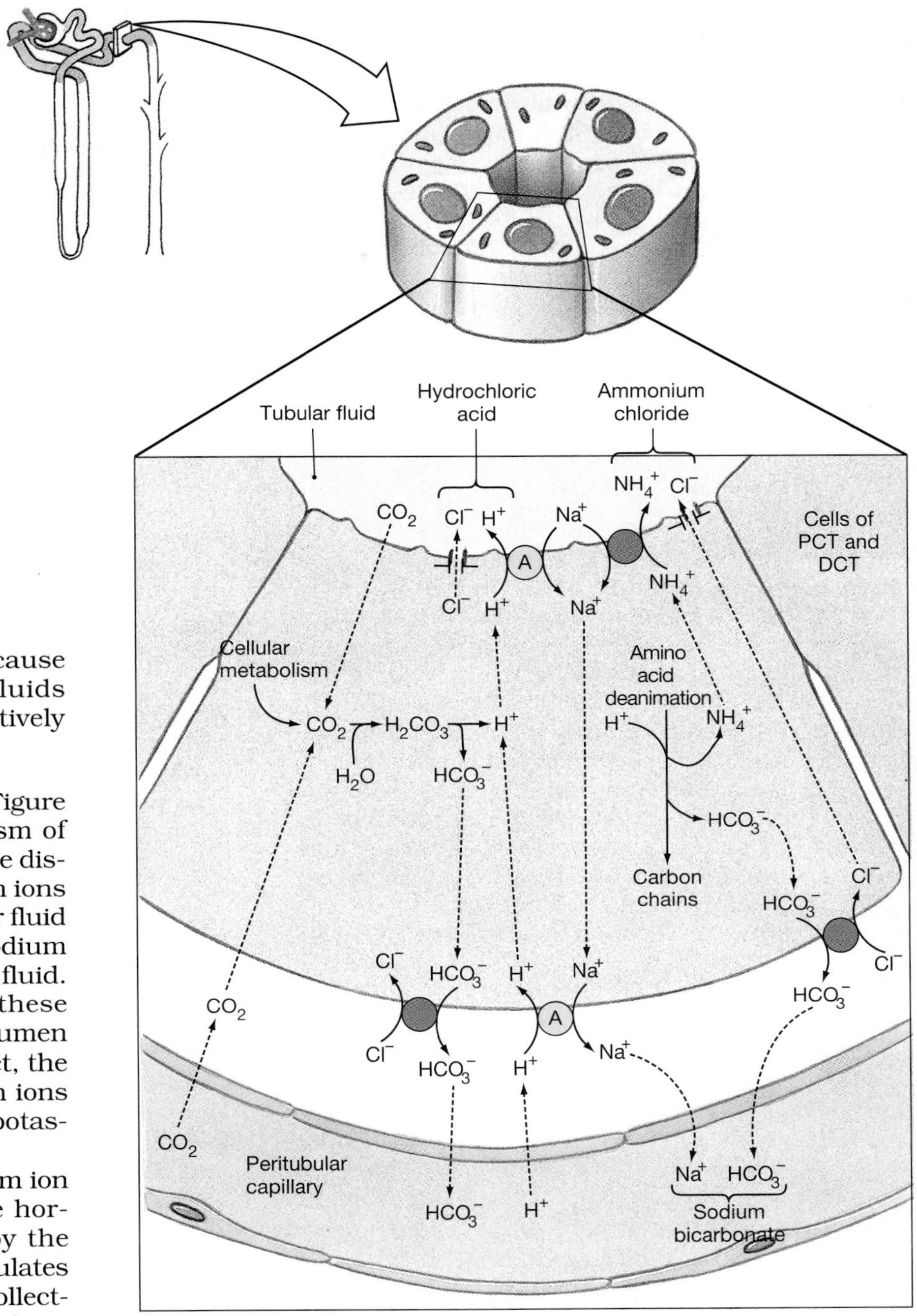

ions merit special attention because their concentration in body fluids must be maintained within relatively narrow limits.

Potassium Ion Secretion. Figure 26-15● diagrams the mechanism of potassium ion secretion along the distal convoluted tubule. Potassium ions are removed from the peritubular fluid by DCT cells in exchange for sodium ions obtained from the tubular fluid. At the surface of the tubule, these potassium ions diffuse into the lumen via potassium channels. In effect, the tubular cells are trading sodium ions in the tubular fluid for excess potassium ions in the body fluids.

The ion pump and the sodium ion channels are controlled by the hormone *aldosterone*, produced by the adrenal cortex. Aldosterone stimulates ion pumps along the DCT, the collecting tubule, and the collecting duct, thus reducing the number of sodium ions lost in the urine. However, sodium ion conservation is associated with potassium ion loss. Prolonged aldosterone stimulation can therefore produce *hypokalemia*, a dangerous reduction in the plasma concentration of potassium ions.

Hydrogen Ion Secretion. Hydrogen ion secretion is also associated with the reabsorption of sodium. Two secretory routes are shown in Figure 26-16●. Both involve a familiar reaction sequence, the generation of carbonic acid by the enzyme *carbonic anhydrase*. Hydrogen ions generated by the dissociation of the carbonic acid are secreted by means of sodium-linked countertransport, in exchange for sodium ions in the tubular fluid. The bicarbonate ions then diffuse into the peritubular fluids and into the bloodstream, where they help prevent changes in plasma pH.

Hydrogen ion secretion acidifies the urine while elevating the pH of the blood. Hydrogen ion secretion accelerates when the pH of the blood falls, as in lactic acidosis (Chapter 10) or ketoacidosis (Chapter 25). ∞ *[pp. 308, 960]* The combination of hydrogen ion removal and bicarbonate ion production at the kidneys plays an important role in the control of blood pH.

Chapter 25 noted that the production of lactic acid and ketone bodies during the postabsorptive state can cause acidosis. Under these conditions the PCT and DCT will deaminate amino acids in reactions that strip off the amino groups (—NH_2). The reaction sequence ties up hydrogen ions and yields **ammonium ions,** NH_4^+, and bicarbonate ions. As indicated in Figure 26-16•, the ammonium ions are then pumped into the tubular fluid through sodium-linked countertransport, and the bicarbonate ions enter the peritubular fluid.

Tubular deamination thus has two major benefits: (1) It provides carbon chains suitable for catabolism; and (2) it generates bicarbonate ions that add to the buffering capabilities of the plasma.

The Collecting System

Figures 26-15, 26-16

The collecting ducts receive tubular fluid, now called urine, from many nephrons and carry it toward the renal sinus through the concentration gradient in the medulla. The normal amount of water and solute loss in the urine is regulated in two ways: (1) by control of the ion pumps that remove sodium ions from the urine, via aldosterone secretion, and (2) by control of the permeability of the collecting system to water, via ADH secretion. The primary site of sodium ion reabsorption is the DCT and the cortical portion of the collecting duct. The primary site of water reabsorption is the medullary portion of the collecting duct, as the collecting duct passes through the concentration gradient in the medulla.

SOLUTE REABSORPTION

- ***Sodium ion reabsorption.*** The cortical region of the collecting system contains aldosterone-sensitive ion pumps that exchange sodium ions in the urine for potassium ions in the peritubular fluid. The mechanism was detailed in Figure 26-15•.
- ***Bicarbonate reabsorption.*** Bicarbonate ions are reabsorbed in exchange for chloride ions in the peritubular fluid, as detailed in Figure 26-16•.
- ***Urea reabsorption.*** The concentration of urea in the urine entering the collecting duct is relatively high. The epithelia lining the thick ascending limb, the distal convoluted tubule, and collecting tubules and ducts are relatively impermeable to urea, and as water reabsorption occurs, the urea concentration continues to rise. The urine entering the papillary duct is usually of the same osmotic concentration as the interstitial fluid of the medulla—around 1200 mOsm—but it contains a much higher concentration of urea. As a result, urea tends to diffuse out of the tubular fluid and into the interstitial fluid of the deepest portion of the medulla. The urea concentration in this region usually averages 450 mOsm, roughly one-third of the total osmolarity.

WATER REABSORPTION Water reabsorption in the collecting system is facultative and requires ADH. At normal levels of ADH the collecting duct will reabsorb roughly 16.9 liters per day, or around 9.4 percent of the original volume of filtrate. In the absence of ADH, the collecting tubules and collecting ducts are impermeable to water and solutes, and all of the water leaving the nephrons will be lost in the urine. This is what happens in *diabetes insipidus* (Chapter 18), when urinary water losses may reach 24 liters per day. ∞ *[p. 614]* Under extreme ADH stimulation, the urinary water losses may fall to 500 ml/day.

Atrial natriuretic peptide (ANP) is released by cardiac muscle fibers in the atrial walls when they are excessively stretched. ANP (1) elevates glomerular pressures and increases the GFR, as noted on 995, (2) suppresses sodium ion reabsorption along the collecting system, (3) blocks the release of aldosterone and ADH, and (4) blocks the responses of the DCT and collecting ducts to aldosterone and ADH already in the circulation. The net result is that ANP stimulates the production of a large volume of relatively dilute urine. This fluid loss soon restores normal plasma volume.

❑ The Function of the Vasa Recta

Figure 26-17

The vasa recta maintains the concentration gradient of the medulla by removing the solutes and water reabsorbed by the loop of Henle and collecting duct. Blood flow through the vasa recta and the movement of tubular fluid through the loop of Henle occur in opposite directions (Figure 26-17•).

Blood enters the vasa recta with an osmolarity of approximately 300 mOsm. The blood flowing into the medulla increases in osmolarity through the gain of solutes; water losses are limited by the presence of plasma proteins. Blood flowing toward the cortex gradually decreases in osmolarity until it departs with an osmolarity of 325 mOsm. This decrease in osmolarity occurs primarily through gain of water. The net result is that:

1. Fewer solute molecules diffuse *out* of the vasa recta as it ascends to the cortex than diffused *in* when it descended into the medulla.
2. The plasma volume leaving the vasa recta is larger than the volume that entered this capillary loop.

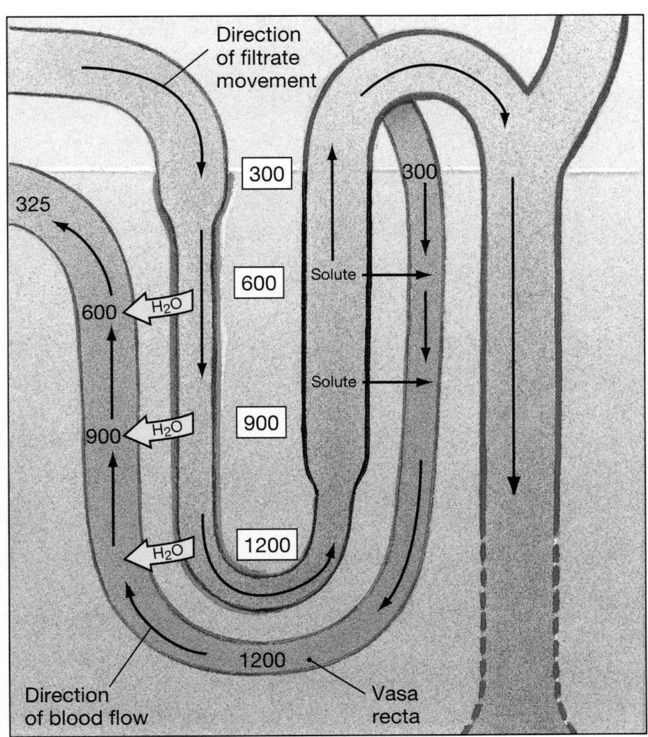
Direction
of filtrate
movement
300
325
300
600
Solute
600
H_2O
Solute
900
900
H_2O
1200
H_2O
1200
Direction
of blood flow
Vasa
recta

● **FIGURE 26-17**

The Vasa Recta. Capillaries of the vasa recta maintain the concentration gradient in the medulla by removing the water and solutes reclaimed from the tubular fluid in the nephron and collecting duct.

Under normal conditions the removal of solutes and water by the vasa recta precisely balances the rates of solute reabsorption and osmosis in the medulla. This mechanism effectively maintains the osmotic gradient and returns the reabsorbed water and solutes to the general circulation.

❑ A Summary of Renal Function

Figure 26-18

Figure 26-18● provides a functional overview that summarizes the major steps involved in the reabsorption of water and the production of hypertonic urine. Table 26-4 details the functions of each segment of the nephron and collecting system.

Step 1. The filtrate produced at the renal corpuscle has the same osmolarity as the plasma, about 300 mOsm.

Step 2. In the proximal convoluted tubule, the active removal of ions and other solutes produces a continual osmotic flow of water out of the tubular fluid. This process reduces the volume of filtrate, but keeps the solutions inside and outside of the tubule isotonic. Between 60 and 70 percent of the filtrate volume has been reabsorbed before the tubular fluid reaches the descending limb of the loop of Henle.

Step 3. In the proximal convoluted tubule and descending limb of the loop of Henle, water moves into the surrounding peritubular fluids, leaving a small volume (roughly 20 percent of the original filtrate) of highly concentrated tubular fluid. The volume reduction has occurred through obligatory water reabsorption.

Step 4. The thick ascending limb is impermeable to water and solutes. The tubular cells actively transport sodium and chloride ions out of the tubular fluid. This transport lowers the osmotic concentration of the tubular fluid *without affecting the volume of the tubular fluid.* The tubular fluid reaching the distal convoluted tubule is hypotonic relative to the peritubular fluid, with an osmolarity of only about 100 mOsm. Because only sodium and chloride ions are removed, urea now accounts for a significantly higher proportion of the total osmotic concentration at the end of the loop than it did at the start.

Step 5. The final composition and concentration of the tubular fluid will be determined by the events under way in the DCT and the collecting system. Although the DCT, collecting tubule, and collecting duct are generally impermeable to solute molecules, the osmolarity of the tubular fluid can be adjusted through active transport. Sodium-chloride cotransport and the aldosterone-stimulated reabsorption of sodium ions are shown in Figure 26-18●.

Step 6. The osmotic concentration of the urine is controlled by variations in the water permeabilities of the distal portions of the DCT, the collecting tubules, and the collecting ducts. These segments are impermeable to water unless exposed to antidiuretic hormone (ADH) from the posterior pituitary. At normal concentrations of ADH, the distal portions of the DCT and the collecting tubules and ducts are somewhat permeable to water, and there is an osmotic flow of water out of the urine as it passes along the collecting ducts. Under these conditions the urine entering the minor calyx has an osmolarity approaching 1200 mOsm.

Step 7. As the tubular fluid becomes increasingly concentrated, the urea concentration rises accordingly. The papillary ducts are permeable to urea, and urea molecules diffuse out of the urine and into the peritubular fluid. The urea entering in this way contributes to the concentration gradient of the medulla.

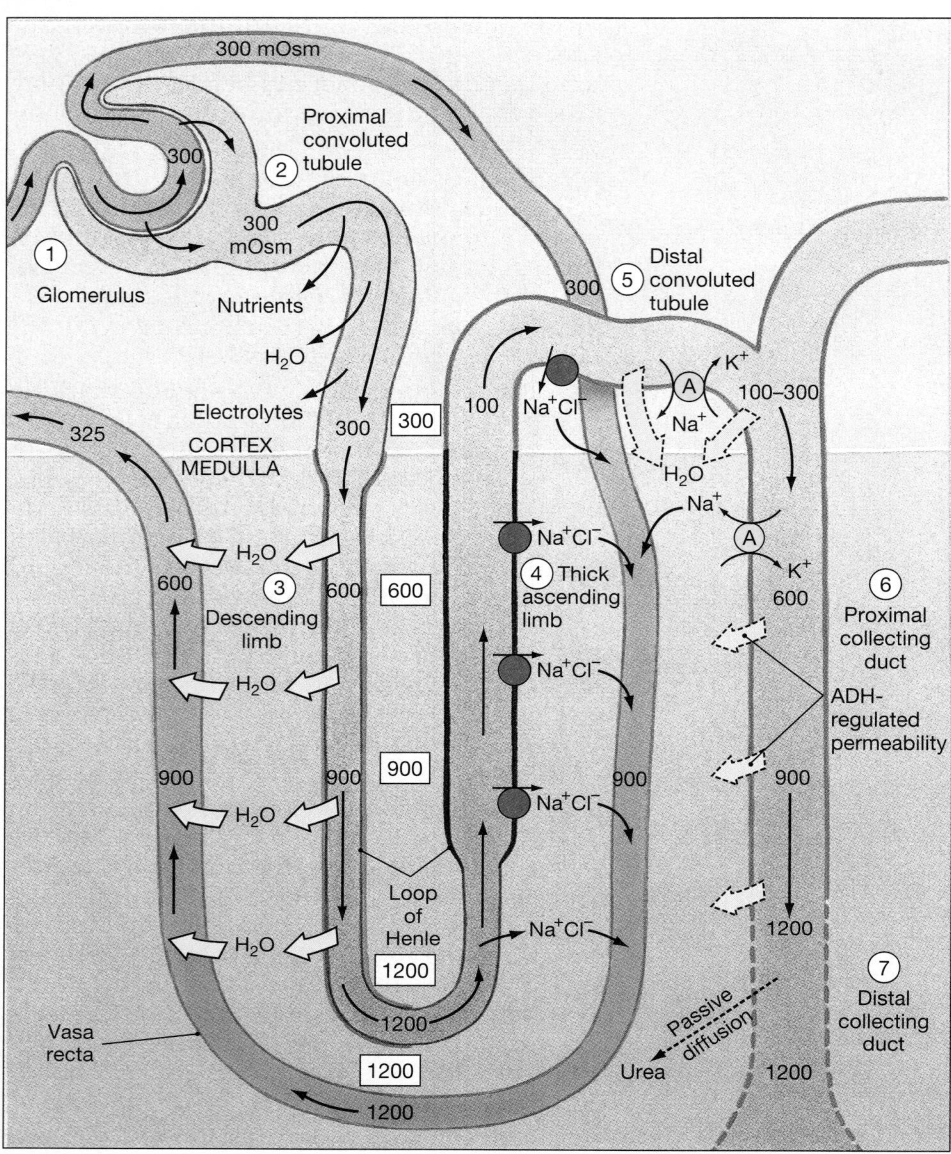
300 mOsm
Proximal convoluted tubule
2
300
300 mOsm
1
Glomerulus
Nutrients
H_2O
Electrolytes
325
CORTEX
MEDULLA
300
300
100
Na^+Cl^-
Distal convoluted tubule
5
300
K^+
A
Na^+
100–300
H_2O
Na^+
A
K^+
600
H_2O
600
3
Descending limb
600
600
Na^+Cl^-
4 Thick ascending limb
6
Proximal collecting duct
H_2O
Na^+Cl^-
ADH-regulated permeability
900
H_2O
900
900
Na^+Cl^-
900
900
Loop of Henle
H_2O
Na^+Cl^-
1200
1200
7
Distal collecting duct
Vasa recta
1200
Passive diffusion
Urea
1200
1200
1200

● **FIGURE 26-18**
Major Steps in Urine Production

KEY:

Membrane permeabilities:

=	Water- and solute-impermeable; Example: thick ascending limb
=	Variable permeability to water; impermeable to urea; Example: DCT
=	Water- permeable, solute-impermeable; Example: thin and thick descending limbs
=	Water- and solute-permeable; Example: thin ascending limb, PCT
=	Variable water permeability

Transport mechanisms:

(A) =	Aldosterone-regulated countertransport carrier
● =	Cotransport pump

DIURETICS **Diuretics** (dī-ū-RET-iks) are drugs that promote the loss of water in the urine. The usual goal in diuretic administration is a reduction of blood volume, blood pressure, or both. The ability to control renal water losses with relatively safe and effective diuretics has saved the lives of many patients, especially those with high blood pressure or congestive heart failure.

Diuretics have many different mechanisms of action, but all affect transport activities or water reabsorption along the nephron and collecting system. For example, a class of diuretics called *thiazides* (THĪ-a-zīdz) promotes water loss by reducing sodium and chloride transport in the proximal and distal convoluted tubules. For a detailed discussion of the major classes of diuretics and their modes of action, refer to the *Applications Manual.* AM *Diuretics*

Diuretic use for nonclinical reasons is currently on the rise. For example, large doses of diuretics may be taken by bodybuilders to improve muscle definition and by fashion models or jockeys to reduce body weight. This is an extremely dangerous practice that has caused several deaths due to severe dehydration and cardiac arrest.

✓ What effect would increased amounts of aldosterone have on the potassium ion concentration of the urine?

✓ What effect would a decrease in the sodium ion concentration of the filtrate have on the pH of the filtrate?

✓ If the nephrons lacked a loop of Henle, how would this lack affect the volume and osmolarity of the urine they produced?

✓ Why does a decrease in the amount of sodium ions in the distal convoluted tubule lead to an increase in blood pressure?

TABLE 26-4 Renal Segments and Their Functions

Segment	General Functions	Specific Functions	Mechanisms
Renal corpuscle	*Filtration* of plasma; generates approximately 180 l/day of filtrate similar in composition to blood plasma without plasma proteins	*Filtration* of water, inorganic and organic solutes from plasma; retention of plasma proteins and blood cells	Glomerular hydrostatic (blood) pressure working across capillary endothelium, lamina densa, and filtration slits
Proximal convoluted tubule	*Reabsorption* of 60–70% of the water (approximately 108 l/day), 99–100% of the organic substrates, and 60–70% of the sodium and chloride ions present in the original filtrate	*Reabsorption:* Active: glucose, other simple sugars, amino acids, vitamins, ions (including sodium, potassium, calcium, magnesium, and bicarbonate) Passive: urea, chloride ions, lipid-soluble materials, water *Secretion:* Hydrogen ions, ammonium ions, creatinine (as at DCT), and phosphate	Carrier-mediated transport, including facilitated transport (glucose, amino acids), cotransport (glucose, ions), or countertransport (with secretion of H^+) Diffusion (solutes) or osmosis (water) Countertransport with sodium ions
Loop of Henle	*Reabsorption* of 25% of the water (45 l/day) and 20–25% of the sodium and chloride ions present in the original filtrate. Creation of the concentration gradient in the medulla	*Reabsorption:* Sodium and chloride ions Water	Active transport via $Na^+ - K^+/2Cl^-$ transporter Osmosis
Distal convoluted tubule	*Reabsorption* of a variable amount of water (usually 5% or 9 l/day), under ADH stimulation, and a variable amount of sodium ions, under aldosterone stimulation	*Reabsorption:* Sodium and chloride ions Sodium ions (variable) Water (variable) Bicarbonate ions *Secretion:* Hydrogen ions, ammonium ions Creatinine, drugs, toxins	Cotransport Countertransport with potassium ions, aldosterone-regulated Osmosis, ADH-regulated Diffusion or generation within tubular cells Countertransport with sodium ions Carrier-mediated transport
Collecting system	*Reabsorption* of a variable amount of water (usually 9.4% or 16.9 l/day), under ADH stimulation, and a variable amount of sodium ions, under aldosterone stimulation	*Reabsorption*: Sodium ions (variable) Bicarbonate ions (variable) Water (variable) Urea (distal portions only) *Secretion:* Potassium and hydrogen ions (variable)	Countertransport with potassium or hydrogen ions, aldosterone-regulated Diffusion, generated within tubular cells Osmosis, ADH-regulated Diffusion Carrier-mediated transport
Vasa recta	*Redistribution* of water and solutes reabsorbed in medulla, and stabilization of the concentration gradient of the medulla	*Reabsorption:* Solutes and water reabsorbed by nephrons and collecting system	Diffusion and osmosis

❑ The Composition of Normal Urine

Out of the 180 liters or more of filtrate produced each day at the glomerulus, more than 99 percent is absorbed before reaching the renal pelvis. General characteristics of normal urine are listed in Table 26-5. However, the composition of the excreted 1.2 liters of urine varies, depending upon the metabolic and hormonal events under way.

The composition and concentration of the excreted urine are two integrated but distinct properties. The *composition* of the urine reflects the filtration, absorption, and secretion activities of the nephrons. Some compounds, such as urea, are neither actively excreted nor reabsorbed along the nephron. In contrast, nutrients are usually

TABLE 26-5 General Characteristics of Normal Urine

pH:	6.0 (range: 4.5–8)
Specific gravity:	1.003–1.030
Osmolarity:	855–1335 mOsm
Water content:	93–97 percent
Volume:	1200 ml/day
Color:	clear yellow
Odor:	varies depending on composition
Bacterial content:	sterile

completely absorbed, and other compounds, such as creatinine, that are missed by the filtration process are actively secreted into the tubular fluid.

These processes determine the identities and amounts of materials eliminated in the urine. The *concentration* of these components in a given urine sample depends on the osmotic movement of water across the walls of the tubules and collecting ducts. Because the composition and concentration of the urine vary independently, an individual can produce a small quantity of concentrated urine or a large quantity of dilute urine and still excrete the same amount of dissolved materials. As a result, physicians often request a 24-hour urine collection rather than a single sample. This practice enables them to assess both quantity and composition accurately.

URINALYSIS Normal urine is a clear, sterile solution with a yellow color. The color results from the presence of urobilin, generated in the kidneys from the urobilinogens produced by intestinal bacteria and absorbed in the colon (see Figure 19-5●, p. 658). The evaporation of small molecules accounts for the characteristic odor of urine. Other substances not normally present, such as ammonia or ketone bodies, may also impart a distinctive smell. The analysis of a urine sample is a diagnostic tool of considerable importance even in modern high-technology medicine. A standard **urinalysis** involves an assessment of urine color and appearance, two characteristics that can be determined without specialized equipment. In the seventeenth century, physicians classified the taste of the urine as sweet, salty, and so on, but quantitative chemical tests have long since replaced the taste bud assay. Average values for urinalysis are presented in Table 26-6. AM *Urinalysis*

Urine Transport, Storage, and Elimination

Figure 26-19

Filtrate modification and urine production end when the fluid enters the renal pelvis. The remaining parts of the urinary system (the *ureters, urinary bladder,* and *urethra)* are responsible for the transport, storage, and elimination of urine. Figure 26-19●, a pyelogram of the urinary system, provides an orientation to the relative sizes and positions of these organs. The sizes of the minor and major calyces, the renal pelvis, the ureters, the urinary bladder, and the proximal portion of the urethra are somewhat variable because these regions are lined by a *transitional epithelium* that can tolerate cycles of distension and contraction without damage. [p. 116]

We will now examine these components of the urinary system and consider their functional interactions.

The Ureters

Figures 26-3, 26-19, 26-21

The ureters are a pair of muscular tubes that extend inferiorly from the kidneys for about 30 cm (12 in.)

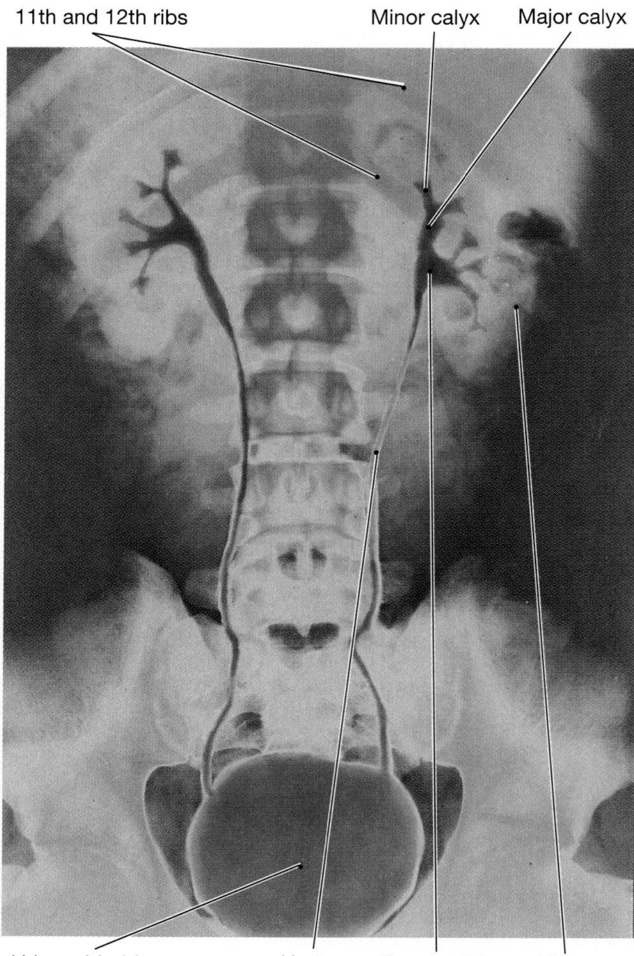
11th and 12th ribs
Minor calyx
Major calyx
Urinary bladder
Ureter
Renal pelvis
Kidney

● **FIGURE 26-19**
A Radiographic View of the Urinary System. This is a color-enhanced view known as a *pyelogram.* D

TABLE 26-6 Typical Values from Standard Urine Testing

Compound	*Primary Source*	*Daily Excretion*[a]	*Concentration*	*Remarks*
Nitrogenous wastes				
Urea	Deamination of amino acids at liver and kidneys	21 g	1.8 g/dl	Rises if negative nitrogen balance exists
Creatinine	Breakdown of creatine phosphate in skeletal muscle	1.8 g	150 mg/dl	Proportional to muscle mass; decreases during atrophy or muscle disease
Ammonia	Deamination by liver and kidney, absorption from intestinal tract	0.68 g	60 mg/dl	
Uric acid	Breakdown of purines	0.53 g	40 mg/dl	Increases in gout, liver diseases
Hippuric acid	Breakdown of dietary toxins	4.2 mg	350 μg/dl	
Urobilin	Urobilinogens absorbed at colon	1.5 mg	125 μg/dl	Gives urine its yellow color
Bilirubin	Hemoglobin breakdown product	0.3 mg	20 μg/dl	Increase may indicate problem with liver excretion or excess production; causes yellow skin color in jaundice
Nutrients and metabolites				
Carbohydrates		0.11 g	9 μg/dl	Primarily glucose; *glycosuria* if T_m is exceeded
Ketone bodies		0.21 g	17 μg/dl	Ketonuria may occur during postabsorptive state
Lipids		0.02 g	1.6 μg/dl	May increase in some kidney diseases
Amino acids		2.25 g	287.5 μg/dl	Note relatively high loss compared with other metabolites—low T_m, no large storage pools exist; excess *(aminoaciduria)* indicates T_m problem
Ions				
Sodium		4.0 g	333 mg/dl	Varies with diet, urine pH, hormones, etc.
Chloride		6.4g	533 mg/dl	
Potassium		2.0 g	166 mg/dl	Varies with diet, urine pH, hormones, etc.
Calcium		0.2 g	17 mg/dl	Hormonally regulated (PTH/CT)
Magnesium		0.15 g	13 mg/dl	
Blood cells[b]				
RBCs		130,000/day	100/ml	Excess *(hematuria)* indicates vascular damage
WBCs		650,000/day	500/ml	Excess *(pyuria)* indicates infection, inflammation

[a]Representative values for a 70-kg male.
[b]These are usually estimated by counting the numbers in a sample of sediment after urine centrifugation.

Advances in the Treatment of Renal Failure

Many different conditions can result in renal failure. Management of chronic renal failure typically involves restricting water and salt intake and reducing protein intake to a minimum, with few dietary proteins. This combination reduces strain on the urinary system by (1) minimizing the volume of urine produced and (2) preventing the generation of large quantities of nitrogenous wastes. Acidosis, a common problem in patients with renal failure, can be countered with ingestion of bicarbonate ions.

If drugs and dietary controls cannot stabilize the composition of the blood, more drastic measures are taken. In one technique, called **hemodialysis** (hē-mō-dī-AL-i-sis), an artificial membrane is used to regulate the composition of the blood. The basic principle involved in this process, called **dialysis,** involves passive diffusion across a semipermeable membrane. The patient's blood flows past an artificial *dialysis* membrane that contains pores large enough to permit the diffusion of small ions, but small enough to prevent the loss of plasma proteins. On the other side of the membrane flows a special **dialysis fluid.**

TABLE 26-7 The Composition of Dialysis Fluid

Constituent	*Normal Plasma*	*Dialyzing Fluid*
ELECTROLYTES (mEq/1)		
Potassium	5	3
Bicarbonate	27	36
Phosphate	3	0
Sulfate	0.5	0
NUTRIENTS (mg/dl)		
Glucose	100	125
NITROGENOUS WASTES (mg/dl)		
Urea	26	0
Creatinine	1	0
Uric acid	0.3	0

Note: Only the significant variations are noted; values for other electrolytes are usually similar. Although these values are representative, the precise composition can be tailored to meet the specific clinical needs. For example, if plasma potassium levels are too low, the dialyzing fluid concentration can be elevated to remedy the situation. Changes in the osmolarity of the dialyzing fluid can also be used to adjust an individual's blood volume, usually by adjusting the glucose content of the dialyzing fluid.

The composition of typical dialysis fluid is indicated in Table 26-7. As diffusion takes place across the membrane, the composition of the blood changes. Potassium ions, phosphate ions, sulfate ions, urea, creatinine, and uric acid diffuse across the membrane into the dialyzing fluid. Bicarbonate ions and glucose diffuse into the bloodstream. In effect, diffusion across the dialysis membrane takes the place of normal glomerular filtration, and the characteristics of the dialysis fluid ensure that important metabolites remain in the circulation, rather than diffusing across the membrane.

In practice, silastic tubes, called *shunts*, are inserted into a medium-sized artery and vein. (The usual location is in the forearm, although the lower leg is sometimes used.) The two shunts are then connected as shown in Figure 26-20b●. The connection acts like a "short circuit" that does not impede blood flow, and the shunts can be used like taps in a wine barrel, to draw a blood sample or to connect the individual to a *dialysis machine* (Figure 26-20a,b●). For long-term dialysis, a surgically created arteriovenous anastomosis provides access.

When connected to the dialysis machine, the individual sits quietly while blood circulates from the arterial shunt, through the machine, and back via the venous shunt. Inside the machine, the blood flows across a dialyzing membrane, where diffusion occurs.

Use of a dialysis machine is suggested when a patient's *BUN (blood urea nitrogen)* exceeds 100 mg/dl (the normal value is up to 30 mg/dl). Dialysis techniques are useful because they can maintain patients awaiting a transplant or those whose kidney function has been temporarily disrupted. Hemodialysis does have a number of drawbacks, however: (1) The patient must sit by the machine about 15 hours per week; (2) between treatments the symptoms of uremia will gradually develop; (3) hypotension can develop as a result of fluid loss during dialysis; (4) air bubbles in the tubing can cause embolism formation in the bloodstream; (5) anemia often develops; and (6) the shunts can serve as sites for recurring infections.

One alternative to the use of a dialysis machine is *peritoneal dialysis.* In **peritoneal dialysis** the peritoneal lining is used as a dialysis membrane. Dialyzing fluid is introduced into the peritoneum through a catheter in the abdominal wall, and at intervals the fluid is removed and replaced. For example, one procedure involves cycling 2 liters of fluid in an hour—15 minutes for infusion, 30 minutes for exchange, and 15 minutes for fluid reclamation. This process is usually performed in a hospital. An interesting variation on this procedure is called **continuous ambulatory peritoneal dialysis (CAPD).** In this procedure the patient administers 2 liters of dialyzing fluid through the catheter and then continues with life as usual until 4–6 hours later, when the fluid is removed and replaced.

Probably the most satisfactory solution, in terms of overall quality of life, is *kidney transplantation.* This procedure involves the implantation of a new kidney obtained from a living donor or a cadaver. One-third of the approximately 8000 kidneys transplanted in 1994 were obtained from living, related donors. The damaged kidney is usually removed and its blood supply is connected to the transplant. When the original kidney is left in place, an arterial graft is inserted to carry blood from the iliac artery or the aorta to the transplant, which is placed in the pelvis or lower abdomen.

The success rate for this procedure varies, depending on how aggressively the recipient's T cells attack the donated organ and whether an infection develops. The 1-year success rate for implantation is now 85–95 percent. The use of kidneys taken from close relatives significantly improves the chances for a successful transplant. Immunosuppressive drugs are administered to reduce tissue rejection, but unfortunately this treatment also lowers the individual's resistance to infection.

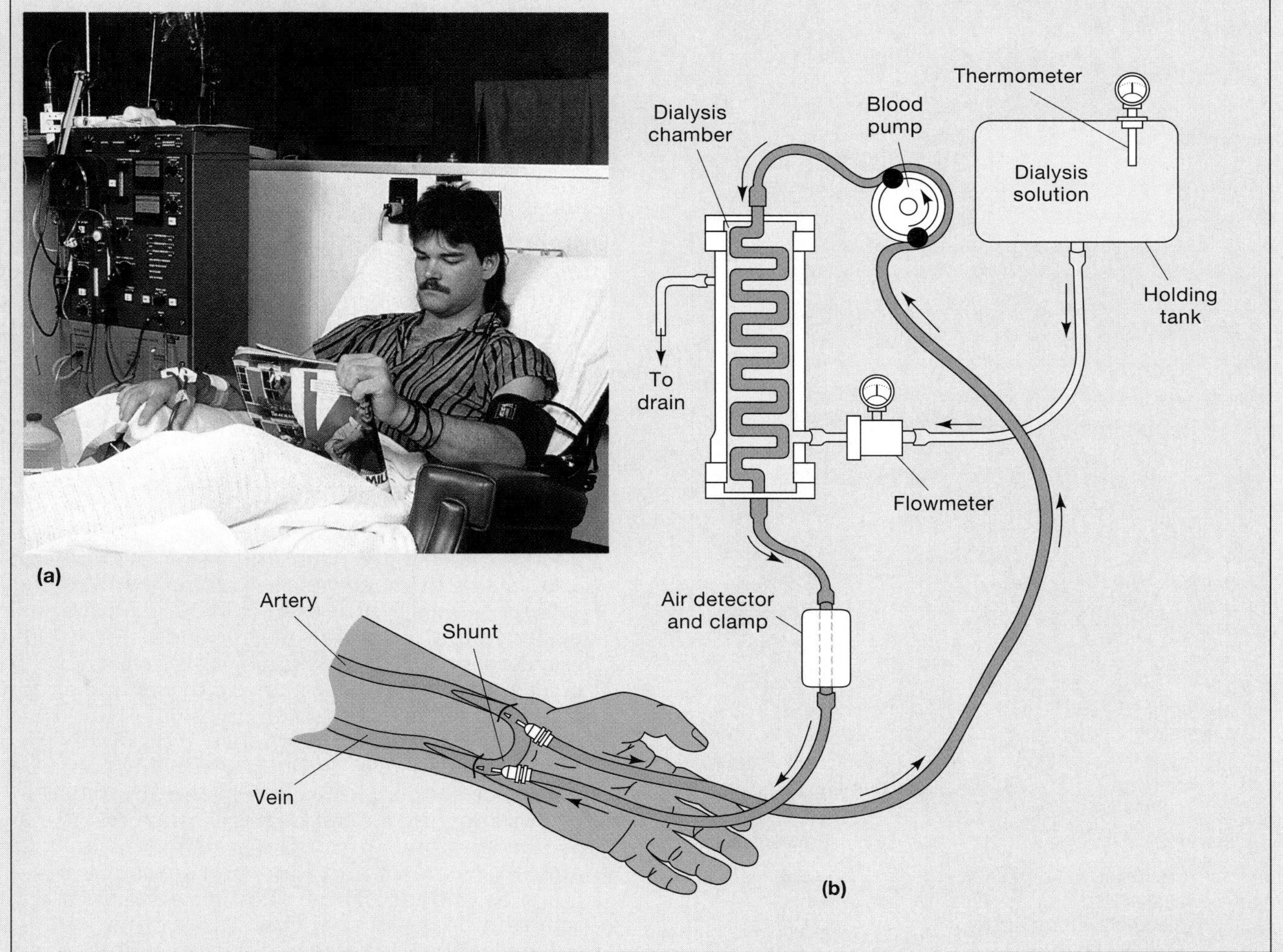
Thermometer
Dialysis chamber
Blood pump
Dialysis solution
Holding tank
To drain
Flowmeter
(a)
Air detector and clamp
Artery
Shunt
Vein
(b)

● **FIGURE 26-20**

Hemodialysis. (a) Preparation for hemodialysis typically involves implantation of a pair of shunts connected by a loop that permits normal blood flow when the patient is not hooked up to the dialysis machine. **(b)** A diagrammatic view of the dialysis procedure.

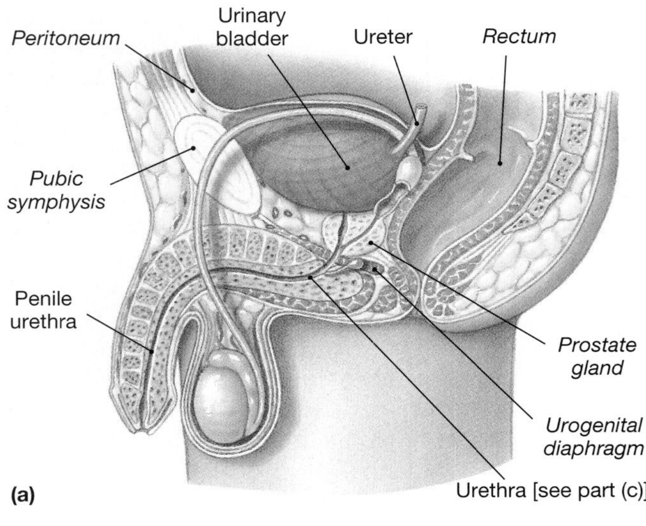
Peritoneum
Urinary bladder
Ureter
Rectum
Pubic symphysis
Penile urethra
Prostate gland
Urogenital diaphragm
(a)
Urethra [see part (c)]

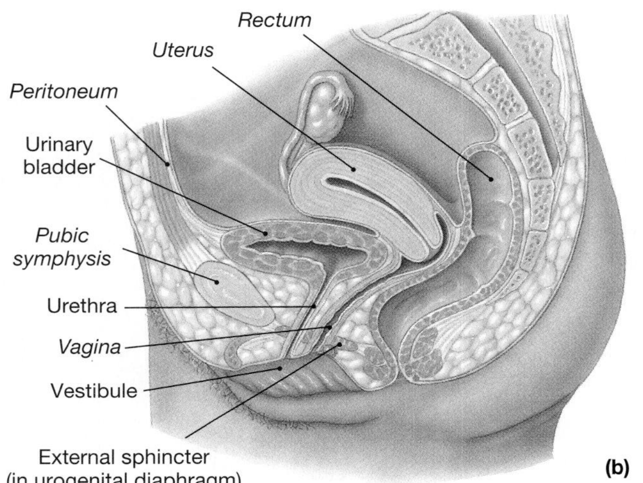
Rectum
Uterus
Peritoneum
Urinary bladder
Pubic symphysis
Urethra
Vagina
Vestibule
External sphincter (in urogenital diaphragm)
(b)

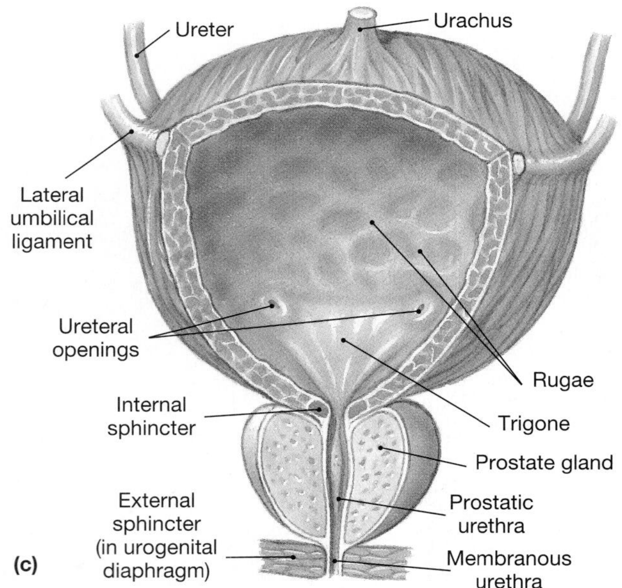
Ureter
Urachus
Lateral umbilical ligament
Ureteral openings
Rugae
Internal sphincter
Trigone
Prostate gland
External sphincter (in urogenital diaphragm)
Prostatic urethra
(c)
Membranous urethra

● **FIGURE 26-21**
Organs Responsible for the Conduction and Storage of Urine. (a) The ureter, urinary bladder, and urethra in the male. **(b)** The same organs in the female. **(c)** The urinary bladder in a male. AM *Plates 6.6, 6.7*

before reaching the urinary bladder. Each ureter begins at the funnel-shaped renal pelvis (Figure 26-19●). As the ureters descend from the kidneys, they pass inferiorly and medially over the psoas muscles. The ureters are retroperitoneal in position, and they are firmly attached to the posterior abdominal wall (Figure 26-3●, p. 982). The paths taken by the ureters in men and women are somewhat different because of variations in the nature, size, and position of the reproductive organs (Figure 26-21a,b●). In the male the base of the urinary bladder lies between the rectum and the symphysis pubis; in the female, the base of the urinary bladder sits inferior to the uterus and anterior to the vagina.

The ureters penetrate the posterior wall of the urinary bladder without entering the peritoneal cavity. They pass through the bladder wall at an oblique angle, and the **ureteral opening** is slit-like, rather than rounded (Figure 26-21c●). This shape helps to prevent backflow of urine toward the ureter and kidneys when the urinary bladder contracts.

Histology of the Ureters

Figure 26-22a

The wall of each ureter consists of three layers: (1) an inner mucosa covered by a transitional epithelium; (2) a middle muscular layer made up of longitudinal and circular bands of smooth muscle; and (3) an outer connective tissue layer that is continuous with the fibrous renal capsule and peritoneum (Figure 26-22a●). Starting at the kidney, about every 30 seconds a peristaltic contraction of the muscular wall serves to "milk" urine out of the renal pelvis and along the ureter to the bladder.

PROBLEMS WITH THE CONDUCTING SYSTEM Local blockages of the collecting tubules, collecting ducts, or ureter may result from the formation of *casts,* small blood clots, epithelial cells, lipids, or other materials. Casts are often excreted in the urine and visible in microscopic analysis of urine samples. **Calculi** (KAL-kū-lī), or "kidney stones," form from calcium deposits, magnesium salts, or crystals of uric acid. This condition is called *nephrolithiasis* (nef-rō-li-THĪ-a-sis). The blockage of the urinary passage by a stone or other factors, such as external compression, results in **urinary obstruction.** Urinary obstruction is a serious problem because, in addition to causing pain, it will reduce or eliminate filtration in the affected kidney by elevating the capsular hydrostatic pressure.

Kidney stones are usually visible on an X-ray, and if peristalsis and fluid pressures are insufficient to dislodge them they must be surgically removed or destroyed. One interesting nonsurgical procedure involves breaking kidney stones apart with a *lithotripter,* the same apparatus used to destroy gallstones. ∞ *[p. 914]* Another nonsurgical approach entails the insertion of a catheter armed with a laser that can shatter kidney stones with intense light beams.

❑ The Urinary Bladder

Figure 26-21c

The urinary bladder is a hollow muscular organ that functions as a temporary "storage reservoir" for urine. The dimensions of the urinary bladder vary, depending on the state of distension, but the full urinary bladder can contain about a liter of urine.

The superior surfaces of the urinary bladder are covered by a layer of peritoneum, and several peritoneal folds assist in stabilizing its position (Figure 26-21c●). The **middle umbilical ligament,** or **urachus** (Ū-ra-kus), extends from the anterior and superior border toward the umbilicus. The **lateral umbilical ligaments** pass along the sides of the bladder and also reach the umbilicus. These fibrous cords contain the vestiges of the two *umbilical arteries* that supplied blood to the placenta during embryonic and fetal development. [p. 767] The bladder's posterior, inferior, and anterior surfaces lie outside the peritoneal cavity, and in these areas tough ligamentous bands anchor the bladder to the pelvic and pubic bones.

In sectional view, the mucosa lining the urinary bladder is usually thrown into folds, or **rugae,** that disappear as the bladder fills. The triangular area bounded by the ureteral openings and the entrance to the urethra constitutes the **trigone** (TRĪ-gōn) of the urinary bladder. The mucosa here is smooth and very thick, and the trigone acts as a funnel that channels urine into the urethra when the urinary bladder contracts.

The urethral entrance lies at the apex of the trigone, at the most inferior point in the bladder. The region surrounding the urethral opening, known as the **neck** of the urinary bladder, contains a muscular **internal urethral sphincter.** The smooth muscle fibers of the internal urethral sphincter provide involuntary control over the discharge of urine from the bladder.

Histology of the Urinary Bladder

Figure 26-22b

The wall of the urinary bladder contains mucosa, submucosa, and muscularis layers (Figure 26-22b●). The muscularis layer consists of inner and

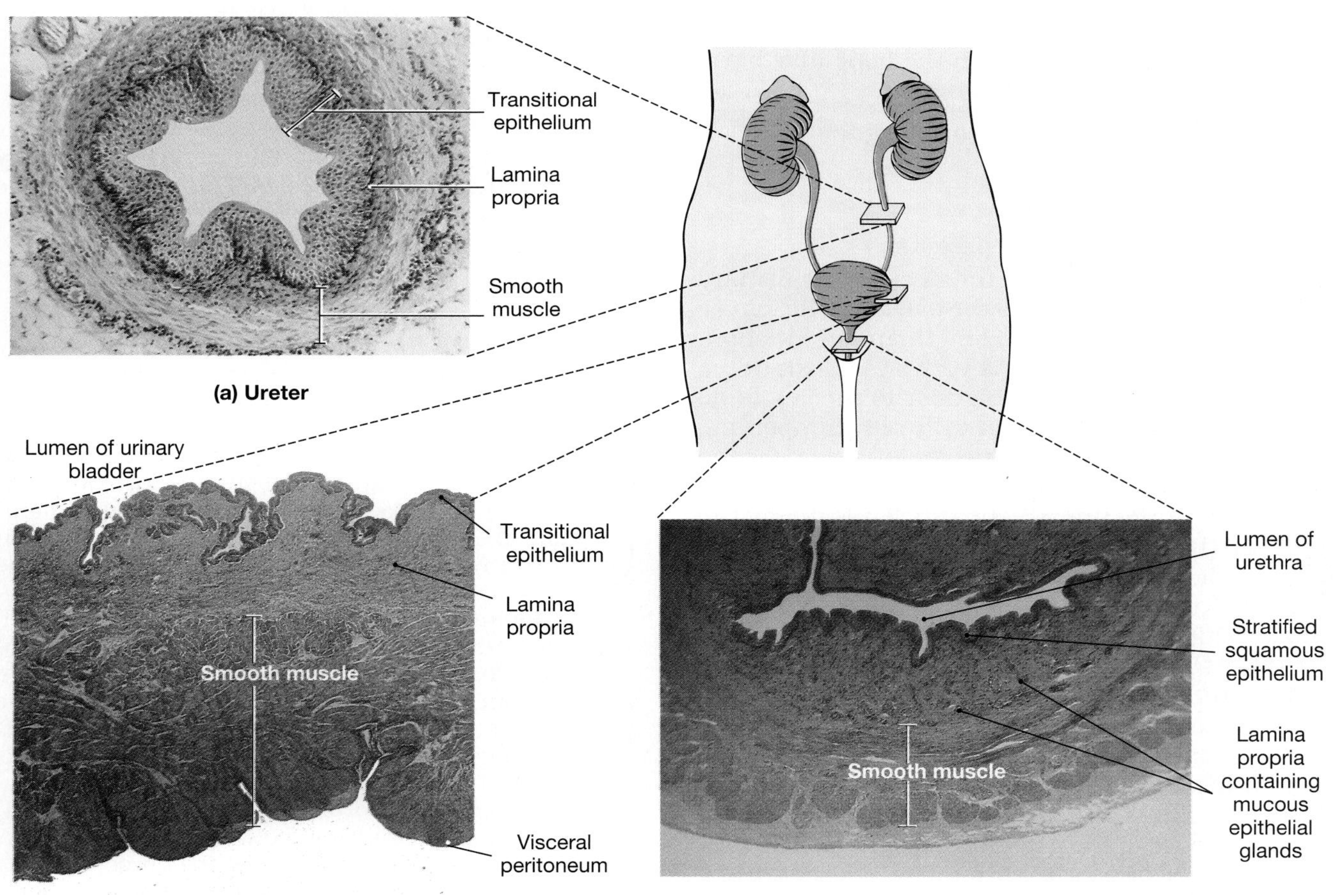
Transitional epithelium
Lamina propria
Smooth muscle
(a) Ureter
Lumen of urinary bladder
Transitional epithelium
Lamina propria
Smooth muscle
Visceral peritoneum
(b) Urinary bladder
Lumen of urethra
Stratified squamous epithelium
Lamina propria containing mucous epithelial glands
Smooth muscle
(c) Urethra

● **FIGURE 26-22**
Histology of the Collecting and Transport Organs. (a) A ureter seen in transverse section. Note the thick layer of smooth muscle surrounding the lumen. (For a close-up of transitional epithelium, review Figure 4-5c●, p. 117.) (LM × 53) **(b)** The wall of the urinary bladder. (LM × 29) **(c)** A transverse section through the urethra. (LM × 49)

outer longitudinal smooth muscle layers, with a circular layer sandwiched between. Collectively, these layers form the powerful **detrusor** (de-TROO-sor) muscle of the urinary bladder. Contraction of this muscle compresses the urinary bladder and expels its contents into the urethra.

BLADDER CANCER Each year approximately 52,000 new cases of bladder cancer are diagnosed in the United States, and there are 9500 deaths from this condition. The incidence among males is three times that among females, and most patients are older males, age 60–70. Environmental factors, especially exposure to *2-naphthylamine* or related compounds, are responsible for most bladder cancers. For this reason, the bladder cancer rate is highest among cigarette smokers and employees of chemical and rubber companies. The mechanism appears to involve damage to tumor suppressor genes, such as p53, that regulate cell division. The prognosis is reasonably good for localized superficial cancers (88 percent 5-year survival), but poor for those with severe metastatic bladder cancer (9 percent 5-year survival). Treatment of metastatic bladder cancer is very difficult because it spreads rapidly via adjacent lymphatics and through the bone marrow of the pelvis.

The Urethra

Figure 26-21

The urethra extends from the neck of the urinary bladder (Figure 26-21c●) to the exterior. The female and male urethrae differ in length and in function. In the female the urethra is very short, extending 3–5 cm (1–2 in.) from the bladder to the vestibule (Figure 26-21b●). The external urethral opening, or **external urethral meatus,** is situated near the anterior wall of the vagina.

In the male the urethra extends from the neck of the urinary bladder to the tip of the penis, a distance that may be 18–20 cm (7–8 in.). The male urethra can be subdivided into three portions (Figure 26-21a●): (1) the *prostatic urethra,* (2) the *membranous urethra,* and (3) the *penile urethra.*

The **prostatic urethra** passes through the center of the prostate gland (Figure 26-21c●). The **membranous urethra** includes the short segment that penetrates the urogenital diaphragm, the muscular floor of the pelvic cavity. The **penile** (PĒ-nīl) **urethra** extends from the distal border of the urogenital diaphragm to the external urethral meatus at the tip of the penis (Figure 26-21a●). The functional differences between these regions will be considered in Chapter 28.

In both sexes, as the urethra passes through the urogenital diaphragm, a circular band of skeletal muscle forms the **external urethral sphincter.** The contractions of both the external and internal urethral sphincters are controlled by branches of the *hypogastric plexus.* [p. 541] Only the external urethral sphincter is under voluntary control, via the perineal branch of the pudendal nerve. [p. 439]

Histology of the Urethra

Figure 26-22c

The urethral lining consists of a stratified epithelium that varies from transitional at the neck of the urinary bladder, through stratified columnar at the midpoint, to stratified squamous near the external urethral meatus. The lamina propria is thick and elastic, and the mucous membrane is thrown into longitudinal folds (Figure 26-22c●). Mucus-secreting cells are found in the epithelial pockets, and in the male the epithelial mucous glands may form tubules that extend into the lamina propria. Connective tissues of the lamina propria anchor the urethra to surrounding structures. In the female, the lamina propria contains an extensive network of veins, and the entire complex is surrounded by concentric layers of smooth muscle.

URINARY TRACT INFECTIONS **Urinary tract infections,** or **UTIs,** result from the colonization of the urinary tract by bacterial or fungal invaders. The intestinal bacterium *Escherichia coli* is most often involved, and women are particularly susceptible to urinary tract infections because of the proximity of the external urethral orifice to the anus. Sexual intercourse may also push bacteria into the urethra, and, since the female urethra is relatively short, the urinary bladder may become infected.

The condition may be asymptomatic (without symptoms), but it can be detected by the presence of bacteria and blood cells in the urine. If inflammation of the urethral wall occurs, the condition may be termed *urethritis,* while inflammation of the lining of the bladder is called *cystitis.* Many infections, including sexually transmitted diseases such as gonorrhea, cause a combination of urethritis and cystitis. These conditions cause painful urination, a symptom known as *dysuria* (dis-Ū-rē-a), and the bladder becomes tender and sensitive to pressure. Despite the discomfort produced, the individual feels the urge to urinate frequently. Urinary tract infections usually respond to antibiotic therapies, although subsequent reinfections may occur.

In untreated cases the bacteria may proceed along the ureters to the renal pelvis. The resulting inflammation of the walls of the renal pelvis produces *pyelitis* (pī-e-LĪ-tis), and if the bacteria invade the renal cortex and medulla as well, *pyelonephritis* (pī-e-lō-nef-RĪ-tis) results. Signs and symptoms include a high fever, intense pain on the affected side, vomiting, diarrhea, and the presence of blood cells and pus in the urine.

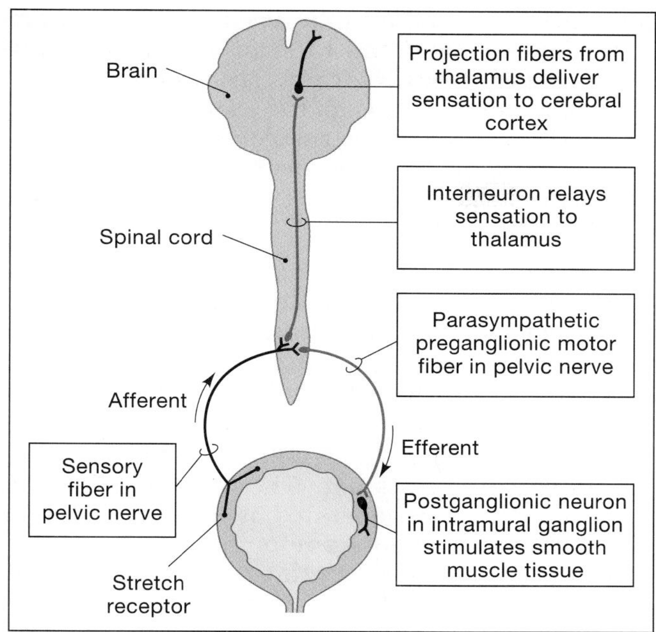
Brain
Projection fibers from thalamus deliver sensation to cerebral cortex
Interneuron relays sensation to thalamus
Spinal cord
Parasympathetic preganglionic motor fiber in pelvic nerve
Afferent
Efferent
Sensory fiber in pelvic nerve
Postganglionic neuron in intramural ganglion stimulates smooth muscle tissue
Stretch receptor

● **FIGURE 26-23**

The Micturition Reflex. The components of the reflex arc that stimulates smooth muscle contractions in the urinary bladder. Micturition occurs after voluntary relaxation of the external sphincter.

❑ The Micturition Reflex and Urination

Figure 26-23

Urine reaches the urinary bladder by the peristaltic contractions of the ureters. The process of urination is coordinated by the **micturition reflex.** Components of this reflex are diagrammed in Figure 26-23●.

Stretch receptors in the wall of the urinary bladder are stimulated as it fills with urine. Afferent fibers in the pelvic nerves carry the impulses generated to the sacral spinal cord. Their increased level of activity (1) facilitates parasympathetic motor neurons in the sacral spinal cord and (2) stimulates interneurons that relay sensations to the cerebral cortex. As a result, we become consciously aware of the fluid pressure in the urinary bladder.

The urge to urinate usually appears when the bladder contains about 200 ml of urine. The micturition reflex begins to function when the stretch receptors have provided adequate stimulation to the parasympathetic motor neurons. At this time activity in the motor neurons generates action potentials that reach the smooth muscle in the bladder wall. These efferent impulses travel over the pelvic nerves, producing a sustained contraction of the urinary bladder.

This contraction elevates fluid pressures in the urinary bladder, but urine ejection does not occur unless both the internal and external sphincters are relaxed. Relaxation of the external sphincter occurs under voluntary control. When the external sphincter relaxes, so does the internal sphincter. If the external sphincter does not relax, the internal sphincter remains closed. Stimulation of the smooth muscles in the urinary bladder soon subsides, and the bladder gradually relaxes.

A further increase in bladder volume begins the cycle again, usually within an hour. Each increase in urinary volume leads to an increase in stretch receptor stimulation that makes the sensation more acute. Once the volume of the urinary bladder exceeds 500 ml, the micturition reflex may generate enough pressure to force open the internal sphincter. This opening leads to a reflexive relaxation in the external sphincter, and urination occurs despite voluntary opposition or potential inconvenience. At the end of a normal micturition, less than 10 ml of urine remains in the bladder.

Infants lack voluntary control over urination because the necessary corticospinal connections have yet to be established. Toilet training before age 2 usually involves training the parent to anticipate the timing of the reflex, rather than training the child to exert conscious control.

INCONTINENCE **Incontinence** (in-KON-ti-nens) is the inability to control urination voluntarily. Incontinence may develop in otherwise normal adults because of trauma to the internal or external sphincters. For example, childbirth can stretch and damage the sphincter muscles, and some women then develop *stress incontinence.* In this condition elevated intra-abdominal pressures, caused for example by a cough or sneeze, can overwhelm the sphincter muscles, causing urine to leak out. Incontinence may also develop in older individuals, because of a general loss of muscle tone.

Damage to the CNS, the spinal cord, or the nerve supply to the bladder or external sphincter may also produce incontinence. For example, incontinence often accompanies Alzheimer's disease, and it may also result from a stroke or spinal cord injury. In most cases the affected individual develops an *automatic bladder.* The micturition reflex remains intact, but voluntary control of the external sphincter is lost and the individual cannot prevent the reflexive emptying of the bladder. Damage to the pelvic nerves can eliminate the micturition reflex entirely, because these nerves carry both afferent and efferent fibers of this reflex arc. In this case the bladder becomes greatly distended with urine. It remains filled to capacity, and the excess trickles into the urethra in an uncontrolled stream. Insertion of a catheter is often required to facilitate the discharge of urine.

Aging and the Urinary System

In general, aging is associated with an increased incidence of kidney problems. Examples, such as *nephrolithiasis* (kidney stones), were described earlier in the chapter. Other age-related changes in the urinary system include:

1. ***A decline in the number of functional nephrons.*** The total number of kidney nephrons drops by 30–40 percent between ages 25 and 85.
2. ***A reduction in the GFR.*** This reduction results from decreased numbers of glomeruli, cumulative damage to the filtration apparatus in the remaining glomeruli, and reductions in renal blood flow.
3. ***Reduced sensitivity to ADH.*** With age the distal portions of the nephron and collecting system become less responsive to ADH. Less reabsorption of water and sodium ions occurs, and more sodium ions are lost in the urine.
4. ***Problems with the micturition reflex.*** Several factors are involved in such problems.

   - The sphincter muscles lose muscle tone and become less effective at voluntarily retaining urine. Problems with *incontinence* result, often involving a slow leakage of urine.
   - The ability to control micturition is often completely lost following a stroke, Alzheimer's disease, or other CNS problems affecting the cerebral cortex or hypothalamus.
   - In males, **urinary retention** may develop secondary to enlargement of the prostate gland *(prostatic hypertrophy)*. In this condition swelling and distortion of surrounding prostatic tissues compress the urethra, restricting or preventing the flow of urine.

Integration with Other Systems

Figure 26-24

The urinary system is not the only organ system concerned with excretion. The urinary, integumentary, respiratory, and digestive systems are sometimes considered to form an anatomically diverse **excretory system.** The components of this system perform all of the excretory functions of the body that affect the composition of body fluids. Examples of excretory activities discussed in earlier chapters include:

1. ***Integumentary system.*** Water and electrolyte losses in sensible perspiration can affect plasma volume and composition. The effects are most apparent when losses are extreme, as in maximal sweat production. Small amounts of metabolic wastes, including urea, are also excreted in perspiration.
2. ***Respiratory system.*** The lungs excrete the carbon dioxide generated by living cells. Small amounts of other compounds, such as acetone and water, evaporate into the alveoli and are eliminated during exhalation.
3. ***Digestive system.*** Small amounts of metabolic waste products are excreted by the liver in bile, and a variable amount of water is lost in feces.

These excretory activities have an impact on the composition of body fluids. The respiratory system, for example, is the primary site of carbon dioxide excretion. But the excretory functions of these systems are not regulated as closely as are those of the kidneys, and under normal circumstances the effects of integumentary and digestive excretory activities are minor compared with those of the urinary system.

Figure 26-24● summarizes the functional relationships between the urinary system and other systems. Many of these relationships will be explored further in the next chapter, which considers major aspects of fluid, pH, and electrolyte balance.

What effect would a high-protein diet have on the composition of urine?

An obstruction of a ureter by a kidney stone would interfere with the flow of urine between what two points?

The ability to control the micturition reflex depends on one's ability to control what muscle?

INTEGUMENTARY SYSTEM

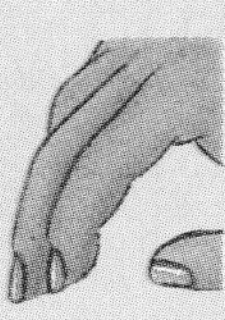

Sweat glands assist in elimination of water and solutes, especially sodium and chloride ions; keratinized epidermis prevents excessive fluid loss through skin surface; epidermis produces vitamin D_3, important for the renal production of calcitriol

URINARY SYSTEM

FOR ALL SYSTEMS

Excretes waste products, maintains normal body fluid pH and ion composition

SKELETAL SYSTEM

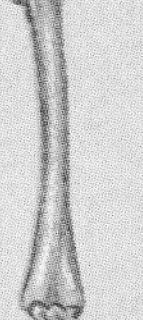

Axial skeleton provides some protection for kidneys and ureters; pelvis protects urinary bladder and proximal urethra

Conserves calcium and phosphate needed for bone growth

MUSCULAR SYSTEM

Sphincter controls urination by closing urethral opening; muscle layers of trunk provide some protection for urinary organs

Removes waste products of protein metabolism; assists in regulation of calcium and phosphate concentrations

NERVOUS SYSTEM

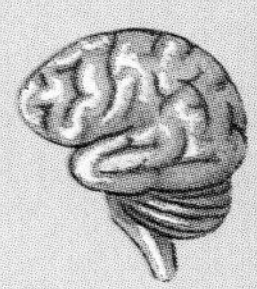

Adjusts renal blood pressure; monitors distension of bladder and controls urination

ENDOCRINE SYSTEM

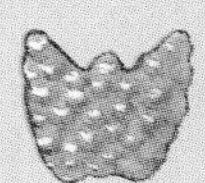

Aldosterone and ADH adjust rates of fluid and electrolyte reabsorption in kidneys

Kidney cells release renin when local blood pressure declines and erythropoietin (EPO) when renal O_2 levels decline

CARDIOVASCULAR SYSTEM

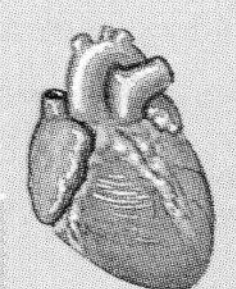

Delivers blood to capillaries when filtration occurs; accepts fluids and solutes reabsorbed during urine production

Releases renin to elevate blood pressure and erythropoietin to accelerate red blood cell production

LYMPHATIC SYSTEM

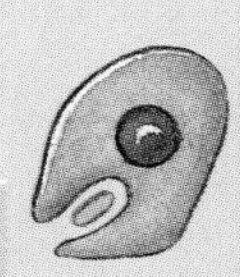

Provides specific defenses against urinary tract infections

Eliminates toxins and wastes generated by cellular activities; acid pH of urine provides nonspecific defense against urinary tract infections

RESPIRATORY SYSTEM

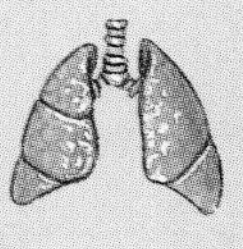

Assists in the regulation of pH by eliminating carbon dioxide

Assists in elimination of CO_2; provides bicarbonate buffers that assist in pH regulation

REPRODUCTIVE SYSTEM

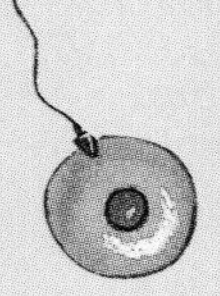

Accessory organ secretions may have antibacterial action that helps prevent urethral infections in males

Urethra in males carries semen to the exterior

DIGESTIVE SYSTEM

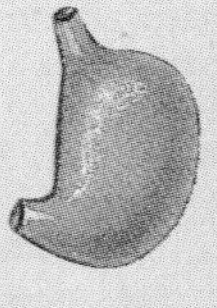

Absorbs water needed to excrete wastes at kidneys; absorbs ions needed to maintain normal body fluid concentrations; liver removes conjugated bilirubin

Excretes toxins, such as urobilinogen, absorbed by the digestive epithelium; excretes some conjugated bilirubin and nitrogenous wastes produced by the liver

FIGURE 26-24
Functional Relationships between the Urinary System and Other Systems

EMBRYOLOGY SUMMARY

Development of the Urinary System

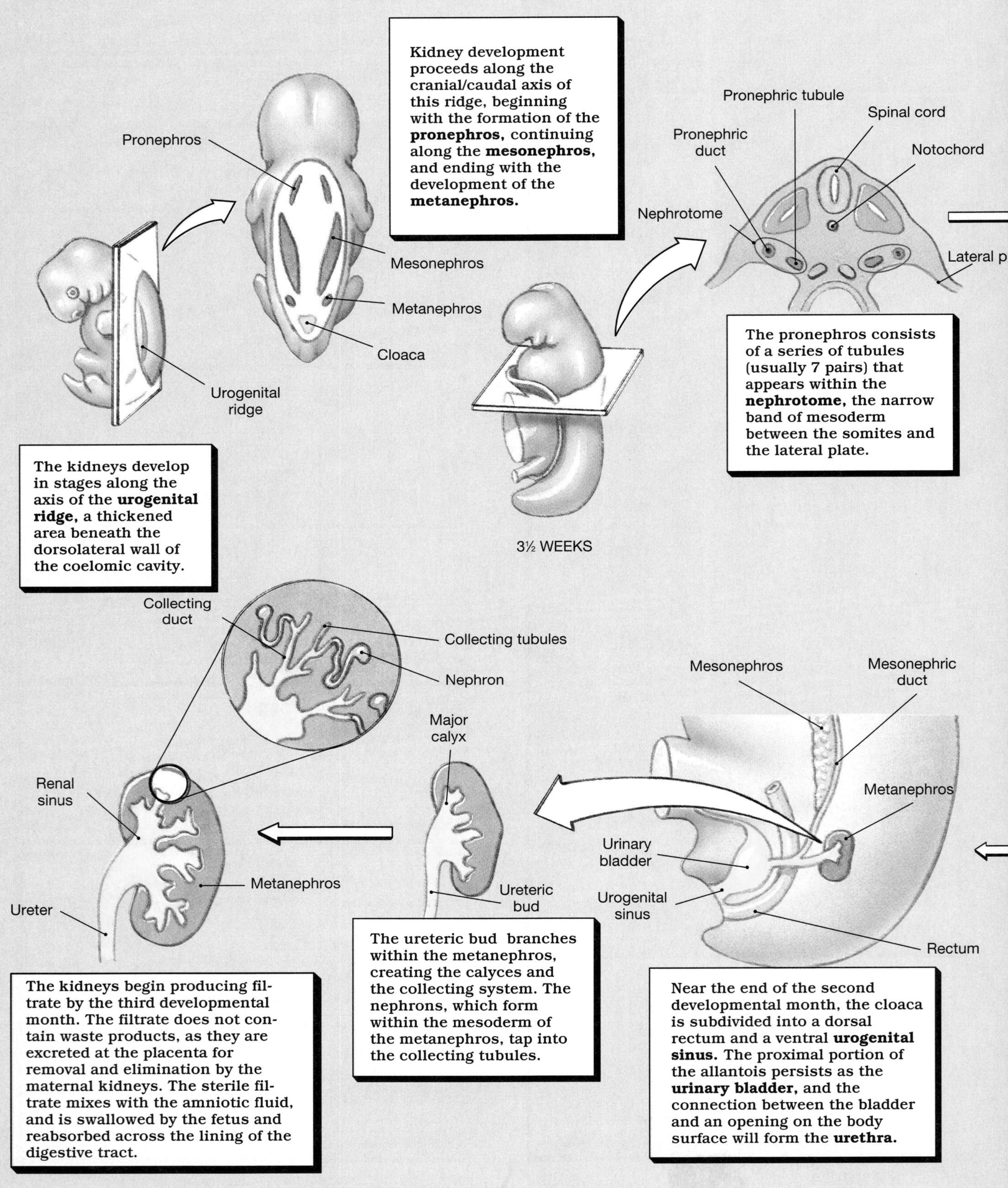

Kidney development proceeds along the cranial/caudal axis of this ridge, beginning with the formation of the **pronephros,** continuing along the **mesonephros,** and ending with the development of the **metanephros.**
Pronephros
Mesonephros
Metanephros
Cloaca
Urogenital ridge
Pronephric tubule
Spinal cord
Pronephric duct
Notochord
Nephrotome
Lateral p
3½ WEEKS
The pronephros consists of a series of tubules (usually 7 pairs) that appears within the **nephrotome,** the narrow band of mesoderm between the somites and the lateral plate.
The kidneys develop in stages along the axis of the **urogenital ridge,** a thickened area beneath the dorsolateral wall of the coelomic cavity.
Collecting duct
Collecting tubules
Nephron
Major calyx
Mesonephros
Mesonephric duct
Metanephros
Renal sinus
Metanephros
Urinary bladder
Ureteric bud
Urogenital sinus
Ureter
Rectum
The kidneys begin producing filtrate by the third developmental month. The filtrate does not contain waste products, as they are excreted at the placenta for removal and elimination by the maternal kidneys. The sterile filtrate mixes with the amniotic fluid, and is swallowed by the fetus and reabsorbed across the lining of the digestive tract.
The ureteric bud branches within the metanephros, creating the calyces and the collecting system. The nephrons, which form within the mesoderm of the metanephros, tap into the collecting tubules.
Near the end of the second developmental month, the cloaca is subdivided into a dorsal rectum and a ventral **urogenital sinus.** The proximal portion of the allantois persists as the **urinary bladder,** and the connection between the bladder and an opening on the body surface will form the **urethra.**

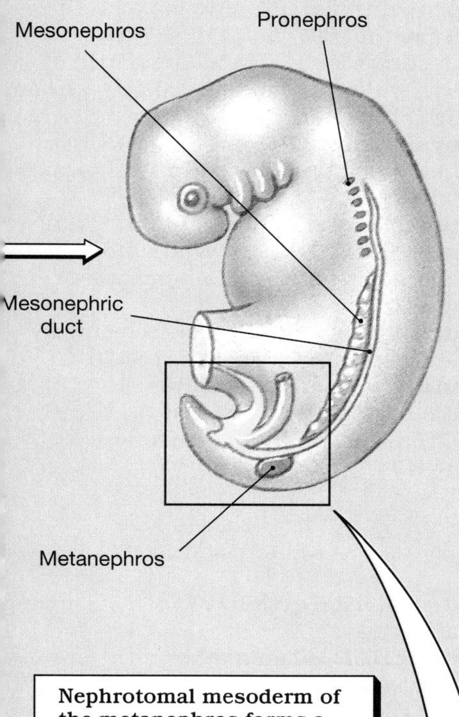

Mesonephros
Pronephros
Mesonephric duct
Metanephros

The pronephric kidneys are very small and nonfunctional, and they disintegrate almost at once. The only significant contribution of the pronephros is the formation of a pair of **pronephric ducts** that grow caudally until they connect to the *cloaca* (p. 887).

Duct

Tubule

After approximately 4 weeks of development, the mesoderm midway along the urogenital ridge begins organizing into the mesonephros. On either side of the midline, approximately 70 tubules develop within these segments. These tubules grow toward the adjacent pronephric duct and fuse with it. From this moment on, the duct will be called the **mesonephric duct.**

Aorta

Nephrotomal mesoderm of the metanephros forms a dense mass without a trace of segmental organization. This will become the functional adult kidney.

4 WEEKS

In each segment, a branch of the aorta grows toward the nephrotome, and the tubules form large nephrons with enormous glomeruli. Like the pronephros, the mesonephros does not persist, and when the last segments of the mesonephros are forming, the first are already beginning to degenerate.

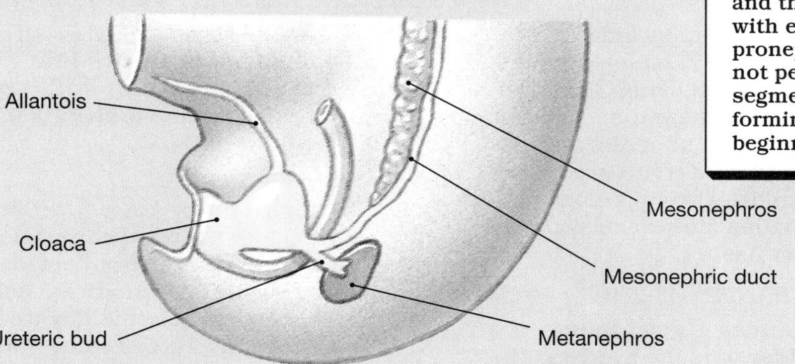

Allantois
Cloaca
Ureteric bud
Mesonephros
Mesonephric duct
Metanephros

Most of the metabolic wastes produced by the developing embryo are passed across the placenta to enter the maternal circulation. The small amount of urine produced by the kidneys accumulates within the cloaca and the *allantois* (p. 887), an endodermal sac that extends into the umbilical stalk.

A **ureteric bud** forms in the wall of each mesonephric duct, and this blind tube elongates and branches within the adjacent metanephros. Tubules developing within the metanephros then connect to the terminal branches.

Selected Clinical Terminology

Terms Discussed in This Chapter

aminoaciduria: Amino acid loss in the urine; the most common form is *cystinuria*. *(p. 998* and *AM)*
calculi (KAL-kū-lī): Insoluble deposits that form within the urinary tract from calcium salts, magnesium salts, or uric acid. *(p. 1014)*
clearance test: A test that permits calculation of the GFR by monitoring plasma and renal concentrations of a specific solute, such as creatinine. *(p. 994* and *AM)*
cystitis: Inflammation of the urinary bladder lining. *(p. 1016)*
diuretics (dī-ū-RET-iks): Drugs that promote fluid loss in the urine. *(p. 1008* and *AM)*
dysuria (dis-Ū-re-a): Painful urination. *(p. 1016)*
glomerulonephritis (glo-mer-ū-lō-nef-RĪ-tis): Inflammation of the renal cortex. *(p. 986)*
hematuria: Blood loss in the urine. *(p. 997* and *AM)*
hemodialysis (hē-mō-dī-AL-i-sis): A technique in which an artificial membrane is used to regulate the composition of blood. *(p. 1012)*
incontinence (in-KON-ti-nens): An inability to control urination voluntarily. *(p. 1017)*
polycystic kidney disease: An inherited abnormality that affects the development and structure of kidney tubules. *(p. 988)*
proteinuria: Protein loss in the urine. *(p. 997* and *AM)*
pyelitis (pī-e-LĪ-tis): Inflammation of the walls of the renal pelvis. *(p. 1016)*
pyelonephritis (pī-e-lō-nef-RĪ-tis): Inflammation of the kidney tissues. *(p. 1016)*
renal failure: An inability of the kidneys to excrete wastes in sufficient quantities to maintain homeostasis. *(p. 997)*
urinalysis: A physical and chemical assessment of urine. *(p. 1010)*
urinary obstruction: Blockage of the urinary tract. *(p. 1014)*
UTI: A urinary tract infection. *(p. 1016)*

AM *Additional Terms Discussed in the Applications Manual*

nephritis: Inflammation of the kidneys.
pyelogram (PĪ-el-o-gram): An image obtained by taking an X-ray of the kidneys after a radiopaque compound has been administered.
uremia (ū-RĒ-mē-a): A change in the composition of the blood indicating that all renal excretory functions are abnormal.
urethritis: Inflammation of the urethral wall.

CHAPTER REVIEW

STUDY OUTLINE

INTRODUCTION, p. 980

1. The functions of the urinary system include (1) eliminating organic waste products; (2) regulating plasma concentrations of ions; (3) regulating blood volume and pressure by adjusting the volume of water lost and releasing erythropoietin and renin; (4) helping to stabilize blood pH; (5) conserving nutrients; and (6) assisting the liver in detoxifying poisons and, during starvation, deaminating amino acids so they can be catabolized by other tissues.

ORGANIZATION OF THE URINARY SYSTEM, p. 981

1. The urinary system includes the **kidneys,** the **ureters,** the **urinary bladder,** and the **urethra.** The kidneys produce **urine** (a fluid containing water, ions, and soluble compounds). During **urination** urine is forced out of the body. (*Figure 26-1*)

THE KIDNEYS, p. 981

1. The left kidney extends superiorly slightly more than the right kidney. Both kidneys and their adrenal gland caps lie in a retroperitoneal position. (*Figure 26-2*)

Superficial Anatomy of the Kidney, p. 982

2. The **hilus,** a medial indentation, provides entry for the renal artery and exit for the renal vein and ureter. (*Figure 26-3*)

Sectional Anatomy of the Kidney, p. 982

3. The **ureter** communicates with the **renal pelvis.** This chamber branches into two **major calyces,** each connected to four or five **minor calyces** that enclose the **renal papillae.** (*Figure 26-4*)

The Nephron, p. 983

4. The **nephron** (the basic functional unit in the kidney) includes the **renal corpuscle** and a **renal tubule** that empties into the **collecting system** via a **collecting tubule,** a tributary of a **collecting duct.** From the **renal corpuscle** the **filtrate** travels through the **proximal convoluted tubule,** the **loop of Henle,** and the **distal convoluted tubule.** (*Figure 26-5*)
5. Nephrons are responsible for (1) production of filtrate, (2) reabsorption of nutrients, and (3) reabsorption of water and ions. (*Table 26-1*)
6. Roughly 85 percent of the nephrons are **cortical nephrons** found within the cortex; the **juxtamedullary nephrons** are closer to the medulla, with their loops of Henle extending deep into the renal pyramids.
7. The renal tubule begins at the renal corpuscle. It includes a knot of intertwined capillaries called the **glomerulus** surrounded by **Bowman's capsule.** Blood arrives at the glomerulus via the **afferent arteriole** and departs in the **efferent arteriole.** (*Figure 26-6a,b,c*)

8. At the glomerulus, **podocytes** cover the **lamina densa** of the capillaries that project into the **capsular space.** The **pedicels** of the podocytes are separated by narrow **filtration slits.** (*Figure 26-6d,e*)
9. The **proximal convoluted tubule (PCT)** actively reabsorbs nutrients, plasma proteins, and electrolytes from the filtrate. The loop of Henle includes a **descending limb** and an **ascending limb;** each limb contains a **thick segment** and a **thin segment.** The ascending limb delivers fluid to the **distal convoluted tubule (DCT),** which actively secretes ions and reabsorbs sodium ions from the urine. (*Figure 26-7*)

The Blood Supply to the Kidneys, p. 988

10. The vasculature of the kidneys includes the **interlobar, arcuate,** and **interlobular arteries,** and the **interlobar, arcuate,** and **interlobular veins.** Blood travels from the efferent arteriole to the **peritubular capillaries** and the **vasa recta.** Diffusion occurs between the capillaries of the vasa recta and the tubular cells through the **peritubular fluid** that surrounds the nephron. (*Figures 26-8, 26-9*)

Innervation of the Kidneys, p. 991

11. The **renal nerves** that innervate the kidneys and ureters contain mostly sympathetic postganglionic fibers.

RENAL PHYSIOLOGY, p. 991

Basic Principles of Urine Formation, p. 991

1. Urine formation involves **filtration, reabsorption,** and **secretion.** (*Table 26-2*)

Filtration, p. 992

2. **Glomerular filtration** occurs as fluids move across the wall of the glomerulus into the capsular space, in response to the hydrostatic (blood) pressure in the glomerular capillaries **(GHP).** This movement is opposed by the **capsular hydrostatic pressure (CHP)** and by the colloid osmotic pressure of the blood **(BCOP).** The **net filtration pressure (NFP)** at the glomerulus is the difference between the blood pressure and the opposing capsular and osmotic pressures. (*Figure 26-10*)
3. The **glomerular filtration rate (GFR)** is the amount of filtrate produced in the kidneys each minute. Any factor that alters the filtration pressure acting across the glomerular capillaries will change the GFR and affect kidney function (*Figure 26-11*)
4. Dropping filtration pressures stimulate the juxtaglomerular apparatus to release renin and erythropoietin. ADH production and release occur after stimulation by angiotensin II or after stimulation of hypothalamic **osmoreceptors.**
5. Sympathetic activation (1) produces a powerful vasoconstriction of the afferent arterioles, decreasing the GFR and slowing the production of filtrate, (2) alters the GFR by changing the regional pattern of blood circulation, and (3) stimulates the release of renin by the juxtaglomerular apparatus.

Reabsorption and Secretion, p. 997

6. Four types of **carrier-mediated transport** (*facilitated diffusion, active transport, cotransport,* and *countertransport*) are involved in modifying the filtrate. The saturation limit of a carrier protein is its **transport maximum.** The transport maximum determines the **renal threshold,** the plasma concentration at which various compounds will appear in the urine. (*Table 26-3*)
7. Glomerular filtration produces a filtrate with a composition similar to blood plasma, but with few, if any, plasma proteins.
8. The cells of the PCT normally reabsorb 60 percent of the volume of filtrate produced in the renal corpuscle. The PCT generally reabsorbs sodium and other ions, water, and almost all of the organic nutrients in the filtrate. It also secretes various substances. (*Figure 26-12*)
9. Water and ions are reclaimed from the filtrate by the loop of Henle. An **osmotic gradient** in the medulla encourages the osmotic flow of water out of the filtrate. The **countercurrent multiplication** between the ascending and descending limbs of the loop of Henle helps create the osmotic gradient in the medulla. As water is lost by osmosis and the filtrate volume decreases, the urea concentration rises. (*Figures 26-13, 26-14*)
10. The DCT performs final adjustments by actively secreting or absorbing materials. Sodium ions are actively absorbed, in exchange for potassium or hydrogen ions discharged into the filtrate. Aldosterone secretion increases the rate of sodium reabsorption and potassium loss. (*Figures 26-15, 26-16*)
11. The amount of water and solutes in the urine of the collecting ducts is further regulated by aldosterone and ADH secretions.

The Function of the Vasa Recta, p. 1006

12. Normally the removal of solutes and water by the vasa recta precisely balances the rates of reabsorption and osmosis in the medulla. (*Figure 26-17*)

A Summary of Renal Function, p. 1007

13. Each segment of the nephron and collecting system contributes to the production of hypertonic urine. (*Figure 26-18; Table 26-4*)

The Composition of Normal Urine, p. 1009

14. More than 99 percent of the filtrate produced each day is reabsorbed before reaching the renal pelvis. (*Table 26-5*)
15. **Hemodialysis** is a technique used to regulate the composition of blood artificially. (*Figure 26-20; Table 26-6*)

URINE TRANSPORT, STORAGE, AND ELIMINATION, p. 1010

1. Filtrate modification and urine production end when the fluid enters the renal pelvis. The rest of the urinary system is responsible for transporting, storing, and eliminating the urine. (*Figures 26-19, 26-21*)

The Ureters, p. 1010

2. The ureters extend from the renal pelvis to the urinary bladder without entering the peritoneal cavity. Peristaltic contractions by smooth muscles move the urine. (*Figure 26-22a*)

The Urinary Bladder, p. 1015

3. The bladder is stabilized by the **urachus (middle umbilical ligament)** and the **lateral umbilical ligaments.** Internal features include the **trigone**, the **neck,** and the **internal sphincter.** The mucosal lining contains prominent **rugae** (folds). Contraction of the **detrusor** muscle compresses the bladder and expels the urine into the urethra. (*Figure 26-22b*)

The Urethra, p. 1016

4. In both sexes, as the urethra passes through the urogenital diaphragm a circular band of skeletal muscles forms the **external sphincter,** which is under voluntary control. (*Figure 26-22c*)

The Micturition Reflex and Urination, p. 1017

5. The process of urination is coordinated by the **micturition reflex,** which is initiated by stretch receptors in the bladder wall. Voluntary urination involves coupling this reflex with the voluntary relaxation of the external sphincter, which allows the opening of the **internal sphincter.** (*Figure 26-23*)

AGING AND THE URINARY SYSTEM, p. 1018

1. Aging is usually associated with increased kidney problems. Age-related changes in the urinary system include (1) declining number of functional nephrons, (2) reduced GFR, (3) reduced sensitivity to ADH, (4) problems with the micturition reflex (**urinary retention** may develop in men whose prostate glands are enlarged).

INTEGRATION WITH OTHER SYSTEMS, p. 1018

1. The urinary, integumentary, respiratory, and digestive systems are sometimes considered as an anatomically diverse **excretory system.** The system's components work together to perform all of the excretory functions that affect the composition of body fluids. (*Figure 26-24*)

REVIEW QUESTIONS

LEVEL 1 Reviewing Facts and Terms

1. The point of entry for the renal artery and exit for the renal vein and ureter is a medial indentation called the:
   (a) renal column (b) medulla
   (c) hilus (d) renal cortex

2. The glomerulus is located within the:
   (a) renal corpuscle (b) renal tubule
   (c) renal pelvis (d) renal column

3. Specialized cells that engulf organic materials that might otherwise clog the filter at the lamina densa are:
   (a) mesangial cells (b) juxtamedullary cells
   (c) podocytes (d) macula densa

4. Large cells with complex processes, or "feet," that wrap around the glomerular capillaries are called:
   (a) vasa recta (b) podocytes
   (c) astrocytes (d) mesangial cells

5. After the filtrate leaves the glomerulus it empties into the:
   (a) distal convoluted tubule
   (b) loop of Henle
   (c) proximal convoluted tubule
   (d) collecting duct

6. The distal convoluted tubule (DCT) is an important site for:
   (a) active secretion of ions
   (b) active secretion of acids and other materials
   (c) selective reabsorption of sodium ions from the tubular fluid
   (d) a, b, and c are correct

7. The endocrine structure that secretes renin and erythropoietin is the:
   (a) juxtaglomerular apparatus
   (b) vasa recta
   (c) Bowman's capsule
   (d) adrenal gland

8. The primary purpose of the collecting system is to:
   (a) transport urine from the bladder to the urethra
   (b) selectively reabsorb sodium ions from tubular fluid
   (c) transport urine from the renal pelvis to the ureters
   (d) make final adjustments to the osmotic concentration and volume of urine

9. Pickup or delivery of substances that are reabsorbed or secreted by the PCT and the DCT is provided by the:
   (a) afferent arteriole (b) peritubular capillaries
   (c) renal artery (d) efferent arteriole

10. The most abundant *organic* waste in urine is:
   (a) uric acid (b) creatinine
   (c) urea (d) creatine phosphate

11. The removal of water and solute molecules from the filtrate after it enters the renal tubules is:
   (a) filtration (b) secretion
   (c) reabsorption (d) excretion

12. The transport of solutes from the peritubular fluid into the tubular fluid is:
   (a) reabsorption (b) filtration
   (c) excretion (d) secretion

13. The force that tends to drive water and solutes into the interstitial fluid is the:
   (a) glomerular hydrostatic pressure
   (b) net hydrostatic pressure
   (c) capsular hydrostatic pressure
   (d) net colloid osmotic pressure

14. Inflammation of the lining of the urinary bladder is called:
   (a) urethritis (b) cystitis
   (c) dysuria (d) pyelitis

15. What is the *primary* function of the urinary system?

16. What structures are included as component parts of the urinary system?

17. Trace the pathway of the protein-free filtrate from the time it is produced in the renal corpuscle until it drains into the renal pelvis in the form of urine. (Use arrows to indicate the direction of flow.)

18. Name the segments of the nephron distal to the renal corpuscle and state the function(s) of each.

19. What role does the lamina densa play in the renal corpuscle?

20. What is the function of the juxtaglomerular apparatus?

21. Using arrows, trace a drop of blood from the time it enters the renal artery until it exits via a renal vein.

22. Name and define the three distinct processes involved in the production of urine.

23. What are the primary effects of angiotensin II on kidney function and regulation?

24. What parts of the urinary system are responsible for the transport, storage, and elimination of urine?

LEVEL 2 Reviewing Concepts

25. The urinary system regulates blood volume and pressure by:
   (a) adjusting the volume of water lost in the urine
   (b) releasing erythropoietin
   (c) releasing renin
   (d) a, b, and c are correct

26. The balance of solute and water reabsorption in the medulla is maintained by the:
   (a) segmental arterioles and veins
   (b) lobar arteries and veins
   (c) vasa recta
   (d) arcuate arteries

27. Sympathetic activation of the nerve fibers in the nephron causes:
   (a) regulation of glomerular blood flow and pressure
   (b) stimulation of renin release from the juxtaglomerular apparatus
   (c) direct stimulation of water and sodium ion reabsorption
   (d) a, b, and c are correct

28. An increase in the capsular colloid osmotic pressure caused by damage to the glomerulus would:
   (a) decrease the amount of plasma delivered to the kidney
   (b) enhance the movement of water and solutes into the capsular space
   (c) cause a decrease in the renal blood pressure
   (d) decrease movement of water and solutes into the capsular space

29. Sodium reabsorption in the DCT and cortical portion of the collecting system is accelerated by secretion of:
   (a) ADH (b) renin
   (c) aldosterone (d) erythropoietin

30. When ADH levels rise:
   (a) the amount of water reabsorbed increases
   (b) the DCT becomes impermeable to water
   (c) the amount of water reabsorbed decreases
   (d) sodium ions are exchanged for potassium ions

31. Control of blood pH by the kidneys involves:
   (a) addition of hydrogen ions and removal of bicarbonate ions from the filtrate
   (b) a decrease in the amount of water reabsorbed
   (c) hydrogen ion removal and bicarbonate ion production
   (d) potassium ion secretion

32. How does a cortical nephron differ from a juxtamedullary nephron?

33. How can you determine the net filtration pressure (NFP) at the glomerulus?

34. What interacting controls operate to stabilize the glomerular filtration rate (GFR)?

35. What is the renal threshold?

36. What primary changes occur in the composition and concentration of the filtrate as a result of activity in the proximal convoluted tubule?

37. What two functions does the countercurrent mechanism perform for the kidney?

38. What events in the DCT and collecting system determine the final composition and concentration of the filtrate?

39. Describe the micturition reflex.

LEVEL 3 Critical Thinking and Clinical Applications

40. Why do long-haul trailer truck drivers frequently experience kidney problems?

41. Doctors often ask for urine samples collected over a 24-hour period rather than a single sample. Why?

42. Randy enjoys his four or five cups of coffee a day and his six or seven beers a few nights a week. What effects does his caffeine and alcohol consumption have on the urinary system?

43. For the past week Susan has felt a burning sensation in the urethral area when she urinates. She checks her temperature and finds that she has a low-grade fever. What unusual substances are likely to be present in her urine?

44. *Lasix* is a diuretic that acts by decreasing the amounts of sodium and chloride ions actively transported by the ascending limb of the loop of Henle. Why would this medication be given to someone suffering from high blood pressure (hypertension)?

45. Carlos suffers from advanced arteriosclerosis. An analysis of his blood indicates elevated levels of aldosterone and decreased levels of ADH. Explain.

46. *Mannitol* is a sugar that is filtered but not reabsorbed by the kidneys. What effect would drinking a solution of mannitol have on the volume of urine produced?

47. A drug known as *Diamox* is sometimes used in the treatment of epilepsy. It functions by inhibiting the action of carbonic anhydrase in the proximal convoluted tubule. Polyuria (the elimination of an unusually large volume of urine) is a side effect associated with this medication. Why would this symptom occur?

CHAPTER 27

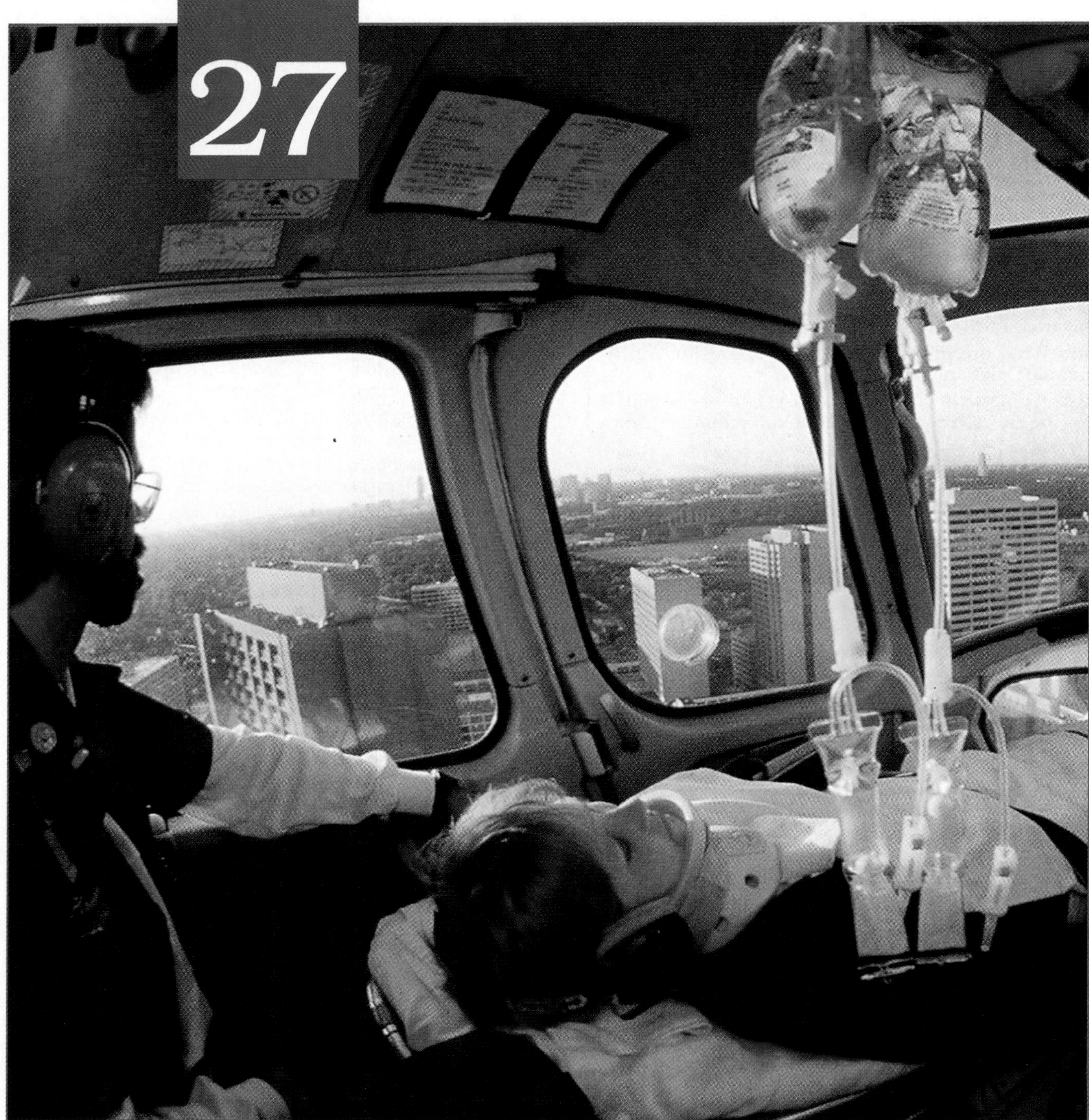

After a severe accident, getting to a medical facility as quickly as possible often means the difference between life and death. But even when the trip is made by Medivac helicopter, emergency measures usually can't wait until the patient arrives at a hospital. After serious trauma, the body's regulatory mechanisms may be unable to maintain homeostasis without assistance. The patient's internal conditions must therefore be monitored and stabilized while in transit. In virtually all medical emergencies this involves adjusting fluid and electrolyte levels, usually by administering intravenous solutions containing water, electrolytes, and buffers. Although the therapy is relatively simple, it can have multiple benefits: replacing lost blood, providing nutrients, and controlling the pH of body fluids. This chapter examines the mechanisms responsible for the maintenance of water, electrolyte, and acid-base balance under normal and abnormal conditions.

Fluid, Electrolyte, and Acid-Base Balance

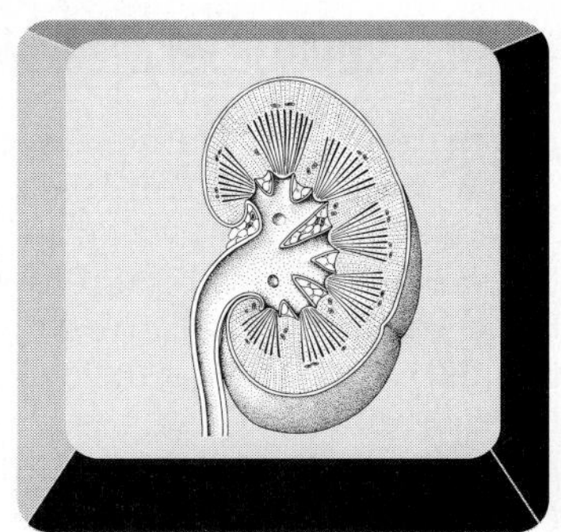

Chapter Outline and Objectives

1 Compare the composition of the intracellular and extracellular fluids.

2 Explain the basic concepts involved in fluid and electrolyte regulation.

3 Identify the hormones that play important roles in regulating fluid and electrolyte balance, and describe their effects.

4 Discuss the mechanisms by which sodium, potassium, calcium, and chloride concentrations are regulated to maintain electrolyte balance in the body.

5 Explain the buffering systems that balance the pH of the intracellular and extracellular fluid.

6 Describe the compensatory mechanisms involved in the maintenance of acid-base balance.

7 Identify the most frequent threats to acid-base balance, and explain how the body responds when the pH of body fluids varies outside normal limits.

The next time you see a small pond, take a moment to think about the fish it contains. They live out their lives totally dependent on the quality of that isolated environment. Polluting the pond with toxic substances will, of course, kill the fish, but more subtle changes can have equally grave effects. Changes in the volume of the pond, for example, can be quite important. If evaporation removes too much of the pond water, the fish become overcrowded; oxygen and food supplies run out, and the fish suffocate or starve. The ionic concentration of the pond water is also crucial. Most of the fish in a freshwater pond will die if the water becomes too salty; those in a saltwater pond will die if their environment becomes too dilute. The pH of the pond water, too, is a vital factor—that is why acid rain is such a problem.

The cells of our bodies live in a pond whose shores are the exposed surfaces of the skin. Most of the weight of the human body is water. Water accounts for up to 99 percent of the volume of extracellular fluid (ECF), and it is an essential ingredient of cytoplasm. All of a cell's operations rely on water as a diffusion medium for the distribution of gases, nutrients, and waste products. If the water content of the body changes, cellular activities are jeopardized. For example, when the water content of the body reaches very low levels, proteins denature, enzymes cease functioning, and cells die.

The ionic concentrations and pH of the body water are as important as its absolute quantity. If concentrations of calcium or potassium ions in the ECF become too high, cardiac arrhythmias develop, and the individual's life is in jeopardy. A pH outside the normal range can lead to a variety of dangerous problems. Low pH is especially dangerous because hydrogen ions break chemical bonds, change the shapes of complex molecules, disrupt cell membranes, and impair tissue functions.

This chapter considers the dynamics of exchange among the various body fluids and between the body and the external environment. Maintenance of normal volume and composition in the extracellular and intracellular fluids is essential to our survival. Stabilizing the volumes, solute concentrations, and pH of the ECF and ICF involves three interrelated processes:

- A person is in **fluid balance** when the amount of water gained each day is equal to the amount lost to the environment. The maintenance of normal fluid balance involves regulating body water *content* and *distribution* in the ECF and ICF. The digestive system is the primary source for water gains. A small amount of water is generated through metabolic activity. The urinary system is the primary route for water loss under normal conditions, although again as noted in Chapter 25, sweat gland activity can become important when body temperature is elevated. ∞ *[p. 970]* Because our cells and tissues cannot transport water, fluid balance reflects primarily the control of *electrolyte balance.*
- **Electrolyte balance** exists when there is neither a net gain nor a net loss of any ion in body fluids. This condition means primarily balancing the rates of absorption across the digestive tract with rates of loss at the kidneys, although excretion at sweat glands and other excretory sites plays a secondary role.
- **Acid-base balance** exists when the production of hydrogen ions precisely offsets their loss. When acid-base balance exists, the pH of body fluids remains within normal limits. Preventing a reduction in pH is the primary problem here, because a variety of acids are generated within the body in the course of normal metabolic operations. Once again, the kidneys play a major role by secreting hydrogen ions into the urine and generating buffers. As we saw in Chapter 26, this secretion occurs primarily in the distal segments of the distal convoluted tubule (DCT) and along the collecting system. ∞ *[p. 1005]* As we shall see, the lungs also play an important role through the elimination of carbon dioxide.

Much of the material in this chapter was introduced in earlier chapters that focused on aspects of fluid, electrolyte, or acid-base balance that affected specific systems. This chapter provides an overview that integrates those discussions to highlight important functional patterns. Few other chapters have such wide-ranging clinical importance: *Treatment of any serious illness affecting the nervous, cardiovascular, respiratory, urinary, or digestive system must include steps to restore normal fluid, electrolyte, and acid-base balance.* Because this chapter builds on information provided in earlier chapters, you will encounter frequent references to relevant discussions and figures that should be used when a quick review is needed.

Fluid and Electrolyte Balance

Figures 27-1, 27-2

Figure 27-1 • provides an overview of the composition of the body of a 70-kg person with a minimum of body fat. Water accounts for roughly 60 percent of the total body weight. The largest share of the total body water (roughly 25 liters) is found inside living cells, in the *intracellular fluid* **(ICF).** The *extracellular fluid* **(ECF)** contains the rest of the body water (rough-

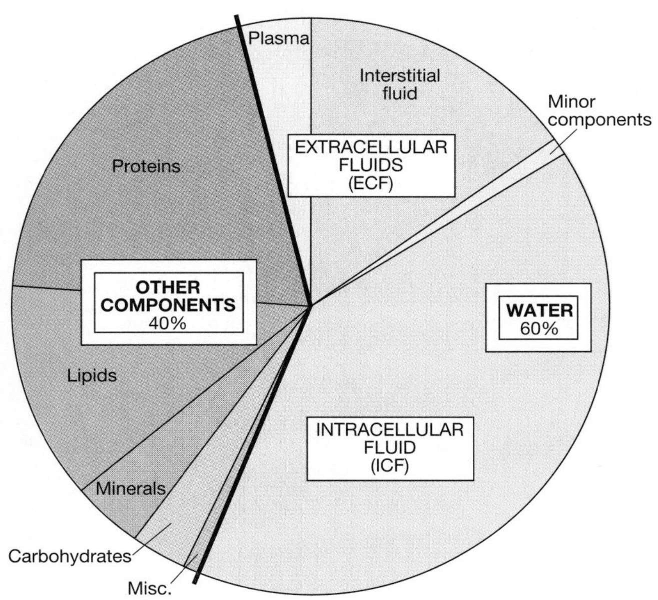
Plasma
Interstitial fluid
Minor components
EXTRACELLULAR FLUIDS (ECF)
Proteins
OTHER COMPONENTS 40%
WATER 60%
Lipids
INTRACELLULAR FLUID (ICF)
Minerals
Carbohydrates
Misc.

● **FIGURE 27-1**
Composition of the Human Body. The body composition and major body fluid compartments in a representative individual.

ly 17 liters). Exchange between the ICF and the ECF occurs across cell membranes, via mechanisms detailed in Chapter 3. (For a review of the mechanisms involved, see Table 3-2, p. 82.)

The largest subdivisions of the ECF are (1) the *interstitial fluid* of peripheral tissues (14 liters) and (2) the *plasma* of the circulating blood (2.8 liters). Minor components of the ECF include lymph, cerebrospinal fluid (CSF), synovial fluid, serous fluids (pleural, pericardial, and peritoneal fluids), aqueous humor, perilymph, and endolymph.

Exchange among the subdivisions of the ECF occurs primarily across the endothelial lining of capillaries. Fluid may also travel from the interstitial spaces to the plasma through lymphatic vessels (Chapter 22) that drain into the venous system. ∞ [pp. 781-783] There are regional variations in the identity and quantity of dissolved electrolytes, proteins, nutrients, and waste products within the ECF. (For a chemical analysis of the composition of ECF compartments, see Appendix VII.) Yet the variations among the segments of the ECF seem minor when compared with the major differences between the ECF and the ICF.

The ECF and ICF are often called **fluid compartments** because they often behave as distinct entities. The presence of a cell membrane and active transport at the membrane surface enable cells to maintain internal environments with a composition different from that of their surroundings. The principal ions in the ECF are sodium, chloride, and bicarbonate. The ICF contains an abundance of potassium, magnesium, and phosphate ions, plus large numbers of negatively charged proteins. (A comparison between the ICF and the two major subdivisions of the ECF can be found in Figure 27-2●.) Chapter 2 introduced the various means used to report solute concentrations in body fluids. ∞ [p. 43]

If the cell membrane were freely permeable, diffusion would continue until these ions were evenly distributed across the membrane. But it does not, because cell membranes are selectively permeable, and ions can enter or leave the cell only via specific membrane channels. In addition, there are active transport mechanisms that move specific ions into or out of the cell.

Despite the differences in the concentration of specific substances, the intracellular and extracellular osmolarities are identical. Osmosis eliminates any minor concentration differences almost at once, because cell membranes are usually freely permeable to water. (The only noteworthy exceptions are the ascending thick segment of the loop of Henle, the DCT, and the collecting system, as detailed in Chapter 26.) ∞ [pp. 1003–1006] Because changes in solute concentrations will lead to immediate changes in water distribution, the regulation of fluid balance and that of electrolyte balance are tightly intertwined.

Physiologists and clinicians pay particular attention to ionic distributions across membranes and to the electrolyte composition of body fluids. Appendix VII, which contains normal values for the analysis of body fluids, reports values in the units most often encountered in clinical reports. For a discussion of the methods of reporting solute concentrations, see the *Applications Manual.* AM *Solutions and Concentrations*

❑ Basic Concepts Pertaining to Fluid and Electrolyte Regulation

Four basic principles must be understood before we can proceed to a discussion of fluid and electrolyte balance.

- ***Concept 1: All of the homeostatic mechanisms that monitor and adjust the composition of body fluids respond to changes in the ECF, not in the ICF.*** Receptors monitoring the composition of two key components of the ECF, the plasma and the cerebrospinal fluid, detect significant changes in composition or volume and trigger appropriate neural and endocrine responses.

  This arrangement makes functional sense because a change in one component of the ECF will spread rapidly throughout the extracellular compartment and will affect all the cells in the body. In contrast, the ICF is contained within trillions of individual cells that are phys-

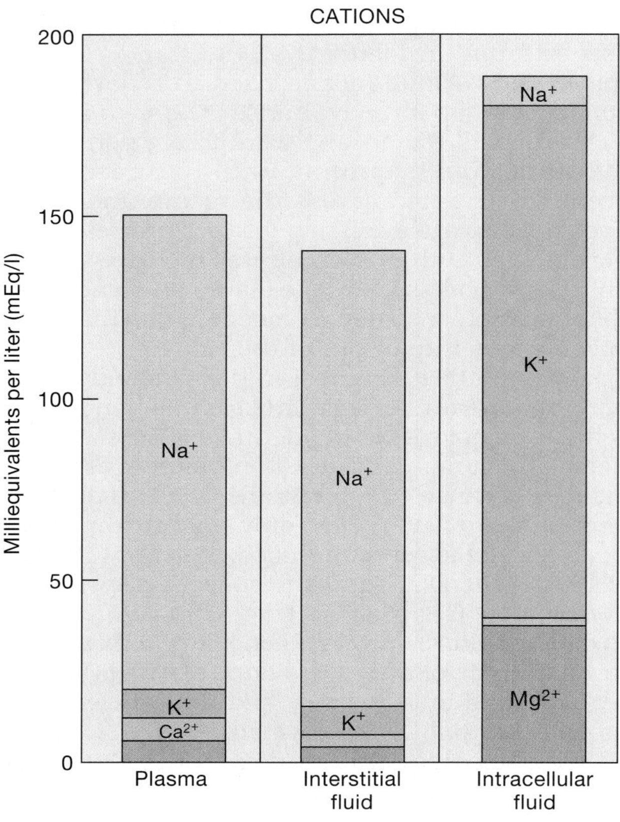
CATIONS
200
150
100
50
0
Milliequivalents per liter (mEq/l)
Na+
K+
Ca2+
Na+
K+
Na+
K+
Mg2+
Plasma
Interstitial fluid
Intracellular fluid

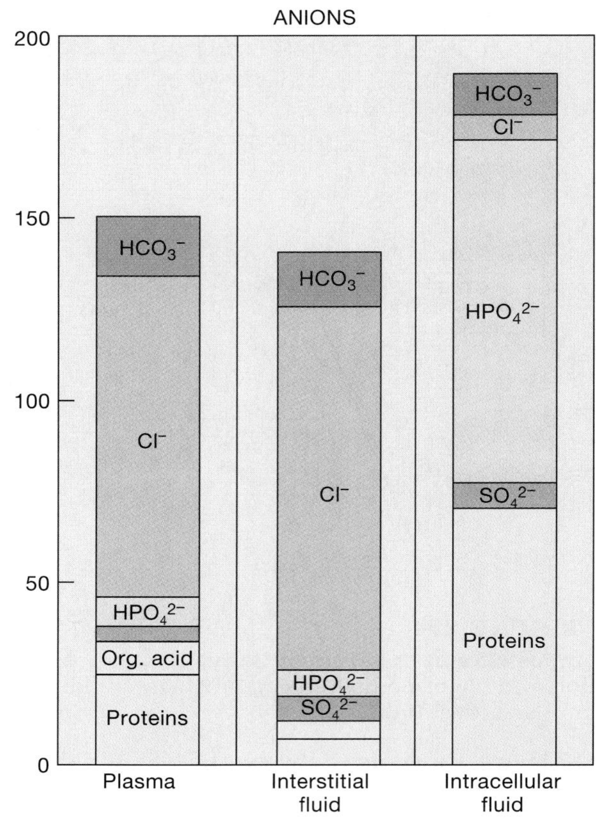
ANIONS
200
150
100
50
0
HCO3−
Cl−
HPO42−
Org. acid
Proteins
HCO3−
Cl−
HPO42−
SO42−
HCO3−
Cl−
HPO42−
SO42−
Proteins
Plasma
Interstitial fluid
Intracellular fluid

● **FIGURE 27-2**
Cations and Anions in Body Fluids. Note the differences in cation and anion concentrations in the various body fluid compartments. For information concerning the composition of other body fluids, see Appendix VII.

ically and chemically isolated from one another by their cell membranes. Thus changes in the intracellular fluid in one cell will have no direct effect on the composition of the intracellular fluid in distant cells and tissues unless those changes affect the extracellular fluid.

- ***Concept 2: There are no receptors that can directly monitor fluid and electrolyte balance.*** Receptors cannot detect how much water, sodium, chloride, or potassium is present in the entire body. Homeostatic adjustments occur in response to changes in *plasma volume* or *osmolarity* (the total concentration of solutes). Minor alterations in the composition or volume of the interstitial fluids are ignored until they affect plasma osmolarity or volume. However, because fluid continually circulates between the interstitial fluid and plasma, significant alterations in the composition or volume of the interstitial fluid will very quickly affect the plasma.

- ***Concept 3: Our cells are unable to move water molecules by active transport.*** All water movement across cell membranes and epithelia occurs passively, in response to osmotic gradients. These gradients can be established by the active transport of specific ions, such as sodium and chloride. You may find it useful to remember the simple phrase *water follows salt.* As we have seen in earlier chapters, when sodium and chloride ions (or other solutes) are actively transported across a membrane or epithelium, water follows by osmosis. This basic principle accounts for water absorption across the digestive epithelium and water conservation in the kidneys.

- ***Concept 4: The body content of water or electrolytes will rise if intake exceeds outflow and fall if losses exceed gains.*** This simple fact must be kept in mind when considering the mechanics of fluid and electrolyte balance. Homeostatic adjustments usually affect the balance between dietary absorption and urinary excretion. These responses may involve both changes in behavior and physiological changes in specific systems.

For example, a person may experience a sensation of thirst that stimulates the drinking of fluids or a craving for heavily salted foods. The rates of water and electrolyte uptake are directly proportional to the dietary supply. The more you consume, the more you absorb. Thus voluntary changes in the diet have a direct effect on fluid and electrolyte balance. The physiological adjustments are regulated primarily by circulating hormones.

An Overview of the Primary Regulatory Hormones

Major physiological adjustments affecting fluid and electrolyte balance are mediated by three hormones: *antidiuretic hormone* (ADH), *aldosterone*, and *atrial natriuretic peptide* (ANP). Interactions among these hormones were shown in Figures 18-16, 18-17, and 21-17 (pp. 627, 628, and 744).

Antidiuretic Hormone

The hypothalamus contains special cells known as *osmoreceptors* that monitor the osmotic concentration of the ECF. These cells are sensitive to subtle changes in osmolarity: A 2 percent change (approximately 6 mOsm) is sufficient to alter osmoreceptor activity.

The population of osmoreceptors includes neurons that secrete ADH. These neurons are located in the anterior hypothalamus, and their axons release ADH near fenestrated capillaries in the posterior pituitary gland. The rate of ADH release varies directly with the osmolarity: the higher the osmolarity, the greater the amount of ADH released.

Increased release of ADH has two important effects: (1) *It stimulates water conservation at the kidneys, reducing urinary water losses and concentrating the urine*, and (2) *it stimulates the thirst center to promote the drinking of fluids*. As we saw in Chapter 18, the combination of decreased water loss and increased water gain gradually restores normal plasma osmolarity. ∞ [p. 614]

Aldosterone

The secretion of aldosterone by the adrenal cortex plays a major role in determining the rate of sodium absorption along the DCT and collecting system of the kidneys. The mechanism was detailed in Chapters 18, 21, and 26. ∞ [pp. 624, 743, 1003–1004] The higher the plasma concentration of aldosterone, the more efficiently the kidneys will conserve sodium ions. Because "water follows salt," conservation of sodium ions also stimulates water retention.

Secretion of aldosterone is not directly influenced by the sodium ion concentration of the plasma. Instead, aldosterone release occurs in response to activation of the renin-angiotensin system by (1) a fall in plasma volume or blood pressure at the juxtaglomerular apparatus of the nephron, (2) a decline in filtrate osmolarity at the DCT, or, as we will see later in this chapter, by (3) rising potassium ion concentrations.

Atrial Natriuretic Peptide

Atrial natriuretic peptide is released by cardiac muscle fibers in response to abnormal stretching of the atrial walls caused by elevated blood pressure or an increase in blood volume. The many effects of ANP were considered in Chapters 18, 21, and 26. ∞ [pp. 629, 743, 995] Among its other effects, this hormone (1) reduces thirst and (2) blocks the release of ADH and aldosterone that might otherwise lead to water and salt conservation. The resulting diuresis lowers both blood pressure and plasma volume, eliminating the source of the stimulation.

The Interplay between Fluid and Electrolyte Balance

At first glance, it can be very difficult to distinguish between water balance and electrolyte balance. For example, when an individual loses body water, plasma volume decreases and electrolyte concentrations rise. Conversely, when excess electrolytes are gained or lost, there is an associated water gain or loss due to osmosis. However, because the regulatory mechanisms involved are quite different, it is often useful to consider fluid balance and electrolyte balance as distinct entities. This distinction is absolutely vital in a clinical setting, where problems with fluid balance and electrolyte balance must be identified and corrected promptly.

Fluid Balance

Figure 27-3

Water circulates freely within the extracellular fluid compartment. At capillary beds throughout the body, hydrostatic pressure forces water out of the plasma and into interstitial spaces. Some of that water is reabsorbed along the distal portion of the capillary bed, and the rest circulates into lymphatic vessels for transport to the venous circulation. ∞ [p. 735] This circulation is diagrammed in Figure 27-3•, which indicates two additional important relationships between components of the ECF:

1. Water moves back and forth across the mesothelial surfaces lining the peritoneal, pleural, and pericardial cavities and through the synovial membranes lining joint capsules. The flow rate is significant; for example, roughly 7 liters of peritoneal fluid are produced and reabsorbed each day.
2. Water also moves between the blood and the cerebrospinal fluid, the aqueous and vitreous

humors of the eye, and the perilymph and endolymph of the inner ear. The volumes involved are very small, and the volume and composition of these fluids are closely regulated. For this reason, we will largely ignore them in the discussion that follows.

Fluid Movement within the ECF

The basic principles determining fluid movement between the divisions of the ECF were considered in Chapter 21 in the discussion of capillary dynamics. [p. 733] The exchange between the plasma and the interstitial fluid, by far the largest components of the ECF, is determined by the relationship between the net hydrostatic pressure, which tends to push water out of the plasma and into the interstitial fluid, and the net colloid osmotic pressure, which tends to draw water out of the interstitial fluid and into the plasma. The interaction between these opposing forces, diagrammed in Figure 21-13● (p. 735), results in the continual filtration of fluid from the capillaries into the interstitial fluid. This volume of fluid is then redistributed, returning to the venous system after passing through the channels of the lymphatic system. At any given moment, the interstitial fluid and minor fluid compartments contain roughly 80 percent of the ECF volume and the plasma contains the other 20 percent.

Any factor that affects the net hydrostatic or colloid osmotic pressures will alter the distribution of fluid within the ECF. The movement of abnormal amounts of water from the plasma into the interstitial fluids is called *edema.* The general causes of edema were detailed in Chapter 21. [p. 736] For example, pulmonary edema can result from an increase in the blood pressure in pulmonary capillaries, and a generalized edema can result from a decrease in blood colloid osmotic pressure, as in advanced starvation, when plasma protein concentrations decline. [p. 742] Localized edema can result from damage to capillary walls, as in bruising, constriction of the regional venous circulation, or blockage of the lymphatic drainage, as in *lymphedema,* discussed in Chapter 22. [p. 783]

Fluid Exchange with the Environment

Figure 27-3

Figure 27-3● and Table 27-1 indicate the major routes of exchange with the environment.

- ***Water losses.*** Roughly 2500 ml of water are lost each day through urine, feces, and *insen-*

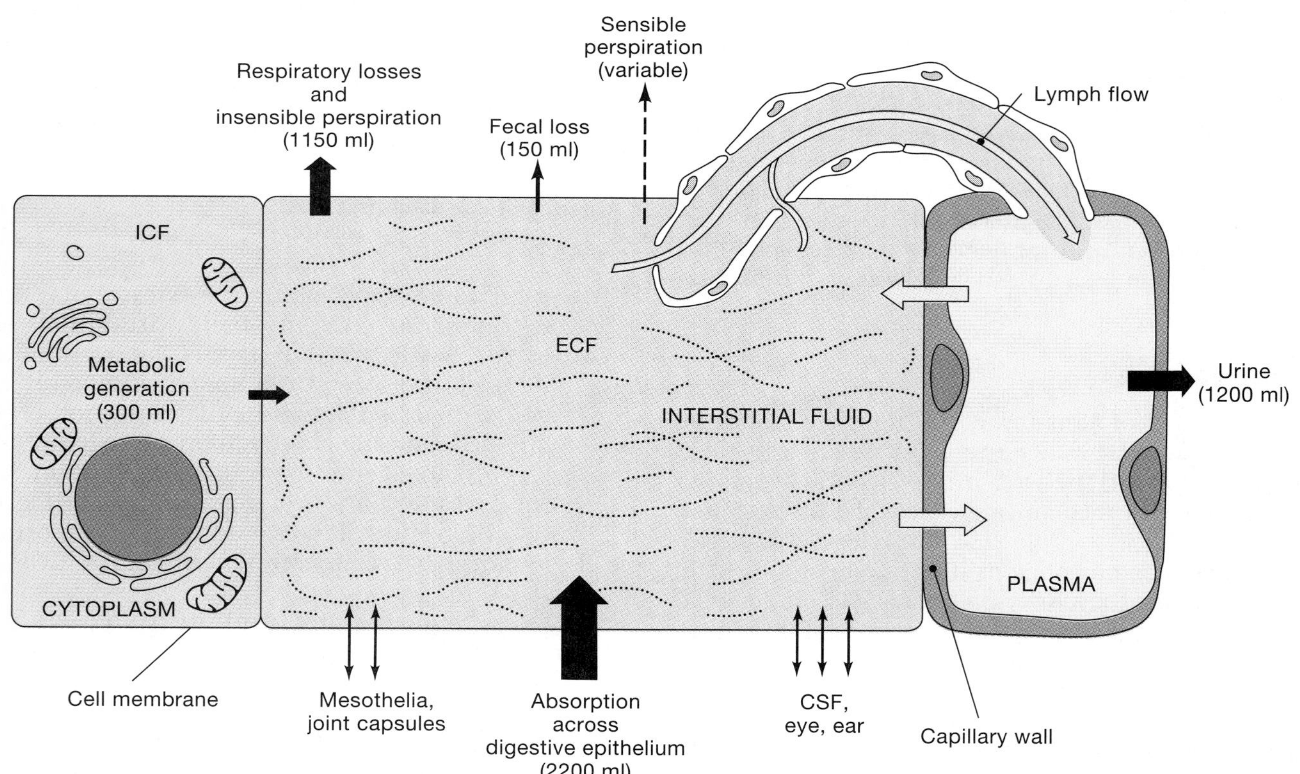
Sensible
perspiration
(variable)
Respiratory losses
and
insensible perspiration
(1150 ml)
Fecal loss
(150 ml)
Lymph flow
ICF
Metabolic
generation
(300 ml)
CYTOPLASM
ECF
INTERSTITIAL FLUID
Urine
(1200 ml)
PLASMA
Cell membrane
Mesothelia,
joint capsules
Absorption
across
digestive epithelium
(2200 ml)
CSF,
eye, ear
Capillary wall

● **FIGURE 27-3**

Fluid Exchanges. A diagrammatic representation of fluid movement between the ICF and ECF, and between the ECF and the environment. The volumes are not drawn to scale.

sible perspiration, the gradual movement of water across the epithelia of the skin and respiratory tract. The losses due to *sensible perspiration,* the secretory activities of the sweat glands, vary depending on the activities undertaken. Sensible perspiration can cause significant water deficits, with maximum perspiration rates reaching well over 4 liters per hour. The relationships between sensible and insensible perspiration were detailed in Chapter 25. [p. 970]

The temperature rise accompanying a fever can also increase water losses. For each degree the temperature rises above normal the daily insensible water loss increases by 200 ml. Thus the advice "drink plenty of fluids" when one is sick has a definite physiological basis.

- ***Water gains.*** A water gain of roughly 2500 ml/day is required to balance average water losses. This value amounts to roughly 40 ml/kg body weight per day. Water is obtained through eating (1000 ml), drinking (1200 ml), and *metabolic generation* (300 ml).

Metabolic generation of water is the production of water within cells, primarily as a result of oxidative phosphorylation within mitochondria. (The synthesis of water at the end of the electron transport system was detailed in Chapter 25.) [p. 943] When a cell breaks down 1g of lipid, it generates 1.7 ml of water. Breaking down proteins or carbohydrates results in much lower values (0.41 ml/g protein, 0.55 ml/g carbohydrate). A typical mixed diet in the United States contains 46 percent carbohydrates, 40 percent lipids, and 14 percent protein. This diet would produce roughly 300 ml of water per day, about 12 percent of the average daily requirement.

TABLE 27-1 Water Balance

Daily Input	
Water content of food	1000
Water consumed as liquid	1200
Metabolic water during catabolism	300
Total	2500 ml
Daily Output	
Urine	1200
Evaporation at skin	750
Evaporation at lungs	400
Lost in feces	150
Total	2500 ml

WATER AND WEIGHT LOSS The safest way to lose weight is to reduce the intake of food while ensuring that all dietary essentials are available in adequate quantities. Water must be included on the list of essentials along with the amino acids, fatty acids, vitamins, and minerals. Because nearly half of our normal water intake comes from food, a person who eats less becomes more dependent on drinking fluids and whatever water is generated metabolically. At the start of a diet, the body conserves water and catabolizes lipids. That is why the first week of dieting may seem rather unproductive. Over that week, the level of circulating ketone bodies gradually increases. During subsequent weeks, the rate of water loss at the kidneys increases in order to excrete waste products, such as the hydrogen ions released by ketone bodies and the urea and ammonia generated during protein catabolism. The rate of fluid intake must also increase or the individual risks dehydration. Dehydration under these conditions is a major problem, because as water is lost, the solute concentration in the ECF climbs, further increasing the concentration of waste products and acids. This effect can be opposed by fluid movement from the ICF into the ECF, but only up to a point.

Fluid Shifts

Water movement between the ECF and ICF is called a **fluid shift.** Fluid shifts occur relatively rapidly, reaching equilibrium within a period of minutes to hours. These shifts occur in response to changes in the osmolarity of the extracellular fluid. The basic relationships were explored in Chapter 3, especially in Figure 3-8•, p. 76, which considered the effects of *hypertonic* and *hypotonic* solutions on cells.

- If the osmolarity of the ECF increases, it will become hypertonic with respect to the ICF. Water will then move from the cells into the ECF until osmotic equilibrium is restored. The osmolarity of the ECF will increase if the individual loses water but retains electrolytes.
- If the osmolarity of the ECF decreases, it will become hypotonic with respect to the ICF. Water will then move from the ECF into the cells, and the volume of the ICF will increase. The osmolarity of the ECF will decrease if the individual gains water without a corresponding gain of electrolytes.

In summary, if the osmolarity of the ECF changes, a fluid shift between the ICF and ECF will tend to oppose the change. Because the volume of the ICF is much greater than that of the ECF, the ICF acts as a "water reserve." In effect, instead of a large change in the osmolarity of the ECF there are smaller changes in both the ECF and ICF. Two examples will demonstrate the dynamic exchange of water between the ECF and ICF.

ALLOCATION OF WATER LOSSES When water is lost but electrolytes are retained, the osmolarity of the ECF rises. Osmosis then moves water out of the ICF and into the ECF until the two solutions are again isotonic. At this point both the ECF and ICF will be somewhat more concentrated than normal, and both volumes will be lower than they were before the fluid loss occurred. Because the ICF has roughly twice the volume of the ECF, the net change in the ECF is kept relatively small.

Conditions that cause severe water losses include excessive perspiration (exercising in hot weather), inadequate water consumption (being lost in a desert), repeated vomiting, and diarrhea. These disorders promote water losses far in excess of electrolyte losses, so body fluids become increasingly concentrated. Responses that attempt to restore homeostasis include ADH and renin secretion and (as soon as possible) an increase in fluid intake.

DISTRIBUTION OF WATER GAINS When a person drinks a glass of pure water, or when hypotonic solutions are administered through intravenous infusion, the water content of the body increases without a corresponding increase in the concentration of electrolytes. As a result, the extracellular fluid increases in volume but becomes hypotonic with respect to the ICF. A fluid shift then occurs, and the volume of the ICF increases at the expense of the ECF. Once again, the larger volume of the ICF enables it to limit the amount of osmotic change. After the fluid shift the ECF and ICF have slightly larger volumes and slightly lower osmolarities than they did originally.

Under normal circumstances this situation will be promptly corrected. The reduced plasma osmolarity depresses the secretion of ADH, discouraging fluid intake and increasing water losses in the urine. If the situation is *not* corrected, a variety of clinical problems develop.

WATER EXCESS AND WATER DEPLETION The body's water content cannot easily be determined. However, the concentration of sodium, the most abundant ion in the ECF, provides useful clues to the state of water balance. When the body water content rises enough to reduce the sodium concentration of the ECF below 130 mEq/l, the state of *hyponatremia* (*natrium*, sodium) exists. When the body water content declines, the sodium concentration rises; when it exceeds 150 mEq/l, *hypernatremia* exists.

Hyponatremia is a sign of **overhydration,** or *water excess.* Hyponatremia may result from (1) ingestion of a large volume of fresh water or infusion of a hypotonic solution, (2) an inability to eliminate excess water in the urine caused by *chronic renal failure* (p. 997), *congestive heart failure* (p. 742), *cirrhosis* (p. 912), or other disorders, or (3) endocrine disorders, such as excessive ADH production. The reduction in sodium ion concentrations leads to a fluid shift into the ICF. The first signs are the effects on CNS function. In the early stages of hyponatremia the individual behaves as if drunk on alcohol. This condition, called *water intoxication,* sounds humorous, but it is extremely dangerous. Untreated cases can rapidly progress from confusion to hallucinations, convulsions, coma, and death. Treatment of severe water intoxication usually involves diuretics and infusing a concentrated salt solution that elevates sodium levels to near normal levels.

Dehydration, or *water depletion,* develops when water losses outpace water gains. Plasma osmolarity gradually increases, and hypernatremia results. The loss of body water is associated with severe thirst, a dryness and wrinkling of the skin, and a fall in plasma volume and blood pressure. Eventually, circulatory shock develops, usually with fatal consequences. Treatment for dehydration entails administering hypotonic fluids by mouth or via intravenous infusion. This treatment increases ECF volume and restores normal electrolyte concentrations.

❑ Electrolyte Balance

An individual is in electrolyte balance when the rates of gain and loss are equal for each of the individual electrolytes in the body. Electrolyte balance is important because:

- ***Total electrolyte concentrations have a direct effect on water balance,*** as detailed above.
- ***The concentrations of individual electrolytes can have an effect on a variety of cell functions.*** Many examples were encountered in earlier chapters, including the effect of abnormal sodium ion concentrations on neuron activity and the effects of high or low calcium and potassium ion concentrations on cardiac muscle tissue.

Two cations, sodium and potassium, merit special attention because (1) they are major contributors to the osmolarities of the ECF and ICF, and (2) they have direct effects on the normal functioning of all cells.

Sodium is the dominant cation within the extracellular fluid. Because more than 90 percent of the osmolarity of the ECF results from the presence of sodium salts, principally sodium chloride (NaCl) and sodium bicarbonate ($NaHCO_3$), alterations in the osmolarity of body fluids usually reflect changes in the concentration of sodium ions. Normal sodium ion concentrations in the ECF average 136–142 mEq/l, versus 10 mEq/l or less in the ICF.

Potassium is the dominant cation in the intracellular fluid, where concentrations reach 160 mEq/l. Extracellular potassium concentrations are usually very low, ranging from 3.8 to 5.0 mEq/l.

Two general rules about sodium balance and potassium balance are worth noting:

Athletes and Salt Loss

There are a number of unfounded myths and rumors about water and salt requirements during exercise. Sweat is a hypotonic solution that contains sodium ions in lower concentration than the ECF. As a result, in prolonged sweating a person loses more water than salt, and this loss leads to a rise in the sodium ion concentration of the ECF. The water content of the ECF decreases as the water loss occurs, and the blood volume declines. This condition is known as *volume depletion.* Because this volume depletion occurs at the same time that blood is being shunted away from the kidneys, waste products accumulate in the blood and kidney function is impaired, an effect considered in Chapter 26. ∞ *[pp. 996-997]*

To prevent volume depletion from occurring, athletes should pause at short intervals to drink liquids. The primary problem in volume depletion is water loss, and research has shown no basis for the rumor that cramps will result from drinking while exercising. Salt pills and the various sports beverages promoted for "faster absorption" and "better electrolyte balance" have no apparent benefits, despite their relatively high cost. Body reserves of electrolytes are sufficient to tolerate extended periods of strenuous activity, and problems with sodium balance are extremely unlikely except during ultramarathons or other activities that involve maximal exertion for more than 6 hours.

Some of the sports beverages contain sugars and vitamins as well as electrolytes. During endurance events (marathons, ultramarathons, distance cycling), solutions containing less than 10 g/dl of glucose may improve performance when consumed late in the competition, as metabolic reserves are exhausted. However, high sugar concentrations (above 10 g/dl) may cause cramps, diarrhea, and other problems. The benefit of "glucose polymers" in sports drinks has yet to be proved. Drinking beverages "fortified" with vitamins is actively discouraged. Vitamins are not lost during exercise, and the consumption of these beverages could lead to hypervitaminosis.

- ***The most common problems with electrolyte balance are caused by an imbalance between gains and losses of sodium ions.***
- ***Problems with potassium balance are less common but significantly more dangerous than those related to sodium balance.***

Sodium Balance

Figures 27-4, 27-5

The total amount of sodium in the ECF represents a balance between two factors:

1. ***Sodium ion uptake across the digestive epithelium.*** Sodium ions enter the ECF by crossing the digestive epithelium via diffusion and carrier-mediated transport. The rate of absorption varies directly with the amount in the diet.
2. ***Sodium ion excretion at the kidneys and other sites.*** Sodium losses occur primarily in the urine and through perspiration. The kidneys are the most important sites of sodium ion regulation. The mechanisms for sodium reabsorption at the kidneys were detailed in Chapter 26. ∞ *[p. 1001–1006]*

When sodium gains exceed sodium losses, the total sodium ion content of the ECF goes up; when losses exceed gains, the sodium ion content declines. However, a change in the sodium ion *content* of the ECF does not produce a change in sodium ion *concentration.* Whenever the rate of sodium intake or output changes, there is a corresponding gain or loss of water that tends to keep the sodium concentration constant.

For example, eating a heavily salted meal will not elevate the osmolarity of the ECF. When sodium ions are pumped out of the digestive epithelium and into the interstitial fluid of the lamina propria, the solute concentration in that portion of the ECF increases and that of the intestinal contents decreases. Osmosis then occurs, and additional water enters the ECF from the digestive tract. Thus:

- Whenever sodium gains exceed sodium losses, the volume of the ECF increases, but its osmolarity remains the same. This is why individuals with high blood pressure are advised to restrict the amount of salt in their diets. If consumed with a normal fluid volume, excess salt intake will result in an elevated blood volume and blood pressure as water follows salt from the digestive tract into the plasma. If salt is consumed *without* adequate fluid, as when eating a bowl of heavily salted potato chips, the plasma sodium concentration will rise temporarily. A change in ECF volume will soon follow, however, because osmoreceptors will respond by secreting ADH, restricting water loss and stimulating thirst to promote additional water consumption.
- Whenever sodium losses exceed gains, the volume of the ECF decreases. Once again, this

reduction occurs without a significant change in the osmolarity of the ECF. Thus, if you perspire heavily but consume only pure water, you will lose sodium, and the osmolarity of the ECF will decline briefly. However, as soon as the osmolarity drops by 2 percent or more, ADH secretion will decrease and water losses at the kidneys will increase. As water leaves the ECF, the osmolarity will return to normal.

THE REGULATION OF SODIUM BALANCE As we have seen, under normal circumstances alterations in sodium balance result in the expansion or contraction of the extracellular fluid compartment. The regulatory mechanism, diagrammed in Figure 27-4●, changes the ECF volume but keeps the sodium ion concentration relatively stable.

Minor changes in the ECF volume do not matter because they do not cause adverse physiological effects. If the regulation of sodium ion concentrations results in a large change in the ECF volume, the situation will be corrected by the same homeostatic mechanisms responsible for regulating blood pressure and blood volume that were detailed in Chapters 18 and 21. [pp. 614, 628–629, 743]

Whenever the ECF volume changes, so does plasma volume, and when plasma volume changes it affects blood volume. If the ECF volume rises, blood volume goes up; if the ECF volume declines, blood volume goes down. As you will recall from Chapter 21, blood volume has a direct effect on blood pressure. A rise in blood volume elevates blood pressure, and a decline in blood volume leads to a fall in blood pressure. The net result is that homeostatic mechanisms can monitor ECF volume indirectly, by monitoring blood pressure.

The receptors involved are *baroreceptors* at the carotid sinus, the aortic sinus, and the right atrium. The regulatory steps are detailed in Figure 27-5●. If the ECF volume is inadequate, blood volume and blood pressure decline:

- The fall in blood pressure along the afferent arterioles of the kidneys stimulates the release of renin by the juxtaglomerular apparatus of each nephron.
- Renin initiates a chain of events leading to the activation of angiotensin II.
- Angiotensin II produces a coordinated elevation in the extracellular fluid volume by stimulating thirst, causing the release of ADH, and triggering the adrenal production and secretion of aldosterone.
- Thirst and ADH secretion lead to increased ingestion and retention of water, while aldosterone promotes the retention of sodium ions along the distal portions of the kidney nephrons.

As a result of this mechanism, water and sodium ion losses are reduced, gains are increased, and the volume of the ECF increases. The concentration of sodium ions, however, remains unchanged.

● **FIGURE 27-4**
Homeostatic Regulation of Normal Sodium Ion Concentrations in Body Fluids

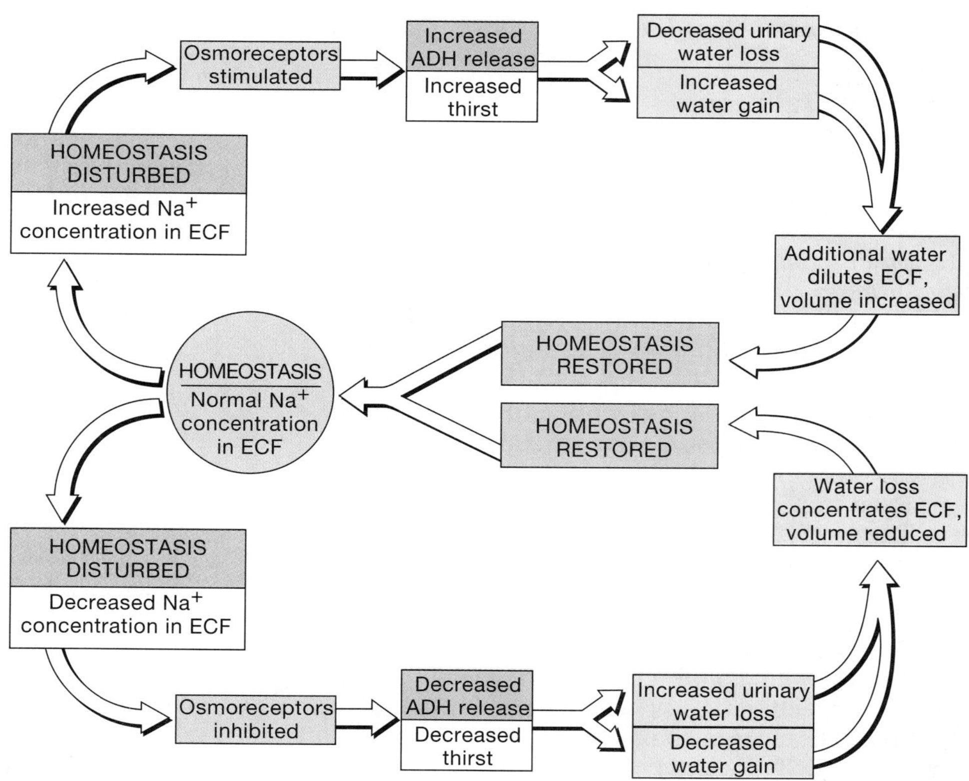
Osmoreceptors stimulated
Increased ADH release
Increased thirst
Decreased urinary water loss
Increased water gain
HOMEOSTASIS DISTURBED
Increased Na+ concentration in ECF
Additional water dilutes ECF, volume increased
HOMEOSTASIS RESTORED
HOMEOSTASIS
Normal Na+ concentration in ECF
HOMEOSTASIS RESTORED
Water loss concentrates ECF, volume reduced
HOMEOSTASIS DISTURBED
Decreased Na+ concentration in ECF
Osmoreceptors inhibited
Decreased ADH release
Decreased thirst
Increased urinary water loss
Decreased water gain

If the plasma volume becomes abnormally large:

- Venous return increases, stretching the atrial wall and stimulating the release of atrial natriuretic peptide.
- ANP reduces thirst and blocks the secretion of ADH and aldosterone, which together would promote water or salt conservation. This mechanism causes increased salt and water loss at the kidneys, and the volume of the ECF declines.

Potassium Balance

Roughly 98 percent of the potassium content of the body is in the ICF. Living cells expend energy to recover potassium ions as they diffuse out of the cytoplasm and into the ECF. The potassium concentration outside of the cell is relatively low, and the concentration in the ECF at any given moment represents a balance between (1) the rate of entry across the digestive epithelium and (2) the rate of loss into the urine. As noted in Chapter 26, potassium loss in the urine is regulated by controlling the activities of ion pumps along the distal portions of the nephron and collecting system. [p. 1005] Whenever a sodium ion is reabsorbed from the tubular fluid, it usually is exchanged for a cation in the peritubular fluid, typically a potassium ion.

Urinary losses are usually limited to the amount gained by absorption across the digestive epithelium, typically 50–150 mEq per day. (The potassium losses in feces and perspiration are negligible by comparison.) The concentration of potassium in the ECF is controlled by adjustments in the rate of active secretion along the DCT of the nephron.

The rate of tubular secretion of potassium ions changes in response to:

1. ***Alterations in the potassium ion concentration in the ECF.*** In general, the higher the extracellular concentration of potassium, the higher the rate of secretion along the DCT.
2. ***Changes in pH.*** When the pH of the ECF falls, so does the pH of the peritubular fluid. The rate of potassium ion secretion then declines, because hydrogen ions, rather than potassium ions, are secreted in exchange for sodium ions in the tubular fluid. The mechanisms for H^+ secretion were summarized in Figure 26-16, p. 1005.

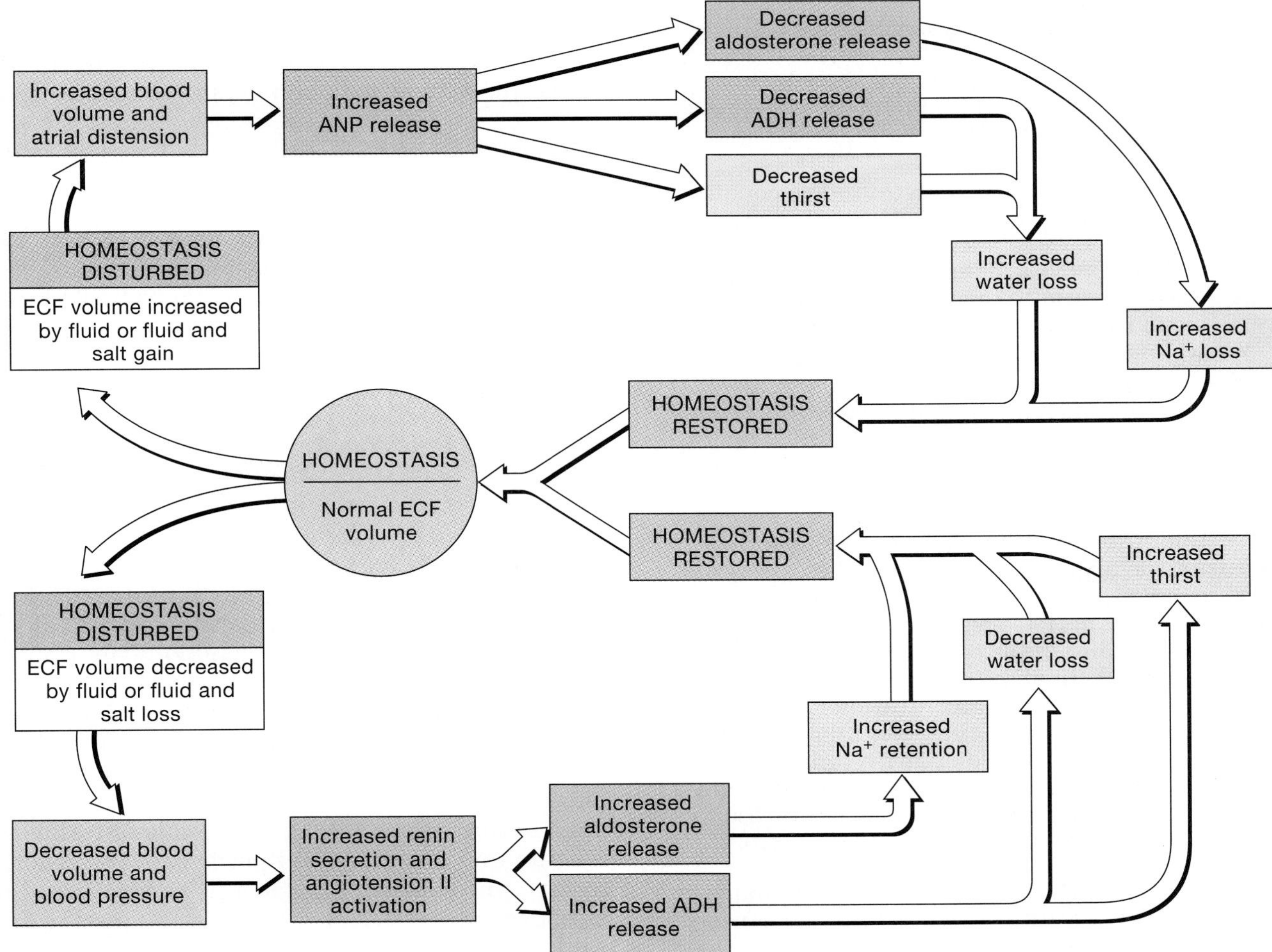
Increased blood volume and atrial distension
Increased ANP release
Decreased aldosterone release
Decreased ADH release
Decreased thirst
Increased water loss
Increased Na+ loss
HOMEOSTASIS DISTURBED
ECF volume increased by fluid or fluid and salt gain
HOMEOSTASIS RESTORED
HOMEOSTASIS
Normal ECF volume
HOMEOSTASIS RESTORED
Increased thirst
Decreased water loss
HOMEOSTASIS DISTURBED
ECF volume decreased by fluid or fluid and salt loss
Increased Na+ retention
Decreased blood volume and blood pressure
Increased renin secretion and angiotension II activation
Increased aldosterone release
Increased ADH release

FIGURE 27-5
The Integration of Fluid Volume Regulation and Sodium Ion Concentrations in Body Fluids

3. ***Aldosterone levels.*** The rate at which potassium is lost in the urine is strongly affected by aldosterone, because the ion pumps sensitive to this hormone reabsorb sodium ions from the filtrate in exchange for potassium ions from the peritubular fluid. Aldosterone secretion is stimulated by angiotensin II as part of the regulation of blood volume. High plasma potassium concentrations also stimulate aldosterone secretion. Either way, under the influence of aldosterone there is a direct relationship between the amount of sodium conserved and the amount of potassium excreted in the urine.

HYPOKALEMIA AND HYPERKALEMIA When the plasma concentration of potassium falls below 2 mEq/l, extensive muscular weakness develops, followed by eventual paralysis. This condition, called *hypokalemia* (*kalium*, potassium), was discussed in Chapter 20 in connection with ion effects on cardiac function. [p. 708] Causes of hypokalemia include:

1. ***Inadequate dietary intake.*** If potassium gains from the diet do not keep pace with the rate of potassium loss in the urine, potassium concentrations in the ECF will decline.
2. ***Administration of diuretic drugs.*** Several diuretics, including Lasix, can produce hypokalemia by increasing the volume of urine produced. Although the concentration of potassium in the urine is low, the greater the total volume, the larger the amount of potassium lost.
3. ***Excessive aldosterone secretion.*** The condition of aldosteronism, characterized by excessive aldosterone secretion, results in hypokalemia because the reabsorption of sodium ions is tied to the secretion of potassium. Aldosteronism was discussed in Chapter 18. [p. 624]
4. ***An increase in the pH of the extracellular fluids.*** When the hydrogen ion concentration in the ECF declines, cells exchange intracellular hydrogen ions for extracellular potassium ions. This ion swap helps stabilize the extracellular pH but gradually lowers the potassium ion concentration in the ECF.

Treatment for hypokalemia usually includes (1) increasing dietary intake, by salting food with potassium salts (KCl) or by taking potassium tablets, such as *Slow-K*, and in severe cases (2) the infusion of a solution containing potassium ions at a concentration of 40–60 mEq/l.

High potassium ion concentrations in the ECF produce an equally dangerous condition known as *hyperkalemia*. Severe cardiac arrhythmias appear when the potassium ion concentration exceeds 8 mEq/l; the mechanism was discussed in Chapter 20. [p. 708] Hyperkalemia may result from:

1. ***Renal failure.*** Kidney failure due to damage or disease will prevent normal potassium ion secretion and so produce hyperkalemia.
2. ***The administration of diuretic drugs that block sodium reabsorption.*** Potassium secretion is linked to sodium reabsorption. When sodium reabsorption slows down, so does potassium secretion, and hyperkalemia can result.
3. ***A decline in the pH of the ECF.*** When the pH in the ECF declines, hydrogen ions move into the ICF in exchange for intracellular potassium ions. In addition, potassium secretion at the kidney tubules slows down because hydrogen ions are secreted instead of potassium ions. The combination of increased potassium entry into the ECF and decreased potassium secretion can produce a dangerous hyperkalemia very rapidly.

Treatment for hyperkalemia includes (1) elevation of ECF volume with a solution low in potassium, (2) stimulation of urinary potassium loss using appropriate diuretics, such as *Lasix*, (3) administration of buffers (usually sodium bicarbonate) that can control the pH of the ECF, (4) restriction of dietary potassium intake, and (5) administration of enemas or laxatives containing compounds, such as *kayexolate*, that promote potassium loss across the digestive lining. In cases resulting from renal failure, kidney dialysis may be required as well.

Other Electrolytes

The ECF concentrations of other electrolytes are regulated as well. We will briefly note the most important ions involved. Additional information concerning these ions can be found in Table 27-2.

CALCIUM BALANCE Calcium is the most abundant mineral in the human body. A typical human body contains 1–2 kg (2.2–4.4 lb) of calcium, with 99 percent of it deposited in the skeleton. In addition to forming the crystalline component of bone, calcium ions play key roles in the control of muscular and neural activity, in blood clotting, as cofactors for enzymatic reactions, and as second messengers in a variety of cell types.

As noted in Chapters 6 and 18, calcium homeostasis reflects primarily an interplay between the reserves in bone, the rate of absorption across the digestive tract, and the rate of loss at the kidneys. The hormones parathyroid hormone, calcitriol, and, to a lesser degree, calcitonin are responsible for maintaining calcium homeostasis in the ECF. Parathyroid hormone and calcitriol elevate calcium ion concentrations, and their actions are opposed by calcitonin. [pp. 189, 621, 628]

A relatively small amount of calcium is lost in the bile, and under normal circumstances very little calcium escapes in the urine or feces. To keep pace with biliary, urinary, and fecal calcium losses an adult must absorb only 0.8–1.2 g/day of calcium. That represents only around 0.03 percent of the calcium reserve in the skeleton. Calcium absorption at the digestive tract and reabsorption along the proximal convoluted tubule (PCT) are stimulated by parathyroid hormone from the parathyroid glands and calcitriol from the kidneys.

TABLE 27-2 Electrolyte Balance

Ion and Normal ECF Range (mEq/l)	*Disorder*	*Symptoms*	*Causes*	*Treatment*
Sodium (136–142)	Hypernatremia (>147)	Thirst, dryness and wrinkling of skin, reduced blood volume and blood pressure, eventual circulatory collapse	Dehydration, loss of hypotonic fluid	Ingestion of water or intravenous infusion of hypotonic solution
	Hyponatremia (<130)	Disturbed CNS function ("water intoxication"): confusion, hallucinations, convulsions, coma, death in severe cases	Infusion or ingestion of large volumes of hypotonic solution	Diuretic use and infusion of hypertonic salt solution
Potassium (3.8–5.0)	Hyperkalemia (>8)	Severe cardiac arrhythmias	Renal failure, diuretic use, chronic acidosis	Infusion of hypotonic solution, selection of different diuretics, infusion of buffers, dietary restrictions
	Hypokalemia (<2)	Muscular weakness and paralysis	Low potassium diet, diuretic drugs, hypersecretion of aldosterone, chronic alkalosis	Increase dietary K^+ content, ingestion of K^+ tablets or solutions, infusion of potassium solution
Calcium (4.5–5.3)	Hypercalcemia (>11)	Confusion, muscle pain, cardiac arrhythmias, kidney stones, calcification of soft tissues	Hyperparathyroidism, cancer, vitamin D toxicity, calcium supplement overdose	Infusion of hypotonic fluid to lower calcium level, surgery to remove parathyroid gland, calcitonin administration
	Hypocalcemia (<4)	Muscle spasms, convulsions, intestinal cramps, weak heartbeats, cardiac arrhythmias, osteoporosis	Poor diet, lack of vitamin D, renal failure, hypoparathyroidism, hypomagnesemia	Calcium supplements, vitamin D administration
Magnesium (1.5–2.5)	Hypermagnesemia (>4)	Confusion, lethargy, respiratory depression, hypotension	Overdose of magnesium supplements or antacids (rare)	Infusion of hypotonic solution to lower plasma concentration
	Hypomagnesemia (<0.8)	Hypocalcemia, muscle weakness, cramps, cardiac arrhythmias, hypertension	Poor diet, alcoholism, severe diarrhea, kidney disease, malabsorption syndromes, ketoacidosis	Intravenous infusion of solution high in magnesium ions
Phosphate (1.8–2.6)	Hyperphosphatemia (>6)	No immediate symptoms; chronic elevation leads to calcification of soft tissues	High dietary intake, hypoparathyroidism	Dietary reduction
	Hypophosphatemia (<1)	Anorexia, dizziness, muscle weakness, cardiomyopathy, osteoporosis	Poor diet, kidney disease, malabsorption syndrome, hyperparathyroidism, vitamin D deficiency	Dietary improvement, hormone or vitamin supplements
Chloride (100–108)	Hyperchloremia (>112)	Acidosis, hyperkalemia	Dietary excess, increased retention	Infusion of hypotonic solution to lower plasma concentration
	Hypochloremia (<95)	Alkalosis, anorexia, muscle cramps, apathy	Vomiting, hypokalemia	Diuretic use and influsion of hypertonic salt solution

HYPERCALCEMIA AND HYPOCALCEMIA *Hypercalcemia* exists when the ECF calcium ion concentration is greater than 11 mEq/l. The primary cause of hypercalcemia in adults is *hyperparathyroidism*, followed by malignant cancers involving the breast, lung, kidneys, or bone marrow. [p. 622] Severe hypercalcemia (12–13 mEq/l) causes a variety of symptoms, including fatigue, confusion, cardiac arrhythmias, and calcification of the kidneys and soft tissues throughout the body.

The condition of *hypocalcemia* (under 4 mEq/l) is much less common than hypercalcemia. Hypoparathyroidism, vitamin D deficiency, or chronic renal failure is often responsible for hypocalcemia. Symptoms include muscle spasms, sometimes including generalized convulsions, weak heartbeats, cardiac arrhythmias, and osteoporosis.

MAGNESIUM BALANCE The adult body contains around 29 g of magnesium, with almost 60 percent of it deposited in the skeleton. The magnesium in body fluids is contained primarily in the ICF, where the concentration averages around 26 mEq/l. Magnesium is required as a cofactor for several important enzymatic reactions, including the phosphorylation of glucose within cells and the use of ATP by contracting muscle fibers. Magnesium is also important as a structural component of bone.

The concentration of magnesium in the ECF averages 1.5–2.5 mEq/l, considerably lower than levels in the ICF. The PCT reabsorbs magnesium very effectively; to keep pace with the daily urinary loss the minimum dietary requirement for magnesium is only 0.3–0.4 g/day.

PHOSPHATE BALANCE Phosphate ions are required for bone mineralization, and roughly 744 g are bound up in the mineral salts of the skeleton. In body fluids the most important functions of phosphate ions involve the ICF, where they are required for the formation of high-energy compounds, the activation of enzymes, and the synthesis of nucleic acids.

The phosphate concentration of the plasma is usually 1.8–2.6 mEq/l. Phosphate ions are reabsorbed from the tubular fluid along the PCT; urinary and fecal losses of phosphate ions total 0.8–1.2 g/day. Phosphate ion reabsorption along the PCT are inhibited by parathyroid hormone and stimulated by calcitriol.

CHLORIDE BALANCE Chloride ions are the most abundant anions in the ECF. The plasma concentration ranges from 100 to 108 mEq/l. Chloride concentrations in the ICF are usually low (3 mEq/l). Chloride ions are absorbed across the digestive tract in company with sodium ions, and several carrier proteins along the renal tubules reabsorb chloride ions with sodium ions. [pp. 1001–1005] The rate of chloride ion loss is small—a gain of 1.7–5.1 g/day will keep pace with losses in urine and perspiration.

 What effect would drinking a pitcher of distilled water have on your level of ADH?

 How would eating a meal high in salt affect the amount of fluid in the ICF compartment?

 What effect would being lost in the desert for a day without water have on your plasma osmolarity?

Acid-Base Balance

Chapter 2 introduced the topic of pH and the chemical nature of acids, bases, and buffers. Table 27-3 reviews key terms important to the discussion that follows; if you need a more detailed review, refer to the appropriate sections of Chapter 2 before proceeding. [pp. 44–45]

The pH may be altered by the introduction of either acids or bases into body fluids. A general classification of acids and bases sorts them into *strong* versus *weak*.

- ***Strong acids and bases.*** Strong acids and strong bases dissociate completely in solution. For example, hydrochloric acid (HCl) is a strong acid introduced in Chapter 2. The reaction in solution is diagrammed as:

$$HCl \rightarrow H^+ + Cl^-$$

- ***Weak acids and bases.*** When **weak acids** and **weak bases** enter a solution, a significant number of molecules remain intact. This means that if you place molecules of a weak acid in one solution and the same number of molecules of a strong acid in another, the weak acid will liberate fewer hydrogen ions and have less of an impact on the pH of the solution than the strong acid will. Carbonic acid is an example of a weak acid encountered in previous chapters. At the normal pH of the ECF, roughly 95 percent of the carbonic acid has dissociated, releasing bicarbonate ions and hydrogen ions. An equilibrium state exists, and the reaction can be diagrammed as:

$$H_2CO_3 \rightleftharpoons H^+ + HCO_3^-$$

The Importance of pH Control

The pH of body fluids reflects interactions among all of the acids, bases, and salts in solution. The pH of the ECF normally remains within relatively narrow limits, usually from 7.35 to 7.45. Any deviation outside the normal range is extremely dangerous because changes in hydrogen ion concentrations disrupt the stability of cell membranes, alter protein structure, and change the activities of important enzymes. An individual cannot survive with an extracellular fluid pH below 6.8 or above 7.7.

When the pH of the plasma falls below 7.35, **acidemia** exists. The physiological state that results is called **acidosis.** When the pH of the blood rises above 7.45, **alkalemia** exists. The physiological state that results is called **alkalosis.** Acidosis and alkalosis affect virtually all body systems, but the nervous system and cardiovascular system are particularly sensitive to pH fluctuations. For example, severe acidosis (pH below 7.0) can be deadly because (1) CNS function deteriorates, and the individual becomes comatose; (2) cardiac contractions grow weak and irregular, and symptoms of heart failure develop; and (3) peripheral vasodilation produces a dramatic drop in blood pressure, and circulatory collapse may occur.

The control of pH is therefore an extremely important homeostatic process of great physiological and clinical significance. Although acidosis and alkalosis are both dangerous, in practice *problems with acidosis are much more common than problems with alkalosis.* This is the case because several different types of acids, including carbonic acid, are generated by normal cellular activities.

❑ Types of Acids in the Body

There are three general categories of acids in the body: *volatile acids, fixed acids,* and *organic acids.*

Volatile Acids

Figure 27-6

A **volatile acid** is an acid that can leave solution and enter the atmosphere. Carbonic acid (H_2CO_3) is an important volatile acid found in body fluids. At the lungs, carbonic acid breaks down into carbon dioxide and water, and the carbon dioxide diffuses into the alveoli, a process described in Chapter 23. ∞ *[p. 861]* Carbonic acid forms in peripheral tissues as carbon dioxide in solution interacts with water to form molecules of carbonic acid, which dissociates to release hydrogen ions and bicarbonate ions. The complete reaction sequence is:

$$\underset{\text{carbon dioxide}}{CO_2} + \underset{\text{water}}{H_2O} \rightleftharpoons \underset{\text{carbonic acid}}{H_2CO_3} \rightleftharpoons H^+ + \underset{\text{bicarbonate ion}}{HCO_3^-}$$

This reaction occurs spontaneously in body fluids, but it occurs most rapidly in the presence of *carbonic anhydrase,* an enzyme found in the cytoplasm of red blood cells, liver and kidney cells, parietal cells of the stomach, and in many other cell types.

Because most of the carbon dioxide in solution is converted to carbonic acid, and most of the carbonic acid dissociates, there is a direct relationship between the P_{CO_2} and the pH (Figure 27-6•). When carbon dioxide levels rise, additional hydrogen ions and bicarbonate ions are released, and the pH goes down. (Remember that the pH is a *negative exponent,* so when the concentration of hydrogen ions goes up, the pH goes down.) The P_{CO_2} is the most important factor affecting the pH in body tissues.

At the alveoli, carbon dioxide diffuses into the atmosphere, the number of hydrogen ions and bicarbonate ions declines, and the pH rises. This process, which effectively removes hydrogen ions from solution, will be considered in more detail later in the chapter.

Fixed Acids

Fixed acids are acids that do not leave solution; once produced, they remain in body fluids until excreted at the kidneys. **Sulfuric acid** and **phosphoric acid** are the most important fixed acids. They are generated in small amounts during the catabolism of amino acids and compounds containing phosphate groups, including phospholipids and nucleic acids.

Organic Acids

Organic acids are acid participants in or byproducts of cellular metabolism. Lactic acid, produced

TABLE 27-3 A Review of Important Terms Relating to Acid-Base Balance

Term	Definition
pH	The negative exponent (negative log) of the hydrogen ion concentration
Neutral	A solution with a pH of 7, which contains equal numbers of hydrogen and hydroxyl ions
Acidic	A solution with a pH below 7, in which hydrogen ions predominate
Basic or alkaline	A solution with a pH above 7, in which hydroxyl ions predominate
Acid	A substance that dissociates to release hydrogen ions, shifting the pH toward acidity
Base	A substance that dissociates to release hydroxide ions or tie up hydrogen ions and increase pH
Salt	An ionic compound consisting of a cation other than hydrogen and an anion other than a hydroxide ion
Buffer	A substance that tends to oppose changes in the pH of a solution by removing or replacing hydrogen ions; in body fluids, buffers maintain pH within normal limits (7.35–7.45)

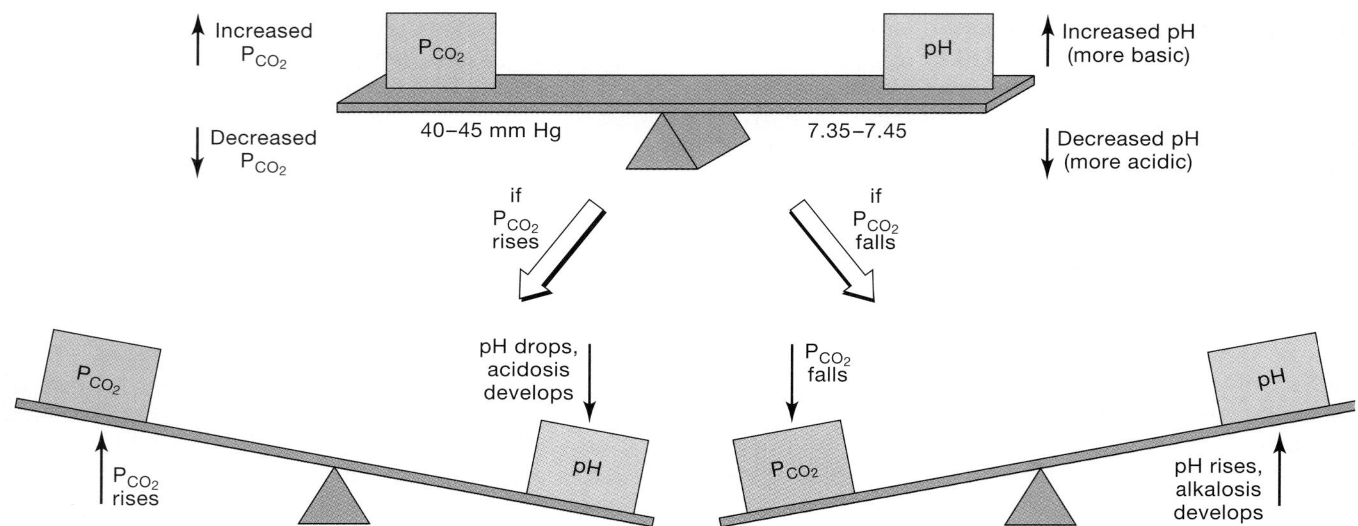
Increased
P_{CO_2}
P_{CO_2}
pH
Increased pH
(more basic)
Decreased
P_{CO_2}
40–45 mm Hg
7.35–7.45
Decreased pH
(more acidic)
if
P_{CO_2}
rises
if
P_{CO_2}
falls
P_{CO_2}
pH drops,
acidosis
develops
P_{CO_2}
falls
pH
pH
P_{CO_2}
rises
P_{CO_2}
pH rises,
alkalosis
develops

● **FIGURE 27-6**

Basic Relationships between P_{CO_2} and Plasma pH. The pH is inversely related to the P_{CO_2}; when the P_{CO_2} increases, the pH declines.

by the anaerobic metabolism of pyruvate, and ketone bodies, synthesized from acetyl-CoA, are important examples of organic acids considered in Chapters 10 and 25. [pp. 307, 961] Under normal conditions, most organic acids are metabolized rapidly, and significant accumulations do not occur. But relatively large amounts of organic acids are produced (1) during periods of anaerobic metabolism, because of the rapid buildup of lactic acid, and (2) during starvation or excessive lipid catabolism, because of the accumulation of ketone bodies.

❑ Mechanisms of pH Control

For acid-base balance to be maintained over long periods of time, there must be a balance between gains and losses of hydrogen ions. Hydrogen ions are gained at the digestive tract and through metabolic activities within our cells. These hydrogen ions must be excreted at the kidneys, through the secretion of hydrogen ions into the urine, and at the lungs, through the formation of water and carbon dioxide from hydrogen ions and bicarbonate ions. The sites of excretion are far removed from the sites of acid production. *Buffer systems* within body fluids neutralize the hydrogen ions as they travel from their points of origin to the lungs or kidneys, where they will be excreted through *compensation mechanisms* described in a later section.

Buffers and Buffer Systems

Figures 27-7, 27-8, 27-9

The acids produced in the course of normal metabolic operations are temporarily neutralized by buffers and buffer systems in body fluids. *Buffers*, introduced in Chapter 2, are dissolved compounds that can provide or remove hydrogen ions and thereby stabilize the pH of a solution. [p. 45] Buffers include weak acids that can donate hydrogen ions and weak bases that can absorb them. A **buffer system** consists of a combination of a weak acid or weak base and the salt of that weak acid or weak base.

There are three major buffer systems, each with slightly different characteristics and distribution.

- ***Protein buffer systems.*** Protein buffer systems contribute to the regulation of pH in the ECF and ICF. These buffer systems interact extensively with the other buffer systems.
- ***The carbonic acid–bicarbonate buffer system.*** The carbonic acid buffer system is most important in the ECF.
- ***The phosphate buffer system.*** The phosphate buffer system has an important role in buffering the pH of the ICF.

The functional relationships between these buffer systems are diagrammed in Figure 27-7●.

PROTEIN BUFFER SYSTEMS Protein buffer systems depend on the ability of amino acids to respond to alterations in pH by accepting or releasing hydrogen ions. The underlying mechanism can be seen in Figure 27-8●.

- If the pH climbs, the carboxyl group (–COOH) of the amino acid can dissociate, releasing a hydrogen ion and becoming a carboxylate ion ($–COO^-$).
- If the pH drops, the amino group ($–NH_2$) can accept an additional hydrogen ion, forming an amino ion ($–NH_3^+$).

Plasma proteins contribute to the buffering capabilities of the blood. In the interstitial fluids, there are extracellular protein fibers and dissolved amino acids that also assist in the regulation of pH. In the intracellular fluids of active cells, structural and other proteins provide an extensive buffering capability that prevents destructive pH changes when organic acids, such as lactic acid or pyruvic acid, are produced by cellular metabolism.

Because there is exchange between the ECF and ICF, the protein buffer system can help stabilize the pH of the extracellular fluid. For example:

- When the pH of the ECF declines, cells pump hydrogen ions out of the ECF and into the ICF, where they can be buffered by intracellular proteins.
- When the pH of the ECF rises, exchange pumps in cell membranes exchange hydrogen ions in the ICF for potassium ions in the ECF.

These mechanisms can assist in buffering the pH of the ECF. but the process is slow because hydrogen ions must be individually transported across the cell membrane. As a result, the protein buffer system in most cells cannot make rapid, large-scale adjustments in the pH of the ECF.

The Hemoglobin Buffer System. The situation is somewhat different when one considers red blood cells. Red blood cells, which contain roughly 5.5 percent of the ICF, are normally suspended in the plasma. These cells are densely packed with hemoglobin, and the cytoplasm contains large amounts of carbonic anhydrase. These cells have a significant effect on the pH of the ECF because they absorb carbon dioxide from the plasma and convert it to carbonic acid. CO_2 can diffuse across the RBC membrane very quickly, and a transport mechanism is not needed. As the carbonic acid dissociates, the bicarbonate ions move into the plasma, via the *chloride shift,* and the hydrogen ions are buffered by hemoglobin molecules. At the lungs the entire reaction sequence, diagrammed in Figure 23-24●, p. 861, proceeds in reverse. This is known as the **hemoglobin buffer system.**

This is the only intracellular buffer system that can have an immediate effect on the pH of the ECF. *The hemoglobin buffer system helps prevent drastic alterations in pH when the plasma* P_{CO_2} *is rising or falling.*

THE CARBONIC ACID–BICARBONATE BUFFER SYSTEM With the exception of red blood cells, some cancer cells, and tissues temporarily deprived of oxygen, our cells generate carbon dioxide virtually 24 hours a day. As detailed above, most of the carbon dioxide is converted to carbonic acid, which then dissociates into a hydrogen ion and a bicarbonate ion. The carbonic acid and its dissociation products form the **carbonic acid–bicarbonate buffer system.**

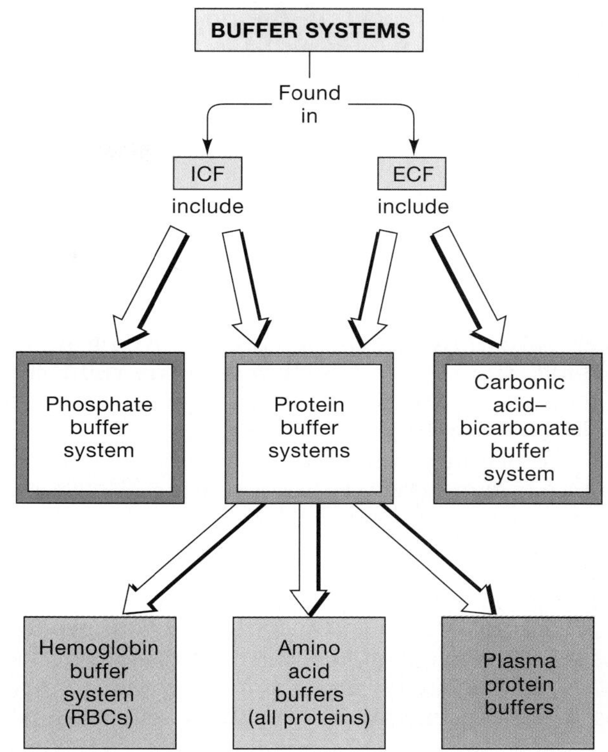
BUFFER SYSTEMS
Found in
ICF
include
ECF
include
Phosphate buffer system
Protein buffer systems
Carbonic acid–bicarbonate buffer system
Hemoglobin buffer system (RBCs)
Amino acid buffers (all proteins)
Plasma protein buffers

● **FIGURE 27-7**
Buffer Systems in Body Fluids. Phosphate buffers are found primarily in the ICF, whereas the carbonic acid–bicarbonate buffer system is found primarily in the ECF. Protein buffer systems are found in both the ICF and the ECF. Extensive interactions occur between these buffer systems.

This buffer system consists of a weak acid, carbonic acid, and *sodium bicarbonate,* the salt of that weak acid (Figure 27-9a●). The buffer system consists of the reaction:

$$CO_2 + H_2O \rightleftharpoons H_2CO_3 \rightleftharpoons H^+ + HCO_3^-$$

Because the reaction is freely reversible, changing the concentrations of any one of the participants will affect the concentrations of all others. For example, if hydrogen ions are added, most will be removed by interactions with bicarbonate ions,

● **FIGURE 27-8**
Amino Acid Buffers. Amino acids can either accept a hydrogen ion or donate one, depending on the pH of their surroundings.

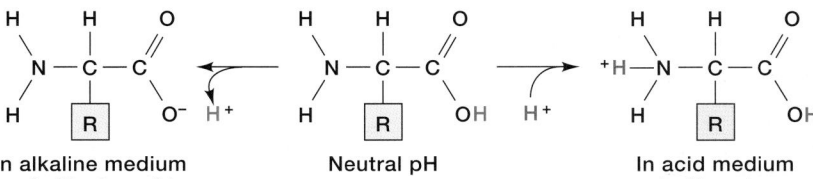

In alkaline medium acts like an acid

Neutral pH

In acid medium acts like a base

forming carbonic acid. In the process, the bicarbonate ion acts as a weak base that buffers the excess hydrogen ions. The carbonic acid formed in this way will in turn dissociate into CO_2 and water (Figure 27-9b●), and the CO_2 can then be excreted at the lungs. In effect, this reaction takes the hydrogen ion released by a strong organic or fixed acid and generates a volatile acid that can be easily eliminated.

This buffer system can also protect against increases in pH, although such changes are relatively rare. If hydrogen ions are removed from the plasma, the reaction is driven to the right: The P_{CO_2} declines, and the dissociation of carbonic acid replaces the missing hydrogen ions.

The primary role of the carbonic acid–bicarbonate buffer system is in preventing pH changes caused by organic acids and fixed acids in the ECF. The most important limitations of this buffer system are that:

- ***It cannot protect the ECF from pH changes that result from elevated or depressed levels of CO_2.*** A buffer system cannot protect against changes in the concentration of the weak acid or its salt. In the case above, the addition of excess H^+ drove the reaction to the left. But consider what would happen if we had added excess CO_2 instead of excess H^+. The elevated CO_2 would drive the reaction to the right, forming additional H_2CO_3 that dissociates into H^+ and HCO_3^-. This reaction would reduce the pH of the plasma.
- ***It can function only when the respiratory system and the respiratory control centers are working normally.*** Under normal circumstances the elevation in P_{CO_2} that occurs when fixed or organic acids are buffered will stimulate an increase in the respiratory rate. This increase accelerates the removal of CO_2 at the lungs. If the respiratory passageways are blocked, circulation to the lungs is impaired, or if the respiratory centers do not respond normally, the efficiency of the buffer system will be reduced.
- ***The ability to buffer acids is limited by the availability of bicarbonate ions.*** Every time a hydrogen ion is removed from the plasma, a bicarbonate ion goes with it. When all of the bicarbonate ions have been tied up, buffering capabilities are lost. Such problems are rare because:

  1. Body fluids contain a large reserve of bicarbonate ions, primarily in the form of dissolved molecules of a weak base, *sodium bicarbonate* ($NaHCO_3$). This readily available supply of bicarbonate ions is known as the **bicarbonate reserve.** The reaction sequence involved is:

  $$Na^+ + HCO_3^- \rightleftharpoons NaHCO_3$$

  When hydrogen ions enter the ECF, the bicarbonate ions tied up in carbonic acid molecules are replaced by bicarbonates from the bicarbonate reserve. This relationship is diagrammed in Figure 27-9b●.
  2. Additional bicarbonate ions can be generated at the kidneys, through mechanisms introduced in Chapter 26 (Figure 26-16●, p. 1005). In the DCT and collecting system, carbonic anhydrase converts CO_2 within tubular cells into carbonic acid, which then dissociates. The hydrogen ion is pumped into the tubular fluid in exchange for a sodium ion, and the bicarbonate ion is transported into the peritubular fluid in exchange for a chloride ion. In effect, the tubular cell removes HCl from the peritubular fluid in exchange for sodium bicarbonate.

THE PHOSPHATE BUFFER SYSTEM The **phosphate buffer system** consists of an anion, $H_2PO_4^-$, that is a weak acid. The basic operation of the phosphate buffer system resembles that of the carbonic acid–bicarbonate buffer system. The reversible reaction involved is:

$$H_2PO_4^- \rightleftharpoons H^+ + HPO_4^{2-}$$

The weak acid is *dihydrogen phosphate,* and the anion released is *monohydrogen phosphate.* In the ECF the phosphate buffer system plays only a supporting role in the regulation of pH, primarily because the concentration of bicarbonate ions far exceeds that of monohydrogen phosphate. However, the phosphate buffer system is quite important in buffering the pH of the intracellular fluids. In addition, cells contain a *phosphate reserve* in the form of the weak base *sodium monohydrogen phosphate* (Na_2HPO_4). The phosphate buffer system is also important in stabilizing the pH of the urine.

The dissociation of this compound provides additional monohydrogen phosphate ions for use by the buffer system:

$$2Na^+ + HPO_4^{2-} \rightleftharpoons Na_2HPO_4$$

❑ Maintenance of Acid-Base Balance

Although buffer systems can tie up excess hydrogen ions, they provide only a temporary solution. The hydrogen ions have not been eliminated, merely rendered harmless. For homeostasis to be preserved, the captured hydrogen ions must eventually be excreted.

The problem here is that the supply of buffer molecules is limited. For example, suppose that a buffer molecule prevents a pH change by binding a hydrogen ion that enters the ECF. That buffer molecule will then be tied up, and this reduces the capacity of the

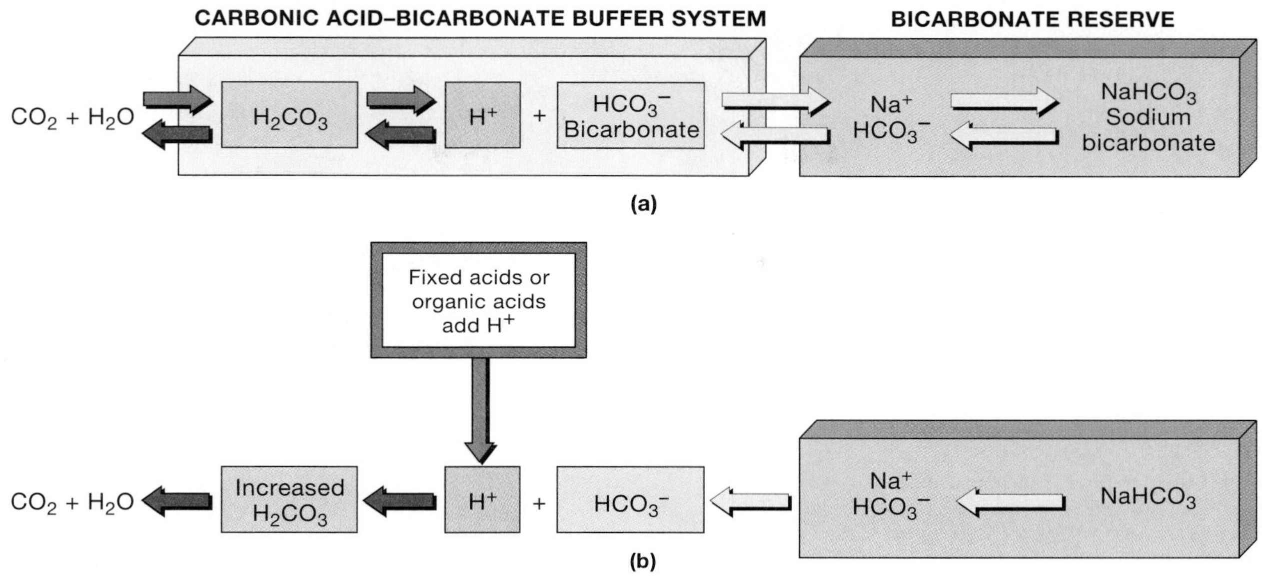
CARBONIC ACID–BICARBONATE BUFFER SYSTEM
BICARBONATE RESERVE
CO2 + H2O
H2CO3
H+
+
HCO3–
Bicarbonate
Na+
HCO3–
NaHCO3
Sodium
bicarbonate
(a)
Fixed acids or
organic acids
add H+
CO2 + H2O
Increased
H2CO3
H+
+
HCO3–
Na+
HCO3–
NaHCO3
(b)

● **FIGURE 27-9**

The Carbonic Acid–Bicarbonate Buffer System. (a) Basic components of the carbonic acid–bicarbonate buffer system, showing their relationships to carbon dioxide and the bicarbonate reserve. **(b)** Response of the carbonic acid–bicarbonate buffer system to hydrogen ions generated by fixed or organic acids in body fluids.

ECF to cope with additional hydrogen ions. Eventually, all the buffer molecules will be bound to hydrogen ions, and pH control will be impossible.

The situation can be resolved only by removing the hydrogen ions from the ECF, thereby freeing the buffer molecules, or by replacing the buffer molecules. Similarly, if a buffer provides a hydrogen ion to maintain normal pH, either another hydrogen ion must be obtained or the buffer must be replaced.

The maintenance of acid-base balance thus includes balancing hydrogen ion gains and losses. This balancing act involves coordinating the actions of buffer systems with pulmonary mechanisms and renal mechanisms. The pulmonary and renal mechanisms support the buffer systems by (1) secreting or absorbing hydrogen ions, (2) controlling the excretion of acids and bases, and, when necessary, (3) generating additional buffers. It is the *combination* of buffer systems and these pulmonary and renal mechanisms that maintains pH within narrow limits.

Respiratory Compensation

Respiratory compensation is a change in the respiratory rate that helps stabilize the pH of the ECF. Respiratory compensation occurs whenever the pH strays outside of normal limits. It is effective because respiratory activity has a direct effect on the carbonic acid–bicarbonate buffer system. Increasing or decreasing the rate of respiration alters pH by lowering or raising the P_{CO_2}. When the P_{CO_2} rises, the pH falls; when the P_{CO_2} falls, the pH rises because the removal of CO_2 drives the carbonic acid-bicarbonate system to the left:

$$CO_2 + H_2O \leftarrow H_2CO_3 \leftarrow H^+ + HCO_3^-$$

Mechanisms responsible for the control of respiratory rate were detailed in Chapter 23, and only a brief summary will be presented here. (Readers seeking a more detailed review should refer to Figures 23-27, p. 865, and 23-28●, p. 867.)

1. Chemoreceptors of the carotid and aortic bodies are sensitive to the P_{CO_2} of the circulating blood. Other receptors located on the ventrolateral surfaces of the medulla oblongata monitor the P_{CO_2} of the cerebrospinal fluid.
2. A rise in P_{CO_2} stimulates the receptors; a fall in the P_{CO_2} inhibits them.
3. Stimulation of the chemoreceptors leads to an increase in the respiratory rate. As the rate of respiration increases, more CO_2 is lost at the lungs, and the P_{CO_2} returns to normal levels.
4. When the P_{CO_2} of the blood or CSF declines, respiratory activity becomes depressed and the breathing rate decreases. This causes an elevation of the P_{CO_2} in the ECF.

Renal Compensation

Figure 27-10

Renal compensation is a change in the rates of hydrogen and bicarbonate ion secretion or reabsorption in response to changes in plasma pH. Under normal conditions, the body generates enough organic and fixed acids to add around 100 mEq of hydrogen ions to the extracellular fluids each day. An equivalent number of hydrogen ions

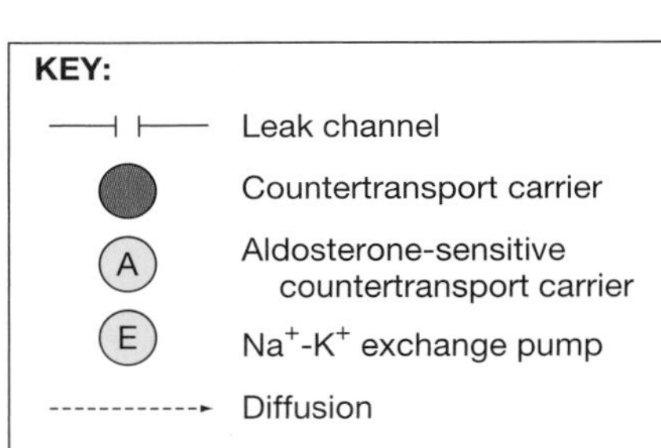
KEY:
Leak channel
Countertransport carrier
Aldosterone-sensitive countertransport carrier
Na+-K+ exchange pump
Diffusion

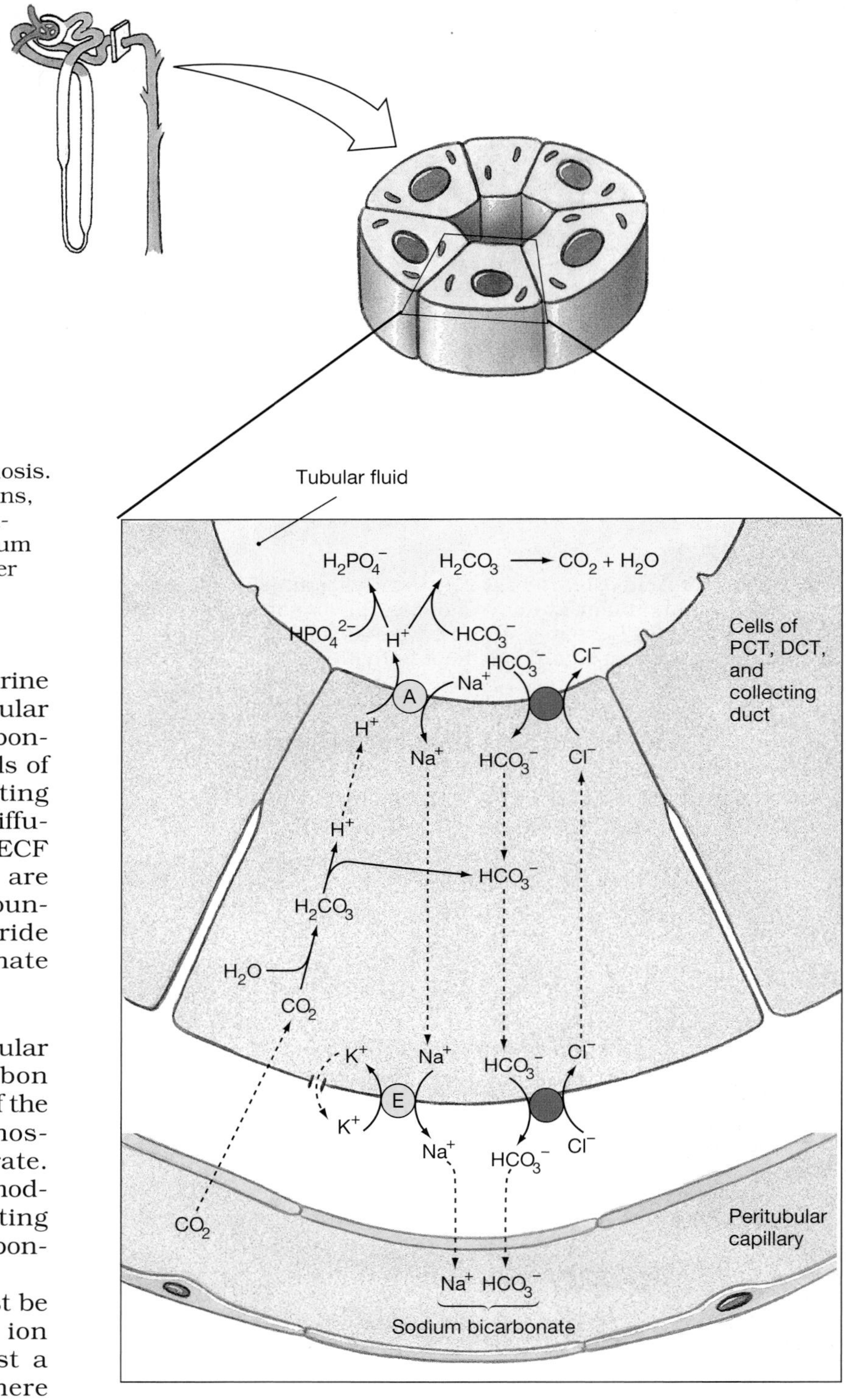

FIGURE 27-10
Kidney Tubules and pH Regulation. Response of the kidney tubules to acidosis. Note (1) active secretion of hydrogen ions, (2) removal of carbon dioxide from peritubular fluids, (3) reabsorption of sodium bicarbonate, and (4) the activity of other buffers in the tubular fluid.

must therefore be excreted in the urine to maintain acid-base balance. Tubular hydrogen ion secretion involves carbonic anhydrase activity within the cells of the PCT, the DCT, and the collecting ducts. This process results in the diffusion of sodium bicarbonate into the ECF (Figure 27-10•). Bicarbonate ions are also recovered from the filtrate by countertransport in exchange for chloride ions. In this way urinary bicarbonate losses are minimized.

BUFFERS IN THE URINE Glomerular filtration puts hydrogen ions, carbon dioxide, and the other components of the carbonic acid–bicarbonate and phosphate buffer systems into the filtrate. The kidney tubules subsequently modify the pH of the filtrate by secreting hydrogen ions or reabsorbing bicarbonate ions.

The pH of the tubular fluid must be kept above 4.5 because hydrogen ion secretion cannot continue against a large concentration gradient. If there were no buffers in the filtrate, the kidney tubules could secrete less than 1 percent of the acid produced each day before the pH reached this limit, and the kidneys would have to produce about 1000 liters of urine just to excrete the excess hydrogen ions. Buffers in the filtrate are therefore important, because they keep the pH within the acceptable range.

THE RENAL RESPONSE TO ACIDOSIS AND ALKALOSIS Acidosis develops when the normal plasma buffer mechanisms are stressed by excessive numbers of hydrogen ions. The kidney tubules do not distinguish among the various acids that may cause acidosis. Whether the fall in pH results from the production of volatile, fixed, or organic acids, the renal contribution remains limited to (1) secretion of hydrogen ions, (2) activity of buffers in the tubular fluid, (3) removal of CO_2, and (4) reabsorption of sodium bicarbonate.

The tubular cells thus bolster the capabilities of the bicarbonate buffer system by increasing the concentration of bicarbonate ions in the ECF, replacing those already used to remove hydrogen ions from solution. During starvation, the tubular cells break down amino acids, yielding carbon chains for catabolism and bicarbonates to help buffer ketone bodies in the blood (see Figure 26-16●, p. 1005).

When alkalosis develops, the rate of hydrogen ion secretion at the kidneys declines, and the tubular cells do not reclaim the bicarbonates entering the filtrate. As a result, the concentration of bicarbonate ions in the plasma declines. This decline promotes the dissociation of carbonic acid molecules, and the hydrogen ions released help to return the pH to normal levels.

Disturbances of Acid-Base Balance

Figure 27-11

Figure 27-11● summarizes the interactions among buffer systems, respiration, and renal function in normal acid-base balance. In combination, these mechanisms are usually able to control pH very precisely, so that the pH of the extracellular fluids seldom varies more than 0.1 pH units, from 7.35 to 7.45. When buffering mechanisms are severely stressed, the pH wanders outside of these limits, producing symptoms of alkalosis or acidosis.

Readers considering a career in any health-related field will find that understanding acid-base dynamics is essential for clinical diagnosis and management of a wide variety of conditions. Temporary alterations in the pH of body fluids occur frequently. Rapid and complete recovery occurs through a combination of buffer system activity and the respiratory and renal responses. More serious and prolonged disturbances of acid-base balance can result from any factor affecting one of the principal regulatory mechanisms.

- Any disorder affecting circulating buffers, respiratory performance, or renal function will potentially disrupt acid-base balance. Several conditions described in earlier chapters, including *emphysema* (Chapter 23) and *renal failure* (Chapter 26), are associated with dangerous changes in pH. ∞ [pp. 863, 997]
- Cardiovascular conditions, such as *heart failure* or *hypotension*, can also affect the pH of internal fluids by causing fluid shifts, changing glomerular filtration rates, and altering respiratory efficiency.
- Conditions affecting the CNS will also disrupt normal acid-base balance. For example, neural damage or CNS disease will affect the respiratory and cardiovascular reflexes that are essential to normal pH regulation.

Serious abnormalities in acid-base balance usually show an initial *acute phase*, in which the pH moves rapidly away from the normal range. If the condition persists, physiological adjustments occur and the individual enters the *compensated phase*. Unless the underlying problem is corrected, compensation cannot be complete, and blood chemistry will remain abnormal. The pH often remains outside normal limits, even after compensation has occurred. Even if the pH is within the normal range, the P_{CO_2} or HCO_3^- concentrations may be abnormal.

The primary source of the problem is usually indicated by the name given to the resulting condition.

- ***Respiratory acid-base disorders*** result from a mismatch between carbon dioxide generation in peripheral tissues and carbon dioxide excretion at the lungs. When a respiratory acid-base disorder is present, the carbon dioxide levels in the ECF are abnormal.
- ***Metabolic acid-base disorders*** are caused by the generation of organic or fixed acids or by conditions affecting the concentration of bicarbonate ions in the extracellular fluids.

Pulmonary compensation alone can often restore normal acid-base balance in individuals suffering from respiratory disorders. In contrast, compensation mechanisms for metabolic disorders may be able to stabilize pH, but other aspects of acid-base balance (buffer system function, bicarbonate levels, and P_{CO_2}) remain abnormal until the underlying metabolic cause is corrected.

The respiratory and metabolic categories can be subdivided, creating four major classes of acid-base disturbances: *respiratory acidosis*, *respiratory alkalosis*, *metabolic acidosis*, and *metabolic alkalosis*. Each of these will be discussed separately.

Respiratory Acidosis

Figure 27-12

Respiratory acidosis develops when the respiratory system is unable to eliminate all of the carbon dioxide generated by peripheral tissues. The primary symptom is low plasma pH due to **hypercapnia,** an elevated plasma P_{CO_2}. The usual cause is *hypoventilation*, an abnormally low respiratory rate. (Hypoventilation and hyperventilation, and their effects on P_{CO_2}, were discussed in Chapter 23.) ∞ [pp. 866-867] When the P_{CO_2} in the ECF rises, hydrogen and bicarbonate ion concentrations also begin rising as carbonic acid forms and dissociates. Other buffer systems can tie up some of the hydrogen ions, but once the combined buffering capacity has been exceeded the pH begins to fall rapidly. The effects are diagrammed in Figure 27-12●, p. 1050.

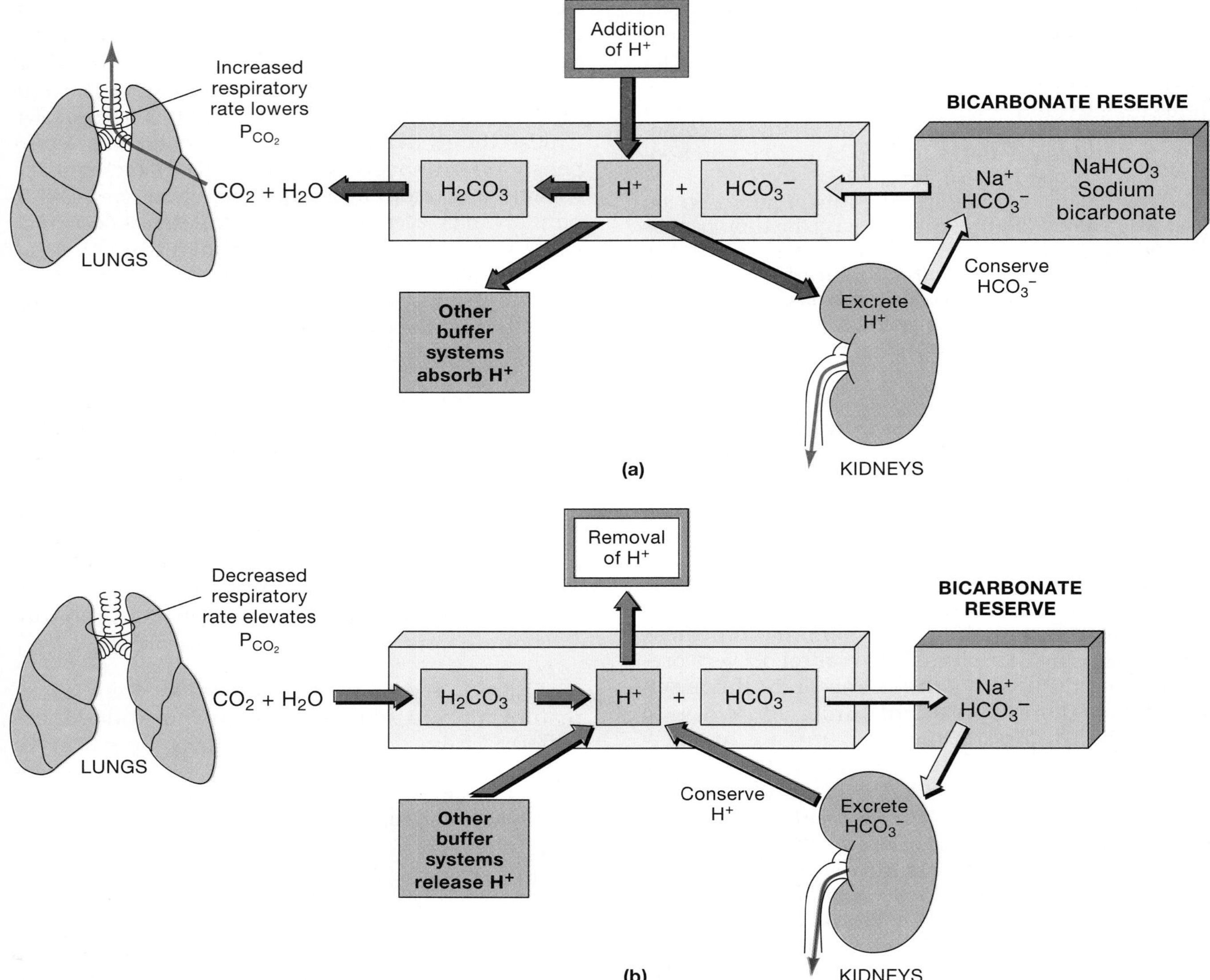

Addition of H+
Increased respiratory rate lowers P_{CO_2}
BICARBONATE RESERVE
$CO_2 + H_2O$
H_2CO_3
H^+
+
HCO_3^-
Na^+
HCO_3^-
$NaHCO_3$ Sodium bicarbonate
LUNGS
Conserve HCO_3^-
Excrete H^+
Other buffer systems absorb H^+
(a)
KIDNEYS
Removal of H+
Decreased respiratory rate elevates P_{CO_2}
BICARBONATE RESERVE
$CO_2 + H_2O$
H_2CO_3
H^+
+
HCO_3^-
Na^+
HCO_3^-
LUNGS
Conserve H^+
Excrete HCO_3^-
Other buffer systems release H^+
(b)
KIDNEYS

● **FIGURE 27-11**
The Central Role of the Carbonic Acid–Bicarbonate Buffer System in the Regulation of Plasma pH. This diagram summarizes the interactions between this buffer system and other buffer systems and compensation mechanisms. Part **(a)** shows the response to a fall in pH caused by the addition of hydrogen ions, part **(b)** the response to a rise in pH caused by the removal of hydrogen ions.

Respiratory acidosis represents the most frequent challenge to acid-base equilibrium. Our tissues generate carbon dioxide at a rapid rate, and even a few minutes of hypoventilation can cause acidosis, reducing the pH of the ECF to as low as 7.0. Under normal circumstances the chemoreceptors monitoring the P_{CO_2} of the plasma and CSF will eliminate the problem by calling for an increase in pulmonary ventilation rates.

Acute Respiratory Acidosis

If the chemoreceptors fail to respond, if pulmonary ventilation cannot be increased, or if the circulatory supply to the lungs is inadequate, the pH continues to decline. If the decline is severe, **acute respiratory acidosis** will develop. Acute respiratory acidosis is an immediate, life-threatening condition. It is especially dangerous in persons whose tissues are generating large amounts of carbon dioxide or who are incapable of normal respiratory activity. For this reason reversing acute respiratory acidosis is probably the major goal in the resuscitation of cardiac arrest or drowning victims. Thus first aid, CPR, and life-saving courses always stress the "ABCs" of emergency care: *Airway, Breathing,* and *Circulation.*

Chronic Respiratory Acidosis

Chronic respiratory acidosis develops when normal respiratory function has been compromised but the compensatory mechanisms have not failed completely. Individuals suffering from central nervous system injuries, or those whose respiratory centers have been desensitized by drugs such as alcohol or barbiturates, do not respond to the warning signals from the chemoreceptors. As a result, these people are more likely to develop acidosis due to chronic hypoventilation.

Even with intact and functional respiratory centers, increased pulmonary exchange may be prevented by damage to some components of the respiratory system. Examples of conditions fostering chronic respiratory acidosis include emphysema, congestive heart failure, and pneumonia (in which alveolar damage or blockage typically occurs). Pneumothorax and respiratory muscle paralysis have a similar effect, for they too limit the ability to maintain adequate pulmonary ventilation rates.

When a normal pulmonary response does not occur, the kidneys respond by increasing the rate of hydrogen ion secretion into the filtrate. This response slows the rate of pH change, but renal mechanisms alone cannot return the pH to normal until the underlying respiratory or circulatory problems are corrected.

Treatment of Respiratory Acidosis

The primary problem in respiratory acidosis is that the rate of pulmonary exchange is inadequate to keep the arterial P_{CO_2} within normal limits. The efficiency of pulmonary ventilation can often be improved temporarily by inducing bronchodilation or using mechanical aids that provide air under positive pressure. If breathing has ceased altogether, artificial respiration or a mechanical ventilator will be required. Because of the nature of the problem, these measures may be sufficient to restore normal pH if the respiratory acidosis was neither severe nor prolonged. Treatment of acute respiratory acidosis is complicated by the fact that it causes a complementary metabolic acidosis because of the generation of lactic acid in oxygen-starved tissues (see below).

❑ Respiratory Alkalosis

Figure 27-12

Problems with **respiratory alkalosis** (Figure 27-12●) are relatively uncommon. Respiratory alkalosis develops when respiratory activity lowers plasma P_{CO_2} to below normal levels, a condition called **hypocapnia.** A temporary hypocapnia can be produced by hyperventilation, when increased respiratory activity leads to a reduction in the arterial P_{CO_2}. Continued hyperventilation can elevate the pH to levels as high as 7.8–8. This condition usually corrects itself, for the reduction in P_{CO_2} removes the stimulation for the chemoreceptors, and the urge to breathe fades until carbon dioxide levels have returned to normal. *Respiratory alkalosis caused by hyperventilation seldom persists long enough to cause a clinical emergency.*

Common causes of hyperventilation include physical or psychological stresses, such as extreme anxiety. Sometimes people will hyperventilate deliberately to prepare for holding their breath; this can sometimes be dangerous for reasons noted in Chapter 23. ∞ *[p. 867]* Hyperventilation gradually elevates the pH of the CSF, and CNS function becomes affected. The initial symptoms involve tingling sensations in the hands, feet, and lips. A light-headed feeling may also be noted, and if the hyperventilation continues the individual may lose consciousness. When this occurs, any contributing psychological stimuli are removed, and the breathing rate declines. The P_{CO_2} then rises until the pH returns to normal.

A simple treatment for this form of respiratory alkalosis consists of having the victim breathe and rebreathe the air contained in a small paper bag. As the P_{CO_2} in the bag rises, so do the alveolar and arterial carbon dioxide concentrations. This change eliminates the problem and restores the pH to normal levels. Other problems with respiratory alkalosis are rare and involve primarily (1) individuals adapting to high altitudes, where the low P_{O_2} promotes hyperventilation, (2) patients on mechanical respirators, or (3) individuals with brain stem injuries who are incapable of responding to alterations in plasma CO_2 concentrations.

❑ Metabolic Acidosis

Figure 27-13

Metabolic acidosis is the second most common type of acid-base imbalance. It has three major causes:

1. The most frequent cause of metabolic acidosis is the production of a large number of fixed or organic acids. As diagrammed in Figure 27-13a●, the hydrogen ions liberated by these acids overload the bicarbonate buffer system, and the pH begins to decline. Two important examples of metabolic acidosis were introduced in earlier chapters.

   - **Lactic acidosis** may develop following severe exercise or prolonged tissue hypoxia (oxygen starvation) as active cells rely upon anaerobic respiration. (For a more detailed discussion, see Chapter 10, especially Figures 10-15, p. 308, and 10-16●, p. 309.)

   - **Ketoacidosis** results from the generation of large quantities of ketone bodies during the postabsorptive state. Ketoacidosis is a problem in starvation and a potential complication of poorly controlled diabetes mellitus. In either case, peripheral tissues are unable to obtain adequate glucose from the circulation and begin metabolizing lipids and ketone bodies. (The ketone bodies are generated at the liver during the catabolism of lipids and ketogenic amino acids.) For additional details, see the related discussion in Chapter 25. [p. 960]

2. A less common cause of metabolic acidosis is an impaired ability to excrete hydrogen ions at the kidneys. For example, conditions marked by severe kidney damage, such as *glomerulonephritis* (Chapter 26), often result in a severe metabolic acidosis. [p. 986] Metabolic acidosis may also be caused by diuretics that turn off the sodium-hydrogen transport system in the kidney tubules. The secretion of hydrogen ions is linked to the reabsorption of sodium, and when sodium reabsorption stops so does hydrogen secretion.
3. Metabolic acidosis can occur after a severe bicarbonate loss (Figure 27-13b●). The carbonic acid–bicarbonate buffer system relies on bicarbonate ions to balance hydrogen ions that threaten pH balance. A decline in the bicarbonate concentration in the ECF thus reduces the effectiveness of this buffer system, and acidosis soon develops. The most common cause of bicarbonate depletion is chronic diarrhea. Under normal conditions most of the bicarbonate ions secreted into the digestive tract in pancreatic, hepatic, and mucous secretions are reabsorbed before the feces are eliminated. In diarrhea these bicarbonates are lost, and the bicarbonate concentration in the ECF drops accordingly.

The proper treatment of a case of metabolic acidosis depends on an understanding of the nature of the problem. Because the potential causes are so varied, a clinician must piece together relevant clues to make a diagnosis. Sometimes the diagnosis is very straightforward—for example, a patient in metabolic acidosis after a bicycle race probably has lactic acidosis. At other times, one must do some detective work. When questions exist, the calculation of an *anion gap* can be very helpful.

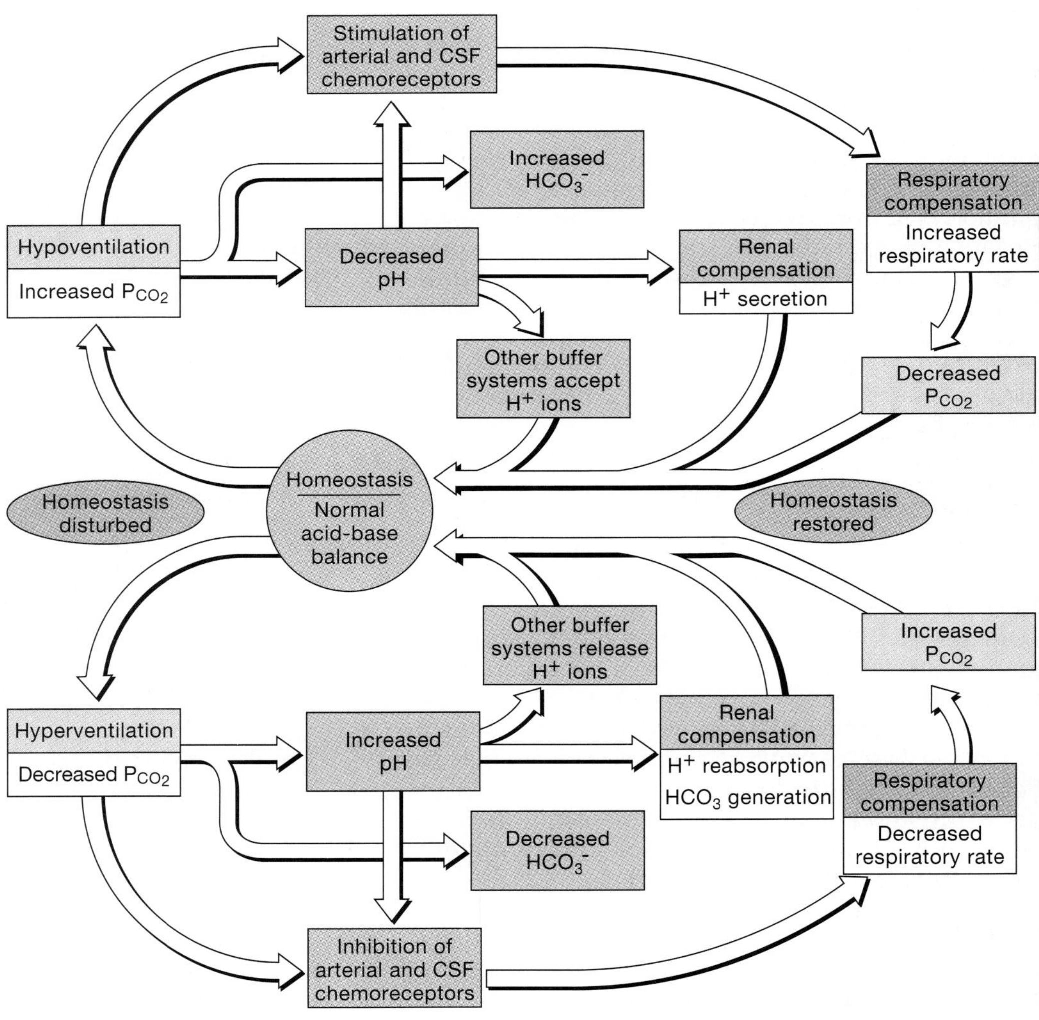

Stimulation of arterial and CSF chemoreceptors
Increased HCO_3^-
Respiratory compensation
Increased respiratory rate
Hypoventilation
Increased P_{CO_2}
Decreased pH
Renal compensation
H^+ secretion
Other buffer systems accept H^+ ions
Decreased P_{CO_2}
Homeostasis
Normal acid-base balance
Homeostasis disturbed
Homeostasis restored
Other buffer systems release H^+ ions
Increased P_{CO_2}
Hyperventilation
Decreased P_{CO_2}
Increased pH
Renal compensation
H^+ reabsorption
HCO_3 generation
Respiratory compensation
Decreased respiratory rate
Decreased HCO_3^-
Inhibition of arterial and CSF chemoreceptors

● **FIGURE 27-12 Respiratory Acid-Base Regulation.** Respiratory acidosis (top) and alkalosis (bottom) result from inadequate or excessive pulmonary ventilation. In normal individuals respiratory responses, combined with those of other buffer systems and the kidneys, are usually able to restore normal acid-base balance.

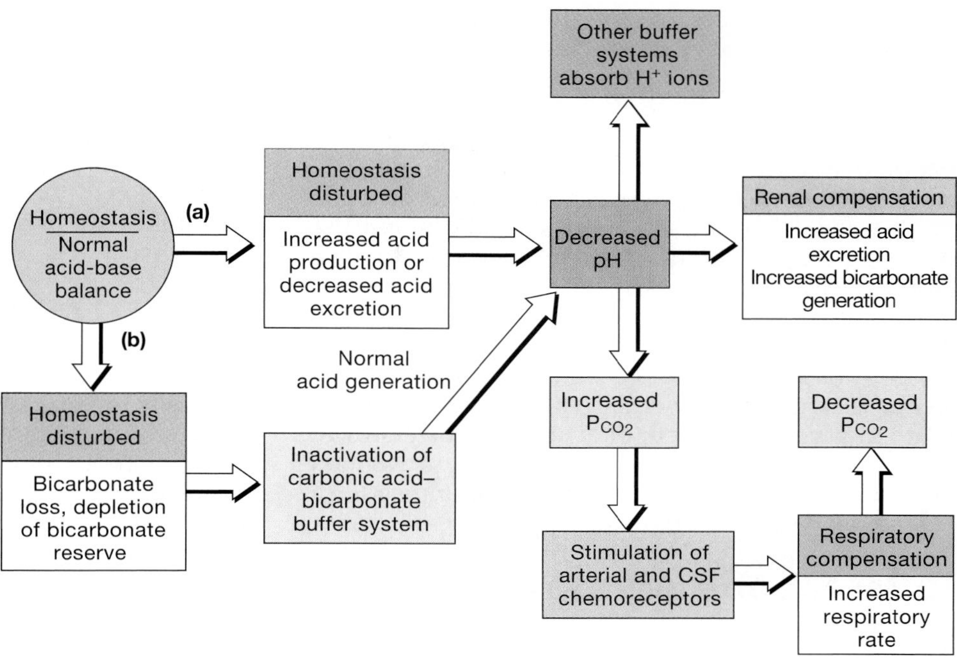

Other buffer systems absorb H+ ions
Homeostasis disturbed
Increased acid production or decreased acid excretion
Homeostasis
Normal acid-base balance
(a)
(b)
Decreased pH
Renal compensation
Increased acid excretion
Increased bicarbonate generation
Normal acid generation
Increased PCO2
Decreased PCO2
Homeostasis disturbed
Bicarbonate loss, depletion of bicarbonate reserve
Inactivation of carbonic acid–bicarbonate buffer system
Stimulation of arterial and CSF chemoreceptors
Respiratory compensation
Increased respiratory rate

● **FIGURE 27-13**
Factors Causing Metabolic Acidosis. (a) Increased acid production or decreased acid secretion leads to a buildup of H^+ in body fluids and a fall in pH. Pulmonary and renal compensation mechanisms can stabilize pH, but the blood chemistry remains abnormal until acid production or secretion returns to normal levels. **(b)** A loss of bicarbonate ions makes the carbonic acid–bicarbonate buffer system incapable of preventing a fall in pH.

THE ANION GAP Under normal circumstances sodium ions are the primary cations in the ECF, and their positive charges are roughly balanced by the negative charges on the major anions (chloride and bicarbonate), minor anions (phosphate and sulfate), and plasma proteins. The concentrations of sodium, chloride, and bicarbonate are relatively easy to determine. The concentration of sodium ions is greater than the concentration of chloride plus bicarbonate ions, and the difference is called the *anion gap.*

$$\text{anion gap} = Na^+ - (HCO_3^- + Cl^-)$$

The anion gap in normal individuals is 10–12 mEq/l. The calculation of the anion gap is useful in diagnosis because it can be used to distinguish among different types of metabolic acidosis. The ECF of an individual in metabolic acidosis will have low pH, and the P_{CO_2} and HCO_3^- will be reduced as the carbonic acid–bicarbonate system attempts to buffer the excess hydrogen ions.

- If the anion gap is normal, the problem results from either (1) the generation or ingestion of HCl or (2) the loss of bicarbonate.
  1. HCl is a strong acid, and it dissociates completely into H^+ and Cl^-. The chloride ion gained is balanced by the loss of a bicarbonate ion, which buffers the hydrogen ion. The net result is that the anion gap remains unchanged, although the bicarbonate level drops.
  2. In chronic diarrhea, the body loses the bicarbonate ions contained in the buffers secreted into the intestinal tract. As you may recall from Chapters 19 and 26, the movement of a bicarbonate ion across a cell membrane involves a countertransport mechanism that exchanges it for a chloride ion. Thus for every bicarbonate ion secreted into the digestive tract and lost, a chloride ion is absorbed and retained. Once again, the net result is that the anion gap remains unchanged, although the bicarbonate level drops.
- If the individual in metabolic acidosis has an increased anion gap, the problem must be either (1) the production or ingestion of organic acids or toxins or (2) renal failure.
  1. Conditions such as *ketoacidosis* and *lactic acidosis* are caused by organic acids that on dissociation release hydrogen ions and anions that are not considered in the calculation of the anion gap. ∞ [pp. 308, 960] As a result, when the hydrogen ions are buffered by bicarbonate ions, bicarbonate levels decline. Because this decline is not accompanied by an elevation in chloride ion concentrations, the anion gap increases.
  2. In renal failure, acids that are normally excreted in the urine, such as sulfuric acid and phosphoric acid, are retained. The anions released by the dissociation of these acids are not considered in the calculation of the anion gap. Once again, bicarbonate levels decline as they buffer the hydrogen ions, and the anion gap increases.

Compensation for Metabolic Acidosis

Compensation for metabolic acidosis usually involves a combination of respiratory and renal mechanisms. Hydrogen ions interacting with bicarbonate ions form carbon dioxide molecules that are eliminated at the lungs, while the kidneys excrete additional hydrogen ions into the urine and generate bicarbonate ions that are released into the ECF.

Combined Respiratory and Metabolic Acidosis

Metabolic and respiratory acidosis are frequently associated because sustained hypoventilation leads to decreased arterial P_{O_2}, and oxygen-starved tissues generate large quantities of lactic acid. The problem can be particularly serious in cases of near-drowning, in which the body fluids have high P_{CO_2}, low P_{O_2}, and large amounts of lactic acid generated by the muscles of the struggling victim. Prompt emergency treatment is essential, and the usual procedure involves some form of artificial or mechanical respiratory assistance coupled with the intravenous infusion of an isotonic solution containing sodium lactate, sodium gluconate, or sodium bicarbonate.

❑ Metabolic Alkalosis

Figure 27-14

Metabolic alkalosis occurs when bicarbonate ion concentrations become elevated (Figure 27-14●). The bicarbonate ions then interact with hydrogen ions in solution, forming carbonic acid. The reduction in H^+ concentrations then causes symptoms of alkalosis.

Cases of metabolic alkalosis are relatively rare, but one particularly interesting cause was noted in Chapter 24. ∞ [p. 897] The secretion of hydrochloric acid by the gastric mucosa is associated with the influx of large numbers of bicarbonate ions into the extracellular fluid. This phenomenon, known as the *alkaline tide*, temporarily elevates the bicarbonate concentration in the ECF at mealtimes. An individual who begins vomiting repeatedly will continue to generate stomach acids to replace those lost, and the ECF bicarbonate concentration will therefore continue to rise.

Compensation for metabolic alkalosis involves a reduction in pulmonary ventilation, coupled with the increased loss of bicarbonates in the urine. Treatment of mild cases addresses the primary cause, usually by controlling vomiting and may involve the administration of solutions containing NaCl or KCl.

Treatment of acute cases of metabolic alkalosis may involve the administration of ammonium chloride, NH_4Cl. Metabolism of the ammonium ion in the liver results in the liberation of a hydrogen ion, so in effect the introduction of ammonium chloride leads to the internal generation of hydrochloric acid, HCl, a strong acid. As the HCl diffuses into the bloodstream, the pH falls toward normal levels.

❑ The Detection of Acidosis and Alkalosis

Virtually anyone with a problem affecting the cardiovascular, respiratory, urinary, digestive, or nervous systems may develop potentially dangerous problems with acid-base balance. For this reason diagnostic blood tests usually include tests designed to provide information on pH and buffer function. Standard tests include blood pH, P_{CO_2}, and bicarbonate levels. These data make the recognition of acidosis or alkalosis and the classification of a particular condition as respiratory or metabolic in nature relatively straightforward. Table 27-4 indicates the patterns that characterize the four major categories of acid-base disorders. Additional steps, such as determination of the anion gap, plotting blood test results on a graph, or *nomogram*, or the use of a diagnostic chart may help in identifying possible causes for the problem and in the diagnosis of compensated versus uncompensated conditions. Details and diagnostic exercises are included in the *Applications Manual.* AM *Diagnostic Classification of Acid-Base Disorders.*

- ✔ What effect would a decrease in the pH of the body fluids have on the respiratory rate?
- ✔ Why does the tubular fluid in nephrons need to be buffered?
- ✔ How would a prolonged fast affect the body's pH?
- ✔ Why can prolonged vomiting produce alkalosis?

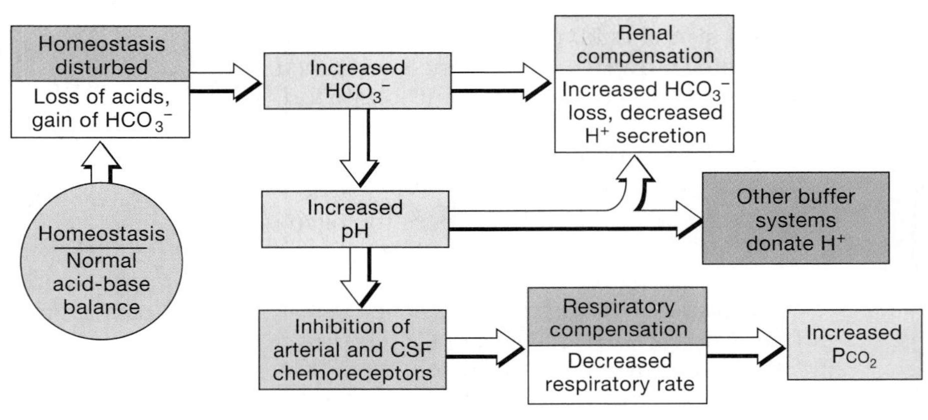
Homeostasis disturbed
Loss of acids, gain of HCO_3^-
Increased HCO_3^-
Renal compensation
Increased HCO_3^- loss, decreased H^+ secretion
Homeostasis
Normal acid-base balance
Increased pH
Other buffer systems donate H^+
Inhibition of arterial and CSF chemoreceptors
Respiratory compensation
Decreased respiratory rate
Increased P_{CO_2}

● **FIGURE 27-14**
Metabolic Alkalosis. Metabolic alkalosis most often results from the loss of acids, especially stomach acid lost through vomiting. During generation of replacement gastric acids, the alkaline tide introduces large numbers of bicarbonate ions into the bloodstream, and pH increases.

TABLE 27-4 Changes in Blood Chemistry Associated with the Major Classes of Acid-Base Disorders

Disorder	*pH (normal = 7.35–7.45)*	*HCO_3^- (normal = 24–28)*	*P_{CO_2} (mm Hg) (normal = 35–45)*	*Remarks*	*Treatment*
Respiratory acidosis	decreased (below 7.35)	*acute:* normal *compensated:* increased (above 28)	increased (above 50)	Usually caused by hypoventilation and CO_2 buildup in tissues and blood	Improve ventilation, sometimes with bronchodilation and mechanical assistance
Metabolic acidosis	decreased (below 7.35)	decreased (below 24)	*acute:* normal *compensated:* decreased (below 35)	Caused by organic or fixed acid buildup, impaired H^+ excretion at kidneys, or bicarbonate loss in urine or feces	Bicarbonate administration (gradual) with other steps as needed to correct primary cause
Respiratory alkalosis	increased (above 7.45)	*acute:* normal *compensated:* decreased (below 24)	decreased (below 35)	Usually caused by hyperventilation and reduction in plasma CO_2 levels	Reduce respiratory rate, allow rise in P_{CO_2}
Metabolic alkalosis	increased (above 7.45)	increased (above 28)	increased (above 45)	Usually caused by prolonged vomiting and associated acid loss	pH below 7.55, no treatment; pH above 7.55 may require ammonium chloride administration

Selected Clinical Terminology

Terms Discussed in This Chapter

hyperkalemia: Plasma potassium levels above 8 mEq/l. *(p. 1038)*
hypernatremia: Plasma sodium levels above 150 mEq/l. *(p. 1034)*
hypokalemia: Plasma potassium levels below 2 mEq/l. *(p. 1038)*
hyponatremia: Plasma sodium levels below 130 mEq/l. *(p. 1034)*
metabolic acidosis: A type of acidosis caused by the inability to excrete hydrogen ions, the production of numerous fixed and/or organic acids, or a severe bicarbonate loss. *(p. 1049* and *AM)*
metabolic alkalosis: A rare form of alkalosis resulting from high concentrations of bicarbonate ions in body fluids. *(p. 1052* and *AM)*
nomogram: A graph that can be used to assess acid-base balance in a clinical setting. *(p. 1052* and *AM)*
respiratory acidosis: Acidosis resulting from inadequate respiratory activity, characterized by elevated levels of carbon dioxide (hypercapnia) in body fluids. *(p. 1047* and *AM)*
respiratory alkalosis: Alkalosis due to excessive respiratory activity, which depresses carbon dioxide levels and elevates the pH of body fluids. *(p. 1049* and *AM)*

CHAPTER REVIEW

STUDY OUTLINE

INTRODUCTION, p. 1028

1. Maintenance of normal volume and composition in the extracellular and intracellular fluids is vital to life. Three types of homeostasis are involved: **fluid balance, electrolyte balance,** and **acid-base balance.**

FLUID AND ELECTROLYTE BALANCE, p. 1028

1. The **intracellular fluid (ICF)** contains nearly two-thirds of the total body water; the **extracellular fluid (ECF)** contains the rest. Exchange occurs between the ICF and ECF, but the two **fluid compartments** retain their distinctive characteristics. *(Figures 27-1, 27-2)*

Basic Concepts Pertaining to Fluid and Electrolyte Regulation, p. 1029

2. Homeostatic mechanisms that monitor and adjust the composition of body fluids respond to changes in the ECF.
3. Receptors involved in fluid and electrolyte balance respond to changes in plasma volume or osmolarity.

4. All water movements across cell membranes and epithelia occur passively in response to osmotic gradients.

An Overview of the Primary Regulatory Hormones, p. 1031

5. ADH encourages water reabsorption at the kidneys and stimulates thirst. Aldosterone increases the rates of sodium reabsorption at the kidneys. ANP opposes these actions and promotes fluid and electrolyte losses in the urine.

The Interplay between Fluid and Electrolyte Balance, p. 1031

Fluid Balance, p. 1031

6. Water circulates freely within the ECF compartment.
7. Water losses are normally balanced by gains through eating, drinking, and **metabolic generation.** *(Figure 27-3; Table 27-1)*
8. If the ECF becomes hypertonic relative to the ICF, water will move from the ICF into the ECF until osmotic equilibrium has been restored. If the ECF becomes hypotonic compared with the ICF, water will move from the ECF into the cells, and the volume of the ICF will increase.

Electrolyte Balance, p. 1034

9. Problems with electrolyte balance generally result from an imbalance between sodium gains and losses. Problems with potassium balance are less common but more dangerous.
10. The rate of sodium uptake across the digestive epithelium is directly proportional to the amount of sodium in the diet. Sodium losses occur mainly in the urine and through perspiration. *(Figure 27-4)*
11. Alterations in sodium balance result in the expansion or contraction of the ECF. Large variations in the ECF volume are corrected by the homeostatic mechanisms triggered by changes in blood volume. If the volume becomes too low, ADH and aldosterone are secreted; if the volume becomes too high, ANP is secreted. *(Figure 27-5)*
12. Potassium ion concentrations in the ECF are very low and not as closely regulated as sodium ion concentrations. Potassium excretion increases as ECF concentrations rise, under aldosterone stimulation, and when the pH rises. Potassium retention occurs when the pH falls.
13. The ECF concentrations of other electrolytes such as calcium, magnesium, phosphate, and chloride are also regulated. *(Table 27-2)*

ACID-BASE BALANCE, p. 1040

1. Acids and bases may be regarded as strong or weak. *(Table 27-3)*

The Importance of pH Control, p. 1040

2. The pH of normal body fluids ranges from 7.35 to 7.45; variations outside this relatively narrow range produce symptoms of **acidosis** or **alkalosis.**

Types of Acids in the Body, p. 1041

3. **Volatile acids** can leave solution and enter the atmosphere; **fixed acids** remain in body fluids until excreted at the kidneys; **organic acids** are acid participants in or byproducts of cellular metabolism.
4. Carbonic acid is the most important factor affecting the pH of the ECF. In solution CO_2 reacts with water to form carbonic acid. An inverse relationship exists between the concentration of CO_2 and pH. *(Figure 27-6)*
5. **Sulfuric acid** and **phosphoric acid,** the most important fixed acids, are generated during the catabolism of amino acids and compounds containing phosphate groups.
6. Organic acids include metabolic products such as lactic acid and ketone bodies.

Mechanisms of pH Control, p. 1042

7. A **buffer system** consists of a weak acid or weak base and its dissociation products. There are three major buffer systems: (1) **protein buffer systems** in the ECF and ICF; (2) the **carbonic acid–bicarbonate buffer system,** most important in the ECF; and (3) the **phosphate buffer system** in the intracellular fluids and urine. *(Figure 27-7)*
8. In protein buffer systems, the component amino acids respond to changes in pH by accepting or releasing H^+. The **hemoglobin buffer system** is a protein buffer system that helps prevent drastic changes in pH when the P_{CO_2} is rising or falling. *(Figure 27-8)*
9. The carbonic acid–bicarbonate buffer system prevents pH changes due to organic acids and fixed acids in the ECF. *(Figure 27-9)*
10. The phosphate buffer system plays a supporting role in regulating the pH of the ECF but is important in buffering the pH of the ICF.

Maintenance of Acid-Base Balance, p. 1044

11. The lungs help regulate pH by affecting the carbonic acid–bicarbonate buffer system; changing the respiratory rate can raise or lower the P_{CO_2} of body fluids, affecting the buffering capacity. This process is called **respiratory compensation.**
12. In the process of **renal compensation** the kidneys vary their rates of hydrogen ion secretion and bicarbonate ion reabsorption depending on the pH of extracellular fluids. *(Figure 27-10)*

DISTURBANCES OF ACID-BASE BALANCE, p. 1047

1. Interactions among buffer systems, respiratory, and renal functions normally maintain pH of the extracellular fluids within a range of 0.1 pH units (7.35–7.45). *(Figure 27-11)*
2. **Respiratory disorders** result when abnormal respiratory function causes an extreme rise or fall in CO_2 levels in the ECF. **Metabolic disorders** are caused by the generation of organic or fixed acids or by conditions affecting the concentration of bicarbonate ions in the ECF.

Respiratory Acidosis, p. 1047

3. **Respiratory acidosis** results from excessive levels of carbon dioxide in body fluids. **Chronic respiratory acidosis** develops when compensatory mechanisms have not completely failed. If normal homeostatic adjustments do not occur, **acute respiratory acidosis** develops. *(Figure 27-12)*

Respiratory Alkalosis, p. 1049

4. **Respiratory alkalosis** is a relatively uncommon condition associated with hyperventilation. *(Figure 27-12)*

Metabolic Acidosis, p. 1049

5. **Metabolic acidosis** results from depletion of the bicarbonate reserve. It is caused by an inability to excrete hydrogen ions at the kidneys, the production of large numbers of fixed and organic acids, or the bicarbonate loss that accompanies chronic diarrhea. *(Figure 27-13)*

Metabolic Alkalosis, p. 1052

6. **Metabolic alkalosis** occurs when bicarbonate ion concentrations become elevated, as from extended periods of vomiting. *(Figure 27-14)*

The Detection of Acidosis and Alkalosis, p. 1052

7. Standard diagnostic blood tests such as blood pH, P_{CO_2}, and bicarbonate are used to recognize and classify acidosis and alkalosis conditions as respiratory or metabolic in nature. *(Figures 27-15, 27-16; Table 27-4)*

REVIEW QUESTIONS

LEVEL 1 Reviewing Facts and Terms

Match each item in Column A with the most closely related item in Column B. Use letters for answers in the space provided.

	Column A	Column B
___	1. ECF	a. reduces thirst
___	2. ADH	b. dominant cation in ECF
___	3. aldosterone	c. hyponatremia
___	4. atrial natriuretic peptide	d. weak acid
___	5. oxidative phosphorylation	e. baroreceptors
___	6. overhydration	f. promotes calcium loss
___	7. dehydration	g. respiratory alkalosis
___	8. sodium	h. stimulates water conservation
___	9. potassium	i. sulfuric acid
___	10. carotid sinus	j. volatile acid
___	11. calcitonin	k. ketone bodies
___	12. hydrochloric acid	l. interstitial fluid
___	13. carbonic acid	m. respiratory acidosis
___	14. carbon dioxide	n. metabolic acidosis
___	15. fixed acid	o. hypernatremia
___	16. organic acid	p. metabolic alkalosis
___	17. hypoventilation	q. dominant cation in ICF
___	18. hyperventilation	r. conservation of sodium ions
___	19. elevated bicarbonate ion concentration	s. strong acid
___	20. severe bicarbonate loss	t. metabolic generation of water

21. A person is in fluid balance when:
(a) the ECF and ICF are isotonic
(b) there is no fluid movement between compartments
(c) the amount of water gained each day is equal to the amount lost to the environment
(d) a, b, and c are correct

22. The primary components of the extracellular fluid are:
(a) lymph and cerebrospinal fluid
(b) blood plasma and serous fluids
(c) interstitial fluid and plasma
(d) a, b, and c are correct

23. The principal ions in the ECF are:
(a) sodium and chloride
(b) sodium and potassium
(c) phosphate and chloride
(d) phosphate and bicarbonate

24. All the homeostatic mechanisms that monitor and adjust the composition of body fluids respond to changes:
(a) in the ICF (b) in the ECF
(c) inside the cell (d) a, b, and c are correct

25. Homeostatic adjustments to monitor fluid and electrolyte balance occur in response to changes in:
(a) pH of body fluids
(b) autonomic nervous system activity
(c) concentration of dissolved gases
(d) plasma volume or osmolarity

26. A rise in body temperature accompanying a fever:
(a) decreases water losses
(b) increases water losses
(c) does not affect water gain or loss
(d) causes the body to retain water

27. The most common problems with electrolyte balance are caused by an imbalance between gains and losses of:
(a) calcium ions (b) chloride ions
(c) potassium ions (d) sodium ions

28. Angiotensin II produces a coordinated elevation in the ECF volume by:
(a) stimulating thirst
(b) causing the release of ADH
(c) triggering production and secretion of aldosterone
(d) a, b, and c are correct

29. The rate of tubular secretion of potassium ions changes in response to:
(a) changes in pH
(b) aldosterone levels
(c) alterations in the potassium ion concentration in the ECF
(d) a, b, and c are correct

30. Respiratory acidosis develops when there is:
(a) elevated plasma pH due to a decreased plasma P_{CO_2}
(b) decreased plasma pH due to an elevated plasma P_{CO_2}
(c) elevated plasma pH due to an elevated plasma P_{CO_2}
(d) decreased plasma pH due to a decreased plasma P_{CO_2}

31. Metabolic alkalosis occurs when:
   (a) bicarbonate ion concentrations become elevated
   (b) there is a severe bicarbonate loss
   (c) the kidneys fail to excrete hydrogen ions
   (d) ketone bodies are generated in large quantities

32. What are fluid shifts? What is their function, and what factors can cause them?

33. What three major hormones mediate major physiological adjustments that affect fluid and electrolyte balance? What are the primary effects of each hormone?

34. What influence does aldosterone have on the relationship between sodium and potassium?

35. Define and give an example of a volatile acid, a fixed acid, and an organic acid. Which represent(s) the greatest threat to acid-base balance? Why?

LEVEL 2 Reviewing Concepts

36. The higher the plasma concentration of aldosterone, the more efficiently the kidney will:
   (a) conserve sodium ions
   (b) retain potassium ions
   (c) stimulate urinary water loss
   (d) secrete greater amounts of ADH

37. If the ECF is *hypertonic* with respect to the ICF:
   (a) water will move from the ECF into the cells until osmotic equilibrium is reached
   (b) ions will move from the ECF into the cells until osmotic equilibrium is reached
   (c) ions will move from the cells into the ICF until osmotic equilibrium is reached
   (d) water will move from the cells into the ECF until osmotic equilibrium is reached

38. The osmolarity of the ECF decreases if the individual gains water without a corresponding:
   (a) gain of electrolytes
   (b) loss of water
   (c) fluid shift from the ECF to the ICF
   (d) a, b, and c are correct

39. Chronic renal failure and congestive heart failure ultimately result in:
   (a) hypernatremia (b) hyponatremia
   (c) hyperkalemia (d) hypokalemia

40. When pure water is consumed:
   (a) the ECF becomes hypertonic with respect to the ICF
   (b) the ECF becomes hypotonic with respect to the ICF
   (c) the ICF becomes hypotonic with respect to the plasma
   (d) water moves from the ICF into the ECF

41. When the plasma concentration of potassium decreases to a dangerous level, the result is:
   (a) kidney failure
   (b) extensive muscular weakness, followed by paralysis
   (c) decreased blood-clotting capability
   (d) cessation of enzyme activity in cells

42. In a protein buffer system, if the pH rises:
   (a) the protein acquires a hydrogen ion from carbonic acid
   (b) hydrogen ions are buffered by hemoglobin molecules
   (c) a hydrogen ion is released and a carboxyl ion is formed
   (d) a chloride shift occurs

43. Increasing or decreasing the rate of respiration alters pH by:
   (a) lowering or raising the partial pressure of carbon dioxide
   (b) lowering or raising the partial pressure of oxygen
   (c) lowering or raising the partial pressure of nitrogen
   (d) a, b, and c are correct

44. Differentiate among fluid balance, electrolyte balance, and acid-base balance, and explain why each is important to homeostasis.

45. A sample of plasma contains chloride ions at a concentration of 250 mg/dl. How many mmol/l of chloride ions are in the sample of plasma?

46. Why should a person with a fever drink plenty of fluids?

47. Describe the effects of losses and gains of sodium on the ECF.

48. What are the three major buffer systems found in body fluids? How does each system work?

49. How do pulmonary and renal mechanisms support the buffer systems?

50. Differentiate between respiratory compensation and renal compensation.

51. Distinguish between respiratory and metabolic disorders that disturb acid-base balance.

52. What is the difference between metabolic and respiratory alkalosis? What can cause these conditions?

53. The most recent advice from the medical and nutritional experts is to decrease the intake of salt so that it does not exceed the amount needed to maintain a constant ECF composition. What effect does excessive salt and water ingestion have on (a) urine volume, (b) urine concentration, and (c) blood pressure?

54. Exercise physiologists recommend that adequate amounts of fluid be ingested before, during, and after exercise. Why is adequate fluid replacement during conditions of extensive sweating important?

LEVEL 3 Critical Thinking and Clinical Applications

55. After falling into an abandoned stone quarry filled with water, a nearly drowned young boy is rescued. His rescuers assess his condition and find that his body fluids have high P_{CO_2} and low P_{O_2} and that large amounts of lactic acid had been generated by the muscles of the victim as he struggled. As a clinician, diagnose the condition of the boy and recommend the necessary treatment to restore his body to homeostasis.

56. A patient is brought into the emergency room hyperventilating and disoriented. Blood analysis indicates that the patient has elevated levels of potassium (hyperkalemia). Is the patient most likely suffering from acidosis or alkalosis? Explain.

57. Willy suffers from chronic emphysema. A blood analysis indicates that his plasma pH is 7.4, although his plasma bicarbonate level is significantly elevated. How can this be?

58. While visiting a foreign country, Milly inadvertently drinks the water, even though she had been advised not to, and contracts an intestinal disease that causes severe diarrhea. How would you expect her condition to affect her blood pH, urine pH, and pattern of ventilation?

59. Tess is suffering from dehydration, and her physician prescribes intravenous fluids. The nurse caring for Tess becomes distracted and erroneously gives her a hypertonic glucose solution instead of normal saline. What effect will this mistake have on Tess's plasma levels of ADH and urine volume?

60. Referring to the diagnostic flow chart below (Figure 27-15), use information from the blood test results to categorize the acid-base disorders affecting these patients.

	Patient 1	*Patient 2*	*Patient 3*	*Patient 4*
pH	7.5	7.2	7.0	7.7
P_{CO_2}	32	45	60	50
Na^+	138	140	140	136
HCO_3^-	22	20	28	34
Cl^-	106	102	101	91
Anion gap	10	18	12	11

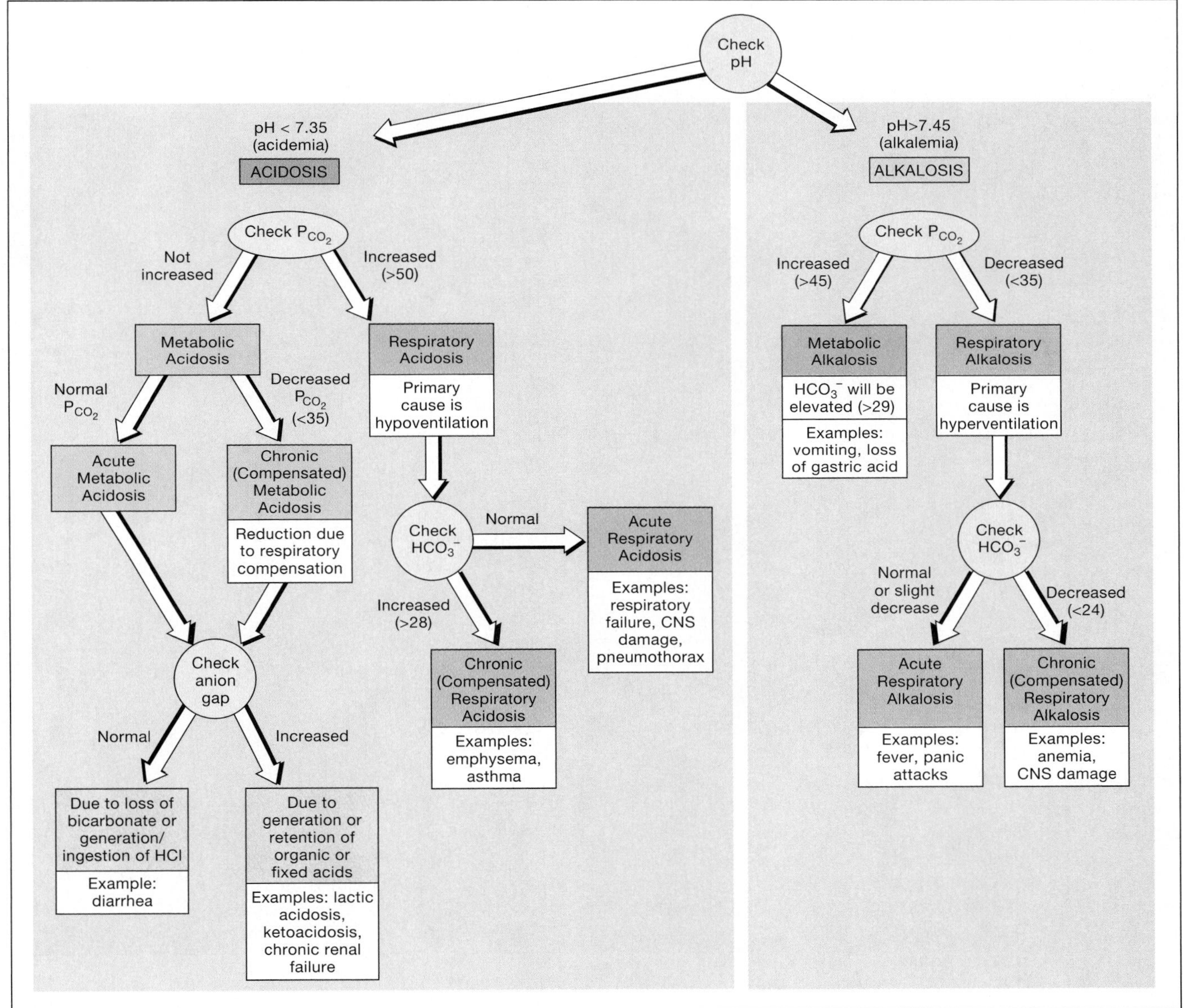
Check pH
pH < 7.35 (acidemia)
ACIDOSIS
pH>7.45 (alkalemia)
ALKALOSIS
Check P_{CO_2}
Not increased
Increased (>50)
Metabolic Acidosis
Respiratory Acidosis
Primary cause is hypoventilation
Normal P_{CO_2}
Decreased P_{CO_2} (<35)
Acute Metabolic Acidosis
Chronic (Compensated) Metabolic Acidosis
Reduction due to respiratory compensation
Check HCO_3^-
Normal
Acute Respiratory Acidosis
Examples: respiratory failure, CNS damage, pneumothorax
Increased (>28)
Chronic (Compensated) Respiratory Acidosis
Examples: emphysema, asthma
Check anion gap
Normal
Increased
Due to loss of bicarbonate or generation/ ingestion of HCl
Example: diarrhea
Due to generation or retention of organic or fixed acids
Examples: lactic acidosis, ketoacidosis, chronic renal failure
Check P_{CO_2}
Increased (>45)
Decreased (<35)
Metabolic Alkalosis
HCO_3^- will be elevated (>29)
Examples: vomiting, loss of gastric acid
Respiratory Alkalosis
Primary cause is hyperventilation
Check HCO_3^-
Normal or slight decrease
Decreased (<24)
Acute Respiratory Alkalosis
Examples: fever, panic attacks
Chronic (Compensated) Respiratory Alkalosis
Examples: anemia, CNS damage

● **FIGURE 27-15**
A Diagnostic Chart for Acid-Base Disorders

CHAPTER 28

Of all our organ systems, only one isn't needed for survival. Indeed, it doesn't even seem to confer any direct benefit on us—at least physically. In a sense, our reproductive systems don't exist for us at all—they exist for our kind, for the human race. But how can we measure the value to us—to the part of us that isn't just cells and tissues and organs—of being able to have children? To contribute something to the next generation of human beings, and in the process make a little part of ourselves immortal?

In this chapter we'll study the system that has enabled human beings to exist on this earth for a very long time—and, with luck, to stick around for at least a while longer.

The Reproductive System

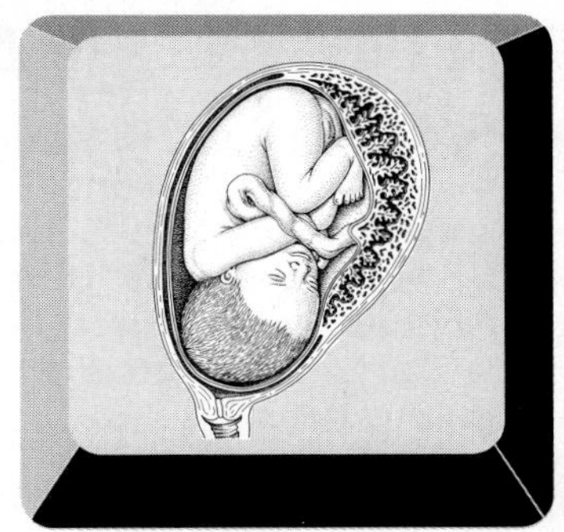

Chapter Outline and Objectives

The reproductive system is the only system that is not essential to the life of the individual. Instead, this system ensures the continued existence of the human species. The entire process of reproduction seems almost magical; many primitive societies even failed to discover the basic link between sexual activity and childbirth and assumed that cosmic forces were responsible for producing new individuals. Although our society has a much clearer view of the reproductive process, a sense of wonder remains. Sexually mature males and females produce individual reproductive cells that are brought together by the sexual act. Their fusion starts a chain of events leading to the appearance of an infant that will mature as part of the next generation.

The next two chapters will consider the mechanics of this remarkable process. We will begin by examining the anatomy and physiology of the reproductive system in this chapter. The human reproductive system produces, stores, nourishes, and transports functional male and female reproductive cells, or **gametes** (GAM-ēts). The next chapter begins with **fertilization,** the fusion of a **sperm** contributed by the father and an **ovum** (Ō-vum) from the mother. All the cells in a human body are the mitotic descendants of a single **zygote** (ZĪ-gōt), the cell created by the fusion of a sperm and an ovum. The gradual transformation of that single cell into a functional adult, over a span of 15–20 years, is the process of *development.* Over this period, hormones produced by the reproductive system direct gender-specific patterns of development. The mechanics of development will be considered in Chapter 29.

An Overview of the Reproductive System

The reproductive system includes:

- Reproductive organs, or **gonads** (GŌ-nadz), that produce gametes and hormones.
- Ducts that receive and transport the gametes.
- Accessory glands and organs that secrete fluids into these or other excretory ducts.
- Perineal structures associated with the reproductive system. These perineal structures are collectively known as the **external genitalia** (jen-i-TĀ-lē-a).

The male and female reproductive systems are functionally quite different. In the adult male the gonads, or **testes** (TES-tēz; singular *testis*) secrete *androgens,* principally testosterone, and produce one-half billion sperm each day. After storage, mature sperm travel along a lengthy duct system, where they are mixed with the secretions of accessory glands. The mixture created is known as **semen** (SĒ-men). During *ejaculation* the semen is expelled from the body.

The female gonads, or **ovaries,** typically release only one immature ovum a month. This gamete travels along short **uterine tubes** (*oviducts*) that end in the muscular **uterus** (Ū-ter-us). A short passageway, the **vagina** (va-JĪ-na), connects the uterus with the exterior. During the sexual act, male ejaculation introduces semen into the vagina, and the sperm cells ascend the female reproductive tract. If fertilization occurs, the uterus will enclose and support a developing embryo as it grows into a fetus and prepares for eventual delivery.

We will now examine the anatomy of the male and female reproductive systems in detail and consider the physiological and hormonal mechanisms responsible for the regulation of reproductive function. The anatomical reference points used in the following discussions were introduced in earlier chapters, and you may find it helpful to review the figures on the pelvic girdle (Figures 8-8 and 8-10•, pp. 248, 249), perineal musculature (Figure 11-13•, p. 349), pelvic innervation (Figure 13-10•, p. 440), and the regional blood supply (Figures 21-28 and 21-34•, pp. 759, 765).

The Reproductive System of the Male

Figure 28-1

The principal structures of the male reproductive system are shown in Figure 28-1•. Proceeding from the testes, the sperm cells, or **spermatozoa** (sper-ma-tō-ZŌ-a), travel along the **epididymis** (ep-i-DID-i-mus), the **ductus deferens** (DUK-tus DEF-e-renz), or *vas deferens,* the **ejaculatory** (ē-JAK-ū-la-tō-rē) **duct,** and the **urethra** before leaving the body. Accessory organs, notably the **seminal** (SEM-i-nal) **vesicles,** the **prostate** (PROS-tāt) **gland,** and the **bulbourethral** (bul-bō-ū-RĒ-thral) **glands,** secrete into the ejaculatory ducts and urethra. The external genitalia include the **scrotum** (SKRŌ-tum), which encloses the testes, and the **penis** (PĒ-nis), an erectile organ that surrounds the distal portion of the urethra.

The Testes

Figures 28-1, 28-2

Each testis has the shape of a flattened oval roughly 5 cm (2 in.) long and 2.5 cm (1 in.) wide, with a weight of 10–15 g (0.35–0.53 oz). The testes hang within the scrotum, a fleshy pouch suspended below the perineum anterior to the anus (Figures 28-1 and 28-2•).

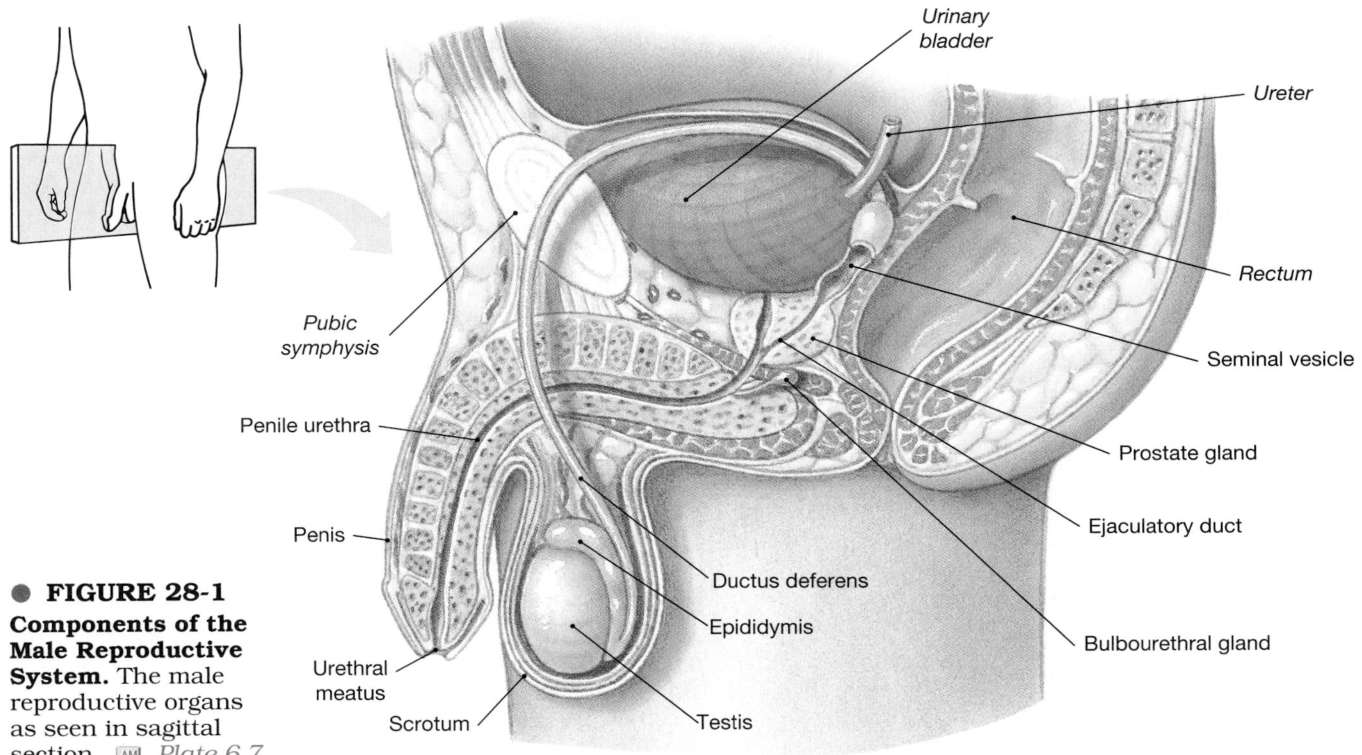

Urinary bladder
Ureter
Rectum
Seminal vesicle
Prostate gland
Ejaculatory duct
Bulbourethral gland
Pubic symphysis
Penile urethra
Penis
Ductus deferens
Epididymis
Urethral meatus
Scrotum
Testis

● **FIGURE 28-1 Components of the Male Reproductive System.** The male reproductive organs as seen in sagittal section. AM *Plate 6.7*

The Spermatic Cord

Figure 28-2

The **spermatic cords** consist of layers of fascia, tough connective tissue, and muscle enclosing the blood vessels, nerves, and lymphatics supplying the testes. Each spermatic cord contains a ductus deferens, a testicular artery, a **pampiniform** (pam-PIN-i-form; *pampinus*, tendril + *forma*, form) **plexus** of a testicular vein, and **ilioinguinal** and *genitofemoral nerves* from the lumbar plexus. These components accompany each testis as it descends into the scrotum during development. Each spermatic cord begins at the deep inguinal ring, extends along the inguinal canal, exits at the superficial inguinal ring, and descends into the scrotum (Figure 28-2●).

The narrow canals linking the scrotal chambers with the peritoneal cavity are called the *inguinal canals*. These passageways usually close, but the persistence of the spermatic cords creates weak points in the abdominal wall that remain throughout life. As a result *inguinal hernias*, discussed in Chapter 11, are relatively common in males. *[p. 347]* The inguinal canals in females are very small, containing only the ilioinguinal nerves and the round ligaments of the uterus. The abdominal wall is nearly intact, and inguinal hernias in women are very rare.

Descent of the Testes

Figure 28-3

During development, the testes form inside the body cavity adjacent to the kidneys. A bundle of connective tissue fibers extends from each testis to the posterior wall of a small, inferior pocket of the peritoneum. These fibers constitute the **gubernaculum** of the testis. As growth proceeds, the gubernacula do not elongate, and the testes are held in position (Figure 28-3●). As a result, the testes gradually move inferiorly and anteriorly toward the anterior abdominal wall as the fetus enlarges. During the seventh developmental month, (1) growth continues at a rapid pace, and (2) circulating hormones stimulate a contraction of the gubernaculum. Over this period the relative position of the testes changes further, and they move through the abdominal musculature accompanied by small pockets of the peritoneal cavity. This process is called the **descent of the testes** (Figure 28-3●).

As it moves through the body wall, each testis is accompanied by the ductus deferens and the testicular blood vessels, nerves, and lymphatics. Together these structures form the body of the *spermatic cord*.

CRYPTORCHIDISM In **cryptorchidism** (kript-OR-ki-dizm; *crypto*, hidden) one or both of the testes have not descended into the scrotum by the time of birth. Typically the testes are lodged in the abdominal cavity or within the inguinal canal. This condition occurs in about 3 percent of full-term deliveries and in roughly 30 percent of premature births. In most instances normal descent occurs a few weeks later, but the condition can be surgically corrected if it persists. Corrective

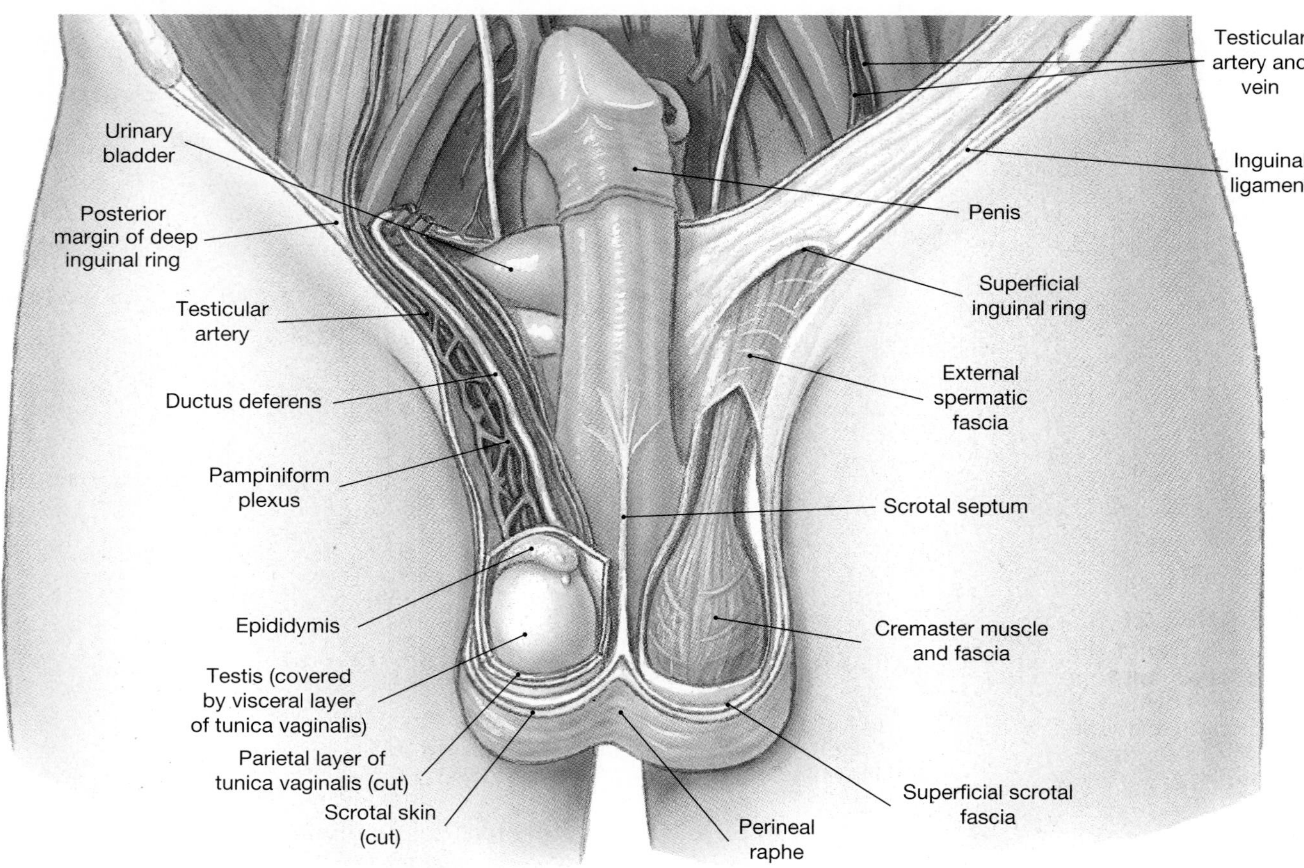
Testicular artery and vein
Inguinal ligament
Urinary bladder
Posterior margin of deep inguinal ring
Testicular artery
Ductus deferens
Pampiniform plexus
Epididymis
Testis (covered by visceral layer of tunica vaginalis)
Parietal layer of tunica vaginalis (cut)
Scrotal skin (cut)
Penis
Superficial inguinal ring
External spermatic fascia
Scrotal septum
Cremaster muscle and fascia
Superficial scrotal fascia
Perineal raphe

FIGURE 28-2
The Male Reproductive System in Frontal View

measures are usually taken before puberty because cryptorchid (abdominal) testes will not produce sperm, and the individual will be sterile. If the testes cannot be moved into the scrotum, they will usually be removed, because about 10 percent of those with uncorrected cryptorchid testes eventually develop testicular cancer. This surgical procedure is called a *bilateral orchiectomy* (ōr-kē-EK-to-mē; *orchis*, testis).

The Scrotum and the Position of the Testes

Figures 28-2, 28-4

The scrotum is divided internally into two separate chambers, and the partition between the two is marked by a raised thickening in the scrotal surface known as the **perineal raphe** (RĀ-fē) (Figures 28-2 and 28-4a●). Each testis lies in a separate compartment, or **scrotal cavity.** Because the scrotal cavities are separated by a partition, infection or inflammation of one testis does not ordinarily spread to affect the other. A narrow space separates the inner surface of the scrotum from the outer surface of the testis. A serous membrane, the **tunica vaginalis,** lines the scrotal cavity and reduces friction between the opposing parietal (scrotal) and visceral (testicular) surfaces. The tunica vaginalis is an isolated portion of the peritoneum that lost its connection with the peritoneal cavity during the descent of the testes.

The scrotum consists of a thin layer of skin and the underlying superficial fascia. The dermis contains a layer of smooth muscle, the **dartos** (DAR-tōs), and tonic contraction of the dartos causes the characteristic wrinkling of the scrotal surface. A layer of skeletal muscle, the **cremaster** (krē-MAS-ter) **muscle,** lies beneath the dermis. Contraction of the cremaster tenses the scrotum and pulls the testes closer to the body. Contraction occurs during sexual arousal and in response to changes in temperature. Normal sperm development in the testes requires temperatures around 1.1° C (2° F) lower than those found elsewhere in the body. The cremaster moves the testes away from or toward the body as needed to maintain acceptable testicular temperatures. When air or body temperatures rise, the cremaster relaxes and the testes then move away from the body. Cooling the scrotum, as when entering a cold swimming pool, results in cremasteric contractions that pull the testes closer to the body and keep testicular temperatures from falling.

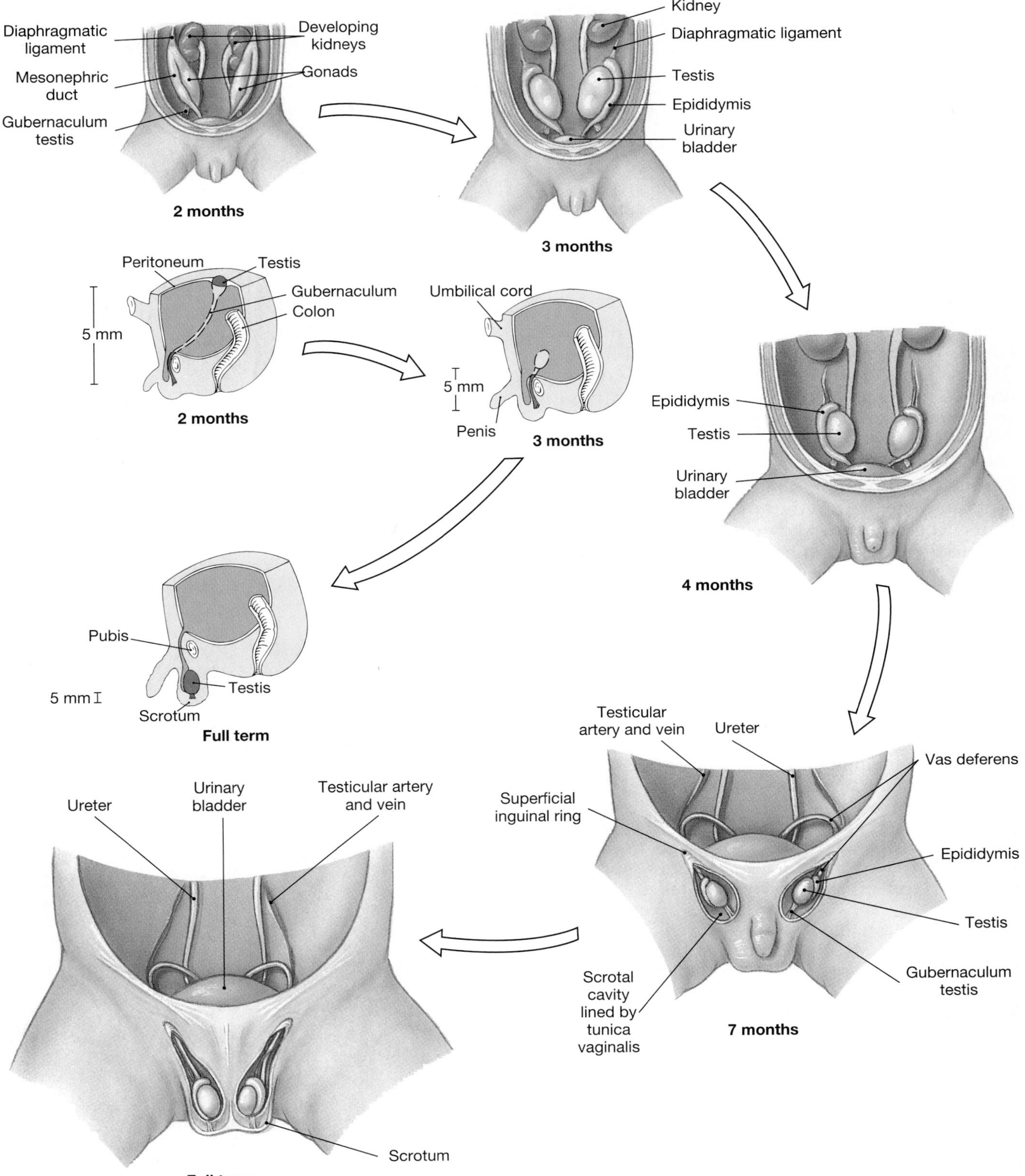
Diaphragmatic ligament
Developing kidneys
Mesonephric duct
Gonads
Gubernaculum testis
2 months
Kidney
Diaphragmatic ligament
Testis
Epididymis
Urinary bladder
3 months
Peritoneum
Testis
Gubernaculum
Colon
5 mm
2 months
Umbilical cord
5 mm
Penis
3 months
Epididymis
Testis
Urinary bladder
4 months
Pubis
Testis
5 mm
Scrotum
Full term
Testicular artery and vein
Ureter
Vas deferens
Superficial inguinal ring
Epididymis
Testis
Gubernaculum testis
Scrotal cavity lined by tunica vaginalis
7 months
Ureter
Urinary bladder
Testicular artery and vein
Scrotum
Full term

● **FIGURE 28-3**

Descent of the Testes. Diagrammatic frontal and sagittal sectional views of the positional changes involved in the descent of the testes and the formation of the spermatic cord. Note the scale bar at the left of each of the sagittal views.

The scrotum is richly supplied with sensory and motor nerves from the *hypogastric plexus* and branches of the *ilioinguinal nerves*, the *genitofemoral nerves*, and the *pudendal nerves*. *[pp. 440–441, 541]* The vascular supply to the scrotum includes the **internal pudendal arteries** (from the internal iliac arteries), the **external pudendal arteries** (from the femoral arteries), and the *cremasteric branch* of the **inferior epigastric arteries** (from the external iliac arteries). *[pp. 758–759]* The names and distributions of the veins follow those of the arteries.

Structure of the Testes

Figure 28-4

Beneath the serous membrane covering the testis lies the **tunica albuginea** (TŪ-ni-ka al-bū-JIN-ē-a), a dense layer of connective tissue rich in collagen fibers. The fibers of this network are continuous with those surrounding the adjacent epididymis. The collagen fibers of the tunica albuginea also extend into the substance of the testis, forming fibrous partitions, or *septa* (Figure 28-4●). These septa converge toward the area adjacent to the proximal portion of the epididymis. This region is called the **mediastinum** of the testis. The connective tissues of the mediastinum support the blood vessels and lymphatics supplying the testis and the ducts transporting sperm to the epididymis.

TESTICULAR TORSION Because the testes are relatively loosely attached to the scrotal walls, they may become twisted within the scrotal cavity. This condition, called *testicular torsion*, usually occurs in children or adolescents. Symptoms include pain in the groin and inguinal region, local inflammation, and swelling of the scrotum on the affected side. Treatment involves prompt external or surgical manipulation of the testis to relieve the twisting. Because any kinks in the spermatic cord severely restrict the arterial supply to the testis, corrective measures must be taken within 4–6 hours, before the testicular tissues become permanently damaged. If a testis is deprived of circulation for longer periods, the damage will be irreversible, and the affected testis will have to be removed.

Histology of the Testes

Figures 28-4, 28-5, 28-6

The septa subdivide the testis into a series of **lobules.** Roughly 800 slender, tightly coiled **seminiferous** (se-mi-NIF-e-rus) **tubules** are distributed among the lobules (Figures 28-4 and 28-5●). Each tubule averages around 80 cm (31 in.) in length, and a typical testis contains nearly one-half mile of seminiferous tubules. Sperm production occurs within these tubules.

Each seminiferous tubule has the form of a U connected to a single **straight tubule** (*tubuli recti*) connected to a maze of passageways known as the **rete** (RĒ-tē; *rete*, a net) **testis** (Figure 28-4a●).

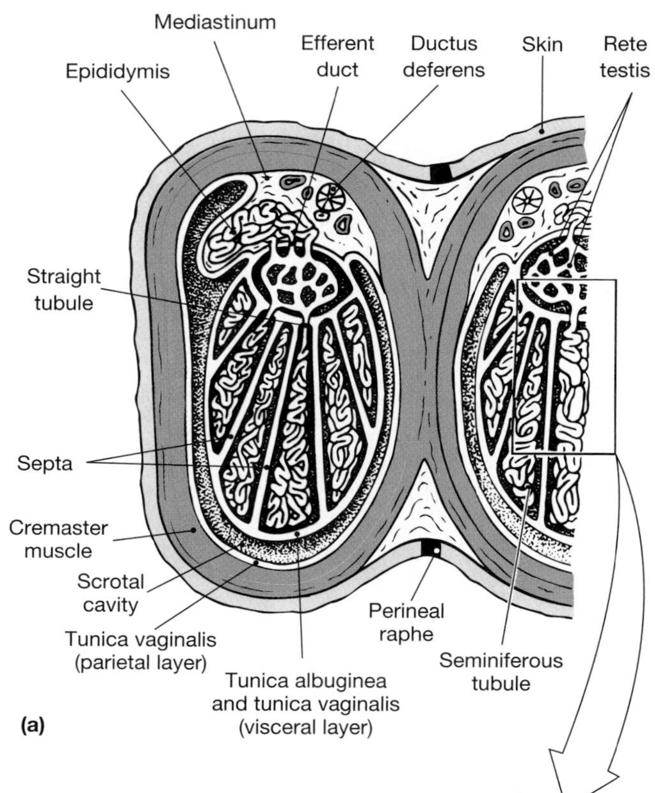
Mediastinum
Epididymis
Efferent duct
Ductus deferens
Skin
Rete testis
Straight tubule
Septa
Cremaster muscle
Scrotal cavity
Tunica vaginalis (parietal layer)
Tunica albuginea and tunica vaginalis (visceral layer)
Perineal raphe
Seminiferous tubule
(a)

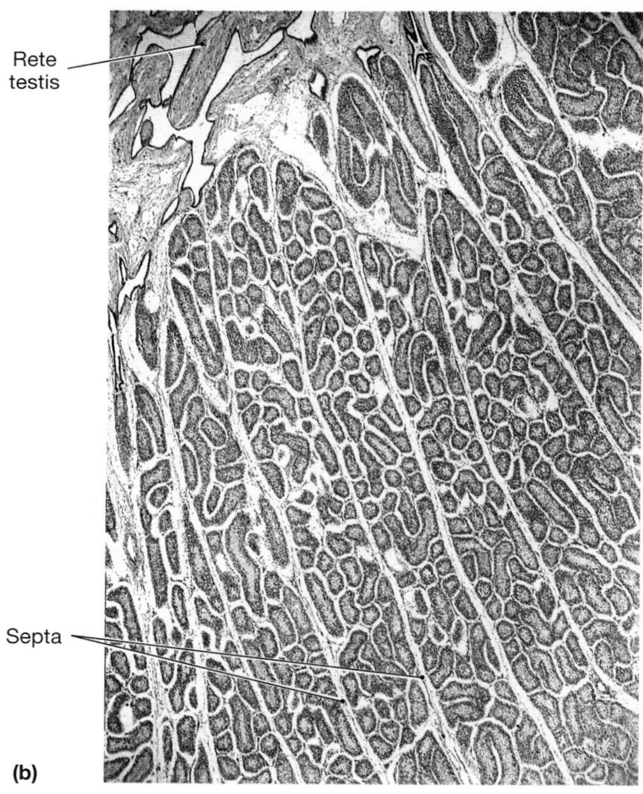
Rete testis
Septa
(b)

● **FIGURE 28-4**

Structure of the Testes. (a) Diagrammatic sketch and anatomical relationships of the testes. **(b)** Light micrograph of a section through a testis. (LM × 26)

Fifteen to twenty large **efferent ducts** (*efferent ductules*) connect the rete testis to the epididymis.

Because the seminiferous tubules are tightly coiled, histological preparations most often show them in transverse section (Figure 28-5a●). Each tubule is surrounded by a delicate capsule, and loose connective tissue fills the spaces between the tubules. Within those spaces are found numerous blood vessels and large **interstitial cells** *(cells of Leydig)* (Figure 28-5●). Interstitial cells are responsible for the production of sex hormones, called *androgens. Testosterone* is the most important androgen. (Testosterone and other sex hormones were introduced in Chapter 18.) ∞ [pp. 634–635]

Sperm cells, or spermatozoa, are produced through the process of **spermatogenesis** (sper-ma-tō-JEN-e-sis). Spermatogenesis begins at the outermost layer of cells in the seminiferous tubules (Figure 28-5b●) and proceeds toward the tubular lumen. Stem cells called *spermatogonia* produce generations of daughter cells, some of which differentiate into *spermatocytes.* Through *meiosis,* a specialized form of cell division involved only in the production of gametes (sperm in the male, ova in the female), spermatocytes give rise to *spermatids.* At each step in this process, the daughter cells move closer to the tubular lumen. The spermatids subsequently differentiate into spermatozoa. This differentiation process, called *spermiogenesis,* ends as the physically mature spermatozoa lose contact with the tubular wall and enter the tubular fluid.

Each seminiferous tubule contains spermatogonia, spermatocytes at various stages of meiosis, spermatids, spermatozoa, and large **sustentacular** (sus-ten-TAK-ū-lar) **cells** *(Sertoli cells).* Sustentacular cells are attached to the tubular capsule and extend toward the lumen between the other cell types (Figures 28-5b and 28-6a●).

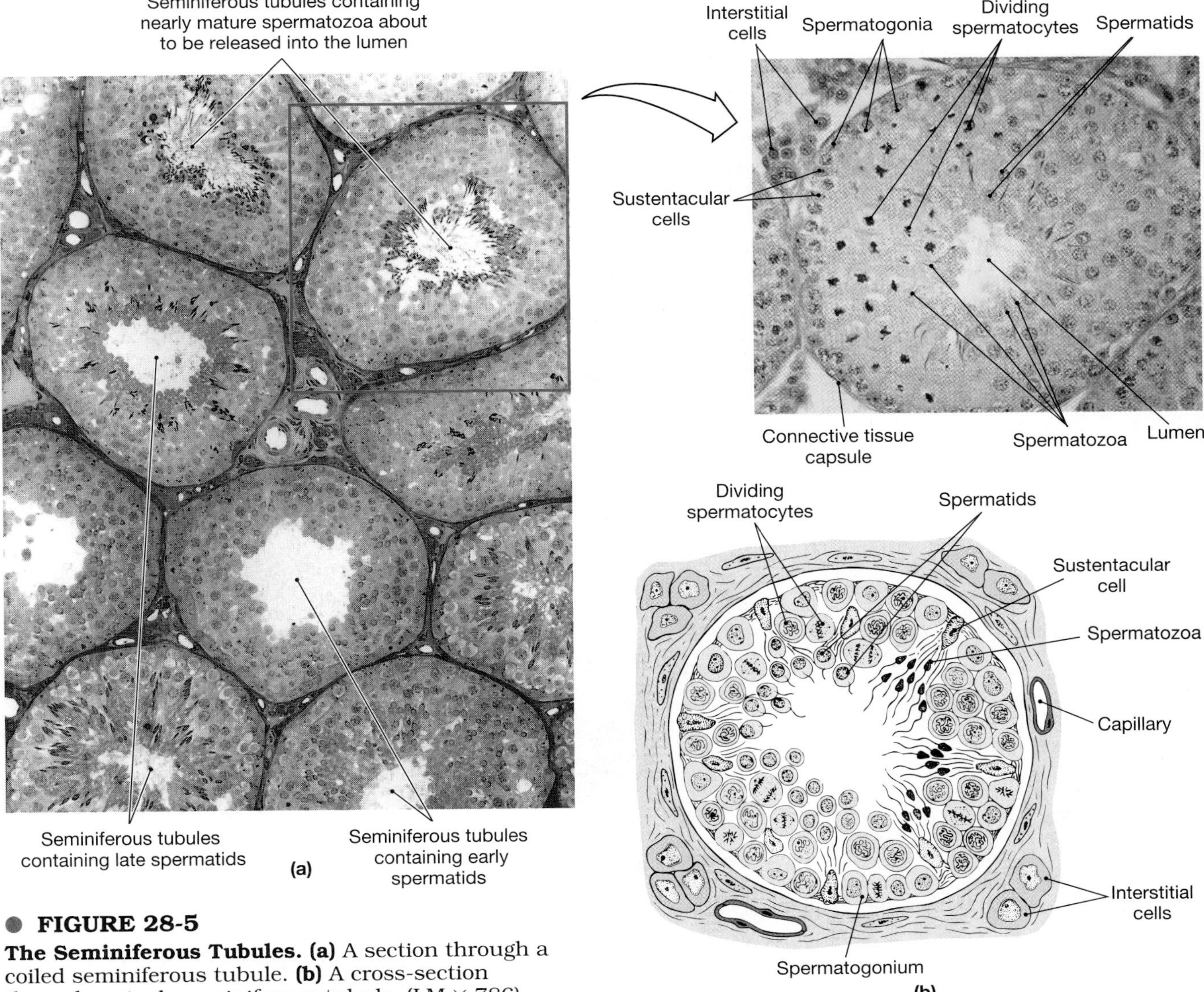
Seminiferous tubules containing nearly mature spermatozoa about to be released into the lumen
Seminiferous tubules containing late spermatids
(a)
Seminiferous tubules containing early spermatids
Interstitial cells
Spermatogonia
Dividing spermatocytes
Spermatids
Sustentacular cells
Connective tissue capsule
Spermatozoa
Lumen
Dividing spermatocytes
Spermatids
Sustentacular cell
Spermatozoa
Capillary
Interstitial cells
Spermatogonium
(b)

● **FIGURE 28-5**

The Seminiferous Tubules. (a) A section through a coiled seminiferous tubule. **(b)** A cross-section through a single seminiferous tubule. (LM × 786)

Spermatogenesis

Figures 28-6, 28-7

Spermatogenesis involves three integrated processes:

- *Mitosis.* Stem cells called **spermatogonia** (sper-ma-to-GŌ-nē-a) undergo cell divisions throughout adult life. (You may wish to review the description of mitosis and cell division in Chapter 3.) [pp. 99–102] The cell divisions produce daughter cells that are pushed toward the lumen of the tubule. These cells differentiate into *spermatocytes* (sper-MA-tō-sīts) that prepare to begin *meiosis.*
- *Meiosis.* **Meiosis** (mī-Ō-sis) is a special form of cell division involved in gamete production. Gametes produced through meiosis contain half of the normal chromosome complement. As a result, the fusion of a sperm and an ovum produces a cell with the normal number of chromosomes. In the seminiferous tubules, the meiotic divisions of spermatocytes produce immature gametes called *spermatids.*
- *Spermiogenesis.* Spermatids are small, relatively unspecialized cells. In the process of **spermiogenesis** spermatids differentiate into physically mature spermatozoa. Spermatozoa are among the most highly specialized cells in the body, and spermiogenesis involves major changes in the internal and external structure of the spermatid.

MITOSIS AND MEIOSIS Mitosis and meiosis differ significantly in terms of the events that take place in the nucleus. Mitosis is part of the process of somatic cell division, which produces two daughter cells, each containing 23 *pairs* of chromosomes. Each pair consists of one chromosome provided by the male parent and another by the female parent at the time of fertilization. Because they contain

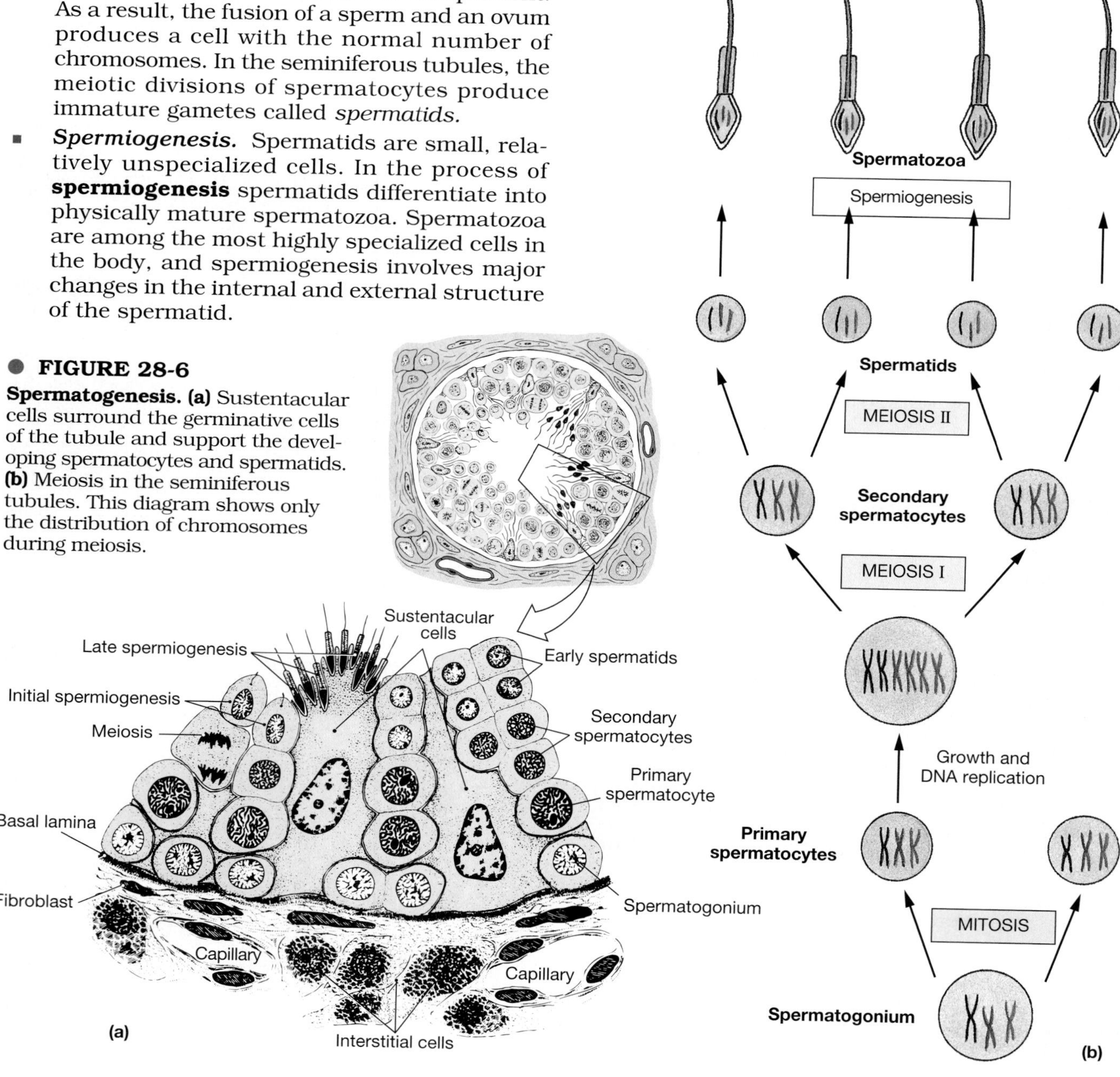

Late spermiogenesis
Initial spermiogenesis
Meiosis
Basal lamina
Fibroblast
Capillary
Sustentacular cells
Early spermatids
Secondary spermatocytes
Primary spermatocyte
Spermatogonium
Capillary
Interstitial cells
(a)
Spermatozoa
Spermiogenesis
Spermatids
MEIOSIS II
Secondary spermatocytes
MEIOSIS I
Growth and DNA replication
Primary spermatocytes
MITOSIS
Spermatogonium
(b)

FIGURE 28-6

Spermatogenesis. (a) Sustentacular cells surround the germinative cells of the tubule and support the developing spermatocytes and spermatids. **(b)** Meiosis in the seminiferous tubules. This diagram shows only the distribution of chromosomes during meiosis.

both members of each chromosome pair, they are called **diploid** (DIP-loyd; *diplo,* double) cells. Meiosis involves a pair of divisions and produces *four* cells, each of which contains 23 *individual chromosomes.* Because these cells contain only one member of each pair of chromosomes, they are called **haploid** (HAP-loyd; *haplo,* single) cells.

The mitotic divisions of spermatogonia produce **primary spermatocytes.** As a primary spermatocyte prepares to begin meiosis, all of the chromosomes within the nucleus replicate themselves as if the cell were about to undergo mitosis. As the prophase of the first meiotic division, **meiosis I,** arrives, the chromosomes condense and become visible. As in mitosis, each chromosome consists of two duplicate *chromatids* (KRŌ-ma-tidz). The events in the nucleus are the same whether you consider the formation of sperm or ova; we will examine spermatogenesis now and *oogenesis* in a later section.

As meiosis begins, the primary spermatocyte contains 46 individual chromosomes. The corresponding maternal and paternal chromosomes now come together, an event known as **synapsis** (sin-AP-sis). Synapsis produces 23 pairs of chromosomes, each member of the pair consisting of two chromatids. A matched set of four chromatids is called a **tetrad** (TET-rad; *tetras,* four). Some exchange of genetic material can occur between the chromatids at this stage of meiosis. This exchange, called *crossing-over,* increases genetic variation among offspring; it will be detailed in Chapter 29.

As metaphase I begins, the nuclear envelope disappears and the tetrads line up along the metaphase plate. At the end of metaphase I the tetrads break up, and the maternal and paternal chromosomes separate. This phase is a major difference between mitosis and meiosis: In mitosis, each daughter cell receives one of the two copies of every chromosome, maternal and paternal, whereas in meiosis each daughter cell receives both copies of *either* the maternal chromosome *or* the paternal chromosome from each tetrad (Figure 28-6b•).

As anaphase proceeds, the maternal and paternal components are randomly distributed. As a result, telophase I ends with the formation of two daughter cells containing unique combinations of maternal and paternal chromosomes. In the testes, the daughter cells produced by the first meiotic division (meiosis I) are called **secondary spermatocytes.**

Each secondary spermatocyte contains 23 chromosomes. Because the first meiotic division reduces the number of chromosomes from 46 to 23, it is called a **reductional division.** Each of these chromosomes still consists of two duplicate chromatids. These duplicates will be separated during **meiosis II.**

The interphase separating meiosis I and meiosis II is very brief, and the secondary spermatocyte soon enters prophase II. The completion of metaphase II, anaphase II, and telophase II produces four **spermatids** (SPER-ma-tidz), each containing 23 chromosomes. Because the chromosome number has not changed, meiosis II represents an **equational division.**

For each primary spermatocyte that enters meiosis, four spermatids are produced (Figure 28-6b•). Because cytokinesis is not completed in meiosis I or meiosis II, the four spermatids initially remain interconnected by cytoplasmic bridges. These connections assist in the transfer of nutrients and hormonal messages between the cells, thus helping to ensure that the cells develop in synchrony.

SPERMIOGENESIS Each spermatid matures into a single **spermatozoon** (sper-ma-tō-ZŌ-on), or sperm cell. This maturation process is called spermiogenesis (Figure 28-7•). Developing spermatocytes undergoing meiosis and spermatids undergoing spermiogenesis are not free in the seminiferous tubules. Instead, they are surrounded by the cytoplasm of the sustentacular cells. As spermiogenesis proceeds, the spermatids gradually develop the appearance of mature spermatozoa. At *spermiation* a spermatozoon loses its attachment to the sustentacular cell and enters the lumen of the seminiferous tubule. The entire process, from spermatogonial division to spermiation, takes approximately 9 weeks.

SPERMATOGENESIS AND SUSTENTACULAR CELLS Sustentacular cells play a key role in the process of spermatogenesis. These cells have six important functions that directly or indirectly affect mitosis, meiosis, and spermiogenesis within the seminiferous tubules:

1. ***Maintenance of the blood-testis barrier.*** The seminiferous tubules are isolated from the general circulation by a **blood-testis barrier** comparable to the blood-brain barrier. ∞ *[pp. 467, 741]* Extensions of sustentacular cells form a layer that surrounds the seminiferous tubule between the spermatogonia and the tubular capsule. Transport across the sustentacular cells is tightly regulated so that conditions inside the tubule remain very stable. The tubular fluid within the lumen of a seminiferous tubule is secreted by sustentacular cells, which also regulate its composition. Tubular fluid is very different from the surrounding interstitial fluid. For example, tubular fluid is high in androgens, estrogens, potassium, and amino acids. The blood-testis barrier is essential to preserving the differences between tubular fluid and interstitial fluid. In addition, developing spermatozoa contain sperm-specific antigens in their cell membranes. These antigens, not found in somatic cell membranes, would be attacked by the immune system if the blood-testis barrier did not prevent their detection.

2. ***Support of mitosis and meiosis.*** Spermatogenesis depends on the stimulation of sustentacular cells by circulating follicle-stimulating hormone (FSH) and testosterone. Stimulated sustentacular cells then in some way promote the division of spermatogonia and the meiotic divisions of spermatocytes.
3. ***Support of spermiogenesis.*** Spermiogenesis requires the presence of sustentacular cells. These cells surround and enfold the spermatids, providing nutrients and chemical stimuli that promote their development.
4. ***Secretion of inhibin.*** Sustentacular cells secrete a peptide hormone, **inhibin** (in-HIB-in), in response to factors released by developing sperm. Inhibin, introduced in Chapter 18, depresses the pituitary production of FSH and perhaps the hypothalamic secretion of gonadotropin-releasing hormone (GnRH) as well. [p. 635] The faster the rate of sperm production, the greater the amount of inhibin secreted. By regulating FSH and GnRH secretion, sustentacular cells provide feedback control of spermatogenesis.
5. ***Secretion of androgen-binding protein.*** **Androgen-binding protein (ABP)** binds androgens (primarily testosterone) in the fluid contents of the seminiferous tubules. This protein is thought to be important in elevating the concentration of androgens within the tubules and stimulating spermiogenesis. ABP production is stimulated by FSH.
6. ***Secretion of Müllerian-inhibiting factor.*** **Müllerian-inhibiting factor (MIF)** is secreted by sustentacular cells in the developing testes. This hormone causes regression of the fetal *Müllerian ducts,* passageways that in a female participate in the formation of uterine tubes and uterus. Inadequate MIF production during development leads to retention of these ducts and failure of the testes to descend into the scrotum.

TESTICULAR CANCER Testicular cancer occurs at a relatively low rate of around 3 cases per 100,000 males per year. Although there are only about 6600 cases each year in the United States, testicular cancer is the most common cancer in males age 15–35. The incidence among Caucasian males has more than doubled since the 1930s, but the incidence among African-Americans has remained unchanged. The reason for this difference is not known.

More than 95 percent of testicular cancers are the result of abnormal spermatogonia or spermatocytes, rather than sustentacular cells, interstitial cells, or other cells of the testes. Treatment usually consists of a combination of orchiectomy and chemotherapy. The survival rate has increased from around 10 percent in 1970 to over 90 percent in 1994, primarily as a result of earlier diagnosis and improved treatment protocols.

❑ Anatomy of a Spermatozoon

Figure 28-7

There are three distinct regions to each spermatozoon: the *head,* the *middle piece,* and the *tail.*

- The **head** is a flattened oval containing densely packed chromosomes. The tip of the head is covered by the **acrosomal** (ak-rō-SŌ-mal) **cap,** a membranous compartment containing enzymes essential to the process of fertilization. The acrosomal cap forms through the fusion of the Golgi saccules during spermiogenesis (Figure 28-7●).
- A short **neck** attaches the head to the **middle piece.** The neck contains both centrioles of the original spermatid. The microtubules of the distal centriole are continuous with those of the middle piece and tail. Mitochondria in the middle piece are arranged in a spiral around the microtubules. Mitochondrial activity provides ATP that is needed to move the tail.
- The **tail** is the only example of a flagellum in the human body. A **flagellum,** an organelle introduced in Chapter 3, moves a cell from one place to another. [p. 85] Although cilia beat in a predictable, waving fashion, the flagellum of a spermatozoon has a complex, corkscrew motion. The tail can be divided into a long **principal piece** and a tapering **end piece.** Microtubules within the principal piece are arranged in the same 9 + 2 pattern seen in cilia. [p. 86] These microtubules are surrounded by a dense, fibrous sheath that thickens and fuses around the microtubules in the end piece.

Unlike other, less specialized cells, a mature spermatozoon lacks an endoplasmic reticulum, Golgi apparatus, lysosomes, peroxisomes, inclusions, and many other intracellular structures. Because the cell does not contain glycogen or other energy reserves, it must absorb nutrients (primarily fructose) from the surrounding fluid.

❑ The Male Reproductive Tract

The testes produce physically mature spermatozoa that are, as yet, incapable of fertilizing an ovum. The other portions of the male reproductive system are concerned with the functional maturation, nourishment, storage, and transport of spermatozoa.

The Epididymis

Figures 28-1, 28-4, 28-8

Late in their development the spermatozoa become detached from the sustentacular cells and lie within the lumen of the seminiferous tubule. Although they have most of the physical characteristics of

● **FIGURE 28-7**
Spermiogenesis and Spermatozoon Structure. (a) Differentiation of a spermatid into a spermatozoon. **(b)** SEM of human spermatozoa (× 1688).

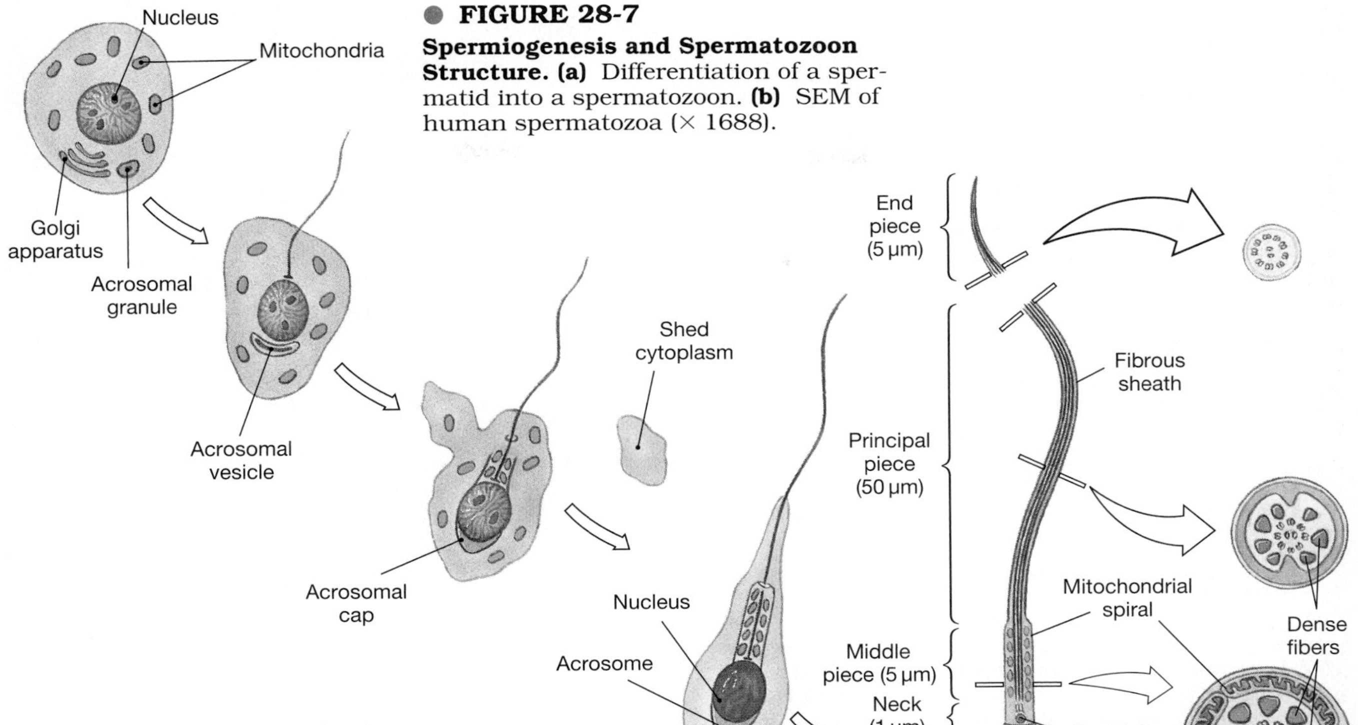

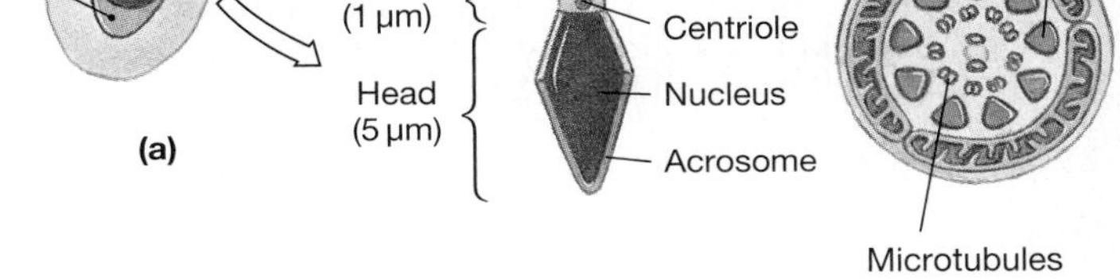

Nucleus
Mitochondria
Golgi apparatus
Acrosomal granule
Acrosomal vesicle
Shed cytoplasm
Acrosomal cap
Nucleus
Acrosome
End piece (5 μm)
Principal piece (50 μm)
Middle piece (5 μm)
Neck (1 μm)
Head (5 μm)
Fibrous sheath
Mitochondrial spiral
Centriole
Nucleus
Acrosome
Dense fibers
Microtubules
(a)

mature sperm cells, they are still functionally immature and incapable of coordinated locomotion or fertilization. Fluid currents then transport the cell along the straight tubule, through the rete testis (Figure 28-4a●, p. 1064), and into the epididymis.

The epididymis lies along the posterior border of the testis (Figures 28-1, p. 1061, 28-4, p. 1064, and 28-8a●). It has a firm texture and can be felt through the skin of the scrotum. The epididymis consists of a tubule almost 7 meters (23 ft) long, coiled and twisted so as to take up very little space. The epididymis has a *head*, a *body*, and a *tail*.

- The superior **head** is the portion of the epididymis proximal to the mediastinum of the testis. The head receives spermatozoa from the efferent ducts that connect the rete testis to the epididymis.
- The **body** begins distal to the last efferent duct and extends inferiorly along the posterior margin of the testis.
- Near the inferior border of the testis the number of convolutions decreases, marking the start of the **tail.** The tail recurves, and as it ascends its histological organization changes until it closely resembles the structure of the ductus deferens. The tail of the epididymis is the principal region involved with sperm storage.

The epididymis has the following functions:

1. ***It monitors and adjusts the composition of the tubular fluid.*** The columnar epithelial lining of the epididymis (Figure 28-8b,c●) bears distinctive *stereocilia* that increase the surface area available for absorption and secretion into the tubular fluid. ∞ [p. 111]

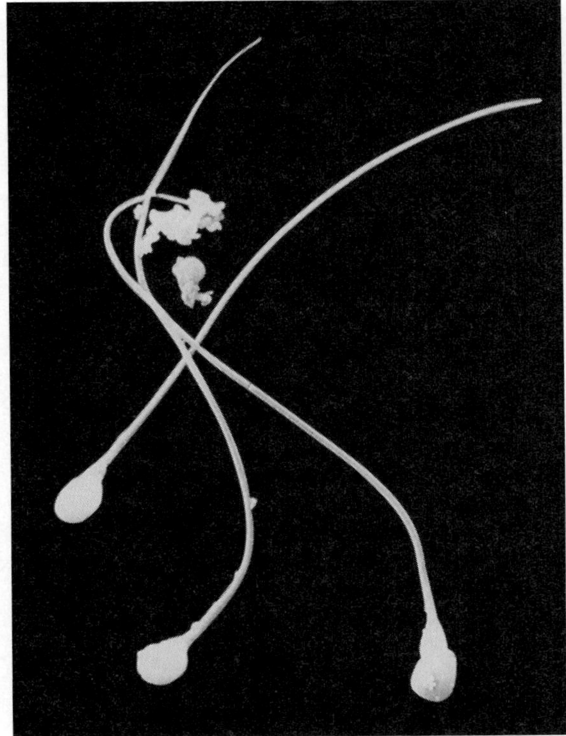

(b)

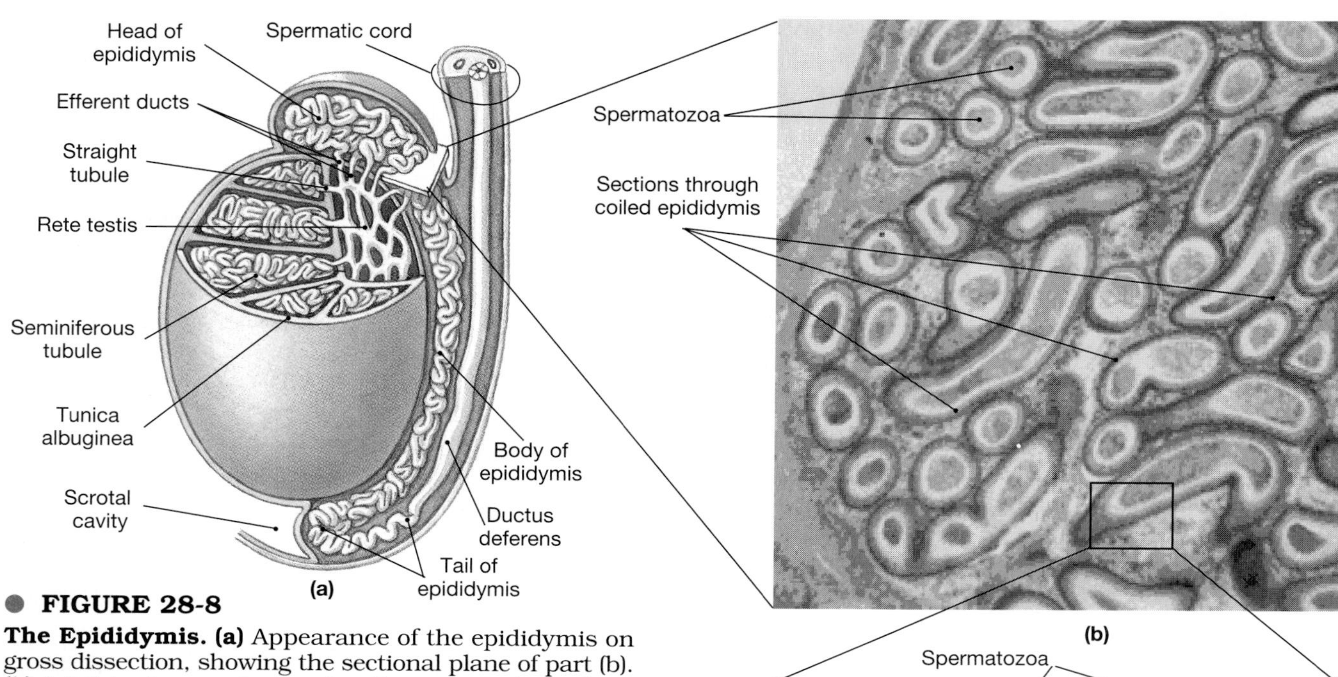
Head of epididymis
Spermatic cord
Efferent ducts
Straight tubule
Rete testis
Seminiferous tubule
Tunica albuginea
Scrotal cavity
Body of epididymis
Ductus deferens
Tail of epididymis
(a)
Spermatozoa
Sections through coiled epididymis
(b)

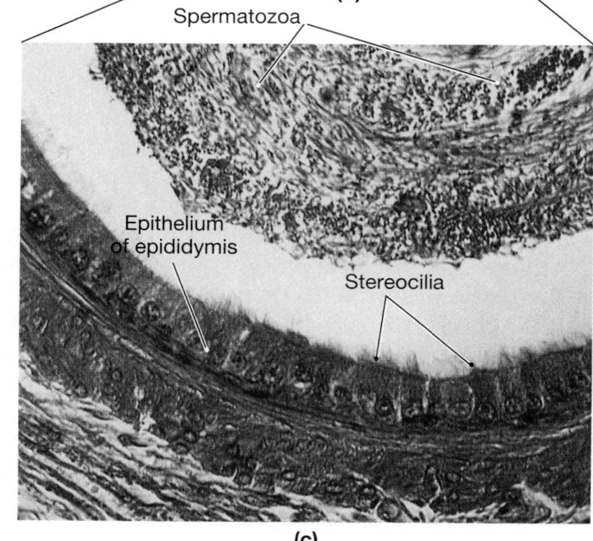
Spermatozoa
Epithelium of epididymis
Stereocilia
(c)

● **FIGURE 28-8**

The Epididymis. (a) Appearance of the epididymis on gross dissection, showing the sectional plane of part (b). **(b)** A light micrograph showing the organization of tubules and the surrounding connective tissues. (LM × 49) **(c)** A micrograph showing epithelial characteristics, especially the elongate stereocilia characteristic of the epididymis. (LM × 1239)

2. ***It acts as a recycling center for damaged spermatozoa.*** Cellular debris and damaged spermatozoa are absorbed, and the products of enzymatic breakdown are released into the surrounding interstitial fluids for pickup by the epididymal circulation.
3. ***It stores spermatozoa and facilitates their functional maturation.*** It takes about 2 weeks for a spermatozoon to pass through the epididymis, and during this period it completes its functional maturation. Although spermatozoa leaving the epididymis are mature, they remain immobile. To become active, motile, and fully functional the spermatozoa must undergo **capacitation,** and the epididymis secretes a substance that prevents premature capacitation. Capacitation normally occurs in two steps: (1) Spermatozoa become motile when mixed with secretions of the seminal vesicles, and (2) they become capable of successful fertilization when exposed to conditions inside the female reproductive tract.

Transport along the epididymis involves some combination of fluid movement and peristaltic contractions of smooth muscle. After passing along the tail of the epididymis, the spermatozoa enter the ductus deferens.

The Ductus Deferens

Figures 28-1, 28-8, 28-9

The ductus deferens, or *vas deferens,* is 40–45 cm (16–18 in.) long. It begins at the tail of the epididymis (Figure 28-8a●) and ascends through the inguinal canal in the spermatic cord (Figure 28-1●, p. 1061). Inside the abdominal cavity, the ductus deferens passes posteriorly, curving downward along the lateral surface of the urinary bladder toward the superior and posterior margin of the prostate gland. Just before it reaches the prostate and seminal vesicles, the ductus deferens becomes enlarged, and the expanded portion is known as the **ampulla** (am-PYŪL-ah) (Figure 28-9●).

The wall of the ductus deferens contains a thick layer of smooth muscle (Figure 28-9e●). Peristaltic contractions in this layer propel spermatozoa and fluid along the duct, which is lined by a pseudostratified ciliated columnar epithelium. In addition to transporting sperm, the ductus deferens can store spermatozoa for several months. During this time the spermatozoa remain in a state of suspended animation, with low metabolic rates.

The junction of the ampulla with the base of the seminal vesicle marks the start of the ejaculatory duct. This relatively short passageway (2 cm, or less than 1 in.) penetrates the muscular wall of the prostate gland and empties into the urethra (Figure 28-1•, p. 1061) near the ejaculatory duct from the opposite side.

The Urethra

The urethra of the male extends from the urinary bladder to the tip of the penis, a distance of 15–20 cm (6–8 in.). It is divided into *prostatic, membranous,* and *penile regions.* These subdivisions were considered in Chapter 26. ∞ [p. 1016] The urethra in the male is a passageway used by both the urinary and reproductive systems.

❑ The Accessory Glands

Figures 28-1, 28-9

The fluids contributed by the seminiferous tubules and the epididymis account for only about 5 percent of the final volume of semen. The seminal fluid is a mixture of the secretions of many different glands, each with distinctive biochemical characteristics. Important glands include the *seminal vesicles,* the *prostate gland,* and the *bulbourethral glands* (Figures 28-1, p. 1061, and 28-9•). Major functions of these glandular organs include: (1) activating the spermatozoa; (2) providing the nutrients spermatozoa need for motility; (3) propelling spermatozoa and fluids along the reproductive tract, primarily through peristaltic contractions; and (4) producing buffers that counteract the acidity of the urethral and vaginal contents.

The Seminal Vesicles

Figures 28-1, 28-9, 28-10

The ductus deferens on each side ends at the junction between the ampulla and the duct draining the seminal vesicle. The seminal vesicles are embedded in connective tissue on either side of the midline, sandwiched between the posterior wall of the urinary bladder and the rectum. Each seminal vesicle is a tubular gland, with a total length of around 15 cm (6 in.). The body of the gland has many short side branches, and the entire assemblage is coiled and folded into a compact, tapered mass roughly 5 cm × 2.5 cm (2 in. × 1 in.). The location of the seminal vesicles can be seen in Figures 28-1, p. 1061 (lateral view), 28-9 and 26-21, p.1014 (posterior/inferior views), and 28-10•, p. 1075 (anterior view).

The seminal vesicles are extremely active secretory glands, with an epithelial lining that contains extensive folds (Figure 28-9d•). The seminal vesicles contribute about 60 percent of the volume of semen, and, although the vesicular fluid usually has the same osmotic concentration as blood plasma, the composition is quite different. In particular, the secretion of the seminal vesicles contains (1) relatively high concentrations of fructose, which is easily metabolized by spermatozoa, (2) prostaglandins, which may stimulate smooth muscle contractions along the male and female reproductive tracts, and (3) fibrinogen, which after ejaculation will form a temporary clot within the vagina. The secretions of the seminal vesicles are slightly alkaline, and this alkalinity helps neutralize acids in the prostatic secretions and within the vagina. When mixed with the secretions of the seminal vesicles, previously inactive but functional spermatozoa begin beating their flagella, becoming highly mobile.

The secretions of the seminal vesicles are discharged into the ampulla of the ductus deferens at *emission,* when peristaltic contractions are under way in the ductus deferens, seminal vesicles, and prostate gland. These contractions are under control of the sympathetic nervous system.

The Prostate Gland

Figures 28-1, 28-9, 28-10

The prostate gland is a small, muscular, rounded organ with a diameter of about 4 cm (1.6 in.). The prostate gland encircles the proximal portion of the urethra as it leaves the urinary bladder (Figures 28-1, p. 1061, sagittal view; 28-9a, posterior view; and 26-21c, p. 1014, and 28-10•, p.1075, anterior views). The glandular tissue of the prostate consists of a cluster of 30–50 compound tubuloalveolar glands (Figure 28-9b•). ∞ [p. 121] These glands are surrounded and wrapped in a thick blanket of smooth muscle fibers.

The prostatic glands produce **prostatic fluid,** a slightly acidic solution that contributes 20–30 percent of the volume of semen. In addition to several other compounds of uncertain significance, prostatic secretions contain **seminalplasmin** (sem-i-nal-PLAZ-min), an antibiotic that may help prevent urinary tract infections in males. These secretions are ejected into the prostatic urethra by peristaltic contractions of the muscular wall.

PROSTATITIS Prostatic inflammation, or **prostatitis** (pros-ta-TĪ-tis), can occur at any age, but it most often afflicts older men. Prostatitis may result from bacterial infections, but the condition may also develop in the apparent absence of pathogens. Individuals with prostatitis complain of pain in the lower back, perineum, or rectum, sometimes accompanied by painful urination and the discharge of mucous secretions from the urethral meatus. Antibiotic therapy is usually effective in treating cases resulting from bacterial infection, but in other cases antibiotics may not provide relief. Prostatitis is taken seriously because the symptoms can resemble those of *prostate cancer* (p. 1072). Prostate cancer is the second most common cancer in men, and it is the second leading cause of cancer deaths in males.

Prostatic Hypertrophy and Prostate Cancer

Prostatic enlargement, or **benign prostatic hypertrophy,** usually occurs spontaneously in men over age 50. The increase in size occurs at the same time that hormonal changes are under way within the testes. Androgen production by the interstitial cells decreases over this period, and at the same time these endocrine cells begin releasing small quantities of estrogens into the circulation. The combination of lower testosterone levels and the presence of estrogen probably stimulates prostatic growth. In severe cases, prostatic swelling can constrict and block the urethra and even the rectum. The urinary obstruction can cause permanent kidney damage if not corrected. Partial surgical removal is the most effective treatment at present. In the procedure known as a **TURP** (*transurethral prostatectomy*), an instrument pushed along the urethra restores normal function by cutting away the swollen prostatic tissue. Most of the prostate remains in place, and there are no external scars.

Prostate cancer is the second most common cancer in men, and it is the second most common cause of cancer deaths in males. In 1994 approximately 135,000 new cases of prostate cancer were diagnosed in the United States, and there were approximately 37,000 deaths. Most patients are elderly (average age 72 at diagnosis). There are racial differences in susceptibility that are poorly understood. At age 50–54 the prostate cancer rates are twice as high for African-Americans as for Caucasian Americans. (The rates at all ages are about one-third higher for African-Americans.) The prostate cancer rates for Asian males are relatively low compared with either Caucasian Americans or African-Americans. For all age groups and all races, the rates of prostate cancer are rising sharply. The rate of increase has been as high as 16 percent per year (1989–1990). The reason for the increase is not known.

Prostate cancer usually originates in one of the secretory glands, and as it progresses it produces a nodular lump or swelling on the prostatic surface. Palpation of the prostate gland through the rectal wall, a procedure known as a *digital rectal exam*, or DRE, is the easiest diagnostic screening procedure. *Transrectal prostatic ultrasound* (TRUS) can be used to obtain more detailed information about the status of the prostate, but at significantly higher cost to the patient.

If the condition is detected before the cancer cells have spread to other organs, the usual treatment is either localized radiation or the surgical removal of the prostate gland. This operation, called a **prostatectomy** (prosta-TEK-to-mē), is often effective in controlling the condition, but undesirable side effects may include a loss of sexual function and urinary incontinence. Modified surgical procedures can reduce these risks and maintain normal sexual function in almost three out of four patients.

One recent screening method involves a blood test for *prostate-specific antigen (PSA).* Elevated levels of this antigen, normally present in low concentrations, may indicate the presence of prostate cancer. This test is more sensitive than the serum enzyme assay previously used for screening purposes. That enzyme test, which checks levels of the isozyme *prostatic acid phosphatase*, detects prostate cancer in comparatively late stages of development. Screening with periodic PSA tests is now being recommended for men over age 50.

Early detection is important because metastasis from the prostate soon involves the lymphatic system, lungs, bone marrow, liver, or adrenal glands. The survival rates at this stage become relatively low. Potential treatments for metastatic prostate cancer include more intensive radiation dosage, hormonal manipulation, lymph node removal, and aggressive chemotherapy. Because the cancer cells are stimulated by testosterone, treatment may involve castration or hormones that depress GnRH or LH production. Until recently the usual hormone selected was *diethylstilbestrol* (DES), an estrogen. There are now two other options: (1) *Drugs that mimic GnRH:* These drugs are given in high doses, producing a surge in LH production followed by a sharp decline to very low levels, presumably as the endocrine cells adapt to the excessive stimulation. (2) *Drugs that block the action of androgens:* Several new drugs, including *flutamide* and *finasateride*, prevent stimulation of the cancer cells by testosterone. Despite these interesting advances in treatment, however, the average survival time for patients diagnosed with advanced prostatic cancer is only 2.5 years.

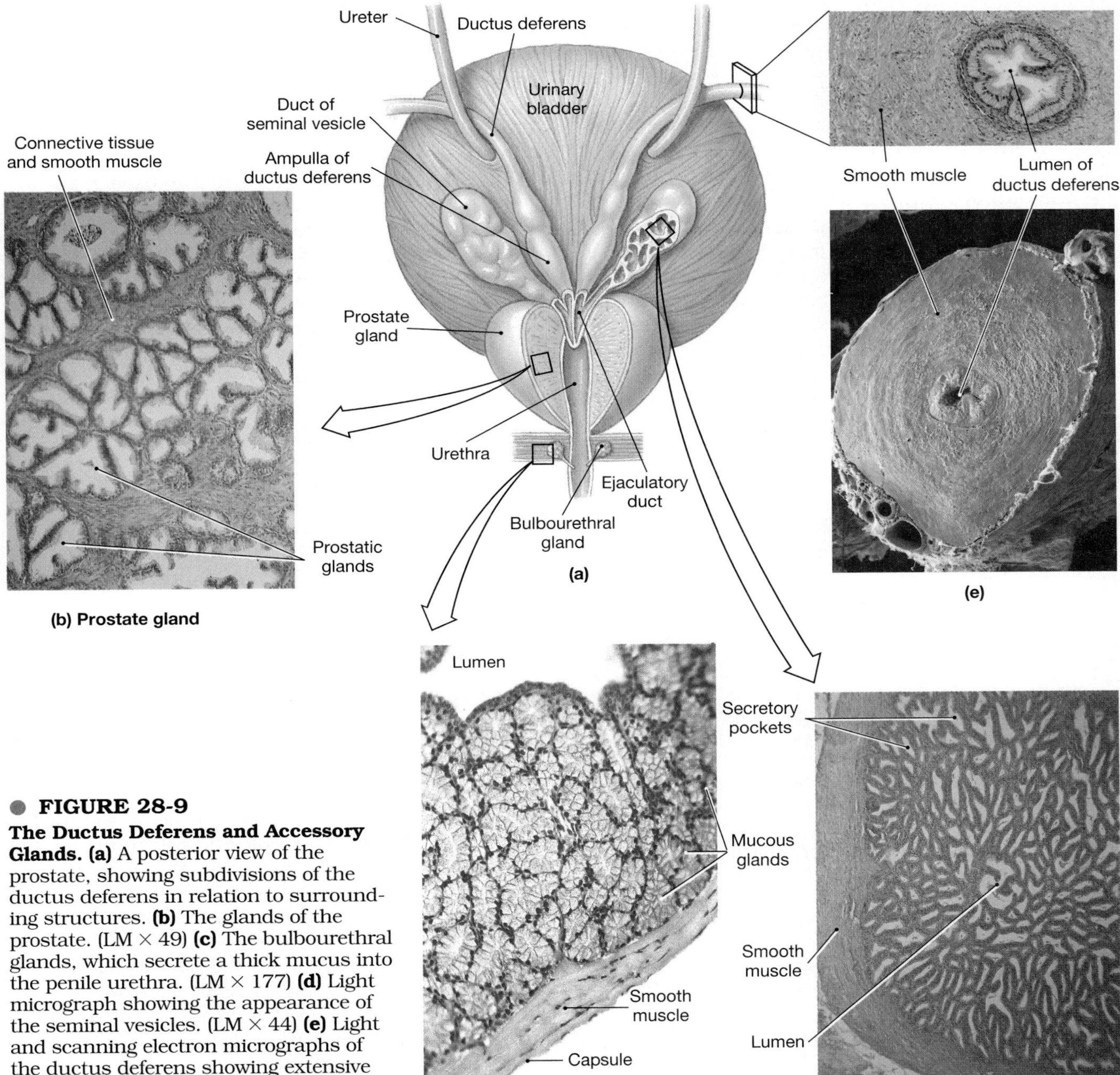

Ureter
Ductus deferens
Urinary bladder
Duct of seminal vesicle
Ampulla of ductus deferens
Connective tissue and smooth muscle
Prostate gland
Urethra
Ejaculatory duct
Bulbourethral gland
(a)
Prostatic glands
(b) Prostate gland
Smooth muscle
Lumen of ductus deferens
(e)
Lumen
Mucous glands
Smooth muscle
Capsule
(c) Bulbourethral gland
Secretory pockets
Smooth muscle
Lumen
(d) Seminal vesicle

● **FIGURE 28-9**
The Ductus Deferens and Accessory Glands. (a) A posterior view of the prostate, showing subdivisions of the ductus deferens in relation to surrounding structures. **(b)** The glands of the prostate. (LM × 49) **(c)** The bulbourethral glands, which secrete a thick mucus into the penile urethra. (LM × 177) **(d)** Light micrograph showing the appearance of the seminal vesicles. (LM × 44) **(e)** Light and scanning electron micrographs of the ductus deferens showing extensive layering with smooth muscle around the lumen. (SEM × 42, LM × 34)

The Bulbourethral Glands

Figures 28-1, 28-9, 28-10

The paired bulbourethral glands, or *Cowper's glands*, are situated at the base of the penis, covered by the fascia of the urogenital diaphragm (Figures 28-1●, p. 1061, sagittal view; 28-9a●, posterior view; and 28-10●, anterior view). The bulbourethral glands are round, with diameters approaching 10 mm (less than 0.5 in.). The duct of each gland travels alongside the penile urethra for 3–4 cm (1.2–1.6 in.) before emptying into the urethral lumen. These are compound, tubuloalveolar mucous glands (Figure 28-9c●) that secrete a thick, alkaline mucus. This secretion helps to neutralize any urinary acids that may remain in the urethra and provides lubrication for the *glans*, or distal tip of the penis.

❑ Semen

A typical ejaculation releases 2–5 ml of semen. This volume of fluid, called an **ejaculate,** contains:

1. ***Spermatozoa.*** A normal **sperm count** ranges from 20 to 100 million spermatozoa per milliliter.

2. *Seminal fluid.* **Seminal fluid,** the fluid component of semen, is a mixture of glandular secretions with a distinctive ionic and nutrient composition. In terms of total volume, a typical sample of seminal fluid contains the combined secretions of the seminal vesicles (60 percent), the prostate (30 percent), the sustentacular cells and epididymis (5 percent), and the bulbourethral glands (<5 percent).
3. *Enzymes.* Several important enzymes are present in the seminal fluid, including (1) a protease that may help to dissolve mucous secretions in the vagina and (2) seminalplasmin, an antibiotic enzyme from the prostate gland that kills a variety of bacteria, including *Escherichia coli,* (3) a prostatic enzyme that converts fibrinogen to fibrin after ejaculation, and (4) *fibrinolysin,* which subsequently liquefies the clotted semen.

SEMEN ANALYSIS Semen analysis is often performed to assess male fertility. A sample is donated after a 36-hour period of sexual abstinence. Standard tests include:

- ***Volume.*** Normal ejaculate volume is 2–5 ml. An abnormally low volume may indicate problems with the prostatic glands or seminal vesicles.
- ***Sperm count.*** The total number of spermatozoa in an ejaculate should be greater than 60 million. The concentration of spermatozoa in semen should be over 20 million per milliliter. Individuals with lower sperm counts are usually infertile. A low sperm count may reflect inflammation of the epididymis, ductus deferens, or prostate.
- ***Motility.*** At least 60 percent of the spermatozoa in the sample should be beating their flagella and swimming actively.
- ***Morphology.*** At least 60 percent of the spermatozoa should have normal shapes. Common abnormalities are malformed heads and "twin" spermatozoa that did not separate at the time of spermiation.
- ***Liquefaction.*** Within a few minutes after ejaculation semen coagulates, liquefying again after a variable period. The function of this clotting is unknown. Normal semen liquefies after 15–30 minutes. An extended liquefaction time may indicate problems with the secretions of accessory glands.

Determination of male fertility problems in the absence of abnormal semen analysis results may require additional tests. In what is often called the "hamster test," a semen sample is placed on a slide with the ovum of a hamster. Normal human spermatozoa will fertilize the ovum, although further development is impossible. If fertilization does not occur, it may indicate problems with the enzymes in the acrosome.

A complete chemical analysis of semen can be found in Appendix VII.

The Penis

Figures 28-1, 28-2, 28-10

The penis is a tubular organ that surrounds the urethra (Figures 28-1, p. 1061, and 28-2●, p. 1062). It conducts urine to the exterior and introduces semen into the female vagina during sexual intercourse. The penis (Figure 28-10●) is divided into three regions:

- The **root** of the penis is the fixed portion that attaches the penis to the body wall. This connection occurs within the urogenital triangle immediately inferior to the pubic symphysis.
- The **body (shaft)** of the penis is the tubular, movable portion. Masses of *erectile tissue* are found within the body.
- The **glans** of the penis is the expanded distal end that surrounds the external urethral meatus.

The skin overlying the penis resembles that of the scrotum. The dermis contains a layer of smooth muscle, and the underlying loose connective tissue allows the thin skin to move without distorting underlying structures. The subcutaneous layer also contains superficial arteries, veins, and lymphatics.

A fold of skin, the **prepuce** (PRĒ-pyūs), or *foreskin,* surrounds the tip of the penis. The prepuce attaches to the relatively narrow **neck** of the penis and continues over the glans. There are no hair follicles on the opposing surfaces, but **preputial** (prē-PYŪ-shal) **glands** in the skin of the neck and the inner surface of the prepuce secrete a waxy material known as **smegma** (SMEG-ma). Unfortunately, smegma can be an excellent nutrient source for bacteria. Mild inflammation and infections in this region are common, especially if the area is not washed thoroughly and frequently. One way of avoiding trouble is to perform a **circumcision** (ser-kum-SIZH-un) and surgically remove the prepuce. In Western societies (especially the United States) this procedure is usually performed shortly after birth. Although it remains controversial, strong cultural biases and epidemiological evidence[1] suggest that this practice will continue.

Beneath the loose connective tissue, a dense network of elastic fibers encircles the internal structures of the penis. Most of the body of the penis consists of three cylindrical columns of **erectile tissue** (Figure 28-10b,c●). Erectile tissue consists of a three-dimensional maze of vascular channels incompletely separated by partitions of elastic connective tissue and smooth muscle fibers. In the resting state, the arterial branches are constricted, and the muscular partitions are tense. This combination restricts blood flow into the erectile tissue. The parasympathetic innervation of the penile arteries involves neu-

[1]Uncircumsized males have a higher incidence of urinary tract infections and are at greater risk for penile cancer than are circumsized males.

● FIGURE 28-10

The Penis. (a) Anterior and lateral view of the penis, showing positions of the erectile tissues. **(b)** Frontal section through the penis and associated organs. **(c)** Sectional view through the penis.

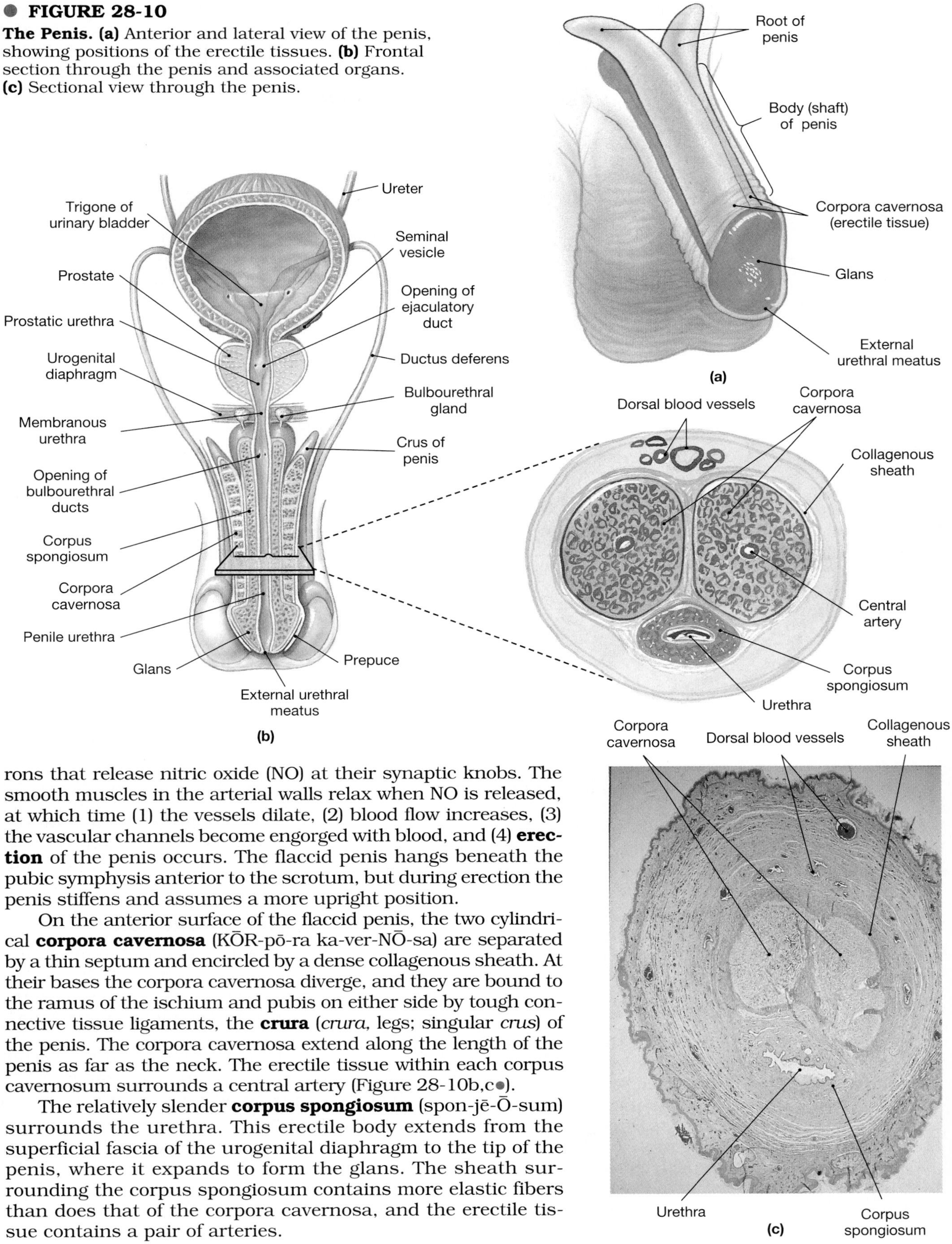

Root of penis
Body (shaft) of penis
Corpora cavernosa (erectile tissue)
Glans
External urethral meatus
(a)

Ureter
Trigone of urinary bladder
Seminal vesicle
Prostate
Opening of ejaculatory duct
Prostatic urethra
Ductus deferens
Urogenital diaphragm
Bulbourethral gland
Membranous urethra
Crus of penis
Opening of bulbourethral ducts
Corpus spongiosum
Corpora cavernosa
Penile urethra
Prepuce
Glans
External urethral meatus
(b)

Dorsal blood vessels
Corpora cavernosa
Collagenous sheath
Central artery
Corpus spongiosum
Urethra

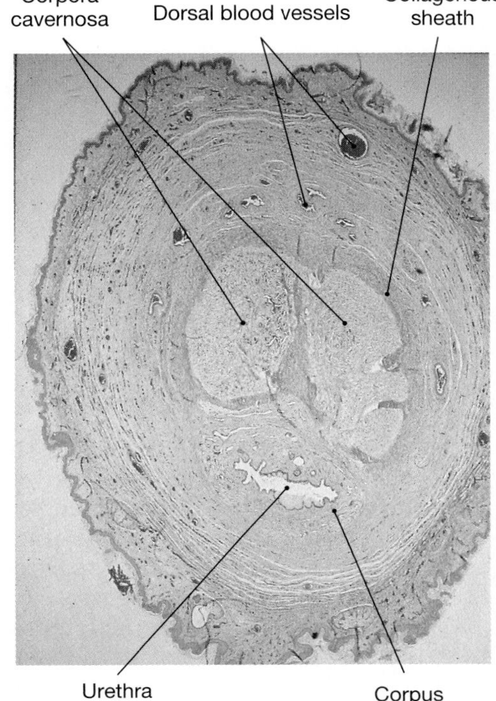
Corpora cavernosa
Dorsal blood vessels
Collagenous sheath
Urethra
Corpus spongiosum
(c)

rons that release nitric oxide (NO) at their synaptic knobs. The smooth muscles in the arterial walls relax when NO is released, at which time (1) the vessels dilate, (2) blood flow increases, (3) the vascular channels become engorged with blood, and (4) **erection** of the penis occurs. The flaccid penis hangs beneath the pubic symphysis anterior to the scrotum, but during erection the penis stiffens and assumes a more upright position.

On the anterior surface of the flaccid penis, the two cylindrical **corpora cavernosa** (KŌR-pō-ra ka-ver-NŌ-sa) are separated by a thin septum and encircled by a dense collagenous sheath. At their bases the corpora cavernosa diverge, and they are bound to the ramus of the ischium and pubis on either side by tough connective tissue ligaments, the **crura** (*crura*, legs; singular *crus*) of the penis. The corpora cavernosa extend along the length of the penis as far as the neck. The erectile tissue within each corpus cavernosum surrounds a central artery (Figure 28-10b,c●).

The relatively slender **corpus spongiosum** (spon-jē-Ō-sum) surrounds the urethra. This erectile body extends from the superficial fascia of the urogenital diaphragm to the tip of the penis, where it expands to form the glans. The sheath surrounding the corpus spongiosum contains more elastic fibers than does that of the corpora cavernosa, and the erectile tissue contains a pair of arteries.

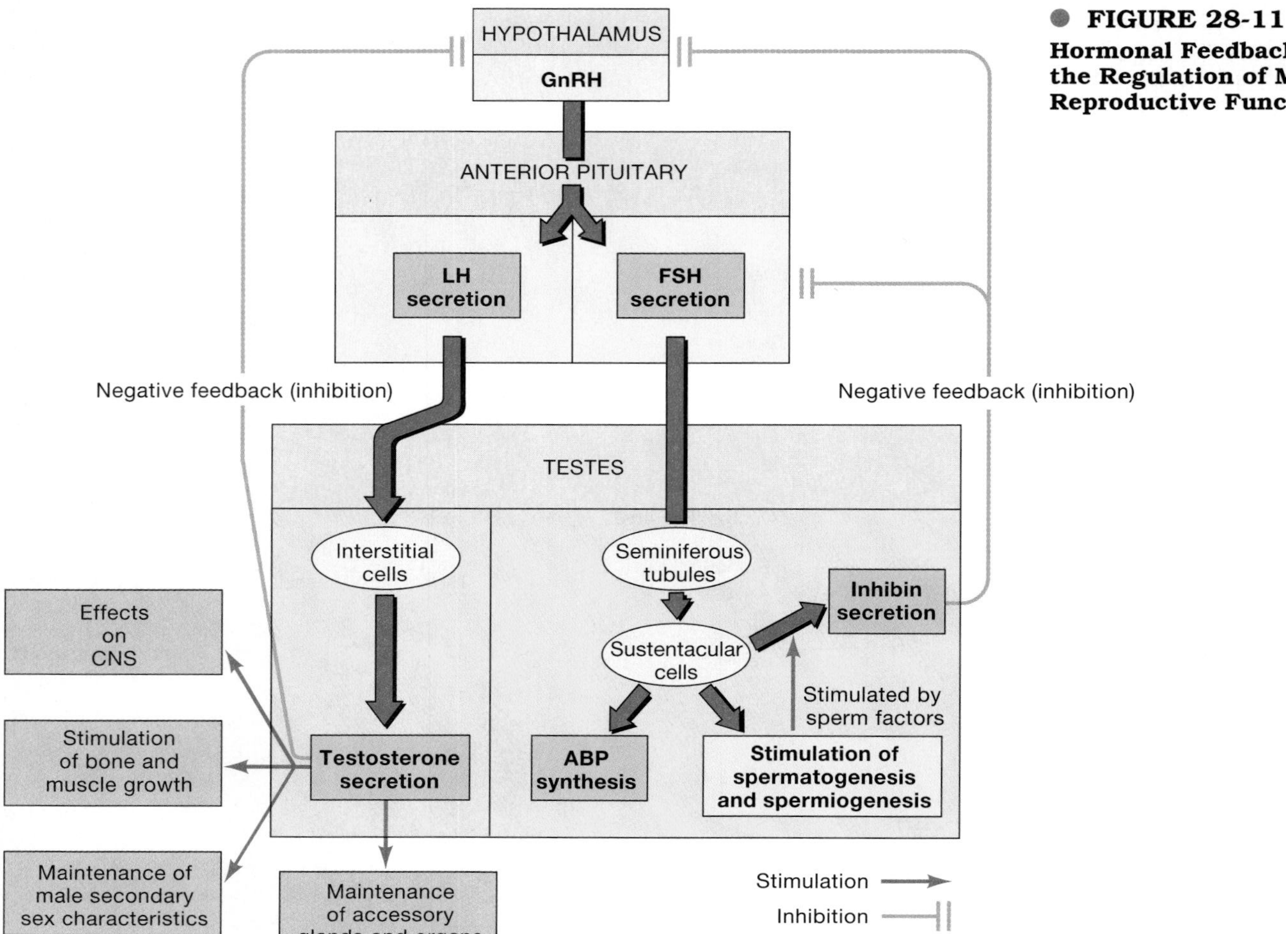
HYPOTHALAMUS
GnRH
ANTERIOR PITUITARY
LH secretion
FSH secretion
Negative feedback (inhibition)
Negative feedback (inhibition)
TESTES
Interstitial cells
Seminiferous tubules
Inhibin secretion
Sustentacular cells
Stimulated by sperm factors
Effects on CNS
Stimulation of bone and muscle growth
Testosterone secretion
ABP synthesis
Stimulation of spermatogenesis and spermiogenesis
Maintenance of male secondary sex characteristics
Maintenance of accessory glands and organs
Stimulation
Inhibition

FIGURE 28-11
Hormonal Feedback and the Regulation of Male Reproductive Function

Hormones and Male Reproductive Function

Figure 28-11

The hormonal interactions in the male are diagrammed in Figure 28-11●, and major reproductive hormones were introduced in Chapter 18. [p. 634] The anterior pituitary releases *follicle-stimulating hormone* **(FSH)** and *luteinizing hormone* **(LH).** The pituitary release of these hormones occurs in the presence of *gonadotropin-releasing hormone* **(GnRH),** a peptide synthesized in the hypothalamus and carried to the anterior pituitary by the hypophyseal portal system.

FSH and Spermatogenesis

In the male, FSH targets primarily the sustentacular cells of the seminiferous tubules. Under FSH stimulation, and in the presence of testosterone from the interstitial cells, sustentacular cells (1) promote spermatogenesis and spermiogenesis and (2) secrete androgen-binding protein.

The rate of spermatogenesis is regulated by a negative feedback mechanism involving GnRH, FSH, and inhibin. Under GnRH stimulation, FSH promotes spermatogenesis along the seminiferous tubules. As spermatogenesis accelerates, however, so does the rate of inhibin secretion by the sustentacular cells of the testes. Inhibin inhibits FSH production in the anterior pituitary and may also suppress secretion of GnRH at the hypothalamus.

The net effect is that when FSH levels become elevated, inhibin production increases until the FSH levels return to normal. If FSH levels decline, inhibin production falls, and the rate of FSH production accelerates.

LH and Androgen Production

In the male, LH causes the secretion of testosterone and other androgens by the interstitial cells of the testes. Testosterone, the most important androgen, has numerous functions that include:

1. Stimulating spermatogenesis and promoting the functional maturation of spermatozoa.
2. Maintaining the accessory organs of the male reproductive tract.
3. Determining the secondary sexual characteristics such as the distribution of facial hair, increased muscle mass and body size, and the quantity and location of characteristic adipose tissue deposits.

4. Stimulating metabolism throughout the body, especially pathways concerned with protein synthesis and muscle growth.
5. Affecting libido (sexual drive) and related behaviors.

Testosterone functions like other steroid hormones. [p. 634] It diffuses across the cell membrane and binds to an intracellular receptor, and the hormone-receptor complex then binds to the DNA in the nucleus. In many target tissues, some of the arriving testosterone is converted to **dihydrotestosterone (DHT).** A small amount of DHT diffuses back out of the cell and into the circulation, and DHT levels are usually around 10 percent of circulating testosterone levels. DHT can also enter peripheral cells and bind to the same hormone receptors targeted by testosterone. In addition, some tissues, notably those of the external genitalia, respond to DHT rather than testosterone, and other tissues, including the prostate gland, are more sensitive to DHT than testosterone.

Testosterone production begins around the seventh week of embryonic development and reaches a peak after roughly 6 months of development. Over this period, secretion of MIF by developing sustentacular cells causes the regression of the Müllerian ducts. The early surge in testosterone levels stimulates the differentiation of the male duct system and accessory organs and affects CNS development. The best-known CNS effects occur within the developing hypothalamus, where testosterone apparently programs the hypothalamic centers involved with (1) GnRH production and the regulation of pituitary FSH and LH secretion, (2) sexual behaviors, and (3) sexual drive. As the result of this prenatal exposure to testosterone, the hypothalamic centers will respond appropriately when the individual becomes sexually mature. The factors responsible for regulating the fetal production of testosterone are not known.

Testosterone levels are low at birth, but from birth to puberty, background testosterone levels are higher in males than in females. Testosterone secretion accelerates markedly at puberty, initiating sexual maturation and the appearance of secondary sexual characteristics. In the adult, the level of testosterone production is controlled by negative feedback. Above-normal testosterone levels inhibit the release of GnRH by the hypothalamus. This inhibition causes a reduction in LH secretion and lowers testosterone levels.

 On a warm day would the cremaster muscle be contracted or relaxed? Why?

 How would the lack of an acrosomal cap affect the ability of a spermatozoon to fertilize an ovum?

 What will occur if the arteries serving the penis dilate?

 What effect would low levels of FSH have on sperm production?

The Reproductive System of the Female

Figures 28-12, 28-13

A woman's reproductive system must produce sex hormones and functional gametes and also protect and support a developing embryo and nourish the newborn infant. The principal organs of the female reproductive system (Figure 28-12•) include the *ovaries*, the *uterine tubes* (*Fallopian tubes* or *oviducts*), the *uterus*, the *vagina*, and the components of the external genitalia. As in the male, a variety of accessory glands secrete into the reproductive tract.

The ovaries, uterine tubes, and uterus are enclosed within an extensive mesentery known as the **broad ligament** (Figure 28-13•). The uterine tubes run along the superior border of the broad ligament and open into the pelvic cavity lateral to the ovaries. A thickened fold of mesentery, the **mesovarium** (mes-ō-VAR-ē-um), supports and stabilizes the position of each ovary. The broad ligament attaches to the sides and floor of the pelvic cavity, where it becomes continuous with the parietal peritoneum. The broad ligament thus subdivides the pelvic cavity. The pocket formed between the posterior wall of the uterus and the anterior surface of the rectum is the **rectouterine** (rek-tō-Ū-te-rin) **pouch;** that between the uterus and the posterior wall of the bladder is the **vesicouterine** (ves-i-kō-Ū-ter-in) **pouch.** These subdivisions are most apparent in sagittal section (Figure 28-12•).

Several other ligaments assist the broad ligament in supporting and stabilizing the position of the uterus and associated reproductive organs. These ligaments travel within the mesentery sheet of the broad ligament on the way to the ovaries or uterus. The broad ligament limits side-to-side movement and rotation, and the other ligaments (described with the ovaries and uterus) prevent superior-inferior movement.

The Ovaries

Figures 28-12, 28-13

The ovaries are small, almond-shaped organs located near the lateral walls of the pelvic cavity (Figures 28-12 and 28-13•). The ovaries are responsible for (1) the production of female gametes, or *ova*, (2) the secretion of female sex hormones, including *estrogens* and *progestins*, and (3) the secretion of inhibin, involved in the feedback control of pituitary FSH production.

The position of each ovary is stabilized by the mesovarium and by a pair of supporting ligaments: the *ovarian ligament* and the *suspensory ligament*. The **ovarian ligament** extends from the uterus, near the attachment of the uterine tube, to the medial surface of the ovary. The **suspensory lig-**

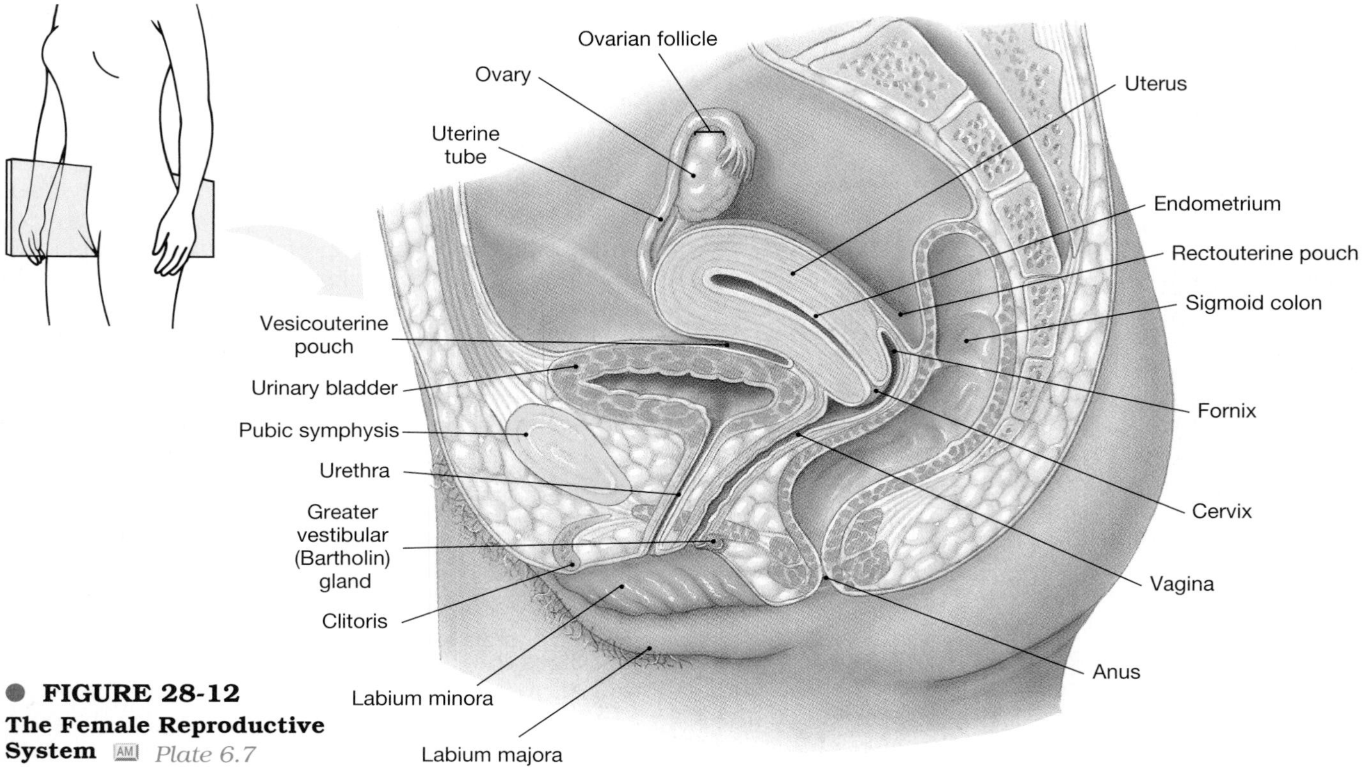

Ovarian follicle
Ovary
Uterine tube
Uterus
Endometrium
Rectouterine pouch
Sigmoid colon
Vesicouterine pouch
Urinary bladder
Pubic symphysis
Urethra
Greater vestibular (Bartholin) gland
Clitoris
Fornix
Cervix
Vagina
Anus
Labium minora
Labium majora

● **FIGURE 28-12**
The Female Reproductive System *Plate 6.7*

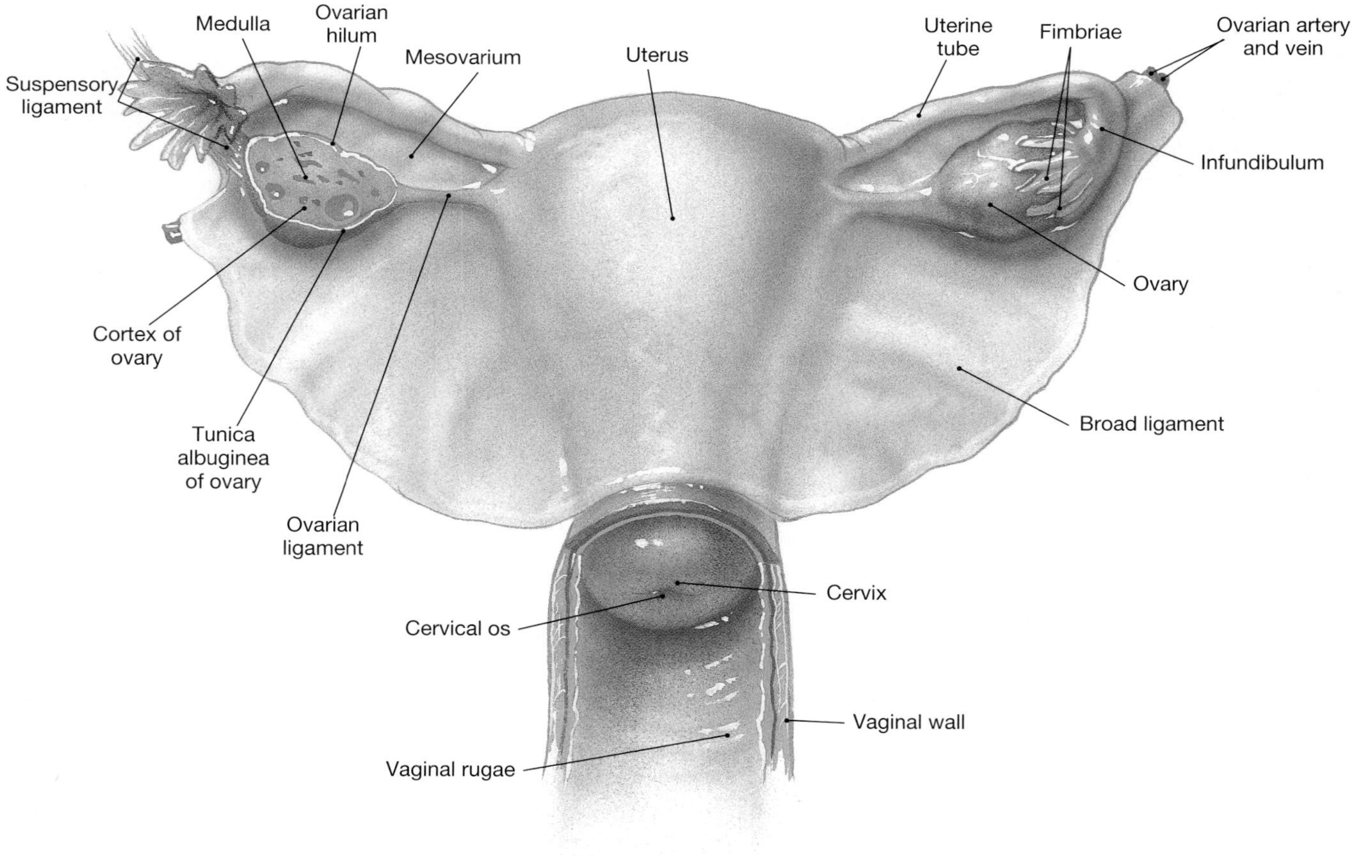

Medulla
Ovarian hilum
Mesovarium
Uterus
Uterine tube
Fimbriae
Ovarian artery and vein
Suspensory ligament
Infundibulum
Ovary
Cortex of ovary
Tunica albuginea of ovary
Ovarian ligament
Broad ligament
Cervix
Cervical os
Vaginal wall
Vaginal rugae

● **FIGURE 28-13**
The Ovaries and Their Supporting Ligaments. Anterior view of the uterus, uterine tubes, and ovaries.

ament extends from the lateral surface of the ovary past the open end of the uterine tube to the pelvic wall. The major blood vessels, the **ovarian artery** and **vein,** travel to and from the ovary within the suspensory ligament. These vessels are connected to the ovary at the **ovarian hilum,** where the ovary attaches to the mesovarium.

A typical ovary measures approximately 5 c × 2.5 cm (2 in. × 1 in.) and weighs 6–8 g (roug 0.25 oz). It has a pink or yellowish coloratio a nodular consistency. The visceral perit or *germinal epithelium,* covering the surface o ovary overlies a layer of dense connective tiss called the **tunica albuginea.** The interior tissues, or **stroma,** of the ovary, can be divided into a superficial *cortex* and a deeper *medulla.* The production of gametes occurs in the cortex.

Oogenesis

Figure 28-14

Ovum production, or **oogenesis** (ō-o-JEN-e-sis), begins before birth, accelerates at puberty, and ends at *menopause.* Between puberty and menopause, oogenesis occurs on a monthly basis, as part of the *ovarian cycle.*

Unlike the situation in the male gonads, the stem cells, or **oogonia** (ō-ō-GŌ-nē-a), of the female complete their mitotic divisions before birth. Between the third and seventh months of fetal development, the daughter cells, or **primary oocytes** (Ō-ō-sīts), prepare to undergo meiosis. They proceed as far as prophase of meiosis I, but at that time the process comes to a halt. The primary oocytes then remain in a state of suspended development until puberty. At that time, rising levels of FSH trigger the start of the ovarian cycle, and each month thereafter some of the primary oocytes will be stimulated to undergo further development. Not all of the primary oocytes produced in development survive until puberty. There are roughly 2 million primordial follicles in the ovaries at birth; by the time of puberty that number has declined to around 400,000. The rest of the primary oocytes degenerate, a process called **atresia** (a-TRĒ-zē-a).

Although the nuclear events under way during meiosis in the ovary are the same as those in the testes, the process differs in two important details:

1. The cytoplasm of the primary oocyte is not evenly distributed during the two meiotic divisions. Oogenesis produces one functional ovum, which contains most of the original cytoplasm, and three nonfunctional *polar bodies* that later disintegrate (Figure 28-14●).
2. The ovary releases a **secondary oocyte** rather than a mature ovum. The secondary oocyte is suspended in metaphase of meiosis II, and meiosis will not be completed unless and until fertilization occurs.

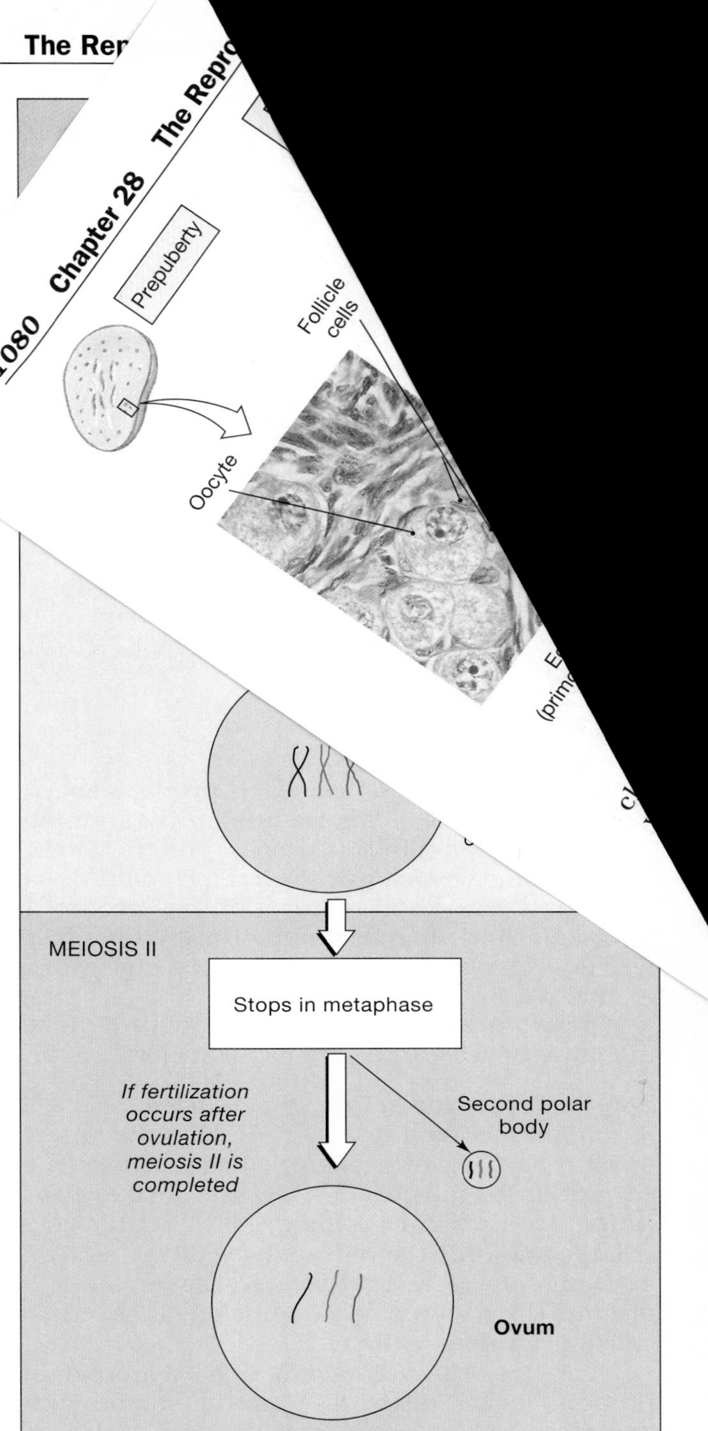
MEIOSIS II
Stops in metaphase
If fertilization occurs after ovulation, meiosis II is completed
Second polar body
Ovum

● **FIGURE 28-14**

Oogenesis. In oogenesis a single primary oocyte produces an ovum and three nonfunctional polar bodies. Compare with Figure 28-6, p. 1066.

The Ovarian Cycle

Figure 28-15

Ovarian follicles are specialized structures where oocyte growth and meiosis I occur. The ovarian follicles are located in the cortex of the ovaries. Primary oocytes are found in the outer portion of the ovarian cortex near the tunica albuginea, in

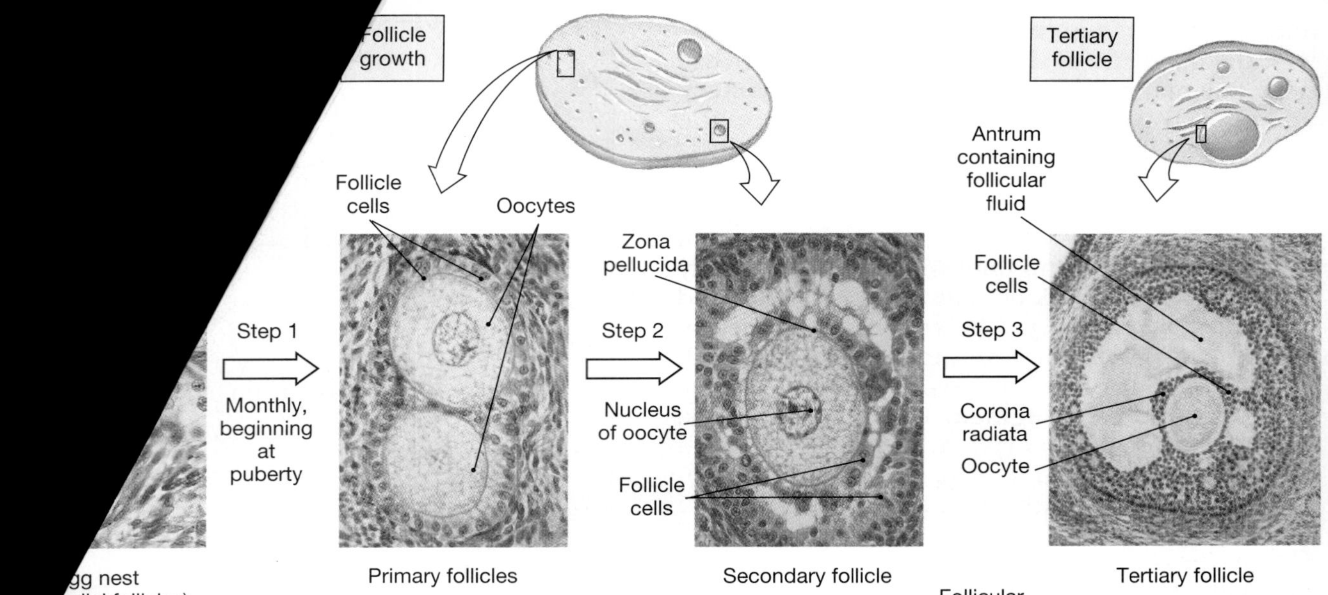

Follicle growth
Tertiary follicle
Follicle cells
Oocytes
Antrum containing follicular fluid
Zona pellucida
Follicle cells
Step 1
Monthly, beginning at puberty
Step 2
Step 3
Nucleus of oocyte
Corona radiata
Oocyte
Follicle cells
gg nest
rdial follicles)
Primary follicles
Secondary follicle
Tertiary follicle
Follicular fluid in antrum
Zona pellucida
Corona radiata
Nucleus of oocyte
Follicle cells
Oocyte in tertiary follicle

FIGURE 28-15
The Ovarian Cycle

usters called *egg nests.* Each primary oocyte within an egg nest is surrounded by a simple squamous layer of follicular cells, and the combination is known as a **primordial** (prī-MŌR-dē-al) **follicle.** After sexual maturation has occurred, each month a different group of primordial follicles is activated. This monthly process is known as the **ovarian cycle.**

Important steps in the ovarian cycle are shown in Figure 28-15● and summarized below.

Step 1: ***Formation of Primary Follicles.*** Follicle formation is stimulated by FSH from the anterior pituitary. The ovarian cycle begins as activated primordial follicles develop into **primary follicles.** In a primary follicle the follicular cells enlarge and undergo repeated cell divisions. This division creates several layers of follicular cells around the oocyte. These follicle cells are now called **granulosa cells.**

As layers of granulosa cells develop around the primary oocyte, microvilli from the surrounding granulosa cells intermingle with those of the primary oocyte. The microvilli are surrounded by a layer of glycoproteins, and the entire region is called the **zona pellucida** (ZŌ-na pel-LŪ-si-da). The microvilli increase the surface area available for the transfer of materials from the granulosa cells to the rapidly enlarging oocyte.

The conversion from primordial to primary follicles and subsequent follicular development occurs under FSH stimulation. As the granulosa cells enlarge and multiply, adjacent cells in the ovarian stroma form a layer of **thecal cells** around the follicle. Thecal cells and granulosa cells work together to produce female sex hormones called *estrogens.*

Step 2: ***Formation of Secondary Follicles.*** Although many primordial follicles develop into primary follicles, only a few will proceed to the next step. The transformation begins as the wall of the follicle thickens and the deeper follicular cells begin secreting small amounts of fluid. This **follicular fluid,** or *liquor folliculi,* accumulates in small pockets that gradually expand and separate the inner and outer layers of the follicle. At this stage the complex is known as a **secondary follicle.** Although the primary oocyte continues to grow at a steady pace, the follicle as a whole now enlarges rapidly because of the accumulation of follicular fluid.

Step 3: ***Formation of a Tertiary Follicle.*** Eight to ten days after the start of the ovarian cycle, the ovaries usually contain only a single secondary follicle destined for further development. By the tenth to fourteenth day of the cycle it has formed a **tertiary follicle,** or *mature Graafian* (GRAF-ē-an) *follicle,* roughly 15 mm in diameter. This complex spans the entire width of the ovarian cortex and distorts the ovarian capsule, creating a prominent bulge in the surface of the ovary. The oocyte projects into the expanded central chamber, or **antrum** (AN-trum), surrounded by a mass of granulosa cells.

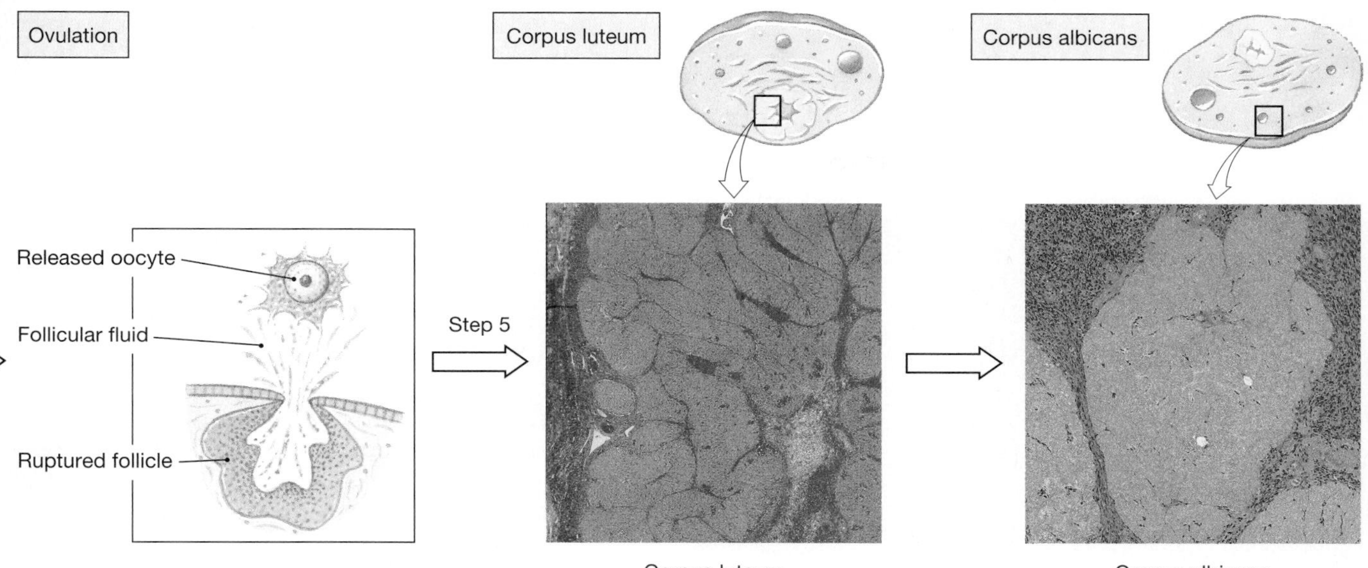
Ovulation
Corpus luteum
Corpus albicans
Released oocyte
Follicular fluid
Ruptured follicle
Step 5
Corpus luteum
Corpus albicans

Until this time, the primary oocyte has been suspended in prophase of the first meiotic division, a state it achieved before birth. As the tertiary follicle completes its development, LH levels begin rising and the primary oocyte now completes meiosis I. Instead of producing two secondary oocytes, the first meiotic division yields a secondary oocyte and a small, nonfunctional polar body. The secondary oocyte then enters meiosis II, but stops once again upon reaching metaphase. Meiosis II will not be completed unless fertilization occurs.

Step 4: ***Ovulation.*** At **ovulation** the tertiary follicle releases the secondary oocyte. As the time of ovulation approaches, the secondary oocyte and the surrounding granulosa cells lose their connections with the follicular wall and drift free within the antrum. This event usually occurs on day 14 of a 28-day cycle. The granulosa cells surrounding the secondary oocyte are now known as the **corona radiata** (ko-RŌ-na rā-dē-A-ta). At ovulation the distended follicular wall ruptures, releasing the follicular contents, including the secondary oocyte and corona radiata, into the pelvic cavity. The sticky follicular fluid usually keeps the corona radiata attached to the surface of the ovary, where direct contact with the fimbriae or fluid currents established by their ciliated epithelium can transfer the secondary oocyte to the uterine tube.

Step 5: ***Formation and Degeneration of the Corpus Luteum.*** The empty tertiary follicle initially collapses, and ruptured vessels bleed into the atrial chamber. The remaining granulosa cells then invade the area, proliferating to create an endocrine structure known as the **corpus luteum** (LŪ-tē-um), named for its yellow color (*lutea*, yellow). This process occurs under LH stimulation.

The lipids contained in the corpus luteum are used to manufacture steroid hormones known as **progestins** (prō-JES-tinz), principally the steroid **progesterone** (prō-JES-ter-ōn). Although moderate amounts of estrogens are also secreted by the corpus luteum, progesterone is the principal hormone of the postovulatory period. Its primary function is to prepare the uterus for pregnancy by stimulating the growth of the uterine lining and the secretion of uterine glands.

Unless pregnancy occurs, the corpus luteum begins to degenerate roughly 12 days after ovulation. Progesterone and estrogen levels then fall markedly. Fibroblasts invade the nonfunctional corpus luteum, producing a knot of pale scar tissue called a **corpus albicans** (AL-bi-kanz). The disintegration, or *involution*, of the corpus luteum marks the end of the ovarian cycle. A new cycle then begins with the activation of another group of primordial follicles.

Age and Oogenesis

Although many primordial follicles may have developed into primary follicles, and several primary follicles converted to secondary follicles, usually only a single oocyte will be released into the pelvic cavity at ovulation. The rest undergo atresia. At puberty there are approximately 200,000 primordial follicles in each ovary. Forty years later few if any follicles remain, although only around 500 will have been ovulated over the interim. AM *Ovarian Cancer*

❑ The Uterine Tubes

Figures 28-12, 28-13, 28-16

Each uterine tube is a hollow, muscular tube measuring roughly 13 cm (5 in.) in length (Figures 28-12•, sagittal view; 28-13 and 28-16•, anterior view. Each uterine tube is divided into three regions:

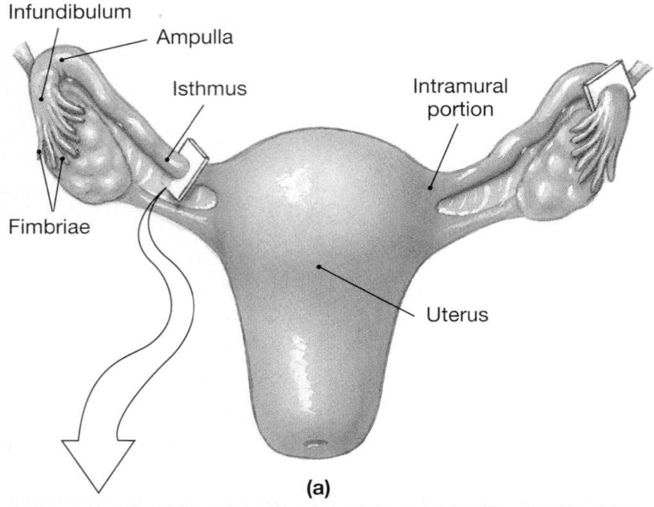
Infundibulum
Ampulla
Isthmus
Intramural portion
Fimbriae
Uterus
(a)

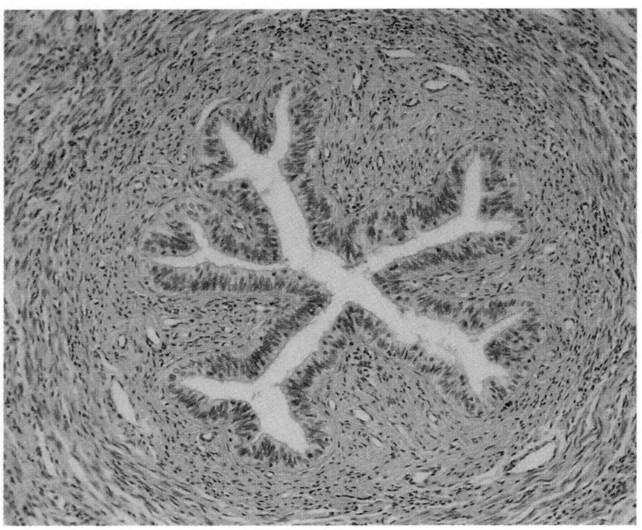
(b)

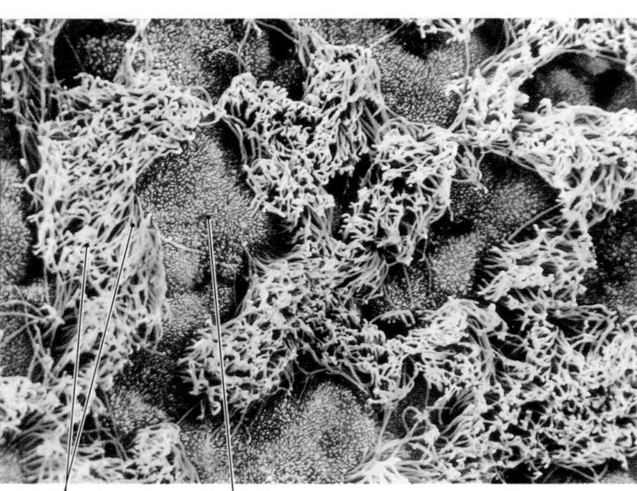
Cilia
Microvilli on mucous cells
(c)

● **FIGURE 28-16**

The Uterine Tubes. (a) Regions of the uterine tubes. **(b)** A sectional view of the isthmus. (LM × 122) **(c)** The ciliated lining of the uterine tube.

1. ***The infundibulum.*** The end closest to the ovary forms an expanded funnel, or **infundibulum,** with numerous fingerlike projections that extend into the pelvic cavity. The projections are called **fimbriae** (FIM-brē-ē). The inner surfaces of the infundibulum are lined with cilia that beat toward the entrance to the expanded initial segment of the uterine tube, the *ampulla.*
2. ***The ampulla.*** As the uterine tube works its way to the uterus the diameter becomes enlarged. This enlarged or expanded middle portion is the **ampulla.**
3. ***The isthmus.*** The enlarged ampulla leads to the **isthmus** (IS-mus), a short segment adjacent to the uterine wall. The **intramural** (in-tra-MŪ-ral) **portion** of the uterine tube passes through the wall of the uterus and opens into the uterine cavity.

Histology of the Uterine Tube

Figure 28-16

The epithelium lining the uterine tube is composed of ciliated columnar epithelial cells with scattered mucus-secreting cells (Figure 28-16c●). The mucosa is surrounded by concentric layers of smooth muscle (Figure 28-16b●). Ovum transport involves a combination of ciliary movement and peristaltic contractions in the walls of the uterine tube. A few hours before ovulation, sympathetic and parasympathetic nerves from the hypogastric plexus "turn on" this beating pattern. The uterine tubes transport a mature oocyte for final maturation and fertilization. It normally takes 3–4 days for an ovum to travel from the infundibulum to the uterine chamber. *If fertilization is to occur, the ovum must encounter spermatozoa during the first 12–24 hours of its passage.* Fertilization typically occurs near the boundary between the ampulla and isthmus of the uterine tube.

Along with its transport function, the uterine tube also provides a rich, nutritive environment containing lipids and glycogen. This mixture provides nutrients to both spermatozoa and a developing pre-embryo. Unfertilized eggs degenerate in the terminal portions of the uterine tubes or within the uterus.

PELVIC INFLAMMATORY DISEASE (PID)

Pelvic inflammatory disease (PID) is a major cause of sterility in women. This condition, an infection of the uterine tubes, affects an estimated 850,000 women each year in the United States alone. Sexually transmitted pathogens are often involved, and as many as 50–80 percent of all first cases may be due to infection by *Neisseria gonorrhoeae,* the organism responsible for symptoms of **gonorrhea** (gon-ō-RĒ-a), a sexually transmitted disease. PID may also result from invasion of the region by bacteria normally found within the vagina. Symptoms of pelvic inflammatory disease include fever, lower abdominal pain, and elevated white blood cell counts. In severe cases the infection may spread to other visceral organs or produce a generalized peritonitis.

Sexually active women in the 15–24 age group have the highest incidence of PID. Although oral contraceptive use decreases the risk of infection, the presence of an intrauterine device (IUD) may increase the risk by 1.4–7.3 times. Treatment with antibiotics may control the condition, but chronic abdominal pain may persist. In addition, damage and scarring of the uterine tubes may cause infertility by preventing the passage of a zygote to the uterus. Recently, another sexually transmitted bacterium, *Chlamydia,* has been identified as the probable cause of up to 50 percent of all cases of PID. Despite the fact that women with this infection may develop few if any symptoms, scarring of the uterine tubes may still produce infertility.

❑ The Uterus

Figure 28-12

The uterus provides mechanical protection, nutritional support, and waste removal for the developing embryo (weeks 1–8) and fetus (from week 9 to delivery). In addition, contractions in the muscular wall of the uterus are important in ejecting the fetus at the time of birth.

The uterus is a small, pear-shaped organ about 7.5 cm (3 in.) long with a maximum diameter of 5 cm (2 in.). It weighs 30–40 g (1–1.4 oz). In its normal position the uterus bends anteriorly near its base, a condition known as **anteflexion** (an-tē-FLEK-shun). In this position the body of the uterus lies across the superior and posterior surfaces of the urinary bladder (Figure 28-12•, p. 1078). If instead the uterus bends backward toward the sacrum, the condition is termed **retroflexion** (re-trō-FLEK-shun). Retroflexion, which occurs in about 20 percent of adult women, has no clinical significance. (A retroflexed uterus usually becomes anteflexed spontaneously during the third month of pregnancy.)

Suspensory Ligaments of the Uterus

Figure 28-17a

In addition to the mesenteric sheet of the broad ligament, three pairs of suspensory ligaments stabilize the position of the uterus and limit its range of movement (Figure 28-17a•). The **uterosacral** (ū-te-rō-SĀ-kral) **ligaments** extend from the lateral surfaces of the uterus to the anterior face of the sacrum, keeping the body of the uterus from moving down and forward. The **round ligaments** arise on the lateral margins of the uterus just below the uterine tubes. They extend anteriorly, passing through the inguinal canal before ending in the connective tissues of the external genitalia. These ligaments restrict primarily posterior movement of the uterus. The **lateral** (*cardinal*) **ligaments** extend from the base of the uterus and vagina to the lateral walls of the pelvis. These ligaments also tend to prevent the downward movement of the uterus. Additional mechanical support is provided by the skeletal muscles and fascia of the pelvic floor. ∞ *[pp. 347–349]*

Internal Anatomy of the Uterus

Figure 28-17c

The uterus is divided into two anatomical regions (Figure 28-17c•): the *body* and the *cervix.* The uterine **body,** or *corpus,* is the largest division of the uterus. The **fundus** is the rounded portion of the body superior to the attachment of the uterine tubes. The body ends at a constriction known as the uterine **isthmus.** The **cervix** (SER-viks) is the inferior portion of the uterus that extends from the isthmus to the vagina.

The tubular cervix projects about 1.25 cm (0.5 in.) into the vagina. Within the vagina, the distal end of the cervix forms a curving surface surrounding the **external orifice** of the uterus, also known as the *cervical os* (*os,* an opening or mouth). The cervical os leads into the *cervical canal,* a constricted passageway that opens into the **uterine cavity** of the body at the **internal orifice,** or *internal os* (Figure 28-17c•).

Each region of the uterus is supplied with branches of the uterine and ovarian arteries and veins. Numerous lymphatic vessels also supply each portion of the uterus. The uterus is innervated by autonomic fibers from the hypogastric plexus and sacral segments S_3 and S_4. Sensory afferents from the uterus enter the spinal cord via the eleventh and twelfth thoracic nerves.

The Uterine Wall

Figures 28-17c, 28-18

The dimensions of the uterus are highly variable. In adult women of reproductive age who have not borne children, the uterine wall is about 1.5 cm (0.5 in.) thick. The uterine wall has an outer, muscular **myometrium** (mī-ō-MĒ-trē-um; *myo-,* muscle + *metra,* uterus) and an inner glandular **endometrium** (en-dō-MĒ-trē-um), or *mucosa.* The fundus and posterior surface of the uterine body are covered by a serous membrane continuous with the peritoneal lining. This incomplete serosal layer is called the **perimetrium** (Figure 28-17c•).

The endometrium contributes about 10 percent to the mass of the uterus. The glandular and vascular tissues of the endometrium support the physiological demands of the growing fetus. Vast numbers of uterine glands open onto the endometrial surface. These glands extend deep into the lamina propria almost all the way to the myometrium. Under the influence of estrogen the uterine glands, blood vessels, and endothelium change with the various phases of the monthly *uterine cycle.*

The myometrium is the thickest portion of the uterine wall, and it forms almost 90 percent of the mass of the uterus. Smooth muscle in the myometrium is arranged into longitudinal, circular, and oblique layers. The smooth muscle tissue of the myometrium provides much of the force needed to move a large fetus out of the uterus and into the vagina.

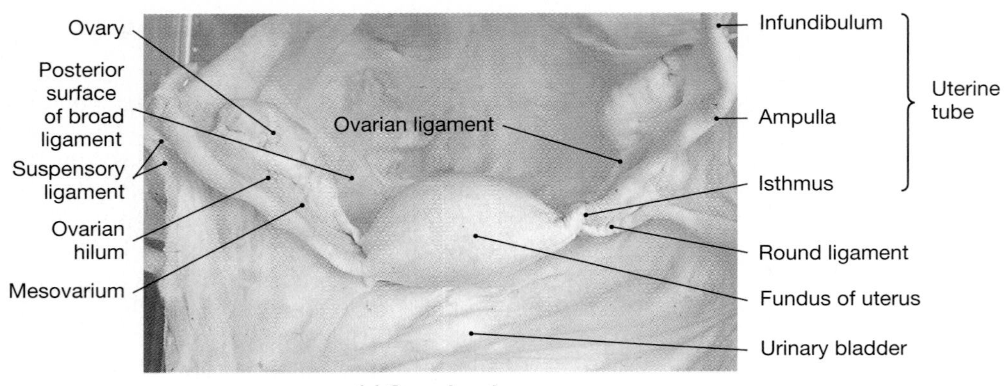
Ovary
Posterior surface of broad ligament
Suspensory ligament
Ovarian hilum
Mesovarium
Ovarian ligament
Infundibulum
Ampulla
Isthmus
Uterine tube
Round ligament
Fundus of uterus
Urinary bladder

(a) Superior view

FIGURE 28-17

The Uterus. (a) The pelvic cavity as seen from above, showing the position of the uterus relative to other structures. **(b)** A diagrammatic view showing ligaments that stabilize the position of the uterus in the pelvic cavity. **(c)** Anterior view with right uterus, uterine tube, and ovary shown in section. For histological details, see Figures 28-18 and 28-19.

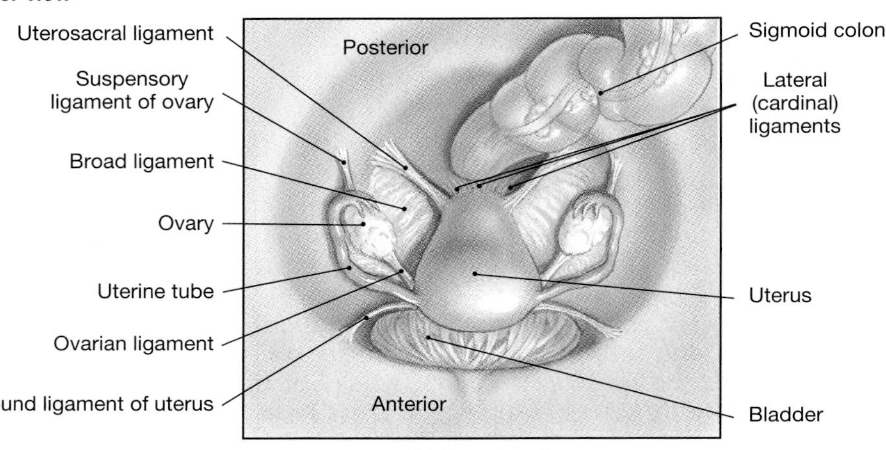
Uterosacral ligament
Suspensory ligament of ovary
Broad ligament
Ovary
Uterine tube
Ovarian ligament
Round ligament of uterus
Posterior
Anterior
Sigmoid colon
Lateral (cardinal) ligaments
Uterus
Bladder

(b) Superior view

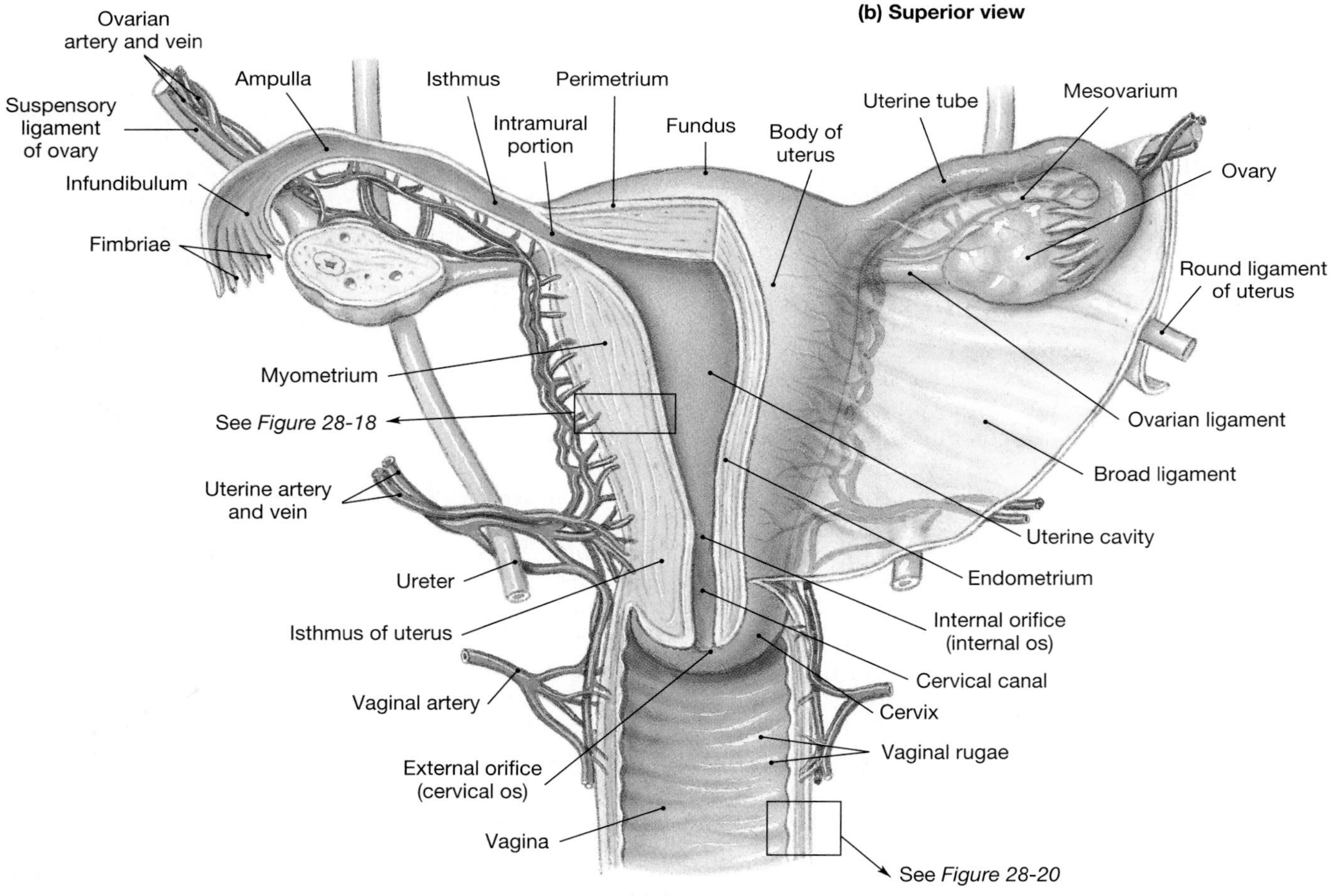

Ovarian artery and vein
Suspensory ligament of ovary
Infundibulum
Fimbriae
Ampulla
Isthmus
Intramural portion
Perimetrium
Fundus
Body of uterus
Uterine tube
Mesovarium
Ovary
Round ligament of uterus
Myometrium
See *Figure 28-18*
Ovarian ligament
Broad ligament
Uterine artery and vein
Uterine cavity
Endometrium
Ureter
Isthmus of uterus
Internal orifice (internal os)
Cervical canal
Vaginal artery
Cervix
Vaginal rugae
External orifice (cervical os)
Vagina
See *Figure 28-20*

(c) Anterior view

BLOOD SUPPLY TO THE UTERUS The uterus receives blood from branches of the **uterine arteries,** which arise from branches of the *internal iliac arteries,* and the *ovarian arteries,* which arise from the abdominal aorta inferior to the renal arteries. There are extensive interconnections among the arteries to the uterus. This arrangement helps ensure a reliable flow of blood to the organ, despite changes in position and the changes in uterine shape that accompany pregnancy.

INNERVATION OF THE UTERUS The uterus is innervated by autonomic nerves from the *hypogastric plexus* (sympathetic) and from the third and fourth sacral nerves (parasympathetic). Sensory innervation reaches the CNS within the dorsal roots of spinal nerves T_{11} and T_{12}. The most delicate anesthetic procedures used during labor and delivery, known as *segmental blocks,* target only spinal nerves T_{10}–L_1.

CERVICAL CANCER **Cervical cancer** is the most common reproductive system cancer in women age 15–34. Roughly 13,000 new cases of invasive cervical cancer are diagnosed each year in the United States, and approximately 40 percent of them will eventually die of this condition. Another 50,000 patients are diagnosed with a less aggressive form of cervical cancer. Cervical and other uterine tumors and cancers are discussed in the *Applications Manual.* AM *Uterine Tumors and Cancers*

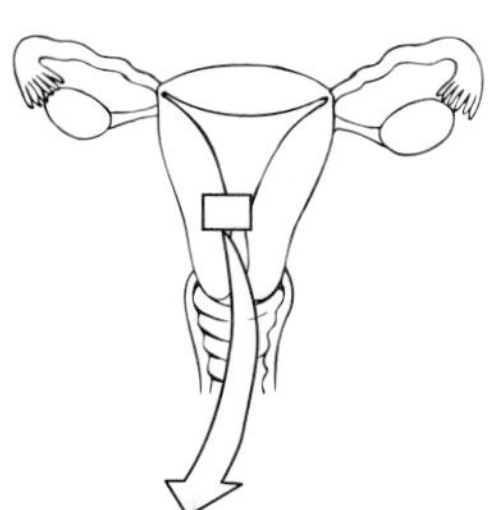

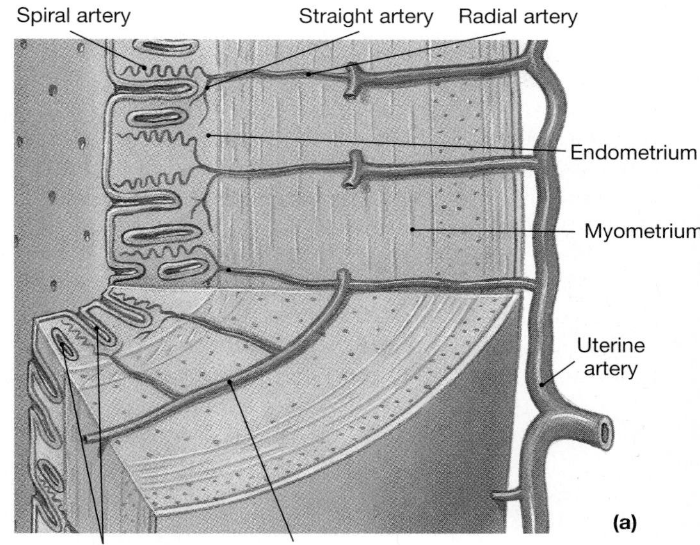
Spiral artery
Straight artery
Radial artery
Endometrium
Myometrium
Uterine artery
(a)
Uterine glands
Arcuate artery

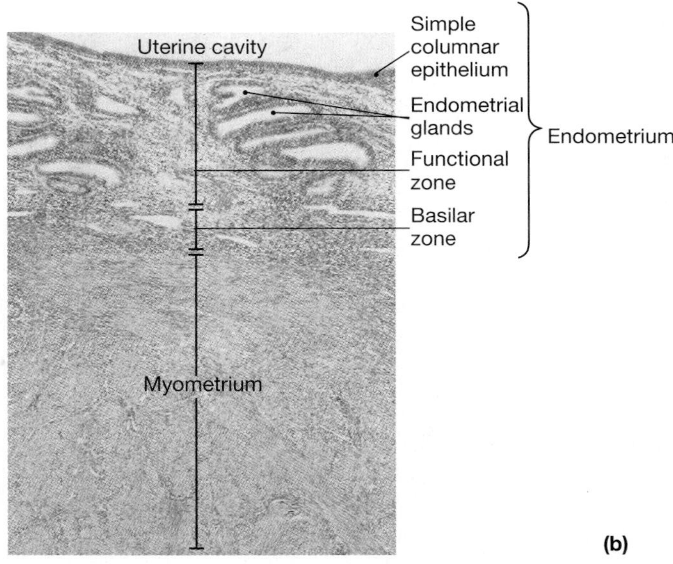
Uterine cavity
Simple columnar epithelium
Endometrial glands
Functional zone
Basilar zone
Endometrium
Myometrium
(b)

● **FIGURE 28-18**
The Uterine Wall.
(a) A sectional view of the uterine wall, showing the endometrial regions and the circulatory supply to the endometrium. **(b)** The basic structure of the endometrium. (LM × 32)

HISTOLOGY OF THE UTERUS The endometrium can be divided into a **functional zone,** the layer closest to the uterine cavity, and an outer **basilar zone** adjacent to the myometrium. The functional zone contains most of the uterine glands and contributes most of the endometrial thickness. The basilar zone attaches the endometrium to the myometrium and contains the terminal branches of the tubular glands (Figure 28-18a●).

Within the myometrium, branches of the uterine arteries form **arcuate arteries** that encircle the endometrium. **Radial arteries** supply **straight arteries** that deliver blood to the basilar zone of the endometrium, and **spiral arteries** that supply the functional zone (Figure 28-18b●).

The structure of the basilar zone remains relatively constant over time, but that of the functional zone undergoes cyclical changes in response to sexual hormone levels. These alterations produce the characteristic histological features of the *uterine cycle.*

The Uterine Cycle

Figure 28-19

The **uterine cycle,** or *menstrual* (MEN-strū-al) *cycle,* is a repeating series of changes in the structure of the endometrium. The uterine cycle averages 28 days in length, but it can range from 21 to 35 days in normal individuals. The cycle can be divided into three phases: (1) *menses,* (2) the *proliferative phase,* and (3) the *secretory phase.* The histological appearance of the endometrium during each phase is shown in Figure 28-19●. The phases occur in response to hormones associated with the regulation of the ovarian cycle, and the regulatory mechanism will be considered in a later section.

MENSES The uterine cycle begins with the onset of **menses** (MEN-sēz), a period marked by the degeneration of the functional zone of the endometrium. The deterioration occurs in patches. It is caused by the constriction of coiled arteries, which reduces blood flow to areas of endometrium. Deprived of oxygen and nutrients, the secretory glands and other tissues in the functional zone begin to deteriorate. Eventually the weakened arterial walls rupture, and blood pours into the connective tissues of the functional zone. Blood cells and degenerating tissues then break away and enter the uterine lumen, to be lost by passage through the external orifice and into the vagina. Only the functional zone is affected, because the deeper, basilar zone is provided with blood from the straight arteries, which remain unconstricted.

The sloughing of tissue is gradual, and at each site repairs begin almost at once. Nevertheless, before menses has ended the entire functional zone has been lost (Figure 28-19a●). This process of endometrial sloughing is called **menstruation** (men-stru-Ā-shun). Menstruation usually lasts from 1 to 7 days, and over this period roughly 35 to 50 ml of blood are lost. The process is often relatively painless. Painful menstruation, or **dysmenorrhea,** may result from uterine inflammation and contraction or from conditions involving adjacent pelvic structures.

THE PROLIFERATIVE PHASE The basilar zone, including the basal portions of the uterine glands, survives menses intact. In the days following the completion of menses, the epithelial cells of the glands multiply and spread across the endometrial surface, restoring the integrity of the uterine epithelium (Figure 28-19b●). Further growth and vascularization will result in the complete restoration of the functional zone. As this reorganization proceeds, the endometrium is said to be in the **proliferative phase.** Because this restoration occurs at the same time as the enlargement of primary and secondary follicles in the ovary, it is also known as the *preovulatory phase* or *follicular phase* of the uterine cycle. The proliferative phase is stimulated and sustained by estrogens secreted by the developing ovarian follicles.

By the time ovulation occurs, the functional zone is several millimeters thick, and prominent mucous glands extend to the border with the basilar zone. At this time the endometrial glands are manufacturing a mucus rich in glycogen. The entire functional zone is highly vascularized, with small arteries spiraling toward the inner surface from larger trunks in the myometrium.

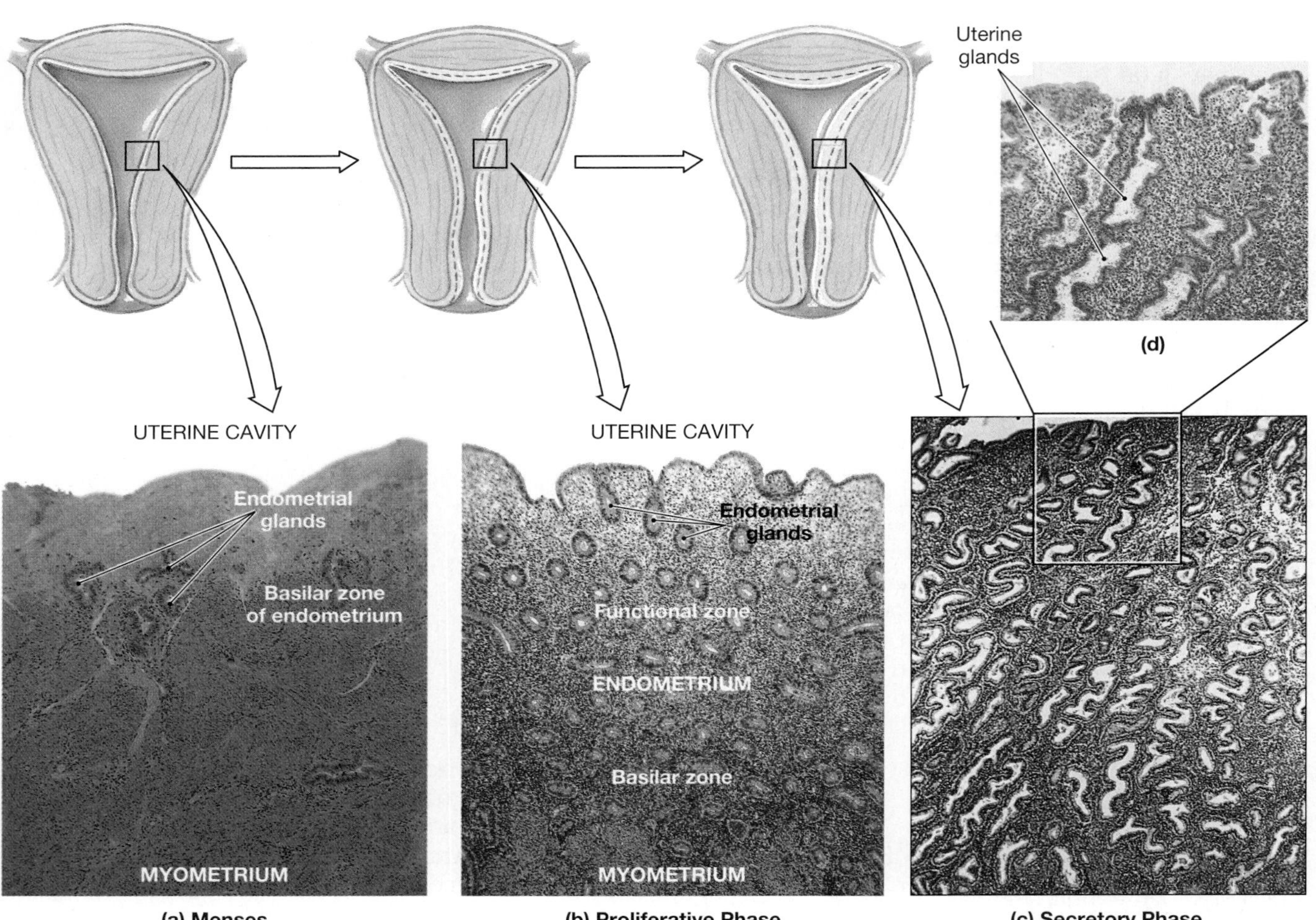
Uterine glands
(d)
UTERINE CAVITY
Endometrial glands
Basilar zone of endometrium
MYOMETRIUM
UTERINE CAVITY
Endometrial glands
Functional zone
ENDOMETRIUM
Basilar zone
MYOMETRIUM
(a) Menses
(b) Proliferative Phase
(c) Secretory Phase

● **FIGURE 28-19**

The Uterine Cycle. These micrographs show the appearance of the endometrium at menses and during the proliferative and secretory phases of the uterine cycle. [LMs: (a) × 63, (b) × 66, (c) × 52, and (d) × 50]

THE SECRETORY PHASE During the secretory phase of the menstrual cycle the endometrial glands enlarge, accelerating their rates of secretion, and the arteries elongate and spiral through the tissues of the functional zone (Figure 28-19c●). This activity occurs under the combined stimulatory effects of progestins and estrogens from the corpus luteum. Because this phase begins at the time of ovulation, and it persists as long as the corpus luteum remains intact, it can also be called the *postovulatory phase*, or *luteal phase*, of the uterine cycle.

Secretory activities peak about 12 days after ovulation. Over the next day or two the glandular activity declines, and the menstrual cycle comes to a close as the corpus luteum stops producing stimulatory hormones. A new cycle then begins with the onset of menses and the disintegration of the functional zone. The secretory phase usually lasts 14 days. As a result, the date of ovulation can be determined after the fact by counting backward 14 days from the first day of menses.

ENDOMETRIOSIS In *endometriosis* (en-dō-mē-trē-Ō-sis) an area of endometrial tissue begins to grow outside the uterus. The severity of the condition depends on the size of the abnormal mass and its location. Abdominal pain, bleeding, pressure on adjacent structures, and infertility are common symptoms. As the island of endometrial tissue enlarges, the symptoms become more severe.

Diagnosis can usually be made by using a laparoscope inserted through a small opening in the abdominal wall. Using this device a physician can inspect the outer surfaces of the uterus and uterine tubes, the ovaries, and the lining of the pelvic cavity. Treatment of endometriosis may involve hormonal therapy or surgical removal of the endometrial mass. If the condition is widespread, a hysterectomy or *oophorectomy* (removal of the ovaries) may be required.

MENARCHE AND MENOPAUSE The menstrual cycle of events begins with the **menarche** (me-NAR-kē), or first menstrual period at puberty, typically at age 11–12. The cycles continue until age 45–50 when **menopause** (MEN-ō-paws), the last menstrual cycle, occurs. Over the intervening three and a half to four decades the regular appearance of menstrual cycles will be interrupted only by unusual circumstances, such as illness, stress, starvation, or pregnancy.

AMENORRHEA If menarche does not appear by age 16, or if the normal menstrual cycle of an adult becomes interrupted for 6 months or more, the condition of **amenorrhea** (ā-men-ō-RĒ-a) exists. *Primary amenorrhea* is the failure to initiate menses. This condition may indicate developmental abnormalities, such as nonfunctional ovaries or even the absence of a uterus, or an endocrine or genetic disorder. Transient *secondary amenorrhea* may be caused by severe physical or emotional stresses. In effect, the reproductive system gets "switched off" under these conditions. Examples of factors that can cause amenorrhea include drastic weight reduction programs, anorexia nervosa, and severe depression or grief. Amenorrhea has also been observed in marathon runners and others engaged in training programs that require sustained high levels of exertion and severely reduce body lipid reserves.

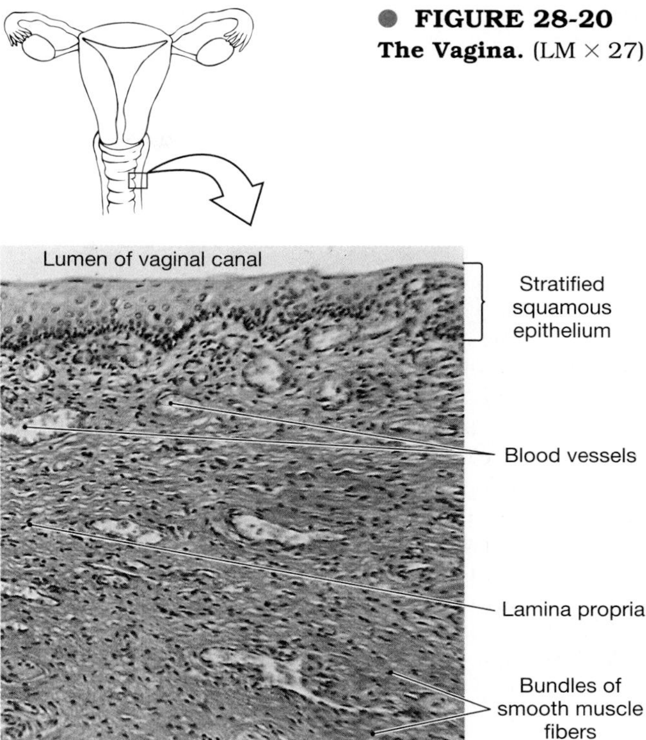
Lumen of vaginal canal
Stratified squamous epithelium
Blood vessels
Lamina propria
Bundles of smooth muscle fibers

● **FIGURE 28-20**
The Vagina. (LM × 27)

The Vagina

Figures 28-12, 28-13, 28-20

The **vagina** is an elastic, muscular tube extending between the cervix of the uterus and the vaginal opening into the *vestibule* (Figures 28-12 and 28-13, p. 1078, and 28-20●). The vagina has an average length of 7.5–9 cm (3–3.5 in.), but because the vagina is highly distensible, its length and width are quite variable.

At the proximal end of the vagina, the cervix projects into the **vaginal canal.** The shallow recess surrounding the cervical protrusion is known as the **fornix** (FŌR-niks). The vagina lies parallel to the rectum, and the two are in close contact posteriorly. Anteriorly, the urethra travels along the superior wall of the vagina as it goes from the urinary bladder to its opening into the vestibule. The primary blood supply of the vagina is via the **vaginal branches** of the internal iliac (or uterine) arteries and veins. Innervation is from the hypogastric plexus, sacral nerves S_2–S_4, and branches of the pudendal nerve.

The vagina has three major functions:

1. It serves as a passageway for the elimination of menstrual fluids.

2. It receives the penis during sexual intercourse and holds spermatozoa prior to their passage into the uterus.
3. It forms the lower portion of the birth canal through which the fetus passes during delivery.

Histology of the Vagina

Figure 28-20

In sectional view the lumen of the vagina appears constricted, forming a rough H. The vaginal walls contain a network of blood vessels and layers of smooth muscle, and the lining is moistened by the secretions of the cervical glands and by the movement of water across the permeable epithelium. The vagina and vestibule are separated by an elastic epithelial fold, the **hymen** (HĪ-men), which may partially or completely block the entrance to the vagina. The two bulbocavernosus muscles pass on either side of the vaginal orifice, and their contractions constrict the entrance. These muscles cover the *vestibular bulbs,* masses of erectile tissue that pass on either side of the vaginal entrance. These are the equivalents of the corpora cavernosa of the male.

The vaginal lumen (Figure 28-20•) is lined by a stratified squamous epithelium (nonkeratinized) that in the relaxed state is thrown into folds, called *rugae.* The underlying lamina propria is thick and elastic, and it contains small blood vessels, nerves, and lymph nodes. The vaginal mucosa is surrounded by an elastic **muscularis** layer, with layers of smooth muscle fibers arranged in circular and longitudinal bundles continuous with the uterine myometrium. The portion of the vagina adjacent to the uterus has a serosal covering continuous with the pelvic peritoneum; over the rest of the vagina the muscularis layer is surrounded by an *adventitia* of fibrous connective tissue.

The vagina contains a normal population of resident bacteria, supported by the nutrients found in the cervical mucus. The metabolic activity of these bacteria creates an acid environment, which restricts the growth of many pathogenic organisms. An inflammation of the vaginal canal, known as *vaginitis* (va-jin-Ī-tis), may be caused by fungi, bacteria, or parasites. In addition to any discomfort that may result, the condition may affect the survival of sperm and thereby reduce fertility. AM *Vaginitis* An acid environment also inhibits sperm motility, and for this reason the buffers contained in seminal fluid are important to successful fertilization.

The hormonal changes associated with the ovarian cycle also have an effect on the vaginal epithelium. A *vaginal smear* is a sample of epithelial cells shed at the vaginal surface. By examination of these cells, the corresponding stage in the ovarian and uterine cycles can be estimated. This is an example of *exfoliative cytology,* a diagnostic procedure introduced in Chapter 4. *[p. 119]*

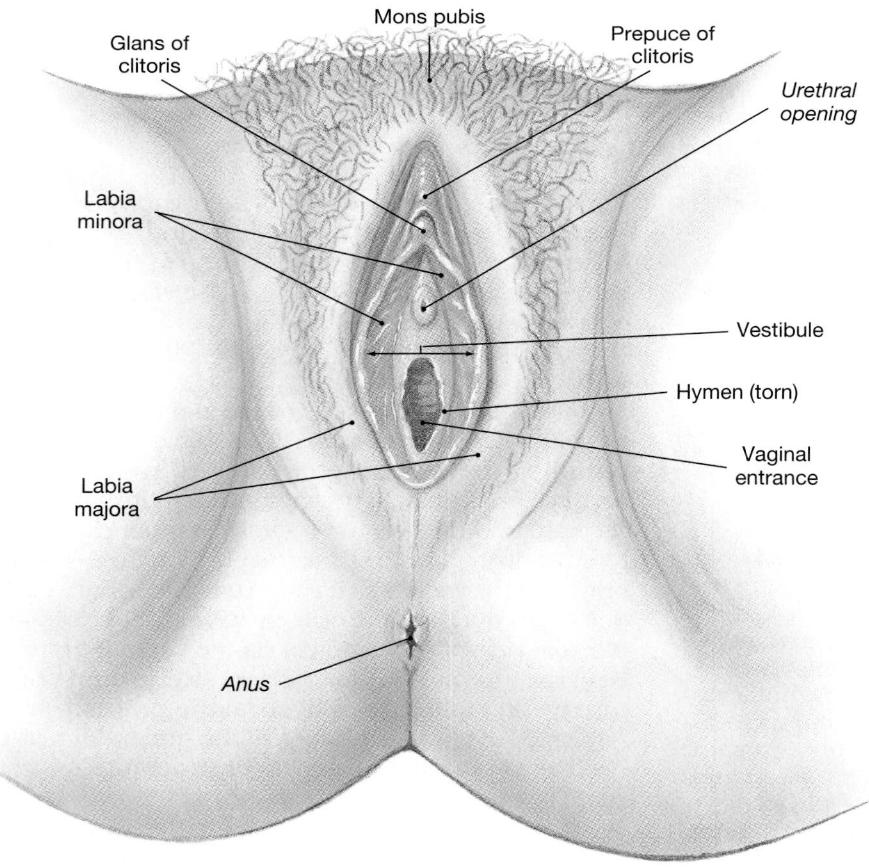
Mons pubis
Glans of clitoris
Prepuce of clitoris
Urethral opening
Labia minora
Vestibule
Hymen (torn)
Vaginal entrance
Labia majora
Anus

• **FIGURE 28-21**
The Female External Genitalia

The External Genitalia

Figure 28-21

The region enclosing the female external genitalia is usually called the **vulva** (VUL-va), or *pudendum* (Figure 28-21●). The vagina opens into the **vestibule,** a central space bounded by the **labia minora** (LĀ-bē-a mi-NOR-a; singular *labium*). The labia minora are covered with a smooth, hairless skin. The urethra opens into the vestibule just anterior to the vaginal entrance. Anterior to the urethral opening, the **clitoris** (KLI-tō-ris) projects into the vestibule. The clitoris is the female equivalent of the penis, derived from the same embryonic structures (see the Embryology Summary later in this chapter). Internally the clitoris contains erectile tissue comparable to the corpus spongiosum of the penis. The clitoris becomes engorged with blood during sexual arousal. A small erectile *glans* sits atop the organ, and extensions of the labia minora encircle the body of the clitoris, forming the **prepuce,** or *hood,* of the clitoris.

A variable number of small **lesser vestibular glands** discharge their secretions onto the exposed surface of the vestibule, keeping it moistened. During arousal a pair of ducts discharges the secretions of the **greater vestibular glands** (*Bartholin's glands*) into the vestibule near the posterolateral margins of the vaginal entrance. These mucous glands resemble the bulbourethral glands of the male.

The outer limits of the vulva are established by the *mons pubis* and the *labia majora.* The prominent bulge of the **mons pubis** is created by adipose tissue beneath the skin anterior to the pubic symphysis. Adipose tissue also accumulates within the fleshy **labia majora** that encircle and partially conceal the labia minora and vestibular structures. The outer margins of the labia majora are covered with the same coarse hair that covers the mons pubis, but the inner surfaces are relatively hairless. Sebaceous glands and scattered apocrine sweat glands secrete onto the inner surface of the labia majora, moistening them and providing lubrication.

The Mammary Glands

Figure 28-22

At birth the newborn infant cannot fend for itself, and several key systems have yet to complete their development. Over the initial period of adjustment to an independent existence, the infant gains nourishment from the milk secreted by the maternal **mammary glands.** Milk production, or **lactation** (lak-TĀ-shun), occurs in these glands, which are specialized organs of the integumentary system of

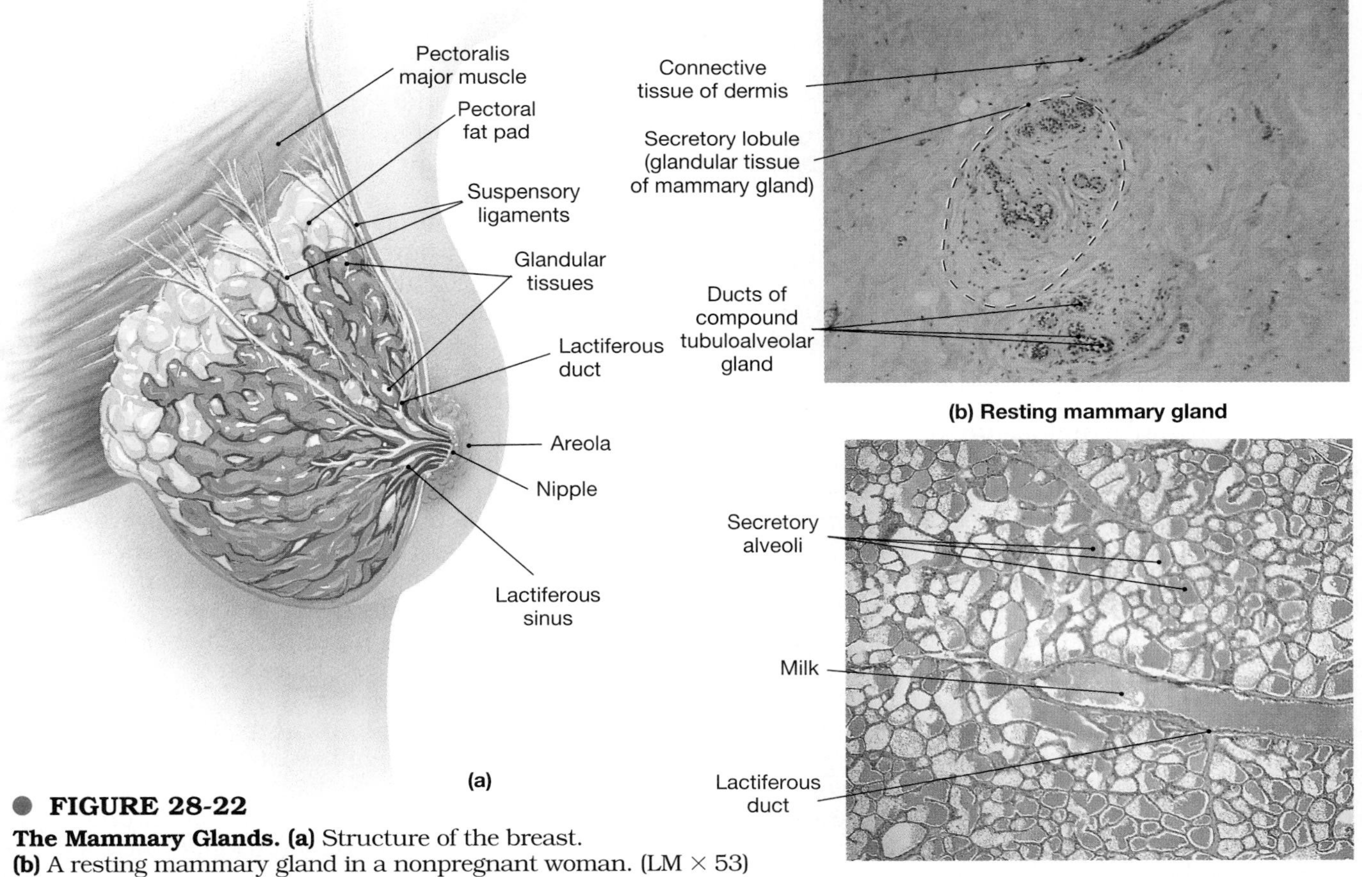
Pectoralis major muscle
Pectoral fat pad
Suspensory ligaments
Glandular tissues
Lactiferous duct
Areola
Nipple
Lactiferous sinus
(a)
Connective tissue of dermis
Secretory lobule (glandular tissue of mammary gland)
Ducts of compound tubuloalveolar gland
(b) Resting mammary gland
Secretory alveoli
Milk
Lactiferous duct
(c) Active mammary gland

● **FIGURE 28-22**
The Mammary Glands. (a) Structure of the breast. **(b)** A resting mammary gland in a nonpregnant woman. (LM × 53) **(c)** An active mammary gland in a pregnant woman. (LM × 119)

Breast Cancer

The mammary glands are cyclically stimulated by the changing levels of circulating reproductive hormones that accompany the menstrual cycle. Usually the effects go unnoticed, but there can be occasional discomfort and even inflammation of mammary gland tissues late in the cycle. If inflamed lobules become walled off with scar tissue, **cysts** are created. Clusters of cysts can be felt in the breast as discrete masses, a condition known as **fibrocystic disease.** Because the symptoms are similar, biopsies may be needed to distinguish between this benign condition and breast cancer.

Breast cancer is the leading cause of death for women between the ages of 35 and 45. It is most common, however, in women over age 50. There were approximately 46,000 deaths in the United States from breast cancer in 1994, and approximately 183,000 new cases reported. An estimated 12 percent of women in the United States will develop breast cancer at some point in their lifetimes, and the rate is steadily rising. The incidence is highest among Caucasians, somewhat lower in African-Americans, and lowest in Asians and American Indians. Notable risk factors include (1) a family history of breast cancer, (2) a pregnancy after age 30, and (3) early menarche (first menstrual period) or late menopause (last menstrual period). Breast cancers in males are very rare; there are only around 300 deaths among males due to breast cancer each year in the United States.

Despite repeated studies, there are no proven links between oral contraceptive use, estrogen therapy, fat consumption, or alcohol use and breast cancer. It appears likely that multiple factors are involved; most women never develop breast cancer, even women in families with a history of this disease. Adequate amounts of nutrients and vitamins, and a diet rich in fruits and vegetables, appear to offer some protection against the development of breast cancer. Women who breast-fed their babies have a 20 percent lower incidence of breast cancer after menopause than mothers who did not. The reason for this effect is not known. (Adding to the mystery, nursing does not appear to affect the incidence of premenopausal breast cancer.)

Early detection of breast cancer is the key to reducing mortalities. *Most breast cancers are found through self-examination,* but the use of clinical screening techniques has increased in recent years. **Mammography** involves the use of X-rays to examine breast tissues; the radiation dosage can be restricted because only soft tissues must be penetrated. This procedure gives the clearest picture of conditions within the breast tissues. Ultrasound can provide some information, but the images lack the detail of standard mammograms.

For treatment to be successful the cancer must be identified while it is still relatively small and localized. Once it has grown larger than 2 cm (0.78 in.) the chances for long-term survival worsen. A poor prognosis also follows if the cancer cells have spread through the lymphatic system to the axillary lymph nodes. If the nodes are not yet involved, the chances of 5-year survival are about 82 percent, but if four or more nodes are involved, the survival rate drops to 21 percent.

Treatment of breast cancer begins with the removal of the tumor. Because the cancer cells usually begin spreading before the condition is diagnosed, surgical treatment involves the removal of part or all of the affected breast.

- In a **segmental mastectomy,** or "lumpectomy," only a portion of the breast is removed.
- In a **total mastectomy** the entire breast is removed, but other tissues are left intact.
- In a **radical mastectomy** the pectoralis muscles, the breast, and the axillary lymph nodes are removed. In a *modified radical mastectomy,* the most common operation, the breast and nodes are removed but the muscular tissue remains intact.

A combination of chemotherapy, radiation treatments, and hormone treatments may be used to supplement the surgical procedures. *Tamoxifen* is a controversial drug that may be used to treat breast cancer. It is more effective than conventional chemotherapy for treating breast cancer in women over 50, and it has fewer unpleasant side effects. It can also be used in addition to regular chemotherapy when treating advanced-stage disease. As an added bonus, tamoxifen prevents and even reverses the osteoporosis of aging. There are risks, however. When given to premenopausal women, tamoxifen can cause amenorrhea and hot flashes similar to those of menopause. Tamoxifen has also been linked to an increased risk of endometrial cancer and perhaps liver cancer as well. It has been proposed that this drug be used to *prevent* breast cancer, rather than treat it. Large-scale trials are under way to determine whether the benefits of chronic low-dose tamoxifen therapy outweigh the risks of complications.

the female that are controlled primarily by hormones of the reproductive system and placenta.

The mammary glands lie in the subcutaneous tissue of the **pectoral fat pad** beneath the skin of the chest (Figure 28-22a●). Each breast bears a small conical projection, the **nipple,** where the ducts of underlying mammary glands open onto the body surface. The skin surrounding each nipple has a reddish-brown coloration, and this region is known as the **areola** (a-RĒ-ō-la). Large sebaceous glands beneath the areolar surface give it a granular texture.

The glandular tissue of the mammary gland consists of separate lobes, each containing several secretory lobules. Ducts leaving the lobules converge, giving rise to a single **lactiferous** (lak-TIF-e-rus) **duct** in each lobule (Figure 28-22c●). Near the nipple, that lactiferous duct expands, forming an expanded chamber called a **lactiferous sinus.** Usually 15–20 lactiferous sinuses open onto the surface of each nipple.

Dense connective tissue surrounds the duct system and forms partitions that extend between the lobes and lobules. These bands of connective tissue, known as the **suspensory ligaments of the breast,** originate in the dermis of the overlying skin. A layer of loose connective tissue separates the mammary complex from the underlying pectoralis muscles. Branches of the *internal thoracic artery* supply blood to each mammary gland (Figure 21-23●, p. 752).

Development of the Mammary Glands during Pregnancy

Figure 28-22

Figure 28-22b,c● compares the histological organization of the inactive and active mammary glands. The resting mammary gland is dominated by a duct system, rather than by active glandular cells. The size of the mammary glands in a nonpregnant woman reflects primarily the amount of adipose tissue present, rather than the amount of glandular tissue. The secretory apparatus does not complete its development until pregnancy occurs. The hormonal mechanisms involved will be discussed in Chapter 29.

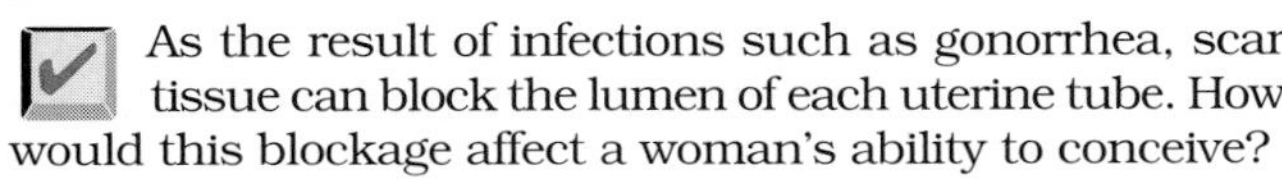
As the result of infections such as gonorrhea, scar tissue can block the lumen of each uterine tube. How would this blockage affect a woman's ability to conceive?

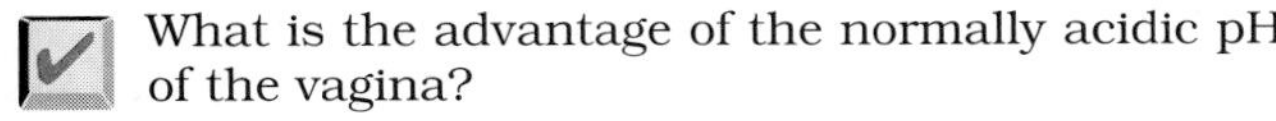
What is the advantage of the normally acidic pH of the vagina?

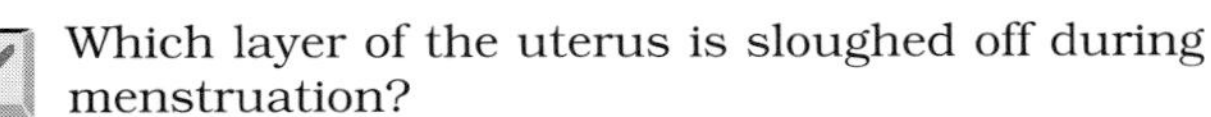
Which layer of the uterus is sloughed off during menstruation?

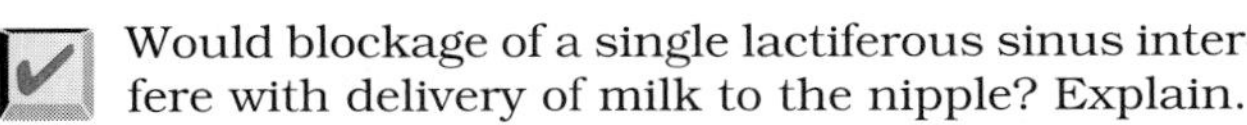
Would blockage of a single lactiferous sinus interfere with delivery of milk to the nipple? Explain.

❑ Hormones and the Female Reproductive Cycle

Figure 28-23

The activity of the female reproductive tract falls under hormonal control involving an interplay between pituitary and gonadal secretions. But the regulatory pattern in females is much more complicated than in males, because it must coordinate the ovarian and uterine cycles. Circulating hormones control the **female reproductive cycle,** coordinating the ovarian and uterine cycles to ensure proper reproductive function. If the two cycles cannot be coordinated normally, infertility results. For example, it is obvious that a woman who fails to ovulate will be unable to conceive, even if her uterus is perfectly normal. A woman who ovulates normally, but whose uterus isn't ready to support an embryo, will be just as infertile. Because the processes are complex and difficult to study, many of the biochemical details of the female reproductive cycle still elude us, but the general patterns are reasonably clear.

Changes in circulating estrogen concentration are the primary mechanism for coordinating the female reproductive cycle, and the relationships are summarized in Figure 28-23●. The upper portion of this figure summarizes the hormonal regulation of the ovarian cycle. The pattern of hormonal regulation of the ovarian cycle differs depending on whether one considers the *preovulatory period* or the *postovulatory period.*

Hormones and the Preovulatory Period

Figure 28-24

Follicular development begins under FSH stimulation, and each month some of the primordial follicles begin their development into primary follicles. As the follicles enlarge, thecal cells begin producing a steroid, *androstenedione,* that is absorbed by the granulosa cells and converted to *estrogens.* Small quantities of estrogens are also contributed by *interstitial cells* scattered throughout the ovarian stroma. The hormone **estradiol** (es-tra-DĪ-ol) is the most important estrogen, and it is the dominant hormone prior to ovulation.

Estrogens have multiple functions, affecting the activities of many different tissues and organs throughout the body. Important general functions include (1) stimulating bone and muscle growth, (2) maintaining female secondary sex characteristics such as body hair distribution and the location of adipose tissue deposits, (3) affecting CNS activity, (4) maintaining functional accessory reproductive glands and organs, and (5) initiating repair and growth of the endometrium.

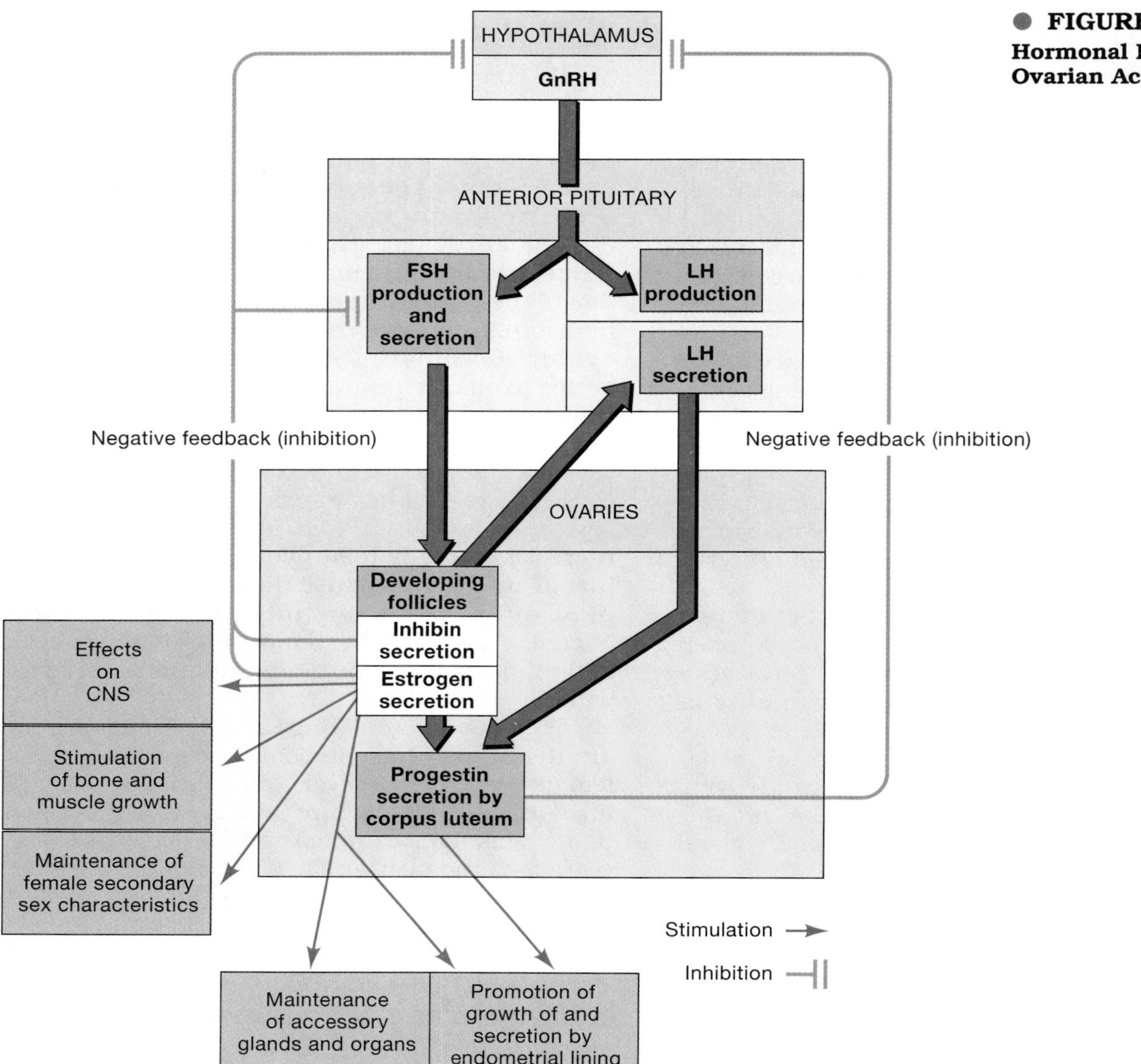
HYPOTHALAMUS
GnRH
ANTERIOR PITUITARY
FSH production and secretion
LH production
LH secretion
Negative feedback (inhibition)
Negative feedback (inhibition)
OVARIES
Developing follicles
Inhibin secretion
Estrogen secretion
Progestin secretion by corpus luteum
Effects on CNS
Stimulation of bone and muscle growth
Maintenance of female secondary sex characteristics
Maintenance of accessory glands and organs
Promotion of growth of and secretion by endometrial lining
Stimulation
Inhibition

● **FIGURE 28-23**
Hormonal Regulation of Ovarian Activity

The upper portion of Figure 28-24● follows the hormonal events associated with the ovarian cycle. The following points should be noted:

- As follicular development proceeds the concentration of circulating estrogens rises, for the follicular cells are increasing in number and secretory activity.
- Granulosa cells of the ovarian follicles secrete inhibin. As the primary follicles enlarge, becoming secondary follicles, the number of granulosa cells increases, and the level of circulating inhibin rises.
- Rising estrogen levels, combined with rising inhibin production, inhibit both the hypothalamic secretion of GnRH and the pituitary production and release of FSH.
- Estrogen also has an effect on the rate of LH production and secretion. LH production occurs under GnRH stimulation, and the effect is enhanced by estrogens. The rate of LH release into the bloodstream is thus linked to the circulating concentration of estrogens. Thus as the follicles develop and estrogen concentrations rise, the pituitary output of LH gradually increases.
- Over the first 14 days of the cycle, LH-secreting cells of the pituitary become more sensitive to GnRH. This increased sensitivity further stimulates LH production as the cycle proceeds.
- The combination of estrogens, FSH, and LH continues to support follicular development and maturation despite a gradual decline in FSH levels.
- Estrogen concentrations take a sharp upturn in the second week of the ovarian cycle, as this month's tertiary follicle enlarges in preparation for ovulation.

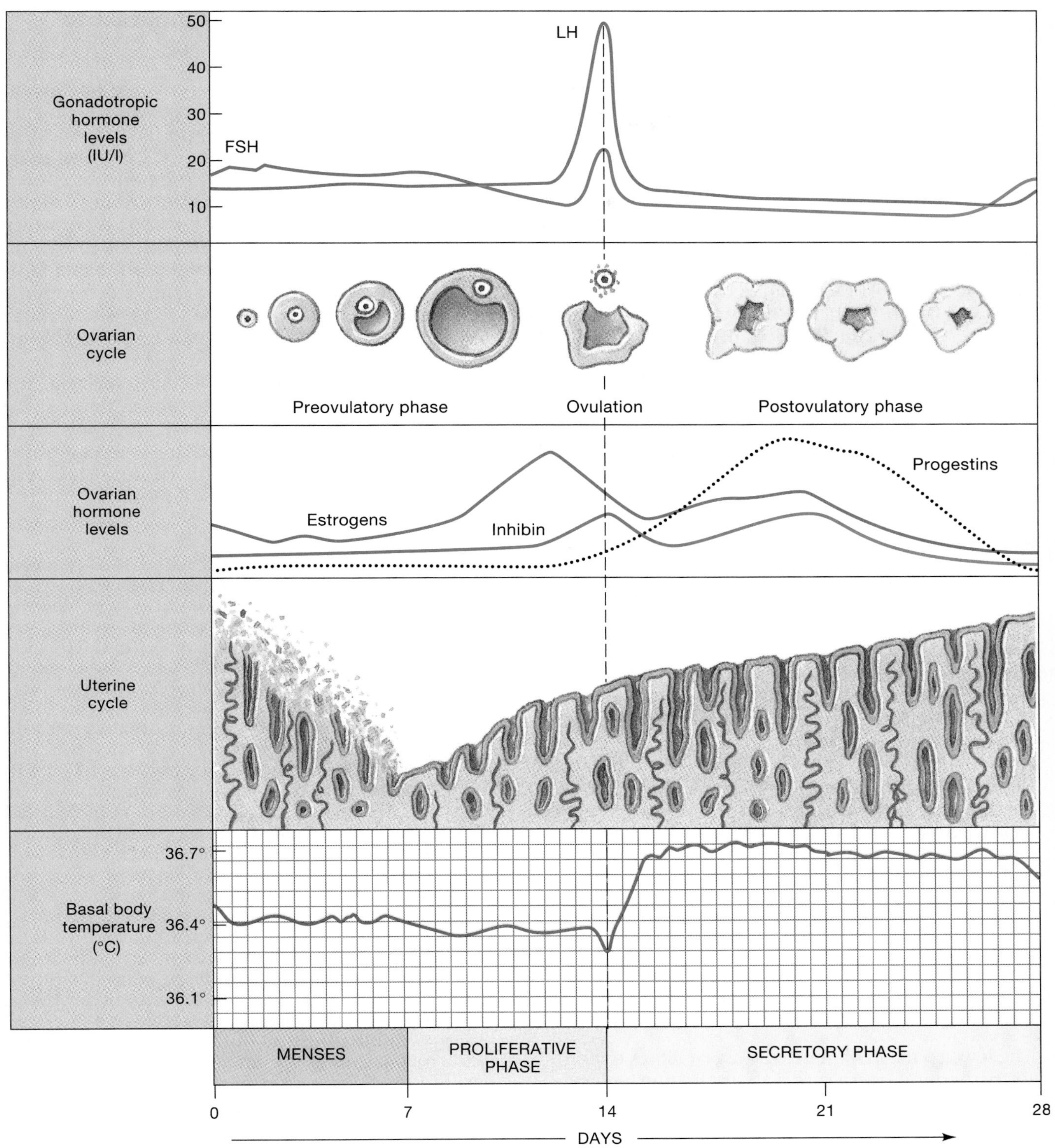
Gonadotropic hormone levels (IU/l)
50
40
30
20
10
LH
FSH
Ovarian cycle
Preovulatory phase
Ovulation
Postovulatory phase
Ovarian hormone levels
Estrogens
Inhibin
Progestins
Uterine cycle
Basal body temperature (°C)
36.7°
36.4°
36.1°
MENSES
PROLIFERATIVE PHASE
SECRETORY PHASE
0
7
14
21
28
DAYS

● **FIGURE 28-24**
Hormonal Regulation of the Female Reproductive Cycle

- At about day 14 estrogen levels peak, accompanying the maturation of that follicle, and the pituitary LH cells become maximally sensitive to GnRH. The combination results in a massive outpouring of LH from the anterior pituitary. That sudden surge in LH concentration causes the completion of meiosis I, the rupture of the follicular wall, and ovulation.

Hormones and the Postovulatory Period

After ovulation, LH stimulates the remaining granulosa cells to form the corpus luteum, which secretes progestins, principally the steroid progesterone. Although moderate amounts of estrogens are also secreted by the corpus luteum, progesterone is the principal hormone of the postovulatory period. Its primary function is to prepare the uterus for pregnancy by stimulating the growth and development of the blood supply and secretory glands of the functional zone.

Luteinizing hormone levels remain elevated for only 2 days, but that is long enough to stimulate the formation of the functional corpus luteum. Progesterone secretion continues at relatively high levels for the next week, but unless pregnancy occurs the corpus luteum then begins to degenerate. Roughly 12 days after ovulation, the corpus luteum becomes nonfunctional, and progesterone and estrogen levels fall markedly.

The decline in progesterone and estrogen levels stimulates the hypothalamic receptors, and GnRH production increases. This increase leads to an increase in FSH and LH production in the anterior pituitary, and the entire cycle begins again.

The hormonal changes involved with the ovarian cycle in turn affect the activities of other reproductive tissues and organs. At the uterus, the hormonal changes are responsible for the maintenance of the uterine cycle.

Hormones and the Uterine Cycle

Figure 28-24

The lower portion of Figure 28-24• follows changes in the endometrium during a single uterine cycle. The sudden declines in progesterone and estrogen levels that accompany the degeneration of the corpus luteum result in menses. The sloughing of endometrial tissue continues for several days, until rising estrogen levels stimulate the repair and regeneration of the functional zone of endometrium. The preovulatory phase continues until rising progesterone levels mark the arrival of the postovulatory phase. The combination of estrogen and progesterone then causes the enlargement of the endometrial glands and an increase in their secretory activities.

Hormones and Body Temperature

Figure 28-24

The hormonal fluctuations also cause physiological changes that affect the core body temperature. During the preovulatory period, when estrogen is the dominant hormone, the resting body temperature, or *basal body temperature*, measured upon awakening in the morning is about 0.3°C lower than it is during the postovulatory period, when progesterone dominates the endocrine picture. At the time of ovulation, basal temperature declines sharply, making the temperature rise over the following day even more noticeable (Figure 28-24•). As a result, by keeping records of body temperature over a few menstrual cycles a woman can often determine the precise day of ovulation. This information can be very important for individuals wishing to avoid or promote a pregnancy, for such an event can occur only if an ovum becomes fertilized, and this usually occurs within a day of ovulation.

PREMENSTRUAL SYNDROME Several physical and physiological changes occur in women 7–10 days before the start of menses. Fluid retention, breast enlargement, headaches, pelvic pain, and an uncomfortable "bloated" feeling are common symptoms. These sensations may be associated with psychological changes producing irritability, anxiety, and depression. This combination has been called **premenstrual syndrome (PMS).**

The mechanism responsible for PMS has yet to be determined. Changes in sex hormone levels may be involved directly, by action on peripheral organ systems, or indirectly, by modifying neurotransmitter release in the CNS. There are no laboratory tests or procedures to diagnose PMS, but tracking the appearance of symptoms over a 2–3 month period can reveal characteristic patterns. Treatment at present is symptomatic and may involve exercise, dietary change, or medication, depending on the nature of the primary symptom. For example, if headache is the major problem, analgesics are prescribed; diuretics may be used to combat bloating and fluid retention. For severe PMS, drugs can be administered that block GnRH secretion and stop uterine cycles completely for 6 months or more. Over the interim, estrogens can be administered to prevent symptoms of premature menopause.

What changes would you expect to observe in the ovarian cycle if the LH surge did not occur?

What effect would blockage of progesterone receptors in the uterus have on the endometrium?

What event occurs in the menstrual cycle when the levels of estrogens and progesterone decline?

EMBRYOLOGY SUMMARY

Development of the Reproductive System: Indifferent Stages

DEVELOPMENT OF THE GONADS

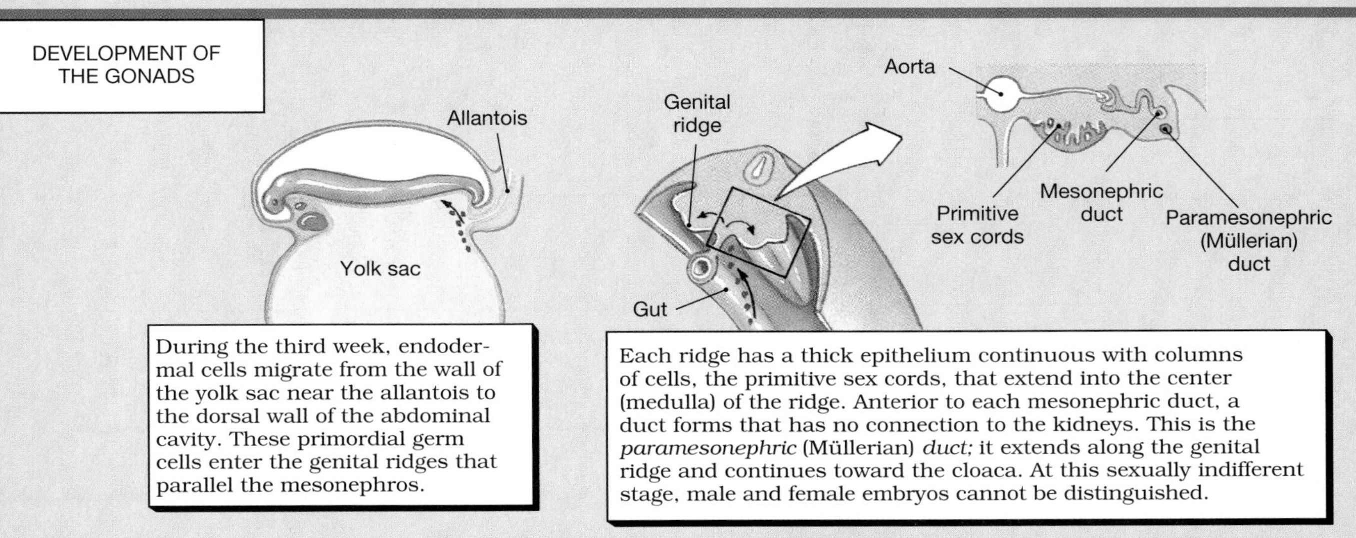
Allantois
Yolk sac
Genital ridge
Gut
Aorta
Primitive sex cords
Mesonephric duct
Paramesonephric (Müllerian) duct

During the third week, endodermal cells migrate from the wall of the yolk sac near the allantois to the dorsal wall of the abdominal cavity. These primordial germ cells enter the genital ridges that parallel the mesonephros.

Each ridge has a thick epithelium continuous with columns of cells, the primitive sex cords, that extend into the center (medulla) of the ridge. Anterior to each mesonephric duct, a duct forms that has no connection to the kidneys. This is the *paramesonephric* (Müllerian) *duct;* it extends along the genital ridge and continues toward the cloaca. At this sexually indifferent stage, male and female embryos cannot be distinguished.

DEVELOPMENT OF DUCTS AND ACCESSORY ORGANS

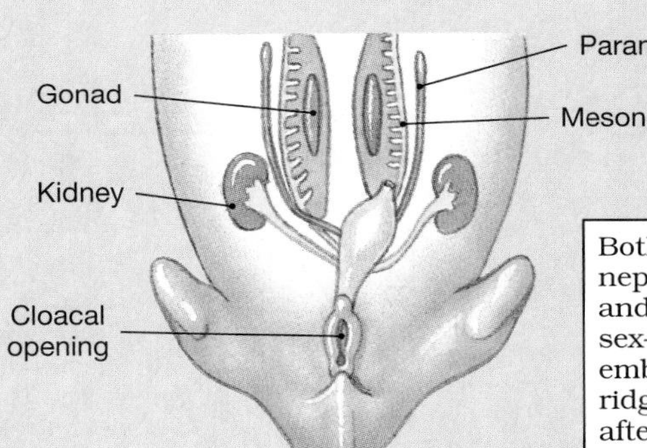
Paramesonephric duct
Gonad
Mesonephric duct
Kidney
Cloacal opening

Both sexes have mesonephric and paramesonephric ducts at this stage. Unless exposed to androgens, the embryo—regardless of its genetic sex—will develop into a female. In a normal male embyro, cells in the core (medulla) of the genital ridge begin producing testosterone sometime after week 6. This triggers the changes in the duct system and external genitalia that are detailed on the following page.

DEVELOPMENT OF EXTERNAL GENITALIA

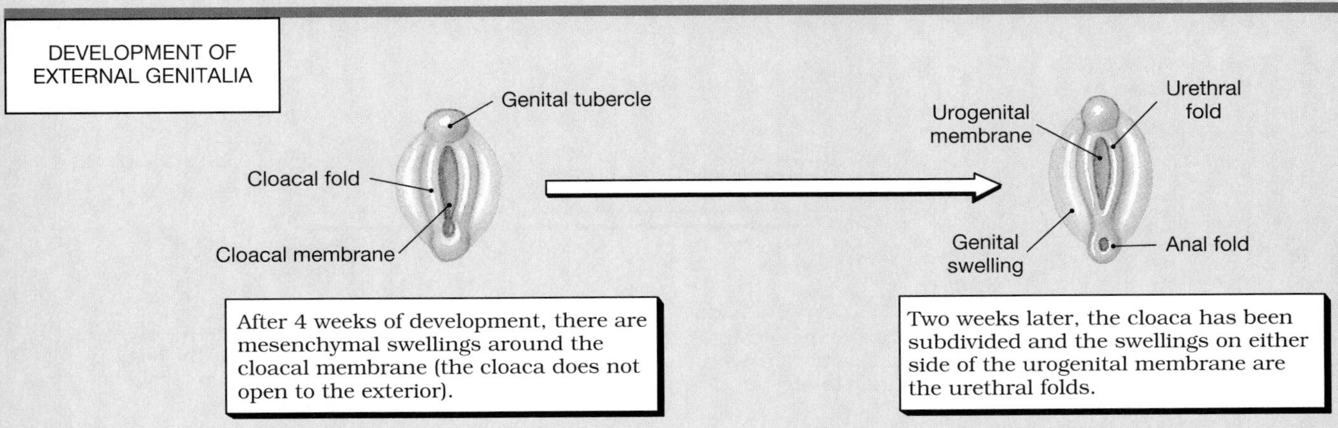
Genital tubercle
Cloacal fold
Cloacal membrane
Urogenital membrane
Urethral fold
Genital swelling
Anal fold

After 4 weeks of development, there are mesenchymal swellings around the cloacal membrane (the cloaca does not open to the exterior).

Two weeks later, the cloaca has been subdivided and the swellings on either side of the urogenital membrane are the urethral folds.

EMBRYOLOGY SUMMARY

Development of the Male Reproductive System

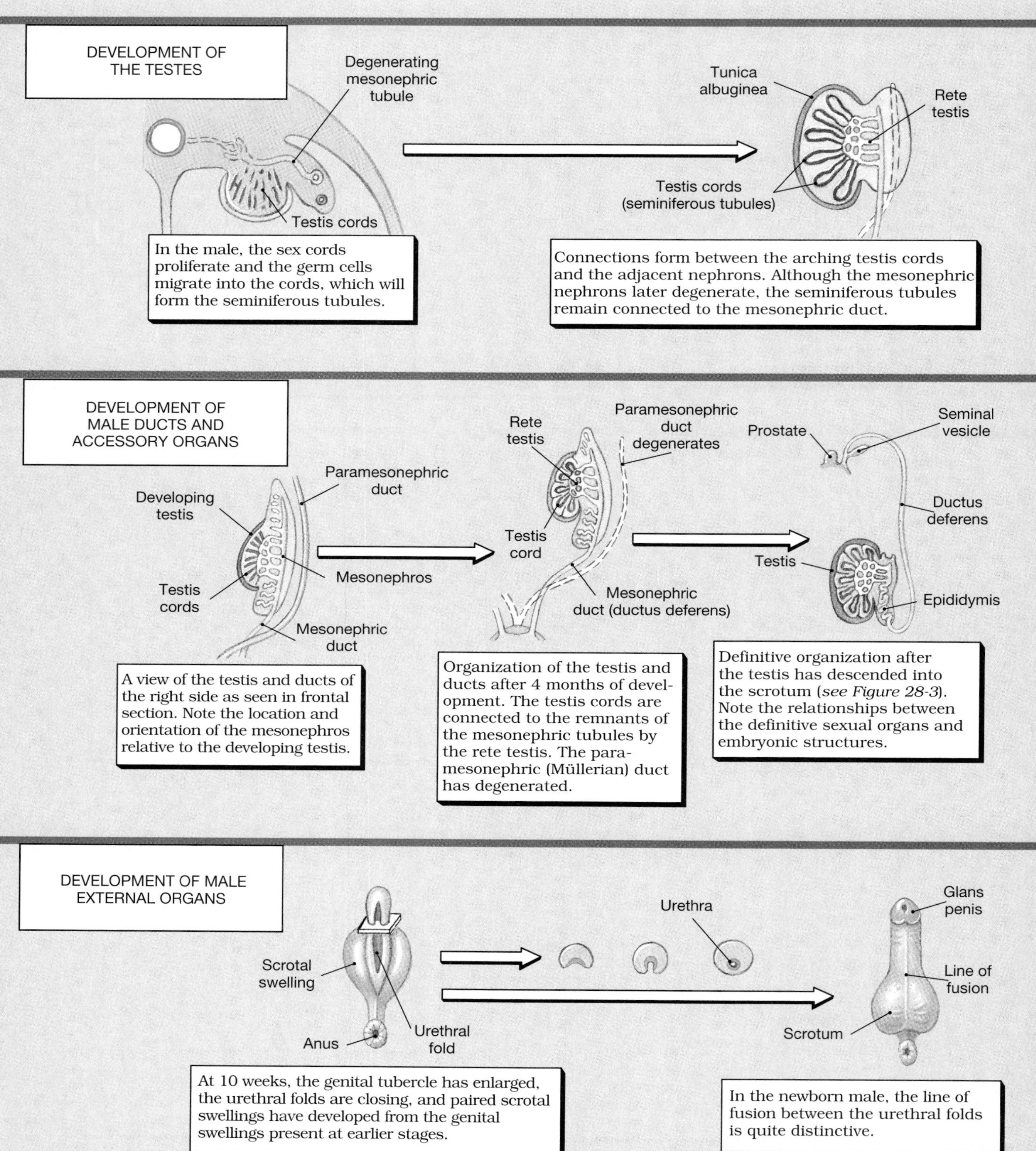

DEVELOPMENT OF THE TESTES
Degenerating mesonephric tubule
Testis cords
Tunica albuginea
Rete testis
Testis cords (seminiferous tubules)
DEVELOPMENT OF MALE DUCTS AND ACCESSORY ORGANS
Developing testis
Paramesonephric duct
Testis cords
Mesonephros
Mesonephric duct
Rete testis
Paramesonephric duct degenerates
Testis cord
Mesonephric duct (ductus deferens)
Prostate
Seminal vesicle
Ductus deferens
Testis
Epididymis
DEVELOPMENT OF MALE EXTERNAL ORGANS
Scrotal swelling
Anus
Urethral fold
Urethra
Glans penis
Line of fusion
Scrotum

In the male, the sex cords proliferate and the germ cells migrate into the cords, which will form the seminiferous tubules.

Connections form between the arching testis cords and the adjacent nephrons. Although the mesonephric nephrons later degenerate, the seminiferous tubules remain connected to the mesonephric duct.

A view of the testis and ducts of the right side as seen in frontal section. Note the location and orientation of the mesonephros relative to the developing testis.

Organization of the testis and ducts after 4 months of development. The testis cords are connected to the remnants of the mesonephric tubules by the rete testis. The paramesonephric (Müllerian) duct has degenerated.

Definitive organization after the testis has descended into the scrotum (*see Figure 28-3*). Note the relationships between the definitive sexual organs and embryonic structures.

At 10 weeks, the genital tubercle has enlarged, the urethral folds are closing, and paired scrotal swellings have developed from the genital swellings present at earlier stages.

In the newborn male, the line of fusion between the urethral folds is quite distinctive.

EMBRYOLOGY SUMMARY

Development of the Female Reproductive System

DEVELOPMENT OF THE OVARIES

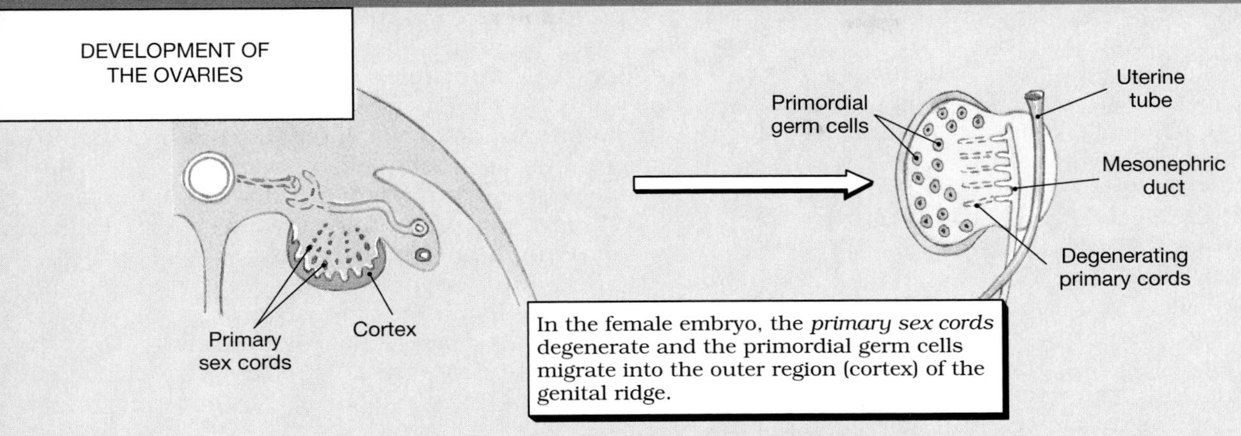
Primary
sex cords
Cortex
Primordial
germ cells
Uterine
tube
Mesonephric
duct
Degenerating
primary cords

In the female embryo, the *primary sex cords* degenerate and the primordial germ cells migrate into the outer region (cortex) of the genital ridge.

DEVELOPMENT OF FEMALE DUCTS AND ACCESSORY ORGANS

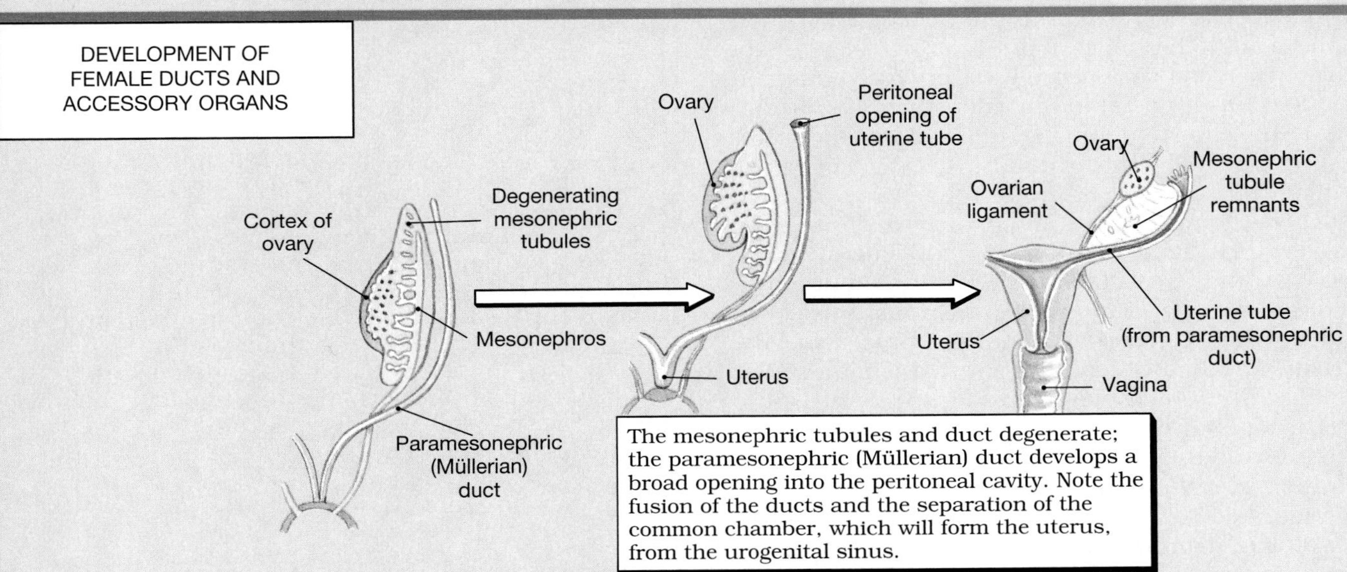
Cortex of
ovary
Degenerating
mesonephric
tubules
Mesonephros
Paramesonephric
(Müllerian)
duct
Ovary
Peritoneal
opening of
uterine tube
Uterus
Ovary
Ovarian
ligament
Mesonephric
tubule
remnants
Uterus
Uterine tube
(from paramesonephric
duct)
Vagina

The mesonephric tubules and duct degenerate; the paramesonephric (Müllerian) duct develops a broad opening into the peritoneal cavity. Note the fusion of the ducts and the separation of the common chamber, which will form the uterus, from the urogenital sinus.

DEVELOPMENT OF FEMALE EXTERNAL GENITALIA

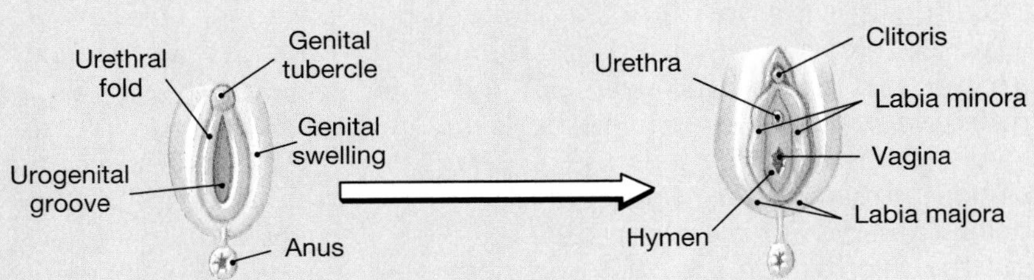
Urethral
fold
Genital
tubercle
Genital
swelling
Urogenital
groove
Anus
Urethra
Clitoris
Labia minora
Vagina
Labia majora
Hymen

In the female, the urethral folds do not fuse; they develop into the labia minora. The genital swellings will form the labia majora. The genital tubercle, which in the male forms the glans of the penis, develops into the clitoris. The urethra opens to the exterior immediately posterior to the clitoris. The hymen remains as an elaboration of the urogenital membrane.

The Physiology of Sexual Intercourse

Sexual intercourse, or *coitus* (KŌ-i-tus), introduces semen into the female reproductive tract. We will now consider the process as it affects the reproductive systems of males and females.

Male Sexual Function

Sexual function in the male is coordinated by complex neural reflexes that are incompletely understood. The reflex pathways utilize the sympathetic and parasympathetic divisions of the autonomic nervous system. During **arousal,** erotic thoughts or the stimulation of sensory nerves in the genital region leads to an increase in the parasympathetic outflow over the pelvic nerves. This outflow leads to erection of the penis. The integument covering the glans of the penis contains numerous sensory receptors, and erection tenses the skin and increases their sensitivity. Subsequent stimulation may initiate the secretion of the bulbourethral glands, lubricating the penile urethra and the surface of the glans.

During intercourse the sensory receptors of the penis are rhythmically stimulated. This stimulation eventually results in the coordinated processes of *emission* and *ejaculation*. **Emission** occurs under sympathetic stimulation. The process begins with the peristaltic contractions of the ampulla, pushing fluid and spermatozoa into the prostatic urethra. The seminal vesicles then begin contracting, and the contractions increase in force and duration over the next few seconds. Peristaltic contractions also appear in the walls of the prostate gland. The combination moves the seminal mixture into the membranous and penile portions of the urethra. While the contractions are proceeding, sympathetic commands also cause the contraction of the sphincter guarding the entrance to the urinary bladder. This contraction effectively prevents the passage of semen into the bladder.

Ejaculation (ē-jak-ū-LĀ-shun) occurs as powerful, rhythmic contractions appear in the *ischiocavernosus* and *bulbocavernosus* muscles, two superficial skeletal muscles of the pelvic floor. The ischiocavernosus muscles insert along the sides of the penis, and their contractions serve primarily to stiffen that organ. The bulbocavernosus muscle wraps around the base of the penis, and its contraction pushes semen toward the external urethral orifice. These contractions are controlled by somatic motor neurons in the lower lumbar and upper sacral segments of the spinal cord. (The positions of these muscles can be seen in Figure 11-13•, p. 349.)

Ejaculation is associated with intensely pleasurable sensations, an experience known as male **orgasm** (ŌR-gazm). Several other noteworthy physiological changes occur at this time, including pronounced but temporary increases in heart rate and blood pressure. After ejaculation has been completed, blood begins to leave the erectile tissue, and the erection begins to subside. This subsidence, called **detumescence** (de-tū-MES-ens), is mediated by the sympathetic nervous system.

In summary, arousal, erection, emission, and ejaculation are mediated by a complex interplay between the sympathetic and parasympathetic divisions of the autonomic nervous system. Higher centers, including the cerebral cortex, can facilitate or inhibit many of the important reflexes, thereby modifying the patterns of sexual function. Any physical or psychological factor that affects a single component of the system can result in male sexual dysfunction, also called **impotence.**

IMPOTENCE Impotence is an inability to achieve or maintain an erection. There may be a variety of possible physical causes, for erection involves vascular changes as well as neural commands. For example, low blood pressure in the arteries servicing the penis, due to a circulatory blockage such as a plaque, will affect the ability to obtain an erection. Drugs, trauma, or illnesses that affect the autonomic nervous system or the CNS may have the same effect. But male sexual performance can also be strongly affected by the psychological state of the individual, and the majority of clinical cases of impotence probably reflect psychological rather than anatomical problems. Temporary periods of impotence are relatively common in normal individuals experiencing severe stresses or emotional problems. Depression, anxiety, and fear of impotence are examples of emotional factors that may result in sexual dysfunction.

Female Sexual Function

The phases of female sexual function are comparable to those of a male. During sexual arousal, parasympathetic activation leads to an engorgement of the erectile tissues of the clitoris and increased secretion of cervical mucous glands and the greater vestibular glands. Clitoral erection increases its sensitivity to stimulation, and the cervical and vestibular glands provide lubrication for the vaginal walls. A network of blood vessels in the vaginal walls becomes filled with blood at this time, and the vaginal surfaces are also moistened by the transudation of fluid from underlying connective tissues. (This process accelerates during intercourse as the result of mechanical stimulation.) Paraympathetic stimulation also causes engorgement of blood vessels at the nipples, making them more sensitive to touch and pressure.

During intercourse, rhythmic contact with the clitoris and vaginal walls, reinforced by touch sensations from the breasts and other stimuli (visual, olfactory, and auditory), provides stimulation that eventually leads to orgasm. Female orgasm is accompanied by peristaltic contractions of the uterine and vaginal walls and, via impulses over the pudendal nerves, rhythmic contractions of the bulbocavernosus and ischiocavernosus muscles. The latter contractions give rise to the sensations of orgasm.

SEXUALLY TRANSMITTED DISEASE **Sexually transmitted diseases (STDs)** are transferred from individual to individual, usually or exclusively by sexual intercourse. A variety of bacterial, viral, and fungal infections are included in this category. At least two dozen different STDs are currently recognized. All are unpleasant. *Chlamydia*, noted on page 1083, can cause PID and infertility. Other types of STDs are quite dangerous, and a few, including AIDS, are deadly. *[p. 810]* The incidence of STDs has been increasing in the United States since 1984, primarily in urban centers and in urban minority populations. Acute poverty, coupled with drug use, prostitution, and the appearance of drug-resistant pathogens all contribute to the problem. The *Applications Manual* contains a detailed discussion of the most common forms of STD, including: *gonorrhea, syphilis, herpes,* and *chancroid.* AM *Sexually Transmitted Diseases*

Aging and the Reproductive System

The aging process affects the reproductive systems of men and women. The most striking age-related changes in the female reproductive system occur at menopause, whereas changes in the male reproductive system occur more gradually and over a longer period of time.

Menopause

Menopause is usually defined as the time that ovulation and menstruation cease. Menopause typically occurs at age 45–55, but in the years preceding it the ovarian and menstrual cycles become irregular. A shortage of primordial follicles is the underlying cause of these developments. It has been estimated that almost 7 million potential oocytes are found in fetal ovaries after 5 months of development, but the number drops to about 2 million at birth, and to a few hundred thousand at puberty. By age 50, there are no primordial follicles left to respond to FSH; in **premature menopause** this depletion occurs before age 40.

Menopause is accompanied by a sharp and sustained rise in the production of GnRH, FSH, and LH, while circulating concentrations of estrogen and progesterone decline. The decline in estrogen levels leads to reductions in the size of the uterus and breasts, accompanied by a thinning of the urethral and vaginal walls. The reduced estrogen concentrations have also been linked to the development of osteoporosis, presumably because bone deposition proceeds at a slower rate. A variety of neural effects are also reported, including "hot flashes," anxiety, and depression, but the hormonal mechanisms involved are not well understood. In addition, the risk of atherosclerosis and other forms of cardiovascular disease increases after menopause.

The majority of women experience only mild symptoms, but some individuals experience acutely unpleasant symptoms either during or after menopause. For those individuals, hormone replacement therapies involving a combination of estrogens and progestins can often prevent osteoporosis and the neural and vascular changes associated with menopause. The hormones may be administered by injection or via "estrogen patches" that rely on transdermal delivery. A synthetic hormone, *etidronate*, inhibits osteoporosis by suppressing osteoclast activity. When given at intervals over a 2-year period, it increases bone mass and reduces the rate of fracture incidence in postmenopausal women.

The Male Climacteric

Changes in the male reproductive system occur more gradually, over a period known as the **male climacteric.** Circulating testosterone levels begin to decline between ages 50 and 60, coupled with increases in circulating levels of FSH and LH. Although sperm production continues (men well into their eighties can father children) there is a gradual reduction in sexual activity in older men. This decrease may be linked to declining testosterone levels, and some clinicians are now tentatively suggesting the use of testosterone replacement therapy to enhance libido (sexual drive) in elderly men.

Integration with Other Systems

Figure 28-25

Figure 28-25● summarizes the relationships between the reproductive system and other physiological systems. Normal human reproduction is a complex process that requires the participation of multiple systems. Hormones play a major role in coordinating these events, and Table 28-1 reviews the hormones discussed in this chapter. The reproductive process depends on a variety of physical, physiological, and psychological factors, many of which require intersystem cooperation. For example, the male's sperm count must be adequate, the semen must have the correct pH and nutrients, and erection and ejaculation must occur in the proper sequence. For these steps to occur, the reproductive, digestive, endocrine, nervous, cardiovascular, and urinary systems must all be functioning normally.

Even when all else is normal and fertilization occurs at the proper time and place, a normal infant will not result unless the zygote, a single cell the size of a pinhead, manages to develop into a full-term fetus weighing 3–4 kg. Chapter 29 considers the process of development, focusing on the mechanisms that determine both the structure of the body and the distinctive characteristics of each individual.

Inability to contract the ischiocavernosus and bulbocavernosus muscles would interfere with what part of the male sex act?

What changes occur in the female during sexual arousal as the result of increased parasympathetic stimulation?

Why does the level of FSH rise and remain high during menopause?

TABLE 28-1 Hormones of the Reproductive System

Hormone	*Source*	*Regulation of Secretion*	*Primary Effects*
GONADOTROPIN-RELEASING HORMONE (GnRH)	Hypothalamus	*Male:* inhibited by testosterone *Female:* inhibited by estrogens and/or progestins	Stimulates FSH secretion, LH synthesis
FOLLICLE-STIMULATING HORMONE (FSH)	Anterior pituitary	*Male:* stimulated by GnRH, inhibited by inhibin *Female:* stimulated by GnRH, inhibited by estrogens and/or progestins	*Male:* stimulates spermatogenesis and spermiogenesis through effects on sustentacular cells *Female:* stimulates follicle development, estrogen production, and egg maturation
ESTROGENS (primarily estradiol)	Follicular and interstitial cells of ovaries	Stimulated by FSH	Stimulates LH secretion, maintains secondary sex characteristics and sexual behavior, stimulates repair of endometrium, inhibits secretion of GnRH
INHIBIN	Sustentacular cells of testes and granulosa cells of ovaries	Stimulated by factors released by developing sperm (male) and developing follicles (female)	Inhibits secretion of FSH and possibly GnRH
LUTEINIZING HORMONE (LH)	Anterior pituitary	*Male:* stimulated by GnRH *Female:* production stimulated by GnRH, secretion by estrogens	*Male:* stimulates interstitial cells *Female:* stimulates follicular and interstitial cells
PROGESTINS (primarily progesterone)	Corpus luteum	Stimulated by LH	Stimulates endometrial growth and glandular secretion, inhibits GnRH secretion
ANDROGENS (primarily testosterone and dihydrotestosterone)	Interstitial cells of testes	Stimulated by LH	Maintains secondary sex characteristics and sexual behavior, promotes maturation of spermatozoa, inhibits GnRH secretion

INTEGUMENTARY SYSTEM

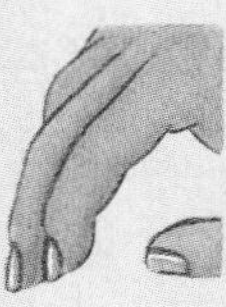

Covers external genitalia; provides sensations that stimulate sexual behaviors; mammary gland secretions provide nourishment for newborn

Reproductive hormones affect distribution of body hair and subcutaneous fat deposits

SKELETAL SYSTEM

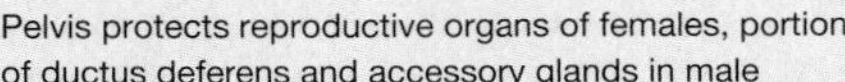

Pelvis protects reproductive organs of females, portion of ductus deferens and accessory glands in male

Sexual hormones stimulate growth and maintenance of bones; sex hormones at puberty accelerate growth and closure of epiphyseal plates

MUSCULAR SYSTEM

Contractions of skeletal muscles eject semen from male reproductive tract; muscle contractions during sexual act produce pleasurable sensations in both sexes

Reproductive hormones, especially testosterone, accelerate skeletal muscle growth

REPRODUCTIVE SYSTEM

NERVOUS SYSTEM

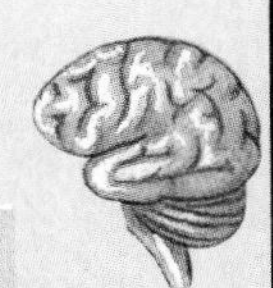

Controls sexual behaviors and sexual function

Sexual hormones affect CNS development and sexual behaviors

ENDOCRINE SYSTEM

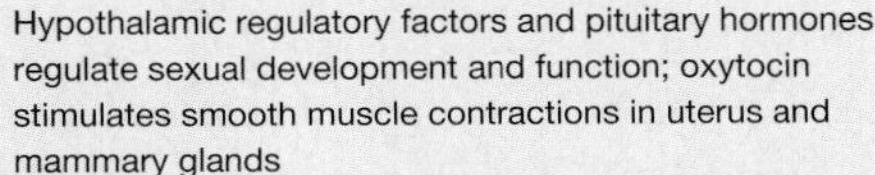

Hypothalamic regulatory factors and pituitary hormones regulate sexual development and function; oxytocin stimulates smooth muscle contractions in uterus and mammary glands

Steroid sex hormones and inhibin inhibit secretory activities of hypothalamus and pituitary

CARDIOVASCULAR SYSTEM

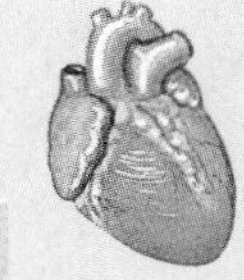

Distributes reproductive hormones; provides nutrients, oxygen, and waste removal for fetus; local blood pressure changes responsible for physical changes during sexual arousal

Estrogens may maintain healthy vessels and slow development of atherosclerosis

LYMPHATIC SYSTEM

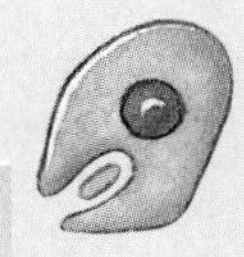

Provides IgA for secretion by epithelial glands; assists in repairs and defense against infection

Lysozymes and bactericidal chemicals in secretions provide nonspecific defense against reproductive tract infections

FOR ALL SYSTEMS

Secretion of hormones with effects on growth and metabolism

RESPIRATORY SYSTEM

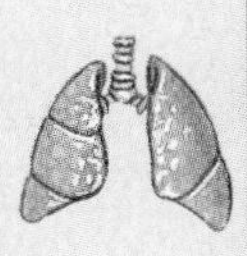

Provides oxygen and removes carbon dioxide generated by tissues of reproductive system

Changes in respiratory rate and depth occur during sexual arousal, under control of the nervous system

URINARY SYSTEM

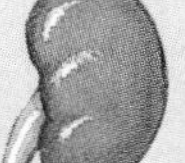

Urethra in males carries semen to exterior; kidneys remove wastes generated by reproductive tissues

Accessory organ secretions may have antibacterial action that helps prevent urethral infections in males

DIGESTIVE SYSTEM

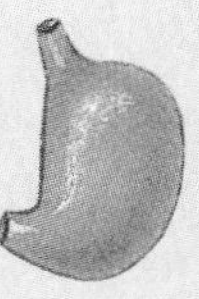

Provides additional nutrients required to support gamete production and (in pregnant women) embryonic and fetal development

● **FIGURE 28-25**
Functional Relationships between the Reproductive System and Other Systems

Birth Control Strategies

For physiological, logistical, financial, or emotional reasons most adults practice some form of conception control during their reproductive years. When the simplest and most obvious method, sexual abstinence, is unsatisfactory for some reason, another method of contraception must be used to avoid unwanted pregnancies. The selection process can be quite involved, for there are many methods available. Because each has specific strengths and weaknesses, the potential risks and benefits must be carefully analyzed on an individual basis.

Well over 50 percent of U.S. women age 15–44 are practicing some method of contraception; in 1994 an estimated 50 million American women were taking oral birth control pills. There are many different methods of contraception; only a few will be considered here.

Sterilization makes one unable to provide functional gametes for fertilization. Either sexual partner may be sterilized with the same net result. In a **vasectomy** (vaz-EK-to-mē), a segment of the ductus deferens is removed, making it impossible for spermatozoa to pass from the epididymis to the distal portions of the reproductive tract. The surgery can be performed in a physician's office in a matter of minutes. The spermatic cords are located as they ascend from the scrotum on either side, and after each cord is opened the ductus deferens is severed. After a 1-cm section is removed, the cut ends are usually tied shut. With the section removed, the cut ends do not reconnect; in time, scar tissue forms a permanent seal. A more recent vasectomy procedure often makes it possible to restore fertility at a later date. In this procedure the cut ends of the ductus deferens are blocked with silicone plugs that can later be removed. After a vasectomy the man experiences normal sexual function, for the epididymal and testicular secretions normally account for only around 5 percent of the volume of the semen. Spermatozoa continue to develop, but they remain within the epididymis until they degenerate. The failure rate for this procedure is 0.08 percent (a failure is defined as a resulting pregnancy).

In the female the uterine tubes can be blocked through a surgical procedure known as a **tubal ligation.** The failure rate for this procedure is estimated at 0.45 percent. Since the surgery involves entering the abdominopelvic cavity, complications are more likely than with vasectomy. As in a vasectomy, attempts may be made to restore fertility after a tubal ligation.

Oral contraceptives manipulate the female hormonal cycle so that ovulation does not occur. The contraceptive pills produced in the 1950s contained relatively large amounts of progestins. These concentrations were adequate to suppress pituitary production of GnRH, so FSH was not released and ovulation did not occur. Unpleasant side effects included endometrial bleeding, and most of the oral contraceptive products developed subsequently added small amounts of estrogens. Current combination pills differ significantly from the earlier products because the hormonal doses are much lower, with only one-tenth the progestins and less than half the estrogens. The hormones are administered in a cyclic fashion, beginning 5 days after the start of menses and continuing for the next 3 weeks. Over the fourth week the woman takes placebo pills or no pills at all. Low-dosage combination pills are sometimes prescribed for women experiencing irregular menstrual cycles, for they create a 28-day cycle.

There are now at least 20 different brands of combination oral contraceptives available, and over 200 million women are using them worldwide. In the United States, 25 percent of women under age 45 use the combination pill to prevent conception. The failure rate for the combination oral contraceptives, when used as prescribed, is 0.24 percent over a 2-year period. Birth control pills are not without their risks, however. For example, women with severe hypertension, diabetes mellitus, epilepsy, gallbladder disease, heart trouble, or acne may find that their problems worsen when taking the combination pills. Women taking oral contraceptives are also at increased risk for venous thrombosis, strokes, pulmonary embolism, and (for women over 35) heart disease.

Two progesterone-only forms of birth control are now available. *Depo-provera* is injected every 3 months. The silastic tubes of the Norplant system are saturated with progesterone and inserted under the skin. This method provides birth control for a period of approximately 5 years, but to date the relatively high cost has limited the use of this contraceptive method. Both Depo-provera and the Norplant system can cause irregular menstruation and temporary amenorrhea, but they are easy to use and extremely convenient.

The **condom,** also called a *prophylactic,* or "rubber," covers the body of the penis during intercourse and keeps spermatozoa from reaching the female reproductive tract. Condoms are also used to prevent transmission of sexually transmitted diseases, such as syphilis, gonorrhea, or AIDS. The reported condom failure rate varies from 6 to 17 percent depending on the study criteria. **Vaginal barriers** such as the *diaphragm, cervical cap,* or *vaginal sponge* rely on similar principles. A diaphragm, the most popular form of vaginal barrier in use at the moment, consists of a dome of latex rubber with a small metal hoop supporting the rim. Because vaginas vary in size, women choosing this method must be individually fitted. Before intercourse the diaphragm is inserted so that it covers the cervical os, and it is usually coated with a small amount of spermicidal jelly or cream, adding to the effectiveness of the barrier. The failure rate for a properly fitted diaphragm is estimated at 5–6 percent. The cervical cap is smaller and lacks the metal rim. It, too, must be fitted carefully, but unlike the diaphragm it may be left in place for several days. The failure rate (8 percent) is higher than that for diaphragm use. The vaginal sponge consists of a small synthetic sponge saturated with a *spermicide,* a sperm-killing foam or jelly. The failure rate for a contraceptive sponge is estimated at 6–10 percent.

An **intrauterine device (IUD)** consists of a small plastic loop or a T that can be inserted into the uterine chamber. The mechanism of action remains uncertain, but it is known that IUDs stimulate prostaglandin production in the uterus. The net result is an alteration in the chemical composition of uterine secretions, and the changes in the intrauterine environment lower the chances for fertilization and subsequent implantation. IUDs are in limited use today in the United States, but they remain popular in many other countries. The failure rate is estimated at 5–6 percent.

The **rhythm method** involves abstaining from sexual activity on the days ovulation might be occurring. The timing is estimated on the basis of previous patterns of menstruation and sometimes by following changes in basal body temperature and cervical mucus. The failure rate for the rhythm method is very high, approaching 25 percent.

Sterilization, oral contraceptives, condoms, and vaginal barriers are the primary contraception methods for all age groups. But the relative proportion of the population using a particular method changes with age. Sterilization is most popular in older women, who may already have had children. Relative availability may also play a role. For example, a sexually active female under age 18 can buy a condom more easily than she can obtain a prescription for an oral contraceptive. But many of the observed changes occur because the relationship between risks and benefits varies for each age group.

When attempting to make a decision concerning the use and selection of contraceptives, many people simply examine the list of potential complications and make the "safest" choice. For example, media coverage of the potential risks associated with oral contraceptives made many women reconsider their use. But complex decisions should not be made on such a simplistic basis, and the risks associated with contraceptive use must be considered in light of their relative efficiencies. Although pregnancy is a natural phenomenon, it has its risks, and the mortality rate for pregnant women in the United States averages around 8 deaths per 100,000 pregnancies. That average incorporates a broad range; the rate is 5.4 per 100,000 in women under 20, and 27 per 100,000 for women over 40. Although these risks are small, for pregnant women over age 35 the chances of dying from complications related to pregnancy are almost twice as great as the chances of being killed in an automobile accident and many times greater than the risks associated with the use of oral contraceptives. For women in Third World countries, the comparison is even more striking. The mortality rate for pregnant women in parts of Africa is approximately 1 per 150 pregnancies. In addition to preventing pregnancy, combination birth control pills have also been shown to reduce the risks of ovarian and endometrial cancers and fibrocystic breast disease.

Before age 35, *any contraceptive method is safer than pregnancy.* In general over this period the risks are proportional to the failure rates of each method. The notable exception involves individuals taking the pill who also smoke cigarettes. Younger women are more fertile, so despite a lower mortality rate for each pregnancy they are likely to have more pregnancies. As a result birth control failures imply a higher risk in the younger age groups.

After age 35 the risks of complications associated with oral contraceptive use increases, while those of other methods remain relatively stable. Women over age 35 (smokers) or 40 (nonsmokers) are therefore often advised to seek other forms of contraception. Each contraceptive method has advantages and disadvantages. As a result, research on contraception control continues. AM *Experimental Contraceptive Methods*

Selected Clinical Terminology

Terms Discussed in This Chapter

amenorrhea: The failure of menarche to appear before age 16, or a cessation of menstruation for 6 months or more in an adult female of reproductive age. *(p. 1087)*

breast cancer: A malignant, metastasizing cancer of the mammary gland that is the primary cause of death for women age 35–45. *(p. 1090)*

cryptorchidism (kript-ŌR-ki-dizm): Failure of one or both testes to descend into the scrotum by the time of birth. *(p. 1061)*

endometriosis (en-dō-mē-trē-Ō-sis): Growth of endometrial tissue outside of the uterus. *(p. 1087)*

fibrocystic disease: Clusters of lobular cysts within the tissues of the mammary gland. *(p. 1090)*

gonorrhea (gon-o-RĒ-a): A sexually transmitted disease. *(p. 1082 and AM)*

impotence: An inability to acheive or maintain an erection. *(p. 1098)*

mammography: The use of X-rays to examine breast tissue. *(p. 1090)*

mastectomy: Surgical removal of part or all of a cancerous breast. *(p. 1090)*

orchiectomy (ōr-kē-EK-to-mē): Surgical removal of a testis. *(p. 1062)*

pelvic inflammatory disease (PID): An infection of the uterine tubes. *(p. 1082)*

prostate cancer: A malignant, metastasizing cancer that is the second most common cause of cancer deaths in males. *(p. 1072)*

prostatectomy (pros-ta-TEK-to-mē): Surgical removal of the prostate gland. *(p. 1072)*

prostate-specific antigen (PSA): An antigen whose concentration in the blood increases in prostate cancer patients. *(p. 1072)*

sexually transmitted diseases (STDs): Diseases transferred from one individual to another primarily or exclusively through sexual contact. Examples include gonorrhea, syphilis, herpes, and AIDS. *(p. 1099 and AM)*

testicular torsion: Twisting of the spermatic cord resulting from rotation of the testis inside the scrotal cavity. *(p. 1064)*

vaginitis (va-jin-Ī-tis): Infection of the vaginal canal by fungal or bacterial pathogens. *(p. 1088 and AM)*

vasectomy (vaz-EK-to-mē): Surgical removal of a segment of the ductus deferens, making it impossible for spermatozoa to reach the distal portions of the reproductive tract. *(p. 1102)*

AM Additional Terms Discussed in the Applications Manual

endometrial polyps: Benign epithelial tumors of the uterine lining.

leiomyomas (lī-ō-mī-Ō-maz), or **fibroids:** Benign myometrial tumors that are the most common tumors in women.

CHAPTER REVIEW

STUDY OUTLINE

INTRODUCTION, p. 1060

1. The human reproductive system produces, stores, nourishes, and transports functional **gametes** (reproductive cells). **Fertilization** is the fusion of a **sperm** from the father and an **ovum** from the mother to create a **zygote** (fertilized egg).

AN OVERVIEW OF THE REPRODUCTIVE SYSTEM, p. 1060

1. The reproductive system includes **gonads,** ducts, accessory glands and organs, and the **external genitalia.**
2. In the male the **testes** produce sperm, which are expelled from the body in **semen** during **ejaculation.** The **ovaries** (gonads) of a sexually mature female produce an egg that travels along **uterine tubes** to reach the **uterus.** The **vagina** connects the uterus with the exterior.

THE REPRODUCTIVE SYSTEM OF THE MALE, p. 1060

1. The **spermatozoa** travel along the **epididymis,** the **ductus deferens,** the **ejaculatory duct,** and the **urethra** before leaving the body. Accessory organs (notably the **seminal vesicles, prostate gland,** and **bulbourethral glands**) secrete into the ejaculatory ducts and urethra. The **scrotum** encloses the testes, and the **penis** is an erectile organ. (*Figure 28-1*)

The Testes, p. 1060

2. The **descent of the testes** through the **inguinal canals** occurs during development. The testes remain connected to internal structures via the **spermatic cord.** The **raphe** marks the boundary between the two chambers in the scrotum. (*Figures 28-2, 28-3*)
3. The **dartos** muscle gives the scrotum a wrinkled appearance; the **cremaster muscle** pulls the testes closer to the body. The **tunica albuginea** surrounds each testis. Septa extend from the tunica albuginea to the **mediastinum** of the testis, creating a series of **lobules.** (*Figure 28-4*)
4. **Seminiferous tubules** within each lobule are the sites of sperm production. From there sperm pass through a **straight tubule** to the **rete testis. Efferent ducts** connect the rete testis to the epididymis. Between the seminiferous tubules there are **interstitial cells** that secrete sex hormones. (*Figure 28-5*)
5. Seminiferous tubules contain **spermatogonia,** stem cells involved in **spermatogenesis,** and **sustentacular cells,** which sustain and promote the development of spermatozoa. (*Figure 28-6*)

Anatomy of a Spermatozoon, p. 1068

6. Each spermatozoon has a **head, middle piece,** and **tail.** (*Figure 28-7*)

The Male Reproductive Tract, p. 1068

7. From the testis the spermatozoa enter the **epididymis,** an elongate tubule with **head, body,** and **tail** regions. The epididymis monitors and adjusts the composition of the tubular fluid and serves as a recycling center for damaged spermatozoa. (*Figure 28-8*)
8. The **ductus deferens,** or *vas deferens,* begins at the epididymis and passes through the inguinal canal as one component of the spermatic cord. Near the prostate it enlarges to form the **ampulla.** The junction of the base of the seminal vesicle and the ampulla creates the **ejaculatory duct,** which empties into the urethra. (*Figure 28-9a,b*)
9. The urethra extends from the urinary bladder to the tip of the penis. It can be divided into three regions: **prostatic urethra, membranous urethra,** and **penile urethra.**

The Accessory Glands, p. 1071

10. Each **seminal vesicle** is an active secretory gland that contributes about 60 percent of the volume of semen; its secretions contain fructose that is easily metabolized by spermatozoa. The **prostate gland** secretes alkaline fluids that help neutralize the acids normally found in the urethra and vagina. Alkaline mucus secreted by the **bulbourethral glands** has lubricating properties. (*Figure 28-9c,d,e*)

Semen, p. 1073

11. A typical ejaculation releases 2–5 ml of semen (an **ejaculate**), which contains 20–100 million sperm per milliliter. (*Appendix VII*)

The Penis, p. 1074

12. The skin overlying the penis resembles that of the scrotum. Most of the body of the penis consists of three masses of **erectile tissue.** Beneath the superficial fascia there are two **corpora cavernosa** and a single **corpus spongiosum** that surrounds the urethra. Dilation of the erectile tissue with blood produces an **erection.** (*Figure 28-10*)

Hormones and Male Reproductive Function, p. 1076

13. Important regulatory hormones include **FSH** (follicle-stimulating hormone), **LH** (luteinizing hormone), and **GnRH** (gonadotropin-releasing hormone). Testosterone is the most important androgen. (*Figure 28-11*)

THE REPRODUCTIVE SYSTEM OF THE FEMALE, p. 1077

1. Principal organs of the female reproductive system include the ovaries, uterine tubes, uterus, vagina, and external genitalia. (*Figure 28-12*)
2. The ovaries, uterine tubes, and uterus are enclosed within the **broad ligament** (an extensive mesentery). The **mesovarium** supports and stabilizes each ovary.

The Ovaries, p. 1077

3. The ovaries are held in position by the **ovarian ligament** and the **suspensory ligament.** Major blood vessels enter the ovary at the **ovarian hilum.** Each ovary is covered by a **tunica albuginea.** (*Figure 28-13*)
4. **Oogenesis** (ovum production) occurs monthly in **ovarian follicles** as part of the **ovarian cycle.** As development proceeds one finds **primordial, primary, secondary,** and **tertiary follicles.** At **ovulation** an **oocyte** and the surrounding follicular walls of the **corona radiata** are released through the ruptured ovarian wall. (*Figures 28-14, 28-15*)

The Uterine Tubes, p. 1081

5. Each **uterine tube** has an **infundibulum** with **fimbriae** (projections), an **ampulla,** an **isthmus,** and an **intramural portion** that opens into the uterine cavity. For fertilization to occur, the ovum must encounter spermatozoa during the first 12–24 hours of its passage from the infundibulum to the uterus. (*Figure 28-16*)

The Uterus, p. 1083

6. The uterus provides mechanical protection and nutritional support to the developing embryo. Normally the uterus bends anteriorly near its base **(anteflexion).** It is stabilized by the broad ligament, **uterosacral ligaments, round ligaments,** and the **lateral ligaments.** (*Figure 28-17*)
7. Major anatomical landmarks of the uterus include the **body, isthmus, cervix, external orifice, uterine cavity, cervical canal,** and **internal orifice.** The uterine wall can be divided into an inner **endometrium,** a muscular **myometrium,** and a superficial **serosa (perimetrium).** (*Figure 28-18*)
8. A typical 28-day **uterine,** or **menstrual, cycle** begins with the onset of **menses** and the destruction of the functional zone of the endometrium. This process of **menstruation** continues from 1 to 7 days. (*Figure 28-19a*)
9. After menses, the **proliferative phase** begins and the functional zone undergoes repair and thickens. Menstrual activity begins at **menarche** and continues until **menopause.** (*Figure 28-19b,c*)

The Vagina, p. 1087

10. The **vagina** is a muscular tube extending between the uterus and external genitalia. Before the onset of sexual activity a thin epithelial fold, the **hymen,** partially blocks the entrance to the vagina. (*Figures 28-20, 28-21*)

The External Genitalia, p. 1089

11. The components of the **vulva** include the **vestibule, labia minora, clitoris, labia majora,** and the **lesser** and **greater vestibular glands.** (*Figure 28-21*)

The Mammary Glands, p. 1089

12. At birth a newborn infant gains nourishment from milk secreted by maternal **mammary glands.** (*Figure 28-22*)

Hormones and the Female Reproductive Cycle, p. 1091

13. Hormonal regulation of the female reproductive system involves coordinating the ovarian and uterine cycles.
14. **Estradiol,** one of the estrogens, is the dominant hormone of the preovulatory period. Ovulation occurs in response to peak levels of estrogen and LH. (*Figure 28-23*)

15. The follicular cells remaining within the ovary form the **corpus luteum** that later degenerates into a **corpus albicans** of scar tissue. The hypothalamic secretion of GnRH triggers the pituitary secretion of FSH and the synthesis of LH. FSH initiates follicular development, and activated follicles and ovarian interstitial cells produce estrogens. **Progesterone,** one of the steroid hormones called **progestins,** is the principal hormone of the postovulatory period. Hormonal changes are responsible for the maintenance of the menstrual cycle. (*Figure 28-24*)

THE PHYSIOLOGY OF SEXUAL INTERCOURSE, p. 1098

Male Sexual Function, p. 1098

1. During **arousal** in the male erotic thoughts or sensory stimulation or both lead to parasympathetic activity that produces erection. Stimuli accompanying **coitus (intercourse** or **copulation)** lead to **emission** and **ejaculation.** Strong muscle contractions are associated with **orgasm.**

Female Sexual Function, p. 1098

2. The phases of female sexual function resemble those of the male, with parasympathetic arousal and skeletal muscle contractions associated with orgasm.

AGING AND THE REPRODUCTIVE SYSTEM, p. 1099

Menopause, p. 1099

1. Menopause (the time that ovulation and menstruation cease in women) typically occurs around age 50. Production of GnRH, FSH, and LH rise, while circulating concentrations of estrogen and progesterone decline.

The Male Climacteric, p. 1099

2. During the **male climacteric,** between ages 50 and 60, circulating testosterone levels decline, while levels of FSH and LH rise.

INTEGRATION WITH OTHER SYSTEMS, p. 1099

1. Hormones play a major role in coordinating reproduction. (*Table 28-1*)
2. In addition to the endocrine and reproductive systems, reproduction requires the normal functioning of the digestive, nervous, cardiovascular, and urinary systems. (*Figure 28-25*)

REVIEW QUESTIONS

LEVEL 1 Reviewing Facts and Terms

1. Perineal structures associated with the reproductive system are collectively known as:
   (a) gonads (b) sex gametes
   (c) external genitalia (d) accessory glands

2. Interstitial cells are responsible for:
   (a) sperm production
   (b) maintenance of the blood-testis barrier
   (c) support of spermiogenesis
   (d) production of androgens

3. Sustentacular cells are responsible for the secretion of:
   (a) inhibin
   (b) androgen-binding protein
   (c) Müllerian-inhibiting factor
   (d) a, b, and c are correct

4. During the process of meiosis, when synapsis occurs, corresponding maternal and paternal chromosomes come together to produce:
   (a) 46 pairs of chromosomes
   (b) 23 chromosomes
   (c) 23 pairs of chromosomes
   (d) the haploid number of chromosomes

5. The completion of the meiotic process in the male produces four spermatids, each containing:
   (a) 23 chromosomes
   (b) 23 pairs of chromosomes
   (c) the diploid number of chromosomes
   (d) 46 pairs of chromosomes

6. Erection of the penis occurs when:
   (a) sympathetic activation of penile arteries occurs
   (b) arterial branches are constricted, and muscular partitions are tense
   (c) the vascular channels become engorged with blood
   (d) a, b, and c are correct

7. In the male, the primary target of FSH is the:
   (a) sustentacular cells of the seminiferous tubules
   (b) interstitial cells of the seminiferous tubules
   (c) cells of Leydig
   (d) epididymis

8. Testosterone and other androgens are secreted by the:
   (a) hypothalamus (b) anterior pituitary gland
   (c) sustentacular cells (d) interstitial cells

9. The ovaries in the female are responsible for:
   (a) the production of female gametes
   (b) the secretion of female sex hormones
   (c) the secretion of inhibin
   (d) a, b, and c are correct

10. The production of gametes occurs in the ________ of the ovary.
   (a) germinal epithelium (b) medulla
   (c) cortex (d) tunica albuginea

11. The process of ovum production, or oogenesis, begins:
   (a) before birth (b) after birth
   (c) at puberty (d) after puberty

12. In the female, the process of meiosis is not completed:
   (a) until birth
   (b) until puberty
   (c) unless and until fertilization occurs
   (d) until uterine implantation occurs

13. If fertilization is to occur, the ovum must encounter spermatozoa during the first ________ of its passage.
(a) 1 to 5 hours (b) 6 to 11 hours
(c) 12 to 24 hours (d) 25 to 36 hours

14. The part of the endometrium that undergoes cyclical changes in response to sexual hormonal levels is the:
(a) serosa
(b) basilar zone
(c) muscular myometrium
(d) functional zone

15. The secretory phase of the uterine cycle is influenced primarily by the stimulatory effects of progestins and estrogens from the:
(a) anterior pituitary (b) hypothalamus
(c) endometrium (d) corpus luteum

16. A sudden surge in LH concentration causes:
(a) the onset of menses
(b) the rupture of the follicular wall and ovulation
(c) the beginning of the proliferative phase
(d) the end of the uterine cycle

17. The principal hormone of the postovulatory period is:
(a) progesterone (b) estradiol
(c) estrogen (d) luteinizing hormone

18. At the time of ovulation, the basal body temperature:
(a) is not affected
(b) increases noticeably
(c) declines sharply
(d) may increase or decrease a few degrees

19. Impotence is the inability to:
(a) produce sufficient amounts of sperm for fertilization
(b) achieve or maintain an erection
(c) achieve orgasm
(d) ejaculate

20. Clitoral erection occurs because of the influence of:
(a) parasympathetic activation
(b) sympathetic activation
(c) increased output of estrogen
(d) increased output of progesterone

21. Menopause is accompanied by:
(a) sustained rises in GnRH, FSH, and LH
(b) declines in circulating levels of estrogen and progesterone
(c) thinning of the urethral and vaginal walls
(d) a, b, and c are correct

22. What reproductive structures are common to both male and female?

23. Trace the duct system that the sperm traverses from the site of production to the exterior.

24. What accessory organs and glands contribute to the composition of semen? What are the functions of each?

25. What two primary cell populations in the testes are responsible for functions related to reproductive activity? What are the functions of these cells?

26. What are the three primary functions of the epididymis in the male?

27. Identify the three regions of the male urethra.

28. Describe the composition of a typical sample of ejaculate.

29. Enumerate the functions of testosterone in the male.

30. What are the primary functions of the ovaries in the female?

31. List and summarize the important steps in the ovarian cycle.

32. Describe the histological composition of the uterine wall.

33. What are the three major functions of the vagina?

34. What is the functional significance of the normally acidic pH of the vagina?

35. What is the role of the clitoris in the female reproductive system?

36. What is the function of the lesser and greater vestibular glands in the female?

37. Trace the route taken by a sample of milk from its site of production to the outside of the body.

LEVEL 2 Reviewing Concepts

38. How does the human reproductive system differ functionally from all other systems in the body?

39. How are the male and female reproductive systems functionally different?

40. How is the process of meiosis involved in the development of the spermatozoon and the ovum?

41. Describe the erectile tissues of the penis. How does erection occur?

42. Using an average uterine cycle of 28 days, describe each of the three phases of the menstrual cycle.

43. Describe the hormonal events associated with the ovarian cycle.

44. Describe the hormonal events associated with the uterine cycle.

45. Summarize how arousal and orgasm are mediated in the male and female. Do these processes differ between the sexes?

46. How does the aging process affect the reproductive systems of men and women?

47. How do birth control pills prevent conception?

LEVEL 3 Critical Thinking and Clinical Applications

48. Diane has an inflammation of the peritoneum (peritonitis), which her doctor says resulted from a urinary tract infection. Why could this situation occur in females but not in males?

49. Rod suffers an injury to the sacral region of his spinal cord. Will he still be able to achieve an erection? Explain.

50. A seven-year-old girl develops an ovarian tumor involving granulosa cells. What symptoms would you expect to observe?

51. If the enzyme required for the production of testosterone were blocked in females, how would this blockage affect a woman's ovarian cycle?

52. Women body builders and women suffering from eating disorders such as anorexia nervosa cease having menstrual cycles, a condition known as amenorrhea. What does this relationship suggest about the role of body fat and menstruation? What benefit might there be in the discontinuance of menstruation under such circumstances?

CHAPTER 29

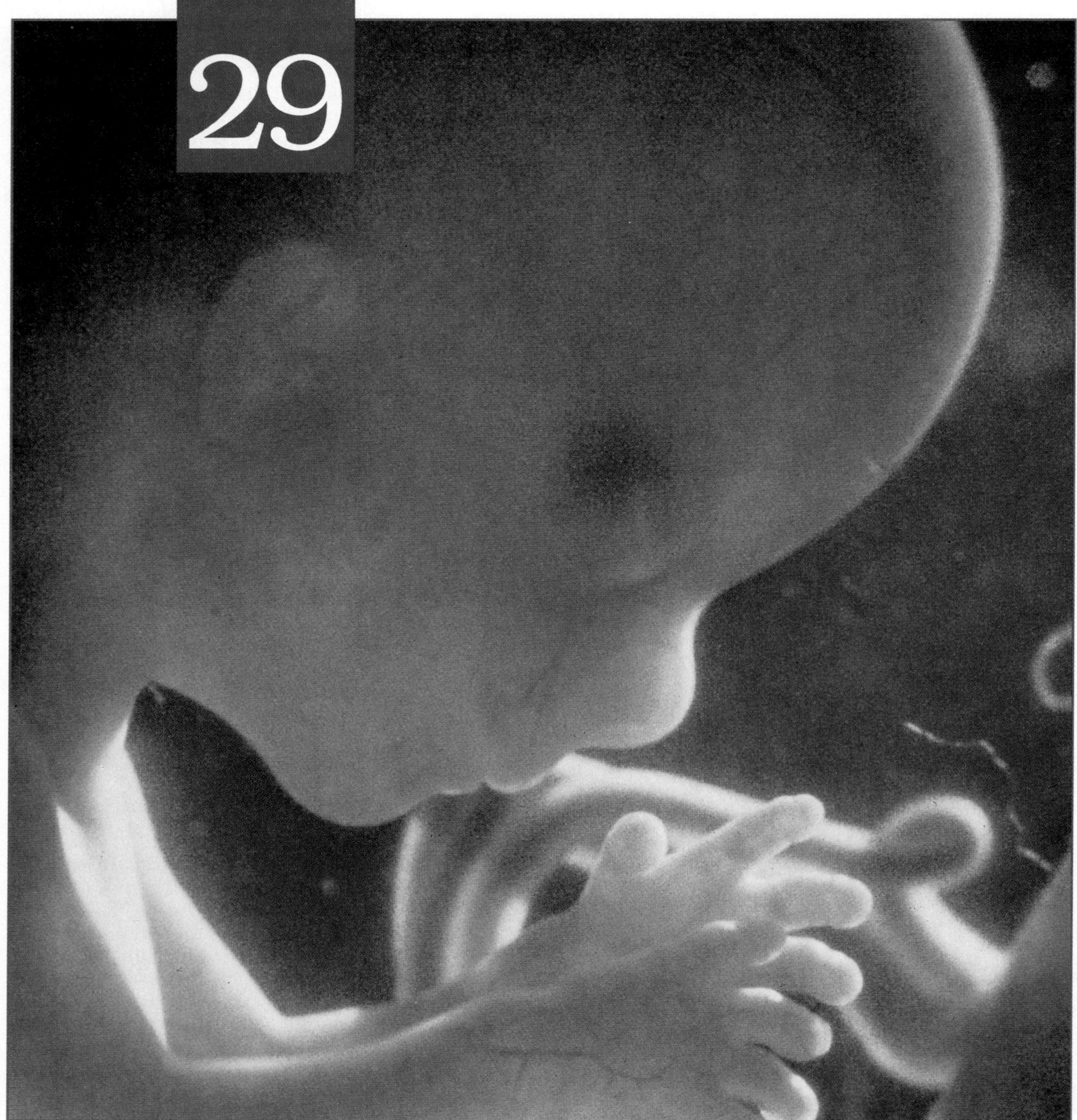

The physiological processes we have studied thus far have been brief ones. Many last only a fraction of a second; others may take hours at most. But some important processes of life are measured in months, years, or decades. A human being develops in the womb for nine months, grows to maturity in 15 or 20 years, and may live the better part of a century. During that whole span of time, he or she will never cease to change. Birth, growth, maturation, aging, and death are all parts of a single, continuous process. And that process does not end with the individual, for human beings can pass at least some of their characteristics on to their offspring. Thus each generation gives rise to a new generation that will repeat the same cycle.... In this chapter we will explore the continuity of life, from conception to death.

Development and Inheritance

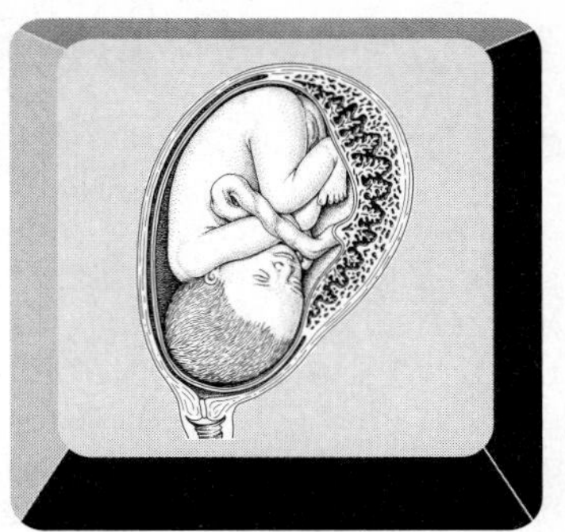

Chapter Outline and Objectives

1 Describe the process of fertilization.

2 Explain how developmental processes are regulated.

3 List the three prenatal periods and describe the major events associated with each period.

4 Explain how the germ layers participate in the formation of extraembryonic membranes.

5 Discuss the importance of the placenta as an endocrine organ.

6 Describe the interplay between the maternal organ systems and the developing fetus.

7 Discuss the structural and functional changes in the uterus during gestation.

8 List and discuss the events that occur during labor and delivery.

9 Identify the features and functions associated with the various life stages.

10 Relate basic principles of genetics to the inheritance of human traits.

Time refuses to stand still; today's infant will be tomorrow's adult. The gradual modification of anatomical structures during the period from conception to maturity is called **development.** The changes that occur during development are truly remarkable—what begins as a single cell slightly larger than the period at the end of this sentence becomes an individual whose body contains trillions of cells organized into highly specialized tissues, organs, and organ systems. The creation of different cell types required by this process is called **differentiation.** Differentiation occurs through selective changes in genetic activity. As development proceeds, some genes are turned off and others turned on. The identities of these genes vary from one cell type to another.

A basic understanding of human development provides important insights into anatomical structures. In addition, many of the mechanisms of development and growth are similar to those responsible for the repair of injuries. This chapter will focus on major aspects of development and consider highlights of the developmental process rather than describe the events in great detail. We will also consider the regulatory mechanisms and how developmental patterns can be modified—for good or ill. Few topics in the biological sciences hold such fascination, and fewer still confront the investigator with so dazzling an array of technological, moral, and logistical challenges.

An Overview of Topics in Development

Development involves (1) the division and differentiation of cells and (2) the changes that produce and modify anatomical structures. Development begins at fertilization, or **conception,** and can be divided into periods characterized by specific anatomical changes. **Embryological development** comprises those events that occur during the first 2 months after fertilization. The study of these events is called **embryology** (em-brē-OL-ō-jē). **Fetal development** begins at the start of the ninth week and continues up to the time of birth. Embryological and fetal development are sometimes referred to collectively as **prenatal development,** the primary focus of this chapter. **Postnatal development** commences at birth and continues to maturity.

Although all human beings go through the same developmental stages, their differences in genetic makeup produce distinctive individual characteristics. **Inheritance** refers to the transfer of genetically determined characteristics from generation to generation. **Genetics** is the study of the mechanisms responsible for inheritance. This chapter considers basic genetics as it applies to the appearance of inherited characteristics such as sex, hair color, and various diseases.

Fertilization

Fertilization involves the fusion of two haploid gametes, producing a zygote containing the normal somatic number of chromosomes, 46. ∞ [p. 1067] The functional roles and contributions of the spermatozoon and the ovum are very different. The spermatozoon simply delivers the paternal chromosomes to the site of fertilization. It is the ovum that must provide all of the nourishment and genetic programming to support the embryonic development for nearly a week after conception. The volume of the ovum is therefore much greater than that of the spermatozoon. At the time of fertilization, the diameter of the ovum is over twice the entire length of the spermatozoon. The relationship between the egg and sperm volumes is even more striking, on the order of 2000:1.

The sperm arriving in the vagina are already motile, but they cannot fertilize an egg until they have undergone **capacitation** (ka-pas-i-TĀ-shun), discussed in Chapter 28. ∞ [p. 1070] It appears that a substance secreted by the epididymis prevents capacitation while spermatozoa are within the male reproductive tract. However, the precise mechanism of capacitation within the female reproductive tract remains uncertain.

Normal fertilization occurs near the junction between the ampulla and isthmus of the uterine tube, usually within a day following ovulation. Over this period of time, the oocyte has traveled a few centimeters, but the spermatozoa must cover the distance between the vagina and the isthmus. An individual spermatozoon can propel itself at speeds of only about 34 μm per second, roughly equivalent to 12.5 cm (5 in.) per hour, so in theory it should be several hours before any spermatozoa arrive in the upper portions of the uterine tubes. The actual passage time, however, ranges from 2 hours to as little as 30 minutes. Contractions of the uterine musculature and ciliary currents in the uterine tubes have been suggested as likely mechanisms for accelerating the movement of spermatozoa from the vagina to the fertilization site.

Even with transport assistance and available nutrients, this is not an easy passage. Of the roughly 200 million spermatozoa introduced into the vagina in a typical ejaculation, only around 10,000 enter the uterine tube, and fewer than 100 actually reach the isthmus. A male with a sperm count below 20 million per milliliter is functionally **sterile** because too few spermatozoa will survive to reach the oocyte. Large numbers of spermatozoa are required for successful fertilization, because of the condition of the oocyte at ovulation.

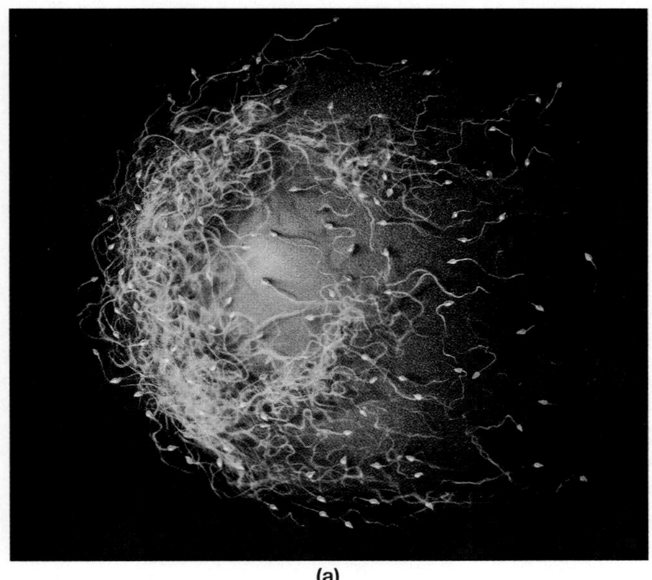

(a)

❑ The Oocyte at Ovulation

Figure 29-1

Ovulation occurs before the oocyte is completely mature: The secondary oocyte leaving the follicle is in metaphase of the second meiotic division. The cell's metabolic operations have been discontinued, and the oocyte drifts in a sort of suspended animation, awaiting the stimulus for further development. If fertilization does not occur, the oocyte disintegrates without completing meiosis.

Fertilization is complicated by the fact that when it leaves the ovary, the oocyte is surrounded by a layer of follicle cells, the *corona radiata.* ∞ [p. 1081] The events that follow are diagrammed in Figure 29-1●. The cells of the corona radiata protect the secondary oocyte as it passes through the ruptured follicular wall and into the infundibulum of the uterine tube. Although the physical

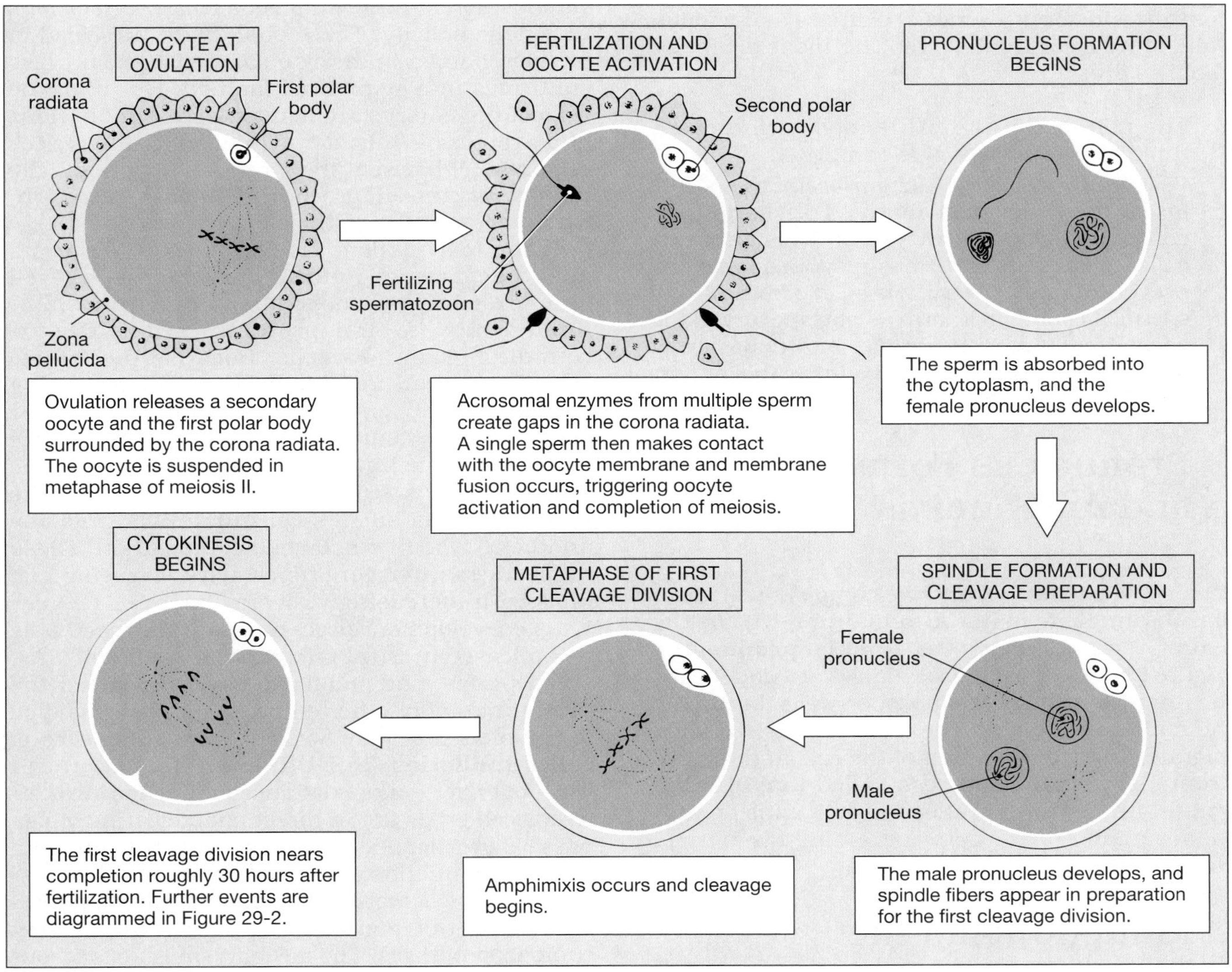

OOCYTE AT OVULATION

Corona radiata

First polar body

Zona pellucida

Ovulation releases a secondary oocyte and the first polar body surrounded by the corona radiata. The oocyte is suspended in metaphase of meiosis II.

FERTILIZATION AND OOCYTE ACTIVATION

Second polar body

Fertilizing spermatozoon

Acrosomal enzymes from multiple sperm create gaps in the corona radiata. A single sperm then makes contact with the oocyte membrane and membrane fusion occurs, triggering oocyte activation and completion of meiosis.

PRONUCLEUS FORMATION BEGINS

The sperm is absorbed into the cytoplasm, and the female pronucleus develops.

SPINDLE FORMATION AND CLEAVAGE PREPARATION

Female pronucleus

Male pronucleus

The male pronucleus develops, and spindle fibers appear in preparation for the first cleavage division.

METAPHASE OF FIRST CLEAVAGE DIVISION

Amphimixis occurs and cleavage begins.

CYTOKINESIS BEGINS

The first cleavage division nears completion roughly 30 hours after fertilization. Further events are diagrammed in Figure 29-2.

(b)

● **FIGURE 29-1**

Fertilization. (a) An oocyte at the time of fertilization; note the difference in size between the gametes. **(b)** Fertilization and the preparations for cleavage.

process of fertilization requires only a single sperm in contact with the oocyte membrane, that spermatozoon must first penetrate the corona radiata. The acrosomal cap of the sperm contains **hyaluronidase** (hī-al-u-RON-a-dāz), an enzyme that breaks down the intercellular cement between adjacent follicle cells. At least a hundred spermatozoa must release hyaluronidase before the connections between the follicular cells break down enough to permit fertilization.

No matter how many spermatozoa slip through the gap, normally only a single spermatozoon will accomplish fertilization and activate the oocyte. When that spermatozoon contacts the secondary oocyte, the plasma membranes of the two cells fuse and the sperm enters the **ooplasm,** or cytoplasm of the oocyte. This event alters the chemical environment inside the oocyte, and **oocyte activation** follows.

Oocyte Activation

Activation involves a series of changes in the metabolic activity of the oocyte. The most dramatic changes are:

1. The metabolic rate of the oocyte increases rapidly, and meiosis II is completed.
2. Vesicles just beneath the surface of the oocyte fuse with the cell membrane and discharge their contents through exocytosis. This process, called a *cortical reaction,* is responsible for the prevention of penetration by more than one sperm, a condition known as **polyspermy.** If the oocyte reaction is abnormal, and polyspermy does occur, the zygote will be incapable of normal development.

❑ Pronucleus Formation and Amphimixis

Figure 29-1

After oocyte activation and the completion of meiosis, the nuclear material remaining within the ovum reorganizes as the **female pronucleus** (Figure 29-1•). While these changes are under way, the nucleus of the spermatozoon swells, becoming the **male pronucleus.** The male pronucleus then migrates toward the center of the cell, and the two pronuclei fuse in a process called **amphimixis** (am-fi-MIK-sis). Fertilization is now complete, with the formation of a zygote containing the normal complement of 46 chromosomes.

❑ Induction and the Regulation of Development

During prenatal development a single cell forms a 3–4 kg infant that in postnatal development grows through adolescence and maturity toward old age and eventual death. One of the most fascinating aspects of development is its apparent order and simplicity. A continuity exists at all levels and at all times. Nothing leaps into existence, unheralded and without apparent precursors; differentiation and increasing structural complexity occur hand in hand.

Differentiation involves changes in the genetic activity of some cells and not others. Chapter 3 considered the exchange of information between the nucleus and cytoplasm in a cell. ∞ [p. 98] Activity in the nucleus varies in response to chemical messages arriving from the surrounding cytoplasm. In turn, ongoing nuclear activity will alter conditions within the cytoplasm by directing the synthesis of specific proteins. In this way the nucleus can affect enzyme activity, cell structure, and membrane properties.

In development, differences in the cytoplasmic composition of individual cells trigger alterations in genetic activity. These changes in turn lead to further changes in the cytoplasm, and the process continues in a sequential fashion. But if all the cells of the embryo are derived from cell divisions of a zygote, how do the cytoplasmic differences originate? What sets this process in motion? The important first step occurs before fertilization, while the oocyte is in the ovary.

Prior to ovulation, the growing oocyte accepts amino acids, nucleotides, and glucose, as well as more complex materials such as phospholipids, mRNAs, and proteins, from the surrounding granulosa cells. Because the follicle cells are not all manufacturing and delivering the same nutrients and instructions to the oocyte, the contents of the oocyte cytoplasm are not evenly distributed. After fertilization, subsequent divisions subdivide the cytoplasm of the zygote into ever smaller cells that differ from one another in their cytoplasmic composition. These differences alter genetic activity, creating cell lines with increasingly diverse fates.

As development proceeds, some of these cells will release chemical substances, such as RNAs, polypeptides, and small proteins, that affect the differentiation of other embryonic cells. This type of chemical interplay between developing cells is called **induction** (in-DUK-shun). Induction can work over very short distances, as when two different cell types are in direct contact. It may also operate over longer distances, with the inducing chemicals functioning as hormones.

This type of regulation, which involves an integrated series of interacting steps, can control very complex processes. The mechanism is not without risk, however, and the appearance of an abnormal or inappropriate inducer can throw the entire development plan off course. AM *Teratogens and Abnormal Development*

❑ A Preview of Prenatal Development

The time spent in prenatal development is known as the period of **gestation** (jes-TĀ-shun). For convenience, the gestation period is usually considered as three integrated **trimesters,** each 3 months in duration:

- The **first trimester** is the period of embryonic and early fetal development. During this period the rudiments of all the major organ systems appear.
- In the **second trimester** the organs and organ systems complete most of their development. The body proportions change, and by the end of the second trimester the fetus looks distinctively human.
- The **third trimester** is characterized by rapid fetal growth. Early in the third trimester most of the major organ systems become fully functional, and an infant born 1 month or even 2 months prematurely has a reasonable chance of survival.

The First Trimester

At the moment of conception, the fertilized ovum is a single cell with a diameter of around 0.135 mm (0.005 in.). In the first month, its weight increases by a factor of 140,000. By the end of the first trimester (twelfth developmental week) the fetus is almost 75 mm (3 in.) long and weighs perhaps 14 g (0.5 oz).

Many important and complex developmental events occur during the first trimester. We will focus on four general processes: *cleavage, implantation, placentation,* and *embryogenesis.*

1. **Cleavage** (KLĒV-ij) is a sequence of cell divisions that begins immediately after fertilization and ends at the first contact with the uterine wall. Over this period the zygote becomes a **pre-embryo** that develops into a multicellular complex known as a *blastocyst.* (Cleavage and blastocyst formation were introduced in the Embryology Summary in Chapter 4.) [p. 124]
2. **Implantation** begins with the attachment of the blastocyst to the endometrium and continues as the blastocyst invades the maternal tissues. During the time implantation is under way, other important events take place that set the stage for the formation of vital embryonic structures.
3. **Placentation** (plas-en-TĀ-shun) occurs as blood vessels form around the periphery of the blastocyst and the **placenta** appears. The placenta is a vital link between maternal and embryonic systems that will support the fetus during the second and third trimesters.
4. **Embryogenesis** (em-brē-ō-JEN-e-sis) is the formation of a viable embryo. This process establishes the foundations for all major organ systems.

These processes are both complex and vital to the survival of the embryo. Perhaps because the events in the first trimester are so complex, this is the most dangerous period in prenatal life. Only about 40 percent of conceptions produce embryos that survive the first trimester. For this reason pregnant women are usually warned to take great care to avoid drugs or other disruptive stresses during the first trimester, in the hopes of preventing an error in the delicate processes under way.

❑ Cleavage and Blastocyst Formation

Figure 29-2

Cleavage (Figure 29-2●) is a series of cell divisions that subdivides the cytoplasm of the zygote. The first cleavage division produces a pre-embryo consisting of two identical cells, called **blastomeres** (BLAS-tō-mērz). The first division is completed roughly 30 hours after fertilization, and subsequent cleavage divisions occur at intervals of 10–12 hours. During the initial cleavage divisions, all the blastomeres undergo mitosis simultaneously, but as the number of blastomeres increases, the timing becomes less predictable.

After 3 days of cleavage, the pre-embryo is a solid ball of cells, resembling a mulberry. This stage is called the **morula** (MOR-ū-la; *morula,* mulberry). The morula usually reaches the uterus on day 4. Over the next 2 days, the blastomeres form a hollow ball, the **blastocyst,** with an inner cavity known as the **blastocoele** (BLAS-tō-sēl). The blastomeres are no longer identical in size and shape. The outer layer of cells, which separates the outside world from the blastocoele, is called the **trophoblast** (TRŌ-fō-blast). The function is implied by the name: *trophos,* food + *blast,* precursor. These cells will be responsible for providing food to the developing embryo. A second group of cells, the **inner cell mass,** lies clustered at one end of the blastocyst. These cells are exposed to the blastocoele but insulated from contact with the outside environment by the trophoblast. In time the inner cell mass will form the embryo.

❑ Implantation

Figure 29-3

At fertilization the zygote is still 4 days away from the uterus. It arrives in the uterine cavity as a morula, and over the next 2–3 days blastocyst formation occurs. Over this period the blastomeres

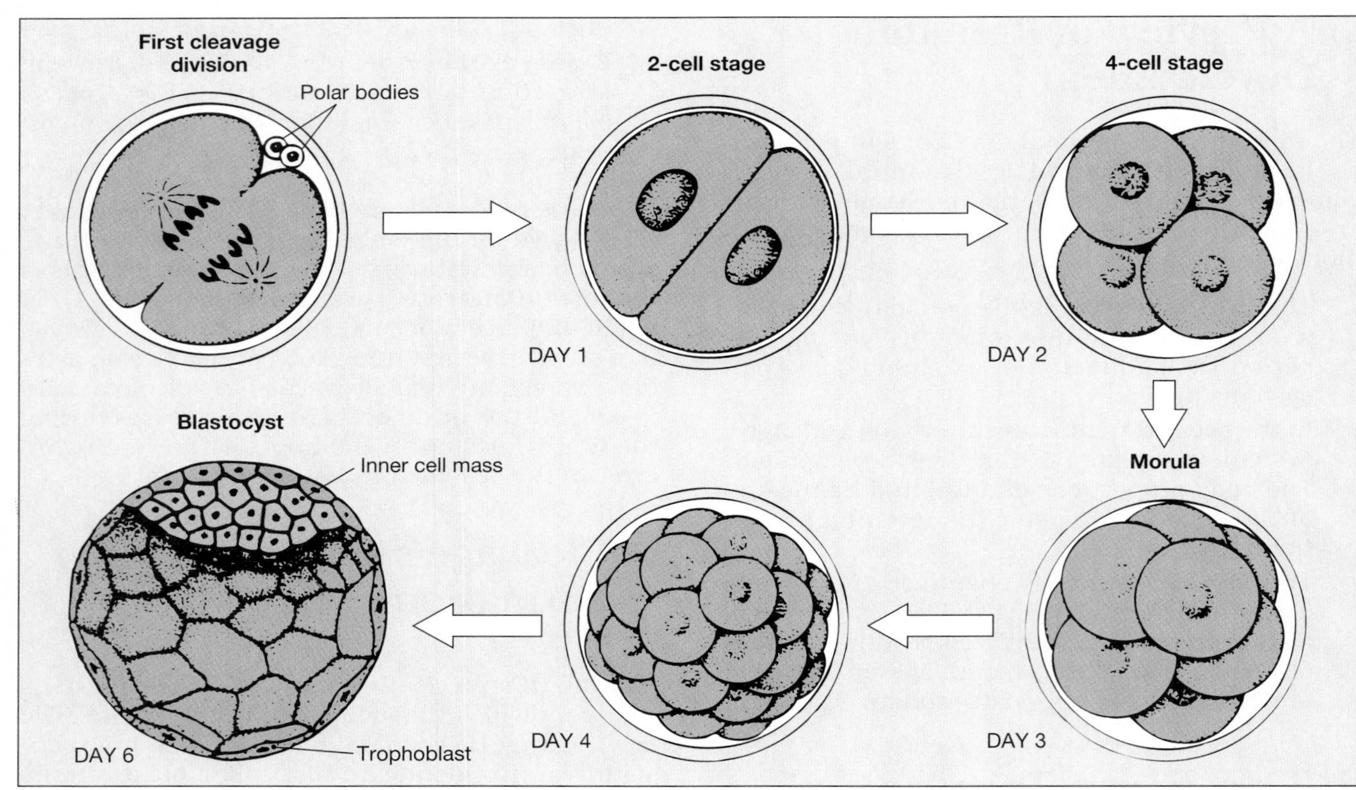
First cleavage division
Polar bodies
2-cell stage
4-cell stage
DAY 1
DAY 2
Morula
Blastocyst
Inner cell mass
DAY 6
Trophoblast
DAY 4
DAY 3

● **FIGURE 29-2**
Cleavage and Blastocyst Formation

are gaining nutrients from the fluid within the uterine cavity. This fluid, rich in glycogen, is secreted by the endometrial glands. When fully formed, the blastocyst contacts the endometrium, usually in the fundus or body of the uterus, and implantation occurs. Stages in the implantation process are illustrated in Figure 29-3●.

Implantation begins as the surface of the blastocyst closest to the inner cell mass touches and adheres to the uterine lining (see *day 7*, Figure 29-3●). At the point of contact the trophoblast cells divide rapidly, making the trophoblast several layers thick. The cells closest to the interior of the blastocyst remain intact, forming a layer of **cellular trophoblast,** or *cytotrophoblast.* Near the endometrial wall, the cell membranes separating the trophoblast cells disappear, creating a layer of cytoplasm containing multiple nuclei (*day 8*). This outer layer is called the **syncytial** (sin-SISH-al) **trophoblast,** or *syncytiotrophoblast.*

The syncytial trophoblast erodes a path through the uterine epithelium by secreting the enzyme *hyaluronidase.* This enzyme dissolves the intercellular cement between adjacent epithelial cells, just as the hyaluronidase released by spermatozoa dissolves the connections between cells of the corona radiata. At first this erosion creates a gap in the uterine lining, but the migration and divisions of epithelial cells soon repair the surface. When the repairs are completed, the blastocyst loses contact with the uterine cavity, and development occurs entirely within the *functional zone* of the endometrium. [p. 1085]

Implantation usually occurs at the endometrial surface lining the uterine cavity. The precise location within the uterus varies, although most often implantation occurs in the body of the uterus. This is not an ironclad rule, and in an **ectopic pregnancy** implantation occurs somewhere other than within the uterus. The incidence of such problems is relatively low, approximately 0.6 percent. AM *Ectopic Pregnancies*

As implantation proceeds, the syncytial trophoblast continues to enlarge and spread into the surrounding endometrium (*day 9*). The digestion of uterine glands releases nutrients that are absorbed by the syncytial trophoblast and distributed by diffusion across the underlying cellular trophoblast to the inner cell mass. These nutrients provide the energy needed to support the early stages of embryo formation. Trophoblastic extensions grow around endometrial capillaries, and as the capillary walls are destroyed, maternal blood begins to percolate through trophoblastic channels known as **lacunae.** Fingerlike **villi** extend away from the trophoblast into the surrounding endometrium, and these extensions gradually increase in size and complexity until around day 21. Beginning around day 9, the syncytial trophoblast begins breaking down larger endometrial veins and arteries, and blood flow through the lacunae accelerates.

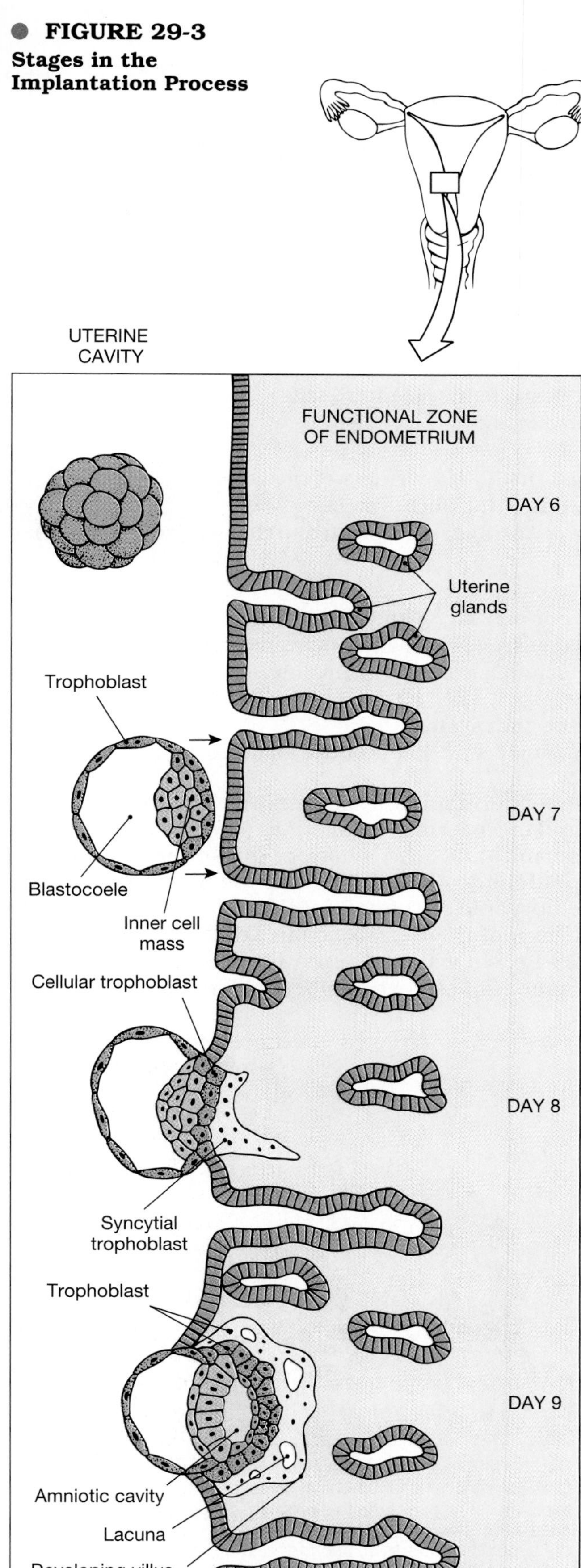
UTERINE
CAVITY
FUNCTIONAL ZONE
OF ENDOMETRIUM
DAY 6
Uterine
glands
Trophoblast
DAY 7
Blastocoele
Inner cell
mass
Cellular trophoblast
DAY 8
Syncytial
trophoblast
Trophoblast
DAY 9
Amniotic cavity
Lacuna
Developing villus

● **FIGURE 29-3**
Stages in the Implantation Process

GESTATIONAL NEOPLASMS The trophoblast undergoes repeated nuclear divisions, shows extensive and rapid growth, has a very high demand for energy, invades and spreads through adjacent tissues, and fails to activate the maternal immune system. In short, the trophoblast has many of the characteristics of cancer cells. In around 0.1 percent of pregnancies, something goes wrong with the regulatory mechanisms, and a normal placenta does not develop. Instead, the syncytial trophoblast behaves like a malignant cancer, forming a **gestational neoplasm,** or *hydatidiform* (hī-da-TID-i-form) *mole.* Prompt surgical removal of the mass is essential, sometimes followed by chemotherapy, for about 20 percent of hydatidiform moles will metastasize, invading other tissues, with potentially fatal results.

Formation of the Blastodisc

Figures 29-3, 29-4

In the early blastocyst stage the inner cell mass has little visible organization. Yet by the time of implantation, the inner cell mass is separating from the trophoblast. The separation gradually increases, creating a fluid-filled chamber called the **amniotic** (am-nē-OT-ik) **cavity.** The trophoblast will later be separated from the amniotic cavity by layers of cells that originate at the inner cell mass. These layers, which line the amniotic cavity, form the *amnion.* The amniotic cavity can be seen in *day 9* of Figure 29-3●, and additional details from *days 10 to 12* are shown in Figure 29-4●. At the time the amniotic cavity first appears, the cells of the inner cell mass are organized into an oval sheet that is two cell layers thick. This oval, called a **blastodisc** (BLAS-tō-disk), initially consists of an epithelial layer, or **epiblast** (EP-i-blast), facing the amniotic cavity and an underlying **hypoblast** (HĪ-pō-blast) exposed to the fluid contents of the blastocoele.

Gastrulation and Germ Layer Formation

Figure 29-4

A few days later, a third layer begins forming through the process of **gastrulation** (gas-troo-LĀ-shun) (*day 12,* Figure 29-4●). During gastrulation, cells in specific areas of the epiblast move toward the center of the blastodisc, toward a line known as the **primitive streak.** At the primitive streak these migrating cells leave the surface and move between the epiblast and hypoblast. This movement creates three distinct embryonic layers with markedly different fates. Once gastrulation begins, the layer remaining in contact with the amniotic cavity is called the **ectoderm,** the hypoblast is known as the **endoderm,** and the intervening, poorly organized layer is the **mesoderm.** The formation of mesoderm and the developmental fates of these **germ layers** were introduced in an Embryology Summary in Chapter 4. ∞ *[p. 124]* Table 29-1 contains a more comprehensive listing of the contributions each germ layer makes to the body systems described in earlier chapters.

TABLE 29-1 The Fates of the Primary Germ Layers

Ectodermal contributions

Integumentary system: epidermis, hair follicles and hairs, nails, and glands communicating with the skin (apocrine and merocrine sweat glands, mammary glands, and sebaceous glands)
Skeletal system, muscular system: pharyngeal cartilages and associated muscles
Nervous system: all neural tissue, including brain and spinal cord
Endocrine system: posterior pituitary gland and the adrenal medullae
Respiratory system: mucous epithelium of nasal passageways
Digestive system: mucous epithelium of mouth and anus, salivary glands

Mesodermal contributions

Skeletal system: all components except some pharyngeal derivatives
Muscular system: all components except some pharyngeal derivatives
Endocrine system: adrenal cortex, endocrine tissues of heart, kidneys, and gonads
Cardiovascular system: all components, including bone marrow
Lymphatic system: all components
Urinary system: the kidneys, including the nephrons and the initial portions of the collecting system
Reproductive system: the gonads and the adjacent portions of the duct systems
Miscellaneous: the lining of the body cavities (thoracic, pericardial, peritoneal) and the connective tissues supporting all organ systems

Endodermal contributions

Endocrine system: thymus, thyroid, pancreas, and anterior pituitary gland
Respiratory system: respiratory epithelium (except nasal passageways) and associated mucous glands
Digestive system: mucous epithelium (except mouth and anus), exocrine glands (except salivary glands), liver, and pancreas
Urinary system: urinary bladder and distal portions of the duct system
Reproductive system: distal portions of the duct system, stem cells that produce gametes

The Formation of Extraembryonic Membranes

Figures 29-4, 29-5, 29-6

Germ layers also participate in the formation of four **extraembryonic membranes:** the *yolk sac* (endoderm and mesoderm), the *amnion* (ectoderm and mesoderm), the *allantois* (endoderm and mesoderm), and the *chorion* (mesoderm and trophoblast). Although these membranes support embryonic and fetal development, they leave few traces of their existence in adult systems. Figure 29-5● shows representative stages in the development of the extraembryonic membranes.

● **FIGURE 29-4**
Blastodisc Organization and Gastrulation

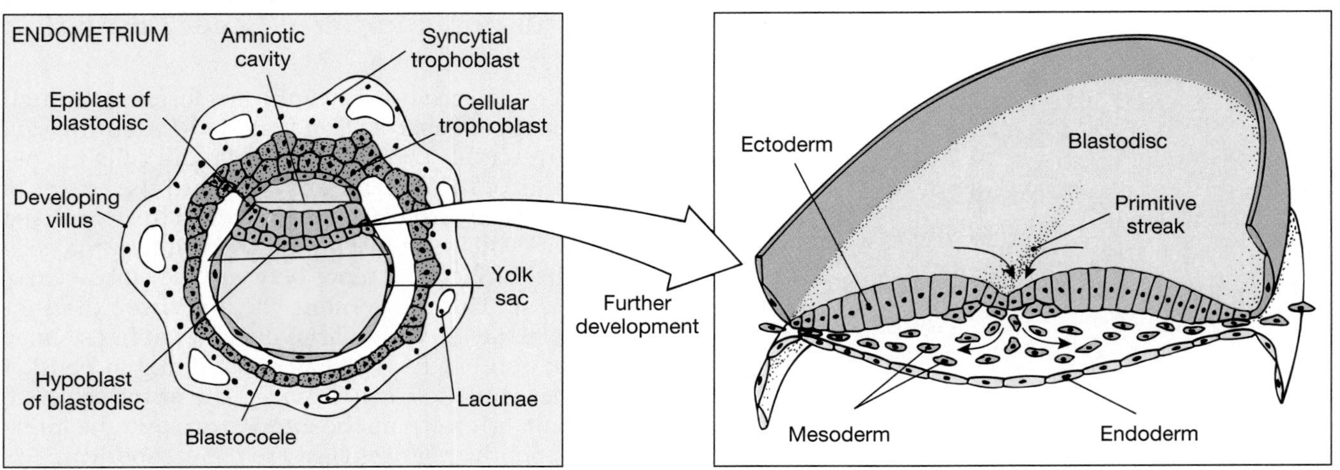
ENDOMETRIUM
Amniotic cavity
Syncytial trophoblast
Epiblast of blastodisc
Cellular trophoblast
Developing villus
Yolk sac
Hypoblast of blastodisc
Lacunae
Blastocoele
Further development
Ectoderm
Blastodisc
Primitive streak
Mesoderm
Endoderm

Day 10: The blastodisc begins as two layers, an epiblast facing the amniotic cavity and the hypoblast exposed to the blastocoele. Migration of epiblast cells around the amniotic cavity is the first step in the formation of the amnion. Migration of hypoblast cells creates a sac that hangs below the blastodisc. This is the first step in yolk sac formation.

Day 12: Migration of epiblast cells into the region between epiblast and hypoblast gives the blastodisc a third layer. From the time this process, called gastrulation, begins, the epiblast is called *ectoderm,* the hypoblast *endoderm,* and the migrating cells *mesoderm.*

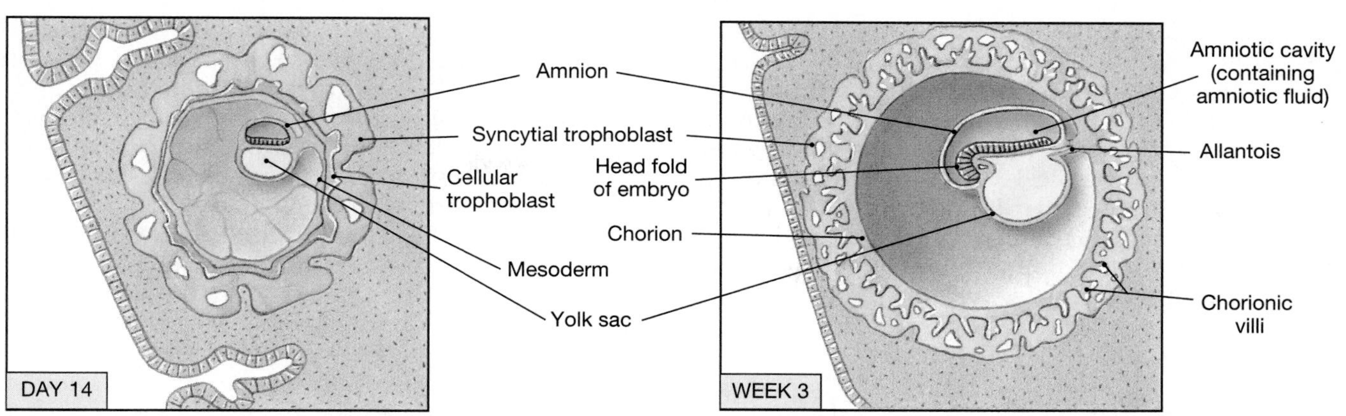
Amnion
Syncytial trophoblast
Cellular trophoblast
Head fold of embryo
Chorion
Mesoderm
Yolk sac
DAY 14
Amniotic cavity (containing amniotic fluid)
Allantois
Chorionic villi
WEEK 3

(a) Migration of mesoderm around the inner surface of the trophoblast creates the chorion. Mesodermal migration around the outside of the amniotic cavity, between the ectodermal cells and the trophoblast, creates the amnion. Mesodermal migration around the endodermal pouch below the blastodisc creates the definitive yolk sac.

(b) The embryonic disc bulges into the amniotic cavity at the head fold. The allantois, an endodermal extension surrounded by mesoderm, extends toward the trophoblast.

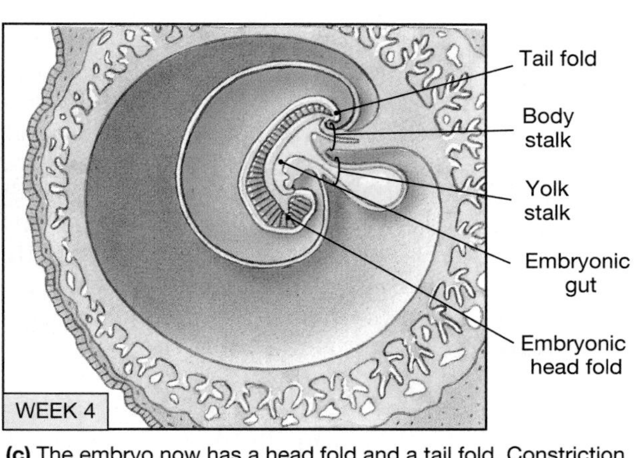
Tail fold
Body stalk
Yolk stalk
Embryonic gut
Embryonic head fold
WEEK 4

(c) The embryo now has a head fold and a tail fold. Constriction of the connection between the embryo and the surrounding trophoblast constricts the yolk stalk and body stalk.

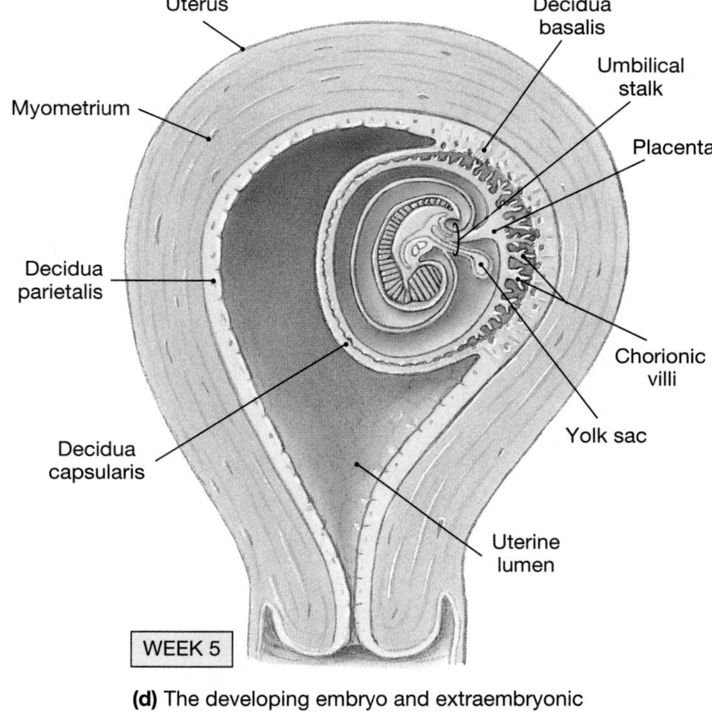
Uterus
Decidua basalis
Umbilical stalk
Myometrium
Placenta
Decidua parietalis
Chorionic villi
Yolk sac
Decidua capsularis
Uterine lumen
WEEK 5

(d) The developing embryo and extraembryonic membranes bulge into the uterine cavity. The trophoblast pushing out into the uterine lumen remains covered by endometrium, but no longer participates in nutrient absorption and embryo support. The embryo moves away from the placenta, and the body stalk and yolk stalk fuse to form an umbilical stalk.

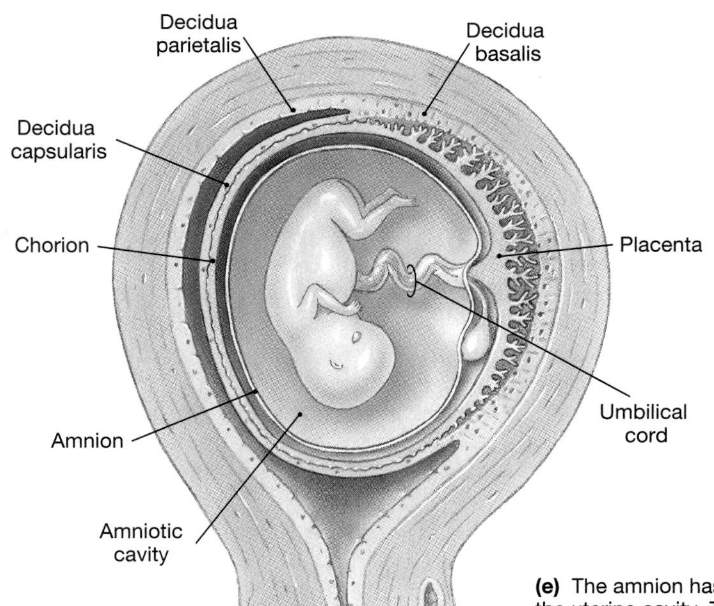
Decidua parietalis
Decidua basalis
Decidua capsularis
Chorion
Placenta
Umbilical cord
Amnion
Amniotic cavity
WEEK 10

(e) The amnion has expanded greatly, filling the uterine cavity. The fetus is connected to the placenta by an elongate umbilical cord that contains a portion of the allantois, blood vessels, and the remnants of the yolk stalk.

● FIGURE 29-5
Embryonic Membranes and Placenta Formation

THE YOLK SAC The first of the extraembryonic membranes to appear is the **yolk sac.** The yolk sac begins as the hypoblast cells spread out around the outer edges of the blastocoele to form a complete pouch suspended below the blastodisc. This pouch is already visible 10 days after fertilization (Figure 29-4•). As gastrulation proceeds, mesodermal cells migrate around this pouch and complete the formation of the yolk sac (Figure 29-5a•). Blood vessels soon appear within the mesoderm, and the yolk sac becomes an important site of blood cell formation.

THE AMNION The ectodermal layer also undergoes an expansion, and ectodermal cells spread over the inner surface of the amniotic cavity. Mesodermal cells soon follow, creating a second, outer layer (Figure 29-5a•). This combination of mesoderm and ectoderm is the **amnion** (AM-nē-on). As development proceeds, this membrane continues to expand, increasing the size of the amniotic cavity. The amnion encloses **amniotic fluid** that surrounds and cushions the developing embryo or fetus (Figure 29-5c,e•).

THE ALLANTOIS The third extraembryonic membrane begins as an outpocketing of the endoderm near the base of the yolk sac (Figure 29-5b,c•). The free endodermal tip then grows toward the wall of the blastocyst, surrounded by a mass of mesodermal cells. This sac of endoderm and mesoderm is the **allantois** (a-LAN-tō-is). The base of the allantois later gives rise to the urinary bladder. (The formation of the allantois and its relationship to the urinary bladder were detailed in the Embryology Summary in Chapter 26, p. 1020.)

THE CHORION The mesoderm associated with the allantois spreads until it extends completely around the inside, forming a mesodermal layer underneath the trophoblast. This combination of mesoderm and trophoblast is the **chorion** (KOR-ē-on) (Figure 29-5a,b•).

When implantation first occurs, the nutrients absorbed by the trophoblast can easily reach the blastodisc by simple diffusion. But as the embryo and the trophoblastic complex enlarge, the distance between the two increases, and diffusion alone can no longer keep pace with the demands of the developing embryo. Blood vessels now begin to develop within the mesoderm of the chorion, creating a rapid-transit system linking the embryo with the trophoblast.

The appearance of blood vessels in the chorion is the first step in the creation of a functional placenta. By the third week of development (Figures 29-5b and 29-6•), the mesoderm extends along the core of each of the trophoblastic villi, forming **chorionic villi** in contact with maternal tissues. These villi continue to enlarge and branch, creating an intricate network within the endometrium. Embryonic blood vessels develop within each of the villi, and circulation through those chorionic vessels begins early in the third week of development, when the heart starts beating. The blood supply to the chorionic villi arises from the allantoic arteries and veins.

As the chorionic villi enlarge, more maternal blood vessels are eroded, and maternal blood slowly percolates through lacunae lined by the syncytial trophoblast. Chorionic blood vessels pass close by, and exchange between the embryonic and maternal circulations occurs by diffusion across the syncytial and cellular trophoblast layers.

❑ Placentation

Figure 29-5

At first the entire blastocyst is surrounded by chorionic villi. The chorion continues to enlarge, expanding like a balloon within the endometrium, and by the fourth week the embryo, amnion, and yolk sac are suspended within an expansive, fluid-filled chamber (Figure 29-5c•). The connection between embryo and chorion, known as the **body stalk,** contains the distal portions of the allantois and blood vessels carrying blood to and from the placenta. The narrow connection between the endoderm of the embryo and the yolk sac is called the **yolk stalk.** (The formation of the yolk stalk and body stalk was detailed in the Embryology Summary in Chapter 24, p. 886.)

The placenta does not continue to enlarge indefinitely. Regional differences in placental organization begin to develop as placental expansion creates a prominent bulge in the endometrial surface. This relatively thin portion of the endometrium, the **decidua capsularis** (dē-SID-ū-a kap-sū-LA-ris; *deciduus*, a falling off), no longer participates in nutrient exchange, and the chorionic villi in this region disappear (Figure 29-5d•). Placental functions are now concentrated in a disc-shaped area situated in the deepest portion of the endometrium, a region called the **decidua basalis** (ba-SA-lis). The rest of the uterine endometrium, which has no contact with the chorion, is called the **decidua parietalis.**

As the end of the first trimester approaches, the fetus moves farther away from the placenta (Figure 29-5d,e•). It remains connected by the **umbilical cord,** or **umbilical stalk,** which contains the allantois, the placental blood vessels, and the yolk stalk.

As the fetus enlarges, the amniotic cavity expands and the space between the chorion and amnion decreases. Eventually the mesoderm on the outer surface of the amnion contacts the mesoderm on the inner surface of the chorion. These layers fuse, creating a compound *amniochorionic membrane.*

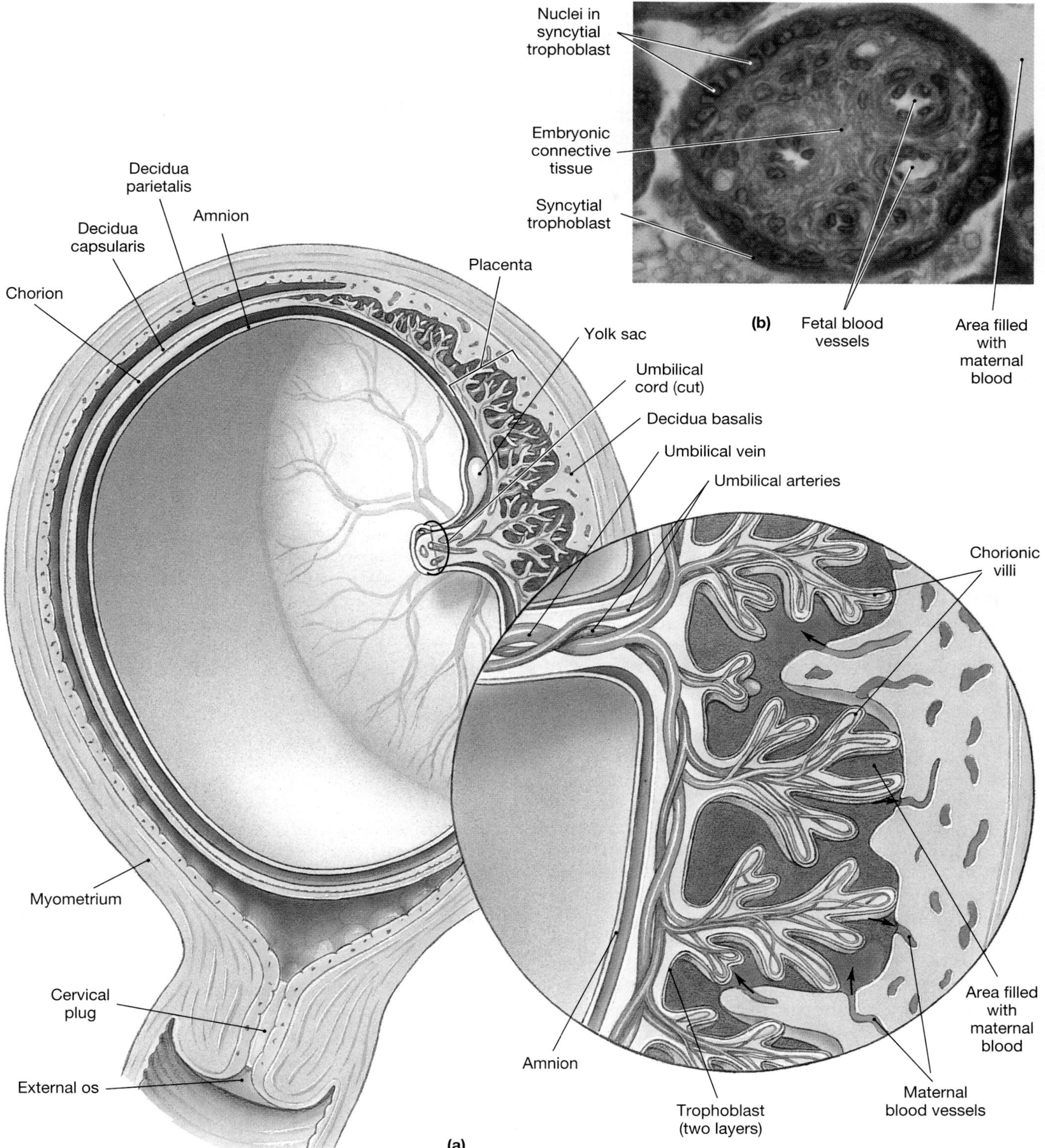
Nuclei in syncytial trophoblast
Embryonic connective tissue
Syncytial trophoblast
(b)
Fetal blood vessels
Area filled with maternal blood
Decidua parietalis
Amnion
Decidua capsularis
Chorion
Placenta
Yolk sac
Umbilical cord (cut)
Decidua basalis
Umbilical vein
Umbilical arteries
Chorionic villi
Myometrium
Cervical plug
External os
Amnion
Area filled with maternal blood
Maternal blood vessels
Trophoblast (two layers)
(a)

FIGURE 29-6

A Three-Dimensional View of Placental Structure. (a) For clarity the uterus is shown after the embryo has been removed and the umbilical cord cut. Blood flows into the placenta through ruptured maternal blood arteries. It then flows around chorionic villi that contain fetal blood vessels. Fetal blood arrives over paired umbilical arteries and leaves over a single umbilical vein. Maternal blood reenters the venous system of the mother through the broken walls of small uterine veins. Note that no actual mixing of maternal and fetal blood occurs. **(b)** A cross section through a chorionic villus, showing the syncytial trophoblast exposed to the maternal blood space. (LM × 2045)

Placental Circulation

Figure 29-6

Figure 29-6a● diagrams circulation at the placenta near the end of the first trimester. Blood flows to the placenta through the paired **umbilical arteries** and returns in a single **umbilical vein.** The chorionic villi (Figure 29-6b●) provide the surface area for active and passive exchange between the fetal and maternal bloodstreams. *[p. 767]*

The placenta places a considerable demand on the maternal cardiovascular system, and blood flow to the uterus and placenta is extensive. If the placenta is torn or otherwise damaged, the consequences may prove fatal for both fetus and mother. AM *Problems with Placentation*

The Endocrine Placenta

In addition to its role in the nutrition of the fetus, the placenta acts as an endocrine organ. The hormones are synthesized by the syncytial trophoblast and released into the maternal circulation. The hormones produced include *human chorionic gonadotropin, human placental lactogen, placental prolactin, relaxin, progestins,* and *estrogens.*

HUMAN CHORIONIC GONADOTROPIN The hormone **human chorionic** (kō-rē-ON-ik) **gonadotropin (hCG)** appears in the maternal bloodstream soon after implantation has occurred. The presence of hCG in blood or urine samples provides a reliable indication of pregnancy, and kits sold for the early detection of pregnancy are sensitive to the presence of this hormone.

In function hCG resembles LH, for it maintains the integrity of the corpus luteum and promotes the continued secretion of progesterone. As a result, the endometrial lining remains perfectly functional, and menses does not occur. If it did, the pregnancy would end, for the functional zone of the endometrium would disintegrate.

In the presence of hCG, the corpus luteum persists for 3–4 months before gradually decreasing in size and secretory function. The decline in luteal function does not trigger the return of menstrual periods, because by the end of the first trimester the placenta is actively secreting both estrogens and progestins.

HUMAN PLACENTAL LACTOGEN AND PLACENTAL PROLACTIN **Human placental lactogen (hPL),** or *human chorionic somatomammotropin* (hCS), helps prepare the mammary glands for milk production. It also has a stimulatory effect on other tissues comparable to that of growth hormone. At the mammary glands, the conversion from resting to active status requires the presence of placental hormones (hPL, **placental prolactin,** estrogen, and progestins) as well as several maternal hormones (GH, PRL, and thyroid hormones). The hormonal control of mammary gland function will be considered in a later section (p. 1124).

RELAXIN **Relaxin** is a peptide hormone that is secreted by the placenta as well as by the corpus luteum during pregnancy. Relaxin (1) increases the flexibility of the symphysis pubis, permitting expansion of the pelvis during delivery; (2) causes dilation of the cervix, making it easier for the fetus to enter the vaginal canal; and (3) suppresses the release of oxytocin by the hypothalamus and delays the onset of labor contractions.

PROGESTINS AND ESTROGENS After the first trimester, the placenta produces sufficient amounts of progestins to maintain the endometrial lining and continue the pregnancy. As the end of the third trimester approaches, estrogen production by the placenta accelerates. As we shall see in a later section, the rising estrogen levels play a role in stimulating labor and delivery.

❑ Embryogenesis

Figures 29-5c, 29-7

Shortly after gastrulation begins, folding and differential growth of the embryonic disc produce a bulge that projects into the amniotic cavity (Figure 29-5b●). This projection is known as the **head fold,** and similar movements lead to the formation of a **tail fold** (Figure 29-5c●). The embryo is now physically as well as developmentally separated from the blastodisc and the extraembryonic membranes. The definitive orientation of the embryo can now be seen, complete with dorsal and ventral surfaces and left and right sides.

These changes in proportions and appearance that occur between the fourth developmental week and the end of the first trimester are visually summarized in Figure 29-7●. The developmental history of structures labeled in Figure 29-7a● can be followed in more detail by referring to the Embryology Summaries in earlier chapters (see Table 29-2).

The first trimester is a critical period for development because events in the first 12 weeks establish the basis for organ formation, a process called **organogenesis.** Embryology Summaries in earlier chapters described major features of organogenesis in each organ system. Important developmental milestones are indicated in Table 29-2; readers interested in additional details should refer to the Embryology Summaries cross-referenced in the table. AM *Complexity and Perfection*

What is the fate of the inner cell mass of the blastocyst?

Improper development of which of the extraembryonic membranes would affect the circulatory system?

Her doctor tells Sue that her pregnancy test indicates elevated levels of the hormone hCG (human chorionic gonadotropin). Is she pregnant or not?

What are two important functions of the placenta?

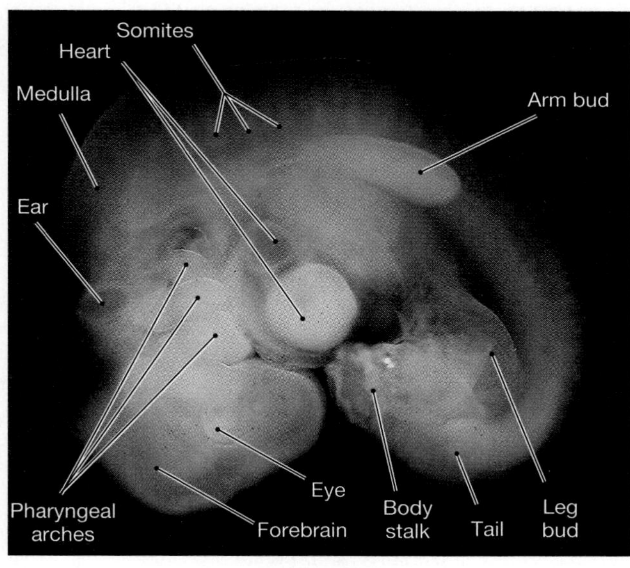
Somites
Heart
Medulla
Arm bud
Ear
Eye
Pharyngeal arches
Forebrain
Body stalk
Tail
Leg bud

(a) Week 4

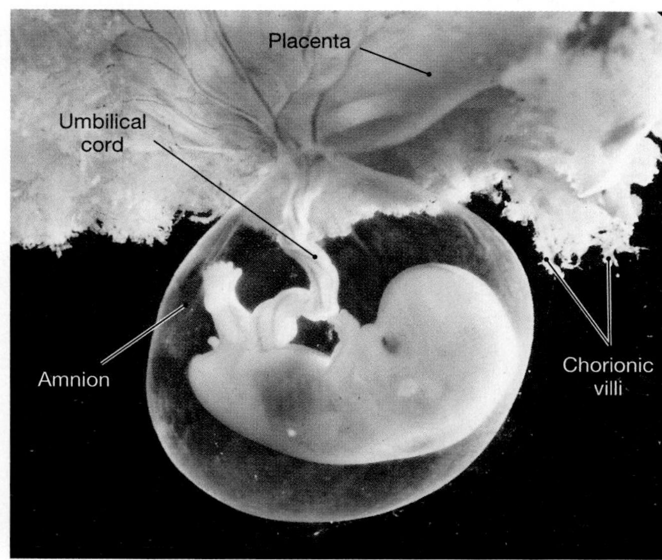
Placenta
Umbilical cord
Amnion
Chorionic villi

(b) Week 8

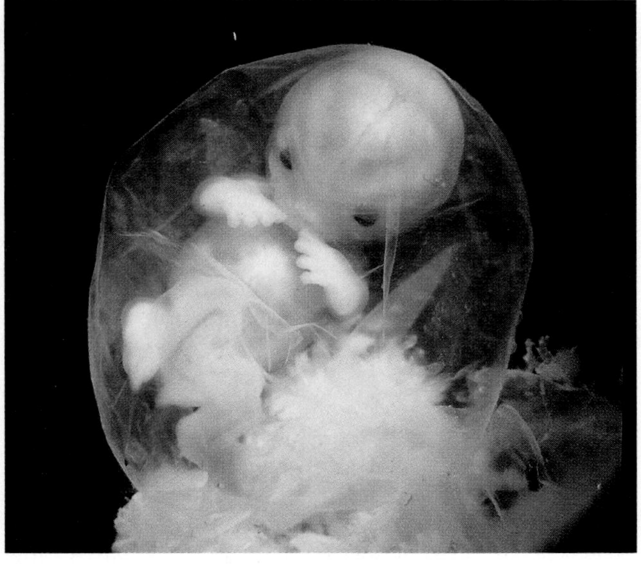

(c) Week 12

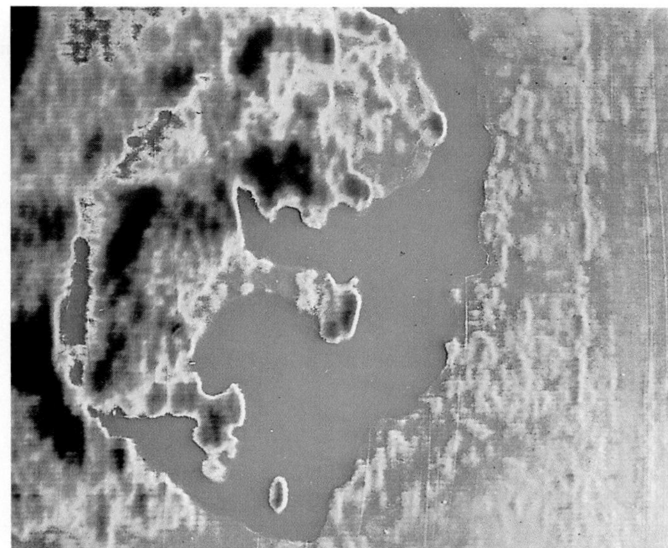

(d)

● **FIGURE 29-7**
The First Trimester. (a,b,c) Fiber-optic views of actual human embryos. **(d)** An ultrasound image that should be compared with part (c). For actual sizes, see Figure 29-14, p. 1133.

The Second and Third Trimesters

Figures 29-7, 29-8, 29-9

By the start of the second trimester (Figure 29-7c,d●) the rudiments of all the major organ systems have formed. Over the next 3 months, the fetus will grow to a weight of around 0.64 kg (1.4 lb). During this period, the fetus, encircled by the amnion, grows faster than the surrounding placenta. Soon the mesodermal outer covering of the amnion fuses with the inner lining of the chorion. Figure 29-8●, p. 1124, shows a 4-month fetus as viewed with a fiber-optic endoscope and a 6-month fetus as seen in ultrasound.

During the third trimester, the basic components of all the organ systems appear, and most become ready to fulfill their normal functions. The rate of growth starts to decrease, but in absolute terms this trimester sees the largest weight gain. In 3 months the fetus puts on around 2.6 kg (5.7 lb), reaching a full-term weight of somewhere near 3.2 kg (7 lb). Important events in organ system development in the second and third trimesters are detailed in the Embryology Summaries in earlier chapters, and highlights are noted in Table 29-2.

At the end of gestation a typical uterus will undergo a tremendous increase in size. Figure

TABLE 29-2 An Overview of Prenatal and Early Postnatal Development

Gestational Age (Months)	Length and Weight	Integumentary System 	Skeletal System 	Muscular System 	Nervous System 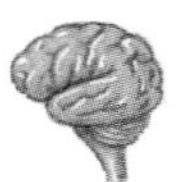	Special Sense Organs 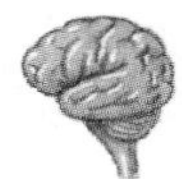
1	5 mm 0.02 g		(b) Somite formation	(b) Somite formation	(b) Neural tube	(b) Eye and ear formation
2	28 mm 2.7 g	(b) Nail beds, hair follicles, sweat glands	(b) Axial and appendicular cartilage formation	(c) Rudiments of axial musculature	(b) CNS, PNS organization, growth of cerebrum	(b) Taste buds, olfactory epithelium
3	78 mm 26 g	(b) Epidermal layers appear	(b) Ossification centers spreading	(c) Rudiments of appendicular musculature	(c) Basic spinal cord and brain structure	
4	133 mm 150 g	(b) Hair, sebaceous glands (c) Sweat glands	(b) Articulations (c) Facial and palatal organization	Fetus starts moving	(b) Rapid expansion of cerebrum	(c) Basic eye and ear structure (b) Peripheral receptor formation
5	185 mm 460 g	(b) Keratin production, nail production			(b) Myelination of spinal cord	
6	230 mm 823 g			(c) Perineal muscles	(b) CNS tract formation (c) Layering of cortex	
7	270 mm 1492 g	(b) Keratinization, nail formation, hair formation				(c) Eyelids open, retina sensitive to light
8	310 mm 2274 g		(b) Epiphyseal plate formation			Taste receptors functional
9	346 mm 2912 g					
Postnatal development		Hair changes in consistency and distribution	Formation and growth of epiphyseal plates continue	Muscle mass and control increase	Myelination, layering, CNS tract formation continue	
Chapter containing relevant Embryology Summary		5 (p. 164)	7, 8 (pp. 220, 232, 256)	11 (p. 362)	12, 13, 14 (pp. 391, 444, 460)	17 (p. 592)

Note: (b) = begin formation; (c) = complete formation.

	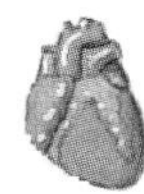 	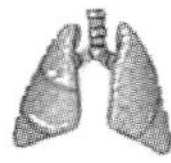			
Endocrine System	*Cardiovascular and Lymphatic Systems*	*Respiratory System*	*Digestive System*	*Urinary System*	*Reproductive System*
	(b) Heartbeat	(b) Trachea and lung formation	(b) Intestinal tract, liver, pancreas (c) Yolk sac	(c) Allantois	
(b) Thymus, thyroid, pituitary, adrenal glands	(c) Basic heart structure, major blood vessels, lymph nodes and ducts (b) Blood formation in liver	(b) Extensive bronchial branching into mediastinum (c) Diaphragm	(b) Intestinal subdivisions, villi, salivary glands	(b) Kidney formation (adult form)	(b) Mammary glands
(c) Thymus, thyroid gland	(b) Tonsils, blood formation in bone marrow		(c) Gallbladder, pancreas		(b) Definitive gonads, ducts, genitalia
	(b) Migration of lymphocytes to lymphatic organs, blood formation in spleen			(b) Degeneration of embryonic kidneys	
	(c) Tonsils	(c) Nostrils open	(c) Intestinal subdivisions		
(c) Adrenal glands	(c) Spleen, liver, bone marrow	(b) Alveolar formation	(c) Epithelial organization, glands		
(c) Pituitary gland			(c) Intestinal plicae		(b) Testes descend
		Complete pulmonary branching and alveolar formation		Complete nephron formation at birth	Descent complete at or near time of delivery
	Cardiovascular changes at birth; immune system becomes operative thereafter				
18	20, 21	23	24	26	28
(p. 632)	(pp. 690, 770)	(p. 844)	(p. 886)	(p. 1020)	(pp. 1095–1097)

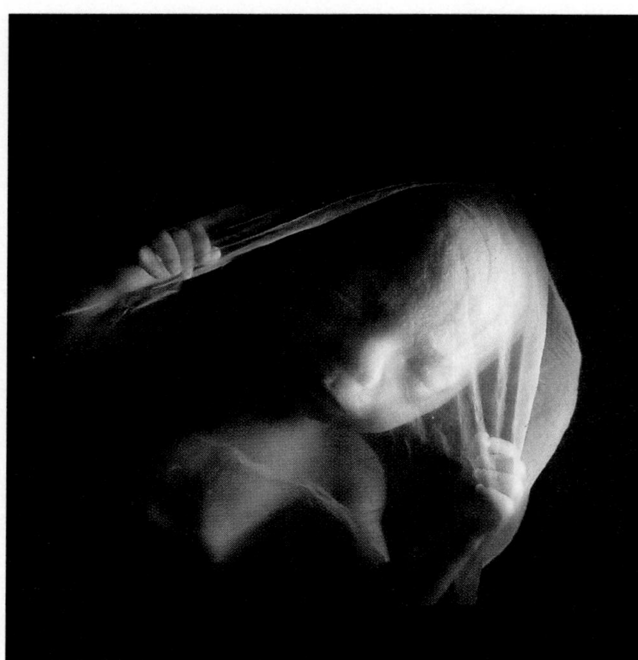

(a)

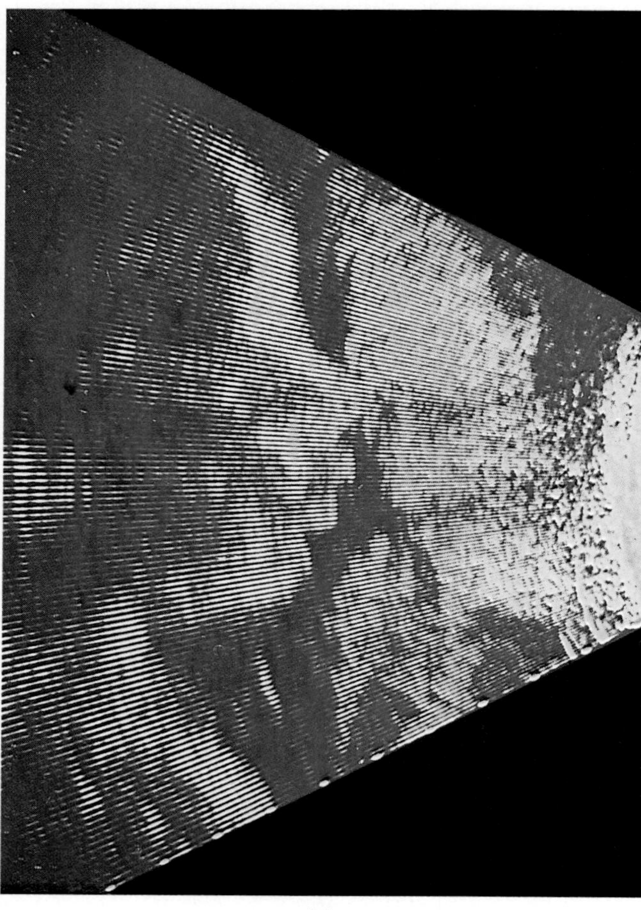

(b)

● **FIGURE 29-8**
The Second and Third Trimesters. (a) A 4-month fetus seen through a fiber-optic endoscope. **(b)** Head of a 6-month fetus as seen using ultrasound equipment.

29-9● shows the position of the uterus, fetus, and placenta from 16 weeks to full term. When the pregnancy is at term, the uterus and fetus push many of the maternal abdominal organs out of their normal positions (Figure 29-9c●).

❑ Pregnancy and Maternal Systems

The developing fetus is totally dependent on maternal organ systems for nourishment, respiration, and waste removal. These functions must be performed by maternal systems in addition to their normal operations. For example, the mother must absorb enough oxygen, nutrients, and vitamins for herself *and* her fetus, and she must eliminate all of the generated wastes. Although this is not a burden over the initial weeks of gestation, the demands placed upon the mother become significant as the fetus grows larger. To survive under these conditions the maternal systems must make major compensatory adjustments. In practical terms the mother must breathe, eat, and excrete for two. AM *Problems with the Maintenance of a Pregnancy*

- ***The respiratory rate goes up and the tidal volume increases.*** As a result, the lungs obtain the extra oxygen required and remove the excess carbon dioxide generated by the fetus.
- ***The maternal blood volume increases.*** This increase occurs because (1) blood flowing into the placenta reduces the volume in the rest of the systemic circuit, and (2) fetal activity lowers the blood P_{O_2} and elevates the P_{CO_2}. The combination stimulates the production of renin and erythropoietin, leading to an increase in maternal blood volume through mechanisms detailed in Chapter 18 (see Figure 18-16b●, p 627). By the end of gestation the maternal blood volume has increased by almost 50 percent.
- ***The maternal requirements for nutrients and vitamins climb 10–30 percent.*** Pregnant women, who must "eat for two," are often hungry.
- ***The glomerular filtration rate increases by roughly 50 percent.*** This increase, which corresponds to the increase in blood volume, accelerates the excretion of metabolic wastes generated by the fetus. Because the volume of urine produced increases and the weight of the uterus presses down on the urinary bladder, pregnant women need to urinate frequently.
- ***The uterus undergoes a tremendous increase in size.*** Structural and functional changes in the expanding uterus are so important that we will discuss them in a separate section.
- ***The mammary glands increase in size and secretory activity begins.*** Mammary gland

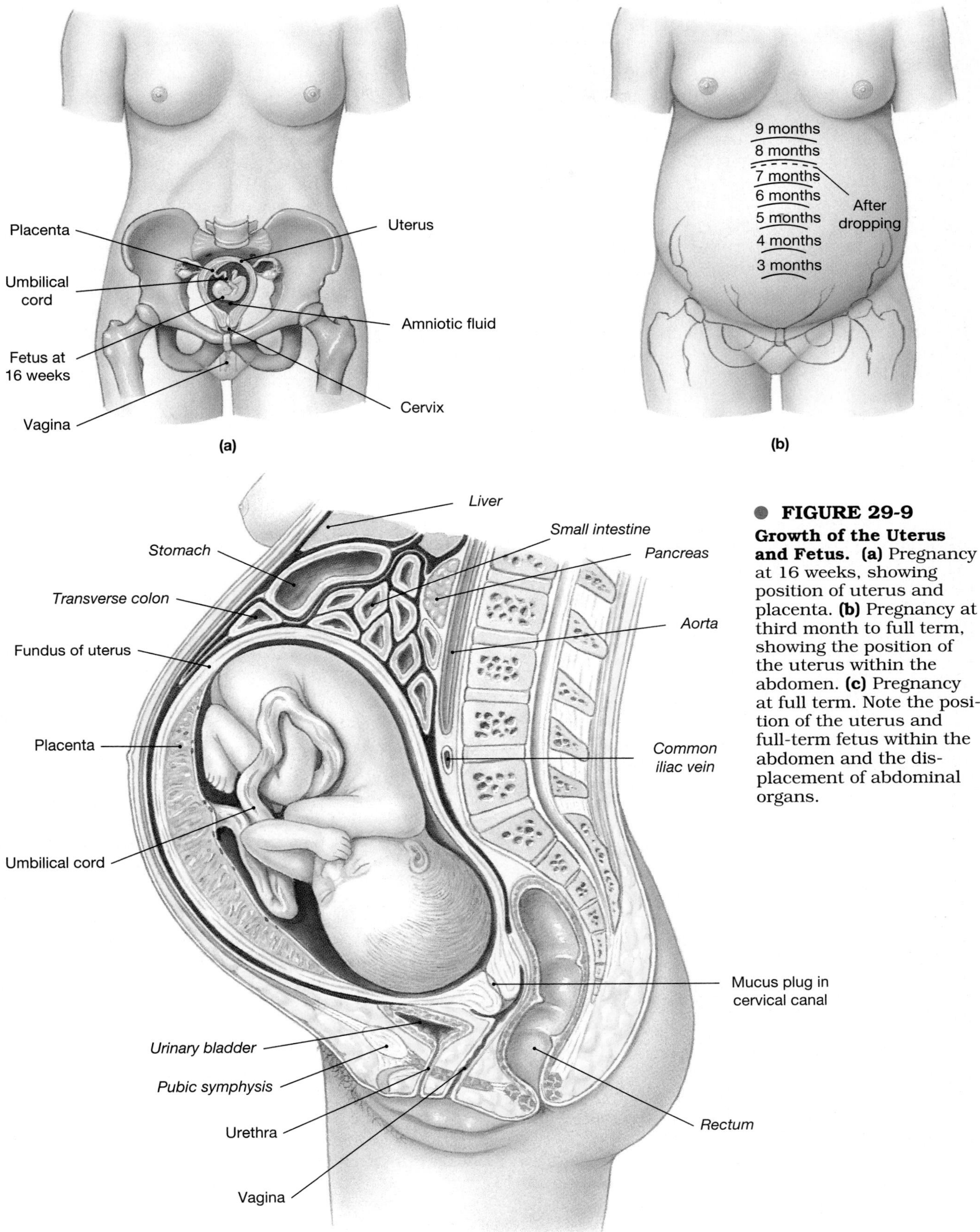
Placenta
Uterus
Umbilical cord
Amniotic fluid
Fetus at 16 weeks
Cervix
Vagina
(a)
9 months
8 months
7 months
6 months
5 months
4 months
3 months
After dropping
(b)
Liver
Small intestine
Stomach
Pancreas
Transverse colon
Aorta
Fundus of uterus
Placenta
Common iliac vein
Umbilical cord
Mucus plug in cervical canal
Urinary bladder
Pubic symphysis
Urethra
Rectum
Vagina
(c)

FIGURE 29-9 Growth of the Uterus and Fetus. **(a)** Pregnancy at 16 weeks, showing position of uterus and placenta. **(b)** Pregnancy at third month to full term, showing the position of the uterus within the abdomen. **(c)** Pregnancy at full term. Note the position of the uterus and full-term fetus within the abdomen and the displacement of abdominal organs.

Technology and the Treatment of Infertility

Infertility, or the inability to have children, has been the focus of media attention in recent years. The reason is simple: Problems with infertility are relatively common. An estimated 10–15 percent of U.S. married couples are infertile, and another 10 percent are unable to have as many children as they desire. It is thus not surprising that reproductive physiology has become a popular field, and the treatment of infertility has become a major medical industry.

An infertile, or *sterile,* woman is unable to produce functional eggs or support a developing embryo. An infertile man is incapable of providing a sufficient number of motile sperm for successful fertilization. Because sterility of either sexual partner will have the same result, diagnosis and treatment of infertility must involve evaluation of both sexual partners. Approximately 60 percent of infertility cases can be attributed to problems with the female reproductive system.

Recent advances in our understanding of reproductive physiology are providing new solutions to fertility problems. These approaches, called **assisted reproductive technologies (ART)** are diagrammed in Figure 29-10●.

- ***Low sperm count.*** In cases of male infertility due to low sperm counts, semen from several ejaculates can be pooled, concentrated, and introduced into the female reproductive tract. This technique, known as *artificial insemination,* may lead to normal fertilization and pregnancy. In special cases, where an individual's spermatozoa are unable to accomplish oocyte penetration, single-sperm fertilization has been accomplished with micromanipulation of the oocyte and corona radiata.
- ***Abnormal spermatozoa.*** If the man cannot produce functional sperm, sperm can be obtained from a "sperm bank" that stores donor sperm.
- ***Hormonal problems.*** If the problem involves the woman's inability to ovulate because of low gonadotropin or estrogen levels or her inability to maintain adequate progesterone levels after ovulation, these hormones can be provided.

  So-called *fertility drugs,* such as clomiphene (*Clomid*), stimulate ovarian ovum production. Clomiphene works by blocking the feedback inhibition of the hypothalamus and pituitary gland by estrogens. As a result, circulating FSH levels rise, and more follicles are stimulated to complete their development. Injected purified gonadotropins, such as *Pergonal* (FSH and LH) and *Metrodine* (FSH), are also used to accelerate ovum development. The chance that a single egg will be fertilized through normal sexual intercourse is around 1 in 3. Increasing the number of eggs released increases the odds of a pregnancy. Unfortunately, it is not easy to determine just how much ovarian stimulation will be needed, so multiple births have often resulted from treatment with fertility drugs. Careful monitoring of follicle development somewhat reduces the chances of a multiple birth.
- ***Problems with oocyte transport.*** When there are problems with the transport of the oocyte from the ovary to the uterine tube, due to scarring of the fimbriae or other problems, a procedure called **GIFT** can be used. GIFT is short for *gamete intrafallopian tube transfer.* (*Fallopian tube* is another name for the uterine tube.) In this procedure the ovaries are stimulated with injected hormones, and a large "crop" of mature oocytes is removed from tertiary follicles. The individual eggs are examined for defects, inserted into the uterine tubes, and exposed to high concentrations of sperm from the husband or donor. The success rate for this procedure is less than that of natural fertilization (33 percent), and not every pregnancy produces an infant. The cost of a single procedure (successful or not) averages $5000.
- ***Blocked uterine tubes.*** In the GIFT procedure, fertilization occurs in its normal location, within the uterine tube. This site is not essential, and fertilization can also take place in a test tube or petri dish. This process is called **in vitro fertilization,** or **IVF** (*vitro,* glass). If a carefully controlled fluid environment is provided, early development will proceed normally. One variation on the GIFT procedure, called **ZIFT** (*zygote intrafallopian tube transfer*), exposes selected eggs to sperm outside the body and inserts zygotes or early cleavage-stage embryos, rather than oocytes, into the uterine tubes. If multiple zygotes are available, some can be frozen and stored for later insertion in case the initial procedure fails to produce a successful pregnancy. The cost for a single ZIFT procedure ranges between $8000 and $10,000.

  Alternatively, the zygote can be maintained in an artificial environment through the first 2–3 days of development. This procedure is often selected if the uterine tubes are damaged or blocked. The cleavage-stage embryo is then placed directly into the uterus, rather than into one of the uterine tubes. The cost of this procedure is comparable to that for ZIFT.

- ***Abnormal oocytes.*** If the oocytes released by the ovaries are abnormal in some way, or if menopause has already occurred, viable oocytes can be obtained from a suitable donor. The donor may be anonymous or known; if anonymous, the donor usually receives a fee for the donation. After treatment with fertility drugs, the donor's ovaries are stimulated to produce a large crop of oocytes. These are collected and fertilized in vitro, usually by the husband's sperm. After cleavage has begun, the pre-embryo is placed in the recipient's uterus, which has been "primed" by progesterone therapy. The pregnancy rate for this procedure is roughly 33 percent for women over age 40, using oocytes donated by women in their early twenties. Oocyte donation has a much higher success rate for these women than ZIFT or GIFT, with either of which the odds of a successful pregnancy are only about 4 percent. This difference suggests that age-related changes in the characteristics and quality of the oocytes, rather than changes in hormone levels or uterine responsiveness, are often the primary cause of infertility in older women.
- ***Abnormal uterine environment.*** If fertilization and transport occur normally but the uterus cannot maintain a pregnancy, the problem may involve low levels of progestin secretion by the corpus luteum. Hormone therapy may solve this problem. If the maternal uterus simply cannot support development, the zygote or cleavage-stage embryo can be introduced into the uterus of a *surrogate mother.* If the embryo survives and makes contact with the endometrium, development will proceed normally despite the fact that the mother has no genetic relationship with the embryo.

Surrogate motherhood, which sounds relatively simple and straightforward, has proven to be one of the most explosive solutions in terms of ethics and legality. Since 1990, several court cases have resulted from disputes over surrogate motherhood and who merits legal custody of the infant. Legal battles have also broken out over a variety of complex questions, and some of them will take years to sort out. To understand the problem, just take a moment and consider:

- Do parents share property rights over frozen and stored zygotes? Can a husband have any of the stored zygotes implanted into the uterus of his second wife without the consent of his first wife, who provided the eggs?
- If both donor egg and donor sperm are used, do adoption laws apply?
- If the husband provided the sperm that fertilized the egg of a donor who is not his wife, for implantation into a surrogate mother, can the wife, the surrogate mother, or the egg donor sue for custody of the child after a divorce?

If you use your imagination, you can probably think of even more complex problems, many of which will probably be debated in a courtroom within the next decade.

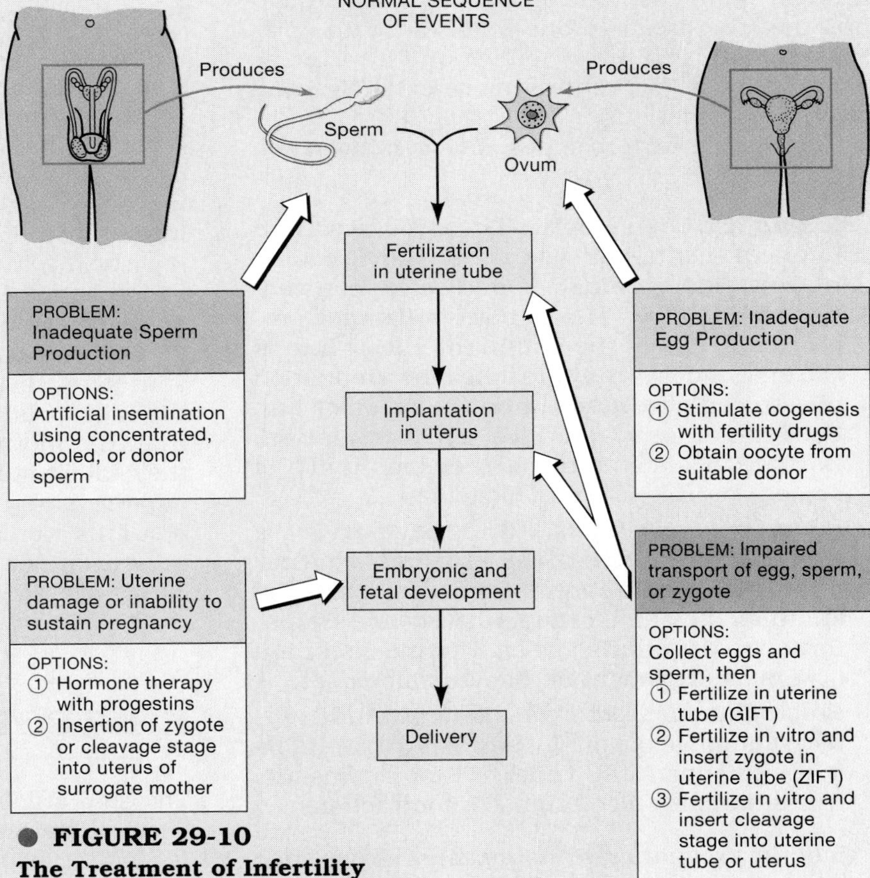
NORMAL SEQUENCE OF EVENTS
Produces
Produces
Sperm
Ovum
Fertilization in uterine tube
Implantation in uterus
Embryonic and fetal development
Delivery
PROBLEM: Inadequate Sperm Production
OPTIONS: Artificial insemination using concentrated, pooled, or donor sperm
PROBLEM: Inadequate Egg Production
OPTIONS: ① Stimulate oogenesis with fertility drugs ② Obtain oocyte from suitable donor
PROBLEM: Uterine damage or inability to sustain pregnancy
OPTIONS: ① Hormone therapy with progestins ② Insertion of zygote or cleavage stage into uterus of surrogate mother
PROBLEM: Impaired transport of egg, sperm, or zygote
OPTIONS: Collect eggs and sperm, then ① Fertilize in uterine tube (GIFT) ② Fertilize in vitro and insert zygote in uterine tube (ZIFT) ③ Fertilize in vitro and insert cleavage stage into uterine tube or uterus

● **FIGURE 29-10**
The Treatment of Infertility

development requires a combination of hormones, including human placental lactogen and prolactin from the placenta, and prolactin (PRL), estrogens, progestins, growth hormone (GH), and thyroxine from maternal endocrine organs. By the end of the sixth month of pregnancy the mammary glands are fully developed, and the glands begin producing secretions that are stored in the duct system.

Structural and Functional Changes in the Uterus

Figure 29-11

At the end of gestation a typical uterus will have grown from 7.5 cm (3 in.) in length and 60 g (2 oz) in weight to 30 cm (12 in.) in length and 1100 g (2.4 lb) in weight. It may then contain almost 5 liters of fluid, giving the organ with contents a total weight of roughly 10 kg (22 lb). This remarkable expansion occurs through the enlargement (hypertrophy) and elongation of existing cells, especially smooth muscle fibers, rather than by an increase in the total number of cells in the uterus.

The tremendous stretching of the myometrium is associated with a gradual increase in the rates of spontaneous smooth muscle contractions. In the early stages of pregnancy the contractions are weak, painless, and brief. There are indications that the progesterone released by the placenta has an inhibitory effect on the uterine smooth muscle, preventing more extensive and powerful contractions.

Three major factors oppose the calming action of progesterone:

1. ***Rising estrogen levels.*** Estrogens produced by the placenta increase the sensitivity of the uterine smooth muscles and make contractions more likely. Throughout pregnancy progesterone exerts the dominant effect, but as the time of delivery approaches the production of estrogen accelerates, and the myometrium becomes more sensitive to stimulation. Estrogens also increase the sensitivity of smooth muscle fibers to oxytocin.
2. ***Rising oxytocin levels.*** Rising oxytocin levels stimulate an increase in the force and frequency of uterine contractions. Oxytocin release is stimulated by high estrogen levels and by distortion of the uterine cervix. Uterine distortion occurs as the weight of the fetus increases.
3. ***Prostaglandin production.*** Estrogens and oxytocin stimulate the production of prostaglandins in the endometrium. These prostaglandins further stimulate smooth muscle contractions.

Late in pregnancy some mothers experience occasional spasms in the uterine musculature, but the contractions are neither regular nor persistent. These contractions are called **false labor. True labor** begins when the biochemical and mechanical factors reach the point of no return. After 9 months of gestation, multiple factors interact to initiate true labor. Once the **labor contractions** have begun in the myometrium, positive feedback ensures that the contractions continue until delivery has been completed.

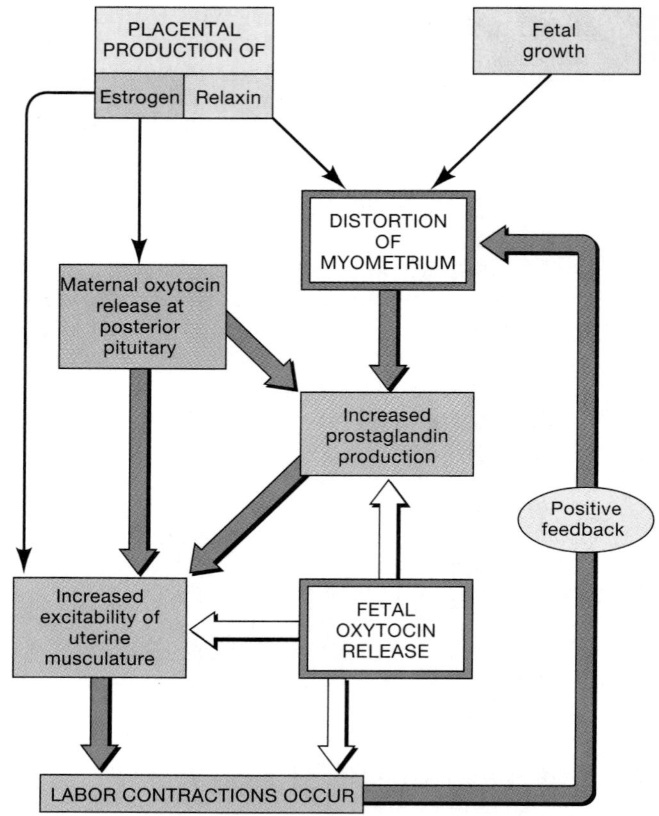
PLACENTAL PRODUCTION OF
Estrogen
Relaxin
Fetal growth
DISTORTION OF MYOMETRIUM
Maternal oxytocin release at posterior pituitary
Increased prostaglandin production
Positive feedback
Increased excitability of uterine musculature
FETAL OXYTOCIN RELEASE
LABOR CONTRACTIONS OCCUR

FIGURE 29-11
Factors Involved in the Initiation of Labor and Delivery

Figure 29-11 diagrams important factors that stimulate and sustain labor. The actual trigger for the onset of labor may be events in the fetus rather than the mother. At the time labor commences, the fetal pituitary secretes oxytocin that is released into the maternal bloodstream at the placenta. The resulting increase in myometrial contractions and prostaglandin production, on top of the priming effects of estrogens and maternal oxytocin, may be the "last straw."

Labor and Delivery

The goal of labor is the forcible expulsion of the fetus, a process known as **parturition** (par-tū-RISH-un), or birth. During true labor, each labor contraction begins near the top of the uterus and sweeps in a wave toward the cervix. These contractions are strong and occur at regular intervals. As parturition approaches the contractions increase in force and frequency, changing the position of the fetus and moving it toward the cervical canal.

❑ Stages of Labor

Figure 29-12

Labor has traditionally been divided into three stages (Figure 29-12●): the *dilation stage*, the *expulsion stage*, and the *placental stage*.

The Dilation Stage

Figure 29-12a

The **dilation stage** (Figure 29-12a●) begins with the onset of true labor, as the cervix dilates completely and the fetus begins to move into the cervical canal. This stage is highly variable in length, but typically lasts 8 or more hours. At the start of this period the labor contractions occur at intervals of once every 10–30 minutes, and their frequency increases steadily. Late in the process, the amnion usually ruptures, an event sometimes referred to as "having the water break."

The Expulsion Stage

Figure 29-12b

The **expulsion stage** (Figure 29-12b●) begins as the cervix dilates completely, pushed open by the approaching fetus. During this stage, contractions reach maximum intensity; they may occur at 2- or 3-minute intervals and last a full minute. Expulsion continues until the fetus has completed its emergence from the vagina, a period usually lasting less than 2 hours. The arrival of the newborn infant into the outside world represents birth, or **delivery.**

If the vaginal canal is too small to permit the passage of the fetus and there is acute danger of perineal tearing, the passageway may be temporarily enlarged by making an incision through the perineal musculature. After delivery, this **episiotomy** (e-pēz-ē-OT-o-mē) can be repaired with sutures, a much simpler procedure than dealing with the bleeding and tissue damage associated with a potentially extensive perineal tear. If unexpected complications arise during the dilation or expulsion stages, the infant may be removed by **cesarean section,** or "C-section." In such cases an incision is made through the abdominal wall, and the uterus is opened just enough to allow passage of the infant's head. This procedure is performed during 15–25 percent of the deliveries in the United States—more often than necessary, according to some studies. Efforts are now being made to reduce the frequency of both episiotomies and cesarian sections.

The Placental Stage

Figure 29-12c

During the **placental stage** of labor (Figure 29-12c●), muscle tension builds in the walls of the partially empty uterus and the organ gradually decreases in size. This uterine contraction tears the connections between the endometrium and the placenta. Usually within an hour after delivery the placental stage ends with the ejection of the placenta, or "afterbirth." The disruption of the placenta is accompanied by a loss of blood (as much as 500–600 ml), but because the maternal blood volume has increased greatly during pregnancy the loss can be tolerated without difficulty. AM *Common Problems with Labor and Delivery*

❑ Premature Labor

Premature labor occurs when true labor begins before the fetus has completed normal development. The chances of newborn survival are directly related to body weight at delivery. Even with massive supportive efforts, infants born weighing less than 400 g (14 oz) will not survive, primarily because the respiratory, cardiovascular, and urinary systems are unable to support life without the aid of maternal systems. As a result, the dividing line between *spontaneous abortion* and **immature delivery** is usually set at 500 g (17.6 oz), the normal weight near the end of the second trimester.

Infants delivered before completing 7 months of gestation (weight under 1 kg) have less than a 50:50 chance of survival, and most survivors suffer from severe developmental abnormalities. A **premature delivery** produces a newborn weighing over 1 kg (35.2 oz). Its chances of survival range from fair to excellent, depending on the individual circumstances. AM *Complexity and Perfection*

❑ Multiple Births

Multiple births (twins, triplets, quadruplets, and so forth) may occur for several reasons. The ratio of twin to single births in the U.S. population is roughly 1:89. In "fraternal," or **dizygotic** (dī-zī-GOT-ik), twins two separate eggs were ovulated and subsequently fertilized. Because of the shuffling of chromosomes that occurs during meiosis, the odds against any two zygotes from the same parents having identical genes exceeds 1:8.4 million. Seventy percent of twins are dizygotic.

"Identical," or **monozygotic,** twins may result from the separation of blastomeres early in cleavage or by splitting of the inner cell mass prior to gastrulation. In either event, the genetic makeup of the pair will be identical, for both formed from the same pair of gametes. Triplets, quadruplets, and larger multiples can result from multiple ovulations, blastomere splitting, or some combination of the two.

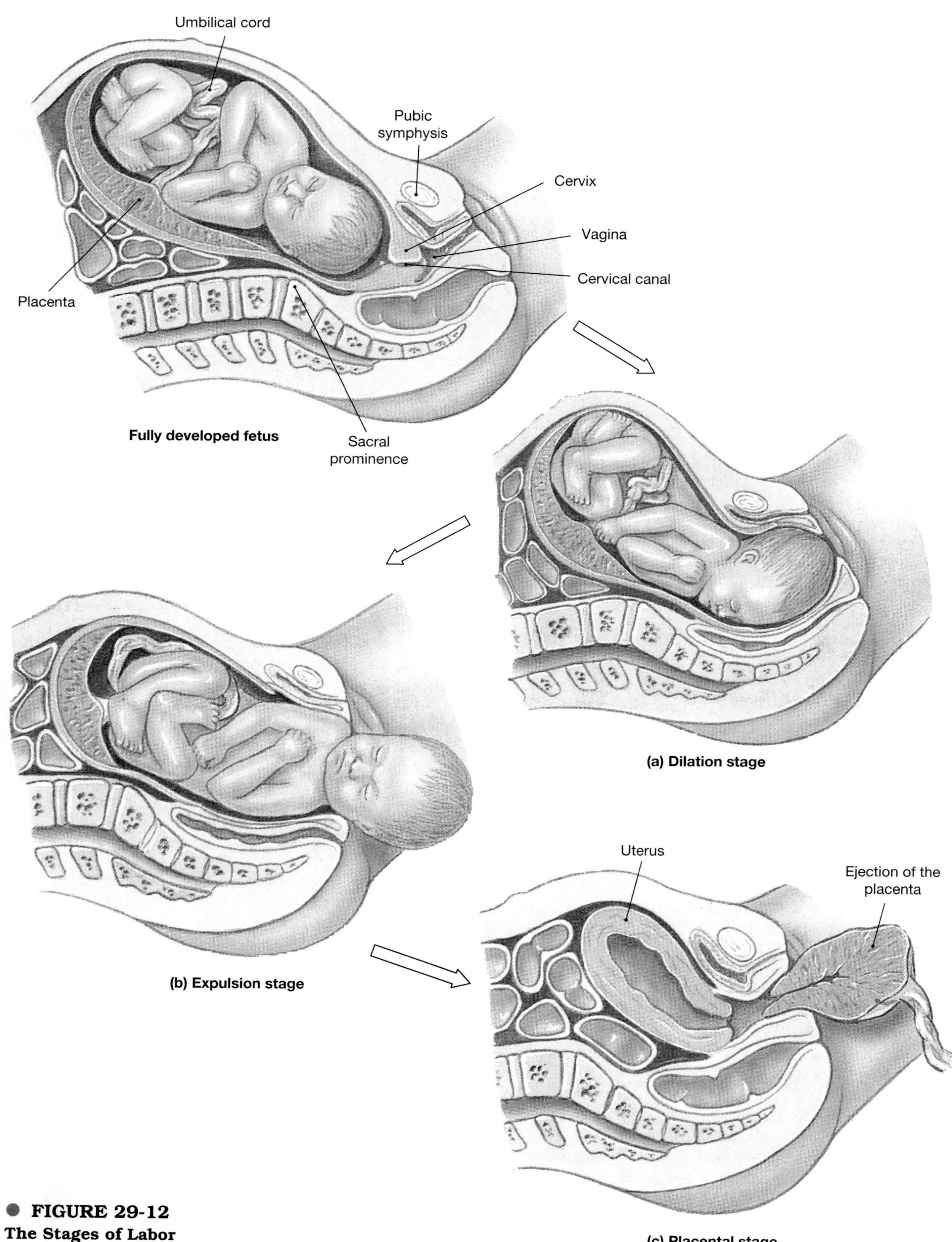
Umbilical cord
Pubic symphysis
Cervix
Vagina
Cervical canal
Placenta
Fully developed fetus
Sacral prominence
(a) Dilation stage
(b) Expulsion stage
Uterus
Ejection of the placenta
(c) Placental stage

● **FIGURE 29-12**
The Stages of Labor

For unknown reasons the statistics for naturally occurring multiple births fall into a pattern: twins occur at a rate of 1:89, triplets at a rate of $1{:}89^2$, quadruplets at $1{:}89^3$, and so forth. The incidence of multiple births can be increased by exposure to fertility drugs that stimulate the maturation of abnormally large numbers of oocytes (see the discussion of fertility and infertility on pp. 1126–1127).

Complete splitting of the embryonic portion of the blastodisc can produce identical twins. If the separation is not complete, **conjoined** (*Siamese*) **twins** may develop. These infants often share a portion of the liver, some skin, and perhaps other internal organs as well. When the fusion is minor the infants can be surgically separated with some success, but those with more extensive fusions usually fail to survive delivery.

Even normal multiple pregnancies pose special problems for maternal systems, for the strains are multiplied proportionately. The chances of premature labor are increased, and the risks to the mother are higher than for single births. The risks for the fetus are also increased, both during gestation and after delivery, for even at full term the newborn infants have a lower average birth weight. They are also more likely to have problems during delivery. For example, in more than half of twin deliveries, one or both fetuses enter the vaginal canal in an abnormal position.

Postnatal Development

Developmental processes do not cease at delivery, for the newborn infant has few of the anatomical, functional, or physiological characteristics of the mature adult. In the course of postnatal development every individual passes through a number of **life stages,** each typified by a distinctive combination of characteristics and abilities.

These stages are a familiar part of human experience. You could probably identify the features and functions associated with *infancy, childhood, adolescence,* and *maturity*. Although each stage has distinctive features, the transitions between them are gradual and the boundaries often indistinct. Once maturity has arrived, development ends and the process of aging, or **senescence,** begins.

The Neonatal Period, Infancy, and Childhood

The **neonatal period** extends from the moment of birth to 1 month thereafter. **Infancy** then continues to 2 years of age, and **childhood** lasts until puberty commences. Two major events are under way during these developmental stages.

1. The major organ systems other than those associated with reproduction become fully operational and gradually acquire the functional characteristics of adult structures.
2. The individual grows rapidly, and there are significant changes in body proportions.

Pediatrics is a medical specialty focusing on postnatal development from infancy through adolescence. Because infants and young children cannot clearly describe the problems they are experiencing, pediatricians and parents must be skilled observers. A number of standardized testing procedures are used to assess an individual's developmental progress relative to normal values. AM *Monitoring Postnatal Development*

The Neonatal Period

Figure 29-13

A variety of physiological and anatomical alterations occur as the fetus completes the transition to the status of a newborn infant, or **neonate.** Prior to delivery, transfer of dissolved gases, nutrients, waste products, hormones, and immunoglobulins occurred across the placental interface. At birth the newborn infant must become relatively self-sufficient, with the processes of respiration, digestion, and excretion performed by its own specialized organs and organ systems. The transition from fetus to neonate may be summarized as follows:

1. The lungs at birth are collapsed and filled with fluid. Filling them with air involves a massive and powerful inspiratory movement. ∞ *[p. 844]*
2. When the lungs expand, the pattern of cardiovascular circulation changes because of alterations in blood pressure and flow rates. The ductus arteriosus closes, isolating the pulmonary and systemic trunks. Closure of the foramen ovale separates the atria of the heart, completing separation of the pulmonary and systemic circuits. These circulatory changes were discussed in Chapters 20 and 21. ∞ *[pp. 685–686, 690, 768]*
3. Typical heart rates of 120–140 beats per minute and respiratory rates of 30 breaths per minute in neonates are considerably higher than those of adults.
4. Prior to birth the digestive system remains relatively inactive, although it does accumulate a mixture of bile secretions, mucus, and epithelial cells. This collection of debris is excreted in the first few days of life. Over that period the newborn infant begins to nurse.
5. As waste products build up in the arterial blood they are filtered into the urine at the kidneys. Glomerular filtration is normal, but the urine cannot be concentrated to any significant degree. As a result, urinary water losses are

high, and neonatal fluid requirements are much greater than those of adults.

6. The neonate has little ability to control body temperature, particularly in the first few days after delivery. As the infant grows larger and increases the thickness of its insulating subcutaneous adipose "blanket," its metabolic rate also rises. Daily and even hourly alterations in body temperature continue throughout childhood. [p. 972]

Throughout the neonatal period the newborn is dependent on nutrients contained in the milk secreted by the maternal mammary glands.

LACTATION AND THE MAMMARY GLANDS By the end of the sixth month of pregnancy the mammary glands are fully developed, and the gland cells begin producing a secretion known as **colostrum** (ko-LOS-trum). Colostrum, which will be provided to the infant during the first 2 or 3 days of life, contains relatively more proteins and far less fat than milk. Many of the proteins are immunoglobulins that may help the infant ward off infections until its own immune system becomes fully functional. In addition, glycoproteins known as *mucins,* present in both colostrum and milk, can inhibit the replication of a family of viruses *(rotaviruses)* that can cause dangerous forms of gastroenteritis and infant diarrhea.

As colostrum production declines, the mammary glands convert to milk production. Milk consists of a mixture of water, proteins, amino acids, lipids, sugars, and salts. It also contains large quantities of *lysozymes,* enzymes with antibiotic properties. Human milk provides roughly 750 Calories per liter. The secretory rate varies, depending on the demand, but a 5–6 kg (11–13 lb) infant usually requires around 850 ml of milk per day.

The actual secretion of the mammary glands is triggered when the infant begins to suck on the nipple. Stimulation of tactile receptors at that site leads to the release of oxytocin at the posterior pituitary. When oxytocin reaches the mammary gland it causes the contraction of myoepithelial cells in the walls of the lactiferous ducts and sinuses. The result is the ejection of milk, and this milk ejection, or *milk let-down,* is diagrammed in Figure 29-13●.

The milk let-down reflex continues to function until weaning occurs, typically 1–2 years after birth. Milk production ceases soon after, and the mammary glands gradually return to a resting state.

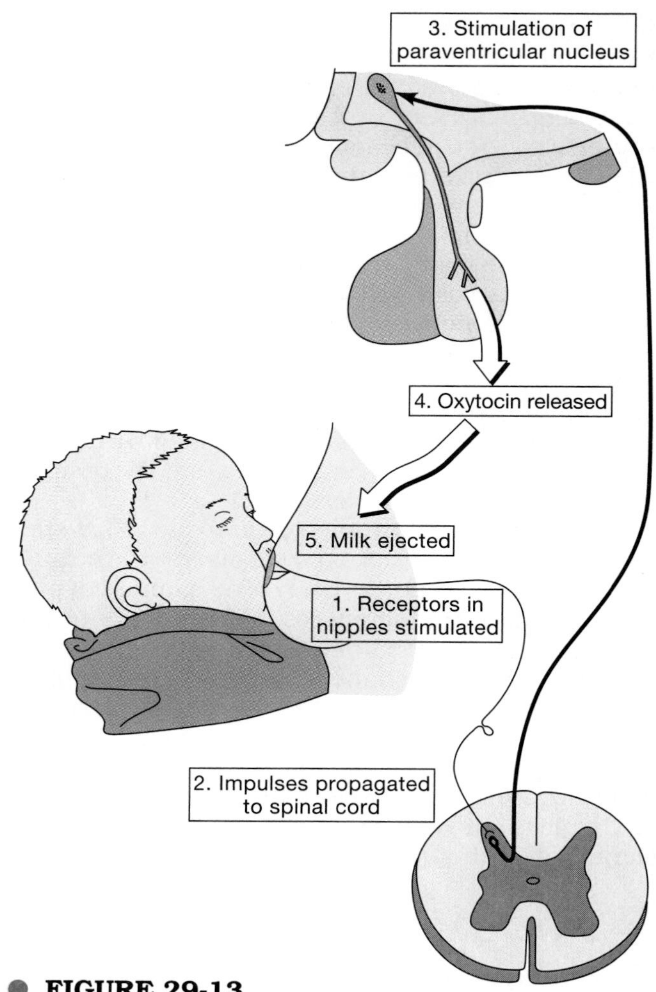
3. Stimulation of paraventricular nucleus
4. Oxytocin released
5. Milk ejected
1. Receptors in nipples stimulated
2. Impulses propagated to spinal cord

● **FIGURE 29-13**
The Milk Let-Down Reflex

Infancy and Childhood

Figure 29-14

The most rapid growth occurs during prenatal development, and after delivery the relative rate of growth continues to decline. Postnatal growth during infancy and childhood occurs under the direction of circulating hormones, notably growth hormone from the pituitary, adrenal steroids, and thyroid hormones. These hormones affect each tissue and organ in specific ways, depending on the sensitivities of the individual cells. As a result, growth does not occur uniformly, and the body proportions gradually change (Figure 29-14●).

❑ Adolescence and Maturity

Adolescence begins at puberty, when three events interact to promote increased hormone production and sexual maturation:

1. The hypothalamus increases its production of gonadotropin-releasing hormone (GnRH).
2. The anterior pituitary becomes more sensitive to the presence of GnRH, and there is a rapid elevation in the circulating levels of FSH and LH.
3. Ovarian or testicular cells become more sensitive to FSH and LH. These changes initiate gametogenesis and the production of male or female sex hormones that stimulate the appearance of secondary sexual characteristics and behaviors.

In the years that follow, the continued background secretion of estrogens or androgens maintains these sexual characteristics. In addition, the combination of sex hormones and growth hormone, adrenal steroids, and thyroxine leads to a sudden acceleration in the growth rate. The timing of the increase in size varies between the sexes, corresponding to different ages at the onset of puberty. In girls the growth rate is maximum between ages 10 and 13, whereas boys grow most rapidly between ages 12 and 15. Growth continues at a slower pace until ages 18–21, when most of the epiphyseal plates become ossified.

The boundary between adolescence and maturity is very hazy, for it has physical, emotional, behavioral, and legal aspects. Maturity is often associated with the end of growth in the late teens or early twenties. Although growth normally ceases at maturity, physiological changes continue. These changes are part of the process of senescence, or aging. Aging reduces the efficiency and capabilities of the individual, and even in the absence of other factors senescence will ultimately lead to death.

Table 29-3 summarizes the age-related changes in physiological systems that have been discussed in earlier chapters. Taken together, these alterations reduce the functional abilities of the individual. They also affect homeostatic mechanisms. As a result, the elderly are less able to make homeostatic adjustments in response to internal or environmental stresses. As immune function deteriorates, the risks of contracting a variety of bacterial or viral diseases are proportionately increased. This deterioration leads to drastic physiological alterations that affect all internal systems. Death finally occurs when some combination of stresses cannot be countered by existing homeostatic mechanisms.

Physicians attempt to forestall death by adjusting homeostatic mechanisms or removing the sources of stress. **Geriatrics** is a medical specialty concerned with the mechanics of the aging process, and physicians trained in geriatrics are known as **geriatricians.** Problems commonly encountered by geriatricians include infections, cancers, heart disease, strokes, arthritis, and anemia; these conditions can be directly related to the age-induced changes in vital systems. AM *Death and Dying*

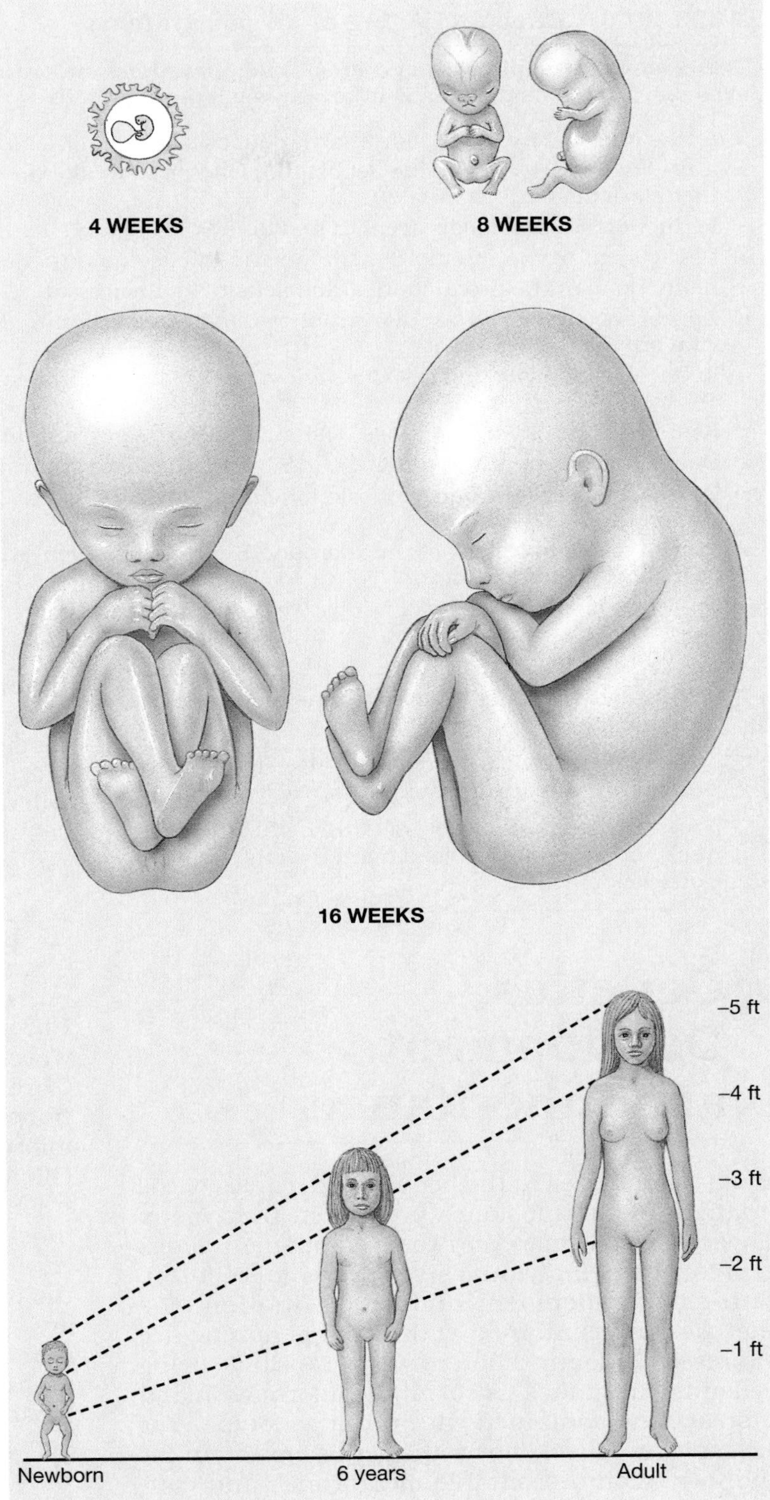
4 WEEKS
8 WEEKS
16 WEEKS
–5 ft
–4 ft
–3 ft
–2 ft
–1 ft
Newborn
6 years
Adult

● **FIGURE 29-14**
Growth and Changes in Body Form. The views at 4, 8, and 16 weeks are presented at actual size. Notice the changes in body form and proportions as development proceeds. These changes do not stop at birth. For example, the head, which contains the brain and sense organs, is relatively large at birth.

TABLE 29-3 Effects of Aging on Organ Systems

The characteristic physical and functional alterations under way as part of the aging process affect all organ systems. Examples discussed in previous chapters include:

- A loss of elasticity in the skin that produces sagging and wrinkling. [p. 168]
- A decline in the rate of bone deposition, leading to weak bones, and degenerative changes in joints that make them less mobile. [pp. 191, 264]
- Reductions in muscular strength and ability. [p. 314]
- Impairment of coordination, memory, and intellectual function. [pp. 521–522]
- Reductions in the production of and sensitivity to circulating hormones. [p. 639]
- Appearance of cardiovascular problems and a reduction in peripheral blood flow that can affect a variety of vital organs. [p. 768]
- Reduced sensitivity and responsiveness of the immune system, leading to problems with infection and cancer. [p. 817]
- Reduced elasticity in the lungs, leading to decreased respiratory function. [p. 869]
- Decreased peristalsis and muscle tone along the digestive tract. [p. 927]
- Decreased peristalsis and muscle tone in the urinary system, coupled with a reduction in the glomerular filtration rate. [p. 1018]
- Functional impairment of the reproductive system, which eventually becomes inactive when the menopause or male climacteric occurs. [p. 1099)

Why does a mother's blood volume increase during pregnancy?

What effect would a decrease in progesterone have on the uterus during late pregnancy?

An increase in the levels of GnRH, FSH, LH, and sex hormones in the blood mark the onset of what stage of development?

Genetics, Development, and Inheritance

Every somatic cell in the body carries copies of the original 46 chromosomes present in the zygote. Those chromosomes and their component genes represent the individual's **genotype** (JĒN-ō-tīp). Through development and differentiation, the instructions contained within the genotype are expressed in many different ways. No single living cell or tissue makes use of all the information and instructions contained within the genotype. For example, in muscle fibers the genes important for the formation of excitable membranes and contractile proteins are active, while a different set of genes is operating in the cells of the pancreatic islets. Collectively, however, the instructions contained within the genotype determine *the anatomical and physiological characteristics of each individual.* Those characteristics are known as the individual's **phenotype** (FĒN-ō-tīp).

Your genotype is derived from those of your parents, but not in a simple way. You are not an exact copy of either parent, nor are you an easily identifiable mixture of their characteristics. Our discussion of genetics will begin with the basic patterns of inheritance and their implications. We will then examine the mechanisms responsible for regulating the activities of the genotype during subsequent prenatal development.

Genes and Chromosomes

Figures 29-15, 29-16, 29-17

Chromosome structure and the functions of genes were introduced in Chapter 3. [pp. 93, 98] Chromosomes contain DNA, and genes are functional segments of DNA. Each gene carries the information needed to direct the synthesis of a specific polypeptide.

Every somatic cell contains 23 pairs of chromosomes. One member of each pair was contributed by the spermatozoon at fertilization and the other by the ovum. The members of each pair are known as **homologous** (hō-MOL-o-gus) **chromosomes.** Twenty-two of those pairs are known as **autosomal** (aw-to-SŌ-mal) **chromosomes.** Most of the genes of these chromosomes affect somatic characteristics, such as hair color or skin pigmentation. The chromosomes of the twenty-third pair are called the **sex chromosomes** because they differ in the two sexes. Figure 29-15• shows the chromosomes of a normal male individual.

Autosomal Chromosomes

Both chromosomes in an autosomal pair have the same structure and carry genes that affect the same traits. If one member of the pair contains three

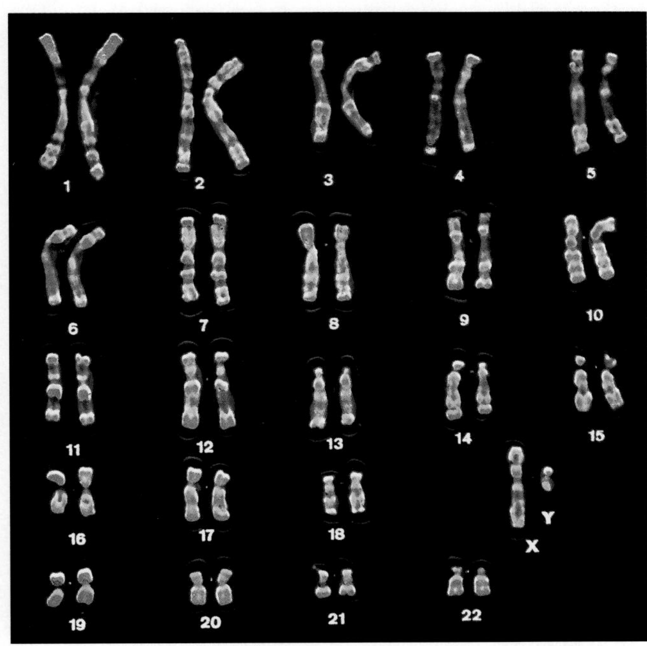
1
2
3
4
5
6
7
8
9
10
11
12
13
14
15
16
17
18
X
Y
19
20
21
22

FIGURE 29-15
Human Chromosomes

genes in a row, with number 1 determining hair color, number 2 eye color, and number 3 skin pigmentation, the other chromosome will carry genes affecting the same traits and in the same sequence.

The various forms of any one gene are called **alleles** (a-LĒLZ). It is these *alternate forms* that determine the precise effect of the gene on the individual's phenotype. If both chromosomes of a homologous pair carry the same allele of a particular gene, the individual is **homozygous** (hō-mō-ZĪ-gus) for that trait. For example, if an individual receives a gene for curly hair from the spermatozoon and one for curly hair from the ovum, the individual will be homozygous for curly hair. Usually about 80 percent of an individual's genetic complement consists of homozygous alleles.

INTERACTIONS BETWEEN ALLELES Because the chromosomes of a homologous pair have different origins, one paternal and the other maternal, they do not necessarily carry the same alleles. When an individual has two different alleles of the same gene, the individual is **heterozygous** (het-er-ō-ZĪ-gus) for that trait. In the case of a heterozygous genotype, the phenotype will be determined by the interactions between the corresponding alleles.

- If an allele is **dominant** it will be expressed in the phenotype *regardless of any conflicting instructions carried by the other allele.*
- If an allele is **recessive,** it will be expressed in the phenotype only if it is present on *both* chromosomes of a homologous pair. For example, Chapter 6 described the albino condition, characterized by an inability to synthesize a yellow-brown pigment, *melanin.* The presence of one normal allele will result in normal coloration. Two recessive alleles must be present to produce an albino individual.
- There can be many different alleles for a single gene, some dominant and others recessive. If an individual receives two dominant but different alleles, the resulting phenotype will depend on how the two dominant genes interact. In *suppression* one allele suppresses the other, and the phenotype is the same as if the second allele was recessive. In *complementary gene action* the two dominant alleles interact to produce a phenotype different from that resulting from either allele by itself.

PENETRANCE AND EXPRESSIVITY Differences in genotype lead to distinctive variations in phenotype, but the relationships are not always easily predictable. The presence of a particular pair of alleles does not affect the phenotype in the same way in every individual. **Penetrance** is the percentage of individuals with a particular genotype that show the expected phenotype. The effects of the genotype in the other individuals may be overridden by the activity of other genes or by environmental factors. For example, *emphysema,* a respiratory disorder discussed in Chapter 23, has been linked to a specific abnormal genotype. [p. 863] However, roughly 20 percent of the individuals with the genotype do not develop emphysema. The penetrance of this genotype is therefore approximately 80 percent.

If a given genotype *does* affect the phenotype, it can do so to varying degrees, again depending on the activity of other genes or environmental stimuli. For example, identical twins do not have the same fingerprints, even though they have the same genotype. The extent to which a particular allele is expressed when it is present is termed its **expressivity.**

PREDICTING INHERITANCE Not every allele can be neatly characterized as dominant or recessive. Some that can are included in Table 29-4. If you restrict attention to these alleles it is possible to predict the characteristics of individuals on the basis of those of their parents.

In such calculations, dominant alleles are traditionally indicated by capitalized abbreviations and recessives are abbreviated in lower case letters. For a given trait, the possibilities are indicated by *AA* (homozygous dominant), *Aa* (heterozygous), or *aa* (homozygous recessive). The gametes involved in fertilization each contribute a single allele for a given trait. That allele must be one of the two contained by all cells in the parental body. Consider, for example, the offspring of an albino mother and a normal father. Because albinism is a recessive trait, the maternal alleles can be abbreviated *aa*. No matter which of her ova gets fertilized, it will carry the recessive *a* allele. The father has normal coloration, a dominant trait. He may therefore be homozygous *or* heterozygous for this trait, since *AA* or *Aa* will both give rise to the same phenotype. A homozygous father will produce sperm that all carry the *A* allele. A heterozygous father, however, will produce two different kinds of sperm; Some will carry the dominant allele *A* and others will carry the recessive allele *a*.

A simple box diagram known as a **Punnett square** lets us predict the probabilities that the children will have particular characteristics by showing us the possible combinations of parental alleles that they can inherit. In the square shown in Figure 29-16•, the maternal alleles are listed along the horizontal axis and the paternal ones along the vertical axis. The possible combinations are indicated in the small boxes. Figure 29-16a• considers the possible offspring of an *aa* mother and an *AA* father. All of the children must have the genotype *Aa*, and so all will have normal skin coloration. Compare these results with those of Figure 29-16b•, for a heterozygous father (*Aa*). The heterozygous individual produces two types of gametes, *A* and *a*, and the egg may be fertilized by either one. As a result, there is a 50 percent probability that a child of such a father will inherit the genotype *Aa*, and so have normal skin color. The probability of inheriting the genotype *aa*, and thus having the albino phenotype, is also 50 percent. The Punnett square can also be used in reverse, to draw conclusions about the identity and genotype of a parent. For example, a man with the genotype *AA* cannot be the father of an albino child (*aa*).

SIMPLE INHERITANCE In **simple inheritance** phenotypic characters are determined by interactions between a single pair of alleles. The frequency of appearance of an inherited disorder resulting from simple inheritance can be predicted using a Punnett square. Although they are rare disorders

TABLE 29-4 The Inheritance of Selected Phenotypic Characters

DOMINANT TRAITS
One allele determines phenotype, and the other is suppressed
normal skin coloration
brachydactyly (short fingers)
ability to taste phenylthiocarbamate (PTC)
free earlobes
curly hair
presence of Rh factor on red blood cell membranes
Both dominant alleles may be expressed (codominance)
presence of A or B antigens on red blood cell membranes
structure of serum proteins (albumins, transferrins)
structure of hemoglobin molecule
RECESSIVE TRAITS
albinism
blond hair
red hair (expressed only if individual is also homozygous for blond hair)
lack of A, B agglutinogens (Type O blood)
inability to roll the tongue into a U-shape
SEX-LINKED TRAITS
color blindness
POLYGENIC TRAITS
eye color
hair colors other than pure blond or red

Note: For a listing of inherited clinical conditions, see Table 29-5.

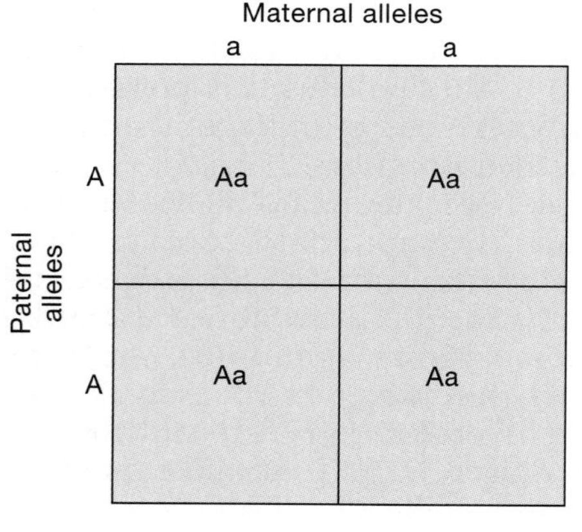
Maternal alleles
a
a
Paternal alleles
A
A
Aa
Aa
Aa
Aa
(a)

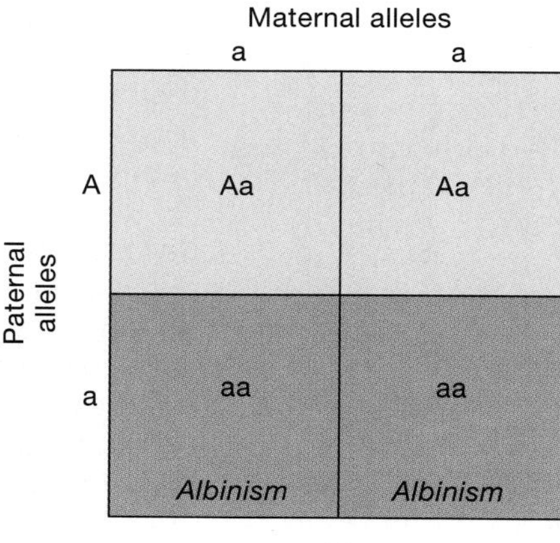
Maternal alleles
a
a
Paternal alleles
A
a
Aa
Aa
aa
aa
Albinism
Albinism
(b)

• **FIGURE 29-16**
Predicting Phenotypic Characteristics. (a) The offspring of a homozygous-dominant father and a homozygous-recessive mother will all be heterozygous for that trait. Their phenotype will be the same as that of the father. **(b)** The offspring of a heterozygous father and a homozygous-recessive mother will be either heterozygous or homozygous for the recessive trait. In this example, half of the offspring will have normal skin coloration and the other half will be albinos.

in terms of overall numbers, more than 1200 different inherited conditions have been identified that reflect the presence of one or two abnormal alleles for a single gene. A partial listing is included in Table 29-5, along with the location where additional information can be obtained.

POLYGENIC INHERITANCE **Polygenic inheritance** involves interactions among the alleles of several different genes. Because multiple alleles are involved, the frequency of occurrence cannot easily be predicted using a simple Punnett square. Several important adult disorders, including hypertension and coronary artery disease, fall into this category.

Many of the developmental disorders responsible for fetal mortalities and congenital malformations result from multiple genetic interactions. In these cases the particular genetic composition of the individual does not by itself determine the onset of the disease. Instead, the conditions regulated by these genes establish a susceptibility to particular environmental influences. This means that not every individual with the genetic tendency for a particular condition will actually develop it. It is therefore difficult to track polygenic conditions through successive generations. However, because many inherited polygenic conditions are *likely* but not *guaranteed* to occur, steps can be taken to prevent a crisis. For example, hypertension may be prevented or reduced by controlling diet and fluid volume, and coronary artery disease may be prevented by lowering serum cholesterol concentrations.

TABLE 29-5 Relatively Common Inherited Disorders

Disorder	*Page in Text or Applications Manual*
Autosomal dominants	
Adult polycystic kidney disease	*AM*
Marfan's syndrome	p. 125 and *AM*
Huntington's disease	p. 521 and *AM*
Autosomal recessives	
Deafness	p. 594
Albinism	pp. 154, 1135
Sickle cell anemia	p. 656 and *AM*
Cystic fibrosis	p. 829 and *AM*
Phenylketonuria	p. 952 and *AM*
Tay-Sachs disease	p. 418 and *AM*
X-linked	
Duchenne's muscular dystrophy	p. 293 and *AM*
Myotonic muscular dystrophy	*AM*
Hemophilia (one form)	p. 675
Testicular feminization	*AM*
Color blindness	p. 577

Note: The genetic basis for the diseases in blue has been identified.

SOURCES OF INDIVIDUAL VARIATION You are not a copy of either parent, nor are you a 50:50 mixture of their characteristics. One reason was noted in Chapter 28: During meiosis, maternal and paternal chromosomes are randomly assorted, so each gamete contains a unique blend of maternal and paternal chromosomes.

In addition, parts of chromosomes can become rearranged during synapsis (Figure 29-17•). When tetrads are formed, adjacent chromatids may overlap, a process called *crossing-over.* Under these conditions they may break and reattach in a process that switches a portion of one chromatid for the corresponding segment on another. This process is called *genetic recombination.* Genetic recombination, which occurs during meiosis in both males and females, greatly increases the range of possible variation among gametes, and thus among members of successive generations, whose genotypes are formed by the combination of gametes in fertilization. It can also complicate the tracing of inheritance of genetic disorders.

Variations at the level of the individual gene may also appear as a result of mutations. **Spontaneous mutations** are the result of random errors in the DNA replication process. Such incidents are relatively common. At least 10 percent of fertilizations produce zygotes with abnormal chromosomes, and because spontaneous mutations usually fail to produce visible defects, the actual number of mutations must be far larger. Many of the changes may have no effect on phenotype, and each of us has unique, individual characteristics. Given the complexity of the interactions between genotype and phenotype, some degree of individual variation is inevitable.

AM *Complexity and Perfection*

Zygotes with more significant abnormalities usually produce embryos that die before completing development, and only about 0.5 percent of newborn infants show *chromosomal abnormalities* resulting from spontaneous mutations. Individuals with **chromosomal abnormalities** have damaged, broken, missing, or extra copies of chromosomes. The normal human chromosomal complement, or *karyotype,* is shown in Figure 29-15•. Chromosomal abnormalities produce a variety of serious clinical conditions, in addition to contributing to prenatal mortality. *Few individuals with chromosomal abnormalities survive to full term.* The high mortality rate and severity of the problems reflect the fact that large numbers of genes have been added or deleted.

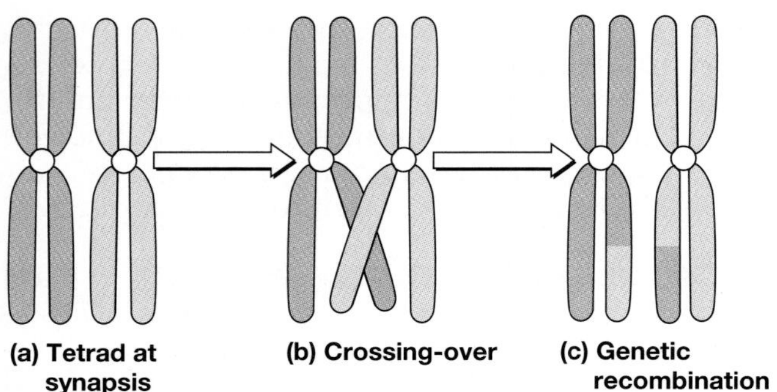
(a) Tetrad at synapsis
(b) Crossing-over
(c) Genetic recombination

● **FIGURE 29-17**
Genetic Recombination. (a) Synapsis, with the formation of tetrads during meiosis. **(b)** Crossing-over of homologous portions of two chromosomes. **(c)** Genetic recombination by the breakage and exchange of corresponding sections on the two chromosomes.

Sex Chromosomes

Figure 29-18

Unlike other chromosomal pairs, the sex chromosomes are not necessarily identical in appearance and gene content. There are two different sex chromosomes, an **X chromosome** and a **Y chromosome.** The *Y* chromosomes are considerably smaller than *X* chromosomes and contain fewer genes, but among those genes are dominant alleles that specify that an individual with that chromosome will be a male. The normal male sex chromosome complement is *XY.* Females do not have a *Y* chromosome, and their sex chromosome pair is *XX.*

The ova produced by a woman will always carry *X,* whereas sperm may carry *X* or *Y.* Thus, if you make a Punnett square you can show that the male:female sex ratio in the offspring should be 1:1. The birth statistics differ slightly from that prediction, with 106 males born for every 100 females. It has been suggested that more males are born because the sperm carrying the *Y* chromosome can reach the oocyte first, since they do not have to carry the extra weight of the larger *X* chromosome.

The *X* chromosome also carries genes that affect somatic structures. These characteristics are called **X-linked** because in most cases there are no corresponding alleles on the *Y* chromosome. The inheritance of characteristics regulated by these genes does not follow the pattern of alleles on autosomal chromosomes. The best known of these single-allele characters are those associated with noticeable diseases or functional deficits.

The inheritance of color blindness, a condition discussed in Chapter 17, exemplifies the differences between sex-linked and autosomal inheritance. ∞ [p. 577] A relatively common form of color blindness is associated with the presence of a dominant or recessive gene on the *X* chromosome. Normal color vision is determined by the presence of a dominant gene, *C,* and color blindness results from the presence of the recessive gene, *c.* A woman, with her two *X* chromosomes, can be either homozygous, *CC,* or heterozygous, *Cc,* and still have normal color vision. She will be color-blind only if she carries two recessive alleles, *cc.* But a male has only one *X* chromosome, so whatever that chromosome carries will determine whether he has normal color vision or is color-blind. A Punnett square for an *X*-linked trait, as in Figure 29-18●, reveals that the sons produced by a normal father and a heterozygous mother will have a 50 percent chance of being color-blind, while the daughters will all have normal color vision.

Paternal alleles \ Maternal alleles	X^C	X^c
X^C	$X^C X^C$ Normal female	$X^C X^c$ Normal female (carrier)
Y	$X^C Y$ Normal male	$X^c Y$ Color blind male

● **FIGURE 29-18**
Sex-Linked Inheritance

A number of other clinical disorders noted earlier in the text are *X*-linked traits, including certain forms of *hemophilia, diabetes insipidus,* and *muscular dystrophy.* In several instances, advances in molecular genetic techniques have permitted the localization of the specific genes on the *X* chromosome. This technique provides a relatively direct method of screening for the presence of a particular condition before the symptoms appear and even before birth.

❑ The Human Genome Project

Figures 29-15, 29-19

Few of the genes responsible for inherited disorders have been identified or even localized to a specific chromosome. That situation is changing rapidly, however, because of the attention devoted to the **Human Genome Project.** This project, funded by the National Institutes of Health and the Department of Energy, is attempting to transcribe the entire human genome, chromosome by chromosome and gene by gene. The project began in October 1990 and was expected to take 10–15 years. Progress has been more rapid than expected, and it may actually take considerably less time.

The first step in understanding the human genome is to prepare a map of the individual chromosomes. **Karyotyping** (KAR-ē-ō-tīp-ing; *karyon,* nucleus + *typos,* mark) is the determination of an individual's chromosomal complement. Figure 29-15●, p. 1135, shows a set of normal human chromosomes. Each chromosome has characteristic banding patterns, and segments can be stained

Chromosomal Abnormalities and Genetic Analysis

There are two types of abnormalities in autosomal chromosomes that do not invariably kill the individual before birth. These are *translocation defects* and *trisomy*.

- In a **translocation defect,** crossing-over occurs between different chromosome pairs. For example, a piece of chromosome 8 may become attached to chromosome 14. The genes moved to their new position may function abnormally, becoming inactive or overactive. A translocation between chromosomes 8 and 14 is responsible for *Burkitt's lymphoma*, a type of lymphatic system cancer.
- In **trisomy** something goes wrong in meiosis II, so that the gamete that contributes to the zygote carries an extra copy of one chromosome. Because there are three copies of this chromosome rather than two, the condition is termed *trisomy*. The nature of the trisomy is indicated by the number of the chromosome involved. Zygotes with extra copies of chromosomes seldom survive. Individuals with trisomy 13 and trisomy 18 may survive until delivery, but rarely live longer than a year. The notable exception is trisomy 21.

Trisomy 21, or **Down syndrome,** is the most common viable chromosomal abnormality. Estimates of the frequency of appearance range from 1.5 to 1.9 per 1000 births for the U.S. population. The affected individual suffers from mental retardation and characteristic physical malformations, including a facial appearance that gave rise to the term *mongolism* once used to describe this condition. The degree of mental retardation ranges from moderate to severe, and few individuals with this condition lead independent lives. Anatomical problems affecting the cardiovascular system often prove fatal during childhood or early adulthood. Although some individuals survive to moderate old age, many develop Alzheimer's disease while still relatively young (before age 40).

For unknown reasons there is a direct correlation between maternal age and the risk of having a child with trisomy 21. Below maternal age 25 the incidence of Down syndrome approaches 1 in 2000 births, or 0.05 percent. For maternal ages 30–34, the odds increase to 1:900, and over the next decade they go from 1:290 to 1:46, or more than 2 percent. These statistics are becoming increasingly significant, for many women have delayed childbearing until their mid-30s or later.

Abnormal numbers of sex chromosomes do not produce effects as severe as those induced by extra or missing autosomal chromosomes. In **Klinefelter syndrome** the individual carries the sex chromosome pattern *XXY*. The phenotype is male, but the extra *X* chromosome causes reduced androgen production. As a result, the breasts are slightly enlarged, the testes fail to mature, and affected individuals are sterile. The incidence of this condition among newborn males averages 1:750.

Individuals with **Turner syndrome** have only a single female sex chromosome; their sex chromosome complement is abbreviated *XO*. This kind of chromosomal deletion is known as **monosomy.** The incidence of this condition at delivery has been estimated as 1:10,000. At birth the condition may not be recognized, for the phenotype is normal female. But maturational changes do not appear at puberty. The ovaries are nonfunctional, and estrogen production occurs at negligible levels.

Chromosomal Analysis. In **amniocentesis** a sample of amniotic fluid is removed and the fetal cells that it contains are analyzed. This procedure permits the identification of over 20 different congenital conditions, including trisomy 21. The needle inserted to obtain a fluid sample is guided into position using ultrasound. Unfortunately, amniocentesis has two major drawbacks:

1. Because the sampling procedure represents a potential threat to the health of the fetus and mother, amniocentesis is performed only when there are known risk factors present. Examples of risk factors would include a family history of specific conditions or, in the case of Down syndrome, a maternal age over 35.
2. Sampling cannot safely be performed until the volume of amniotic fluid is large enough that the fetus will not be injured during the sampling process. The usual time for amniocentesis is at a gestational age of 14–15 weeks. It may take several weeks to obtain results once samples have been collected, and by the time the results are received, the option of therapeutic abortion may no longer be available.

An alternative procedure known as **chorionic villus sampling** analyzes cells collected from the villi during the first trimester. Although it can be performed at an earlier gestational age, this technique has largely been abandoned because of an associated increased risk of spontaneous abortion.

with special dyes. The banding patterns are useful as reference points when more detailed genetic maps are prepared. The banding patterns themselves can be useful, as abnormal banding patterns are characteristic of some genetic disorders and several cancers, including a form of leukemia.

As of 1994, a progress report includes the following:

- Two chromosomes—chromosome 21 and the *Y* chromosome—have been mapped completely, and preliminary maps have been made for all other chromosomes.
- Over 2200 genes have been identified. Although a significant number, this is only a small fraction of the estimated 100,000 genes in the human genome.
- The specific genes responsible for 27 inherited disorders have been identified, including several of the disorders listed in Table 29-5. Genetic screening can now be done for many of these conditions, as indicated in Figure 29-19●.

The Human Genome Project is attempting to determine the normal genetic composition of a "typical" human being. Yet we all are variations on a basic theme. As we improve our abilities to manipulate our own genetic foundations, we will face troubling ethical and legal decisions. For example, few people object to the insertion of a "correct" gene into somatic cells to cure a specific disease. But what if we could insert that modified gene into a gamete, and change not only that individual but all his or her descendants? And what if the gene did not correct or prevent a disorder, but "improved" the individual by increasing intelligence, height, or vision, or altering some other phenotypic characteristic? These and other difficult questions will not go away, and in the years to come we will have to find answers we all can live with.

Curly hair is an autosomal dominant trait. What would be the phenotype of a person who is heterozygous for this trait?

Why are children not identical copies of their parents?

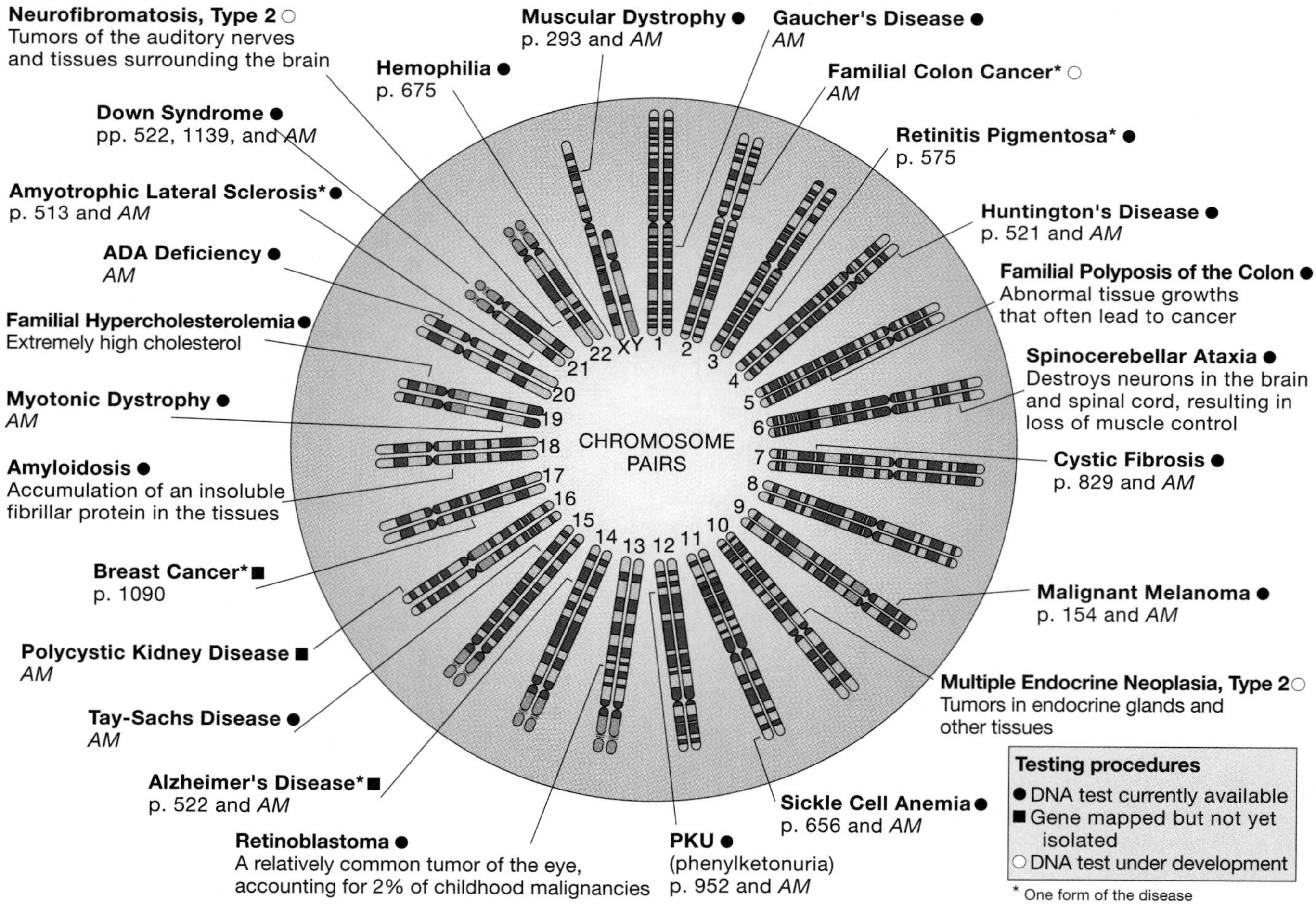

Neurofibromatosis, Type 2 ○
Tumors of the auditory nerves and tissues surrounding the brain

Down Syndrome ●
pp. 522, 1139, and *AM*

Amyotrophic Lateral Sclerosis* ●
p. 513 and *AM*

ADA Deficiency ●
AM

Familial Hypercholesterolemia ●
Extremely high cholesterol

Myotonic Dystrophy ●
AM

Amyloidosis ●
Accumulation of an insoluble fibrillar protein in the tissues

Breast Cancer* ■
p. 1090

Polycystic Kidney Disease ■
AM

Tay-Sachs Disease ●
AM

Alzheimer's Disease* ■
p. 522 and *AM*

Retinoblastoma ●
A relatively common tumor of the eye, accounting for 2% of childhood malignancies

Hemophilia ●
p. 675

Muscular Dystrophy ●
p. 293 and *AM*

Gaucher's Disease ●
AM

Familial Colon Cancer* ○
AM

Retinitis Pigmentosa* ●
p. 575

Huntington's Disease ●
p. 521 and *AM*

Familial Polyposis of the Colon ●
Abnormal tissue growths that often lead to cancer

Spinocerebellar Ataxia ●
Destroys neurons in the brain and spinal cord, resulting in loss of muscle control

Cystic Fibrosis ●
p. 829 and *AM*

Malignant Melanoma ●
p. 154 and *AM*

Multiple Endocrine Neoplasia, Type 2 ○
Tumors in endocrine glands and other tissues

Sickle Cell Anemia ●
p. 656 and *AM*

PKU ●
(phenylketonuria)
p. 952 and *AM*

CHROMOSOME PAIRS
1 2 3 4 5 6 7 8 9 10 11 12 13 14 15 16 17 18 19 20 21 22 XY

Testing procedures
● DNA test currently available
■ Gene mapped but not yet isolated
○ DNA test under development

* One form of the disease

● **FIGURE 29-19**

A Map of the Human Chromosomes. This diagram shows the banding patterns of typical chromosomes in a male individual, and indicates the locations of the genes responsible for specific inherited disorders. The chromosomes are not drawn to scale.

Selected Clinical Terminology

Terms Discussed in This Chapter

amniocentesis: An analysis of fetal cells taken from a sample of amniotic fluid. *(p. 1139)*
chorionic villus sampling: An analysis of cells collected from the chorionic villi during the first trimester. *(p. 1139)*
ectopic pregnancy: A pregnancy in which implantation occurs somewhere other than the uterus. *(p. 1114* and *AM)*
gestational neoplasm: Tumor formed by undifferentiated, rapid growth of the syncytial trophoblast; if untreated, the neoplasm may become malignant. *(p. 1115)*
infertility: The inability to have children. *(p. 1126)*
in vitro fertilization: Fertilization outside the body, usually in a petri dish. *(p. 1126)*

AM Additional Terms Discussed in the Applications Manual

abruptio placentae (ab-RUP-shē-ō pla-SEN-tē): Tearing of the placenta sometime after the fifth gestational month.
Apgar rating: A method of evaluating newborn infants; a test for developmental problems and neurological damage.
breech birth: A delivery during which the legs or buttocks of the fetus enter the vaginal canal first.
congenital malformation: A severe structural abnormality, present at birth, that affects major systems.
DDST: The Denver Developmental Screening Test, given to infants and children up to age 5 to monitor their development.
fetal alcohol syndrome (FAS): A neonatal condition resulting from maternal alcohol consumption; characterized by developmental defects often involving the skeletal, nervous, and/or cardiovascular system.
placenta previa: A condition resulting from implantation in or near the cervix, where the placenta covers the cervix and prevents normal birth.
pseudohermaphrodite (su-dō-her-MAF-ro-dīt): An individual who has the genotype of one sex but has some or most of the phenotypic characteristics of the opposite sex.
teratogens (TER-a-tō-jenz): Stimuli that disrupt normal development by damaging cells, altering chromosome structure or altering the chemical environment of the embryo.
testicular feminization syndrome: A form of male pseudohermaphroditism resulting from an inability to respond to testosterone.
toxemia (tok-SĒ-mē-a) **of pregnancy:** Disorders affecting the maternal cardiovascular system; includes *eclampsia* and *preeclampsia.*

CHAPTER REVIEW

STUDY OUTLINE

INTRODUCTION, p. 1110

1. **Development** is the gradual modification of physical and physiological characteristics from conception to maturity. The creation of different cell types is **differentiation.**

AN OVERVIEW OF TOPICS IN DEVELOPMENT, p. 1110

1. **Inheritance** refers to the transfer of genetically determined characteristics from generation to generation. **Genetics** is the study of the mechanisms of inheritance. **Prenatal development** occurs before birth; **postnatal development** begins at birth and continues to maturity, when **senescence** (aging) begins.

FERTILIZATION, p. 1110

1. Fertilization normally occurs in the uterine tube within a day after ovulation. Sperm cannot fertilize an egg until they have undergone **capacitation.**

The Oocyte at Ovulation, p. 1111

2. The acrosomal caps of the spermatozoa release **hyaluronidase,** an enzyme that separates cells of the corona radiata and exposes the oocyte membrane. When a single spermatozoon contacts that membrane, fertilization occurs and **activation** follows.
3. During activation the oocyte completes meiosis II, and a **cortical reaction** occurs that prevents the penetration of additional sperm.

Pronucleus Formation and Amphimixis, p. 1112

4. After activation the **female pronucleus** and **male pronucleus** fuse in a process called **amphimixis.** *(Figure 29-1)*

Induction and the Regulation of Development, p. 1112

5. During development, differences in the cytoplasmic composition of individual cells trigger changes in genetic activity. The chemical interplay between developing cells is **induction.**

A Preview of Prenatal Development, p. 1113

6. The 9-month **gestation** period can be divided into three **trimesters.**

THE FIRST TRIMESTER, p. 1113

1. **Cleavage** subdivides the cytoplasm of the zygote in a series of mitotic divisions; the zygote becomes a **blastocyst.** During **implantation** the blastocyst burrows into the uterine endometrium. **Placentation** occurs as blood vessels form around the blastocyst and the **placenta** appears.

Cleavage and Blastocyst Formation, p. 1113

2. The blastocyst consists of an outer **trophoblast** and an **inner cell mass.** (*Figure 29-2*)

Implantation, p. 1113

3. Implantation occurs about 7 days after fertilization as the blastocyst adheres to the uterine lining. (*Figure 29-3*)
4. As the trophoblast enlarges and spreads, maternal blood flows through open **lacunae.** After **gastrulation,** the **blastodisc** becomes an embryo composed of **endoderm, ectoderm,** and an intervening **mesoderm.** It is from these **germ layers** that the body systems differentiate. (*Figure 29-4; Table 29-1*)
5. Germ layers help form four **extraembryonic membranes:** the yolk sac, amnion, allantois, and chorion. (*Figure 29-5*)
6. The **yolk sac** is an important site of blood cell formation. The **amnion** encloses fluid that surrounds and cushions the developing embryo. The base of the **allantois** later gives rise to the urinary bladder. Circulation within the vessels of the **chorion** provides a rapid-transit system linking the embryo with the trophoblast.

Placentation, p. 1118

7. **Chorionic villi** extend outward into the maternal tissues, forming an intricate, branching network through which maternal blood flows. As development proceeds the **umbilical cord** connects the fetus to the placenta. The trophoblast synthesizes hCG, estrogens, progestins, hCS (hPL), and relaxin. (*Figure 29-6*)

Embryogenesis, p. 1120

8. The first trimester is critical because events in the first 12 weeks establish the basis for **organogenesis** (organ formation). (*Figure 29-7; Table 29-2*)

THE SECOND AND THIRD TRIMESTERS, p. 1121

1. In the second trimester the organ systems near functional completion. During the third trimester many of the organ systems become fully functional. (*Figure 29-8*)
2. The fetus undergoes its largest weight gain in the third trimester. At the end of gestation the fetus and enlarged uterus push many abdominal organs out of their positions. (*Figure 29-9*)
3. **Infertility** is relatively common, but recent advances are providing new solutions. (*Figure 29-10*)

Pregnancy and Maternal Systems, p. 1124

4. The developing fetus is totally dependent on maternal organs for nourishment, respiration, and waste removal. Maternal adaptations include increased blood volume, respiratory rate, tidal volume, nutrient intake, and glomerular filtration.

Structural and Functional Changes in the Uterus, p. 1128

5. Progesterone produced by the placenta has an inhibitory effect on uterine muscles; its calming action is opposed by estrogens, oxytocin, and prostaglandins. At some point multiple factors interact to produce **labor contractions** in the uterine wall. (*Figure 29-11*)

LABOR AND DELIVERY, p. 1128

1. The goal of labor is **parturition** (forcible expulsion of the fetus).

Stages of Labor, p. 1129

2. Labor can be divided into three stages: the **dilation stage, expulsion stage,** and **placental stage.** (*Figure 29-12*)

Premature Labor, p. 1129

3. Premature labor may result in **immature delivery** or **premature delivery.**

Multiple Births, p. 1129

4. Twin births may be **dizygotic** ("fraternal") or **monozygotic** ("identical").

POSTNATAL DEVELOPMENT, p. 1131

1. Postnatal development involves a series of **life stages,** including **infancy, childhood, adolescence,** and **maturity. Senescence** begins at maturity and ends in the death of the individual.

The Neonatal Period, Infancy, and Childhood, p. 1131

2. In the transition from fetus to **neonate** the respiratory, circulatory, digestive, and urinary systems begin functioning independently. The newborn must also begin thermoregulation.
3. Mammary gland cells produce protein-rich **colostrum** during the infant's first few days of life and then convert to milk production. These secretions are released as a result of the *milk let-down reflex.* (*Figure 29-13*)
4. Body proportions gradually change during infancy and childhood. (*Figure 29-14*)

Adolescence and Maturity, p. 1132

5. Adolescence begins at puberty when (1) the hypothalamus increases its production of GnRH; (2) circulating levels of FSH and LH rise rapidly; and (3) ovarian or testicular cells become more sensitive to FSH and LH. These changes initiate gametogenesis, production of sex hormones, and a sudden acceleration in growth rate. (*Table 29-3*)

GENETICS, DEVELOPMENT, AND INHERITANCE, p. 1134

1. Every somatic cell carries copies of the original 46 chromosomes in the zygote; these represent the individual's **genotype.** The physical expression of the genotype is the **phenotype** of the individual.

Genes and Chromosomes, p. 1134

2. Every somatic human cell contains 23 pairs of chromosomes; each pair consists of **homologous** chromosomes. Twenty-two pairs are **autosomal** chromosomes. The twenty-third pair of chromosomes are the **sex chromosomes** because they differ between the sexes. (*Figure 29-15*)
3. Chromosomes contain DNA, and genes are functional segments of DNA. The various forms of a gene are called **alleles.** If both homologous chromosomes carry the same allele of a particular gene, the individual is **homozygous;** if they carry different instructions, the individual is **heterozygous.**

4. Alleles are considered **dominant** or **recessive** depending upon how their traits are expressed. (*Table 29-4*)
5. Combining maternal and paternal alleles in a **Punnett square** diagram allows us to predict the characteristics of offspring. (*Figure 29-16*)
6. In **simple inheritance** phenotypic characters are determined by interactions between a single pair of alleles. **Polygenic (multifactorial) inheritance** involves interactions among alleles on several genes.
7. Genetic recombination, the crossing-over of chromosomes during meiosis, increases the genetic variation of male and female gametes. (*Figure 29-17*)
8. There are two different sex chromosomes, an ***X* chromosome** and a ***Y* chromosome.** The normal male sex chromosome complement is *XY*; that of females is *XX*. The *X* chromosome carries **X-linked genes** that affect somatic structures but have no corresponding alleles on the *Y* chromosome. (*Figure 29-18*)

The Human Genome Project, p. 1138

9. The Human Genome Project has identified 2200 of our estimated 100,000 genes, including some of those responsible for inherited disorders. (*Figure 29-19; Table 29-5*)

REVIEW QUESTIONS

LEVEL 1 Reviewing Facts and Terms

1. The gradual modification of anatomical structures during the period from conception to maturity is:
(a) development (b) differentiation
(c) embryogenesis (d) capacitation

2. Human fertilization involves the fusion of two haploid gametes, producing a zygote containing:
(a) 23 chromosomes
(b) 46 chromosomes
(c) the normal haploid number of chromosomes
(d) 46 pairs of chromosomes

3. The secondary oocyte leaving the follicle is in:
(a) interphase
(b) metaphase of the first meiotic division
(c) telophase of the second meiotic division
(d) metaphase of the second meiotic division

4. Chemical interplay between developing cells is called:
(a) gestation (b) induction
(c) capacitation (d) placentation

5. The stage of development that follows cleavage is the:
(a) blastocyst (b) morula
(c) trophoblast (d) blastodisc

6. The process that establishes the foundation of all major organ systems is:
(a) cleavage (b) implantation
(c) placentation (d) embryogenesis

7. The zygote arrives in the uterine cavity as a:
(a) morula (b) trophoblast
(c) lacuna (d) blastomere

8. The first extraembryonic membrane to appear during development is the:
(a) allantois (b) yolk sac
(c) amnion (d) chorion

9. The primitive streak appears during:
(a) cleavage (b) implantation
(c) gastrulation (d) placentation

10. The part of the uterine endometrium that has no contact with the chorion is the:
(a) decidua capsularis (b) decidua basalis
(c) decidua parietalis (d) allantois

11. The surface that provides for active and passive exchange between the fetal and maternal bloodstreams is the:
(a) yolk stalk (b) chorionic villi
(c) umbilical veins (d) umbilical arteries

12. The placental hormone that appears in the maternal bloodstream soon after implantation is:
(a) human chorionic gonadotropin
(b) human placental lactogen
(c) placental prolactin
(d) relaxin

13. The placental stage of labor ends with:
(a) the afterbirth
(b) rupture of the amniotic sac
(c) delivery of the newborn
(d) parturition

14. Milk let-down is associated with:
(a) events occurring in the uterus
(b) placental hormonal influences
(c) circadian rhythms
(d) reflex action triggered by suckling

15. When an individual has two different alleles for the same gene, the individual is __________ for that trait
(a) heterozygous (b) homozygous
(c) dominant (d) recessive

16. If an allele must be present on both the maternal and paternal chromosomes in order to affect the phenotype, the allele is said to be:
(a) dominant (b) recessive
(c) complementary (d) heterozygous

17. The percentage of individuals with a particular genotype that show the expected phenotype reflects:
(a) dominance
(b) polygenic inheritance
(c) spontaneous mutation
(d) penetrance

18. For a trait "A," the genotype of a homozygous recessive individual would be represented as:
(a) *AA* (b) *Aa*
(c) *aa* (d) *Ab*

19. The normal male sex chromosome composition is:
   (a) *XX* (b) *XY*
   (c) *XXY* (d) *YY*

20. Describe the changes that occur in the sperm and egg immediately after fertilization.

21. What is induction? Explain its significance.

22. Summarize the developmental changes that occur during the first, second, and third trimesters.

23. What are the four extraembryonic membranes, how do they form, and what are their functions?

24. Indicate when each of the following appears during development, and describe each: blastodisc, blastocyst, morula, zygote.

25. What is the trophoblast, and what are its three major functions?

26. Identify the three stages of labor, and describe the events that characterize each stage.

27. What factors are involved in initiating labor contractions?

28. Identify the three life stages that occur between birth and approximately age 10. Describe the characteristics of each stage and when it occurs.

29. What hormonal events are responsible for puberty? What life stage does it initiate?

30. What occurs during the life stage of maturity?

31. What is senescence? Give some examples of how it affects organ systems throughout the human body.

LEVEL 2 Reviewing Concepts

32. Relaxin is a peptide hormone that:
   (a) increases the flexibility of the symphysis pubis
   (b) causes dilation of the cervix
   (c) suppresses the release of oxytocin by the hypothalamus
   (d) a, b, and c are correct

33. The production of prostaglandins in the endometrium:
   (a) initiates release of oxytocin for parturition
   (b) stimulates smooth muscle contractions
   (c) initiates secretory activity in the mammary glands
   (d) a, b, and c are correct

34. During adolescence, the events that interact to promote increased hormone production and sexual maturation result from activity of the:
   (a) hypothalamus
   (b) anterior pituitary
   (c) ovaries and testicular cells
   (d) a, b, and c are correct

35. What activity during oocyte activation prevents penetration by additional sperm?

36. In addition to its role in the nutrition of the fetus, what are the primary endocrine functions of the placenta?

37. Discuss the changes that occur in maternal systems during pregnancy. Why are these changes functionally significant?

38. Discuss the changes that occur in the uterus during pregnancy. How do these changes affect uterine tissues, and what hormones are involved?

39. Why are uterine contractions in the early stages of pregnancy weak, painless, and brief?

40. During true labor, what physiological mechanisms ensure that uterine contractions continue until delivery has been completed?

41. To what does the phrase "having the water break" refer during the process of labor?

42. What physiological adjustments must an infant make during the neonatal period in order to survive?

43. Distinguish between the following paired terms:
   (a) monozygotic and dizygotic
   (b) genotype and phenotype
   (c) heterozygous and homozygous
   (d) simple and polygenic inheritance

44. What would you conclude about a trait in each of the following situations?
   (a) Children who exhibit this trait have at least one parent who exhibits the same trait.
   (b) Children exhibit this trait even though neither of the parents exhibits it.
   (c) The trait is expressed more frequently in sons than in daughters.
   (d) The trait is expressed equally in both daughters and sons.

45. Explain why more men than women are color-blind. What type of inheritance is involved?

46. Explain the impact of genetic recombination on the production of gametes and the traits of individuals in future generations.

47. Explain the goals and possible benefits of the Human Genome Project.

LEVEL 3 Critical Thinking and Clinical Applications

48. Hemophilia A, a condition in which the blood does not clot properly, is a recessive trait located on the *X* chromosome (*Xh*). A woman heterozygous for the trait marries a normal male. What is the probability that this couple will have hemophiliac daughters? What is the probability that this couple will have hemophiliac sons?

49. Joan is a 27-year-old nurse who is in labor with her first child. She remembers from her anatomy and physiology class that calcium ions can increase the force of smooth muscle contractions, and since the labor is prolonged, she asks her physician for a calcium injection. The surprised physician informs her that that is definitely out of the question. Why?

50. A new mother confides to you that when she nurses her baby, she feels as if she is having menstrual cramps. How would you explain this phenomenon?

51. Explain why the normal heart and respiratory rates of neonates are so much higher than those of adults, even though adults are so much larger.

52. Women who are pregnant are advised against taking aspirin during their last trimester. Why?

53. Sally gives birth to a baby with a congenital deformity of the stomach. She swears that it is the result of a viral infection that she suffered during her third trimester of pregnancy. Do you think this is a possibility? Explain.

APPENDIX I

Answers to Concept Check Questions

CHAPTER 1

Page 6, Set 1

1. The ability of a cell or organism to respond to changes in its environment is called responsiveness or irritability.
2. A *histologist* investigates the structure and properties of tissues.
3. Study of the physiology of specific organs is called *special physiology*. In this particular case the field of study is *cardiac physiology* (study of heart function). Since heart failure is often caused by disease, this specialty would overlap or be closely related to *pathological physiology*.

Page 6, Set 2

1. The ability to move around and produce heat are functions of the muscular system.
2. Defense against infection and disease is a function of the lymphatic system.

Page 16

1. Physiological systems can function normally only under carefully controlled conditions. Homeostatic regulation serves to prevent potentially disruptive changes in the body's internal environment.
2. When homeostasis fails, organ systems function less efficiently or begin to malfunction. The result is the state that we call *disease*. If the situation is not corrected, death may result.
3. Positive feedback is useful in processes that must move quickly to completion once they have begun, such as blood clotting. It is harmful in situations where a stable condition must be maintained, because it will serve to increase any departure from the desired condition. For example, positive feedback in the regulation of body temperature would cause a slight fever to spiral out of control, with fatal results. For this reason physiological systems usually exhibit negative feedback, which tends to oppose any departure from the norm.

Page 24

1. The two eyes would be separated by a *midsagittal section*.
2. The body cavity inferior to the diaphragm is the *abdominopelvic* (or *peritoneal*) *cavity*.

CHAPTER 2

Page 37

1. Atoms combine with each other so as to gain a complete set of eight electrons in their outer energy levels. Oxygen atoms do not have a full outer energy level and so will readily react with many other elements to attain this stable arrangement. Neon already has a full other energy level and thus has little tendency to combine with other elements.
2. Hydrogen can exist as three different isotopes: hydrogen-1, with a mass of 1; deuterium, with a mass of 2; and tritium, with a mass of
3. The heavier sample must contain a higher proportion of one or both of the heavier isotopes.
4. A water molecule is formed by polar covalent bonds. Water molecules are attracted to one another by hydrogen bonds.

Page 40

1. Since this reaction involves a large molecule being broken down into two smaller ones, it is an example of a *decomposition* reaction. Because energy is released in the process, the reaction can also be classified as *exergonic*.
2. Removing the product of a reversible reaction would keep its concentration low compared to the concentration of the reactants. Thus the formation of product molecules would continue but the reverse reaction would slow down, resulting in a shift in the equilibrium toward the product.

Page 45

1. When salt dissolves in water it dissociates into charged particles called ions that are capable of conducting an electric current. Sugar molecules are held together by covalent bonds and do not dissociate in solution; thus there are no ions to carry a current.
2. Stomach discomfort is often the result of excess stomach acidity ("acid indigestion"). Antacids contain a weak base that neutralizes the excessive acid.

Page 52

1. A C:H:O ratio of 1:2:1 would indicate that the molecule is a *carbohydrate*. The body uses carbohydrates chiefly as an energy source.
2. When two monosaccharides undergo a dehydration synthesis type of reaction they form a disaccharide.
3. The most abundant lipid in a sample taken from beneath the skin would be a triglyceride.
4. Analysis of the lipid content of human cell membranes would indicate the presence of mostly phospholipids with small amounts of cholesterol and glycolipids.

Page 56

1. Proteins are composed of chains of small molecules called amino acids.
2. An agent that disrupts hydrogen bonding would affect the secondary level of protein structure.
3. The heat of boiling will break bonds that maintain the protein's tertiary & /or quaternary structure. The resulting change in shape will affect the ability of the protein molecule to perform its normal biological functions. These alterations are known as *denaturation*.
4. If the active site of an enzyme changes so that it better fits its substrate, the change will increase the level of enzyme activity. On the other hand, if the change alters the active site to the extent that the enzyme's substrate can no longer bind or binds poorly, the enzyme's activity will decrease or be inhibited.

Page 60

1. The nucleic acid RNA (ribonucleic acid) contains the sugar ribose. DNA (deoxyribonucleic acid) contains the sugar deoxyribose.
2. Phosphorylation of an ADP molecule would yield a molecule of ATP.

CHAPTER 3

Page 71

1. The phospholipid bilayer is primarily responsible for forming a physical barrier between the internal environment of the cell and the external environment.
2. Channel proteins are integral proteins that allow water, some ions, and small molecules to pass through the cell membrane.

Page 77

1. The process of diffusion is driven by a concentration gradient. The greater the concentration difference, the faster the rate of diffusion. The smaller the concentration difference, the slower the rate of diffusion. If the concentration of oxygen in the lungs decreased, the concentration difference between oxygen in the lungs and oxygen in the blood would decrease (assuming that the oxygen level of the blood remained constant). Thus oxygen would diffuse more slowly into the blood.
2. The 10 percent salt solution would be hypertonic with respect to the cells of the nasal lining, because it contains a higher concentration of salt than do the cells. The hypertonic solution would draw water out of the cells, causing the cells to shrink and loosening the mucus, thus relieving the congestion.

Page 82

1. In order to transport H^+ ions against their concentration gradient—that is, from a region where they are less concentrated (the cells lining the stomach) to a region where they are more concentrated (the interior of the stomach)—energy must be expended. An *active transport* process must be involved.
2. If the cell membrane were freely permeable to sodium ions, more of these positively charged ions would move into the cell and the transmembrane potential would move closer to zero.
3. This is an example of phagocytosis.

Page 87

1. The finger-like projections on the surface of the intestinal cells are *microvilli*. They serve to increase the cells' surface area so that they can absorb nutrients more efficiently.
2. Since the flagellum is an organelle of locomotion, sperm cells lacking a flagellum would be unable to move. (As a result, they would probably be unable to reach and fertilize an egg.)

Page 92

1. The function of mitochondria is to produce energy for the cell in the form of ATP molecules. A large number of mitochondria in a cell would indicate a high demand for energy.

2. SER functions in the synthesis of lipids such as steroids. Ovaries and testes would be expected to have a great deal of SER because they produce large amounts of steroid hormones.

Page 98

1. The nucleus of a cell contains DNA that codes for production of all of the cell's polypeptides and proteins. Some of these proteins are structural proteins that are responsible for the shape and other physical characteristics of the cell. Other proteins are enzymes that govern cellular metabolism, direct the production of cell proteins, and control all of the cell's activities.
2. If a cell lacked the enzyme RNA-polymerase it would not be able to transcribe RNA from DNA.
3. The deletion of a base from a coding sequence of DNA would alter the entire base sequence after the deletion point. This would result in different codons on the messenger RNA that was transcribed from the affected region, and this, in turn, would result in the incorporation of a different series of amino acids into the polypeptide. Almost certainly the polypeptide product would be not functional.

Page 104

1. This cell would be in the G_1 phase of its life cycle.
2. If spindle fibers failed to form during mitosis, the cell would not be able to separate the chromosomes into two sets. If cytokinesis occurred, the result would be one cell with two sets of chromosomes and one cell with none.

CHAPTER 4

Page 113

1. The presence of many microvilli on the free surface of epithelial cells greatly increases the surface area for absorption.
2. Gap junctions are especially prominent in cardiac and smooth muscle.
3. All of these regions are subject to mechanical trauma and abrasion—by food (pharynx and esophagus), feces (anus), and intercourse or childbirth (vagina).

Page 121

1. No. A simple squamous epithelium does not provide enough protection against infection, abrasion, and dehydration and is not found in the skin surface.
2. The process described is *holocrine secretion*.
3. The gland is an endocrine gland.

Page 128

1. Collagen fibers add strength to connective tissue. We would therefore expect vitamin C deficiency to result in the production of connective tissue that is weaker and more prone to damage.
2. Antihistamines act against the molecule histamine which is released from mast cells.
3. The tissue is adipose (fat) tissue.

Page 134

1. Cartilage lacks a direct blood supply which is necessary for proper healing to occur. Materials that are needed to repair damaged cartilage must diffuse from the blood to the chondroblasts, a process that takes a long time and retards the healing process.
2. Intervertebral disks are composed of fibrocartilage.
3. The two connective tissues that contain a liquid matrix are blood and lymph.

Page 136

1. The pleural, peritoneal, and pericardial cavity are all lined by serous membranes.
2. This is an example of a *mucous membrane*.
3. The tissue is probably *fascia*, a type of dense regular connective tissue that attaches muscles to skin and bones.

Page 138

1. Since cardiac and skeletal muscle are both striated (banded), this must be *smooth muscle* tissue.
2. The cells are neurons.
3. New muscle fibers can be produced through the division of satellite cells, which are mesenchymal cells that persist in adult skeletal muscle tissue.

CHAPTER 5

Page 153

1. Cells are constantly shed from the outer layers of the *stratum corneum*.
2. The splinter is lodged in the *stratum granulosum*.
3. Because fresh water is hypotonic with respect to skin cells, water will move into the cells by osmosis, causing them to swell.
4. Sanding the tips of one's fingers will not permanently remove fingerprints. Since the ridges of the fingerprints are formed in layers of the skin that are constantly regenerated, they will eventually reappear. The actual pattern of the ridges is determined by arrangement of tissue in the dermis, which is not affected by sanding.

Page 155

1. When exposed to the ultraviolet radiation in sunlight or tanning lamps, melanocytes in the epidermis and dermis synthesize the pigment melanin, darkening the color of the skin.
2. When skin is warm, more blood is diverted to the superficial dermis for the purpose of eliminating heat. Since the blood is red in color it imparts a red cast to the skin.
3. The hormone cholecalciferol (Vitamin D_3) is needed to form strong bones and teeth. The first step in this hormone's production involves the skin where cholesterol is exposed to specific wavelengths of ultraviolet light. By covering the body surface, the ultraviolet light is not able to penetrate to the blood in the skin to begin vitamin D_3 production, resulting in fragile bones.

Page 157

1. The capillaries and sensory neurons of the skin are located in the papillary layer of the dermis.
2. The presence of elastic fibers and the resilience of skin turgor account for dermal elasticity.

Page 161

1. Contraction of the arrector pili muscles pulls the hair follicles erect, depressing the area at the base of the hair and making the surrounding skin appear higher. The result is known as "goose bumps" or "goose pimples."
2. Hair is a derivative of the epidermis, and if the epidermis is destroyed by the injury there would be no hair follicles to produce new hair.

Page 163

1. Sebaceous glands produce a secretion called sebum. Sebum lubricates and protects the keratin of the hair shaft, lubricates and conditions the surrounding skin, and inhibits the growth of bacteria.
2. Apocrine sweat glands produce a secretion containing several kinds of organic compounds. Some of these have an odor and others produce an odor when metabolized by skin bacteria. Deodorants are used to mask the odor of these secretions.
3. Apocrine sweat glands enlarge and increase secretion in response to the increase in sex hormones that occurs at puberty.

Page 166

1. The name given to the combination of fibrin clot, fibroblasts, and the extensive network of capillaries that is found in tissue that is healing is granulation tissue.
2. Skin can still regenerate effectively even after considerable damage has occurred because stem cells persist in both the epithelial and connective tissue components of skin. When injury occurs, cells of the stratum germinativum replace epithelial cells while mesenchymal cells replace cells lost from the dermis.
3. As a person ages, the blood supply to the dermis decreases and merocrine sweat glands become less active. These changes make it more difficult for the elderly to cool themselves in hot weather.

CHAPTER 6

Page 180

1. If the ratio of collagen to hydroxyapatite in a bone increased, the bone would be more flexible and less strong.
2. Concentric layers of bone around a central canal is indicative of a Haversian system. Haversian systems make up compact bone. Since the ends (epiphyses) of long bones are primarily cancellous (spongy) bone, this sample most likely came from the shaft (diaphysis) of a long bone.

3. Since osteoclasts function in breaking down or demineralizing bone, the bone would have less mineral content and as a result it would be weaker.

Page 185

1. During the process of intramembranous ossification, fibrous connective tissue is replaced by bone.
2. In endochondral ossification, mesenchymal cells associated with the inner layer of the perichondrium differentiate into osteoblasts.
3. Long bones of the body, like the femur, have a plate of cartilage, called the epiphyseal plate, that separates the epiphysis from the diaphysis as long as the bone is still growing lengthwise. An X-ray would indicate whether or not the epiphyseal plate was still present. If it was, then growth was still occurring, and if not the bone had reached its adult length.

Page 188

1. The larger arm muscles of the weight lifter will apply more mechanical stress to the bones of the arms. In response to the stress, the bones will grow thicker. We would expect the jogger to have heavier thigh bones for similar reasons.
2. Growth continues throughout childhood. At puberty there is a growth spurt followed by epiphyseal closure. The later puberty begins, the taller the child will be when the growth spurt begins, and the taller the individual will be when growth is completed.
3. Increased levels of growth hormone (GH) prior to puberty will result in excessive bone growth leading to taller stature.

Page 190

1. Children suffering from rickets have bones that are poorly mineralized and as a result are quite flexible. Under the weight of the body, the leg bones bend. The instability makes bending and walking difficult and can lead to other problems of the legs and feet.
2. Parathyroid hormone (PTH) stimulates osteoclasts to release calcium ions from bone. This would result in an increase in the level of calcium ions in the blood.
3. Calcitonin acts to lower blood calcium levels by (1) inhibiting osteoclast activity and (2) increasing the rate of calcium excretion at the kidneys.

Page 195

1. An external callus forms early in the healing process when cells from the endosteum and periosteum migrate to the area of the fracture and form an enlarged collar (external callus) that circles the bone in the area of the fracture.
2. The sex hormones known as estrogens play an important role in moving calcium into bones. After menopause, the level of these hormones decreases dramatically and as a result, it is difficult to replace the calcium in bones that is being lost due to normal aging. Males do not show a decrease in sex hormone levels (androgens).
3. Bones that form in the tendons of joints like the knee and wrist are known as sesamoid bones.

CHAPTER 7

Page 210

1. Wormian bones are most often located in the lambdoidal suture between the parietal bones and occipital bone.
2. The foramen magnum is located in the base of the occipital bone.
3. Tom has fractured his right parietal bone.

Page 213

1. The internal jugular veins pass through openings in the temporal bones.
2. The sella turcica contains the pituitary gland and is located in the sphenoid bone.
3. Nerve fibers to the olfactory bulb, which deals with the sense of smell, pass through the cribriform plate from the nasal cavity. If the cribriform plate failed to form, these sensory nerves could not reach the olfactory bulbs and the sense of smell (olfaction) would be lost.

Page 219

1. The paranasal sinuses function to make some of the heavier skull bones lighter and to produce mucus.
2. The most powerful muscles that are involved in closing the mouth attach to the mandible at the coronoid process. A fracture of the coronoid process would make it difficult for these muscles to function properly and close the mouth.
3. Since many muscles that move the tongue and the larynx are attached to the hyoid bone, you would expect a person with a fractured hyoid bone to have difficulty moving the tongue, breathing, and swallowing.

Page 223

1. The adult vertebral column has fewer vertebrae because the 5 sacral vertebrae fuse to form a single sacrum and the 4 coccygeal vertebrae fuse to form a single coccyx.
2. The secondary curves of the spine are important because they allow us to balance the weight of our body on the legs with minimal muscular effort. Without the secondary curves, we would not be able to stand upright for extended periods.
3. When you run your finger along a person's spine you can feel the spinous processes of the vertebrae.

Page 229

1. The odontoid process is found on the second cervical vertebra or axis which is located in the neck.
2. The presence of transverse foramina indicate that this is a cervical vertebra.
3. The lumbar vertebrae must support a great deal more weight than vertebrae more superior in the spinal column. The large bodies allow the weight to be distributed over a larger area.

Page 234

1. True ribs are attached directly to the sternum by their own costal cartilage. False ribs either do not attach to the sternum at all (floating ribs) or attach by means of a common costal cartilage (the fusion of several).
2. Improper compression of the chest during CPR could and frequently does result in a fracture of the sternum or ribs.

CHAPTER 8

Page 242

1. The clavicle attaches the scapula to the sternum and thus restricts the scapula's range of movement. If the clavicle is broken, the scapula will have a greater range of movement and will be less stable.
2. The head of the humerus articulates with the glenoid fossa of the scapula.

Page 246

1. The two rounded prominences on either side of the elbow are parts of the humerus (the lateral and medial epicondyles).
2. The radius is in a lateral position when the arm is pronated and in a medial position when the arm is supinated.
3. The first distal phalanx is located at the tip of the thumb, so Bill has broken his thumb.

Page 249

1. The three bones that make up the coxa are the ilium, ischium, and the pubis.
2. The female pelvis is usually smoother and lighter and has less prominent markings. The pelvic outlet is enlarged and there is less curvature on the sacrum and coccyx. The pelvic inlet is wider and more circular. The pelvis as a whole is relatively broad and low. The ilia project farther laterally and the pubic arch is broader with an inferior angle between the pubic bones greater than 100°. These differences adapt the pelvis for supporting the weight of the developing fetus and the passage of the newborn through the pelvic outlet at the time of delivery.
3. When seated, the weight of the body is borne by the ischial tuberosities of the pelvis.

Page 254

1. Although the fibula is not part of the knee joint nor does it bear weight, it is an important point of attachment for many leg muscles. When the fibula is fractured, these muscles cannot function properly to move the leg and walking is difficult and painful. The fibula also helps stabilize the ankle joint.
2. Joey has most likely fractured the calcaneus (heel bone).
3. The talus transmits the weight of the body from the tibia anterior toward the toes.

4. Adductor muscles attach to the femur at the linea aspera. As these muscles are strengthened, they will apply more stress to to the femur at the linea aspera, and this portion of the femur will get larger and heavier.

CHAPTER 9

Page 266

1. All of these joints (other than synostoses) consist of bony regions separated by fibrous or cartilaginous connective tissue.
2. Originally, the joint is a type of syndesmosis. When the bones interlock, they form sutural joints.
3. The articular cartilages lack a blood supply and rely on synovial fluid to supply nutrients and eliminate wastes. If the circulation of synovial fluid were impaired, it would have the same effect as impairing a tissue's blood supply. Nutrients would not be delivered to meet the tissue's needs and wastes would accumulate. This would lead to damage and ultimately death of the cells in the tissue.

Page 270

1. (a) abduction; (b) supination; (c) flexion
2. In performing "jumping jacks," when the lower limbs move away from the midline of the body, the movement is abduction. When the lower limbs are brought back together, the movement is adduction.
3. Flexion and extension are the movements associated with hinge joints.

Page 272

1. Intervertebral discs are not found between the first and second cervical vertebrae, between sacral vertebrae in the sacrum, or between coccygeal vertebrae in the coccyx. An intervertebral disc between the first and second cervical vertebrae would prohibit rotation. The vertebrae in the sacrum and coccyx are fused together.
2. In a herniated disk, the deformed nucleus pulposus forces the annulus fibrosus into the vertebral foramen. This causes pressure on sensory nerves and accounts for the pain.
3. (a) flexion; (b) lateral flexion; (c) rotation

Page 277

1. Ligaments and muscles provide most of the stability for the shoulder joint.
2. Since the subscapular bursa is located in the shoulder joint, inflammation of this structure (bursitis) would be found in the tennis player. The condition is associated with repetitive motion that occurs at the shoulder, such as swinging a tennis racket. The jogger would be more at risk for injuries to the knee joint.
3. A shoulder separation is an injury involving partial or complete dislocation of the acromioclavicular joint.
4. Terry has most likely damaged his annular ligament.

Page 280

1. The iliofemoral, pubofemoral, and ischiofemoral ligaments would all be found in the hip joint.
2. Damage to the menisci in the knee joint would result in a decrease in the joint's stability. The individual would have a harder time locking the knee in place while standing and would have to use muscle contractions to stabilize the joint. When standing for long periods, the muscles would fatigue and the knee would "give out." We would also expect the individual to experience pain.
3. "Clergyman's knees" is a common bursitis found in members of the clergy due to excessive kneeling during prayer or other religious rituals. The work of carpet layers and roofers necessitates kneeling and sliding along on their knees causing a similar inflammation of the bursae in the knees.

CHAPTER 10

Page 293

1. Since tendons attach muscles to bones, severing the tendon would disconnect the muscle from the bone so that when the muscle contracted nothing would happen.
2. Skeletal muscle appears striated when viewed under the microscope because it is composed of the myofilaments actin and myosin which are arranged in such a way as to produce a banded appearance in the muscle.
3. You would expect to find the greatest concentration of calcium ions in the cisternae of the sarcoplasmic reticulum of the muscle.

Pages 300

1. Since the ability of a muscle to contract depends upon the formation of cross bridges between the myosin and actin myofilaments, a drug that would interfere with cross bridge formation would prevent the muscle from contracting.
2. Because the amount of cross bridge formation is proportional to the amount of available calcium ions, increased permeability of the sarcolemma to calcium ions would lead to an increased intracellular concentration of calcium and a greater degree of contraction. In addition, since relaxation depends on decreasing the amount of calcium in the sarcoplasm, an increase in the permeability of the sarcolemma to calcium could result in a situation in which the muscle would not be able to relax completely.
3. Without acetylcholinesterase, the motor endplate would be continuously stimulated by the acetylcholine and the muscle would be locked into contraction.

Page 305

1. The ability of the muscle to contract depends on the ability to form cross bridges between the actin and myosin. If the myofilaments overlap very little, then very few cross bridges are formed and the contraction is weak. If the myofilaments to not overlap at all, then no cross bridges form and the muscle cannot contract.
2. During the phenomenon known as treppe, there is not enough time between successive contractions to reabsorb all of the calcium ions that were released during the prior contraction event. As a result, calcium ions accumulate in the sarcoplasm at higher than normal levels, allowing more cross bridges to form and tension to increase.
3. The muscle may shorten (isotonic, concentric), elongate (isotonic, eccentric), or remain the same length (isometric) depending on the relationship between the resistance and the tension produced by actin and myosin interactions.

Page 315

1. The sprinter requires large amounts of energy for a relatively short burst of activity. To supply this demand for energy, the muscles switch to anaerobic metabolism. Anaerobic metabolism is not as efficient in producing energy as aerobic metabolism and the process also produces acidic waste products. The combination of less energy and the waste products contributes to fatigue. Marathon runners, on the other hand, derive most of their energy from aerobic metabolism, which is more efficient and does not produce the level of waste products that anaerobic respiration does.
2. We would expect activities that require short periods of strenuous activity to produce a greater oxygen debt because this type of activity relies heavily on energy production by anaerobic respiration. Since lifting weights is more strenuous over the short term we would expect this type of exercise to produce a greater oxygen debt than swimming laps, which is an aerobic activity.
3. Individuals who are naturally better at endurance types of activities such as cycling or marathon running have a higher percentage of slow muscle fiber which are physiologically better adapted to this type of activity than the fibers which are less vascular and fatigue faster.

Page 320

1. Cardiac muscle cells are joined by gap junctions which allow ions and small molecules to flow directly from one cell to another. This type of junction allows for action potentials generated in one cell to spread rapidly to adjacent cells. Thus all the cells will contract simultaneously, as if they were one single unit (a syncytium).
2. Cardiac muscle and smooth muscle are more affected by changes in the concentration of calcium ions in the extracellular fluid than skeletal muscle because in cardiac and smooth muscle, the majority of the calcium ions that trigger a contraction come from the extracellular fluid. In skeletal muscle, most of the calcium ions come from the sarcoplasmic reticulum.
3. The actin and myosin filaments of smooth muscle are not as rigidly organized as they are in skeletal muscle. This allows smooth muscle to contract over a relatively large range of resting lengths.

CHAPTER 11

Page 329

1. A pennate muscle will contain more muscle fibers than a parallel muscle of the same size. A muscle that has more muscle fibers has more myofibrils and sarcomeres, and as a result, the contraction of the pennate muscle generates more tension than a parallel muscle of the same size.
2. The opening between the stomach and the small intestine would be guarded by a circular muscle known as a sphincter muscle. The concentric circles of muscle fibers found in sphincter muscles are ideally suited for opening and closing openings or acting as valves in the body.
3. The joint between the occipital bone and the first cervical vertebra would be an example of a first-class lever system. The joint between the two bones is the fulcrum, which lies between the skull which is the resistance, and the neck muscles which are the force.

Page 331

1. The origin of a muscle is the end that remains stationary during an action. Since the gracilis muscle moves the tibia, the origin must be on the pelvis (pubis and ischium).
2. Muscles A and B are antagonists since they perform opposite actions.
3. The name flexor carpi radialis longus tells you that this is a long muscle that lies next to the radius and functions to flex the hand.

Page 341

1. Contraction of the masseter muscle raises the mandible, while the mandible depresses when the muscle is relaxed. These movements are important in the process of chewing or mastication.
2. You would expect the buccinator muscle which forms the mouth for blowing to be well-developed in a trumpet player.
3. Swallowing involves contractions of the palatal muscles that raise the soft palate as well as portions of the superior pharyngeal wall. Elevation of the superior portion of the pharynx enlarges the opening to the pharyngotympanic (Eustachian) tube that permits air flow to the middle ear and the inside of the eardrum. Making this opening larger facilitates air flow into or out of the middle ear cavity.

Page 348

1. Damage to the external intercostal muscles would interfere with the process of breathing.
2. A blow to the rectus abdominis would cause the muscle to contract forcefully resulting in flexion of the torso. In other words, you would "double up."
3. The sore muscles are most likely the sacrospinalis muscles composed of the longissimus and the iliocostalis of the lumbar region. These muscles would have to contract harder to counterbalance increased anterior weight, as when carrying heavy boxes.

Page 360

1. When you shrug your shoulders you are contracting your levator scapulae muscles.
2. The rotator cuff muscles include the supraspinatus, infraspinatus, subscapularis, and teres minor. The tendons of these muscles help to enclose and stabilize the shoulder joint.
3. Injury to the flexor carpi ulnaris would impair the ability to flex and adduct the hand.

Page 373

1. Injury to the obturator muscle would interfere with your ability to rotate your leg laterally.
2. The hamstring refers to a group of five muscles that collectively function in flexing the leg. These muscles are the biceps femoris, semimembranous, semitendinosus, gracilis, and sartorius.
3. The achilles (calcaneal) tendon attaches the soleus and gastrocnemius muscles to the calcaneus (heel bone). When these muscles contract, they cause extension of the foot. A torn achilles tendon would make extension of the foot difficult and the opposite action, flexion, would be more pronounced as a result of less antagonism from the soleus and gastrocnemius.

CHAPTER 12

Page 390

1. The afferent division of the nervous system is composed of nerves that carry sensory information to the brain and spinal cord. Damage to this division would interfere with a person's ability to experience a variety of sensory stimuli.
2. Sensory neurons of the peripheral nervous system are usually unipolar; thus this tissue is most likely associated with a sensory organ.
3. Microglial cells are small phagocytic cells that are found in increased number in damaged and diseased areas of the CNS.

Page 407

1. Depolarization of the neuron membrane involves the opening of the sodium channels and the rapid influx of sodium ions into the cell. If the sodium channels were blocked, a neuron would not be able to depolarize and conduct an action potential.
2. If the extracellular concentration of potassium ions decreased, more potassium ions would leave the cell and the electrical difference across the membrane (transmembrane potential) would be greater. This condition is called hyperpolarization.
3. Action potentials are propagated along myelinated axons by saltatory propagation at speeds much higher than those seen along unmyelinated axons. An axon with a propagation speed of 50 msec must be myelinated.

Page 413

1. When an action potential reaches the presynaptic terminal of a cholinergic synapse, calcium channels are opened and the influx of calcium triggers the release of acetylcholine into the synapse to stimulate the next neuron. If the calcium channels were blocked, the acetylcholine would not be released and transmission across the synapse would cease.
2. Because of synaptic delay, the pathway with fewer neurons (3) will conduct impulses faster.
3. The effect of a neurotransmitter on postsynaptic membranes depends upon the receptor molecule that binds it. In this case although the neurotransmitter and target tissues are the same, the smooth muscle in blood vessels of skeletal muscle have different NE receptors than those in the blood vessels of intestines; thus the opposite effects.

Page 419

1. This pattern represents a convergent circuit.
2. In order for a severed axon to heal, it must come into contact with and grow into the new cord of Schwann cells that forms distal to the site of the injury. If the axon fails to make this connection, the column of Schwann cells degenerates and the axon stops growing and the connection will not be reestablished. By closely aligning the two ends of the axons after an injury, there is a better chance that the connection will be made and innervation reestablished.
3. Interneurons are found in the CNS and are responsible for analyzing sensory input and coordinating motor output. Without the interneurons, the nervous system would not be able to process sensory information or make appropriate motor responses.

CHAPTER 13

Page 432

1. The ventral root of spinal nerves is composed of visceral and somatic motor fibers. Damage to this root would interfere with motor function.
2. The cerebrospinal fluid that surrounds the spinal cord is found in the subarachnoid space which lies beneath the epithelium of the arachnoid layer and on top of the pia mater.

Page 433

1. Since the polio virus would be located in the somatomotor neurons, we would find it in the anterior gray horns of the spinal cord where the cell bodies of these neurons are located.
2. A disease that damages myelin sheaths would affect the columns of the spinal cord, since this part of the cord is composed of bundles of myelinated axons.

Page 439

1. The dorsal rami of spinal nerves innervates the skin and muscles of the back. In this case we would expect the skin and muscles of the back of the neck and shoulders to be affected.
2. The phrenic nerves that innervate the diaphragm originate in the brachial plexus. Damage to this plexus or more specifically to the phrenic nerves would greatly interfere with the ability to breathe and possibly result in death.

3. Compression of the sciatic nerve produces the characteristic "pins and needles" sensation that one perceives when one's leg falls asleep.

Page 449

1. The minimum number of neurons required for a reflex arc is two. One must be a sensory neuron to bring impulses to the central nervous system, and the other a motor neuron that can bring about a response to the sensory input.
2. The suckling reflex is an example of an innate reflex.
3. When the stretch receptors are stimulated by the gamma motor neurons, the spindles become narrower and less sensitive to stretch. As a result it would take more force to get the muscles of the leg to contract for the knee jerk reflex, and as a result the reflex would be slower.

Page 452

1. This response is the tendon reflex.
2. During the withdrawal reflex, the limb on the opposite side is extended. This response is called a cross-extensor reflex.
3. A positive Babinski reflex is abnormal for an adult and indicates possible damage of descending tracts in the spinal cord.

CHAPTER 14

Page 462

1. The three primary brain vesicles are the prosencephalon, mesencephalon, and rhombencephalon. The prosencephalon gives rise to the cerebrum and diencephalon; the mesencephalon does not subdivide further; the rhombencephalon develops into the cerebellum, pons, and medulla oblongata.
2. The response is controlled by the inferior colliculi of the midbrain (mesencephalon).

Page 468

1. If one of the interventricular foramina became blocked, cerebrospinal fluid would not be able to flow from the first or second ventricle into the third. Since cerebrospinal fluid would continue to be formed, the blocked ventricle would swell with fluid, a condition known as hydrocephalus.
2. Diffusion across the arachnoid villi is the means by which cerebrospinal fluid re-enters the blood stream. If this process decreased, then excess fluid would start to accumulate in the ventricles and the volume of fluid in the ventricles would increase.
3. The blood-brain barrier consisting of capillary endothelium and the associated astrocytes restricts and regulates the movement of water-soluble molecules from the blood to the extracellular fluid of the brain.

Page 472

1. The primary motor cortex is located in the precentral gyrus of the frontal lobe of the cerebrum.
2. Damage to the temporal lobe of the cerebrum would interfere with the processing of olfactory (smell) and auditory (sound) impulses.
3. The stroke has damaged Jake's motor speech center in the motor cortex of the frontal lobe.

Page 475

1. The extrapyramidal system is composed of pathways that control muscle tone and coordinate learned movement patterns. A person suffering an injury to these tracts would exhibit difficulty in walking and in fluid precise movements of the arms and hands.
2. Paul is probably having problems associated with the temporal lobe of the cerebrum, specifically the hippocampus and the amygdala. His problems may also involve other parts of the limbic system that act as a gate for loading and retrieving long-term memories.
3. Terri's symptoms imply injury to the mammillary nuclei of the hypothalamus. She may also have sustained injury to motor areas of the brainstem that control the muscles involved in these processes.

Page 479

1. The lateral geniculate nuclei are involved with processing visual information. Damage to these nuclei would interfere with the sense of sight.
2. Changes in body temperature would stimulate the preoptic area of the hypothalamus, a division of the diencephalon.

Page 483

1. The vermis and arbor vitae are structures associated with the cerebellum.
2. Even though the medulla oblongata is small, it contains many vital reflex centers including those that control breathing and regulate the heart and blood pressure. Damage to the medulla oblongata can result in a cessation of breathing, or changes in heart rate and blood pressure that are incompatible with life.
3. Cells of the substantia nigra release the neurotransmitter dopamine at synapses with cerebral neurons. The dopamine produces IPSPs and decreases the activity of motor neurons. Decreased secretion would result in the motor neurons being more active, producing a variety of symptoms including twitches, tremors, other types of involuntary contractions, and possibly tetany.

CHAPTER 15

Page 508

1. The fasciculus gracilis in the posterior column of the spinal cord is responsible for carrying information about touch and pressure from the lower part of the body to the brain.
2. Since nociceptors are stimulated by pain, the action potentials generated by these receptors would be carried by the lateral spinothalamic tracts.
3. Impulses carried along the right fasciculus gracilis are destined for the primary sensory cortex of the left cerebral hemisphere.

Page 513

1. The anatomical basis for opposite-side motor control is that crossing over (decussation) occurs, and the pyramidal motor fibers innervate lower motor neurons on the opposite side of the body.
2. The superior portion of the motor cortex exercises control over the hand, arm, and upper portion of the leg. An injury to this area would affect the ability to control the muscles in those regions of the body.
3. Motor neurons of the red nucleus help control the muscle tone of skeletal muscles. Increased stimulation of these neurons would increase stimulation of the skeletal muscles producing increased muscle tone.

Page 519

1. We would expect Tina's brain waves to be beta waves, which are characteristic of adults who are experiencing stress and/or tension.
2. An inability to comprehend the written or spoken word indicates a problem with the general interpretive area of the brain, which in most individuals is located in the left temporal lobe of the cerebrum.
3. Recalling information for an A & P test involves fact memory, specifically secondary memory (long-term memory).

Page 522

1. The reticular activating system (RAS) is responsible for rousing the cerebrum to a state of consciousness. If a sleeping individual's RAS were stimulated, she would certainly wake up.
2. We would expect a drug that increases serotonin levels to produce a heightened perception of certain sensory stimuli such as auditory or visual, and hallucinations.
3. Some possible reasons for slower recall and loss of memory in the elderly include a loss of neurons (possibly those involved in specific memories), changes in synaptic organization of the brain, changes in the neurons themselves, and decreased blood flow which would affect the metabolic rate of neurons and perhaps slow the retrieval of information from memory.

CHAPTER 16

Page 530

1. It would require two neurons to carry an action potential from the spinal cord to the smooth muscle of the intestine. One neuron is required to carry the action potential from the spinal cord to the autonomic ganglion and a second to carry the action potential from the autonomic ganglion to the smooth muscle.
2. The sympathetic division of the autonomic nervous system is responsible for the physiological changes that occur in response to stress and increased activity.
3. In the sympathetic division of the autonomic nervous system the preganglionic fiber is relatively short and the postganglionic fiber is relatively long. In the parasympathetic division. The preganglionic fiber is relatively long and the postganglionic fiber is relatively short.

Page 537

1. The neurons that synapse in the collateral ganglia originate in the lower thoracic and upper lumbar portion of the spinal cord and pass through the chain ganglia to the collateral ganglia.
2. Since acetylcholine is the neurotransmitter that is released by all of the preganglionic fibers of the sympathetic nervous system, a drug that stimulates acetylcholine receptors would stimulate the postganglionic fibers of the sympathetic nerves, resulting in increased sympathetic activity.
3. Blocking the beta-receptors on cells would decrease or prevent sympathetic stimulation of those tissues. This would result in decreased heart rate and force of contraction and relaxation of the smooth muscle in the walls of blood vessels. These changes would contribute to lowering a person's blood pressure.

Page 540

1. The vagus nerve (N X) carries the parasympathetic fibers that innervate the lungs, heart, stomach, liver, pancreas, and parts of the small and large intestine as well as several other visceral organs.
2. Muscarinic receptors are a type of acetylcholine receptors found in the postganglionic synapse of the parasympathetic nervous system. Stimulation of these receptors at the heart would cause an opening of more potassium channels resulting in hyperpolarization of the membrane and a decreased heart rate.
3. The parasympathetic division is sometimes referred to as the anabolic system because parasympathetic stimulation leads to a general increase in the nutrient content of the blood. Cells throughout the body respond to the increase by absorbing the nutrients and using them to support growth and other anabolic activities.

Page 546

1. Since most blood vessels receive sympathetic stimulation, a decrease in sympathetic tone would lead to a relaxation of the muscles in the walls of the vessels and vasodilation (the vessels would increase in diameter). This in turn would result in increased blood flow to the tissue.
2. A patient who is anxious about impending root canal would probably exhibit some or all of the following changes: a dry mouth, increased heart rate, increased blood pressure, increased rate of breathing, cold sweats, an urge to urinate or defecate, change in motility of the digestive tract (i.e. "butterflies in the stomach"), and dilated pupils. These changes would be the result of anxiety or stress causing an increase in sympathetic stimulation.
3. A brain tumor that interferes with hypothalamic function would be expected to interfere with autonomic function as well. Centers in the posterior and lateral hypothalamus coordinate and regulate sympathetic function while centers in the anterior and medial hypothalamus control parasympathetic function.

CHAPTER 17

Page 558

1. The receptor with the smaller receptor field will provide the most precise sensory information, thus receptor A.
2. Since nociceptors are pain receptors, if they are stimulated, you would perceive a painful sensation in your affected hand.
3. Proprioceptors relay information about limb position and movement to the central nervous system, especially the cerebellum. Lack of this information would result in uncoordinated movements and the individual probably would not be able to walk.

Page 562

1. By the end of the lab period adaptation has occurred. In response to the constant level of stimulation, the receptor neurons have become less active, partially as the result of synaptic fatigue.
2. The taste receptors (taste buds) are only sensitive to molecules and ions that are in solution. If you dry the surface of the tongue, there is no moisture for the sugar molecules or salt ions to dissolve in and they will not stimulate the taste receptors.
3. You would tell your grandfather that the difference in the taste of his food is the result of several age-related factors. The number of taste buds declines dramatically after age 50 and those that remain are not as sensitive as they once were. In addition the loss of olfactory receptors contributes to the perception of less flavor in foods.

Page 573

1. The first layer of the eye to be affected by inadequate tear production would be the conjunctiva. Drying of this layer would produce an irritated, scratchy feeling.
2. When the lens is round you are looking at something closer to you.
3. Renee will probably not be able to see at all. The fovea of the eye contains cones but no rods. Rods respond to light of low intensity but cones need high light intensities to be stimulated. In a dimly lit room, the light would not be strong enough to stimulate these photoreceptors and as a result, Renee would not be able to see.
4. If the canal of Schlemm is blocked, aqueous humor will not be able to drain and glaucoma will develop. As the quantity of fluid increases, the pressure within the eye increases, distorting soft tissues and interfering with vision. If untreated, the condition will ultimately cause blindness.

Page 581

1. Even with a congenital lack of cone cells in the eye you would still be able to see as long as you had functioning rod cells. Since cone cells function in color vision, you would see only black and white.
2. A deficiency or lack of vitamin A in the diet would affect the quantity of retinal that the body could produce and thus interfere with night vision.
3. A decrease in phosphodiesterase activity would lead to higher levels of intracellular cGMP. This in turn would keep the gated sodium ion channels open and decrease the ability of receptor neurons to respond to photons of light.

Page 595

1. Without the movement of the round window, the perilymph would not be moved by the vibration of the stapes at the oval window, and there would be little or no perception of sound.
2. Loss of stereocilia (as a result of constant exposure to loud noises for instance) would reduce hearing sensitivity and could eventually result in deafness.
3. The role of the pharyngotympanic tube (Eustachian tube) is to allow for equalizing pressure on both sides of the tympanic membrane (eardrum). If this tube is blocked, there will be greater pressure on the inside of the tympanic membrane, forcing it outward and producing pain.

CHAPTER 18

Page 609

1. Neural responses occur within fractions of a second and do not last long (short duration). Endocrine responses, on the other hand, require 30 seconds or more before the response is noted and will last for minutes to hours or more (long duration).
2. Adenylate cyclase is the enzyme that converts ATP to cAMP. A molecule that blocks this enzyme would block action of any hormone that required cAMP for a second messenger.
3. A cell's hormonal sensitivity is determined by the presence or absence of the necessary receptor complex.

Page 616

1. Dehydration increases the osmotic pressure of the blood. The increase in blood osmotic pressure would stimulate the neurohypophysis to release more ADH.
2. Somatomedins are the mediators of growth hormone action. If the level of somatomedins is elevated, we would expect to see the level of growth hormone elevated as well.
3. Increased levels of cortisol would inhibit the cells that control ACTH release from the pituitary; therefore the level of ACTH would decrease. This is an example of a negative feedback mechanism.

Page 627

1. If an individual lacked iodine in her diet, she would not be able to form the hormone thyroxine. As a result we would expect to see the symptoms associated with thyroxins deficiency, such as decreased rate of metabolism, decreased body temperature, poor response to physiological stress, and an increase in the size of the thyroid gland (goiter).
2. Most of the thyroid hormone in the blood is bound to a protein called thyroid-binding globulins. This represents a large reservoir of thyroxine that guards against rapid fluctuations in the level of this important

hormone. Because there is such a large amount stored in this way, it takes several days to deplete the supply of hormone, even after the thyroid gland has been removed.
3. Removal of the parathyroid glands would result in a decrease in the blood levels of calcium ion. This could be counteracted by increasing the amount of vitamin D and calcium in the diet.
4. One of the functions of cortisol is to decrease the cellular use of glucose while increasing the available glucose by promoting the breakdown of glycogen and the conversion of amino acids to carbohydrates. The net result is an elevation in the level of glucose in the blood.

Page 629

1. One of the functions of the hormone atrial natriuretic peptide is to increase the rate of sodium excretion in the kidneys. As a result the sodium excreted in the urine would increase.

Page 636

1. An individual with Type I diabetes has such elevated levels of glucose in the blood that the kidney cannot reabsorb all of the glucose; some is lost in the urine. The water lost with the glucose elevates blood osmotic pressure and promotes the thirst.
2. Glucagon stimulates the conversion of glycogen to glucose in the liver. Increased amounts of glucagon would then lead to decreased amounts of liver glycogen.
3. The pineal gland receives neural input from the optic tracts and its secretion, melatonin, is influenced by light-dark cycles. Increased amounts of light inhibit the production and release of melatonin from the pineal gland.

Page 639

1. The type of hormonal interaction exemplified by the insulin and glucagon is antagonism. In this type of hormonal interaction, two hormones have opposite effects on their target tissues.
2. The hormones growth hormone, thyroid hormone, parathyroid hormone, and the gonadal hormones all play a role in formation and development of the skeletal system.
3. During the resistance phase of GAS, there is a high demand for glucose, especially by the nervous system. The GH-RH and CRH increase the levels of growth hormone and ACTH respectively. Growth hormone mobilizes fat reserves and promotes the catabolism of protein. ACTH increases cortisol which stimulates the conversion of glycogen to glucose as well as the catabolism of fat and protein.

CHAPTER 19

Page 653

1. Venipuncture is a popular sampling technique because (1) superficial veins are easy to locate, (2) the walls of veins are thinner than arteries and (3) blood pressure in veins is relatively low so the puncture wound seals quickly.
2. A decrease in the amount of plasma proteins in the blood may cause (1) a decrease in plasma osmotic pressure, (2) a decreased ability to fight infection, and (3) a decrease in the transport and binding of some ions, hormones, and other molecules.
3. During a viral infection, you would expect to see an increase in the level of gamma-globulins (antibodies) in the blood.

Page 660

1. The hematocrit measures the amount of formed elements (mostly red blood cells) as a percentage of the total blood. In hemorrhage the loss of blood, especially red blood cells, would cause the hematocrit to be less.
2. A decreased blood flow to the kidneys would trigger the release of erythropoietin. The elevated erythropoietin would lead to an increase in erythropoiesis (red blood cell formation). Thus, Dave's hematocrit should increase.
3. The liver conjugates bilirubin, that is makes it more soluble by combining it with glucouronic acid so that it can be excreted more easily. Diseases that damage the liver such as hepatitis or cirrhosis would impair the liver's ability to perform this function. As a result, the bilirubin would accumulate in the blood, producing a condition known as jaundice.

Page 664

1. A person with Type AB blood can accept Type A, Type B, Type AB, or Type O) blood.
2. If a person with Type A blood receives a transfusion of Type B blood, the red cells would clump or agglutinate, potentially blocking blood flow to various organs and tissues.
3. Surface antigens on RBCs are glycoproteins and glycolipids in the cell membrane.

Page 667

1. In an infected cut we would expect to find a large number of neutrophils. Neutrophils are phagocytic white cells that are usually the first to arrive at the site of an injury and which specialize in dealing with infectious bacteria.
2. The type of white blood cell that produces circulating antibodies is the B lymphocyte, and these would be found in increased numbers.
3. During an inflammatory response, basophils release a variety of chemicals such as histamine and heparin that exaggerate the inflammation and attract other types of white blood cells.

Page 676

1. Megakaryocytes are the precursors of platelets which play an important role in hemostasis and the clotting process. A decreased number of megakaryocytes would result in fewer platelets, which in turn would interfere with the ability to clot properly.
2. The use of broad-spectrum antibiotics would lower the number of intestinal bacteria, and thus the amount of vitamin K produced. This decrease in vitamin K would lead to a decrease in the production of several clotting factors, most notably prothrombin. As a result, clotting time increases.
3. Activation of the Hagemann factor, factor XII, initiates the series of events known as the intrinsic system.

CHAPTER 20

Page 692

1. The semilunar valves on the right side of the heart guard the opening to the pulmonary artery. Damage to these valves would interfere with the blood flow through this vessel.
2. When the ventricles begin to contract, they force the AV valves to close which in turn pulls on the chordae tendineae which then pull on the papillary muscles. The papillary muscles respond by contracting, counteracting the force that is pushing the valves upward.
3. The wall of the left ventricle is more muscular than that of the right ventricle because the left ventricle has to generate enough force to propel blood throughout all of the body's systems except the lungs. The right ventricle only has to generate enough force to propel the blood a few centimeters to the lungs. Since the left ventricle is so muscular it takes more force to push blood into the chamber against the normal tension of the muscle; this in turn requires that the left atrium be more muscular than the right atrium.

Page 700

1. If these cells were not functioning, the heart would still continue to beat but at a slower rate.
2. If the impulses from the atria were not delayed at the AV node, they would be conducted through the ventricles so quickly by the bundle branches and Purkinje fibers that the ventricles would begin contracting immediately before the atria had finished their contraction. As a result the ventricles would not be as full of blood as they could be and the pumping of the heart would not be as efficient, especially during activity.
3. The cardioinhibitory center of the medulla oblongata is part of the parasympathetic branch of the autonomic nervous system. Damage to this center would result in fewer parasympathetic action potentials to the heart and an increase in heart rate due to sympathetic dominance.
4. An increase in extracellular calcium ion would result in an increased force of cardiac contraction. More calcium ions would enter heart cells when they were stimulated because of the increased concentration difference. The increased amount of intracellular calcium would lead to increased binding of troponin, more myosin heads binding to actin, and ultimately a stronger contraction.

Page 706

1. When pressure in the left ventricle begins rising, the heart is contracting but no blood is leaving the heart. During this initial phase of contraction, the AV valves and semilunar valves are both closed. The increase in pressure is the result of increased tension as the muscle contracts.

When the pressure in the ventricle exceeds the pressure in the aorta, the aortic semilunar valves are forced open and the blood is rapidly ejected from the ventricle.

2. An increase in the size of the QRS complex would indicate a larger than normal amount of electrical activity during ventricular depolarization. One possible cause would be an increase in the size of the heart (left ventricular hypertrophy). Since there is more muscle depolarizing, the magnitude of the electrical event would be greater.

3. If the heart beats too quickly (tachycardia), there is not sufficient time for it to fill completely between the beats. Since the heart pumps blood proportionately to what enters, the less blood that enters, the less it will be able to pump. If it beats too fast, very little blood will enter circulation and tissues will suffer damage from lack of blood supply.

Page 708

1. Stimulating the acetylcholine receptors of the heart would cause the heart to slow down. Since the cardiac output is the product of stroke volume times the heart rate, if the heart rate decreases so will the cardiac output (assuming no change in the stroke volume).
2. The venous return fills the heart with blood, stretching the heart muscle. According to Starling's law, the more the heart muscle is stretched, the more forcefully it will contract (to a point). The more forceful the contraction the more blood the heart will eject with each beat (stroke volume). Therefore, increased venous return will increase the stroke volume, assuming all other factors are constant.
3. Increased sympathetic stimulation of the heart will result in increased heart rate and increased force of contraction. The ESV represents the amount of blood that remains in a ventricle after a contraction (systole). The more forcefully the heart contracts the more blood it will eject and the lower the ESV will be. Therefore, increased sympathetic stimulation should result in a lower ESV.
4. SV = EDV − ESV
   SV = 125 ml − 40 ml = 85 ml

Page 712

1. Caffeine acts directly on the conducting system and contractile cells of the heart, increasing the rate at which they depolarize. The net result would then be an increased heart rate.
2. Bradycardia (low heart rate) would likely result in a decreased cardiac output, since CO = HR × SV. It is possible that the stroke volume would increase enough to balance the decreased heart rate, but this is not likely since factors that decrease heart rate also tend to decrease force of myocardial contraction and thus stroke volume as well.
3. A drug that increases the time required for pacemaker cells to repolarize would decrease the heart rate, since the pacemaker cells would generate fewer action potentials per minute.

CHAPTER 21

Page 728

1. The blood vessels are veins. Arteries and arterioles have a relatively large amount of smooth muscle tissue in a thick, well-developed tunica media.
2. Blood pressure in the arterial system pushes blood into the capillaries. Blood pressure on the venous side is very low, and other forces help keep the blood moving. Valves prevent the blood from flowing backward whenever the venous pressure drops.
3. You would expect to find fenestrated capillaries in organs and tissues where large polypeptides or small proteins move freely in and out of the blood. These would include endocrine glands, the choroid plexus of the brain, absorptive areas of the intestine, and filtration areas of the kidneys.

Page 737

1. In a normal individual, the pressure should be greatest in the aorta and least in the venae cavae. Blood, like other fluids, moves along a pressure gradient from high pressure to low pressure. If the pressure were higher in the inferior vena cava, the blood would flow backwards.
2. While standing for periods of time, blood tends to pool in the lower extremities. This decreases the venous return to the heart, and in turn the cardiac output decreases, sending less blood to the brain, causing light-headedness and fainting. A hot day adds to the effect, due to a loss of body water through sweating.
3. The nurse would have first heard the Korotkoff's sounds when the pressure in the cuff reached 125 mm Hg. At this point the pressure in the vessel during systole is just enough to overcome the pressure in the cuff. Turbulent flow produced in the constricted vessel then produces the audible sounds.
4. Patients suffering from congestive heart failure (a condition in which the heart is a weak and inefficient pump) frequently have swollen feet and ankles due to edema. Since the heart cannot generate enough force to properly circulate the blood, it pools in the extremities, especially those in the inferior regions of the body. The pooling increases hydrostatic pressure, promoting the movement of fluid into the interstitial spaces but reducing capillary reabsorption. The lymphatics cannot drain the excess effectively, so it accumulates and produces the noticeable, edematous swelling.

Page 749

1. In exercise (1) blood flow to muscles increases, (2) cardiac output increases and (3) resistance in visceral tissues increases.
2. Pressure at this site would decrease blood pressure at the carotid sinus, where the carotid baroreceptors are located. This causes a decreased frequency of action potentials along the glossopharyngeal nerve (IX) to the medulla, and more sympathetic impulses will be sent to the heart. The net result will be an increase in the heart rate.
3. Vasoconstriction of the renal artery will decrease both blood flow and blood pressure at the kidney. In response the kidney will increase the amount of renin that it releases, which in turn will lead to an increase in the level of angiotensin II. The angiotensin II will bring about increased blood pressure and increased blood volume.
4. In circulatory shock, there is a decreased venous return to the heart. As a result of the decrease in venous return, the cardiac output is decreased, which accounts for the weak pulse. Because the cardiac output is decreased, the baroreceptors are stimulated and in turn there is increased sympathetic stimulation to the heart, causing the rapid heart rate. Although the heart is beating faster, there is less blood to pump, and pulse pressure remains low.

Page 760

1. The left subclavian artery is the branch of the aorta that sends blood to the left shoulder and arm.
2. The common carotid arteries carry blood to the head. A compression of one of the common carotid arteries would result in decreased blood flow to the brain and loss of consciousness or even death.
3. Organs served by the celiac artery include the stomach, spleen, liver, and pancreas.

Page 768

1. The vein that is bulging is the external jugular vein.
2. Blockage of the popliteal vein would interfere with blood flow in the tibial and peroneal veins (which form the popliteal vein) and the small saphenous vein (which joins the popliteal vein).
3. This blood sample must have come from the umbilical vein, which carries oxygenated, nutrient-rich blood from the placenta to the fetus.

CHAPTER 22

Page 791

1. The thoracic duct drains lymph from the area beneath the diaphragm and the left side of the head and thorax. Most of the lymph enters the venous blood by way of this duct. A blockage of this duct would not only impair circulation of lymph through most of the body, it would also promote accumulation of fluid in the extremities (lymphedema).
2. The thymosins from the thymus play a role in the differentiation of stem lymphocytes into T lymphocytes. A lack of these hormones would result in an absence of T lymphocytes.
3. During an infection, the lymphocytes and phagocytes in the lymph nodes in the affected region undergo cell division to better deal with the infectious agent. This increase in the number of cells in the node causes the node to become enlarged or swollen.

Page 798

1. A decrease in the number of monocyte forming cells in the bone marrow would result in a decreased number of macrophages in the body, since all of the different macrophages are derived from the monocytes. This would include the microglia of the CNS, the Kuppfer's cells of the

liver, Langerhans cells in the skin and digestive tract, alveolar macrophages as well as others.
2. A rise in interferon would indicate a viral infection. Interferon is released from cells that are infected with viruses. It does not help the infected cell, but "interferes" with the virus' ability to infect other cells.
3. Pyrogens stimulate the temperature control area of the preoptic nucleus of the hypothalamus. The result is an increase in body temperature or fever.

Page 808

1. Abnormal peptides in the cytoplasm of a cell can become attached to MHC (major histocompatibility complex) proteins and displayed on the surface of the cell's membrane. Peptides presented in this manner are then recognized by T cells which can initiate an immune response.
2. Cytotoxic T cells function in cell-mediated immunity. A decrease in the number of cytotoxic T cells would interfere with the ability to kill foreign cells and tissues as well as cells infected by viruses.
3. Helper T cells promote B cell division, the maturation of plasma cells and the production of antibody by the plasma cells. Without the helper T cells the humoral immune response would take much longer to occur and would not be as efficient.
4. Since plasma cells produce and secrete antibodies, we would expect to see increased levels of circulating antibodies in the blood if the number of plasma cells were increased.

Page 819

1. The secondary response would be affected by the lack of memory B cells for a specific antigen. The ability to produce a secondary response depends upon the presence of memory B cells and T cells that are formed during the primary response to an antigen. These cells are not involved in the primary response, but are held in reserve against future contact with the same antigen.
2. The developing fetus is protected primarily by passive immunity, the product of IgG antibodies that cross the placenta from the mother's circulation. In addition, the fetus may show some degree of active cellular immunity by the third month of development.
3. Stress can interfere with the immune response by depressing the inflammatory response, reducing the number and activity of phagocytes, and inhibiting interleukin secretion.

CHAPTER 23

Page 834

1. The rich blood supply to the nose brings in large amounts of heat energy that is used to warm the air as it passes through the nasal cavity. The heat also evaporates moisture from the epithelium to humidify the incoming air. The moisture is derived from the blood supply as well.
2. The nasopharynx only receives air from the nasal cavity. The oropharynx and laryngopharynx receive air from the nasal cavity and food from the oral cavity. Ingested solids and liquids can be damaging to delicate cells, thus in the areas in contact with food we find a highly protective stratified squamous epithelium, like that of the exterior skin. The lining of the nasopharynx is the same as the nasal cavity, a pseudostratified columnar epithelium.
3. Increased tension in the vocal cord will cause a higher pitch in the voice.

Page 842

1. The tracheal cartilages are C-shaped to allow room for esophageal expansion when large portions of food or liquid are swallowed.
2. Without surfactant, surface tension in the thin layer of water that moistens their surfaces would cause the alveolar to collapse.
3. Chronic smoking damages the lining of the air passageways. Cilia are seared off the surface of the cells by the heat, and the large number of particles that escape filtering in the nose are trapped in the excess mucus that is secreted to protect the irritated lining. This combination of circumstances creates a situation in which there is a large amount of thick mucus that is difficult to clear from the passages. The cough reflex is an attempt to remove this material from the airways.

Page 854

1. Since the rib penetrates the chest wall, atmospheric air will enter the thoracic cavity. This condition is called a pneumothorax. Pressure within the pleural cavity is normally lower than atmospheric pressure. When air enters the pleural cavity, the natural elasticity of the lung may cause it to collapse. The resulting condition is called atelectasis, or a collapsed lung.
2. Since the fluid produced in pneumonia takes up space that would normally be occupied by air, the vital capacity will be decreased.

Page 862

1. On a hot, humid day, the air that we breathe contains more water vapor than on a cool, dry day, and this means that the partial pressure of oxygen in the air is less. Since gases expand when heated, on a hot day the same volume of gas would contain fewer molecules. Thus an individual must breathe deeper or faster (or both) to gain the same amount of oxygen as compared to a cool, dry day.
2. As skeletal muscles become more active they generate more heat and more acid waste products which lowers the pH of surrounding fluid. The combination of lower pH and higher temperature causes the hemoglobin to release more oxygen than it would under conditions of lower temperature and higher pH.
3. An obstruction of the airways would interfere with the body's ability to gain oxygen and eliminate carbon dioxide. Since most carbon dioxide is carried in the blood as bicarbonate ion that is formed from the dissociation of carbonic acid, an inability to eliminate carbon dioxide would result in an excess of hydrogen ions thus lowering the body's pH.

Page 869

1. The pneumotaxic center inhibits the inspiratory center and the apneustic center. Exciting this center in the brainstem would result in shorter breaths and a more rapid rate of breathing.
2. Chemoreceptors are more sensitive to carbon dioxide. When this gas dissolves it produces hydrogen ions that lower pH and alter cell or tissue activity.
3. Johnny's mother shouldn't worry. When Johnny holds his breath, the level of carbon dioxide in his blood will increase. This will lead to increased stimulation of the inspiratory center, forcing Johnny to breathe again.

CHAPTER 24

Page 883

1. The mesenteries are double layers of serous membrane that support and stabilize the positions of organs in the abdominopelvic cavity and provide a route for the associated blood vessels, nerves, and lymphatics.
2. Peristalsis would be more efficient in propelling intestinal contents. Segmentation is essentially a churning action that mixes intestinal contents with digestive fluids. Peristalsis consists of waves of contractions that not only mix the contents, but also propel them along the digestive tract.
3. Parasympathetic stimulation increases muscle tone and motility in the digestive tract. A drug that blocks this activity would decrease the rate of peristalsis.

Page 892

1. The oral cavity is lined by a stratified squamous epithelium. This type of lining is found in areas of the body that receive a great deal of friction or abrasion, and is very protective.
2. Since the parotid glands secrete amylase, an enzyme that digests starch, damage to these glands would interfere with the digestion of carbohydrates.
3. The incisors are the type of tooth best suited for chopping, cutting, or shearing pieces of relatively rigid food, such as raw vegetables.

Page 895

1. The muscularis externa of the esophagus is unusual because (1) it contains skeletal muscle fibers along most of the length of the esophagus and (2) it is surrounded by an adventitia rather than a serosa.
2. The fauces is the opening between the oral cavity and the pharynx.
3. The process that is being described is deglutition or swallowing.

Page 903

1. The larger a meal (especially in terms of protein) the more stomach acid there is secreted. The hydrogen ions for the acid come from the

blood that enters the stomach; therefore, the blood leaving the stomach will have fewer than normal hydrogen ions and be decidedly alkaline, that is, have a higher pH. This is referred to as the alkaline tide.
2. The vagus nerve contains parasympathetic motor fibers that can stimulate gastric secretions. This can occur even if food is not present in the stomach (caphalic phase of gastric digestion). Cutting the branches of the vagus that supply the stomach would prevent this type of secretion from occurring and decrease the chance of ulcer formation.

Page 915

1. The small intestine has several adaptations that increase surface area to increase its absorptive capacity. First walls of the small intestine are thrown into folds called the plicae circularis. The tissue that covers the plicae forms fingerlike projections, the villi. The cells that cover the villi have an exposed surface that is covered by small fingerlike projections called the microvilli. In addition, the small intestine has a very rich blood and lymphatic supply to transport the nutrients that are absorbed.
2. It would increase.
3. The hormone secretin, among other things, stimulates the pancreas to release fluid high in bicarbonate ion to neutralize the chyme entering the duodenum from the stomach. If the intestine did not secrete secretin, we would expect the pH of the intestinal contents to be lower than normal.
4. Damage to the exocrine pancreas would most affect the digestion of fats (lipids). Enzymes for carbohydrate digestion are produced by salivary glands and the small intestine as well as the pancreas. Enzymes for protein digestion are produced by the stomach and the small intestine as well as the pancreas. Even though digestion of carbohydrates and proteins would not be as complete as with the pancreas functioning, some digestion would still take place. Since the pancreas is the primary source of lipases, lipid digestion would be most interfered with.

Page 927

1. Chylomicrons are formed from the fats that are digested in a meal. A meal that is high in fat would increase the number of chylomicrons in the lacteals.
2. Removal of the upper portion of the stomach would interfere with the absorption of vitamin B_{12}. This vitamin requires a glycoprotein, called intrinsic factor, that is produced by the parietal cells in the stomach.
3. Diarrhea is potentially life-threatening because a person could lose fluid and electrolytes faster than they can be replaced. This would result in dehydration and possibly death. Constipation is not potentially life-threatening, although it can be quite uncomfortable, since it does not interfere with any major body process that supports life. The few toxic waste products that are normally eliminated by way of the digestive system can move into the blood and be eliminated by the kidneys.

CHAPTER 25

Page 945

1. The primary role of the TCA cycle in ATP production is to transfer electrons from substrates to coenzymes. These electrons carry energy that can then be used as an energy source for the production of ATP by the electron transport system.
2. The NADH produced by glycolysis cannot enter the mitochondria where the enzymes of the electron transport chain are located. A carrier molecule in the mitochondrial membrane can, however, transfer the electrons from the NADH to a coenzyme within the mitochondria. In skeletal muscle, the intermediate transfers the electrons to FAD, whereas in cardiac muscle a different intermediate is found which transfers the electrons to another NAD. This accounts for the difference in ATP yield.
3. A decrease in cytoplasmic NAD would lead to a decrease in the amount of ATP produced by the mitochondria. The mitochondria depends on a supply of pyruvate from glycolysis. Glycolysis in turn requires NAD. A decrease in NAD would decrease the available pyruvate for the TCA cycle and thus a decrease in overall ATP production.

Page 954

1. Vitamin B_6 (pyridoxine) is an important coenzyme in the processes of deamination and transamination, the first step in processing amino acids in the cell. A deficiency in this vitamin would interfere with the ability to metabolize proteins.
2. Uric acid is the product of purine degradation in the body. The macromolecules that contain purines are the nucleic acids. An increase in uric acid levels could indicate increased breakdown of nucleic acids.
3. HDLs are beneficial because they reduce the amount of fat (including cholesterol) in the bloodstream by transporting it back to the liver for storage or excretion in the bile.

Page 961

1. After a meal that is high in carbohydrate, we would expect increased glycogenesis (the formation of glycogen) to occur in the liver.
2. Urea is formed from by-products of protein metabolism. During the postabsorptive state, many amino acids are being metabolized and the ammonia produced by deamination is converted to urea in the liver, thus the amount of urea in the blood increases.
3. Excess acetyl-CoA is generally converted into acetoacetate and related compounds collectively known as ketone bodies.

Page 966

1. We would expect a person adding muscle mass to be in positive nitrogen balance.
2. Bile salts are necessary for the digestion and absorption of fats and fat soluble vitamins. Vitamin A is a fat soluble vitamin. A decrease in the amount of bile salts in the bile would result in a decreased ability to absorb the vitamin A from food and result in a vitamin A deficiency.

Page 973

1. The BMR of a pregnant woman should be higher than the BMR of the woman in a nonpregnant state because of increased metabolism associated with support of the fetus as well as the added effect of fetal metabolism.
2. Vasoconstriction of peripheral vessels would decrease blood flow to the skin and decrease the amount of heat that the body can lose. As a result, the body temperature would increase.
3. Infants have higher surface area to volume ratios than adults, and the temperature-regulating mechanisms of the body are not fully functional at birth. As a result they must expend more energy to maintain body temperature and they get cold more easily than a normal adult.

CHAPTER 26

Page 991

1. The renal corpuscle, proximal convoluted tubule, distal convoluted tubule, and the proximal portions of the loop of Henle and collecting duct are all located within the renal cortex.
2. The slit pores created by the podocytes will only allow substances less than 6-9 nm in size to pass into the capsular space.
3. Damage to the juxtaglomerular apparatus portion of the nephrons would interfere with the normal control of blood pressure.

Page 999

1. The absorption of calcium ion by the kidney is an example of a countertransport mechanism in which sodium ions are traded for calcium. As more calcium is reabsorbed more sodium enters the filtrate, increasing the sodium ion concentration of the filtrate.
2. When the plasma concentration of a substance exceeds its tubular maximum, the excess is not reabsorbed and is excreted in the urine.
3. Decreases in blood pressure would reduce the blood hydrostatic pressure within the glomerulus and decrease the GFR.

Page 1008

1. Aldosterone promotes sodium retention and potassium secretion at the kidneys. In response to increases in aldosterone levels, the potassium ion concentration of the urine would increase.
2. The secretion of hydrogen ions by the nephron involves a countertransport mechanism with sodium. If the concentration of sodium in the filtrate decreased, fewer hydrogen ions could be secreted. This would result in a urine with a higher pH.
3. If the nephrons lacked a loop of Henle, the kidneys would not be able to form a concentrated urine.
4. When the number of sodium ions in the filtrate passing through the DCT is low, the cells of the macula densa are stimulated to release renin. Renin activates angiotensin and this brings about an increase in blood pressure.

Page 1018

1. A nitrogenous waste, urea, is formed during metabolism of amino acids that come from proteins. Thus we would expect an increased

amount of urea in an individual that is on a high protein diet. We might also observed an increased volume, the result of more fluid needed to flush the excess urea.
2. An obstruction of the ureters would interfere with the passage of urine from the renal pelvis to the urinary bladder.
3. In order to control the micturition reflex, one must be able to control the external urinary sphincter, a ring of skeletal muscle that acts as a valve.

CHAPTER 27

Page 1040

1. Drinking a pitcher of distilled water would temporarily lower the blood osmolarity (osmotic pressure). Since ADH release is triggered by increases in osmolarity, a decrease in osmolarity would lead to a decrease in the level of ADH in the blood.
2. Consuming a meal high in salt would temporarily increase the osmolarity of the ECF. As a result some of the water in the ICF would shift to the ECF.
3. Fluid loss through perspiration, urine formation, and respiration would increase the osmolarity of body fluids.

Page 1052

1. A decrease in the pH of body fluids would have a stimulating effect on the respiratory center in the medulla. The result would be an increase in the rate of breathing. This would lead to an elimination of more carbon dioxide which would tend to cause the pH to increase.
2. Hydrogen ion secretion ceases at a pH of 4.5 to 5.0. Since the body usually needs to eliminate more hydrogen ions than the amount necessary to lower the pH to that critical range, the filtrate needs to be buffered. The buffers allow the filtrate to take more hydrogen ions without decreasing the pH below the critical level.
3. In a prolonged fast, fatty acids are mobilized and large numbers of ketone bodies are formed. These molecules are acids that lower the body's pH. This would eventually lead to a condition known as ketoacidosis.
4. In vomiting, large amounts of stomach acid are lost from the body. This acid is formed by the parietal cells of the stomach by taking hydrogen ions from the blood. Excessive vomiting would lead to excessive removal of hydrogen ions from the blood to produce the acid, thus raising the body's pH and creating a condition called metabolic alkalosis.

CHAPTER 28

Page 1077

1. The cremaster muscle as well as the dartos muscle would be relaxed on a warm day, so that the scrotal sac could descend away from the warmth of the body and cool the testes.
2. The acrosomal cap contains enzymes necessary for penetrating the cell layers around the ovum. Without these enzymes, fertilization would not occur.
3. Relaxation of the arteries serving the penis will result in erection.
4. FSH is required for production of androgen-binding protein, a protein that binds testosterone and keeps a high level of the hormone available to support spermatogenesis. Low levels of FSH would lead to low levels of testosterone in the seminiferous tubules and thus a lower rate of sperm production and a low sperm count.

Page 1091

1. A blockage of the uterine tube would cause sterility.
2. The acidic pH of the vagina helps to prevent bacterial, fungal, and protozoal infections in this area.
3. The functional layer of the endometrium is sloughed off during menstruation.
4. Blockage of a single lactiferous sinus would not interfere with milk moving to the nipple because each breast usually has between 15 and 20 lactiferous sinuses.

Page 1094

1. If the LH surge did not occur during an ovulatory cycle, ovulation and corpus luteum formation would not occur.
2. Progesterone is responsible for the functional maturation and secretion of the endometrial lining. Blocking progesterone receptors would inhibit endometrial development and make it unsuitable for implantation.
3. A sudden decline in the levels of estrogen and progesterone during the menstrual cycle signals the beginning of the menses.

Page 1100

1. Inability to contract the ischiocavernosus and bulbocavernosus muscles would interefere with a male's ability to ejaculate and experience orgasm.
2. As the result of parasympathetic stimulation in females during sexual arousal, there is engorgement of the erectile tissues of the clitoris, increased secretion of cervical and vaginal glands, increased blood flow to the walls of the vagina, and engorgement of the blood vessels in the nipples.
3. At menopause, circulating estrogen levels begin to drop. Estrogen has an inhibitory effect on GN-RH and FSH and as the level of estrogen declines the levels of these two hormones rise and remain elevated.

CHAPTER 29

Page 1120

1. The inner cell mass of the blastocyst eventually develops into the embryo.
2. The yolk sac would thus affect the development and function of the circulatory system.
3. After fertilization, the developing trophoblasts and then later on the placenta produce and release the hormone hCG. She is pregnant.
4. Placental functions include (1) supplying the developing fetus with a route for gas exchange, nutrient transfer, and waste product elimination, and (2) producing hormones that affect maternal systems.

Page 1134

1. Because blood flow through the placenta reduces the volume of blood in the systemic circuit, and this stimulates an increase in maternal blood volume.
2. Progesterone reduces uterine contractions. A decrease in progesterone at any time during the pregnancy can lead to uterine contractions and in late pregnancy, labor.
3. An increase in blood levels of GnRH, FSH, LH, and sex hormones would signal the onset of puberty.

Page 1140

1. A person who is heterozygous for curly hair would have one dominant gene and one recessive gene. The person's phenotype would be "curly hair."
2. One reason that children are not identical copies of their parents is during meiosis, parental chromosomes randomly assorted such that each gamete has a unique set of chromosomes. Also the crossing-over that occurs during meiosis and mutation introduces new variations.

Accurate descriptions of physical objects would be impossible without a precise method of reporting the pertinent data. Dimensions such as length and width are reported in standardized units of measurement, such as inches or centimeters. These values can be used to calculate the volume of an object, a measurement of the amount of space it fills. **Mass** is another important physical property. The mass of an object is determined by the amount of matter it contains; on earth the mass of an object determines its weight.

Most U.S. readers describe length and width in terms of inches, feet, or yards; volumes in pints, quarts, or gallons; and weights in ounces, pounds, or tons. These are units of the **U.S. system** of measurement. Table 1 summarizes the familiar and unfamiliar terms used in the U.S. system. For reference purposes, this table also includes a definition of the "household units," popular in recipes and cookbooks. The U.S. system can be very difficult to work with, because there is no logical relationship between the various units. For example, there are 12 inches in a foot, 3 feet in a yard, and 1760 yards in a mile. Without a clear pattern of organization, converting feet to inches or miles to feet can be confusing and time consuming. The relationships between ounces, pints, quarts and gallons, or ounces, pounds, and tons are no more logical.

In contrast, the **metric system** has a logical organization based on powers of 10, as indicated in Table 2. For example, a **meter** (m) represents the basic unit for the measurement of size. When measuring larger objects, data can be reported in terms of **dekameters** (*deka*, ten), **hectometers** (*hekaton*, hundred), or **kilometers** (**km**; *chilioi*, thousand); for smaller objects, data can be reported in **decimeters** (0.1 m; *decem*, ten), **centimeters** (**cm** = 0.01 m; *centum, hundred*), **millimeters** (**mm** = 0.001 m; *mille*, thousand), and so forth. Notice that the same prefixes are used to report weights, based on the **gram (g)**, and volumes, based on the **liter (l)**. This text reports data in metric units, usually with U.S. equivalents. You should use this opportunity to become familiar with the metric system, because most technical sources report data only in metric units, and most of the rest of the world uses the metric system exclusively. Conversion factors are included in Table 2.

The U.S. and metric systems also differ in their methods of reporting temperatures; in the U.S., temperatures are usually reported in degrees Fahrenheit (°F), whereas scientific literature and individuals in most other countries report temperatures in degrees Centigrade or Celsius (°C). The relationship between temperatures in degrees Fahrenheit and those in degrees Centigrade has been indicated at the bottom of Table 2.

TABLE 1 The U.S. System of Measurement

Physical Property	*Unit*	*Relationship to Other U.S. Units*	*Relationship to Household Units*
Length	inch (in.)	1 in. = 0.083 ft	
	foot (ft)	1 ft = 12 in.	
		= 0.33 yd	
	yard (yd)	1 yd = 36 in.	
		= 3 ft	
	mile (mi)	1 mi = 5.280 ft	
		= 1,760 yd	
Volume	fluidram (fl dr)	1 fl dr = 0.125 fl oz	
	fluid ounce (fl oz)	1 fl oz = 8 fl dr	= 6 teaspoons (tsp)
		= 0.0625 pt	= 2 tablespoons (tbsp)
	pint (pt)	1 pt = 128 fl dr	= 32 tbsp
		= 16 fl oz	= 2 cups (c)
		= 0.5 qt	
	quart (qt)	1 qt = 256 fl dr	= 4 c
		= 32 fl oz	
		= 2 pt	
		= 0.25 gal	
	gallon (gal)	1 gal = 128 fl oz	
		= 8 pt	
		= 4 qt	
Mass	grain (gr)	1 gr = 0.002 oz	
	dram (dr)	1 dr = 27.3 gr	
		= 0.063 oz	
	ounce (oz)	1 oz = 437.5 gr	
		= 16 dr	
	pound (lb)	1 lb = 7000 gr	
		= 256 dr	
		= 16 oz	
	ton (t)	1 t = 2000 lb	

TABLE 2 The Metric System of Measurement

Physical Property	*Unit*	*Relationship to Standard Metric Units*	*Conversion to U.S. Units*	
Length	nanometer (nm)	1 nm = 0.000000001 m (10^{-9})	= 4×10^{-8} in.	25,000,000 nm = 1 in.
	micrometer (µm)	1 µm = 0.000001 m (10^{-6})	= 4×10^{-5} in.	25,000 mm = 1 in.
	millimeter (mm)	1 mm = 0.001 m (10^{-3})	= 0.0394 in.	25.4 mm = 1 in.
	centimeter (cm)	1 cm = 0.01 m (10^{-2})	= 0.394 in.	2.54 cm = 1 in.
	decimeter (dm)	1 dm = 0.1 m (10^{-1})	= 3.94 in.	0.25 dm = 1 in.
	meter (m)	standard unit of length	= 39.4 in.	0.0254 m = 1 in.
			= 3.28 ft	0.3048 m = 1 ft
			= 1.09 yd	0.914 m = 1 yd
	dekameter (dam)	1 dam = 10 m		
	hectometer (hm)	1 hm = 100 m		
	kilometer (km)	1 km = 1000 m	= 3280 ft	
			= 1093 yd	
			= 0.62 mi	1.609 km = 1 mi
Volume	microliter (µl)	1 µl = 0.000001 l (10^{-6})		
		= 1 cubic millimeter (mm^3)		
	milliliter (ml)	1 ml = 0.001 l (10^{-3})	= 0.03 fl oz	5 ml = 1 tsp
		= 1 cubic centimeter (cm^3 or cc)		15 ml = 1 tbsp
				30 ml = 1 fl oz
	centiliter (cl)	1 cl = 0.01 l (10^{-2})	= 0.34 fl oz	3 cl = 1 fl oz
	deciliter (dl)	1 dl = 0.1 l (10^{-1})	= 3.38 fl oz	0.29 dl = 1 fl oz
	liter (l)	standard unit of volume	= 33.8 fl oz	0.0295 l = 1 fl oz
			= 2.11 pt	0.473 l = 1 pt
			= 1.06 qt	0.946 l = 1 qt
Mass	picogram (pg)	1 pg = 0.000000000001 g (10^{-12})		
	nanogram (ng)	1 ng = 0.000000001 g (10^{-9})		
	microgram (µg)	1 µg = 0.000001 g (10^{-6})	= 0.000015 gr	66,666 mg = 1 gr
	milligram (mg)	1 mg = 0.001 g (10^{-3})	= 0.015 gr	66.7 mg = 1 gr
	centigram (cg)	1 cg = 0.01 g (10^{-2})	= 0.15 gr	6.7 cg = 1 gr
	decigram (dg)	1 dg = 0.1 g (10^{-1})	= 1.5 gr	0.67 dg = 1 gr
	gram (g)	standard unit of mass	= 0.035 oz	28.35 g = 1 oz
			= 0.0022 lb	453.6 g = 1 lb
	dekagram (dag)	1 dag = 10 g		
	hectogram (hg)	1 hg = 100 g		
	kilogram (kg)	1 kg = 1000 g	= 2.2 lb	0.453 kg = 1 lb
	metric ton (kt)	1 mt = 1000 kg	= 1.1 t	
			= 2205 lb	0.907 kt = 1 t

Temperature		*Centigrade*		*Fahrenheit*
Freezing point of pure water		0°		32°
Normal body temperature		36.8°		98.6°
Boiling point of pure water		100°		212°
Conversion	°C → °F:	°F = (1.8 × °C) + 32	°F → °C:	°C = (°F - 32) × 0.56

The figure below spans the entire range of measurements that we will consider in this book. Gross anatomy traditionally deals with structural organization as seen with the naked eye or with a simple hand lens. A microscope can provide higher levels of magnification and reveal finer details. Before the 1950s most information was provided by *light microscopy*. A photograph taken through a light microscope is called a **light micrograph (LM)**. Light microscopy can magnify cellular structures about 1000 times and show details as fine as 0.25 μm. The symbol μm stands for *micrometer*; 1 μm = 0.001 mm, or 0.00004 inches. With a light microscope one can identify cell types, such as muscle cells or neurons, and see large structures within the cell. Because individual cells are relatively transparent, thin sections taken through a cell are treated with dyes that stain specific structures, making them easier to see.

Although special staining techniques can show the general distribution of proteins, lipids, carbohydrates, and nucleic acids in the cell, many fine details of intracellular structure remained a mystery until investigators began using electron microscopy. This technique uses a focused beam of electrons, rather than a beam of light, to examine cell structure. In *transmission electron microscopy*, electrons pass through an ultrathin section to strike a photographic plate. The result is a **transmission electron micrograph (TEM)**. Transmission electron microscopy shows the fine structure of cell membranes and intracellular structures. In *scanning electron microscopy*, electrons bouncing off exposed surfaces create a **scanning electron micrograph (SEM)**. Although scanning microscopy cannot achieve as much magnification as transmission microscopy, it provides a three-dimensional perspective on cell structure.

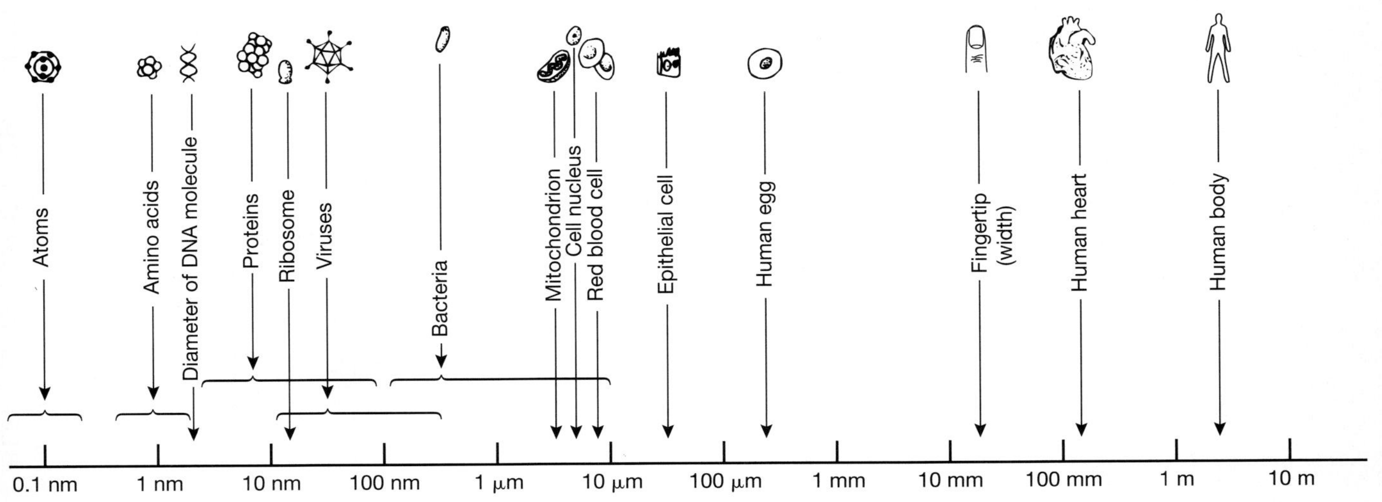

Atoms
Amino acids
Diameter of DNA molecule
Proteins
Ribosome
Viruses
Bacteria
Mitochondrion
Cell nucleus
Red blood cell
Epithelial cell
Human egg
Fingertip (width)
Human heart
Human body
0.1 nm
1 nm
10 nm
100 nm
1 μm
10 μm
100 μm
1 mm
10 mm
100 mm
1 m
10 m
Transmission electron microscope
Scanning electron microscope
Light microscope
Unaided human eye

APPENDIX III

PERIODIC TABLE

The periodic table presents the known elements in order of their atomic weights. Each horizontal row represents a single electron shell; the number of elements in that row is determined by the maximum number of electrons that can be stored at that energy level. The element at the left end of each row contains a single electron in its outermost electron shell; the last element in the row has a filled outer electron shell. Organizing the elements in this fashion highlights similarities that reflect the compositiion of the outer electron shell, and these relationships are evident when you examine the vertical columns. Helium, neon, argon, krypton, xenon, and radon all have full electron shells, they are all gases at normal atmospheric temperature and pressure, and they do not react readily with other elements. These elements, highlighted in green, are known as the noble or inert gases. In contrast, lithium, sodium, potassium, and so forth are silvery, soft metals that are so highly reactive that pure forms cannot be found in nature. The fourth and fifth electron levels can hold 18 electrons: table inserts are used to save space, as higher levels may store up to 32 electrons. Elements of particular importance to our discussion of human anatomy and physiology are highlighted in yellow.

Atomic number — 1 H — Chemical symbol
Hydrogen — Element name
Atomic weight — 1.01

1 H Hydrogen 1.01																	2 He Helium 4.00
3 Li Lithium 6.94	4 Be Beryllium 9.01											5 B Boron 10.81	6 C Carbon 12.01	7 N Nitrogen 14.01	8 O Oxygen 16.00	9 F Fluorine 19.00	10 Ne Neon 20.18
11 Na Sodium 22.99	12 Mg Magnesium 24.31											13 Al Aluminum 26.98	14 Si Silicon 28.09	15 P Phosphorus 30.97	16 S Sulfur 32.07	17 Cl Chlorine 35.45	18 Ar Argon 39.95
19 K Potassium 39.10	20 Ca Calcium 40.08	21 Sc Scandium 44.96	22 Ti Titanium 47.88	23 V Vanadium 50.94	24 Cr Chromium 52.00	25 Mn Manganese 54.94	26 Fe Iron 55.85	27 Co Cobalt 58.93	28 Ni Nickel 58.69	29 Cu Copper 63.55	30 Zn Zinc 65.39	31 Ga Gallium 69.72	32 Ge Germanium 72.61	33 As Arsenic 74.92	34 Se Selenium 78.96	35 Br Bromine 79.90	36 Kr Krypton 83.80
37 Rb Rubidium 85.47	38 Sr Strontium 87.62	39 Y Yttrium 88.91	40 Zr Zirconium 91.22	41 Nb Niobium 92.91	42 Mo Molybdenum 95.94	43 Tc Technetium (98)	44 Ru Ruthenium 101.07	45 Rh Rhodium 102.91	46 Pd Palladium 106.42	47 Ag Silver 107.87	48 Cd Cadmium 112.41	49 In Indium 114.82	50 Sn Tin 118.71	51 Sb Antimony 121.76	52 Te Tellurium 127.60	53 I Iodine 126.90	54 Xe Xenon 131.29
55 Cs Cesium 132.91	56 Ba Barium 137.33	57 La Lanthanum 138.91 *	72 Hf Hafnium 178.49	73 Ta Tantalum 180.95	74 W Tungsten 183.85	75 Re Rhenium 186.21	76 Os Osmium 190.2	77 Ir Iridium 192.22	78 Pt Platinum 195.08	79 Au Gold 196.97	80 Hg Mercury 200.59	81 Tl Thallium 204.38	82 Pb Lead 207.2	83 Bi Bismuth 208.98	84 Po Polonium (209)	85 At Astatine (210)	86 Rn Radon (222)
87 Fr Francium (223)	88 Ra Radium 226.03	89 Ac Actinium 227.03 †	104 Db Dubnium (261)	105 Jl Joliotium (262)	106 Rf Ruther-fordium (263)	107 Bh Bohrium (262)	[108] Hn Hahnium (265)	[109] Mt Meitnerium (266)									

*Lanthanide series	58 Ce Cerium 140.12	59 Pr Praseo-dymium 140.91	60 Nd Neodymium 144.24	61 Pm Promethium (145)	62 Sm Samarium 150.36	63 Eu Europium 151.96	64 Gd Gadolinium 157.25	65 Tb Terbium 158.93	66 Dy Dysprosium 162.50	67 Ho Holmium 164.93	68 Er Erbium 167.26	69 Tm Thulium 168.93	70 Yb Ytterbium 173.04	71 Lu Lutetium 174.97
†Actinide series	90 Th Thorium 232.04	91 Pa Protactinium 231.04	92 U Uranium 238.03	93 Np Neptunium 237.05	94 Pu Plutonium (244)	95 Am Americium (243)	96 Cm Curium (247)	97 Bk Berkelium (247)	98 Cf Californium (251)	99 Es Einsteinium (252)	100 Fm Fermium (257)	101 Md Mendelevium (258)	102 No Nobelium (259)	103 Lr Lawrencium (260)

APPENDIX IV Condensed Structural Formulas for the Amino Acids

Amino acids with acidic or basic side chain functional groups

Aspartic acid (asp)

Glutamic acid (glu)

Tyrosine (tyr)

Lysine (lys)

Arginine (arg)

Histidine (his)

Amino acids with unchanged but polar side chains

Serine (ser)

Threonine (thr)

Methionine (met)

Cysteine (cys)

Amino acids with hydrocarbon side chains

Glycine (gly)

Alanine (ala)

Valine (val)

Tryptophan (trp)

Asparagine (asn)

Glutamine (gln)

Leucine (leu)

Isoleucine (ile)

Phenylalanine (phe)

Proline (pro)

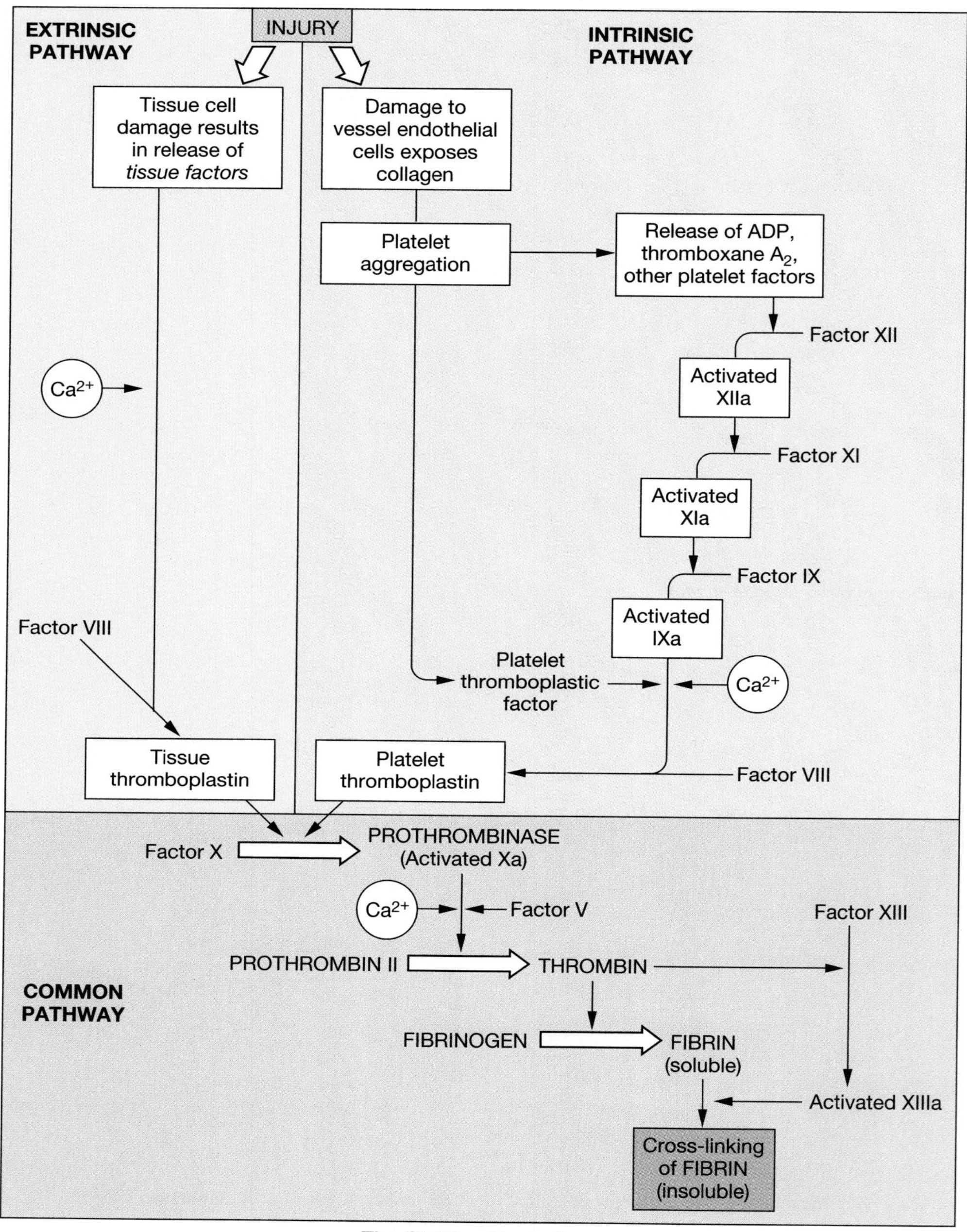
EXTRINSIC PATHWAY
INJURY
INTRINSIC PATHWAY
Tissue cell damage results in release of tissue factors
Damage to vessel endothelial cells exposes collagen
Platelet aggregation
Release of ADP, thromboxane A2, other platelet factors
Factor XII
Activated XIIa
Factor XI
Activated XIa
Factor IX
Activated IXa
Ca2+
Factor VIII
Platelet thromboplastic factor
Ca2+
Tissue thromboplastin
Platelet thromboplastin
Factor VIII
Factor X
PROTHROMBINASE (Activated Xa)
Ca2+
Factor V
Factor XIII
PROTHROMBIN II
THROMBIN
COMMON PATHWAY
FIBRINOGEN
FIBRIN (soluble)
Activated XIIIa
Cross-linking of FIBRIN (insoluble)

The Clotting Cascade

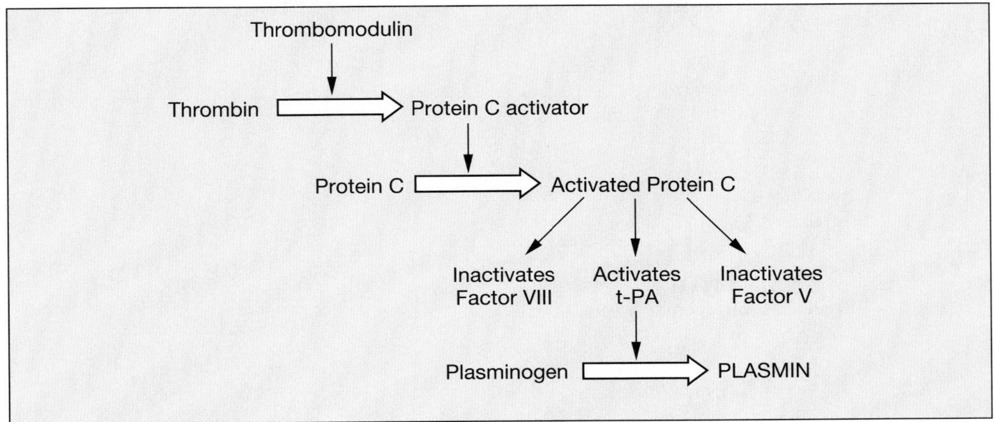
Thrombomodulin
Thrombin
Protein C activator
Protein C
Activated Protein C
Inactivates Factor VIII
Activates t-PA
Inactivates Factor V
Plasminogen
PLASMIN

The Fibrinolytic System

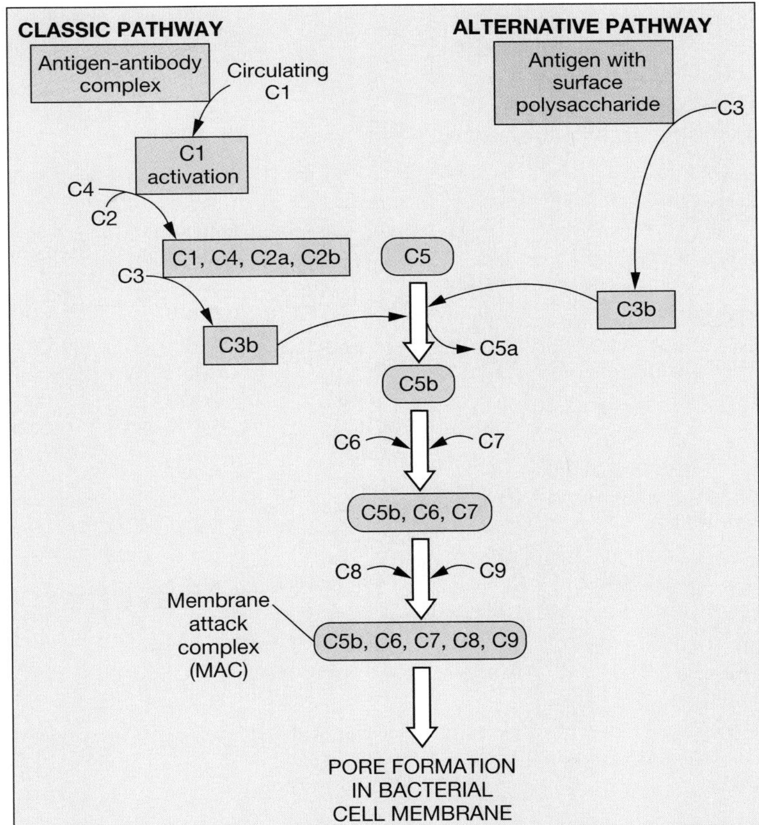
CLASSIC PATHWAY
ALTERNATIVE PATHWAY
Antigen-antibody complex
Circulating C1
Antigen with surface polysaccharide
C3
C1 activation
C4
C2
C1, C4, C2a, C2b
C5
C3
C3b
C3b
C5a
C5b
C6
C7
C5b, C6, C7
C8
C9
Membrane attack complex (MAC)
C5b, C6, C7, C8, C9
PORE FORMATION IN BACTERIAL CELL MEMBRANE

The Complement Pathways

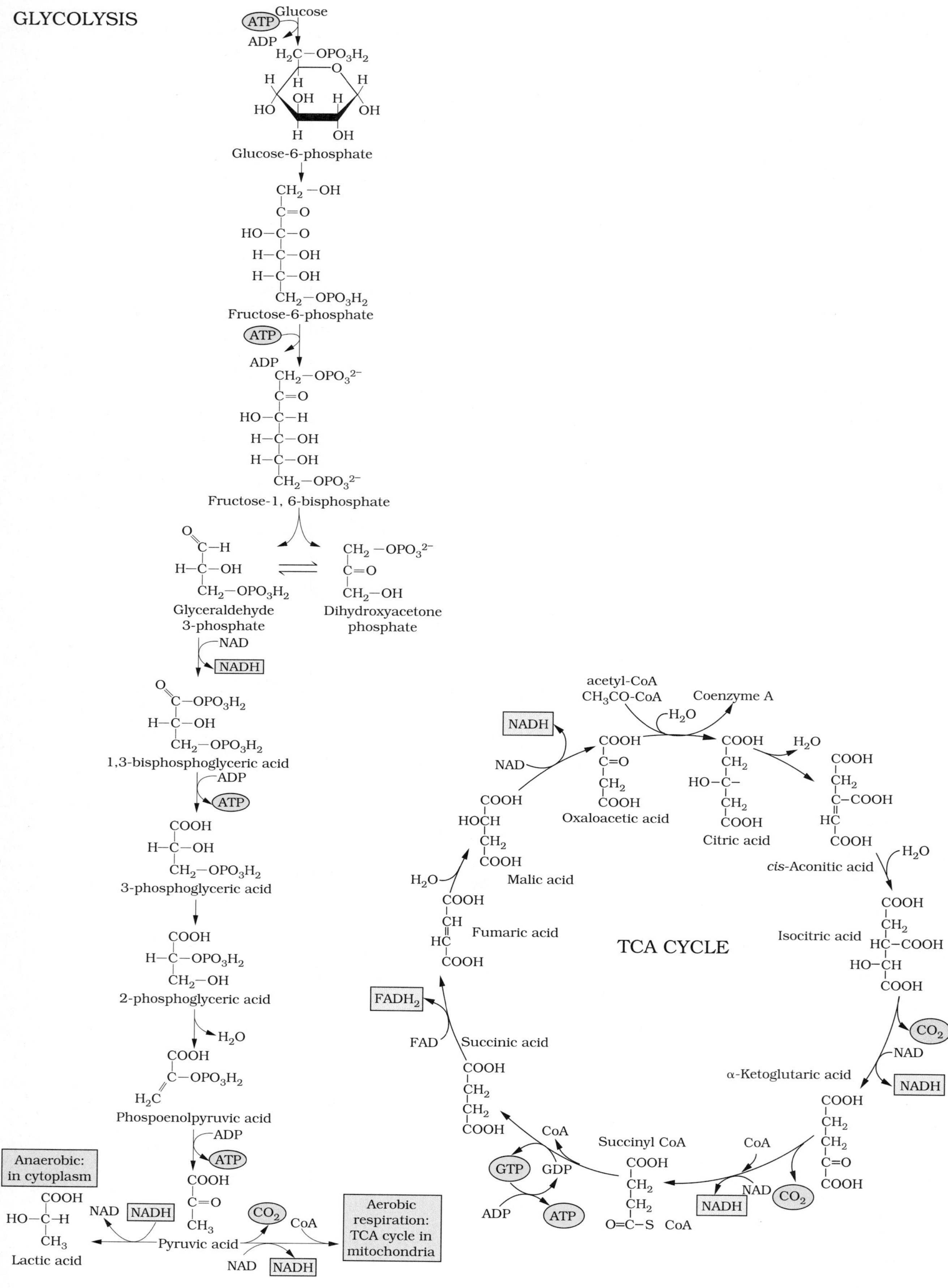
GLYCOLYSIS
Glucose
ATP
ADP
Glucose-6-phosphate
Fructose-6-phosphate
ATP
ADP
Fructose-1, 6-bisphosphate
Glyceraldehyde 3-phosphate
Dihydroxyacetone phosphate
NAD
NADH
1,3-bisphosphoglyceric acid
ADP
ATP
3-phosphoglyceric acid
2-phosphoglyceric acid
H2O
Phospoenolpyruvic acid
ADP
ATP
Anaerobic: in cytoplasm
Lactic acid
NAD NADH
Pyruvic acid
CO2
CoA
NAD NADH
Aerobic respiration: TCA cycle in mitochondria
acetyl-CoA
CH3CO-CoA
Coenzyme A
H2O
NADH
NAD
Oxaloacetic acid
Citric acid
H2O
cis-Aconitic acid
H2O
Isocitric acid
TCA CYCLE
CO2
NAD
NADH
α-Ketoglutaric acid
CoA
NAD
CO2
NADH
Succinyl CoA
CoA
GTP
GDP
ADP
ATP
Succinic acid
FAD
FADH2
Fumaric acid
H2O
Malic acid

APPENDIX VII — NORMAL PHYSIOLOGICAL VALUES

Tables 3 and 4 present normal averages or ranges for the chemical composition of body fluids. These should be considered approximations, rather that absolute values, as test results vary from laboratory to laboratory due to differences in procedures, equipment, normal solutions, and so forth. Blanks in the tabular data appear where data were not available; sources used in the preparation of these tables are indicated below. Additional information concerning body fluid analysis can be found at the following locations in the text:

Table 19-3 (p. 668) presents data on the cellular composition of whole blood.

Table 26-2 (p. 992) compares the average compositions of plasma and urine.

Tables 26-5 and 26-6 (pp. 1010-1011) give the general characteristics of normal urine.

Sources

Ballenger, John Jacob. 1977. *Diseases of the Nose, Throat, and Ear.* Philadelphia, Pa.: Lea and Febiger.

Braunwauld, Eugene, Kurt J. Isselbacher, Robert G. Petersdorf, Jean D. Wilson, Joseph B. Martin, and Anthony S. Fauci, eds. 1987. *Harrison's Principles of Internal Medicine*, 11th ed. New York: McGraw-Hill.

Lentner, Cornelius, ed. 1981. *Geigy Scientific Tables*, 8th ed. Basel, Switzerland: Ciba-Geigy Limited.

Davidsohn, Israel, and John Bernard Henry, eds. 1969. *Todd-Sanford Clinical Diagnosis by Laboratory Methods*, 14th ed. Philadelphia, Pa.: W. B. Saunders Company.

Diem, K., and C. Lenter, eds. 1970. *Scientific Tables*, 7th ed. Basel, Switzerland: Ciba-Geigy Limited.

Halsted, James A. 1976. *The Laboratory in Clinical Medicine: Interpretation and Application.* Philadelphia, Pa.: W.B. Saunders Company.

Harper, Harold A. 1987. *Review of Physiological Chemistry.* Los Altos, Calif.: Lange Medical Publications.

TABLE 3 The Composition of Minor Body Fluids

	Normal Averages or Ranges					
Test	*Perilymph*	*Endolymph*	*Synovial Fluid*	*Sweat*	*Saliva*	*Semen*
pH			7.4	4—6.8	6.4[a]	7.19
Specific gravity			1.008–1.015	1.001–1.008	1.007	1.028
Electrolytes (mEq/l)						
Potassium	5.5–6.3	140–160	4.0	4.3–14.2	21	31.3
Sodium	143–150	12–16	136.1	0–104	14[a]	117
Calcium	1.3–1.6	0.05	2.3–4.7	0.2–6	3	12.4
Magnesium	1.7	0.02		0.03–4	0.6	11.5
Bicarbonate	17.8–18.6	20.4–21.4	19.3–30.6		6[a]	24
Chloride	121.5	107.1	107.1	34.3	17	42.8
Proteins(mg/dl)						
Total	200	150	1.72 g/dl	7.7	386[b]	4.5 g/dl
Metabolites (mg/dl)						
Amino acids				47.6	40	1.26 g/dl
Glucose	104		70–110	3.0	11	224 (fructose)
Urea				26–122	20	72
Lipids, total	12		20.9	[d]	25–500[c]	188

[a] Increases under salivary stimulation.
[b] Primarily alpha-anylase, with some lysozomes.
[c] Cholesterol.
[d] Not present in eccrine secretions.

TABLE 4 The Chemistry of Blood, Cerebrospinal Fluid, and Urine

	Normal Ranges		
Test	*Blood*[a]	*CSF*	*Urine*
pH	S: 7.38—7.44	7.31—7.34	4.6—8.0
Osmolarity (mOsm/l)	S: 280—295	292—297	500—800
Electrolytes	(mEq/l unless noted)		(urinary loss per 24-hour period[b])
Bicarbonate	P: 21—28	20—24	
Calcium	S: 4.5—5.3	2.1—3.0	6.5—16.5 mEq
Chloride	S: 100—108	116—122	120—240 mEq
Iron	S: 50—150 μg/l	23—52 μg/l	40—150 μg
Magnesium	S: 1.5—2.5	2—2.5	4.9—16.5 mEq
Phosphorus	S: 1.8—2.6	1.2—2.0	0.8—2 g
Potassium	P: 3.8—5.0	2.7—3.9	35—80 mEq
Sodium	P: 136—142	137—145	120—220 mEq
Sulfate	S: 0.2—1.3		1.07—1.3 g
Metabolites	(mg/dl unless noted)		(urinary loss per 24-hour period[c])
Amino acids	P/S: 2.3—5.0	10.0—14.7	41—133 mg
Ammonia	P: 20—150 μg/dl	25—80 μg/dl	340—1200 mg
Bilirubin	S: 0.5—1.2	<0.2	0.02—1.9 mg
Creatinine	P/S: 0.6—1.2	0.5—1.9	1.01—2.5
Glucose	P/S: 70—110	40—70	16—132 mg
Ketone bodies	S: 0.3—2.0	1.3—1.6	10—100 mg
Lactic acid	WB: 5—20[d]	10—20	100—600 mg
Lipids (total)	S: 400—1000	0.8—1.7	0—31.8 mg
Cholesterol (total)	S: 150—300	0.2—0.8	1.2—3.8 mg
Triglycerides	S: 40—150	0—0.9	
Urea	P/S: 23—43	13.8—36.4	12.6—28.6
Uric Acid	S: 2.0—7.0	0.2—0.3	80—976 mg
Proteins	(g/dl)	(mg/dl)	(urinary loss per 24-hour period[c])
Total	S: 6.0—7.8	20—4.5	47—76.2 mg
Albumin	S: 3.2—4.5	10.6—32.4	10—100 mg
Globulins (total)	S: 2.3—3.5	2.8—15.5	7.3 mg (average)
Immunoglobulins	S: 1.0—2.2	1.1—1.7	3.1 mg (average)
Fibrinogen	P: 0.2—0.4	0.65 (average)	

[a] S = serum, P = plasma, WB = whole blood

[b] Because urinary output averages just over 1 liter per day, these electrolyte values are comparable to mEq/l.

[c] Because urinary metabolite and protein data approximate mg/l or g/l, they must be divided by 10 for comparison with CSF or blood concentrations.

[d] Venous blood sample

GLOSSARY OF KEY TERMS

abdomen: Region of trunk bounded by the diaphragm and pelvis.

abdominopelvic cavity: portion of the ventral body cavity that contains abdominal and pelvic subdivisions.

abducens (ab-DŪ-senz): Cranial nerve VI, innervates the lateral rectus muscle of the eye.

abduction: Movement away from the midline.

abortion: Premature loss or expulsion of an embryo or fetus.

abruptio placentae (ab-RUP-shē-ō pla-SEN-tē): Premature loss of placental connection to uterus, leading to maternal hemorrhaging and shock.

abscess: A localized collection of pus within a damaged tissue.

absorption: The active or passive uptake of gases, fluids, or solutes.

accommodation: Alteration in the curvature of the lens to focus an image on the retina; decrease in receptor sensitivity or perception following chronic stimulation.

acetabulum (a-se-TAB-ū-lum): Fossa on lateral aspect of pelvis that accommodates the head of the femur.

acetyl group: $CH_3C{=}O$.

acetyl-CoA: An acetyl group bound to Coenzyme A, a participant in the anabolic and catabolic pathways for carbohydrates, lipids, and many amino acids.

acetylcholine (ACh) (as-ē-til-KŌ-lēn): Chemical neurotransmitter in the brain and PNS; dominant neurotransmitter in the PNS, released at neuromuscular junctions and synapses of the parasympathetic division.

acetylcholinesterase (AChE): Enzyme found in the synaptic cleft, bound to the postsynaptic membrane, and in tissue fluids; breaks down and inactivates ACh molecules.

achalasia (āk-a-LĀ-zē-a): Condition that develops when the lower esophageal sphincter fails to dilate, and ingested materials cannot enter the stomach.

Achilles tendon: Calcaneal tendon.

acid: A compound whose dissociation in solution releases a hydrogen ion and an anion; an acid solution has a pH below 7.0 and contains an excess of hydrogen ions.

acidosis (a-sid-Ō-sis): An abnormal physiological state characterized by a plasma pH below 7.35.

acinus/acini (AS-i-nī): Histological term referring to a blind pocket, pouch, or sac.

acne: Condition characterized by inflammation of sebaceous glands and follicles, commonly affects adolescents and most often involves the face.

acoustic: Pertaining to sound or the sense of hearing.

acquired immunodeficiency syndrome (AIDS): A disease caused by the **human immunodeficiency virus (HIV),** characterized by destruction of helper T cells and a resulting severe impairment of the immune response.

acromegaly: Condition caused by overproduction of growth hormone in the adult, characterized by thickening of bones and enlargement of cartilages and other soft tissues.

acromion (a-KRŌ-mē-on): Continuation of the scapular spine that projects above the capsule of the scapulohumeral joint.

acrosomal cap (ak-rō-SŌ-mal): Membranous sac at the tip of a sperm cell that contains hyaluronic acid.

actin: Protein component of microfilaments; form thin filaments in skeletal muscles and produce contractions of all muscles through interaction with thick (myosin) filaments; *see* **sliding filament theory.**

action potential: A conducted change in the transmembrane potential of excitable cells, initiated by a change in the membrane permeability to sodium ions; *see also* **nerve impulse.**

activation energy: The energy required to initiate a specific chemical reaction.

active transport: The ATP-dependent absorption or excretion of solutes across a cell membrane.

acute: Sudden in onset, severe in intensity, and brief in duration.

adaptation: Alteration of pupillary size in response to changes in light intensity; in CNS often used as a synonym for accommodation; physiological responses that produce acclimatization.

Addison's disease: Condition resulting from hyposecretion of glucocorticoids, characterized by lethargy, weakness, hypotension, and increased skin pigmentation.

adduction: Movement toward the axis or midline of the body as viewed in the anatomical position.

adenine: A purine, one of the nitrogen bases in the nucleic acids RNA and DNA.

adenohypophysis (ad-e-nō-hī-POF-i-sis): The anterior portion of the pituitary gland, also called the **anterior pituitary** or the **pars distalis.**

adenoids: The pharyngeal tonsil.

adenosine: A nucleoside consisting of adenine and a 5-carbon sugar.

adenosine diphosphate (ADP): Adenosine with two phosphate groups attached.

adenosine phosphate (AMP): A nucleotide consisting of adenine plus a phosphate group (PO_4^{3-}); also known as adenosine monophosphate.

adenosine triphosphate (ATP): A high-energy compound consisting of adenosine with three phosphate groups attached; the third is attached by a high-energy bond.

adenylate cyclase: An enzyme bound to the inner surfaces of cell membranes that can convert ATP to cyclic AMP. Also called *adenyl cyclase* and *adenylyl cyclase.*

adhesion: Fusion of two mesenterial layers following damage or irritation of their opposing surfaces.

adipocyte (AD-i-pō-sīt): A fat cell.

adipose tissue: Loose connective tissue dominated by adipocytes.

adrenal cortex: Superficial portion of adrenal gland that produces steroid hormones.

adrenal gland: Small endocrine gland secreting steroids and catecholamines, located superior to each kidney.

adrenal medulla: Core of the adrenal gland; a modified sympathetic ganglion that secretes catecholamines into the blood following sympathetic activation.

adrenergic (ad-ren-ER-jik): A synaptic terminal that releases norepinephrine when stimulated.

adrenocortical hormone: Any of the steroids produced by the adrenal cortex.

adrenocorticotrophic hormone (ACTH): Hormone that stimulates the production and secretion of glucocorticoids by the zona fasciculata of the adrenal cortex; released by the anterior pituitary in response to CRH.

adventitia (ad-ven-TISH-a): Superficial layer of connective tissue surrounding an internal organ; fibers are continuous with those of surrounding tissues, providing support and stabilization.

aerobic: Requiring the presence of oxygen.

aerobic respiration: The complete breakdown of organic substrates into carbon dioxide and water, via pyruvic acid; a process that yields large amounts of ATP but requires mitochondria and oxygen.

afferent: Toward.

afferent arteriole: An arteriole bringing blood to a glomerulus of the kidney.
afferent fiber: Axons carrying sensory information to the CNS.
afterbirth: The distal portions of the umbilical cord and placenta that are ejected from the uterus during the placental stage of labor.
agglutination (a-gloo-ti-NĀ-shun): Aggregation of red blood cells due to interactions between surface agglutinogens and plasma agglutinins.
agglutinins (a-gloo-TĪ-ninz): Immunoglobulins in plasma that react with antigens on the surfaces of foreign red blood cells when donor and recipient differ in blood type.
agglutinogens (a-gloo-TIN-ō-jenz): Antigens on the surfaces of red blood cells whose presence and structure are genetically determined.
agonist: A muscle responsible for a specific movement.
agranular: Without granules; *agranular leukocytes* are monocytes and lymphocytes; the *agranular reticulum* is an intracellular organelle that synthesizes and stores carbohydrates and lipids.
AIDS: *See* **Acquired immunodeficiency syndrome.**
alba, albicans, albuginea (AL-bi-kanz) (al-bū-JIN-ē-a): White.
albinism: Absence of pigment in hair and skin caused by inability of melanocytes to produce melanin.
aldosterone: A mineralocorticoid produced by the zona glomerulosa of the adrenal cortex; stimulates sodium and water conservation at the kidneys; secreted in response to the presence of angiotensin II.
aldosteronism: Condition caused by the oversecretion of aldosterone, characterized by fluid retention, edema, and hypertension.
alkalosis (al-kah-LŌ-sis): Condition characterized by a plasma pH of greater than 7.45, and associated with relative deficiency of hydrogen ions or an excess of bicarbonate ions.
allantois (a-LAN-tō-is): One of the extraembryonic membranes; it provides vascularity to the chorion and is therefore essential to placenta formation; the proximal portion becomes the urinary bladder.
alleles (a-LĒLZ): Alternate forms of a particular gene.
allergen: An antigenic compound that produces a hypersensitivity response.
alpha-blockers: Drugs that prevent stimulation of *alpha receptors*.
alpha cells: Cells in the pancreatic islets that secrete glucagon.
alpha receptors: Membrane receptors sensitive to norepinephrine or epinephrine; stimulation usually results in excitation of the target cell.
alveolar sac: An air-filled chamber that supplies air to several alveoli.
alveolus/alveoli (al-VĒ-o-lī): Blind pockets at the end of the respiratory tree, lined by a simple squamous epithelium and surrounded by a capillary network; gas exchange with the blood occurs here.
Alzheimer's disease: Disorder resulting from degenerative changes in populations of neurons in the cerebrum, causing dementia characterized by problems with attention, short-term memory, and emotions.
amacrine cells (AM-a-krīn): Modified neurons in the retina that facilitate or inhibit communication between bipolar and ganglion cells.
amenorrhea (ā-men-ō-RĒ-a): Failure to commence menstruation at adolescence or the cessation of menstruation prior to menopause.
amination: The attachment of an amine group to a carbon chain; performed by a variety of cells and important in the synthesis of *amino acids*.
amino acids: Organic compounds whose chemical structure can be summarized as $R\text{-}CHNH_2COOH$.
amino group: NH_2.
amnesia: Temporary or permanent memory loss.
amniocentesis: Sampling of amniotic fluid for analytical purposes; used to detect certain forms of genetic abnormalities.
amnion (AM-nē-on): One of the extraembryonic membranes; surrounds the developing embryo/fetus.
amniotic fluid (am-nē-OT-ik): Fluid that fills the amniotic cavity; provides cushioning and support for the embryo/fetus.
amphiarthrosis (am-fē-ar-THRŌ-sis): An articulation that permits a small degree of independent movement.
amphicytes (AM-fi-sīts): Supporting cells that surround neurons in the PNS; also called *satellite cells*.
amphimixis (am-fi-MIK-sis): The fusion of male and female pronuclei following fertilization.
ampulla/ampullae (am-PYŪL-la): A localized dilation in the lumen of a canal or passageway.
amygdala/amygdaloid nucleus (ah-MIG-da-loyd): A cerebral nucleus that is a component of the limbic system and acts as an interface between that system, the cerebrum, and sensory systems.
amylase: An enzyme that breaks down polysaccharides, produced by the salivary glands and pancreas.
anabolism (a-NAB-ō-lizm): The synthesis of complex organic compounds from simpler precursors.
anaerobic: Without oxygen.
analgesia: Relief from pain.
anal triangle: The posterior subdivision of the perineum.
anamnestic response (an-am-NES-tic): Sudden and exaggerated production of antibodies following second exposure to a specific antigen, due to the activation of memory B cells.
anaphase (AN-a-fāz): Mitotic stage in which the paired chromatids separate and move toward opposite ends of the spindle apparatus.
anaphylaxis (a-na-fi-LAK-sis): Hypersensitivity reaction due to antigen binding to immunoglobulins (IgE) on the surfaces of mast cells; mast cell release of histamine, serotonin, and prostaglandins then causes widespread inflammation; a sudden decline in blood pressure may occur, producing a condition known as anaphylactic shock.
anastomosis (a-nas-to-MŌ-sis): The joining of two tubes, usually referring to a connection between two peripheral vessels without an intervening capillary bed.
anatomical position: An anatomical reference position, the body viewed from the anterior surface with the palms facing forward; supine.
anatomy (a-NAT-ō-mē): The study of the structure of the body.
anaxonic neuron (an-ak-SON-ik): A CNS neuron that has many processes but no apparent axon.
androgen (AN-drō-jen): A steroid sex hormone primarily produced by the interstitial cells of the testis, and manufactured in small quantities by the adrenal cortex in either sex.
anemia (a-NĒ-mē-ah): Condition marked by a reduction in the hematocrit and/or hemoglobin content of the blood.
anencephaly (an-en-SEF-a-lē): Developmental defect characterized by incomplete development of cerebral hemispheres and cranium.

anesthesia: Total or partial loss of sensation from a region of the body.
aneurysm (AN-ū-rizm): A weakening and localized dilation in the wall of a blood vessel.
angiogram (AN-jē-ō-gram): An X-ray image of circulatory pathways.
angiography: X-ray examination of vessel distribution following the introduction of radiopaque substances into the bloodstream.
angiotensin I, II: Angiotensin II is a hormone that causes an elevation in systemic blood pressure, stimulates secretion of aldosterone, promotes thirst, and causes the release of ADH; a converting enzyme in the pulmonary capillaries converts angiotensin I to angiotensin II.
angiotensinogen: Blood protein produced by the liver that is converted to angiotensin I by the enzyme renin.
angstrom (Å): Unit of measure equivalent to 0.1 nanometers.
anion (AN-ī-on): An ion bearing a negative charge.
ankyloglossia (ang-kī-lō-GLOS-ē-a): Condition characterized by an overly robust and restrictive lingual frenulum.
annulus (AN-ū-lus): A cartilage or bone shaped like a ring.
anorectal canal (ā-nō-REC-tal): The distal portion of the rectum that contains the rectal columns and ends at the anus.
anorexia nervosa: An eating disorder marked by a loss of appetite and pronounced weight loss.
anoxia (a-NOKS-ē-a): Tissue oxygen deprivation.
antagonist: A muscle that opposes the movement of an agonist.
antebrachium: The forearm.
anteflexion (an-te-FLEK-shun): Normal position of the uterus, with the superior surface bent forward.
anterior: On or near the front or ventral surface of the body.
anterior pituitary: *See* **pituitary gland.**
anterograde amnesia: Inability to store memories of events that occur after a specific incident or time.
anthracosis (an-thra-KŌ-sis): "Black lung disease," deterioration of respiratory exchange efficiency due to chronic inhalation of coal dust.
antibiotic: Chemical agent that selectively kills pathogenic microorganisms.
antibody (AN-ti-bod-ē): A globular protein produced by plasma cells that will bind to specific antigens and promote their destruction or removal from the body.
anticholinesterase: Chemical compound that blocks the action of acetylcholinesterase and causes prolonged and intensive stimulation of postsynaptic membranes.
anticoagulant: Compound that slows or prevents clot formation by interfering with the clotting system.
anticodon: Triplet of nitrogenous bases on a tRNA molecule that interacts with an appropriate codon on a strand of mRNA.
antidiuretic hormone (ADH) (an-tī-dī-ū-RET-ik): Hormone synthesized in the hypothalamus and secreted at the posterior pituitary; causes water retention at the kidneys, and an elevation of blood pressure.
antigen: A substance capable of inducing the production of antibodies.
antigen-antibody complex: The combination of an antigen and a specific antibody.
antigenic determinant site: A portion of an antigen that can interact with an antibody molecule.
antihistamine (an-ti-HIS-ta-mēn): A chemical agent that blocks the action of histamine on peripheral tissues.
antipyretic agents: Chemicals that reduce fever.
antrum (AN-trum): A chamber or pocket.
anuria (a-NŪ-rē-a): Cessation of urine production.
anus: External opening of the anorectal canal.
aorta: Large, elastic artery that carries blood away from the left ventricle, and into the systemic circuit.
aortic reflex: Baroreceptor reflex triggered by increased aortic pressures; leads to a reduction in cardiac output and a fall in systemic pressure.
Apgar: A test used to assess the neurological status of a newborn infant.
aphasia: Inability to speak.
apnea (AP-nē-a): Cessation of breathing.
apneustic center (ap-NŪ-stik): Respiratory center whose chronic activation would lead to apnea at full inspiration.
apocrine secretion: Mode of secretion where the glandular cell sheds portions of its cytoplasm.
aponeurosis/aponeuroses (ap-ō-nū-RŌ-sēz): A broad tendinous sheet that may serve as the origin or insertion of a skeletal muscle.
apoplexy: A stroke (cerebrovascular accident).
appendicitis: Inflammation of the appendix.
appendicular: Pertaining to the upper or lower limbs.
appendix: A blind tube connected to the cecum of the large intestine.
appositional growth: Enlargement by the addition of cartilage or bony matrix to the outer surface.
aqueous humor: Fluid similar to perilymph or CSF that fills the anterior chamber of the eye.
arachidonic acid: One of the essential fatty acids.
arachnoid (a-RAK-noyd): The middle meninges that encloses CSF and protects the central nervous system.
arachnoid villi: Processes of the arachnoid that project into the superior sagittal sinus; sites where CSF enters the venous circulation.
arbor vitae: Central, branching mass of white matter inside the cerebellum.
arcuate (AR-kū-āt): Curving.
areflexia (ā-re-flek-sē-a): Absence of normal reflex responses to stimulation.
areola (a-RĒ-ō-la): Pigmented area that surrounds the nipple of a breast.
areolar: Containing minute spaces, as in areolar connective tissue.
arrector pili (ar-REK-tor PI-li): Smooth muscles whose contractions cause piloerection.
arrhythmias (a-RITH-mē-as): Abnormal patterns of cardiac contractions.
arteriole (ar-TE-rē-ol): A small arterial branch that delivers blood to a capillary network.
artery: A blood vessel that carries blood away from the heart and toward a peripheral capillary.
arthritis (ar-THRĪ-tis): Inflammation of a joint.
arthroscope: Fiberoptic device intended for visualizing the interior of joints; may also be used for certain forms of joint surgery.
articular: Pertaining to a joint.
articular capsule: Dense collagen fiber sleeve that surrounds a joint and provides protection and stabilization.
articular cartilage: Cartilage pad that covers the surface of a bone inside a joint cavity.
articulation (ar-tik-ū-LĀ-shun): A joint; formation of words.
arytenoid cartilages (ar-i-TĒ-noyd): A pair of small cartilages in the larynx.

ascending tract: A tract carrying information from the spinal cord to the brain.
ascites (a-SĪ-tēz): Overproduction and accumulation of peritoneal fluid.
aseptic: Free from pathogenic contamination.
asphyxia: Unconsciousness due to oxygen deprivation at the CNS.
aspirate: To remove or obtain by suction; to inhale.
association areas: Cortical areas of the cerebrum responsible for integration of sensory inputs and/or motor commands.
association neuron: *See* **interneuron.**
asthma (AZ-ma): Reversible constriction of smooth muscles around respiratory passageways, frequently caused by an allergic response.
astigmatism: Visual disturbance due to an irregularity in the shape of the cornea.
astrocyte (AS-trō-sīt): One of the glial cells in the CNS; responsible for the blood-brain barrier.
ataxia: Failure to coordinate muscular activities normally.
atelectasis (at-e-LEK-ta-sis): Collapse of a lung or a portion of a lung.
atherosclerosis (ath-er-ō-skle-RŌ-sis): Formation of fatty plaques in the walls of arteries, leading to circulatory impairment.
atom: The smallest stable unit of matter.
atomic number: The number of protons in the nucleus of an atom.
atomic weight: Roughly, the average total number of protons and neutrons in the atoms of a particular element.
atresia (a-TRĒ-zē-a): Closing of a cavity, or its incomplete development; used in the reproductive system to refer to the degeneration of developing ovarian follicles.
atria: Thin-walled chambers of the heart that receive venous blood from the pulmonary or systemic circuits.
atrial natriuretic peptide (nā-trē-ū-RET-ik): Hormone released by specialized atrial cardiocytes when they are stretched by an abnormally large venous return; promotes fluid loss and reductions in blood pressure and venous return.
atrial reflex: Reflexive increase in heart rate following an increase in venous return; due to mechanical and neural factors; also called the **Bainbridge reflex.**
atrioventricular (AV) node (ā-trē-ō-ven-TRIK-ū-lar): Specialized cardiocytes that relay the contractile stimulus to the bundle of His, the bundle branches, the Purkinje fibers, and the ventricular myocardium; located at the boundary between the atria and ventricles.
atrioventricular valve: One of the valves that prevent backflow into the atria during ventricular systole.
atrophy (AT-rō-fē): Wasting away of tissues from lack of use, ischemia, or nutritional abnormalities.
auditory: Pertaining to the sense of hearing.
auditory ossicles: The bones of the middle ear: *malleus, incus, and stapes.*
autoantibodies: Antibodies that react with antigens on the surfaces of one's own cells and tissues.
autodigestion: The digestion of tissues by digestive acids or enzymes from the stomach or pancreas.
autoimmunity: Immune system sensitivity to normal cells and tissues, resulting in the production of autoantibodies.
autolysis: Destruction of a cell due to the rupture of lysosomal membranes in its cytoplasm.
automatic bladder: Reflex micturition following stimulation of stretch receptors in the bladder wall; seen in patients who have lost motor control of the lower body.
automaticity: Spontaneous depolarization to threshold, a characteristic of cardiac pacemaker cells.
autonomic ganglion: A collection of visceral motor neurons outside the CNS.
autonomic nerve: A peripheral nerve consisting of preganglionic or postganglionic autonomic fibers.
autonomic nervous system (ANS): Centers, nuclei, tracts, ganglia, and nerves involved in the unconscious regulation of visceral functions; includes components of the CNS and PNS.
autopsy: Detailed examination of a body after death, usually performed by a pathologist.
autoregulation: Alterations in activity that maintain homeostasis in direct response to changes in the local environment; does not require neural or endocrine control.
autosomal (aw-to-SŌ-mal): Chromosomes other than the X or Y chromosomes.
avascular (ā-VAS-kū-lar): Without blood vessels.
avitaminosis (ā-vī-ta-min-Ō-sis): Condition caused by inadequate intake of one or more essential vitamins.
avulsion: An injury involving the violent tearing away of body tissues.
axilla: The armpit.
axolemma: The cell membrane of an axon, continuous with the cell membrane of the soma and dendrites and distinct from any glial cell coverings.
axon: Elongate extension of a neuron that conducts an action potential.
axon hillock: In a multipolar neuron, the portion of the neural soma adjacent to the initial segment.
axoplasm (AK-so-plazm): Cytoplasm within an axon.
azotemia (a-zo-TĒ-mē-a): Condition resulting from impaired kidney function and the retention of nitrogenous wastes, especially urea.
Babinski sign: Reflexive dorsiflexion of the toes following stroking of the plantar surface of the foot; positive reflex (Babinski sign) is normal up to age 1.5 yrs; thereafter a positive reflex indicates damage to descending tracts.
bacteria: Single-celled microorganisms, some pathogenic, that are common in the environment.
Bainbridge reflex: *See* **atrial reflex.**
baroreception: Ability to detect changes in pressure.
baroreceptor reflex: A reflexive change in cardiac activity in response to changes in blood pressure.
baroreceptors (bar-ō-rē-SEP-torz): Receptors responsible for baroreception.
basal metabolic rate: The resting metabolic rate of a normal fasting subject under homeostatic conditions.
base: A compound whose dissociation releases a hydroxide ion (OH^-) or removes a hydrogen ion from the solution.
basement membrane: A layer of filaments and fibers that attach an epithelium to the underlying connective tissue.
basilar membrane: Membrane that supports the organ of Corti and separates the cochlear duct from the scala tympani in the inner ear.
basophils (BĀ-sō-filz): Circulating granulocytes (WBCs) similar in size and function to tissue mast cells.
B cells: Lymphocytes capable of differentiating into the plasma cells that produce antibodies.
benign: Not malignant.
beta cells: Cells of the pancreatic islets that secrete insulin in response to elevated blood sugar concentrations.
beta oxidation: Fatty acid catabolism that produces molecules of acetyl-CoA.

beta receptors: Membrane receptors sensitive to epinephrine; stimulation may result in excitation or inhibition of the target cell.
bicarbonate ions: HCO_3^-; anion components of the carbonic acid–bicarbonate buffer system.
bicuspid (bī-KUS-pid): A sharp, conical tooth, also called a canine tooth.
bicuspid valve: The left AV valve, also known as the **mitral valve.**
bifurcate: To branch into two parts.
bile: Exocrine secretion of the liver that is stored in the gallbladder and ejected into the duodenum.
bile salts: Steroid derivatives in the bile, responsible for the emulsification of ingested lipids.
bilirubin (bil-ē-ROO-bin): A pigment, byproduct of hemoglobin catabolism.
bioenergetics: The analysis of energy production and utilization by living cells.
biofeedback: Using artificial signals to provide feedback about unconscious, visceral motor activities.
biopsy: The removal of a small sample of tissue for pathological analysis.
bipennate: A muscle whose fibers are arranged on either side of a common tendon.
bladder: A muscular sac that distends as fluid is stored, and whose contraction ejects the fluid at an appropriate time; used alone, the term usually refers to the urinary bladder.
blastocoele (BLAS-tō-sēl): Fluid-filled cavity within a blastocyst.
blastocyst (BLAS-tō-sist): Early stage in the developing embryo, consisting of an outer trophoblast and an inner cell mass.
blastodisc (BLAS-tō-disk): Later stage in the development of the inner cell mass; it includes the cells that will form the embryo.
blastomere (BLAS-tō-mēr): One of the cells in the morula, a collection of cells produced by the division of the zygote.
blockers/blocking agents: Drugs that block membrane pores or prevent binding to membrane receptors.
blood-brain barrier: Isolation of the CNS from the general circulation; primarily the result of astrocyte regulation of capillary permeabilities.
blood clot: A network of fibrin fibers and trapped blood cells.
blood pressure: A force exerted against the vascular walls by the blood, as the result of the push exerted by cardiac contraction and the elasticity of the vessel walls. It is usually measured along one of the muscular arteries, with systolic pressure measured during ventricular systole, and diastolic pressure during ventricular diastole.
blood-testis barrier: Isolation of the seminiferous tubules from the general circulation, due to the activities of the sustentacular (Sertoli) cells.
Bohr effect: Increased oxygen release by hemoglobin in the presence of elevated carbon dioxide levels.
boil: An abscess of the skin, usually involving a sebaceous gland.
bolus: A compact mass; usually refers to compacted ingested material on its way to the stomach.
bone: *See* **osseous tissue**.
botulinus toxin (bot-ū-LĪ-nus): A toxin produced by the anaerobic bacterium *Clostridium botulinum* that can cause severe food poisoning.
bowel: The intestinal tract.
Boyle's law: The principle that in a gas, pressure and volume are inversely related.
brachial: Pertaining to the arm.
brachial plexus: Network formed by branches of spinal nerves C_5-T_1 en route to innervate the upper limb.
brachium: The arm.
bradycardia (brad-ē-KAR-dē-a): Slow heart rate, below 50 bpm.
brain stem: The brain minus the cerebrum, diencephalon, and cerebellum.
brevis: Short.
Broca's center: The speech center of the brain, usually found on the neural cortex of the left cerebral hemisphere.
bronchial tree: The trachea, bronchi, and bronchioles.
bronchitis (brong-KĪ-tis): Inflammation of the bronchial passageways.
bronchodilation: Dilation of the bronchial passages; may be caused by sympathetic stimulation.
bronchodilators (brong-kō-dī-LĀ-torz): Drugs that produce bronchodilation; some are used clinically in treating asthma.
bronchoscope: A fiberoptic instrument used to examine the bronchial passageways.
bronchus/bronchi: One of the branches of the bronchial tree between the trachea and bronchioles.
buccal (BUK-al): Pertaining to the cheeks.
buffer: A compound that stabilizes the pH of a solution by removing or releasing hydrogen ions.
buffer system: Interacting compounds that prevent increases or decreases in the pH of body fluids; includes the carbonic acid–bicarbonate buffer system, the phosphate buffer system, and the protein buffer system.
bulbar: Pertaining to the brain stem.
bulbourethral glands (bul-bō-ū-RĒ-thral): Mucous glands at the base of the penis that secrete into the penile urethra; also called *Cowper's glands.*
bundle branches: Specialized conducting cells in the ventricles that carry the contractile stimulus from the bundle of His to the Purkinje fibers.
bundle of His (hiss): Specialized conducting cells in the interventricular septum that carry the contracting stimulus from the AV node to the bundle branches and thence to the Purkinje fibers.
bursa: A small sac filled with synovial fluid that cushions adjacent structures and reduces friction.
bursectomy: The surgical removal of an inflamed bursa.
bursitis: Painful inflammation of one or more bursae.
Caesarian section: Surgical delivery of an infant via an incision through the lower abdominal wall and uterus.
calcaneal tendon: Large tendon that inserts on the calcaneus; tension on this tendon produces plantar flexion of the foot; also called the *Achilles tendon.*
calcaneus (kal-KĀ-nē-us): The heelbone, the largest of the tarsal bones.
calcification: The deposition of calcium salts within a tissue.
calcitonin (kal-si-TŌ-nin): Hormone secreted by C cells of the thyroid when calcium ion concentrations are abnormally high; restores homeostasis by increasing the rate of bone deposition and the renal rate of calcium loss.
calculus/calculi (KAL-kū-lī): Concretions of insoluble materials that form within body fluids, especially the gallbladder, kidneys, or urinary bladder.
callus: A localized thickening of the epidermis due to chronic mechanical stresses; a thickened area that forms at the site of a bone break as part of the repair process.

calorie (c) (KAL-o-rē): The amount of heat required to raise the temperature of one gram of water 1°C.
Calorie (C): The amount of heat required to raise the temperature of one kilogram of water 1°C.
calorigenic effect: The stimulation of energy production and heat loss by thyroid hormones.
calvaria (kal-VAR-ē-a): The skullcap, formed of the frontal, parietal, and occipital bones.
calyx/calyces (KĀL-i-sēz): A cup-shaped division of the renal pelvis.
canaliculi (kan-a-LIK-ū-lī): Microscopic passageways between cells; bile canaliculi carry bile to bile ducts in the liver; in bone, canaliculi permit the diffusion of nutrients and wastes to and from osteocytes.
cancellous bone (KAN-sel-us): Spongy bone, composed of a network of bony struts.
cancer: A malignant tumor that tends to undergo metastasis.
cannula: A tube that can be inserted into the body; often placed in blood vessels prior to transfusion or dialysis.
canthus, medial and lateral (KAN-thus): The angles formed at either corner of the eye between the upper and lower eyelids.
capacitation (ka-pas-i-TĀ-shun): Activation process that must occur before a spermatozoon can successfully fertilize an oocyte; occurs in the vagina following ejaculation.
capillary: Small blood vessels, interposed between arterioles and venules, whose thin walls permit the diffusion of gases, nutrients, and wastes between the plasma and interstitial fluids.
capitulum (ka-PIT-ū-lum): General term for a small, elevated articular process; used to refer to the rounded distal surface of the humerus that articulates with the radial head.
caput: The head.
carbaminohemoglobin (kar-bam-ē-nō-hē-mō-GLŌ-bin): Hemoglobin bound to carbon dioxide molecules.
carbohydrase (kar-bō-HĪ-drāz): An enzyme that breaks down carbohydrate molecules.
carbohydrate (kar-bō-HĪ-drāt): Organic compound containing carbon, hydrogen, and oxygen in a ratio that approximates 1:2:1.
carbon dioxide: CO_2, a compound produced by the decarboxylation reactions of aerobic metabolism.
carbonic anhydrase: An enzyme that catalyzes the reaction $H_2O + CO_2 \rightleftharpoons H_2CO_3$; important in carbon dioxide transport, gastric acid secretion, and renal pH regulation.
carboxyl group (kar-BOKS-il): -COOH, an acid group found in fatty acids, amino acids, etc.
carboxypeptidase (kar-bok-sē-PEP-ti-dāz): A protease that breaks down proteins and releases amino acids.
carcinogenic (kar-SIN-ō-jen-ik): Stimulates cancer formation in affected tissues.
cardia (KAR-dē-a): The area of the stomach surrounding its connection with the esophagus.
cardiac: Pertaining to the heart.
cardiac cycle: One complete heartbeat, including atrial and ventricular systole and diastole.
cardiac glands: Mucous glands characteristic of the cardia of the stomach.
cardiac output: The amount of blood ejected by the left ventricle each minute; normally about 5 liters.
cardiac reserve: The potential percentage increase in cardiac output above resting levels.
cardiac tamponade: Compression of the heart due to fluid accumulation in the pericardial cavity.
cardiocyte (KAR-dē-ō-sīt): A cardiac muscle cell.
cardiomyopathy (kar-dē-ō-mī-OP-a-thē): A progressive disease characterized by damage to the cardiac muscle tissue.
cardiopulmonary resuscitation: Method of artificially maintaining respiratory and circulatory function.
cardiovascular: Pertaining to the heart, blood, and blood vessels.
cardiovascular centers: Poorly localized centers in the reticular formation of the medulla of the brain; includes cardioacceleratory, cardioinhibitory, and vasomotor centers.
cardium: The heart.
carina (ka-RĪ-na): A ridge on the inner surface of the base of the trachea that runs anteroposteriorly, between two primary bronchi.
carotene (KAR-ō-tēn): A yellow-orange pigment found in carrots and in green and orange leafy vegetables; a compound that the body can convert to vitamin A.
carotid artery: The principal artery of the neck, servicing cervical and cranial structures; one branch, the internal carotid, represents a major blood supply for the brain.
carotid body: A group of receptors adjacent to the carotid sinus that are sensitive to changes in the carbon dioxide levels, pH, and oxygen concentrations of the arterial blood.
carotid sinus: A dilated segment at the base of the internal carotid artery whose walls contain baroreceptors sensitive to changes in blood pressure.
carotid sinus reflex: Reflexive changes in blood pressure that maintain homeostatic pressures at the carotid sinus, stabilizing blood flow to the brain.
carpus/carpal: The wrist.
cartilage: A connective tissue with a gelatinous matrix containing an abundance of fibers.
castration: Removal of the testes, also called *bilateral orchiectomy*.
catabolism (ka-TAB-ō-lizm): The breakdown of complex organic molecules into simpler components, accompanied by the release of energy.
catalyst (KAT-ah-list): A substance that accelerates a specific chemical reaction, but that is not altered by the reaction.
cataract: A reduction in lens transparency that causes visual impairment.
catecholamine (kat-e-KŌL-am-in): Epinephrine, norepinephrine, dopamine, and related compounds.
cathepsins (ka-THEP-sinz): Enzymes present in the sarcoplasm of skeletal muscle cells that can break down contractile proteins, providing amino acids that can act as a supplemental energy source.
catheter (KATH-e-ter): Surgical instrument, a tube inserted into a body cavity or along a blood vessel or excretory passageway for the collection of body fluids, blood pressure monitoring, or the introduction of medications or radiographic dyes.
cation (KAT-ī-on): An ion that bears a positive charge.
cauda equina (KAW-da ek-WĪ-na): Spinal nerve roots distal to the tip of the adult spinal cord; they extend caudally inside the vertebral canal en route to lumbar and sacral segments.
caudal/caudally: Closest to or toward the tail (coccyx).
caudate nucleus (KAW-dāt): One of the cerebral nuclei of the extrapyramidal system, involved with the unconscious control of muscular activity.
cavernous tissue: Erectile tissue that can be engorged with blood; found in the penis and clitoris.

cecum (SĒ-kum): An expanded pouch at the start of the large intestine.
cell: The smallest living unit in the human body.
cell-mediated immunity: Resistance to disease through the activities of sensitized T cells that destroy antigen-bearing cells by direct contact or through the release of lymphotoxins; also called *cellular immunity.*
cellulitis (sel-ū-LĪ-tis): Diffuse inflammation, usually involving areas of loose connective tissue, such as the subcutaneous layer.
cementum (se-MEN-tum): Bony material covering the root of a tooth, not shielded by a layer of enamel.
center of ossification: Site in a connective tissue where bone formation begins.
central canal: Longitudinal canal in the center of an osteon that contains blood vessels and nerves, also called the *Haversian canal;* a passageway along the longitudinal axis of the spinal cord that contains cerebrospinal fluid.
central nervous system (CNS): The brain and spinal cord.
central sulcus: Groove in the surface of a cerebral hemisphere, between the primary sensory and primary motor areas of the cortex.
centriole: A cylindrical intracellular organelle composed of 9 groups of microtubules, 3 in each group; functions in mitosis or meiosis by organizing the microtubules of the spindle apparatus.
centromere (SEN-trō-mēr): Localized region where two chromatids remain connected following chromosome replication; site of spindle fiber attachment.
centrosome: Region of cytoplasm containing a pair of centrioles oriented at right angles to one another.
centrum: The vertebral body.
cephalic: Pertaining to the head.
cerebellum (ser-e-BEL-um): Posterior portion of the metencephalon, containing the cerebellar hemispheres; includes the arbor vitae, cerebellar nuclei, and cerebellar cortex.
cerebral cortex: An extensive area of neural cortex covering the surfaces of the cerebral hemispheres.
cerebral hemispheres: Expanded portions of the cerebrum covered in neural cortex.
cerebral nuclei: Nuclei of the cerebrum that are important components of the extrapyramidal system.
cerebral palsy: Chronic condition resulting from damage to motor areas of the brain during development or at delivery.
cerebral peduncle: Mass of nerve fibers on the ventrolateral surface of the mesencephalon; contains ascending tracts that terminate in the thalamus and descending tracts that originate in the cerebral hemispheres.
cerebrospinal fluid: Fluid bathing the internal and external surfaces of the CNS; secreted by the choroid plexus.
cerebrovascular accident (CVA): A stroke; occlusion of a blood vessel supplying a portion of the brain, resulting in damage to the dependent neurons.
cerebrum (SER-ē-brum): The largest portion of the brain, composed of the cerebral hemispheres; includes the cerebral cortex, the cerebral nuclei, and the internal capsule.
cerumen: Waxy secretion of integumentary glands along the external auditory canal.
ceruminous glands (se-RŪ-mi-nus): Integumentary glands that secrete cerumen.
cervical enlargement: Relative enlargement of the cervical portion of the spinal cord due to the abundance of CNS neurons involved with motor control of the upper limbs.
cervix: The lower part of the uterus.
chalazion (kah-LA-zē-on): An inflammation and distension of a Meibomian gland on the eyelid; also called a *sty.*
chancre (SHANG-ker): A skin lesion that develops at the primary site of a syphilis infection.
charleyhorse: Soreness and stiffness in a strained muscle, usually involving the quadriceps group.
chemoreception: Detection of alterations in the concentrations of dissolved compounds or gases.
chemotaxis (kē-mō-TAK-sis): The attraction of phagocytic cells to the source of abnormal chemicals in tissue fluids.
chemotherapy: Treatment of illness through the administration of specific chemicals.
chloride shift: Movement of plasma chloride ions into RBCs in exchange for bicarbonate ions generated by the intracellular dissociation of carbonic acid.
cholecystitis (kō-lē-sis-TĪ-tis): Inflammation of the gallbladder.
cholecystokinin (CCK) (kō-lē-sis-tō-KĪ-nin): Duodenal hormone that stimulates the contraction of the gallbladder and the secretion of enzymes by the exocrine pancreas; also called *pancreozymin.*
cholelithiasis (kō-lē-li-THI-a-sis): The formation or presence of gallstones.
cholesterol: A steroid component of cell membranes and a substrate for the synthesis of steroid hormones and bile salts.
choline: Chemical compound, a breakdown product or precursor of acetylcholine.
cholinergic synapse (kō-lin-ER-jik): Synapse where the presynaptic membrane releases ACh on stimulation.
cholinesterase (kō-li-NES-te-rās): Enzyme that breaks down and inactivates ACh.
chondrocyte (KON-drō-sīt): Cartilage cell.
chondroitin sulfate (kon-DROI-tin): The predominant proteoglycan in cartilage, responsible for the gelatinous consistency of the matrix.
chordae tendineae (KOR-dē TEN-di-nē-ē): Fibrous cords stabilize the position of the AV valves in the heart, preventing back-flow during ventricular systole.
chorion/chorionic (KOR-ē-on) (ko-rē-ON-ik): An extraembryonic membrane, consisting of the trophoblast and underlying mesoderm, that forms the placenta.
choroid: Middle, vascular layer in the wall of the eye.
choroid plexus: The vascular complex in the roof of the third and fourth ventricles of the brain, responsible for CSF production.
chromatid (KRŌ-ma-tid): One complete copy of a DNA strand.
chromatin (KRŌ-ma-tin): Histological term referring to the grainy material visible in cell nuclei during interphase; the appearance of the DNA content of the nucleus when the chromosomes are uncoiled.
chromosomes: Dense structures, composed of tightly coiled DNA strands and associated histones, that become visible in the nucleus when a cell prepares to undergo mitosis or meiosis; normal human somatic cells contain 46 chromosomes apiece.
chronic: Habitual or long term.
chylomicrons (kī-lō-MĪ-kronz): Relatively large droplets that may contain triglycerides, phospholipids, and cholesterol in association with proteins; synthesized and released by intestinal cells and transported to the venous blood via the lymphatic system.
chyme (kīm): A semifluid mixture of ingested food and digestive secretions that is found in the stomach and small intestine as digestion proceeds.

chymotrypsin (kī-mō-TRIP-sin): A protease found in the small intestine.

chymotrypsinogen: Inactive proenzyme secreted by the pancreas that is subsequently converted to chymotrypsin.

ciliary body: A thickened region of the choroid that encircles the lens of the eye; it includes the ciliary muscle and the ciliary processes that support the suspensory ligaments of the lens.

cilium/cilia: A slender organelle that extends above the free surface of an epithelial cell, and usually undergoes cycles of movement; composed of a basal body and microtubules in a 9×2 array.

circulatory system: The network of blood vessels that are components of the cardiovascular system.

circumduction (sir-kum-DUK-shun): A movement at a synovial joint where the distal end of the bone describes a circle, but the shaft does not rotate.

circumvallate papilla (sir-kum-VAL-āt pa-PIL-la): One of the large, dome-shaped papillae on the dorsum of the tongue that form the V that separates the body of the tongue from the root.

cirrhosis (sir-RŌ-sis): A liver disorder characterized by the degeneration of hepatocytes and their replacement by fibrous connective tissue.

cisterna (sis-TUR-na): An expanded chamber.

citric acid cycle: *See* **TCA cycle.**

cleavage (KLĒ-vij): Mitotic divisions that follow fertilization of the ovum and lead to the formation of a blastocyst.

cleavage lines: Stress lines in the skin that follow the orientation of major bundles of collagen fibers in the dermis.

climacteric: Age-related cessation of gametogenesis in the male or female due to reduced sex hormone production.

clitoris (KLI-to-ris): A small erectile organ of the female that is the developmental equivalent of the male penis.

clone: The production of genetically identical cells.

clonus (KLŌ-nus): Rapid cycles of muscular contraction and relaxation.

clot: A network of fibrin fibers and trapped blood cells; also called a *thrombus* if it occurs within the circulatory system.

clotting factors: Plasma proteins synthesized by the liver that are essential to the clotting response.

clotting response: Series of events that result in the formation of a clot.

coccygeal ligament: Fibrous extension of the dura mater and filum terminale; provides longitudinal stabilization to the spinal cord.

coccyx (KOK-siks): Terminal portion of the spinal column, consisting of relatively tiny, fused vertebrae.

cochlea (KŌK-lē-a): Spiral portion of the bony labyrinth of the inner ear that surrounds the organ of hearing.

cochlear duct (KOK-lē-ar): Membranous tube within the cochlea that is filled with endolymph and contains the organ of Corti; also called the *scala media.*

codon (KŌ-don): A sequence of three nitrogenous bases along an mRNA strand that will specify the location of a single amino acid in a peptide chain.

coelom (SĒ-lom): The ventral body cavity, lined by a serous membrane and subdivided during development into the pleural, pericardial, and abdominopelvic (peritoneal) cavities.

coenzymes (kō-EN-zīmz): Complex organic cofactors, usually structurally related to vitamins.

cofactor: Ions or molecules that must be attached to the active site before an enzyme can function; examples include mineral ions and several vitamins.

colectomy (ko-LEK-tō-mē): Surgical removal of part or all of the colon.

colitis: Inflammation of the colon.

collagen: Strong, insoluble protein fiber common in connective tissues.

collateral ganglion (kō-LAT-er-al): A sympathetic ganglion situated in front of the spinal column and separate from the sympathetic chain.

Colles' fracture (KOL-lēz): Fracture of the distal end of the radius and possibly the ulna, with posterior and dorsal displacement of the distal bone fragments.

colliculus/colliculi (kol-IK-ū-lus): A little mound; in the brain, used to refer to one of the thickenings in the roof of the mesencephalon; the superior colliculus is associated with the visual system, and the inferior colliculi with the auditory system.

colloid/colloidal suspension: A solution containing large organic molecules in suspension.

colon: The large intestine.

colonoscope (kō-LON-ō-skōp): A fiberoptic device for examining the interior of the colon.

colostomy (kō-LOS-to-mē): The surgical connection of a portion of the colon to the body wall, sometimes performed after a colectomy to permit the discharge of fecal materials.

colostrum (ko-LOS-trum): Secretion of the mammary glands at the time of childbirth and for a few days thereafter; contains more protein and less fat than the milk secreted later.

coma (kō-ma): An unconscious state from which the individual cannot be aroused, even by strong stimuli.

comedo (kō-MĒ-dō): An inflamed sebaceous gland.

comminuted: Broken or crushed into small pieces.

commissure: A crossing over from one side to another.

common bile duct: Duct formed by the union of the cystic duct from the gallbladder and the bile ducts from the liver; terminates at the duodenal ampulla, where it meets the pancreatic duct.

common pathway: In the clotting response, the events that begin with the appearance of thromboplastin and end with the formation of a clot.

compact bone: Dense bone containing parallel osteons.

compensation curves: The cervical and lumbar curves that develop to center the body weight over the legs.

complement: 11 plasma proteins that interact in a chain reaction following exposure to activated antibodies or the surfaces of certain pathogens, and which promote cell lysis, phagocytosis, and other defense mechanisms.

compliance: Distensibility; the ability of certain organs to tolerate changes in volume; a property that reflects the presence of elastic fibers and smooth muscles.

compound: A molecule containing two or more elements in combination.

concentration: Amount (in grams) or number of atoms, ions, or molecules (in moles) per unit volume.

concentration gradient: Regional differences in the concentration of a particular substance.

conception: Fertilization.

concha/conchae (KONG-kē): Three pairs of thin, scroll-like bones that project into the nasal cavities; the superior and medial conchae are part of the ethmoid, and the inferior are separate bones.

concussion: A violent blow or shock; loss of consciousness due to a violent blow to the head.

conducted change: An action potential; a change in the transmembrane potential that spreads across an excitable membrane.

condyle: A rounded articular projection on the surface of a bone.
cone: Retinal photoreceptor responsible for color vision.
congenital (kon-JEN-i-tal): Already present at the birth of an individual.
congestive heart failure (CHF): Failure to maintain adequate cardiac output due to circulatory problems or myocardial damage.
conjunctiva (kon-junk-TĪ-va): A layer of stratified squamous epithelium that covers the inner surfaces of the lids and the anterior surface of the eye to the edges of the cornea.
conjunctivitis: Inflammation of the conjunctiva.
connective tissue: One of the four primary tissue types; provides a structural framework for the body that stabilizes the relative positions of the other tissue types; includes connective tissue proper, cartilage, bone, and blood; always has cell products, cells, and ground substance.
contractility: The ability to contract, possessed by skeletal, smooth, and cardiac muscle cells.
contracture: A permanent contraction of an entire muscle following the atrophy of individual muscle cells.
contralateral reflex: A reflex that affects the opposite side of the body from the stimulus.
conus medullaris: Conical tip of the spinal cord that gives rise to the filum terminale.
convergence: In the nervous system, the term indicates that the axons from several neurons innervate a single neuron; this is most common along motor pathways.
coracoid process (ko-RA-kōyd): A hook-shaped process of the scapula that projects above the anterior surface of the capsule of the shoulder joint.
Cori cycle: Metabolic exchange of lactic acid from skeletal muscle for glucose from the liver; performed during the recovery period following muscular exertion.
cornea (KOR-nē-a): Transparent portion of the fibrous tunic of the anterior surface of the eye.
corniculate cartilages (kor-NIK-ū-lāt): A pair of small laryngeal cartilages.
cornification: The production of keratin by a stratified squamous epithelium; also called *keratinization.*
cornu: Shaped like a horn.
corona radiata (ko-RŌ-na rā-dē-A-ta): A layer of follicle cells surrounding a secondary oocyte at ovulation.
coronoid (kō-RŌ-nōyd): Hooked or curved.
corpora quadrigemina (KOR-po-ra quad-ri-JEM-i-na): The superior and inferior colliculi of the mesencephalic tectum (roof) in the brain.
corpus/corpora: Body.
corpus callosum: Bundle of axons linking centers in the left and right cerebral hemispheres.
corpus cavernosum (KOR-po-ra ka-ver-NŌ-sum): Masses of erectile tissue within the body of the penis (male) or clitoris (female).
corpus luteum (LOO-tē-um): Progestin-secreting mass of follicle cells that develops in the ovary after ovulation.
corpus spongiosum (spon-JĒ-ō-sum): Mass of erectile tissue that surrounds the urethra in the male penis, and expands distally to form the glans.
cortex: Outer layer or portion of an organ.
Corti, organ of: Receptor complex in the scala media of the cochlea that includes the inner and outer hair cells, supporting cells and structures, and the tectorial membrane; provides the sensation of hearing.
corticobulbar tracts (kor-ti-kō-BUL-bar): Descending tracts that carry information/commands from the cerebral cortex to nuclei and centers in the brain stem.
corticospinal tracts: Descending tracts that carry motor commands from the cerebral cortex to the anterior gray horns of the spinal cord.
corticosteroid: A steroid hormone produced by the adrenal cortex.
corticosterone (kor-ti-KOS-te-rōn): One of the corticosteroids secreted by the zona fasciculata of the adrenal cortex; a glucocorticoid.
corticotropin: *See* **adrenocorticotropic hormone (ACTH).**
corticotropin-releasing hormone (CRH): Releasing hormone secreted by the hypothalamus that stimulates secretion of ACTH by the anterior pituitary.
cortisol (KOR-ti-sōl): One of the corticosteroids secreted by the zona fasciculata of the adrenal cortex; a glucocorticoid.
costa/costae: A rib.
cotransport: Membrane transport of a nutrient, such as glucose, in company with the movement of an ion, usually sodium; transport requires a carrier protein but does not involve direct ATP expenditure and can occur regardless of the concentration gradient for the nutrient.
countercurrent exchange: Diffusion between two solutions traveling in opposite directions.
countercurrent multiplication: Active transport between two limbs of a loop that contains a fluid moving in one direction; responsible for the concentration of the urine in the kidney tubules.
covalent bond (kō-VĀ-lent): A chemical bond between atoms that involves the sharing of electrons.
coxa/coxae: The bones of the hip.
cranial: Pertaining to the head.
cranial nerves: Peripheral nerves originating at the brain.
craniosacral division (krā-nē-ō-SAK-ral): *See parasympathetic division.*
craniostenosis (krā-nē-ō-sten-Ō-sis): Skull deformity caused by premature closure of the cranial sutures.
cranium: The brainbox; the skull bones that surround the brain.
creatine: A nitrogenous compound synthesized in the body that can bind a high-energy phosphate and serve as an energy reserve.
creatine phosphate: A high-energy compound present in muscle cells; during muscular activity the phosphate group is donated to ADP, regenerating ATP. Also called *phosphocreatine* and *phosphorylcreatine.*
creatinine: A breakdown product of creatine metabolism.
crenation: Cellular shrinkage due to an osmotic movement of water out of the cytoplasm.
cribriform plate: Portion of the ethmoid bone of the skull that contains the foramina used by the axons of olfactory receptors en route to the olfactory bulbs of the cerebrum.
cricoid cartilage (KRĪ-koyd): Ring-shaped cartilage forming the inferior margin of the larynx.
crista/cristae: A ridge-shaped collection of hair cells in the ampulla of a semicircular canal; the crista and cupula form a receptor complex sensitive to movement along the plane of the canal.
cross-bridge: Myosin head that projects from the surface of a thick filament, and that can bind to an active site of a thin filament in the presence of calcium ions.
cruciate ligaments: A pair of intracapsular ligaments (anterior and posterior) in the knee.
cryosurgery: A surgical technique that involves freezing and killing cells in a localized area.
cryptorchid testis: An undescended testis that is in the abdominopelvic cavity rather than the scrotum.

cuneiform cartilages (kū-NĒ-i-form): A pair of small cartilages in the larynx.
cupula (KYŪ-pū-la): A gelatinous mass that sits in the ampulla of a semicircular canal in the inner ear, and whose movement stimulates the hair cells of the crista.
curare: A toxin that prevents neural stimulation of neuromuscular junctions.
Cushing's disease: Condition caused by oversecretion of adrenal steroids.
cutaneous membrane: The epidermis and papillary layer of the dermis.
cuticle: Layer of dead, cornified cells surrounding the shaft of a hair; for nails, *see eponychium.*
cutis: The skin.
cyanosis: Bluish coloration of the skin due to the presence of deoxygenated blood in vessels near the body surface.
cyst: A fibrous capsule containing fluid or other material.
cystic duct: A duct that carries bile between the gallbladder and the common bile duct.
cystitis: Inflammation of the urinary bladder.
cytochrome (Sī-tō-krōm): A pigment component of the electron transport system; a structural relative of heme.
cytokinesis (sī-tō-ki-NĒ-sis): The cytoplasmic movement that separates two daughter cells at the completion of mitosis.
cytology (sĪ-TOL-ō-jē): The study of cells.
cytoplasm: The material between the cell membrane and the nuclear membrane.
cytosine: One of the nitrogenous base components of nucleic acids.
cytoskeleton: A network of microtubules and microfilaments in the cytoplasm.
cytosol: The fluid portion of the cytoplasm.
cytotoxic: Poisonous to living cells.
cytotoxic T cells: Lymphocytes of the cellular immune response that kill target cells by direct contact or through the secretion of lymphotoxins; also called *killer T cells* and T_C *cells.*
daughter cells: Genetically identical cells produced by somatic cell division.
deamination (dē-am-i-NĀ-shun): The removal of an amino group from an amino acid.
decarboxylation (dē-kar-boks-i-LĀ-shun): The removal of a molecule of CO_2.
decerebrate: Lacking a cerebrum.
decomposition reaction: A chemical reaction that breaks a molecule into smaller fragments.
decubitis ulcers: Ulcers that form where chronic pressure interrupts circulation to a portion of the skin.
decussate: To cross over to the opposite side, usually referring to the crossover of the pyramidal tracts on the ventral surface of the medulla oblongata.
defecation (def-e-KĀ-shun): The elimination of fecal wastes.
deglutition (deg-loo-TISH-un): Swallowing.
degradation: Breakdown, catabolism.
dehydration: A reduction in the water content of the body that threatens homeostasis.
dehydration synthesis: The joining of two molecules associated with the removal of a water molecule.
delta cell: A pancreatic islet cell that secretes somatostatin.
dementia: Loss of mental abilities.
demyelination: The loss of the myelin sheath of an axon, usually due to chemical or physical damage to Schwann cells or oligodendrocytes.
denaturation: Irreversible alteration in the three-dimensional structure of a protein.
dendrite (DEN-drĪt): A sensory process of a neuron.
denticulate ligaments: Supporting fibers that extend laterally from the surface of the spinal cord, tying the pia mater to the dura mater and providing lateral support for the spinal cord.
dentin (DEN-tin): Bonelike material that forms the body of a tooth; it differs from bone in lacking osteocytes and osteons.
deoxyribonucleic acid (dē-ok-sē-rĪ-bo-nū-KLĀ-ik): DNA strand: a nucleic acid consisting of a chain of nucleotides containing the sugar deoxyribose and the nitrogen bases adenine, guanine, cytosine, and thymine; **DNA molecule:** two DNA strands wound in a double helix and held together by weak bonds between complementary nitrogen base pairs.
deoxyribose: A 5-carbon sugar resembling ribose but lacking an oxygen atom.
depolarization: A change in the transmembrane potential that moves it from a negative value toward 0 mV.
depression: Inferior (downward) movement of a body part.
dermatitis: Inflammation of the skin.
dermatome: A sensory region monitored by the dorsal rami of a single spinal segment.
dermis: The connective tissue layer beneath the epidermis of the skin.
detrusor muscle (de-TROO-sor): Smooth muscle in the wall of the urinary bladder.
detumescence (dē-tū-MES-ens): Loss of a penile erection in the male.
development: Growth and the acquisition of increasing structural and functional complexity; includes the period from conception to maturity.
diabetes insipidus: Polyuria due to inadequate production of ADH.
diabetes mellitus (mel-LI-tus): Polyuria and glycosuria, most often due to inadequate production of insulin with resulting elevation of blood glucose levels.
diabetogenic effect: Elevation in blood sugar concentrations following the secretion of growth hormone or glucagon.
dialysis: Diffusion between two solutions of differing solute concentrations across a semipermeable membrane containing pores that permit the passage of some solutes and not others.
diapedesis (dī-a-pe-DĒ-sis): Movement of white blood cells through the walls of blood vessels by migration between adjacent endothelial cells.
diaphragm (Dī-a-fram): Any muscular partition; often used to refer to the respiratory muscle that separates the thoracic from the abdominopelvic cavities.
diaphysis (dī-A-fi-sis): The shaft of a long bone.
diarrhea (dī-a-RĒ-a): Abnormally frequent defecation, associated with the production of unusually fluid feces.
diarthrosis (dī-ar-THRO-sis): A synovial joint.
diastolic pressure: Pressure measured in the walls of a muscular artery when the left ventricle is in diastole.
diencephalon (dī-en-SEF-a-lon): A division of the brain that includes the epithalamus, thalamus, and hypothalamus.
differential count: The determination of the relative abundance of each type of white blood cell, based on a random sampling of 100 WBCs.
differentiation: The gradual appearance of characteristic cellular specializations during development, as the result of gene activation or repression.

diffusion: Passive molecular movement from an area of relatively high concentration to an area of relatively low concentration.

digestion: The chemical breakdown of ingested materials into simple molecules that can be absorbed by the cells of the digestive tract.

digestive system: The digestive tract and associated glands.

digestive tract: An internal passageway that begins at the mouth and ends at the anus.

dilate: To increase in diameter; to enlarge or expand.

diploid (DIP-lōyd): Having a complete somatic complement of chromosomes (23 pairs in human cells).

disaccharide (di-SAK-ah-rīd): A compound formed by the joining of two simple sugars by dehydration synthesis.

dislocation: Forceful displacement of an articulating bone to an abnormal position, usually accompanied by damage to tendons, ligaments, the articular capsule, or other structures.

dissociation (di-sō-sē-Ā-shun): *See* **ionization**.

distal: Movement away from the point of attachment or origin; for a limb, away from its attachment to the trunk.

distal convoluted tubule: Portion of the nephron closest to the collecting tubule and duct; an important site of active secretion.

diuresis: Fluid loss at the kidneys; the production of urine.

divergence: In neural tissue, the spread of excitation from one neuron to many neurons; an organizational pattern common along sensory pathways of the CNS.

diverticulitis (dī-ver-tik-ū-LĪ-tis): Inflammation of a diverticulum.

diverticulosis (dī-ver-tik-ū-LŌ-sis): The formation of diverticula.

diverticulum: A sac or pouch in the wall of the colon or other organ.

dizygotic twins (dī-zī-GOT-ik): Twins that result from the fertilization of two different ova.

dominant gene: A gene whose presence will determine the phenotype, regardless of the nature of its allelic companion.

dopamine (DŌ-pa-mēn): An important neurotransmitter in the CNS.

dorsal: Toward the back, posterior.

dorsal root ganglion: PNS ganglion containing the cell bodies of sensory neurons.

dorsiflexion: Elevation of the superior surface of the foot.

Down syndrome: A genetic abnormality resulting from the presence of three copies of chromosome 21; individuals with this condition have characteristic physical and intellectual deficits.

duct: A passageway that delivers exocrine secretions to an epithelial surface.

ductus arteriosus (ar-te-rē-Ō-sus): Vascular connection between the pulmonary trunk and the aorta that functions throughout fetal life; normally closes at birth or shortly thereafter, and persists as the ligamentum arteriosum.

ductus deferens (DUK-tus DEF-e-renz): A passageway that carries sperm from the epididymis to the ejaculatory duct.

duodenal ampulla: Chamber that receives bile from the common bile duct and pancreatic secretions from the pancreatic duct.

duodenal glands: *See* **submucosal glands.**

duodenal papilla: Conical projection from the inner surface of the duodenum that contains the opening of the duodenal ampulla.

duodenum (dū-A-dē-num): The proximal 25 cm of the small intestine that contains short villi and submucosal glands.

dura mater (DŪ-ra MĀ-ter): Outermost component of the meninges that surround the brain and spinal cord.

dynamic equilibrium: Maintenance of normal body orientation as sudden changes in position (rotation, acceleration, etc.) occur.

dynorphin (dī-NOR-fin): A powerful neuromodulator produced in the CNS that blocks pain perception by inhibiting pain pathways.

dyslexia: Impaired ability to comprehend written words.

dysmenorrhea: Painful menstruation.

dysmetria (dis-MET-rē-a): Difficulty in performing movements due to problems with the interpretation and anticipation of the distance to be covered.

dysuria (dis-Ū-rē-a): Painful urination.

eccrine glands (EK-rin): Sweat glands of the skin that produce a watery secretion.

echocardiography (ek-ō-kar-dē-OG-ra-fē): Examination of the heart using modified ultrasound techniques.

ectoderm: One of the three primary germ layers; covers the surface of the embryo and gives rise to the nervous system, the epidermis and associated glands, and a variety of other structures.

ectopic (ek-TOP-ik): Outside of its normal location.

effector: A peripheral gland or muscle cell innervated by a motor neuron.

efferent: Away from.

efferent arteriole: An arteriole carrying blood away from a glomerulus of the kidney.

efferent fiber: An axon that carries impulses away from the CNS.

ejaculation (e-jak-ū-LĀ-shun): The ejection of semen from the penis as the result of muscular contractions of the bulbocavernosus and ischiocavernosus muscles.

ejaculatory duct (e-JAK-ū-la-to-rē): Short ducts that pass within the walls of the prostate and connect the ductus deferens with the prostatic urethra.

elastase (ē-LAS-tāz): A pancreatic enzyme that breaks down elastin fibers.

elastin: Connective tissue fibers that stretch and rebound, providing elasticity to connective tissues.

electrical coupling: A connection between adjacent cells that permits the movement of ions and the transfer of graded or conducted changes in the transmembrane potential from cell to cell.

electrocardiogram (ECG, EKG) (e-lek-trō-KAR-dē-ō-gram): Graphic record of the electrical activities of the heart, as monitored at specific locations on the body surface.

electroencephalogram (EEG): Graphic record of the electrical activities of the brain.

electrolytes (ē-LEK-trō-līts): Soluble inorganic compounds whose ions will conduct an electric current in solution.

electron: One of the three fundamental particles; a subatomic particle that bears a negative charge and normally orbits around the protons of the nucleus.

electron transport system: Cytochrome system responsible for most of the energy production in living cells; a complex bound to the inner mitochondrial membrane.

eleidin (el-Ē-i-din): A protein that forms as a precursor of keratin.

element: All of the atoms with the same atomic number.

elephantiasis (el-e-fan-TĪ-a-sis): A lymphedema caused by infection and blockage of lymphatics by mosquito-borne parasites.

elevation: Movement in a superior, or upward, direction.
embolism (EM-bō-lizm): Obstruction or closure of a vessel by an embolus.
embolus (EM-bo-lus): An air bubble, fat globule, or blood clot drifting in the circulation.
embryo (EM-brē-o): Developmental stage beginning at fertilization and ending at the start of the third developmental month.
embryology (em-brē-OL-o-jē): The study of embryonic development, focusing on the first 2 months after fertilization.
emesis (EM-e-sis): Vomiting.
emmetropia: Normal vision.
emulsification (ē-mul-si-fi-KĀ-shun): The physical breakup of fats in the digestive tract, forming smaller droplets accessible to digestive enzymes; normally the result of mixing with bile salts.
enamel: Crystalline material similar in mineral composition to bone, but harder and without osteocytes, that covers the exposed surfaces of the teeth.
encephalitis: Inflammation of the brain.
endocarditis: Inflammation of the endocardium of the heart.
endocardium (en-dō-KAR-dē-um): The simple squamous epithelium that lines the heart and is continuous with the endothelium of the great vessels.
endochondral ossification (en-dō-KON-dral): The conversion of a cartilaginous model to bone; the characteristic mode of formation for skeletal elements other than the bones of the cranium, the clavicles, and sesamoid bones.
endocrine gland: A gland that secretes hormones into the blood.
endocrine system: The endocrine glands of the body.
endocytosis (EN-dō-sī-tō-sis): The movement of relatively large volumes of extracellular material into the cytoplasm via the formation of a membranous vesicle at the cell surface; includes pinocytosis and phagocytosis.
endoderm: One of the three primary germ layers; the layer on the undersurface of the embryonic disc, that gives rise to the epithelia and glands of the digestive system, the respiratory system, and portions of the urinary system.
endogenous: Produced within the body.
endolymph (EN-dō-limf): Fluid contents of the membranous labyrinth (the saccule, utricle, semicircular canals, and cochlear duct) of the inner ear.
endometrial glands: Secretory glands of the endometrium.
endometrium (en-dō-MĒ-trē-um): The mucous membrane lining the uterus.
endomysium (en-dō-MĪS-ē-um): A delicate network of connective tissue fibers that surrounds individual muscle cells.
endoneurium: A delicate network of connective tissue fibers that surrounds individual nerve fibers.
endoplasmic reticulum (en-dō-PLAZ-mik re-TIK-ū-lum): A network of membranous channels in the cytoplasm of a cell that function in intracellular transport, synthesis, storage, packaging, and secretion.
endorphins (en-DOR-finz): Neuromodulators produced in the CNS that inhibit activity along pain pathways.
endosteum: An incomplete cellular lining found on the inner (medullary) surfaces of bones.
endothelium (en-dō-THĒ-lē-um): The simple squamous epithelium that lines blood and lymphatic vessels.
enkephalins (en-KEF-a-linz): Neuromodulators produced in the CNS that inhibit activity along pain pathways.
enteritis (en-ter-Ī-tis): Inflammation of the intestinal tract.
enterocrinin: A hormone secreted by the duodenal lining when exposed to acid chyme; stimulates the secretion of the duodenal glands.
enteroendocrine cells (en-ter-ō-EN-dō-krin): Endocrine cells scattered among the epithelial cells lining the digestive tract.
enterogastric reflex: Reflexive inhibition of gastric secretion initiated by the arrival of acid chyme in the small intestine.
enterohepatic circulation: Excretion of bile salts by the liver, followed by absorption of bile salts by intestinal cells for return to the liver via the hepatic portal vein.
enterokinase: An enzyme in the lumen of the small intestine that activates the proenzymes secreted by the pancreas.
enzyme: A protein that catalyzes a specific biochemical reaction.
eosinophils (ē-ō-sin-ō-filz): A granulocyte (WBC) with a lobed nucleus and red-staining granules; participates in the immune response, and is especially important during allergic reactions.
ependyma (ep-EN-di-mah): Layer of cells lining the ventricles and central canal of the CNS.
epicardium: Serous membrane covering the outer surface of the heart; also called the visceral pericardium.
epidermis: The epithelium covering the surface of the skin.
epididymis (ep-i-DID-i-mus): Coiled duct that connects the rete testis to the ductus deferens; site of functional maturation of spermatozoa.
epidural block: Anesthesia caused by the elimination of sensory inputs from dorsal nerve roots following the introduction of drugs into appropriate regions of the epidural space.
epidural space: Space between the spinal dura mater and the walls of the vertebral foramen; contains blood vessels and adipose tissue; a frequent site of injection for regional anesthesia.
epiglottis (ep-i-GLOT-is): Blade-shaped flap of tissue, reinforced by cartilage, that is attached to the dorsal and superior surface of the thyroid cartilage; it folds over the entrance to the larynx during swallowing.
epimysium (ep-i-MĪS-ē-um): A dense investment of collagen fibers that surrounds a skeletal muscle, and is continous with the tendons/aponeuroses of the muscle, and with the perimysium.
epineurium: A dense investment of collagen fibers that surrounds a peripheral nerve.
epiphyseal plate (e-pi-FI-sē-al): Cartilaginous region between the epiphysis and diaphysis of a growing bone.
epiphysis (e-PIF-i-sis): The head of a long bone.
epistaxis (ep-i-STAK-sis): Nosebleed.
epithelium (e-pi-THĒ-lē-um): One of the four primary tissue types; a layer of cells that forms a superficial covering or an internal lining of a body cavity or vessel.
eponychium (ep-ō-NIK-ē-um): A narrow zone of stratum corneum that extends across the surface of a nail at its exposed base; also called the **cuticle.**
equational division: The second meiotic division.
equilibrium (ē-kwi-LIB-rē-um): A dynamic state, where two opposing forces or processes are in balance.
erection: Stiffening of the penis due to the engorgement of the erectile tissues of the corpora cavernosa and the corpus spongiosum.
erythema (er-i-THĒ-ma): Redness and inflammation at the surface of the skin.
erythrocyte (e-RITH-rō-sīt): A red blood cell; an anucleate blood cell containing large quantities of hemoglobin.

erythrocytosis (e-rith-rō-sī-TŌ-sis): An abnormally large number of erythrocytes in the circulating blood.
erythropoietin (e-rith-rō-POI-ē-tin): Hormone released by tissues, especially the kidneys, exposed to low oxygen concentrations; stimulates erythropoiesis in bone marrow.
Escherichia coli: Normal bacterial resident of the large intestine.
esophagus: A muscular tube that connects the pharynx to the stomach.
essential amino acids: Amino acids that cannot be synthesized in the body in adequate amounts, and must be obtained from the diet.
essential fatty acids: Fatty acids that cannot be synthesized in the body, and must be obtained from the diet.
estrogens (ES-tro-jenz): A class of steroid sex hormones that includes estradiol.
eupnea (ŪP-nē-a): Normal quiet breathing.
evaporation: Movement of molecules from the liquid to the gaseous state.
eversion (ē-VER-shun): A turning outward.
excitable membranes: Membranes that conduct action potentials, a characteristic of muscle and nerve cells.
excitatory postsynaptic potential: The depolarization of a postsynaptic membrane by a chemical neurotransmitter released by the presynaptic cell.
excretion: Elimination from the body.
exocrine gland: A gland that secretes onto the body surface or into a passageway connected to the exterior.
exocytosis (EK-sō-sī-tō-sis): The ejection of cytoplasmic materials by fusion of a membranous vesicle with the cell membrane.
expiration: Exhalation; breathing out.
expiratory reserve: The amount of additional air that can be voluntarily moved out of the respiratory tract after a normal tidal expiration.
extension: An increase in the angle between two articulating bones; the opposite of flexion.
external auditory canal: Passageway in the temporal bone that leads to the tympanic membrane of the inner ear.
external ear: The pinna, external auditory meatus, external auditory canal, and tympanic membrane.
external nares: The nostrils; the external openings into the nasal cavity.
external respiration: Diffusion of gases between the alveolar air and the alveolar capillaries.
exteroceptors: Sensory receptors in the skin, mucous membranes, and special sense organs that provide information about the external environment and our position within it.
extracellular fluid: All body fluid other than that contained within cells; includes plasma and interstitial fluid.
extraembryonic membranes: The yolk sac, amnion, chorion, and allantois.
extrafusal fibers: Contractile muscle fibers, as opposed to the sensory intrafusal fibers (muscle spindles).
extrapyramidal system: Nuclei and tracts associated with the involuntary control of muscular activity.
extremities: The limbs.
extrinsic pathway: Clotting pathway that begins with damage to blood vessels or surrounding tissues and ends with the formation of tissue thromboplastin.
fabella: A sesamoid bone often found in the gastrocnemius muscle just behind the knee.
facilitated diffusion: Passive movement of a substance across a cell membrane via a protein carrier.
facilitation: Depolarization of a nerve cell membrane toward threshold, or making the cell more sensitive to depolarizing stimuli.
falciform ligament (FAL-si-form): A sheet of mesentery that contains the *ligamentum teres*, the fibrous remains of the umbilical vein of the fetus.
falx (falks): Sickle-shaped.
falx cerebri (falks ser-E-brē): Curving sheet of dura mater that extends between the two cerebral hemispheres; encloses the superior sagittal sinus.
fascia (FASH-a): Connective tissue fibers, primarily collagenous, that form sheets or bands beneath the skin to attach, stabilize, enclose, and separate muscles and other internal organs.
fasciculus (fa-SIK-ū-lus): A small bundle, usually referring to a collection of nerve axons or muscle fibers.
fatty acids: Hydrocarbon chains ending in a carboxyl group.
fauces (FAW-sēz): The passage from the mouth to the pharynx, bounded by the palatal arches, the soft palate, and the uvula.
febrile: Characterized by or pertaining to a fever.
feces: Waste products eliminated by the digestive tract at the anus; contains indigestible residue, bacteria, mucus, and epithelial cells.
fenestra: An opening.
fertilization: Fusion of egg and sperm to form a zygote.
fetus: Developmental stage lasting from the start of the third developmental month to delivery.
fibrillation (fi-bri-LĀ-shun): Uncoordinated contractions of individual muscle cells that impair or prevent normal function.
fibrin (FĪ-brin): Insoluble protein fibers that form the basic framework of a blood clot.
fibrinogen (fi-BRIN-o-jen): Plasma protein, soluble precursor of the fibrous protein fibrin.
fibrinolysis (fi-bri-no-LI-sis): The breakdown of the fibrin strands of a blood clot by a proteolytic enzyme.
fibroblasts (FĪ-brō-blasts): Cells of connective tissue proper that are responsible for the production of extracellular fibers and the secretion of the organic compounds of the extracellular matrix.
fibrocartilage: Cartilage containing an abundance of collagen fibers; found around the edges of joints, in the intervertebral discs, the menisci of the knee, etc.
fibrous tunic: The outermost layer of the eye, composed of the sclera and cornea.
fibula: The lateral, relatively small bone of the leg.
filariasis (fil-a-RĪ-a-sis): Condition resulting from infection by mosquito-borne parasites; may cause *elephantiasis*.
filiform papillae: Slender conical projections from the dorsal surface of the anterior two-thirds of the tongue.
filtrate: Fluid produced by filtration at a glomerulus in the kidney.
filtration: Movement of a fluid across a membrane whose pores restrict the passage of solutes on the basis of size.
filtration pressure: Hydrostatic pressure responsible for the filtration process.
filum terminale: A fibrous extension of the spinal cord that extends from the conus medullaris to the coccygeal ligament.
fimbriae (FIM-brē-ē): A fringe; used to describe the fingerlike processes that surround the entrance to the uterine tube.
fissure: An elongate groove or opening.
fistula: An abnormal passageway between two organs or from an internal organ or space to the body surface.

flaccid: Limp, soft, flabby; a muscle without muscle tone.
flagellum/flagella (fla-JEL-ah): An organelle structurally similar to a cilium, but used to propel a cell through a fluid.
flatus: Intestinal gas.
flavin adenine dinucleotide: A coenzyme important in oxidative phosphorylation; cycles between the oxidized ($FADH_2$) and reduced (FAD) states.
flavin adenine mononucleotide: A coenzyme important in oxidative phosphorylation; cycles between the oxidized ($FMNH_2$) and reduced (FMN) states.
flexion (FLEK-shun): A movement that reduces the angle between two articulating bones; the opposite of extension.
flexor: A muscle that produces flexion.
flexor reflex: A reflex contraction of the flexor muscles of a limb in response to an unpleasant stimulus.
flexure: A bending.
fluoroscope: An instrument that permits the examination of the body with X-rays in real-time, rather than via fixed images on photographic plates.
folia (FŌ-lē-ah): Leaf-like folds; used in reference to the slender folds in the surface of the cerebellar cortex.
follicle (FOL-i-kl): A small secretory sac or gland.
follicle-stimulating hormone (FSH): A hormone secreted by the anterior pituitary; stimulates oogenesis (female) and spermatogenesis (male).
folliculitis (fo-lik-ū-LĪ-tis): Inflammation of a follicle, such as a hair follicle of the skin.
fontanel (fon-tah-NEL): A relatively soft, flexible, fibrous region between two flat bones in the developing skull.
foramen: An opening or passage through a bone.
forearm: Distal portion of the arm between the elbow and wrist.
forebrain: The cerebrum.
fornix (FOR-niks): An arch, or the space bounded by an arch; in the brain, an arching tract that connects the hippocampus with the mamillary bodies; in the eye, a slender pocket found where the epithelium of the ocular conjunctiva folds back upon itself as the palpebral conjunctiva.
fossa: A shallow depression or furrow in the surface of a bone.
fourth ventricle: An elongate ventricle of the metencephalon (pons and cerebellum) and the myelencephalon (medulla) of the brain; the roof contains a region of choroid plexus.
fovea (FŌ-vē-a): Portion of the retina providing the sharpest vision, with the highest concentration of cones; also called the *macula lutea.*
fracture: A break or crack in a bone.
frenulum (FREN-ū-lum): A bridle; *see* **lingual frenulum.**
frontal plane: A sectional plane that divides the body into anterior and posterior portions; also called *coronal plane.*
fructose: A hexose (simple sugar containing 6 carbons) found in foods and in semen.
fundus (FUN-dus): The base of an organ.
fungiform papillae: Mushroom-shaped papillae on the dorsal and dorsolateral surfaces of the tongue.
furuncle (FŪ-rung-kl): A boil, resulting from the invasion and inflammation of a hair follicle or sebaceous gland.
gallbladder: Pear-shaped reservoir for the bile secreted by the liver.
gametes (GAM-ēts): Reproductive cells (sperm or eggs) that contain one-half of the normal chromosome complement.
gametogenesis (ga-mē-tō-JEN-e-sis): The formation of gametes.
gamma aminobutyric acid (GABA) (GAM-ma a-MĒ-nō-bū-TIR-ik): A neurotransmitter of the CNS whose effects are usually inhibitory.
gamma motor neurons: Motor neurons that adjust the sensitivities of muscle spindles (intrafusal fibers).
ganglion/ganglia: A collection of neuron cell bodies outside of the CNS.
gangliosides: Glycolipids that are important components of cell membranes in the CNS.
gap junctions: Connections between cells that permit electrical coupling.
gaster (GAS-ter): The stomach; the body or belly of a skeletal muscle.
gastrectomy (gas-TREK-to-mē): Partial or total surgical removal of the stomach.
gastric: Pertaining to the stomach.
gastric glands: Tubular glands of the stomach whose cells produce acid, enzymes, intrinsic factor, and hormones.
gastrin (GAS-trin): Hormone produced by enteroendocrine cells of the stomach, when exposed to mechanical stimuli or vagal stimulation, and the duodenum, when exposed to chyme containing undigested proteins.
gastritis (gas-TRĪ-tis): Inflammation of the stomach.
gastroenteric reflex (gas-trō-en-TER-ik): An increase in peristalsis along the small intestine triggered by the arrival of food in the stomach.
gastroileal reflex (gas-trō-IL-ē-al): Peristaltic movements that shift materials from the ileum to the colon, triggered by the arrival of food in the stomach.
gastrointestinal (GI) tract: An internal passageway that begins at the mouth, ends at the anus, and is lined by a mucous membrane; also known as the *digestive tract.*
gastroscope: A fiberoptic instrument that permits visual inspection of the stomach lining.
gastrulation (gas-troo-LĀ-shun): The movement of cells of the inner cell mass that creates the three primary germ layers of the embryo.
gene: A portion of a DNA strand that functions as a hereditary unit and is found at a particular locus on a specific chromosome.
genetic engineering: Term used to describe research and experiments involving the manipulation of the genetic makeup of an organism.
genetics: The study of mechanisms of heredity.
geniculate (je-NIK-ū-lāt): Like a little knee; the medial geniculates and the lateral geniculates are thalamic nuclei in the walls of the thalamus of the brain.
genitalia (jen-i-TĀ-lē-a): Reproductive organs.
genotype (JĒN-ō-tīp): The genetic complement of a particular individual.
germinal centers: Pale regions in the interior of lymphoid tissues or nodules, where cell divisions are under way.
gestation (jes-TĀ-shun): The period of intrauterine development.
gingivae (JIN-ji-vē): The gums.
gingivitis: Inflammation of the gums.
gland: Cells that produce exocrine or endocrine secretions.
glans penis: Expanded tip of the penis that surrounds the urethra meatus; continuous with the corpus spongiosum.
glaucoma: Eye disorder characterized by rising intraocular pressures due to inadequate drainage of aqueous humor at the canal of Schlemm.
glenoid fossa: A rounded depression that forms the articular surface of the scapula at the shoulder joint.
glial cells (GLĒ-al): Supporting cells in the neural tissue of the CNS and PNS.

globular proteins: Proteins whose tertiary structure makes them rounded and compact.
glomerular capsule: Expanded initial portion of the nephron that surrounds the glomerulus.
glomerular filtration rate: The rate of filtrate formation at the glomerulus.
glomerulonephritis (glo-mer-ū-lō-nef-RĪ-tis): Inflammation of the glomeruli of the kidneys.
glomerulus (glo-MER-ū-lus): A ball or knot; in the kidneys, a knot of capillaries that projects into the enlarged, proximal end of a nephron; the site where filtration occurs, the first step in the production of urine.
glossopharyngeal nerve (glos-ō-fa-RIN-jē-al): Cranial nerve IX.
glottis (GLOT-is): The passage from the pharynx to the larynx.
glucagon (GLOO-ka-gon): Hormone secreted by the alpha cells of the pancreatic islets; elevates blood glucose concentrations.
glucocorticoids: Hormones secreted by the zona fasciculata of the adrenal cortex to modify glucose metabolism; cortisol, cortisone, and corticosterone are important examples.
glucogenic amino acids: Amino acids that can be broken down, converted to pyruvic acid, and used in the synthesis of glucose (gluconeogenesis).
gluconeogenesis (gloo-kō-nē-ō-JEN-e-sis): The synthesis of glucose from protein or lipid precursors.
glucose (GLOO-kōs): A 6-carbon sugar, $C_6H_{12}O_6$, the preferred energy source for most cells and the only energy source for neurons under normal conditions.
glucose-dependent insulinotrophic hormone (GIP): A duodenal hormone released when the arriving chyme contains large quantities of carbohydrates; triggers the secretion of insulin and a slowdown in gastric activity.
glycerides: Lipids composed of glycerol bound to 1–3 fatty acids.
glycogen (GLĪ-kō-jen): A polysaccharide that represents an important energy reserve; a polymer consisting of a long chain of glucose molecules.
glycogenesis: The synthesis of glycogen from glucose molecules.
glycogenolysis: Glycogen breakdown, and the liberation of glucose molecules.
glycolipids (glī-cō-LIP-idz): Compounds created by the combination of carbohydrate and lipid components.
glycolysis (glī-KOL-i-sis): The cytoplasmic breakdown of glucose into lactic acid via pyruvic acid, with a net gain of 2 ATP.
glycoprotein (glī-kō-PRŌ-tēn): A compound containing a relatively small carbohydrate group attached to a large protein.
glycosuria (glī-cō-SŪ-rē-a): The presence of glucose in the urine.
goblet cell: A goblet-shaped, mucus-producing, unicellular gland found in certain epithelia of the digestive and respiratory tracts.
goiter: Enlargement of the thyroid gland.
Golgi apparatus (gol-jē): Cellular organelle consisting of a series of membranous plates that give rise to lysosomes and secretory vesicles.
Golgi tendon organ (gol-jē): *See tendon organ.*
gomphosis (gom-FŌ-sis): A fibrous synarthrosis that binds a tooth to the bone of the jaw; *see* **periodontal ligament.**
gonadotrophic hormones: FSH and LH, hormones that stimulate gamete development and sex hormone secretion.
gonadotropin-releasing hormone (GnRH) (gō-nad-ō-TRŌ-pin): Hypothalamic releasing hormone that causes the secretion of FSH and LH by the anterior pituitary gland.
gonadotropins: Gonadotropic hormones.
gonads (GŌ-nads): Organs that produce gametes and hormones.
gout: Clinical condition resulting from elevated uric acid concentrations in the blood and peripheral tissues.
granulocytes (GRAN-ū-lō-sīts): White blood cells containing granules visible with the light microscope; includes eosinophils, basophils, and neutrophils; also called *granular leukocytes.*
gray matter: Areas in the CNS dominated by neuron bodies, glial cells, and unmyelinated axons.
gray ramus: A bundle of postganglionic sympathetic nerve fibers that go to a spinal nerve for distribution to effectors in the body wall, skin, and extremities.
greater omentum: A large fold of the dorsal mesentery of the stomach that hangs in front of the intestines.
greater vestibular glands: Mucous glands in the vaginal walls that secrete into the vestibule; the equivalent of the bulbourethral glands of the male.
greenstick fracture: A fracture where a bone cracks and bends, most often affecting the long bones of young children.
groin: The inguinal region.
gross anatomy: The study of the structural features of the human body without the aid of a microscope.
growth hormone (GH): Anterior pituitary hormone that stimulates tissue growth and anabolism when nutrients are abundant, and restricts tissue glucose dependence when nutrients are in short supply.
guanine: A purine, one of the nitrogenous bases found in nucleic acids.
gustation (GUS-tā-shun): Taste.
gynecologists (gī-ne-KOL-o-jists): Physicians specializing in the pathology of the female reproductive system.
gyrus (JĪ-rus): A prominent fold or ridge of neural cortex on the surfaces of the cerebral hemispheres.
hair: A keratinous strand produced by epithelial cells of the hair follicle.
hair cells: Sensory cells of the inner ear.
hair follicle: An accessory structure of the integument; a tube lined by a stratified squamous epithelium that begins at the surface of the skin and ends at the hair papilla.
hair root: A thickened, conical structure consisting of a connective tissue papilla and the overlying matrix, a layer of epithelial cells that produces the hair shaft.
hallux: The big toe.
haploid (HAP-lōyd): Possessing one-half of the normal number of chromosomes; a characteristic of gametes.
hapten: A partial antigen; one that can bind to an antibody but not stimulate antibody production; a foreign compound that has only one antigenic-determinant site.
hard palate: The bony roof of the oral cavity, formed by the maxillary and palatine bones.
Hassall's corpuscles: Aggregations of epithelial cells in the thymus whose functions are unknown.
haustra (HAWS-truh): Saclike pouches along the length of the large intestine that result from tension in the *taenia coli.*
Haversian system: *See* **osteon.**
heart block: A cardiac arrhythmia due to conduction delays that affect communication between the atria and ventricles.

heat exhaustion: Condition characterized by excessive perspiration and resulting in dangerous fluid and salt losses.

heat stroke: Condition resulting from failure of the normal temperature control mechanisms, characterized by a cessation of sweating and a potentially fatal elevation in body temperature.

Heimlich maneuver (HĪM-lik): A technique for removing an airway blockage by external compression of the abdomen and forceful elevation of the diaphragm.

helper T cells: Lymphocytes (T_H cells) whose secretions and other activities coordinate the cellular and humoral immune responses.

hematocrit (hē-MAT-ō-krit): Percentage of the volume of whole blood contributed by cells; also called the packed cell volume (PCV) or the volume of packed red cells (VPRC).

hematoma: A tumor or swelling filled with blood.

hematuria (hē-ma-TŪ-rē-a): The presence of abnormal numbers of red blood cells in the urine.

heme (hēm): A porphyrin ring containing a central iron atom that can reversibly bind oxygen molecules; a component of the hemoglobin molecule.

hemiplegia: Paralysis affecting one side of the body (arm, trunk, and leg).

hemocytoblasts: Stem cells whose divisions produce all of the various populations of blood cells.

hemodialysis (hē-mō-dī-AL-i-sis): Dialysis of the blood.

hemoglobin (hē-mō-GLŌ-bin): Protein composed of four globular subunits, each bound to a single molecule of heme; the protein found in red blood cells that gives them the ability to transport oxygen in the blood.

hemolysis: Breakdown (lysis) of red blood cells.

hemophilia (hē-mo-FIL-ē-a): A congenital condition resulting from the inadequate synthesis of one of the clotting factors.

hemopoiesis (hē-mō-poi-Ē-sis): Blood cell formation and differentiation.

hemorrhage: Blood loss.

hemorrhoids (HEM-o-rōydz): Swollen, varicosed veins that protrude from the walls of the rectum and/or anorectal canal.

hemostasis: The cessation of bleeding.

hemothorax: The entry of blood into one of the pleural cavities.

heparin (HEP-a-rin): An anticoagulant released by activated basophils and mast cells.

hepatic duct: Duct carrying bile away from the liver lobes and toward the union with the cystic duct.

hepatic portal vein: Vessel that carries blood between the intestinal capillaries and the sinusoids of the liver.

hepatitis (hep-a-TĪ-tis): Inflammation of the liver, resulting from exposure to toxic chemicals, drugs, or viruses.

hepatocyte (he-PAT-ō-sit): A liver cell.

hernia: The protrusion of a loop or portion of a visceral organ through the abdominopelvic wall or into the thoracic cavity.

herniated disc: Rupture of the connective tissue sheath of the nucleus pulposus of an intervertebral disc.

heterotopic: Ectopic; outside of its normal location.

heterozygous (het-er-ō-ZĪ-gus): Possessing two different alleles at corresponding loci on a chromosome pair; the individual's phenotype may be determined by one or both of the alleles.

hexose: A 6-carbon simple sugar.

hiatus (hī-Ā-tus): A gap, cleft, or opening.

high-density lipoprotein: A lipoprotein with a relatively small lipid content, thought to be responsible for the movement of cholesterol from peripheral tissues to the liver.

hilum or **hilus** (HĪ-lus): A localized region where blood vessels, lymphatics, nerves, and/or other anatomical structures are attached to an organ.

hippocampus: A region beneath the floor of a lateral ventricle involved with emotional states and the conversion of short-term to long-term memories.

hirsutism (HER-sut-izm): Excessive hair growth in women that follows the distribution pattern typical of adult males; sometimes caused by the overproduction of androgens.

histamine (HIS-ta-min): Chemical released by stimulated mast cells or basophils to initiate or enhance an inflammatory response.

histology (his-TOL-ō-jē): The study of tissues.

histones: Proteins associated with the DNA of the nucleus, and around which the DNA strands are wound.

holocrine (HŌ-lō-krin): Form of exocrine secretion where the secretory cell becomes swollen with vesicles and then ruptures.

homeostasis (hō-mē-ō-STĀ-sis): The maintenance of a relatively constant internal environment.

homologous chromosomes (hō-MOL-o-gus): The members of a chromosome pair, each containing the same gene loci.

homozygous (hō-mō-ZĪ-gus): Having the same allele for a particular character on two homologous chromosomes.

hormone: A compound secreted by one cell that travels through the circulatory system to affect the activities of cells in another portion of the body.

human chorionic gonadotrophin (hCG): Placental hormone that maintains the corpus luteum for the first 3 months of pregnancy.

human immunodeficiency virus (HIV): The infectious agent that causes *acquired immunodeficiency syndrome (AIDS).*

human leukocyte antigen (HLA): Antigens on cell surfaces important to foreign antigen recognition and that play a role in the coordination and activation of the immune response. Also called *MHC proteins.*

human placental lactogen (HPL): Placental hormone that stimulates the functional development of the mammary glands.

humoral immunity: Immunity resulting from the presence of circulating antibodies produced by plasma cells.

hyaluronic acid: A proteoglycan in the matrix of many connective tissues that gives the matrix a viscous consistency; also functions as intercellular cement.

hyaluronidase: An enzyme that breaks down hyaluronic acid; produced by some bacteria and found in the acrosomal cap of a sperm cell.

hydrocephalus: Condition resulting from excessive production or inadequate drainage of cerebrospinal fluid.

hydrogen bond: Weak interaction between the hydrogen atom on one molecule and a negatively charged portion of another molecule.

hydrolysis (hī-DROL-i-sis): The breakage of a chemical bond through the addition of a water molecule; the reverse of dehydration synthesis.

hydrophilic (hī-drō-PHIL-ik): Freely associating with water; readily entering into solution.

hydrophobic: Incapable of freely associating with water molecules; insoluble.

hydrostatic pressure: Fluid pressure.

hydroxyl group (hī-DROK-sil): OH^-.

hypercapnia (hī-per-KAP-nē-a): High plasma carbon dioxide concentrations, often the result of hypoventilation or inadequate tissue perfusion.

hyperglycemia: Elevated plasma glucose concentrations.
hyperkalemia (hī-per-kā-LĒ-mē-a): Abnormally high potassium concentrations in the extracellular fluid.
hypernatremia: Abnormally high sodium concentrations in the extracellular fluid.
hyperopia: The farsighted condition, characterized by an inability to focus on objects close by.
hyperplasia: Abnormal enlargement of an organ due to an increase in the number of cells.
hyperpnea (hī-perp-NĒ-a): Abnormal increases in the rate and depth of respiration.
hyperpolarization: Movement of the transmembrane potential away from the normal resting potential and farther from 0 mV.
hyperreflexia: Abnormally exaggerated reflex responses to stimulation.
hypersecretion: Overactivity of glands that produce exocrine or endocrine secretions.
hypersensitivity: Overreaction to an allergen that results in tissue damage and inflammation.
hypertension: Abnormally high blood pressure.
hyperthermia: Excessively high body temperature.
hyperthyroidism: Excessive production of thyroid hormones.
hypertonic: A term used when comparing two solutions, to refer to the solution with the higher osmolarity.
hypertrophy (HĪ-per-trō-fē): Increase in the size of tissue without cell division.
hyperventilation (hī-per-ven-ti-LĀ-shun): A rate of respiration sufficient to reduce the plasma P_{CO_2} to levels below normal.
hypervitaminosis (hī-per-vī-ta-min-Ō-sis): Clinical condition caused by the excessive ingestion and uptake of vitamins.
hypesthesia: Abnormally decreased sensitivity to stimuli.
hypoblast (HĪ-pō-blast): The undersurface of the inner cell mass that faces the blastocoel of the early embryo.
hypocapnia: Abnormally low plasma P_{CO_2}, usually the result of hyperventilation.
hypodermic needle: A needle inserted through the skin to introduce drugs into the subcutaneous layer.
hypodermis: The subcutaneous layer, a region of loose connective tissue also called the *superficial fascia.*
hypokalemia (hī-pō-kā-LĒ-mē-a): Abnormally low plasma potassium concentrations.
hyponatremia: Abnormally low plasma sodium concentrations.
hyponychium (hī-pō-NIK-ē-um): A thickening in the epidermis beneath the free edge of a nail.
hypophyseal portal system (hī-po-FI-sē-al): Network of vessels that carry blood from capillaries in the hypothalamus to capillaries in the anterior pituitary gland (hypophysis).
hypophysis (hī-POF-i-sis): The anterior pituitary gland, which can be further subdivided into the pars distalis and the pars intermedia.
hyporeflexia: Abnormally depressed reflex responses to stimuli.
hyposecretion: Abnormally low rates of exocrine or endocrine secretion.
hypothalamus: The floor of the diencephalon; region of the brain containing centers involved with the unconscious regulation of visceral functions, emotions, drives, and the coordination of neural and endocrine functions.
hypothermia (hī-pō-THER-mē-a): Abnormally low body temperature.
hypothesis: A prediction that can be subjected to scientific analysis and review.
hypotonic: When comparing two solutions, used to refer to the one with the lower osmolarity.
hypoventilation: A respiratory rate insufficient to keep plasma P_{CO_2} within normal levels.
hypovitaminosis: Clinical condition resulting from inadequate vitamin ingestion and uptake; vitamin deficiency.
hypovolemic (hī-pō-vo-LĒ-mik): An abnormally low blood volume.
hypoxia (hī-POKS-ē-a): Low tissue oxygen concentrations.
ileum (IL-ē-um): The last 2.5 m of the small intestine.
ileocecal valve (il-ē-ō-SĒ-kal): A fold of mucous membrane that guards the connection between the ileum and the cecum.
ileostomy (il-ē-OS-to-mē): Surgical creation of an opening into the ileum; the opening created when the ilium is surgically attached to the abdominal wall.
ilium (IL-ē-um): The largest of the three bones whose fusion creates a coxa.
immunity: Resistance to injuries and diseases caused by foreign compounds, toxins, and pathogens.
immunization: Developing immunity by the deliberate exposure to antigens under conditions that prevent the development of illness but stimulate the production of memory B cells.
immunodeficiency: An inability to produce normal numbers and types of antibodies and sensitized lymphocytes.
immunoglobulin (i-mū-nō-GLOB-ū-lin): A circulating antibody.
immunosuppression (i-mū-nō-su-PRE-shun): The suppression of immune responses by the administration of drugs or exposure to toxic chemicals, radiation, or infection.
implantation (im-plan-TĀ-shun): The erosion of a blastocyst into the uterine wall.
impotence: Inability to obtain or maintain an erection in the male.
inclusions: Aggregations of insoluble pigments, nutrients, or other materials in the cytoplasm.
incontinence (in-KON-ti-nens): Inability to voluntarily control micturition (or defecation).
incus (IN-kus): The central auditory ossicle, situated between the malleus and the stapes in the middle ear cavity.
inducer: A stimulus that promotes the activity of a specific gene.
inexcitable: Incapable of conducting an action potential.
infarct: An area of dead cells resulting from an interruption of circulation.
infection: Invasion and colonization of body tissues by pathogenic organisms.
inferior: A directional reference meaning below.
inferior vena cava: The vein that carries blood from the parts of the body below the heart to the right auricle.
infertility: Inability to conceive.
inflammation: A nonspecific defense mechanism that operates at the tissue level, characterized by swelling, redness, warmth, pain, and some loss of function.
inflation reflex: A reflex mediated by the vagus nerve that prevents overexpansion of the lungs.
infundibulum (in-fun-DIB-ū-lum): A tapering, funnel-shaped structure; in the nervous system, refers to the connection between the pituitary gland and the hypothalamus; the infundibulum of the uterine tube is the entrance bounded by fimbriae that receives the ova at ovulation.

ingestion: The introduction of materials into the digestive tract via the mouth.
inguinal canal: A passage through the abdominal wall that marks the path of testicular descent, and that contains the testicular arteries, veins, and ductus deferens.
inguinal region: The area near the junction of the trunk and the thighs that contains the external genitalia.
inhibin (in-HIB-in): A hormone produced by the sustentacular cells that inhibits the pituitary secretion of FSH.
inhibitory postsynaptic potential (IPSP): A hyperpolarization of the postsynaptic membrane following the arrival of a neurotransmitter.
initial segment: The proximal portion of the axon where an action potential first appears.
injection: Forcing of fluid into a body part or organ.
inner cell mass: Cells of the blastocyst that will form the body of the embryo.
inner ear: *See* **internal ear.**
innervation: The distribution of sensory and motor nerves to a specific region or organ.
insertion: Point of attachment of a muscle that is more movable.
inspiration: Inhalation, the movement of air into the respiratory system.
inspiratory reserve: The maximum amount of air that can be drawn into the lungs over and above the normal tidal volume.
insoluble: Incapable of dissolving in solution.
insomnia: Sleep disorder characterized by an inability to fall asleep.
insulin (IN-su-lin): Hormone secreted by the beta cells of the pancreatic islets; causes a reduction in plasma glucose concentrations.
integument (in-TEG-ū-ment): The skin.
intercalated discs (in-TER-ka-lā-ted): Regions where adjacent cardiocytes interlock and where gap junctions permit electrical coupling between the cells.
intercellular cement: Proteoglycans, especially hyaluronic acid, found between adjacent epithelial cells.
intercellular fluid: *See* **interstitial fluid.**
interdigitate: To interlock.
interferons (in-ter-FĒR-ons): Peptides released by virally infected cells, especially lymphocytes, that make other cells more resistant to viral infection and slow viral replication.
interleukins (in-ter-LOO-kins): Peptides released by activated monocytes and lymphocytes that assist in the coordination of the cellular and humoral immune responses.
internal capsule: Term given to the appearance of the white matter of the cerebral hemispheres on gross dissection of the brain.
internal ear: The membranous labyrinth that contains the organs of hearing and equilibrium.
internal nares: The entrance to the nasopharynx from the nasal cavity.
internal respiration: Diffusion of gases between the blood and interstitial fluid.
interneuron: An association neuron; neurons inside the CNS that are interposed between sensory and motor neurons.
interoceptors: Sensory receptors monitoring the functions and status of internal organs and systems.
interosseous membrane: Fibrous connective tissue membrane between the shafts of the tibia and fibula or the radius and ulna; an example of a fibrous amphiarthrosis.
interphase: Stage in the life of a cell during which the chromosomes are uncoiled and all normal cellular functions except mitosis are under way.
intersegmental reflex: A reflex that involves several segments of the spinal cord.
interstitial cell-stimulating hormone: An alternative name for LH in the male.
interstitial fluid (in-ter-STISH-al): Fluid in the tissues that fills the spaces between cells.
interstitial growth: Form of cartilage growth through the growth, mitosis, and secretion of chondrocytes inside the matrix.
interventricular foramen: The opening that permits fluid movement between the lateral and third ventricles.
intervertebral disc: Fibrocartilage pad between the centra of successive vertebrae that acts as a shock absorber.
intestinal crypt: A tubular epithelial pocket lined by secretory cells and opening into the lumen of the digestive tract; also called an intestinal gland.
intestine: Tubular organ of the digestive tract.
intracellular fluid: The cytosol.
intrafusal fibers: Muscle spindle fibers.
intramembranous ossification (in-tra-MEM-bra-nus): The formation of bone within a connective tissue without the prior development of a cartilaginous model.
intramuscular injection: Injection of medication into the bulk of a skeletal muscle.
intraocular pressure: The hydrostatic pressure exerted by the aqueous humor of the eye.
intrapleural pressure: The pressure measured in a pleural cavity; also called the *intrathoracic pressure.*
intrapulmonary pressure (in-tra-PUL-mo-ner-ē): The pressure measured in an alveolus; also called the *intraalveolar pressure.*
intrauterine: Within the uterus; used to refer to the period of prenatal development.
intrinsic factor: Glycoprotein secreted by the parietal cells of the stomach that facilitates the intestinal absorption of vitamin B_{12}.
intrinsic pathway: A pathway of the clotting system that begins with the activation of platelets and ends with the formation of platelet thromboplastin.
inversion: A turning inward.
in vitro: Outside of the body, in an artificial environment.
in vivo: In the living body.
involuntary: Not under conscious control.
ion: An atom or molecule bearing a positive or negative charge due to the acceptance or donation of an electron.
ionic bond (ī-ON-ik): Molecular bond created by the attraction between ions with opposite charges.
ionization (ī-on-i-ZĀ-shun): Dissociation; the breakdown of a molecule in solution to form ions.
ipsilateral: A reflex response affecting the same side as the stimulus.
iris: A contractile structure made up of smooth muscle that forms the colored portion of the eye.
ischemia (is-KĒ-mē-a): Inadequate blood supply to a region of the body.
ischium (IS-kē-um): One of the three bones whose fusion creates the coxa.
islets of Langerhans: *See* **pancreatic islets.**
isometric contraction: A muscular contraction characterized by rising tension production but no change in length.
isotonic: A solution having an osmolarity that does not result in water movement across cell membranes; of the same contractive strength.

isotonic contraction: A muscular contraction during which tension climbs and then remains stable as the muscle shortens.

isotope: Forms of an element whose atoms contain the same number of protons but different numbers of neutrons (and thus differ in atomic weight).

isthmus (IS-mus): A narrow band of tissue connecting two larger masses.

jaundice (JAWN-dis): Condition characterized by yellowing of connective tissues due to elevated tissue bilirubin levels; usually associated with damage to the liver or biliary system.

jejunum (je-JŪ-num): The middle portion of the small intestine.

joint: An area where adjacent bones interact; an articulation.

juxtaglomerular apparatus: The macula densa and the juxtaglomerular cells; a complex responsible for the release of renin and erythropoietin.

juxtaglomerular cells: Modified smooth muscle cells in the walls of the afferent and efferent arterioles adjacent to glomerulus and the macula densa.

karyotyping (KAR-ē-ō-tī-ping): The determination of the chromosomal characteristics of an individual or cell.

keratin (KER-a-tin): Tough, fibrous protein component of nails, hair, calluses, and the general integumentary surface.

keratinization (KER-a-tin-ī-za-shun): The production of keratin by epithelial cells.

keratohyalin (ker-a-tō-HĪ-a-lin): A protein precursor of keratin.

keto acid: A molecule that ends in -COCOOH; the carbon chain that remains after the deamination or transamination of an amino acid.

ketoacidosis (kē-tō-as-i-DŌ-sis): A reduction in the pH of body fluids due to the presence of large numbers of ketone bodies.

ketogenic amino acids: Amino acids whose catabolism yields ketone bodies rather than pyruvic acid.

ketone bodies: Keto acids produced during the catabolism of lipids and ketogenic amino acids; specifically acetone, acetoacetate, and beta-hydroxybutyrate.

ketonemia (kē-to-NĒ-mē-a): Abnormal concentrations of ketone bodies in the blood.

ketonuria (kē-to-NŪ-rē-a): Abnormal concentrations of ketone bodies in the urine.

ketosis (kē-TŌ-sis): Condition characterized by the abnormal production of ketone bodies.

kidney: A component of the urinary system; an organ functioning in the regulation of plasma composition, including the excretion of wastes and the maintenance of normal fluid and electrolyte balance.

killer T cells: *See* **cytotoxic T cells.**

kilocalorie (KIL-o-kal-o-rē): The amount of heat required to raise the temperature of a kilogram of water 1°C.

Krebs cycle: *See* **TCA cycle.**

Kupffer cells (KOOP-fer): Stellate reticular cells of the liver; phagocytic cells of the liver sinusoids.

kyphosis (kī-FŌ-sis): Exaggerated thoracic curvature.

labia (LĀ-bē-a): Lips; labia majora and minora are components of the female external genitalia.

labrum: A lip or rim.

labyrinth: A maze of passageways; usually refers to the structures of the inner ear.

lacrimal gland (LAK-ri-mal): Tear gland on the dorsolateral surface of the eye.

lactase: An enzyme that breaks down milk proteins.

lactation (lak-TĀ-shun): The production of milk by the mammary glands.

lacteal (LAK-tē-al): A terminal lymphatic within an intestinal villus.

lactic acid: Compound produced from pyruvic acid under anaerobic conditions.

lactiferous duct (lak-TIF-e-rus): Duct draining one lobe of the mammary gland.

lactiferous sinus: An expanded portion of a lactiferous duct adjacent to the nipple of a breast.

lacuna (la-KŪ-nē): A small pit or cavity.

lambdoidal suture (lam-DŌYD-al): Synarthrotic articulation between the parietal and occipital bones of the cranium.

lamellae (la-MEL-lē): Concentric layers of bone within an osteon.

lamina (LA-min-a): A thin sheet or layer.

lamina propria (LA-mi-na PRO-prē-a): Loose connective tissue that underlies a mucous epithelium and forms part of a mucous membrane.

laminectomy: Removal of the spinous processes of a vertebra to gain access and treat a herniated disc.

Langerhans cells (LAN-ger-hanz): Cells in the epithelium of the skin and digestive tract that participate in the immune response by presenting antigens to T cells.

laparoscope (LAP-a-ro-skōp): Fiberoptic instrument used to visualize the contents of the abdominopelvic cavity.

large intestine: The terminal portions of the intestinal tract, consisting of the colon, the rectum, and the anorectal canal.

laryngopharynx (la-rin-gō-FAR-inks): Division of the pharynx inferior to the epiglottis and superior to the esophagus.

larynx (LAR-inks): A complex cartilaginous structure that surrounds and protects the glottis and vocal cords; the superior margin is bound to the hyoid bone and the inferior margin is bound to the trachea.

latent period: The time between the stimulation of a muscle and the start of the contraction phase.

lateral: Pertaining to the side.

lateral apertures: Openings in the roof of the fourth ventricle that permit the circulation of CSF into the subarachnoid space.

lateral ventricle: Fluid-filled chamber within one of the cerebral hemispheres.

laxatives: Compounds that promote defecation via increased peristalsis or an increase in the water content and volume of the feces.

lens: The transparent body lying behind the iris and pupil and in front of the vitreous humor.

lesion: A localized abnormality in tissue organization.

lesser omentum: A small pocket in the mesentery that connects the lesser curvature of the stomach to the liver.

leukemia (loo-KĒ-mē-ah): A malignant disease of the blood-forming tissues.

leukocyte (LOO-kō-sīt): A white blood cell.

leukocytosis (loo-kō-sī-TŌ-sis): Abnormally high numbers of circulating white blood cells.

leukopenia (loo-kō-PĒ-nē-ah): Abnormally low numbers of circulating white blood cells.

ligament (LI-ga-ment): Dense band of connective tissue fibers that attach one bone to another.

ligamentum arteriosum: The fibrous strand found in the adult that represents the remains of the ductus arteriosus of the fetus.

ligamentum teres: The fibrous strand in the falciform ligament that represents the remains of the umbilical vein of the fetus.

ligate: To tie off.
limbic system (LIM-bik): Group of nuclei and centers in the cerebrum and diencephalon that are involved with emotional states, memories, and behavioral drives.
limbus (LIM-bus): The edge of the cornea, marked by the transition from the corneal epithelium to the ocular conjunctiva.
liminal stimulus: A stimulus sufficient to depolarize the transmembrane potential of an excitable membrane to threshold, and produce an action potential.
linea alba: Tendinous band that runs along the midline of the rectus abdominis.
lingual: Pertaining to the tongue.
lingual frenulum: An epithelial fold that attaches the inferior surface of the tongue to the floor of the mouth.
lipase (LI-pāz): A pancreatic enzyme that breaks down triglycerides.
lipemia (lip-Ē-mē-a): Elevated concentration of lipids in the circulation.
lipid: An organic compound containing carbons, hydrogens, and oxygens in a ratio that does not approximate 1:2:1; includes fats, oils, and waxes.
lipofuscin (li-po-FŪ-shun): A pigment inclusion of uncertain significance that is found in aging nerve cells.
lipogenesis (li-pō-JEN-e-sis): Synthesis of lipids from nonlipid precursors.
lipoids: Prostaglandins, steroids, phospholipids, glycolipids, etc.
lipolysis: The catabolism of lipids as a source of energy.
lipoprotein (li-po-PRŌ-tēn): A compound containing a relatively small lipid bound to a protein.
liver: An organ of the digestive system with varied and vital functions that include the production of plasma proteins, the excretion of bile, the storage of energy reserves, the detoxification of poisons, and the interconversion of nutrients.
lobule (LOB-ūl): The basic organizational unit of the liver at the histological level.
long-term memories: Memories that persist for an extended period.
loose connective tissue: A loosely organized, easily distorted connective tissue containing several different fiber types, a varied population of cells, and a viscous ground substance.
lordosis (lor-DŌ-sis): An exaggeration of the lumbar curvature.
lumbar: Pertaining to the lower back.
lumen: The central space within a duct or other internal passageway.
lungs: Paired organs of respiration, situated in the left and right pleural cavities.
luteinizing hormone (LH) (LOO-tē-in-ī-zing): Anterior pituitary hormone that in the female assists FSH in follicle stimulation, triggers ovulation, and promotes the maintenance and secretion of the endometrial glands; in the male, stimulates spermatogenesis; formerly known as *interstitial cell-stimulating hormone* in males.
luxation (luks-Ā-shun): Dislocation of a joint.
lymph: Fluid contents of lymphatic vessels, similar in composition to interstitial fluid.
lymphadenopathy (lim-fad-e-NOP-a-thē): Pathological enlargement of the lymph nodes.
lymphatics: Vessels of the lymphatic system.
lymphedema (lim-fe-DĒ-ma): Swelling of peripheral tissues due to excessive lymph production or inadequate drainage.
lymph nodes: Lymphatic organs that monitor the composition of lymph.
lymphocyte (LIM-fō-sīt): A cell of the lymphatic system that participates in the immune response.
lymphokines: Chemicals secreted by activated lymphocytes.
lymphopoiesis: The production of lymphocytes.
lymphotoxin (lim-fō-TOK-sin): A secretion of lymphocytes that kills the target cells.
lysis (LĪ-sis): The destruction of a cell through the rupture of its cell membrane.
lysosome (LĪ-so-sōm): Intracellular vesicle containing digestive enzymes.
lysozyme: An enzyme present in some exocrine secretions that has antibiotic properties.
macrophage: A phagocytic cell of the monocyte-macrophage system.
macula (MAK-ū-la): A receptor complex in the saccule or utricle that responds to linear acceleration or gravity.
macula densa (MAK-ū-la DEN-sa): A group of specialized secretory cells in a portion of the distal convoluted tubule adjacent to the glomerulus and the juxtaglomerular cells; a component of the juxtaglomerular apparatus.
macula lutea (LOO-tē-a): The fovea.
malignant cancer: A form of cancer characterized by rapid cellular growth and the spread of cancer cells throughout the body.
malleus (MAL-ē-us): The first auditory ossicle, bound to the tympanic membrane and the incus.
malnutrition: An unhealthy state produced by inadequate dietary intake of nutrients, calories, and/or vitamins.
mamillary bodies (MAM-i-lar-ē): Nuclei in the hypothalamus concerned with feeding reflexes and behaviors; a component of the limbic system.
mammary glands: Milk-producing glands of the female breast.
manus: The hand.
marrow: A tissue that fills the internal cavities in a bone; may be dominated by hemopoietic cells (red marrow) or adipose tissue (yellow marrow).
mass peristalsis: Powerful peristaltic contraction that moves fecal materials along the colon and into the rectum.
mass reflex: Hyperreflexia in an area innervated by spinal cord segments distal to an area of injury.
mast cell: A connective tissue cell that when stimulated releases histamine, serotonin, and heparin, initiating the inflammatory response.
mastectomy: Surgical removal of part or all of a mammary gland.
mastication (mas-ti-KĀ-shun): Chewing.
mastoid sinus: Air-filled spaces in the mastoid process of the temporal bone.
matrix: The ground substance of a connective tissue.
maxillary sinus (MAK-si-ler-ē): One of the paranasal sinuses; an air-filled chamber lined by a respiratory epithelium that is located in a maxillary bone and opens into the nasal cavity.
meatus (mē-Ā-tus): An opening or entrance into a passageway.
mechanoreception: Detection of mechanical stimuli, such as touch, pressure, or vibration.
medial: Toward the midline of the body.
mediastinum (mē-dē-as-TĪ-num): Central tissue mass that divides the thoracic cavity into two pleural cavities; includes the aorta and other great vessels, the esophagus, trachea, thymus, the pericardial cavity and heart, and a host of nerves, small vessels, and

lymphatics; area of connective tissue attaching a testis to the epididymus, proximal portion of ductus deferens, and associated vessels.
medulla: Inner layer or core of an organ.
medulla oblongata: The most caudal of the five brain regions, also known as the *myelencephalon.*
medullary cavity: The space within a bone that contains the marrow.
medullary rhythmicity center: Center in the medulla that sets the background pace of respiration; includes inspiratory and expiratory centers.
megakaryocytes (meg-a-KAR-ē-ō-sīts): Bone marrow cells responsible for the formation of platelets.
meiosis (mī-Ō-sis): Cell division that produces gametes with half of the normal somatic chromosome complement.
melanin (ME-la-nin): Yellow-brown pigment produced by the melanocytes of the skin.
melanocyte (me-LAN-ō-sīt): Specialized cell found in the deeper layers of the stratified squamous epithelium of the skin, responsible for the production of melanin.
melanocyte-stimulating hormone (MSH) (me-LAN-ō-sit): Hormone of the pars intermedia of the anterior pituitary that stimulates melanin production.
melanomas (mel-a-NŌ-mas): Dangerous malignant skin cancers that involve melanocytes.
melatonin (mel-a-TŌ-nin): Hormone secreted by the pineal gland; inhibits secretion of MSH and GnRF.
membrane: Any sheet or partition; a layer consisting of an epithelium and the underlying connective tissue.
membrane flow: The movement of sections of membrane surface to and from the cell surface and components of the endoplasmic reticulum, the Golgi apparatus, and vesicles.
membrane potential: *See transmembrane potential.*
membranous labyrinth: Endolymph-filled tubes of the inner ear that enclose the receptors of the inner ear.
memory: The ability to recall information or sensations; can be divided into short-term and long-term memories.
menarche (me-NAR-kē): The beginning of menstrual function.
meninges (men-IN-jēz): Three membranes that surround the surfaces of the CNS; the dura mater, the pia mater, and the arachnoid.
meningitis: Inflammation of the spinal or cranial meninges.
meniscectomy: Removal of a meniscus.
meniscus (mēn-IS-kus): A fibrocartilage pad between opposing surfaces in a joint.
menses (MEN-sēz): The first menstrual period that normally occurs at puberty.
merocrine (MER-o-krin): A method of secretion where the cell ejects materials through exocytosis.
mesencephalic aqueduct: Passageway that connects the third ventricle (diencephalon) with the fourth ventricle (meten cephalon).
mesencephalon (mez-en-SEF-a-lon): The midbrain.
mesenchyme: Embryonic/fetal connective tissue.
mesentery (MEZ-en-ter-ē): A double layer of serous membrane that supports and stabilizes the position of an organ in the abdominopelvic cavity and provides a route for the associated blood vessels, nerves, and lymphatics.
mesoderm: The middle germ layer that lies between the ectoderm and endoderm of the embryo.
mesothelium (mez-ō-THĒ-lē-um): A simple squamous epithelium that lines one of the divisions of the ventral body cavity.
messenger RNA (mRNA): RNA formed at transcription to direct protein synthesis in the cytoplasm.
metabolic turnover: The continual breakdown and replacement of organic materials within living cells.
metabolism (me-TAB-ō-lizm): The sum of all of the biochemical processes underway within the human body at a given moment; includes anabolism and catabolism.
metabolites (me-TAB-ō-līts): Compounds produced in the body as the result of metabolic reactions.
metacarpals: The five bones of the palm of the hand.
metalloproteins (met-al-ō-PRO-tēnz): Plasma proteins that transport metal ions.
metaphase (MET-a-faz): A stage of mitosis wherein the chromosomes line up along the equatorial plane of the cell.
metaphysis (me-TA-fi-sis): The region of a long bone between the epiphysis and diaphysis, corresponding to the location of the epiphyseal plate of the developing bone.
metarteriole (met-ar-TĒ-rē-ōl): A vessel that connects an arteriole to a venule and that provides blood to a capillary plexus.
metastasis (me-TAS-ta-sis): The spread of a disease from one organ to another.
metatarsal: One of the five bones of the foot that articulate with the tarsals (proximally) and the phalanges (distally).
metencephalon (met-en-SEF-a-lon): The pons and cerebellum of the brain.
micelle (mī-SEL): A spherical aggregation of bile salts, monoglycerides, and fatty acids in the lumen of the intestinal tract.
microcephaly (mī-krō-SEF-a-lē): An abnormally small cranium, due to premature closure of one or more fontanelles.
microfilaments: Fine protein filaments visible with the electron microscope; components of the cytoskeleton.
microglia (mī-KROG-lē-a): Phagocytic glial cells in the CNS, derived from the monocytes of the blood.
microphages: Neutrophils and eosinophils.
microtubules: Microscopic tubules that are part of the cytoskeleton, and are found in cilia, flagella, the centrioles, and spindle fibers.
microvilli: Small, fingerlike extensions of the exposed cell membrane of an epithelial cell.
micturition (mīk-tū-RI-shun): Urination.
midbrain: The mesencephalon.
middle ear: Space between the external and internal ear that contains auditory ossicles.
midsagittal plane: A plane passing through the midline of the body that divides it into left and right halves.
mineralocorticoid: Corticosteroids produced by the zona glomerulosa of the adrenal cortex; steroids such as aldosterone, that affect mineral metabolism.
miscarriage: Spontaneous abortion.
mitochondrion (mī-tō-KON-drē-on): An intracellular organelle responsible for generating most of the ATP required for cellular operations.
mitosis (mī-TŌ-sis): The division of a single cell that produces two identical daughter cells; the primary mechanism of tissue growth.
mitral valve (MĪ-tral): The left AV, or bicuspid, valve of the heart.
mixed gland: A gland that contains exocrine and endocrine cells, or an exocrine gland that produces serous and mucous secretions.
mixed nerve: A peripheral nerve that contains sensory and motor fibers.

mole: A quantity of an element or compound having a mass in grams equal to its atomic or molecular weight.
molecular weight: The sum of the atomic weights of the atoms in a molecule.
molecule: A compound containing two or more atoms that are held together by chemical bonds.
monoclonal antibodies (mo-nō-KLŌ-nal): Antibodies produced by genetically identical cells under laboratory conditions.
monocytes (MON-o-sīts): Phagocytic agranulocytes (white blood cells) in the circulating blood.
monoglyceride (mo-nō-GLI-se-rīd): A lipid consisting of a single fatty acid bound to a molecule of glycerol.
monokines: Secretions released by activated cells of the monocyte/macrophage system to coordinate various aspects of the immune response.
monosaccharide (mon-ō-SAK-ah-rīd): A simple sugar, such as glucose or ribose.
monosynaptic reflex: A reflex where the sensory afferent synapses directly on the motor efferent.
monozygotic twins: Twins produced through the splitting of a single fertilized egg (zygote).
morula (MOR-ū-la): A mulberry-shaped collection of cells produced through the mitotic divisions of a zygote.
motor unit: All of the muscle cells controlled by a single motor neuron.
mucins (MŪ-sins): Proteoglycans responsible for the lubricating properties of mucus.
mucosa (mū-KŌ-sa): A mucous membrane; the epithelium plus the lamina propria.
mucous: An adjective referring to the presence or production of mucus.
mucous membrane: *See* **mucosa.**
mucus: Lubricating secretion produced by unicellular and multicellular glands along the digestive, respiratory, urinary, and reproductive tracts.
multifactorial trait: A phenotypic character that reflects the interactions of many different genes.
multipennate: A muscle whose internal fibers are organized around several different tendons.
multipolar neuron: A neuron with many dendrites and a single axon, the typical form of a motor neuron.
multiunit smooth muscle: Smooth muscle tissue whose muscle cells are innervated in motor units.
muriatic acid: Hydrochloric acid (HCl).
muscarinic receptors (mus-kar-IN-ik): Membrane receptors sensitive to acetylcholine (ACh) and to muscarine, a toxin produced by certain mushrooms; found at all parasympathetic neuroeffector junctions and at a few sympathetic neuroeffector junctions.
muscle: A contractile organ composed of muscle tissue, blood vessels, nerves, connective tissues, and lymphatics.
muscle tissue: A tissue characterized by the presence of cells capable of contraction; includes skeletal, cardiac, and smooth muscle tissue.
muscularis externa (mus-kū-LAR-is): Concentric layers of smooth muscle responsible for peristalsis.
muscularis mucosae: Layer of smooth muscle beneath the lamina propria; responsible for moving the mucosal surface.
mutagens (MŪ-ta-jenz): Chemical agents that induce mutations and may be carcinogenic.
myalgia (mī-AL-jē-a): Muscle pain.
myasthenia gravis (mī-as-THĒ-nē-a GRA-vis): Muscular weakness due to a reduction in the number of ACh receptor sites on the sarcolemmal surface; suspected to be an autoimmune disorder.
myelencephalon (mī-el-en-SEF-a-lon): The medulla oblongata.
myelin (MĪ-e-lin): Insulating sheath around an axon consisting of multiple layers of glial cell membrane; significantly increases conduction rate along the axon.
myelination: The formation of myelin.
myenteric plexus (mī-en-TER-ik): Parasympathetic motor neurons and sympathetic postganglionic fibers located between the circular and longitudinal layers of the muscularis externa.
myocardial infarction (mī-ō-KAR-dē-al): Heart attack; damage to the heart muscle due to an interruption of regional coronary circulation.
myocarditis: Inflammation of the myocardium.
myocardium: The cardiac muscle tissue of the heart.
myofibril: Organized collections of myofilaments in skeletal and cardiac muscle cells.
myofilaments: Fine protein filaments, composed of the proteins actin (thin filaments) and myosin (thick filaments).
myoglobin (MĪ-ō-glō-bin): An oxygen-binding pigment especially common in slow skeletal and cardiac muscle fibers.
myogram: A recording of the tension produced by muscle fibers on stimulation.
myometrium (mī-ō-MĒ-trē-um): The thick layer of smooth muscle in the wall of the uterus.
myopia: Nearsightedness, an inability to accommodate for distant vision.
myosepta: Connective tissue partitions that separate adjacent skeletal muscles.
myosin: Protein component of the thick myofilaments.
myositis (mī-ō-SĪ-tis): Inflammation of muscle tissue.
nail: Keratinous structure produced by epithelial cells of the nail root.
narcolepsy: A sleep disorder characterized by falling asleep at inappropriate moments.
nares, external (NA-rēz): The entrance from the exterior to the nasal cavity.
nares, internal: The entrance from the nasal cavity to the nasopharynx.
nasal cavity: A chamber in the skull bounded by the internal and external nares.
nasolacrimal duct: Passageway that transports tears from the nasolacrimal sac to the nasal cavity.
nasolacrimal sac: Chamber that receives tears from the lacrimal ducts.
nasopharynx (nā-zō-FAR-inks): Region posterior to the internal nares, superior to the soft palate, and ending at the oropharynx.
N compound: An organic compound containing nitrogen atoms.
necrosis (nek-RŌ-sis): Death of cells or tissues from disease or injury.
negative feedback: Corrective mechanism that opposes or negates a variation from normal limits.
neonate: A newborn infant, or baby.
neoplasm: A tumor, or mass of abnormal tissue.
nephritis (nef-RĪ-tis): Inflammation of the kidney.
nephrolithiasis (nef-rō-li-THĪ-a-sis): Condition resulting from the formation of kidney stones.
nephron (NEF-ron): Basic functional unit of the kidney.
nerve impulse: An action potential in a neuron cell membrane.
neural cortex: An area where gray matter is found at the surface of the CNS.
neurilemma (nū-ri-LEM-ma): The outer surface of a glial cell that encircles an axon.

neuroeffector junction: A synapse between a motor neuron and a peripheral effector, such as a muscle, gland cell, or fat cell.
neurofibrils: Microfibrils in the cytoplasm of a neuron.
neurofilaments: Microfilaments in the cytoplasm of a neuron.
neuroglandular junction: A specific type of neuroeffector junction.
neuroglia (nū-ROG-lē-a): Cells of the CNS and PNS that support and protect the neurons.
neurohypophysis (NŪ-rō-hī-pof-i-sis): The posterior pituitary, or pars nervosa.
neuromodulator (nū-rō-MOD-ū-la-tor): A chemical that adjusts the sensitivities of another neuron to specific neurotransmitters.
neuromuscular junction: A specific type of neuroeffector junction.
neuron (NŪ-ron): A cell in neural tissue specialized for intercellular communication via (1) changes in membrane potential and (2) synaptic connections.
neurotransmitter: Chemical compound released by one neuron to affect the transmembrane potential of another.
neurotubules: Microtubules in the cytoplasm of a neuron.
neurulation: The embyrological process responsible for the formation of the CNS.
neutron: A fundamental particle that does not carry a positive or negative charge.
neutropenia: An abnormally low number of neutrophils in the circulating blood.
neutrophil (NŪ-tro-fil): A phagocytic microphage that is very numerous and usually the first of the mobile phagocytic cells to arrive at an area of injury or infection.
nicotinic receptors (nik-o-TIN-ik): ACh receptors found on the surfaces of sympathetic and parasympathetic ganglion cells, that will also respond to the compound nicotine.
nipple: An elevated epithelial projection on the surface of the breast, containing the openings of the lactiferous sinuses.
Nissl bodies: The ribosomes, Golgi, RER, and mitochondria of the perikaryon of a typical nerve cell.
nitrogenous wastes: Organic waste products of metabolism that contain nitrogen, such as urea, uric acid, and creatinine.
nociception (nō-sē-SEP-shun): Pain perception.
node of Ranvier: Area between adjacent glial cells where the myelin covering of an axon is incomplete.
nodose ganglion (NŌ-dōs): A sensory ganglion of cranial nerve X.
noradrenaline: Catecholamine secreted by the adrenal medulla, released at most sympathetic neuroeffector junctions, and at certain synapses inside the CNS; also called *norepinephrine.*
norepinephrine (nor-ep-i-NEF-rin): A catecholamine neurotransmitter in the PNS and CNS, and a hormone secreted by the adrenal medulla; also called *noradrenaline.*
nucleic acid (nū-KLĒ-ik): A polymer of nucleotides containing a pentose sugar, a phosphate group, and one of four nitrogenous bases that regulate the synthesis of proteins and make up the genetic material in cells.
nucleolus (nū-KLĒ-ō-lus): Dense region in the nucleus that represents the site of RNA synthesis.
nucleoplasm: Fluid content of the nucleus.
nucleoproteins: Proteins of the nucleus that are generally associated with the DNA.
nucleoside: A nitrogenous base plus a simple sugar.
nucleotide: Compound consisting of a nitrogenous base, a simple sugar, and a phosphate group.
nucleus: Cellular organelle that contains DNA, RNA, and proteins; a mass of gray matter in the CNS.
nucleus pulposus (pul-PO-sus): Gelatinous central region of an intervertebral disc.
nutrient: An organic compound that can be broken down in the body to produce energy.
nystagmus: Involuntary, continual movement of the eyes as if to adjust to constant motion.
obesity: Body weight 10–20 percent above standard values as the result of body fat accumulation.
occlusal surface (o-KLŪ-sal): The opposing surfaces of the teeth that come into contact when processing food.
ocular: Pertaining to the eye.
oculomotor nerve (ok-ū-lō-MŌ-ter): Cranial nerve III, that controls the extrinsic oculomotor muscles other than the superior oblique and the lateral rectus.
olecranon: The proximal end of the ulna that forms the prominent point of the elbow.
olfaction: The sense of smell.
olfactory bulb (ol-FAK-tor-ē): Expanded ends of the olfactory tracts; the sites where the axons of NI synapse on CNS interneurons that lie beneath the frontal lobe of the cerebrum.
oligodendrocytes (o-li-gō-DEN-drō-sīts): CNS glial cells responsible for maintaining cellular organization in the gray matter and providing a myelin sheath in areas of white matter.
oligopeptide (ol-i-gō-PEP-tīd): A short chain of amino acids.
oncogene (ON-kō-jēn): A gene that can turn a normal cell into a cancer cell.
oncologists (on-KOL-ō-jists): Physicians specializing in the study and treatment of tumors.
oocyte (Ō-ō-sīt): A cell whose meiotic divisions will produce a single ovum and three polar bodies.
oogenesis (ō-ō-JEN-e-sis): Ovum production.
oogonia (ō-ō-GŌ-nē-a): Stem cells in the ovaries whose divisions give rise to oocytes.
oophorectomy (ō-of-ō-REK-to-mē): Surgical removal of the ovaries.
oophoritis (ō-of-ō-RĪ-tis): Inflammation of the ovaries.
ooplasm: The cytoplasm of the ovum.
opsin: A protein, one structural component of the visual pigment rhodopsin.
opsonization: An effect of coating an object with antibodies; the attraction and enhancement of phagocytosis.
optic chiasm (OP-tik KI-asm): Crossing point of the optic nerves.
optic nerve: Nerve that carries signals from the eye to the optic chiasm.
optic tract: Tract over which nerve impulses from the retina are transmitted between the optic chiasm and the thalamus.
orbit: Bony cavity of the skull that contains the eyeball.
orchiectomy (or-kē-EK-tō-mē): Surgical removal of one or both testes; also called *orchidectomy.*
orchitis: Inflammation of the testes.
organelle (or-gan-EL): An intracellular structure that performs a specific function or group of functions.
organic compound: A compound containing carbon, hydrogen, and usually oxygen.
organogenesis: The formation of organs during embryological and fetal development.
organs: Combinations of tissues that perform complex functions.

origin: Point of attachment of a muscle that is less movable.

oropharynx: The middle portion of the pharynx, bounded superiorly by the nasopharynx, anteriorly by the oral cavity, and inferiorly by the laryngopharynx.

osmolarity (oz-mo-LAR-i-tē): The total concentration of dissolved materials in a solution, regardless of their specific identities, expressed in terms of moles.

osmoreceptor: A receptor sensitive to changes in the osmolarity of the plasma.

osmosis (oz-MŌ-sis): The movement of water across a semipermeable membrane toward a solution containing a relatively high solute concentration.

osmotic pressure: The force of osmotic water movement; the pressure that must be applied to prevent osmotic movement across a membrane.

osseous tissue: A strong connective tissue containing specialized cells and a mineralized matrix of crystalline calcium phosphate and calcium carbonate.

ossicles: Small bones.

ossification: The formation of bone.

osteoblast: A cell that produces the fibers and matrix of bone.

osteoclast (OS-tē-ō-klast): A cell that dissolves the fibers and matrix of bone.

osteocyte (OS-tē-ō-sīt): A bone cell responsible for the maintenance and turnover of the mineral content of the surrounding bone.

osteogenic layer (os-tē-ō-JEN-ik): The inner, cellular layer of the periosteum that participates in bone growth and repair.

osteolysis (os-tē-ŌL-i-sis): The breakdown of the mineral matrix of bone.

osteon (OS-tē-on): The basic histological unit of compact bone, consisting of osteocytes organized around a central canal and separated by concentric lamellae.

otic: Pertaining to the ear.

otitis media: Inflammation of the middle ear cavity.

otoconia (otoliths) (ō-tō-KŌ-nē-a): Aggregations of calcium carbonate crystals in a gelatinous membrane that sits above one of the maculae of the vestibular apparatus.

oval window: Opening in the bony labyrinth where the stapes attaches to the membranous wall of the scala vestibuli.

ovarian cycle (ō-VAR-ē-an): Monthly chain of events that leads to ovulation.

ovary: Female reproductive gland that produces gametes.

ovulation (ōv-ū-LĀ-shun): The release of a secondary oocyte, surrounded by cells of the corona radiata, following the rupture of the wall of a tertiary follicle.

ovum/ova (Ō-vum): The functional product of meiosis II, produced after fertilization of a secondary oocyte.

oxidation: The loss of electrons or hydrogen atoms, or the acceptance of an oxygen atom.

oxidative phosphorylation: The capture of energy as ATP during a series of oxidation/reduction reactions; a reaction sequence that occurs in the mitochondria and involves coenzymes and the electron transport system.

oxytocin (oks-i-TŌ-sin): Hormone produced by hypothalamic cells and secreted into capillaries at the posterior pituitary; stimulates smooth muscle contractions of the uterus or mammary glands in the female, but has no known function in males.

pacemaker cells: Cells of the SA node that set the pace of cardiac contraction.

Pacinian corpuscle (pa-SIN-ē-an): Receptor sensitive to vibration.

palate: Horizontal partition separating the oral cavity from the nasal cavity and nasopharynx; can be divided into an anterior bony (hard) palate and a posterior fleshy (soft) palate.

palatine: Pertaining to the palate.

palpate: To examine by touch.

palpebrae (pal-PĒ-brē): Eyelids.

pancreas: Digestive organ containing exocrine and endocrine tissues; exocrine portion secretes pancreatic juice, endocrine portion secretes hormones, including insulin and glucagon.

pancreatic duct: A tubular duct that carries pancreatic juice from the pancreas to the duodenum.

pancreatic islets: Aggregations of endocrine cells in the pancreas.

pancreatic juice: A mixture of buffers and digestive enzymes that is discharged into the duodenum under the stimulation of the enzymes secretin and cholecystokinin.

pancreatitis (pan-krē-a-TĪ-tis): Inflammation of the pancreas.

Papanicolaou (Pap) test: Test for the detection of malignancies based on cytological appearance of epithelial cells, especially those of the cervix and uterus.

papilla (pa-PIL-la): A small, conical projection.

paralysis: Loss of voluntary motor control over a portion of the body.

paranasal sinuses: Bony chambers lined by respiratory epithelium that open into the nasal cavity; includes the frontal, ethmoidal, sphenoidal, and maxillary sinuses.

parasagittal: A section or plane that parallels the midsagittal plane but that does not pass along the midline.

parasympathetic division: One of the two divisions of the autonomic nervous system; also known as the craniosacral division; generally responsible for activities that conserve energy and lower the metabolic rate.

parasympathomimetic drugs: Drugs that mimic the actions of parasympathetic stimulation.

parathyroid glands: Four small glands embedded in the posterior surface of the thyroid; responsible for parathyroid hormone secretion.

parathyroid hormone: Hormone secreted by the parathyroid gland when plasma calcium levels fall below the normal range; causes increased osteoclast activity, increased intestinal calcium uptake, and decreased calcium ion loss at the kidneys.

parenchyma (pa-RENG-kī-ma): The cells of a tissue or organ that are responsible for fulfilling its functional role.

paresthesia: Sensory abnormality that produces a tingling sensation.

parietal: Referring to the body wall or outer layer.

parietal cell: Cells of the gastric glands that secrete HCl and intrinsic factor.

Parkinson's disease: Progressive motor disorder due to degeneration of the cerebral nuclei.

parotid glands (pa-ROT-id): Large salivary glands that secrete a saliva containing high concentrations of salivary (alpha) amylase.

pars distalis (dis-TAL-is): The large, anterior portion of the anterior pituitary gland.

pars intermedia (in-ter-MĒ-dē-a): The portion of the anterior pituitary immediately adjacent to the posterior pituitary and the infundibulum.

pars nervosa: The posterior pituitary gland.

parturition (par-tū-RISH-un): Childbirth, delivery.

patella (pa-TEL-ah): The sesamoid bone of the kneecap.

pathogenic: Disease-causing.

pathologist (pa-THO-lo-jist): An M.D. specializing in the identification of diseases based on characteristic structural and functional changes in tissues and organs.
pedicel (PED-i-sel): A slender process of a podocyte that forms part of the filtration apparatus of the kidney glomerulus.
pedicles (PE-di-kels): Thick bony struts that connect the vertebral body with the articular and spinous processes.
pelvic cavity: Inferior subdivision of the abdominopelvic (peritoneal) cavity; encloses the urinary bladder, the sigmoid colon and rectum, and male or female reproductive organs.
pelvis: A bony complex created by the articulations between the coxae, the sacrum, and the coccyx.
penis (PĒ-nis): Component of the male external genitalia; a copulatory organ that surrounds the urethra, and that serves to introduce semen into the female vagina.
pepsin: Proteolytic enzyme secreted by the chief cells of the gastric glands in the stomach.
peptidases: Enzymes that split peptide bonds and release amino acids.
perforating canal: A passageway in compact bone that runs at right angles to the axes of the osteons, between the periosteum and endosteum.
perfusion: The blood flow through a tissue.
pericardial cavity (per-i-KAR-dē-al): The space between the parietal pericardium and the epicardium (visceral pericardium) that covers the outer surface of the heart.
pericarditis: Inflammation of the pericardium.
pericardium (per-i-KAR-dē-um): The fibrous sac that surrounds the heart, and whose inner, serous lining is continuous with the epicardium.
perichondrium (per-i-KON-drē-um): Layer that surrounds a cartilage, consisting of an outer fibrous and an inner cellular region.
perikaryon (per-i-KAR-ē-on): The cytoplasm that surrounds the nucleus in the soma of a nerve cell.
perilymph (PER-ē-limf): A fluid similar in composition to cerebrospinal fluid; found in the spaces between the bony labyrinth and the membranous labyrinth of the inner ear.
perimysium (pe-ri-MĪS-ē-um): Connective tissue partition that separates adjacent fasciculi in a skeletal muscle.
perineum (pe-ri-NĒ-um): The pelvic floor and associated structures.
perineurium: Connective tissue partition that separates adjacent bundles of nerve fibers in a peripheral nerve.
periodontal ligament (per-ē-ō-DON-tal): Collagen fibers that bind the cementum of a tooth to the periosteum of the surrounding alveolus.
periosteum (per-ē-OS-tē-um): Layer that surrounds a bone, consisting of an outer fibrous and inner cellular region.
peripheral nervous system (PNS): All neural tissue outside of the CNS.
peripheral resistance: The resistance to blood flow primarily caused by friction with the vascular walls.
peristalsis (per-i-STAL-sis): A wave of smooth muscle contractions that propels materials along the axis of a tube such as the digestive tract, the ureters, or the ductus deferens.
peritoneal cavity: *See abdominopelvic cavity.*
peritoneum (per-i-tō-NĒ-um): The serous membrane that lines the peritoneal (abdominopelvic) cavity.
peritonitis (per-i-to-NĪ-tis): Inflammation of the peritoneum.
peritubular capillaries: A network of capillaries that surrounds the proximal and distal convoluted tubules of the kidneys.
peroxisome: A membranous vesicle containing enzymes that break down hydrogen peroxide (H_2O_2).
permeability: Ease with which dissolved materials can cross a membrane; if freely permeable, any molecule can cross the membrane; if impermeable, nothing can cross; most biological membranes are selectively permeable.
perspiration, insensible: Evaporative water loss by diffusion across the epithelium of the skin or evaporation across the alveolar surfaces of the lungs.
perspiration, sensible: Water loss due to sweat gland secretion.
pes: The foot.
petrosal ganglion: Sensory ganglion of the glossopharyngeal nerve (N IX).
petrous: Stony, usually used to refer to the thickened portion of the temporal bone that encloses the inner ear.
Peyer's patches (PĪ-erz): Lymphoid nodules beneath the epithelium of the small intestine.
pH: The negative exponent of the hydrogen ion concentration, in moles per liter.
phagocyte: A cell that performs phagocytosis.
phagocytosis (FA-gō-sī-tō-sis): The engulfing of extracellular materials or pathogens; movement of extracellular materials into the cytoplasm by enclosure in a membranous vesicle.
phalanx/phalanges (fa-LAN-gez): A bone of the fingers or toes.
pharmacology: The study of drugs, their physiological effects, and their clinical uses.
pharyngotympanic tube: A passageway that connects the nasopharynx with the middle ear cavity; also called the *Eustachian* or *auditory tube.*
pharynx: The throat; a muscular passageway shared by the digestive and respiratory tracts.
phasic response: A pattern of response to stimulation by sensory neurons that are normally inactive; stimulation causes a burst of neural activity that ends when the stimulus either stops or stops changing in intensity.
phenotype (FĒN-ō-tīp): Physical characteristics that are genetically determined.
phonation (fō-NĀ-shun): Sound production at the larynx.
phosphate group: PO_4^{3-}.
phospholipid (fos-fō-LIP-id): An important membrane lipid whose structure includes hydrophilic and hydrophobic regions.
phosphorylation (fos-for-i-LĀ-shun): The addition of a high-energy phosphate group to a molecule.
photoreception: Sensitivity to light.
physiology (fiz-ē-OL-o-jē): Literally the study of function; considers the ways living organisms perform vital activities.
pia mater: The tough, outer meningeal layer that surrounds the CNS.
pigment: A compound with a characteristic color.
piloerection: "Goosebumps" effect produced by the contraction of the arrector pili muscles of the skin.
pineal gland: Neural tissue in the posterior portion of the roof of the diencephalon, responsible for the secretion of melatonin.
pinna: The expanded, projecting portion of the external ear that surrounds the external auditory meatus.

pinocytosis (PI-nō-sī-tō-sis): The introduction of fluids into the cytoplasm by enclosing them in membranous vesicles at the cell surface.

pituitary gland: Endocrine organ situated in the sella turcica of the sphenoid bone and connected to the hypothalamus by the infundibulum; includes the posterior pituitary (pars nervosa) and the anterior pituitary (pars intermedia and pars distalis).

placenta (pla-SENT-a): A complex structure in the uterine wall that permits diffusion between the fetal and maternal circulatory systems; also called the *afterbirth.*

plantar: Referring to the sole of the foot.

plasma (PLAZ-mah): The fluid ground substance of whole blood; what remains after the cells have been removed from a sample of whole blood.

plasma cell: Activated B cells that secrete antibodies.

plasmalemma (plaz-ma-LEM-a): Cell membrane.

platelets (PLĀT-lets): Small packets of cytoplasm that contain enzymes important in the clotting response; manufactured in the bone marrow by cells called *megakaryocytes.*

pleura (PLOO-ra): The serous membrane lining the pleural cavities.

pleural cavities: Subdivisions of the thoracic cavity that contain the lungs.

pleuritis (ploor-Ī-tis): Inflammation of the pleura.

plexus (PLEK-sus): A network or braid.

plica (PLĪ-ka): A permanent transverse fold in the wall of the small intestine.

pneumotaxic center (nū-mō-TAKS-ik): A center in the reticular formation of the pons that regulates the activities of the apneustic and respiratory rhythmicity centers to adjust the pace of respiration.

pneumothorax (nū-mō-THO-raks): The introduction of air into the pleural cavity.

podocyte (PŌ-do-sīt): A cell whose processes surround the glomerular capillaries and assist in the filtration process.

polar bond: A form of covalent bond in which there is an unequal sharing of electrons.

polar body: A nonfunctional packet of cytoplasm containing chromosomes eliminated from an oocyte during meiosis.

polarized: Histological use indicates cells that have regional differences in organelle distribution or cytoplasmic composition along a specific axis, such as between the basement membrane and free surface of an epithelial cell.

pollex: The thumb.

polymer: A large molecule consisting of a long chain of subunits.

polymorph: Polymorphonuclear leukocyte; a neutrophil.

polypeptide: A chain of amino acids strung together by peptide bonds; those containing over 100 peptides are called proteins.

polyribosome: Several ribosomes linked by their translation of a single mRNA strand.

polysaccharide (pol-ē-SAK-ah-rīd): A complex sugar, such as glycogen or a starch.

polysynaptic reflex: A reflex with interneurons interposed between the sensory fiber and the motor neuron(s).

polyunsaturated fats: Fatty acids containing carbon atoms linked by double bonds.

polyuria (pol-ē-Ū-rē-a): Excessive urine production.

pons: The portion of the metencephalon anterior to the cerebellum.

popliteal (pop-LIT-ē-al): Pertaining to the back of the knee.

porphyrins (POR-fi-rinz): Ring-shaped molecules that form the basis for important respiratory and metabolic pigments, including heme and the cytochromes.

porta hepatis: A region of mesentery between the duodenum and liver that contains the hepatic artery, the hepatic portal vein, and the common bile duct.

positive feedback: Mechanism that increases a deviation from normal limits following an initial stimulus.

postabsorptive state: A period that begins 4 hours after a meal, characterized by falling blood glucose concentrations and the mobilization of metabolic reserves.

postcentral gyrus: The primary sensory cortex, where touch, vibration, pain, temperature, and taste sensations arrive and are consciously perceived.

posterior: Toward the back; dorsal.

postganglionic neuron: An autonomic neuron in a peripheral ganglion, whose activities control peripheral effectors.

postovulatory phase: The secretory phase of the menstrual cycle.

potential difference: The separation of opposite charges; requires a barrier that prevents ion migration.

precentral gyrus: The primary motor cortex on a cerebral hemisphere, located rostral to the central sulcus.

prefrontal cortex: Rostral portion of each cerebral hemisphere thought to be involved with higher intellectual functions, predictions, calculations, and so forth.

preganglionic neuron: Visceral motor neuron inside the CNS whose output controls one or more ganglionic motor neurons in the PNS.

premolars: Bicuspids; teeth with flattened occlusal surfaces located anterior to the molar teeth.

premotor cortex: Motor association area between the precentral gyrus and the prefrontal area.

preoptic nucleus: Hypothalamic nucleus that coordinates thermoregulatory activities.

preovulatory phase: A portion of the menstrual cycle; period of estrogen-induced repair of the functional zone of the endometrium through the growth and proliferation of epithelial cells in the glands not lost during menses.

prepuce (PRĒ-pūs): Loose fold of skin that surrounds the glans penis (males) or the clitoris (females).

preputial glands (prē-PŪ-shal): Glands on the inner surface of the prepuce that produce a viscous, odorous secretion, called *smegma.*

presbyopia: Farsightedness; an inability to accommodate for near vision.

presynaptic membrane: The synaptic surface where neurotransmitter release occurs.

prevertebral ganglion: *See* **collateral ganglion.**

prime mover: A muscle that performs a specific action.

proenzyme: An inactive enzyme secreted by an epithelial cell.

progesterone (prō-JES-ter-ōn): The most important progestin secreted by the corpus luteum following ovulation.

progestins (prō-JES-tinz): Steroid hormones structurally related to cholesterol.

prognosis: A prediction concerning the possibility or time course of recovery from a specific disease.

projection fibers: Axons carrying information from the thalamus to the cerebral cortex.

prolactin (prō-LAK-tin): Hormone that stimulates functional development of the mammary gland in females; a secretion of the anterior pituitary gland.

prolapse: The abnormal descent or protrusion of a portion of an organ, such as the vagina or anorectal canal.

proliferative phase: *See* **preovulatory phase.**

pronation (prō-NĀ-shun): Rotation of the forearm that makes the palm face posteriorly.
pronucleus: Enlarged egg or sperm nucleus that forms after fertilization but before amphimixis.
properdin: Complement factor that prolongs and enhances nonantibody dependent complement binding to bacterial cell walls.
prophase (PRŌ-fāz): The initial phase of mitosis, characterized by the appearance of chromosomes, breakdown of the nuclear membrane, and formation of the spindle apparatus.
proprioception (prō-prē-ō-SEP-shun): Awareness of the positions of bones, joints, and muscles.
prostaglandin (pros-tah-GLAN-din): Lipoid secreted by one cell that alters the metabolic activities or sensitivities of adjacent cells; sometimes called "local hormones."
prostatectomy (pros-ta-TEK-tō-mē): Surgical removal of the prostate.
prostate gland (PROS-tāt): Accessory gland of the male reproductive tract, contributing roughly one-third of the volume of semen.
prostatitis (pros-ta-TĪ-tis): Inflammation of the prostate.
prosthesis: An artificial substitute for a body part.
protease: *See* **proteinase**.
protein: A large polypeptide with a complex structure.
proteinase: An enzyme that breaks down proteins into peptides and amino acids.
proteinuria (prō-tēn-ŪR-ē-a): Abnormal amounts of protein in the urine.
proteoglycan (prō-tē-ō-GLĪ-kan): Compound containing a large polysaccharide complex attached to a relatively small protein; examples include hyaluronic acid and chondroitin sulfate.
prothrombin: Circulating proenzyme of the common pathway of the clotting system; converted to thrombin by the enzyme thromboplastin.
proton: A fundamental particle bearing a positive charge.
protraction: To move anteriorly in the horizontal plane.
proximal: Toward the attached base of an organ or structure.
proximal convoluted tubule: The portion of the nephron between Bowman's capsule and the loop of Henle; the major site of active reabsorption from the filtrate.
pruritis (prū-RĪ-tus): Itching.
pseudopodia (sū-dō-PŌ-dē-a): Temporary cytoplasmic extensions typical of mobile or phagocytic cells.
pseudostratified epithelium: An epithelium containing several layers of nuclei, but whose cells are all in contact with the underlying basement membrane.
psoriasis (sō-RĪ-a-sis): Skin condition characterized by excessive keratin production and the formation of dry, scaly patches on the body surface.
psychosomatic condition: An abnormal physiological state with a psychological origin.
puberty: Period of rapid growth, sexual maturation, and the appearance of secondary sexual characteristics; usually occurs at ages 10–15.
pubic symphysis: Fibrocartilaginous amphiarthrosis between the pubic bones of the coxae.
pubis (PŪ-bis): The anterior, inferior component of the coxa.
pudendum (pū-DEN-dum): The external genitalia.
pulmonary circuit: Blood vessels between the pulmonary semilunar valve of the right ventricle and the entrance to the left atrium; the blood circulation through the lungs.
pulmonary ventilation: Movement of air in and out of the lungs.
pulp cavity: Internal chamber in a tooth, containing blood vessels, lymphatics, nerves, and the cells that maintain the dentin.
pulpitis (pul-PĪ-tis): Inflammation of the tissues of the pulp cavity.
pupil: The opening in the center of the iris through which light enters the eye.
purine: An N compound with a ring-shaped structure; examples include adenine and guanine, two nitrogen bases common in nucleic acids.
Purkinje cell (pur-KIN-jē): Large, branching neuron of the cerebellar cortex.
Purkinje fibers: Specialized conducting cardiocytes in the ventricles.
pus: An accumulation of debris, fluid, dead and dying cells, and necrotic tissue.
putamen (pū-TĀ-men): Thalamic nucleus involved in the integration of sensory information prior to projection to the cerebral hemispheres.
P wave: Deflection of the ECG corresponding to atrial depolarization.
pyelogram (PI-el-ō-gram): A radiographic image of the kidneys and ureters.
pyelonephritis (pī-e-lō-nef-RĪ-tis): Inflammation of the kidneys.
pyloric sphincter (pi-LOR-ic): Sphincter of smooth muscle that regulates the passage of chyme from the stomach to the duodenum.
pylorus (pī-LOR-us): Gastric region between the body of the stomach and the duodenum; includes the pyloric sphincter.
pyrexia (pī-REK-sē-a): A fever.
pyrimidine: An N compound with a ring-shaped structure; examples include cytosine, thymine, and uracil, nitrogenous bases common in nucleic acids.
pyrogen (PĪ-ro-jen): A compound that promotes a fever.
pyruvic acid (pī-RŪ-vik): 3-carbon compound produced by glycolysis.
quadriplegia: Paralysis of the upper and lower limbs.
quaternary structure: Three-dimensional protein structure produced by interactions between individual protein sub-units.
radiodensity: Relative resistance to the passage of X-rays.
radiographic techniques: Methods of visualizing internal structures using various forms of radiational energy.
radiopaque: Having a relatively high radiodensity.
rami communicantes: Axon bundles that link the spinal nerves with the ganglia of the sympathetic chain.
ramus: A branch.
raphe (RĀ-fē): A seam.
receptor field: The area monitored by a single sensory receptor.
recessive gene: An allele that will affect the phenotype only when the individual is homozygous for that trait.
recombinant DNA: DNA created by splicing together a specific gene from one organism into the DNA strand of another organism.
rectal columns: Longitudinal folds in the walls of the anorectal canal.
rectouterine pouch (rek-tō-Ū-te-rin): Peritoneal pocket between the anterior surface of the rectum and the posterior surface of the uterus.
rectum (REK-tum): The last 15 cm (6 in.) of the digestive tract.
rectus: Straight.
red blood cell: *See* **erythrocyte.**
reduction: The gain of hydrogen atoms or electrons, or the loss of an oxygen molecule.

reductional division: The first meiotic division, which reduces the chromosome number from 46 to 23.
reflex: A rapid, automatic response to a stimulus.
reflex arc: The receptor, sensory neuron, motor neuron, and effector involved in a particular reflex; interneurons may or may not be present, depending on the reflex considered.
refraction: The bending of light rays as they pass from one medium to another.
refractory period: Period between the initiation of an action potential and the restoration of the normal resting potential; over this period the membrane will not respond normally to stimulation.
relaxation phase: The period following a contraction when the tension in the muscle fiber returns to resting levels.
relaxin: Hormone that loosens the pubic symphysis; a hormone secreted by the placenta.
renal: Pertaining to the kidneys.
renal corpuscle: The initial portion of the nephron, consisting of an expanded chamber that encloses the glomerulus.
renin: Enzyme released by the juxtaglomerular cells when renal blood pressure or pO_2 declines; converts angiotensinogen to angiotensin I.
rennin: Gastric enzyme that breaks down milk proteins.
replication: Duplication.
repolarization: Movement of the transmembrane potential away from + mV values and toward the resting potential.
residual volume: Amount of air remaining in the lungs after maximum forced expiration.
respiration: Exchange of gases between living cells and the environment; includes pulmonary ventilation, external respiration, internal respiration, and cellular respiration.
respiratory minute volume: The amount of air moved in and out of the respiratory system each minute.
resting potential: The transmembrane potential of a normal cell under homeostatic conditions.
rete (RĒ-tē): An interwoven network of blood vessels or passageways.
reticular activating center: Mesencephalic portion of the reticular formation responsible for arousal and the maintenance of consciousness.
reticular formation: Diffuse network of gray matter that extends the entire length of the brain stem.
reticulospinal tracts: Descending tracts that carry involuntary motor commands issued by neurons of the reticular formation.
retina: The innermost layer of the eye, lining the vitreous chamber; also known as the neural tunic.
retinal (RET-i-nal): Visual pigment derived from vitamin A.
retraction: Movement posteriorly in the horizontal plane.
retroflexion (re-trō-FLEK-shun): A posterior tilting of the uterus that has no clinical significance.
retrograde flow (RET-rō-grād): Transport of materials from the telodendria to the soma of a neuron.
retroperitoneal (re-trō-per-i-to-NĒ-al): Situated behind or outside of the peritoneal cavity.
reverberation: Positive feedback along a chain of neurons, so that they remain active once stimulated.
rheumatism (ROO-ma-tizm): A condition characterized by pain in muscles, tendons, bones, or joints.
Rh factor: Agglutinogen that may be present (Rh-positive) or absent (Rh-negative) from the surfaces of red blood cells.
rhizotomy: Surgical transection of a dorsal root, usually performed to relieve pain.
rhodopsin (rō-DOP-sin): The visual pigment found in the membrane disks of the distal segments of rods.
rhythmicity center: Medullary center responsible for the basic pace of respiration; includes inspiratory and expiratory centers.
ribonucleic acid (rī-bō-nū-KLĀ-ik): A nucleic acid consisting of a chain of nucleotides that contain the sugar ribose and the nitrogen bases adenine, guanine, cytosine, and uracil.
ribose: A 5-carbon sugar that is a structural component of RNA.
ribosome: An organelle containing rRNA and proteins, that is essential to mRNA translation and protein synthesis.
right lymphatic duct: Lymphatic vessel delivering lymph from the right side of the head, neck, and chest to the venous system via the right subclavian vein.
rigor mortis: Extended muscular contraction and rigidity that occurs after death, as the result of calcium ion release from the SR and the exhaustion of cytoplasmic ATP reserves.
rod: Photoreceptor responsible for vision under dimly lit conditions.
rostral: Toward the nose; used when referring to relative position inside the skull.
rough endoplasmic reticulum (RER): A membranous organelle that is a site of protein synthesis and storage.
round window: An opening in the bony labyrinth of the inner ear that exposes the membranous wall of the scala tympani to the air of the middle ear cavity.
rubrospinal tracts: Descending tracts that carry involuntary motor commands issued by the red nucleus of the mesencephalon.
Ruffini corpuscles (rū-FĒ-nē): Receptors sensitive to tension and stretch in the dermis of the skin.
rugae (ROO-gē): Mucosal folds in the lining of the empty stomach that disappear as gastric distension occurs.
saccule (SAK-ūl): A portion of the vestibular apparatus of the inner ear, responsible for static equilibrium.
sagittal plane: Sectional plane that divides the body into left and right portions.
salivatory nucleus (SAL-i-va-tō-rē): Medullary nucleus that controls the secretory activities of the salivary glands.
salt: An inorganic compound consisting of a cation other than H^+ and an anion other than OH^-.
saltatory conduction: Relatively rapid conduction of an action potential between successive nodes of a myelinated axon.
sarcolemma: The cell membrane of a muscle cell.
sarcoma (sar-KŌ-ma): A tumor of connective tissues.
sarcomere: The smallest contractile unit of a striated muscle cell.
sarcoplasm: The cytoplasm of a muscle cell.
scala media: The central, endolymph-filled chamber of the inner ear; *see* **cochlear duct.**
scala tympani: The perilymph-filled chamber of the inner ear below the basilar membrane; pressure changes here distort the round window.
scala vestibuli: The perilymph-filled chamber of the inner ear above the vestibular membrane; pressure changes here result from distortions of the oval window.
scar tissue: Thick, collagenous tissue that forms at an injury site.

Schlemm, canal of: Passageway that delivers aqueous humor from the anterior chamber of the eye to the venous circulation.
Schwann cells: Glial cells responsible for the neurilemma that surrounds axons in the PNS.
sciatic nerve (sī-A-tik): Nerve innervating the posteromedial portions of the thigh and leg.
sciatica (sī-AT-i-ka): Pain felt along the peripheral distribution of the sciatic nerve.
sclera (SKLER-a): The fibrous, outer layer of the eye forming the white area of the anterior surface; a portion of the fibrous tunic of the eye.
sclerosis: A hardening and thickening that often occurs secondary to tissue inflammation.
scoliosis (skō-lē-Ō-sis): An abnormal, exaggerated lateral curvature of the spine.
scrotum (SKRŌ-tum): Loose-fitting, fleshy pouch that encloses the testes of the male.
sebaceous glands (sē-BĀ-shus): Glands that secrete sebum, usually associated with hair follicles.
sebum (SĒ-bum): A waxy secretion that coats the surfaces of hairs.
secondary sex characteristics: Physical characteristics that appear at puberty in response to sex hormones, but that are not involved in the production of gametes.
secretin (sē-KRĒ-tin): Duodenal hormone that stimulates pancreatic buffer secretion and inhibits gastric activity.
semen (SĒ-men): Fluid ejaculate containing spermatozoa and the secretions of accessory glands of the male reproductive tract.
semicircular canals: Tubular components of the vestibular apparatus responsible for dynamic equilibrium.
semilunar valve: A three-cusped valve guarding the exit from one of the cardiac ventricles; includes the pulmonary and aortic valves.
seminal vesicles (SEM-i-nal): Glands of the male reproductive tract that produce roughly 60 percent of the volume of semen.
seminiferous tubules (se-mi-NIF-e-rus): Coiled tubules where sperm production occurs in the testis.
senescence: Aging.
septae (SEP-tē): Partitions that subdivide an organ.
serosa: *See* **serous membrane.**
serotonin (ser-ō-TO-nin): A neurotransmitter in the CNS; a compound that enhances inflammation, released by activated mast cells and basophils.
serous cell/secretion: A cell that produces a watery secretion containing high concentrations of enzymes.
serous membrane: A squamous epithelium and the underlying loose connective tissue; the lining of the pericardial, pleural, and peritoneal cavities.
serum: Blood plasma from which clotting agents have been removed.
sesamoid bone: A bone that forms within a tendon.
sigmoid colon (SIG-mōyd): The S-shaped 18 cm portion of the colon between the descending colon and the rectum.
sign: A clinical term for visible evidence of the presence of a disease.
simple epithelium: An epithelium containing a single layer of cells above the basement membrane.
sinus: A chamber or hollow in a tissue; a large, dilated vein.
sinusoid (SĪ-nus-ōyd): An extensive network of vessels found in the liver, adrenal cortex, spleen, and pancreas; similar in histological structure to capillaries.
sinusitis: Inflammation of a nasal sinus.
skeletal muscle: A contractile organ of the muscular system.
skeletal muscle tissue: Contractile tissue dominated by skeletal muscle fibers; characterized as striated, voluntary muscle.
sliding filament theory: The concept that a sarcomere shortens as the thick and thin filaments slide past one another.
small intestine: The duodenum, jejunum, and ileum; the digestive tract between the stomach and large intestine.
smegma (SMEG-ma): Secretion of the preputial glands of the penis or clitoris.
smooth endoplasmic reticulum: Membranous organelle where lipid and carbohydrate synthesis and storage occur.
smooth muscle tissue: Muscle tissue found in the walls of many visceral organs; characterized as nonstriated, involuntary muscle.
soft palate: Fleshy posterior extension of the hard palate, separating the nasopharynx from the oral cavity.
sole: The inferior surface of the foot.
solute: Materials dissolved in a solution.
solution: A fluid containing dissolved materials.
solvent: The fluid component of a solution.
soma (SŌ-ma): Body.
somatic (sō-MAT-ik): Pertaining to the body.
somatic nervous system: The efferent division of the nervous system that innervates skeletal muscles.
somatomedins: Compounds stimulating tissue growth, released by the liver following GH secretion.
somatostatin: GH-IH, a hypothalamic regulatory hormone that inhibits GH secretion by the anterior pituitary.
somatotrophin: Growth hormone, produced by the anterior pituitary in response to GH-RH.
sperm: *See* **spermatozoa.**
spermatic cord: Spermatic vessels, nerves, lymphatics, and the ductus deferens, extending between the testes and the proximal end of the inguinal canal.
spermatids (SPER-ma-tidz): The product of meiosis in the male, cells that differentiate into spermatozoa.
spermatocyte (sper-MA-to-sīt): Cells of the seminiferous tubules that are engaged in meiosis.
spermatogenesis (sper-ma-to-JEN-e-sis): Sperm production.
spermatogonia (sper-ma-to-GŌ-nē-a): Stem cells whose mitotic divisions give rise to other stem cells and spermatocytes.
spermatozoa (sper-ma-to-ZŌ-a): Sperm cells; singular form = spermatozoon.
spermicide: Compound toxic to sperm cells, sometimes used as a contraceptive method.
spermiogenesis: The process of spermatid differentiation that leads to the formation of physically mature spermatozoa.
sphincter (SFINK-ter): Muscular ring that contracts to close the entrance or exit of an internal passageway.
spinal nerve: One of 31 pairs of nerves that originate on the spinal cord from anterior and posterior roots.
spindle apparatus: A muscle spindle (intrafusal fibers) and its sensory and motor innervation.
spinocerebellar tracts: Ascending tracts carrying sensory information to the cerebellum.
spinothalamic tracts: Ascending tracts carrying poorly localized touch, pressure, pain, vibration, and temperature sensations to the thalamus.
spinous process: Prominent posterior projection of a vertebra, formed by the fusion of two laminae.
spleen: Lymphoid organ important for red blood cell phagocytosis, immune response, and lymphocyte production.

splenectomy (sple-NEK-tō-mē): Surgical removal of the spleen.
sprain: Forceful distortion of an articulation that produces damage to the capsule, ligaments, or tendons but not dislocation.
sputum (SPŪ-tum): Viscous mucus ejected from the mouth after transport to the pharynx by the mucus escalator of the respiratory tract.
squama: A broad, flat surface.
squamous (SKWĀ-mus): Flattened.
squamous epithelium: An epithelium whose superficial cells are flattened and platelike.
stapedius (stā-PĒ-dē-us): A muscle of the middle ear whose contraction tenses the auditory ossicles and reduces the forces transmitted to the oval window.
stapes (STĀ-pez): The auditory ossicle attached to the tympanic membrane.
stenosis (ste-NŌ-sis): A constriction or narrowing of a passageway.
stereocilia: Elongate microvilli characteristic of the epithelium of the epididymis, portions of the ductus deferens, and the inner ear.
steroid: A ring-shaped lipid structurally related to cholesterol.
stimulus: An environmental alteration that produces a change in cellular activities; often used to refer to events that alter the transmembrane potentials of excitable cells.
stratified: Containing several layers.
stratum (STRĀ-tum): Layer.
stretch receptors: Sensory receptors that respond to stretching of the surrounding tissues.
stroma: The connective tissue framework of an organ, as distinguished from the functional cells (parenchyma) of that organ.
subarachnoid space: Meningeal space containing CSF; the area between the arachnoid membrane and the pia mater.
subclavian (sub-CLĀ-vē-an): Pertaining to the region under the clavicle.
subcutaneous layer: The layer of loose connective tissue below the dermis; also called the *hypodermis* or *superficial fascia.*
sublingual glands (sub-LING-gwal): Mucus-secreting salivary glands situated under the tongue.
submandibular glands: Salivary glands nestled in depressions on the medial surfaces of the mandible; salivary glands that produce a mixture of mucins and enzymes (salivary amylase).
submucosa (sub-mū-KŌ-sa): Region between the muscularis mucosae and the muscularis externa.
submucosal glands: Mucous glands in the submucosa of the duodenum.
subserous fascia: Loose connective tissue layer beneath the serous membrane lining the ventral body cavity.
substrate: A participant (product or reactant) in an enzyme-catalyzed reaction.
sulcus (SUL-kus): A groove or furrow.
summation: Temporal or spatial addition of stimuli.
superficial fascia: *See* **subcutaneous layer.**
superior: Directional reference meaning *above.*
superior vena cava: The vein that carries blood from the parts of the body above the heart to the right atrium.
supination (su-pi-NĀ-shun): Rotation of the forearm so that the palm faces anteriorly.
supine (SU-pīn): Lying face up, with palms facing anteriorly.
suppressor T cells: Lymphocytes that inhibit B cell activation and plasma cell secretion of antibodies.
suprarenal gland (sū-pra-RĒ-nal): *See* **adrenal gland.**
surfactant (sur-FAK-tant): Lipid secretion that coats alveolar surfaces and prevents their collapse.
sustentacular cells (sus-ten-TAK-ū-lar): Supporting cells of the seminiferous tubules of the testis, responsible for the differentiation of spermatids, the maintenance of the blood-testis barrier, and the secretion of inhibin, ABP, and MIH.
sutural bones: Irregular bones that form in fibrous tissue between the flat bones of the developing cranium; also called *Wormian bones.*
suture: Fibrous joint between flat bones of the skull.
sympathectomy (sim-path-EK-to-mē): Transection of the sympathetic innervation to a region.
sympathetic division: Division of the autonomic nervous system responsible for "fight or flight" reactions; primarily concerned with the elevation of metabolic rate and increased alertness.
sympathomimetic drugs: Drugs that mimic the actions of sympathetic stimulation.
symphysis: A fibrous amphiarthrosis, such as those between adjacent vertebrae or between the pubic bones of the coxae.
symptom: Clinical term for an abnormality of function due to the presence of disease.
synapse (SIN-aps): Site of communication between a nerve cell and some other cell; if the other cell is not a neuron, the term neuroeffector junction is often used.
synaptic delay (sin-AP-tik): The period between the arrival of an impulse at the presynaptic membrane and the initiation of an action potential in the postsynaptic membrane.
synarthrosis (sin-ar-THRŌ-sis): A joint that does not permit relative movement between the articulating elements.
synchondrosis (sin-kon-DRŌ-sis): A cartilaginous synarthrosis, such as the articulation between the epiphysis and diaphysis of a growing bone.
syncope: A sudden, transient loss of consciousness; a faint.
syncytial trophoblast: Multinucleate cytoplasmic layer that covers the blastocyst; the layer responsible for uterine erosion and implantation.
syncytium: A multinucleate mass of cytoplasm, produced by the fusion of cells or repeated mitoses without cytokinesis.
syndesmosis (sin-dez-MŌ-sis): A fibrous amphiarthrosis.
syndrome: A discrete set of symptoms that occur together.
syneresis (si-NER-e-sis): Clot retraction.
synergist (SIN-er-jist): A muscle that assists a prime mover in performing its primary action.
synostosis (sin-os-TŌ-sis): A synarthrosis formed through the fusion of the articulating elements.
synovial cavity (si-NŌ-vē-ul): Fluid-filled chamber in a diarthrodial joint.
synovial fluid (sī-NŌV-ē-ul): Substance secreted by synovial membranes that lubricates joints.
synovial membrane: An incomplete layer of fibroblasts confronting the synovial cavity, plus the underlying loose connective tissue.
synthesis (SIN-the-sis): Manufacture; anabolism.
system: An interacting group of organs that performs one or more specific functions.
systemic circuit: Vessels between the aortic semilunar valve and the entrance to the right atrium; the circulatory system other than vessels of the pulmonary circuit.
systole (SIS-to-lē): The period of cardiac contraction.
systolic pressure: Peak arterial pressures measured during ventricular systole.

tachycardia (tak-ē-KAR-dē-a): Abnormally rapid heart rate.
tactile: Pertaining to the sense of touch.
taenia coli (TĒ-nē-a KŌ-lī): Three longitudinal bands of smooth muscle in the muscularis externa of the colon.
tarsus: The ankle.
TCA cycle: Aerobic reaction sequence occurring in the mitochondrial matrix; in the process organic molecules are broken down, carbon dioxide molecules are released, and hydrogen molecules are transferred to coenzymes that deliver them to the electron transport system.
T cells: Lymphocytes responsible for cellular immunity, and for the coordination and regulation of the immune response; includes regulatory T cells (helpers and suppressors) and cytotoxic (killer) T cells.
tears: Fluid secretion of the lacrimal glands that bathes the anterior surfaces of the eyes.
tectorial membrane (tek-TŌR-ē-al): Gelatinous membrane suspended over the hair cells of the organ of Corti.
tectospinal tracts: Descending extrapyramidal tracts carrying involuntary motor commands issued by the colliculi.
tectum: The roof of the mesencephalon of the brain.
telencephalon (tel-en-SEF-a-lon): The forebrain or cerebrum, including the cerebral hemispheres, the internal capsule, and the cerebral nuclei.
telodendria (te-lō-DEN-drē-a): Terminal axonal branches that end in synaptic knobs.
telophase (TEL-o-fāz): The final stage of mitosis, characterized by the disappearance of the spindle apparatus, the reappearance of the nuclear membrane and the disappearance of the chromosomes, and the completion of cytokinesis.
temporal: Pertaining to time (temporal summation) or pertaining to the temples (temporal bone).
tendinitis: Painful inflammation of a tendon.
tendon: A collagenous band that connects a skeletal muscle to an element of the skeleton.
tendon organ: Receptor sensitive to tension in a tendon.
tentorium cerebelli (ten-TŌR-ē-um ser-e-BEL-ē): Dural partition that separates the cerebral hemispheres from the cerebellum.
teratogen (TER-a-to-jen): Stimulus that causes developmental defects.
teres: Long and round.
terminal: Toward the end.
tertiary follicle: A mature ovarian follicle, containing a large, fluid-filled chamber.
tertiary structure: Protein structure that develops as the result of interactions between distant portions of the same molecule.
testes (TES-tēz): The male gonads, sites of gamete production and hormone secretion.
testosterone (tes-TOS-te-rōn): The principal androgen produced by the interstitial cells of the testes.
tetanic contraction: Sustained skeletal muscle contraction due to repeated stimulation at a frequency that prevents muscle relaxation.
tetanus: A tetanic contraction; also used to refer to a disease state resulting from the stimulation of muscle cells by bacterial toxins.
tetany: A tetanic contraction; also used to refer to abnormally prolonged contractions resulting from disturbances in electrolyte balance.
tetrad (TET-rad): Paired, duplicated chromosomes visible at the start of meiosis I.
tetraiodothyronine (tet-ra-ī-ō-dō-THĪ-rō-nēn): T_4, or thyroxine, a thyroid hormone.
thalamus: The walls of the diencephalon.
thalassemia (thal-ah-SĒ-mē-ah): A hereditary disorder affecting hemoglobin synthesis and producing anemia.
theory: A hypothesis that makes valid predictions, as demonstrated by evidence that is testable, unbiased, and repeatable.
therapy: Treatment of disease.
thermogenesis (ther-mō-JEN-e-sis): Heat production.
thermography: Diagnostic procedure involving the production of an infrared image.
thermoreception: Sensitivity to temperature changes.
thermoregulation: Homeostatic maintenance of body temperature.
thick filament: A myosin filament in a skeletal or cardiac muscle cell.
thin filament: An actin filament in a skeletal or cardiac muscle cell.
thoracoabdominal pump (tho-ra-kō-ab-DOM-i-nal): Changes in the intrapleural pressures during the respiratory cycle that assist the venous return to the heart.
thoracolumbar division (thō-ra-kō-LUM-bar): The sympathetic division of the ANS.
thorax: The chest.
threshold: The transmembrane potential at which an action potential begins.
thrombin (THROM-bin): Enzyme that converts fibronogen to fibrin.
thrombocyte (THROM-bo-sīt): *See* **platelets.**
thrombocytopenia (throm-bō-sī-tō-PĒ-nē-ah): Abnormally low platelet count in the circulating blood.
thromboembolism (throm-bō-EM-bō-lizm): Occlusion of a blood vessel by a drifting blood clot.
thromboplastin: Enzyme that converts prothrombin to thrombin; enzyme formed by the intrinsic or extrinsic clotting pathways.
thrombus: A blood clot.
thymine: A pyrimidine found in DNA.
thymosins (thī-MO-sinz): Thymic hormones essential to the development and differentiation of T cells.
thymus: Lymphoid organ, site of T cell formation.
thyroglobulin (thī-rō-GLOB-ū-lin): Circulating transport globulin that binds thyroid hormones.
thyroid gland: Endocrine gland whose lobes sit lateral to the thyroid cartilage of the larynx.
thyroid hormones: Thyroxine (T_4) and triiodothyronone (T_3) hormones of the thyroid gland; hormones that stimulate tissue metabolism, energy utilization, and growth.
thyroid stimulating hormone (TSH): Anterior pituitary hormone that triggers the secretion of thyroid hormones by the thyroid gland.
thyroxine (TX) (thī-ROKS-in): A thyroid hormone (T_4).
tidal volume: The volume of air moved in and out of the lungs during a normal quiet respiratory cycle.
tissue: A collection of specialized cells and cell products that perform a specific function.
titer: Plasma antibody concentration.
tolerance: Failure to produce antibodies against antigenic compounds normally present in the body.
tonic response: An increase or decrease in the frequency of action potentials by sensory receptors that are chronically active.
tonsil: A lymphoid nodule beneath the epithelium of the pharynx; includes the palatine, pharyngeal, and lingual tonsils.
topical: Applied to the body surface.
toxic: Poisonous.
trabecula (tra-BEK-ū-la): A connective tissue partition that subdivides an organ.

trabeculae carneae (tra-BEK-ū-lē CAR-nē-ē): Muscular ridges projecting from the walls of the ventricles of the heart.
trachea (TRĀ-kē-a): The windpipe, an airway extending from the larynx to the primary bronchi.
tracheal ring: C-shaped supporting cartilage of the trachea.
tracheostomy (trā-kē-OS-to-mē): Surgical opening of the anterior tracheal wall to permit airflow.
trachoma: An infectious disease of the conjunctiva and cornea.
tract: A bundle of axons inside the CNS.
tractotomy: The surgical transection of a tract, sometimes used to relieve pain.
transamination (trans-am-i-NĀ-shun): The enzymatic transfer of an amine group from an amino acid to another carbon chain.
transcription: The encoding of genetic instructions on a strand of mRNA.
transdermal medication: Administration of medication by absorption through the skin.
transection: To sever or cut in the transverse plane.
transfusion: Transfer of blood from a donor directly into the bloodstream of another person.
transient ischemic attack: A temporary loss of consciousness due to the occlusion of a small blood vessel in the brain.
translation: The process of peptide formation using the instructions carried by an mRNA strand.
transmembrane potential: The potential difference, in millivolts, measured across the cell membrane; a potential difference that results from the uneven distribution of positive and negative ions across a cell membrane.
transudate (TRANS-ū-dāt): Fluid that diffuses across a serous membrane and lubricates opposing surfaces.
treppe (TREP-e): "Staircase" increase in tension production following repeated stimulation of a muscle, even though the muscle is allowed to complete each relaxation phase.
triad (liver): The combination of branches of the hepatic duct, the hepatic portal vein, and the hepatic artery, found at each corner of a liver lobule.
triad (muscle cell): The combination of a T tubule and two cisternae of the sarcoplasmic reticulum.
tricarboxylic acid cycle (trī-kar-bok-SIL-ik): *See* **TCA cycle.**
tricuspid valve (trī-KUS-pid): The right atrioventricular valve that prevents backflow of blood into the right atrium during ventricular systole.
trigeminal nerve (trī-GEM-i-nal): Cranial nerve V, responsible for providing sensory information from the lower portions of the face, including the upper and lower jaws, and delivering motor commands to the muscles of mastication.
triglyceride (trī-GLIS-e-rīd): A lipid composed of a molecule of glycerol attached to three fatty acids.
trigone (TRĪ-gōn): Triangular region of the urinary bladder bounded by the exits of the ureters and the entrance to the urethra.
triiodothyronine: T_3, one of the thyroid hormones.
trisomy: The abnormal possession of three copies of a chromosome; trisomy 21 is responsible for the Down syndrome.
trochanter (trō-KAN-ter): Large processes near the head of the femur.
trochlea (TRŌK-le-a): A pulley.
trochlear nerve (TRŌK-lē-ar): Cranial nerve IV, controlling the superior oblique muscle of the eye.
trophoblast (TRŌ-fō-blast): Superficial layer of the blastocyst that will be involved with implantation, hormone production, and placenta formation.
troponin/tropomyosin (trō-PŌ-nin) (trō-pō-MĪ-ō-sin): Proteins on the thin filaments that mask the active sites in the absence of free calcium ions.
trunk: The thoracic and abdominopelvic regions.
trypsin (TRIP-sin): One of the pancreatic proteases.
trypsinogen: The inactive, proenzyme secreted by the pancreas and converted to trypsin in the duodenum.
T tubules: Transverse, tubular extensions of the sarcolemma that extend deep into the sarcoplasm to contact cisternae of the sarcoplasmic reticulum.
tuberculum (tū-BER-kū-lum): A small, localized elevation on a bony surface.
tuberosity: A large, roughened elevation on a bony surface.
tubulin: Protein subunit of microtubules.
tumor: A tissue mass formed by the abnormal growth and replication of cells.
tunica (TŪ-ni-ka): A layer or covering; in blood vessels: t. externa, the outermost layer of connective tissue fibers that stabilizes the position of the vessel; t. intima, the innermost layer, consisting of the endothelium plus an underlying elastic membrane; t. media, a middle layer containing collagen, elastin, and smooth muscle fibers in varying proportions.
turbinates: *See* **concha/conchae.**
T wave: Deflection of the ECG corresponding to ventricular repolarization.
twitch: A single contraction/relaxation cycle in a skeletal muscle.
tympanic membrane (tim-PAN-ik): Membrane that separates the external auditory canal from the middle ear; membrane whose vibrations are transferred to the auditory ossicles and ultimately to the oval window; the "eardrum."
type A axons: Large myelinated axons.
type B axons: Smaller myelinated axons.
type C axons: Small unmyelinated axons.
ulcer: An area of epithelial sloughing associated with damage to the underlying connective tissues and vasculature.
ultrasound: Diagnostic visualization procedure that uses high-frequency sound waves.
umbilical cord (um-BIL-i-kal): Connecting stalk between the fetus and the placenta; contains the allantois, the umbilical arteries, and the umbilical vein.
umbilicus: The navel.
unicellular gland: Goblet cells.
unipennate muscle: A muscle whose fibers are all arranged on one side of the tendon.
unipolar neuron: A sensory neuron whose soma lies in a dorsal root ganglion or a sensory ganglion of a cranial nerve.
unmyelinated axon: Axon whose neurilemma does not contain myelin, and where continuous conduction occurs.
uracil: One of the pyrimidines characteristic of RNA.
uremia (ū-RĒ-mē-a): Abnormal condition caused by impaired kidney function, characterized by the retention of wastes and the disruption of many other organ systems.
ureters (ū-RĒ-terz): Muscular tubes, lined by transitional epithelium, that carry urine from the renal pelvis to the urinary bladder.
urethra (ū-RĒ-thra): A muscular tube that carries urine from the urinary bladder to the exterior.
urethritis: Inflammation of the urethra.
urinalysis: Analysis of the physical and chemical characteristics of the urine.
urinary bladder: Muscular, distensible sac that stores urine prior to micturition.

urination: The voiding of urine; micturition.
urobilinogen: A compound derived from the bilirubin excreted in the bile.
uterus (Ū-ter-us): Muscular organ of the female reproductive tract where implantation, placenta formation, and fetal development occur.
utricle (Ū-trē-k'l): The largest chamber of the vestibular apparatus; contains a macula important for static equilibrium.
uvea: The vascular tunic of the eye.
uvula (Ū-vū-la): A dangling, fleshy extension of the soft palate.
vagina (va-JĪ-na): A muscular tube extending between the uterus and the vestibule.
varicose veins (VAR-i-kōs): Distended superficial veins.
vasa vasorum: Blood vessels that supply the walls of large arteries and veins.
vascular: Pertaining to blood vessels.
vascularity: The blood vessels in a tissue.
vascular spasm: Contraction of the wall of a blood vessel at an injury site, a process that may slow the rate of blood loss.
vasoconstriction: A reduction in the diameter of arterioles due to contraction of smooth muscles in the tunica media; an event that elevates peripheral resistance, and that may occur in response to local factors, through the action of hormones, or from stimulation of the vasomotor center.
vasodilation (vaz-ō-dī-LĀ-shun): An increase in the diameter of arterioles due to the relaxation of smooth muscles in the tunica media; an event that reduces peripheral resistance, and that may occur in response to local factors, through the action of hormones, or following decreased stimulation of the vasomotor center.
vasomotion: Alterations in the pattern of blood flow through a capillary bed in response to changes in the local environment.
vasomotor center: Medullary center whose stimulation produces vasoconstriction and an elevation in peripheral resistance.
vein: Blood vessel carrying blood from a capillary bed toward the heart.
venae cavae (VĒ-na CĀ-va): The major veins delivering systemic blood to the right atrium.
ventilation: Air movement in and out of the lungs.
ventilatory rate: The respiratory rate.
ventral: Pertaining to the anterior surface.
ventricle (VEN-tri-k'l): One of the large, muscular pumping chambers of the heart that discharges blood into the pulmonary or systemic circuits.
ventricular escape: The initiation of ventricular contractions after a pause caused by impaired conduction of the contractile stimulus from the AV node.
ventricular folds: Mucosal folds in the laryngeal walls that do not play a role in sound production; the false vocal cords.
venule (VEN-ūl): Thin-walled veins that receive blood from capillaries.
vermis (VER-mis): Midsagittal band of neural cortex on the surface of the cerebellum.
vertebral canal: Passageway that encloses the spinal cord, a tunnel bounded by the neural arches of adjacent vertebrae.
vertebral column: The cervical, thoracic, and lumbar vertebrae, the sacrum, and the coccyx.
vertebrochondral ribs: Ribs 8–10, false ribs connected to the sternum by shared cartilaginous bars.
vertebrosternal ribs: Ribs 1–7, true ribs connected to the sternum by individual cartilaginous bars.
vertigo: Dizziness.
vesicle: A membranous sac in the cytoplasm of a cell.
vestibular membrane: The membrane that separates the scala media from the scala vestibuli of the inner ear.
vestibular nucleus: Processing center for sensations arriving from the vestibular apparatus; located near the border between the pons and medulla.
vestibule (VES-ti-būl): A chamber; in the inner ear, the term refers to the utricle, saccule, and semicircular canals.
vestibulospinal tracts: Descending tracts of the extrapyramidal system, carrying involuntary motor commands issued by the vestibular nucleus to stabilize the position of the head.
villus: A slender projection of the mucous membrane of the small intestine.
virus: A pathogenic microorganism.
viscera: Organs in the ventral body cavity.
visceral: Pertaining to viscera or their outer coverings.
visceral smooth muscle: Smooth muscle tissue forming sheets or layers in the walls of visceral organs; the cells may not be innervated, and the layers often show automaticity (rhythmic contractions).
viscosity: The resistance to flow exhibited by a fluid due to molecular interactions within the fluid.
viscous: Thick, syrupy.
vital capacity: The maximum amount of air that can be moved in or out of the respiratory system; the sum of the inspiratory reserve, the expiratory reserve, and the tidal volume.
vitamin: An essential organic nutrient that functions as a coenzyme in vital enzymatic reactions.
vitreous humor: Gelatinous mass in the vitreous chamber of the eye.
vocal folds: Folds in the laryngeal wall containing elastic ligaments whose tension can be voluntarily adjusted; the true vocal cords, responsible for phonation.
voluntary: Controlled by conscious thought processes.
vulva (VUL-va): The female pudendum (external genitalia).
Wallerian degeneration: Disintegration of an axon and its myelin sheath distal to an injury site.
white blood cells: Leukocytes; the granulocytes and agranulocytes of the blood.
white matter: Regions inside the CNS that are dominated by myelinated axons.
white ramus: A nerve bundle containing the myelinated preganglionic axons of sympathetic motor neurons en route to the sympathetic chain or a collateral ganglion.
Wormian bones: *See* **sutural bones.**
xiphoid process (ZĪ-fōid): Slender, inferior extension of the sternum.
Y chromosome: The sex chromosome whose presence indicates that the individual is a genetic male.
yolk sac: One of the three extraembryonic membranes, composed of an inner layer of endoderm and an outer layer of mesoderm.
Zeis, glands of (ZĪS): Enlarged sebaceous glands on the free edges of the eyelids.
zona fasciculata (ZŌ-na) (fa-sik-ū-LA-ta): Region of the adrenal cortex responsible for glucocorticoid secretion.
zona glomerulosa (glo-mer-u-LŌ-sa): Region of the adrenal cortex responsible for mineralocorticoid secretion.
zona pellucida (pel-LŪ-si-da): Region between a developing oocyte and the surrounding follicular cells of the ovary.
zona reticularis (re-ti-kū-LAR-is): Region of the adrenal cortex responsible for androgen secretion.
zygote (ZĪ-gōt): The fertilized ovum, prior to the start of cleavage.

INDEX

F

G

J

Q

T

U

ILLUSTRATION CREDITS

Photographs

Page xxvi Hawaiian paddlers: Don King/The Image Bank; ice fishing: Robert Semeniuk/The Stock Market

Page xxxvii scan of pelvis: Dr. Eugene C. Wasson, III; muscles of the thigh: Ralph Hutchings; pelvis and lower extremity: Custom Medical Stock Photo; major arteries of the leg: Ralph Hutchings

Chapter 1 **CO** Carol L. Newsom **1-14a,b** Science Source/Photo Researchers, Inc. **1-15a** CNRI/Photo Researchers, Inc. **1-15b** Photo Researchers, Inc. **1-15c** Diagnostic Radiological Helath Sciences Learning Lab

Chapter 2 **CO** David Spears/Science Photo Library **2-20c** PH Archives

Chapter 3 **CO** P. Motta/Dept. of Anatomy/University La Sapienza/Photo Researchers, Inc. **3-8a,c** David M. Phillips/Visuals Unlimited **3-8b** Dana-Farber Cancer Institute **3-13b** Lennart Nilsson/Bonnier Fakta **3-15a** Fawcett/Hirokawa/Heuser/Science Source/ Photo Researchers, Inc. **3-16a** PH Archives **3-17a** CNRI/Science Source/Photo Researchers, Inc. **3-18b** PH Archives **3-19a** Biophoto Associates/Photo Researchers, Inc. **3-19d** Dr. Birgit Satir, Yeshiva University **3-21a** Don Fawcett/Harvard Medical School **3-21b** Biophoto Associates/Photo Researchers, Inc. **3-29** Ed Reschke/ Peter Arnold, Inc.

Chapter 4 **CO** David M. Phillips/Photo Researchers, Inc. **4-1b** Custom Medical Stock Photo **4-2e** M.G. Farquhar and J.E. Palade, "Junctional Complexes in Epithelia" *Journal of Cell Biology*, 17: 379 (1963) **4-3c** Dr. C.P. LeBlond, McGill University **4-4a** Wards Natural Science Establishment, Inc. **4-4b** Frederic H. Martini **4-5a** PH Archives **4-5b,c** Frederic H. Martini **4-6a,b,c** Frederic H. Martini **4-7a** J.J. Head/Carolina Biological Supply **4-9a** Wards Natural Science Establishment, Inc. **4-9b,c** L.V. Bergman & Associates, Inc. **4-10a** Frederic H. Martini **4-10b** Wards Natural Science Establishment, Inc. **4-11a,b** John D. Cunningham/Visuals Unlimited **4-11c** Frederic H. Martini **4-13a** Robert Brons/Biological Photo Service/Terraphotographics **4-13b** Science Source/ Photo Researchers, Inc. **4-13c** Ed Reschke/Peter Arnold, Inc. **4-14** Frederic H. Martini **4-17a** Eric Grave/Phototake **4-17b** Phototake **4-17c & 4-18** Frederic H. Martini

Chapter 5 **CO** Paul Trummer/The Image Bank **5-2b,c** Frederic H. Martini **5-3** Wehr/Custom Medical Stock Photo **5-4** PH Archives **5-6a** Elias-Paul's & Peters, Amenta, *Histology and Microanatomy, 5/E,* Puccin Publishers, 1987 **5-6b** Manfred Kage/Peter Arnold, Inc. **5-7b** John D. Cunningham/Visuals Unlimited **5-8** Frederic H. Martini **5-9a** Carolina Biological Supply/Phototake **5-9b** Frederic H. Martini

Chapter 6 **CO** Superstock **6-1a,b** Dr. Richard Kessel and Dr. Randy H. Kardon **6-2b,c** Biophoto Associates/ Photo Researchers, Inc. **6-4** Frederic H. Martini **6-5a,b** Ralph Hutchings **6-7b** Frederic H. Martini **6-7c,d** L.V. Bergman & Associates, Inc. **Focus** comminuted and epiphyseal fractures: Visuals Unlimited; transverse fracture: Grace Moore/Medichrome; spiral fracture: Peter Arnold, Inc; Colles' fracture: Scott Camazine/ Photo Researchers, Inc.; compression fracture: L.V. Bergman & Associates, Inc.; displaced fracture: Custom Medical Stock Photo

Chapter 7 **CO** Martha Cooper/Peter Arnold, Inc. **7-1b** Ralph Hutchings **7-7** Michael J. Timmons **7-21a** Ralph Hutchings **Table 7-2** Ralph Hutchings

Chapter 8 **CO** Roy Morsch/The Stock Market **8-2** Bates/Custom Medical Stock Photo **8-3 & 8-12** Ralph Hutchings **8-14c** Photo Researchers, Inc.

Chapter 9 **CO** The Stock Market

Chapter 10 **CO** PHOTRI **10-2** Wards Natural Science Establishment, Inc. **10-3** Don Fawcett/Photo Researchers, Inc. **10-6** J.J. Head/Carolina Biological Supply **10-7a** Fred Hossler/Visuals Unlimited **10-7b** Don Fawcett/Science Source/Photo Researchers, Inc. **10-17a** Frederic H. Martini **10-17b** Cormack, D. (ed.), *Ham's Histology, 9/e.* Philadelphia: J.B. Lippincott Publishers. 1987. Fig. 15-11. **10-20a** G.W. Willis/ Biological Photo Service/Terraphotographics **10-21a** Frederic H. Martini

Chapter 11 **CO** Anne Marie Weber/The Stock Market

Chapter 12 **CO** Synapier/Science Photo Library/Photo Researchers, Inc. **12-2** Frederic H. Martini **12-3** Michael J. Timmons **12-4a** Biophoto Associates/Photo Researchers, Inc. **12-4b** Photo Researchers, Inc. **12-5a** PH Archives **12-6** David Scott/Phototake

Chapter 13 **CO** Globus Brothers/The Stock Market **13-2c** Ralph Hutchings **13-3b** Michael J. Timmons **13-4b** R.G. Kessel and R.H. Kardon, Tissues and Organs: A Text-Atlas of Scanning Electron Microscopy. Copyright C 1979 W.H. Freeman & Co. Used with permission.

Chapter 14 **CO** Will & Deni McIntyre/Photo Researchers, Inc. **14-1a,b** Ralph Hutchings **14-6a** Michael J. Timmons **14-8b** Michael J. Timmons **14-8c** Pat Lynch/Photo Researchers, Inc. **14-9, 14-13b, 14-15a,b lower** Ralph Hutchings **14-15b upper** Wards Natural Science Establishment, Inc. **14-17a** Ralph Hutchings

Chapter 15 **CO** Max Hilaire/The Image Bank **15-9e** Larry Mulvehill/Photo Researchers, Inc.

Chapter 16 **CO** Gary Gladstone/The Image Bank

Chapter 17 **CO** M.C. Escher/Art Resource **17-5b** Frederic H. Martini **17-5c** G.W. Willis/Biological Photo Service/Terraphotographics **17-7a** Ralph Hutchings **17-9a** Ed Reschke/Peter Arnold, Inc. **17-9c** Custom Medical Stock Photo **17-19b** Richmond Products **17-23b** Lennart Nilsson/Bonnier Fakta **17-27d** Wards Natural Science Establishment, Inc.

Chapter 18 **CO** Focus on Sports **18-7b** Manfred Kage/ Peter Arnold, Inc. **18-11b,c & 18-13b** Frederic H. Martini **18-13c** Martin Rotker/Phototake **18-15c & 18-18b** Wards Natural Science Establishment, Inc. **18-18c,d** Pfizer, Inc. **18-22a,b** L.V. Bergman & Associates, Inc. **18-22c** John Paul Kay/Peter Arnold, Inc. **18-22d** Custom Medical Stock Photo **18-22e** Biophoto Associates/Science Source/Photo Researchers, Inc.

Chapter 19 **CO** Lester Lefkowitz/Tony Stone Images **19-1** Martin M. Rotker **19-2a** Frederic H. Martini

19-2b Ed Reschke/Peter Arnold, Inc. **19-2c** David Scharf/Peter Arnold, Inc. **19-4a,b** Stanley Flegler/Visuals Unlimited **19-9a-e** Ed Reschke/Peter Arnold, Inc. **19-11** Frederic H. Martini **19-13** Custom Medical Stock Photo

Chapter 20 **CO** Lunagrafix, Inc./Photo Researchers, Inc. **20-3a** Ralph Hutchings **20-4b** Bonnier Fakta **20-4c** Ralph Hutchings **20-6c left** Science Photo Library/Photo Researchers, Inc. **20-6c right** Biophoto Associates/Science Source/Photo Researchers, Inc. **20-7c** Peter Arnold, Inc. **20-8b,d** Ralph Hutchings **20-13** St. Bartholomew's Hospital/Science Photo Library/Photo Researchers, Inc.

Chapter 21 **CO** Biophoto Associates/Photo Researchers, Inc. **21-2** Michael J. Timmons **21-4a** B & B Photo/Custom Medical Stock Photo **21-4b** William Ober/Visuals Unlimited **21-6** Biophoto Associates/Photo Researchers, Inc. **21-11** PH Archives

Chapter 22 **CO** Jeffrey Reed/Medichrome **22-3b** Frederic H. Martini **22-6a** Wards Natural Science Establishment, Inc. **22-6b** Biophoto Associates/Photo Researchers, Inc. **22-8c,d & 22-9c** Frederic H. Martini **22-19b** Ken Eward/Biografx-Science Source/Photo Researchers, Inc.

Chapter 23 **CO** D. Kirkland/Sygma **23-2a** Photo Researchers, Inc. **23-2b** Frederic H. Martini **23-4d** Phototake **23-5c** John D. Cunningham/Visuals Unlimited **23-7** Ralph Hutchings **23-8a** University of Toronto **23-8b** Ralph Hutchings **23-10a & 23-11b,c** Adapted from Junqueira, Carneira, & Long, *Basic Histology, 5/e.* Norwalk, CT: Appleton-Century-Crofts, 1986. **23-10b** Wards Natural Science Establishment, Inc. **23-10c** Bloom & Fawcett, *Textbook of Histology, 11/e.* Copyright 1986, W.B. Saunders Publishers

Chapter 24 **CO** Erich Lessing/Art Resource **24-7b** Frederic H. Martini **24-10a** Alfred Pasieka/Peter Arnold, Inc. **24-10b** Wards Natural Science Establishment, Inc. **24-10c** Astrid & Hanns Friedler-Michler/Science Photo Library/Photo Researchers, Inc. **24-12a** Ralph Hutchings **24-13a** Reproduced from R.G. Kessel & R.H. Kardon, *Tissues and Organs: A Text-Atlas of Scanning Electron Microscopy,* W.H. Freeman & Co., 1979 **24-13d** Wards Natural Science Establishment, Inc. **24-16a** Phototake **24-13e** Michael J. Timmons **24-18c** Wards Natural Science Establishment, Inc. **24-20b,c** Michael J. Timmons **24-23c** Wards Natural Science Establishment, Inc.

Chapter 25 **CO left** Don King/The Image Bank **CO right** Robert Semenlik/The Stock Market

Chapter 26 **CO** Kavaler/Art Resource **26-3a & 26-4b** Ralph Hutchings **26-6b** PH Archives **26-6d** David M. Phillips/Visuals Unlimited **26-7a,b,c** PH Archives **26-9a** Ralph Hutchings **26-9b** Reproduced from R.G. Kessel & R.H. Kardon, *Tissues and Organs: A Text-Atlas of Scanning Electron Microscopy,* W.H. Freeman & Co., 1979 **26-19** Photo Researchers, Inc. **26-20a** Peter Arnold, Inc. **26-22a** Wards Natural Science Establishment, Inc. **26-22b,c** Frederic H. Martini

Chapter 27 **CO** Jim Olive/Peter Arnold, Inc.

Chapter 28 **CO** Elyse Lewin/The Image Bank **28-4b** Frederic H. Martini **28-5a** Bloom & Fawcett, *Textbook of Histology, 11/e.* Copyright 1986, W.B. Saunders Publishers **28-5b** Wards Natural Science Establishment, Inc. **28-7b** David M. Phillips/Visuals Unlimited **28-8b** Wards Natural Science Establishment, Inc. **28-8c & 28-9b,c,d** Frederic H. Martini **28-9e upper** Wards Natural Science Establishment, Inc. **28-9e lower** Dr. Richard Kessel and Dr. Randy H. Kardon **28-10c** Wards Natural Science Establishment, Inc. **28-15(1,2,3)** Frederic H. Martini **28-15(4)** Wards Natural Science Establishment, Inc. **28-15(5)** PH Archives **28-15(6,7)** G.W. Willis/Biological Photo Service **28-16b** K.R. Porter **28-16c** Custom Medical Stock Photo **28-17a** Ralph Hutchings **28-18b** Wards Natural Science Establishment, Inc. **28-19a,b,d** Frederic H. Martini **28-19c & 28-20** Michael J. Timmons **28-22b** Wards Natural Science Establishment, Inc. **28-22c** Frederic H. Martini

Chapter 29 **CO** Petit Format/Nestle/Photo Researchers, Inc. **29-1a** Francis Leroy, Biocosmos/Science Photo Library/Custom Medical Stock Photo **29-6b** Frederic H. Martini **29-7a,b,c** Lennart Nilsson/Bonnier Fakta **29-7d** Photo Researchers, Inc. **29-8a** Lennart Nilsson/Bonnier Fakta **29-8b & 29-15** Photo Researchers, Inc.

Art

All embryology summaries, system integrators, and anatomical paintings were rendered by William Ober and Claire Garrison, with the exception of the paintings noted below.

Illustrations by Ron Ervin

3-19b, 6-13, 7-1a, 7-2, 7-3, 7-4, 7-5c, 7-6, 7-7a,d, 7-8, 7-9, 7-10, 7-11b, 7-13, 7-15, 7-16, 7-20, 7-21b,c, 7-23a,b, 8-1, 8-4, 8-5, 8-8, 8-10, 8-11, 8-13, 8-14, 9-8, 9-10, 9-11, 11-4a, 11-6, 11-7, 11-13, 11-21a,b, 11-22a, 11-23, 13-18

Illustrations by Tina Sanders

7-11a, 7-12, 7-14b, 8-9, 13-4a, 14-5b, 14-7, 14-14, 15-8, 16-3, 16-7, 16-10, 17-8a, 18-11c, 24-22, 25-20

Illustrations by Craig Luce

1-15, 20-3, 20-4, 20-8, 20-11a, 23-3a, 28-10a,b (right), 28-13, 28-21, 28-22

Credits

Figures 2-7, 2-9, 22-19a, 22-21, 22-26, and **A-1** from Jacquelyn C. Black, *Microbiology: Principles and Applications, 2/e.* Englewood Cliffs, NJ: Prentice-Hall, Inc., 1993 (Figures 2.4, 2.6a, 2.7, 18.7, 18.8, 19.1, and 3.2)

Table 21-3 Data Sources: Asmussen and Kielson, *Acta Physiogogica Scandinavia,* 27:217 (1952) and Reindell et. al., *Schweiz, Zeitschr. f. Sportmed.,* 1:97 (1953)

Figure 29-16 Adapted from *Time,* January 17, 1994, pp. 50–51, graphic by Nigel Holmes, research by Leslie Dickstein, source: Dr. Victor A. McKusick, Johns Hopkins University

ABBREVIATIONS USED IN THIS TEXT

ACh	acetylcholine
AChE	acetylcholinesterase
ACTH	adrenocorticotrophic hormone
ADH	antidiuretic hormone
ADP	adenosine diphosphate
AIDS	acquired immunodeficiency syndrome
ALS	amyotrophic lateral sclerosis
AMP	adenosine monophosphate
ANP	atrial natriuretic peptide
ANS	autonomic nervous system
AP	arterial pressure
ARDS	adult respiratory distress syndrome
atm	atmospheric pressure
ATP	adenosine triphosphate
ATPase	adenosine triphosphatase
AV	atrioventricular
BCDF	B cell differentiation factor
BCGF	B cell growth factor
BMR	basal metabolic rate
BCOP	blood colloid osmotic pressure
bpm	beats per minute
BUN	blood urea nitrogen
C	kilocalorie; centigrade
CABG	coronary artery bypass graft
CAD	coronary artery disease
cAMP	cyclic-AMP
CAPD	continuous ambulatory peritoneal dialysis
CCK	cholecystokinin
CCOP	capsular colloid osmotic pres sure
CF	cystic fibrosis
CHF	congestive heart failure
C_{hp}	capillary hydrostatic pressure
CNS	central nervous system
CO	cardiac output; carbon monoxide
CoA	coenzyme A
COMT	cathecol-O-methyltransferase
COPD	chronic obstructive pulmonary disease
CP	creatine phosphate
CPK or CK	creatine phosphokinase
CPM	continual passive motion
CPR	cardiopulmonary resuscitation
CRF	chronic renal failure
CRH	corticotropin-releasing hormone
CSF	cerebrospinal fluid; colony-stimulating factors
CT	computed tomography; calcitonin
CVA	cerebrovascular accident
CVS	cardiovascular system
DAG	diacylglycerol
DC	Doctor of Chiropractory
DCT	distal convoluted tubule
DDST	Denver Developmental Screening Test
DIC	disseminated intravascular coagulation
DJD	degenerative joint disease
DMD	Duchenne's muscular dystrophy
DNA	deoxyribonucleic acid
DO	Doctor of Osteopathy
DPG	diphosphoglycerate
DPM	Doctor of Podiatric Medicine
ECF	extracellular fluid
ECG	electrocardiogram
EDV	end-diastolic volume
EEG	electroencephalogram
EKG	electrocardiogram
ELISA	enzyme-linked immunoabsorbent assay
EPSP	excitatory postsynaptic potential
ESV	end-systolic volume
ETS	electron transport system
FAD	flavin adenine dinucleotide
FAS	fetal alcohol syndrome
FES	functional electrical stimulation
FMN	flavin mononucleotide
FSH	follicle-stimulating hormone
GABA	gamma aminobutyric acid
GAS	general adaptation syndrome
GC	glucocorticoids
GFR	glomerular filtration rate
GH	growth hormone
GH-IH	growth hormone-inhibiting hormone
G_{hp}	glomerular hydrostatic pressure
GH-RH	growth hormone-releasing hormone
GIP	glucose-dependent insulinotropic hormone
GnRH	gonadotropin-releasing hormone
GTP	guanosine triphosphate
Hb	hemoglobin
hCG	human chorionic gonadotropin
HCl	hydrochloric acid
HDL	high-density lipoprotein
HDN	hemolytic disease of the newborn
hGh	human growth hormone
HIV	human immunodeficiency virus
HLA	human leukocyte antigen
HMD	hyaline membrane disease
HP	hydrostatic pressure
HPL	human placental lactogen
HR	heart rate
HTLV-III	human T cell lymphotrophic virus, type III (= HIV)
Hz	Hertz
ICF	intracellular fluid
ICSH	interstitial cell-stimulating hormone, also called LH
IH	inhibiting hormone
IM	intramuscular
IP_3	inositol triphosphate
IPSP	inhibitory postsynaptic potential
ISF	interstitial fluid
IUD	intrauterine device
IVF	in vitro fertilization
kc	kilocalorie
LDH	lactate dehydrogenase
LDL	low-density lipoprotein
L-DOPA	levodopa
LH	luteinizing hormone
LM	light micrograph
LSD	lysergic acid diethylamide
MAO	monoamine-oxidase
MAP	mean arterial pressure
MC	mineralocorticoids
MD	Doctor of Medicine
mEq	millequivalents
MHC	major histocompatibility complex
MI	myocardial infarction
mm Hg	millimeters of mercury
mmol	millimoles
mOsm	milliosmoles
MRI	magnetic resonance imaging
mRNA	messenger RNA
MS	multiple sclerosis
MSH	melanocyte-stimulating hormone
MSH-IH	melanocyte-stimulating hormone–inhibiting hormone
NAD	nicotinamide adenine dinucleotide
NE	norepinephrine
NFP	net filtration pressure
NRDS	neonatal respiratory distress syndrome
OP	osmotic pressure
Osm	osmoles
OT	oxytocin
PAC	premature atrial contractions
PAT	paroxysmal atrial tachycardia
PCT	proximal convoluted tubule
PCV	packed cell volume
PEEP	positive end-expiratory pressure
PET	positron-emission tomography
PFC	perfluorochemical emulsions
PG	prostaglandin
PID	pelvic inflammatory disease
PIH	prolactin-inhibiting hormone
PIP	phosphatidylinositol
PKC	protein kinase C
PKU	phenylketonuria
PLC	phospholipase C
PMN	polymorphonuclear leukocyte
PNS	peripheral nervous system
PR	peripheral resistance
PRL	prolactin
psi	pounds per square inch
PT	prothrombin time
PTA	post-traumatic amnesia; plasma thromboplastin antecedent
PTC	phenylthiocarbamide
PTH	parathyroid hormone
PVC	premature ventricular contraction
RAS	reticular activating system
RBC	red blood cell
RDA	recommended daily allowance
RDS	respiratory distress syndrome
REM	rapid eye movement
RER	rough endoplasmic reticulum
RH	releasing hormone
RHD	rheumatic heart disease
RLQ	right lower quadrant
RNA	ribonucleic acid
rRNA	ribosomal RNA
SA	sinoatrial
SCA	sickle cell anemia
SCID	severe combined immunodeficiency disease
SEM	scanning electron micrograph
SER	smooth endoplasmic reticulum
SGOT	serum glutamic oxaloacetic transaminase
SIADH	syndrome of inappropriate ADH secretion
SIDS	sudden infant death syndrome
SLE	systemic lupus erythematosus
SNS	somatic nervous system
STD	sexually transmitted diseases
SV	stroke volume
SVC	superior vena cava
T_3	triiodothyronine
T_4	tetraiodothyronine, also called thyroxine
TB	tuberculosis
TBG	thyroid-binding globulin
TEM	transmission electron micrograph
TIA	transient ischemic attack
T_m	transport (tubular) maximum
TMJ	temporomandibular joint
t-PA	tissue plasminogen activator
TRH	thryotropin-releasing hormone
tRNA	transfer RNA
TSH	thyroid stimulating hormone
TSS	toxic shock syndrome
TX	thyroxine
U.S.	United States
UTI	urinary tract infection
UTP	uridine triphosphate
UV	ultraviolet
VF	ventricular fibrillation
VLDL	very low density lipoprotein
VPRC	volume of packed red cells
VT	ventricular tachycardia
WBC	white blood cell